Y0-BSD-750

The chemistry of

peroxides

THE CHEMISTRY OF FUNCTIONAL GROUPS

A series of advanced treatises under the general editorship of Professor Saul Patai

The chemistry of alkenes (2 volumes)
The chemistry of the carbonyl group (2 volumes)
The chemistry of the ether linkage
The chemistry of the amino group
The chemistry of the nitro and nitroso groups (2 parts)
The chemistry of carboxylic acids and esters
The chemistry of the carbon–nitrogen double bond
The chemistry of amides
The chemistry of the cyano group
The chemistry of the hydroxyl group (2 parts)
The chemistry of the azido group
The chemistry of acyl halides
The chemistry of the carbon–halogen bond (2 parts)
The chemistry of the quinonoid compounds (2 parts)
The chemistry of the thiol group (2 parts)
The chemistry of the hydrazo, azo and azoxy groups (2 parts)
The chemistry of amidines and imidates
The chemistry of cyanates and their thio derivatives (2 parts)
The chemistry of diazonium and diazo groups (2 parts)
The chemistry of the carbon–carbon triple bond (2 parts)
The chemistry of ketenes, allenes and related compounds (2 parts)
The chemistry of the sulphonium group (2 parts)
Supplement A: The chemistry of double-bonded functional groups (2 parts)
Supplement B: The chemistry of acid derivatives (2 parts)
Supplement C: The chemistry of triple-bonded functional groups (2 parts)
Supplement D: The chemistry of halides, pseudo-halides and azides (2 parts)
Supplement E: The chemistry of ethers, crown ethers, hydroxyl groups and their sulphur analogues (2 parts)
Supplement F: The chemistry of amino, nitroso and nitro compounds and their derivatives (2 parts)
The chemistry of the metal–carbon bond—Volume 1
The chemistry of peroxides

—O—O—

The chemistry of
peroxides

Edited by

SAUL PATAI

The Hebrew University, Jerusalem

1983

JOHN WILEY & SONS

CHICHESTER – NEW YORK – BRISBANE – TORONTO – SINGAPORE

An Interscience ® Publication

Library of Congress Cataloging in Publication Data:

Main entry under title:

The Chemistry of peroxides.

(The Chemistry of functional groups)
'An Interscience publication.'
Includes indexes.
1. Peroxides. I. Patai, Saul. II. Series.
QD181.01C46 1983 546′.7212 83-14844
ISBN 0 471 10218 0

British Library Cataloguing in Publication Data:

The chemistry of peroxides.
1. Peroxides
I. Patai, Saul
547′.23 QD181.01

ISBN 0 471 10218 0

Filmset by Speedlith Photo Litho Ltd., Manchester and Printed by The Pitman Press, Bath, Avon

Contributing Authors

W. Adam	Institut für Organische Chemie, Universität Würzburg, Am Hubland, D-8700 Würzburg, BRD and Departamento de Quimica, Universidad de Puerto Rico, Rio Piedras, Puerto Rico 00931, U.S.A.
A. C. Baldwin	Department of Chemical Kinetics, SRI International, Menlo Park, California 94025, U.S.A.
L. Batt	Department of Chemistry, University of Aberdeen, Aberdeen AB9 2UE, Scotland
G. Bouillon	Fachrichtung 14.1—Organische Chemie, Universität des Saarlandes, D-6600 Saarbrücken, W. Germany
P. B. Brindley	Kingston Polytechnic, Kingston upon Thames KT1 2EE, England
R. Ceresa	Polymer Consultant, Pepys Cottage, 13 High Street, Cottenham, Cambridge CB4 4SA, England
D. Cremer	Lehrstuhl für Theoretische Chemie, Universität Köln, Köln, W. Germany
O. Exner	Institute of Organic Chemistry and Biochemistry, Czechoslovak Academy of Sciences, 16610 Prague 6, Czechoslovakia
A. A. Frimer	Department of Chemistry, Bar-Ilan University, Ramat-Gan, Israel
K. Fujimori	Department of Chemistry, the University of Tsukuba, Sakura-mura, Niihari-gun, Ibaraki-ken, 305 Japan
K. Furuta	Department of Applied Chemistry, Faculty of Engineering, Nagoya University, Nagoya, Japan
J. Z. Gougoutas	Department of Chemistry, University of Minnesota, Minneapolis, MN 55455, U.S.A.
R. V. Hoffman	Department of Chemistry, New Mexico State University. Las Cruces, New Mexico 88003, U.S.A.
J. A. Howard	Division of Chemistry, National Research Council of Canada, Ottawa K1A OR9, Canada
E. Keinan	Department of Organic Chemistry, The Weizmann Institute of Science, Rehovot, 76100, Israel
M. Lazár	Polymer Institute of the Slovak Academy of Sciences, 80934 Bratislava, Czechoslovakia

C. Lick — Fachrichtung 14.1—Organische Chemie, Universität des Saarlandes, D-6600 Saarbrücken, W. Germany

M. T. H. Liu — Department of Chemistry, University of Prince Edward Island, Charlottetown, Prince Edward Island, C1A 4P3, Canada

H. Mimoun — Laboratoire d'Oxydation, Institut Français du Pétrole, 92506 Rueil-Malmaison, France

S. S. Nittala — Department of Synthetic Chemistry, Faculty of Engineering, Kyoto University, Kyoto, Japan

S. Oae — Department of Chemistry, The University of Tsukuba, Sakura-mura, Niihari-gun, Ibaraki-ken, 305 Japan

Y. Ogata — Department of Applied Chemistry, Faculty of Engineering, Nagoya University, Nagoya, Japan

B. Plesničar — University of Ljubljana, Ljubljana, Yugoslavia

W. H. Richardson — San Diego State University, San Diego, California 92182, U.S.A.

I. Saito — Department of Synthetic Chemistry, Faculty of Engineering, Kyoto University, Kyoto, Japan

K. Schank — Fachrichtung 14.1—Organische Chemie, Universität des Saarlandes, D-6600 Saarbrücken, W. Germany

H.-M. Schiebel — Institut für Organische Chemie, Technische Universität Braunschweig, D-3300 Braunschweig, W. Germany

H. Schwarz — Institut für Organische Chemie, Technische Universität Berlin, D-1000 Berlin 12, W. Germany

R. A. Sheldon — Océ-Andeno B.V., Venlo, The Netherlands

K. Tomizawa — Department of Applied Chemistry, Faculty of Engineering, Nagoya University, Nagoya, Japan

H. T. Varkony — Department of Organic Chemistry, The Weizmann Institute of Science, Rehovot, 76100, Israel

Foreword

The present volume in "The Chemistry of Functional Groups" series deals with functional groups which include the —O—O— bond, i.e., organic peroxides, hydroperoxides, acyl peroxides, peroxy acids and esters and ozonides. None of these groups have been treated in the volumes which appeared up to now in the series.

Unfortunately, out of the originally planned 31 chapters, seven did not materialize. These should have treated the following subjects: NMR; ESR and CIDNP; Analytical methods; Electrochemistry; Chemiluminescence; Biological formation and reactions; Safety and toxicity. We hope that these missing chapters will be incorporated into one of the future supplementary volumes of the Series.

The Editor will be very grateful to readers who would communicate to him mistakes and omissions relating to this volume as well as to other volumes in the Series.

SAUL PATAI

Jerusalem, June 1983

The Chemistry of Functional Groups
Preface to the series

The series 'The Chemistry of Functional Groups' is planned to cover in each volume all aspects of the chemistry of one of the important functional groups in organic chemistry. The emphasis is laid on the functional group treated and on the effects which it exerts on the chemical and physical properties, primarily in the immediate vicinity of the group in question, and secondarily on the behaviour of the whole molecule. For instance, the volume *The Chemistry of the Ether Linkage* deals with reactions in which the C—O—C group is involved, as well as with the effects of the C—O—C group on the reactions of alkyl or aryl groups connected to the ether oxygen. It is the purpose of the volume to give a complete coverage of all properties and reactions of ethers in as far as these depend on the presence of the ether group but the primary subject matter is not the whole molecule, but the C—O—C functional group.

A further restriction in the treatment of the various functional groups in these volumes is that material included in easily and generally available secondary or tertiary sources, such as Chemical Reviews, Quarterly Reviews, Organic Reactions, various 'Advances' and 'Progress' series as well as textbooks (i.e. in books which are usually found in the chemical libraries of universities and research institutes) should not, as a rule, be repeated in detail, unless it is necessary for the balanced treatment of the subject. Therefore each of the authors is asked *not* to give an encyclopaedic coverage of his subject, but to concentrate on the most important recent developments and mainly on material that has not been adequately covered by reviews or other secondary sources by the time of writing of the chapter, and to address himself to a reader who is assumed to be at a fairly advanced post-graduate level.

With these restrictions, it is realized that no plan can be devised for a volume that would give a *complete* coverage of the subject with *no* overlap between chapters, while at the same time preserving the readability of the text. The Editor set himself the goal of attaining *reasonable* coverage with *moderate* overlap, with a minimum of cross-references between the chapters of each volume. In this manner, sufficient freedom is given to each author to produce readable quasi-monographic chapters.

The general plan of each volume includes the following main sections:

(a) An introductory chapter dealing with the general and theoretical aspects of the group.

(b) One or more chapters dealing with the formation of the functional group in question, either from groups present in the molecule, or by introducing the new group directly or indirectly.

(c) Chapters describing the characterization and characteristics of the functional groups, i.e. a chapter dealing with qualitative and quantitative methods of determination including chemical and physical methods, ultraviolet, infrared, nuclear magnetic resonance and mass spectra: a chapter dealing with activating and directive effects exerted by the group and/or a chapter on the basicity, acidity or complex-forming ability of the group (if applicable).

(d) Chapters on the reactions, transformations and rearrangements which the functional group can undergo, either alone or in conjunction with other reagents.

(e) Special topics which do not fit any of the above sections, such as photo-chemistry, radiation chemistry, biochemical formations and reactions. Depending on the nature of each functional group treated, these special topics may include short monographs on related functional groups on which no separate volume is planned (e.g. a chapter on 'Thioketones' is included in the volume *The Chemistry of the Carbonyl Group*, and a chapter on 'Ketenes' is included in the volume *The Chemistry of Alkenes*). In other cases certain compounds, though containing only the functional group of the title, may have special features so as to be best treated in a separate chapter, as e.g. 'Polyethers' in *The Chemistry of the Ether Linkage*, or 'Tetraaminoethylenes' in *The Chemistry of the Amino Group*.

This plan entails that the breadth, depth and thought-provoking nature of each chapter will differ with the views and inclinations of the author and the presentation will necessarily be somewhat uneven. Moreover, a serious problem is caused by authors who deliver their manuscript late or not at all. In order to overcome this problem at least to some extent, it was decided to publish certain volumes in several parts, without giving consideration to the originally planned logical order of the chapters. If after the appearance of the originally planned parts of a volume it is found that either owing to non-delivery of chapters, or to new developments in the subject, sufficient material has accumulated for publication of a supplementary volume, containing material on related functional groups, this will be done as soon as possible.

The overall plan of the volumes in the series 'The Chemistry of Functional Groups' includes the titles listed below:

The Chemistry of Alkenes (*two volumes*)
The Chemistry of the Carbonyl Group (*two volumes*)
The Chemistry of the Ether Linkage
The Chemistry of the Amino Group
The Chemistry of the Nitro and Nitroso Groups (*two parts*)
The Chemistry of Carboxylic Acids and Esters
The Chemistry of the Carbon–Nitrogen Double Bond
The Chemistry of the Cyano Group
The Chemistry of Amides
The Chemistry of the Hydroxyl Group (*two parts*)
The Chemistry of the Azido Group
The Chemistry of Acyl Halides
The Chemistry of the Carbon–Halogen Bond (*two parts*)
The Chemistry of Quinonoid Compounds (*two parts*)
The Chemistry of the Thiol Group (*two parts*)
The Chemistry of Amidines and Imidates
The Chemistry of the Hydrazo, Azo and Azoxy Groups (*two parts*)
The Chemistry of Cyanates and their Thio Derivatives (*two parts*)
The Chemistry of Diazonium and Diazo Groups (*two parts*)

The Chemistry of the Carbon–Carbon Triple Bond (*two parts*)
Supplement A: The Chemistry of Double-bonded Functional Groups (*two parts*)
The Chemistry of Ketenes, Allenes and Related Compounds (*two parts*)
Supplement B: The Chemistry of Acid Derivatives (*two parts*)
Supplement C: The Chemistry of Triple-Bonded Groups (*two parts*)
Supplement D: The Chemistry of Halides, Pseudo-halides and Azides (*two parts*)
Supplement E: The Chemistry of Ethers, Crown Ethers, Hydroxyl Groups and their Sulphur Analogues (*two parts*)
The Chemistry of the Sulphonium Group (*two parts*)
Supplement F: The Chemistry of Amino, Nitroso and Nitro Groups and their Derivatives (*two parts*)
The Chemistry of the Metal–Carbon Bond, Vol. 1.
The Chemistry of Peroxides

Titles in press:

The Chemistry of the Metal–Carbon Bond, Vol. 2.
The Chemistry of Organic Se and Te Compounds

Advice or criticism regarding the plan and execution of this series will be welcomed by the Editor.

The publication of this series would never have started, let alone continued, without the support of many persons. First and foremost among these is Dr Arnold Weissberger, whose reassurance and trust encouraged me to tackle this task, and who continues to help and advise me. The efficient and patient cooperation of several staff-members of the Publisher also rendered me invaluable aid (but unfortunately their code of ethics does not allow me to thank them by name). Many of my friends and colleagues in Israel and overseas helped me in the solution of various major and minor matters, and my thanks are due to all of them, especially to Professor Z. Rappoport. Carrying out such a long-range project would be quite impossible without the non-professional but none the less essential participation and partnership of my wife.

The Hebrew University
Jerusalem, ISRAEL

SAUL PATAI

Contents

1. General and theoretical aspects of the peroxide group — 1
   D. Cremer
2. Stereochemical and conformational aspects of peroxy compounds — 85
   O. Exner
3. Thermochemistry of peroxides — 97
   A. C. Baldwin
4. Mass spectrometry of organic peroxides — 105
   H. Schwarz and H.-M. Schiebel
5. Acidity, hydrogen bonding and complex formation — 129
   W. H. Richardson
6. Synthesis and uses of alkyl hydroperoxides and dialkyl peroxides — 161
   R. A. Sheldon
7. Singlet oxygen in peroxide chemistry — 201
   A. A. Frimer
8. Free-radical reaction mechanisms involving peroxides in solution — 235
   J. A. Howard
9. Organic sulphur and phosphorus peroxides — 259
   R. V. Hoffman
10. Diacyl peroxides, peroxycarboxylic acids and peroxy esters — 279
    G. Bouillon, C. Lick and K. Schank
11. Endoperoxides — 311
    I. Saito and S. S. Nittala
12. Structural aspects of organic peroxides — 375
    J. Z. Gougoutas
13. Polymeric peroxides — 417
    R. Ceresa
14. Organic reactions involving the superoxide anion — 429
    A. A. Frimer
15. Transition-metal peroxides—synthesis and use as oxidizing agents — 463
    H. Mimoun

16. Organic polyoxides — 483
B. Plesničar

17. Polar reaction mechanisms involving peroxides in solution — 521
B. Plesničar

18. Preparation and uses of isotopically labelled peroxides — 585
S. Oae and K. Fujimori

19. Ozonation of single bonds — 649
E. Keinan and H. T. Varkony

20. Pyrolysis of peroxides in the gas phase — 685
L. Batt and M. T. H. Liu

21. Photochemistry and radiation chemistry of peroxides — 711
Y. Ogata, K. Tomizawa and K. Furuta

22. Solid-state reactions of peroxides — 777
M. Lazár

23. Organometallic peroxides — 807
P. B. Brindley

24. Four-membered ring peroxides: 1,2-dioxetanes and α-peroxylactones — 829
W. Adam

Author index — 921

Subject index — 981

The Chemistry of Functional Groups, Peroxides
Edited by S. Patai

CHAPTER 1

General and theoretical aspects of the peroxide group

DIETER CREMER
Lehrstuhl für Theoretische Chemie, Universität Köln, Köln, West Germany

I. INTRODUCTION . . . 2
II. STRUCTURE . . . 4
A. Topology of Atomic Assemblies XO_2 and X_2O_2 . . . 4
B. The Configuration Space of XO_2 and X_2O_2 . . . 5
III. ORBITAL DESCRIPTION . . . 5
A. Qualitative Valence Bond Treatment . . . 5
1. Molecular oxygen . . . 6
2. Radical, biradical and ionic states of XO_2 . . . 7
3. X_2O_2 geometries with chain, Y or bridged structures . . . 8
B. Qualitative Molecular Orbital Treatment . . . 8
1. MO description of O_2 . . . 8
2. The hydrogenperoxyl radical . . . 13
3. XO_2: ozone . . . 13
4. Hydrogen peroxide . . . 13
5. X_2O_2:F_2O_2 . . . 21
IV. PROPERTIES OF XO_2 AND X_2O_2 PROTOTYPES . . . 21
A. Stationary Points on the Potential Hypersurface . . . 21
B. The Conformational Subspace of H_2O_2 . . . 36
C. Total Energies, Heats of Formation and Bond Dissociation Enthalpies . . . 40
D. Orbital Energies and Ionization Potentials . . . 44
E. Geometry and Vibrational Analysis . . . 47
F. Charge Density and One-electron Properties . . . 53
G. Excited States . . . 58
V. SUBSTITUENT EFFECTS . . . 61
A. General Trends . . . 61
1. Peroxy compounds XO_2 . . . 61

2. Peroxides XOOH and XOOX . . . 63
B. Special Compounds . . . 66
1. Peroxy acids and acyl peroxides . . . 66
2. Polyoxides . . . 68
3. Ozonides and other cyclic peroxides . . . 69
VI. ABBREVIATIONS, SYMBOLS, CONSTANTS AND CONVERSION FACTORS 73
A. List of Abbreviations . . . 73
B. List of Symbols . . . 74
C. Constants and Conversion Factors . . . 77
VII. ACKNOWLEDGEMENTS . . . 77
VIII. REFERENCES . . . 78

I. INTRODUCTION

Compounds containing the OO linkage are key species in oxidation reactions. Knowledge of their chemical properties is essential in the elucidation of atmospheric and stratospheric chemistry, the chemistry of combustion and flames, pollution, polymerization, biochemical synthesis and metabolism. This has been shown in previous monographs and review articles on hydrogen peroxide[1,2] and organic peroxides[3–7], oxidation reactions[7–9], especially those involving singlet oxygen[10,11] and ozone[12], combustion[13], decomposition of peroxides[7,14–16], smog reactions[17,18], degradation of polymers[19], oxidation in biochemical and biological systems[11,20,21] and metal–dioxygen complexes[22,23].

In most cases where a peroxo compound is formed its precursor has been molecular oxygen. Since O_2 is the second most abundant molecule in the atmosphere, one might ask why only a vanishing small amount is converted to per- or poly-oxides. What force prevents O_2 from polymerizing in chains and rings held together by O—O single bonds?

The presentation given here is an attempt to answer this, and related questions, by providing an insight into the electronic features of molecules possessing OO bonds. In order to establish the scenario of per- and poly-oxide chemistry, it seems appropriate to first compare and contrast the OO group with other groups of chemical importance.

Table 1 contains some data relevant to the question of the stability of the O—O single bond. The average bond energies[24] listed in Table 1 indicate that oxygen prefers bonding to H, C, N or F rather than to another O atom. Actually, this tendency has been traced to the difference in electronegativities of singly bonded atoms X and O[25]. The larger this difference, the more ionic the X—O bond (Table 1). Since bond strength is always enhanced by ionic character, X—O bond energies are generally larger than the 34 kcal mol^{-1} of the O—O bond.

TABLE 1. Bond parameters of molecules containing a X—O single bond

Parameter	H—O	C—O	N—O	O—O	F—O	Ref.
Bond energy (kcal mol^{-1})	110	84	53	34	44	24
Electronegativity difference $\varepsilon_O - \varepsilon_X$[a]	1.4	1	0.5	0	−0.5	25
Ionic character of bond (%)	18	15	7	0	6	24

[a]According to the Pauling scale the electronegativity of oxygen is 3.5.

The average O—O single-bond energy, however, is also smallest when compared with values for X—X single bonds where X is a neighbouring atom of the same period (Table 2) or the same group (Table 3). For example, the C—C bond energy is 50 kcal mol^{-1} larger than that of the O—O bond while the S—S bond energy is more than 20 kcal mol^{-1} larger. Table 3 reveals that single-bond energies of Group VI elements do not decrease with atomic number as double-bond energies do. Both the S—S and Se—Se bond are stronger than O—O and Te—Te bonds, the latter being comparable in strength. In this respect, O is similar to N and F, both of which also form weaker homonuclear single bonds than their higher homologues. This anomaly of N, O and F also becomes apparent when looking at average X—X bond lengths. These are larger by 5–12% than bond lengths predicted from covalent radii, which have been derived from C—X bond lengths.

Both these anomalies of the O—O bond are indicative of the weakening effect of lone-pair–lone-pair repulsion. If destabilization resulting from lone-pair repulsion is lowered, a strengthening of the OO linkage occurs. This is best achieved in the O_2 molecule where two of the four lone pairs are no longer localized at one atom. By delocalization they gain bonding character (see Section III.A.1). This explains the high bond energy of O_2 (119.2 kcal mol^{-1}, Table 3).

Lone-pair–lone-pair repulsion also causes a weakening of SS or SeSe bonds. However, due to the larger covalent radii of these atoms (Table 3) and the corresponding increase of the bond lengths the effect is much smaller than for the OO linkage. This difference constitutes the source of the anomaly of Group VI single-bond energies discussed above.

The atomization energy of O_2 is 59.6 kcal mol^{-1} of atoms to be compared with 34 kcal mol^{-1} of atoms for an oxygen polymer. This means that polymerization of O_2 would be endothermic by 26 kcal mol^{-1} of atoms, which corresponds to a change in the free enthalpy larger than 26 kcal mol^{-1} as polymerization would be accompanied by a decrease of entropy. Therefore, oxygen polymers are not likely to occur in nature.

The fact that the O—O single bond can easily be broken is responsible for the unusual reactivity of peroxo compounds.

TABLE 2. Bond parameters of homonuclear single bonds X—X[a]

Parameter	H—H	C—C	N—N	O—O	F—F
Bond energy (kcal mol^{-1})	104[b]	84	38	34	38[b]
Bond length (Å)	0.71	1.54	1.47	1.46	1.43
Covalent radius r_c of X (Å)	—	0.77	0.74	0.73	0.71
			(0.70)[c]	(0.66)[c]	(0.64)[c]

[a]All values from Ref. 24.
[b]Dissociation enthalpy of X—X.
[c]Evaluated from X—C bond lengths.

TABLE 3. Average bond energies and covalent radii of Group VI elements[a]

Parameter	OO	SS	SeSe	TeTe
Single-bond energy (kcal mol^{-1})	34	58	44	34
Double-bond energy (kcal mol^{-1})	119	102	63	53
Covalent radius r_c (Å)	0.73	1.04	1.16	1.35

[a]All values from Ref. 24.

II. STRUCTURE

Molecules with the stoichiometric formulas XO_2 or X_2O_2 will be considered in this chapter provided that they contain the subunit OO, i.e. that there exists some kind of bonding interaction between the O atoms. At the moment, it suffices to indicate these interactions by a string ∿∿ tying the atoms together. Then, the following question has to be answered: What are the possible structures of molecules XO_2 and X_2O_2?

A. Topology of Atomic Assemblies XO_2 and X_2O_2

If the number of bonding interactions involving X and O is not limited, two topologically different structures can be expected for XO_2 and ten for X_2O_2. These are shown in Figure 1.

To distinguish between them we may term them chain (**1**, **3**, **4**), Y or branched (**5**), cyclic or polycyclic (**2**, **6**–**12**) forms. Since X,X interactions are of secondary interest, we can consider **4**, **6** and **9** as special cases of **1** and **2** and, similarly, **7**, **11** and **12** as being special cases of **5**, **8** and **10**. If we assume that both atoms X interact in a similar manner with the OO moiety, then **8** can be dropped as an unlikely candidate for a peroxide structure*. That leaves us with the chain structures **1**, **3**, the branched or Y structure **5**, the cyclic structure **2** and the bicyclic or bridged one (**10**).

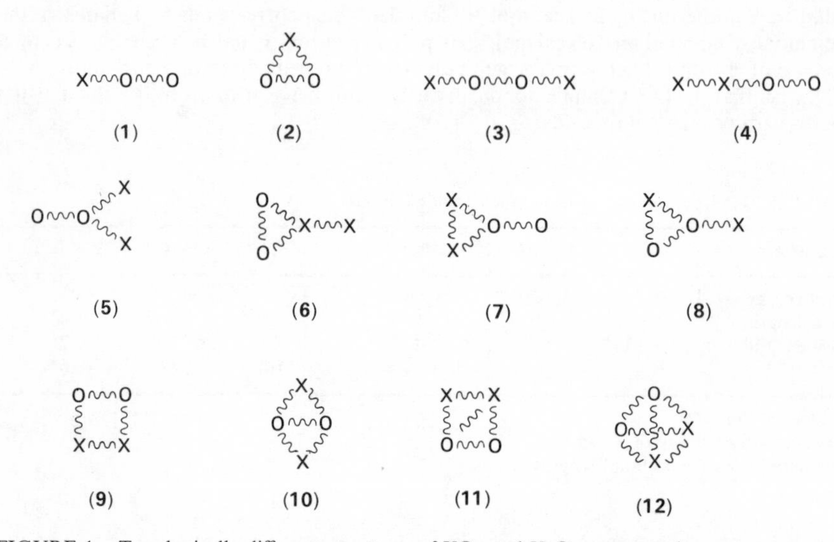

FIGURE 1. Topologically different structures of XO_2 and X_2O_2 compounds.

* Actually, this line of reasoning can only be followed if the stability of **8** is compared with those of the other peroxide structures. We anticipate the result of such a stability analysis in order to simplify the topological analysis. However, if two different substituents X and Y are attached to the OO moiety, structure **8** may very well correspond to a stable peroxide form (see, e.g. $\overline{CH_2OOH^+}$) and, therefore, cannot be dropped.

B. The Configuration Space of XO_2 and X_2O_2

In order to get a better understanding of the topological arrangements **1**, **2**, **3**, **5** and **10**, we shall now discuss specific geometrical forms generated from these structures. The configuration space of XO_2 can be spanned by the three internal coordinates R, R' and α, where R and R' denote the OO and OX bond length, respectively, and α is the OOX bond angle. In Figure 2 the interconversion of **1** to **2** is depicted. It involves linear, bent and cyclic geometries of XO_2, which are related by the angle α. In this respect interconversion can be viewed as corresponding to a movement approximately parallel to the α axis of the XO_2 space. This is indicated in Figure 2. Movements roughly parallel to the R or R' axis ultimately lead to dissociation of XO_2.

As for X_2O_2, the chain structure **3** is certainly the one most familiar to chemists. Geometries generated from **3** comprise linear–linear, linear–bent and bent–bent forms. The latter can be further distinguished by the dihedral angle τ. Characteristic geometries are obtained for $\tau = 0°$ (*cis* form), $\tau = 180°$ (*trans* form) or $0° < \tau < 180°$ (skewed forms). They are shown in Figure 3.

The Y structure (**5**) was historically one of the first discussed in connection with the elucidation of the hydrogen peroxide structure[1]. It can be either planar or pyramidal as shown in Figure 3. A similar distinction can be made for the bicyclic or bridged geometries, which have hardly been considered in peroxide chemistry.

The main interconversional modes together with some dissociative paths of X_2O_2 molecules are sketched in Figure 3. There the total six-dimensional space of the four-atom system has been projected onto a four-dimensional subspace by keeping the two R' and two α coordinates equal. Again, movements in geometrical space have been constrained to occur roughly parallel to one of the four axes, defined by R, R', α and τ.

III. ORBITAL DESCRIPTION

A. Qualitative Valence Bond Treatment

In order to understand the bonding situation in XO_2 and X_2O_2 it is helpful to 'synthesize' them in a step-by-step manner from atoms O and X via the 'intermediate' O_2.

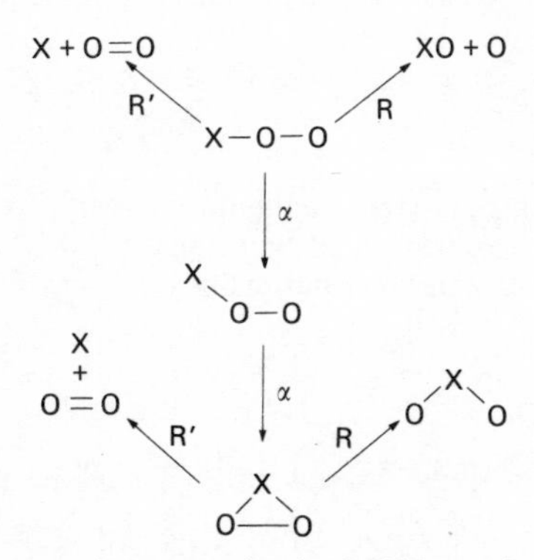

FIGURE 2. Possible interconversions occurring in the XO_2 configuration space approximately parallel to the space axes (R, R', α). Note that lines between atoms symbolize bonding interactions rather than electron-pair bonds.

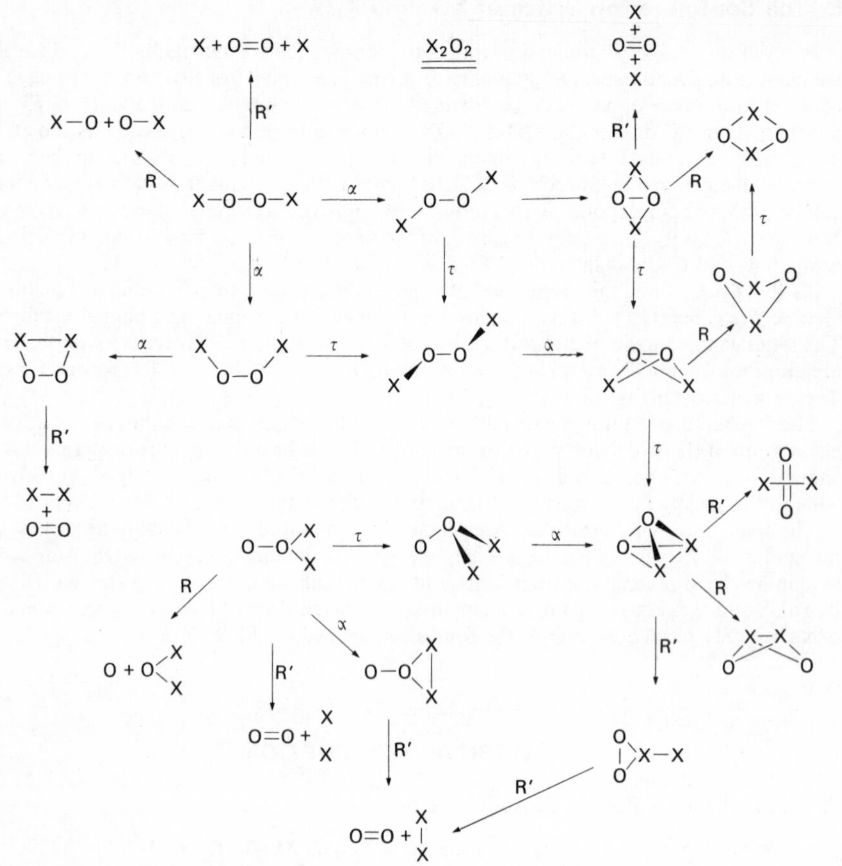

FIGURE 3. Possible interconversions occurring in the X_2O_2 configuration space. See caption of Figure 2.

1. Molecular oxygen

Atomic oxygen possesses the electron configuration $(1s)^2\ (2s)^2\ (2p)^4$, which leads to a ^3P ground state and ^1D and ^1S excited states. Ignoring the low-lying and doubly occupied 1s and 2s AOs, the $O(^3P)$ state can be visualized as:

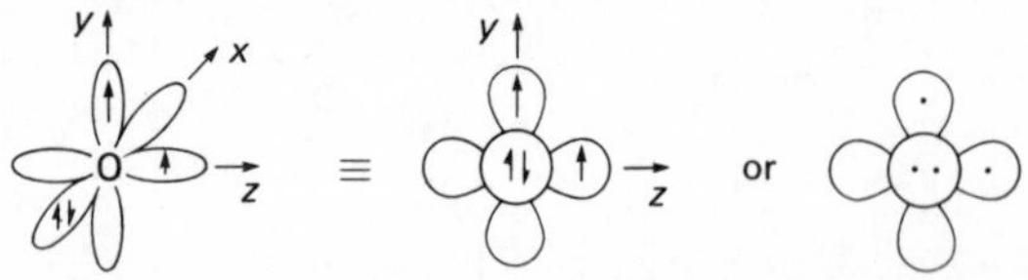

There are 81 different ways of combining two $O(^3P)$ atoms leading to a total of 81 O_2 states. If a strong σ bond is formed by coupling of the $2p_z$ AOs, essentially two possibilities a and b remain to combine the other 2p electrons. Each combination mode splits into a singlet (S) and a triplet (T) state of O_2:

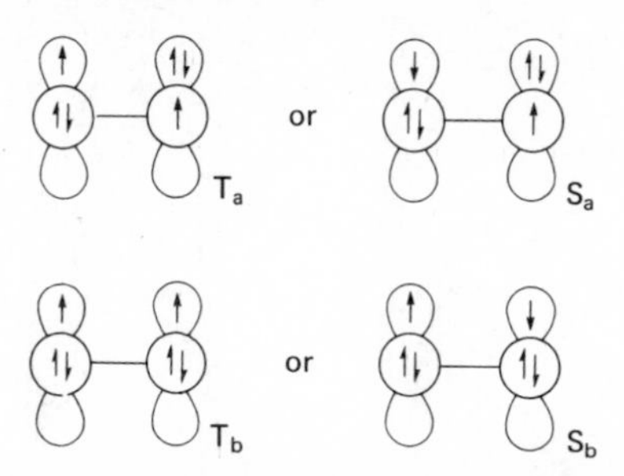

Combination mode a is clearly preferred, since it allows the electron pair on one O atom to delocalize onto the second O atom, thus reducing destabilizing Coulomb repulsions typical for combination mode b. The delocalization effect involving three $2p\pi$ electrons has been estimated to account for about 30 kcal mol^{-1} of the bond strength[26]. Accordingly, π bonding should contribute 60 kcal mol^{-1} to the total bond strength of O_2. The singly occupied orbitals are orthogonal in case a. Exchange interactions stabilize the triplet but destabilize the singlet state, just as would be predicted if Hund's rule of maximum multiplicity would be applied. Hence T_a represents the ground state (GS) of O_2, while S_a, S_b and T_b describe excited states.

2. *Radical, biradical and ionic states of XO_2*

If an atom X, with a single electron in a 1s, 2s or $2p\sigma$ AO, approaches O_2, it can form a σ bond with O via one of the singly occupied 2p orbitals of the T_a state. A bent XO_2 radical results. This should be the GS if X is a monovalent atom or group, like H, Li, CH_3, NH_2, etc. The GS is characterized by the term symbol $1^2\pi(3\pi)$ where the plane containing the three atoms serves to classify the symmetry of the singly occupied orbital (σ or π) and the total number of π electrons is given in parentheses (see Figure 4).

Excited states of XO_2 are derived from the $1^2\pi(3\pi)$ state by $p\sigma \rightarrow p\pi$ promotion or a $p\pi \rightarrow p\pi$ charge transfer. Thus, a covalent (cov) $1^2\sigma(4\pi)$ state with the single electron occupying the $p\sigma$ orbital and a ionic $2^2\pi(3\pi)$ state of XO_2, both with bent geometries, are obtained. They are shown in Figure 4.

A cyclic XO_2 state becomes possible when X has a second unpaired electron available for bonding. If X = O, F^+, S etc., there is in addition an electron lone pair and the configuration at atom X may be either $(p\sigma)^1(p\pi)^2$ or $(p\sigma)^2(p\pi)^1$. As indicated in Figure 5, the latter is more favourable since it avoids the destabilizing pair–pair Coulomb repulsion between π electrons at adjacent atomic centres of XO_2. Accordingly, the 4π states of XO_2 should be more stable than its 5π states.

If the $p\pi$ orbitals at atom X and the terminal oxygen did not overlap, the biradical XO_2 would be more stable in the T state $^3\pi\pi(4\pi)$ according to Hund's rule. But $p\pi$ orbitals separated by more than 2 Å still have a finite overlap. This brings the S (4π) state below the T state[26].

Excited 5π and 6π states are generated from the 4π states by $p\sigma \rightarrow p\pi$ promotion(s). The $^1\sigma\sigma(6\pi)$ state, characterized by bad Coulomb repulsions, can stabilize itself by decreasing α and forming a three membered ring. Ionic states of XO_2 are obtained by $p\pi \rightarrow p\pi$ or $p\sigma \rightarrow p\pi$ charge transfer to one of the terminal atoms (Figure 5). They correspond to resonance descriptions of XO_2 in terms of Lewis structures.

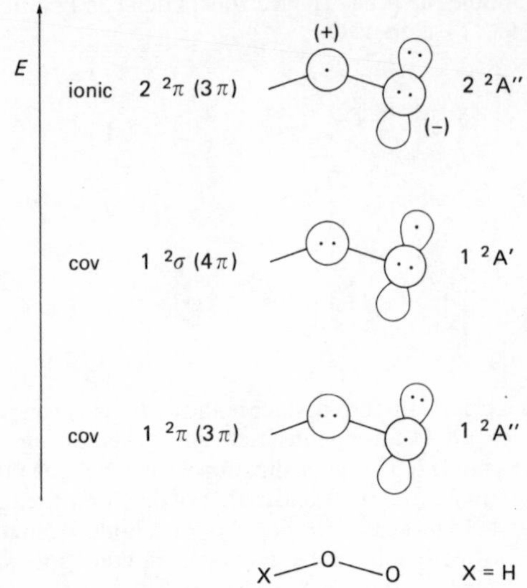

FIGURE 4. Schematic representation of low-lying states of XO_2 with monovalent X. Sigma bonds are denoted by solid lines, σ orbitals by lobes, π orbitals by circles and electrons by dots. Appropriate state symbols are given for HO_2. The covalent (cov) or ionic nature of each state is indicated.

3. X_2O_2 geometries with chain, Y or bridged structures

In Figure 6, four different geometries of X_2O_2 are rationalized by adding another monovalent atom X to each of the three lowest states of XO_2. Assuming that the energy content of XO_2 is carried over to X_2O_2, the orthogonal bent–bent geometry should be more stable than both *trans* and *cis* forms, which in turn should be more stable than the pyramidal Y form.

Some of the other possible X_2O_2 geometries can only be derived from high-lying XO_2 states (Figure 6). Accordingly, their energies should be considerably higher than those of the bent–bent geometries. This can be verified by counting the number of electron-pair–electron-pair repulsions.

Bridged geometries of X_2O_2 cannot be rationalized in this way. However, one can predict that the planar bridged form is also destabilized because of pair–pair repulsions.

B. Qualitative Molecular Orbital Treatment

1. MO description of O_2

One of the early triumphs of MO theory was the explanation of the paramagnetism of molecular oxygen. In Figure 7, the MOs of O_2 are schematically shown[27]. With respect to their shape and energy (Figure 7, left-hand side), they differ from those which one obtains by a simple pairwise combination of oxygen 2s and 2p AOs (Figure 7, right-hand side and middle). This is due to mixing of valence MOs with the same symmetry as indicated by the interaction lines of Figure 7.

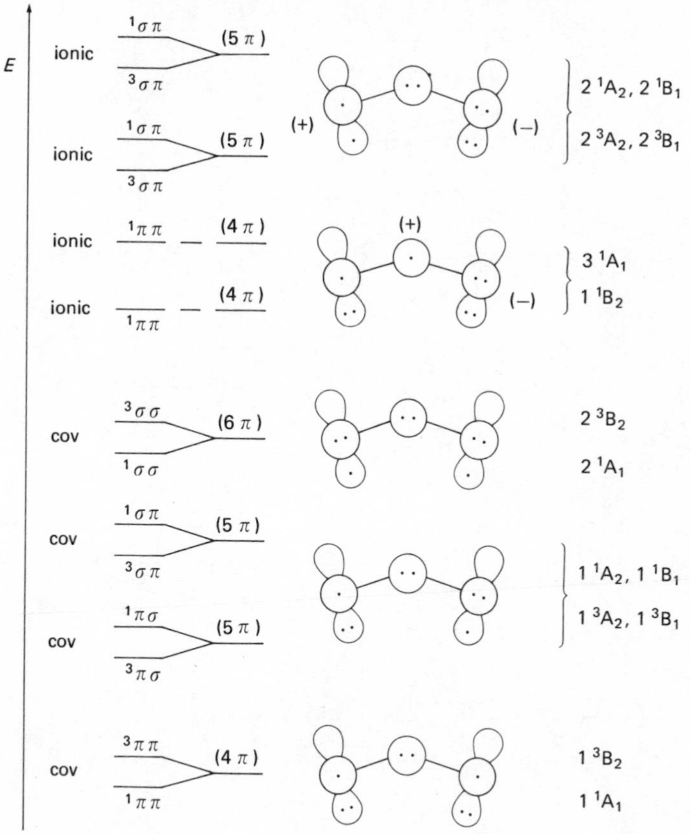

E

ionic $^1\sigma\pi$ $^3\sigma\pi$ (5 π) (+) (−) 2 $^1A_2$, 2 $^1B_1$; 2 $^3A_2$, 2 $^3B_1$

ionic $^1\sigma\pi$ $^3\sigma\pi$ (5 π)

ionic $^1\pi\pi$ (4 π) (+) (−) 3 $^1A_1$; 1 $^1B_2$

ionic $^1\pi\pi$ (4 π)

cov $^3\sigma\sigma$ $^1\sigma\sigma$ (6 π) 2 $^3B_2$; 2 $^1A_1$

cov $^1\sigma\pi$ $^3\sigma\pi$ (5 π) 1 $^1A_2$, 1 $^1B_1$; 1 $^3A_2$, 1 $^3B_1$

cov $^1\pi\sigma$ $^3\pi\sigma$ (5 π)

cov $^3\pi\pi$ $^1\pi\pi$ (4 π) 1 $^3B_2$; 1 $^1A_1$

X—O—O or O—O—X X = O

FIGURE 5. Schematic representation of low-lying states of XO_2 with divalent X. Note that the order of states depends on the nature of X. Appropriate state symbols are given for O_3. See caption of Figure 4.

Assigning 18 electrons to the MOs of O_2, the electron configuration

$$O_2{:}(1\sigma_g)^2(1\sigma_u)^2(2\sigma_g)^2(2\sigma_u)^2(3\sigma_g)^2(1\pi_u)^4(1\pi_g)^2 = [N_2](1\pi_g)^2$$

results. Ten electrons occupy bonding σ_g and π_u MOs, while six are in antibonding σ_u and π_g MOs. Hence, O_2 possesses two bonds, one formed by the $3\sigma_g$ and one by a $1\pi_u$ electron pair.

There are only two electrons occupying the $1\pi_g$ set, which can hold a total of four electrons. When spin is considered, there exist

$$\binom{4}{2} = \frac{4 \times 3}{2} = 6$$

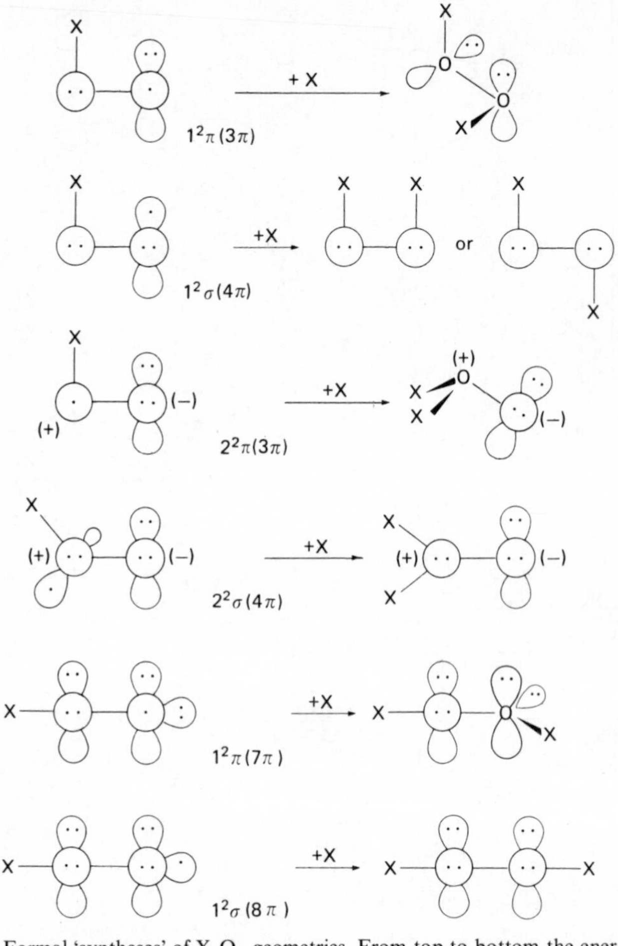

FIGURE 6. Formal 'syntheses' of X_2O_2 geometries. From top to bottom the energy of the XO_2 'precursor' increases. Compare with Figure 4.

possible assignments of the two electrons to the four $1\pi_g$ spin orbitals. They are given in Table 4. By writing for each assignment the corresponding Slater determinant, six state functions are obtained, which are depicted in Table 5 in terms of both real and complex spin orbitals[28].

The real state functions gain physical significance if O_2 is approached by a reacting molecule. For the free molecule, however, the distinction between the x and y directions is completely arbitrary. Then, the state functions have to be expressed in terms of complex spin orbitals $\pi_g^{m_l}$. The cylindrical symmetry of the latter complies with the requirements of the cylindrical point group $D_{\infty h}$ of the O_2 molecule.

The six state functions describe the three electronic states $^3\Sigma_g^-$, $^1\Delta_g$ and $^1\Sigma_g^+$ of O_2 (Tables 4 and 5). According to Hund's rule of maximum multiplicity and the orbital diagrams shown in Table 5 for the real state functions of O_2, these states should

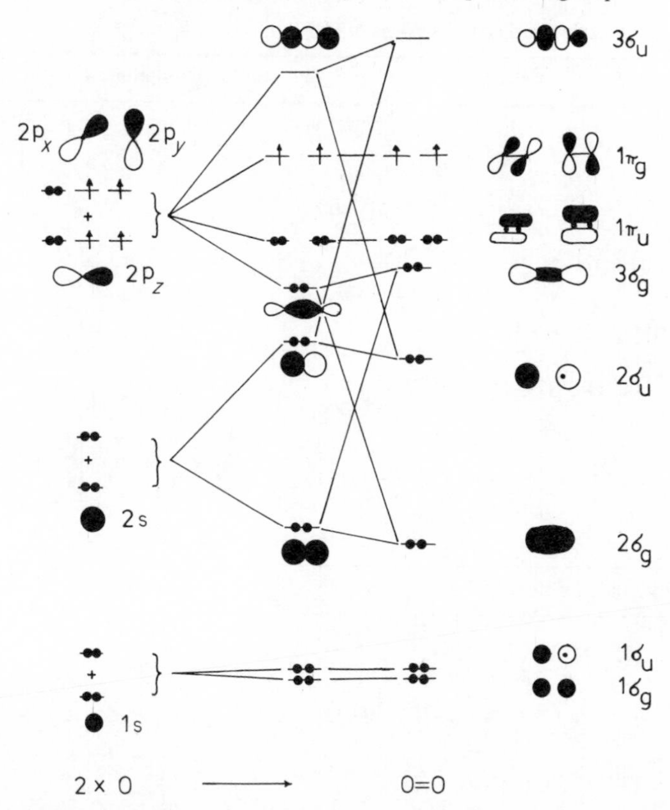

FIGURE 7. Qualitative MO correlation diagram for O_2. The shape of the MOs before (middle) and after (left-hand side) mixing is indicated. Solid lines between different MO levels denote orbital mixing.

TABLE 4. Assignments of the π_g electrons of O_2[a,b]

Assignment	$\pi_g^1\alpha$	$\pi_g^1\beta$	$\pi_g^{-1}\alpha$	$\pi_g^{-1}\beta$	M_L	M_S	Term
1	1	1	0	0	2	0	$^1\Delta_g$
2	1	0	1	0	0	1	$^3\Sigma_g^-$
3[c]	1	0	0	1	0	0	$^3\Sigma_g^-$, $^1\Sigma_g^+$
4[c]	0	1	1	0	0	0	$^3\Sigma_g^-$, $^1\Sigma_g^+$
5	0	1	0	1	0	-1	$^3\Sigma_g^-$
6	0	0	1	1	-2	0	$^1\Delta_g$

[a] M_L and M_S are the eigenvalues of the total orbital and spin angular momentum operator $\hat{L}_Z$ and $\hat{S}_Z$.
[b] Assignments are given for complex spin orbitals $\pi_g^1 = \frac{1}{\sqrt{2}}(\pi_g^x + i\pi_g^y)$ and $\pi_g^{-1} = \frac{1}{\sqrt{2}}(\pi_g^x - i\pi_g^y)$ where the superscript ± 1 corresponds to the eigenvalue m_l of the operator $\hat{l}_z$. The complex spin orbitals $\pi_g^{m_l}$ are more easily obtained when starting from complex atomic orbitals $2p_{m_l} = f(r, \theta) \cdot e^{m_l i\phi}$ expressed in terms of spherical polar coordinates r, θ, ϕ. For a more detailed description, see, for example, Ref. 28.
[c] Assignments are degenerate. To obtain the correct state functions, in-phase and out-of-phase combinations of the corresponding Slater determinants have to be taken.

TABLE 5. Complex and real state functions of O_2[a]

State	Complex state functions	Description
$^1\Sigma_g^+$	$\frac{1}{\sqrt{2}}[\pi_g^1(1)\pi_g^{-1}(2) + \pi_g^{-1}(1)\pi_g^1(2)] \times \frac{1}{\sqrt{2}}[\alpha(1)\beta(2) - \beta(1)\alpha(2)]$	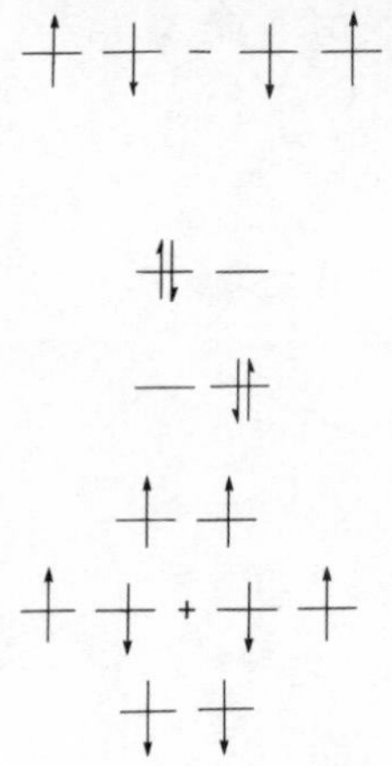
$^1\Delta_g$	$\left.\begin{matrix}[\pi_g^1(1)\pi_g^1(2)] \\ [\pi_g^{-1}(1)\pi_g^{-1}(2)]\end{matrix}\right\} \times \frac{1}{\sqrt{2}}[\alpha(1)\beta(2) - \beta(1)\alpha(2)]$	
$^3\Sigma_g^-$	$\frac{1}{\sqrt{2}}[\pi_g^1(1)\pi_g^{-1}(2) - \pi_g^{-1}(1)\pi_g^1(2)] \times \left\{\begin{matrix}\alpha(1)\alpha(2) \\ \frac{1}{\sqrt{2}}[\alpha(1)\beta(2) + \beta(1)\alpha(2)] \\ \beta(1)\beta(2)\end{matrix}\right.$	

State	Real state functions	Description
$^1\Sigma_g^+$	$\frac{1}{\sqrt{2}}[\pi_g^x(1)\pi_g^x(2) + \pi_g^y(1)\pi_g^y(2)] \times \frac{1}{\sqrt{2}}[\alpha(1)\beta(2) - \beta(1)\alpha(2)]$	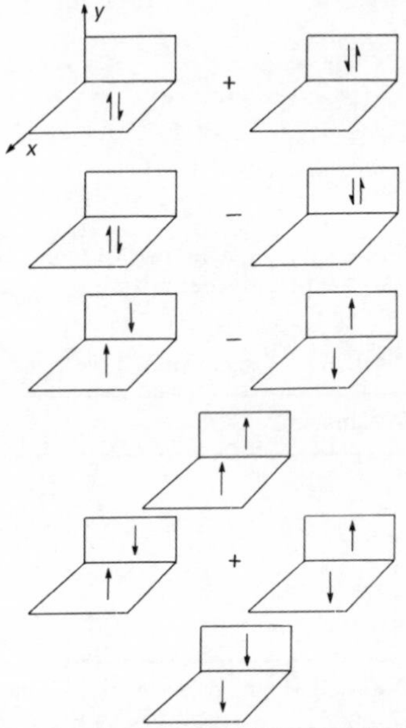
$^1\Delta_g$	$\left.\begin{matrix}\frac{1}{\sqrt{2}}[\pi_g^x(1)\pi_g^x(2) - \pi_g^y(1)\pi_g^y(2)] \\ \frac{1}{\sqrt{2}}[\pi_g^x(1)\pi_g^y(2) + \pi_g^y(1)\pi_g^x(2)]\end{matrix}\right\} \times \frac{1}{\sqrt{2}}[\alpha(1)\beta(2) - \beta(1)\alpha(2)]$	
$^3\Sigma_g^-$	$\frac{1}{\sqrt{2}}[\pi_g^x(1)\pi_g^y(2) - \pi_g^y(1)\pi_g^x(2)] \times \left\{\begin{matrix}\alpha(1)\alpha(2) \\ \frac{1}{\sqrt{2}}[\alpha(1)\beta(2) + \beta(1)\alpha(2)] \\ \beta(1)\beta(2)\end{matrix}\right.$	

[a]Complex state functions have been obtained by expanding the Slater determinants derived from Table 4. Their form is schematically represented by orbital diagrams. The two linear combinations correspond to degenerate electron assignments (Table 4, footnote *c*). The real state functions have been obtained using the relation between complex and real MOs (Table 4, footnote *b*). In case of the two $^1\Delta_g$ functions linear combinations of the resulting functions have to be taken in order to cancel imaginary terms.

correspond to the GS and the first and second excited state of O_2, i.e.

$$E(^3\Sigma_g^-) < E(^1\Delta_g) < E(^1\Sigma_g^+),$$

which is experimentally confirmed (Section IV.A).

2. *The hydrogenperoxyl radical*

The MOs of the HO_2 radical are closely related to those of molecular oxygen, as can be judged from a comparison of Figures 7 and 8. Figure 8 contains contour-line representations of the actual MOs of HO_2, calculated with HF theory for the linear (**13**), bent (**14**) and bridged (**15**) geometries. For each MO the appropriate symmetry notation is given. They should be used to understand Figure 9 where a qualitative MO correlation diagram for O_2 and HO_2, the latter with varying bond angle α, is given.

Figure 9 can be analysed in terms of increasing or decreasing bonding overlap[29]. Since the 2π–$7a'$ MO is stabilized for $90° < \alpha < 180°$, bent geometries of HO_2 should be the most stable ones, irrespective of the occupation of the 2π–$2a''$ MO. This means that the HO_2 cation, radical and anion should all prefer geometry **14** rather than **15** or **13**.

3. *XO_2: ozone*

If X disposes of suitable $2p\pi$ AOs three degenerate pairs of π MOs determine the electronic features of the linear XO_2 form. They possess OO bonding, nonbonding and antibonding character (Figure 10). In case of bending of the molecule, degeneracy is removed and the MO levels split (Figure 11). Both the in-plane (σ) and out-of-plane (π) nonbonding MOs are destabilized, while the two other σ,π pairs, bonding and antibonding, become more stable. This is due to developing 1,3 bonding or antibonding interactions in the bent form as can be seen from inspection of the corresponding MOs of ozone depicted in Figure 10.

Depending on the occupancy of these MOs XO_2 prefers the bent rather than the linear structure. This is demonstrated in Table 6 where predictions with regard to the most probable GS geometry of XO_2 systems with 14–20 valence electrons are given. These are based on Figures 10 and 11 and suggest that XO_2 peroxides with 14–16 valence electrons are linear, while those with 17–20 valence electrons adopt bent geometries with $100 < \alpha < 130°$ [30–32].

As indicated in Figure 11 conversion of bent to cyclic ozone is symmetry forbidden and, therefore, should be characterized by a relatively high energy barrier. The orbital diagram of Figure 11 suggests that the cyclic state of O_3 should be more stable than the bent one. A quantitative analysis of O_3, however, reveals that configuration interaction (CI), especially between the GS electron configuration and the ... $(1a_2)^0(4b_2)^2(6a_1)^2(2b_1)^2$ configuration leads to stabilization of the bent form below the cyclic structure. So far only one XO_2 system has been observed experimentally in a cyclic form, namely dioxirane ($X = CH_2$)[33].

4. *Hydrogen peroxide*

Linear H_2O_2 (**16**) possesses degenerate π_u and π_g MOs, similar to those of O_2, but fully occupied. The $^1\Sigma_g^+$ state is the GS of **16**. In the Y (**17**) and the bridged form (**18**), the in-plane components of the π MOs gain OH bonding character (Figure 12), thus leading to a lowering of the corresponding orbital energies (Figure 13) and an overall stabilization of these forms. According to the qualitative MO diagram of Figure 13, one can expect **17** to

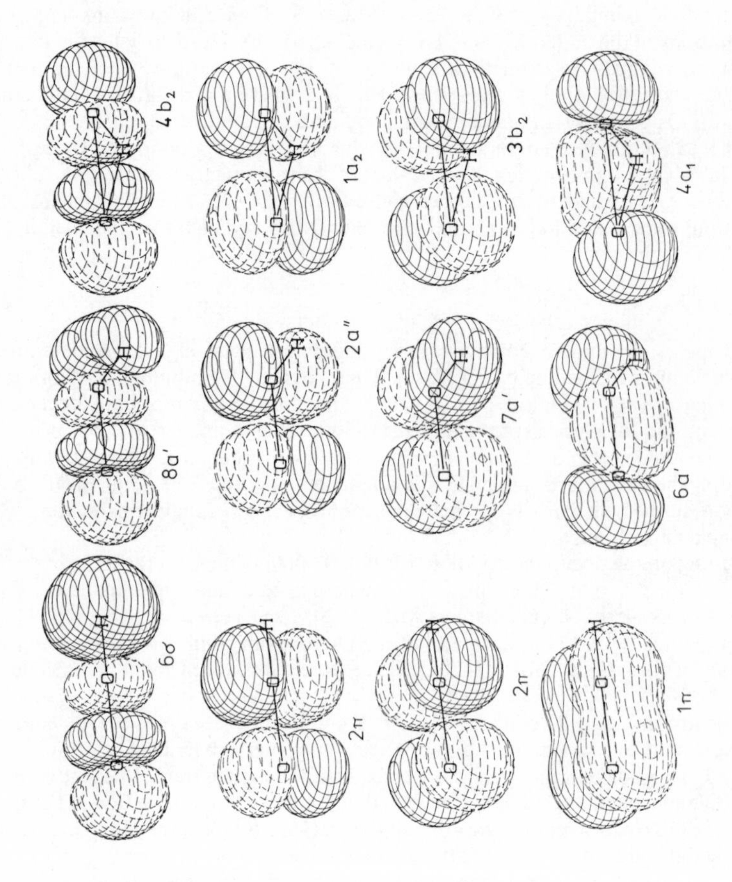
4b2
1a2
3b2
4a1
8a'
2a''
7a'
6a'
6σ
2π
2π
1π

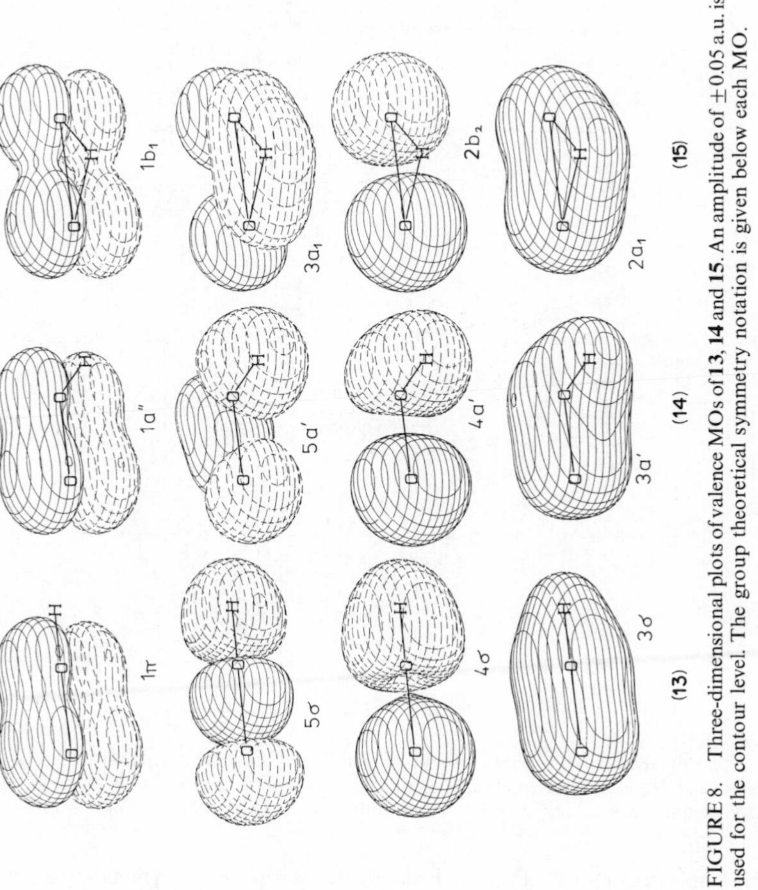

FIGURE 8. Three-dimensional plots of valence MOs of **13**, **14** and **15**. An amplitude of ± 0.05 a.u. is used for the contour level. The group theoretical symmetry notation is given below each MO.

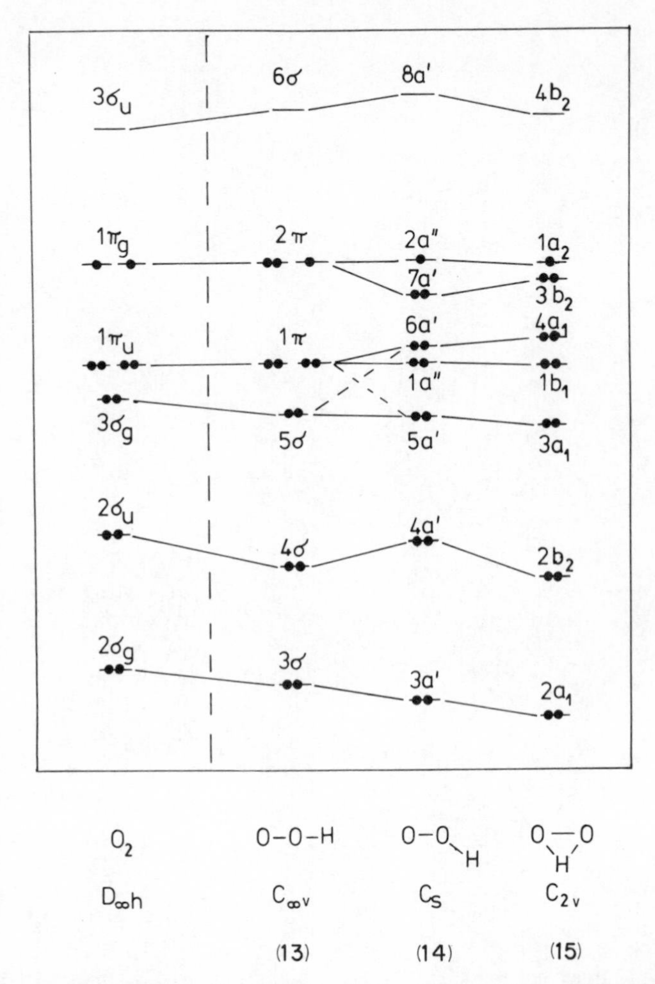

FIGURE 9. Qualitative MO correlation diagram for O_2 and linear (**13**), bent (**14**) and bridged (**15**) HO_2 based on UHF calculations and experimental ionization potentials. A crossing of the 5a′ and 6a′ MO, indicated by dashed lines, is symmetry forbidden (noncrossing rule). Compare with Figure 8.

be more stable than **18**. In both cases the nonplanar forms, i.e. the pyramidal Y form (**19**) and the puckered bridged form (**20**), are characterized by additional OH bonding as one (**20**) or both (**19**)[29] π MOs can mix with the 1s(H) orbitals. Hence, Figures 12 and 13 suggest the following ordering of total energies:

$$E(\mathbf{16}) > E(\mathbf{18}) > E(\mathbf{20}) > E(\mathbf{17}) > E(\mathbf{19}).$$

Undoubtedly the argument of increased stabilization due to developing OH bonding in the highest occupied MOs applies even more strongly to the bent–bent forms with $\tau = 0°$ (**21**), 120° (**22**) or 180° (**23**). This is documented by the shape of the $3b_1$–$4b$–$1b_g$ and $1a_2$–$5a$–$4a_g$ MOs depicted in Figure 14. (A quantitative MO correlation diagram for the

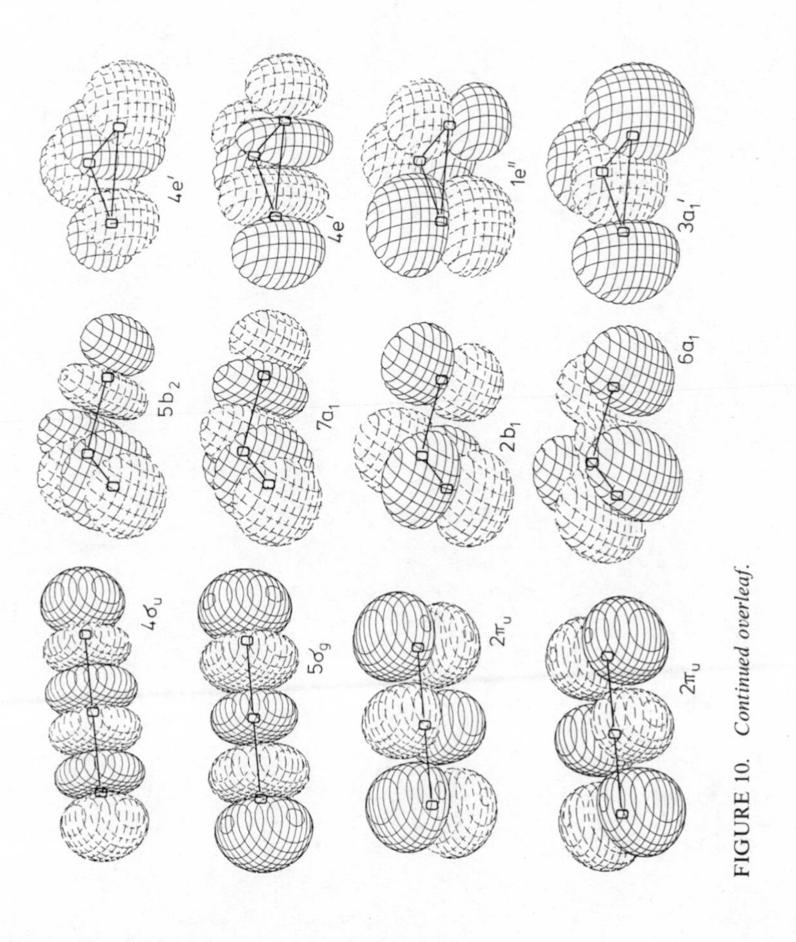

FIGURE 10. *Continued overleaf.*

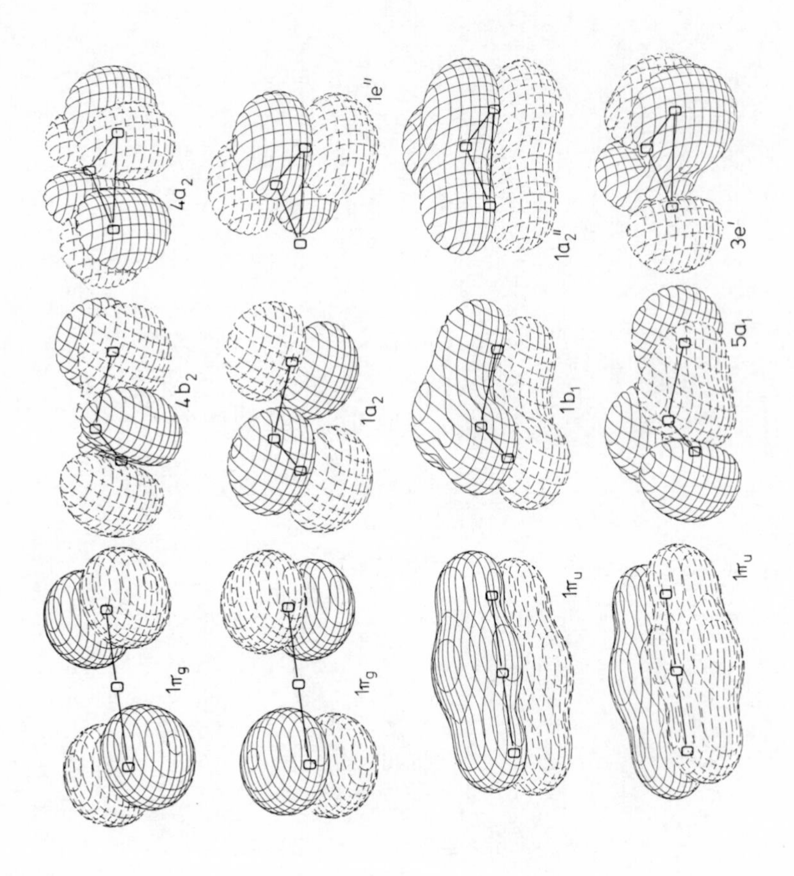
$4a_2$
$1e''$
$1a_2''$
$3e'$
$4b_2$
$1a_2$
$1b_1$
$5a_1$
$1\pi_g$
$1\pi_g$
$1\pi_u$
$1\pi_u$

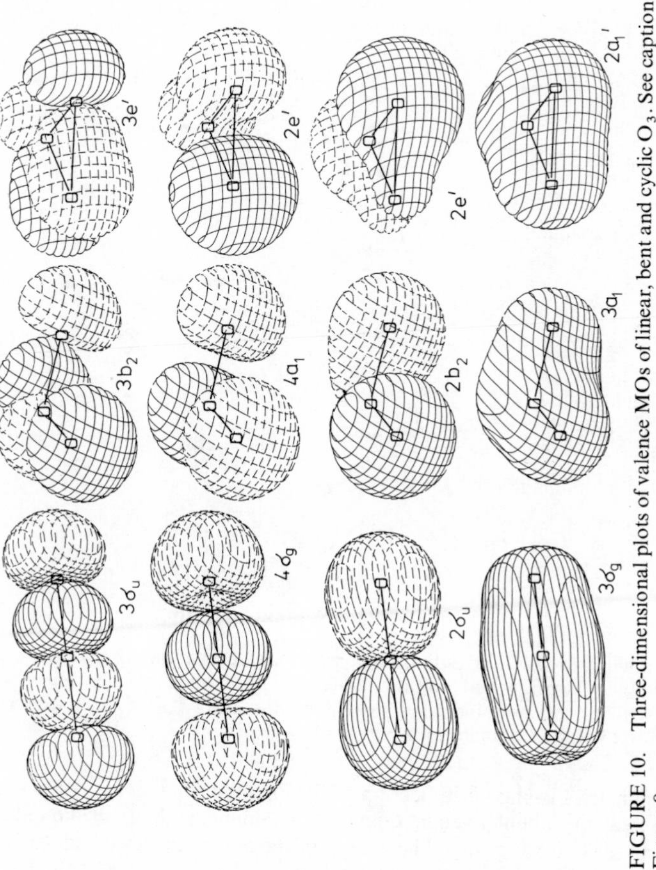

FIGURE 10. Three-dimensional plots of valence MOs of linear, bent and cyclic O_3. See caption of Figure 8.

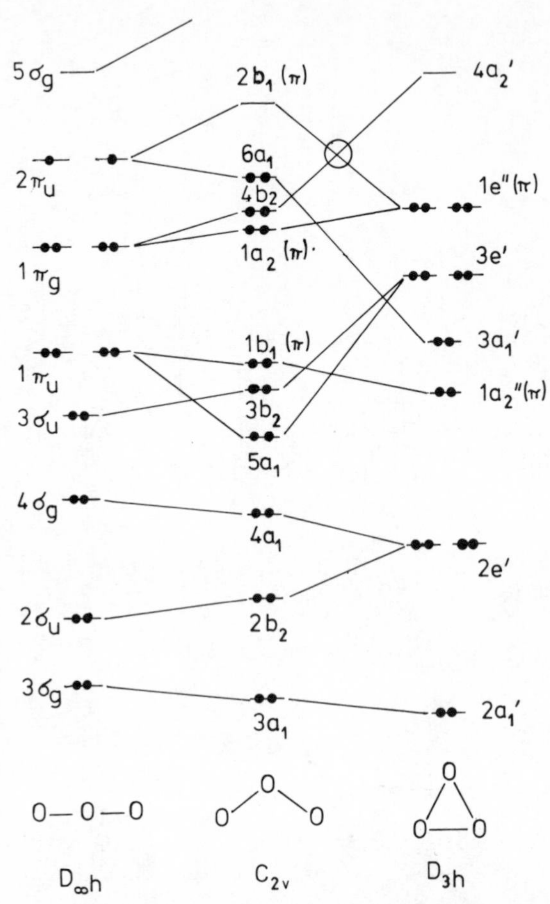

FIGURE 11. Qualitative MO correlation diagram for linear, bent and cyclic O_3 based on UHF and RHF calculations and experimental ionization potentials. The symmetry-forbidden crossing of the $2b_1$–le'' and $4b_2$–$4a_2'$ MOs is indicated. Note that the $2b_1$ MO—if occupied—possesses a considerably lower energy. Compare with Figure 10.

bent–bent geometries is shown in Section IV.B, Figure 19).

According to an argument given by Gimarc[34], the higher stability of skewed H_2O_2 can be explained in the following way: The change of the orbital energy of the 4b and 5a MO for τ increasing from 0° to 180° is approximately parallel to the change in the 1s(H)–2p(O) orbital overlap, which in turn depends on $\cos\tau$. For simplicity the $2p_x$ and $2p_y$ AOs are kept fixed and interconversion from **21** to **23** is considered to comprise clockwise and counterclockwise rotation of the OH groups by $\tau' = 90°$. At $\tau' = 45°$ ($\tau = 90°$) the 4b and 5a MO cross, both possessing then some OH bonding character (Figure 14). Overlap in the orthogonal form is larger than for **21** and **23** by a factor of $2\cos 45° = 1.7$. Hence skewed H_2O_2 with τ close to 90° should be the most stable bent–bent form. Actually, the lowest energy is found for **22**, since τ depends on a delicate balance among various electronic factors (see Section IV.B).

TABLE 6. Ground-state geometry of XO_2 peroxides

Valence electrons	Molecule	Electron configuration[a]	State[a]	Geometry	α(deg.)[b]
14	BeOO	$\ldots(1\pi_u)^4(1\pi_g)^2$	$^3\Sigma_g^-$	Linear	180[c]
15	BOO	$\ldots(1\pi_u)^4(1\pi_g)^3$	$^2\Pi_g$	Linear	
16	BOO^-	$\ldots(1\pi_u)^4(1\pi_g)^4$	$^1\Sigma_g^+$	Linear	
	HBOO			Linear	180[c]
	NOO^+			Linear	
17	NOO	$\ldots(1a_2)^2(4b_2)^2(6a_1)^1$	$^2A_1$	Bent	122
	OOO^+			Bent	132
18	H_2COO	$\ldots(1a_2)^2(4b_2)^2(6a_1)^2$	$^1A_1$	Bent	120[c]
	HNOO			Bent	119
	NOO^-			Bent	118
	OOO			Bent	118
	FOO^+			Bent	113
19	OOO^-	$\ldots(1a_2)^2(4b_2)^2(6a_1)^2(2b_1)^1$	$^2B_1$	Bent	116
	FOO			Bent	109
20	FOO^-	$\ldots(1a_2)^2(4b_2)^2(6a_1)^2(2b_1)^2$	$^1A_1$	Bent	

[a]Electron configuration and appropriate state symbol are given for the isoelectronic ozone ion (O_3^{4+}, $\ldots O_3^{2-}$). Compare with Figures 10 and 11.
[b]Sources of α values are given in Section V.A.1.
[c]Cyclic form is more stable; see Table 36, Section V.A.1.

5. X_2O_2:F_2O_2

X_2O_2 compounds with 26 valence electrons are best known in peroxide chemistry. They prefer bent–bent geometries[35] as is revealed by Figures 15 and 16, which depict the valence MOs and the corresponding orbital correlation diagram of F_2O_2. Overlap arguments similar to those used in the H_2O_2 case suggest the existence of a stable skewed form.

Knowledge of X_2O_2 compounds with 16–24 valence electrons is scarce. Some of these peroxides can be formed as diradical intermediates by a homolytic X–X cleavage reaction of cyclic peroxides X–O–O–X. As can be inferred from studies on dioxetanes[15], decomposition to X=O fragments (X = Be, BH, CH_2, NH, O) should be rapid in all cases.

According to theory, stable X_2O_2 systems with 18 (X = BeH) or 22 (X = BH_2) valence electrons should exist[32]. Their orbital diagrams differ considerably from that of F_2O_2. Predictions with regard to their geometry are difficult to make without a complete MO analysis (see Section V.A.2).

IV. PROPERTIES OF XO_2 AND X_2O_2 PROTOTYPES

A. Stationary Points on the Potential Hypersurface

Few molecules have been studied as extensively, both theoretically and experimentally, as the O_2 molecule. The vast literature on O_2 through early 1971 has been reviewed by Krupenie[36]. Since then several very accurate calculations of GS and excited states of O_2 and its ions have been carried out[37–44]. They confirm the qualitative ordering of the three

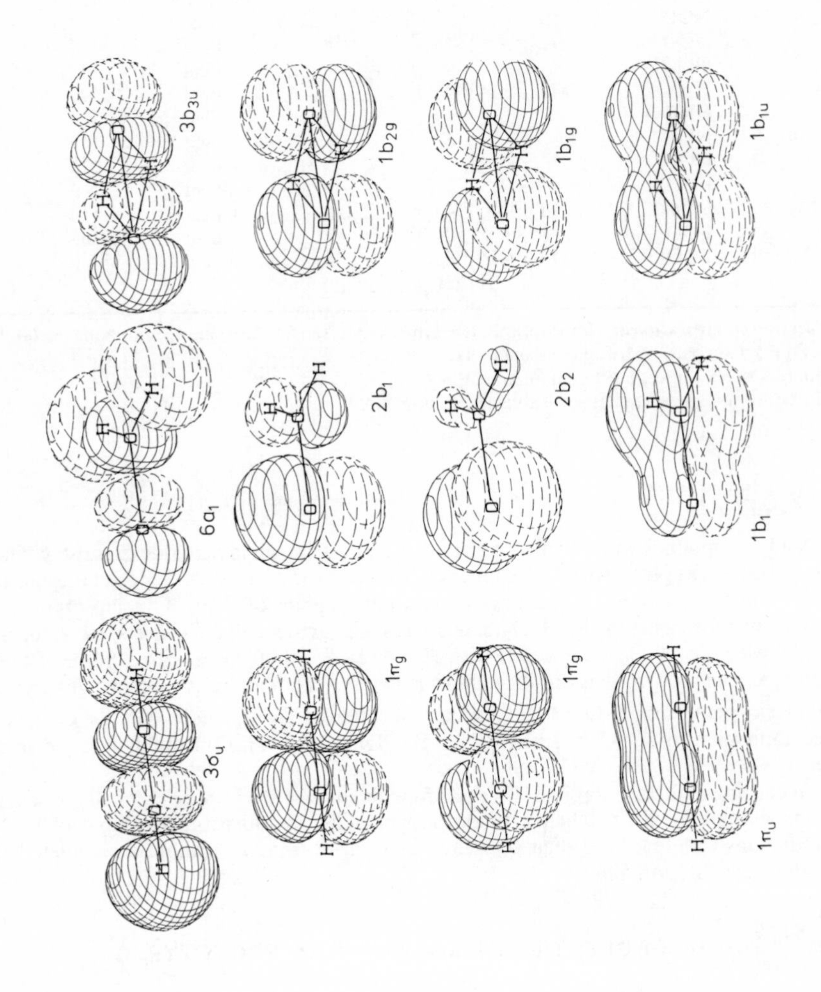
3b3u
1b2g
1b1g
1b1u
6a1
2b1
2b2
1b1
3σu
1πg
1πg
1πu

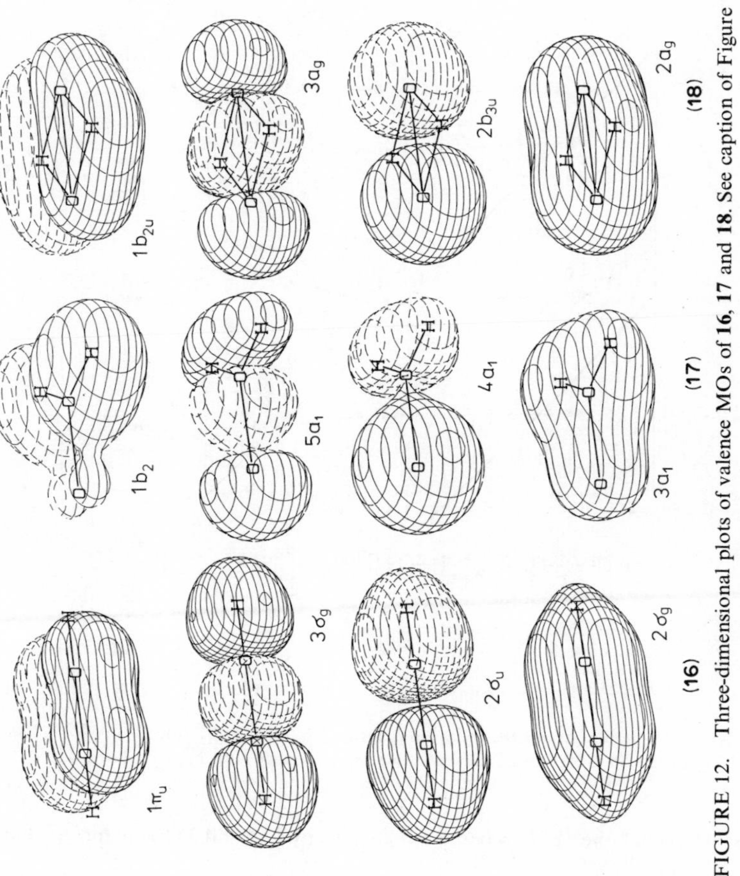

FIGURE 12. Three-dimensional plots of valence MOs of **16**, **17** and **18**. See caption of Figure 8.

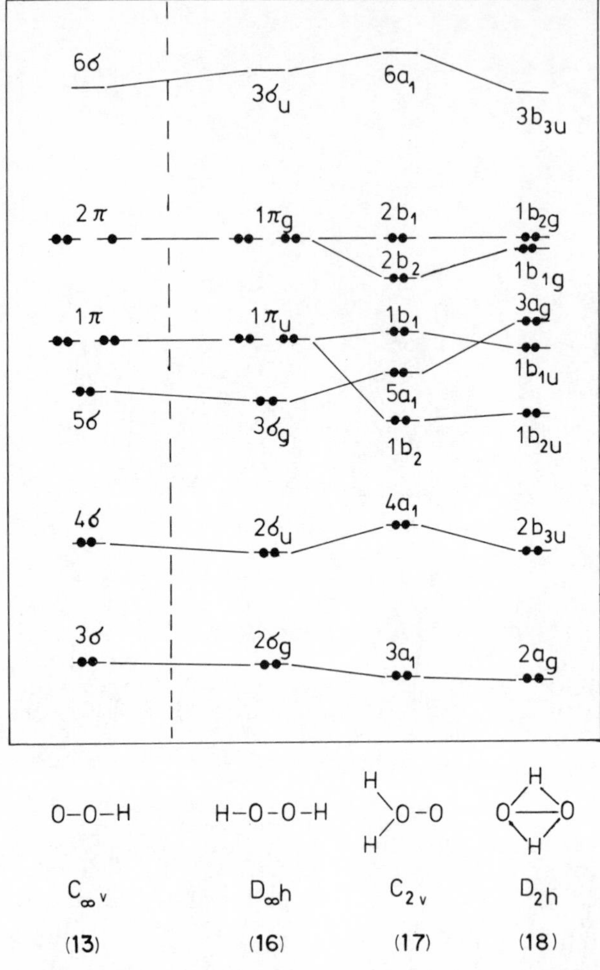

FIGURE 13. Qualitative MO correlation diagram for linear HO_2, linear (**16**), Y (**17**) and bridged (**18**) H_2O_2 based on UHF and RHF calculations. Compare with Figures 8 and 12.

lowest states with the $^3\Sigma_g^-$ GS being more stable by 22.5 and 37.5 kcal mol^{-1} than the $^1\Delta_g$ and $^1\Sigma_g^+$ states[36].

Also available are detailed theoretical data on special features of the O_3 hypersurface[45–61], some of which are summarized in Table 7. They show that calculations which go beyond the HF level of theory predict the bent 4π state to be more stable by 5–40 kcal mol^{-1} than the cyclic state of O_3 with 6π electrons. More recent calculations seem to suggest a value of about 23–28 kcal mol^{-1} [51,60]. Bent ozone is separated from its cyclic form by a barrier of about 30–40 kcal mol^{-1} [52,57,62].

Wright has suggested that cyclic O_3 with C_{2v} symmetry may be an intermediate on the decomposition path leading to $O_2(^3\Sigma_g^-)$ and $O(^3P)$[46]. His assumption is based on the discovery of an ozone precursor in radiolysis experiments. An experimental estimate of the activation energy of O_3 decomposition (24 kcal mol^{-1}), reported by Benson and Axworthy[63], excludes this possibility by describing O_3 decomposition as an endothermic process with no activation barrier. A recent theoretical evaluation of the decomposition surface of the bent form is in line with this estimate[60]. (For a different view see Reference 56.)

Widening of the angle α is accompanied by an increase of the energy. Linear O_3 is less stable than the bent form by 77[49] to 89 kcal mol^{-1} [59]. It adopts a triplet GS, $^3\Sigma_g^-$ (see Figure 11), which correlates with the first $^3B_1$ state of bent ozone (Figure 5). The $^1A_1$ GS of ozone, however, correlates with a degenerate $^1\Delta_g$ state of the linear configuration lying about 13 kcal mol^{-1} above the $^3\Sigma_g^-$ state[49].

In the past, the GS of ozone has been mostly described by a zwitterionic structure in order to explain the observed reactivity of O_3. Recent calculations carried out with different methods unanimously find a relative high biradical character for this and related XO_2 species in the gas phase[26,64–68] (Table 8). Harding and Goddard have shown that biradical character is consistent with the electrophilic nature observed for ozone and that there is no need to postulate zwitterionic structures[69]. The latter can become important in solution-phase reactions of XO_2 systems, especially when X bears a π-donating substituent R(X = NR, CR_2) [58,69].

Several theoretical investigations on the HO_2 radical in its most stable GS configuration have been published recently[71–81]. They describe HO_2 as possessing C_s symmetry with an equilibrium angle α close to 104° which is in accordance with qualitative MO arguments (Section III.B.2, Figure 9) and experiment[82]. Specific results are compared in Table 9.

In recent studies by Melius and Blint[78] and Langhoff and Jaffe[79] large portions of the HO_2 potential energy surface have been computed employing CI methods and augmented basis sets. Contour plots of various sections of the theoretical surface[78] are presented in Figure 17. They indicate that either widening or closing of the angle α causes an increase of the total energy with the linear form being more destabilized than the bridged one. The lowest linear HO_2 state, $^2\Pi$, correlates with the $1^2A''$ GS and the $1^2A'$ excited state. The barrier to linearity is computed to be 60–70 kcal mol^{-1} [79,83]. It is interesting to note that CI calculations describe the linear state to be ionic because of a transfer of the H electron to the π_g MO of O_2 [79].

Unfortunately, only a C_s geometry (R' = 0.968 and 1.198 Å at $\alpha = 60°$) of bridged HO_2 has been computed[79]. It lies about 40 kcal mol^{-1} above the GS of bent HO_2. Geometry optimization should lead to a value of about 35 kcal mol^{-1}. An early estimate of the energy of the bridged form[84] suggesting a minimum is unreliable because it is based on *ab initio* calculations of HO_2^+ and HO_2^- rather than a direct calculation of HO_2.

The theoretical analysis of the HO_2 surface suggests a small barrier (≤ 2 kcal mol^{-1}) at $\alpha = 120°$, $R' = 1.99$ and $R = 1.23$ Å (Figure 17) for the reaction $H + O_2 \rightarrow HO_2$ due to partial breaking of the π bond of O_2. HO_2 is more stable by 44 kcal mol^{-1} than the reactants, which has to be compared with an experimental value of 46 kcal mol^{-1} [85]. Breaking of the O—O bond of HO_2 requires 56 kcal mol^{-1} (63 kcal mol^{-1}, obs.[86]). The corresponding reaction channel proceeds uphill directly towards the products HO and O, i.e. there is no activation barrier for the reverse process leading to HO_2.

Although H_2O_2 has been the subject of numerous quantum-chemical calculations, only the conformational subspace of its bent–bent form (see Section IV.B) has been explored so far. Therefore, we have carried out *ab initio* calculations on forms **16**–**20** at various levels of theory[87]. Some of our results are listed in Table 10. They confirm the order of stabilities given in Section III.B.4. Thus, inversion at one of the O atoms of **23** is an unlikely process (ΔE = 71 kcal mol^{-1}, Table 10). Equally unstable are the bridged forms **18** and **20**. The Y form, however, may occur under certain conditions. Depending on the level of theory

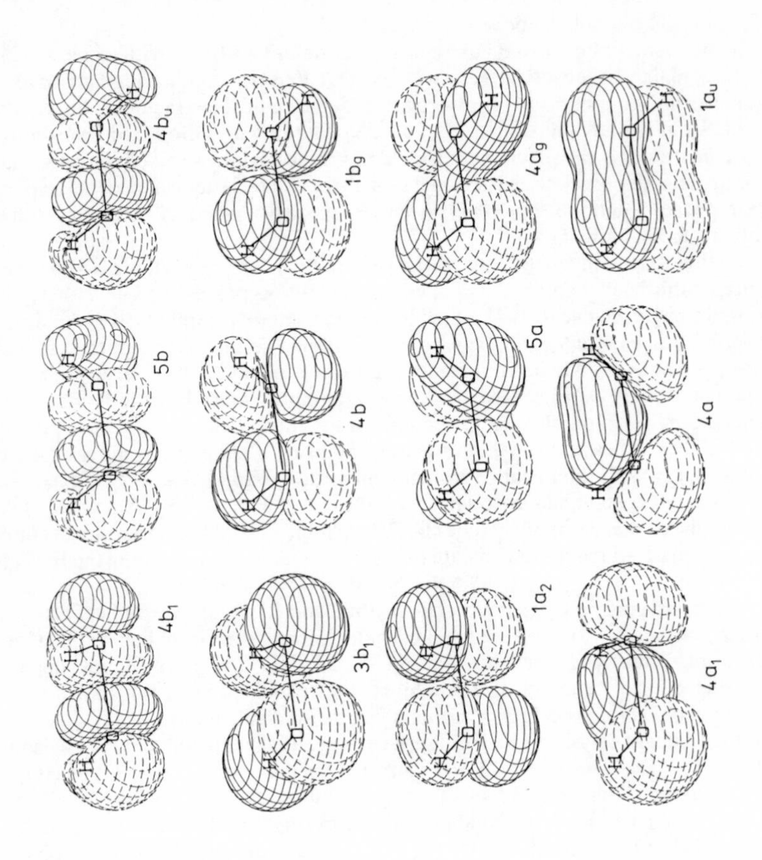
4b_u
1b_g
4a_g
1a_u
5b
4b
5a
4a
4b_1
3b_1
1a_2
4a_1

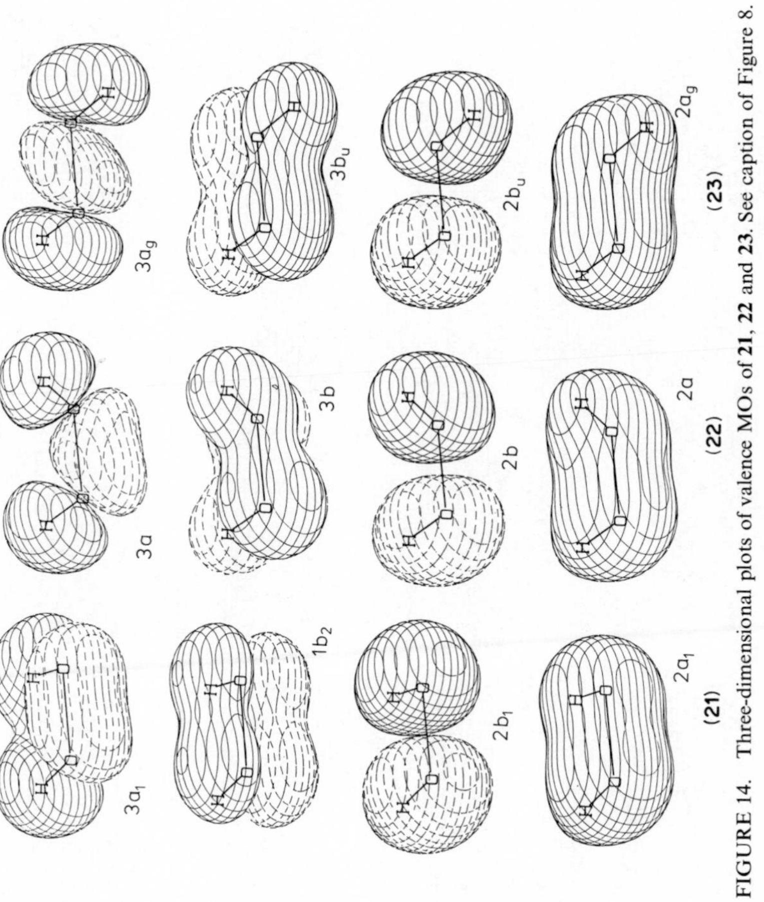

FIGURE 14. Three-dimensional plots of valence MOs of **21**, **22** and **23**. See caption of Figure 8.

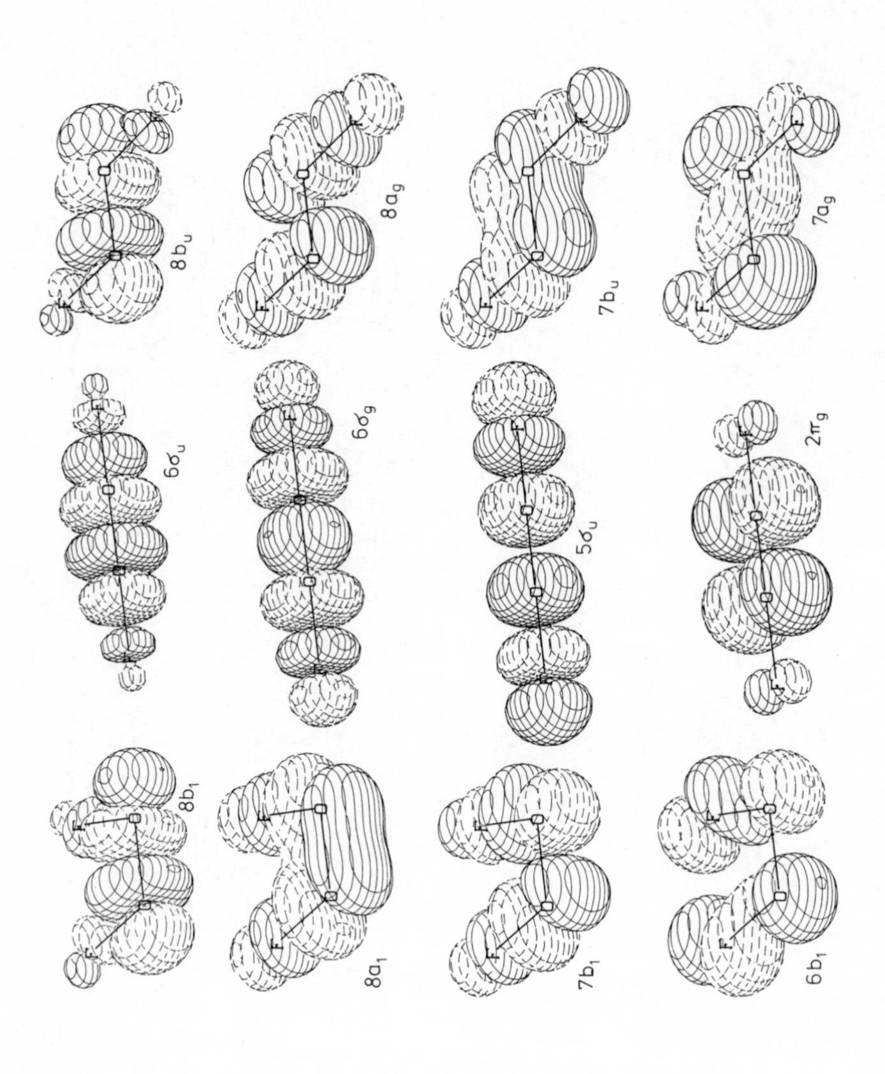
$8b_u$
$8a_g$
$7b_u$
$7a_g$
$6\sigma_u$
$6\sigma_g$
$5\sigma_u$
$2\pi_g$
$8b_1$
$8a_1$
$7b_1$
$6b_1$

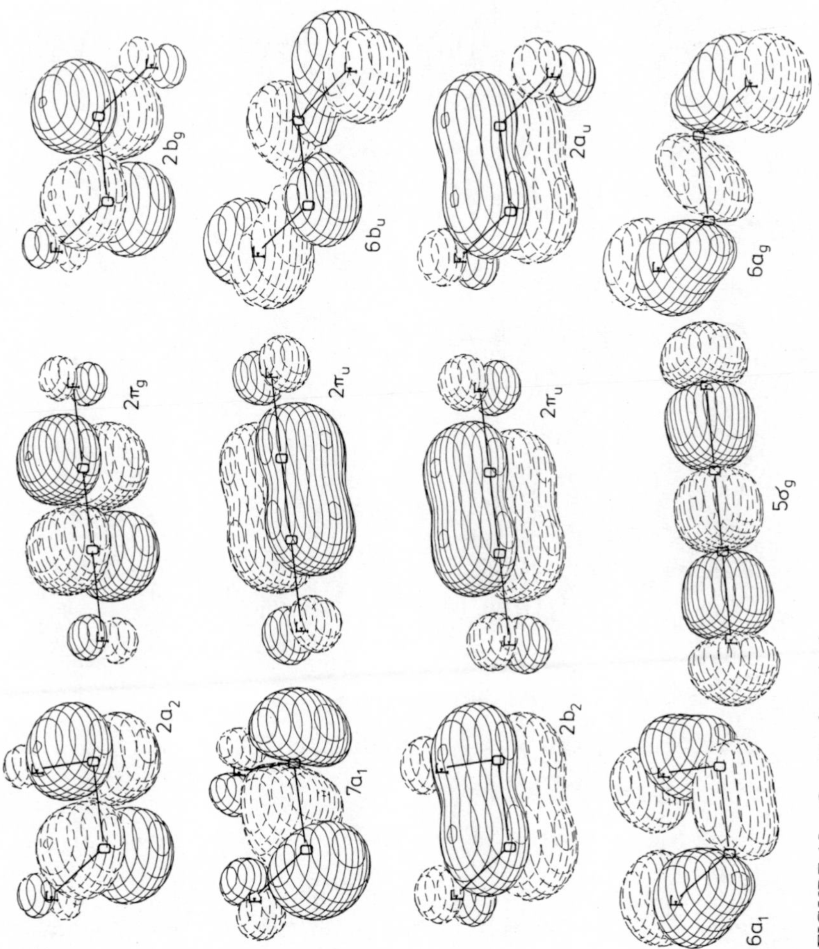

FIGURE 15. *Continued overleaf.*

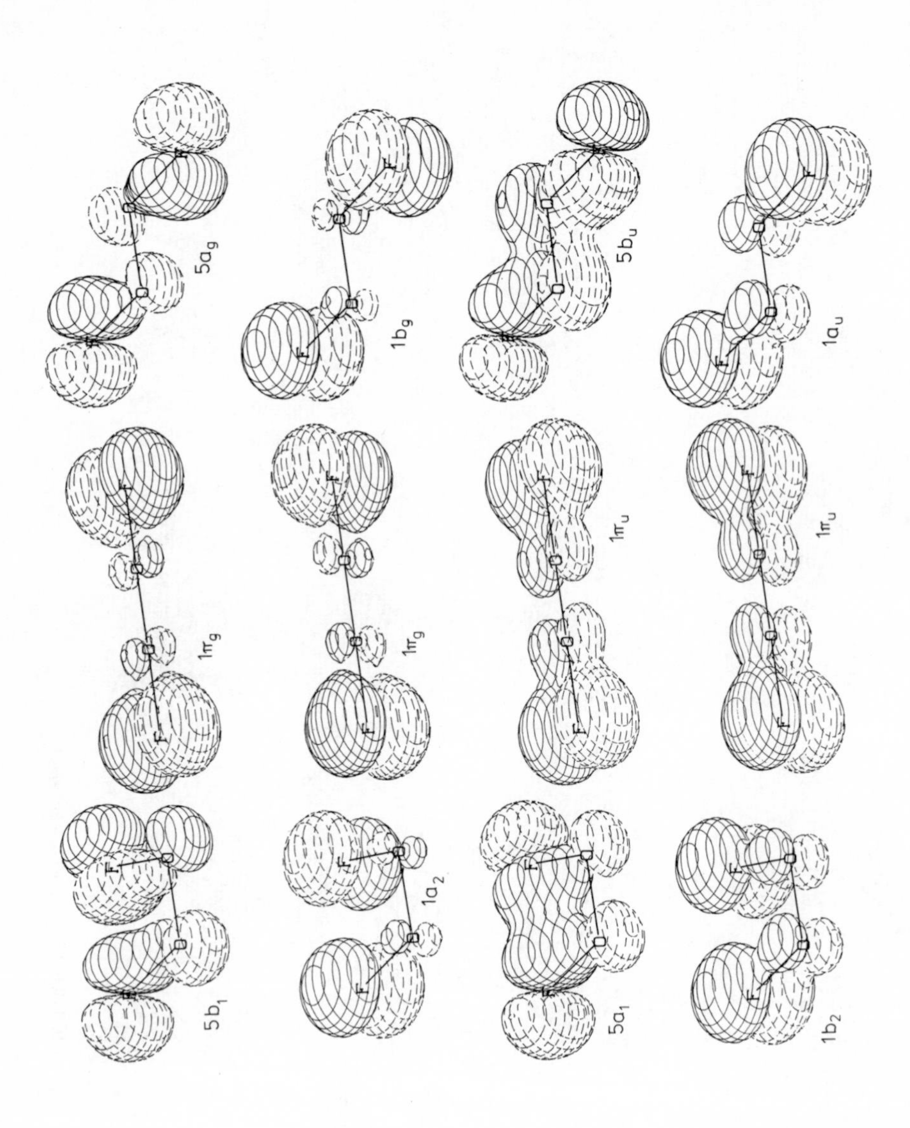
$5a_g$
$1b_g$
$5b_u$
$1a_u$
$1\pi_g$
$1\pi_g$
$1\pi_u$
$1\pi_u$
$5b_1$
$1a_2$
$5a_1$
$1b_2$

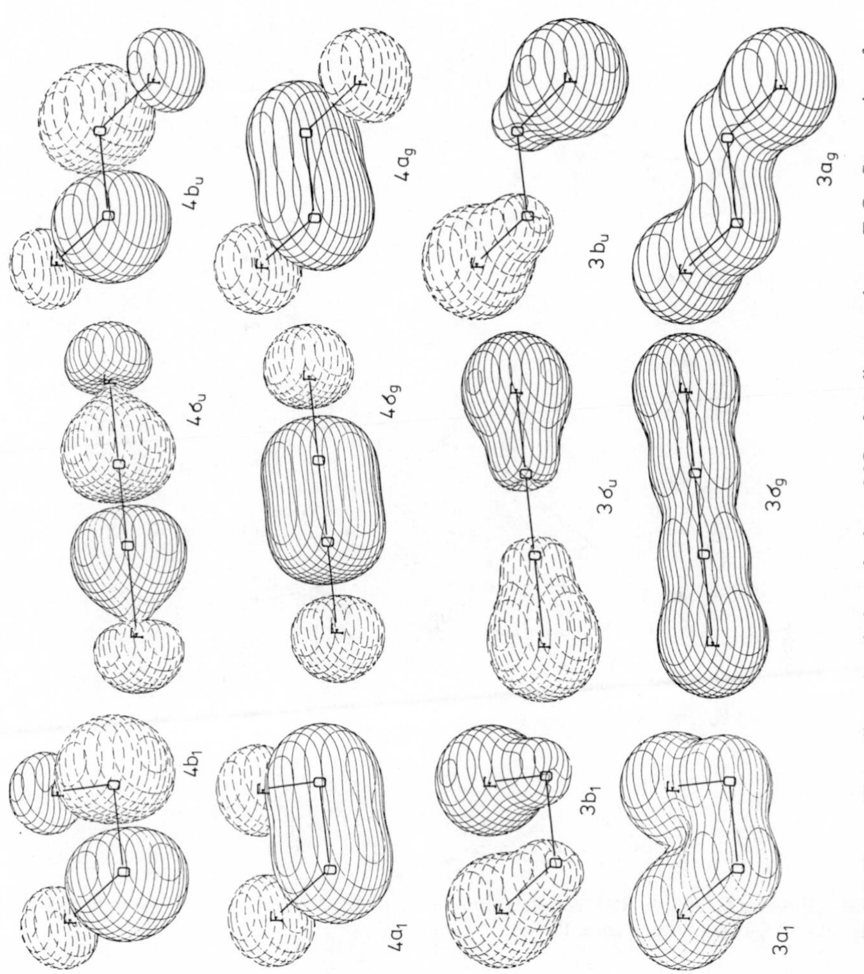

FIGURE 15. Three-dimensional plots of valence MOs of *cis*, linear and *trans* F_2O_2. See caption of Figure 8.

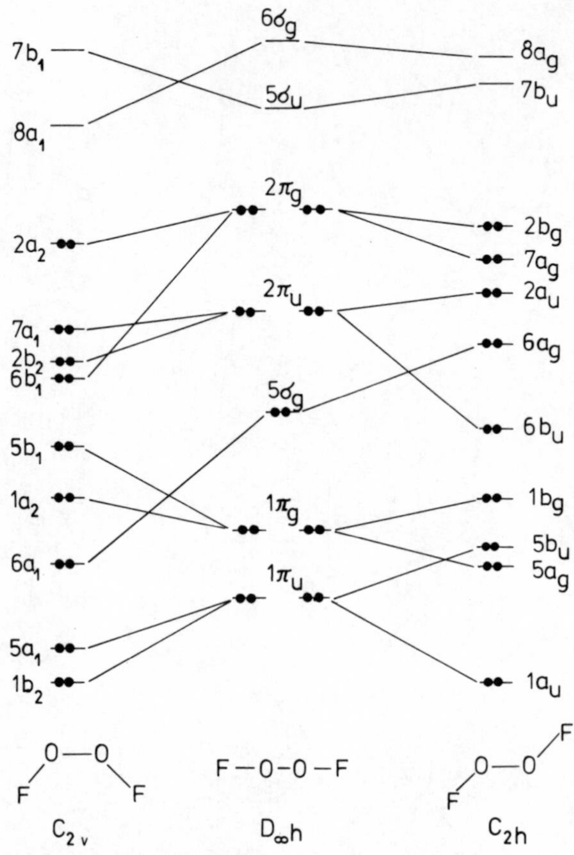

FIGURE 16. Qualitative MO correlations for *cis*, linear and *trans* F_2O_2 based on RHF calculations. Compare with Figure 15.

TABLE 7. Energies and geometries of bent (C_{2v}) and cyclic ozone[a]

Method	Absolute energy (bent)	Relative energy (cyclic)	Geometry: Bent R	Bent α	Cyclic R	Cyclic α	Ref.
HF/DZb	−224.2386	−7.0	1.244	118	1.397	(60)	52
INO CI/DZb	−224.4226	16.1	1.322	115	1.482	62	52
CEPA/DZb	−224.7710	4.6[b]	1.264	117.3	1.435	(60)	56
GVB-CI/DZd	−224.78578[c]	28.1	1.299	116	1.449	(60)	50
HF-CI/DZd	−224.80065	21.0	(1.278)	(116.8)	(1.44)	(60)	55
RSMP/DZd	−225.05309	38.6	1.289	116.8	1.450	(60)	58
Exp.[d]	−225.557	23–28	1.272	117.8	1.45	(60)	70

[a]Absolute energies in hartree, relative energies in kcal mol^{-1}, distances in Å, angles in deg.; values in parentheses are assumed.
[b]In a more recent study, Burton proposes a value of 12 kcal mol^{-1} [56].
[c]Calculated at experimental geometries.
[d]Experimental r_e geometry of ozone (C_{2v}). Absolute energy from Table 14, Section IV.C. Estimates of relative energy from Refs. 60 and 51.

TABLE 8. Biradical character χ of some XO_2 compounds (%)

Molecule	RHF/CI (Ref. 64)	GVB[b] (Ref. 26)	UHF/CI[b] (Ref. 66)	VB[c] (Ref. 67)	MC SCF-CI (Ref. 68)
·OOO·	30[a]	48	55	59	23[a]
·NHOO·				55	
$\cdot CH_2OO\cdot$			42	43	
·OO·	100	100	100	100	100

[a]Calculated from coefficients of $\Psi = 1/\sqrt{2}[C_1\Phi(\ldots(1a_2)^2(2b_1)^0) + C_2\Phi(\ldots(1a_2)^0(2b_1)^2)]$; $\chi = 100C_2^2/(1/\sqrt{2})^2 = 200C_2^2$.
[b]Calculated from overlap S in the highest occupied orbital set: $\chi = [1 - 2S/(1 + S^2)] \times 100$; $S = 0.28$ (GVB) and 0.24 (UHF/CI) for O_3.
[c]Calculated by expanding Ψ_{HF} in terms of VB functions.

TABLE 9. Energies and geometries for the $^2A''$ state of the HO_2 radical[a]

Method	Absolute energy	Geometry: R	R'	α	Ref.
UHF/MBS	−148.1967	1.357	1.004	104.1	74
UHF/DZ	−150.1579	1.384	0.968	106.8	71
UHF/DZ	−150.2360	1.315	0.948	105.7	79
CI/DZ	−150.2448	1.458	0.973	104.6	75
MC SCF-CI/DZdp[b]	−150.2998	1.365	0.995	104.2	78
GVB-CI/DZdp	−150.4271	1.369	0.991	103.3	81
Exp.		1.335	0.977	104.1	82

[a]Energies in hartree, distances in Å, angles in deg.
[b]MCSCF energy given.

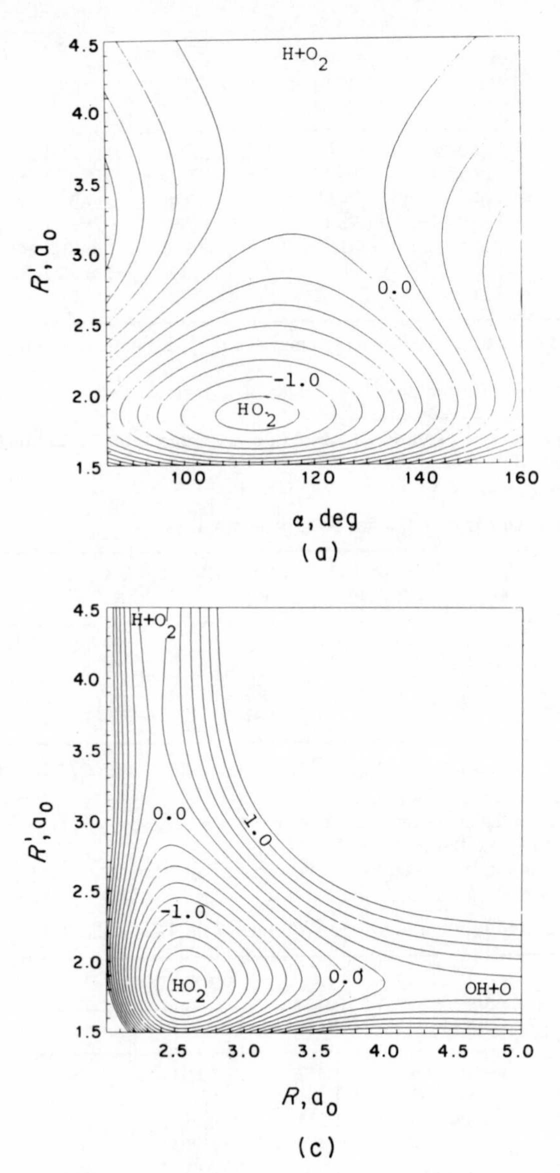

FIGURE 17. Equal potential energy contour plots of the HO_2 potential surface: (a) α versus R' ($R = 1.233$ Å), (b) R versus R' ($\alpha = 120°$), (c) R versus R' ($\alpha = 104°$) and (d) α versus R($R' = 0.979$ Å). The contour spacings are 0.2 eV. The zero-energy contour is taken with respect to the H + O_2

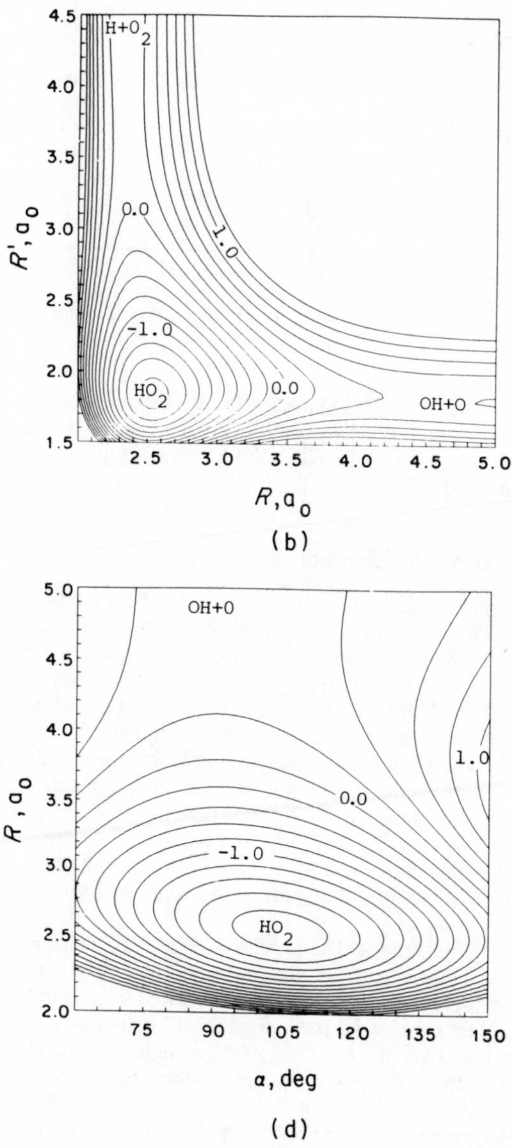

reactants. Contour levels greater than 1.2 eV are not included. $a_0 = 0.52918$ Å. Reproduced by permission of North-Holland Publishing Company from C. F. Melius and R. J. Blint, *Chem. Phys. Letters*, **64**, 183 (1979).

TABLE 10. Theoretical energies and geometries of various H_2O_2 structures[a]

Structure	Form	R	R'	α	Energy
Chain	Bent–bent (**23**)	1.476	0.967	97.2	−151.15613
	Linear (**16**)	1.333	0.941	180.0	155.4
	Linear–bent	1.406	0.945	180.0	70.7
			0.971	99.6	
Y	Planar (**17**)	1.487	0.959	119.8	63.2
	Pyramidal (**19**)	1.521	0.969	100.7	52.8
				108.7	
Bridged	Planar (**18**)	1.720	1.167	42.5	71.4
	Puckered (**20**)[b]	1.662	1.179	45.2	68.3

[a]RSMP/SVdp calculations, Ref. 87. Absolute energy of **23** in hartree, relative energies in kcal mol^{-1}, bond lengths in Å, angles in deg.
[b]The puckering angle δ is 57°. It corresponds to $\tau = 123°$ ($\delta = 180 - \tau$). The puckering amplitude q is 0.40 Å.

employed, the relative energy of **17** ranges from 26–63 kcal mol^{-1}[87]. Form **17** can gain about 10 kcal mol^{-1} by pyramidalization. Since the dipole moment of the pyramidal geometry **19** is rather high (4.3 D, RHF/SVdp), solvation in polar solvents will lead to further stabilization of the Y form.

B. The Conformational Subspace of H_2O_2

One of the benchmark tests in quantum chemistry is the computation of the rotational potential of H_2O_2. The pros and cons of newly developed methods and techniques have been scrutinized by comparing computed and observed barrier data[88–107]. In addition, attempts to explain the origin of the H_2O_2 barriers have revealed merits and limitations of interpretative models[108–113]. Aspects relevant to this work have been discussed in several reviews on the quantum-chemical treatment of internal rotation in molecules[114–117].

Despite the fact that H_2O_2 is the simplest molecule to show internal rotation, it was not until the early seventies that a reasonable account of the rotational barriers could be provided by *ab initio* calculations of the RHF type.

In Figure 18 15 selected H_2O_2 barriers and the corresponding RHF molecular energies are plotted, where the latter may be considered as roughtly reflecting the size of the basis set employed. It is obvious that only with elaborate basis sets are reliable barrier values obtained.

In contrast to the situation for ethane where RHF/MBS calculations performed for a rigid rotor model are satisfactory, *ab initio* calculations of H_2O_2 must fulfill at least two criteria: (1) The basis set employed has to be augmented by polarization functions. (2) All geometrical parameters have to be optimized for all values of τ to be considered.

Cremer has demonstrated that rescaling of the basis set functions during rotation leads to a further improvement of the barrier values[107]. Inclusion of correlation effects into the theoretical approach does not lead to more accurate results[107]. This is in line with purely theoretical considerations by Freed who has shown that correlation effects should contribute little to rotation or inversion barriers[118].

In Table 11 experimental[119–122] and *ab initio* barriers[101,107] are compared. There exists no ambiguity with regard to the stability of the skew form at $\tau = 120°$ and its *trans* barrier. A value of 1.1 kcal mol^{-1} has been widely accepted. With regard to the *cis* barrier, reported barrier data, both experimental and theoretical, are less conclusive (Table 11). Ewig and

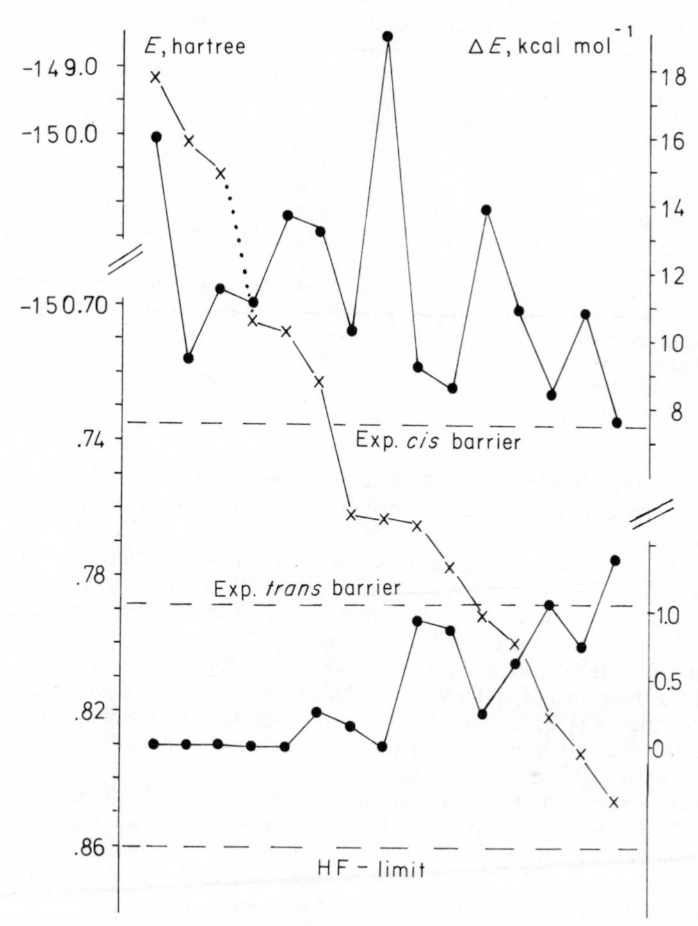

FIGURE 18. Total molecular energies E(crosses: ×) and barrier values ΔE (dots: •) of H_2O_2 according to 15 selected RHF calculations taken from References 101 and 107.

Harris[122] have demonstrated that small changes in the torsional frequencies of H_2O_2 increase the *cis* barrier height from 7.6 to 14.4 kcal mol^{-1}, while the *trans* barrier and the shape of the torsional potential between 140° and 220° remain unchanged. A value of 7.4 kcal mol^{-1}, predicted on the basis of large basis set calculations[107], seems to be the best estimate of the *cis* barrier available at present.

The preference of the skewed conformation has also been observed in gas-phase investigations of substituted peroxides[123–127]. Depending on the size and the electronic features of the group X replacing H, the dihedral angle τ may vary from 90 to 170°. Smaller angles τ are observed for persulphides[128–130].

Various hypotheses have been put forward concerning the origin of the rotational barriers of ethane-like molecules. A critical review, published by Payne and Allen[117], compares no less than 14 distinctly different models for explaining conformational behaviour.

TABLE 11. Comparison of experimental and *ab initio* barriers of H_2O_2

Authors	Year	Reference	ΔE (*trans*) (kcal mol^{-1})	ΔE (*cis*) (kcal mol^{-1})	τ (deg.)
Redington, Olson and Cross[a]	1962	119	0.85	3.71	109.5
Hunt and coworkers[a]	1965	120	1.10	7.03	111.5
Oelfke and Gordy[b]	1969	121	1.1	7.0	120
Ewig and Harris[a]	1969	122	1.10	7.57	112.8
Dunning and Winter[c]	1975	101	1.10	8.35	113.7
Cremer[c]	1978	107	0.94	7.69	119.3[d]

[a] From infrared spectrum.
[b] From millimetre-wave spectrum.
[c] From RHF calculations.
[d] From RSMP calculations.

Despite the appealing character of some of the proposed models, their quantitative verification turns out to be especially difficult in the case of H_2O_2. Rather inconclusive have been attempts to trace the origin of the *cis* and *trans* barriers back to orbital orthogonality or exchange contributions imposed by the Pauli Principle[112,131], interference effects between the weak vicinal 'tails' of OH or lone-pair (n) LMOs[111,132], bond–antibond interactions[133], the dominance of attractive or repulsive energy terms[109,116,134,135] or the prevailing role of special MOs.

In Figure 19, RHF/SVd orbital energies ε_i for the five highest occupied MOs of H_2O_2 are plotted as functions of τ. They reveal that reliable predictions with regard to the relative stabilities of **21**, **22** and **23** cannot be made with the aid of ε_i values. This holds for the two HOMOs discussed in Section III.B.4 as well as for the total orbital energy $2\sum_i^{occ}\varepsilon_i$ as was first shown by Fink and Allen[89].

An elegant way to avoid these difficulties has been pursued by Radom, Hehre and Pople[100]. These authors have expanded the *ab initio* rotational potential of H_2O_2 in form of a truncated Fourier series (equation 1) and have used the constants V_i^c to analyse the *cis* and *trans* barriers. Figure 20 illustrates this procedure. In Table 12, the corresponding V_i^c constants of H_2O_2 are compared with those of some other peroxides.

$$
\begin{aligned}
V(\tau) &= V_1(\tau) + V_2(\tau) + V_3(\tau) \\
&= \tfrac{1}{2}V_1^c(1 - \cos\tau) + \tfrac{1}{2}V_2^c(1 - \cos 2\tau) + \tfrac{1}{2}V_3^c(1 - \cos 3\tau) \qquad (1)
\end{aligned}
$$

The $V_1(\tau)$ term can be considered as indicating repulsive ($V_1^c < 0$, H_2O_2, Table 12) or attractive ($V_1^c > 0$, HOOF, Figure 21) interactions between OH or OX bond dipole moments. The V_2^c term has been connected with the degree of lone-pair (n) delocalization. According to Pople and coworkers[100], there seems to be a general tendency of n orbitals to become coplanar with adjacent polar bonds, thus guaranteeing an overall stabilization of N or O containing rotors in the corresponding conformation[100]. Hence, a negative V_2^c is indicative of maximum n delocalization at $\tau = 90°$. Actually, this description is related to the explanation of the anomeric effect given by Altona and coworkers[137]. Both ways of interpreting $V_2(\tau)$ are illustrated in Figure 21.

Figure 20 as well as Table 12 reveal that the rotational minimum at $\tau = 120°$ results from a delicate balance of $V_1(\tau)$ and $V_2(\tau)$, which clearly dominate the conformational behaviour of H_2O_2 and other peroxides. The V_3^c term is relatively small and negative suggesting a slight preference for staggering of bonds.

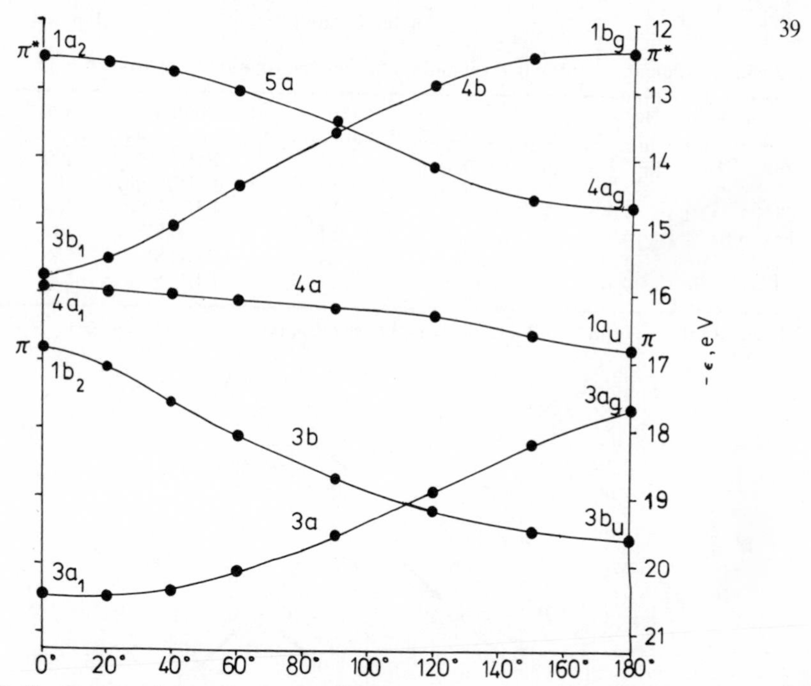

FIGURE 19. Functional dependence of RHF/SVd orbital energies on τ calculated for H_2O_2 (D. Cremer, unpublished results).

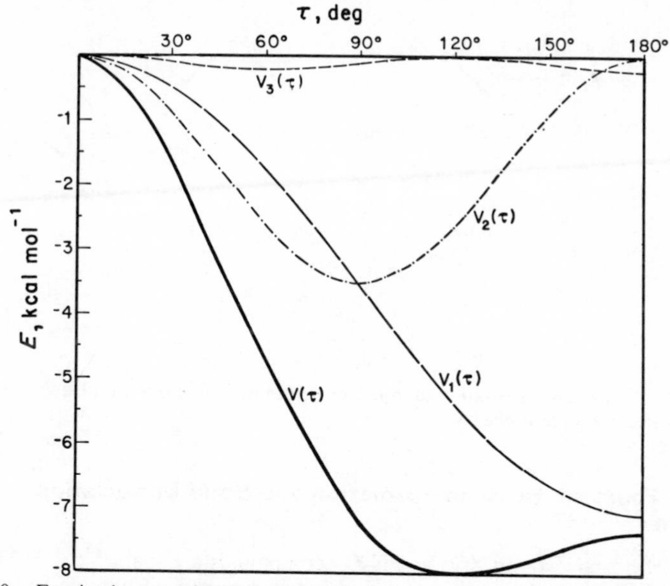

FIGURE 20. Fourier decomposition of potential function $V(\tau)$ for H_2O_2. Adapted from L. Radom, W. J. Hehre and J. A. Pople, *J. Amer. Chem. Soc.*, **94**, 2371 (1972), by permission of the American Chemical Society.

TABLE 12. Potential constants (kcal mol^{-1}) for internal rotation in peroxides[a]

Molecule	V_1^c	V_2^c	V_3^c	Method	Ref.
HO—OH	−7.1	−3.5	−0.2	RHF/SV	100
	−8.0	−3.7	−0.3	RHF/SVdp	114
CH_3O—OH	−7.5	−2.9	−0.4	RHF/SV	100
FO—OH	4.2	−5.2	−0.1	RHF/SV	100
CF_3O—OH	−5.6	−3.9	−0.4	RHF/SV	136
CF_3O—OF	−4.1	−6.1	−0.7	RHF/SV	136

[a]Energy of the *cis* form ($\tau = 0°$) is taken as the reference point.

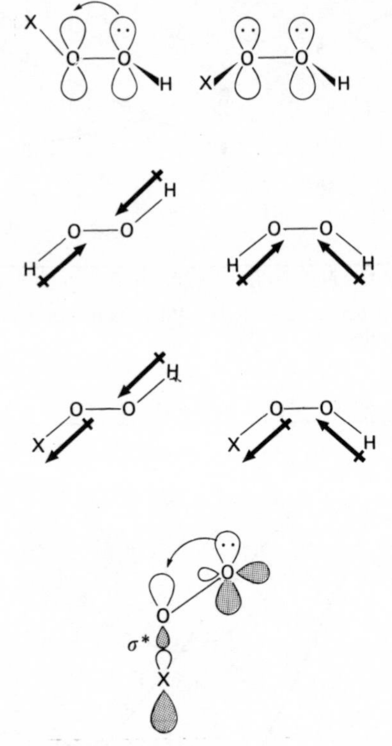

FIGURE 21. Schematic illustration of lone-pair delocalization, interaction between bond dipole moments, and the anomeric effect.

C. Total Energies, Heats of Formation and Bond Dissociation Enthalpies

Thermochemical data on XO_2 and X_2O_2 compounds are sparse[138–141]. That is why Benson and Shaw in their review on the thermochemistry of organic per- and polyoxides[142] have dwelt on empirical methods to estimate heats of formation $\Delta H_f^0(298)$ and bond dissociation enthalpies $DH^0(298)$. Subsequent work of Benson and coworkers[143–148], based on group additivity principles, has led to an improvement and extension of ΔH_f^0 estimates for polyoxides and polyoxide radicals. Some of these values are listed in Table 13[143,148].

A theoretical determination of ΔH_f^0 is only possible with recourse to an appropriate reference state and its experimental ΔH_f^0 value (equation 2). By calculating molecular (MOL) and reference state (REF) energies and using known $\Delta H_f^0\,(0)^{REF}$ data, $\Delta H_f^0\,(0)^{MOL}$ can be determined from equation (2). The crucial point is the evaluation of the 'experimental' energy E(EXP)[149]. As is illustrated in Figure 22, this requires the knowledge of (*a*) HF limit energies E(HF), (*b*) their correlation corrections E(CORR) to obtain Schrödinger energies E(S), (*c*) relativistic corrections E(REL) to obtain true theoretical energies E(THEO) and (*d*) vibrational corrections E(VIB), which primarily comprise zero-point vibrational energies *ZPE*.

$$E(\mathrm{EXP})^{\mathrm{MOL}} - \sum^{k} E(\mathrm{EXP})^{\mathrm{REF}} = \Delta H_f^0\,(0)^{\mathrm{MOL}} - \sum^{k} \Delta H_f^0\,(0)^{\mathrm{REF}} \quad (2)$$

For polyatomic molecules none of these energies can be accurately determined by theory. However, it is possible to obtain estimates of the molecular energies E(HF), E(S), E(THEO) and E(EXP) if *ab initio* and experimental data are combined[149]. In Table 14 E(EXP) values as well as some other characteristic molecular energies, obtained in this way, are given for O_2, H_2O_2, O_3, H_2O_3, MeO_2H and MeO_2Me [149]. The theoretical estimates lead to ΔH_f^0 values, generally not more accurate than ± 5 kcal mol^{-1}. This is also true when differences ΔE(EXP) are approximated by computed SCF energies E(X) (X: SV or DZ basis)[150,151,153] or estimates of E(HF) [152] using closed-shell molecules (H_2, H_2O, H_2O_2, XH_n, XOH, etc.) or ions (O_2^{2+}) as reference states[150–153].

Due to the relatively large uncertainties of theoretical ΔH_f^0 values, the data of Table 13 are used to discuss dissociation enthalpies DH^0 of peroxo compounds. A cleavage of the O—O bond of H_2O_2 requires[142]:

$$DH^0(\mathrm{HO{-}OH}) = 2\,\Delta H_f^0\,(\mathrm{HO\cdot}) - \Delta H_f^0\,(\mathrm{H_2O_2}) = 51\ \mathrm{kcal\,mol^{-1}}$$

TABLE 13. Estimated ΔH_f^0 (298) values (kcal mol^{-1}) for polyoxides and polyoxide radicals[143,148 b]

	X¹, X²				
	H, H	Me, Me	*t*-Bu, *t*-Bu	Me, H	*t*-Bu, H
Polyoxides					
$X^1O_2X^2$	−32.5[a]	−30.0[a]	−83.4[a]	−31.3	−58.0[a]
$X^1O_3X^2$	−15.7	−13.2	−66.6	−14.5	−41.2
$X^1O_4X^2$	1.1	3.6	−49.8	2.3	−24.4
$X^1O_5X^2$	17.9	20.4	−33.0	19.1	−7.6
Polyoxide Radicals					
$X^1O\cdot$	9.4[a]	3.9	−22.8		
$X^1O_2\cdot$	3.0	6.2	−22.5		
$X^1O_3\cdot$	17.8	23.0	−7.7		
$X^1O_4\cdot$	32.6	39.8	7.1		
$X^1O_5\cdot$	47.4	56.0	21.9		

[a]Experimental values.
[b]Adapted with permission from P. S. Nangia and S. W. Benson, *J. Phys. Chem.*, **83**, 1138 (1979).

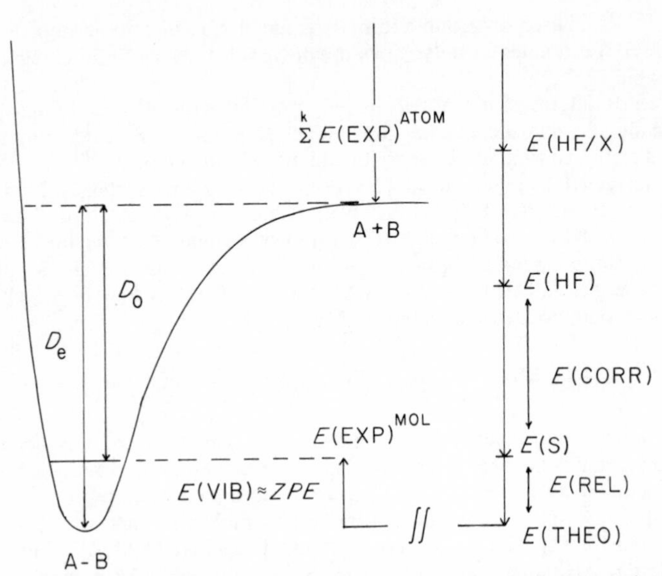

FIGURE 22. Theoretical determination of dissociation energies D_e and D_0 for a molecule AB. (See text for an explanation of the various energies.) Note that different energy scales are used on the right- and left-hand sides of the drawing.

TABLE 14. Theoretical energies (hartree) and heats of formation (kcal mol^{-1}) of some compounds containing the OO moiety[a]

Energy	O_2	H_2O_2	MeOOH	MeOOMe	O_3	H_2O_3
E(HF)	−149.670	−150.860	−189.901	−228.956	−224.391	−225.678
E(CORR)	−0.647	−0.693	−0.946	−1.199	−1.032	−1.024
E(REL)	−0.100	−0.100	−0.128	−0.150	−0.150	−0.150
E(THEO)	−150.417	−151.653	−190.961	−230.283	−222.573	−226.852
ZPE	0.004	0.025	0.054	0.083	0.008	0.030
E(EXP)	−150.408	−151.621	−190.898	−230.188	−225.557	−226.814
$\Delta H_f^0(0)$	0	−31.1	−23.1	−23.2	34.8	−24.2
ΔH_f^0 (298)	0	−32.6	−26.5	−28.5	34.1	−26.5

[a]Taken from Ref. 149. For O_2, H_2O_2 and O_3 experimental ΔH_f^0 values have been used to obtain correlation energy increments upon which the estimation of *E*(CORR) energies of higher peroxides is based. *ZPE* values have to be enlarged by corrections for the nuclear motion relative to the centre of mass when calculating *E*(VIB); $ZPE = \frac{1}{2} h N_A \sum_i^{3k-6} \nu_i$, where *k* is the number of atoms and the ν_i are the experimental frequencies. Reference states are the atoms H, O and C.

and that of the O—H bond[143]:

$$DH^0(HO_2{-}H) = \Delta H_f^0\,(HO_2\cdot) + \Delta H_f^0\,(H\cdot) - \Delta H_f^0\,(H_2O_2)$$
$$= 88\ \text{kcal mol}^{-1}.$$

Thus, DH^0 (O—O) nicely fits into the series of DH^0 (X—X) values of isoelectronic X_2H_{2n} molecules as can be seen from Table 15 (compare also with Table 2, Section I).

TABLE 15. Dissociation enthalpies DH^0 (kcal mol^{-1}) of isoelectronic X_2H_{2n} molecules

Molecule	DH^0(X—X)	DH^0(X—H)	Reference
C_2H_6	88	99	141
N_2H_4	69	104[a]	141
O_2H_2	51	88	142
F_2	38	—	141

[a]Estimated value.

It is interesting to compare O—O, C—O and O—H dissociation enthalpies for organic polyoxides. According to the data listed in Table 16, the O—O bond is considerably weakened if H is replaced by Me or an additional O atom is inserted into the O—O linkage. Delocalization of the 3π electrons of a peroxyl radical (see Section III.A.2) strengthens the O—O bond by about 15 kcal mol^{-1}. The same effect makes the O—H bond of H_2O_2 more susceptible to bond rupture than that of an alcohol. This holds also for C—O bonds (Table 16).

The π delocalization energy of O_3 can be estimated by the formal reaction[145]

$$\text{HOOOH} \rightarrow \cdot\text{OOO}\cdot + 2\,\text{H}\cdot$$

where the unpaired electrons of the ·OOO· fragment are thought not to interact and the enthalpy change is taken as twice the DH^0 (O—H) value of H_2O_2. Thus, a ΔH_f^0 value for the hypothetical ·OOO· species with 100% biradical character (i.e. when the overlap (S) equals zero, see Table 8, Section IV.A) can be evaluated and compared with the experimental heat of formation of ozone. Depending on the value of ΔH_f^0 (H_2O_3), the 4π delocalization energy has been predicted to be 17–19 kcal mol^{-1} [69,145]. Since the actual dissociation enthalpy of O_3 is 25 kcal mol^{-1} (Table 16), only 6–8 kcal mol^{-1} can be assigned to the O—O σ bond.

A heterolytic cleavage of the O—O bond

$$X^1O{-}OX^2 \rightarrow X^1O^+ + {}^-OX^2$$

requires a considerably higher amount of energy if X^1 and X^2 cannot stabilize the emerging ions. The reaction energy ΔE can be estimated utilizing the dissociation enthalpy for homolytic cleavage (equation 3)[144]. From experimental ionization potentials I and

TABLE 16. Dissociation enthalpies DH^0 (kcal mol^{-1}) of molecules containing O—O, C—O and O—H bonds[141,143,148,154]

Molecule	DH^0(O—H)	Molecule	DH^0(O—O)	Molecule	DH^0(C—O)
H—O·	102	O=O	119	CH_3—OH	91
H—OH	119	HO—O·	66	CH_3—O_2·	28
H—OCH_3	104	CH_3O—O·	59	CH_3—O_2H	70
H—O_2·	49	HO—OH	51	HCO—OH	107
H—O_2H	88	CH_3O—OH	45	HCO—O_2H	86
H—O_2CH_3	90	CH_3O—OCH_3	38		
H—O_3·	68	$O{=}\overset{+}{O}{-}\overset{-}{O}$	25		
		HO_2—OH	30		
		CH_3O_2—OCH_3	23		

electron affinities EA[146,147], the energy of heterolytic O—O cleavage in the gas phase is predicted to be 5–8 times larger than homolytic cleavage (Table 17).

$$\Delta E \approx DH^0(X^1O{-}OX^2) + I(X^1O\cdot) + EA(X^2O\cdot) \quad (3)$$

However, heterolytic cleavage needs less energy if (*a*) an ion pair is formed at a distance r_{ip} separating the effective charge centres, and (*b*) formation of the ion pair occurs in solution. Then the energy of heterolytic cleavage is given by equation (4). Benson and coworkers[143,147] have estimated r_{ip} to be 2.65 ± 0.05 Å. This leads to a Coulomb attraction energy of 124 ± 2 kcal mol^{-1}. Accordingly, a dialkyl trioxide can undergo heterolytic cleavage, provided the solvation energy ΔE_{solv} of the ion pair compensates for a difference of about 20 kcal mol^{-1}. The energy ΔE_{solv} can be approximated by Kirkwood's formula[155] (equation 5), where ε is the dielectric constant of the solvent, μ the dipole moment ($\mu = 2.65 \times 4.8 = 12.72$ D) and a the radius of a spherical cavity formed by solvent molecules surrounding the ion pair. With $a = 3.5$ Å [147] the solvation energy of a typical hydrocarbon solvent ($\varepsilon = 2$) is predicted to be 11 kcal mol^{-1}. This energy will increase to 20 kcal mol^{-1} if a solvent with $\varepsilon = 5.2$ is used.

$$\Delta E_{ip}(\text{solvent}) = \Delta E - e^2/r_{ip} - \Delta E_{solv} \quad (4)$$

$$\Delta E_{solv} \approx 14.39 \cdot \frac{\varepsilon - 1}{2\varepsilon + 1} \cdot \frac{\mu^2}{a^3} \quad (5)$$

D. Orbital Energies and Ionization Potentials

According to Koopmans' theorem[156] the values $-\varepsilon_i$ of UHF orbital energies provide reasonable approximations to vertical ionization potentials (*IP*s), I_{vert}. As can be seen from Table 18, magnitudes of the energies ε_i obtained with UHF theory for O_2 [157] are of the same order as the experimental *IP*s measured with ESCA [158–160]. An exception occurs in the case of the $3\sigma_g$ and $1\pi_u$ MOs where the experimentally observed order is reversed. This failure of UHF theory results from the neglect of (*a*) Coulomb correlation of electrons and (*b*) MO relaxation effects upon ionization.

The correct MO sequence of O_2 has been obtained by the ΔE_{SCF} approach, i.e. by separately calculating the GS of O_2 and the ground and excited 'hole states' of O_2^+ listed in Table 18[161]. In this way relaxation effects are accounted for. Electron propagator calculations, which consider in addition correlation effects, provide the best agreement between experiment and theory (Table 18)[157].

A similar discrepancy between Koopmans' values and experimental *IP*s has been observed in the case of ozone. The experimental PE spectrum[162–164] reveals considerable

TABLE 17. Energy for heterolytic O—O cleavage estimated according to equation (3)[a]

Products	I(eV)	EA(eV)	ΔE(kcal mol^{-1})
$HO^+ + {}^-OH$	13.2	1.83	313
$CH_3O^+ + {}^-OH$	8.3	1.83	194
$CH_3O^+ + {}^-OCH_3$	8.3	1.57	193
$HO^+ + {}^-O_2H$	13.2	1.85	292
$HO_2{}^+ + {}^-OH$	11.5	1.83	253
$CH_3O_2{}^+ + {}^-OCH_3$	6.75	1.57	142

[a] *I* and *E* values from Refs. 146 and 147.

TABLE 18. Theoretical and experimental ionization potentials I_{vert} (eV) of O_2

Spin orbital	Ion state	UHF[a,d]	ΔE_{SCF}[b,e]	Electron propagator[a,e]	Expt.[c]
$1\pi_g\alpha$	$^2\Pi_g$	15.3	13.1	11.8	12.1
$1\pi_u\beta$	$^4\Pi_u$	15.8	14.3	17.0	16.1
$1\pi_u\alpha$	$^2\Pi_u$	22.8	15.6	17.4	17.0
$3\sigma_g\beta$	$^4\Sigma_g^-$	19.3	17.3	18.0	18.2
$3\sigma_g\alpha$	$^2\Sigma_g^-$	20.8	21.0	19.5	20.3
$2\sigma_u\beta$	$^4\Sigma_u^-$	27.5	26.0	24.1	24.6
$2\sigma_u\alpha$	$^2\Sigma_u^-$	33.0	33.5	26.7	27.9
$2\sigma_g\beta$	$^4\Sigma_g^-$	43.3	41.0	39.0	39.6
$2\sigma_g\alpha$	$^2\Sigma_g^-$	46.6	45.9	40.2	41.6
$1\sigma_u\beta$	$^4\Sigma_u^-$	563.4	554.4	542.7	543.1
$1\sigma_g\beta$	$^4\Sigma_g^-$	563.5		542.8	
$1\sigma_u\alpha$	$^2\Sigma_u^-$	564.9	556.6	544.5	544.2
$1\sigma_g\alpha$	$^2\Sigma_g^-$	565.0		544.5	

[a]Ref. 157.
[b]Ref. 161.
[c]Ref. 158 (above 28 eV) and Ref. 159 (below 28 eV).
[d]Koopmans' values.
[e]From calculation of the ion states.

vibrational structure for the first ionization. This is not consistent with depopulation from the nonbonding $1a_2$ MO as suggested by HF calculations (see Table 19). Investigations, which go beyond the HF level of theory, agree on the assignment of the first *IP* as resulting from $6a_1$ ionization[49,53,62,165–167]. In addition, they provide sufficient evidence for the MO sequence $6a_1$, $4b_2$, $1a_2$, $1b_1$. This has been taken into consideration when summarizing the relevant data in Table 19 (see also Figure 11).

Available data on the *IP*s of the HO_2 radical are sparse. Foner and Hudson[85] deduced a preliminary value for the first *IP*, which has been confirmed theoretically by Shih and coworkers (Table 20)[77]. It corresponds to ionization of a 7a′ electron resulting in a triplet state, namely the $^3A''$ GS of the HO_2^+ ion[74]. Shih and coworkers have tentatively assigned the second *IP* to 2a″ ionization. Again, Koopmans' values lead to a different order of *IP*s[77].

TABLE 19. Theoretical and experimental ionization potentials I_{vert} (eV) of O_3

MO	Ion state	RHF[a] (Ref. 59)	MBPT[b] (Ref. 166)	CI[b] (Ref. 53)	GVB[b] (Ref. 49)	Expt. (Refs. 162–164)
$6a_1$	$^2A_1$	15.2	12.9	12.5	12.9	12.75
$4b_2$	$^2B_2$	15.5	13.3	12.6	13.0	13.02
$1a_2$	$^2A_2$	13.3	13.2	13.0	13.6	13.57
$1b_1$	$^2B_1$	21.1				
$3b_2$	$^2B_2$	21.7				20.1
$5a_1$	$^2A_1$	22.6				
$4a_1$	$^2A_1$	30.0				

[a]Koopmans' values.
[b]From calculations of the ion states.

TABLE 20. Ionization potentials I_{vert} (eV) of the HO_2 radical

MO	Ion state	SCF[a] (Ref. 77)	MRD-CI[a] (Ref. 77)	Exp. (Refs. 85, 168)
7a′	$^3A''$[b]	10.9	11.6	11.5
2a″	$^1A'$	12.9	12.3	12.2
7a′	$^1A''$	12.2	12.6	

[a]From calculations of the ion states.
[b]Note that a 'synthesis' of HO_2^+ from O_2 and H^+ also leads to a triplet GS:

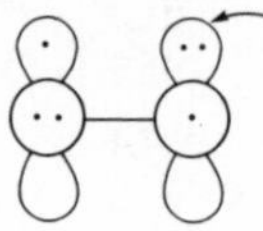

$+ \ H^+$

PE spectra of H_2O_2 are obscured due to decomposition of the sample to H_2O and O_2 in the electron source[169–171]. Because of the contaminants, the uncertainty of the experimental *IP*s is rather large (0.2–1 eV)[171]. The Koopmans' values obtained with augmented basis sets[107] are 8–10% larger than the observed *IP*s (Table 21), which is in accordance with Robin's 8% rule[172].

The first band of the PE spectrum of H_2O_2 reveals some vibrational fine structure ($\nu'' = 1050\,cm^{-1}$). Brown[170] has argued that ionization from an O—O antibonding MO should lead to a strengthening of the bond and, hence, to an increase of the O—O stretching frequency ν_3 of H_2O_2 ($\nu_3 = 863\,cm^{-1}$, see Section IV.E). Comparison with Figure 14 shows that both the 4b and 5a MO possess O—O antibonding character. With the auxiliary information about τ being larger than 90° and, hence, 5a below 4b (Figure 19), an assignment of the first and second *IP* to the 4b and 5a MO is straightforward.

The third *IP* resulting from ionization of the O – O bonding 4a electron (Figure 14) also exhibits a vibrational progression ($\nu'' = 1100\,cm^{-1}$). Arguments have been given, which connect this progression with an excitation of the symmetric OOH bending vibration ν_2 ($1393\,cm^{-1}$)[171]. If correct, the same reasoning, of course, could apply to the first PE band.

TABLE 21. Theoretical and experimental ionization potentials I_{vert} (eV) of H_2O_2

MO	$-\varepsilon_i$[a] RHF/SVd	$-\varepsilon_i$[b] RHF/SVd	Exp. I_{vert}	
			Ref. 169	Ref. 171
4b	12.9	11.9	11.7	11.7
5a	14.1	13.0	12.7	13.0
4a	16.3	15.0	15.3	15.4
3a	18.9	17.4	17.4	17.5
3b	19.2	17.7	17.4	18.5
2b	32.9	30.3		
2a	39.8	36.6		
1b 1a	561.5	516.6		

[a]Koopmans' values obtained with an augmented SV basis. See Section IV.B, Figure 19.
[b]Corrected according to Robins 8% rule, Ref. 172.

Assignment of 3a and 3b ionization on the basis of theoretical Koopmans' values depends very much on the use of the correct τ value ($\tau = 120°$) of skewed H_2O_2. Figure 19 reveals that the 3a and 3b MOs cross at $\tau \approx 116°$. Hence, any calculation with $\tau < 116°$ [173] leads to a wrong assignment.

For a series of organic peroxides PE spectra have been recorded[170,171,174–177]. The spacing of the first two *IP*s has been used to determine the conformation of a peroxide. According to Figure 19 the splitting $\Delta\varepsilon_1 = \varepsilon(5a) - \varepsilon(4b) = -\Delta I_1$ varies with τ, which can be described analytically by a truncated Fourier expansion (Figure 23) (equation 6). Rademacher and Elling[176] have calibrated equation (6) experimentally by utilizing known τ values of organic peroxides in conjunction with PE measurements. The function $\Delta I_1(\tau)$ thus obtained is depicted in Figure 23. It has been used to estimate the dihedral angle τ of cyclic peroxides (Table 22), for which $\tau < 90°$ and, hence, the sign of ΔI_1 is known[176]. Since the magnitude of both I_{vert} and ΔI_1 is influenced by the substituents attached to the peroxo group (compare with Figure 23), τ values determined with equation (6) on the basis of PE investigations are rather inaccurate. This becomes obvious when applying the analytic form of $\Delta I_1(\tau)$ given in Reference 176 to alicyclic organic peroxides ($\Delta I_1 = 0.45$ eV for ROOH; R = pentyl, hexyl or heptyl[171]).

$$\Delta I_1(\tau) = A\cos\tau + B\cos 2\tau + C \tag{6}$$

E. Geometry and Vibrational Analysis

Pertinent to a discussion of the O—O bond strength measured by the depth of the potential function (see, for example, Figure 22) is the analysis of the O—O bond length (location of the potential minimum) and its stretching frequency and force constant (width

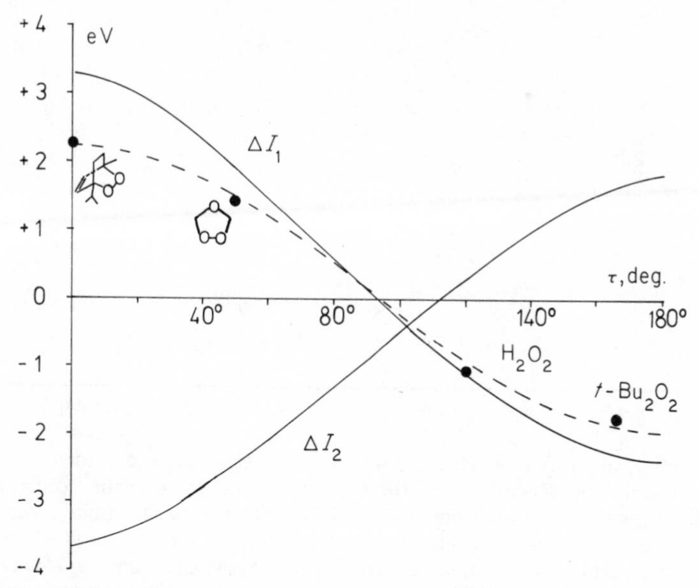

FIGURE 23. Functional dependence of the spacings $\Delta I_1 = I(4b) - I(5a)$ and $\Delta I_2 = I(3b) - I(3a)$ on the dihedral angle τ of H_2O_2 as obtained with RHF/SVd calculations [$\Delta I_1(\tau)$: $A = 2.7$, $B = 0.2$, $C = 0.3$ eV]. Experimental measurements of ΔI_1 are denoted by dots. They lead to $\Delta I_1(\tau) = 2.08\cos\tau + 0.15$ (dashed line)[176].

TABLE 22. Determination of the dihedral angle τ from measured ionization potentials I_{vert} [176]

Molecule[a]	I(b) (eV)	I(a) (eV)	ΔI (eV)	τ (deg.)	Ref.
HO—OH*	11.51	12.56	−1.05	120	170
MeO—OMe	9.71	11.61	−1.90	170	177
t-BuO—OBu-*t**	8.83	10.57	−1.74	166	176
O—O	10.94	8.98	1.96	30	170
O—O	11.13	9.86	1.27	57	175
O—O	10.40	9.25	1.15	61	170, 174
O, O—O *	12.4	10.96	1.44	50	175
O—O	10.35	10.17	0.18	89	176
O—O	9.76	9.35	0.31	86	170, 174
O, O *	10.71	8.42	2.29	10	170
O, O	10.36	8.50	1.86	35	170

[a]Starred molecules have been used for determining the functional dependence $\Delta I(\tau)$.

of the potential function). In Table 23 some relevant data for O_2 and its ions[36,151,178–182] are summarized. The dependence of these properties on the electron configuration is evident. Depopulation of the antibonding π_g MOs of O_2 increases the bond order (Section III.B.1, Figure 7) and bond strength. The strengthening of the OO bond is reflected by lower R_e and higher ν_e and D_e values. Conversely, if the partially vacant π_g MOs of O_2 are filled, thus lowering the bond order to 1.5 (superoxide ion) and 1 (peroxide ion), the OO distance increases, while stretching frequency and dissociation energy decrease.

The $\mathbf{r}_e$ and ν parameter of ozone[70,183–189] suggest that its OO bonds resemble more O_2 than O_2^{2-} despite its low dissociation energy[188] (Table 24). Even its anion, O_3^-,

TABLE 23. Electronic configuration and properties of O_2 and its ions

Molecule	Configuration	State	Bond order P^a	R_e (Å)	ν_e (cm^{-1})	D_e (eV)	T_e (eV)	Ref.
O_2^{2+}	$\ldots(1\pi_g)^0(3\sigma_u)^0$	$^1\Sigma_g^+$	3	1.034^b				151
O_2^+	$\ldots(1\pi_g)^1(3\sigma_u)^0$	$^2\Pi_g$	2.5	1.123	1876.4	6.55		36
O_2	$\ldots(1\pi_g)^2(3\sigma_u)^0$	$^3\Sigma_g^-$	2	1.207	1580.2	5.21	0	36
		$^1\Delta_g$	2	1.216	1509.3	4.23	0.98	36
		$^1\Sigma_g^+$	2	1.227	1432.7	3.58	1.63	36
O_2^-	$\ldots(1\pi_g)^3(3\sigma_u)^0$	$^2\Pi_g$	1.5	1.341	1089	4.09^c		178, 179
O_2^{2-}	$\ldots(1\pi_g)^4(3\sigma_u)^0$	$^1\Sigma_g^+$	1	1.50^d	848^e			180, 182
$(O^{2-})_2$	$\ldots(1\pi_g)^4(3\sigma_u)^2$	$^1\Sigma_g^+$	0	Large		0		

$^a P = P_b - P_a$; P_b, P_a: number of electron pairs occupying bonding or antibonding MOs.
^{b}RHF/DZ value; see also Ref. 181 for RHF/SV calculations.
$^c D_0$ value.
^{d}Distance observed for alkali peroxides, Ref. 180.
^{e}Average of the A_g vibrational frequencies computed for Li_2O_2, Ref. 182.

TABLE 24. Electron configuration and molecular properties of ozone and some of its ions

Molecule	Configuration	State	Bond order P^a	R_e (Å)	α_e (deg.)	ν (cm^{-1})	D_0 (eV)	Ref.
O_3^+	$\ldots(6a_1)^1(2b_1)^0$	$^2A_1$	1.72	1.26^b	131.7		1.85	185, 189
O_3	$\ldots(6a_1)^2(2b_1)^0$	$^1A_1$	1.66	1.272	117.8	1103	1.05^c	70, 184
O_3^-	$\ldots(6a_1)^2(2b_1)^1$	$^2B_1$	1.22	1.38 (2)	116 (2)	982	1.39^d	186, 188
							2.41^e	
				1.35^b	114.1			189

^{a}Evaluated with the aid of Pauling's bond-order relationship[25]: $R_e = (a - b)\ln P$, where $a = 1.452$ Å [$R_e(H_2O_2)$ for $P = 1$] and $b = -0.353$ Å from $R_e = 1.207$ Å for $P(O_2) = 2$.
^{b}Theoretical value of Ref. 189, corrected with the aid of exp. and theoret. ozone parameters.
^{c}Hiller and Vestal[188] suggest a value of ≤ 0.75 eV on the basis of photodissociation measurements on O_3^-.
$^d D_0(O_2{-}O^-)$.
$^e D_0(O{-}O_2^-)$.

possesses an OO distance and stretching frequency[186] closer to the superoxide than the peroxide ion. Noteworthy is the reduction of the angle α in the series O_3^+, O_3, O_3^-, which is parallel to the stepwise occupation of 1,3 bonding MOs (Figure 10).

A first direct measurement of R for H_2O_2 was obtained by Giguère and Shomaker as early as 1943[190]. Subsequent work on the $\mathbf{r}_0$ structure of H_2O_2 by Redington, Olson and Cross[119] led to a complete set of geometrical data. However, the experimental determination of a $\mathbf{r}_0$ structure of H_2O_2 has to cope with the dilemma of extracting four internal parameters out of three rotational B_0 constants, since no accurate spectroscopic data on D_2O_2 are available. This problem has been solved by assuming an O—H length. Recently, convincing evidence has been gathered from neutron diffraction results of H_2O_2 [191,192], D_2O_2 [193] and D_2O [194], from the microwave study of HOF[195] and an elaborate *ab initio* study[107], which suggest $R_e(OH) \approx R_0(OH) \approx 0.965$ Å. With this parameter the $\mathbf{r}_0$ structure of H_2O_2 published by Redington and coworkers[119] has been revised[107,196,197]. In addition, $\mathbf{r}_e$ parameters have been derived utilizing published vibrational–rotational constants[196]. In Table 25, the $\mathbf{r}_0$ and $\mathbf{r}_e$ geometries are compared. Both theory and experiment support a R_e value of 1.452 Å for H_2O_2. The corresponding difference $R_0 - R_e$ obtained by Cremer and Christen[197] is rather large, which has been criticized by Giguère and Srinivasan[198].

Out of the wealth of computed *ab initio* geometries of H_2O_2 a rather confusing picture emerges as is illustrated in Figure 24. Some theoretical R_e distances cluster around 1.39–1.40 Å, while the others are scattered between 1.40 and 1.56 Å, i.e. most *ab initio* distances are clearly outside the range of error of spectroscopic H_2O_2 geometries. This indicates that the calculation of the O—O bond length of H_2O_2 is very sensitive to basis set and correlation errors of the HF approach. HF calculations with extended basis sets

TABLE 25. Geometrical parameters of H_2O_2 as determined by experiment and theory

Parameter	$\mathbf{r}_0$ [a] IR (Ref. 197)	$\mathbf{r}_e$ [b] IR, MW (Ref. 196)	$\mathbf{r}_e$ RHF/SV (Ref. 107)	$\mathbf{r}_e$ RSMP/DZdp (Ref. 107)
R'(Å)	0.965[c]	0.965[c]	0.965	0.967
R(Å)	1.464	1.452	1.460	1.451
α (deg.)	99.4	100.0	102.3	99.3
τ (deg.)	120.2	119.1	120.0[c]	119.3

[a] Reinterpretation of infrared data of Ref. 119.
[b] Deduced from infrared and microwave data of Refs. 119 and 121.
[c] Assumed value.

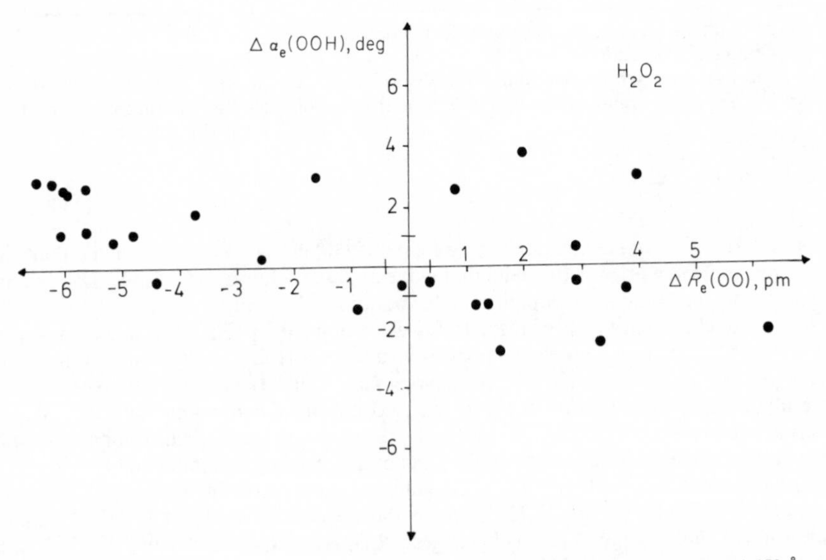

FIGURE 24. Deviation of *ab initio* geometries for skewed H_2O_2. Origin at $R_e = 1.452$ Å and $\alpha_e = 100°$ [196]. ΔR_e measured in picometre. Uncertainty of experimental $\mathbf{r}_e$ geometry is indicated.

severely underestimate R_e. This is the result of an artificial accumulation of electron charge close to the O nuclei, which increases the stabilizing Coulomb interactions between electrons and nuclei. The latter are shielded by the surrounding electron charge. Coulomb repulsion between the O nuclei is considerably reduced, which explains the short O—O bond lengths. As soon as electron correlation is considered, accumulation of charge in one area is no longer possible. Removal of electron charge from the inner to the outer valence sphere of the O nuclei causes a lengthening of the theoretical O—O bond towards the true R_e value[107].

Because of a distinctively different description of the inner and outer valence spheres of O and the polarity of the OH bond with MBS, SV, DZ or augmented DZ basis sets[107,199], theoretical R_e values of H_2O_2 depend strongly on the size of the basis set. This dependence is qualitatively described in Figure 25. It is responsible for the scattering of *ab initio* values of R_e reflected by Figure 24. In addition, it indicates that calculations carried out with relatively small basis sets can lead to reasonable R_e and R_o values due to a fortuitous cancellation of basis set and correlation errors[107].

Crystallographic data on the O—O bond length in H_2O_2 [191,192], D_2O_2 [193] and perhydrates[200–206] vary between 1.44 and 1.47 Å. It has been noted[200] that R of H_2O_2 molecules in solids is generally smaller than for peroxides due to the presence of water. Correcting for this effect and the thermal motion in the crystal, Pedersen has predicted the average R value of H_2O_2 to be 1.456 Å [200], which is in line with $R_e = 1.452$ Å.

The six normal modes of vibrational motion of H_2O_2 are sketched in Figure 26. Harmonized frequencies have been published by Khachkuruzov and Przhevalskii[207], who examined the available spectroscopic data of H_2O_2 and D_2O_2. Recent Raman measurements of H_2O_2 vapour[208] provide evidence for an O—O stretching frequency, significantly different from liquid- or solid-phase values (Table 26). However, these

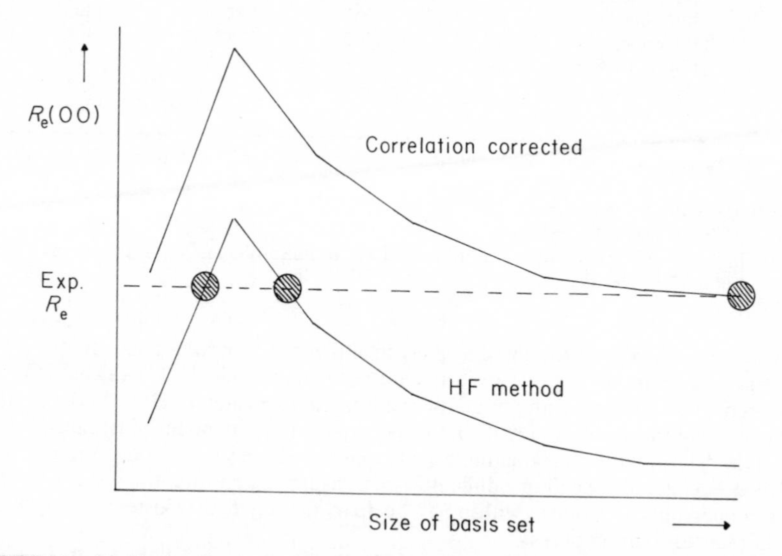

FIGURE 25. Qualitative illustration of the dependence of the theoretical O—O bond length R_e on the size of the basis set and the method. Two 'Pauling points' found for HF/small-basis-set calculations are indicated.

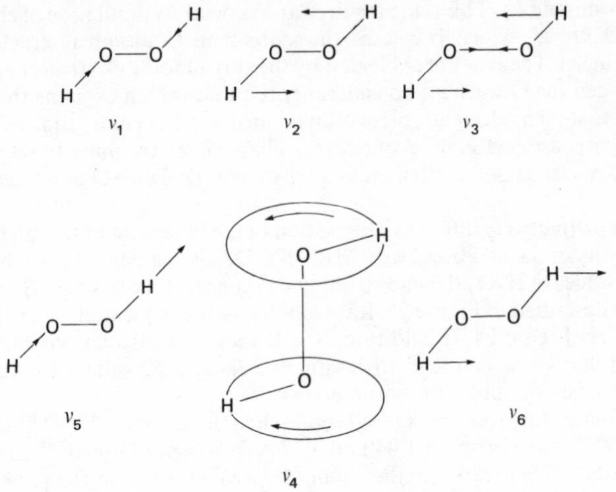

FIGURE 26. Normal modes of vibrational motion for H_2O_2.

TABLE 26. Vibrational frequencies (cm^{-1}) of HO_2 and H_2O_2

Frequency	Character	HO_2 Exp.[a]	HO_2 UHF[b]	HO_2 GVB-CI[c]	H_2O_2 Exp.[d]
ν_1	OH stretch	3414	3488	3655	3607
ν_2	OOH bend	1389	1357	1416	1393
ν_3	OO stretch	1101	1083	1181	863
ν_4	HOOH torsion				317
ν_5[e]	OH stretch				3608
ν_6[e]	OOH bend				1266

[a]From matrix isolation studies, Ref. 209. Recently, $\nu_3 = 1097\ cm^{-1}$ has been found in the gas phase[210].
[b]Scaled UHF/DZdp calculations, Ref. 80.
[c]GVB-CI/DZdp values from Ref. 81.
[d]Ref. 208. In the infrared spectra of liquid and solid H_2O_2 a value of 880 cm^{-1} has been observed for ν_3.
[e]Asymmetric modes of H_2O_2 (Figure 26).

differences are still within the margin of error of theoretical ν values as is revealed by a comparison of *ab initio*[80,81] and experimental HO_2 frequencies[209,210] (Table 26).

Theoretical attempts at evaluating absolute infrared intensities of HO_2 (via calculation of derivatives of the dipole moment with respect to the normal modes of vibration)[80] and H_2O_2 (via determination of a suitable hydrogen atomic tensor)[211] have recently been published. Because of the extreme difficulties of measuring these quantities, the theoretical data, although only accurate to within 50%, help to investigate the existence of H_2O_2 and HO_2 in planetary atmospheres.

A general harmonic *ab initio* force field of H_2O_2 has been calculated by Botschwina, Meyer and Semkow at the HF level of theory[102]. Theoretical values of diagonal quadratic force constants are considerably overestimated, e.g. stretching force constants by 10–45%.

Correcting for correlation effects proves as important as in the case of $\mathbf{r}_e$ values. Alternatively, diagonal force constants can be adjusted empirically with the aid of observed frequencies. Force constants obtained in this way for H_2O_2 [102] are compared in Table 27 with experimentally based values published by Khachkuruzov and Przhevalskii[212].

Trends in theoretical quadratic and cubic force constants of FOH, H_2O_2, NH_2OH and CH_3OH are extensively discussed by Meyer and coworkers[102]. Absolute values of f_{rr}, f_{RR}, f_{rrr} and f_{RRR} increase monotonically from FOH to CH_3OH. The diagonal cubic stretching force constants are negative ($f_{rrr} = -60.8$, $f_{RRR} = -36.7$ aJ Å^{-3}) and dominate the anharmonicity of the potential energy function.

The OO stretching force constants, either experimentally or theoretically determined, clearly indicate the weakening of the OO bond along the series O_2^+, O_2, O_2^-, O_2^{2-} or O_3, O_3^- or O_2, HO_2, H_2O_2. Evidence for these trends is summarized in Table 28, which complements Tables 23 and 24.

F. Charge Density and One-electron Properties

The electron density distribution ρ of H_2O_2 has been computed by *ab initio* methods and analysed with the aid of a Mulliken population analysis[92,93,96,101,107,215]. Gross atomic charges q and overlap populations p obtained in this way reflect changes of the electron density distribution during rotation around the O—O bond as can be seen from Table 29. They can be used to substantiate the qualitative models discussed in Section IV.B. On the other hand, q and p values have to be interpreted with care since the Mulliken population analysis suffers from serious drawbacks[216,217]. For certain basis sets p(OO)

TABLE 27. Quadratic force constants of H_2O_2 at its equilibrium geometry

Force constant[a]	Exp. (Ref. 212)[b]	*Ab initio*/empirical (Ref. 102)[c]
f_{rr}	8.311	8.009
f_{RR}	4.493	4.322
$f_{\alpha\alpha}$	0.696	0.894
$f_{rr'}$	−0.045	−0.020
f_{rR}	0.069	−0.083
$f_{r\alpha}$	−0.370	−0.040
$f_{r\alpha'}$	−0.002	−0.001
$f_{R\alpha}$	0.385	0.605
$f_{\alpha\alpha'}$	0.074	0.079

[a]For reasons of simplicity the symbols designated in the text as R(OH) and R(OO) are abbreviated here to r and R. All values in aJ Å^{-n} where n is the number of stretching coordinates involved in the partial differential quotient of the potential energy. 1 aJ (atto joule) = 10^{-18} J = 1 mdyn Å = 0.2294 hartree = 6.24 eV. See I. M. Mills in *Theoretical Chemistry*, Vol. I, The Chemical Society, London, 1974.

[b]Based on $\nu_2 = 1390$ cm^{-1} and $\nu_3 = 880$ cm^{-1}.

[c]Least squares adjustment to ν values of Table 26; OH vibrations harmonized (ν_1: +176; ν_5: +188 cm^{-1}) with anharmonicity constants of H_2O and D_2O: K. Kuchitsu and Y. Morino, *Bull. Chem. Soc. Japan*, **38**, 814 (1965).

TABLE 28. Experimental and theoretical OO stretching force constants of molecules containing the O_2 unit

Molecule	State	f_{RR} (aJ Å^{-2}) Exp.	Theory	Ref.
O_2^+	$^2\Pi_g$	16.5		213
O_2	$^3\Sigma_g^-$	11.8[a]	15.2	213, 73
	$^1\Delta_g$	10.7[a]	15.1	213, 73
O_2^-	$^2\Pi_g$	5.6		214
O_2^{2-}	$^1\Sigma_g^+$		4.0	182[b]
O_3	$^1A_1$	6.2	10.6	184, 73
O_3^-	$^2B_1$	3.8[c]		186
HO_2	$^2A''$	5.9	7.4	209, 73
H_2O_3	^1A		6.7	73
H_2O_2	^1A	4.5	6.3	212, 73

[a] $f_e = 5.8883 \times 10^{-7}\ \mu_a \nu_e$ [aJ Å^{-2}]; μ_a = reduced mass.
[b] Calculated for Li_2O_2.
[c] Assumed.

TABLE 29. Mulliken population analysis of RHF/SVdp calculations on H_2O_2 [a,b]

Parameter	*cis*	skew	*trans*
q(O)	8.348	8.365	8.371
q(H)	0.652	0.635	0.629
p(OO)	0.123	0.138	0.128
p(OH)	0.599	0.616	0.620
p(HH)	−0.019	0.003	0.007

[a] All values in atomic units.
[b] From Ref. 107.

may even become negative[93], which is in clear conflict with the chemical picture of the O—O bond.

Another way of analysing the electron density distribution at the O—O bond is to evaluate the deformation density function (equation 7) as suggested by Daudel and coworkers[218]. In equation (7) $\rho(\mathbf{r})$ is the electron density at a point $\mathbf{r}$ and $\Sigma\rho^A(\mathbf{r})$ that which would result if the atoms forming the molecule could be added together without perturbing each other. In the case of the O atom the function ρ is not spherical and it is not self-evident how to form ρ^A. One eludes this problem by averaging the electron density of O(^{3}P) over all orientations in space, thus reintroducing spherical symmetry.

$$\Delta\rho(\mathbf{r}) = \rho(\mathbf{r}) - \sum_{\text{ATOM}} \rho^A(\mathbf{r}) \qquad (7)$$

Applying this method to O_2, the O lone-pair electrons can be pictured. But at the same time a negative $\Delta\rho(\mathbf{r})$ is found in the internuclear region[218]. This has been interpreted as a result of strong Coulomb repulsion between bonding electrons of O_2, which forces electron density to a region outside the space surrounding the bond axis.

A similar result has been obtained for H_2O_2 by Coppens and Stevens[219] using the RHF wave function published by Dunning and Winter[101]. The computed deformation density $\Delta\rho$ is negative in the O—O bond region. This has been verified by X-ray and neutron diffraction studies on H_2O_2 [192]. On the other hand, a theoretical determination of $\Delta\rho$ for H_2S_2 [220] leads to $\Delta\rho > 0$ along the bond axis. Obviously the interpretative value of density difference descriptions is poor in the case of bonds between valence-electron-rich atoms like oxygen.

A more appealing way of analysing ρ has been worked out by Bader and coworkers[221–224]. It involves the evaluation of the gradient vector field $\nabla\rho(\mathbf{r})$ from *ab initio* (or experimental) electron density functions and the determination of its critical points at which the field vanishes. In Figures 27 and 28 a contour line diagram and the corresponding gradient vector field of **23** are shown[225]. All the gradient paths terminating at one of the four nuclei define a subspace of the total molecular space, which can be assigned to the atom in question. There is a saddle point of ρ, at location $\mathbf{r}_c$ between each pair of bonded atoms. This is an O—O or O—H bond critical point of ρ, which serves as the origin for two gradient paths connecting neighbouring nuclei. Together they define the bond path, along which the charge density is a maximum with regard to a lateral displacement. Gradient paths terminating at $\mathbf{r}_c$ form the interatomic surfaces between the O and H atoms[221].

In Table 30, $\rho(\mathbf{r}_c)$, $\nabla^2\rho(\mathbf{r}_c)$ and the eigenvalues $\lambda_i (i = 1, 2, 3)$ of the Hessian matrix of ρ at $\mathbf{r}_c$(O—O) are listed for H_2O_2 [225]. The negative sign of $\nabla^2\rho(\mathbf{r}_c)$ is indicative of O—O bonding. The curvature of ρ along the internuclear axis is positive ($\lambda_1 > 0$), while it is negative perpendicular to the axis ($\lambda_2, \lambda_3 < 0$). Normally, accumulation of electron density between the nuclei, characteristic of a strengthening of the bond, reduces the

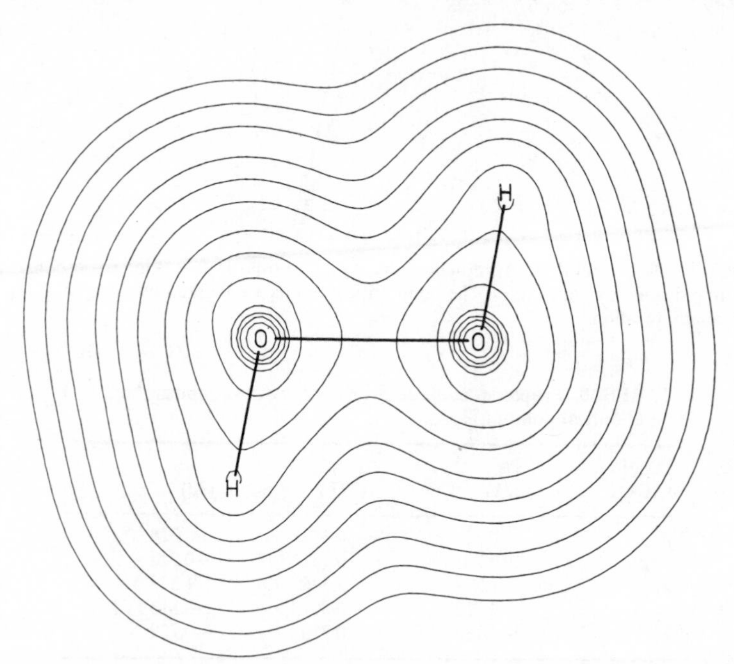

FIGURE 27. Contour plot of $\rho(\mathbf{r})$ for *trans* H_2O_2. (Wave function and geometry from Reference 101.)

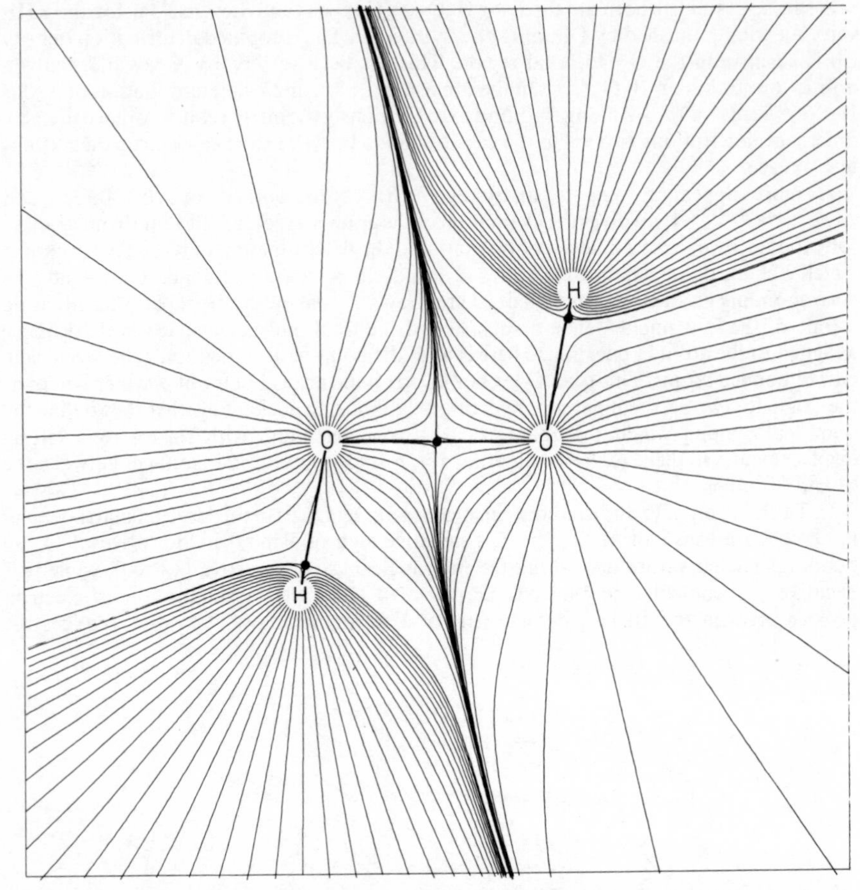

FIGURE 28. Representation of gradient paths of $\rho(\mathbf{r})$ for *trans* H_2O_2. Bond paths connecting the nuclei are indicated by heavy lines. The bond critical points are marked by dots. (R. F. W. Bader, private communication.)

TABLE 30. Properties of the electron density distribution at the O—O bond critical point of H_2O_2 [a]

Position $\mathbf{r}_c$(OO)	*cis* (**21**)	skew (**22**)	*trans* (**23**)
ρ	0.329	0.352	0.328
$\nabla^2\rho$	−0.225	−0.231	−0.218
λ_1	1.328	1.346	1.323
λ_2	−0.822	−0.807	−0.813
λ_3	−0.730	−0.770	−0.729

[a] R. F. W. Bader, unpublished results; geometry and wave function from Ref. 101.

magnitude of λ_1, thus making $\nabla^2\rho(\mathbf{r}_c)$ more negative. Although both $\rho(\mathbf{r}_c)$ and $\nabla^2\rho(\mathbf{r}_c)$ suggest a maximum of the O—O bond strength for the skewed form, the eigenvalues of the Hessian matrix reveal that the decrease of λ_3 rather than λ_1 influences the value of $\nabla^2\rho(\mathbf{r}_c)$.

In the planar forms $|\lambda_2| > |\lambda_3|$, i.e. a lower curvature perpendicular to the molecular plane signalizes a 'π-like' nature of the charge distribution in this direction. The fact that at $\tau = 120°$ the value of λ_3 adjusts to that of λ_2, thus causing the decrease of $\nabla^2\rho(\mathbf{r}_c)$ in skewed H_2O_2, indicates that 'π-like' charge arranges more uniformly around the O—O bond. This is in line with the MO description of n delocalization discussed in Section IV.B.

Noteworthy is the computed deviation of the O—O bond path from the internuclear axis for $\tau < 180°$. At $\mathbf{r}_c$ a displacement of 0.012 Å for **22** and 0.032 Å for **21** is computed. This is strongly suggestive of the bent-bond picture of strained molecules and, therefore, may be interpreted as increasing strain for τ going to 0° [223].

RHF calculations close to the HF limit[107,226] predict a molecular dipole moment μ for skewed H_2O_2, which is 0.3–0.6 D lower than the experimental value of 2.26 D [227]. Although RHF dipole moments cannot directly be compared with vibrationally averaged values, such a large difference is indicative of a sizeable contribution to the theoretical value of μ due to correlation effects. The importance of correlation corrections has been demonstrated in calculations on O_3 [49]. In this case, RHF theory leads to an overestimation of μ by 0.3 D[45] while GVB-CI/DZd calculations yield $\mu = -0.54$ D[49], in line with an experimental value of 0.53 D[228]. Recently, a value of 2.0 D has been predicted for HO_2 [80].

One-electron properties of H_2O_2 have been determined at various levels of theory[92,98,106,215]. At present the only quantity which can be compared with an experimental value seems to be the ^{17}O nuclear quadrupole coupling constant K observed in the ^{17}O nuclear quadrupole resonance spectrum of a 90% H_2O_2 solution[229]. The calculated K values are 5–15% too large, while the theoretical values of the asymmetry parameter η, which describes how the electric field gradient $\mathbf{q}$ departs from cylindrical symmetry ($0 < \eta < 1$; $\eta = 0$ corresponds to axial symmetry around the principal axis z', see footnote *a* of Table 31), exceed the observed value by 20–40% (Table 31).

In Table 32 some selected one-electron properties of skewed H_2O_2 are listed[98,106]. Although the comparison of values obtained by different methods reveals no dramatic changes, the accuracy of one-electron properties may vary considerably. At least this is suggested by RHF/DZd calculations on O_3. Rothenberg and Schaefer[45] have found surprisingly good agreement between experimental and RHF second moments of the electronic charge distribution while, for example, the computed quadrupole moment tensor elements bear little resemblance to experimental values.

TABLE 31. Theoretical and experimental quadrupole coupling constants (MHz) at ^{17}O of H_2O_2

Property[a]	RHF/DZ (Ref. 215)	RHF/MBS[b] (Ref. 98)	RHF/DZdp (Ref. 106)	Exp. (Ref. 229)
$K_{z'z'} = eq_{z'z'}Q/h$	−17.11	−18.23	−18.70	16.31(7)
η	0.814	0.930	0.953	0.687(11)

[a] $Q(^{17}O) = -0.0256$ barn from H. F. Schaefer, R. A. Klemm and F. E. Harris, *Phys. Rev.*, **176**, 49 (1968). $\eta = (q_{y'y'} - q_{x'x'})/q_{z'z'}$ with $|q_{y'y'}| < |q_{x'x'}| < |q_{z'z'}|$; $q_{x'y'}$, etc. are zero in the principal axes system of H_2O_2, which is defined in footnote *a* of Table 32. See T. D. Das and E. L. Hahn, *Nuclear Quadrupole Resonance Spectroscopy*, Academic Press, New York, 1958.
[b] MBS calculations with STFs.

TABLE 32. Selected one-electron properties for skewed H_2O_2

Property[a]		RHF/MBS (Ref. 98)	RHF/DZdp (Ref. 106)	APSG[c] (Ref. 106)
Quadrupole moment[b]	$\theta_{x'x'}$	5.48	5.83	5.64
(10^{-26} esu cm^2)	$\theta_{y'y'}$	−1.35	−1.34	−1.39
	$\theta_{z'z'}$	−4.12	−4.49	−4.25
	ψ	32.5	36.1	36.7
Electric field gradient	$q_{x'x'}$	0.95	0.98	0.93
at 0	$q_{y'y'}$	0.03	0.02	−0.01
(10^{-16} esu cm^{-3})	$q_{z'z'}$	−0.98	−1.01	−0.92
at H	$q_{x'x'}$	0.10	0.09	0.09
	$q_{y'y'}$	0.07	0.06	0.06
	$q_{z'z'}$	−0.17	−0.15	−0.15
Asymmetry parameter	η^H	0.19	0.20	0.21

[a]The molecular x and z axes are parallel to the C_2 symmetry axis and the O—O bond, respectively. Primed coordinates denote the principal axes of the tensor. The eulerian angles (ψ, $\varphi = 90°$, $\vartheta = -90°$) relate these axes to the molecular axes.
[b]Calculated with the centre of mass as origin.
[c]Calculations with the antisymmetrized product of strongly orthogonal geminals (APSG) only consider intrapair electron correlation.

G. Excited States

Because of the importance of O_2 and O_3 in atmospheric chemistry and photochemistry, their excited states have been theoretically studied by various groups[36,38–40,42,48–50,53,54,230,231]. A detailed discussion of these investigations would go beyond the scope of this chapter. Therefore, just some of the results for ozone are cited here.

Table 33 contains computed vertical transition energies to the excited states shown in Figure 5 (Section III.A.2). MRD-CI results of Thunemann, Peyerimhoff and Buenker[53] agree quite well with observed spectral features, while HF calculations lead to a false order of states and an underestimation of excitation energies. Hay, Dunning and Goddard[48–50] have reported state diagrams, adiabatic excitation energies, geometries, force constants, frequencies and dipole moments for excited states of bent, cyclic and linear O_3. In a recent MCSCF-CI investigation the potential surface of the $^3B_2$ state has been explored[62]. This state is found to be bound with an O_2—O binding energy of 0.4 eV.

The importance of the HO_2 radical in atmospheric chemistry has triggered elaborate studies on its excited states only recently. In Table 34, vertical excitation energies and oscillator strengths taken from MRD-CI calculations of Shih, Peyerimhoff and Buenker[77] and an extensive CI investigation of Langhoff and Jaffe[79] are compared with the available experimental data[232–235]. Adiabatic potential energy curves for the covalent $1^2A'$, the ionic $2^2A''$ (Figure 4, Section III.A.2) and the ionic $2^2A'$ state, can be found in Reference 79. Covalent or ionic character is reflected by theoretical dipole moments of 4 D ($2^2A''$) and 3 D ($2^2A'$) as compared with 2.3 D ($1^2A''$) and 2 D ($1^2A'$)[79].

In the $1^2A'$ (4π) state the O—O bond is elongated to about 1.41 Å, which causes a decrease of the O—O stretching frequency to 968 cm^{-1}[79] (exp. 951[232], 881 cm^{-1}[233]). This is indicative of the higher O—O antibonding character of the 2a″ MO (Figure 8, Section III.B.2). Shih and coworkers[77] have calculated a radiation lifetime of the $1^2A'$ state of $\tau_{1/2} = 7.6 \times 10^{-3}$ s as compared with 1.1×10^{-3} s found by Langhoff and Jaffe[79] and Buenker and Peyerimhoff[75].

TABLE 33. Calculated vertical transition energies (eV) of ozone[a]

No.	State	Excitation	HF	MRD-CI	Exp.[b]
			DZb + diff		
1	$1^1A_1(4\pi)$	$\ldots 1a_2^2, 4b_2^2, 6a_1^2$	0	0	0
2	$1^3B_2(4\pi)$	$1a_2 \rightarrow 2b_1$	−2.27	1.20	Peaks at 1.29, 1.43, 1.55, 1.67, 1.80, 1.92 eV (states 2–5)
3	$1^3A_2(5\pi)$	$4b_2 \rightarrow 2b_1$	0.73	1.44	
4	$1^3B_1(5\pi)$	$6a_1 \rightarrow 2b_1$	0.69	1.59	
5	$1^1A_2(5\pi)$	$4b_2 \rightarrow 2b_1$	1.18	1.72	
6	$1^1B_1(5\pi)$	$6a_1 \rightarrow 2b_1$	1.51	1.95	2.1 (Chappuis)
7	$2^3B_2(6\pi)$	$4b_2, 6a_1 \rightarrow 2b_1^2$	−0.55	3.27	
8	$2^1A_1(6\pi)$	$36\%\ 4b_2^2 \rightarrow 2b_1^2$ $+45\%\ 6a_1^2 \rightarrow 2b_1^2$		3.60	3.5–4.2 (Huggins)
9	$1^1B_2(4\pi)$	$1a_2 \rightarrow 2b_1$	3.73	4.97	4.86 (Hartley)
10	$3^1A_1(4\pi)$	$50\%\ 1a_2^2 \rightarrow 2b_1^2$ $+23\%\ 1b_1 \rightarrow 2b_1$		7.60	
11	$2^3A_2(5\pi)$	$6a_1, 1a_2 \rightarrow 2b_1^2$	4.38	5.58	
12	$2^3B_1(5\pi)$	$4b_2, 1a_2 \rightarrow 2b_1^2$	4.98	6.50	
13	$2^1A_2(5\pi)$	$6a_1, 1a_2 \rightarrow 2b_1^2$	5.10	6.37	
14	$2^1B_1(5\pi)$	$4b_2, 1a_2 \rightarrow 2b_1^2$	6.14	7.26	7.18

[a]Ref. 53. Calculated at $R = 1.277$ Å and $\alpha = 116.8°$ with a DZ basis augmented by bond functions and diffuse Rydberg functions.
[b]For quotations of the experimental work see Ref. 53.

TABLE 34. Vertical excitation energies (eV) and oscillator strengths for HO_2

State	Excitation	Vertical excitation energies			Oscillator strength
		Ref. 79	Ref. 77	Exp.	(Ref. 77)
$1^2A''(3\pi)$	$\ldots 7a'^2 2a''^1$	0	0	0	0
$1^2A'(4\pi)$	$7a' \rightarrow 2a''$	1.02	0.93	0.88[a], 0.87[b]	3.9×10^{-6}
$2^2A''(3\pi)$	$1a'' \rightarrow 2a''$	6.26	5.90	5.9–6.2[c]	0.065
$2^2A'(4\pi)$	$6a' \rightarrow 2a''$	6.73	6.49		0.0012

[a]Ref. 232.
[b]Ref. 233.
[c]Refs. 234 and 235.

According to Table 34 the UV spectrum of HO_2 is dominated by a single continuous feature corresponding to the $2^2A'' \leftarrow 1^2A''$ transition with a peak near 2100 Å. The $2^2A' \leftarrow 1^2A''$ transition is far too weak to be observed. Theory suggests that if the $1^2A'$ state is appreciably populated, it may be possible to observe photoabsorption at about 2500 Å corresponding to the $2^2A' \leftarrow 1^2A'$ transition.

Some excited states of H_2O_2 have been investigated by Rauk and Barriel[236] with the aid of perturbative CI calculations. Only singly excited configurations were considered and an empirical correction for correlation and orbital relaxation effects applied. This was based on the assumption that computed Rydberg state energies of H_2O_2 may suffer from an error similar in magnitude to that found for the Koopmans' value $-\varepsilon_i$ of the occupied MO ϕ_i from which excitation takes place[236] (equation 8), where I^{exp} is the experimentally observed vertical IP.

$$\Delta E_{cor} = \Delta E_{calc} + I_i^{exp} + \varepsilon_i \tag{8}$$

In Table 35 results of Rauk and Barriel[236] are summarized. The experimental UV spectrum of H_2O_2 lacks any absorption bands below 6.7 eV (1850 Å)[237]. The absorption increases towards 10.3 eV (1200 Å) where the spectrum becomes obscured due to H_2O contamination. There is a single broad maximum at 7.5 eV (1650 Å) and the suggestion of a shoulder at 7 eV (1770 Å). Corrected excitation energies for the third and fourth singlet states, ^1B and ^1A, are of comparable magnitude (Table 35). These states arise from excitation to a bonding (a symmetry) combination of the 3s(O) orbitals and, hence, should be bound states.

Excitation to the corresponding antibonding combination (b symmetry) from the n MOs yields two states in the 4–6 eV region, which are probably dissociative. Rauk and Barriel[236] assume on the basis of the higher oscillator strength of the ^1B state that photolytic decomposition of H_2O_2 occurs via this state. Rupture of the O—O bond of alkyl peroxides has been observed in the first absorption region, 3100–2500 Å, while below 2300 Å C—O rupture appears as a new primary dissociative mode[238].

Rupture of the O—O bond in peroxides has been classified by Dauben, Salem and Turro[239] as being of the tetratopic $(\sigma\pi)$ $(\sigma\pi)$ type, thus yielding the four pairs of diradical states shown in Figure 29. In the case of H_2O_2, only the GS is bonding while the three excited S states and all the T states are repulsive in nature. These states correlate with the GS of two OH radicals as has been confirmed by Evleth on the basis of CNDO-CI calculations[240,241]. The key to this correlation lies in the doubly degenerate character of the $^2\Pi$ ground state of OH, which leads in double combination to four S and four T states. This eightfold energetic degeneracy is an essential feature of O—O bond rupture. It is responsible for an extreme complexity of surfaces in O—O dissociation processes of larger peroxides.

TABLE 35. Vertical excitation energies and oscillator strengths for H_2O_2[a]

State	Excitation[b]	Energy (eV)			Oscillator strength
		ΔE_{calc}	ΔE_{cor}[c]	ΔE_{exp}	
^1A	n(4b) → 3sσ*	6.2	4.0		0.0013
^1B	81% n(5a) + 14% n(4a) } → 3sσ*	7.5	5.6		0.0207
^1B	n(4b) → 3sσ	9.1	6.9	7.0	0.0054
^1A	73% n(5a) + 16% n(4b) } → 3sσ	9.7	7.8	7.5	0.0078
^1A	15% n(5a) + 72% n(4b) } → 3sσ*	10.5	8.3		0.0131
^1B	71% n(4b) + 20% n(4a) } → 3pπ	11.1	8.9		0.1371
^1B	n(5a) → 3pπ*	11.2	9.3		0.0388

[a]Ref. 236. Computed at $R = 1.475$ Å, $R' = 0.95$ Å, $\alpha = 94.8°$, $\tau = 111.5°$ with a DZ basis augmented by diffuse Rydberg functions.

[b]Fractional excitations out of more than one occupied MO are given in percent. Rydberg character of excited states is indicated. Note that valence or Rydberg character of computed states also depends on the inclusion of double excitations. Compare with Ref. 241.

[c]Correction for n(4b) calculated with $\varepsilon = -13.71$, $I_{Vert} = 11.4$ eV and for n(5a) with $\varepsilon = -14.47$ and $I_{Vert} = 12.56$ eV (Ref. 236).

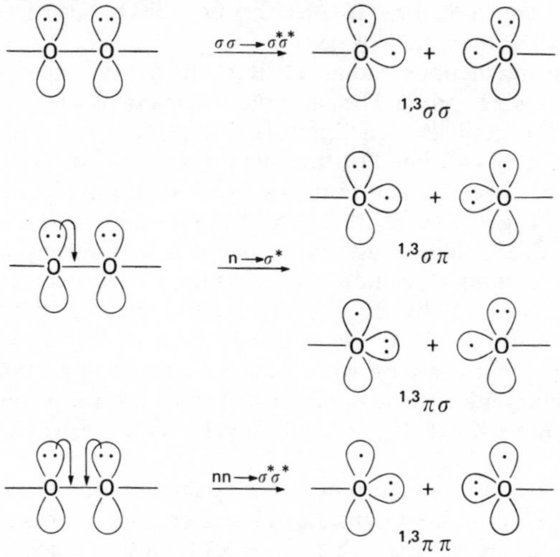

FIGURE 29. Photoexcitation and dissociation of H_2O_2. Mode of excitation and diradical states of OH fragments are given.

V. SUBSTITUENT EFFECTS

A. General Trends

1. Peroxy compounds XO_2

Apart from the parent compounds HO_2 and O_3 only scattered data on XO_2 peroxides are available. Experimental and/or theoretical investigations on structure and bonding in LiO_2 [182,214,242–246], MeO_2 [247], NH_2O_2 [248], HO_3 (X = HO) [73,249], FO_2 [189,250–252] and ClO_2 [253] with monovalent X have been reported. As for XO_2 systems with divalent X, most attention has been focused on carbonyl oxide and dioxirane[58,67,69,254–259] because of their important role in the ozonolysis[12] and other oxidation reactions of hydrocarbons[13]. Dioxirane has recently been detected by microwave spectroscopy in the low-temperature reaction of O_3 with ethylene[33]. Other theoretical investigations considered the bent or cyclic form of NOO[260,261], NOO^- [262] and HNOO[67,263,264].

When varying the monovalent substituent X from F to Li, the O—O bond strength decreases. This is reflected by corresponding changes in the bond distance, stretching frequency and force constant (e.g. $f_{RR} = 10.5$, 9.7, 5.6 aJ Å^{-2} for X = F[250], Cl[253] and Li[182]). Spratley and Pimentel[265] have suggested that—depending on the electronegativity of X—electron charge is either withdrawn from or donated to the antibonding π_g MO of O_2, thus stengthening or weakening the OO bond. This description has been corroborated by McCain and Palke[252], who have investigated trends in electron spin *g* values for peroxy radicals on the basis of *ab initio* calculations on HO_2 and FO_2. They consider bonding in XOO to result from Lewis acid–base reactions between a diamagnetic group X and a O_2^+ or O_2^- radical. The unoccupied π_g MO in O_2^+ is an electron acceptor which acts as a Lewis σ acid, whereas the filled level of O_2^- is an electron donor or Lewis σ

base. In addition, the open shell π_g MO on either O_2^+ or O_2^- can act as a Lewis π acid or base toward a pπ orbital on X.

Strong acid–base interactions lead to relatively strong XO bonding. Conversely, weak interactions lead to ionic bonding. The latter situation obtains when X is a weak σ/π donor ($X = BF_4^-$, AsF_6^-) or a weak σ/π acceptor ($X = Li^+$, Na^+). Accordingly, ionicity of the XO bond is revealed by OO bond features typical for O_2^+ or O_2^-.

This prediction has been verified in the case of LiO_2. Bonding between Li and O_2 is essentially ionic with at least 0.77 e transferred from the alkali metal to the O_2 moiety[246]. In order to maximize Coulomb attraction between a positively charged Li and the negatively charged O atoms, the molecule adopts the C_{2v} symmetrical cyclic structure **2**. This has been confirmed by matrix IR measurements[214,242,243] and *ab initio* calculations[244–246] ($R = 1.30$ Å, $R' = 1.77$ Å, $\alpha = 68.5°$, $\alpha' = 43°$ [246]; ν_1(O—O stretch) = 1097 cm^{-1} [214]; compare with Tables 9 and 23 of Section IV). Alkali-metal superoxides all seem to prefer structure **2** since the ionic character varies only slightly for X = Li, Na (maximum), K, Rb, Cs, as is indicated by the corresponding O—O stretching frequencies[243].

For X = BeH or BH_2 the equilibrium geometry should also correspond to an isosceles triangle, yet with less ionic X—O bonding character. This, at least, is suggested by the relative energies of XO_2 peroxides with divalent X (Table 36), which we have calculated in order to compare OO bonding in these compounds at a consistent level of theory[266]. From Be to F^+ the energy difference ΔE between bent (linear) and cyclic XO_2 increases steadily from −80 to 80 kcal mol^{-1}, i.e. for X = Be, BH and CH_2 structure **2** is more stable than **1**, while for X = NH, O and F^+ the reverse is true (see Table 6, Section III.B.3). Cyclic NO_2^- with $\Delta E = 27$ kcal mol^{-1} ($R = 1.47$ Å, $R' = 1.50$ Å, $\alpha' = 59°$)[262] nicely fits into this trend. Parallel to the increase in ΔE, the O—O bond length R and the angle α of structures **1** and **2** decrease. Again, this is indicative of a stepwise depopulation of the π_g MOs of O_2^- and a smooth change from ionic to covalent XO bonding. For example, the charge of the O_2 moiety changes from a surplus of 0.55–0.65 e for Be and BH to a lack of 0.13 e for F^+.

It is interesting to note that peroxynitrene, HNO_2, prefers the *syn* form by about 2 kcal mol^{-1}, probably due to Coulomb attraction between H and the terminal oxygen atom[266]. A similar effect has been found for alkyl-substituted carbonyl oxides[58,267].

TABLE 36. Energies and geometries of some XO_2 peroxides calculated at the RSMP/SVd level of theory[58,266]

No. of valence electrons	No. of π electrons[a]	Molecule	Abs. energy[b] (hartree)	Geometry[b] R (Å)	R' (Å)	α (deg.)	ΔE^c (kcal mol^{-1})	Geometry[c] R (Å)	R' (Å)	α' (deg.)
20	8/4	BeOO	−164.5230	1.34	1.33	180	−82	1.66	1.44	70
22	8/4	HBOO	−175.2611	1.36	1.21	180	−70	1.62	1.37	72
24	4/6	H_2COO^d	−189.0528	1.29	1.30	120	−34	1.53	1.40	66
24	4/6	HNOO	−205.0625	1.27	1.38	119	10	1.49	1.45	62
24	4/6	OOO^e	−224.8768	1.31	1.31	116	35	1.48	1.48	60
24	4/6	FOO^+	−248.9809	1.28	1.33	113	78	1.43	1.59	54

[a]Number of π electrons in the chain/cyclic state.
[b]Absolute energy and geometry (linear or bent) of chain structure **1**.
[c]Relative energy and geometry of cyclic structure **2**. The angle OXO is denoted by α'.
[d]$\mathbf{r}_s$ geometry of dioxirane: $R = 1.516$ Å, $R' = 1.388$, $\alpha' = 66.2°$ [33].
[e]$\mathbf{r}_e$ geometry of ozone: $R = 1.2716$ Å, $\alpha = 117.79°$ [70].

There, through-space interactions between a pseudo-π orbital of a methylene group and the $2p\pi$ AO of the terminal O atom (homoaromatic 6π system) can lead to additional stabilization of the *syn* forms. In this respect, the configurational and conformational preferences of carbonyl oxides may be considered to represent examples of the *cis* effect[267].

Rupture of the XO or OO bond in the acyclic GS of XO_2 leads to $XO(^3\pi\sigma)$ and $O(^3P)$ or $X(^3\pi\sigma)$ and $O_2(^3\Sigma_g^-)$ (compare with Figure 5, Section III.A.2). In the case of carbonyl oxide, these processes have been calculated to require about 43 and 56 kcal mol^{-1}, respectively[254]. Hence, the 4π state of H_2CO_2 is stable with respect to dissociation although it is actually higher in energy than $H_2CO(^1A_1) + O(^3P)$. Because of the high reactivity of carbonyl oxide in the presence of electrophilic, nucleophilic or dipolarophilic agents, there is only indirect evidence for its existence[12].

2. Peroxides XOOH and XOOX

A number of theoretical investigations have been carried out in order to establish equilibrium geometry and conformational behaviour of closed-shell peroxides. To be mentioned are *ab initio* studies on the hydroperoxides LiOOH [268], MeOOH [100,150,269], EtOOH[270], PhOOH[271], CF_3OOH[136], NH_2OOH[150], HOOOH[73,150,272,273], FOOH [100,150] and the peroxides LiOOLi [182,246,267,274,275], NaOONa [275], KOOK [275], BH_2OOBH_2 [276], MeOOMe [277], CF_3OOCF_3 [277], CF_3OOF [136], HOOOOH [272] and FOOF [278]. We have supplemented these investigations by RHF/SV calculations on XOOH and XOOX varying X systematically from Li to F [266]. Our results are condensed into Table 37.

Almost all hydroperoxides adopt a bent–bent form. Exceptions are LiOOH and HBeOOH which seem to prefer a linear–bent form with a positively charged X($\sim +0.7$ e) and a bent OOH moiety with some anionic character. However, the stability of the bridged forms may be underestimated by as much as 20 kcal mol^{-1} (compare with RSMP/SVdp results for H_2O_2, Table 10, Section IV.A) due to basis set and correlation errors at the RHF/SV level. Accordingly, the bridged form of LiOOH is likely to be the most stable one as has been suggested by Peslak[268].

If both hydrogens are replaced by Li or BeH, the stability of the bridged form increases. This is in line with experimental[180,214] and theoretical results[182,246,268,274,275] on alkali-metal peroxides. For Li, Na and K, the planar rhombic form is the most stable one since it minimizes Coulomb repulsion between the positively charged metal atoms. Puckering leads to an energy increase[266,274]. Obviously, the electrostatic factor outweighs stabilizing orbital interactions found for the puckered form of H_2O_2. As soon as the ionic character of the X—O bond is reduced, puckering will lead to stabilization. This is true for the persulphide analogue of Li_2O_2 [274], namely Li_2S_2, which is more stable in the puckered geometry (puckering angle $\delta = 53°$ [274]; compare with Table 10). According to RHF/SV calculations the planar form is destabilized by 1.7 kcal mol^{-1} [274].

On the other hand, increased covalent bonding between X and O leads to destabilization of the bridged forms. For X = NH_2, OH and F the bicyclic forms open to monocyclic forms with very weak O—O interactions. Their relative energy is considerably larger than the usual O—O dissociation energies (Table 16, Section IV.C). For example, ΔE of $\overline{OFOF}$ is 2.5 times larger than D_0 (FO—OF) (ca. 62 kcal mol^{-1} [279]). This holds also for the linear geometries, which represent the most unstable peroxide forms of Table 37. Their relative energy increases steadily from X = Li towards X = F, probably because of enhanced repulsion between electron lone pairs.

The data of Table 37 suggest that Y forms are stable under certain experimental conditions. The relative energies of the planar geometries represent an upper limit of their

TABLE 37. Absolute and relative energies (in hartree and kcal mol^{-1}) of hydroperoxides XOOH and peroxides XOOX calculated at the RHF/SV level of theory for optimized geometries[266]

Structure (geometry)	H	Li	BeH	BH_2	CH_3	NH_2	OH	F
XOOH								
Bent–bent (*trans*)		−157.4631[a]	−165.2282[a]	−175.8255	−189.5311	−205.4516	−225.2052	−249.1634
Bridged (planar)		17	23	59	115	60	81	61
Linear		59	67	110	143	167	180	210
XOOX								
Bent–bent (*trans*)	−150.5599	−164.3484[b]	−179.9047[c]	−201.0896	−228.5023	−260.3441	−299.8534	−347.7629
Y (planar)	26	Nonstable[d]		29	28	49	34	39
Bridged (planar)	88	−48	−33	36	155	177	133	155
Linear	144	0	3	78	143	190	209	271

[a] Energy of linear–bent form.
[b] Energy of linear form.
[c] Energy of *cis* form.
[d] Geometry collapses to bridged form.

actual stability since pyramidalization and solvent effects will decrease this energy. The first experimental evidence for the existence of Y forms was found in persulphide chemistry. Kuczkowski[280] identified by microwave and mass-spectrometric studies two stable F_2S_2 isomers, namely FSSF and F_2SS, the latter with pyramidal geometry. Recent *ab initio* calculations of Hinchliffe[281] suggest that a H_2SS isomer may also exist.

In connection with the reaction of $^1O_2$ with alkenes, the intermediacy of the peroxirane **24** has been discussed[10,11]. Dewar and Thiel[282] have predicted the formation of **24** on the basis of MINDO/3 calculations. However, more recent GVB/CI[283] and HF investigations[284–286] indicate that **24** is much higher in energy than other possible reaction intermediates.

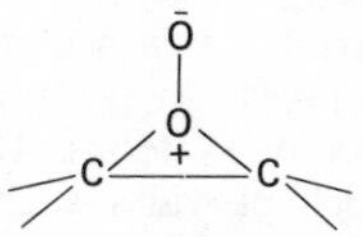

(24)

Unambiguous evidence for the existence of a peroxide with Y structure has recently been given by Atwood and coworkers[287]. They have synthesized the stable complex [K·dibenzo-18-crown-6] $[Al_2Me_6O_2]$, which according to X-ray measurements contains the Y structure **25** with normal AlO single bonds and a long O—O bond. Since ν(O—O) of **25** (851 cm^{-1}) is similar to the O—O stretching frequency found for hemerythrins (844 cm^{-1})[288], a group of oxygen-carrying proteins, it is likely that **25** models the bonding situation in these compounds[287].

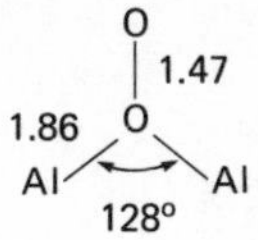

1.47 1.86 128°

(25)

Additional information about the influence of the group X is provided by the RHF/SV bond separation energy (*BSE*) of formal reactions leading to H_2O_2 and XOH (Table 38). Positive *BSE*s are indicative of stabilizing bond interactions, probably via an electron transfer of the type

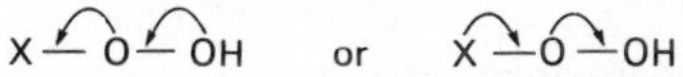

X—O—OH or X—O—OH

TABLE 38. RHF/SV bond separation energies (kcal mol^{-1}) of the formal reactions (1), (2) and (3)[a]

Bond separation reaction for X =	Li	BeH	BH_2	CH_3	NH_2	OH	F
(1) $XOOH + H_2O \rightarrow H_2O_2 + XOH$	−3.2	−19.2	−1.6	5.7	5.3	4.7	3.1
(2) $XOOX + 2H_2O \rightarrow H_2O_2 + 2XOH$	30.5	0.5	−4.1	11.3	11.1	4.8	1.0
(3) $XOOX + H_2O_2 \rightarrow 2XOOH$	36.9	38.9	−0.9	−0.1	0.5	−4.6	−5.2

[a] Geometries of molecules XOH and XOOX have been completely optimized at the RHF/SV level[266]. BSE values at standard geometries are 6.0, 7.8, 9.3 and 6.3 kcal mol^{-1} for XOOH with X = CH_3, NH_2, OH and F, respectively[150].

involving π-type donation and σ- or π-type acceptance of electrons as shown in Figure 21 of Section IV.B. These stabilizing interactions are smaller for X^1OX^2 than for $X^1CH_2X^2$ or X^1NHX^2 [150]. An electropositive substituent like Li, BeH or BH_2 leads to destabilization, which, however, is partially offset by attractive Coulomb interactions, especially if a second Li or BeH substituent is attached to the O—O group. Disproportionation of XOOX (reaction **3** of Table 38) becomes more likely with increasing electronegativity of the substituent X. In this case the *BSE*s decrease rather than double upon going from XOOH to XOOX (reactions 1 and 2 of Table 38).

The σ,π interactions discussed in Section IV.B and illustrated in Figure 21 are primarily responsible for stabilization of skewed forms of peroxides. The equilibrium angle τ is close to 90° if X is both a π donor and a strong σ acceptor like OH or F (dominance of V_2 term in equation 1, Section IV.B). However, if X is a σ electron donor with weak π acceptor property, repulsion between bond dipoles $^+X{-}O^-$ leads to a shift of the conformational minimum towards 180° (dominance of V_1 term). These trends are confirmed by the geometrical data of Table 39 which provides a comparison between peroxide and persulphide geometries determined by gas-phase measurements or *ab initio* calculations.

Noteworthy are the changes in R for increasing electronegativity of the group X. The O—O and S—S bond lengths in the difluoro derivatives are abnormally short, being close to bond lengths in O_2 (1.207 Å, Table 23) and S_2 (1.888 Å, Reference 281). A simple explanation in descriptive VB terms is that structures like $F^-O{=}O^+F$ and $F^-S{=}S^+F$ are very important. *Ab initio* theory is presently unable to reproduce the experimental *R* values of F_2O_2 and F_2S_2 (see Table 39).

If the substituent X possesses a low-lying unoccupied π^* or pseudo-π^* orbital, secondary overlap with the occupied π^* MO of the peroxo group will increase stabilizing two-electron interactions. This explains why conformation **26** rather than **27** is more stable for methyl- or amino-peroxides[266].

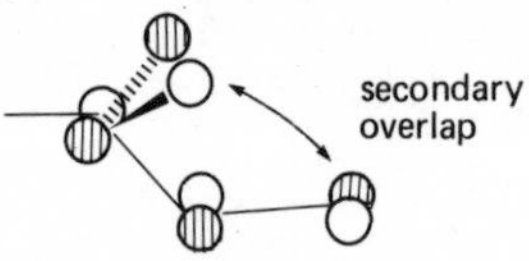

secondary overlap

(**26**)

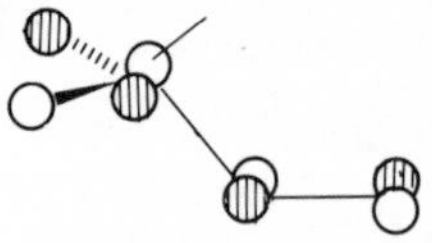

(**27**)

B. Special Compounds

1. Peroxy acids and acyl peroxides

The structure of organic peroxy acids poses some interesting questions. IR measurements have led to the proposal of a planar conformation of the —C(O)OOH moiety with a *cis–cis* (**28**) rather than a *cis–trans* (**29**) geometry, the former being stabilized by an intramolecular hydrogen bond. Recent RHF/SV calculations of the harmonic and anharmonic force field and the fundamental frequencies of performic acid published by Bock, Trachtman and George[293] add support to this argument as do *ab initio*

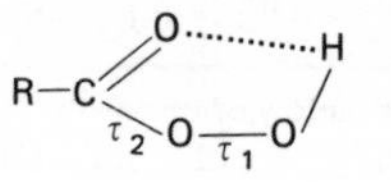

(**28**)

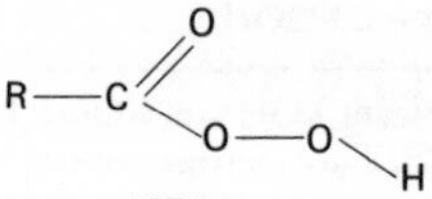

(**29**)

TABLE 39. Comparison between experimental and theoretical geometries (distances in Å, angles in deg.) and barriers for internal rotation (kcal mol^{-1}) obtained for some peroxides and persulphides

Molecule	Geometry				Barriers		
	R	R'	α	τ	(*trans*; *cis*)	Method[a]	Ref.
HOOH	1.464	0.965	99.4	120.0	1.1; 7.4	IR,MW	197,122
CH_3OOH	1.46[b]	1.43[b]	109.5[b]	140	0.1; 8.0	RHF/SV	100
CF_3OOH	1.447	1.376; 0.974	107.6; 100[b]	95[b]		ED	124
	1.436	1.404; 0.96[b]	109.5; 109.5[b]	90[b]	0.9; 6.9	RHF/SV	136
HOOOH	1.442	0.98	106.1; 100.3	78.5	9.0; 19.6	RSMP/SVd	272
FOOH	1.419	1.441; 0.960	104.5; 104.2	81.8	5.4; 4.6	RHF/SV	266,100
CMe_3OOCMe_3	1.480[b]	1.460	103.9	165.8		ED	123
CF_3OOCF_3	1.419	1.399	107.2	123.3		ED	185
CF_3OOF	1.366	1.419; 1.449	108.2; 104.5	97.1		ED	124
	1.417	1.423; 1.440	110.0; 105.0	97	8.4; 14.6	RHF/SV	136
CF_3OOCl	1.447	1.372; 1.699	108.1; 110.8	93.2		ED	124
$SiMe_3OOSiMe_3$	1.480	1.681	106.6	143.5		ED	123
	1.447	1.667	101.2	136.5		RHF/SV	289
SF_5OOSF_5	1.47	1.66	105	107		ED	127
FOOF	1.217	1.575	109.5	87.5		MW	126
	1.395	1.432	104.7	83.6	8.1; 11.8	RHF/SV	278
	1.29	1.49	109.5[b]	87.5[b]		CI/DZd[d]	278
HSSH	2.055	1.327	91.3	90.6	6–12(?)	MMW	128
	1.922	1.298	95.9	90.7	7.6; 12.7	RHF/FSGO	290
CH_3SSCH_3	2.022	1.806	104.1	83.9		ED	129
	2.029	1.816	103.2	85.3		ED	130
$CH_3SSC_2H_5$	2.031	1.817[c]	103.2[c]	84.4		ED	130
	2.030	1.821[c]	104.0[c]	83.5		MM	291
FSSF	1.888[e]	1.635	71.7	87.9		MW	280
ClSSCl	1.97	2.07	107	82.5		ED	292
BrSSBr	1.98	2.24	105	83.5		ED	292

[a] IR = infrared, ED = electron diffraction, MW = microwave, MMW = millimetre-wave, MM = molecular mechanics.
[b] Assumed values.
[c] Averaged values.
[d] Only O-centred polarization functions used.
[e] RHF/DZd: 1.957 Å, Ref. 281.

investigations on peroxyacetic acid[294,295] and peroxytrifluoroacetic acid[295]. Another RHF/SV study[296] on performic acid, however, has described **29** to be more stable than **28** by about 1 kcal mol^{-1}, the two forms being separated by a rotational barrier of less than 2 kcal mol^{-1}.

Since none of these studies has employed an augmented basis set, care has to be taken when referring to the published *ab initio* results. On the other hand, both theory and experiment clearly establish the *cis* conformation of the O=C—O—O fragment. The *cis* arrangement is stabilized by secondary overlap effects between the π^* orbitals of C=O and O—O, similar to those encountered in methyl- or amino-peroxides (Section V.A). These overlap effects are also responsible for an equilibrium value of τ_1 equal to 0 or 180°.

Organic peroxy acids convert alkenes to oxiranes by an electrophilic attack on the double bond. The *cis–cis* form (**28**) plays an important part in the epoxidation reaction. Exploratory RHF/MBS calculations by Plesničar and coworkers[297,298] on the oxidation of ethylene and methylenimine with performic acid suggest that the reaction is characterized by an asymmetric but highly ordered transition state and an intramolecular transfer of the proton. Since the O atoms of **28** all bear negative charges, it has been argued[295] that the electrophilic attack is overlap- rather than charge-controlled. The availability of a low-lying peroxide σ^* MO, especially in compounds like peroxytrifluoroacetic acid, is in line with this reasoning.

A low-lying σ^* MO of the peroxo group seems to play a similar role in the radical-induced decomposition of dibenzoyl peroxide:

$$\mathrm{Ph{-}C(=O){-}OO{-}C(=O){-}Ph + R\cdot \longrightarrow Ph{-}C(=O){-}OR + Ph{-}C(=O){-}O\cdot}$$

Semiempirical MINDO-CI calculations[299] on the decomposition of diformyl peroxide (DFP) reveal that a charge transfer from the SOMO of the radical to the LUMO of DFP (σ^*_{OO}) is very important in the TS of the reaction. Therefore, an electron-withdrawing substituent at the acyl group and an electron-donating group at the radical enlarge the charge transfer and, hence, speed up the reaction.

Some of the attention, which acyl peroxides and acylperoxy radicals have received in the past years, has stemmed from their role in the chemistry of polluted atmospheres[17,18,300]. In photochemical smog, the latter are formed in a rapid reaction between O_2 and an acyl radical. This can lead to the $^2A'$ excited state rather than the $^2A''$ GS of the peroxy radical (Figure 4, Section III.A.2). HF/DZ calculations on the formylperoxy radical (FPR)[301] show that the reaction

$$\mathrm{HCO(^2A') + O_2(^3\Sigma_g^-) \rightarrow HC(O)O_2(^2A'')}$$

is exothermic by 36 kcal mol^{-1} whereas the $^2A'' \rightarrow {}^2A'$ excitation energy is lower than 20 kcal mol^{-1}. In the $^2A'$ state the SOMO is in the right position to facilitate H migration (Figure 4) and, hence, the decomposition to CO_2 and an OH radical. This mechanism is in line with the observed generation of methoxy radicals from acetyl radicals via $MeC(O)O_2(^2A')$[302].

2. *Polyoxides*

The structural and conformational features of the polyoxides H_2O_n and their F, Me and CF_3 derivatives are affected by *n* electron-pair interactions[272,273,277,303]. The determining electronic factor is the tendency of a n electron pair to delocalize into a coplanar vicinal bond (Section IV.B). This leads to stable *helix* conformations of H_2O_n with dihedral angles of 80–90° (Figure 30), as has been demonstrated by *ab initio* calculations[272,273,277].

For H_2O_3 the theoretically determined conformational surface[273], spanned by two rotational angles τ_1 and τ_2, is shown in Figure 31 in form of a contour-line diagram. Least-energy paths connecting the potential minima (two global minima, GMIN, at $\tau_1 = \tau_2 = 78°$ and $-78°$ corresponding to helix forms of H_2O_3; two local minima, LMIN, at $\tau_1 = \pm 92°$, $\tau_2 = \mp 92°$ corresponding to forms with both OH bonds either above or below the heavy-atom plane) are shown by dashed lines. There are barriers of 6.5 kcal mol^{-1} (S points in Figure 31), which have to be surmounted to convert one GMIN form into the other. This interconversion corresponds to successive rotations of the OH bonds through the heavy-atom plane (*flip-flop* rotation, see Figure 32). It needs 16 kcal mol^{-1} less energy than synchronous rotation of the OH bonds. Flip-flop rotations are the preferred conformational modes of geminal double rotors since they involve only smooth changes of the electronic stucture and, hence, the geometry of the rotor molecule as has been demonstrated extensively for H_2O_3 (see Figures 12, 13 and 14 in Reference 273). If flip-flop rotations of adjacent bonds of a cyclic compound are coupled, ring pseudorotation results. Pseudorotation generally requires less energy than ring-inversion[304], which can be understood by inspection of Figures 31 and 32[273].

Tetroxides are probably intermediates in the self-reaction of peroxyl radicals. Experimental observations suggest that the gas-phase reaction between HO_2 radicals:

$$HO_2 + HO_2 \rightarrow H_2O_2 + O_2$$

proceeds via a H_2O_4 conformer with an intramolecular hydrogen bond[305]. (Actually, a double hydrogen-bonded association complex could not be excluded by experiment.) Such a conformer is probably 6 kcal mol^{-1} less stable than the helix conformation of H_2O_4[272]. Both ^{18}O labelling experiments[305] and semiempirical CI calculations[306] exclude a four-centre TS involving a H_2O_4 conformer with strong n pair repulsions.

The observation of the isotopic exchange reaction between $^{16}O_2$ and $^{18}O_2$ has led to the proposal of a four-membered oxygen ring[307,308]. According to *ab initio* calculations[47] the formation of O_4 is endothermic ($\Delta E > 30$ kcal mol^{-1}). This holds also for the hypothetical O_5 ring formed from O_3 and $^1O_2$[266]. The average O—O bond length in O_5 would be 1.46 Å while the actual bond lengths range from 1.43 to 1.48 Å. The ring is expected to be strongly puckered and, like cyclopentane, a free pseudorotor[266].

Knowledge on HO_n ($n > 3$) molecules is meagre. Recent experiments have revealed the existence of HO_{2n}^+ ions[309]. UHF/SV calculations[310] show that these are clusters of O_2 molecules sharing a common proton.

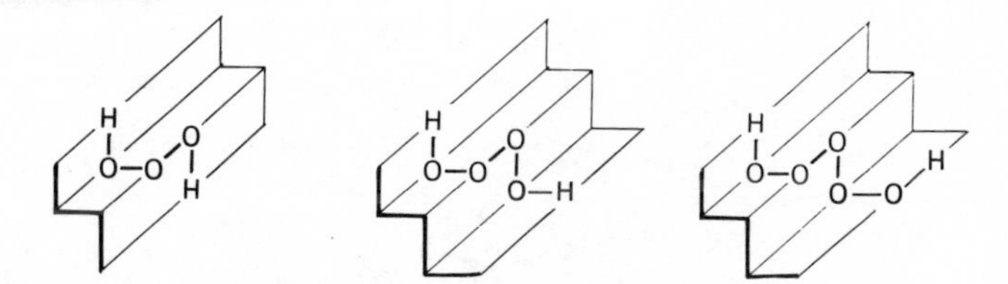

FIGURE 30. Helix conformation of H_2O_3, H_2O_4 and H_2O_5.

3. *Ozonides and other cyclic peroxides*

Interactions between the n electron pairs of oxygen influence the geometry and conformation of cyclic peroxides. If the size of the ring implies a small value of τ, the O—O bond turns out to be rather long. For example, dioxirane contains one of the longest O—O bonds so far observed[33]. With increasing ring size, the cyclic peroxide can pucker more strongly. Accordingly n-pair delocalization as described for alicyclic peroxides becomes a

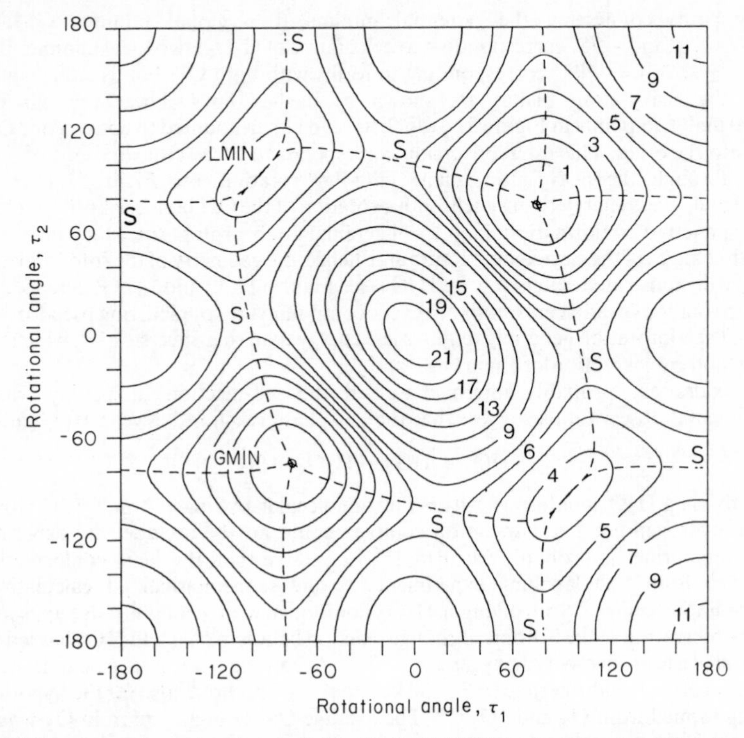

FIGURE 31. Internal rotational potential for H_2O_3 (RSMP/C calculations) as a function of the dihedral angles τ_1 and τ_2. Contours indicate kcal mol^{-1} above the energy of the global minimum GMIN. The dashed lines represent the steepest descent and ascent paths to and from the saddle-points S. Adapted by permission of the American Institute of Physics from D. Cremer, *J. Chem. Phys.*, **69**, 4456 (1978).

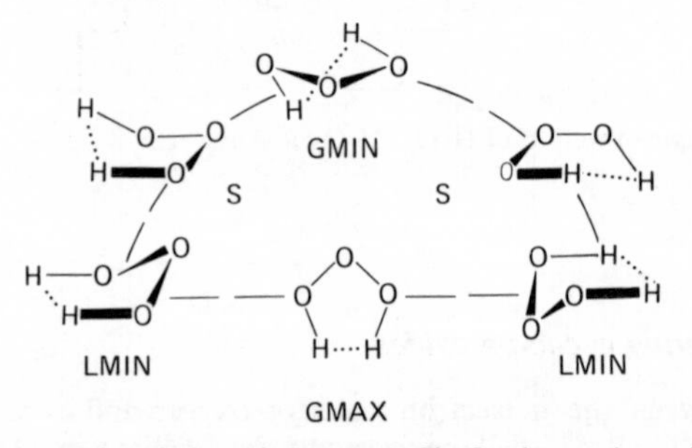

FIGURE 32. Illustration of the relationship between a flip-flop internal rotation of the double rotor H_2O_3 and the pseudorotation of a five-membered ring. GMAX: global maximum; GMIN: global minimum; LMIN: local minimum; S: saddlepoint. Reproduced by permission of the American Institute of Physics from D. Cremer, *J. Chem. Phys.*, **69**, 4456 (1978).

stabilizing factor. This is reflected by the *ab initio* and experimental geometrical data compiled in Table 40.

In a ring of given size, the tendency for puckering is stronger the more O atoms are incorporated into the ring framework. Thus, 1,2,3-trioxolane (primary ozonide) and tetroxolane with 3 and 4 adjacent O atoms are more strongly puckered than 1,2-dioxolane or 1,2,4-trioxolane (final ozonide). Parallel to this trend increases the barrier to inversion (Table 40), which is determined by the energy of the planar ring form.

The mode of puckering is also influenced by n,n interactions. We have shown this by analysing the HOMOs of the trioxolanes, which are primarily out-of-phase combinations of the 2pπ(O) orbitals[270,316]. Antibonding overlap of the HOMOs is reduced if the rotational angle of the O—O rather than the C—O or C—C bond becomes large. This leads to stable C_2-symmetrical twist (T) forms for **32**, **33** and **35** but a C_s-symmetrical envelope (E) form for **34** (Table 40). Along the same lines stabilizing or destabilizing substituent effects can be explained[270,316].

TABLE 40. Geometries and conformational barriers or cyclic peroxides as determined by experiment or theory

Molecule	R (Å)	R' (Å)	τ (deg.)	q^a (Å)	ϕ^a (deg.)	$\Delta E_{PR}{}^b$ (kcal mol^{-1})	$\Delta E_{IV}{}^c$ (kcal mol^{-1})	Method	Ref.
(**30**)	1.516	1.388	0					MW	33
	1.529	1.398	0					RSMP/DZd	58
(**31**)	1.491	1.475[d]	22.1	0.28				X-ray[e]	311
	1.497	1.473	0	0			0	RHF/SV	285
(**32**)	1.483	1.451[d]	20					X-ray[f]	312
	1.461	1.439	50.2	0.45	90; 270	2.2	5.7	RHF/SVd	313
(**33**)	1.461	1.415	49.4	0.46	90; 270			MW	314
	1.467	1.433	47.4	0.45	90; 270	3.3	6.3	RHF/SVd	269
(**34**)	1.454	1.437	48.9	0.47	0; 180	3.0	7.9	RHF/SVd	269
(**35**)	1.450[d]	1.441	52.3	0.49	90; 270	2.5	10.9	RHF/SVd	313
(**36**)	1.45	1.46	60.2					X-ray[g]	315
			68.3					PE[h]	176

[a]Puckering amplitude q and pseudorotational phase angle ϕ of most stable conformer; $\phi = 0°$ or $180°$ corresponds to envelope, $\phi = 90°$ or $270°$ to twist forms. See Refs. 269 and 317.
[b]Pseudorotational barrier.
[c]Inversion barrier of most stable conformer.
[d]Averaged value.
[e]X-ray analysis of dispiro(adamantane-2,3′-(1,2)dioxetane-4′2″-adamantane) ('adamantylidene-adamantane peroxide').
[f]X-ray analysis of 10,10-dimethyl-3,4-dioxatricyclo[5.2.1.0^{1,5}]decane-2-spiro-2′-adamantane. See also Table 22, Section IV.D.
[g]X-ray analysis of 3,3,6,6-tetra(bromomethyl)-1,2,4,5-tetroxane. Average value of R' is given. Ideal chair form assumed: the q value corresponds to q_3, q_2 is zero (see Refs. 317 and 318).
[h]PE analysis of 3,3,6,6-tetramethyl-1,2,4,5-tetroxane.

In Figure 33 the theoretically determined conformational surface of the final ozonide (**33**) is shown in the form of a contour-line diagram[269]. There the conformational space of the five-membered ring is spanned by the puckering amplitude q and the phase angle ϕ ($0° \leq \phi < 360°$)[317,318]. The dashed line indicates the energetically most favourable psuedorotation itinerary. The energy difference between E and T forms determines the pseudorotational barriers. For compounds **32**–**35** these are $\leq 3\,\mathrm{kcal\,mol^{-1}}$ (Table 40), which means that five-membered ring peroxides and ozonides are rather flexible in spite of relatively large barriers to ring inversion. Again, this is due to relatively small changes of the electronic structure during pseudorotation.

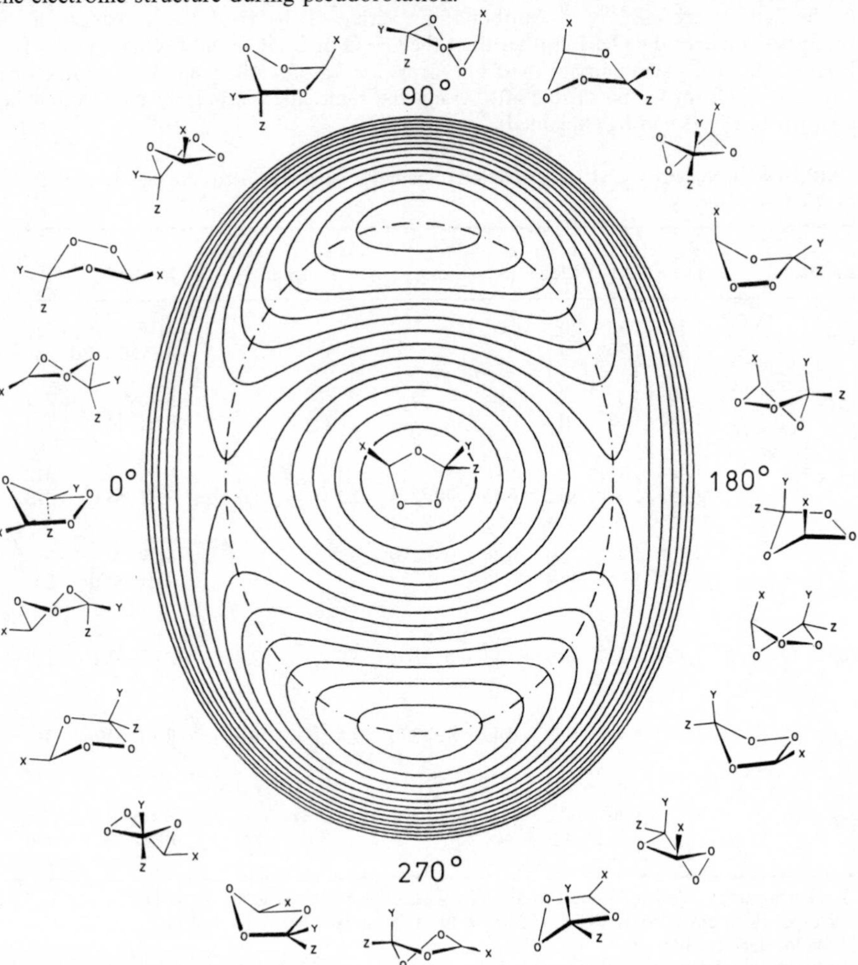

FIGURE 33. Pseudorotational surface of 1,2,3-trioxolane (RHF/C calculations). The potential is zero at the centre of the (q, ϕ) diagram, the innermost contour line corresponds to $-0.5\,\mathrm{kcal\,mol^{-1}}$. The vertical spacing of two contour lines is $0.5\,\mathrm{kcal\,mol^{-1}}$. The dashed line indicates the energetically most favourable pseudorotation path. Conformers are shown along this path in intervals of 18°. Substituents X, Y, Z correspond to hydrogen. Reproduced by permission of the American Institute of Physics from D. Cremer, *J. Chem. Phys.*, **70**, 1898 (1979).

Since the ozonides are important intermediates in the ozonolysis reaction, their conformational properties have been extensively discussed on both experimental[12] and theoretical grounds[319,320].

VI. ABBREVIATIONS, SYMBOLS, CONSTANTS AND CONVERSION FACTORS

A. List of Abbreviations

AO	Atomic Orbital
APSG	Antisymmetrized Product of Strongly-orthogonal Geminals
BSE	Bond Separation Energy
CEPA	Coupled Electron Pair Approximation
CI	Configuration Interaction
DZ	Double Zeta (Basis with 2 GTFs or STFs per AO)
DZb	Double Zeta basis augmented by bond functions
DZb + diff	Double Zeta basis augmented by bond functions and diffuse Rydberg functions
DZd	Double Zeta basis augmented by 3d functions in the heavy-atom part
DZdp	Double Zeta basis augmented by 3d functions in the heavy-atom part and 2p functions in the H part
ED	Electron Diffraction spectroscopy
ESCA	Electron Spectroscopy for Chemical Analysis
FSGO	Floating Spherical Gaussian Orbital
GS	Ground State
GTF, GTO	Gaussian Type Function, GT Orbital
GVB	Generalized Valence Bond
HF	Hartree–Fock
HOMO	Highest Occupied Molecular Orbital
INDO	Intermediate Neglect of Differential Overlap
IP	Ionization Potential
IR	Infrared spectroscopy
INO	Iterative Natural Orbital
LMO	Localized Molecular Orbital
LUMO	Lowest Unoccupied Molecular Orbital
MB, MBS	Minimal Basis Set (1 GTF or STF per AO; e.g. STO-3G)
MBPT	Many-Body Perturbation Theory
MCSCF	MultiConfiguration Self-Consistent Field
MINDO	Modified Intermediate Neglect of Differential Overlap
MM	Molecular Mechanics
MMW	MilliMetre-Wave spectroscopy
MO	Molecular Orbital
MOL	Molecule
MRD-CI	Multi-Reference Double-excitation Configuration Interaction
MW	Microwave spectroscopy
PE	Photoelectron spectroscopy
REF	Reference State
RHF	Restricted Hartree–Fock
RSMP	Rayleigh–Schrödinger Møller–Plesset perturbation theory
SCF	Self-Consistent Field

SOMO	Singly Occupied Molecular Orbital
STF, STO	Slater Type Function, ST Orbital
SV	Split-Valence basis (basis with two functions per AO of the valence shell; e.g. Pople's 4-31G or 6-31G basis sets)
SVd	Split-valence basis augmented by 3d functions in the heavy-atom part (e.g. Pople's 6-31G* basis)
SVdp	Split-valence basis augmented by 3d functions in the heavy-atom part and 2p functions in the H part (e.g. Pople's 6-31G** basis)
TS	Transition State
UHF	Unrestricted Hartree–Fock
VB	Valence Bond
ZPE	Zero-Point vibrational Energy

B. List of Symbols

A	Arbitrary atom
A,B	Term symbols for nonlinear molecules
Å	Ångstrom; 1 Å = 10^{-10} m
a	Radius of spherical cavity (Å) (Section IV.C)
aJ	Atto Joule
a_0	Bohr radius: Atomic unit of length (see conversion factors)
a.u.	Atomic unit
B_0	Rotational constant of lowest vibrational level of ground state
C_i	Coefficient of linear combination
D	Atomic term symbol
D	Debye; unit of dipole moment
D_e	Dissociation energy measured relative to the minimum of the potential energy function ($D_e = D_0 + ZPE$).
D_0	Dissociation energy measured relative to the lowest vibrational level
DH^0, $DH^0(T)$	Dissociation enthalpy measured at temperature T
E	Envelope form of five-membered ring (Section V.B)
E	Energy
E(EXP)	Absolute energy (hartree) at lowest vibrational level of molecular ground state
E(CORR)	Correlation energy (hartree)
E(HF)	Hartree–Fock limit energy (hartree)
E(HF/X)	SCF energy (hartree) obtained with basis set X
E(REL)	Relativistic energy (hartree)
E(S)	Schrödinger energy (hartree)
E(THEO)	Theoretical molecular energy (hartree) for fixed nuclei
E(VIP)	Vibrational energy. (hartree)
E(X)	SCF energy (hartree) obtained with basis set X
EA	Electron affinity
e	Electron charge (see conversion factors)
f	Force constant (aJ Å^{-n}; see conversion factors)
f_e	Equilibrium molecular force constant
f_{RR}, f_{rr}	Quadratic OO and HO stretching force constants (aJ Å^{-2})
$f_{\alpha\alpha}$	Quadratic HOO bending force constant (aJ).
f_{rR}, $f_{rr'}$	HO,OO and HO,OH stretch–stretch coupling constants (aJ Å^{-2})

$f_{R\alpha}, f_{r\alpha}, f_{r\alpha'}$	OO,HOO, HO,HOO and HO,OOH stretch–bend coupling constants (aJ Å^{-1})
$f_{\alpha\alpha'}$	HOO,OOH bend–bend coupling constant (aJ)
f_{RRR}, f_{rrr}	Cubic OO and HO stretching constants (aJ Å^{-3})
$f(r, \theta)$	Common radial function of 2p AOs
g	Electron spin g tensor (Section V.A)
g	Subscript used to denote a 'gerade' function
h	Planck's constant (see conversion factors)
I	Ionization potential (eV) (Section IV.D)
I_{vert}	Vertical ionization potential (eV)
I_i^{exp}	Experimentally observed vertical ionization potential (eV)
i	Subscript used to denote molecular orbitals, orbital energies, etc.
i	$\sqrt{-1}$
$K, K_{z'z'}$	Nuclear quadrupole coupling constant (MHz) (Section IV.F)
k	Number of atoms
$\hat{L}_z$	z component of total orbital angular momentum operator
$\hat{l}_z$	z component of orbital angular momentum operator for a specific electron
M_L	Eigenvalue of $\hat{L}_z$ operator
M_S	Eigenvalue of $\hat{S}_z$ operator
m_l	Eigenvalue of $\hat{l}_z$ operator
m_s	Eigenvalue of $\hat{s}_z$ operator
N_A	Avogadro number
n	Electron lone pair
occ	Occupied orbitals (summation limit)
P	Atomic term symbol
P, P(AB)	Bond order of bond AB
p, p(AB)	Overlap population between atoms A and B
pm	Picometer; 1 pm = 10^{-12} m
Q, Q(A)	Electric quadrupole moment of nucleus A (barn, see conversion factors) (Section IV.F)
q	Puckering amplitude (Å) of a nonplanar ring compound; if not otherwise denoted q corresponds to q_2
q_2	Puckering amplitude (Å) of four- and five-membered rings
q_3	Puckering amplitude (Å) of chair form of six-membered ring
q, q(A)	Charge at atom A (e)
$q_{x'x'}$, etc.	Diagonal elements of electric field gradient tensor measured in principal axes system (esu cm^{-3}) (Section IV.F)
R, R(OO)	Interatomic OO distance
R_e	Equilibrium distance between O atoms of a peroxide at the minimum of the potential energy function
R_0	Effective OO distance derived directly from ground-state rotational constants
R', R(OH), R(OX)	Interatomic OH or OX distance
R'_e, R'_0	See corresponding definitions of R_e and R_0
r_c	Covalent radius of an atom A
$\mathbf{r}_c$(AB)	AB bond critical point of ρ
$\mathbf{r}_e$	Equilibrium geometry determined at the minimum of the potential energy function
$\mathbf{r}_0$	Effective geometry derived directly from ground-state rotational constants

$\mathbf{r}_s$	Effective geometry derived from rotational constants via Kraitchman's equations for a sequence of isotopic substitutions
r_{ip}	Distance between the centres of charge in an ion pair
r, θ, ϕ	Polar coordinates
S	Singlet state
S	Overlap integral
S	Atomic term symbol
$\hat{S}_z$	z component of total spin angular momentum operator
$\hat{s}_z$	z component of spin angular momentum operator for a specific electron
T	Triplet state
T	Temperature (Kelvin)
T_e	Energy of excited state relative to the minimum of the ground-state potential energy function
u	Subscript used to denote an 'ungerade' function
V_i^c	Fourier constant
$V_i(\tau)$	Fourier term
X	Arbitrary basis set
x, y, z	Cartesian coordinates (arbitrary axes system)
x', y', z'	Cartesian coordinates (principal axes system)
α, α(OOH), α(OOX)	Bond angle OOH or OOX
α_e	Equilibrium OOH or OOX bond angle at the minimum of the potential energy function
α'	Bond angle OXO
$\alpha, \beta; \alpha(i), \beta(i)$	Spin functions with $m_s = \frac{1}{2}$ and $m_s = -\frac{1}{2}$
Δ	Term symbol for linear molecules
ΔE	Difference between various energy levels
ΔE_{solv}	Solvation energy
ΔH_f^0	Enthalpy (heat) of formation
$\Delta H_f^0(0)$, $\Delta H_f^0(T)$	Enthalpy of formation at 0° and T° Kelvin
ΔI	Difference between succeeding ionization potentials
ΔR_e	Deviation from equilibrium distance R_e
$\Delta\alpha_e$	Deviation from equilibrium bond angle α_e
$\Delta\rho(\mathbf{r})$	Deformation (difference) density function (ea_0^{-3})
$\nabla\rho(\mathbf{r})$	Gradient vector field of electron density distribution
δ	Puckering angle of a four-membered ring
ε	Dielectric constant
ε_A	Electronegativity of atom A
ε_i	Energy of orbital ϕ_i
η, η^A	Asymmetry parameter of atom A (dimensionless) (Section IV.F)
$\theta_{x'x'}$, etc.	Diagonal elements of molecular quadrupole moment tensor in principal axes system (esu cm^2) (Section IV.F)
λ_i	Eigenvalue i of Hessian matrix of ρ (matrix of second derivatives) (Section IV.F)
μ	Dipole moment (Debye)
μ_a	Reduced mass (atomic-weight units)
v_i	Fundamental vibrational frequency i (cm^{-1})
v_e	Equilibrium vibrational frequency (cm^{-1})
Π	Term symbol for linear molecules
π	Orbitals being antisymmetrical with respect to the molecular plane

π^*	Antibonding π orbitals
ρ,ρ^{A}	Electron density of atom A (ea_0^{-3})
$\rho(\mathbf{r})$	Electron density distribution at point **r** (ea_0^{-3})
Σ	Term symbol for linear molecules
$\sum$	Summation symbol
σ	Orbital being symmetrical with respect to molecular plane or specified bond axis
σ^*	Antibonding σ orbital
τ	HOOH or XOOX dihedral angle
τ_1, τ_2, etc.	Dihedral angles in a polyoxide
τ'	Angle between HOO plane and plane defined by OO bond and C_2 axis
$\tau_{1/2}$	Radiation lifetime of excited state(s) (Section IV.G)
Φ	Slater determinant
ϕ_i	Molecular orbital *i*
φ, ϑ, ψ	Eulerian angles (Section IV.F)
χ	Biradical character given in percent (Section IV.A)
Ψ	Molecular wave function

C. Constants and Conversion Factors

$a_0 = 0.52918 \times 10^{-8}$ cm
$e = 4.803 \times 10^{-10}$ esu $= 1.6022 \times 10^{-19}$ C
$h = 6.6256 \times 10^{-27}$ erg s $= 6.6256 \times 10^{-34}$ Js
$N_A = 6.0225 \times 10^{23}$ mol^{-1}
1 eV = 23.06 kcal mol^{-1}
1 hartree = 27.211 eV = 627.525 kcal mol^{-1}
1 kcal mol^{-1} = 4.184 kJ mol^{-1} = 349.74 cm^{-1}
1 aJ = 10^{-18} J = 1 mdyn Å
= 0.2294 hartree = 6.24 eV
1 barn = 10^{-24} cm^2
1 Debye = 10^{-18} esu cm
1 ea_0 = 2.54158 Debye
1 $ea_0^2 = 1.34492 \times 10^{-26}$ esu cm^2 = 1.34492 Buckingham
1 $ea_0^{-2} = 17.1524 \times 10^{6}$ esu cm^{-2}
1 $ea_0^{-3} = 32.4140 \times 10^{14}$ esu cm^{-3}

VII. ACKNOWLEDGEMENTS

The author owes a great debt of gratitude to his wife Susi for her continuous help and patience in the preparation of the manuscript. Useful discussions were held with Prof. R. F. W. Bader, Prof. S. W. Benson, Dr. J. Bull, Dr. H. Freund, Dr. S. L. Manatt, Dr. R. Pachter, Prof. P. Rademacher, Dr. S. Razumovskii, Dr. P. v. Royen, Prof. K. Schank, Prof. P. N. Skancke and Prof. U. Wahlgren. Prof. W. Jorgensen kindly provided an early version of his MO plot program. Technical assistance by J. Normann helped to prepare the MO drawings. The manuscript was partially written during a stay at the Council for Scientific and Industrial Research, Pretoria, SAR. The author thanks the members of the NCRL for accommodating him during this time. Support of the Fonds der Chemischen Industrie, the Deutschen Forschungsgemeinschaft and the Rechenzentrum der Universität Köln is gratefully acknowledged.

VIII. REFERENCES

1. W. C. Schumb, C. N. Satterfield and R. L. Wentworth, *Hydrogen Peroxide*, Reinhold, New York, 1955.
2. W. M. Weigert (Ed.), *Wasserstoffperoxid und seine Derivate, Chemie und Anwendungen*, Hüthig Verlag, Heidelberg, 1978.
3. A. V. Tobolsky and R. B. Mesrobian, *Organic Peroxides*, Interscience, New York, 1954.
4. A. G. Davies, *Organic Peroxides*, Butterworths, London, 1961.
5. E. G. E. Hawkins, *Organic Peroxides*, Van Nostrand, Princeton, 1961.
6. O. L. Mageli and C. S. Sheppard in *Encyclopedia of Chemical Technology*, 2nd ed., Vol. 14 (Eds. K. E. Kirk and D. F. Othmer), John Wiley and Sons, New York, 1967, pp. 746–820.
7. D. Swern (Ed.), *Organic Peroxides*, Volumes I, II and III, Wiley–Interscience, New York–London, 1971.
8. W. A. Waters, *Mechanisms of Oxidation of Organic Compounds*, Methuen, London, 1964.
9. K. B. Wiberg (Ed.), *Oxidation in Organic Chemistry*, Part A. Academic Press, New York, 1965.
10. D. R. Kearns, *Chem. Rev.*, **71**, 395 (1971); A. A. Frimer, *Chem. Rev.*, **79**, 359 (1979); M. Balci, *Chem. Rev.*, **81**, 91 (1981).
11. H. H. Wasserman and R. W. Murray (Eds.), *Singlet Oxygen*, Academic Press, New York, 1979.
12. P. S. Bailey, *Ozonation in Organic Chemistry*, Vol. I, *Olefinic Compounds*, Academic Press, New York, 1978.
13. S. W. Benson and P. S. Nangia, *Acc. Chem. Res.*, **12**, 223 (1979).
14. O. E. Edwards and R. Curci in *Encyclopedia of Chemical Technology*, 2nd ed., Vol. 14 (Eds. K. E. Kirk and D. F. Othmer), John Wiley and Sons, New York, 1967, pp. 820–839.
15. N. J. Turro and P. Lechtken, *Pure Appl. Chem.*, **33**, 363 (1973).
16. W. Adam, *Acc. Chem. Res.*, **12**, 390 (1979).
17. K. L. Demerjian, J. A. Kerr and J. G. Calvert, *Advan. Environ. Sci. Technol.*, **4**, 1 (1973).
18. J. H. Seinfeld, *Air Pollution, Physical and Chemical Fundamentals*, McGraw–Hill, New York, 1975, Chap. 4, pp. 142–217.
19. M. L. Kaplan and A. M. Trozzolo in *Singlet Oxygen* (Eds. H. H. Wasserman and R. W. Murray), Academic Press, New York, 1979, pp. 575–595.
20. M. A. J. Rodgers and E. L. Powers (Eds.), *Oxygen and Oxy-Radicals in Chemistry and Biology*, Academic Press, New York, 1981.
21. R. D. Jones, D. D. Summerville and F. Basolo, *Chem. Rev.*, **79**, 139 (1979).
22. L. Vaska, *Acc. Chem. Res.*, **9**, 175 (1976).
23. A. B. P. Lever and H. B. Gray, *Acc. Chem. Res.*, **11**, 348 (1978).
24. R. T. Sanderson, *Chemical Bonds and Bond Energy*, Academic Press, New York, 1971.
25. L. Pauling, *The Nature of the Chemical Bond*, Cornell University Press, Ithaca, 1969.
26. W. A. Goddard, III, T. H. Dunning, Jr., W. J. Hunt and P. J. Hay, *Acc. Chem. Res.*, **6**, 368 (1973).
27. W. L. Jorgensen and L. Salem, *The Organic Chemist's Book of Orbitals*, Academic Press, New York, 1973, p. 88. There three-dimensional MO drawings of O_2 are shown.
28. M. Kasha and D. E. Brabham in *Singlet Oxygen* (Eds. H. H. Wasserman and R. W. Murray), Academic Press, New York, 1979, pp. 1–33.
29. B. M. Gimarc, *J. Amer. Chem. Soc.*, **93**, 815 (1971); *Acc. Chem. Res.*, **7**, 384 (1974).
30. R. J. Buenker and S. D. Peyerimhoff, *Chem. Rev.*, **74**, 127 (1974).
31. B. M. Gimarc, *Molecular Structure and Bonding, The Qualitative Molecular Orbital Approach*, Academic Press, New York, 1979, Chap. 7, pp. 153–169.
32. D. Cremer, unpublished results.
33. F. J. Lovas and R. D. Suenram, *Chem. Phys. Letters*, **51**, 453 (1977); R. D. Suenram and F. J. Lovas, *J. Amer. Chem. Soc.*, **100**, 5117 (1978).
34. B. M. Gimarc, *J. Amer. Chem. Soc.*, **92**, 266 (1970); see also Reference 31, Chap. 6, pp. 127–131.
35. B. M. Gimarc, *J. Amer. Chem. Soc.*, **100**, 1996 (1978); see also Reference 31, Chap. 8, pp. 194–200.
36. P. H. Krupenie, *J. Phys. Chem. Ref. Data*, **1**, 423 (1972).
37. H. F. Schaefer, III, *J. Chem. Phys.*, **54**, 2207 (1971).
38. B. J. Moss and W. A. Goddard, III, *J. Chem. Phys.*, **63**, 3523 (1975).
39. B. J. Moss, F. W. Bobrowicz and W. A. Goddard, III, *J. Chem. Phys.*, **63**, 4632 (1975).
40. N. H. F. Beebe, E. W. Thulstrup and A. Andersen, *J. Chem. Phys.*, **64**, 2080 (1976).
41. S. L. Guberman, *J. Chem. Phys.*, **67**, 1125 (1977).
42. R. P. Saxton and B. Liu, *J. Chem. Phys.*, **67**, 5432 (1977).

43. W. T. Zemke, G. Das and A. C. Wahl, *Chem. Phys. Letters*, **14**, 310 (1972).
44. G. Das, W. T. Zemke and W. C. Stwalley, *J. Chem. Phys.*, **72**, 2327 (1980).
45. S. Rothenberg and H. F. Schaefer, III, *Mol. Phys.*, **21**, 317 (1971).
46. J. S. Wright, *Can. J. Chem.*, **51**, 139 (1973).
47. J. S. Wright, *Theoret. Chim. Acta*, **36**, 37 (1974).
48. P. J. Hay, T. H. Dunning, Jr. and W. A. Goddard, III, *Chem. Phys. Letters*, **23**, 457 (1973).
49. P. J. Hay, T. H. Dunning, Jr. and W. A. Goddard, III, *J. Chem. Phys.*, **62**, 3912 (1975).
50. P. J. Hay and T. H. Dunning, Jr., *J. Chem. Phys.*, **67**, 2290 (1977).
51. L. B. Harding and W. A. Goddard, III, *J. Chem. Phys.*, **67**, 2377 (1977).
52. S. Shih, R. J. Buenker and S. D. Peyerimhoff, *Chem. Phys. Letters* **28**, 463 (1974).
53. K. H. Thunemann, S. D. Peyerimhoff and R. J. Buenker, *J. Mol. Spectroy*, **70**, 432 (1978).
54. D. Grimbert and A. Devaquet, *Mol. Phys.*, **27**, 831 (1974).
55. R. R. Lucchese and H. F. Schaefer, III, *J. Chem. Phys.*, **67**, 848 (1977).
56. P. G. Burton, *Intern. J. Quant. Chem., Symp.*, **11**, 207 (1977); *J. Chem. Phys.*, **77**, 961 (1979).
57. G. Karlström, S. Engström and B. Jönsson, *Chem. Phys. Letters*, **57**, 390 (1978).
58. D. Cremer, *J. Amer. Chem. Soc.*, **101**, 7199 (1979).
59. D. Cremer, to be published.
60. J. S. Wright, S. Shih and R. J. Buenker, *Chem. Phys. Letters*, **75**, 513 (1980).
61. C. W. Wilson, Jr. and D. G. Hopper, *J. Chem. Phys.*, **74**, 595 (1981).
62. M. J. S. Dewar, S. Olivella and H. S. Rzepa, *Chem. Phys. Letters*, **47**, 80 (1977).
63. S. W. Benson and A. E. Axworthy Jr., *J. Chem. Phys.*, **26**, 1718 (1957).
64. E. F. Hayes and A. K. Q. Siu, *J. Amer. Chem. Soc.*, **93**, 2090 (1971).
65. K. Yamaguchi, K. Ohta and T. Fueno, *Chem. Phys. Letters*, **50**, 266 (1977).
66. K. Yamaguchi, *Intern. J. Quant. Chem.*, **28**, 101 (1980).
67. P. C. Hiberty and C. Leforestier, *J. Amer. Chem. Soc.*, **100**, 2012 (1978).
68. W. D. Laidig and J. F. Schaefer, III, *J. Chem. Phys.*, **74**, 3411 (1981).
69. L. B. Harding and W. A. Goddard, III, *J. Amer. Chem. Soc.*, **100**, 7180 (1978).
70. J. C. Depannemaecker and J. Bellet, *J. Mol. Spectry*, **66**, 106 (1977).
71. D. H. Liskow, H. F. Schaefer, III and C. F. Bender, *J. Amer. Chem. Soc.*, **93**, 6734 (1971).
72. J. L. Gole and E. F. Hayes, *J. Chem. Phys.*, **57**, 360 (1972).
73. R. J. Blint and M. D. Newton, *J. Chem. Phys.*, **59**, 6220 (1973).
74. W. A. Lathan, L. A. Curtiss, W. J. Hehre, J. B. Lisle and J. A. Pople, *Progr. Phys. Org. Chem.*, **2**, 175 (1974).
75. R. J. Buenker and S. D. Peyerimhoff, *Chem. Phys. Letters*, **37**, 208 (1976).
76. S. K. Shih and S. D. Peyerimhoff, *Chem. Phys. Letters*, **28**, 299 (1978).
77. S. K. Shih, S. D. Peyerimhoff and R. J. Buenker, *Chem. Phys.*, **28**, 299 (1978).
78. C. F. Melius and R. J. Blint, *Chem. Phys. Letters*, **64**, 183 (1979).
79. S. R. Langhoff and R. L. Jaffe, *J. Chem. Phys.*, **71**, 1475 (1979).
80. A. Komornicki and R. L. Jaffe, *J. Chem. Phys.*, **71**, 2150 (1979).
81. T. H. Dunning, Jr., S. P. Walch and M. M. Goodgame, *J. Chem. Phys.* **74**, 3482 (1981).
82. Y. Beers and C. J. Howard, *J. Chem. Phys.*, **64**, 1541 (1976).
83. B. R. De and A. B. Sannigrahi, *J. Comp. Chem.*, **1**, 334 (1980).
84. M. E. Boyd, *J. Chem. Phys.*, **37**, 1317 (1962).
85. S. N. Foner and R. L. Hudson, *J. Chem. Phys.*, **36**, 2681 (1962).
86. P. Gray, *Trans. Faraday Soc.*, **55**, 408 (1959).
87. D. Cremer, to be published.
88. U. Kaldor and I. Shavitt, *J. Chem. Phys.*, **44**, 1823 (1966).
89. W. H. Fink and L. C. Allen, *J. Chem. Phys.*, **46**, 2261 (1967); **46**, 2276 (1967).
90. L. Pedersen and K. Morukuma, *J. Chem. Phys.*, **46**, 3941 (1967).
91. W. E. Palke and R. M. Pitzer, *J. Chem. Phys.*, **46**, 3948 (1967).
92. P. F. Franchini and C. Vergani, *Theoret. Chim. Acta*, **13**, 46 (1969).
93. A. Veillard, *Chem. Phys. Letters*, **4**, 51 (1969); *Theoret. Chim. Acta*, **18**, 21 (1970).
94. R. M. Stevens, *J. Chem. Phys.*, **52**, 1397 (1970).
95. I. H. Hillier, V. R. Saunders and J. F. Wyatt, *Trans. Faraday Soc.*, **66**, 2665 (1970).
96. R. B. Davidson and L. C. Allen, *J. Chem. Phys.*, **55**, 519 (1971).
97. J. P. Ranck and J. Johansen, *Theoret. Chim. Acta*, **24**, 334 (1972).
98. C. Guidotti, U. Lamanna, M. Maestro and R. Moccia, *Theoret. Chim. Acta.*, **27**, 55 (1972).

99. T. H. Dunning, Jr. and N. W. Winter, *Chem. Phys. Letters*, **11**, 194 (1971).
100. L. Radom, W. J. Hehre and J. A. Pople, *J. Amer. Chem. Soc.*, **94**, 2371 (1972).
101. T. H. Dunning, Jr. and N. W. Winter, *J. Chem. Phys.*, **63**, 1847 (1975).
102. P. Botschwina, W. Meyer and A. M. Semkow, *Chem. Phys.*, **15**, 25 (1976).
103. R. E. Howard, M. Levy, H. Shull and S. Hagstrom, *J. Chem. Phys.*, **66**, 5181 (1977); **66**, 5189 (1977).
104. P. B. Ryan and J. D. Todd, *J. Chem. Phys.*, **67**, 4787 (1977).
105. P. G. Burton and B. R. Markey, *Australian J. Chem.*, **30**, 231 (1977).
106. W. R. Rodwell, N. R. Carlsen and L. Radom, *Chem. Phys.*, **31**, 177 (1978).
107. D. Cremer, *J. Chem. Phys.*, **69**, 4440 (1978).
108. M. P. Melrose and R. G. Parr, *Theoret. Chim. Acta*, **8**, 150 (1967).
109. W. H. Fink and L. C. Allen, *J. Chem. Phys.*, **46**, 3270 (1967).
110. G. F. Musso and V. Magnasco, *J. Chem. Phys.*, **60**, 3754 (1974).
111. W. England and M. S. Gorden, *J. Amer. Chem. Soc.*, **94**, 4818 (1972).
112. P. A. Christiansen and W. E. Palke, *J. Chem. Phys.*, **67**, 57 (1977).
113. E. Lombardi, G. Tarantini, L. Pirola and P. Torsellini, *J. Chem. Phys.*, **64**, 5229 (1976).
114. J. A. Pople in *The World of Quantum Chemistry* (Ed. R. Daudel and B. Pullman), Reidel Publishing Co., Dordrecht, Holland, 1974, p. 49.
115. A. Golebiewski and A. Parczewski, *Chem. Rev.*, **74**, 519 (1974).
116. A. Veillard in *Internal Rotation in Molecules* (Ed. W. J. Orville-Thomas), John Wiley and Sons, London, 1974, p. 385.
117. P. W. Payne and L. C. Allen in *Modern Theoretical Chemistry*, Vol. 4, *Applications of Electronic Structure Theory* (Ed. H. F. Schaefer, III), Plenum Press, New York, 1977, p. 29.
118. K. F. Freed, *Chem. Phys. Letters*, **2**, 255 (1968).
119. R. L. Redington, W. B. Olson and P. C. Cross, *J. Chem. Phys.*, **36**, 1311 (1962).
120. R. H. Hunt, R. A. Leacock, C. W. Peters and K. T. Hecht, *J. Chem. Phys.*, **42**, 1931 (1965).
121. W. C. Oelfke and W. Gordy, *J. Chem. Phys.*, **51**, 5336 (1969).
122. C. S. Ewig and D. O. Harris, *J. Chem. Phys.*, **52**, 6268 (1970).
123. D. Käss, H. Oberhammer, D. Brandes and A. Blaschette, *J. Mol. Struct.*, **40**, 65 (1977).
124. C. J. Marsden, D. D. DesMarteau and L. S. Bartell, *Inorg. Chem.*, **16**, 2359 (1977).
125. C. J. Marsden, L. S. Bartell and F. P. Diodati, *J. Mol. Struct.*, **39**, 253 (1977).
126. R. H. Jackson, *J. Chem. Soc.*, 4585 (1962).
127. R. B. Harvey and S. H. Bauer, *J. Amer. Chem. Soc.*, **76**, 859 (1954).
128. G. Winnewisser, M. Winnewisser and W. Gordy, *J. Chem. Phys.*, **49**, 3465 (1968).
129. B. Beagley and K. T. McAlloon, *Trans. Faraday Soc.*, **67**, 3216 (1972).
130. A. Yokozeki and S. H. Bauer, *J. Phys. Chem.*, **80**, 618 (1976).
131. O. J. Sovers, C. W. Kern, R. M. Pitzer and M. Karplus, *J. Chem. Phys.*, **49**, 2592 (1968).
132. K. Ruedenberg, *Rev. Mod. Phys.*, **34**, 326 (1962).
133. T. K. Brunck and F. Weinhold, *J. Amer. Chem. Soc.*, **101**, 1700 (1979).
134. L. C. Allen, *Chem. Phys. Letters*, **2**, 597 (1968).
135. A. Liberles, B. O'Leary, J. E. Eilers and D. R. Whitman, *J. Amer. Chem. Soc.*, **94**, 6894 (1972).
136. J. F. Olsen, *J. Mol. Struct.*, **49**, 361 (1978).
137. C. Romers, C. Altona, H. R. Buys and E. Havinga in *Topics in Stereochemistry*, Vol. 4 (Ed. E. L. Eliel and N. L. Allinger), Wiley–Interscience, New York–London, 1969, p. 39.
138. J. D. Cox and G. Pilcher, *Thermochemistry of Organic and Organometallic Compounds*, Academic Press, London, 1970.
139. D. D. Wagman, W. H. Evans, V. B. Parker, I. Halow, S. M. Bailey and R. H. Schumm, '*Selected values of chemical thermodynamic properties*', *National Bureau of Standards Technical Note*, No. 270–3, Washington D.C., 1968; D. R. Stull and H. Prophet, *Natl. Stand. Ref. Data Ser., Natl. Bur. Stand.*, No. 37 (1971).
140. S. W. Benson, F. R. Cruickshanck, D. M. Golden, G. R. Hangen, H. E. O'Neal, A. S. Rodgers, R. Shaw and R. Walsh, *Chem. Rev.*, **69**, 279 (1969).
141. S. W. Benson, *Thermochemical Kinetics*, 2nd ed., John Wiley and Sons, New York, 1976.
142. S. W. Benson and R. Shaw in *Organic Peroxides*, Vol. 1 (Ed. D. Swern), Wiley–Interscience, New York–London, 1971, p. 105.
143. P. S. Nangia and S. W. Benson, *J. Phys. Chem.*, **83**, 1138 (1979).
144. S. W. Benson and P. S. Nangia, *Acc. Chem. Res.*, **12**, 223 (1979).

145. P. S. Nangia and S. W. Benson, *Intern. J. Chem. Kinet.*, **12**, 29, 43, 169 (1980).
146. S. W. Benson and P. S. Nangia, *J. Amer. Chem. Soc.*, **102**, 2843 (1980).
147. P. S. Nangia and S. W. Benson, *J. Amer. Chem. Soc.*, **102**, 3105 (1980).
148. S. W. Benson, private communication.
149. D. Cremer, *J. Comp. Chem.*, in press.
150. L. Radom, W. J. Hehre and J. A. Pople, *J. Amer. Chem. Soc.*, **93**, 289 (1971).
151. A. C. Hopkinson, K. Yates and I. G. Csizmadia, *Theoret. Chim. Acta*, **23**, 369 (1972).
152. A. C. Hurley, *Advan. Quant. Chem.*, **7**, 315 (1973).
153. H. v. Hirschhausen and K. Wenzel, *Theoret. Chim. Acta*, **25**, 293 (1974).
154. J. G. Calvert and J. N. Pitts, Jr., *Photochemistry*, John Wiley and Sons, New York, 1967, pp. 824–826.
155. J. G. Kirkwood, *J. Chem. Phys.*, **2**, 351 (1934).
156. T. Koopmans, *Physica*, **1**, 104 (1934).
157. G. D. Purvis and Y. Öhrn, *J. Chem. Phys.*, **62**, 2045 (1975).
158. K. Siegbahn, C. Nordling, G. Johansson, J. Hedman, P. F. Heden, K. Hamrin, U. Gelius, T. Bergmark, L. O. Werme, R. Manne and Y. Baer, *ESCA Applied to Free Molecules*, North-Holland, Amsterdam, 1969.
159. O. Edqvist, E. Lindholm, L. E. Selin and L. Åsbrink, *Phys. Scripta*, **1**, 25 (1970).
160. J. L. Gardner and J. A. R. Samson, *J. Chem. Phys.*, **62**, 4460 (1975).
161. P. S. Bagus and H. F. Schaefer, III, *J. Chem. Phys.*, **56**, 224 (1972); P. S. Bagus, M. Schrenck, D. W. Davis and D. A. Shirley, *Phys. Rev.*, **9A**, 1090 (1974).
162. D. C. Frost, S. T. Lee and C. A. McDowell, *Chem. Phys. Letters*, **24**, 149 (1974).
163. C. R. Brundle, *Chem. Phys. Letters*, **26**, 25 (1974).
164. J. M. Dyke, L. Golob, N. Jonathan, A. Morris and M. Okuda, *J. Chem. Soc. Faraday Trans.* 2, 1828 (1974).
165. H. Basch, *J. Amer. Chem. Soc.*, **97**, 6047 (1975).
166. L. S. Cederbaum, W. Domcke, W. v. Niessen and W. P. Kramer, *Mol. Phys.*, **34**, 381 (1977).
167. H. Sambe and R. H. Felton, *J. Chem. Phys.*, **61**, 3862 (1974); but see also R. P. Messmer and D. R. Salahub, *J. Chem. Phys.*, **65**, 779 (1976).
168. A. J. B. Robertson, *Trans. Faraday. Soc.*, **48**, 228 (1952).
169. K. Osafune and K. Kimura, *Chem. Phys. Letters*, **25**, 47 (1974).
170. R. S. Brown, *Can. J. Chem.*, **53**, 3439 (1975).
171. F. S. Ashmore and A. R. Burgess, *J. Chem. Soc., Faraday Trans.* 2, **73**, 1247 (1977).
172. C. R. Brundle, M. B. Robin and H. Basch, *J. Chem. Phys.*, **53**, 2196 (1970).
173. D. W. Davies, *Chem. Phys. Letters*, **28**, 520 (1974).
174. C. Batich and W. Adam, *Tetrahedron Letters* 1467 (1974).
175. R. S. Brown and R. W. Marcinko, *J. Amer. Chem. Soc.*, **100**, 5584 (1978).
176. P. Rademacher and W. Elling, *Liebigs Ann. Chem.*, 1473 (1979).
177. K. Kimura and K. Osafune, *Bull. Chem. Soc. Japan*, **48**, 2421 (1975).
178. M. J. W. Boness and G. J. Schulz, *Phys. Rev.*, **A2**, 2182 (1970).
179. R. J. Celotta, R. A. Bennett, J. L. Hell, M. W. Siegel and J. Levine, *Phys. Rev.*, **A6**, 631 (1972).
180. H. Föppl, *Z. Anorg. Allgem. Chem.*, **291**, 12 (1957).
181. N. L. Summers and J. Tyrrell, *J. Amer. Chem. Soc.*, **99**, 3960 (1977).
182. J. H. Yates and R. M. Pitzer, *J. Chem. Phys.*, **66**, 3592 (1977).
183. T. Tanaka and Y. Morino, *J. Mol. Spectry*, **33**, 552 (1970).
184. A. Barbe, C. Secroun and P. Jouve, *J. Mol. Spectry*, **49**, 171 (1974).
185. M. L. Vestal and G. H. Mauclaire, *J. Chem. Phys.*, **67**, 3767 (1977).
186. S. E. Novick, P. C. Engelking, P. L. Jones, J. H. Futrell and W. C. Lineberger, *J. Chem. Phys.*, **70**, 2652 (1979).
187. G. P. Smith and L. C. Lee, *J. Chem. Phys.*, **71**, 2323 (1979).
188. J. F. Hiller and M. L. Vestal, *J. Chem. Phys.*, **74**, 6096 (1981).
189. D. C. McCain and W. E. Palke, *J. Chem. Phys.*, **56**, 4957 (1972).
190. P. A. Giguère and V. Schomaker, *J. Amer. Chem. Soc.*, **65**, 2025 (1943).
191. W. R. Busing and H. A. Levy, *J. Chem. Phys.*, **42**, 3054 (1965).
192. J. M. Savariault and M. S. Lehmann, *J. Amer. Chem. Soc.*, **102**, 1298 (1980).
193. S. F. Trevino, C. S. Choi and M. K. Farr, *J. Chem. Phys.*, **63**, 2620 (1975).
194. S. W. Peterson and H. A. Levy, *Acta Cryst.*, **10**, 70 (1957).

195. H. Kim, E. F. Pearson and E. H. Appelman, *J. Chem. Phys.*, **56**, 1 (1972).
196. I. N. Przhevalskii and G. A. Khachkuruzov, *Opt. Spectry*, **36**, 172 (1974).
197. D. Cremer and D. Christen, *J. Mol. Spectry*, **74**, 480 (1979).
198. P. A. Giguère and T. K. K. Srinivasan, *J. Mol. Spectry*, **66**, 168 (1977).
199. D. Cremer, *Fresenius Z. Anal. Chem.*, **304**, 261 (1980).
200. B. F. Pedersen, *Acta Cryst.*, **B28**, 1014 (1972).
201. B. F. Pedersen, *Acta Chem. Scand.*, **23**, 1871 (1969).
202. B. F. Pedersen and B. Pedersen, *Acta Chem. Scand.*, **18**, 1454 (1964).
203. B. F. Pedersen, *Acta Chem. Scand.*, **21**, 779 (1967).
204. J. M. Adams, V. Ramdas and A. W. Hewat, *Acta Cryst.*, **B36**, 570 (1980).
205. J. M. Adams and R. G. Pritchard, *Acta Cryst.*, **B32**, 2438 (1976).
206. J. M. Adams and V. Ramdas, *Acta Cryst.*, **B34**, 2150 (1978).
207. G. A. Khachkuruzov and I. N. Przhevalskii, *Opt. Spectry*, **33**, 127, 434 (1972).
208. P. A. Giguère and T. K. K. Srinivasan, *J. Raman Spectry*, **2**, 125 (1974).
209. D. W. Smith and L. Andrews, *J. Chem. Phys.*, **60**, 81 (1974).
210. J. W. Johns, A. R. W. McKellar and M. Riggin, *J. Chem. Phys.*, **68**, 3957 (1978).
211. J. D. Rogers and J. J. Hillman, *J. Chem. Phys.*, **75**, 1085 (1981).
212. G. A. Khachkuruzov and I. N. Przhevalskii, *Opt. Spectry*, **36**, 175 (1974).
213. G. Herzberg, *Molecular Spectra and Molecular Structure: I. Spectra of Diatomic Molecules*, Van Nostrand, New York, 1950.
214. L. Andrews, *J. Chem. Phys.*, **50**, 4288 (1969).
215. L. C. Snyder and H. Basch, *Molecular Wave Functions and Properties*, Wiley–Interscience, New York–London, 1972, p. T–88.
216. R. Janoschek, *Z. Naturforsch.*, **25a**, 311 (1970).
217. R. E. Christoffersen and K. A. Baker, *Chem. Phys. Letters*, **8**, 4 (1971).
218. S. Bratož, R. Daudel, M. Roux and M. Allavena, *Rev. Mod. Phys.*, **32**, 412 (1960).
219. P. Coppens and E. D. Stevens, *Advan. Quantum Chem.*, **10**, 1 (1977).
220. J. Rys and M. Dupuis, results cited in P. Coppens, Y. W. Yang, R. H. Blessing, W. F. Cooper and F. K. Larsen, *J. Amer. Chem. Soc.*, **99**, 760 (1977).
221. R. F. W. Bader, S. G. Anderson and A. J. Duke, *J. Amer. Chem. Soc.*, **101**, 1389 (1979).
222. R. F. W. Bader, T. T. Nguyen-Dang and Y. Tal, *J. Chem. Phys.*, **70**, 4316 (1979).
223. R. F. W. Bader, *J. Chem. Phys.*, **73**, 2871 (1980).
224. R. F. W. Bader, T. H. Tang, Y. Tal and F. W. Biegler-König, *J. Amer. Chem. Soc.*, to be published.
225. R. F. W. Bader, private communication.
226. L. C. Snyder, *J. Chem. Phys.*, **61**, 747 (1974).
227. J. T. Massey and D. R. Bianco, *J. Chem. Phys.*, **22**, 442 (1954).
228. M. Lichtenstein, J. J. Gallagher and S. A. Clough, *J. Mol. Spectry*, **40**, 10 (1971).
229. O. Lumpkin and W. T. Dixon, *J. Chem. Phys.*, **71**, 3550 (1979).
230. R. J. Buenker and S. D. Peyerimhoff, *Chem. Phys. Letters*, **34**, 225 (1975).
231. R. P. Messmer and D. R. Salahub, *J. Chem. Phys.*, **65**, 779 (1976).
232. H. E. Hunziker and H. R. Wendt, *J. Chem. Phys.*, **60**, 4622 (1974).
233. K. H. Becker, E. H. Fink, P. Langen and U. Schurath, *J. Chem. Phys.*, **60**, 4623 (1974).
234. T. T. Paukert and H. S. Johnston, *J. Chem. Phys.*, **56**, 2824 (1972).
235. C. J. Hochanadel, J. A. Ghormley and P. J. Ogrem, *J. Chem. Phys.*, **56**, 4426 (1972).
236. A. Rauk and J. M. Barriel, *Chem. Phys.*, **25**, 409 (1977).
237. M. Schürgers and K. H. Welge, *Z. Naturforsch.*, **23a**, 1508 (1968).
238. J. G. Calvert and J. N. Pitts Jr., *Photochemistry*, Wiley–Interscience, London–New York, 1967, pp. 443, 447–450.
239. W. G. Dauben, L. Salem and N. J. Turro, *Acc. Chem. Res.*, **8**, 41 (1975).
240. E. M. Evleth, *J. Amer. Chem. Soc.*, **98**, 1637 (1976).
241. E. M. Evleth and E. Kassab, *J. Amer. Chem. Soc.*, **100**, 7859 (1978).
242. D. A. Hatzenbuhler and L. Andrews, *J. Chem. Phys.*, **56**, 3398 (1972).
243. L. Andrews and R. R. Smardzewski, *J. Chem. Phys.*, **58**, 2258 (1973).
244. F. B. Billingsley and C. Trindle, *J. Phys. Chem.*, **76**, 2295 (1972).
245. S. V. O'Neil, H. F. Schaefer, III and C. F. Bender, *J. Chem. Phys.*, **59**, 3608 (1973).
246. D. T. Grow and R. M. Pitzer, *J. Chem. Phys.*, **67**, 4019 (1977).
247. K. Ohkubo, T. Fujita and H. Sato, *J. Mol. Struct.*, **36**, 101 (1977).

248. K. Yamaguchi and S. Iwata, *Chem. Phys. Letters*, **76**, 375 (1980).
249. M. M. L. Chen, R. W. Wetmore and H. F. Schaefer, III, *J. Chem. Phys.*, **74**, 2938 (1981).
250. P. N. Noble and G. C. Pimentel, *J. Chem. Phys.*, **44**, 3641 (1966).
251. J. L. Gole and E. F. Hayes, *Intern. J. Quantum Chem.*, **3S**, 519 (1970).
252. D. C. McCain and W. E. Palke, *J. Magn. Res.*, **20**, 52 (1975).
253. A. Arkell and I. Schwager, *J. Amer. Chem. Soc.*, **89**, 5999 (1967).
254. W. R. Wadt and W. A. Goddard, III, *J. Amer. Chem. Soc.*, **97**, 3004 (1975).
255. P. C. Hiberty, *J. Amer. Chem. Soc.*, **98**, 6088 (1976).
256. C. A. Hull, *J. Org. Chem.*, **43**, 2780 (1978).
257. K. Yamaguchi, K. Ohta, S. Yabushita and T. Fueno, *J. Chem. Phys.*, **68**, 4323 (1978).
258. G. Karlström, S. Engström and B. Jönsson, *Chem. Phys. Letters*, **67**, 343 (1979).
259. K. Yamaguchi, S. Yabushita, T. Fueno, S. Kato and K. Morokuma, *Chem. Phys. Letters*, **71**, 563 (1980).
260. H. F. Schaefer, III, C. F. Bender and J. H. Richardson, *J. Phys. Chem.*, **80**, 2035 (1976).
261. P. A. Benioff, G. Das and A. C. Wahl, *J. Chem. Phys.*, **67**, 2449 (1977); G. Das and P. A. Benioff, *Chem. Phys. Letters*, **75**, 519 (1980).
262. P. K. Pearson, H. F. Schaefer, III, J. H. Richardson, L. M. Stephenson and J. I. Brauman, *J. Amer. Chem. Soc.*, **96**, 6778 (1974).
263. K. Yamaguchi, S. Yabushita and T. Fueno, *J. Chem. Phys.*, **71**, 2321 (1979).
264. W. A. Lathan, L. Radom, P. C. Hariharan, W. J. Hehre and J. A. Pople, *Topics Curr. Chem.*, **40**, 1 (1973).
265. R. D. Spratley and G. C. Pimentel, *J. Amer. Chem. Soc.*, **88**, 2394 (1966).
266. D. Cremer, to be published.
267. E. Block, R. E. Penn, A. A. Bazzi and D. Cremer, *Tetrahedron Letters*, **22**, 29 (1981).
268. J. Peslak, Jr., *J. Mol. Struct.*, **12**, 235 (1972).
269. D. Cremer, *J. Chem. Phys.*, **70**, 1898 (1979).
270. D. Cremer, *J. Chem. Phys.*, **70**, 1911 (1979).
271. W. J. Hehre, L. Radom and J. A. Pople, *J. Amer. Chem. Soc.*, **94**, 1486 (1972).
272. B. Plesničar, S. Kaiser and A. Ažman, *J. Amer. Chem. Soc.*, **95**, 5476 (1973).
273. D. Cremer, *J. Chem. Phys.*, **69**, 4456 (1978).
274. J. Kao, *J. Mol. Struct.*, **56**, 147 (1979).
275. M. Allavena, E. Blaisten-Barojas and B. Silvi, *J. Chem. Phys.*, **75**, 787 (1981).
276. O. Gropen, *J. Mol. Struct.*, **32**, 85 (1976).
277. B. Plesničar, D. Kocjan, S. Murovec and A. Ažman, *J. Amer. Chem. Soc.*, **98**, 3143 (1976).
278. R. R. Luchese, J. F. Schaefer, III, W. R. Rodwell and L. Radom *J. Chem. Phys.*, **68**, 2507 (1978).
279. A. D. Kirshenbaum, A. V. Grosse and J. G. Aston, *J. Amer. Chem. Soc.*, **81**, 6398 (1959).
280. R. L. Kuczkowski, *J. Amer. Chem. Soc.*, **85**, 3047 (1963); **86**, 3617 (1964).
281. A. Hinchliffe, *J. Mol. Struct.*, **55**, 127 (1979).
282. M. J. S. Dewar and W. Thiel, *J. Amer. Chem. Soc.*, **97**, 3978 (1975); **99**, 2338 (1977).
283. L. B. Harding and W. A. Goddard, III, *J. Amer. Chem. Soc.*, **99**, 4520 (1977); **102**, 439 (1980).
284. K. Yamaguchi, T. Fueno, I. Saito and T. Matsuura, *Tetrahedron Letters*, **21**, 4087 (1980).
285. K. Yamaguchi. S. Yabushita and T. Fueno, *Chem. Phys. Letters*, **78**, 572 (1981).
286. K. Yamaguchi, S. Yabushita, T. Fueno and K. N. Houk, *J. Amer. Chem. Soc.*, **103**, 5043 (1981).
287. D. C. Hrncir, R. D. Rogers and J. L. Atwood, *J. Amer. Chem. Soc.*, **103**, 4277 (1981).
288. D. M. Kurtz, Jr., D. F. Shriver and I. M. Klotz, *J. Amer. Chem. Soc.*, **98**, 5033 (1976).
289. H. Oberhammer and J. E. Boggs, *J. Amer. Chem. Soc.*, **102**, 7241 (1980).
290. P. H. Blustin, *Theoret. Chim. Acta*, **48**, 1 (1978).
291. N. L. Allinger, J. Kao, H.-M. Chang and D. B. Boyd, *Tetrahedron*, **32**, 2867 (1976).
292. E. Hirota, *Bull. Chem. Soc. Japan*, **31**, 130 (1958).
293. C. W. Bock, M. Trachtman and P. George, *J. Mol. Spectry*, **84**, 256 (1980); *J. Mol. Struct.*, **71**, 327 (1981).
294. L. M. Hjelmeland and G. H. Loew, *Chem. Phys. Letters*, **32**, 309 (1975).
295. L. M. Hjelmeland and G. Loew, *Tetrahedron*, **33**, 1029 (1977).
296. C. Petrongolo, *Chem. Phys.*, **26**, 243 (1977).
297. B. Plesničar, M. Tasevski and A. Ažman, *J. Amer. Chem. Soc.*, **100**, 743 (1978).
298. A. Ažman, J. Koller and B. Plesničar, *J. Amer. Chem. Soc.*, **101**, 1107 (1979).
299. O. Kikuchi, K. Suzuki and K. Tokumaru, *Bull. Chem. Soc. Japan*, **52**, 1086 (1979).

300. K. Ohkubo and J. Sato, *Bull. Chem. Soc. Japan*, **52**, 1525 (1979).
301. N. W. Winter, W. A. Goddard, III and C. F. Bender, *Chem. Phys. Letters*, **33**, 25 (1975).
302. G. R. McMillan and J. G. Calvert, *Oxidation Combustion Rev.*, **1**, 83 (1965).
303. P. Brant, J. A. Hashmall, F. L. Carter, R. DeMarco and W. B. Fox, *J. Amer. Chem. Soc.*, **103**, 329 (1981).
304. D. Cremer and J. A. Pople, *J. Amer. Chem. Soc.*, **97**, 1358 (1975).
305. H. Niki, P. D. Maker, C. M. Savage and L. P. Breitenbach, *Chem. Phys. Letters*, **73**, 43 (1980).
306. T. Minato, S. Yamabe, H. Fujimoto and K. Fukui, *Bull. Chem. Soc. Japan*, **51**, 682 (1978).
307. H. F. Caroll and S. H. Bauer, *J. Amer. Chem. Soc.*, **91**, 7727 (1969).
308. A. Bar-nun and A. Lifshitz, *J. Chem. Phys.*, **47**, 2878 (1967).
309. K. Hiraoka, P. P. S. Saluja and P. Kebarle, *Can. J. Chem.*, **57**, 2159 (1979).
310. S. Yamabe and K. Hirao, *J. Amer. Chem. Soc.*, **103**, 2176 (1981).
311. J. Hess and A. Voss, *Acta Cryst.*, **B33**, 3527 (1977).
312. P. B. Hitchcock and I. Beheshti, *J. Chem. Soc., Perkin* 2, 126 (1979).
313. D. Cremer, *J. Amer. Chem. Soc.*, to be published.
314. C. W. Gillies and R. L. Kuczkowski, *J. Amer. Chem. Soc.*, **94**, 6337 (1972); R. L. Kuczkowski, C. W. Gillies and K. L. Gallaher, *J. Mol. Spectry*, **60**, 361 (1976); U. Mazur and R. L. Kuczkowski, *J. Mol. Spectry*, **65**, 84 (1977).
315. M. Schulz, K. Kirschke and E. Hohne, *Chem. Ber.*, **100**, 2242 (1967).
316. D. Cremer, *J. Chem. Phys.*, **70**, 1928 (1979).
317. D. Cremer and J. A. Pople, *J. Amer. Chem. Soc.*, **97**, 1354 (1975).
318. D. Cremer, *Israel J. Chem.*, **20**, 12 (1980).
319. D. Cremer, *J. Amer. Chem. Soc.*, **103**, 3619, 3627, 3633 (1981).
320. D. Cremer, *Angew. Chem.*, **93**, 934 (1981).

The Chemistry of Functional Groups, Peroxides
Edited by S. Patai

CHAPTER 2

Stereochemical and conformational aspects of peroxy compounds

OTTO EXNER
Institute of Organic Chemistry and Biochemistry, Czechoslovak Academy of Sciences, 16610 Prague 6, Czechoslovakia

I. INTRODUCTION . . . 85
II. CONFORMATION OF HYDROGEN PEROXIDE AND GENERAL THEORY . 86
III. ALKYL PEROXIDES AND HYDROPEROXIDES . . . 89
IV. ACYL PEROXIDES . . . 91
V. PEROXY ACIDS . . . 92
VI. PEROXY ESTERS . . . 93
VII. REFERENCES . . . 94

1. INTRODUCTION

The stereochemistry of peroxy compounds is dominated by the conformation around the O—O bond which is more or less influenced by the neighbouring groups. A characteristic feature of this conformation is a relatively flat potential energy curve, at least in the part near to the antiperiplanar arrangement. There are two important consequences. Firstly, the conformers of a given molecule may interconvert rather rapidly, giving rise to some problems, e.g. in NMR spectroscopy and in dipole-moment studies. Secondly, the conformation varies from one molecule to another and in different states under the influence of intramolecular and intermolecular forces, viz. hydrogen bonds, substituent effects, solvent effects and crystal packing forces. This makes even the exact results of X-ray analysis or of neutron diffraction less telling to the structure of an isolated molecule.

An additional complicating feature is of experimental origin and connected with the inherent instability of all peroxy compounds. It becomes evident even in X-ray crystallography and makes the results less accurate; it further almost prevents work in the gas phase (electron diffraction, microwave spectroscopy) giving thus room to less precise solution methods, like dipole moments and infrared spectroscopy. It is for all these reasons that the conformation of many peroxy compounds is known only approximately. The general problems and basic features of the conformation have been clearly stated in a 1971 review[1], reporting in particular crystal structures; however, some results concerning isolated molecules have been later revised. The present survey is focused on experimental conformations in the gas phase and in solution and theoretical predictions; some crystal structures are quoted for comparison.

II. CONFORMATION OF HYDROGEN PEROXIDE AND GENERAL THEORY

All the problems involved are best seen on the parent compound, hydrogen peroxide, which has been given more attention than all organic peroxides put together, especially as far as theoretical work is concerned. It is why the experimental results on organic peroxides are often compared to calculations carried out for H_2O_2. The early investigations have been reviewed in a book[2], and some new aspects discussed recently[3]. There is an early classic paper by Penney and Sutherland[4] in which the nonplanar conformation **1** with a dihedral angle τ of approximately 100° was suggested, the controlling factor being reuplsion of lone electron pairs assumed to have pure p_y character. Although the reasoning seems more than primitive in the light of contemporary quantum chemistry, it was correct in the qualitative sense. Its merit was of drawing attention to the predictive power of quantum theory and also to the stereochemistry about single bonds. It was thus shown that even simple molecules need not possess the simplest conformation with the highest symmetry, as had been always written in common structural formulae.

(1) **(2)**

Penney and Sutherland's prediction challenged both experimental and theoretical work but it lasted 40 years before some degree of agreement was reached. The first qualitative confirmation was seen in the crystal structure[5] of the crystalline complex $CO(NH_2)_2 \cdot H_2O_2$ ($\tau = 106°$). In its quantitative aspect, however, this result was depreciated by studies of further complexes[6–8] (Table 1) which revealed τ ranging from 102° up to 180° (structure **2**) compared to 90° in the crystal of pure hydrogen peroxide[9–11]. From the sensitivity to molecular environment the conclusion was drawn that the potential curve must be rather flat near its minimum. Therefore, the crystallographic data in the field of peroxides may be considered only as a crude approximation of the minimum energy conformation. Later examples of crystalline complexes are numerous, e.g. References 12–19 (Table 1); they reveal among others the values[13,19] of $\tau = 180°$, two different conformations in a single crystal[12–14], disordered molecules[15,16] and lowered symmetry[12]; the main subject of investigation has been intermolecular hydrogen bonds. Besides hydrogen bonds and packing forces, the dihedral angle depends also on substitution [see the effect of electronegativity in fluorine peroxide[20] and the steric effect in di-*t*-butyl peroxide[21] (Table 1).] It follows that a discussion of conformation must not blend results obtained on different compounds.

Returning to hydrogen peroxide itself, a more exact experimental determination of the gas-phase equilibrium geometry is a rather complex matter[3,22]. The most accurate primary data are rotational constants, but four internal coordinates cannot be derived from three constants without an assumption based on some less dependable reasoning. The dihedral angle τ is particularly sensitive to these assumptions so that its estimates are in a strict sense not purely experimental values. Therefore, even the good agreement between microwave[23] and infrared[24] spectral results ($\tau = 120°$) is depreciated by the common assumption concerning the O—H bond length. Further infrared studies[25,26] of H_2O_2 and D_2O_2 have claimed the values to be near 111° and only recent very precise analyses[22,27] have again preferred the values near to 120°. In exact considerations it is necessary to distinguish the geometry at minimum potential energy (τ_e) from the average geometry of the vibrational ground state (τ_0). In H_2O_2 the latter angle is about 1° higher; the difference does not seem important but some confusion in the results arises[3,22].

If an attempt is made to calculate τ from dipole moments using the H—O bond moment derived from water molecule, one obtains 83° (from gas-phase data[28]) or 93° (from measurement in dioxan[29]). Since the uncertainty in the H—O—O angle[23] is of no consequence, the bond-moment scheme[30] is evidently not accurate enough for such molecules. It may follow that even the dihedral angles of organic peroxides, determined from dipole moments, are somewhat small.

Regarding the quantum-chemical calculations, the H_2O_2 geometry has been a hard nut for theoreticians. Almost all possible kinds of improvement and refinement have been tested on this simple molecule till an agreement with experiment has finally been reached. It was the equilibrium dihedral angle which presented the greatest problems; remarkably enough the results of more sophisticated methods were at first worse than those of semiempirical and still simpler approaches. In Table 1 some representative calculated values are arranged approximately according to the complexity of the procedure. Striking is the relative success of very simple ion-pair[4] and electrostatic[31] calculations. CNDO[32], INDO[33] and MINDO[34,35] were apparently also successful provided that the geometry was completely optimized; however, the acceptable values of τ were redeemed by unrealistic remaining parameters. Surprising was the failure of early *ab initio* calculations[36,37], even on the Hartree–Fock limit[38–41]; the results have been summarized and the reasons discussed in some detail[42]. Besides some good results (maybe fortuitous) with the 4-31G basis[43], mostly very complicated calculations[44–47] were necessary to meet the experimental findings. They included a very extended basis, in particular polarization functions[40,41,44,46–48], full optimization of geometry[39,44,46,47] for each value of τ, electron correlation[47,49] and Rayleigh–Schrödinger perturbation theory[47]; even so a survey of results[46] revealed that some parameters may be poorly predicted. Note that even certain errors in calculations[37,44] were later disclosed[38,46]. The best possible agreement with experiment has probably already been reached[47], and recent studies[48–53] focus attention merely to the importance of individual named improvements with respect to the economy of calculation, and suggest simplified procedures[50,53]; in the extreme the angle τ is simply assigned a fixed value[54].

In conclusion it seems that too much emphasis should not be given to the exact value of an equilibrium parameter of such a flexible molecule. On the other hand, many attempts have been made to describe the conformation in a simplified manner and/or to explain it in terms of structural chemistry. The original model[4,31] is still widely accepted[21]; according to this one lone electron pair on either oxygen has largely s character, the other is essentially p_y, perpendicular to the OOH plane. The repulsion between the latter pairs would be at minimum for $\tau = 90°$, a larger value experimentally found is attributed to repulsion of hydrogen atoms. In alternative models the lone electron pairs may be described as largely delocalized and opposite to the O—H bonds[33], or as nearly sp^3-

TABLE 1. A survey of dihedral angles, calculated and experimental, in some peroxides and peroxy compounds

Compound	Dihedral angle, X—O—O—X		
	Calculated	Experimental (gas or solution)	Experimental (X-ray in crystal)
HO—OH	~100 (estim.[4,31])	120 (MW[23])	90.2[9,a]
	84–109 (semiemp.[32–35])	120 (IR[24])	93.5[10]
	120 (4-31G[43])	(111.5) (IR[25])	96–139[b]
	180 (*ab initio*[36–41])	119.1 (IR[27])	180[c]
	114–132 (large basis[42,44,46])		
	117.3 (APSG[49])		
	119.3 (el. correl.[47])		
DO—OD		(110.8) (IR[26])	90.8[11,a]
FO—OH	75 (4-31G[43])		
FO—OF	86.5 (CNDO/2[32])	87.5 (MW[20])	
RO—OH	140 (4-31G[43])	~100 (μ[60,61])	90,92[d]
			119[d]
CF_3O—OH	~120 (4-31G[76])	~95 (ED[71])	
MeO—OMe	96.5 (MINDO[35])	170 (PES[65])	
	110.1 (MINDO/3[67])	<180 (ED[68])	
t-BuO—OBu-*t*	130 (MM[66])	166 (ED[21])	180[74,e]
	180 (CNDO/2[21])	123–126 (μ[60–62])	
		155 (PES[65])	
CF_3O—OCF_3		123, 71 (ED[69,70])	
CF_3O—OHal	97 (4-31G[76])	93–97 (ED[71])	
Me_3SiO—$OSiMe_3$		~144 (ED[21])	
F_5SO—OSF_5		107 (ED[72])	
BzO—OBz		(100) (μ[61])	91[80]
		115 (μ[87])	81[f]
			86.6[f]
HCOO—OH	0 (INDO[102])	0 (IR[94])	
	0 (STO-2G[103])		
	180 (4(5)-31G[102,104])		
RCOO—OH	0 (EHT[89])	0 (MW[95])	133[96]
	0 (STO-3G[105])	0 (IR[90,94])	
		(72) (μ[99])	
BzO—OH		0 (IR[93])	146[g]
		0 (μ[100])	170[g]
RCOO—OMe	~90 (EHT[89])	~130 (μ[110])	
RCOO—OBu-*t*		~170 (μ[110])	144[108]
		~130 (μ[109])	<180[107]

[a]Neutron diffraction data.
[b]Various crystalline complexes of hydrogen peroxide[5–7,12–18].
[c]Crystalline complexes $NaOCOCOONa{\cdot}H_2O_2$ [8], $Na_2CO_3{\cdot}\frac{3}{2}H_2O_2$ [13], $NH_4F{\cdot}H_2O_2$ [19].
[d]α-Substituted peroxides $(CH_2)_{11}C(OOH)_2$[77] and 2-Ph-1-OOH-*c*-C_6H_9—N=NC_6H_4OMe-2[78], respectively.
[e]Triphenylmethyl peroxide.
[f]4-Chlorobenzoyl peroxide[81] and acetyl benzoyl peroxide[84], respectively.
[g]2-Nitroperoxybenzoic[97] and 4-nitroperoxybenzoic[98] acid, respectively.

hybridized and tetragonally arranged[12,55,56]. In the latter case a stabilizing interaction must occur either between two adjacent lone pairs[55] or between a lone pair and the antibond pertinent to the opposite bond[56].

Another analysis[57] attributes the conformation of H_2O_2 to maximum electron–nuclear attraction in spite of electron–electron repulsion which is also at maximum. The most recent description has been given in terms of group orbitals[47] corresponding to the concept of two interacting OH radicals. Wolfe has coined the term *gauche* effect for the general preference of a conformation with the maximum number of *gauche* interactions between adjacent electron pairs and/or polar bonds[55]. This is merely a label rather than an explanation and in complex cases the preferred conformation can be barely predicted; probably the preference is better interpreted in terms of local symmetry[58].

Any static description of conformation in terms of equilibrium geometry is rather incomplete because the values of rotational barriers may be of deciding importance. Note, for example, that the conformation **1** is chiral and it is only due to the low barriers that enantiomers cannot be isolated. Experimental and theoretical studies agree that there are two barriers around the O—O bond in H_2O_2 (Table 2). The higher *cis* barrier corresponds to eclipsing of the two hydrogen atoms and the much lower *trans* barrier allows a rapid degenerate rearrangement into the enantiomer via the conformation **2**. As to the actual values, difficulties similar to those found in the calculations of equilibrium angles have been encountered. Even here the 'experimental' values also depend on the underlying theory and have been several times improved[24–27,59]; the uncertainty is more marked with the *cis* barrier, less important in chemical terms. In some recent papers the agreement between theory[46,47] and experiment[25,59] is apparently good, but the most reliable *trans* barrier is higher[22,27]. Compared for example to ethane the *trans* barrier is less than one half, the *cis* barrier is more than twofold. One can guess that the barriers are rather sensitive to substitution, although there are only sporadic data available (Table 2).

III. ALKYL PEROXIDES AND HYDROPEROXIDES

Alkyl derivatives of hydrogen peroxide exhibit a similar conformation, some variations of the dihedral angle τ being understandable in terms of steric and polar effects. Experimental difficulties make themselves felt for many compounds of this class, hence most evidence concerns the relatively stable *t*-butyl peroxide. Due to strong steric repulsion, its dihedral angle is enlarged[21] to 166° and the potential curve is probably very flat. This may be partly responsible for the variance between ED[21] and dipole moments[60–62] (Table 1): If a continuous series of conformations is present, the effective mean value may be defined differently for different methods. In addition, the dipole-moment approach might yield somewhat biased values as shown by the example of hydrogen peroxide. Also, the apparent independence of ^{1}H-NMR[62] and IR[63] spectra of *t*-butyl peroxide of temperature, as well as a dielectric loss curve with a single relaxation time[64] could be connected with a special form of the potential curve. A reasonable dihedral angle was predicted from empirical correlation of PES data[65], but *t*-butyl peroxide was just one of the compounds on which the correlation was established. Similarly, the Kerr constant could not be exploited for conformational analysis since the polarizability of the O—O bond was determined on the same compound[62]. Theoretical calculations for *t*-butyl peroxide have been restricted to very simple methods: CNDO/2[21] and molecular mechanics[66], so more than qualitative results cannot be expected.

Of the other alkyl peroxides, dimethyl peroxide has been studied more theoretically[35,54,67] but the experimental evidence[65,68] is poor. There is a serious disagreement in the literature[69,70] concerning the conformation of trifluoromethyl

TABLE 2. Some values of rotational barriers about the O—O bond

Compound	Rotational barrier[a] (kcal mol^{-1})			
	Calculated		Experimental	
	cis	*trans*	*cis*	*trans*
HO—OH	5	3–4[32,33]	~4	0.9[24]
	9–14	0[38–42,53]	7.0	1.10[25]
	10.9	0.7[44,45]	7.57	1.10[59]
	8.35	1.10[46]	7.420	1.3410[27]
	9.30	1.15[51]		
	7.4	1.1[47]		
DO—OD			7.1	1.08[26]
			8.80	1.08[59]
			7.51	1.32[27]
FO—OH	7.39	3.33[43]		
FO—OF	6.05	4.50[32]		
CF_3O—OH		~2.4[76]		
CF_3O—OF	14.6	8.4[76]		
HCOO—OH	−3.7	0[102]		
	−0.4	−1.55[102]		
	−0.3	−1.8[104]		
AcO—OH	−3.7	−0.6[105]		
	−1.8[89]			
CF_3CO—OH	−2.8	−0.8[105]		

[a] If there is an energy minimum between $\tau = 0°$ and $\tau = 180°$, the barriers are positive; if there is a maximum they are negative.

peroxide; however, in any case, the angle τ is much less than in *t*-butyl peroxide. The same effect is clearly also seen in the molecules of FOOF[20], CF_3OOHal[71] and SF_5OOSF_5[72] (Table 1) and may be rationalized by the fact that the dilution of electrons on the O—O bond relieves the repulsion between electron pairs. Several substituted peroxides have been studied by solution dipole moments[73] and molecular mechanics[66]; due to the complexity of the molecules the results are not particularly reliable. Crystal structure analysis is available for triphenylmethyl peroxide[74]; the angle $\tau = 180°$ is not unexpected. Extensive crystallographic work has been devoted to cyclic peroxides (see Reference 1) but is not relevant to the conformation of open-chain compounds. ED has revealed that even the cyclic derivative **3** (ozonide) is not planar[75] ($\tau = 50°$). The conformation of other ozonides has been predicted from PES spectral correlation[65].

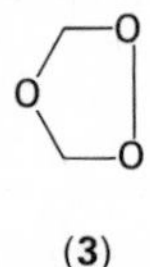

(**3**)

Alkyl hydroperoxides have been less investigated than alkyl peroxides, and the results are very similar (Table 1). The values of τ from dipole moments[60,61] could be somewhat underestimated and the effect of electron-attracting substituents is less expressed[71]. Theoretical calculations[43,76] (4-31G) have yielded a qualitative agreement with experiments. Crystal structures of some complex hydroperoxides are available[77,78] but the angle τ has sometimes not been explicitly evaluated[77].

IV. ACYL PEROXIDES

With acyl peroxides two new problems emerge. Firstly, the acyl group is conjugated with the oxygen atom as expressed in formula **4** (an electron-attracting mesomeric effect as compared to the inductive effect in the molecule CF_3OOCF_3). The hybridization on oxygen is thus changed and the dihedral angle τ may be affected. Secondly, two additional axes of rotation enter the molecule, viz. the C—O bonds. Hence two main conformations may be taken into consideration: sp (carbonyl oxygen inward, facing the O—O bond, formula **5**), or ap (carbonyl oxygen outward, formula **6**), in addition to the unsymmetrical combination and to further conformations in which the planarity of arrangement of each O=C—O—O group is more or less distorted.

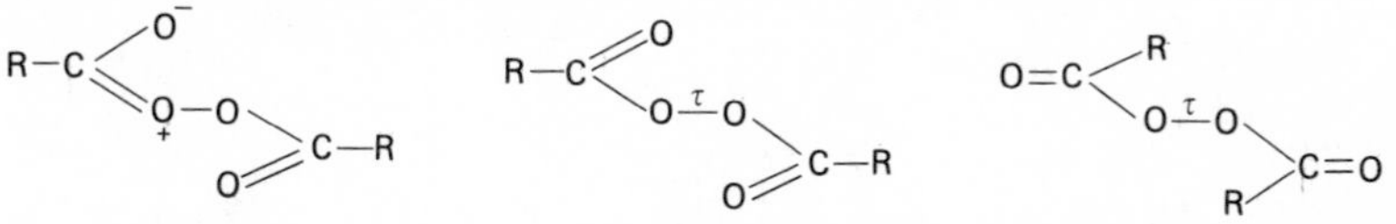

(4) **(5)** **(6)**

Earlier discussions concentrated on the latter point[1]: Conformation **5** was preferred on the basis of dipole moments[61], or at least a C_2 conformation (**5** or **6**) on the basis of IR spectra[79]; the angle τ remained in fact undetermined. Most of the further arguments concerned the common compound benzoyl peroxide. Crystallographic analysis[80] confirmed for it the form **5** with $\tau = 91°$; for its 4,4′-dichloroderivative[81] τ equals 81°: Results on *ortho* derivatives[82,83] as well as on acetyl benzoyl peroxide[84] suggest that the conformation is again dependent on molecular environment but probably less than in the case of hydrogen peroxide. An early X-ray study[85] was evidently in error; however, all the determinations[80–85] agree as far as the conformation on the C—O bonds is concerned.

Gas-phase data being apparently not available, the conformation in solution must essentially be deduced from dipole moments. For this method, benzoyl peroxide is more favourable than alkyl peroxides in two respects. The most important bond moments are more distant from the axis of rotation, and suitable polar substituents can be introduced into the benzene rings. The second point is of utmost importance in dipole-moment work[86]. A systematic study of substituted dibenzoyl peroxides[87] has confirmed in a convincing and independent manner the conformation **5** rather than the alternative **6**. The dihedral angle has been determined less accurately: With standard bond moments[30] $\tau = 115°$ would result, while the crystallographic value of 91° was compatible with dipole moments if some plausible modifications of bond moments were admitted. Anyhow this angle is reduced as compared with hydrogen peroxide, in accord with the electron-attracting effect shown by the mesomeric formula **4**. On the other hand, some details of conformations can be inferred[87]: i.e. the nonexact parallelism of the two R—C bonds in **5** and the statistically distributed positions of the two benzene nuclei with respect to the O=C—O planes.

The sp conformation about the C—O bond in formula **5** is in accord with ample evidence[88] obtained uniformly on esters and ester-like compounds with the grouping CO—O, and has been indirectly corroborated by EHT calculations on methyl peroxyacetate[89]. Quantum-chemical calculations on acyl peroxides themselves are very scarce; in a semiempirical procedure[54] the dihedral angle $\tau = 110°$ has been simply adopted. In conclusion the conformation of acyl peroxides is relatively well proven and essentially the same in solution and in crystal. It is also in accord with the conformation of simple model compounds, viz. H_2O_2 and esters.

V. PEROXY ACIDS

With peroxy acids a new factor enters into consideration, viz. the intramolecular hydrogen bond possible in the sp conformation (**7**). It would not only relieve the *cis* barrier but also stabilize the sp conformation by an additional energy contribution; as a result the potential curve can change profoundly. If once proven, the hydrogen bond would become the deciding factor in the conformation. From the experimental point of view its existence extends the number of methodic tools, in particular by infrared spectroscopy; on the other hand, some methods are difficult to apply due to the instability of the compounds.

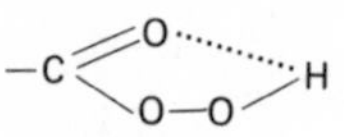

(**7**) 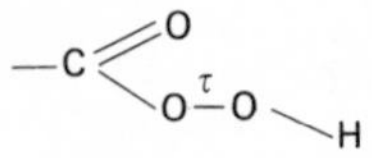 (**8**) 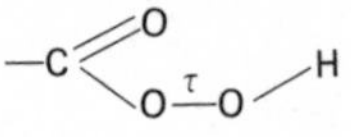 (**9**)

The hydrogen bond has already been anticipated on the basis of boiling points and other observations; the early work is summarized in References 1 and 90. The present evidence is based on the IR spectra of aliphatic[90–92] and aromatic[91,93] peroxy acids in various nonpolar and slightly polar solvents; particularly convincing is a complete normal coordinate analysis[94]. Recently, the planar conformation **7** was proven in the vapour of peroxyacetic acid by MW spectroscopy[95]. In crystal several peroxy acids possess the conformation **8** with $\tau = 133–170°$ (Table 1) and the hydrogen atoms are engaged in intermolecular bonds[96–98]; this difference is clearly displayed also by IR spectra in crystal and in solution[92,93]. Hence the crystal structures cannot tell anything about the conformation of a free molecule and the form **7** is to be held for proven in the gas phase and in solution. Only two possible refinements could be still considered. Firstly, the nonbonded form **8** could exist in minute amounts in equilibrium, but it has not been detected by spectroscopy. Secondly, even the bonded form need not be exactly planar, a small angle τ as in **9** is possible with a certain weakening of the hydrogen bond. This latter possibility (with $\tau = 72°$), contradicting the MW results[95], has been advocated on the basis of dipole moments[99]; however, only unsubstituted aliphatic peroxy acids have been studied and the contribution of the hydrogen bond itself to the resulting dipole moment has not been taken into account. A new analysis based on substituted benzoic acids[100] has assumed the planar form **7** and estimated the contribution of the hydrogen bond (μ_H) to 2.5 D. Its direction is in accord with similar hydrogen-bonded molecules, in particular with α-hydroxyketones[101]. In Figure 1 are shown the gross dipole moment of peroxybenzoic acid, its resolution into bond moments and the resulting vector μ_H. Although the resolution is only approximate and formal, the latter vector can give some idea about the strength of the hydrogen bond and the charge transfer connected with it. Accordingly, the hydrogen bond is rather strong, in aliphatic peroxy acids possibly somewhat weaker; the direction of μ_H is *a priori* unexpected.

Theoretical calculations on peroxy acids are rather inconsistent; possibly some problems encountered with hydrogen peroxide reappear, but these have not been followed in such detail. All the calculations[89,102–105] agree that there are two minima of energy in

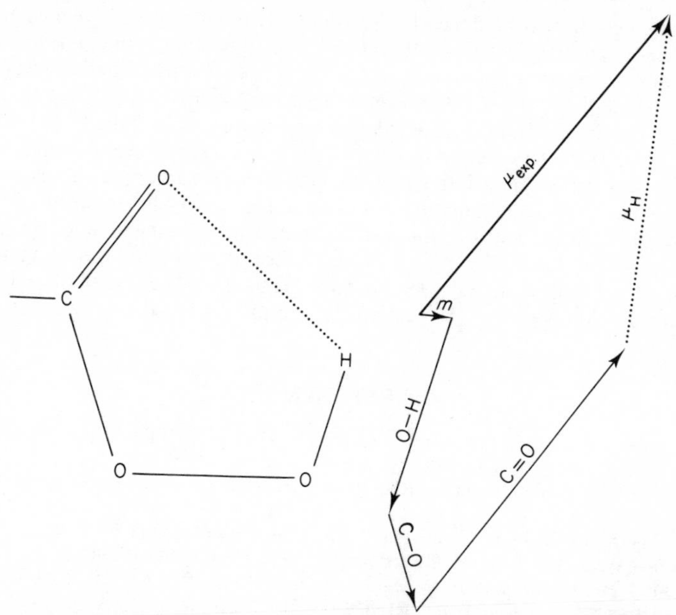

FIGURE 1. The experimental dipole moment of peroxybenzoic acid and its resolution into bond moments, a mesomeric correction m and the contribution from the hydrogen bond μ_H. Based on data of Reference 100.

the two planar conformations, the barrier between them being rather high (in contradistinction to the quoted dipole moment study[99]). The shape of the potential curve is just reverse compared to H_2O_2. However, the respective calculations differ in attributing the absolute minimum either to $\tau = 0$ (structure **7**) in accord with experimental evidence, or to $\tau = 180°$ at variance with it (Table 1). Again we encounter the strange fact that simple methods (EHT[89], INDO[102] and STO-2G[103]) yield better results than more sophisticated ones: 4-31G[104] or 5-31G[102]. The rotational barriers are of reasonable magnitude (Table 2) and the expected energy maximum [102,105] is between 35 and 100°. Sometimes the calculated energy difference between sp and ap rotamers is too low[102], so that an equilibrium should be observable. All the results quoted also throw some doubts on the calculations of peroxyformidic acid[106] (not yet isolated).

VI. PEROXY ESTERS

Of all the compounds mentioned, this group has been least investigated. When the hydrogen bond is no more present, the sp conformation is impossible for steric reasons and one can anticipate that the angle τ in formula **10** will be rather large, as large as it is, for example, in *t*-butyl peroxide. This is confirmed by some X-ray data[107,108], which, however, are insufficient to estimate the effect of crystal forces.

$R^1—C(=O)—O—O—R^2$ (τ about the O—O bond)

(10)

The solution conformation depends only on dipole-moment measurements, mostly on *t*-butyl esters. From the experimental point of view those compounds represent a more difficult problem compared to symmetrical molecules of alkyl peroxides or acyl peroxides (**5**) as the interval of calculated values is more narrow. An early study[109] was restricted to unsubstituted derivatives and considered also the rotation around the C—O bond. This meant that two parameters were to be determined from one experimental quantity; hence the decision was rather speculative and the predicted distortion of the carboxyl part improbable (see the conformation of ester and ester-like molecules[88]). In a reinvestigation[110] polar substitution was systematically exploited and the effects of conjugation and of the bulky *t*-butyl group taken into account referring to model compounds. The dihedral angles found are large (Table 1); in the case of *t*-butyl esters the planar conformation with $\tau = 180°$ cannot be actually excluded.

VII. REFERENCES

1. L. S. Silbert in *Organic Peroxides*, Vol. 2 (Ed. D. Swern), John Wiley and Sons, New York–London, 1971, pp. 637–798.
2. W. C. Schumb, C. N. Satterfield and R. L. Wentworth, *Hydrogen Peroxide*, Reinhold, New York, 1955.
3. P. A. Giguère and T. K. K. Srinivasan, *J. Mol. Spectroy.*, **66**, 168 (1977).
4. W. G. Penney and G. B. B. M. Sutherland, *J. Chem. Phys.*, **2**, 492 (1934).
5. C. S. Lu, E. W. Hughes and P. A. Giguère, *J. Amer. Chem. Soc.*, **63**, 1507 (1941).
6. B. F. Pedersen, *Acta Chem. Scand.*, **21**, 779 (1967).
7. I. Olovsson and D. H. Templeton, *Acta Chem. Scand.*, **14**, 1325 (1960).
8. B. F. Pedersen and B. Pedersen, *Acta Chem. Scand.*, **18**, 1454 (1964).
9. W. R. Busing and H. A. Levy, *J. Chem. Phys.*, **42**, 3045 (1965).
10. S. C. Abrahams, R. L. Collin and W. N. Lipscomb, *Acta Cryst.* **4**, 15 (1951).
11. E. Prince, S. F. Trevino, C. S. Choi and M. K. Farr, *J. Chem. Phys.*, **63**, 2620 (1975).
12. V. A. Sarin, V. Ya. Dudarev, T. A. Dobrynina, L. E. Fykin and V. E. Zavodnik, *Kristallografiya*, **21**, 929 (1976); **22**, 982 (1977).
13. M. A. A. F. de C. T. Carrondo, W. P. Griffith, P. D. Jones and A. C. Skapski, *J. Chem. Soc., Dalton Trans.*, 2323 (1977).
14. J. M. Adams and V. Ramdas, *Acta Cryst.* (B), **34**, 2150, 2781 (1978).
15. J. M. Adams and R. G. Pritchard, *Acta Cryst.* (B), **33**, 3650 (1977).
16. J. M. Adams and R. G. Pritchard, *Acta Cryst.* (B), **34**, 1428 (1978).
17. T. C. W. Mak and Y.-S. Lam, *Acta Cryst.* (B), **34**, 1732 (1978).
18. D. Thierbach, F. Huber and H. Preut, *Acta Cryst.* (B), **36**, 974 (1980).
19. V. A. Sarin, V. Ya. Dudarev, T. A. Dobrynina and V. E. Zavodnik, *Kristallografiya*, **24**, 825 (1979).
20. R. H. Jackson, *J. Chem. Soc.*, 4585 (1962).
21. D. Käss, H. Oberhammer, D. Brandes and A. Blaschette, *J. Mol. Struct.*, **40**, 65 (1977).
22. D. Cremer and D. Christen, *J. Mol. Spectroy.*, **74**, 480 (1979).
23. W. C. Oelfke and W. Gordy, *J. Chem. Phys.*, **51**, 5336 (1969).
24. R. L. Redington, W. B. Olson and P. C. Cross, *J. Chem. Phys.*, **36**, 1311 (1962).
25. R. H. Hunt, R. A. Leacock, C. W. Peters and K. T. Hecht, *J. Chem. Phys.*, **42**, 1931 (1965).
26. R. H. Hunt and R. A. Leacock, *J. Chem. Phys.*, **45**, 3141 (1966).
27. G. A. Khachkuruzov and I. N. Przhevalskii, *Opt. Spektrosk.*, **36**, 299 (1974); **44**, 194 (1978).
28. J. T. Massey and D. R. Bianco, *J. Chem. Phys.*, **22**, 442 (1954).
29. E. P. Linton and O. Maass, *Can. J. Res.*, **7**, 81 (1932).
30. O. Exner, *Dipole Moments in Organic Chemistry*, Thieme, Stuttgart, 1975, Chap. 3.3.
31. F. N. Lassettre and L. B. Dean, *J. Chem. Phys.*, **17**, 317 (1949).
32. M. S. Gordon, *J. Amer. Chem. Soc.*, **91**, 3122 (1969).
33. W. England and M. S. Gordon, *J. Amer. Chem. Soc.* **94**, 4818 (1972).
34. R. F. Nalewajski, *Z. Naturforsch.* (*A*), **32**, 276 (1977).

35. K. Ohkubo, T. Fujita and H. Sato, *J. Mol. Struct.*, **36**, 101 (1977).
36. Y. Amako and P. A. Giguère, *Can. J. Chem.*, **40**, 765 (1962).
37. U. Kaldor and I. Shavitt, *J. Chem. Phys.*, **44**, 1823 (1966).
38. W. E. Palke and R. M. Pitzer, *J. Chem. Phys.*, **46**, 3948 (1967).
39. R. M. Stevens, *J. Chem. Phys.*, **52**, 1397 (1970).
40. L. Pedersen and K. Morokuma, *J. Chem. Phys.*, **46**, 3941 (1967).
41. W. H. Fink and L. C. Allen, *J. Chem. Phys.*, **46**, 2261, 2276 (1967).
42. R. B. Davidson and L. C. Allen, *J. Chem. Phys.*, **55**, 519 (1971).
43. L. Radom, W. J. Hehre and L. A. Pople, *J. Amer. Chem. Soc.*, **94**, 2371 (1972).
44. A. Veillard, *Theoret. Chim. Acta*, **18**, 21 (1970).
45. C. Guidotti, V. Lamanna, M. Maestro and R. Moccia, *Theoret. Chim. Acta*, **27**, 55 (1972).
46. T. H. Dunning and N. W. Winter, *J. Chem. Phys.*, **63**, 1847 (1975).
47. D. Cremer, *J. Chem. Phys.*, **69**, 4440 (1978).
48. J. P. Ranck and H. Johansen, *Theoret. Chim. Acta*, **24**, 334 (1972).
49. W. R. Rodwell, N. R. Carlsen and L. Radom, *Chem. Phys.*, **31**, 177 (1978).
50. A. H. Pakiari and J. W. Linnett, *J. Chem. Soc., Faraday Trans.* 2, 1590 (1975).
51. P. G. Burton and B. R. Markey, *Australian J. Chem.*, **30**, 231 (1977).
52. P. B. Ryan and H. D. Todd, *J. Chem. Phys.*, **67**, 4787 (1977).
53. R. E. Howard, M. Levy, H. Shull and S. Hagstrom, *J. Chem. Phys.*, **66**, 5181, 5189 (1977).
54. A. O. Litinskii, A. I. Rakhimov, D. B. Shatkovskaya and A. B. Bolotin, *Zh. Strukt. Khim.*, **19**, 393 (1978).
55. S. Wolfe, *Acc. Chem. Res.*, **5**, 102 (1972).
56. T. K. Brunck and F. Weinhold, *J. Amer. Chem. Soc.*, **101**, 1700 (1979).
57. A. M. Semkov and J. W. Linnett, *J. Chem. Soc., Faraday Trans.* 2, 1595 (1975).
58. O. Exner and J. B. F. N. Engberts, *Collect. Czech. Chem. Commun.*, **44**, 3378 (1979).
59. C. S. Ewig and D. O. Harris, *J. Chem. Phys.*, **52**, 6268 (1970).
60. M. T. Rogers and T. W. Campbell, *J. Amer. Chem. Soc.*, **74**, 4742 (1952).
61. W. Lobunez, J. R. Rittenhouse and J. G. Miller, *J. Amer. Chem. Soc.*, **80**, 3505 (1958).
62. M. J. Aroney, R. J. W. LeFèvre and R. K. Pierens, *Australian J. Chem.*, **20**, 2251 (1967).
63. D. C. McKean, J. L. Duncan and R. K. M. Hay, *Spectrochim. Acta* (*A*), **23**, 605 (1967).
64. M. J. Aroney, H. Chio, R. J. W. LeFèvre and D. V. Radford, *Australian J. Chem.*, **23**, 199 (1970).
65. P. Rademacher and W. Elling, *Justus Liebigs Ann. Chem.* 1473 (1979).
66. A. A. Turovskii and N. A. Turovskii, *Zh. Strukt. Khim.*, **20**, 937 (1979).
67. R. C. Bingham, M. J. S. Dewar and D. H. Lo, *J. Amer. Chem. Soc.*, **97**, 1302 (1975).
68. W. Shand, *Ph.D. Thesis*, California Institute of Technology, 1946.
69. C. J. Marsden, L. S. Bartell and F. P. Diodati, *J. Mol. Struct.*, **39**, 253 (1977).
70. A. H. Lowrey, G. George and P. D'Antonio, Personal communication, quoted according to Reference 21.
71. C. J. Marsden, D. D. DesMarteau and L. S. Bartell, *Inorg. Chem.*, **16**, 2359 (1977).
72. R. B. Harvey and S. H. Bauer, *J. Amer. Chem. Soc.*, **76**, 859 (1954).
73. R. V. Kucher, A. A. Turovskii, N. A. Turovskii, N. V. Dzumedzei and A. F. Dmitruk, *Dokl. Akad. Nauk SSSR*, **210**, 598 (1973).
74. C. Glidewell, D. C. Liles, D. J. Walton and G. M. Sheldrick, *Acta Cryst.* (*B*), **35**, 500 (1979).
75. C. W. Gillies and R. L. Kuczkowski, *J. Amer. Chem. Soc.*, **94**, 6337 (1972).
76. J. F. Olsen, *J. Mol. Struct.*, **49**, 361 (1978).
77. P. Groth, *Acta Chem. Scand.* (*A*), **29**, 840 (1975).
78. S. Bozzini, S. Gratton, A. Risaliti, A. Stener, M. Calligaris and G. Nardin, *J. Chem. Soc. Perkin Trans.* 1, 1377 (1977).
79. C. Fayat, *Compt. Rend., Ser. C*, **265**, 1406 (1967).
80. M. Sax and R. K. McMullan, *Acta Cryst.*, **22**, 281 (1967).
81. S. Caticha-Ellis and S. C. Abrahams, *Acta Cryst.* (*B*), **24**, 277 (1968).
82. J. Z. Gougoutas and J. C. Clardy, *Acta Cryst.* (*B*), **26**, 1999 (1970).
83. J. Z. Gougoutas and K. H. Chang, *Cryst. Struct. Commun.*, **8**, 977 (1979).
84. N. J. Karch, E. T. Koh, B. L. Whitsel and J. M. McBride, *J. Amer. Chem. Soc.*, **97**, 6729 (1975).
85. V. Kasatochkin, S. Perlina and K. Ablezova, *Dokl. Akad. Nauk SSSR*, **47**, 36 (1945).
86. Reference 30, Chap. 5, 6.

87. V. Jehlička, B. Plesničar and O. Exner, *Collect. Czech. Chem. Commun.*, **40**, 3004 (1975).
88. O. Exner in *The Chemistry of Functional Groups, Supplement A* (Ed. S. Patai), John Wiley and Sons, London–New York, 1976, pp. 1–92.
89. T. Yonezawa, H. Kato and O. Yamamoto, *Bull. Chem. Soc. Japan*, **40**, 307 (1967).
90. D. Swern and L. S. Silbert, *Anal. Chem.*, **35**, 880 (1963).
91. W. H. T. Davison, *J. Chem. Soc.*, 2456 (1951).
92. P. A. Giguère and A. W. Olmos, *Can. J. Chem.*, **30**, 821 (1952).
93. R. Kavčič, B. Plesničar and D. Hadži, *Spectrochim. Acta* (*A*), **23**, 2483 (1967).
94. (a) W. F. Brooks and C. M. Haas, *J. Phys. Chem.*, **71**, 650 (1967). (b) J. Cugley, R. Meyer and H. H. Günthard, *Chem. Phys.*, **18**, 281 (1976).
95. J. A. Cugley, W. Bossert, A. Bauder and H. H. Günthard, *Chem. Phys.*, **16**, 229 (1976).
96. D. Belitskus and G. A. Jeffrey, *Acta Cryst.*, **18**, 458 (1965).
97. M. Sax, P. Beurskens and S. Chu, *Acta Cryst.*, **18**, 252 (1965).
98. H. S. Kim and G. A. Jeffrey, *Acta Cryst.* (*B*), **26**, 896 (1970).
99. J. R. Rittenhouse, W. Lobunez, D. Swern and J. G. Miller, *J. Amer. Chem. Soc.*, **80**, 4850 (1958).
100. O. Exner and B. Plesničar, *Collect. Czech. Chem. Commun.*, **43**, 3079 (1978).
101. B. Plesničar, J. Smolíková, V. Jehlička and O. Exner, *Collect. Czech. Chem. Commun.*, **43**, 2754 (1978).
102. R. D. Bach, C. L. Willis and T. J. Lang, *Tetrahedron*, **35**, 1239 (1979).
103. B. Plesničar, M. Tasevski and A. Ažman, *J. Amer. Chem. Soc.*, **100**, 743 (1978).
104. C. Petrongolo, *Chem. Phys.*, **26**, 243 (1977).
105. L. M. Hjelmeland and G. Loew, *Tetrahedron*, **33**, 1029 (1977).
106. T. J. Lang, G. J. Wolber and R. D. Bach, *J. Amer. Chem. Soc.*, **103**, 3275 (1981).
107. L. S. Silbert, L. P. Witnauer, D. Swern and C. Ricciuti, *J. Amer. Chem. Soc.*, **81**, 3244 (1959).
108. R. S. Miller, Personal communication, quoted according to Reference 84.
109. F. D. Verderame and J. G. Miller, *J. Phys. Chem.*, **66**, 2185 (1962).
110. B. Plesničar and O. Exner, *Collect. Czech. Chem. Commun.*, **46**, 490 (1981).

The Chemistry of Functional Groups, Peroxides
Edited by S. Patai

CHAPTER **3**

Thermochemistry of peroxides

ALAN C. BALDWIN

Department of Chemical Kinetics, SRI International, Menlo Park, California 94025, U.S.A.

I. INTRODUCTION . . . 97
II. MEASURED THERMOCHEMICAL DATA . . . 98
A. Heats of Formation . . . 98
B. Bond Strengths . . . 99
C. Entropies and Heat Capacities . . . 100
III. GROUP ADDITIVITY SCHEME FOR THERMOCHEMICAL DATA . . . 100
A. Introduction to Group Additivity Method . . . 100
B. Heats of Formation . . . 100
1. Peroxides and hydroperoxides . . . 100
2. Polyoxides . . . 101
3. Peroxy and polyoxy radicals . . . 101
4. Peroxynitrates . . . 101
5. Summary . . . 102
C. Entropies and Heat Capacities . . . 102
IV. CONCLUSIONS . . . 102
V. ACKNOWLEDGEMENTS . . . 102
VI. REFERENCES . . . 102

I. INTRODUCTION

This chapter reviews the available experimental data, and methods for the estimation of the thermochemistry of peroxides. Here the term peroxides covers organic and organometallic peroxides, hydroperoxides, polyoxides, peroxynitrates and the radicals derived from these compounds. In keeping with previous chapters in this series by Shaw, the method of group additivity will be used to correlate and compare measured data, and to develop a consistent scheme for the estimation of unmeasured values. The group additivity method was developed by Benson and coworkers[1–3], and has been discussed in previous volumes in this series[4,5]. It considers the thermochemical parameters of a molecule to be the sum of contributions from groups, each defined to be a polyvalent atom in the molecule together with its ligands. Only groups unique to peroxides will be discussed here; Reference 3 contains the most recent compilation of group values for other types of compounds.

The thermochemistry of organic peroxides was reviewed by Benson and Shaw[6] in 1970, and a basic group additivity scheme was determined. I have drawn heavily on their work in preparing this chapter, updating their values in the light of recent measurements and including some new classes of compounds. The basic sources for new thermochemical data have been the IUPAC *Bulletin of Thermochemistry and Thermodynamics*, Cox and Pilcher's[7] *Thermochemistry of Organic and Organometallic Compounds* and the Sussex-NPL[8] *Computer Analyzed Thermochemical Data.*

The thermochemical properties to be discussed are the heat of formation (ΔH_f^0), the entropy (S^0) and the heat capacity (C_p^0) for the ideal gas state at a temperature of 298.15 K (25°C) and a standard state of 1 atmosphere. For convenience, both Joule and calorie units will be used. Thus, ΔH_f^0 is given in kJ mol^{-1} and kcal mol^{-1}, and S^0 and C_p^0 are given in J mol^{-1} K^{-1} and cal mol^{-1} K^{-1}.

II. MEASURED THERMOCHEMICAL DATA

A. Heats of Formation

The experimentally determined heats of formation for peroxides and hydroperoxides are listed in Tables 1 and 2. In Section III, these values will be used to derive the individual group contributions which will form the basis of a scheme for estimating unmeasured heats of formation.

The heat of formation of one peroxy radical, HO_2, has been determined experimentally. Foner and Hudson[9] have obtained ΔH_f^0 (HO_2) = 20.9 ± 8.4 kJ mol^{-1} (5 ± 2 kcal mol^{-1}) from appearance potential measurements. Recently, Howard[10] has obtained ΔH_f^0 (HO_2) = 10.5 ± 2.5 kJ mol^{-1} (2.5 ± 0.6 kcal mol^{-1}) from measurements of the equilibrium constant for reaction (1) from 452 K to 1115 K. Howard's value is less susceptible to experimental problems (appearance potential measurements really yield upper limits for heats of formation) and is used for all subsequent discussion.

$$OH + NO_2 \rightleftharpoons HO_2 + NO \quad (1)$$

TABLE 1. Measured heats of formation for peroxides

Compound	ΔH_f^0		Reference
	kJ mol^{-1}	kcal mol^{-1}	
Hydrogen peroxide	−135.9	−32.5	11
Dimethyl peroxide	−125.8	−30.1	12
Diethyl peroxide	−192.7	−46.1	12
Di-*t*-butyl peroxide	−340.7	−81.5	12
Diacetyl peroxide	−497.4	−119.0	13
Dipropionyl peroxide	−580.6	−138.9	13
Dibutyryl peroxide	−627.0	−150.0	13
Bis(hydroxymethyl) peroxide	−575.2	−137.6	14
$Me_3SiOOBu$-*t*	−483.6	−115.7	15
$Et_3GeOOBu$-*t*	−460.2	−110.1	15
$Et_3SnOOBu$-*t*	−371.2	−88.8	15

TABLE 2. Measured heats of formation for hydroperoxides

Compound	ΔH_f^0 kJ mol^{-1}	ΔH_f^0 kcal mol^{-1}	Reference
Hydrogen peroxide	−135.9	−32.5	11
Ethyl hydroperoxide	[−199.9][a]	[−47.6][a]	7
1-Propyl hydroperoxide	[−271.7][a]	[−65.0][a]	7
t-Butyl hydroperoxide	−245.8	−58.8	16
Cyclohexyl hydroperoxide	−229.9	−55.0	17
1-Methylcyclohexyl hydroperoxide	−263.3	−63.0	17
1-Hydroperoxyhexane	−259.2	−62.0	13
2-Hydroperoxyhexane	−267.5	−64.0	13
3-Hydroperoxyhexane	−263.3	−63.0	13

[a] Quoted in Cox and Pilcher[7] but clearly inconsistent with other data on hydroperoxides.

B. Bond Strengths

The kinetics of the thermal decomposition of a number of peroxides have been studied. The bond strengths (bond dissociation enthalpies) derived from the measured activation energies for decomposition are listed in Table 3. As can be readily seen, the strength of the O—O bond in dialkyl peroxides is about 155 ± 5 kJ mol^{-1} (37 ± 1 kcal mol^{-1}) independent of the alkyl group. Similarly the strength of the O—O bond in diacyl peroxides is about 125 ± 4 kJ mol^{-1} (30 ± 1 kcal mol^{-1}). The strength of the O—O bond

TABLE 3. Bond strengths (D_{298}^0)

Bond	D_{298}^0 kJ mol^{-1}	D_{298}^0 kcal mol^{-1}	Reference
MeO—OMe	155.0	37.1	19, 20
EtO—OEt	158.6	37.9	21
n-PrO—OPr-*n*	155.2	37.1	21
i-PrO—OPr-*i*	157.7	37.7	21
s-BuO—OBu-*s*	152.3	36.4	22
t-BuO—OBu-*t*	152.0	36.4	23
neoC_5H_{11}O—OneoC_5H_{11}	152.3	36.4	24
MeC(O)O—O(O)CMe	125.8	30.1	25, 26
EtC(O)O—O(O)CEt	127.9	30.6	25, 26
n-PrC(O)O—O(O)CPr-*n*	126.2	30.2	25, 26
$(CF_3)_3CO$—$OC(CF_3)_3$	149	35.7	27
CF_3O—OCF_3	193	46.2	28, 29
SF_5O—OSF_5	156	37.2	30
SF_5OO—OSF_5	105	25	31
CF_3OO—OCF_3	127	30.3	32
Me_3SiO—OBu-*t*	197	47	15
Et_3GeO—OBu-*t*	192	46	15
Et_3SnO—OBu-*t*	205	49	15
HO_2—NO_2	96 ± 8	23 ± 2	33
$CH_3C(O)O_2$—NO_2	109 ± 4	26 ± 1	34

in trioxides is reduced by 50–70 kJ mol^{-1} (12–16 kcal mol^{-1}) compared to the corresponding peroxide for the two cases listed. A considerable strengthening of the bond results from substitution of one of the carbon atoms adjacent to the breaking bond with Si, Ge or Sn.

Experimental data in the pyrolysis of hydroperoxides are not consistent with any reasonable thermochemical analysis due to the occurrence of complex wall-sensitive chain reactions[18], and are not included here.

C. Entropies and Heat Capacities

Entropy and heat capacity data are available for only one peroxide, H_2O_2[6]. Fortunately, entropies and heat capacities may be estimated to good accuracy by comparison with other model compounds, and from statistical mechanical calculations on assigned structures. A group additivity scheme for entropies and heat capacities is developed in Section III.

III. GROUP ADDITIVITY SCHEME FOR THERMOCHEMICAL DATA

A. Introduction to Group Additivity Method

Group additivity represents the thermochemical properties of a molecule as the sum of contributions due to all the groups in the molecule, where a group is any polyvalent atom in the molecule together with its ligands. For example, *t*-BuOOH contains the following groups*:

3 [C—(C)(H)$_3$]	3 carbons bonded to another carbon and three hydrogens
1 [C—(C)$_3$(O)]	1 carbon bonded to 3 other carbons and an oxygen
1 [O—(C)(O)]	1 oxygen bonded to a carbon and an oxygen
1 [O—(O)(H)]	1 oxygen bonded to an oxygen and a hydrogen

By summing the values of these groups for heat of formation, entropy or heat capacity, we would obtain the appropriate property of *t*-butyl hydroperoxide. Group values, derived from measured properties, are available for many common groups[3]. In this section, the unique groups for peroxides, hydroperoxides, polyoxides and peroxynitrates will be derived.

B. Heats of Formation

1. *Peroxides and hydroperoxides*

In order to estimate the heat of formation of a peroxide or hydroperoxide, we need values for the groups [O—(O)(C)] and [O—(O)(H)]. The latter is obtained directly from the heat of formation of hydrogen peroxide which is composed of 2 [O—(O)(H)]. Thus, [O—(O)(H)] = −135.9/2 = −68.0 kJ mol^{-1} (−16.3 kcal mol^{-1}). The group [O – (O)(C)] is obtained by subtracting the known hydrocarbon groups from the

* Note that there are some small correction factors for non-next-nearest-neighbour interactions which are described in Reference 3. *t*-BuOOH should be corrected for the '*gauche*' interaction.

measured heats of formation of alkyl peroxides. Benson and Shaw[6] have derived $[O-(O)(C)] = -18.8\,kJ\,mol^{-1}$ ($-4.5\,kcal\,mol^{-1}$) and shown that these group values reproduce most of the measured heats of formation of the organic peroxides and hydroperoxides in Tables 1 and 2 to within $\pm 8\,kJ\,mol^{-1}$ ($\pm 2\,kcal\,mol^{-1}$).

2. Polyoxides

The polyoxides H_2O_n[35] ($n = 3,4$) and R_2O_n[36] ($n = 3,4$) have been observed in solution, and $CF_3O_3CF_3$[32] and $SF_5O_3SF_5$[31] have been studied in the gas phase. Nangia and Benson[37] have shown that the unique polyoxide group $[O-(O)(O)]$ can be derived from an analysis of the heats of reaction of equilibria of the type:

$$R_2O_4 \rightleftharpoons 2\,RO_2\cdot$$

$$R_2O_3 \rightleftharpoons RO_2\cdot + RO\cdot$$

$$R_2O_2 \rightleftharpoons 2\,RO\cdot$$

if the heats of formation of $RO\cdot$ and $RO_2\cdot$ are known. ΔH_f^0 ($MeO\cdot$) is known to be $16.3\,kJ\,mol^{-1}$, ($3.9\,kcal\,mol^{-1}$) and ΔH_f^0 ($MeO_2\cdot$) can be estimated from ΔH_f^0 (CH_3OOH) which can be calculated as $-129.2\,kJ\,mol^{-1}$ ($-30.9\,kcal\,mol^{-1}$) from Section III.B.1. Taking the known heats of formation of H_2O_2 and H with the Howard[10] value for ΔH_f^0 (HO_2) gives a bond strength D_{298}^0 (HO_2-H)[10] of $364.1\,kJ\,mol^{-1}$ ($87.1\,kcal\,mol^{-1}$). Assuming that D_{298}^0 (MeO_2-H) is the same as D_{298}^0 (HO_2-H), the heat of formation of methylperoxy is ΔH_f^0 ($MeO_2\cdot$) $= 17.1\,kJ\,mol$ ($4.1\,kcal\,mol^{-1}$). Thus from Nangia and Benson[37] the heat of formation of the group $[O-(O)(O)]$ is $61.5\,kJ\,mol^{-1}$ ($14.7\,kcal\,mol^{-1}$).

3. Peroxy and polyoxy radicals

Peroxy radicals require the group $[O-(\dot{O})(C)]$. MeO_2 contains the groups $[C-(H)_3(O)]$ and $[O-(\dot{O})(C)]$; taking ΔH_f^0 ($MeO_2\cdot$) $= 17.1\,kJ\,mol^{-1}$ ($4.1\,kcal\,mol^{-1}$), as above, with the known $[C-(H)_3(O)]$ gives the heat of formation of $[O-(\dot{O})(C)]$ as $59.4\,kJ\,mol^{-1}$ ($14.2\,kcal\,mol^{-1}$).

Polyoxy radicals require the group $[O-(\dot{O})(O)]$, which can be estimated from ΔH_f^0 (CH_3OOOH), calculated as $-67.7\,kJ\,mol^{-1}$ ($-16.2\,kcal\,mol^{-1}$) from the groups given above. Assuming that the ROOO—H bond strength is unchanged from hydrogen peroxide, ΔH_f^0 ($CH_3OOO\cdot$) $= 78.6\,kJ\,mol^{-1}$ ($18.8\,kcal\,mol^{-1}$), and subtraction of the known groups leaves the heat of formation of $[O-(O)(\dot{O})]$ as $139.6\,kJ\,mol^{-1}$ ($33.4\,kcal\,mol^{-1}$).

4. Peroxynitrates

The bond strengths measured for HO_2NO_2 and $CH_3C(O)O_2NO_2$ enable us to calculate the group $[O-(O)(NO_2)]$. Taking the bond strength from Table 3 with the known heats of formation of NO_2 and HO_2 gives ΔH_f^0 (HO_2NO_2) $= -52.7\,kJ\,mol^{-1}$ ($-12.5\,kcal\,mol^{-1}$). Thus, $[O-(O)(NO_2)] = 15.5\,kJ\,mol^{-1}$ ($3.7\,kcal\,mol^{-1}$). Using the groups from Section III.B.3, ΔH_f^0 ($CH_3C(O)O_2\cdot$) is $-114.1\,kJ\,mol^{-1}$ ($-27.3\,kcal\,mol^{-1}$), which with the measured bond strength gives ΔH_f^0

$(CH_3C(O)O_2NO_2) = -189.8$ kJ mol^{-1} (-45.4 kcal mol^{-1}). Subtraction of the known groups gives the heat of formation of [O—(O)(NO_2)] as 18.4 kJ mol^{-1} (4.4 kcal mol^{-1}). Thus, a value of 17 ± 4 kJ mol^{-1} (4 ± 1 kcal mol^{-1}) for [(O)—(O)(NO_2)] is established.

5. Summary

The derived heats of formation for the various groups are listed in Table 4.

TABLE 4. Derived heats of formation for various groups

	ΔH_f^0	
Group	kJ mol^{-1}	kcal mol^{-1}
[O—(O)(C)]	−18.8	−4.5
[O—(O)(H)]	−68.0	−16.3
[O—(O)(O)]	61.5	14.7
[O—($\dot{O}$)(C)]	59.4	14.2
[O—($\dot{O}$)(O)]	139.6	33.4
[O—(O)NO_2)]	17	4

C. Entropies and Heat Capacities

There have been no measurements of entropies of heat capacities of peroxides since the review of Shaw and Benson[6], who estimated the necessary groups based on the known data for hydrogen peroxide and from standard corrections to model compounds. The values they derived are listed in Table 5.

IV. CONCLUSIONS

The group values listed in Tables 4 and 5, together with the standard values[3], will enable the calculation of the thermochemical properties of most common peroxide-type compounds. Although based on somewhat sparse data in some areas, these estimates will probably be as accurate as most measured values. At present, the data base is insufficient to derive groups for the halogenated and organometallic compounds. Hopefully, more data will soon be available.

V. ACKNOWLEDGEMENTS

It is a pleasure to acknowledge the many contributions of my colleagues at SRI International, and, in particular, David M. Golden and Elaine Adkins.

VI. REFERENCES

1. S. W. Benson and J. H. Buss, *J. Chem. Phys.*, **29**, 546 (1958).
2. S. W. Benson, F. R. Cruickshank, D. M. Golden, G. R. Haugen, H. E. O'Neal, A. S. Rodgers, R. Shaw and R. Walsh, *Chem. Rev.*, **69**, 279 (1969).
3. S. W. Benson, *Thermochemical Kinetics*, 2nd ed., John Wiley and Sons, New York–London, 1976.

TABLE 5. Group values for calculating entropies and heat capacities

Group	S^0		C_p^0 (300 K)		C_p^0 (500 K)		C_p^0 (1000 K)	
	J mol^{-1} K^{-1}	cal mol^{-1} K^{-1}	J mol^{-1} K^{-1}	cal mol^{-1} K^{-1}	J mol^{-1} K^{-1}	cal mol^{-1} K^{-1}	J mol^{-1} K^{-1}	cal mol^{-1} K^{-1}
[O—(O)(C)]	39.3	9.4	16.3	3.9	19.2	4.6	24.2	5.8
[O—(O)(H)]	116.6	27.9	21.7	5.2	26.3	6.3	31.4	7.5
[O—(O)(O)]	39.3	9.4	9.2	2.2	17.6	4.2	20.5	4.9
[O—($\dot{O}$)(C)]	150.5	36.0[a]	29.7	7.1	32.6	7.8	40.6	9.7
[O—($\dot{O}$)(O)]	154.7	37.0[a]	17.6	4.2	23.4	5.6	33.4	8.0

[a]Radical entropies are intrinsic. They include the contribution from the unpaired electron.

4. R. Shaw in *Chemistry of Hydrazo, Azo and Azoxy Groups* (Ed. S. Patai), John Wiley and Sons, New York–London, 1974.
5. R. Shaw in *Chemistry of Amidines and Imidates* (Ed. S. Patai), John Wiley and Sons, New York–London, 1974.
6. S. W. Benson and R. Shaw in *Organic Peroxides*, Vol. I (Ed. D. Swern), Wiley–Interscience, New York–London, 1970).
7. J. D. Cox and G. Pilcher, *Thermochemistry of Organic and Organometallic Compounds*, Academic Press, London, 1970.
8. J. B. Pedley and J. Rylance, *Computer-analyzed Thermochemical Data*, University of Sussex, England, 1977.
9. S. N. Foner and R. L. Hudson, *J. Chem. Phys.*, **36**, 2681 (1962).
10. C. J. Howard, *J. Amer. Chem. Soc.*, **102**, 6937 (1980).
11. *JANAF Thermochemical Tables*, Dow Chemical Company, Midland, Mich.
12. G. Baker, J. H. Littlefair, R. Shaw and J. C. J. Thynne, *J. Chem. Soc.*, 6970 (1965).
13. J. Jaffé, E. J. Prosen and M. Szwarc, *J. Chem. Phys.*, **27**, 416 (1957).
14. A. D. Jenkins and D. W. G. Style, *J. Chem. Soc.*, 2337 (1953).
15. I. B. Rabinovich, E. G. Kiparisova and Yu. A. Alexksandrov, *Dokl. Akad. Nauk. SSSR*, **20**, 1116 (1971).
16. N. A. Kozolov and I. B. Rabinovich, *Tr. po Khim. i Khim. Tekhnol.*, **2**, 189 (1964); *Chem. Abstr.*, **63**, 6387 (1965).
17. W. Pritzkow and K. A. Muller, *Chem. Ber.*, **89**, 2321 (1956).
18. S. W. Benson and H. E. O'Neal, 'Kinetic Data on Gas Phase Unimolecular Reactions', *National Standard Reference Data Service*, NBS 21 (1970).
19. L. Batt and R. D. McCullough, *Intern. J. Chem. Kinet.* **8**, 911 (1976).
20. D. H. Shaw and H. O. Pritchard, *Can. J. Chem.*, **46**, 2722 (1968).
21. L. Batt, K. Christie, R. T. Milne and A. J. Summers, *Intern. J. Chem. Kinet.*, **6**, 877 (1974).
22. R. F. Walker and L. Phillips, *J. Chem. Soc. (A)*, 2103 (1968).
23. D. K. Lewis, *Can. J. Chem.*, **54**, 581 (1976).
24. M. J. Perona and D. M. Golden, *Intern. J. Chem. Kinet.* **5**, 55 (1973).
25. J. A. Kerr, *Chem. Rev.*, **66**, 465 (1966).
26. A. Rembaum and M. Szwarc, *J. Amer. Chem. Soc.*, **76**, 5975 (1954).
27. R. Ireton, A. S. Gordon and D. C. Tardy, *Intern. J. Chem. Kinet.* **9**, 769 (1977).
28. R. C. Kennedy and J. B. Levy, *J. Phys. Chem.*, **76**, 3480 (1972).
29. J. Czarnowski and H. J. Schumacher, *Z. Physik. Chem. (Frankfurt)*, **92**, 329 (1974).
30. J. Czarnowski and H. J. Schumacher, *Intern. J. Chem. Kinet.*, **10**, 111 (1978).
31. J. Czarnowski and H. J. Schumacher, *Intern. J. Chem. Kinet.*, **11**, 613 (1979).
32. J. Czarnowski and H. J. Schumacher, *Intern. J. Chem. Kinet.*, **13**, 639 (1981).
33. A. C. Baldwin and D. M. Golden, *J. Phys. Chem.*, **82**, 644 (1978).
34. D. G. Hendry and R. Kenley, *J. Amer. Chem. Soc.*, **99**, 3198 (1977).
35. P. A. Giguere and K. Herman, *Can. J. Chem.*, **48**, 3473 (1970).
36. P. D. Barlett and P. Günther, *J. Amer. Chem. Soc.*, **88**, 3288 (1966).
37. P. S. Nangia and S. W. Benson, *J. Phys. Chem.*, **83**, 1179 (1979).

The Chemistry of Functional Groups, Peroxides
Edited by S. Patai

CHAPTER **4**

Mass spectrometry of organic peroxides

HELMUT SCHWARZ

Institut für Organische Chemie, Technische Universität Berlin, D-1000 Berlin 12, W. Germany

HANS-MARTIN SCHIEBEL

Institut für Organische Chemie, Technische Universität Braunschweig, D-3300 Braunschweig, W. Germany

I. PERESTERS AND PEROXYLACTONES 106
II. ALKYL HYDROPEROXIDES 113
III. ACYCLIC AND CYCLIC PEROXIDES 116
IV. OZONIDES 123
V. ACKNOWLEDGEMENTS 126
VI. REFERENCES 126

Although mass spectrometry has been applied to most groups of organic compounds and detailed investigations concerning the unimolecular decay of various cation radicals have been carried out over the last two decades, there have been comparatively few reports in the chemical literature dealing with systematic investigations of the mass spectral behaviour of organic peroxides. Whether this situation is caused by the well-known thermal instability of many of these compounds is open to question, because the mass spectra of other classes of thermally quite unstable organic compounds, e.g. azides[1] or polyacetylenes[2], have been reported. Moreover, in most of the published work ionization of the peroxides has been achieved by electron impact, whereas 'soft ionization' techniques, e.g. field desorption (FD) and chemical ionization (CI) mass spectrometry have been scarcely used. Detailed mechanistic investigations of the unimolecular decay of cation radicals by means of extensive isotopic labelling or by employing more recent techniques for ion structure elucidations, e.g. collisional activation (CA) or metastable ion characteristics, do not seem to have been carried out. In most publications on mass spectrometry of peroxides we have studied, the method has been exclusively used to provide analytical data concerning the molecular weight and gross structural features. To our knowledge there are only a small number of reports dealing with systematic investigations on the mass spectral behaviour of organic peroxides. We shall refer to these studies, covering the literature from 1965 to early 1980.

I. PERESTERS AND PEROXYLACTONES

Systematic investigations of the electron impact mass spectra of mono- and di-*t*-butyl peresters have been carried out by Krull and Mandelbaum[3]. They have suggested mass spectrometry as a method of choice for the general characterization of these compounds, provided special care is taken when the spectra are run (e.g. using the direct insertion probe and keeping the ion source the lowest possible temperature, ideally with external cooling). In all of the peresters examined in the monocarboxylic acid series (**1**–**7**, R = *t*-Bu) a distinct molecular ion peak is observed. In the case of *t*-butyl perbenzoate (**1**) the relative intensity of $M^{+\cdot}$ is 5.5% of the most abundant ion (*m/z* 105, $PhCO^+$). In all other cases, the molecular ion is of much lower abundance (0.1–1.7%), but in each case it is easily detected. The peresters of dicarboxylic acids **8**–**13** (R = *t*-Bu) exhibit a rather general fragmentation behaviour. Distinct molecular ions are formed for the di-*t*-butyl esters of **8** and **9**. There is some speculation that it is the presence of a rigid cyclic structure which prevents the interaction of the two perester groups and thus allows the formation of detectable molecular ions. This requirement is met in the ground states of **8** and **9**, but not in those of the other di-*t*-butyl esters of dicarboxylic esters **10**–**13**. However, it should be borne in mind that there is now ample evidence for the fact that cation radicals of substituted three-and four-membered rings are prone to undergo ring-opening at molecular ion lifetimes as short as $t \leq 10^{-10}$s[4]. Thus it is possible that a temperature effect (higher inlet temperatures are used due to the relatively low volability) may have been partially or fully responsible for the absence of a $M^{+\cdot}$ion in **10**–**13**.

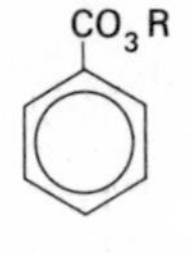

CO_3R (benzene) **(1)**

$CH_3(CH_2)_nCO_3R$
(**2**) $n = 5$
(**3**) $n = 6$
(**4**) $n = 7$

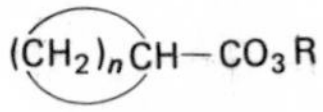

$(CH_2)_nCH—CO_3R$
(**5**) $n = 3$
(**6**) $n = 4$
(**7**) $n = 5$

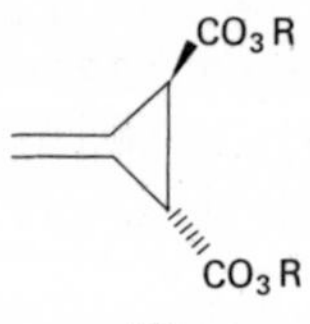

(8) 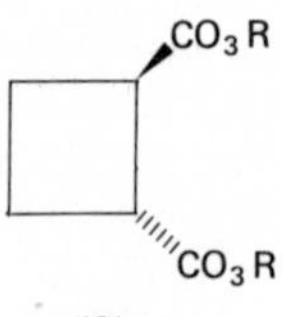 **(9)**

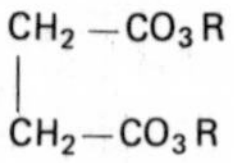

$CH_2—CO_3R$
$CH_2—CO_3R$
(10)

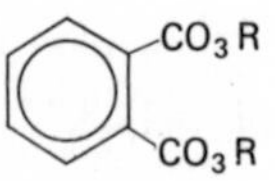

(11) 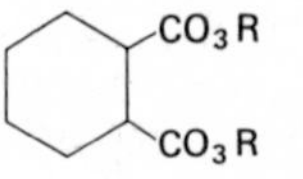 **(12)** 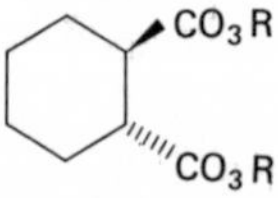 **(13)**

The fragmentation processes are also of interest for the application of mass spectrometry as an analytical method for characterizing peresters in general. The unimolecular loss of a *t*-butyl peroxy radical from the molecular ion, giving rise to the formation of very abundant $[M - 89]^+$ ions, is a general decomposition pathway in all of the examples studied by Krull and Mandelbaum, with the exception of the *cis* ester **12**. The *m/z* 57 $(C_4H_9)^+$ and *m/z* 73 $(C_4H_9O)^+$ peaks are also of relatively high intensity in the mass spectra of **1–13**.

An interesting difference has been observed between the fragmentation processes for the cation radicals of the stereoisomers **12a** and **13a**. The *trans* compound (**13a**) exhibits an $[M - 89]^+$ peak, corresponding to the loss of a *t*-butylperoxy radical from $M^{+\cdot}$. This ion is practically absent in the mass spectrum of the *cis* isomer **12a**, in which a relatively prominent $[M - 88]^+$ peak is present. This ion is generated from $M^{+\cdot}$ via elimination of $C_4H_8O_2$ (H transfer). It appears that in the *cis* compound an interaction between the two functional groups facilitates an intramolecular hydrogen migration from one *t*-butyl group to the other CO_3R function. In the *trans* isomer such an interaction is less likely or even impossible due to the separation of both functional groups. The most important fragmentation processes of ionized **12a** and **13a**, substantiated by high-resolution mass spectrometry and metastable transitions, are illustrated in Scheme 1.

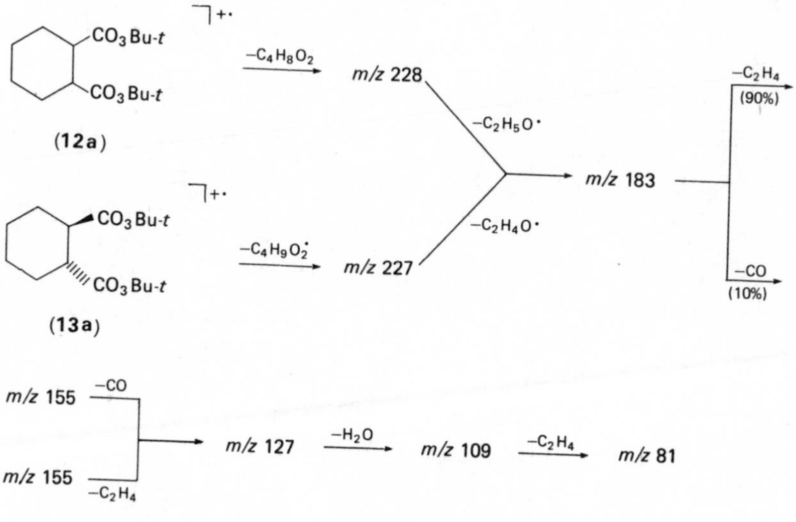

(12a) CO_3Bu-*t* / CO_3Bu-*t* $\xrightarrow{-C_4H_8O_2}$ *m/z* 228 $\xrightarrow{-C_2H_5O\cdot}$ *m/z* 183

(13a) CO_3Bu-*t* / CO_3Bu-*t* $\xrightarrow{-C_4H_9O_2^\cdot}$ *m/z* 227 $\xrightarrow{-C_2H_4O\cdot}$ *m/z* 183

m/z 183 $\xrightarrow{-C_2H_4}$ (90%) ; $\xrightarrow{-CO}$ (10%)

m/z 155 $\xrightarrow{-CO}$ / *m/z* 155 $\xrightarrow{-C_2H_4}$ → *m/z* 127 $\xrightarrow{-H_2O}$ *m/z* 109 $\xrightarrow{-C_2H_4}$ *m/z* 81

SCHEME 1

The mass spectral fragmentation of the peroxy esters *t*-butyl 5-phenyl-Δ^2-isoxazoline-3-peroxycarboxylate (**14a**) and *t*-butyl 5-phenylisoxazole-3-peroxycarboxylate (**15a**) has been found[5] to parallel the oxidative fragmentation to give benzaldehyde as one of the products. The peroxy esters decompose by processes which correspond to the loss of isobutene followed by $O_2H\cdot$ (or $O_2 + H\cdot$) elimination, but the molecular and $[M - C_4H_8]^{+\cdot}$ ions of **14a** are considerably less abundant than these ions in the 70 eV mass spectrum of the peroxy ester **15a**.

The most important processes believed to occur in the peroxy esters **14a** and **15a** and their related ethyl esters (**14b** and **15b**) are given in Scheme 2. Distinct similarities may be seen between the thermal decomposition of **14a** and the electron-impact-induced decays,

(14a) ($M^{+\cdot}$ < 1%)

m/z 57 (82)

m/z 59 (100)

$-C_4H_8, O_2$

m/z 175

$-H^{\cdot}$

m/z 174 (11)

m/z 147

m/z 147

m/z 106 (53)

m/z 107 (42)

Ph—C≡O^+

m/z 105 (76)

(14b) (46%)

$-OEt^{\cdot}$

m/z 174 (8)

$-CO$

m/z 146 (15)

Ph—C≡O^+

m/z 105 (44)

$-H^{\cdot}$

m/z 218 (30)

$-C_2H_4$

m/z 109 (5)

$-H_2O$

m/z 172 (5)

Ph—CH_2—CH $^{+\cdot}$

m/z 104 (100)

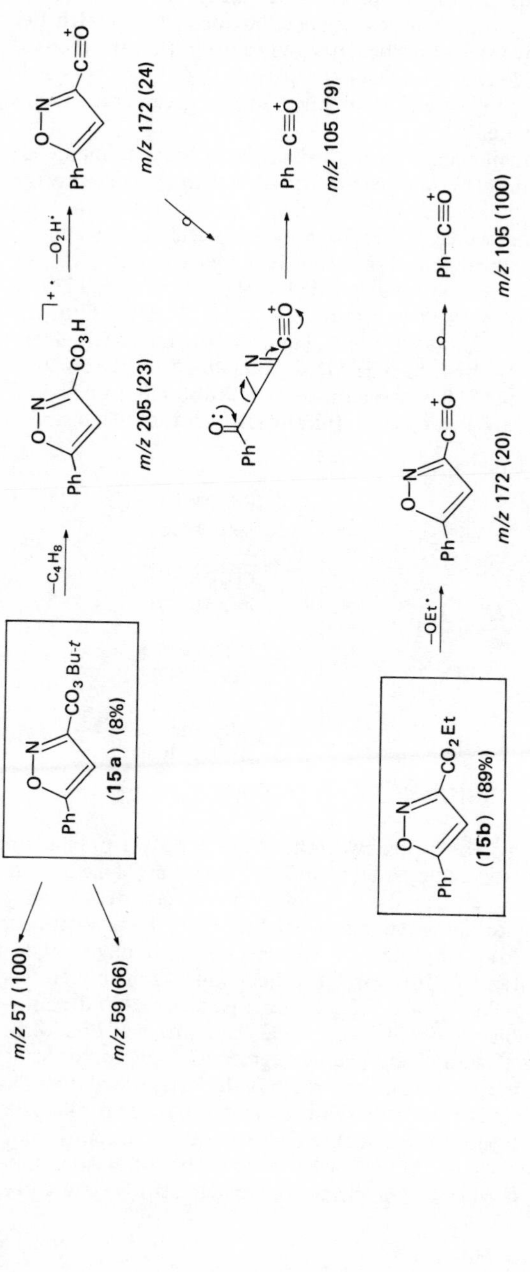

m/z 57 (100) ← (15a) (8%) → m/z 59 (66)

Ph–isoxazole–CO_3Bu-t (15a) (8%) $\xrightarrow{-C_4H_8}$ Ph–isoxazole–CO_3H $]^{+\cdot}$ m/z 205 (23) $\xrightarrow{-O_2H^{\cdot}}$ Ph–isoxazole–$C\equiv O^+$ m/z 172 (24) → Ph–CO–C(=N)–$C\equiv O^+$ → Ph–$C\equiv O^+$ m/z 105 (79)

Ph–isoxazole–CO_2Et (15b) (89%) $\xrightarrow{-OEt^{\cdot}}$ Ph–isoxazole–$C\equiv O^+$ m/z 172 (20) → Ph–$C\equiv O^+$ m/z 105 (100)

SCHEME 2

especially with regard to the ion at *m/z* 146, which under thermolysis is generated as 3-hydroxy-3-phenylpropionitrile. Comparison of the fragmentations of the cation radicals of the ethyl ester **14b** and the peroxy ester **14a** shows that the *t*-butoxy group has a strong directing effect for fragmentation along two quite similar pathways. Of particular interest are the suppression of the formal 'retro' 1,3-dipolar processes (giving rise to the formation of *m/z* 104, 105). The appearance of very intensive peaks at *m/z* 106 (ionized benzaldehyde) and *m/z* 107 (protonated benzaldehyde) is of particular significance in that the thermal decomposition of **14a** gives benzaldehyde as one of the products. For the 5-phenylisoxazole derivatives **15a** and **15b** the similarity of the fragmentation processes rather than their pronounced differences is striking.

Peroxyacetyl nitrate (**16**, PAN), a major component of urban photochemical smog[6], has been extensively investigated by means of mass spectrometry. All attempts to measure the molecular weight of PAN using conventional and time-of-flight mass spectrometers, both equipped with electron impact ion sources, have been unsuccessful in detecting a molecular ion. However, chemical ionization mass spectrometry employing either CH_4 or $i\text{-}C_4H_{10}$ as reactant gases have provided very abundant $[M + H]^+$ ions at *m/z* 122 (3.5 and 100%, respectively) and a small peak at *m/z* 150 $[M + C_2H_5]^+$ in the $CI(CH_4)$ spectrum[7]. Therefore, the molecular weight of PAN is 121 amu. Moreover, the fragment ions at *m/z* 43 (CH_3CO^+), *m/z* 46 (NO_2^+), *m/z* 61 (CH_3ONO^+) and *m/z* 77 ($CH_3ONO_2^+$) in the $CI(CH_4)$ spectrum of **16** (Scheme 3) have been taken as indication that PAN does have the commonly accepted structure of a peroxyacetyl nitrate (**16**)[6] and not that of acetylpernitrate, $CH_3(CO)ONO_3$[8].

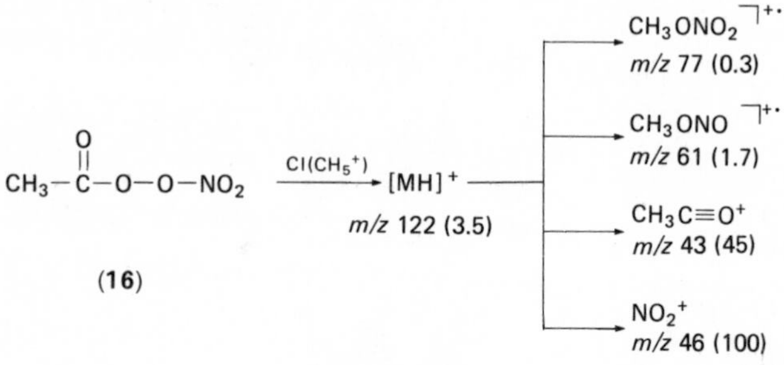

$CH_3-C(=O)-O-O-NO_2$ (**16**) $\xrightarrow{CI(CH_5^+)}$ $[MH]^+$ *m/z* 122 (3.5) →

- $CH_3ONO_2^{+\cdot}$ *m/z* 77 (0.3)
- $CH_3ONO^{+\cdot}$ *m/z* 61 (1.7)
- $CH_3C{\equiv}O^+$ *m/z* 43 (45)
- NO_2^+ *m/z* 46 (100)

SCHEME 3

A detailed mass spectral study of a series β-peroxylactones **17** with α-alkyl and β-alkyl or β-phenyl substitution has been carried out by Adam and Tsai[9]. By means of metastable ion characteristics, isotopic labelling and analysing substituent effects, five primary decomposition pathways of the quite abundant molecular ions have been identified, consisting of (*a*) loss of carbon trioxide, (*b*) α-lactone elimination, (*c*) β-alkyl loss, (*d*) elimination of a hydroperoxy radical and (*e*) CO_2 loss. Of particular mechanistic interest is the electron-impact-induced process $M \rightarrow [M - CO_2]^{+\cdot}$, because thermo- and photodecarboxylation of **17** has been studied in detail. Thus, while thermolysis of **17** affords, predominantly, rearranged ketones (via alkyl and phenyl migration)[10], photolysis leads principally to epoxides[11]. The mass spectral investigation[9] of **17a–h** clearly reveals that the $[M - CO_2]^{+\cdot}$ ions are epoxide-like in structure, thus indicating that the electron-impact behaviour of β-peroxylactones **17a–h** parallels the photolytic behaviour of the compounds (Scheme 4). It should be noted that cyclic carbonates **18**, which are structural isomers of the β-peroxylactones **17**, also fragment on electron impact into epoxide-like $[M - CO_2]^{+\cdot}$ ions[12].

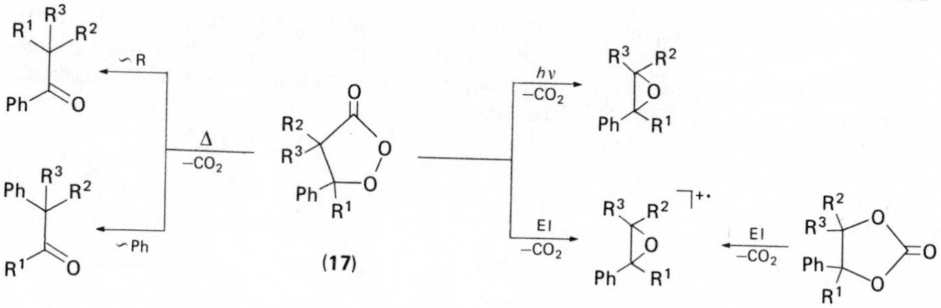

	a	b	c	d	e	f	g	h
R^1	CH_3	CH_3	CD_3	C_2H_5	*i*-Pr	$PhCH_2$	Ph	$PhCH_2$
R^2	H	D	H	H	H	H	CH_3	CH_3
R^3	H	D	H	H	H	H	H	H

SCHEME 4

One convincing piece of structural information concerning the structures of the $[M - CO_2]^{+\cdot}$ ions comes from peroxylactone **17c**, in which the abundant *m/z* 108 ion ($PhCHCD_2$) derives from eliminating a CHO group from the $[M - CO_2]^{+\cdot}$ ion. Such a reaction pathway (Scheme 5) is most characteristic of an epoxide-like $[M - CO_2]^{+\cdot}$ ion[13]. Keto-like structures would be expected to give rise to the elimination of $CD_3\cdot$, $PhCH_2{}^+$ and $CD_3CH_2\cdot$ from $[M - CO_2]^{+\cdot}$, which, however, is not found to be the case.

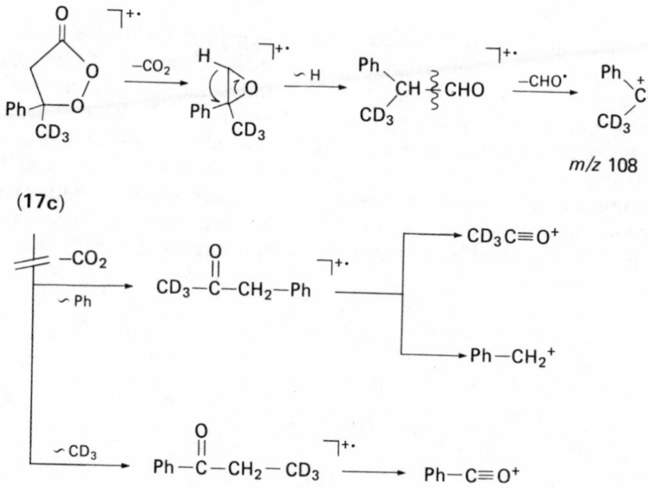

SCHEME 5

A prerequisite for decarboxylation is peroxide bond rupture. The minor abundant $[M - CO_2]^{+\cdot}$ fragments (1–20%) observed in the mass spectra of β-peroxylactones **17a–e** suggest that the peroxide bond is *strengthened* on ionization. This is indicated by the resonance structures in Scheme 6. However, α-methyl and β-benzyl substitution change the fragmentation pattern to favour the decarboxylation process (34–81%).

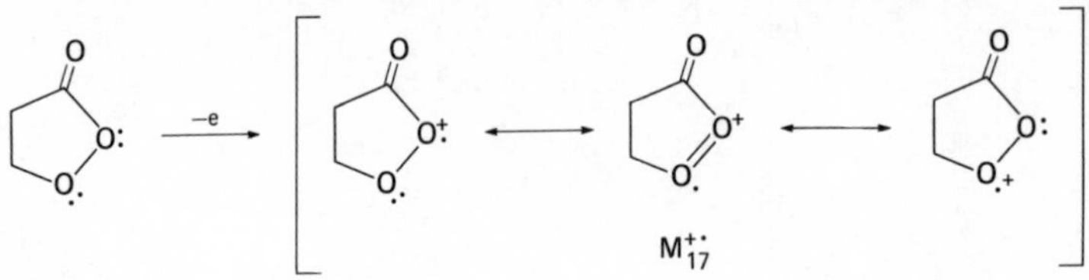

(17)

$M_{17}^{+\cdot}$

SCHEME 6

While direct loss of a CO_3 neutral from $M^{+\cdot}$ has been confirmed by metastable ion transitions (m*) for **17b**, **17c** and **17f**, respectively (Scheme 7), for the β-peroxylactones **17d** and **17h** the $[M - CO_3]^{+\cdot}$ fragments arise from an $[M - CO_2]^{+\cdot}$ ion by further loss of an oxygen atom. The $[M - CO_3]^{+}$ ion subsequently decomposes via the typical mass spectral patterns known for alkenes.

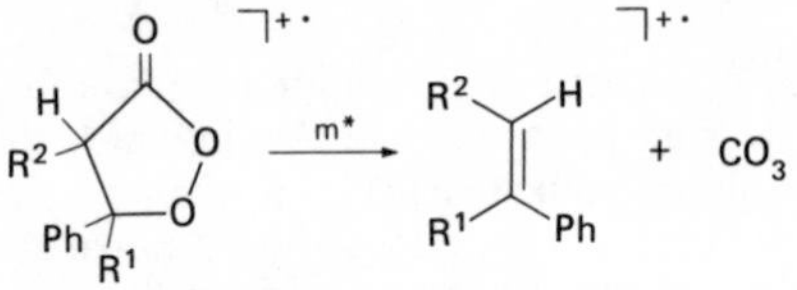

(17)

SCHEME 7

Loss of an α-lactone from $M^{+\cdot}$ (Scheme 8) is most evident for the β-peroxylactone **17g** (60%). The driving force lies probably in the stabilization of the product, the benzophenone cation radical, which subsequently decomposes according to its characteristic fragmentation pattern. For the loss of β-ethyl radicals from $M^{+\cdot}$ the theoretically expected relative abundance of the $[M - R^1]^+$ ions should follow the order of the stability of β-alkyl radicals, e.g. $PhCH_2 > i\text{-}Pr > Et > Me$. Due to the fact that the same β substituents may also kinetically favour other reaction pathways, such as decarboxylation or elimination of an α-lactone, it is not surprising that the β-peroxylactones **17f**, **17g** and **17h** show reduced abundance for the $[M - R^1]^+$ ions (9%, <1% and 2%, respectively).

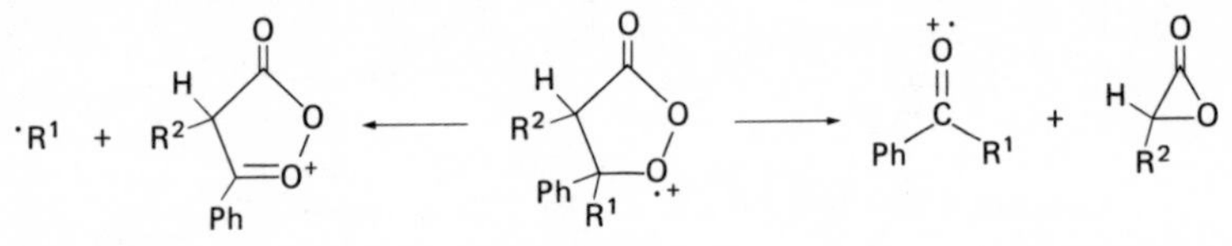

(17)

SCHEME 8

The investigation of the deuterium-labelled β-peroxylactones **17b** and **17c** reveals that the hydrogen involved in the loss of a hydroperoxy radical from **17a**, **17b** and **17c** (Scheme 9) is abstracted from the α position. A [1,2]elimination of $HO_2\cdot$ involving a five-membered transition state has been suggested to account for the experimental results. Although hydrogen atoms from other sources are available for abstraction, e.g. from the α-methyl substituent ([1,3]elimination) in **17g** and **17h**, or from the β-alkyl groups ([1,4]elimination) in all the other β-peroxylactones (except **17g**) the [1,2]elimination mode is clearly preferred.

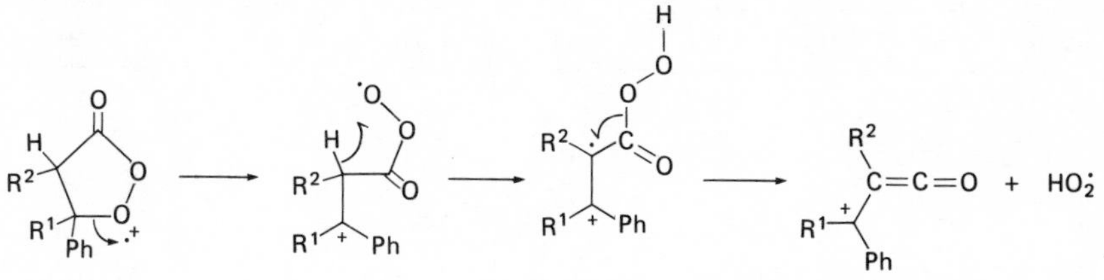

(**17**)

SCHEME 9

II. ALKYL HYDROPEROXIDES

The 70 eV mass spectra of some alkyl hydroperoxides, compounds which are known to be important intermediates in many oxidation reactions, have been recorded by Burgess and coworkers[14]. Signals for the molecular ions for compounds of the general structure **19** (Scheme 10) have been detected for all compounds investigated with moderate to low intensity. The relative abundance decreases substantially with the molecular weight. For isomeric compounds it seems that the relative stabilities of the molecular ions may be summarized as 3- > 2- > 1-compounds, and pentyl > hexyl > heptyl. Signals due to the loss of OH· from $M^{+\cdot}$ (cleavage of the peroxide link) are not very intensive in all the spectra, suggesting that either the peroxide bond is stronger for the cation radicals than for the neutrals (see also Scheme 6 for the discussion of β-peroxylactones) or that the product $R^1R^2CH{-}O^+$ is quite high in energy and thus unlikely to be formed when the molecular ions of **19** undergo decomposition. Elimination of the hydroperoxy radical $HO_2\cdot$ is substantially affected by the substitution pattern, being most favoured when the alkyl cation is tertiary and of less importance when the incipient cation is primary. Unimolecular water loss, which is observed in ionized alcohols, is only important in 3-pentyl hydroperoxide (**19c**). By analogy to water loss from ionized alcohols, the process $\mathbf{19c}^{+\cdot} \rightarrow [\mathbf{19c} - H_2O]^+$ is believed[14] to occur according to Scheme 11. Similarly, it is assumed that elimination of H_2O_2 gives rise to the formation of ionized cycloalkenes. However, these suggestions are neither substantiated by appropriate deuterium labelling nor by the determination of the product ion structures.

The 70 eV electron-impact mass spectra of the naturally occurring 20-, 24- and 25-hydroperoxides of 3β-hydroxycholest-5-ene (**20a–c**)[15] and the peroxy-Y structure **21a** isolated from phenylalanine t-RNA of the plant *Lupimus lutens*[16] contain detectable molecular ions. Moreover, signals are observed for $[M - 16]^{+\cdot}$ ions which have been assigned to unimolecular deoxygenation of $M^{+\cdot}$. However, no metastable ions have been observed (or reported) for the process $M^{+\cdot} \rightarrow [M - O]^+$; therefore it cannot be excluded that this process is—at least partially—thermally induced. For **21a–c** it is found that, irrespective of the substituent R, the fragment at m/z 216 forms the base peak in the 70 eV mass spectrum.

	R^1	R^2	M^+	$[M-HO]^+$	$[M-HO_2]^+$	$[M-H_2O]^+$	$[M-H_2O_2]^+$	$[R^1]^+$	$[R^2]^+$
a	*n*-Bu	H	1.8	0.1	3.2	<0.1	2.7	6.9	–
b	*n*-Pr	CH_3	2.7	0.4	14.5	0.4	2.8	100	10.8
c	C_2H_5	C_2H_5	8.5	0.8	62	8.5	4.7	67.8	67.8
d	*n*-C_5H_{11}	H	<0.1	0.1	1.6	<0.1	1.5	2.4	–
e	*n*-Bu	CH_3	<0.1	0.1	2.4	<0.1	1.6	7.2	62
f	*n*-Pr	C_2H_5	<0.1	0.3	5.5	1.0	0.5	57.6	91
g	*n*-C_6H_{11}	H	<0.1	0.4	0.1	<0.1	0.5	0.4	–
h	*n*-C_5H_9	CH_3	<0.1	0.1	1.1	0.6	1.1	3.9	16.2

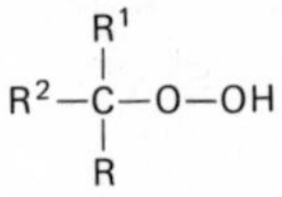

$R^2—C(R^1)(R)—O—OH$

(19)

SCHEME 10

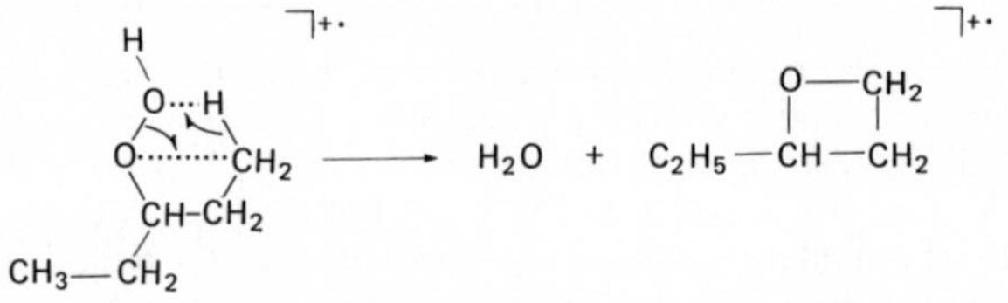

$[CH_3CH_2CH(CH_2CH_2)O—OH]^{+\cdot}$ **(19c)** ⟶ $H_2O + [C_2H_5—CH(CH_2)—O—CH_2]^{+\cdot}$

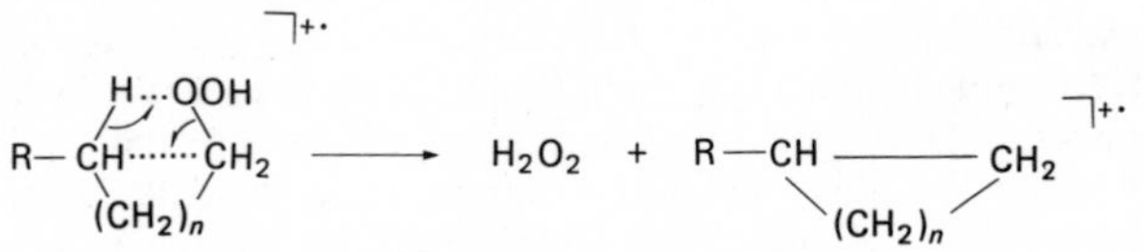

$[R—CH(CH_2)_n CH_2 ... H...OOH]^{+\cdot}$ **(19)** ⟶ $H_2O_2 + [R—CH—(CH_2)_n—CH_2]^{+\cdot}$

SCHEME 11

No molecular ions have been detected in the 70 eV mass spectra of the hydroperoxides of the antitumour agents of various cyclohosphamides **22**[17]. The most pronounced fragment ions are formed by loss of H_2O and $HO_2^{\cdot}$ from $M^{+\cdot}$. Quite surprising is the observation that even in the field desorption mass spectra of **22** signals due to $M^{+\cdot}$ or $[MH]^+$ are not very abundant. The prevailing decomposition mode is, again, H_2O loss. The cytotoxic germacranolide hydroperoxides **23** and **24** have been investigated by means of chemical ionization mass spectrometry[18]. The CI(*i*-C_4H_{10}) spectrum of **23** shows prominent peaks corresponding to $[MH]^+$, $[MH - O]^+$, $[MH - H_2O]^+$ and $[MH - H_2O_2]^+$.

In the CI (NH_3/CH_4) spectrum of **24** abundant signals are observed for the cluster ion $[MNH_4]^+$ and (surprisingly) $[MNH_4 - O]^+$.

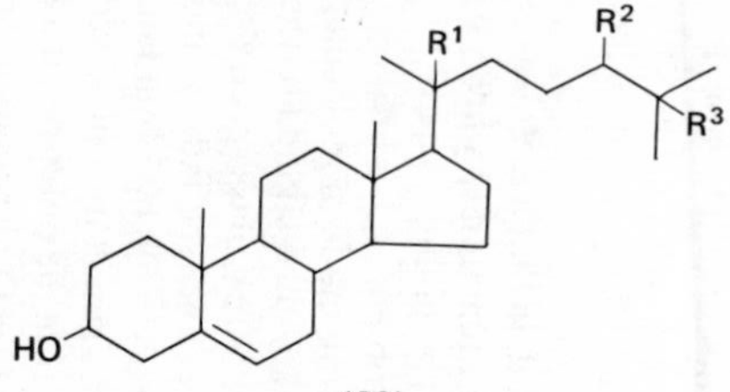

(20)

	R^1	R^2	R^3
a	OOH	H	H
b	H	OOH	H
c	H	H	OOH

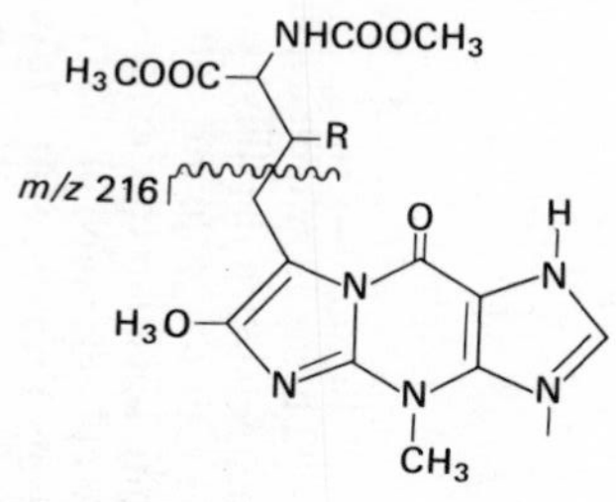

(21)

	R
a	OOH
b	OH
c	CH_3

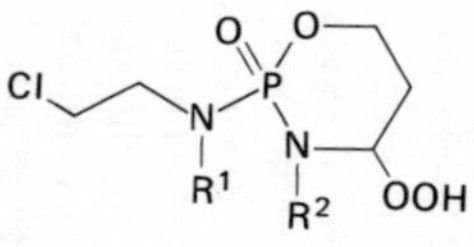

(22)

	R^1	R^2
a	C_2H_4Cl	H
b	H	C_2H_4Cl
c	C_2H_4Cl	C_2H_4Cl

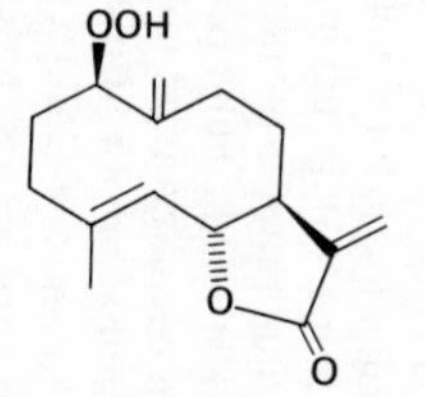

(23)

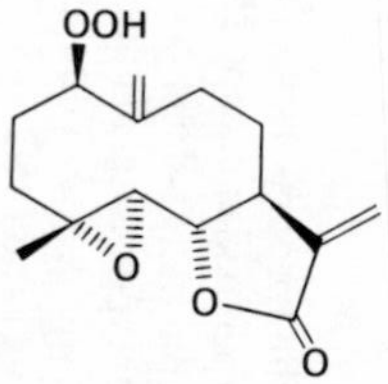

(24)

SCHEME 12

III. ACYCLIC AND CYCLIC PEROXIDES

The mass spectral behaviour of acyclic symmetric dialkyl peroxides of the general structure **25** has been studied by Fraser and coworkers[19] by means of low-resolution 70 eV and 10 eV mass spectra and metastable ion transitions. The relative intensities of the molecular ions are found to be generally much lower than those observed for the corresponding disulphides, and to decrease with increasing molecular weight. Very abundant ion currents are observed for the formal α-cleavage process $ROOR^{+\cdot} \rightarrow R^+ + OOR$; however, metastable ion transitions are not found for this decomposition. For the related hydroperoxides **19** it has been concluded[19] from the presence of appropriate metastable ion transitions that the alkyl ions R^+ are formed at least partially by loss of $HO_2\cdot$ from $M^{+\cdot}$. Other processes prevailing in the mass spectra of **25** are the

ROOR
(**25**)

	R
a	Me
b	Et
c	*n*-Pr
d	*i*-Pr
e	*n*-Bu
f	*i*-Bu

olefin (process a) and hydroperoxide (process b) eliminations (Scheme 13) which occur in all peroxides except **25a**. Pathway b tends to predominate as the size of the alkyl group increases. There is no evidence for the loss of a second olefin group as is found with the corresponding disulphides. Both processes a and b are quite sensitive to the ionizing energy. Lower voltages also favour the formation of $M^{+\cdot}$, $M/2^+$ (cleavage of the O—O bond) and $(M/2 - 1)^+$ ions, the latter by normal loss of an alcohol molecule. At 70 eV, the intensity of the $[M/2]^+$ ions decreases rapidly, and, according to the analysis of metastable ion transitions, the process does *not* follow a simple one-step dissociation $RO{-}OR^{+\cdot} \rightarrow RO^+ + RO\cdot$. Instead, the $[M/2]^+$ ions are formed in a two-step reaction involving the olefin elimination product from which, subsequently, OH· is eliminated associated with hydrogen migration (reaction c, Scheme 13). Preliminary results using deuterium labelling indicate that hydrogen atoms from the 1-position are involved to a very minor extent only in the olefin elimination. Positive evidence concerning the site of the hydrogen source has not been reported yet.

Dimethyl peroxide (**25a**) differs from all the other peroxides studied, in that a metastable peak is found for the process $M^{+\cdot} \rightarrow [M/2]^+$. It is very likely that the reaction (pathway d, Scheme 13) is, again, assisted by 'hidden' hydrogen migration[20] giving rise to the formation of protonated formaldehyde and not the methoxy cation.

For alkyl *t*-butyl peroxides **26** it has been found that the prevailing mode of molecular ion decomposition involves the formation of $[M - 89]^+$ ions, which corresponds to the elimination of a *t*-BuOO· radical[21] (Scheme 14). The low-resolution 70 eV mass spectra of some mixed dialkyl peroxides, **27**, containing the *t*-butyl group and secondary and primary groups have been published by Salomon and coworkers[22]. Very abundant molecular ion peaks are observed, and one of the most prominent fragment ions observed in the spectra of symmetric peroxides **27**, corresponds to $[M/2]^+$. From the published data it cannot be decided whether this fragment is a result of a simple O—O dissociation or, more likely, generated in a two-step process (olefin elimination followed by OH loss in analogy to reaction c, Scheme 13). The low-voltage mass spectrum of the 1,3-bis(*t*-butyl)

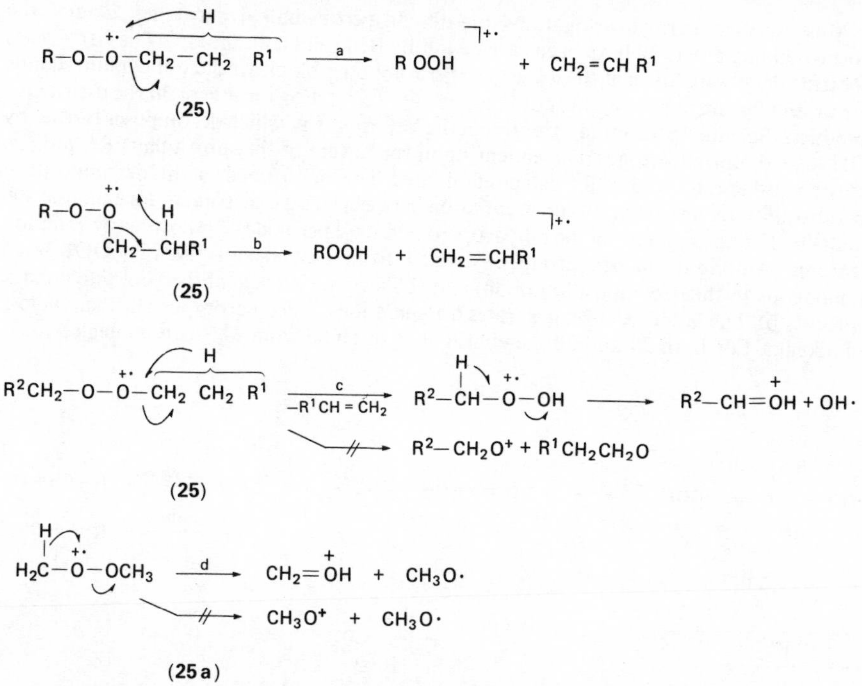

$$R{-}O{-}\overset{+\cdot}{O}{-}CH_2{-}CH_2\ R^1 \xrightarrow{a} [R\,OOH]^{+\cdot} + CH_2{=}CH\,R^1$$

(25)

$$R{-}O{-}\overset{+\cdot}{O}(CH_2{-}CHR^1)(H) \xrightarrow{b} [ROOH + CH_2{=}CHR^1]^{+\cdot}$$

(25)

$$R^2CH_2{-}O{-}\overset{+\cdot}{O}{-}CH_2\ CH_2\ R^1 \xrightarrow[-R^1CH=CH_2]{c} R^2{-}CH(H){-}\overset{+\cdot}{O}{-}OH \longrightarrow R^2{-}CH{=}\overset{+}{O}H + OH\cdot$$

$$\not\longrightarrow R^2{-}CH_2O^+ + R^1CH_2CH_2O$$

(25)

$$H{-}H_2C{-}\overset{+\cdot}{O}{-}OCH_3 \xrightarrow{d} CH_2{=}\overset{+}{O}H + CH_3O\cdot$$

$$\not\longrightarrow CH_3O^+ + CH_3O\cdot$$

(25a)

SCHEME 13

peroxide of cyclopentane (**28**) is dominated by the *m/z* 190 ion (100%); this fragment is formed via loss of *i*-butene (Scheme 14). Stereoisomeric effects were not reported for the *cis* and *trans* isomers of **28**.

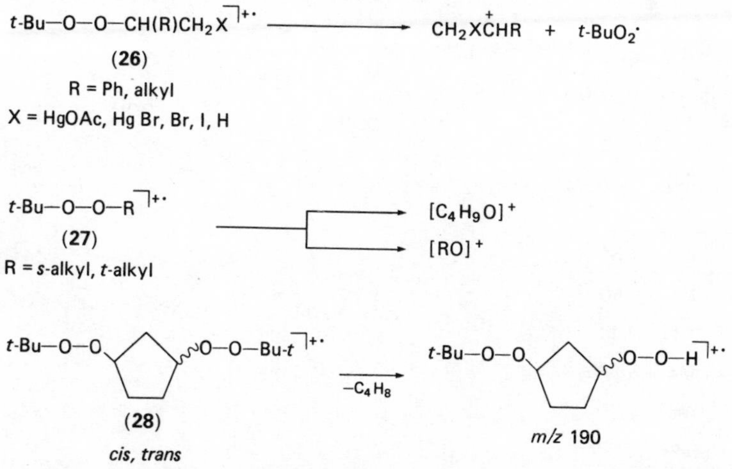

$$[t\text{-Bu}{-}O{-}O{-}CH(R)CH_2X]^{+\cdot} \longrightarrow CH_2X\overset{+}{C}HR + t\text{-BuO}_2^{\cdot}$$

(26)

R = Ph, alkyl

X = HgOAc, Hg Br, Br, I, H

$$[t\text{-Bu}{-}O{-}O{-}R]^{+\cdot} \longrightarrow [C_4H_9O]^+ \quad [RO]^+$$

(27)

R = *s*-alkyl, *t*-alkyl

$$[t\text{-Bu}{-}O{-}O{-}C_5H_8{-}O{-}O{-}Bu\text{-}t]^{+\cdot} \xrightarrow{-C_4H_8} [t\text{-Bu}{-}O{-}O{-}C_5H_8{-}O{-}O{-}H]^{+\cdot}$$

(28) *cis, trans* — *m/z* 190

SCHEME 14

Mass spectrometric investigations on dihydroperoxydialkyl peroxides **29** and the corresponding ethers **30** have been carried out by Belič and colleagues[23]. The 70 eV mass spectra which were taken at the lowest temperature possible either did not contain signals for molecular ions or were of low abundance ($<0.3\%$). For the ethers **30** the α-cleavage products (Scheme 15, reaction a) represent the key fragment, which decomposes further by OH loss to ionized ketones. Dependent upon the nature of the substituents R^1 and R^2, ethylene, alkene, $R^1\cdot$ and/or $R^2\cdot$ can be eliminated. The ions formed via this decomposition pattern allow an unambiguous differentiation between structural isomers, for example **30b** and **30c**. The mass spectra of the dihydroperoxydialkyl peroxides **29** show many common features. Among them are products derived from cleavage of the (C—OO) bond (analogous to the decomposition of **30**) and the formal cleavage of the peroxide linkage (process b). The latter reaction generates fragment ions which decompose further via loss of alkenes. For both **29** and **30** the elimination of $HO_2\cdot$ from $M^{+\cdot}$ can be neglected.

HOO—C(R^1)(R^2)—O—C(R^1)(R^2)—OOH $^{+\cdot}$ —a→ R^1R^2C=$\overset{+}{O}$—OH —(− OH·)→ R^1R^2C=O $^{+\cdot}$ —(− Alkene)→ R—CH=O $^{+\cdot}$; —(− $R^{1(2)}$)→ $R^{1(2)}$—C≡$\overset{+}{O}$

(**30**)

	R^1	R^2
a	Me	Me
b	Et	Et
c	Me	*n*-Bu
d	*n*-Pr	*n*-Pr

HOO—C(R^1)(R^2)—O—O—C(R^1)(R^2)—OOH $^{+\cdot}$ —b→ R^1R^2C(O^+)(OOH) —(− Alkene)→ R—$\overset{+}{C}$(OH)(OOH)

(**29**)

	R^1	R^2
a	Me	Me
b	Et	Et
c	Me	*n*-Pr
d	Me	*i*-Pr
e	*n*-Pr	*n*-Pr

SCHEME 15

Mass spectral data for various cyclic peroxides have been reported occasionally in the literature. The main decomposition processes for cyclic peroxides[24] of the general form **31** are outlined in Scheme 16. For all compounds investigated so far, relatively abundant molecular ions have been recorded. The fragmentation is greatly influenced by the presence of the bromine atoms in **31**. For R = CH_2Br, the loss of R· is the strongly preferred mode of homolytic β-cleavage in $M^{+\cdot}$, and formation of CH_2Br^+ can be envisaged by heterolytic β-scission in either $M^{+\cdot}$ or $[M - CH_2Br]^+$. Quite unusual is the reported[24b] loss of OH· from $M^{+\cdot}$ of **31** (R = H, $n = 2$), with a relative intensity of 21%! For the other peroxides **31** no signals corresponding to the process $M^{+\cdot} \rightarrow [M - OH]^+$ have been reported.

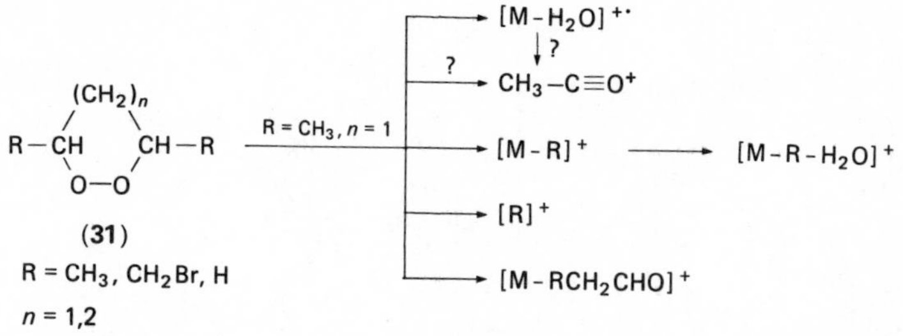

$[M-H_2O]^{+\cdot}$

$CH_3-C\equiv O^+$

$(CH_2)_n$

R—CH CH—R

O—O

R = CH_3, n = 1

$[M-R]^+$ → $[M-R-H_2O]^+$

$[R]^+$

(**31**)

R = CH_3, CH_2Br, H

n = 1,2

$[M-RCH_2CHO]^+$

SCHEME 16

For the thermally stable 1,2-dioxetane **32** *no* molecular ion can be detected in the 70 eV mass spectrum[25]. Instead, losses of O and O_2 (via cycloreversion) are reported. The base peak is found at m/z 150 ($C_{10}H_{14}O$) (Scheme 17).

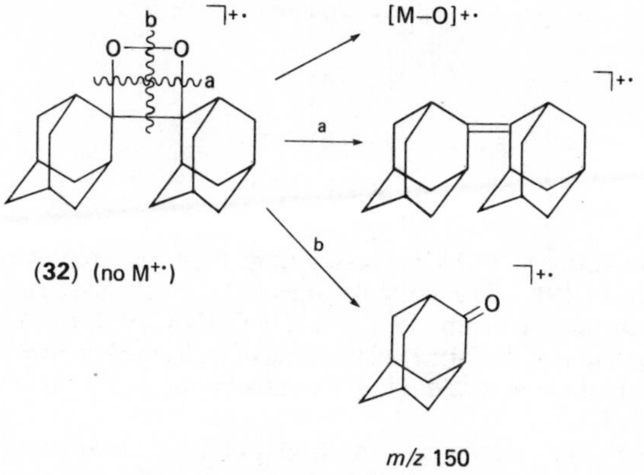

$[M-O]^{+\cdot}$

(**32**) (no $M^{+\cdot}$)

m/z 150

SCHEME 17

The electron impact mass spectra of the six-membered cyclic peroxides of ferrocenophane (**33**) have been reported to depend decisively on temperature effects[26]. The results have been explained as follows: Some of the peroxide molecules **33** undergo thermally induced rearrangement to the isomeric ketal derivatives **34** prior to ionization and decomposition (Scheme 18). From the cation radicals of **34** the phenoxy radical PhO· is unimolecularly eliminated, giving rise to the formation of the quite abundant fragment ions at m/z 329 (for **33a**) and 343 (for **33b**). The unrearranged molecular ions of **33** serve as precursors for the following processes: (1) elimination of oxygen, thus forming the ionized epoxide **35** and (2) ethylene and propene loss, respectively, via a retro-Diels–Alder (RDA) process whereby the ionized diketone is formed (Scheme 18). Of minor importance are the unimolecular eliminations of O_2 and H_2O_2.

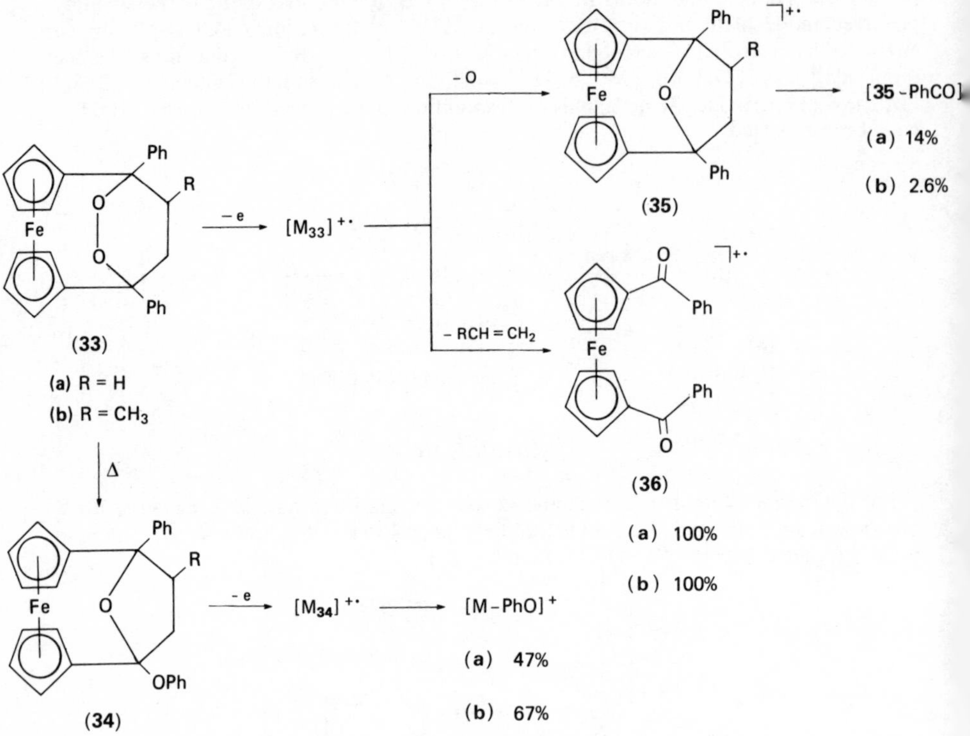

SCHEME 18

Electron-impact-induced RDA processes have been also reported for the bicyclic peroxides **37** (Scheme 19)[27]. The molecular ions of **37**, which are recorded with moderate relative abundance, undergo unimolecular loss of $C_2H_2R_2$, thus generating the fragment ion *m/z* 224. Fragments for the formal elimination of O_2 from **37** are observed; however, no metastable transitions are reported. Thus, it cannot be concluded that this process may be thermally produced.

The base peak in the mass spectrum of the octamethyl-substituted 2,3-epoxynaphthalene 1,4-endoperoxide **38** is the ion *m/z* 202[28]. Unfortunately, no further experimental details have been published, so we have no clues to the understanding of the genesis and structure of this fragment. In the spectra of the tetramethyl-substituted analogue **39** the base peak is formed by the fragment *m/z* 173, for the structure of which again no information is available.

Partial electron-impact mass spectra of the peroxides **40**[29], **41**[30], **42**[31] and **43**[32] have been reported. Whereas **40** gives a very abundant molecular ion peak (22%) and **41** and **42** a minor one (<1%), no molecular ions are detected for **43**. The molecular ion of **40** decomposes via O loss (24%) and the formation of $Ph\overset{+}{C}{=}O$ (100%). Common to **42** and

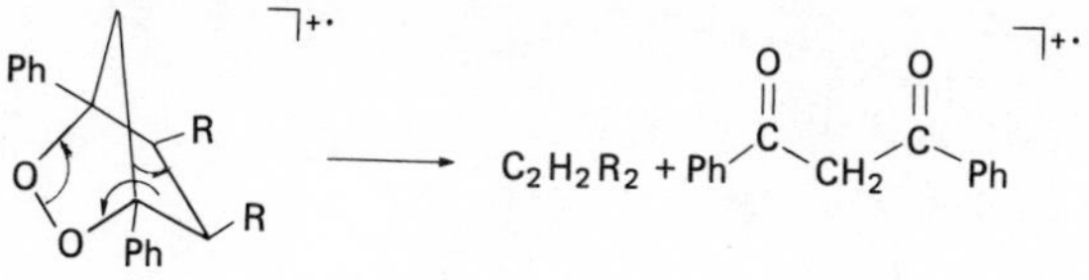

$C_2H_2R_2$ + Ph–CO–CH_2–CO–Ph]+·

(37)

(a) R = H

(b) R = D

(c) R = Cl

m/z 224

(a) 16%

(b) 23%

(c) 24%

SCHEME 19

43 is the elimination of O_2 (via cycloreversion). The macrocyclic peroxide ether **44** under electron impact yields the monomeric molecular ion m/z 274, from which various structure-unspecific fragments are formed[32]. In the mass spectrum of 1,1′-dihydroxy-dicyclohexyl peroxide (**45**) no molecular ion has been detected[33]. Ions of m/z 196, the highest m/z values observed in the spectrum, are believed to be formed by loss of two hydroxyl radicals. The low abundance of the m/z 196 ion (0.16 %) has been interpreted as indication for a rapid further decomposition to yield ionized cyclohexanone. A second breakdown pattern consists of an homolytic scission of the O—O— bond to give an

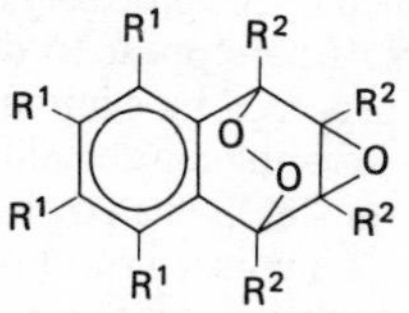

(**38**) $R^1 = R^2$ = Me

(**39**) R^1 = H, R^2 = Me

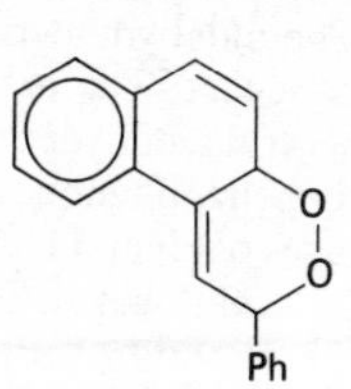

(**40**)

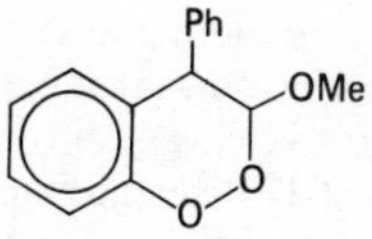

(**41**)

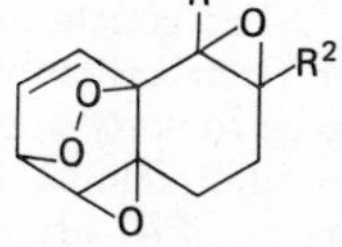

(**42**)

	R^1	R^2
a	H	H
b	Ph	Ph
c	*i*-Pr	H
d	Me	Ph

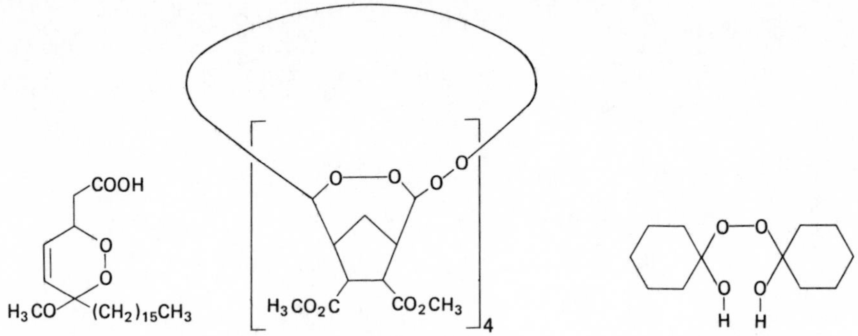

(43) (44) (45)

intermediate at m/z 115. This process may also give rise to the formation of a cyclohexanone cation radical by elimination of a hydroxyl radical.

Cyclic diperoxides **46** have been investigated by low-resolution mass spectrometry and analysis of metastable ion transitions[33]. For the acetone-derived diperoxide **46a** the process outlined in Scheme 20 covers for the most abundant fragment ions ($M^{+\cdot}$ has not been detected). A simultaneous (or consecutive) loss of O_2 and $CH_3\cdot$ generates fragment m/z 101, which decomposes further to the acylium ion, m/z 43, and ionized acetone (pathway a, Scheme 20). In the mass spectrum of the benzophenone-derived diperoxide, **46b**, signals are observed which are likely to be formed by the processes outlined in Scheme 20, pathway b. The aldehyde-derived diperoxides **46c**–**e** appear to decompose via two main reaction sequences. The first one (path c) proceeds by elimination of an oxygen molecule and a neutral aldehyde molecule, thus generating ionized aldehydes (analogous to the formation of benzophenone cation radical from **46b**), which then decompose further by consecutive losses of H· and CO. The second path involves a rupture of the heterocycle to produce $[M/2]^+$ (pathway d, Scheme 20). This splitting can occur in one of two ways. Unfortunately, the method of preparation of these diperoxides does not permit useful isotopic labelling of the oxygen atoms within the molecules **46c**–**e**, and hence further investigation is required to draw any mechanistic conclusions.

Considerably different mass spectra have been reported for the cyclohexanone diperoxide **47**[33,34]. The conclusion[33] that the cyclohexyl ring is less stable than the heterocycle toward electron bombardment has been questioned by Ledaal[34] using high-resolution mass spectrometry and with the temperature of the ionizing chamber well below the decomposition temperature of the peroxide. In fact, Ledaal has demonstrated that the mass spectra of **47** are extremely sensitive to both the temperature and the insertion mode. His results have led him to the conclusion that most of the important fragments observed in the mass spectrum of **47** are due to cleavage of the heterocyclic ring-system followed by consecutive elimination. Spectra of **47** taken at higher temperature or by means of indirect sample introduction are completely different from those obtained at lower temperatures. The differences have been ascribed to thermolysis of **47** prior to ionization.

For the valerophenone-derived diperoxide **48** no molecular ion is detected in the mass spectrum[35]. In close analogy to the decomposition pattern of **46b**, the main fragments are derived from the primary ion m/z 162 (PhCOBu, 33%), which decomposes to m/z 120 (PhCOMe, 93%) and m/z 105 (PhCO, 100%).

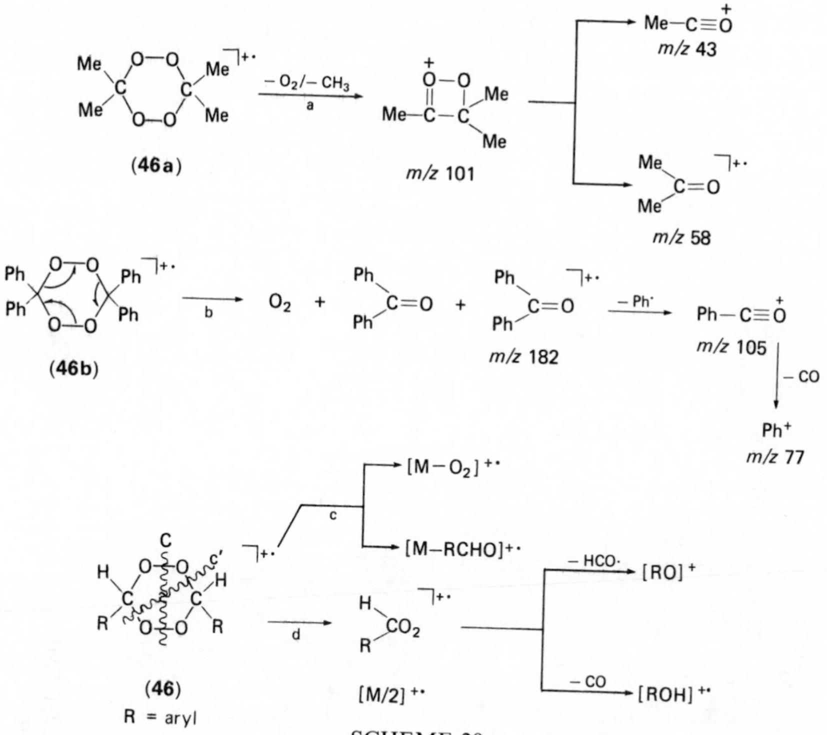

SCHEME 20

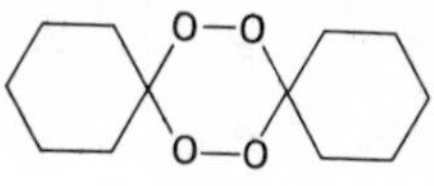

(47)

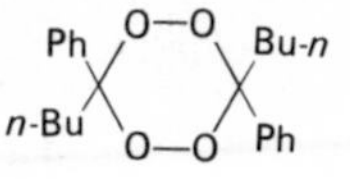

(48)

IV. OZONIDES

The mass spectral behaviour of ozonides **49** has been studied by means of ^{18}O labelling, low-resolution mass spectra and metastable ion transitions[36]. Although in specific cases the substituent groups affect the relative intensity of certain fragments, e.g. aromatic substituents favour the formation of ions containing carbonyl groups, whereas for alkyl substituents such ions are less abundant and more ions due to secondary decompositions are observed, it has been demonstrated that fragmentation of the molecular ions of **49** occurs by two distinct pathways, both of which involve a rupture of the heterocycle (Scheme 21). The first involves the cleavage of an O—O and a C—O bond, whereas the second, which occurs less frequently, proceeds by the scission of the two C—O bonds. The ions reported in Scheme 21 account for most of the ion current recorded. In some cases the elimination of a radical R· from the molecular ion is detected, but this process gives rise to peaks of very low intensity. For stereoisomers, e.g. **49b**, **c**, **e** and **f**, **g**, no differences are observed. With the exception of the $[M - 32]^{+\cdot}$ ion, processes a and b of Scheme 21 give

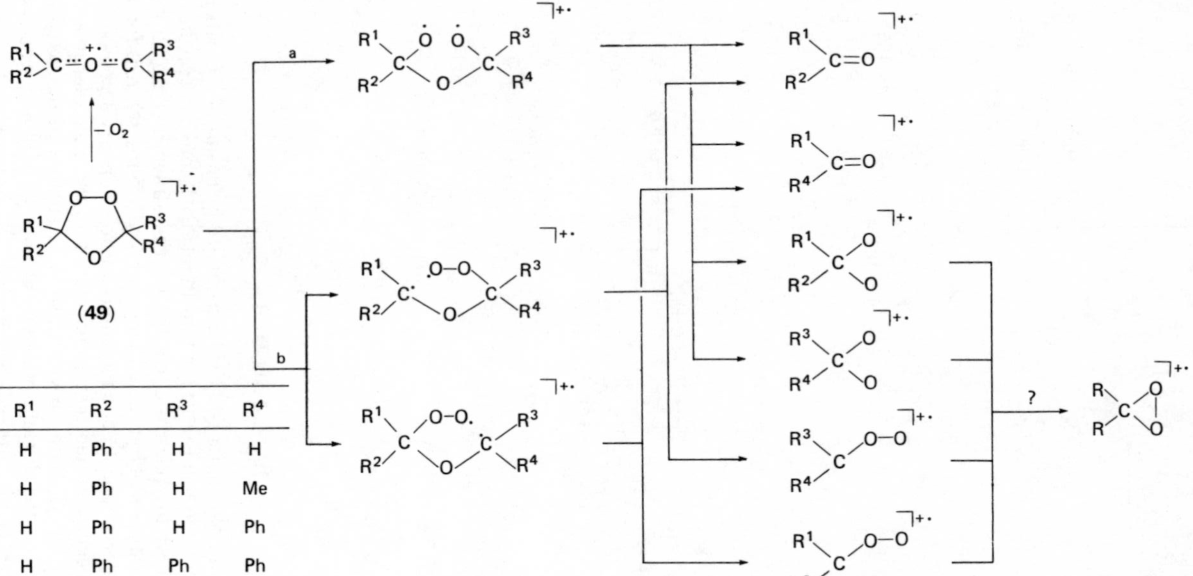

	R^1	R^2	R^3	R^4
a	H	Ph	H	H
b	H	Ph	H	Me
c	H	Ph	H	Ph
d	H	Ph	Ph	Ph
e	H	*n*-Pr	H	*n*-Pr
f	H	$PhCH_2$	H	$PhCH_2$
g	H	$ClCH_2$	H	$ClCH_2$

SCHEME 21

rise to ions having the same elemental compositions but different structures. The spectra of the ozonides, specifically labelled with ^{18}O in the ether position, show clearly that there are, indeed, two distinct pathways for the generation of the $R_2CO_2^+$ ions. One involves cleavage of the O—O and one of the C—O bonds and the other cleavage of the two C—O bonds. Whether the latter is a two-step reaction or a synchronous process (retro-1,3-cycloaddition) is open to question. The same holds for the structure(s) of the $R_2CO_2^+$ ions; it is possible that one or both of the initially formed ions can exist as a dioxacyclopropane derivative, although no definite indication of this has been obtained.

Decomposition patterns as described for **49** are also observed for the ozonides of vinyl fluoride (**50a**)[37] and 1,2-difluoroethylene (**50b**)[38]. From the very abundant molecular ions, among other fragmentations, processes giving rise to the formation of the ions depicted in Scheme 22 are observed. These primary fragments undergo subsequent decompositions to various products. For **50b** no differences are observed for the two diastereomers; the same holds for the diastereomeric D-labelled isotopomers of **50a**. The dissociation of the molecular ions of **50a** and **50b** must be preceded, however, by some skeletal rearrangements; this has to be inferred from the presence of the quite abundant signal at m/z 84 (loss of CO from $M^{+\cdot}$).

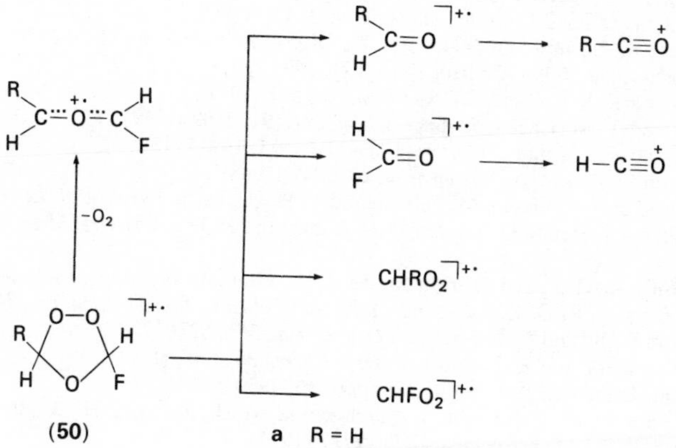

(**50**)

a R = H
b R = F

SCHEME 22

The 70 eV electron-impact mass spectrum of the ozonide **51** is unique in that the base peak is formed by the molecular ion[39]. Other important processes, summarized in Scheme 23, are due to unimolecular losses of O, O_2 and CO. The quite abundant fragment ion at m/z 165 is likely to be a $C_{13}H_9^+$ ion. However, this as well as the genesis and structure (fluorenyl cation?) has yet to be established.

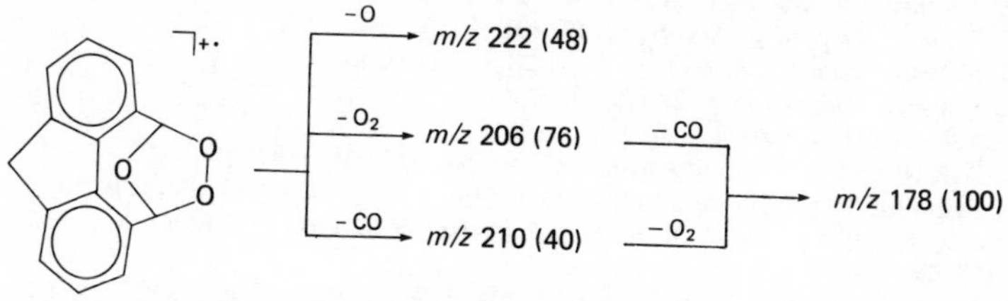

(**51**) ($M^{+\cdot}$ 100%)

SCHEME 23

V. ACKNOWLEDGEMENTS

The financial support from the Fonds der Chemischen Industrie, the Technische Universität Berlin (exchange programm TU Berlin/HU Jerusalem) and the Gesellschaft von Freunden der Technischen Universität Berlin is gratefully acknowledged. H.S. is pleased to thank Churchill College, Cambridge, for an Overseas Fellowship (1981) during the tenure of which this article has been written.

VI. REFERENCES

1. R. A. Abramovitch, E. P. Kyla and E. F. V. Scriven, *J. Org. Chem.*, **36**, 3796 (1971).
2. See for example: F. Bohlmann, C. Zdero, H. Bethke and D. Schumann, *Chem. Ber.*, **101**, 1553 (1968).
3. (a) I. S. Krull, *Org. Prep. Proced. Int.*, **4**, 119 (1972).
   (b) I. S. Krull and A. Mandelbaum, *Proc. Israel Chem. Soc.*, 46 (1972).
   (c) I. S. Krull and A. Mandelbaum, *Org. Mass Spectrom.*, **11**, 504 (1976).
4. H. D. Beckey, *Principles of Field Ionization and Field Desorption Mass Spectrometry*, Pergamon Press, Oxford, 1977.
5. G. S. King, P. G. Magnus and H. S. Rzepa, *J. Chem. Soc., Perkin Trans. 1*, 437 (1972).
6. E. R. Stephans, *Advan. Environ. Sci. Technol.*, **1**, 119 (1969).
7. C. T. Pate, J. L. Sprung and J. N. Pitts, Jr., *Org. Mass Spectrom.*, **11**, 552 (1976).
8. P. L. Hanst, *J. Air Pollut. Control Assoc.*, **21**, 269 (1971).
9. W. Adam and R. S.-C. Tsai, *J. Org. Chem.*, **42**, 537 (1977).
10. W. Adam and Y. M. Cheng, *J. Amer. Chem. Soc.*, **91**, 2109 (1969).
11. W. Adam and G. Santiago, *J. Amer. Chem. Soc.*, **93**, 4300 (1971).
12. P. Brown and C. Djerassi, *Tetrahedron*, **24**, 2949 (1968).
13. (a) H. E. Audier, J. F. Dupin, M. Fétizon and Y. Hoppilliard, *Tetrahedron Letters* 2077 (1966);
    (b) M. Fétizon, Y. Henry, G. Aranda, H. E. Audier and H. D. Lutes, *Org. Mass Spectrom.*, **8**, 201 (1974).
14. (a) A. R. Burgess, D. K. Sen Sharma and M. J. D. White, *Advan. Mass Spectrom.*, **4**, 345 (1968);
    (b) A. R. Burgess, R. D. G. Lane and D. K. Sen Sharma, *J. Chem. Soc.* (*B*), 341 (1969).
15. (a) J. E. van Liehr and L. L. Smith, *J. Org. Chem.*, **35**, 2627 (1970);
    (b) J. E. van Liehr and L. L. Smith, *J. Org. Chem.*, **36**, 1007 (1971);
    (c) J. E. van Liehr and L. L. Smith, *Steroids*, **15**, 385 (1970).
16. A. M. Feinberg, K. Nakanishi, J. Barciszewski, A. J. Rafalski, H. Augustyniak and M. Wiemiórowski, *J. Amer. Chem. Soc.*, **96**, 7797 (1974).
17. M. Przybylski, H. Rinsdorf, U. Lenssen, G. Peter, G. Voelcker, T. Wagner and H. J. Hohorst, *Biomed. Mass Spectrom.*, **4**, 209 (1977).
18. (a) F. S. El-Feraly, Y. M. Chan, E. H. Fairchild and R. W. Doskotch, *Tetrahedron Letters* 1973 (1977).
    (b) R. W. Doskotch, F. S. El-Feraly, E. H. Fairchild and C. T. Huang, *J. Org. Chem.*, **42**, 3614 (1977).
    (c) F. S. El-Feraly, Y. M. Chan, G. A. Capiton, R. W. Doskotch and E. H. Fairchild, *J. Org. Chem.*, **44**, 3952 (1979).
19. R. T. M. Fraser, N. C. Paul and L. Phillips, *J. Chem. Soc.* (*B*), 1278 (1970).
20. For the concept of 'hidden' hydrogen migrations see:
    (a) H. Schwarz, *Org. Mass Spectrom.*, **15**, 491 (1980).
    (b) H. Schwarz, *Nachr. Chem. Tech. Lab.*, **28**, 158 (1980).
    (c) H. Schwarz, *Annali Chim.*, 29 (1981).
    (d) H. Schwarz, *Top. Curr. Chem.*, **97**, 1 (1981).
21. D. H. Ballard and A. J. Bloodworth, *J. Chem. Soc.* (*C*), 945 (1971).
22. M. F. Salomon, R. G. Salomon and R. D. Gleim, *J. Org. Chem.*, **41**, 3983 (1976).
23. I. Bellič, T. Kastelic-Suhadok, R. Kavčič, J. Marsel, V. Kramer and B. Kralj, *Tetrahedron*, **32**, 3045 (1976).
24. (a) A. J. Bloodworth and M. E. Loveitt, *J. Chem. Soc., Perkin Trans 1*, 522 (1978).
    (b) M. F. Salomon and R. G. Salomon, *J. Amer. Chem. Soc.*, **101**, 4290 (1979).

25. J. H. Wieringa, J. Strating, H. Wynberg and W. Adam, *Tetrahedron Letters*, 169 (1972).
26. M. Hisatome and K. Yamakawa, *Org. Mass Spectrom.*, **13**, 1 (1978).
27. D. J. Coughlin, R. S. Brown and R. G. Salomon, *J. Amer. Chem. Soc.*, **101**, 1533 (1979).
28. M. Sasavka and H. Hart, *J. Org. Chem.*, **44**, 368 (1979).
29. M. Matsumoto and K. Kondo, *Tetrahedron Letters*, 3935 (1975).
30. C. S. Foote, S. Mazur, P. A. Burns and D. Lendal, *J. Amer. Chem. Soc.*, **95**, 586 (1973).
31. R. J. Wells, *Tetrahedron Letters*, 2637 (1976).
32. V. N. Odinokov, O. S. Kukovinets and G. A. Tolstikov, *Zh. Org. Khim.*, **14**, 1209 (1978).
33. M. Bertrand, S. Fliszar and Y. Rousseau, *J. Org. Chem.*, **33**, 1931 (1968).
34. T. Ledaal, *Tetrahedron Letters*, 3661 (1969).
35. Y. Ito, M. Konishi and T. Matsuura, *Photochem. Photobiol.*, **30**, 53 (1979).
36. J. Castonguay, M. Bertrand, J. Carles, S. Fliszár and Y. Rouseau, *Coan. J. Chem.*, **47**, 919 (1969).
37. W. Mazur, R. O. Lattimer, A. Lopata and R. L. Kuczkowski, *J. Org. Chem.*, **44**, 3181 (1979).
38. C. W. Gillies, *J. Amer. Chem. Soc.*, **99**, 7239 (1977).
39. M. Yoshida, A. Kadokura, M. Minabe and K. Suzuki, *Tetrahedron*, **35**, 2237 (1979).

The Chemistry of Functional Groups, Peroxides
Edited by S. Patai

CHAPTER 5

Acidity, hydrogen bonding and complex formation

WILLIAM H. RICHARDSON
San Diego State University, San Diego, California 92182, U.S.A.

I. INTRODUCTION . . . 130
II. ACIDITY OF PEROXY ACIDS AND HYDROPEROXIDES . . . 130
A. Acidity of Peroxy Acids . . . 130
B. Acidity of Hydroperoxides . . . 131
C. Transmission Effects . . . 133
III. HYDROGEN BONDING . . . 134
A. Peroxy Acids . . . 134
1. Self-association . . . 134
2. Intermolecular association . . . 135
B. Hydroperoxides . . . 135
1. Self-association . . . 135
2. Intermolecular association . . . 138
a. Arenes . . . 138
b. Alkenes and alkynes . . . 139
c. Ethers . . . 142
d. Alcohols and phenols . . . 143
e. Carboxylic acids . . . 143
f. Amines . . . 144
g. Carbonyls and carboxylic acid derivatives . . . 144
h. Sulphoxides . . . 145
i. Miscellaneous donors . . . 148
j. Summary . . . 148
3. Intramolecular association . . . 150
a. Arenes . . . 150
b. Alkynes, ethers and amines . . . 151
c. Carbonyls . . . 151
d. Peroxides . . . 153
IV. COMPLEXES . . . 155
A. Peroxy Acids . . . 155

B. Hydroperoxides 156
1. Amine–hydroperoxide complexes 156
2. Ammonium–hydroperoxide complexes 157
V. REFERENCES 157

I. INTRODUCTION

The three topics considered in this chapter are closely related, since the acidity of the OOH group can be correlated with hydrogen bonding and complex formation. The complexes considered here will not involve σ bonding or transition-metal ions. This effectively limits complexes to hydrogen-bonded species. The section on complexes arbitrarily considers only isolable complexes. The section on hydrogen bonding considers complexes that are detected in solution, but not isolated.

The section on acidity of peroxides will present data primarily from direct measurements of pK_a values. Correlations between hydrogen bonding and pK_a can be used to estimate acidities of peroxides. These correlations will be presented in the section on hydrogen bonding.

II. ACIDITY OF PEROXY ACIDS AND HYDROPEROXIDES

The two classes of organic peroxides that have ionizable protons and will display acidic properties are peroxy acids and hydroperoxides. The pK_a values for several peroxides of these two types have been reported. In both classes of peroxides, the pK_a values fall into a rather narrow range. With the aid of linear free-energy relationships (LFERs), and due to the narrow pK_a range, good pK_a estimates can be obtained for most peroxy acids and tertiary hydroperoxides.

Acidities of peroxides have been of particular interest in mechanistic studies and to correlate rates of peroxide decompositions[1]. The use of pK_a values in this manner has been one of the motivating forces for obtaining reliable pK_a data of peroxides.

A. Acidity of Peroxy Acids

The pK_a values for aliphatic peroxy acids are given in Table 1. It is seen that the values fall over a rather narrow range. A Taft correlation gives

$$pK_a = (-0.950 \pm 0.233)\sigma^* + 8.00 \pm 0.12$$

with $r = 0.898$ and the standard deviation of pK_a on $\sigma^*(S_{y.x})$ of ± 0.264. Since the correlation coefficient (r) is a function of the slope and decreases with smaller ρ^* values, the $S_{y.x}$ parameter is a better measure of the goodness of fit[5]. For comparison, the ρ^* value for ionization of aliphatic carboxylic acids in water at 25°C is 1.72[6] vs. 0.950 for the peroxy acids. The reduced sensitivity of substituents on pK_as of peroxy acids relative to carboxylic acids is qualitatively understandable. The acidic proton in peroxy acids is removed from the substituents by one additional atom as compared to carboxylic acids. A fall-off in the effect of substituents on the pK_a of peroxy acids is then expected. The decreased acidity of peroxy acids relative to carboxylic acids is also evident from Table 1. For example, peroxyacetic acid is a weaker acid by about 3.5 pK_a units as compared to acetic acid ($pK_a = 4.75$)[7]. Since resonance stabilization of the peroxy acid anion is not possible, the decreased acidity relative to carboxylic acids is understandable.

TABLE 1. pK_a values of aliphatic peroxy acids (RCO_3H) in water

R	pK_a	T(°C)	Reference
H	7.1	19.5	2
Me	8.2	20.0	2
Et	8.1	23.0	2
n-Pr	8.2	21.5	2
$ClCH_2$	7.2	25	3
t-Bu	8.23	25	4

Acidities of arylcarboperoxy acids are given in Table 2. The decreased acidity of the arylcarboperoxy acids relative to the arylcarboxylic acids is similar to that observed in the aliphatic series. For example, benzenecarboperoxy acid is a weaker acid by 3.6 pK_a units than benzoic acid (pK_a = 4.20)[7]. A Hammett correlation of the data in Table 2, expressed as ionization constants, gives $\rho = 0.704 \pm 0.043$ ($r = 0.991$ and $S_{y.x} = \pm 0.037$). This ρ value compared to arylcarboxylic acids ($\rho \equiv 1.00$) again indicates a poorer relay of electronic effects of the substituents.

TABLE 2. pK_a values of arylcarboperoxy acids ($ArCO_3H$)

Ar	pK_a	T(°C)	Solvent	Reference
C_6H_5	7.9	20	H_2O	8
	7.78	25	H_2O	1b
	7.74	35	H_2O	1b
	7.63	45	H_2O	1b
p-$MeOC_6H_4$	8.07	25	H_2O	1b
p-MeC_6H_4	7.95	25	H_2O	1b
	7.80	35	H_2O	1b
	7.73	45	H_2O	1b
p-FC_6H_4	7.76	25	H_2O	1b
p-ClC_6H_4	7.67	25	H_2O	1b
m-ClC_6H_4	7.60	25	H_2O	1b
	7.46	35	H_2O	1b
	7.37	45	H_2O	1b
p-$NO_2C_6H_4$	7.29	25	H_2O	1b
o-$(CO_2H)C_6H_4$	~8.5	20	Aq. dioxan	8

Considering the rather narrow range of pK_a values for peroxy acids and the reasonably good LFER correlations, there appears little need to measure additional pK_a values. In terms of the standard deviation estimate in pK_a, the pK_as of arylcarboperoxy acids can be estimated to ± 0.037 pK_a units, while aliphatic peroxy acid pK_as can be estimated to ± 0.26 pK_a units.

B. Acidity of Hydroperoxides

The pK_a values for hydroperoxides are listed in Table 3 and Figure 1 displays the data in a Taft plot. Reasonably good correlations are obtained if the hydroperoxides are separated into three groups. With this division, data points are sparse for the

TABLE 3. pK_a values of hydroperoxides (ROOH)

R	pK_a	T(°C)	Solvent	Reference
H	11.6	20	H_2O	2
	12.02	25	*a*	9
Me	11.5	20	H_2O	2
Et	11.8	20	H_2O	2
i-Pr	12.1	20	H_2O	2
t-Bu	12.8	20	H_2O	2
	13.27	25	*a*	9
i-BuC(Me)Et	12.8	20	H_2O	2
$PhCH_2CMe_2$	13.25	25	*a*	9
$PhCMe_2$	13.08	25	*a*	9
	12.6	25	H_2O^b	10
Ph_2CEt	13.02	25	*a*	9
Ph_2CMe	12.94	25	*a*	9
Ph_3C	13.07	25	*a*	9
Me_2CCH_2Cl	~13.2	0–30	*c*	1a

[a] 40% aq. methanol, ionic strength = 0.600.
[b] By distribution between benzene and 0.253 M sodium hydroxide solution.
[c] From kinetics of the basic decomposition in 40% aq. methanol, ionic strength = 1.58.

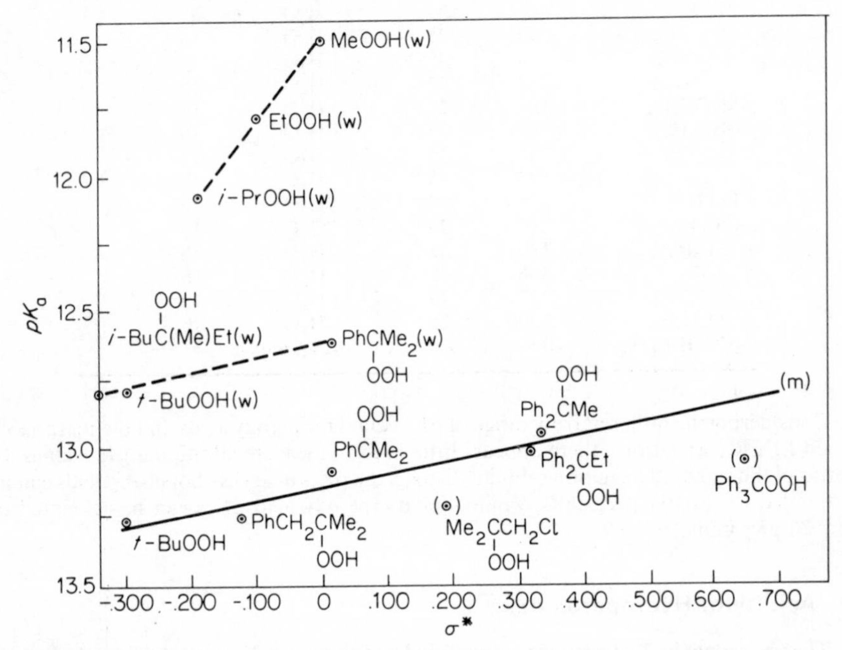

FIGURE 1. An Ingold–Taft correlation of the pK_a values of the hydroperoxides in water (w) and 40% aqueous methanol (m). Reprinted with permission from W. H. Richardson and V. F. Hodge, *J. Org. Chem.*, **35**, 4012 (1970). Copyright (1970), American Chemical Society.

primary–secondary hydroperoxide line and the tertiary hydroperoxide (water) line. The correlation equations for these two lines are, respectively:

$$pK_a = (-3.15 \pm 0.096)\sigma^* + 11.50 \pm 0.01$$

($r = 0.9995$ and $S_{y.x} = \pm 0.01$) and

$$pK_a = (-0.591 \pm 0.061)\sigma^* + 12.61 \pm 0.02$$

($r = 0.995$ and $S_{y.x} = \pm 0.02$). The correlation equation for tertiary hydroperoxides in 40% aq. methanol is: $pK_a = (-0.499 \pm 0.095)\sigma^* + 13.14 \pm 0.02$ ($r = 0.949$ and $S_{y.x} = \pm 0.05$), where chloro-*t*-butyl hydroperoxide and trityl hydroperoxide are excluded. The pK_a value of the former peroxide has been obtained by an indirect kinetic method[1a]. The lower acidity of trityl hydroperoxide, as judged from the correlation line of the remaining peroxides, may result from excessive steric effects.

Although the number of points for the pK_a correlation of tertiary hydroperoxides in water is limited, the ρ^* value (0.591) is similar to that for this type of peroxide in aqueous methanol ($\rho^* = 0.499$). The similarity of these two ρ^* values gives added support to the reliability of these correlations. Translation of the pK_as to higher values in aqueous methanol is expected, based on similar trends in the pK_as of carboxylic acids in aqueous alcoholic solvents[11]. The primary and secondary hydroperoxides appear anomalous in their deviation from tertiary hydroperoxides and their large ρ^* values (3.15). It can be noted that primary and secondary hydroperoxides readily undergo reaction with bases to yield carbonyl products[12]. This elimination reaction did prevent the measurement of the pK_a of benzhydryl hydroperoxide by the standard base titration method[9]. Steric effects are another possible explanation for the deviation in ρ^* values between primary–secondary and tertiary hydroperoxides. A suggestion of the importance of steric effects is seen in the deviation of trityl hydroperoxide from the tertiary hydroperoxide correlation line. Unfortunately, insufficient E_s values do not allow a test of this possibility by a Taft polar–steric multilinear regression analysis.

It can be noted that the pK_a values of *n*-butyl, *s*-butyl and *t*-butyl hydroperoxide have been measured by the base titration–UV method and reported to give a linear correlation with σ^* where $\rho^* = 4.1$[13]. A different correlation line in a plot of pK_a vs. σ^* with 'almost' the same slope has been reported for benzyl, 1-phenylethyl and cumyl hydroperoxide[13].

In summary, reasonably good estimates of the pK_as of tertiary hydroperoxides in water and aqueous methanol can be made. In contrast, the reliability of pK_a predictions for primary and secondary hydroperoxides is uncertain.

C. Transmission Effects

The relative transmission factor (ϕ) is defined as $\phi = \rho^*_A/\rho^*_B$ and is a relative measure of the ability of substituents to transmit polar effects to the equilibrium or reaction site in systems A and B[14]. The relay of polar effects through oxygen vs. CH_2 can be evaluated by comparison of ROOH vs. RCH_2OH, where $\phi(ROOH/RCH_2OH) = 0.35$ ($= 0.499/1.42$[15]). The poorer relay through oxygen has been analysed in terms of the Kirkwood–Westheimer equation and it has been concluded that this effect is due to increasing the effective dielectric between the substituent and the OO^- group by replacement of CH_2 with O[9]. For tertiary hydroperoxides and aliphatic peroxy acids, $\phi(ROOH/RCO_3H) = 0.53$ ($= 0.499/0.950$), which indicates that the relay of polar effects is better with peroxy acids even through an added CO group. In terms of the Kirkwood–Westheimer equation, this means that the effective dielectric constant is less for the peroxy acids. It can be noted that the effective dielectric constant can be minimized with an *anti* conformation of the oxygen anion and the carbonyl group.

III. HYDROGEN BONDING

In this section, hydrogen bonding involving peroxy acids and hydroperoxides is considered. Three types of association are reviewed: self-association and inter- and intramolecular hydrogen bonding. There is a considerable amount of literature concerning hydrogen bonding of hydroperoxides. The interest in part arises from attempts to correlate rates of hydroperoxide decomposition with hydrogen bonding. It is proposed that the rate of decomposition of hydrogen-bonded hydroperoxides to give radicals is faster than homolysis of the uncomplexed hydroperoxide[16]. In fact, equilibrium constants of complex formation can be obtained indirectly from the kinetics of hydroperoxide decomposition. Since the formation and decomposition of hydroperoxides has practical consequences, particularly in autooxidation, the interest in hydroperoxide hydrogen bonding is understandable.

A. Peroxy Acids

1. Self-association

In the crystalline state, peroxy acids are intermolecularly hydrogen-bonded to the oxygen atom of the carbonyl group. X-ray crystallographic studies of *o*-nitroperoxybenzoic acid[17] and peroxypelargonic acid[18] are reported and reviewed[19]. The intermolecular hydrogen-bonding energy of solid peroxydecanoic acid, as estimated from vapour pressure measurements, is reported to be 2.5 kcal mol^{-1} [20]. In comparison, the intermolecular hydrogen-bonding energy of decanoic acid is estimated to be 4.7 kcal mol^{-1} [20].

In contrast to the crystalline state, peroxy acids in the liquid state or in solution are intramolecularly hydrogen-bonded and monomeric. The first suggestion of a monomeric state for peroxy acids as opposed to a dimeric state for carboxylic acids was given by a comparison of boiling points, where the corresponding peroxy acids showed lower boiling points[21]. Analysis of the O—H and C—O infrared absorptions leads to the proposal of the intramolecular hydrogen-bonded structure **1** as the sole species in liquid and vapour

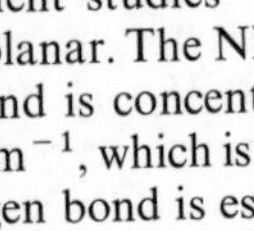

(1)

states[22]. The monomeric intramolecularly hydrogen-bonded structure is supported by other studies which have employed infrared[23–27] NMR[28], molecular weight[24], vapour pressure[29] and electric dipole moment studies[30]. The last study[30] has indicated that structure **1** is puckered, rather than planar. The NMR absorption of the peroxidic proton falls in the range of 10.9 to 11.85 and is concentration-independent[28]. A sharp O—H infrared band occurs at about 3300 cm^{-1}, which is also concentration-independent[24]. The energy of the intramolecular hydrogen bond is estimated to be about 1–2 kcal mol^{-1} [31].

The carbonyl stretching frequencies (1727–1735 cm^{-1}) of a series of substituted peroxybenzoic acids (*m*-Cl, *m*-F, *p*-Br, *p*-Cl, *p*-F, H, *m*-Me, *p*-Me) in carbon tetrachloride have been correlated with the Hammett σ constant to give a negative ρ value[26]. The linear correlation has been contrasted with nonlinear relations with either monomer or dimeric carboxylic acids. It has been proposed that the LFER with the peroxybenzoic acids is the result of electronic effects on both the carbonyl groups and the strength of the intramolecular hydrogen bond.

2. *Intermolecular association*

Equilibrium constants and enthalpies have been determined by calorimetric measurements for equation (1). The results are given in Table 4 along with infrared frequency shifts ($\Delta\nu$) between **1** and the peroxy-acid–donor complex. Complex formation

$$R-C(=O\cdots H-O-O) + D \rightleftharpoons R-C(=O)-OOH\cdots D \qquad (1)$$

(**1**)

increases with increasing acidity of the peroxy acid, as seen from the first three entries in Table 4[32]. The basicity of the donors decrease in the order DMAA > DMF > THF, which follows decreasing complex formation with *p*-chloroperoxybenzoic acid[32]. According to the Badger–Bauer relationship[34], increasing $\Delta\nu$ values correspond to increasing complex formation. With *p*-nitroperoxybenzoic acid, the donors are listed in Table 4 according to increasing complex formation based on $\Delta\nu$. It has been noted that increased complex formation (equation 1) corresponds to a decrease in rate of olefin oxidation by peroxy acids[32,33].

TABLE 4. Intermolecular association of peroxy acids with various donors

Peroxy acid	Donor (D)	$K(25°C)^a$ (M^{-1})	$-\Delta H^a$ (kcal mol^{-1})	$\Delta\nu_{OH}{}^b$ (cm^{-1})	Reference
p-t-$BuC_6H_4CO_3H$	DMAA[c]	4.0 ± 0.5	6.3 ± 0.1	—	32
p-$ClC_6H_4CO_3H$	DMAA[c]	8.5 ± 0.5	6.6 ± 0.1	—	32
m-$ClC_6H_4CO_3H$	DMAA[c]	9.9 ± 0.5	7.1 ± 0.1	—	32
p-$ClC_6H_4CO_3H$	DMF	6.0 ± 0.5	6.0 ± 0.1	—	32
p-$ClC_6H_4CO_3H$	THF	2.3 ± 0.5	5.8 ± 0.1	—	32
p-$NO_2C_6H_4CO_3H$	$MeCO_2Et$	—	—	33	33
p-$NO_2C_6H_4CO_3H$	Dioxane	—	—	100	33
p-$NO_2C_6H_4CO_3H$	Et_2O	—	—	115	33
p-$NO_2C_6H_4CO_3H$	THF	—	—	149	33
p-$NO_2C_6H_4CO_3H$	DMF	—	—	165	33

[a]In carbon tetrachloride, determined by calorimetry.
[b]In methylene chloride, where $\Delta\nu_{OH} = \nu_{OH}$ (free) $- \nu_{OH}$ (hydrogen-bonded).
[c]*N,N*-Dimethylacetamide.

B. Hydroperoxides

The majority of the hydrogen-bonding literature of peroxides deals with hydroperoxides. The discussion of this topic is organized in order of self-association, intermolecular hydrogen bonding to other donors or acceptors and intramolecular hydrogen bonding.

1. *Self-association*

One of the most characteristic physical properties of hydroperoxides is self-association. Of the associated species, it appears that the dimer and trimer predominate, with lesser amounts of more highly associated species at moderate concentrations. One formulation of the self-association equilibria is shown in Scheme 1. A more complex self-associated scheme has been proposed which involves (*i*) the dimer, (*ii*) the dimer with additional hydroperoxide molecules associated with it and (*iii*) linear-associated hydroperoxide molecules[35]. There is no direct evidence for the various self-associated species. A reasonable argument has been made for the six-ring structures of the dimer and trimer based on the preference for alcohols to form trimers[36].

$$2\,ROOH \underset{}{\overset{K_2}{\rightleftharpoons}} (ROOH)_2\ \text{(cyclic dimer)} \underset{}{\overset{K_3,\ ROOH}{\rightleftharpoons}} (ROOH)_3\ \text{(cyclic trimer)}$$

SCHEME 1

Equilibrium constants and thermodynamic parameters for self-association of hydroperoxides are given in Table 5. Variation in hydroperoxide concentration is used to fit the IR or NMR data to dimer or higher-order association, and along with temperature variation the thermodynamic parameters are obtained. The equilibrium constants and thermodynamic parameters for self-association beyond the dimer are calculated by two different methods. In one method, the average equilibrium constant and thermodynamic parameters per subsequent hydroperoxide association are calculated[35,37]. If the higher order associated species beyond the dimer is predominately trimer, then this equilibrium constant will correspond to K_3 of Scheme 1. This assumption has been made in Table 5. By the second method, the data for higher-order associated species beyond the dimer have been fitted to a monomer–trimer equilibrium (equation 2)[38]. The equilibrium constants in Scheme 1 and equation (1) are related as $K_T = K_2K_3$ and the thermodynamic parameters as $\Delta H_T = (\Delta H_2 + \Delta H_3)$ and $\Delta S_T = (\Delta S_2 + \Delta S_3)$. In Table 5, the reported data associated with equation (2) are given in parentheses and values of K_3, ΔH_3 and ΔS_3 (Scheme 1) are calculated from the above relationships.

$$3\,ROOH \overset{K_T}{\rightleftharpoons} (ROOH)_3 \qquad (2)$$

Rather small changes in the thermodynamic parameters (ΔH_2 and ΔS_2) for dimerization are observed with varying substituents. Larger variations in ΔH_3 and ΔS_3 are noted, but this may be due in part to the different methods of calculating these parameters. Dependence of monomer, dimer, trimer and tetramer concentration on the stoichiometric hydroperoxide concentration has been calculated for the cyclooctyl and cyclopentyl hydroperoxides[45].

Replacement of the hydroperoxy proton with deuterium decreases dimer formation as shown by the first and second entries of Table 5. It has been noted that a similar result occurs in the dimerization of phenol and trifluoroacetic acid[39].

It is interesting that trityl hydroperoxide forms a trimer, but not a dimer[42]. This observation can be rationalized in terms of less severe steric effects in the trimer compared to the dimer. Steric effects also appear to alter the distribution of the heptyl hydroperoxide species as seen from Table 6. As steric effects increase, with internal placement of the hydroperoxy group, the % trimer increases at the expense of the dimer.

R	Solvent	Method	$K_2(T, °C)$ (M^{-1})	$K_3(T, °C)$ (M^{-1})	$-\Delta H_2$ (kcal mol^{-1})	$-\Delta S_2$ (e.u.)	$-\Delta H_3$ (kcal mol^{-1})	$-\Delta S_3$ (e.u.)	Ref.
t-Bu	CCl_4	IR	1.90(30)	—	5.95	18.4	—	—	39
t-Bu (OOD)	CCl_4	IR	1.65(30)	—	4.9	15	—	—	39
t-Bu	CCl_4	IR	0.46(40.5)	2.75(40.5)[a]	6.30	21.6[b]	2.80[a]	6.9[a,b]	40
t-Bu	c	NMR	19(21)	180(21)[a]	6.6	17	3[a]	0.3[a]	35
t-Bu	d	NMR	0.67(30)[e]	—	5.7	19.6	—	—	41
t-Bu	CCl_4	NMR	1.98(20)[j]	—	5.7	19.6	—	—	42
t-Bu (av.)	—	—	[1.6(30)][e]	—	[6.1]	[19.2]	—	—	
Me_2CEt	d	NMR	0.95(30)[e]	—	5.4	18	—	—	41
$Me_2C(CH{=}CHMe)$	d	NMR	0.44(30)[e]	—	4.5	17.6	—	—	41
				0.32(30)[f]			−0.7[f]	0[f]	
1-Heptyl	CCl_4	IR	3.1(30)	[1.0(30)][g]	4.9	13.9	(4.2)[g]	(13.9)[g]	38
				3.6(30)[f]			0.6[f]	−0.5[f]	
2-Heptyl	CCl_4	IR	0.70(30)	[2.5(30)][g]	4.7	16.2	(5.3)[g]	(15.7)[g]	38
				3.4(30)[f]			1.0[f]	0.9[f]	
3-Heptyl	CCl_4	IR	0.90(30)	[3.1(30)][g]	4.4	14.7	(5.4)[g]	(15.6)[g]	38
				2.9(30)[f]			1.4[f]	1.9[f]	
4-Heptyl	CCl_4	IR	1.3(30)	[3.7(30)][g]	4.5	14.3	(5.9)[g]	(16.2)[g]	38
1-Methylcyclohexyl-1	Decane	IR	1.3(22)	—	—	—	—	—	37
1-Methylcyclohexyl-1	Decane	NMR	1.1(20)	4.9(20)[a]	5.6	19	2.8[a]	6.4[a,h]	37
1-Methylcyclohexyl-1	i	NMR	1.2(39)	—	—	—	—	—	43
1-Ethynylcyclohexyl-1	—	NMR	—	—	4.2	—	—	—	44
1-Phenylcyclohexyl-1	d	NMR	0.42(30)	—	5.2	20	—	—	41
Cumyl	d	NMR	0.85(30)	—	6.8	23	—	—	41
Cumyl	c	NMR	15(21)	16(21)[a,e]	7.7	21	2.2[a]	2[a]	35
Ph_3C	CCl_4	NMR	—	[3.3(28)][g]	—	—	(4.8)[g]	(13.8)[g]	42

[a]Reported for subsequent association of each additional hydroperoxide beyond the dimer. K is in units of M^{-1}.
[b]Calculated from K_3 and ΔH_3 at 40.5°C.
[c]Reported as 'inert solvent', which is presumably carbon tetrachloride or a saturated hydrocarbon.
[d]Carbon tetrachloride or cyclohexane.
[e]Calculated from ΔH and ΔS.
[f]Calculated $K_3(M^{-1})$, ΔH_3 and ΔS_3 according to Scheme 1 from $K_T(M^{-2})$, ΔH_T and ΔS_T of equation (2); see text.
[g]$K_T(M^{-2})$, ΔH_T and ΔS_T as reported according to equation (2).
[h]Calculated from K_3 and ΔH_3 at 20°C.
[i]Hydrocarbon solvent.
[j]This is the reported value, the calculated value from ΔH_2 and ΔS_2 at 20°C is $K_2 = 0.93\ M^{-1}$.

TABLE 6. Distribution of isomeric heptyl hydroperoxide species in carbon tetrachloride (30°C) at 1 M concentration[42]

ROOH	% Monomer	% Dimer	% Trimer
1-Heptyl	31.1	59.9	9.7
2-Heptyl	38.0	20.2	41.0
3-Heptyl	36.0	23.3	43.3
4-Heptyl	32.8	28.0	39.2

2. *Intermolecular association*

Several different types of functional groups have been complexed with hydroperoxides. This section is organized according to the type of group that is complexed to the hydroperoxide.

a. Arenes. Table 7 presents data for the complex formation of equation (3), where D is the aromatic π electron donor. The equilibrium constants are calculated according to a 1:1

$$\text{ROOH} + \text{D} \overset{K}{\rightleftharpoons} \text{ROOH}\cdots\text{D} \tag{3}$$

complex as indicated by equation (3). However, the situation may be more complicated. In distribution experiments with *t*-butyl hydroperoxide between water (presumably monomeric and hydrogen-bonded to water) and an arene (neat or in CCl_4, where it exists as a monomer, self-associated and associated to the arene), the results could be accomodated to a 1:1 complex of the hydroperoxide with 4 M styrene in carbon tetrachloride[39]. With other aromatic solvent systems, particularly at higher hydroperoxide concentration, an additional term involving association of the hydroperoxide dimer with the arene is required to fit the data[39].

Replacement of the hydroperoxy proton in *t*-butyl hydroperoxide with deuterium increases complexation to benzene and styrene. As seen previously, this substitution decreases dimer formation. It has been noted that a similar trend occurs with phenol[39].

For *t*-butyl hydroperoxide, there is not a large change in K with various substituted arenes. As expected from a cursory survey of Table 7, no satisfactory LFERs have been observed with $\log K$ (30°C) vs. $\Sigma(\sigma_m + \sigma_p)/2$ or $\log K/K_o$ for iodine–arene complexes relative to *p*-xylene[50]. Similarly, no LFE correlations have been observed with the ΔH values. A good correlation has been reported for *t*-butyl hydroperoxide using the difference in frequencies between the free and aromatic-complexed O—H species ($\Delta\nu_{OH}$, cm^{-1}) vs. $\Sigma(\sigma_m + \sigma_p)/2$. The correlation equation is:

$$\Delta\nu_{OH}(\text{ArH}) - \Delta\nu_{OH}\,(\text{benzene}) = -43.7\,\Sigma(\sigma_m + \sigma_p)/2 + 5.67,$$

$r = 0.96$[47]. The value of $\Delta\nu_{OH}$ then increases with electron-releasing groups. According to the Badger–Bauer rule[34], a linear relationship is expected between $\Delta\nu_{OH}$ and $-\Delta H$. The correlation then indicates that complex formation is increased by electron-releasing groups, as would be expected in hydroperoxide–π-arene hydrogen-bonded species. The reason for a lack of correlation between $\log K$ or ΔH vs. $\Sigma(\sigma_m' + \sigma_p)/2$ is uncertain, but it may reflect the lack of accuracy of these values. Reasonably linear correlations of $\Delta\nu_{OH}$ for *t*-butyl hydroperoxide vs. $\Delta\nu_{OH}$ for aromatic complexes with HCl, HCN, phenol and *t*-butyl alcohol have been observed[47].

It is interesting to note that with iodobenzene and *t*-butyl hydroperoxide an additional IR band is observed at 3481 cm^{-1} (CCl_4), which has been assigned to I···HOO hydrogen bonding. A band at 3521 cm^{-1} has been assigned to the π-arene–HOO complex[47]. Two bands have also been observed for anisole and assigned to a π complex (3508 cm^{-1}) and an O···HOO complex (3436 cm^{-1})[47].

Satisfactory LFERs for cumyl hydroperoxide have been obtained by correlating $-\Delta H$ (Table 7)[35] vs. $\Sigma(\sigma_m + \sigma_p)/2$ and log $(K/K_o)_{I_2}$ for iodine–arene complexes relative to *p*-xylene[50]. The correlation equations are:

$$-\Delta H = (-1.22 \pm 0.12)\Sigma(\sigma_m + \sigma_p)/2 + 1.54 \pm 0.05,$$

$r = 0.979$, $S_{y.x} = \pm 0.12$; and

$$-\Delta H = (1.09 \pm 0.11)\log(K/K_o)_{I_2} + 1.82 \pm 0.04,$$

$r = 0.985$, $S_{y.x} = \pm 0.079$. With *t*-amyl hydroperoxide, there are insufficient data available for a correlation with log $(K/K_o)_{I_2}$, but a good correlation has been obtained with $\Sigma(\sigma_m + \sigma_p)/2$;

$$-\Delta H = (-1.11 \pm 0.18)\Sigma(\sigma_m + \sigma_p)/2 + 1.52 \pm 0.07,$$

$r = 0.949$, $S_{y.x} = \pm 0.16$. For both cumyl and *t*-amyl hydroperoxide, electron-releasing groups stabilize the arene–hydroperoxide complex and the ρ values are within experimental error. The positive ρ value of about 1.0, obtained in the $\log(K/K_o)_{I_2}$ correlation with cumyl hydroperoxide, indicates that the arene–hydroperoxide complex responds similarly to electronic effects as does the well-known I_2–π-arene complex. Although data are not available for a thorough test of steric effects on arene–hydroperoxide complexes, it appears that increased methyl substitution to pentamethylbenzene does not introduce excessive steric effects, since that latter arene is satisfactorily correlated with the cumyl hydroperoxide complex formation.

NMR data has also been used to probe hydroperoxide hydrogen bonding to arenes[35]. It is claimed that the chemical shifts of the hydrogen-bonded OOH proton as well as IR frequency shifts ($\Delta\nu_{OH}$) are satisfactorily correlated with the ionization potentials of the aromatic donor[35,48].

b. Alkenes and alkynes. Relative to the studies with arene–hydroperoxide complexes, only a small amount of data is available for alkene– or alkyne–π-hydroperoxide complexes. In Table 8, infrared absorption bands for the free and the π-complexed OH group are given for cyclic olefins along with the difference between these frequencies ($\Delta\nu_{OH}$). It has been noted that the strain energy of the olefin, as measured by the heat of hydrogenation of the olefin, parallels $\Delta\nu_{OH}$[51]. Although, only three data points are available, an excellent correlation is obtained:

$$\Delta\nu_{OH} = (6.01 \pm 0.21)[-\Delta H(H_2)] - (83.7 \pm 5.9),$$

$r = 0.999$, $S_{y.x} = \pm 1.14$. The correlation suggests that complex formation increases with increasing strain in the olefin.

The NMR and IR spectra of *t*-butyl hydroperoxide complexes with cyclohexene, isomeric pentenes and pentadienes have been measured[52]. Unsymmetrical olefins form stronger complexes than symmetrical olefins. It has been claimed that *n*-butyl, *s*-butyl and *t*-butyl hydroperoxides form hydrogen-bonded complexes to both the carbonyl oxygen and the olefin π bond in methyl acrylate[47]. The free O—H stretching frequencies of these hydroperoxides in carbon tetrachloride solution are: 3560, 3560 and 3562 cm^{-1}, respectively. The corresponding $\Delta\nu_{OH}$ values assigned to the olefin–hydroperoxide complexes are: 218, 222 and 224 cm^{-1}, respectively. These $\Delta\nu_{OH}$ values are surprisingly

TABLE 7. Intermolecular association of hydroperoxides with aromatic compounds

$$\mathrm{ROOH} + \mathrm{D} \overset{K}{\rightleftharpoons} \mathrm{ROOH\cdots D}$$

R	D	Solvent	Method	$K(T, ^\circ C)$ (M^{-1})	$-\Delta H$ (kcal mol^{-1})	$-\Delta S$ (e.u.)	Ref.
t-Bu	C_6H_6	CCl_4	IR	0.11(70)	1.58	9.0	39
t-Bu(OOD)	C_6H_6	CCl_4	IR	0.25(70)	—	—	39
t-Bu	C_6H_6	C_6H_6	IR	0.073(20)	1.46	10[a]	46
t-Bu	C_6H_6	CCl_4	IR	0.16(30)[b]	1.96[b]	10.1	47
t-Bu	Toluene	CCl_4	IR	0.27(30)[b]	1.37[b]	7.1	47
t-Bu	Ethylbenzene	CCl_4	IR	0.20(30)	—	—	47
t-Bu	Ethylbenzene	Ethylbenzene	IR	0.095(20)	1.45	9.6[a]	46
t-Bu	Cumene	CCl_4	IR	0.19(30)[b]	2.80[b]	12.5	47
t-Bu	Butylbenzene	CCl_4	IR	0.18(30)	—	—	47
t-Bu	Isobutylbenzene	CCl_4	IR	0.16(30)	—	—	47
t-Bu	*o*-Xylene	CCl_4	IR	0.24(30)	—	—	47
t-Bu	*p*-Xylene	CCl_4	IR	0.25(30)	2.34[b]	10.5	47
t-Bu	*m*-Xylene	—	IR	—	1.75	—	35, 49
t-Bu	*m*-Diethylbenzene	CCl_4	IR	0.25(30)	—	—	47
t-Bu	*m*-Diisopropylbenzene	CCl_4	IR	0.22(30)	—	—	47
t-Bu	1,2,4-Trimethylbenzene	CCl_4	IR	0.26(30)	—	—	47
t-Bu	1,2,4-Trimethylbenzene	—	IR	—	1.95		35, 49
t-Bu	1,2,4,5-Tetramethylbenzene	CCl_4	IR	0.25(30)[b]	2.38[b]	10.6	47
t-Bu	Styrene	CCl_4	IR	0.28(30)[b]	1.79[b]	8.46	47
t-Bu	Styrene	CCl_4	IR	0.27(30)	2.39	10.5	39
t-Bu	Styrene	Styrene	IR	0.15(20)	2.5	12[a]	46
t-Bu (OOD)	Styrene	CCl_4	IR	0.84(30)	2.7	9	39
t-Bu	Biphenyl	CCl_4	IR	0.29(30)	—	—	47
t-Bu	Bibenzyl	CCl_4	IR	0.43(30)	—	—	47
t-Bu	PhCl	CCl_4	IR	0.12(30)[b]	1.81[b]	10.2	47

TABLE 7 *continued*

R	D	Solvent	Method	$K(T, °C)$ (M^{-1})	$-\Delta H$ (kcal mol^{-1})	$-\Delta S$ (e.u.)	Ref.
t-Bu	PhCl	CCl_4	IR	0.10(70)	—	—	39
t-Bu	$PhCH_2Cl$	CCl_4	IR	0.24(30)	—	—	47
t-Bu	Anisole	CCl_4	IR	0.56(30)	—	—	47
t-Bu	Nitrobenzene	CCl_4	IR	0.37(30)	—	—	47
Cumyl	Benzene	CCl_4	IR	—	0.95	—	48
Cumyl	Benzene	CCl_4	IR	—	1.46	—	35
Cumyl	Pentamethylbenzene	CCl_4	IR	—	1.51	—	48
Cumyl	Chlorobenze	*c*	IR	1.22	1.22	—	35
Cumyl	Anisole	CCl_4	IR	0.46(20)	—	—	49
Cumyl	*o*-Dichlorobenzene	*c*	IR	—	0.86	—	35
Cumyl	Chlorobenzene	*c*	IR	—	1.22	—	35
Cumyl	Toluene	*c*	IR	—	1.50	—	35
Cumyl	*m*-Xylene	*c*	IR	—	1.82	—	35
Cumyl	Durene	*c*	IR	—	2.28	—	35
t-Amyl	*o*-Dichlorobenzene	*c*	IR	—	0.80	—	35
t-Amyl	Chlorobenzene	*c*	IR	—	0.98	—	35
t-Amyl	Benzene	*c*	IR	—	1.41	—	35
t-Amyl	Toluene	*c*	IR	—	1.51	—	35
t-Amyl	*m*-Xylene	*c*	IR	—	1.84	—	35
t-Amyl	1,2,4-Trimethylbenzene	*c*	IR	—	1.95	—	35

[a] Calculated from K and ΔH at 20°C.
[b] The ΔH values are given as calculated values, but it is assumed that they are experimental values. The temperature is not reported for K, but the reported values correspond to those calculated at 30°C from ΔH and ΔS.
[c] Solvent was not reported.

TABLE 8. Hydrogen bonding of *t*-butyl hydroperoxide to cyclic olefins in CCl_4 [51]

Olefin	ν_{OH} (cm^{-1})	$\Delta\nu_{OH}$ (cm^{-1})	$-\Delta H(H_2)$ (Olefin) (kcal mol^{-1})
None	3560 (free)	0	—
Cyclopentene	3490	70	25.7
Cyclohexene	3480	80	27.1
Norbornene	3445	115	33.1

large compared to the values given in Table 8, particularly considering that the olefinic bond in methyl acrylate is substituted with an electron-withdrawing group. No olefin–hydroperoxide complex bands have been detected for methyl methacrylate with the same hydroperoxides[47].

Intermolecular hydrogen bonding of *t*-butyl hydroperoxide to the alkyne π bond in ethynylbenzene has been claimed[47]. The infrared spectrum of this complex shows bands at 3518 and 3483 cm^{-1}. The first band has been assigned to a π aromatic complex and the second band to a π alkyne complex.

c. Ethers. Hydrogen-bonding data derived from infrared studies of hydroperoxide–ether complexes are given in Table 9. Compared to the π complexes of hydroperoxides that have previously been considered, the n oxygen complexes are considerably stronger, as seen from the larger K, $\Delta\nu_{OH}$ and $-\Delta H$ values in Table 9. Presumably 1:1 complexes are formed with monofunctional ethers. In the case of dioxane–cumyl hydroperoxide, the data require a 1:2 complex[48], where presumably one hydroperoxide molecule is hydrogen-bonded to each oxygen.

On the basis of the Badger–Bauer relationship[34], increased $\Delta\nu_{OH}$ values correspond to increased values of $-\Delta H$, i.e., stronger hydrogen bonds. If a stronger hydrogen bond corresponds to a more acidic hydroperoxy proton[56], then the Ph_3MOOH hydroperoxides decrease in acidity on the basis of $\Delta\nu_{OH}$ as: M = Si > Ge > C > Sn. It has been noted that this series does not follow the order of electronegativity of M[56], where on this basis a decreasing order of acidity would be expected as: M = C > Ge > Si > Sn. The acidity order deduced from $\Delta\nu_{OH}$ values can be explained by a decrease in d_π–p_π bonding in the M—O bonds in the order Si > Ge > Sn[56], where this effect is outweighed by the electronegativity of carbon in the case of M = Sn. A similar conclusion was reached in a study of hydrogen bonding to THF by Ph_3MOH [57].

TABLE 9. Infrared data for the intermolecular hydrogen bonding of hydroperoxides (ROOH) to ethers

R	Ether	$\Delta\nu_{OH}$ (cm^{-1})	K(20°C) (M^{-1})	$-\Delta H$ (kcal mol^{-1})	Ref.
t-Bu	Pr_2O	220	—	3.7	53
Cumyl	Et_2O	216	—	—	54
Cumyl	Pr_2O	225	1.78	—	54
Cumyl	Dioxane	194	86[a]	—	48
Cumyl	Dioxane	194	—	3.9	55
Ph_3C	THF	330	—	—	56
Ph_3Si	THF	408	—	—	56
Ph_3Ge	THF	335	—	—	56
Ph_3Sn	THF	258	—	—	56

[a]Units of M^{-2}.

Hydrogen-bonding studies with diethyl ether as the donor to *t*-butyl, *t*-amyl and cumyl hydroperoxides have been made with NMR measurements[58].

d. Alcohols and phenols. Complexes of tertiary hydroperoxides with several alcohols have been studied by NMR[35]. With these types of complexes the alcohol can act as a donor or an acceptor. With the alcohol as a donor via the n oxygen electrons and the hydroperoxy proton as the acceptor, the bond energy is about 2.5 kcal mol^{-1} as determined by monitoring the hydroperoxy protons. By observing the alcohol protons, the bond energy is 3.5–4.5 kcal mol^{-1}, where the alcohol is the acceptor. Only 1:1 complexes of alcohol to hydroperoxide have been considered[35]. Since hydroperoxides form six-ring dimers and alcohols have a propensity to form six-ring trimers (cf. Section III.B.1), a likely alternative to the 1:1 complex would be a six-ring 1:2 hydroperoxide to alcohol complex (**2**). Presumably complex **2** could be ruled out by concentration studies,

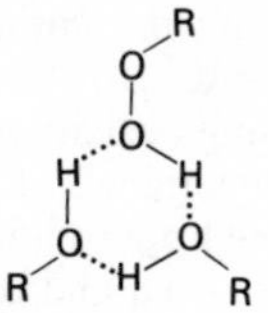

R
O
O
H H
O O
R H R

(**2**)

but this was not mentioned[35]. An NMR study of complex formation between cumyl hydroperoxide and phenol has been reported[59]. The data are fitted to a 1:1 complex, where the equilibrium constant for complex formation is calculated to be $22 \pm 5\,\text{M}^{-1}$ at 20°C. In comparison, the equilibrium constants for cumyl hydroperoxide dimer formation and phenol dimer formation are calculated to be 5.1 ± 0.5 and $3.6 \pm 0.5\,\text{M}^{-1}$ at 20°C, respectively[59]. Thus, complex formation between the hydroperoxide and phenol is greater than between either of the individual components. The large equilibrium constant for the hydroperoxide–phenol complex clearly implicates the oxygen atoms as the donors rather than the π system of phenol. The structure **3** has been proposed for the complex, where phenol acts both as a donor and an acceptor[59].

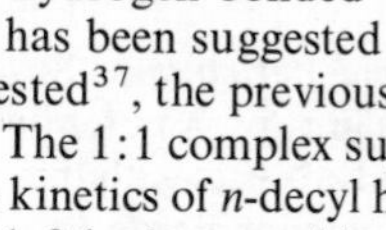

O
PhCMe$_2$—O H
H—O
Ph

(**3**)

e. Carboxylic acids. A 1:1 hydrogen-bonded complex of 1-methylcyclohexyl hydroperoxide and hexanoic acid has been suggested from NMR studies[37]. Although a seven-ring complex has been suggested[37], the previous arguments for six-ring complexes (cf. Section III.B.1) would favour . The 1:1 complex suggested by NMR studies contrasts with an earlier report based on the kinetics of *n*-decyl hydroperoxide decomposition with caproic acid in *n*-decane where a 1:2 hydroperoxide–acid complex was proposed[16].

O···H R^2
R^1—C O—O
O—H

(**4**)

f. Amines. Hydroperoxides can form isolable complexes with amines. These isolable complexes will be considered later. In this section, hydrogen-bonded associations in solution will be considered.

Complex formation between *t*-butyl hydroperoxide and triethylamine has been studied by UV spectroscopy in the range of 258–280 nm[60]. The data obtained by concentration variation could *not* be fitted to equation (4) to give an invariant equilibrium constant. However, an invariant equilibrium constant did result if the equilibrium involved the hydroperoxide dimer (equation 5). At 20°C, $K = 96.0\,\mathrm{M}^{-1}$ with $-\Delta H = 26\,\mathrm{kcal\,mol}^{-1}$ and $-\Delta G = 2.69\,\mathrm{kcal\,mol}^{-1}$, which gives a surprisingly large negative entropy ($-\Delta S = 79.6$ e.u.). The enthalpies of complex formation of *t*-butyl, *t*-amyl and cumyl hydroperoxide to pyridine have been reported to be twice those for complexes with diethyl ether[58]. Since the *t*-butyl hydroperoxide–diethyl ether complex enthalpy is $-\Delta H = 3.7\,\mathrm{kcal\,mol}^{-1}$[53]; this places the hydroperoxide–pyridine complex enthalpy at about $-\Delta H = 8\,\mathrm{kcal\,mol}^{-1}$. Comparing the enthalpies for hydroperoxide complexes with triethylamine ($pK_a = 10.7$)[7] and pyridine ($pK_a = 5.17$)[7], it appears that complexation is increased with increasing basicity of the amine. In contrast to the report with *t*-butyl hydroperoxide and triethylamine[60], the equilibrium constant for complex formation with this amine and cumyl hydroperoxide in aqueous solution is reported[61] to be $K = 0.18\,\mathrm{M}^{-1}$. A 1:1 complex is proposed and the data have been obtained by NMR measurements.

$$2\,Me_3COOH + Et_3N \rightleftharpoons (Me_3COOH)_2NEt_3 \tag{4}$$

$$(Me_3COOH)_2 + Et_3N \rightleftharpoons (Me_3COOH)_2NEt_3 \tag{5}$$

Complex formation between *t*-butyl hydroperoxide and β-ethanolamine has been studied in aqueous solutions by UV (257–259 nm)[62]. The data have been fitted to a 1:1 complex and from IR measurements it has been concluded that the complex is between the peroxide and the amino group. The equilibrium constant at 20° has been reported to be $K = 38.7\,\mathrm{M}^{-1}$ with $-\Delta H = 25.3\,\mathrm{kcal\,mol}^{-1}$ and $-\Delta S = 78$ e.u.

Infrared band assignments have been made where primary and secondary amines are both donors and acceptors in hydrogen bonding with hydroperoxides. Complexes of *t*-butyl, cumyl, diphenylmethylcarbinyl and trityl hydroperoxides with aniline in dilute solution have been studied[63]. Infrared bands at 3470 and 3370 cm^{-1} have been assigned to the asymmetric and symmetric modes of the NH···O hydrogen bond, while the OOH···N hydrogen bond has been associated with a band at 3350 cm^{-1}. Similar assignments have been made for a series of substituted anilines, α-naphthylamine and β-naphthylamine complexed to cumyl hydroperoxide[64]. The asymmetric and symmetric N—H stretching of the NH···O hydrogen bond has been assigned to bands at 3500 and 3400 cm^{-1}. A band at 3300 cm^{-1} (shoulder) has been assigned to the O—H stretching of the OOH···N hydrogen bond.

g. Carbonyls and carboxylic acid derivatives. Several reports appear for hydrogen bonding of hydroperoxides to these groups. A potential difficulty with these measurements is the formation of hemiperoxy ketal adducts with the carbonyl group[65]. It is reported that complexes of 1-methylcyclohexyl hydroperoxide and 2-octanone show no infrared bands characteristic of hemiperoxy ketals[37]. A 1:1 complex has been suggested in this system from NMR measurements of the hydroperoxy proton as a function of concentration[37]. A PMR study of complex formation between cumyl hydroperoxide and acetone has been reported, where a 1:1 complex is proposed with $K = 3.5 \pm 0.3\,\mathrm{M}^{-1}$ at 20°C[59]. In comparison, the equilibrium constant is reported to be 5.4 ± 0.6 at 20°C for the 1:1 complex between acetone and phenol[59]. Equilibrium constants for hydrogen bonding of cumyl hydroperoxide to benzophenone, acetophenone, cyclohexanone and acetone by infrared measurements at 20°C in carbon tetrachloride are reported to 1.67, 2.28, 2.50 and 1.92 M^{-1}, respectively[54]. The IR frequency shift ($\Delta\nu_{OH}$) for the *t*-butyl hydroperoxide–acetone complex is reported to be 172 cm^{-1} at 20°C[66]. Increased temperatures, required to obtain the enthalpy of this hydrogen bond, cause a 'chemical reaction' (presumably hemiperoxy ketal formation), so the enthalpy cannot be obtained. This suggests some caution in the interpretation of these data for hydroperoxide–carbonyl complexes. These carbonyl–hydroperoxide complexes are proposed to involve n-carbonyl oxygen–HOO hydrogen bonding.

Two structures are suggested for hydroperoxide–ester complexes, where **5** is proposed to be the major species with a small amount of **6**[66]. Hydrogen bonding associated with

$$R^1C(=\ddot{O}\cdots HOOR^2)OR^1 \quad \textbf{(5)} \qquad\qquad R^1C(=O)\ddot{O}(R^1)\cdots HOOR^2 \quad \textbf{(6)}$$

structures **5** and **6** is assigned to bands at 3442 and 3380 cm^{-1}, respectively, where $R^1 = Me$ and $R^2 = t\text{-Bu}$. The results of complex formation between esters and hydroperoxides, assuming structure **5**, are given in Table 10. The results with methyl methacrylate (MMA) and methyl acrylate (MA) compared to methyl acetate as donors are surprising. The $\Delta\nu_{OH}$ values suggest that both MMA and MA are poorer donors than methyl acetate. Also the $-\Delta H$ values indicate that MMA is a poorer donor than methyl acetate. Resonance release by the double bond in MA or MMA should increase the basicity of the carbonyl oxygen and it would be expected that the stability of complex **5** should increase with MA or MMA rather than decrease relative to methyl acetate. Also small negative entropies are found with MMA complexes, which contrasts with large negative entropies for other intermolecular hydroperoxide complexes.

As seen from Table 11, stronger complexes are formed between amides and hydroperoxides as compared to ester–hydroperoxide complexes. Hydrogen bonding to the oxygen atom of the amide is favoured[67], but the actual structure of the complex is uncertain. The difference in the equilibrium constants is small between DMF and *N*-*n*-butylacetamide for both *t*-butyl and cumyl hydroperoxides, which suggests little or no involvement of the N—H group as an acceptor in hydrogen bonding.

h. Sulphoxides. Equilibrium constants for hydrogen bonding of hydroperoxides with methyl cyclohexyl sulphoxide as the donor, according to equation (6), have been determined by IR methods. The results are given in Table 12. A correction has been made for intramolecular hydrogen bonding to phenyl in the case of cumyl hydroperoxide, which causes a considerable change in K (equation 6)[68]. Similar corrections have not been made for other substituents bearing π systems.

$$R^1OOH + OS(R^1)(R^2) \overset{K}{\rightleftharpoons} R^1OOH\cdots OS(R^1)(R^2) \qquad (6)$$

For the data in Table 12, increased complex formation, as measured by K, parallels the increase in $\Delta\nu_{OH}$. Presumably, an increase in the equilibrium constant of equation (6) also corresponds to an increase in the acidity of the hydroperoxide. Comparison of Me_3COOH to Me_3SiOOH and Ph_3COOH to Ph_3SiOOH shows that in both cases silicon increases K and presumably the acidity. The same conclusion has been reached from intermolecular

TABLE 10. Hydrogen-bonded complexes to the carbonyl oxygen of esters with hydroperoxides[a]

Hydroperoxide	Ester	Solvent	$\Delta\nu_{OH}$(20°C) (cm^{-1})	K(20°C) (M^{-1})	$-\Delta H$ (kcal mol^{-1})	$-\Delta S$ (e.u.)	Ref.
t-Butyl	$MeCO_2Me$	CCl_4	120	—	3.3	—	66
n-Butyl	$CH_2{=}C(Me)CO_2Me$ (MMA)	MMA	97	—	—	—	46
s-Butyl	MMA	MMA	95	—	—	—	46
t-Butyl	MMA	MMA	95	1.76	1.1	2.6[b]	46
Cumyl	MMA	MMA	97	2.06	0.72	1.0[b]	46
n-Butyl	$CH_2{=}CHCO_2Me$ (MA)	MA	70	—	—	—	46
s-Butyl	MA	MA	72	—	—	—	46
t-Butyl	MA	MA	72	—	—	—	46

[a] Data are from infrared measurements.
[b] Calculated at 20°C from K and ΔH.

TABLE 11. Hydrogen-bonded complex formation between amides and hydroperoxides in carbon tetrachloride[67][a]

Hydroperoxide	Amide	K(20°C) (M^{-1})	$-\Delta H$ (kcal mol^{-1})	$-\Delta S$ (e.u.)
t-Butyl	$MeCONMe_2$	22.9	5.41	12.2
t-Butyl	MeCONHBu-*n*	22.4	5.72	13.3
t-Butyl	MeCONHPr-*i*	16.2	4.71	10.4
Cumyl	$MeCONMe_2$	13.5	5.00	11.9
Cumyl	MeCONHBu-*n*	12.3	4.90	11.7
Cumyl	MeCONHPr-*i*	8.3	5.09	13.3

[a]Data are from infrared measurements.

hydrogen bonding studies to THF[56] (cf. Section III.B.2.c). Again the order of acidity ($R_3SiOOH > R_3COOH$) can be explained by d_π–p_π bonding in the Si—O bond[69].

A LFER was suggested between the pK_a of the hydroperoxide and log K of equation (6)[68]. With pK_a values in aqueous solution for *t*-butyl hydroperoxide (12.8)[2], cumyl hydroperoxide (12.6)[10], phenol (9.95)[68] and thymol (10.49)[68] and the K values in Table 12 (Reference 68) along with those for phenol (314)[68] and thymol (127)[68], one obtains:

$$pK_a = (-3.15 \pm 0.50)\log K + 17.55 \pm 0.98,$$

$r = 0.976$, $S_{y.x} = \pm 0.39$. The generality of this correlation for the prediction of the pK_a values of hydroperoxides and phenols has not been thoroughly tested. Presently, the correlation rests on two hydroperoxides with similar pK_as and two phenols with similar pK_as.

A trend to increasing chemical shift of the peroxy proton with increasing acidity has been noted in hydrogen bonding with DMSO as seen from Table 13[9]. An attempted LFER with all of the points gives a poor correlation ($r = 0.770$, $S_{x.y} = \pm 0.087$). If 1,1-diphenyl-1-propyl hydroperoxide is excluded, a fair correlation results:

$$pK_a = (-0.430 \pm 0.0095)\delta + 17.88 \pm 1.05,$$

TABLE 12. Hydrogen bonding of hydroperoxides with methyl cyclohexyl sulphoxide[a]

Hydroperoxide	K(25°C) (M^{-1})	$\Delta\nu_{OH}$ (cm^{-1})	Ref.
t-Butyl	35	—	68
t-Butyl	25.1	283	69
1-Tetralyl	53	—	68
Cyclohex-2-enyl	43	—	68
Cyclohexyl	39	—	68
Cumyl	41[b](22)[c]	—	68
Ph_3COOH	34.2	304	69
Me_3SiOOH	42.4	331	69
Ph_3SiOOH	132	412	69

[a]From infrared data.
[b]Corrected for intramolecular hydrogen bonding to the phenyl group.
[c]Uncorrected for intramolecular hydrogen bonding.

TABLE 13. Chemical shifts of tertiary hydroperoxides in DMSO and pK_a values in 40% aqueous methanol[9]

Hydroperoxide	δ(ppm)[a]	pK_a
Me_3COOH	10.72	13.27
$PhCH_2C(Me_2)OOH$	10.76	13.25
$Me_2C(CH_2Cl)OOH$	10.96	13.2[b]
$Ph_2C(Et)OOH$	10.89	13.02
$Ph_2C(Me)OOH$	11.34	12.94
Ph_3COOH	11.34	13.07

[a]Relative to TMS as the internal standard.
[b]Obtained indirectly from kinetic data.

$r = 0.915$, $S_{y.x} = \pm 0.057$. The low correlation coefficient is due in part to the small spread in both δ and pK_a. The correlation of pK_a of hydroperoxides with σ^* (cf. Section II.B) is preferable unless σ^* values are unavailable.

i. Miscellaneous donors. Equilibrium constants for complex formation between 4-hydroxy-2,2,6,6-tetramethylpiperidine-1-oxyl (**7**) and *t*-butyl, *t*-amyl and 1-phenylethyl hydroperoxide have been determined by NMR in carbon tetrachloride at 18°C to be: 5.7, 2.8 and 2.0 M^{-1}, respectively[70]. On the basis of the hyperfine coupling constants, it is proposed that the oxygen atom of the nitroxide group is the donor. At moderate concentrations of hydroperoxide, a 1:1 complex is formed, but at high concentrations ([ROOH] > 8 M) a 1:2 (**7**:ROOH) complex is detected.

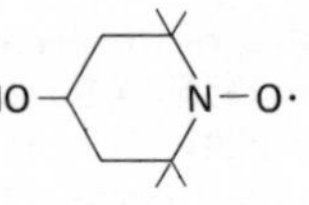

HO—⟨ ⟩N—O·

(**7**)

The cyano group apparently acts as a donor in hydrogen bonding with hydroperoxides. An infrared band at 3417 cm^{-1} ($\Delta\nu_{OH} = 134\ cm^{-1}$) is assigned to this type of complex between acetonitrile and cumyl hydroperoxide[54]. The $-\Delta H$ value for complex formation is reported to be 2.78 kcal mol^{-1} [55].

Both the nitro group and the π aromatic system in nitrobenzene are proposed to act as donors to *t*-butyl hydroperoxide. Infrared bands at 3490 cm^{-1} ($\Delta\nu_{OH} = 68\ cm^{-1}$) and 3522 cm^{-1} ($\Delta\nu_{OH} = 36\ cm^{-1}$), respectively, are assigned to these complexes[47].

The equilibrium constant for *t*-butyl peroxide as the donor with cumyl hydroperoxide is reported to be 0.65 M^{-1} at 20°C in carbon tetrachloride by infrared measurements[54]. The frequency shift ($\Delta\nu_{OH}$)[54] for this complex is 151 cm^{-1}.

j. Summary. A comparison of the effect of various donors on intermolecular hydroperoxide hydrogen bonding is given in Table 14. For simplicity an attempt has been made to restrict the data to *t*-butyl hydroperoxide. When data are not available for this hydroperoxide, results with cumyl hydroperoxide are given. The table is ordered in decreasing complex formation as well as the data permit. Preference in the ordering is given first to equilibrium values, then enthalpy, and finally $\Delta\nu_{OH}$ values.

A word of caution is worthwhile with regard to the application of the Badger–Bauer relation (equation 7)[34] to obtain enthalpies of complex formation. It has been found that the proportionality constant (K_{BB}) varies according to the donor, so that a universal value

TABLE 14. Comparison of hydroperoxide intermolecular hydrogen-bonded complexes

$$ROOH + D \overset{K}{\rightleftharpoons} ROOH \cdots D$$

$ROOH^{a}$	D	$K(20°C)$ (M^{-1})	$-\Delta H$ (kcal mol^{-1})	$\Delta\nu_{OH}$ (cm^{-1})	Ref.
TBHP	Et_3N	96	26	—	60
TBHP	Methyl cyclohexyl sulphoxide	25^{b}	—	280	69
TBHP	Amides	16–23	4.7–5.4	—	67
CHP	PhOH	22	—	—	59
TBHP	HO–(2,2,6,6-tetramethylpiperidine)N–O·	5.7^{c}	—	—	70
TBHP	TBHP	2.2	6.1	154[39]	*d*
ROOH	R^1OH (complex: ROO(H)···HOR^1)	—	3.5–4.5	—	35
TBHP	Pr_2O	est. ~2	3.7	220	53
CHP	MeCOMe	1.9	—	176	54
TBHP	$MeCO_2Me$	—	3.3	120	66
CHP	MeCN	—	2.78	134	54,55
ROOH	R^1OH(complex: ROOH···O(H)R^1)	—	2.5	—	35
CHP	$Me_3COOCMe_3$	0.65	—	151	54
TBHP	Cyclohexene	—	—	80	51
TBHP	$PhNO_2$	—	—	80	47
TBHP	ArH	~0.2	1.4–2.4	$45(C_6H_6)$[39]	*e*

[a]TBHP = *t*-butyl hydroperoxide, CHP = cumyl hydroperoxide.
[b]25°C.
[c]18°C.
[d]Cf. Table 5.
[e]Cf. Table 7.

of K_{BB} cannot be used to evaluate $-\Delta H$ from infrared data[66]. Some values of K_{BB} (mol kcal^{-1}) for hydroperoxide complexes are[66]: $(t\text{-BuOOH})_2$ (1.1×10^{-2}), *t*-BuOOH–styrene (0.64×10^{-2}), *t*-BuOOH–benzene (0.77×10^{-2}), *t*-BuOOH–*o*-dichlorobenzene (0.88×10^{-2}), *t*-BuOOH–Pr_2O (1.65×10^{-2}) and *t*-BuOOH–$MeCO_2Me$ (1.0×10^{-2}).

$$-\Delta H \cdot K_{BB} = \frac{\Delta\nu_{OH}}{\nu_{OH}\,(\text{free})} \tag{7}$$

In general complex formation is increased with n electron donors relative to π electron donors. Within each category of donor type, increasing basicity appears to increase complex formation. If a group possesses both n and π electrons, complexation with the n electrons should be preferable. With carbonyl[66], ester[66] and amide[67] groups, n electron complexes have been proposed. An n electron complex would also be expected with the cyano group.

A weak complex is suggested by the $\Delta\nu$ value for the nitro group as a donor in nitrobenzene. The nitro group is expected to be an n electron donor, which should place it higher in Table 14. As stated above, the use of $\Delta\nu$ values to evaluate complex formation between different types of donors is subject to considerable error.

3. *Intramolecular association*

Hydroperoxides can form intramolecular hydrogen-bonded complexes with n or π electron donors. It can be noted that intramolecular association can complicate the calculation of intermolecular complex equilibrium constants[55,68]. Since concentration variation studies cannot be used to obtain equilibrium constants for intramolecular hydrogen bonding, enthalpy is the only thermodynamic parameter that can be obtained from experimental data. This parameter is obtained by temperature variation studies.

a. Arenes. In carbon tetrachloride solution, cumyl hydroperoxide shows two infrared bands in the region of about 3500–3550 cm^{-1}. The higher frequency band was assigned to the free O—H group and the lower frequency band to the O—H group intramolecularly hydrogen-bonded to the π aromatic system[48,68,71,72]. Intramolecular hydrogen bonds to aromatic groups have been noted with several other hydroperoxides and the data are given in Table 15. The energy of the intramolecular hydrogen bond is somewhat less than that of the intermolecular hydrogen bond to π aromatic systems. For example, the enthalpy of intermolecular association to benzene, toluene and *p*-xylene are: $-\Delta H \cong 1.5$ kcal mol^{-1} (cf. Table 7), 1.4 kcal mol^{-1} [47] and 2.34 kcal mol^{-1} [47]. The $\Delta\nu_{OH}$ values for intramolecular hydrogen bonding to the π aromatic systems (with the exception of Ph_3MOOH hydroperoxides) are also less than that for intermolecular hydrogen bonding to benzene ($\Delta\nu_{OH} = 45$ cm^{-1} with *t*-BuOOH)[39].

With *p*-$ROOC(Me_2)C_6H_4C(Me_2)OOH$, the $\Delta\nu_{OH}$ and ΔH values are similar to other systems which involve π aromatic intramolecular hydrogen bonding. On this basis,

TABLE 15. Infrared frequency shifts ($\Delta\nu_{OH}$) and enthalpy for intramolecular hydrogen bonding of hydroperoxides to aromatic groups in carbon tetrachloride

Hydroperoxide	Free ν_{OH} (cm^{-1})	Intra ν_{OH} (cm^{-1})	$\Delta\nu_{OH}$ (cm^{-1})	$-\Delta H$ (kcal mol^{-1})	Ref.
CHP[a]	3535	3510	25	—	68
CHP[a]	3530	3497	33	0.9 ± 0.1	71
CHP[a]	3551	3519	32	—	48, 72
1-Tetralyl	—	—	28	0.9 ± 0.1	66
$Ph_2C(Me)OOH$	—	—	30	0.96 ± 0.11	66
Ph_3COOH	Not observed	3518	—	—	42
Ph_3COOH	3615	3518	97	—	56
Ph_3SiOOH	3688	3547	141	—	56
Ph_3GeOOH	3655	3556	99	—	56
Ph_3SnOOH	3618	3577	41	—	56
p-$ROOC(Me_2)C_6H_4C(Me_2)OOH$:					
$R = Me_3C$	3556	3528	28	0.98 ± 0.04	73
$R = EtCMe_2$	3561	3528	33	1.0 ± 0.02	73
$R = PhCMe_2$	3548	3527	21	1.1 ± 0.1	73

[a]Cumyl hydroperoxide.

hydrogen bonding to the π aromatic system rather than to the ROO group has been proposed[73]. A linear correlation is noted between $\Delta\nu_{OH}$ (intra-) and $\Delta\nu_{OH}$ (inter-molecular to THF) for M = C, Ge and Sn in Ph_3MOOH, while M = Si deviates slightly from the line ($\Delta\nu_{OH}$ (intra) is less than expected from the correlation line). With M = Si, Ge and Sn, d_π–p_π bonding could effect both the basicity of the aromatic rings and the acidity of the hydroperoxides. Since a linear correlation between intra- and inter-molecular hydrogen bonding is observed with M = C, Ge and Sn, M—O d_π–p_π bonding to effect hydroperoxide acidity appears to dominate with these atoms. With M = Si, M—Ph d_π–p_π bonding apparently contributes as well to decrease the basicity of the aryl group and to cause a deviation from the correlation. This appears reasonable, considering the order of decreasing d_π–p_π bonding: Si > Ge > Sn[56].

b. Alkynes, ethers and amines. Intramolecular hydrogen bonding of hydroperoxides containing alkene and alkyne groups has been largely neglected. In one study, intramolecular hydrogen bonding between the OOH and the alkyne in $Me_2CHC(Me)(OH)C\equiv CC(Me_2)OOH$ has been proposed with $-\Delta H = 0.47$ kcal mol^{-1} [74].

Infrared and NMR studies have been reported for 2-tetrahydropyranyl 1,4-dioxan-2-yl and 2-tetrahydrofuryl hydroperoxides[75]. It has been claimed that none of these oxygen heterocyclic hydroperoxides show intramolecular hydrogen bonding.

Intramolecular hydrogen bonding of the hydroperoxy group to nitrogen in several nitrogen heterocycles has been reported and the data are given in Table 16. Both the low O—H stretching frequency and the down-field NMR shift of the hydroperoxy group suggest strong intramolecular hydrogen bonding in these heterocycles. Also in the second to the last entry in Table 16, the absorption at 3100 cm^{-1} is reported to be invariant on dilution[79]. The absorption at 3100 cm^{-1} seems high relative to the other reports which place the intramolecular hydrogen-bonded OH stretching frequency at 2800–2820 cm^{-1}.

Intramolecular hydrogen bonding is suggested in a series of *N*-(1-aryl-1-hydroperoxymethyl)-3,5-di-*t*-butyl-*p*-benzoquinone monoimines (**8**)[80]. The IR (KBr, cm^{-1}) and NMR (ppm, $CDCl_3$) absorptions of the hydroperoxy group are: R = C_6H_5 (3070, 11.65), *p*-MeC_6H_4 (3100, 11.58), *p*-$MeOC_6H_4$ (3110, 11.20), *p*-ClC_6H_4 (3100, 11.27) and 2-furyl (3080, 10.75).

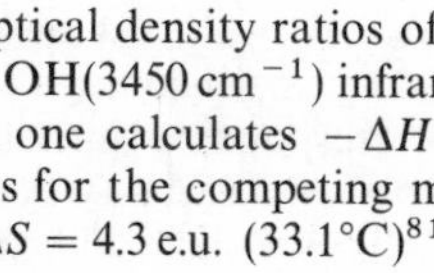

(**8**)

c. Carbonyls. From a plot of optical density ratios of the free OH(3546 cm^{-1}) to the intramolecular hydrogen-bonded OH(3450 cm^{-1}) infrared absorptions, extrapolated to infinite dilution, vs. temperature, one calculates $-\Delta H = 1.86$ kcal mol^{-1} for equation (8)[81]. Thermodynamic parameters for the competing monomer–dimer equilibrium are $-\Delta H = 1.45$ kcal mol^{-1} and $-\Delta S = 4.3$ e.u. (33.1°C)[81]. The structure of the dimer is

$$Me_2CHC(=O)C(OOH)Me_2 \rightleftharpoons Me_2CHC(=O\cdots H\text{–}O\text{–}O)CMe_2 \qquad (8)$$

(**9**)

TABLE 16. Intramolecular hydrogen bonding of five-membered ring-nitrogen heterocycles[a]

Hydroperoxide	Intra ν_{OH} (cm^{-1})	Intra δ_{OH} (ppm)	Ref.
$R^1 = R^2 = Ph$	2820[b]	14.4[b]	76
$R^1 = R^2 = Ph$, OOD	2105[b]	—	76
$R^1 = R^2 = Ph$	2816 (KBr)	12.4($(CD_3)_2CO$)	77
$R^1 = R^2 = R^3 = R^4 = Ph$	2800($CHCl_3$)	—	78
$R^1 = R^2 = Ph$, $R^3 = R^4 = H$	2800($CHCl_3$)	—	78
$R^1 = R^2 = R^3 = t$-Bu, $R^4 = H$	3100(Nujol)	12.65(CCl_4)	79
$R^1 = R^2 = t$-Bu, $R^3 = R^4 = H$	3100(Nujol)[c]	11.7(CCl_4)	79

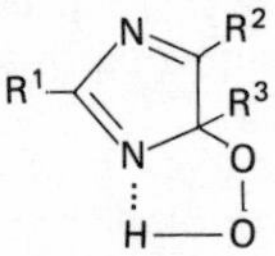

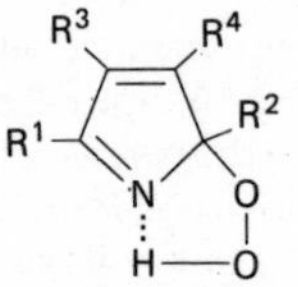

[a]Free O—H stretching frequency is not reported, but see footnote *c*.
[b]Solvent not reported.
[c]ν_{OH} (free) = 3500 cm^{-1}.

presumed to be the six-membered ring species that is proposed for other hydroperoxides (cf. Section III.B.1). The low $-\Delta H$ value for the monomer–dimer equilibrium, as compared to *t*-butyl hydroperoxide ($-\Delta H = 5.95\,kcal\,mol^{-1}$)[39], results from the increased stability of the monomer due to intramolecular hydrogen bonding.

Cyclization of **9** to the dioxetane **10** is not considered to be significant, since a strong carbonyl absorption is observed at 1717 cm^{-1} and the NMR spectrum shows a hydrogen-bonded hydroperoxy proton at 9.39 ppm. In contrast to **9**, in the case shown in equation (10) there is no evidence for the corresponding intramolecular hydrogen-bonded species **11**[82]. The infrared spectrum fails to show a carbonyl band in Nujol or in a 10% solution in dimethoxyethane, which along with other data indicates structure **12**.

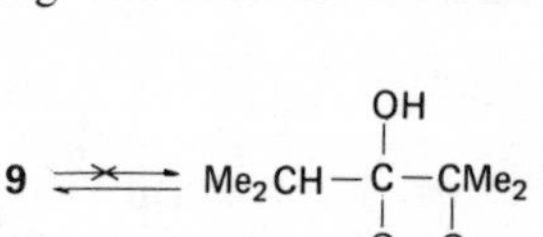

$$\mathbf{9} \rightleftharpoons Me_2CH-C(OH)-CMe_2 \text{ (with O—O ring)} \quad (9)$$

(10)

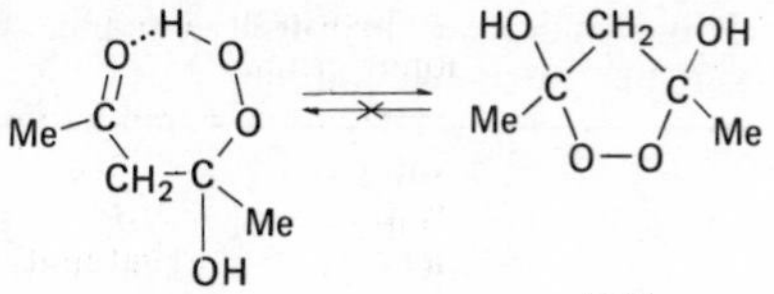

(10)

(11) (12)

d. Peroxides. In Table 17, intramolecular hydrogen-bonding data are presented where the peroxide group acts as the donor. Five-membered ring hydrogen-bonded structures involving the hydroperoxy groups have been proposed[66,83], but alternative structures involving six-membered rings cannot be excluded. Originally the structure of **15** was shown with one free and one hydrogen-bonded hydroperoxy group[66,83]. Considering the similar ΔH values of **13** and **15**, and that there are two hydrogen-bonded OOH groups in **13**, the structure for **15** shown here seems preferable. Also, the alcoholic OH group in **14** was originally shown as free[66,83]. To account for the more negative enthalpy of **14** vs **13** or **15**, the alcoholic OH group is now shown as being hydrogen-bonded.

Intramolecular hydrogen bonding of an alcoholic OH group with a peroxide group (**16**) has been studied in relationship to two model hydroxy ethers (**17** and **18**)[84]. The results from infrared measurements are given in Table 18. The enthalpy of intramolecular

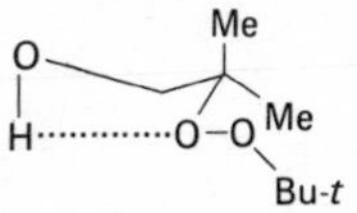

(16) 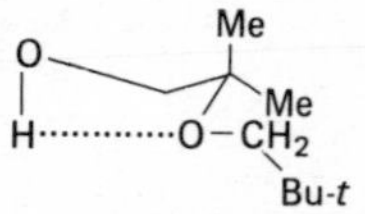 (17) 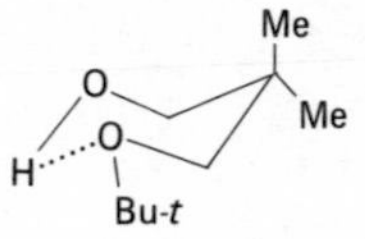 (18)

hydrogen bonding is more favourable for a 1,6- than a 1,5-interaction as seen from **17** vs. **18**. However, the populations of the 1,5- and 1,6-species depend on the free energies of these interactions. Unfortunately, the entropy cannot be obtained experimentally for intramolecular hydrogen bonding, which is needed to obtain the free energy. Empirical correlations suggest that the entropy should be lowered by 4 e.u. per restricted bond rotation[85]. The rotation of one additional bond is restricted in the six-membered ring species as compared with the five-membered ring species. At the average temperature of the measurements (50°C),

$$T\Delta\Delta S_{6\text{-}5} \cong -1.3\,\text{kcal mol}^{-1}\ [= 323(-4 \times 10^{-3})],$$

where $\Delta\Delta S_{6\text{-}5}$ is the estimated entropy difference between the five- and six-membered species. The enthalpy difference between 1,5- and 1,6-interactions is obtained from **17** and **18** to be

$$\Delta\Delta H_{6\text{-}5} = -0.34\,\text{kcal mol}^{-1}\ [= -1.41 - (-1.07)].$$

The free energy difference is then

$$\Delta\Delta G_{6\text{-}5} = 1.0\,\text{kcal mol}^{-1}\ [= -0.34 - (-1.3)],$$

so that the 1,5-interaction is favoured over the 1,6-interaction. On this basis, peroxide **16** is shown as a 1,5- rather than a 1,6-hydrogen-bonded species. It is seen that the difference in free energy between these two types of interactions is small, which points out the difficulty

TABLE 17. Intramolecular hydrogen bonding of hydroperoxides with a peroxide donor group[66 a]

Peroxide	Intra ν_{OH} (cm^{-1})	$-\Delta H$ (kcal mol^{-1})
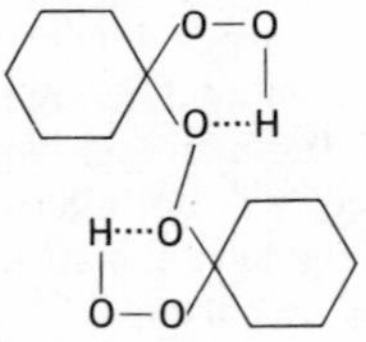 **(13)**	107[b]	7.0 ± 0.5[c]
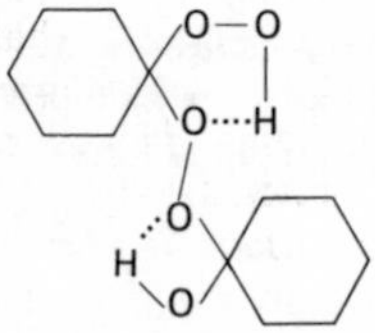 **(14)**	172	11.2 ± 0.8
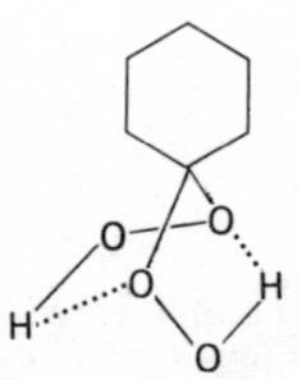 **(15)**	112	7.3 ± 0.5

[a]In carbon tetrachloride.
[b]ν_{OH} (free) = 3542 cm^{-1}.
[c]The van't Hoff plot shows curvature.

TABLE 18. Intramolecular hydrogen bonding of the hydroxy group with peroxide and ether donor groups[84 a]

Compound	Free ν_{OH} (cm^{-1})	Intra ν_{OH} (cm^{-1})	$\Delta\nu_{OH}$ (cm^{-1})	$-\Delta H$ (kcal mol^{-1})
16	3638	3593	45	0.95 ± 0.05
17	3638	3584	54	1.07 ± 0.03
18	3638	3510	128	1.41 ± 0.05

[a]In carbon tetrachloride.

in evaluating the type of hydrogen bonding in more complex structures such as **13**–**15**.

A comparison of **16** with **17**, where both compounds involve 1,5-hydrogen bonding, reveals that the peroxy and ether groups are similar as donors.

IV. COMPLEXES

This section will survey *isolable* complexes of peroxides, where σ bonds are not formed. Transition-metal-ion–peroxide complexes will not be included. Isolable complexes have been reported for two types of peroxides: peroxy acids and hydroperoxides.

A. Peroxy Acids

Isolable 1:1 complexes of substituted peroxybenzoic acids with triphenylphosphine oxide, triphenylarsine oxide, pyridine *N*-oxide and 4-methylpyridine *N*-oxide have been prepared[86]. These appear to be the only isolable complexes of peroxy acids. The adducts are unstable at room temperature, with the pyridine *N*-oxide complexes being least stable (rapid decomposition at 0°C). The complexes are white solids, which are soluble in haloalkanes. Infrared data and melting points are given in Table 19.

TABLE 19. Peroxybenzoic acid complexes with triphenylphosphine oxide, triphenylarsine oxide and pyridine *N*-oxides[86 a]

Complex	m.p.(°C)	ν_{OH}(cm^{-1})	$\nu_{C=O}$, (cm^{-1})
$Ph_3PO \cdots HOOC(=O)C_6H_4X$			
X = 3-NO_2	82–84	2820	1753
4-NO_2	84–86	2750	1758
3-Cl	64–66	2840	1748
$Ph_3AsO \cdots HOOC(=O)C_6H_4X$			
X = 3-NO_2	77–79	(2500, 2370)	1742
4-NO_2	78–81	(2450, 2340)	1750
3-Cl	57–60	(2500, 2380)	(1760, 1730)
$C_5H_5N-O \cdots HOOC(=O)C_6H_4X$			
X = 4-NO_2	77–81	(2400, 2330)	1735
4-Cl	79–82	2590	1728
$Me-C_5H_4N-O \cdots HOOC(=O)C_6H_4NO_2$-*p*	94–95	2660	1745

[a]Infrared spectra measured in Nujol–hexachlorobutadiene mulls.

The OH stretching frequencies of the crystalline peroxybenzoic acids are about 400–500 cm^{-1} higher than observed in the triphenylphosphine oxide adducts[86]. This suggests strong hydrogen bonding in all of the adducts. The carbonyl frequencies are shifted to higher values (20–30 cm^{-1}) in the triphenylphosphine complexes, while the P—O stretching frequency is lowered by 30–40 cm^{-1}.

B. Hydroperoxides

Two types of isolable complexes are reported; amine–hydroperoxide adducts and ammonium–hydroperoxide adducts.

1. Amine–hydroperoxide complexes

These complexes have proved useful in the isolation and purification of hydroperoxides. For example, 1:1 pyridine:hydroperoxyfluorene adducts (**19**) have been used to isolate the hydroperoxides produced in the autooxidation of the fluorenes[87]. The melting points of these adducts are: R = *p*-$MeC_6H_4CH_2$ (76–77°C), *p*-$MeOC_6H_4CH_2$ (82–92°C, decomp.), and Ph (90–91°C). In another example, DABCO (1,4-diazabicyclo-[2.2.2]octane) has been used to purify β-halo hydroperoxides via crystalline adducts[88].

(**19**)

Table 20 lists some representative hydroperoxide–amine complexes. The references in the table may be checked for additional complexes. The adducts are prepared by mixing the reactants in the stoichiometry for the desired type of complex (1:1, 1:2, etc.) in a hydrocarbon solvent. The adduct precipitates (sometimes cooling is required), the precipitate is washed with a hydrocarbon solvent and the adduct is vacuum-dried. It appears that the adducts involve 1:1 hydroperoxide:amino group association, where one or all of the amino groups in polyamines may be complexed. In general, the polyamines appear best suited for high-melting derivatives.

Although these complexes are sometimes referred to as salts, in most instances they are instead strongly bonded adducts. With approximate pK_a values for hydroperoxides and tertiary aliphatic amines of 13 and 10, the equilibrium constant (K_c) for equation (11) is 10^{-3}, which indicates little salt formation. A possible exception to this conclusion may be

$$R^1OOH + R^2_3N \underset{}{\overset{K_c}{\rightleftharpoons}} R^1OO^- + R^2_3\overset{+}{N}H \qquad (11)$$

found with guanidine–hydroperoxide adducts, where the pK_a of guanidine (13.65)[94] is comparable or somewhat greater than pK_a values of hydroperoxides. It has been noted[90] that guanidine adducts of *t*-butyl and cumyl hydroperoxide are insoluble in benzene, while cumyl hydroperoxide adducts of 1,3-diphenylguanidine (pK_a = 10.12)[94] and 1,2,3-

TABLE 20. Representative hydroperoxide–amine complexes

Hydroperoxide	Amine	Ratio HP:Amine	m.p. (°C)	% Yield	Ref.
t-Butyl	DABCO	1:1	52–55	90	89
t-Butyl	DABCO	2:1	74–76	93	89
t-Butyl	$(NH_2)_2C{=}NH$	1:1	91(d)	68	90
t-Butyl	$(NH_2)_2C{=}NH$	3:1	101–102(d)	92	90
Cumyl	$(NH_2)_2C{=}NH$	1:1	82(d)	71	90
Cumyl	$(NH_2)_2C{=}NH$	2:1	83–84(d)	89	90
Cumyl	$(PhNH)_2C{=}NH$	1:1	84(d)	75	90
Cumyl	$(PhNH)_2C{=}NPh$	1:1	101–102(d)	89	90
Cumyl	DABCO	2:1	87.5–88.5	96	89
Cumyl	*p*-$MeC_6H_4NH_2$	1:1	94		91
Cumyl	β-Napthylamine	1:1	64–66		92
Trityl	*p*-$MeC_6H_4NH_2$	1:1	63		91
Trityl	Benzidine	2:1	91		91
BH[a]	DABCO	1:1	128–131	82	89
BH[a]	Acridine	1:2	104–105	92	89
TH[b]	DABCO	2:1	50–52	75	93

[a] $HOOC(Me)_2(CH_2)_2C(Me)_2COOH$.
[b] $PhSC(Me)_2CH{=}CHC(Me)_2OOH$.

triphenylguanidine ($pK_a = 9.10$)[94] are soluble in benzene. The solubility differences may reflect salt formation with guanidine, but not with the lower pK_a phenyl-substituted guanidines. A broad infrared band at about 3400–3300 cm^{-1} is observed for hydroperoxide–amine complexes which is assigned to OOH···N[90–93], while bands at about 3500 and 3400 cm^{-1} are assigned to O—O···H—N[91,92].

2. Ammonium–hydroperoxide complexes

Cumyl[95] and *t*-butyl[96] hydroperoxide form 1:1 complexes with tetra-*n*-butylammonium bromide. A 2:1 (hydroperoxide to ammonium salt) complex is also reported for *t*-butyl hydroperoxide[96]. The cumyl hydroperoxide adduct (m.p. 45–47°C) has been prepared in 95% yield [$\nu_{OH}(cm^{-1})$: 3180 ($\Delta\nu = 370$) in carbon tetrachloride and 3240 ($\Delta\nu = 280$) in benzene][95]. These infrared absorptions have been assigned to Br^-···HOO hydrogen bonding.

Preparation of $R_4\overset{+}{N}\overset{-}{O}OR^2$ salts have been reported where R^1 = Me, Et, *n*-Bu and R^2 = *t*-Bu, $PhCMe_2$[97]. The salts cannot be isolated in a pure state.

V. REFERENCES

1. (a) W. H. Richardson and V. F. Hodge, *J. Amer. Chem. Soc.*, **93**, 3996 (1971).
   (b) J. F. Goodman, P. Robson and E. R. Wilson, *Trans. Faraday Soc.*, **58**, 1846 (1962).
   (c) E. Koubek, M. L. Haggett, C. J. Battaglia, K. M. Ibne-Rasa, H. Y. Pyun and J. O. Edwards, *J. Amer. Chem. Soc.*, **85**, 2263 (1963).
2. A. J. Everett and G. J. Minkoff, *Trans. Faraday Soc.*, **49**, 410 (1953).
3. E. Koubek and J. O. Edwards, *J. Org. Chem.*, **28**, 2157 (1963).
4. E. Koubek and J. E. Welch, *J. Org. Chem.*, **33**, 445 (1968).
5. W. H. Davis and W. A. Pryor, *J. Chem. Ed.*, **53**, 270 (1976).

6. J. E. Leffler and E. Grunwald, *Rates and Equilibria of Organic Reactions*, John Wiley and Sons, New York–London, 1963, p. 226.
7. Y. Yukawa, *Handbook of Organic Structural Analysis*, Benjamin, New York, 1965.
8. R. Wolf, *Bull. Soc. Chim. Fr.*, 644 (1954).
9. W. H. Richardson and V. F. Hodge, *J. Org. Chem.*, **35**, 4012 (1970).
10. I. M. Kolthoff and A. I. Medalia, *J. Amer. Chem. Soc.*, **71**, 3789 (1949).
11. (a) A. Albert and E. P. Serjeant, *Ionization Constants of Acids and Bases*, John Wiley and Sons, New York–London, 1962; p. 66.
(b) R. G. Bates, *Determination of pH, Theory and Practice*, John Wiley and Sons, New York–London, 1964, p. 222.
12. R. Hiatt, *Organic Peroxides*, Vol. II (Ed. D. Swern), Wiley–Interscience, New York–London, 1971; p. 80.
13. S. Kato, T. Ishihara and F. Masuo, *Bull. Jap. Petrol. Inst.*, **12**, 117 (1970); *Chem. Abstr.*, **73**, 65843d (1970).
14. K. Bowden, *Can J. Chem.*, **43**, 3354 (1965).
15. P. Ballinger and F. A. Long, *J. Amer. Chem. Soc.*, **82**, 795 (1960).
16. E. T. Denison, *Comprehensive Chemical Kinetics*, Vol. 16, (Eds. C. H. Bamford and C. F. H. Tipper), Elsevier, New York, 1980, p. 184.
17. M. Sax, P. Beurskens and S. Chu, *Acta Cryst.*, **18**, 252 (1965).
18. D. Belitskus and G. A. Jeffry, *Acta Cryst.*, **18**, 458 (1965).
19. L. S. Silbert, in *Organic Peroxides*, Vol. II (Ed. D. Swern), Wiley–Interscience, New York–London, 1971, pp. 648–656.
20. H. A. Swain, C.-Y. Kwan and H.-N. Sung, *J. Phys. Chem.*, **84**, 1347 (1980).
21. (a) J. D'Ans and A. Kneip, *Ber.*, **48**, 1136 (1915).
(b) D. Swern, *Chem. Rev.*, **45**, 1 (1949).
22. P. A. Giguere and A. W. Olmos, *Can. J. Chem.*, **30**, 821 (1952).
23. Z. K. Maĭzus, G. Ya. Timofeeva and N. M. Emanuel, *Dokl. Akad. Nauk SSSR*, **70**, 655 (1950); Z. K. Maĭzus, V. M. Cherdnickenko and N. M. Emanuel, *Dokl. Akad. Nauk SSSR*, **70**, 855 (1950); B. S. Neporent, T. E. Pavlovskaya, N. M. Emanuel and N. G. Yaroslavkii, *Dokl. Akad. Nauk SSSR*, **70**, 1025 (1950); D. G. Knorre and N. M. Emanuel, *Zh. Fiz. Khim.*, **26**, 425 (1952).
24. D. Swern, L. P. Witnauer, C. R. Eddy and W. E. Parker, *J. Amer. Chem. Soc.*, **77**, 5537 (1955).
25. W. H. T. Davison, *J. Chem. Soc.*, 2456 (1951).
26. R. Kavčič, B. Plesničar and D. Hadži, *Spectrochim. Acta*, **23A**, 2483 (1967).
27. D. Hadži, R. Kavčič and B. Plesničar, *Spectrochim. Acta*, **27A**, 179 (1971).
28. D. Swern, A. H. Clements and T. M. Luong, *Anal. Chem.*, **41**, 412 (1969).
29. A. C. Egerton, W. Emte and G. J. Minkoff, *Disc. Faraday Soc.*, **10**, 278 (1951).
30. J. R. Rittenhouse, W. Lobunez, D. Swern and J. G. Miller, *J. Amer. Chem. Soc.*, **80**, 4850 (1958).
31. S. W. Benson, cited in Reference 19, p. 750.
32. J. Škerjanc, A. Regent and B. Plesničar, *J. Chem. Soc., Chem. Commun.*, 1007 (1980).
33. R. Kavčič and B. Plesničar, *J. Org. Chem.*, **35**, 2033 (1970).
34. (a) R. M. Badger and S. H. Bauer, *J. Chem. Phys.*, **5**, 839 (1937).
(b) R. M. Badger, *J. Chem. Phys.*, **8**, 288 (1940).
(c) K. F. Purcell and R. S. Drago, *J. Amer. Chem. Soc.*, **89**, 2874 (1967).
35. O. P. Yablonskii, V. A. Belyaev and A. N. Vinogradov, *Russ. Chem. Rev.*, **41**, 565 (1972).
36. Reference 19, p. 699.
37. M. N. Puring, V. A. Itskovich, V. M. Potekchin, Yu. F. Sigolaev and V. A. Proskuryakov, *J. Gen. Chem. USSR*, **45**, 1826 (1975).
38. G. Geisler and H. Zimmermann, *Z. Phys. Chem. (Frankfurt)*, **43**, 84 (1964).
39. C. Walling and L. Heaton, *J. Amer. Chem. Soc.*, **87**, 48 (1965).
40. V. V. Zharkov and N. K. Rudnevskii, *Optics and Spectrocopy*, **12**, 264 (1962).
41. O. P. Yablonskii, N. S. Lastochkina and V. A. Belyaev, *Neftekhimiya*, **13**, 851 (1973); *Chem. Abstr.*, **80**, 81796r (1974).
42. I. F. Franchuk and L. I. Kalinina, *J. Appl. Spectry*, **15**, 1510 (1971).
43. P. A. Smirnov, T. N. Timofeeva, A. M. Syroezhko, V. M. Potehkin and V. A. Proskuryakov, *J. Org. Chem. USSR*, **10**, 1903 (1974).
44. A. I. Chirko, I. G. Tishchenko, G. M. Sosnovskii and A. F. Abramov, *Vesti Akad. Navuk B.SSR, Ser. Khim. Navuk*, 122 (1975); *Chem. Abstr.*, **84**, 16358t (1976).

45. O. P. Yablonskii, N. M. Rodionova, T. B. Ushakova, L. F. Lapuka, A. I. Nefedova and G. N. Koshel, *Neftekhimiya*, **16**, 130 (1976); *Chem. Abstr.*, **84**, 163765b (1976).
46. S. S. Ivanchev, Yr. N. Anisimov and B. A. Khomenko, *J. Appl. Spectry*, **16**, 374 (1972).
47. G. L. Bitman, Yu. Z. Karasev and M. E. Élyashberg, *J. Gen. Chem. USSR*, **44**, 1336 (1974).
48. I. I. Shabalin, M. A. Klimchuk and E. A. Kiva, *Optics and Spectry*, **24**, 294 (1968).
49. I. I. Shabalin and E. A. Kiva, *Optics and Spectry*, **24**, 377 (1968).
50. L. J. Andrews and R. M. Keefer, *Molecular Complexes in Organic Chemistry*, Holden–Day, New York, 1964; p. 93.
51. V. V. Voronekov, T. B. Ushakova, N. M. Rodionova and E. A. Lazurin, *Russ. J. Phys. Chem.*, **54**, 279 (1980).
52. O. P. Yablonskii, V. F. Bystrov, V. A. Belyaev, V. V. Voronenkov and Z. A. Pokrovskaya, *Neftekhimiya*, **12**, 577 (1972); *Chem. Abstr.*, **78**, 28697s (1973).
53. V. A. Terent'ev and V. L. Antonovskii, *Russ. J. Phys. Chem.*, **42**, 990 (1968).
54. I. I. Shabalin and E. A. Kiva, *Optics and Spectry*, **24**, 377 (1968).
55. I. I. Shabalin and E. A. Kiva, *J. Appl. Spectry*, **8**, 422 (1968).
56. A. N. Egorochkin, N. S. Vyazankin, S. Ya. Khorshev, S. E. Skobeleva, V. A. Yablokov and A. P. Tarabarina, *Proc. Acad. Sci. USSR*, **194**, 755 (1970).
57. R. West, R. H. Baney and D. L. Powell, *J. Amer. Chem. Soc.*, **82**, 6269 (1960).
58. O. P. Yablonskii, A. K. Kobyakov and V. A. Beolyaev, *Neftekhimiya*, **14**, 446 (1974); *Chem. Abstr.*, **82**, 30763m (1975).
59. M. G. Markarov, A. B. Kudryavstev, D. D. Patlyakevich and N. N. Lebedev, *J. Struct. Chem.*, **16**, 280 (1975).
60. O. A. Chaltykyan, O. A. Vartapetyan and A. G. Aloyan, *Russ. J. Phys. Chem.*, **46**, 357 (1972).
61. S. K. Gregorijan, Sh. A. Margaryan and N. M. Beileryan, *Arm. Khim. Zh.*, **32**, 516 (1979); *Chem. Abstr.*, **93**, 7498x (1980).
62. A. A. Martirosyan, O. A. Vartapotyan and N. M. Beileryan, *Russ. J. Phys. Chem.*, **47**, 1169 (1973).
63. N. N. Vyshinskii, O. S. Morozov and N. K. Rudnevskii, *Khim. Elementoorg. Soedin*, **4**, 75 (1976); *Chem. Abstr.*, **88**, 6116f (1978).
64. Yu. A. Aleksandrov, N. N. Vyshinskii, E. I. Solo'eva, O. S. Morozov and B. V. Sul'din, *J. Gen. Chem. USSR*, **45**, 664 (1975).
65. R. Hiatt, *Organic Peroxides*, Vol. II (Ed. D. Swern), Wiley–Interscience; New York–London, 1971; pp. 62–63.
66. V. A. Terent'ev and V. L. Antonovskii, *Russ. J. Phys. Chem.*, **42**, 990 (1968).
67. T. I. Drozdova, V. T. Varlamov, V. P. Lodygina and A. L. Aleksandrov, *Russ. J. Phys. Chem.*, **52**, 414 (1978).
68. D. Barnard, K. R. Hargrave and G. M. C. Higgins, *J. Chem. Soc.*, 2845 (1956).
69. A. Blaschette and B. Bressel, *Inorg. Nucl. Letters*, **4**, 175 (1968).
70. R. B. Svitych, O. P. Yablonskii and A. L. Buchanchenko, *Dokl. Phys. Chem.*, **204**, 416 (1972).
71. V. V. Zarkov and N. K. Rudnevskii, *Optics and Spectry*, **7**, 497 (1959).
72. I. I. Shabalin and M. A. Klimchuk, *Optics and Spectry*, **24**, 294 (1968).
73. L. P. Mamchur, I. P. Zyat'kov and M. A. Diki, *J. Appl. Spectry*, **22**, 809 (1975).
74. I. P. Mikhailova, I. A. Shingel, A. I. Chirko, I. G. Tischenko and M. V. Sorkina, *Vesti Akad. Navuk B.SSR*, **2**, 62 (1976); *Chem. Abstr.*, **85**, 5035e (1976).
75. K. Itoh, *Nippon Kagaku Kaishi*, 714 (1978); *Chem. Abstr.*, **88**, 106969a (1978).
76. J. Sonnenberg and D. M. White, *J. Amer. Chem. Soc.*, **86**, 5685 (1964).
77. E. H. White and M. J. C. Harding, *Photochem. Photobiol.*, **4**, 1129 (1965); *J. Amer. Chem. Soc.*, **86**, 5686 (1964).
78. G. Rio, A. Ranjon, O. Pouchot and M.-J. Scholl, *Bull. Soc. Chim. Fr.*, 1667 (1969).
79. R. Ramasseul and A. Rassat, *Tetrahedron Letters*, 1337 (1972).
80. A. Nishinaga, T. Shimizu and T. Matsuura, *Tetrahedron Letters*, **21**, 1265 (1980).
81. W. H. Richardson and R. F. Steed, *J. Org. Chem.*, **32**, 771 (1967).
82. N. A. Milas, O. L. Mageli, A. Golubović, R. W. Arndt and J. C. J. Ho, *J. Amer. Chem. Soc.*, **85**, 222 (1963).
83. A. H. M. Cosijn and M. G. J. Ossewold, *Rec. Trav. Chim.*, **87**, 1264 (1968).
84. W. H. Richardson and R. S. Smith, *J. Org. Chem.*, **33**, 3882 (1968).
85. H. E. O'Neal and S. W. Benson, *J. Phys. Chem.*, **71**, 2903 (1967) and references therein.
86. B. Plesničar, R. Kavčič and D. Hadži, *J. Mol. Struct.*, **20**, 457 (1974).

87. Y. Sprinzak, *J. Amer. Chem. Soc.*, **80**, 5449 (1958).
88. (a) K. R. Kopecky and J. A. L. Sastre, *Can. J. Chem.*, **58**, 2089 (1980).
(b) K. R. Kopecky, J. E. Filby, C. Mumford, P. A. Lockwood and J.-Y. Ding, *Can. J. Chem.*, **53**, 1103 (1975).
89. A. A. Oswald, F. Noel and A. J. Stephenson, *J. Org. Chem.*, **26**, 3969 (1961).
90. N. A. Sokolov, L. G. Usova, N. N. Vyshinskii and O. S. Morozov, *J. Gen. Chem. USSR*, **43**, 1346 (1973).
91. N. N. Vyshinskii, O. S. Morozov, E. I. Solov'eva, B. V. Sul'din and Yu. A. Aleksandrov, *J. Gen. Chem. USSR*, **49**, 1866 (1979).
92. Yu. A. Aleksandrov, N. N. Vyshinskii, E. I. Solov'eva, O. S. Morozov and B. V. Sul'din, *J. Gen. Chem. USSR*, **45**, 664 (1975).
93. A. A. Oswald, K. Griesbaum and B. E. Hudson, *J. Gen. Chem. USSR*, **28**, 2351 (1963).
94. N. F. Hall and M. R. Sprinkle, *J. Amer. Chem. Soc.*, **54**, 3469 (1932).
95. N. A. Sokolov, G. Ya. Perchugov and O. S. Morozov, *J. Gen. Chem. USSR*, **49**, 1633 (1979).
96. N. A. Sokolov and G. Ya. Perchugov, *Tr. Khim. Khim. Tekhnol.*, **2**, 29 (1973); *Chem. Abstr.*, **80**, 82294n (1974).
97. N. M. Lapshin, M. F. Erykalova, N. R. Nikulina and N. A. Sokolov, *J. Org. Chem. USSR*, **6**, 2032 (1970).

The Chemistry of Functional Groups, Peroxides
Edited by S. Patai

CHAPTER 6

Synthesis and uses of alkyl hydroperoxides and dialkyl peroxides

ROGER A. SHELDON
Océ-Andeno B. V., Venlo, The Netherlands

I. INTRODUCTION . . . 162
II. SYNTHESIS . . . 163
A. Autoxidation of Hydrocarbons . . . 163
B. Autoxidation of Organometallic Compounds . . . 168
C. Alkyl Hydroperoxides and Dialkyl Peroxides from Hydrogen Peroxide . . . 168
D. α-Alkoxyalkyl Hydroperoxides from Ozonides . . . 168
E. Dialkyl Peroxides from Metal-catalysed reactions of Alkyl Hydroperoxides . . . 169
III. GENERAL CONSIDERATION OF REACTION TYPES AND MECHANISMS . . . 169
IV. METAL-CATALYSED HOMOLYSIS . . . 173
A. Intermolecular Processes . . . 173
B. Intramolecular Processes . . . 174
V. ACID-CATALYSED HETEROLYSIS . . . 176
VI. METAL-CATALYSED HETEROLYSIS . . . 177
A. Epoxidation of Alkenes . . . 177
B. Alkene Hydroxyketonization . . . 184
C. Oxidation of Nitrogen Compounds . . . 184
D. Oxidation of Sulphur Compounds . . . 186
E. Alcohol Oxidations . . . 186
F. Ketone Oxidations . . . 187
G. Alkene Ketonization . . . 187
VII. ALKYL HYDROPEROXIDES AS TERMINAL OXIDANTS . . . 188
A. Allylic Oxidation of Alkenes and Alkynes . . . 189
B. Vicinal Dihydroxylation of Alkenes and Alkynes . . . 189
C. Alkene Epoxidation . . . 191
D. Alcohol Oxidation . . . 192
VIII. REDUCTION OF ALKYL HYDROPEROXIDES TO ALCOHOLS . . . 193

IX. REACTIONS OF α-SUBSTITUTED ALKYL HYDROPEROXIDES 194
X. CONCLUDING REMARKS 195
XI. REFERENCES 196

I. INTRODUCTION

Alkyl hydroperoxides (ROOH) and dialkyl peroxides (ROOR) are derivatives of hydrogen peroxide (HOOH). Their thermal stability generally increases in the order primary < secondary < tertiary alkyl. The lower molecular weight members, methyl, ethyl, propyl and allyl, are liable to explode when in concentrated form and their isolation is not to be recommended. Tertiary alkyl hydroperoxides and di-*t*-alkyl peroxides on the other hand, are quite stable towards thermal decomposition. Indeed *t*-butyl hydroperoxide (TBHP) and di-*t*-butyl peroxide (DTBP) are two of the most stable organic peroxides. When subjected to heat, alkyl hydroperoxides and dialkyl peroxides undergo unimolecular, homolytic dissociation (reactions 1 and 2). A major application of these compounds is, therefore, as initiators in free-radical polymerization processes. They are also used as crosslinking agents for polyolefins, vulcanizing agents for elastomers and as curing agents for polyester resins. In this chapter, however, we shall be concerned only with the uses of alkyl hydroperoxides and dialkyl peroxides in organic synthesis. We shall, of necessity, concentrate mainly on the alkyl hydroperoxides, as these are more reactive, particularly in combination with metal catalysts, and have consequently found wider usage. The uses of alkyl hydroperoxides in organic synthesis may be broadly divided into two types: those involving alkyl hydroperoxides as intermediates and those employing them as reagents, i.e. as oxidizing agents. An example of the former is the conversion of cumene to phenol and acetone via cumene hydroperoxide and an example of the latter is the metal-catalysed epoxidation with alkyl hydroperoxides (*vide infra*).

$$\mathrm{RO{-}OR} \longrightarrow 2\,\mathrm{RO}\cdot \tag{1}$$

$$\mathrm{RO{-}OH} \longrightarrow \mathrm{RO}\cdot + \mathrm{HO}\cdot \tag{2}$$

t-Butyl hydroperoxide (TBHP) was first prepared in 1938 by Milas[1], but remained pretty much a laboratory curiosity until the sixties. The situation changed with the discovery, by several industrial laboratories[2–4], that propylene could be selectively epoxidized by TBHP in the presence of a homogeneous molybdenum catalyst (reaction 3). This process presently accounts for the manufacture of several hundred thousand tonnes per annum of propylene oxide.

$$\mathrm{CH_3CH{=}CH_2} + t\text{-}\mathrm{BuO_2H} \xrightarrow{[\mathrm{Mo^{VI}}]} \mathrm{CH_3\overset{\frown O \frown}{CH{-}CH_2}} + t\text{-}\mathrm{BuOH} \tag{3}$$

TBHP now constitutes a bulk organic chemical and in recent years has found increasing use, generally in combination with a metal catalyst, as a selective oxidant in organic chemistry[5]. However, as Sharpless[5] has recently pointed out, organic chemists have generally failed to fully appreciate the advantages of TBHP as a mild selective oxidant. These advantages, which make TBHP superior to other, more well-known, sources of active oxygen, such as H_2O_2 and peracetic acid, are worth noting:

(1) TBHP has a high thermal stability in dilute organic solutions, e.g. a 0.2 M solution of TBHP in benzene has a half-life of 520 hours at 130°C[6].
(2) TBHP is less sensitive to contamination by metals than H_2O_2 or peracetic acid and is consequently safer to handle.

(3) Perhaps the key advantage of TBHP is its selectivity. In contrast to H_2O_2 and peracetic acid it is unreactive to most organic compounds in the absence of metal catalysts.
(4) TBHP is readily soluble in hydrocarbon solvents. This is an important advantage over H_2O_2 which is soluble only in strongly polar solvents and many of the metal-catalysed oxidations to be discussed later are seriously retarded by such solvents.
(5) Oxidations with TBHP are carried out under essentially neutral conditions and are, consequently, applicable to acid-sensitive substrates.
(6) The co-product, *t*-butanol, has a relatively low boiling point (83°C) and is readily removed by distillation. Other readily available alkyl hydroperoxides, such as ethylbenzene hydroperoxide (EBHP) and cumene hydroperoxide (CHP), produce higher boiling co-products and could be the reagents of choice when the product is low boiling and difficult to separate from *t*-butanol.

The fact that organic chemists, who are accustomed to employing organic peracids in their syntheses, have failed to exploit the advantages of TBHP, can probably be attributed to a lack of familiarity with this reagent. We hope, therefore, that this article will stimulate the further exploitation of this useful oxidant. In more than ten years of working with TBHP and other alkyl hydroperoxides we have not yet experienced a single explosion.

In dividing the material for this chapter we have opted for a categorization based on the type of mechanism involved. We have chosen this approach because we are of the opinion that an understanding of mechanism provides a sound basis for extrapolation to other, as yet undiscovered, applications in organic synthesis. Before proceeding to a discussion of the mechanistic features of these reactions we shall first consider the various methods available for the synthesis of alkyl hydroperoxides and dialkyl peroxides.

The comprehensive review by Hiatt[7] constitutes a useful account of the literature up to 1970 on the synthesis and reactions of alkyl hydroperoxides.

II. SYNTHESIS

A. Autoxidation of Hydrocarbons

The liquid-phase autoxidation of hydrocarbons is the method of choice for industrial-scale synthesis and is used for the manufacture of the three most important alkyl hydroperoxides: TBHP, EBHP and CHP.

Me_3COOH	PhCH(Me)OOH	$PhC(Me_2)OOH$
TBHP	EBHP	CHP

Autoxidations proceed via a free-radical chain mechanism, described by Scheme 1[7].

Initiation

$$In_2 \xrightarrow{R_i} 2\, In\cdot$$

$$In\cdot + RH \longrightarrow InH + R\cdot$$

Propagation

$$R\cdot + O_2 \longrightarrow RO_2\cdot$$

$$RO_2\cdot + RH \xrightarrow{k_p} RO_2H + R\cdot$$

Termination

$$2\, RO_2\cdot \xrightarrow{2k_t} \text{non-radical products} + O_2$$

SCHEME 1

The rate expression for hydroperoxide formation and oxygen consumption is given by equation (4). This, however, represents an ideal situation in which chain lengths are long and the ratio of propagation to termination is high, resulting in high yields of hydroperoxides. In practice this is observed only with reactive tertiary alkanes and aralkanes and certain alkenes (see later). The susceptibility of any particular substrate to autoxidation is determined by the ratio $(k_p/2k_t)^{\frac{1}{2}}$ which is referred to as its *oxidizability*. The rate of propagation (k_p) roughly parallels the ease of breaking of the C—H bond and increases in the series *n*-alkanes < branched alkanes < aralkanes < alkenes.

$$\frac{-\mathrm{d}[O_2]}{\mathrm{d}t} = \frac{\mathrm{d}[RO_2H]}{\mathrm{d}t} = k_p[RH]\left(\frac{R_i}{2k_t}\right)^{\frac{1}{2}} \qquad (4)$$

The rate of termination ($2k_t$) increases in the order primary < secondary < tertiary alkylperoxy radicals. Thus the lower rates of autoxidation of primary and secondary C—H functions compared to their tertiary counterparts are not only due to the lower reactivity of the C—H bonds in the former but also to the significantly higher rate of termination of primary and secondary alkylperoxy radicals. This explains why such a fairly reactive hydrocarbon as toluene undergoes a relatively slow rate of autoxidation (see Table 1).

Chain initiation is readily accomplished by the deliberate addition of initiators, such as aliphatic azo compounds and organic peroxides, which yield free radicals by thermal decomposition. Tertiary alkanes and aralkenes are readily autoxidized in the temperature range 100–150°C even in the absence of added initiators. Such reactions exhibit induction periods, followed by increasingly rapid oxygen uptake and eventual levelling off. The reactions become considerably more complicated at high hydrocarbon conversions due to the accumulation of secondary products, such as carbonyl compounds, formed by the thermal decomposition of the alkyl hydroperoxide. Autoxidative syntheses are thus generally carried out to low conversions (10–20%) and the excess unreacted hydrocarbon, which is easily separated from the hydroperoxide, is recycled. In practice further reactions of the alkyl hydroperoxide may be carried out directly on the solution in unreacted hydrocarbon.

The autoxidation of isobutane to TBHP has been thoroughly studied because of the commercial importance of the product[9–14]. Both liquid[9,11–14] and vapour-phase[10,11,14] processes have been described. The reaction is carried out in the temperature range 100–140°C and, under commercially relevant conditions, produces TBHP in 75% yield at ca. 10% isobutane conversion. The selectivity to TBHP drops to 64% at 20% conversion of isobutane[9]. The main by-products are *t*-butyl alcohol and di-*t*-butyl peroxide (DTBP), together with small amounts of acetone. Commercially available TBHP generally contains ca. 70% wt. TBHP.

TABLE 1. Absolute rate constants for hydrocarbon autoxidations at 30°C[8]

Substrate	$k_p/(2k_t)^{\frac{1}{2}} \times 10^3$ ($M^{-\frac{1}{2}} s^{-\frac{1}{2}}$)	k_p ($M^{-1} s^{-1}$)	$2k_t \times 10^{-6}$ ($M^{-1} s^{-1}$)
Cumene	1.50	0.18	0.015
Tetralin	2.30	6.4	7.6
Ethylbenzene	0.21	1.3	40
Toluene	0.014	0.24	300
Cyclohexene	2.3	5.4	5.6
Heptene-3	0.54	1.4	6.4
Octene-1	0.062	1.0	260

Similarly, other tertiary alkanes can be selectively autoxidized in the liquid phase to produce the corresponding hydroperoxide[15,16]. Examples are shown in Table 2.

Alkanes containing two tertiary C—H bonds afford bishydroperoxides[17,18]. When the two relevant C—H bonds are suitably juxtaposed, the formation of the bishydroperoxide occurs in high yield as a result of efficient intramolecular hydrogen abstraction. For example, 2,4-dimethylpentane afforded the bishydroperoxide in 95% selectivity at 10% conversion (reaction 5)[17].

CH2 H C O—O· → CH2 C· C OOH

$\xrightarrow{O_2}$ CH2 C C OOH OO· $\xrightarrow{RH}$ CH2 C C O_2H O_2H + R·, etc. (5)

TABLE 2. *t*-Alkyl hydroperoxides from the liquid-phase autoxidation of alkanes

Substrate	Temperature (°C)	Major product	Selectivity (%)	Ref.
R-cyclohexane	100	1-R-cyclohexyl O_2H	—[a]	15
pinane	120	pinane O_2H	96[b]	16
p-menthane	120	p-menthane O_2H	95[b]	16
carane	120	carane O_2H	95[b]	16
bornane	120	bornane O_2H	96[b]	16

[a] R = Me, Et, *i*-Pr, *t*-Bu, *c*-C_6H_{11}.

[b] The selectivity refers to total hydroperoxide; isomer distributions were not reported. The tertiary hydroperoxide is presumed to be the major product.

As mentioned earlier the oxidizabilities of aralkanes decrease significantly in the order: tertiary > secondary > primary benzylic C—H bonds. This is largely due to the significant increase in rate of termination in the order: primary > secondary > tertiary alkylperoxy radicals (see Table 1).

Aralkanes containing tertiary benzylic C—H bonds undergo smooth autoxidation at moderate temperatures to afford the corresponding hydroperoxides in high selectivities. The autoxidation of cumene to cumene hydroperoxide (CHP) (reaction 6) is of considerable industrial importance as the first step in the well-known process[19] for the co-production of phenol and acetone (reaction 7), a reaction discovered in 1944 by Hock and Lang[20]. The autoxidation of cumene is carried out with air at 90–110°C, usually in the presence of a small amount of base, such as sodium carbonate, in order to circumvent premature acid-catalysed decomposition of the CHP to the autoxidation inhibitor, phenol. The reaction is usually carried out to 20–30% conversion and the solution concentrated to 80–90% CHP by distillation of unreacted cumene. A continuous process, which affords a 95% selectivity to CHP at ca. 30% cumene conversion, has been described[21].

$$PhCHMe_2 + O_2 \longrightarrow PhC(Me)_2O_2H \quad (6)$$

$$PhC(Me)_2O_2H \xrightarrow{[H^+]} PhOH + Me_2CO \quad (7)$$

Diisopropylbenzenes are similarly autoxidized to give the corresponding bishydroperoxides, which undergo subsequent acid-catalysed decomposition to the corresponding dihydric phenols. Thus industrial processes have been developed for the large-scale manufacture of resorcinol (reaction 8)[22] and hydroquinone (reaction 9)[23]. They are rather complex processes involving separation of the required bishydroperoxide, by crystallization or extraction techniques, and recycling of the monohydroperoxide and unreacted hydrocarbon to the autoxidation step.

$$m\text{-}C_6H_4(CHMe_2)_2 + 2O_2 \longrightarrow m\text{-}C_6H_4(CMe_2O_2H)_2 \xrightarrow{[H^+]} m\text{-}C_6H_4(OH)_2 + 2\,Me_2CO \quad (8)$$

$$p\text{-}C_6H_4(CHMe_2)_2 + 2O_2 \longrightarrow p\text{-}C_6H_4(CMe_2O_2H)_2 \xrightarrow{[H^+]} p\text{-}C_6H_4(OH)_2 + 2\,Me_2CO \quad (9)$$

Ethylbenzene hydroperoxide (EBHP) has achieved commercial importance as the oxidizing agent in the Oxirane and Shell processes for the co-production of propylene oxide and styrene (see Section VI.A). Processes have been described for the production of EBHP by air oxidation of ethylbenzene in the presence of small amounts of sodium pyrophosphate[24] or barium oxide[25]. Thus, oxidation of ethylbenzene with air at 135°C for three hours in the presence of 0.0067% BaO affords EBHP in 95% selectivity at 21% conversion[25]. Autoxidations of methylbenzenes, such as toluene and xylenes, proceed at low rates and give low hydroperoxide selectivities due to efficient termination of the primary benzylperoxy radicals which affords substantial amounts of aldehydes and alcohols as primary products (reaction 10). The rate of termination can, however, be decreased and the oxidizability $[k_p/(2k_t)^{\frac{1}{2}}]$ increased by increasing the dielectric constant of the reaction medium[26,27]. In agreement with this finding the rates and selectivities of autoxidations of relatively unreactive hydrocarbons, such as *p*-xylene and cyclohexane, to the corresponding hydroperoxides, increase considerably in the presence of nitriles[28].

$$2\,ArCH_2O_2\cdot \longrightarrow ArCHO + O_2 + ArCH_2OH \qquad (10)$$

Isopropyltoluenes are selectively autoxidized at the isopropyl group to give hydroperoxides which can be converted to cresols via acid-catalysed decomposition[29], e.g. reaction (11).

$$\text{p-}CH_3C_6H_4CH(CH_3)_2 \xrightarrow{O_2} \text{p-}CH_3C_6H_4C(CH_3)_2O_2H \xrightarrow{[H^+]} \text{p-}CH_3C_6H_4OH \qquad (11)$$

In the autoxidation of alkenes, chain propagation can occur via the usual abstraction mechanism, to form an allylic hydroperoxide as the primary product (reaction 12), or via the addition of the alkylperoxy radical to the double bond (reaction 13)[14]. Reaction via addition leads to the formation of epoxides (reaction 14) and/or polyperoxides (reaction 15). Reasonable yields of allylic hydroperoxides are obtained only with reactive alkenes which possess a favourable abstraction/addition ratio, e.g. cyclohexene. With less reactive alkenes, such as 1-octene, the intermediate hydroperoxides are not stable at the temperatures required for a reasonable rate of reaction. Indeed, with few exceptions the autoxidations of alkenes generally afford complex product mixtures and are of little synthetic value. Many of the high yields of hydroperoxides reported in the early literature are erroneous and can generally be attributed to the inadequate analytical techniques available at the time.

$$RO_2\cdot + -\overset{|}{\underset{H}{C}}-\overset{|}{C}=C< \xrightarrow{\text{abstraction}} RO_2H + -\overset{|}{\underset{\cdot}{C}}-\overset{|}{C}=C< \qquad (12)$$

$$RO_2\cdot + -\overset{|}{\underset{H}{C}}-\overset{|}{C}=C< \xrightarrow{\text{addition}} RO_2-\overset{|}{\underset{|}{C}}-\overset{|}{\underset{|}{C}}\cdot \qquad (13)$$

$$RO_2-\overset{|}{\underset{|}{C}}-\overset{|}{\underset{|}{C}}\cdot \longrightarrow RO\cdot + >C\overset{O}{-}C< \qquad (14)$$

$$RO_2-\overset{|}{\underset{|}{C}}-\overset{|}{\underset{|}{C}}\cdot \xrightarrow{O_2} RO_2-C-C-O_2\cdot \qquad (15)$$

In contrast to the nonselective autoxidation of alkenes, photosensitized oxidation of alkenes with singlet oxygen can lead to the formation of allylic hydroperoxides in high yields[30,31]. The ease of oxidation and yield increase with increasing substitution of the double bond by alkyl groups. High yields of allylic hydroperoxides are generally obtained with tri- and tetra-substituted alkenes, such as 2,3-dimethylbutene-2 and cyclohexylidene-cyclohexane. Terminal alkenes, e.g. 1-octene, and unsubstituted cyclic alkenes, e.g. cyclohexene, in contrast, react very slowly or not at all. The reaction proceeds via a nonradical mechanism and invariably leads to migration of the double bond[32], e.g. reaction (16).

$$\text{(tetramethylethylene)} + {}^1O_2 \longrightarrow \text{(allylic hydroperoxide)}-O_2H \qquad (16)$$

B. Autoxidation of Organometallic Compounds

Organometallic compounds undergo facile autoxidation via a free-radical chain mechanism involving the propagation steps shown in reactions (17) and (18). Thus alkyl derivatives of Group I, II and III metals react vigorously with air to give, initially, alkylperoxymetal compounds which in many cases react further with the alkylmetal to give alkoxymetal compounds[33]. Walling and Buckler[34,35] have shown that this reduction can be minimized by employing the technique of inverse addition, i.e. by adding the alkylmetal to a saturated solution of oxygen. Using this technique the low-temperature autoxidation of Grignard reagents is a useful procedure for preparing primary and secondary alkyl hydroperoxides which are often difficult to obtain by other methods (reaction 19).

$$R\cdot + O_2 \longrightarrow RO_2\cdot \qquad (17)$$

$$RO_2\cdot + RM \longrightarrow RO_2M + R\cdot \qquad (18)$$

$$RMgX + O_2 \longrightarrow RO_2MgX \xrightarrow{H^+} RO_2H \qquad (19)$$

C. Alkyl Hydroperoxides and Dialkyl Peroxides from Hydrogen Peroxide

Tertiary alkyl hydroperoxides are readily prepared by reaction of 30% hydrogen peroxide with alkyl hydrogen sulphates (reaction 20), a reaction discovered by Milas[36,37]. The alkyl hydrogen sulphate is prepared *in situ* from equimolar quantities of an alcohol, or an alkene, and 70% sulphuric acid. When half of the required quantity of hydrogen peroxide is used good yields of dialkyl peroxides are obtained via subsequent reaction of the alkyl hydroperoxide with the alkyl hydrogen sulphate. This reaction can also be employed for the preparation of unsymmetrical dialkyl peroxides (reaction 21). Kochi has provided detailed procedures for the synthesis of a variety of alkyl hydroperoxides using this procedure[38]. It should be emphasized that although the addition of hydrogen peroxide to alcohol/H_2SO_4 mixtures is usually safe, the addition of H_2SO_4 to alcohol/H_2O_2 mixtures frequently results in an explosion.

$$ROSO_3H + H_2O_2 \longrightarrow RO_2H + H_2SO_4 \qquad (20)$$

$$ROSO_3H + R^1O_2H \longrightarrow ROOR^1 + H_2SO_4 \qquad (21)$$

D. α-Alkoxyalkyl Hydroperoxides from Ozonides

The reaction of ozone with olefins in alcohol (usually methanol or ethanol) solvents leads to the formation of α-alkoxyalkyl hydroperoxides via trapping of the zwitterion intermediate[39] (reaction 22).

$$\text{(ozonide)} \longrightarrow {>}C{=}O + {>}\overset{+}{C}{-}O{-}\overset{-}{O} \xrightarrow{ROH} {>}C(O_2H)(OR) \qquad (22)$$

E. Dialkyl Peroxides from Metal-catalysed Reactions of Alkyl Hydroperoxides

In the presence of cobalt or copper salts alkyl hydroperoxides react with various reactive substrates such as alkylarenes, alkenes or ethers to give dialkyl peroxides via reaction (23). The mechanism and scope of reaction (23) are discussed in detail in Sections III and IV.

$$R^1H + R^2O_2H \xrightarrow{Co^{II} \text{ or } Cu^{II}} R^1OOR^2 + H_2O \qquad (23)$$

III. GENERAL CONSIDERATION OF REACTION TYPES AND MECHANISMS

The reactions of alkyl hydroperoxides and dialkyl peroxides can be divided into two types on a mechanistic basis: *homolytic* and *heterolytic*. Both of these types of processes are promoted by metal compounds. Thus, although thermal homolysis of alkyl hydroperoxides occurs readily only at temperatures in excess of 100°C, facile homolytic decomposition occurs at ambient temperatures in the presence of catalytic amounts of metal ions such as manganese (II) and cobalt (II) which readily undergo one-electron redox reactions.

Catalysis by these metals can be explained on the basis of the outer-sphere redox reactions (24)–(26)[40]. Since alkyl hydroperoxides are oxidizing agents reaction (24) is generally much faster than reaction (25) which proceeds readily only with strong oxidants. For catalytic decomposition of the alkyl hydroperoxide both reactions (24) and (25) should proceed at a reasonable rate, which is the case with Co^{II}/Co^{III} and Mn^{II}/Mn^{III} couples. Although the outer-sphere mechanism shown above is probably relevant to reactions in polar solvents (e.g. aqueous solution), in hydrocarbon solvents these reactions are more likely to involve inner-sphere electron transfer reactions of metal–alkyl hydroperoxide complexes (reactions 27–29). It is interesting to note that the homolytic cleavage of an alkylperoxymetal species at the O—O bond affords what is formally an oxometal (M=O) species containing the metal in the next highest oxidation state.

$$RO_2H + M^{(n-1)+} \longrightarrow M^{n+} + RO\cdot + HO^- \qquad (24)$$

$$RO_2H + M^{n+} \longrightarrow M^{(n-1)+} + RO_2\cdot + H^+ \qquad (25)$$

$$\text{Overall reaction: } 2\,RO_2H \xrightarrow{[M]} RO_2\cdot + RO\cdot + H_2O \qquad (26)$$

$$M^{(n-1)+} \leftarrow \underset{\displaystyle H}{\overset{}{O}}-OR \longrightarrow M^{n+}-OH + RO\cdot \qquad (27)$$

$$\text{or } M^{(n-1)+}-O-OR \longrightarrow M^{n+}=O + RO\cdot \qquad (28)$$

$$M^{n+}-O-OR \rightleftharpoons M^{(n-1)+} + RO_2\cdot \qquad (29)$$

In the presence of reactive substrates the intermediate alkoxy radicals formed in reaction (24) or (28) can undergo hydrogen transfer with the substrate (solvent) (reaction 30). The radical derived from the latter can then undergo electron-transfer oxidation to the corresponding carbenium ion (reaction 31) with regeneration of the reduced form of the catalyst. Subsequent reaction of the carbenium ion with the alkyl hydroperoxide affords a dialkyl peroxide (reaction 32). Alternatively, the carbenium ion can undergo either solvolysis by a hydroxylic solvent (reaction 33) or proton elimination (reaction 34).

$$R^1O\cdot + RH \longrightarrow R^1OH + R\cdot \qquad (30)$$

$$R\cdot + M^{n+} \longrightarrow R^+ + M^{(n-1)+} \qquad (31)$$

$$R^+ + R^1O_2H \longrightarrow RO_2R^1 + H^+ \qquad (32)$$

$$R^+ + HOAc \longrightarrow ROAc + H^+ \qquad (33)$$

$$R^+ \longrightarrow R(-H) + H^+ \qquad (34)$$

Reaction (31) constitutes an alternative pathway for the regeneration of the reduced form of the catalyst when regeneration via reaction (25) is not favourable. This is particularly relevant in the case of copper compounds. Thus, alkyl hydroperoxides are not readily decomposed by catalytic amounts of copper (II) compounds since reaction (25) is unfavourable with this weak oxidant. On the other hand facile decomposition occurs in the presence of reactive substrates and copper (I) catalysts since the electron-transfer oxidation of alkyl radicals by copper (II) is extremely facile[41].

Dialkyl peroxides can similarly undergo metal-ion-mediated homolysis with reducing agents such as copper (I) (reaction 35). In this case, however, there is no equivalent reaction to (25) and catalytic decomposition is possible only in the presence of reactive substrates where the reduced catalyst is regenerated via reactions (31) and (33) or (34).

$$RO{-}OR + M^{(n-1)+} \longrightarrow M^{n+} + RO\cdot + RO^- \qquad (35)$$

Alkyl hydroperoxides (and to a lesser extent dialkyl peroxides) can also undergo heterolytic reactions with organic substrates via nucleophilic displacement at oxygen (reaction 36). Since alkyl hydroperoxides are weak electrophiles they undergo facile heterolysis only with relatively strong nucleophiles (reducing agents) such as iodide ion and trivalent phosphorus compounds, e.g. reaction (37). Divalent sulphur and trivalent nitrogen nucleophiles do not react readily with alkyl hydroperoxides in the absence of metal catalysts and where reactions are observed at elevated temperatures they tend to be complex processes involving both homolytic and heterolytic pathways.

$$\ddot{Nu} \curvearrowright \underset{H}{O}{-}O{-}R \longrightarrow \overset{+}{Nu}{-}OH + RO^- \longrightarrow \overset{+}{Nu}{-}O^- + ROH \qquad (36)$$

$$R^1_3P + R^2O_2H \longrightarrow R^1_3PO + R^2OH \qquad (37)$$

The *heterolysis* of alkyl hydroperoxides is catalysed both by acids (see Section V) and by certain metal compounds. Thus, metal complexes in high oxidation states, such as those of molybdenum (VI), tungsten (VI), vanadium (V) and titanium (IV) can facilitate the heterolysis of alkyl hydroperoxides by nucleophiles via the intermediacy of alkylperoxymetal complexes[40]. The superior catalysts are characterized by metals that exhibit a low oxidation potential (i.e. homolytic decomposition via reaction 25 is unfavourable), and a high Lewis acidity, in their highest oxidation state. The Lewis acidity of transition metal oxides decreases in the order: CrO_3, $MoO_3 \gg WO_3 > TiO_2$, V_2O_5 etc., which is in agreement with the high activity observed for molybdenum (VI) compounds. Chromium (VI) compounds should, on the basis of their high Lewis acidity, also be active catalysts but are strong oxidants and cause rapid decomposition of the alkyl hydroperoxide. Certain main-group elements such as selenium (IV), boron (III) and tin (IV) are also catalysts but generally show a lower activity than the best transition metals.

It is significant that the superior catalysts are generally oxometal complexes such as molybdenyl (MoO_2^{2+}) and vanadyl (VO^{3+}). Complexation with the alkyl hydroperoxide can, in principle, involve the displacement of another ligand or addition to the oxometal group, e.g. reactions (38) and (39) for a trialkoxyvanadyl catalyst.

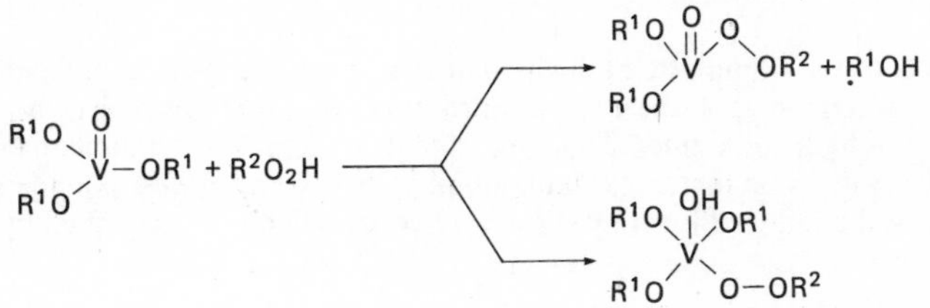

$$(R^1O)_2V(=O)-OR^1 + R^2O_2H \longrightarrow (R^1O)_2V(=O)-O-OR^2 + R^1OH \quad (38)$$

$$(R^1O)_2V(=O)-OR^1 + R^2O_2H \longrightarrow (R^1O)_2V(OH)(OR^1)(O-OR^2) \quad (39)$$

Coordination to the metal increases the electrophilicity of the peroxidic oxygens and facilitates oxygen transfer to nucleophiles such as dialkyl sulphides, amines and alkenes. With alkenes this results in epoxidation which is the most important application of the metal catalyst–alkyl hydroperoxide reagents.

The mechanism of oxygen transfer has been the subject of much discussion in the literature[40,42–45]. The rate-determining step involves the attack of an electrophilic species on the alkene (or other nucleophile) since the rate of epoxidation increases with increasing substitution of the double bond with alkyl groups. Sharpless[44] favours rate-determining attack of the alkene at the peroxidic oxygen coordinated to the M=O group (Scheme 2). Mimoun[45], on the other hand, favours rate-determining coordination of the alkene at the electrophilic metal centre followed by intramolecular addition of the coordinated alkylperoxy ligand to the coordinated alkene to give a peroxyalkylmetal intermediate. The latter decomposes to the epoxide and the alkoxymetal catalyst (Scheme 3). It is, at this juncture, not possible to distinguish unequivocally between these mechanisms.

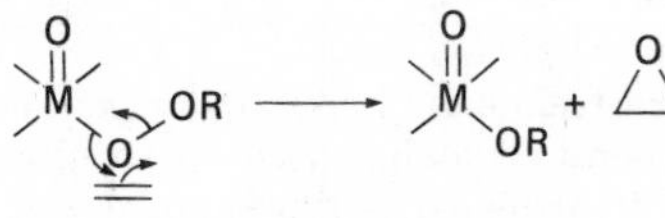

SCHEME 2

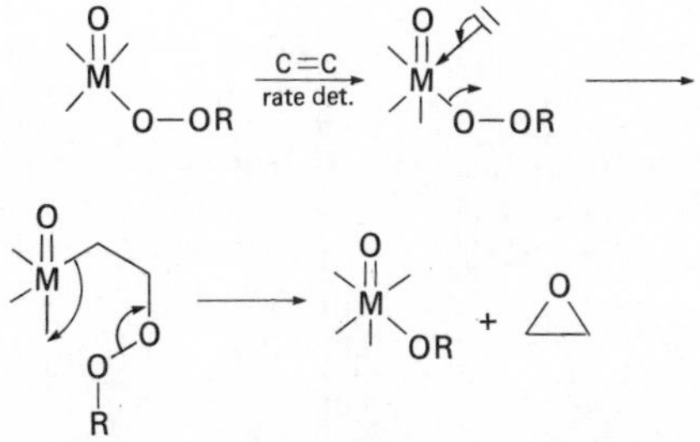

C=C
rate det.

SCHEME 3

Mimoun[45] has similarly postulated the intermediacy of peroxyalkylpalladium(II) species in the palladium(II)-catalysed oxidation of terminal alkenes with TBHP which affords the corresponding methyl ketones (reaction 40) (see Section VI.G).

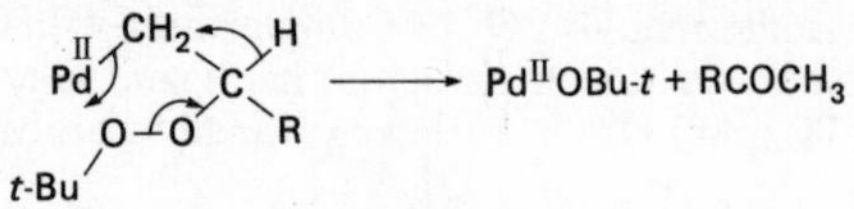

$$\overset{\text{II}}{\text{Pd}}(CH_2)\text{–}C(H)(R)\text{–}O\text{–}O\text{–}t\text{-Bu} \longrightarrow Pd^{II}OBu\text{-}t + RCOCH_3 \qquad (40)$$

In addition to the homolytic pathway and that involving alkylperoxymetal species as electrophilic intermediates there is a third type of mechanism for metal-catalysed oxidations with hydroperoxides. This type of reaction involves the direct oxidation of an organic substrate by a metal oxidant, usually an oxometal compound, followed by reoxidation of the reduced form by the alkyl hydroperoxide, as described by the general Scheme 4.

$$M{=}O + S \longrightarrow M + SO \qquad (41)$$

$$M + RO_2H \longrightarrow M{=}O + ROH \qquad (42)$$

S = substrate, SO = oxidized substrate

SCHEME 4

A typical example of this class of oxidations is the SeO_2-catalysed allylic oxidation of alkenes with TBHP (see Section VII). Reaction (42) is, in principle, favourable for metals which readily undergo two-electron redox reactions and may be envisaged as a concerted two-electron change or as two successive one-electron transfers, e.g. for a divalent metal reaction (43) or (44) would apply.

$$M^{II}{-}O{-}OR \longrightarrow M^{IV}{-}O + RO^- \qquad (43)$$

$$\text{or } M^{II}{-}O{-}OR \longrightarrow M^{III}{=}O + RO\cdot \longrightarrow M^{IV}{=}O + RO^- \qquad (44)$$

In practice metal-catalysed reactions of alkyl hydroperoxides may involve competing homolytic and heterolytic pathways as has been observed for the tin(II)- and nickel(II)-catalysed decomposition of hydroperoxides[46]. Summarizing, alkylperoxymetal species, formed as transient intermediates in metal-promoted oxidations with alkyl hydroperoxides, may decompose via a variety of pathways as illustrated in Scheme 5.

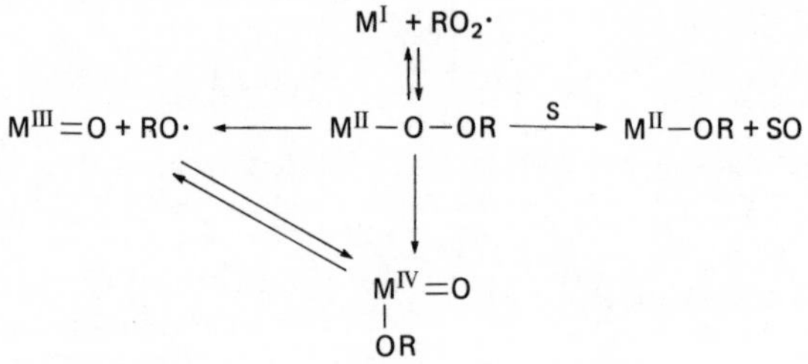

$$M^{I} + RO_2\cdot \rightleftharpoons M^{II}{-}O{-}OR$$

$$M^{III}{=}O + RO\cdot \longleftarrow M^{II}{-}O{-}OR \xrightarrow{S} M^{II}{-}OR + SO$$

$$M^{II}{-}O{-}OR \longrightarrow M^{IV}(=O)(OR); \quad M^{III}{=}O + RO\cdot \rightleftharpoons M^{IV}(=O)(OR)$$

SCHEME 5

IV. METAL-CATALYSED HOMOLYSIS

A. Intermolecular Processes

As mentioned in the preceding section the redox decomposition of alkyl hydroperoxides can be employed for introducing the alkylperoxy group into reactive substrates[47–49]. Copper and cobalt salts are usually the superior catalysts for this reaction. For example, the cuprous chloride-catalysed decomposition of TBHP in cumene at 100–110°C affords *t*-butyl α-cumyl peroxide in 90% yield via steps (45)–(48)[49].

$$t\text{-BuO}_2\text{H} + \text{Cu}^{\text{I}} \longrightarrow \text{Cu}^{\text{II}}\text{OH} + t\text{-BuO}\cdot \qquad (45)$$

$$t\text{-BuO}\cdot + \text{PhCHMe}_2 \longrightarrow t\text{-BuOH} + \text{Ph(Me)}_2\text{C}\cdot \qquad (46)$$

$$\text{Ph(Me)}_2\text{C} + \text{Cu}^{\text{II}} \longrightarrow \text{Ph(Me)}_2\text{C}^+ + \text{Cu}^{\text{I}} \qquad (47)$$

$$\text{Ph(Me)}_2\text{C}^+ + t\text{-BuO}_2\text{H} \longrightarrow \text{Ph(Me)}_2\text{CO}_2\text{Bu-}t + \text{H}^+ \qquad (48)$$

Similarly, reaction in *p*-xylene at 50°C gives *p*-methylbenzyl *t*-butyl peroxide in 85% yield[49] (reaction 49).

$$\text{Me—C}_6\text{H}_4\text{—CH}_3 + 2\ t\text{-BuO}_2\text{H} \xrightarrow{\text{CuCl}} \text{Me—C}_6\text{H}_4\text{—CH}_2\text{O}_2\text{Bu-}t + t\text{-BuOH} + \text{H}_2\text{O} \qquad (49)$$

In the presence of alkenes allylic peroxides are formed via abstraction of an allylic hydrogen, e.g. cyclohexene gives mainly cyclohexenyl *t*-butyl peroxide[50] (reaction 50). When the reaction is carried out with cuprous chloride in acetic acid, cyclohexenyl acetate is formed in quantitative yield via interception of the intermediate cyclohexenyl cation with acetic acid[51] (reactions 51–53).

$$\text{C}_6\text{H}_{10} + t\text{-BuO}_2\text{H} \xrightarrow[\text{or Cu}^{\text{II}}]{\text{Cu}^{\text{I}}} \text{C}_6\text{H}_9\text{—O}_2\text{Bu-}t + t\text{-BuOH} + \text{H}_2\text{O} \qquad (50)$$

$$\text{C}_6\text{H}_{10} + t\text{-BuO}\cdot \longrightarrow \text{C}_6\text{H}_9\cdot + t\text{-BuOH} \qquad (51)$$

$$\text{C}_6\text{H}_9\cdot + \text{Cu}^{\text{II}} \longrightarrow \text{C}_6\text{H}_9^+ + \text{Cu}^{\text{I}} \qquad (52)$$

$$\text{C}_6\text{H}_9^+ + \text{HOAc} \longrightarrow \text{C}_6\text{H}_9\text{—OAc} + \text{H}^+ \qquad (53)$$

This reaction can, in principle, be applied to the α-acetoxylation of a variety of substrates containing reactive C—H bonds[47,48]. Similar reactions are also obtained with dialkyl peroxides. Copper-catalysed oxidation of hydrocarbons with DTBP in the presence of benzamide, for example, affords the corresponding *N*-alkylbenzamide[52] (reaction 54).

$$RH + (t\text{-}BuO)_2 + PhCONH_2 \xrightarrow{Cu^I} PhCONHR + 2\,t\text{-}BuOH \quad (54)$$

B. Intramolecular Processes

The alkoxy radical intermediates formed in the redox decomposition of alkyl hydroperoxides can also undergo intramolecular hydrogen transfer with a suitably juxtaposed C—H bond. This reaction has synthetic utility as a method for the introduction of remote double bonds by $FeSO_4/Cu(OAc)_2$-catalysed decomposition of alkyl hydroperoxides in acetic acid. The reaction proceeds via the sequence shown in Scheme 6[53–56]. The method has been successfully applied in the preparation of unsaturated alcohols from a variety of primary and secondary alkyl hydroperoxides in addition to tertiary alkyl hydroperoxides[55,56].

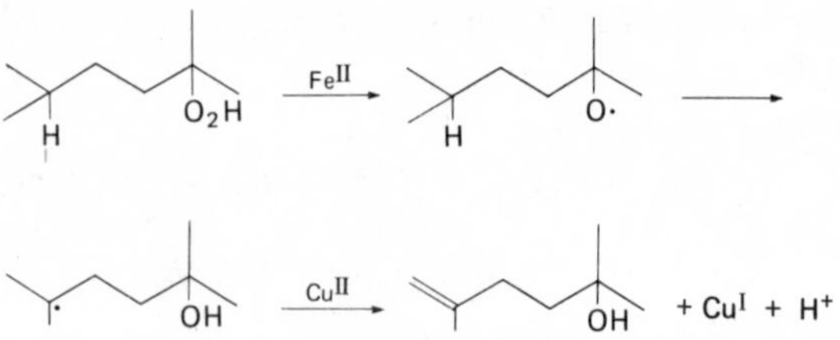

O_2H $\xrightarrow{Fe^{II}}$ O· $\longrightarrow$

OH $\xrightarrow{Cu^{II}}$ OH $+ Cu^I + H^+$

SCHEME 6

Another characteristic reaction of alkoxy radicals is β-fragmentation as illustrated in the general equation (55). Reaction (55) is favoured when R is branched alkyl and when the solvent is inert to hydrogen abstraction by the alkoxy radical. With cycloalkyl hydroperoxides reaction (55) leads to ring-opening. The ferrous-ion-catalysed decomposition of 1-alkylcycloalkyl hydroperoxides, for example, provides a synthetically useful method for preparing long-chain diketones[57,58], e.g. reactions (56) and (57).

$$R(Me)_2CO\cdot \longrightarrow R\cdot + Me_2CO \quad (55)$$

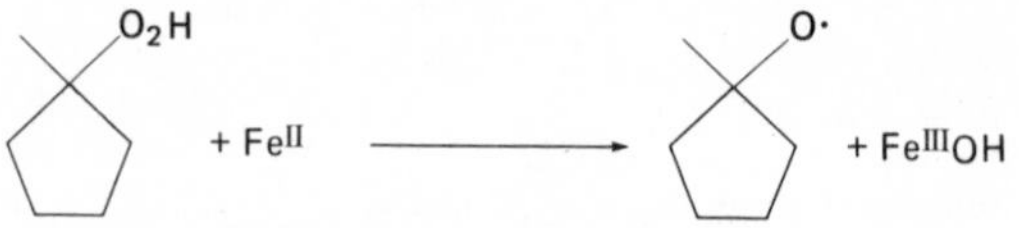

$$(O_2H) + Fe^{II} \longrightarrow (O\cdot) + Fe^{III}OH \quad (56)$$

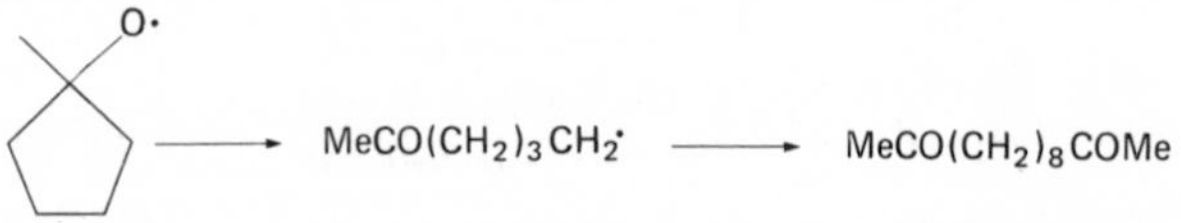

$$(O\cdot) \longrightarrow MeCO(CH_2)_3CH_2\cdot \longrightarrow MeCO(CH_2)_8COMe \quad (57)$$

A related reaction is the palladium(II)-catalysed decomposition of cyclohexyl hydroperoxide, in a two-phase cyclohexane/water system, which affords *trans*-2-hexenal in 51% yield[59]. The mechanism shown in Scheme 7 has been suggested.

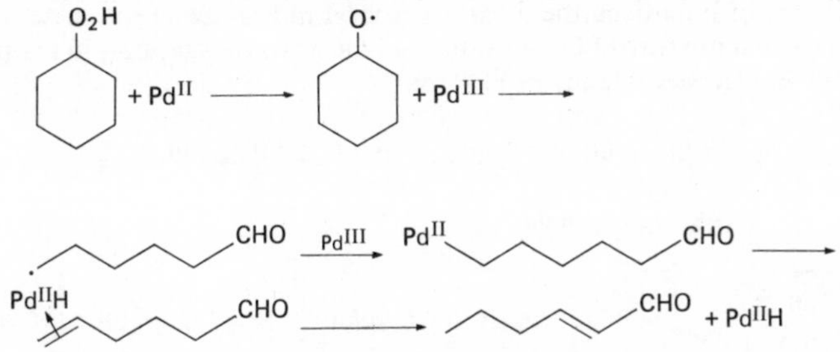

$$\text{C}_6\text{H}_{11}\text{O}_2\text{H} + \text{Pd}^{\text{II}} \longrightarrow \text{C}_6\text{H}_{11}\text{O}\cdot + \text{Pd}^{\text{III}} \longrightarrow$$

$$\cdot\text{CH}_2(\text{CH}_2)_4\text{CHO} \xrightarrow{\text{Pd}^{\text{III}}} \text{Pd}^{\text{II}}(\text{CH}_2)_5\text{CHO} \longrightarrow$$

$$\text{Pd}^{\text{II}}\text{H}\text{–alkene–CHO} \longrightarrow \text{CH}_3\text{CH}_2\text{CH}_2\text{CH=CHCH}_2\text{CHO} + \text{Pd}^{\text{II}}\text{H}$$

SCHEME 7

An interesting series of synthetically useful reactions has been developed based on the redox decomposition of the 1-hydroxycycloalkyl hydroperoxides formed *in situ* by the reaction of hydrogen peroxide with cycloalkanones. For example, 1-hydroxycyclohexyl hydroperoxide reacts with ferrous sulphate in acidic solutions to produce the 5-carboxypentyl radical (reaction 58). In the absence of reactive substrates the latter dimerizes to give dodecanedioic acid in good yield (reaction 59)[60].

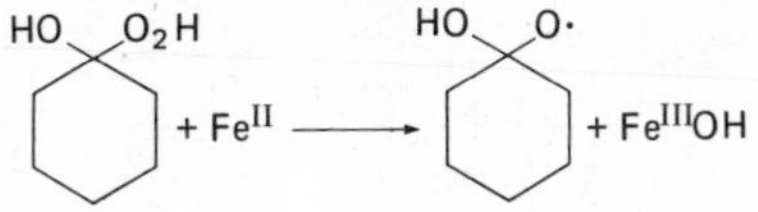

$$\text{1-HO-C}_6\text{H}_{10}\text{-1-O}_2\text{H} + \text{Fe}^{\text{II}} \longrightarrow \text{1-HO-C}_6\text{H}_{10}\text{-1-O}\cdot + \text{Fe}^{\text{III}}\text{OH} \quad (58)$$

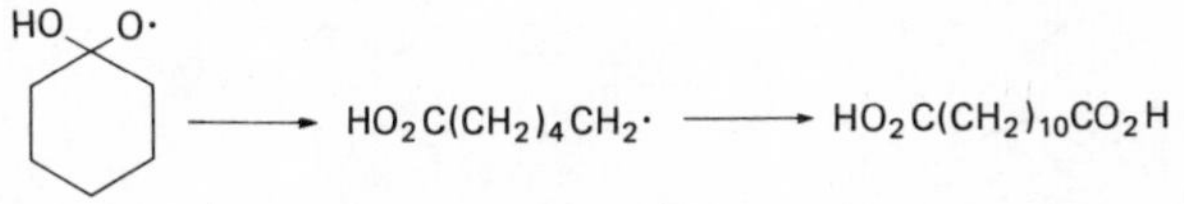

$$\text{1-HO-C}_6\text{H}_{10}\text{-1-O}\cdot \longrightarrow \text{HO}_2\text{C(CH}_2)_4\text{CH}_2\cdot \longrightarrow \text{HO}_2\text{C(CH}_2)_{10}\text{CO}_2\text{H} \quad (59)$$

When the reaction is carried out in the presence of copper(II) sulphate or copper(II) chloride the intermediate 5-carboxypentyl radicals are intercepted to afford 5-hexenoic acid (electron transfer) (reaction 60) or 6-chlorohexanoic acid (ligand transfer) (reaction 61), respectively[61,62].

$$\text{HO}_2\text{C(CH}_2)_4\text{CH}_2\cdot \xrightarrow{\text{Cu}^{\text{II}}} \text{HO}_2\text{C(CH}_2)_3\text{CH=CH}_2 + \text{Cu}^{\text{I}} + \text{H}^+ \quad (60)$$

$$\text{HO}_2\text{C(CH}_2)_4\text{CH}_2\cdot \xrightarrow{\text{Cu}^{\text{II}}\text{Cl}_2} \text{HO}_2\text{C(CH}_2)_4\text{CH}_2\text{Cl} + \text{CuCl} \quad (61)$$

Similarly, the reaction can be carried out in the presence of other anions that can take part in ligand-transfer oxidation of the intermediate alkyl radicals to give a variery of ω-hexanoic acid derivatives. When the reaction is carried out in methanol at acidic pH the corresponding methyl esters are obtained according to the general equation (62)[61–63].

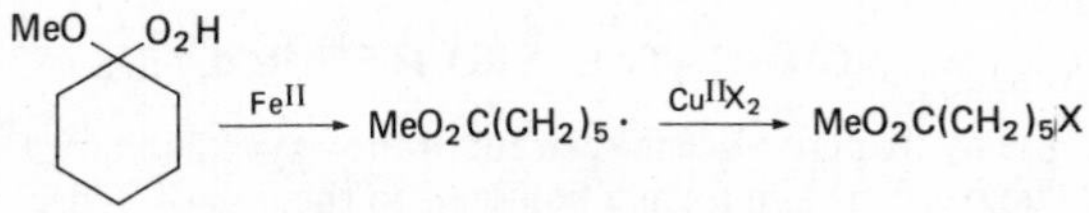

$$\text{1-MeO-C}_6\text{H}_{10}\text{-1-O}_2\text{H} \xrightarrow{\text{Fe}^{\text{II}}} \text{MeO}_2\text{C(CH}_2)_5\cdot \xrightarrow{\text{Cu}^{\text{II}}\text{X}_2} \text{MeO}_2\text{C(CH}_2)_5\text{X} \quad (62)$$

X = Cl, Br, I, CN, SCN, N_3, etc.

In the presence of butadiene the 5-carboxypentyl radical generates an allylic radical which dimerizes to a mixture of C_{20} unsaturated dicarboxylic acids or, in the presence of copper(II) salts, undergoes ligand- or electron-transfer oxidation (Scheme 8)[62].

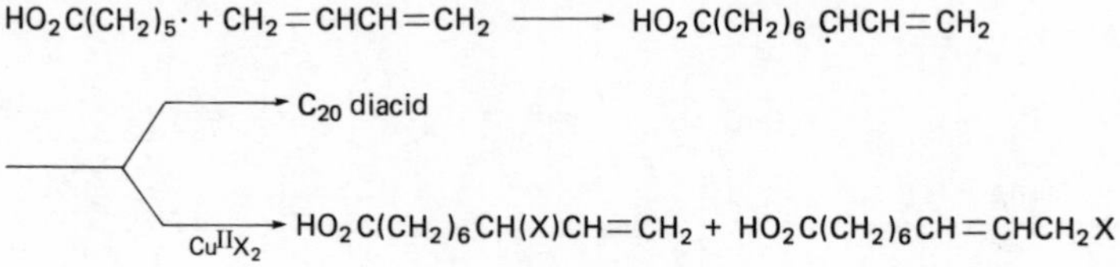

$$HO_2C(CH_2)_5\cdot + CH_2{=}CHCH{=}CH_2 \longrightarrow HO_2C(CH_2)_6\dot{C}HCH{=}CH_2$$

$\longrightarrow C_{20}$ diacid

$\xrightarrow{Cu^{II}X_2} HO_2C(CH_2)_6CH(X)CH{=}CH_2 + HO_2C(CH_2)_6CH{=}CHCH_2X$

SCHEME 8

By variation of the ketone, alkene and the catalyst these reactions have been utilized by Minisci and coworkers[64,65] for the synthesis of a wide variety of polyfunctional long-chain molecules.

An interesting recent application of this type of reaction is the synthesis of the naturally occurring macrolide recifeiolide (**2**), in 96% yield, by $FeSO_4/Cu(OAc)_2$-catalysed decomposition of the alkoxy hydroperoxide (**1**) in methanol (see Scheme 9)[66].

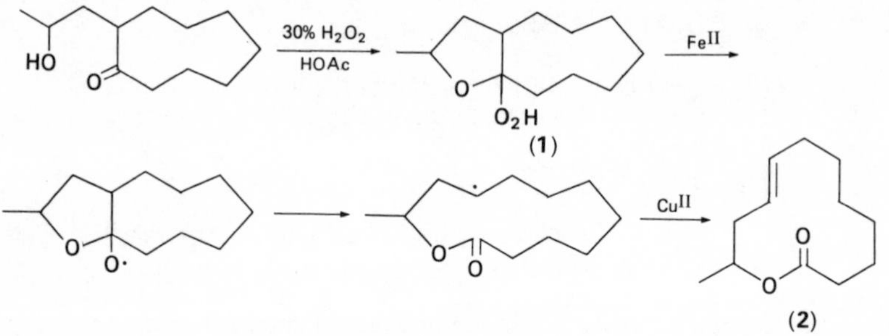

30% H_2O_2, HOAc; Fe^{II}; (**1**); Cu^{II}; (**2**)

SCHEME 9

The reaction has also been applied[66] in a simple synthesis of (+)-6-methylcyclohex-2-enone (**4**) from (−)carvone (**3**) (reaction 63).

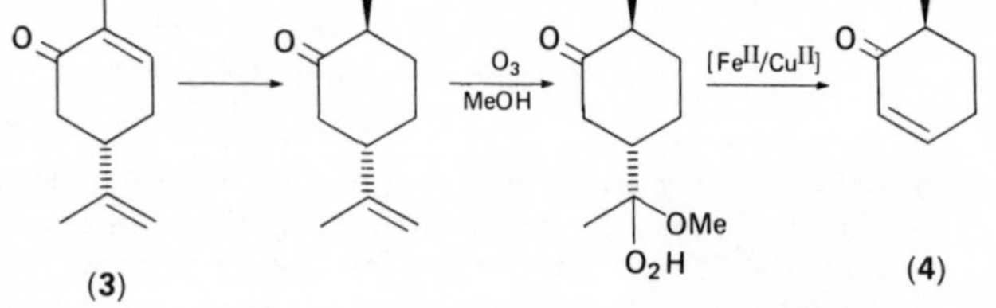

O_3, MeOH; $[Fe^{II}/Cu^{II}]$; OMe; O_2H; (**3**); (**4**) (63)

V. ACID-CATALYSED HETEROLYSIS

Protonation of alkyl hydroperoxides may occur at the oxygen adjacent to the alkyl group or at the hydroxylic oxygen. The former can lead to the loss of hydrogen peroxide from those hydroperoxides which form a stable carbenium ion (reaction 64). Protonation at the hydroxyl group, on the other hand, can lead to O—O heterolysis accompanied by rearrangement (Scheme 10).

$$RO_2H + H^+ \rightleftharpoons R-\overset{+}{O}(H)-OH \rightleftharpoons R^+ + H_2O_2 \quad (64)$$

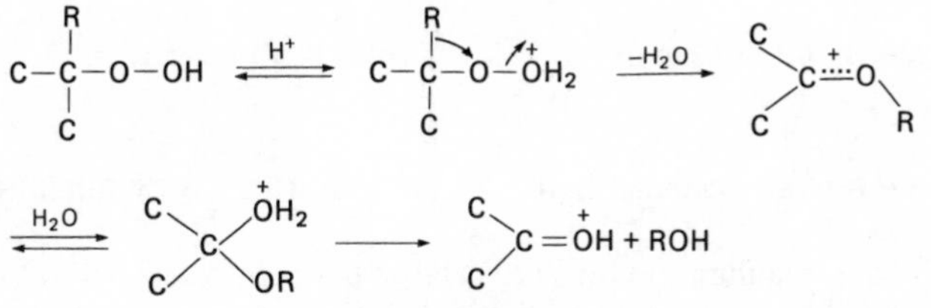

$$C-C(R)(C)-O-OH \overset{H^+}{\rightleftharpoons} C-C(R)(C)-O-\overset{+}{O}H_2 \xrightarrow{-H_2O} C_2C\overset{+}{=}O-R$$

$$\overset{H_2O}{\rightleftharpoons} C_2C(\overset{+}{O}H_2)(OR) \longrightarrow C_2C=\overset{+}{O}H + ROH$$

SCHEME 10

The ease of acid-catalysed heterolysis depends on the ability of the R group to undergo a 1,2-shift from carbon to an incipiently positive oxygen. It is favourable when R is hydrogen, aryl or branched alkyl but not when R is methyl as in *t*-butyl hydroperoxide. The intermediate alkoxycarbenium ions have been observed by proton magnetic resonance spectroscopy in FSO_3H/SbF_5 at $-40°C$[67].

The reaction provides a useful method for the preparation of a wide variety of phenols from the appropriate isopropylbenzene via autoxidation and acid-catalysed heterolysis (see earlier)[68] (reaction 65).

$$ArCHMe_2 \xrightarrow{O_2} Ar(Me)_2C-O-OH \xrightarrow{H^+} ArOH + Me_2CO \quad (65)$$

Highly branched alkyl hydroperoxides undergo facile acid cleavage. For example, the acid-catalysed reaction of hydrogen peroxide with 2,4,4-trimethylpentene-1-(di-isobutylene) provides a useful synthesis of neopentyl alcohol[69] (reaction 66).

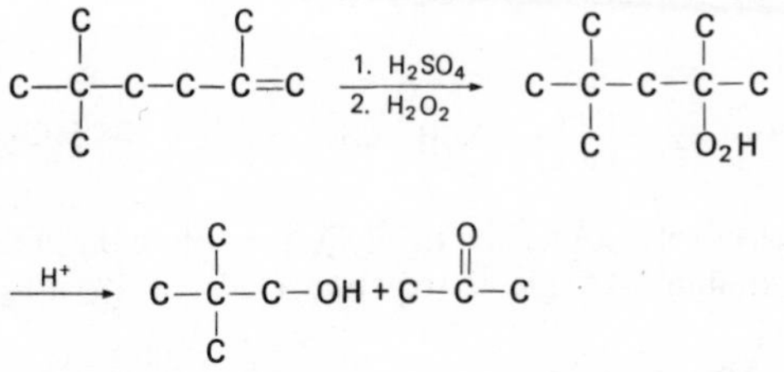

$$C-C(C)(C)-C-C(C)=C \xrightarrow[2.\ H_2O_2]{1.\ H_2SO_4} C-C(C)(C)-C-C(C)(O_2H)-C$$

$$\xrightarrow{H^+} C-C(C)(C)-C-OH + C-C(=O)-C \quad (66)$$

VI. METAL-CATALYSED HETEROLYSIS

A. Epoxidation of Alkenes

The single most important synthetic application of alkyl hydroperoxides is without doubt the metal-catalysed epoxidation of alkenes[42,43] (reaction 67). The reaction is catalysed by certain high-valent metals such as molybdenum (VI), tungsten (VI), vanadium (V) and titanium (IV). Molybdenum (VI) compounds are particularly effective catalysts. Propylene oxide is currently manufactured on a large scale by the molybdenum-catalysed epoxidation of propylene with TBHP (reaction 68) or EBHP (reaction 69)[70]. The latter are prepared by autoxidation of isobutane or ethylbenzene, respectively.

$$\mathrm{C{=}C + RO_2H \longrightarrow \overset{O}{C{-}C} + ROH} \tag{67}$$

$$\mathrm{MeCH{=}CH_2 + \mathit{t}\text{-}BuO_2H \longrightarrow Me\overset{O}{CH{-}CH_2} + \mathit{t}\text{-}BuOH} \tag{68}$$

$$\mathrm{MeCH{=}CH_2 + PhCH(Me)O_2H \longrightarrow Me\overset{O}{CH{-}CH_2} + PhCH(Me)OH} \tag{69}$$

A heterogeneous, titanium (IV)-on-silica catalyst has been developed by Shell[43,71]. The catalyst is prepared by impregnating silica with $TiCl_4$ or an organotitanium compound, followed by calcination. Since the catalyst is insoluble in the reaction medium it can be effectively employed in continuous, fixed-bed operation and propylene oxide selectivities of 93–94% at 96% EBHP conversion have been claimed[43]. By comparison, the homogeneous molybdenum catalyst affords 90% selectivity at 92% EBHP conversion under comparable conditions.

The *t*-butyl alcohol co-product formed in reaction (68) can be recycled via dehydration and subsequent hydrogenation or converted to methyl *t*-butyl ether, an octane booster for gasoline. The methylphenylcarbinol co-product formed in reaction (69) is dehydrated to styrene which is sold as such or recycled.

Alkyl hydroperoxides in combination with homogeneous (Mo, V) and heterogeneous (Ti^{IV}/SiO_2) catalysts form a versatile group of reagents for the epoxidation of alkenes in general. The reactions are performed in hydrocarbon solvents at moderate temperatures (80–120°C) and afford epoxides in very high selectivities. Epoxides are versatile chemical intermediates and many industrial applications have been envisaged for the hydroperoxide process[72–74].

The epoxidation of terminal alkenes followed by catalytic hydrogenation over a nickel catalyst, for example, provides a route to long-chain primary alcohols (reaction 70).

$$\mathrm{RCH{=}CH_2 \xrightarrow[\text{[Mo^{VI}]}]{\text{TBHP}} R\overset{O}{CH{-}CH_2} \xrightarrow[\text{[Ni]}]{H_2} RCH_2CH_2OH} \tag{70}$$

Epoxides can be selectively isomerized to allylic alcohols in the presence of basic lithium phosphate[75] or aluminium alkoxides[76] (reaction 71).

$$\mathrm{RCH_2\overset{O}{CH{-}CH_2} \xrightarrow[\text{or } Al(OR)_3]{Li_2HPO_4} RCH{=}CHCH_2OH + RCH(OH)CH{=}CH_2} \tag{71}$$

Treatment with Brønsted acids or Lewis acids such as $MgBr_2$, on the other hand, generally results in isomerization to ketones[77]. In one instance, the direct conversion of an alkene to a ketone has been reported. Thus, molybdenum–catalysed reaction of $TBHP/H_2O$ with 2-methylbutene-2 at 155°C affords methylisopropyl ketone in high yield[78] (reaction 72). In this case the molybdenum(VI), probably present as molybdic acid, not only catalyzes the epoxidation but also the subsequent isomerization.

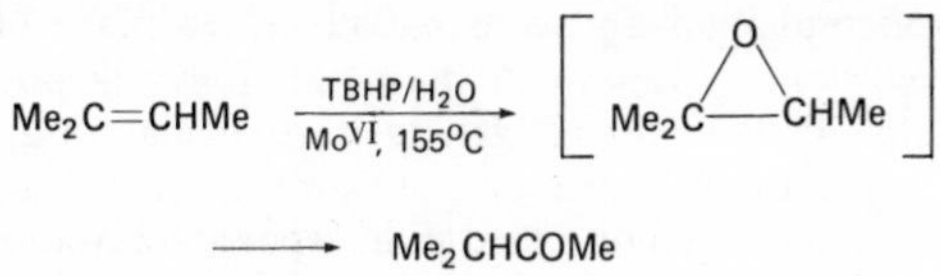

$$\mathrm{Me_2C{=}CHMe} \xrightarrow[\mathrm{Mo^{VI},\ 155^{\circ}C}]{\mathrm{TBHP/H_2O}} \left[\mathrm{Me_2C\overset{O}{—}CHMe}\right] \longrightarrow \mathrm{Me_2CHCOMe} \qquad (72)$$

TBHP offers many advantages over the traditional organic peracids for use in organic synthesis (see earlier). TBHP–metal-catalyst reagents are particularly useful for the epoxidation of acid-sensitive alkenes and those which contain a functional group which reacts with a peracid. An example of the latter is the selective epoxidation of citral with TBHP in the presence of Ti^{IV}/SiO_2 [43] (reaction 73).

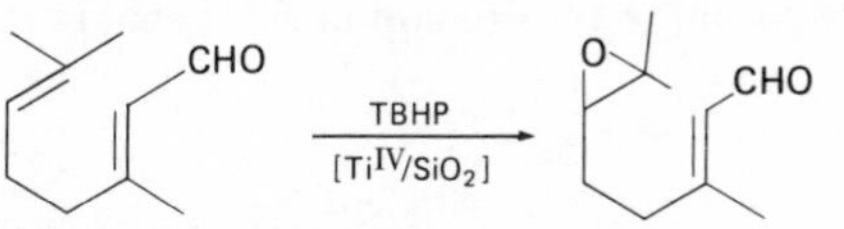

$$\xrightarrow[\mathrm{[Ti^{IV}/SiO_2]}]{\mathrm{TBHP}} \qquad (73)$$

Sharpless and Verhoeven[5] have reported a detailed procedure for epoxidations with $TBHP/Mo(CO)_6$ which affords 85–95% isolated yields with a variety of nonfunctionalized alkenes. In agreement with the electrophilic nature of the epoxidizing agent the rate of epoxidation increases with increasing substitution of the double bond with alkyl groups (ca. a factor of 10 per alkyl group)[79]. This is reflected in the selective monoepoxidation of nonconjugated dienes, e.g. reactions (74) and (75)[4,73].

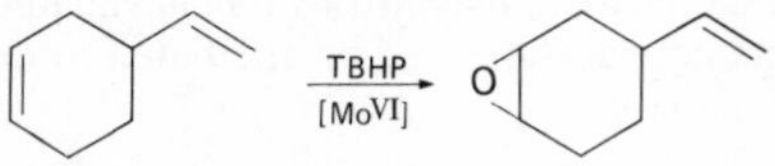

$$\xrightarrow[\mathrm{[Mo^{VI}]}]{\mathrm{TBHP}} \qquad (74)$$

$$\xrightarrow[\mathrm{[Mo^{VI}]}]{\mathrm{TBHP}} \qquad (75)$$

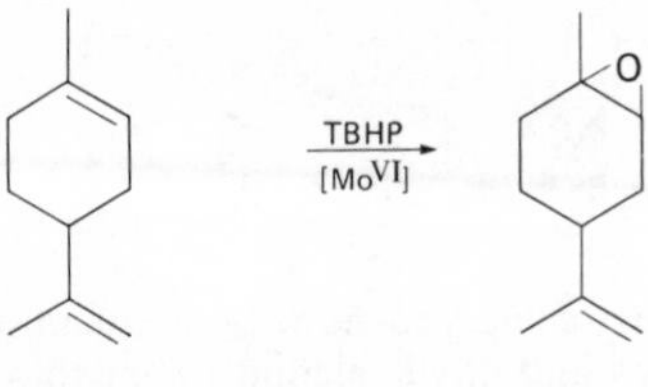

Conjugated dienes are generally less reactive than the analogous compounds containing isolated double bonds (e.g. isoprene is less reactive than 2-methylbutene-2), and they are selectively epoxidized to monoepoxides. Electron-withdrawing groups also retard the rate of epoxidation. Allyl chloride, for example, is about one tenth as reactive as propylene[79]. Acrylic esters and acrylonitrile are unreactive towards these reagents. Epoxidation is not seriously impeded, however, when the electron-withdrawing group is sufficiently removed from the double bond, e.g. 4-cyanocyclohexene is epoxidized in 88% yield[73] (reaction 76).

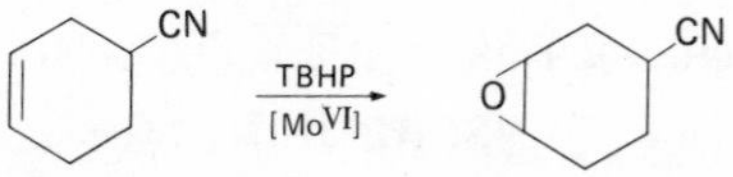

$$\xrightarrow[\mathrm{[Mo^{VI}]}]{\mathrm{TBHP}} \qquad (76)$$

In the epoxidation of nonfunctionalized alkenes only minor differences in regio- and stereo-selectivity are observed, and expected, between the TBHP/metal catalyst reagents and organic peracids. Alkenes containing functional groups, on the other hand, can produce completely different regio- and stereo-selectivities with the TBHP/metal reagents. Orientation of the alkene by coordination to the metal catalyst through a functional group can lead to preferential delivery of oxygen to a particular double bond in a diene (regioselectivity) or to a particular face of a substrate (stereoselectivity). For example, quite remarkable results have been obtained in the epoxidations of unsaturated alcohols with these reagents[5,42,43].

List and Kuhnen[80] were the first to observe that allylic alcohols give excellent yields of epoxides with RO_2H/V_2O_5. Subsequently, Sheng and Zajacek[4] noted, in a study of epoxidations of functionalized alkenes, that allylic alcohols give unexpected results. Thus, with simple alkenes molybdenum-catalysed epoxidation is ca. 10^2 times faster than vanadium-catalysed epoxidation, whereas with allylic alcohols vanadium gives higher rates and better yields than molybdenum. This difference is reflected in the different regioselectivities observed in the epoxidation of the dienol in equation (77)[4].

OH —TBHP [Cat.]→ (epoxide) OH + OH (epoxide) (77)

$VO(acac)_2$	4	:	1
$Mo(CO)_6$	1	:	1

The exceptional reactivity of allylic alcohols towards the vanadium(v)–alkyl hydroperoxide reagent is due to an efficient intramolecular oxygen transfer from the coordinated alkylperoxy group to the double bond of an allylic alcohol coordinated through its alcohol group as shown in reaction (78). That vanadium catalysts in particular are able to cause such rate accelerations can be attributed to the strong coordination of alcohol ligands to vanadium(v).

(78)

The different reactivity ratios of molybdenum- and vanadium- alkyl hydroperoxide reagents towards simple alkenes and allylic alcohols are exploited in the conversion of propylene to glycidol, and glycerol by subsequent hydrolysis (Scheme 11)[81].

$$CH_3CH{=}CH_2 \xrightarrow[\text{[Mo}^{VI}]]{\text{TBHP}} CH_3\overset{O}{CH{-}CH_2} \xrightarrow{Li_2HPO_4}$$

$$CH_2{=}CHCH_2OH \xrightarrow[\text{[V}^{V}]]{\text{TBHP}} \overset{O}{CH_2{-}CHCH_2OH} \xrightarrow{H_2O}$$

$$HOCH_2CH(OH)CH_2OH$$

SCHEME 11

The exceptionally facile epoxidation of allylic alcohols by vanadium–alkyl hydroperoxide reagents has been utilized for the regioselective epoxidation of a wide variety of complex organic molecules[5,42,43]. For example geraniol(**5**) and linalool(**7**) are selectively epoxidized to the previously unknown monoepoxides (**6**) and (**8**), respectively, with TBHP/$VO(acac)_2$ [82] (reactions 79 and 80). Such regioselectivities are not possible with other epoxidation agents. The reactivity of the TBHP/$VO(acac)_2$ reagent is underscored by the fact that reactions (79) and (80) proceed readily at ambient temperature. High regioselectivities have also been observed with homoallylic and bishomoallylic alcohols[5,42,43].

TBHP, [$VO(acac)_2$]: (**5**) → (**6**) 93% (98%) (79)

TBHP, [$VO(acac)_2$]: (**7**) → (**8**) 84% (95%) (80)

The transition metal–alkyl hydroperoxide reagents also exhibit remarkable stereoselectivities resulting from the preferential *syn* transfer of oxygen within the ternary metal–alkene-hydroperoxide complex. A comparison of the stereoselectivities of epoxidation of cyclohexen-3-ol with $Mo(CO)_6$/TBHP, $VO(acac)_2$/TBHP and perbenzoic acid (PBA) reveals that the Mo and V-catalysed reactions are virtually stereospecific[82] (reaction 81). They also show rate enhancements, compared to cyclohexene, of 4.5 and 200, respectively. PBA, on the other hand, shows no rate enhancement and is less stereoselective. Differences in stereoselectivity are even more pronounced with the homoallylic alcohol, cyclohexen-4-ol[82] (reaction 82). The rate enhancements in reaction (82), compared to cyclohexene, are ca. 10 for both Mo and V, making Mo the catalyst of choice because of its much higher rate with cyclohexene. There is evidence that suggests that the stereoselectivities of these reactions decrease with increasing alkene conversion[42,83,84]. This can probably be attributed to competing coordination of the product alcohol to the catalyst hindering coordination of the substrate through its alcohol group. The *syn*-directive effect in these systems results from preferential transfer of oxygen to one face of the substrate within a ternary metal–substrate–hydroperoxide complex. Teranishi and coworkers[85] have studied the epoxidation of cyclic allylic alcohols of

(81)

	syn	*anti*
TBHP/$Mo(CO)_6$	98%	2%
TBHP/$VO(acac)_2$	98%	2%
PBA	92%	8%

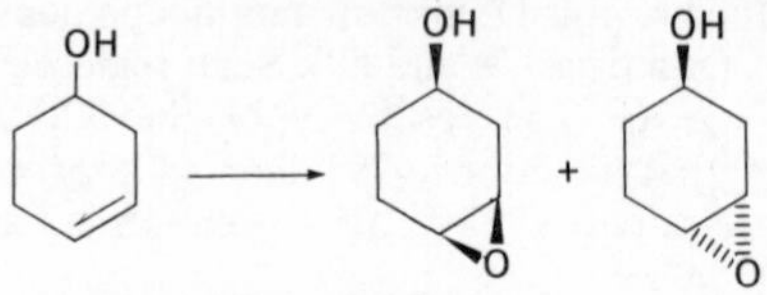

(82)

TBHP/Mo(CO)$_6$	98%	2%
TBHP/VO(acac)$_2$	98%	2%
PBA	60%	40%

varying ring size. They have found that with 5- and 6-membered rings both TBHP/VO(acac)$_2$ and *m*-chloroperbenzoic acid (MCPBA) afford predominantly the *syn* epoxide. With medium-ring alcohols, on the other hand, MCPBA affords predominantly the *anti* epoxide, whilst VO(acac)$_2$/TBHP gives the *syn* isomer. These results have been rationalized on the basis of different transition-state geometries. Epoxidation with VO(acac)$_2$/TBHP and MCPBA involves a 5.5-membered (**9**) and 6.5-membered (**10**) transition state, respectively.

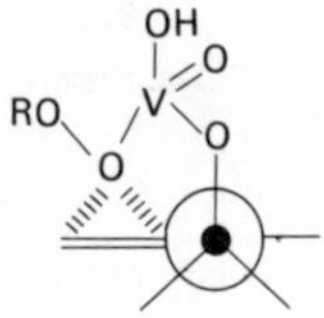

(**9**) 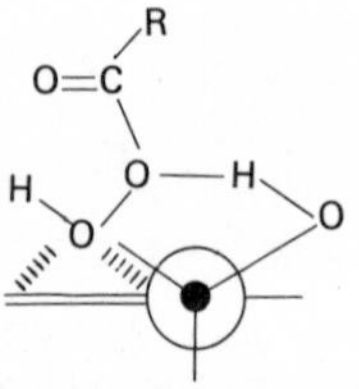 (**10**)

Stereoselectivity is not restricted to cyclic unsaturated alcohols. Thus, the highly stereoselective epoxidations of a series of acyclic allylic alcohols, to give predominantly the *erythro* epoxy alcohol have been reported,[86,87] e.g. reaction (83).

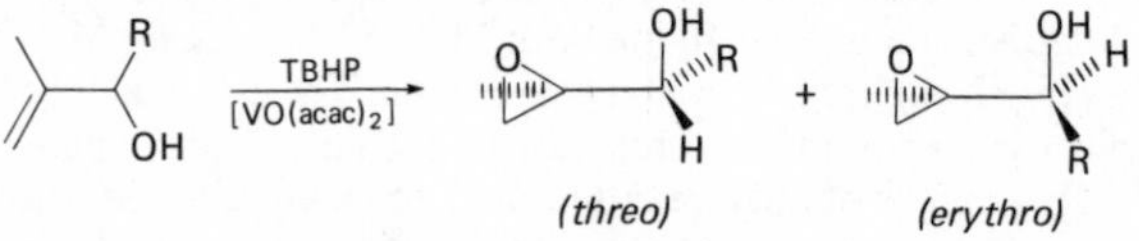

(83)

	(threo)	(erythro)
R = Me	5%	95%
R = *n*-Bu	2%	98%

Allylic alcohols have also proved to be suitable substrates for effecting asymmetric epoxidations with alkyl hydroperoxide/metal catalyst reagents. In initial studies chiral vanadium[88] and molybdenum[89] catalysts have been used to obtain 30–50 % and 13–33 % enantiomeric excesses, respectively, with a variety of allylic alcohols. More recently, Katsuki and Sharpless[90] have reported that epoxidations of allylic alcohols with TBHP–titanium isopropoxide, in the presence of (+)- or (−)-diethyl tartrate (DET) as the chiral ligand, afford enantiomeric excesses consistently greater than 90 %, and in many cases, greater than 95 %. Although the method employs stoichiometric quantities of the titanium isopropoxide and the chiral ligand it has been pointed out that these reagents are commercially available at low to moderate cost. This novel chiral epoxidizing agent exhibits two especially striking features. Firstly, it gives uniformly high asymmetric inductions throughout a range of substitution patterns in the allylic alcohol substrate. Secondly, the oxygen is always delivered from the same enantioface of the alkene when a particular tartrate enantiomer is used, as illustrated in reaction (84).

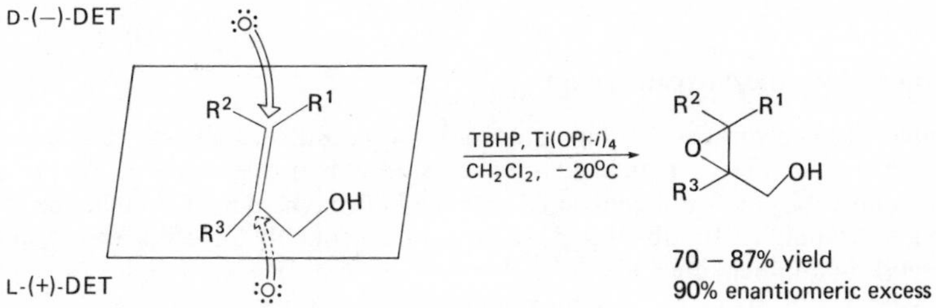

D-(−)-DET

L-(+)-DET

TBHP, $Ti(OPr\text{-}i)_4$

CH_2Cl_2, −20°C

70 − 87% yield
90% enantiomeric excess

(84)

Since chiral epoxides constitute ideal building blocks for asymmetric synthesis of complex molecules this method is likely to be widely applied in the future. The Sharpless group[91] has utilized the method in the synthesis of (+)-disparlure (**11**), the sex attractant of the gypsy moth. (Scheme 12).

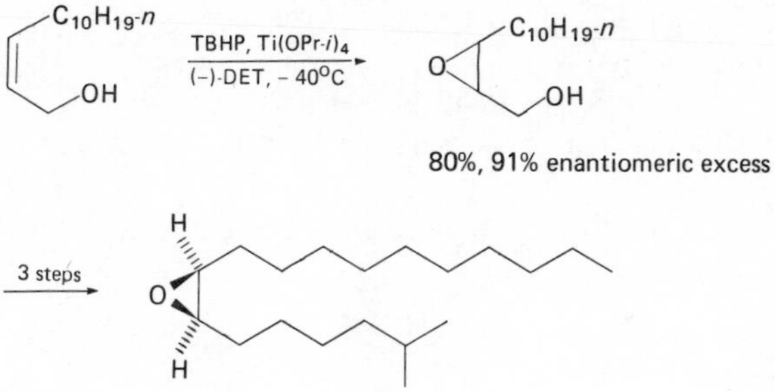

TBHP, $Ti(OPr\text{-}i)_4$

(−)-DET, − 40°C

80%, 91% enantiomeric excess

3 steps

(**11**)

SCHEME 12

Similarly the chiral epoxy alcohols **12**, **13** and **14**, which are key intermediates in the syntheses of methymycin, erythromycin and leukotriene C-1, respectively, have been prepared enantioselectively[91].

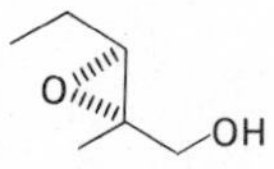

(**12**) 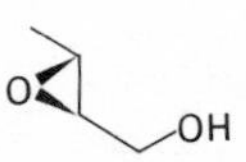 (**13**) 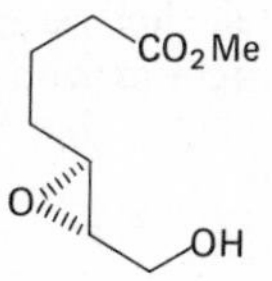 (**14**)

A major drawback of the systems described above is the fact that asymmetric epoxidation is restricted to allylic alcohols as substrates. They are not successful with nonfunctionalized alkenes. Otsuka and coworkers[92] have reported the asymmetric epoxidation of simple alkenes such as 1-methyl cyclohexene, with the TBHP/Mo^{VI}/tartrate system. Enantiomeric excesses of less than 10% have been observed which emphasizes the need for a more efficient system for the asymmetric epoxidation of simple, nonfunctionalized alkenes.

B. Alkene Hydroxyketonization

Tolstikov and coworkers[93,94] have shown that trisubstituted alkenes can be converted into α-hydroxyketones by reaction with excess alkyl hydroperoxide in the presence of molybdenum catalysts, e.g. α-pinene affords the hydroxvketone **15**[93] (reaction 85). The reaction is limited to trisubstituted alkenes and probably involves the sequence of reactions shown in Scheme 13.

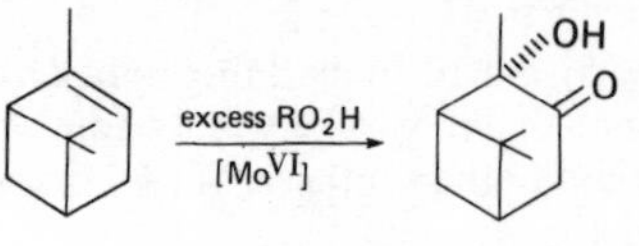

excess RO_2H, $[Mo^{VI}]$ (85)

(15)

$+ RO_2H \xrightarrow{[Mo^{VI}]}$ ... O_2R ... OH

→ ... O· ... OH + RO· → ... O ... OH

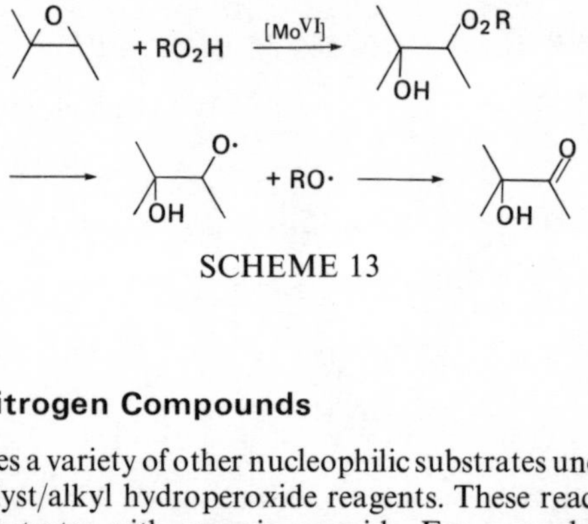

SCHEME 13

C. Oxidation of Nitrogen Compounds

In addition to alkenes a variety of other nucleophilic substrates undergo oxygen transter with these metal catalyst/alkyl hydroperoxide reagents. These reactions closely parallel those of the same substrates with organic peracids. For example, tertiary amines are smoothly oxidized to the corresponding *N*-oxides with TBHP (reaction 86) or *t*-amyl hydroperoxide (TAHP) (reaction 87) in the presence of vanadium or molybdenum catalysts[95–98]. Reaction rates and selectivities are often significantly higher than the corresponding reactions with peracids.

$$R_3N \xrightarrow[{[VO(acac)_2]}]{TBHP} R_3NO \qquad (86)$$

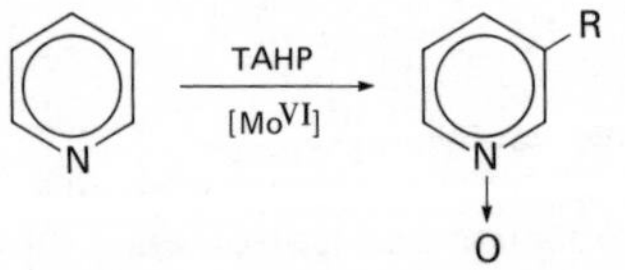

N ($\xrightarrow[{[Mo^{VI}]}]{TAHP}$) R, N → O (87)

Imines are oxidized to the corresponding oxaziridines in 80–95% yield by TAHP in the presence of molybdenum catalysts[96,97] (reaction 88).

$$R^1R^2C{=}N{-}R^3 \xrightarrow[\text{[Mo}^{VI}\text{]}]{\text{TAHP}} R^1R^2C(O)N{-}R^3 \qquad (88)$$

The oxidation of primary aliphatic amines with metal catalyst/hydroperoxide reagents produces the corresponding hydroxylamines or oximes depending on the amount of hydroperoxide used and the temperature[99]. Titanium(IV) compounds are the catalysts of choice for these reactions[99] (reaction 89).

$$>CHNH_2 \xrightarrow[\text{[Ti}^{VI}\text{]}]{RO_2H} >CHNHOH \xrightarrow[\text{[Ti}^{VI}\text{]}]{RO_2H} >C{=}NOH + ROH + H_2O \qquad (89)$$

A process has been envisaged[99,100] for the co-production of cyclohexanone oxime and styrene as shown in reaction (90).

$$C_6H_{11}{-}NH_2 + PhCH(CH_3)O_2H \xrightarrow{\text{[Ti}^{VI}\text{]}} C_6H_{10}{=}NOH + PhCH(CH_3)OH \longrightarrow PhCH{=}CH_2 + H_2O \qquad (90)$$

The product of oxidation of anilines by TBHP depends on the metal catalysts used. Molybdenum and vanadium catalysts give nitrobenzene[101] (reaction 91) whilst titanium catalysts afford azoxybenzene[102] (reaction 92).

$$ArNH_2 + t\text{-}BuO_2H \xrightarrow{\text{[Mo}^{VI}\text{, V}^{V}\text{]}} ArNO_2 \qquad (91)$$

$$ArNH_2 + t\text{-}BuO_2H \xrightarrow{\text{[Ti}^{IV}\text{]}} ArN{=}N(\rightarrow O)Ar \qquad (92)$$

TBHP/VO(acac)$_2$ has been used to effect a mild, oxidative conversion of nitroalkanes to carbonyl compounds via hydroxylation of the nitronate anion (Scheme 14)[103]. This method provides a synthetically useful alternative to the Nef reaction which requires strongly acidic conditions[104].

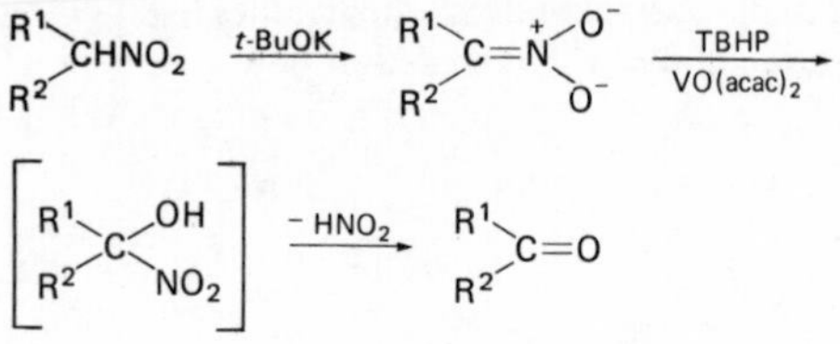

$$\begin{matrix}R^1\\R^2\end{matrix}\!>\!CHNO_2 \xrightarrow{t\text{-BuOK}} \begin{matrix}R^1\\R^2\end{matrix}\!>\!C{=}\overset{+}{N}(O^-)O^- \xrightarrow[VO(acac)_2]{TBHP} \left[\begin{matrix}R^1\\R^2\end{matrix}\!>\!C(OH)(NO_2)\right] \xrightarrow{-HNO_2} \begin{matrix}R^1\\R^2\end{matrix}\!>\!C{=}O$$

SCHEME 14

D. Oxidation of Sulphur Compounds

Sulphides are oxidized to the corresponding sulphoxides with alkyl hydroperoxides in the presence of Mo, W, Ti and V catalysts. In the presence of excess hydroperoxide further oxidation to the sulphone occurs[105–112] (reaction 93).

$$R_2S \xrightarrow[[Mo^{VI},V^{V}]]{TBHP} R_2SO \xrightarrow[[Mo^{VI},V^{V}]]{TBHP} R_2SO_2 \qquad (93)$$

Sulphides are generally oxidized much faster than alkenes. For example, with $TBHP/VO(acac)_2$ in ethanol at 25°C the relative rates decrease in the order

n-Bu_2S (100) > n-BuSPh (58) > n-Bu_2SO (1.7) > cyclohexene (0.2)[107].

This is also reflected in the selective oxidation of unsaturated sulphides at the sulphur atom[106] (reaction 94).

$$CH_3(CH_2)_3SCH_2CH{=}CH_2 \xrightarrow[[MoO_2(acac)_2]]{TBHP} CH_3(CH_2)_3SO_2CH_2CH{=}CH_2 \qquad (94)$$

When the oxidation of unsymmetrical sulphides with $TBHP/VO(acac)_2$ is carried out in a mixture of benzene and a chiral alcohol, such as (−)-menthol, as solvent, asymmetric induction is observed, although enantiomeric excesses are rather low (5–10%)[112].

The oxidation of thiols with TBHP in the presence of molybdenum or vanadium catalysts produces sulphonic acids, presumably via the corresponding sulphenic (RSOH) and sulphinic (RSO_2H) acids as intermediates[113] (reaction 95).

$$RSH + 3\,t\text{-}BuO_2H \xrightarrow{[Mo^{VI}, V^{V}]} RSO_3H + 3\,t\text{-}BuOH \qquad (95)$$

E. Alcohol Oxidations

Tolstikov and coworkers[114] have reported the selective oxidation of secondary alcohols to ketones with TAHP in the presence of molybdenum catalysts at 60–80°C. The mechanism was not discussed. It could involve an intramolecular hydrogen transfer in an alkylperoxymolybdenum(VI) complex as shown in reaction (96). An alternative mechanism involving direct oxidation of the alcohol by oxomolybdenum(VI), analogous to the mechanism proposed for vanadium- catalysed oxidations with TBHP (see Section VII.E), is unlikely under these conditions considering the weak oxidizing power of molybdenum(VI).

The mild oxidation of primary alkoxy alcohols to the corresponding carboxylic acids (reaction 97) with TBHP/NaOH, in the presence of a palladium-on-charcoal catalyst at

$$Mo^{VI}{=}O + RO_2H + \rangle C(H)(OH) \longrightarrow [Mo^{VI}(O)(OOR)\cdots O{-}C(H)\langle] \longrightarrow Mo^{VI}{=}O + ROH + \rangle C{=}O \quad (96)$$

60°C, has been reported[115]. The mechanism of this interesting reaction was not discussed but may be related to the palladium-catalysed oxidation of alkenes with TBHP (see Section VI.G).

$$RO(CH_2CH_2O)_nCH_2CH_2OH \xrightarrow[\text{[Pd/C]}]{\text{TBHP/NaOH}} RO(CH_2CH_2O)_nCH_2CO_2H \quad (97)$$

F. Ketone Oxidations

Cyclohexanone and cyclopentanone are oxidized to the corresponding lactones by TBHP in the presence of boron catalysts, such as B_2O_3 (reaction 98)[116]. The reaction is analogous to the Baeyer–Villiger oxidation of the same ketones with organic peracids and, presumably, involves an alkylperoxyboron(III) intermediate as the active oxidant. In contrast, oxidation of cyclohexanone with TBHP in the presence of formic acid as catalyst is reported to give 2-hydroxycyclohexanone (reaction 99)[117].

$$\text{cyclohexanone} + t\text{-BuO}_2H \xrightarrow{[B_2O_3]} \text{caprolactone} + t\text{-BuOH} \quad (98)$$

$$\text{cyclohexanone} + t\text{-BuO}_2H \xrightarrow{[HCO_2H]} \text{2-hydroxycyclohexanone} + t\text{-BuOH} \quad (99)$$

An alternative approach to the α-hydroxylation of ketones is via epoxidation of the corresponding enol ester with the TBHP/molybdenum reagent (reaction 100). These reactions have been reported to proceed smoothly in high yield[118].

$$\text{cyclohexanone} \longrightarrow \text{1-acetoxycyclohexene} \xrightarrow[\text{[Mo}^{VI}]]{\text{TBHP}} \text{1-acetoxy-1,2-epoxycyclohexane} \xrightarrow{\text{NaOH}} \text{2-hydroxycyclohexanone} \quad (100)$$

G. Alkene Ketonization

Mimoun and coworkers[45,119] have shown that terminal alkenes are selectively oxidized to the corresponding methyl ketones by *t*-butylperoxypalladium(II) complexes at ambient temperatures. For example, the reaction of *t*-butylperoxypalladium(II) trifluoroacetate (**16**) with 1-hexene, in benzene at 20°C for less than 10 minutes, affords 2-hexanone in 98% yield (reaction 101). Since (**16**) can be regenerated by reaction of **17** with TBHP the reaction can be carried out with a catalytic amount of $Pd(O_2CF_3)_2$ (reaction 102). Although reaction (102) bears a formal resemblance to the Wacker oxidation[120] of alkenes

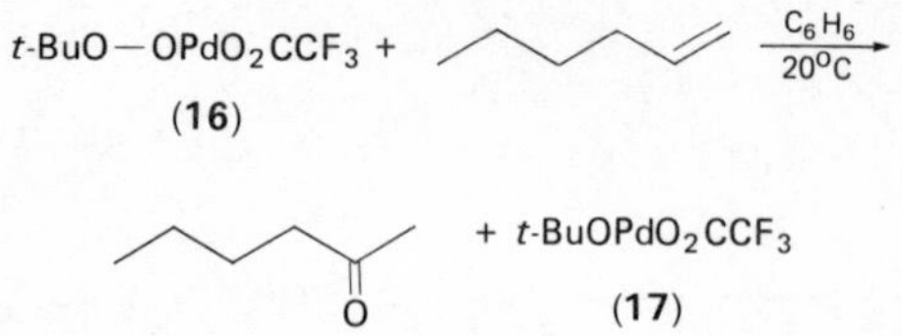

$$t\text{-BuO}-\text{OPdO}_2\text{CCF}_3 + \text{(1-hexene)} \xrightarrow[20^{\circ}C]{C_6H_6} \text{(2-hexanone)} + t\text{-BuOPdO}_2\text{CCF}_3 \quad (101)$$

(16) (17)

$$RCH{=}CH_2 + t\text{-BuO}_2H \xrightarrow{[Pd(O_2CCF_3)_2]} RCOCH_3 + t\text{-BuOH} \quad (102)$$

to ketones with molecular oxygen in the presence of palladium(II) catalysts it involves a completely different mechanism. In the Wacker process the oxygen source is water and oxidation involves external nucleophilic attack of water on the coordinated alkene. Reaction (102), on the other hand, is inhibited by water and other hydroxylic solvents. The results are compatible with a mechanism involving an alkylperoxyalkylpalladium(II) intermediate, formed via nucleophilic attack of the *t*-butylperoxy group on the coordinated alkene (Scheme 15).

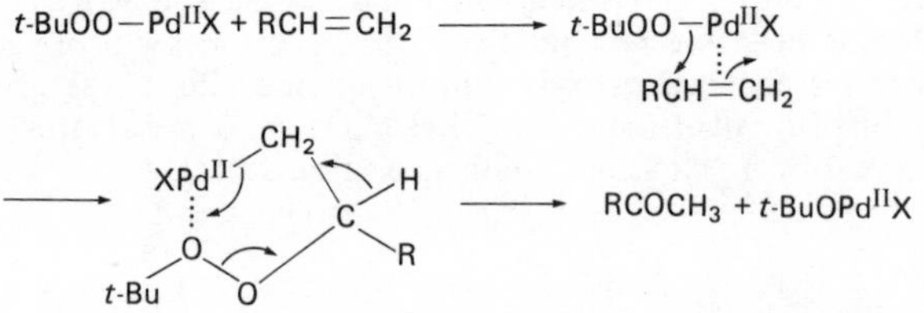

$$t\text{-BuOO}-\text{Pd}^{II}X + RCH{=}CH_2 \longrightarrow t\text{-BuOO}-\text{Pd}^{II}X \cdots RCH{=}CH_2$$

$$\longrightarrow [XPd^{II}-CH_2-C(H)(R)-O-O(t\text{-Bu})] \longrightarrow RCOCH_3 + t\text{-BuOPd}^{II}X$$

$X = CF_3CO_2$

SCHEME 15

Tsuji and coworkers[121] have similarly reported the selective oxidation of α,β-unsaturated ketones and esters to the corresponding 1,3-diketones and β-keto esters, respectively, with TBHP in aqueous acetic acid using Na_2PdCl_4 as the catalyst, e.g. reaction (103). This constitutes a useful synthetic method for converting readily available α,β-unsaturated carbonyl compounds to 1,3-dicarbonyl compounds, reactions which proceed in poor yields under normal Wacker conditions. These authors assumed that TBHP was involved in the reoxidation of Pd^0 to Pd^{II} in a normal Wacker-type process although the Mimoun mechanism (see above) cannot be excluded on the basis of the experimental data.

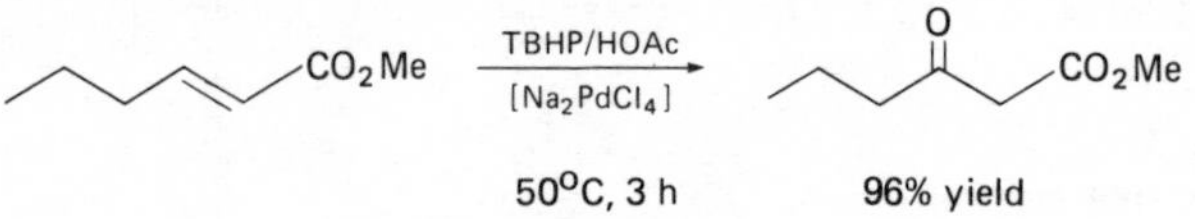

$$CH_3CH_2CH_2CH{=}CHCO_2Me \xrightarrow[\text{[Na}_2\text{PdCl}_4]]{\text{TBHP/HOAc}} CH_3CH_2CH_2COCH_2CO_2Me \quad (103)$$

50°C, 3 h — 96% yield

VII. ALKYL HYDROPEROXIDES AS TERMINAL OXIDANTS

The majority of the reactions discussed in this section involve the oxidation of substrates by oxometal (M=O) oxidants, followed by regeneration of the oxidant with alkyl hydroperoxide (reactions 41 and 42).

A. Allylic Oxidation of Alkenes and Alkynes

Selenium dioxide is the most reliable reagent for the selective allylic oxidation of alkenes[122]. A serious drawback of SeO_2 oxidations in organic syntheses is the formation of colloidal selenium and organoselenium compounds, which are difficult to remove. Umbreit and Sharpless[123] have found that this can be circumvented by using catalytic amounts of SeO_2 in combination with TBHP as the terminal oxidant. The reaction proceeds at ambient temperature in dichloromethane to afford allylic alcohols in moderate to good yields using 2–50% SeO_2 based on TBHP[124]. For example, β-pinene is selectively oxidized in 86% yield[123] (reaction 104).

$$\xrightarrow[\text{TBHP}]{[SeO_2]} \quad \text{OH} \qquad (104)$$

The same group have subsequently found that alkynes undergo a similar oxidation with TBHP in the presence of catalytic amounts of SeO_2 to afford the corresponding acetylenic alcohols[125], e.g. reaction (105).

$$\longrightarrow \quad \text{OH} \quad \text{88\% yield} \qquad (105)$$

Internal alkynes show a pronounced tendency to undergo α,α′-dihydroxylation (reaction 106). With unsymmetrical alkynes the reactivity sequence is $CH_2 \simeq CH > CH_3$ allowing for selective monohdyroxylation in the case of CH_2 or CH vs. CH_3. Relative rate data indicate that alkynes are slightly more reactive than the corresponding alkenes. Alkynes bearing one methylene and one methine substituent afford the enynone as the major product at 80 °C, e.g. reaction (107).

$$RCH_2C{\equiv}CCH_2R \xrightarrow[\text{excess TBHP}]{[SeO_2]} RCH(OH)C{\equiv}CCH(OH)R \qquad (106)$$

$$-C{\equiv}C- \xrightarrow[[SeO_2],\ 80^{\circ}C]{\text{3 mol TBHP}} \text{C}{\equiv}\text{C (OH, OH)} \ (18\%) \; + \; -C{\equiv}C- \text{(=O)} \ (52\%) \qquad (107)$$

B. Vicinal Dihydroxylation of Alkenes and Alkynes

The classical catalytic method for converting alkenes to the corresponding glycols (vicinal dihydroxylation) employs H_2O_2 in the presence of catalytic amounts of osmium tetroxide[126]. OsO_4 reacts with the alkene to form a cyclic osmate ester that undergoes oxidative cleavage with H_2O_2 (reactions 108 and 109).

$$>C=C< + OsO_4 \longrightarrow \text{cyclic osmate ester} \qquad (108)$$

$$\text{cyclic osmate ester} \xrightarrow{H_2O_2} \text{C(OH)-C(OH)} + OsO_4 \qquad (109)$$

A disadvantage of this method is that overoxidation to, *inter alia*, ketols often occurs. This has led to the search for more efficient catalytic procedures and culminated in the development of methods employing TBHP[127–131] or *N*-methylmorpholine oxide[132] as the terminal oxidant. Byers and Hickinbottom[127] were the first to use TBHP in the presence of catalytic amounts of OsO_4. After an interval of more than 25 years, in which this method was completely ignored, Sharpless and coworkers have developed procedures employing OsO_4/TBHP in the presence of Et_4NOH[128] or Et_4NOAc[129], in *t*-butyl alcohol or acetone respectively. It has been suggested[5] that the role of the nucleophile(AcO^- or HO^-) is to increase the turnover rate by facilitating removal of the glycol product from the coordination sphere of the osmium. These new TBHP-based methods are generally more selective than the classical procedure employing H_2O_2. The Et_4NOAc modification is the preferred method with base-sensitive alkenes, e.g. reaction (110).

$$MeCH=CHCO_2Et \xrightarrow[\text{Acetone}]{TBHP/OsO_4/Et_4NOAc} MeCH(OH)CH(OH)CO_2Et \qquad (110)$$

72% yield

Two recent patents[130,131] have similarly described the use of catalytic amounts of OsO_4 in combination with TBHP or other alkyl hydroperoxides for the selective vicinal dihydroxylation of alkenes. The reaction of ethylene or propylene with EBHP in a two-phase aqueous/organic medium in the presence of OsO_4 and an alkali metal hydroxide affords the corresponding glycols in high yield[131] e.g. reaction (111).

$$CH_2{=}CH_2 \xrightarrow{EBHP/OsO_4/CsOH} HOCH_2CH_2OH \qquad (111)$$

84% yield

The analogous OsO_4-catalysed oxidation of alkynes with TBHP affords α-diketones[126] e.g. reaction (112).

$$PhC{\equiv}CMe \xrightarrow{TBHP/OsO_4} PhC(=O)C(=O)Me \qquad (112)$$

62% yield

Cyclic osmate esters have been isolated as intermediates in the reaction of OsO_4L (L = quinuclidine) with alkynes[133] (reaction 113). Hydrolysis of these complexes with aqueous sodium sulphite yields the corresponding diketone from internal alkynes and carboxylic acids, via oxidative cleavage of the initially formed α-keto aldehyde, from terminal alkynes.

$$R^1C{\equiv}CR^2 + 2\,OsO_4L \longrightarrow L{-}Os(=O)_2(O)_2C(R^1)C(R^2)(O)_2Os(=O)_2{-}L \tag{113}$$

C. Alkene Epoxidation

Much attention has been devoted recently to the epoxidation of alkenes with high-valent oxometal porphyrin complexes of iron[134], manganese[135,136] and chromium[137]. These potent oxidants are generated *in situ* by reaction of the trivalent metal salts with an oxygen atom donor, such as iodosyl benzene[134,135,137] (reaction 114) or hypochlorite[136] (reaction 115).

$$(TPP)M^{III} + PhIO \longrightarrow (TPP)M^{V}{=}O + PhI \tag{114}$$

$$(TPP)M^{III} + NaClO \longrightarrow (TPP)M^{V}{=}O + NaCl \tag{115}$$

M = Fe, Mn, Cr; TPP = *meso*-tetraphenylporphyrinato

Interest in the reactions of these oxometal species stems from the fact that an oxoiron(V) porphyrin has been implicated as the active oxidant in reactions mediated by the cytochrome P450-containing monooxygeneases[138]. These enzymes effect a variety of selective oxidations such as alkane hydroxylation and alkene epoxidation[138]. In agreement with the postulated mechanism the model oxometalporphyrin complexes have been shown to selectively epoxidize alkenes at ambient temperature[134–137] (reaction 116).

$$(TPP)M^{V}{=}O + C{=}C \longrightarrow (TPP)M^{III} + C{-}C\ (\text{epoxide, bridging }O) \tag{116}$$

M = Fe, Mn, Cr

Alkyl hydroperoxides should, in principle, also be able to function as oxygen atom donors in a reaction analogous to equations (114) and (115). Indeed, it has been shown that alkyl hydroperoxides can replace the combination of molecular oxygen and hydrogen donor utilized by the enzymes[138]. It is interesting in this context to mention the work of Stautzenberger[139] who has studied the oxidation of cyclohexene with TBHP in the presence of metal phthalocyanine catalysts. He has found, as would be expected, that MoO_2Pc and VOPc (Pc = phthalocyanine) afford high yields of epoxide (see Section VI). An unexpected result is the high selectivity to epoxide (97%) observed with nickel(II) phthalocyanine as catalyst. Nickel(II) falls into the class of metal compounds that generally catalyse the homolytic decomposition of hydroperoxides (Section IV) and not the class of high-valent metal compounds that catalyse epoxidation via a heterolytic mechanism (Section VI). In view of the results obtained with the oxometal porphyrins (*vide supra*) it seems likely that the reaction involves epoxidation by a putative oxonickel(IV) phthalocyanine complex (reactions 117 and 118). Recent preliminary results

$$PcNi^{II} + t\text{-}BuO_2H \longrightarrow PcNi^{IV}{=}O + t\text{-}BuOH \tag{117}$$

$$PcNi^{IV}{=}O + C{=}C \longrightarrow PcNi^{II} + C{-}C\ (\text{epoxide, bridging }O) \tag{118}$$

have revealed that NiPc also catalyses the selective epoxidation of alkenes, such as styrene, with other oxygen atom donors, e.g. NaOCl, PhIO, etc.[140].

It is interesting to note that a mechanism involving an analogous oxopalladium(IV) intermediate as the active oxidant in the palladium(II)-catalysed oxidations of, *inter alia*, alkenes with TBHP (see Section VI) cannot be completely ruled out (Scheme 16).

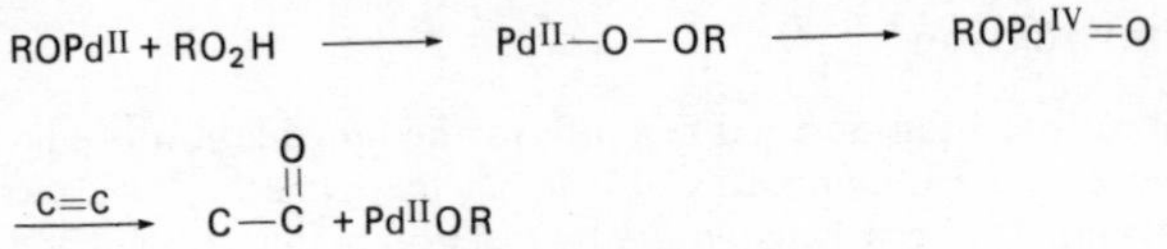

$$ROPd^{II} + RO_2H \longrightarrow Pd^{II}{-}O{-}OR \longrightarrow ROPd^{IV}{=}O$$

$$\xrightarrow{C=C} C{-}\overset{\overset{O}{\|}}{C} + Pd^{II}OR$$

SCHEME 16

D. Alcohol Oxidation

The oxidation of secondary alcohols to ketones has been observed as a side-reaction in the vanadium-catalysed epoxidation of allylic alcohols with TBHP (see Section VI.A)[85]. Oxidation to the enone becomes the predominant reaction with conformationally rigid allylic alcohols that are unable to attain the transition state for intramolecular oxygen transfer to the double bond in the vanadium(V)/TBHP/substrate complex (see Section VI.A). For example, *cis*-5-*t*-butylcyclohex-2-en-1-ol affords the corresponding enone as the major product in 74% yield[85] (reaction 119).

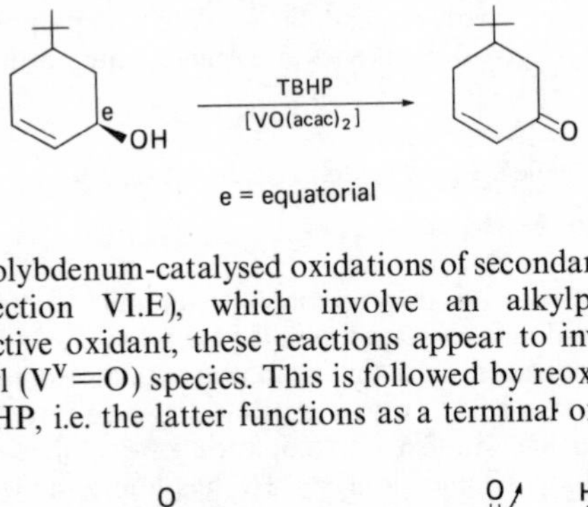

$$\text{cis-5-}t\text{-butylcyclohex-2-en-1-ol (e, OH)} \xrightarrow[\text{[VO(acac)}_2]]{\text{TBHP}} \text{5-}t\text{-butylcyclohex-2-en-1-one} \quad (119)$$

e = equatorial

In contrast to the molybdenum-catalysed oxidations of secondary alcohols with TBHP described earlier (Section VI.E), which involve an alkylperoxymolybdenum(VI) intermediate as the active oxidant, these reactions appear to involve oxidation of the alcohol by the vanadyl ($V^{V}{=}O$) species. This is followed by reoxidation of the reduced form (V^{III}) by the TBHP, i.e. the latter functions as a terminal oxidant (Scheme 17).

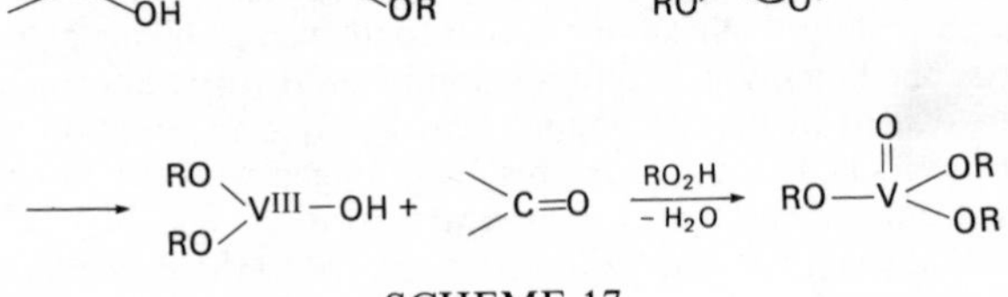

$$\gt C(H)(OH) + RO{-}V(=O)(OR)_2 \xrightarrow{-ROH} (RO)_2V(=O){-}O{-}C(H)\lt$$

$$\longrightarrow (RO)_2V^{III}{-}OH + \gt C{=}O \xrightarrow[-H_2O]{RO_2H} RO{-}V(=O)(OR)_2$$

SCHEME 17

Similarly the recently reported[141] oxidations of primary and secondary alcohols to aldehydes and ketones, respectively, by H_2O_2 or RO_2H in the presence of $RuCl_3$ as catalyst may involve high-valent oxoruthenium compounds as the active oxidants [compare also the palladium(II)-catalysed oxidations of alcohols with TBHP described in

Section VI.E]. Vicinal glycols have been similarly reported to undergo oxidative cleavage[142] (reaction 120).

$$>C(OH)-C(OH)< \xrightarrow[H_2O_2 \text{ or } RO_2H]{[RuCl_3]} 2 >C{=}O \quad (120)$$

VIII. REDUCTION OF ALKYL HYDROPEROXIDES TO ALCOHOLS

We have already mentioned (Section II.A) that alkanes and aralkanes containing tertiary C—H bonds can be selectively autoxidized, at moderate temperatures, to give the corresponding *t*-alkyl hydroperoxides. The alkyl hydroperoxides can be selectively reduced to the corresponding alcohol by aqueous sodium sulphite[143] or by catalytic hydrogenation[144]. This provides a method for the conversion of saturated hydrocarbons to the corresponding alcohols. For example, pinane has been converted to 2-pinanol by autoxidation (to 5% pinane conversion) and subsequent reduction of the pinane-2-hydroperoxide with sulphite[145] (reaction 121).

$$\text{pinane} \xrightarrow{O_2} \text{pinane-2-}O_2H \xrightarrow{Na_2SO_3} \text{2-pinanol (OH)} \quad (121)$$

The conversion of alkanes to alcohols should, in principle, be accomplished in one step by carrying out the autoxidation at elevated temperatures where the alkyl hydroperoxide is decomposed *in situ* via the sequence shown in reactions (122)–(124). For long kinetic chain lengths the overall reaction is shown in reaction (125). However, reaction under these conditions generally leads to substantial by-product formation via β-scission of intermediate alkoxy radicals (see Section IV.B). Selectivities to alcohols can be significantly improved by carrying out the autoxidation in the presence of stoichiometric amounts of boric acid (H_3BO_3), metaboric acid (HBO_2) or boric anhydride (B_2O_3), an effect discovered by Bashkirov[146]. The boron compound converts the intermediate hydroperoxide to the corresponding alkyl borate, dioxygen and water[147] (reaction 126). The alkyl borate is hydrolysed, in a subsequent step, to the corresponding alcohol and boric acid, which is recycled to the oxidation reactor. In the oxidation of cyclopentadecane (reaction 127), for example, the selectivity to cyclopentadecanol (at ca. 20% conversion) increased from 17% in the absence of boric acid to 84% in its presence[148].

$$RO_2H \longrightarrow RO\cdot + HO\cdot \quad (122)$$

$$RO\cdot + RO_2H \longrightarrow RO_2\cdot + ROH \quad (123)$$

$$2RO_2\cdot \longrightarrow 2RO\cdot + O_2 \quad (124)$$

$$2RO_2H \longrightarrow 2ROH + O_2 \quad (125)$$

$$6RO_2H + B_2O_3 \longrightarrow 2(RO)_3B + 3H_2O + 3O_2 \quad (126)$$

$$\text{cyclododecane} \xrightarrow[\text{2. } H_2O,\ 95^{\circ}C]{\text{1. } O_2/H_3BO_3/160^{\circ}C} \text{cyclododecanol (84\%)} \quad (127)$$

The catalytic decomposition of TBHP by organoselenium compounds, at 75°C in benzene, is also reported to yield *t*-butanol and dioxygen as the sole products[149]. The scope of this method in the direct conversion of hydrocarbons to the corresponding alcohols does not appear to have been investigated.

IX. REACTIONS OF α-SUBSTITUTED ALKYL HYDROPEROXIDES

As mentioned earlier, alkyl hydroperoxides are generally unreactive towards organic compounds in the absence of metal catalysts. This contrasts with organic peracids ($RCO.O_2H$) in which the electron-withdrawing carbonyl group renders the peroxidic oxygens more electrophilic compared to those in alkyl hydroperoxides. Thus, organic peracids oxidize a variety of organic functional groups in the absence of catalysts. In this section we shall be concerned with a group of alkyl hydroperoxides in which substitution by electron-withdrawing groups at the α-C—H bond produces oxidants with properties comparable to the organic peracids. For example, the hydroperoxide (**18**), produced by the addition of H_2O_2 to hexafluoracetone at ambient temperature[150] (reaction 128) is capable of epoxidizing alkenes under mild conditions[151]. The cyclic mechanism shown in reaction (129) has been suggested for the epoxidation of alkenes by (**18**)[43].

$$CF_3COCF_3 + H_2O_2 \longrightarrow (CF_3)_2C(OH)(O_2H) \quad (128)$$

(**18**)

$$(CF_3)_2C(OH)OOH + C{=}C \longrightarrow (CF_3)_2CO + \underset{\text{epoxide}}{C{-}C} + H_2O \quad (129)$$

This reagent has also been utilized for the oxidation of aldehydes to carboxylic acids[152] and the oxidation of tertiary amines and dialkyl sulphides to the corresponding amine oxides and sulphoxides, respectively[153]. Rebek and coworkers[154] have similarly shown that the α-substituted hydroperoxides (**19**)–(**22**), obtained from the autoxidation of ketones, esters, amides and nitriles are able to epoxidize fairly reactive alkenes, such as 2,3-dimethylbutene-2 and 2-methylstyrene, in high yield. The highest yields have been observed with **20** and **22** in chloroform at 60°C. The same group have shown that α-substituted hydroperoxides, derived from the reaction of ortho esters and imines with H_2O_2 (reactions 130 and 131), are also able to epoxidize alkenes[155].

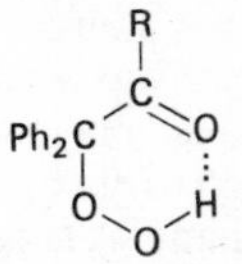

(**19**) R = $CHMe_2$

(**20**) R = OMe

(**21**) R = NMe_2

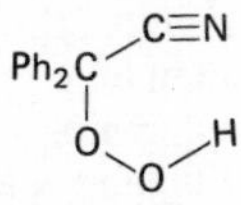

(**22**)

$$MeC(OEt)_3 + H_2O_2 \longrightarrow MeC(OEt)_2O_2H + EtOH \quad (130)$$

$$>C{=}N{-}R + H_2O_2 \longrightarrow >C(O_2H){-}NHR \quad (131)$$

Many of these reagents bear a structural resemblance to the flavin hydroperoxides (**23**) which have been implicated as the active oxidants in oxidations mediated by flavin-containing enzymes[156,157].

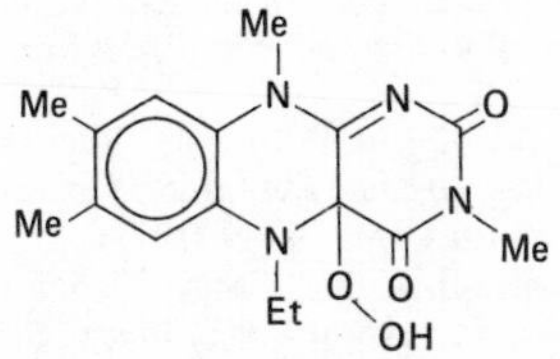

(**23**)

Bruice and coworkers[157,158] have compared the relative rates of oxidation of thioxane to the corresponding sulphoxide (reaction 132) with H_2O_2, TBHP and the model flavin hydroperoxide (**23**). The latter has been found to react ca. 10^5 times as fast as TBHP.

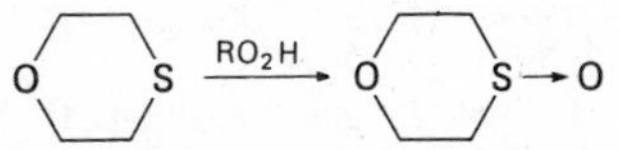

$$\text{thioxane} \xrightarrow{RO_2H} \text{thioxane } S\rightarrow O \quad (132)$$

RO_2H	k^{rel}
TBHP	1
H_2O_2	24
(**23**)	1.8×10^5

X. CONCLUDING REMARKS

In the foregoing discussion we have attempted to show that alkyl hydroperoxides, readily prepared via autoxidation of appropriate hydrocarbons, constitute an important class of mild oxidants in organic synthesis. In combination with a variety of transition-metal catalysts they are able to effect diverse selective oxidative transformations. Many of these reactions have been developed in the last five years. Selective coordination of the substrate to the metal catalyst, through functional groups not directly involved in the oxidation, can result in a high degree of regio- and/or stereo-selectivity (e.g. in the epoxidation of unsaturated alcohols). Furthermore, the use of chiral ligands offers the additional possibility of effecting asymmetric induction in these reactions. We confidently expect, therefore, that the scope of metal-catalysed oxidations with alkyl hydroperoxides will be further expanded in the near future. The mechanisms of many of these reactions are generally poorly understood. In particular, the question of alkylperoxymetal vs. oxometal species as the putative oxygen-atom-transfer agents is a problem that warrants further investigation.

XI. REFERENCES

1. N. A. Milas and S. A. Harris, *J. Amer. Chem. Soc.*, **60**, 2434 (1938).
2. N. Indictor and W. F. Brill, *J. Org. Chem.*, **30**, 2074 (1965).
3. J. Kollar, *U.S. Patent*, No. 3,350,422 (1967), *Chem. Abstr.*, **68**, 2922e (1967); *U.S. Patent*, No. 3,351,635 (1968), *Chem. Abstr.*, **68**, 21821n (1968).
4. M. N. Sheng and J. G. Zajacek, *J. Org. Chem.*, **33**, 588 (1968).
5. K. B. Sharpless and T. R. Verhoeven, *Aldrichim. Acta*, **12**, 63 (1979).
6. Technical Data Bulletin, *Evaluation of Organic Peroxides from Half Life Data*, Lucidol Division, Pennwalt Corporation, 1979.
7. R. Hiatt in *Organic Peroxides*, Vol. 2 (Ed. D. Swern), Wiley–Interscience, New York–London, 1971, p. 1; see also A. G. Davies, *Organic Peroxides*, Butterworths, London, 1961.
8. J. A. Howard, *Advan. Free Radical Chem.*, **4**, 49 (1972).
9. D. E. Winkler and G. W. Hearne, *Ind. Eng. Chem.*, **53**, 655 (1961); D. E. Winkler and G. W. Hearne, *U.S. Patent*, No. 2,845,461 (1958); *Chem. Abstr.*, **53**, 3057e (1959).
10. E. R. Bell, F. H. Dickey, J. H. Raley, F. F. Rust and W. E. Vaughan, *Ind. Eng. Chem.*, **41**, 2597 (1949).
11. D. L. Allara, T. Mill, D. G. Hendry and F. R. Mayo, *Advan. Chem. Ser.*, **76**, 40 (1968).
12. H. R. Grane, *U.S. Patent*, No. 3,478,108 (1969); *Chem. Abstr.*, **69**, 76635f (1968).
13. R. J. Harvey, *U.S. Patent*, No. 3,449,217 (1969); *Chem. Abstr.*, **71**, 60704h (1969).
14. F. R. Mayo, *Acc. Chem. Res.*, **1**, 193 (1968).
15. W. Pritzkow and K. H. Grobe, *Chem. Ber.*, **93**, 2156 (1960).
16. W. Ester and A. Sommer, *U.S. Patent*, No. 3,259,661 (1966); *Chem. Abstr.*, **57**, 1190e (1962).
17. F. F. Rust, *J. Amer. Chem. Soc.*, **79**, 4000 (1957).
18. R. Criegee and P. Ludwig, *Erdol Kohle Erdgas Petrochem.*, **15**, 523 (1962).
19. W. T. Reichle, F. M. Konrad and J. R. Brooks in *Benzene and its Industrial Derivatives* (Ed. E. G. Hancock), Benn, London, 1975).
20. H. Hock and S. Lang, *Chem. Ber.*, **77**, 257 (1944).
21. R. L. Feder, R. Fuhrmann, J. Pisanchijn, S. Elishewitz, T. H. Insinger and C. T. Mathew, *German Patent*, No. 2,035,496 (1972); *Chem. Abstr.*, **76**, 85551x (1972).
22. Mitsui Petrochemical Industries, *German Patent*, No. 2,737,302 (1978), *Chem. Abstr.*, **88**, 152240h (1978); *German Patent*, No. 2,843,754 (1979) *Chem. Abstr.*, **91**, 39120t (1979); *German Patent*, No. 2,846,892 (1979); *Chem. Abstr.*, **91**, 5627f (1979); *German Patent*, No. 2,921,503 (1979); *Chem. Abstr.*, **92**, 94059a (1980).
23. D. W. Cooper, W. J. Burkholder, R. C. Hays and D. H. May, *German Patent*, No. 2,036,937 (1972), *Chem. Abstr.*, **76**, 99336t (1972); *British Patent*, No. 1,342,506 (1974), *Chem. Abstr.*, **80**, 133033m (1974).
24. M. Becker, *German Patent*, No. 2,631,016 (1977); *Chem. Abstr.*, **86**, 189508m (1977).
25. C. Y. Wu, H. E. Swift and J. E. Bozik, *U.S. Patent*, No. 4,158,022 (1979); *Chem. Abstr.*, **91**, 91360k (1979).
26. G. E. Zaikov and Z. K. Maizus, *Bull. Acad. Sci. USSR, Div. Chem. Sci.*, 267 (1969).
27. J. A. Howard and K. U. Ingold, *Can. J. Chem.*, **42**, 1250 (1964).
28. K. Tanaka and J. Imamura, *Chem. Letters*, 1347 (1974).

29. H. Suda, I. Dohgane, S. Yamamoto, K. Tanimoto, T. Chinuki and N. Kotera, *German Patent*, No. 2,325,354 (1973); *Chem. Abstr.*, **80**, 70535d (1974).
30. G. O. Schenck, *Angew. Chem.*, **69**, 579 (1957).
31. K. Gollnick and G. O. Schenck, *Pure Appl. Chem.*, **9**, 507 (1964).
32. For a recent review of mechanistic aspects see L. M. Stephenson, M. J. Grdina and M. Orfanopoulos, *Acc. Chem. Res.*, **13**, 419 (1980).
33. For a review see A. G. Davies in *Organic Peroxides*, Vol. 2 (Ed. D. Swern), Wiley–Interscience, New York–London, 1971, p. 337.
34. C. Walling and S. A. Buckler, *J. Amer. Chem. Soc.*, **75**, 4372 (1953).
35. C. Walling and S. A. Buckler, *J. Amer. Chem. Soc.*, **77**, 6032 (1955).
36. N. A. Milas and D. M. Surgenor, *J. Amer. Chem. Soc.*, **68**, 205 (1946); N. A. Milas, *U.S. Patent*, No. 2,223,807 (1940); *Chem. Abstr.*, **35**, 1802^9 (1940).
37. N. A. Milas and D. M. Surgenor, *J. Amer. Chem. Soc.*, **68**, 643 (1946).
38. J. K. Kochi, *J. Amer. Chem. Soc.*, **84**, 1193 (1962).
39. For reviews see: P. S. Bailey, *Chem. Rev.*, **58**, 925 (1958); A. T. Menyailo and M. V. Pospelov, *Russ. Chem. Rev.*, **36**, 284 (1967).
40. R. A. Sheldon and J. K. Kochi, *Metal Catalyzed Oxidations in Organic Chemistry*, Academic Press, New York, 1981.
41. J. K. Kochi in *Free Radicals*, Vol. 1, Wiley–Interscience, New York–London, 1973.
42. R. A. Sheldon in *Aspects of Homogeneous Catalysis*, Vol. 4 (Ed. R. Ugo), D. Reidel, Dordrecht, 1981, p. 1.
43. R. A. Sheldon, *J. Mol. Catal.*, **7**, 107 (1980).
44. A. O. Chong and K. B. Sharpless, *J. Org. Chem.*, **42**, 1587 (1977).
45. H. Mimoun, *J. Mol. Catal.*, **7**, 1 (1980).
46. L. I. Matiénko, I. P. Skibida and Z. K. Maizus, *Kinet. Catal.*, **12**, 525 (1971).
47. G. Sosnovsky and D. J. Rawlinson in *Organic Peroxides*, Vol. 2 (Ed. D. Swern), Wiley–Interscience, New York–London, 1970, p. 153.
48. D. J. Rawlinson and G. Sosnovsky, *Synthesis*, 1 (1972).
49. M. S. Kharasch and A. Fono, *J. Org. Chem.*, **24**, 72 (1959); **23**, 325 (1958).
50. M. S. Kharasch, P. Pauson and W. Nudenberg, *J. Org. Chem.*, **18**, 322 (1953).
51. C. Walling and A. Zavitsas, *J. Amer. Chem. Soc.*, **85**, 2084 (1963).
52. A. B. Evnin and A. V. Lam, *Chem. Commun.*, 1184 (1968).
53. B. Acott and A. L. J. Beckwith, *Australian J. Chem.*, **17**, 1342 (1964).
54. J. K. Kochi, *J. Amer. Chem. Soc.*, **85**, 1958 (1963).
55. Z. Cekovic and M. M. Green, *J. Amer. Chem. Soc.*, **96**, 3000 (1974).
56. Z. Cekovic, J. Dimitrijevic, G. Djokic and T. Srnic, *Tetrahedron*, **35**, 2021 (1979).
57. E. G. E. Hawkins and D. P. Young, *J. Chem. Soc.*, 2804 (1950); E. G. E. Hawkins, *Nature (London)*, **166**, 69 (1950).
58. J. O. Punderson, *U.S. Patent*, No. 2,822,399 (1958), *Chem. Abstr.*, **52**, 10159g (1958); *U.S. Patent*, No. 2,700,057 (1955), *Chem. Abstr.*, **49**, 15953a (1955).
59. K. Formanek, J. P. Aune, M. Jouffret and J. Metzger, *Nouveau J. Chim.*, **3**, 311 (1979).
60. M. S. Kharasch and W. Nudenberg, *J. Org. Chem.*, **19**, 1921 (1954).
61. H. E. De La Mare, J. K. Kochi and F. F. Rust, *J. Amer. Chem. Soc.*, **83**, 2013 (1961); **85**, 1437 (1963).
62. J. K. Kochi and F. F. Rust, *J. Amer. Chem. Soc.*, **84**, 3946 (1962).
63. F. Minisci and R. Galli, *Tetrahedron Letters*, 357 (1963).
64. F. Minisci, *Acc. Chem. Res.*, **8**, 165 (1975).
65. F. Minisci, M. Cecere, R. Galli and A. Selva, *Org. Prep. Proc.*, **1**, 11 (1969).
66. S. L. Schreiber, *J. Amer. Chem. Soc.*, **102**, 6163 (1980).
67. R. A. Sheldon and J. A. van Doorn, *Tetrahedron Letters*, 1021 (1973).
68. *Houben–Weyl: Methoden der Organischen Chemie*, Vol. 6, Part 1C, Georg Thieme Verlag, Stuttgart, 1976, pp. 117–139.
69. J. Hoffman, *Org. Synth.*, **40**, 76 (1960).
70. R. Landau, G. A. Sullivan and D. Brown, *Chem. Tech.*, 602 (1979).
71. Shell Oil, *British Patent*, No. 1,249,079 (1971); *Chem. Abstr.*, **74**, 12981m (1971).
72. M. I. Farberov, *Sov. Chem. Ind. (USSR)*, **11**, (1), 7 (1979).
73. G. A. Tolstikov, V. P. Yurev and U. M. Dzhemilev, *Russ. Chem. Rev.*, **44**, 319 (1975).

74. B. N. Bobylev, M. I. Farberov and S. A. Kesarev, *Sov. Chem. Ind.* (*USSR*), **11** (3), 153 (1979).
75. W. I. Denton, *U.S. Patent*, No. 2,986,585 (1961); *Chem. Abstr.*, **55**, 20956i (1961).
76. E. H. Eschinasi, *J. Org. Chem.*, **35**, 1598 (1970).
77. H. O. House, *J. Amer. Chem. Soc.*, **77**, 5083 (1955); B. Rickborn and R. M. Gerkin, *J. Amer. Chem. Soc.*, **93**, 1693 (1971).
78. J. K. Cox, *U.S. Patent*, No. 4,000,200 (1976); *Chem. Abstr.*, **86**, 89180e (1977).
79. R. A. Sheldon, *Rec. Trav. Chim.*, **92**, 253 (1973).
80. F. List and L. Kuhnen, *Erdöl Kohle Erdgas Petrochem.*, **20**, 192 (1967).
81. B. N. Bobylev, L. V. Melnik, M. I. Farberov, L. I. Bobyleva and I. V. Subbotina, *J. Appl Chem. USSR*, **50**, 589 (1977); M. I. Farberov, L. V. Melnik, B. A. Bobylev and V. A. Podgornova, *Kinet. Catal.* (*USSR*), **12**, 1144 (1971).
82. K. B. Sharpless and R. C. Michaelson, *J. Amer. Chem. Soc.*, **95**, 6136 (1973).
83. J. E. Lyons, *Advan. Chem. Ser.*, **132**, 64 (1974).
84. J. E. Lyons in *Catalysis in Organic Synthesis* (Eds. P. N. Rylander and H. Greenfield), Academic Press, New York, 1976, pp. 235–255.
85. T. Itoh, K. Jitsukawa, K. Kaneda and S. Teranishi, *J. Amer. Chem. Soc.*, **101**, 159 (1979); see also R. B. Daniel and G. H. Whitham, *J. Chem. Soc., Perkin Trans. 1*, 953 (1979).
86. S. Tanaka, H. Yamamoto, H. Nozaki, K. B. Sharpless, R. C. Michaelson and J. D. Cutting, *J. Amer. Chem. Soc.*, **96**, 5254 (1974); B. E. Rossiter, T. R. Verhoeven and K. B. Sharpless, *Tetrahedron Letters*, 4733 (1979).
87. E. D. Mihelich, *Tetrahedron Letters*, 4729 (1979).
88. R. C. Michaelson, R. E. Palermo and K. B. Sharpless, *J. Amer. Chem. Soc.*, **99**, 1990 (1977).
89. S. Yamada, T. Mashiko and S. Terashima, *J. Amer. Chem. Soc.*, **99**, 1988 (1977).
90. T. Katsuki and K. B. Sharpless, *J. Amer. Chem. Soc.*, **102**, 5974 (1980).
91. B. E. Rossiter, T. Katsuki and K. B. Sharpless, *J. Amer. Chem. Soc.*, **103**, 464 (1981).
92. K. Tani, M. Hanafusa and S. Otsuka, *Tetrahedron Letters*, 3017 (1979).
93. G. A. Tolstikov, U. M. Dzhemilev and V. P. Yurev, *J. Org. Chem. USSR*, **8**, 1204 (1971); *J. Gen. Chem. USSR*, **43**, 2058 (1973).
94. U. M. Dzhemilev, V. P. Yurev, G. A. Tolstikov, F. B. Gershanov and S. R. Rafikov, *Proc. Acad. Sci. USSR, Chem. Sect.*, **196**, 79 (1971).
95. M. N. Sheng and J. G. Zajacek, *J. Org. Chem.*, **33**, 588 (1968).
96. G. A. Tolstikov, U. M. Jemilev, V. P. Jurjev, F. B. Gershanov and S. R. Rafikov, *Tetrahedron Letters*, 2807 (1971).
97. G. A. Tolstikov, U. M. Dzhemilev and V. P. Yurev, *J. Org. Chem. USSR*, **8**, 1200 (1971).
98. G. A. Tolstikov, U. M. Dzhemilev, V. P. Yurev, A. A. Pozdeeva and F. G. Gerchikova, *J. Gen. Chem. USSR*, **43**, 1350 (1973).
99. J. L. Russell and J. Kollar, *British Patent*, No. 1,100,672 (1968), *Chem. Abstr.*, **64**, 6527d (1966); *British Patent*, No. 1,111,892 (1968), *Chem, Abstr.*, **65**, 8792b (1966).
100. G. N. Koshel, M. I. Farberov, L. L. Zalygin and G. A. Krushinskaya, *J. Appl. Chem. USSR*, **44**, 885 (1971).
101. G. R. Howe and R. R. Hiatt, *J. Org. Chem.*, **35**, 4007 (1970).
102. K. Kosswig, *Justus Liebigs Ann. Chem.*, **749**, 206 (1971).
103. P. A. Bartlett, F. R. Green and T. R. Webb, *Tetrahedron Letters*, 331 (1977).
104. W. E. Noland, *Chem. Rev.*, **55**, 137 (1955).
105. L. Kuhnen, *Angew. Chem.*, **78**, 957 (1966).
106. W. F. List and L. Kuhnen, *Erdöl Kohle Erdgras Petrochem.*, **20**, 192 (1967).
107. R. Curci, F. DiFuria, R. Testi and G. Modena, *J. Chem. Soc., Perkin Trans. 2*, 752 (1974).
108. R. Curci, F. DiFuria and G. Modena, *J. Chem. Soc., Perkin Trans. 2*, 576 (1977).
109. S. Cenci, F. DiFuria, G. Modena, R. Curci and J. O. Edwards, *J. Chem. Soc., Perkin Trans. 2*, 979 (1978).
110. G. A. Tolstikov, U. M. Dzhemilev, N. N. Novitskaya, V. P. Yurev and R. G. Kantyukova, *J. Gen. Chem. USSR*, **41**, 1896 (1971).
111. G. A. Tolstikov, U. M. Dzhemilev, N. N. Novitskaya and V. P. Yurev, *Bull. Acad. Sci. USSR, Div. Chem. Sci.*, **21**, 2675 (1972).
112. F. DiFuria, G. Modena and R. Curci, *Tetrahedron Letters*, 4637 (1976).
113. M. N. Sheng and J. G. Zajacek, *U.S. Patent*, No. 3,670,002 (1972); *Chem. Abstr.*, **77**, 88089j (1972).

114. G. A. Tolstikov, U. M. Dzhemilev and V. P. Yurev, *J. Gen. Chem. USSR*, **42**, 1602 (1972).
115. C. L. Willis and L. H. Slaugh, *European Patent Appl.*, No. 18,681 (1980); *Chem. Abstr.*, **94**, 174350v (1981).
116. G. S. P. Field, *British Patent*, No. 1,413,475 (1975); *Chem. Abstr.*, **84**, 73672q (1976).
117. R. Rosenthal and G. A. Bonetti, *U.S. Patent*, No. 3,755,453 (1973); *Chem. Abstr.*, **79**, 125957e (1973).
118. G. A. Tolstikov, V. P. Yurev and I. A. Gailyunas, *Bull. Acad. Sci. USSR, Div. Chem. Sci.*, **22**, 1395 (1973).
119. H. Mimoun, R. Charpentier, A. Mitschler, J. Fischer and R. Weiss, *J. Amer. Chem. Soc.*, **102**, 1047 (1980).
120. R. Jira and W. Freiesleben in *Organometallic Reactions* (Eds. E. Becker and M. Tsutsui), Vol. III, Wiley–Interscience, New York–London, 1972, p. 1.
121. J. Tsuji, H. Nagashima and K. Hori, *Chem. Letters*, 257 (1980).
122. N. Rabjohn, *Org. Reactions*, **24**, 261 (1976).
123. M. A. Umbreit and K. B. Sharpless, *J. Amer. Chem. Soc.*, **99**, 5526 (1977).
124. See also O. Campos and J. M. Cook, *Tetrahedron Letters*, 1025 (1979); F. W. Sum and L. Weiler, *J. Amer. Chem. Soc.*, **101**, 4401 (1979); H. E. Paaren, D. E. Hamer, H. K. Schoes and H. F. De Luca, *Proc. Nat. Acad. Sci., U.S.A.*, **75**, 2080 (1978).
125. B. Chabaud and K. B. Sharpless, *J. Org. Chem.*, **44**, 4202 (1979).
126. For a recent review see M. Schroder, *Chem. Rev.*, **80**, 187 (1980).
127. A. Byers and W. J. Hickinbottom, *J. Chem. Soc.*, 1328 (1948).
128. K. B. Sharpless and K. Akashi, *J. Amer. Chem. Soc.*, **98**, 1986 (1976).
129. K. Akashi, R. E. Palermo and K. B. Sharpless, *J. Org. Chem.*, **43**, 2063 (1978).
130. M. N. Sheng and W. A. Mameniskis, *U.S. Patent*, No. 4,049,724 (1977); *Chem. Abstr.*, **88**, 22130m (1978).
131. T. P. Kobylinski and C. Y. Wu, *U.S. Patent*, No. 4,203,926 (1980); *Chem. Abstr.*, **93**, 167618q (1980).
132. V. van Rheenen, R. C. Kelly and D. Y. Cha, *Tetrahedron Letters*, 1973 (1976).
133. M. Schroder, A. J. Nielson and W. P. Griffith, *J. Chem. Soc., Dalton*, 1607 (1979).
134. J. T. Groves, T. E. Nemo and R. S. Myers, *J. Amer. Chem. Soc.*, **101**, 1032 (1979).
135. J. T. Groves, W. J. Kruper and R. C. Haushalter, *J. Amer. Chem. Soc.*, **102**, 6375 (1980).
136. E. Guilmet and B. Meunier, *Tetrahedron Letters*, **21**, 4449 (1980).
137. J. T. Groves and W. J. Kruper, *J. Amer. Chem. Soc.*, **101**, 7613 (1979).
138. For leading references see J. T. Groves in *Metal Ion Activation of Dioxygen* (Ed. T. G. Spiro), Wiley–Interscience, New York–London, 1980; V. Ullrich, *Top. Curr. Chem.*, **83**, 68 (1979).
139. A. L. Stautzenberger, *U.S. Patent*, No. 3,931,249 (1976); *Chem. Abstr.*, **84**, 105379t (1976).
140. R. A. Sheldon, unpublished results.
141. Mitsui Petrochemical Company, *Japanese Patent*, No. 80,102,527 (1980); *Chem. Abstr.*, **94**, 46783b (1981).
142. Mitsui Petrochemical Company, *Japanese Patent*, No. 80,102,528 (1980); *Chem. Abstr.*, **94**, 46774z (1981).
143. H. Hock and S. Lang, *Chem. Ber.*, **75**, 313 (1942).
144. A. G. Davies, *J. Chem. Soc.*, 3474 (1958).
145. C. Filliatre and R. Lalande, *Bull. Soc. Chim. Fr.*, 4141 (1968).
146. A. N. Bashkirov, V. V. Kamzolkin, K. M. Sokova and T. P. Andreyeva, in *The Oxidations of Hydrocarbons in the Liquid Phase* (Ed. N. M. Emanuel), Pergamon, Oxford, 1965, p. 183.
147. H. Sakaguchi, Y. Kamiya and N. Ohta, *Bull. Japan Petr. Inst.*, **14**, 71 (1972).
148. J. R. Sanderson, K. Paul, R. J. Wilterdink and J. A. Alford, *J. Org. Chem.*, **41**, 3767 (1976).
149. D. T. Woodbridge, *J. Chem. Soc., Phys. Org.*, 50 (1966).
150. R. D. Chambers and M. Clark, *Tetrahedron Letters*, 2741 (1970).
151. L. Kim, *British Patent*, No. 1,399,639 (1975); *Chem. Abstr.*, **78**, 159400n (1973). See also R. O. Heggs and B. Ganem, *J. Amer. Chem. Soc.*, **101**, 2484 (1979).
152. B. Ganem, R. P. Heggs, A. J. Biloski and D. R. Schwarts, *Tetrahedron Letters*, **21**, 685 (1980).
153. B. Ganem, A. J. Biloski and R. P. Heggs, *Tetrahedron Letters*, **21**, 689 (1980).
154. J. Rebek, R. McCready and R. Wolak, *J. Chem. Soc., Chem. Commun.*, 705 (1980).
155. J. Rebek and R. McCready, *J. Amer. Chem. Soc.*, **102**, 5602 (1980); J. Rebek and R. McCready, *Tetrahedron Letters*, **21**, 2491 (1980).

156. T. C. Bruice, *Acc. Chem. Res.*, **13**, 256 (1980).
157. T. C. Bruice, *Advan. Chem. Ser.*, **191**, 89 (1980).
158. C. Kemal, T. W. Chan and T. Bruice, *Proc. Natl. Acad. Sci. U.S.A.*, **74**, 405 (1977).

The Chemistry of Functional Groups, Peroxides
Edited by S. Patai

CHAPTER 7

Singlet oxygen in peroxide chemistry

ARYEH A. FRIMER

Department of Chemistry, Bar-Ilan University, Ramat-Gan, Israel

I. INTRODUCTION . . . 202
II. THEORETICAL DESCRIPTION . . . 202
III. METHODS OF PREPARATION . . . 204
A. Photosensitization . . . 204
B. Oxidation of H_2O_2 . . . 205
C. Decomposition of Trialkyl and Triaryl Phosphite Ozonides . . . 206
D. Thermal Decomposition of Endoperoxides . . . 207
E. Microwave Discharge . . . 208
IV. DIAGNOSTIC TESTS . . . 208
V. MODES OF REACTION . . . 209
A. Introduction . . . 209
B. Singlet Oxygen Diels–Alder Reaction . . . 210
1. Nature of the substrate . . . 210
2. Mechanism . . . 212
3. Reactions of endoperoxides . . . 212
a. Rearrangement to diepoxides . . . 213
b. Transformation to 4-hydroxy-2-en-1-ones and furans . . . 214
c. Reductions . . . 216
d. Thermal rearrangements . . . 216
e. Solvolysis . . . 217
C. Singlet Oxygen Ene Reaction . . . 218
1. Nature of the substrate . . . 218
2. Mechanism . . . 219
3. Reactions of allylic hydroperoxides . . . 220
a. Reduction to allylic alcohols . . . 220
b. Transformation to two carbonyl fragments or a divinyl ether . . . 221
c. Homolysis of the peroxy linkage yielding enones, enols and epoxides . . . 224
d. Miscellaneous reactions . . . 225
D. Singlet Oxygen Dioxetane-forming Reaction . . . 226
1. Nature of the substrate . . . 226

2. Mechanism 226
3. Reactions of dioxetanes 227
a. Cleavage 227
b. Rearrangement to α-hydroxyketones and α-diketones 227
c. Nucleophilic addition 228
d. Reduction 228
VI. CONCLUSION 229
VII. REFERENCES 229

I. INTRODUCTION

The photoinduced damage or destruction of aerobic species in the presence of natural or adventitious dyes has piqued the curiosity of scientists for the past century[1]. In recent years it has been shown that the first excited state of molecular oxygen, singlet oxygen ($^1O_2$) is one of the active species involved in this 'photodynamic action'. Keen interest has developed in understanding fully the mechanistic details of this destructive effect and in uncovering other possible biosynthetic functions of singlet oxygen.

Concomitant with the growing intrigue with singlet oxygen chemistry, there has developed an interest in the resulting products which include three major classes of peroxides: allylic hydroperoxides, 1,2-dioxacyclobutanes (dioxetanes) and 4,5-dioxacyclohexenes (endoperoxides). Often, secondary reactions of these labile primary products occur spontaneously or during work-up. Hence, a knowledge of some of the chemistry of endoperoxides, allylic hydroperoxides and dioxetanes is crucial to an understanding of the course of a reaction. Noteworthy as well is that these secondary reactions can, to some extent, be controlled and selectively induced. Synthetically, therefore, $^1O_2$ has proven to be a powerful tool which allows for ready access to a plethora of functionalities including allylic alcohols, enones, epoxyenones, β-diepoxides and diones.

It is the goal of this chapter to discuss the organic chemistry of $^1O_2$ and particularly its role in peroxide chemistry. To that end we shall also present various aspects of the chemistry of the primary $^1O_2$ products. We will confine ourselves, however, to those transformations which commonly occur spontaneously during the course of a singlet oxygen reaction or are induced thereafter for synthetic purposes. Above all, we trust that the reader will become fascinated with the versatility of this petite but mighty little molecule.

II. THEORETICAL DESCRIPTION

Ever since the discovery of oxygen over two centuries ago, mankind has invested a good deal of time and resources in attempting to understand the exact role this life-supporting molecule plays in autoxidative, photooxidative and metabolic processes. Since it is the electronic makeup of a molecule which determines its reactivity, it was to molecular orbital theory and electronic excitation spectroscopy that scientists turned in order to get an exact description of the configuration of the various electronic states of molecular oxygen[2a]. We shall limit our discussion to the structure of the lowest three electronic states of O_2 which differ primarily in the manner in which the two electrons of highest energy occupy the two degenerate π^*_{2p} molecular orbitals. Following Hund's rule, in the ground state of O_2 these two electrons will have parallel spins and be located one each in the two degenerate π^*_{2p} orbitals (Figure 1). Such an electronic configuration corresponds to a triplet $^3\Sigma_g^-$ state and we shall henceforth refer to ground-state molecular oxygen as triplet oxygen, $^3O_2$.

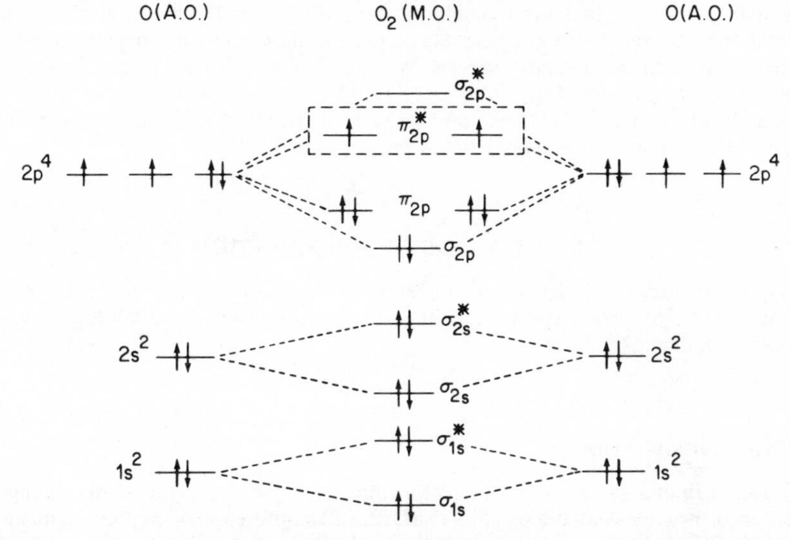

FIGURE 1. Schematic energy level diagram showing how the atomic orbitals (A.O.) of two atoms of elemental oxygen interact to form the molecular orbitals (M.O.) of molecular oxygen. The electron distribution is, according to Hund's rule, yielding ground-state molecular oxygen ($^3\Sigma_g^-$).

This triplet character is responsible for the paramagnetism and diradical-like properties of $^3O_2$. More importantly, this triplet electronic configuration only permits reactions involving one-electron steps. Thus, despite the exothermicity of oxygen reactions, a spin barrier prevents $^3O_2$ from reacting indiscriminately with the plethora of singlet ground-state organic compounds surrounding it. One could well argue that it is this spin barrier that permits life to be maintained.

The two lowest excited states are both singlets in which the two highest energy electrons have antiparallel spins. Thus no spin barrier should exist for their reaction with organic substrates. In the first ($^1\Delta_g$) state, which lies 22.5 kcal mol^{-1} above the ground state, both of the highest energy electrons occupy the same π^*_{2p} orbital. In the second, a $^1\Sigma_g^+$ state lying 15 kcal mol^{-1} higher, each of the π^*_{2p} orbitals is half full (Table 1).

In the gas phase the lifetimes of $^1\Delta$ and $^1\Sigma$ oxygen are 45 minutes and 7 seconds respectively[2b]. However, in solution these lifetimes are dramatically reduced through

TABLE 1. The three lowest electronic states of molecular oxygen and selected properties

Electronic state	Configuration of π^*_{2p}	Relative energy (kcal mol^{-1})	Lifetime (s)[2b,c] Gas phase	Liquid phase	Valence bond representation
$^1\Sigma_g^+$	↑ ↓	37.5	7–12	10^{-9}	
$^1\Delta_g$	↑↓ —	22.5	2700	10^{-3}	O=O
$^3\Sigma_g^-$	↑ ↑	0	∞	∞	

collisional deactivation to approximately 10^{-3} and 10^{-9} seconds respectively[2b,c]. Because the reactions that concern us are generally carried out in solution, it is the longer lived $^1\Delta$ O_2 that is involved as the active species. We shall henceforth refer to this longer lived species as singlet oxygen, $^1O_2$.

A simplified picture of the three lowest electronic states of molecular oxygen and a comparison of some of their properties is presented in Table 1.

III. METHODS OF PREPARATION

An impressive variety of physical and chemical sources of $^1O_2$ are now available for laboratory-scale purposes. However, in the following discussion we shall focus only on those most commonly in use.

A. Photosensitization

By the beginning of the twentieth century there were several reports[3] describing the oxidation of organic and biological substrates in the presence of oxygen, light and a photosensitizer. It has become apparent during the last two decades that there are in fact two general classes of photooxidations[4]. In the first, called Type I, the sensitizer serves as a photochemically activated free-radical initiator. In its excited state the sensitizer reacts with a molecule of a substrate, resulting in either hydrogen atom abstraction or electron transfer. The radicals thus formed react further with $^3O_2$ or other molecules. In the second class of reactions, dubbed Type II, the sensitizer triplet (sens^3), formed via intersystem crossing (ISC) of the excited singlet state sensitizer (sens^{1*}), interacts with oxygen, most commonly by transferring excitation, to produce $^1O_2$ (equations 1 and 2). The direct absorption of light by $^3O_2$ to produce $^1O_2$ is a spin-forbidden process. Type II generally predominates with coloured sensitizers (dyes) which absorb visible light while Type I processes are favoured by high-energy UV absorbing sensitizers. Table 2 includes several common sensitizers used to produce $^1O_2$. Note that their triplet energies (E_T)[5,6] range from 30–46 kcal mol^{-1}.

$$\text{sens}^1 \xrightarrow{h\nu} \text{sens}^{1*} \xrightarrow{\text{ISC}} \text{sens}^3 \tag{1}$$

$$\text{sens}^3 + {}^3O_2 \longrightarrow \text{sens}^1 + {}^1O_2 \tag{2}$$

TABLE 2. Triplet energy[5,6] of sensitizers commonly used to produce $^1O_2$

Sensitizer	Triplet energy, E_T (kcal mol^{-1})
Fluorescein	45.2–48.1
Eosin	43.2–46.0
Erythrosin	43.1–45.8
Rhodamin B	43.0
Rose Bengal	39.5–42.2
Tetraphenylporphyrin	34.0
Methylene blue	34.0

A variety of photochemical apparatus and procedures have been described[7,8]. In a typical reaction, the substrate and the sensitizer (10^{-3}–10^{-5} M) are dissolved in an appropriate solvent (see Table 3) and photolysed (250–1000 W) while oxygen is bubbled through the reaction mixture. Alternatively the solution is rapidly stirred under an oxygen atmosphere with the uptake of oxygen followed by means of a gas buret. A UV cut-off filter is often placed between the light source and the reaction vessel to prevent the initiation of free-radical reactions.

As is clear from Table 3, solubility properties differ from sensitizer to sensitizer according to their chemical structure. Boden[9] reports that the problem of solubilizing anionic dyes (such as Rose Bengal and Eosin Y) in aprotic solvents can be circumvented by the use of phase-transfer catalysts such as crown ethers and tetramethylammonium salts.

Recently the use of polymer-based or adsorbant-bound sensitizers[10–15] has become quite popular and several products are commercially available. Problems such as solubility, removal, recovery and bleaching, often confronted with unbound sensitizers, are eliminated by using this heterogenous photooxygenation method. The polymer-based sensitizer need simply be suspended in any (mostly organic) solvent which will 'wet' the polymer. Upon conclusion of the photolysis, the sensitizer may be filtered off, washed and reused if so desired.

Of all the techniques available for generating $^1O_2$, photosensitization is clearly the most convenient. It is applicable to a large spectrum of reaction temperatures, solvents and sensitizers. Most importantly for unreactive substrates this physical method, unlike the many chemical methods discussed below, requires no additional reagents, merely longer photolysis times. Nevertheless, the possible intervention of Type I and other free-radical processes requires independent nonphotochemical sources of $^1O_2$.

B. Oxidation of H_2O_2

In 1960 Seliger[16] tested the ability of a number of different oxidizing agents to stimulate the chemiluminescence of luminol. He noted that upon mixing the two oxidants H_2O_2 and HOCl he obtained a red luminescent flash. The reaction proved more efficient when alkaline solutions of the two reactants were used. Seliger reported further that the spectrum of the emission consisted of a rather narrow band centred at 634 nm which we now know[17] corresponds to the dimole emission of $^1O_2$. The accepted mechanism[18] for

TABLE 3. Solubility properties of common $^1O_2$ sensitizers[a]

Solvent	Eosin	Methylene Blue	Rose Bengal	Tetraphenylporphyrin
Acetone	S	S	S	SS
CH_3OH	S	VS	S	I
$CHCl_3$	SS	VS	I	S
CH_2Cl_2	SS	S	I	S
CH_3CN	S	S	S	I
C_6H_6	SS	I	I	S
CS_2	SS	I	I	S
CCl_4	SS	I	I	S
Pyridine	S	S	S	S
Ether	SS	I	SS	SS
THF	S	SS	S	S
H_2O	S	S	S	I

[a]VS, very soluble; S, soluble; SS, slightly soluble; I, insoluble.

this process is based on the early work of Connick[19] and Cahill and Taube[20] and involves the intermediacy of a chlorohydroperoxy anion[21] as the active species, formed as shown in equations (3)–(6).

$$HOOH + {}^{-}OCl \rightleftharpoons HOO^{-} + Cl{-}OH \quad (3)$$

$$HOO^{-} + Cl{-}OH \rightleftharpoons HOO{-}Cl + HO^{-} \quad (4)$$

$$HO^{-} + H{-}OOCl \rightleftharpoons H_2O + {}^{-}OOCl \quad (5)$$

$$^{-}O{-}O{-}Cl \longrightarrow {}^{1}O_2 + Cl^{-} \quad (6)$$

Foote and Wexler[22,23] have reported that when sodium hypochlorite is added dropwise to chilled and stirred alkaline alcoholic (MeOH, EtOH, MeOH/*t*-BuOH) solutions containing the substrate (~0.1 M) and excess hydrogen peroxide, yields of up to 80% can be obtained.

Despite the utility of this method for reactive $^1O_2$ acceptors, unreactive substrates require large quantities of reagents which further complicate the solubility problems. Furthermore, this system produces various free radicals which may induce autoxidative processes. Functional groups on the starting material or products may be sensitive to either H_2O_2 or aqueous base.

To circumvent the solubility problem, McKeown and Waters[24] suggested a two-phase system. In this procedure the upper layer contains the substrate dissolved in a suitable organic solvent while the lower layer contains aqueous KOH and H_2O_2. Br_2 is added dropwise while the lower layer is slowly stirred and $^1O_2$ bubbles up through the upper layer. The reaction occurs well with Br_2 and Cl_2 but not I_2. Although this method allows for some diversity in the choice of solvents it is not recommended for general synthetic use because of the low $^1O_2$ yields, competing radical and halogenation reactions, and possible sensitivity of the substrate or products to the alkaline oxidizing conditions.

C. Decomposition of Trialkyl and Triaryl Phosphite Ozonides

Alkyl and aryl phosphites react with ozone to give the corresponding phosphates and molecular oxygen. In 1961, Thompson[25] found that when the reaction is carried out at −70°C, some triaryl phosphites first form a metastable 1:1 adduct with ozone, which then decomposes upon warming to phosphate and oxygen. Murray and Kaplan[26–28] later demonstrated that in the case of triphenyl phosphite ozonide, the oxygen evolved displayed the reactivity of $^1O_2$ generated by other procedures (equation 7). Bartlett and coworkers[29–31] have, however, shown that with highly reactive substrates, triphenyl phosphite ozonide reacts slowly and directly far below its decomposition temperature. These results have been corroborated by the thermochemical and kinetic studies of Koch[32,33].

$$(PhO)_3P + O_3 \xrightarrow[CH_2Cl_2]{-70^\circ C} (PhO)_3P\langle O_3 \rangle \xrightarrow{>-35^\circ C} (PhO_3)P{=}O + {}^{1}O_2 \quad (7)$$

Nevertheless, for most substrates the phosphite ozonide method is a convenient source of $^1O_2$ which is susceptible to temperature control. Thus ozonide can be added to the acceptor solution at temperatures below −35°C and reaction will take place only upon warming. Bartlett and colleagues have further shown that the decomposition of the ozonide to $^1O_2$ is very susceptible to base catalysis[34]. For example, the use of pyridine in methanol as solvent permits the controlled thermal generation of singlet oxygen at temperatures as low as −100°C without significant diminution in the yield.

The major drawback of the phosphite ozonide technique is that large quantities of triphenyl phosphate are produced, particularly in the case of unreactive substrates. The phosphate is difficult to separate from the peroxidic products.

A number of other phosphite–ozone adducts have been used to carry out oxygenation reactions. Thus the tricyclic phosphites 4-ethyl-2,6,7-trioxa-1-phosphabicyclo[2,2,2]octane (**1**)[35a] and 1-phospha-2,8,9-trioxaadamantane (**2**)[35b] readily form ozone adducts which efficiently produce $^1O_2$ at temperatures above 0°C. Phosphite **2** is of particular interest since its ozonide is water-soluble and may therefore be used as a source of $^1O_2$ in aqueous solutions. Furthermore, the corresponding phosphate is insoluble in CCl_4. Thus CCl_4 solutions of pure products are easily obtainable.

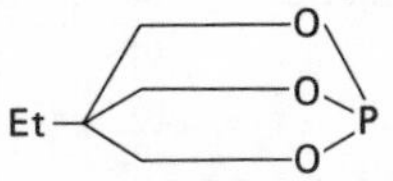

(**1**) 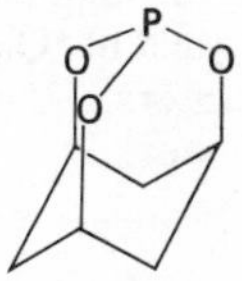 (**2**)

D. Thermal Decomposition of Endoperoxides

Polynuclear aromatic hydrocarbons are among the best and oldest acceptors of singlet oxygen[36] (see Section V.B.1). In many cases, the resulting transannular peroxides (also called endoperoxides or epidioxides) have the interesting property of regenerating the original arene and singlet oxygen when heated. Wasserman and coworkers[37,38] have reported that 9,10-diphenylanthracene endoperoxide has a half-life of 16 hours at 80°C and 8 hours at 90°C. In a typical experiment the substrate is heated with two equivalents of peroxide in refluxing benzene for 2–5 days depending on substrate reactivity. Aprotic solvents other than benzene (e.g. toluene, chloroform or dimethyl sulphoxide) may be used as well. Alcoholic solvents generally react with endoperoxides and are hence not suitable.

Clearly the thermal stability of the peroxides will depend on the structure of the parent arene. Thus while the endoperoxides of 1,2,4-trimethylnaphthalene ($\tau_{25^\circ} = 70$ h)[39], 1,4,5-trimethylnaphthalene ($\tau_{25^\circ} = 290$ h, $\tau_{34.6^\circ} = 34.5$ h, $\tau_{51^\circ} = 2.7$ h)[39] and 1,2,3,4-tetramethylnaphthalene ($\tau_{51^\circ} = 47$ h)[39] are stable at room temperature, those of 1-methylnaphthalene[39], 1,5-dimethylnaphthalene[39], acenaphthene[39], 5-methylacenaphthene[39], 1-methoxynaphthalene[39], 1,4-dimethoxy-9,10-diphenylanthracene[40] and 2,5-diphenyl-2,5-dihydrofuran[41] are not. The latter may be prepared at −78°C and decomposed at room temperature[41]. The endoperoxide of 3-(4-methyl-1-naphthyl)-propionic acid[42] is water-soluble and decomposes readily at 35°C ($\tau = 23$ min).

The major disadvantage of this method is that large amounts of peroxide are required for unreactive substrates. This only complicates the difficulty of separating the arene from the reaction products. Rosenthal and Acher[43] have got round this problem by preparing a polymer-based 9,10-diphenylanthracene derivative which can be removed after the decomposition of its endoperoxide by simple filtration and then reused if desired. However, the decomposition of the endoperoxide of the polymer like that of the free 9,10-diphenylanthracene requires relatively high temperatures (>80°C) and long reaction times. It would seem then that the endoperoxide decomposition method, though well suited for mechanistic studies, is inconvenient for preparative use.

E. Microwave Discharge

In the generation of singlet oxygen in a stream of gaseous oxygen by microwave discharge[44], only about 10% of the oxygen stream is in the $^1\Delta$ state. Another 10% is oxygen atoms which can be removed by a variety of methods. In order to avoid the formation of ozone only a few millimetres of oxygen pressure are used. The $^1O_2$-rich gas stream can then be passed into a solution of acceptor.

This physical method has had widespread importance for gas-phase studies. However, because of the short lifetime of $^1O_2$ in solution, reaction occurs only where the stream initially comes into contact with the liquid phase making it unsuitable for unreactive substrates. Compare this, for example, with photosensitization where the sites of $^1O_2$ formation are dispersed throughout the solution. If we take into consideration as well the relatively low absolute yield of $^1O_2$, because of the low pressures required, as well as the cost of microwave generators, it is not surprising that this technique has had only a minor importance in solution work.

IV. DIAGNOSTIC TESTS

We have noted above that free-radical processes can be inadvertantly initiated during the course of $^1O_2$ generation, particularly when photosensitization is the method used. As a result, a series of diagnostic tests are required to determine whether the products observed result from a singlet oxygen reaction or a free-radical autoxidation. Below are listed a few such tests which have proven quite useful.

(1) 1,4-Diazabicyclo[2.2.2]octane (DABCO, **3**) is chemically inert to $^1O_2$ and is particularly efficient in quenching it to the ground state[45]. It should be borne in mind though that DABCO in concentrations above 0.05 M may quench the excited singlet state of the sensitizing dye[46]. Thus, a sharp reduction in the rate of oxidation as a result of the presence of <0.05 M DABCO is strong evidence for a $^1O_2$ process.

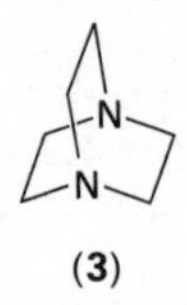

(3)

(2) On the other hand, 2,6-di-*t*-butylphenol is an effective free-radical inhibitor. A dramatic reduction of oxygen uptake upon addition of this phenol to the reaction mixture suggests that the process is free-radical in nature. Foote[47] has pointed out that one must be careful in interpreting negative results since this phenol is not completely inert to $^1O_2$ and does react slowly. However, at concentrations below 0.01 M, its influence on the $^1O_2$ reaction is small in methanol, though it will be larger for most aprotic solvents in which singlet oxygen has a longer lifetime.

(3) Because of the extremely low activation energies required for singlet oxygen processes[48–51], little, if any, dependence of the rate on temperature should be observed. On the other hand, the initiation of free-radical oxidation is well known to be slowed or inhibited at low temperatures[52]. Hence, if a free-radical process is suspected the oxygenation should be repeated at $-78°C$.

(4) In the case of photosensitization, the rate of reaction and product distribution in a given solvent should be independent of the sensitizing dye utilized. This is because the sensitizer is not involved in the oxidation of the substrate but rather only in the generation of the $^1O_2$. The latter should react in the same fashion independent of the source. For the same reason photosensitization results should be comparable to those obtained using chemically generated $^1O_2$ (e.g. from the thermal decomposition of phosphite ozonides or polycyclic arene transannular peroxides). Large variance in product distribution as a result of changes in the sensitizer or $^1O_2$ source is symptomatic of the involvement of a Type I process.

V. MODES OF REACTION

A. Introduction

Unlike $^3O_2$, which displays a biradical character, all the electrons in $^1O_2$ are paired. Hence, the type of reactions it undergoes are expected to involve electron pairs. What's more, it is convenient to think of $^1O_2$ as the oxygen analogue of ethylene. Indeed, each of the three modes in which $^1O_2$ reacts with unsaturated compounds finds precedent in one of the reaction pathways of ethylene.

The first of the singlet oxygen reaction modes is a [2 + 2]cycloaddition to a double bond to form a 1,2-dioxacyclobutane or dioxetane (equation 8). These cyclic peroxides are sometimes of moderate stability but readily cleave thermally or photochemically into two carbonyl-containing fragments. The cleavage is quite often accompanied by chemiluminescence.

(8)

The second mode bears a striking resemblence to the Alder 'ene' reaction[53,54]. In the $^1O_2$ ene reaction, olefins containing an allylic hydrogen are oxidized to the corresponding allylic hydroperoxides in which the double bond has shifted to a position adjacent to the original double bond (equation 9).

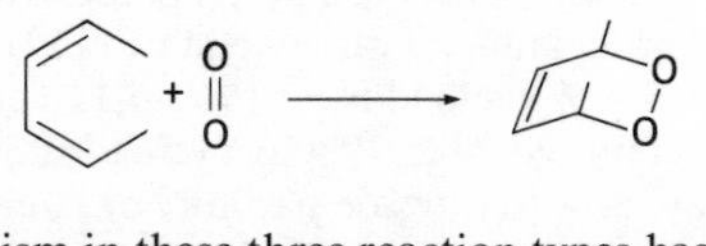

(9)

The third and final mode involves a [4 + 2] Diels–Alder-type addition of singlet oxygen to a diene producing endoperoxides (equation 10).

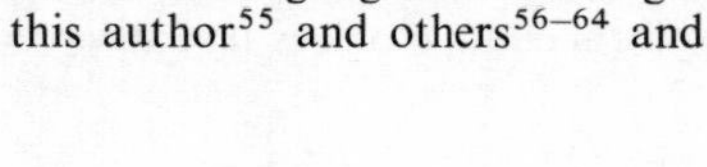

(10)

The question of mechanism in these three reaction types has been the subject of much heated debate over the past decade. The highlights of this long-standing controversy have been recently reviewed by this author[55] and others[56–64] and a detailed discussion is beyond the scope of this chapter. For the purpose of completeness, however, we shall briefly summarize the various positions as we consider each reaction mode in turn. Let us simply note that despite extensive research, the question of mechanism of these three modes has yet to be resolved.

A variety of factors have been shown to control all singlet oxygen reactions[65a]. The *rate* of reaction within a homologous series of compounds is generally inversely proportional to their ionization potential. This suggests that singlet oxygen is mildly electrophilic and sensitive to the nucleophilicity of the olefinic bond. Thus as a rule, alkyl substitution increases the reactivity of olefins 10–100-fold per group. Solvent has only a minimal effect on the rate of reaction; changes in rate are commonly due to solvent effects on the lifetime of singlet oxygen. Because of the low activation energy for singlet oxygen processes (1–5 kcal)[48–51] little if any temperature effect on the rate of reaction is observed. Regarding the *mode* of reaction, electron-rich olefins (such as vinyl sulphides, enol ethers and enamines) as well as sterically hindered alkenes (such as 2,2-biadamantylidene[65b] and 7,7-binorbornylidene[65c]) tend to prefer dioxetane formation, though two modes often compete. Finally, the *direction* of singlet oxygen attack is predominantly, if not exclusively, from the less hindered side of the molecule. Other mode-specific factors will be described as we discuss each reaction type, a task to which we now turn.

B. Singlet Oxygen Diels–Alder Reaction

1. Nature of the substrate

We have already pointed out that the [2 + 4]cycloaddition of $^1O_2$ to dienes[66] is analogous to the well-known Diels–Alder reaction in which oxygen serves as the dienophile. The *cisoid* 1,3-diene functionality requisite for such a cycloaddition commonly resides in nonaromatic systems. Thus Schenck and Ziegler[67] succeeded in the early fifties in synthesizing the pharmacologically active and stable endoperoxide ascaridole (**5**) via the photosensitized oxidation of α-terpinene (**4**). In the absence of sensitizer, free-radical processes intervene and polyperoxide (**6**) is obtained (equation 11).

(**6**) $\xleftarrow{O_2/h\nu}$ (**4**) $\xrightarrow[\text{sensitizer}]{O_2/h\nu}$ (**5**) (11)

A more recent and fascinating example is the photooxidation of cycloheptatriene[68]. In view of its facile valence isomerization between the tropilidene (**7**) and the norcaradiene (**8**) forms, its dienic reactivity is abundant and varied (equation 12). Both forms react with $^1O_2$ to give the corresponding [2 + 4]cycloadducts **10** and **11** respectively. However **7** also undergoes an uncommon, yet precedented, [2 + 6]cycloaddition yielding endoperoxide **9**.

The 1,3-diene moiety may, however, reside partially or completely within an aromatic system. Thus, vinyl aromatic systems such as indenes[69–71], 1,2-dihydronaphthalenes[72], substituted styrenes[73–77], vinylnaphthalenes[76b,78], 2-vinylthiophenes[79] and phenylpyruvic acid[80] all react to form endoperoxides (equation 13).

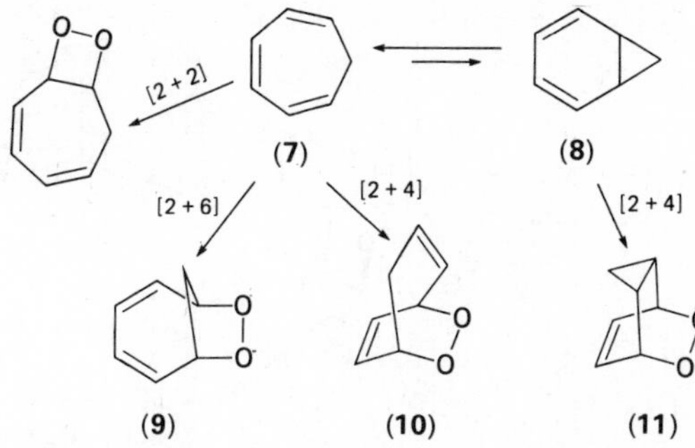

(12)

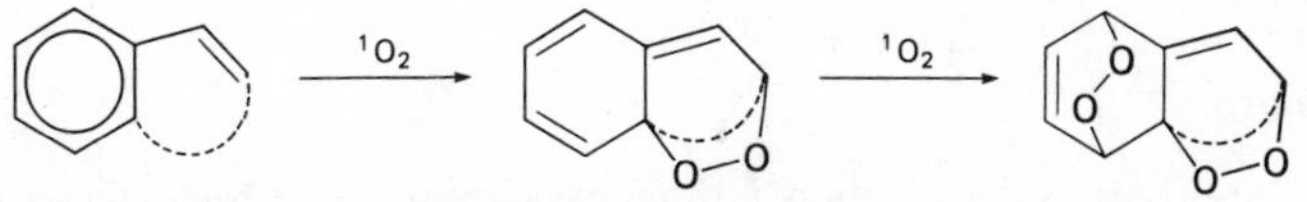

(13)

Polynuclear systems are likewise susceptible to the [2 + 4] mode of attack. For example, it has been known for more than half a century[36] that the red rubrene **12** can be converted into its colourless transannular peroxide **13** by a self-sensitized photooxidation. Like the endoperoxides of other polynuclear arenes (see Section III.D), this endoperoxide reverts thermally to $^1O_2$ and the coloured starting material (equation 14).

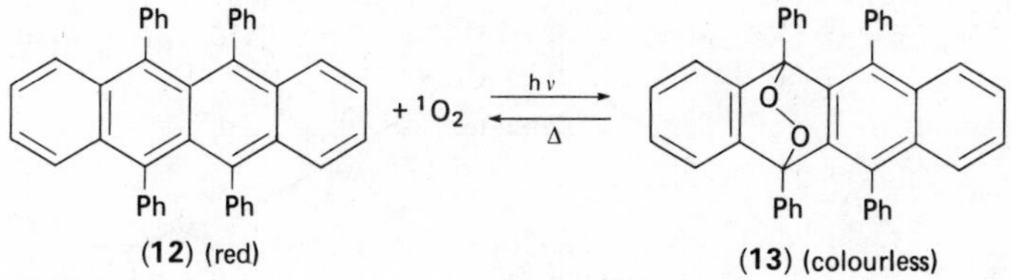

(14)

The 1,4-cycloaddition of $^1O_2$ to electron-rich monocyclic aromatic systems has also been reported. Thus methoxy-[81], dimethylamino-[82] and polymethyl-benzenes[83,84] all react with $^1O_2$ affording endoperoxides as the initial products. Similarly a very large number of diverse heterocycles such as furans, pyrroles, indoles, imidazoles, purines, oxazoles, thiazoles and thiophenes react by this mode[85,86]. Interestingly, the reaction of one mole of $^1O_2$ with a furan in a [2 + 4] fashion yields what may be viewed as a cyclobutadiene monoozonide. Not surprisingly, the latter is quite labile but, like other ozonides, can be reduced with triphenylphosphine (equation 15)[87].

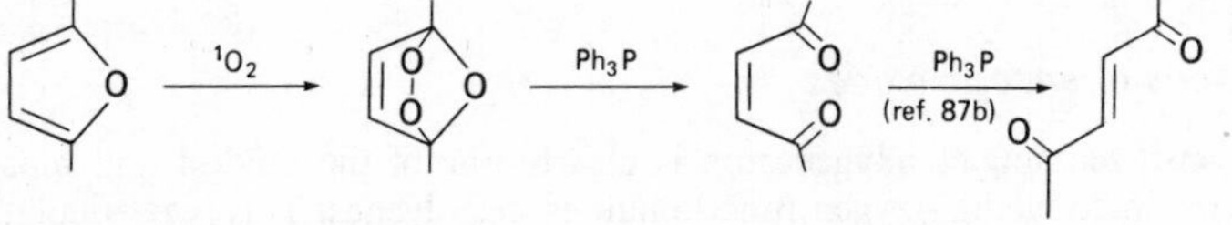

(15)

We close this section by citing a fascinating homo-Diels–Alder reaction which has been reported by Takeshita and coworkers[88] in the photooxidation of spiro[2.4]hepta-4,6-diene (equation 16).

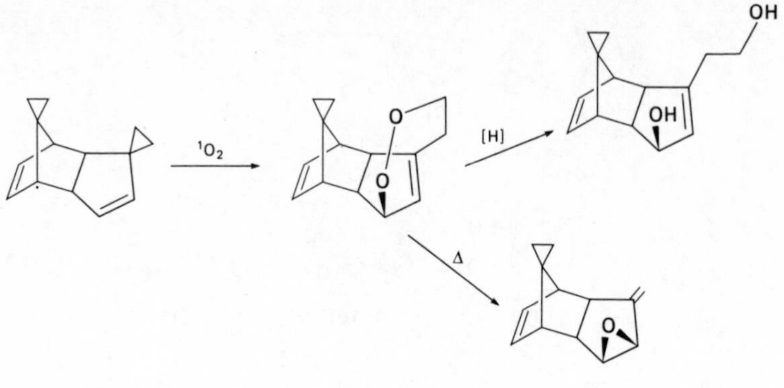

$^1O_2$ [H] Δ OH (16)

2. *Mechanism*

For the $^1O_2$ Diels–Alder reaction[55] three mechanisms have been suggested and are summarized in Scheme 1. The mechanism most commonly invoked is a concerted [2 + 4]cycloaddition[66b,c]; however, two dissenting views have recently been posited. Ogryzlo[63] has presented evidence suggesting the formation of an initial charge-transfer complex. Theoretical calculations by Dewar and Thiel[64] and certain experimental data[55] are consistent with the intermediacy of an allylic perepoxide[64].

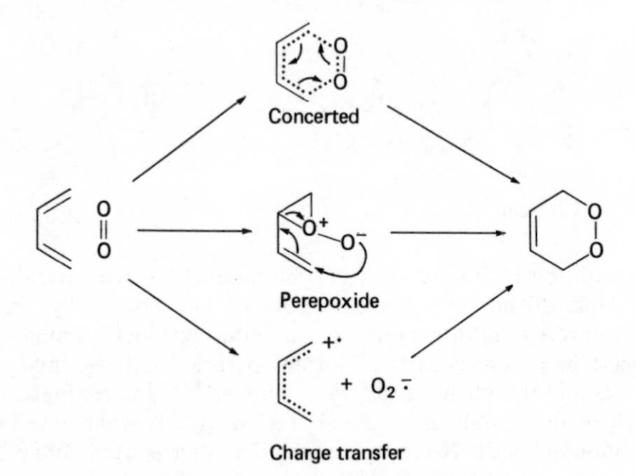

SCHEME 1. Possible mechanisms for the singlet oxygen Diels–Alder reaction.

3. *Reactions of endoperoxides*

Photosensitized singlet oxygenation is clearly one of the mildest and most efficient methods for introducing oxygen functionalities in polyenes. This is particularly true in light of the selective and useful transformations which the labile endoperoxide oxygen–oxygen bond undergoes. Little wonder then that endoperoxide chemistry has long been of interest to the synthetic organic chemist[66d,e,g]. Most recently, however, the discovery of the pivotal role endoperoxides play in the biosynthesis of prostaglandins[89] has spurred scientific exploration of the modes and mechanism of endoperoxide rearrangements and reactions. Scheme 2 illustrates several of the transformations observed for ascaridole (**14**).

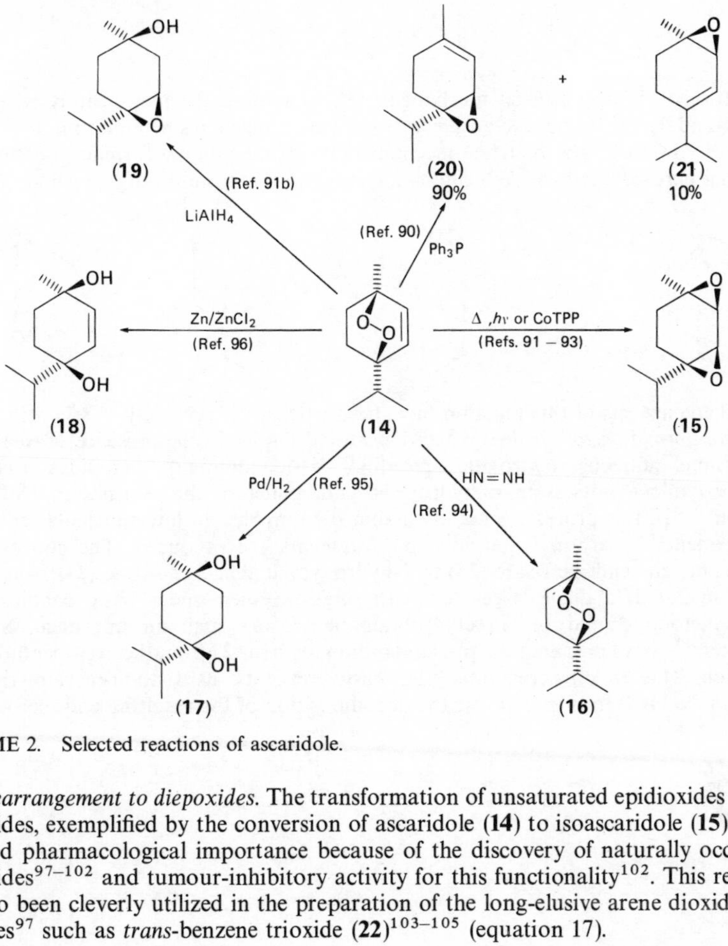

SCHEME 2. Selected reactions of ascaridole.

a. Rearrangement to diepoxides. The transformation of unsaturated epidioxides to 1,3-diepoxides, exemplified by the conversion of ascaridole (**14**) to isoascaridole (**15**)[91], has assumed pharmacological importance because of the discovery of naturally occurring diepoxides[97–102] and tumour-inhibitory activity for this functionality[102]. This reaction has also been cleverly utilized in the preparation of the long-elusive arene dioxides and trioxides[97] such as *trans*-benzene trioxide (**22**)[103–105] (equation 17).

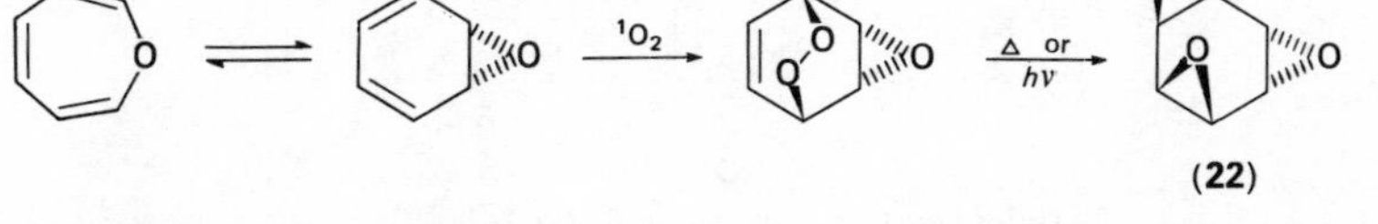

(17)

(**22**)

The rearrangement can be induced thermally[97], photochemically[92] or via metal catalysis[93,106]. The mechanism generally proposed for the thermal and photolytic processes involves homolytic fission of the oxygen–oxygen linkage followed by attack of the oxy radicals on the double bond (equation 18). (The thermal process, however, might

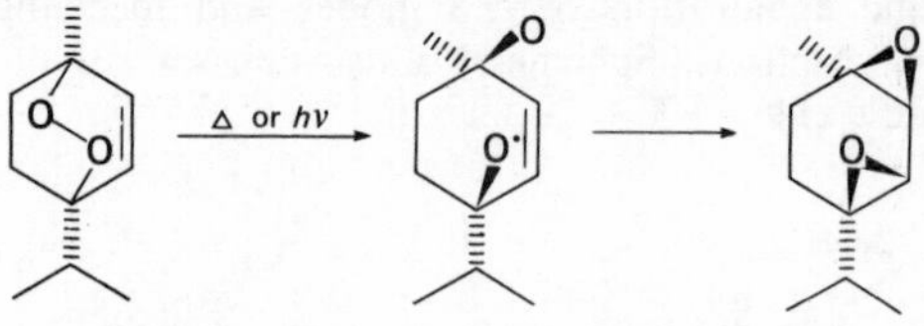

Δ or $h\nu$

(18)

also proceed via a concerted mechanism)[104]. Regarding the metal-catalysed process, Turner and Herz[106] suggest an electron-exchange mechanism (equation 19). This is rejected by Foote[93] who prefers a mechanism involving complex formation between the oxidizing peroxide and the reducing catalyst without separation of ion pairs.

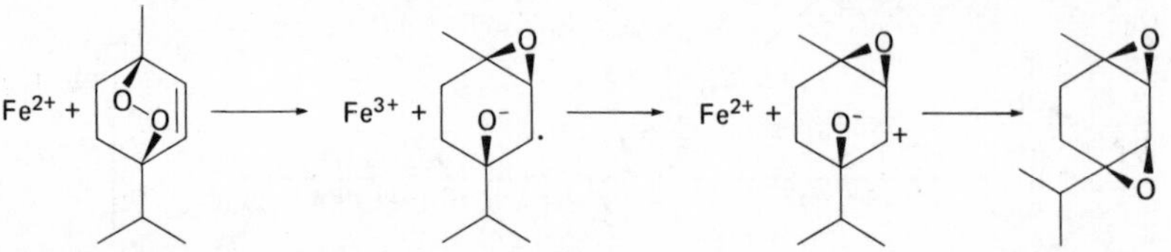

Fe^{2+} + ⟶ Fe^{3+} + ⟶ Fe^{2+} + ⟶ (19)

Synthetic aspects of this reaction have been extensively reviewed[66d–g,97].

b. Transformation to 4-*hydroxy*-2-*en*-1-*ones and furans*. In the presence of weak bases (e.g. dilute aqueous hydroxide, pyridine, basic alumina), peroxides (including hydroperoxides) possessing α-hydrogens can undergo the Kornblum–DeLaMare reaction[107]. In this process, generally assumed to involve an intramolecular carbanion displacement[108], carbonyl and hydroxy fragments are produced. The conversion of cyclopentadiene endoperoxide (**23**) to 1-hydroxycyclopent-2-en-4-one (**24**)[109] is typical (equation 20). If a diene is reacted with singlet oxygen under basic conditions the hydroxyketone should be directly obtainable. Using such an approach, Sih and coworkers[110] have prepared the prostaglandin precursor **27** from the cyclopentadiene **25** (equation 21). In this reaction, the basic reagents used to generate the $^1O_2$ (H_2O_2 + NaOCl) in turn catalyse the decomposition of the resulting endoperoxide **26**.

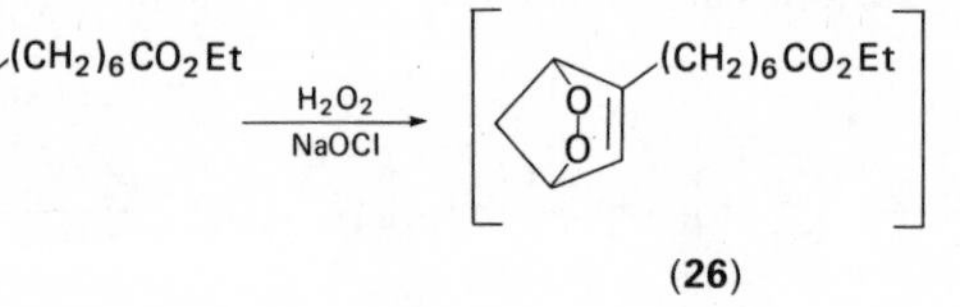

$^1O_2$ (20)

(**23**) (**24**)

$(CH_2)_6CO_2Et$ $\xrightarrow{H_2O_2,\ NaOCl}$ [$(CH_2)_6CO_2Et$] ⟶

(**25**) (**26**)

$(CH_2)_6CO_2Et$ + $(CH_2)_6CO_2Et$ (21)

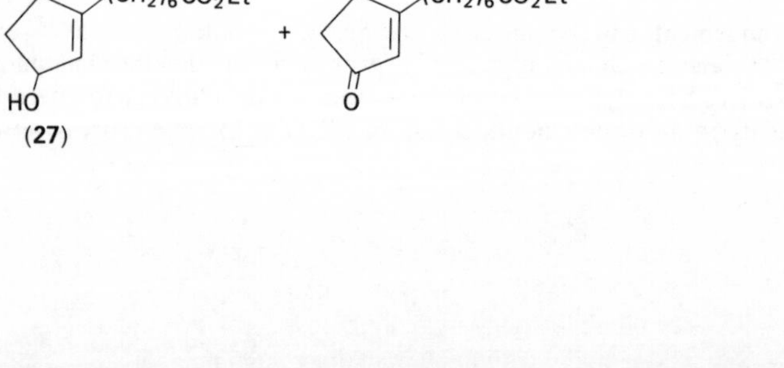

(**27**)

Hydroxyketones are often formed in the thermolysis or photolysis of endoperoxides in the absence of added base, though in this case they are rarely the exclusive product. For example, the endoperoxide of cyclohexadiene (**28**) yields both diepoxide (**29**) and hydroxyketone (**30**) (equation 22).

(22)

	(**28**) → (**29**)	+ (**30**)
Δ (140°C)	36%	45%
$h\nu$ (366 nm)	27%	22%

This transformation has also been catalysed by metals. Hagenbuch and Vogel[111] report that endoperoxide **31** can be converted to **32** in moderate yields (60–75%) through the agency of fluoride ion (as base)[112], or the metal cations Fe^{II} and Rh^{I} (equation 23). Compound **32** is also obtained in low yields when **30** is thermolysed at 130°C. The corresponding diepoxide is the major product.

(23)

(**31**) → CsF, $FeSO_4$ or $Rh_2(CO)_4Cl_2$ → (**32**) CHO, CH_2OH; 60–75% yield

(**31**) → 130°C → diepoxide 65% + 32 <15%

(**31**) → MeCOOH 80°C → (**33**) 60%

In the case of endoperoxides (**35**) formed from acyclic conjugated dienes (**34**), the resulting γ-hydroxy carbonyl compounds (**36**) can close to a cyclic hemiacetal (**37**) and then dehydrate yielding a furan system (**38**) (equation 24). This overall process is generally acid-catalysed. Thus endoperoxide **31** is converted to furan **33** by heating it in acetic acid (equation 23)[111]. Metal catalysis has also been reported[113] Often, however, this reaction occurs spontaneously. For example, in the photooxidation of the sex pheromone 9,11-tetradecadienyl acetate [**34**; R^1 = Et, R^2 = $(CH_2)_8OAc$] only the corresponding furan can be isolated in a 70% yield[114].

Synthetically this reaction has been utilized for the preparation of various furanoterpenes[115–119].

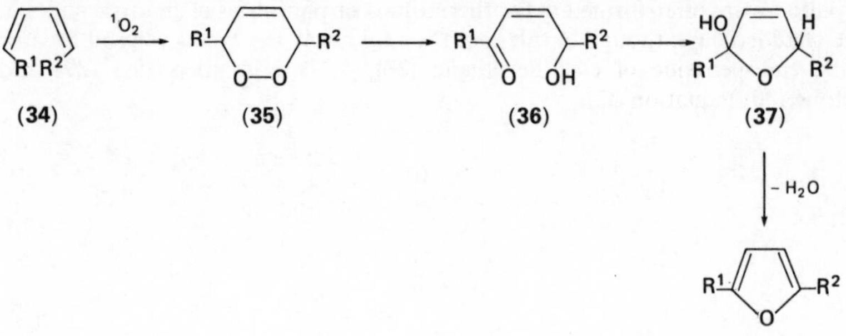

(34) (35) (36) (37) (38)

(24)

c. Reductions. The peroxide linkage is one of the most susceptible bonds towards reductive cleavage by a variety of reductants. It is, therefore, not surprising that in catalytic hydrogenation of an endoperoxide both the peroxide bond and the double bond are reduced yielding *cis*-1,4-diols. One classic example is the conversion of ascaridole to *cis*-*p*-menthane-1,4-diol (**17**, Scheme 2). This sequence suggests a convenient synthetic method of converting 1,3-dienes to 1,4-diols[120,121] (equation 25).

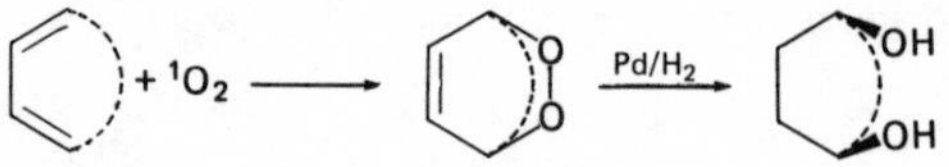

(25)

Selective reduction of the double bond has been accomplished in a variety of endoperoxides [e.g. the transformation of ascaridole (**14**) to its dihydro analogue **16** as shown in Scheme 2] with diimide[94,122,123]. This method has been used recently for the preparation of prostaglandin endoperoxide model compounds[122].

Selective reductions of the oxygen–oxygen bond using $Zn/ZnCl_2$ [96] or $LiAlH_4$ [124] generally yield 2-ene-1,4-diols. Thus ascaridole (**14**) is converted by $Zn/ZnCl_2$ to 1,4-dihydroxy-*p*-menth-2-ene (**18**)[96]. Somewhat surprisingly, with lithium aluminium hydride *cis*-1-hydroxy-3,4-epoxy-*p*-menthane (**19**) is formed[91b]. Trivalent phosphorus compounds in general, and triphenylphosphine in particular, reduce 1,4-endoperoxides to the corresponding 3,4-unsaturated 1,2-epoxide[90,105,124a,125]. The reaction sequence is exemplified by the reduction of naphthalene endoperoxide (**39**, equation 26)[124a].

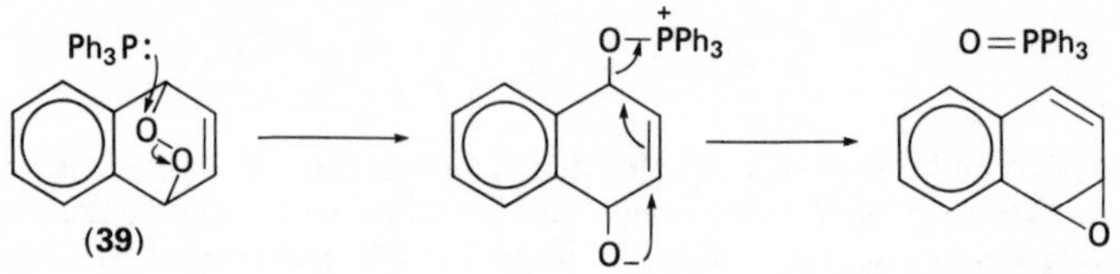

(39)

(26)

d. Thermal rearrangements. We have already noted previously in this section the thermal rearrangement of endoperoxides to diepoxides and/or hydroxyenones. Indeed, the formation of these two rearrangement products is quite general with various endoperoxides of acyclic and cyclic dienes. However, when the diene moiety resides in a five-membered ring the resulting endoperoxides rearrange to epoxy aldehydes as well as bisepoxides (equation 27)[66d].

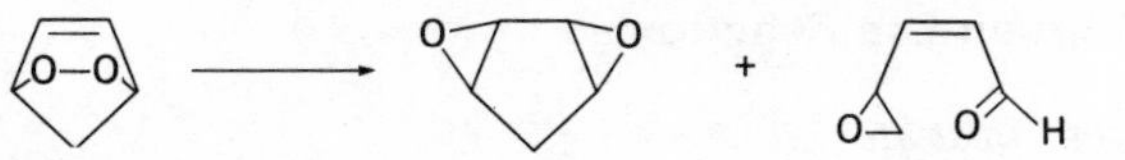

(27)

Polycyclic arenes which undergo [2 + 4]cycloaddition with $^1O_2$ suffer a loss of substantial resonance stabilization energy. Reattainment of this energy is undoubtedly the driving force in the frequent retroreversion of such transannular endoperoxides to $^1O_2$ and the starting aromatic hydrocarbon (see Section III.D).

Endoperoxides are also known to rearrange to dioxetanes[55]. Goto and Nakamura[126] report that endoperoxide **40** rearranges to dioxetane **41** (equation 28). Note that despite the isolation of dioxetane cleavage products, initial $^1O_2$ attack involves not [2 + 2]- but [2 + 4]-cycloaddition. This is merely one example of many which demonstrate the caution required in determining the mode of reaction simply based on product analysis[55].

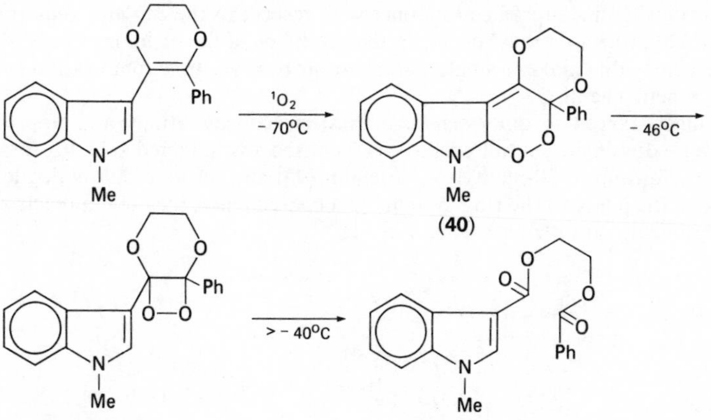

(**40**)

(**41**) (**42**) (28)

e. Solvolysis. The reaction of endoperoxides with nucleophilic solvents leads to hydroperoxides. For example, the photooxidation of furans (**43**) in methanol[87] yields 2-methoxy-5-hydroperoxyfurans (**44**) (equation 29). Similarly, the endoperoxides of 9,10-disubstituted anthracenes undergo hydrolysis or methanolysis upon addition of dilute acid (equation 30)[127].

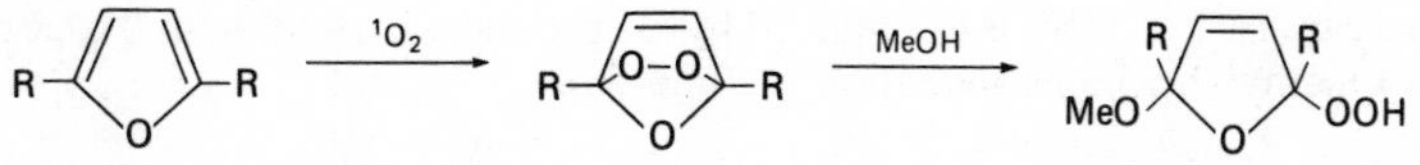

(**43**) (**44**) (29)

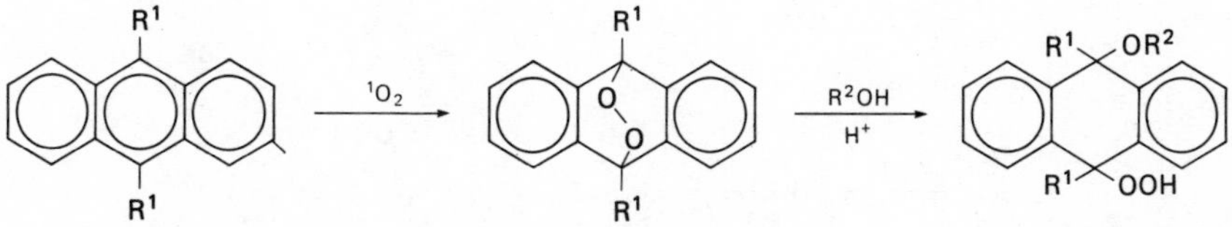

(30)

C. Singlet Oxygen Ene Reaction

1. Nature of the substrate

As we have mentioned already, the 1,3-cycloaddition of $^1O_2$ to olefins bearing allylic hydrogens yields allylic hydroperoxides in which the double bond has shifted to a neighbouring position (equation 9). Because of the facility with which hydroperoxides can be converted to a variety of other functional groups (*vide infra*) this oxidative process permits allylic functionalization, an important synthetic tool. However, a variety of complications may set in when substrates contain more than one allylic hydrogen and/or more than one double bond. We shall briefly review, therefore, the various factors which determine the regio- and stereoselectivity of this reaction and set it apart from free-radical autoxidation.

(*a*) The singlet oxygen ene reaction, in sharp contrast to autoxidation, proceeds stereospecifically in a suprafacial manner with respect to the ene unit such that oxygen attack and hydrogen removal occur on the same side of the olefin molecule[128]. It is this stereospecificity that makes a singlet oxygen route to allylic functionalization so attractive to the synthetic chemist.

(*b*) Singlet oxygen is quite sensitive to steric considerations and approaches the substrate predominantly if not exclusively from the less hindered side. For example, the axial methyl group in 10-methyl-$\Delta^{1(9)}$-octalin (**45**) and its steroidal analogues, inhibits reaction on the β face of the ring-system. As a consequence, the octalin reacts practically stereospecifically on its α face (equation 31)[4b].

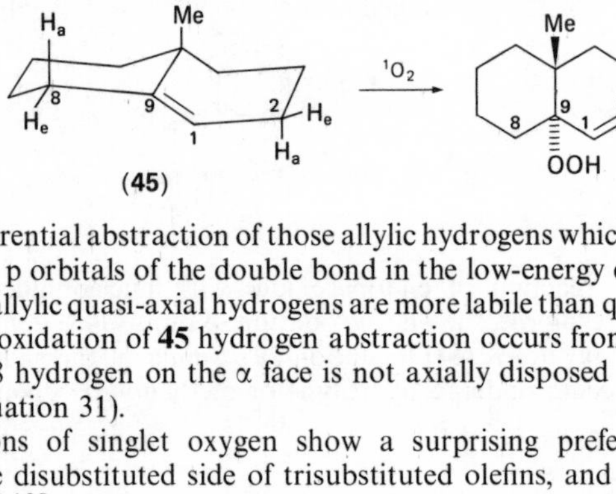

H$_a$ Me H$_e$ 8 9 1 2 H$_e$ H$_a$ (**45**) $\xrightarrow{^1O_2}$ Me 8 9 1 2 OOH (31)

(*c*) There is preferential abstraction of those allylic hydrogens which are aligned parallel to the plane of the p orbitals of the double bond in the low-energy conformations of the olefin. As a result, allylic quasi-axial hydrogens are more labile than quasi-equatorial ones. Thus in the photooxidation of **45** hydrogen abstraction occurs from the α face but only from C-2. The C-8 hydrogen on the α face is not axially disposed in the lowest energy conformation (equation 31).

(*d*) The reactions of singlet oxygen show a surprising preference for hydrogen abstraction on the disubstituted side of trisubstituted olefins, and for *cis* disubstituted olefins over *trans*[60,128]. This is exemplified by the product distribution in the cases of (*E*)- and (*Z*)-3-methyl-2-pentene (equations 32 and 33).

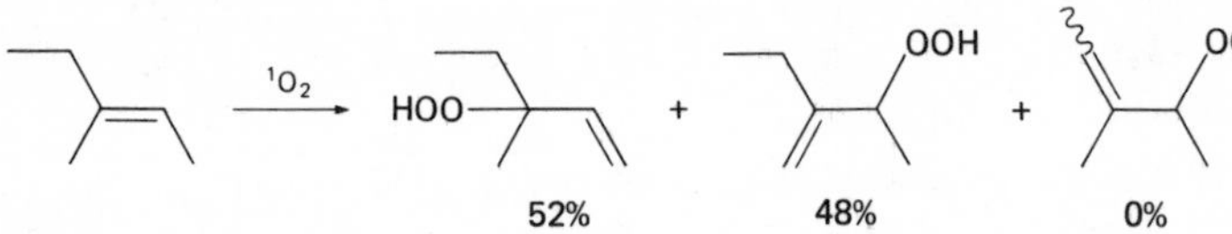

$\xrightarrow{^1O_2}$ HOO— (52%) + OOH (48%) + OOH (0%) (32)

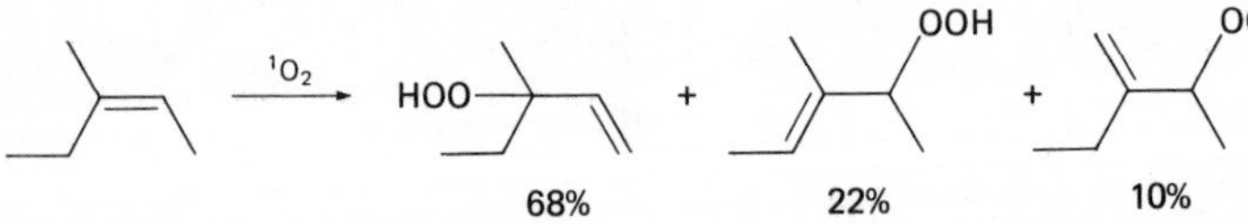

$\xrightarrow{^1O_2}$ HOO— (68%) + OOH (22%) + OOH (10%) (33)

(*e*) In polyolefinic systems, singlet oxygen attack occurs preferentially at the more highly substituted double bond, i.e. at the double bond of lowest ionization potential[113,129]. Thus in the photooxidation of (+)-dihydromyrcene (**46**) only the trisubstituted double bond reacts (equation 34).

1. $^1O_2$ 2. Na_2SO_3 HO HO

(**46**) (**47**) 51% (**48**) 49% (34)

(*f*) In contradistinction to autoxidation, the susceptibility of a C—H bond to abstraction in singlet oxygenations is not inherently related to whether it is primary, secondary or tertiary[130]. Not surprising then is the observation that **47** and **48** are formed in equal amounts in the photooxidation of **46** (equation 34). Furthermore, the thermodynamic stability of the final double bond has little effect on the reaction[130,131]. Indeed, Frimer and Roth[132] have reported that in the photooxidation of 1,1-dicyclopropylpropene (**49**) in benzene, the more highly strained alkylidene cyclopropane product **50** predominates over **51** by a ratio of 6:1 (equation 35). This is despite the required investment of more than 11 kcal in additional strain energy. Finally, there is a lack of a strong Markownikoff directing influence on the product distribution in the singlet oxygen ene reaction. This has been clearly demonstrated by Foote and Denny[133] for a series of phenyl-substituted 2-methyl-3-phenyl-2-butenes (**52**). The ratio of the two products **53** and **54** remained virtually ($\pm 2\%$) unchanged over a large series of substituents (equation 36).

$^1O_2$ C_6H_6 OOH HOO

(**49**) (**50**) 86% (**51**) 14% (35)

$^1O_2$ MeOH OOH HOO R R R

(**52**) (**53**) 73% (**54**) 27% (36)

R = *p*-MeO, *m*-MeO, *p*-Me, *m*-Me, H, *p*-Cl, *m*-Cl, *p*-CN, *m*-CN, *p*-NMe_2

(g) Frimer and coworkers have argued that the interatomic distance between the α-olefinic carbon and the γ-allylic hydrogen may play a role in determining whether the latter is abstractable[131,134]. Similarly Jefford and Rimbault[135] have recently suggested that the regioselectivity of hydroperoxidation of 1-alkylcycloalkenes can be rationalized in terms of the interatomic distance between the terminal oxygen of a supposed zwitterionic intermediate (see Section V.C.2) and the γ hydrogen. These concepts will still require substantial testing to prove their generality.

2. *Mechanism*

For the singlet oxygen 'ene' reaction, five mechanisms are presently under consideration and these are summarized in Scheme 3. According to the first two proposals, $^1O_2$ attacks one end of the olefinic linkage generating either a biradical[55,56,58] or a zwitterion[55,56,59].

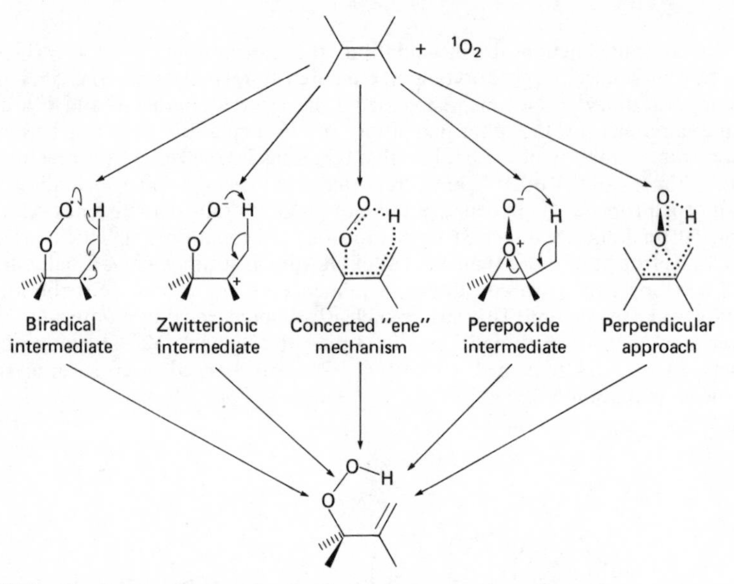

SCHEME 3. Possible mechanisms for the singlet oxygen 'ene' reaction.

In a subsequent step these intermediates collapse to the observed allylic hydroperoxides. In the concerted ene mechanism, a six-centre transition state is involved in which attack of one end of the singlet dioxygen molecule occurs at the α-olefinic carbon while the other end abstracts the γ-allylic hydrogen. In the two remaining proposals, approach of the $^1O_2$ is along the perpendicular bisector of the plane of the p orbitals. They differ however on whether a discrete perepoxide (peroxirane) intermediate is formed[55,56] or whether the initial interaction proceeds smoothly to product[60–62].

3. *Reactions of allylic hydroperoxides*

a. Reduction to allylic alcohols. Allylic hydroperoxides can be conveniently reduced to the corresponding alcohols by a variety of reagents including Ph_3P, $(PhO)_3P$, $LiAlH_4$, $NaBH_4$, Na_2SO_3 and Me_2S. Because of the general instability of hydroperoxides, their reduction prior to work-up and product isolation is common practice in singlet oxygen studies. The synthetic use of singlet oxygen in the preparation of allylic alcohols both as final products and as intermediates has been recently reviewed[66d,e].

An interesting variation of this reaction has been developed by Conia and coworkers[136] for the synthesis of α,β-unsaturated carbonyl compounds (Scheme 4). For this purpose ketones bearing an α-hydrogen are converted to methyl enol ethers with methoxymethylenetriphenylphosphorane. Photooxygenation of the enol ether in benzene produces primarily the ene product, a peroxy hemiacetal. Reduction of the latter with triphenylphosphine yields the hemiacetal which spontaneously eliminates methanol generating the desired enone. Alternatively, treatment of the peroxy hemiacetal with acetyl chloride in pyridine results in the net elimination of the elements of water, yielding an α,β-unsaturated ester. This latter reaction probably occurs via *in situ* generation of the peroxyacetate followed by a Kornblum DeLaMare elimination[107,108].

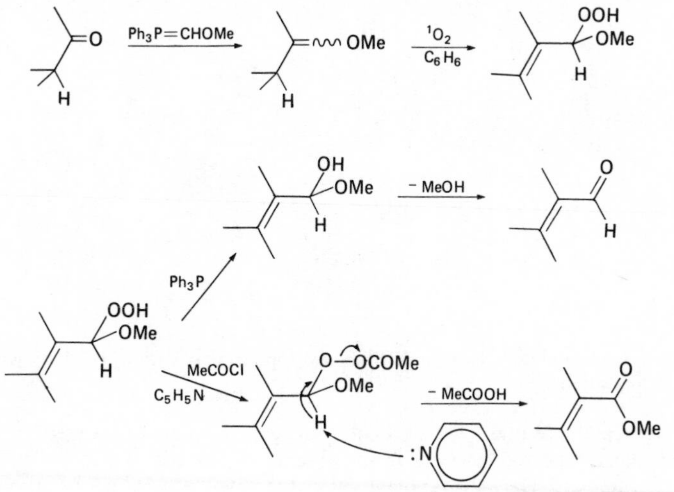

SCHEME 4. Scheme for the synthesis of α,β-unsaturated aldehydes and esters using singlet oxygen.

b. Transformation to two carbonyl fragments or a divinyl ether. In principle heterolysis of the peroxide bond should generate both a negative and a positive oxygen fragment. The instability of the latter with respect to a carbocation would then initiate skeletal changes in the carbon framework resulting from migration of groups to the electron-deficient oxygen. Such heterolyses and ensuing rearrangements have indeed been observed with hydroperoxides and are generally acid-catalysed. One classic example is the acid-catalysed cleavage of a hydroperoxide to an alcoholic and ketonic fragment[137–139], for which the accepted mechanism is outlined in Scheme 5. Relative migratory aptitudes have been determined for this reaction and their qualitative order is as follows[90]:

$$\text{cyclobutyl} > \text{aryl} > \text{vinyl} > \text{hydrogen} > \text{cyclopentyl} \simeq \text{cyclohexyl} \gg \text{alkyl}.$$

In the particular case of allylic hydroperoxides the migrating group is generally vinylic. In such cases the resulting fragments will both be ketonic (Scheme 6, path a). Because of this fundamental difference in the make-up of the products, this transformation of allylic hydroperoxides to *two* carbonyl fragments, called Hock cleavage, has for a long time been

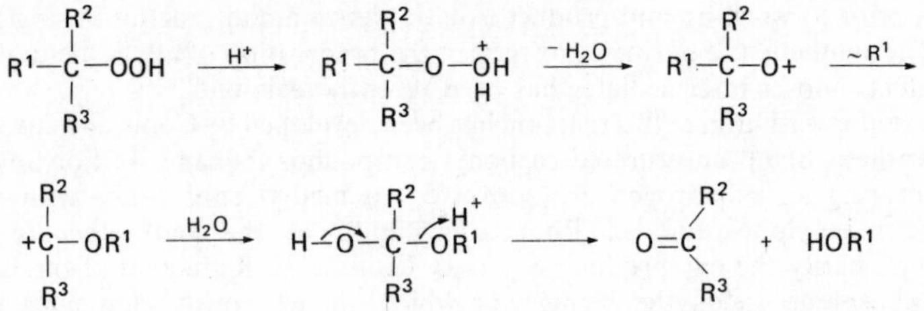

$$R^1-C(R^2)(R^3)-OOH \xrightarrow{H^+} R^1-C(R^2)(R^3)-O-\overset{+}{O}H_2 \xrightarrow{-H_2O} R^1-C(R^2)(R^3)-O^+ \xrightarrow{\sim R^1}$$

$$^+C(R^2)(R^3)-OR^1 \xrightarrow{H_2O} H-\overset{+}{O}H-C(R^2)(R^3)-OR^1 \longrightarrow O=C(R^2)(R^3) + HOR^1$$

SCHEME 5. Mechanism for the acid-catalysed cleavage of hydroperoxides.

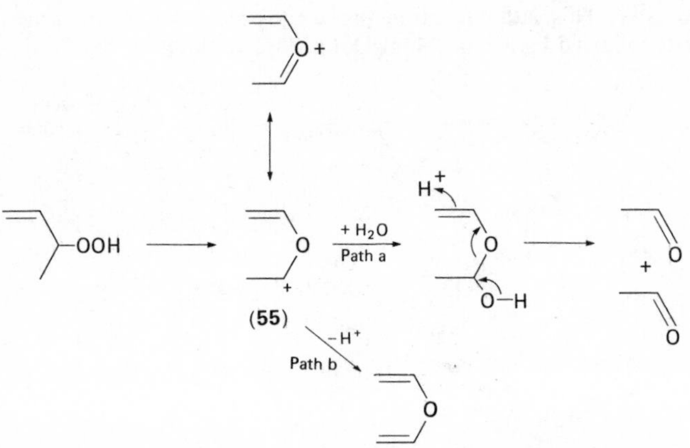

O+

OOH

O

+ H_2O

Path a

H^+

O

O–H

O

+

O

(**55**)

$-H^+$

Path b

O

SCHEME 6. Acid-catalysed cleavage of allylic hydroperoxides. Path a: Hock cleavage, path b: divinyl ether formation.

classified separately. While such cleavages are generally acid-catalysed[66d,134,140–145], several have been reported to occur in the absence of any added acid[46,132,146–151]. For example, Turner and Herz[147] report that in the low-temperature photooxidation of dihydrohexamethyl(Dewar benzene) (**56**) the resulting hydroperoxide **57** is stable below 0°C and can be reduced to alcohol **58**. Above 0°C the hydroperoxide undergoes Hock cleavage to diketone **59** (equation 37).

It should be pointed out in passing that carbonyl fragments also result from the decomposition of a dioxetane. Hence it is crucial that one be able to distinguish between the two modes. Low-temperature reduction of the labile hydroperoxide to the corresponding alcohol (**58**) is one common solution (equation 37). A discussion of several other techniques has been presented elsewhere[55] (see also Section V.D.3.c).

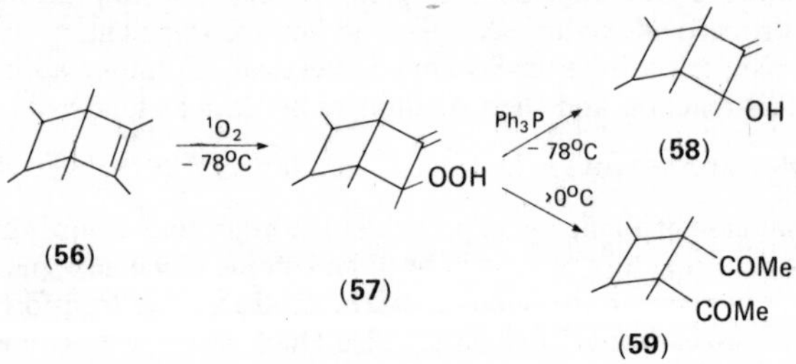

$^1O_2$, −78°C

OOH

Ph_3P, −78°C

OH

>0°C

COMe

COMe

(**56**) (**57**) (**58**) (**59**) (37)

A further example of the principles thus far delineated can be found in the photooxidation of methylenecyclopropanes **60a** and **b** (Scheme 7)[134]. Allylic hydroperoxide **61**, formed as the ene reaction product, is thermally quite labile but can be reduced to the corresponding alcohol **62** at −78°C. As the temperature is raised heterolysis of the oxygen–oxygen linkage occurs. If vinyl migration occurs in a typical Hock cleavage (Scheme 7, path a), this leads to ketones **63** and **64**. The reader is reminded however that the migratory aptitude of a cyclobutyl bond[90] is of the same order of magnitude as that of a vinyl group. It should not be surprising then that products (such as **65**) resulting from the shift of one side of the cyclopropyl ring to positive oxygen (path b) are also observed.

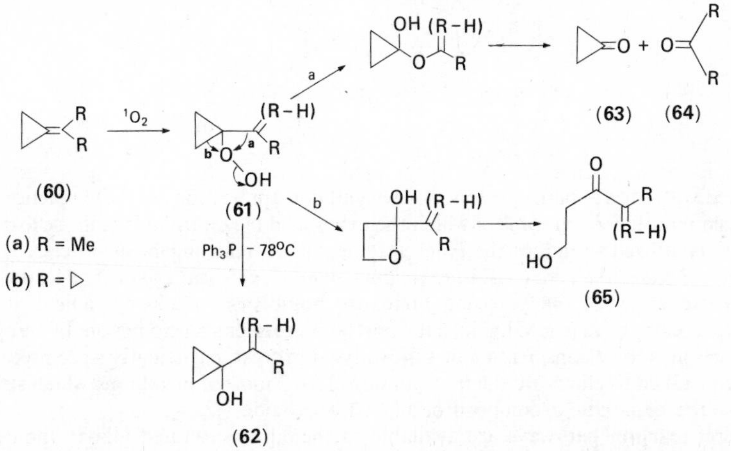

SCHEME 7. Allylic hydroperoxide rearrangement products in the photooxidation of α,α-dialkylmethylenecyclopropanes. Path a: vinyl group migration (Hock cleavage), path b: cyclopropyl migration.

An interesting variation on the Hock cleavage theme is shown in Scheme 6 (path b). In this variant a proton is eliminated α to the oxycarbonium ion **55** yielding a divinyl ether. For example 1,2-dihydronaphthalene 2-hydroperoxide (**66**) rearranges thermally to 3-benzoxepin (**67**) (equation 38)[152]. Similarly, in the photooxidation of 3β-acetoxylanost-8-ene (**68**) divinyl ether **70** has been isolated, presumably also a rearrangement product of the corresponding hydroperoxide **69** (equation 39)[153]. A biological analogy for this reaction is the enzymic conversion of 9-hydroperoxylinoleic acid to the divinyl ether colneleic acid[154]. Hock cleavage[155] and divinyl ether formation[156] have also been

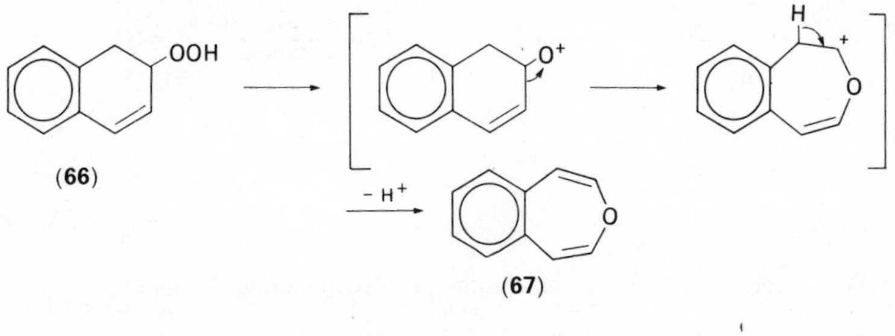

(38)

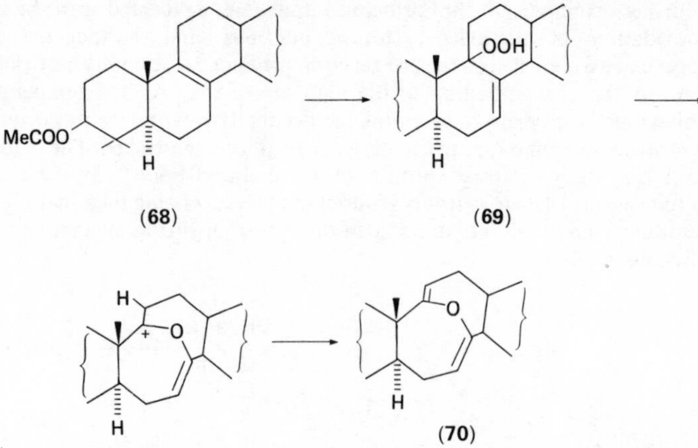

(39)

observed with several peroxy esters. A few synthetic applications of these reactions have also been reported[157], in some of which the Lewis acid boron trifluoride in the form of its etherate is utilized to induce the Hock fragmentation–rearrangement process.

c. Homolysis of the peroxy linkage yielding enones, enols and epoxides. Because of the relative weakness of the peroxide bond, its homolysis to alkoxy radical at room temperature or above (e.g. GLC injector port) is a prevalent phenomenon. In many cases this reaction is to be considered a metal-catalysed process, particularly since precautions are rarely taken to eliminate the trace amount (10^{-8} mol) of metal ions which suffice to catalyse the homolytic decomposition of hydroperoxides[158].

Several reaction pathways are available to the α,β-unsaturated alkoxy radical thus generated (Scheme 8)[159–162]. Firstly an allylic alcohol can be formed via hydrogen abstraction. Alternatively, β cleavage of a neighbouring β hydrogen, alkyl or alkoxy group would lead to an α,β-unsaturated carbonyl compound. In the case of primary and secondary hydroperoxides loss of a hydrogen atom is quite prevalent. In sum total, this corresponds to the elimination of the elements of water from the hydroperoxide, a process commonly called Hock dehydration. For tertiary hydroperoxides carbonyl formation

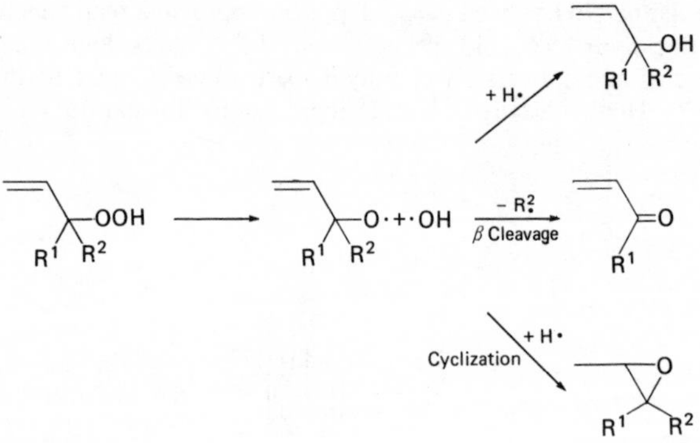

SCHEME 8. Reaction pathways of α,β-unsaturated alkoxy radicals.

requires carbon–carbon bond scission, while for α-hydroperoxy ethers or esters carbon–oxygen cleavage often results.

One interesting case, which demonstrates the various pathways discussed above, is the photooxidation of 1-methoxycyclohexene[163,164] (Scheme 9) which produces two hydroperoxides (**72** and **73**) and a dioxetane (**74**) as primary products. Thermolysis of allylic hydroperoxide **72** in the GLC injector port generates allylic alcohol **75** and Hock dehydration product enone **76**. The former is the sole product when **72** is treated with triphenylphosphine. Peroxyhemiacetal **73** is thermolysed to cyclohexenone **77** (via β cleavage) and to aldehydo ester **78** (via Hock cleavage). When **73** is reduced with Ph_3P only **77** is obtained. Dione **78** is, of course, also the dioxetane cleavage product.

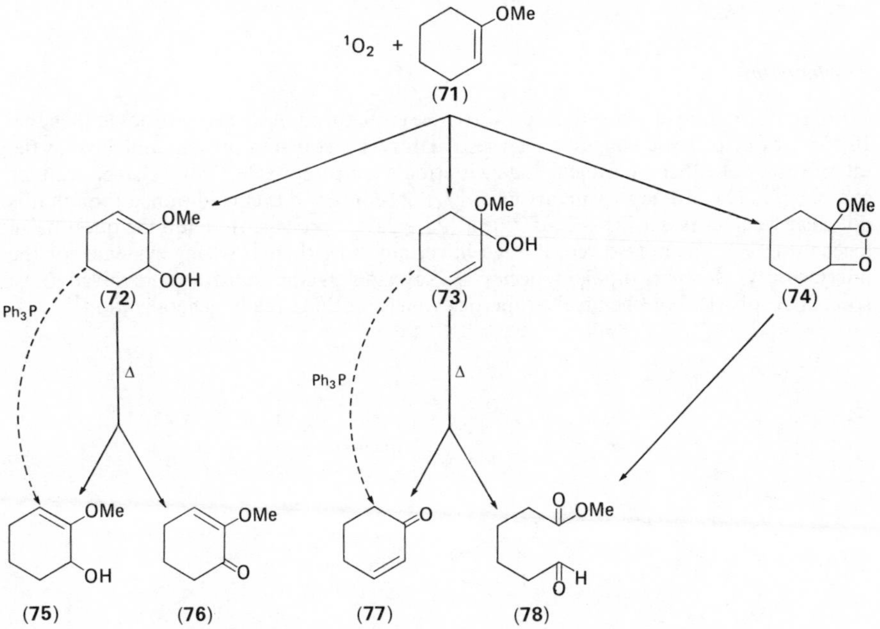

$^1O_2$ + (71) → (72), (73), (74); (72) —Ph_3P→ (75); (72) —Δ→ (75) + (76); (73) —Ph_3P→ (77); (73) —Δ→ (77) + (78); (74) → (78)

SCHEME 9. Photooxidation products of 1-methoxycyclohexene.

There is a third pathway for the allylic alkoxy radical and that is to cyclize to an epoxide (Scheme 8). While this course is less common, a few examples[165–168] exist in the literature and should be kept in mind.

d. Miscellaneous reactions. Primary and secondary hydroperoxides undergo Kornblum–DeLaMare dehydration[107,108] in the presence of bases. To prevent competing reactions, the transformation is often carried out in the presence of acetyl chloride[136] or acetic anhydride[156]. These presumably convert the hydroperoxides to peroxy esters which then readily eliminate acetic acid yielding the desired enone. As noted above, Conia and coworkers[136] have used this method to prepare α,β-unsaturated esters from α,β-unsaturated peroxyhemiacetals (see Scheme 4). Allylic hydroperoxides are also reported to undergo a 1,3- and 1,5-hydroperoxide shift[169].

C. Singlet Oxygen Dioxetane-forming Reaction

1. Nature of the substrate

The [2 + 2]cycloaddition of $^1O_2$ to olefins occurs most commonly with heteroatom-activated double bonds such as are found in enol ethers, vinyl sulphides and enamines[57]. However, a variety of other substrates undergo this reaction including those containing double bonds which are either sterically hindered[65b,c,170–173], strained[174,175] or cumulated (allenes[176], ketenes[177–180], sulphines[181–183] and thioketenes[183b]). Carbon–heteroatom double bonds (sulphines [181–183], oximes[184], thiones[185], thioketenes[183b] and phosphorous ylides[186]), 1-methylene-2,5-cyclohexadienes[187] and certain conformationally rigid vinyl cyclopropanes[131] also react by this mode.

2. Mechanism

For the formation of a dioxetane product, five mechanisms have been proposed (Scheme 10)[55,57]. Four of these suggest that the reaction is a two-step process and invoke the intermediacy of either a biradical[58], a zwitterion[188], a perepoxide[189] or a charge-transfer complex[188]. The fifth argues in favour of [2 + 2] concerted cycloaddition, although it is not clear whether it is a $[2_S + 2_A]$[190] or a $[2_S + 2_S]$[189] process. Here too the question of mechanism is far from resolved, though in certain cases there is strong evidence for the intermediacy of some dipolar species[55]. Nevertheless the substrates involved have specialized physical or chemical properties which preclude ready generalization[55].

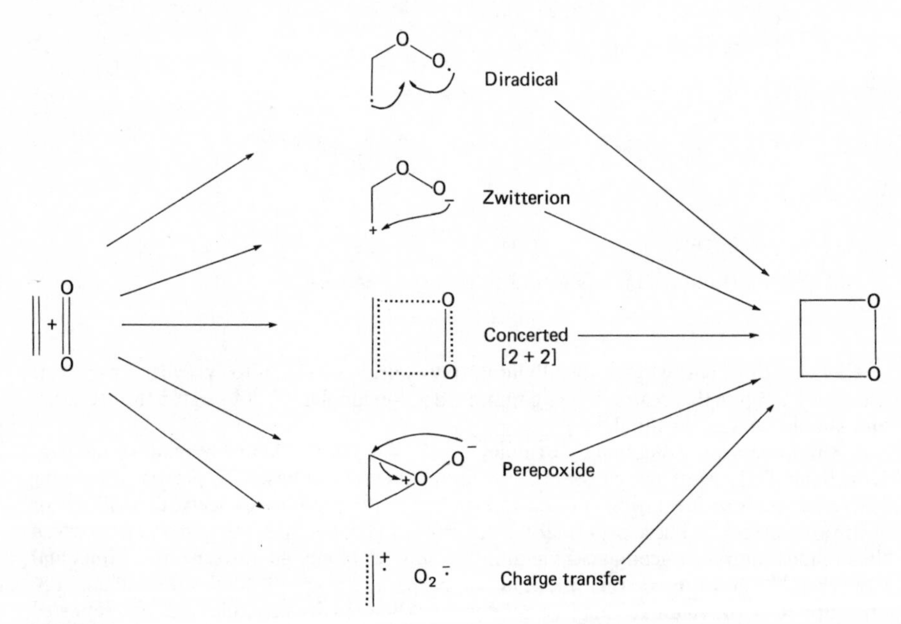

SCHEME 10. Possible mechanisms for the singlet oxygen dioxetane-forming reaction.

3. Reactions of dioxetanes

a. Cleavage. Dioxetanes cleave thermally[191,192] or photochemically[193,194] producing carbonyl fragments and chemiluminescence[195,196]. Bulky rigid groups render the dioxetane considerably more stable. Thus while most dioxetanes cleave somewhere around 50°C ($E_a \simeq 25$ kcal), adamantylideneadamantanedioxetane (**79**)[65b] and norbornylidenenorbornanedioxetane (**80**)[65c] are thermally stable up to 240°C and 200°C, respectively ($E_a \simeq 35$ kcal).

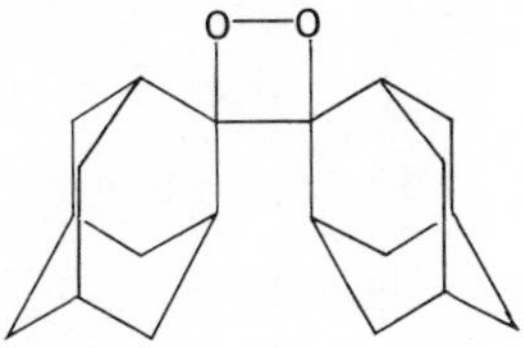

(**79**) 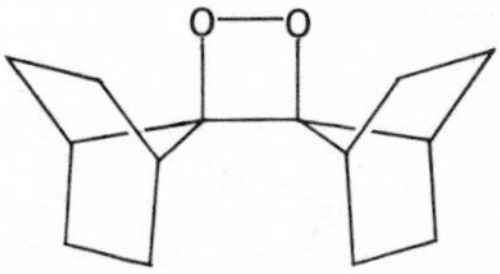 (**80**)

The rate of decomposition of dioxetanes is catalysed by traces of metals[197], silica gel[198] and electron donors such as enol ethers and amines[197,199]. The mechanism of this catalysis in the case of electron donors is not at all clear. Schuster discounts a role for electron-transfer-initiated reactions in the case of dioxetanes[200] based on pulsed laser spectroscopic investigations[201]. He finds no evidence for electron transfer even when the donor is excited singlet pyrene which is some 60 kcal mol^{-1} easier to oxidize than triethylamine. Some other mechanism must be responsible for the observed catalytic effect in this case[201].

b. Rearrangement to α-hydroxyketones and α-diketones. While the carbon–carbon bond cleavage discussed above predominates for most dioxetanes, those bearing sulphur and nitrogen substituents have shown additional modes of fragmentation. Wasserman[202–205] reports that enamines of cyclic ketones react with singlet oxygen, forming dioxetanes which are isolable at low temperatures. At room temperature, however, these cleave (probably via a Kornblum–DeLaMare mechanism[107]) almost exclusively to keto aminals. The latter are unstable and readily expel the amine group, thereby generating α-diketones in high overall yield (equation 40). Cyclic keto aminals are stable and have been isolated in the photooxidation of pyrroles[206,207] and 4-azaandrostenones[208]. In all these cases little carbon–carbon cleavage has been observed.

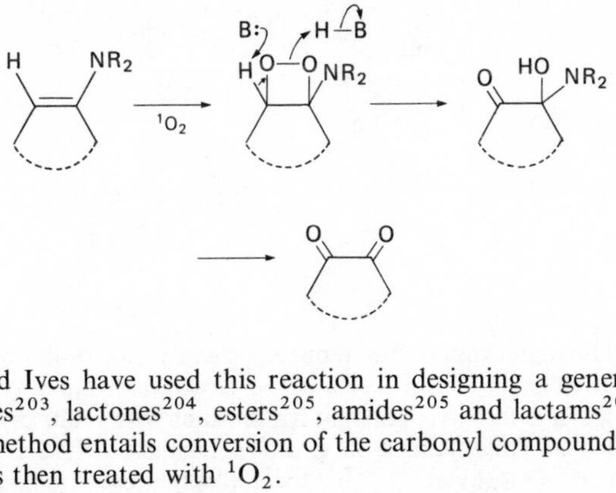

H NR$_2$ $\xrightarrow{^1O_2}$ B: H–B HO–O NR$_2$ ⟶ O HO NR$_2$ ⟶ O O (40)

Wasserman and Ives have used this reaction in designing a general procedure for converting ketones[203], lactones[204], esters[205], amides[205] and lactams[205] to their α-keto congeners. This method entails conversion of the carbonyl compound to its α-enamino analogue which is then treated with $^1O_2$.

In the photooxidation of vinyl sulphides and the enamines of acyclic ketones[209–213] the usual C–C cleavage products are observed as well as diketones and hydroxyketones. The latter are formed even in the absence of an α-hydrogen; hence, a Kornblum–DeLaMare elimination[107] is precluded. A likely mechanism (Scheme 11) for the transformations observed here would involve initial O—O bond fission followed by either C—C or C—X cleavage. In the latter case, hydrogen abstraction by the resulting α-ketoalkoxy radical would lead to a hydroxyketone while β-cleavage would generate a diketone.

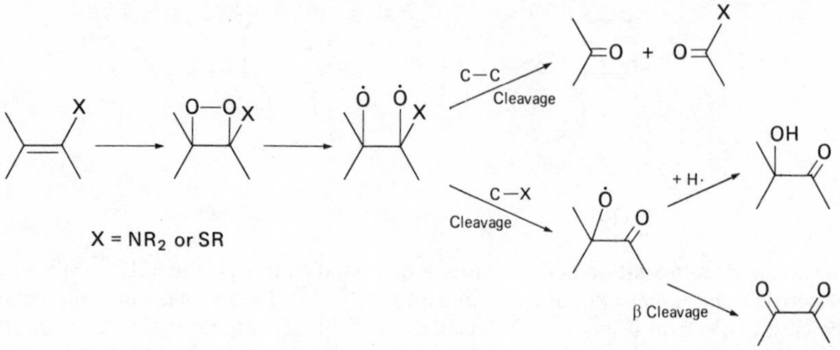

X = NR$_2$ or SR

C—C Cleavage; C—X Cleavage; + H·; β Cleavage

SCHEME 11. Photooxidation of vinyl sulphides and enamines.

c. Nucleophilic addition. Alkyl- and aryl-dioxetanes do not generally undergo nucleophilic attack at carbon. For example, tetramethyldioxetane is quite insensitive to acidic, basic or neutral methanol at 25°C[191]. However, there have been several reports of solvolysis where dioxetanes with heteroatom substituents are involved[208–211,214]. This is illustrated by the photooxidation of 1-ethoxycyclohexene (**81**) (equation 41)[209]. When the reaction is carried out in acetone the expected dioxetane cleavage product (**82**) is obtained. The aldehydo ester **82** is, however, essentially absent when the reaction is performed in methanol. In its place appears the dioxetane solvolysis product 2-methoxycyclohexanone (**83**).

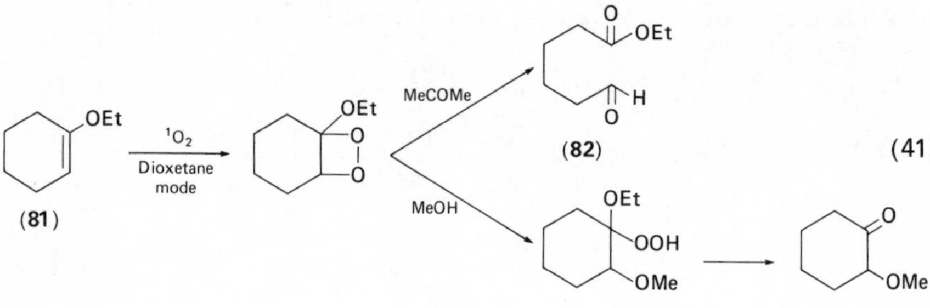

(**81**) $\xrightarrow[\text{Dioxetane mode}]{^1O_2}$ dioxetane; MeCOMe → (**82**); MeOH → OEt/OOH/OMe intermediate → (**83**) (41)

d. Reduction. Dioxetanes can be reduced cleanly to diols by the action of $LiAlH_4$ [191,215,216] or $NaBH_4$ [168]. Takeshita and coworkers report that dioxetanes are convertible to *cis*-1,2-glycols by visible light irradiations with relatively large amounts of xanthene dyes, such as Rose Bengal in protic solvents[217,218]. Trivalent phosphorus compounds (such as phosphines[191,197,219,220], phosphites[197,221]), bisulphite ion[191], sulphoxylates[222] and sulphides[222,223] react with acyclic dioxetanes to yield O—O bond insertion products which collapse to epoxides. Cyclic dioxetanes generally yield allylic alcohols[191]. Since diphenyl sulphide is inert to $^1O_2$, endoperoxides and hydroperoxides[223b], yet readily reacts with dioxetanes[223], it can be used to discern which carbonyl fragments result from dioxetane cleavage and which from Hock cleavage[55] (see also Section V.C.3.b).

VI. CONCLUSION

The primary emphasis of this chapter has been on the organic chemistry of $^1O_2$ and its primary products in the liquid phase. It has become increasingly clear over the past decade that $^1O_2$ is of importance in a wide range of fields including industrial polymer reactions[224], atmospheric chemistry[225], biology and medicine[1,226] just to mention a few. We are confident that the dizzying pace of two hundred papers a year[227], that has characterized this field since the mid-1970s, will continue. More importantly we trust that many new horizons lie yet ahead for the creative and insightful singlet oxygen chemist.

VII. REFERENCES

1. C. S. Foote in *Free Radicals in Biology*, Vol. II (Ed. W. A. Pryor), Academic Press, New York, 1972, Chap. 3, pp. 85–133.
2. (a) J. C. Slater, *Quantum Theory of Molecules and Solids*, Vol. 1, McGraw-Hill, New York, 1963.
   (b) O. Raab, *Chem. Zentr.*, **1**, 618 (1900); C. W. Hausman, *Biochem. Z.* **7**, **14**, 275 (1908).
   (c) P. B. Merkel and D. R. Kearns, *J. Amer. Chem. Soc.*, **94**, 1029 (1972).
3. See for example:
   (a) M. Fritzche, *Compt. Rend.*, **64**, 1035 (1867).
   (b) O. Raab, *Chem. Zentr.*, **1**, 618 (1900); C. W. Hausman, *Biochem. Z.* **7**, **14**, 275 (1908).
4. (a) G. O. Schenck and E. Koch, *Z. Elektrochem.*, **64**, 170 (1960).
   (b) K. Gollnick, *Advan. Photochem.*, **6**, 1 (1968).
5. W. Adam, *Chem. Z.*, **99**, 142 (1975).
6. K. Gollnick, T. Franken, G. Schaade and G. Dörhöfer, *Ann. N. Y. Acad. Sci.*, **171**, 89 (1970).
7. A. A. Frimer, P. D. Bartlett, A. F. Boschung and J. E. Jewett, *J. Amer. Chem. Soc.*, **99**, 7977 (1977).
8. W. R. Adams in *Oxidation*, Vol. II (Ed. R. L. Augustine and D. J. Trecker), Marcel Dekker, New York, 1971, Chap. 2, p. 65.
9. R. M. Boden, *Synthesis*, 738 (1975).
10. J. R. Williams, G. Orton and L. R. Unger, *Tetrahedron Letters*, 4603 (1973).
11. E. C. Blossey, D. C. Neckers, A. L. Thayer and A. P. Schaap, *J. Amer. Chem. Soc.*, **95**, 5820 (1973).
12. R. Nilsson and D. R. Kearns, *Photochem. Photobiol.*, **19**, 181 (1974).
13. A. P. Schaap, A. L. Thayer, E. C. Blossey and D. C. Neckers, *J. Amer. Chem. Soc.*, **97**, 3741 (1975).
14. C. Lewis and W. H. Scouten, *Biochem. Biophys. Acta*, **444**, 326 (1976).
15. A. P. Schaap, A. L. Thayer, K. A. Zaklika and P. C. Valenti, *J. Amer. Chem. Soc.*, **101**, 4016 (1979).
16. H. H. Seliger, *Anal. Biochem.*, **1**, 60 (1960).
17. T. Wilson and J. W. Hastings, *Photophysiology*, **5**, 49 (1970).
18. B. Z. Shakhashiri and L. G. Williams, *J. Chem. Ed.*, **53**, 358 (1976).
19. R. E. Connick, *J. Amer. Chem. Soc.*, **69**, 1509 (1947).
20. A. E. Cahill and H. Taube, *J. Amer. Chem. Soc.*, **74**, 2312 (1952).
21. A. U. Khan and M. Kasha, *J. Amer. Chem. Soc.*, **92**, 3293 (1970).
22. C. S. Foote and S. Wexler, *J. Amer. Chem. Soc.*, **86**, 3879, 3880 (1964).
23. C. S. Foote, S. Wexler, W. Ando and R. Higgens, *J. Amer. Chem. Soc.*, **90**, 975 (1968).
24. E. McKeown and W. A. Waters, *J. Chem. Soc. (B)*, 1040 (1966).
25. Q. E. Thompson, *J. Amer. Chem. Soc.*, **83**, 845 (1961).
26. R. W. Murray and M. L. Kaplan, *J. Amer. Chem. Soc.*, **90**, 537, 4161 (1968).

27. E. Wasserman, R. W. Murray, M. L. Kaplan and W. A. Yager, *J. Amer. Chem. Soc.*, **90**, 4160 (1968).
28. R. W. Murray and M. L. Kaplan, *J. Amer. Chem. Soc.*, **91**, 5358 (1969).
29. P. D. Bartlett and G. D. Mendenhall, *J. Amer. Chem. Soc.*, **92**, 210 (1970).
30. A. P. Schaap and P. D. Bartlett, *J. Amer. Chem. Soc.*, **92**, 6055 (1970).
31. P. D. Bartlett and H.-K. Chu, *J. Org. Chem.*, **45**, 3000 (1980).
32. E. Koch, *Tetrahedron*, **26**, 3503 (1970).
33. E. Koch, *Anal. Chem.*, **45**, 2120 (1973).
34. P. D. Bartlett, G. D. Mendenhall and D. L. Durham, *J. Org. Chem.*, **45**, 4269 (1980).
35. (a) M. E. Brennan, *Chem. Commun.*, 957 (1970).
(b) A. P. Schaap, K. Kees and A. L. Thayer, *J. Org. Chem.*, **40**, 1185 (1975).
36. C. Moreau, C. Dufraisee and P. M. Dean, *Compt. Rend.*, **182**, 1440 (1926).
37. H. H. Wasserman and J. R. Scheffer, *J. Amer. Chem. Soc.*, **89**, 3073 (1967).
38. H. H. Wasserman, J. R. Scheffer and J. L. Cooper, *J. Amer. Chem. Soc.*, **94**, 4991 (1972).
39. H. H. Wasserman and D. L. Larsen, *Chem. Commun.*, 253 (1972).
40. T. Wilson, *Photochem. Photobiol.*, **10**, 441 (1969).
41. A. M. Trozzolo and S. R. Fahrenholtz, *Ann. N. Y. Acad. Sci.*, **171**, 61 (1970).
42. I. Saito, T. Matsuura and K. Inoue, *J. Amer. Chem. Soc.*, **103**, 188 (1981).
43. I. Rosenthal and A. J. Acher, *Israel J. Chem.*, **12**, 897 (1974).
44. For a recent review see E. A. Ogryzlo in *Singlet Oxygen* (Ed. H. H. Wasserman and R. W. Murray), Academic Press, New York, 1979, Chap. 2, pp. 35–58.
45. C. Ouannes and T. Wilson, *J. Amer. Chem. Soc.*, **90**, 6527 (1968).
46. R. S. Davidson and K. R. Trethewey, *J. Amer. Chem. Soc.*, **98**, 4008 (1976).
47. C. S. Foote in Reference 44, Chap. 5, footnote to p. 159.
48. G. O. Schenck and E. Koch, *Z. Electrochem.*, **64**, 170 (1960).
49. G. O. Schenck, H. Mertens, W. Muller, E. Koch and G. P. Scheiraenz, *Angew. Chem.*, **68**, 303 (1956).
50. E. Koch, *Tetrahedron*, **24**, 6295 (1968).
51. R. D. Ashford and E. A. Ogryzlo, *J. Amer. Chem. Soc.*, **91**, 5358 (1969).
52. K. U. Ingold, *Chem. Rev.*, **61**, 563 (1961).
53. K. Alder, F. Pascher and A. Schmitz, *Ber. Dtsch. Chem. Ges.*, **76**, 27 (1943).
54. H. M. R. Hoffman, *Angew. Chem.* (*Intern. Ed. Engl.*), **8**, 556 (1969).
55. A. A. Frimer, *Chem. Rev.*, **79**, 359 (1979) and references cited therein.
56. K. Gollnick and H. J. Kuhn, in Reference 44, Chap. 8, pp. 287–427.
57. A. P. Schaap and K. A. Zaklika in Reference 44, Chap. 6, pp. 173–242.
58. L. B. Harding and W. A. Goddard, III, *J. Amer. Chem. Soc.*, **102**, 439 (1980).
59. C. W. Jefford and C. G. Rimbault, *Tetrahedron Letters*, **22**, 91 (1981), and references cited therein.
60. L. M. Stephenson, M. J. Grdina and M. Orfanopoulos, *Acc. Chem. Res.*, **13**, 419 (1980).
61. K. Yamaguchi, T. Fueno, I. Saito, T. Matsuura and K. N. Houk, *Tetrahedron Letters*, **22**, 749 (1981).
62. K. N. Houk, J. C. Williams Jr., P. A. Mitchell and K. Yamaguchi, *J. Amer. Chem. Soc.*, **103**, 949 (1981).
63. R. D. Ashford and E. A. Ogryzlo, *Can. J. Chem.*, **52**, 3544 (1974).
64. M. J. S. Dewar and W. Thiel, *J. Amer. Chem. Soc.*, **99**, 2338 (1977).
65. (a) R. W. Denny and A. Nickon, *Org. Reactions*, **20**, 133 (1973).
(b) J. H. Wieringa, J. Strating and H. Wynberg, *Tetrahedron Letters*, 169 (1972).
(c) P. D. Bartlett and M. S. Ho, *J. Amer. Chem. Soc.*, **96**, 627 (1974).
66. For leading reviews on the singlet oxygen Diels–Alder reaction, see:
(a) Y. A. Arbuzov, *Russ. Chem. Rev.*, **34**, 558 (1965).
(b) K. Gollnick and G. O. Schenck in *1,4-Cycloaddition Reactions* (Ed. J. Hamer), Academic Press, New York, 1967, p. 255.
(c) R. W. Denny and A. Nickon, *Org. Reactions*, **20**, 133 (1973).
(d) G. Ohloff, *Pure Appl. Chem.*, **43**, 481 (1975).
(e) H. H. Wasserman and J. L. Ives, *Tetrahedron*, **37**, 1825 (1981).
(f) W. Adam, *Chem. Z.* **99**, 142 (1975).

(g) M. Balci, *Chem. Rev.*, **81**, 91 (1981).
67. (a) G. O. Schenck and K. Ziegler, *Naturwissenschaften*, **32**, 157 (1954).
(b) G. O. Schenck, *Angew. Chem.*, **64**, 12 (1952).
68. (a) W. Adam and M. Balci, *J. Amer. Chem. Soc.*, **101**, 7537 (1979).
(b) W. Adam and H. Rebollo, *Tetrahedron Letters*, **22**, 3049 (1981).
69. P. A. Burns, C. S. Foote and S. Mazur, *J. Org. Chem.*, **41**, 899 (1976).
70. J. D. Boyd and C. S. Foote, *J. Amer. Chem. Soc.*, **101**, 6758 (1979).
71. J. D. Boyd, C. S. Foote and D. K. Imagawa, *J. Amer. Chem. Soc.*, **102**, 3641 (1980).
72. P. A. Burns and C. S. Foote, *J. Org. Chem.* **41**, 908 (1976).
73. G. Rio and J. Berthelot, *Bull. Soc. Chim. Fr.*, 3610 (1969).
74. M. Matsumoto, S. Dobashi and K. Kondo, *Tetrahedron Letters*, 2329 (1977).
75. M. Matsumoto, S. Dobashi and K. Kuroda, *Tetrahedron Letters*, 3361 (1977).
76. (a) M. Matsumoto and K. Kuroda, *Tetrahedron Letters*, 1607 (1979).
(b) M. Matsumoto, K. Kuroda and Y. Suzuki, *Tetrahedron Letters*, **22**, 3253 (1981).
77. D. S. Steichen and C. S. Foote, *Tetrahedron Letters*, 4363 (1979).
78. M. Matsumoto and K. Kondo, *Tetrahedron Letters*, 3935 (1975).
79. M. Matsumoto, S. Dobashi and K. Kondo, *Tetrahedron Letters*, 4471 (1975).
80. H. Kotsuki, I. Saito and T. Matsuura, *Tetrahedron Letters*, 469 (1981).
81. I. Saito, M. Imuta and T. Matsuura, *Tetrahedron*, **28**, 5307 (1972).
82. I. Saito, S. Abe, Y. Takahashi and T. Matsuura, *Tetrahedron Letters*, 4001 (1974).
83. C. J. M. Van den Heuvel, A. Hofland, H. Steinberg and Th. J. de Boer, *Rec. Trav. Chim.*, **99**, 275 (1980).
84. K. Onodera, H. Sakuragi and K. Tokumaru, *Tetrahedron Letters*, **21**, 2831 (1980).
85. H. H. Wasserman and B. H. Lipshutz in Reference 44, Chap. 9, pp. 429–502.
86. M. V. George and V. Bhat, *Chem. Rev.*, **79**, 447 (1979).
87. (a) C. S. Foote, M. T. Wuesthoff, S. Wexler, I. G. Burstain, R. Denny, G. O. Schenck and K. H. Schulte-Elte, *Tetrahedron*, **23**, 2583 (1967).
(b) K. H. Schulte-Elte, B. Willhalm and G. Ohloff, *Angew Chem.* (*Inter. Ed.*), **8**, 985 (1969); these authors describe the *cis–trans* isomerization of 4,5-epoxy-*cis*-pentenal in the presence of triphenylphosphine.
88. (a) H. Takeshita and T. Hatsui, *J. Org. Chem.*, **43**, 3080 (1978);
(b) H. Takeshita, T. Hatsui, R. Iwabuchi and S. Itoh, *Bull. Chem. Soc. Japan*, **51**, 1257 (1978).
(c) H. Takeshita, T. Hatsui and H. Kanamori, *Tetrahedron Letters*, 1697 (1973).
89. K. H. Gibson, *Chem. Soc. Rev.*, **7**, 489 (1978).
90. G. O. Pierson and O. A. Runquist, *J. Org. Chem.*, **34**, 3654 (1969).
91. (a) J. Boche and O. Runquist, *J. Org. Chem.*, **33**, 4285 (1968);
(b) J. Hudee and R. S. A. Kelley, *Tetrahedron Letters*, 3175 (1967).
92. K. K. Maheshwari, P. de Mayo and D. Wiegand, *Can. J. Chem.*, **48**, 3265 (1970).
93. J. D. Boyd, C. S. Foote and D. K. Imagawa, *J. Amer. Chem. Soc.*, **102**, 3641 (1980).
94. (a) W. Adam and H. J. Eggelte, *Angew. Chem.* (*Intern. Ed. Engl.*), **16**, 713 (1977).
(b) W. Adam and H. J. Eggelte, *J. Org. Chem.*, **42**, 3987 (1977).
95. O. Wallach, *Justus Liebigs Ann. Chem.*, **392**, 49 (1912).
96. G. O. Schenck, K. G. Kinkel and H. J. Mertens, *Justus Liebigs Ann. Chem.*, **584**, 125 (1953).
97. For a recent review on various aspects of cyclic polyepoxides see W. Adam and M. Balci, *Tetrahedron*, **36**, 833 (1980).
98. S. M. Kupchan, R. J. Hemingway, P. Coggon, A. T. McPhail and G. A. Sim, *J. Amer. Chem. Soc.*, **90**, 2982 (1968).
99. S. M. Kupchan, R. J. Hemingway and R. M. Smith, *J. Org. Chem.*, **34**, 3898 (1969).
100. D. B. Borders, P. Shu and J. E. Lancaster, *J. Amer. Chem. Soc.*, **94**, 2540 (1972).
101. D. B. Borders and J. E. Lancaster, *J. Org. Chem.*, **39**, 435 (1974).
102. S. M. Kupchan, W. A. Court, R. G. Dailey, Jr., J. Gilmore and R. F. Bryan, *J. Amer. Chem. Soc.*, **94**, 7194 (1972).
103. C. H. Foster and G. A. Berchtold, *J. Amer. Chem. Soc.*, **94**, 7939 (1972).
104. C. H. Foster and G. A. Berchtold, *J. Org. Chem.*, **40**, 3743 (1975).
105. E. Vogel, H. J. Altenbach and C. D. Sommerfeld, *Angew. Chem.* (*Intern. Ed. Engl.*), **11**, 939 (1972).
106. J. A. Turner and W. Herz, *J. Org. Chem.*, **42**, 1895 (1977).

107. N. Kornblum and H. E. DeLaMare, *J. Amer. Chem. Soc.*, **73**, 880 (1951).
108. R. Hiatt in *Organic Peroxides* (Ed. D. Swern), Wiley–Interscience, New York–London, Vol. II, 1971, pp. 79–80; Vol. III, 1972, p. 23.
109. K. H. Schulte-Elte, B. Willhalm and G. Ohloff, *Angew. Chem.* (*Intern. Ed. Engl.*), **8**, 985 (1969).
110. C. J. Sih, R. G. Solomon, P. Price, R. Sood and G. Peruzzotti, *J. Amer. Chem. Soc.*, **97**, 857 (1975).
111. J. P. Hagenbuch and P. Vogel, *Tetrahedron Letters*, 561 (1979).
112. (a) J. P. Hagenbuch and P. Vogel, *Chimia*, **31**, 136 (1977).
(b) J. H. Clark, *Chem. Rev.*, **80**, 429 (1980).
113. J. A. Turner and W. Herz, *J. Org. Chem.*, **42**, 1900 (1977).
114. A. Shani and J. T. Klug, *Tetrahedron Letters*, **21**, 1563 (1980).
115. E. Demole, C. Demole and D. Berthet, *Helv. Chim. Acta*, **56**, 265 (1973).
116. K. Kondo and M. Matsumoto, *Chem. Letters*, 701 (1974).
117. M. Matsumoto and K. Kondo, *J. Org. Chem.*, **40**, 2259 (1975).
118. K. Kondo and M. Matsumoto, *Tetrahedron Letters*, 391, 4363 (1976).
119. B. Harirchia and P. D. Magnus, *Synth. Commun.*, **7**, 119 (1977).
120. G. O. Schenck, *Angew. Chem.*, **69**, 579 (1957).
121. W. A. Ayer, L. M. Browne and S. Fung, *Can. J. Chem.*, **54**, 3276 (1976).
122. D. J. Coughlin, R. S. Brown and R. G. Salomon, *J. Amer. Chem. Soc.*, **101**, 1533 (1979).
123. (a) W. Adam, A. J. Bloodworth, H. J. Eggelte and M. E. Loveitt, *Angew. Chem.* (*Intern. Ed. Engl.*), **17**, 209 (1978).
(b) W. Adam and I. Erden, *Angew. Chem.* (*Intern. Ed. Engl.*), **17**, 210, 211 (1978).
124. (a) M. Schafer-Rider, U. Brocker and E. Vogel, *Angew. Chem.* (*Intern. Ed. Engl.*), **15**, 228 (1976).
(b) M. Oda, Y. Kayama and Y. Kitahara, *Tetrahedron Letters*, 2019 (1974).
125. Y. Ito, M. Oda and Y. Kitahara, *Tetrahedron Letters*, 239 (1975).
126. T. Goto and H. Nakamura, *Chem. Commun.*, 781 (1978).
127. J. Rigaudy and D. Sparfel, *Bull. Soc. Chim. Fr.*, 3441 (1972).
128. L. M. Stephenson, *Tetrahedron Letters*, **21**, 1005 (1980).
129. L. A. Paquette and D. C. Liotta, *Tetrahedron Letters*, 2681 (1976).
130. A. Nickon, V. T. Chuang, P. J. L. Daniels, R. W. Denny, J. B. DiGiorgio, J. Tsunetsugu, H. G. Vilhuber and E. Werstiuk, *J. Amer. Chem. Soc.*, **94**, 5517 (1972).
131. A. A. Frimer, *Israel J. Chem.*, **21**, 194 (1981) and references cited therein.
132. A. A. Frimer and D. Roth, *J. Org. Chem.*, **44**, 3882 (1979).
133. C. S. Foote and R. W. Denny, *J. Amer. Chem. Soc.*, **93**, 5162 (1971).
134. A. A. Frimer, T. Farkash and M. Sprecher, *J. Org. Chem.*, **44**, 989 (1979).
135. C. W. Jefford and C. G. Rimbault, *Tetrahedron Letters*, **22**, 91 (1981).
136. G. Rousseau, P. Le Perchec and J. M. Conia, *Synthesis*, 67 (1978).
137. R. Hiatt in *Organic Peroxides* (Ed. D. Swern), Wiley–Interscience, New York–London, Vol. II, 1971, p. 67.
138. J. B. Lee and B. C. Uff, *Quart. Rev. Chem. Soc.*, **21**, 429 (1967).
139. J. Hoffman, *Org. Synth.*, **40**, 76 (1960).
140. (a) H. Hock and O. Schrader, *Angew. Chem.*, **49**, 595 (1936).
(b) H. Hock and K. Ganicke, *Ber.*, **71**, 1430 (1938).
141. (a) H. Hock and S. Lang, *Ber.*, **75**, 300 (1942).
(b) H. Hock and S. Lang, *Ber.*, **77**, 257 (1944).
142. H. Hock and H. Kropf, *Angew. Chem.*, **69**, 313 (1957).
143. E. H. Farmer and A. Sundralingam, *J. Chem. Soc.*, 121 (1942).
144. G. O. Schenck, *Angew. Chem.*, **69**, 579 (1957).
145. G. O. Schenck and K. H. Schulte-Elte, *Justus Liebigs Ann. Chem.*, **618**, 185 (1958).
146. M. S. Kharasch and J. G. Burt, *J. Org. Chem.*, **16**, 150 (1951).
147. J. A. Turner and W. Herz, *J. Org. Chem.*, **42**, 1657 (1977).
148. P. D. Bartlett and A. A. Frimer, *Heterocycles*, **11**, 419 (1978).
149. G. Ohloff and E. Klein, *Tetrahedron*, **18**, 37 (1962), footnote 22.
150. E. Koerner von Gustorf, F. W. Grevels and G. O. Schenck, *Justus Liebigs Ann. Chem.*, **719**, 1, 6 (1968).
151. G. Ohloff, H. Shuckler, B. Willhalm and M. Hinder, *Helv. Chim. Acta*, **53**, 623, 628 (1970) and references cited therein.

152. A. M. Jeffrey and D. M. Jerina, *J. Amer. Chem. Soc.*, **94**, 4048 (1972).
153. J. E. Fox, A. I. Scott and P. W. Young, *Chem. Commun.*,4 (1967).
154. (a) T. Galliard and D. Phillips, *Biochem. J.*, **129**, 743 (1972).
(b) T. Galliard, D. A. Wardale and J. A. Matthew, *Biochem. J.*, **138**, 23 (1974).
(c) T. Galliard and J. A. Matthew, *Biochim. Biophys. Acta*, **398**, 1 (1975).
155. (a) A. Nishinaga, K. Nakamura and T. Matsuura, *Tetrahedron Letters*, **21**, 1269 (1980).
(b) A. Nishinaga, K. Nakamura, T. Matsuura, A. Rieker, D. Koch and R. Griesshammer, *Tetrahedron*, **35**, 2493 (1979).
(c) J. A. M. Peters, N. P. van Vliet and F. J. Zeelen, *Rec. Trav. Chim.*, **100**, 226 (1981).
156. R. E. Ireland, K. Huthmacher, P. A. Aristoff and R. Farr, unpublished results cited in Reference 66e.
157. (a) B. Maurer, M. Fracheboud, A. Grieder and G. Ohloff, *Helv. Chim. Acta*, **55**, 2371 (1972).
(b) G. Ohloff, J. Becker and K. H. Schulte-Elte, *Helv. Chim. Acta*, **50**, 705 (1967).
(c) A. F. Thomas, *Helv. Chim. Acta*, **55**, 2429 (1972).
158. See for example Reference 137, top of p. 80 and the footnotes to pp. 87 and 96.
159. W. F. Brill, *Advan. Chem. Ser.*, **75**, 93 (1968).
160. A. D. Walsh, *Trans. Faraday Soc.*, **42**, 99, 269 (1946).
161. C. E. Frank, *Chem. Rev.*, **46**, 155, 161 (1950).
162. L. Bateman and H. Hughes, *J. Chem. Soc.*, 4594 (1952).
163. A. A. Frimer, *J. Org. Chem.*, **42**, 3194 (1977).
164. P. D. Bartlett and A. A. Frimer, *Heterocycles*, **11**, 419 (1978).
165. C. W. Jefford and C. G. Rimbault, *J. Org. Chem.*, **43**, 1908 (1978).
166. Reference 4b, footnote to p. 48.
167. (a) H. W. Gardner, *Lipids*, **14**, 208 (1979).
(b) T. A. Dix and L. G. Marnett, *J. Amer. Chem. Soc.*, **103**, 6744 (1981).
168. R. M. Kellog and J. K. Kaiser, *J. Org. Chem.*, **40**, 2575 (1975), footnote 6.
169. For discussion and sources see Reference 55; see also N. A. Porter, J. Logan and V. Kontoyiannidou, *J. Org. Chem.*, **44**, 3177 (1979).
170. H. Takeshita, T. Hatsui and O. Jinnai, *Chem. Letters*, 1059 (1976).
171. F. McCapra and I. Beheshti, *Chem. Commun.*, 517 (1977).
172. L. A. Paquette and R. V. C. Carr, *J. Amer. Chem. Soc.*, **102**, 7553 (1980).
173. (a) C. W. Jefford and A. F. Boschung, *Tetrahedron Letters*, 4771 (1976).
(b) C. W. Jefford and A. F. Boschung, *Helv. Chim. Acta*, **60**, 2673 (1977).
174. N. J. Turro, V. Ramamurthy, K.-C. Liu, A. Krebs and R. Kemper, *J. Amer. Chem. Soc.*, **98**, 6758 (1976).
175. Y. Inoue and N. J. Turro, *Tetrahedron Letters*, **21**, 4327 (1980).
176. T. Greibrokk, *Tetrahedron Letters*, 1663 (1973).
177. L. J. Bollyky, *J. Amer. Chem. Soc.*, **92**, 3230 (1970).
178. A. Aoyagi, T. Tsuyuki, T. Takahashi and R. Stevenson, *Tetrahedron Letters*, 3397 (1972).
179. W. Adam, N. Duran and G. A. Simpson, *J. Amer. Chem. Soc.*, **97**, 5464 (1975).
180. N. J. Turro, Y. Ito, M.-F. Chow, W. Adam, O. Rodriguez and F. Yang, *J. Amer. Chem. Soc.*, **99**, 5836 (1977).
181. B. Zwanenburg, A. Wagenaar and J. Strating, *Tetrahedron Letters*, 4683 (1970).
182. S. Tamagaki and K. Hotta, *Chem. Commun.*, 598 (1980).
183. Di-*t*-butylsulphine and di-*t*-butylthioketene *S*-oxide are perfectly stable to $^1O_2$, see:
(a) S. Tamagaki, A. Akatsuka, M. Nakamura and S. Kozuka, *Tetrahedron Letters*, 3665 (1979).
(b) V. Jayathirtha Rao and V. Ramamurthy, *Chem. Commun.*, 638 (1981).
184. C. C. Wamser and J. W. Herring, *J. Org. Chem.*, **41**, 1476 (1976).
185. (a) R. Rajee and V. Ramamurthy, *Tetrahedron Letters*, 5127 (1978).
(b) N. Ishibe, M. Odani and M. Sunami, *Chem. Commun.*, 118 (1971).
(c) See, however, References 182 and 183.
186. C. W. Jefford and G. Bachietto, *Tetrahedron Letters*, 4531 (1977).
187. H. E. Zimmerman and G. E. Keck, *J. Amer. Chem. Soc.*, **97**, 3527 (1975).
188. C. S. Foote, *Pure Appl. Chem.*, **27**, 639 (1971).
189. D. R. Kearns, *Chem. Rev.*, **71**, 395 (1971).
190. P. D. Bartlett, *Pure Appl. Chem.*, **27**, 597 (1971).

191. K. R. Kopecky, J. E. Filby, C. Mumford, P. A. Lockwood and J. Y. Ding, *Can. J. Chem.*, **53**, 1103 (1975).
192. W. H. Richardson, F. C. Montgomery, M. B. Yelvington and H. E. O'Neal, *J. Amer. Chem. Soc.*, **96**, 7525 (1974) and references cited therein.
193. P. A. Burns and C. S. Foote, *J. Amer. Chem. Soc.*, **96**, 4339 (1974).
194. N. J. Turro and W. H. Waddel, *Tetrahedron Letters*, 2069 (1975).
195. T. Wilson, D. E. Golan, M. S. Harris and A. L. Baumstark, *J. Amer. Chem. Soc.*, **98**, 1086 (1976) and references cited therein.
196. J. W. Hastings and T. Wilson, *Photochem. Photobiol.*, **23**, 461 (1976).
197. P. D. Bartlett and M. E. Landis in Reference 44, Chap. 7, p. 243.
198. K. A. Zaklika, P. A. Burns and P. A. Schaap, *J. Amer. Chem. Soc.*, **100**, 318 (1978).
199. D. C. Lee and T. Wilson in *Chemiluminescence and Bioluminescence* (Eds. M. J. Cormier, D. M. Hercules and J. Lee), Plenum Press, New York, 1973, p. 625.
200. Compare:
(a) K. A. Horn and G. A. Schuster, *J. Amer. Chem. Soc.*, **101**, 7097 (1979).
(b) J. J. Zupancic, K. A. Horn and G. B. Schuster, *J. Amer. Chem. Soc.*, **102**, 5279 (1980).
201. G. A. Schuster, personal communication, September 16, 1980.
202. H. H. Wasserman and S. Terao, *Tetrahedron Letters*, 1735 (1980).
203. H. H. Wasserman and J. L. Ives, *J. Amer. Chem. Soc.*, **98**, 7868 (1976).
204. H. H. Wasserman and J. L. Ives, *J. Org. Chem.*, **43**, 3238 (1978).
205. J. L. Ives, *Ph.D. Thesis*, Yale University, 1978.
206. D. A. Lightner, G. S. Bisacchi and R. D. Norris, *J. Amer. Chem. Soc.*, **98**, 802 (1976).
207. D. A. Lightner and C.-S. Pak, *J. Org. Chem.*, **40**, 2724 (1975).
208. F. Abello, J. Bois, J. Gomez, J. Morell and J. J. Bonet, *Helv. Chim. Acta*, **58**, 2549 (1975).
209. W. Ando, T. Saiki and T. Migita, *J. Amer. Chem. Soc.*, **97**, 5028 (1975).
210. W. Ando, J. Suzuki, T. Arai and T. Migita, *Tetrahedron*, **29**, 1507 (1973).
211. W. Ando, K. Watanabe, J. Suzuki and T. Migita, *J. Amer. Chem. Soc.*, **96**, 6766 (1974).
212. (a) W. Ando, K. Watanabe and T. Migita, *Chem. Commun.*, 961 (1975);
(b) W. Ando, K. Watanabe and T. Migita, *Tetrahedron Letters*, 4127 (1975).
213. W. Adam and J.-C. Liu, *J. Amer. Chem. Soc.*, **94**, 1206 (1972).
214. D. A. Lightner and L. K. Low, *J. Heterocycl. Chem.*, **12**, 793 (1975).
215. G. Bachi and H. Wüest, *J. Amer. Chem. Soc.*, **100**, 294 (1978).
216. P. R. Story, E. A. Whited and J. A. Alford, *J. Amer. Chem. Soc.*, **94**, 2143 (1972).
217. H. Takeshita and T. Hatsui, *J. Org. Chem.*, **43**, 3083 (1978).
218. H. Takeshita, T. Hatsui and I. Shimooda, *Tetrahedron Letters*, 2889 (1978).
219. P. D. Bartlett, A. L. Baumstark and M. E. Landis, *J. Amer. Chem. Soc.*, **95**, 6486 (1973).
220. P. D. Bartlett, A. L. Baumstark and M. J. Shapiro, *J. Org. Chem.*, **42**, 1661 (1977).
221. P. D. Bartlett, A. L. Baumstark, M. E. Landis and C. L. Lerman, *J. Amer. Chem. Soc.*, **96**, 5267 (1974).
222. B. S. Campbell, D. B. Denney, D. Z. Denney and L. S. Shih, *J. Amer. Chem. Soc.*, **97**, 3850 (1975).
223. (a) H. H. Wasserman and I. Saito, *J. Amer. Chem. Soc.*, **97**, 905 (1975).
(b) Ph_2S reduces α-alkoxy hydroperoxides; see H.-S. Ryang and C. S. Foote, *J. Amer. Chem. Soc.*, **103**, 4951 (1981).
224. (a) B. Ranby and J. F. Rabeck (Eds.), *Singlet Oxygen Reactions with Organic Compounds and Polymers*, Wiley–Interscience, New York–London, 1978.
(b) B. Ranby and J. F. Rabeck, *Photodegradation, Photooxidation and Photostabilization of Polymers*, Wiley–Interscience, New York–London, 1975.
225. J. N. Pitts, *Advan. Environ. Sci. Technol.*, **1**, 289 (1969).
226. N. I. Krinsky in Reference 44, Chap. 12, pp. 597–641.
227. M. Kasha in Reference 44, Introductory Remarks, pp. xiii–xviii.

The Chemistry of Functional Groups, Peroxides
Edited by S. Patai

CHAPTER 8

Free-radical reaction mechanisms involving peroxides in solution[1]

J. A. HOWARD
Division of Chemistry, National Research Council of Canada, Ottawa K1A OR9, Canada

I. INTRODUCTION . . . 235
II. DIALKYL PEROXIDES . . . 236
III. DIACYL PEROXIDES . . . 240
IV. DIAROYL PEROXIDES . . . 242
V. CYCLIC PEROXIDES . . . 243
VI. TRIOXIDES . . . 245
VII. TETROXIDES . . . 247
VIII. HYDROPEROXIDES . . . 247
IX. PERACIDS . . . 252
X. HYDROTRIOXIDES . . . 252
XI. PEROXY ESTERS . . . 253
XII. β-PEROXYLACTONES . . . 255
XIII. CONCLUSION . . . 256
XIV. ACKNOWLEDGEMENT . . . 256
XV. REFERENCES . . . 256

I. INTRODUCTION

This chapter deals with the mechanisms that are involved in the liquid-phase homolytic formation and decomposition of compounds which contain the peroxide function. Compounds which contain this function, e.g., hydroperoxides, peroxides, trioxides, tetroxides and peracids are most commonly found in organic compounds which have been in contact with atmospheric oxygen. A knowledge of their free-radical chemistry is, therefore, vital to a complete understanding of the autoxidation of organic and organometallic compounds. Ozone also reacts with organic compounds, especially olefins, to give the peroxide function in the form of cyclic trioxides and peroxides. Because of this, reaction of plastics and rubbers with ozone leads to their breakdown and loss of usefulness.

Other peroxidic compounds such as diacyl peroxides and peroxy esters are important photochemical and thermal sources of free radicals for initiation of vinyl polymerizations and for electron spin resonance spectroscopic studies of lifetimes and structures of transient species.

The chemistry of the peroxide function has been the subject of several books and review articles[2–7] and the free-radical chemistry discussed in these articles will receive cursory treatment here. This chapter will, therefore, concentrate mainly on those aspects of free-radical peroxide chemistry that have not been recently reviewed.

II. DIALKYL PEROXIDES

Dialkyl peroxides are formed homolytically by the reaction of an alkyl radical ($R\cdot$) with an alkylperoxyl ($RO_2\cdot$), by the self-reaction of tertiary alkylperoxyls and by the addition of an alkylperoxyl to an olefin[8] (reactions 1–3).

$$R\cdot + RO_2\cdot \longrightarrow ROOR \quad (1)$$

$$2t\text{-}RO_2\cdot \longrightarrow t\text{-}ROOR\text{-}t + O_2 \quad (2)$$

$$RO_2\cdot + R_2C{=}CR_2 \longrightarrow ROOCR_2\dot{C}R_2 \quad (3)$$

Reaction of alkyls with alkylperoxyls occurs during autoxidation of hydrocarbons at low oxygen pressures and the importance of this reaction depends on the resonance stabilization energy of the radical $R\cdot$. Thus reaction of triphenylmethyl with dissolved oxygen gives high yields of triphenylmethyl peroxide. Another example of this reaction is found in the inhibition of hydrocarbon autoxidation by 2,6-di-*t*-butyl-4-methylphenol where 2,6-di-*t*-butyl-4-methylphenoxyl reacts rapidly with a chain-carrying peroxyl to give a 4-alkylperoxy-4-methyl-2,6-di-*t*-butyl-2,5-cyclohexadiene-1-one[8] (reaction 4).

$$\text{(2,6-di-}t\text{-butyl-4-methylphenoxyl, O}\cdot) + RO_2\cdot \longrightarrow \text{(4-methyl-2,6-di-}t\text{-butyl-2,5-cyclohexadien-1-one, 4-OOR)} \quad (4)$$

The self-reaction of tertiary alkylperoxyls to give dialkyl peroxide is an important mode of chain-termination for autoxidation of hydrocarbons containing labile tertiary hydrogen atoms and occurs via the intermediacy of a tetroxide[9]. Rate constants for the overall formation of peroxide are quite small (Table 1) and this is one of the reasons for the ease with which the parent hydrocarbons undergo autoxidation.

Addition of an alkylperoxyl to an olefin is the rate-controlling propagation step for autoxidation of vinyl monomers and the final reaction product is a polyperoxide. Rate

TABLE 1. Rate constants for the formation of tertiary alkyl peroxides from tertiary alkylperoxyls[9]

Peroxyl	$2k_2(\mathrm{M}^{-1}\,\mathrm{s}^{-1}) \times 10^{-3a}$	$\log [A_2(\mathrm{M}^{-1}\,\mathrm{s}^{-1})]$	$E_2(\mathrm{kJ\,mol}^{-1})$
$(CH_3)_3CO_2\cdot$	1.2	9.2	35.5
$C_6H_5C(CH_3)_2O_2\cdot$	6.0	10.7	39.7
$(C_6H_5)_2C(CH_3)O_2\cdot$	64	—	—

[a] At 303 K.

constants for this reaction (Table 2) depend principally on the nature of $RO_2\cdot$ and the stability of the incipient β-alkylperoxyalkyl radical, although polar and steric effects can be of some influence[8].

At low oxygen pressures and elevated temperatures β-alkylperoxyalkyls undergo an intramolecular S_H2 reaction at the peroxide function to give an epoxide and an alkoxyl[10] (reaction 5).

$$ROOCR_2\dot{C}R_2 \longrightarrow R_2C\overset{O}{\frown}CR_2 + RO\cdot \qquad (5)$$

The weakest bond in a dialkyl peroxide is the O—O bond ($D_{O-O} = 159\,\mathrm{kJ\,mol^{-1}}$)[11] and these compounds thermolyse and photolyse to give alkoxyls (reaction 6).

$$ROOR \xrightarrow{\Delta \text{ or } h\nu} 2RO\cdot \qquad (6)$$

Rates of thermolysis follow first-order kinetics and true rate constants are virtually independent of the nature of the alkyl moiety R (Table 3).

Di-*t*-butyl peroxide decomposes to give *t*-butoxyl with almost 100% efficiency while primary and secondary alkoxyls can undergo a cage disproportionation reaction (reaction 7).

$$[R_2CHO\cdot + R_2CHO\cdot]_{cage} \longrightarrow R_2C{=}O + R_2CHOH \qquad (7)$$

The alkoxyls which escape the solvent cage do not usually encounter another alkoxyl. Instead they either abstract a hydrogen atom from the solvent, add to the solvent or undergo β scission, e.g. reactions (8) and (9).

$$(CH_3)_3CO\cdot + C_6H_5CH_3 \longrightarrow (CH_3)_3COH + C_6H_5CH_2\cdot \qquad (8)$$

$$(CH_3)_3CO\cdot \longrightarrow (CH_3)_2C{=}O + CH_3\cdot \qquad (9)$$

The overall rate constant for reaction of *t*-butoxyl with toluene is $2.3 \times 10^5\,\mathrm{M^{-1}\,s^{-1}}$ at 303 K and for other solvents there is an approximate correlation with the strength of the C—H bond that is broken[12]. Alkoxyls add quite rapidly to double bonds and the rate constant for addition to norbornene is $1.1 \times 10^6\,\mathrm{M^{-1}\,s^{-1}}$ at 303 K[13]. Rates of β scission of alkoxyls depend on the stability of the alkyl radical and carbonyl compound produced in the reaction. The rate constant for β scission of *t*-butoxyl in CCl_4 can be represented by $\log(k_9/s^{-1}) = 12.5 - 58.1/\theta$, where $\theta = 2.303\,RT\,\mathrm{kJ\,mol^{-1}}$ and at 303 K, $k_9 = 3 \times 10^2\,\mathrm{s^{-1}}$[14].

TABLE 2. Rate constants for the addition of alkylperoxyls to olefins at 303 K[8]

Olefin	$k_3^{rr}(\text{M}^{-1}\,\text{s}^{-1})^a$	$k_3^{br}(\text{M}^{-1}\,\text{s}^{-1})^b$
Styrene	41	1.3
α-Methylstyrene	10	2.8
Vinyl acetate	2.5	0.002
Methyl methacrylate	4.5	0.094

[a] Homopropagation rate constant, i.e. reaction of substrate (superscript r) with its own peroxyl (superscript r).
[b] Absolute rate constant for addition of $(CH_3)_3CO_2\cdot$ (superscript b) to the substrate (superscript r).

Alkoxyls can rearrange to give carbon-centred radicals. The best known example of this reaction is the rearrangement of triphenylmethoxyl to give diphenylphenoxymethyl[15] (reaction 10).

$$(C_6H_5)_3CO\cdot \longrightarrow (C_6H_5)_2\dot{C}OC_6H_5 \qquad (10)$$

Dialkyl peroxides often homolyse faster than would be predicted from Arrhenius parameters for true unimolecular decomposition because of induced decomposition which occurs either by hydrogen-atom abstraction from the alkyl moiety or S_H2 displacement at the O—O bond (reactions 11–13).

$$R_2CHOOCHR_2 + R_2CHO\cdot \longrightarrow R_2\dot{C}OOCHR_2 + R_2CHOH \qquad (11)$$

$$R_2\dot{C}OOCHR_2 \longrightarrow R_2C{=}O + R_2CHO\cdot \qquad (12)$$

$$R_2\dot{C}HOOCHR_2 + R^1\cdot \longrightarrow R_2CHO\cdot + R^1OCHR_2 \qquad (13)$$

The susceptibility of dialkyl peroxides to induced decomposition via hydrogen-atom abstraction increases in the order *t*-alkyl < *p*-alkyl < *s*-alkyl as might be expected from the absence of α hydrogens in di-*t*-alkyl peroxides and the presence of secondary and tertiary hydrogens in primary and secondary dialkyl peroxides.

An S_H2 reaction at the peroxide function has been invoked to explain the formation of epoxides from peroxide decomposition[16]. Thus decomposition of neat di-*t*-butyl peroxide at 383 K gives a substantial yield of isobutylene oxide (reactions 14 and 15).

$$(CH_3)_3CO\cdot + (CH_3)_3COOC(CH_3)_3 \longrightarrow (CH_3)_3COH + \cdot CH_2(CH_3)_2COOC(CH_3)_3 \qquad (14)$$

$$\cdot CH_2(CH_3)_2COOC(CH_3)_3 \longrightarrow \underset{\diagdown O \diagup}{CH_2 \text{———} C(CH_3)_2} + (CH_3)_3CO\cdot \qquad (15)$$

Di-*t*-butyl peroxide is also susceptible to intermolecular radical attack at the O—O function by nucleophilic radicals which have a hydroxy group or a primary or secondary amino group attached directly to the radical centre[16]. Because of this, rates of peroxide

TABLE 3. Rate constants and Arrhenius parameters for thermolysis of some dialkyl peroxides[11]

Peroxide	$10^3 k_6 (s^{-1})^a$	$\log [A_6 (s^{-1})]$	$E_6 (kJ\,mol^{-1})$
CH_3OOCH_3	4	15.6	154.2
$CH_3CH_2OOCH_2CH_3$	7.9	16.1	156
$(CH_3)_3COOC(CH_3)_3$	2.5	15.9	159

^{a}At 448 K.

decomposition are appreciably faster in primary and secondary alcohols and amines than they are in hydrocarbon solvents (reactions 16 and 17).

$$(CH_3)_3CO\cdot + R_2CHOH \longrightarrow (CH_3)_3COH + R_2\dot{C}OH \qquad (16)$$

$$R_2\dot{C}OH + (CH_3)_3COOC(CH_3)_3 \longrightarrow R_2C{=}O + (CH_3)_3COH + (CH_3)_3CO\cdot \qquad (17)$$

Attack on the peroxide function by α-hydroxy- and α-amino-alkyls apparently occurs through the hydroxy and amino hydrogen[16], e.g. reaction (18).

$$\overset{\delta+}{R_2C}\cdots\cdots O\cdots\cdots H\cdots\cdots \begin{matrix}\overset{\delta}{O}C(CH_3)_3\\ |\\ \dot{O}C(CH_3)_3\end{matrix} \qquad (18)$$

Peroxide decomposition is also faster in ethers[16] and in this case α-alkoxyalkyl attack at the O—O function occurs (reaction 19).

$$\begin{matrix}R_2C\cdot\\ |\\ OR\end{matrix} + \begin{matrix}O{-}R^1\\ |\\ O{-}R^1\end{matrix} \longrightarrow \begin{matrix}R_2COR^1\\ |\\ OR\end{matrix} + R^1O\cdot \qquad (19)$$

Allyl *t*-butyl peroxide[17] decomposes in toluene with $\log(k/s^{-1}) = 13.8 - 138.8/\theta$, parameters somewhat lower than the values in Table 3. This is because the reaction is a mixture of homolysis and radical-induced decomposition, the latter involving both solvent-derived radical addition to C=C and alkyl hydrogen abstraction by alkoxy radicals (reactions 20–22). Interestingly there is no *t*-butoxyl addition to the unsaturated function. Computer modelling[17] of this system indicates that $k_{20} \sim 700\,M^{-1}\,s^{-1}$ and $k_{22} \sim 4.4 \times 10^6\,M^{-1}\,s^{-1}$ at 393 K.

$$C_6H_5CH_2\cdot + CH_2{=}CHCH_2OOC(CH_3)_3 \longrightarrow C_6H_5CH_2CH_2\dot{C}HCH_2OOC(CH_3)_3 \qquad (20)$$

$$C_6H_5CH_2CH_2\dot{C}HCH_2OOC(CH_3)_3 \longrightarrow C_6H_5CH_2CH_2\overset{O}{CH{-}CH_2} + (CH_3)_3CO\cdot \qquad (21)$$

$$(CH_3)_3CO\cdot + CH_2{=}CHCH_2OOC(CH_3)_3 \longrightarrow (CH_3)_3COH + CH_2{=}CH\dot{C}HOOC(CH_3)_3$$

$$\longrightarrow CH_2{=}CHCHO + (CH_3)_3CO\cdot \qquad (22)$$

Steric effects are extremely important in the induced decomposition of peroxides by the S_H2 mechanism. Thus trialkyltin radicals do not react with di-*t*-butyl peroxide whereas the rate constant for reaction of *n*-$Bu_3Sn\cdot$ with $C_2H_5OOC_2H_5$ in benzene at 263 K is $7.5 \times 10^4\,M^{-1}\,s^{-1}$[16].

Metal ions, such as Fe^{2+}, Cu^+ and Co^{2+}, accelerate the decomposition of dialkyl peroxides by a one-electron transfer mechanism[18]. Thus the ferrous-ion-catalysed decomposition of diethyl peroxide in aqueous solution proceeds rapidly to give ethanol and acetaldehyde (reaction 23).

$$CH_3CH_2OOCH_2CH_3 + Fe^{2+} \longrightarrow CH_3CH_2O\cdot + CH_3CH_2O^- + Fe^{3+} \longrightarrow CH_3CHO + CH_3CH_2OH \tag{23}$$

III. DIACYL PEROXIDES

Diacyl peroxides might be expected to be formed in the termination reaction of aldehyde autoxidation by a mechanism analagous to the formation of peroxides in tertiary hydrocarbon autoxidation (reaction 24). There is, however, no experimental evidence for this reaction since it has been shown that acetyl peroxide is not formed during autoxidation of acetaldehyde. Termination for this compound in fact occurs by reaction of acetylperoxyl with methylperoxyl and by self-reaction of methylperoxyl[19].

$$2RC(O)O_2\cdot \rightleftharpoons RC(O)OOOOC(O)R \longrightarrow RC(O)OOC(O)R + O_2 \tag{24}$$

Diacyl peroxides readily homolyse to give alkyls either by a one-step or concerted mechanism, with the concerted mechanism increasing in importance as the stability of the alkyl moiety, R, increases[20] (reaction 25). In this respect solid acetyl peroxide can be handled, albeit with extreme caution, whereas di-*t*-alkyl acyl peroxides are too unstable to isolate.

$$RC(O)OOC(O)R \longrightarrow \left[\begin{array}{c} 2RC(O)O\cdot \\ R\cdots C(O)\cdots O\cdots OC(O)R \\ R\cdots C(O)\cdots O\cdots O\cdots C(O)\cdots R \end{array}\right] \longrightarrow 2R\cdot + 2CO_2 \tag{25}$$

Diacyl peroxides in dilute solution usually decompose with first-order kinetics and Arrhenius parameters and rate constants for unimolecular decomposition of several of these compounds are given in Table 4. It is clear from a comparison of this data with the

TABLE 4. Rate constants and Arrhenius parameters for thermolysis of acyl and aroyl peroxides[20]

Acyl peroxide	$10^5 k_{25}(s^{-1})^a$	$\log[A_{25}(s^{-1})]$	$E_{25}(kJ\,mol^{-1})$
$CH_3C(O)OOC(O)CH_3$	2.5	15.8	133.8
$C_2H_5C(O)OOC(O)C_2H_5$	4	15.4	130
$C_6H_5C(O)OOC(O)C_6H_5$	63	15.9	125

[a] At 343 K.

data in Table 3 that diacyl peroxides are less stable than dialkyl peroxides because of a difference of 20–30 kJ mol^{-1} in the activation energy for decomposition, a difference which reflects the difference in the stabilization energies of acyloxyls and alkoxyls.

Diacyl peroxides in high concentrations and in polar solvents are significantly more susceptible to induced decomposition than dialkyl peroxides. This generally involves an S_H2 reaction on the peroxidic oxygen by a solvent-derived radical, analogous to reaction (13), although in certain cases decomposition may be induced by abstraction of a hydrogen atom from the β position of the alkyl moiety[16], e.g. reaction (26).

$$R^1\cdot + H{-}\overset{R^2}{\overset{|}{C}}HCH_2\overset{O}{\overset{\|}{C}}O\overset{O}{\overset{\|}{O}}CR^3 \longrightarrow R^1H + R^2CH{=}CH_2 + CO_2 + R^3CO_2\cdot \qquad (26)$$

As with peroxides the decomposition of acyl peroxides is catalysed by metal ions of variable valency by a one-electron transfer mechanism[18].

The products of the homolysis of acyl peroxides are alkanes, R_2, formed by combination of two alkyls, alkanes, RH and alkenes, R(−H), produced by the disproportionation reaction of two alkyls, and products, such as alkyl halides, produced by reaction of alkyls which escape the solvent cage, with a halo compound added as a radical scavenger (reaction 27). Combination and disproportionation can of course occur inside or outside the solvent cage.

$$R\cdot + R\cdot \longrightarrow \begin{cases} R_2 \\ RH + R(-H) \\ \xrightarrow{SX} RX + S\cdot \end{cases} \qquad (27)$$

Time-resolved nuclear magnetic resonance spectroscopy has revealed that several of these products initially possess abnormal NMR spectra which contain negative peaks and absorption signals of unusually high intensity[21]. That is, the products have nuclear spin state populations which are perturbed from equilibrium in either direction. Diacyl peroxides have in fact proved to be extremely useful radical sources for the study of chemically induced dynamic nuclear polarization because of their convenient half-lives.

Diacyl peroxides are also useful photochemical and thermal sources of alkyl radicals for electron spin resonance spectroscopic studies of the structure and lifetimes of alkyl radicals in solution[22]. In particular di-(6-heptenoyl) peroxide has been used as a source of 5-hexenyl which cyclizes to cyclopentylmethyl (reactions 28 and 29).

$$(CH_2{=}CH(CH_2)_4C(O){-}O)_2 \longrightarrow 2\,CH_2{=}CH(CH_2)_3CH_2\cdot + 2CO_2 \qquad (28)$$

$$CH_2{=}CH(CH_2)_3CH_2\cdot \longrightarrow c\text{-}C_5H_9CH_2\cdot \qquad (29)$$

The rate constant for this radical rearrangement has been measured by kinetic electron spin resonance spectroscopy[23] and can be represented by $\log(k_{29}/s^{-1}) = (9.5 \pm 1.1) - (25.5 \pm 4.6)/\theta$. A knowledge of this Arrhenius equation enables rate constants for the reaction of 5-hexenyl with a variety of substrates to be estimated by measuring the ratio of 5-hexenyl to cyclopentylmethyl produced in the presence of a known concentration of the substrate[24].

IV. DIAROYL PEROXIDES

Benzoyl peroxide is by far the most thoroughly studied diaroyl peroxide[20] and its thermal decomposition is a complex process in most solvents because of the ease with which it undergoes radical displacement by solvent-derived radicals, R·, on the peroxidic oxygen as shown in reaction (31). This S_H2-induced decomposition is particularly facile for nucleophilic radicals such as α-ethoxyethyl derived by hydrogen-atom abstraction from diethyl ether.

$$C_6H_5C(=O)OOC(=O)C_6H_5 \longrightarrow 2C_6H_5C(=O)O\cdot \tag{30}$$

$$R\cdot + C_6H_5C(=O)OOC(=O)C_6H_5 \longrightarrow C_6H_5C(=O)OR + C_6H_5C(=O)O\cdot \tag{31}$$

Decomposition of benzoyl peroxide is further complicated by induced decomposition involving radical addition to the *para* position of the peroxide (reaction 32). Thus decomposition of benzoyl peroxide in cyclohexane gives *p*-cyclohexylbenzoic acid. Several mechanisms have been offered to explain the absence of *meta* products including a suggestion that *para* substitution is enhanced by α-lactone formation concerted with addition, a transannular hydrogen-atom transfer involving reversible free-radical addition and formation of a radical intermediate stabilized by an adjacent carboxylate function.

$$R\cdot + C_6H_5C(=O)OOC(=O)C_6H_5 \longrightarrow \text{[4-R-4-H-cyclohexadiene spiro-}\alpha\text{-lactone (O, C=O)]} + C_6H_5CO_2\cdot \longrightarrow R\text{-}C_6H_4\text{-}CO_2H \tag{32}$$

Nucleophilic compounds such as amines, sulphides, phosphines and olefins often react instantaneously with benzoyl peroxide by an ionic mechanism. Free radicals are, however, often produced in low percentages in these reactions by a mechanism which involves either electron transfer or formation of an intermediate which is more susceptible to homolysis than benzoyl peroxide[25].

Ring-substitution has a marked influence on the stability of aroyl peroxides and electron-withdrawing substituents in the *meta* and *para* positions retard decomposition while the reverse is true for electron-donating substituents[20]. The rate constants fit a Hammett $\rho\sigma$ plot with $\rho = -0.38$. This has been rationalized in terms of an inductive effect removing or adding to the excess of electron density on the peroxidic oxygens thereby stabilizing or destabilizing it with respect to cleavage.

Ortho substituents accelerate the decomposition of aroyl peroxides because of steric effects, although certain *ortho*-substituted aroyl peroxides, e.g. iodo and vinyl, exhibit anchimerically assisted decomposition[20].

Aroyloxyls such as $C_6H_5C(O)O\cdot$ are more long-lived than acetoxyls. Thus the rate constant for decomposition of $CH_3C(O)O\cdot$ has been estimated to be $1.6 \times 10^9\, s^{-1}$ at 333 K[14] whereas $C_6H_5C(O)O\cdot$ lives long enough to be trapped by phenyl *t*-butylnitrone[26].

Because of its susceptibility to induced decomposition homolysis of benzoyl peroxide does not show simple first-order kinetics but exhibits a rate law of the form

$$\frac{-d[P]}{dt} = k_{30}[P] + k_{31}[P]^{\frac{1}{2}}$$

where [P] is the peroxide concentration.

V. CYCLIC PEROXIDES

Cyclic peroxides are primary products of the autoxidation of polyunsaturated compounds such as natural and synthetic rubbers, oils and fats and are formed by the intramolecular addition of a peroxy radical to a double bond[27]. For instance, squalene absorbs two moles of oxygen to give one mole of the hydroperoxy cyclic peroxides, **1** and **2**[28] (Scheme 1).

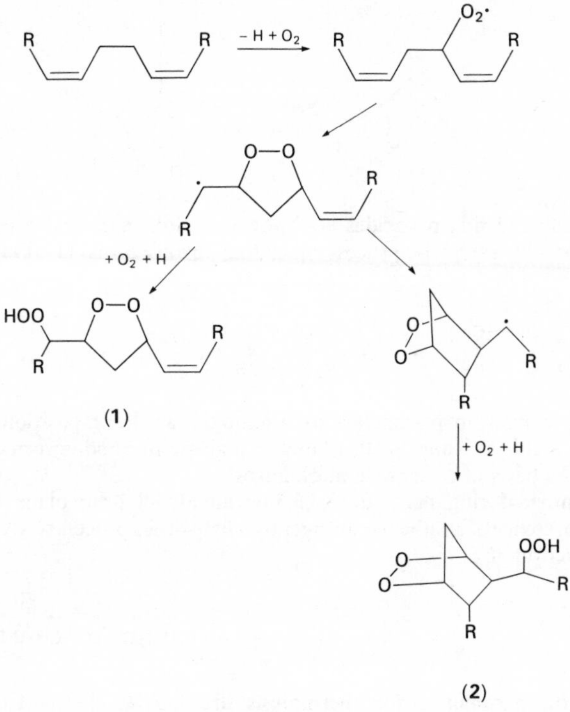

SCHEME 1

Similarly, a cyclic peroxide is the primary product of α-farnesene autoxidation[29] (reaction 33).

$$\xrightarrow[\text{R}\cdot]{\text{O}_2} \qquad (33)$$

Bicyclic endoperoxides analogous to **2** are intermediates in prostagladin biosynthesis and two members of this important class of natural products, PGG and PGH, have bicyclic endoperoxide functionally incorporated in their structure[27].

Much simpler cyclic peroxides can be prepared from unsaturated hydroperoxides by the reaction sequence (34).

$$\xrightarrow{\text{R}\cdot} \qquad (34)$$

The four-membered ring peroxides are known as dioxetanes and decompose to give carbonyl products by a stepwise process which may involve initial O—O bond homolysis followed by C—C bond scission[30] (reaction 35).

$$R_2\overset{\text{O—O}}{\text{C—CR}_2} \longrightarrow R_2\overset{\text{O}^\bullet\ \ ^\bullet\text{O}}{\text{C—CR}_2} \longrightarrow 2\ R_2C{=}O \qquad (35)$$

Experimental activation parameters for unimolecular decomposition of substituted dioxetanes are $\sim 10^{12}\,s^{-1}$ and $\sim 105\,kJ\,mol^{-1}$ and are in good agreement with values calculated on the basis of a stepwise mechanism.

The five-membered ring peroxide 3,3,5,5-tetramethyl-1,2-dioxolane decomposes in benzene in the presence of a radical scavenger by a first-order process to give acetone and a diradical[31] (reaction 36).

$$\rightleftharpoons \longrightarrow (CH_3)_2C{=}O + \cdot CH_2\overset{\cdot O}{C}(CH_3)_2 \qquad (36)$$

The activation parameters for thermolysis are $\log(A_{36}/s^{-1}) = 15.85 \pm 0.42$ and $E_{36} = 186 \pm 4\,kJ\,mol^{-1}$, values which are consistent with the stepwise mechanism shown in reaction (36). Activation parameters in the absence of a radical scavenger are lower than these values because this peroxide is susceptible to induced decomposition by an S_H2 mechanism because the peroxidic oxygens are exposed to radical attack.

Cyclic peroxides of the type **3**, where R^1 and R^2 are aryl and alkyl groups, undergo initial O—O bond homolysis to give a diradical which subsequently breaks down by C—O bond scission to give two molecules of ketone plus oxygen or C—C bond scission to give radical products[32,33] (reaction 37).

$$\mathbf{(3)} \longrightarrow 2R^1C(O)R^2 + O_2 \quad \text{or} \quad 2R^1\cdot + (R^2CO_2)_2 \qquad (37)$$

The proportion of radical products increases with the stability of the radical $R^1\cdot$. Thus if $R^2 = R^1 = C_6H_5$, benzophenone is formed in 95–96% yield in benzene at 423 K whereas if $R^1 = C_6H_5CH_2$ and $R^2 = C_6H_5$, ketone is only formed in 12–14% yield[32].

The cyclic peroxide from cyclohexanone can either lose oxygen to give cyclohexanone, one molecule of CO_2 to give a cyclic lactone or two molecules of CO_2 to give a cyclic hydrocarbon[34] (reaction 38).

(38)

2 cyclohexanone $+ O_2$; lactone $+ CO_2$; cyclic hydrocarbon $+ 2\ CO_2$

It should perhaps be noted here that cyclic peroxides are more stable than acyclic peroxides because the reversal of O—O bond cleavage is much more favourable.

VI. TRIOXIDES

Di-*t*-alkyl trioxides are produced by the combination of *t*-alkoxy and *t*-alkylperoxy radicals at low temperatures[8]. For instance *t*-butoxyl and *t*-butylperoxyl combine to give di-*t*-butyl trioxide which dissociates above about 240 K to regenerate the radicals (reaction 39).

$$(CH_3)_3CO\cdot + (CH_3)_3CO_2\cdot \rightleftharpoons (CH_3)_3COOOC(CH_3)_3 \qquad (39)$$

Despite their instability di-*t*-butyl trioxide and dicumyl trioxide have been prepared, and decomposition of the latter in $CFCl_3$ has been followed by NMR spectroscopy[35]. This trioxide has half-lives of 3.2×10^3 s and 3×10^2 s at 248 and 265.5 K, respectively, and an activation energy for thermolysis of 78.4 kJ mol^{-1} which gives some indication of the O—O bond strength in these compounds.

There is ESR spectroscopic evidence[36] that unsymmetric secondary (or primary) and tertiary alkyl trioxides are formed at low temperatures by combination of *t*-butoxyl and a secondary (or primary) alkylperoxyl (reaction 40).

$$(CH_3)_3CO\cdot + s\text{-}RO_2\cdot \rightleftharpoons (CH_3)_3COOOR\text{-}s \longrightarrow (CH_3)_3CO_2\cdot + s\text{-}RO\cdot \quad (40)$$

These crossed trioxides, as well as dissociating back to the original radicals, decompose to give *t*-butylperoxyl and a *s*-alkoxyl. Thus, if the *s*-alkylperoxyl is prepared from oxygen enriched in ^{17}O, *t*-butylperoxyl with the terminal oxygen specifically enriched in ^{17}O is prepared[37].

These unsymmetric trioxides are less stable than di-*t*-alkyl trioxides and decompose to give *t*-butylperoxyl at 173 K. A satisfactory explanation has, however, not been advanced for the marked influence of the nature of the alkyl group on the stability of dialkyl trioxides apart from the possible involvement of induced decomposition.

Perfluorination of the substituents increases the stability of trioxides dramatically and bis-trifluoromethyl trioxide can be isolated at ambient temperatures[38,39].

1,2,3-Trioxolanes are the cyclic trioxides that are the initial products of the reaction of ozone with olefins[40] (reaction 41).

$$R^1HC{=}CHR^2 + O_3 \longrightarrow \text{1,2,3-trioxolane (H, R}^1\text{; R}^2\text{, H)} \quad (41)$$

These trioxides are unstable and decompose to give 1,2,4-trioxolanes, carbonyl products and ketone peroxides by a mechanism which involves the intermediacy of a diradical or zwitterion (reaction 42).

$$\text{1,2,3-trioxolane} \longrightarrow H(R^1)C^{+}{-}O{-}O^{-} \text{ or } H(R^1)C\cdot{-}O{-}O\cdot + H(R^2)C{=}O \longrightarrow \text{1,2,4-trioxolane} + \text{1,2,4,5-tetroxane} \quad (42)$$

Decomposition of several 1,2,3-trioxolanes have been followed by low-temperature infrared spectroscopy[40] and have been found to follow first-order kinetics. Arrhenius parameters for a series of these trioxides are presented in Table 5.

TABLE 5. Rate constants and Arrhenius parameters for decomposition of 1,2,3-trioxolanes

Parent alkene	Solvent	T(K)	$\log[k(s^{-1})]$	$\log[A(s^{-1})]$	$E(kJ\,mol^{-1})$
trans-Diethylethylene	CS_2	173.5	−3.8	5.3 ± 1.3	29 ± 6
	CCl_2F_2	173	−3.27	6.7	33
trans-Diisopropylethylene	CS_2	174	−4.22	8.4 ± 1.5	40 ± 5
trans-Di-*t*-butylethylene	CS_2	190.5	−5.00	11.0 ± 2	60 ± 7
cis-Diisopropylethylene	Ethane	95	−5	~3	20
	CS_2	168	2		
Hex-1-ene	Ethane	95	−5	~3	20
	CS_2	168	−2		

VII. TETROXIDES

Dialkyl tetroxides are produced at low temperatures (193 K) by dimerization of alkylperoxyls[8,9] and the equilibrium shown in (43) has been demonstrated by kinetic ESR spectroscopy for several tertiary and secondary alkylperoxyls[41–43].

$$RO_2\cdot + RO_2\cdot \rightleftharpoons ROOOOR \tag{43}$$

Di-*t*-butyl tetroxide is the most stable dialkyl tetroxide[9] and estimates of the equilibrium constant K_{43} at different temperatures has enabled values of $\Delta H^0_{43} \sim -35\,\text{kJ mol}^{-1}$ and $\Delta S^0_{43} \sim -125\,\text{J deg}^{-1}\,\text{mol}^{-1}$ to be evaluated. Thermodynamic parameters for other tetroxides indicate that K_{43} is independent of the nature of the alkyl moiety R.

Di-*t*-alkyl tetroxides decompose irreversibly to give two *t*-alkoxyls and oxygen and the alkoxyls may either combine in the solvent cage to give di-*t*-alkyl peroxide or escape from the solvent cage and undergo typical alkoxy radical chemistry (reaction 44).

$$R_3COOOOCR_3 \longrightarrow \begin{cases} R_3COOCR_3 + O_2 \\ 2\,R_3CO\cdot + O_2 \end{cases} \tag{44}$$

It is not yet clear whether tetroxides decompose by a two-step or a concerted mechanism[9] because kinetic evidence, i.e. high and low *A* factors, have been presented in support of both mechanisms[42,44,45]. It has, however, been argued on thermochemical grounds that $RO_3\cdot$ is an unlikely intermediate at any temperature and that decomposition must be concerted[46].

Di-*s*-alkyl- and di-*p*-alkyl tetroxides decompose at ambient temperatures and below primarily by the nonradical Russell mechanism[47] (reaction 45).

$$R_2C\!\left(\begin{matrix}O{-}O\\ \diagdown\ \ O \\ H\cdots O{-}CHR_2\end{matrix}\right) \longrightarrow R_2C{=}O + O_2 + R_2CHOH \tag{45}$$

At higher temperatures (>373 K), however, decomposition to give alkoxyls becomes increasingly important because the radical process has a higher activation energy than the nonradical process.

Diacyl tetroxides have been proposed as intermediates in aldehyde autoxidation and in the radical-induced decomposition of peracids[19,48]. These tetroxides decompose to give acyloxy radicals which rapidly decarboxylate to give alkyl radicals.

VIII. HYDROPEROXIDES

Hydroperoxides are the principal reaction products of the autoxidation of organic compounds with abstractable hydrogen atoms (RH) at temperatures below 423–473 K and are key intermediates in many autoxidations at higher temperatures[8]. They are produced in the rate-controlling propagation reaction (46).

$$RO_2\cdot + RH \longrightarrow ROOH + R\cdot \tag{46}$$

Homo- and crossed-propagation rate constants for some hydroperoxide-forming reactions are summarized in Table 6. Values of these rate constants depend principally on the strength of the C—H bond that is broken and on the nature of the alkylperoxyl with polar and steric effects playing important but minor roles. Thus, primary, secondary and tertiary aliphatic hydrogens have relative reactivities of ca. 1 : 35–70 : 3000[49,50]. Similarly, primary, secondary and tertiary benzylic hydrogens have relative reactivities of 1:8.3:13.3[8]. Although most of this difference in reactivity is reflected in differences in activation energies there are small but real differences in the preexponential factors. Thus, the A factor for cumene is about an order of magnitude lower than the A factor for the tertiary hydrogen of 3-methylpentane[51]. This is because the transition state for the formation of a resonance-stabilized substituted benzyl radical has fewer degrees of freedom than the transition state for formation of an alkyl radical.

The reactivity of a peroxy radical toward hydrogen-atom abstraction must depend on the nature of the alkyl moiety because cumene has about the same reactivity towards cumylperoxyl as ethylbenzene has towards 1-phenylethylperoxyl. This difference in peroxyl reactivity is, however, best illustrated by the pronounced difference in reactivity of benzoylperoxyl and *t*-butylperoxyl of $\sim 10^4:1$ [8].

The transition state for the transfer of a hydrogen atom to a peroxyl, although dominated by the nature of the incipient radical must contain a contribution from the dipolar structure $ROO^{\delta -}:H\cdot R^{\delta +}$ because rate constants for reaction of *t*-butylperoxyl with ring-substituted toluenes can be correlated by the Hammett equation using σ^+ substituent constants with a ρ^+ value of -0.53[52].

The influence of steric effects on the rate of the hydroperoxide-forming reaction is best illustrated by the reactivities of alkanes towards *t*-butylperoxyl[49]. Thus, 2,2,4-trimethylpentane is relatively unreactive because of steric hindrance to attack by the peroxyl while cyclopentane exhibits enhanced reactivity because of relief of steric strain upon removal of a hydrogen atom.

The O—O bond is the weakest bond in a hydroperoxide[11] ($R_{RO-OH} \sim 175.5\,\mathrm{kJ\,mol^{-1}}$) and homolysis (either thermal or photochemical) gives an alkoxyl and a hydroxyl (reaction 47).

$$ROOH \longrightarrow RO\cdot + \cdot OH \qquad (47)$$

TABLE 6. Absolute rate constants for the hydroperoxide-forming reaction in hydrocarbon autoxidation[a]

Substrate	$k_p^{rr}(M^{-1}s^{-1})$	$k_p^{br}(M^{-1}s^{-1})$	$\log [A_p^{br}(M^{-1}s^{-1})]$	$E_p^{br}(kJ\,mol^{-1})$
$(CH_3)_3CCH_2\underline{CH_3}$	—	0.00002	—	—
$CH_3CH_2CH(CH_3)\underline{CH_2}CH_3$	—	0.0002	8.9	73.1
$CH_3CH_2\underline{CH}(CH_3)CH_2CH_3$	—	0.007	9.5	68.1
$C_6H_5CH_3$	0.08	0.012	—	—
$C_6H_5CH_2CH_3$	0.25	0.1	—	—
$C_6H_5C(CH_3)_2H$	0.18	0.16	8.7	55.2
C_6H_5CHO	33,000	0.85	—	—
$CH_2\text{-}CH_2\text{-}CH_2\text{-}CH_2\text{-}O$ (ring)	1.1	0.085	—	—

[a] Per active hydrogen at 303 K, where subscript p refers to the rate-controlling propagation reaction.

At moderate temperatures (373 K) and in inert solvents hydroperoxides are stable and hydrocarbons and other organic compounds such as ethers and ketones can be autoxidized to give high yields of hydroperoxides. Thus, cumene gives cumene hydroperoxide, tetrahydrofuran gives α-hydroperoxytetrahydrofuran, and methyl ethyl ketone gives α-hydroperoxyethyl methyl ketone.

The stability of a hydroperoxide is, however, very dependent on its environment and there is a complex dependence on the medium and the hydroperoxide concentration[53]. At low hydroperoxide concentrations, the reaction is first order with respect to the hydroperoxide concentration while at high concentrations, the reaction tends to second order. This is because the bimolecular reaction (48) is thermodynamically more favourable than unimolecular homolysis. It has, however, been noted that well-documented cases of bimolecular reaction are limited to initiation of olefin autoxidation and thermal decomposition of allylic hydroperoxides[54].

$$\begin{matrix} R-O-O-H \\ \vdots \quad \vdots \\ H-O-O-R \end{matrix} \longrightarrow RO\cdot + H_2O + RO_2\cdot \tag{48}$$

The thermal stability of tertiary alkyl hydroperoxides does not depend to any great extent on the nature of the alkyl moiety[55], thus $(CH_3)_3COOH$ and $C_6H_5CH_2C(CH_3)_2OOH$ have very similar half-lives in benzene at 427.5 K (1.6×10^5 and $1.4 \times 10^5\ s^{-1}$, respectively). The thermodynamic parameters for decomposition of ROOH are $\Delta H^{\neq} \sim 125\ kJ\ mol^{-1}$ and $\Delta S^{\neq} \sim -50\ J\ deg^{-1}\ mol^{-1}$ and the values for $(CH_3)_3COOH$($\Delta H^{\neq} \sim 170\ kJ\ mol^{-1}$ and $\Delta S^{\neq} = +51\ J\ deg^{-1}$) are considered anomalous.

The hydroperoxidic hydrogen of an alkyl hydroperoxide is relatively weak, $D_{ROO-H} \sim 368\ kJ\ mol^{-1}$ [56], and reaction with many radicals is either exothermic or thermoneutral. In addition, rate constants for removal of this hydrogen by alkoxyl and alkylperoxyl are large because of low activation energies[12,57]. Hydroperoxides are, therefore, very susceptible to induced decomposition by a mechanism which involves abstraction of the hydroperoxidic hydrogen. In the case of tertiary butyl hydroperoxide, the rate of hydroperoxide decomposition is very much faster than the rate of free-radical initiation because most of the self-reactions of *t*-butylperoxyls give radical rather than nonradical products. Kinetic chain lengths are, therefore, greater than 1. The di-*t*-butyl peroxyoxalate-initiated decomposition of this hydroperoxide (reactions 49–52) provides an example of this mode of induced decomposition[58].

$$(CH_3)_3COOC(=O)C(=O)OOC(CH_3)_3 \longrightarrow 2(CH_3)_3CO\cdot + 2CO_2 \tag{49}$$

$$(CH_3)_3CO\cdot + (CH_3)_3COOH \longrightarrow (CH_3)_3COH + (CH_3)_3CO_2^{\cdot} \tag{50}$$

$$2(CH_3)_3CO_2^{\cdot} \longrightarrow 2(CH_3)_3CO\cdot + O_2 \tag{51}$$

$$2(CH_3)_3CO_2^{\cdot} \longrightarrow (CH_3)_3COOC(CH_3)_3 + O_2 \tag{52}$$

The rate of decomposition of *t*-butyl hydroperoxide is given by

$$\frac{-d[(CH_3)_3COOH]}{dt} = R_i\left(1 + \frac{k_{51}}{k_{52}}\right)$$

where R_i is the rate of chain initiation and k_{51} and k_{52} are the rate constants for reactions (51) and (52). The chain length for hydroperoxide decomposition depends on the ratio of rate constants k_{51}/k_{52}, i.e. on the relative rates of nonterminating and terminating interactions of *t*-butylperoxyl. For *t*-butyl hydroperoxide the chain length is ~9 at 303 K[58].

All tertiary alkyl hydroperoxides undergo an induced decomposition analogous to reactions (49)–(52) and chain lengths are independent of the nature of the alkyl moiety[59]. The fate of the alkoxy radicals produced in reaction (51) is, however, very dependent on the nature of R.

As discussed above in the section on cyclic peroxides, abstraction of the hydroperoxidic hydrogen from unsaturated hydroperoxides is an important route to model compounds for prostaglandin biosynthesis.

Tertiary allylic hydroperoxides such as 3-hydroperoxy-2,3-dimethylbut-1-ene are more susceptible to radical-induced decomposition than nonolefinic hydroperoxides because peroxyls add to the carbon–carbon double bond and the addition product decomposes with a relatively low activation energy[54] (reactions 53 and 54).

$$RO_2\cdot + CH_2{=}C(CH_3)C(CH_3)_2OOH \longrightarrow ROOCH_2\dot{C}(CH_3)C(CH_3)_2OOH \quad (53)$$

$$ROOCH_2\dot{C}(CH_3)C(CH_3)_2OOH \longrightarrow \begin{cases} RO\cdot + \overset{O}{\overbrace{CH_2-C}}(CH_3)C(CH_3)_2OOH \\ ROOCH_2-\overset{O}{\overbrace{C(CH_3)-C}}(CH_3)_2 + HO\cdot \end{cases} \quad (54)$$

Primary and secondary alkyl hydroperoxides do not undergo a chain-induced decomposition at ambient temperatures because primary and secondary alkylperoxy radicals undergo a nonradical mutual reaction[9] (reaction 55).

$$2R_2CHO_2\cdot \rightleftharpoons [R_2C(H)-O-O-O-O-CHR_2 \text{ cyclic intermediate}] \longrightarrow R_2C{=}O + O_2 + R_2CHOH \quad (55)$$

At higher temperatures reactions analagous to (53) and (54) increase in importance relative to (55) and chain lengths greater than one are observed[9].

The alkyl moieties of tertiary alkyl hydroperoxides are reasonably inert to free-radical attack because they contain only unactivated primary hydrogens. Secondary alkyl hydroperoxides however, contain tertiary hydrogens α to the hydroperoxide function and this hydrogen is readily abstracted by a free radical (equation 56).

$$R^1\cdot + R^2CHOOH \longrightarrow R^1H + R^2\dot{C}OOH \quad (56)$$

The α-hydroperoxyalkyl radical $R_2\dot{C}OOH$ produced by this reaction is unstable and decomposes, even in the presence of oxygen, to give ketone and hydroxyl[49] (reaction 57).

$$R_2\dot{C}OOH \longrightarrow R_2C{=}O + HO\cdot \quad (57)$$

Primary alkyl hydroperoxides are less susceptible to induced decomposition at the alkyl moiety because they contain secondary hydrogens α to the hydroperoxide function.

Metal ions of variable valency such as Co and Mn, in catalytic concentrations, decompose hydroperoxides rapidly at ambient temperatures to give alkoxyls and alkylperoxyls by the sequence shown in reactions (58) and (59)[60]. The alkoxyls and alkylperoxyls produced in these two reactions may then undergo a variety of reactions such as reaction with uncomplexed hydroperoxide or metal ion, solvent or themselves.

$$\mathrm{ROOH + M^{n+} \rightleftharpoons [ROOH\,M^{n+}] \longrightarrow RO\cdot + M^{(n+1)} + HO^-} \quad (58)$$

$$\mathrm{ROOH + M^{(n+1)} \rightleftharpoons [ROOHM^{(n+1)}] \longrightarrow RO_2\cdot + H^+ + M^{n+}} \quad (59)$$

The reaction products in the case of *t*-butyl hydroperoxide are *t*-butyl alcohol, di-*t*-butyl peroxide and oxygen. Other hydroperoxides such as cumene hydroperoxide give significant yields of products derived from alkoxyl β-scission.

Rates of reaction depend on the nature of the transition-metal ion and the ligand. Reaction kinetics are complex because reactions (58) and (59) involve reversible association of the metal ion and the hydroperoxide before electron transfer takes place. Furthermore, the reaction is retarded by reaction products such as alcohol and water. Interestingly, reaction can be enhanced by a low concentration of amines[61].

Metal ions such as iron and titanium give alkoxyls almost exclusively because reaction (59) is slow (reaction 60).

$$\mathrm{ROOH + Fe^{2+} \longrightarrow RO\cdot + HO^- + Fe^{3+}} \quad (60)$$

Reactions (58) and (59) constitute a catalytic cycle because M^{n+} or M^{n+1} are not destroyed. In practice, however, the complex is destroyed by reaction with $RO_2\cdot$ or $RO\cdot$ by a reaction which usually takes place at the metal centre[62] (reaction 61).

$$\mathrm{RO_2\cdot + M^{n+} \longrightarrow ROO^- + M^{(n+1)+}} \quad (61)$$

Reaction of alkyl hydroperoxides with complexes of transition-metal ions with only one readily accessible valency state such as nickel and zinc occurs with the production of free radicals[63]. Thus dialkyldithiophosphates $[(RO)_2PS_2]_2M$ and dialkyldithiocarbamates $(R_2NCS_2)_2M$ of these two metal ions react with various alkyl hydroperoxides to give free radicals which are capable of initiating hydrocarbon autoxidation, inducing decomposition of the hydroperoxide and destroying radical scavengers such as aromatic amines and stable free radicals[63].

Lead tetraacetate reacts with alkyl hydroperoxides to produce alkylperoxyls via a single electron-transfer mechanism (reactions 62 and 63).

$$\mathrm{Pb(OAc)_4 + ROOH \longrightarrow Pb(OAc)_3 + HOAc + RO_2\cdot} \quad (62)$$

$$\mathrm{Pb(OAc)_3 + ROOH \longrightarrow Pb(OAc)_2 + HOAc + RO_2\cdot} \quad (63)$$

Alternatively, in view of the nonionic character of lead tetraacetate, the hydroperoxide may enter the coordination shell of the lead either in addition to the acetate or with partial displacement of it (reaction 64).

$$\mathrm{Pb(OAc)_4 + ROOH \rightleftharpoons Pb(OAc)_3OOR + HOAc} \quad (64)$$

IX. PERACIDS

Peracids are the initial products of aldehyde autoxidation although they subsequently react with excess aldehyde to form carboxylic acids (reactions 65 and 66).

$$\mathrm{RCHO + O_2 \longrightarrow RC(=O)OOH} \quad (65)$$

$$\mathrm{RC(=O)OOH + RC(=O)H \longrightarrow RC(OH)(H){-}O{-}O{-}C(=O)R \longrightarrow 2\,RC(=O)OH} \quad (66)$$

Although peracids have strong (25–29 kJ mol^{-1}) internal hydrogen bonds, thermally generated *t*-butoxyls do abstract the hydroperoxyl hydrogen to give acylperoxyls, e.g. reaction (67). Consequently there is a chain (although a very short chain) induced decomposition of peracetic acid[48].

$$\mathrm{(CH_3)_3CO\cdot + CH_3C(O)OOH \longrightarrow (CH_3)_3COH + CH_3C(O)O_2\cdot} \quad (67)$$

Photolysis of peracetic acid in xylenes gives more ethyltoluenes from side-chain methylation at 2537 Å than at 2900 Å, but more methylbenzyl alcohols at 2900 Å[64]. Peracids decompose to give alkyl radicals which either react with the peracid in an S_H2 process to give alcohol or with the solvent to give hydrocarbons (reactions 68–71). The proportion of alcohol formed increases with increasing nucleophilic character of the radical[65].

$$\mathrm{RC(O)OOH \longrightarrow RCO_2\cdot + \cdot OH} \quad (68)$$

$$\mathrm{RCO_2\cdot \longrightarrow R\cdot + CO_2} \quad (69)$$

$$\mathrm{R\cdot + RC(O)OOH \longrightarrow ROH + RCO_2\cdot} \quad (70)$$

$$\mathrm{R\cdot + SH \longrightarrow RH + S\cdot} \quad (71)$$

X. HYDROTRIOXIDES

Saturated compounds such as alcohols, ethers, acetals and hydrocarbons react with ozone at low temperatures (195 K) to give oxygenated products by a reaction mechanism which is believed to involve the intermediacy of hydrotrioxides, RO_3H[66]. These trioxides are formed by a heterolytic rather than a homolytic process[66].

Hydrotrioxides are believed to decompose as shown in reaction (72) for cyclohexane trioxide with reactions (72a) and (72b) occurring within the solvent cage.

$$\text{C}_6\text{H}_{11}\text{OOOH} \rightleftharpoons [\text{C}_6\text{H}_{11}\text{O}\cdot + \cdot\text{OOH}] \begin{cases} \xrightarrow{a} \text{C}_6\text{H}_{10}{=}\text{O} + \text{H}_2\text{O}_2 \\ \xrightarrow{b} \text{C}_6\text{H}_{11}\text{OH} + {}^1\text{O}_2 \\ \xrightarrow{c} \text{C}_6\text{H}_{11}\text{O}\cdot + \text{HO}_2\cdot \end{cases} \quad (72)$$

The weakest bond in a hydrotrioxide (RO—OOH) has a dissociation energy of $\sim 94\,\text{kJ mol}^{-1}$ and decomposition should be rapid above 233 K[66]. Reaction products and their yields will of course depend on the efficiency with which the caged radicals can escape the solvent cage and the possible reactions which can occur within the solvent cage.

XI. PEROXY ESTERS

These peroxides decompose somewhat more slowly than diacyl peroxides and considerably faster than dialkyl peroxides. The rate-determining step for homolysis can be a single or multiple bond-cleavage process depending on the nature of the alkyl moiety[67] (reactions 73 and 74).

$$CH_3C(O)OOC(CH_3)_3 \longrightarrow CH_3C(O)O\cdot + (CH_3)_3CO\cdot \quad (73)$$

$$(C_6H_5)_3CC(O)OOC(CH_3)_3 \longrightarrow (C_6H_5)_3C\cdot + CO_2 + (CH_3)_3CO\cdot \quad (74)$$

Rate constants increase and enthalpies and entropies of activation decrease as the stability of R increases (Table 7). This has been attributed to an increase in the importance of concerted decomposition. Thus, although concerted decomposition produces a decrease in $\Delta H^{\neq}$ it also produces a decrease in $\Delta S^{\neq}$ because of the increased importance of restricted rotations in the transition state.

Peroxy esters are subject to induced decomposition at the O—O function by a variety of free radicals. For instance the chain length for induced decomposition of 0.2–0.4 M *t*-butyl peracetate at 348 K is 250–300 in secondary alcohols[68] (reactions 75 and 76).

$$(CH_3)_3CO\cdot + CH_3CH_2CH(CH_3)OH \longrightarrow (CH_3)_3COH + CH_3CH_2\dot{C}(CH_3)OH \quad (75)$$

$$CH_3CH_2\dot{C}(CH_3)OH + CH_3C(O)OOC(CH_3)_3 \longrightarrow CH_3CH_2COCH_3 + CH_3C(O)OH + (CH_3)_3CO\cdot \quad (76)$$

TABLE 7. Rate constants and Arrhenius parameters for thermolysis of some peroxy esters

Peroxy ester	$10^9 k(s^{-1})$	$\log [A(s^{-1})]$	$E(\text{kJ mol}^{-1})$
$CH_3CO_3C(CH_3)_3$	0.0113	17.61	165.5
$(CH_3)_3CCO_3C(CH_3)_3$	37.55	15.66	128
$C_6H_5CH_2CO_3C(CH_3)_3$	13.7	13.67	119
$(C_6H_5)_3CCO_3C(CH_3)_3$	345,274	14.31	103

Peroxy esters containing an α hydrogen atom are sensitive to induced decomposition via hydrogen atom abstraction to give α-lactones[16] (reaction 77).

$$R^1R^2CHC(=O)OOBu\text{-}t + t\text{-}BuO\cdot \longrightarrow t\text{-}BuOH + R^1R^2\dot{C}C(=O)OOBu\text{-}t \longrightarrow R^1R^2\overset{\frown O \frown}{C{-}C}{=}O + t\text{-}BuO\cdot \qquad (77)$$

Large rate enhancements have been observed in the decomposition of *t*-butyl peroxybenzoates substituted in the *ortho* position by substituents which can accomodate one additional electron beyond their ground-state complement such as iodo, vinyl and thio groups[69]. Anchimerically accelerated bond homolysis has been attributed to extra stabilization of the transition state leading to homolytic dissociation by participation of bridged radicals, i.e. reaction (78).

$$o\text{-}X\text{-}C_6H_4C(=O)O{-}OC(CH_3)_3 \longrightarrow \text{(cyclic } \dot{X}\text{-bridged radical)} + (CH_3)_3CO\cdot \qquad (78)$$

The largest acceleration results from participation by sulphide sulphur while a neighbouring sulphinyl group exhibits much smaller acceleration and the sulphonyl substituent shows no acceleration. Neighbouring iodine and vinyl substituents give substantial accelerations while the bulky *ortho* *t*-butyl substituent shows very little acceleration. Acceleration due to steric influences can, therefore, be ruled out (Table 8).

Decomposition of *t*-butyl *o*-phenylthioperbenzoate involves the intermediacy of a sulphuranyl radical and such three-coordinate sulphur species have recently been detected by electron spin resonance spectroscopy[70] (reaction 79). The sulphuranyl radical then either reacts with *t*-butoxyl to give a sulphurane which is long-lived enough to be detected by ^{1}H-NMR (reaction 80) or escapes from the solvent cage to undergo reaction (81).

$$o\text{-}(C_6H_5S)C_6H_4C(=O)OOC(CH_3)_3 \longrightarrow \text{(cyclic } C_6H_5\dot{S}\text{-O radical)} + (CH_3)_3CO\cdot \qquad (79)$$

TABLE 8. Kinetic data for *ortho*-substituted *t*-butyl perbenzoates in chlorobenzene[69]

Substituent	k^a	$\log[A(\text{M}^{-1}\,\text{s}^{-1})]$	$E(\text{kJ mol}^{-1})$
SC_6H_5	2.78×10^4	12.60	96.92
SOC_6H_5	72.7	14.33	123.7
$CH{=}C(C_6H_5)_2$	67.0	14.44	110.7
I	54.1	13.2	118.3
$C(CH_3)_3$	3.8	16.08	143.7
H	1.0	15.53	143.3

[a]Relative to *t*-butylperbenzoate at 393 K.

C_6H_5 S• O + $(CH_3)_3CO\cdot$ ⟶ C_6H_5 $OC(CH_3)_3$ S O O ⟶ C_6H_5 S=O COOH (80)

C_6H_5 S. O O ⟶ S• COC_6H_5 O ⟶ dimer; ⟶ S—C_6H_5 COOH (81)

Di-*t*-butyl peroxyoxalate is widely used as a source of *t*-butoxy radicals at ambient temperatures and since there is no evidence for cage production of di-*t*-butyl monoperoxycarbonate this peroxy ester possibly decomposes by a three-bond homolysis[67]. (reaction 82)

$$(CH_3)_3COOC(O)C(O)OOC(CH_3)_3 \longrightarrow \begin{cases} \not\longrightarrow [(CH_3)_3COOC(O)OC(CH_3)_3]_{cage} \\ \longrightarrow [(CH_3)_3COOC(CH_3)_3]_{cage} \\ \longrightarrow 2(CH_3)_3CO\cdot + 2CO_2 \end{cases} \quad (82)$$

The activation parameters for homolysis, $\log [A(s^{-1})] = 14.5$ and $E = 109\,kJ\,mol^{-1}$, are, however, similar to the values for other peroxy esters which are believed to decompose by two-bond homolysis[67].

Although the products of the decomposition of peresters are reliably free radical the transition state is quite polar, as is demonstrated by a large rate dependence on solvent polarity and the influence of electron-donating and -withdrawing ring-substituents on the stability of *t*-butyl perbenzoates[67].

XII. β-PEROXYLACTONES

Thermolysis of β-peroxylactones gives ketones as the principal reaction products by a mechanism which involves the intermediacy of a diradical[7] (reaction 83).

$$\text{(cyclic peroxylactone, } R^1, R^2\text{)} \longrightarrow \text{(1,5-diradical)} \longrightarrow R^2\overset{O}{\overset{\|}{C}}CH_2R^1 + R^1\overset{O}{\overset{\|}{C}}CH_2R^2 + CO_2 \quad (83)$$

The ketone is produced by migration of R^1 or R^2 and interestingly when R^2 is phenyl alkyl radical migration is preferred to the more usual phenyl migration. Furthermore epoxide is only formed in trace amounts. Stereolabelling experiments have indicated that the 1,5-diradical undergoes a fully concerted 1,2-shift, i.e. concurrent β scission and decarboxylation.

Photolysis of β-peroxylactones gives mainly epoxide along with low yields of ketone. In this case the 1,5-diradical must lose carbon dioxide to give the 1,3-diradical which ring-closes to give the epoxide (reaction 84).

$$\text{(1,5-diradical)} \longrightarrow R^1R^2\dot{C}CH_2O\cdot \longrightarrow \text{(epoxide } R^1R^2C\text{—}CH_2\text{—}O\text{)} \quad (84)$$

XIII. CONCLUSION

It is clear from the examples given above that the peroxide function provides a rich source of free-radical reaction mechanisms and the continuing interest in this area of organic chemistry indicates that the well is not yet dry.

XIV. ACKNOWLEDGEMENT

The author thanks Drs. D. C. Nonhebel and J. Warkentin for reading and correcting an earlier version of this chapter.

XV. REFERENCES

1. Issued as NRCC No. 19975.
2. A. G. Davies, *Organic Peroxides*, Butterworths, London, 1961.
3. E. G. E. Hawkins, *Organic Peroxides–Their Formation and Reactions*, Van Nostrand, Princeton, N. J., 1961.
4. D. Swern (Ed.), *Organic Peroxides*, Vols. I, II and III, Wiley–Interscience, New York–London, 1970–1973.
5. E. G. E. Hawkins, *MTP Int. Rev. Sci.: Org. Chem., Ser. 1*, **10**, 95 (1973).
6. R. R. Hiatt in *Frontiers of Free-Radical Chemistry* (Ed. W. A. Pryor), 1980, p. 225.
7. W. Adam, *Acc. Chem. Res.*, **12**, 390 (1979).
8. J. A. Howard in *Free Radicals*, Vol. II (Ed. J. K. Kochi), Wiley–Interscience, New York–London, 1973, Chap. 12.
9. J. A. Howard, *Amer. Chem. Soc. Symp. Ser.*, **69**, 413 (1978).
10. F. R. Mayo, *Acc. Chem. Res.*, **1**, 193 (1968).
11. S. W. Benson and R. Shaw in *Organic Peroxides*, Vol. I (Ed. D. Swern), Wiley–Interscience, New York–London, 1970, Chap. 2.
12. H. Paul, R. D. Small, Jr. and J. C. Scaiano, *J. Amer. Chem. Soc.*, **100**, 4520 (1978).

13. P. C. Wong, D. Griller and J. C. Scaiano, private communication.
14. K. U. Ingold in *Free Radicals*, Vol. I (Ed. J. K. Kochi), Wiley–Interscience, New York–London, 1973, Chap. 2.
15. H. Wieland, *Ber.*, **44**, 2553 (1911).
16. K. U. Ingold and B. P. Roberts, *Free-radical Substitution Reactions*, Wiley–Interscience, New York–London, 1971.
17. R. Hiatt and V. G. K. Nair, *Can. J. Chem.*, **58**, 450 (1980).
18. G. Sosnovsky and D. J. Rawlinson in *Organic Peroxides*, Vol. I (Ed. D. Swern), Wiley–Interscience, New York–London, 1970, Chap. 9.
19. N. A. Clinton, R. A. Kenley and T. G. Traylor, *J. Amer. Chem. Soc.*, **97**, 3746, 3752, 3757 (1975).
20. R. Hiatt in *Organic Peroxides*, Vol. II (Ed. D. Swern), Wiley–Interscience, New York–London, 1971, Chap. 8.
21. H. R. Ward in *Free Radicals*, Vol. I (Ed. J. K. Kochi), Wiley–Interscience, New York–London, 1973, Chap. 6.
22. J. K. Kochi and P. J. Krusic, *Chemical Society Special Publication*, No. 24, 147 (1970).
23. P. Schmid, D. Griller and K. U. Ingold, *Int. J. Chem. Kinet.*, **11**, 333 (1979).
24. D. Griller and K. U. Ingold, *Acc. Chem. Res.*, **13**, 317 (1980).
25. J. A. K. Harmony, *Methods in Free Rad. Chem.*, **5**, 101 (1974).
26. E. G. Janzen, C. A. Evans and Y. Nishi, *J. Amer. Chem. Soc.*, **94**, 8236 (1972).
27. N. A. Porter, J. R. Nixon and D. W. Gilmore, *Am. Chem. Soc. Symp. Ser.*, **69**, 89 (1978).
28. R. Bateman, *Quart. Rev.*, **8**, 164 (1954).
29. E. F. Anet, *Australian J. Chem.*, **22**, 2403 (1969).
30. W. H. Richardson, J. H. Anderegg, M. E. Price, W. A. Tappen and H. E. O'Neal, *J. Org. Chem.*, **43**, 2236 (1978).
31. W. H. Richardson, R. McGinness and H. E. O'Neal, *J. Org. Chem.*, **46**, 1887 (1981).
32. K. J. McCullough, A. R. Morgan, D. C. Nonhebel and P. L. Pauson, *J. Chem. Res.* (*S*), 35 (1980).
33. K. J. McCullough, A. R. Morgan, D. C. Nonhebel and P. L. Pauson, *J. Chem. Res.* (*S*), 36 (1980).
34. M. Schutz and K. Kuschke in *Organic Peroxides*, Vol. III (Ed. D. Swern), Wiley–Interscience, New York–London, 1973, Chap. 2.
35. P. D. Barlett and M. Lahav, *Israel J. Chem.*, **10**, 101 (1972).
36. J. E. Bennett, G. Brunton and R. Summers, *J. Chem. Soc., Perkin 2*, 981 (1980).
37. J. A. Howard, *Can. J. Chem.*, **50**, 1981 (1972).
38. P. G. Thompson, *J. Amer. Chem. Soc.*, **89**, 4316 (1967).
39. L. R. Anderson and W. B. Fox, *J. Amer. Chem. Soc.*, **89**, 4313 (1967).
40. B. Mile, G. W. Morris and W. G. Alcock, *J. Chem. Soc., Perkin 2*, 1644 (1979).
41. J. E. Bennett, D. M. Brown and B. Mile, *Trans. Faraday Soc.*, **66**, 386 (1970).
42. K. Adamic, J. A. Howard and K. U. Ingold, *Can. J. Chem.*, **47**, 3803 (1969).
43. E. Furimsky, J. A. Howard and J. Selwyn, *Can. J. Chem.*, **58**, 677 (1980).
44. P. D. Bartlett and G. Guaraldi, *J. Amer. Chem. Soc.*, **89**, 4799 (1967).
45. T. Mill and R. S. Stringham, *J. Amer. Chem. Soc.*, **90**, 1062 (1968).
46. P. S. Nangia and S. W. Benson, *Int. J. Chem. Kinet.*, **12**, 29 (1980).
47. G. A. Russell, *J. Amer. Chem. Soc.*, **79**, 3871 (1957).
48. R. A. Kenley and T. G. Traylor, *J. Amer. Chem. Soc.*, **97**, 4700 (1975).
49. J. H. B. Chenier, S. B. Tong and J. A. Howard, *Can. J. Chem.*, **56**, 3047 (1978).
50. J. A. Howard and J. H. B. Chenier, *Can. J. Chem.*, **58**, 2808 (1980).
51. J. A. Howard, J. H. B. Chenier and D. A. Holden, *Can. J. Chem.*, **56**, 170 (1978).
52. J. A. Howard and J. H. B. Chenier, *J. Amer. Chem. Soc.*, **95**, 3054 (1973).
53. R. Hiatt in *Organic Peroxides*, Vol. II (Ed. D. Swern), Wiley–Interscience, New York–London, 1971, Chap. 1.
54. R. Hiatt and T. McCarrick, *J. Amer. Chem. Soc.*, **97**, 5234 (1975).
55. R. Hiatt and W. M. J. Strachan, *J. Org. Chem.*, **28**, 1893 (1963).
56. L. R. Mahoney and M. A. DaRooge, *J. Amer. Chem. Soc.*, **92**, 4063 (1970).
57. J. H. B. Chenier and J. A. Howard, *Can. J. Chem.*, **53**, 623 (1975).
58. R. Hiatt, J. Clipsham and T. Visser, *Can. J. Chem.*, **42**, 2754 (1964).
59. J. A. Howard and K. U. Ingold, *Can. J. Chem.*, **47**, 3797 (1969).
60. J. F. Black, *J. Amer. Chem. Soc.*, **100**, 527 (1978).

61. T. Yamada and Y. Kamiya, *Bull. Chem. Soc. Japan*, **52**, 474 (1979).
62. J. A. Howard and S. B. Tong, *Can. J. Chem.*, **58**, 1962 (1980).
63. J. A. Howard in *Frontiers of Free-radical Chemistry* (Ed. W. Pryor), Academic Press, New York, 1980, p. 237.
64. Y. Ogata and K. Tomizawa, *J. Org. Chem.*, **45**, 785 (1980).
65. J. Fossey and D. Lefort, *Tetrahedron*, **36**, 1023 (1980).
66. P. S. Nangia and S. W. Benson, *J. Amer. Chem. Soc.*, **102**, 3105 (1980).
67. L. A. Singer in *Organic Peroxides*, Vol. I (Ed. D. Swern), Wiley–Interscience, New York–London, 1970, Chap. 5.
68. C. Walling and J. C. Azar, *J. Org. Chem.*, **33**, 3888 (1968).
69. J. C. Martin, *Amer. Chem. Soc. Symp. Ser.*, **69**, 71 (1978).
70. C. W. Perkins, J. C. Martin, A. J. Arduengo, W. Lau, A. Alegria and J. K. Kochi, *J. Amer. Chem. Soc.*, **102**, 7753 (1980).

The Chemistry of Functional Groups, Peroxides
Edited by S. Patai

CHAPTER 9

Organic sulphur and phosphorus peroxides

ROBERT V. HOFFMAN

Department of Chemistry, New Mexico State University, Las Cruces, New Mexico 88003, U.S.A.

I. INTRODUCTION . . . 259
II. SULPHONYL PEROXIDES . . . 260
A. Introduction . . . 260
B. Synthesis of Sulphonyl Peroxides . . . 261
C. Reactions of Sulphonyl Peroxides . . . 261
1. Decomposition reactions . . . 261
2. Reactions with electron donors . . . 266
a. π Donors . . . 266
b. n Donors . . . 269
c. σ Donors . . . 273
III. ORGANOPHOSPHORUS PEROXIDES . . . 274
IV. REFERENCES . . . 276

I. INTRODUCTION

In the last ten years or so, there has been increasing interest shown in peroxides in which the O—O bond is flanked by elements other than carbon or hydrogen[1a,b]. This discussion will summarize the chemistry of peroxides in which one or both ends of the oxygen–oxygen bond are flanked by organosulphur or organophosphorus groups.

The known examples of these peroxides are largely restricted to bis-sulphonyl peroxides (**1**), sulphonyl acyl peroxides (**2**), sulphonyl alkyl peroxides (**3**) and bis-diphenylphosphinyl peroxide (**5**). Sulphonic peracids (**4**) are not well known[2a] but are among the products formed in the sulphoxidation of alkanes[2b] and may be formed from the reaction of sulphonic acids with hydrogen peroxide[3].

$RSO_2{-}O{-}O{-}SO_2R$ (1)

$R^1SO_2{-}O{-}O{-}C(=O){-}R^2$ (2)

$R^1SO_2{-}O{-}OR^2$ (3)

$Ph_2P(=O){-}O{-}O{-}P(=O)Ph_2$ (5)

RSO_2OOH (4)

II. SULPHONYL PEROXIDES

A. Introduction

In general the attachment of sulphonyl groups to the ends of the oxygen–oxygen bond enhances the thermal stability of the peroxide link due to inductive and/or resonance withdrawal of electrons from the peroxide oxygens. The thermal stability is further enhanced by attachment of arenesulphonyl groups in which the aryl group is substituted with an electron-withdrawing group (**1**; R = *p*-$NO_2C_6H_4$, *m*-$NO_2C_6H_4$, *m*-$CF_3C_6H_4$, 3,5-$(CF_3)_2C_6H_3$), and these compounds represent the most stable examples. The same effect has been noted in aroyl peroxides where substitution of the aromatic ring by electron-withdrawing groups stabilizes the peroxide towards thermal decomposition[4].

Thermal stability does not necessarily imply greater O—O bond strength, however, since thermal stability is a descriptive term which depends on the relative efficiencies of several modes of decomposition[5]. The three principal modes for thermal decomposition of peroxides are (*a*) O—O homolysis, (*b*) radical-induced decomposition and (*c*) polar decomposition. Only the first is directly dependent on the peroxide bond strength. Most likely the thermal stability of sulphonyl peroxides derives from a resistance to radical-induced decomposition and a preference for nonchain, polar decomposition pathways. They can be kept indefinitely at −20°C without measurable decomposition and some are stable at room temperature for many days.

In addition to their good thermal stability, sulphonyl peroxides (**1**) have rather low active oxygen content, typically 4–8%, so that the pure materials decompose exothermically but not violently. They are thus easily handled and stored and can be routinely purified by crystallization and analysed by standard iodometry.

Mixed sulphonyl acyl peroxides (**2**) are qualitatively less stable than bis-sulphonyl peroxides (**1**) but they can nevertheless be prepared, handled at room temperature for short periods, stored, and analysed routinely. Sulphonyl alkyl peroxides (**3**) are somewhat less stable and decompose violently at room temperature after a few minutes. They can be stored in a freezer for a few weeks. Sulphinyl peroxides have not been reported in the literature but a mixed alkyl sulphenyl peroxide $PhC(Me_2)OOSPh$ from cumyl hydroperoxide and phenylsulphenyl chloride is reported as an intermediate which rapidly rearranges to the cumyl sulphinate ester $PhC(Me_2)OS(O)Ph$[6].

Peracid analogues of organosulphur acids are not well known[2a]. Recently Nielson and coworkers[3] have reported that pertrifluoromethanesulphonic acid (**4**, R = CF_3), generated *in situ* from trifluoromethanesulphonic acid and 98% hydrogen peroxide, is one of the most powerful peracid oxidants known, but other reports of organosulphonic peracids are rather scarce[2]. Kice and collaborators[7] have postulated a sulphenic peracid intermediate, PhSOOH, in the reaction of phenyl benzenethiosulphonate with the hydroperoxide anion, but this material rearranges immediately to benzenesulphinic acid.

Based on this behaviour, sulphinic peracids, RS(O)OOH might likewise be expected to isomerize to sulphonic acids (RSO_3H).

B. Synthesis of Sulphonyl Peroxides

There are two general routes available for the preparation of bis-sulphonyl peroxides, **1**. The first utilizes the base-promoted condensation of hydrogen peroxide with sulphonyl chlorides (equation 1) and is the method of choice for the preparation of many bis-arenesulphonyl peroxides (**1**, R = Ar). The best results are obtained when an aqueous alcohol mixture is used in which both the sulphonyl chloride and the hydrogen peroxide are soluble[8], or when a two-phase mixture of aqueous alcohol containing the hydrogen peroxide and a second organic phase (typically chloroform) containing the sulphonyl chloride are stirred vigorously[9]. The solid product is collected and purified by recrystallization to give sulphonyl peroxides of high purity (>98%).

$$2\,ArSO_2Cl + H_2O_2 \xrightarrow{K_2CO_3} ArSO_2OOSO_2Ar \qquad (1)$$

Persulphonic acids resulting from the initial condensation of hydrogen peroxide with the sulphonyl chloride are undoubtedly formed[2a], but they have never been detected in this preparation (equation 2), even when large excesses of hydrogen peroxide are used. Perhaps the greater acidity of the peracid causes it to be deprotonated, and hence sulphonylated, faster than hydrogen peroxide itself; or perhaps any peracid not converted to the bis-peroxide is unstable in the basic reaction mixture.

$$ArSO_2Cl + H_2O_2 \longrightarrow ArSO_2OOH \xrightarrow{ArSO_2Cl} ArSO_2OOSO_2Ar \qquad (2)$$

A second method for the preparation of bis-sulphonyl peroxides uses the electrolysis of the sulphonic acid. Methanesulphonyl peroxide[10] and trifluoromethanesulphonyl peroxide[11] have been prepared by this method, which requires a very concentrated (10 M) aqueous solution of the acid. Table 1 lists bis-sulphonyl peroxides.

Unsymmetrical sulphonyl acyl peroxides (**2**) and sulphonyl alkyl peroxides (**3**) are prepared by the base-promoted condensation of a carboxylic peracid or alkyl hydroperoxide, respectively, with a sulphonyl chloride. Table 2 lists mixed sulphonyl peroxides.

C. Reactions of Sulphonyl Peroxides

The reactions of sulphonyl peroxides can be loosely grouped into two categories. The first is the thermolytic decomposition reactions of sulphonyl peroxides which are usually first-order processes. The second category is the intermolecular reactions of sulphonyl peroxides with electron donors which are generally second-order reactions and which are generally thought to be ionic processes.

1. Decomposition reactions

Of the three principal modes of peroxide decomposition described earlier—homolytic, polar and free-radical-induced decomposition—only the first two are important in the thermal decompositions of sulphonyl peroxides. When bis-arenesulphonyl peroxides (**1**, R = Ar) are allowed to decompose in chloroform at near ambient temperatures

TABLE 1. Bis-sulphonyl peroxides (**1**), RSO_2OOSO_2R

R	m.p.(°C)[a]	Stability[b]	Method of prep.[c]	Reference
C_6H_5	66	Poor	A	9a
$4\text{-}CH_3C_6H_4$	50	Poor	A	8
$4\text{-}ClC_6H_4$	75	Moderate	A	8
$4\text{-}BrC_6H_4$	76	Moderate	A	9a
$3,4\text{-}Cl_2C_6H_3$	72	Moderate	A	9a
$3\text{-}CF_3C_6H_4$	82	Good	A	12
$2\text{-}NO_2C_6H_4$	97	Good	A	9a
$3\text{-}NO_2C_6H_4$	112	Good	A	9a
$4\text{-}NO_2C_6H_4$	128	Good	A	9a
$3,5\text{-}(CF_3)_2C_6H_3$	72	Good	A	13
CH_3	79	Good	B	10
CF_3	liq.	Very poor	B	11

[a]All of these peroxides decompose explosively upon melting.
[b]Definitions of stability are: very poor—decomposes within minutes at 25°C; poor—decomposes within hours at 25°C; Moderate—can be handled at 25°C but decomposes after several days at room temperature; good—can be easily handled at 25°C and does not appreciably decompose in a week or more at 25°C.
[c]Method A—condensation of the sulphonyl chloride with hydrogen peroxide; Method B—electrolysis of the sulphonic acid.

TABLE 2. Unsymmetrical acyl sulphonyl peroxides (**2**) and alkyl sulphonyl peroxides (**3**)

(a) **2**, $R^1SO_2OOC(O)R^2$

R^1	R^2	m.p.(°C)[a]	Stability[b]	Reference
$4\text{-}MeC_6H_4$	Ph	59	Moderate	14
$4\text{-}MeC_6H_4$	$4\text{-}ClC_6H_4$	73	Moderate	15
$c\text{-}C_6H_{11}$	Me	35	Very poor	16
Me	Ph	54	Poor	17
Et	Ph	46.5	Poor	17
n-Pr	Ph	24	Poor	17
i-Pr	Ph	49	Poor	17
$PhCH_2$	Ph	—	Poor	17

(b) **3**, $R^1SO_2OOR^2$

R^1	R^2	m.p.(°C)[a]	Stability[b]	Reference
$4\text{-}MeOC_6H_4$	*t*-Bu	47	Very poor	18
$4\text{-}MeC_6H_4$	*t*-Bu	37	Very poor	18
Ph	*t*-Bu	liq.	Very poor	18
$4\text{-}ClC_6H_4$	*t*-Bu	30	Very poor	18
$4\text{-}BrC_6H_4$	*t*-Bu	40	Very poor	18

[a]All of these materials decompose explosively upon melting.
[b]See footnote b, Table 1, for definitions of stability.

(25–40°C), the corresponding arylsulphonic acid is produced quantitatively in addition to products derived from the trichloromethyl radical[19,20]. These results are consistent with homolytic O—O bond fission followed by hydrogen abstraction from the solvent (equation 3). The kinetics are strictly first order suggesting that radical-induced decomposition is not important, and the activation energy, $E_a = 24.5\,\text{kcal mol}^{-1}$, which approximates the O—O bond energy, is indicative of a labile O—O bond. (The O—O bond energy in benzoyl peroxide is $30\,\text{kcal mol}^{-1}$ [5].) Earlier it had been reported that at relatively low temperatures (25–50°C), benzenesulphonyl peroxide initiated polymerization of methyl methacrylate more effectively than benzoyl peroxide, again attesting to the ready homolysis of the O—O bond in solution[21].

$$(m\text{-}NO_2C_6H_4SO_2O)_2 \longrightarrow 2\,m\text{-}NO_2C_6H_4SO_2O\cdot + CHCl_3 \longrightarrow 2\,m\text{-}NO_2C_6H_4SO_3H + CCl_3\cdot \qquad (3)$$

Arenesulphonyl free radicals produced from sulphonyl peroxides efficiently abstract hydrogen from solvent[22]. They have not been observed to split out sulphur trioxide and give carbon-centred radicals analogous to decarboxylation in acyloxy radicals (equation 4). When generated in the presence of aromatic compounds, tosyloxy radicals can also undergo aromatic substitution in low yield. The addition to the aromatic system is a recombination and addition reactions[5,24]. Activation energies on the order of tosyloxy radicals are quite electrophilic[22b,23] which is in keeping with the strong electron-withdrawing capacity of the sulphonyl group.

$$X\text{-}C_6H_4\text{-}SO_3\cdot \not\longrightarrow X\text{-}C_6H_4\cdot + SO_3 \qquad (4)$$

$$TsO\cdot + C_6H_5X \rightleftharpoons [TsO(H)C_6H_5X]\cdot \longrightarrow TsO\text{-}C_6H_4X \qquad (5)$$

Mixed acyl sulphonyl peroxides undergo more complicated thermal decomposition in solution. Simple homolysis of the peroxide bond gives sulphonoxy radicals and acyloxy radicals (equation 6). The sulphonoxy radicals undergo those processes that have been discussed above, and the acyloxy radicals undergo the usual decarboxylation, recombination and addition reactions[5,24]. Activation energies of the order of $25\,\text{kcal mol}^{-1}$ are observed in the decompositions of a variety of mixed peroxides[14,15,22,25–27]. Mixed acyl sulphonyl peroxides are therefore used commercially as polymerization initiators since they provide a source of free radicals at relatively low temperatures. Acetyl cyclohexanesulphonyl peroxide and acetyl *s*-heptanesulphonyl peroxide are routinely used in the polymerization of vinyl chloride. Many formulations can be found in the patent literature[28].

$$R^1SO_2OOC(=O)\text{—}R^2 \longrightarrow R^1SO_3\cdot + \cdot O_2CR^2 \longrightarrow CO_2 + R^1\cdot \qquad (6)$$

Since acyl sulphonyl peroxides (**2**) are unsymmetrical, a polar decomposition mode competes with O—O bond homolysis. The heterolytic cleavage of the O—O bond in unsymmetric peroxides is expected, considering the classic work of Leffler concerning the carboxy inversion reaction of unsymmetrical aroyl peroxides in polar solvents (equation 7)[29,30]. Acyl sulphonyl peroxides might be expected to undergo heterolysis of the O—O bond even more readily than unsymmetric aroyl peroxides, since they might be thought of as materials having an excellent sulphonoxy leaving group attached to oxygen. They have been shown to undergo polar rearrangement by a first-order reaction (equation 8). This rearrangement is important mainly for mixed peroxides (**2**) in which R^2 = aryl, indicating that the ease of polar decomposition depends on the migratory aptitude of the R^2 group. Unlike unsymmetric aroyl peroxides, however, the extent of rearrangement is largely insensitive to solvent polarity but is subject to acid catalysis[14,31].

$$NO_2-C_6H_4-C(=O)-O-O-C(=O)-C_6H_4-OMe \xrightarrow{\circ} NO_2-C_6H_4-C(=O)-O-C(=O)-O-C_6H_4-OMe \quad (7)$$

$$\underset{(\mathbf{2})}{R^1SO_2OOC(=O)-R^2} \xrightarrow{\circ} \underset{(\mathbf{6})}{R^1SO_2OC(=O)-OR^2} \quad (8)$$

Thus the overall rate of decomposition of benzoyl *p*-toluenesulphonyl peroxide (**2**; R^1 = *p*-tolyl, R^2 = Ph), in several solvents, increases steadily as the reaction progresses, due to formation of benzenesulphonic acid which catalyses the carboxy inversion. The rearrangement product, phenyl *p*-toluenesulphonyl carbonate (**6**; R^1 = *p*-tolyl, R^2 = Ph) comprises 48% of the products. The addition of acid catalysts results in a much faster decomposition, and the proportion of rearrangement increases to 71%. If a heterogeneous base such as magnesium oxide is suspended in the reaction mixture, the rate of decomposition becomes a constant first-order rate and the proportion of rearrangement decreases to 22%. These observations indicate that acids catalyse the carboxyl inversion, but that it occurs by an uncatalysed route also. Kinetic studies of acetyl cyclohexanesulphonyl peroxide decompositions give rate constants of $13.9 \times 10^{-5} s^{-1}$ and $9.59 \times 10^{-5} s^{-1}$ at 20°C in isopropanol and cyclohexane, respectively, showing that solvent polarity exerts a minor influence on the rate[25]. Furthermore the partitioning of reaction products is the same in several solvents[14].

The mechanism of the carboxy inversion in acyl sulphonyl peroxides has been determined by ^{18}O labelling experiments. When the carbonyl oxygen in benzoyl *p*-toluenesulphonyl peroxide is labelled with ^{18}O, there is no scrambling of the label, whereas labelling the sulphonyl oxygens with ^{18}O leads to complete scrambling of the label (equations 9 and 10)[23]. These results are well accomodated by the mechanism shown in equation (11) which has a phenyl-bridged ion pair as an intermediate stage, and which is

$$C_6H_5-C(=O^*)-O-O-S(=O)_2-C_6H_4-Me \xrightarrow{\circ} C_6H_5-O-C(=O^*)-O-S(=O)_2-C_6H_4-Me \quad (9)$$

$$C_6H_5-C(=O)-O-O-S(=O^*)_2-C_6H_4-Me \xrightarrow{\circ} C_6H_5-O-C(=O)-O^*-S(=O^*)_2-C_6H_4-Me \quad (10)$$

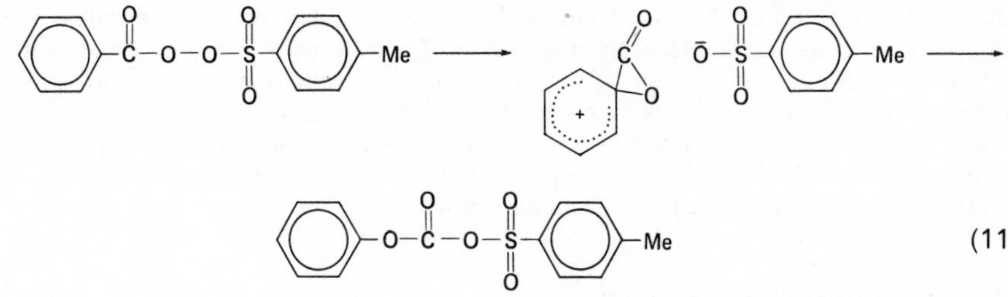

(11)

mechanistically similar to the carboxy inversion in acyl peroxides[30,32]. The bridging of the phenyl ring may be an important feature of the transition state leading to the ion-pair intermediate. Consequently the ability of acyl substituents to interact with the electron-deficient oxygen (migratory aptitude) is an important factor in determining the energy of that transition state, and the extent to which ionic rearrangement competes with O—O bond homolysis. Thus for mixed acyl sulphonyl peroxides (**2**), where the acyl group is aromatic, ionic rearrangement is a significant decomposition pathway[14,31]; for those where the acyl group is alkyl with low migratory aptitude, polar rearrangement is inconsequential[25]. Acid catalysis observed for polar rearrangement of these peroxides presumably results from protonation of the sulphonyl group in the precursor peroxide.

Alkyl arenesulphonyl peroxides (**3**) are unstable materials which decompose in alcohol solution without the production of free radicals. In absolute methanol, acetone dimethyl ketal is produced along with the arylsulphonic acid (equation 12)[33]. A Hammett study of several substituted arylsulphonyl groups has given $\rho = +1.36$, consistent with ionization of the O—O bond, in which electron density is increased on the sulphonoxy group. It follows then that a methyl group migrates to an electron-deficient oxygen. The rate of rearrangement is sensitive to solvent polarity, Grunwald–Winstein $m = 0.59$[33], as is the ionic rearrangement of 9-decalyl perbenzoate, $m = 0.57$ (equation 13) which has a similar mechanism.

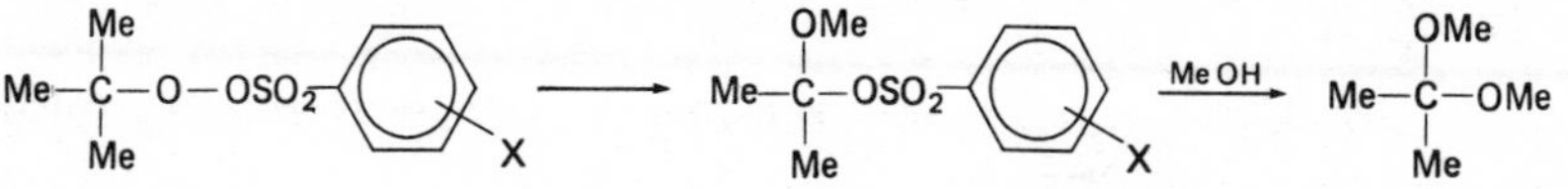

(12)

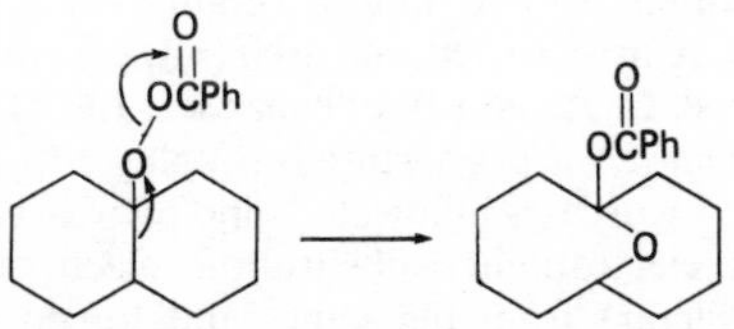

(13)

It is interesting that the rearrangement of alkyl sulphonyl peroxides is reported to be sensitive to solvent polarity and not subject to acid catalysis[33]. This behaviour is opposite to that discussed above for acyl sulphonyl peroxides. It is quite likely that the very different solvent systems used for the respective decompositions can account for this apparent discrepancy. The work with alkyl sulphonyl peroxides has utilized aqueous methanol solvent mixtures, while the work with acyl sulphonyl peroxides has used nonbasic solvents like benzene, chloroform and carbon tetrachloride. Acid catalysis is more likely in the latter solvents which are not levelling solvents and which tend to enhance the acidity of dissolved acids. On the other hand, aqueous methanol is a basic levelling solvent so that the peroxide competes ineffectively with solvent for protons. Furthermore the former group of nonprotic solvents span a narrow range of low polarities and should be very inefficient in solvating ionic intermediates. The polar protic nature of aqueous methanol might be more indicative of normal solvent effects in ionic reactions. Further experiments utilizing similar solvents for both peroxide types are needed to clarify differences in solvent and catalyst effects in these polar decompositions.

2. *Reactions with electron donors*

Although bis-sulphonyl peroxides can decompose by homolysis of the O—O bond (*vide supra*), in the presence of electron donors they react as electrophiles. Conversely, given the electron-deficient nature of the peroxide bond in these peroxides, electron donors can be thought of as attacking the O—O bond nucleophilically (equation 14). Bis-sulphonyl peroxides can thus be considered pseudo-halogens which give products from heterolytic cleavage of the peroxide bond. The reactions of several types of electron donors (π, n and σ donors) with sulphonyl peroxides have been investigated, and all can promote the heterolysis of the peroxide bond to give donor–peroxide adducts.

$$RSO_2{-}O{-}O{-}SO_2R + :D \longrightarrow RSO_2O^- \; D^+ + RSO_3^- \longrightarrow \text{products} \qquad (14)$$

a. π Donors. Based on the ability of benzenesulphonyl peroxide to initiate the polymerization of methyl methacrylate, the formation of phenyl benzenesulphonate from the decomposition of benzenesulphonyl peroxide in benzene (equation 15) was thought to arise from a free-radical aromatic substitution of benzene by the benzenesulphonoxy radical[21]. Dannley's very thorough study of the reactions of arenesulphonyl peroxides with aromatic substrates has revealed, however, that the attachment of the sulphonoxy group to the aromatic ring occurs by electrophilic aromatic substitution[9,12,35,36,38,40,42].

$$(PhSO_2O)_2 + C_6H_6 \longrightarrow PhSO_2O{-}C_6H_5 + PhSO_3H \qquad (15)$$

When *m*-nitrobenzenesulphonyl peroxide is decomposed in the presence of a variety of alkylbenzenes, reactivities relative to benzene itself (k_{Ar}/k_B) are found to be: toluene, 19.0; ethylbenzene, 17.6; cumene, 13.5; and *t*-butylbenzene, 11.9. The product orientations are predominantly *ortho* and *para*; the proportion is roughly 30% *ortho* and 65% *para*. Thus the Baker–Nathan reactivity order is followed[34], and relative reactivities and orientations are consistent with an electrophilic substitution mechanism. The data for these substitutions are very different from the same kinds of data obtained in free-radical aromatic substitutions[9b]. Furthermore in the reactions with cumene, no bicumyl is detected. Bicumyl is often produced in free-radical reactions of cumene by abstraction of the methine hydrogen of the isopropyl side-chain, followed by dimerization of the cumyl radical. The absence of side-chain attack argues against the involvement of free radicals in the substitution reaction, as does the lack of detectable ESR signals during the reaction[9b].

Relative reactivity and orientation studies on a wide variety of aromatic substrates have been carried out. Electron-donating substituents make the aromatic substrate more reactive than benzene and give *ortho*, *para*-orientation, while electron-withdrawing substituents decrease the reactivity and give mostly *meta* orientation. Partial rate factors derived from these data have been correlated by a Hammett plot using Brown's σ^+ values[37] and give $\rho = -4.4$, similar to other known electrophilic substitutions like mercuration (−4.0), bromination (−5.8) and nitration (−6.2), but much larger in magnitude than free-radical substitutions (~ −1 to −2)[35,36].

The mechanism of aromatic substitution by sulphonyl peroxides can best be pictured as rate-determining formation of a Wheland intermediate (equation 16). Several additional studies support this mechanism. There is no kinetic deuterium isotope effect ($k_H/k_D = 1$), indicating that proton loss from the cationic intermediate is not rate-determining[38]. The possibility that π complex formation is rate-determining is discounted by the relative reactivities (k_{Ar}/k_B) of *p*-xylene (340) and mesitylene (2400) which are much greater than those for reactions where π complexation is rate-determining[38]. There is evidence, however, that π complexation does occur as a preequilibrium step. The half-life for decomposition of *m*-nitrobenzenesulphonyl peroxide in ethyl acetate at room temperature is 20 hours, however, in ethyl acetate that is 1 M in ethyl benzoate, the same peroxide decomposes with a half-life of 50 hours and gives high yields of aromatic substitution[35]. The formation of peroxide–aromatic π complexes can be used to explain this behaviour. Finally the reactions of sulphonyl peroxides with arenes are second-order reactions, first-order in both peroxide and substrate, which argues against predissociation of the peroxide[38].

RSO2—O / RSO2—O + C6H5X ⇌ [RSO2—O / RSO2—O]←C6H5X —slow→ RSO2—O(H)C6H5X(+) —(−H+)→ RSO3—C6H4X (16)

The detailed mechanism of the formation of the σ complex has been studied by oxygen-18 labelling of the sulphonyl oxygens in *p*-nitrobenzenesulphonyl peroxide. In methylene chloride solution, the labelled peroxide reacts with benzene to give phenyl *p*-nitrobenzenesulphonate with no label in the phenolic oxygen (equation 17). Thus direct interaction of benzene with the peroxide oxygens leads to product. *p*-Xylene behaves similarly. In some solvents partial label scrambling occurs, which has been interpreted to result from competitive attack on the peroxide and sulphonyl oxygens[19,39]. Due to the fact that π complexation with aromatics is highly likely, and considering that the stability of sulphonyl peroxides in various solvents varies widely[9b], it is possible that π complexation by some solvents may compete with that by the aromatic substrate. If such π complexation by solvent involves the peroxidic oxygens, then the aromatic substrate may be forced to attack the sulphonyl oxygens.

$NO_2C_6H_4S(O^*)_2OOS(O^*)_2C_6H_4NO_2 + C_6H_6 \longrightarrow NO_2C_6H_4S(O^*)_2OC_6H_5$ (17)

The ionic mechanism for aromatic substitution by sulphonyl peroxides has been further substantiated by Kovacic and coworkers, who have carried out aromatic substitutions with *m*-nitrobenzenesulphonyl peroxide in the presence of redox catalysts such as copper and cobalt salts, and the Lewis acid, aluminium trichloride. Little difference in rates or products has been found in the presence or absence of these additives, and it has been concluded that the ionic mechanism satisfactorily accounts for these results[41]. Since a variety of sulphonyl peroxides give similar products with aromatic substrates[10a,12,35,40,42], electrophilic aromatic substitution seems to be an inherent property of these reagents.

In addition to aromatic π systems, olefinic π electron donors react readily with bis-sulphonyl peroxides. Kergomard and collaborators[8] have shown that *p*-toluenesulphonyl peroxide reacts with *cis*- and *trans*-stilbene to give *meso*-1,2-ditosyloxy-1,2-diphenylethane (equation 18). Experiments with norbornene yield products of substitution and addition whose structures are indicative of a Wagner–Meerwein rearrangement (equation 19)[43]. It is thus proposed that ionic addition to the double bond gives an intermediate cation which proceeds to products.

$$(TsO)_2 + PhCH{=}CHPh \longrightarrow \text{TsO(H)(Ph)C–C(H)(Ph)OTs} \quad (18)$$

cis or *trans* 20 – 30%

$$\text{norbornene} + ArSO_2O{-}OSO_2Ar \longrightarrow \ \longrightarrow \ \xrightarrow{-H^+} ArSO_3\text{-nortricyclene/norbornene} \qquad (19)$$

OSO$_2$Ar, +, ArSO$_2$O, +, −H$^+$, ArSO$_3$, ArSO$_2$O$^-$, ArSO$_3$, ArSO$_3$

Several other studies have substantiated the ionic nature of the reaction of sulphonyl peroxides with olefins. When β-alkylstyrenes are reacted with arenesulphonyl peroxides in methanol, solvent capture of the intermediate cation is the major process (equation 20)[44]. The orientation of the products is Markownikoff in that the peroxide acts as an electrophile and the nucleophile is incorporated on the most stable carbenium ion centre.

$$PhCH{=}CHR + (ArSO_2O)_2 \xrightarrow{MeOH} PhCH(OMe){-}CHR(O_3SAr) \qquad (20)$$

The structure of the intermediate cation is of interest since the sulphonoxy function is known to destabilize adjacent carbenium ions[45] yet the sulphonoxy group could undergo bridging interactions as in **7a** or **7b** (equation 21). A study of the addition or *p*-nitrobenzenesulphonyl peroxide to *cis*- or *trans*-stilbene, contrary to earlier reports, has been found to give different ratios of *d,l* and *meso* adducts[46]. *Syn* addition to the double bond is a major contributor to the product stereochemistry. This argues against

$$RCH{=}CHR + (ArSO_2O)_2 \longrightarrow R\overset{+}{C}H{-}CH(OSO_2Ar)R \text{ or } RCH{-}CHR \text{ bridged by } {}^{+}O{-}SO_2Ar \ (\mathbf{7a}) \text{ or } RCH{-}CHR \text{ bridged by } O{-}S(O)(Ar){-}O^{+} \ (\mathbf{7b}) \quad (21)$$

sulphonoxy bridging and suggests instead that phenyl bridging helps control the stereochemistry of addition (equation 22). The additions of sulphonyl peroxides to β-alkylstyrenes, in which phenyl bridging is not possible, give the same mixture of *threo* and *erythro* adducts from either *cis* or *trans* precursors[44]. The same cation is formed from either isomer and sulphonoxy bridging cannot be an important feature of the cation structure. Similar studies need to be undertaken with aliphatic olefins which do not give benzylic cations and which therefore might be more prone to sulphonoxy bridging.

$$PhCH{=}CHPh + (ArSO_2O)_2 \longrightarrow ArSO_3^- \ PhCH{-}CH(O_3SAr)Ph \ (\text{phenyl-bridged cation}) \longrightarrow PhCH(ArSO_3){-}CH(O_3SAr)Ph \ (syn) \quad (22)$$

Electron-rich olefins such as enols react readily with sulphonyl peroxides to give α-sulphonoxy carbonyls. Deoxybenzoin, in the presence of boron trifluoride etherate, reacts with arenesulphonyl peroxides to give the α-sulphonoxycarbonyl product in good yield (equation 23)[44]. Other carbonyl compounds, such as cyclohexanone and acetophenone, show similar behaviour.

$$PhCH_2\overset{O}{\overset{\|}{C}}Ph \xrightarrow[MeOH]{BF_3} PhCH{=}C(OH)Ph \xrightarrow{(ArSO_2O)_2} PhCH(ArSO_3){-}\overset{O}{\overset{\|}{C}}Ph \ (83\%) \quad (23)$$

Mixed acyl sulphonyl peroxides do not seem to react ionically with π electron donors, but this question has not been addressed to any great extent in the literature.

b. n Donors. Until recently little was known of the reactions of sulphonyl peroxides with n electron donors. Triphenylphosphine was used to analyse *m*-nitrobenzenesulphonyl peroxide, and by labelling experiments was found to attack the peroxidic oxygens exclusively (equation 24)[19]. Furthermore dimethyl sulphoxide was known to rapidly decompose sulphonyl peroxides[9b], suggestive of attack by sulphur on the peroxide bond. Recent work on the reactions of n electron donors with sulphonyl peroxides has utilized amines and amine derivatives as donors.

$$Ar{-}S({=}O^*)_2{-}O{-}O{-}S({=}O^*)_2{-}Ar + Ph_3P{:} \longrightarrow Ph_3P{=}O + Ar{-}S({=}O^*)_2{-}O{-}S({=}O^*)_2{-}Ar \quad (24)$$

Primary and secondary amines react readily with arenesulphonyl peroxides at −78°C. Hydrolytic work-up affords good yields of carbonyl products according to the stoichiometry shown in equation (25)[47]. The initial oxidation product is an amine as shown by its isolation in several cases. Table 3 is a representative sample of amine oxidations with *p*-nitrobenzenesulphonyl peroxide.

$$3R^1CH_2NHR^2 + (ArSO_2O)_2 \longrightarrow R^1{-}CH{=}N{-}R^2 + 2ArSO_3^-\,R^1CH_2N^+H_2R^2 \xrightarrow{H_2O} R^1CHO + R^2NH_2 \qquad (25)$$

The above oxidative deamination is of limited synthetic value since two extra equivalents of amine are required to neutralize the two equivalents of arenesulphonic acid produced from the peroxide. In order to be synthetically useful, the sulphonic acid must be removed by a base other than the amine being oxidized. A heterogeneous base suspended in the reaction mixture is useful in this regard. When one equivalent of amine is oxidized with one equivalent of an arenesulphonyl peroxide in the presence of five equivalents of powdered potassium hydroxide, good yields of carbonyl product are obtained (Table 4)[47].

This oxidation is very interesting since there are several viable and quite different mechanistic alternatives. Based on analogy to π electron donors, the amine could attack the O—O bond nucleophilically to give an *O*-sulphonylhydroxylamine (**8**), which by elimination gives the imine (equation 26). Alternatively the *O*-sulphonylhydroxylamine could undergo homolysis of the N—O bond and yield the imine by a radical or radical-cation process (equation 27). Finally the amine could undergo an electron-transfer

TABLE 3. The oxidation of amines with *p*-nitrobenzenesulphonyl peroxide in ethyl acetate at −78°C

Amine	Product	Yield (%)
$(PhCH_2)_2NH$	PhCHO	96
$PhCH_2NH_2$	PhCHO	84
$(c\text{-}C_6H_{11})_2NH$	Cyclohexanone	86
$c\text{-}C_6H_{11}NH_2$	Cyclohexanone	83
$(n\text{-}Bu)_2NH$	*n*-PrCHO	73
$n\text{-}BuNH_2$	*n*-PrCHO	39
$PhCH(NH_2)Me$	PhCOMe	66
$PhCH_2NHBu\text{-}t$	PhCHO	100
$c\text{-}C_5H_9NH_2$	Cyclopentanone	25

TABLE 4. The oxidation of amines (1 equiv.) with *p*-nitrobenzenesulphonyl peroxide (1 equiv.) with powdered potassium hydroxide (5 equiv.)

Amine	Product	Yield (%)
$PhCH_2NH_2$	PhCHO	63
$(PhCH_2)_2NH$	PhCHO	91
$PhCH_2NHMe$	PhCHO	69
$c\text{-}C_6H_{11}NH_2$	Cyclohexanone	66
$(c\text{-}C_6H_{11})_2NH$	Cyclohexanone	53
$PhCH(NH_2)Me$	Acetophenone	28

reaction with the peroxide to give a nitrogen cation radical and hence the imine (equation 28).

$$R^1CH_2\ddot{N}HR^2 + \begin{matrix} O-SO_2Ar \\ | \\ O-SO_2Ar \end{matrix} \longrightarrow \underset{(8)}{R^1CH_2\overset{OSO_2Ar}{\overset{|}{\ddot{N}}}-R^2} + ArSO_3H \xrightarrow{:B} R^1CH=\ddot{N}R^2 + \overset{+}{B}HArSO_3^- \quad (26)$$

$$R^1CH_2\ddot{N}HR^2 + \begin{matrix} O-SO_2Ar \\ | \\ O-SO_2Ar \end{matrix} \longrightarrow R^1CH_2\overset{OSO_2Ar}{\overset{|}{\underset{+}{N}}}HR^2 \longrightarrow R^1CH_2\overset{+\cdot}{N}HR^2 \longrightarrow R^1CH=\ddot{N}R^2$$

$$R^1CH_2\overset{OSO_2Ar}{\overset{|}{N}}R^2 \longrightarrow R^1CH_2\dot{N}R^2 \longrightarrow R^1CH=\ddot{N}R^2 \quad (27)$$

$$R^1CH_2\ddot{N}HR^2 + \begin{matrix} O-SO_2Ar \\ | \\ O-SO_2Ar \end{matrix} \longrightarrow R^1CH_2\overset{+\cdot}{N}R^2 + \begin{matrix} \bar{O}SO_2Ar \\ \dot{O}SO_2Ar \end{matrix} \longrightarrow R^1CH=\ddot{N}R^2 \quad (28)$$

There are precedents in the literature for each of these mechanistic possibilities. In the first place amines attack acyl peroxides nucleophilically to give *N*-acylhydroxylamines[48]. Furthermore *N*-substituted amines such as chloramines[49] and *N*-acylhydroxylamines[50] undergo base-promoted elimination to imines. Thus both steps of equation (26) have literature precedents. Secondly, *N*-substituted amines such as chloramines[51] and *N*-acylhydroxylamines[52] can also undergo N—X bond homolysis to give nitrogen radicals and radical cations as in equation (27). Finally, it has been asserted that amines can react with acyl peroxides by electron transfer as in equation (28)[53]. Furthermore since nitrogen radicals and nitrogen cation radicals have been shown to give imines as stable products[54], equations (27) and (28) also account for the observed products.

A number of studies have shown that amines react with sulphonyl peroxides by the two-step, two-electron process of equation (26) and that unpaired-electron species are not important in these reactions. When substituted benzylamines are reacted with *p*-nitrobenzenesulphonyl peroxide, a Hammett treatment gives $\rho = -0.53$ for the nucleophilic attack by the amine on the peroxide[55]. Kinetic studies with the same amine series show that $\rho = 0.57$ for the formation of the imine. The two-step process of equation (26) predicts an electron deficiency on the benzylic carbon in the first step (negative ρ), and an increase in electron density at the same position during the elimination step (positive ρ), in agreement with the observed data.

A series of unsymmetric amines have been oxidized with sulphonyl peroxides and the regiochemistry of the elimination reaction measured (equation 29). Increased branching of

$$C_6H_5-CH_2NHCHR^1R^2 \xrightarrow{(ArSO_2O)_2} \underset{(9)}{C_6H_5-CH=NCHR^1R^2} + \underset{(10)}{C_6H_5-CH_2N=CR^1R^2}$$

$$R^1 = R^2 = H;\ R^1 = H,\ R^2 = Me;\ R^1 = R^2 = Me \quad (29)$$

the N-alkyl substituent causes increased elimination towards the alkyl substituent (**10** > **9**). This regioselectivity is opposite to that observed for imine formation in similar amines from radical-cation intermediates produced electrochemically[55].

When α protons are lacking or when good migrating groups are present in the precursor amine, carbon to nitrogen rearrangement becomes important. The oxidation of tritylamine with arenesulphonyl peroxides gives benzophenone as the only product (55% yield) after hydrolysis (equation 30)[56,57]. Steiglitz rearrangement in the N-sulphonoxy intermediate (**11**) accounts for the observed products. Benzhydryl amines give products of both rearrangement and elimination (equation 31).

$$Ph_3CNH_2 \xrightarrow{(ArSO_2O)_2} \underset{(\mathbf{11})}{Ph_3CNHOSO_2Ar} \xrightarrow{\varnothing} Ph_2C{=}NPh \xrightarrow{H_3O^+} Ph_2C{=}O \quad (30)$$

$$Ph_2CHNH_2 \xrightarrow{(ArSO_2O)} Ph_2CHNHOSO_2Ar \xrightarrow{elimin.} Ph_2C{=}NH \xrightarrow{H_3O^+} Ph_2C{=}O$$
$$Ph_2CHNHOSO_2Ar \xrightarrow{\varnothing} PhCH{=}NPh \xrightarrow{H_3O^+} PhCHO \quad (31)$$

Finally, kinetic studies dramatically illustrate the two-step nature of the oxidation. Since two equivalents of ammonium salt are produced in the oxidation (equation 25), the rate of reaction can be conveniently monitored by measuring an increase in conductivity of the reaction mixture. When benzylamine is mixed with arenesulphonyl peroxides at $-10°C$, there is an immediate conductivity increase corresponding to the formation of one equivalent of the ammonium salt. Then a second equivalent of the salt is formed in a second, slower, base-dependent step[55].

Taken together, the above data rule out one-electron processes for the oxidation of amines by sulphonyl peroxides. The electron transfer of equation (28) gives two equivalents of the salt from the rate-determining step of electron transfer. Furthermore the regioselectivity of imine formation from radical cations is opposite to that observed. Homolytic cleavage of the N—O bond of the N-sulphonoxy intermediate as in equation (27) is not consistent with the migratory aptitude data in the Steiglitz rearrangement or the base-dependent kinetics of the imine-forming step. One type of mechanism which cannot be specifically excluded is initial electron transfer from the amine to the peroxide followed by rapid collapse to the O-sulphonylhydroxylamine (equation 32). Collapse would have to occur in the solvent cage and would have to be faster than proton loss from the nitrogen radical cation which is known to be very fast[54]. There is presently no concrete reason to invoke this mechanism.

$$R\ddot{N}H_2 + \begin{matrix} O-SO_2Ar \\ | \\ O-SO_2Ar \end{matrix} \xrightarrow{slow} \left[R\overset{+\cdot}{N}H_2 \quad \begin{matrix} \dot{O}-SO_2Ar \\ \\ \bar{O}-SO_2Ar \end{matrix} \right]_{\text{solvent cage}} \xrightarrow{fast} \underset{(\mathbf{8})}{R-NH-OSO_2Ar} \quad (32)$$

It is clear that O-sulphonylhydroxylamines (**8**) can be produced easily from amines and sulphonyl peroxides. These adducts have been very useful in studying base-promoted, imine-forming eliminations[58]. They are also potential sources of nitrenium ions by solvolysis[59] and might serve as aminating agents[60]. In the absence of base, which promotes imine-forming elimination, N-alkyl-O-sulphonylhydroxylamines are sufficiently stable to be isolated and purified at low temperatures (<0°C). Several of these compounds **8** have been prepared[61].

$$\mathrm{R{-}NH{-}OSO_2C_6H_4NO_2}$$

(8)

R = Me, *t*-Bu, $PhCH_2$, *m*-$ClC_6H_4CH_2$

The reactions of several n electron donors with acyl sulphonyl peroxides (**2**) have been reported in the literature[62]. The unsymmetrical character of these peroxides leads to a greater diversity of reaction type than for bis-sulphonyl peroxides. Treatment of benzoyl *p*-toluenesulphonyl peroxide with triphenylphosphine yields triphenylphosphine oxide and benzoic *p*-toluenesulphonic anhydride (equation 33). Oxygen-18 labelling studies indicate that phosphorus attacks the peroxide oxygen bonded to the carbonyl function. On the other hand diphenyl sulphide is oxidized to diphenyl sulphoxide and the mixed anhydride, but label randomization occurs. The means by which label scrambling occurs is not known[62].

$$\mathrm{PhC(=O){-}O{-}O{-}S(=O)_2{-}C_6H_4Me + Ph_3P \longrightarrow Ph_3P{=}O + PhC(=O){-}O{-}S(=O)_2{-}C_6H_4Me} \qquad (33)$$

p-Tolylmagnesium bromide reacts with benzoyl *p*-toluenesulphonyl peroxide to give benzoic acid and *p*-tolyl *p*-toluenesulphonate, which is formed exclusively by attack of the Grignard reagent on the peroxide oxygen bonded to the sulphonyl group. Coordination of magnesium to the carbonyl oxygen of the peroxide apparently directs such specificity of attack (equation 34).

$$\mathrm{PhC(=O){-}O{-}O{-}S(=O)_2{-}C_6H_4Me + Me{-}C_6H_4{-}MgBr \rightleftharpoons [\text{cyclic Mg-coordinated complex: Br, Mg, Me}{-}C_6H_4, \text{Ph}, C_6H_4Me]}$$

$$\longrightarrow \mathrm{PhC(=O){-}OMgBr + Me{-}C_6H_4{-}O{-}S(=O)_2{-}C_6H_4Me} \qquad (34)$$

Finally, hydrazine and sodium methoxide attack the carbonyl carbon of the mixed peroxide to yield benzoyl hydrazide and methyl benzoate respectively[62]. Thus acyl sulphonyl peroxides are attacked at both peroxide oxygens and at the carbonyl carbon by various n electron donors. The factors which influence these various modes of attack by nucleophiles are not known with any certainty, but their elucidation would be very useful. If nucleophilic attack on the peroxide oxygen next to the carbonyl group could be controlled, then mixed acyl sulphonyl peroxides could be used to attach the acyloxy function electrophilically to substrates. This synthetic capability is not generally possible at present.

c. σ Donors. Electron-rich carbon–metal σ bonds can act as electron donors to sulphonyl peroxides and the products are those of carbon–metal bond cleavage (equation

35)[63]. Organomercury compounds, upon treatment with methanesulphonyl peroxide, give good yields of cleavage products (equation 36). When diphenyl- or dibenzyl-mercury react with two equivalents of the peroxide both carbon–metal bonds are cleaved; however, the second cleavage is slower than the first (equation 37).

$$\text{R}-\overset{|}{\underset{|}{\text{C}}}-\text{M} + \begin{matrix}\text{O}-\text{SO}_2\text{R}\\ |\\ \text{O}-\text{SO}_2\text{R}\end{matrix} \longrightarrow \text{R}-\overset{|}{\underset{|}{\text{C}}}-\text{OSO}_2\text{R} + \text{MOSO}_2\text{R} \qquad (35)$$

$$\text{RHgX} + (\text{MeSO}_2\text{O})_2 \xrightarrow[25^{\circ}\text{C}]{\text{CH}_2\text{Cl}_2} \underset{75-98\%}{\text{ROSO}_2\text{Me}} + \text{XHgOSO}_2\text{Me} \qquad (36)$$

$$\text{R} = \text{Ph, PhCH}_2, p\text{-MeC}_6\text{H}_4, p\text{-MeOC}_6\text{H}_4, p\text{-MeC}_6\text{H}_4\text{CH}_2, p\text{-ClC}_6\text{H}_4$$
$$\text{X} = \text{Cl, Br}$$

$$\text{R}_2\text{Hg} + 2(\text{MeSO}_2\text{O})_2 \xrightarrow{\text{faster}} \text{ROSO}_2\text{Me} + \text{RHgOSO}_2\text{Me} \xrightarrow{\text{slower}} \text{ROSO}_2\text{Me} + \text{Hg(OSO}_2\text{Me)}_2 \qquad (37)$$
$$\text{R} = \text{Ph, PhCH}_2$$

Tetraalkyl- and tetraaryl-tins are also cleaved by methanesulphonyl peroxide, but somewhat less efficiently (equation 38). Organo-lead and -silicon compounds give little or no cleavage with methanesulphonyl peroxide. Perhaps the more electrophilic arenesulphonyl peroxides would be more effective in cleaving these latter materials.

$$\text{R}_4\text{Sn} + (\text{MeSO}_2\text{O})_2 \longrightarrow \underset{10-83\%}{\text{ROSO}_2\text{Me}} + \text{R}_3\text{SnOSO}_2\text{Me} \qquad (38)$$
$$\text{R} = \text{Me, Ph}, m\text{-MeC}_6\text{H}_4, n\text{-Bu}$$

A Hammett study of carbon–metal bond cleavage in substituted arylmercuric chlorides has given $\rho = -1.97$ which is indicative of an electrophilic cleavage mechanism. More work is needed on factors influencing the reactivity and the stereochemistry of these reactions in order to completely understand their mechanism.

Sulphonyl peroxides have been shown to react effectively with π, n and σ electron donors by two-electron pathways. This reactivity seems inherent to this peroxide type. It is quite likely that other donor functions (sulphur, phosphorus) will give similar chemistry; thus, a wide range of interesting and useful transformations using sulphonyl peroxides is possible.

III. ORGANOPHOSPHORUS PEROXIDES

Only one example of an organophosphorus peroxide has been studied to any extent. Bis-diphenylphosphinyl peroxide (**5**) has been prepared by the condensation of diphenylphosphinyl chloride with sodium peroxide (equation 39)[64]. It is quite interesting that no peroxides were obtained when substituted arenephosphinyl chlorides were used in the same procedure. Hydrolysis to the corresponding phosphinic acids was the only

$$\text{Ph}_2\overset{\overset{\text{O}}{\|}}{\text{P}}-\text{Cl} + \text{Na}_2\text{O}_2 \longrightarrow \underset{\underset{(\mathbf{5})}{57\%}}{\text{Ph}_2\overset{\overset{\text{O}}{\|}}{\text{P}}-\text{O}-\text{O}-\overset{\overset{\text{O}}{\|}}{\text{P}}\text{Ph}_2} \qquad (39)$$

observable reaction. Diphenylphosphinyl peroxide **5** is stable indefinitely at $-80°C$ but decomposes exothermically upon warming to room temperature. Apparently the arenephosphinyl groups do not confer thermal stability to an attached peroxide link to the extent that arenesulphonyl groups do.

The decomposition of **5** in several solvents occurs by a first-order process and yields, after hydrolysis, diphenylphosphinic acid (**12**) and phenyl hydrogen phenyl phosphonate (**13**). The unsymmetrical anhydride **14** has been postulated as an intermediate (equation 40). The product results from a phenyl migration from phosphorus to oxygen and is analogous to a carboxy inversion process. The anhydride **14** could not be isolated but has been identified in the mass spectrum of the crude product and trapped by reaction with methanol.

$$\underset{(\mathbf{5})}{\mathrm{Ph_2P(=O)-O-O-P(=O)Ph_2}} \longrightarrow \underset{(\mathbf{14})}{\mathrm{Ph_2P(=O)-O-P(=O)(OPh)-Ph}} \xrightarrow{\mathrm{H^2O}} \underset{(\mathbf{12})}{\mathrm{Ph_2P(=O)-OH}} + \underset{(\mathbf{13})}{\mathrm{PhP(=O)(OPh)-OH}} \quad (40)$$

The mechanism of the rearrangement has been elucidated by oxygen-18 labelling experiments. When the phosphinyl oxygen is labelled with oxygen-18, there is no scrambling of the label in the rearranged product (equation 41)[65]. Phenyl migration occurs exclusively to the peroxide oxygen, and dissociation processes, radical or ionic, are ruled out. An ionic mechanism analogous to that found for mixed acylsulphonyl peroxides and unsymmetrical acyl peroxides is favoured, which involves concerted phenyl migration and O—O bond heterolysis as in equation (42). This formulation is substantiated by the observation that photolysis of the same labelled peroxide affords the same products but gives a randomization of the label. Photolysis apparently promotes homolysis of the O—O bond, and phenyl migration in the radical occurs to either of the equivalent oxygens (equation 43).

$$\mathrm{Ph_2P(=O^*)-O-O-P(=O^*)Ph_2} \longrightarrow \mathrm{PhOP(Ph)(=O^*)-O-P(=O^*)Ph_2} \quad (41)$$

$$\mathrm{(C_6H_5)_2P(=O)-O-O-P(=O)(C_6H_5)_2} \longrightarrow \left[\mathrm{C_6H_5P(=O)(C_6H_5^{+})O}\right] \; {}^{-}\mathrm{O_2P(C_6H_5)_2} \longrightarrow \mathrm{C_6H_5P(=O)(OC_6H_5)-O-P(=O)(C_6H_5)_2} \quad (42)$$

$$(Ph_2P{-}O)_2 \text{ (with } =O^*) \xrightarrow{h\nu} Ph_2P{-}O^* \longrightarrow PhP(^*OPh)(={}^*O)\cdot \;\; {}^*O{-}PPh_2(={}^*O) \longrightarrow PhP(^*OPh)(=O^*){-}\overset{*}{O}{-}PPh_2(={}^*O) \qquad (43)$$

In view of the tendency of phosphinyl peroxide (**5**) to undergo ionic rearrangement, it may show electrophilic behaviour towards electron donors analogous to sulphonyl peroxides. As yet reactions of **5** with donor functions have not been pursued, but they may provide fruitful methods for attaching oxidized phosphorus groups to electron donors.

IV. REFERENCES

1. (a) Reference to organometallic peroxide chapter in this volume entitled
   (b) G. Sosnovsky and J. H. Brown, *Chem. Rev.*, **66**, 529 (1966).
2. (a) D. Swern, *Chem. Rev.*, **45**, 1 (1949).
   (b) Reports of sulphonic peracids from sulphoxidation are largely restricted to the patent literature. For example, J. A. Manner, *U.S. Patent*, No. 761221 (1977); *Chem. Abstr.*, **86**, (19), 139498N (1977).
3. A. T. Nielsen, R. L. Atkins, W. P. Novis, C. L. Coon and M. E. Sitzmann, *J. Org. Chem.*, **45**, 2341 (1980).
4. C. G. Swain, W. H. Stockmayer and J. T. Clark, *J. Amer. Chem. Soc.*, **72**, 5426 (1950).
5. R. Hiatt in *Organic Peroxides*, Vol. II (Ed. D. Swern), Wiley–Interscience, New York–London, Chap. 8, pp. 812–813.
6. H. Kropf, M. Ball and A. Gilberz, *Justus Liebigs Ann. Chem.*, 1013 (1974).
7. J. L. Kice, T. E. Rogers and A. C. Warheit, *J. Amer. Chem. Soc.*, **96**, 8020 (1974).
8. J. Bolte, A. Kergomard and S. Vincent, *Tetrahedron Letters* 1529 (1965).
9. (a) R. L. Dannley and G. E. Corbett, *J. Org. Chem.*, **31**, 153 (1966).
   (b) R. L. Dannley, J. E. Gagen and O. J. Stewart, *J. Org. Chem.*, **35**, 3076 (1970).
10. (a) R. W. Hazeldine, R. B. Heslop and J. W. Lethbridge, *J. Chem. Soc.*, 4901 (1964).
    (b) C. J. Myall and D. Pletcher, *J. Chem. Soc., Perkin Trans. 1*, 953 (1975).
11. R. E. Noftle and G. H. Cady, *Inorg. Chem.*, **4**, 1010 (1965).
12. R. L. Dannley and P. K. Tornstrom, *J. Org. Chem.*, **40**, 2278 (1975).
13. R. V. Hoffman and E. L. Belfoure, *J. Amer. Chem. Soc*, in press.
14. R. Hisada, H. Minato and M. Kobayashi, *Bull. Chem. Soc. Japan*, **44**, 2541 (1971).
15. R. Hisada, M. Kobayashi and H. Minato, *Bull. Chem. Soc. Japan*, **45**, 564 (1972).
16. R. Graf, *Justus Liebigs Ann. Chem.*, **578**, 50 (1952).
17. V. R. Likhterov, V. S. Etlis, G. A. Razuvaev and A. V. Gorelik, *Vysokomolecul. Soed.*, **4**, 357 (1962); *Chem. Abstr.*, **57**, 12691i (1962).
18. P. D. Bartlett and B. T. Storey, *J. Amer. Chem. Soc.*, **80**, 4954 (1958).
19. Y. Yokoyama, H. Wada, M. Kobayashi and H. Minato, *Bull. Chem. Soc. Japan*, **44**, 2479 (1971).
20. J. Bolte, G. Dauphin and A. Kergomard, *Bull. Soc. Chim. Fr.*, 2291 (1970).
21. L. W. Crovatt and R. L. McKee, *J. Org. Chem.*, **24**, 2031 (1959).
22. (a) R. Hisada, N. Kamigata, H. Minato and M. Kobayashi, *Bull. Chem. Soc. Japan*, **44**, 3475 (1971).
    (b) M. Kobayashi, M. Sekiguchi and H. Minato, *Chem. Letters* 393 (1972).
23. R. Hisada, M. Kobayashi and H. Minato, *Bull. Chem. Soc. Japan*, **45**, 2902 (1972).
24. (a) E. G. E. Hawkins, *Organic Peroxides*, Van Nostrand Co. Inc., Princeton, N.J., 1961, pp. 311–316.
    (b) M. Szwarc in *Peroxide Reaction Mechanisms* (Ed. J. O. Edwards), Interscience, New York, 1962, pp. 153–174.
25. G. A. Razuvaev, V. R. Likhterov and V. S. Etlis, *Zh. Obshch. Khim.*, **31**, 274 (1961).
26. G. A. Razuvaev, L. M. Terman, V. R. Likhterov and V. S. Etlis, *J. Poly. Sci.*, **52**, 123 (1961).
27. G. A. Razuvaev, V. R. Likhterov and V. S. Etlis, *Zh. Obshch. Khim.*, **32**, 2033 (1962).

28. See for example R. L. Friedman and R. N. Lewis, *U. S. Patent*, No. 4151339 (1979), *Chem, Abstr.*, **91**, 21448W (1979); J. A. Manner, *U. S. Patent*, No. 3998888 (1977), *Chem. Abstr.*, **86**, 139498N (1977); S. Aruga and H. Fujii, *Japanese Patent*, No. 7674078 (1976), *Chem, Abstr.*, **85**, 124671U (1976).
29. J. E. Leffler, *J. Amer. Chem. Soc.*, **72**, 67 (1950).
30. D. B. Denney, *J. Amer. Chem. Soc.*, **78**, 590 (1956).
31. G. A. Rasuwajew, V. R. Likhterov and V. S. Etlis, *Tetrahedron Letters*, 527 (1961).
32. D. B. Denney and D. Z. Denney, *J. Amer. Chem. Soc.*, **79**, 4806 (1957).
33. P. D. Bartlett and T. G. Traylor, *J. Amer. Chem. Soc.*, **83**, 856 (1961).
34. J. A. March, *Advanced Organic Chemistry*, 2nd ed., McGraw-Hill, New York, 1977, pp. 70–71.
35. R. L. Dannley and W. R. Knipple, *J. Org. Chem.*, **38**, 6 (1973).
36. F. N. Keeney, *Ph.D. Thesis*, Case-Western Reserve University, 1969.
37. L. M. Stock and H. C. Brown, *Advan. Phys. Org. Chem.*, **1**, 35 (1963).
38. R. L. Dannley, J. E. Gagen and K. Zak, *J. Org. Chem.*, **38**, 1 (1973).
39. R. L. Dannley, R. V. Hoffman, P. K. Tornstrom, R. L. Walker and R. B. Srivastava, *J. Org. Chem.*, **39**, 2543 (1974).
40. N. B. Kurnath, *Ph.D. Thesis*, Case Western Reserve University, 1970.
41. E. M. Levi, P. Kovacic and J. F. Gormish, *Tetrahedron*, **26**, 4537 (1970).
42. P. K. Tornstrom, *Ph.D. Thesis*, Case Western Reserve University, 1970.
43. J. Bolte, A. Kergomard and S. Vincent, *Bull. Soc. Chim. Fr.*, 301 (1972).
44. R. D. Bishop and R. V. Hoffman, unpublished results.
45. J. B. Lambert, H. W. Mark, A. G. Holcomb and E. S. Magyar, *Acc. Chem. Res.*, **12**, 317 (1979).
46. R. V. Hoffman and R. D. Bishop, *Tetrahedron Letters*, 33 (1976).
47. R. V. Hoffman, *J. Amer. Chem. Soc.*, **98**, 6702 (1976).
48. C. Walling and N. Indictor, *J. Amer. Chem. Soc.*, **80**, 5814 (1858); R. Huisgen and F. Bayerlein, *Justus Liebigs Ann. Chem.*, **630**, 138 (1960); C. L. Horner and B. Anders, *Chem. Ber.*, **95**, 2470 (1962); D. B. Denney and D. Z. Denney, *J. Amer. Chem. Soc.*, **82**, 1389 (1960); G. Zinner, *Arch. Pharm.* (*Weinheim. Ger.*), **296**, 57 (1963); Reference 5, pp. 870–875.
49. R. A. Bartsch and B. R. Cho, *J. Amer. Chem. Soc.*, **101**, 3587 (1979); W. E. Bachmann, M. P. Cava and A. S. Dreiding, *J. Amer. Chem. Soc.*, **76**, 5554 (1954); S. Dayagi and Y. Degani in *The Chemistry of the Carbon–Nitrogen Double Bond* (Ed. S. Patai), Wiley–Interscience, New York–London 1970, pp. 117–127.
50. S. Oae and T. Sakurai, *Bull. Chem. Soc. Japan*, **49**, 730 (1976).
51. M. E. Wolff, *Chem. Rev.*, **62**, 55 (1962); O. E. Edwards, D. Vocelle and J. W. ApSimon, *Can. J. Chem.*, **50**, 1167 (1972).
52. R. G. Roy and G. A. Swann, *Chem. Commun.*, 427 (1966).
53. C. Walling and J. P. Sloan, *J. Amer. Chem. Soc.*, **101**, 7679 (1979); W. A. Pryor and W. H. Hendrickson, Jr., *J. Amer. Chem. Soc.*, **97**, 1580, 1582 (1975).
54. L. A. Hull, G. T. Davis, D. H. Rosenblatt and C. K. Mann, *J. Phys. Chem.*, **73**, 2124 (1969) and references therein; C. A. Audeh and J. R. L. Smith, *J. Chem. Soc.* (*B*), 1741 (1971).
55. R. V. Hoffman and R. Cadena, *J. Amer. Chem. Soc.*, **99**, 8226 (1977).
56. R. V Hoffman, R. Cadena and D. J. Poelker, *Tetrahedron Letters*, 203 (1978).
57. R. V. Hoffman and D. J. Poelker, *J. Org. Chem.*, **44**, 2364 (1979).
58. R. V. Hoffman and E. L. Belfoure, *J. Amer. Chem. Soc.*, **101**, 5687 (1979).
59. P. G. Gassman, *Acc. Chem. Res.*, **3**, 26 (1970); P. G. Gassman and G. D. Hartman, *J. Amer. Chem. Soc.*, **95**, 449 (1973).
60. Y. Tamura, J. Muramikava and M. Ikeda, *Synth.*, 1 (1977); G. Boche, N. Mayer, M. Bernheim and K. Wagner, *Angew. Chem.* (*Intern. Ed. Engl.*), **17**, 687 (1978).
61. E. L. Belfoure, *Ph.D. Thesis*, New Mexico State University, 1981.
62. R. Hisada, M. Kobayashi and H. Minato, *Bull. Chem. Soc. Japan*, **45**, 2035 (1972).
63. F. U. Ahmed, *Ph.D. Thesis*, New Mexico State University, 1979.
64. R. L. Dannley and K. Kabre, *J. Amer. Chem. Soc.*, **87**, 4805 (1965).
65. R. L. Dannley, R. L. Waller, R. V. Hoffman and R. F. Hudson, *J. Org. Chem.*, **37**, 418 (1972); R. L. Dannley, R. L. Waller, R. V. Hoffman and R. F. Hudson, *Chem. Commun.*, 1362 (1971).

The Chemistry of Functional Groups, Peroxides
Edited by S. Patai

CHAPTER **10**

Diacyl peroxides, peroxycarboxylic acids and peroxy esters

GÜNTER BOUILLON, CARLO LICK AND KURT SCHANK*

Fachrichtung 14.1–Organische Chemie, Universität des Saarlandes, D-6600 Saarbrücken, W. Germany

I. DIACYL PEROXIDES . . . 280
A. General Syntheses . . . 280
1. Symmetrical diacyl peroxides . . . 280
2. Unsymmetrical diacyl peroxides . . . 280
3. Cyclic diacyl peroxides . . . 281
4. Peroxycarbonates . . . 281
B. Stability of Diacyl Peroxides . . . 281
C. Reactions of Diacyl Peroxides with Nucleophiles . . . 282
1. Solvolysis . . . 282
2. Reactions with carbanions . . . 282
3. Reactions with enamines . . . 283
4. Reactions with aromatic, heteroaromatic and olefinic hydrocarbons . . . 285
5. Reactions with other nucleophiles . . . 285
II. PEROXYCARBOXYLIC ACIDS . . . 287
A. General Syntheses . . . 287
B. Stability and Structure of Peroxycarboxylic Acids . . . 288
C. Reactions of Peroxycarboxylic Acids with Olefins . . . 288
1. Mechanism of epoxidation . . . 288
2. Stereospecifity and stereoselectivity of epoxidation . . . 290
3. Reaction rates of epoxidation . . . 290
4. Special epoxidation methods . . . 290
5. Epoxidation by related peroxycarbon acids . . . 292
D. Reactions of Peroxycarboxylic Acids with Acetylenes and Allenes . . . 292
E. Reactions of Peroxycarboxylic Acids with Other Reaction Partners . . . 295
1. Hydroxylation of alkanes . . . 295
2. Oxidation of alcohols . . . 295

* Correspondence author.

3. Oxidation of aldehydes, ketones and acetals . . . 295
4. Oxidation of sulphur compounds . . . 296
5. Oxidation of nitrogen compounds . . . 297
6. Oxidation of miscellaneous compounds . . . 298
III. PEROXYCARBOXYLATES (PEROXY ESTERS) . . . 299
A. General Syntheses . . . 299
B. Stability of Alkyl Peroxycarboxylates . . . 299
C. Reactions of *t*-Alkyl Peroxycarboxylates . . . 300
1. Reactions of *t*-butyl peroxycarboxylates with reactive nucleophiles (transfer of alkoxy groups) . . . 300
2. Reactions of *t*-butyl peroxycarboxylates with CH bonds, forming stabilized radicals after hydrogen abstraction (introduction of acyloxy groups) . . . 300
a. Substitution of allylic (and related) hydrogen . . . 301
b. Substitution of α-H in ethers . . . 302
c. Substitution of α-H in thio ethers . . . 303
d. Miscellaneous CH substitutions . . . 303
IV. REFERENCES . . . 303

I. DIACYL PEROXIDES

Recent results in the chemistry of diacyl peroxides have been summarized in comprehensive reviews[1–3]. The following chapter includes fundamental principles as well as a selection of newer results.

A. General Syntheses

1. Symmetrical diacyl peroxides

The reaction of sodium peroxide (or H_2O_2) with acyl chlorides represents the most convenient method of synthesizing diacyl peroxides[4–22] (equation 1a). Acyl chlorides may be replaced by acid anhydrides[19–21] (equation 1b).

$$2\,RCOCl + Na_2O_2\ (H_2O_2) \xrightarrow{-2\,NaCl\ (-2\,HCl)} RCO{-}O{-}O{-}COR \qquad (1a)$$

$$2\,(RCO)_2O + Na_2O_2 \xrightarrow{-2\,RCO_2Na} RCO{-}O{-}O{-}COR \qquad (1b)$$

2. Unsymmetrical diacyl peroxides

Diacyl peroxides with different acyl groups have been similarly synthesized by reaction of peroxycarboxylic acids with acyl chlorides or acyl anhydrides[19–21,23–25] (equation 2). Autoxidation of aldehydes in presence of acetic anhydride leads to an '*in situ*' formation of peroxycarboxylic acids which are immediately acetylated by the added anhydride[25–34] (equation 3).

$$R^1COX + R^2CO_3H \xrightarrow[-HX]{} R^1CO{-}O{-}O{-}COR^2 \qquad (2)$$

$$X = Cl,\ OCOR^1$$

$$RCHO + (CH_3CO)_2O + O_2 \xrightarrow[- CH_3CO_2H]{} RCO{-}O{-}O{-}COCH_3 \quad (3)$$

A very specific method uses the ozonolysis of 4,5-diphenyl-1,3-dioxol-4-en-2-one (**1**) to yield **2**[35] (equation 4).

$$(\mathbf{1}) \xrightarrow[75\%]{O_3} PhCO{-}O{-}O{-}CO_2COPh \; (\mathbf{2}) \quad (4)$$

3. *Cyclic diacyl peroxides*

Starting from dicarboxylic acids or from their acyl chlorides conversion with sodium peroxide leads to five- or six-membered cyclic peroxides[13,14,36,37] (equations 5 and 6).

$$\text{1,1-cyclohexanedicarboxylic acid} \xrightarrow[CH_3SO_3H]{Na_2O_2} \text{cyclic peroxide (75\%)} \quad (5)$$

$$C_6H_4(COX)_2 \xrightarrow{Na_2O_2} \text{cyclic phthaloyl peroxide} \quad (6)$$

X = OH, Cl

4. *Peroxycarbonates*

Recently, peroxycarbonates obtained importance in connection with acyloxylations of carbon nucleophiles. Dibenzyl peroxydicarbonate in particular has been used frequently, due to its high reactivity in solution on the one hand and its relative stability for storage on the other[38–43]. Peroxycarbonates are easily prepared by reaction of sodium peroxide with chlorocarbonates in a two-phase reaction[44] (equation 7).

$$2ClCO_2R + Na_2O_2 \xrightarrow[- 2NaCl]{0^oC} RO_2C{-}O{-}O{-}CO_2R \quad (7)$$

(dissolved in CH_2Cl_2) (dissolved in H_2O)

B. Stability of Diacyl Peroxides

Diacyl peroxides are cleaved homolytically under the influence of energy or catalysts. Accordingly, they can be used as initiators of polymerization reactions[1,45,46] (equation 8). Diacyl peroxides containing Ph, p-$NO_2C_6H_4$ and $OC(CH_3)_2CCl_3$[47] groups are preferentially used because of their sufficient stability at ambient temperature. *Caution:*

$$\mathrm{RCO{-}O{-}O{-}COR} \xrightarrow[\text{or catalyst}]{\Delta E} \left[2\,\mathrm{RCO_2}^{\cdot} \right] \xrightarrow{-2\,\mathrm{CO_2}} \left[2\,\mathrm{R}^{\cdot} \right] \qquad (8)$$

NEVERTHELESS, ALL LABORATORY EXPERIMENTS WITH PEROXY COMPOUNDS MUST BE CARRIED OUT BEHIND A SAFETY SHIELD IN A HOOD! The smaller the molecular weight of R the higher will be the danger of explosions at ambient temperature and at high concentrations[1,48,49].

Besides serving as sources of free radicals, diacyl peroxides behave as ambident electrophiles; they react either with 'hard bases' at the carbonyl carbon, or with 'soft bases' at the peroxide oxygen.

C. Reactions of Diacyl Peroxides with Nucleophiles

1. Solvolysis

Hydrolysis of diacyl peroxides has been used as one method of preparing perbenzoic acids; similarly, cleavage of diacyl peroxides by sodium alcoholate/alcohol yields esters and sodium peroxycarboxylates[50–52] (equation 9).

$$\mathrm{RCO{-}O{-}O{-}COR} + \mathrm{R^1O^-} \xrightarrow[-\,\mathrm{RCO_2R^1}]{} \mathrm{RCO_3}^- \xrightarrow{\mathrm{H^+}} \mathrm{RCO_3H} \qquad (9)$$

2. Reactions with carbanions

In 1962, Lawesson and coworkers[53–57] investigated the nucleophilic attack of carbanions at the peroxy oxygen of benzoyl peroxide yielding benzoyloxylated products (equation 10).

$$-\overset{|}{\underset{|}{\mathrm{C}}}{}^{-} + \underset{\mathrm{COPh}}{\mathrm{O}}{-}\mathrm{O{-}COPh} \longrightarrow -\overset{|}{\underset{|}{\mathrm{C}}}{-}\mathrm{OCOPh} + \mathrm{PhCO_2}^- \qquad (10)$$

In the case of open-chain α-unsubstituted β-dicarbonyl compounds mono- and bis-substitution occurs simultaneously, yielding monoacylated reductones and acylals of *vic*-triketones.

5,5-Disubstituted derivatives of 1,3-cyclohexanedione react initially in the same way; however, the acylals **5** have proved to be highly sensitive to bases, and show rearrangements in the course of which derivatives **6** of 2,5-cyclohexadienone are formed[39] (equation 11).

The importance of the direct acyloxylation lies in the fact that halogen substituents in *open-chain* β-dicarbonyl compounds are easily substituted by carboxylate nucleophiles, whereas *cyclic* β-dicarbonyl compounds don't react under similar conditions[38,58]

Figure 1 shows the composition of the reaction mixture in the conversion of the sodium salt of dimedone (**3a**) with benzoyl peroxide versus time.

$$\mathbf{(3)} \xrightarrow[-\,R^2CO_2H]{+\,(R^2CO_2)_2} \mathbf{(4)} \xrightarrow[-\,R^2CO_2Na]{+\,(R^2CO_2)_2} \mathbf{(5)} \xrightarrow[-\,R^2CO_2H]{[R^2CO_2Na]} \mathbf{(6)} \qquad (11)$$

(**a**) R^1, R^1 = Me, Me

(**b**) $R^1, R^1 = (CH_2)_5$

R^2 = X–C_6H_4– (X = H, MeO, Cl, NO_2); OCH_2Ph

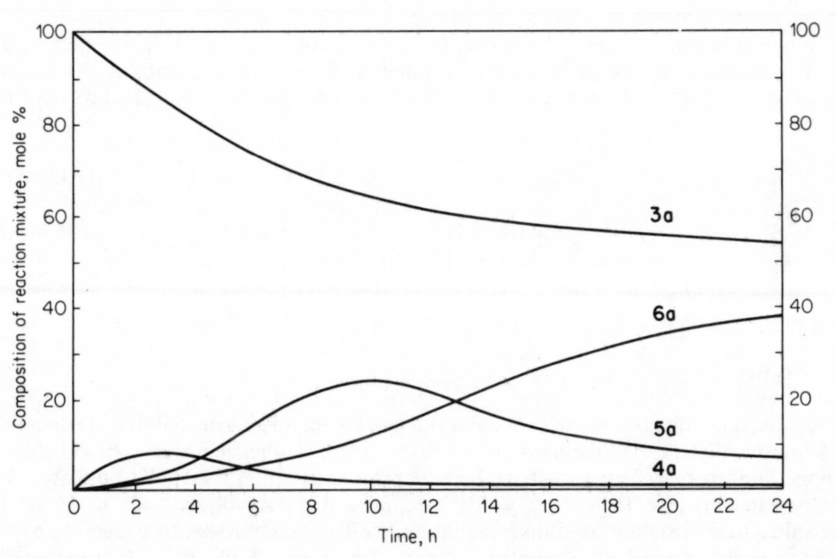

FIGURE 1. Reaction of dimedone (as sodium salt **3a**) with dibenzoyl peroxide in acetonitrile at 20–25°C[39].

3. Reactions with enamines

Like carbanions, enamines from secondary amines show equally high reactivity towards diacyl peroxides[59,60]. Similarly, enamines derived from *open-chain* β-dicarbonyl compounds ('β-acyl enamines') have been monoacyloxylated[61] (equation 12).

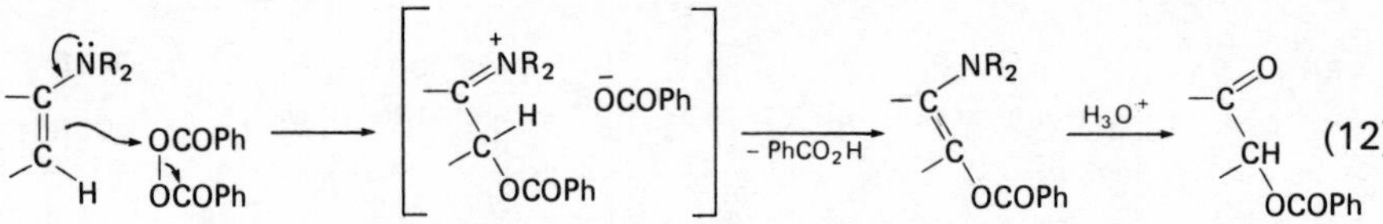

$$\text{(12)}$$

In the case of cyclic acyl secondary enamines obtained from dimedone (and related 1,3-cyclanediones) acyloxylation leads to a mixture of monoacyloxylated acylenamines **8** and diacyloxylated sterically hindered enamines **9**[43] (equation 13). The reactivity of enamines **9** is so poor that only traces of trisacyloxylation products are obtained.

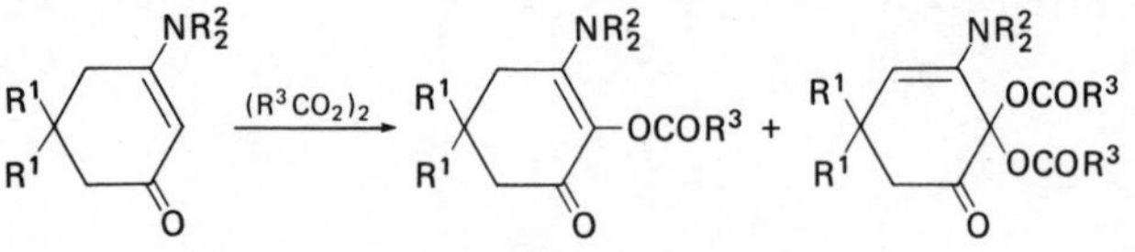

$$\text{(13)}$$

(7) (8) (9)

The enamine **10** from dimedone and cyclohexylamine is correspondingly mono-acyloxylated and obviously also bisacyloxylated; however, the bisacyloxylation product **12** cannot be isolated because of its immediate rearrangement to **13**[62] (equation 14). Although this rearrangement resembles that of the triketone acylals **5**, the mechanism must be different. In the presence of two different R^2 groups, **5** eliminates the *weaker* carboxylic acid to yield **6** whereas **12** eliminates the *stronger* carboxylic acid to yield **13**.

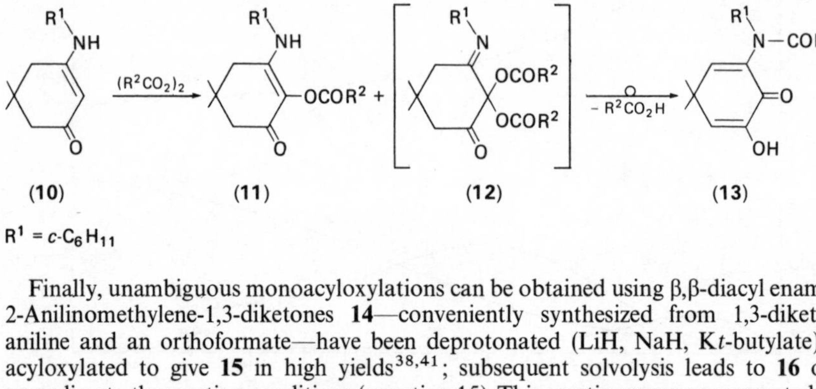

$$\text{(14)}$$

(10) (11) (12) (13)

$R^1 = c\text{-}C_6H_{11}$

Finally, unambiguous monoacyloxylations can be obtained using β,β-diacyl enamines. 2-Anilinomethylene-1,3-diketones **14**—conveniently synthesized from 1,3-diketones, aniline and an orthoformate—have been deprotonated (LiH, NaH, K*t*-butylate) and acyloxylated to give **15** in high yields[38,41]; subsequent solvolysis leads to **16** or **17** according to the reaction conditions (equation 15). This reaction sequence seems to be the most convenient method to synthesize pure 2-acyloxy-1,3-diketones **16** from easily available β-diketones. With R^1 = benzyloxy particularly, smooth hydrogenations of **16** yield the corresponding reductones **17**.

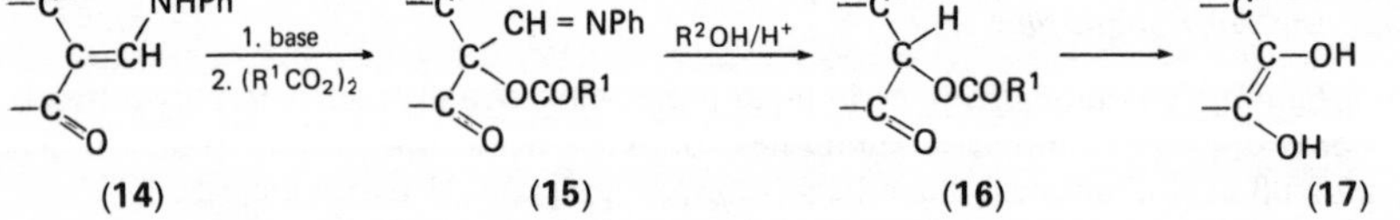

$$\text{(15)}$$

(14) (15) (16) (17)

4. *Reactions with aromatic, heteroaromatic and olefinic hydrocarbons*

Under catalysis of iodine, benzoyl peroxide is found to introduce the benzoyloxy group into aromatic compounds[63] (equation 16).

$$R\text{-}C_6H_5 + (PhCO_2)_2 \xrightarrow{[I_2]} R\text{-}C_6H_4\text{-}OCOPh \qquad (16)$$

R = H, Me, Cl, OMe

Pyrroles and indoles[64] behave as enamines, and accordingly mono- and bis-benzoyloxylations[65] have been observed (equations 17 and 18).

$$\text{2-phenylindole} \xrightarrow[\text{3. } H_2O]{\text{1. NaH; 2. } (PhCO_2)_2} \text{3-OCOPh-2-phenylindole} \qquad (17)$$

$$\text{N-R-pyrrole} \xrightarrow[C_6H_6 \text{ or } MeCN]{(PhCO_2)_2} \text{2-OCOPh-N-R-pyrrole} + \text{2,5-}(PhCO_2)_2\text{-N-R-pyrrole} \qquad (18)$$

Unlike enamines, olefins react with diacyl peroxides by pathways including radical intermediates. Conversions with stilbenes[66] and with tetramethylethylene[67] (equation 19) have been described.

$$(RCO_2)_2 + Me_2C{=}CMe_2 \xrightarrow[45^{\circ}C]{C_6H_6} Me_2C\underset{O}{—}CMe_2 + (RCO)_2O \qquad (19)$$

R = 3-BrC$_6$H$_4$ (72%)

β-Substituted crotonates are acyloxylated at the α carbon by benzoyl peroxide[68] (equation 20).

$$CH_3C(R^1){=}CHCO_2R^2 \xrightarrow{(PhCO_2)_2} CH_2{=}C(R^1)CH(OCOPh)CO_2R^2 \qquad (20)$$

70% (R^1 = R^2 = Me)

5. *Reactions with other nucleophiles*

Like those of enols, acyloxylations of phenols are facile. Electron-donating substituents in the aromatic nucleus accelerate the conversions[69,70] (equation 21). Double acyloxylation at one carbon has also been observed[71] (equation 22).

$$R\text{-}C_6H_4OH \xrightarrow{(PhCO_2)_2} R\text{-}C_6H_3(OH)\text{-}2\text{-}OCOPh \qquad (21)$$

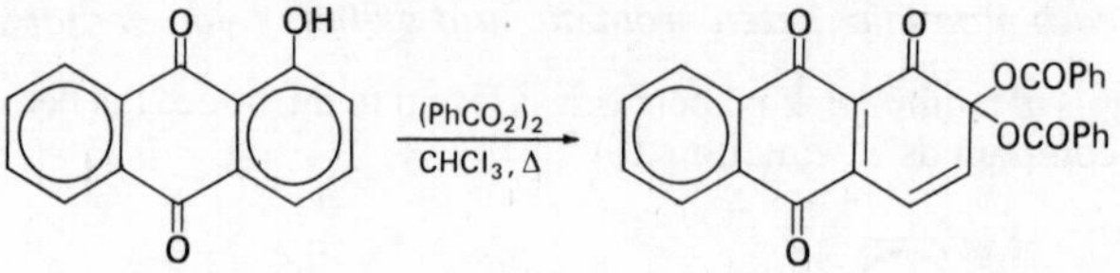

$$\xrightarrow[\text{CHCl}_3,\,\Delta]{(\text{PhCO}_2)_2} \qquad (22)$$

β-Naphthol reacts similarly[72] whereas 2,4-dibromo-α-naphthol (**18**) yields the quinone **19** as a condensation product of **18** with an intermediate oxidation product[73] (equation 23).

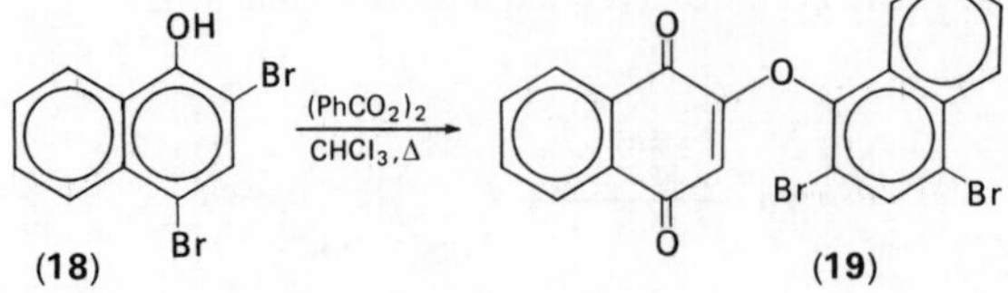

$$(18) \xrightarrow[\text{CHCl}_3,\Delta]{(\text{PhCO}_2)_2} (19) \qquad (23)$$

Aldehydes and ketones have been isolated by oxidation of primary and secondary alcohols with benzoyl peroxide in the presence of $NiBr_2$ [74] (equations 24 and 25).

$$\text{RCH}_2\text{OH} \xrightarrow[\text{[NiBr}_2\text{]}]{(\text{PhCO}_2)_2} \underset{34-96\%}{\text{RCHO}} + \underset{0-81\%}{\text{RCO}_2\text{CH}_2\text{R}} \qquad (24)$$

$$\text{R}_2\text{CHOH} \longrightarrow \underset{81-98\%}{\text{R}_2\text{CO}} \qquad (25)$$

Reactions of amines with diacyl peroxides usually follow radical reaction paths yielding very different products depending on the nature of the amine[1]. DABCO (**20**) has been found to be oxidized to its monoxide **21** by benzoyl peroxide[75] (equation 26). In this case, an *inter*molecular acyl transfer from the intermediate acyloxylation product takes place. *Intra*molecular acyl migrations have been observed in the course of acyloxylations of secondary aromatic amines[76–78] (equation 27). Primary aromatic amines yield small amounts of the corresponding products among other reaction products[78].

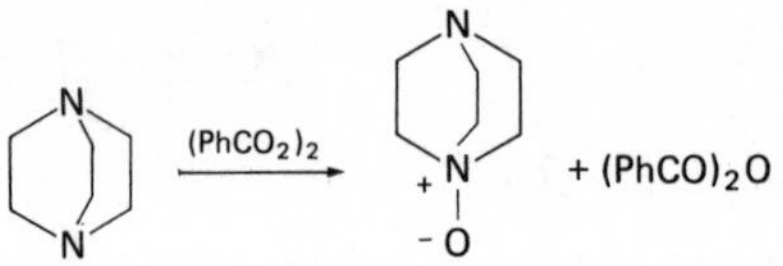

$$(20) \xrightarrow{(\text{PhCO}_2)_2} (21) + (\text{PhCO})_2\text{O} \qquad (26)$$

$$\text{PhNHR} \xrightarrow{(\text{PhCO}_2)_2} o\text{-HOC}_6\text{H}_4\text{N(R)COPh} \qquad (27)$$

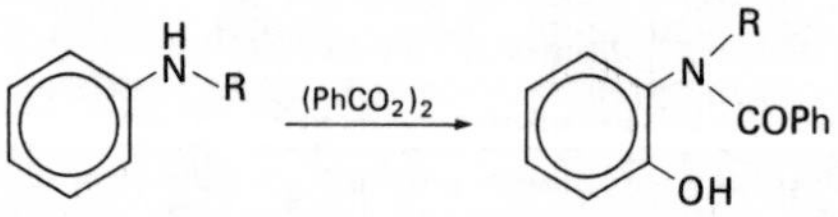

Ethers are acyloxylated in the α position with high yields[79,80] (equation 28). Thio ethers react in a more complex manner with diacyl peroxides. On the one hand the sulphur is attacked in analogy to the attack on nitrogen with tertiary amines, but on the other they behave like ethers with acyloxylation α to the sulphur atom[81] (equation 29). Mechanistic aspects have been reported in detail[81–83].

$$CH_3CH_2OCH_2CH_3 \xrightarrow{(PhCO_2)_2} CH_3CH_2OCH(OCOPh)CH_3 \quad (84\%) \tag{28}$$

$$R^1SCH_2R^2 \xrightarrow{(PhCO_2)_2} \begin{cases} R^1SOCH_2R^2 + (PhCO)_2O \\ R^1SCH(OCOPh)R^2 + PhCO_2H \end{cases} \tag{29}$$

II. PEROXYCARBOXYLIC ACIDS

A. General Syntheses

Peroxycarboxylic acids are prepared in analogy to diacyl peroxides by reaction of H_2O_2 (Na_2O_2) with carboxylic acids, their halides or anhydrides[22,84,85] (equations 30–32). Frequently, formation of diacyl peroxides occurs as a side reaction. In order to avoid the formation of complex reaction mixtures, alkaline solvolysis of diacyl peroxides has been proved to be a favourable method of preparing pure peroxycarboxylic acids[50,86,87] (equation 33).

$$RCO_2H + H_2O_2 \underset{}{\overset{[H^+]}{\rightleftharpoons}} RCO_3H + H_2O \tag{30}$$

$$RCOCl + H_2O_2 \longrightarrow RCO_3H + HCl \tag{31}$$

$$(RCO)_2O + H_2O_2 \longrightarrow RCO_3H + RCO_2H \tag{32}$$

$$(RCO_2)_2 + R^1O^- \xrightarrow[-RCO_2R^1]{} RCO_3^- \xrightarrow{H^+} RCO_3H \tag{33}$$

Alkoxycarbonyl hydroperoxides—which are particularly mild epoxidizing reagents—have been synthesized similarly by solvolysis of the corresponding peroxycarbonates with H_2O_2 [88,89] (equation 34).

$$(ROCO_2)_2 + H_2O_2 \xrightarrow{NaOH/CH_3OH} 2\,ROCO_3H \tag{34}$$

Furthermore, photooxidation of aromatic aldehydes has been used to prepare peroxycarboxylic acids[90] (equation 35).

$$MeO_2C\text{-}C_6H_4\text{-}C(=O)H \xrightarrow{O_2/h\nu} MeO_2C\text{-}C_6H_4\text{-}C(=O)O_2H \tag{35}$$

As further methods, reactions of H_2O_2 with ketenes[91], acyl imidazoles[92] and acyl dialkyl phosphates[93] have been described.

B. Stability and Structure of Peroxycarboxylic Acids

Pure hydroperoxy derivatives of lower fatty acids are explosives even at low temperatures. Nevertheless, distilled 40% peroxyacetic acid in acetic acid/water is commercially available. The stability of peroxycarboxylic acids increases with rising chain length, i.e. with rising molecular weight. At present, the fairly stable *m*-chloroperbenzoic acid (MCPB, m.p. 92°C, commercially available) is mostly preferred in laboratory experiments.

X-ray structure analyses have confirmed the presence of intramolecular as well as intermolecular hydrogen bonds[94,95] (**22**). The latter can be detected only in inert solvents (benzene, CCl_4)[96–98] and disappear in polar solvents[84] (equation 36). Solvents with electron-donating functions (ethers, alcohols, amides, ketones) show nearly exclusive formation of intermolecular hydrogen bonds with peroxycarboxylic acids[97,99,100]. Hydrogen bonds as well as reduced resonance stabilization reduce the acidities of peroxycarboxylic acids compared with the corresponding carboxylic acids (e.g. CH_3CO_3H: pK_a 7.6; CH_3CO_2H: pK_a 4.7)[3].

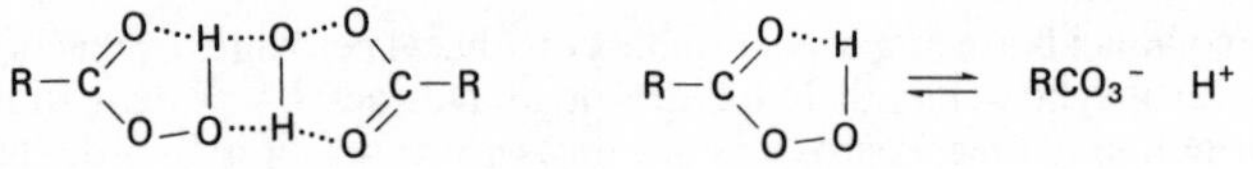

(22)

$$\text{R-C(=O)-O-OH (cyclic intramolecular H-bond)} \rightleftharpoons RCO_3^- \quad H^+ \qquad (36)$$

C. Reactions of Peroxycarboxylic Acids with Olefins

Olefins are epoxidized by peroxycarboxylic acids to yield oxiranes. Reactions are carried out in polar as well as in apolar solvents[101,102]. Peroxyacetic acid is used for large-scale oxidations. Other frequently used peracids are MCPB, perbenzoic acid, monoperphthalic acid and the highly reactive trifluoroperacetic acid[101] (equation 37).

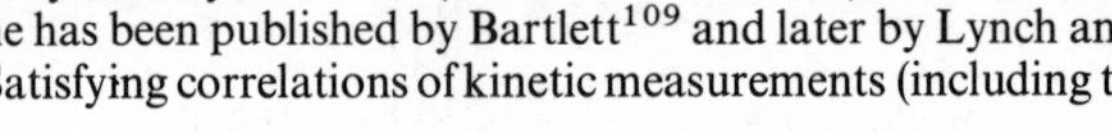

$$>C{=}C< \; + \; RCO_3H \longrightarrow >C\overset{O}{—}C< \; + \; RCO_2H \qquad (37)$$

1. Mechanism of epoxidation

Several publications have appeared concerning the mechanism of epoxidation of olefins by peroxycarboxylic acids (Prilezhaev reaction)[101,103–108]. The commonly accepted scheme has been published by Bartlett[109] and later by Lynch and Pausacker [110] (equation 38). Satisfying correlations of kinetic measurements (including the conversion of

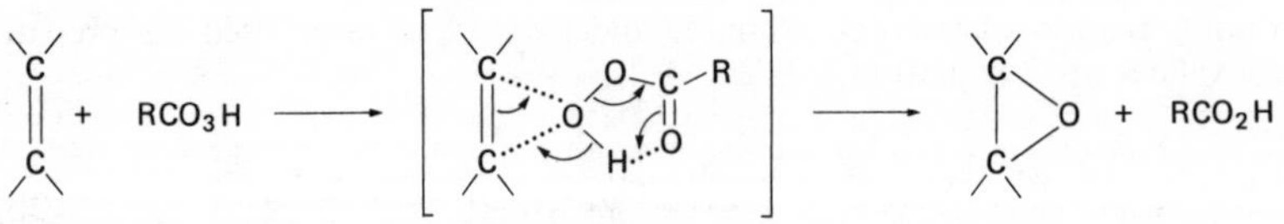

$$>C{=}C< \; + \; RCO_3H \longrightarrow [\text{Activated complex}] \longrightarrow >C\overset{O}{—}C< \; + \; RCO_2H \qquad (38)$$

Activated complex

peracetic acid with open-chained olefins[107] as well as with ring-substituted styrenes[108]) with the three- and five-parameter equations of Taft indicate that the epoxidation ensues through an activated (nearly) nonpolar complex. Therefore the epoxidation is of S_N2 type with highly negative activation entropy.

Important facts concerning epoxidation rates (Table 1) are the following: Reaction rates rise (*a*) with increasing electron density of the double bond, (*b*) with decreasing electronegativity of R and (*c*) with increasing dielectric constant of the solvents (supposing that the intramolecular hydrogen bond remains intact[106]).

A certain parallel of the reaction parameters and kinetic data of the epoxidation reaction compared with 1,3-dipolar cycloaddition has led to the proposal of a closely related mechanism[111] (equation 39).

$$\text{R—C(=O}\cdots\text{H)(O—O)} \overset{?}{\rightleftharpoons} \text{R—C}^{+}\text{(OH)(O—O}^{-}\text{)} \xrightarrow{\text{C=C}} [\ \text{transition state}\ ] \longrightarrow \text{RCO}_2\text{H} + \text{epoxide} \qquad (39)$$

Calculations[112] concerning the transition states of the 'normal' epoxidation scheme (Lynch and Pausacker, Bartlett; usually designed as the 'generally accepted' scheme) and the Kwart mechanism[111] have given no decision as to which mechanism is correct.

Determination of the rate constants[113] of the epoxidation of cyclohexene ($1.92 \times 10^{-2}\,l\,mol^{-1}\,s^{-1}$) and of norbornene ($2.28 \times 10^{-2}\,l\,mol^{-1}\,s^{-1}$) has given no spectacular differences (in 1,3-dipolar cycloaddition norbornene is known to be much more reactive than cyclohexene).

However, the results of Schneider and coworkers[114] are in contradiction with the Kwart scheme which demands a distinct positive charge at the peroxycarboxylic acid carbon. Their investigations concerning polar and steric effects of oxidations with aliphatic peracids include a satisfying correlation between epoxidation rates and the corresponding Taft constants σ^*. The experimental reaction constant $\rho^* = +2.0$ contradicts a cycloaddition of a 1,3-dipolar species bearing a positive charge at the carbonyl carbon

TABLE 1. Epoxidation rates of olefins

Olefin	$k \times 10^{3a}$
$CH_2{=}CH_2$	0.19
$CH_2{=}CHMe$	4.2
$CH_2{=}CMe_2$	92
$MeCH{=}CHMe$	93
$MeCH{=}CMe_2$	1240
	129
$CH_2{=}CHPh$	11.2
$CH_2{=}CPh_2$	48

[a] Peroxyacetic acid, 25.8°C: D. Swern, *J. Amer. Chem. Soc.*, **69**, 1692 (1947).

atom. Further investigations are necessary to elucidate remaining uncertainties. Although epoxidations with peroxycarboxylic acids are very fast, it has been found that rates can be further enhanced by catalysis with MoO_2 $(acac)_2$ [115].

2. *Stereospecifity and stereoselectivity of epoxidation*

Usually, epoxidations of olefins by peroxycarboxylic acids proceed stereospecifically; high stereoselectivity has been found in the course of epoxidations of *cis*- or *trans*-stilbenes yielding the corresponding 2,3-diphenyloxiranes[110]. Further examples are epoxidations of *Z,E* isomers of 2-butene[116] and those of cycloalkenes with 8-, 9- and 10-membered rings[117a–c,118a–c]. A recent review[119] reports in detail on the stereochemistry of epoxidations by peroxycarboxylic acids. Generally, sterically hindered olefins are attacked from the least hindered part of the molecule; however, in cases of interactions of appropriate substituents with the epoxidizing reagent, this rule can be inverted[120].

3. *Reaction rates of epoxidation*

Since open-chain *cis* olefins are of higher energy than their *trans* isomers, they are epoxidized more quickly (oleic acid/elaidic acid[121,122], *cis*/*trans*-stilbene[123,125]). Ratios of k_{cis}/k_{trans} are ranging between 1 and 2. Olefins of medium-sized rings show inverse reactivities. *Trans* isomers of cyclononene[117b] and cyclodecene[124] are epoxidized eight times faster than their *cis* isomers due to higher ring tension. In the case of *cis,trans,trans*-1,5,9-cyclododecatriene, epoxidation of *cis* and *trans* double bonds could be studied in the same molecule[126] (equation 40). Epoxidation of a *trans* double bond was 6.6 times faster than that of the *cis*. Analogously, MCPB epoxidizes *cis,trans*-1,5-cyclodecadiene yielding a 9:1 mixture of *trans* and *cis*-oxirane[127].

CH_3CO_3H (92%) (7%) (40)

4. *Special epoxidation methods*

Generally, two different epoxidation techniques are used[128–130]. On the larger scale *in situ* generations of peroxycarboxylic acids are preferred, whereas in laboratory experiments isolated peroxycarboxylic acids are mostly used.

Peracetic acid (in presence of a mineral acid) and performic acid are frequently prepared *in situ* by reaction of the carboxylic acids with hydrogen peroxide[84,131,132]; technical epoxidations are generally carried out with such epoxidation mixtures. In many cases the use of strong acids is unfavourable owing to side-reactions. One possibility is to use less than stoichiometric amounts of formic acid and of hydrogen peroxide in order to epoxidize oils and fats[133,134] as well as polybutadienes[135]. In other cases, peracetic acid prepared *in situ* is buffered by sodium carbonate, sodium acetate or a stoichiometric amount of sodium hydroxide; peracetic acid can also be utilized after distillation in vacuum. 9,10-Epoxystearic acid can be obtained in high yield from oleic acid by oxidation with perbenzoic acid prepared *in situ* from benzaldehyde and molecular oxygen[136].

Recently, particularly mild epoxidations have been carried out utilizing *in situ*-generated ethoxycarbonyl hydroperoxide (**23**)[137] (equations 41 and 42).

$$\text{EtOCOCl} \xrightarrow{H_2O_2/\text{buffer}} \text{EtO–C}(\text{=O}\cdots\text{H})\text{–O–O} \quad (\mathbf{23}) \qquad (41)$$

(23) + norbornene → exo-2,3-epoxynorbornane, 94%; + 1-methylcyclohexene → 1-methylcyclohexene oxide, 84%; + $CH_2{=}CHC_7H_{15}$ → 1,2-epoxynonane (C_7H_{15}-oxirane), 33% (42)

Aryloxiranes are obtained from styrenes by simple oxidations with peracids; however, care must be taken to avoid their contact with carboxylic acids[138]. Therefore, a new method[139] has described the stereospecific synthesis of acid-sensitive oxiranes (**24**) using a PTC epoxidation technique (PTC: Phase Transfer Catalysis) (equation 43).

$$\text{PhC(Me)=C(Me)R} \xrightarrow[\text{Phosphate buffer, pH 8, } 0\text{–}20^{\circ}\text{C}]{\text{MCPB–}CH_2Cl_2/} \text{oxirane } (\mathbf{24}) \quad 90-95\% \qquad (43)$$

R = Me, Et, H

Another method has used the system monoperphthalic acid–finely powdered disodiumphthalate in order to synthesize the highly sensitive oxirane ethers **25**; without buffer, yields drop to half[140].

Generally, peroxycarboxylic acids are unstable at higher temperatures, and therefore they are unsuitable for epoxidations of unreactive olefins. Investigations on the stability of MCPB solutions in the presence of radical inhibitors have exhibited no noticeable decomposition at temperatures up to 90°C; under these conditions, even unreactive olefins like 1-dodecene or methyl methacrylate can be epoxidized in quantitative yield[141].

$n = 6-9$ (52 – 70%)

(**25**)

5. *Epoxidation by related peroxycarbon acids*

Carbonitriles and hydrogen peroxide react in methanol solution at pH 8 to yield *in situ* imidoperoxycarboxylic acids **26**[142] (equation 44). With excess hydrogen peroxide the latter decompose to yield carbamides and molecular oxygen. In the presence of olefins this reaction is suppressed, and oxiranes are generated. Because of the pH this method is particularly suitable for the epoxidization of acid-sensitive olefins like **27**[143] (equation 45). Further examples are the successful epoxidations of 2-allylcyclohexanone[142b], ergocalciferol[144] and methylenecyclohexane[142b].

The relative reactivities of some olefins towards peroxyacetic acid and peroxybenzimidic acid have been compared[145].

$$RC{\equiv}N + H_2O_2 \xrightarrow{\text{buffer}} \left[RC(=NH)\text{–O–O–H} \right] \xrightarrow{+H_2O_2} RC(=O)NH_2 + O_2 + H_2O \quad (44)$$

R = Me, Ph

(**26**)

$$CH_2{=}CHCH(OEt)_2 \xrightarrow{+26} H_2C\overset{O}{—}CH—CH(OEt)_2 \quad (45)$$

(**27**)

In analogy to imidoperacids (**26**), peroxycarbamic acids (**28**) are obtained by reaction of hydrogen peroxide with isocyanates or carbonyl azolides; they are suitable reagents for the epoxidation of acid-sensitive olefins (as well as azomethines) (equation 46). Some results are summarized in Table 2.

$$R^1NCO \text{ or (triazolyl)}_2C{=}O \xrightarrow{+H_2O_2} R^1R^2N\text{–}C(=O)\text{–O–OH} \xrightarrow{>C=Y} \text{oxirane } >C\overset{O}{—}Y \quad (46)$$

(**28**)

Y = >C, –N

D. Reactions of Peroxycarboxylic Acids with Acetylenes and Allenes

Although epoxidation of acetylenes has been of theoretical and preparative interest for a long time, comparatively few details on chemistry and mechanism were published[149]. More recent investigations are the reactions of di-*t*-butylacetylene (**29**) (equation 47) and of cyclodecyne (**34**) with MCPB in methylene chloride at 25°C (equation 48)[150]. Intermediates in the reactions are oxirenes[151] which can also be obtained independently by photolysis of α-diazoketones.

TABLE 2. Epoxidations with peroxycarbamic acids (**28**)

28 R^1	R^2	Oxirane (oxaziridine)	Yield (%)	Ref.
PhCO	H	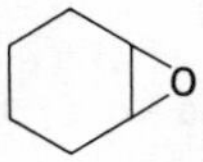 [cyclohexene oxide]	68	146
PhCO	H	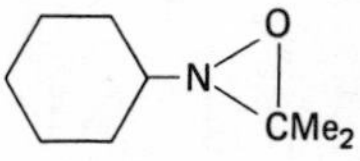 [cyclohexyl–N oxaziridine, O, CMe_2]	72	146
Ph	H	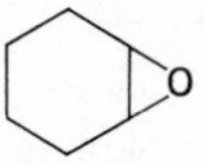 [cyclohexene oxide]	33–69[a]	147
4-ClC_6H_4	H	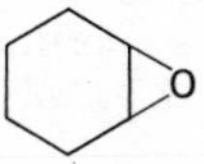 [cyclohexene oxide]	58–75[b]	147
4-ClC_6H_4	H	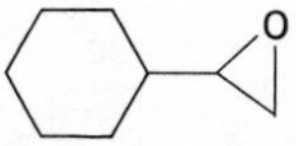 [cyclohexyl oxirane]	25–42[b]	147
[imidazol-1-yl]		[cyclohexene oxide] [c]	—	148
		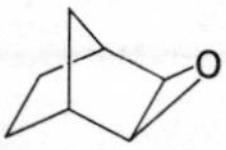 [norbornene oxide] [c]	—	148

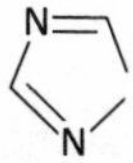

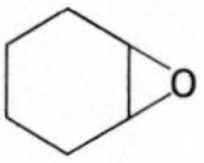

[a]Yields are dependent on the reaction of PhNCO:H_2O_2.
[b]Solvent-dependent yields.
[c]Epoxidation rates are 200 times faster than with peroxybenzoic acid.

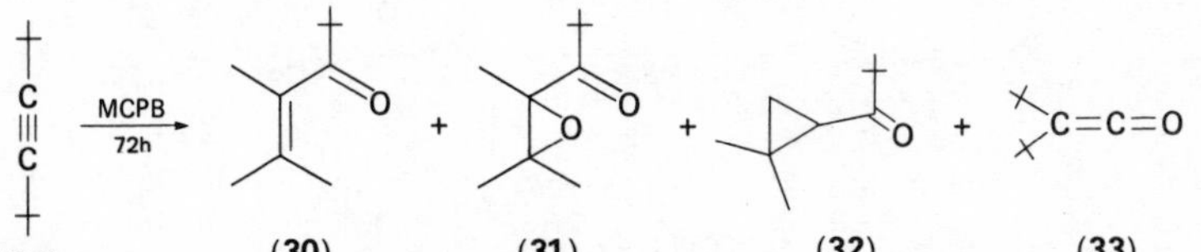

$$\text{(29)} \xrightarrow[\text{72h}]{\text{MCPB}} \text{(30)} + \text{(31)} + \text{(32)} + \text{(33)} \qquad (47)$$

30:**31**:**32** = 21:72:7

The question of the existence of allene oxides (methylene oxiranes) (**39**) as well as that of spirobisoxiranes (**40**) has been the subject of many peracid oxidations of allenes in the earlier literature[84]. This question may have been answered by Crandall and coworkers.

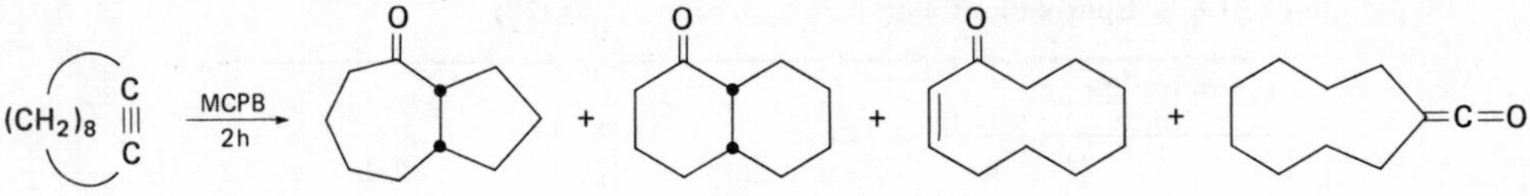

$(CH_2)_8$ C≡C →(MCPB, 2h) (35) + (36) + (37) + (38) C=O

(34) (35) (36) (37) (38)

(48)

35:**36**:**37** = 67:12:21

(39) (40)

Their oxidation of 1,1-dimethylallene (**41**) with peracetic acid yielded a mixture of products the formation of which could be explained by the intermediate presence of three reactive species[152] (equation 49). When large substituents are introduced into the allene system the synthesis of derivatives of type **39**, as well as of type **40**, is made possible[153] (equations 50–52).

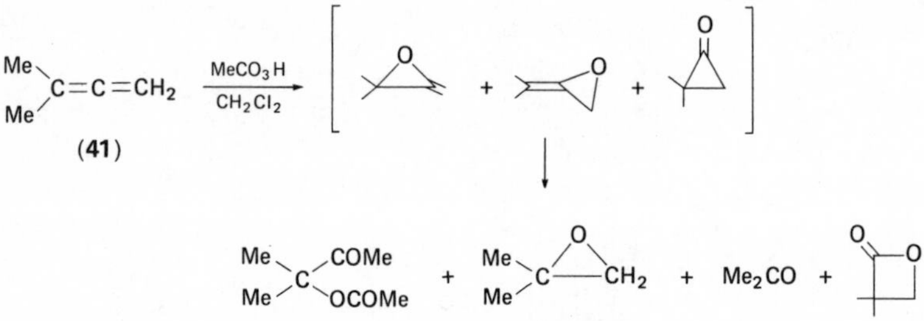

$Me_2C{=}C{=}CH_2$ (**41**) →($MeCO_3H$, CH_2Cl_2) [...] →

$Me_2C(COMe)(OCOMe)$ (**42**) + $Me_2C(O)CH_2$ (**43**) + Me_2CO (**44**) + (**45**) (49)

42:**43**:**44**:**45** = 50:25:17:8

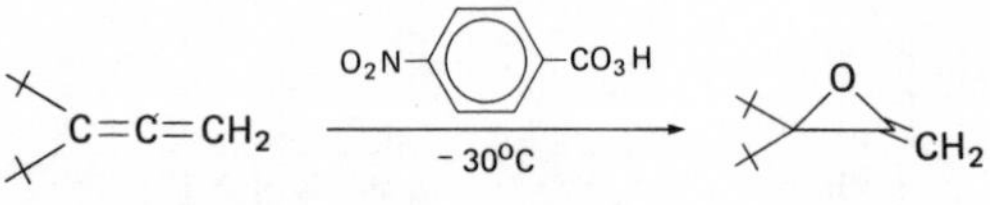

$t\text{-}Bu_2C{=}C{=}CH_2$ →($O_2N\text{-}C_6H_4\text{-}CO_3H$, −30°C) epoxide CH_2 (50)

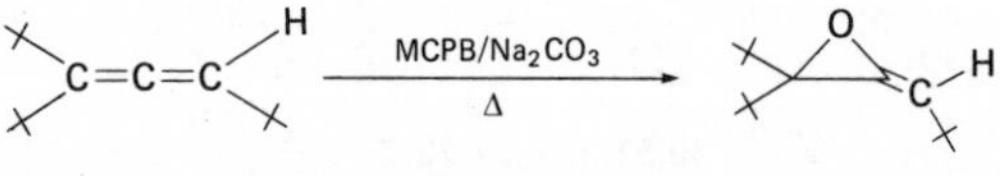

$t\text{-}Bu_2C{=}C{=}C(H)(t\text{-}Bu)$ →($MCPB/Na_2CO_3$, Δ) epoxide $C(H)(t\text{-}Bu)$ (51)

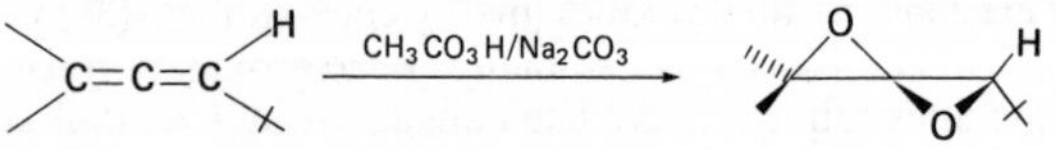

$Me_2C{=}C{=}C(H)(t\text{-}Bu)$ →(CH_3CO_3H/Na_2CO_3) dioxaspiropentane (b.p.$_2$ 40°C) (52)

E. Reactions of Peroxycarboxylic Acids with Other Reaction Partners

1. Hydroxylation of alkanes

Hydroxylations at C(25) and at C(5α) have been observed in the course of irradiations of cholestanes in presence of peroxyacetic acid[154]. Tertiary (71%), secondary (29%) and primary (<0.3%) alcohols were obtained when methylcyclohexane (100-fold excess) was hydroxylated by peroxytrifluoroacetic acid[155]. Later experiments to reproduce this conversion under preparative conditions gave unsatisfying results[156]. Only very poor conversion was observed, and the reaction mixture consisted of 38% 1-methylcyclohexanol and 62% isomeric secondary alcohols, the separation of which presented considerable difficulties. However, hydroxylations of other more suitable alicyclic hydrocarbons (**46**–**48**) took place with *p*-nitroperbenzoic acid[157]. In these cases good yields, high regioselectivity and retention of configuration could be observed.

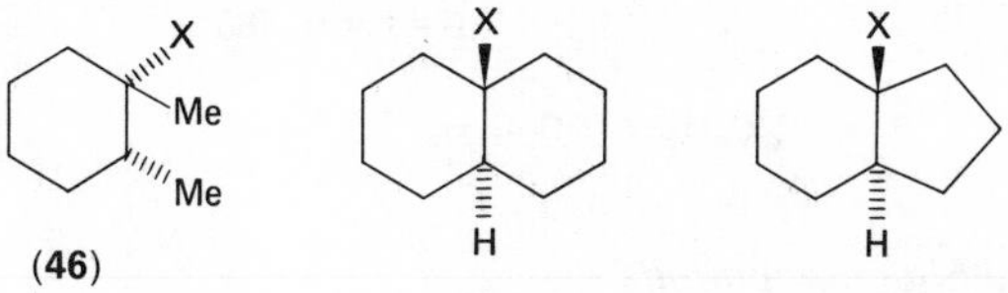

(**46**) (**47**) (**48**)

(X = H ⟶ X = OH)

2. Oxidation of alcohols

Generally, primary and secondary alcohols are not attacked by simple peroxycarboxylic acids. However, in the presence of appropriate catalysts (2,2,6,6-tetramethylpiperidine hydrochloride[158] or hydrogen chloride in THF[159]) oxidations take place yielding aldehydes and ketones (equation 53).

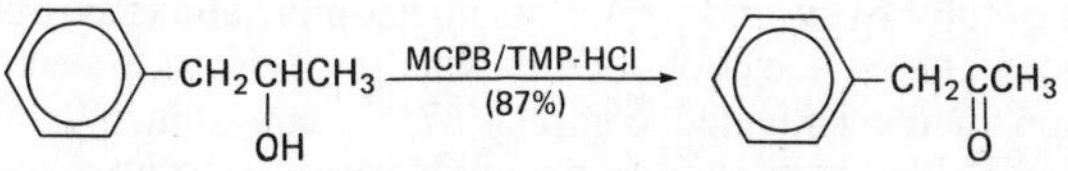

$$C_6H_5-CH_2CH(OH)CH_3 \xrightarrow[(87\%)]{MCPB/TMP\cdot HCl} C_6H_5-CH_2C(=O)CH_3 \qquad (53)$$

3. Oxidation of aldehydes, ketones and acetals

Oxygen insertion into the carbon chain of aldehydes and ketones by means of peroxy acids to yield esters is known as the Baeyer–Villiger oxidation[160–162] (equation 54). Alicyclic ketones yield lactones[163–165]; special efforts have been made to synthesize ε-caprolactone[166–168]. Ketones with different substituents R yield mixtures of esters, whereas aldehydes yield carboxylic acids and alkyl or aryl formates[169] (equation 55). Oxidation of aldehyde acetals has proved to be an excellent method for synthesizing the corresponding esters[170] (equation 56).

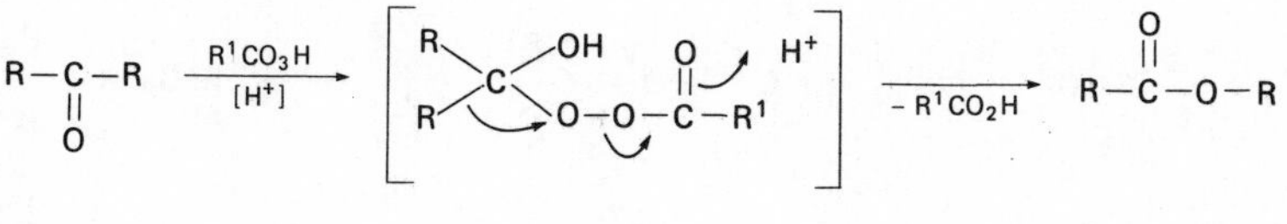

$$R-C(=O)-R \xrightarrow[[H^+]]{R^1CO_3H} \left[R_2C(OH)-O-O-C(=O)-R^1 \quad H^+ \right] \xrightarrow{-R^1CO_2H} R-C(=O)-O-R \qquad (54)$$

$$\mathrm{RCHO} \xrightarrow{\text{MCPB}} \underset{(\mathbf{49})}{\mathrm{ROCHO}} + \underset{(\mathbf{50})}{\mathrm{RCO_2H}} \qquad (55)$$

R	49	50
cyclohexyl	76%	23%
$PhCH_2$	76%	24%
$Ph(CH_2)_2$	5%	84%

$$\mathrm{RCH(OEt)_2} \xrightarrow{\mathrm{R^1CO_3H}} \mathrm{RCO_2Et} \qquad (56)$$

R = cyclohexyl (90%)

R = $PhCH_2$ (80%)

R^1 = 3-ClC_6H_4, 4-$NO_2C_6H_4$

4. Oxidation of sulphur compounds

Peroxycarboxylic acids are able to oxidize thiols as well as disulphides directly yielding sulphonic acids[171]. Sulphides are converted to sulphoxides by a similar mechanism as described with olefins[172] (equation 57). With less reactive peroxycarboxylic acids further oxidation of sulphoxides to sulphones by a similar mechanism[173,174] evidently proceeds much slower than the first step. However, highly reactive peroxycarboxylic acids like trifluoroperoxyacetic acid are able to yield sulphones directly[175]. In these cases the oxidizing power of the peracid is so high that phenylalkyl functions suffer degradation to yield carboxylic acids whereas phenyl groups connected to the sulphonyl functions are protected (equation 58). As n increases, increasing yields of a sulphone acid **51** are obtained. In the course of oxidation of sulphoxides by peroxycarboxylic acids in alkaline medium, the roles of sulphoxide and peracid as nucleophile and electrophile are changed. Now, peroxycarboxylates as nucleophiles attack sulphoxides as *S*-electrophiles followed by a fragmentation of the activated complex **52**[176] (equation 59). As a general rule oxidation rates of sulphides decrease with rising pH values in contrast to those found with sulphoxides.

$$R^1R^2S{:} + HO{-}O{-}C(=O){-}R^3 \longrightarrow R^1R^2SO + R^3CO_2H \qquad (57)$$

$$C_6H_5{-}S{-}(CH_2)_n{-}C_6H_5 \xrightarrow{CF_3CO_3H} \underset{(\mathbf{51})}{C_6H_5{-}SO_2{-}(CH_2)_n{-}CO_2H} \qquad (58)$$

$$RCO_3^- + R^1R^2S{=}O \rightleftharpoons \underset{(\mathbf{52})}{\left[R{-}C(=O){-}O{-}O{-}S(R^1)(R^2){-}O^-\right]} \longrightarrow RCO_2^- + R^1R^2SO_2 \qquad (59)$$

Some selected interesting oxidation reactions are described in the following examples: Sulphoxides have been obtained by means of polymeric peroxycarboxylic acids (polymer resins like polystyrenes bearing the hydroperoxycarbonyl group as substituent in the benzene nucleus) (yields 70–100%)[177]. Methylphenylethynyl sulphide has been oxidized selectively to yield the corresponding sulphoxide (with one mole of MCPB) or sulphone (with two moles of MCPB)[178]. Peroxytrifluoroacetic acid is reactive enough to convert dibenzothiophene to the corresponding sulphone in quantitative yield[179]. Thioketones are known to be oxidized by peroxycarboxylic acids to give sulphines[180]; a second, slower (epoxidation-like) step leads to ketones (equation 60). By this reaction benzophenones have been synthesized in nearly quantitative yields using perbenzoic acid in carbon tetrachloride[181] (equation 61).

$$R_2C{=}S \longrightarrow R_2C{=}S{=}O \longrightarrow R_2C{=}O \qquad (60)$$

$$(RC_6H_4)_2C{=}S{=}O \longrightarrow (RC_6H_4)_2C{=}O \; + \; (RC_6H_4\text{–}O\text{–}C(=O)\text{–}C_6H_4R) \qquad (61)$$

R = H, 4-MeO, 4-Me, 3-Me, 4-Cl, 3-Cl

5. *Oxidation of nitrogen compounds*

Preparation of tertiary amine oxides by the use of hydrogen peroxide[182] in water, acetic acid or acetic anhydride proceeds slowly and often affords low yields. Using MCPB excellent yields are obtained under mild conditions[183] (equation 62). Pyridine *N*-oxide is obtained in 80% yield using peroxyacetic acid at 80°C[184], but in quantitative yield using peroxytrifluoroacetic acid at low temperature[179]. The latter reagent converts even 2,6-dibromopyridine to its *N*-oxide in 75% yield[185].

$$R^1R^2R^3N| \xrightarrow[CHCl_3/0-25^{\circ}C]{MCPB} R^1R^2R^3N^+{-}O^- \qquad (62)$$

86 – 98%

Primary amines are reported to be oxidized by peroxycarboxylic acids to yield aliphatic or aromatic nitro compounds[186]; similarly, oximes yield nitroalkanes[187], whereas oxaziranes are formed by oxidation of imines[188] (equation 63). Ketene imines have also been oxidized[189].

$$C_6H_{10}{=}N{-}C_6H_{11} \xrightarrow[(89\%)]{CH_3CO_3H} \text{(cyclohexane-spiro-oxaziridine)}{-}N{-}C_6H_{11} \qquad (63)$$

Similarly to the conversion of thiobenzophenone yielding benzophenone through sulphines as intermediates, benzophenone hydrazone has been oxidized with peroxyacetic acid (with I_2 as catalyst) to form diphenyldiazomethane[190]; further oxidation of diphenyldiazomethane as well as of other diaryldiazomethanes[191] and cyclic α-diazoketones[192] leads to an exchange of the diazo group for oxygen (equation 64).

$$Ph_2C{=}N{-}NH_2 \xrightarrow[CH_2Cl_2\ (85\%)]{CH_3CO_3H,\ [I_2]} Ph_2C{=}N_2 \xrightarrow[X\,=\,H,\,Cl]{X{-}C_6H_4{-}CO_3H} Ph_2C{=}O \qquad (64)$$

6. Oxidation of miscellaneous compounds

Reactions of peroxycarboxylic acids with arenes yield different reaction products. Under catalysis of Lewis acids, electron-rich aromatic systems are hydroxylated[193] (equation 65). Depending on structure, aromatic rings can be cleaved as soon as a second hydroxyl group has been introduced[194] (equation 66). Finally, alkylbenzenes are degraded by peroxytrifluoroacetic acid to yield fatty acids[195] (equation 67).

$$\text{1,3,5-}Me_3C_6H_3 \xrightarrow[CH_2Cl_2,\ 0^\circ C]{CF_3CO_3H,\ [BF_3]} \text{2,4,6-}Me_3C_6H_2OH\ (88\%) \qquad (65)$$

$$\text{2-naphthol} \xrightarrow[(92-95\%)]{CH_3CO_3H} \text{2-}(HO_2C{-}CH{=}CH){-}C_6H_4{-}CO_2H \qquad (66)$$

$$n\text{-}Pr{-}C_6H_5 \xrightarrow[\Delta]{CF_3CO_3H} n\text{-}PrCO_2H\ (74\%) \qquad (67)$$

Alkyl and aryl iodides are easily oxidized by peracetic acid. Depending on reaction conditions, iodoso or iodoxy compounds are formed[196] (equation 68). Iodoso compounds from iodides containing an electron-withdrawing substituent on the same carbon atom show *cis* eliminations yielding α,β-unsaturated carbonyl, sulphinyl or sulphonyl compounds[197] (equation 69).

$$C_6H_5{-}I \xrightarrow{CH_3CO_3H} C_6H_5{-}I{=}O \; \begin{cases} \underset{H_2O}{\overset{CH_3CO_2H}{\rightleftharpoons}} C_6H_5{-}I(OCOCH_3)_2 \\ \xrightarrow{CH_3CO_3H,\ \Delta} C_6H_5{-}IO_2 \end{cases} \qquad (68)$$

$$\text{(tetralin, 1-H, 1-D, 2-I, 2-SPh)} \xrightarrow[O_2]{\text{MCPB}} \text{(dihydronaphthalene, 1-D, 2-SO}_2\text{Ph)} \tag{69}$$

III. PEROXYCARBOXYLATES (PEROXY ESTERS)

A. General Syntheses

Usually, peroxycarboxylates are prepared by reaction of alkyl hydroperoxides with acylating agents[3,198–207] (equation 70). Ketene has also been used as acylating agent[3,206,207] (equation 71). Peroxycarbonates have been synthesized from chlorocarbonates (equation 72) and *t*-peroxycarbamates have been generated from carbamoyl chlorides[205,208,209] (equation 73). Isocyanates can be utilized instead of ketenes. In these cases, *s*-peroxycarbamates are formed (equation 74).

$$R^1COX + R^2O_2H \xrightarrow[-HX]{} R^1CO_3R^2 \qquad X = Cl, OCOR^3, \text{1-imidazolyl} \tag{70}$$

$$CH_2{=}CO + R^2O_2H \longrightarrow CH_3CO_3R^2 \tag{71}$$

$$R^1OCOCl + R^2O_2H \xrightarrow[-HCl]{} R^1OCO_3R^2 \tag{72}$$

$$R^3R^4NCOCl + R^2O_2H \xrightarrow[-HCl]{} R^3R^4NCO_3R^2 \tag{73}$$

$$R^5NCO + R^2O_2H \longrightarrow R^5NHCO_3R^2 \tag{74}$$

B. Stability of Alkyl Peroxycarboxylates

Esters generated from primary and secondary alkyl hydroperoxides have proved to be particularly unstable. As they decompose very easily to give carbonyl compounds and carboxylic acids, they have been synthesized only rarely, using especially mild procedures[198,210–212] (equation 75).

$$R^1{-}C(=O)O{-}O{-}CHR^2R^3 \longrightarrow R^1CO_2H + R^2COR^3 \tag{75}$$

Tertiary alkyl peroxycarboxylates are comparatively more stable. Compounds with low molecular weights may be distilled at reduced pressure at low temperatures, but even so caution is recommended because an explosion has been reported[213]. A comprehensive summary on alkyl peroxycarboxylates containing physical data has been given[205].

C. Reactions of *t*-Alkyl Peroxycarboxylates

Reactions of peroxycarboxylates have been the subject of several review articles. Sosnovsky and Lawesson[214] have reported on copper-catalysed reactions of peresters and Rüchardt[210] has summarized noncatalysed decompositions of peresters. Acyloxylations by means of peresters have been reviewed by Rawlinson and Sosnovsky[2].

1. *Reactions of* t-*butyl peroxycarboxylates with reactive nucleophiles (transfer of alkoxy groups)*

Aryl Grignard compounds as highly reactive nucleophiles have been shown to react with *t*-butyl perbenzoate yielding aryl *t*-butyl ethers[215]. The reaction mechanism probably includes nucleophilic attack of the carbanion at the peroxy oxygen attached to the *t*-butyl group, whereas carboxylate anion acts as leaving group (equation 76). This reaction sequence has been used to synthesize 4-*t*-butoxybenzenesulphenyl chloride[216] (equation 77).

$$\mathrm{RC(=O)O{-}OBu\text{-}}t + \mathrm{ArMgX} \longrightarrow \mathrm{ArOBu\text{-}}t + \mathrm{RCO_2MgX} \qquad (76)$$

$$t\text{-BuS}\text{-C}_6\text{H}_4\text{-MgBr} \xrightarrow[(54\%)]{\mathrm{PhCO_3Bu\text{-}}t} t\text{-BuS-C}_6\text{H}_4\text{-OBu-}t \xrightarrow[(92\%-96\%)]{+\mathrm{Cl_2}} t\text{-BuO-C}_6\text{H}_4\text{-SCl} + t\text{-BuCl} \qquad (77)$$

Phosphines represent another type of nucleophile which is able to attack alkyl peroxycarboxylates at the alkyl-bearing oxygen. The resulting alkoxyphosphonium ions behave as very efficient alkylating agents transferring carboxylate anions to esters[217] (equation 78). This reaction sequence can be used to generate carboxylic acid esters from the corresponding peresters.

$$\mathrm{Ph_3P| + R^2O{-}OCOR^1} \longrightarrow \left[\mathrm{Ph_3\overset{+}{P}OR^2}\ \ {}^{-}\mathrm{OCOR^1}\right] \longrightarrow \mathrm{Ph_3PO + R^1CO_2R^2} \qquad (78)$$

2. *Reactions of* t-*butyl peroxycarboxylates with CH bonds, forming stabilized radicals after hydrogen abstraction (introduction of acyloxy groups)*

Unlike reactions described in Section III.C.1 which prefer an ionic reaction mechanism, the copper (I/II)-catalysed acyloxylations of CH bonds forming stabilized radicals after hydrogen abstraction by *t*-butyl peroxycarboxylates (Kharasch–Sosnovsky reaction[2]) proceed via a radical mechanism (equation 79). Generally, reaction path (A) yielding acyloxylation products represents the main reaction. In certain cases however, reaction path (B) becomes important, and can domineer over (A), particularly if the reactions require longer reaction periods at elevated temperatures.

$$RCO_3Bu\text{-}t \xrightarrow{+Cu^+} RCO_2Cu^+ + t\text{-}Bu\cdot \quad (79a)$$

$$t\text{-}Bu\cdot + -\overset{|}{\underset{|}{C}}-H \longrightarrow -\overset{|}{\underset{|}{C}}\cdot + t\text{-}BuOH \quad (79b)$$

$$-\overset{|}{\underset{|}{C}}\cdot + RCO_2Cu^+ \longrightarrow \left[-\overset{|}{\underset{|}{C}}^+ + RCO_2Cu\right] \underset{\pm Cu^+}{\overset{(A)}{\rightleftharpoons}} -\overset{|}{\underset{|}{C}}-OCOR \quad (79c)$$

$$\left[-\overset{|}{\underset{|}{C}}^+ + RCO_2Cu\right] \xrightarrow[(B)]{+t\text{-}BuOH,\ -Cu^+} -\overset{|}{\underset{|}{C}}-OBu\text{-}t + RCO_2H$$

a. Substitution of allylic (and related) hydrogen. Although reactions of peroxycarboxylates proceed predominantly in an unspecific manner, olefins with terminal double bonds, like 1-octene or allylbenzene, can be acyloxylated selectively with good yields[218] (equation 80). The main products prove to be the unrearranged olefins whereas allyl-rearranged olefins appear only as by-products[219,220].

$$R^1CH_2CH{=}CH_2 + R^2CO_3Bu\text{-}t \xrightarrow{Cu(I)/(II)} R^1\underset{\displaystyle OCOR^2}{\underset{|}{C}H}CH{=}CH_2 + t\text{-}BuOH \quad (80)$$

(a) $R^1{=}Pr, C_5H_{11}, Ph$, $R^2{=}Ph$
(b) $R^1{=}Pr, C_5H_{11}$; $R^2{=}Me$

In contrast the corresponding 2-olefins give rise to predominantly rearranged acyloxylation products with terminal double bonds[221,222] (equation 81). Mechanistic investigations have shown that no uniform reaction sequence can be derived in these cases[214,218,219,223].

$$CH_3CH{=}CHCH_3 + CH_3CO_3Bu\text{-}t \xrightarrow{Cu(I)/(II)} CH_3\underset{\displaystyle OCOCH_3}{\underset{|}{C}H}CH{=}CH_2 + CH_3CH{=}CH\underset{\displaystyle OCOCH_3}{\underset{|}{C}H_2} \quad (81)$$

9 : 1

Similarly to allylic hydrogens, propargylic hydrogens can be acyloxylated in the same manner[224] (equation 82). In these cases no bond shifting occurs.

$$R^1C{\equiv}\overset{\displaystyle R^2}{\underset{\displaystyle R^3}{C}}H + C_6H_5CO_3Bu\text{-}t \xrightarrow{Cu(I)} R^1C{\equiv}\overset{\displaystyle R^2}{\underset{\displaystyle R^3}{C}}OCOC_6H_5 \quad (82)$$

(a) $R^1 = R^2 = H, R^3 = Pr$
(b) $R^1 = Me, R^2 = H, R^3 = Et$
(c) $R^1 = Et, R^2 = H, R^3 = Me$
(d) $R^1 = Ph, R^2 = R^3 = Me$
(e) $R^1 = SiMe_3, R^2 = H, R^3 = Pr$

Tetramethylallene reacts under the same conditions which are used for acyloxylations of 2-olefins, to give mainly the rearranged derivative of 1,3-butadiene containing a terminal double bond[224] (equation 83).

$$Me_2C{=}C{=}CMe_2 + PhCO_3Bu\text{-}t \xrightarrow{Cu(I)} Me_2C{=}C(OCOPh){-}C(Me){=}CH_2 \qquad (83)$$

In analogy to substitutions of allylic and propargylic hydrogens, benzylic hydrogens can also be replaced by acyloxy groups in the course of the above reaction. However, yields are described to be lower[2].

b. Substitution of α-H in ethers. Ethers containing α-hydrogen are able to be acyloxylated by peroxy esters forming α-acyloxy ethers, which may be regarded as derivatives of aldehydes and ketones[225–228] (equation 84). These acyloxylations also proceed with ethers which are not attacked by molecular oxygen in the course of autoxidation, e.g. with aryl alkyl ethers[226].

$$EtOCH_2CH_3 + RCO_3Bu\text{-}t \xrightarrow{Cu(I)/(II)} EtOCH(OCOR)CH_3 + t\text{-}BuOH \qquad (84)$$

$$R = Me\ (64\%),\ Ph\ (82\%)$$

Cyclic ethers behave similarly to the open-chain ones; however, the resulting α-acyloxy ethers are preferentially converted to give acetals and carboxylic acids in presence of *t*-butanol[225,227,228,230] (equation 85).

$$R^1OCHR^2(OCOCH_3) + t\text{-}BuOH \xrightarrow{Cu(I)} R^1OCHR^2(OBu\text{-}t) + CH_3CO_2H \qquad (85)$$

Cyclic acetals like 1,3-dioxane form cyclic derivatives of orthocarboxylic acids[231] (equation 86).

$$\text{1,3-dioxane} + PhCO_3Bu\text{-}t \xrightarrow{Cu(I)} \text{2-(OCOPh)-1,3-dioxane} + t\text{-}BuOH \qquad (86)$$

In cases where the acetal carbon is substituted by alkyl groups the outer positions are attacked[232] (equation 87).

$$\text{2,2-Me}_2\text{-1,3-dioxolane} + PhCO_3Bu\text{-}t \xrightarrow{CuBr} \text{2,2-Me}_2\text{-4-(OCOPh)-1,3-dioxolane}\ (63\%) \qquad (87)$$

c. Substitution of α-H in thio ethers. Whereas sulphides have proved to be convenient reagents for the reduction of alkyl hydroperoxides in the course of ozonolytic cleavage of olefins[233] yielding sulphoxides, they react with *t*-butyl peroxycarboxylates in analogy to the ethers described above. Accordingly, formation of α-acyloxylated thio ethers of varying types has been observed[234–239]. Copper salts as well as styrene[237] have been used as catalysts (equations 88–90).

$$\mathrm{MeSMe} + \mathrm{PhCO_3Bu\text{-}}t \xrightarrow{\text{Styrene}} \mathrm{MeSCH_2OCOPh} + t\text{-}\mathrm{BuOH} \qquad (88)$$

$$\text{1,3-dithiane} + \mathrm{PhCO_3Bu\text{-}}t \xrightarrow{\text{CuCl}} \text{2-(OCOPh)-1,3-dithiane} + \text{2-(OCOPh)-1,4-dithiane} \qquad (89)$$

$$\text{1,3,5-trithiane} + \mathrm{PhCO_3Bu\text{-}}t \xrightarrow{\text{CuCl}} \text{2-(OCOPh)-1,3,5-trithiane} + \text{bis(OCOPh)-trithiane (Ref. 238)} \qquad (90)$$

Reaction of 1,4-thioxane with *t*-butyl peroxybenzoate takes place selectively at hydrogens α to sulphur[231] (equation 91).

$$\text{1,4-thioxane} + \mathrm{PhCO_3Bu\text{-}}t \xrightarrow{\text{CuCl}} \text{3-(OCOPh)-1,4-thioxane} + t\text{-}\mathrm{BuOH} \qquad (91)$$

d. Miscellaneous CH substitutions[2]. Several other species of CH compounds have been acyloxylated by *t*-butyl peroxycarboxylates (equations 92–94). These reactions have been of no great interest up till now.

$$\mathrm{(EtO_2C)_2CH_2} + \mathrm{PhCO_3Bu\text{-}}t \xrightarrow{\text{Cu(I)/(II)}} \underset{16\%}{\mathrm{(EtO_2C)_2CHOCOPh}} + \underset{26\%}{\mathrm{(EtO_2C)_2CHOBu\text{-}}t} \text{ (Ref. 231)} \qquad (92)$$

$$\mathrm{R{-}CHO} + \mathrm{PhCO_3Bu\text{-}}t \xrightarrow{\text{Cu(I)/(II)}} \mathrm{R{-}C(=O){-}O{-}C(=O){-}Ph} \text{ (Ref. 240)} \qquad (93)$$

R = Pr, Ph (70%)

$$\mathrm{OCH{-}N(Me){-}Me} + \mathrm{RCO_3Bu\text{-}}t \xrightarrow{\text{Cu(I)/(II)}} \mathrm{OCH{-}N(Me){-}CH_2OCOR} \text{ (Refs. 214, 229, 231)} \qquad (94)$$

R = Me (25%), Ph (35%)

IV. REFERENCES

1. R. Hiatt in *Organic Peroxides*, (Ed. D. Swern), Vol. II, Wiley–Interscience, New York–London, 1970, pp. 799–930.
2. D. J. Rawlinson and G. Sosnovsky, *Synthesis*, 1 (1972).
3. A. F. Hegarty in *Comprehensive Organic Chemistry* (Eds. D. H. R. Barton and W. D. Ollis), Vol. 2 (Ed. I. O. Sutherland), Pergamon Press, Oxford, 1979, pp. 1105–1118.

4. D. E. Van Sickle, *J. Org. Chem.*, **34**, 3446 (1969).
5. C. C. Price, R. W. Kell and E. Krebs, *J. Amer. Chem. Soc.*, **64**, 1103 (1942).
6. C. C. Price and E. Krebs, *Org. Synth., Collect. Vol. III*, 649 (1955).
7. C. E. H. Bawn and S. F. Mellish, *Trans. Faraday Soc.*, **47**, 1216 (1951).
8. J. R. Slagle and H. J. Shine, *J. Org. Chem.*, **24**, 107 (1959).
9. C. G. Swain, W. T. Stockmayer and J. T. Clarke, *J. Amer. Chem. Soc.*, **72**, 5426 (1950).
10. R. D. Schuetz, F. M. Gruen, D. R. Byrne and R. L. Brennan, *J. Heterocycl. Chem.*, **3**, 184 (1966).
11. C. C. Price and H. Moritz, *J. Amer. Chem. Soc.*, **75**, 3686 (1953).
12. G. A. Russell, *J. Amer. Chem. Soc.*, **77**, 4814 (1955).
13. H. Kleinfeller and K. Rastädter, *Angew. Chem.*, **65**, 543 (1953).
14. (a) F. D. Greene, *J. Amer. Chem. Soc.*, **78**, 2246 (1956).
    (b) F. D. Greene, *J. Amer. Chem. Soc.*, **78**, 2250 (1956).
    (c) F. D. Greene, *J. Amer. Chem. Soc.*, **81**, 1503 (1959).
15. J. Lichtenberger and F. Weiss, *Bull. Soc. Chim. Fr.*, 915 (1962).
16. H. C. Haas, N. W. Schuler and H. S. Kolesinski, *J. Polymer Sci., Part A-1*, **5**, 2964 (1967).
17. W. Cooper, *J. Chem. Soc.*, 3106 (1951).
18. (a) L. S. Silbert and D. Swern, *J. Amer. Chem. Soc.*, **81**, 2364 (1959).
    (b) L. S. Silbert and D. Swern, *J. Org. Chem.*, **27**, 1336 (1962).
19. A. G. Davies, *Organic Peroxides*, Butterworths, London, 1961, pp. 63–71.
20. E. G. E. Hawkins in *Organic Peroxides*, Van Nostrand, Princeton, N.J., 1961, pp. 300–329.
21. A. V. Tobolsky and R. B. Mesrobian, *Organic Peroxides*, Interscience, New York, 1954.
22. P. R. H. Speakman, *Chem. Ind.* (*London*), 579 (1978).
23. H. Wieland, T. Ploetz and K. Indest, *Justus Liebigs Ann. Chem.*, **532**, 179 (1937).
24. H. Wieland and G. A. Razuvaev, *Justus Liebigs Ann. Chem.*, **480**, 157 (1930).
25. A. Baeyer and V. Villiger, *Chem. Ber.*, **33**, 1581 (1900).
26. Yu. A. Ol'dekop, A. N. Sevchenko, I. P. Zvat'kov, G. S. Bylina and A. P. El'nitskii, *J. Gen. Chem. USSR* (*Engl. Transl.*), **31**, 2706 (1961).
27. Yu. A. Ol'dekop, G. S. Bylina, L. K. Grakovich, Zh. J. Buloichik and Zh. D. Teif, *J. Org. Chem. USSR* (*Engl. Transl.*), **1**, 80 (1965).
28. (a) Yu. A. Ol'dekop and A. P. El'nitskii, *Zh. Org. Khim.*, **1**, 876 (1965).
    (b) Yu. A. Ol'dekop, G. S. Bylina, L. K. Grakovich, Zh. I. Buloichik and Zh. D. Teif, *Zh. Org. Khim.*, **1**, 82 (1965).
    (c) Yu. A. Ol'dekop and A. P. El'nitskii, *Zh. Org. Khim.*, **2**, 1257 (1966).
    (d) Yu. A. Ol'dekop, G. S. Bylina, I. K. Burykina and G. S. Kislyak, *Zh. Org. Khim.*, **2**, 2175 (1966).
    (e) Yu. A. Ol'dekop, G. S. Bylina and S. F. Petrashkevich, *Akad. Nauk SSSR, Otd. Obshch. Tekh. Khim.*, 152 (1967); *Chem. Abstr.*, **68**, 39292 (1968).
    (f) Yu. A. Ol'dekop, G. S. Bylina and L. K. Burykina, *Izv. Akad. Nauk Belorussk. SSR, Ser. Khim. Nauk*, 118 (1968); *Chem. Abstr.*, **70**, 96350 (1969).
29. (a) Yu. A. Ol'dekop, A. P. El'nitskii, S. F. Petrashkevich and A. A. Karaban, *Izv. Akad. Nauk Belorussk SSR, Ser. Khim. Nauk*, 80 (1968); *Chem. Abstr.*, **70**, 11057 (1969).
    (b) Yu. A. Ol'dekop, A. P. El'nitskii and S. I. Budai, *Izv. Akad. Nauk Belorussk SSR, Ser. Khim. Nauk SSSR*, **128**, 1201 (1959); *Chem. Abstr.*, **54**, 7630 (1960).
30. (a) Yu. A. Ol'dekop, A. N. Sevchenko, I. P. Zyat'kov, G. S. Bylina and A. P. El'nitskii, *Dokl. Akad. Nauk SSSR*, **128**, 1201 (1959); *Chem. Abstr.*, **54**, 7630 (1960).
    (b) Yu. A. Ol'dekop, A. N. Sevchenko, I. P. Zyat'kov, G. S. Bylina and A. P. El'nitskii, *Zh. Obshch. Khim.*, **31**, 2904 (1961); *Chem. Abstr.*, **56**, 15400 (1962).
    (c) Yu. A. Ol'dekop, A. N. Sevchenko, I. P. Zyat'kov and A. P. El'nitskii, *Zh. Obshch. Khim.*, **33**, 2771 (1963); *Chem. Abstr.*, **60**, 15726 (1964).
    (d) Yu. A. Ol'dekop and R. F. Sokolova, *Zh. Obshch. Khim.*, **23**, 1159 (1953); *Chem. Abstr.*, **47**, 12226 (1953).
31. A. Baeyer and V. Villiger, *Chem. Ber.*, **33**, 1969 (1900).
32. H. R. Appel, *U.S. Patent*, No. 3,397,245 (1968); *Chem. Abstr.*, **69**, 76954 (1969).
33. N. J. Bunce and D. D. Tanner, *J. Amer. Chem. Soc.*, **91**, 6069 (1969).
34. Imperial Chemical Industries, Ltd., *French Patent*, No. 1,337,986 (1963); *Chem. Abstr.*, **60**, 2787 (1964).

35. J. R. Hurst and G. B. Schuster, *J. Org. Chem.*, **45**, 1053 (1980).
36. K. E. Russell, *J. Amer. Chem. Soc.*, **77**, 4814 (1955).
37. A. H. Alberts, H. Wynberg and J. Strating, *Synth. Commun.*, **3**, 237 (1973).
38. G. Bouillon and K. Schank, *Chem. Ber.*, **112**, 2332 (1979).
39. K. Schank, R. Blattner, V. Schmidt and H. Hasenfratz, *Chem. Ber.*, **114**, 1938 (1981).
40. K. Schank, R. Blattner and G. Bouillon, *Chem. Ber.*, **114**, 1951 (1981).
41. G. Bouillon and K. Schank, *Chem. Ber.*, **113**, 2630 (1980).
42. M. Adler, K. Schank and V. Schmidt, *Chem. Ber.*, **112**, 2324 (1979).
43. M. Adler, K. Schank and V. Schmidt, *Chem. Ber.*, **112**, 2314 (1979).
44. F. Strain, W. E. Bissinger, W. R. Dial, H. Rudolph, B. J. DeWitt, H. C. Stevens and J. H. Langston, *J. Amer. Chem. Soc.*, **72**, 1254 (1950).
45. C. Walling, *Free Radicals in Solution*, Wiley, New York, 1958.
46. F. G. Edwards and F. R. Mayo, *J. Amer. Chem. Soc.*, **72**, 1265 (1950).
47. K. Schank and G. Bouillon, unpublished results, 1980; m.p. 148°C (dec.).
48. J. C. Martin and E. H. Drew, *J. Amer. Chem. Soc.*, **83**, 1232 (1961).
49. J. C. Martin and E. Hargis, *J. Amer. Chem. Soc.*, **91**, 5399 (1969).
50. G. Braun, *Org. Synth., Collect. Vol. I*, 431 (1941).
51. H. von Pechmann and L. Vanino, *Chem. Ber.*, **27**, 1510 (1894).
52. L. S. Silbert and D. Swern, *J. Org. Chem.*, **27**, 1336 (1962).
53. S.-O. Lawesson, P. G. Jönsson and J. Taipale, *Arkiv. Kemi*, **17**, 441 (1961).
54. S.-O. Lawesson and E. H. Larson, *Org. Synth.*, **45**, 37 (1965).
55. S.-O. Lawesson and C. Frisell, *Arkiv. Kemi*, **17**, 409 (1961).
56. S.-O. Lawesson, M. Andersson and C. Berglund, *Arkiv. Kemi*, **17**, 429 (1961).
57. S.-O. Lawesson, C. Frisell, D. Z. Denney and D. B. Denney, *Tetrahedron*, **19**, 1229 (1963).
58. H. De Pooter and N. Schamp, *Bull. Soc. Chim. Belg.*, **77**, 377 (1968).
59. R. L. Augustine, *J. Org. Chem.*, **28**, 581 (1963).
60. S.-O. Lawesson, H. J. Jakobsen and E. H. Larsen, *Acta Chem. Scand.*, **17**, 1188 (1963).
61. H. J. Jakobsen, E. H. Larsen, P. Madsen and S.-O. Lawesson, *Arkiv Kemi*, **24**, 519 (1965).
62. K. Schank and M. Adler, *Chem. Ber.*, **114**, 2019 (1981).
63. P. Kovacic, C. G. Reid and M. J. Brittain, *J. Org. Chem.*, **35**, 2152 (1970).
64. T. Nishio, M. Yuyama and Y. Omote, *Chem. Ind.* (*London*), 480 (1975).
65. M. Aiura and Y. Kanaoka, *Chem. Pharm. Bull.*, **23**, 2835 (1975).
66. F. D. Green, W. Adam and J. E. Cantrill, *J. Amer. Chem. Soc.*, **83**, 3461 (1961).
67. F. D. Green and W. W. Rees, *J. Amer. Chem. Soc.*, **82**, 890 (1960).
68. P. R. Ortiz de Montellano and C. K. Hsu, *Tetrahedron Letters*, 4215 (1976).
69. J. J. Batten and M. F. R. Mulcahy, *J. Chem. Soc.*, 2949 (1956).
70. C. Walling and R. B. Hodgdon, *J. Amer. Chem. Soc.*, **80**, 228 (1958).
71. J. S. Chauhan and K. B. L. Mathur, *Indian J. Chem.* (*B*), **15**, 51 (1977).
72. V. P. Bhatia and K. B. L. Mathur, *Tetrahedron Letters*, 4057 (1966).
73. K. N. Sawhney and K. B. L. Mathur, *Indian J. Chem.* (*B*), **17**, 511 (1979).
74. M. P. Dryle, W. J. Patrie and S. B. Williams, *J. Org. Chem.*, **44**, 2955 (1979).
75. R. Huisgen and W. Kolbeck, *Tetrahedron Letters*, 783 (1965).
76. S. Gambarjan, *Chem. Ber.*, **42**, 4003 (1909).
77. J. T. Edward, *J. Chem. Soc.*, 1464 (1954).
78. R. B. Roy and G. A. Swan, *J. Chem. Soc.* (*C*), 80 (1968).
79. O. C. Musgrave, *Chem. Rev.*, **69**, 499 (1969).
80. W. E. Cass, *J. Amer. Chem. Soc.*, **69**, 500 (1947).
81. L. Horner and E. Jürgens, *Justus Liebigs Ann. Chem.* **602**, 135 (1957).
82. L. Horner and B. Anders, *Chem. Ber.*, **95**, 2470 (1962).
83. D. I. Davies, D. H. Hey and B. Summers, *J. Chem. Soc.* (*C*), 2653 (1970).
84. D. Swern in *Organic Peroxides* (Ed. D. Swern), Vol. I, Wiley–Interscience, New York–London, 1970, pp. 313–516.
85. J. Y. Nedelec, J. Sorba and D. Lefort, *Synthesis*, 821 (1976).
86. J. d'Ans, J. Mattner and W. Busse, *Angew. Chem.*, **65**, 7 (1953).
87. R. Yoshizawa and T. Inukai, *Bull. Chem. Soc. Japan*, **42**, 3238 (1969).
88. R. M. Coates and J. W. Williams, *J. Org. Chem.*, **39**, 3054 (1974).

89. R. D. Bach, W. M. Klein, R. A. Ryntz and J. W. Holubka, *J. Org. Chem.*, **44**, 2569 (1979).
90. N. Kawabe, K. Okada and M. Ohmo, *J. Org. Chem.*, **37**, 4210 (1972).
91. J. d'Ans and W. Frey, *Chem. Ber.*, **45**, 1845 (1912).
92. U. Folli and D. Jarossi, *Bull. Sci. Fac. Chim. Ind. Bologna*, **20**, 61 (1958).
93. D. A. Konen and L. S. Silbert, *J. Org. Chem.*, **36**, 2162 (1971).
94. D. Belitskus and G. A. Jeffrey, *Acta Cryst.*, **18**, 458 (1965).
95. L. M. Hjelmeland and G. Loew, *Tetrahedron*, **33**, 1029 (1977).
96. B. S. Neporent, T. E. Pavlovskaya, N. M. Emanuel and N. G. Jaroslavskii, *Dokl. Akad. Nauk SSSR*, **70**, 1025 (1950).
97. P. A. Gigere and A. W. Olmos, *Can. J. Chem.*, **30**, 821 (1952).
98. D. Swern and L. P. Witnauer, *J. Amer. Chem. Soc.*, **77**, 5557 (1955).
99. R. Kavic and B. Plesnicar, *J. Org. Chem.*, **35**, 2033 (1970).
100. R. Curci, R. A. Diprete, J. Edwards and G. Modena, *J. Org. Chem.*, **35**, 740 (1970).
101. D. Swern in *Organic Peroxides* (Ed. D. Swern), Vol. II, Wiley–Interscience, New York–London 1971, pp. 355–534.
102. H. O. House in *Modern Synthetic Reactions* (Ed. H. O. House), Benjamin, New York, 1972, pp. 293–337.
103. H. Krauch and W. Kunz, *Reaktionen der Organischen Chemie*, 5th ed. Hüthig, Heidelberg, 1976 (references therein till 1972), p. 560.
104. E. N. Prilezhaev, *Prilezhaev Reaction: Electrophilic Oxidation*, Nauka, Moscow, 1974.
105. V. G. Dryuk, *Tetrahedron*, **32**, 2855 (1976).
106. N. N. Schwartz and J. H. Blumbergs, *J. Org. Chem.*, **29**, 1976 (1964).
107. M. M. Khalil and W. Pritzkow, *J. Prakt. Chem.*, **315**, 58 (1973).
108. B. Füllbier, M. Hampel and W. Pritzkow, *J. Prakt. Chem.*, **319**, 693 (1977).
109. P. D. Bartlett, *Rec. Chem. Progr.*, **11**, 47 (1950).
110. B. M. Lynch and K. H. Pausacker, *J. Chem. Soc.*, 1525 (1955).
111. H. Kwart and D. M. Hoffman, *J. Org. Chem.*, **31**, 419 (1966).
112. A. Azman, B. Borstnik and B. Plesnicar, *J. Org. Chem.*, **34**, 971 (1969).
113. K. D. Bingham, G. D. Meakins and G. H. Witham, *Chem. Commun.*, 445 (1966).
114. H.-J. Schneider, N. Becker and K. Philippi, *Chem. Ber.*, **114**, 1562 (1981).
115. D. Schnurpfeil, *Z. Chem.*, **20**, 445 (1980).
116. D. J. Pasto and C. C. Cumbo, *J. Org. Chem.*, **30**, 1271 (1965).
117. (a) V. Prelog and V. Boarland, *Helv. Chim. Acta*, **38**, 1776 (1955).
(b) V. Prelog, K. Schenker and W. Kung, *Helv. Chim. Acta*, **36**, 471 (1953).
(c) V. Prelog and M. Speck, *Helv. Chim. Acta*, **38**, 1786 (1955).
118. (a) A. C. Cope, S. W. Fenton and C. F. Spencer, *J. Amer. Chem. Soc.*, **74**, 5884 (1952).
(b) A. C. Cope, A. Fournier and H. E. Simmons, *J. Amer. Chem. Soc.*, **79**, 3905 (1957).
(c) A. C. Cope, A. H. Keaugh, P. E. Peterson, H. E. Simmons and G. W. Wood, *J. Amer. Chem. Soc.*, **79**, 3900 (1957).
119. B. Plesnicar in *Oxidation in Organic Chemistry* (Ed. W. S. Trahanovsky), Vol. 5C, Academic Press, New York, 1978, p. 211–294.
120. H. B. Henbest and R. A. L. Wilson, *J. Chem. Soc.*, 1958 (1957).
121. W. C. Smit, *Rec. Trav. Chim.* **49**, 686 (1930).
122. D. Swern, *J. Amer. Chem. Soc.*, **69**, 1692 (1947).
123. J. Stuurman, *Kon. Akad. Wetensch. Amsterdam, Proc.*, **38**, 450 (1935); *Chem. Abstr.*, **29**, 4657 (1935).
124. V. Prelog, K. Schenker and H. H. Günthard, *Helv. Chim. Acta*, **35**, 1598 (1952).
125. J. Boeseken and J. S. P. Blumberger, *Rec. Trav. Chim. Pays-Bas*, **44**, 90 (1925).
126. W. Stumpf and K. Rombusch, *Justus Liebigs Ann. Chem.*, **687**, 136 (1965).
127. J. G. Traynham, G. R. Franzen, G. A. Knesel and D. J. Northington, *J. Org. Chem.*, **32**, 3285 (1967).
128. W. Weigert, A. Kleemann and H. Offermanns in *Methodicum Chimicum* (Ed. F. Korte), Vol. 5, Georg Thieme Verlag, Stuttgart, 1975, pp. 164–180.
129. (a) S. N. Lewis in *Oxidation* (Ed. R. L. Augustine), Vol. I, Dekker, New York, 1969, p. 223.
(b) J. G. Wallace, 'Epoxidation' in *Kirk-Othmer: Encyclopedia of Chemical Technology*, 2nd ed., Vol. III, Interscience, New York, 1965, p. 238.

130. R. Wegler and R. Schmitt-Josten in *Houben-Weyl: Methoden der Organischen Chemie* (Ed. E. Müller), 4th ed., Vol. XIV/2, Georg Thieme Verlag, Stuttgart, 1963, p. 481.
131. O. Weiberg in *Ullmanns Encyclopädie der technischen Chemie, Ergänzungsband*, Urban und Schwarzenberg, München, 1970, p. 181.
132. D. Swern, *Encyclop. Polym. Sci. Technol*, **6**, 83 (1967).
133. Rohm and Haas, *US Patent*, No. 2485100 (1949); *Chem. Abstr.*, **44**, 7346 (1950).
134. Rohm and Haas, *DBP*, No. 1042565 (1955); *Chem. Abstr.*, **54**, 12620 (1960).
135. W. Dittmann and K. Hamann, *Chemiker Ztg.*, **95**, 857 (1971).
136. D. Swern, T. W. Findley and J. T. Scanlan, *J. Amer. Chem. Soc.*, **66**, 1925 (1944).
137. R. D. Bach, M. W. Klein, R. A. Ryntz and J. W. Holubka, *J. Org. Chem.*, **44**, 2569 (1979).
138. N. C. Jang, W. Chiang, D. Leonov, I. Bilyk and B. Kim, *J. Org. Chem.*, **43**, 3425 (1978).
139. M. Imuta and H. Ziffer, *J. Org. Chem.*, **44**, 1351 (1979).
140. K. Schank, J.-H. Felzmann and M. Kratzsch, *Chem. Ber.*, **102**, 388 (1969).
141. J. Kishi, M. Aratani, H. Tanino, T. Fukuyama and T. Goto, *J. Chem. Soc., Chem. Commun.*, 64 (1972).
142. (a) G. B. Payne, P. H. Denning and P. H. Williams, *J. Org. Chem.*, **26**, 659 (1961).
(b) G. B. Payne, *Tetrahedron*, **18**, 763 (1962).
143. K. Schank and D. Wessling, *Justus Liebigs Ann. Chem.*, **710**, 137 (1967).
144. N. L. Boulch, J. Raoul and G. Ourisson, *Bull. Soc. Chim. Fr.*, 646 (1965).
145. R. G. Carlson, N. S. Behn and C. Cowles, *J. Org. Chem.*, **36**, 3832 (1971).
146. E. Höft and S. Ganschow, *J. Prakt. Chem.*, **314**, 145 (1972).
147. N. Matsumura, N. Sonoda and S. Tsutsumi, *Tetrahedron Letters*, 2029 (1970).
148. J. Rebek, Jr., S. F. Wolf and A. B. Mossman, *J. Chem. Soc., Chem. Commun.*, 711 (1974).
149. (a) V. Frantzen, *Chem. Ber.*, **88**, 717 (1955).
(b) R. N. McDonald and P. A. Schwab, *J. Amer. Chem. Soc.*, **86**, 4866 (1964).
(c) J. K. Stille and D. D. Whitehurst, *J. Amer. Chem. Soc.*, **86**, 4871 (1964).
150. J. Ciabattoni, R. A. Campbell, C. A. Renner and P. W. Concannon, *J. Amer. Chem. Soc.*, **92**, 3826 (1970).
151. K.-P. Zeller, *Chem. Ber.*, **112**, 678 (1979).
152. J. K. Crandall, W. H. Machleder and S. A. Sojka, *J. Org. Chem.*, **38**, 1149 (1973).
153. J. K. Crandall, W. W. Conover, J. B. Komin and W. H. Machleder, *J. Org. Chem.*, **39**, 1723 (1974).
154. A. Rotman and Y. Mazur, *J. Chem. Soc., Chem. Commun.*, 15 (1974).
155. U. Frommer and V. Ullrich, *Z. Naturforsch.*, **26B**, 322 (1971).
156. W. Müller, *Diplomarbeit*, Universität Saarbrücken, 1979.
157. W. Müller and H.-J. Schneider, *Angew Chem.*, **91**, 438 (1979).
158. (a) E. G. Rozantzer and M. B. Neiman, *Tetrahedron*, **20**, 131 (1964).
(b) B. Ganem, *J. Org. Chem.*, **40**, 1998 (1975).
159. (a) J. A. Cella and J. P. McGrath, *Tetrahedron Letters*, 4115 (1975).
(b) J. A. Cella, J. A. Kelley and E. F. Kenehan, *J. Org. Chem.*, **40**, 1860 (1975).
160. A. Baeyer and V. Villiger, *Chem. Ber.*, **32**, 3625 (1899).
161. P. A. S. Smith in *Molecular Rearrangements* (Ed. P. de Mayo), Part I, Interscience, New York, 1963, p. 457.
162. C. H. Hassall, *Org. Reactions*, **9**, 73 (1957).
163. S. L. Friess, *J. Amer. Chem. Soc.*, **71**, 2571 (1949).
164. S. L. Friess and P. E. Frankenburg, *J. Amer. Chem. Soc.*, **74**, 2679 (1952).
165. L. Ruzicka and M. Stoll, *Helv. Chim. Acta*, **11**, 1159 (1928).
166. (a) Union Carbide Corporation, *DAS*, No. 1086686 (1955).
(b) Degussa, *DAS*, No. 1258858 (1965).
167. (a) Société d' Electro-Chimie, d'Electro-Métallurgie et des Acieries Electriques d'Ugine, *DAS*, No. 1216283 (1963); *Chem. Abstr.*, **62**, 1571 (1965).
(b) F. Weiss, *Chim. Ind. Genie Chim.*, **103**, 1083 (1970).
168. E. G. E. Hawkins, *J. Chem. Soc. (C)*, 2691 (1969).
169. J. Royer and M. Beugelmans-Verrier, *Compt. Rend., Ser. C*, **272**, 1818 (1971).
170. J. Royer and M. Beugelmans-Verrier, *Compt. Rend., Ser. C*, **279**, 1049 (1974).
171. J. S. Showell, J. R. Russell and D. Swern, *J. Org. Chem.*, **27**, 2853 (1962).
172. F. Montanari and M. Cinquini, *Mech. React. Sulfur Comp.*, **3**, 121 (1968).

173. (a) A. Cerniani and G. Modena, *Gazz. Chim.*, **89**, 843 (1959).
(b) A. Greco, G. Modena and P. E. Todesco, *Gazz. Chim.*, **90**, 671 (1960).
174. G. Kresze, W. Schramm and G. Cerc, *Chem. Ber.*, **94**, 2060 (1961).
175. C. G. Venier, T. G. Squires, J. Shei, Y. Y. Chen and G. Hussman, *Abstr.* **263**, ACS Las Vegas Meeting, 1980.
176. R. Curci, A. Giovine and G. Modena, *Tetrahedron*, **22**, 1235 (1966).
177. C. R. Harrison and P. Hodge, *J. Chem. Soc., Perkin Trans. 1*, 2252 (1976).
178. G. A. Russell and L. A. Ochrymowycz, *J. Org. Chem.*, **35**, 2106 (1970).
179. R. Liotta and W. S. Hoff, *J. Org. Chem.*, **45**, 2887 (1980).
180. (a) B. Zwanenburg and J. Strating, *Quart. Rep. Sulfur Chem.* **5**, 79 (1970).
(b) B. Zwanenburg, *Rec, Trav. Chim.*, **101**, 1 (1982).
181. (a) A. Battaglia, A. Dondoni, P. Giorgianni, G. Maccagnani and G. Mazzanti, *J. Chem. Soc. (B)*, 1547 (1971).
(b) A. Battaglia, A. Dondoni. G. Maccagnani and G. Mazzanti, *J. Chem. Soc., Perkin Trans. 2*, 609 (1974).
182. (a) J. Meisenheimer and K. Bratring, *Justus Liebigs Ann. Chem.*, **397**, 286 (1913).
(b) D. Jerchel and G. Jung, *Chem. Ber.*, **85**, 1130 (1952).
(c) M. Izumi, *Pharm. Bull. (Tokyo)*, **2**, 279 (1954).
183. J. C. Craig and K. K. Purushothaman, *J. Org. Chem.*, **35**, 1721 (1970).
184. H. S. Mosher, L. Turner and A. Carlsmith, *Org. Synth., Collect. Vol. IV*, 828 (1963).
185. R. F. Evans, M. van Ammers and H. J. den Hertog, *Rec. Trav. Chim.*, **78**, 408 (1959).
186. (a) W. D. Emmons, *J. Amer. Chem. Soc.*, **79**, 5528 (1957).
(b) R. W. White and W. D. Emmons, *Tetrahedron* **17**, 31 (1962).
187. W. D. Emmons and A. S. Pagano, *J. Amer. Chem. Soc.*, **77**, 4557 (1955).
188. W. D. Emmons, *J. Amer. Chem. Soc.*, **79**, 5739 (1957).
189. J. K. Crandall and L. C. Crawley, *J. Org. Chem.*, **39**, 489 (1974).
190. Glaxo, DOS No. 2 335 107 (1974).
191. R. Curci, F. DiFuria and F. Marcuzzi, *J. Org. Chem.*, **36**, 3774 (1971).
192. R. Curci, F. DiFuria, J. Ciabattoni and P. W. Concannon, *J. Org. Chem.*, **39**, 3295 (1974).
193. (a) C. A. Buehler and H. Hart, *J. Amer. Chem. Soc.*, **85**, 2177 (1963).
(b) M. E. Kurz and G. J. Johnson, *J. Org. Chem.*, **36**, 3184 (1971).
(c) For mechanistic aspects of related systems see: C. Walling, D. M. Camaioni and S. S. Kim, *J. Amer. Chem. Soc.*, **100**, 4814 (1978).
194. (a) A. Wacek and R. Fiedler, *Monatsh. Chem.*, **80**, 170 (1949).
(b) G. A. Page and D. S. Tarbell, *Org. Synth.*, **34**, 8 (1954).
195. N. C. Deno, B. A. Greigger, L. A. Messer, M. D. Meyer and S. G. Stroud, *Tetrahedron Letters*, 1703 (1977).
196. J. G. Sharefkin and H. Salzman, *Org. Synth.*, **43**, 65 (1963).
197. H. J. Reich and S. L. Peake, *J. Amer. Chem. Soc.*, **100**, 4888 (1978).
198. R. Criegee, *Chem. Ber.*, **77**, 22 (1944).
199. (a) N. A. Milas and D. M. Surgenov, *J. Amer. Chem. Soc.*, **68**, 642 (1946).
(b) N. A. Milas and D. G. Orphanos and R. J. Klein, *J. Org. Chem.*, **29**, 3099 (1964).
200. P. D. Bartlett and R. R. Hiatt, *J. Amer. Chem. Soc.*, **80**, 1398 (1958).
201. L. S. Silbert and D. Swern, *J. Amer. Chem. Soc.*, **81**, 1364 (1959).
202. J. P. Lorand and P. D. Bartlett, *J. Amer. Chem. Soc.*, **88**, 3294 (1966).
203. R. Hecht and C. Rüchardt, *Chem. Ber.*, **96**, 1281 (1963).
204. H. A. Staab, W. Rohr and F. Graf, *Chem. Ber.*, **98**, 1122 (1965).
205. O. L. Mageli and C. S. Sheppard in *Organic Peroxides* (Ed. D. Swern), Vol. I, Wiley–Interscience, New York–London, 1970, pp. 1–92.
206. R. Naylor, *J. Chem. Soc.*, 244 (1945).
207. D. Harman, *US Patent*, No. 2,608,570 (1952), *Chem. Abstr.*, **48**, 3387 (1954).
208. J. Pederson, *J. Org. Chem.*, **23**, 252, 255 (1958).
209. (a) E. L. O'Brien, F. M. Beringer and R. B. Mesrobian, *J. Amer. Chem. Soc.*, **79**, 6238 (1957).
(b) E. L. O'Brien, F. M. Beringer and R. B. Mesrobian, *J. Amer. Chem. Soc.*, **81**, 1506 (1959).
210. C. Rüchardt, *Fortschr. Chem. Forsch.*, **6**, 251 (1966).
211. A. Rieche and F. Hitz, *Chem. Ber.*, **63**, 2504 (1930).

212. A. G. Davies, *Organic Peroxides*, Butterworths, London, 1961, p. 58.
213. R. C. Schnur, *Chem. Engng. News*, **59**, 19 (1981).
214. G. Sosnovsky and S.-O. Lawesson, *Angew Chem.*, **76**, 218 (1964).
215. S.-O. Lawesson and N. C. Yang, *J. Amer. Chem. Soc.*, **81**, 4320 (1959).
216. F. Marcuzzi and G. Melloni, *Synthesis*, 451 (1976).
217. M. A. Greenbaum, D. B. Denney and A. K. Hoffmann, *J. Amer. Chem. Soc.*, **78**, 2563 (1956).
218. M. S. Kharasch, G. Sosnovsky and N. C. Yang, *J. Amer. Chem. Soc.*, **81**, 5819 (1959).
219. D. B. Denney, D. Z. Denney and G. Feig, *Tetrahedron Letters*, 19 (1959).
220. A. L. J. Beckwith and G. W. Evans, *Proc. Chem. Soc.* (*London*), 63 (1962).
221. J. K. Kochi, *J. Amer. Chem. Soc.*, **84**, 774 (1962).
222. D. Z. Denney, A. Appelbaum and D. B. Denney, *J. Amer. Chem. Soc.*, **84**, 4969 (1962).
223. J. K. Kochi, *Tetrahedron*, **18**, 483 (1962).
224. H. Kropf, R. Schröder and R. Fölsing, *Synthesis*, 894 (1977).
225. G. Sosnovsky, *J. Org. Chem.*, **25**, 874 (1960).
226. S.-O. Lawesson, C. Berglund and S. Grönwall, *Acta Chem. Scand.*, **15**, 249 (1961).
227. G. Sosnovsky, *Tetrahedron*, **13**, 241 (1961).
228. S.-O. Lawesson and C. Berglund, *Arkiv Kemi*, **17**, 465 (1961).
229. G. Sosnovsky, *Tetrahedron*, **21**, 871 (1965).
230. S.-O. Lawesson and C. Berglund, *Acta Chem. Scand.*, **14**, 1854 (1960).
231. C. Berglund and S.-O. Lawesson, *Arkiv Kemi*, **20**, 225 (1963).
232. H.-D. Scharf and E. Wolters, *Angew. Chem.*, **88**, 718 (1976).
233. (a) J. J. Pappas, W. P. Keavenay, E. Gancher and M. Berger, *Tetrahedron Letters*, 4273 (1966).
(b) P. L. Stotter and J. B. Eppner, *Tetrahedron Letters*, 2417 (1973).
(c) C. Lick and K. Schank, *Chem. Ber.*, **111**, 2461 (1978).
234. G. Sosnovsky, *J. Org. Chem.*, **26**, 281 (1961).
235. G. Sosnovsky, *Tetrahedron*, **18**, 15 (1962).
236. S.-O. Lawesson and C. Berglund, *Acta Chem. Scand.*, **15**, 36 (1961).
237. W. A. Pryor and W. H. Hendrickson, Jr., *J. Amer. Chem. Soc.*, **97**, 1580 (1975).
238. T. Sugawara, H. Iwamura and M. Oki, *Bull. Chem. Soc. Japan*, **47**, 1496 (1974).
239. C. Berglund and S.-O. Lawesson, *Acta Chem. Scand.*, **16**, 773 (1962).
240. G. Sosnovsky and N. C. Yang, *J. Org. Chem.*, **25**, 899 (1960).

The Chemistry of Functional Groups, Peroxides
Edited by S. Patai

CHAPTER **11**

Endoperoxides

ISAO SAITO and S. SARMA NITTALA
Department of Synthetic Chemistry, Faculty of Engineering, Kyoto University, Kyoto, Japan

I. INTRODUCTION . . . 312
II. NATURALLY OCCURRING ENDOPEROXIDES . . . 312
III. SYNTHETIC METHODS FOR ENDOPEROXIDES . . . 314
A. Nucleophilic Displacement Reaction . . . 314
B. Singlet Oxygen Reaction of 1,3-Dienes . . . 317
1. Carbocyclic and acyclic dienes . . . 318
2. Aromatic compounds . . . 324
3. Heterocyclic compounds . . . 332
C. Oxidation with Triplet Oxygen . . . 338
IV. REACTIONS OF ENDOPEROXIDES . . . 342
A. Reduction . . . 342
1. Diimide . . . 343
2. Lithium aluminium hydride and thiourea . . . 343
3. Triphenylphosphine and triphenyl phosphite . . . 344
B. Thermolysis . . . 346
1. Release of molecular oxygen . . . 346
2. Cleavage of O—O bond and fragmentation . . . 348
C. Photolysis . . . 350
D. Base- and Acid-catalysed Reaction . . . 352
E. Metal-catalysed Reaction . . . 356
F. Reaction of Prostaglandin Endoperoxides and Their Model Compounds . . . 359
G. Endoperoxides in Natural-product Synthesis . . . 365
V. REFERENCES . . . 369

I. INTRODUCTION

Much of the recent interest in endoperoxide chemistry stems from the recognition that endoperoxides have an important and varied role in biological systems. The establishment of the intervention of prostaglandin endoperoxide in the metabolism of arachidonic acid has been one of the most important findings in the field of endoperoxide chemistry. The identification of endoperoxides as key intermediates or metabolites in biological systems has stimulated the development of new synthetic methodology for a variety of complex and labile endoperoxides.

In the area of synthetic organic chemistry, endoperoxides have served as important precursors for many types of oxygen-functionalized organic molecules. A number of synthetically useful intermediates including epoxides are derived from endoperoxides with stereoregulation.

In this chapter, various aspects of the synthesis and reaction of endoperoxides will be reviewed by selecting recent representative examples from among innumerable references up to late 1981. In the strict sense of the term 'endoperoxide', monocyclic peroxides should not be included in this category. However, in view of the structural similarity and biological importance inherent with monocyclic peroxides, we have not limited ourselves to bicyclic endoperoxides, and some of the important monocyclic peroxides are included in this chapter. We have also attempted to illustrate the varied use of endoperoxides in organic synthesis, mostly in the natural product field.

II. NATURALLY OCCURRING ENDOPEROXIDES

It is often difficult to establish that a particular endoperoxide is of natural origin, being formed within the organism by enzymatic oxygenation. Adventitious oxygenation can take place with molecular oxygen or its related species such as hydrogen peroxide, singlet oxygen and superoxide ion, during isolation of the compound. Also, the aerial parts of a plant containing large amounts of pigments may be involved in sensitized oxygenations, making it difficult to realize the distinction between such and enzymatic processes for the origin of the endoperoxide within the organism. This problem has been partly responsible for the tardy growth of chemistry in this area. Ergosterol endoperoxide (**1**) first prepared in 1928[1] was claimed as a natural product in 1947 from the mycelium *Aspergillus fumigatus* cultured in the dark[2]. Later, it was isolated from several fungi, the early reports being from *Trichophyton schonleini*[3], *Daedalea quercina*[4] and *Penicillium sclerotigenum*[5]. However, it was suggested that **1** could be an artifact obtained by photosensitized oxygenation of ergosterol since fungal extracts contain pigments[6]. More recently, its isolation has been reported from several fungi[7], lichens[8], plants[9] and marine sponges[10].

The first discovery of a naturally occurring endoperoxide was ascaridole (**2**) characterized in 1911[11], and isolated as a natural product in 1949 from chenopodium oil[12].

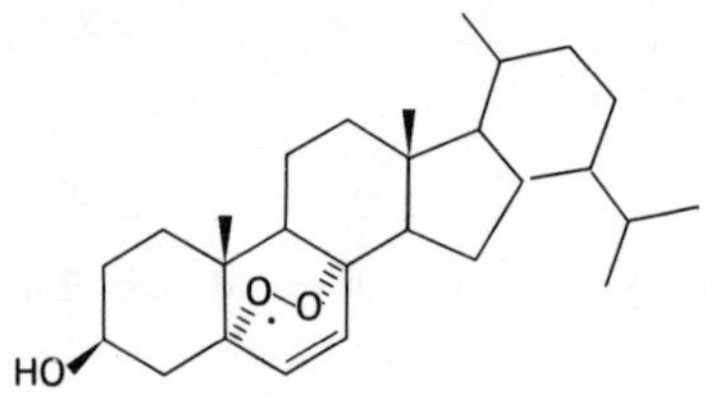

HO

(**1**)

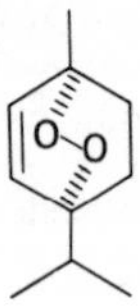

(**2**)

The formation of endoperoxides in the prostanoid biosynthesis was first suggested in 1965 by Samuelsson[13]. Accordingly in 1973, the unstable prostanoid endoperoxide PGH_2 (**3**) having 100–450 times more activity than PGE_2 (**4**) on the superfused aorta strip was isolated[14]. The incubation of arachidonic acid with a microsomal fraction of a homogenate of the vesicular gland of sheep produced PGH_2 and PGG_2 [15]. The former has since been recognized to serve as the biogenetic precursor to many potent prostaglandins, PGE_2 (**4**), $PGF_{2\alpha}$ (**5**), thrombaxane (**6**) and prostacycline (**7**)[16].

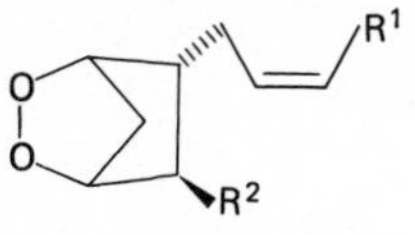

(**3**) 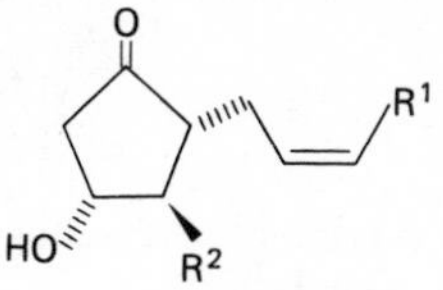 (**4**) 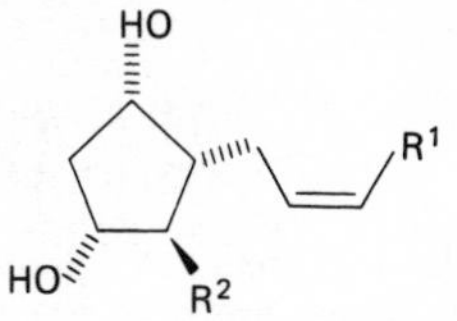 (**5**)

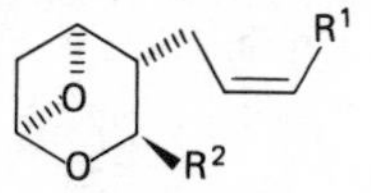

(**6**) 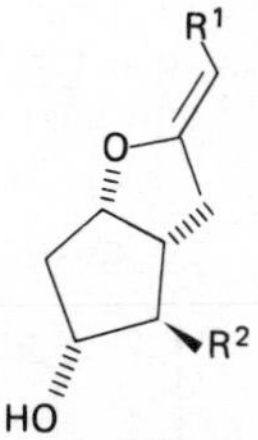 (**7**)

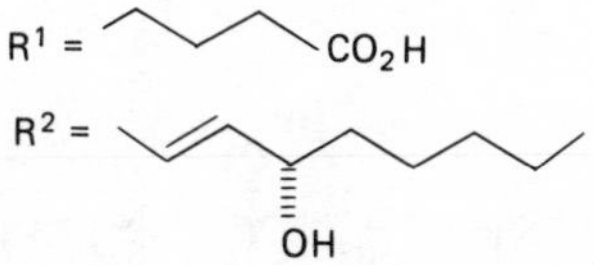

Apart from these bicyclic endoperoxides, several monocyclic endoperoxides have been isolated recently from marine organisms and terrestrial plants. A sponge of the genus *Chondrilla* is the source of the peroxyketal, chondrilline (**8**)[17], while the antimicrobial agent plakortin (**9**)[18] is isolated from *Plakortis halichondriodes*. The red seaweed *Chondria oppositinlada* gives the unique vinyl endoperoxide **10**[19]. In the leaves of the plant *Eucalyptus grandis* has been revealed the endoperoxide **11** which has the property of inhibiting root formation[20].

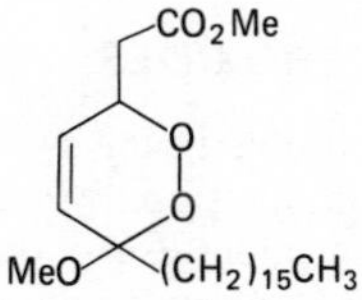

(**8**) 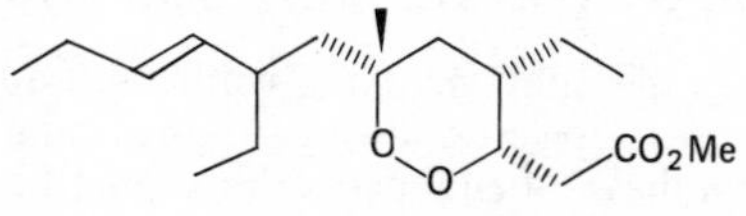 (**9**)

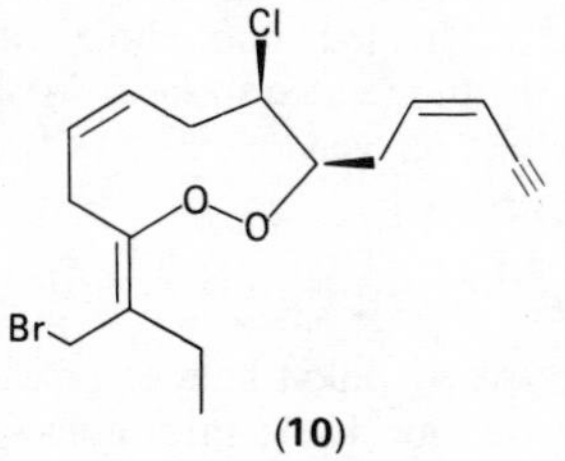

(**10**) 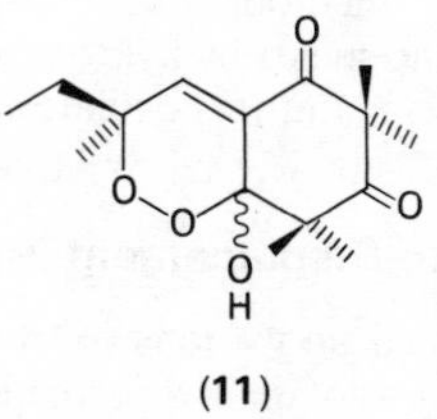 (**11**)

The antibacterial endoperoxides **12–15** of the marine sponge *Chondrosia collectrix* origin can be isolated only from a fresh extract. These are not found on storage of the ethanolic extract but products resulting from rearrangements have been isolated[21].

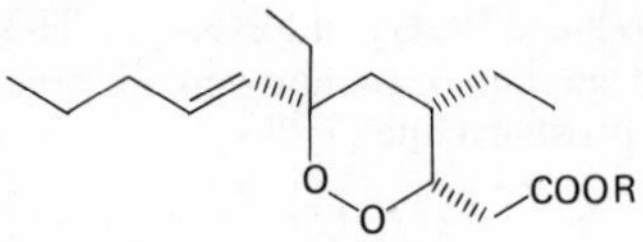

(**12**) R = Me
(**13**) R = H

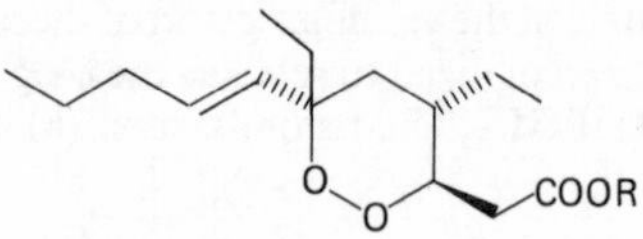

(**14**) R = Me
(**15**) R = H

Novel natural peroxides of plant origin are verruculogen (**16a**) and fumitremorgin A (**16b**) both of which are tremorgenic agents[22,23].

In addition to these naturally occurring peroxides, there has been known a class of natural products which are thought to be derived via metabolic transformations of the intermediate endoperoxides. Some of these possess structural units of either *cis* diepoxides or furans as typically seen in crotepoxide (**17**)[24] and the fungal antibiotic **18**[25].

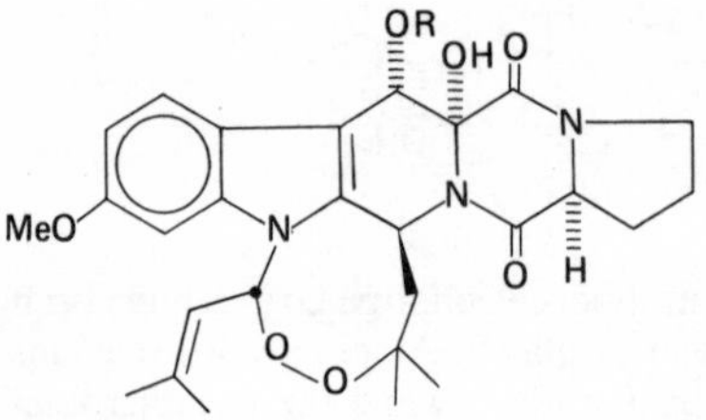

(**16**)
(**a**) R = H
(**b**) R =

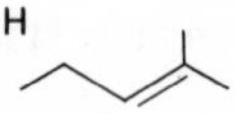

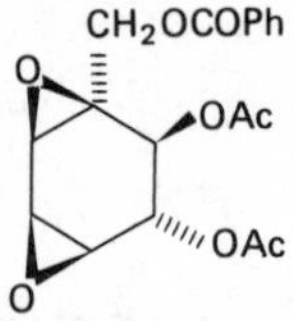

(**17**)

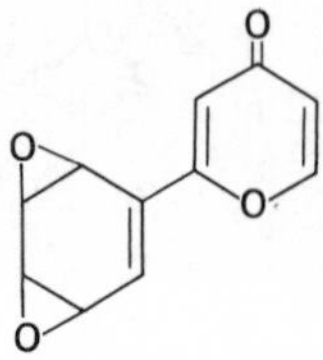

(**18**)

III. SYNTHETIC METHODS FOR ENDOPEROXIDES

The methodology of dialkyl peroxide synthesis is directly applicable to the synthesis of endoperoxides[26,27]. However, owing to their marked thermal instability and high reactivity, the synthesis of endoperoxides should be conducted under exceptional mild conditions. Strong acid or basic media and long reaction times at elevated temperatures may often cause extensive decomposition of the desired endoperoxides. In recent years, a number of significant advances have been made for the development of new methods for the synthesis of labile endoperoxides including prostaglandin endoperoxides and related compounds. The methods are mechanistically divided into three categories, (*i*) nucleophilic displacements by hydroperoxy groups and related species, (*ii*) singlet oxygen reaction of 1,3-diene and (*iii*) oxidation with triplet oxygen.

A. Nucleophilic Displacement Reaction

Silver-salt-assisted alkylations of hydroperoxides by alkyl halides provide the main route to the synthesis of dialkyl peroxides[28,29] (equation 1). An intramolecular variation

$$R^1OOH + R^2X \xrightarrow{AgOCOCF_3} R^1OOR^2 + AgX + CF_3CO_2H \quad (1)$$

of this reaction constitutes a promising method for the preparation of cyclic peroxides. 1,2-Dioxacyclopentane (**19**) is prepared from cyclopropane by the route indicated in equation (2)[30]. A prostaglandin endoperoxide model, 2,3-dioxabicyclo[2.2.1]heptane (**20**) is obtained from the silver trifluoroacetate reaction with *trans*-3-bromocyclopentyl hydroperoxide (**21**) which is prepared either from bicyclopentane (**22**) or *cis*-1,3-cyclopentandiol (**23**) (equation 3)[31]. The bicyclic peroxide **24** can also be synthesized by a similar route (equation 4)[32]. It should be noted here that the S_N2-type displacement proceeds with predominant inversion of configuration at alkyl halides. It is also advisable to carry out the reaction in the dark to avoid formation of metallic silver which can often catalyse peroxide decomposition. Silver nitrate, silver oxide and silver trifluoromethanesulphonate are also used in these reactions.

Ar, R — NBS/H_2O_2 → Ar, R, Br, OOH — Ag_2O → Ar, R, O—O (**19**) (2)

(**22**) — 1. H_2O_2 2. NBS → HOO, Br (**21**) ← 1. Ph_3PBr 2. H_2O_2–AgX — HO, HO (**23**) (3)

(**21**) — $AgOCOCF_3$, CH_2Cl_2 → O, O (**20**)

OOH — Br_2 → Br, Br, OOH (**24**) — $AgOCOCF_3$ → Br, O, O (4)

Reaction of alkyl trifluoromethanesulphonates (triflates) with hydroperoxy anions is known to provide dialkyl peroxides[33]. This method has been applied to the synthesis of **25**[34] and **20**[35] utilizing the combination of the triflate leaving group and the bis(tributylstannyl) peroxide nucleophile (equations 5 and 6).

TfO—$(CH_2)_n$—OTf — $Bu_3SnOOSnBu_3$ → $(CH_2)_n$, O—O (**25**) (5)

$$\text{TfO-cyclopentane-OTf} \xrightarrow{Bu_3SnOOSnBu_3} \textbf{(20)} \qquad (6)$$

Corey and coworkers have demonstrated the utility of crown-ether-complexed potassium superoxide for the synthesis of a 1,2-dioxacyclopentane **26** (equation 7)[36]. By the modified method the methyl ester of prostaglandin endoperoxide (PGH_2–Me) (**27**) is prepared, albeit in low yield (3%)[37]. Porter and collaborators have achieved the same conversion with a sevenfold increase in yield by using the silver trifluoroacetate and hydrogen peroxide reaction (equation 8)[38]. The same procedure has also been applied to the synthesis of the biologically important prostaglandin G_2 (PGG_2) (**28**) having both endoperoxide and hydroperoxy groups in the same molecule (equation 9)[39].

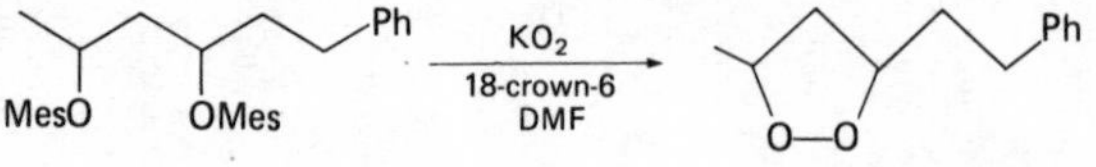

$$\text{MesO, OMes, Ph} \xrightarrow[\text{18-crown-6, DMF}]{KO_2} \textbf{(26)} \qquad (7)$$

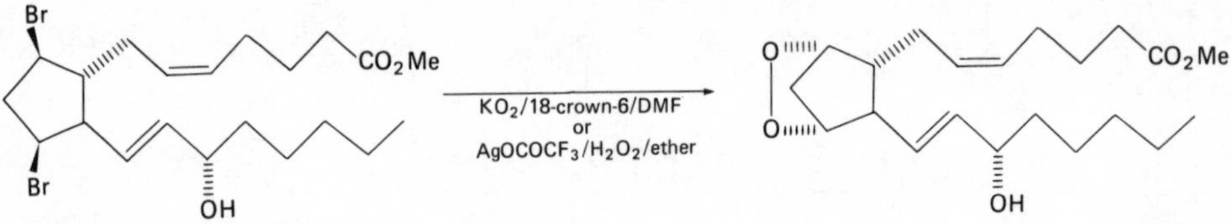

$$\text{Br, Br, } CO_2Me\text{, OH} \xrightarrow[AgOCOCF_3/H_2O_2/\text{ether}]{KO_2/\text{18-crown-6/DMF or}} \textbf{(27)} \qquad (8)$$

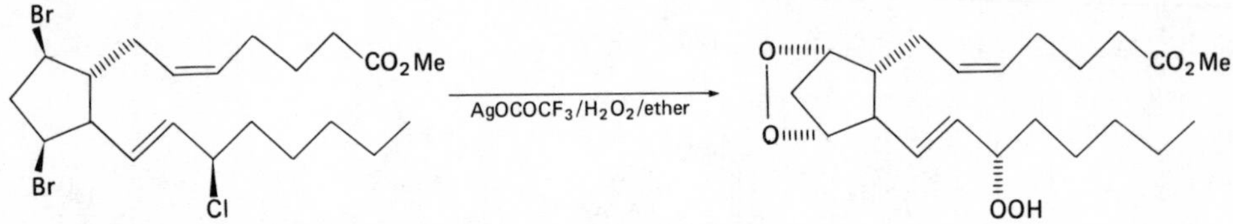

$$\text{Br, Br, } CO_2Me\text{, Cl} \xrightarrow{AgOCOCF_3/H_2O_2/\text{ether}} \textbf{(28)} \qquad (9)$$

Peroxymercuration where the electrophile is a mercury(II) salt has proven to be a versatile method for the preparation of dialkyl peroxides[40,41]. Demercurations occur rapidly under mild conditions with sodium borohydride or halogens as shown in equation (10). By using hydrogen peroxide and suitable dienes, mercury-free cyclic peroxides **29** are

$$R^1CH{=}CHR^2 + R^3OOH + HgX_2 \longrightarrow R^1CH(OOR^3)CH(HgX)R^2 + HX \qquad (10)$$

$$R^1CH(OOR^3)CH(HgX)R^2 \xrightarrow{Br_2} R^1CH(OOR^3)CHBrR^2$$

$$R^1CH(OOR^3)CH(HgX)R^2 \xrightarrow{NaBH_4} R^1CH(OOR^3)CH_2R^2$$

X = halgoen

obtained in moderate yield (equation 11)[41]. The synthesis of bicyclic endoperoxides **30** and **31** can be achieved by this peroxymercuration as given in equations (12) and (13) where each reaction is regiospecific[42,43]. By analogy with peroxymercuration, the reaction of epoxides with a hydroperoxy group under acidic conditions affords cyclic peroxide **32** in good yield (equation 14)[44].

$(CH_2)_n$ — H_2O_2, $2\,Hg(CF_3CO_2)_2$ → $NaBH_4$ or Br_2 → **(29)**, $n = 1, 2$; Y = H, Br (11)

$2\,Hg(CF_3CO_2)_2$ → $NaBH_4$ or Br_2 → **(30)** Y = H, Br (12)

H_2O_2, $2\,Hg(CF_2CO_2)_2$ → $NaBH_4$ or Br_2 → **(31)** Y = H, Br (13)

HOO — cat. CCl_3CO_2H, $0°C$ → **(32)** (OH) (14)

B. Singlet Oxygen Reaction of 1,3-Dienes

Singlet oxygen ($^1O_2$) is the first electronically excited state ($^1\Delta_g$) of molecular oxygen lying 22.4 kcal mol^{-1} above the ground-state triplet oxygen. Singlet oxygen reacts with various types of conjugated dienes and aromatic substrates by the Diels–Alder mode of addition. This stereoselective oxygenation of the terminal carbons of a 1,3-diene system has had widespread application in the synthesis of endoperoxides. As described later, the finding that diimide reduces the double bond of singlet–oxygen–diene adducts provides an excellent opportunity to prepare saturated endoperoxides as well.

There are a number of methods for generating singlet oxygen including reaction of hydrogen peroxide with sodium hypochlorite, thermolysis of triaryl phosphite ozonides or arene endoperoxides and dye-sensitized photooxygenation. The latter technique is the most efficient method and has been employed in the majority of singlet oxygen reactions. Use of an appropriate filter or a sodium lamp as a light source largely eliminates the problem of thermal and UV-induced decomposition of the peroxidic products. An important feature of the reaction is that it often proceeds satisfactorily at temperatures as low as −78°C thereby facilitating the preparation of unstable endoperoxides under mild conditions. The field of singlet oxygen chemistry has been extensively reviewed up to 1978[45] and its importance to organic peroxides is described in another chapter.

1. Carbocyclic and acyclic dienes

Singlet oxygen reacts with many acyclic and cyclic 1,3-dienes[45]. As will be discussed in Section IV, the resulting 1,4-endoperoxides provide an efficient route to a variety of 1,4-oxygenated systems as shown in equation (15). The addition of singlet oxygen occurs in a

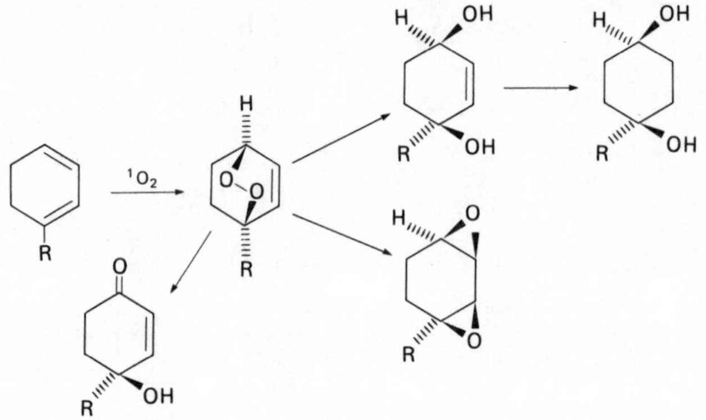

(15)

stereospecific fashion as typically exemplified in equation (16)[45a]. As shown in Table 1, a number of 1,3-butadiene derivatives give the corresponding 1,2-dioxenes **33** under photosensitized oxygenation[46,47]. Generally, diene systems having electron-donating substituents are more reactive toward singlet oxygen, whereas acyclic 1,3-dienes are usually less reactive than cyclic 1,3-dienes. Since the subject has already been reviewed[45], only some typical examples are listed in Table 1.

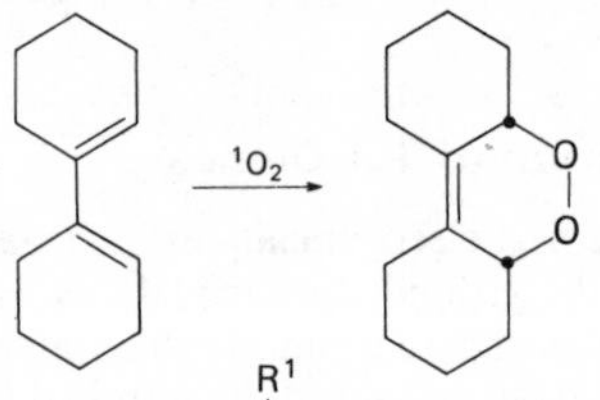

(16)

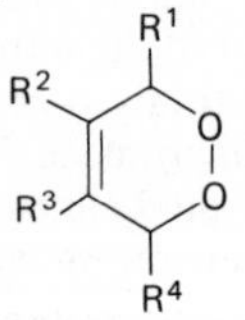

(**33**)

TABLE 1. 1,4-Cycloaddition of singlet oxygen with conjugated dienes[46]

Diene	Yield of endoperoxide (%)	Reference
Cyclopentadiene	86	48
1,1-Dicyclohexenyl	51	45a
1,3-Butadienes:		
Unsubstituted	20	46
trans-1-Methyl	31	46
2-Methyl	51	49
trans,trans-1,4-Dimethyl	56	49
trans-1,3-Dimethyl	58	46
2-Benzyl	84	46
2-Phenyl	57	46
trans-1-Phenyl	57	46
trans,trans-1,4-Diphenyl	92	50
1,2-Diphenyl	77	46
1-Acetoxy	22	46
2-Hydroxymethyl	50	46
2-Chloro	Polymeric products	46
2,3-Diphenyl	No reaction	46

The stereochemistry of the addition of singlet oxygen is greatly influenced by steric hindrance as seen in the example shown in equation (17)[51]. In addition, control of stereoselectivity by remote electronic effects has also been observed in certain cases. For example, the reaction of propellane **34** containing two cyclohexadiene rings with singlet oxygen takes the course of exclusive *syn* attack with respect to the hetero ring containing two CO groups (equation 18)[52]. When only one or no CO group is found in the hetero ring, both *syn* and *anti* attack occur. The former course is interpreted in terms of secondary orbital interactions which may stabilize the transition state **35** for *syn* attack.

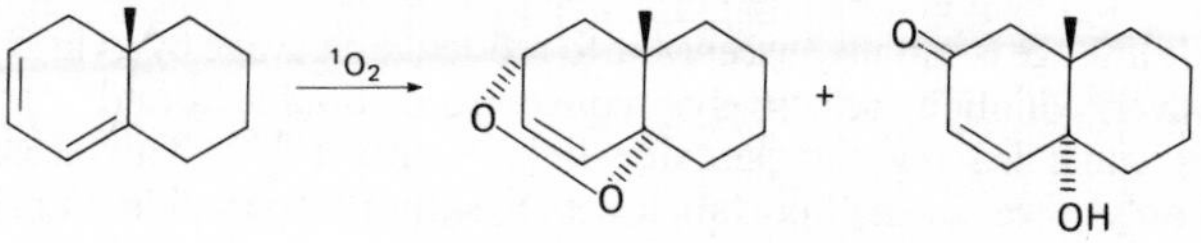

$^1O_2$ (17)

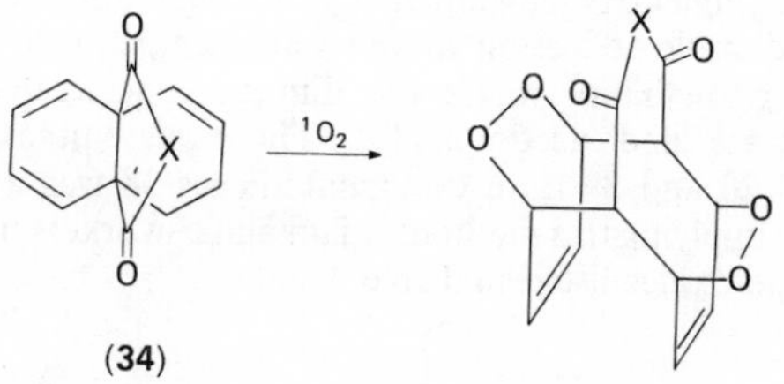

$^1O_2$ (18)

(34)
X = NMe

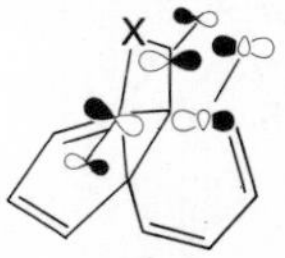

(35)

Singlet oxygenation of **36** and **37** proceeds only with moderate *endo* stereoselectivity, in contrast to the behaviour of a wide range of conventional dienophiles to form Diels–Alder adducts arising exclusively from *endo* attack (equation 19)[53]. The loss of stereochemical control is attributed to energetic factors arising from the ionization potential of singlet oxygen and the π_1(S) energies of the dienes **36** and **37**[53].

$^1O_2$ → endoperoxides →(Δ)

(**36**) 44% + 16% + CHO 23% (19)

(**37**)

The peroxide linkage is usually susceptible to reductive cleavage by a variety of reducing agents. However, diimide selectively reduces the double bond of unsaturated endoperoxides while leaving the peroxide linkage intact[54,55]. This selective diimide reduction not only serves as an important access to saturated peroxides but also facilitates the trapping of the unstable singlet oxygen adducts by running photooxygenation at low temperature followed by diimide reduction at the same temperature. Normally, the reduction is carried out in methanol where the diimide is generated *in situ* from dipotassium azodicarboxylate and acetic acid[56]. The first synthesis of the 2,3-dioxabicyclo[2.2.1] system **20** and **39** from cyclopentadienes **38** was achieved by this technique (equation 20). By applying this method, Adam and coworkers have synthesized a number of bicyclic endoperoxides listed in Table 2.

$^1O_2$ → [ ] →(HN = NH, CH_2Cl_2) (20)

(**38**)

(**39**)[54] R = Ph
(**20**)[57] R = H

TABLE 2. Synthesis of endoperoxides by diimide reduction of singlet-oxygen–diene adducts

Singlet oxygen acceptor	Singlet-oxygen–diene adduct	Product (Yield, %)	m.p. (°C)	Ref.
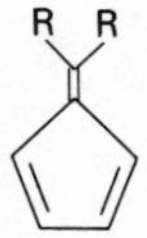	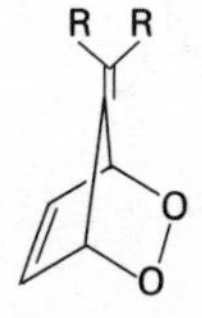 (NMR at − 70°C)	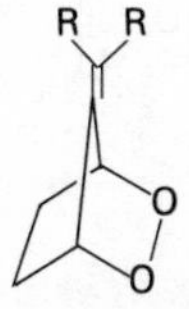 R = Me (63) R = Ph (88)	53 77–83	58
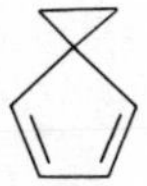	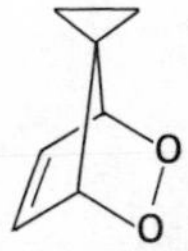	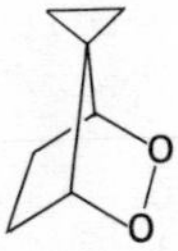 (68)	30	59
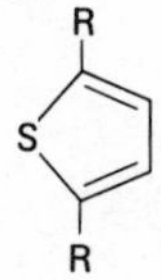	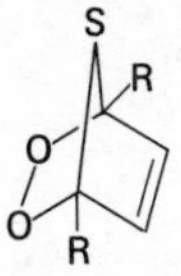	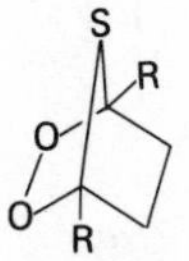 R = Me (46) R = t-Bu (50)	oil 68–69	60
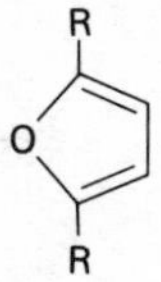	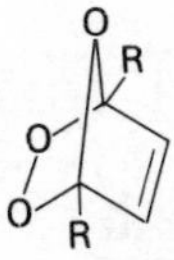	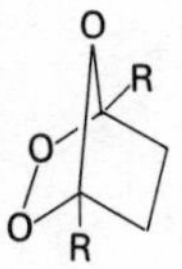 R = H (54) R = Me (65) R = Ph (45)	 oil 100	61
	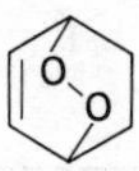	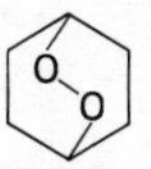 (48)	117–118	55

TABLE 2. (*cont.*)

Singlet oxygen acceptor	Singlet-oxygen–diene adduct	Product (Yield, %)	m.p. (°C)	Ref.
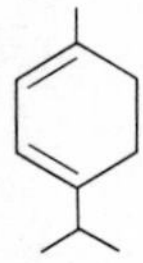 Me	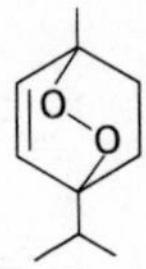 Me	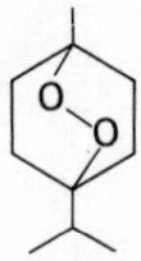 (100)		55
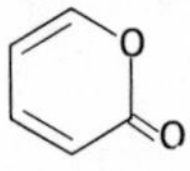 O, O	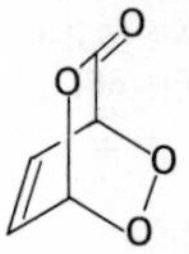 O, O, O, O	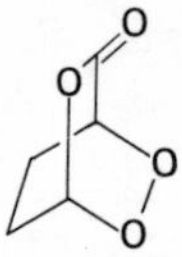 (25)	30	62
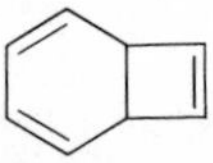	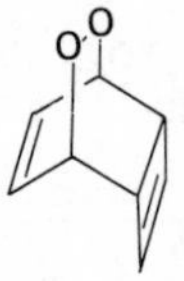 O, O	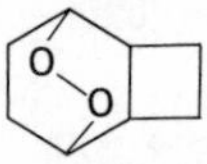 O, O (80)	120–121	63
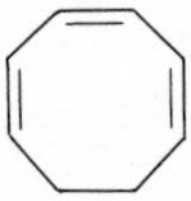	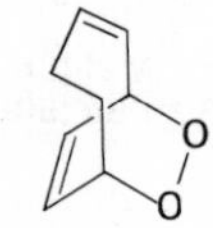 O, O	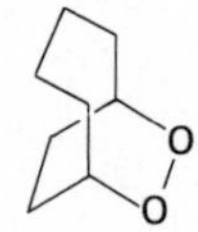 O, O (86)	95–96	64
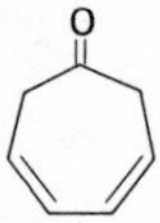 O	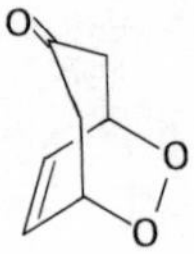 O, O, O	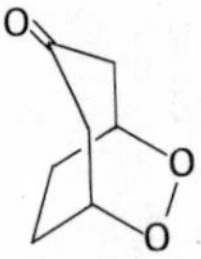 O, O, O (92)	89–96	65
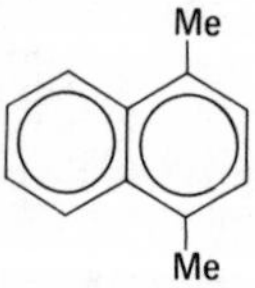 Me, Me	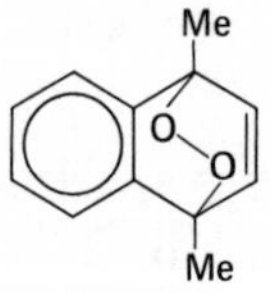 Me, O, O, Me	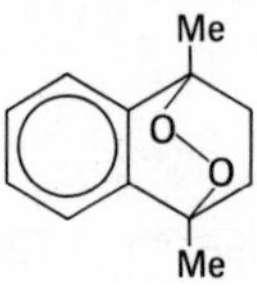 Me, O, O, Me		55

In view of the facile valence isomerization between tropilidene (T) and norcaradiene (N) forms, cycloaddition of singlet oxygen with 1,3,5-cycloheptatriene (**40**) is particularly interesting. Photosensitized oxygenation of **40** in carbon tetrachloride at 0°C gives all of four possible cycloadducts in the yields shown in equation (21)[66,67]. The resulting endoperoxides are reduced with diimide to the corresponding saturated endoperoxides in good yields.

(**40**) T N $^1O_2$ $^1O_2$

[4 + 2] 40% [6 + 2] 37% [2 + 2] 9% [4 + 2] 3.5%

HNNH HNNH HNNH

(21)

Singlet oxygenation of 7-substituted cycloheptatriene derivatives **41** also gives a mixture of cycloadducts as given in equation (22)[68,69]. Substituents at C(7) with π-electron acceptor ability, such as CHO, COOR and CN, tend to favour the norcaradiene adduct

(**41**) $^1O_2$ T T T N

(22)

(N), while substituents with π-electron donor ability such as OR favour the tropilidene-type adducts (T)[68]. All of these endoperoxides are convenient and useful precursors to oxygen-functionalized cycloheptane derivatives[70]. Various types of cyclic polyepoxides including **42**, **43** and **44** can be prepared from 1,3,5-cycloheptatriene endoperoxides[71,72].

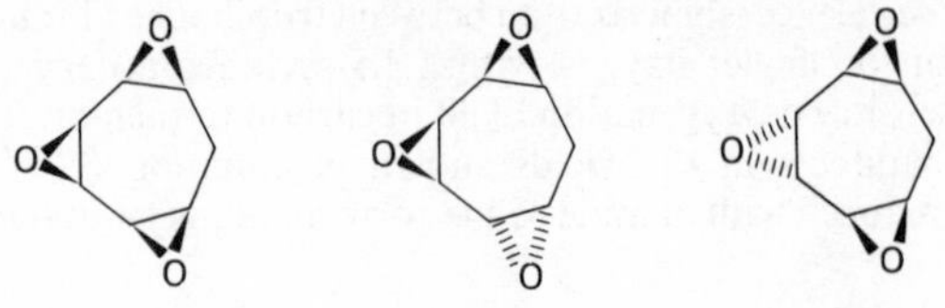

(42) (43) (44)

The synthesis of benzene trioxide (**45**) is achieved from the benzene oxide **46** via the thermal rearrangement of endoperoxide **47** (equation 23)[73,74]. Vitamin D_2 contains an *s-cis* diene function and gives a 1:1 mixture of the expected epimeric peroxide **48** of potential biological interest[75].

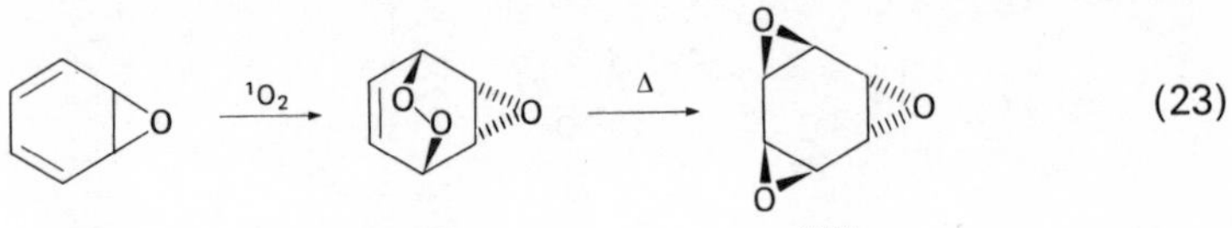

$$\xrightarrow{^1O_2} \quad \xrightarrow{\Delta} \tag{23}$$

(46) (47) (45)

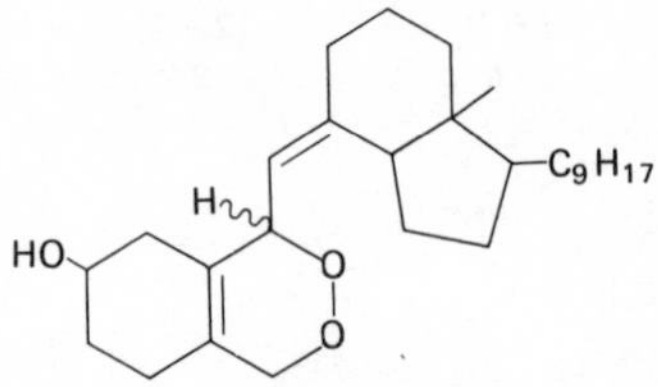

(48)

2. Aromatic compounds

The formation of endoperoxides has long been observed during self-sensitized photooxygenation of polycyclic aromatic hydrocarbons. A large number of polycyclic aromatic systems including anthracenes, pentacenes, hexacenes and azaanthracenes undergo 1,4-cycloaddition with singlet oxygen to give stable endoperoxides. Since many of the earlier examples of arene photooxidation have appeared in recent reviews[45d,76,77], we shall focus only on the recent results which are important for the synthesis of arene endoperoxides.

In the 1,4-cycloaddition of singlet oxygen to polycyclic aromatic systems, one can alter regioselectivity by introducing electron-donating groups such as methyl, methoxy and dimethylamino groups at appropriate positions on an aromatic nucleus. As seen from typical examples of anthracene derivatives **49** given in equation (24), the predominant or exclusive formation of 1,4-endoperoxide **51** over 9,10-endoperoxide **50** requires that the substituents be located at the 1- and 4-positions and not at the 2- and 3-positions[77,78]. Substitution at the 9- and 10-positions favours the formation of 9,10-endoperoxides **50**.

In contrast to the above-mentioned polycondensed aromatic hydrocarbons, benzene, naphthalene and phenanthrene show practically no reactivity toward singlet oxygen. However, the introduction of electron-donating substituents into suitable positions causes singlet oxygen addition. Photooxygenations of di-[79], tri-[79], tetra-[79] and octa-methylated[80] naphthalenes, 1,4-dimethoxynaphthalene[77] and 1,4-dimethylphenanthrene[81] all afford the corresponding endoperoxides in high yields. A remarkable example is the singlet oxygenation of 1,4-dimethylnaphthalene (**52**). Photosensitized oxygenation of **52** at

(49) (50) (51) (24)

	(50)		(51)
X = H, Y = Ph, Z = H	100	:	0
X = H, Y = OMe, Z = H	0	:	100
X = Ph, Y = H, Z = H	100	:	0
X = Ph, Y = OMe, Z = H	50	:	50
X = H, Y = H, Z = OMe	100	:	0

ambient temperature does not give the corresponding endoperoxide **53** because of its thermal instability (half-life at 25°C, 5 h), whereas the same photooxygenation at below 0°C rapidly produces the endoperoxide **53** in high yield (equation 25)[79,81].

(52) + $^1O_2$ ⇌ (Δ) (53) (25)

Photooxygenation of the naphthalene analogue of [2.2]paracyclophane **54** in methanol gives **55**, probably through the formation of endoperoxide **56** which undergoes internal Diels–Alder reaction followed by solvolysis (equation 26)[82].

(54) —$^1O_2$→ [(56) →] —MeOH→ (55) (26)

The naphthalene 1,4-endoperoxide **57**, which is not formed by singlet oxygenation of the parent naphthalene, can be prepared from the endoperoxide **58** derived from 1,6-imino[10]annulene (**59**) (equation 27)[83]. Treatment of **58** with nitrosyl chloride and triethylamine at −78°C provides **57** which readily decomposes to naphthalene and singlet oxygen with a half-life of 303 min at 20°C.

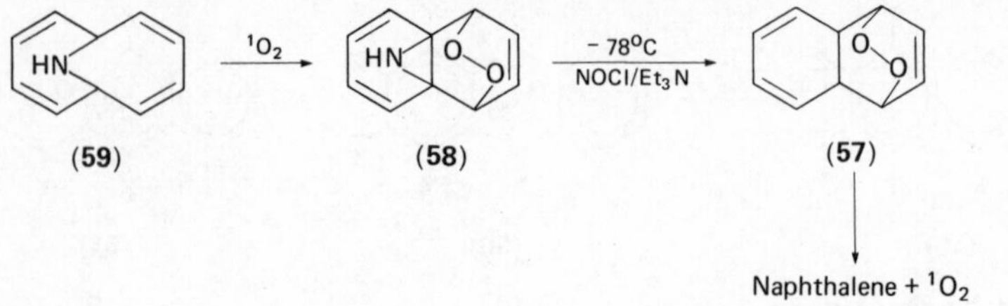

(59) → $^1O_2$ → (58) → −78°C, NOCl/Et_3N → (57) → Naphthalene + $^1O_2$ (27)

Even sufficiently activated benzenes can undergo 1,4-cycloaddition with singlet oxygen to give products which are believed to be derived from 1,4-endoperoxides **60** as indicated in equation (28)[84,85].

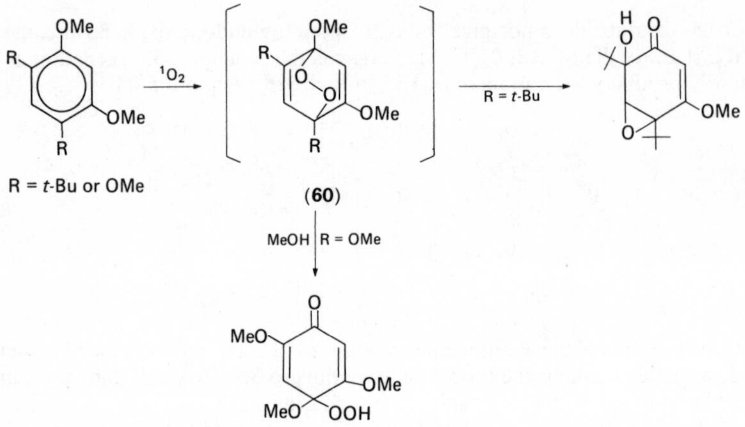

R = *t*-Bu or OMe; $^1O_2$; (**60**); R = *t*-Bu; MeOH, R = OMe (28)

The strained benzene ring in [2.2.2.2]-(1,2,2,5)cyclophane (**61**) readily undergoes 1,4-addition with singlet oxygen to give an isolable benzene endoperoxide (**62**) (equation 29)[86].

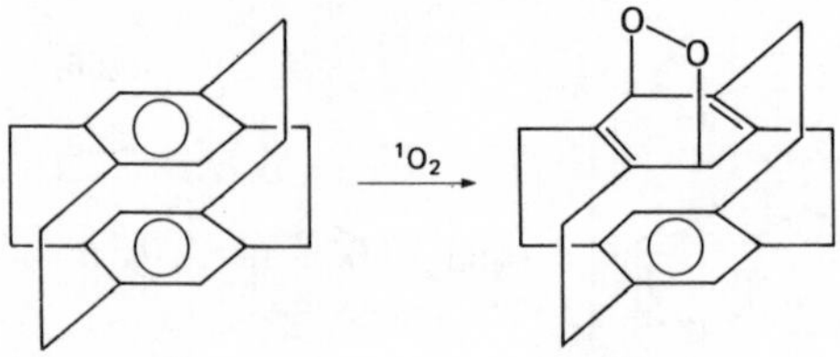

(**61**) → $^1O_2$ → (**62**) (29)

Hexamethylbenzene does not undergo direct 1,4-cycloaddition with singlet oxygen but gives the product derived from an ene-type reaction followed by 1,4-cycloaddition of another molecule of singlet oxygen (equation 30)[87]. In an attempt to obtain a benzene-1,4-endoperoxide **63**, the peroxide **64** is treated with base at −5°C. However, **63** could not be characterized but decomposed to *p*-terphenyl and singlet oxygen as shown in equation (31)[88].

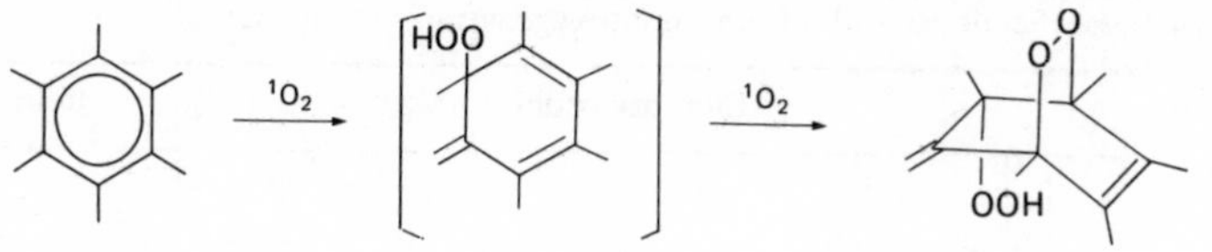

(30)

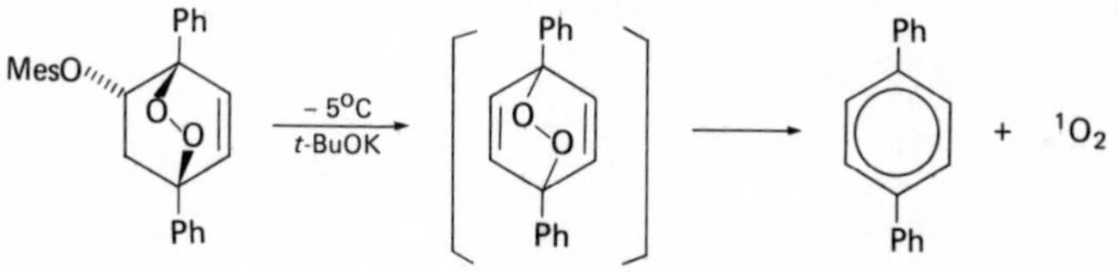

(64) (63)

(31)

Only relatively recently could it be demonstrated that vinylarenes undergo 1,4-cycloaddition with singlet oxygen. A number of vinylarenes give the corresponding cyclic peroxides by stereospecific addition of singlet oxygen. These reactions proceed with complete retention of stereochemistry, and constitute an efficient method for the synthesis of various types of cyclic peroxides. One of the examples is given in equation (32). Low-temperature (-70°C) photooxygenation of *cis*-α-methoxystyrene (**65**) gives the *cis* endoperoxide **66**, together with other secondary products including the diendoperoxide **67**, whereas the *trans* isomer **68** produces the *trans* endoperoxide **69** exclusively. Thermolysis of **66** produces formylmethide quinone (**70**) which can be trapped with electron-rich olefins (equation 32)[89].

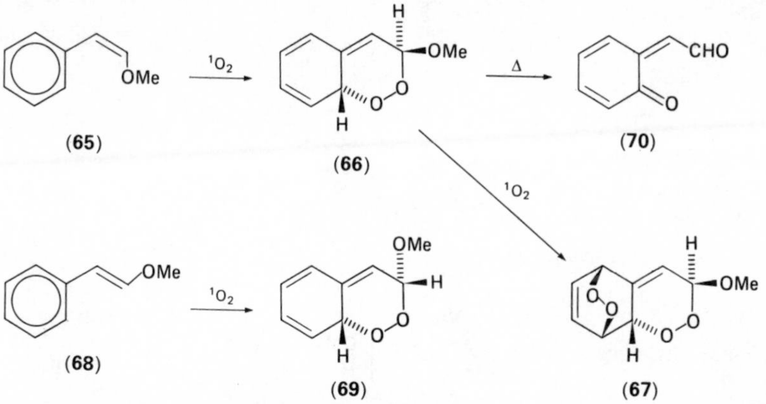

(65) (66) (70)

(68) (69) (67)

(32)

This type of singlet oxygen addition has been first observed by Foote and coworkers in the photooxygenation of 1,1-diphenyl-2-methoxyethylene and indenes[90]. Some examples are listed in Table 3. The resulting endoperoxides have received wide application in organic synthesis as described in Section IV.G.

In many cases in Table 3, the photooxidation products have a strong dependency on the photosensitizers and reaction conditions such as solvent, temperature and light intensity. For example, low-temperature (-78°C) photooxygenation of 2,3-diphenylindene in methanol gives dioxetane **71** exclusively, whereas in acetone the rose-bengal-sensitized

TABLE 3. Synthesis of endoperoxides from photooxygenation of vinylarenes

Vinylarene	Endoperoxide	Reference
$(XC_6H_4)_2C{=}CHOMe$	X = H, *p*-OMe	90
	R^1 = Me, R^2 = *t*-Bu; R^1 = R^2 = Me; R^1 = Me, R^2 = H	91
	R = H, Me	92
$CH{=}C(OMe)CO_2Me$	R = H, OMe	93
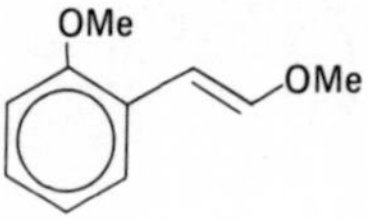	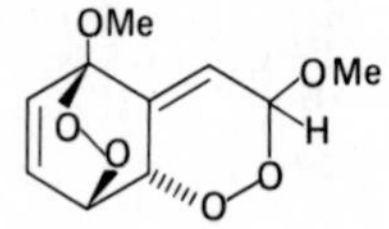	94
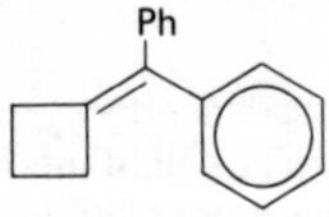	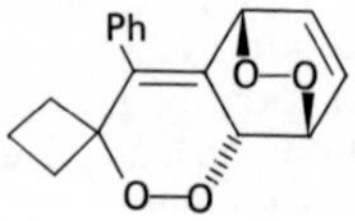	95

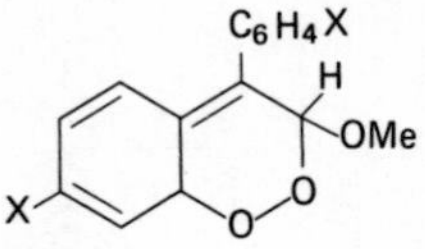

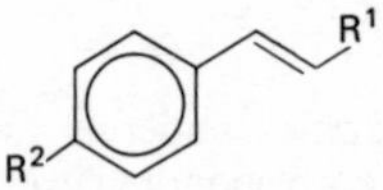

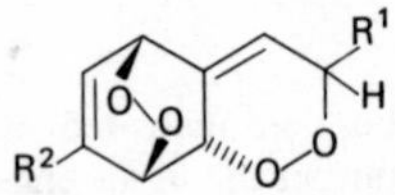

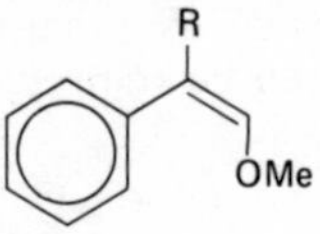

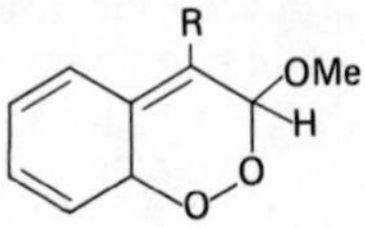

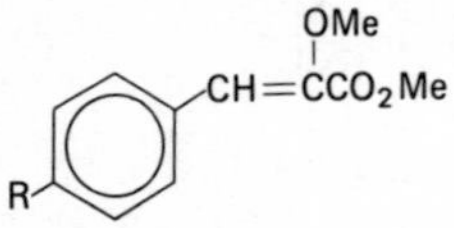

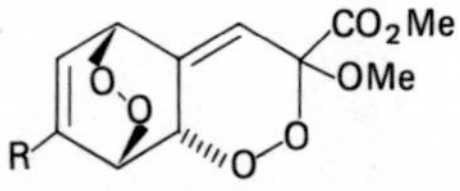

TABLE 3. (*cont.*)

Vinylarene	Endoperoxide	Reference
R^1 = H, R^2 = Ph R^1 = R^2 = Ph R^1 = Me, R^2 = Ph R^1 = R^2 = H R^1 = Me, R^2 = H		96
R^1 = R^2 = H R^1 = Me, R^2 = H R^1 = H, R^2 = Me		97
		98
R^1 = R^2 = Me or H R^1 = Me, R^2 = H		99

photooxygenation produces **72** as a sole product[100]. Tetraphenylporphine-sensitized photooxygenation in Freon-11 provides a different product, the diendoperoxide **73**. High light intensity promotes the formation of **73** over **72**. Likewise, the solvents Freon-11 and acetone-d_6, in which singlet oxygen has an unusually long lifetime, favour the formation of **73** (equation 33). Thus, high steady singlet-oxygen concentration would trap **74** more efficiently and increase the amount of **73**[101]. Photooxygenation of silyl enol ethers **75** in Freon-11 at 0°C affords the diendoperoxide **76**, whereas at −78°C the same photooxygenation produces the monoendoperoxide **77** exclusively (equation 34)[102].

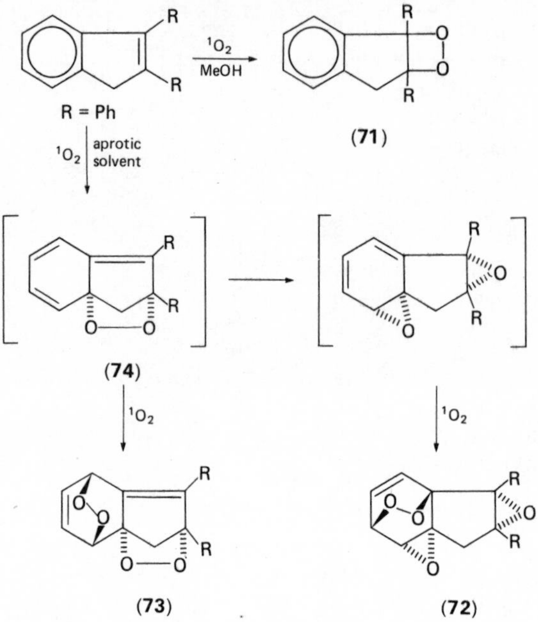

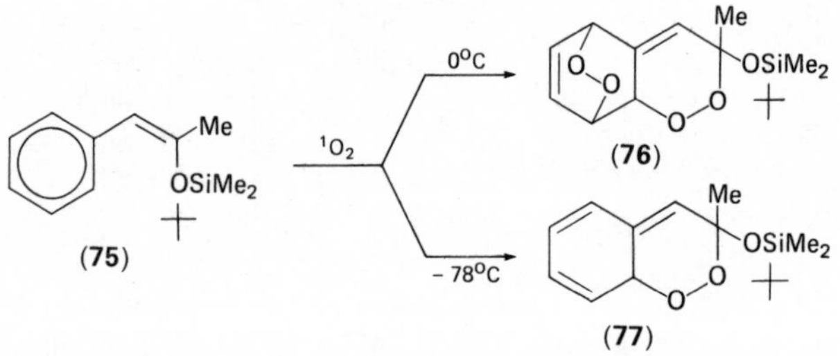

Reaction of singlet oxygen with annulenes has been studied extensively by Vogel and coworkers. Photooxygenation of 1,6-methano[10]annulene (**78**) gives the 1,4-endoperoxide **79** which on warming rapidly rearranges to the diepoxide **80**. Continued photooxygenation of **78** under the same conditions provides the tetraepoxide **81**[103]. A similar result is observed in the photooxygenation of 1,6-oxido[10]annulene (**82**) (equation 35)[104].

(78) X = CH2
(82) X = O
(79)
(80)
(81)

(35)

Photooxygenation of the bridged [14]annulene **83** gives an isolable endoperoxide **84** which, however, decomposes over silica gel (equation 36)[103].

(83)
(84)

(36)

An interesting photochromic system which is based on the photoreversible addition of singlet oxygen into an aromatic nucleus has been observed. Self-sensitized photooxygenation of red–violet heterocoerdianthrone (**85**) leads to the formation of the colourless endoperoxide **86**. Upon UV irradiation of **86** at 313 nm, the parent **85** and singlet oxygen are reformed with a maximum quantum yield of 0.26 (equation 37)[105].

(85) Red-violet
(86) Colourless

(37)

Photooxygenation of 10,10′-disubstituted phenafulvenes **87** gives crystalline epiperoxides **88**[106], whereas 15,16-dimethyldihydropyrene **89** provides diepoxide **90** by way of endoperoxide **91** by singlet oxygenation (equation 38 and 39)[107].

R = Me or Ph

(87) (88) (38)

(89) (91) (90) (39)

3. Heterocyclic compounds

The reaction of singlet oxygen with heterocyclic compounds usually gives rise to a complex mixture of products. The diverse transformations are primarily attributed to the multitude of pathways that are available for the decomposition of the primary peroxidic intermediates such as endoperoxides. Except for reactions at very low temperature, these endoperoxides have not been isolated in normal photooxygenations. Therefore, there are not so many examples in which endoperoxides have been isolated. Exhaustive coverage of the photooxidation of heterocyclic compounds has been made[108]. Consequently, we shall focus only on the recent results in which endoperoxides are experimentally confirmed.

Generally, addition of singlet oxygen to a heterocyclic system occurs by one of the following three modes: (*a*) 1,4-cycloaddition to the 1,3-diene system leading to endoperoxides as frequently observed in many heterocycles including furans, pyrroles, oxazoles, thiophenes, imidazoles and purines; (*b*) dioxetane formation resulting in double-bond cleavage; (*c*) hydroperoxide formation by typical ene-type reactions. Some of these processes may be preceded by the initial formation of a zwitterionic peroxide species **92**, when highly electron-rich double-bond systems are exposed to photosensitized oxygenation in polar solvents[109] (equation 40).

X = NR, O, S

(92) (40)

In the decompositions of initially formed endoperoxides, the effects of solvent, temperature, substituents and geometry all play an important role in determining the nature of the products obtained. Oxazoles have been chosen as illustrating typical pathways for the decomposition of endoperoxides. The reaction of oxazoles with singlet oxygen most generally results in formation of triacylamides **93**[110]. The oxidation proceeds through an initially formed endoperoxide **94** which rearranges to an imino anhydride **95**[111]. Subsequent *O* to N acyl migration then yields the triacylamide, as outlined in equation (41). The imino anhydride **96** is formed as the main product when the fused-ring oxazole **97** is oxidized by singlet oxygen. In this case the restrictions imposed by the geometry of the ring system inhibit the rearrangement to triacylamides (equation 42)[112].

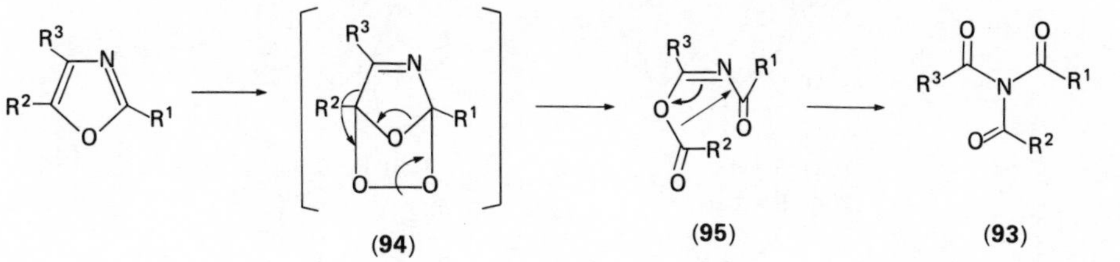

(94) (95) (93)

(41)

(97) (96)

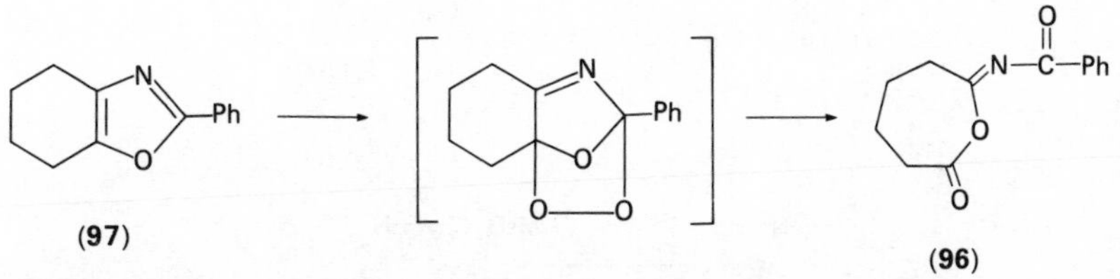

(42)

Solvent effects may also play an important role in governing the outcome of singlet oxygen reaction with oxazoles. Thus, during photooxygenation of **98** in nonpolar solvents like methylene chloride the intermediate endoperoxide **99** undergoes fragmentation, affording cyano anhydride **100** which may undergo hydrolysis or loss of carbon monoxide to ω-cyanocarboxylic acid **101** (equation 43)[113].

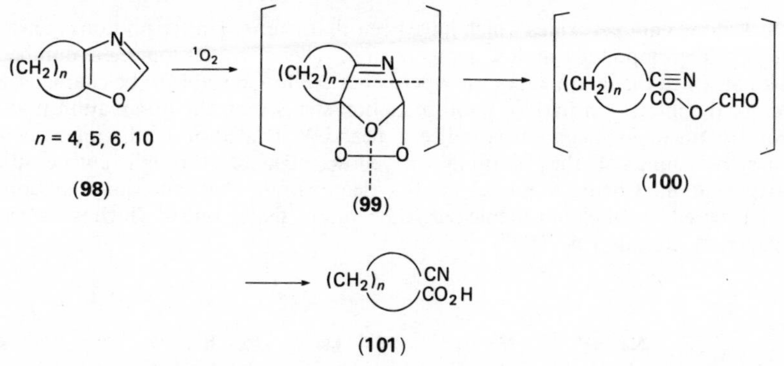

n = 4, 5, 6, 10

(98) (99) (100)

(43)

(101)

The intermediate endoperoxides derived from oxazoles can be trapped by an intramolecular reaction with a carboxylic group. Thus, photooxygenation of **102** in chloroform at 0°C gives spirolactone hydroperoxide **103**. When the photooxygenation of the trimethylsilyl ester **104** is carried out at 0°C, the corresponding endoperoxide **105** is indeed detectable by NMR, and its treatment with methanol immediately produces **103** (equation 44)[114].

In contrast to the above cases, methylene-blue(MB)-sensitized photooxygenation of 5-methoxyoxazoles **106** gives the dioxazoles **107** (equation 45)[115]. The yields of **107** are remarkably improved when the reaction is carried out in the presence of 1,4-diazabicyclo[2.2.2]octane (DABCO).

(**102**) R = H
(**104**) R = $SiMe_3$
(**105**) R = $SiMe_3$
(**103**)
(44)

(**106**)
R = Me, Ph
(**107**)
(45)

Table 4 shows endoperoxides which have been characterized in the photooxygenation of heterocyclic compounds at low temperatures. There are, of course, a number of examples in which the intermediate endoperoxides are too unstable to be characterized.

Much of the interest in pyrrole photooxidation stems from the observation that the common treatment for neonatal jaundice is near-UV irradiation of the infant, which promotes bleaching of the bilirubin, a yellow pigment, through autosensitized photooxygenation. Chemical model studies demonstrate that photooxygenation of bilirubin proceeds through unisolable peroxidic intermediates, one of which is suggested to be the endoperoxide **108**[120,130].

(**108**)

$R^1 = -CH{=}CH_2$, $R^2 = -CH_2CH_2CO_2H$

TABLE 4. Endoperoxides obtained in the photooxygenation of heterocyclic compounds

Singlet oxygen acceptor	Endoperoxide	Reference
	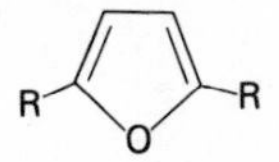 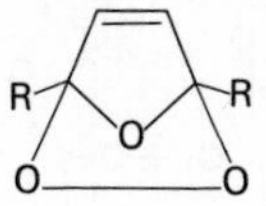 R = Me R = t-Bu	116 117
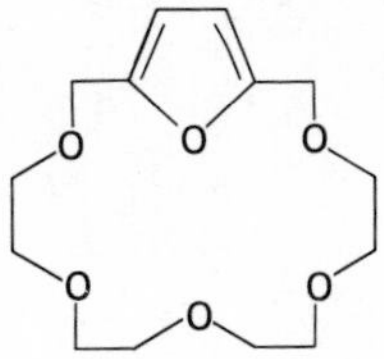	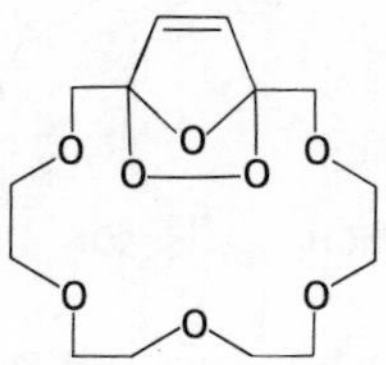	118
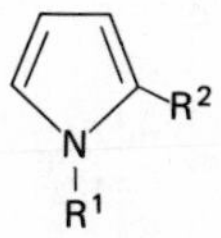 R^1 = Me, t-Bu; R^2 = H R^1 = H; R^2 = Me, t-Bu	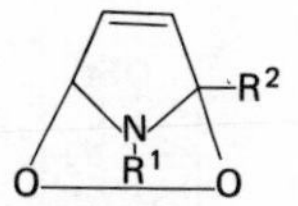 (NMR at − 80°C)	119 120
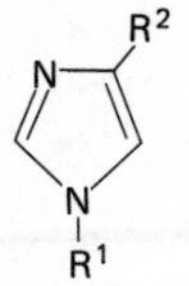 R^1 = R^2 = Me R^1 = H, R^2 = $CH_2CH(NHCO_2Bu\text{-}t)CO_2Me$	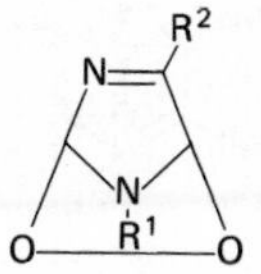	121
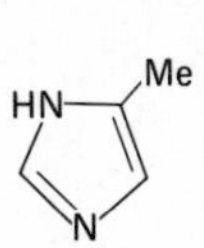	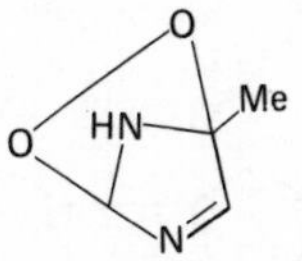 (NMR at − 80°C)	121
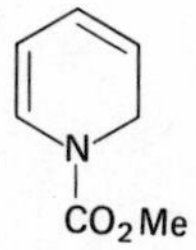	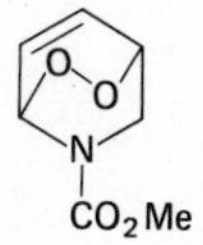	122

TABLE 4. (*cont.*)

Singlet oxygen acceptor	Endoperoxide	Reference
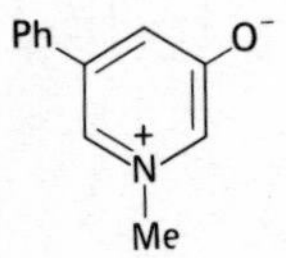 Ph, O^-, $\overset{+}{N}$, Me	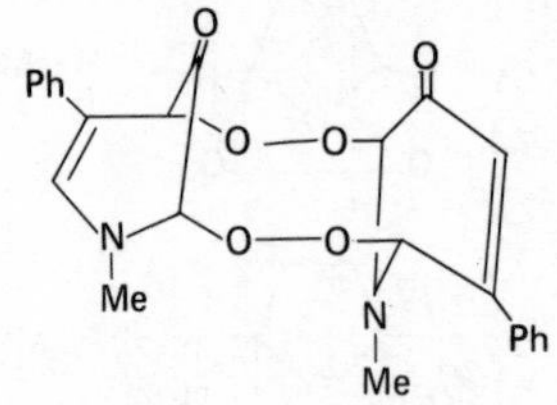 O, Ph, O—O, N, Me, O—O, O, N, Me, Ph	123
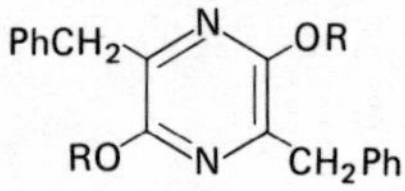 $PhCH_2$, N, OR, RO, N, CH_2Ph	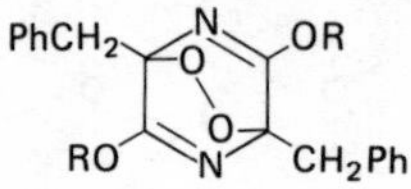 $PhCH_2$, N, OR, O, O, RO, N, CH_2Ph	124
R = H, Me, Ph		
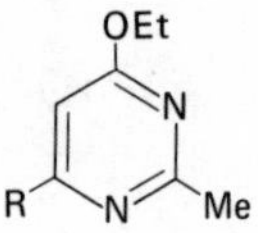 OEt, N, R, N, Me	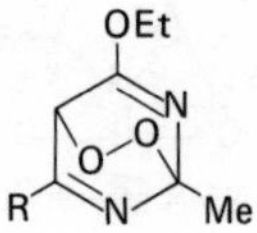 OEt, N, O, O, R, N, Me	124
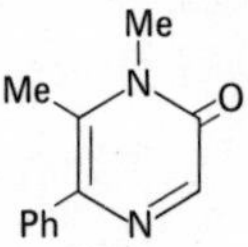 Me, Me, N, O, Ph, N	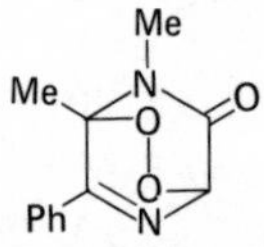 Me, Me, N, O, O, O, Ph, N	125
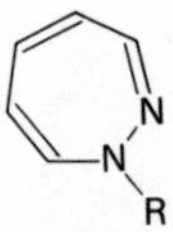 N, N, R	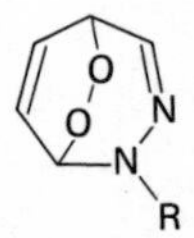 O, O, N, N, R	126
$R = CO_2Et$		
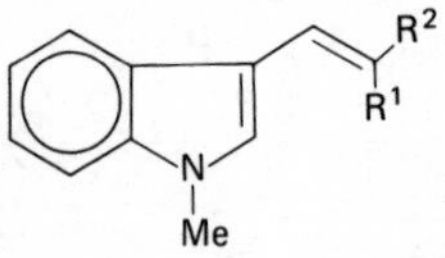 R^2, R^1, N, Me	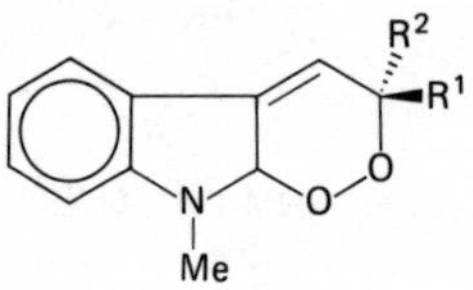 R^2, R^1, O, O, N, Me	127
$R^1 = H$, $R^2 = Ph$ $R^1 = R^2 = H$ or Me $R^1 = OMe$, $R^2 = OSiMe_2(Bu\text{-}t)$		

TABLE 4. (*cont.*)

Single oxygen acceptor	Endoperoxide	Reference
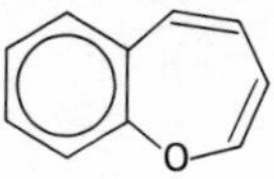 O	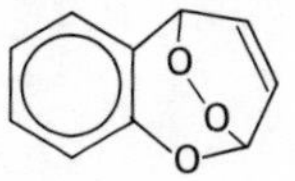 O O O	128
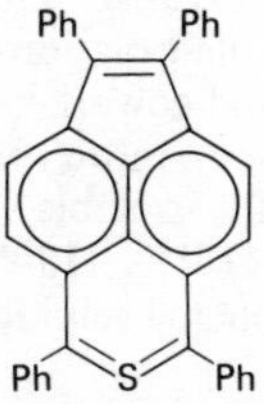 Ph Ph Ph S Ph	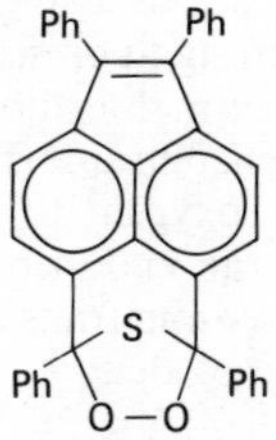 Ph Ph S Ph Ph O—O	129

Smith and Schuster have demonstrated that photooxygenation of **109** produces the stable endoperoxide **110** which upon thermal decomposition gives *o*-dibenzoylbenzene and phenyl *o*-benzoylbenzoate[131]. A novel class of *o*-xylylene peroxide (**111**) is suggested as an intermediate. Isolation of **112** from the thermolysis of **110** in the presence of maleic anhydride is convincing support for this proposal. Thermolysis of **110** in the presence of easily oxidizable aromatic hydrocarbons, such as rubrene, produces detectable chemiluminescence by the chemically initiated electron-exchange chemiluminescence (CIEEC) mechanism (equation 46). In a related work, Adam and Erden have synthesized endoperoxides **114** from some α-pyrones **113**[132]. In this case, however, all efforts to trap the expected *o*-dioxin **115** with dienophiles have failed. The endoperoxides **114** also undergo fluorescer-enhanced chemiluminescence (equation 47).

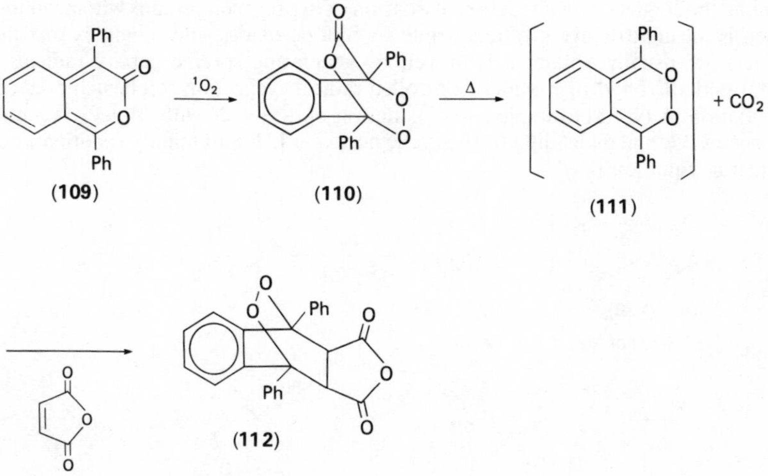

Ph O O Ph (**109**) $\xrightarrow{^1O_2}$ O O Ph O O Ph (**110**) $\xrightarrow{\Delta}$ [Ph O O Ph] (**111**) + CO_2

(46)

$\longrightarrow$ (maleic anhydride: O O O) O O Ph O O O Ph (**112**)

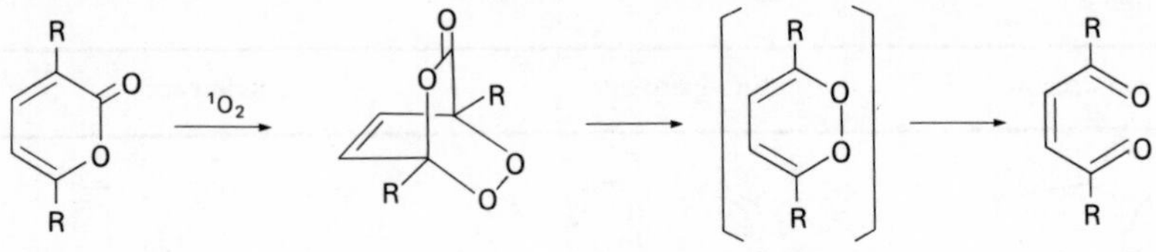

(113) (114) (115) (47)

1,3-Diphenylisobenzofuran is widely used as a singlet oxygen trap because of its high reactivity ($k_r = 7 \times 10^8\ \text{M}^{-1}\text{s}^{-1}$) and strong fluorescence. However, the isolation of the endoperoxide is not easy owing to its thermal instability[133]. Saito and coworkers have demonstrated that 1,3-di-*t*-butylisobenzofuran (**116**) reacts even more rapidly ($k_r = 2.8 \times 10^9\ \text{M}^{-1}\text{s}^{-1}$) with singlet oxygen to give the easily isolable stable endoperoxide **117** (equation 48)[134]. Furan endoperoxides such as **117**[134], **118**[135] and **119**[135] are capable of undergoing oxygen atom transfer to olefins, diphenyl sulphide and naphthalene probably through carbonyl oxides.

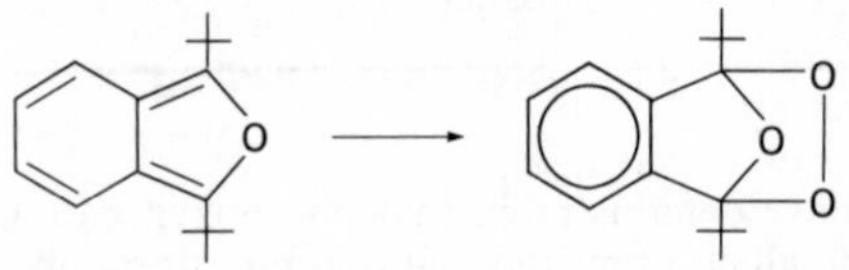

(116) (117) (48)

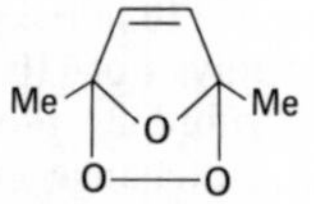

(118) 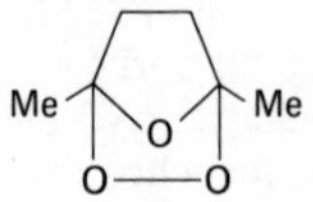 (119)

C. Oxidation with Triplet Oxygen

Unlike the singlet oxygen reaction, oxidation of organic compounds with ground-state oxygen is an unattractive synthetic route to endoperoxides since complex mixtures of products are usually obtained. However, by generating specific peroxy radicals from unsaturated alkyl hydroperoxides, controlled radical cyclization can be achieved during autoxidation. A typical example is the oxidation of diene **120** with triplet oxygen in the presence of thiophenol leading to the cyclic peroxide **121** with mainly *cis* orientation, as outlined in equation (49)[136].

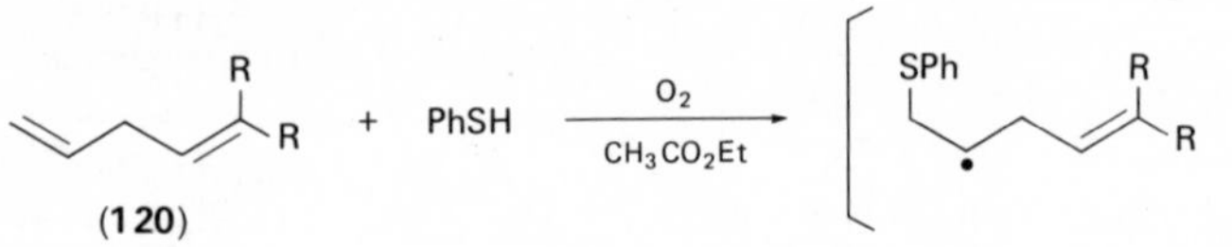

(120) + PhSH $\xrightarrow[\text{CH}_3\text{CO}_2\text{Et}]{\text{O}_2}$

R = H or Me

$\xrightarrow{\text{O}_2}$

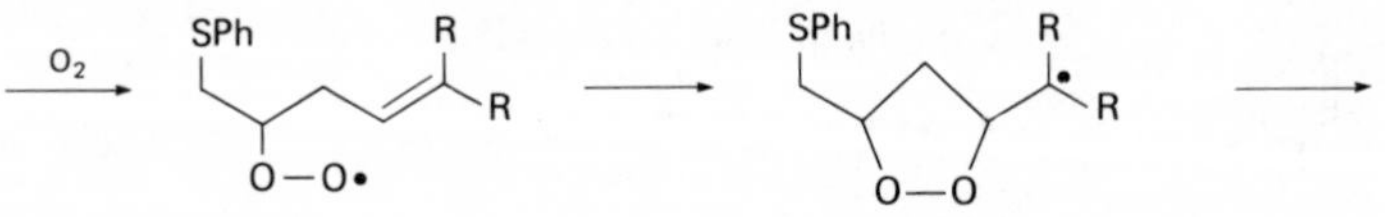

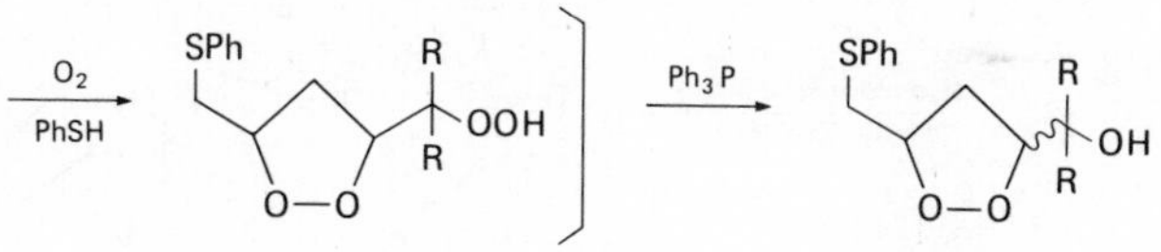

(49)

(121)
cis/trans = 4

Such a mechanism involving alkenylperoxy radical cyclization has been proposed as taking place during the biosynthesis of prostaglandin endoperoxides[16,137]. In accomplishing mechanistically similar transformations, alkenylperoxy radical cyclizations have been studied in detail (equations 50[44] and 51[138]). As a model of prostaglandin biosynthesis, radical-initiated autooxidation of **122** is carried out. The PGF_1-like product **123** is obtained in extremely low yield (equation 52)[139].

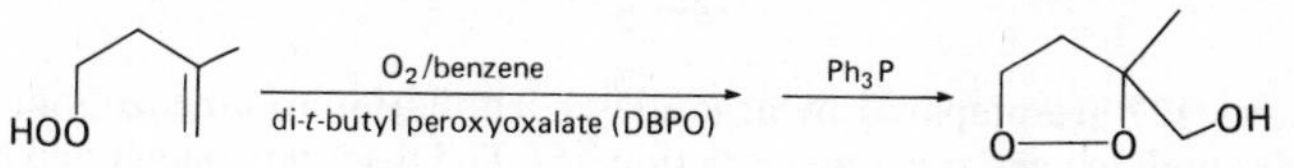

(50)

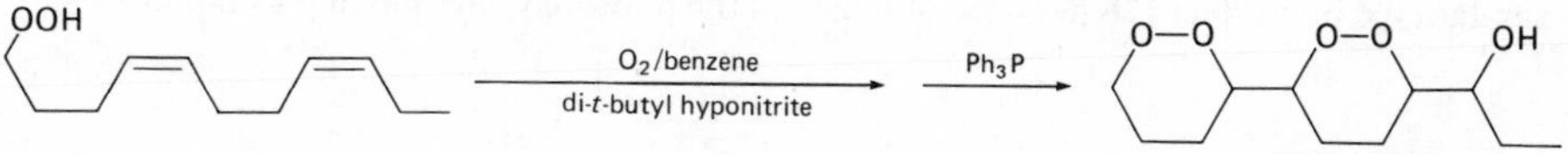

(51)

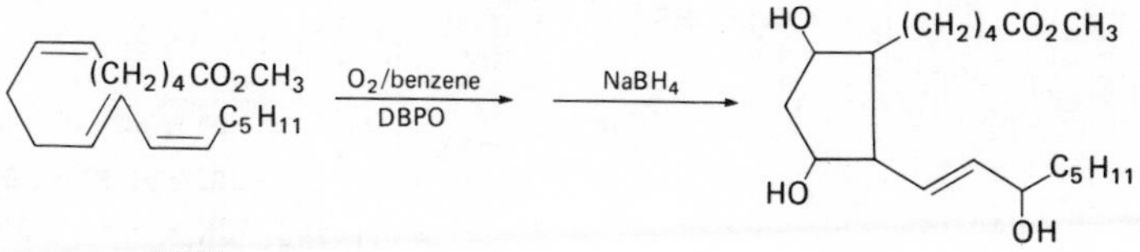

(52)

(122) (123)

Bicyclic endoperoxides are prepared by a route involving generation of triplet biradicals via benzophenone-sensitized decomposition of the corresponding azoalkanes, using argon laser light, and trapping these species with triplet oxygen under high pressure. Bicyclo[2.2.1]endoperoxides (**20**, **124**, **125**) can be synthesized from the corresponding azoalkanes in moderate yields by this method (equations 53 and 54)[140]. The synthesis of bicyclo[3.2.1]endoperoxide **126** can be achieved by a similar procedure (equation 55)[141]. In these reactions, it is essential to irradiate only the benzophenone chromophore. Oxygen pressure and reaction time must also be carefully regulated to obtain optimum yield of endoperoxides.

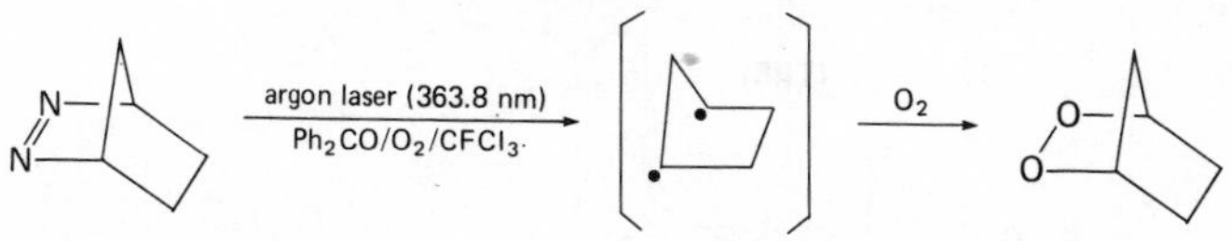

(53)

(20)

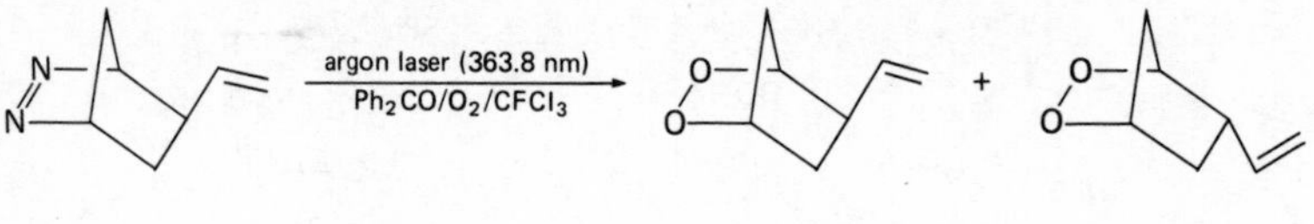

argon laser (363.8 nm), $Ph_2CO/O_2/CFCl_3$ (54)

(124) (125)

(1:1)

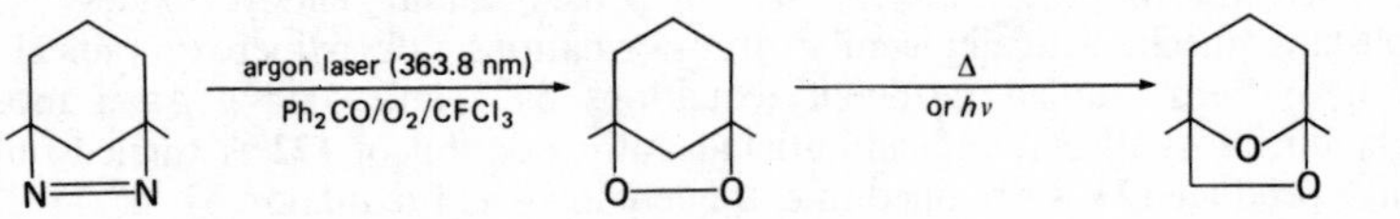

argon laser (363.8 nm), $Ph_2CO/O_2/CFCl_3$; Δ or $h\nu$ (55)

(126)

The trioxanes **127** are prepared by argon laser irradiation of *p*-benzoquinone–olefin mixtures under high oxygen pressure (equation 56). The reaction is suggested to proceed through a charge-transfer-like intermediate which is trapped by triplet oxygen[142,143]. A similar type of product **128** has been obtained in the photooxygenation of plastoquinone-1[144].

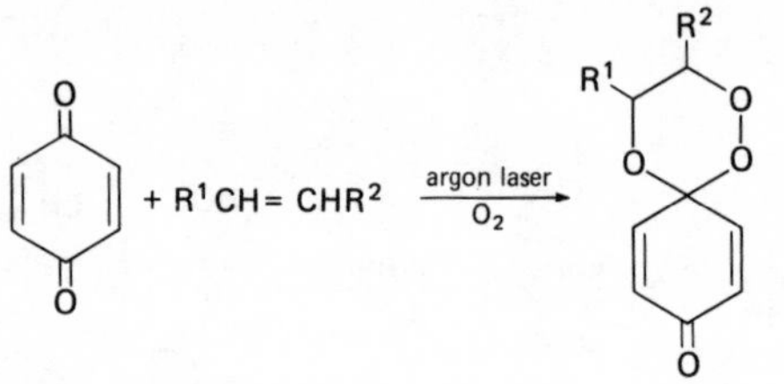

+ $R^1CH{=}CHR^2$ —argon laser, O_2→ (56)

(127)

$R^1 = R^2 = -(CH_2)_4-$
$R^1 = H$, $R^2 = t$-Bu
$R^1 = H$, $R^2 = OAc$

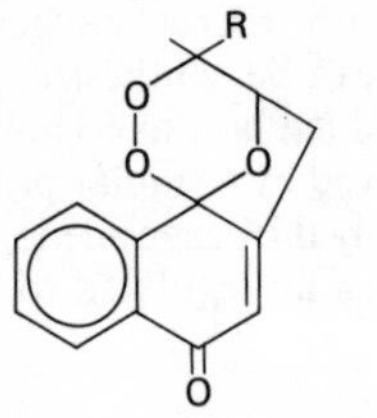

(128)

R =

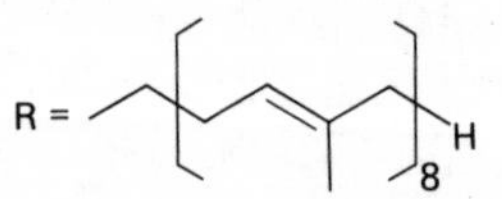

Lewis-acid-catalysed oxidation of cyclic conjugated dienes with triplet oxygen provides bicyclic endoperoxides. This type of reaction represents an alternative to singlet oxygenation of 1,3-dienes. The first example is the oxygenation of ergosteryl acetate (**129**) in the presence of a catalytic amount of trityl cation at $-78°C$ leading to the corresponding endoperoxide (**130**) in high yield (equation 57)[145]. The catalysts BF_3,

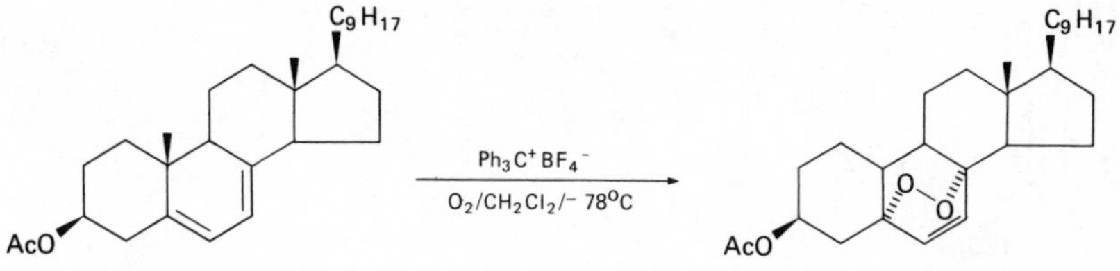

$$\textbf{(129)} \xrightarrow[O_2/CH_2Cl_2/-78°C]{Ph_3C^+BF_4^-} \textbf{(130)} \qquad (57)$$

$SnCl_4$, $SnBr_4$, SbF_5, $SnCl_5$, WF_6, I_2 and Ph_3C^+ BF_4^- require simultaneous irradiation with ordinary or UV light to be effective, whereas with $VOCl_3$, $FeCl_3$, $MoCl_5$, WCl_6 and $(p\text{-}BrC_6H_4)_3N^{\dot{+}}$ BF_4^- the reaction proceeds smoothly in the dark[146,147]. Electron-rich *cisoid* dienes such as **131**, **132** and **133** give the corresponding endoperoxides or their decomposition products as in the case of singlet oxygenation, whereas **134**, **135** and **136**, well-known singlet oxygen acceptors, are inert toward this triplet oxygen oxidation[148].

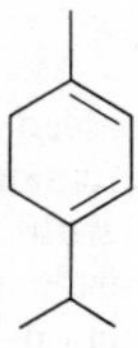

(**131**) 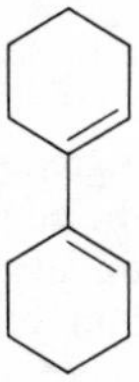 (**132**) 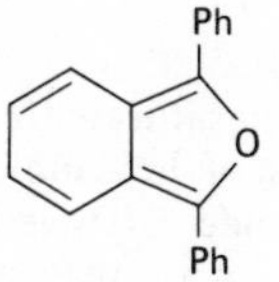 (**133**)

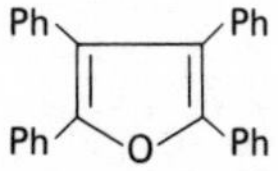

(**134**) 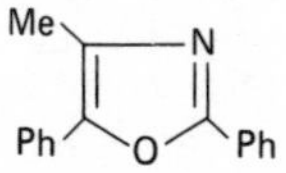 (**135**) 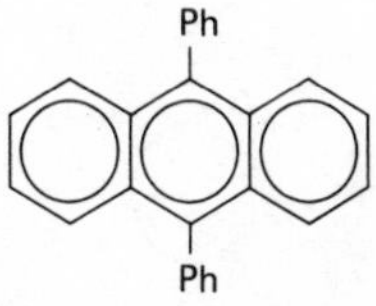 (**136**)

The mechanism currently favoured involves one-electron transfer from dienes to catalysts to generate substrate radical cations **137** which are followed by radical cation chain autoxidation with triplet oxygen, as outlined in equation (58)[149]. A similar type of triplet oxygenation initiated by one-electron transfer has been proposed in the 9,10-dicyanoanthracene-sensitized photooxidation of 1,3-dienes leading to endoperoxides[150]. Electrochemical reaction in the presence of triplet oxygen also induces one-electron transfer type oxygenation of certain olefins[151].

129 $\xrightarrow{Ar_3N^{+\cdot}BF_4^-}$ (137) BF_4^- + Ar_3N $\xrightarrow{O_2}$ [... or ...] BF_4^- (58)

$\xrightarrow{129}$ (130) + 137

IV. REACTIONS OF ENDOPEROXIDES

The potential for synthetic utility of endoperoxides is enormous. The reactions of bicyclic endoperoxides has been recently reviewed[152]. The route has also been used in the synthesis of natural products[153], of polyepoxides[72] and in prostanoid biosynthesis[154]. It has also been realized that endoperoxides are probably involved in metabolic processes of certain natural products. Thus their chemistry might also shed light on the complex processes of biological oxidations.

A. Reduction

In the case of unsaturated endoperoxides, reduction can proceed in four ways: (*i*) Selective reduction of the double bond with the peroxide linkage remaining intact—with reagents like diimide, (*ii*) selective reduction of the peroxide with the double bond remaining intact—with reagents like lithium aluminium hydride or thiourea, (*iii*) reduction of both the double bond and the peroxide in one reaction—e.g. by catalytic hydrogenation and (*iv*) extrusion of one peroxide oxygen—with reagents like triphenylphosphine or triphenyl phosphite. These four reaction courses are represented in equation (59).

PPh$_3$ or P(OPh)$_3$; HN=NH; cat/H$_2$; LiAlH$_4$ or thiourea (59)

1. Diimide

The reaction of unsaturated endoperoxides with diimide to produce their saturated analogues has already been mentioned (see Table 2). The reactivity of the double bond is governed by various factors such as its substitution pattern and stereochemistry[155]. The endoperoxides of substituted furans[61] and thiophenes[60] are very unstable and could not be isolated. However, by using diimide, these peroxides could be trapped *in situ*. Thus diimide reduction provides a useful method for the preparation of saturated endoperoxides as mentioned before (see Table 2).

2. Lithium aluminium hydride and thiourea

For controlled reductions, thiourea is usually preferred over lithium aluminium hydride; it also has the advantage of being used *in situ* without isolation of the endoperoxide. Hence, rigorous low-temperature reaction for the production of the endoperoxide can be avoided. Cyclopentene endoperoxide (**138**) was reduced to the *cis* diol **139** by using thiourea[48] or by using this reagent without isolation of the endoperoxide (equation 60)[156].

$^1O_2$ – MeOH, – 100°C

O–O

(**138**)

thiourea

OH

$^1O_2$ – MeOH

thiourea

OH

(**139**)

(60)

The reaction of spiro[2,4]hepta-4,6-diene (**140**) with singlet oxygen produces the *bis*-epoxide **142** and the keto epoxide **143**, products originating from thermal rearrangement (*vide infra*) (equation 61)[157]. The origin of **142** and **143** from the common endoperoxide intermediate **141** has been postulated but cannot be supported since no trace of **141** is isolated due to its rapid rearrangement to **142** and **143** at 0–5°C. On the other hand, singlet oxygen reaction in the presence of thiourea produces the *cis* diol **144** exclusively, confirming the above postulate (equation 61)[59].

(**140**) $^1O_2$ [O–O (**141**)] Δ O O (**142**) + O O (**143**)

thiourea

OH

(**144**) OH

(61)

Lithium aluminium hydride reduction of **145** gives the diol **146** (equation 62)[66]. The endoperoxide **147** upon thiourea reduction also gives the diol **148** (equation 63)[156]. The latter, which is a *bis*-allylic keto alcohol, suffers further elimination to the tropolone **149**. The endoperoxide **150** gives under similar conditions the dihydroxyketone **151** which produces the α-diketone **153** by the rearrangement of the unstable 7-hydroxy-2,3-homotropone (**152**) (equation 64)[158].

(145) $\xrightarrow{LiAlH_4}$ (146) (62)

(147) $\xrightarrow{\text{thiourea}}$ (148) $\xrightarrow{-H_2O}$ (149) (63)

(150) $\xrightarrow{\text{thiourea}}$ (151) $\xrightarrow{-H_2O}$ (152) → (153) (64)

3. *Triphenylphosphine and triphenyl phosphite*

These two reagents are versatile in the reduction of the endoperoxides where extrusion of one of the peroxide oxygens takes place with the production of unsaturated epoxides. As a typical example, the deoxygenation of naphthalene 1,4-endoperoxide (**57**) gives the naphthalene monoepoxide **154** (equation 65)[159]. The tropone-2,5-endoperoxide (**147**) gives a mixture of the two isomeric tropone oxides **155** and **156** (equation 66)[160].

(57) $\xrightarrow{PPh_3}$ [$Ph_3\overset{+}{P}$–O] ⟶ (154) (65)

(147) $\xrightarrow{PPh_3}$ (155) + (156) (66)

Valence tautomerism allows the synthesis of arene oxides which are otherwise very difficult to obtain. Thus, *anti*-benzene dioxide (**157**) has been prepared from the endoperoxide **47** resulting from singlet oxygenation of benzene oxide **46**, a tautomer of the oxepin **158** (equation 67)[161]. The occurrence of two epoxide groups in *trans* fashion on adjacent sites makes the cyclohexene ring in **157** planar and the double bond is stabilized due to a 'locked-in' configuration. A stable *anti* diepoxide **159** has been obtained similarly from the epoxyendoperoxide **160** (equation 68)[161].

(158) ⇌ (46) $\xrightarrow{^1O_2}$ (47) $\xrightarrow{PPh_3}$ (157) (67)

(160) $\xrightarrow{P(OPh)_3}$ (159) (68)

In the cycloheptatriene (T)-norcaradiene(N) tautomerism, the presence of an electron-withdrawing group favours the latter species. On triphenylphosphine reduction, the easily obtained endoperoxides **161** from cycloheptratrienes give the stable unsaturated *anti* epoxides **162** (equation 69)[162].

T ⇌ N $\xrightarrow{^1O_2}$ (161) ⟶ (162) (69)

X = H, COOMe, CH (*endo* or *exo*) and Ph

Triphenylphosphine reduction has been used in the elegant synthesis of the cyclobutene annelated α-diketone **163** starting from the endoperoxide **164** (equation 70)[163].

PPh3 (70)

(164) (163)

When the desired stereochemical arrangement for a S_N2' displacement of the intermediate zwitterion (see equation 65) cannot be achieved, side-reactions will determine the products. Reaction of ascaridole (**2**) with triphenylphosphine gives a complex mixture among which the epoxide **165** is the major product (equation 71)[164]. The triphenylphosphine deoxygenation of the endoperoxide **166** does not give the expected unsaturated epoxide **167** but produces the cyclic ether **168** (equation 72)[64].

PPh3 + + + OH (71)

(2) (165)

PPh3 PPh3 (72)

(167) (166) (168)

B. Thermolysis

Like reduction, thermolysis can also be carried out under several conditions in polar and nonpolar solvents and the nature of the products depends upon the structural features of the substrates. Thermolysis of endoperoxides is most intriguing in view of the chemistry of peroxide rearrangements and their use for biomimetic-type synthesis of certain natural products.

1. Release of molecular oxygen

While many unsaturated endoperoxides are prepared by 1,4-cycloaddition of 1,3-dienes with singlet oxygen, the latter two can be regarded as the dissociation products of the endoperoxides in retro-Diels–Alder fashion. In cases where endoperoxides gain significant resonance stabilization upon loss of oxygen, this dissociation indeed takes place under thermolytic conditions (equation 73). Thus, heating of certain arene endoperoxides generates singlet oxygen with high efficiency. When substrates cannot be oxidized with singlet oxygen under photolytic conditions due to their photochemical instability, this controlled thermolytic generation of singlet oxygen can be used as a preparative tool. Several aromatic endoperoxides suitable for singlet oxygenation under various reaction conditions have been synthesised[76]. These include the water-soluble naphthalene endoperoxide **169** which is useful in the singlet oxygenation of polar biological substrates in aqueous systems[165]. A polymeric naphthalene endoperoxide **170** is formed below 0°C and liberates singlet oxygen at room temperature ($\simeq$25°C)[166]. This oxygen carrier **170** can be used with or without solvent and has the potential for application in insect-repellent systems because of its mild reversible toxicity[166].

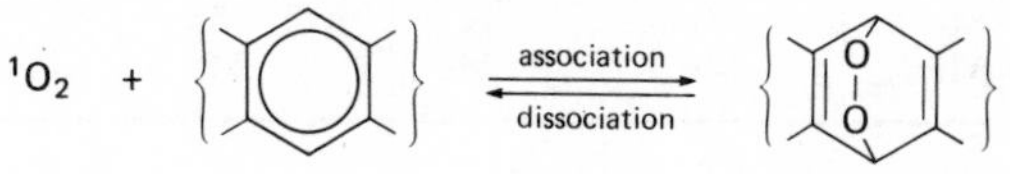

$$^1O_2 + \text{arene} \underset{\text{dissociation}}{\overset{\text{association}}{\rightleftharpoons}} \text{endoperoxide} \qquad (73)$$

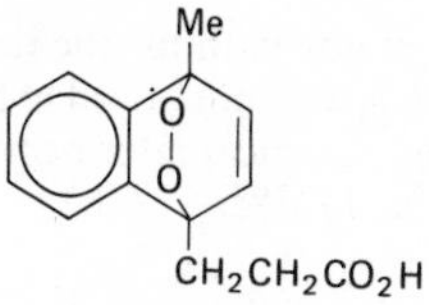

(**169**)

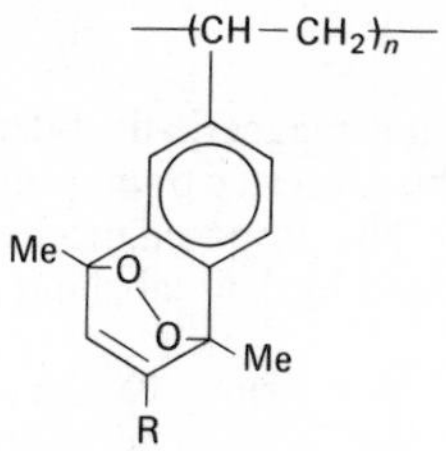

(**170**) R = H or Me

The thermal conditions necessary for the liberation of singlet oxygen from arene endoperoxides highly depend on their structures, particularly on the extent and nature of the nuclear substitutions. Wasserman and Larsen, in their study of the thermal decomposition of polymethylnaphthalene 1,4-endoperoxides **171**, have concluded from the data of half-lives (Table 5) that the more the relief of the *ortho* and *peri* steric strains in the parent naphthalene by endoperoxide formation, the more stable are the endoperoxides[79,81].

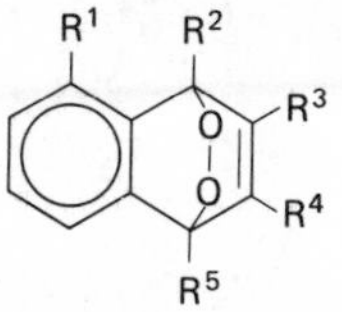

(**171**)

In spite of the principle of spin conservation, singlet oxygen released from endoperoxides under normal conditions is associated with triplet oxygen (equation 74). Turro and collaborators[167] have studied the thermal decomposition of the anthracene 9,10-(**136**, **172**) and 1,4-(**173**, **174**)endoperoxides. The thermal process to generate singlet oxygen approaches ca. 100% for **173** and **174**, whereas **136** and **172** give only 35% and 50% singlet oxygen, respectively. They have also studied the effect of a magnetic field on the thermolysis of these endoperoxides[168]. The rate of singlet–triplet conversion is expected to increase proportionally to the strength of the magnetic field. Singlet oxygen yield from the 1,4-endoperoxide **173** has been found to be unchanged by application of an external magnetic field, whereas under the same conditions a substantial change in the singlet oxygen yield has been observed for the 9,10-endoperoxide **136**. Turro has suggested that in an initial step **136** forms a singlet diradical (^{1}D) which competitively fragments to singlet oxygen ($^1O_2$) or intersystem-crosses to ^{3}D which releases triplet oxygen ($^3O_2$)[168]. The ^{1}D

TABLE 5. Half-life ($\tau_{1/2}$) of polymethylated naphthalene 1,4-endoperoxides (**171**)[76,81]

R^1	R^2	R^3	R^4	R^5	$\tau_{1/2}$ (h)
H	Me	H	H	Me	5[a]
Me	Me	H	H	H	30[a]
Me	Me	H	H	Me	290[a]
H	Me	Me	H	Me	70[a]
H	Me	Me	Me	Me	47[b]

[a] At 25°C.
[b] At 50°C.

to ^{3}D conversion is magnetic-field-dependent. The major path for the decomposition of **173** or **174** may be concerted or may involve ^{1}D which is too short-lived to be influenced by a magnetic field. The proposed mechanism involving diradical intermediates ^{1}D and ^{3}D has been confirmed by ^{17}O labelling experiments on **172**[168].

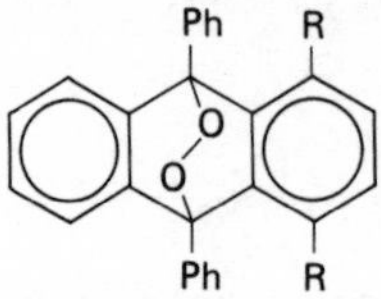

(**136**) R = H
(**172**) R = Me

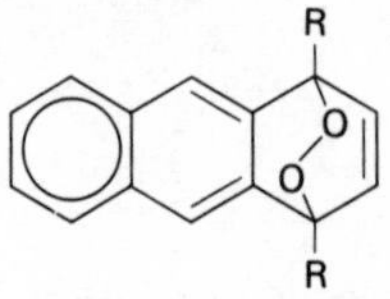

(**173**) R = Me
(**174**) R = OMe

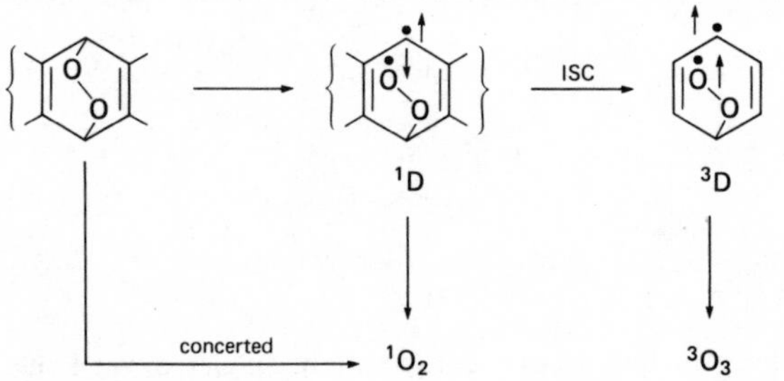

(74)

2. Cleavage of O—O bond and fragmentation

The most common reaction of unsaturated endoperoxides is the cleavage of the weak O—O bond followed by addition of the oxygen radicals to adjacent double bonds leading to *syn* bis-epoxides (equation 75). This method has been widely used for the synthesis of a number of cyclic bis-epoxides including natural products (*vide infra*). Since extensive reviews[45e,152,153] have already appeared, examples for epoxide formation will not be discussed here in detail. A typical example may be found in the thermolytic reaction of the endoperoxide obtained from β-damascenone (**175**) producing the diepoxyvinyl ketone **176** (equation 76)[169]. The endoperoxide **177** with *syn* configuration is preferentially formed from its diene procursor by singlet oxygenation and its thermolysis gives **178** exclusively. Interestingly, the diepoxyacetate **178** has all-*cis* configuration (equation 77)[64].

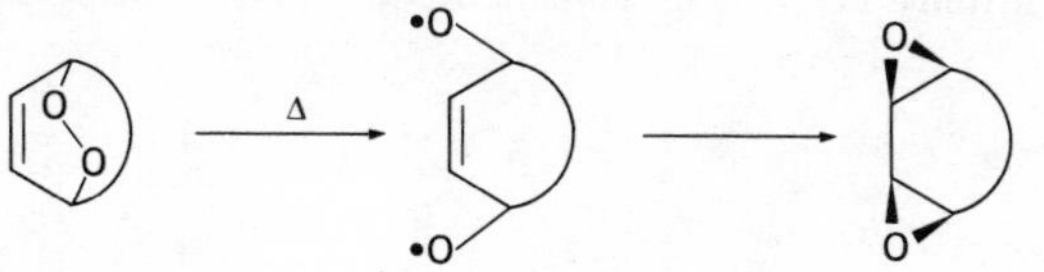

(75)

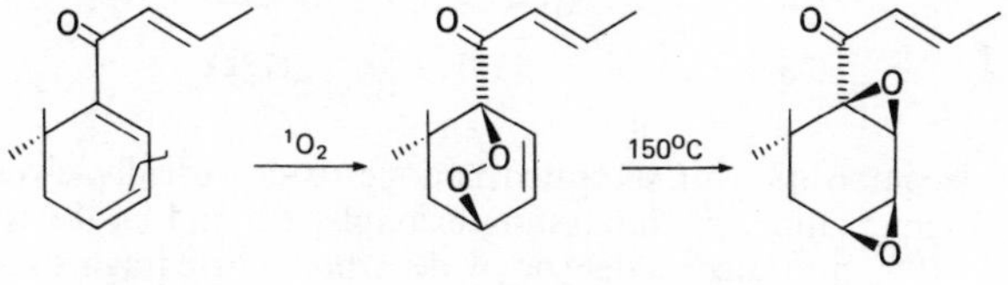

$^1O_2$ 150°C

(76)

(175) (176)

toluene 160°C

(77)

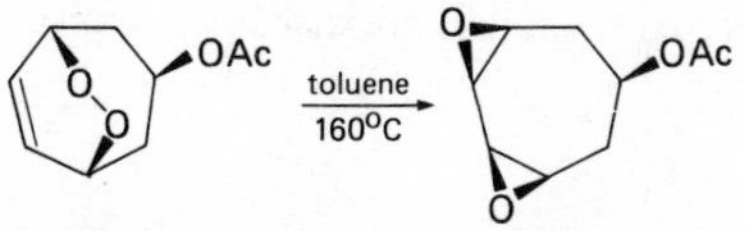

(177) (178)

Rearrangements of arene endoperoxides have also been noted to produce diepoxides as exemplified in equation (78)[170]. Thermolysis of **50b** obtained from 9,10-dimethoxyanthracene gives products of deep-seated rearrangement by way of a diepoxide as indicated in equation (79)[77,171].

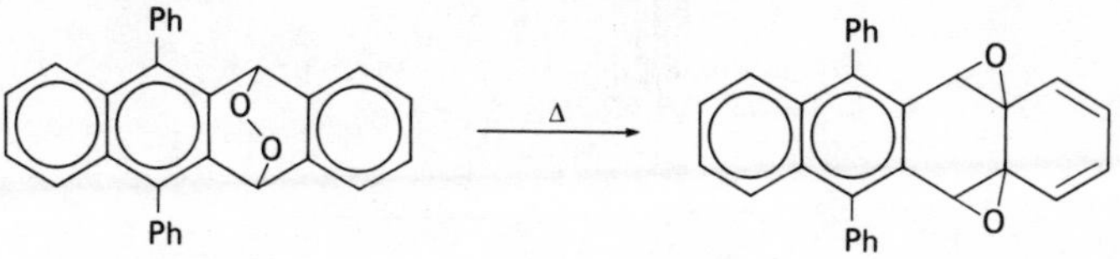

(78)

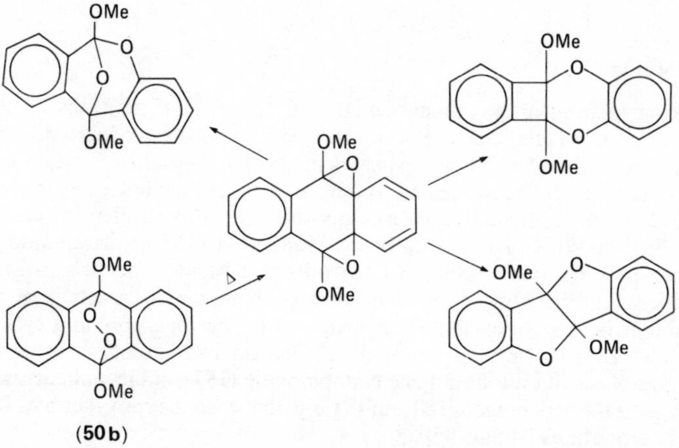

(79)

(50b)

In cases where endoperoxides are highly strained or perturbed by double bonds or by epoxide groups, other side-reactions become more important. Thermolysis of **145** gives the bis-epoxide **179** in only 11% yield, with the major product being the epoxyenone **180** (equation 80)[70].

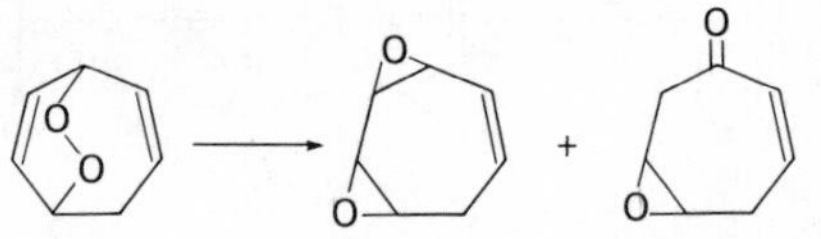

(145) (179) (180) (80)

Thermolytic decomposition of saturated endoperoxides usually gives the products resulting from fragmentation. An interesting example is found in **181** which is cleanly transformed at -10°C into succinaldehyde by decarbonylative fragmentation (equation 81)[62]. Decomposition of **182** gives succinaldehyde and ethylene (equation 82)[172]. Other examples for fragmentation of endoperoxides are shown in equations (46) and (47). Reactions of prostaglandin endoperoxide models are discussed in more detail in Section IV.F.

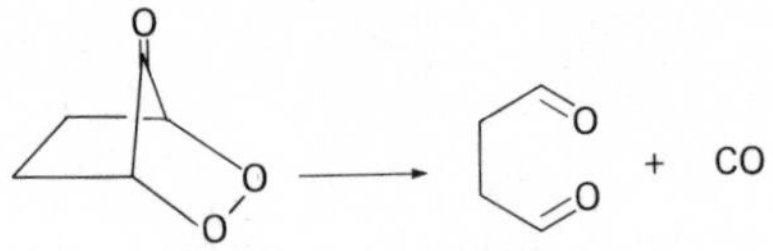

(181) + CO (81)

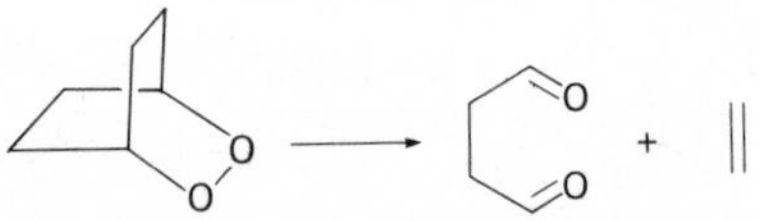

(182) (82)

C. Photolysis

By making theoretical assignments of the electronic configurations of peroxides in excited states, Kearns predicted that long-wavelength photolysis of peroxides should lead to the cleavage of the O—O bond, whereas at short wavelengths C—O bond cleavage should be observed[173]. The two reaction courses are indeed confirmed by the photolysis of ascaridole (**2**) giving a different product composition at 366 nm and at 185 nm as shown in equation (83)[174]. With high energy being available at 185 nm irradiation, the retro-Diels–Alder products (**183**, **185**, **186**) are obtained besides the isoascaridole **184**[174]. Oxygen liberated, probably in the excited state, immediately reacts with the solvent cyclohexane giving rise to cyclohexyl hydroperoxide, cyclohexanol and cyclohexanone. Isoascaridole (**184**) is the major product at 366 nm irradiation[175]. Long-wavelength (366 nm) photolysis of cyclohexadiene endoperoxide (**187**) and levopimaric acid methyl ester (**190**) gives the bis-epoxides **188** and **191** together with the epoxyketones **189** and **192**, respectively (equations 84 and 85)[175].

(2)

185 nm

366 nm

(83)

(183) (184) (185) (186)

+ O_2^*

366 nm

(84)

(187) (188) (189)

OH

366 nm

(85)

(190) (191) (192)

CO_2Me

As mentioned earlier, thermolysis of some arene endoperoxides gives preferentially the parent arenes rather than the corresponding bis-epoxides. In such cases, photolysis at long wavelength serves as a useful method for the preparation of bis-epoxides. Endoperoxide **193** is cleanly transformed into naphthalene diepoxide **194** upon photolysis, whereas heating the endoperoxide leads to oxygen extrusion to give 1,4-dimethoxynaphthalene (equation 86)[176]. Rigaudy and coworkers have recently succeeded in isolating the anthracene diepoxide **195** by irradiation of anthracene 9,10-endoperoxide (**196**) at wavelengths greater than 435 nm (equation 87)[177]. Upon photolysis at shorter wavelengths or by thermolysis, **195** rearranges to dioxan **197** by way of **198**.

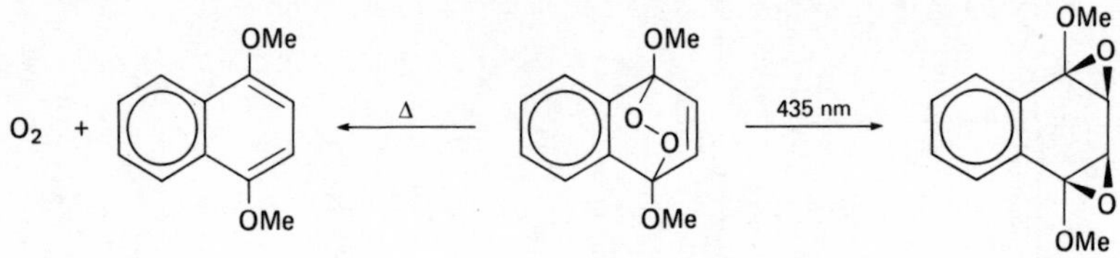

(86)

(193) (194)

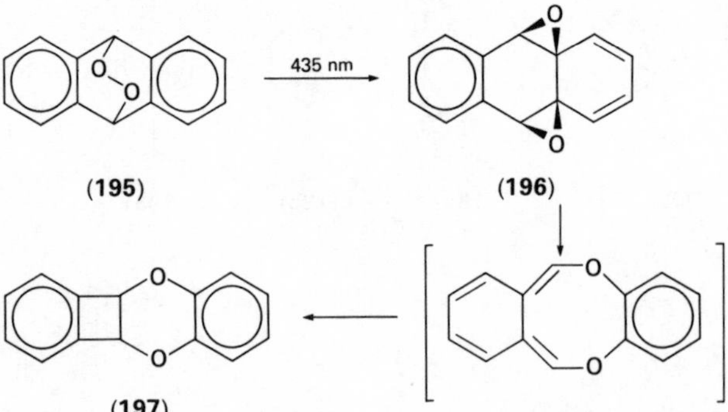

(87)

(195) (196)

(197) (198)

D. Base- and Acid-catalysed Reaction

Base-catalysed rearrangements of peroxides proceed via a general type of β-elimination mechanism to give hydroxyketones (equation 88)[178]. The application of this reaction to bicyclic endoperoxides and the oxidation of the resulting monocyclic hydroxyketones provide a useful method for the synthesis of 1,4-diketones from cyclic 1,3-dienes. Triethylamine-catalysed rearrangement of 1,3-cyclooctadiene endoperoxide (**199**) produces the hydroxyketone which upon oxidation with manganese dioxide gives 1,4-cyclooctenedione (**200**) in high yield (equation 89)[179]. Substituted *p*-homobenzoquinones (**201**) can be synthesized efficiently by a similar route (equation 90)[180]. Treatment of tropone endoperoxide (**147**) with triethylamine results in cleavage of the peroxide bond to give 5-hydroxytropolone (**202**) in quantitative yield (equation 91)[181].

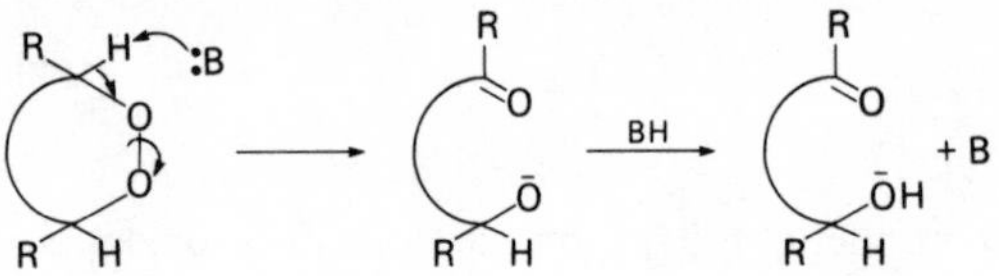

(88)

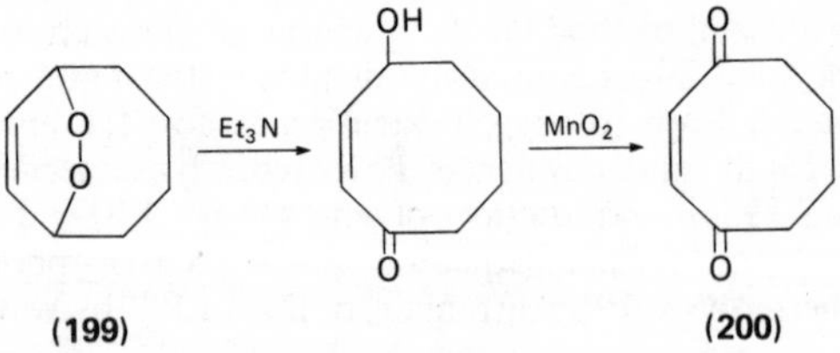

(89)

(199) (200)

$$\xrightarrow{Et_3N} \xrightarrow{MnO_2} \quad (90)$$

(201)

X = H, CHO, CO_2Me

$$\xrightarrow{Et_3N} \longrightarrow \quad (91)$$

(147) (202)

An interesting quinone cyclophane (**203**) can be prepared in high yield by treatment of the endoperoxide **62** (cf. equation 29) with methanolic potassium hydroxide followed by manganese dioxide oxidation (equation 92)[86].

1. MeOH–KOH
2. MnO_2 (92)

(62) (203)

The endoperoxide **204** obtained by addition of singlet oxygen to 2,3-bis(methylene)-7-oxanorbornane (**205**) is rearranged to the chiral γ-hydroxy α,β-unsaturated aldehyde **206** by catalytic amounts of various natural bases such as (+)-quinidine, (−)-cinchonidine and (−)-ephedrine with an enantiomeric excess ranging up to 46%[182]. It is envisaged that the hydroxy group in a base could lead to better asymmetric induction because of the possibility of hydrogen bonding between the oxygen bridge of **204** and the catalyst (equation 93). Some examples of the base-catalysed rearrangements of saturated endoperoxides are described in Section IV.F.

$$\xrightarrow{^1O_2} \xrightarrow{base} \quad (93)$$

(205) (204) (206)

X = O

Acid-catalysed reaction of unsaturated endoperoxides is more complicated. In the case of arene endoperoxides the products resulting from 1,4-endoperoxide–dioxetane rearrangement are often observed. For example, 1,4-dimethoxyanthracene 1,4-endoperoxide under acidic conditions gives two sets of products: in aqueous acid the *p*-quinone **207** is obtained exclusively, whereas the reaction in anhydrous acidic media gives rise to the *o*-quinone **208** and the aldehyde ester **209** (equation 94)[183]. In a similar way the isolated bridged [14]annulene endoperoxide **84** rearranges over silica gel to give **210, 211** and **212** (equation 95)[103]. The formation of both **211** and **212** has been explained in terms of 1,4-endoperoxide–dioxetane rearrangement.

OMe OMe (178) aq.HCl anhydrous H^+ (207) MeO OMe (94)

(208) OMe + OMe CHO OMe (209)

(84) SiO_2 (210) (95)

(211) CHO CHO (212)

Le Roux and Goasdoue reported the isolation of the dioxetane **214** in the acid-catalysed rearrangement of the 1,4-endoperoxide **213** obtained by singlet oxygenation of a tetraarylfulvene (equation 96)[184]. More recently, Schaap and collaborators have demonstrated that the isolated endoperoxide **215** is quantitatively converted to the dioxetane **216** by means of silica gel (equation 97)[98].

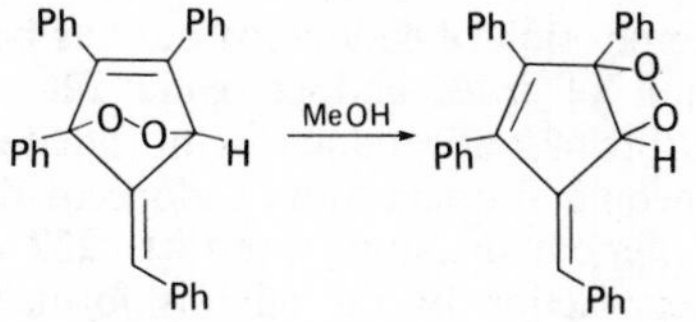

(213) —MeOH→ (214) (96)

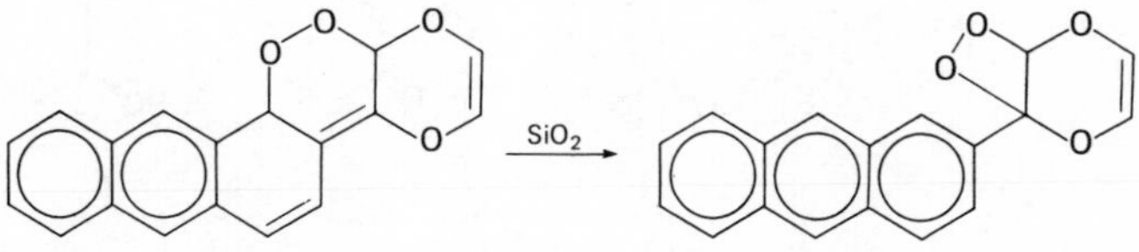

(215) —SiO_2→ (216) (97)

Acid-catalysed transformations of acene *meso* peroxides have been studied extensively by Rigaudy and his coworkers. For example, treatment of 9,10-diphenylanthracene 9,10-endoperoxide (**136**) with strong acid in aqueous media gives **217** and **218**, while acid treatment under anhydrous conditions produces dibenzo[*b*,*e*]oxepin **219** (equation 98)[185].

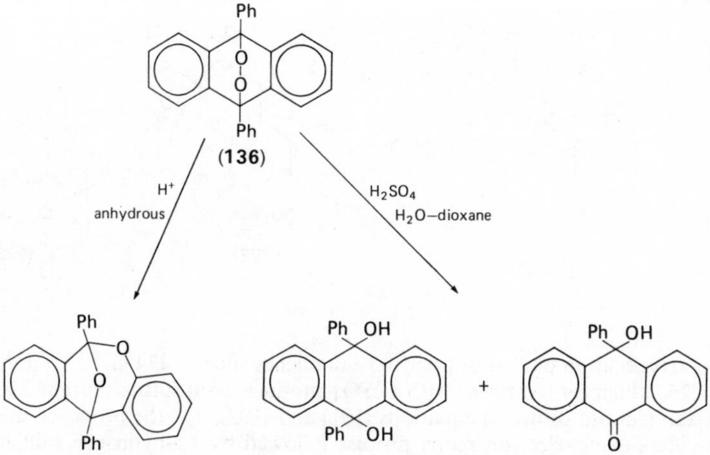

(136) —H^+, anhydrous→ (219); (136) —H_2SO_4, H_2O–dioxane→ (217) + (218) (98)

E. Metal-catalysed Reaction

The reactions of alkyl peroxides with organometallic compounds are extremely varied and depend both on the character of the peroxide and of the metal. Such processes have been reviewed[186] and are also discussed in other chapters of this volume. Hence, we have limited ourselves to selected recent examples of the metal-catalysed reactions of bicyclic endoperoxides.

The Fe(II)-induced decomposition of endoperoxides has been studied extensively by Turner and Herz. Reaction of *trans* endoperoxide **220** with $FeSO_4$ in aqueous tetrahydrofuran gives a mixture of four products as indicated in equation (99)[187]. Fe(II)-induced decomposition of levopimaric acid epoxy endoperoxide (**221**), however, results in the unusual formation of remote oxidation products **222** and **223** as the result of intramolecular hydrogen abstraction by the initially formed anion radical (equation 100)[188]. A general scheme for the initial step of the reaction of endoperoxides with Fe(II) is proposed, involving a one-electron redox process[187,189].

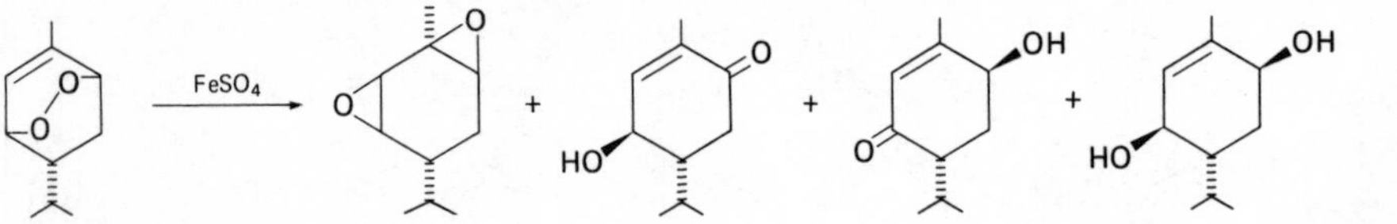

(**220**) (99)

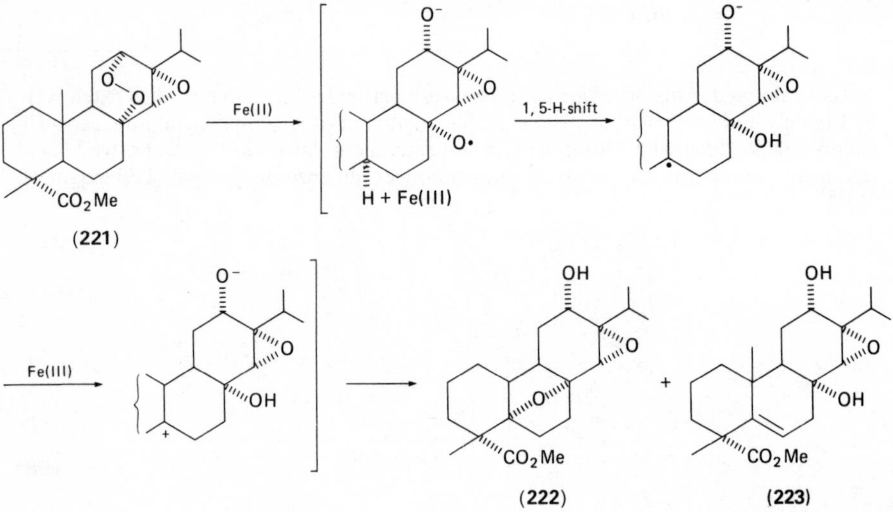

(**221**) (**222**) (**223**) (100)

Singlet oxygenation of 2-substituted 1,3-butadienes such as **224** gives 3,6-dihydro-1,2-dioxins **225**, which on treatment with $FeSO_4$ produce 3-substituted furans **226** in high yields. Examples are shown in equations (101) and (102), and the proposed mechanism again involves a one-electron redox process followed by 1,5-hydrogen shift (equation 103)[190]. The overall sequence constitutes a model for the biogenesis of naturally occurring 3-alkylfurans.

(224) → (225) $\xrightarrow{FeSO_4 \text{, aq.THF}}$ (226) (101)

R =

$\xrightarrow{^1O_2}$ $\xrightarrow{FeSO_4}$ (102)

Fe(II), + Fe(III), ~1,5-H-shift, Fe(III), H^+ (103)

Catalytic rearrangement of unsaturated endoperoxides to *syn* diepoxides also proceeds with cobalt *meso*-tetraphenylporphine (CoTPP)[191]. The rearrangement occurs in a stereospecific fashion under mild conditions. Foote and coworkers have demonstrated that endoperoxide **227** gives tetraepoxide **228** in 99% yield with 5 mol % CoTPP in toluene even at −78°C (equation 104)[191]. Neither zinc *meso*-tetraphenylporphine nor *meso*-tetraphenylporphine, both of which are weaker reductants than CoTPP, promote the rearrangement. A mechanism involving complex formation between the endoperoxide and the catalyst rather than an electron-exchange mechanism has been proposed.

(227) $\xrightarrow{CoTPP}$ (228) (104)

Noyori and collaborators have studied the $Ru(II)Cl_2(PPh_3)_3$-catalysed decomposition of saturated 1,4-endoperoxides[192]. The reaction is proposed to proceed through an inner-sphere radical such as **229** to give a mixture of products as exemplified in equation (105)[192]. Palladium (0)-catalysed reaction of bicyclic 1,4-endoperoxides has also been reported[193]. The reaction of **230** in dichloromethane in the presence of a catalytic amount of $Pd(PPh_3)_4$ (5 mol %) gives **231**, **232** and **233** (equation 106)[193]. The formation of **231** is suggested as involving the insertion of Pd(0) into the O—O linkage (path a) or a back-side S_N2 displacement by Pd(0) to generate the zwitterion **234** (path b). Subsequent hydrogen reorganization, leading to **231**, occurs via a palladium hydride species formed by β-elimination. The diol **233** is thought to arise from an inner-sphere radical **235** formed by Pd(0)/Pd(I) one-electron redox reaction (equation 107). The reaction with prostaglandin endoperoxides will be discussed in the next section.

5% Ru(II) (105)

(229)

Pd(o) (106)

(230) (231) (232) (233)

230 → (a: Pd(o)) → HPdO … → −Pd(o) → 231 (107)

230 → (b: Pd(o)) → $-\overset{+}{Pd}H$ → 231

(234) * = +, −

(235) * = •

(233)

In the presence of $SnCl_2$, endoperoxide **236**, derived from singlet oxygenation of dihydropyridine, reacts with carbon nucleophiles such as trimethylsilyl enol ethers, enamines, vinyl ethers, indoles, pyrroles and furans to give products **237**, as exemplified in equation (108)[122,194]. The reaction is believed to proceed through the intermediate **238**. Pyrrole endoperoxide **239** also undergoes similar reactions with carbon nucleophiles to produce 2-substituted pyrroles as typically seen in equation (109)[119].

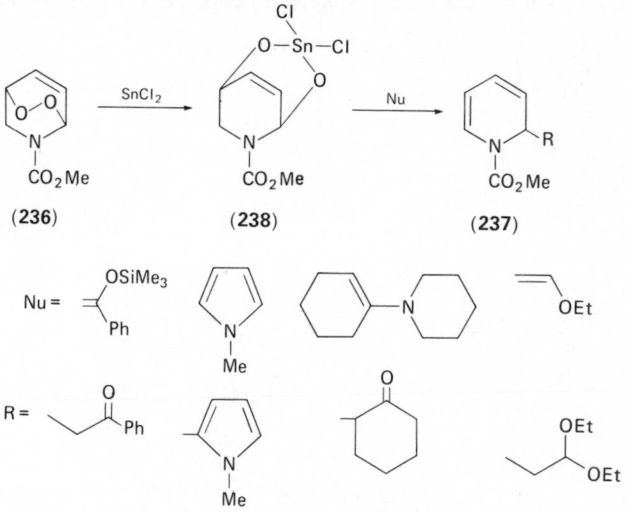

(236) (238) (237) (108)

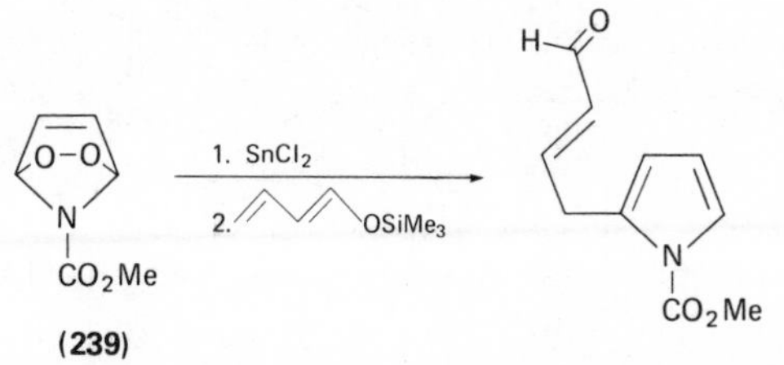

(239) (109)

F. Reaction of Prostaglandin Endoperoxides and Their Model Compounds

In the beginning of this chapter it has been mentioned that the prostaglandin endoperoxides like PGH_2 (**3**) and PGG_2 (**28**) serve as biological precursors for other prostanoids such as PGE_2 (**4**), $PGF_{2\alpha}$ (**5**), thromboxane (**6**) and prostacycline (**7**). In connection with these biological transformations, extensive studies have been devoted to the chemical reactions of prostaglandin endoperoxides and their model compounds. Many earlier examples have been covered in recent reviews[16,195]. The thermolysis at 150–200°C and the photolysis at 310 nm of substituted 1,2-dioxolanes (**240**) have been investigated by Adam and Duran as the simplest models of prostaglandin endoperoxides[196]. On the basis of product analysis, the mechanism of the reaction is postulated to proceed via the radical **241** as a key intermediate to give the products of cyclization **242** and of fragmentation **243** and **244** (equation 110). The 2,3-dioxabicyclo[2.2.1]heptane system has been recognized as a more suitable model of prostaglandin endoperoxides, and much interest has been devoted to its preparation and reactions. The thermal reactions of the endoperoxides of the 2,3-diphenyl (**245**) and the 2,3-diphenyl-4,5-dichloro(**246**) derivatives give products which are believed to arise from the dioxy diradical **247**[54]. The products in the thermolysis of **245** in an NMR tube at 78°C in C_6D_6 are **248** and ethylene. Under the same conditions, **246** gives acetophenone and **249**. Although different types of products are obtained from **245** and **246**, a similar diradical intermediate **247** has been postulated in both cases (equation 111)[54].

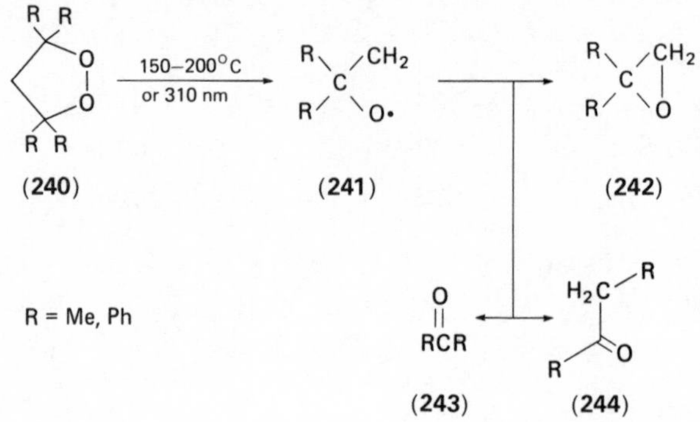

(110)

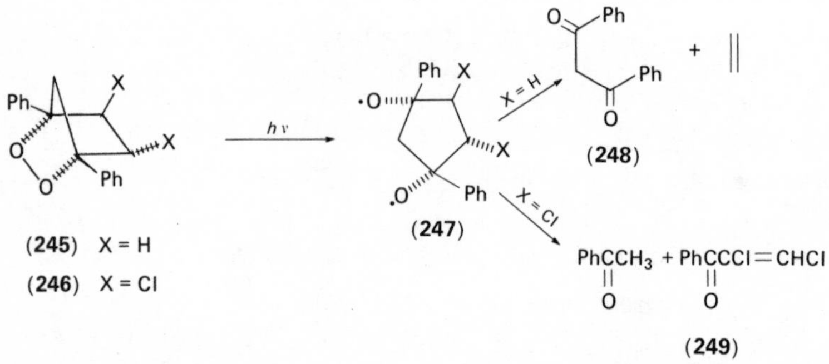

(111)

The thermal decomposition of the unsubstituted 2,3-dioxabicyclo[2.2.1]heptane **20**, however, leads to an unusual product mixture of **250**–**253** (equation 112)[197]. The relative yields of these products are strongly solvent-dependent. Further, the rate of decomposition increases with solvent polarity, protic solvents being exceptionally efficient (Table 6). The data obtained by NMR analyses suggest that there might be two mechanisms operative: (*i*) a nonpolar mechanism giving **250** and **252** slowly, and (*ii*) a polar mechanism producing **251** and **253** fast. The epoxy aldehyde **250** is believed to arise from simple homolysis of the peroxide bond of **20** followed by ring-opening to nonpolar diradical **254a**. Cyclization would produce the observed product **250** (equation 113)[197]. For the formation of **251** and **253**, a polar rearrangement mechanism involving **254b** rather than a diradical mechanism has been suggested, since they are favoured in polar and protic solvents. In connection with this, Zagorski and Salomon have studied the kinetics and mechanism of the amine-catalysed fragmentation of **20**[198]. They suggest that the amine-catalysed fragmentation leading to **251** and the disproportionation to give **253** are closely related mechanistically. Rate-determining cleavage of the bridgehead C—H bond of **20** generates a keto alkoxide **255** which undergoes either retro-aldol cleavage leading to **251** or protonation to **253** (equation 114)[198]. Imidazole, which is an efficient proton abstractor as well as a proton donor, produces a 1:1 mixture of **251** and **253** from **20**[198], whereas the aprotic bicyclic trimethylene diamine, DABCO, produces **251** exclusively[35].

(20) —Δ→ (250) + (251) + (252) + (253) (112)

20 → (254a) → 250

20 → (254b) → 251 + 253 (113)

(20) → (255) ⇌ H^+ (253)

(255) → H^+ → (251) (114)

TABLE 6. Thermal decomposition of 2,3-dioxabicyclo[2.2.1]heptane (**20**) in various solvents[197]

Solvent	Dielectric constant (73°C)	Reaction temp. (°C)	$\tau_{1/2}$ (h)	Products, mol % **250**	**251**	**252**	**253**
Cyclohexane	1.94	73	5.7	97	2	1	0
CCl_4	2.13	73	4.1	87	10	3	0
Benzene-d_6	2.18	73	2.9	86	11	3	0
Chlorobenzene	4.85	73	2.4	85	15	0	0
CD_3CO_2D	6.63	73	0.22	0	100	0	0
2-Butanone	14.35	73	1.5	63	35	2	0
CD_3CN	28	73	1.3	40	59	1	0
D_2O	74	40	0.12	0	72	0	28

The decomposition of yet another model, 1,5-dimethyl-6,7-dioxabicyclo[3.2.1]octane (**126**), has also been studied[141]. In contrast to the [2.2.1] system, the [3.2.1] model suffers cleavage of the one-carbon bridge giving **256, 257** and **258** (path a) and of the three-carbon bridge giving **259** (path b). A nonconcerted biradical mechanism, as shown in equation (115), has been proposed. The study of the decomposition of **126** under various conditions has led to the establishment of the conditions for the preferential formation of the monocyclic ether **259**, the keto epoxide **256** or the bicyclic ether, frontalin (**258**), which is the naturally occurring pine beetle pheromone (Table 7)[141].

Metal-catalysed decomposition of prostaglandin endoperoxide models has also been investigated. As mentioned above **20** decomposes in the presence of Ru(II) or Pd(0).

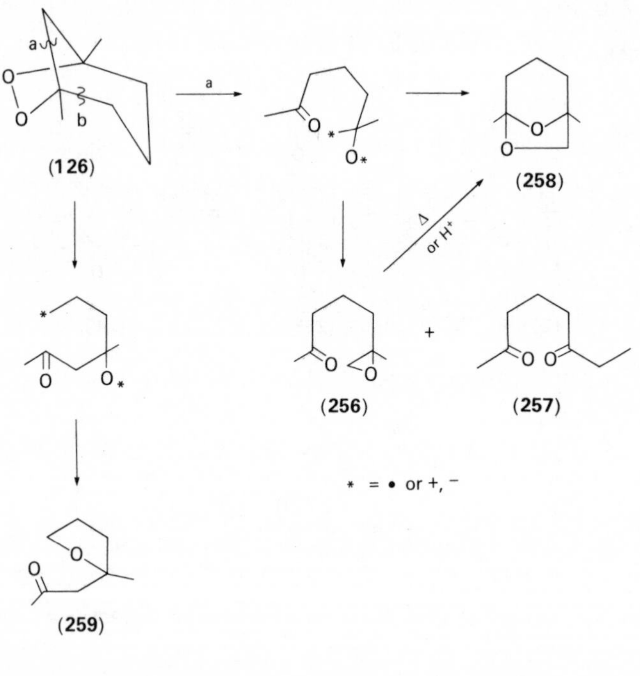

(115)

TABLE 7. Decomposition of 1,5-dimethyl-6,7-dioxabicyclo[3.2.1]octane (**126**)[54]

Decomposition conditions	Yield, %			
	256	**257**	**258**	**259**
1. Thermal: 400°C vapour phase		21	40	5
2. *hv* (argon laser): direct irradiation, $CFCl_3$, −20°C	100			
3. *hv*: Ph_2CO sensitized, $CFCl_3$, −20°C			100	
4. $TiCl_4$, CH_2Cl_2, 0°C		95		
5. $AlCl_4$, $CDCl_3$, 43°C		45		
6. TsOH, THF, reflux		5	5	50
7. $FeCl_3$, THF, H_2O, reflux		5	9	34

Reaction of **20** with $RuCl_2(PPh_3)_3$ (5 mol %) in CD_2Cl_2 at 0°C gives five products as indicated in equation (116)[192]. However, when PGH_2 methyl ester (**27**) is decomposed under the same conditions, methyl (5*Z*, 8*E*, 10*E*, 12*S*)-12-hydroxy-5,8,10-heptadecatrienoate (HHT methyl ester **261**) and malondialdehyde are obtained exclusively (equation 117)[192].

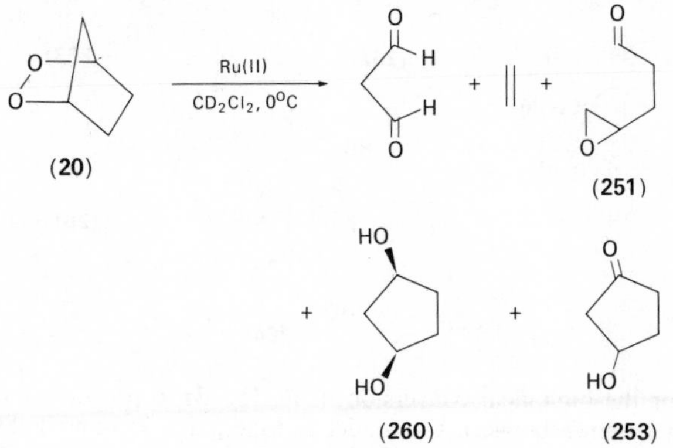

(116)

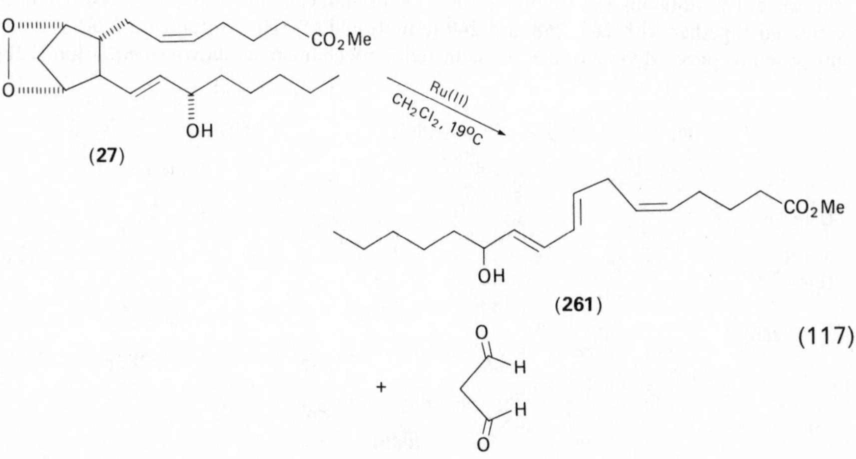

(117)

Similarly, the decomposition of **20** in the presence of $Pd(PPh_3)_4$ (5 mol %) in dichloromethane gives three products, **251**, **253** and **260** (equation 118)[193]. The former two are also obtained in the thermolysis of **20**[197]. Noteworthy is the fact that a similar type of reaction has been observed in the reaction of PGH_2 methyl ester (**27**). When **27** is exposed to 10 mol % of $Pd(PPh_3)_4$ in dichloromethane at 19°C, a mixture of methyl esters of PGD_2 (**262**) (17 %), PGE_2 (**263**) (11 %), $PGF_{2\alpha}$ (**264**) (41 %) and HHT (**261**) (4 %) is produced (equation 119)[193].

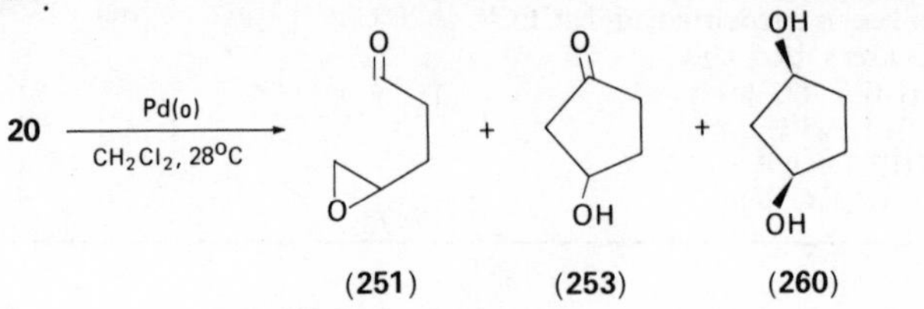

20 —Pd(o), CH_2Cl_2, 28°C→ (**251**) + (**253**) + (**260**) (118)

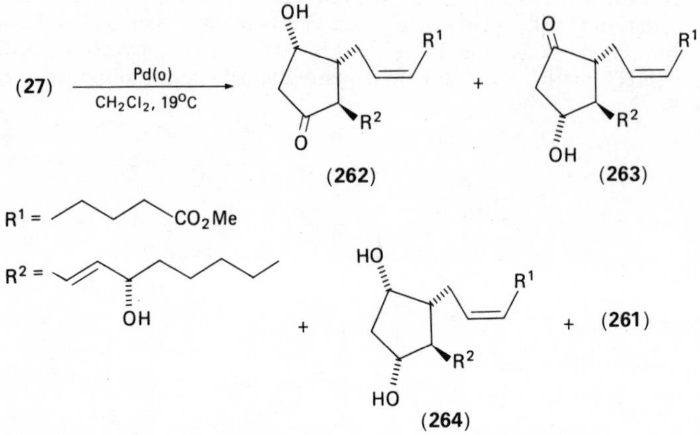

(**27**) —Pd(o), CH_2Cl_2, 19°C→ (**262**) + (**263**) + (**264**) + (**261**) (119)

R^1 = ~~~CO_2Me

R^2 = ~~~ (OH)

As a model for the biological conversion of PGH_2 (**3**) to prostacycline (**7**), a one-equivalent redox reaction between the model endoperoxide **265** and $FeSO_4$ has been carried out in aqueous tetrahydrofuran[154]. A prostacycline-type product **266** has been obtained together with **267**, **268** and **269** (equation 120). The formation of **266** is again proposed to proceed via a one-equivalent redox mechanism as shown in equation (121).

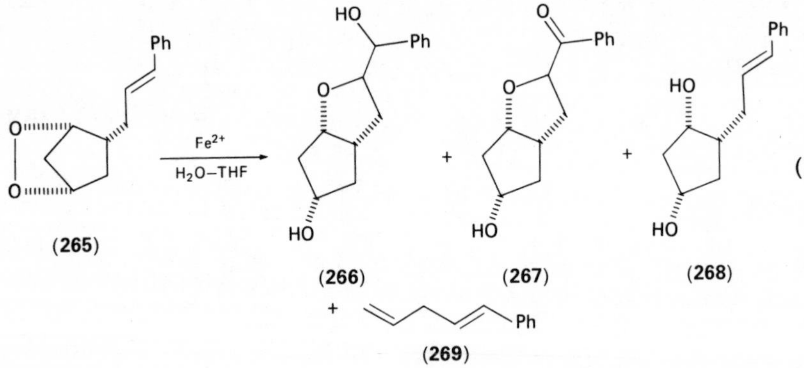

(**265**) —Fe^{2+}, H_2O–THF→ (**266**) + (**267**) + (**268**) + (**269**) (120)

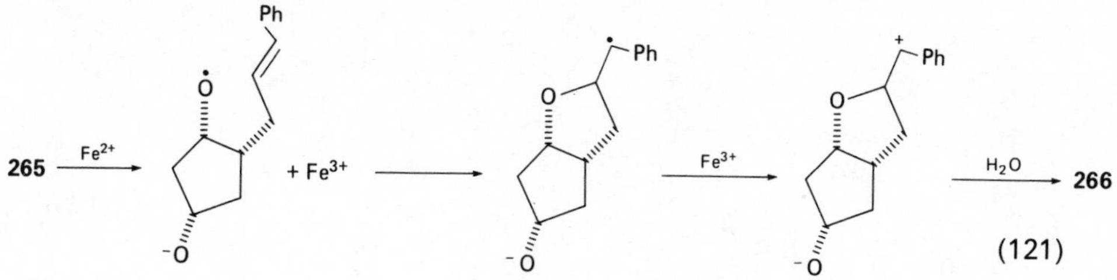

(121)

G. Endoperoxides in Natural-product Synthesis

Endoperoxides have served as important precursors for many types of oxygen-functionalized molecules. As mentioned earlier, the easy accessibility of unsaturated endoperoxides from singlet oxygenation of 1,3-dienes enables the synthesis of a variety of 1,4-oxygenated systems. In this section the utility of endoperoxide reactions as a key step for natural product synthesis is briefly described[153].

The carbon–carbon bond formation effected by $SnCl_2$-catalysed reaction of endoperoxide **270** with indole has been used as a key step in the total synthesis of an indole alkaloid, 3-epiuleine (**271**) (equation 122)[199].

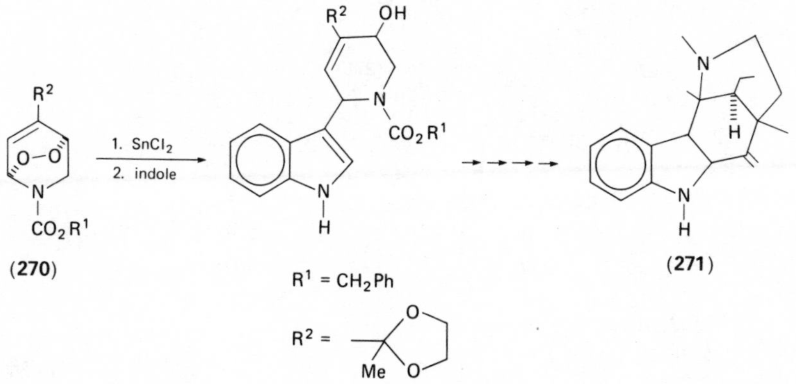

(122)

Acyclic 1,3-dienes react with singlet oxygen to afford endoperoxides which can be rearranged and dehydrated to furans. Demole and coworkers[200] have first illustrated the application of this sequence in their biomimetic conversion of solanone (**272**) to solanofuran (**273**) (equation 123). The endoperoxide **274** undergoes a β-elimination on treatment with alumina to form a hydroxy aldehyde, which on dehydration gives **273**. Matsumoto and Kondo[201] have utilized this method for converting dienes to furans in the synthesis of a variety of furanoterpenes such as perillene[202], neotorreyol[202] and ipomeamarone[203].

(272) $\xrightarrow{^1O_2}$ (274) $\xrightarrow{\text{alumina}}$ → (273) (123)

Base-catalysed rearrangement of endoperoxides to 4-hydroxyenones has been applied in the synthesis of α-agarofuran (**279**) as a means of introducing a bridgehead hydroxy group stereoselectively[204]. Singlet oxygenation of the triene **275** gives mainly the endoperoxide **276** which on treatment with basic alumina is converted to **277** bearing the desired hydroxy group in the α configuration. Subsequent cyclization to **278** followed by reduction affords **279** (equation 124).

(275) $\xrightarrow{^1O_2}$ (276) $\xrightarrow{\text{alumina}}$ (277) → (278)

1. $NaBH_4$
2. $SOCl_2$
3. $LiAlH_4$

(279) (124)

Photosensitized oxygenation has been used for the biomimetic conversion of berberine alkaloids[205]. For example, rose-bengal-sensitized photooxygenation of 13-oxidoberberine (**280**) gives the stable endoperoxide **281** which on photolysis provides berberal (**282**) together with a small amount of **283** (equation 125)[206]. The reductive transformation of endoperoxides has had direct application in the synthesis of cybullol (**286**)[207].

Photooxygenation of **284** gives stereoselectively the endoperoxide **285** due to the steric effect of the angular 10-methyl group. The peroxide is then reduced to give **286** (equation 126).

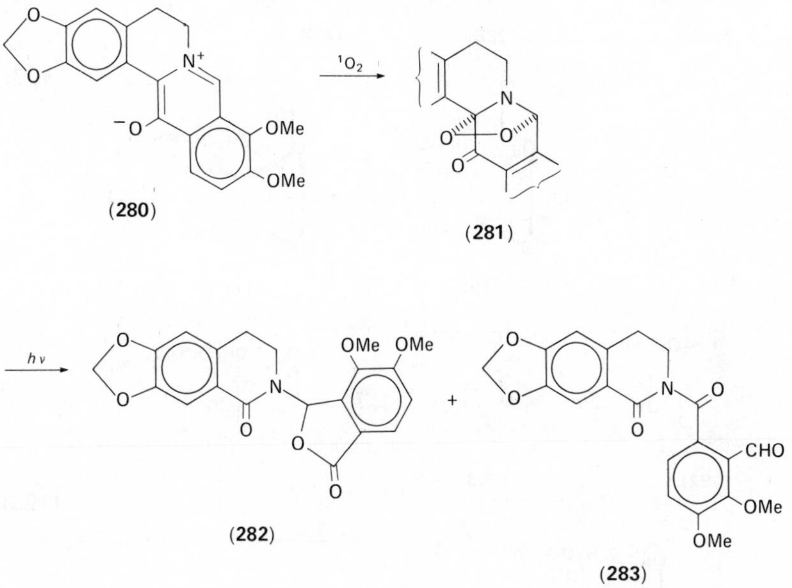

(280) $\xrightarrow{^1O_2}$ (281) $\xrightarrow{h\nu}$ (282) + (283) (125)

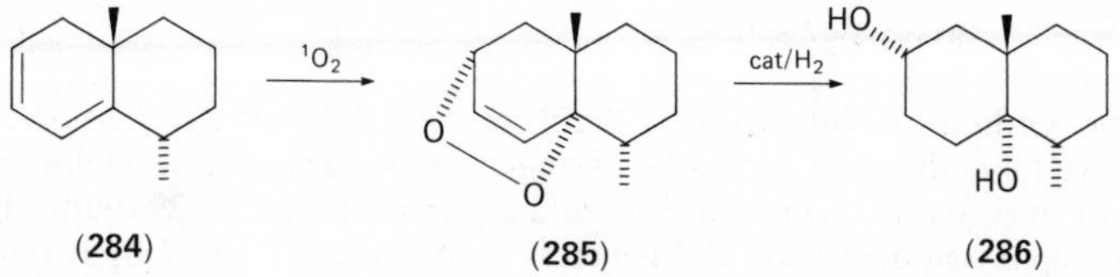

(284) $\xrightarrow{^1O_2}$ (285) $\xrightarrow{cat/H_2}$ (286) (126)

The thermal rearrangement of unsaturated 1,4-endoperoxides to *syn* bis-epoxides has been used in the synthesis of the crotepoxide (**17**) family of naturally occurring 1,3-diepoxides which exhibit tumour-inhibiting or antibiotic activity. White and coworkers[208], have demonstrated that photooxygenation of **287** gives a mixture of the unstable endoperoxides **288** and **289**. Acetylation and thermolysis give the diepoxide **290** resulting from the rearrangement of **288**. Hydrogenolysis and benzoylation of **290** furnish the desired **17** (equation 127). In the synthesis of senepoxide (**291**), Ganem and collaborators[209] have observed that photooxygenation of **292** affords a single crystalline *anti* endoperoxyepoxide **293** which is reduced regioselectively and then hydrolysed to give a mixture of isomeric diols. Acetylation of the desired isomer **294** gives senepoxide **291** (equation 128).

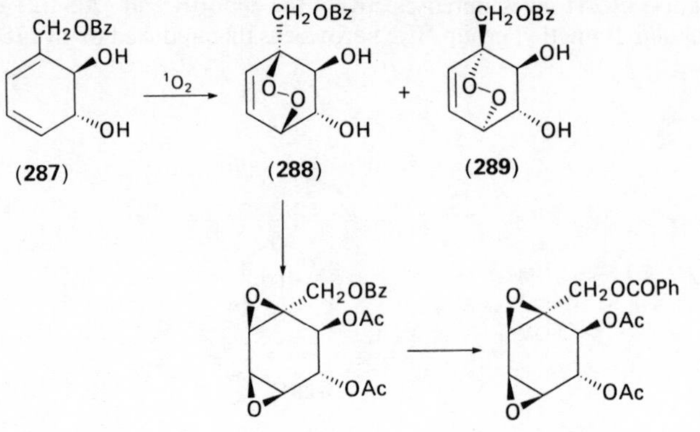

(127)

(287) (288) (289) (290) (17)

(128)

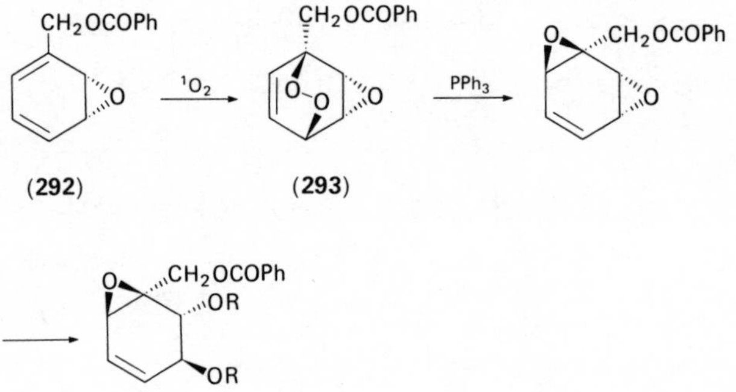

(292) (293)

(294) R = H
(291) R = Ac

An alternative synthesis of crotepoxide (**17**) utilizes the β,β-dimethylstyrene derivative **295**[210]. Successive 1,4-addition of two moles of singlet oxygen gives the bis-endoperoxide **296** along with its epimer. On heating in 1,2-dichloropropane **296** affords the corresponding diepoxyendoperoxide **297** which is then converted to **17** by ozonolysis, reduction and acetylation (equation 129).

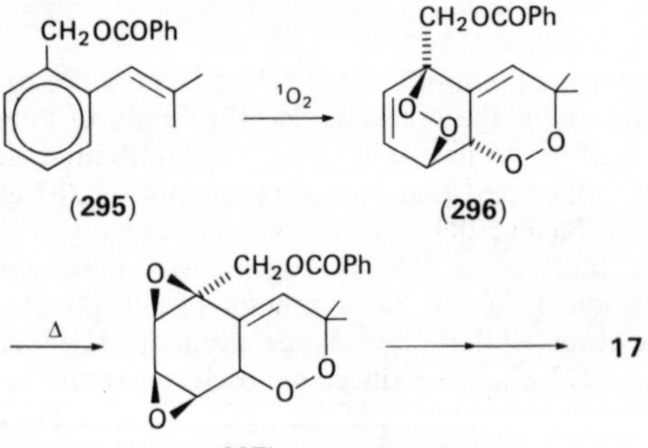

(129)

(295) (296) (297)

The synthetic route to the antileukaemic diterpene diepoxide, stemolide (**299**), involves thermal rearrangement of an endoperoxide to a *syn* diepoxide in the last step[211]. Heating the β,β-endoperoxide **298** under reflux in toluene gives **299** (equation 130).

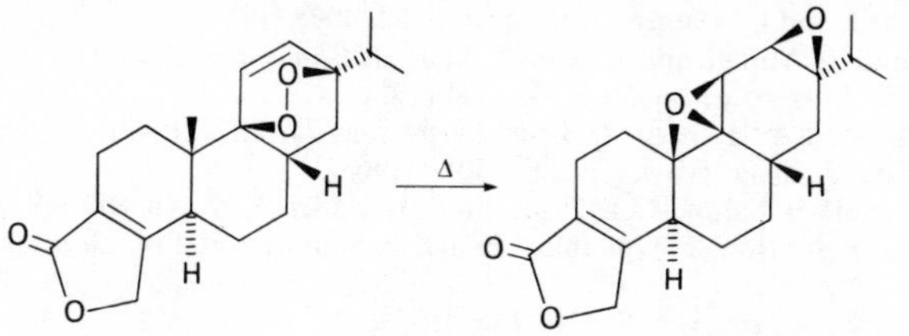

(298) (299) (130)

As indicated in equation (41), the reaction of oxazoles with singlet oxygen leads to triacylamide formation by way of unstable endoperoxides. The oxazole–triacylamide rearrangement provides a means of generating activated carboxylates in the form of triacylamides. Wasserman and coworkers have applied this methodology in the synthesis of macrolides and polyether lactones[212]. A typical example is the synthesis of di-*O*-methylcurvularin (**300**) as outlined in equation (131)[213].

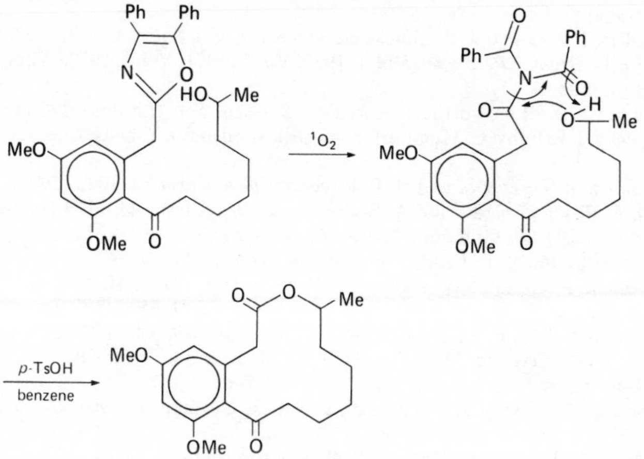

(300) (131)

V. REFERENCES

1. A. Windaus and J. Brunken, *Justus Liebigs Ann. Chem.*, **460,** 225 (1928).
2. P. Wieland and V. Prelog, *Helv. Chim. Acta*, **30**, 1028 (1947).
3. G. Bauslaugh, G. Just and F. Blank, *Nature*, **202**, 1218 (1964).
4. Y. Tanahashi and T. Takahashi, *Bull. Chem. Soc. Japan*, **39**, 848 (1966).
5. S. M. Clarke and M. McKenzie, *Nature*, **213**, 504 (1967).
6. H. K. Adam, I. M. Campbell and N. J. McCorkindale, *Nature*, **216**, 397 (1967).
7. (a) A. N. Alvin, *Phytochemistry*, **15**, 2002 (1976).
   (b) H. K. Bhat, G. N. Qazi, C. L. Chopra, K. L. Dhar and C. K. Atal, *J. Indian Chem. Soc.*, **56**, 934 (1979).

8. V. N. Sviridonov and L. I. Strigina, *Khim. Prir. Soedin.*, 669 (1976).
9. (a) J. Diak, *Planta Med.*, **32**, 181 (1977).
   (b) B. Achari, P. C. Majumder and S. C. Pakrashi, *Indian J. Chem.*, **17B**, 215 (1979).
10. (a) E. Fattorusso, S. Magno, C. Santacrose and D. Sica, *Gazz. Chim. Ital.*, 409 (1974).
   (b) Y. M. Sheik and C. Djerassi, *Tetrahedron*, **30**, 4095 (1974).
   (c) A. Marloni, L. Minale and R. Riccio, *Nouv. J. Chim.*, **2**, 351 (1978).
11. E. K. Nelson, *J. Amer. Chem. Soc.*, **33**, 1404 (1911).
12. H. H. Szmant and A. Halpern, *J. Amer. Chem. Soc.*, **71**, 1133 (1949).
13. B. Samuelsson, *J. Amer. Chem. Soc.*, **87**, 3011 (1965).
14. M. Hamberg and B. Samuelsson, *Proc. Natl. Acad. Sci. N. Y.*, **70**, 899 (1973).
15. M. Hamberg, J. Svenson, T. Wakabayashi and B. Samuelsson, *Proc. Nat. Acad. Sci. N. Y.*, **71**, 345 (1974).
16. (a) K. H. Gibson, *Chem. Soc. Rev.*, **6**, 489 (1977).
   (b) K. C. Nicolaou, G. P. Gasi and W. E. Barnette, *Angew. Chem. (Intern. Ed.)*, **17**, 293 (1978).
17. R. J. Wells, *Tetrahedron Letters*, 2637 (1976).
18. M. D. Higgs and D. J. Faulkner, *J. Org. Chem.*, **43**, 3454 (1978).
19. W. Fenical, *J. Amer. Chem. Soc.*, **96**, 5580 (1974).
20. W. D. Crow, W. Nicholls and M. Sterns, *Tetrahedron Letters*, 1353 (1971).
21. D. B. Stierle and D. J. Faulkner, *J. Org. Chem.*, **44**, 964 (1979).
22. J. Fayos, D. Lokensgard, J. Clardy, R. J. Cole and J. W. Kirksey, *J. Amer. Chem. Soc.*, **96**, 6785 (1974).
23. N. Eickman, J. Clardy, R. J. Cole and J. W. Kirksey, *Tetrahedron Letters*, 1051 (1975).
24. S. M. Kupchan, R. J. Hemingway, P. Coggon, A. J. McPhail and G. A. Sim, *J. Amer. Chem. Soc.*, **90**, 2983 (1968).
25. D. B. Borders, P. Shu and J. E. Lancaster, *J. Amer. Chem. Soc.*, **94**, 2540 (1972).
26. D. Swern (Ed.), *Organic Peroxides*, Vol. 1, 1970; Vol. 2, 1971; Vol. 3, 1972, Wiley–Interscience, New York–London.
27. W. Adam and A. J. Bloodworth, *Ann. Reports B, Chem. Soc. (London)*, 342 (1978).
28. K. R. Kopecky, J. E. Filby, C. Mumford, P. A. Lockwood and J. Y. Ding, *Can. J. Chem.*, **53**, 1103 (1975).
29. P. G. Cookson, A. G. Davies and B. P. Roberts, *Chem. Commun.*, 1022 (1976).
30. W. Adam, A. Birke, C. Cádiz and A. Rodriquez, *J. Org. Chem.*, **43**, 1154 (1978).
31. N. A. Porter and D. W. Gilmore, *J. Amer. Chem. Soc.*, **99**, 3503 (1977).
32. A. J. Bloodworth and B. P. Leddy, *Tetrahedron Letters*, 729 (1979).
33. M. F. Salomon, R. G. Salomon and R. D. Gleim, *J. Org. Chem.*, **41**, 3983 (1976).
34. M. F. Salomon and R. G. Salomon, *J. Amer. Chem. Soc.*, **99**, 3500 (1977).
35. R. G. Salomon and M. F. Salomon, *J. Amer. Chem. Soc.*, **99**, 3501 (1977).
36. E. J. Corey, K. C. Nicolaou, M. Shibasaki, Y. Machida and C. S. Shiner, *Tetrahedron Letters*, 3183 (1975).
37. R. A. Johnson, E. G. Nidy, L. Baczynskyj and R. R. Gorman, *J. Amer. Chem. Soc.*, **99**, 7738 (1977).
38. N. A. Porter, J. D. Byers, R. C. Mebane, D. W. Gilmore and J. R. Nixon, *J. Org. Chem.*, **43**, 2088 (1978).
39. N. A. Porter, J. D. Byers, A. E. Ali and J. E. Eling, *J. Amer. Chem. Soc.*, **102**, 1183 (1980).
40. A. J. Bloodworth and I. M. Griffin, *J. Chem. Soc. Perkin Trans. 1*, 195 (1975).
41. A. J. Bloodworth and M. E. Loveitt, *J. Chem. Soc. Perkin Trans. 1*, 522 (1972).
42. E. Adam, A. J. Bloodworth, H. J. Eggelte and M. E. Loveitt, *Angew. Chem. (Intern. Ed.)*, **17**, 209 (1978).
43. A. J. Bloodworth and J. A. Khan, *Tetrahedron Letters*, 3075 (1978).
44. N. A. Porter, M. O. Funk, D. Gilmore, R. Isaac and J. Nixon, *J. Amer. Chem. Soc.*, **98**, 6000 (1976).
45. (a) K. Gollnick, *Advan. Photochem.*, **6**, 1 (1968).
   (b) C. S. Foote, *Acc. Chem. Res.*, **1**, 104 (1968).
   (c) D. R. Kearns, *Chem. Rev.*, **71**, 395 (1971).
   (d) R. W. Denny and A. Nichon, *Org. Reactions*, **20**, 133 (1973).
   (e) H. H. Wasserman and R. W. Murray (Eds.) *Singlet Oxygen*, Academic Press, New York, 1979.

46. M. Matsumoto and K. Kondo, *J. Synth. Org. Chem. Japan*, **35**, 188 (1977).
47. E. Demole, C. Demole and D. Berthet, *Helv. Chim. Acta*, **56**, 265 (1973).
48. G. O. Schenck and D. E. Dunlap, *Angew. Chem.*, **68**, 248 (1956).
49. K. Kondo and M. Matsumoto, *J. Chem. Soc., Chem. Commun.*, 1332 (1972).
50. G. Rio and J. Berthelot, *Bull. Soc. Chim. Fr.*, 1664 (1969).
51. W. A. Ayer, L. M. Browne and S. Fung, *Can. J. Chem.*, **54**, 3276 (1976).
52. I. Landheer and D. Ginsberg, *Tetrahedron*, **37**, 133 (1981).
53. L. A. Paquette, R. V. C. Carr, E. Arnold and J. Clardy, *J. Org. Chem.*, **45**, 4907 (1980).
54. D. J. Coughlin and R. G. Salomon, *J. Amer. Chem. Soc.*, **99**, 655 (1977).
55. W. Adam and H. J. Eggelte, *Angew. Chem. (Intern. Ed.)*, **16**, 713 (1977).
56. E. E. van Tamelen, R. S. Dewey and R. J. Timmons, *J. Amer. Chem. Soc.*, **83**, 3725 (1961).
57. W. Adam and H. J. Eggelte, *J. Org. Chem.*, **42**, 3987 (1977).
58. W. Adam and H. J. Eggelte, *Angew. Chem. (Intern. Ed.)*, **17**, 210 (1978).
59. W. Adam and I. Erden, *J. Org. Chem.*, **43**, 2737 (1978).
60. W. Adam and H. J. Eggelte, *Angew Chem. (Intern. Ed.)*, **17**, 765 (1978).
61. W. Adam, H. J. Eggelte and A. Rodriquez, *Synthesis*, 383 (1979).
62. W. Adam and I. Erden, *Angew. Chem. (Intern. Ed.)*, **17**, 211 (1978).
63. W. Adam, O. Cueto and O. De Lucchi, *J. Org. Chem.*, **45**, 5220 (1980).
64. W. Adam and I. Erden, *Tetrahedron Letters*, 2781 (1979).
65. W. Adam and I. Erden, *Tetrahedron Letters*, 1975 (1979).
66. W. Adam and M. Balci, *J. Amer. Chem. Soc.*, **101**, 7537 (1979).
67. W. Adam and H. Rebollo, *Tetrahedron Letters*, 3049 (1981).
68. W. Adam, M. Balci and B. Pietrzak, *J. Amer. Chem. Soc.*, **101**, 6289 (1979).
69. W. Adam, M. Balci, B. Pietrzak and H. Rebollo, *Synthesis*, 820 (1980).
70. W. Adam and M. Balci, *J. Amer. Chem. Soc.*, **101**, 7542 (1979).
71. W. Adam and M. Balci, *J. Amer. Chem. Soc.*, **102**, 1961 (1980).
72. W. Adam and M. Balci, *Tetrahedron*, **36**, 833 (1980).
73. C. H. Foster and G. H. Berchthold, *J. Amer. Chem. Soc.*, **94**, 7939 (1972).
74. E. Vogel, H.-J. Altenbach and C. D. Sommerfeld, *Angew. Chem.*, **84**, 986 (1972).
75. S. Yamada, K. Nakayama and H. Takayama, *Tetrahedron Letters*, 4895 (1978).
76. I. Saito and T. Matsuura, Reference 45e, p. 511.
77. J. Rigaudy, *Pure Appl. Chem.*, **16**, 169 (1968).
78. O. Chalvet, R. Dandel, G. M. Schmid and J. Rigaudy, *Tetrahedron*, **26**, 365 (1970).
79. H. H. Wasserman and D. L. Larsen, *J. Chem. Soc., Chem. Commun.*, 253 (1972).
80. H. Hart and A. Oku, *J. Chem. Soc., Chem. Commun.*, 254 (1972).
81. D. L. Larsen, *Dissertation*, Yale University, 1973.
82. H. H. Wasserman and P. M. Keehn, *J. Amer. Chem. Soc.*, **94**, 278 (1972).
83. M. Schäfer-Ridder, V. Brocker and E. Vogel, *Angew. Chem.*, **88**, 262 (1976).
84. I. Saito, S. Kato and T. Matsuura, *Tetrahedron Letters*, 239 (1970).
85. I. Saito, M. Imuta and T. Matsuura, *Tetrahedron*, **28**, 5313 (1972).
86. R. Gray and V. Boekelheide, *J. Amer. Chem. Soc.*, **101**, 2128 (1979).
87. C. J. M. Van Den Heuvel, H. Steinberg and T. J. De Boer, *Rec. Trav. Chim.*, **96**, 157 (1977).
88. I. Saito, K. Tamoto, A. Katsumura and T. Matsuura, *Chem. Letters*, 127 (1978).
89. M. Matsumoto and K. Kondo, *Tetrahedron Letters*, 1607 (1979).
90. C. S. Foote, S. Mazur, B. A. Burns and D. Lerdal, *J. Amer. Chem. Soc.*, **95**, 586 (1973).
91. M. Matsumoto, S. Dobashi and K. Kondo, *Tetrahedron Letters*, 2329 (1977).
92. D. Lerdal and C. S. Foote, *Tetrahedron Letters*, 3227 (1978).
93. H. Kotsuki, I. Saito and T. Matsuura, *Tetrahedron Letters*, 469 (1981).
94. M. Matsumoto, K. Kuroda and Y. Suzuki, *Tetrahedron Letters*, 3253 (1981).
95. G. Rio, D. Bricout and M. Lacombe, *Tetrahedron*, **29**, 3553 (1973).
96. P. A. Burns and C. S. Foote, *J. Org. Chem.*, **41**, 908 (1976).
97. M. Matsumoto and K. Kondo, *Tetrahedron Letters*, 3935 (1975).
98. A. P. Schaap, P. A. Burns and K. A. Zaklika, *J. Amer. Chem. Soc.*, **99**, 1270 (1977).
99. M. Matsumoto, S. Dobashi and K. Kondo, *Tetrahedron Letters*, 4471 (1975).
100. P. A. Burns, C. S. Foote and S. Mazur, *J. Org. Chem.*, **41**, 899 (1976).
101. J. D. Boyd and C. S. Foote, *J. Amer. Chem. Soc.*, **101**, 6758 (1979).
102. I. Saito, R. Nagata, H. Kotsuki and T. Matsuura, *Tetrahedron Letters*, **23**, 1717 (1982).

103. E. Vogel, A. Alscher and K. Wilms, *Angew. Chem.*, **86**, 407 (1974).
104. E. Vogel, H.-H. Klug and M. Schäfer-Ridder, *Angew. Chem.*, **88**, 268 (1976).
105. R. Schmidt, W. Drens and H.-D. Brauer, *J. Amer. Chem. Soc.*, **102**, 2791 (1980).
106. N. Morita and T. Asao, *Chem. Letters*, 71 (1975).
107. D. Kamp and V. Bockelheide, *J. Org. Chem.*, **43**, 3475 (1978).
108. (a) T. Matsuura and I. Saito in *Photochemistry of Heterocyclic Compounds* (Ed. O. Buchardt), Wiley–Interscience, New York–London, 1976, p. 456.
(b) M. V. George and V. Bhat, *Chem. Rev.*, **79**, 447 (1979).
(c) H. H. Wasserman and B. H. Lipshutz, Reference 45e, p. 429.
109. (a) I. Saito, S. Matsugo and T. Matsuura, *J. Amer. Chem. Soc.*, **101**, 7332 (1979).
(b) K. Yamaguchi, T. Fueno, I. Saito and T. Matsuura, *Tetrahedron Letters*, 3433 (1979).
(c) C. W. Jefford and C. G. Rimbault, *J. Amer. Chem. Soc.*, **100**, 6515 (1978).
(d) E. W. H. Asveld and K. M. Kellogg, *J. Amer. Chem. Soc.*, **102**, 3644 (1980).
110. H. H. Wasserman and M. B. Floyd, *Tetrahedron Suppl.*, **7**, 441 (1966).
111. H. H. Wasserman, F. J. Vinick and Y. C. Chang, *J. Amer. Chem. Soc.*, **94**, 7181 (1972).
112. H. H. Wasserman and G. R. Lenz, *Heterocycles*, **4**, 409 (1976).
113. H. H. Wasserman and E. Druckrey, *J. Amer. Chem. Soc.*, **90**, 2440 (1968).
114. H. H. Wasserman, J. F. Pickett and F. J. Vinick, *Heterocycles*, **15**, 1069 (1981).
115. M. L. Graziano, M. R. Iesce, A. Carotenuto and R. Scarpati, *Synthesis*, 572 (1977).
116. W. Adam and K. Takayama, *J. Org. Chem.*, **44**, 1727 (1979).
117. I. Saito, A. Nakata and T. Matsuura, unpublished results.
118. B. L. Feringa, *Tetrahedron Letters*, 1443 (1981).
119. H. Natsume and H. Muratake, *Tetrahedron Letters*, 3477 (1979).
120. D. A. Lightner, G. S. Bisacchi and R. D. Norris, *J. Amer. Chem. Soc.*, **98**, 802 (1976).
121. H.-S. Ryang and C. S. Foote, *J. Amer. Chem. Soc.*, **101**, 6683 (1979).
122. M. Natsume, Y. Sekine and H. Soyagimi, *Chem. Pharm. Bull. Japan*, **26**, 2188 (1978).
123. J. Banerji, N. Dennis, A. R. Katritzky, R. L. Harlow and S. H. Simonsen, *J. Chem. Res. (S)*, 38 (1977)
124. J. L. Markham and P. G. Sammes, *J. Chem. Soc., Chem. Commun.*, 417 (1976).
125. T. Nishio, N. Nakajima and Y. Omote, *Tetrahedron Letters*, 753 (1981).
126. T. Tsuchiya, H. Arai, H. Hasegawa and H. Igeta, *Tetrahedron Letters*, 4103 (1974).
127. M. Matsumoto and K. Kondo, *J. Amer. Chem. Soc.*, **99**, 2393 (1977).
128. J. E. Baldwin and D. W. Lever, Jr., *J. Chem. Soc., Chem. Commun.*, 344 (1973).
129. J. M. Hoffman and R. H. Schlessinger, *Tetrahedron Letters*, 797 (1970).
130. D. A. Lightner and G. B. Quistad, *F.E.B.S. Letters*, **25**, 94 (1972).
131. J. P. Smith and G. Schuster, *J. Amer. Chem. Soc.*, **100**, 2564 (1978).
132. W. Adam and I. Erden, *J. Amer. Chem. Soc.*, **101**, 5692 (1979).
133. G. Rio and M. J. Scholl, *J. Chem. Soc., Chem. Commun.*, 474 (1975).
134. I. Saito, A. Nakata and T. Matsuura, *Tetrahedron Letters*, 1697 (1981).
135. W. Adam and A. Rodriquez, *J. Amer. Chem. Soc.*, **102**, 404 (1980).
136. A. L. J. Beckwith and R. D. Wagner, *J. Amer. Chem. Soc.*, **101**, 7099 (1979).
137. F. J. van der Ouderaa, M. Buytenhek, D. H. Nugteren and D. A. van Dorp, *Biochim. Biophys. Acta*, **487**, 315 (1977).
138. N. A. Porter, A. N. Roe and A. T. McPhail, *J. Amer. Chem. Soc.*, **102**, 4574 (1980).
139. N. A. Porter and M. O. Funk, *J. Org. Chem.*, **40**, 3614 (1975).
140. R. M. Wilson and F. Geisen, *J. Amer. Chem. Soc.*, **100**, 2225 (1978).
141. R. M. Wilson and J. W. Rekers, *J. Amer. Chem. Soc.*, **103**, 206 (1981).
142. R. M. Wilson, E. J. Elden, R. H. Squire and L. R. Florian, *J. Amer. Chem. Soc.*, **96**, 2955 (1974).
143. R. W. Wilson and S. W. Wunderly, *J. Amer. Chem. Soc.*, **96**, 7350 (1974).
144. D. Greed, H. Webrin and E. T. Strom, *J. Amer. Chem. Soc.*, **93**, 502 (1971).
145. D. H. R. Barton, G. LeClerc, P. D. Magnus and I. D. Menzies, *J. Chem. Soc., Chem. Commun.*, 447 (1972).
146. D. H. R. Barton, G. LeClerc, P. D. Magnus and I. D. Menzies, *J. Chem. Soc., Perkin Trans. 1*, 2055 (1975).
147. D. H. R. Barton and R. K. Haynes, *J. Chem. Soc., Perkin Trans. 1*, 2065 (1975).
148. R. K. Haynes, *Australian J. Chem.*, **31**, 131 (1978).
149. R. Tang, H. J. Yue, J. F. Wolf and F. Mares, *J. Amer. Chem. Soc.*, **100**, 5248 (1978).

150. J. Eriksen, C. S. Foote and T. L. Parker, *J. Amer. Chem. Soc.*, **99**, 6455 (1977).
151. S. F. Nelsen and R. Akaba, *J. Amer. Chem. Soc.*, **103**, 2096 (1981).
152. M. Balci, *Chem. Rev.*, **81**, 91 (1981).
153. H. H. Wasserman and J. L. Ives, *Tetrahedron*, **37**, 1825 (1981).
154. N. A. Porter, *Free Radicals Biol.*, **4**, 216 (1980).
155. A. Gabrische, Jr., S. M. Schildcrout, D. P. Patterson and C. M. Sprecher, *J. Amer. Chem. Soc.*, **87**, 2932 (1965).
156. C. Kanoko, A. Sugimoto and S. Tanaka, *Synthesis*, 876 (1974).
157. H. Takeshita, H. Kanamori and T. Hatsui, *Tetrahedron Letters*, 3139 (1973).
158. Y. Ito, M. Oda and Y. Kitahara, *Tetrahedron Letters*, 239 (1975).
159. M. Schäfer-Ridder, U. Brocker and E. Vogel, *Angew. Chem.*, **88**, 262 (1976).
160. R. Miyamoto, T. Tezuka and T. Mukai, *Tetrahedron Letters*, 891 (1975).
161. C. H. Forster and G. A. Berchtold, *J. Org. Chem.*, **40**, 3743 (1975).
162. W. Adam, M. Balci and B. Pietrzak, *J. Amer. Chem. Soc.*, **101**, 6285 (1975).
163. M. Oda, M. Oda and Y. Kitahara, *Tetrahedron Letters*, 839 (1976).
164. G. O. Pierson and O. A. Rinquist, *J. Org. Chem.*, **34**, 3654 (1969).
165. I. Saito, T. Matsuura and K. Inoue, *J. Amer. Chem. Soc.*, **103**, 188 (1981).
166. I. Saito, R. Nagata and T. Matsuura, *Tetrahedron Letters*, **22**, 4231 (1981).
167. N. J. Turro, M. F. Chow and J. Rigaudy, *J. Amer. Chem. Soc.*, **101**, 1300 (1979).
168. (a) N. J. Turro and M. F. Chow, *J. Amer. Chem. Soc.*, **101**, 3701 (1979).
(b) N. J. Turro and M. F. Chow, *J. Amer. Chem. Soc.*, **102**, 1191 (1980).
169. K. H. Schulte-Elte, M. Gadola and G. Ohloff, *Helv. Chim. Acta*, **56**, 2028 (1973).
170. J. Rigaudy and D. Sparfel, *Tetrahedron*, **34**, 2263 (1978).
171. J. Rigaudy, N. C. Cohen, N. K. Cuong and M. C. Duffraisse, *Compt. Rend.*, **264**, 1851 (1967).
172. D. J. Coughlin and R. G. Salomon, *J. Amer. Chem. Soc.*, **101**, 2761 (1979).
173. D. R. Kearns, *J. Amer. Chem. Soc.*, **91**, 6554 (1969).
174. R. Srinivasan, K. H. Brown, J. A. Oars, L. S. White and W. Adam, *J. Amer. Chem. Soc.*, **101**, 7424 (1979).
175. K. K. Maheshwari, P. de Mayo and D. Wiegand, *Can. J. Chem.*, **48**, 3265 (1970).
176. J. Rigaudy, C. Deletang and J. J. Basselier, *Compt. Rend.*, **268**, 344 (1969).
177. (a) J. Rigaudy, A. Defion and J. B. Lefort, *Angew. Chem.* (*Intern. Ed. Engl.*), **18**, 413 (1979).
(b) J. Rigaudy, C. Brelière and P. Scribe, *Tetrahedron Letters*, 687 (1978).
178. N. Kornblum and H. E. de la Mare, *J. Amer. Chem. Soc.*, **73**, 880 (1951).
179. M. Oda, Y. Kayama, H. Miyazaki and Y. Kithara, *Angew. Chem.*, **87**, 414 (1975).
180. W. Adam, M. Balci and J. Rivera, *Synthesis*, 807 (1979).
181. M. Oda and Y. Kithara, *Tetrahedron Letters*, 3295 (1969).
182. J. P. Hagenbuch and P. Vogel, *J. Chem. Soc., Chem. Commun.*, 1062 (1980).
183. J. E. Baldwin, H. H. Basson and H. Krauss, *J. Chem. Soc., Chem. Commun.*, 984 (1968).
184. J. P. Le Roux and C. Goasdoue, *Tetrahedron*, **31**, 2761 (1975).
185. J. Rigaudy and C. Brelière, *Bull. Soc. Chim. Fr.*, 1390 (1972); 1399 (1972).
186. G. A. Razuvaev, V. A. Schushunov, V. A. Dodonov and T. G. Brilkina in *Organic Peroxides* (Ed. D. Swern), Vol. 3, Wiley–Interscience, New York–London, 1972, Chap. III.
187. J. A. Turner and W. Herz, *J. Org. Chem.*, **42**, 1895 (1977).
188. W. Herz, R. C. Ligon, J. A. Turner and J. F. Blount, *J. Org. Chem.*, **42**, 1885 (1977).
189. O. Brown, B. T. Davis, T. G. Halsall and A. P. Hands, *J. Chem. Soc.*, 4492 (1962).
190. J. A. Turner and W. Herz, *J. Org. Chem.*, **42**, 1900 (1977).
191. J. D. Boyd, C. S. Foote and D. K. Imagawa, *J. Amer. Chem. Soc.*, **102**, 3641 (1980).
192. M. Suzuki, R. Noyori and N. Hamanaka, (a) *24th Symposium on the Chemistry of Natural Products, Osaka*, 1981, p. 418; (b) *J. Amer. Chem. Soc.*, **104**, 2024 (1982).
193. M. Suzuki, R. Noyori and N. Hamanaka, *J. Amer. Chem. Soc.*, **103**, 5606 (1981).
194. M. Natsume, Y. Sekine and M. Ogawa, *Tetrahedron Letters*, 3473 (1979).
195. D. A. van Dorp in *Chemistry, Biochemistry and Pharmacological Activity of Prostanoids* (Eds. S. M. Roberts and F. Scheinmann), Pergamon, New York, 1979, p. 233.
196. W. Adam and N. Duran, *J. Amer. Chem. Soc.*, **99**, 2729 (1977).
197. R. G. Salomon, M. F. Salomon and D. F. Coughlin, *J. Amer. Chem. Soc.*, **100**, 660 (1978).
198. M. G. Zagorski and R. G. Salomon, *J. Amer. Chem. Soc.*, **102**, 2501 (1980).
199. M. Natsume and Y. Kitagawa, *Tetrahedron Letters*, **21**, 839 (1980).

200. E. Demole, C. Demole and D. Berthet, *Helv. Chim. Acta*, **56**, 265 (1973).
201. M. Matsumoto and K. Kondo, *J. Org. Chem.*, **40**, 2259 (1975).
202. K. Kondo and M. Matsumoto, *Tetrahedron Letters*, 391 (1976).
203. K. Kondo and M. Matsumoto, *Tetrahedron Letters*, 4363 (1976).
204. H. C. Barrett and G. Büchi, *J. Amer. Chem. Soc.*, **89**, 5665 (1967).
205. (a) M. Hanaoka, C. Mukai and Y. Arata, *Heterocycles*, **6**, 895 (1977).
(b) M. Schamma, D. M. Hindenlang, T.-T. Wu and J. L. Moniot, *Tetrahedron Letters*, 4285 (1977).
(c) Y. Kondo, H. Inoue and J. Imai, *Heterocycles*, **6**, 953 (1977).
206. Y. Kondo, J. Imai and H. Inoue, *J. Chem. Soc., Perkin Trans. 1*, 911 (1980).
207. W. A. Aher, L. M. Browne and S. Fung, *Can. J. Chem.*, **54**, 3276 (1976).
208. M. R. Demuth, P. E. Garret and J. D. White, *J. Amer. Chem. Soc.*, **98**, 634 (1976).
209. B. Ganem, G. W. Holbert, L. B. Weiss and K. Ishizumi, *J. Amer. Chem. Soc.*, **100**, 6483 (1978).
210. M. Matsumoto, S. Dobashi and K. Kuroda, *Tetrahedron Letters*, 3361 (1977).
211. E. E. van Tamelen and E. G. Taylor, *J. Amer. Chem. Soc.*, **102**, 1202 (1980).
212. H. H. Wasserman, R. J. Gambale and M. J. Pulwer, *Tetrahedron Letters*, 1737 (1981).
213. H. H. Wasserman and R. J. Gambale, *Tetrahedron Letters*, 4849 (1981).

The Chemistry of Functional Groups, Peroxides
Edited by S. Patai

CHAPTER 12

Structural aspects of organic peroxides

J. Z. GOUGOUTAS
Department of Chemistry, University of Minnesota, Minneapolis, MN 55455, U.S.A.

I. INTRODUCTION . . . 376
A. Background . . . 376
B. Caveat . . . 376
C. The Geometry of the Peroxide Link in Organic Peroxides . . . 377
II. DIALKYL PEROXIDES . . . 381
A. Acyclic Structures . . . 381
B. Cyclic Structures . . . 384
1. Six-membered rings . . . 384
2. Five-membered rings . . . 384
3. Four-membered rings . . . 385
III. PEROXY KETALS AND ACETALS . . . 387
A. Cyclic Structures . . . 387
1. Five-membered rings . . . 387
2. Six-membered rings . . . 391
3. Eight-membered rings . . . 394
4. Nine-membered rings . . . 396
B. Acyclic Structures . . . 397
IV. HYDROPEROXIDES . . . 398
A. Alkyl Hydroperoxides . . . 400
B. Hydroperoxy Acetals and Ketals . . . 400
V. PEROXY ACIDS AND THEIR ESTERS . . . 403
VI. DIACYL PEROXIDES . . . 405
A. Cyclic Structures . . . 405
B. Acyclic Structures . . . 406
VII. ISOSTERIC FUNCTIONAL GROUPS . . . 409
VIII. ACKNOWLEDGEMENTS . . . 412
IX. REFERENCES . . . 412

I. INTRODUCTION

A. Background

Many of the structural aspects of organic peroxides can be understood in terms of the much studied molecular structure of H_2O_2. On the basis of Raman studies, Penney and Sutherland[1] first described its salient conformational features: eclipsed (*cis*, $\phi = 0°$) conformation several kilocalories less stable than *anti* (*trans*, $\phi = 180°$) conformation; minimum energy at $\phi \sim 100°$. Subsequently, a detailed far-infrared analysis[2] confirmed the extreme energy conformations of the earlier model, but found that the magnitude of the rotational barriers had been overestimated. The revised torsional potential function (equation 1), has a minimum at 111.5°, which is 7 kcal mol^{-1} below the eclipsed conformational maximum, but only 1.1 kcal mol^{-1} below the other maximum which corresponds to the *trans* conformation.

$$V(\phi)(\text{kcal mol}^{-1}) = 2.26 + 2.84\cos\phi + 1.82\cos 2\phi + 0.12\cos 3\phi \qquad (1)$$

In the solid state, H_2O_2 has been examined by diffraction methods, in its anhydrous crystal structure (m.p. −0.89°C)[3,4], its dihydrated structure (m.p. −51°C)[5] and, most often, in perhydrated salts (alkali metal[6–8], ammonium[9] or guanidinium[10–13] cations) of simple acids (oxalic, pyromellitic, pyrophosphoric). The H_2O_2 in perhydrates is 'solvent of crystallization' analogous to the H_2O in crystalline hydrates.

The early X-ray analysis[14] of urea perhydrate ('Hyperol') revealed the great capacity of H_2O_2 to become involved in hydrogen bonds—it is a donor for two and an acceptor for four. The same number, though often fewer, of hydrogen bonds have since been found in other perhydrates. The HOOH torsional angle* shows considerable variation in these structures. Thus, the 90° angle in anhydrous crystalline H_2O_2 increases to 130° in its crystalline dihydrate. It is not uncommon for H_2O_2 to occupy crystallographic sites of inversion symmetry and thereby have the *trans* planar conformation. Indeed, the crystal structure of guanidinium pyromellitate triperhydrate[13] shows internal variation. Its unit cell contains two crystallographically independent molecules. One sits on a centre, while the other occupies a general position and has a skew ($\phi = 154°$) conformation.

These variations are not surprising in light of the above torsional potential function. Even an unrealistically conservative estimate of the stabilization gained through hydrogen bonding in the solid is sufficient to account for all of the observed conformations of H_2O_2.

B. Caveat

The O—O bond length of H_2O_2 in many perhydrated crystal structures has been underestimated. In a survey of the then known peroxide structures[15], Pedersen noted that the apparent shortening (relative to 1.453(7) Å for anhydrous H_2O_2) for the most part could be attributed to substitutional disorder in which H_2O sometimes occupies lattice sites of H_2O_2. The mixed hydrate–perhydrate crystal structures appear to be particularly prone to such disorder (Adams and Ramdas[12] so accounted for the apparent O—O bond length of 0.98 Å which they observed in crystals of guanidinium pyromellitate trihydrate monoperhydrate. A value of 1.41 Å was recently reported for the perhydrate of hexamethylenetetramine *N*-oxide).[16] Anhydrous perhydrate crystal structures are also susceptible, particularly when the corresponding hydrates are crystallographically

* Defined as the O···O—O···O torsional angle, ϕ_{acc}, when the donor hydrogens of H_2O_2 are not observed. ϕ and ϕ_{acc} will be identical only if the O—H···O angle is 180°.

isostructural. Thus, even when sealed in a capillary, ammonium oxalate perhydrate undergoes topotactic decomposition to the corresponding hydrate[9].

Organic peroxy acids and hydroperoxides may show similar behaviour. Thus, the gradual topotactic decomposition of *o*-chloro- and *o*-bromo-peroxybenzoic acids to the corresponding halobenzoic acids has been observed in these laboratories. While such decompositions are usually detectable in X-ray photographic studies, they may be missed entirely during diffractometric measurements of intensity data. As the following example illustrates, crystal instability may also lead to an *over*estimation of the OO bond length.

In the extreme, electronic effects occasion drastic changes in bond order and length: O_2^+(1.123)[17], O_2(1.208)[18], O_3(1.278)[19], O_2^-(1.28)[20], F_2O_2(1.217)[21], H_2O_2(1.453)[4], $O_2^=$(1.49 Å)[22]. Some of these effects are evident in the many X-ray analyses of dioxygen–organometallic complexes recently reviewed by Vaska[23]. Both superoxo and peroxo structures are well represented. The OO bond in the superoxo complexes is characteristically short (1.302 Å [24]), while that of the peroxo group normally falls in the range 1.45–1.50 Å.

There are, in addition, several reports of relatively precise but unusually long (1.6–1.7 Å) OO bond lengths in metal complexes. Intensity data from one structure which was reinvestigated[25] were found to suffer from systematic errors introduced by crystal instability. The monoclinic angle β was observed to change by 0.7°, gradually during continuous exposure to X-rays over a period of 62 h, and marked intensity variations were noted. Redetermination of the structure gave a normal OO bond length.

More recent reports of long OO bonds[26] are similarly suspect in light of these findings.

C. The Geometry of the Peroxide Link in Organic Peroxides

Approximately fifty crystal structures of organic peroxides are surveyed below, with emphasis on the geometric bonding parameters unique to the peroxide link: the OO and CO bond lengths, the COO angle and the peroxide torsional angle, ϕ. The structures are presented in two groups—first alkyl, then acyl derivatives of hydrogen peroxide, with hydroperoxides included in the former, and peroxy acids in the latter group.

Table 1 serves as a compendium of the geometric bonding parameters found in each structure.

General observations on trends in the geometric parameters of the compendium are presented here.

(1) *The O—O bond.* A 'standard' bond length for dialkyl peroxides can be set at 1.48 Å. Longer bonds (1.49–1.50 Å) signify cyclic strain, as in 1,2-dioxetans. Somewhat shorter bonds (1.46–1.47 Å) are generally observed in hydroperoxides. Electron withdrawal allows considerable bond compression [1.45 Å in diacyl peroxides, 1.42 Å in $(CF_3O)_2$].

In general there appears to be no systematic lengthening of the OO bond as the peroxide torsional geometry approaches planarity, $\phi \rightarrow 0, 180°$. This is in contrast to disulphides where increasing repulsion between the lone electron pairs as $\phi \rightarrow 0$ lengthens by ~0.07 Å the SS bond (2.03 Å) of orthogonal conformations[27].

(2) *The C—O bond.* This bond length shows considerable variation (1.34–1.47 Å). Short bonds (1.35–1.38 Å) are expected when the carbon atom is sp^2-hybridized as in acyl peroxides and peroxyacids. Marked variation (1.41–1.47 Å) remains in bonds to sp^3-hybridized carbon atoms. A length of 1.44 Å can be expected for CO bonds to *alkyl* carbon, as opposed to ketalic or acetalic carbons which bear another oxygen (or nitrogen). Larger values signify strain.

Variation is still present in bonds involving ketalic or acetalic carbons (1.41–1.44 Å), with somewhat longer values in strained ozonides, and it appears that conformational effects akin to the anomeric effects in carbohydrates may be important.

TABLE 1. A compendium of the geometric parameters of the organic peroxide link

Structure	Bond lengths (Å)				Angles	
	O—O	X—O	C—O	C—O	COO	ϕ^a
F_2O_2	1.217	1.575	—	—	109.5	85.5
CF_3OOF	1.366	1.449	1.419	—	108.2	97
$(CF_3O)_2$	1.419	—	1.399	—	107	123
CF_3OOH	1.447	0.97	1.376	—	107.6	95
CF_3OOCl	1.447	1.699	1.372	—	108.1	93
H_2O_2	1.453	—	—	—	—	90
Acyclic alkyl peroxides						
$(MeO)_2$	—	—	1.44	—	105	—
$(t\text{-BuO})_2$	—	—	1.46	—	103.9	166,123
(**2**) Bis(triphenylmethyl)peroxide	1.480	—	1.461	—	107.5	180
(**3**) 1-(*p*-Methylphenyl)ethyldioxybis(dimethylglyoximato)pyridinecobalt	1.455	—	1.45	—	107.2	114
(**4**) Cumylperoxybis(dimethylglyoximato)pyridinecobalt	1.455	—	1.462	—	110.3	132
(**5**) Oxabis(*t*-butylperoxytriphenylantimony)	1.47	—	1.45	—	107	145
threo-1-Bromomercuri-2-*t*-butylperoxy-1,2-diphenylethane	—	—	—	—	—	135
Cyclic alkyl peroxides						
2-β-(Bromomercurymethyl)-3,4-dioxabicyclo[4.4.0]decane	—	—	—	—	—	75
8α,10α,Epidioxy-8,14-dihydro-14β-nitrothebaine	—	—	—	—	—	70.4
1,4-Diphenyl-2,3-dioxabicyclo[2.2.1]heptane	1.501	—	1.459	1.459	104.2	0
(**6**) 10,10-Dimethyl-3,4-dioxatricyclo[5,2,1,$0^{1,5}$]decane-2-spiro-2′-adamantane	1.483	—	1.45	—	107.2	20
(**7**) Dispiro[adamantane-2,3′-(1,2)dioxetane-4′,2″-adamantane]	1.491	—	1.475	—	89	21
(1,4-β)-(2,3-β)-Bis(1,1,4,4-tetramethylcycloheptane)-5,6-dioxabicyclo[2.2.0]hex-2-ene	1.488	—	1.492	1.488	91.2	0
Peroxy ketals and acetals						
Ethylene ozonide	1.47(1.487)	—	1.395	—	99.2	50
(**16**) Gilvanol *p*-bromobenzoate	1.485	—	1.446	1.463	104.5	—
(**17a**) 4-(4-Nitrobenzyl)-1-phenyl-1,4-epoxy-1*H*-2,3-benzodioxepin-5(4*H*)-one						
(**17b**) 4-(4-Bromobenzyl)-1,4-epoxy-1-phenyl-1-*H*-2,3-benzodioxepin-5(4*H*)-one						
(**18**) 8-Acetyl-1,2,6,7,8-pentamethyl-2,4,5,9-tetraoxatricyclo-[4.2.1.$0^{3,7}$]nonane	1.476	—	1.418	—	106.8	0.2

TABLE 1. (*cont.*)

Structure	Bond lengths (Å)				Angles	
	O—O	X—O	C—O	C—O	COO	ϕ^a
(**19**) Dimeric cyclohexanone peroxide	1.476	—	1.441	—	107.8	63.9
(**20**) Dimeric cycloheptanone peroxide	1.467	—	1.435	—	108.1	63.5
(**21**) Dimeric cyclooctanone peroxide	1.470	—	1.440	—	108.3	64.2
(**22**) Dimeric dibromoacetone peroxide	1.45	—	1.46	—	110	61
(**23**) Dimeric benzaldehyde peroxide	1.48	—	1.42	—	106	67
(**25**) 3-Isopropyl-6,6-dimethyl-5-(1-naphthylamino)-1,2,4-trioxane	1.469	—	1.419	1.458	107.6	68.6
(**26**) 3-Bromo-4′a,10′a-dihydrospiro[2,5-cyclohexadiene-1,3′-cycloocta-*as*-trioxin]-4-one	1.49	—	1.47	—	108	67
(**27**) 4-Ethyl-1-hydroxy-4,8,8,10,10-pentamethyl-7,9-dioxo-2,3-dioxabicyclo[4.4.0]dec-5-ene	1.479	—	1.423	1.446	107.6	77
(**30**) 3,8-Dimethoxy-4,5,6,7-dibenzo-1,2-dioxacyclooctane	1.45	—	1.42	—	108.5	120
(**31**) Verruculogen	1.51	—	1.43	1.47	107.5	137
(**33**) Trimeric acetone peroxide	1.477	—	1.417	—	107.6	135.6
(**37**) 4-Peroxycyclophosphamide	1.47	—	1.43	—	106	133
Hydroperoxides						
(**38**) Neoconcinndiol hydroperoxide	1.461	—	1.438	—	108.1	103
(**39**) 5-(1-Perhydroxy-1-isopropyl)-1-*p*-tolyl-1,2,3,4-tetrazole	1.457	—	1.446	—	107.8	109*
(**24**) *p*-Dioxanyl hydroperoxide	1.48	—	1.39	—	106.4	99
(**40**) 4-Hydroperoxycyclophosphamide	1.459	—	1.440	—	107.2	104
(**41**) *cis*-4-Hydroperoxyisophosphamide	1.462	—	1.442	—	107.4	112.5
(**42**) *trans*-4-Hydroperoxyisophosphamide	1.454	—	1.413	1.417	108.0	103,117*
(**43**) 4-Hydroperoxytrophosphamide	1.446	—	1.464	—	107.0	109
(**44**) 1-Hydroperoxy-1-(*o*-methoxyphenylazo)-2-phenylcyclohexane	1.48	—	1.42.	—	110.7	61,55*
(**45**) 1,1′-Di(hydroperoxycyclohexanyl)-1,1′-peroxide	1.472(1.486)	—	1.433	1.450	109.1	92*,99*,126
(**46**) 1,1′-Di(hydroperoxycyclododecanyl)-1,1′-peroxide	1.473(1.476)	—	1.425	1.429	108.0	92*,93*,124.3
(**47**) 1,1-Di(hydroperoxy)cyclododecane	1.465	—	1.410	1.442	109.5	90,116
Peroxy acids and esters						
(**48**) Peroxypelargonic acid	1.44	—	1.35	—	112	133*,72
(**49**) *o*-Nitroperoxybenzoic acid	1.478	—	1.337	—	108.9	146
(**50**) *p*-Nitroperoxybenzoic acid	1.48	—	1.37	—	107	170*
Di-*t*-butyl diperoxyoxalate	—	—	—	—	—	143.6

TABLE 1. (*cont.*)

Structure	Bond lengths (Å)				Angles	
	O—O	X—O	C—O	C—O	COO	ϕ^a
Di-*t*-butyl diperoxyadipate	—	—	—	—	—	121.2
Diacyl peroxides[b]						
(**51**) Phthaloyl peroxide	1.473	—	1.356	—	120.0	11.4
(**52**) Acetyl benzoyl peroxide	1.445	—	1.382	—	110.9	86.6
(**53**) Dibenzoyl peroxide	1.435	—	—	—	—	91.3
(**54**) 4,4′-Dichlorodibenzoyl peroxide	1.48	—	1.32	—	—	81
(**56**) 2,2′-Dichlorodibenzoyl peroxide	1.45	—	1.41	—	107	106
(**57**) 2,2′-Dibromodibenzoyl peroxide	1.45	—	1.32	—	111	112
(**63**) Bis(3,3,3-triphenylpropanoyl)peroxide	1.476	—	1.276	—	110.0	180
(**66**) 2,2′-Diiododibenzoyl peroxide	1.52	—	1.33	—	106	110
(**71**) 2-Iodo-3′-chlorodibenzoyl peroxide	1.46	—	1.32	—	111	117
(**73β**) 2-Iodo-3′-bromodibenzoyl peroxide	1.47	—	1.38	—	111	86

[a] An asterisk indicates the value is ϕ_{app}, defined as the torsional angle C—O—O···O, where the dotted line represents a hydrogen bond.
[b] Torsional angles for several other diacyl peroxides are given in Table 4.

(3) $\langle COO$. The angle has been found to vary from 89° to 120°. Except for the roughly square 1,2-dioxetan structures, values near 107–108° can be expected for both cyclic and acyclic dialkyl peroxides. Nonbonded interactions may increase the angle by several degrees in acyclic peroxide links. Angles several degrees smaller are found in ozonides. In acyclic diacyl peroxides, the angle is expanded to 111°, while in cyclic diacyl peroxides the angle is expected to depend on ring size. The largest angle (120°) was observed in phthaloyl peroxide.

(4) *The torsional angle* ϕ. In the absence of hydrogen bonding or cyclic constraints, disubstituted peroxides have torsional angles between 80° and 180°, the lower bound being attributable to general steric factors. On the other hand, ϕ for cyclic peroxides can probably be adjusted to any value through appropriate choice of ring size and substituent. Values in the range 0–135° have been observed.

Peroxy acids and hydroperoxides similarly exhibit large torsional variation as a consequence of hydrogen bonding effects.

The smallest angle ($\phi = 60°$) yet observed for any acyclic peroxide link obtains as a consequence of *intra*molecular hydrogen bonding in hydroperoxide structure **44**.

II. DIALKYL PEROXIDES

Structural studies of dialkyl peroxides constitute a relatively small fraction of the reported crystallographic investigations of organic peroxides. The earliest justifiably emphasized the fascinating topotactic solid-state behaviour of the epidioxide of anthracene ('anthracene photooxide'), **1**, and not its detailed molecular structure. Although the latter was determined and illustrated in packing diagrams of the crystal structure, neither the atomic coordinates nor details of the molecular geometry were given ($a = 15.94$, $b = 5.863$, $c = 11.43$ Å, $\beta = 108.2°$, $P2_1/a$, $z = 4$).

The conformation shown in Figure 1 for **1** is that depicted in the original report of that classic study[28] of the crystallographic mechanism of the transformation of **1** to a single-crystal phase, solid solution of anthrone and anthraquinone.

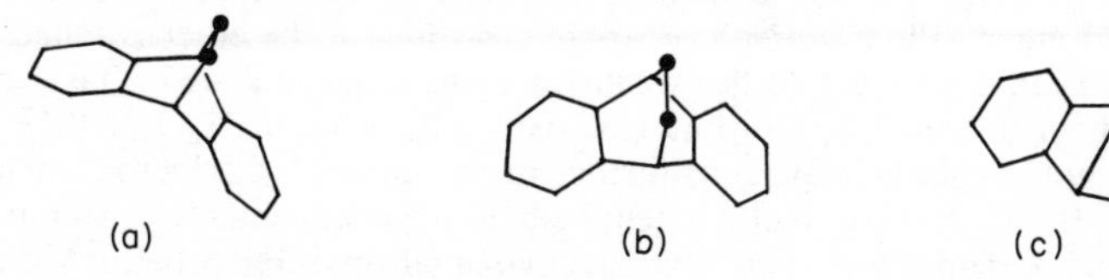

FIGURE 1. The conformation of **1**: (left) [010] projection; (centre) [001] projection; (right) [100] projection. Reproduced with permission from J. Z. Gougoutas, *Pure Appl. Chem.*, **27**, 305 (1971).

A. Acyclic Structures

Most of the experimental structural information on simple acyclic alkyl peroxides comes from gas-phase electron diffraction studies. Bartell and his coworkers[29,30] have reported geometric parameters for several derivatives of trifluoromethyl hydroperoxide, $XOOCF_3$ (X = H, F, Cl, CF_3) which fill in some of the gap separating the OO bond lengths of O_2F_2 and O_2H_2. The OO and CO bond lengths (1.45 and 1.37 Å) in $HOOCF_3$ and $ClOOCF_3$ are those in diacyl peroxides, and it is clear that the highly electron-withdrawing fluorine atom places these derivatives, and particularly the dialkyl peroxide $(CF_3O)_2$ (CO, OO: 1.40, 1.42 Å) in a class apart.

The gas-phase structure of di-*t*-butyl peroxide[31] has been analysed assuming an OO bond length of 1.48 Å and C_{3v} symmetry for the methyl and *t*-butyl groups. The electron diffraction data indicate a somewhat skew conformation, $\phi = 166°$, with the *t*-butyl group ~10° rotated from a staggered conformation ($\phi_{OOCC} = 170°$). The long CO bond length (1.46 Å) has also been found in the strained peroxides **2**–**4**.

Its nearly *trans* gas-phase peroxide conformation is supported by photoelectron spectroscopic measurements[32], although dipole moment studies[33] ($\mu = 0.94$ D at 30°C) suggest a more nearly orthogonal ($\phi = 123°$) conformation in benzene solutions.

The recently reported[34] crystal structure of bis(triphenylmethyl)peroxide, **2** (Figure 2), is remarkable in that it is one of the few known organic peroxides which occupy a site of inversion symmetry in the crystal ($P\overline{1}$, $z = 1$, m.p. 184°C). It has therefore adopted a symmetry-defined conformation in which the peroxide torsional angle is at a maximum ($\phi = 180°$). The extended conformation is no doubt a consequence of steric factors, which

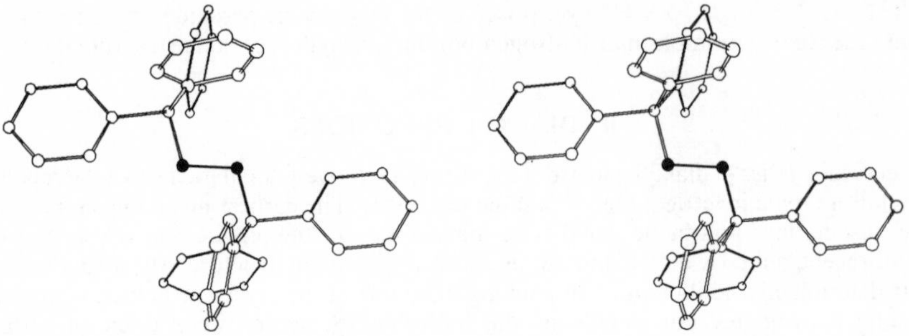

FIGURE 2. A stereoscopic drawing of the centrosymmetric conformation of **2**.

also account for the long CO and CC_{ar} bonds [1.461(2), 1.533(8) Å], but have no effect on the OO bond [1.480(2) Å]. The same long CO bond length has been found in the even more crowded ether, $(Ph_3C)_2O$[35]. The severe crowding in the ether is relieved mainly through expansion of ⟨COC to 128°. While the COO angle in **2** (107.5°) is expanded in comparison with the angle in di-*t*-butyl peroxide (103.9°) and H_2O_2 (102.7°, neutron study[4]), it is not atypical. It expands further in the unusual alkylperoxycobaloxime **4** (Figure 3). Both **3** [1-(*p*-methylphenyl)ethyldioxybis(dimethylglyoximato)pyridinecobalt][36], and **4** [cumylperoxybis(dimethylglyoximato)pyridinecobalt][37] have similar bond distances and angles in the chain $Co{-}O{-}O{-}C{-}C_{ar}$, [(for **4**) CoO: 1.909(3) Å, ⟨CoOO: 111.7(2)°, OO: 1.455(3) Å, OC: 1.462(5) Å] and in both the peroxide bond is essentially *anti* to the CC_{ar} bond. However, the added steric interactions in **4** are relieved through expansion of the peroxide torsional angle, and ⟨OOC [114° and 107.2(5)° in **3**; 132° and 110.3(2)° in **4**].

The novel centrosymmetric peroxyantimony structure **5** [oxabis(*t*-butylperoxytriphenylantimony)][38a] (Figure 4) combines some features from all of the previously described structures in this section. Unfortunately, its structure has not been completely refined. The reported bond distances and angles include Sb—OO(2.09 Å), Sb—OSb(1.97 Å), OO(1.47 Å), CO(1.45 Å), ⟨SbOO(110°), ⟨COO(107°).

Rotation about the metal–peroxide bond is restricted by the phenyl rings, which more or less have a propellor-blade arrangement about the antimony atom. As in **3** and **4**, the alkyl group of **5**, staggered with respect to the OO bond, sits above the plane of one ring which is more nearly perpendicular to the metal–peroxide bond. In this conformation, close

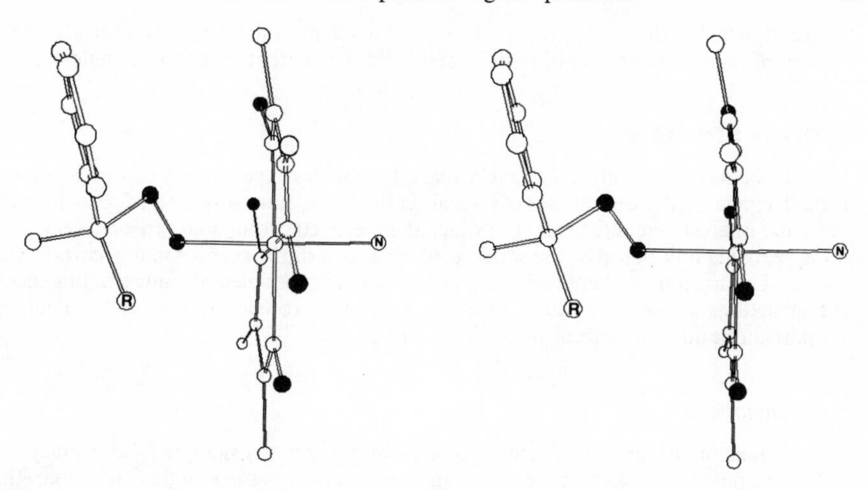

FIGURE 3. A stereoscopic drawing of the structure of **4** (R = Me). **3** (R = H) has a similar conformation. Only the nitrogen atom of the *trans* pyridine ring is shown.

contacts between a methyl group and the phenyl ring ($C \cdots C_{ar} = 3.5$ Å) appear to define the minimum permissible peroxide torsional angle ($\phi = 145°$).

Two other organometallic structures containing the *t*-butylperoxo group have recently been examined by X-ray diffraction: the peroxide–metal bonding in the tetrameric, nearly D_{2d} symmetric structure tetra-μ-(trichloro-acetato)-tetra-μ-(*t*-butylperoxy)-tetra-palladium[38b], $(CCl_3CO_2PdOO$-*t*-$Bu)_4$, differs from that in **3**, **4** and **5** in that the terminal oxygen of each *t*-butyl peroxide anion forms equivalent bridge bonds to *two* palladium atoms (Pd—O = 1.994 Å, ⟨PdOPd = 94°, ⟨PdOO = 110.5°, O—O = 1.49 Å).

In contrast to the above alkyl peroxidic transition-metal complexes, *threo*-1-bromomercuri-2-*t*-butylperoxy-1,2-diphenylethane contains a dialkylperoxy moiety, which is only indirectly associated with the metal atom. The OCCHg torsion angle is 52° and there is a strong *intra*molecular interaction [2.68(4) Å] between the metal and the

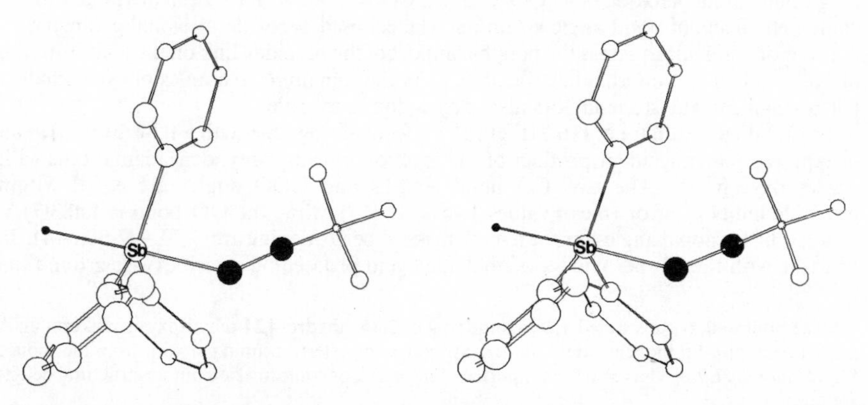

FIGURE 4. A stereoscopic drawing of half of the centrosymmetric structure of **5**. The bridging oxygen atom (left) sits on a crystallographic inversion centre.

oxygen attached to the *t*-butyl group (the sum of covalent radii for Hg and O is 2.10 while the sum of their van der Waals' radii is 2.90 Å)[38c]. The peroxide torsional angle is 135°.

B. Cyclic Structures

Experimental structural data on cyclic dialkyl peroxides have recently become available through crystal structure analyses of several synthetic and naturally derived peroxides. Of particular interest here is the extent to which the preferred orthogonal torsional geometry of the peroxide link perturbs the stable conformations of the corresponding carbocyclic systems. Unfortunately, there have been no systematic experimental studies of this effect. The structures chosen for study differ widely, and present little basis for detailed comparison besides the size of the peroxidic ring.

1. Six-membered rings

The *trans*-fused peroxidic ring in 2-β-(bromomercurymethyl)-3,4-dioxabicyclo-[4.4.0]decane[38d] is more puckered (average endocyclic torsional angle = 64°) than the other, cyclohexyl ring (average angle = 58°), as a consequence of the relatively large peroxide torsional angle [−75(3)°]. A comparable peroxide torsional angle (−70.4°) has been found in the multiply fused 1,2-dioxane ring of 8α,10α-epidioxy-8,14-dihydro-14β-nitrothebaine[38e].* Similar distortions attributable to the presence of a peroxide link were found for the six-membered rings of the peroxyketals **25** and **26** ($\phi = 68°$) and even greater pucker is evident in the dihydro-1,2-dioxin ring† of **27** ($\phi = 77°$, see Section III.A.2). It is of interest to note that less pucker is introduced through the incorporation of *two* peroxide links in a six-membered ring ($\phi = 64°$ in the 'dimeric ketone peroxides' **19**–**21**, Section III.A.2).

2. Five-membered rings

The torsional energetics of the peroxide link clearly have little effect on the conformation of relatively rigid rings. The five-membered peroxide ring in 1,4-diphenyl-2,3-dioxabicyclo[2.2.1]heptane[38h] has nearly perfect mirror symmetry with $\phi \sim 0°$. The long bonds of its peroxide link (OO = 1.501, CO = 1.459 Å) have been attributed to the combined effects of bond-angle strain and the eclipsed peroxide torsional geometry.

It is more difficult to assess the possible import of the peroxide link on the conformation of cyclic systems of intermediate flexibility. As the remaining examples of cyclic dialkyl peroxides illustrate, steric factors also play a dominant role.

The 1,2-dioxolan, **6** (m.p. 116°) (Figure 5), is formed (together with a 1,2-dioxetan) as an unexpected rearrangement product of the reaction of camphenylideneadamantane with singlet oxygen[39,40]. The two CO bond lengths and COO angles are equal within probable limits of error (mean values 1.45 Å, 107.2°), while the OO bond is 1.483(7) Å. Endocyclic torsional angles for the five-membered peroxide ring are −20, −7, 30, −41, 38 (starting with the 20° peroxide torsional angle and proceeding anticlockwise around the

* The nonfused, trisubstituted 1,2-dioxane ring of 2,24-dihydro-4,24-dihydroxysigmosceptillin A methyl ester 24-*p*-bromobenzoate[38f] (a derivative of a norsesterterpenoid peroxide from the sponge *Sigmosceptrella Laevis*) has also been reported to have a chair conformation but no structural details are given.

† No structural parameters have been reported for the dihydro-1,2-dioxin ring in the crystal structure of (6*R*)-6,19-epidioxy-9,10-*seco*-5(10),7,22-ergostatriene-3β-ol benzoate[38g].

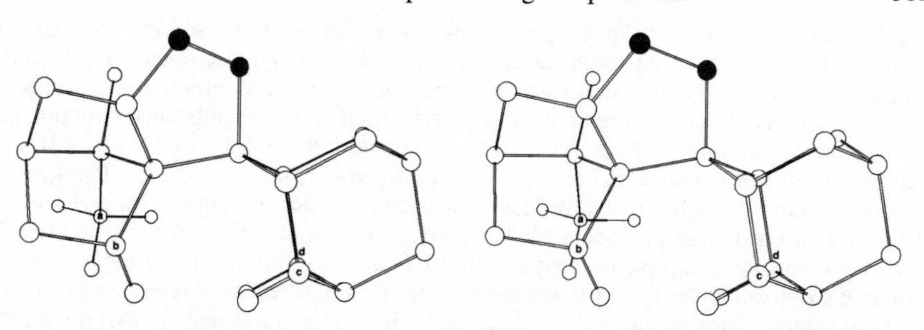

FIGURE 5. A stereoscopic drawing of the conformation of **6**. The three hydrogens of methyl group a, and methylene hydrogens on carbons b, c and d are shown.

ring). The repulsive electronic effects of the peroxide link are probably of little consequence in defining the preferred conformation of the 1,2-dioxolan ring in **5** (*exo*-2-substituted bornanes typically have a torsional angle X—C(2)—C(1)—C(10) ~ 35° (cf. 30° in **6**). In addition to the rigidity imposed by the fused bornane moiety, remote steric factors in the crowded structure of **6** are also important. Thus, when hydrogen atoms are introduced at carbons a, b, c and d, the calculated H · · H distances (2.3–2.4 Å) suggest close contacts and little conformational flexibility.

3. *Four-membered rings*

A model for the 1,2-dioxetan isomer of **6** is found in the crystal structure[41] of the unusually stable dioxetan **7**. While crystalline dioxetans of simple olefins melt near room temperature (−8°C for tetramethoxydioxetan[42]) crystals of **7** melt at 163°C. Molten **7**, at 240°C, decomposes vigorously into the monomer, adamantanone. Thermal decomposition to the monomer under less persuasive conditions (gentle heating in ethylene glycol) is accompanied by very bright chemiluminescence[43]. Even at room temperature, however, crystals slowly decompose into adamantanone; 35% crystal decomposition was observed upon exposure to X-rays during data acquisition.

(**7**)

The structural aspects of crystalline **7** are no less unusual. At room temperature, the unit cell parameters and measured density indicate the presence of six molecules per unit cell. The symmetry and systematic absences on Weissenberg photographs are consistent with space group $P2_1/c$. At −160°C, however, reflections from a supercell become apparent together with diffuse streaks at non-Bragg points. Only the room temperature structure has been examined.

Four of the six molecules occupy general lattice positions, and are ordered, while the other two are disordered. The centre of inertia of each of the two disordered molecules is nearly coincident with a space group inversion centre. Least-squares refinements converged to the relatively high value $R_w = 0.11$ for 2164 reflections. Despite these limitations, the molecular structure, gleaned only from the ordered molecules, is reasonably well defined (Figure 6). Within experimental error, the published structure has C_2 symmetry—all bonds of the dioxetan are stretched (OO: 1.491(7), CO: 1.475(8), CC: 1.549(9) Å) in comparison with corresponding bond types of other peroxides. This feature, however, is not unique to dioxetans. The elongation of cyclic bonds in several cyclobutane derivatives was noted several years ago[44]. More recently, and in light of the structure of **7**, Hitchcock and Behesti have pointed out that CO and OO bonds, to a greater or lesser extent, also show elongation with decreasing ring size (all three bond types are shortened again in three-membered rings)[40]. Endocyclic bond angles at oxygen and carbon are 89° and 87° respectively.

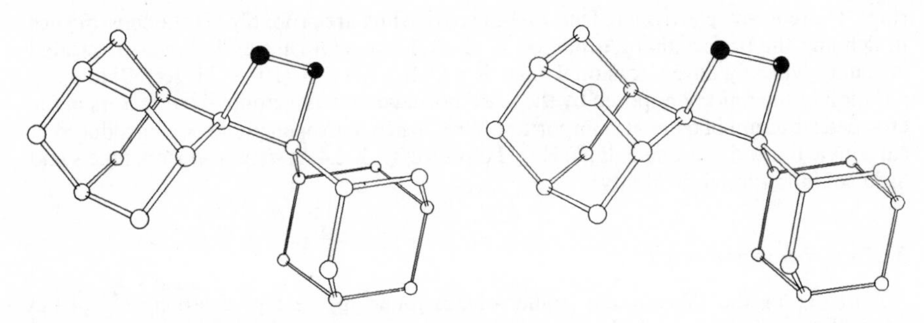

FIGURE 6. A stereoscopic drawing of the conformation of **7**.

As in the case of **6**, but to a much greater degree, the conformation of the peroxide ring is governed by remote steric factors. Very severe crowding is evident between several pairs of hydrogen atoms which are forced together during dioxetan ring formation. The crowding is alleviated partly through the observed twist of the dioxetan ring (ring torsional angles are ±21°), and partly through a distortion of the exocyclic bond angles at the dioxetan carbon atoms (CCC angles are ~12° larger than exocyclic OCC angles). Even so, several distances (2.0–2.2 Å) between hydrogens of the two adamantyl rings are considerably less than twice the effective van der Waals' radius of hydrogen (1.17 Å)[45a].

Although the bond lengths in the dioxetan ring of (1,4-β)-(2,3-β)-bis(1,1,4,4-tetramethylcycloheptane)-5,6-dioxabicyclo[2.2.0]hex-2-ene (m.p. 123–124°C), the minor product from the addition of singlet oxygen to the diene[45b], are not significantly different from those of **7**, the dioxetan ring is essentially flat*.

Several relatively recent analyses of oxetane structures provide an indication of the range of torsional angles which can be expected in four-membered oxygen heterocycles (Table 2).

* The ring torsional angles are not given; however, the reported bond distances and endocyclic angles of the dioxetan ring are only consistent with a virtually planar geometry.

Not surprisingly, large torsional angles are found in strained polycyclic structures such as **13** and **14** (20° and 32°), while trimethylene oxide and **8** have planar rings. McGandy and Fasiska suggest that the buckle in **12** (torsional angles ~18°) cannot be rationalized on steric grounds alone[51]. In this connection it is of interest to note that while *trans*-1,3-cyclobutane dicarboxylic acid has ring torsional angles near 18°, its dianion has a planar ring[44].

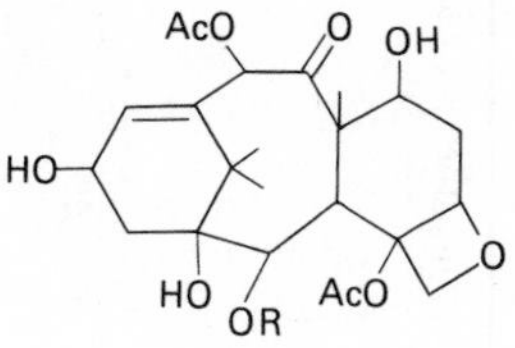

(**8**)

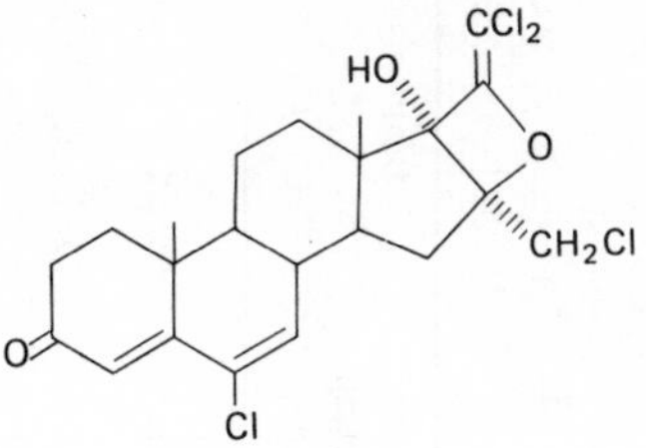

(**9**)

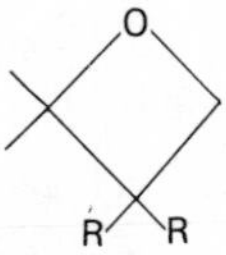

(**10**)
R = 4-$MeOC_6H_4$

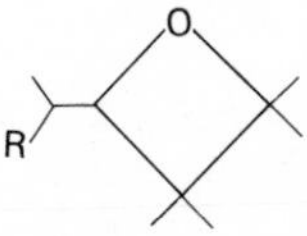

(**11**)
R = 3,5-di$(NO_2)C_6H_3CO_2$

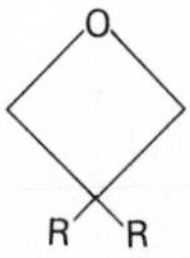

(**12**)
R = $Me_3\overset{+}{N}CH_2$

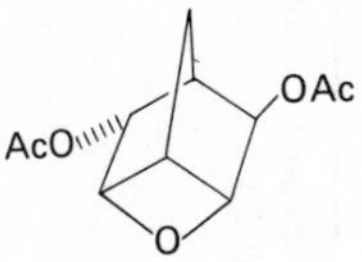

(**13**)

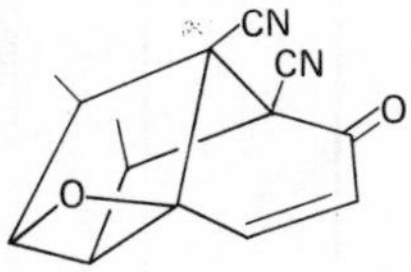

(**14**)

III. PEROXY KETALS AND ACETALS

The disubstituted peroxides in this section have been placed in a separate group since in each example, at least one of the carbon atoms bears another oxygen (or nitrogen) atom, and is therefore ketalic or acetalic. Cyclic peroxides are considered first, in order of increasing ring size.

A. Cyclic Structures

1. Five-membered rings

Ethylene ozonide—an explosive oily liquid at room temperature—was first examined experimentally by gas-phase electron diffraction[54], in light of force-field energy calculations[55] which suggested a lower energy (0.1–1.3 kcal mol^{-1}) for the C_2 half-chair

TABLE 2. The geometry of oxetanes and 1,2-dioxetanes

Structure	Typical torsional angle (deg.)	Buckling angle (deg.)[a]	Bond distances (Å)					Ref.
			C—C	C—C	C—O	C—O	O—O	
Oxetane	0	0	—	—	-	—	—	46
(**8**) Baccatin V	0	0	1.58	1.58	1.46	1.42	—	47
(**9**) 6,21,21-Trichloro-16α-chloromethyl-16β,20-oxido-17α-hydroxypregna-4,6,20-trien-3-one	9	13	1.59	1.52	1.54	1.40	—	48
(**10**) 2,2-Di(*p*-ethoxyphenyl)-3,3-dimethyloxetane	11	16	—	—	1.48	1.47	—	49
(**11**) *threo*-3,3,4,4,α-Pentamethyl-2-oxetanemethanol 3,5-dinitrobenzoate	16	23	—	—	—	—	—	50
(**12**) 3,3-Di(*N*-trimethylammoniummethyl)oxetane di(methanesulphonate)	18	26	—	—	—	—	—	51
(**13**) 2,8-Diacetoxy-4-oxatricyclo[$4.1.1^{1,5}.0^{3,6}$]octane	20	28	—	—	—	—	—	52
(**14**) 5,10-Dicyano-6,9-dimethyl-11-oxatetracyclo[$6.2.1.0^{1,7}.0^{5,10}$]undec-2-en-4-one	32	46	1.537	1.548	1.500	1.443	—	53
(7) Dispiro(adamantane-2,3′-(1,2)dioxetane-4′,2″-adamantane)	21	30	1.549	—	1.475	—	1.491	41
(1,4-β)-(2,3-β)-Bis(1,1,4,4-tetramethylcycloheptane)-5,6-dioxabicyclo[2.2.0]hex-2-ene	0	0	1.552	—	1.492	1.488	1.488	45b

[a] Some authors report the 'buckling angle', θ, while others give ring torsional angles, ϕ. In a D_{2d} symmetric structure, they are related as $2\cos\phi = 1 + \cos\theta$, in which case the bond angle ψ is given by $\cos\psi = (1 - \cos\theta)/(3 + \cos\theta)$.

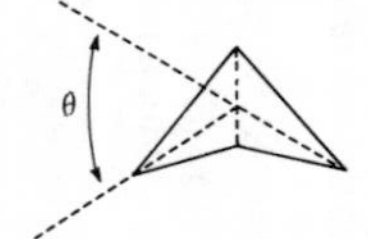

form than for the mirror symmetric C_S envelope conformation. Later studies[56] of its microwave spectrum placed on firm ground the conformational preference for the C_2 half-chair, and, further, found no evidence for free- or hindered pseudo-rotation. In addition, the microwave analysis demonstrated a difference in CO bond lengths which had not been resolved in the diffraction study. Subsequent microwave studies[57] of propylene and *trans*-2-butene ozonides gave similar geometric parameters and demonstrated the conformational preference for equatorial substituents in the C_2 half-chair. The geometry (microwave) for ethylene ozonide is summarized below. The diffraction analysis suggested a longer OO bond, 1.487(6) Å.

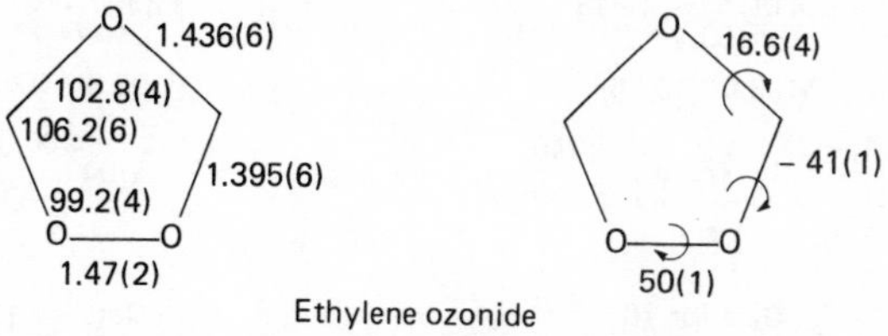

Ethylene ozonide

In an attempt to identify the *solid state* conformational preference of ozonides, Groth examined the racemic ozonide of methyl *trans*-*p*-methoxycinnamate (**15**) and in 1970 published the first X-ray data for an ozonide[58]. Unfortunately, the crystal structure (Pbca)

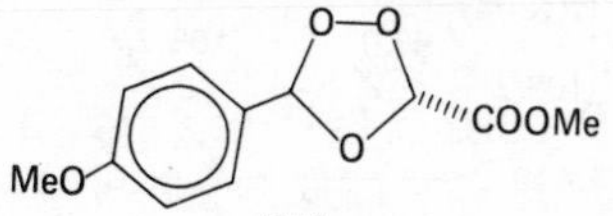

(**15**)
m.p. = 62°C

proved to be disordered, such that a given site appeared to be occupied by either enantiomer, in one of two different conformations. The better-resolved image suggested a peroxide dihedral angle of 51° and OO bond length of 1.48 Å. The substitueents clearly were *trans*, but no further conformational conclusions could be drawn.

In the three other ozonides presented here (**16**, **17a** and **17b**) the ozonide ring is incorporated in a relatively rigid 6,7,8-trioxa[3,2,1]bicyclic framework. The triterpene gilvanol **16** (R = H, m.p. 212–215°C) is the first-known *naturally occurring* ozonide. It was acylated (R = 4-BrC_6H_4CO, m.p. 212°C) in order to simplify the X-ray analysis[59]. Ozonide **17a** (m.p. 106°C) results from ozonolysis of 2-(4-nitrophenylmethyl)-3-phenylindenone[60]. These ozonides, together with **18**, the ozonolysis product of *cis*-3,4-diacetyl-1,2,3,4-tetramethylcyclobutene (m.p. 156°C)[61], and **6** give presage of the variation in CO bond lengths frequently found in the peroxy ketal and acetal structures described below.

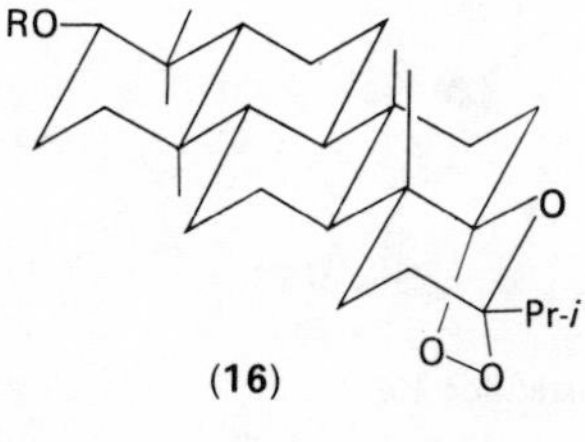

(**16**)

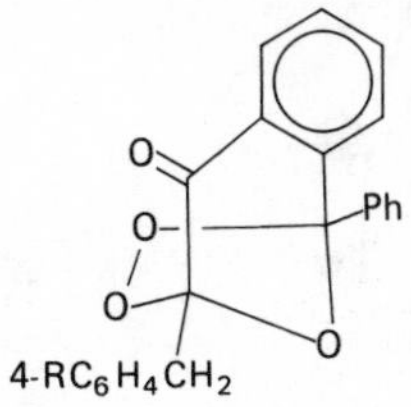

(**17**)
(**a**) R = NO_2
(**b**) R = Br

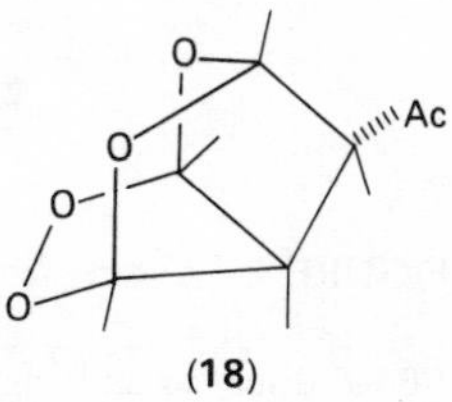

(**18**)

In contrast to ethylene ozonide and to **6**, the five-membered peroxide rings of **17a** and **18** (and apparently **16** also*) are forced to adopt the least stable[62a] mirror symmetric C_s envelope conformation. Some flexibility of the bicyclic nucleus is indicated by the larger peroxide torsional angle ($\phi = 15°$) of the bromo derivative **17b**[62b]. The three different torsional angles of the five-membered peroxide ring, starting at the peroxide link, in **18** are: -0.2, 26.7 and -39.9. Essentially the same angles have been found in the less precisely defined structure of **17a** (Figure 7). Striking differences in the distances and angles of ethylene ozonide, **16**, **17a** and **18** are clearly outside the limits of probable errors.

O
1.418 1.418
103.2
103.0 103.5
1.463 104.2 1.446
104.9
O—O
1.485

Data for **16**

O
1.44 104.6 1.41
103.3 101.8
1.47 106.3 1.47
104.8
O—O
1.46

Data for **17a**

1.439
1.417 O
O 106.0 1.543 1.434
106.0
1.476 96.0 108.2
106.7
O 106.2 1.543 1.435
1.419 O
1.429

Data for **18**

The OO bond lengths, however, are not unusually long, despite the perfectly eclipsed peroxide torsional geometry. Although the apparent shortening of the OO bond in **17a** [and **17b**, OO = 1.46(1) Å] is of questionable statistical significance, it is interesting to speculate on the extent to which transannular electron delocalization can occur into the bridging carbonyl group.

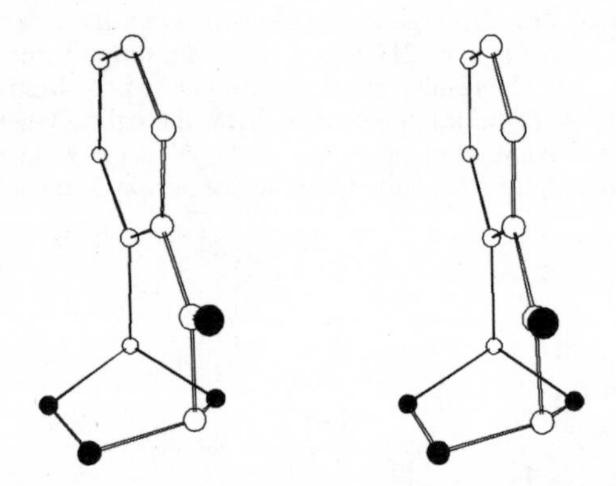

FIGURE 7. A stereoscopic drawing of the 6,7,8-trioxa[3,2,1]bicyclic nucleus of **17**.

* Bond distances and angles, but not coordinates, have been reported for **16**.

2. Six-membered rings

'Dimeric ketone peroxides' constitute the most systematically surveyed members of this group. Groth has reported relatively precise crystal structure analyses for several 'dimeric peroxides' of cyclic ketones **19**[63], **20**[64] and **21**[65]. Crystal structures of the corresponding peroxides of dibromoacetone (**22**)[66] and benzaldehyde (**23**)[67] have also been described but their atomic parameters are known with less precision. All crystallize in a centrosymmetric

	n	m.p.(°C)
(**19**)	5	132
(**20**)	6	103
(**21**)	7	98

(**22**) R = CH_2Br
m.p. = 104°C

(**23**)
m.p. = 202°C

chair conformation with the molecular inversion centre coincident with a space group ($P\bar{1}$ or $P2_1/c$) inversion centre. Dimeric cyclododecanone peroxide is reported[63] to crystallize in space group $P\bar{1}$ with one molecule per cell; presumably it has a similar centrosymmetric structure in the solid state. The crystal structure of dimeric acetone peroxide has not been reported; Groth apparently could obtain only twinned crystals[63]. However, proton magnetic resonance studies[68] indicate that, in solvents, it exists in a chair conformation with an energy of activation of 12.3 kcal mol^{-1} for conformational interconversion of the axial and equatorial methyl groups.

The structure of the 1,2,4,5-tetraoxanes is exemplified by the solid state conformation of **19** (Figure 8). In the similar structure of **22**, the axial bromine is essentially *anti* to C(2) and therefore directed over the ring, while the equatorial bromine is essentially *anti* to 0(1). The phenyl groups of **23** are equatorial and oriented such that the plane of the phenyl ring is perpendicular to the plane of the four oxygen atoms.

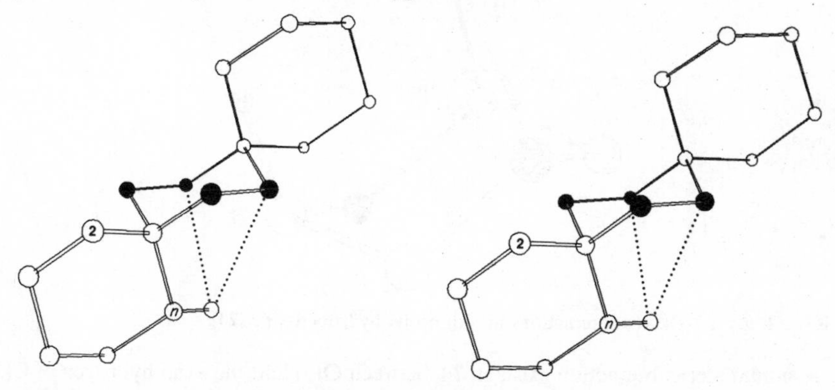

FIGURE 8. A stereoscopic drawing of the conformation of **19**. An equatorial hydrogen on C(n) is shown.

There are no statistically significant variations in the distances and angles involving the six atoms of the heterocyclic rings of these structures. In each, the three independent dihedral angles of the peroxide ring are virtually identical and there is but little torsional variation among the structures. The conformation of the 1,2,4,5-tetraoxane ring is only slightly more puckered than an idealized chair conformation. The averages of symmetric bonds, angles and dihedral angles in **19**–**21** are summarized below.

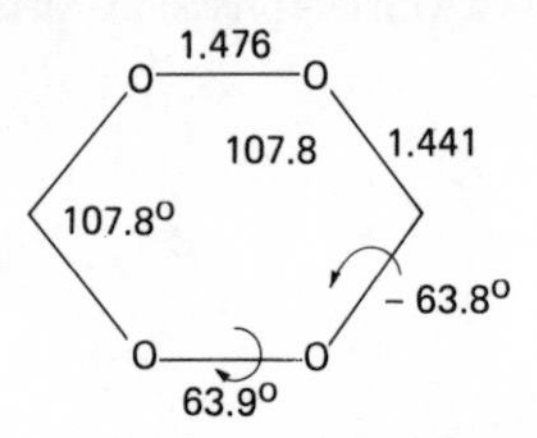

Bond angles involving a ring bond and an axial substituent [e.g. O(1)—C(1)—C(*n*)] are consistently larger by ~4° than the ideal tetrahedral angle of 109.5°, while those involving a ring bond and an equatorial substituent are consistently smaller by ~4° than the tetrahedral angle. Identical angular distortions are evident in the structure of **22**. Groth[63] has ascribed these distortions to repulsive nonbonded interactions which occur across the face of the peroxide ring, and involve the peroxide oxygens and a hydrogen on the axial carbon atom, *C*(*n*).

Some conformational features of the structure of *p*-dioxanyl hydroperoxide **24**[69] (Figure 9) lend support to Groth's explanation. (The peroxidic features of **24** are considered in Section IV.B.) The dioxane ring exists in a perfect chair conformation in which the hydroperoxide group is axial. The O(3)···O(1) distance (2.84 Å) is identical to the C(*n*)···O(1) distance in **19**. While this separation in **24** is realized with *no* distortion of the C(1)—C(2)—O(3) [or C(1)—C(2)—O(2)] angle (109°), it is attained in **19** only after an expansion of the angle OCC(*n*) to 113°. Steric interaction between oxygen and an equatorial hydrogen on *C*(*n*) of **19** is thus reduced [O(1)···H = 2.5 Å].

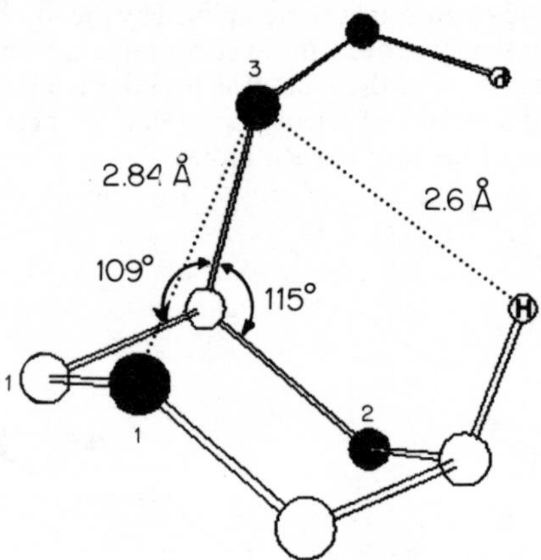

FIGURE 9. 1,3-Diaxial interactions in *p*-dioxanyl hydroperoxide (**24**).

A similar steric interaction exists in **24**, between O(3) and the axial hydrogen of C(3). The O(3)—C(2)—O(2) angle accordingly is expanded to 115° and the resulting O(3)···H distance is 2.6 Å.

There are no unusual intermolecular interactions in the crystal structures of **19–23**.

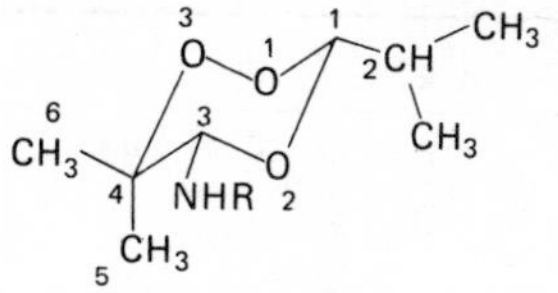

(**25**)
R = l-naphthyl

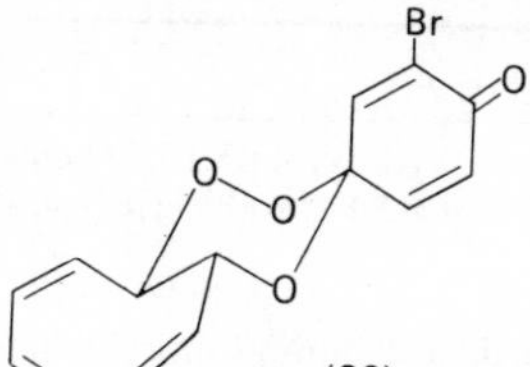

(**26**)

X-ray analysis of the chemiluminescent **25**[70] (m.p. 80°C) has confirmed that the products from the reaction of arylamines and aldehydes or ketones in the presence of oxygen, possess the 1,2,4-trioxane structure. The products had been initially formulated as aminodioxetanes[71].

The structure of **26**[72] (m.p. 139°C) has been elucidated in connection with studies of the laser-initiated photoaddition of *p*-benzoquinone to cyclooctatetraene in the presence of air. The geometry of **26** is known less precisely than that of **25**, primarily because the crystal used for analysis contains ~8% of the diastereomeric bromide in which the bromine is bonded to the other α-carbon atom of the dienone. Nevertheless, it is clear that the peroxide rings of **25** and **26** have similar, somewhat distorted, chair conformations in which a maximum number of substituents occupy equatorial positions. Their conformations differ in detail, primarily because the unusual *trans*-fused cyclooctatriene moiety imparts considerable conformational rigidity to the peroxide ring. Their peroxide torsional angles, however, are identical* within experimental error, and ~4° larger than that of **19**.

In comparison with the CO bonds in **19** (1.44 Å) there is a marked decrease in the CO bonds to the acetalic carbon C(1) (1.42 Å) and a lengthening of the (alkyl) C(4)—O(3) bond in **25** (1.46 Å). By contrast, the CO bond length involving the other aldehydic carbon, C(3), and the peroxide bond length are those in **19**.

In the absence of an axial substituent at C(1) of **25**, the O(1)—C(1)—C(2) (108.5°) and O(2)—C(1)—C(2) (110.3°) angles to the equatorial isopropyl group are more nearly tetrahedral than in the *geminally* disubstituted ketone dimers.

The angles at C(4) [O(3)—C(4)—C(6) = 101.2°, C(3)—C(4)—C(5) = 113.8°] are consistent with the angular contraction for equatorial substituents, and the angular expansion for axial substituents found in **19**. However, angles C(3)—C(4)—C(6) and O(3)—C(4)—C(5) are essentially equal (110.7°C). A better analysis of these distortions at C(4) should properly take into account the influence of the bulky arylamine substituent at C(3).

Somewhat surprisingly, there are no hydrogen bonds in the crystal structure of **25**.

The crystal structure of **27**[73] has been determined in order to establish its molecular structure. It occurs together with **28** and **29** in adult tissue of *Eucalyptus* species *E. grandis*.

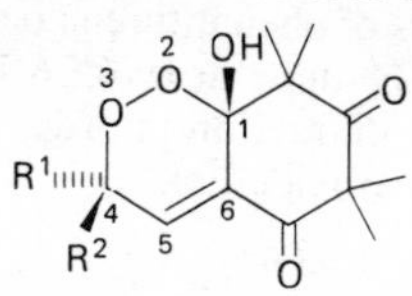

(**27**) R^1 = Me, R^2 = Et (m.p. = 98°C)
(**28**) R^2 = Me, R^1 = Et (m.p. = 127°C)
(**29**) R^1 = R^2 = Me (m.p. = 169°C)

* The torsional angle of 111(2)° reported for **26** is the supplement of the conventionally defined peroxide angle.

TABLE 3. Torsional angles of some six-membered rings

Structure	Angles	
(**27**) 4-Ethyl-1-hydroxy-4,8,8,10,10-pentamethyl-7,9-dioxo-2,3-dioxabicyclo[4.4.0]-dec-5-ene	C(1)—O(2)—O(3)—C(4) −77	O(2)—O(3)—C(4)—C(5) 58
3-Cyanomethylsulphonyl-2-morpholinocyclohexene	C(3)—C(4)—C(5)—C(6) −58	C(4)—C(5)—C(6)—C(1) 39
4-Androsten-17β *p*-bromobenzenesulphonate[a]	C(6)—C(1)—C(2)—C(3) −60	C(1)—C(2)—C(3)—C(4) 46
(**44**) 1-Hydroperoxy-1-(*o*-methoxyphenylazo)-2-phenylcyclohexane	HOOC −61	OOCN 71
(**51**) Phthaloyl peroxide	C(1)—O(2)—O(3)—C(4) −11.4	O(2)—O(3)—C(4)—C(10) 8.2
1,4-Naphthoquinone	C(1)—C(2)—C(3)—C(4) −3	C(2)—C(3)—C(4)—C(10) 2

[a]The *A*-ring torsional angles.

All three inhibit root formation in cuttings. The naturally occurring *racemic* modifications, **27** and **28**, are interconvertible under weakly alkaline conditions, evidently via the keto hydroperoxide form[74].

The presence of the relatively rigid double bond in the heterocyclic ring of **27** presents an interesting variation on the previously described structures. The peroxide ring adopts a distorted half-chair conformation with the hydroxyl group and methyl substituent occupying pseudo-axial positions. The peroxide torsional angle of 77° is remarkably large for a six-membered ring. This and the other cyclic torsional angles of **27** can be compared (Table 3) with corresponding angles in 3-cyanomethylsulphonyl-2-morpholinocyclohexene[75] (like the hydroxyl of **27**, the sulphur atom is pseudo-axial), and the *A*-ring torsional angles of 4-androsten-17β-yl *p*-bromobenzenesulphonate[76].

Short CO bonds to the ketalic carbon, C(1), are again evident in **27**, with the shortest (1.407 Å) involving the pseudo-axial hydroxyl group. The C(4)—O(3) bond length (1.447 Å), however, is not unusual.

An *inter*molecular hydrogen bond is formed between the hydroxyl group and the nonconjugated ketone oxygen (O···O = 2.85 Å).

3. *Eight-membered rings*

Acid-catalysed methanolysis of phenanthrene ozonide resulted in the formation of the relatively stable (m.p. 180°C) 3,8-dimethoxy-4,5,6,7-dibenzo-1,2-dioxacyclooctane (**30**)[77] which was examined by X-ray diffraction in order to determine the stereochemistry and geometric parameters of the peroxide link.

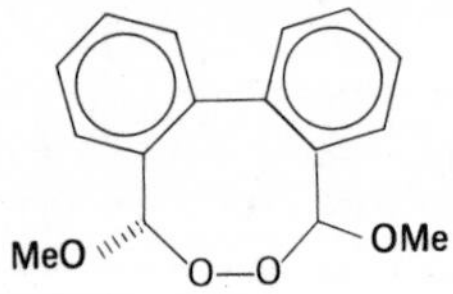

MeO O—O OMe

(**30**)

(deg.)			
O(3)—C(4)—C(5)—C(6)	C(4)—C(5)—C(6)—C(1)	C(5)—C(6)—C(1)—O(2)	C(6)—C(1)—O(2)—O(3)
−21	−2	−14	51
C(5)—C(6)—C(1)—C(2)	C(6)—C(1)—C(2)—C(3)	C(1)—C(2)—C(3)—C(4)	C(2)—C(3)—C(4)—C(5)
−10	−2	−11	46
C(2)—C(3)—C(4)—C(1)	C(3)—C(4)—C(5)—C(10)	C(4)—C(5)—C(10)—C(1)	C(5)—C(10)—C(1)—C(2)
−18	−1	−9	40
OCNN	CNNH	NNHO	NHOO
−14	−24	42	20
O(3)—C(4)—C(10)—C(9)	C(4)—C(10)—C(9)—C(1)	C(10)—C(9)—C(1)—O(2)	C(9)—C(1)—O(2)—O(3)
−2.0	−1.5	−1.6	7.7
C(3)—C(4)—C(10)—C(9)	C(4)—C(10)—C(9)—C(1)	C(10)—C(9)—C(1)—C(2)	C(9)—C(1)—C(2)—C(3)
1	−2	1	2

It crystallizes in a symmetric C_2 conformation (Figure 10) with the molecular twofold symmetry axis coincident with a crystallographic twofold axis of space group C2/c ($z = 4$). The symmetry-independent torsional angles around the bonds of the eight-membered ring, starting with the OO bond, are: 120, −31, −52, 3, 65°. The dihedral angle between the planes of the phenyl rings is 62°. The reported bond lengths along the chain COCOO [1.434(10), 1.396(12), 1.416(17), 1.452(18) Å respectively] were described as alternately long and short. The authors proposed a possible explanation for these trends, based on the methanediol model used convincingly to explain bond length trends among anomers in carbohydrate structures[78].

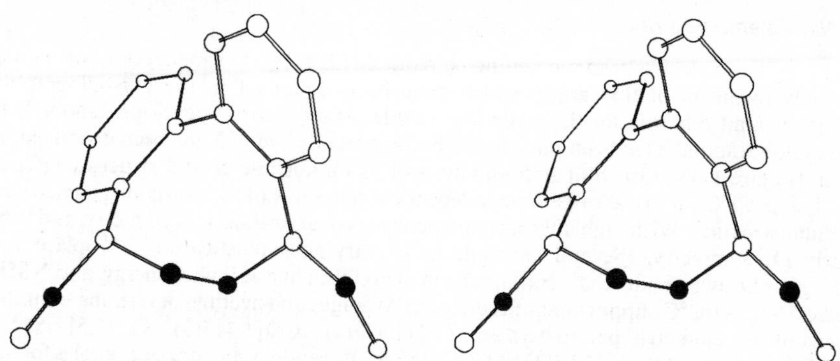

FIGURE 10. A stereoscopic drawing of the C_2 symmetric conformation of **30**.

Unless the reported errors have been greatly overestimated, the alternate long–short characterization for **30** can hardly be justified on statistical grounds[79]. While the least-squares errors do seem somewhat pessimistic in view of the internal consistency of aromatic CC bond lengths and angles [mean values and average deviations: 1.398(6) Å, 120.0(9)°], it is not uncommon that some atomic parameters are more precisely defined than others in the same crystal structure. Noting the reported anisotropic crystal decomposition (corrected by an isotropic linear correction of 40%), it is surprising that the geometric parameters agree as well as they do with values from other determinations.

Whatever the case, the above suggestion that bond lengths in organic peroxides may show conformational dependence certainly should be considered further in light of the numerous examples of variable CO bond lengths.

An asymmetric eight-membered peroxide ring was discovered in the crystallographic elucidation of the structures of two closely related tremorgenic metabolites isolated from fungi: Verruculogen, **31**[80] (m.p. 233°C), and Fumitremorgin A, **32**[81].

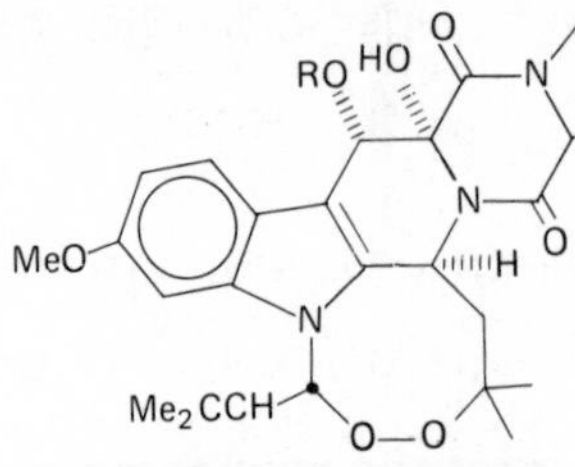

(**31**) R = H
(**32**) R = $CH_2CH{=}CMe_2$

Geometric details and atomic coordinates have been published only in the case of **31**; presumably **32** has a similar structure. The conformation of the peroxide ring is understandably different in **30** and **31**, as the former structure contains *two* groups of four, contiguous, nearly planar atoms, while the latter contains only *one* such group. Endocyclic torsional angles* and bond distances, starting with the OO bond and proceeding clockwise around the ring, are: 137(1.51), −65(1.43), −26(1.46), 11(1.40), 80(1.51), −90(1.54), 55(1.53), −77°(1.47 Å).

4. *Nine-membered rings*

The highly explosive 'trimeric ketone peroxides' (1,2,4,5,7,8-hexaoxacyclononanes are the only members of this group which have been described in the crystallographic literature. Unit cell data for the trimeric peroxides of acetone (**33**), cyclopentanone (**34**) and cyclohexanone have been reported[63], but apparently only **33** has been examined in detail (Figure 11)[82]. Groth's careful analysis of its monoclinic crystal structure ($P2_1/c$, z = 4, m.p. 97°C) nicely affords these independent measures of the peroxide geometry in medium-size rings. With high internal consistency, its structure has C_3 symmetry, and very nearly D_3 symmetry. (Several calculations of varying sophistication[83–86] place the analogous D_3 twist boat–chair conformation of cyclononane at lowest energy, and NMR studies[87] at −162°C support their prediction.) Averages and average deviations from the means of the geometric parameters are: OO[1.477(4)], CO[1.417(3)], CC[1.517(5) Å], ⟨COO[107.6(2)], ⟨OCO[112.3(3)], ⟨CCC[113.8(1)°]; endocyclic torsional angles for C_3 symmetry: ⟨COOC[135.6(3)], ⟨OOCO[−56.3(8)], ⟨OCOO[−59.2(8)°]. The latter two torsional angles would be equal in a D_3 conformation.

Each methyl group is *anti* (178°) to an oxygen of one peroxide group, and *gauche* (58°) to an oxygen of the other peroxide group bonded to the same carbon atom. Consistent

* Calculated from the published coordinates. A peroxide dihedral angle of 155(3)° is given in the original paper.

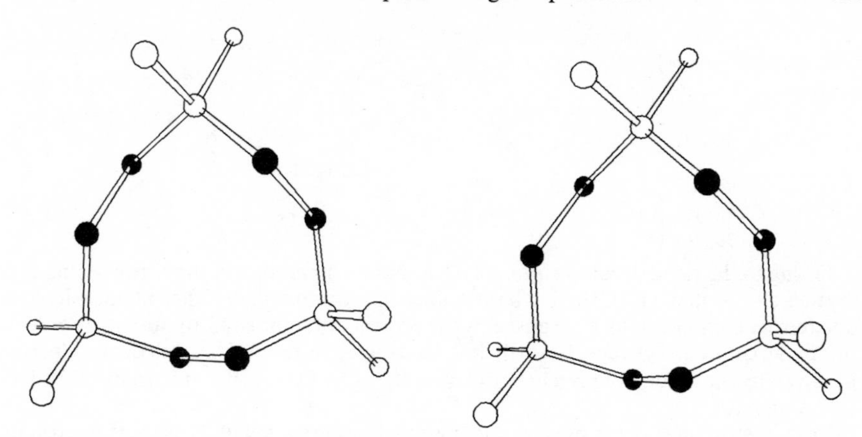

FIGURE 11. A stereoscopic drawing of the conformation of **33**.

expansions of the CCO angle [112.5(5)°] involving the *gauche* link, and contractions of the CCO angle [103.1(3)°] involving the *anti*- link are attributable to steric interaction between the hydrogen and oxygen atoms.

Although the OO bond length is equal to that in the 'dimeric ketone peroxides', (**19**–**22**), the CO bond length is significantly shorter in the 'trimer', (**33**). The molecular threefold axis is directed along the crystallographic direction $\bar{u} = -0.016\,\bar{a} + 0.086\,\bar{b} + 0.035\,\bar{c}$, which is inclined by 22° to the *b* monoclinic symmetry axis.

The isostructural crystal structure of the corresponding hexamethylcyclotrisilaperoxane has recently been studied at −120°C[87b]. Each of its unit cell lengths is ~6% longer than the corresponding length in **33**, and the difference in their molecular volumes ($\Delta V_{cell}/Z$) is 55 Å^3. The molecular expansion reflects the fact that the bonds to silicon (SiO = 1.674, SiC = 1.834 Å) are ~20% longer than the corresponding bonds to carbon in **33** (CO = 1.417, CC = 1.517 Å). The peroxide bond length is ~1% longer (1.492 Å) than in **33** [1.501(1) Å in the centrosymmetric molecular structure of bis(dimethylbenzylsilyl) peroxide at −120°C[87c]].

The cyclotrisilaperoxane also adopts the nearly D_3 symmetric twist-boat–chair conformation having an average peroxide torsional angle (Si—O—O—Si) identical to that of **33**.

The 'peroxide trimer' of cyclopentanone (**34**) reportedly[63] crystallizes in the hexagonal space group $P6_322$ with $z = 2/3$. Something appears to be wrong here, for barring disorder or polymeric structures, at least two molecules related by a screw axis must be present in a unit cell of this symmetry.

B. Acyclic Structures

Much interest has been focused on the structure of 4-hydroxycyclophosphamide (**35**), an intermediate thought to be produced *in vivo* early in the biological activation of the widely used antitumour drug, cyclophosphamide (**36**)[88–91]. Since the carbinolamine **35** is relatively unstable, various synthetic (Fenton) oxidation products of **36** have been studied in order to obtain structural evidence pertaining to the configuration at the carbinolamine carbon, C(4). The structure of **37**[92] is considered here. Four hydroperoxides closely related to **37** are considered in a separate section (IV.B) since the torsional geometry of the hydroperoxide group is intimately linked to hydrogen-bonding effects. Rather unexpectedly, hydrogen bonding also plays a leading role in defining the torsional geometry of **37**.

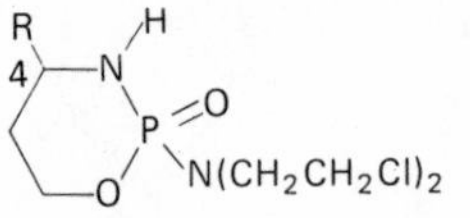

(35) R = OH
(36) R = H

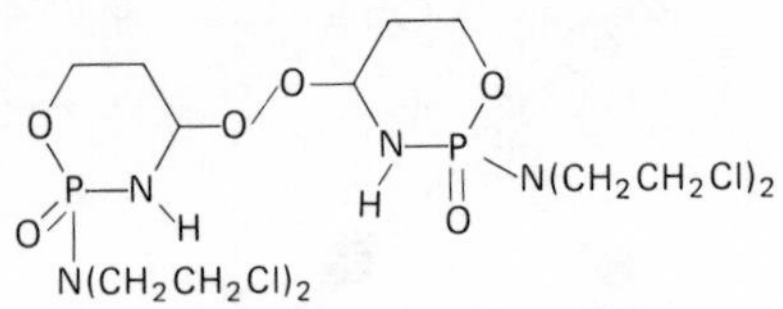

(37)

Owing to the rapid decomposition of **37** at room temperature in the X-ray beam, seven crystals ($P2_1/c$, m.p. 112°C) had to be examined in order to obtain sufficient intensity data, and further complications were posed by the presence of some conformational disorder in one of the chlorethyl groups. Nevertheless, its structure is reasonably well defined (peroxy group *cis* to phosphoryl oxygen in both halves; C—O—O bond distances and angles 1.43, 1.47 Å and 106°).

From a structural point of view, the observed *conformation* of **37** is most interesting. Each heterocyclic ring has the same absolute configuration, and somewhat distorted chair conformation with axial oxygen substituents (axial O···O distance = 3.62 Å). Except for a few atoms of the chloromethyl groups, the entire structure has C_2 molecular symmetry, within experimental error. The conformation is folded, rather than extended, through rotations about the three bonds of the peroxide link, such that the amide nitrogen of each ring approaches the phosphoryl oxygen of the other ring (N···O = 2.96 Å). As no short intermolecular approaches occur, there appear to be two *intra*molecular interactions. (In view of the low-field chemical shift of the amide protons (δ_{TMS}^{DMSO} 6.26), the hydrogen bonds probably persist in dispersed phases[93].) Introduction of trigonal amide hydrogen atoms (assumed NH = 1.00 Å, ⟨CNH = 119°) leads to the geometry: ⟨PO···H = 122°, O···H = 2.3 Å, ⟨NH···O = 122°.

If the hydrogen atoms are regarded as cyclic members, the structure of **37** has a heterobicyclic [4.3.3] nucleus (Figure 12). The peroxide link, which in this extension is contained in two nine-membered rings, has virtually the same COOC torsional angle (133°) observed in the nine-membered ring of the trimeric peroxide **33**. Endocyclic torsional angles for **33** and **37** are presented below.

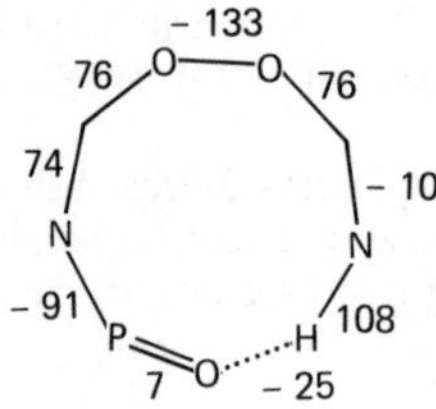

Data for **37**

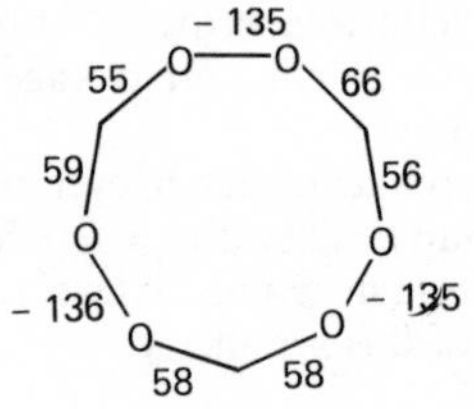

Data for **33**

IV. HYDROPEROXIDES

Whereas a 'standard' OO bond length of 1.48 Å with somewhat longer bonds in highly strained cyclic peroxides is indicated for disubstituted structures, it is likely that 1.48 Å is an upper limit for the OO bond in hydroperoxides. Usually, it is found to be somewhat shorter, 1.46–1.47 Å, but it is not clear whether the apparent shortening is real or the result of systematic effects[15]. The 'standard' CO bond length (1.44 Å) in alkyl hydroperoxides again shows variation in hydroperoxy acetals and ketals.

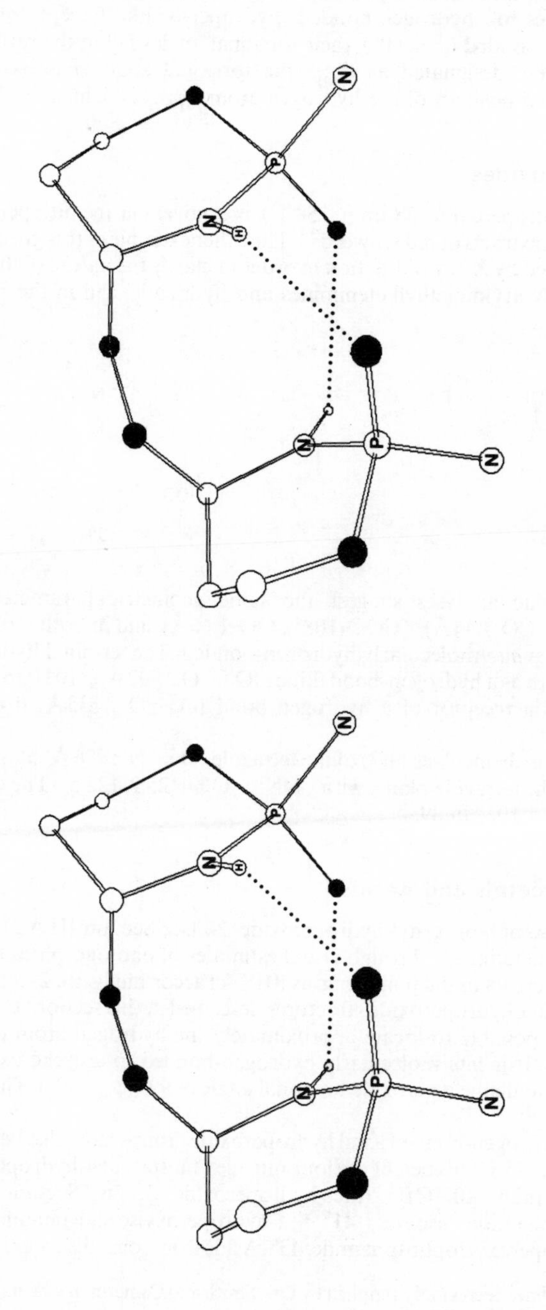

FIGURE 12. A stereoscopic drawing of the two *intra*molecular hydrogen bonds (dotted lines) in the structure of **37**. The 2-chloroethyl groups are not shown.

The large torsional variation seen in perhydrate crystal structures has not been found in these twelve examples of hydrogen-bonded hydroperoxides. Except when intramolecularly hydrogen-bonded ($\phi = 60°$), their torsional angles fall in the rather narrow range 103 ± 13. (Unless designated as ϕ_{app}, the torsional angle ϕ is based on the experimentally observed position of the hydrogen atom.) ϕ_{app} is defined in Table 1.

A. Alkyl Hydroperoxides

Neoconcinndiol hydroperoxide, **38** (m.p. 158°C), is a novel marine diterpenoid which has been isolated from extracts of red seaweed[94]. The other example of this group, **39** (m.p. 128°C), has been studied by X-ray diffraction in order to clarify the nature of the products from the reaction of *N*-aryldimethylketenimines and hydrazoic acid in the presence of oxygen[95].

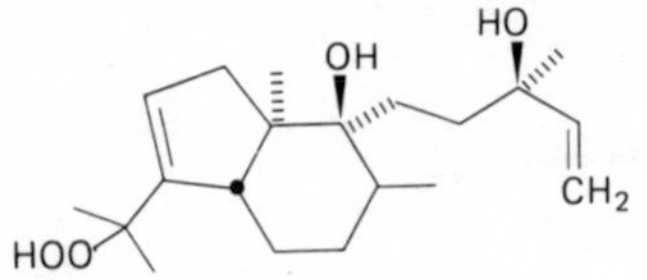

(38) 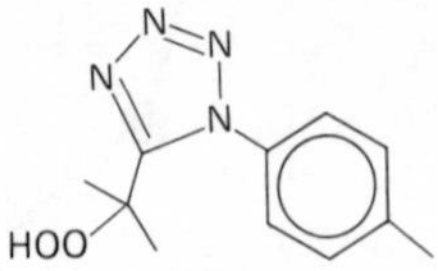 (39)

Both crystallographic analyses suggest the same geometric parameters for the hydroperoxide group: CO(1.44 Å), ⟨COO(108°), OO(1.46 Å), and in both structures, the hydroperoxide group is *inter*molecularly hydrogen-bonded. The terminal hydroperoxide oxygen of **38** serves both as a hydrogen-bond donor (O···O 2.792 Å, ϕ 103°) to a hydroxyl oxygen atom, and as a receptor of a hydrogen bond (O···O 2.833 Å) from another hydroxyl group.

Hydrogen bonding in **39** involves N(3) of the tetrazole (O···N 2.796 Å, ϕ_{app} 109°). The donor oxygen lies in the tetrazole plane, with ⟨NN···O = 125.9, 122.5°. The O—O···X angle is 118° in **38** and 102° in **39**.

B. Hydroperoxy Acetals and Ketals

The crystal structure of *p*-dioxanyl hydroperoxide, **24** (see Section III.A.2), at −40°C (m.p. 56°C) has been determined through visual estimates of photographically recorded intensities. Estimated errors in the bond lengths (0.01 Å) accordingly are 2–3 times larger than in most of the other hydroperoxide structures described in this section. Despite these limitations, it proved possible to locate approximately the hydrogen atom of the axial hydroperoxide group. It is *inter*molecularly hydrogen-bonded to a cyclic oxygen atom (O···O 2.80 Å) so as to define a peroxide torsional angle of 99° (ϕ_{app} 90°). The OO···O angle is 115°.

*Inter*molecularly hydrogen-bonded axial hydroperoxide groups have also been found in the crystal structures (−5°C) of each of the four nitrogen mustards: 4-hydroperoxycyclophosphamide, **40**[96] (m.p. 107°C)[90] (the hydroperoxide of **36**, Section III.B); 4-hydroperoxyisophosphamide—isomer I, **41**[97]*; 4-hydroperoxyisophosphamide—isomer II, **42**[97]*; and 4-hydroperoxytrophosphamide, **43***. Aside from some disorder in one of the

* Atomic coordinates have been kindly supplied by Drs. Smith and Camerman who have also made available results from their unpublished analysis of **43**.

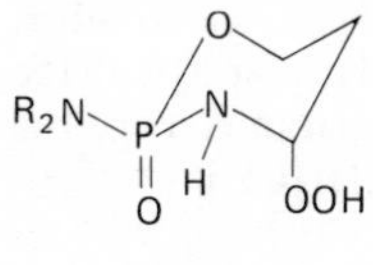

(40) 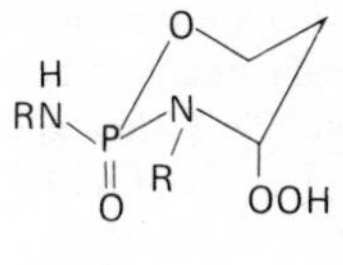 **(41)** 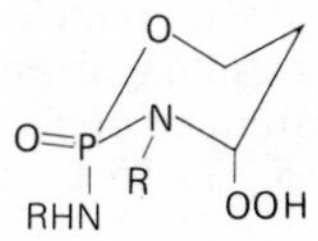 **(42)** R = CH_2CH_2Cl 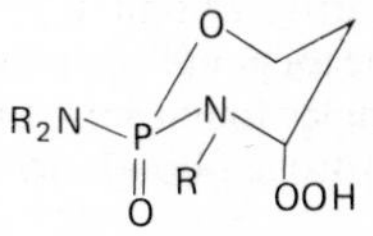 **(43)**

chloroethyl groups of **43** (similar to the disorder in **37**), the structures are well defined. The heterocyclic ring in each adopts a distorted chair conformation, flattened somewhat due to phosphamide resonance. The hydroperoxide hydrogen atom and others were located in each case.

There is no significant variation of the OO bond length and ⟨COO in this group [average values and deviations from the mean are 1.455(5) Å 107.5(4)°]. On the other hand, the (axial) CO bond length is appreciably longer in **40**, **41** and **43** [average = 1.45(1) Å] than in **42** [two independent molecules are present in the crystal structure of **42** ($P2_1/c$, $z = 8$), and both have a shorter CO length: 1.414(3) Å].

Another, probably related, difference is evident in the angular distortions at C(4): angle O—C(4)—C(5) (111°) greater than angle O—C(4)—N (105°) in **40**, **41** and **43**, while in **42** the distortions are of the opposite sense. Similar distortions are evident in **37** and **24** (Section III.A.2). Conformational analysis of the various dipole and 1,3-diaxial steric interactions might shed light on the observed variation of CO bond lengths.

The hydrogen bonds from the hydroperoxide group in **40**–**43** invariably involve the acyclic phosphoryl oxygen of a neighbouring molecule, with O···O distances in the range 2.62–2.75 Å. Angular ranges for ⟨OO···O and ϕ_{app} are 95–106° and 98–119° respectively.

The remaining hydroperoxide structures in this group, **44**–**47**, share a potential for *intra*molecular hydrogen bonding. **44** is formed through autoxidation of either the *syn* or

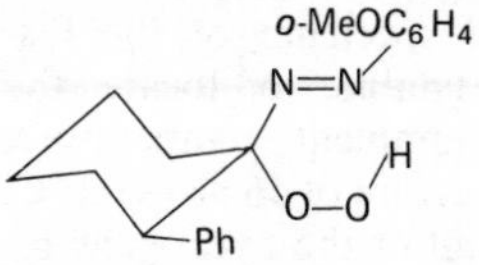

(44) 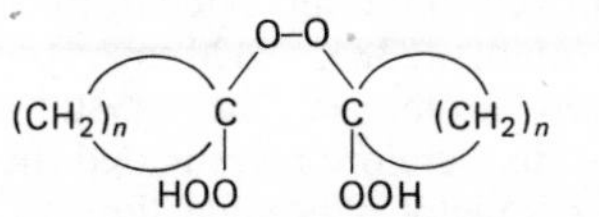 **(45)** $n = 5$ **(46)** $n = 11$ 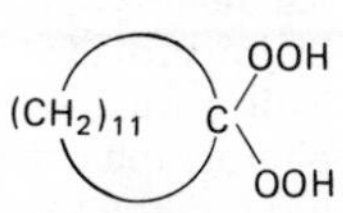 **(47)**

anti *o*-methoxyphenylhydrazone of 2-phenylcyclohexanone. Since the deep-yellow crystals of **44** are thermally unstable and low-melting (28°C), their structure was examined at −30°C[98]. While the results accordingly are not very precise (errors in bond lengths ~0.01 Å, OO 1.48, CO 1.42, CN 1.47, NN 1.28), the conformational aspects of structure are clear. The observed stereochemistry (hydroperoxide *trans* to phenyl) is a result of sterically favoured attack on a radical intermediate in the synthesis. This, together with the equatorial preference of the phenyl ring, accounts for the observed *equatorial* disposition of the hydroperoxide group.

Both the planar *trans* diazo and the hydroperoxide groups are rotated away from the phenyl substituent at C(2) (torsional angles: C(2)—C(1)—N(1)—N(2) = 111°, C(2)—C(1)—O(1)—O(2) = −53°) such that the terminal O(2)···N(2) distance is 2.66 Å. The observation of electron density between these atoms, together with the absence of any close *inter*molecular approaches to O(2), is evidence in support of an *intra*molecular hydrogen bond (Figure 13). If the (unrefined) hydrogen atom is regarded as a cyclic member, the torsional angles of the resulting six-membered ring may be compared (Table 3) with the angles found in the unsaturated cyclic peroxide **27**. Other geometric parameters of the hydrogen bond include ϕ_{app} (−55°), ⟨OOH(85°), ⟨NHO(148°), ⟨NNH(105°).

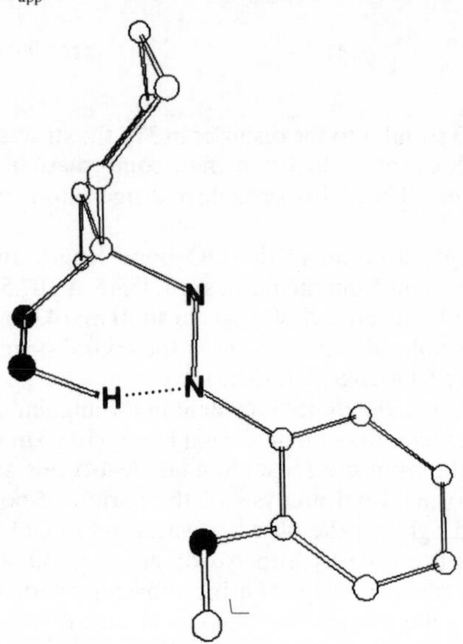

FIGURE 13. The *intra*molecular hydrogen bonding in **44**. Only the first atom of the equatorial phenyl group at C(2) is shown.

Groth had earlier[99] attempted to characterize the geometric parameters of the two different types of hydrogen bonding evident in the infrared spectrum of the bis-hydroperoxide, **45** [ν(nujol) 3385, 3420 cm^{-1}], but was unable to locate the hydroperoxide hydrogen atoms in the crystal structure (ambient temperature, photographic methods). He therefore investigated the crystal structure of **46** at −160°C, using diffractometric methods of data collection[100]. Although all of the hydrogens on carbon were visible, once again the hydroperoxide hydrogens were not.

The bridging peroxo group of **45** is equatorial with respect to both (chair) cyclohexane rings, while the hydroperoxide groups are axial. The 'square' conformation of the 12-membered ring of **46** has D_4 symmetry with eight *syn*-clinal and four *anti*-periplanar endocyclic torsional angles.

Despite these differences, every OO bond in both structures is *anti*-periplanar to one cyclic bond and +*syn*-clinal (−*syn*-clinal in the enantiomer) with respect to another. Moreover, the two structures have virtually identical geometric parameters along the eight-membered chain containing the six oxygen atoms. Mean values for the C_2 symmetric chain of **46** are shown below. The central peroxide torsional angle is $\phi = 124.3°$, while the torsional angles about CO bonds are ~ −66°.

108.3° 110.2° 109.6°

$O_{(1)}—O_{(2)}—C_{(3)}—O_{(4)}—O_{(5)}—C_{(6)}—O_{(7)}—O_{(8)}$

1.473 Å 1.425 Å 1.429 Å 1.476 Å

Close *intra*molecular approach (2.765–2.895 Å) of each terminal oxygen to the penultimate oxygen at the other end of the C_2 symmetric chain raises the possibility of *two* equivalent *intra*molecular hydrogen bonds. As in the structure of **44**, the angles about H cannot be linear since $\langle OO \cdot\cdot O = 80°$ (ϕ_{app} −92°).

Since no *inter*molecular close approaches to the terminal oxygens occur in crystalline **46**, the hydrogen bonding is probably *intra*molecular (Figure 14). However, *one* of the terminal oxygens of **45** is packed near a penultimate oxygen of a neighbouring molecule (2.824 Å, $\langle OO \cdot\cdot O$ 152°, ϕ_{app} −99°). In view of the infrared absorption, Groth concluded that one *intra*- and one *inter*-molecular hydrogen bond is present in crystalline **45**.

By contrast, both experimentally observed hydroperoxide hydrogen atoms of **47** are involved in endless double chains of *inter*molecular hydrogen bonds in the ($P2_12_12_1$) crystal structure at −160°C[101] (O···O 2.764, 2.843 Å; ϕ 90, 116°; $\langle OOH$ ~100°). The two oxygen atoms of *one* of the hydroperoxide groups are the receptor atoms for the hydrogen bonds.

All sixteen carbon and oxygen atoms of the roughly C_2 symmetric structure of **47** can be rigidly mapped on the corresponding atoms of **46**, with a positional RMS deviation of ~0.15 Å. Despite their very similar conformations, there is a difference in their CO bond lengths which lies outside the limits of experimental error: whereas all four CO bonds in **46** are equal [mean = 1.427(4) Å], the two in **47** are not [1.442(3), 1.410(3) Å]. Groth has suggested that the difference in **47** may be related to the fact that the shorter bond involves the oxygen which is not a hydrogen-bond acceptor. In any case, it cannot obviously be attributed to gross conformational differences.

V. PEROXY ACIDS AND THEIR ESTERS

The few crystal structures of organic peroxy acids which have been analysed were studied primarily in order to compare the molecular and crystallographic features with those of their extensively studied parents—the carboxylic acids. Peroxy acids generally melt at a considerably lower temperature than the corresponding carboxylic acid[102]. The well-known 'saw-toothed' variations of melting point and long crystal lattice spacing, with total number of carbon atoms, in straight-chain fatty acids, are not observed in the corresponding peroxy acids. Instead, the even and odd members of the homologous series of peroxy acids (C_9 through C_{16}) have similar crystal structures. Consequently, the long lattice spacing increases linearly by ~2.11 Å from one member to the next, and there are no discontinuities in the variation of their melting points[103]. In the aromatic realm, the relatively large increase in effective molecular volume (unit cell volume/z) in going from parent acid to the corresponding peroxy acid (12–18 $Å^3$ for *m*-Cl*, *o*-NO_2- and *p*-NO_2-peroxybenzoic acid (*vide infra*) suggests that the latter are less efficiently packed in the crystalline state.

The above differences are attributable to differences in modes of hydrogen bonding. While carboxylic acids show a strong tendency to associate pairwise, through two hydrogen bonds, into characteristic dimers which persist even in dispersed phases, the generally less acidic peroxy acids are essentially monomeric. Early infrared studies indicated that in noncrystalline phases the peroxide hydrogen atom of fatty peroxy acids is *intra*molecularly hydrogen-bonded to the carbonyl oxygen so as to define a five-membered ring[103,105]. Subsequent measurements of the electric dipole moment (2.32 D) of peroxypelargonic (**48**, $CH_3(CH_2)_7CO_3H$) and other fatty acids demonstrated that the peroxide torsional

* Although he did not analyse the unstable crystal structure of *m*-chloroperoxybenzoic acid, Lessinger[104] measured the cell parameters: $a = 4.035(1)$, $b = 5.969(1)$, $c = 30.552(2)$ Å, $\beta = 91.70(2)°$. Space group $P2_1/c$, $z = 4$. $\Delta v/z = 12$ $Å^3$

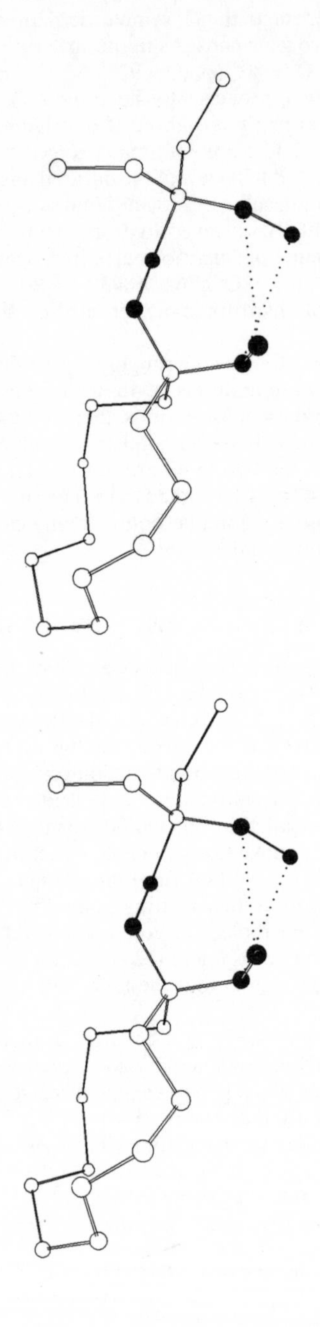

FIGURE 14. A stereoscopic drawing of the two *intra*molecular hydrogen bonds (dotted lines, hydrogens not shown) in **46**. Only five carbon atoms of the other cyclododecane ring are shown.

geometry was skewed ($\phi = 72°$) in the cyclic intramolecular hydrogen bond[106] (cf. $\phi = 61°$ in **44**).

However, the hydrogen bonding is *inter*molecular in the solid state, at least in the crystal structures of **48** ($-30°C$)[107], and the *ortho* and *para* isomers of nitroperoxybenzoic acid, **49**[108] and **50**[109] respectively ($-15°C$). In each case, glide or screw-related molecules are joined through hydrogen bonds (O··O 2.74 Å) to form infinite molecular chains, instead of the discrete 'dimers' of the corresponding acids (O··O = 2.65 Å for *o*-[110] and *p*-[111]nitrobenzoic acids). The peroxide torsional angles in **48–50** show considerable variation (ϕ_{app} 133–170°)

Of further interest in peroxy acids is the OOCO torsional angle, ω. As in the definition of ϕ, ω is measured from the eclipsed conformation, $\omega = 0°$. When sighting down the bond in question, a positive value for ϕ or ω indicates that the forwardmost atom must be rotated clockwise in order to eclipse the back atom. For $0° < \omega, \phi < 180°$, the carbonyl oxygen and peroxide substituent (here hydrogen) lie on opposite sides of the COO plane. Resonance effects are expected to keep ω near 0° (or 180°) in peroxy acids (0° in **48**, **50**; +5° in **49**) and diacyl peroxides, but steric factors in peroxy esters may result in appreciable values for ω. Thus, electric dipole moment studies indicate that the preferred conformations of *t*-butyl peroxy esters have $\omega = 30–45°$, $\phi = 100–150°$ in dispersed phases[112]*.

Within experimental error, the analyses of **48–50** give consistent results for the other geometric parameters: (for **49**) C=O 1.214(7) Å, C_{ar}—C 1.495(7) Å, OO 1.478(7) Å, ⟨OOC 108.9(4)°, ⟨OCO and ⟨CC=O 124.9(5)°, CCO 109.9(4)°. The above-mentioned resonance effects associated with electron withdrawal by the carbonyl group result in bonds from (sp^2) carbon to the peroxide group which are considerably shorter [1.337(6) Å] than the (sp^3) C—O bonds in other peroxides.

VI. DIACYL PEROXIDES

A. Cyclic Structures

Although the chemical behaviour of phthaloyl peroxide[113], **51** (m.p. 126°C), continues to attract much attention[114–120], its structural features have been virtually ignored since Greene[121] first proposed an *upper limit* (38°) for its peroxide torsional angle, on the basis of geometric considerations. The relatively low activation energy (24 kcal mol^{-1}) for radical cleavage of its peroxide bond[122] is indicative of the strain expected upon incorporation of the peroxide link in a six-membered ring containing four sp^2-hybridized carbon atoms.

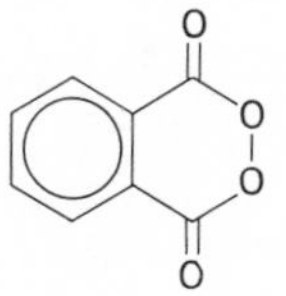

O
O
O
O

(51)

* Professor J. M. McBride of Yale University has kindly informed me that the peroxide torsional angles, ϕ, in the yet unpublished crystal structures of the bis-*t*-butyl peroxy esters of oxalic and adipic acids are 144° and 121° respectively.

We have determined its crystal structure* and here summarize the molecular results. Bond lengths for the heterocyclic ring of the nearly C_2 symmetric structure are OO 1.473(3) Å, CO 1.356(4) Å, $C_{ar}C$ 1.462(4) Å, $C_{ar}C_{ar}$ 1.396(3) Å. All endocyclic angles of *both* rings are 120°, and accordingly there is an unusual pattern of bond angles about the planar carboxy groups. Whereas in acyclic diacyl peroxides, e.g. acetyl benzoyl peroxide **52**[124], the angle between the single bonds (109°) is considerably smaller than the two angles involving the double bond [roughly 1/2 (360 − 109)°], it is the exocyclic OCO angle in **51** which is small. The three angles are essentially equal in 1,4-naphthoquinone[125].

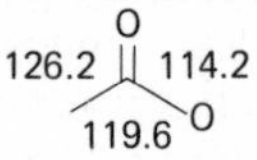

126.2 | 114.2 | 119.6 (O, O)

Data for **51**

127.9 | 122.7 | 109.4 (O, O)

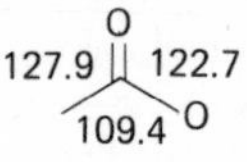

Data for **52**

The benzene ring is planar (rms deviation = 0.003 Å) but atoms in the chain OCOOCO are slightly displaced from the plane by 0.011, 0.012, 0.072, −0.061, −0.012, −0.026 Å respectively, so as to define the relatively small peroxide torsional angle $\phi = 11.4°$, and $\omega = 174.0°$ (ω is defined in Section V). Torsional angles about CC bonds in the heterocyclic ring are 1.5°.

Comparisons with the torsional angles (Table 3) in the dihydro-1,2-dioxin ring of **27** suggest that the comparatively flat geometry of **51** is primarily attributable to resonance effects which restrict rotations about the bonds to the carbonyl carbons. In general, somewhat larger values for the $C_{ar}C$ bond length and the torsional angle about this bond are observed in acyclic diacyl peroxides, and in phthalate salts where resonance delocalization of π electrons into the carboxylate group is not important, the $C_{ar}C$ bond length and torsional angle increase to 1.51 Å and 30–80° respectively[126] [1.518(2) Å and 6.6° in potassium terephthalate[127]].

The OO bond in the relatively flat structure of **51** is no longer than in the more nearly orthogonal peroxide geometry of **52**; further studies of other cyclic diacyl peroxides are needed to assess the possible significance of this observation.

The ground-state structure of **51** may accordingly be a good model for the structure of its electronically excited states.

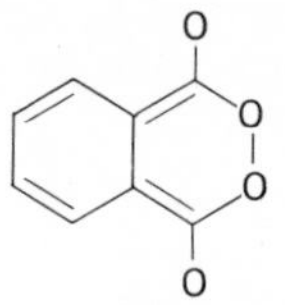

(51)

B. Acyclic Structures

A large number of acyclic diacyl peroxide crystal structures, **52**–**74** have been examined, but relatively few precise structures are described in the literature. Substituents R^1 and R^2

* The orthorhombic crystals have $a = 8.826(2)$, $b = 5.609(1)$, $c = 14.274(3)$ Å, $z = 4$ in space group $Pna2_1$. Refinements based on 551 observed intensities converged to $R = 0.038$. All hydrogens were located[123].

are defined in Table 4, which also presents the observed values of the peroxide torsional angle, ϕ, and the torsional angles ω_1, ω_2 about the CO bonds.

(52 – 74)

McBride's elegant X-ray and electron paramagnetic resonance study[124] of the structure (−95°C) and photolysis of acetyl benzoyl peroxide **52** (m.p. 38°C) has provided the most reliable measure of the acyclic diacyl peroxide linkage (estimated errors in bond lengths and angles are 0.003 Å and 0.2° respectively): OO 1.445 Å, CO 1.382, 1.190 Å, $C_{ar}C$ 1.477 Å, ⟨COO 110.9°. There is no significant difference in the geometries of the alkyl and aryl carboxy groups. The latter are twisted by 4.6° relative to the phenyl plane, presumably to alleviate steric interactions between the oxygens and *ortho* hydrogens. (Much larger angles of twist are of course necessary to overcome such steric interactions when bulky *ortho* substituents are present. Twist angles in the range 18–55° have been observed in the *o*-halodibenzoyl peroxides.) The methyl group of **52** is oriented such that one of the methyl hydrogen atoms defines a torsional angle, HCC=O, of −90°.

TABLE 4. Acyclic diacyl peroxides, $R^1C(=O)OOC(=O)R^2$

Structure	R^1	R^2	ω_1(deg.)	ω_2(deg.)	ϕ(deg.)	Reference
52	Me	Ph	−3.6	−6.3	86.6	124
53	Ph	Ph	−3	−4	91.3	128,129
54	4-ClC_6H_4	4-ClC_6H_4	−2	−5	81	130
55	4-BrC_6H_4	4-BrC_6H_4	—	—	~81	130
56	2-ClC_6H_4	2-ClC_6H_4	8	8	106	131
57	2-BrC_6H_4	2-BrC_6H_4	—	—	112	131
58	2-Naphthyl	2-Naphthyl	—	—	91.2	129
59	4-*t*-BuC_6H_4	4-*t*-BuC_6H_4	—	—	96.3	132
60	2-MeC_6H_4	2-MeC_6H_4	—	—	~90	132
61	PhCCl(Me)CH_2	3-ClC_6H_4	—	—	83	132
62	3,5-$Me_2C_6H_3$	3,5-$Me_2C_6H_3$	—	—	180[a]	132
63	Ph_3CCH_2	Ph_3CCH_2	−3.4	3.4	180[a]	129,141
64	PhC(Me_2)CH_2	PhC(Me_2)CH_2	—	—	~90	132
65	*n*-Decyl	*n*-Decyl	—	—	90.1	132
66	2-IC_6H_4	2-IC_6H_4	—	—	110	131
67	2-IC_6H_4	2-BrC_6H_4	—	—	~110	131
68	2-IC_6H_4	2-ClC_6H_4	—	—	~110	131,133
69	2-IC_6H_4	Ph	—	—	—	134
70	2-IC_6H_4	2-FC_6H_4	—	—	80	135,136
71	2-IC_6H_4	3-ClC_6H_4	0	0	117	137
72	2-IC_6H_4	3-FC_6H_4	—	—	—	136
73α	2-IC_6H_4	3-BrC_6H_4	—	—	—	138
73β	2-IC_6H_4	3-BrC_6H_4	2	6	86	138
74	2-IC_6H_4	4-$NO_2C_6H_4$	—	—	—	104

[a] Centrosymmetrically constrained in the crystal.

With few exceptions, ranges for the OO, C_{ar}—C(O), and C—C(O) bond lengths of the diacyl peroxide groups of **53–65** are 1.460 ± 0.015, 1.485 ± 0.015 and 1.51 ± 0.01 Å respectively*.

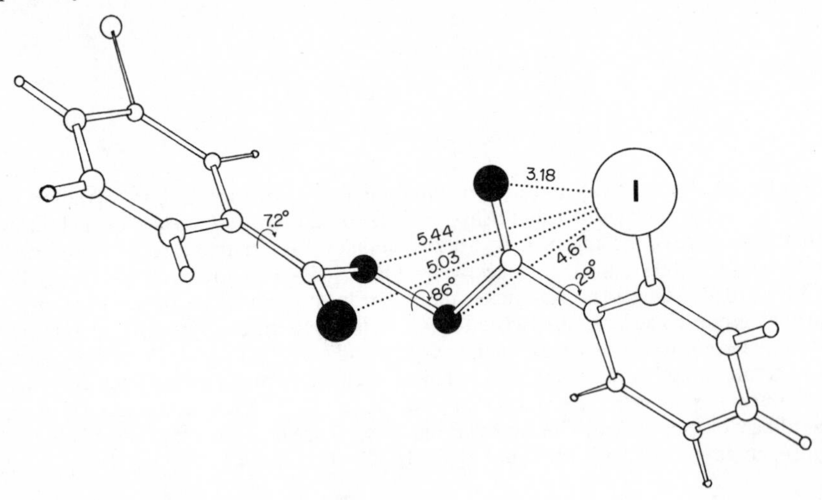

FIGURE 15. The solid-state conformation of **73β**.

It is of interest to note that the *intra*molecular distance between the two carbonyl oxygen atoms (3.136 Å) in **52** is greater than twice the van der Waals' radius of oxygen (1.52 Å)[45a], despite the *opposite* sense of the torsional angles: ($\phi = +86.6$) ($\omega_1, \omega_2 < 0$). Indeed, if $\omega_1 = \omega_2 = 0$ in the otherwise same structure of **52**, a minimum torsional angle $\phi = 78°$ can be realized before the distance between the carbonyl oxygens becomes that of van der Waals' contact. (Positive rotations, $\omega_1, \omega_2 > 0$, allow further reduction of ϕ; in the extreme, one formally arrives at a torsional variant of the dimeric ketone peroxide structure.)

A torsional angle near 80° appears to be the lower limit for acyclic diacyl peroxides. While values near 90° usually prevail, crystallographically constrained centrosymmetric conformations with $\phi = 180°$ are known (**62, 63**). The structures of **57** and **66–68** are unusual in this regard. Although crystal packing forces favour a centrosymmetric ($P2_1/c$) conformation for the halobenzene rings, the preferred skew conformation of the diacyl peroxide link reduces the crystal symmetry to Pc.

Dipole moment studies of the conformations of several dibenzoyl peroxide derivatives in solvents have been described[139].

The 2-iododibenzoyl peroxide derivatives **66–74** are very unstable (Figure 15). Their structures have been qualitatively examined in order to gain insight into their remarkable topotactic solid-state behaviour[133]. Nearly all crystallize in one of two types of layered structures exemplified in Figure 16 for **66** (type A) and **71** (type B). Translational stacking of the layers defines the short ~4 Å monoclinic symmetry axis. Single crystals of type A structures smoothly transform to (twinned) single crystals of the corresponding benzoxiodole isomers (Figure 17) during storage at ambient temperatures for several days. Under the same conditions, most type B structures are degraded to mutually aligned

* The as yet unpublished results from crystal structure analyses of several diacyl peroxides have been kindly made available by Prof. J. M. McBride of Yale University.

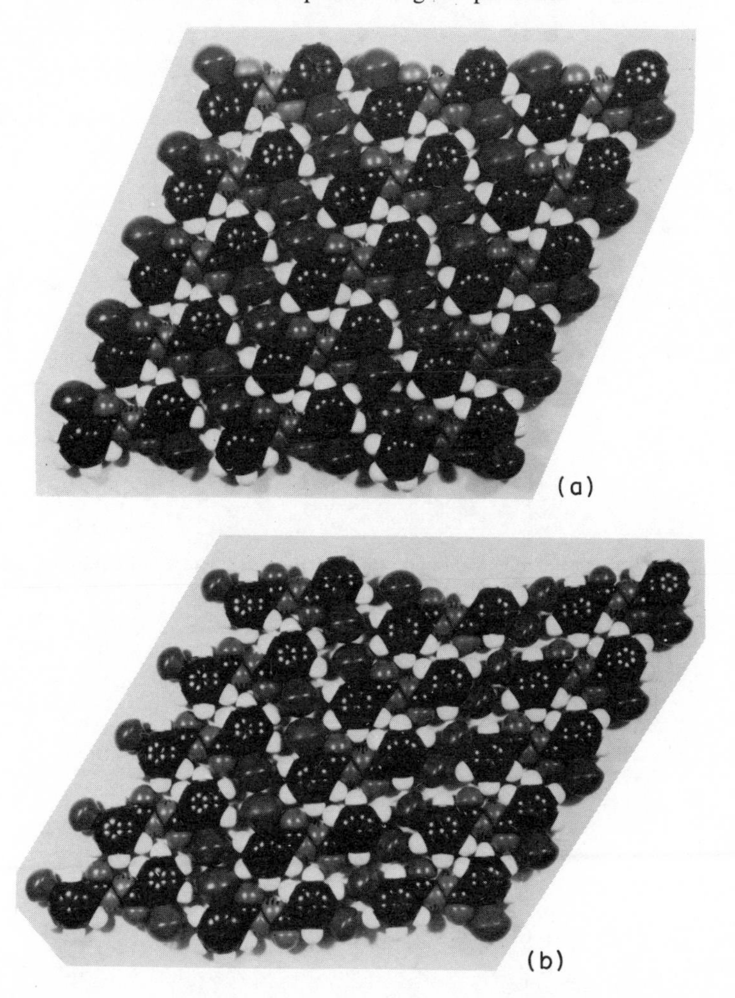

FIGURE 16. Molecular packing in layers of the 2-iododibenzoyl peroxides which undergo topotactic transformations. (a) The type A structure of **66**, (b) the type B structure of **71**. The *c*-glide direction of both is inclined to the vertical direction by $\sim 30°$.

single-crystal phases of *o*-iodoso- and *o*-halo-benzoic acids. Under different conditions, type A structures undergo several competitive and sequential topotactic transformations as a consequence of isomerization, hydrolysis and photochemical reduction (Scheme 1).

VII. ISOSTERIC FUNCTIONAL GROUPS

The crystallographic literature contains numerous examples of two or more closely related chemical compounds which crystallize in virtually the same crystal structure (crystallographic isomorphism). In its most widely recognized form isomorphism is usually associated with the substitution of one atom in the chemical structure by another

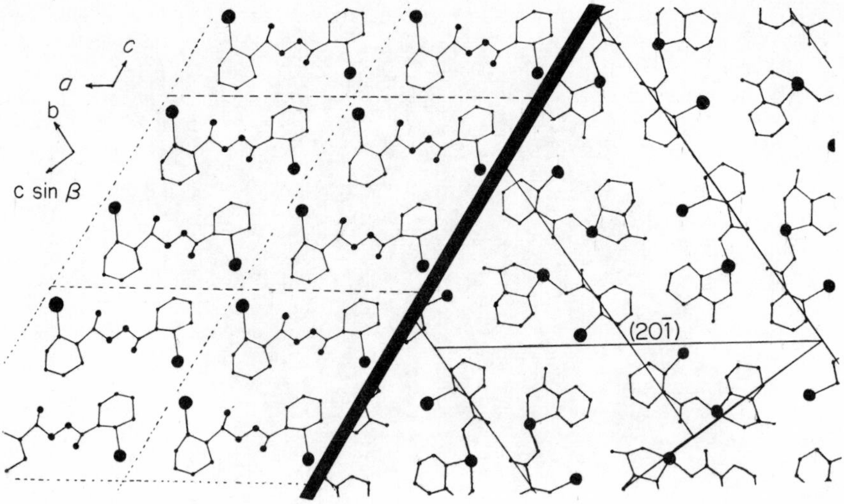

FIGURE 17. The observed alignment of crystal structures in the topotactic transformation of **66** (left) to its benzoxiodole isomer (right).

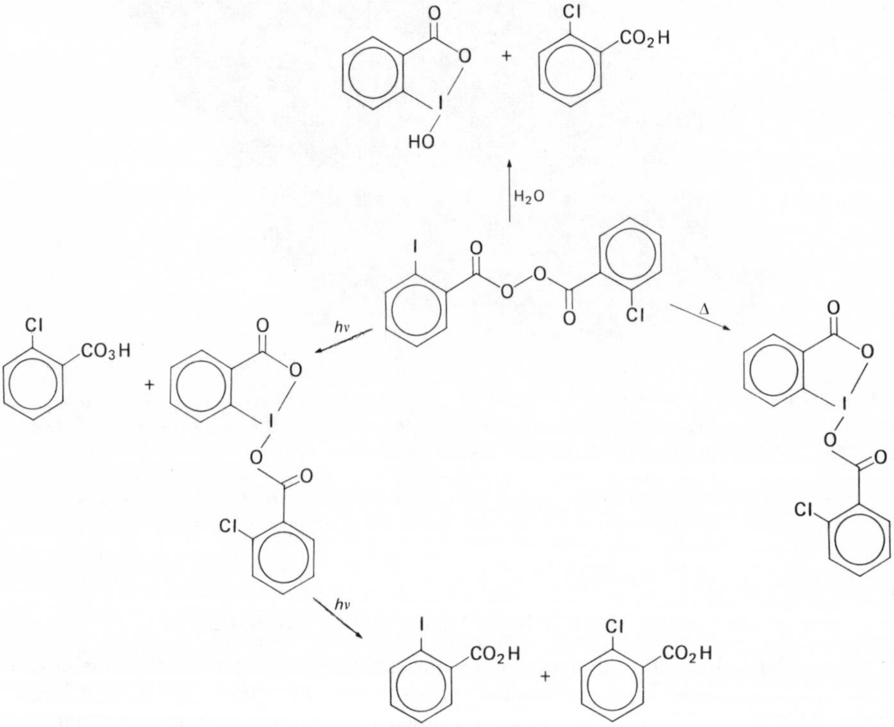

SCHEME 1

having similar chemical and spatial properties, e.g. one ion replaced by another of slightly different ionic radius, or one covalently bound halogen replaced by another (peroxides **54**, **55**, **66**–**68** and **71**–**73**).

The author and Professor D. Britton of these laboratories have for some time been interested in an extension of this notion to include entire groups of atoms—*isosteric functional groups*: e.g. (—CN and —NC), or (—OCN and —NCO and —NNN), or ($-\overset{\overset{\Large O^-}{|}}{N^+}=N-$ and $-\overset{\overset{\Large O}{||}}{C}-O-$) or ($-NO_2$ and COF). *Isosteric molecules*, that is, those containing isosteric functional groups, by virtue of their potentially equivalent spatial requirements, represent a powerful probe into the interdependence of molecular topology, electronic structure and chemical, biochemical and physical properties. Our studies are focused on several questions, many of which bear directly on the structure and properties of organic peroxides:

(*a*) Are isosteric molecules crystallographically isomorphous?
(*b*) Do they form limited or continuous solid solutions?
(*c*) To what extent are isosteric functional groups present in the same molecule (e.g. O_2N—C_6H_4—COF) crystallographically disordered?
(*d*) What is the variation in physical properties (e.g. conductivity, colour) of solid solutions of isosteric molecules?
(*e*) How do isosteric diluents effect the distribution of products of various solid-state reactions?
(*f*) What are the effects of isosteric diluents in explosive solids?

In looking for an isosteric surrogate for the peroxide group, we retain the restrictions most likely to result in isosteric molecules:

(*a*) The surrogate functional group should contain the same number of atoms in an identical connectivity scheme, with a 'close' correspondence of bond distances, angles and torsional angles. The latter geometric restrictions ordinarily would preclude the alkyne functional group as an isostere of the peroxide link ($-C\equiv C- \neq -O-O-$).
(*b*) Replacement of one atom by another in the same group of the Periodic Table should be considered only for atoms of comparable van der Waals' radii. It is thus unlikely that peroxides and the corresponding disulphides or sulphenic esters will be isosteric ($-S-S- \neq -S-O- \neq -O-O-$).
(*c*) Hydrogen-bonding characteristics should be comparable for isosteric functional groups. Therefore a methyl group and a protonated primary amine are unlikely bedfellows ($R-CH_3 \neq R-\overset{+}{N}H_3$), but $-N=N-H$ may be isosteric with the hydroperoxide group.

It remains to be seen whether the ~13% reduction in bond length and small angular differences preclude the potentially isosteric relationship between the diazo group (N=N = 1.28 Å in **44**) and the peroxide link. A more serious obstacle is posed by the preference for a skew torsional geometry in peroxides. However, in light of the centrosymmetric structures of **2**, **62** and **63**, the isosterism $-N=N- \equiv -O-O-$ cannot be discounted, even for acyclic links. With appropriate modification, oximes and oximino ethers offer a compromise ($=N-OR \equiv -O-OR$).

Cyclic peroxides with small torsional angles ϕ (e.g. **6**, **7**, **16**, **17**, **18** and **51**) would seem most likely to be isosteric with the corresponding diazo compounds. The last example **51** $\overset{?}{\equiv}$ **75** raises several interesting possibilities, particularly in light of the properties and solid-state behaviour of the emerald-green crystals of the diazanaphthaquinone, **75**[140].

(51) (75)

VIII. ACKNOWLEDGEMENTS

Most of the structures described in this chapter were uncovered through a connectivity search of the Cambridge Data File (through September 1981), the access to which was kindly provided by the Squibb Institute for Medical Research, Princeton, N.J. The gargantuan efforts of all those involved in the preparation of that extraordinary resource are very gratefully acknowledged. I would also like to express warm appreciation of Ms. Sharon Haugen's patience and skill in typing the manuscript.

IX. REFERENCES

1. W. G. Penney and G. B. B. M. Sutherland, *Trans. Faraday Soc.*, **30**, 898 (1934).
2. R. H. Hunt, R. A. Leacock, C. W. Peters and K. T. Hecht, *J. Chem. Phys.*, **42**, 1931 (1965).
3. S. C. Abrahams, R. L. Collin and W. N. Lipscomb, *Acta Cryst.*, **4**, 15 (1951).
4. W. R. Busing and H. A. Levy, *J. Chem. Phys.*, **42**, 3054 (1965).
5. I. Olovsson and D. H. Templeton, *Acta Chem. Scand.*, **14**, 1325 (1960).
6. B. F. Pedersen and B. Pedersen, *Acta Chem. Scand.*, **18**, 1454 (1964).
7. B. F. Pedersen, *Acta Chem. Scand.*, **21**, 779 (1967).
8. B. F. Pedersen, *Acta Chem. Scand.*, **23**, 1871 (1969).
9. B. F. Pedersen, *Acta Cryst.*, **B28**, 746 (1972).
10. J. M. Adams and R. G. Pritchard, *Acta Cryst.*, **B32**, 2438 (1976).
11. J. M. Adams and V. Ramdas, *Acta Cryst.*, **B34**, 2150 (1978).
12. J. M. Adams and V. Ramdas, *Acta Cryst.*, **B34**, 2781 (1978).
13. J. M. Adams and V. Ramdas, *Inorg. Chim. Acta*, **34**, L225 (1979).
14. C.-S. Lu, E. W. Hughes and P. A. Giguère, *J. Amer. Chem. Soc.*, **63**, 1507 (1941).
15. B. F. Pedersen, *Acta Cryst.*, **B28**, 1014 (1972).
16. T. C. W. Mak and Y.-S. Lam, *Acta Cryst.*, **B34**, 1732 (1978).
17. L. E. Sutton (Ed.), *Interatomic Distances*, The Chemical Society, London, 1958, p. M69.
18. I. L. Karle, *J. Chem. Phys.*, **23**, 1739 (1955).
19. R. H. Hughes, *J. Chem. Phys.*, **24**, 131 (1956).
20. S. C. Abrahams and J. Kalnajs, *Acta Cryst.*, **8**, 503 (1955).
21. R. H. Jackson, *J. Chem. Soc.*, 4585 (1962).
22. R. L. Tallman, J. L. Margrave and S. W. Bailey, *J. Amer. Chem. Soc.*, **79**, 2979 (1957).
23. L. Vaska, *Accounts Chem. Res.*, **9**, 175 (1976).
24. A. Avdeef and W. P. Schaefer, *J. Amer. Chem. Soc.*, **98**, 5153 (1976).
25. M. J. Nolte, E. Singleton and M. Laing, *J. Amer. Chem. Soc.*, **97**, 6396 (1975).
26. D. C. Bradley, J. S. Ghotra, F. A. Hart, M. B. Hursthouse and P. R. Raithby, *J. Chem. Soc. (Dalton)*, 1166 (1977).
27. A. Hordvik, *Acta Chem. Scand.*, **20**, 1885 (1966).
28. K. Lonsdale, E. Nave and J. F. Stephens, *Phil. Trans. Roy. Soc.*, **A261**, 1 (1966).
29. C. J. Marsden, D. D. DesMarteau and L. S. Bartell, *Inorg. Chem.*, **16**, 2359 (1977).
30. C. J. Marsden, L. S. Bartell and F. P. Diodati, *J. Mol. Struct.*, **39**, 253 (1977).
31. D. Käss, H. Oberhammer, D. Brandes and A. Blaschette, *J. Mol. Struct.*, **40**, 65 (1977).
32. C. Batich and W. Adam, *Tetrahedron Letters*, 1467 (1974).
33. W. Lobunez, J. Rittenhouse and J. Miller, *J. Amer. Chem. Soc.*, **80**, 3505 (1958).

34. C. Glidewell, D. C. Liles, D. J. Walton and G. M. Sheldrick, *Acta Cryst.*, **B35**, 500 (1979).
35. C. Glidewell and D. C. Liles, *Acta Cryst.*, **B34**, 696 (1978).
36. A. Chiaroni and C. Pascard-Billy, *Bull. Soc. Chim. Fr.*, 781 (1973).
37. C. Giannotti, C. Fontaine, A. Chiaroni and C. Riche, *J. Organomet. Chem.*, **113**, 57 (1976).
38. (a) Z. A. Starikova, T. M. Shchegoleva, V. K. Trunov and I. E. Pokrovskaya, *Kristallografiya*, **23**, 969 (1978).
(b) H. Mimoun, R. Charpentier, A. Mitschler, J. Fischer and R. Weiss, *J. Amer. Chem. Soc.*, **102**, 1047 (1980).
(c) J. Halfpenny and R. W. H. Small, *J. Chem. Soc., Chem. Commun.*, 879 (1979).
(d) N. A. Porter, M. A. Cudd, R. W. Miller and A. T. McPhail, *J. Amer. Chem. Soc.*, **102**, 414 (1980).
(e) R. M. Allen, C. J. Gilmore, G. W. Kirby and D. J. McDougall, *J. Chem. Soc., Chem. Commun.*, 22 (1980).
(f) M. Albericci, M. Collart-Lempereur, J. C. Braekman, D. Dalcze, B. Tursch, J. P. Declercq, G. Germain and M. Van Meerssche, *Tetrahedron Letters*, 2687 (1979).
(g) S. Yamada, K. Nakayama, H. Takayama, A. Itai and Y. Iitaka, *Chem. Pharm. Bull.*, **27**, 1949 (1979).
(h) D. A. Langs, M. G. Erman, G. T. Detitta, D. J. Coughlin and R. G. Salomon, *J. Cryst. Mol. Struct.*, **8**, 239 (1978).
39. F. McCapra and I. Beheshti, *J. Chem. Soc., Chem. Commun.*, 517 (1977).
40. P. B. Hitchcock and I. Beheshti, *J. Chem. Soc., Perkin Trans. 2*, 126 (1979).
41. J. Hess and A. Vos, *Acta Cryst.*, **B33**, 3527 (1977).
42. C. S. Foote and S. Mazur, *J. Amer. Chem. Soc.*, **92**, 3225 (1970).
43. J. H. Wieringa, J. Strating, W. Adam and H. Wynberg, *Tetrahedron Letters* 169 (1972).
44. T. N. Margulis and E. Adman, *J. Amer. Chem. Soc.*, **90**, 4517 (1968).
45. (a) A. I. Kitaigorodsky, *Molecular Crystals and Molecules*, Academic Press, New York, 1973, Chap. 1, p. 11.
(b) A. Krebs, H. Schmalstieg, O. Jarchow and K.-H. Klaska, *Tetrahedron Letters*, **21**, 3171 (1980).
46. S. I. Chan, T. R. Borgers, J. W. Russell, H. L. Strauss and W. D. Gwinn, *J. Chem. Phys.*, **44**, 1103 (1966).
47. E. E. Castellano and O. J. R. Hodder, *Acta Cryst.*, **B29**, 2566 (1973).
48. E. L. Shapiro, L. Weber, S. Polovsky, J. Morton, A. T. McPhail, K. D. Onan and D. H. R. Barton, *J. Org. Chem.*, **41**, 3940 (1976).
49. G. Holan, C. Kowala and J. A. Wunderlich, *J. Chem. Soc., Chem. Commun.*, 34 (1973).
50. M. Hospital, F. Leroy, J. P. Bats and J. Moulines, *Cryst. Struct. Commun.*, **7**, 309 (1978).
51. E. L. McGandy and E. J. Fasiska, *Amer. Cryst. Assoc.* (abstract of summer meeting), 53 (1970).
52. A. K. Saksena, P. Mangiaracina, R. Brambilla, A. T. McPhail and K. D. Onan, *Tetrahedron Letters*, 1729 (1978).
53. S. E. V. Phillips and J. Trotter, *Acta Cryst.*, **B33**, 1602 (1977).
54. A. Almenningen, P. Kolsaker, H. M. Seip and T. Willadsen, *Acta Chem. Scand.*, **23**, 3398 (1969).
55. H. M. Seip, *Acta Chem. Scand.*, **23**, 2741 (1969).
56. C. W. Gillies and R. L. Kuczkowski, *J. Amer. Chem. Soc.*, **94**, 6337 (1972).
57. R. P. Lattimer, R. L. Kuczkowski and C. W. Gillies, *J. Amer. Chem. Soc.*, **96**, 348 (1974).
58. P. Groth, *Acta Chem. Scand.*, **24**, 2137 (1970).
59. H. Itokawa, Y. Tachi, Y. Kamano and Y. Iitaka, *Chem. Pharm. Bull.*, **26**, 331 (1978).
60. J. Karban, J. L. McAtee, Jr., J. S. Belew, D. F. Mullica, W. O. Milligan and J. Korp, *J. Chem. Soc., Chem. Commun.*, 729 (1978).
61. H. Henke and H. Keul, *Cryst. Struct. Commun.*, **4**, 451 (1975).
62. (a) R. A. Rouse, *J. Amer. Chem. Soc.*, **95**, 3460 (1973).
(b) J. D. Oliver, D. F. Mullica, W. O. Milligan, J. Karban, J. L. McAtee, Jr. and J. S. Belew, *Acta Cryst.*, **B35**, 2273 (1979).
63. P. Groth, *Acta Chem. Scand.*, **21**, 2608 (1967).
64. P. Groth, *Acta Chem. Scand.*, **21**, 2631 (1967).
65. P. Groth, *Acta Chem. Scand.*, **21**, 2695 (1967).
66. M. Schulz, K. Kirschke and E. Hohne, *Chem. Ber.*, **100**, 2242 (1967).
67. P. Groth, *Acta Chem. Scand.*, **21**, 2711 (1967).

68. R. W. Murray and M. L. Kaplan, *Tetrahedron*, **23**, 1575 (1967).
69. A. G. Nord and B. Lindberg, *Acta Chem. Scand.*, **27**, 1175 (1973).
70. A. Takenaka, Y. Sasada and H. Yamamoto, *Acta. Cryst.*, **B33**, 3564 (1977).
71. H. Yamamoto, H. Aoyama, Y. Omote, M. Akutagawa, A. Takenaka and Y. Sasada, *J. Chem. Soc., Chem. Commun.*, 63 (1977).
72. R. M. Wilson, E. J. Gardner, R. C. Elder, R. H. Squire and L. R. Florian, *J. Amer. Chem. Soc.*, **96**, 2955 (1974).
73. M. Sterns, *J. Cryst. Mol. Struct.*, **1**, 373 (1971).
74. W. D. Crow, W. Nicholls and M. Sterns, *Tetrahedron Letters*, 1353 (1971).
75. M. P. Sammes, R. L. Harlow and S. H. Simonsen, *J. Chem. Soc., Perkin Trans. 2*, 1126 (1976).
76. J. Bordner, S. G. Levine, Y. Mazur and L. R. Morrow, *Cryst. Struct. Commun.*, **2**, 59 (1973).
77. J. N. Brown, R. L. R. Towns, M. J. Kovelan and A. H. Andrist, *J. Org. Chem.*, **41**, 3756 (1976).
78. G. A. Jeffrey, J. A. Pople and L. Radom, *Carbohydr. Res.*, **25**, 117 (1972).
79 D. W. J. Cruickshank and A. P. Robertson, *Acta Cryst.*, **6**, 698 (1953).
80. J. Fayos, D. Lokensgard, J. Clardy, R. J. Cole and J. W. Kirksey, *J. Amer. Chem. Soc.*, **96**, 6785 (1974).
81. N. Eickman, J. Clardy, R. J. Cole and J. W. Kirksey, *Tetrahedron Letters*, 1051 (1975).
82. P. Groth, *Acta Chem. Scand.*, **23**, 1311 (1969).
83. J. B. Hendrickson, *J. Amer. Chem. Soc.*, **83**, 4537 (1961); **86**, 4854 (1964); **89**, 7036 (1967).
84. K. B. Wiberg, *J. Amer. Chem. Soc.*, **87**, 1070 (1965).
85. M. Bixon and S. Lifson, *Tetrahedron*, **23**, 769 (1967).
86. J. Dale, *Acta Chem. Scand.*, **27**, 1115 (1973).
87. (a) F. A. L. Anet and J. J. Wagner, *J. Amer. Chem. Soc.*, **93**, 5266 (1971).
(b) V. E. Shklover, P. Ad Yaasuren, I. Tsinker, V. A. Yablokov, A. V. Ganyushkin and Yu. T. Struchkov, *Zh. Strukt. Khim.*, **21**, 112 (1980).
(c) V. E. Shklover, A. V. Ganyushkin and Yu. T. Struchkov, *Cryst. Struct. Commun.*, **8**, 869 (1979).
88. H. J. Hohorst, A. Ziemann and N. Brock, *Arzneim. Forsch.*, **21**, 1254 (1971).
89. D. W. Hill, W. R. Laster, Jr. and R. F. Struck, *Cancer Res.*, **32**, 658 (1972).
90. A. Takamizawa, S. Matsumoto, T. Iwata, K. Katagiri, Y. Tochino and K. Yamaguchi, *J. Amer. Chem. Soc.*, **95**, 985 (1973).
91. A. Takamizawa, S. Matsumoto, T. Iwata, Y. Tochino, K. Katagiri, K. Yamaguchi and O. Shiratori, *J. Med. Chem.*, **18**, 376 (1975).
92. H. Sternglanz, H. M. Einspahr and C. E. Bugg, *J. Amer. Chem. Soc.*, **96**, 4014 (1974).
93. R. F. Struck, M. C. Thorpe, W. C. Coburn, Jr. and W. R. Laster, Jr., *J. Amer. Chem. Soc.*, **96**, 313 (1974).
94. B. M. Howard, W. Fenical, J. Finer, K. Hirotsu and J. Clardy, *J. Amer. Chem. Soc.*, **99**, 6440 (1977).
95. G. L'abbé, J.-P. Dekerk, A. Verbruggen, S. Toppet, J. P. Declercq, G. Germain and M. Van Meerssche, *J. Org. Chem.*, **43**, 3042 (1978).
96. A. Camerman, H. W. Smith and N. Camerman, *Acta Cryst.*, **B33**, 678 (1977).
97. A. Camerman, H. W. Smith and N. Camerman, *Acta. Cryst. Assoc., Ser. 2, 1*, 31 (1977).
98. S. Bozzini, S. Gratton, A. Risaliti, A. Stener, M. Calligaris and G. Nardin, *J. Chem. Soc., Perkin Trans. 1*, 1377 (1977).
99. P. Groth, *Acta Chem. Scand.*, **23**, 2277 (1969).
100. P. Groth, *Acta Chem. Scand.*, **29**, 783 (1975).
101. P. Groth, *Acta Chem. Scand.*, **29**, 840 (1975).
102. L. S. Silbert, E. Siegel and D. Swern, *J. Org. Chem.*, **27**, 1336 (1962).
103. D. Swern, L. P. Witnauer, C. R. Eddy and W. E. Parker, *J. Amer. Chem. Soc.*, **77**, 5537 (1955).
104. L. Lessinger, *Ph.D. Thesis*, Harvard University, 1971.
105. P. A. Giguère and A. Weingartshofer Olmos, *Can. J. Chem.*, **30**, 821 (1952).
106. J. R. Rittenhouse, W. Lobunez, D. Swern and J. G. Miller, *J. Amer. Chem. Soc.*, **80**, 4850 (1958).
107. D. Belitskus and G. A. Jeffrey, *Acta Cryst.*, **18**, 458 (1965).
108. M. Sax, P. Beurskens and S. Chu, *Acta Cryst.*, **18**, 252 (1965).
109. H. S. Kim, S.-C. Chu and G. A. Jeffrey, *Acta Cryst.*, **B26**, 896 (1970).
110. T. D. Sakore, S. S. Tavale and L. M. Pant, *Acta Cryst.*, **22**, 720 (1967).
111. T. D. Sakore and L. M. Pant, *Acta Cryst.*, **21**, 715 (1966).

112. F. D. Verderame and J. G. Miller, *J. Phys. Chem.*, **66**, 2185 (1962).
113. M. Schulz and K. Kirschke in *Organic Peroxides*, Vol. III (Ed. D. Swern), Wiley–Interscience, New York–London, 1972, p. 114.
114. O. L. Chapman, C. L. McIntosh, J. Pacansky, G. V. Calder and G. Orr, *J. Amer. Chem. Soc.*, **95**, 4061 (1973).
115. R. T. Luibrand and R. W. Hoffmann, *J. Org. Chem.*, **39**, 3887 (1974).
116. G. A. Russell, V. Malatesta, D. E. Lawson and R. Steg, *J. Org. Chem.*, **43**, 2242 (1978).
117. G. B. Schuster, *Accounts Chem. Res.*, **12**, 366 (1979).
118. K. A. Horn and G. B. Schuster, *J. Amer. Chem. Soc.*, **101**, 7097 (1979).
119. J. J. Zupancic, K. A. Horn and G. B. Schuster, *J. Amer. Chem. Soc.*, **102**, 5279 (1980).
120. C. W. Bird, *Tetrahedron*, **36**, 535 (1980).
121. F. D. Greene, *J. Amer. Chem. Soc.*, **81**, 1503 (1959).
122. K. E. Russell, *J. Amer. Chem. Soc.*, **77**, 4814 (1955).
123. J. Z. Gougoutas and M. Malley, in press.
124. N. J. Karch, E. T. Koh, B. L. Whitsel and J. M. McBride, *J. Amer. Chem. Soc.*, **97**, 6729 (1975).
125. J. Gaultier and C. Hauw, *Acta Cryst.*, **18**, 179 (1965).
126. J. Z. Gougoutas, W. H. Ojala and J. A. Miller, *Cryst. Struct. Commun.*, **9**, 519 (1980).
127. J. Z. Gougoutas, *Cryst. Struct. Commun.*, in press.
128. M. Sax and R. K. McMullan, *Acta Cryst.*, **22**, 281 (1967).
129. M. W. Vary and J. M. McBride, *Mol. Cryst. Liq. Cryst.*, **52**, 133 (1979).
130. S. Caticha-Ellis and S. C. Abrahams, *Acta Cryst.*, **B24**, 277 (1968).
131. J. Z. Gougoutas and J. C. Clardy, *Acta Cryst.*, **B26**, 1999 (1970).
132. J. M. McBride, private communication.
133. J. Z. Gougoutas and D. G. Naae, *J. Phys. Chem.*, **82**, 393 (1978).
134. J. Z. Gougoutas, *Pure. Appl. Chem.*, **27**, 305 (1971).
135. J. Z. Gougoutas, K. H. Chang and M. C. Etter, *J. Solid State Chem.*, **16**, 283 (1976).
136. K. H. Chang, *Ph.D. Thesis*, University of Minnesota, 1977.
137. J. Z. Gougoutas and L. Lessinger, *J. Solid State Chem.*, **7**, 175 (1973).
138. J. Z. Gougoutas and K. H. Chang, *Cryst. Struct. Commun.*, **8**, 977 (1979).
139. V. Jehlička, B. Plesničar and O. Exner, *Collect. Czech. Chem. Commun.*, **40**, 3004 (1975).
140. T. J. Kealy, *J. Amer. Chem. Soc.*, **84**, 966 (1962).
141. D. W. Walter and J. M. McBride, *J. Amer. Chem. Soc.*, **103**, 7074 (1981).

The Chemistry of Functional Groups, Peroxides
Edited by S. Patai

CHAPTER **13**

Polymeric peroxides

RAY CERESA
Polymer Consultant, Pepys Cottage, 13 High Street, Cottenham, Cambridge CB4 4SA, England

I. INTRODUCTION 417
II. POLYMERIC HYDROPEROXIDES 419
A. Polymers with Terminal Hydroperoxide Groups 419
B. Polymers with Pendant Hydroperoxide Groups 420
C. Polymers with Peroxide Groups within the Backbone Chain 422
D. Polymers with Terminal Peroxide Groups 424
E. Polymers with Pendant Peroxide Groups 425
F. Polymers with Perester Groups 425
III. CONCLUSION 426
IV. REFERENCES 426

I. INTRODUCTION

In this chapter consideration in detail will only be given to peroxidic compounds of a truly polymeric nature with molecular weights of several thousand or more. Lower molecular weight prepolymers, the so-called oligomers, which exist in oxidized forms, will only be considered in passing and without detail. The term 'peroxide' is taken to encompass simple peroxides, hydroperoxides, peresters and ozonides which decompose to a peroxide or hydroperoxide, but not epoxides or other oxidized forms which may be produced via a peroxide or ozonide grouping but which cannot be isolated at the 'peroxide' stage.

There are basically three types of monomeric 'peroxy' compounds: ROOH, ROOR and RCOOOR′, where R and R′ are aliphatic or aromatic hydrocarbon groups or substituted groups. By replacing the monomeric R and R′ groups by polymeric chains we increase the number of distinct possible structures to ten although to my knowledge not all of these have been described to date in the literature. We can classify our structures as follows:

(1) ∿∿∿∿∿∿∿∿∿∿∿OOH and (2) HOO∿∿∿∿∿∿∿∿∿∿∿OOH

where we have hydroperoxide groups at one or at both ends of the macromolecule;

```
(3) ~~~~~~~~~~~~~~~~~~~~~~~~~
          O       O       O
          O       O       O
          H       H       H
```

where the hydroperoxide groups are at selected or random sites along the backbone of the polymer;

```
(4) ~~~~~~~~~~~~~~~~OOR   and   (5)  ROO~~~~~~~~~~~~~~~~OOR
```

where the end-group or -groups are peroxy groups, R being aliphatic or aromatic, and when there are groups at both ends, they may be identical or different);

```
(6) ~~~~~~OO~~~~~~OO~~~~~~
```

where the peroxy groups are present within the backbone of the macromolecule;

```
(7) ~~~~~~~~~~~~~~~~~~~~~~~~~
          O       O       O
          O       O       O
          R       R       R
```

where the peroxy groups are present in side-chains which are pendant to the macromolecular backbone;

```
(8) ~~~~~~~~~~~~~~~~~~~~~~~~COOR    and   (9) ROOC ~~~~~~~~~~~~~~~~~~~~COOR
                            ||                  ||                      ||
                            O                   O                       O
```

where the perester groups are at one end or at both ends of the macromolecule, R being aliphatic or aromatic;

```
(10) ~~~~~~~~~~~~~~~~~~~~~~~~~~
          |       |       |
         C=O     C=O     C=O
         O       O       O
         O       O       O
         R       R       R
```

where the perester groups are in pendant positions at selected or random sites along the backbone of the macromolecule.

Each of these types of polymeric peroxides requires a different synthesis and for some of these types a number of methods have been described whilst for others new syntheses are required.

The major application for polymeric peroxides has been as the first step in the synthesis of block or graft copolymers[1–3] and specific references will be given in the text where appropriate. In a number of applications for peroxidic materials described in other chapters of this book it would be advantageous if the peroxidic material could be easily separated from the reaction zone or contained within a defined area or volume of the reaction. By attaching the peroxidic groups to macromolecules which themselves can be further confined by cross-linking, it is possible to influence peroxidation reactions.

In this way Greig and coworkers[4] have used polymer resins based on cross-linked polystyrene to support peroxy acids in the oxidation of tetrahydrothiophene. This is a new field of activity for polymer chemists which should show rewarding technological applications in the 1980s[78].

II. POLYMERIC HYDROPEROXIDES

A. Polymers with Terminal Hydroperoxide Groups

Polystyrene molecules with hydroperoxy end-groups have been obtained by several different methods. During the polymerization of styrene monomer which was initiated by the redox reaction of *p*-diisopropylbenzene dihydroperoxide in the presence of ferrous ion in an emulsion system, polymeric peroxide molecules were formed by the mutual termination of the growing chains which become oil-soluble rapidly enough that no significant reaction of the remaining terminal groups took place in the aqueous phase[5–7]. Fairly pure difunctional polystyrene hydroperoxide was obtained and used to initiate block copolymerization of methyl methacrylate and other monomers (equations 1–3). Emulsions of poly(methyl methacrylate) and of copolymers of butadiene with styrene and butadiene with methyl methacrylate have also been prepared by this method to yield polymers and copolymers with terminal hydroperoxide groups[7]. In the syntheses where methyl methacrylate was used we can assume that some termination by disproportionation occurs such that a fraction of the polymeric hydroperoxide molecules would be monofunctional (equation 4).

$$\text{HOOROOH(aq.)} + \text{Fe}^{2+} \longrightarrow \text{HOORO}\bullet\text{(aq.)} + \text{Fe}^{3+} + \bullet\text{OH} \quad (1)$$

$$\text{HOORO}\bullet\text{(aq.)} + \text{M(aq.)} \longrightarrow \text{HOOROM}\bullet\text{(aq.)} \longrightarrow \text{HOOROM}_n^\bullet\text{(oil)} \quad (2)$$

$$\text{HOOROM}_n^\bullet\text{(oil)} + \text{HOOROM}_m^\bullet\text{(oil)} \longrightarrow \text{HOOROM}_{(m+n)}\text{(oil)} \quad (3)$$

$$\text{HOOROM}_n^\bullet\text{(oil)} + \text{HOOROM}_m^\bullet\text{(oil)} \longrightarrow \text{HOOROM}_n\text{(oil)} + \text{HOOROM}_m\text{(oil)} \quad (4)$$

A different approach was used by Unwin[8] in that he used *m*-diisopropylbenzene monohydroperoxide to initiate the bulk polymerization of styrene monomer in the presence of cumenyl mercaptan as a transfer agent. In this way the molecular weight was reduced from 83,000 to 28,000 so that five-sixths of the polymer chains formed had cumenyl end-groups. The isolated polymer was then oxidized in cumene solution by molecular oxygen to the polymeric dihydroperoxide which was then used to initiate block copolymerization of a number of monomers.

A measure of success was achieved in free-radical activation of hydroperoxidation by degrading polystyrene in solution with AZBN in the presence of oxygen to produce a hydroperoxidized polymer in which the hydroperoxide groups were mainly at terminal positions[9]. On the whole, attempts to use this technique with other polymers have led to more complex reactions including side-group hydroperoxidation.

As part of the process for synthesizing block copolymers with alternating 'hard' and 'soft' segments, Baysal and coworkers[10] reacted a commercially available low-molecular-weight dihydroxyl-terminated prepolymer with an aliphatic diisocyanate and with 2,5-dimethyl-2,5-bis(hydroperoxy)hexane to produce polymeric peroxycarbamates (equations 5 and 6), where n is dependent on the molecular weight of the prepolymer and R^1 represents the prepolymer residue, R^2 the hydrocarbon chain and R^3 the diisocyantate residue.

Established commercial procedures were used to choose exact stoichometric quantities in the reactions in order to retain terminal hydroperoxy groups.

$$\left[HOO-\underset{CH_3}{\overset{CH_3}{\underset{|}{\overset{|}{C}}}}-CH_2CH_2\underset{CH_3}{\overset{CH_3}{\underset{|}{\overset{|}{C}}}}-OO-\overset{O}{\overset{||}{C}}-NH-C_{13}H_{22}-NH\overset{O}{\overset{||}{C}}-(-R^1-)_{n/2}-CH_2- \right] \qquad (5)$$

$$-\!\!\left[-O-O-R^2-O-O-R^3-O-O-R^2-O-O-R^3-O-O-\right]_m\!- \qquad (6)$$

A sophisticated technique for introducing hydroperoxide groups into polymers in terminal positions makes use of living polymers prepared by anionic polymerization. A number of different complex reactions can take place when oxygen is allowed to react with the living ends of the polymer chains, but it has been shown that quantitative hydroperoxidation can be achieved by the slow addition of the living polymer solution to a polar solvent saturated by oxygen, provided that there is a low concentration of living ends, that bulky terminal groups are used and that the temperature is kept as low as possible[11]. Agouri and others approached the problem in a different way, polymerizing ethylene in the presence of diethyl zinc to obtain polyethylene with terminal zinc atoms. By controlled oxidation the latter were converted to terminal hydroperoxide groups which were used for the initiation of block copolymerization of styrene[12].

B. Polymers with Pendant Hydroperoxide Groups

When neither an exact number of pendant hydroperoxide groups nor an exact positioning of the groups along the macromolecular backbone is required, then there are many more ways of introducing the hydroperoxide functions. One of the simplest methods is by the direct oxidation of suitable side-groups along the polymer chain. Whilst polystyrene gives very few hydroperoxide groups when it is oxidized with benzoyl peroxide, simple air oxidation is claimed to introduce 6–12 hydroperoxide groups per 100 styrene units[13]. However, the partial alkylation of polystyrene introducing isopropyl groups allows the oxidation to be carried out in a more controlled manner to give, for instance, 2.37 moles of hydroperoxide groups per 100 moles of monomer units[14,15]. These polymeric initiators have been used to synthesize graft copolymers with methyl methacrylate, reducing the undesirable homopolymerisation due to hydroxyl free radicals by carrying out the second step of the synthesis in the latex phase. By using a redox initiation system, e.g. ferrous ion in conjunction with the hydroperoxidized polymer, graft copolymers can be obtained free of contaminating homopolymer[16] (equation 7).

$$\underset{OOH\qquad\qquad OOH}{\sim\!\sim\!\sim\!\sim\!\sim\!\sim\!\sim} \xrightarrow{Fe^{2+}} \underset{O^{\bullet}\qquad\qquad OOH}{\sim\!\sim\!\sim\!\sim\!\sim\!\sim\!\sim} + {}^{-}OH + Fe^{3+} \qquad (7)$$

$$+\ \text{Vinyl monomer} \longrightarrow \text{Grafted copolymer}$$

The introduction of oxidizable groups into acrylic polymers and copolymers is more difficult, but some of the ester groups can be converted by reaction with phosphorus pentachloride into acid chloride groups, which in turn may react with *t*-butyl hydroperoxide (equation 8). Saigusa and coworkers used this technique to produce base polymers onto which they grafted styrene, acrylonitrile and vinyl acetate, starting with poly(methyl acrylate)[17,18]. Because the hydroperoxidation step of the syntheses is complicated by side-reactions, later workers have preferred to copolymerize acryl chloride directly into the poly(vinyl ester) and to react the acid chloride groups with perbenzoic acid to form peranhydrides which could be used directly for the synthesis of graft copolymers or reduced to the hydroperoxide state[19] (equation 9).

$$\sim CH_2-CH(C(=O)OMe)-CH_2-CH(C(=O)OMe)\sim \xrightarrow{PCl_5} \sim CH_2-CH(C(=O)OMe)-CH_2-CH(C(=O)Cl)\sim \xrightarrow{t\text{-BuOOH}} \sim CH_2-CH(C(=O)OMe)-CH_2-CH(C(=O)OOBu\text{-}t)\sim \quad (8)$$

$$\sim CH_2-CH(C(=O)Cl)-CH_2-CH(C(=O)Cl)-CH_2\sim \xrightarrow{\text{perbenzoic acid}} \sim CH_2-CH-CH_2-CH-CH_2\sim \text{ (the two C=O groups linked by } -O-O-) \quad (9)$$

A copolymer of methyl methacrylate with isopropenyl acetate was partially hydrolysed before peroxidation was effected using hydrogen peroxide. The same workers treated a styrene/maleic anhydride copolymer with hydrogen peroxide to introduce hydroperoxide groups at random along the backbone chains in order to synthesize graft copolymers from a range of monomers[19].

Hydroperoxide groups can be introduced randomly into cellulosic derivatives in similar ways. Thus, Jahn[20] reacted *o*-chlorobenzylcellulose with hydrogen peroxide to introduce a mixture of hydroperoxide and peroxide groups along the cellulose backbone as grafting sites for monomers. More recently, working with cellulose itself, Ogiwara and Kubota used three different techniques to preoxidize the cellulose prior to grafting via ceric ion redox systems. The methods they used were oxidation by periodic acid, oxidation by reduction of ceric salts and direct oxidation with hydrogen peroxide.[21] Early stages of peroxidation produced hydroperoxide side-groups along the cellulose backbone. These methods were later extended to synthesizing poly(vinyl alcohol) with pendant hydroperoxide groups[22] following their work with aldehyde and carbonyl group containing celluloses[23,24]. These techniques seemed to be an improvement on the much earlier work of oxidizing the polymer with AZBN and oxygen (air) in benzene/cumene solutions at 60°C which was successfully used by Gleason and Stannett to introduce hydroperoxide groups at random sites along the macromolecular chains of benzyl and ethyl celluloses[25].

The presence of unsaturation in the backbone polymer assists in the activation process that allows oxidation to hydroperoxide groups and advantage was taken of this by Morris and Sekhar in 1959[26]. The natural oxidation process of raw rubber field latex results in the formation of hydroperoxide groups at an early stage of oxidation, (aldehyde and ketone groups being formed later in the oxidation process), and by adding methyl methacrylate monomer to stabilized fresh-field latex together with a trace of ferrous sulphate and tetraethylenepentamine (to complete the redox system); natural *cis*-1,4-poly(isoprene-g-methyl methacrylate) was obtained in good yield, a reaction which was later adopted commercially on a limited scale.

Poly-α-olefins are more readily oxidized than the polyethylenes and Natta and others found that atactic polypropylene could be readily oxidized in isopropylbenzene solution, containing a little methanol, by air at atmospheric pressure. After 240 minutes at 70°C a hydroperoxide content of 1.98 % was obtained and after 460 min this had increased to 4.03 %[27,28]. This reaction was used to initiate the graft copolymerization of methyl methacrylate onto polypropylene. Atactic polystyrene was oxidised in a similar way in cumene solution to give 6–12 hydroperoxide groups per 100 styrene units[11]. Degtayareva and coworkers have recently studied the kinetics of the oxidative degradation of polypropylene, using chlorobenzene as an inert solvent, by measuring the oxygen absorption and the formation of hydroperoxide groups[29], following earlier work on the kinetics of the oxidation of polyethylene[30]. It was found that the rate of formation of hydroperoxide groups at random points along the chain was considerably lower than the rate of oxygen absorption, confirming the complexity of the intramolecular reactions.

A completely different method of introducing random hydroperoxide groups was used by Smets and coworkers[31] in which poly(methacrylic acid) and copolymers of methacrylic acid were hydrolysed in aqueous solution. The hydroperoxide groups were produced by decarboxylation followed by oxidation at the anode (equation 10). Hydroperoxides of poly(methacrylic acid) containing 0.4 % of active groups were used to synthesize graft copolymers with acrylamide, vinyl pyrrolidone and acrylonitrile. This technique should be capable of extension into a wider field of hydroperoxidation reactions.

$$\sim CH_2-C(R)(C{=}O)(O^-)-CH_2-C(R)(C{=}O)(OH)\sim \xrightarrow{\text{electrolysis}} \sim CH_2-C(R)(C{=}O)(O^{\bullet})-CH_2-C(R)(C{=}O)(OH)\sim \xrightarrow{-CO_2} \sim CH_2-\dot{C}(R)-CH_2-C(R)(C{=}O)(OH)\sim \xrightarrow{\text{oxidation}} \sim CH_2-C(R)(O{-}OH)-CH_2-C(R)(C{=}O)(OH)\sim \qquad (10)$$

C. Polymers with Peroxide Groups within the Backbone Chain

Although the fact that oxygen copolymerizes with many monomers during free-radical polymerizations has been known for many years[32,33], it is only in more recent years that such polymeric peroxides have been studied in detail. Cais and Bovey[34] used NMR techniques to study the microstructure and molecular dynamics of polystyrene peroxide and showed that copolymerization at atmospheric pressure produced an equimolar copolymer, i.e. poly(styrene oxide), with a molecular weight of 8000 as measured by light-scattering techniques, corresponding to a degree of polymerization of 30. Earlier work in this field has been reviewed by Mayo[35]. Optically active polystyrene peroxides were the subject of a recent short communication by Nukui and coworkers[36] who carried out the copolymerization of styrene and oxygen in the presence of cobalt (II) Schiff bases as oxygen carriers which orientated the oxygen molecule during the copolymerization step. This is an interesting extension to the work in this field which has important possibilities in related synthetic systems.

The rates of photodegradation of polystyrene peroxides on irradiation with UV light were found to be very much dependent upon the solvent used and solvent transfer and diffusion effects were found to play important roles in the kinetics of degradation[37]. From this work it can be concluded that UV degradation of polystyrene peroxide could be used as a route to the synthesis of block copolymers following the lines of the classic method of Smets and Woodward[38,39], who synthesized polymeric phthaloyl peroxide and used the peroxypolymer to prepare polymers with residual internal peroxy groups which were then used in the second step of a block copolymerization process.

The presence of fortuitous peroxy groups within polymers formed by copolymerization with oxygen as an impurity has been used as a synthetic approach to block copolymerization[40,41], although the presence of such groups was not claimed in the earlier reference but in a later review[42a].

The ozonization of polymers is a very complex series of reactions which may result in the formation of stable ozonides by reaction with double bonds within the backbone polymer chain, or within side-groups to the main chain, or both; or it may result in almost simultaneous degradation of the initial entities with the consequent formation of hydroperoxides, peroxides and free oxygen. The reactivity of these groups and that of the free oxygen is such that further side-reactions may be promoted. It must also be borne in mind that the degradation of ozonides is a free-radical process, so that the overall reactions are further complicated by chain-reactions initiated by hydrogen abstractions. Because of this it is not feasible to give meaningful equations outlining the schematic reactions of ozonization.

Landler and Lebel[43,44] were the first to use ozonization to prepare graft copolymers with substrates of poly(terafluoroethylene), polystyrene, polyethylene and poly(vinyl chloride) using residual unsaturation within the polymer chain as the oxidation sites, or, in the case of polystyrene, the unsaturation of the benzene ring of the styryl groups. Russian workers continued in this field using ozonized starch, cellulose, amylose and isotactic polystyrene onto which they grafted both styrene and methyl methacrylate[45–47]. Korshak and his group extended the field still further by using ozonized polyamides and polyesters for grafting experiments[48,49]. Despite their complexities, ozonization reactions do offer a great deal of scope when using up-to-date methods of ozonolysis. Ozonolysis itself has been used in characterizing some block and graft copolymers, where unsaturated segments can be destroyed leaving the untreated saturated segments to be characterized by viscometry or osmometry, etc.[50]. There is no reason why these same techniques should not be used to prepare ozonides of unsaturated rubbers, or, by choice of the right solvents, the ozonides of polymers with limited numbers of unsaturated groups along the backbone or in side-chains (introduced by copolymerization, for example). Cross-linking of the rubber molecules still leaving some unsaturation offers a general method of immobilizing peroxidized systems for catalytic and other reactions. In nonrubber systems cross-links can be introduced by the use of difunctional monomers such as divinylbenzene or ethylene dimethacrylate.

The irradiation of polymers in the presence of oxygen (or the reaction of irradiated polymers with oxygen as a second-step reaction), is also a very complex process leading to either hydroperoxides or peroxides within the macromolecule according to the structure of the polymer, the irradiation conditions, the temperature and other factors affecting diffusion within the system. In this way Chapiro and his coworkers introduced peroxidic groups into polyethylene, polytetrafluoroethylene, copolymers of polyethylene and polypropylene and polystyrene as base polymers for grafting[51–55], but the generalized method should be capable of extension to other applications of 'peroxidized' polymers.

Mechanochemical degradation of high-molecular-weight polymers proceeds via the shearing of carbon–carbon bonds in the polymer backbone with the formation of free radicals at the point of rupture. When such degradations are carried out in the presence of atmospheric oxygen terminal hydroperoxide groups are formed initially. If 'hot spots' develop, or if the viscoelastic state for degradation is achieved thermally, then secondary reactions result in the formation of peroxy linkages within the degraded chains. Some of these reactions are reviewed in Reference 42b. At higher temperatures than those normally used for effecting mechanochemical reactions, thermal oxidation of polymer chains takes place with the rapid transformation of hydroperoxide to peroxide linkages and this method of introducing peroxide groups within a polymer chain has been used to initiate block copolymerization[56]. Isotactic polypropylene can be oxidized by treatment with oxygen at 100°C after previous treatment with ozone at room temperature to give a polymer containing peroxy groups within the chain as well as some terminal hydroperoxide groups[57–59]. Polypropylenes prepared in this way contained 7–23 peroxy linkages per macromolecule and were used for the synthesis of block copolymers with vinyl chloride in an emulsion system.

D. Polymers with Terminal Peroxide Groups

As has been indicated in the previous section, a number of the reactions discussed may produce terminal hydroperoxide or peroxide groups in addition to peroxy groups at intervals within the backbone chain of the polymer macromolecule. However, peroxide groups can be introduced exclusively at the ends of the chain by polymerizing the monomer with *t*-butyl hydroperoxide in the presence of a small quantity of a divalent copper salt, such as copper octoate[60]. The reaction can be represented by equations (11) and (12). Equimolecular amounts of *t*-butyloxy- and *t*-butyl-peroxy radicals are formed which initiate, for example, the polymerization of styrene and, since in this case the termination step is largely by a combination of growing radicals, some 75% of the chains would be expected to have one or more terminal peroxide groups. Block copolymers of styrene with methyl methacrylate have been synthesized in this way.

$$t\text{-BuOOH} + Cu^{2+} \dashrightarrow t\text{-BuOO}\cdot + Cu^{+} + H^{+} \quad (11)$$

$$t\text{-BuOOH} + Cu^{+} \dashrightarrow t\text{-BuO}\cdot + Cu^{2+} + {}^{-}OH \quad (12)$$

By using a peroxidic initiator with a highly active chain-transfer group, transfer to initiator can be promoted during polymerization so that a high proportion of the polymer chains formed have unreacted peroxy groups in terminal positions. A suitable initiator is *p,p*-bisbromomethyl benzoyl peroxide (**1**)[61], which gives end-groups with the structure **2**. A

$$BrCH_2-C_6H_4-\underset{\underset{O}{\|}}{C}OO\underset{\underset{O}{\|}}{C}-C_6H_4-CH_2Br$$

(**1**)

polystyrene with peroxy end-groups obtained in this way was used as a macromolecular initiator for block copolymerization.

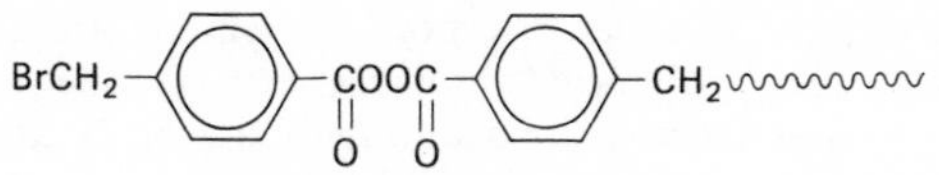

$$BrCH_2-C_6H_4-\underset{\underset{O}{\|}}{C}OO\underset{\underset{O}{\|}}{C}-C_6H_4-CH_2\sim\sim\sim$$

(2)

E. Polymers with Pendant Peroxide Groups

If peroxy groups can be introduced into a vinyl monomer to give a stable monomeric entity that is inert to intramolecularly induced decomposition then a way is open to introduce pendant peroxy groups into a macromolecule by copolymerization with such a peroxy monomer. Dalton and Tidwell[62] used this approach to synthesize *t*-butyl *p*-vinylperbenzoate by reacting *p*-vinylbenzoyl chloride with sodium *t*-butyl hydroperoxide (equation 13). This monomer can be polymerized on its own to a d.p. of 33 without loss of any of the peroxy groups or can be copolymerized with styrene to give copolymers with an average structure of 8 peroxy monomer groups to 24 of styrene. Earlier workers obtained similar results using *t*-butyl *p*-vinylbenzene peroxide[63,64]. More recently Russian workers have copolymerized alkene–alkyne peroxide monomers with a variety of monomeric and copolymeric (e.g. butadiene–styrene) systems to produce cooligomers with functional peroxidic groups capable of cross-linking reactions[65–68], but the structure of these oligomers is less well defined. This general type of copolymerization reaction which leaves unreacted peroxy groupings is an important synthetic method which is capable of further extension.

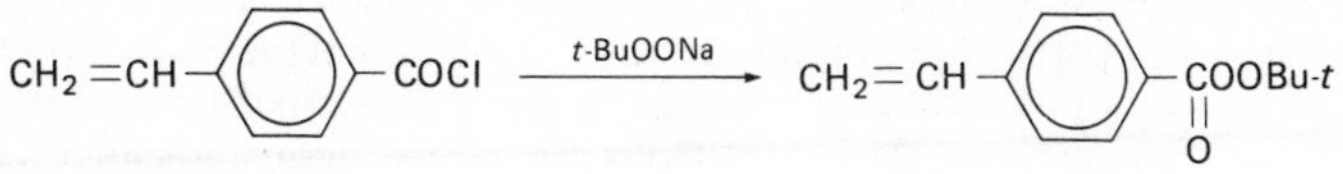

$$CH_2{=}CH-C_6H_4-COCl \xrightarrow{t\text{-BuOONa}} CH_2{=}CH-C_6H_4-\underset{\underset{O}{\|}}{C}OOBu\text{-}t \qquad (13)$$

The use of hydrogen peroxide as the oxidizing system has been used to produce random peroxy groups on a variety of polymers and polymeric systems such as ion-exchange resins[69,70], carboxymethylcellulose[71] and liquid rubbers[72] and the application of such polymeric reagents has been reviewed by Manecke and Reuter[73].

F. Polymers with Perester Groups

In general there are only passing references to the synthesis of polymeric peresters and this is a very much neglected field. When polyacetaldehyde is prepared according to the method of Delzenne and Smets[74,75] it has a polyacetalic structure containing 1–4% of hydroperoxide groups and of the order of 1% of peracetate groups, due to the presence of traces of peracetic acid during the preparative step. The presence of these mixed groups was used to initiate the graft copolymerization of methyl methacrylate.

Peroxy acid groups have been introduced into cross-linked polystyrene resins by using chloromethylation, then formylation, followed by oxidation with hydrogen peroxide[76]. In this way oxidation reactions were confined to the vicinity of the modified resin, a technique which will probably be adopted by other workers in the field.

III. CONCLUSION

The preceding sections of this chapter have dealt with the syntheses available for the different types of peroxy polymers as defined in the introduction. The majority of these syntheses were carried out in order to produce polymeric precursors for block or graft copolymerization and the copolymers so prepared were identified following the description of the synthesis. To describe the applications of the block and graft copolymers so synthesized is outside the scope of this book but the reader is referred to several general references which review the properties and applications of block and graft copolymers[1–3,77]. Apart from the synthesis of 'ordered copolymers' there have been very few published applications for peroxide polymers and those of which I am aware of have already been mentioned following the description of the synthesis. However, the importance of these applications is such that they can be restated in this section.

By fixing peroxy groups on the surface of a polymeric resin it is possible to carry out oxidation reactions with controlled concentration of peroxy groups at the site of the reaction[4]. The insoluble nature of the resin enables the oxidant to be rapidly brought into contact with the system to be oxidized and equally rapidly removed at any desired time of reaction. It is feasable in many instances to use a peroxidized polymer which is not cross-linked, but then one is faced with the problem of isolating the polymer at the required point in the reaction. This could be effected by making use of temporary bonding by salt formation, e.g. by addition of a diamine or triamine where the peroxy polymer contains acid groups, as in a peroxidized acrylic acid copolymer. In this way the peroxidized polymer could be rapidly brought out of solution and isolated and the isolated polymer could then be readily converted back to its soluble form for characterization or further reaction. In other cases it may be feasible to isolate the oxidizing polymer by the addition of methanol or ethanol (or other nonsolvent) to the system, followed by resolvation. As a physical chemist, it seems to me that these alternative techniques could give a greater degree of flexibility than that obtained by using cross-linked systems.

Peroxides have been used to cross-link natural and synthetic rubbers for many years, so it is not surprising that cross-linking of liquid rubber systems has been attempted using blends of 'normal' and peroxidized liquid rubbers[72]. Isolation of the peroxidized polymer from the other phase or phases of the peroxidizing synthesis is usually a problem and this has hindered developments in this field. However, as has been indicated, controlled thermal oxidation can be used to degrade many rubbers and at the same time introduce peroxy groups with a measure of control. By carrying out the degradation to an extent that the polymer is liquid at room temperature it then becomes possible to mix into the system reactants that would cross-link via the peroxy groups when the temperature is elevated.

IV. REFERENCES

1. R. J. Ceresa, *Block and Graft Copolymers*, Butterworths, London, 1962.
2. R. J. Ceresa, *Block and Graft Copolymerization*, John Wiley and Sons, New York–London, Vol. I, 1973; Vol. II, 1976.
3. W. J. Burlant and A. S. Hoffman, *Block and Graft Polymers*, Reinhold, New York, 1960.
4. J. A. Greig, R. D. Hancock and D. C. Sherrington, *Eur. Polym. J.*, **16**, 293 (1980).
5. P. Allen, J. Downer, G. Hastings, H. Melville, P. Molyneux and T. Unwin, *Nature*, **177**, 910 (1956).
6. R. Orr and S. H. Williams, *J. Amer. Chem. Soc.*, **78**, 3237 (1956).
7. R. Orr and S. H. Williams, *J. Amer. Chem. Soc.*, **79**, 3137 (1957).
8. J. R. Unwin, *J. Polym. Sci.*, **27**, 580 (1958).
9. G. Riess and A. Banderet, *Bull. Soc. Chim. Fr.*, **51**, 733 (1959).
10. B. M. Baysal, E. H. Orhan and I. Yilgor, *J. Polym. Sci.*, *Symposium No. 46*, 237 (1974).

11. J. M. Catala, G. Riess and J. Brossas, *Makromol. Chem.*, **178**, 1249 (1977).
12. E. Agouri, R. Laputte, Y. Philardeau and J. Rideau, *Inf. Chim.*, **144**, 167 (1975).
13. G. Natta, *J. Polym. Sci.*, **34**, 531 (1959).
14. D. Metz and R. Mesrobian, *J. Polym. Sci.*, **16**, 345 (1955).
15. J. Manson and L. Cragg, *Can. J. Chem.*, **36**, 858 (1958).
16. W. Hahn and H. Lechtenbohmer, *Makromol. Chem.*, **16**, 50 (1955).
17. R. Oda and T. Saigusa, *Bull. Inst. Chem. Res. Kyoto Univ.*, **32**, 32 (1954).
18. T. Saigusa, T. Nozaki and R. Oda, *J. Chem. Soc. Japan* (*Ind. Chem. Sect.*), **57**, 233, 243 (1954).
19. N. G. Gaylord, *Interchem. Rev.*, **15**, 91 (1957).
20. A. Jahn, *Angew. Chem.*, **66**, 177 (1954).
21. Y. Ogiwara and H. Kubota, *J. Appl. Polym. Sci.*, **17**, 2427 (1973).
22. H. Kubota and Y. Ogiwara, *J. Appl. Polym. Sci.*, **23**, 2271 (1979).
23. H. Kubota and Y. Ogiwara, *J. Appl. Polym. Sci.*, **22**, 3363 (1978).
24. Y. Ogiwara, T. Umasaka and H. Kubota, *J. Appl. Polym. Sci.*, **23**, 837 (1979).
25. E. Gleason and V. Stannett, *J. Polym. Sci.*, **44**, 183 (1960).
26. J. E. Norris and B. C. Sokhar, *Proc. Intern. Rubb. Conf., Washington*, 277 (1959).
27. G. Natta, E. Beati and F. Severini, *J. Polym. Sci.*, **34**, 685 (1959).
28. Solvic. Soc., *Belgian Patent*, No. 571,542 (1959); *Chem. Abstr.*, **53**, 16599 (1959).
29. T. G. Degtayareva, N. F. Trofimova and V. V. Kharitonov, *Vysokomol. Soed.*, **A20**, 1873 (1978).
30. P. A. Ivanchenko, V. V. Kharitonov and Ye. T. Denisev, *Vysokomol. Soed.*, **A11**, 1622 (1969).
31. G. Smets, A. Poot, M. Mullier and J. P. Bex, *J. Polym. Sci.*, **34**, 287 (1959).
32. A. A. Miller and F. R. Mayo, *J. Amer. Chem. Soc.*, **78**, 1017 (1956).
33. F. R. Mayo, *J. Amer. Chem. Soc.*, **80**, 2465 (1958).
34. R. E. Cais and F. A. Bovey, *Macromolecules*, **10**, 169 (1977).
35. F. R. Mayo, *Acc. Chem. Res.*, **1**, 193 (1968).
36. M. Nukui, K. Yoshino, Y. Ohkatsu and T. Tsuruta, *Makromol. Chem.*, **180**, 523 (1979).
37. N. A. Weir and T. H. Milkie, *Makromol. Chem.*, **179**, 1989 (1978).
38. G. Smets and A. E. Woodward, *J. Polym. Sci.*, **14**, 126 (1954).
39. A. E. Woodward and G. Smets, *J. Polym. Sci.*, **17**, 51 (1955).
40. H. W. Melville and W. F. Watson, *J. Polym. Sci.*, **11**, 299 (1953).
41. R. J. Ceresa, *Polymer*, **1**, 397 (1960).
42. (a) Reference 1, p. 67
    (b) Reference 1, pp. 65–101.
43. Y. Landler and P. Lebel, *Makromol. Khim., Dokl., Moscow, Sektsiya*, **3**, 191 (1960).
44. Y. Landler and P. Lebel, *J. Polym. Sci.*, **48**, 477 (1960).
45. V. A. Kargin, P. V. Kozlov, N. A. Platé and I. I. Konoreva, *Vysokomol. Soed.*, **1**, 114 (1959).
46. V. A. Kargin, Kh. V. Ustanov and V. I. Atkhodzhaev, *Vysokomol. Soed.*, **1**, 149 (1959).
47. N. A. Platé, V. P. Shybnev and V. A. Kargin, *Vysokomol. Soed.*, **1**, 1853 (1959).
48. V. V. Korshak and K. K. Musgova, *Bull. Acad. Sci., USSR*, 656 (1958).
49. V. V. Korshak and M. A. Shkolina, *Dokl. Akad. Nauk. SSSR, Chem. Sci. Sect.*, **122**, 609 (1958).
50. D. Barnard, *J. Polym. Sci.*, **22**, 213 (1956).
51. A. Chapiro, *J. Polym. Sci.*, **29**, 321 (1958).
52. A. Chapiro, *J. Polym. Sci.*, **34**, 439 (1959).
53. A. Chapiro, *Chim. Ind. Plast. mod.*, **9**, 34 (1957).
54. A. Chapiro, A.-M. Jendrychowska-Bonamour and J.-P. Leca, *Eur. Polym. J.*, **8**, 1301 (1972).
55. A.-M. Jendrychowska-Bonamour and C. Michot, *Eur. Polym. J.*, **13**, 241 (1977).
56. R. J. Ceresa in *Science and Technology of Rubber* (Ed. F. R. Eirich), Academic Press, New York, 1978, Chap. 11.
57. G. Smets, G. Weinand and S. Deguchi, *J. Polym. Sci., Pol. Chem. Ed.*, **16**, 3077 (1978).
58. G. Weinand and G. Smets, *J. Polym. Sci., Pol. Chem. Ed.*, **16**, 3091 (1978).
59. N. Overbergh, G. Weinand and G. Smets, *J. Polym. Sci., Pol. Chem. Ed.*, **16**, 3107 (1978).
60. W. Kern, M. A. Achon, G. Schroder and R. Schulz, *Z. Elektrochem*, **60**, 309 (1956).
61. Z. Nicolova-Nankova, F. Palacin, F. Raviola and G. Riess, *Eur. Polym. J.*, **11**, 301 (1975).
62. A. J. Dalton and T. T. Tidwell, *J. Polym. Sci., Pol. Chem. Ed.*, **12**, 2957 (1974).
63. W. Kawai and S. Tsutsumi, *Kogyu Kagaku Zashi*, **62**, 1048 (1959).
64. O. L. Mageli, R. E. Light, Jr. and R. B. Gallagher, *J. Polym. Sci.*, **24**, 57 (1968).
65. L. S. Chuiko, N. M. Grinenko and T. I. Yurzhenko, *Vysokomol. Soed.*, **A17**, 91 (1975).

66. S. S. Ivanchev, O. N. Romantsova and O. S. Romanova, *Vysokomol. Soyd.*, **A17**, 2401 (1975).
67. V. A. Puchin, S. A. Voronov, Yu. A. Lastukhin and S. P. Prokopchuk, *Vysokomol. Soyd.*, **A18**, 107 (1976).
68. S. S. Ivanchev, V. P. Budtov, O. N. Romantsova, V. M. Belyayev, O. S. Romanova, N. G. Podosenova and G. A. Otradina, *Vysokomol. Soed.*, **A18**, 1005 (1976).
69. T. Takagi, *Kogyo Kagaku Zasshi*, **69**, 298 (1966).
70. J. S. Bates and R. A. Shanks, *J. Macromol. Sci.*, **A14**, 137 (1980).
71. H. Kubota and Y. Ogiwara, *J. Appl. Polym. Sci.*, **22**, 3363 (1978).
72. L. S. Chuiko, N. M. Grinenko and A. V. Krat, *Vysokomol. Soed.*, **A18**, 494 (1976).
73. G. Manecke and P. Reuter, *J. Polym. Sci., Polymer Symposium*, **62**, 227 (1978).
74. G. Delzenne and G. Smets, *Makromol. Chem.*, **18**, 82 (1956).
75. G. Delzenne and G. Smets, *Makromol. Chem.*, **23**, 16 (1957).
76. J. A. Creig, R. D. Hancock and D. Sherington, *Eur. Polym. J.*, **16**, 293 (1980).
77. E. Matuselli, R. Palumbo and M. Kryszewski (Eds.), *Polymer Blends*, Plenum Press, New York, 1979.
78. P. Hodge and D. C. Sherrinton (Eds.), *Polymer-Supported Reactions in Organic Syntheses*, John Wiley and Sons, Chichester, 1980.

The Chemistry of Functional Groups, Peroxides
Edited by S. Patai

CHAPTER **14**

Organic reactions involving the superoxide anion

ARYEH A. FRIMER
Department of Chemistry, Bar-Ilan University, Ramat-Gan, Israel

I. INTRODUCTION . . . 429
II. DEPROTONATION VS. HYDROGEN ATOM ABSTRACTION . . . 430
A. Kinetic and Thermodynamic Data . . . 430
B. O—H bonds . . . 432
C. S—H bonds . . . 435
D. N—H bonds . . . 436
E. C—H bonds . . . 437
III. NUCLEOPHILIC REACTIONS . . . 441
A. Introduction . . . 441
B. Alkyl Halides and Sulphonates . . . 441
C. Acid Chlorides, Anhydrides and Esters . . . 442
D. Diacyl Peroxides . . . 443
E. Aldehydes and Ketones . . . 444
F. Amides and Nitriles . . . 445
G. Sulphur Compounds . . . 445
H. Cations and Cation Radicals . . . 446
IV. ONE-ELECTRON REDUCTIONS . . . 448
A. Introduction . . . 448
B. Conjugated Ketones . . . 448
C. Olefins . . . 449
D. Aryl Systems . . . 451
E. Peroxides . . . 452
F. Free Radicals . . . 454
G. Generation of Singlet Oxygen ($^1O_2$) . . . 455
V. REFERENCES . . . 456

I. INTRODUCTION

Despite the omnipresence of one-electron processes in nature, free-radical damage presents a serious and constant threat to living organisms[1–3]. One available source of radicals in the body is the superoxide anion radical, $O_2^{\cdot-}$, which is formed in a large number of reactions of biological importance in both enzymic and nonenzymic processes[4]. It follows then that it is of great value to understand the organic chemistry of $O_2^{\cdot-}$, for as Fridovich[5] has poignantly noted: 'If we are going to know how it does its dirty work, we have to know what it is capable of doing'. Nevertheless, had convenient methods not been found for generating $O_2^{\cdot-}$ in aprotic organic solvents, progress in this direction would have undoubtedly been slow and tedious.

Two basic approaches have been developed and are presently in use. The first involves *in situ* generation of $O_2^{\cdot-}$ by the electrolytic reduction of molecular oxygen[6,7]. This method permits the controlled generation of low concentrations ($<10^{-2}$M) of pure $O_2^{\cdot-}$ and is well suited for mechanistic studies. This is particularly true for cyclic voltametry which allows the researcher to follow the course of the reaction and detect unstable intermediates. Efficient product studies, however, require greater $O_2^{\cdot-}$ levels[8].

An alternate approach utilizes superoxide salts as well-defined sources of $O_2^{\cdot-}$. The inorganic salts, such as the commercially available potassium superoxide (KO_2), are generally insoluble in aprotic organic solvents, though they are slightly soluble in those of high polarity like DMSO. Nevertheless, solutions of KO_2 have been conveniently prepared in benzene, toluene, acetonitrile, DMSO, pyridine, triethylamine, THF, etc. through the agency of phase-transfer catalysts such as crown ethers[9]. Tetramethylammonium superoxide has also been synthesized and, in contrast to its alkali metal analogues, is quite soluble in a number of aprotic solvents[10,11].

With the introduction of the KO_2–crown ether reagent[9] as a convenient source of $O_2^{\cdot-}$, the organic chemistry of the radical anion was pursued with renewed vigour. It is now clear that $O_2^{\cdot-}$ displays four basic modes of action including deprotonation, H-atom abstraction, nucleophilic attack and electron transfer. It is important to note, however, that a variety of autoxidative processes can take over following the initial $O_2^{\cdot-}$ reaction; hence, one must proceed with due caution in any attempt to determine the mechanism of reaction simply based on product analysis. Let us turn now to a discussion of the various modes of reaction and the fascinating organic chemistry of $O_2^{\cdot-}$.

II. DEPROTONATION VS. HYDROGEN ATOM ABSTRACTION

A. Kinetic and Thermodynamic Data

We have repeatedly referred to the use of aprotic solvents for carrying out organic reactions of $O_2^{\cdot-}$. This is simply because $O_2^{\cdot-}$, in the presence of a proton or proton source, rapidly disproportionates to molecular oxygen and hydroperoxy anion (equation 1). This process involves primarily two steps (equations 2 and 3) for which kinetic and thermodynamic data have been evaluated by pulse radiolysis[12]. Two other reactions (equations 4 and 5) are of lesser importance.

$$H^+ + 2O_2^{\cdot-} \rightleftharpoons HO_2^- + O_2 \tag{1}$$

$$H^+ + O_2^{\cdot-} \rightleftharpoons HO_2^{\cdot} \qquad pK_a\,(HO_2^{\cdot}) = 4.69 \tag{2}$$

$$HO_2^{\cdot} + O_2^{\cdot-} \longrightarrow HO_2^- + O_2 \qquad k_3 = 1 \times 10^8\ \text{M}^{-1}\,\text{s}^{-1} \tag{3}$$

$$HO_2^{\cdot} + HO_2^{\cdot} \longrightarrow H_2O_2 + O_2 \qquad k_4 = 8.6 \times 10^5\ \text{M}^{-1}\,\text{s}^{-1} \tag{4}$$

$$O_2^{\cdot-} + O_2^{\cdot-} \xrightarrow{H^+} O_2 + HO_2^- \qquad k_5 < 0.3\ \text{M}^{-1}\,\text{s}^{-1} \tag{5}$$

The low pK_a of the hydroperoxy radical (see equation 2) suggests that $O_2^{\cdot-}$ is not a strong Brønsted base and is comparable in strength to acetate. Sawyer[6,7,13,14] has noted, however, that the initial deprotonation step (equation 2) is driven far to the right by the subsequent electron-transfer step (equation 3). This in turn raises the effective basicity of $O_2^{\cdot-}$, i.e. the efficiency with which $O_2^{\cdot-}$ can effect proton transfer.

To obtain a more quantitative idea as to the magnitude of the effective basicity of $O_2^{\cdot-}$, let us consider briefly the disproportionation of $O_2^{\cdot-}$ in water[13,14]. In this case H_2O becomes our proton source (equation 6). The standard potential for this process was derived by summing the two half-cell reactions equations (7) and (8)[13] and is 0.53 V vs. NHE at pH 14. The equilibrium constant for equation (6), K_6, can be calculated from E_6^0 by utilizing the well-known electrochemical relationship $\ln K = nFE^0/RT$. At 298 K, $K_6 = 9.1 \times 10^8$ (equation 9).

$$2O_2^{\cdot-} + H_2O \rightleftharpoons O_2 + HO_2^- + HO^- \qquad E_6^0 = 0.53 \text{ V} \tag{6}$$

$$O_2^{\cdot-} + H_2O + e^- \longrightarrow HO_2^- + HO^- \qquad E_7^0 = 0.20 \text{ V} \tag{7}$$

$$O_2^{\cdot-} \longrightarrow O_2 + e^- \qquad E_8^0 = 0.33 \text{ V} \tag{8}$$

$$K_6 = \frac{[O_2][HO_2^-][HO^-]}{[O_2^{\cdot-}]^2[H_2O]} = 9.1 \times 10^8 \tag{9}$$

For the purpose of measuring the effective basicity of $O_2^{\cdot-}$, let us view equation (6) as the hydrolysis of the hypothetical base ($2\,O_2^{\cdot-}$) which generates the corresponding conjugate acid ($HO_2^- + O_2$). The K_b for this hypothetical base and the pK_a for the corresponding conjugate acid can be calculated as shown in equation (10).

$$\begin{aligned} K_b(2\,O_2^{\cdot-}) &= K_6[H_2O] = (9.1 \times 10^8)(55.5) = 5.1 \times 10^{10} \\ pK_a &= pK_w - pK_b = 14 - (-10.7) = 24.7 \end{aligned} \tag{10}$$

The value of pK_a indicates that $O_2^{\cdot-}$ can promote proton transfer from substrates and solvents to an extent equivalent to that of a conjugate base of an acid with a pK_a of approximately 25. It is not at all surprising then that the addition of $O_2^{\cdot-}$ to an aqueous solution results in the formation of HO_2^- and HO^-, both strong Brønsted bases. We should expect therefore that even some weakly acidic organic compounds can be deprotonated efficiently by $O_2^{\cdot-}$. More importantly we may conclude that only those modes of reaction that can compete with rapid dismutation will be observed in protic media. In aprotic media, however, the lifetime of $O_2^{\cdot-}$ is expected to be long because *solvent-induced* acid-catalysed disproportionation (equation 1) is precluded, while simple disproportionation (equation 11) is energetically unfavourable[15].

$$2O_2^{\cdot-} \longrightarrow {}^3O_2 + O_2^{2-} \qquad \Delta G > 28 \text{ kcal mol}^{-1} \tag{11}$$

In contradistinction to its effectiveness as a base, $O_2^{\cdot-}$ is expected to be a poor hydrogen-atom abstractor. A simple thermochemical calculation proves this point. The heat of reaction for a hydrogen abstraction by $O_2^{\cdot-}$ (equation 12) can be estimated from the net bond energy difference[16a] between the R—H bond broken and the H—O bond in HOO^- formed, as shown in equation (13). Since the H—O bond dissociation energy for HOO^- has

$$R-H + O_2^{\cdot-} \longrightarrow R^{\cdot} + HOO^- \qquad (12)$$

$$\Delta H_{12} = D(R-H) = D(H-OO^-) \qquad (13)$$

been evaluated at 63.4 kcal[16b], a hydrogen abstraction would only be an exothermic process if the R—H bond energy were less than 63.4 kcal. A quick scan of any table of bond dissociation energies[17] reveals that only a handful of substrates bear R—H bonds that are that weak. Even the labile allylic (89 kcal), benzylic (85) kcal and aldehydic (86 kcal) C—H bonds lie substantially above this figure. The reader is reminded that the above calculation is strictly applicable only to the gas phase. In solution a variety of other considerations come into play. Nevertheless, these results do suggest that hydrogen abstraction is not expected to be a primary reaction pathway while proton transfer is.

B. O—H Bonds

Alcohols (and hydroperoxides — see Section IV.E) serve as excellent disproportionation catalysts for $O_2^{\cdot-}$ and are converted to the corresponding alkoxides[18–20]. Stanley[20] reports that steric considerations seem to control the rate of reaction. Primary alcohols, even those as weakly acidic as *n*-butanol[13] ($pK_a = 33$ in DMF)[21] apparently cause the instantaneous disproportionation of $O_2^{\cdot-}$. Isopropanol, on the other hand, requires several minutes for complete reaction while *t*-butanol reacts at appreciable rates only at relatively high concentrations.

The reaction of $O_2^{\cdot-}$ with aromatic hydroxylic substrates such as phenols[22], *o*- and *p*-dihydroxyarenes[14,21–27] and α-tocopherol and related 6-hydroxychroman compounds[21,28–31] have been studied and substrate oxidation products result. The observation of semiquinones in the case of dihydroxyarenes, and chromanoxyl radicals in the case of 6-hydroxychroman compounds, suggests that the initial steps of these oxidations involve net hydrogen-atom transfer from ArOH to $O_2^{\cdot-}$ as outlined in equation (14). Semiquinone formation merely entails subsequent proton transfer to base.

$$O_2^{\cdot-} + HO\text{-}C_6H_4\text{-}OH \longrightarrow HO_2^- + HO\text{-}C_6H_4\text{-}O^{\cdot} \xrightarrow[-\overset{+}{B}H]{B:} {}^-O\text{-}C_6H_4\text{-}O^{\cdot} \qquad (14)$$

It should be noted, however, that a true H-atom transfer mechanism should in no way lead to the formation of molecular oxygen. Nevertheless, Sawyer and coworkers[21] have strong evidence for the transient formation of O_2 in a reversible equilibrium in the cases of 3,5-di-*t*-butylcatechol and α-tocopherol. Molecular oxygen formation is generally symptomatic of an acid-catalysed disproportionation process (equation 1). Indeed, when the reaction mixture is continuously purged of O_2 (by vigorous argon bubbling through the solution), quantitative yields of the substrate anion are obtained without significant oxidation. Based on this and other experimental electrochemical evidence, Sawyer and coworkers[21] suggest that the oxidation of catechols, tocopherol and other acidic substrates by superoxide ion involves an initial rate-determining proton transfer from the substrate to $O_2^{\cdot-}$ (equation 15). This is followed by rapid disproportionation to give peroxide and molecular oxygen (equation 16), with the latter oxidizing the substrate anion (equation 17).

$$O_2^{\cdot-} + ROH \longrightarrow HO_2^{\cdot} + RO^- \qquad (15)$$

$$HO_2^{\cdot} + O_2^{\cdot -} \longrightarrow HO_2^{-} + O_2 \quad (16)$$

$$O_2 + RO^{-} \longrightarrow RO^{\cdot} + O_2^{\cdot -} \quad (17)$$

Following the formation of RO· a variety of free-radical oxidative processes can take over. In the case of *o*-dihydroxyarenes[24], *o*-quinones and dicarboxylic acids (or their cyclization products) are formed from semiquinone **1** perhaps by the mechanism shown in Scheme 1.

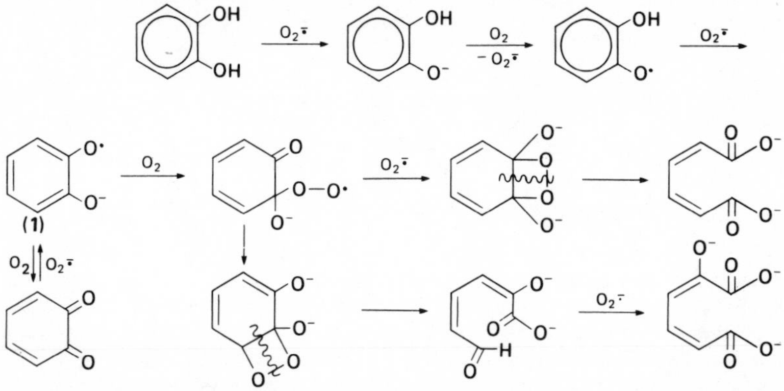

SCHEME 1. Suggested mechanism for the $O_2^{\cdot -}$-induced oxidative cleavage of *o*-hydroquinone.

In the case of α-tocopherol model compounds (see Scheme 2), chroman **2a** (R = COOH) reacts with enzymatically generated $O_2^{\cdot -}$ in aqueous media to yield quinone **3**[30] via 8a-hydroxychroman-6-one **4**, while chroman **2b** (R = Me) yields 6-hydroxychroman-5-one **5** when reacted with KO_2 suspended in THF[29]. When chroman **2b** (R = Me) is reacted with KO_2/18-crown-6 in acetonitrile, diepoxide **6** results[31]. A possible mechanism for these reactions are shown in Scheme 2.

The oxidation of ascorbic acid[21,32,33] has also been studied in some detail but the exact mechanistic details are far from clear[34].

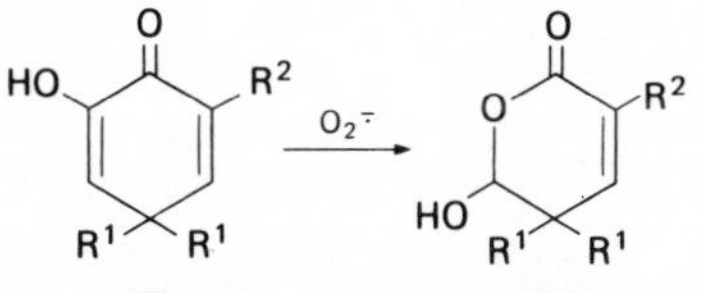

(7) $\xrightarrow{O_2^{\cdot -}}$ (8) (18)

(a) – (d) R^1 = Me; R^2 = H, Me, OMe, OEt

(e) R^1, R^1 = $-(CH_2)_5-$ · R^2 = OMe

(f) 2-Hydroxycholesta-1,4-dien-3-one

(g) 2-Hydroxy-1,2-dehydrotestosterone

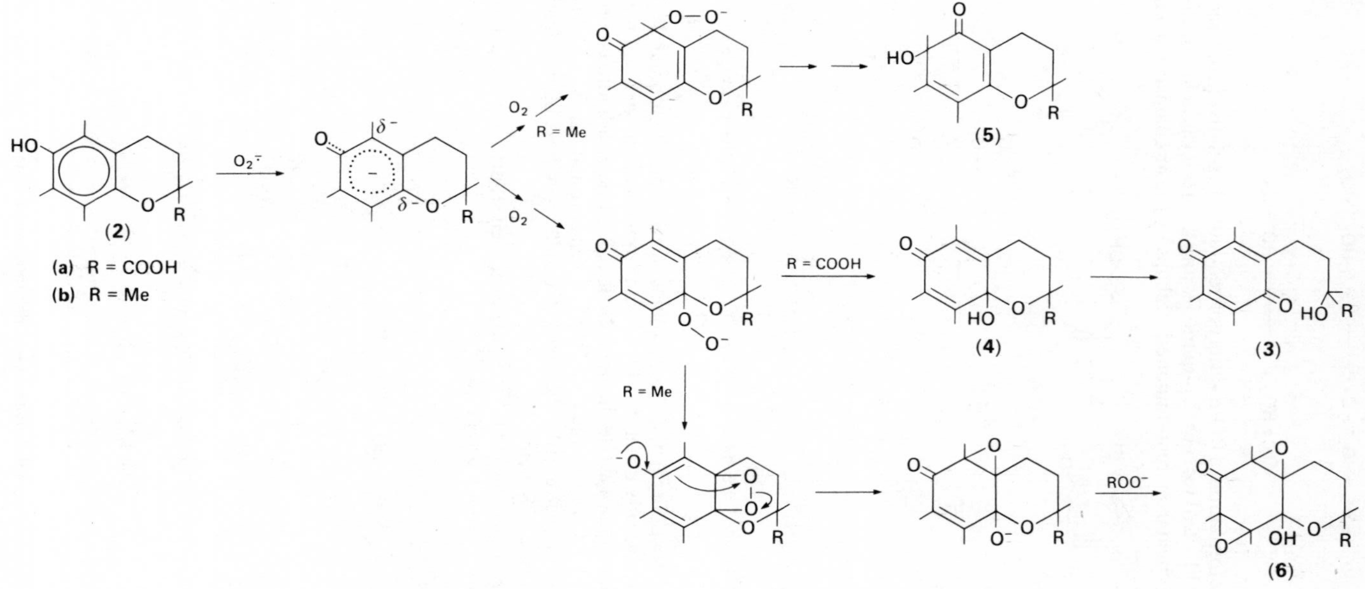

SCHEME 2. Reaction mechanism for $O_2^{\cdot -}$-induced oxidation of α-tocopherol model compounds.

Frimer and Gilinsky have studied the reaction of $O_2^{\cdot -}$ with a variety of enols[35–37], in particular 2-hydroxycyclohexa-2,5-dien-1-ones (**7**) (equation 18). Interestingly, these enols (**7**) were oxidized to the corresponding lactols (**8**). Considering that the reaction proceeded even more rapidly with *t*-butoxide and oxygen, these authors consider the $O_2^{\cdot -}$ process to be a base-catalysed autoxidation. A possible mechanism[35,37a,38,39] is outlined in Scheme 3 and involves initial proton removal as suggested by Sawyer[21].

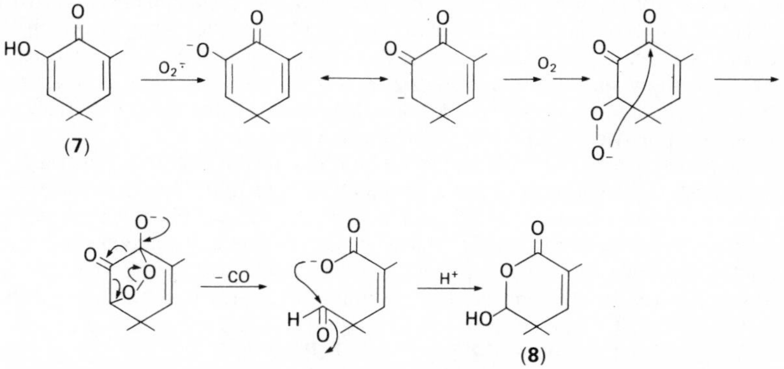

SCHEME 3. Mechanism for $O_2^{\cdot -}$-induced base-catalysed autoxidation of enols **7**.

There has been an interesting report[40] that $O_2^{\cdot -}$, produced from KO_2 and Pd(II) catalysts, reacts with ketoximes and CH_2Cl_2 to produce methylene dioximes (equation 19). Here too it is likely that the initial step involves oximate anion formation.

$$R_2C{=}N{-}OH \xrightarrow[Pd(II)]{KO_2/CH_2Cl_2} (R_2C{=}N{-}O)_2CH_2 \qquad (19)$$

C. S—H Bonds

A variety of thiols[41–43,126c] have been reacted with $O_2^{\cdot -}$ and generally disulphides result. Considering the greater acidity of S—H bonds as compared to the corresponding O—H bonds, the mechanism of these reactions probably involves initial deprotonation, oxidation of the resulting anion followed by radical coupling (equation 20, path a). Alternatively (path b), a sulphur radical and anion might couple, in a step typical of an $S_{RN}1$ mechanism[44,45], yielding a disulphide anion radical. Electron transfer to oxygen generates $O_2^{\cdot -}$ and the desired disulphide.

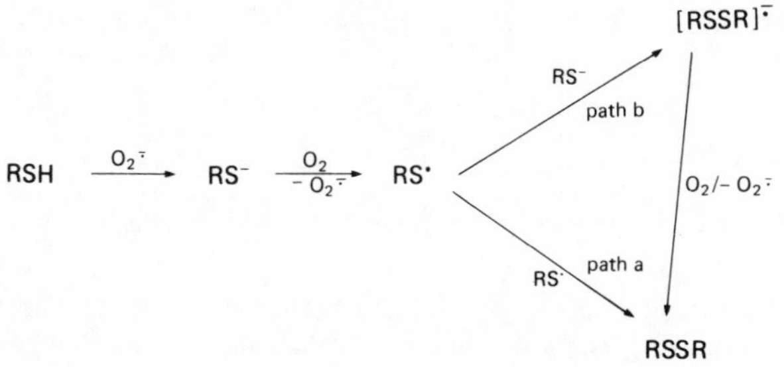

$$RSH \xrightarrow{O_2^{\cdot -}} RS^- \xrightarrow[-O_2^{\cdot -}]{O_2} RS^{\cdot} \quad \begin{cases} \xrightarrow[\text{path b}]{RS^-} [RSSR]^{\cdot -} \xrightarrow{O_2/-O_2^{\cdot -}} RSSR \\ \xrightarrow[\text{path a}]{RS^{\cdot}} RSSR \end{cases} \qquad (20)$$

D. N—H Bonds

One readily observes upon perusal of standard pK_a tables that as a general rule the pK_a values of N—H bonds are 15–20 units higher than those of the corresponding O—H analogues. Hence, proton-catalysed dismutation of O_2 is not expected to be a general phenomenon with nitrogen compounds. On the other hand, since the average bond energy of an N—H bond is substantially lower than that of a O—H or C—H bond[46], one might expect H-atom abstraction by $O_2^{\cdot-}$ to be more prevalent. Hydrophenazines, hydrazine, reduced flavins, hydroxylamine and related substrates bear labile hydrogens which, though not acidic, are readily abstractable via free-radical processes. The electrochemical evidence[47] strongly suggests that oxidation by $O_2^{\cdot-}$ occurs via initial H-atom transfer. Similarly Hussey and coworkers[48] have shown that proton abstraction is not involved in the oxidation of *o*-phenylenediamine by electrogenerated $O_2^{\cdot-}$.

Poupko and Rosenthal[49] report that based on ESR data it is clear that dialkylamines are instantaneously oxidized to dialkyl nitroxides by KO_2 in DMSO even in the absence of atmospheric oxygen (equation 21). Tertiary amines such as triphenylamine seem inert. Nevertheless, Frimer and coworkers[50] have failed to isolate any reaction products when $O_2^{\cdot-}$ (KO_2/18-crown-6) is contacted with diethylamine, suggesting that the dialkyl nitroxides observed by the sensitive ESR technique are formed in very low yields.

$$R_2NH + O_2^{\cdot-} \longrightarrow R_2NO\cdot + HO^- \qquad (21)$$

The reaction of $O_2^{\cdot-}$ with aromatic amines[41,48–51] yields symmetrical azobenzenes which are also obtained in similar yields starting with the corresponding 1,2-disubstituted hydrazines[51,52]. It is likely, therefore, that the hydrazine, formed by the coupling of two anilinyl radicals, is a reactive intermediate in this process (equation 22). Further evidence for the intermediacy of the anilinyl radical comes from the work of Frimer, Ziv and Aljadeff[50] on the reaction of aniline with KO_2/crown ether (equation 23). These researchers found in addition to azobenzene (34% yield) and nitrobenzene (4% yield), a 24% yield of *N*-(*p*-nitrophenyl)aniline. The latter undoubtedly results from the trapping of the anilinyl radical by nitrobenzene.

$$ArNH_2 \xrightarrow{O_2^{\cdot-}} ArNH\cdot \longrightarrow ArNH{-}NHAr \xrightarrow{O_2^{\cdot-}} ArN{=}NAr \qquad (22)$$

$$PhNH_2 \xrightarrow{O_2^{\cdot-}} PhNH\cdot \begin{cases} \xrightarrow{PhNH\cdot} Ph{-}N{=}N{-}Ph \\ \longrightarrow \xrightarrow{O_2} PhNO_2 \\ \xrightarrow{PhNO_2} Ph{-}NH{-}C_6H_4{-}NO_2 \end{cases} \qquad (23)$$

Although the above data are most consistent with an initial H-atom abstraction, a base-catalysed autoxidative process has not been ruled out. Indeed Balogh-Hergovich and coworkers[51] report that azobenzenes can be obtained (although at a slower rate) when arylamines are oxidized with molecular oxygen catalysed by 'naked' hydroxide ion (KOH/18-crown-6).

Superoxide anion radical reacts with hydrazo compounds and related substances in a variety of ways[52,53]. 1,2-Diarylhydrazines are converted to the corresponding azo compounds (equation 24), though 1,2-dialkylhydrazines seem to be unreactive. Monosubstituted alkyl- and aryl-hydrazines are oxidized in a reaction which appears to involve free alkyl or aryl radicals (equation 25). 1,1-Disubstituted hydrazines are oxidized to *N*-nitrosamines (equation 26) while the hydrazines of aryl ketones yield the corresponding azine in high yield (equation 27). However, the hydrazine of the alkyl ketone cyclohexanone was unreactive.

$$\mathrm{ArNHNHAr} \xrightarrow{O_2^{\bar{\cdot}}} \mathrm{ArN{=}NAr} \qquad (24)$$

$$\mathrm{RNHNH_2} \xrightarrow{O_2^{\bar{\cdot}}} \mathrm{RH} \qquad (25)$$

$$\mathrm{R_2NNH_2} \xrightarrow{O_2^{\bar{\cdot}}} \mathrm{R_2NNO} \qquad (26)$$

$$\mathrm{Ar(R)C{=}NNH_2} \xrightarrow{O_2^{\bar{\cdot}}} \mathrm{Ar(R)C{=}N{-}N{=}C(R)Ar} \qquad (27)$$

The mechanisms for these processes are unknown. Chern and San Filippo[52] suggest that the ability of $O_2^{\bar{\cdot}}$ to effect the oxidation of these substrates is related to their acidity and that the initial step involves proton abstraction by $O_2^{\bar{\cdot}}$. However, in light of the above-mentioned work of the research groups of Sawyer[47] and Hussey[48] on related compounds, it would seem likely that a H-atom abstraction is involved here as well[54].

Nitroaromatic amines such as *o*- and *p*-nitroaniline with acidic N—H bonds ($pK_a = 17.9$ and 18.4, respectively) have been shown by Hussey[55] to react with $O_2^{\bar{\cdot}}$ by proton transfer. The failure of *m*-nitroaniline to react with $O_2^{\bar{\cdot}}$ is not surprising since it should be much less acidic than the *ortho* or *para* isomers.

Purines, such as purine, adenine and N-benzyladenine, also bear an acidic proton at the 7-position. It is not surprising, therefore, that upon reaction with $O_2^{\bar{\cdot}}$, the corresponding salt is formed which readily regenerates the starting material in acid (equation 28). Superoxide does not react appreciably with diethyl- or triethyl-amine which are convenient solvents for this reaction[50].

(28) [purine (R at C-6, N–H at position 7) $\underset{H^+}{\overset{O_2^{\bar{\cdot}}}{\rightleftharpoons}}$ purine 7-anion]

E. C—H Bonds

C—H linkages with low pK_a values undoubtedly react with $O_2^{\bar{\cdot}}$ via initial proton transfer. Thus Stanley[56] reports that cyclopentadiene ($pK_a = 16$)[57] and diethyl malonate ($pK_a = 13$)[57] induce instantaneous disproportionation of $O_2^{\bar{\cdot}}$. Dibenzoylmethane and 1,3-cyclohexadione are rapidly deprotonated by $O_2^{\bar{\cdot}}$, though the resulting anions are stable to O_2 and $O_2^{\bar{\cdot}}$ [37,66]. Similarly, the oxidation of benzoylacetonitrile[8], malononitrile[58,59] ($pK_a = 11.2$)[57], benzyl cyanide[58,59] and assorted carbonyl compounds[60–62] ($pK_a = 20$)[57]

at the α-position is expected to proceed via the base-catalysed autoxidation outlined in equations (29)–(36). It should be pointed out that unlike Russel and coworkers[60–62], we have not included in this sequence of reactions a radical coupling between R· and superoxide anion (equation 37), a process with the same outcome as equations (33) and (34) combined. This is simply because electron transfer (see equations 32 and 34) rather than radical coupling is generally observed with $O_2^{\cdot -}$ (a point we shall discuss at length in Section IV.F).

$$RH + O_2^{\cdot -} \longrightarrow R^- + HOO^\cdot \tag{29}$$

$$HOO^\cdot + O_2^{\cdot -} \longrightarrow HOO^- + O_2 \tag{30}$$

$$HOO^\cdot + R^- \longrightarrow HOO^- + R^\cdot \tag{31}$$

$$R^- + O_2 \rightleftharpoons R^\cdot + O_2^{\cdot -} \tag{32}$$

$$R^\cdot + O_2 \longrightarrow RO_2^\cdot \tag{33}$$

$$RO_2^\cdot + O_2^{\cdot -} \longrightarrow RO_2^- + O_2 \tag{34}$$

$$RO_2^- + RH \longrightarrow ROOH + R^- \tag{35}$$

$$ROOH \longrightarrow \text{ketones, alcohols, carboxylic acids, etc.} \tag{36}$$

$$R^\cdot + O_2^{\cdot -} \longrightarrow RO_2^- \tag{37}$$

In accordance with the above prediction α oxidation has been cited in numerous reports regarding the reaction of $O_2^{\cdot -}$ with ketones. Thus ketones are converted to either α-diketones or acids[38,63–66] and α-hydroxyketones yield acids[67,68]. In the case of enone systems often two acidic protons are present, positioned on the α′ and γ carbons. In instances where both are available abstraction of the latter should be thermodynamically preferred since the dienolate anion formed is more stable than its cross-conjugated isomer[69]. Thus Frimer and Gilinsky[35,36] have found that both 4,4- and 5,5-disubstituted cyclohex-2-en-1-ones (**9** and **10** respectively) yield the corresponding 2-hydroxycyclohexa-2,5-dien-1-ones (**7**) (equation 38). In the case of **9** it is the α′ hydrogen that is removed since the γ position is blocked and oxygenation yields the diketone **11** which in turn enolizes to **7**. In the case of **10** the γ hydrogen is preferentially removed. The resulting dienolate anion is oxygenated α to the carbonyl, leading again to diketone **11** (equation 39).

We have previously mentioned (Section II.B) that enols such as **7** can be further oxidized to lactols **8** (equation 18 and Scheme 3). Indeed Frimer and Gilinsky[35,37] have been able to convert enones directly to lactols in a one-pot reaction in overall yields of 50–90%. These researchers[35,37] have also utilized this method in a two-step preparation of 2-oxa-Δ^4-steroids from the corresponding parent compounds (equation 40).

(9) or (10) $\xrightarrow{O_2^{\cdot-}}$ (11) $\rightleftharpoons$ (7) (38)

(9):
(a) R^1 = Me, R^2 = H
(b) R^1 = Ph, R^2 = H
(c) R^1 = Me, R^2 = OEt
(d) R^1 = Me, R^2 = OMe
(e) Cholest-4-en-3-one
(f) Testosterone

(10):
(a) R^1 = Me, R^2 = H
(b) R^1 = R^2 = Me
(c) R^1 = Me, R^2 = OEt
(d) R^1,R^2 = –$(CH_2)_5$–, R^2 = OMe

(10) $\xrightarrow{O_2^{\cdot-}}$ $\xrightarrow{O_2}$ $\longrightarrow$ **11** $\longrightarrow$ **7** (39)

(9e,f) $\xrightarrow{O_2^{\cdot-}}$ (7e,f) $\xrightarrow{O_2^{\cdot-}}$ (8e,f) $\xrightarrow{NaBH_4}$ (12) (40)

In the case of 6,6-disubstituted cyclohex-2-en-1-ones (**13**), epoxides and dimers result by a mechanism which most likely involves initial γ-proton removal as outlined in equation (41). Condensation of the resulting anion with starting material yields a dimer, while oxygenation generates a peroxy anion. Either the latter or HOO^- (formed by the disproportionation of $O_2^{\cdot-}$ can epoxidize the substrate. Epoxides are also isolated when the γ hydrogen is exocyclic. Thus 3,4,4-trimethylcyclohex-2-en-1-one (**14**)) reacts with $O_2^{\cdot-}$ yielding primarily acid **15** but also epoxide **16** (equation 42)[70].

(13) $\xrightarrow{O_2^{\cdot-}}$ $\longleftrightarrow$ $\xrightarrow{13}$ Dimers and trimers; $\xrightarrow{O_2}$ $\xrightarrow{13}$ (41)

One indication of the fact that $O_2^{\cdot-}$ is acting as a base in the above oxidations of enones **9**, **10**, **13** and **14** is that the same products are obtained with 'naked' hydroxide and *t*-butoxide[35,36,70] with the order of decreasing rates *t*-butoxide > superoxide > hydroxide.

$$\mathbf{(14)} \xrightarrow{O_2^{\bar{\cdot}}} \mathbf{(15)} + \mathbf{(16)} \quad (42)$$

Additional evidence comes from the observation that the rate of the $O_2^{\bar{\cdot}}$ reaction is essentially the same whether carried out in air or under argon (after carefully degassing the solvent via five freeze-thaw cycles). As seen from equation (1) for every molecule of substrate deprotonated a molecule of O_2 is formed. Hence the autoxidation may proceed even in the absence of an oxygen atmosphere. It should also be noted that epoxides have been invoked as intermediates in the base-catalysed autoxidations of enones with available γ hydrogens, but these have never been isolated[71].

The mechanistic picture may not be as simple as we have thus far presented. Sawyer and coworkers[8] report that they have observed no loss of electrogenerated $O_2^{\bar{\cdot}}$ in the presence of cyclohexanone, acetone or 2-butanone. These authors suggest that the KO_2/crown ether autoxidations of ketones may well be catalysed by $O_2^{\bar{\cdot}}$-derived oxidants/bases (perhaps HO_2^-). It is clear, therefore, that more mechanistic work is required.

In the case of substrates with pK_a values greater than 20 the question of mechanism is not always clear. Sagae and coworkers[72,73] have reported that *o*- and *p*- (but not *m*-) nitrotoluenes are oxidized by electrogenerated $O_2^{\bar{\cdot}}$ to the corresponding benzoic acids. These authors suggest an initial H-atom abstraction by $O_2^{\bar{\cdot}}$. However, nitrotoluene also undergoes base-catalysed autoxidation[60,61].

Various diarylmethanes have been oxidized by $O_2^{\bar{\cdot}}$ to the corresponding ketones including anthrone, 9,10-dihydroanthracene, fluorene, xanthine, diphenylmethane and distyrylmethane[63,74–77]. Although H-atom abstraction has been proposed as the initial step in these $O_2^{\bar{\cdot}}$-induced processes, *bona fide* base-catalysed autoxidations have also been reported for each of these compounds with other bases[60–62,77–79]. It is interesting to note, however, that the C—H bond dissociation energy [D(R—H)] in each of these cases is substantially above the 63.4 kcal mol^{-1} borderline we calculated at the end of Section II.A for the onset of a H-abstraction mechanism.

Although 1,4- and 1,3-cyclohexadiene [D(R—H) = 70 kcal mol^{-1}][17] are converted to benzene[74], most benzylic and allylic hydrogens are inert to the action of $O_2^{\bar{\cdot}}$. Thus, 9,10-dihydrophenanthrene, acenaphthene, tetralin, cyclohexene, trimethyl- and tetramethyl-ethylene, 2-methyl-2-pentene and cholesterol[74,80,81,86] are unaffected by $O_2^{\bar{\cdot}}$. The benzyl hydrogen may be activated by an adjacent amino group and as a result benzyl-(**17**, R = H) and furfuryl-amines are oxidized to the corresponding amides[50,82,83] (equation 43). The nature of the substituents on the nitrogen of compound **17** seems to play a crucial role in controlling its reactivity. When R = alkyl or aryl, the reaction proceeds as expected[50,82], while when R = SO_2Ph or C(O)NHPh only starting material is recovered[63]. Quite surprisingly ring hydroxylation products are observed when R = C(O)Ar (equation 44)[63]. The mechanism for this latter process is described in Section III. F.

$$\text{PhCH}_2\text{NHR}\ \mathbf{(17)} \xrightarrow{O_2^{\bar{\cdot}}} \text{PhC(=O)NHR} \quad (43)$$

$$\text{C}_6\text{H}_5\text{CH}_2\text{NHC(O)Ar} \xrightarrow{O_2^{\cdot -}} \text{HO-C}_6\text{H}_4\text{-CH}_2\text{NHC(O)Ar (40–70\%)} + \text{(2-HO)C}_6\text{H}_4\text{CH}_2\text{NHC(O)Ar (2–5\%)} \quad (44)$$

Methylpyridines and methylpyridine-*N*-oxides are oxidized by electrogenerated $O_2^{\cdot -}$ to the corresponding carboxylic acids in low yields. Here too an H-atom abstraction is assumed to be the first step[84].

$O_2^{\cdot -}$ can also induce E2 eliminations producing olefins from the corresponding quaternary ammonium salts[85] and halides[18,19,65,86–89], imines from *N*-chloramines[90] and diphenylacetylene from bromostilbene[91].

III. NUCLEOPHILIC REACTIONS

A. Introduction

In Section II.A we mentioned the low pK_a of the hydroperoxy radical HOO·. Valentine[92] notes that this low value should in fact not be very surprising. After all, $O_2^{\cdot -}$, like fluoride, is a small nonpolarizable anion which is expected to be particularly stabilized in protic media by a tightly bound solvation sphere of hydrogen-bonded solvent. Indeed, in the gas phase both $O_2^{\cdot -}$ and F^- have large hydration affinities[93]. The solvent envelope renders fluoride a weak base (pK_a of HF $= 3.2$) and a relatively unreactive nucleophile[94–98]. In aprotic solvents, however, the situation is expected to be radically different, since, with the hydrogen bonds now absent, the dissolved anions are essentially 'naked'[98,99]. Fluoride, for example, proves to be strongly nucleophilic[94,95,97] and quite basic[98,99] in aprotic solvents.

We must remember that in protic solutions any nucleophilic attack by $O_2^{\cdot -}$ must compete with rapid disproportionation (equation 1). It should not be surprising then that in aqueous solution little evidence for nucleophilic reactivity has been found. In aprotic solvents, on the other hand, uncatalyzed disproportionation is essentially absent and the lifetime of $O_2^{\cdot -}$ is quite long (see Section II.A). In the absence of hydrogen bonding with solvent, the 'naked' $O_2^{\cdot -}$ anion should now show a substantial degree of nucleophilic reactivity as does F^-. Indeed $O_2^{\cdot -}$ has been shown to be one of the most potent S_N2 nucleophiles yet studied[100,101]. Thus the second-order rate constant for the reaction of KO_2 with alkyl bromides in DMSO is of the order of $10^2\ \text{M}^{-1}\,\text{s}^{-1}$, while for nearly all other nucleophiles this value is 10^{-2}–$10^1\ \text{M}^{-1}\,\text{s}^{-1}$. This supernucleophilicity has been rationalized[100,101] in terms of an α effect[102,103]. Alternatively it may be attributed to a significant electron-transfer contribution in the transition state which should be particularly important for $O_2^{\cdot -}$, an excellent electron donor (see Section IV)[100,101].

B. Alkyl Halides and Sulphonates

Aliphatic halides and sulphonates undergo rapid nucleophilic substitution with $O_2^{\cdot -}$ to produce peroxides, hydroperoxides, alcohols, aldehydes or acids depending on reaction conditions and work-up[18,19,65,67,86,87,89,100,101,104–114]. The reaction bears all the characteristics of an S_N2 mechanism. The usual reactivity order is present: $1° > 2° > 3°$ is the order of reactivity in substrate while I > Br > OTs > Cl is the order in leaving group. The reaction is highly stereoselective and Walden inversion is observed. While substitution predominates with primary substrates, substantial elimination occurs with secondary and tertiary systems. Remembering of course that $O_2^{\overline{\cdot}}$ is a strongly basic nucleophile, this is not unexpected.

As noted above, product distribution is dependent on the reaction conditions. Interestingly, in most aprotic solvents, primary and secondary alkyl halides and tosylates react with $O_2^{\overline{\cdot}}$ to yield dialkyl peroxides as the major product. In DMSO, however, alcohols predominate. The explanation for this phenomenon is that the peroxide anion, formed via equations (45) and (46), is reduced by DMSO (equation 48) before it has the opportunity to react with starting material (equation 47)[20,105,115].

$$RX + O_2^{\overline{\cdot}} \longrightarrow RO_2^{\cdot} + X^- \qquad (45)$$

$$RO_2^{\cdot} + O_2^{\overline{\cdot}} \longrightarrow RO_2^- + O_2 \qquad (46)$$

$$RO_2^- + RX \longrightarrow RO_2R + X^- \qquad (47)$$

$$RO_2^- + CH_3SOCH_3 \longrightarrow RO^- + CH_3SO_2CH_3 \qquad (48)$$

C. Acid Chlorides, Anhydrides and Esters

Acid chlorides[19,88,116] and anhydrides[19,20,88] react rapidly with $O_2^{\overline{\cdot}}$ in aprotic solvents to produce diacyl peroxides (50–74% yield)[116] and small amounts of peracids[20,88]. Approximately one mole of oxygen gas is evolved for every two moles of $O_2^{\overline{\cdot}}$ consumed and in the case of anhydrides the corresponding acid is formed as a primary product. The mechanism suggested for this reaction[88] (equations 49–51) is analogous to that proposed for the formation of peroxides from alkyl halides and tosylates (equations 45–47). Short reaction times (1–3 hours) are usually sufficient for these reactions even when slurries of potassium superoxide in benzene (without crown ether) are used as the $O_2^{\overline{\cdot}}$ source. Excess $O_2^{\overline{\cdot}}$ is not only not required to drive the reaction to completion, but is deleterious, since it converts the diacyl peroxide efficiently to two moles of carboxylic acid (see Section III.D)[116,117].

$$RC(=O)X + O_2^{\overline{\cdot}} \longrightarrow RC(=O)OO\cdot + X^- \qquad (49)$$

$$RC(=O)OO\cdot + O_2^{\overline{\cdot}} \longrightarrow RC(=O)OO^- + O_2 \qquad (50)$$

$$RC(=O)OO^- + RC(=O)X \longrightarrow RC(=O)OOC(=O)R + X^- \qquad (51)$$

By comparison, the reaction of *carboxylic esters* with $O_2^{\overline{\cdot}}$ is in general[118] quite slow (1–3 days, excess KO_2/crown ether). Nevertheless, good to excellent yields of the corresponding acid and alcohol are obtained subsequent to aqueous work-up[18,19,88,117–119]. The mechanistic details of this reaction have been investigated to some extent[8,117,119]. Considering first of all the substantial nucleophilic character of $O_2^{\overline{\cdot}}$, one mechanistic possibility[117] would be $O_2^{\overline{\cdot}}$ attack at the alcohol carbon resulting in displacement of a carboxylate anion (equation 52). Such a mechanism requires Walden inversion at R; however, 99% net retention of configuration at the chiral carbon has been observed in the case of the acetate ester of (*l*)-(*R*)-2-octanol[117].

$$O_2^{\cdot-} + R^1{-}O\overset{\displaystyle O}{\overset{\|}{C}}R^2 \longrightarrow \cdot OOR^1 + {}^-O\overset{\displaystyle O}{\overset{\|}{C}}R^2 \qquad (52)$$

An alternative mechanism[8,117] might involve nucleophilic attack at the carbonyl carbon in an addition–elimination process (equations 53 and 54). For such a situation, k_{54} would be highly dependent upon the stability of the leaving group (i.e. pK_a of R^2OH). The overall rate, on the other hand, should depend upon the competition between the loss of $O_2^{\cdot-}$ (k_{-53}) and the loss of R^2O^- (k_{54}). Such a mechanism is therefore consistent with the observation that the structure of the leaving group has a strong effect on the overall rate of reaction. In particular it has been observed[8,117,119] that the relative rate of reaction decreases in the order R^2 = acyl > *p*-chlorophenyl > phenyl > 1° alkyl > 2° alkyl > 3° alkyl.

$$R^1\overset{\displaystyle O}{\overset{\|}{C}}OR^2 + O_2^{\cdot-} \underset{k_{-53}}{\overset{k_{53}}{\rightleftharpoons}} R^1\underset{\displaystyle OO\cdot}{\overset{\displaystyle O^-}{C}}OR^2 \qquad (53)$$

$$R^1\underset{\displaystyle OO\cdot}{\overset{\displaystyle O^-}{C}}OR^2 \underset{k_{-54}}{\overset{k_{54}}{\rightleftharpoons}} R^1\overset{\displaystyle O}{\overset{\|}{C}}OO\cdot + R^2O^- \qquad (54)$$

Once the peracid radical is formed (equation 54) it is suggested[6–8,88,117,119] that reactions such as equations (50) and (51) follow, with the ultimate cleavage of the diacyl peroxide by excess $O_2^{\cdot-}$ resulting in the observed carboxylic acid (see Section III.D).

D. Diacyl Peroxides

Until recently it was assumed that the conversion of alkoyl and aroyl peroxides to carboxylic acids proceeds by a simple two-step electron transfer from $O_2^{\cdot-}$ to the peroxy linkage[6–8,88,117] (equations 55 and 56). Several pieces of evidence speak against this

$$O_2^{\cdot-} + R\overset{\displaystyle O}{\overset{\|}{C}}OO\overset{\displaystyle O}{\overset{\|}{C}}R \longrightarrow R\overset{\displaystyle O}{\overset{\|}{C}}O\cdot + {}^-O\overset{\displaystyle O}{\overset{\|}{C}}R + O_2 \qquad (55)$$

$$R\overset{\displaystyle O}{\overset{\|}{C}}O\cdot + O_2^{\cdot-} \longrightarrow R\overset{\displaystyle O}{\overset{\|}{C}}O^- + O_2 \qquad (56)$$

mechanism. As will be discussed in Section IV.E, electron transfer to the oxygen–oxygen bond has yet to be observed for any other peroxide. More importantly, Stanley[20] has observed that an epoxidizing species is formed in the reaction of $O_2^{\cdot-}$ with diacyl peroxides and anhydrides. For example, when acetic anhydride or lauroy peroxide are reacted with $O_2^{\cdot-}$ in the presence of tetramethylethylene, a 35% yield of the corresponding epoxide is formed. Other research groups have reported that action of $O_2^{\cdot-}$ on benzoyl chloride leads to a species which can efficiently (40% yield) epoxidize chalcones and stilbenes[120,121]. The exact nature of the epoxidizing species is not known although it may be $RC(=O)OO\cdot$[120,121]. This would lead one to suggest that the initial step in the conversion of diacyl peroxides to carboxylic acids is essentially the same as proposed for acyl chlorides, anhydrides and esters, i.e. nucleophilic attack at the carbonyl carbon[20] (equations 57 and 58). Decomposition of the peroxycarboxylate anion[122] to the corresponding carboxylate requires hydrogen-ion catalysis which may occur to some extent during work-up. Alternatively, a nucleophilic attack on the peroxy oxygen may be involved (equations 59–61) in which case $RC(=O)OOO\cdot$ may well be the epoxidizing species. In any case, the question of mechanism still remains a matter of speculation.

$$RC(=O)OOC(=O)R + O_2^{\cdot-} \longrightarrow RC(=O)OO\cdot + RC(=O)OO^- \quad (57)$$

$$RC(=O)OO + O_2^{\cdot-} \longrightarrow RC(=O)OO^- + O_2 \quad (58)$$

$$RC(=O)OOC(=O)R + O_2^{\cdot-} \longrightarrow RC(=O)OOO + RC(=O)O^- \quad (59)$$

$$RC(=O)OOO\cdot + O_2^{\cdot-} \longrightarrow RC(=O)OOO^- + O_2 \quad (60)$$

$$RC(=O)OOO^- \longrightarrow RC(=O)O^- + O_2 \quad (61)$$

E. Aldehydes and Ketones

There have been many reports indicating that aldehydes are oxidized by $O_2^{\cdot-}$ to carboxylic acids[18,19,64,88,117,119]. Sawyer and colleagues[8] have investigated the reaction of electrogenerated $O_2^{\cdot-}$ with benzaldehyde in some detail. When impure aldehyde is used, a rapid decomposition of $O_2^{\cdot-}$ results. However, when the aldehyde is freshly and carefully purified, there is no decomposition of the $O_2^{\cdot-}$ beyond that observed in the absence of substrate. Using the purified benzaldehyde after it has been stored under argon for even one day results in increased rates of $O_2^{\cdot-}$ decomposition. Thus it would seem that not $O_2^{\cdot-}$ but some derived oxidant is responsible for the conversion of benzaldehyde to benzoic acid.

This research group also reports that when benzaldehyde is reacted with $O_2^{\cdot-}$ (generated from KO_2/18-crown-6) benzyl alcohol is formed at a rate one-half that of the disappearance rate of substrate. The benzyl alcohol is in turn oxidized to benzoic acid.

These results suggest that the oxidation of benzaldehyde involves some base B^- which is not $O_2^{\cdot-}$. The reaction occurs via a Cannizzaro-type process yielding benzyl alcohol and an oxidized benzaldehyde species. The nature of the base B^- is not clear but may well be HOO^- formed as described in equations (1)–(3).

Ketones lacking acidic α hydrogens such as benzophenone[8,35] are unreactive toward $O_2^{\cdot-}$. Those bearing acidic protons are oxidized by the base-catalysed autooxidative processes described in Section II.E. Regarding the oxidative cleavage of α-diketones, α-keto acids and α-hydroxyketones see Section IV.B.

F. Amides and Nitriles

Amides as a general rule are essentially inert[50,82,83,117] although with large excesses of $O_2^{\cdot-}$ (10-fold) and long reaction times (7 days) some hydrolysis product is observed[63]. Galliani and Rindone[63] report that *N*-aroylbenzylamides undergo ring hydroxylation (equation 44) and propose a mechanism (equation 62) involving initial nucleophilic attack of $O_2^{\cdot-}$ on the carbonyl carbon atom.

$$C_6H_5CH_2NHC(=O)Ar \longrightarrow C_6H_5CH(H)NHC(O^-)(OO^{\cdot})Ar \longrightarrow C_6H_5\dot{C}HNHC(O^-)(OOH)Ar \tag{62}$$

$$\longrightarrow [C_6H_5\cdots CHNHC(=O)Ar]^{\cdot} \longrightarrow \longrightarrow HO{-}C_6H_4CH_2NHC(=O)Ar$$

Nitriles, too, seem to be inert to superoxide[8,117], despite several reports in which nitriles were converted to amides. However, these reactions appear to involve not $O_2^{\cdot-}$ but rather ROO^- generated *in situ*[20,105,115,118] (see also Section IV.E). Recently Sawaki and Ogata[123] have reexamined the reaction of $O_2^{\cdot-}$ with acetonitrile and suggested that the acetamide obtained results from a direct nucleophilic attack of $O_2^{\cdot-}$ on the nitrile group. It should be noted, however, that the yield of amide was less than 2% despite the fact that more than 78% of the KO_2 decomposed. Thus the decomposition of $O_2^{\cdot-}$ may well involve acid catalysis generating HOO^- anion *in situ*.

G. Sulphur Compounds

Several sulphides have been reported to be inert to $O_2^{\cdot-}$. Included in this list are diphenyl sulphide[124], di-*t*-butyl sulphide[120,125], di-*t*-butyl disulphide[126a] and thianthrene[124]. Presumably other sulphides are oxidized. Thus Takata and coworkers[126a,c] report that $O_2^{\cdot-}$ (KO_2/18-crown-6) converts aryl and alkyl disulphides, thiosulphinates, thiosulphonates and sodium thiolates to the corresponding sulphinic and sulphonic acids (equation 63). In the case of thiosulphinates substantial yields (35–45%) of the symmetrical disulphide R^2SSR^2 has also been isolated. The authors posit that the latter results from the coupling of two thiyl radicals formed in the nucleophilic attack of $O_2^{\cdot-}$ on the sulphinyl sulphur of the

$$R^1SSR^2,\ R^1S(=O)SR^2,\ R^1S(=O)_2SR^2\ \text{or}\ RS^-Na^+ \xrightarrow[\text{pyridine}]{O_2^{\cdot-}} RSO_2^- + RSO_3^- \tag{63}$$

thiosulphinate. This fact is supported by the exclusive formation of the acids from the sulphinyl side (equation 64).

$$R^1S(=O)SR^2 + O_2^{\overline{\cdot}} \longrightarrow R^1S(\dot{O})(O{-}O^-)SR^2 \longrightarrow R^1SO_2 + R^2S\cdot \longrightarrow R^2SSR^2 \quad (64)$$

In the reaction of thiosulphonates a small amount of symmetrical disulphate was also detected, though the main products were the acids. In this case, attack of $O_2^{\overline{\cdot}}$ occurs at both the sulphenyl and sulphonyl sulphur atoms of the thiosulphonates.

Nagano and coworkers have reported[126b] that a mixture of symmetrical disulphides react with electrogenerated $O_2^{\overline{\cdot}}$ to yield disulphides in which there has been an interchange of substituents. Oxidation products were not observed in this instance and the reason for the discrepancy is not clear[126c].

Sulphonyl chlorides, sulphinyl chlorides and thiosulphonates react with $O_2^{\overline{\cdot}}$ generating *in situ* peroxysulphenate and peroxysulphinate[120,121,126c]. The latter epoxidize chalcones, stilbene and acenaphthylene presumably via nucleophilic attack on the double bond.

H. Cations and Cation Radicals

The nucleophilic attack of $O_2^{\overline{\cdot}}$ at positively charged centres has been reported by several research groups. Thus the reaction of tropylium ion with $O_2^{\overline{\cdot}}$ (and O_2^{2-}) leads to benzaldehyde, benzene, cycloheptatriene and carbon monoxide in a 2:2:2:1 ratio[127,128]. The likely intermediate is tropyl peroxide equation (65). Oxonium ($C{=}\overset{+}{O}R$)[129] and immonium ($C{=}\overset{+}{N}R_2$)[130] cations (equation 66) as well as nitrones ($C{=}\overset{+}{N}\langle^{R}_{O^-}$)[130–136] react as expected at the carbon end of the double bond. The final product obtained, however, depends to some extent on the exact nature of the substrate and the work-up conditions.

Although radical coupling is not generally observed with $O_2^{\overline{\cdot}}$ (see Section IV.F), reaction between cation radicals and $O_2^{\overline{\cdot}}$ has been suggested as the mechanism of photoinduced electron-transfer oxidations[137–147]. These are photooxidations sensitized by an electron-deficient compound such as 9,10-dicyanoanthracene. The experimental evidence would seem to be consistent with an electron-transfer mechanism involving a donor radical cation ($D^{+\cdot}$) and a sensitizer radical anion ($Sens^{\overline{\cdot}}$) which subsequently reduces oxygen to $O_2^{\overline{\cdot}}$. The coupling of the donor radical cation and superoxide anion radical leads to oxidation products (equation 67). Other researchers[68,145,148], however, posit an alternative mechanism not involving $O_2^{\overline{\cdot}}$. In this proposal, following electron transfer from substrate to sensitizer, the substrate radical cation reacts with molecular oxygen[149–152]. This is followed by back-transfer of an electron from the sensitizer anion radical (equation 68).

Mayeda and Bard[153] report that the reaction between the electrogenerated cation radicals of either 1,3-diphenylisobenzofuran or ferrocene with electrogenerated $O_2^{\overline{\cdot}}$ results not in coupling but in electron transfer with concomitant formation of molecular oxygen, which these researchers indicate is of singlet multiplicity (equation 69; see Section IV.G). Nishinaga and colleagues[154] also report that stable 2,5-dialkyl-1,4-dimethoxybenzene radical cations oxidize KO_2 to $^1O_2$ in 82–84% yield.

On the other hand, several groups report coupling between $O_2^{\overline{\cdot}}$ and a cation radical. Thus Ando and coworkers[124] report that while electron transfer is indeed observed in the

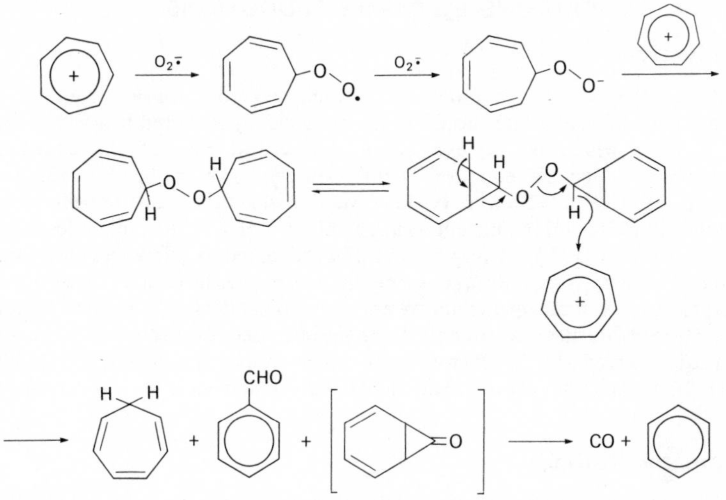

$$\xrightarrow{O_2^{\cdot-}} \qquad \xrightarrow{O_2^{\cdot-}} \qquad \longrightarrow \qquad \rightleftharpoons \qquad \longrightarrow \; + \; + \; \longrightarrow CO + \quad (65)$$

$$\xrightarrow{O_2^{\cdot-}} \quad (66)$$

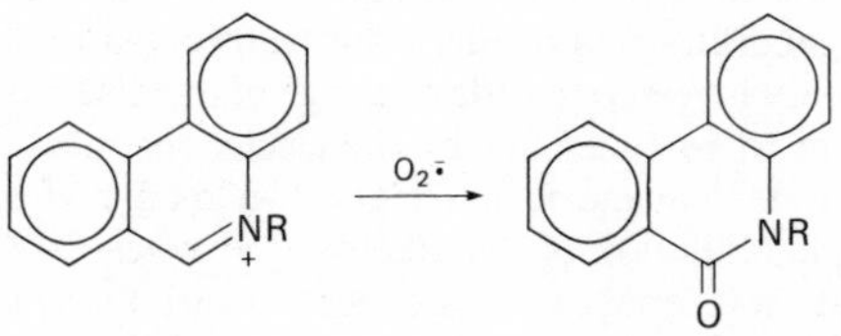

$$\text{Sens} \xrightarrow{h\nu} {}^1\text{Sens}^* + \text{D} \longrightarrow \text{Sens}^{\cdot-} + \text{D}^{+\cdot} \xrightarrow{O_2} \text{Sens} + \text{D}^{+\cdot} + O_2^{\cdot-} \longrightarrow \text{DO}_2 \quad (67)$$

$$\text{Sens} \xrightarrow{h\nu} {}^1\text{Sens}^* \longrightarrow \text{Sens}^{\cdot-} + \text{D}^{+\cdot} \xrightarrow{O_2} \text{Sens}^{\cdot-} + [\text{DO}_2]^{+\cdot} \longrightarrow \text{Sens} + \text{DO}_2 \quad (68)$$

$$\text{D}^{+\cdot} + O_2^{\cdot-} \longrightarrow \text{D} + {}^1O_2 \quad (69)$$

reaction of thianthrene cation radical perchlorate and KO_2 in acetonitrile, coupling also occurs leading to sulphinyl oxides. Similarly, Sawyer and Nanni[155] report that when DMF solutions of reduced methyl viologen ($MV^{+\cdot}$) and of $O_2^{\cdot-}$ are combined in 1:1 stoichiometry, the reversible formation of a dioxygen adduct [MVO_2] results. The reaction chemistry of [MVO_2] conforms with what one would expect for a dioxetane.

IV. ONE-ELECTRON REDUCTIONS

A. Introduction

Considering the stability of molecular oxygen, electron transfer from $O_2^{-\cdot}$ to an appropriate reducible substrate should be an energetically favoured process. Indeed it is electron transfer which is the essence of the second step of the acid-catalysed disproportionation reaction (equation 3). Interestingly, however, the reduction potential for oxygen is sensitive to the nature of the solvent in which the measurement is made. In aprotic solvents E^0 (vs NHE) has an average value[156] of -0.57 V, much less than that observed for water (-0.33 V). This gap cannot be attributed to differences in the dielectric constants of the media, since the reduction potential of oxygen is relatively insensitive to the differing dielectric constants of a variety of aprotic solvents (CH_2Cl_2, acetone, acetonitrile, DMF and DMSO). Rather it is probably due to the increased solvation (and hence stability) of $O_2^{-\cdot}$ in aqueous media[157,158]. As we have noted previously, the data indicate that $O_2^{-\cdot}$ is expected to be a much stronger reducing agent in aprotic solvents than it is in protic media.

B. Conjugated Ketones

Diketones and keto acids are cleaved by $O_2^{-\cdot}$ to carboxylic acids[67,159,160]. Three mechanisms are possible (equation 70). One involves nucleophilic attack on the carbonyl group (path a). The second and third alternatives (paths b and c) involve initial ketyl formation. This radical ion may then be scavenged by oxygen (path b) or $O_2^{-\cdot}$ (path c). Mechanism a is the one generally presumed for this transformation[6,13,92] and finds strong precedent in the alkaline hydrogen peroxide cleavage of these substrates[161,162]. However, this pathway would seem to be ruled out by the electrochemical studies of Boujlel and Simonet[160], who report that cleavage occurs in good yield only when the fixed potential of the cathode allows for the reduction of the diketone, i.e. when $O_2^{-\cdot}$ and [diketone]$^{-\cdot}$ are simultaneously present at the interface (and later in solution). These authors therefore prefer path c. However, radical coupling with $O_2^{-\cdot}$ is not observed (Section IV.F) and this should be all the more true in this case where the coupling would have to be between two negatively charged radicals. Thus path b seems the most likely mechanistic route.

$O_2 + R^1\text{—}\dot{C}(O^-)\text{—}C(=O)\text{—}R^2 \xrightarrow[O_2^{-\cdot}]{\text{Path c}} R^1\text{—}C(O\text{—}O^-)(O^-)\text{—}C(=O)\text{—}R^2$

Paths b and c (upward from $O_2^{-\cdot} + R^1\text{—}C(=O)\text{—}C(=O)\text{—}R^2$); Path b ($O_2$); $O_2^{-\cdot}$ (upward)

$O_2^{-} + R^1\text{—}C(=O)\text{—}C(=O)\text{—}R^2 \xrightarrow{\text{Path a}} R^1\text{—}C(O\text{—}O\cdot)(O^-)\text{—}C(=O)\text{—}R^2$

$\longrightarrow R^1C(=O)O^- + {}^-OC(=O)R^2$ (70)

α-Hydroxy- and α-halo-ketones, esters and carboxylic acids are also oxidatively cleaved to carboxylic acids[67], presumably via the corresponding α-diketones and α-keto acids.

The simple enone moiety *per se* is unreactive to $O_2^{-\cdot}$ be it via electron transfer or Michael addition. Thus, 4,4,6,6-tetrasubstituted cyclohex-2-en-1-ones have proved totally inert to $O_2^{-\cdot}$ even after being in contact for several days[36]. However, as discussed previously, this moiety does labilize the adjacent hydrogens towards proton abstraction (see Section II.E). If the π system is extended then electron transfer from $O_2^{\cdot -}$ to substrates may be observed. Thus several research groups report that anion radicals can be detected in the reaction of $O_2^{\cdot -}$ with *N*-methylacridone[163] and various benzoquinones[49,164]

Frimer and Rosenthal[80,165,166] have studied the oxidative cleavage of tetracyclone and chalcones by the action of $O_2^{\cdot -}$. Carboxylic acids were obtained as the final products and no intermediate epoxide formation could be detected. The mechanism suggested involves initial electron transfer (equation 71). A Michael-type addition to the enone system was excluded on the basis of $K^{18}O_2$ experiments.

$$\mathrm{Ar^1C(=O)CH{=}CHAr^2} \xrightarrow{O_2^{\cdot -}} \left[\mathrm{Ar^1C(=O)CH{=}CHAr^2}\right]^{\cdot -} \xrightarrow{O_2} \mathrm{Ar^1C(O^-)(OO\cdot)CH{=}CHAr^2} \longrightarrow$$

$$\mathrm{Ar^1C(O^-)(O{-}O)\dot{C}HCHAr^2} \longrightarrow \mathrm{Ar^1CO_2^-} + \mathrm{Ar^2\dot{C}HCHO}$$

$$\mathrm{Ar^2\dot{C}HCHO} \xrightarrow{O_2} \mathrm{Ar^2CH{-}CH(O^-)(O{-}O)} \longrightarrow \mathrm{Ar^2CHO} \xrightarrow{O_2^{\cdot -}} \mathrm{Ar^2CO_2^-}$$

$$\mathrm{Ar^2\dot{C}HCHO} \xrightarrow{RH} \mathrm{Ar^2CH_2CHO} \xrightarrow{O_2^{\cdot -}} \mathrm{Ar^2CH_2CO_2^-} \qquad (71)$$

Saito and colleagues[167] report that 2,3-dimethyl-1,4-naphthaquinone and other vitamin-K-related compounds react with KO_2/18-crown-6 to give the corresponding oxirane and its secondary oxidation product phthalic acid in a 25–35% yield. The remaining products are unidentified. The mechanistic details are unclear, but based on the reactions of other benzoquinones[49,163,164] initial electron transfer is likely. Alternatively a base-catalysed autoxidative process may be invoked, entailing initial γ-proton abstraction analogous to that proposed for $O_2^{\cdot -}$-induced oxidation of enones **13** and **14** which also yields epoxyketones (see Section II.E).

Kobayashi and coworkers[127] report that tropone reacts with $O_2^{\cdot -}$ in DMSO generating salicylaldehyde. Here, too, electron transfer is proposed as the initial step. Surprisingly, however, no reaction occurs in either DMF, benzene or acetonitrile which leads the authors to conclude that the oxidation of DMSO by a reversibly formed intermediate is a crucial step in this reaction (Scheme 4).

C. Olefins

Unlike many other free radicals, $O_2^{\cdot -}$ does not add directly to carbon–carbon double bonds and this has recently been rationalized on thermodynamic grounds[34]. Thus simple olefins such as tetramethyl- and trimethyl-ethylene, 2-methyl-2-pentene, cyclohexene and its 1,2-dimethyl analogue and cholesterol are totally unreactive to $O_2^{\cdot -}$ [80,81,86]. Tetraphenylethylene[80] and styrene[18,19] yield only traces of polar compounds. Feroci and Roffia[168] have studied the reaction of electrogenerated $O_2^{\cdot -}$ with styrene in detail and have isolated only small amounts of ethylbenzene and 1,4-diphenylbutane. The absence of oxygenated compounds supports the hypothesis that the only reaction between $O_2^{\cdot -}$ and

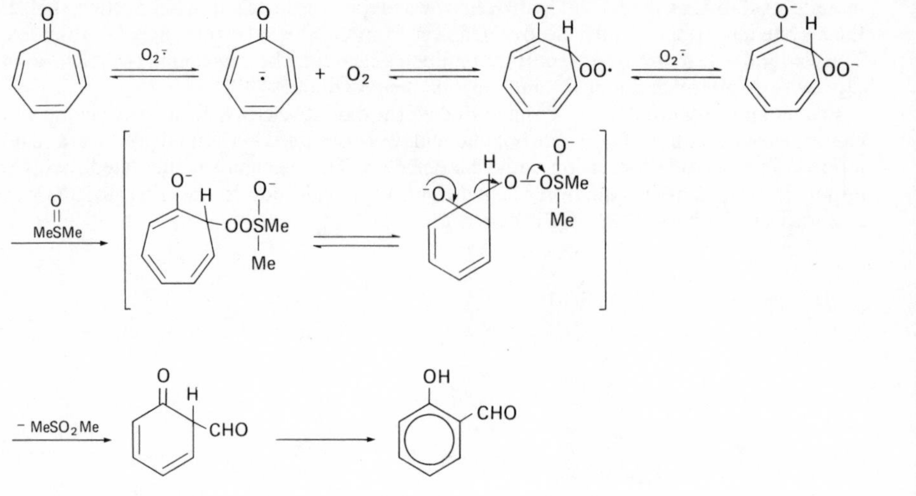

SCHEME 4. Proposed mechanism for the reaction of $O_2^{\overline{\cdot}}$ with tropone.

styrene is electron transfer. Protonation of the styryl radical ion yields ethylbenzene while dimerization produces 1,4-diphenylbutane.

Interestingly, electron-poor nitro and cyano olefins are oxidatively cleaved by $O_2^{\overline{\cdot}}$ [10,137]. When $^{18}O_2^{\overline{\cdot}}$ is used only very little of the label is actually incorporated into the product. Hence electron transfer would seem to be the reaction pathway here as well. A likely mechanism is outlined in equation (72). Cinnamonitrile, PhCH=CHCN, is reported[8] to be stable to KO_2/18-crown-6. It would seem then that one nitro or *two* cyano groups are required to obtain proper electron-transfer conditions.

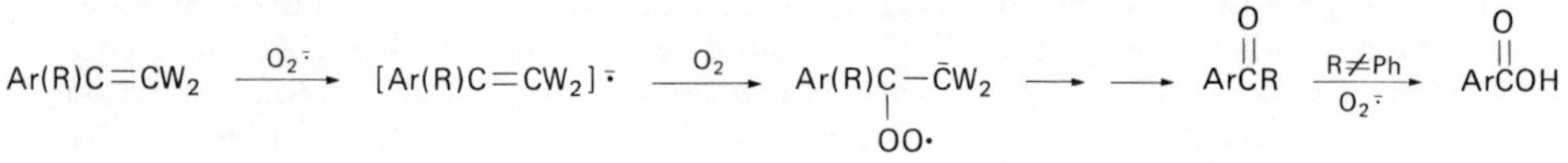

$$\mathrm{Ar(R)C{=}CW_2} \xrightarrow{O_2^{\overline{\cdot}}} [\mathrm{Ar(R)C{=}CW_2}]^{\overline{\cdot}} \xrightarrow{O_2} \mathrm{Ar(R)C(OO\cdot){-}\bar{C}W_2} \longrightarrow \longrightarrow \mathrm{ArC(O)R} \xrightarrow[O_2^{\overline{\cdot}}]{R \neq Ph} \mathrm{ArC(O)OH} \qquad (72)$$

R = H, Me, Et, Ph

W = NO_2, CN

Dietz and colleagues[86] have suggested that the conversion of benzylidenefluorene (BF) to fluorenone by $O_2^{\overline{\cdot}}$ is a result of initial nucleophilic attack. However, in light of the work described above on styrene and electron-poor olefins, an initial electron transfer to the extended π system would appear to be more plausible. The resulting radical ion would then be scavenged by molecular oxygen (equation 73).

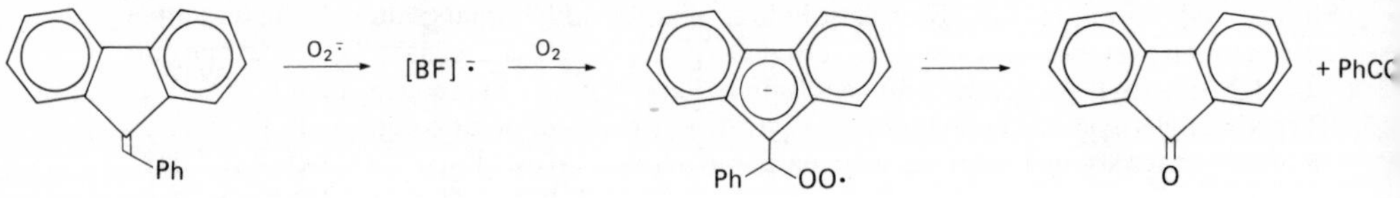

$$\mathrm{(BF)} \xrightarrow{O_2^{\overline{\cdot}}} [\mathrm{BF}]^{\overline{\cdot}} \xrightarrow{O_2} \longrightarrow \text{fluorenone} + \mathrm{PhC}\ldots \qquad (73)$$

D. Aryl Systems

Unsubstituted aromatic hydrocarbons are inert to $O_2^{\overline{\cdot}}$. Thus, benzene, toluene, pyridine, 2,5-diphenylfuran, naphthalene, anthracene and rubrene are unreactive[80], as are nitro-, chloro- and bromo-benzene[169]. Even the highly reactive and readily oxidized 1,3-diphenylisobenzofuran is reported by several groups to be unreactive to $O_2^{\overline{\cdot}}$ [154,170–173]. However, rings activated by one or two electron-withdrawing groups readily undergo nucleophilic aromatic substitutions[166,169,174,175]. Thus 1-halo-2,4-dinitrobenzenes react rapidly ($F > Br \sim I > Cl$) to yield the corresponding 2,4-dinitrophenols; *o*- and *p*-bromonitrobenzene as well as *o*-, *m*- and *p*-dinitrobenzene react with $O_2^{\overline{\cdot}}$ to yield the corresponding nitrophenols.

Three mechanisms are possible for the above reactions and are outlined in Scheme 5. Path a involves direct addition of $O_2^{\overline{\cdot}}$ to the aromatic ring according to a typical nucleophilic aromatic substitution mechanism[176]. Paths b and c involve initial electron transfer from $O_2^{\overline{\cdot}}$ to the substituted benzene yielding an anion radical. In path b bond scission between the nucleofugic substituent and the aryl ring occurs, a process typical of nucleophilic radical aromatic substitution[177]. The resulting aryl radical is scavenged by oxygen. In path c oxygen scavenges the radical anion directly and only subsequently does the nucleofugic group depart.

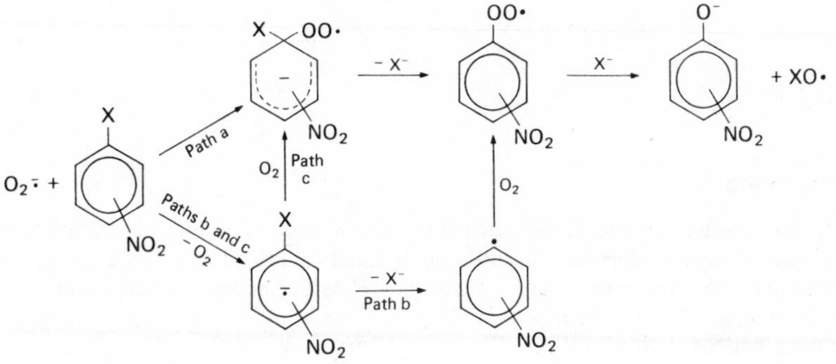

SCHEME 5. Possible reaction mechanisms for the reaction of $O_2^{\overline{\cdot}}$ with nitrobenzenes (X = halogen or nitro group).

The fact that upon reaction with $O_2^{\overline{\cdot}}$ well-resolved EPR spectra were obtained for the radical anions of 1,3- and 1,4-dinitrobenzenes as well as for several other nitrobenzenes tends to affirm the intermediacy of an electron-transfer step. Additional evidence against path a was provided by the use of $K^{18}O_2$ [166,169,178] which indicated that only a small amount of label entered the product. [It should be noted in passing that there has been a growing number of reports suggesting that for at least some substrates (e.g. *o*- and *p*-dinitrobenzenes) nucleophilic aromatic substitution proceeds not through the conventional Meisenheimer-type intermediate, but rather via a substrate anion radical generated by initial electron transfer[179,180]].

Path b has been rejected by several authors[166,175,181] because no coupling product is observed, as might have been expected had a phenyl radical been formed[182]. However, biaryls are not observed in *bona fide* $S_{RN}1$ reactions either (probably as a result of a very rapid reaction between the aryl radical and the attacking nucleophile[183]). Perhaps more convincing evidence comes from the work of Sagae and colleagues[175] and Behar and Neta[184a], who report that in inert atmosphere the radical anions of halonitroaromatic compounds simply do not undergo a dehalogenation reaction (i.e. C—X scission) at an appreciable rate[184a] (equation 74). Hence path c seems to operate in the $O_2^{\cdot -}$ reaction of nitrobenzenes. Nevertheless, Savéant and coworkers[184b] have reported kinetic data which they argue favours path a.

$$[X\text{-}C_6H_4\text{-}NO_2]^{\cdot -} \longrightarrow {}^{\cdot}C_6H_4\text{-}NO_2 + X^- \quad k < 1\ s^{-1} \tag{74}$$

A related reaction has been observed in the case of halogenoquinolines by Yamaguchi and van der Plas[181]. These researchers have observed the conversion of 3-bromoquinoline, 2-chloroquinoline and 1-bromoisoquinoline to the corresponding hydroxy analogue, upon reaction with $O_2^{\cdot -}$ (equation 75). It should be noted that 3-halogenoquinolines are generally unreactive in S_NAr substitution reactions, thus ruling out a path a type process. Although the formation of heteroaryl radicals from radical anions of heteroaryl halides has ample precedent, these authors prefer path c because of the absence of any coupling product. As before, this argument is not conclusive and the issue of mechanism in this case remains moot.

$$\text{3-bromoquinoline} \xrightarrow{O_2^{\cdot -}} \text{3-quinolinolate } (O^-) \tag{75}$$

E. Peroxides

Extensive studies on the biological role of superoxide have focused attention on the possible occurrence of the reaction outlined in equation (76). The essence of this reaction, dubbed the Haber–Weiss reaction[185] when R = H, is an electron transfer from $O_2^{\cdot -}$ to the peroxide linkage. Such a process would induce scission of the oxygen–oxygen bond, generating hydroxide ion and an alkoxy (or hydroxy, when R = H) radical.

$$O_2^{\cdot -} + ROOH \longrightarrow O_2 + RO^{\cdot} + HO^- \tag{76}$$

In the case of H_2O_2, the generation of hydroxy radicals by the Haber–Weiss reaction could explain the crucial role $O_2^{\cdot -}$ plays in a variety of biological hydroxylations as well as the reason for its toxicity[186]. However, a series of kinetic studies [156,186–190] indicate that the Haber–Weiss reaction (equation 76, R = H) is much too slow ($k < 1.3 \pm 0.7 \times 10^{-1}$)[187] to compete with the rapid disproportionation of $O_2^{\cdot -}$ (equation 1) induced by protic solvent and/or the slightly acidic hydroperoxide itself. A theoretical justification for the absence of an electron-transfer mode has been suggested by Koppenol and coworkers[190].

The wave of interest in the Haber–Weiss reaction and its possible role in biological processes brought in its wake a series of studies on the reactions of alkyl hydroperoxides with $O_2^{\cdot -}$. Le Berre and Berguer[18,19] studied the reaction of triphenylmethyl hydroperoxide with dispersed sodium superoxide, NaO_2. They reported that while only simple disproportionation was observed in benzene (with the concomitant formation of a peroxy anion), a 15% yield of reductive cleavage was observed after 24 hours when THF served as solvent. In the latter case, however, oxygen yields are far below that predicted by electron transfer suggesting that the solvent is slowly attacked by the $O_2^{\cdot -}$. It is not clear what other species are produced in this latter process but they may well be responsible for the reductive cleavage observed.

More recently Peters and Foote[11] reported that tetramethylammonium superoxide reacts with either *t*-butyl or *t*-amyl hydroperoxide in acetonitrile yielding the corresponding alcohols as the major product. Similarly Pryor and coworkers[191,192] found that *t*-butyl and linoleic hydroperoxides are reductively cleaved by KO_2/18-crown-6 polyether complex in acetonitrile–hexane or acetonitrile–DMSO solvent mixtures. These authors suggest that a Haber–Weiss-type process is at play.

Gibian and Ungerman[105,115] and Stanley[20], however, found that the peroxy linkage of *t*-butyl hydroperoxide is inert to $O_2^{\cdot -}$ when the reaction is carried out in pyridine, benzene or toluene. In these cases, only acid- catalysed disproportionation of $O_2^{\cdot -}$ is observed with concomitant formation of a stable peroxy anion. Furthermore, they report[115,120] that the results observed by Foote[11] and Pryor[191,192] can also be obtained in acetonitrile when KOH replaces KO_2. More importantly, the isolation of acetamide as a product indicates that the solvent plays a crucial role in the course of the reaction (see Section III.F). It is probable, therefore, that the mechanism of the acetonitrile reactions involves initial deprotonation of the hydroperoxides by $O_2^{\cdot -}$ as observed in toluene, benzene and pyridine. In a well-precedented process[193], the resulting peroxy anion reacts with nitrile solvent to form a perimidic ester which decomposes generating acetamide, alcohol and β-cleavage products. Gibian and Underman[115] conclude, therefore, that no Haber–Weiss-type reaction occurs with alkyl hydroperoxides in aprotic solvents.

Similar reactions have been carried out in aqueous solutions in the absence of acetonitrile. Pryor[191,192] has reported that enzymatically (xanthine–xanthine oxidase) generated $O_2^{\cdot -}$ does indeed convert hydroperoxides to alcohols. Bors and coworkers[194] have, however, generated $O_2^{\cdot -}$ by radiolytic techniques and conclude that superoxide anions do not react with hydroperoxides. They argue that, in contradistinction to their pulse radiolysis technique wherein $O_2^{\cdot -}$ is generated cleanly, the enzymatic system utilized by Pryor does not produce $O_2^{\cdot -}$ exclusively, and presumably the other species are causing the effects observed.

Thus the data suggest that superoxide reacts with the protic peroxides primarily as a base; other potential reactions (i.e. the Haber–Weiss reaction) do not compete effectively.

Dialkyl peroxides show varying reactivity with $O_2^{\cdot -}$. In general, tertiary dialkyl peroxides (such as di-*t*-butyl peroxide[87,100,105] and biadamantylidene dioxetane[195]) are essentially inert. This is of course consistent with the absence of a Haber–Weiss-type reaction. Nevertheless, Lee-Ruff[106] has reported that upon reaction with KO_2 in DMF for 5 days the endoperoxide of 9,10-diphenylanthracene yields the corresponding diol in high yield. Primary and secondary peroxides[86,87,106] react slowly producing alcohols as the major isolable product, along with small amounts of carbonyl compounds.

The latter cases can be explained in terms of a Kornblum–De La Mare[122,196,197] reaction (equation 77). In this base-catalysed process, a peroxide bearing an α proton is decomposed directly to alcohol and carbonyl functionalities. Such a mechanism[86,87,115] nicely explains the formation of both alcohol and carbonyl compounds as primary products in the reaction of $O_2^{\cdot -}$ with primary and secondary dialkyl peroxides. The observed low yield of carbonyl-containing product undoubtedly results from its further base-catalysed autoxidation to

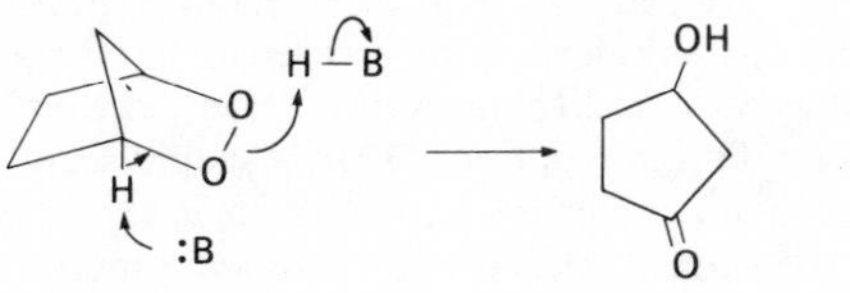

(77)

acid (see Sections II.E and III.E). Furthermore, it would explain the general inertness of tertiary peroxides as well as the greater than six-fold increase in the reaction half-life of di-*n*-hexyl-1,1,1',1'-d_4 peroxide as compared to the undeuterated analogue[87].

Nevertheless, the fact that alcohol is produced in excess of 50% yield in the case of di-*n*-hexyl peroxide requires that an additional pathway be partially operative. Such a secondary pathway is also required by the observation that the tertiary peroxide 9,10-diphenylanthracene-9,10-peroxide is cleaved to diol after being stirred for five days with KO_2 in DMF. What may be involved in these cases is a nucleophilic displacement at the peroxide oxygen[122]. Indeed when di-*n*-hexyl peroxide is reacted with KOH/18-crown-6 in benzene, a 60% yield of 1-hexanol is obtained[87]. Similarly nucleophilic displacement on oxygen has been invoked to explain the production of diol in the reaction of trimethyldioxetane with alkali[198]. In any case, the role of trace metals[12b] in catalysing the homolytic scission of the peroxide linkage should not be overlooked[197,199].

F. Free Radicals

One reaction which is quite typical of free radicals (though not necessarily of radical anions)[200] is radical–radical combination which in the case of $O_2^{\cdot-}$ would lead to peroxide anion formation (equation 78). Radical combination is the essence of the 'termination' step in free-radical chain-reactions and the rates of such radical couplings approach the diffusion-controlled limit[201]. Indeed a mechanism involving the combination of $O_2^{\cdot-}$ with organic radicals has been postulated for the oxidation of phenols[24,74,202], carbanions[60–62] and diketones[160]. In these cases, however, coupling of the generated radical with molecular oxygen is equally plausible (equations 79 and 80). The peroxy radical formed would then be reduced by $O_2^{\cdot-}$ to the corresponding peroxide anion.

$$R^{\cdot} + O_2^{\cdot-} \longrightarrow RO_2^{-} \tag{78}$$

$$R^{-} + O_2 \longrightarrow R^{\cdot} + O_2^{\cdot-} \tag{79}$$

$$R^{\cdot} + O_2 \longrightarrow RO_2^{\cdot} \xrightarrow{O_2^{\cdot-}} RO_2^{-} \tag{80}$$

There are several documented cases where it is clear that $O_2^{\cdot-}$ does not react with free radicals by coupling. For example, the reaction of $O_2^{\cdot-}$ with the hydroperoxy radicals during the acid-catalysed disproportionation of $O_2^{\cdot-}$ yields hydroperoxy anion and molecular oxygen (equation 3) via electron transfer.

A second case has been discussed by Le Berre and Berguer[18,19] who report that $O_2^{\cdot-}$ reacts only extremely sluggishly with triphenylmethyl radical. Since the triphenylmethyl-peroxy anion is stable[18], this sluggishness cannot be attributed to the reversibility of the reaction.

Nishinaga and colleagues[154,170] have found that $O_2^{\cdot-}$ does not couple with phenoxy radicals, but reduces them to the corresponding phenolates. Similarly the radical cations of 1,3-diphenylisobenzofuran[153], ferrocene[153] and 2,5-dialkyl-1,4-dimethoxybenzenes[154] do not couple with $O_2^{\cdot-}$ but are reduced by it (see Section III.H).

Radical–radical coupling has indeed been observed in the reaction of $O_2^{\cdot-}$ with the cation radicals of methyl viologen[155] and thianthrene[124] (see Section III.H). These couplings, however, are really anion–cation reactions. Thus it would seem that the radical–radical combination may be ruled out as a possible mechanism in $O_2^{\cdot-}$ reactions. Instead, electron transfer from $O_2^{\cdot-}$ to the radical species is the observed mode.

G. Generation of Singlet Oxygen ($^1O_2$)

Molecular oxygen (O_2) has two electrons in its highest occupied molecular orbitals, the two degenerate π^*_{2p} orbitals[203,204]. Following Hund's rule, in the ground state of O_2, these two electrons have parallel spins and are located one each in the two π^*_{2p} orbitals (Figure 1). Such an electronic configuration corresponds to a triplet ($^3\Sigma_g^-$) state and we shall henceforth refer to ground-state molecular oxygen as triplet oxygen, $^3O_2$. The first excited state of O_2 lies only 22.5 kcal mol^{-1} above the ground state and has both electrons paired with antiparallel spins in the same π^*_{2p} orbital. Such a state is a $^1\Delta_g$ state and we shall refer to molecular oxygen in its first excited state as singlet oxygen or $^1O_2$.

$O_2^{\cdot-}$ differs from $^3O_2$ and $^1O_2$ in that the former has three — not two — electrons in its π^*_{2p} orbitals. This leads to a situation in which one of the two degenerate π^*_{2p} orbitals is totally occupied while the second is only half-full. (The electronic configuration of the π^*_{2p} orbitals for these three species is outlined in equation 81). It should be noted that no Jahn–Teller splitting can occur with diatomic molecules[205]; hence, all three of the π^*_{2p} electrons in $O_2^{\cdot-}$ are of equal energy.

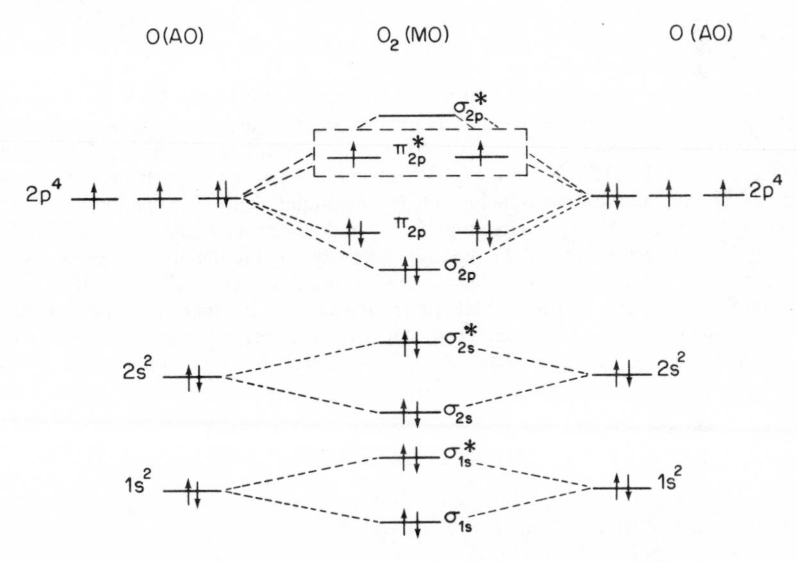

FIGURE 1. Schematic energy level diagram shows how the atomic orbitals (AO) of two atoms of elemental oxygen interact to form the molecular orbitals (MO) of molecular oxygen. The electron distribution is according to Hund's rule yielding ground-state molecular oxygen ($^3\Sigma_g^-$).

$$O_2^{\cdot-}\ \uparrow\downarrow\ \ \uparrow\ \xrightarrow{-e}\ ?\ \begin{cases}\uparrow\downarrow\ \ —\ \ {}^1O_2\\ \uparrow\ \ \uparrow\ \ {}^3O_2\end{cases} \qquad (81)$$

Throughout this chapter we have seen a variety of instances in which an electron is transferred from $O_2^{\cdot-}$ to a reducible substrate. The question[206] that arises is whether the molecular oxygen generated is $^3O_2$ or $^1O_2$ (equation 81)?

Because of its biological importance a great deal of effort has been invested in the study of the acid-catalysed disproportionation of $O_2^{\overline{\cdot}}$ (equations 1–3). Theoretical studies[190,191,206–209] suggest that $^1O_2$ may indeed be formed in the crucial second electron-transfer step of this process (equation 3), which under acid conditions can be modified as shown in equation (82). Nevertheless there has been no real experimental substantiation for its occurrence. Indeed the fact that the well-known $^1O_2$ scavengers, 2,3-methyl-2-butene[49], 2-methyl-2-pentene[49], cyclohexene[49] and its 1,2-dimethyl analogue[210], potassium rubrene-2,3,8,9-tetracarboxylate[211], 1,3-diphenylisobenzofuran (DPBF)[173] and cholesterol[212], are unaffected by the dismutating $O_2^{\overline{\cdot}}$ suggests that $^1O_2$ is in fact not formed.

$$HOO\cdot + O_2^- + H^+ \longrightarrow HOOH + {}^1O_2 \qquad (82)$$

Foote and coworkers[212] have been able to demonstrate that if $^1O_2$ indeed results from $O_2^{\overline{\cdot}}$ acid-catalysed disproportionation the upper limit for its production is no more than 0.2%. This value is noteworthy in that it includes corrections for trapping efficiency and for any quenching of the $^1O_2$ down to the triplet state that may have been induced by the $O_2^{\overline{\cdot}}$ supposedly generated. In light of all this evidence, the report of Mayeda and Bard[172] that $^1O_2$ can be trapped by DPBF from the disproportionation of electrochemically generated $O_2^{\overline{\cdot}}$ deserves reexamination.

No $^1O_2$ can be detected in the electron-transfer reaction between $O_2^{\overline{\cdot}}$ and phenoxy radical[154,170,213] or nitrohalobenzenes[91]. Nevertheless, there are several superoxide reactions in which $^1O_2$ is clearly generated. We have seen for example (Section III.H) that the cation radicals of DPBF[153], ferrocene[153], 2,5-dialkyl-1,4-dimethoxybenzene[154], and thianthrene[124] react with $O_2^{\overline{\cdot}}$ producing $^1O_2$ in substantial yields. The production of $^1O_2$ has been verified by DPBF scavenging[153,154], and spectroscopically in the case of thianthrene[124]. The generation of $^1O_2$ has also been observed in the reaction of $O_2^{\overline{\cdot}}$ with diacyl peroxides[210]. The presence of $^1O_2$ in this system has been verified by DPBF, 1,2-dimethylcyclohexene and tetramethylethylene scavenging. Furthermore the $^1O_2$ was quenched by β-carotene and $O_2^{\overline{\cdot}}$ as expected. More research is clearly necessary before the conditions required for the conversion of $O_2^{\overline{\cdot}}$ to $^1O_2$ can be clearly delineated.

V. REFERENCES

1. W. A. Pryor, *Photochem. Photobiol.*, **28**, 787 (1978).
2. L. Parker and J. Walton, *Chem. Tech.*, **7**, 278 (1977).
3. J. Bland, *J. Chem. Ed.*, **55**, 151 (1978).
4. I. Fridovich, *Science*, **201**, 875 (1978).
5. I. Fridovich, as quoted in J. D. Spikes and H. M. Swartz, *Photochem. Photobiol.*, **28**, 921, 930 (1978).
6. D. T. Sawyer and M. J. Gibian, *Tetrahedron*, **35**, 1471 (1979).
7. J. Wilshire and D. T. Sawyer, *Acc. Chem. Res.*, **12**, 105 (1979).
8. M. J. Gibian, D. T. Sawyer, T. Ungermann, R. Tangpoonpholvivat and M. M. Morrison, *J. Amer. Chem. Soc.*, **101**, 640 (1979).
9. J. S. Valentine and A. B. Curtis, *J. Amer. Chem. Soc.*, **97**, 224 (1975).
10. A. D. McElroy and J. S. Hasman, *Inorg. Chem.*, **40**, 1798 (1964).
11. J. W. Peters and C. S. Foote, *J. Amer. Chem. Soc.*, **98**, 873 (1976).
12. (a) B. H. J. Bielski, *Photochem. Photobiol.*, **28**, 645 (1978).
    (b) B. H. J. Bielski and A. O. Allen, *J. Phys. Chem.*, **81**, 1048 (1977).
13. D. T. Sawyer and E. J. Nanni, Jr. in *Oxygen and Oxy-Radicals in Chemistry and Biology* (Eds. M. A. J. Rodgers and E. L. Powers), Academic Press, New York, 1981, pp. 15–44.
14. D. T. Sawyer, M. J. Gibian, M. M. Morrison and E. T. Seo, *J. Amer. Chem. Soc.*, **100**, 627 (1978); see footnote 20 therein.
15. A. D. Goolsby and D. T. Sawyer, *Anal. Chem.*, **40**, 83 (1968).

16. (a) B. E. Knox and H. B. Palmer, *Chem. Rev.*, **61**, 247 (1961).
(b) J. S. Valentine, personal communication, 1981, regarding Reference 92.
17. See for example, *Handbook of Chemistry and Physics*, 61st ed. (Ed. R. C. Weast), Chemical Rubber Co., Boca Raton, Florida, 1980, F-233.
18. A. Le Berre and Y. Berguer, *Bull. Soc. Chim. Fr.*, 2363 (1966).
19. A. Le Berre and Y. Berguer, *Compt. Rend.*, **260**, 1995 (1965).
20. J. P. Stanley, *J. Org. Chem.*, **45**, 1413 (1980).
21. E. J. Nanni, M. D. Stallings, and D. T. Sawyer, *J. Amer. Chem. Soc.*, **102**, 4481 (1980).
22. H. J. James and R. F. Broman, *J. Phys. Chem.*, **75**, 4019 (1971).
23. R. Dietz, A. E. J. Forno, B. E. Larcombe and M. E. Peover, *J. Chem. Soc. (B)*, 816 (1970).
24. Y. Moro-Oka and C. S. Foote, *J. Amer. Chem. Soc.*, **98**, 1510 (1976).
25. E. Lee-Ruff, A. B. P. Lever and J. Rigaudy, *Can. J. Chem.*, **54**, 1837 (1976).
26. I. B. Afanas'ev and N. I. Polozova, *J. Org. Chem. USSR*, **12**, 1799 (1976).
27. I. B. Afanas'ev and N. I. Polozova, *J. Org. Chem. USSR*, **14**, 947 (1978).
28. M. Nishikimi and L. J. Machlin, *Arch. Biochem. Biophys.*, **170**, 648 (1975).
29. (a) S. Matsumoto and M. Matsuo, *Tetrahedron Letters*, 1999 (1977).
(b) S. Matsumoto and M. Matsuo, *Tetrahedron Letters*, **22**, 3649 (1981).
(c) S. Matsumoto and T. Matsuo, *Chem. Commun.*, 1267 (1981).
30. T. Ozawa, A. Hanaki, S. Matsumoto and M. Matsuo, *Biochem. Biophys. Acta*, **531**, 72 (1978).
31. M. Matsuo, S. Matsumoto, Y. Titaka, A. Hanaki and T. Ozawa, *Chem. Commun.*, 105 (1979).
32. M. Nishikimi, *Biochem. Biophys. Res. Commun.*, **63**, 463 (1975).
33. E. F. Elstner and R. Kramer, *Biochim. Biophys. Acta*, **314**, 340 (1973).
34. A. A. Frimer in *Superoxide Dismutase*, Vol. II (Ed. L. W. Oberley), Chemical Rubber Co., Boca Raton, Florida, 1982, in press.
35. A. A. Frimer and P. Gilinsky in *Oxygen and Oxy-Radicals in Chemistry and Biology* (Eds. E. L. Power and M. A. J. Rogers), Academic Press, New York, 1981, pp. 639–640.
36. A. A. Frimer and P. Gilinsky, *Tetrahedron Letters*, 4331 (1979).
37. (a) A. A. Frimer P. Gilinsky-Sharon and G. Aljadeff, *Tetrahedron Letters*, **23**, 1301 (1982).
(b) A. A. Frimer, P. Gilinsky-Sharon, J. Hameiri and G. Aljadeff, *J. Org. Chem.*, in press.
38. M. Utaka, S. Matsushita, H. Yamasaki and A. Takula, *Tetrahedron Letters*, **21**, 1063 (1980).
39. R. Hanna and G. Ourisson, *Bull. Soc. Chim. Fr.*, 587 (1961).
40. T. Hosokawa, T. Ohta, Y. Okamoto and S.-I. Murahashi, *Tetrahedron Letters*, **21**, 1259 (1980).
41. G. Crank and M. I. H. Makin, *Tetrahedron Letters*, 2169 (1979).
42. K. Assada and S. Kanematsu, *Agr. Biol. Chem.*, **40**, 1891 (1976).
43. D. A. Armstrong and J. D. Buchanan, *Photochem. Photobiol.*, **28**, 743 (1978).
44. J. F. Bunnett, *Acc. Chem. Res.*, **11**, 413 (1978).
45. N. Kornblum, *Angew. Chem. (Intern. Ed. Engl.)*, **14**, 734 (1975).
46. J. D. Roberts and M. J. Caserio, *Basic Principles of Organic Chemistry*, 2nd ed., W. A. Benjamin, California, 1977, p. 77, Table 4-3.
47. E. J. Nanni and D. T. Sawyer, *J. Amer. Chem. Soc.*, **102**, 7591 (1980).
48. C. L. Hussey, T. M. Laker and J. M. Achord, *J. Electrochem. Soc.*, **127**, 1865 (1980).
49. R. Poupko and I. Rosenthal, *J. Phys. Chem.*, **77**, 1722 (1973).
50. A. A. Frimer, Y. Ziv and G. Aljadeff, unpublished results.
51. E. Balogh-Hergovich, G. Speier and E. Winkelmann, *Tetrahedron Letters*, 3541 (1979).
52. C. I. Chern and J. San Filippo, Jr., *J. Org. Chem.*, **42**, 178 (1977).
53. H. P. Misra and I. Fridovich, *Biochemistry*, **15**, 681 (1976).
54. D. T. Sawyer, personal communication, 1981.
55. C. L. Hussey, T. M. Laker and J. M. Achord, *J. Electrochem. Soc.*, **127**, 1484 (198).
56. Reference 20, footnote 20.
57. A. Streitweiser, Jr. and C. H. Heathcock, *Introduction to Organic Chemistry*, Macmillan, New York, 1976, Appendix IV, p. 1191.
58. M. Tezuka, H. Hamada, Y. Ohkatsu and T. Osa, *Denki Kagaku*, **44**, 17 (1976); *Chem. Abstr.*, **85**, 26672a (1976).
59. S. A. Dibiase, R. P. Wolak, Jr., D. M. Dishong and G. W. Gokel, *J. Org. Chem.*, **45**, 3630 (1980).
60. G. A. Russel, E. G. Janzen, A. G. Bemis, E. J. Geels, A. J. Moye, S. Mak and E. T. Strom, *Advan. Chem. Ser.*, **51**, 112 (1965).
61. G. A. Russel, *Pure Appl. Chem.*, **15**, 185 (1967).

62. G. A. Russel, A. G. Bemis, E. J. Geels, E. G. Janzen and A. J. Moye, *Advan. Chem. Ser.*, **75**, 174 (1968).
63. G. Galliani and B. Rindone, *Tetrahedron*, **37**, 2313 (1981).
64. A. A. Frimer, I. Rosenthal and S. Hoz, *Tetrahedron Letters*, 4631 (1977).
65. E. J. Corey, K. C. Nicolaou, M. Shibasaki, Y. Machida and C. S. Shiner, *Tetrahedron Letters*, 3183 (1975).
66. M. Lissel and E. V. Dehmlow, *Tetrahedron Letters*, 3689 (1978).
67. J. San Fillipo, Jr., C. I. Chern and J. S. Valentine, *J. Org. Chem.*, **41**, 1077 (1976).
68. M. P. L. Caton, G. Darnbrough and T. Parker, *Tetrahedron Letters*, **21**, 1685 (1980).
69. R. A. Lee, C. McAndrews, K. M. Patel and W. Reusch, *Tetrahedron Letters*, 965 (1973).
70. A. A. Frimer and J. Hameiri, unpublished results.
71. G. Sosnovsky and E. H. Zaret in *Organic Peroxides*, Vol. 1 (Ed. D. Swern), John Wiley and Sons, New York–London, 1970, Chap. 8, Sect. IV.C.
72. H. Sagae, M. Fujihira, H. Lund and T. Osa, *Bull. Chem. Soc. Japan*, **53**, 1537 (1980).
73. H. Sagae, M. Fujihira, T. Osa and H. Lund, *Chem. Letters*, 793 (1977).
74. Y. Moro-oka, P. J. Chung, H. Arakawa and T. Ikawa, *Chem. Letters*, 1293 (1976).
75. M. Tezuka, Y. Ohkatsu and T. Osa, *Bull. Chem. Soc. Japan*, **48**, 1471 (1975).
76. E. Lee-Ruff and N. Timms, *Can. J. Chem.*, **58**, 2138 (1980).
77. S. Top, G. Jaoven and M. McGlinckey, *Chem. Commun.*, 643 (1980).
78. J. O. Hawthorne, K. A. Schowalter, A. W. Simon, M. H. Wilt and M. S. Morgan, *Advan. Chem. Ser.*, **75**, 203 (1968).
79. S. A. DiBiase and G. W. Gokel, *J. Org. Chem.*, **43**, 447 (1978).
80. I. Rosenthal and A. Frimer, *Tetrahedron Letters*, 3731 (1975).
81. L. L. Smith, M. J. Kulig and J. I. Teng, *Chem. Phys. Lipids*, **20**, 211 (1977).
82. Y. Leshem, S. Grossman, A. Frimer and J. Ziv in *Biochemistry and Physiology of Plant Lipids* (Eds. L. A. Appelqvist and C. Liljenberg), Elsevier North-Holland Biomedical Press, Netherlands, 1979, pp. 193–198.
83. Y. Y. Leshem, Y. Liftman, S. Grossman and A. A. Frimer in *Oxygen and Oxy-Radicals in Chemistry and Biology* (Eds. E. L. Powers and M. A. J. Rodgers) Academic Press, New York, 1981, pp. 676–678.
84. H. Sagae, M. Fujihira, H. Lund and T. Osa, *Heterocycles*, **13**, 321 (1979).
85. G. Feroci and S. Roffia, *J. Electroanal. Chem.*, **71**, 191 (1976).
86. R. Dietz, A. E. J. Forno, B. E. Larcombe and M. E. Peover, *J. Chem. Soc.* (*B*), 816 (1970).
87. C. I. Chern, R. DiCosimo, R. DeJesus and J. San Filippo, Jr., *J. Amer. Chem. Soc.*, **100**, 7317 (1978).
88. A. Le Berre and Y. Berguer, *Bull. Soc. Chim. Fr.*, 2368 (1966).
89. J. San Filippo, Jr., C. I. Chern and J. S. Valentine, *J. Org. Chem.*, **40**, 1678 (1975).
90. F. E. Scully, Jr. and R. C. Davis, *J. Org. Chem.*, **43**, 1467 (1978).
91. A. A. Frimer and I. Rosenthal, unpublished results.
92. J. S. Valentine in *Biochemical and Clinical Aspects of Oxygen* (Ed. W. S. Caughey) Academic Press, New York, 1979, pp. 659–677.
93. R. Yamadagni, J. D. Payzant and P. Kebarle, *Can. J. Chem.*, **51**, 2507 (1973).
94. A. J. Parker, *J. Chem. Soc.*, 1328 (1961).
95. A. J. Parker, in *Advances in Physical Organic Chemistry*, Vol. V (Ed. V. Gold) Academic Press, New York, 1967, p. 173.
96. A. J. Parker, *Chem. Rev.*, **69**, 1 (1969).
97. W. A. Sheppard and C. M. Sharts, *Organic Fluorine Chemistry*, Benjamin, New York, 1969, pp. 14–17.
98. J. H. Clark, *Chem. Rev.*, **80**, 429 (1980).
99. C. L. Liotta and H. P. Harris, *J. Amer. Chem. Soc.*, **96**, 2250 (1974).
100. W. C. Danen and R. J. Warner, *Tetrahedron Letters*, 989 (1977).
101. W. C. Danen, R. J. Warner and R. L. Arudi in *Organic Free Radicals: ACS Symposium Series*, Vol. 69 (Ed. W. A. Pryor), American Chemical Society, Washington, D.C., 1978, pp. 244–257.
102. (a) J. O. Edwards and G. O. Pearson, *J. Amer. Chem. Soc.*, **84**, 16 (1962).
(b) J. O. Edwards, *J. Chem. Ed.*, **45**, 386 (1968).
103. A. P. Grekov and V. Y. Veselov, *Russ. Chem. Rev.*, **47**, 631 (1978).
104. R. A. Johnson, E. G. Nidy and M. W. Merrit, *J. Amer. Chem. Soc.*, **100**, 7960 (1978).
105. M. J. Gibian and T. Ungerman, *J. Org. Chem.*, **41**, 2500 (1976).

106. L. H. Dao, A. C. Hopkinson, E. Lee-Ruff and J. Rigaudy, *Can. J. Chem.* **55**, 3791 (1977).
107. M. V. Merritt and D. T. Sawyer, *J. Org. Chem.*, **35**, 2157 (1970).
108. R. A. Johnson and E. G. Nidy, *J. Org. Chem.*, **40**, 1680 (1975).
109. E. J. Corey, K. C. Nicolaou and M. Shibasaki, *Chem. Commun.*, 658 (1975).
110. J. Divisek and B. Kastening, *J. Electroanal. Chem.*, **65**, 603 (1978).
111. M. V. Merrit and R. A. Johnson, *J. Amer. Chem. Soc.*, **99**, 3713 (1977).
112. Z. Osawa, H. Nakano, E. Mitsui and M. Nakano, *J. Polymer Sci., Polymer Chem. Ed.*, **17**, 139 (1979).
113. W. H. Perkle and P. L. Renaldi, *J. Org. Chem.*, **44**, 1025 (1979).
114. J. L. Roberts, Jr. and D. T. Sawyer, *J. Amer. Chem. Soc.*, **103**, 712 (1981).
115. M. J. Gibian and T. Ungerman, *J. Amer. Chem. Soc.*, **100**, 1291 (1978).
116. R. A. Johnson, *Tetrahedron Letters*, 331 (1976).
117. J. San Filippo, Jr., L. J. Romano, C.-I. Chern and J. S. Valentine, *J. Org. Chem.*, **41**, 586 (1976).
118. N. Kornblum and S. Singaram, *J. Org. Chem.*, **44**, 4727 (1979).
119. F. Magno and G. Bontempelli, *J. Electroanal. Chem.*, **68**, 337 (1976).
120. S. Oae and T. Takata, *Tetrahedron Letters*, **21**, 3689 (1980).
121. T. Nagano, K. Arakane and M. Hirobe, *Chem. Pharm. Bull.*, **28**, 3719 (1980).
122. R. Curci and J. O. Edwards in *Organic Peroxides*, Vol. I, (Ed. D. Swern), John Wiley and Sons, New York–London, 1970, Chap. 4.
123. Y. Sawaki and Y. Ogata, *Bull. Chem. Soc. Japan*, **54**, 793 (1981).
124. W. Ando, Y. Kabe, S. Kabayashi, C. Takyu, A. Yamagishi and H. Inaba, *J. Amer. Chem. Soc.*, **102**, 4526 (1980).
125. P. D. Bartlett in *Organic Free Radicals: ACS Symposium Series*, Vol. 69 (Ed. W. A. Pryor), American Chemical Society, Washington, D.C., 1978, pp. 15–32.
126. (a) T. Takata, Y. H. Kim and S. Oae, *Tetrahedron Letters*, 821 (1979).
(b) T. Nagano, K. Arakane and M. Hirobe, *Tetrahedron Letters*, **21**, 5021 (1980).
(c) S. Oae, T. Takata and Y. H. Kim, *Bull. Chem. Soc. Japan*, **54**, 2712 (1981).
127. S. Kobayashi, T. Tezuka and W. Ando, *Chem. Commun.*, 508 (1979).
128. S. Kobayashi, T. Tezuka and W. Ando, *Tetrahedron Letters*, 261 (1979).
129. S. Kobayashi and W. Ando, *Chem. Letters*, 1159 (1978).
130. A. Picot, P. Milliet, M. Cherest and X. Lusinchi, *Tetrahedron Letters*, 3811 (1977).
131. C. A. Evan, *Aldrichim. Acta.*, **12**, 23 (1979).
132. E. Finkelstein, G. M. Rosen and E. J. Rauchmann, *J. Amer. Chem. Soc.*, **102**, 4994 (1980).
133. D. G. Sanderson and M. R. Chedekel, *Photochem. Photobiol.*, **32**, 573 (1980).
134. G. R. Beuttner, L. W. Oberley and S. W. H. Chan-Leuthauser, *Photochem. Photobiol.*, **28**, 693 (1978).
135. B. Kalyanaraman, E. Perez-Reyes and P. P. Mason, *Biochim. Biophys. Acta*, **630**, 119 (1980).
136. G. VanGinkel and J. K. Raison, *Photochem. Photobiol.*, **32**, 793 (1980).
137. J. Eriksen, C. S. Foote and T. L. Parker, *J. Amer. Chem. Soc.*, **99**, 6455 (1977).
138. J. Eriksen and C. S. Foote, *J. Phys. Chem.*, **82**, 2659 (1978).
139. L. T. Spada and C. S. Foote, *J. Amer. Chem. Soc.*, **102**, 391 (1980).
140. J. Eriksen and C. S. Foote, *J. Amer. Chem. Soc.*, **102**, 6083 (1980).
141. L. E. Mauring, J. Eriksen and C. S. Foote, *J. Amer. Chem. Soc.*, **102**, 4275 (1980).
142. D. S. Steichen and C. S. Foote, *J. Amer. Chem. Soc.*, **103**, 1855 (1981).
143. A. P. Schaap, K. A. Zaklika, B. Kashar and L. W.-M. Fung, *J. Amer. Chem. Soc.*, **102**, 389 (1980).
144. S. L. Maltes and S. Farid, *Chem. Commun.*, 456 (1980).
145. N. Berenjian, P. de Mayo, F. H. Phoenix and A. C. Weedon, *Tetrahedron Letters*, 4179 (1979).
146. K. A. Brown-Winsley, S. L. Maltes and S. Farid, *J. Amer. Chem. Soc.*, **100**, 4162 (1978).
147. D. G. Whilten, *Acc. Chem. Res.*, **13**, 83 (1980).
148. I. Saito, K. Tamoto and T. Matsuura, *Tetrahedron Letters*, 2889 (1979).
149. R. Tang, H. J. Yue, J. F. Wolf and F. Mares, *J. Amer. Chem. Soc.*, **100**, 5248 (1978).
150. M. E. Landis and D. C. Madoux, *J. Amer. Chem. Soc.*, **101**, 5106 (1979).
151. S. F. Nelsen and R. Akaba, *J. Amer. Chem. Soc.*, **103**, 2096 (1981).
152. E. L. Clennan, W. Simmons and C. W. Almgren, *J. Amer. Chem. Soc.*, **103**, 2098 (1981).
153. E. A. Mayeda and E. J. Bard, *J. Amer. Chem. Soc.*, **95**, 6223 (1973).
154. A. Nishinaga, T. Shimizu and T. Matsuura, *Chem. Letters*, 547 (1977).
155. E. J. Nanni, Jr., C. T. Angelis, J. Dickson and D. T. Sawyer, *J. Amer. Chem. Soc.*, **103**, 4268 (1981).

156. J. A. Fee and J. S. Valentine in *Superoxide and Superoxide Dismutase* (Eds. A. M. Michelson, J. M. McCord and I. Fridovich), Academic Press, New York, 1977, pp. 19–60.
157. J. Chevalet, F. Rouelle, L. Greist and J. P. Lambert, *J. Electroanal. Chem.*, **39**, 201 (1972).
158. R. Dietz, M. E. Peover and P. Rothbaum, *Chem. Ing. Techn.*, **42**, 185 (1970).
159. E. Lee-Ruff, *Chem. Soc. Rev.*, **6**, 195 (1977).
160. K. Boujlel and J. Simonet, *Tetrahedron Letters*, 1063 (1979).
161. Y. Ogata, Y. Sawaki and M. Shiroyama, *J. Org. Chem.*, **42**, 4061 (1977).
162. Y. Sawaki and C. S. Foote, *J. Amer. Chem. Soc.*, **101**, 6292 (1979).
163. I. Rosenthal and T. Bercovici, *Chem. Commun.* 200 (1973).
164. K. B. Palet and R. L. Wilson, *J. Chem. Soc., Faraday Trans. 1*, **69**, 814 (1973).
165. I. Rosenthal and A. A. Frimer, *Tetrahedron Letters*, 2805 (1976).
166. A. A. Frimer and I. Rosenthal, *Photochem. Photobiol.*, **28**, 711 (1978).
167. I. Saito, T. Otsuki and T. Matsuura, *Tetrahedron Letters*, 1693 (1979).
168. G. Feroci and S. Roffia, *J. Electroanal. Chem.*, **81**, 387 (1977).
169. A. A. Frimer and I. Rosenthal, *Tetrahedron Letters*, 2809 (1976).
170. A. Nishinaga, T. Itahara, T. Shimizu, H. Tomita, K. Nishizawa and T. Matsuura, *Photochem. Photobiol.*, **28**, 687 (1978).
171. E. A. Mayeda and A. J. Bard, *J. Amer. Chem. Soc.*, **95**, 6223 (1973).
172. E. A. Mayeda and A. J. Bard, *J. Amer. Chem. Soc.*, **96**, 4023 (1974).
173. R. Nilsson and D. R. Kearns, *J. Phys. Chem.*, **78**, 1681 (1974).
174. P. F. Levonowich, H. P. Tannenbaum and R. C. Doughtery, *Chem. Commun.*, 597 (1975).
175. H. Sagae, M. Fujihara, K. Komazawa, H. Lund and T. Osa, *Bull. Chem. Soc. Japan*, **53**, 2188 (1980).
176. (a) C. K. Ingold, *Structure and Mechanism in Organic Chemistry*, 2nd ed., G. Bell and Sons, London, 1969, p. 389.
(b) G. Bartoli and P. E. Todesco, *Acc. Chem. Res.*, **10**, 125 (1977), and references cited therein.
177. J. F. Bunnett, *Acc. Chem. Res.*, **11**, 413 (1978).
178. I. Rosenthal, *J. Labelled Comp. Radiopharm.*, **12**, 317 (1976).
179. T. Abe and Y. Ikegami, *Bull. Chem. Soc. Japan*, **49**, 3227 (1976), and references cited therein.
180. T. Abe and Y. Ikegami, *Bull. Chem. Soc. Japan*, **51**, 196 (1978).
181. T. Yamaguchi and H. C. van der Plas, *Rec. Trav. Chim.*, **96**, 89 (1977).
182. G. A. Russel and R. F. Bridger, *J. Amer. Chem. Soc.*, **85**, 3765 (1963).
183. J. F. Bunnett, personal communication, 1981.
184. (a) D. Behar and P. Neta, *J. Phys. Chem.*, **85**, 690 (1981).
(b) M. Gareil, J. Pinson and J. M. Savéant, *Nouveau. J. Chim.*, **5**, 311 (1981).
185. F. Haber and J. Weiss, *Proc. Roy. Soc. (London)*, **A147**, 332 (1934).
186. C. Ferradni, J. Foos, C. Houce and J. Pucheault, *Photochem. Photobiol.*, **28**, 697 (1978).
187. J. Weinstein and B. H. J. Bielski, *J. Amer. Chem. Soc.*, **101**, 58 (1979).
188. W. H. Melhuish and H. C. Sulton, *Chem. Commun.*, 970 (1978).
189. G. Czapski and Y. A. Ilan, *Photochem. Photobiol.*, **28**, 651 (1978).
190. W. H. Koppenol, J. Butler and J. W. Van Leeuwen, *Photochem. Photobiol.*, **28**, 655 (1978).
191. M. J. Thomas, K. S. Mehl and W. A. Pryor, *Biochem. Biophys. Res. Commun.*, **83**, 927 (1978).
192. W. A. Pryor, *Photochem. Photobiol.*, **28**, 787 (1978).
193. D. B. Denney and J. D. Rosen, *Tetrahedron*, **20**, 271 (1964), and references cited therein.
194. W. Bors, C. Michel and M. Saran, *FEBS Letters*, **107**, 403 (1979).
195. C. W. Jefford and A. F. Boschung, *Helv. Chim. Acta*, **60**, 2673 (1977).
196. N. Kornblum and H. E. DeLaMare, *J. Amer. Chem. Soc.* **73**, 880 (1951).
197. R. Hiatt in *Organic Peroxides* (Ed. D. Swern), John Wiley and Sons, New York–London, Vol. II, 1971, Chap. 1; Vol. III, 1972, Chap. 1.
198. C. Mumford, *Chem. Brit.*, **11**, 402 (1975).
199. G. Sosnovsky and D. I. Rawlinson in *Organic Peroxides*, Vol. I (Ed. D. Swern), John Wiley and Sons, New York–London, 1970, Chap. 5.
200. J. F. Garst in *Free Radicals*, Vol. 1 (Ed. J. K. Kochi), John Wiley and Sons, London–New York, 1973, p. 503.
201. R. W. Alder, R. Baker and J. M. Brown, *Mechanism in Organic Chemistry*, John Wiley and Sons, London–New York, 1971, p. 158.
202. R. K. Kaynes and H. Musso, *Chem. Ber.*, **107**, 3723 (1974).

203. J. C. Slater, *Quantum Theory of Molecules and Solids*, Vol. 1, McGraw-Hill, New York, 1963.
204. P. E. Cade, R. F. W. Bader and J. Pelletier, *J. Chem. Phys.*, **54**, 3517 (1971).
205. Cf. Reference 156, p. 22 which errs on this point.
206. A. U. Kahn, *Science*, **168**, 476 (1970).
207. A. U. Kahn, *J. Amer. Chem. Soc.*, **99**, 370 (1977).
208. W. H. Koppenol and J. Butler, *FEBS Letters*, **83**, 1 (1977).
209. A. U. Kahn, *Photochem. Photobiol.*, **28**, 615 (1978).
210. W. C. Danen and R. L. Arudi, *J. Amer. Chem. Soc.*, **100**, 3944 (1978).
211. J. M. Aubry, J. Rigaudy, C. Ferrandini and J. Pucheault, *J. Amer. Chem. Soc.*, **103**, 4965 (1981).
212. C. S. Foote, F. C. Shook and R. A. Abakerli, *J. Amer. Chem. Soc.*, **102**, 2503 (1980).
213. D. T. Sawyer, unpublished results reported in S. Muto and T. C. Bruice, *J. Amer. Chem. Soc.*, **102**, 7559 (1980), footnote 17.

The Chemistry of Functional Groups, Peroxides
Edited by S. Patai

CHAPTER **15**

Transition-metal peroxides—synthesis and use as oxidizing agents

HUBERT MIMOUN
Laboratoire d'Oxydation, Institut Français du Pétrole, 92506 Rueil-Malmaison, France

I. INTRODUCTION . . . 464
II. PEROXO COMPLEXES . . . 465
A. General Properties and Synthesis . . . 465
B. General Reactivity . . . 467
1. Reaction with electrophiles . . . 467
2. Reaction with nucleophiles . . . 467
C. Transition-metal peroxo complexes . . . 468
1. Titanium, zirconium, hafnium . . . 468
2. Vanadium, niobium, tantalum . . . 469
3. Chromium, molybdenum, tungsten . . . 469
4. Manganese, technecium, rhenium . . . 471
5. Iron, ruthenium, osmium . . . 471
6. Cobalt, rhodium, iridium . . . 473
7. Nickel, palladium, platinum . . . 474
8. Uranium and actinides . . . 474
III. μ-PEROXO COMPLEXES . . . 475
A. General Properties and Synthesis . . . 475
B. Reactivity . . . 476
1. Group VIII metals . . . 476
2. Early transition metals . . . 478
IV. CONCLUSION . . . 478
V. REFERENCES . . . 479

Abbreviations

acac	Acetylacetonate
Bipy	2,2′-Bipyridyl
COD	1,5-Cyclooctadiene
Cp	Cyclopentadienyl
dipic	Pyridine-2,6-dicarboxylate
DMF	*N*,*N*-Dimethylformamide
DMG	Dimethylglyoxime
(S)DML	(S)-*N*,*N*-Dimethyllactamide
dppe	1,2-Bis(diphenylphosphino)ethane
HL	2,2′-Bipyridylamine
HMPA	Hexamethylphosphoric triamide
OEP	Octaethylporphyrin
Pic	Pyridine-2-carboxylate
Phen	Phenantroline
Py	Pyridine
PyH	Pyridinium cation
2=phos	1,2-Bis(diphenylphosphino)ethylene
Salen	Salicylaldehyde ethyleneimine
Salpr, Salptr	Five-coordinate Schiff base (see References 147 and 157)
tetars	$[Me_2As(CH_2)_3As(Ph)CH_2]_2$
TPP	Tetraphenylporphyrin

I. INTRODUCTION

Transition metals play an essential role as active centres in biological oxidation processes[1,2] and have been extensively used in industry as catalysts for the selective oxidation of hydrocarbons[3]. Transition-metal peroxides represent an important class of reactive intermediates in catalytic oxidations, since they can be obtained upon interaction of molecular oxygen or inorganic and organic peroxides with metal salts or complexes.

According to the rationalization made by Vaska[4], transition-metal peroxides involve covalently bound dioxygen resembling O_2^{2-} in the peroxo configuration (I), or the μ-peroxo configuration (II) comprising bimetallic peroxides (IIa), hydroperoxides (IIb) and alkyl peroxides (IIc).

M(O)(O)	M–O–O–M	M–O–O–H	M–O–O–R
(I) Peroxo complexes	(IIa) Bimetallic	(IIb) Hydroperoxide	(IIc) Alkyl peroxide

μ-Peroxo complexes

A common characteristic of these complexes is the O—O distance, which is between 1.4 and 1.52 Å (1.49 Å for O_2^{2-}) and the corresponding infrared frequency ν(O—O) which is between 800 and 950 cm^{-1} (802 cm^{-1} for O_2^{2-}).

Several review articles have appeared on dioxygen–metal complexes[4–14] and inorganic peroxy compounds[15,16] and their reactivity toward hydrocarbons[17–20]. However, no systematic attempt has been made to consider the whole problem of transition-metal peroxides associated with oxygen transfer to organic substrates.

In this nonexhaustive review, we shall describe the most representative and well-defined transition-metal peroxo and μ-peroxo complexes together with their reactivity toward hydrocarbons. Owing to space limitation, we shall not consider the theoretical aspects of this problem, the studies in solution, and the mineral and nonmolecular aspects, for which we refer to previous reviews[4–16].

II. PEROXO COMPLEXES

A. General Properties and Synthesis

Peroxo complexes, also referred to as side-on dioxygen complexes, are the most widely known peroxides, and have been prepared from a large number of transition metals. A list of the most representative and well-characterized examples is given in Table 1.

TABLE 1. Representative transition-metal peroxo complexes[a]

Peroxo compounds	Structure	Distances (Å) O—O	Distances (Å) M—O	ν(O—O) (cm⁻¹)	ν(M(O)(O))s,as (cm⁻¹)	Oxygen source	Reactivity and/or references
Ti, Zr, Hf							
$(Dipic)TiO_2, 2H_2O$	BP	1.46	1.85	—	—	H_2O_2	32
$(Pic)_2TiO_2$, HMPA	BP	1.42	1.85	895	575,615	H_2O_2	A,B[26]
$(OEP)TiO_2$		1.44	1.82	895	595,635	H_2O_2,O_2	33
V, Nb, Ta							
$NH_4[VO(O_2)dipic]$	BP	1.44	1.87	—	575,590	H_2O_2	34,35
$NH_4[VO(O_2)_2NH_3]$	BP	1.47	1.88	880	—	H_2O_2	36
$K[VO(O_2)_2Phen], 3H_2O$	BP	—	—	870,854	590,635	H_2O_2	37,38
$Cp_2Nb(O_2)Cl$	PT	1.47	1.97	870	525,550	H_2O_2	40,A,C[39]
$(NH_4)[Nb(O_2)_2(C_2O_4)_2],H_2O$	Dh	1.48	1.97	—	—	H_2O_2	41
$K[Nb(O_2)_3(Phen)], 3H_2O$	Dh	1.50	1.99	865	—	H_2O_2	42
Cr, Mo, W							
$CrO(O_2)_2Py$	BP	1.40	1.81	875	—	H_2O_2	46,D[51],E[50]
$K_3Cr(O_2)_4$	Dh	1.49	1.84	875	—	H_2O_2	49
$MoO(O_2)(dipic), H_2O$	BP	1.45	1.90	900	575,592	H_2O_2	54,53
$MoO(O_2)_2,HMPA,H_2O$	BP	1.44	1.94	865,875	540,590	H_2O_2	57,61,A[62], F[62–65],E[70]
$MoO(O_2)_2$,(S)DML	BP	1.45	1.93	—	—	H_2O_2	F[66],67
$(TPP)Mo(O_2)_2$	—	1.4	1.96	970	—	H_2O_2	76
$WO(O_2)$dipic, HMPA	BP	—	—	880	569,594	H_2O_2	53
$WO(O_2)_2$, HMPA, H_2O	BP	—	—	835,850	535,560	H_2O_2	57,E[80]
$K_2[WO(O_2)_2]_2O$	BP	1.5	1.93	—	—	H_2O_2	79
Fe, Ru, Os							
$Ru(O_2)Cl(NO)(PPh_3)_2$	BT	—	—	875	—	O_2	86,A[90]
$Os(O_2)(CO)_2(PPh_3)_2$	BT	—	—	820	—	O_2	88
Co, Rh, Ir							
$[Co(O_2)(2{=}phos)_2]BF_4$	BT	1.42	1.87	909	—	O_2	95
$[Co(O_2)(HL)_2]ClO_4$	—	—	—	892	542,565	H_2O_2	97
$Co_2(CN)_4(PMe_2Ph)_5(O_2)$	BT	1.44	1.89	881	—	O_2	98,A[98]
$[Co(O_2)(tetars)]ClO_4$	—	1.42	—	—	—	O_2,H_2O_2	22

TABLE 1. (*cont.*)

Peroxo compounds	Structure	Distances (Å) O—O	Distances (Å) M—O	ν(O—O) (cm^{-1})	ν(M<O/O)$_{s,as}$ (cm^{-1})	Oxygen source	Reactivity and/or references
$[Rh(O_2)(AsMe_2Ph)_4]ClO_4$	BT	1.46	2.03	892	540	O_2	103,104, A[106],G[106]
$Rh(O_2)Cl(PPh_3)_3$	BT	1.41	2.0	850	—	O_2	100
$Rh(O_2)Cl(t\text{-}BuNC)(PPh_3)_2$	BT	—	—	892	576	O_2	102
$[IrO_2(PMe_2Ph)_4]BF_4$	BT	1.49	2.04	838	—	O_2	104
$[IrO_2(dppe)_2]PF_6$	BT	1.52	2.05	—	—	O_2	114
$Ir(O_2)Cl(CO)(PPh_2Et)_2$	BT	1.46	2.03	—	—	O_2	110
$Ir(O_2)I(CO)(PPh_3)_2$	BT	1.51	2.06	—	—	O_2	111, C[112,116], H[117]
Ni, Pd, Pt							
$Ni(O_2)(t\text{-}BuNC)_2$	T	1.45	1.81	898	—	O_2	120,121, A[120],I[125]
$Pd(O_2)[PPh(t\text{-}Bu)_2]_2$	T	1.37	2.05	915	—	O_2	122
$Pd(O_2)(t\text{-}BuNC)_2$	T	—	—	893	—	O_2	120
$Pt(O_2)[PPh(t\text{-}Bu)_2]_2$	T	1.43	2.02	835	—	O_2	122
$Pt(O_2)(PPh_3)_2$	T	1.45	2.01	830	—	O_2	123,A[124], C[128,129], H[30],B[23]
Actinides							
$[UO_2(O_2)_3]Na_4$	D_{3h}	1.51	2.28	850–900	—	H_2O_2	139
$UO_2(O_2)_2Ph_3PO$	—	—	—	808	—	H_2O_2	137

[a]Abbreviations for Table 1:
Structure: BP = Bipyramidal pentagonal, PT = pseudotetrahedral, Dh = dodecahedral, BT = bipyramidal trigonal, T = trigonal, D_{3h} = symmetry D_{3h}.
Reactivity: A = oxidation of phosphines to phosphine oxides, B = peroxometallocyclic adduct formation with tetracyanoethylene, C = oxidation of SO_2 to coordinated $SO_4^{=}$, D = oxidative cleavage of olefins, E = oxidation of alcohols to carbonyl compounds, F = epoxidation of olefins, G = oxidation of terminal olefins to methyl ketones, H = metalloozonide formation with carbonyl compounds, I = oxidation of isocyanides to isocyanates.
Infrared: s = symmetric, as = antisymmetric.

The triangular peroxo group is characterized for all the metals by a peroxidic O—O distance between 1.4 and 1.5 Å and three infrared vibrations of the C_{2v} structure at 800–950 cm^{-1} [ν(O—O)], and 500–650 cm^{-1} [ν(M—O) sym and asym].

The bipyramidal pentagonal structure is most frequently found for early transition metals (Groups IVb, Vb and VIb), while the trigonal bipyramidal and the trigonal structure exist very often for Group VIII metal complexes.

Peroxo–metal complexes can be synthesized by two main methods:

(*a*) For early metals, from the reaction of hydrogen peroxide with high-valent metal–oxo complexes (equation 1).

$$M{=}O + H_2O_2 \longrightarrow M\langle{}^{O}_{O}| + H_2O \qquad (1)$$

(*b*) For Group VIII metals, from the interaction of dioxygen with reduced two-electron donor metal complexes (equation 2).

$$M^n + O_2 \longrightarrow M^{n+2}\!<\!\begin{smallmatrix}O\\|\\O\end{smallmatrix} \quad (2)$$

Despite these different methods of preparation, and contrary to previous thoughts based on erroneous X-ray structure determinations[21], there is no indication that dioxygen carriers, even reversibly formed, are different from true peroxo compounds synthesized from H_2O_2[4]. In fact, there are several examples of peroxo complexes obtained from both oxygen sources, e.g. $[Co^{III}(O_2)(tetars)]ClO_4$ (equation 3)[20,22].

$$[Co^{I}(tetars)]^+ + O_2 \longrightarrow [Co^{III}(O_2)(tetars)]^+ \longleftarrow [Co^{III}(H_2O)_2(tetars)]^{3+} + H_2O_2 \quad (3)$$

B. General Reactivity

Peroxo complexes are potential oxygen donors to organic substrates. However, only a few of them are reactive. One important criterion is solubility in organic solvents. Another even more important point is the geometry of the complex and the availability of a vacant, or releasable, site for the coordination of a nucleophilic substrate on the metal.

Peroxo–metal complexes generally act as 1,3-dipolar reagents $M^+{-}O{-}O^-$, in which the positive charge is localized on the metal, and the negative charge on the terminal oxygen atom of the opened peroxo group. They can react with both electrophilic and nucleophilic substrates.

1. Reaction with electrophiles

This reaction generally occurs in a bimolecular fashion. 1,3-Dipolar cycloaddition of peroxo complexes with electrophilic dipolarophiles such as cyano olefins or hexafluoroacetone results in the formation of cyclic five-membered peroxidic adducts (equation 4)[23–27].

$$M\!<\!\begin{smallmatrix}O\\|\\O\end{smallmatrix} + (NC)_2C{=}C(CN)_2 \longrightarrow \begin{smallmatrix}M^+ \;\; O \;\; O^-\\ >C{=}C<\end{smallmatrix} \longrightarrow \begin{smallmatrix}M \;\; O \;\; O\\ (NC)_2C{-}C(CN)_2\end{smallmatrix} \quad (4)$$

The reaction with alkyl halides results in the formation of alkyl peroxidic complexes, presumably from an S_N2 nucleophilic attack of the terminal oxygen on the carbon atom, and occurs with inversion of configuration (equation 5)[27,28].

$$L_2Pt\!<\!\begin{smallmatrix}O\\|\\O\end{smallmatrix} + R{-}Br \longrightarrow L_2Pt\!<\!\begin{smallmatrix}O{-}OR\\Br\end{smallmatrix} \quad (5)$$

2. Reaction with nucleophiles

Peroxo complexes can oxidize a large variety of nucleophilic substrates, such as phosphines, sulphides, alcohols, olefins, enolates, imines, amides, etc., as summarized in

Table 1. Oxidation of phosphine to phosphine oxide is common behaviour for most of the peroxo complexes.

In contrast, the oxidation of olefins, which has been the most studied reaction, is limited to a few complexes (Cr, Mo, W, U, Rh). It largely depends upon the nature of the metal and the geometry of the complex. Oxygen transfer from peroxo complexes to nucleophilic olefins requires the coordination of the substrate on the metal, probably on an adjacent and coplanar position with respect to the peroxo moiety. This is followed by the intramolecular 1,3-dipolar cycloaddition of the peroxo group to the olefin which becomes an electrophilic dipolarophile upon coordination to the metal. The resulting five-membered peroxometallocyclic adduct then decomposes to give an oxo complex and the oxygenated olefin which can be either an epoxide if the metal is an early transition metal with a high-valent d(0) state, or a methyl ketone if the olefin is terminal and if the metal belongs to Group VIII (equations 6 and 7)[20,29].

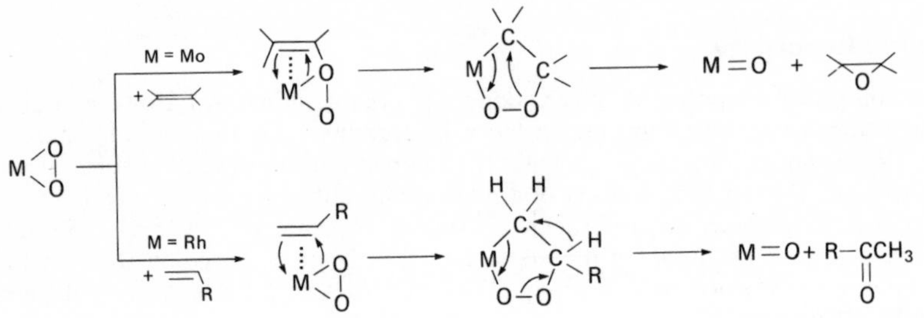

(6)

(7)

This mechanism, called cyclic peroxymetalation, has also been shown to occur in the reaction of platinum peroxo complexes with nucleophilic ketones forming a stable metalloozonide (equation 8)[30]. It obeys the general π–σ rearrangement procedure occurring in most transition-metal-catalysed transformations of olefins[31].

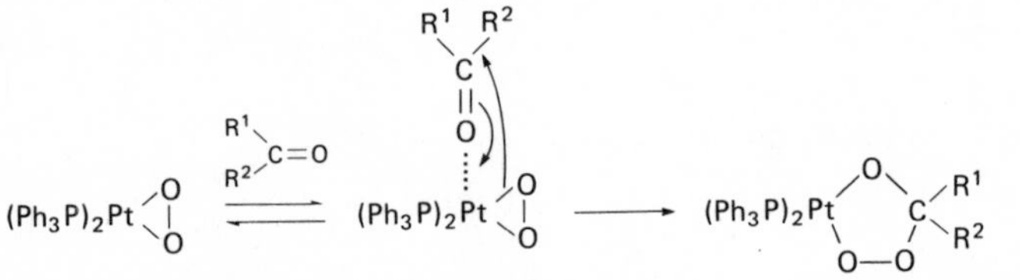

(8)

C. Transition-metal Peroxo Complexes

1. *Titanium, zirconium, hafnium*

Although the formation of an orange colour from the reaction of Ti(IV) with H_2O_2 has been noted for over 100 years[15], the synthesis of well-defined molecular Ti(IV) peroxo complexes such as (dipic)Ti(O_2), 2 H_2O (**1**) is recent[32].

We have also recently prepared a series of unusually stable Ti(IV) peroxo compounds of general formula [Ti(O_2)(A—B)$_2$,L] where A—B is a bidendate monoanionic ligand such as pyridine-2-carboxylate or 8-hydroxy quinolinate and L a basic ligand such as HMPA[26]. Ti(O_2)(Pic)$_2$, HMPA (**2**) has been shown by X-ray crystallography to exhibit a bipyramidal pentagonal geometry, similar to that of complex **1**. Complex **2** oxidizes PPh_3

to PPh_3O and forms a peroxometallocyclic adduct with tetracyanoethylene, but it does not react with sulphides and unactivated olefins which do not coordinate on the metal[26].

Peroxotitanium porphyrins such as (OEP)Ti(O_2) (**3**) have been prepared from both H_2O_2 and O_2, and have also been found to be unreactive toward olefins[33].

No well-characterized molecular peroxo complexes of Zr(IV) and Hf(IV) have yet been prepared.

2. *Vanadium, niobium, tantalum*

Vanadium peroxo complexes exist in the form of anionic monoperoxo complexes, e.g. $NH_4[VO(O_2)dipic]$[34,35], and diperoxo complexes, e.g. $NH_4[VO(O_2)_2NH_3]$[36] and $K[VO(O_2)_2Phen], 3\,H_2O$[37,38], with a bipyramidal pentagonal structure. However, the reactivity of such complexes is not yet known.

Niobium monoperoxo complexes $Cp_2Nb(O_2)Cl$(**4**)[39,40], diperoxo complexes e.g. $(NH_4)_3\,[Nb(O_2)_2(C_2O_4)_2], H_2O$(**5**)[41] and triperoxo complexes $K\,[Nb(O_2)_3(Phen)],3\,H_2O$ (**6**)[42,43] have been prepared and characterized by X-ray structure analysis. The structure of **4** is pseudo-tetrahedral, while **5** and **6** are dodecahedral.

The complex $Cp_2Nb(O_2)Cl$(**4**) is one of the rare organometallic peroxo complexes with a metal–carbon π bond. It oxidizes PPh_3 to PPh_3O, and SO_2 to coordinated sulphate, but is unreactive towards olefins, probably because of the 18 electrons existing in the metal valence orbitals which prevent coordination of the substrate[40].

Vanadium peroxo complexes are also probably involved as intermediates in the $VO(acac)_2$-catalysed oxidation of sulphides to sulphoxides and allylic alcohols to epoxides by H_2O_2, but have not yet been fully characterized[44].

3. *Chromium, molybdenum, tungsten*

Peroxo complexes of Group VIb are perhaps the most numerous and the best characterized complexes.

a. Chromium (VI) *peroxo complexes.* These exist in the form of crystalline bipyramidal pentagonal diperoxo compounds $CrO(O_2)_2$,L (L = Py[45,46], Phen[47], Bipy[48], Et_2O[48]) and dodecahedral tetraperoxo compounds $K_3Cr(O_2)_4$[49].

Chromium oxo diperoxo complexes CrO_5L have been used for the selective oxidation of alcohols to carbonyl compounds (L = Py)[50], and for the oxidative cleavage of tetracyclone (L = Et_2O)[51], presumably via intermediate formation of an epoxide[52].

b. Molybdenum peroxo complexes. Molybdenum salts are known to react with H_2O_2 to form a large variety of complexes in which the peroxide to metal ratio is 1:1, 2:1, 3:1 and 4:1[15].

Representative monoperoxo complexes are $MoO(O_2)$ (dipic), H_2O[53,54], $MoO(O_2)$ $[PhCON(Ph)O]_2$[55] and $[MoO(O_2)F_4](NH_4)_3$[56], which all exhibit a bipyramidal pentagonal structure.

Covalent molybdenum(VI) compounds of general formula $MoO(O_2)_2$,L,L′ have been prepared with a great variety of basic ligands L and L′, including tertiary amide[57], phosphoramides[57], amine oxides[57,58], phosphine and arsine oxide[57–59], aromatic amines[57,60] and pyridine-2-carboxylic acid[53]. These complexes are easily prepared from addition of the ligand to a solution of MoO_3 in H_2O_2 (permolybdic acid $Mo_2O_{11}H_2$)[57].

The X-ray structure of MoO_5, HMPA, H_2O (**7**), shown in Figure 1[61], exhibits a bipyramidal pentagonal structure with HMPA occupying the equatorial position on the same plane of the two peroxo triangles, and H_2O the axial position *trans* to the oxo group.

This compound **7** can be easily dehydrated, producing MoO_5, HMPA (**8**), and its high solubility in organic solvents has helped to study its reactivity toward organic substrates.

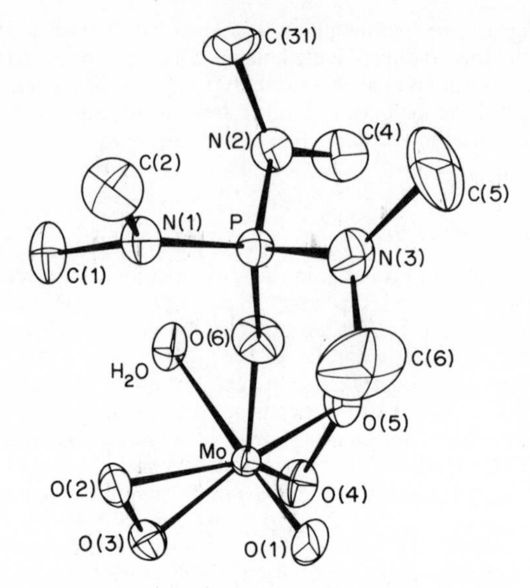

FIGURE 1. X-ray structure of MoO_5, HMPA, H_2O (7).

MoO_5,HMPA has been found to oxidize stoichiometrically olefins to epoxides at room temperature and in aprotic solution[52,62–65]. The reaction is stereospecific, and the reactivity of olefins increases with increasing nucleophilicity. The following mechanism, based on kinetic and NMR measurement, and on the inhibitory effect of added basic ligands on the reaction, has been proposed (equation 9)[62].

$$\mathrm{MoO(O_2)_2L} \underset{}{\overset{+\ \text{olefin},\ -L}{\rightleftharpoons}} \mathrm{MoO(O_2)_2(olefin)} \longrightarrow [\mathrm{Mo\text{–}O\text{–}O\text{–}C\text{–}C}] \longrightarrow \mathrm{MoO_2(O_2)} + \text{epoxide} \qquad (9)$$

Asymmetric epoxidation of light olefins (e.g. propene, *trans*-2-butene) has been achieved using molybdenum peroxo complexes bearing a chiral bidentate ligand, i.e. MoO_5,(*S*)DML[66,67].

Molybdenum diperoxo compounds have also been used to hydroxylate enolizable ketones on an adjacent position[68], to cause Baeyer–Villiger lactonization of cyclic ketones[69], to transform alcohols to carbonyl compounds[70], to oxidize *n*-butyllithium[71], alkylboron and magnesium compounds to alcohols[72,73] and to convert amides to hydroxamic acids[74,75]. The common characteristic of most oxidations by MoO_5 compounds is high selectivity.

It is worth pointing out that the reactivity of molybdenum peroxo compounds toward organic nucleophilic substrates is strongly dependent upon availability of coordination sites on the metal. Saturated complexes such as $MoO(O_2)$(dipic),H_2O[70], MoO-$(O_2)_2$, Bipy[62] and TPP,$Mo(O_2)_2$[76] are inactive.

The monoperoxo complex $MoO(O_2)$ $[PhCON(Ph)O]_2$ is also unreactive toward simple olefins, but allylic alcohols which are able to displace one hydroxamate ligand, are easily converted to α,β-epoxy alcohols[55]. Primary and secondary alcohols are also oxidized to the corresponding carbonyl compounds by this complex[55].

Molybdenum peroxo complexes are also involved as reactive intermediates in the catalysed oxidation of sulphides to sulphoxides and olefins to epoxides by H_2O_2 (equation 10)[77]. Continuous removal of water formed in this reaction has been found highly beneficial for the selectivity of epoxide formation[78].

$$\text{>C=C<} + H_2O_2 \xrightarrow{\text{Mo}} \text{epoxide} + H_2O \qquad (10)$$

c. *Tungsten peroxo complexes.* The chemistry of tungsten peroxo complexes appears very similar to that of molybdenum complexes, with the existence of monoperoxo species, e.g. $WO(O_2)_2$ dipic, HMPA[53], and diperoxo species, e.g. $WO(O_2)_2$,L,L′[57] and $K_2[WO(O_2)_2]_2O$[79]. However, the reactivity of tungsten complexes has not been very much studied, probably because of their low solubility. $WO(O_2)_2$,HMPA,H_2O has been found to epoxidize olefins[80], and $H^+[WO(O_2)_2\ (Pic)]^-$ converts stoichiometrically secondary alcohols to ketones, and catalytically in the presence of excess H_2O_2[70].

4. *Manganese, technecium, rhenium*

Although manganese peroxidic complexes have been detected on interaction of dioxygen with Mn(II) porphyrins[12,81] there are still no examples of manganese peroxo compounds well-characterized by crystal X-ray structure determinations.

Technecium and rhenium peroxo compounds have not yet been reported, except in solution studies[15].

5. *Iron, ruthenium, osmium*

In contrast to end-on iron superoxo complexes formed in the interaction of haemoglobin and myoglobin with dioxygen, and characterized by X-ray structure of porphyrin model compounds[82], no well-defined iron(III) peroxo complexes have been yet reported.

However, porphyrin iron(III) peroxo species are probably involved as intermediates in enzymatic cytochrome P-450 monooxygenases, upon one-electron reduction of the primary superoxo complex[83,84]. They have been recently identified by spectroscopial methods and by the ν(O—O) band at 806 cm^{-1} upon reaction of O_2^- with Fe^{II}(TPP)[85].

Ruthenium(III) peroxo complexes are well-characterized crystalline compounds, obtained upon interaction of dioxygen with Ru(I) complexes. However, no X-ray crystal structure of these compounds is available to date. Known $Ru(O_2)$ complexes include $RuX(NO)(O_2)(PPh_3)_2$($X = OH$, Cl, Br, I, NCS...)[86], $Ru(CO)(CNR)(O_2)(PPh_3)_2$[87], $Ru(CO)_2(O_2)(PPh_3)_2$[88] and the paramagnetic $RuCl_2(O_2)(AsPh_3)_2$[89].

Most of these complex catalyse the oxidation of PPh_3 to PPh_3O by dioxygen[90]. However, the oxidation of olefins such as cyclohexene produces radical-chain oxidation products[91,92]. A porphyrin (DMF)Ru(OEP)(O_2) complex has been formed from molecular O_2, but the nature of the coordinated oxygen remains uncertain[93,94]. This compound was found inactive for the oxidation of phosphines and olefins.

Osmium peroxo complexes exist in the form of $Os(CO)_2(PPh_3)_2(O_2)$ and $OsCl(NO)(PPh_3)_2O_2$[88], but no reactivity has been reported for these compounds.

6. *Cobalt, rhodium, iridium*

a. Cobalt(III) *peroxo complexes.* These have been prepared by the reaction of cobalt(I) complexes with dioxygen (equation 11)[95,96] or by the reaction of H_2O_2 with Co(II) complexes (equation 12)[97].

$$[\mathrm{Co^{I}(2{=}phos)_2}]\mathrm{BF_4} + \mathrm{O_2} \longrightarrow [(2{=}\mathrm{phos})_2\mathrm{Co^{III}}(\mathrm{O_2})]\mathrm{BF_4} \quad (\mathbf{9}) \tag{11}$$

$$\mathrm{Co^{II}(HL)_3(ClO_4)_2} + \mathrm{H_2O_2} \longrightarrow [(\mathrm{HL})_2\mathrm{Co^{III}}(\mathrm{O_2})]\mathrm{ClO_4} \tag{12}$$

HL = 2,2′-bipyridylamine

The X-ray crystal structure of **9** indicates a trigonal bipyramidal structure with bond distances O—O = 1.42 Å and Co—O = 1.87 and 1.90 Å[95].

Other reported examples involve $L_3(CN)_2Co(NC)Co(CN)L_2(O_2)$ (**10**) (L = PMe_2Ph) obtained from the reaction of $Co(CN)_2(PMe_2Ph)_3$ with O_2 [98], and $[Co(O_2)(tetars)]ClO_4$ previously shown in equation (3)[22].

The bimolecular complex **10** oxidizes PMe_2Ph to Me_2PhPO, and $CoCl_2(PEt_3)_2$ catalyses the oxidation by O_2 of phosphines to phosphine oxides[99].

b. Rhodium (III) *peroxo complexes.* These compounds have all been prepared from the reaction of Rh(I) complexes with molecular oxygen. They exist in both covalent and ionic form.

Covalent complexes involve $Rh(O_2)Cl(PPh_3)_3$ [100], $[RhCl(O_2)(PPh_3)_2]$ [101] and Rh $X(O_2)(PPh_3)_2(RNC)$(X = Cl, Br, I; R = *t*-Bu, C_5H_{11}, *p*-$CH_3C_6H_4$)[102]. Cationic peroxo complexes have the general formula

$$[\mathrm{Rh(O_2)L_4}^+\mathrm{A}^- \ (\mathrm{L} = \mathrm{AsPh_3},\ \mathrm{AsPhMe_2};\ \mathrm{A} = \mathrm{ClO_4},\ \mathrm{PF_6},\ \mathrm{BF_4})$$ [103,104].

Both types of complexes have a bipyramidal trigonal structure with an O—O distance between 1.4 and 1.5 Å [104].

Rhodium complexes such as $RhCl(PPh_3)_3$ have been shown to promote the cooxygenation of terminal olefins to methyl ketones and PPh_3 to PPh_3O (equation 13)[105].

$$\mathrm{RCH{=}CH_2} + \mathrm{PPh_3} + \mathrm{O_2}, \xrightarrow{\mathrm{RhCl(PPh_3)_3}} \mathrm{RCOCH_3} + \mathrm{Ph_3PO} \tag{13}$$

Rhodium(III) peroxo complexes act as reactive intermediates in this catalytic transformation, as indicated by the selective formation of methyl ketone in high yield from the reaction of the ^{18}O-labelled $[(AsPh_3)_4Rh(O_2)]^+ClO_4^-$ complex (**11**) with terminal olefins under anhydrous conditions (equation 14)[106,107].

Reaction (14), which is somewhat similar to the epoxidation of olefins by molybdenum peroxo complexes, requires the coordination of the olefin to the metal displacing $AsPh_3$, prior to its insertion to the rhodium–oxygen bond, forming the five-membered peroxometallocyclic adduct. However, the decomposition of this adduct occurs differently from that of the Mo analogue, and produces a methyl ketone instead of an epoxide, owing to the β-hydride abstraction ability of rhodium.

The cationic dioxygen complex $[RhO_2(Ph_2P(CH_2)_2PPh_2)_2]^+ClO_4^-$ is completely inactive toward terminal olefins, due to the nonavailability of coordination sites on the metal for the complexation of the substrate.

$$\left[(AsPh_3)_4Rh\langle{}^{O}_{O}\right]^+ClO_4^- \ (\mathbf{11}) \xrightarrow[-AsPh_3]{+\ R\text{-}CH{=}CH_2} \left[(AsPh_3)_4Rh(\text{O–O–CH(R)–CH}_2)\right]^+ClO_4^- \ (\mathbf{12}) \longrightarrow R-\underset{\|}{C}(=O)-Me \ \ 85\% \qquad (14)$$

$$(\mathbf{11}) \xrightarrow[-2AsPh_3]{+TCNE} \left[(AsPh_3)_2Rh(\text{O–O–C(CN)}_2\text{–C(CN)}_2)\right]^+ClO_4^- \ (\mathbf{13}) \qquad (15)$$

In further support for this mechanism, the stable peroxometallocyclic adduct **13** has been isolated from the reaction of **11** with tetracyanoethylene. Peroxometallocyclic intermediates such as **12** can also result from the reaction of rhodium(I) π-olefinic complexes with molecular oxygen. The reaction of $[Rh(1,7\text{-octadiene})_2]^+BF_4^-$ with O_2 produces 1-octene-7-one in quantitative yield, and this reaction can be made catalytic in the presence of a strong acid[106]. This catalytic oxidation of terminal olefins to methyl ketones by O_2, which also occurs in the presence of $Rh(ClO_4)_3$ or $RhCl_3 + Cu(ClO_4)_2$ in alcoholic solvents, has been shown to result from the coupling of an oxygen activation path (peroxymetalation) consuming the first oxygen atom, with a Wacker-type hydroxymetalation path consuming the second protonated oxygen atom[108].

c. Iridium peroxo complexes. The reversible reaction of $IrCl(CO)(PPh_3)_2$ with molecular oxygen was first reported by Vaska in 1963[109]. Since that time, the syntheses and structures of many iridium dioxygen complexes have been reported[4,6,8,11].

Iridium peroxo complexes are similar to those of rhodium. They exist in the form of covalent complexes, e.g. $Ir(O_2)X(CO)L_2$ (X = Cl, Br, I, etc.; L = phosphine, arsine)[110,111] and $Ir(O_2)(PPh_3)_3$[112], and cationic complexes, e.g. $[Ir(O_2)L_4]^+A^-$ (L = phosphine, arsine; A = BF_4, PF_6, ClO_4)[104,113] and $[Ir(O_2)L_2]^+A^-$ (L = diphosphine; A = BPh_4, PF_6, ClO_4)[104,114,115].

Iridium dioxygen complexes $Ir(O_2)X(CO)(PPh_3)_2$ react with SO_2 to give the sulphato complex[112,116] and form a five-membered iridium(III) ozonide with hexafluoroacetone[117].

In contrast to rhodium cationic dioxygen complexes[106], iridium analogues such as $[(AsPhMe_2)_4Ir(O_2)]ClO_4$ do not undergo an oxygen transfer to terminal olefins. Also, no ketone is formed from the reaction of $[Ir(cyclooctene)]^+BF_4^-$ with O_2, although it has been noted that one mole of O_2 is absorbed per mole of iridium[118]. This lack of reactivity is attributed to the fact that O_2 and olefin coordinate to the metal on positions unsuitable for oxygen transfer[29]. In support of this hypothesis, a stable iridium complex having both O_2 and ethylene coordinated to the same metal has been synthesized, without any oxygen transfer occurring in this case (equation 16)[118].

$$[IrCl(C_8H_{14})_2]_2 + 2PPh_3 + C_2H_4 + O_2 \longrightarrow Ir(O_2)Cl(CH_2{=}CH_2)(PPh_3)_2 \qquad (16)$$

7. Nickel, palladium, platinum

A great number of peroxo complexes of the nickel triad, of the general formula $L_2M(O_2)$ and trigonal structure, have been reported:

(*a*) M = Ni; L = PPh_3[119], RNC[120,121]

(*b*) M = Pd; L = PPh_3[119], $PPh(t\text{-}Bu)_2$[122]

(*c*) M = Pt; L = PPh_3[119,123], $PPh(t\text{-}Bu)_2$[122], *t*-BuNC[120]

All these complexes are irreversibly formed from interaction of molecular oxygen with the corresponding metal(O) compounds, except complexes with bulky *t*-phosphine ligands, such as $PPh(t\text{-}Bu)_2$, which are reversible[122].

These peroxo complexes undergo a series of reactions with reactive substrates[18]: oxidation of phosphine to phosphine oxide[119,124], isonitriles to isocyanates[120,125], CO to coordinated carbonate (for Pt)[126], CO_2 to coordinated percarbonate (for Ni[125] and Pt[127]), SO_2 to coordinated sulphate (for Pd, Pt)[128,129], NO to coordinated nitrites (Pd, Pt)[129,130] and NO_2 to coordinated nitrates (Pd, Pt)[128,129] Five-membered peroxometallocyclic adducts have been isolated from the reaction of $(Ph_3P)_2M(O_2)$ (M = Pd, Pt) with aldehydes and ketones (equation 8)[30], hexafluoroacetone[131] and electrophilic cyano olefins[23].

The peroxidic nature of coordinated oxygen in $(Ph_3P)_2MO_2$ (M = Pd, Pt) has been demonstrated by the formation of H_2O_2 upon acid hydrolysis[132,133] and dialkyl peroxides by reaction with alkyl halides (equation 17)[27].

$$PtO_2L_2 \xrightarrow{+\,RBr} L_2PtBr(OOR) \xrightarrow{+\,RBr} cis\text{-}PtBr_2L_2 + ROOR \qquad (17)$$

The reaction of $Pt(O_2)L_2$ with electrophilic acetylenes occurs rather unusually with cleavage of the O—O bond (equation 18)[27,134].

$$L_2PtO_2 + RC{\equiv}CR \longrightarrow L_2Pt\begin{matrix} O-C-R \\ \| \\ O-C-R \end{matrix} \qquad (18)$$

Peroxo complexes of nickel, palladium and platinum are reluctant to react with unactivated olefins, probably due to the lack of complexation of these olefins to the metal. However, such a transfer occurs from palladium dioxygen complexes to terminal olefins, producing methyl ketones, in the presence of strong acids or alkylating agents. The active species in this case are hydroperoxidic or alkylperoxidic palladium compounds (see below)[135].

8. Uranium and actinides

Although inorganic actinide peroxo compounds have been known for a long time[15], few molecular complexes have been described. Covalent peroxo compounds involve $UO_2(O_2)$, $4\,H_2O$[136] and $UO_2(O_2)L_2$(L = Ph_3PO, Ph_3AsO, pyridine *N*-oxide)[137].

Peroxouranium oxide $UO_2(O_2),4\,H_2O$ has been found to oxidize olefins mainly to epoxides and oxidative cleavage products[138]. The suggested mechanism is similar to that previously shown for molybdenum peroxo complexes (equation 9)[137]. The ionic sodium peruranate Na_2UO_8 is found to be unreactive[137].

III. μ-PEROXO COMPLEXES

A. General Properties and Synthesis

Well-identified μ-peroxo complexes are much less numerous than peroxo complexes, and exist in the three structures IIa–c shown in the introduction to this chapter.

Table 2 lists some of the most characteristic μ-peroxo complexes synthesized to date. They exhibit a peroxidic character, with a O—O distance between 1.4 and 1.5 Å and an infrared band (ν(O—O) at 800–950 cm^{-1} (except for symmetrical MOOM species for which this absorption is only Raman-active).

TABLE 2. Representative transition-metal μ-peroxo complexes[a]

μ-Peroxo compound	Distances (Å) O—O	Distances (Å) M—O	ν(O—O) (cm^{-1})	Oxygen source	References and/or reactivity
$[MoO(O_2)_2OOH]_2(PyH)_2$	1.46	2.04 2.39	—	H_2O_2	155
$PyCo(DMG)_2(OOCMe_2Ph)$	1.45	1.91	—	ROOH	156
(Salpr)Co—OO—Quinol	1.50	1.85	—	O_2	147
(Salptr)Co—OO—Co(Salptr)	1.45	1.93	—	O_2	157
$K_3[Rh(OOH)(CN)_4]$	—	—	825	O_2	143
(COD)RhOORh(COD)	—	—	—	KO_2	A[140]
$IrCl(CO)(OOBu\text{-}t)_2$	—	—	885	ROOH	146
$[CCl_3CO_2Pd(OOBu\text{-}t)]_4$	1.49	1.99	855	ROOH	G[144]
$[(PPh_3)_2Pt(O_2)OHPt(PPh_3)_2]ClO_4$	1.54	1.99	880	O_2	133
$PtBr(OOCPh_3)(PPh_3)_2$	—	—	831	O_2	27
$[(CF_3CO_2)_2Pt(OOBu\text{-}t)(t\text{-}BuOH)]_2$	—	—	870	ROOH	G[145]
$(2{=}phos)Pt(CF_3)OOH$	—	—	825	H_2O_2	142
$[UO_2Cl_3(O_2)Cl_3UO_2]^{4-}$	1.49	2.30	905	O_2	158
$La_2[N(SiMe_3)_2]_4(O_2)(Ph_3PO)_2$	1.65	2.33	—	H_2O_2	159

[a]Abbreviations: A = oxidation of phosphines to phosphine oxides, G = oxidation of terminal olefins to methyl ketones.

a. Bimetallic complexes IIa. These can result from the reaction of a cobalt superoxo complex with a reduced metal (equation 19)[12,13], from the acid hydrolysis of platinum peroxo complexes (equation 20)[133] or from the reaction of potassium superoxide with rhodium (equation 21)[140] or palladium complexes[141].

$$LnCo^{II} \xrightarrow{+O_2} LnCo^{III}{-}O{-}O^{\bullet} \xrightarrow{+LnCo^{II}} LnCo^{III}{-}O{-}O{-}Co^{III}Ln \quad (19)$$

Ln = CN, NH_3, Salen, etc.

$$2(Ph_3P)_2PtO_2 + HClO_4 \xrightarrow{H_2O} [(Ph_3P)_2Pt\langle(O{-}O)(O)\rangle Pt(Ph_3P)_2]\,ClO_4 + H_2O_2 \quad (20)$$

$$[(COD)RhCl]_2 + 2KO_2 \longrightarrow (COD)Rh{-}O{-}ORh(COD) + O_2 + 2KCl \quad (21)$$

b. Transition-metal hydroperoxides IIb. These can be prepared from the reaction of H_2O_2 with platinum hydroxo complexes (equation 22)[142], from the insertion of molecular oxygen into a rhodium–hydride bond (equation 23)[143] or from the hydrolysis of a platinum peroxo complex (equation 24)[132].

$$L_2Pt(CF_3)OH + H_2O_2 \longrightarrow L_2Pt(CF_3)OOH + H_2O \qquad (22)$$

$$[CoH(CN)_5]^{3-} + O_2 \longrightarrow [Co(OOH)(CN)_5]^{3-} \qquad (23)$$

$$(Ph_3P)_2PtO_2 + CF_3CO_2H \longrightarrow (Ph_3P)_2Pt(CF_3CO_2)OOH \qquad (24)$$

c. Alkyl peroxide complexes IIc. These can be obtained from the reaction of hydroperoxides with palladium (equation 25)[144] and platinum carboxylates[145] or iridium complexes[146].

$$4Pd(RCO_2)_2 + 4t\text{-BuOOH} \longrightarrow [RCO_2PdOOBu\text{-}t]_4 + 4RCO_2H \qquad (25)$$

They can also by synthesized from the reaction of alkyl or acyl halides with platinum peroxo complexes (equation 26)[27] or from the reaction of a cobalt superoxo complex with a substituted phenol (equation 27)[147].

$$(Ph_3P)_2PtO_2 + Ph_3CBr \longrightarrow (Ph_3P)_2PtBr(OOCPh_3) \qquad (26)$$

$$\text{2,4,6-}t\text{-Bu}_3C_6H_2OH \xrightarrow{Co^{III}(salpr)O_2} \text{2,4,6-}t\text{-Bu}_3\text{-4-}(Co^{III}(salpr)OO)\text{cyclohexa-2,5-dienone} \qquad (27)$$

B. Reactivity

Only a few reports deal with the study of oxygen transfer from μ-peroxo transition-metal complexes to organic substrates[20,29]. However, alkyl peroxidic and hydroperoxidic species play an important role as reactive intermediates in numerous metal-catalysed oxidations of hydrocarbons by O_2, ROOH and H_2O_2.

1. Group VIII metals

a. Palladium(II) t-butyl peroxide carboxylates. These carboxylates, of the general formula $[RCO_2PdOOBu\text{-}t]_4$, selectively oxidize terminal olefins to metylketones with a high selectivity (equation 28)[144].

$$RCO_2PdOOBu\text{-}t + RCH{=}CH_2 \longrightarrow RCO_2PdOBu\text{-}t + RCOMe \qquad (28)$$

The mechanism proposed for this reaction involves complexation of the olefin to the metal, followed by its insertion into the palladium–oxygen bond, forming a pseudocyclic five-membered peroxidic intermediate which decomposes into methyl ketone and the palladium *t*-butoxy complex (equation 29).

$$R^1CO_2PdOOBu\text{-}t \ (\mathbf{14}) \xrightarrow{+\ CH_2{=}CHR^2} R^1CO_2PdOOBu\text{-}t\cdots(CH_2{=}CHR^2) \longrightarrow R^1CO_2Pd[CH_2CH(R^2)OO(t\text{-Bu})] \ (\mathbf{15}) \longrightarrow R^1CO_2PdOBu\text{-}t \ (\mathbf{16}) + R^2C(O)CH_3 \quad (29)$$

This mechanism has been called pseudocyclic peroxymetalation. The decomposition of the pseudocyclic intermediate **15** in equation (29) is similar to that of the cyclic intermediate formed in the oxidation of terminal olefins to methyl ketones by rhodium(III) peroxo complexes previously shown in equations (7) and (14).

Methyl ketones can be obtained on a catalytic scale (equation 30) when an excess of *t*-BuOOH is used for regenerating the initial *t*-butyl peroxidic complex **14** from the *t*-butoxy complex **16** in equation (29)[148].

$$t\text{-BuOOH} + RCH{=}CH_2 \xrightarrow{Pd(RCO_2)_2} t\text{-BuOH} + RC(O)CH_3 \quad (30)$$

Palladium carboxylates are effective catalysts for the selective oxidation of terminal olefins to methyl ketones by hydrogen peroxide (equation 31)[149]. A mechanism similar to that of equation (29), but involving hydroperoxidic palladium(II) complexes RCO_2PdOOH as reactive intermediates, has been suggested for this catalytic oxidation.

$$RCH{=}CH_2 + H_2O_2 \xrightarrow{Pd(RCO_2)_2} RC(O)CH_3 + H_2O \quad (31)$$

Palladium hydroperoxidic or alkyl peroxidic species have been generated from protonation or alkylation of palladium dioxygen complexes $(Ph_3P)_2PdO_2$, and shown to oxidize selectively terminal olefins to methyl ketones (equation 32)[135]. ^{18}O-labelling studies have indicated that the oxygen source for ketone formation is molecular oxygen, and not water[135].

$$L_2Pd(O_2) \xrightarrow[+S]{+H^+A^-} [L_2Pd(OOH)(S)]^+A^- \xrightarrow[-S]{+\ CH_2{=}CHR} [L_2Pd(OOH)(CH_2{=}CHR)]^+ \longrightarrow [L_2Pd(CH_2CHR\text{-}O\text{-}OH)]^+$$

$$\xrightarrow[+S]{-RCOCH_3} [L_2Pd(OH)(S)]^+ \longrightarrow [L_2Pd(\mu\text{-}OH)_2PdL_2]^{++}\ 2A^- \quad (32)$$

$L = PPh_3$; $S = CH_2Cl_2$, $A = CH_3SO_3^-$, BF_4^-

b. Platinum hydroperoxidic complexes. These complexes, $[(Ph_3P)_2Pt\text{-}OOH]^+$, generated from protonation of platinum peroxo complexes, have been found to be inactive for the oxidation of terminal olefins[29].

However, the *t*-butyl peroxidic complex, $[(CF_3CO_2)_2Pt(OOBu\text{-}t)(t\text{-}BuOH)]_2$, having strong electron-acceptor trifluoroacetate substituents, is found to oxidize selectively terminal olefins to methyl ketones[145].

The acyl peroxidic complex, $(PPh_3)_2PtX(OOC(O)Ph)(X = Cl, Br)$, obtained from the reaction of PhCOX with $(PPh_3)_2PtO_2$, selectively oxidizes olefins such as norbornene to the corresponding *exo*-norbornene epoxide[27,150].

c. Iridium hydroperoxidic species. These have been recently assumed to be the reactive intermediates in the oxidation of cyclooctene by O_2 and H_2 in the presence of $[IrCl(C_8H_{14})_2]_2$ (equation 33)[94].

$$\text{cyclooctene} + O_2 + H_2 \; [Ir] \text{cyclooctanone} + H_2O \tag{33}$$

Cyclooctanone has also been obtained from the reaction of $[Ir(C_8H_{14})_4]^+ BF_4^-$ with O_2 in the presence of HBF_4[29]. IrOOH species, presumably obtained from insertion of O_2 into IrH species, have been assumed to intervene as oxidants in this reaction.

2. *Early transition metals*

Despite many attempts, well-characterized alkyl peroxidic complexes of Groups IVb, Vb and VIb transition metals have not been isolated to date.

However, such species are widely considered to be involved as reactive oxidants in the epoxidation of olefins by alkyl hydroperoxides in the presence of molybdenum, vanadium and titanium catalysts (equation 34)[18,151–153],

$$\text{ROOH} + \text{olefin} \xrightarrow{\text{Mo,V,Ti}} \text{ROH} + \text{epoxide} \tag{34}$$

The molybdenum-catalysed epoxidation of propylene by ethylbenzene or *t*-butyl hydroperoxide has been developed on a large scale by the Halcon company[154].

Owing to the high selectivity and stereospecificity of this epoxidation, and its general characteristics, which are very similar to epoxidation of olefins by molybdenum peroxo complexes, the mechanism shown in equation (35) has been suggested[20].

$$\text{Mo}^{VI}\text{OOR} \xrightarrow{+\text{olefin}} \text{MoOOR}\cdots\text{olefin} \longrightarrow \text{Mo–C–C–O–O(R)}\cdots\text{Mo} \longrightarrow \text{MoOR} + \text{epoxide}; \quad \text{MoOR} \xrightarrow[-\text{ROH}]{+\text{ROOH}} \text{Mo}^{VI}\text{OOR} \tag{35}$$

IV. CONCLUSION

The synthesis of transition-metal peroxides and the study of their oxidizing properties toward organic substrates afford an heuristic approach for the understanding of catalytic oxidation reactions. The mechanism of oxygen-transfer reactions to nucleophilic substrates such as olefins can be rationalized in terms of a cyclic peroxymetalation of the substrate by peroxometal complexes, and a pseudocyclic peroxymetalation by alkyl peroxidic or hydroperoxidic species. The nature of the resulting oxygenated substrate strongly depends upon the nature of the metal. However, much yet remains to be done in the synthesis of catalysts, particularly in the design of their shape and ligand surroundings for the selective oxidation of organic substrates. The recent use of titanium isopropoxide–diethyl tartrate for the asymmetric epoxidation of allylic alcohols by *t*-BuOOH[160] is an example of the usefulness of transition-metal peroxides in effecting a variety of selective transformations that are unattainable with any other reagent.

V. REFERENCES

1. O. Hayaishi (Ed.), *Molecular Mechanisms of Oxygen Activation*, Academic Press, New York–London, 1974.
2. V. Ullrich, *Angew. Chem. (Intern. Ed. Engl.)*, **11**, 701 (1972).
3. J. E. Lyons, *Hydroc. Process*, 107 (1980).
4. L. Vaska, *Acc. Chem. Res.*, **9**, 175 (1976).
5. E. Bayer and P. Stretzmann, *Struct. Bonding*, **2**, 181 (1967).
6. J. S. Valentine, *Chem. Rev.*, **73**, 235 (1973).
7. J. A. McGinetty in *International Review of Sciences: Inorganic Chemistry*, Vol. 15 (Ed. R. Sharp), Butterworths, London, 1972, p. 229.
8. V. J. Choy and C. J. O'Connor, *Coord. Chem. Rev.*, **9**, 145 (1972).
9. G. Henrici Olivé and S. Olivé, *Angew. Chem. (Intern. Ed. Engl.)*, **13**, 29 (1974).
10. A. V. Savitskii and V. I. Nelyubin, *Russ. Chem. Rev.*, **44**, 110 (1975).
11. R. W. Erskine and B. O. Field, *Struct. Bonding*, **28**, 1 (1976).
12. R. D. Jones, D. A. Summerville and F. Basolo, *Chem. Rev.*, **79**, 139 (1979).
13. G. McLendon and A. E. Martell, *Coord. Chem. Rev.*, **19**, 1 (1976).
14. D. A. Summerville, R. D. Jones, B. M. Hoffman and F. Basolo, *J. Chem. Ed.*, **56**, 157 (1979).
15. J. A. Connor and E. A. V. Ebsworth, *Advan. Inorg. Chem. Radiochem.*, **6**, 279 (1964).
16. I. I. Volnov, *Russ. Chem. Rev.*, **41**, 314 (1972).
17. H. Mimoun, *Rev. Inst. Fr. Petr.*, **33**, 259 (1979).
18. J. E. Lyons, Aspects Homogeneous Catalysis, Vol. 3 (Ed. R. Ugo), Reidel, Dordrecht, 1977, p. 3.
19. H. Mimoun in *Chemical and Physical Aspects of Catalytic Oxidation* (Eds. J. L. Portefaix and F. Figueras), CNRS, Paris, 1980, p. 15.
20. H. Mimoun, *J. Mol. Catal.*, **7**, 1 (1980).
21. J. A. McGinetty, N. C. Payne and J. A. Ibers, *J. Amer. Chem. Soc.*, **91**, 6301 (1969).
22. B. Bosnich, W. G. Jackson, S. T. D. Lo and J. W. McLaren, *Inorg. Chem.*, **13**, 2605 (1974).
23. R. A. Sheldon and J. A. Van Doorn, *J. Organomet. Chem.*, **94**, 115 (1975).
24. W. B. Beaulieu, G. D. Mercer and D. M. Roundhill, *J. Amer. Chem. Soc.*, **100**, 1147 (1978).
25. F. Igersheim and H. Mimoun, *Nouveau J. Chim.*, **4**, 161 (1980).
26. H. Mimoun, M. Postel, F. Casabianca, A. Mitschler and J. Fisher, *Inorg. Chem.*, **21**, 1303 (1982).
27. Y. Tatsuno and S. Otsuka, *J. Amer. Chem. Soc.*, **103**, 5832 (1981).
28. S. Otsuka, A. Nakamura, Y. Tatsuno and M. Miki, *J. Amer. Chem. Soc.*, **94**, 3761 (1972).
29. H. Mimoun, *Pure Appl. Chem.*, **53**, 2389 (1981).
30. R. Ugo, G. M. Zanderighi, A. Fusi and D. Carreri, *J. Amer. Chem. Soc.*, **102**, 3745 (1980), and references therein.
31. M. Tsutsui and A. Courtney, *Advan Organometal. Chem.*, **16**, 241 (1977).
32. D. Schwarzenbach, *Helv. Chim. Acta*, **55**, 2990 (1972); *Inorg. Chem.*, **9**, 2371 (1970).
33. J. C. Marchon, J. M. Latour and C. J. Boreham, *J. Mol. Catal.*, **7**, 227 (1980); *Inorg. Chem.*, **17**, 1228, 2024 (1978); *J. Amer. Chem. Soc.*, **101**, 3974 (1979).
34. R. E. Drew and F. W. B. Einstein, *Inorg. Chem.*, **12**, 829 (1973).
35. K. Wieghardt, *Inorg. Chem.*, **17**, 57 (1978).
36. R. E. Drew and F. W. B. Einstein, *Inorg. Chem.*, **11**, 1079 (1972).
37. N. Vuletic and C. Djordjevic, *J. Chem. Soc., Dalton Trans.*, 1137 (1973).
38. J. Sala-Pala and J. E. Guerchais, *J. Chem. Soc. (A)*, 1132 (1971).
39. J. Sala-Pala, J. Roue and J. E. Guerchais, *J. Mol. Catal.*, **7**, 141 (1980).

40. A. Bkouche-Waksman, C. Bois, J. Sala-Pala and J. E. Guerchais, *J. Organometal. Chem.*, **195**, 307 (1980).
41. G. Mathern and R. Weiss, *Acta Cryst.*, **B27**, 1572 (1971).
42. G. Mathern and R. Weiss, *Acta Cryst.*, **B27**, 1582 (1971).
43. C. Djordjevic and N. Vuletic, *Inorg. Chem.*, **7**, 1864 (1968).
44. O. Bortolini, F. di Furia, P. Scrimin and G. Modena, *J. Mol. Catal.*, **7**, 59 (1980), and references therein.
45. O. F. Wiede, *Ber.*, **30**, 2178 (1897).
46. R. Stomberg, *Arkiv Kemi*, **22**, 29 (1964).
47. R. Stomberg and I. B. Ainalem, *Acta Chem. Scand.*, **22**, 1439 (1968).
48. R. Stomberg, *Arkiv Kemi*, **24**, 111 (1965).
49. R. Stomberg, *Acta Chem. Scand.*, **17**, 1563 (1963).
50. G. W. J. Fleet and W. Little, *Tetrahedron Letters*, 3749 (1977).
51. J. E. Baldwin, J. C. Swallow and H. W. S. Chan, *J. Chem. Soc., Chem. Commun.*, 1407 (1971).
52. A. A. Frimer, *J. Chem. Soc., Chem. Commun.*, 207 (1977).
53. S. E. Jacobson, R. Tang and F. Mares, *Inorg. Chem.*, **17**, 3055 (1978).
54. D. Westlake, R. Kergoat and J. E. Guerchais, *Compt. Rend. (C)*, **280**, 113 (1975).
55. H. Tomioka, K. Takai, K. Oshima and H. Nozaki, *Tetrahedron Letters*, **21**, 4843 (1980).
56. D. Grandjean and R. Weiss, *Bull. Soc. Chim. Fr.*, 3044 (1967).
57. H. Mimoun, I. Séree de Roch and L. Sajus, *Bull. Soc. Chim. Fr.*, 1481 (1969).
58. A. D. Westland, F. Haque and J. M. Bouchard, *Inorg. Chem.*, **19**, 2255 (1980).
59. J. Lewis and R. Whyman, *J. Chem. Soc. (A)*, 211 (1966).
60. S. Sarkar and R. C. Maurya, *J. Indian Chem. Soc.*, **47**, 341 (1978).
61. J. M. Le Carpentier, A. Mitschler and R. Weiss, *Acta Cryst.*, **B28**, 1288 (1972).
62. H. Mimoun, I. Sérée de Roch and L. Sajus, *Tetrahedron*, **26**, 37 (1970).
63. K. B. Sharpless, J. M. Townsend and D. R. Williams, *J. Amer. Chem. Soc.*, **94**, 295 (1972).
64. H. Arakawa, Y. Moro-Aka and A. Nozaki, *Bull. Chem. Soc. Japan*, **47**, 2958 (1974).
65. A. A. Achrem, T. A. Timoschtschuck and D. I. Metelitza, *Tetrahedron*, **30**, 3165 (1974).
66. H. B. Kagan, H. Mimoun, C. Mark and V. S. Schurig, *Angew. Chem. (Intern. Ed. Engl.)*, **6**, 485 (1979).
67. W. Winter, C. Mark and V. Schurig, *Inorg. Chem.*, **19**, 2045 (1980).
68. E. Vedejs, D. A. Engler and J. E. Tolschow, *J. Org. Chem.*, **43**, 188 (1978).
69. S. E. Jacobson, R. Tang and F. Mares, *J. Chem. Soc., Chem. Commun.*, 888 (1978).
70. S. E. Jacobson, D. A. Muccigrosso and F. Mares, *J. Org. Chem.*, **44**, 921 (1979).
71. S. L. Regen and G. M. Whitesides, *J. Organometal. Chem.*, **59**, 293 (1973).
72. G. Schmitt and B. Olbertz, *J. Organometal. Chem.*, **152**, 271 (1978).
73. M. M. Midland and S. B. Preston, *J. Org. Chem.*, **45**, 4514 (1980).
74. S. A. Matlin, P. G. Sammes and R. M. Upton, *J. Chem. Soc., Perkin Trans. 1*, 2481 (1979).
75. G. A. Brewer and E. Sinn, *Inorg. Chem.*, **20**, 1823 (1981).
76. B. Chevrier, Th. Diebold and R. Weiss, *Inorg. Chim. Acta*, **19**, L57 (1976).
77. O. Bortolini, F. di Furia, G. Modena, C. Scardellato and P. Scrimin, *J. Mol. Catal.*, **11**, 107 (1981).
78. M. Pralus, J. C. Lecoq and J. P. Shirmann in *Fundamental Research in Homogeneous Catalysis*, Vol. 3 (Ed. M. Tsutsui), Plenum Press, New York, 1979, p. 327, and references therein.
79. F. W. B. Einstein and P. R. Penfold, *Acta Cryst.*, **17**, 1127 (1964).
80. H. Mimoun, unpublished results.
81. B. M. Hoffman, C. J. Wechsler and F. Basolo, *J. Amer. Chem. Soc.*, **98**, 5473 (1976).
82. J. P. Collman, T. R. Halpert and K. S. Suslick in *Metal Ion Activation of Dioxygen* (Ed. T. G. Spiro), John Wiley and Sons, London–New York, 1980, p. 1, and references therein.
83. M. J. Coon and R. E. White in Reference 82, p. 73.
84. J. T. Groves in Reference 82, p. 125.
85. E. M. C. Candlish, A. M. Miksztal, M. Nappa, A. Q. Sprenger, J. S. Valentine, J. D. Strong and T. G. Spiro, *J. Amer. Chem. Soc.*, **102**, 4268 (1980).
86. B. W. Graham, K. R. Laing, C. J. O'Connor and W. R. Roper, *J. Chem. Soc., Chem. Commun.*, 1272 (1970); 1556, 1558 (1968).
87. D. F. Christian and W. R. Roper, *J. Chem. Soc., Chem. Commun.*, 1271 (1971).

88. B. E. Cavit, K. R. Grundry and W. R. Roper, *J. Chem. Soc., Chem. Commun.*, 60 (1972).
89. M. M. Taqui Khan, R. K. Andal and P. T. Manoharan, *J. Chem. Soc., Chem. Commun.*, 561 (1971).
90. B. W. Graham, K. R. Laing, C. J. O'Connor and W. R. Roper, *J. Chem. Soc., Dalton Trans.*, 1237 (1942).
91. S. Cenini, A. Fusi and G. Capparella, *Inorg. Nucl. Chem. Letters*, **8**, 127 (1972).
92. S. Muto and K. Kamiya, *J. Catal.*, **41**, 148 (1976).
93. N. P. Farrell, D. H. Dolphin and B. R. James, *J. Amer. Chem. Soc.*, **100**, 324 (1978).
94. B. R. James, *Advan. Chem. Ser.*, **191**, 253 (1980).
95. L. Vaska, L. C. Chen and W. H. Miller, *J. Amer. Chem. Soc.*, **93**, 667 (1971); N. W. Terry, E. L. Amma and L. Vaska, *J. Amer. Chem. Soc.*, **94**, 653 (1972).
96. V. M. Miskowski, J. L. Robbins, G. S. Hammond and H. B. Gray, *J. Amer. Chem. Soc.*, **98**, 2447 (1976).
97. V. L. Johnson and J. F. Geldard, *Inorg. Chem.*, **17**, 1675 (1978).
98. J. Halpern, B. L. Goodall, G. P. Khare, H. S. Lim and J. J. Pluth, *J. Amer. Chem. Soc.*, **97**, 2301 (1975).
99. D. D. Schmidt and J. T. Yoke, *J. Amer. Chem. Soc.*, **93**, 637 (1971).
100. M. J. Bennett and P. B. Donaldson, *Inorg. Chem.*, **16**, 1583 (1977), and references therein.
101. M. J. Bennett and P. B. Donaldson, *Inorg. Chem.*, **16**, 1585 (1977).
102. A. Nakamura, Y. Tatsuno and S. Otsuka, *Inorg. Chem.*, **11**, 2058 (1972).
103. L. M. Haines, *Inorg. Chem.*, **10**, 1685 (1971).
104. M. Laing, M. J. Nolte and E. Singleton, *J. Chem. Soc., Chem. Commun.*, 660 (1975).
105. C. Dudley, G. Read and P. J. C. Walker, *J. Chem. Soc., Dalton Trans.*, 883 (1977), and references therein.
106. F. Igersheim and H. Mimoun, *Nouveau J. Chim.*, **4**, 161 (1980); *J. Chem. Soc., Chem. Commun.*, 559 (1978).
107. R. Tang, F. Mares, N. Neary and D. E. Smith, *J. Chem. Soc., Chem. Commun.*, 274 (1979).
108. H. Mimoun, M. M. Perez-Machirant and I. Sérée de Roch, *J. Amer. Chem. Soc.*, **100**, 5437 (1978).
109. L. Vaska, *Science*, **140**, 809 (1963).
110. M. S. Weininger, I. F. Taylor and E. L. Amma, *J. Chem. Soc., Chem. Commun.*, 1172 (1971).
111. J. A. McGinetty, R. J. Doedens and J. A. Ibers, *Inorg. Chem.*, **6**, 2243 (1967).
112. J. Valentine, D. Valentine and J. P. Collman, *Inorg. Chem.*, **10**, 219 (1971).
113. L. M. Haines and E. Singleton, *J. Chem. Soc., Dalton Trans.*, 1891 (1972).
114. M. J. Nolte, E. Singleton and M. Laing, *J. Amer. Chem. Soc.*, **97**, 6396 (1975).
115. M. J. Nolte, E. Singleton and M. Laing, *J. Chem. Soc., Dalton Trans.*, 1976 (1979).
116. R. W. Horn, E. Weissberger and J. P. Collman, *Inorg. Chem.*, **9**, 2367 (1970).
117. W. B. Beaulieu, G. D. Mercer and D. M. Roundhill, *J. Amer. Chem. Soc.*, **100**, 1147 (1978).
118. H. Van Gaal, H. G. M. Cuppers and A. Van der Ent, *J. Chem. Soc., Chem. Commun.*, 1594 (1970).
119. G. Wilke, H. Schott and P. Heimbach, *Angew Chem.*, **79**, 62 (1967).
120. S. Otsuka, A. Nakamura and S. Tatsuno, *J. Amer. Chem. Soc.*, **91**, 6994 (1969).
121. M. Matsumoto and N. Nakatsu, *Acta Cryst.*, **B31**, 2711 (1975).
122. Y. Yoshida, K. Tatsumi, M. Matsumoto, K. Nakatsu, A. Nakamura, T. Fueno and S. Otsuka, *Nouveau J. Chim.*, **3**, 761 (1979).
123. T. Kashigawi, N. Yasuoka, N. Kasai, M. Kakudo, S. Takahashi and N. Hagihara, *J. Chem. Soc., Chem. Commun.*, 743 (1969).
124. A. Sen and J. Halpern, *J. Amer. Chem. Soc.*, **99**, 8337 (1977), and references therein.
125. S. Otsuka, A. Nakamura, Y. Tatsuno and M. Niki, *J. Amer. Chem. Soc.*, **94**, 3761 (1972).
126. J. P. Collman, *Acc. Chem. Res.*, **1**, 136 (1968).
127. F. Cariati, R. Mason, G. B. Robertson and R. Ugo, *J. Chem. Soc., Chem. Commun.*, 408 (1967).
128. C. D. Cook and G. S. Jauhal, *J. Amer. Chem. Soc.*, **89**, 3066 (1967).
129. J. J. Levison and S. D. Robinson, *J. Chem. Soc. (A)*, 762 (1971).
130. J. P. Collman, M. Kubota and J. W. Hosking, *J. Amer. Chem. Soc.*, **89**, 4809 (1967).
131. P. J. Hayward and C. J. Nyman, *J. Amer. Chem. Soc.*, **93**, 617 (1971).
132. S. Muto, H. Ogata and Y. Kamiya, *Chem. Letters*, 809 (1975).
133. S. Badhuri, L. Casella, R. Ugo, R. Routhby, C. Zuccaro and M. B. Husthouse, *J. Chem. Soc., Dalton Trans.*, 1624 (1979).

134. H. C. Clark, A. B. Goel and C. S. Wong, *J. Amer. Chem. Soc.*, **100**, 6241 (1978).
135. F. Igersheim and H. Mimoun, *Nouveau J. Chim.*, **4**, 711 (1980).
136. G. Gordon and H. Taube, *J. Inorg. Nucl. Chem.*, **16**, 268 (1961).
137. R. G. Bhattacharya, *J. Indian Chem. Soc.*, L3, 1166 (1976).
138. G. A. Olah and J. Welch, *J. Org. Chem.*, **43**, 2830 (1978).
139. N. W. Alcock, *J. Chem. Soc. (A)*, 1588 (1968).
140. F. Sakurai, H. Suzuki, Y. Moro-Oka and T. Ikawa, *J. Amer. Chem. Soc.*, **102**, 1749 (1980).
141. H. Suzuki, K. Mizutani, Y. Moro-Oka and T. Ikawa, *J. Amer. Chem. Soc.*, **101**, 748 (1979); *Chem. Letters*, 63 (1980).
142. R. A. Michelin, R. Ros and G. Strukul, *Inorg. Chim. Acta*, **37**, L491 (1979).
143. H. L. Roberts and W. R. Symes, *J. Chem. Soc. (A)*, 1450 (1968).
144. H. Mimoun, R. Charpentier, A. Mischler, J. Fischer and R. Weiss, *J. Amer. Chem. Soc.*, **102**, 1047 (1980).
145. J. M. Brégeault and H. Mimoun, *Nouveau J. Chim.*, **5**, 287 (1981).
146. B. Booth, R. Haszeldine and G. Neuss, *J. Chem. Soc., Chem. Commun.*, 1074 (1972); *J. Chem. Soc., Dalton Trans.*, 511 (1982).
147. A. Nishinaga, H. Tomita, K. Nishizawa, T. Matsuura, S. Ooi and K. Hirotsu, *J. Chem. Soc., Dalton Trans.*, 1504 (1981), and references therein.
148. H. Mimoun, R. Charpentier and M. Roussel, *German Patent, No.* 2,949,847 (to IFP) (1980); *Chem. Abstr.*, **93**, 185771 p (1980).
149. M. Roussel and H. Mimoun, *J. Org. Chem.*, **45**, 5387 (1980).
150. M. J. W. Chen and J. K. Kochi, *J. Chem. Soc., Chem. Commun.*, 204 (1977).
151. R. A. Sheldon in *Aspects of Homogeneous Catalysis*, Vol. 4 (Ed. R. Ugo), Reidel, Dordrecht, 1981, p. 3; *J. Mol. Catal.*, **7**, 107 (1980).
152. K. B. Sharpless and T. R. Verhoeven, *Aldrichim. Acta*, **12**, 63 (1979).
153. J. Sobczak and J. J. Ziolkowski, *J. Mol. Catal.*, **13**, 11 (1981).
154. R. Landau, G. A. Sullivan and D. Brown, *Chemtech.*, 602 (1979).
155. J. M. Le Carpentier, A. Mischler and R. Weiss, *Acta Cryst.*, **B28**, 1288 (1972).
156. C. Gianotti, C. Fontaine, A. Chiaroni and C. Riche, *J. Organomet. Chem.*, **113**, 57 (1976).
157. L. A. Lindblom, W. P. Shaefer and R. E. Marsh, *Acta Cryst.*, **B27**, 1461 (1971).
158. J. C. A. Boeyens and R. Haegele, *Inorg. Chim. Acta*, **20**, L7 (1976).
159. D. C. Bradley, I. C. Ghotra, F. A. Hart, M. B. Hursthouse and P. R. Raithby, *J. Chem. Soc., Dalton Trans.*, 1166 (1977).
160. T. Katsuki and K. B. Sharpless, *J. Amer. Chem. Soc.*, **102**, 5974 (1980).

The Chemistry of Functional Groups, Peroxides
Edited by S. Patai

CHAPTER **16**

Organic polyoxides

BOŽO PLESNIČAR
University of Ljubljana, Ljubljana, Yugoslavia

I. INTRODUCTION . . . 483
II. HYDROGEN POLYOXIDES . . . 484
III. DIALKYL POLYOXIDES . . . 486
A. Tetroxides . . . 486
1. Tertiary tetroxides . . . 486
2. Primary and secondary tetroxides . . . 490
B. Trioxides . . . 494
1. Di-*t*-butyl and dicumyl trioxides . . . 494
2. Bis(trifluoromethyl) trioxide . . . 496
3. Bis(pentafluorosulphur) trioxide . . . 497
4. Methyl *t*-butyl trioxide . . . 497
5. Miscellaneous trioxides . . . 498
C. Structure of Dialkyl Polyoxides . . . 499
IV. DIACYL TETROXIDES . . . 499
V. ALKYL HYDROTRIOXIDES . . . 501
A. Mechanism of Hydrotrioxide Formation . . . 501
B. Generation, Identification and Decomposition of Hydrotrioxides . . . 503
1. Ether hydrotrioxides . . . 503
2. Acetal hydrotrioxides . . . 507
3. Aldehyde hydrotrioxides . . . 509
4. Hydrocarbon and alcohol hydrotrioxides . . . 513
C. Structure of Hydrotrioxides . . . 515
VI. ACKNOWLEDGEMENT . . . 517
VII. REFERENCES . . . 517

I. INTRODUCTION

The present chapter deals with the formation, identification and decomposition of polyoxides, i.e. compounds with the general formula $R^1O_nR^2$, where R^1 and R^2 stand for hydrogen or other atoms or groups and $n \geqslant 3$. These compounds may be regarded as higher homologues of hydrogen peroxide, alkyl hydroperoxides, dialkyl peroxides and diacyl peroxides.

$$\begin{array}{ll} H-(O)_n-H & \text{Hydrogen polyoxides} \\ R-(O)_n-H & \text{Alkyl hydropolyoxides} \\ R-(O)_n-R & \text{Dialkyl polyoxides} \\ R-C(=O)-(O)_n-C(=O)-R & \text{Diacyl polyoxides} \end{array}$$

Considerable interest has been devoted in recent years to these classes of compounds. Namely, polyoxides and their radicals have been found to be key intermediates in low-temperature oxidation, atmospheric and stratospheric chemistry, combustion, flames and biochemical oxidative processes.

II. HYDROGEN POLYOXIDES

The hydrogen polyoxides, HO_nH, the higher homologues of hydrogen peroxide, were proposed as possible metastable intermediates long ago but it was only recently that direct spectroscopic evidence for their existence became available. Since the early history of these compounds has been well documented[1], only the main points on the way to their discovery will be mentioned here.

It appears that Berthelot[2] was the first to propose hydrogen trioxide as an intermediate in the decomposition of hydrogen peroxidé. Mendeleev[3] and Bakh[4] suggested that hydrogen tetroxide was formed in aging aqueous solutions of hydrogen peroxide. However, all these hypotheses were never confirmed.

Approximately 60 years ago it was discovered that water vapour could be dissociated in an electrical discharge. When the discharged vapour stream was quickly chilled to −190°C, a glassy solid was obtained. This material crystallized at about −115°C, and then began to melt with subsequent exothermic decomposition to produce oxygen and hydrogen peroxide. Much controversy developed to accommodate these phenomena. Ohara[5] observed that the decomposition of the glassy material yielded constant ratios of evolved oxygen to residual hydrogen peroxide. He suggested the reaction sequence (1)–(3) to take place at low temperatures.

$$H + O_2 \longrightarrow HO_2^{\bullet} \quad (1)$$

$$2HO_2^{\bullet} \longrightarrow H_2O_4 \quad (2)$$

$$H_2O_4 \longrightarrow H_2O_2 + O_2 \quad (3)$$

Nekrasov and coworkers[6] studied the low-temperature (−195°C) reaction of liquid ozone with atomic hydrogen. They obtained a glassy solid, similar to that reported by the above-mentioned authors. The infrared spectra of this material showed bands which they attributed to H_2O_3 and H_2O_4. Similar results were reported by Wojtowicz and coworkers[7].

Czapski and Bielsky[8,9] reported the evidence for the formation of H_2O_3 by exposing a flow of oxygen-saturated water to an intense electron beam.

After several unsuccessful attempts[10,11], Giguère and Herman[12] in 1970 succeeded in observing unambiguously for the first time fundamental 'skeletal' vibrations of D_2O_3 and

D_2O_4 molecules by studying infrared spectra of products obtained after the condensation of the effluent of deuterated water–hydrogen peroxide mixtures, which were subjected to electrical discharges. The absorption at $760\,cm^{-1}$ was assigned to D_2O_3 and the one at $820\,cm^{-1}$ to D_2O_4, respectively. Further confirmation of these bands was achieved by laser Raman spectroscopy[13].

Giguère and Herman[12] found that H_2O_4 begins to decompose at $-100°C$ and H_2O_3 at about $-55°C$. They reported an apparent first-order kinetic process for the decomposition of H_2O_3 ($k = 10^{-4.4}\,s^{-1}$) in the temperature range between -55 and $-45°C$ with an activation energy of about $15\,kcal\,mol^{-1}$. Recently, Nangia and Benson[14] suggested that decomposition of H_2O_3 is most likely a chain-reaction involving reactions (4)–(9).

$$H_2O_3 \rightleftharpoons HO_2^{\bullet} + {}^{\bullet}OH \tag{4}$$

$$^{\bullet}OH + H_2O_3 \longrightarrow (H_2O + HO_3^{\bullet}) \longrightarrow H_2O + {}^{\bullet}OH + O_2 \tag{5}$$

$$HO_2^{\bullet} + H_2O_3 \longrightarrow (H_2O_2 + HO_3^{\bullet}) \longrightarrow H_2O_2 + {}^{\bullet}OH + O_2 \tag{6}$$

$$HO^{\bullet} + H_2O_2 \longrightarrow H_2O + HO_2^{\bullet} \tag{7}$$

$$HO^{\bullet} + HO_2^{\bullet} \longrightarrow H_2O + O_2 \tag{8}$$

$$2HO_2^{\bullet} \longrightarrow H_2O_2 + O_2 \tag{9}$$

Little is known about the mechanism of decomposition of H_2O_4 except that it yields H_2O_2 and oxygen as the final products.

By applying the concept of group additivity scheme and thermochemical group properties, Benson[15] estimated the standard enthalpy of formation, ΔH_f^0, of -15.7 kcal and $+1.1$ kcal for H_2O_3 and H_2O_4, respectively. The estimated value for H_2O_5 is 17.9 kcal. These predictions indicate a rather great instability of H_2O_4 and only a moderate stability of H_2O_3. Thus, neither of these polyoxides can be prepared in the free state. It was suggested that both species are stabilized through intermolecular hydrogen bonds in a frozen matrix of water and hydrogen peroxide at temperatures below $-170°C$. Under these conditions, their enthalpy of formation from free radicals was estimated to be ca. -40 and $-25\,kcal\,mol^{-1}$, respectively, for H_2O_3 and H_2O_4[14].

Spectroscopic evidence for the existence of hydrogen polyoxides raises an interesting question of the possible existence of higher homologues in the series with $n \geqslant 5$. Although sulphur atoms are capable of forming stable chains of almost any length, it is obvious from the previous discussion that hydrogen pentoxide, H_2O_5, would be too unstable to exist.

It is obvious from the above discussion that no experimental data are available on the structure of hydrogen polyoxides. However, several *ab initio* molecular orbital studies, employing minimal[16–20] and extended basis sets[21,22], indicate a zig-zag skew chain structure for H_2O_3 and H_2O_4. The dihedral angle, Φ, which is of principal chemical interest in all peroxides, is lower in both compounds (H_2O_3, 94.5°[18], 91°[21], 78.1°[22]; H_2O_4, 91°[18]) than in H_2O_2 (111.5°). The variation of the dihedral angle in going from H_2O_2 to polyoxides most probably reflects (among other factors) the lowered electrostatic interaction between OH bonds with increasing number of oxygen atoms in the molecule.

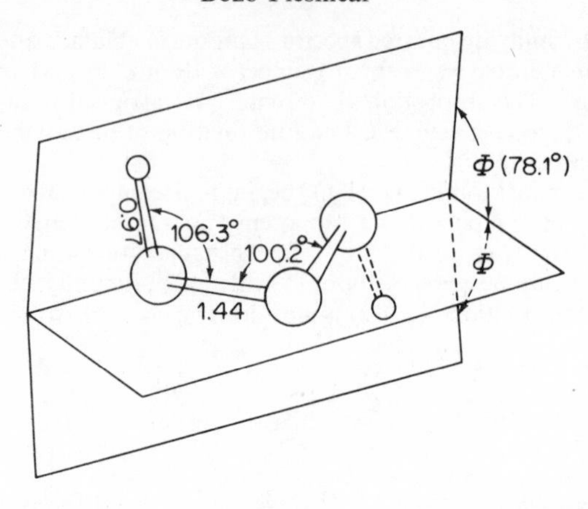

FIGURE 1. Geometry of hydrogen trioxide from an *ab initio* study[22]. Bond lengths are in Å.

Cremer has recently reported[22] (calculations at the level of Rayleigh–Schrödinger–Møller–Plesset perturbation theory) that simultaneous rotation of both OH bonds in H_2O_3 is hindered by relatively large barriers of 22.5 and 11.5 kcal mol^{-1}, and that the actual barriers (saddle points on the internal rotational surface) of this rotational process are 6.5 and 5.4 kcal mol^{-1}, respectively.

Although only limited theoretical data on the structure of H_2O_4 are available[18], it seems safe to predict that this compound also possesses a zig-zag skew chain structure, and that several conformations of similar energy are possible, among them some with intramolecular hydrogen bonds (*syn* forms):

Syn form *Anti* form

III. DIALKYL POLYOXIDES

The known role of dialkyl polyoxides, $R(O)_nR$, goes back to the kinetic analysis of autoxidation, which showed that kinetic chains are terminated by a bimolecular reaction between two alkylperoxy radicals to produce molecular products[23–31]. It was suggested that tetroxides, RO_4R, were formed as short-lived intermediates in these reactions[25].

A. Tetroxides

1. *Tertiary tetroxides*

Bartlett and Guaraldi[32] were the first to demonstrate the equilibrium between *t*-butylperoxy radicals and di-*t*-butyl tetroxide by the ESR technique (reactions 10 and 11). They generated *t*-butylperoxy radicals by irradiation of di-*t*-butyl peroxycarbonate in frozen methylene chloride at $-196°$ with 2540 Å light, or by the oxidation of *t*-butyl hydroperoxide with lead tetraacetate at $-90°C$, and observed a reversible change in their concentration below $-85°C$. They concluded that below this temperature the *t*-butylperoxy radicals were in equilibrium with di-*t*-butyl tetroxide.

$$ROO\cdot + ROO\cdot \rightleftharpoons [ROOOOR] \tag{10}$$

$$ROOOOR \longrightarrow \boxed{RO\cdot + O_2 + \cdot OR} \longrightarrow \begin{cases} ROOR + O_2 \\ 2RO\cdot + O_2 \end{cases} \tag{11}$$

The criterion of the temperature of oxygen evolution has also been used to characterize tetroxides. Milas and Plesničar[33,34] prepared di-*t*-butyl tetroxide by the reaction of *t*-butyl hydroperoxide with iodosobenzene or (diacetyloxy)iodobenzene in methylene chloride below $-80°C$. On warming to ca. $-75°C$, a strong evolution of oxygen was observed, which indicated the decomposition temperature of the tetroxide.

Di-*t*-butyl tetroxide has also been prepared by photolysis of azoisobutane, azoisobutyronitrile and azocyclohexylnitrile in the presence of oxygen[35].

Confirmation for the existence of tetroxides in equilibrium with *t*-butylperoxy radicals at low temperatures has been further provided independently by two groups, both using the ESR technique. Bennett and coworkers studied this equilibrium by photolysing *t*-butyl peroxide in oxygenated alkanes[36–38]. Adamic, Howard and Ingold[39–42] used photolysis of azo compounds in dichlorodifluoromethane and photolysis of hydroperoxides as source of the peroxy radicals. The variation of the equilibrium constant, K, with temperature can be described by an integrated form of the van't Hoff isochore:

$$\ln K = \frac{\Delta S^0}{R} - \frac{\Delta H^0}{RT}$$

where

$$K = \frac{[ROO\cdot]^2}{[ROOOOR]}$$

and ΔS^0 and ΔH^0 are the changes in the standard entropy and enthalpy, respectively. Combination of these two equations leads to the equation:

$$2\ln [ROO\cdot] - \ln [ROOOOR] = \frac{\Delta S^0}{R} - \frac{\Delta H^0}{RT}$$

Maximum peroxy radical concentration at complete dissociation below $-115°C$ where no irreversible decay occurs, is given by the expression:

$$[ROO\cdot]_{max} = 2\,[ROOOOR] + [ROO\cdot]$$

Combination of the last two equations gives:

$$2\ln [ROO\cdot] - \ln([ROO\cdot]_{max} - [ROO\cdot]) - \ln 2 = \frac{\Delta S^0}{R} - \frac{\Delta H^0}{RT}$$

The values of $[ROO\cdot]_{max}$, ΔH^0 and ΔS^0 in this equation were determined computationally as the best fit between experimental and predicted curves for the variation of $[ROO\cdot]$ with temperature. $[ROO\cdot]_{max}$ could not be determined experimentally since at temperatures where dissociation of the tetroxide was nearly complete, the irreversible decay of the peroxy radicals was appreciable. Values of K, ΔH^0 and ΔS^0 for several tertiary peroxy radical–tetroxide equilibria are summarized in Table 1. It is evident from Table 1 that changes in ΔH^0 and ΔS^0 show little dependence on the structure of *t*-peroxy radicals. The thermodynamic parameters for the equilibrium constant, K, for tertiary peroxy radicals can be described as $\Delta H^0 = -(8.8 \pm 1.0)\,\text{kcal mol}^{-1}$ and $\Delta S^0 = -(34 \pm 6)\,\text{cal deg}^{-1}\,\text{mol}^{-1}$ (standard state being 1 M). Similar values were reported for the equilibrium between the tertiary peroxy radicals of germanium (the trimethyl- and triphenyl-germylperoxy radicals), and the corresponding tetroxides, i.e. $\Delta H^0 = -11.5\,\text{kcal mol}^{-1}$ and $\Delta S^0 = -32\,\text{cal deg}^{-1}\,\text{mol}^{-1}$[43].

TABLE 1. Thermodynamic parameters for the equilibrium between tertiary alkylperoxy radicals and their corresponding tetroxides

Peroxy radical	$K \times 10^5$ (M) at $-120°C$	$-\Delta H^0$ (kcal mol^{-1})	$-\Delta S^0$ (cal deg^{-1} mol^{-1})	References
t-Butyl[a]	—	8.8 ± 0.4	34 ± 1	40
t-Butyl[b]	—	8.4 ± 0.4	30	40
t-Butyl[c]	—	8.0 ± 0.2	31 ± 1	40
Cumyl[d]	0.31	11.2 ± 0.9	48 ± 7	37
Cumyl[b]	—	10.6 ± 0.8	—	40
Cumyl[a]	—	9.2 ± 0.4	32 ± 1	40
2-Methylpentyl-2[d]	6.7	8.9 ± 0.7	39 ± 6	37
2,2,3-Trimethylbutyl-3[e]	7.7	8.7 ± 1.0	38 ± 7	37

[a] From irradiation of RN = NR and oxygen in CF_2Cl_2.
[b] From irradiation of RO_2H in isopentane.
[c] From irradiation of RO_2H in CF_2Cl_2.
[d] From irradiation of *t*-butyl peroxide in oxygenated 2-methylpentane ($t\text{-BuOOBu-}t \rightarrow 2\text{-}t\text{-BuO}\cdot$; $t\text{-BuO}\cdot + RH \rightarrow t\text{-BuOH} + R\cdot$; $R\cdot + O_2 \rightarrow ROO\cdot$; abstraction of tertiary H-atoms).
[e] From irradiation of *t*-butyl peroxide in oxygenated 2,2,3-trimethylbutane.

Thermodynamic parameters for tertiary tetroxide–peroxy radical equilibria can be used, together with kinetic constants for the termination reaction, to calculate activation parameters for irreversible tetroxide decomposition (Scheme 1). The rate of termination is

$$2ROO\cdot \underset{}{\overset{K}{\rightleftharpoons}} [ROOOOR] \xrightarrow{k_1} [RO\cdot + O_2 + \cdot OR] \xrightarrow{k_2} ROOR + O_2$$

$$[RO\cdot + O_2 + \cdot OR] \xrightarrow{k_3} 2RO\cdot + O_2 \xrightarrow{k_4} R\cdot + \text{ketone}$$

$$2RO\cdot + O_2 \xrightarrow[\text{or ROOH}]{1.\ RH\quad 2.\ O_2} 2ROO\cdot$$

SCHEME 1

given by:

$$\frac{d\,[ROO\cdot]}{dt} = 2fk_1\,[ROOOOR] = \frac{2fk_1}{K}\,[ROO\cdot]^2$$

where $f = k_2/(k_2 + k_3)$, the fraction of alkoxy radicals that escape the solvent cage. The measured rate constant, k_{ESR}, is equivalent to $2fk_1/K$, and

$$\ln k_{ESR} = \ln 2 + \ln f + \ln k_1 - \ln K$$

$$\ln A - \frac{E}{RT} = \ln 2 + \ln f - \frac{\Delta S^0}{R} + \frac{\Delta H^0}{RT} + \ln A_1 - \frac{E_a}{RT}$$

The A factors and activation parameters for irreversible decay of tertiary tetroxides (the k_1 process) are given in Table 2.

TABLE 2. Activation parameters for the irreversible decomposition of some tetroxides

Peroxy radical	E_a (kcal mol^{-1})	log A	References
t-Butyl	17.5	16.6	40
Cumyl	16.5	17.1	40
2,2,3-Trimethylbutyl-3	16.2 ± 2.0	17.5 ± 2.4	37
2-Methylpentyl-2	18.2 ± 1.7	19.6 ± 2.2	37

Rate constants for termination of tertiary peroxy radicals are in the range $10^{2.5}$–$10^4\,M^{-1}\,s^{-1}$ at room temperature (thus at a rate which is much less than the diffusion-controlled limit), and are sensitive to variation of radical structure and solvent. Reaction order ranges from pure first order to pure second order[44].

Although polar effects do not appear to play an important role in the self-reaction of peroxy radicals derived from ring-substituted cumenes and styrenes as well as α-monosubstituted toluenes, replacement of one of the methyl groups of *t*-butylperoxy by an electron-withdrawing or electron-donating group (CN, OR) causes an increase of the termination rate constant by a factor of 10^3–10^6, most probably as a result of decreased stabilities of the corresponding tetroxides.

Radicals which do not give thermodynamically stable products generally decay rather slowly. For example, *t*-butylperoxy radicals undergo a relatively slow self-reaction, leading to nonradical products ($2k_t = 1.2 \times 10^3\,M^{-1}\,s^{-1}$ at ambient temperature). This is a consequence of the low equilibrium concentration of the tetroxide dimer and the relatively high activation energy for its irreversible decomposition. Howard, Ingold and their coworkers[45–47] recommended the following parameters for the overall termination rate constants for *t*-butylperoxy radicals:

$$\log k_t\,(M^{-1}\,s^{-1}) = 8.9 - \frac{8.5}{2.303\,RT}$$

Recently, Nangia and Benson[48] after reanalysing and correcting the existing data on self-reaction of tertiary peroxy radicals recommended for termination (k_t) and nontermination

(k_{nt}) pathways the following values:

$$\log k_t\,(\mathrm{M}^{-1}\,\mathrm{s}^{-1}) = 7.1 - \frac{7.0}{2.303\,RT}$$

$$\log k_{nt}\,(\mathrm{M}^{-1}\,\mathrm{s}^{-1}) = 9.4 - \frac{9.0}{2.303\,RT}$$

which seem to be in better agreement with those reported by Fukuzumi and Ono[49].

The study of the isotopic composition of O_2 evolved during the self-reactions of t-Bu$^{18}O^{18}O$ and t-Bu$^{16}O^{16}O$ radicals, present in equilibrium concentration, showed that there is head-to-head collision[50], and not head-to-tail as suggested earlier by Thomas[51]. The relative amounts of $^{32}O_2$, $^{34}O_2$ and $^{36}O_2$ were found in amounts expected from statistical scrambling.

The suggestion that tertiary tetroxides decompose to form RO· and ROOO· radicals (reaction 12) was rejected on thermochemical grounds[48]. Namely,

$$\mathrm{ROOOOR} \rightarrow \mathrm{ROOO\cdot} + \mathrm{RO\cdot} \qquad (12)$$

the formation of ROOO· from RO· and O_2 is endothermic by ca. 15 kcal mol^{-1} (R = t-Bu, so that trioxide radical is not expected to be formed in this way at any temperature[47,48]. All attempts to detect t-BuOOO· spectroscopically (ESR, at −196°C) also failed[40].

2. *Primary and secondary tetroxides*

In a pioneering study reported in 1956, Russell[25] suggested that the self-reaction of secondary and primary peroxy radicals involves tetroxides which are formed rapidly and reversibly with subsequent decomposition to molecular products via a cyclic transition state (reaction 13). The formation of ketone, alcohol and oxygen together with the observation of a deuterium isotope effect ($k_H/k_D = 1.37 \pm 0.14$, at 30°C)[52] when the α hydrogen of the alkylperoxy radical is replaced by deuterium, would appear to exclude a mechanism analogous to the one proposed for the self-reaction of tertiary alkylperoxy radicals. An alternative mechanism (reaction 14), involving alkoxy radicals as intermediates, was also discounted on the basis of the stoichiometry of the reaction.

$$2\,-\overset{|}{\underset{\mathrm{H}}{\mathrm{C}}}\mathrm{OO\cdot} \underset{}{\overset{(1)}{\rightleftharpoons}} -\overset{|}{\underset{\mathrm{H}}{\mathrm{C}}}\mathrm{OOOO}\overset{|}{\underset{\mathrm{H}}{\mathrm{C}}}- \xrightarrow{(2)} \left[\text{cyclic transition state}\right]^{\ddagger} \longrightarrow \;\rangle\mathrm{C{=}O} + \mathrm{O_2} + \mathrm{HO}-\underset{\mathrm{H}}{\mathrm{C}}\langle \; + 99\ \mathrm{kcal\ mol^{-1}} \qquad (13)$$

$$2\,-\overset{|}{\underset{\mathrm{H}}{\mathrm{C}}}\mathrm{OO\cdot} \xrightarrow{(1)} -\overset{|}{\underset{\mathrm{H}}{\mathrm{C}}}\mathrm{OOOO}\overset{|}{\underset{\mathrm{H}}{\mathrm{C}}}- \xrightarrow{(2)} \underbrace{\left[\;\rangle\underset{\mathrm{H}}{\mathrm{C}}-\mathrm{O\cdot} + \mathrm{O_2} + \cdot\mathrm{O}-\underset{\mathrm{H}}{\mathrm{C}}\langle\;\right]}_{\text{cage}} \longrightarrow \;\rangle\underset{\mathrm{H}}{\mathrm{C}}-\mathrm{OH} + \mathrm{O_2} + \mathrm{O{=}C}\langle \qquad (14)$$

The Russell mechanism provided an attractive explanation for the observed low activation energy and rapid reaction of primary and secondary peroxy radicals. Namely, in contrast to the tertiary peroxy radicals, bimolecular self-reaction of primary and secondary peroxy radicals is a second-order reaction with the absolute values of $2k_t$ (the overall termination rate constant) in the range 10^6–10^8 $M^{-1}s^{-1}$ [37,49,53–56]. These values are about 10^3–10^5 higher than the $2k_t$ values for tertiary peroxy radicals.

The reported termination rate constants of secondary peroxy radicals are approximately 10^6–10^7 $M^{-1}s^{-1}$ at 30°C [37,52,53]. For example, the rate constant for the termination of *s*-heptylperoxy radicals in heptane (temperature range investigated, −60 to 0°C) is given by:

$$\log k_t\,(M^{-1}s^{-1}) = (7.7 \pm 1.0) - \frac{(1.9 \pm 3)}{2.303\,RT}$$

Activation energies of relatively low accuracy in the range 1–4 kcal mol^{-1} have been reported for the termination of some other secondary peroxy radicals[37,53].

The overall rate constant for the second-order decay of $CH_3OO\cdot$ radicals in the gas phase has recently been reported as 3×10^8 $M^{-1}s^{-1}$, suggesting that there is no significant solvent effect in these terminations[54–57]. It is interesting to note that a ratio of termination to nontermination steps of 2:1 was reported for these radicals. This ratio is in contrast with the ratio reported for the same two reactions of tertiary peroxy radicals, which is 1:10 at room temperature.

The observation of relatively high values for the rate constants for the cross-reactions of tertiary peroxy radicals and peroxy radicals containing α hydrogen is also indicative (*t*-BuOO· and MeOO·, 6×10^7 $M^{-1}s^{-1}$)[55].

An ESR spectroscopic study of the isotopic distribution of the oxygen evolved from the self-reaction of a number of primary and secondary alkylperoxy radicals ($R^{16}O^{16}O\cdot$ and $R^{18}O^{18}O\cdot$) in an inert solvent indicates unambiguously a head-to-head reaction[50].

The Wigner spin-conservation rule requires that oxygen evolved in the self-reaction of secondary peroxy radicals should be formed in an excited singlet state if this reaction proceeds via the tetroxide as an intermediate rather than merely a transition state. A small amount of oxygen evolved from secondary peroxy radicals is indeed trapped with 9,10-diphenylanthracene, i.e. a singlet oxygen acceptor[52]. On the other hand, no transannular peroxide can be detected in the self-reaction of *t*-butylperoxy radicals. The reaction is also sufficiently exothermic to produce an electronically excited ketone. The generation of both singlet states of oxygen, $^1\Delta_g$ and $^1\Sigma_g^+$, has been confirmed recently by spectroscopic methods[58].

The ESR spectroscopic evidence for the reversible formation of tetroxides from secondary peroxy radicals not withstanding, these species are too unstable to allow spectroscopical study of the variations of the equilibrium constant with temperature[53].

The observation that secondary peroxy radicals behave similarly to tertiary peroxy radicals in forming reversibly tetroxides below −100°C, has recently been taken as an evidence against the Russell mechanism. Benson[59] has argued that the Russell mechanism would require for step (2) an activation energy of $E_2 \leqslant 2$ kcal mol^{-1} to accommodate the observed rate parameters, and that with such a low activation energy the tetroxide would always react to form termination products rather than redissociate into free peroxy radicals (irreversible decomposition). Although the calculated activation energy is unusually low for a six-centred 1,5-hydrogen shift reaction, it cannot by itself rule out the cyclic mechanism because of the high exothermicity (60–70 kcal mol^{-1}, assuming that oxygen is formed completely in its first excited state) of the self-reaction of secondary peroxy radicals[60]. On the other hand, it has also been pointed out that such a low activation energy for irreversible tetroxide decomposition would lead to an apparent negative activation energy for the overall reaction[60]. However, this seems to be not in accord with the observed small but positive activation energies (1–4 kcal mol^{-1})[37,53].

All the above-mentioned mechanistic dilemmas, together with reports of various products formed in the decay of methylperoxy radicals (methanol, formaldehyde, formic acid[61]; methyl hydroperoxide, methanol, ozone[62]; methyl hydroperoxide, dimethyl peroxide[63]) led Benson to propose a new mechanism to account for the self-reaction of peroxy radicals containing α hydrogens[64a]. Scheme 2 shows the proposed reaction steps, where $R\overline{CHOO}$ is the zwitterion, first proposed by Criegee[64b]. Thermochemical calculations showed that all steps in the proposed mechanism are exothermic. Singlet oxygen (Δ^1O_2) found in these reactions can be accounted for by step (2), involving formation and decomposition of the $RCH_2OOO\cdot$ radical.

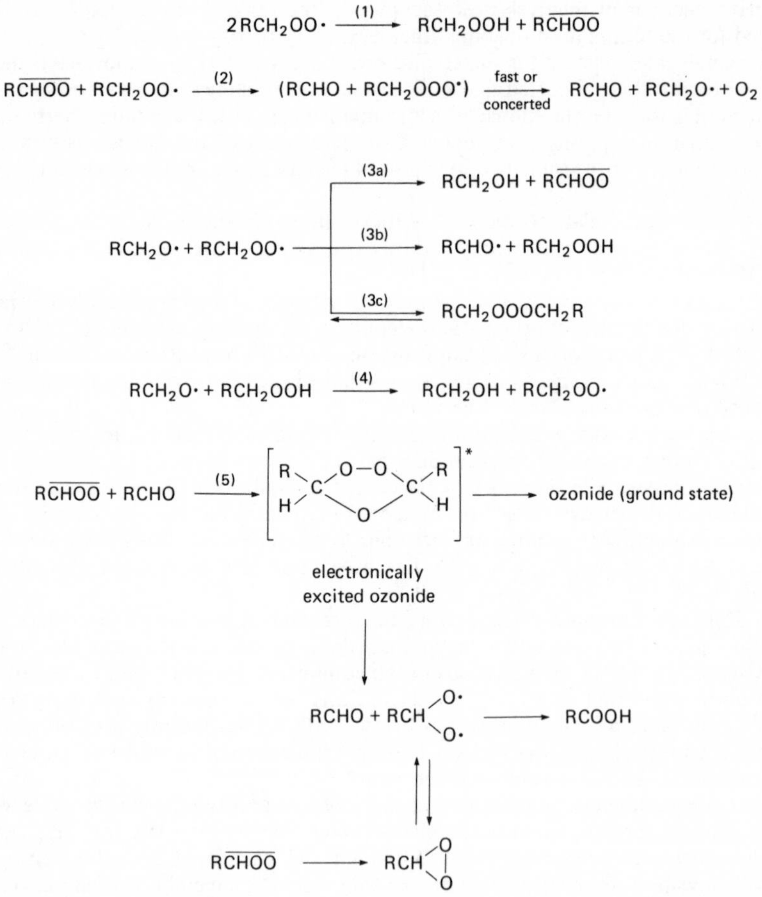

$$2RCH_2OO\cdot \xrightarrow{(1)} RCH_2OOH + R\overline{CHOO}$$

$$R\overline{CHOO} + RCH_2OO\cdot \xrightarrow{(2)} (RCHO + RCH_2OOO^{\cdot}) \xrightarrow[\text{concerted}]{\text{fast or}} RCHO + RCH_2O\cdot + O_2$$

$$RCH_2O\cdot + RCH_2OO\cdot \xrightarrow{(3a)} RCH_2OH + R\overline{CHOO}$$

$$RCH_2O\cdot + RCH_2OO\cdot \xrightarrow{(3b)} RCHO\cdot + RCH_2OOH$$

$$RCH_2O\cdot + RCH_2OO\cdot \underset{}{\overset{(3c)}{\rightleftharpoons}} RCH_2OOOCH_2R$$

$$RCH_2O\cdot + RCH_2OOH \xrightarrow{(4)} RCH_2OH + RCH_2OO\cdot$$

$$R\overline{CHOO} + RCHO \xrightarrow{(5)} [\text{RHC}(\text{O–O})(\text{O})\text{CHR}]^{*} \longrightarrow \text{ozonide (ground state)}$$

electronically excited ozonide

$$\downarrow$$

$$RCHO + RCH(O\cdot)(O\cdot) \longrightarrow RCOOH$$

$$\updownarrow$$

$$R\overline{CHOO} \longrightarrow RCH\langle\text{O–O}\rangle \text{ (dioxirane)}$$

SCHEME 2

Benson estimated that the formation of the zwitterion has a low activation energy of about 1–2 kcal mol^{-1} with an A factor of ca. 10^9. Trioxide is believed to be important at temperatures below 30°C with a short lifetime (< 10 s) at room temperature.

It appears that the Benson mechanism for the first time explains the presence of small amounts of products, such as carboxylic acids, in the termination reaction of primary and secondary peroxy radicals. Although the origin of these products is still not completely settled, it seems appropriate to reanalyse the data reported by Russell in terms of this newer mechanism.

Russell found that in α,α′-azobisisobutyronitrile-catalysed autoxidation of ethylbenzene one mole of acetophenone was formed and one mole of oxygen was present in nonperoxidic products for each pair of peroxy radicals destroyed in the termination process[25]. Under the cited autoxidation conditions (i.e. in the presence of RCH_2OOH) the Benson termination process would have to involve a rapid reaction in step (4), and steps (3a) and (3b) of this termination could be ignored. The Benson termination would thus become as shown in reactions (15)–(18). This process cannot be distinguished from the Russell mechanism on the basis of products, stoichiometry, oxygen-18 labelling or α-deuterium isotope effect. At low concentration of the hydroperoxide, RCH_2OOH, the $RCH_2O\cdot$ in the Benson mechanism would be expected to attack RCH_3 (an alkoxy radical is considerably more reactive than a peroxy radical). This would lead to the termination process (reactions 19–23), which is again indistinguishable from the previously mentioned processes.

$$2RCH_2OO\cdot \longrightarrow RCH_2OOH + R\overline{CHOO} \tag{15}$$

$$R\overline{CHOO} + RCH_2OO\cdot \longrightarrow (RCHO + RCH_2OOO\cdot) \longrightarrow RCHO + RCH_2O\cdot + O_2 \tag{16}$$

$$RCH_2O\cdot + RCH_2OOH \longrightarrow RCH_2OH + RCH_2OO\cdot \tag{17}$$

$$2RCH_2OO\cdot \longrightarrow RCHO + RCH_2OH + O_2 \tag{18}$$

$$2RCH_2OO\cdot \longrightarrow RCH_2OOH \quad R\overline{CHOO} \tag{19}$$

$$R\overline{CHOO} + RCH_2OO\cdot \longrightarrow RCHO + RCH_2O\cdot + O_2 \tag{20}$$

$$RCH_2O\cdot + RCH_3 \longrightarrow RCH_2OH + RCH_2^{\cdot} \tag{21}$$

$$RCH_2^{\cdot} + O_2 \longrightarrow RCH_2OO\cdot \tag{22}$$

$$2RCH_2OO\cdot + RCH_3 \longrightarrow RCHO + RCH_2O\cdot + RCH_2OOH \tag{23}$$

It is obvious from the above discussion that further work is needed to clarify the role of solvent polarity on the decay of primary and secondary peroxy radicals and to obtain accurate activation parameters of these processes. Also, detailed product studies in various solvents have yet to be carried out. In particular, studies in protic nucleophilic solvents, such as alcohols, would hopefully allow the transformation of the zwitterion intermediates (presumably involved in these reactions) into much more persistent α-oxyalkyl hydroperoxide.

B. Trioxides

1. Di-t-butyl and dicumyl trioxides

The history of the preparation of di-*t*-butyl trioxide has been rather devious. Milas and Djokic[65] studied the low-temperature reaction of ozone with potassium *t*-butyl peroxide and observed the formation of an oxygen-rich intermediate, which decomposed on warming with evolution of oxygen (> -30°C). This intermediate was believed to be di-*t*-butyl tetroxide. Bartlett and Günther[66] have later shown, on the basis of oxygen balance and oxygen evolution temperature, that this intermediate was in fact the corresponding trioxide.

Oxygen-evolution temperature has been used independently by Bartlett and Guaraldi[32] and by Milas and Plesničar[33,34] to characterize di-*t*-butyl tetroxide and trioxide present in a solution formed by oxidation of *t*-butyl hydroperoxide with lead tetraacetate, and (diacetyloxy)iodobenzene or iodosobenzene, respectively.

When present in large excess, *t*-butyl hydroperoxide is decomposed in a chain-reaction initiated by *t*-butoxy radicals present after the decomposition of the corresponding tetroxide (reactions 24–26; R = *t*-Bu). Some of the *t*-butoxy radicals are trapped by *t*-butylperoxy radicals at temperatures below -30°C to produce the trioxide. Trioxide can thus be regarded as the product of decomposition of tetroxide, or of oxidation of *t*-butyl hydroperoxide at temperatures between -75 and -30°C. A high yield of trioxide will be expected when *t*-butoxy and *t*-butylperoxy radicals are generated in close vicinity of each other, as in the case of the photolysis of diperoxymonocarbonates. Bartlett and Lahav[67] photolysed solid di-*t*-butyl diperoxymonocarbonate at -78°C (reaction 27) and obtained the solid trioxide in a yield of 20% after 20 h of irradiation. To minimize involatile impurities these authors used the reaction of ozone in Freon 12 or methyl chloride with either dissolved *t*-butyl hydroperoxide or suspended hydrated sodium *t*-butyl hydroperoxide. By using a special low-temperature procedure for freeing the obtained trioxide from accompanying *t*-butyl alcohol and other impurities, they obtained crystalline di-*t*-butyl trioxide. Dicumyl trioxide was also prepared by an analogous procedure.

$$\mathrm{ROOOOR} \longrightarrow 2\,\mathrm{RO}\cdot + \mathrm{O_2}; \quad \mathrm{ROOOOR} \longrightarrow \mathrm{ROOR} + \mathrm{O_2} \tag{24}$$

$$\mathrm{ROOH} + \mathrm{RO}\cdot \longrightarrow \mathrm{ROO}\cdot + \mathrm{ROH} \tag{25}$$

$$\mathrm{RO}\cdot + \mathrm{ROO}\cdot \underset{< -30^{\circ}\mathrm{C}}{\rightleftharpoons} \mathrm{ROOOR} \tag{26}$$

$$t\text{-BuOOC(=O)OOBu-}t \longrightarrow \mathrm{CO_2} + t\text{-BuOOOBu-}t \tag{27}$$

Both trioxides decompose with vigorous evolution of gas on warming in the solid state. In solution, these compounds showed the properties already observed previously[66]. Both trioxides are unstable at temperatures above -30°C. Spectral characteristics (NMR) and kinetic parameters for their decomposition are summarized in Table 3.

TABLE 3. Spectral (NMR) data of di-*t*-butyl and dicumyl trioxides, and kinetic and activation parameters for their decomposition

Trioxide	Solvent	δ (ppm)	T(°C)	k (s^{-1})	E_a (kcal mol^{-1})	$-\Delta S^{\neq}$ (cal deg^{-1} mol^{-1})	References
Di-*t*-butyl	CH_2Cl_2	1.25[a] (*t*-Bu)	−24.8	4.6×10^{-4}	23	20	63
			−33.3	7.2×10^{-5}			
Dicumyl	$CFCl_3$	1.67 (Me)	−25	0.93×10^{-4}	—	—	65
			−17.5	2.1×10^{-4}	24	—	
			−15	6.2×10^{-4}	—	—	
Dicumyl	CH_2Cl_2		−24.8	2.1×10^{-4}	—	—	65

[a]$CFCl_3$ at −70°C[35].

The kinetics of decomposition of di-*t*-butyl trioxide and dicumyl trioxide were measured originally by following the evolution of oxygen from solutions of either *t*-butyl or cumyl hydroperoxide and lead tetraacetate at approximately the same temperature range as that used in NMR spectroscopy. Although it is difficult to separate the decomposition of the trioxide from the chain decomposition of the hydroperoxides with alkoxy radicals by this method, it is nevertheless evident that comparable rate constants were obtained in both cases.

Nangia and Benson[14] have recently reanalysed the data of Bartlett and Günther by taking into account the fact that decomposition of both the trioxide and the hydroperoxide contributes to the rate of oxygen evolution. They have reported $E_a = 20.1$ kcal mol^{-1} for the decomposition of di-*t*-butyl trioxide, i.e. a value in acceptable agreement with that originally reported. For the preexponential factor, A, a value of $10^{13.2}$ was suggested.

2. *Bis(trifluoromethyl) trioxide*

Anderson and Fox[68] isolated CF_3OOOCF_3 in yields up to 84% from the reaction of F_2O with CF_2O over CsF catalyst. Thompson[69] independently prepared this trioxide in low yields (<5%) by the direct fluorination of salts of trifluoroacetic acid. The isolation of three other perfluorotrioxides has also been reported ($CF_3OOOC_2F_5$, $C_2F_5OOOC_2F_5$, $CF_3OOOCF_2OOCF_3$). The structure of these compounds has been determined by elemental analysis, molecular weights, ^{19}F-NMR spectra, IR and mass spectra.

Bis(trifluoromethyl) trioxide, having a melting point of −138°C and a normal boiling point of −16°C, is a surprisingly stable compound if one takes into account that FO_2F decomposes readily above −100°C.

The thermal decomposition yields primarily bis(perfluoromethyl) peroxide and oxygen in almost quantitative yields (reaction 28). The half-life for this reaction is approximately 65 weeks at 25°C. DesMarteau and coworkers[70,71] reported that decomposition of CF_3OOOCF_3 is first-order with an activation energy about 30 kcal mol^{-1}, in close agreement with the value proposed by Benson and Shaw[15b] (31 kcal mol^{-1}). The latter authors estimated the maximum value for $D(CF_3OO{-}OCF_3)$ of about 27 kcal mol^{-1}. The mechanism of decomposition is most probably the same as in other trioxides (reactions 29 and 30). As expected, perfluoroalkyl trioxides are oxidizing agents. They liberate iodine from acidified aqueous potassium iodide.

$$CF_3OOOCF_3 \longrightarrow CF_3OOCF_3 + \tfrac{1}{2}O_2 \qquad (28)$$

Bis(trifluoromethyl) trioxide was shown to be a convenient source of compounds containing $CF_3O{-}$ and $CF_3OO{-}$ groups. The syntheses of new compounds, $CF_3OOSO_2OCF_3$, *cis*-$CF_3OOSF_4OCF_3$ and $CF_3OOC(O)OCF_3$, and improved syntheses of previously reported compounds, were reported[71].

$$CF_3OOOCF_3 \longrightarrow CF_3OO\cdot + CF_3O\cdot \qquad (29)$$

$$2CF_3OO\cdot \longrightarrow 2CF_3O\cdot + O_2 \longrightarrow \begin{cases} CF_3OOCF_3 + O_2 \\ 2CF_3O\cdot + O_2 \end{cases} \qquad (30)$$

3. *Bis(pentafluorosulphur) trioxide*

The synthesis of SF_5OOOSF_5, still another relatively stable trioxide, has been reported by Czarnowski and Schumacher[72,73]. This inorganic trioxide is a colourless substance with a vapour pressure of 55.6 torr at 0°C. The thermal decomposition between 5 and 25°C yields in the presence of sufficiently high pressure of oxygen, bis(pentafluorosulphur) dioxide and oxygen as the only products (reaction 31). In the absence of oxygen and at higher temperatures, the decomposition of the trioxide according to equation (32) is favoured. The homogenous decomposition of the trioxide, in the presence of over 100 torr oxygen is strictly first order with respect to the trioxide pressure and independent of the total pressure, inert gases and the reaction products. The mechanism of decomposition (reactions 33–35) is believed to be parallel to that for *t*-BuOOOBu-*t*, except for the cage effect.

$$SF_5OOOSF_3 \longrightarrow SF_5OOSF_5 + 1/2O_2 \tag{31}$$

$$SF_5OOOSF_5 \longrightarrow SF_5OSF_5 + O_2 \tag{32}$$

$$SF_5OOOSF_5 \longrightarrow SF_5OO\cdot + SF_5O^{\cdot} \tag{33}$$

$$2SF_5OO\cdot \longrightarrow 2SF_5O\cdot + O_2 \tag{34}$$

$$2SF_5O\cdot \longrightarrow SF_5OOSF_5 \tag{35}$$

The following rate expression was reported for the decomposition:

$$\log k(\mathrm{s}^{-1}) = (16.06 \pm 0.37) - \frac{(26.0 \pm 0.5)}{2.303\,RT}$$

4. *Methyl* t-*butyl trioxide*

Recently, Brunton and coworkers[74] have reported ESR evidence for the formation of thermally unstable methyl *t*-butyl trioxide. UV photolysis of the solution of di-*t*-butyl peroxide in oxygenated cyclopropane or Freon 12 at ca. −130°C and below, produces no detectable amounts of free radicals. When irradiation stopped, a single-line ESR spectrum, assigned to the *t*-butylperoxy radical grows and in the absence of further photolysis reaches a steady intensity within a few minutes. Raising the temperature, after the signal has reached its maximum in the dark, from −130°C to ca. −100°C, causes a further increase in intensity, and after cooling back to −130°C, the intensity decreases to its original value. Brunton and coworkers have explained these observations by proposing the formation of methyl *t*-butyl trioxide during photolysis. This trioxide is unstable even at −130°C. The most plausible explanation for its formation is by the cross-termination of the methylperoxy radical with the *t*-butoxy radical. Both radicals are produced according to the reactions (36)–(39). The trioxide can decompose in two ways (reaction

$$t\text{-BuOOBu-}t \longrightarrow 2t\text{-BuO}\cdot \tag{36}$$

$$t\text{-BuO}\cdot \longrightarrow \text{Me}\cdot + \text{Me}_2\text{C}{=}\text{O} \tag{37}$$

$$\text{Me}\cdot + O_2 \longrightarrow \text{MeOO}^{\cdot} \tag{38}$$

$$MeOO\cdot + t\text{-}BuO\cdot \rightleftharpoons MeOOOBu\text{-}t \quad (39)$$

$$MeOOOBu\text{-}t \longrightarrow \begin{cases} MeOO\cdot + t\text{-}BuO\cdot \\ MeO\cdot + t\text{-}BuOO\cdot \end{cases} \quad (40)$$

40). The absence of an ESR signal of methylperoxy radical is not surprising since their self-reaction is several orders of magnitude faster than that for the *t*-butylperoxy radical. At the same time, thermochemical calculations predict that cleavage of the trioxide to produce *t*-butylperoxy and methoxy radicals is energetically more favourable, i.e. the bond dissociation energy $D(t\text{-BuOO—OMe}) = 20\,\text{kcal mol}^{-1}$ and $D(t\text{-BuO—OOMe}) = 21\,\text{kcal mol}^{-1}$.

The observation that MeOOOBu-*t* is thermally unstable even at very low temperatures is surprising. As mentioned before, di-*t*-butyl trioxide is stable below −30°C and perfluoration of both substituents increases the stability still further, since di(trifluoromethyl) trioxide is stable even at ambient temperature.

Evidence for the formation of other unsymmetrically substituted trioxides has also been obtained by the same authors by studying the photolysis of di-*t*-butyl peroxide in other oxygenated hydrocarbons. For instance, photolysis of di-*t*-butyl peroxide in oxygenated toluene below −83°C initially produces the ESR spectrum of the benzylperoxy radical but with continued photolysis the spectrum of the *t*-butylperoxy radical becomes dominant. This observation seems to indicate the formation of the dialkyl trioxide, $PhCH_2OOOBu\text{-}t$, which decomposes to the *t*-butylperoxy radical (reactions 41–43).

$$t\text{-}BuO\cdot + PhCH_3 \longrightarrow t\text{-}BuOH + PhCH_2^{\cdot} \quad (41)$$

$$PhCH_2^{\cdot} + O_2 \longrightarrow PhCH_2OO\cdot \quad (42)$$

$$PhCH_2OO\cdot + t\text{-}BuO\cdot \longrightarrow PhCH_2OOOBu\text{-}t \longrightarrow PhCH_2O\cdot + t\text{-}BuOO\cdot \quad (43)$$

5. *Miscellaneous trioxides*

There are some reports in the literature to indicate that trioxides might be also formed by the direct insertion of ozone into the C—C bond. The isolation of 8% acetone as the only product of the ozonation of neopentane has been explained by such a mechanism (reaction 44)[75]. The suggested dipolar intermediate, which collapses to the dialkyl trioxide, is analogous to the one already proposed in the reactions of protonated ozone with alkanes[76].

$$-\overset{|}{\underset{|}{C}}-\overset{|}{\underset{|}{C}}- + O_3 \longrightarrow \left[-\overset{|}{\underset{|}{C}}\cdots\overset{+}{O}-O-O^{-} \;\; \cdots C\!\!< \right] \longrightarrow -\overset{|}{\underset{|}{C}}OOO\overset{|}{\underset{|}{C}}- \quad (44)$$

The formation of a trioxide by direct insertion of ozone was also postulated to explain the cleavage of the C—C bond in bicyclo[*n*.1.0]alkanes[77] as well as for the formation of ketones from 3,7-dimethyloctyl acetate[78].

Trioxide intermediates have been proposed in the reaction of hydroperoxides with tetranitromethane. However, no direct proof for their involvement in these reactions has been given[79].

$$\mathrm{R{-}O{-}O{-}O{-}\underset{\displaystyle O^-}{\underset{|}{N}}{-}C(NO_2)_3}$$

C. Structure of Dialkyl Polyoxides

Very little of a definitive nature is known about the structure of dialkyl trioxides. The infrared spectrum of the most stable of trioxides, bis(trifluoromethyl) trioxide, suggests that the C—O and O—O bonds are normal single bonds.

An *ab initio* MO SCF study of dimethyl trioxide by using the minimal basis set (STO-3G) reveals a zig-zag skew chain structure[20]. A staggered conformation has been found to have the lowest energy. The dihedral angle, Φ, is lower than that in dimethyl peroxide in the gas phase ($\sim 120°$) reflecting the same trend of closing of this bond angle as observed in going from hydrogen peroxide to hydrogen trioxide.

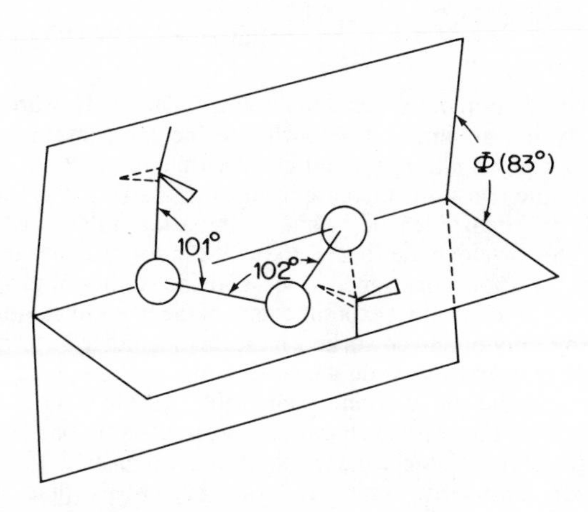

FIGURE 2. Geometry of dimethyl trioxide from an *ab initio* study[21].

There are no reports in the literature on the structure of tetroxides. It seems rather surprising that bis(trifluoromethyl) tetroxide has not yet been prepared in view of the relatively great stability of the corresponding trioxide. At present, a theoretical study is needed to specify the three dihedral angles in a chain of six atoms in simple tetroxides.

IV. DIACYL TETROXIDES

Although no direct spectroscopic evidence has been given till now for the existence of acyl tetroxides, there is strong indication that such intermediates are formed in the autoxidation of various aliphatic and aromatic aldehydes[80–84]. In the liquid phase, a

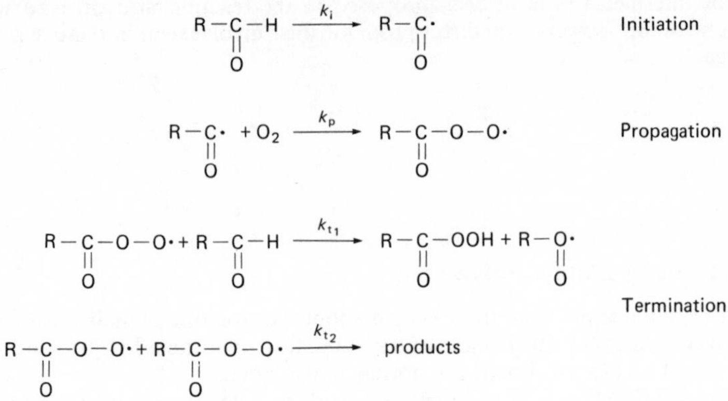

$$\text{R-C(=O)-H} \xrightarrow{k_i} \text{R-C(=O)}\cdot \qquad \text{Initiation}$$

$$\text{R-C(=O)}\cdot + O_2 \xrightarrow{k_p} \text{R-C(=O)-O-O}\cdot \qquad \text{Propagation}$$

$$\text{R-C(=O)-O-O}\cdot + \text{R-C(=O)-H} \xrightarrow{k_{t_1}} \text{R-C(=O)-OOH} + \text{R-C(=O)}\cdot$$

$$\text{R-C(=O)-O-O}\cdot + \text{R-C(=O)-O-O}\cdot \xrightarrow{k_{t_2}} \text{products} \qquad \text{Termination}$$

SCHEME 3

chain-reaction leads first to a peroxy acid according to the Scheme 3. The initiated autoxidation of aldehydes follows the kinetic rate law:

$$\frac{d[O_2]}{dt} = k_p\left(\frac{k_i}{2k_t}\right)^{1/2}[\text{Initiator}]^{1/2}[\text{RCHO}]$$

if a sufficient concentration of oxygen is maintained. Zaikov, Howard and Ingold[80] found that aldehydes oxidize at similar rates owing to the compensating changes in the rate constants for chain propagation, k_p, and chain termination, $2k_t$ $[k_p/(2k_t)^{1/2} = \text{constant}]$. The propagation rate constants increase from ca. $1 \times 10^3\,\text{M}^{-1}\,\text{s}^{-1}$ for pivalaldehyde to $1 \times 10^4\,\text{M}^{-1}\,\text{s}^{-1}$ for benzaldehyde, while termination rate constants increase from $7 \times 10^6\,\text{M}^{-1}\,\text{s}^{-1}$ for pivaldehyde to $2 \times 10^9\,\text{M}^{-1}\,\text{s}^{-1}$ for cyclohexanecarboxaldehyde.

The results of the above-mentioned study and the studies of products[81] and carbon dioxide evolution[82], and labelling experiments[83] of the free-radical-initiated autoxidation of acetaldehyde by Traylor and coworkers indicate that aldehyde termination is preceded by the formation of acetyl tetroxide which decomposes completely to methyl radicals, carbon dioxide and oxygen without appreciable cage collapse. A nearly concerted cleavage of all bonds in the tetroxide has been suggested on the basis of its very exothermic decomposition (reaction 45). Since the termination rate constant for benzaldehyde is close to the diffusion-controlled limit, it was suggested that benzoyl tetroxide, $PhC(O)O_4C(O)Ph$ is formed irreversibly and that the majority of benzoyloxy radicals combine in the cage[80]. The slower termination of acylperoxy radicals may arise either from the rapid decarboxylation of the acyloxy radicals or the reversible formation of the acyl tetroxide, which decomposes more slowly than aroyl tetroxide.

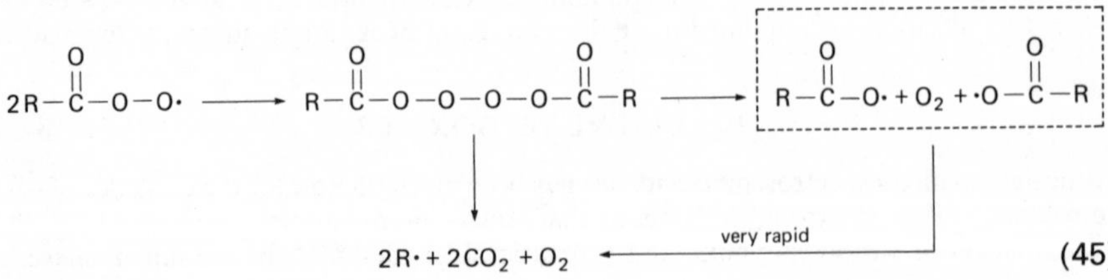

$$2\text{R-C(=O)-O-O}\cdot \longrightarrow \text{R-C(=O)-O-O-O-O-C(=O)-R} \longrightarrow [\text{R-C(=O)-O}\cdot + O_2 + \cdot\text{O-C(=O)-R}] \xrightarrow{\text{very rapid}} 2\text{R}\cdot + 2CO_2 + O_2$$

$$\text{R-C(=O)-O-O-O-O-C(=O)-R} \longrightarrow 2\text{R}\cdot + 2CO_2 + O_2 \qquad (45)$$

The induced decomposition of peroxyacetic acid with *t*-butoxy radicals in acetic acid[84] has also been shown to involve the formation of acetyl tetroxide as a result of nonterminating interaction of acetylperoxy radicals (reactions 46 and 47). No evidence for the decomposition of the tetroxide to acetylperoxy radicals has been found in this case either.

$$\mathrm{Me{-}C({=}O){-}OOH} + t\text{-}\mathrm{BuO\cdot} \longrightarrow \mathrm{Me{-}C({=}O){-}OO\cdot} + t\text{-}\mathrm{BuOH} \qquad (46)$$

$$\mathrm{2Me{-}C({=}O){-}OO\cdot} \longrightarrow \mathrm{Me{-}C({=}O){-}O{-}O{-}O{-}O{-}C({=}O){-}Me} \longrightarrow \mathrm{2Me\cdot} + 2CO_2 + O_2 \qquad (47)$$

V. ALKYL HYDROTRIOXIDES

Alkyl hydrotrioxides, ROOOH, have been proposed as unstable intermediates in the low-temperature ozonation of various saturated organic compounds, i.e. ethers[85,86], aldehydes[87,88], hydrocarbons[89,90], alcohols[89], silanes[91], amines[92–94], diazo compounds[95] and acetals[96–99,120]. Nevertheless, it is only recently that direct spectroscopic evidence for their existence has become available[100–103]. Much remains to be learned about the chemistry of this class of compounds, but already several interesting observations have been reported.

A. Mechanism of Hydrotrioxide Formation

Several mechanisms can be envisaged for the oxidation of the C—H bond of organic saturated substrates to form the corresponding hydrotrioxides. These are outlined in Scheme 4. Most of the presently available evidence indicates that either dipolar insertion or hydride ion transfer are probably involved in the formation of hydrotrioxides.

Taillefer, Fliszar and their coworkers[104] reported the results of a systematic study of the reaction of ozone with acetals. They found that the stoichiometry of the reaction is 1:1 in each reactant and that the reaction is first order in acetal and first order in ozone. A Hammett ρ value between -1.10 and -1.58 was found for these reactions. The solvent polarity has little effect on the rate, which would tend to eliminate highly polar transition states. Relatively large and negative entropies of activation indicate a high degree of orientation in the transition state. On the basis of these results, a 1,3-dipolar insertion mechanism was suggested[120].

Taillefer and coworkers[105] also demonstrated that an isokinetic relationship exists for the ozonation of acyclic acetals of heptaldehyde with the isokinetic temperature below the experimental temperature range, i.e. in a domain of temperatures where entropy factors control the reactivity. In cyclic acetals of the same aldehyde the isokinetic temperature is above the experimental temperatures, i.e. where the reactivity depends mainly on enthalpy factors. These results were interpreted in terms of conformational changes before ozonation in acyclic acetals, supporting the previous findings of Deslongchamps and coworkers[96–99]. In an important series of papers these authors demonstrated that in order for ozonation of acetals to proceed, one nonbonded electron pair on each oxygen atom has to be oriented antiperiplanar to the C—H bond of the acetal function.

Recently, Nangia and Benson[106] argued on thermochemical grounds that the concerted insertion of ozone into the C—H bond would require an activation energy of about 20–26 kcal mol^{-1} since it involves a five-membered transition state with a pentavalent

Proton transfer

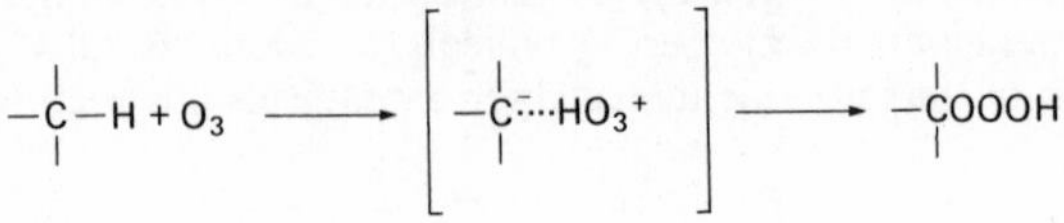

$$-\overset{|}{\underset{|}{C}}-H + O_3 \longrightarrow \left[-\overset{|}{\underset{|}{C}}^{-}\cdots HO_3^{+}\right] \longrightarrow -\overset{|}{\underset{|}{C}}OOOH$$

Hydride ion transfer

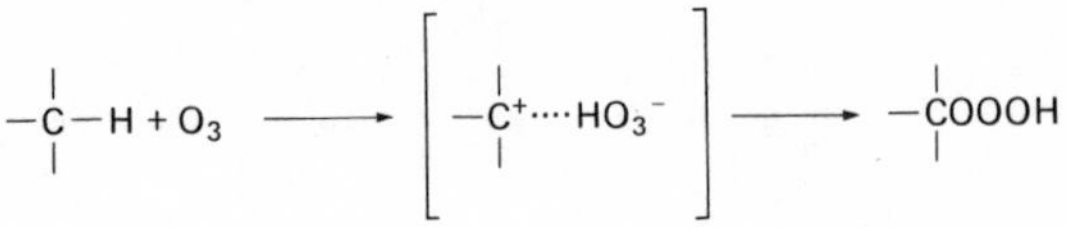

$$-\overset{|}{\underset{|}{C}}-H + O_3 \longrightarrow \left[-\overset{|}{\underset{|}{C}}^{+}\cdots HO_3^{-}\right] \longrightarrow -\overset{|}{\underset{|}{C}}OOOH$$

Electron transfer

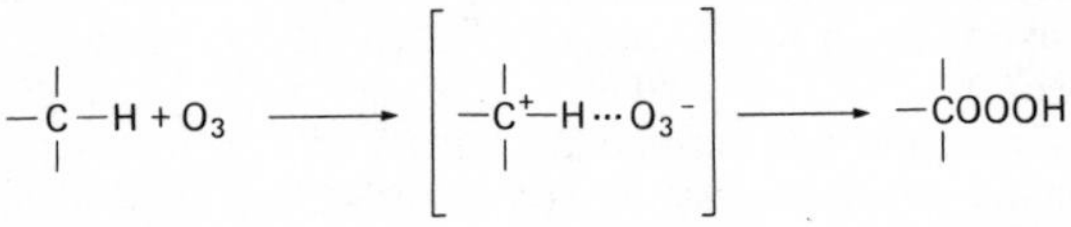

$$-\overset{|}{\underset{|}{C}}-H + O_3 \longrightarrow \left[-\overset{|}{\underset{|}{C}}^{+}-H\cdots O_3^{-}\right] \longrightarrow -\overset{|}{\underset{|}{C}}OOOH$$

1,3-*Dipolar insertion*

$$-\overset{|}{\underset{|}{C}}-H + O_3 \longrightarrow \left[\text{cyclic C}\cdots O-O-O\cdots H\text{ transition state}\right]^{\ddagger} \longrightarrow -\overset{|}{\underset{|}{C}}OOOH$$

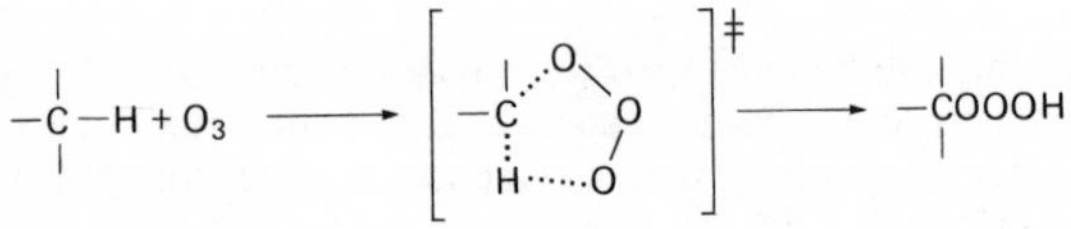

Radical H-atom abstraction

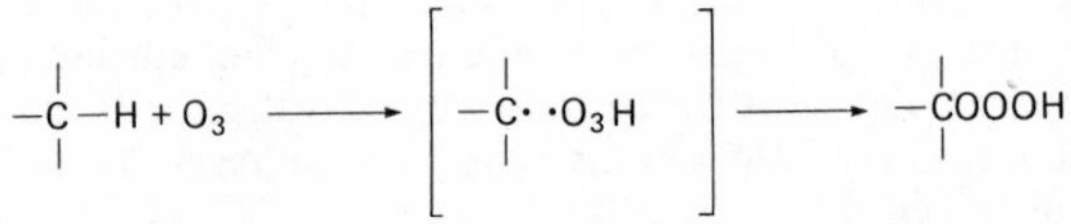

$$-\overset{|}{\underset{|}{C}}-H + O_3 \longrightarrow \left[-\overset{|}{\underset{|}{C}}\cdot\ \cdot O_3H\right] \longrightarrow -\overset{|}{\underset{|}{C}}OOOH$$

SCHEME 4

carbon. Relatively low activation energies were indeed found in the above-mentioned study of acetal ozonation (5–9 kcal mol^{-1})[105]. The hydrogen atom abstraction was also shown to be too endothermic at -78°C to proceed with a measurable rate except for R—H bonds strengths less than 80 kcal mol^{-1}. On the basis of these conclusions, Nangia

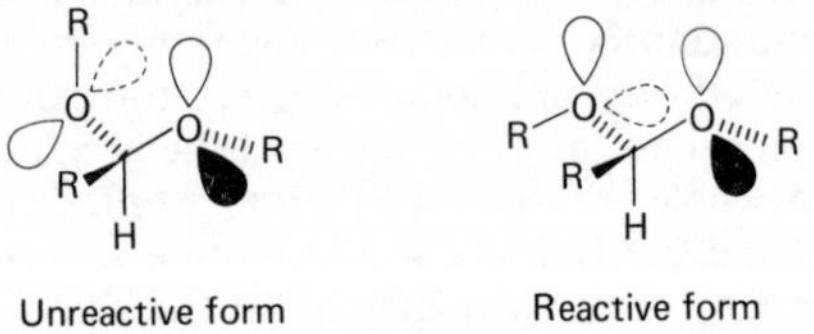

Unreactive form

Reactive form

and Benson suggested the hydride ion transfer as the mechanism of ozonation saturated organic compounds. The ion pair which is formed in a solvent cage collapses to the hydrotrioxide in an exothermic process.

Olah and coworkers[107] proposed on the basis of product analysis of the reaction of ozone with a series of alkanes in superacid (FSO_3H–SbF_5–SO_2ClF) solution at $-78°C$ that the reaction pathway involves electrophilic attack by protonated ozone (ozonium ion, H^+O_3) into the C—H and C—C bonds through two-electron three-centre-bonded pentacoordinated carbonium ions[108]. These could yield, after the loss of proton, the neutral hydrotrioxide or cleave in other ways as shown in Scheme 5.

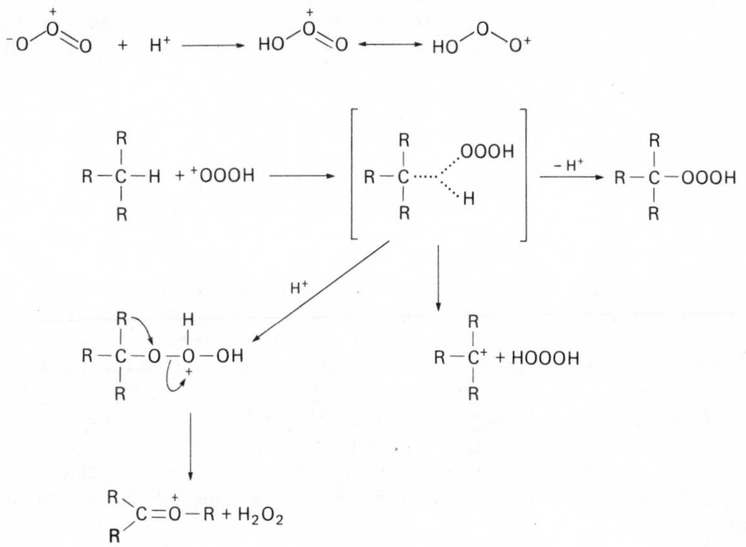

$$^-O-\overset{+}{O}=O + H^+ \longrightarrow HO-\overset{+}{O}=O \longleftrightarrow HO-O-O^+$$

$$R_3C-H + {}^+OOOH \longrightarrow [R_3C\cdots(OOOH)(H)]^+ \xrightarrow{-H^+} R_3C-OOOH$$

$$[R_3C\cdots(OOOH)(H)]^+ \longrightarrow R_3C^+ + HOOOH$$

$$[R_3C\cdots(OOOH)(H)]^+ \xrightarrow{H^+} R_3C-O-\overset{+}{O}(H)-OH \longrightarrow R_2C=\overset{+}{O}-R + H_2O_2$$

SCHEME 5

B. Generation, Identification and Decomposition of Hydrotrioxides

1. Ether hydrotrioxides

The idea of hydrotrioxides being reactive intermediates in the low-temperature ozonation of ethers has been around for quite a while. Price and Tumolo[85] demonstrated that ethers are attacked by ozone only at C—H bonds which are α to the ether oxygen, and that the less acidic these are, the faster the reaction. They proposed an ozone insertion mechanism for the formation of hydrotrioxide derivatives.

Erickson and coworkers[86] observed evolution of gas when isopropyl ether was ozonized at $-78°C$ and the resulting solutions were allowed to warm up to ambient temperature. These authors suggested that the hydrotrioxide is stable up to ca. 0°C.

Murray and coworkers[101] studied the low-temperature ozonation of isopropyl ether, 2-methyltetrahydrofuran and methyl isopropyl ether. The NMR spectral analysis of ozonized ethers showed the presence of a greatly deshielded absorption at ca. δ 13 ppm (downfield from Me_4Si). The chemical shift of these absorptions changed very little with dilution; thus, it was tentatively assigned to the intramolecularly hydrogen-bonded hydrotrioxides. Decomposition of these ether hydrotrioxides was investigated by following the decay of OOOH absorptions and was found to obey first-order kinetics.

Decomposition of ether hydrotrioxides gave singlet oxygen, which was characterized by its reaction with typical singlet oxygen acceptors. For example, 80% of the absorbed ozone was available to react with 1,3-diphenylisobenzofuran in the case of 2-methyltetrahydrofuran and 30% in the case of methyl isopropyl ether. Although no detailed analysis of other decomposition products was reported, it was suggested on the basis of kinetics, the singlet oxygen determination, and the NMR spectroscopic confirmation of acetone among decomposition products of methyl isopropyl ether hydrotrioxide that decomposition of this oxygen-rich intermediate (as well as other hydrotrioxides investigated) involves a concerted process with a cyclic six-membered ring transition state (reaction 48). Nevertheless, the overall efficiency of singlet oxygen formation suggested that other reactions also occur.

$$\text{R(Me)}_2\text{C–O–O–O–H} \longrightarrow [\ldots]^{\ddagger} \longrightarrow \text{ROH} + \text{Me}_2\text{C=O} + \Delta\ {}^1\text{O}_2 \qquad (48)$$

Bailey and Lerdal[109] found evidence for the formation of hydrotrioxides in the low-temperature ozonation of ethyl isopropyl ether in Freon 11. They observed gas evolution (presumably oxygen) when the ozonized solution was allowed to warm up. The only products isolated were acetone, ethanol, ethyl acetate, isopropyl formate and isopropyl acetate. On the basis of quantitative analysis of products they concluded that the major ozone attack occurred at the isopropyl rather than ethyl group (i-Pr/Et = 7.0 at −78°C), and that the 1,3-dipolar insertion is a pathway that leads to the formation of the hydrotrioxide in this case. Concerted decomposition of the hydrotrioxide mentioned above, and other radical pathways shown in Scheme 6 were suggested to accommodate the obtained products.

The results of ozonation of 4-oxa-2-heptanone at 0°C are just the opposite to that obtained with ethyl isopropyl ether. Bailey and Lerdal suggested the internal oxidation mechanism (Scheme 7) for the formation of the hydrotrioxide (or hemiacetal) derivative. Namely, ozone attacks more strongly at the acetomethyl group than at the propyl group as indicated by the formation of propyl formate, 1-propanol, propyl piruvate, acetomethyl formate and acetomethyl propionate (for the explanation of the formation of these products, see Scheme 6).

It is interesting to mention that ozonation of 4-oxy-2-heptanone at −78°C yields as the sole product propyl-2-hydroperoxy-2-hydroxypropanoate. This observation seems to indicate that the major intermediate in the low-temperature ozonation of this ether is the hydrotrioxide, presumably formed according to Scheme 8.

Low-temperature NMR evidence has recently been obtained for the formation of hydrotrioxides in the ozonation of α-methylbenzyl alkyl ethers at −78°C in various solvents (reaction 49)[110,111]. Typical OOOH absorption occurs at ca δ 13 ppm, which is similar to the OOOH absorptions reported for other hydrotrioxides. Since it does not change much with dilution, it was tentatively assigned to the intramolecularly hydrogen-bonded form of the hydrotrioxide.

Kinetic and activation parameters of the first-order decay of α-methylbenzyl alkyl ether hydrotrioxides are summarized in Table 4. Electron-withdrawing substituents accelerate the rate of decomposition of these hydrotrioxides, which decompose in the temperature range −40 to 0°C to produce, among other products, singlet oxygen. For instance,

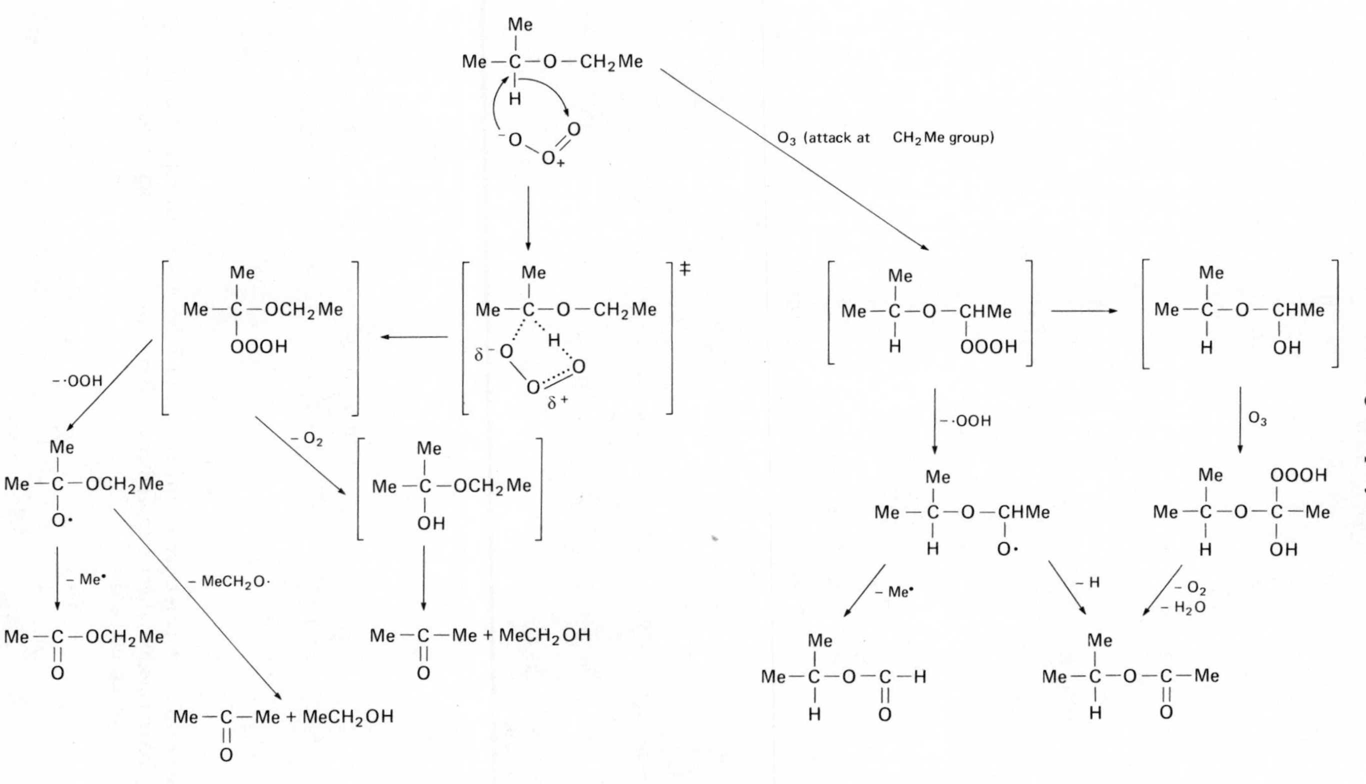
O3 (attack at CH2Me group)
−·OOH
−O2
−Me•
−MeCH2O·
O3
−H
−O2
−H2O
Me—C—Me + MeCH2OH

SCHEME 6

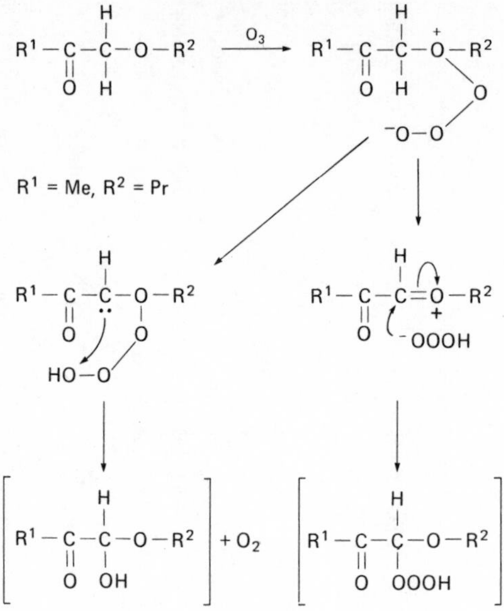

$R^1 = Me$, $R^2 = Pr$

SCHEME 7

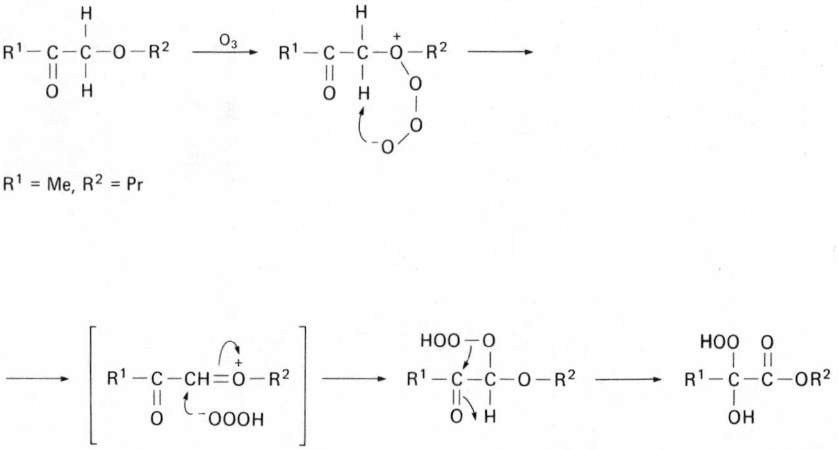

$R^1 = Me$, $R^2 = Pr$

SCHEME 8

50–60% of absorbed ozone is available to react with 1,3-diphenylisobenzofuran in the case of α-methylbenzyl methyl ether (diethyl ether as solvent). In all cases, a vigorous evolution of gas is observed around −10°C.

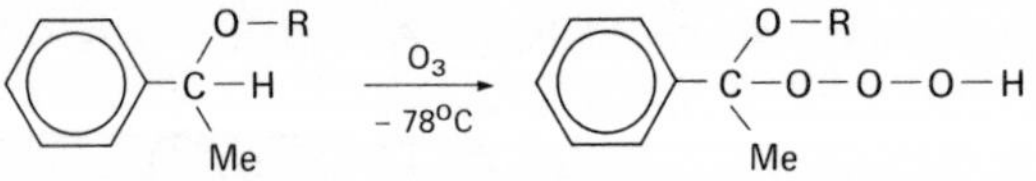

$$PhCH(OR)Me \xrightarrow[-78^\circ C]{O_3} PhC(OR)(Me)-O-O-O-H \quad (49)$$

TABLE 4. Activation parameters for the decomposition of ether hydrotrioxides

Hydrotrioxide	Solvent	E_a (kcal mol^{-1})	log A	References
Me OMe C Me OOOH	Neat	16.6 ± 0.6	11.5 ± 0.5	101
Me O OOOH	2-MeTHF	8.04 ± 0.2	4.1 ± 0.3	101
	Et_2O[a]	17.4 ± 0.4	11.7 ± 0.4	101
OR X—C—OOOH Me				
X = H, R = Me	$MeCO_2Et$[b]	15.2	11.4	110
	Et_2O[b]	17.7	14.9	110
X = Cl, R = Me	Et_2O[b]	16.7	14.1	110
X = Br, R = Me	Et_2O[b]	16.9	14.1	110
X = Me, R = Me	Et_2O[b]	18.1	16.0	110
X = Br; R = Et	Et_2O[b]	17.6	14.7	110

[a]2-MeTHF, 15.4%; Et_2O, 84.6%.
[b]α-Methylbenzyl alkyl ether, 25%; solvent 75%.

The decomposition of α-methylbenzyl methyl ether hydrotrioxide yields acetophenone, methanol and methyl benzoate as the main nongaseous products (ethyl acetate or Freons as solvents, VPC analysis) and relatively smaller amounts of water and hydrogen peroxide. The NMR spectroscopic (δOOH ~ 9 ppm) and chemical evidence has also been given for the presence of a hydroperoxide in the decomposition mixture although this peroxide could not be isolated.

Although acetophenone, methanol and singlet oxygen could be accommodated by a pericyclic mechanism already proposed by Murray and coworkers[101], the presence of the other decomposition products indicates the importance of alternative homolytic decomposition pathways as shown in Scheme 9. It is interesting to note that singlet oxygen can be formed in, at least, two homolytic processes presumably involved in the decomposition of these hydrotrioxides. The first is the cage disproportionation of the corresponding alkoxy and hydroperoxy radicals as already proposed by Benson[106]; the second is the decomposition of the ROOO· to RO· and O_2 (Δ^1O_2 and/or Σ^3O_2).

2. *Acetal hydrotrioxides*

In 1971, Deslongchamps and Moreau[96] first demonstrated that aldehyde acetals react with ozone to produce esters in high yields. Deslongchamps and coworkers also showed (see Section V.A.) that for the reaction to proceed, one nonbonded electron pair on each oxygen atom of the acetal has to lie antiperiplanar with respect to the C—H bond of the acetal functional group[97–99]. It was also assumed that the reaction proceeds via a

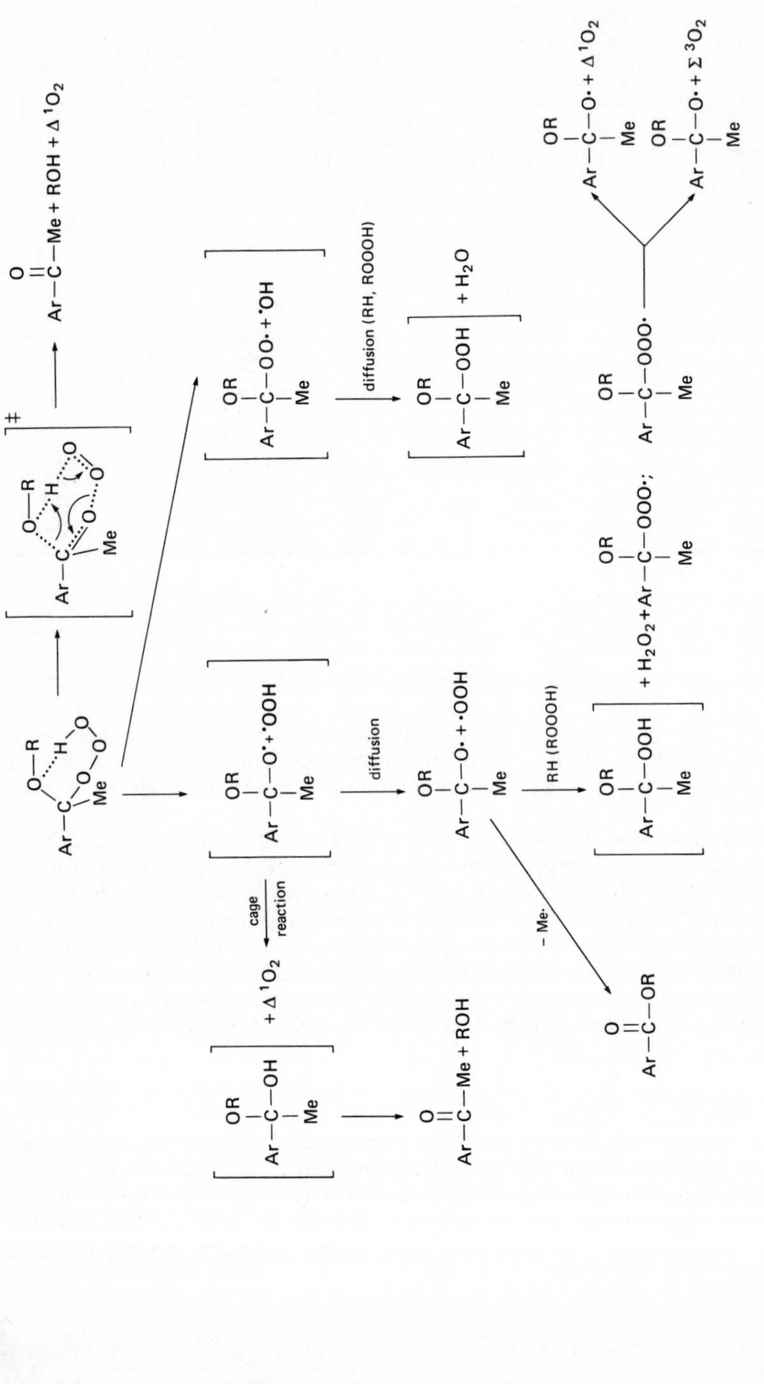

SCHEME 9

hydrotrioxide intermediate (1,3-dipolar insertion mechanism) which then breaks down to the corresponding ester and alcohol in a heterolytic process (reaction 50).

$$R^1-CH(OR^2)_2 + O_3 \longrightarrow R^1-C(OR^2)_2-OOOH \longrightarrow R^1-C(=O)-OR^2 + R^2OH + O_2 \qquad (50)$$

The intermediacy of hydrotrioxides in these reactions has recently been verified by observing the characteristic OOOH absorptions at δ 13 ppm[102,103]. The decomposition of these intermediates was investigated by following the decay of OOOH absorptions in several solvents and was found to obey first-order kinetics. The activation parameters for the decomposition of some representative examples are collected in Table 5. It is seen from Table 5 that the effect of solvents on the decomposition is rather low. Although only a limited number of solvents were suitable for the decomposition studies, it is evident that comparable rate constants for decomposition were obtained in the parent acetal, or in diethyl ether as well as in methylene chloride as a solvent.

Electron-withdrawing groups in the hydrotrioxide derivatives of benzaldehyde dimethyl acetals accelerate decomposition while electron-releasing groups retard it. A Hammett ρ value of 1.2 ± 0.2 was obtained in diethyl ether. A similar trend of the substituent effect on decomposition of substituted derivatives of cyclic acetals investigated has been observed.

Acetal hydrotrioxides investigated decompose in the temperature range -45 to -10°C to produce singlet oxygen as determined by singlet oxygen acceptors (Table 6).

The decomposition of the hydrotrioxide derivative of acetaldehyde diethyl acetal yields ethyl acetate (0.80–0.85 mol/mol of ozone absorbed), ethanol (0.75–0.80 mol), water, acetaldehyde, acetic acid, ethyl formate, diethyl carbonate (0.06–0.12 mol) and a mixture of gases (presumably methane and ethane). α-Hydroperoxydiethyl ether and α,α-diethoxydiethyl peroxide as well as hydrogen peroxide were also found among the decomposition products (organic peroxides, 5–8% of all products). The same types of peroxides were also isolated from decomposition mixtures of aromatic acyclic acetal hydrotrioxides.

On the basis of the above-mentioned results, Kovač and Plesničar[103] concluded that the formation of a large part of ethyl acetate, ethanol and singlet oxygen resulted from a nonradical ('pericyclic') process while the presence of the other decomposition products indicates alternate free-radical processes. Among the latter, an induced decomposition of the hydrotrioxide to produce the alkyltrioxy radical, ROOO•, plays an important role. It is thus obvious that the E_a and $\log A$ values reported in Table 5 are deduced from the observed rate of decomposition which is proceeding by several simultaneous first-order pathways as shown in Scheme 10.

Nangia and Benson[106] have recently reanalysed the kinetic data for the decomposition of acetal hydrotrioxides and suggested that the major reaction leading to singlet oxygen is cage disproportionation of alkoxy and hydroperoxy radicals (step 2). As mentioned before, an induced decomposition of the hydrotrioxide (step 3) cannot be ruled out since it can also explain the formation of singlet oxygen.

3. Aldehyde hydrotrioxides

White and Bailey[87] were the first to propose hydrotrioxides as intermediates in the ozonation of aromatic aldehydes. They studied the ozonation of benzaldehyde in various solvents at 0–25°C. The observed order of reactivity, i.e. anisaldehyde > benzaldehyde > *p*-nitrobenzaldehyde, indicates that nucleophilic attack of

TABLE 5. Activation parameters for the decomposition of some acetal hydrotrioxides[102,103]

Hydrotrioxide		Solvent[a]	E_a (kcal mol^{-1})	log A	k (s^{-1})[b] [T(°C)]
$R^1-C(OR^2)_2-OOOH$	R^1 = Me, R^2 = Me	Et_2O	16.1	11.8	2.9×10^{-2} [−10]
		CH_2Cl_2	—	—	4.0×10^{-2} [−10]
	R^1 = Me, R^2 = Et	Et_2O	13.2	9.8	7.6×10^{-2} [−10]
		Neat	14.5	11.1	1.1×10^{-2} [−10]
$C_6H_5-C(O_3H)(O-CH_2)_2$ (cyclic)		Et_2O	15.9	13.1	6.6×10^{-2} [−30]
$X-C_6H_4-C(O_3H)(O-R)_2$	X = H, R = Et	Et_2O	19.8	16.7	7.6×10^{-2} [−30]
	X = H, R = Me	Et_2O	19.0	15.9	6.6×10^{-2} [−30]
	X = 4-F, R = Me	Et_2O	19.5	16.4	6.3×10^{-2} [−30]
	X = 4-Cl, R = Me	Et_2O	18.4	15.4	3.2×10^{-2} [−35]
	X = 4-OMe, R = Me	Et_2O	17.2	13.7	2.0×10^{-2} [−30]

[a]Acetal (20 mmol), 30%; solvent, 70% (by weight).
[b]Standard deviation, ±8%.

TABLE 6. Singlet oxygen determination in the decomposition of some acetal hydrotrioxides[103]

Hydrotrioxide		Tetraphenylcyclopentadienone (%)[a]	1,3-Diphenylisobenzofuran (%)[a]
$MeC(OEt)_2O_3H$		20 ± 5[b]	54 ± 5[d]
2-Phenyl-1,3-dioxolan-2-yl hydrotrioxide ($PhC(O_3H)(OCH_2CH_2O)$)		55 ± 3[c]	85 ± 8[d,e]
$XC_6H_4C(OR)_2O_3H$	X = H; R = Et	—	46 ± 5[d]
	X = 4-Cl; R = Me	30 ± 5	55 ± 8[d]

[a]Percent of absorbed ozone available to react with the reagent.
[b]Neat or in methylene chloride.
[c]Methylene chloride.
[d]Ethyl acetate.
[e]20–30% of the absorbed ozone was available to react with 1,2-dimethylcyclohexene to form 1-methyl-2-methylenecyclohexanol (after sodium sulphite reduction of the corresponding hydroperoxide).

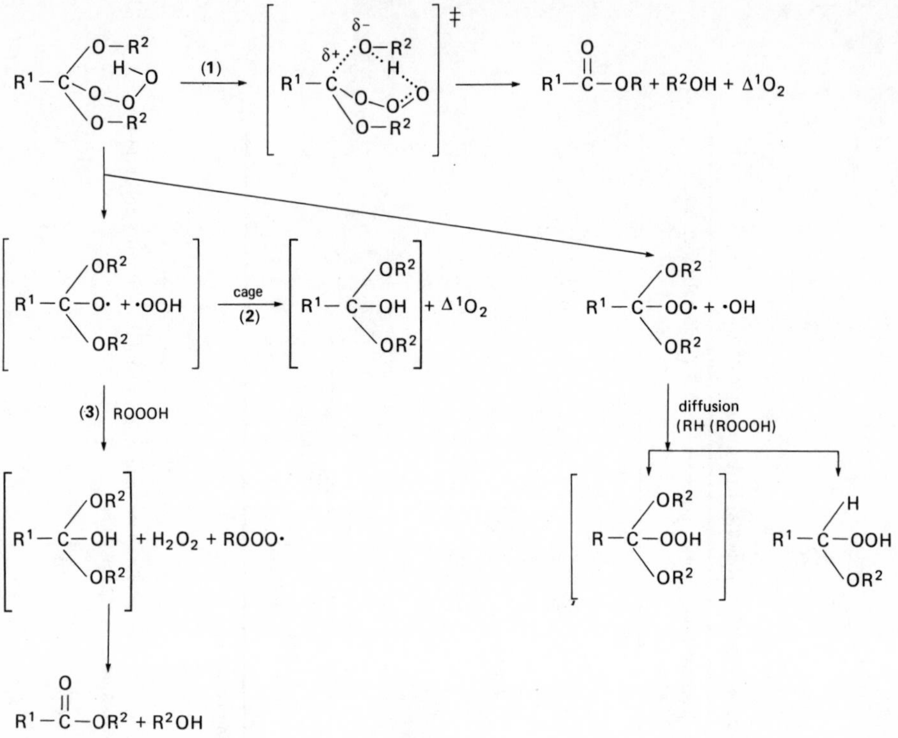

$R^1-C(O-R^2)(O-R^2)(O-O-O-H)$ —(1)→ $[R^1-C \cdots O(\delta-)-R^2 \cdots H \cdots O-O-O]^{\ddagger}$ (δ+, δ−) → $R^1-C(=O)-OR + R^2OH + \Delta^1O_2$

$[R^1-C(OR^2)_2-O\cdot + \cdot OOH]$ —cage (2)→ $[R^1-C(OR^2)_2-OH] + \Delta^1O_2$

$R^1-C(OR^2)_2-OO\cdot + \cdot OH$

(3) ROOOH

$[R^1-C(OR^2)_2-OH] + H_2O_2 + ROOO\cdot$

$R^1-C(=O)-OR^2 + R^2OH$

diffusion (RH (ROOOH)

$[R-C(OR^2)_2-OOH]$, $R^1-C(H)(OR^2)-OOH$

SCHEME 10

$Ar-C(=O)-O-O-H$

$HO\cdot + Ar-C(=O)-O-O\cdot$ → $Ar-C(=O)-OOH$

$Ar-C(=O)-O\cdot + \cdot OOH$ → $Ar-C(=O)-OH + O_2$

$Ar-C(=O)\cdot + O_2 + HO\cdot$ → $Ar-C(=O)-H + CO + C_6H_5\cdot$

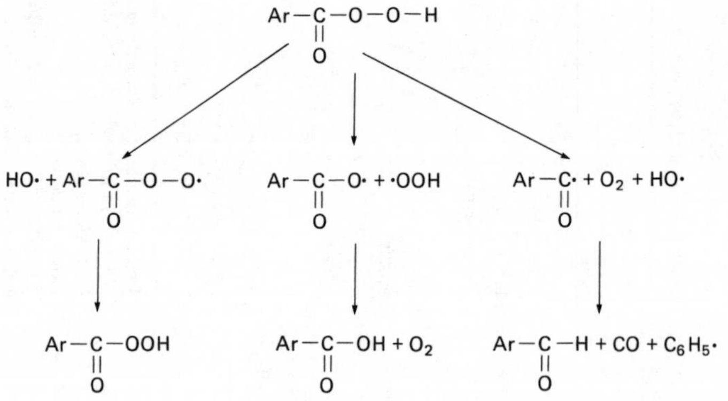

SCHEME 11

ozone on the carbonyl group of the aldehyde is not an important reaction step. The hydrotrioxide intermediate (presumably formed in a concerted 1,3-dipolar insertion of ozone into the C—H bond) was suggested to decompose homolytically to produce predominantly benzoylperoxy and hydroxy radicals. Ozonations with 0.5 mol equivalents of ozone yielded mostly peroxybenzoic acid in the early stages of reaction. Relatively smaller amounts of other products, beside benzoic acid, are explained by the alternative homolytic pathways (Scheme 11).

Murray and coworkers have recently confirmed the involvement of the hydrotrioxide in the low-temperature ozonation (-50°C) of benzaldehyde by observing an OOOH absorption in the NMR spectra (δ OOOH = 13.1 ppm at -50°C). This absorption was assigned to the intramolecularly hydrogen-bonded form of the hydrotrioxide. Benzoic acid and singlet oxygen (96%) were reported to be the main decomposition products of benzaldehyde hydrotrioxide. A vigorous evolution of gas was observed in the vicinity of 10°C. Activation parameters of a first-order decay of the hydrotrioxide ($E_a = 10.7 \pm 0.2$ kcal mol^{-1}, log $A = 5.4 \pm 0.4$ in benzaldehyde as solvent; $E_a = 16.6 \pm 0.6$ kcal mol^{-1}, log $A = 11.5 \pm 0.5$ in diethyl ether) were interpreted in terms of a concerted mechanism involving some charge separation in the transition state (reaction 51).

$$\text{Ar—C(=O}\cdots\text{H—O—O—O)} \longrightarrow \left[\text{Ar—C}\cdots\text{O}\cdots\text{H}^{\delta+}\cdots\text{O}^{\delta-}\cdots\text{O}\cdots\text{O}\right]^{\ddagger} \longrightarrow \text{Ar—C(=O)OH} + \Delta^1\text{O}_2 \qquad (51)$$

4. Hydrocarbon and alcohol hydrotrioxides

Although there is now good evidence that hydrotrioxides are formed in the ozonation of ethers, acetals and aldehydes, there is still no direct indication for hydrotrioxides being reactive intermediates in alkane ozonation. Nevertheless, several studies seem to suggest their participation in these reactions.

Durland and Adkins[112] and later Whiting and coworkers[89] found that *cis*- and *trans*-decalin react with ozone even at -78°C to give the corresponding alcohols and ketones with over 90% retention of configuration.

Hamilton and coworkers[113,114] reported that cyclohexane reacts with ozone to produce cyclohexanol and cyclohexanone in a 3.5:1 ratio. They found that the reaction proceeds with considerable retention of configuration (60–70% for tertiary alcohol formation). Their other findings in a detailed study of the reaction of ozone with alkanes at ambient temperature can be summarized as follows: (1) Considerable C—H bond-breaking in the transition state for the reaction is indicated by a large kinetic isotope effect ($k_H/k_D = 4.0$–4.5). (2) The relative reactivity of primary, secondary and tertiary hydrogens is 1:13:110. This is somewhat greater than for typical radical reactions. (3) The Hammett ρ value of -2.07 for the ozonation of substituted toluenes suggests a considerable charge separation in the transition state. (4) Some steric requirements of the reaction are indicated by the fact that equatorial tertiary hydrogens react seven times more readily than axial tertiary hydrogens.

On the basis of the above-mentioned evidence, Hamilton and coworkers[113,114] proposed a mechanism in which alkane and ozone react to form a radical pair and/or an ion pair as an intermediate (or transition state) without the involvement of the hydrotrioxide (Scheme 12).

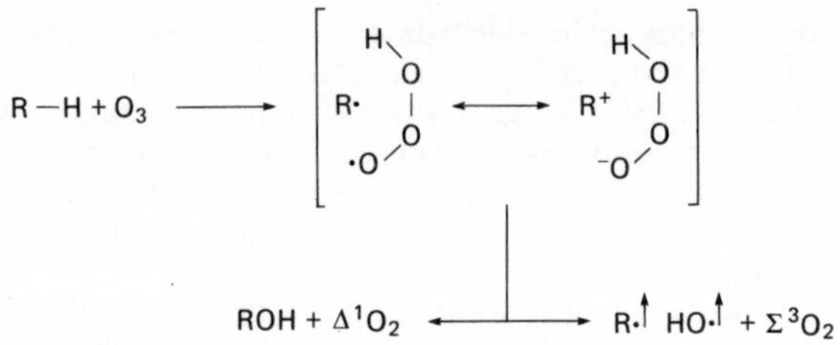

$$\mathrm{R{-}H + O_3 \longrightarrow \left[R\cdot \; {}^{H}\!O{-}O{-}O\cdot \longleftrightarrow R^+ \; {}^{H}\!O{-}O{-}O^- \right]}$$

$$\mathrm{ROH + \Delta^1O_2 \longleftarrow \;\; \longrightarrow R\cdot\!\uparrow \; HO\cdot\!\uparrow + \Sigma^3O_2}$$

SCHEME 12

Nangia and Benson[106] suggested (based on thermochemical calculations) that an ion pair is indeed most probably formed in the reaction of cyclohexane with ozone which then collapses to the hydrotrioxide. The radical decomposition of the latter would produce a singlet caged radical pair which could disproportionate in two ways; either to produce singlet oxygen and alcohol or ketone in the ground state and hydrogen peroxide (Scheme 13). The radicals which escape the cage initiate a chain reaction. Since both disproportionation reactions are very exothermic (by 55 and 75.5 kcal mol^{-1}, respectively), it would be helpful to determine experimentally whether electronically excited states are formed in these reactions.

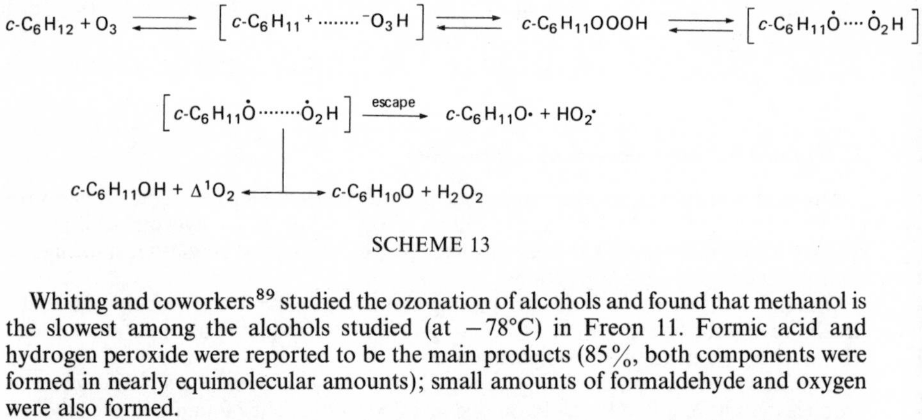

$$c\text{-}\mathrm{C_6H_{12} + O_3 \rightleftharpoons \left[\mathit{c}\text{-}C_6H_{11}^+ \cdots\cdots {}^-O_3H \right] \rightleftharpoons \mathit{c}\text{-}C_6H_{11}OOOH \rightleftharpoons \left[\mathit{c}\text{-}C_6H_{11}\dot{O} \cdots \dot{O}_2H \right]}$$

$$\mathrm{\left[\mathit{c}\text{-}C_6H_{11}\dot{O} \cdots\cdots \dot{O}_2H \right] \xrightarrow{escape} \mathit{c}\text{-}C_6H_{11}O\cdot + HO_2\cdot}$$

$$c\text{-}\mathrm{C_6H_{11}OH + \Delta^1O_2 \longleftarrow \;\; \longrightarrow \mathit{c}\text{-}C_6H_{10}O + H_2O_2}$$

SCHEME 13

Whiting and coworkers[89] studied the ozonation of alcohols and found that methanol is the slowest among the alcohols studied (at -78°C) in Freon 11. Formic acid and hydrogen peroxide were reported to be the main products (85%, both components were formed in nearly equimolecular amounts); small amounts of formaldehyde and oxygen were also formed.

The oxidation of ethanol was found to be much faster and gave acetic acid and hydrogen peroxide as the main products (70%), and acetaldehyde, formaldehyde, peroxyacetic acid and oxygen as the minor ones.

Whiting suggested the formation of a radical pair (or an ion pair) with subsequent formation of the hydrotrioxide. The minor products were explained by a base-catalysed decomposition of the hydrotrioxide, i.e. by a reaction analogous to the Kornblum–De La Mare carbonyl-forming elimination mechanism (Scheme 14). Recent evidence[115] indicates that oxygen bases bond to polar O—H groups, forming intermolecularly hydrogen-bonded adducts, rather than attack C—H hydrogen atoms. It seems, therefore, that the mechanistic explanation advanced by Whiting is unlikely. The hydroxymethyl hydrotrioxide should rather decompose in a radical process as outlined in Scheme 15. Radical decomposition can also account for the minor products being formed.

Further studies of low-temperature ozonation of alkanes and alcohols are needed to clarify the role of hydrotrioxides in these reactions.

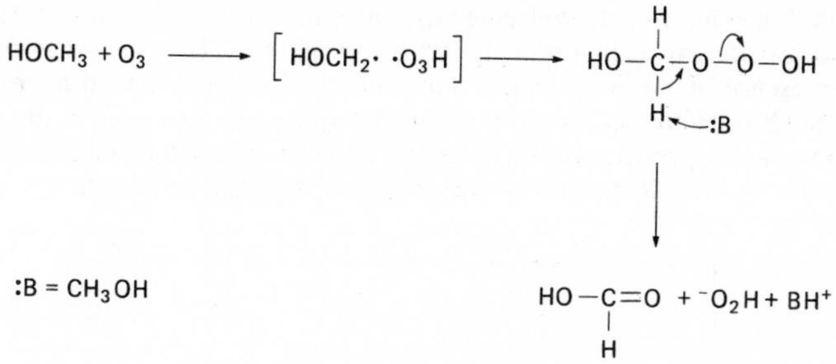

$$HOCH_3 + O_3 \longrightarrow [HOCH_2\cdot \ \cdot O_3H] \longrightarrow HO-CH_2-O-O-OH \longrightarrow HO-CH{=}O + {}^-O_2H + BH^+$$

$:B = CH_3OH$

SCHEME 14

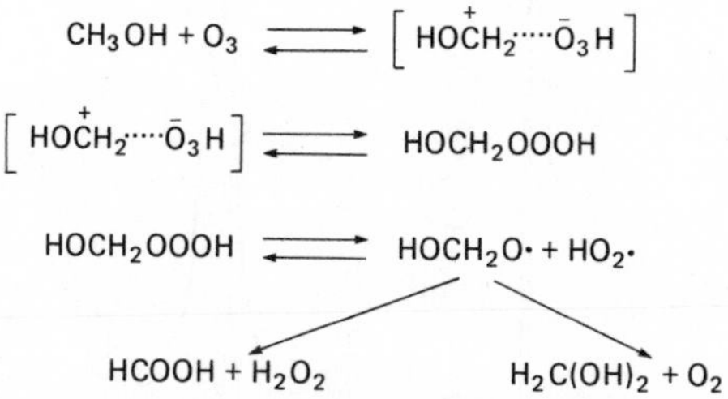

$$CH_3OH + O_3 \rightleftharpoons [HO\overset{+}{C}H_2 \cdots \bar{O}_3H]$$

$$[HO\overset{+}{C}H_2 \cdots \bar{O}_3H] \rightleftharpoons HOCH_2OOOH$$

$$HOCH_2OOOH \rightleftharpoons HOCH_2O\cdot + HO_2\cdot$$

$$HOCH_2O\cdot + HO_2\cdot \longrightarrow HCOOH + H_2O_2 \quad \text{or} \quad H_2C(OH)_2 + O_2$$

SCHEME 15

C. Structure of Hydrotrioxides

The low-temperature NMR spectra of hydrotrioxides (as already mentioned in the previous sections) show absorptions at δ 13 ppm downfield from Me_4Si. The position of these proton absorptions shows little change with dilution and they are thus tentatively assigned to the OOOH absorptions of the intramolecularly hydrogen-bonded six-membered rings of the hydrotrioxides. The observed change with dilution (<0.15 ppm) is in accord with that observed for peroxy acids (<0.10 ppm) which are known to exist in 'inert' solvents exclusively in the intramolecularly hydrogen-bonded form[115].

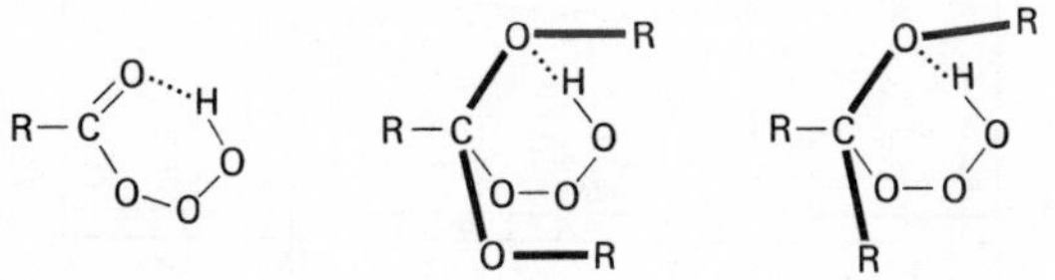

Further confirmation of the structure of hydrotrioxides in solution comes from the low-temperature ozonation of the dimethyl acetal of deuterated benzaldehyde; an oxygen-rich intermediate, which does not show any absorption around δ 13 ppm, is formed in the reaction.

It is interesting to mention that hydrotrioxides of ethers and acetals show two (or three) OOOH absorptions in the low-temperature NMR spectra in Freon 11, methylene chloride, ethyl acetate and diethyl ether as solvents. Both peaks show a tendency to merging and broadening at higher temperatures (Figure 3). This phenomenon might be due to the presence of two loosely, intermolecularly hydrogen-bonded forms of hydrotrioxides with OOOH bonded to either of the nonequivalent lone pairs of the tetrahedral oxygen atom or to the presence of two (or more) conformational forms (chair–boat) of the xix-membered ring with R groups on the oxygen atom axial or equatorial.

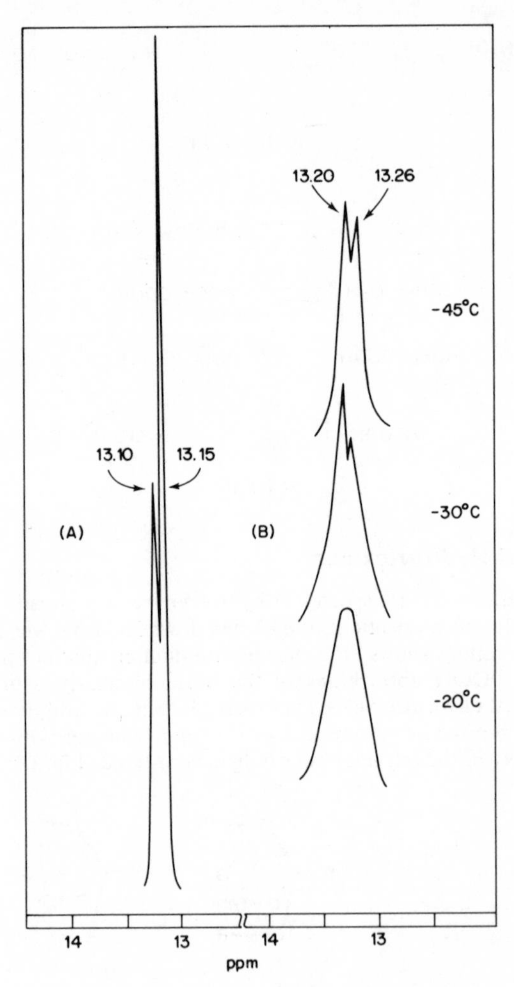

FIGURE 3. Segments of NMR spectra of the ozonized dimethyl acetal of benzaldehyde: (A) in diethyl ether at $-50°C$; (B) in methylene chloride. Adapted from Reference 103.

Addition of *N*,*N*-dimethylacetamide, a relatively strong oxygen acceptor base, to ozonized α-methylbenzyl methyl ether in diethyl ether (hydrotrioxide/base ratio, 1:0.5–2) at $-50°C$ causes a downfield shift of the OOOH absorption up to 1.5 ppm, depending on the amount of base added[110]. A downfield shift of OOOH absorptions was tentatively assigned to the opening of the six-membered chelate ring of the hydrotrioxide with subsequent formation of intermolecularly hydrogen-bonded adducts, since it is analogous to the recently studied complexation of peroxy acids with oxygen bases (equation 52)[116].

$$\text{Ph}-\overset{\text{Me}}{\underset{\text{Me}}{\text{C}}}(\text{O}\cdots\text{H}-\text{O}-\text{O}-\text{O}) + (:\text{B}) \rightleftharpoons \text{Ph}-\overset{\text{O}-\text{Me}}{\underset{\text{Me}}{\text{C}}}-\text{O}-\text{O}-\text{O}-\text{H}\cdots\cdots(\text{B}) \quad (52)$$

Decomposition of the adduct is faster than that of the chelated hydrotrioxide. Disappearance of the OOOH absorption is accompanied by the simultaneous appearance of the absorption at δ 10 ppm, assigned to the corresponding hydroperoxide. It appears that intermolecular hydrogen bonding in the adduct weakens the ROO—OH bond more than the RO—OOH bond (reported to be the weakest one in the monomeric species, amounting to 22–23 kcal mol^{-1}), thus favouring homolytic cleavage to produce ROO· radicals, which then abstract hydrogen to form hydroperoxides.

On the basis of the above-mentioned evidence it can be concluded that the relatively great stability of hydrotrioxides in 'inert' solvents is the result of the presence of intramolecular hydrogen bonds in these species.

Although the existence of higher alkyl hydropolyoxides ($n = 4,5$) is at least theoretically possible at temperatures below -80°C[117,118], it will certainly be a challenging task to detect them unambiguously by the known spectroscopic methods. It is interesting, however, to mention that a thermochemical analysis of the reaction of *t*-butyl hydroperoxide with ozone has already shown that alkyl hydropentoxide, RO_5H, cannot be a transient intermediate in this reaction ('insertion' mechanism)[119].

VI. ACKNOWLEDGEMENT

The author wishes to thank Professor Glen A. Russell for some very helpful comments on parts of the manuscript.

VII. REFERENCES

1. (a) M. Venugopalan and R. G. Jones, *Chemistry of Dissociated Water and Related Systems*, John Wiley and Sons, New York–London, 1968, Chap. 7.
   (b) P. A. Giguère, *Trans. N. Y. Acad. Sci.*, **34**, 334 (1972).
   (c) P. A. Giguère, *Peroxyde d'Hydrogene*, Vol. 4, Complements au Nouveau Traite de Chimie Minerale, Manon, 1975.
2. M. Berthelot, *Compt. Rend.*, **90**, 656 (1880).
3. D. I. Mendeleev, *Osnovy Khimii*, 228 (1895).
4. A. N. Bakh, *Zh. Russ. Fiz.-Khim. Obshch.*, **29**, 373 (1897).
5. E. Ohara, *J. Chem. Soc. Japan*, **61**, 569 (1940).
6. T. V. Yagodovskaya and L. I. Nekrasov, *Russ. J. Phys. Chem.*, **44**, 965 (1970), and references cited therein.
7. J. A. Wojtowicz, F. Martinez and J. A. Zaslowski, *J. Phys. Chem.*, **67**, 849 (1963).
8. G. Czapski and B. H. J. Bielski, *J. Phys. Chem.*, **67**, 2180 (1963).
9. B. H. J. Bielski and H. A. Schwartz, *J. Phys. Chem.*, **72**, 3836 (1968).
10. P. A. Giguère and D. Chin, *J. Phys. Chem.*, **51**, 1685 (1959).
11. K. Herman and P. A. Giguère, *Can. J. Chem.*, **46**, 2649 (1968).
12. P. A. Giguère and K. Herman, *Can. J. Chem.*, **48**, 3473 (1970).
13. X. Deglise and P. A. Giguère, *Can. J. Chem.*, **49**, 2242 (1971).

14. P. S. Nangia and S. W. Benson, *J. Phys. Chem.*, **83**, 1138 (1979).
15. (a) S. W. Benson, *J. Amer. Chem. Soc.*, **86**, 3922 (1964).
(b) S. W. Benson and R. Shaw, *Organic Peroxides*, Vol. 1 (Ed. D. Swern), John Wiley and Sons, New York–London, 1970, Chap. 2.
16. L. Radom, W. J. Hehre and J. A. Pople, *J. Amer. Chem. Soc.*, **93**, 289 (1971).
17. L. Radom, W. J. Hehre and J. A. Pople, *J. Chem. Soc.(A)*, 2299 (1971).
18. B. Plesničar, S. Kaiser and A. Ažman, *J. Amer. Chem. Soc.*, **95**, 5476 (1973).
19. J. L. Arnau and P. A. Giguère, *J. Chem. Phys.*, **60**, 270 (1974).
20. R. J. Blint and M. D. Newton, *J. Chem. Phys.*, **59**, 6220 (1973).
21. B. Plesničar, D. Kocjan, S. Murovec and A. Ažman, *J. Amer. Chem. Soc.*, **98**, 3143 (1976).
22. D. Cremer, *J. Chem. Phys.*, **69**, 4456 (1978).
23. L. Bateman, *Quart. Rev. Chem. Soc.*, **8**, 147 (1954).
24. H. S. Blanchard, *J. Amer. Chem. Soc.*, **81**, 4548 (1959).
25. G. A. Russell, *Chem. Ind. (London)*, 1483 (1956).
26. G. A. Russell, *J. Amer. Chem. Soc.*, **79**, 3871 (1957).
27. T. G. Traylor and P. D. Bartlett, *Tetrahedron Letters*, 30 (1960).
28. P. D. Bartlett and T. G. Traylor, *J. Amer. Chem. Soc.*, **85**, 2407 (1963).
29. T. G. Traylor, *J. Amer. Chem. Soc.*, **85**, 2411 (1963).
30. R. Hiatt and T. G. Traylor, *J. Amer. Chem. Soc.*, **87**, 3766 (1965).
31. J. R. Thomas, *J. Amer. Chem. Soc.*, **89**, 4872 (1967).
32. P. D. Bartlett and G. Guaraldi, *J. Amer. Chem. Soc.*, **89**, 4799 (1967).
33. N. A. Milas and B. Plesničar, *International Symposium on Organic Peroxides, Book of Abstracts*, Deutsche Akademie der Wissenschaften zu Berlin, Berlin–Adlershof, 1967, p. 68.
34. N. A. Milas and B. Plesničar, *J. Amer. Chem. Soc.*, **90**, 4450 (1968).
35. T. Mill and R. S. Stringham, *J. Amer. Chem. Soc.*, **90**, 1062 (1968).
36. J. E. Bennett, D. M. Brown and B. Mile, *Chem. Commun.*, 504 (1969).
37. J. E. Bennett, D. M. Brown and B. Mile, *Trans. Faraday Soc.*, **66**, 397 (1970).
38. J. E. Bennett, B. Mile, A. Thomas and B. Ward, *Advan. Phys. Org. Chem.*, Vol. 8 (Ed. V. Gold), Academic Press, London, 1970, Chap. 1.
39. K. Adamic, J. A. Howard and K. U. Ingold, *Chem. Commun.*, 505 (1969).
40. K. Adamic, J. A. Howard and K. U. Ingold, *Can. J. Chem.*, **47**, 505 (1969).
41. K. U. Ingold, *Acc. Chem. Res.*, **2**, 1 (1969).
42. J. A. Howard, *Advan. Free-Radical Chem.*, **4**, 49 (1972).
43. J. A. Howard and J. C. Tait, *Can. J. Chem.*, **54**, 2669 (1976).
44. S. Fukuzumi and Y. Ono, *J. Chem. Soc., Perkin Trans. 2*, 622 (1977).
45. J. A. Howard and K. U. Ingold, *Can. J. Chem.*, **47**, 3797 (1969).
46. S. Korcek, J. H. B. Chenier, J. A. Howard and K. U. Ingold, *Can. J. Chem.*, **50**, 2285 (1972).
47. J. A. Howard, *Amer. Chem. Soc. Symp. Ser.*, **69**, 413 (1978).
48. P. S. Nangia and S. W. Benson, *Intern. J. Chem. Kinet.*, **12**, 29 (1980).
49. S. Fukuzumi and Y. Ono, *J. Phys. Chem.*, **81**, 1895 (1977).
50. J. E. Bennett and J. A. Howard, *J. Amer. Chem. Soc.*, **95**, 4008 (1973).
51. J. R. Thomas, *J. Amer. Chem. Soc.*, **87**, 3935 (1965).
52. J. A. Howard and K. U. Ingold, *J. Amer. Chem. Soc.*, **90**, 1058 (1968).
53. J. A. Howard and J. E. Bennett, *Can. J. Chem.*, **50**, 2375 (1972).
54. C. J. Hochanadel, J. A. Ghormley, J. W. Boyle and P. J. Ogren, *J. Phys. Chem.*, **81**, 3 (1977).
55. D. A. Parkes, *Intern. J. Chem. Kinet.*, **9**, 451 (1977).
56. E. Sanhueza, R. Simonaitis and J. Heicklen, *Intern. J. Chem. Kinet.*, **11**, 907 (1979).
57. C. S. Kan, R. D. McQuigg, M. R. Whitebeck and J. G. Calvert, *Intern. J. Chem. Kinet.*, **11**, 921 (1979).
58. M. Nakano, K. Takayama, Y. Shimizu, Y. Tsuji, H. Inaba and T. Migita, *J. Amer. Chem. Soc.*, **98**, 1974 (1976).
59. S. W. Benson, 'Symposium on the Mechanisms of Pyrolysis, Oxidation and Burning of Organic Materials', *NBS Special Publication*, **357**, 121 (1972). S. W. Benson and P. S. Nangia, *Acc. Chem. Res.*, **12**, 223 (1979).
60. D. F. Bowman, T. Gillan and K. U. Ingold, *J. Amer. Chem. Soc.*, **93**, 6555 (1971).
61. D. F. Dever and J. G. Calvert, *J. Amer. Chem. Soc.*, **84**, 1362 (1962).
62. N. R. Subbaratnam and J. G. Calvert, *J. Amer. Chem. Soc.*, **84**, 1113 (1962).

63. R. Shortridge and J. Heicklen, *Can. J. Chem.*, **51**, 2251 (1973).
64. (a) P. S. Nangia and S. W. Benson, *Intern. J. Chem. Kinet.*, **12**, 43 (1980).
(b) R. Criegee, *Justus Liebigs Ann. Chem.*, **583**, 1 (1953).
65. N. A. Milas and S. M. Djokic, *J. Amer. Chem. Soc.*, **84**, 3098 (1962).
66. P. D. Bartlett and P. Günther, *J. Amer. Chem. Soc.*, **88**, 3288 (1966).
67. P. D. Bartlett and M. Lahav, *Israel J. Chem.*, **10**, 101 (1972).
68. L. R. Anderson and W. B. Fox, *J. Amer. Chem. Soc.*, **89**, 4313 (1967).
69. P. G. Thompson, *J. Amer. Chem. Soc.*, **89**, 4316 (1967).
70. D. D. DesMarteau, *Inorg. Chem.*, **9**, 2179 (1970).
71. F. A. Hohorst, D. D. DesMarteau, L. R. Anderson, D. E. Gould and W. B. Fox, *J. Amer. Chem. Soc.*, **95**, 3866 (1973).
72. J. Czarnowski and H. J. Schumacher, *J. Fluorine Chem.*, **12**, 497 (1978).
73. J. Czarnowski and H. J. Schumacher, *Intern. J. Chem. Kinet.*, **11**, 613 (1979).
74. J. E. Bennett, G. Brunton and R. Summers, *J. Chem. Soc., Perkin Trans. 2*, 981 (1980).
75. D. Tal, E. Keinan and Y. Mazur, *J. Amer. Chem. Soc.*, **101**, 502 (1979).
76. N. Yoneda and G. A. Olah, *J. Amer. Chem. Soc.*, **99**, 3113 (1977).
77. T. Preuss, E. Proksch and A. deMeijere, *Tetrahedron Letters*, 833 (1978).
78. A. L. J. Beckwith, C. L. Bodkin and T. Duong, *Australian J. Chem.*, **30**, 2177 (1977).
79. W. F. Sager and J. C. Hoffsommer, *J. Phys. Chem.*, **73**, 4155 (1969).
80. G. E. Zaikov, J. A. Howard and K. U. Ingold, *Can. J. Chem.*, **47**, 3017 (1969).
81. N. A. Clinton, R. A. Kenley and T. G. Traylor, *J. Amer. Chem. Soc.*, **97**, 3746 (1975).
82. N. A. Clinton, R. A. Kenley and T. G. Traylor, *J. Amer. Chem. Soc.*, **97**, 3752 (1975).
83. N. A. Clinton, R. A. Kenley and T. G. Traylor, *J. Amer. Chem. Soc.*, **97**, 3757 (1975).
84. R. A. Kenley and T. G. Traylor, *J. Amer. Chem. Soc.*, **97**, 4700 (1975).
85. C. C. Price and A. L. Tumolo, *J. Amer. Chem. Soc.*, **86**, 4691 (1964).
86. R. E. Erickson, R. T. Hansen and J. Harkins, *J. Amer. Chem. Soc.*, **90**, 6777 (1968).
87. H. M. White and P. S. Bailey, *J. Org. Chem.*, **30**, 3037 (1965).
88. R. E. Erickson, D. Bakalik, C. Richards, M. Scanion and D. Huddleston, *J. Org. Chem.*, **31**, 461 (1966).
89. M. C. Whiting, A. J. N. Bolt and J. H. Parish, *Advan. Chem. Ser.*, **77**, 4 (1968).
90. J. E. Batterbee and P. S. Bailey, *J. Org. Chem.*, **32**, 3899 (1967).
91. J. D. Austin and L. Spialter, *Advan. Chem. Ser.*, **77**, 26 (1968).
92. P. S. Bailey, D. A. Mitchard and A.-I. Khashab, *J. Org. Chem.*, **33**, 2675 (1968).
93. P. S. Bailey and J. E. Keller, *J. Org. Chem.*, **33**, 2680 (1968).
94. P. S. Bailey, T. P. Carter, Jr. and L. M. Southwick, *J. Org. Chem.*, **37**, 2997 (1972).
95. P. S. Bailey, A. M. Reader, P. Kolsaker, H. M. White and J. C. Barborak, *J. Org. Chem.*, **30**, 3042 (1965).
96. P. Deslongchamps and C. Moreau, *Can. J. Chem.*, **49**, 2465 (1971).
97. P. Deslongchamps, C. Moreau, D. Frehel and P. Atlani, *Can. J. Chem.*, **50**, 3402 (1972).
98. P. Deslongchamps, P. Atlani, D. Frehel and C. Moreau, *Can. J. Chem.*, **52**, 3651 (1974).
99. P. Deslongchamps, C. Moreau, D. Frehel and R. Chenevert, *Can. J. Chem.*, **53**, 1204 (1975).
100. R. W. Murray, W. C. Lumma, Jr. and J. W.-P. Lin, *J. Amer. Chem. Soc.*, **92**, 3205 (1970).
101. F. E. Stary, D. E. Emge and R. W. Murray, *J. Amer. Chem. Soc.*, **98**, 1880 (1976).
102. F. Kovač and B. Plesničar, *J. Chem. Soc., Chem. Commun.*, 122 (1978).
103. F. Kovač and B. Plesničar, *J. Amer. Chem. Soc.*, **101**, 2677 (1979).
104. R. J. Taillefer, S. E. Thomas, Y. Nadeau, S. Fliszar and H. Henry, *Can. J. Chem.*, **58**, 1138 (1980).
105. R. J. Taillefer, S. E. Thomas, Y. Nadeau and H. Beierbeck, *Can. J. Chem.*, **57**, 3041 (1979).
106. P. S. Nangia and S. W. Benson, *J. Amer. Chem. Soc.*, **102**, 3105 (1980).
107. G. A. Olah, N. Yoneda and D. G. Parker, *J. Amer. Chem. Soc.*, **98**, 5261 (1976); *Angew. Chem.*, **17**, 909 (1978).
108. H. Varkony, S. Pass and Y. Mazur, *J. Chem. Soc., Chem. Commun.*, 437 (1974).
109. P. S. Bailey and D. A. Lerdal, *J. Amer. Chem. Soc.*, **100**, 5820 (1978).
110. (a) B. Plesničar and F. Kovač, *Fifth IUPAC Conference on Physical Organic Chemistry, Book of Abstracts*, University of California at Santa Cruz, California, 1980, p. 132.
(b) B. Plesničar and F. Kovač, *Third International Symposium on Organic Free Radicals, Book of Abstracts*, Federation of European Chemical Societies (Gesellschaft Deutscher Chemiker), Freiburg, 1981, p. 10.

111. F. Kovač and B. Plesničar in preparation.
112. J. D. Durland and H. Adkins, *J. Amer. Chem. Soc.*, **61**, 429 (1939).
113. G. A. Hamilton, B. S. Ribner and T. M. Hellman, *Advan. Chem. Ser.*, **77**, 15 (1968).
114. T. M. Hellman and G. A. Hamilton, *J. Amer. Chem. Soc.*, **96**, 1530 (1974).
115. W. H. T. Davison, *J. Chem. Soc.*, 2456 (1951); R. Kavčič, B. Plesničar and D. Hadži, *Spectrochim. Acta*, **A23**, 2483 (1967); B. Plesničar, R. Kavčič and D. Hadži, *J. Mol. Struct.*, **20**, 457 (1974).
116. J. Škerjanc, A. Regent and B. Plesničar, *J. Chem. Soc., Chem. Commun.*, 1007 (1980).
117. S. W. Benson, *Thermochemical Kinetics*, 2nd ed., John Wiley and Sons, New York–London, 1976.
118. Y. R. Luo, *J. Chem. Educ.*, **58**, 26 (1981).
119. M. E. Kurz and W. A. Pryor, *J. Amer. Chem. Soc.*, **100**, 7953 (1978).
120. B. M. Brudkin, S. S. Zlotski, U. B. Imashev and D. L. Rakhmankulov, *Dokl. Akad. Nauk SSSR*, **241**, 129 (1978); B. M. Brudkin, U. B. Imashev, S. S. Zlotski and D. L. Rakhmankulov, *Zh. Org. Khim.*, **17**, 700 (1981).

The Chemistry of Functional Groups, Peroxides
Edited by S. Patai

CHAPTER 17

Polar reaction mechanisms involving peroxides in solution

BOŽO PLESNIČAR
University of Ljubljana, Ljubljana, Yugoslavia

I. INTRODUCTION . . . 522
II. INTERMOLECULAR SUBSTITUTION AT THE PEROXIDE OXYGEN . 522
A. General Characteristics . . . 522
B. Oxidation of Olefins . . . 525
1. Oxidation with peroxy acids (Prilezhaev reaction) . . . 525
a. Kinetics and substituent effects . . . 525
b. Stereochemistry . . . 525
c. Solvent effects . . . 527
d. Catalysis . . . 528
e. Mechanism . . . 528
f. Theoretical studies . . . 530
2. Oxidation with hydroperoxides . . . 533
C. Hydroxylation . . . 534
D. Oxidation of Acetylenes . . . 536
E. Oxidation of Organic Nitrogen Compounds . . . 539
1. Amines . . . 539
2. Oximes . . . 540
3. Azo and diazo compounds . . . 541
F. Oxidation of Organic Sulphur Compounds . . . 542
1. Sulphides and sulphoxides . . . 542
2. Thiobenzophenones . . . 545
G. Oxidation of Organic Iodine Compounds . . . 546
H. Reactions of Diacyl Peroxides with some Nucleophiles . . . 546
1. General characteristics . . . 546
2. Oxidation of amines . . . 547
3. Oxidation of olefins . . . 549

III. INTRAMOLECULAR NUCLEOPHILIC REARRANGEMENTS . . . 551
A. General Characteristics . . . 551
B. Peroxy Esters (Criegee Rearrangement) . . . 551
C. Hydroperoxides and Dialkyl Peroxides . . . 554
D. Diacyl Peroxides . . . 557
E. Oxidation of Ketones and Aldehydes with Peroxy Acids (Baeyer–Villiger Oxidation) . . . 559
1. Ketones . . . 559
a. Polar reaction mechanisms involving the formation of the peroxy-acid–ketone adduct . . . 559
b. Mechanisms involving ionic and/or radical intermediates . . . 562
2. Aldehydes . . . 564
IV. OXIDATIONS INVOLVING PEROXIDES AND THEIR ANIONS AS NUCLEOPHILES . . . 566
A. General Characteristics . . . 566
B. Oxidations with Alkaline Hydrogen Peroxide and other Peroxy Anions . 567
C. Oxidation of Imines (Schiff Bases) . . . 570
V. CARBONYL-FORMING ELIMINATIONS . . . 573
A. Base-catalysed Decompositions . . . 573
1. Hydroperoxides and dialkyl peroxides . . . 573
2. Peroxy esters . . . 577
VI. REFERENCES . . . 578

I. INTRODUCTION

The subject of polar reaction mechanisms involving peroxides in solution has been covered prior to 1970 in a number of reviews in which surveys of most important early data can be found[1]. Thus, this chapter is not exhaustive in coverage but rather stresses the general principles and emphasizes more recent developments in the field. An attempt has been made to evaluate the status of the most important areas along with the outlook for future research.

The division of the material is according to the type of reaction mechanism, introduced 20 years ago by Davies in his, by now, classic treatise on the subject[1a], and later developed by Edwards[1b,c].

The coverage of the literature is through 1981. Since organometallic peroxides are treated separately in this volume, the reaction mechanisms of these compounds are not discussed in this review. The discussion of ionization of peroxides, which has been comprehensively reviewed previously, has also been omitted.

In recent years it has become clear that a number of reactions of peroxides with various organic substrates, traditionally believed to proceed by a 'polar' mechanism, actually involve nonradical and/or radical processes. A number of such reactions are discussed in the chapter, and it is quite probable that additional examples will be found in the future.

II. INTERMOLECULAR SUBSTITUTION AT THE PEROXIDE OXYGEN

A. General Characteristics

A large number of reactions of peroxides with reducing agents may be regarded as nucleophilic displacements on the 'electrophilic' oxygen as shown in Scheme 1, where R^2 is usually hydrogen, and N: denotes a nucleophile. The following general characteristics of these reactions might be expected on the basis of this mechanistic scheme: (*a*) The reaction

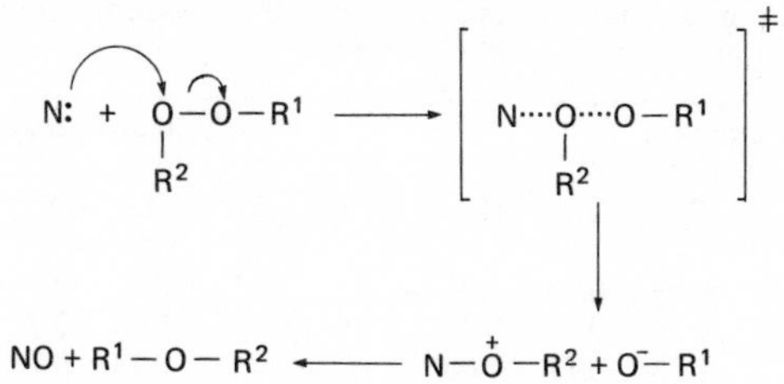

$$\mathrm{N{:}} + \underset{\displaystyle R^2}{\overset{|}{O}}-O-R^1 \longrightarrow \left[\ \mathrm{N}\cdots\underset{\displaystyle R^2}{\overset{|}{O}}\cdots O-R^1\ \right]^{\ddagger}$$

$$\downarrow$$

$$\mathrm{NO} + R^1-O-R^2 \longleftarrow \mathrm{N}-\overset{+}{O}-R^2 + \bar{O}-R^1$$

SCHEME 1

should follow second-order kinetics; first order each in nucleophile and peroxide; (*b*) electron-donating groups on N: will accelerate the reaction (increased nucleophilicity), while just the opposite effect is expected when such groups are located on $R^1-O-O-R^2$ (decreased electrophilicity); (*c*) increased solvent polarity should increase the reaction rates; (*d*) general or specific acid catalysis may be observed (lone electron pairs on the oxygen atoms); (*e*) the activation entropies should be negative because of considerable orientation of the transition state; (*f*) the activation energies will be relatively low (ca. 12–16 kcal mol^{-1}). Higher values would indicate the involvement of free radicals in these reactions[1b,c].

Before we go into discussion of mechanistic details of some reactions of nucleophiles with peroxides, the question may be posed as to what aspects of ground-state structure of peroxides reflect the electrophilic behaviour of these compounds, and what is the major driving force for the heterolytic cleavage of the O—O bond in peroxides in their reaction with nucleophiles.

Peroxy acids are best characterized of all systems which are capable of delivering an electrophilic oxygen[2,3]. The following discussion will, therefore, concentrate on these compounds only. The results of an *ab initio* MO study on peroxyacetic and

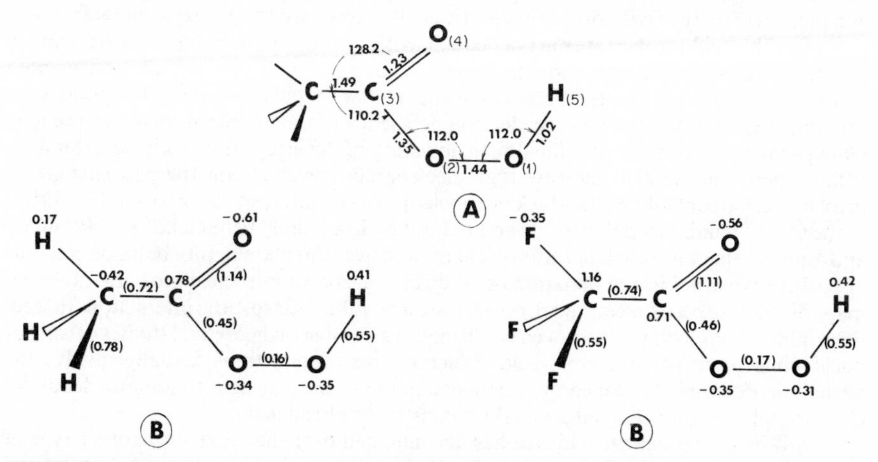

FIGURE 1. (A) The geometry of the peroxycarboxyl group. Bond lengths and angles are in Ångstroms and degrees, respectively. (B) Calculated net atomic charges of peroxyacetic and peroxytrifluoroacetic acid. Calculated overlap populations are in parentheses.

TABLE 1. Calculated atomic orbital charges on the peroxide oxygens, and frontier orbitals of atoms in peroxyacetic and trifluoroperoxyacetic acid (6-31G**)[2]

Peroxyacetic acid							Trifluoroperoxyacetic acid						
Energy[a]		O(1)	O(2)	C(3)	O(4)	H(5)	Energy[a]		O(1)	O(2)	C(3)	O(4)	H(5)
σ^*	0.192	0.91	0.40	0.03	0.06	0.00	σ^*	0.157	0.92	0.40	0.05	0.07	0.49
π^*	0.159	0.00	0.20	1.16	0.58	0.00	π^*	0.097	0.00	0.22	1.00	0.60	0.00
π	−0.464	0.70	0.75	0.03	0.51	0.00	π	−0.497	0.73	0.76	0.03	0.47	0.00
σ	−0.475	0.02	0.16	0.11	1.45	0.01	σ	−0.526	0.01	0.20	0.09	1.46	0.01
π	−0.526	0.56	0.01	0.27	0.81	0.00	π	−0.578	0.76	0.01	0.35	0.84	0.00
σ	−0.561	0.87	0.21	0.13	0.03	0.05	σ	−0.608	1.19	0.21	0.16	0.08	0.09

Peroxyacetic acid O(1)[b]	Peroxyacetic acid O(2)[b]	Trifluoroperoxyacetic acid O(1)[b]	Trifluoroperoxyacetic acid O(2)[b]
2s 1.82	2s 1.84	2s 1.82	2s 1.86
$2p_x$ 0.92	$2p_x$ 1.07	$2p_x$ 0.88	$2p_x$ 1.10
$2p_y$ 1.57	$2p_y$ 1.51	$2p_y$ 1.57	$2p_y$ 1.49
$2p_z$ 1.98	$2p_z$ 1.85	$2p_z$ 1.98	$2p_z$ 1.84

[a]Energies are given in atomic units and distribution of orbitals in fraction of electrons.
[b]Atomic orbital charges.

trifluoroperoxyacetic acid[2], presented in Figure 1 and Table 1, indicate that in terms of the perturbation treatments, overlap control must be more significant than charge interaction control. Namely, it is evident from Figure 1 and Table 1 that the oxygen atom involved in reactions is not electrophilic in the sense of charge interaction since both peroxidic oxygens bear negative charges. There are also regions of low potential associated with the axis of the peroxide bond, as evident from the potential maps. Previously, it was suggested that the enhanced reactivity of trifluoroperoxyacetic acid, as compared to peroxyacetic acid, is due to the fact that trifluoroacetate ion is a better leaving group or that stronger polarization of the peroxide bond is present in this peroxy acid. Similar charge distribution in both compounds as well as small differences in the overlap populations do not support such presumptions. Rather, the relative electron deficiency of the oxygen p_x orbitals, the low-lying peroxide σ^* orbitals and the drastic lowering of the eigenvalues of these orbitals in trifluoroperoxyacetic acid indicate that 'electrophilic' reactions on the peroxide group involve overlap control, i.e. the attack of a nucleophile on the lowest vacant orbital (LUMO) of the O—O bond. Nevertheless, as pointed out by Loew and Hjelmeland[2] and Plesničar and Azman[3], the complete separation of charge and overlap effects cannot be made. Regions of positive potential in the peroxide bond direction, shown in the electrostatic potential maps of trifluoroperoxyacetic acid, do indicate a possible electrostatic interaction. Indeed, these interactions have recently been confirmed in a number of theoretical studies of various nucleophiles with peroxyformic acid. Whether these interactions actually involve the formation of charge-transfer complexes in reactions of peroxy acids with some nucleophiles (for example, unsaturated substrates) remains to be elucidated.

As will be shown below, evidence has accumulated over the years for another type of mechanism in the reaction of peroxides, R^1OOR^2 (where R^2 is not hydrogen), with various nucleophiles capable of serving as one- or two-electron donors. These reactions are generally classified as electron-transfer-initiated transformations.

B. Oxidation of Olefins

1. Oxidation with peroxy acids (Prilezhaev reaction)

The reaction of olefins with peroxy acids to produce epoxides (oxiranes) has been known for almost 80 years[4] (equation 1). In spite of considerable work on the mechanism of this reaction[1,5], a number of important questions have remained unanswered. Let us briefly review the experimental facts concerning the mechanism of this reaction.

$$>C{=}C< + RCO_3H \longrightarrow >C\overset{O}{—}C< + RCO_2H \quad (1)$$

a. Kinetics and substituent effects. The reaction is a second-order process, first order in olefin and first order in peroxy acid[6–11].

$$v = k\,[\text{peroxy acid}]\,[\text{olefin}]$$

The substituent effect studies of epoxidation with peroxybenzoic acid reveal the Hammett ρ value of $-1,2(\sigma)$ for substituted *trans*-stilbenes in benzene, and $+1.3$ for substituted peroxybenzoic acids[7–9]. The importance of σ^+ contributions to the correlation have been reported for the oxidation of *para*-substituted styrenes with peroxybenzoic acid ($\rho = -1.3$, $\sigma + 0.48\,\Delta\sigma^+$)[10]. Activation enthalpies are decreased by electron-withdrawing groups in *trans*-stilbene; electron-withdrawing groups in peroxybenzoic acids decrease this activation parameter while electron-releasing groups increase it[7]. Thus, alkyl groups on ethylene increase the rate of epoxidation. A linear plot of log k_2 vs. the number of methyl groups on ethylene was obtained[12]. *Cis* double bonds in straight-chain olefins are epoxidized faster than *trans* double bonds; just the opposite is usually observed for *cis–trans* isomers of medium-size cycloalkenes[13,14].

The observed trends of kinetic and activation parameters clearly indicate the nucleophilic nature of olefins, and the electrophilic nature of peroxy acids. The order of 'electrophilicity' of these reagents approximately parallels the pK values[15,16].

Relatively large and negative entropies of activation (-20 to $-40\,\text{cal mol}^{-1}\,\text{deg}^{-1}$) indicate a high degree of orientation in the transition state[7].

The kinetics of epoxidation of strained olefins imply that the rate is independent of the strain[17–19]. For example, 1,2-diphenylcyclopentene reacts 11.7 times faster than 1,2-diphenylcyclopropene with *m*-chloroperoxybenzoic acid in carbon tetrachloride at 0°C[19], and norbornene reacts 2.4 times faster than cyclohexene with peroxylauric acid in chloroform at 25°C[17].

b. Stereochemistry. The stereochemistry of olefin epoxidation has been comprehensively reviewed[20]. The reaction is *syn*-stereospecific. Thus, *cis*- and *trans*-stilbene always give *cis*- and *trans*-stilbene epoxide[7]. Neighbouring functional groups may influence the direction as well as the rate of attack of the peroxy acid. For example, epoxidation of allylic compounds with a bulky group as a neighbouring substituent proceeds with a preferential *anti* attack and is slower than that of cyclohexene. 3-*t*-Butylcyclohexene (**1**) yields a 9:1 ratio of *trans* to

Me

Me Me

(**1**)

cis epoxide (*trans* and *cis* with respect to the *t*-butyl group) when oxidized with *m*-chloroperoxybenzoic acid in dichloromethane[21,22]. Steric influence is, in general, important only in cases where there are not more than two carbon atoms between the substituent and the reaction centre. Nevertheless, preference for *anti* epoxidation has also been reported for systems with a polar group too far from the double bond to exhibit any steric influence[23]. On the other hand, the epoxidation of the allylic alcohol 2-cyclohexene-1-ol (**2**) gives a 9:91 ratio of *trans* to *cis* epoxide (equation 2); the *cis* epoxide is also formed at a significantly faster rate[23,24]. Allylic carboxylic acids[25], and carboxylates[26] are also effective *syn*-directing groups. Contrary to previous findings, an interesting cooperative effect by a hydroxy group and an ether oxygen near the asymmetric centre in directing the steric course of the approaching peroxy acid was recently reported[27]. Namely, epoxidation of **3** with *m*-chloroperoxybenzoic acid in dichloromethane at 0°C affords the epoxide **4** as the main product (ratio >25:1) (equation 3).

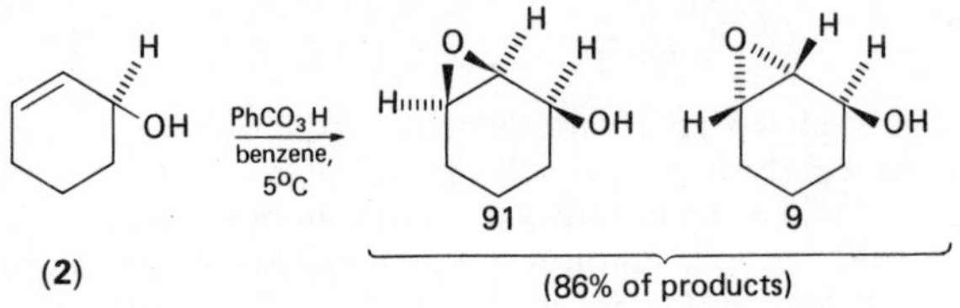

(**2**) $\xrightarrow[\text{benzene, } 5^{\circ}\text{C}]{PhCO_3H}$ 91 : 9 (86% of products) (2)

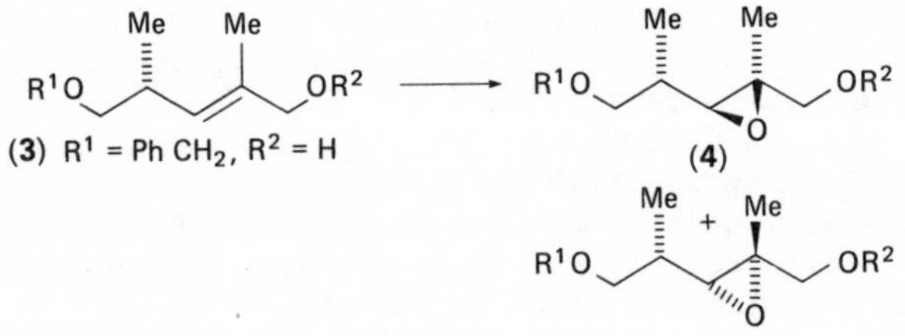

(**3**) $R^1 = PhCH_2$, $R^2 = H$ ⟶ (**4**) + ... (3)

Steric hindrance is most likely operating in reactions of bridged cycloalkenes[28–34]. For example, norbornene (**5**) reacts with *m*-chloroperoxybenzoic acid to give the *exo* epoxide in 99% yield, while 7,7-dimethylnorbornene (**6**) reacts with the same peroxy acid to afford a 9:1 ratio of *endo* to *exo* epoxide (equations 4 and 5)[28,29]. The otherwise unfavourable *endo* attack becomes predominant, most probably due to the shielding to *exo* attack by the *syn*-7-methyl group.

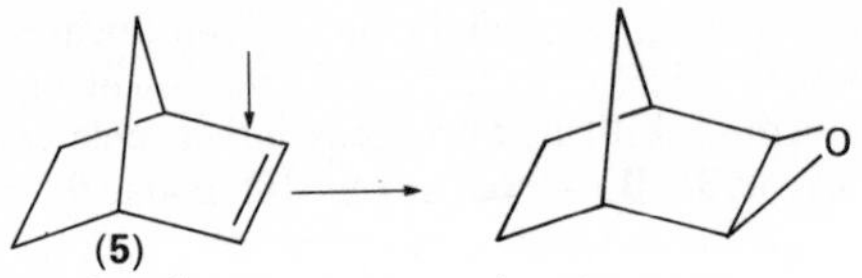

(**5**) ⟶ (4)

(**6**) ⟶ + (5)

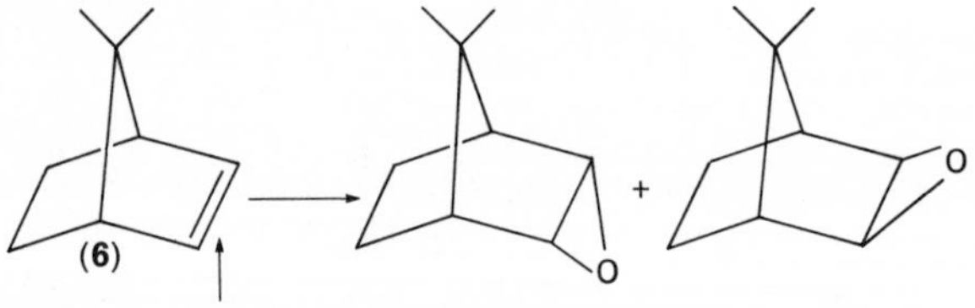

In an important series of papers, Paquette and Gleiter and their coworkers have reported the results of a comparative study of the reaction of 9-isopropylidenebenzonorbornenes (**7**) and snoutanes (**8**) with *m*-chloroperoxybenzoic acid, *N*-bromosuccinimide, singlet oxygen and *N*-methyltriazolinedione (**9**)[32,33] They have reported that all these reagents attack **7a**

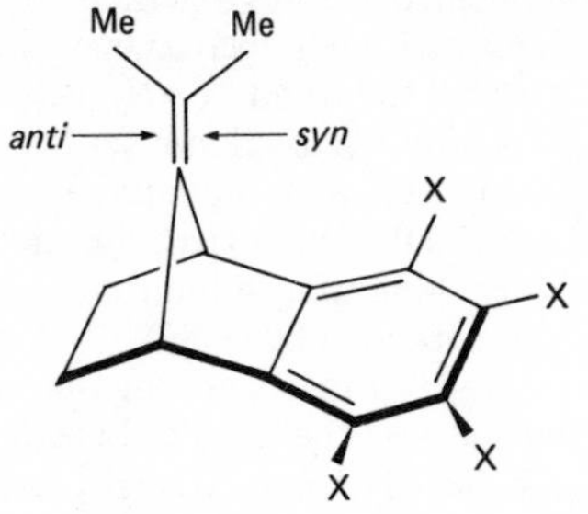

(**7**)

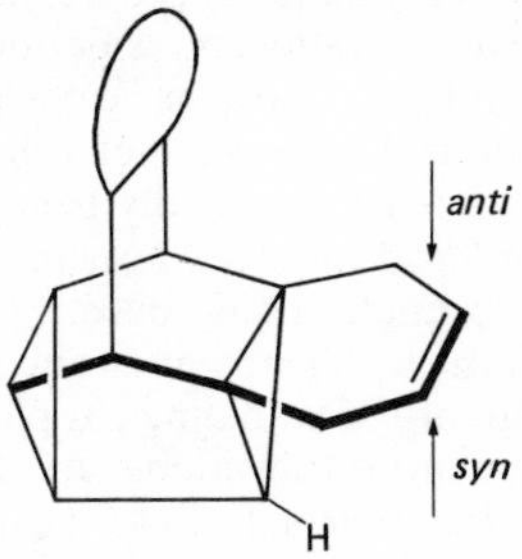

(**8**)

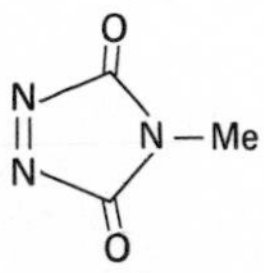

(**9**)

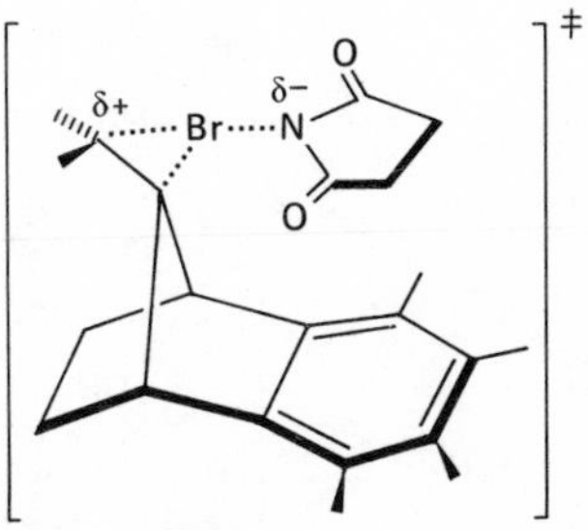

(**10**)

predominantly from the *anti* direction (*syn*:*anti* ratio ≅ 20:80), but that this ratio is appreciably reversed in the case of **7b** and **7c** (*syn*:*anti* ≅ 55:45), except for the reaction with *m*-chloroperoxybenzoic acid (*syn*:*anti* ≅ 35:65)[32]. It is evident that electronic control of stereoselectivity is operative in the reactions of **7b** and **7c** with singlet oxygen, *N*-bromosuccinimide and *N*-methyltriazolinedione. These compounds are believed to react with unsaturated systems through a dipolar transition state (**10**), similar to the one shown above for the attack of *N*-bromosuccinimide. Much less propensity for charge separation in the transition state for the attack of the peroxy acid molecule is thus indicated. The preferential *syn* attack of the peroxy acid on **8** also indicates the predominance of steric factors in these reactions[33].

c. Solvent effects. It appears that epoxidation with peroxy acids is facilitated in solvents of high polarity and low basicity. In aprotic nonbasic solvents, the parallelism between the rate constants and Dimroth–Reichardt E_T values is reported[34].

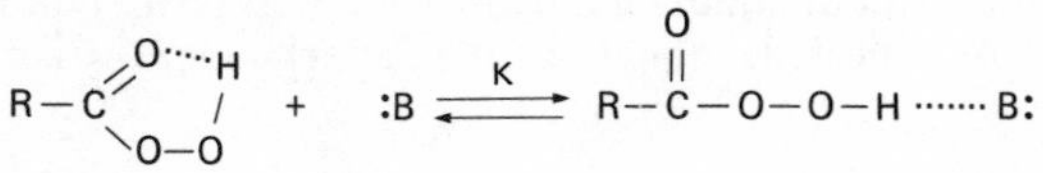

$$\text{R—C(=O}\cdots\text{H)—O—O (chelate)} + \text{:B} \underset{}{\overset{K}{\rightleftharpoons}} \text{R—C(=O)—O—O—H} \cdots\cdots \text{B:} \qquad (6)$$

Considerable reduction of reaction rates has been observed in solvents capable of disrupting the chelated peroxycarboxyl ring to form intermolecularly hydrogen-bonded adducts[35,36] (equation 6). Intermolecular association between the peroxy acid and basic solvent appears to be the major factor influencing the kinetics of epoxidation in these solvents[36,37]. A recent calorimetric study of intermolecular association between both components shows that the calculated enthalpy change per mole of hydrogen-bonded adduct, ΔH_0, is sensitive to the acidity of peroxy acid as well as the basicity of the oxygen base (ethers, esters, amides). A progressive reduction of reaction rates together with increasing enthalpies of activation for oxidation of cyclohexene with *t*-butylperoxybenzoic acid, when going from less to more basic solvents, parallels well the trend in ΔH_0 values[37].

d. Catalysis. The question of acid catalysis of epoxidation is still not settled completely. Although it is now generally believed that epoxidations with organic peroxy acids are usually not acid catalysed (except in the presence of sufficiently large amounts of strong acids, for example, trichloroacetic acid)[38], it is interesting to mention that epoxidation of *trans*-stilbene with peroxymonophosphoric acid is catalysed by H_2SO_4 [39]. A straight line with a slope of 0.87 is obtained by plotting log k_2 vs. $-H_0$, thus suggesting the participation of a proton in the transition state. The attacking species is believed to be either **11** or **12**.

The protonation of the P=O p_π–d_π bonding is believed to proceed easily, resulting in the activation of the peroxy acid as an electrophile (equation 7).

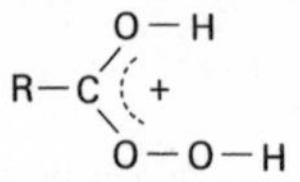

(11) 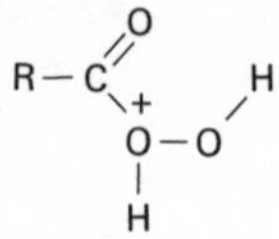 (12)

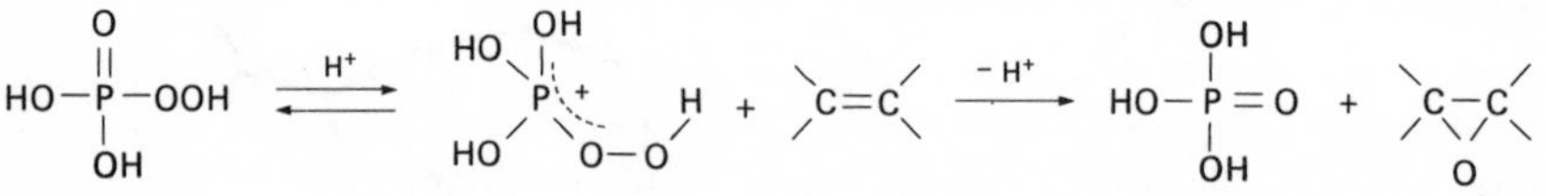

(7)

e. Mechanism. Several mechanisms, which can accommodate the experimental data, have been proposed but a clear-cut experimental substantiation of either of these is still lacking.

Bartlett was the first to propose the so-called 'butterfly' mechanism, which involves nucleophilic attack of the olefin on the peroxy acid[40] (equation 8). The proposed mechanism, which is in accord with the infrared and NMR findings supporting the chelated form of peroxy acids in 'inert' solvents, suggests that proton transfer occurs by a concerted intramolecular process. It also appears to explain adequately large and negative entropies of activation as well as the effect of basic solvents, i.e. the reduction of reaction rates with the reduced 'effective' concentration of peroxy acid in these solvents. Nevertheless, there is some evidence that peroxy acid-base adducts might be involved in the transition state in these cases[5c,37].

Waters suggested the initial attack of a hydroxyl cation (OH^+) on the olefin via the transition state **13**[41]. As indicated in equation (9) the attacking species is not necessarily the free ion.

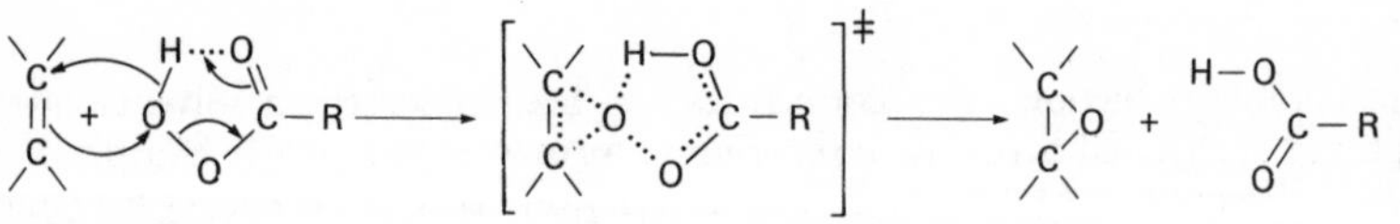

(8)

Kwart and Hoffman proposed an epoxidation mechanism involving 1,3-dipolar addition of a hydroxycarbonyl oxide, derived from the intramolecularly hydrogen-bonded peroxy acid molecule, to the olefinic dipolarophile[42] (equations 10 and 11). To accommodate the fact that the proposed intermediates have actually been isolated and are stable under conditions of epoxidation, a modified transition state **14** was proposed by Kwart and coworkers[43]. The reactivity ratio norbornene:cyclohexene in peroxy acid epoxidation is rather low (2.4; peroxylauric acid). This observation was taken as an evidence against this mechanism[44], since a characteristic feature of reagents believed to react by a cyclic five-membered transition-state mechanism is a high norbornene:cyclohexene ratio (6500, phenyl azide addition)[45]. Therefore, it was argued that a 1,3-dipolar addition mechanism cannot be involved in epoxidation[44]. Nevertheless, according to Huisgen, a 1,3-dipole without a double bond in the sextet (hydroxycarbonyl oxide) should not necessarily show enhanced reactivity with cyclic alkenes[46]. Thus it appears that there is not, as yet, enough experimental evidence to rule out this mechanistic proposal (the '1,3-dipolar addition' approach of peroxy acid to olefin).

(9)

(13)

or

$+ RCO_2H$

(10)

(11)

(14)

Hanzlik and Schearer have reported the results of a study of the secondary kinetic deuterium isotope effect for the epoxidation of *p*-phenylstyrene and three deuterated derivatives with *m*-chloroperoxybenzoic acid in 1,2-dichloroethane[47]. A clear distinction between the α and β carbon atoms of the olefin in the transition state has been found. The k_H/k_D of 0.82 for the β,β-d_2 derivative implies that a significant hybridization change occurs on the β carbon; the α carbon remains essentially sp^2-hybridized. Therefore, the transition state must have a substantial C_β—O bond formation. These data seem to suggest an unsymmetrical transition state **15**. The primary peroxy acid isotope effect (OH/OD) is small, thus indicating that the O—H bond is not broken in the transition state. This result seems to explain why the reaction is not catalysed by acids. Despite the nonlinearity of the O—H···O system, a k_H/k_D of at least 2 would be expected for a process characterized by the symmetrical ('butterfly') transition state. The observed *syn* stereospecificity of epoxidation can only be explained by the retention of some amount of π-bond character of the C—C bond in the olefinic part of the transition state. Rotation around this bond must be slower than the ring-closure.

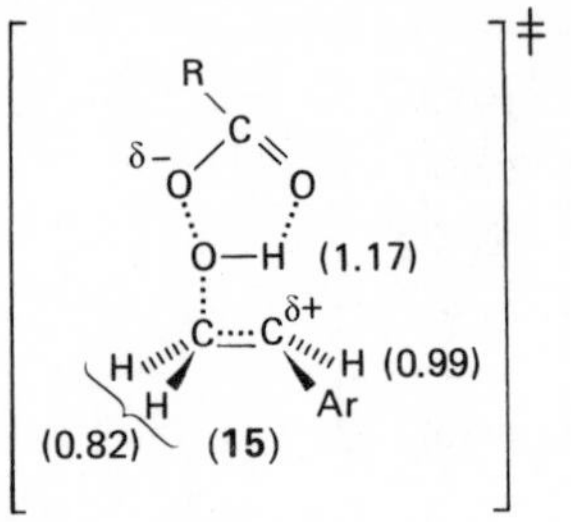

Hanzlik and Schearer have also demonstrated that the unsymmetrical transition state is not a result of asymmetry of the olefin investigated. The relative rates of epoxidation of stilbene, 4-methoxystilbene and 4,4-dimethoxystylbene indicate an equal enhancement of approximately four for each methoxy group. This is expected for statistical and unsymmetrical attack of the peroxy acid at either carbon of the stilbene, regardless to substitution. In addition, *p*-nitrostyrene is oxidized by *m*-chloroperoxybenzoic acid 13 times slower than *p*-phenylstyrene; the isotope effect is still absent ($k_H/k_D = 0.98$, deuteration at C_α) for this olefin.

f. Theoretical studies. The mechanism of epoxidation has been the subject of four theoretical investigations[3,48,49,51]. An extended Hückel study indicated a favourable interaction between the σ orbital on the oxygen atom next to hydrogen in peroxy acid, and the πMO of the olefin. The results were interpreted as supporting the Waters mechanism[48].

A theoretical study of the electronic structure of peroxyformic acid in its intramolecularly hydrogen-bonded and dipolar form (hydroxycarbonyl oxide), using the semiempirical Pariser–Parr–Pople SCF-MO method could not distinguish between the proposed 1,1-addition and 1,3-dipolar cycloaddition mechanism[49].

Detailed *ab initio* (STO-3G and 6-31G) and semiempirical (PCILO, INDO, MINDO/3) molecular orbital calculations have been performed on both peroxytrifluoroacetic and peroxyacetic acid. All these calculations agree that the rotational barriers about the O—O bond are rather low (ca. 3 kcal mol^{-1}), and that the planar *syn* form is the most stable one[50] (equation 12). Recent *ab initio* calculations (5-31G) have shown that the rotational barrier in peroxyformic is indeed low (1.04 kcal mol^{-1}) but that the *anti* planar, rather than the *syn* form, has a minimum energy[52]. Nevertheless, all the available experimental evidence indicates that peroxy acids exist in 'inert' solvents in the chelated form[38]. It seems, therefore, safe to predict that the chelated form is most probably the 'effective' one in the oxidation of unsaturated substrates.

$$\text{R}-\text{C}\overset{\text{O}\cdots\text{H}}{\underset{\text{O}-\text{O}}{}} \rightleftharpoons \text{R}-\text{C}(=\text{O})-\text{O}-\text{O}-\text{H} \tag{12}$$

An *ab initio* molecular orbital theory has also been used to study arbitrarily chosen molecular arrangements believed to be near the transition state for epoxidation of ethylene with peroxyformic acid[3]. It was found that unsymmetric 'transition states' are energetically more favourable than symmetric ones. The inspection of charges (Mulliken population analysis) showed in all 'transition states' under investigation a small but definite shift of electrons from olefin to peroxy acid, indicating a weak electronic interaction between both components. This finding supports the experimentally determined electrophilic nature of peroxy acids. Nevertheless, the reaction is not just simply charge-interaction-controlled but must also involve some overlap control. This is most clearly evident in the 'transition state' with a minimum energy shown in Figure 2. The electron-deficient p_y orbital of O(1) (the oxygen–oxygen bond axes) and the lowest unoccupied (LUMO) antibonding σ^* orbital (oriented in the same direction) of the peroxy acid are aligned nearly perpendicularly to the plane of the olefin part of the 'transition state'. Thus, displacement on the peroxide occurs most favourably from the backside and along the axis of the O—O bond being broken. Theoretical evidence indicates an exothermic reaction of 46 kcal mol^{-1}. Relatively small

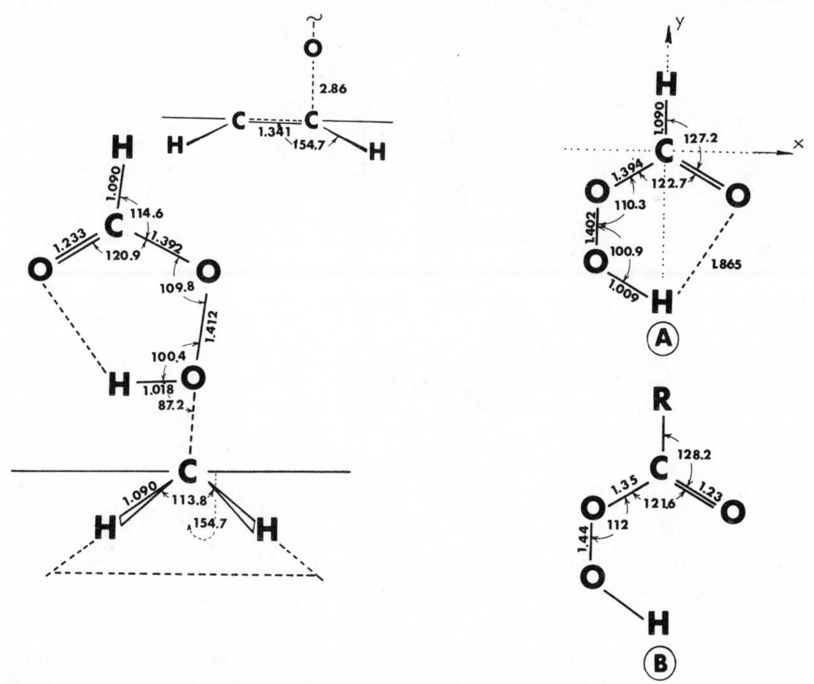

FIGURE 2. Optimized geometry of the transition state for the reaction of peroxyformic acid with ethylene. Bond lengths and angles are in Ångstroms and degrees. (A) Optimized geometry of peroxyformic acid. (B) Crystal-structure geometry of peroxypelargonic acid.

reorganization in both participating molecules indicates an 'early' transition state, which is in accord with the relatively great exothermicity of the reaction (the Hammond postulate). The findings of this study also support the previously reported nonequivalency of both olefinic carbon atoms in the epoxidation of substituted styrenes, which is evidently not a consequence of a choice of an unsymmetric olefin[47].

The *syn*-directing ability of a hydroxyl group in the peroxy acid oxidation of allylic alcohols[23,24] (Section II.B.1.b) has recently been explained by the stereoelectronically favourable orientation of both participating molecules as shown in Figure 3. The nonbonding electron pair *a* is believed to be oriented favourably as to form a hydrogen bond with an allylic hydroxyl group (O—C—C=C dihedral angle, 120°)[53,54]. It appears that this suggestion explains more satisfactorily the directing ability of the hydroxyl group compared to the previous interpretation, which indicated the formation of hydrogen bonding to the carbonyl oxygen of the peroxy acid[24].

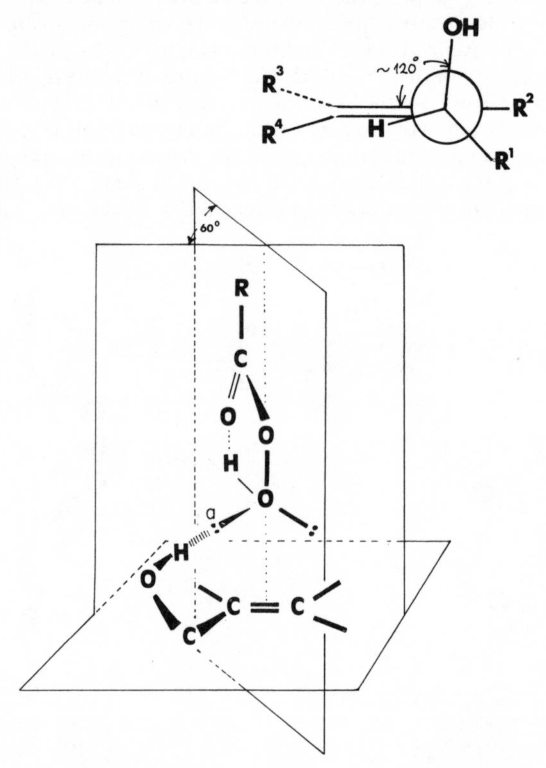

FIGURE 3. The preferred geometry of the transition state for the oxidation of allylic alcohols with peroxy acids suggested by Sharpless and Verhoeven[53].

The observed cooperative effect by a hydroxy group and ether oxygen in directing the steric course of the attacking peroxy acid (Section II.B.1.b), has been rationalized as being due to the energetically favourable possibility of formation of two hydrogen bonds in the transition state for epoxidation of the *trans*-allylic alcohol derivative, as shown in Figure 4[27].

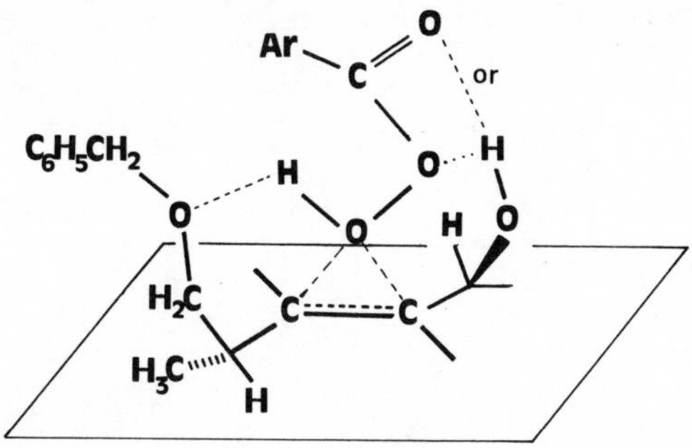

FIGURE 4. The preferred geometry of the transition state for the oxidation of **3** with *m*-chloroperoxybenzoic acid, as suggested by Johnson and Kishi[27].

2. Oxidation with hydroperoxides

Recently, a number of α-substituted hydroperoxides of esters (**16**), amides (**17**) and nitriles (**18**), which can be regarded as homologues of peroxy acids, have been found to epoxidize olefins in a stereospecific manner[55,56]. The reactions are first order with respect to the hydroperoxide and olefin and show the same trend of reduction of reaction rates as observed

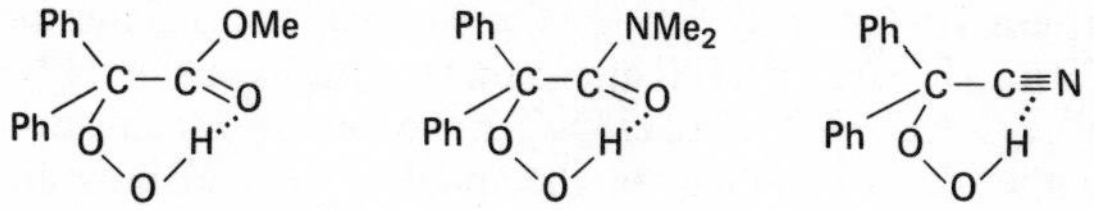

(**16**) (**17**) (**18**)

with peroxy acids when going from dichloromethane to diethyl ether as solvent. A mechanism similar to the one proposed for peroxy acid epoxidation[41] might be operative in these reactions (equation 13). A preference for *syn* epoxidation of 2-cyclohexenen-1-ol has been reported. It is interesting that, although peroxy acids generally show a slight preference for *cis* isomers, **16** attacks more rapidly the *trans* isomer.

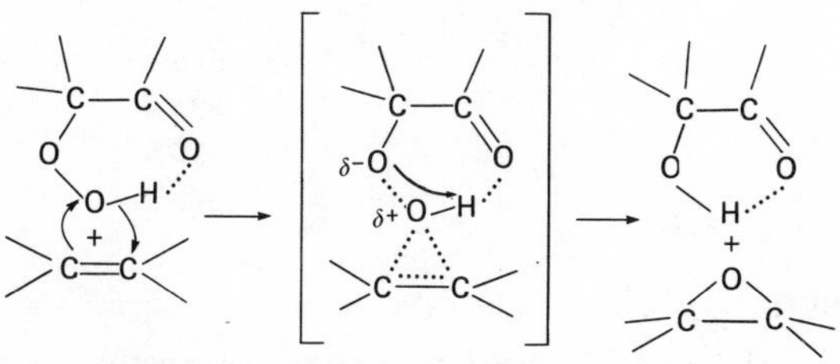

(13)

It is now widely accepted that epoxidations with alkyl hydroperoxides in the presence of 5b and 6b transition metals (Mo, W, V) have characteristics of nucleophilic displacement on the electrophilic oxygen (from the less hindered side) of the O—O bond in the hydroperoxide–catalyst complex[53,57]. Various studies seem to indicate the mechanism shown in equations (14)–(16)[58–60]. Since these reactions have recently been extensively reviewed[53,57–60], only one example will be given here.

$$L_nM + ROOH \longrightarrow L_{(n-1)}MOOR + LH \tag{14}$$

$$M{-}O{-}O{-}R + >C{=}C< \longrightarrow \left[>C{-}C< \cdots O(OR){-}M \right]^{\ddagger} \longrightarrow MOR + \text{epoxide} \tag{15}$$

$$RO{-}O{-}M \;\; >C{=}C< \longrightarrow RO{\cdots}O{-}M \;\; >C{-}C< \longrightarrow MOR + \text{epoxide} \tag{16}$$

We have already discussed the *cis*-directing effect of a hydroxy group in the epoxidation of a suitably constituted allylic alcohol with peroxy acids (Section II.B.1b). This effect is even more pronounced when *t*-butyl hydroperoxide/bis(acetylacetonato)oxovanadium(IV) [$VO(acac)_2$] reagent is used[61,62] (equation 17). The opposite direction of stereoselectivity for both reagents has been found for the eight- and nine-membered-ring allylic alcohols, respectively[61,62]. On the basis of data obtained in the study of cyclic and acyclic allylic alcohols[63], Sharpless has recently proposed a mechanistic scheme for the vanadium-catalysed epoxidation (Figure 5)[53]. It is interesting to note a considerably smaller value for the dihedral angle, ⟨O—C—C=C, of ca. 50°, in the stereoelectronically most favourable conformation of the allyloxy moiety of the roughly trigonal bipyramidal complexes, compared to the one in the epoxidation with peroxy acids (Figure 4).

(a) m-$ClC_6H_4CO_3H$, CH_2Cl_2, 0°C
(b) t-BuOOH/$VO(acac)_2$, benzene, 40°C

(17)

		cis (H, OH / O same side)	trans
$n = 3$	(a)	61%	39%
	(b)	99.6%	0.4%
$n = 4$	(a)	0.2%	99.8%
	(b)	97%	3%

C. Hydroxylation

Aromatic hydrocarbons are oxidized to phenols with peroxy acids. For example, mesitylene (**19**) is easily oxidized by trifluoroperoxyacetic acid, either alone[64] or in the presence of boron trifluororide[65] (equation 18). The reaction is regarded as an electrophilic aromatic substitution, although the nature of the attacking species is still unclear. The hydroxy cation HO^+, or its equivalent, has been suggested as the electrophilic agent[65] (equation 19). Direct nucleophilic attack by the aromatic compound on the electrophilic oxygen of the peroxy acid has also been proposed[66] (equation 20).

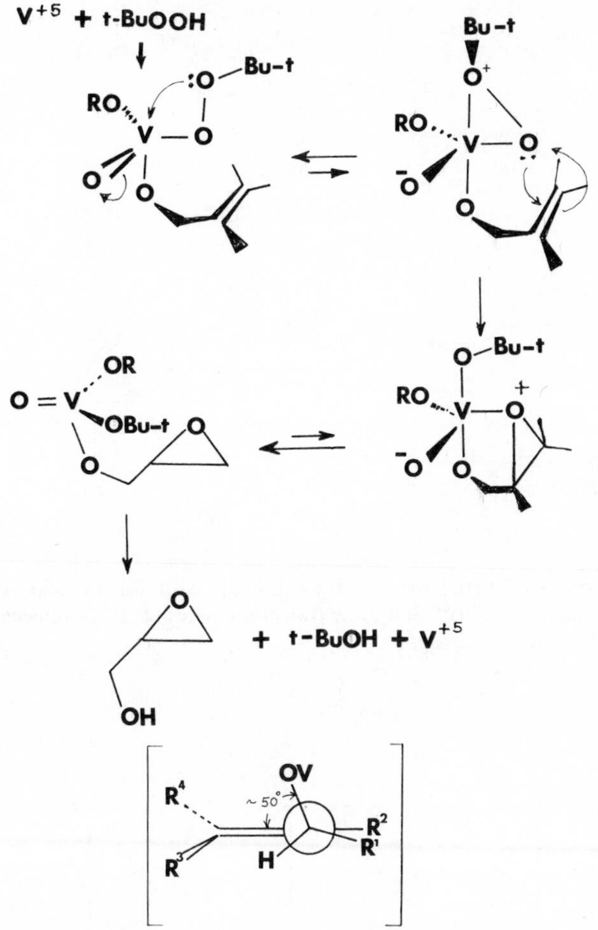

FIGURE 5. The mechanism for the vanadium-catalysed epoxidation of allylic alcohols with *t*-butyl hydroperoxide suggested by Sharpless and coworkers[53,63].

A recent study of the hydroxylation of mesitylene and phenol with peroxymonophosphoric acid, H_3PO_5[67], in acetonitrile reveals that the kinetics is expressed by the equation

$$v = k_2[\mathrm{ArH}][\mathrm{H_3PO_5}]$$

It is interesting that this peroxy acid reacts ca. 100 times faster than peroxyacetic or peroxybenzoic acid and is comparable in activity to trifluoroperoxyacetic acid.

A linear plot of log k_2 vs. $-H_0$ with a slope of 1.17 for mesitylene and 1.26 for phenol has been reported for reactions catalysed by H_2SO_4. A simultaneous attack of unprotonated and protonated H_3PO_5 on the aromatic substrate has been suggested[67].

$$\text{(19) 1,3,5-trimethylbenzene} \xrightarrow[\text{CH}_2\text{Cl}_2]{\text{CF}_3\text{CO}_3\text{H–BF}_3} \text{2,4,6-trimethylphenol} \quad (18)$$

$$RCO_3H \longrightarrow RCO_2^- \; (OH^+)$$

$$CF_3-C(=O^{\delta+}\cdots BF_3)-O-OH \quad \text{or} \quad CF_3-C(=O)-O^{\delta+}(\cdots BF_3)-OH \quad (19)$$

$$ArH + HO-O-C(=O)-R \longrightarrow \overset{+}{Ar}(H)(OH) \xrightarrow{-H^+} Ar-OH \quad (20)$$

Hydroxylation of aliphatic saturated compounds with peroxy acids has also been reported. Both nonionic (**20**)[68] and ionic transition states (**21**)[69] have been suggested as being involved in these reactions.

$$v = k_2 [ArH][H_3PO_5]$$

(**20**) $[CF_3-C(=O\cdots H)-O-O \; / \; R-C-H]^{\ddagger}$ $[CF_3-C(=O)-\overset{+}{O}(H)-O-H \; / \; R-C-H]^{\ddagger}$ (**21**)

The hydroxylation of alkanes can also be accomplished with hydrogen peroxide in the presence of FSO_3H–SbF_5, FSO_3H, H_2SO_4 [69]. Initial electrophilic hydroxylation of the σ bonds in the alkane by the hydroxyl cation, HO^+, formed by cleavage of the hydroperoxonium ion, $H_3O_2^+$, and proceeding through a pentacoordinated carbonium ion, has been suggested[69] (Scheme 2).

D. Oxidation of Acetylenes

Although mechanistic studies on the oxidation of acetylenes are not as abundant as those for olefins, several interesting observations concerning the mechanism of this reaction have been made in recent years[70–76]. At least theoretically, monoepoxidation of acetylenes

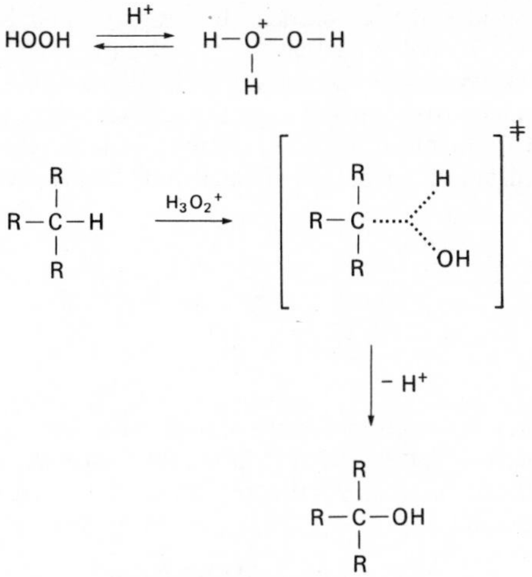

$$\mathrm{HOOH} \underset{}{\overset{H^+}{\rightleftharpoons}} \mathrm{H{-}\overset{+}{O}(H){-}O{-}H}$$

$$\mathrm{R_3C{-}H} \xrightarrow{H_3O_2^+} \left[\mathrm{R_3C\cdots\langle^{H}_{OH}}\right]^{\ddagger} \xrightarrow{-H^+} \mathrm{R_3C{-}OH}$$

SCHEME 2

should first produce oxirenes (**22**), which can be further converted to dioxabicyclo derivatives, i.e. 1,3-dioxabicyclo[1.1.0]butanes (**23**), as shown in equation (21). Neither of these two intermediates have yet been detected. However, oxidation of acetylenes with peroxy acids usually gives a mixture of products which have been explained on the basis of these species[70,71]. Particularly revealing is the conversion of disubstituted acetylenes to disubstituted ketenes in a substituent migration process reported some years ago[72]. It was only recently that a 1,2-hydride shift in both the peroxy acid and enzymatic oxidation of a triple bond was unambiguously demonstrated by studying the fate of the deuterium in the oxidation of monodeuterated biphenylacetylene[76]. It appears that two mechanisms are *a priori* possible for the oxidation of terminal acetylenes. The first pathway (a) involves insertion of oxygen into the carbon–hydrogen bond (the hydrogen becomes an exchangeable hydroxyl proton) and in the pathway (b) the hydrogen is transferred in an intramolecular 1,2-shift (equation 22). Nearly quantitative retention of deuteration in the esterified 2-biphenylacetic acid was found, indicating that this oxidation involves reaction of the reagent with the π electrons (oxirene formation), rather than with the terminal C—H bond of acetylene (hydroxyacetylene formation)[76].

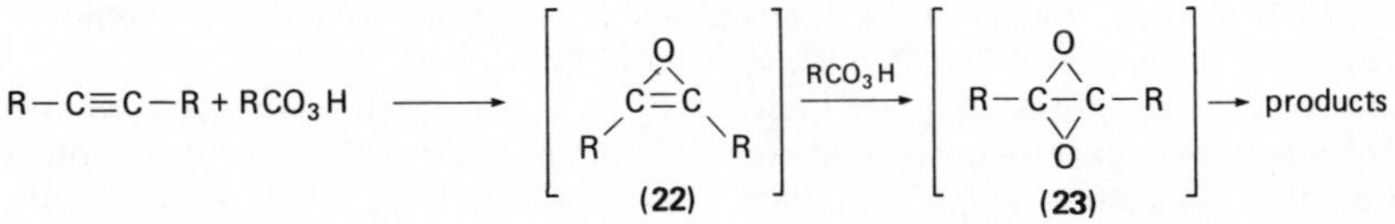

$$\mathrm{R{-}C{\equiv}C{-}R + RCO_3H} \longrightarrow \left[\text{oxirene (RC=CR with O bridge)}\ (\mathbf{22})\right] \xrightarrow{RCO_3H} \left[\mathrm{R{-}C{-}C{-}R}\ \text{(two O bridges)}\ (\mathbf{23})\right] \longrightarrow \text{products} \quad (21)$$

$$\mathrm{R{-}C{\equiv}C{-}H^*} \begin{cases} \xrightarrow{(a)} \mathrm{R{-}C{\equiv}C{-}OH} \longrightarrow \mathrm{R(H)C{=}C{=}O} \xrightarrow{H_2O} \mathrm{R(H^*(H))CH{-}C(=O){-}OH} \\ \xrightarrow{(b)} \text{oxirene (R, H}^*\text{; 1,2-shift)} \longrightarrow \mathrm{R(H^*)C{=}C{=}O} \xrightarrow{H_2O} \mathrm{R(H^*(H))CH{-}C(=O){-}OH} \end{cases} \quad (22)$$

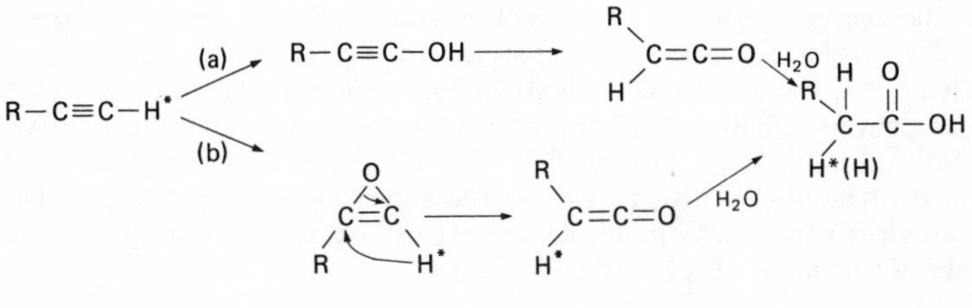

Oxirenes are of considerable theoretical interest as models of four π-electron antiaromatic molecules[77]. It is obvious that the ring strain and electronic destabilization should render oxirenes exceptionally reactive and difficult to stabilize. In principle, oxirenes can exist in equilibrium with oxocarbenes (equation 23), and it is quite possible that both kinds of intermediates are involved in these reactions. Indeed, recent *ab initio* MO SCF calculations show a small activation barrier of 2 kcal mol^{-1} for rearrangement of oxirene to formylacetylene[77d].

$$\text{RC}{=}\text{CR (oxirene, O bridging)} \rightleftharpoons \text{R}-\overset{\text{O}}{\overset{\|}{\text{C}}}-\ddot{\text{C}}-\text{R} \qquad (23)$$

In general, acetylenes react considerably slower with peroxy acids compared to structurally similar olefins[78,79]. It has been reported that, when the acetylene:peroxy acid ratio is 5:1 or greater, the stoichiometry of the reaction is 1:1. In this case, the reaction is first order in each of the reactants:

$$v = k[\text{peroxy acid}][\text{acetylene}]$$

The kinetics of the reaction of substituted phenylacetylenes with peroxybenzoic acid has been measured in benzene at 25°C. The Hammett plot with σ^+ gives a ρ value of -1.40 indicating the electrophilic nature of the reaction[78]. Oxidation of 4-octyne with *m*-chloroperoxybenzoic acid in various solvents capable of intermolecular association with peroxy acids has revealed that peroxy acid–solvent interactions play a similar role in the oxidation of this compound as in epoxidation of cyclohexene. A linear free-energy relationship with a slope of one was obtained by correlating the logarithmic rates of 4-octyne oxidation with those of epoxidation of cyclohexene. The trend of activation parameters in both cases is also the same[79].

All above-mentioned results seem strongly to support oxirene as the first intermediate in the oxidation of acetylenes with peroxy acids. A recent theoretical study[80], using the restricted Hartree–Fock level of *ab initio* theory (STO-4G, 6-31G), has indicated that unsymmetric attacks of peroxyformic acid on acetylene are energetically more favourable than symmetric ones. In the framework of the 1,1-addition mechanism, the unsymmetric attack of the peroxy acid with a trajectory having the H—C—PA angle of ca. 90° was found to be energetically most favourable (Figure 6.) The approach of the 'electrophilic' peroxy acid appears to be dominated by charge-transfer control although some overlap-control is also indicated. The 'transition state' presumably involved in the 1,3-dipolar addition mechanism is energetically slightly more favourable than that involved in the 1,1-addition mechanism. A 'loose' transition state is indicated in both cases.

The unsymmetrical 'transition state' indicated by *ab initio* calculations is in accord with the above-mentioned experimental observation of the importance of the σ^+ contribution to the Hammett correlation in the oxidation of *para*-substituted phenylacetylenes with peroxybenzoic acid, indicating the accumulation of a partial positive charge on one of the carbon atoms in the acetylenic part of the transition state (partially bridged or open transition state). In terms of a recent classification of electrophilic reagents in their reaction toward unsaturated C—C bonds, peroxy acids could be classified as class B reagents[80b].

It is interesting to mention that the second step of the reaction, i.e. the oxidation of oxirene, cannot yield 1,3-dioxabicyclo[1.1.0]butane (**23**). The relatively large bond distance between bridgehead carbons (1.876 Å), as well as the corresponding negative overlap population, indicate clearly that such a molecule cannot exist. Nevertheless, the possibility of a diradical state of this molecule cannot be excluded.

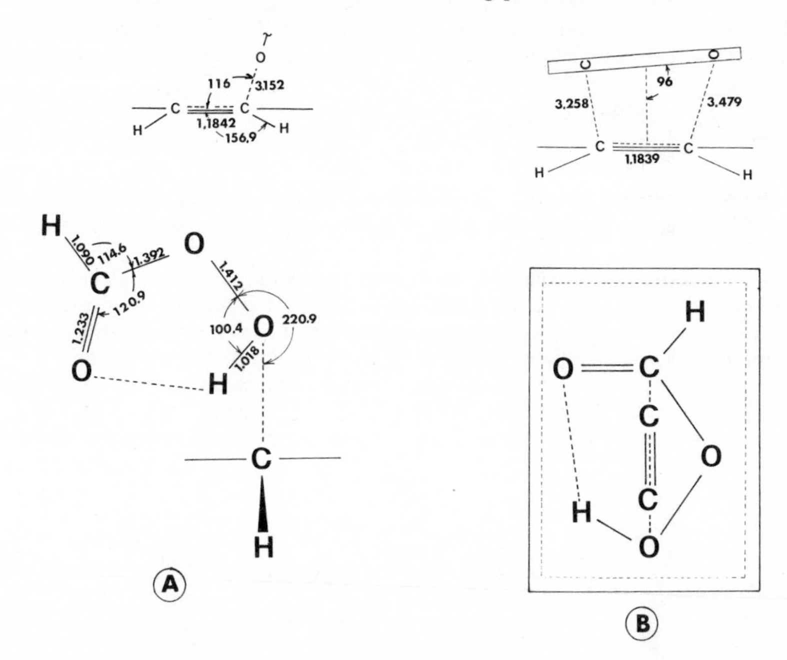

FIGURE 6. Optimized geometry of the 'transition state' for the oxidation of acetylene with peroxyformic acid. (A) 1,1-Addition mechanism; (B) 1,3-dipolar addition.

E. Oxidation of Organic Nitrogen Compounds

1. Amines

Tertiary amines react with peroxy acids to form amine oxides in a process which is believed to be analogous to the epoxidation reaction. Namely, it was reported that an increase in nucleophilicity of the amine accelerates the reaction[81,82] (equation 24). These reactions are of the second order as are oxidations of tertiary amines with hydrogen peroxide. The Hammett ρ value of -2.35 was obtained for the peroxybenzoic acid oxidation of substituted pyridines[82].

$$R_3N\colon + RCO_3H \longrightarrow \left[R_3N\cdots O\overset{H\cdots O}{}\!\!\!\!\overset{}{\underset{O}{}}C{-}R \right]^{\ddagger} \longrightarrow R_3\overset{+}{N}{-}\overset{-}{O} + RCO_2H \qquad (24)$$

Primary amines react with peroxy acids to form nitroso compounds via initially formed hydroxylamines[83] (equation 25). Detailed study of the peroxyacetic acid oxidation of anilines in aqueous ethanol revealed a slow formation of phenylhydroxylamines, and fast conversion of the latter to nitrosobenzenes. The Hammett ρ value of -1.86 was reported for the slow step of this reaction. The unexpected greater reactivity of phenylhydroxylamines compared to anilines was explained as being due to the 'α effect'[83]. The reaction most probably involves the initial formation of an N-oxide which rearranges to the hydroxylamine. Direct insertion of the oxygen atom into the N—H bond seems unlikely.

$$-\overset{|}{\underset{|}{C}}-NH_2 \xrightarrow{[O]} -\overset{|}{\underset{|}{C}}-NHOH \xrightarrow{[O]} -\overset{|}{\underset{|}{C}}-N{=}O \xrightarrow{[O]} -\overset{|}{\underset{|}{C}}-NO_2 \quad (25)$$

Oxidation of nitroso to nitro compounds presumably involves similar nucleophilic attack of nitrogen on electrophilic oxygen of the peroxy acid (the Hammett ρ value, -1.58)[84].

Although no detailed mechanistic studies on the oxidation of secondary amines to nitroxides or nitrones with peroxy acids are available, it is reasonable to assume that intermediate formation of hydroxylamines in these reactions most probably involves the same initial nucleophilic attack of the amine on the electrophilic peroxy acid[84].

A recent mechanistic study of oxidation of tertiary and secondary amines, as well as hydroxylamines with 4a-hydroperoxyflavin (**24**) in t-butanol shows that all these oxidations are second-order reactions; first order in amine and first order in **24**[85,86].

(**24**)

Reactions with secondary amines give quantitatively the corresponding hydroxylamines. The flavin pseudo-base, 4-FlEtOH, is the reduction product in these oxidations. Primary amines react much slower, the reaction is not first-order and a complex mixture of products is obtained. It is interesting that a change of solvent from t-butanol to the aprotic solvent dioxane decreases the second-order rate but does not change the kinetic order. The relative reactivity of amines toward **24** (4a-FlEtOOH) is hydroxylamines > t-amines > s-amines. All these results are consistent with a bimolecular nucleophilic attack of amine nitrogen on the terminal oxygen of **24** with back-donation of the hydroperoxy hydrogen (equation 26). Bruice and coworkers have found that the oxidizing ability of **24** in N-oxidation is more than four orders of magnitude greater than that of hydrogen peroxide or t-butyl hydroperoxide. This fact also seems to explain the lack of a requirement for general acid catalysis in N-oxidations with this reagent.

$$\underset{(\mathbf{24})}{\text{4a-FlEt}-O-O-H} + :NR_3 \longrightarrow \text{4a-FlEtOH} + R_3\overset{+}{N}-\overset{-}{O} \quad (26)$$

2. Oximes

A study of the stereochemistry of oxidation of 2-substituted cyclohexanone oximes with trifluoroperoxyacetic acid in acetonitrile shows a predominant formation of cis-2-substituted nitrocyclohexanes[87,88]. It has been suggested that the formation of intermediate nitronic acids (**25**) involves displacement by sp^2-hybridized nitrogen on the electrophilic oxygen of the peroxy acid[88] (equation 27).

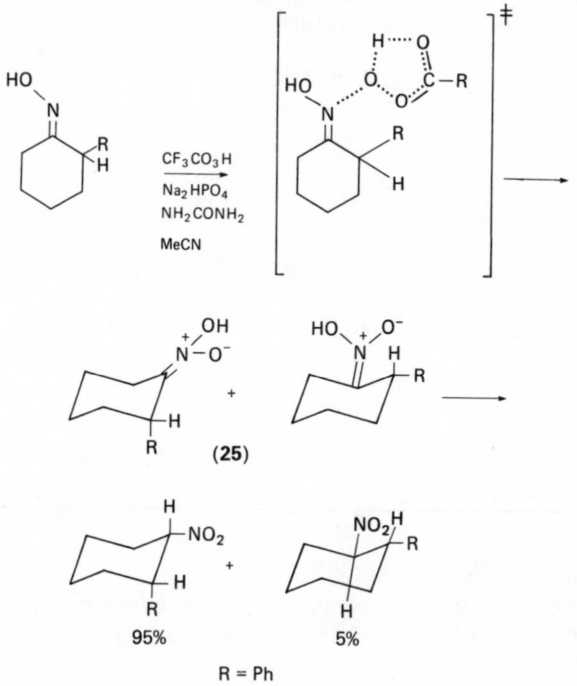

CF_3CO_3H, Na_2HPO_4, NH_2CONH_2, MeCN

(**25**)

(27)

95% 5%

R = Ph

3. *Azo and diazo compounds*

Peroxy acids oxidize azo compounds to azoxy derivatives in a one-step electrophilic attack at either of the lone pairs of the azo nitrogens. The substituent effect on the rate and product isomer ratio in the reaction of *trans*-azobenzene with peroxybenzoic acid seems to exclude the attack of the peroxy acid on the π electrons of the azo double bond to form an oxaziridine-type intermediate (**26**)[89] (Scheme 3).

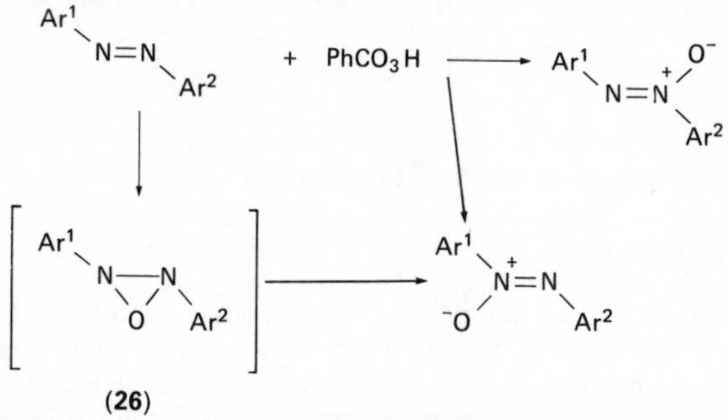

$PhCO_3H$

(**26**)

SCHEME 3

It has been demonstrated that electron-releasing groups on one of the benzene rings of diazoaminobenzenes (**27**) direct the electrophilic attack of the peroxy acid to the nearest nitrogen atom (Scheme 4). The formation of a metastable intermediate (**28**), which collapses to the isomeric *N*-oxides, has been suggested[90].

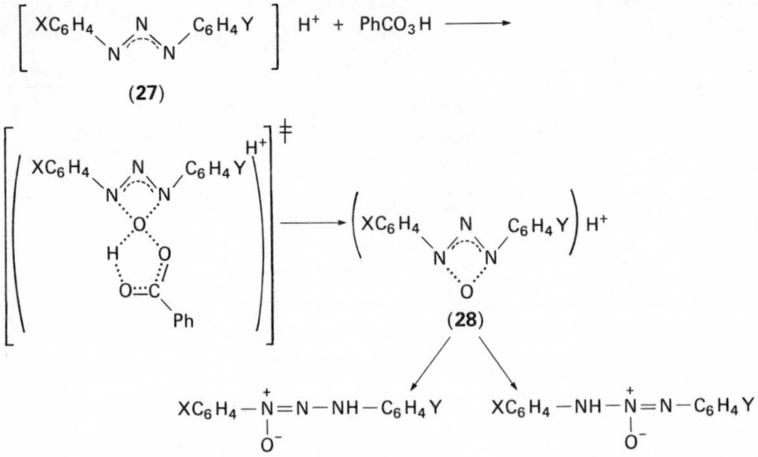

$$[XC_6H_4-N=N-N-C_6H_4Y]H^+ + PhCO_3H \longrightarrow$$

(**27**)

$$[\ldots]^{\ddagger} \longrightarrow (XC_6H_4-N\cdots N\cdots N-C_6H_4Y)H^+$$

(**28**)

$$XC_6H_4-\overset{+}{N}(O^-)=N-NH-C_6H_4Y \qquad XC_6H_4-NH-\overset{+}{N}(O^-)=N-C_6H_4Y$$

SCHEME 4

The direct attack by the diazoalkane nucleophilic carbon atom on the electrophilic oxygen of the peroxy acid has been suggested as being involved in the oxidation of diazodiphenylmethane (**29**) to benzophenone with peroxybenzoic acid in various solvents[91] (equation 28).

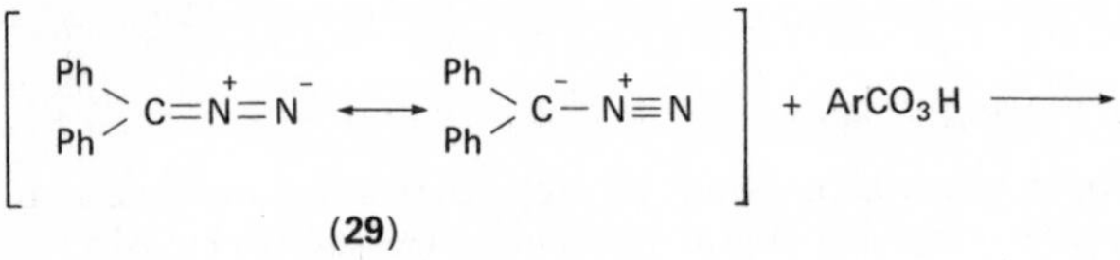

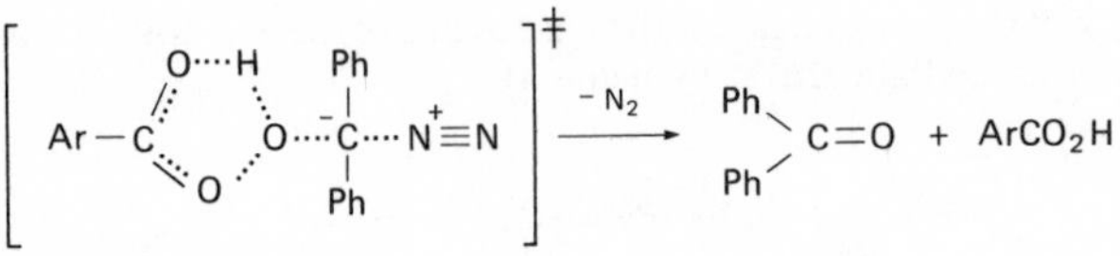

$$\left[Ph_2C=\overset{+}{N}=\overset{-}{N} \longleftrightarrow Ph_2\overset{-}{C}-\overset{+}{N}\equiv N\right] + ArCO_3H \longrightarrow$$

(**29**)

$$\left[Ar-C(=O)-O\cdots H\cdots O\cdots \overset{-}{C}Ph_2\cdots\overset{+}{N}\equiv N\right]^{\ddagger} \xrightarrow{-N_2} Ph_2C=O + ArCO_2H \qquad (28)$$

F. Oxidation of Organic Sulphur Compounds

1. Sulphides and sulphoxides

The peroxy acid oxidation of sulphides to sulphoxides is a second-order process; first order in peroxy acid and first order in sulphide[92,93]. Electron-withdrawing groups in the peroxy acid and electron-donating groups in the sulphides accelerate the reaction, again indicating the electrophilic nature of peroxy acids in these oxidations. A Hammett ρ value of $+1.05$ was found for the oxidation of bis(p-chlorobenzyl)sulphide with substituted peroxybenzoic acids in isopropanol[92]. The reactions are not acid catalysed or subject to salt effects. All these seem to suggest a nucleophilic displacement by the sulphur atom on the electrophilic oxygen in the peroxy acid (equation 29). Specific solute–solvent interactions by hydrogen bonding rather than the polarity of the medium have been found to be the major factor affecting the rates and activation parameters of oxidation of *p*-nitrodiphenyl sulphide with peroxybenzoic acid[93]. Solvents with basic oxygen retard the rates of oxidation by disrupting the chelate ring of the peroxy acid with subsequent formation of the intermolecularly hydrogen-bonded peroxy acid–base adduct, thus lowering the concentration of the 'effective' form of the peroxy acid. Higher activation enthalpies, $\Delta H^{\neq}$, and more positive activation entropies. $\Delta S^{\neq}$, observed in basic solvents compared to 'inert' solvents have been suggested as reflecting desolvation processes on going from reactants to the transition state. Recent studies show that peroxy acids indeed form intermolecularly hydrogen-bonded adducts with bases with the strength of intermolecular association being roughly parallel to the basicity of the base and the acidity of the peroxy acid[37]. Although little is known about the geometry of the transition state in these reactions, it is believed that the above-mentioned interpretation of this phenomenon must be an oversimplification, and that the solvent molecule is most probably a part of the transition state[249].

$$R_2S: + RC(=O)OOH \longrightarrow [R_2S\cdots O \cdots H \cdots O=C(R)\cdots O]^{\ddagger} \longrightarrow R_2S{=}O + RCO_2H \quad (29)$$

The oxidation of sulphides to sulphoxides by hydrogen peroxide and by alkyl hydroperoxides is susceptible to acid catalysis and is believed to involve nucleophilic displacement on the peroxide oxygen[94] (equation 30). A Hammett ρ value of -0.98 is reported for the oxidation of substituted diphenyl sulphides (XC_6H_4SPh) with hydrogen peroxide[94]. The acid catalyst HA is believed to assist the fission of the O—O bond, and to form the transition state which allows proton transfer. In protic solvents which can serve as acids (alcohols, carboxylic acids, water), the reaction is second order, first order in both peroxide and sulphide. The rate of oxidation of sulphides by hydroperoxides can thus be given by the equation

$$v = k[\text{sulphide}][\text{hydroperoxide}][\text{general acid catalyst}]$$

In aprotic solvents (for example, dioxane), the alkyl hydroperoxide can serve as a proton donor and the reaction becomes second order in hydroperoxide.

$$R_2S: + ROOH \longrightarrow [R_2S\cdots O(H\cdots A)\cdots O(R)\cdots H]^{\ddagger} \longrightarrow R_2S{=}O + ROH + HA \quad (30)$$

It is interesting that the oxidation of sulphides by 4a-hydroperoxyflavin, (**24**) (4a-FlEtOOH), in dioxane is first order in hydroperoxide and not second order as expected[95,96]. The lack of the requirement for general acid catalysis in *S*-oxidation by this compound is attributed to the much greater oxidizing potential compared to hydrogen peroxide or alkyl hydroperoxide. The following ratio of second-order rate constants for the oxidation of dioxane in methanol is found:

$$\text{4a-FlEtOOH}:H_2O_2:t\text{-BuOOH} = 2 \times 10^5:20:1.$$

Further oxidation of sulphoxides to sulphones with peroxy acids is much slower compared to the oxidation of sulphides[97]. Electron-donating groups in sulphoxides (the Hammett $\rho = -1.06$, oxidation of substituted diphenyl sulphoxides, $(XPh)_2SO$, with peroxybenzoic acid in benzene at 20°C)[97c] and electron-withdrawing groups in peroxy acids ($\rho = 0.75$, oxidation of *p*-tolyl methyl sulphoxide with substituted peroxybenzoic acids)[97b] accelerate the reaction.

Two mechanisms have been proposed for this oxidation. The first one involves the nucleophilic displacement on the peroxidic oxygen, similar to the one proposed in the oxidation of sulphides ('electrophilic' oxidation)[97]. Lower rates of oxidation of sulphoxides compared to sulphides have been rationalized as being due to the lower nucleophilicity of sulphoxides.

Solvent effect data, on the other hand, appear to favour a two-step process[84,93] (equation 31). In the framework of the two-step mechanism, a positive Hammett ρ value, obtained in the oxidation with substituted peroxybenzoic acids, could be explained by the occurrence of the O—O cleavage in the rate-determining step which should be favoured by the withdrawal of electrons from this bond.

$$R_2S{=}O + RCO_3H \rightleftharpoons R_2\overset{+}{S}{-}OH + RCO_3^- \xrightarrow{\text{fast}} R_2S(OH)(O{-}O{-}C(=O){-}R) \xrightarrow{\text{slow}} R_2SO_2 + RCO_2H \qquad (31)$$

An analogous mechanism, involving nucleophilic attack of the peroxycarboxylate anion on the sulphur atom, has been proposed in the alkaline oxidation of sulphoxides with peroxy acids[97] (equation 32). The present author advocates still another possibility, i.e. the

$$RCO_3^- + R_2S{=}O \longrightarrow \left[R{-}C(=O){-}O{-}O{-}S(R_2){-}O^- \right] \longrightarrow RCO_2^- + R_2SO_2 \qquad (32)$$

formation of a sulphur bis(acyldioxy)derivative (**30**), which could easily break down to the corresponding sulphone. Recent evidence for the existence of diperoxysulphuranes (**32**) in the reaction of sulphuranes (**31**) with *t*-butyl hydroperoxide[98] appears to substantiate such a proposal (equation 33).

$$\left[R_2S(O{-}O{-}C(=O){-}R)_2 \right] \qquad \textbf{(30)}$$

$$Ph_2S(OR)_2 \ (\mathbf{31}) + t\text{-BuOOH} \longrightarrow [Ph_2S(OOBu\text{-}t)(OH)] \xrightarrow{t\text{-BuOOH}} [Ph_2S(OOBu\text{-}t)_2] \ (\mathbf{32}) \qquad (33)$$

$R = -C(CF_3)_2Ph$

It is interesting to mention that the hydroperoxy anion, HOO^-, oxidizes sulphoxides to sulphones in aprotic solvents[97b], while this reaction does not proceed in aqueous media[97a].

2. *Thiobenzophenones*

The oxidation of thiobenzophenones (**33**) to sulphines (**34**) with peroxybenzoic acid is a second-order reaction[99]. The second-order rate constants at 25°C in carbon tetrachloride for the oxidation of substituted thiobenzophenones fit the Hammett equation with the ρ value of -0.88 (equation 34). The second step of the reaction, the oxidation of sulphines to carbonyl compounds, is a slower process[100]. The negative Hammett ρ value of -1.16 ($\sigma^+ + \sigma$) indicates that sulphines also act as nucleophiles in these reactions. Since the effect of solvent is similar to that found for the oxidation of sulphides or olefins, a nucleophilic attack of the sulphur atom on the peroxy acid has been proposed for the first step of the reaction, and an electrophilic attack of the peroxy acid on the π-electron system of sulphines for the second step (equation 35). The importance of σ^+ in the Hammett correlation indicates an unsymmetric transition state.

$$(RC_6H_4)_2C{=}S \ (\mathbf{33}) \longrightarrow (RC_6H_4)_2C{=}SO \ (\mathbf{34}) \longrightarrow (RC_6H_4)_2C{=}O + RCO_2H + 1/2\ SO_2 + 1/2\ S \qquad (34)$$

$$Ar_2C{=}SO + RC(O)OOH \longrightarrow [\text{transition state}]^{\ddagger} \longrightarrow [Ar_2C\text{—}SO \text{ (with bridging O)}] + RCO_2H \longrightarrow Ar_2C{=}O \qquad (35)$$

G. Oxidation of Organic Iodine Compounds

Oxidation of aromatic iodine compounds with peroxy acids to produce the corresponding iodosyl compounds, ArIO, is believed to involve a nucleophilic attack of the iodine compound on the electrophilic peroxy acid[1d,101–107] (equation 36). Further oxidation of iodosyl compounds to iodyl derivatives presumably proceeds via *I,I*-bis(acyldioxy)iodobenzenes (**35**), which have actually been isolated in the reaction of iodosylbenzene with substituted peroxybenzoic acids[104] (equation 37). *I,I*-bis(benzoyldioy)iodobenzenes decompose at higher temperatures to produce, among other products, the corresponding iodylbenzenes[105].

$$\mathrm{Ar{-}\ddot{I}{:}} + \mathrm{RC(O)OOH} \longrightarrow \mathrm{Ar{-}\overset{+}{I}{-}\overset{-}{O}} + \mathrm{RCO_2H} \quad (36)$$

$$\mathrm{C_6H_5{-}\overset{+}{I}{-}\overset{-}{O}} + 2\,\mathrm{XCH_4CO_3H} \longrightarrow \mathrm{C_6H_5{-}I(O{-}O{-}C(O){-}C_6H_4X)_2} \quad (37)$$

(**35**)

Relatively little has been done in the past to elucidate the mechanism of oxidation of aliphatic iodides with peroxides[107–109]. Indirect evidence has recently been reported for the involvement of iodosyl compounds in the peroxy acid oxidation of these compounds[109]. Thus, primary alkyl iodides are converted in good yields to the corresponding alcohols when treated with *m*-chloroperoxybenzoic acid in dichloromethane or carbon tetrachloride. Only small amounts of *m*-chlorobenzoate esters are found. Both compounds are regarded as displacement products of the hypervalent iodine by water and *m*-chlorobenzoic acid, respectively.

Secondary iodides give a mixture of products resulting from displacement, elimination and α-carbon oxidation, while tertiary iodides yield products of displacement and elimination. Scheme 5 has been proposed for the formation of products from peroxy acid oxidation of alkyl iodides. The overall rate of conversion of alkyl iodides to products is dependent on the alkyl group: *t*-butyl > *s*-octyl > methyl > *n*-heptyl. A polarized transition state (**36**), or a significant shift of electrons from carbon to iodine as the reaction progresses, has been suggested[109].

$$\left[\;\rangle\overset{\delta+}{C}{-}\overset{\delta-}{I}\cdots\cdots\overset{\delta+}{O}(H\cdots O{=})\,C{-}R\;(\overset{\delta-}{O})\right]^{\ddagger}$$

(**36**)

H. Reactions of Diacyl Peroxides with some Nucleophiles

1. General characteristics

Considerable interest has been devoted in recent years to the study of the reactions of diacyl peroxides with reagents capable of serving as one- and two-electron donors[110]. Among the compounds investigated are amines[111–118], olefins[119–122], trivalent phosphorus compounds[123–127], sulphides[128], carbanions[129], phenols[113,130,131], ethers[116] and aromatic hydrocarbons[116].

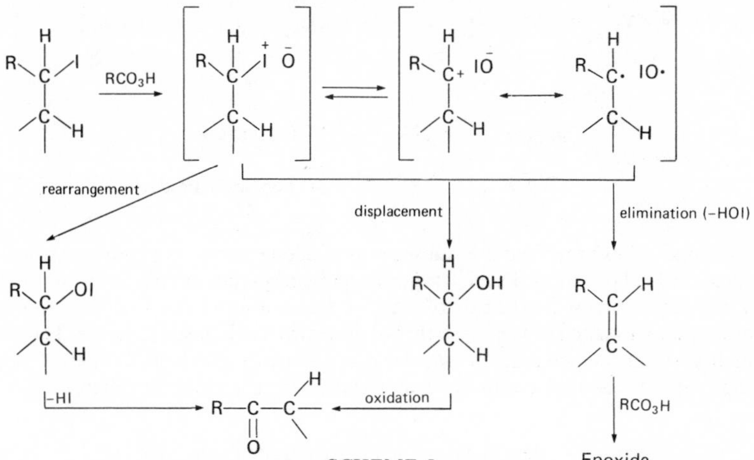

RCO$_3$H

rearrangement

displacement

elimination (−HOI)

−HI

oxidation

RCO$_3$H

Epoxide

SCHEME 5

In general, two broad mechanistic schemes have been suggested at various times for these reactions. In the first scheme, the transformation of the reacting molecules is conceptualized as the result of the interaction of a nucleophile with an electrophile in a process analogous to an S_N2 mechanism. The first-formed ion pair subsequently decomposes by diverse pathways to various products in ionic and/or radical processes, including the possibility of a one-electron transfer after the rate-determining step (equation 38).

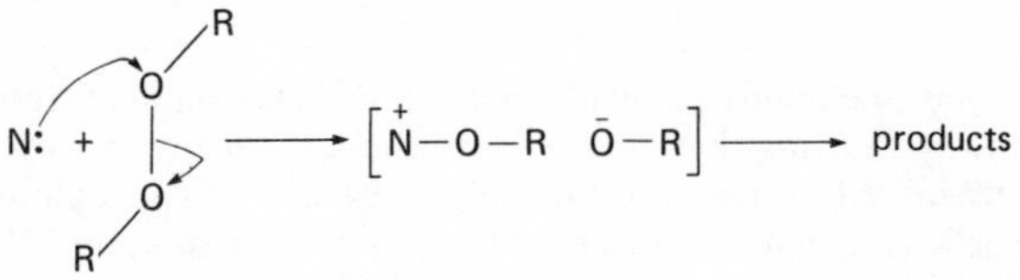

$$\text{N:} + \text{R-O-O-R} \longrightarrow \left[\overset{+}{\text{N}}\text{-O-R} \quad \overset{-}{\text{O}}\text{-R}\right] \longrightarrow \text{products} \tag{38}$$

Although it appears that, in most cases, the observed products can most easily be accounted for by this simple nucleophilic displacement, there are cases where another mechanism is operative. This involves as an initiating process a one-electron transfer to produce odd-electron intermediates, which react further to give (by coupling, disproportionation or a second electron transfer) nonradical products (equation 39).

$$\text{N:} + \text{R-O-O-R} \longrightarrow \left[\dot{\text{N}}^{+} \quad \text{R}\dot{\text{O}} \quad \bar{\text{O}}\text{R}\right] \longrightarrow \text{products} \tag{39}$$

Since earlier work has been comprehensively reviewed[110], only a few typical examples will be presented here to illustrate the difficulties involved in distinguishing between the two- and one-electron processes presumably involved in these reactions.

2. Oxidation of amines

Amines are well known to react nucleophilically with diacyl peroxides[111–116]. Horner and coworkers originally proposed a mechanism involving a one-electron transfer from dimethylaniline to dibenzoyl peroxide[111] (equations 40 and 41). Walling[112], on the other hand, suggested that tertiary amines react with dibenzoyl peroxide to form quaternary hydroxylamine derivatives in a two-electron nucleophilic attack of the amine on the

$$ArNMe_2 + (PhCO_2)_2 \longrightarrow [ArNMe_2^{+\cdot}\ PhCO_2^{\cdot}\ ^{-}O_2CPh] \quad (40)$$

$$\left[ArNMe_2^{+\cdot}\ \begin{matrix} PhCO_2^{\cdot} \\ PhCO_2^{-} \end{matrix}\right] \begin{cases} \xrightarrow{\text{diffusion}} \text{free radicals} \\ \xrightarrow{\text{cage reaction}} \text{nonradical products} \end{cases} \quad (41)$$

peroxide bond. This intermediate subsequently decomposes to produce radical- and nonradical-derived products. The S_N2 type mechanism was favoured mainly on the basis of the effect of substitutents on the rate of reaction (equations 42 and 43). The same type of kinetic evidence was taken to support either of these two mechanistic schemes: The reaction is of the first order in each reagent (second order overall), and is facilitated by electron-donating groups in the amine, and electron-withdrawing groups in the dibenzoyl peroxide.

$$ArNMe_2 + (PhCO_2)_2 \longrightarrow [Ar\overset{+}{N}(Me_2)OCOPh\quad PhCO_2^{-}] \quad (42)$$

$$Ar\overset{+}{N}(Me)_2OCOPh \begin{cases} \xrightarrow{\text{homolysis}} Ar\overset{+\cdot}{N}Me_2 + PhCO_2^{\cdot} \\ \xrightarrow{\text{elimination}} [ArN(Me){=}CH_2]^{+} + PhCO_2H \end{cases} \quad (43)$$

Denney and Denney seemingly clarified the dilemma in favour of the initial nucleophilic attack of the amine by showing that dibenzylamine and dibenzoyl peroxide, labelled with oxygen-18 in the carbonyl group, react to product 94% of *O*-benzoyldibenzylhydroxylamine with the labelled oxygen in the carbonyl group[113,114] (equation 44). On the basis of the observation that solvent polarity has only a small effect on the reaction rate, a concerted proton transfer via a cyclic transition state **37** has been suggested.

$$(PhCH_2)_2\ddot{N}H + (Ph{-}\overset{^{18}O}{\overset{\|}{C}}{-}O)_2 \longrightarrow \left[(PhCH_2)_2\overset{\delta+}{N}(H)\cdots O(\cdots\overset{\delta-}{O}{-}\overset{^{18}O}{\overset{\|}{C}}{-}Ph)\cdots Ph{-}C{=}O^{18}\right]^{\ddagger} \longrightarrow$$

$$(PhCH_2)_2\overset{+}{N}(H){-}O{-}\overset{^{18}O}{\overset{\|}{C}}{-}Ph + {}^{-}O{-}\overset{^{18}O}{\overset{\|}{C}}{-}Ph \longrightarrow (PhCH_2)_2NO\underset{^{18}O}{\underset{\|}{C}}Ph + PhCO_2H \quad (44)$$

Schuster and coworkers[115,116] have recently argued that the lack of scrambling of the label can be accommodated with the electron transfer as well as with the nucleophilic displacement mechanism. McBride and coworkers[117] have recently suggested on the basis of an ESR study of the benzoyl radical in a crystalline matrix at low temperature that the unpaired electron is in an orbital of σ symmetry. Therefore, it is quite possible that the two oxygen atoms of the radical ion pair, during its short lifetime, do not become chemically equivalent.

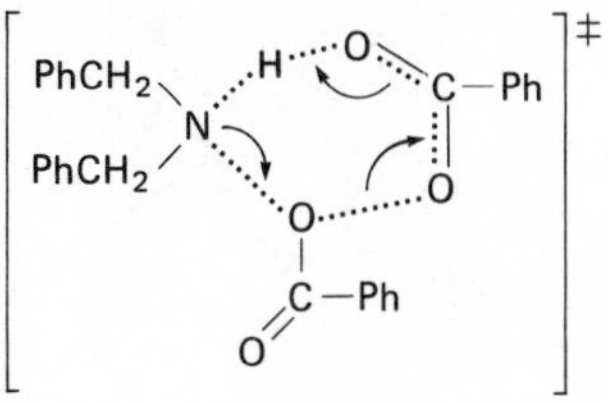

(37)

Similar mechanistic dilemmas appear to be involved in the oxidation of benzylamines with arylsulphonyl peroxides, although a detailed study of kinetics, substituent effects and stereochemistry of the reaction of several substituted benzylamines with *p*-nitrobenzoyl peroxide seems to favour a two-step, two-electron mechanism, with the first step involving nucleophilic attack by the amine on the peroxide to form a hydroxylamine-*O*-*p*-nitrobenzenesulphonate adduct, and the second step involving base-catalysed elimination from the adduct[118] (equations 45 and 46).

$$RCH_2NH_2 + (PhSO_2O)_2 \longrightarrow RCH_2NHOSO_2Ph + PhSO_3H \qquad (45)$$

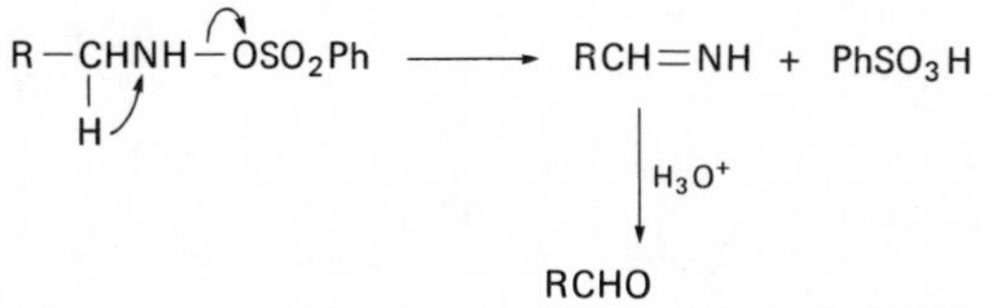

$$R{-}CH(H)NH{-}OSO_2Ph \longrightarrow RCH{=}NH + PhSO_3H \xrightarrow{H_3O^+} RCHO \qquad (46)$$

3. Oxidation of olefins

Greene and coworkers[119] made an extensive investigation of the reaction of phthaloyl peroxide with a variety of olefins. For example, the reaction with *trans*-stilbene in carbon tetrachloride at 80°C was found to give two adducts, the cyclic phthalate **38** and the phthalide **39** (equation 47). The reaction was found to be stereospecific and to obey second-order kinetics. Oxygen-18 labelling studies showed that about half of the oxygen of one carbonyl group of the peroxide appeared in the ether oxygens of the phthalate. On the basis

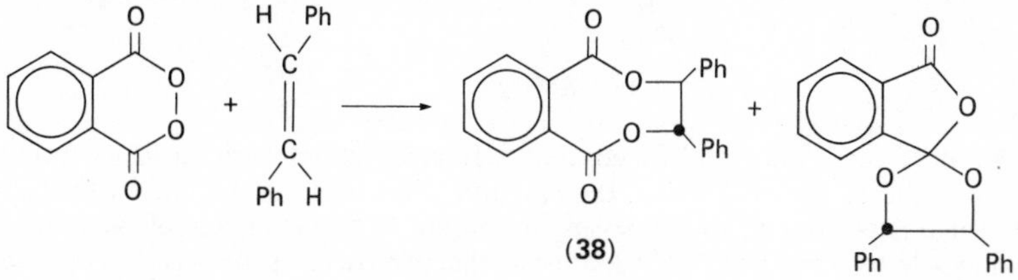

(47)

(38) **(39)**

of these results, Greene and coworkers proposed a two-electron pathway involving a transition state (**40**) resembling a charge-transfer complex between peroxide and olefin. The complex then rearranges through additional intermediates to the products.

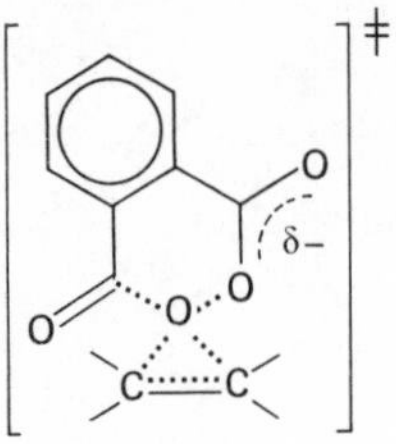

(**40**)

Strong support for a one-electron transfer mechanism in the reaction of various nucleophiles, i.e. olefins, amines, ethers, and aromatic hydrocarbons, with phthaloyl peroxide, has recently been presented by Schuster and coworkers[115,116]. They detected odd-electron intermediates in the reaction of the peroxide with ground- and excited-state electron donors by laser pulse spectrometry. The dependence of the observed bimolecular rate constants on the one-electron oxidation potential for electronically excited donors as well as ground-state donors also speak in favour of such a process. The proposed mechanistic scheme is shown in Scheme 6. Nevertheless, as pointed out by Schuster, there is a danger in extrapolating these conclusions to the reactions of various nucleophiles with other diacyl peroxides. The peroxide bond of the planar, constrained six-membered ring in phthaloyl peroxide is expected to be more easily reduced than the one in the skew dibenzoyl peroxide, and would thus seem to favour odd-electron processes.

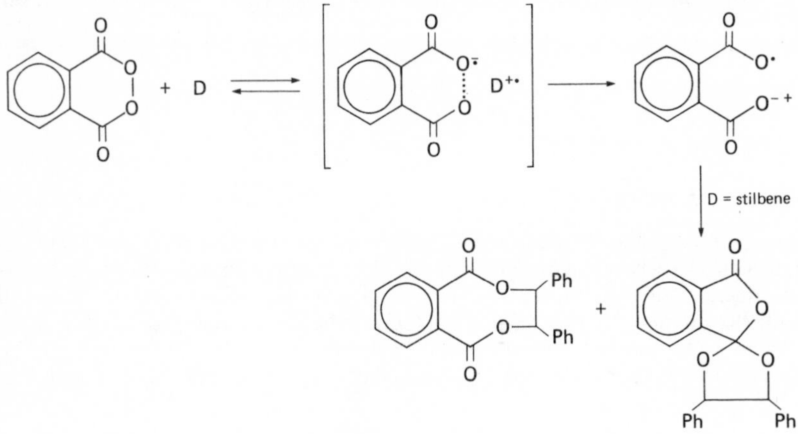

SCHEME 6

Although it is now well established that odd-electron intermediates, as well as scavengeable radicals, are involved in the reaction of various nucleophiles with diacyl peroxides, there is at present no consensus of opinion as to which step of the reaction involves single-electron transfer[131,132]. It appears that in most cases the observed products can most easily be accounted for by a simple nucleophilic displacement with an 'early', polarized transition state and that, at least in some cases, single-electron transfer is involved in processes after the rate-determining step. Clearly, further work is needed to clarify these points.

III. INTRAMOLECULAR NUCLEOPHILIC REARRANGEMENTS

A. General Characteristics

A large group of reactions in which oxygen behaves electrophilically involves rearrangement of peroxides where an O—O bond is cleaved heterolytically and an alkyl or aryl group migrates from carbon to oxygen (equation 48). Although it is generally believed that in most cases heterolysis of the O—O bond and the alkyl or aryl group shift are highly concerted, there is also indication that, at least in some cases, an 'intimate' ion pair is formed first (equation 49). Since the alkoxy cation, RO^+, has never been observed as a long-lived species, a bridged positively charged transition state with a pentacoordinate central carbon atom of the R group is expected to be energetically preferable over the alkoxy cation.

(48)

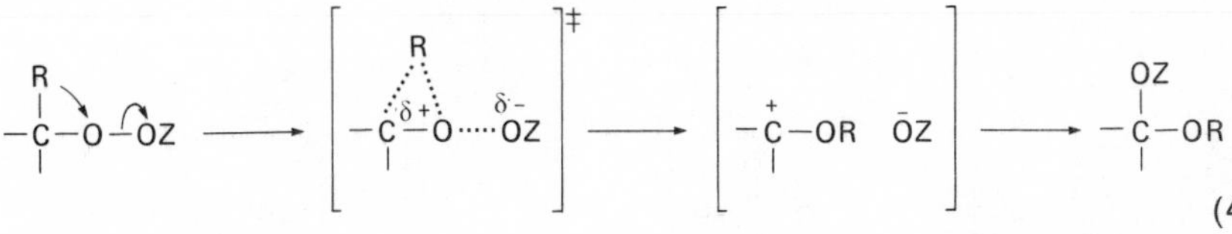

(49)

In general, these reactions are expected to be more sensitive to polar substituent effects and polar solvents than radical reactions, and should also be susceptible to acid catalysis.

B. Peroxy Esters (Criegee Rearrangement)

Acetate and benzoate esters of *trans*-9-decalyl hydroperoxide (**41**) rearrange on standing to produce isomeric esters which are easily hydrolysed to 6-hydroxycyclododecanone (equation 50)[133,134]. Rearrangements of this type received considerable attention in the 1940s and 1950s, but little since. Below are some characteristic features of the Criegee rearrangement.

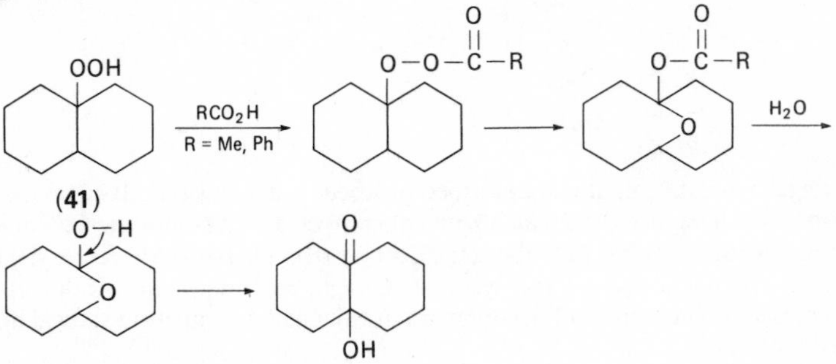

(50)

The reaction is facilitated by electron-withdrawing groups in R ($R = C_6H_4X$, $\rho = 1.35$)[135–137], by groups of high migratory ability, and is strongly influenced by the ionizing power of the solvent. All these suggest an ionic reaction.

The reaction is first order and is subject to slight acid catalysis. Foreign anions are not incorporated into the product, indicating the intramolecular nature of the process. Almost complete retention (98%) of an ^{18}O label in the carbonyl group was observed in the rearrangement of *trans*-9-decalyl peroxybenzoate labelled in the carbonyl position[138,139].

All the above-mentioned evidence indicates that the Criegee rearrangement involves a highly concerted heterolysis of the O—O bond with simultaneous carbon-to-oxygen migration via a bicyclic, highly oriented transition state (**42**), or through a tightly oriented intimate ion pair (**43**). This ion pair then collapses to the product so that the acyl oxygen retains its identity[140]. The five-membered transition state (**44**) seems to be ruled out on the bases of ^{18}O tracer studies. Due to the lack of consensus of opinion on the definition of the intimate ion pair, it is difficult, if not impossible, to differentiate between intermediates and transition states in such cases, and it is quite probable that such ion pairs have energies that are similar to that of the transition states or are transition states themselves.

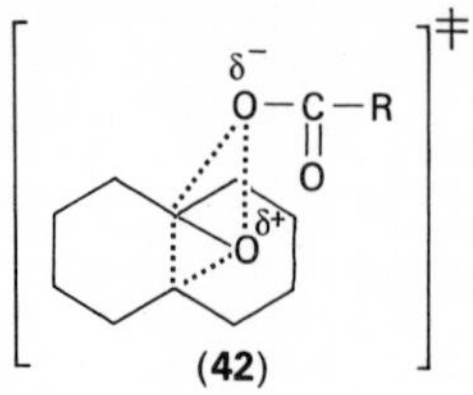

(**42**)

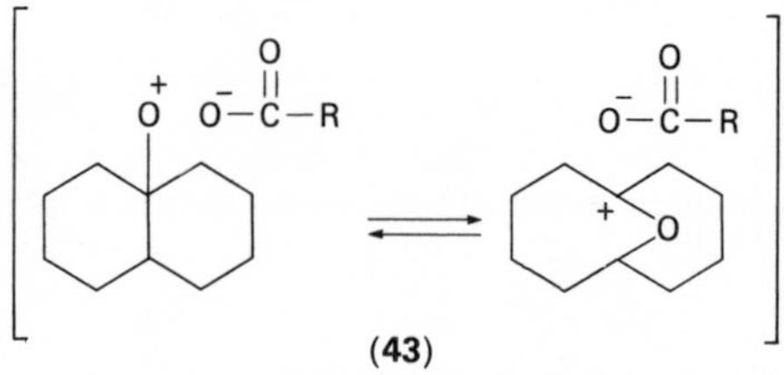

(**43**)

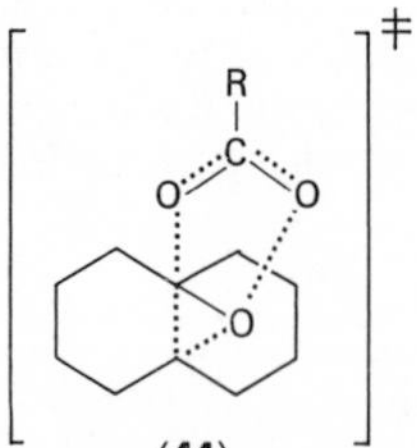

(**44**)

The Criegee rearrangement just described proceeds much less readily in simple *t*-alkyl peroxy esters. In these cases, the rearrangement competes with homolytic decomposition. For example, in polar solvents (nitrobenzene, acetic acid), cumyl peroxyacetate yields acetic acid, acetone and phenol after hydrolysis (the Criegee rearrangement products). On the other hand, the same compound decomposes in toluene by competing radical and ionic pathways[141].

A completely ionic form of decomposition has been proposed for the decomposition of *t*-butyl arylperoxysulphonates in methanol. Aceton and the corresponding sulphonic acid are formed quantitatively by rearrangement and methanolysis of the ion pair and subsequent hydrolysis of the ketal according to Scheme 7[142,143].

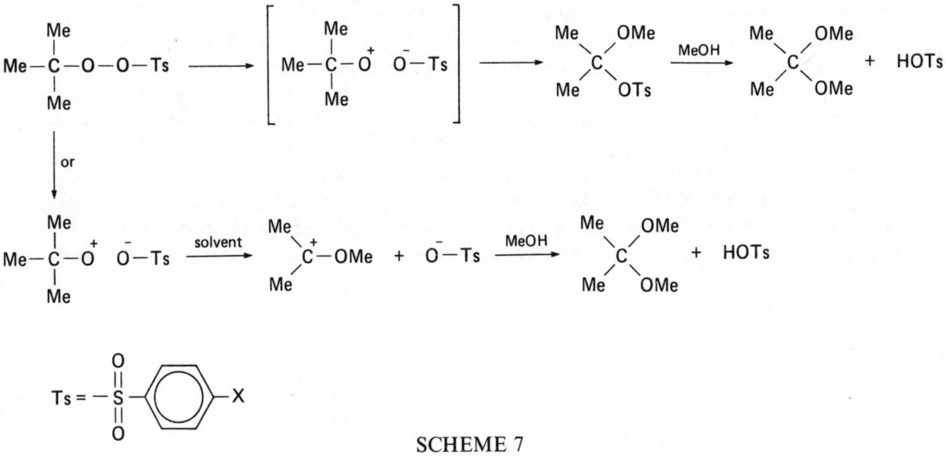

$$Me_3C{-}O{-}O{-}Ts \longrightarrow \left[Me_3C{-}\overset{+}{O}\ \ \overset{-}{O}{-}Ts \right] \longrightarrow Me_2C(OMe)(OTs) \xrightarrow{MeOH} Me_2C(OMe)_2 + HOTs$$

or

$$Me_3C{-}\overset{+}{O}\ \ \overset{-}{O}{-}Ts \xrightarrow{solvent} Me_2\overset{+}{C}{-}OMe + \overset{-}{O}{-}Ts \xrightarrow{MeOH} Me_2C(OMe)_2 + HOTs$$

$$Ts = {-}SO_2{-}C_6H_4{-}X$$

SCHEME 7

Winstein and Hedaya have studied the decomposition of a series of 2-substituted-2-propyl *p*-nitroperoxybenzoates (**45**) in methanol, ethanol and acetic acid[144]. They have found that the relative rate order is very sensitive to the nature of the migrating group, R, strongly indicating the importance of the anchimeric assistance by the migrating group (Scheme 8). A nonclassical-type, bridged, positively charged transition state is suggested to

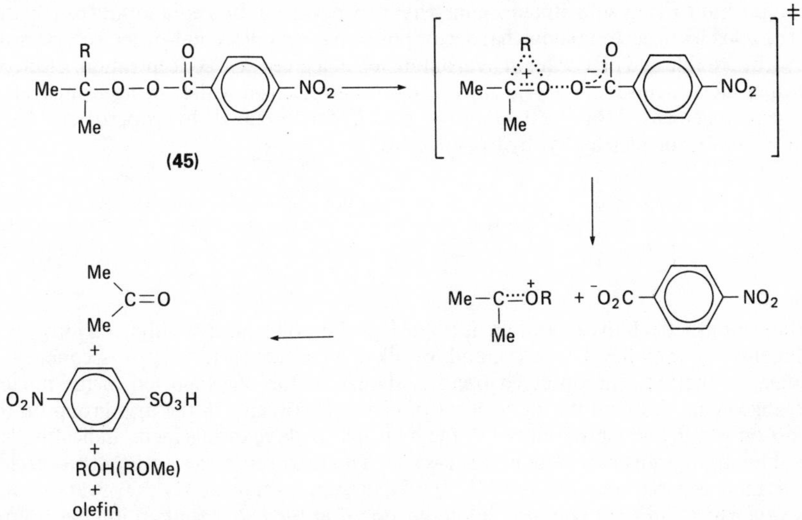

$$Me_2C(R){-}O{-}O{-}C(=O){-}C_6H_4{-}NO_2 \longrightarrow \left[Me_2C\overset{+}{\cdots}O(R)\cdots O\overset{-}{=}C(=O){-}C_6H_4{-}NO_2 \right]^{\ddagger}$$

(**45**)

$$\longrightarrow Me{-}C(Me)\overset{+}{\cdots}OR + {}^{-}O_2C{-}C_6H_4{-}NO_2$$

$$\longrightarrow Me_2C{=}O + O_2N{-}C_6H_4{-}SO_3H + ROH(ROMe) + olefin$$

SCHEME 8

be formed first, collapsing to an α-alkoxy carbonium ion intermediate. The observed products are believed to be derived from the cation $Me_2\overset{+}{C}OR$, except in cases where the alkyl cation, R^+, is stable enough to become a leaving group itself (R = *t*-Bu, *i*-Pr). The relative reactivity order, which ranges over five powers of ten, is *t*-butyl > phenyl > 4-camphyl > *p*-methoxybenzyl > benzyl > secondary > primary > methyl, thus reflecting the ability of these groups to sustain a positive charge.

The Hammett ρ value (−2.1) of the relative migration of substituted benzyl groups also implies that R is carrying an appreciable positive charge in the transition state of the rearrangement[145]. On the other hand, the relatively small migration aptitude of the benzyl group (Et:benzyl = 1:36), together with the small difference in migratory aptitude of 2-*exo*- and 2-*endo*-norbornyl groups, suggest relatively small geometrical changes during the migration[145]. By studying the migratory aptitude of a number of other cyclic and polycyclic bridgehead groups, Rüchardt and coworkers have suggested a nonclassical penta-coordinated central C atom in the transition state as shown in **46**. Nevertheless, the hyperconjugation component in this rearrangement, as indicated by the transition state **47** cannot be completely ruled out[145].

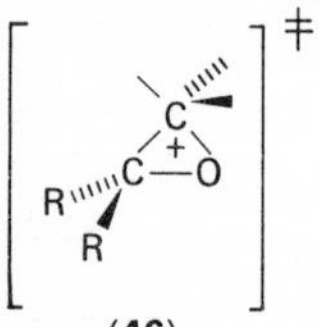

(46) 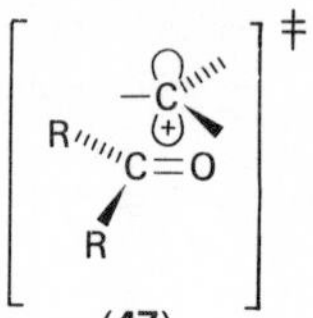 **(47)**

C. Hydroperoxides and Dialkyl Peroxides

The ionic nature of the acid-catalysed cleavage of hydroperoxides to alcoholic and ketonic fragments is now well established[146] (equation 51). The reaction is mechanistically similar to the Criegee rearrangement. Detailed studies of acid-catalysed ($HClO_4$, H_2SO_4) decomposition of ring-substituted α-cumyl hydroperoxides in 50% ethanol and benzhydryl hydroperoxides in acetone show that decomposition is overall second-order, i.e. first order in hydroperoxide and first order in hydronium ion at a given acid concentration. Electron-releasing substituents in the aryl group accelerate and electron-withdrawing groups retard the rearrangement [Hammett $\rho(\sigma^+) = -4.57$, for α-cumyl hydroperoxides[147]; $\rho(\sigma^+) = -3.78$, for benzhydryl hydroperoxides[148]].

$$R_3C{-}O{-}O{-}H \longrightarrow R_2C{=}O + ROH \qquad (51)$$

The established relative migratory aptitude is cyclobutyl > aryl > vinyl > hydrogen > cyclopentyl > cyclohexyl > alkyl, and for alkyl substituents: tertiary > secondary > primary > methyl. The observed trend is similar to the one reported in the pinacol rearrangement, although the steric effect of *ortho* substituents in the aryl group on the migration is much less pronounced in the hydroperoxide rearrangement, indicating less crowding in the transition state in this case[147]. The rearrangement is subject to specific, rather than general, acid catalysis[148]. Partial decomposition in $H_2{}^{18}O$ gives residual peroxide which does not contain ^{18}O, indicating that the alkyloxonium ion has no free existence[149,150].

The results of all the above-mentioned kinetic, isotope and migratory aptitude studies imply an acidity-dependent equilibrium between the protonated and unprotonated hydroperoxide which is followed by a rate-determining concerted rearrangement of the protonated hydroperoxide (equation 52).

$$R_3C{-}O{-}O{-}H \underset{}{\overset{H^+}{\rightleftharpoons}} R_3C{-}O{-}\overset{+}{O}H_2 \longrightarrow \left[R_2C\overset{\delta+}{\cdots}(R)\,{-}O\cdots\overset{\delta+}{O}H_2 \right]^{\ddagger} \longrightarrow$$

$$R_2\overset{+}{C}{-}OR + H_2O \longrightarrow R_2C(OH){-}OR \longrightarrow R_2C{=}O + ROH \qquad (52)$$

Electrophilic attack by a proton may also take place at the other oxygen atom to produce the second possible conjugate acid (equation 53). Although rearrangement is not possible in this case, a reversible process to produce hydrogen peroxide and a carbonium ion may occur, particularly with triarylmethyl hydroperoxides. Protonation at either oxygen atom has recently been confirmed by studying the magic-acid-catalysed cleavage–rearrangement reactions of various *t*-alkyl hydroperoxides by ^{13}C-NMR spectroscopy[151]. Olah and coworkers have found that careful addition of an equimolar amount of *t*-butyl hydroperoxide in SO_2ClF at -78°C to magic acid in SO_2ClF gives the corresponding dimethylmethoxycarbonium ion which is hydrolysed by the water formed to produce acetone and methanol (equation 54). On the other hand, a fivefold excess of magic acid shows formation of trimethylcarbonium ion as well as dimethylmethoxycarbonium ion, indicating both O—O (85%) and C—O (15%) cleavage reactions.

$$R_3C{-}O{-}O{-}H + H^+ \rightleftharpoons R_3C{-}\overset{+}{O}(H){-}O{-}H \rightleftharpoons R_3\overset{+}{C} + H_2O_2 \qquad (53)$$

$$Me_3C{-}O{-}O{-}H \overset{H^+}{\rightleftharpoons} Me_3C{-}O{-}\overset{+}{O}H_2 \longrightarrow Me_2C{=}\overset{+}{O}Me + H_2O \qquad (54)$$

Two carbonyl fragments are formed in the acid-catalysed cleavage of allylic hydroperoxides (the Hock cleavage)[152,153]. The reaction is believed to proceed according to the Criegee rearrangement mechanism (migration of a vinyl group)[152–157] (equation 55), although suggestions have been made that this reaction might involve a dioxetane intermediate since it proceeds sometimes without any acid catalysis[158–161] (equation 56).

$$>C{=}C<{-}C(<){-}O{-}OH \longrightarrow [\text{cyclic intermediate}] \longrightarrow H{-}C(<){-}C(<){=}O + O{=}C< \qquad (55)$$

$X = CR_2$, allylic hydroperoxides[158]
$X = O$, α-ketohydroperoxides[159, 160]
$X = -N$, α-cyanohydroperoxides[161]

(56)

Acid-catalysed decomposition of α-hydroperoxyketones in benzene affords carboxylic acids and ketones (equation 57). Sawaki and Ogata[162] have reported that the decomposition obeys the rate equation

$$v = k_2\ [\text{hydroperoxide}][\text{acid}].$$

$$R^1-C(=O)-C(OOH)R^2R^3 \xrightarrow[\text{benzene}]{CF_3CO_2H} R^1-C(=O)-OH + R^2R^3C=O \qquad (57)$$

Weak bases (benzophenone, acetonitrile, dioxane) retard the reaction. A general acid catalysis by the free acid has been suggested on the basis of the effect of added amine. The relative migratory aptitude of acyl groups is in the order PhC=O < MeC=O < *i*-PrC=O. The substituent effect in the migratory benzoyl group gives $\rho = -2.23$ (σ). On the other hand, the substituent effect in the nonmigrating phenyl group (R^3) gives the ρ value of -0.82 with σ^+. All the evidence indicates that the rate of the acyl migration is determined by the electron-donating power of R^1, and that the positive charge is stabilized mostly by the interaction with the carbonyl group (see **48**). The mechanistic scheme shown in Scheme 9 has been suggested by these authors. Aryl (R^3 = *p*-anisyl, *p*-tolyl) migration begins to compete with benzoyl migration with the increasing acidity of the catalyst. Similar observations have also been reported for the ratio of aryl:hydride shift in the decomposition of benzhydryl hydroperoxides[148], the Baeyer–Villiger oxidation of benzaldehydes[163] and the pinacol rearrangement[164]. Sawaki and Ogata have explained this observation by suggesting that the aryl shift is more sensitive to acidity than the benzoyl shift (different extent of charge separation in the transition state for migration).

Dialkyl peroxides have been much less studied[165]. In strong acid media the equilibria shown in Scheme 10 exist. Acid-catalysed reactions of dialkyl peroxides in nonionizing media usually involve the Criegee rearrangement[166]. On the other hand, trapping of carbonium ions has also been reported in some cases[167].

$$\left[R^1-\overset{+}{C}=\ddot{O}: \longleftrightarrow R^1-C\equiv\overset{+}{O} \longleftrightarrow \overset{+}{R^1}=C=\ddot{O}:\right]$$

(**48**)

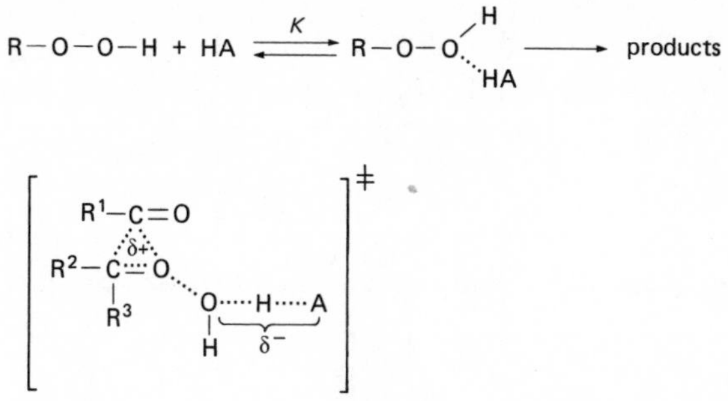

$$R{-}O{-}O{-}H + HA \underset{}{\overset{K}{\rightleftharpoons}} R{-}O{-}O\overset{H}{\underset{HA}{}} \longrightarrow \text{products}$$

$$\left[\begin{array}{c} R^1{-}C{=}O \\ \overset{\delta+}{R^2{-}C{\cdots}O}\cdots O\cdots H\cdots A \\ R^3 \quad\quad H \quad \delta^- \end{array}\right]^{\ddagger}$$

SCHEME 9

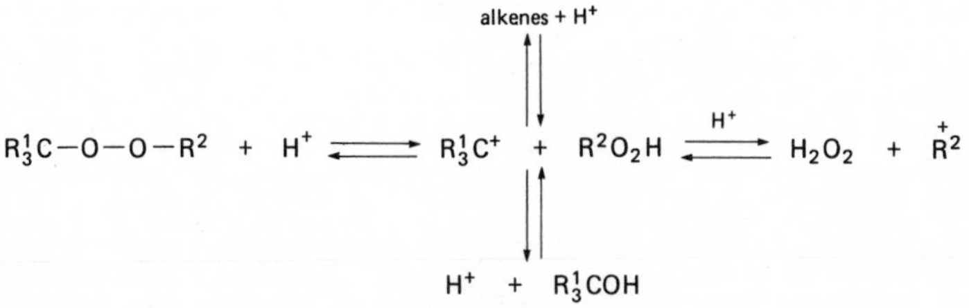

$$R^1_3C{-}O{-}O{-}R^2 + H^+ \rightleftharpoons R^1_3C^+ + R^2O_2H \overset{H^+}{\rightleftharpoons} H_2O_2 + \overset{+}{R}{}^2$$

$$R^1_3C^+ \rightleftharpoons \text{alkenes} + H^+$$

$$R^1_3C^+ \rightleftharpoons H^+ + R^1_3COH$$

SCHEME 10

D. Diacyl Peroxides

It has long been known that unsymmetrically substituted benzoyl peroxides and mixed diacyl peroxides, in which one or both Rs are secondary or tertiary alkyl groups as well as some symmetrical diacyl peroxides, decompose thermally to give, beside radical products, also the 'Leffler carboxy inversion' product, i.e. carbonic anhydride[168–171] (equation 58). The 'polar' reaction is favoured by polar solvents and electron-withdrawing groups in the potentially anionic fragment, and is subject to acid catalysis. Assuming that both types of decomposition, i.e. ionic and radical, are independent, the product ratio has been used to obtain the separate rate constants for the disappearance of substituted benzoyl isobutyryl peroxides. By analysing the obtained data in terms of competing reactions, Lamb and Sanderson found that the apparent k_{rad} (obtained by measuring the yield of scavengeable radicals) give a Hammett ρ value of -0.10, while the rates of rearrangement, $k_{rear.}$ (obtained by measuring the yield of the carboxy inversion product) give a positive ρ value ($+0.70$)[172].

$$R^1{-}\overset{O}{\overset{\|}{C}}{-}O{-}O{-}\overset{O}{\overset{\|}{C}}{-}R^2 \longrightarrow R^1{-}\overset{O}{\overset{\|}{C}}{-}O{-}\overset{O}{\overset{\|}{C}}{-}OR^2 \qquad (58)$$

There is mounting evidence in favour of simultaneous ionic and radical decomposition of these compounds[172–180,250].

By studying the effect of solvents on the rates and products of decomposition of several diacyl peroxides, Walling and coworkers have provided evidence that a common rate-determining transition state is involved in both polar and radical reaction pathways[175,176]. They have proposed a mechanistic scheme which is outlined in Scheme 11. The results of an

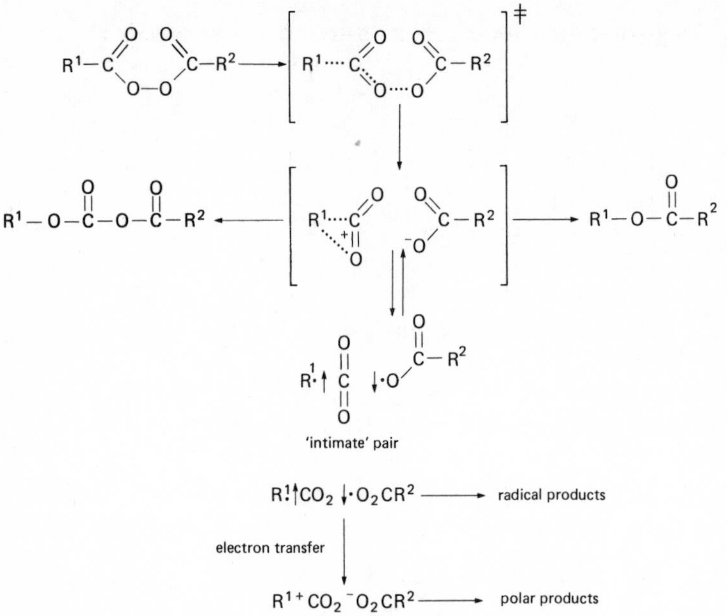

'intimate' pair

$R^{1}\cdot\uparrow CO_2 \downarrow\cdot O_2CR^2$ ⟶ radical products

electron transfer

$R^{1+} CO_2\ {}^{-}O_2CR^2$ ⟶ polar products

SCHEME 11

investigation of the stereochemistry of products obtained by the capture of the ion pairs with nucleophilic solvents (acetonitrile, acetic acid, 2-butanol) indicate that partitioning into tight ion and radical pairs occurs 'late' on the reaction coordinate.

Taylor and coworkers have found that cyclobutyl and cyclopropylmethyl groups are very sensitive structural probes for following posttransition-state events in the thermal decomposition of a number of alkyl–aryl (R^1 = *m*-ClC_6H_4, R_2 = cyclobutyl, cyclopentylmethyl, 3-butenyl) and symmetric diacyl peroxides ($R^1 = R^2$ = cyclobutyl, cyclopropylmethyl, 3-butenyl)[177]. Their results may be summarized as follows: (1) formation of carbonic anhydride involves, at least in part, a concerted pathway, (2) ester formation is in large part due to ion-pair and not to radical-pair collapse and (3) differentiation into ion-radical processes occurs 'early' on the reaction coordinate. Taylor and coworkers favour Leffler's unbridged structure of a common intermediate (**49**)[178] as the most plausible transition-state model. This intermediate is believed to be formed by single electron transfer from the weak C—C bond to the peroxide linkage, with alkyl and carbon dioxide still weakly bonded.

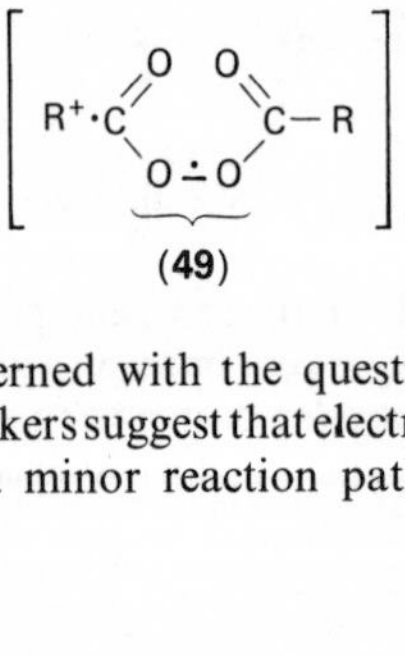

(**49**)

Much current interest is concerned with the question of electron transfer in these decompositions. Taylor and coworkers suggest that electron transfer between separated ion pairs and radical pairs is only a minor reaction pathway[177]. Electron transfer from neopentyl to *m*-chlorobenzoyl radical to form the neopentyl carbocation and *m*-chlorobenzoate has been reported by Lawler and coworkers, who have found CIDNP evidence for such a process[179]. More recently, similar observations have been reported for other alkyl–aryl diacyl peroxides (R^1 = neopentyl, isobutyl, *n*-propyl; R^2 = *m*-ClC_6H_4, 4-$NO_2C_6H_4$). It is surprising that the solvent and substituent effects on the electron transfer are found to be small[180].

Clearly, further work is needed to clarify the sequence of events occurring along the reaction coordinate of the thermal decomposition of these interesting compounds.

E. Oxidation of Ketones and Aldehydes with Peroxy Acids (Baeyer–Villiger Oxidation)

1. Ketones

The oxidation of acyclic ketones to esters (equation 59) and cyclic ketones to lactones (equation 60) with peroxy acids is known as the Baeyer–Villiger oxidation[181,182].

$$R^1-\overset{O}{\overset{\|}{C}}-R^2 + RCO_3H \longrightarrow R^1-\overset{O}{\overset{\|}{C}}-OR^2 \text{ or } R^1O-\overset{O}{\overset{\|}{C}}-R^2 + RCO_2H \qquad (59)$$

$$(CH_2)_n\left<\begin{matrix}CH_2\\CH_2\end{matrix}\right>C=O \longrightarrow (CH_2)_n\left<\begin{matrix}CH_2-O\\CH_2\end{matrix}\right>C=O \qquad (60)$$

a. Polar reaction mechanisms involving the formation of the peroxy-acid–ketone adduct. Several mechanistic studies indicate that these oxidations proceed via a transient tetrahedral intermediate, the peroxy-acid–ketone adduct (Criegee adduct) **50**, with subsequent migration of an electron-rich substituent to an electrophilic oxygen[183–187] (equation 61). As will be shown below, aliphatic acyclic ketones with groups of poor migratory ability do not follow this reaction pathway.

$$R-\overset{O}{\overset{\|}{C}}-R + RCO_3H \longrightarrow \left[HO-\underset{R}{\overset{R}{C}}-O-O-\overset{O}{\overset{\|}{C}}-R\right]\ (\mathbf{50}) \longrightarrow$$

$$R-\overset{^+OH}{\overset{\|}{C}}-OR + {}^-O-\overset{O}{\overset{\|}{C}}-R \longrightarrow R-\overset{O}{\overset{\|}{C}}-OR + HO-\overset{O}{\overset{\|}{C}}-R \qquad (61)$$

Peroxy acid oxidations of ketones are, usually second-order reactions, i.e. first-order in each of the reactants[188–192]:

$$v = k_{obs}\,[\text{peroxy acid}][\text{ketone}]$$

Oxidations with trifluoroperoxyacetic acid are usually of third order and sometimes of pseudo-first order and second order[192].

The activation entropies of the Baeyer–Villiger reactions are large and negative (−25 to −45 cal mol^{-1} deg^{-1})[186]. The reaction is catalysed by acids and facilitated by polar solvents[187,189].

Electron-withdrawing groups in the *para* and *meta* position retard and electron-releasing groups facilitate the migration of the phenyl group.[187,188] *Ortho*-substituted phenyl groups migrate less readily than *para*-substituted analogues[188].

The rearrangement in the adduct must be intramolecular since it proceeds with complete retention of configuration[193]. Isomerization may occur with unsaturated compounds.

Benzophenone containing isotopically labelled oxygen gives phenyl benzoate with the same carbonyl oxygen content as the starting material (equation 62), indicating that no free electron-deficient oxygen species are formed during the reaction. This observation also rules out dioxirane (**51**) as an intermediate[185].

$$Ph-C(={}^{18}O)-Ph + ArCO_3H \longrightarrow Ph-C(={}^{18}O)-OPh + ArCO_2H \quad (62)$$

(51)

The relative migratory aptitude of various groups in unsymmetrical ketones has been extensively studied and the following order has been found: *t*-alkyl > cyclohexyl > *s*-alkyl > benzyl > phenyl > *p*-alkyl > cyclopentyl > methyl[187,188,194,195].

An unusual remote substituent effect on the regioselectivity of the peroxy acid oxidation of 8-oxabicyclo[3.2.1]octan-3-one derivatives (**52**) has recently been reported[196] (equation 63). The rigid system with a plane of symmetry, which is not distorted upon introduction of substituents, allowed the study of through-bond electronic effects. It was found that electron-withdrawing groups γ to carbonyl favour the α′-methylene migration over α-methylene migration. The following relative order of α′-directing abilities was reported: $OSO_2R > OCOR > OR > XCH_2 = Ph$. Through-bond electronic effects, caused by an electronegative substituent X, rather than steric and conformational factors, are thus strongly indicated.

(52) $\xrightarrow{CF_3CO_3H}$ [HO OOCOCF$_3$ adduct] ⟶ lactone + lactone (63)

A similar, remote substituent effect has also been reported in the peroxy acid oxidation of a number of β-trimethylsilyl ketones (**53**)[197]. This reaction is directed by the trimethylsilyl group to give esters of β-hydroxysilanes (equation 64). Oxidation of **53a** gives a 2:1 ratio of esters **54** and **55** (80% yield). Ketone **53b** gives a 1:2 ratio of **54** and **55** (79% yield). The migratory aptitude of the $Me_3SiCH_2CH_2$ group in this system is thus intermediate between that of secondary and tertiary alkyl groups.

Me_3Si–(53) $\xrightarrow[CH_2Cl_2]{m\text{-}ClC_6H_4CO_3H}$ (54) + (55)

(53) (54) (55)

(a) R = H
(b) R = Me

(64)

The nature of the leaving group also influences the migratory ability. Peroxytrifluoroacetic acid has been found to be more sensitive to steric effects of various groups than monoperoxymaleic acid in the oxidation of simple ketones. It has been suggested that this phenomenon might be due to the altered stabilities of the corresponding transition states leading to altered product ratios[195].

Steric effects have also received considerable attention in the Baeyer–Villiger oxidation of bicyclic systems[198] and *ortho*-substituted acetophenones[194].

All the above-mentioned studies seem to indicate that the size of the migrating group, its ability to support a positive charge in the transition state of the migration, and the reactivity and steric requirements of the peroxy acid, are the major factors influencing the rearrangement.

Rate-determining migration in the peroxy-acid–ketone adduct has been shown by the negative Hammett ρ value of $-1.45\ (\sigma)$[192] and $-1.3\ (\sigma^+)$[191], and even more conclusively by the ^{14}C isotope-effect study. Palmer and Fry studied the oxidation of ^{14}C-labelled *para*-substituted acetophenones (MeO, Me, H, Cl, CN) and found a significant ^{14}C isotope effect for all acetophenones except in the case of *p*-methoxyacetophenone[191] (equation 65). Rate-determining formation of the peroxy-acid–ketone adduct would not show an isotope effect; this step does not involve bond alteration at the labelled site.

$$X\text{-}C_6H_4\text{-}C(=O)\text{-}Me + m\text{-}ClC_6H_4CO_3H \longrightarrow X\text{-}C_6H_4\text{-}O\text{-}C(=O)\text{-}Me \quad (65)$$

Although the rate of the Baeyer–Villiger reaction in the presence of a suitable catalyst is governed by the migration step, carbonyl addition can become rate-determining under some conditions in ketones substituted with electron-releasing groups. Ogata and Sawaki[199] made an extensive investigation of the peroxybenzoic acid oxidation of acetophenones in various organic solvents and 40% aqueous ethanol. They found that acid catalysis by perchloric acid in 40% ethanol is general rather than specific, since the reaction is not correlated with the acidity function, H_0. Acetic acid is able to catalyse only addition to the C=O group while trifluoroacetic acid catalyses migration as well as addition. The species involved in the catalysis are believed to be the intermolecularly hydrogen-bonded adducts **56** and **57**. An approximately estimated Hammett ρ value of -3 was reported for the migration step in the oxidation of ring-substituted acetophenones (*p*-Me, *p*-Cl, H, *p*-OMe). This value is comparable with those reported for other peroxide rearrangements[144,147,148].

Rate-determining carbonyl addition was observed in the oxidation of hydroxyacetophenones. The rates of addition of peroxybenzoate ion to *o*- and *p*-hydroxyacetophenone at pH 5–12 satisfies the Taft equation[248] with $\rho^* = 1.3\text{–}1.6\ (\sigma^*)$ and $\delta = 0.6\text{–}0.8$, reflecting the importance of electronic and steric effects. Ogata and Sawaki[199] reported that migration is the rate-determining step in the oxidation of *p*-methoxyacetophenone at pH > 5.5, while

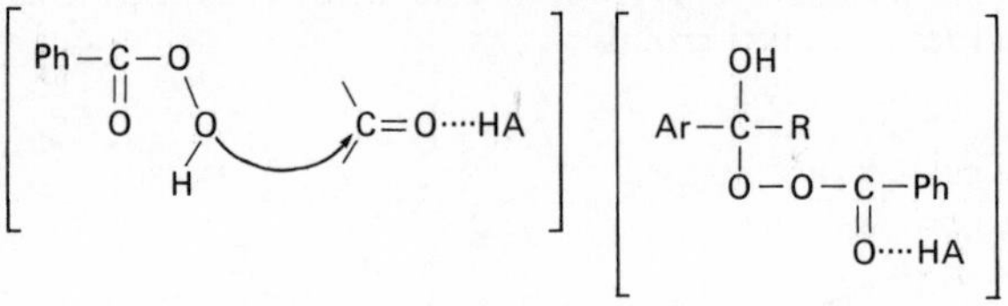

(56) (57)

both migration and addition are important at pH 0.7–5.5. At even higher acidity, migration becomes predominant. The change of the rate-determining step from migration to addition in the latter case appears to explain the lack of ^{14}C isotope effect in the oxidation of this ketone reported by Palmer and Fry[191].

Similar results were obtained by studying the kinetics of the Baeyer–Villiger oxidation of acetophenones with peroxymonophosphoric acid, H_3PO_5, in acetonitrile. The rates were of second order:

$$v = k\ [\mathrm{PhCOMe}][\mathrm{H_3PO_5}]$$

The reaction is subject to catalysis by H_2SO_4. A plot of log k_2 vs. $-H_0$ gave a straight line with a slope of 0.75, indicating proton involvement in the transition state. The migration step is rate-determining as indicated by the Hammett ρ value of -2.55 (σ). The observed better correlation of log k_2 vs. σ rather than σ^+ implies a relatively low positive charge formation in the migration step, which is believed to be concerted with the leaving of phosphoric acid according to Scheme 12[200].

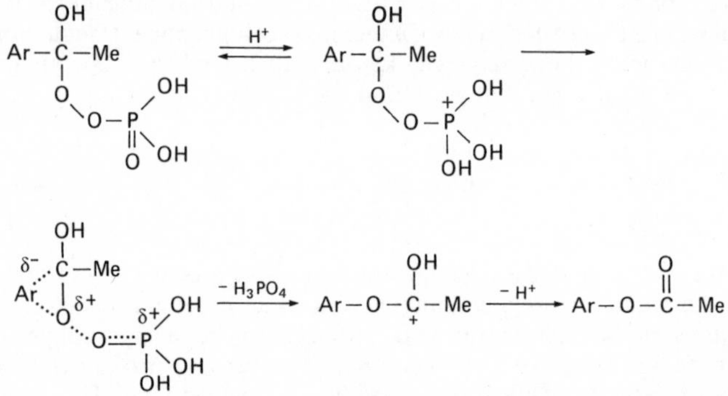

SCHEME 12

The question of the timing of the migration and the loss of the leaving group has also been discussed in other studies[201–203]. Considerable carbonyl formation in the transition state with little reorganization in the migrating group (**58**) are implied on the basis of *ab initio* and CNDO/2 MO calculations[202]. These results are also in accord with the small α- and relatively large β-secondary deuterium isotope effect observed iun the rearrangement of deuterated phenyl 2-propanones[203]. An ion-pair intermediate **59** seems to be ruled out on the basis of these calculations.

b. Mechanisms involving ionic and/or radical intermediates. As mentioned before the Baeyer–Villiger reaction of aliphatic acyclic ketones with groups of poor migratory aptitude yields ketone peroxides instead of esters.

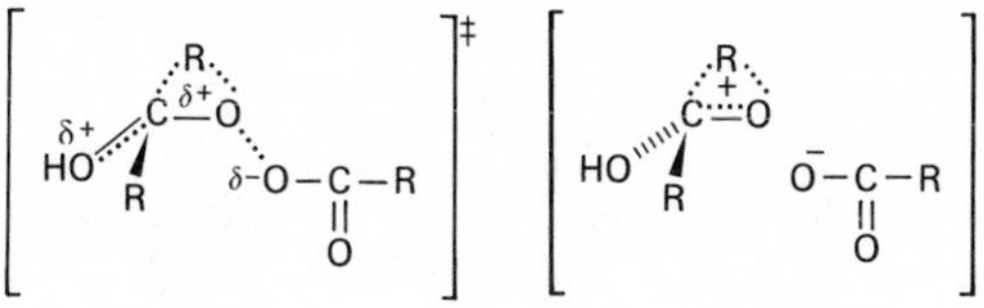

(58) (59)

The oxidation of acetone under Baeyer–Villiger conditions with peroxysulphuric or peroxyacetic acid has been reexamined by Edwards and coworkers[204] and Murray and Ramachandran[205]. Their conclusion is that an intermediate species is formed in these reactions which is capable of epoxidizing olefins. The carbonyl oxide (**60**) and/or the isomeric dioxirane (**61**) have been suggested as being the potential epoxidizing agents (Scheme 13). The remarkable reactivity of the intermediate was demonstrated by the oxidation of phenylpropiolic acid, which is otherwise unreactive with *m*-chloroperoxybenzoic acid.

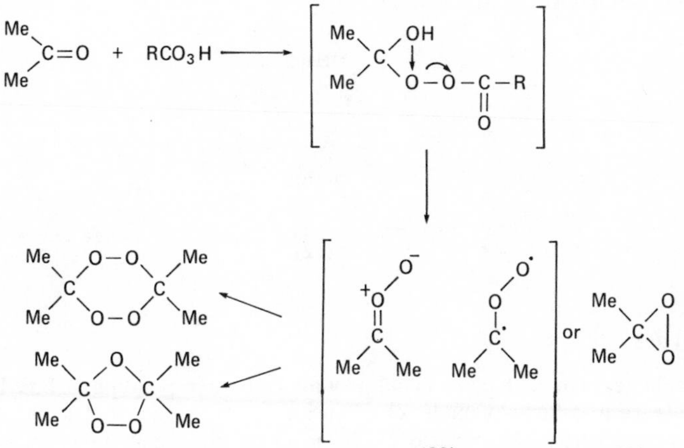

(60) (61)

SCHEME 13

It is not clear at present whether the cleavage of the O—O bond in the peroxyacid–ketone adduct in these cases is heterolytic or homolytic, or both. With peroxyacetic acid, the relatively small inductive effect of the acetyl group could favour the homolytic scission leading to the formation of dioxirane.

Homolytic cleavage of the O—O bond in the adduct has been suggested previously in the reaction of peroxyacetic acid with benzophenones containing electron-withdrawing groups[206]. It has been reported that such groups enhance migration of phenyl rings, which is opposite to the 'normal' order of migration observed when the same benzophenones are oxidized with trifluoroperoxyacetic acid. Formation of the end-products, according to this free-radical mechanism, involves 1,2-shift of one of the phenyl groups from carbon to oxygen, and an 'in-cage' abstraction of hydrogen by the acetoxy radical (Scheme 14).

Still another mechanistic possibility for the Baeyer–Villiger oxidation of ketones with peroxy acids should be mentioned at this point. Kwart and coworkers suggested a 1,3-

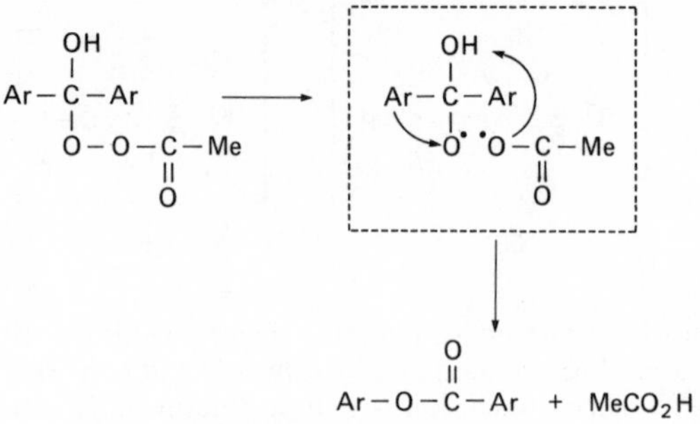

SCHEME 14

dipolar addition of the peroxy acid molecule to the ketone involving a transition state **62**, with more or less expressed dipolar character[43]. However, direct experimental evidence to support this mechanism is still lacking.

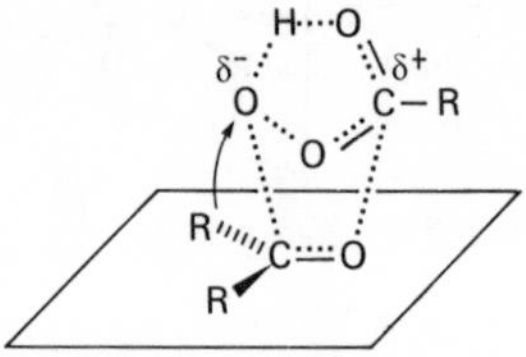

(**62**)

2. *Aldehydes*

Benzaldehydes react with peroxy acids to afford the corresponding carboxylic acids and phenols. Ogata and Sawaki made an extensive kinetic study of the reaction of these compounds with peroxybenzoic acid[163]. The mechanistic scheme shown in equation (66), analogous to the one already proposed for ketones, has been suggested. It is interesting to mention that the intermediate adduct has actually been detected in the reaction of aldehydes with peroxy acids[207–209]. It has been suggested that **63** exists in solution in an equilibrium of three forms. Still another structure (**63a**) for the adduct has been proposed to explain the formation of acetic anhydride in the autoxidation of acetaldehyde at ambient temperature[210]. Assuming that the adduct actually has structure **63**, either the aryl group or the hydrogen atom may migrate.

It has been found that selectivity for migration of an aryl group changes sharply with substituent and pH of the reaction medium[163]. The aryl migration is catalysed by acids via a hydrogen-bonded complex similar to the one involved in the oxidation of benzophenones. It is interesting to note that electron-releasing groups in the *ortho* position accelerate the migration by a factor of approximately ten compared to *para*-substituted phenyl rings. This is just the opposite to the findings reported in the Baeyer–Villiger oxidation of benzophenones, where *ortho* substituents retard the reaction[188]. The aryl migration is insensitive to hydroxide ion catalysis. It is also characterized by the absence of a kinetic deuterium isotope effect, but is quite strongly influenced by the solvent polarity (Grunwald–Winstein m value $= 0.5$). Thus, significant charge separation in the transition

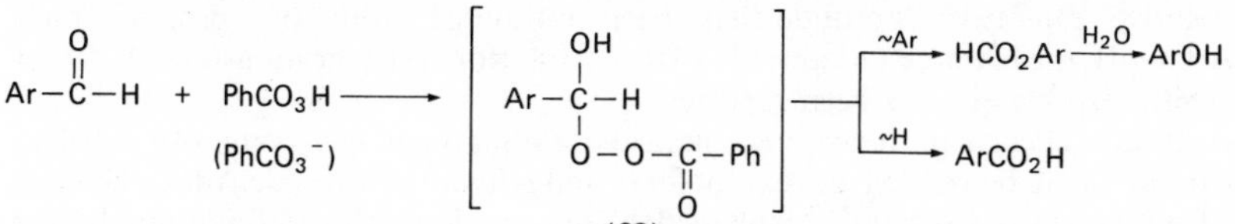

$$\mathrm{Ar{-}C({=}O){-}H} + \mathrm{PhCO_3H}\ (\mathrm{PhCO_3^-}) \longrightarrow \left[\mathrm{Ar{-}CH(OH){-}O{-}O{-}C({=}O){-}Ph}\right]\ (\mathbf{63}) \begin{cases} \xrightarrow{\sim \mathrm{Ar}} \mathrm{HCO_2Ar} \xrightarrow{\mathrm{H_2O}} \mathrm{ArOH} \\ \xrightarrow{\sim \mathrm{H}} \mathrm{ArCO_2H} \end{cases} \tag{66}$$

$$\left[\mathrm{R^1{-}C(O\cdots HO)(O{-}O)C(H){-}R^2} \rightleftharpoons \mathrm{R^1{-}C({=}O){-}O{-}O{-}CH(OH){-}R^2} \underset{}{\overset{\mathrm{R^3CHO}}{\rightleftharpoons}} \mathrm{R^1{-}C({=}O){-}O{-}O{-}CH(R^2){-}OH\cdots O{=}C(H){-}R^3}\right]$$

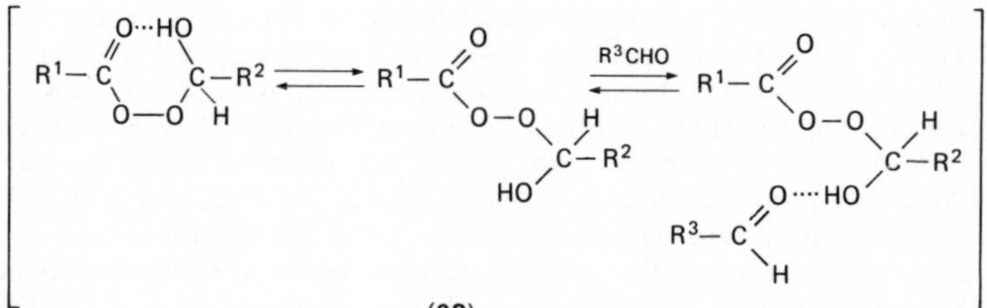

(**63**)

state for migration (**64**) has been suggested. Hydride migration is, on the other hand, characterized by a relatively large deuterium isotope effect ($k_H/k_D = 1.4$–3.0). It is rather insensitive to solvent polarity ($m \approx 0.1$) as well as to proton and hydroxide ion catalysis. A hydride shift presumably proceeds via adduct anion **65** at pH 8 rather than via the neutral adduct **63**. The migration ratio (Ar/H) changes at pH > 8.5, due to the change of the intermediate from an anion to a neutral adduct, and then increases with increasing acidity of the medium since aryl migration is acid-catalysed.

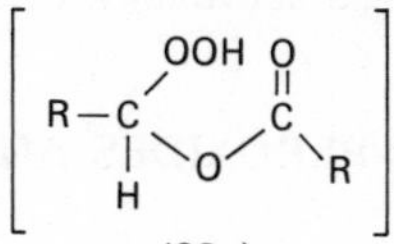

$$\left[\mathrm{R{-}CH(OOH){-}O{-}C({=}O){-}R}\right]$$

(**63a**)

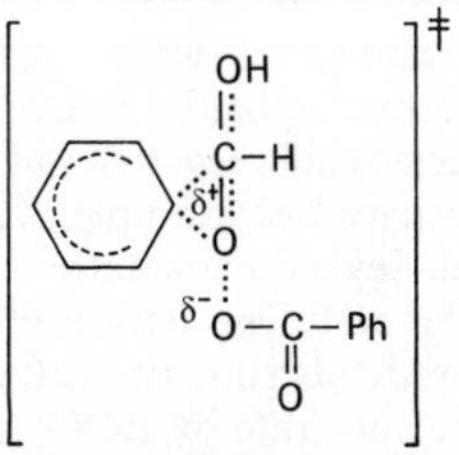

$$\left[\mathrm{Ph}\cdots\mathrm{C(OH)(H)}\cdots\overset{\delta+}{\mathrm{O}}\cdots\overset{\delta-}{\mathrm{O}}{-}\mathrm{C({=}O){-}Ph}\right]^{\ddagger}$$

(**64**)

$$\left[\mathrm{Ar{-}CH(O^-){-}O{-}O{-}C({=}O){-}Ph}\right]$$

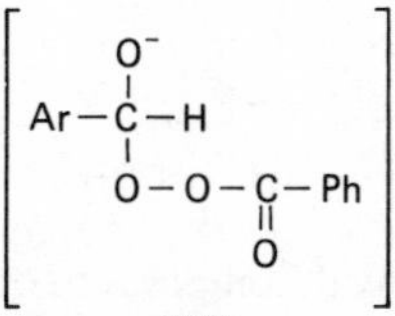

(**65**)

The relative migratory aptitude has been estimated from the product ratio $ArOH/ArCO_2H$; the Hammett ρ value of -4 to -5 (σ^+) for aryl migration and 1.1–1.8 for hydride shift, respectively, has been reported.

The substituent effect in the peroxybenzoic acids is rather small. The ρ value of 0.2–0.4 (σ) has been found for all benzaldehydes except for *o*- and *p*-hydroxybenzaldehydes where ρ is ≈ 0 at pH > 9. Ogata and Sawaki[163] explained this observation with a presumption that the main driving force for the O—O bond heterolysis in the peroxy acid is a weakening of the O—O bond caused by an attack of the nucleophile on the lowest unoccupied orbital (LUMO) of the O—O bond and not the lowering of the energy level of the antibonding orbital of the O—O bond by introducing electron-withdrawing substituents.

On the basis of all the above-mentioned evidence, Ogata and Sawaki concluded that carbonyl addition is the rate-determining step in the oxidation of hydroxy-substituted benzaldehydes, while migration is rate-determining in Me-, H-, and NO_2-substituted benzaldehydes. *p*-Methoxybenzaldehydes appears to be the borderline case.

It is interesting to compare the above-mentioned results with those obtained in the Baeyer–Villiger oxidation of benzaldehydes with peroxymonophosphoric acid. Ogata and coworkers have reported that at pH 1.3, the aryl migration is characterized by the Hammett ρ value of -2.88 (σ)[211]. This correlation is in contrast to the better correlation with σ^+ observed in the reaction with peroxybenzoic acid. This phenomenon has been explained as being due to the relatively weak loosening of the O—O bond in the adduct of benzaldehyde with H_3PO_5 (analogous to **64**) in the transition state of the migration, compared to the one in the oxidation with peroxybenzoic acid. H_3PO_5 is a much stronger acid than peroxybenzoic acid, so that a lower development of cationic charge (δ^+) is expected in the transition state. On the other hand, at pH 4 the hydride shift is characterized by $\rho = 1.74$ and 2.0 at pH $= 7.3$. Similar values of 1.1–1.8 have been obtained at pH 1–12 in the oxidation with peroxybenzoic acid affording hydride shift.

IV. OXIDATIONS INVOLVING PEROXIDES AND THEIR ANIONS AS NUCLEOPHILES

A. General Characteristics

A number of reactions are known in which peroxides or their anions act as nucleophiles[212]. Although much attention has been devoted in recent years to the so-called 'α effect', that is, a far greater nucleophilic reactivity of peroxy anions compared to oxy anions in spite of much lower basicity [for example, $K_b(HO_2^-)/K_b(HO^-) = 10^{-4}$], the problem is far from being solved. Several explanations have been put forward, and the interested reader is referred to the pertinent literature[213].

In general, these reactions can be divided into two categories: (*a*) reactions which involve displacement of a group X in the substrate by ROO^-

$$ROO^- + \geq C{-}X \longrightarrow ROOC{\leq} + X^-$$

(*b*) additions to a multiple bond

$$\underset{(ROOH)}{ROO^-} + \gt C{=}X \longrightarrow \gt\underset{OOR}{\overset{|}{C}}{-}X^- \quad \text{or} \quad \gt\underset{OOR}{\overset{|}{C}}{-}XH$$

The first formed adduct subsequently decomposes to other products, although there are cases where such intermediates have actually been isolated.

We have already encountered in our previous discussion some reactions of this type, i.e. the Baeyer–Villiger oxidation of carbonyl compounds (see Section III.E.1.) and the oxidation of sulphoxides by peroxy acids (see Section II.F.1.). Some further examples are given below.

B. Oxidations with Alkaline Hydrogen Peroxide and other Peroxy Anions

Selective epoxidation of α,β-unsaturated ketones and aldehydes can be accomplished by the sodium salt of hydrogen peroxide or the sodium salt of *t*-butyl hydroperoxide. These reactions are, in contrast to epoxidations with peroxy acids, stereoselective rather than stereospecific. For example, only one isomer is formed by the epoxidation of either isomer of the unsaturated ketone **66**[214] (equation 67). The mechanism of these reactions is believed to involve nucleophilic addition of the hydroperoxide anion at the β carbon of the unsaturated ketone[215] (equation 68).

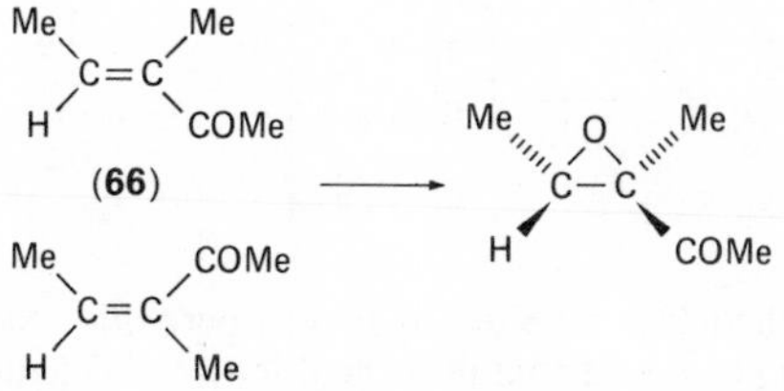

(**66**) (67)

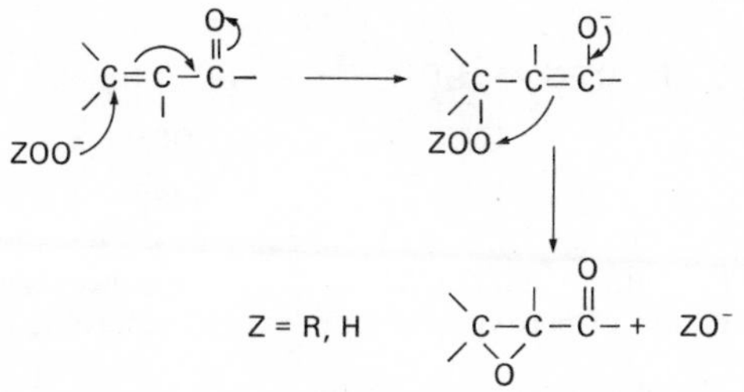

Z = R, H (68)

The Dakin reaction[212,216], i.e. the oxidation of *ortho* and *para* hydroxy- and amino-benzaldehydes to the corresponding phenols with alkaline hydrogen peroxide, is believed to resemble the Baeyer–Villiger-type of reaction. A benzenonium-type transition state **67** is presumably involved[217] (equation 69).

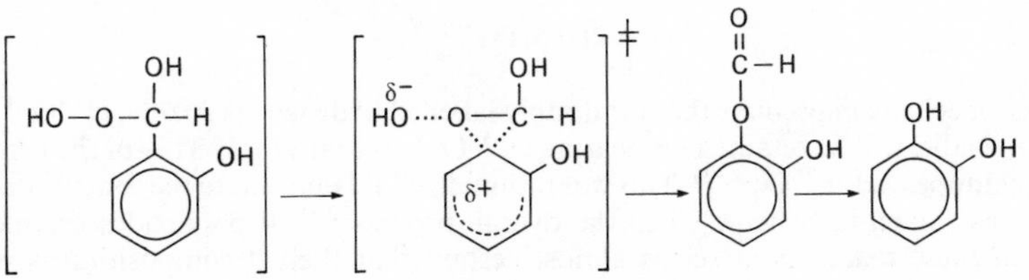

(69)

(**67**)

An acyclic mechanism, involving acyl migration, has recently been proposed by Sawaki and Foote for the reaction of benzil with alkaline hydrogen peroxide in aqueous methanol[218] (equation 70). Alternative 'epoxide' and 'dioxetane' mechanisms were ruled out on the basis of product and ^{18}O tracer studies.

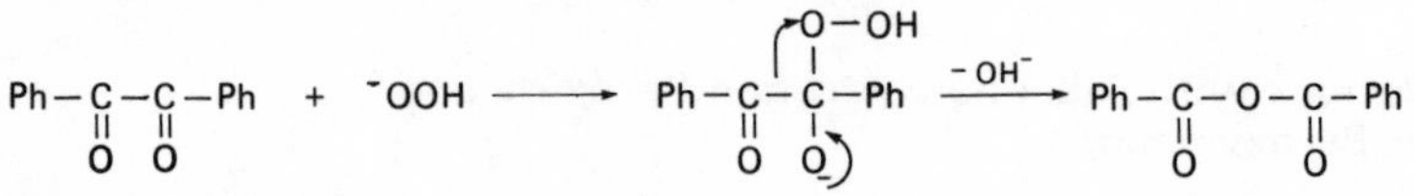

$$Ph-\underset{\underset{O}{\|}}{C}-\underset{\underset{O}{\|}}{C}-Ph + {}^{-}OOH \longrightarrow Ph-\underset{\underset{O}{\|}}{C}-\overset{\overset{O-OH}{|}}{\underset{\underset{O^{-}}{|}}{C}}-Ph \xrightarrow{-OH^{-}} Ph-\underset{\underset{O}{\|}}{C}-O-\underset{\underset{O}{\|}}{C}-Ph \quad (70)$$

A number of other reactions, some of them of preparative value, involve nucleophilic attack of peroxy anions on the electron-deficient acyl carbon of a suitable substrate according to the general scheme shown in equation (71). The preparation of peroxy acids ($Z = H$, $Y = O_2COR$ or $Z = H$, $Y = Cl$) and diacyl peroxides ($Z = RCO$, $Y = Cl$), for example, belong to these nucleophilic displacements, classified as $B_{AC}2$ reactions[212,219–221]. Hydrolysis of peroxy acids in alkaline media also involves attack of the peroxycarboxylate anion on the undissociated peroxy acid[1c,222].

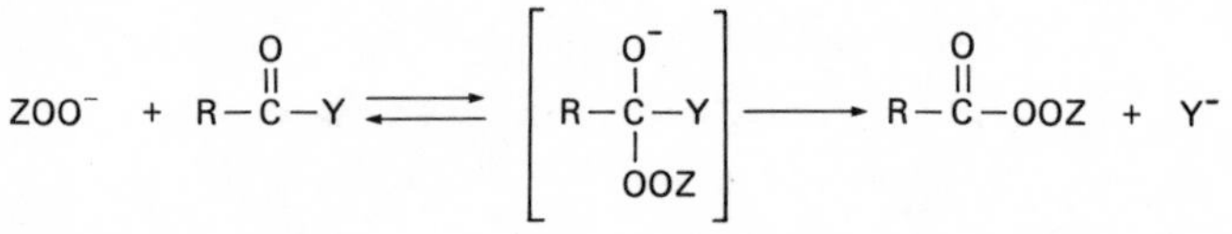

$$ZOO^{-} + R-\overset{\overset{O}{\|}}{C}-Y \rightleftharpoons \left[R-\overset{\overset{O^{-}}{|}}{\underset{\underset{OOZ}{|}}{C}}-Y\right] \longrightarrow R-\overset{\overset{O}{\|}}{C}-OOZ + Y^{-} \quad (71)$$

It is now well established that alkaline hydrogen peroxide reacts with aliphatic and aromatic nitriles to form peroxycarboximidic acid (**68**)[223,224] (equation 72), which is an efficient oxidizing agent[224].

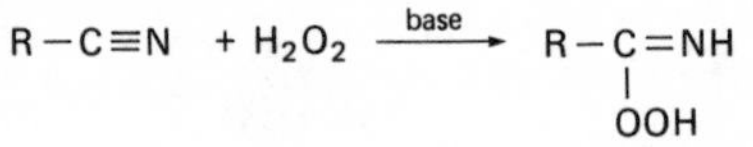

$$R-C\equiv N + H_2O_2 \xrightarrow{\text{base}} \underset{(\mathbf{68})}{R-\underset{\underset{OOH}{|}}{C}=NH} \quad (72)$$

On the basis of kinetic studies of this reaction at pH < 10, Wiberg originally suggested the rate-determining formation of peroxycarboximidic acid according to Scheme 15[223a].

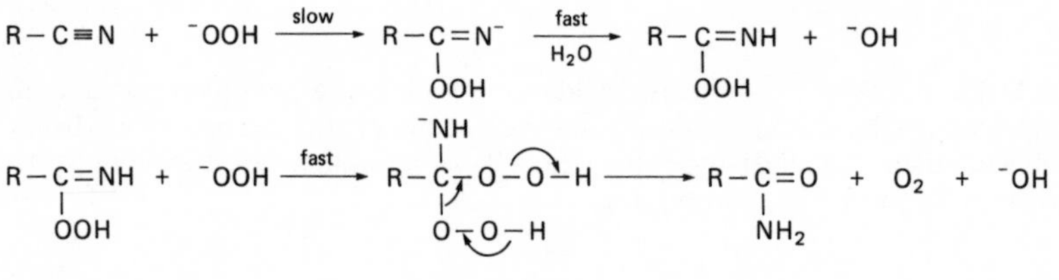

$$R-C\equiv N + {}^{-}OOH \xrightarrow{\text{slow}} R-\underset{\underset{OOH}{|}}{C}=N^{-} \xrightarrow[H_2O]{\text{fast}} R-\underset{\underset{OOH}{|}}{C}=NH + {}^{-}OH$$

$$R-\underset{\underset{OOH}{|}}{C}=NH + {}^{-}OOH \xrightarrow{\text{fast}} R-\overset{\overset{{}^{-}NH}{|}}{\underset{\underset{O-O-H}{|}}{C}}-O-O-H \longrightarrow R-\underset{\underset{NH_2}{|}}{C}=O + O_2 + {}^{-}OH$$

SCHEME 15

Wiberg's mechanism predicts that for doubly labelled hydrogen peroxide, $H^{18}O^{18}OH$, 100% of the label will be retained (appearing as $^{36}O_2$). Actually, only 81% of the original double label appeared as $^{36}O_2$ [223a]. This was rationalized as being due to the contribution of the reactions shown in Scheme 16 to the overall process[223a]. If polyoxides are indeed involved in these reactions, it seems almost certain that their decomposition is more complex than indicated above.

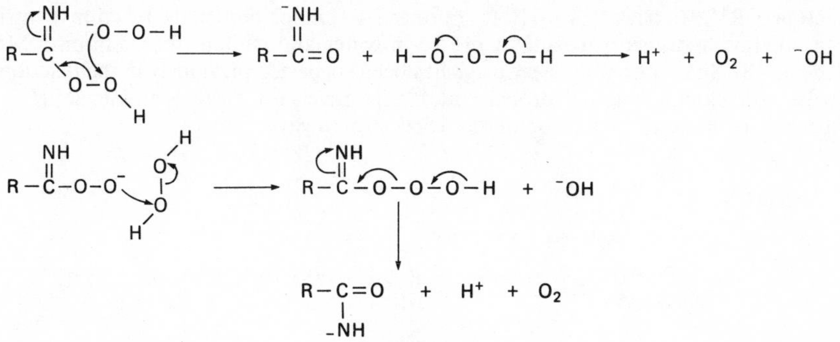

$$R-C(=NH)-O-O^- \ldots \longrightarrow R-C(=O)-NH^- + H-O-O-O-H \longrightarrow H^+ + O_2 + {}^-OH$$

$$R-C(=NH)-O-O^- + HO-OH \longrightarrow R-C(=NH)-O-O-O-H + {}^-OH$$

$$\downarrow$$

$$R-C(=O)-NH^- + H^+ + O_2$$

SCHEME 16

Some doubts have recently been expressed as to the validity of the Wiberg mechanism. Ogata and coworkers have studied the reaction of aliphatic and aromatic nitriles with alkaline hydrogen peroxide and found that k_{-OOH}, calculated from the following rate equation,

$$v = k_{obs.}[RCN][H_2O_2] = k_{-OOH}[RCN][HOO^-]$$

is constant[224]. Thus, the addition of HOO^- to nitrile cannot be the rate-determining step. The yield of amides, based on H_2O_2 consumed, varies from 20 to 60%. The addition of dimethyl sulphoxide (DMSO), which is easily oxidized by peroxycarboximidic acid to the sulphone, accelerates the reaction considerably. In this case, the rate becomes independent of the concentration of DMSO but dependent on $[HO^-]$ or $[HOO^-]$ so that k_{OOH} is constant. Thus the rate is governed by the addition of ^-OOH to the imine. Scheme 17 has been suggested. The k_{-OOH} values for the addition of ^-OOH to nitriles range from 10^{-3} to $3\,l\,mol^{-1}\,s^{-1}$. Aliphatic nitriles are less reactive than aromatic ones by a factor of ca. 10. A Hammett ρ value of 1.54 (σ) implies a nucleophilic attack of HOO^- on the $C{\equiv}N$ bond. The α effect is estimated to be 10^4 (k_{-OOH}/k_{-OH}) for benzonitrile under these conditions.

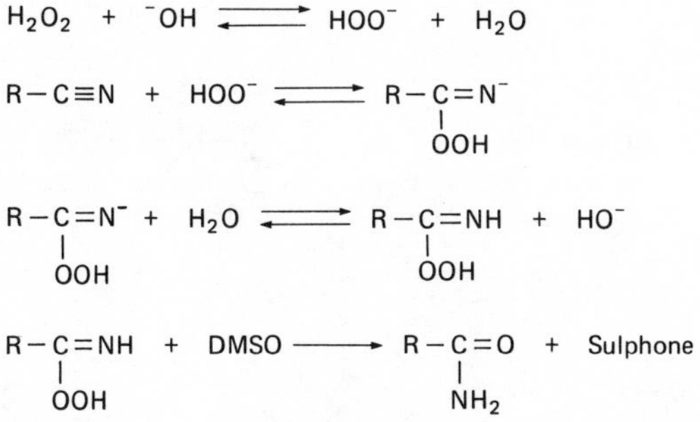

$$H_2O_2 + {}^-OH \rightleftharpoons HOO^- + H_2O$$

$$R-C{\equiv}N + HOO^- \rightleftharpoons R-C(OOH){=}N^-$$

$$R-C(OOH){=}N^- + H_2O \rightleftharpoons R-C(OOH){=}NH + HO^-$$

$$R-C(OOH){=}NH + DMSO \longrightarrow R-C(NH_2){=}O + \text{Sulphone}$$

SCHEME 17

A puzzling question, however, still remains to be answered; that is, the explanation of 81% of unscrambled oxygen observed in the ^{18}O labelling study[223a]. Ogata and coworkers[224] have found evidence for a homolytic cleavage of peroxyimidate ion **69** and subsequent reaction of $R^1O\cdot$ radicals[69a] with H_2O_2 (induced decomposition) in the reaction of nitriles with alkaline hydrogen peroxide at pH 10 in water and without the addition DMSO (Scheme 18). Similar homolytic cleavage has been suggested previously in the reaction of nitriles with alkaline *t*-butyl hydroperoxide[225]. Relatively low yields of amides at pH > 10 appear to be in accord with homolytic reaction pathways.

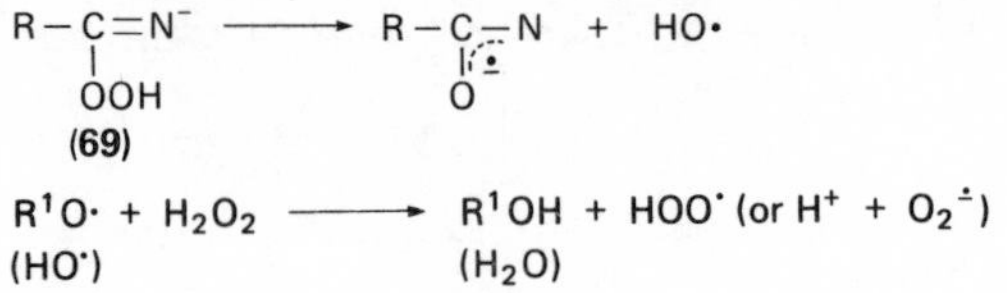

$$\underset{\textbf{(69)}}{\mathrm{R{-}C(OOH){=}N^-}} \longrightarrow \mathrm{R{-}C(O^{\cdot -}){=}N} + \mathrm{HO\cdot}$$

$$\mathrm{R^1O\cdot\ (HO^\cdot)} + \mathrm{H_2O_2} \longrightarrow \mathrm{R^1OH\ (H_2O)} + \mathrm{HOO^\cdot\ (or\ H^+ + O_2^{\cdot -})}$$

SCHEME 18

C. Oxidation of Imines (Schiff Bases)

Peroxy acids oxidize carbon–nitrogen double bonds in imines (Schiff bases) into oxaziridines[226–230]. Two mechanisms have been proposed for this reaction. The first is analogous to the olefin epoxidation mechanism, i.e. a concerted electrophilic attack of the peroxy acid on C=N via a three-membered cyclic transition state (**70**)[227] (equation 73). The second mechanism resembles the Baeyer–Villiger reaction involving the two-step process shown in equation (74)[228]. Earlier kinetic investigations found support for a concerted

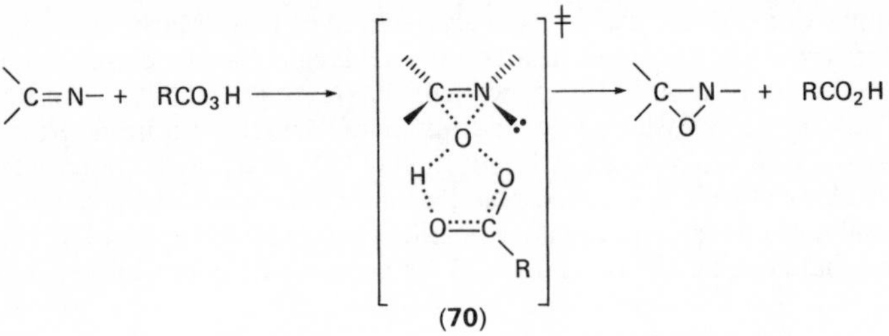

$$\mathrm{{>}C{=}N{-}} + \mathrm{RCO_3H} \longrightarrow [\textbf{(70)}]^{\ddagger} \longrightarrow \mathrm{{>}C{-}N{-}\ (O\ bridged)} + \mathrm{RCO_2H} \qquad (73)$$

$$\mathrm{{>}C{=}N{-}} + \mathrm{RCO_3H} \longrightarrow [\textbf{(71)}] \xrightarrow[-\mathrm{H^+}]{-\mathrm{RCOO^-}} \mathrm{{>}C{-}N{<}\ (O\ bridged)} + \mathrm{RCO_2H} \qquad (74)$$

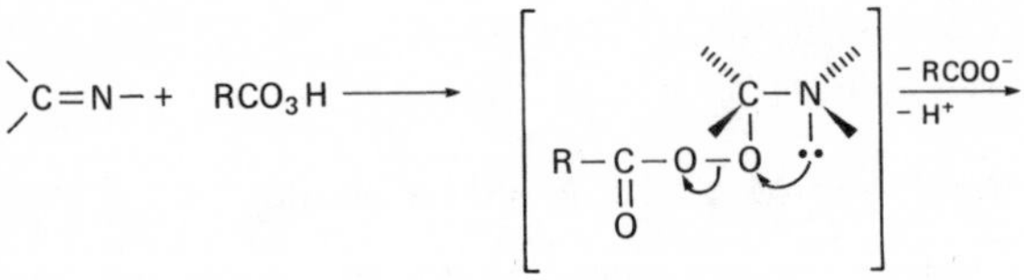

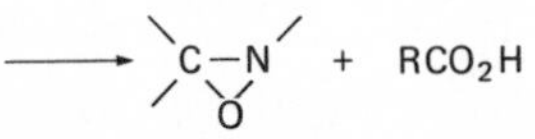

mechanism suggesting that the solvent molecule (protic solvent) or complex of peroxy acid with carboxylic acid or another molecule of peroxy acid is an attacking species[227]. Although a relatively small retardation of imine oxidation in basic solvents was noted, the solvent effect in these reactions bears little resemblance to the trends observed in the oxidation of olefins and sulphides known to react by a concerted mechanism.

$$R^1CO_3H + R^2OH \rightleftharpoons R^1CO_3H\cdots HOR^2$$

$$R^2 = R, RCO, RCO_2$$

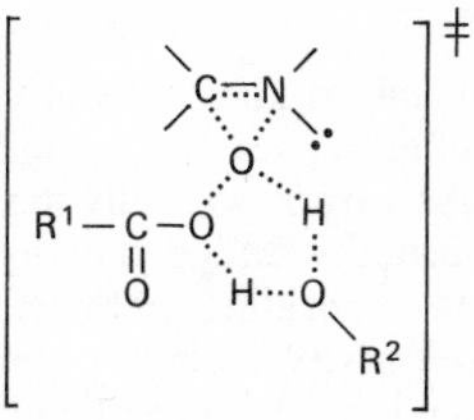

Recent kinetic studies support the oxaziridine formation via a two-step mechanism[228]. The reaction of benzylidene-*t*-butylimine with peroxybenzoic acid in various solvents exhibits complex kinetics since it is catalysed by dilute carboxylic acids and protic solvents and retarded by more concentrated carboxylic acids and basic solvents (alcohols, ethers). It has been suggested that the acid catalysis is due to the formation of an intermolecularly hydrogen-bonded complex of the imine and acid (equation 75) rather than to the formation of a dimer peroxy-acid–acid catalyst. The formation of the adduct **71** is the rate-determining step for acyclic imines while the internal nucleophilic (S_Ni) reaction of the adduct is believed to be the rate-determining step for cyclic imines. The formation of small amounts of nitrones in these oxidations can be rationalized by a mechanism analogous to that proposed for the peroxy acid oxidation of amines, i.e. the nucleophilic site in the imine is the lone pair of electrons at the nitrogen atom (equation 76). Electron-donating groups on the imine and aprotic solvents favour the formation of nitrones.

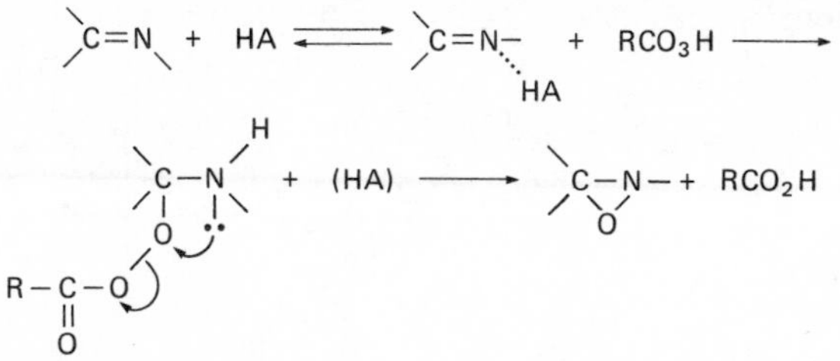

$$>C=N- + HA \rightleftharpoons >C=N-\cdots HA + RCO_3H \longrightarrow$$

$$\text{(71)} + (HA) \longrightarrow >C\text{—}N\text{—}(O) + RCO_2H \quad (75)$$

(**71**)

$$>C=N\text{—} + RCO_3H \longrightarrow >C=N\rightarrow O + RCO_2H \quad (76)$$

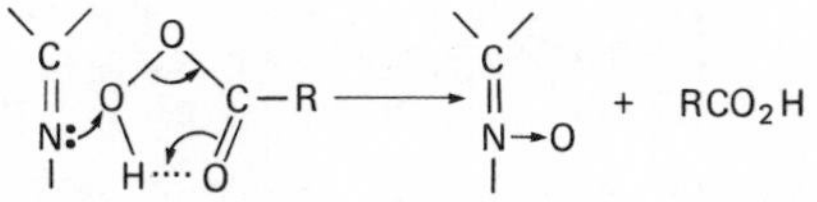

Additional evidence for the two-step mechanism comes from the observation that aliphatic imines react with hydrogen peroxide to form an adduct which can easily be converted to an oxaziridine on heating[231]. The isolation of a typical Baeyer–Villiger oxidation product **73** in the oxidation of sulphonimines **72** with *m*-chloroperoxybenzoic acid in a two-phase system (equation 77) also supports this type of mechanism[232].

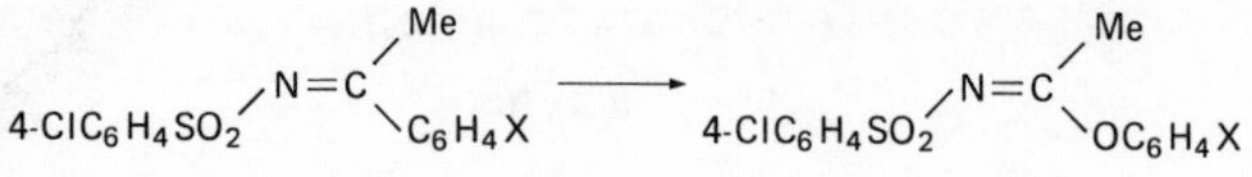

$$4\text{-}ClC_6H_4SO_2N{=}C(Me)C_6H_4X \longrightarrow 4\text{-}ClC_6H_4SO_2N{=}C(Me)OC_6H_4X \quad (77)$$

(72) (73)

The results of stereochemical studies seem to favour slightly the two-step mechanism[229,230]. Oxidation of aldimines and ketimines with achiral and chiral peroxy acids indicate that these reactions are not generally stereospecific[229,230]. For example, the reaction of optically active or racemic *N*-diphenylmethylene-α-methylbenzylamine (**74**)[230] with chiral peroxy acids yielding oxaziridines of known absolute configuration (equation 78) revealed that the diastereoselectivity depends only on the temperature of the reaction while the enantioselectivity depends on the solvent and the reaction temperature as well as the chirality of the peroxy acid[230]. It is believed that the diastereoisomeric ratio is controlled (*a*) by the relative nonbonded interactions between the imine and peroxy acid during the first step of reaction, which may involve rotation about the C—N bond, inversion at nitrogen or still more complex bond-breaking and bond-making phenomena, (*b*) by nonbonded interactions which control the conformational free energy of the intermediate adduct and (*c*) by the difference in the free energy levels of the two transition-state conformations during the ring-closure step. The preferred ground-state geometry of the adduct in the case of *C*-arylaldimines is suggested as possessing a staggered conformation (**75**) on the basis of nonbonded interactions only[229]. The preferred eclipsed conformation of the transition state **76** in the oxidation of these compounds leading to *trans*-oxaziridines is thus expected on steric grounds alone. The considerable amount of *cis*-oxaziridines formed in the oxidation of these imines implies that either other factors in addition to nonbonded interactions must be operative in stabilizing an alternative eclipsed transition state (**77**) required for the formation of *cis*-oxaziridines (polar, hydrogen bonding, solvent effects) or that the bond-breaking and the bond-making phases of the intramolecular acid elimination are not concerted processes. A 'late' transition state (**78**) has been suggested for the second step of the reaction[230].

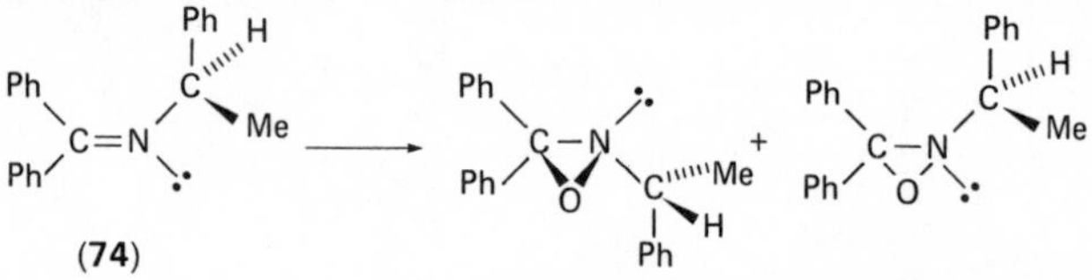

$$Ph_2C{=}N\text{–}CH(Me)Ph \longrightarrow \text{oxaziridine} + \text{oxaziridine} \quad (78)$$

(**74**)

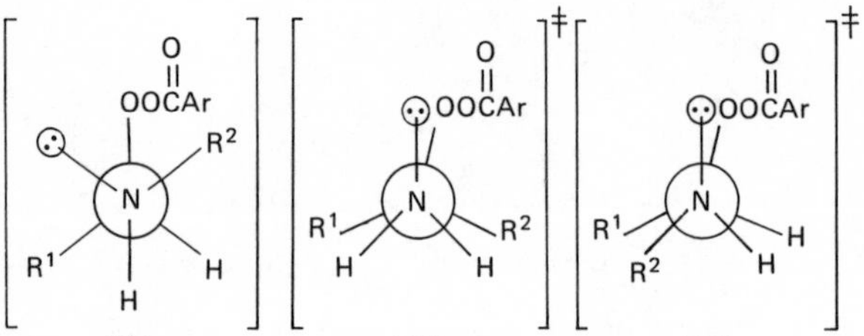

(**75**) (**76**) (**77**)

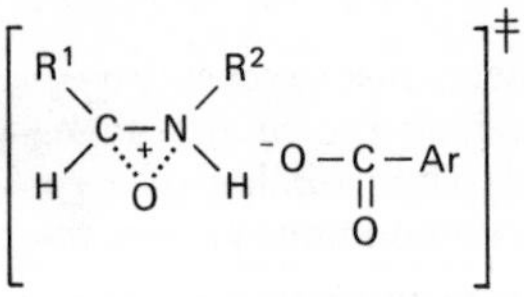

(**78**)

Various plausible trajectories of the peroxy acid approach to methyleneimine, and the geometries of the adduct involved in the two-step mechanism, have recently been investigated by using the *ab initio* theory (STO-4G)[233]. In the context of a concerted 1,1-addition mechanism, unsymmetrical 'transition states' are energetically more favourable than the symmetrical ones. Thus, it seems safe to predict that at least the formation of nitrones in the imine oxidation involves electrophilic attack of peroxy acid on the imine nitrogen, although no convincing evidence has been found that the nitrogen lone-pair electrons rather than π electrons are the nucleophilic centre in the imine. As expected, a staggered conformation of the adduct has been found to have the lowest energy. The energy of this adduct conformation is lower than that of the reactants, indicating the possibility of the involvement of such intermediates in these reactions.

A complete exploration of the energy surface for the reaction of peroxy acids with imines is necessary before one can claim to have shown computationally the detailed features of the mechanism of this reaction. Detection and decomposition studies of the intermediate adduct would also be desirable.

V. CARBONYL-FORMING ELIMINATIONS

A. Base-catalysed Decompositions

1. Hydroperoxides and dialkyl peroxides

Primary and secondary dialkyl peroxides decompose in the presence of base in a process involving a carbonyl-forming elimination (equation 79). The reaction was discovered in 1951 by Kornblum and DeLaMare who found that α-methylbenzyl *t*-butyl peroxide decomposes into acetophenone and *t*-butyl alcohol in the presence of potassium hydroxide, sodium ethoxide and piperidine[234] (equation 80). Tertiary amines also cause decomposition of dialkyl peroxides which carry a hydrogen on the α carbon atom. Bell and McDougall studied the kinetics of decomposition of benzyl *t*-butyl peroxide in chlorobenzene in the presence of amines[235]. They found that the rate law obeys the following expression:

$$v = k[\text{amine}][\text{peroxide}]$$

$$-\overset{|}{\underset{|}{C}}-O-\bar{O}R \;(\text{H on C, attacked by :B}) \longrightarrow \overset{+}{B}H + \rangle C{=}O + \bar{O}R \tag{79}$$

$$\text{Ph}-\overset{\text{Me}}{\underset{\text{H}\cdots:\text{B}}{C}}-O-O-\text{Bu-}t \longrightarrow \text{Ph}-\overset{\text{Me}}{C}{=}O + \bar{O}\text{Bu-}t + \overset{+}{B}H \tag{80}$$

The rate increases with increasing amine basicity (triethylamine > collidine > 2,6-lutidine). The low A factors reported ($A = 10^4$–$10^5\,\text{l}\,\text{mol}^{-1}\,\text{s}^{-1}$) indicate at least a partial separation of charge in the transition state of the rate-determining step of the reaction. It was suggested that proton transfer to form *t*-butanol and regenerate the amine occurs after the rate-determining step.

Alkyl hydroperoxides decompose in the presence of base by a similar process. This was recognized as far back as in 1932 when Medvedev and Aleksejeva reported base-catalysed dehydration of isopropyl hydroperoxide to acetone[236]. Hofmann and coworkers[237] studied the base-catalysed decomposition of tetralin hydroperoxide and α-methylbenzyl hydroperoxide and found that decomposition obeyed the following rate expression:

$$-\frac{d(RO_2H)}{dt} = \frac{k[RO_2H][^-OH]}{1+k[^-OH]}$$

where K and k are defined as follows:

$$^-OH + H-\underset{Me}{\overset{Ar}{C}}-O-OH \overset{K}{\rightleftharpoons} H-\underset{Me}{\overset{Ar}{C}}-O-O^- + H_2O$$

$$^-OH + H-\underset{Me}{\overset{Ar}{C}}-O-OH \xrightarrow{k} Me-\underset{O}{\underset{\|}{C}}-Ar + H_2O + {}^-OH$$

At 30°C, k and K for the tetralin hydroperoxide decomposition are $4.0\,l\,mol^{-1}\,s^{-1}$ and $4.1\,l\,mol^{-1}$, respectively, with $\Delta H^{\neq} = 17.3\,kcal\,mol^{-1}$ and $\Delta S^{\neq} = -17\,cal\,mol^{-1}\,deg^{-1}$. A relatively large isotope effect was reported ($k_H/k_D = 3.9$). Although these data seem to confirm the Kornblum–DeLaMare mechanism they also indicate that this mechanism holds only for peroxides with electron-withdrawing substituents near the peroxide bond, which are capable of facilitating carbanion formation. Indeed, many reported examples are in accord with this requirement. For example, ozonides (**79**)[238,239] and peroxides (**80, 81**)[241] formed during the ozonolysis of olefins in alcohols were reported to react in this way. Some examples are given in equations (81)–(83). Recent evidence indicates that, at least in the case of α-alkoxybenzyl hydroperoxides (**80**), the reaction is more complex than previously reported[242]. A reinvestigation of the decomposition of α-methoxybenzyl

$$(\mathbf{79}) \text{ (ozonide, } R_2C(H)\text{, attacked by MeOH)} \longrightarrow R-C(=O)OH + R_2C=O \qquad (81)$$

$$Ar-\underset{H}{\overset{OR}{C}}-O-OH \ (\mathbf{80}) \xrightarrow{:B} Ar-\underset{O}{\underset{\|}{C}}-OR + \overset{+}{B}H + \overset{-}{O}H \qquad (82)$$

:B = DMSO, amines

$$Ar-\underset{H}{\overset{OR}{C}}-O-O-\underset{H}{\overset{OR}{C}}-Ar \ (\mathbf{81}) \xrightarrow{:B} Ar-\underset{O}{\underset{\|}{C}}-OR + Ar-\underset{O}{\underset{\|}{C}}-H + ROH \qquad (83)$$

:B = amines

hydroperoxide in dimethyl sulphoxide and *N.N*-dimethylacetamide shows that the main decomposition products are benzaldehyde, water and methanol, and that only small amounts ($<5\%$) of methyl benzoate are formed at room temperature. Also a downfield shift of the OOH absorption is observed in these solvents ($\Delta\delta$OOH = 3 ppm) compared to solutions in carbon tetrachloride. This phenomenon has been tentatively assigned to the formation of the intermolecularly hydrogen-bonded hydroperoxide–oxygen base complex. The first-order decay of the complex is rather slow at room temperature as measured by the disappearance of the OOH or C—H absorption in the NMR spectra ($k = 5.1 \times 10^{-3}\,s^{-1}$ at 50°C, ~0.1 M in DMSO). All this evidence suggests a decomposition mechanism which involves homolytic cleavage of the RO—OH bond, which is believed to be weakened due to the intermolecular hydrogen bonding in the complex (Scheme 19). However, the isolation of some dimethyl sulphone in DMSO seems to indicate the involvement of a nonradical contribution to the reaction paths[225].

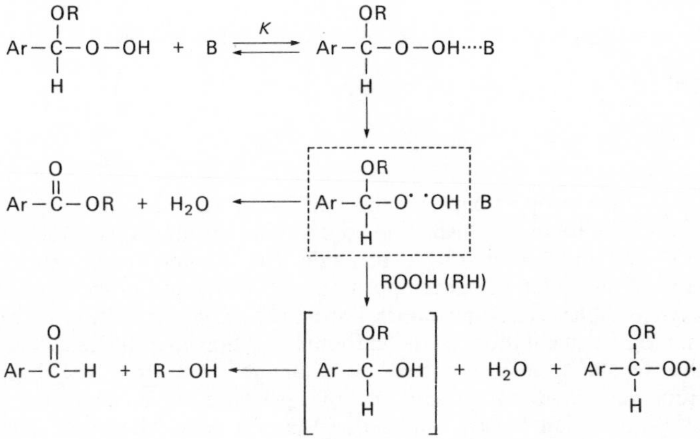

$$\mathrm{Ar{-}CH(OR){-}O{-}OH} + \mathrm{B} \overset{K}{\rightleftharpoons} \mathrm{Ar{-}CH(OR){-}O{-}OH\cdots B}$$

$$\mathrm{Ar{-}C({=}O){-}OR} + \mathrm{H_2O} \longleftarrow [\mathrm{Ar{-}CH(OR){-}O^{\bullet}\ {}^{\bullet}OH}\ \mathrm{B}]$$

$$\mathrm{Ar{-}C({=}O){-}H} + \mathrm{R{-}OH} \longleftarrow [\mathrm{Ar{-}CH(OR){-}OH}] + \mathrm{H_2O} + \mathrm{Ar{-}CH(OR){-}OO^{\bullet}} \quad (\text{via ROOH (RH)})$$

SCHEME 19

Base-catalysed decomposition of α-hydroperoxyketones has been studied extensively in recent years. Two mechanisms have been proposed[159,160,243]. Richardson and coworkers[159] suggested, on the basis of kinetic and product studies, that base-catalysed decomposition of 2,4-dimethyl-2-hydroperoxy-3-pentanone ($R^1 = (Me)_2CH$, $R^2 = R^3 = Me$; **82**) and also of 2,3-diphenyl-2-hydroperoxyvalerophenone proceeds largely through a cyclic or dioxetane mechanism. The production of a carbonyl group in an excited state and chemiluminescence were suggested as being associated with the dioxetane formation. The kinetic results indicated the formation of two dioxetane forms in the case of **82**, i.e. a hydroxy-1,2-dioxetane and its oxy anion (equation 84). It was also recognized that at least part of the reaction proceeds through an acyclic route with the formation of a carbonyl addition intermediate (equation 85). Sodium methoxide-catalysed decomposition in absolute methanol indicated that about 30% of the reaction proceeds by this pathway. Sawaki and Ogata[243], on the other hand, concluded, on the basis of the high yields of ketones (80–100%) and esters (70–100%) obtained in the alkoxy-catalysed decomposition of a number of α-hydroperoxyketones, that the α-cleavage reaction proceeds largely via an acyclic mechanism. The observation, that the pseudo-first-order rate

$$R^1-C(=O)-C(R^2)(R^3)-O-O^- \longrightarrow \left[R^1-C(O^-)-C(R^2)(R^3) \text{ (cyclic O-O)} \longrightarrow R^1-C(OH)-C(R^2)(R^3) \text{ (cyclic O-O)} \right] \longrightarrow$$

(82)

$$\longrightarrow R^1-CO_2^- + R^2R^3C{=}O \qquad (84)$$

$$R^1-C(=O)-C(R^2)(R^3)OOH + {}^-OR \longrightarrow \left[R^1-C(OR)(O^-)-C(R^2)(R^3)OOH \right] \longrightarrow R^1-C(=O)-OR + R^2R^3C{=}O + {}^-OH \qquad (85)$$

constants, k_{obs}, are proportional to the sodium methoxide concentration at low concentrations and become constant at higher concentrations, suggested a reaction between RO^- and undissociated hydroperoxide. The Hammett ρ values of 2.5–3.2 and 0.9–1.7 (σ) were found for the decomposition of benzoyl and α-phenyl-substituted α-hydroperoxy-α,α-diphenylacetophenones, respectively. A 'late' transition state (**83**) for the rate-determining fragmentation of the carbonyl addition intermediate was suggested. Sawaki and Ogata[243] also reported chemiluminescence from the base-catalysed decomposition of these compounds in the presence of a fluorescer (dibromoanthracene ≫ diphenylanthracene) indicating that the cyclic 'dioxetane' mechanism is actually operative in these reactions. Nevertheless, the low quantum yields together with the solvent effects on the chemiluminescence and the rate of decomposition suggest acyclic decomposition as the predominant reaction pathway.

$$\left[R^1-C(OR)(\cdots O)\cdots C(R^2)(R^3)(O\cdots OH) \right]^{\ddagger}$$

(**83**)

On the basis of kinetic and product studies, a cyclic mechanism involving a rate-determining fragmentation of a dioxetaneimine has been suggested as the major mechanistic feature of the base-catalysed decomposition of α-hydroperoxynitriles (**84**) in methanol and water[244] (Scheme 20). Chemiluminescence was observed on addition of dibromoanthracene to the reaction mixture ($\Phi = 4 \times 10^{-8}$ to 3×10^{-6}). As in the case of α-hydroperoxyketones, a predominant formation of triplet ketone was indicated by the inefficiency of diphenylanthracene as a fluorescer.

$$R^1R^2C(OOH)-C\equiv N + RO^- \overset{K}{\rightleftharpoons} R^1R^2C(O-O^-)-C\equiv N + ROH$$

(**84**)

$$R^1R^2C(O-O^-)-C\equiv N \rightleftharpoons R^1R^2C(-O-O-)C=N^- \longrightarrow R^1R^2C=O + {}^-OCN$$

SCHEME 20

2. *Peroxy esters*

Pincock[245] reported that *t*-butyl peroxyformate undergoes a base-catalysed elimination in the presence of pyridine and other bases. The catalytic effect of various amines follows the Brønsted relationship. It is interesting to mention that this fragmentation increases dramatically in rate when the solvent is frozen. Activation parameters and the kinetic isotope effect (pyridine as base, $\Delta H^{\neq} = 15.3\,\text{kcal mol}^{-1}$, $\Delta S^{\neq} = -23\,\text{cal mol}^{-1}\,\text{deg}^{-1}$, $k_H/k_D = 4.1$ at 90°C, in chlorobenzene as solvent) seem to support the suggestion that proton removal by the base is the rate-determining step (equation 86). In the absence of base, *t*-butyl peroxyformate undergoes a one-bond homolytic decomposition ($\Delta H^{\neq} = 38\,\text{kcal mol}^{-1}$, $\Delta S^{\neq} = 15\,\text{cal mol}^{-1}\,\text{deg}^{-1}$, chlorobenzene as solvent).

$$\text{B:} + \text{H}-\overset{\text{O}}{\overset{\|}{\text{C}}}-\text{O}-\text{O}-\text{Bu-}t \longrightarrow \left[\overset{\delta+}{\text{B}}\cdots\text{H}\cdots\overset{\text{O}}{\overset{\|}{\text{C}}}\cdots\text{O}\cdots\overset{\delta-}{\text{O}}-\text{Bu-}t\right]^{\ddagger} \longrightarrow \tag{86}$$

$$\longrightarrow BH^+ + CO_2 + t\text{-BuO}^-$$

Another example, while not base-catalysed, illustrates the dilemmas involved in the study of decomposition of peroxy esters. Hiatt and coworkers[246] have reported that primary alkyl peroxyacetates (*n*-butyl and isobutyl) and secondary peroxyacetates (*s*-butyl and cyclohexyl) decompose in the liquid phase (neat, chlorobenzene or α-methylstyrene as solvent) to produce mainly acetic acid and the aldehyde or ketone corresponding to the alcohol moiety of the peroxy ester. On the basis of product and kinetic data ($E_a = 24$–$28\,\text{kcal mol}^{-1}$, $\log A = 12$–14), the concerted six-centres mechanism shown in equation (87) was suggested. On the other hand, *n*-butyl peroxyacetate decomposes in the vapour phase to give products and kinetics ($E_a = 36\,\text{kcal mol}^{-1}$, $\log A = 16$) indicative of a simple O—O homolysis.

$$\text{>C(H)}-\text{O}-\text{O}-\text{C(=O)}-\text{R} \longrightarrow \left[\text{cyclic six-centre TS: } \overset{\delta+}{\text{C}}\cdots\overset{\delta+}{\text{H}}\cdots\overset{\delta-}{\text{O}}=\overset{\delta+}{\text{C}}(\text{R})-\overset{\delta-}{\text{O}}\cdots\overset{\delta-}{\text{O}}\right]^{\ddagger} \longrightarrow \text{>C=O} + \text{H}-\text{O}-\text{C(=O)}-\text{R} \tag{87}$$

Recent work by Schuster and Dixon seems to argue against this mechanism[247]. These authors have reported that 1-phenylethyl peroxyacetate decomposes in benzene to give quantitatively acetic acid and acetophenone, a small fraction of which is electronically excited. The activation parameters ($\Delta H^{\neq} = 33.2\,\text{kcal mol}^{-1}$, $\Delta S^{\neq} = 11.0\,\text{cal mol}^{-1}\,\text{deg}^{-1}$)

indicate a unimolecular decomposition. Schuster and Dixon suggest that the results are best explained by a stepwise process in which O—O bond homolysis is followed by rapid hydrogen abstraction in the cage (Scheme 21).

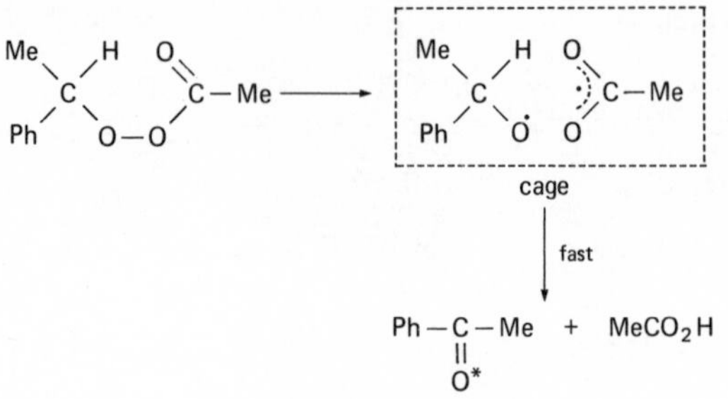

SCHEME 21

In conclusion, it is evident that base-catalysed and other carbonyl-forming eliminations are more complex than previously believed, and it will be a challenging task to confirm the involvement of one-electron processes in these reactions.

VI. REFERENCES

1. (a) A. G. Davies, *Organic Peroxides*, Butterworths, London, 1961.
   (b) J. O. Edwards, *Peroxide Reaction Mechanisms* (Ed. J. O. Edwards), John Wiley and Sons, New York–London, 1962, Chap. 4.
   (c) R. Curci and J. O. Edwards in *Organic Peroxides*, Vol. 1, (Ed. D. Swern), John Wiley and Sons, New York–London, 1970, Chap. 4.
   (d) J. B. Lee and B. C. Uff, *Quart. Rev. Chem. Soc.*, **21**, 431 (1967).
   (e) H. O. House, *Modern Synthetic Reactions*, 2nd ed., Benjamin, Menlo Park, California, 1972, Chap. 6.
   (d) B. Plesničar in *Oxidation in Organic Chemistry*, Part C (Ed. W. S. Trahanovsky), Academic Press, New York, 1978, Chap. 3.
2. L. M. Hjelmeland and G. Loew, *Tetrahedron*, **33**, 1029 (1977).
3. B. Plesničar, M. Tasevski and A. Ažman, *J. Amer. Chem. Soc.*, **100**, 743 (1978).
4. N. Prilezhaev, *Chem. Ber.*, **42**, 4811 (1909).
5. (a) D. Swern, *Organic Peroxides*, Vol. 2 (Ed. D. Swern), John Wiley and Sons, New York–London, 1971, Chap. 5.
   (b) E. N. Prilezhaeva, *Prilezhaev Reaction*, Nauka, Moscow, 1974.
   (c) V. G. Dryuk, *Tetrahedron*, **32**, 2855 (1976);
   (d) G. H. Schmid and D. G. Garatt in *The Chemistry of Double-Bonded Functional Groups* (Ed. S. Patai), John Wiley and Sons, London–New York, 1977, Chap. 9.
6. D. Swern, *J. Amer. Chem. Soc.*, **69**, 1692 (1947).
7. B. M. Lynch and K. H. Pausacker, *J. Chem. Soc.*, 1525 (1955).
8. J. Boeseken and C. J. A. Nonengraaf, *Rec. Trav. Chim.*, **61**, 69 (1942).
9. V. S. Aksenov, Sovrem. *Probl. Org. Khim.*, **58**, 1971; *Chem. Abstr.*, **78**, 3361 (1973).
10. Y. Ishii and Y. Inamoto, *Kogyo Kagaku Zasshi*, **63**, 765 (1960).
11. Y. Ogata and Y. Sawaki, *Bull. Chem. Soc., Japan*, **38**, 194 (1965).
12. M. M. Khalil and W. Pritzkow, *J. prakt. Chemie*, **315**, 58 (1973).
13. V. Prelog, K. Schenker and W. Küng, *Helv. Chim. Acta*, **36**, 471 (1953).
14. F. Asinger, B. Fell, G. Hadik and G. Steffan, *Chem. Ber.*, **97**, 1568 (1964).
15. Y. Ogata and I. Tabushi, *J. Amer. Chem. Soc.*, **83**, 3444 (1961).
16. M. Vilkas, *Bull. Soc. Chim. Fr.*, 1401 (1959).
17. K. D. Bingham, G. D. Meakins and G. H. Whitham, *Chem. Commun.*, 445 (1966).

18. M. S. Newman, N. Gill and D. W. Thomson, *J. Amer. Chem. Soc.*, **89**, 2059 (1967).
19. L. E. Friedrich and R. A. Fiatto, *J. Org. Chem.*, **39**, 416 (1974).
20. G. Berti, *Topics Stereochem.*, **7**, 93 (1973).
21. J. C. Richter and C. Freppel, *Can. J. Chem.*, **46**, 3709 (1968).
22. G. Bellucci, G. Berti, G. Ingrosso and E. Mastrorilli, *Tetrahedron Letters*, 3911 (1973).
23. (a) H. B. Henbest, *Chem. Soc., Spec. Publ.*, **19**, 23 (1964).
    (b) H. B. Henbest, *Proc. Chem. Soc., London*, 159 (1963).
24. P. Chamberlain, M. L. Roberts and G. H. Whitham, *J. Chem. Soc. (B)*, 1374 (1970).
25. B. A. Chiasson and G. A. Berchtold, *J. Amer. Chem. Soc.*, **96**, 2898 (1974).
26. S. A. Cerefice and E. K. Fields, *J. Org. Chem.*, **41**, 355 (1976).
27. M. R. Johnson and Y. Kishi, *Tetrahedron Letters*, 4347 (1979).
28. H. C. Brown, J. H. Kawakami and S. Ikegami, *J. Amer. Chem. Soc.*, **92**, 6914 (1970).
29. H. C. Brown, J. H. Kawakami and K.-T. Liu, *J. Amer. Chem. Soc.*, **95**, 2209 (1973).
30. P. G. Gassman and J.-H. Dygos, *Tetrahedron Letters*, 4749 (1970).
31. J. H. Haywood-Farmer, B. F. Friedlander and L. M. Hammond, *J. Org. Chem.*, **38**, 3145 (1973).
32. L. A. Paquette, L. W. Hertel, R. Gleiter and M. Böhm, *J. Amer. Chem. Soc.*, **100**, 6510 (1978). L. A. Paquette, L. W. Hertel, R. Gleiter, M. C. Böhm, M. A. Beno and G. G. Cristoph, *J. Amer. Chem. Soc.*, **103**, 7106 (1981).
33. L. A. Paquette and G. Kretschmer, *J. Amer. Chem. Soc.*, **101**, 4655 (1979).
34. H. Kropf and M. R. Yazdanbachsch, *Tetrahedron*, **30**, 3455 (1974).
35. P. Renolen and J. Ugelstad, *J. Chim. Phys.*, **57**, 1976 (1960).
36. R. Kavčič and B. Plesničar, *J. Org. Chem.*, **35**, 2033 (1970).
37. J. Škerjanc, A. Regent and B. Plesničar, *J. Chem. Soc., Chem. Commun.*, 1007 (1980).
38. G. Berti and F. Bottari, *J. Org. Chem.*, **25**, 1286 (1960); V. N. Sapunov and N. N. Lebedev, *Zh. Org. Khim.*, **2**, 225 (1966).
39. Y. Ogata, K. Tomizawa and T. Ikeda, *J. Org. Chem.*, **44**, 2362 (1979).
40. P. D. Bartlett, *Rec. Chem. Progr.*, **18**, 111 (1957).
41. W. A. Waters, *Mechanism of Oxidation of Organic Compounds*, John Wiley and Sons, New York–London, 1965.
42. H. Kwart and D. H. Hoffman, *J. Org. Chem.*, **31**, 419 (1966).
43. H. Kwart, P. S. Starcher and S. W. Tinsley. *Chem. Commun.*, 335 (1966).
44. K. D. Bingham, G. D. Meakins and G. H. Whitham, *Chem. Commun.*, 445 (1966).
45. F. Freeman, *Chem. Rev.*, **75**, 439 (1975).
46. R. Huisgen, *Angew. Chem.*, **75**, 604 (1963); R. Huisgen, L. Mobius, G. Müller, H. Stangl, C. Sczeimies and J. Vernon, *Chem. Ber.*, **98**, 3992 (1965).
47. R. P. Hanzlik and G. O. Schearer, *J. Amer. Chem. Soc.*, **97**, 5231 (1975).
48. T. Yonezava, H. Kato and O. Yamamoto, *Bull. Chem. Soc. Japan*, **40**, 307 (1967).
49. A. Ažman, B. Borštnik and B. Plesničar, *J. Org. Chem.*, **34**, 971 (1969).
50. L. M. Hjelmeland and G. H. Lowe, *Chem. Phys. Letters*, **42**, 309 (1975).
51. R. D. Bach, C. L. Willis and J. M. Domagala in *Application of MO Theory in Organic Chemistry*, Vol. 2 (Ed. I. G. Csizmadia), Elsevier, Amsterdam, 1977, p. 221.
52. T. J. Lang, G. J. Wolber and R. D. Bach, *J. Amer. Chem. Soc.*, **103**, 3275 (1981).
53. K. B. Sharpless and T. R. Verhoeven, *Aldrichim. Acta*, **12**, 63 (1979); E. D. Mihelich, K. Daniels and D. J. Eickhoff, *J. Amer. Chem. Soc.*, **103**, 7690 (1981).
54. P. Chautemps and J.-L. Pierre, *Tetrahedron*, **32**, 549 (1976).
55. J. Rebek, Jr., R. McReady and R. Wolak, *J. Chem. Soc., Chem. Commun.*, 705 (1980).
56. J. Rebek, Jr. and R. McCready, *J. Amer. Chem. Soc.*, **102**, 5602 (1980).
57. R. A. Sheldon in *Aspects of Homogenous Catalysis*, Vol. 4 (Ed. C. Manfred), Monotipia Rossi-Santi-Milano, Italy, 1979.
58. R. A. Sheldon and J. K. Kochi, *Metal-Catalyzed Oxidations of Organic Compounds*, Academic Press, New York, 1981.
59. R. Curci, F. Di Furia and G. Modena, *Fundamental Research in Homogenous Catalysis*, (Eds. I. Ishii and M. Tsutsui), Plenum Press, New York, 1978.
60. H. Mimoun, *J. Mol. Catal.*, **7**, 1 (1980).
61. T. Itoh, K. Jitsukawa, K. Kaneda and S. Teranishi, *J. Amer. Chem. Soc.*, **101**, 159 (1979).
62. R. B. Dehnel and G. H. Whitham, *J. Chem. Soc., Perkin Trans. 1*, 953 (1979).
63. B. E. Rossiter, T. R. Verhoeven and K. B. Sharpless, *Tetrahedron Letters*, (1979).

64. R. D. Chambers, P. Goggin and W. K. R. Musgrave, *J. Chem. Soc.*, 1804 (1959).
65. H. Hart, *Acc. Chem. Res.*, **4**, 337 (1971). H.-J. Schneider and W. Müller, *Angew. Chem.*, **94**, 153 (1982).
66. A. J. Davidson and R. O. C. Norman, *J. Chem. Soc.*, 5404 (1964).
67. Y. Ogata, Y. Sawaki, K. Tomizawa and T. Ohno, *Tetrahedron*, **37**, 1485 (1981).
68. U. Frommer and V. Ullrich, *Z. Naturforsch.* (*B*), **26**, 332 (1971).
69. G. A. Olah, N. Yoneda and D. G. Parker, *J. Amer. Chem. Soc.*, **99**, 483 (1977).
70. V. Franzen, *Chem. Ber.*, **88**, 717 (1955).
71. R. N. McDonald and P. A. Schwab, *J. Amer. Chem. Soc.*, **86**, 4866 (1964).
72. J. K. Stille and D. D. Whitehurst, *J. Amer. Chem. Soc.*, **86**, 4871 (1964).
73. J. Ciabattoni, R. A. Campbell, C. A. Renner and P. W. Concannon, *J. Amer. Chem. Soc.*, **92**, 3862 (1970).
74. P. W. Concannon and J. Ciabattoni, *J. Amer. Chem. Soc.*, **95**, 3824 (1973).
75. U. Timm, K. P. Zeller and H. Meier, *Chem. Ber.*, **111**, 1549 (1978).
76. P. R. Ortiz de Montellano and K. L. Kunze, *J. Amer. Chem. Soc.*, **102**, 7375 (1980).
77. (a) O. P. Strausz, R. K. Gosavi, A. S. Denes and I. G. Csizmadia, *J. Amer. Chem. Soc.*, **98**, 4784 (1976).
(b) O. P. Strausz, R. K. Gosavi and H. E. Gunning, *J. Chem. Phys.*, **67**, 3057 (1977).
(c) C. E. Dykstra, *J. Chem. Phys.*, **68**, 4244 (1978).
(d) K. Tanaka and M. Yoshimine, *J. Amer. Chem. Soc.*, **102**, 7655 (1980).
78. Y. Ogata, Y. Sawaki and H. Inoue, *J. Org. Chem.*, **38**, 1044 (1973); Y. Ogata, Y. Sawaki and T. Ohno, *J. Amer. Chem. Soc.*, **104**, 216 (1982).
79. K. M. Ibne-Rasa, R. H. Pater, J. Ciabattoni and J. O. Edwards, *J. Amer. Chem. Soc.*, **95**, 7894 (1973).
80. (a) J. Koller and B. Plesničar, *J. Chem. Soc., Perkin Trans* **2**, 1361 (1982).
(b) G. Melloni, G. Modena and U. Tonelatto, *Acc. Chem. Res.*, **14**, 227 (1981).
81. Y. Ogata and I. Tabushi, *Bull. Chem. Soc. Japan*, **31**, 969 (1958).
82. A. Dondoni, G. Modena and P. E. Todesco, *Gazz. Chim. Ital.*, **91**, 613 (1961).
83. K. B. Ibne-Rasa and J. O. Edwards, *J. Amer. Chem. Soc.*, **84**, 763 (1962).
84. K. B. Ibne-Rasa, J. O. Edwards, M. T. Kost and A. R. Gallopo, *Chem. Ind.* (*London*), 964 (1974).
85. S. Ball and T. C. Bruice, *J. Amer. Chem. Soc.*, **101**, 4018 (1979).
86. S. Ball and T. C. Bruice, *J. Amer. Chem. Soc.*, **102**, 6498 (1980).
87. W. D. Emmons and A. S. Pagano, *J. Amer. Chem. Soc.*, **77**, 4557 (1955).
88. R. J. Sundberg and P. A. Bukowick, *J. Org. Chem.*, **33**, 4098 (1968).
89. T. Mitsuhashi, O. Simamura and Y. Tezuka, *Chem. Commun.*, 1300 (1970).
90. T. Mitsuhashi and O. Simamura, *J. Chem. Soc.* (*B*), 705 (1970).
91. R. Curci, F. Di Furia and F. Marcuzzi, *J. Org. Chem.*, **36**, 3774 (1971).
92. C. G. Overberger and R. W. Cummins, *J. Amer. Chem. Soc.*, **75**, 4250 (1953); V. Marangelli, G. Modena and P. E. Todesco, *Gazz. Chim. Ital.*, **90**, 681 (1960).
93. R. Curci, R. A. Di Prete, J. O. Edwards and G. Modena, *J. Org. Chem.*, **35**, 740 (1970).
94. G. Modena and L. Maiolli, *Gazz. Chim Ital.*, **87**, 1306 (1957); M. A. P. Dankleft, R. Curci, J. O. Edwards and H. Y. Pyun, *J. Amer. Chem. Soc.*, **90**, 3209 (1968).
95. C. Kemal, T. W. Chan and T. C. Bruice, *Proc. Natl. Acad. Sci., U.S.A.*, **74**, 405 (1977).
96. S. Ball and T. C. Bruice, *J. Amer. Chem. Soc.*, **102**, 6498 (1980).
97. (a) A. Cerniani and G. Modena, *Gazz. Chim. Ital.*, **89**, 843 (1959).
(b) R. Curci, A. Giovine and G. Modena, *Tetrahedron*, **22**, 1235 (1966).
(c) Y. Sawaki, H. Kato and Y. Ogata, *J. Amer. Chem. Soc.*, **103**, 3832 (1981).
98. P. D. Bartlett, T. Aida, H.-K. Chu and T.-S. Fang, *J. Amer. Chem. Soc.*, **102**, 3515 (1980).
99. A. Battaglia, A. Dondoni, P. Giorgianni, G. Maccagnani and G. Mazzanti, *J. Chem. Soc.* (*B*), 1547 (1971).
100. A. Battaglia, A. Dondoni, G. Maccagnani and G. Mazzanti, *J. Chem. Soc., Perkin Trans. 2*, 610 (1974).
101. D. Swern, *Chem. Rev.*, **45**, 1 (1949).
102. D. F. Banks, *Chem. Rev.*, **66**, 243 (1966).
103. G. F. Kosser and R. H. Wettach, *J. Org. Chem.*, **45**, 1542 (1980).
104. B. Plesničar and G. A. Russell, *Angew. Chem.* (*Intern. Ed. Engl.*), **9**, 797 (1970).
105. B. Plesničar, *J. Org. Chem.*, **40**, 3267 (1975).

106. Y. Ogata and K. Aoki, *J. Org. Chem.*, **34**, 3974 (1969).
107. R. C. Cambie, B. G. Lindsay, P. C. Rutledge and P. D. Woodgate, *J. Chem. Soc., Chem. Commun.*, 919 (1978); *J. Chem. Soc., Perkin Trans. 1*, 822 (1980).
108. H. J. Reich and S. L. Peake, *J. Amer. Chem. Soc.*, **100**, 4888 (1978).
109. T. L. Mcdonald, N. Narasimhan and L. T. Burka, *J. Amer. Chem. Soc.*, **102**, 7760 (1980).
110. (a) R. Hiatt in *Organic Peroxides*, Vol. 2 (Ed. D. Swern), John Wiley and Sons, New York–London, 1972, Chap. 8.
(b) W. A. Pryor in *Free Radicals in Biology*, Vol. 1 (Ed. W. A. Pryor), Academic Press, New York, 1976, Chap. 1. D. F. Church and W. A. Pryor, *J. Org. Chem.*, **45**, 2866 (1980).
111. (a) L. Horner and E. Schwenk, *Angew. Chem.*, **61**, 411 (1949);
(b) L. Horner and E. Schwenk, *Justus Liebigs Ann. Chem.*, **566**, 35 (1950).
(c) L. Horner and C. Betzel, *Justus Liebigs Ann. Chem.*, **579**, 175 (1953).
(d) L. Horner, *J. Polym. Sci.*, **18**, 438 (1955).
(e) L. Horner and H. Junkerman, *Justus Liebigs Ann. Chem.*, **591**, 53 (1955).
112. (a) C. Walling, *Free Radicals in Solution*, John Wiley and Sons, New York, 1957, pp. 590–595.
(b) C. Walling and N. Indictor, *J. Amer. Chem. Soc.*, **80**, 5815 (1958).
(c) C. Walling and R. B. Hodgdon, *J. Amer. Chem. Soc.*, **80**, 228 (1958).
113. D. B. Denney and D. Z. Denney, *J. Amer. Chem. Soc.*, **82**, 1389 (1960).
114. (a) D. M. Graham and R. B. Mesrobian, *Can. J. Chem.*, **41**, 2945 (1963).
(b) T. Sato, K. Takemoto and M. Imoto, *J. Macromol. Sci., Chem.*, **2**, 69 (1968).
(c) H. Ohta and K. Tokumaru, *J. Chem. Soc. (D)*, 1601 (1970).
(d) H. Hasegawa, S. Kashino and Y. Mukai, *Bull. Soc. Chem. Japan*, **44**, 69 (1971).
115. G. B. Schuster, *Acc. Chem. Res.*, **12**, 366 (1979).
116. J. J. Zupancic, K. A. Horn and G. B. Schuster, *J. Amer. Chem. Soc.*, **102**, 5279 (1980).
117. J. M. McBride and R. A. Merrill, *J. Amer. Chem. Soc.*, **102**, 1725 (1980).
118. (a) R. V. Hoffman and R. Cadena, *J. Amer. Chem. Soc.*, **99**, 8226 (1977).
(b) R. V. Hoffman and E. L. Belfoure, *J. Amer. Chem. Soc.*, **101**, 5687 (1979).
119. (a) F. D. Greene, *J. Amer. Chem. Soc.*, **78**, 2250 (1956).
(b) F. D. Greene and W. W. Rees, *J. Amer. Chem. Soc.*, **80**, 3432 (1958).
(c) F. D. Greene, *J. Amer. Chem. Soc.*, **81**, 1503 (1959).
120. F. D. Greene, W. Adam and J. E. Cantrill, *J. Amer. Chem. Soc.*, **83**, 3461 (1961).
121. T. W. Koenig and J. C. Martin, *J. Org. Chem.*, **29**, 1520 (1964).
122. M. Kobayashi, *Bull. Soc. Chem. Japan*, **43**, 1158 (1970).
123. L. Horner and W. Jurgeleit, *Justus Liebigs Ann. Chem.*, **591**, 138 (1955).
124. (a) D. B. Denney and M. A. Greenbaum, *J. Amer. Chem. Soc.*, **79**, 979 (1957).
(b) D. B. Denney, W. F. Goodyear and B. Goldstein, *J. Amer. Chem. Soc.*, **83**, 1726 (1961).
(c) D. B. Denney and S. T. D. Gough, *J. Amer. Chem. Soc.*, **87**, 138 (1965).
125. D. L. Tullen, W. G. Bentrude and J. C. Martin, *J. Amer. Chem. Soc.*, **85**, 1938 (1963).
126. G. Sosnovsky and D. J. Rawlinson, *J. Org. Chem.*, **34**, 3462 (1969).
127. W. Adam and A. Rios, *J. Org. Chem.*, **36**, 407 (1971).
128. (a) W. A. Pryor and H. T. Bickley, *J. Org. Chem.*, **37**, 2885 (1972);
(b) W. A. Pryor and W. H. Hendrickson, *J. Amer. Chem. Soc.*, **97**, 1580 (1975).
(c) W. A. Pryor and W. H. Hendrickson, *J. Amer. Chem. Soc.*, **97**, 1582 (1975).
129. S. O. Lawesson and N. C. Yang, *J. Amer. Chem. Soc.*, **81**, 4230 (1959).
130. V. P. Bhatia and K. B. L. Mathur, *Tetrahedron Letters*, 4057 (1966).
131. C. Walling, *J. Amer. Chem. Soc.*, **102**, 6855 (1980).
132. F. Scandola, V. Balzani and G. B. Schuster, *J. Amer. Chem. Soc.*, **103**, 2519 (1981).
133. R. Criegee, *Chem. Ber.*, **77**, 722 (1944).
134. R. Criegee and R. Caspar, *Justus Liebigs Ann. Chem.*, **560**, 127 (1948).
135. P. D. Bartlett and J. L. Kice, *J. Amer. Chem. Soc.*, **75**, 5591 (1953).
136. H. L. Goering and A. C. Olsen, *J. Amer. Chem. Soc.*, **75**, 5853 (1953).
137. G. D. Schenk and K. H. Schulte-Elte, *Justus Liebigs Ann. Chem.*, **618**, 185 (1958).
138. D. B. Denney, *J. Amer. Chem. Soc.*, **77**, 1706 (1955).
139. D. B. Denney and D. Z. Denney, *J. Amer. Chem. Soc.*, **79**, 4806 (1957).
140. S. Winstein and G. C. Robinson, *J. Amer. Chem. Soc.*, **80**, 169 (1958).
141. J. E. Leffler and F. E. Scrivener, Jr., *J. Org. Chem.*, **37**, 1794 (1972).
142. P. D. Bartlett and B. T. Storey, *J. Amer. Chem. Soc.*, **80**, 4954 (1958).

143. P. D. Bartlett and T. G. Traylor, *J. Amer. Chem. Soc.*, **83**, 856 (1961).
144. E. Hedaya and S. Winstein, *J. Amer. Chem. Soc.*, **89**, 1661 (1967).
145. E. Wistuba and C. Rüchardt, *Tetrahedron Letters*, **72**, 3389 (1981).
146. R. Hiatt in *Organic Peroxides*, Vol. 2 (Ed. D. Swern), John Wiley and Sons, New York–London, 1971, Chap. 1, p. 65.
147. A. W. Van Stevenick and E. C. Kooyman, *Rec. Trav. Chim.*, **79**, 413 (1960).
148. G. H. Anderson and J. G. Smith, *Can J. Chem.*, **46**, 1553, 1561 (1968).
149. M. Bassey, C. A. Bunton, A. G. Davies, T. A. Lewis and D. Llewellyn, *J. Chem. Soc.*, 2471 (1955).
150. G. Burtzlaff, U. Felber, H. Hubner, W. Pritzkow and W. Rolle, *J. prakt. Chem.*, **28**, 305 (1965).
151. G. A. Olah, D. G. Parker, N. Yoneda and F. Pelizza, *J. Amer. Chem. Soc.*, **98**, 2445 (1976).
152. A. A. Frimer, *Chem. Rev.*, **79**, 359 (1979).
153. H. Hock and O. Schrade, *Naturwissenschaften*, **24**, 159 (1936); H. Hock and H. Kropf, *Angew. Chem.*, **69**, 313 (1957).
154. G. Ohloff, H. Strickler, B. Willhalm and M. Hinder, *Helv. Chim. Acta*, **53**, 623 (1970).
155. J. A. Turner and W. Herz, *J. Org. Chem.*, **42**, 1657 (1977).
156. A. A. Frimer, T. Farkash and M. Sprecher, *J. Org. Chem.*, **44**, 989 (1979).
157. P. D. Bartlett and A. Frimer, *Heterocycles*, **11**, 419 (1978).
158. E. H. Farmer and A. Sundralingam, *J. Chem. Soc.*, 121 (1942).
159. W. H. Richardson, V. F. Hodge, D. L. Stiggall, M. B. Yelvington and F. C. Montgomery, *J. Amer. Chem. Soc.*, **96**, 6652 (1974).
160. Y. Sawaki and Y. Ogata, *J. Amer. Chem. Soc.*, **97**, 6983 (1975).
161. Y. Sawaki and Y. Ogata, *J. Amer. Chem. Soc.*, **99**, 5412 (1977).
162. Y. Sawaki and Y. Ogata, *J. Amer. Chem. Soc.*, **100**, 856 (1978).
163. Y. Ogata and Y. Sawaki, *J. Amer. Chem. Soc.*, **94**, 4189 (1972).
164. C. J. Collins, *J. Amer. Chem. Soc.*, **77**, 5517 (1955); C. J. Collins, W. J. Rainey, W. B. Smith and I. A. Kaye, *J. Amer. Chem. Soc.*, **81**, 460 (1959).
165. R. Hiatt in *Organic Peroxides*, Vol. 3 (Ed. D. Swern), John Wiley and Sons, New York–London, 1972, Chap. 1.
166. V. P. Maslenikov, V. P. Sergeeva and V. A. Shushunov, *Zh. Obshch. Khim.*, **37**, 1727 (1967).
167. A. G. Davies, R. V. Foster and R. Nery, *J. Chem. Soc.*, 2204 (1954).
168. J. E. Leffler, *J. Amer. Chem. Soc.*, **72**, 67 (1950).
169. J. E. Leffler and C. D. Petropoulos, *J. Amer. Chem. Soc.*, **79**, 3068 (1957).
170. H. Hart and R. A. Cipriani, *J. Amer. Chem. Soc.*, **84**, 1962 (1962).
171. F. D. Greene, H. P. Stein, C. C. Chu and F. M. Vane, *J. Amer. Chem. Soc.*, **86**, 2080 (1964).
172. R. C. Lamb and J. R. Sanderson, *J. Amer. Chem. Soc.*, **91**, 5034 (1969).
173. T. Kashiwagi, S. Kozuka and S. Oae, *Tetrahedron*, **26**, 3169 (1970).
174. S. Oae, K. Fujimori and S. Kozuka, *J. Chem. Soc., Perkin Trans. 2*, 1884 (1974).
175. C. Walling, H. P. Waits, J. Milovanovic and C. G. Pappiaonnou, *J. Amer. Chem. Soc.*, **92**, 4927 (1970).
176. C. Walling and J. P. Sloan, *J. Amer. Chem. Soc.*, **101**, 7679 (1979).
177. K. G. Taylor, C. K. Govindan and M. S. Kaelin, *J. Amer. Chem. Soc.*, **101**, 2091 (1979).
178. J. E. Leffler and A. A. More, *J. Amer. Chem. Soc.*, **94**, 2483 (1972).
179. R. G. Lawler, P. F. Barbara and D. Jacobs, *J. Amer. Chem. Soc.*, **100**, 4912 (1978).
180. R. G. Lawler, M. Kline and P. F. Barbara, *Fifth IUPAC Conference on Physical Organic Chemistry, Santa Cruz, Book of Abstracts*, 1980.
181. A. von Baeyer and V. Villiger, *Chem. Ber.*, **32**, 3625 (1899).
182. C. H. Hassall, *Org. React.*, **9**, 73 (1957); P. A. S. Smith in *Molecular Rearrangements*, Vol. 1 (Ed. P. de Mayo), John Wiley and Sons, New York–London, 1963, p. 568.
183. R. Criegee, *Justus Liebigs Ann. Chem.*, **560**, 127 (1948).
184. S. L. Fries and N. Farnham, *J. Amer. Chem. Soc.*, **72**, 5518 (1950).
185. W. von E. Doering and E. Dorfman, *J. Amer. Chem. Soc.*, **75**, 5595 (1953).
186. S. L. Friess and A. H. Soloway, *J. Amer. Chem. Soc.*, **73**, 3968 (1951); S. L. Friess and P. E. Frankenburg, *J. Amer. Chem. Soc.*, **74**, 2679 (1952).
187. W. von E. Doering and L. Speers, *J. Amer. Chem. Soc.*, **72**, 5515 (1950).
188. W. H. Saunders, *J. Amer. Chem. Soc.*, **77**, 4679 (1955).
189. M. F. Hawthorne, W. D. Emmons and K. S. McCallum, *J. Amer. Chem. Soc.*, **80**, 6393 (1958).
190. J. L. Matheos and H. Menchaca, *J. Org. Chem.*, **29**, 2026 (1964).

191. B. M. Palmer and A. Fry, *J. Amer. Chem. Soc.*, **92**, 2580 (1970).
192. M. F. Hawthorne and W. D. Emmons, *J. Amer. Chem. Soc.*, **80**, 6398 (1958).
193. J. A. Berson and S. Suzuki, *J. Amer. Chem. Soc.*, **81**, 4088 (1959); T. F. Gallagher and T. H. Kritchevsky, *J. Amer. Chem. Soc.*, **72**, 882 (1950); K. Mislow and J. Brenner, *J. Amer. Chem. Soc.*, **75**, 2318 (1953).
194. E. E. Smissman, J. P. Li and Z. H. Israili, *J. Org. Chem.*, **30**, 3938 (1965).
195. M. A. Winnik and V. Stoute, *Can. J. Chem.*, **51**, 2788 (1973).
196. R. Noyori, T. Sato and H. Kobayashi, *Tetrahedron Letters*, **21**, 2569 (1980).
197. P. F. Hurdlik, A. M. Hurdlik, G. Nagendrappa, T. Yimenu, E. T. Zellers and E. Chin, *J. Amer. Chem. Soc.*, **102**, 6896 (1980). See also, B. M. Trost, P. Buhlmayer and M. Mao, *Tetrahedron Letters*, **23**, 1443 (1982).
198. A. Rassat and G. P. Ahearn, *Bull. Soc. Chim. Fr.*, 1133 (1959). J. Meinwald and E. Frauenglass, *J. Amer. Chem. Soc.*, **82**, 5235 (1960); R. R. Sauers and J. A. Beisler, *J. Org. Chem.*, **29**, 210 (1964).
199. Y. Ogata and Y. Sawaki, *J. Org. Chem.*, **37**, 2953 (1972).
200. Y. Ogata, K. Tomizava and O. Simamura, *J. Org. Chem.*, **43**, 2417 (1978).
201. T. Mitsuhashi, H. Miyadera and O. Simamura, *Chem. Commun.*, 1301 (1970).
202. V. Stoute, M. A. Winnik and I. G. Csizmadia, *J. Amer. Chem. Soc.*, **96**, 6388 (1974).
203. M. A. Winnik, V. Stoute and P. Fitzgerald, *J. Amer. Chem. Soc.*, **96**, 1977 (1974).
204. J. O. Edwards, R. H. Pater, R. Curci and F. Di Furia, *Photochem. Photobiol.*, **30**, 63 (1979).
205. R. W. Murray and V. Ramachandran, *Photochem. Photobiol.*, **30**, 187 (1979).
206. J. C. Robertson and A. Swelin, *Tetrahedron Letters*, 2871 (1967).
207. B. Phillips, F. C. Frostick, Jr. and P. S. Starcher, *J. Amer. Chem. Soc.*, **79**, 5982 (1957).
208. J. Lemaire and M. Niclause, *Compt. Rend.*, 260 (1965).
209. O. P. Yablonskii, M. G. Vinogradov, R. V. Kereselidze and G. I. Nikishin, *Izv. Akad. Nauk SSSR, Ser. Khim.*, 318 (1969).
210. C. E. H. Bawn and J. B. Williamson, *Trans. Faraday Soc.*, **47**, 735 (1951).
211. Y. Ogata, Y. Sawaki and Y. Tsukamoto, *Bull. Chem. Soc. Japan*, **54**, 2061 (1981).
212. C. A. Bunton in *Peroxide Reaction Mechanisms* (Ed. J. O. Edwards), John Wiley and Sons, New York–London, 1962, Chap. 2.
213. J. O. Edwards and R. G. Pearson, *J. Amer. Chem. Soc.*, **84**, 16 (1962); J. Gerstein and W. P. Jencks, *J. Amer. Chem. Soc.*, **86**, 4655 (1964); J. E. Dixon and T. C. Bruice, *J. Amer. Chem. Soc.*, **93**, 6592 (1971); N. J. Fina and J. O. Edwards, *Intern. J. Chem. Kinet.*, **5**, 1 (1973); J. L. Kice and L. F. Mullan, *J. Amer. Chem. Soc.*, **98**, 4259 (1976); A. P. Grekoc and V. Ya. Veselov, *Russ. Chem. Rev.*, **47**, 661 (1978).
214. H. O. House and R. S. Ro, *J. Amer. Chem. Soc.*, **80**, 2428 (1958).
215. C. A. Bunton and G. O. Minkoff, *J. Chem. Soc.*, 665 (1949).
216. H. D. Dakin, *Organic Syntheses, Collect. Vol. I*, 2nd ed. (Eds. H. Gilman and A. H. Blatt), John Wiley and Sons, New York–London, 1941, p. 759.
217. Y. Ogata and A. Tabushi, *Bull. Chem. Soc. Japan*, **32**, 108 (1959).
218. Y. Sawaki and C. S. Foote, *J. Amer. Chem. Soc.*, **101**, 6292 (1979).
219. H. R. Williams and H. S. Mosher, *J. Amer. Chem. Soc.*, **76**, 2987 (1954).
220. F. D. Greene and J. Kazan, *J. Org. Chem.*, **28**, 2168 (1963).
221. Y. Ogata and Y. Sawaki, *Tetrahedron*, **23**, 3327 (1967).
222. J. F. Goodman, J. F. Robson and E. R. Wilson, *Trans. Faraday Soc.*, **58**, 1846 (1962).
223. (a) K. B. Wiberg, *J. Amer. Chem. Soc.*, **77**, 2519 (1955).
(b) J. E. McIsaac, Jr., R. E. Ball and E. J. Behrman, *J. Org. Chem.*, **36**, 3048 (1971).
224. Y. Sawaki and Y. Ogata, *Bull. Chem. Soc. Japan*, **54**, 793 (1981).
225. M. J. Gibian and T. Ungermann, *J. Amer. Chem. Soc.*, **101**, 1291 (1979).
226. W. D. Emmons, *J. Amer. Chem. Soc.*, **79**, 5739 (1957).
227. V. Madan and L. B. Clapp. *J. Amer. Chem. Soc.*, **92**, 4902 (1970).
228. Y. Ogata and Y. Sawaki, *J. Amer. Chem. Soc.*, **95**, 4687 , 4692 (1973).
229. D. R. Boyd, D. C. Neill, C. G. Watson and G. Torre, *J. Chem. Soc., Perkin Trans. 2*, 1813 (1975).
230. M. Bucciarelli, A. Forni, I. Moretti and G. Torre, *J. Chem. Soc., Perkin Trans. 2*, 1339 (1977).
231. E. Hoft and A. Rieche, *Angew. Chem.*, **77**, 548 (1966).
232. F. A. Davis, J. Lamendola, Jr., U. Nadir, E. W. Kluger, T. C. Sedergran, T. W. Panunto, R. Billmers, R. Jenkins, Jr., I. J. Turchi, W. H. Watson, J. S. Chen and M. Kimura, *J. Amer. Chem. Soc.*, **102**, 2000 (1980).

233. A. Ažman, J. Koller and B. Plesničar, *J. Amer. Chem. Soc.*, **101**, 1107 (1979).
234. N. Kornblum and H. E. DeLaMare, *J. Amer. Chem. Soc.*, **73**, 880 (1951).
235. R. P. Bell and A. O. McDougall, *J. Chem. Soc.*, 1697 (1958).
236. S. S. Medvedev and E. Aleksejeva, *Chem. Ber.*, **65**, 133 (1932).
237. R. Hofmann, H. Hubner, G. Just, L. Kratzsch, A. K. Litkowez, W. Pritzkow, W. Rolle and M. Wahren, *J. prakt. Chem.*, **37**, 1022 (1968).
238. R. M. Ellam and J. M. Padbury, *Chem. Commun.*, 1094 (1971).
239. R. Criegee and H. Korber, *Advan. Chem. Ser.*, **112**, 22 (1972).
240. P. Keaveney, M. G. Berger and J. J. Pappas, *J. Org. Chem.*, **32**, 1537 (1967).
241. R. M. Ellam and J. M. Padbury, *Chem. Commun.*, 1086 (1972).
242. B. Plesničar, F. Kovac and V. Menart, unpublished results.
243. Y. Sawaki and Y. Ogata, *J. Amer. Chem. Soc.*, **99**, 6313 (1977).
244. Y. Sawaki and Y. Ogata, *J. Amer. Chem. Soc.*, **99**, 5412 (1977).
245. R. E. Pincock, *J. Amer. Chem. Soc.*, **86**, 1820 (1964).
246. R. Hiatt, L. C. Glover and H. S. Mosher, *J. Amer. Chem. Soc.*, **97**, 1556 (1975).
247. B. G. Dixon and G. B. Schuster, *J. Amer. Chem. Soc.*, **101**, 3116 (1979); B. G. Dixon and G. B. Schuster, *J. Amer. Chem. Soc.*, **103**, 3068 (1981).
248. W. A. Pavelich and R. W. Taft, *J. Amer. Chem. Soc.*, **79**, 4935 (1957).
249. B. Plesničar and J. Škerjanc, *Abstracts, Fifth IUPAC Conference on Physical Organic Chemistry*, University of California at Santa Cruz, California, 1980, p. 21.
250. T. N. Shatkina, A. N. Lovtsova, K. S. Mazel, T. I. Pekhk, E. T. Lippmaa and O. A. Reutov, *Dokl. Akad. Nauk SSSR*, **237**, 368 (1977); V. A. Yablokov, *Usp. Khim.*, **49**, 833 (1980).

The Chemistry of Functional Groups, Peroxides
Edited by S. Patai

CHAPTER 18

Preparation and uses of isotopically labelled peroxides

SHIGERU OAE and KEN FUJIMORI

Department of Chemistry, The University of Tsukuba, Sakura-mura, Niihari-gun, Ibaraki-ken, 305 Japan

---

I. INTRODUCTION 586

II. PREPARATION OF ISOTOPICALLY LABELLED PEROXIDES . . . 587
A. Hydrogen Peroxide 587
B. Hydroperoxides 587
1. Autoxidation of hydrocarbons 588
2. Ene reaction of alkyl-substituted olefins with singlet oxygen . . . 588
3. Reaction of hydrogen peroxide with alkylating agents 590
4. Addition of hydrogen peroxide to double bonds 591
5. Perester alcoholysis 591
C. Dialkyl Peroxides 591
1. Reactions of hydrogen peroxide or hydroperoxides with alkylating agents . 591
2. Addition of hydrogen peroxide or hydroperoxides to unsaturated compounds 592
3. Cycloadditions of singlet oxygen to olefins 592
a. 1,2-Cycloaddition 592
b. 1,4-Cycloaddition 592
D. Peracids 593
1. Autoxidation 593
2. Reaction of hydrogen peroxide with acylating agents 593
3. Reaction of diacyl peroxides with sodium methoxide 594
E. Peresters 594
F. Diacyl Peroxides 596
1. Reaction between sodium peroxide and acylating agents 596
2. Autoxidation of aldehydes in the presence of acylating agents . . . 597
3. Reaction of peracids with acylating agents 598
4. Diimide synthesis 599

III. USES OF ISOTOPICALLY LABELLED PEROXIDES 600
A. Unimolecular Homolytic Decomposition of Peroxides 600
1. Peresters 600
a. Classification of *O*-acyl *O'*-*t*-butyl hyponitrites, *t*-butyl peresters and *t*-butoxy acyloxy radical pairs 601
b. Formation of peresters by thermal decomposition of *O*-acyl *O'*-*t*-butyl hyponitrites 602
c. Thermal decomposition of *t*-butyl peresters 603
d. [1,3]-Sigmatropic mechanism for ^{18}O scrambling 608
2. Diacyl peroxides 608
a. Classification of diacyl peroxides and acyloxy radical pairs. . . . 610
b. Case I_A diacyl peroxides 614
c. Case I_B diacyl peroxides. 617
d. Case II diacyl peroxides. 617
e. Case III diacyl peroxides 618
3. Thermal decomposition of *t*-butyl cycloalkaneperformates and cycloalkaneformyl peroxides 619
4. Polyoxides 622
5. Polymerization initiated by free radicals generated by unimolecular decomposition of organic peroxides 622
B. Unimolecular Heterolytic Decomposition 623
1. Peresters 623
2. Diacyl peroxides 623
3. Ozonides 632
C. Induced Decomposition 634
1. Peresters 634
2. Diacyl peroxides 636
3. Miscellaneous 641
IV. CONCLUDING REMARKS 642
V. REFERENCES 643

I. INTRODUCTION

While many stable organic peroxides have been used as oxidants and free-radical sources in various organic reactions, there are numerous unstable organic peroxides which have been observed, or suggested, as intermediates in oxidation reactions. Isotopically labelled organic peroxides are generally prepared for investigating the mechanism of reactions involving organic peroxides. They may be used as tracers to follow reaction pathways or to determine kinetic isotope effects. Isotopically labelled peroxides have been used to examine CIDNP or ESR signals during the reaction of organic peroxides. Among the isotopes used to label peroxides, ^{13}C, ^{14}C, ^{2}H, ^{3}H, ^{17}O and ^{18}O are most common. The radioisotope, ^{14}C (β^-, 5570 year) is often used. A useful means of probing the structure and the nature of bonding in carbon compounds is by measuring the NMR spectra of ^{13}C-labelled compounds. By Fourier transform spectroscopy the measurement can be made using compounds with ^{13}C natural abundance (1.11%). Thus, the ^{13}C-NMR chemical shifts are often used to diagnose the polarity of electron densities of various organic compounds while the ^{13}C–H coupling constants are used to estimate the s-character of carbon as we shall see later. Tritium (β^-, 12.4 years) is commercially available to be used as a radioactive tracer. Deuterium is also widely used in mechanistic studies as a tracer, in the measurement of kinetic isotope effects and also in spectroscopic studies, since deuterium is twice as heavy as hydrogen and has a nuclear spin $S = 1$. Oxygen occurs in nature in three stable isotopic species: ^{16}O (99.759 %), and ^{17}O (0.0374 %) and ^{18}O (0.203 %)[1]. ^{18}O of up to 99 atom % is now commercially available and is widely used as a tracer in studies involving oxygen migrations, O—O bond cleavages and others. Since ^{17}O has a nuclear spin $S = 5/2$, ^{17}O labelled compounds are used conveniently for NMR and ESR spectroscopic studies. However, ^{18}O tracer experiments have been the most effective and powerful tool, especially to clarify the mode of O—O bond cleavage in organic peroxides, and much of our attention will be focused on the use of ^{18}O-labelled organic peroxides in elucidating their reaction mechanisms.

II. PREPARATION OF ISOTOPICALLY LABELLED PEROXIDES

A. Hydrogen Peroxide

Hydrogen peroxide labelled with ^{17}O or ^{18}O is an important starting material for the synthesis of labelled organic peroxides. Hydrogen peroxide containing ^{18}O has been prepared from alkali or alkaline earth metal peroxides which may be obtained by treatment of one of these metals with $^{18}O_2$ (equation 1)[2].

$$2Na + {}^{*}O_2 \longrightarrow Na_2{}^{*}O_2 \xrightarrow{H^+} H_2{}^{*}O_2 \qquad (1)$$

Acid-catalysed hydrolysis of ^{18}O-labelled perbenzoic acid[3] (prepared by autoxidation of benzaldehyde with $^{18}O_2$ can also afford ^{18}O-labelled hydrogen peroxide (equation 2).

$$PhC(=O){}^{*}O{}^{*}OH + H_2O \xrightarrow{H^+} PhCOOH + H_2{}^{*}O_2 \qquad (2)$$

Recently Sawaki and Foote[4] prepared ^{18}O-labelled hydrogen peroxide by a convenient base-catalysed autoxidation of benzhydrol with $^{18}O_2$ in a good yield (more than 90% based on $^{18}O_2$) (equation 3).

$$Ph_2CHOH + {}^{*}O_2 \xrightarrow{t\text{-BuOK/benzene}} Ph_2C{=}O + H_2{}^{*}O_2 \qquad (3)$$

Electronic discharge of ^{18}O-enriched water is another way to prepare ^{18}O-labelled hydrogen peroxide[5].

The I. G. Farben industrial process (equation 4) gives H_2O_2 labelled with ^{17}O or ^{18}O, when $^{17}O_2$ or $^{18}O_2$ are used[6]. The anthraquinone can be reduced back to the quinol by catalytic hydrogenation.

$$\text{2-ethyl-9,10-dihydroxyanthracene} + O_2 \longrightarrow \text{2-ethylanthraquinone} + H_2O_2 \qquad (4)$$

B. Hydroperoxides

The general synthetic methods for the preparation of hydroperoxides have been reviewed by Hiatt in detail[7].

1. *Autoxidation of hydrocarbons*

Hydrocarbons with labile hydrogens react with ground-state triplet molecular oxygen. Oxygen itself is too unreactive to abstract even the labile hydrogen; however, in the presence of certain free-radical initiators, metal salts or base catalysts, they can react with oxygen in a chain-reaction, as Russian authors have reported in the preparation of benzylic-^{14}C-labelled cumyl hydroperoxide[8] (equation 5).

$$PhC^{*}(Me)_2-H + O_2 \longrightarrow PhC^{*}(Me)_2-OOH \qquad (5)$$

Alkyl or aryl hydrazones of aldehydes and ketones are readily autoxidized to α-azoalkyl hydroperoxides in good yields[9a]. Recently the latter were revealed to be good hydroxyl radical sources in anhydrous media[9b,c] (equation 6).

$$R^1R^2C{=}N-NHR^3 + O_2 \longrightarrow R^1R^2C(OOH)-N{=}NR^3 \qquad (6)$$

2. *Ene reaction of alkyl-substituted olefins with singlet oxygen*[10]

Treatment of alkyl-substituted olefins with $^1\Delta$ singlet oxygen gives the corresponding allylic hydroperoxides in which the hydroperoxy group is introduced into C(1) with concomitant shift of the double bond[10,11] (equation 7).

$$\text{(7)}$$

Singlet oxygen is conveniently generated by photosensitized excitation of ground-state triplet molecular oxygen with some dyes[10], and can also be generated chemically[12a]. For example, treatment of hydrogen peroxide with hypochlorite also generates singlet oxygen[13]. Both oxygen atoms of the singlet oxygen generated in this reaction were shown, by Cahill and Taube who carried out experiments with ^{18}O-enriched hydrogen peroxide, to originate from hydrogen peroxide and not from hypochlorite or water[14a]. The mechanism shown in equation (8) was proposed by Kasha and Kahn[14b] for the generation of singlet oxygen. The triphenyl-phosphite–ozone adduct is a good source of singlet oxygen without light[12a]. Bartlett and Schaap revealed that triphenyl phosphite ozonide can also react directly with electron-rich olefins to give 1,2-dioxetanes[12b]. Decomposition of transition-metal oxygen complexes[12a] and photo-endoperoxides can also generate singlet

$$OCl^- + H_2{}^{*}O_2 \longrightarrow HOCl + {}^{-}\,{}^{*}O^{*}OH \xrightarrow{H_2O} H^{*}O^{*}OCl \xrightarrow{-HCl} {}^1{}^{*}O_2 \qquad (8)$$

$$\text{9,10-diphenylanthracene 9,10-endoperoxide ({}^{*}O{-}O^{*})} \xrightarrow{\Delta} {}^1{}^{*}O_2 + \text{9,10-diphenylanthracene} \qquad (9)$$

oxygen[12a,15]. One convenient procedure involves thermolysis of 9,10-diphenylanthracene peroxide (equation 9).

The ene reaction of singlet oxygen with olefin has usually been explained in terms of either a concerted or a perepoxide mechanism (equations 10 and 11). Allylic hydroperoxides labelled with either deuterium or tritium have been obtained in order to elucidate the mechanism through determination of k_H/k_D. For example, the reaction of singlet oxygen with olefins **1** and **2** gives mixtures of **3** + **4** and **5** + **6**, respectively[16].

Concerted mechanism

(10)

Perepoxide mechanism

(11)

Kopecky and Sande[16] have suggested the concerted mechanism for the ene reactions (equations 12 and 13), based on the smaller kinetic deuterium isotope effect ($k_H/k_D = 1.4$) than that ($k_H/k_D = 2.2$) for the base-catalysed elimination from a mixture of **7** and **8** (equation 14) and from **9** (equation 15), in which apparent migration of the hydroperoxyl group takes place via a perepoxide intermediate (**10**) as shown in equation (16). Furutachi, Nakadaira and Nakanishi[17] have suggested that the photooxidation shown in equation (17) is another concerted ene reaction based on the fact that 94% of original deuterium label at the 4β position in cholest-5-en-3-one (**11**) is retained in the (photooxidation product, 6α-hydroperoxycholest-4-en-3-one even in the presence of base (equation 17). Frimer, Bartlett and coworkers[18] suggested that the transition state of the ene reaction of 4-methyl-2,3-dihydro-γ-pyrane in the sensitized photooxidation occurs substantially later on the reaction coordinate than has been previously suggested based on the k_H/k_T values shown below (equations 18 and 19). MB and TPP denote methylene blue and *meso*-tetraphenylporphin, respectively.

$$(CD_3)_2C{=}C(CH_3)_2\ (\mathbf{1}) + {}^1O_2 \longrightarrow D_3C{-}C(OOH)(CD_3){-}C(CH_3){=}CH_2\ (\mathbf{3}) + D_2C{=}C(CD_3){-}C(OOD)(CH_3){-}CH_3\ (\mathbf{4}) \qquad (12)$$

$$(H_3C)(D_3C)C{=}C(CH_3)(CD_3)\ (\mathbf{2}) + {}^1O_2 \longrightarrow H_3C{-}C(OOH)(CD_3){-}C(CD_3){=}CH_2\ (\mathbf{5}) + D_2C{=}C(CH_3){-}C(OOD)(CD_3){-}CH_3\ (\mathbf{6}) \qquad (13)$$

$$D_3C{-}C(OOH)(CD_3){-}C(Br)(CH_3){-}CH_3\ (\mathbf{7}) + D_3C{-}C(Br)(CD_3){-}C(OOH)(CH_3){-}CH_3\ (\mathbf{8}) \xrightarrow[-\ HBr]{CD_3O^-} \mathbf{3} + \mathbf{4} \qquad (14)$$

(7) : (8) = 1 : 1.6; 3 : 4 = 2.2 : 1

$$H_3C-C(OOH)(CD_3)-C(Br)(CD_3)-CH_3 \xrightarrow[-HBr]{CD_3O^-} 5 + 6 \quad (2.2:1) \tag{15}$$

(9)

$$-C(OOH)-C(Br)- \xrightarrow[-H^+]{:B} -C(OO^-)-C(Br)- \xrightarrow{-Br^-} (10) \longrightarrow 3+4 \text{ or } 5+6 \tag{16}$$

$$(11) + {}^1O_2 \xrightarrow{Py.} \text{product (D, OOH)} \tag{17}$$

$$+ {}^1O_2 \longrightarrow \quad k_H/k_T = 0.866 \pm 0.003(CH_3CN/MB) = 0.908 \pm 0.023\ (C_6H_6/TPP) \tag{18}$$

$$+ {}^1O_2 \longrightarrow \quad k_H/k_T = 1.067 + 0.10(CH_3CN/MB) = 0.980 + 0.101(C_6H_6/TPP) \tag{19}$$

3. *Reaction of hydrogen peroxide with alkylating agents*

Nucleophilic displacement on aliphatic carbon by H_2O_2 gives alkyl hydroperoxides[7]. Thus, the acid-catalysed conversion of alcohols to the corresponding hydroperoxides often proceeds through an S_N1-type process[7,19], while the reaction between an optically active alkyl methanesulphonate and hydrogen peroxide in the presence of base yields the hydroperoxide of inverted configuration[20] (equation 20). When R is a primary alkyl group, the corresponding alkyl sulphate or methanesulphonate is used under S_N2 conditions[7], while the alcohols themselves are used under acidic (S_N1) conditions, e.g. with sulphuric acid in the preparation of *t*-alkyl hydroperoxides[7,19]. The reaction of acetals or orthoesters with hydrogen peroxide[21] also gives α-alkoxy alkyl hydroperoxides (equation 21).

$$RY + H_2O_2 \longrightarrow ROOH + HY \tag{20}$$

$$Y = OH,\ OSO_2OR,\ \text{halides, etc.}$$

$$(RO)_2CH_2 + H_2O_2 \longrightarrow (HOO)(RO)CH_2 + (HOO)_2CH_2 \tag{21}$$

4. *Addition of hydrogen peroxide to double bonds*

Acid-catalysed addition of hydrogen peroxide to alkenes also gives hydroperoxides (equation 22). In the presence of *t*-butyl hypochlorite or bromine, the initial step is an electrophilic addition of a halonium ion to the C=C bond, followed by nucleophilic addition of a hydroperoxide anion to form a β-halogenated alkyl hydroperoxide[16,22] (equations 23 and 24).

$$\mathrm{>C{=}C<} + H_2O_2 \xrightarrow{H^+} -\overset{OOH}{\underset{|}{\overset{|}{C}}}-\overset{|}{\underset{H}{\underset{|}{C}}}- \qquad (22)$$

$$(D_3C)_2C{=}C(CH_3)_2 + Br_2 + H_2O_2 \longrightarrow D_3C-\overset{OOH}{\underset{CD_3}{C}}-\overset{CH_3}{\underset{Br}{C}}-CH_3 + D_3C-\overset{CD_3}{\underset{Br}{C}}-\overset{OOH}{\underset{CH_3}{C}}-CH_3 \qquad (23)$$

1 : 1.6

$$(D_3C)(H_3C)C{=}C(CH_3)(CD_3) + Br_2 + H_2O_2 \longrightarrow D_3C-\overset{OOH}{\underset{CH_3}{C}}-\overset{CH_3}{\underset{Br}{C}}-CD_3 \qquad (24)$$

Addition of hydrogen peroxide to carbonyl compounds or imines also forms hydroperoxides[7].

5. *Perester alcoholysis*

Koenig and coworkers[23,24] have obtained *t*-butyl hydroperoxide specifically ^{18}O-labelled at the peroxidic OH group by alcoholysis of the perester formed by thermolysis of carbonyl-^{18}O-labelled *N*-acyl-*N*-nitroso-*O*-*t*-butylhydroxylamine (see equation 43).

C. Dialkyl Peroxides

Dialkyl peroxides can be prepared by essentially the same methods as are used for the synthesis of hydroperoxides.

1. *Reactions of hydrogen peroxide or hydroperoxides with alkylating agents*

$$H_2O_2 + 2RY \longrightarrow ROOR \qquad (25)$$

$$RY + ROOH \longrightarrow ROOR \qquad (26)$$

Allen and Bevington[25] have prepared methyl-^{14}C-labelled di-*t*-butyl peroxide from methyl-^{14}C-labelled *t*-butanol (equation 27), while carbonyl-^{14}C-labelled acetone is converted further to *t*-carbon-^{14}C-labelled di-*t*-butyl peroxide.

$$(^*CH_3)_2C{=}O \xrightarrow{CH_3MgI} (^*CH_3)_3COH \xrightarrow{H_2O_2/H_2SO_4} [(^*CH_3)_3C-O-]_2 \qquad (27)$$

Intramolecular nucleophilic displacement of hydroperoxides bearing a good leaving group Y leads to cyclic peroxides. Thus 1,2-dioxetanes have been prepared by Kopecky and Mumford[26] (equation 28).

$$R^1-C(OOH)(R^2)-C(R^3)(R^4)-Y \xrightarrow{B:} R^1-C(R^3)(-O-O-)-C(R^4)-R^3 + BH^+Y^- \quad (28)$$

2. *Addition of hydrogen peroxide or hydroperoxides to unsaturated compounds*

Both hydrogen peroxide and hydroperoxides are excellent nucleophiles and readily add to carbonyl carbon[7] (equations 29 and 30).

$$R^1R^2C{=}O + H_2O_2 \longrightarrow R^1C(OH)(R^2)OOC(OH)(R^2)R^1 + \text{cyclic diperoxide } [R^1R^2C(O-O)_2CR^1R^2] + \text{cyclic triperoxide } [(R^1R^2COO)_3] \quad (29)$$

$$R^1R^2C{=}O + R^3OOH \longrightarrow R^1C(OH)(R^2)OOR^3 + R^1C(OOR^3)(R^2)OOR^3 \quad (30)$$

3. *Cycloadditions of singlet oxygen to olefins*

a. 1,2-Cycloaddition[27]. The reactions of singlet oxygen with highly electron-donating olefins such as enol ethers[28,29] and enamines, or with sterically hindered olefins, like adamantylideneadamantane[30] and 7,7′-norbornylidene[31], give the corresponding 1,2-dioxetanes. Thus, $^{18}O_2$ or $^{17}O_2$ can be directly introduced into 1,2-dioxetanes (equation 31).

$$R^1R^2C{=}CR^3R^4 + {}^1O_2 \longrightarrow R^1-C(R^2)(-O-O-)-C(R^4)-R^3 \quad (31)$$

b. 1,4-Cycloaddition. 1,3-Dienes afford six-membered cyclic endoperoxides with singlet oxygen[10a,b,32] (equation 32). Sensitized photooxidation of 9,10-diphenylanthracene with $^{18}O_2$ produces ^{18}O-labelled endoperoxides which liberate singlet $^{18}O_2$ upon heating[15] (equation 33).

$$\text{1,3-diene} + O{=}O \longrightarrow \text{3,6-dihydro-1,2-dioxin} \quad (32)$$

$$\text{9,10-diphenylanthracene} + {}^{1*}O_2 \longrightarrow \text{9,10-diphenylanthracene 9,10-endoperoxide (*O–*O)} \tag{33}$$

Unsaturated heterocycles react with singlet oxygen to yield heterocyclic endoperoxides. The sequence in equation (34) involving endoperoxide was confirmed by using ^{18}O-enriched singlet oxygen generated according to equation (9)[15].

$$\text{oxazole } (R^1, R^2, R^3) + {}^{1*}O_2 \longrightarrow [\text{endoperoxide } R^2\text{–C(–*O–O*–)C–}R^1 \longrightarrow R^2C(=N–C(=O^*)R^1)–{}^*O–C(=O)R^3] \longrightarrow R^3C(=O)–N(C(=O^*)R^2)–C(=^*O)R^1 \tag{34}$$

D. Peracids

The usual methods of preparing peracids have been reviewed by Swern[33].

1. Autoxidation

Akiba and Simamura[3] have prepared peroxy-^{18}O-labelled perbenzoic acid by autoxidation of benzaldehyde with $^{18}O_2$ (equation 35). Carbonyl-^{18}-O-labelled perbenzoic acid has been prepared by Kobayashi, Minato and Hisada[34] by autoxidation of ^{18}O-labelled benzaldehyde with unlabelled oxygen and allowed to react with acetic anhydride *in situ* yielding benzoyl-^{18}O-labelled acetyl benzoyl peroxide.

$$PhCHO + {}^*O_2 \longrightarrow PhC(=O)^*O^*OH \tag{35}$$

2. Reaction of hydrogen peroxide with acylating agents

The reaction between acyl halides or acid anhydrides and hydrogen peroxide in basic solution provides a convenient method to prepare peracids. Carbonyl-^{18}O-labelled perbenzoic acid has been prepared from benzoyl chloride obtained from the acid formed by hydrolysis of benzotrichloride in ^{18}O-enriched water[35–37] (equation 36). Oae, Kitao and Kitaoka[38] have reported a general convenient method for the synthesis of ^{18}O-labelled acyl chlorides: alkaline hydrolysis of nitriles with a stoichiometric amount of ^{18}O-enriched water and sodium ethoxide in alcohol gives sodium salts of ^{18}O-labelled carboxylic acids which are converted to acyl chlorides by phosphorus pentachloride (equation 37). Labelled acyl chlorides can also be obtained by hydrolysis of acyl halide in

$$PhCCl_3 \xrightarrow[2.\ SOCl_2]{1.\ Na^*OH/H_2{}^*O} PhC(=^*O)Cl \xrightarrow[2.\ H^+]{1.\ H_2O_2/OH^-} PhC(=^*O)OOH \tag{36}$$

$H_2{}^{18}O$, followed by treatment with a chlorinating agent, which regenerates the acyl chloride labelled with ^{18}O. However, in this process only half of the original ^{18}O label of the water is used[39].

$$\mathrm{RCN} \xrightarrow{H_2{}^{*}O/Na^{*}OH} \mathrm{RC^{*}O_2Na} \xrightarrow{PCl_5} \mathrm{R\overset{\overset{*}{O}}{\overset{\|}{C}}Cl} \tag{37}$$

3. Reaction of diacyl peroxides with sodium methoxide

Oae and coworkers[40,41] have obtained carbonyl-^{18}O-labelled perbenzoic acid by the reaction of carbonyl-^{18}O-labelled benzoyl peroxide with sodium methoxide. In this reaction half of the ^{18}O content in benzoyl peroxide is lost (equation 38).

$$(\mathrm{Ph\overset{\overset{*}{O}}{\overset{\|}{C}}O})_2 + \mathrm{NaOCH_3} \longrightarrow \mathrm{Ph\overset{\overset{*}{O}}{\overset{\|}{C}}ONa} + \mathrm{Ph\overset{\overset{*}{O}}{\overset{\|}{C}}OCH_3} \tag{38}$$

E. Peresters

Peresters are usually prepared from hydroperoxides and acylating agents, such as acyl chlorides, acid anhydrides[42–44] or imidazolides[45] in the presence of base. Carbonyl-^{18}O-labelled peresters have been prepared from the corresponding ^{18}O-labelled acyl chlorides (equation 39).

$$\mathrm{R\overset{\overset{*}{O}}{\overset{\|}{C}}Cl} + \mathrm{R'OOH} + \mathrm{Py} \longrightarrow \mathrm{R\overset{\overset{*}{O}}{\overset{\|}{C}}OOR} + \mathrm{Py{\cdot}HCl} \tag{39}$$

R = Me, R′ = *t*-Bu[46]; R = Ph, R′ = *trans*-9-decalyl[36]; R = Ph, R′ = *t*-Bu[24]

Koenig and coworkers have synthesized deuterated *t*-butyl peresters from specifically deuterated acyl chlorides, and *t*-butyl hydroperoxide in the presence of base. In this synthesis, the specifically deuterated carboxylic acid is the key intermediate. *t*-Butyl phenylperacetate-2-d_2 and *t*-butyl 2-methyl-2-phenylperacetate-2-d_1 have been prepared by sodium-deuterioxide-catalysed H–D exchange of phenylacetic acid and 2-methyl-2-phenylacetic acid in deuterium oxide[47]. With *p*-nitrophenylacetic acid, potassium carbonate is a strong enough base to promote this H–D exchange[47]. 2-Dimethyl-d_6-2-phenylacetyl chloride has been synthesized by the reaction of phenyldilithioacetonitrile with commercially available methyl iodide-d_3, followed by hydrolysis of the nitrile group[47] (equation 40). Pivalic acid-d_9 has been prepared by bubbling CO_2 into the Grignard reagent prepared from commercially available *t*-butyl chloride-d_9 [47]. Wolf and coworkers have prepared *t*-butyl alicyclic percarboxylates deuterated specifically at the *t*-carbon from the corresponding alicyclic carboxylic acids-d_1 [48] (equations 41 and 42).

$$\mathrm{PhCH_2CN} \longrightarrow \mathrm{PhCLi_2CN} \xrightarrow{2CD_3I} \mathrm{Ph\underset{CD_3}{\overset{CD_3}{C}}CN} \longrightarrow \mathrm{Ph\underset{CD_3}{\overset{CD_3}{C}}COOH} \tag{40}$$

$$(CH_2)_{n-1}C(H)C(=O)CH_3 \xrightarrow{NaOD/D_2O} (CH_2)_{n-1}C(D)C(=O)CD_3 \xrightarrow[2.\ H^+]{1.\ NaOBr/H_2O} (CH_2)_{n-1}C(D)COOH \tag{41}$$

$$(CH_2)_{n-1}C(COOH)_2 \xrightarrow{1.\ D_2O\quad 3.\ Dry} (CH_2)_{n-1}C(COOD)_2 \xrightarrow{\Delta} (CH_2)_{n-1}C(D)COOD \xrightarrow{SOCl_2} (CH_2)_{n-1}C(D)COCl \xrightarrow{t\text{-BuOOH/NaOH}} (CH_2)_{n-1}C(D)C(=O)\text{-O-O Bu-}t \tag{42}$$

Koenig and coworkers[23,24] have obtained *t*-butyl peresters labelled with ^{18}O at the carbonyl oxygen and the peroxidic oxygen adjacent to carbonyl group (equation 43). The peresters thus formed can be converted to terminal ^{18}O-labelled *t*-butyl hydroperoxide, which in turn can be converted to *t*-butyl *p*-nitroperbenzoates labelled with ^{18}O at the oxygen adjacent to the carbonyl group (equation 43).

$$R\overset{*O}{\overset{\|}{C}}NHOBu\text{-}t \xrightarrow{NOCl} R\overset{*O}{\overset{\|}{C}}N(N{=}O)OBu\text{-}t \longrightarrow R\overset{*O}{\overset{\|}{C}}ON{=}NOBu\text{-}t \xrightarrow[\text{in nujol}]{32^{\circ}C} R\overset{*O}{\overset{\|}{C}}{}^{*}OOBu\text{-}t \quad [R{=}Me(10\%)^{23},\ Ph(38\%)^{24},\ \text{cyclopropyl}\ (46\%)^{49}]$$

$$\xrightarrow{MeO^-} R\overset{*O}{\overset{\|}{C}}OMe + t\text{-BuO}^{*}OH \xrightarrow{p\text{-}O_2NC_6H_4COCl/Py} O_2N\text{-}C_6H_4\text{-}\overset{O}{\overset{\|}{C}}{}^{*}OOBu\text{-}t \tag{43}$$

Treatment of sodium salts of alkyl hydroperoxides with ^{18}O-labelled acyl chloride affords carbonyl-^{18}O-labelled perester[37] (equation 44).

$$R\overset{*O}{\overset{\|}{C}}Cl + R'OONa \longrightarrow R\overset{*O}{\overset{\|}{C}}OOR' + NaCl \tag{44}$$

R = Me, R′ = *t*-Bu; R = Ph, R′ = *t*-Bu; R = R′ = *t*-Bu

An alternative method of preparing isotopically labelled peresters involves the reaction of hydroperoxides with imidazolides which are obtained by condensation of suitable acids with either *N,N'*-sulphinyl or *N,N'*-carbonyl diimide[45]. Carbonyl-^{17}O-labelled *t*-butyl-2-methylthioperbenzoate has been prepared by this procedure[50] (equation 45).

$$2\text{-}(CH_3S)C_6H_4C({=}^{17}O)OH + (C_3H_3N_2)C({=}O)(C_3H_3N_2) \longrightarrow 2\text{-}(CH_3S)C_6H_4C({=}^{17}O)(C_3H_3N_2) \xrightarrow{t\text{-BuOOH}} 2\text{-}(CH_3S)C_6H_4C({=}^{17}O)OOBu\text{-}t \qquad (45)$$

F. Diacyl Peroxides

1. *Reaction between sodium peroxide and acylating agents*

The method most widely used for the preparation of diacyl peroxides involves the reaction of acylating agents such as acid anhydrides and acid chlorides with either sodium peroxide or a combination of hydrogen peroxide and base[51,52]. By this method, several carbonyl-^{18}O-labelled diacyl peroxides have been synthesized (equation 46).

$$2\,R{-}C({=}O^*){-}Cl + Na_2O_2\ (H_2O_2/B{:}) \longrightarrow (R{-}C({=}O^*){-}O{-})_2 + 2NaCl\ \ (2BH^{+}\cdot Cl^{-}) \qquad (46)$$

R = Me[53], Ph[54], PhCH$_2$CH(Me)–[39], Ph(CH$_2$)$_4$–[39], cycloalkyl[55,56], cyclopropyl-CH$_2$–[41], Ph$_3$CCH$_2$–[57]

Benzoyl peroxide-d$_2$ has been prepared by Yoshida[58] while Tokumaru and coworkers[59] have prepared benzoyl peroxide-2,2',4,4',6,6'-d$_6$ by the route shown in equation (47). Perdeuteriobenzoyl peroxide has been prepared by treatment of perdeuteriobenzoyl chloride with sodium peroxide[59].

$$PhNH_3^{+}Cl^{-} \xrightarrow{D_2O/\text{reflux}} 2,4,6\text{-}D_3C_6H_2NH_3^{+}Cl^{-} \xrightarrow{\text{Sandmeyer}} 2,4,6\text{-}D_3C_6H_2CN \qquad (47)$$

$$\xrightarrow{\text{Hydrolysis}} 2,4,6\text{-}D_3C_6H_2COOH \xrightarrow{SOCl_2} 2,4,6\text{-}D_3C_6H_2COCl \xrightarrow{Na_2O_2} (2,4,6\text{-}D_3C_6H_2C({=}O){-}O{-})_2$$

Bevington and Brooks[60] have synthesized phenyl-^{14}C-labelled benzoyl peroxide from commercially available ^{14}C-labelled aniline (equation 48), while carbonyl-^{14}C-labelled benzoyl peroxide has been prepared from commercially available carbonyl-^{14}C-labelled benzoic acid[60] (equation 48). Methyl-^{14}C-labelled *m*- and *p*-anisyl peroxides have also been prepared from the corresponding methyl hydroxybenzoates by Bevington, Toole and Trossavelli[61] (equation 49).

$$^{*}PhNH_2 \longrightarrow {}^{*}PhCN \longrightarrow {}^{*}PhCOOH \longrightarrow {}^{*}PhCOCl \longrightarrow (^{*}PhC(=O)O)_2 \quad (48)$$

$$HO-C_6H_4-CO_2Me \xrightarrow{^{*}CH_3I} {}^{*}CH_3O-C_6H_4-CO_2Me \longrightarrow {}^{*}CH_3O-C_6H_4-CO_2H \longrightarrow (^{*}CH_3O-C_6H_4-CO_2)_2 \quad (49)$$

Greene[62] has synthesized carbonyl-^{18}O-labelled phthaloyl peroxide by treating ^{18}O-labelled phthaloyl chloride with hydrogen peroxide in the presence of sodium carbonate in ether (equation 50). while carbonyl-^{18}O-labelled *trans*-hexahydrophthaloyl peroxide has been prepared by Fujimori, Oshibe and Oae[63a] by the method previously used for the preparation of phthaloyl peroxide by Russell[63b].

$$C_6H_4(C(={}^{*}O)-Cl)_2 + Na_2O_2 \longrightarrow C_6H_4(C(={}^{*}O)-O-)_2 + NaCl \quad (50)$$

Sulphonyl-^{18}O-labelled *m*-nitrobenzenesulphonyl peroxide has been synthesized from the corresponding chloride by Kobayashi and coworkers[64] (equation 51).

$$2\ O_2N-C_6H_4-S^{*}O_2Cl + Na_2O_2 \longrightarrow (O_2N-C_6H_4-S(={}^{*}O)_2-O)_2 + 2NaCl \quad (51)$$

Hyperol, a crystalline complex of hydrogen peroxide with urea, has often been used for the synthesis of diacyl peroxides instead of hydrogen peroxide, since it is easy to handle[65]. Koenig and Cruthoff have prepared both carbonyl-^{18} O-labelled acetyl peroxide and acetyl peroxide-d_6 from the corresponding labelled chlorides and hyperol[66] (equation 52).

$$2CD_3COCl + H_2O_2\cdot Urea + Py \longrightarrow (CD_3C(=O)O)_2 + Urea + Py\cdot HCl \quad (52)$$

2. *Autoxidation of aldehydes in the presence of acylating agents*

Peracids formed during autoxidation of aldehydes can react with acid anhydrides *in situ* to give unsymmetrical diacyl peroxides (equation 53). This method is suitable for preparing labelled stable symmetrical and unsymmetrical diacyl peroxides, if specifically labelled starting materials are available. Kobayashi, Minato and Hisada[67] have modified

$$(CH_3\overset{*O}{\overset{\|}{C}})_2O + PhCHO + O_2 \xrightarrow[MgCO_3]{40^\circ C} CH_3\overset{*O}{\overset{\|}{C}}OO\overset{O}{\overset{\|}{C}}Ph \qquad (53)$$

an earlier procedure[68] for the synthesis of acetyl-^{18}O-labelled and benzoyl-^{18}O-labelled acetyl benzoyl peroxide (equation 54). McBride and coworkers[69–71] have conducted a similar autoxidation of benzaldehyde with $^{18}O_2$, ^{17}O or ^{16}O in the presence of acetic anhydride-d_6 and sodium acetate-d_6 and prepared labelled acetyl benzoyl peroxides as shown in equations 55–58.

$$Ph\overset{H}{\overset{|}{C}}{=}NPh \xrightarrow{H_2{}^*O/H^+} PhCH{}^*O \xrightarrow[MgCO_3,\ 40^\circ C]{Ac_2O/O_2} CH_3\overset{O}{\overset{\|}{C}}OO\overset{*O}{\overset{\|}{C}}Ph \qquad (54)$$

$$(CD_3CO)_2O + \left[C_6D_5CD_3 \xrightarrow{Ce^{4+}/6N\ HClO_4} C_6D_5CDO\right] \xrightarrow{O_2} CD_3\overset{O}{\overset{\|}{C}}OO\overset{O}{\overset{\|}{C}}C_6D_5 \qquad (55)$$

$$(CH_3\overset{O}{\overset{\|}{C}})_2O + C_6H_5CHO + {}^*O_2 \longrightarrow CH_3\overset{O}{\overset{\|}{C}}{}^*O{}^*O\overset{O}{\overset{\|}{C}}C_6H_5 \qquad (56)$$

$$C_6H_5CHBr_2 \xrightarrow{H_2{}^{17}O/{}^{17}OH^-} C_6H_5CH^{17}O \xrightarrow{(CD_3\overset{O}{\overset{\|}{C}})_2O/O_2} CD_3\overset{O}{\overset{\|}{C}}OO\overset{^{17}O}{\overset{\|}{C}}C_6H_5 \qquad (57)$$

$$(CD_3\overset{O}{\overset{\|}{C}})_2O + C_6H_5CHO + {}^{17}O_2 \xrightarrow{CD_3CO_2Na} CD_3\overset{O}{\overset{\|}{C}}-{}^{17}O{}^{17}O-\overset{O}{\overset{\|}{C}}C_6H_5 \qquad (58)$$

Kaptein and coworkers[72] have synthesized hexadeuterioacetyl peroxide by treating commercially available deuterated acetic anhydride with sodium peroxide according to the method of Price and Morita[51], while dideuterioacetyl peroxide has been prepared from acetic anhydride-d_2 obtained from ketene and deuterium oxide (equation 59). The yield of methyl-^{13}C-labelled acetyl peroxide in the reaction of ^{13}C-labelled acetyl chloride and sodium peroxide may reach 80% under optimal conditions[72].

$$2\,CH_2{=}C{=}O + D_2O \longrightarrow (CH_2D\overset{O}{\overset{\|}{C}})_2O \xrightarrow{Na_2O_2} (CH_2D\overset{O}{\overset{\|}{C}}O)_2 \qquad (59)$$

3. Reaction of peracids with acylating agents

The reaction of peracids with acylating agents in the presence of a base usually affords unsymmetrical diacyl peroxides (equations 60 and 61). Since disproportionation of the peroxide is catalysed by base, prolonged reaction times or high temperatures must be avoided.

Sulphonyl-^{18}O-labelled benzoyl *p*-toluenesulphonyl peroxide is prepared as shown in equation (62), while treatment of benzoyl-^{18}O-labelled perbenzoic acid with tosyl chloride gives benzoyl-^{18}O-labelled benzoyl *p*-toluenesulphonyl peroxide[76].

$$R{-}\overset{\overset{*}{O}}{\overset{\|}{C}}{-}Cl + R'CO_3H + Py \longrightarrow R\overset{\overset{*}{O}}{\overset{\|}{C}}OO\overset{O}{\overset{\|}{C}}R' + Py{\cdot}HCl \tag{60}$$

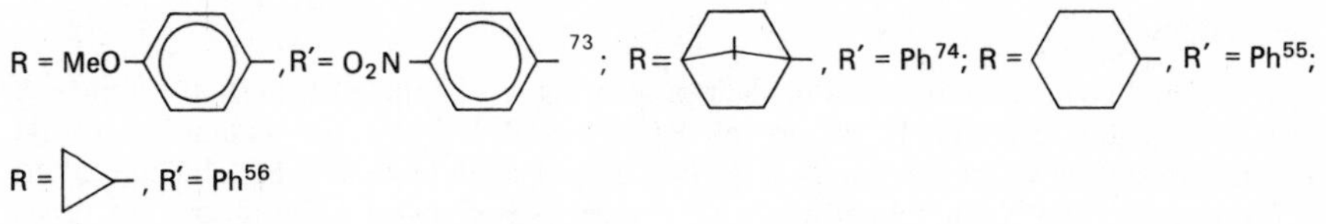

R = MeO–C₆H₄–, R′ = O₂N–C₆H₄–[73]; R = (bicyclic), R′ = Ph[74]; R = cyclohexyl, R′ = Ph[55];

R = cyclopropyl, R′ = Ph[56]

$$R\overset{O}{\overset{\|}{C}}Cl + R'\overset{\overset{*}{O}}{\overset{\|}{C}}OOH + Py \longrightarrow R\overset{O}{\overset{\|}{C}}OO\overset{\overset{*}{O}}{\overset{\|}{C}}R' + Py{\cdot}HCl \tag{61}$$

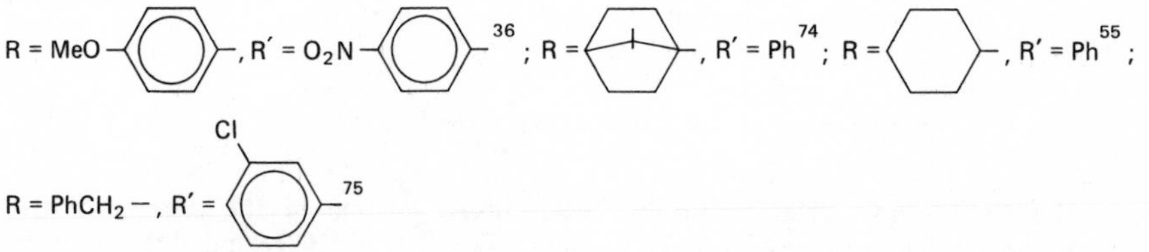

R = MeO–C₆H₄–, R′ = O₂N–C₆H₄–[36]; R = (bicyclic), R′ = Ph[74]; R = cyclohexyl, R′ = Ph[55];

R = PhCH₂–, R′ = 3-Cl–C₆H₄–[75]

$$p\text{-TolSO}_2\text{H} \xrightarrow{H_2{}^{*}O} p\text{-TolS}^{*}\text{O}_2\text{H} \xrightarrow{Cl_2} p\text{-TolS}^{*}\text{O}_2\text{Cl} \xrightarrow{PhCO_3H/B:} p\text{-Tol}\overset{\overset{*}{O}}{\overset{\uparrow}{S}}\underset{\underset{*}{O}}{\underset{\|}{}}OO\overset{O}{\overset{\|}{C}}Ph \tag{62}$$

4. Diimide synthesis

Both symmetrical and unsymmetrical diacyl peroxides can be prepared directly from the corresponding carboxylic acids and hydrogen peroxide in the presence of dicyclohexylcarbodiimide (DCD)[77], for example, as shown in equation (63) for the synthesis of hydrocinnamoyl-β,β-d_2 peroxide[78].

$$PhCO_2Et \xrightarrow{LiAlD_4} PhCD_2OH \xrightarrow[2.\ TsCl]{1.\ NaH} PhCD_2OTs \xrightarrow{NaCH(CO_2Et)_2} PhCD_2CH(CO_2Et)_2 \xrightarrow[2.\ H^+]{1.\ OH^-} PhCD_2CH_2CO_2H \xrightarrow{H_2O_2/DCC} (PhCD_2CH_2\overset{O}{\overset{\|}{C}}{-}O)_2 \tag{63}$$

III. USES OF ISOTOPICALLY LABELLED PEROXIDES

A. Unimolecular Homolytic Decomposition of Peroxides

1. Peresters

Bartlett and coworkers revealed in their pioneering works that the thermal stability of perester depends on the R group of RC(O)OOR′[43,44,79]; for example, *t*-butyl triphenylperacetate which decomposes by two-bond fission (equation 65), due mainly to the formation of the stable trimethylphenyl radical, was shown to decompose 10^6 times more rapidly than *t*-butyl peracetate that undergoes decomposition by one-bond fission (equation 64)[44]. The activation energy of reaction (65) is considerably lowered by the concertedness of the transition complex for the two-bond fission, while the activation entropy becomes small, since more than two bonds are frozen at the transition state[43].

$$\mathrm{RC(=O)OOR'} \longrightarrow [\mathrm{RC(=O)O\cdots OR'}]^{\ddagger} \longrightarrow \mathrm{RC(=O)O\cdot \ \cdot OR'} \quad (64)$$

$$\mathrm{RC(=O)OOR'} \longrightarrow [\mathrm{R\cdots C(=O)\overset{\cdots}{=}O\cdots OR'}]^{\ddagger} \longrightarrow \mathrm{R\cdot \ CO_2 \ \cdot OR'} \quad (65)$$

Thus the $\Delta H^{\neq} - \Delta S^{\neq}$ relationship has been taken as one of the criteria of the concertedness in the decomposition of a particular perester[43,80]. Freezing of the rotation of the R—CO bond in the transition state of the decomposition of *t*-butyl peresters of substituted phenylacetic acids was further demonstrated by the good correlation of the rates with σ^+ constants rather than σ constants[81].

Many criteria have been proposed for the diagnosis of the mechanism of the thermal decomposition of peresters; but only investigations using isotopically labelled peresters will be described here.

Koenig has succeeded in generating various acyloxy–*t*-butoxy radical pairs (**14**) which are the intermediates in the decomposition of one-bond fission in peresters (**13**) as well as in the homolytic denitrogenation of *O*-acyl *O*′-*t*-butyl hyponitrites (**12**) formed by acyl migration of *N*-acyl-*N*-nitroso-*O*-*t*-butylhydroxylamines as shown in Scheme 1[82]. Based on detailed investigations on the chemical behaviour of **14**, generated from either **12** or **13**, by using ^{18}O-tracers, secondary kinetic hydrogen deuterium isotope effects and solvent viscosity effects, equation (66) was introduced, assuming that the rate constant k_3 of the diffusion process of the caged geminate radical pair was the only process sensitive to the viscosity of the medium[83] (f_r = fraction of cage return, k = constant, η = viscosity of the medium, k_2 = rate of decarboxylation of $RCO_2\cdot$ and k_{-1} = rate of recombination of the radical pair **14**).

$$F = \frac{1}{f_r} - 1 = \sqrt{\frac{k}{\eta}} + \frac{k_2}{k_{-1}} \quad (66)$$

Theory predicts that the cage recombination ratio, F, can be correlated with the square root of fluidity, $\eta^{-1/2}$. Indeed, in the thermolysis of hyponitrites **12** in several hydrocarbon solvents of different viscosities, one obtains a line whose intercept gives the value of k_2/k_{-1} at the infinitely viscous solvent, namely the ratio of the rate of decarboxylation of the acyloxy radical over that of the recombination of the radical pair **14** to give the perester **13**[82b,83].

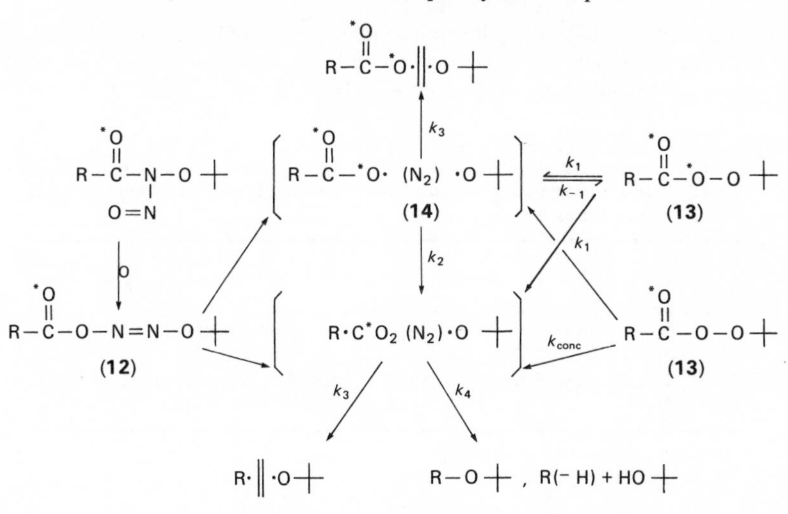

SCHEME 1. Decomposition of *O*-acyl *O'*-*t*-butyl hyponitrites (**12**) and *t*-butyl peresters (**13**), in solution.

a. Classification of O-*acyl* O'-t-*butyl hyponitrites,* t-*butyl peresters and* t-*butoxyacyloxy radical pairs.* The cage recombination, however, depends not only on the viscosity of the medium but also on the nature of the acyloxy radical. Therefore in order to understand the mechanism of the decomposition and recombination, it is desirable to make an educated guess on the chemical behaviour and especially the lifetimes of acyloxy radicals of different structures. On the basis of the stabilities of the geminate acyloxy radicals, we have divided diacyl peroxides (**15**), RCO_2O_2CR', into three classes, Case I (R,R' = alkyl), Case II (R = alkyl, R' = aryl) and Case III (R,R' = aryl)[40]. A similar classification can be made for the hyponitrites and peresters. In this case, the classification is based on the structural effect on the rates of three reactions which start from the caged acyloxy–*t*-butoxy radical pair (**14**) generated primarily from **12** and **13**, i.e. recombination, k_{-1}, decarboxylation of the acyloxy radical, k_2, and diffusion out of solvent cage, k_3 (Scheme 1), whose relative magnitudes determine the amount of cage return (f_r) to form perester,. namely $[f_r = k_{-1}/(k_2 + k_3 + k_{-1})]$.

Thus, both **12** and **13** may be classified as Case I (R = alkyl) and Case II (R = aryl) assuming k_{-1} for $\mathbf{14}_{R=aryl}$ is smaller than k_{-1} for $\mathbf{14}_{R=alkyl}$ [40]; namely the amount of cage return of **14** decreases in the order of Case I > Case II. Meanwhile, since the rate of decarboxylation of acyloxy radicals, $RCO_2\cdot$, is known to decrease in the order R = *s*- or *t*-alkyl, benzyl > primary alkyl, strained cycloalkyl > aryl, we may divide Case I initiators into Case I_A (R = primary alkyl, strained cycloalkyl) and Case I_B (R = *s*- and *t*-alkyl, benzyl) depending upon the lifetime of the acyloxy radical, i.e. if only the lifetime of acyloxy radical were responsible for the cage return, the extent of cage return to give perester, f_r, is anticipated to decrease in the order of Case II > Case I_A > Case I_B. The latter order is opposite to that which was deduced on the assumption that k_{-1} for a Case II radical pair (**14**) is smaller than k_{-1} for a Case I radical pair (**14**). In the usual solvents of fairly low viscosities, k_2 for aliphatic acyloxy radicals is close to k_3. In such a solvent, the effect of structural change appears much stronger on k_{-1} than on k_2 and hence, f_r for Case I would be higher than that for Case II. Increase of viscosity of the medium decreases k_3 and in an infinitely viscous medium ($k_3 = 0$) f_r depends entirely on k_2, and the extent of cage return to form perester, f_r, falls in the following order: Case II > Case I_A > Case I_B. Thus, the extent of cage recombination in connection with the structural change and the solvent viscosity may be summarized in the *a priori* precept shown in Table 1 (see also Section III.2).

TABLE 1. Structural effects on the fates of caged radical pairs (**14**), $RCO_2 \cdot\cdot OBu$-*t*, generated by decomposition of *O*-acyl *O'*-*t*-butyl hyponitrites (**12**) and *t*-butyl peresters (**13**)

Case	R	f_r^a	f_r^b	Solvent viscosity effect on f_r
I_A	Primary alkyl, strained cycloalkyl	Large	Large	Small
I_B	*S*- and *t*-alkyl, benzyl	Small	Small	Very small
II	Aryl	Medium	Very large	Large

[a]In usual solvents of low viscosity.
[b]In infinitely viscous solvent.

Among the reactions of the *t*-butoxy radical, its decomposition to give acetone and methyl radical (equation 67) and its hydrogen abstraction from hydrocarbons[85] (equation 68) are both known to be negligibly slower than its diffusion which is in the order of $10^9\ s^{-1}$ in the usual solvents[86]. Hence the yield of the perester is independent of the rate of the reactions shown in equations (67) and (68).

$$(CH_3)_3CO\cdot \xrightarrow{2 \times 10^3\ s^{-1}/40^\circ C} CH_3C(=O)CH_3 + \cdot CH_3 \qquad (67)$$

$$t\text{-BuO}\cdot + C_6H_{14} \xrightarrow{10^5\ M^{-1}s^{-1}/40^\circ C} t\text{-BuOH} + C_6H_{13}\cdot \qquad (68)$$

b. Formation of peresters by thermal decomposition of O-*acyl* O'-t-*butyl hyponitrites.* Table 2 summarizes the results of thermal decomposition of **12** obtained by Koenig and coworkers[82b], i.e. the values of f_r, the extent of ^{18}O scrambling in the perester **13** formed from carbonyl-^{18}O-labelled hyponitrite **12** and the intercept of the fluidity plots (equation 66). One finds that f_r decreases in the order of R = ▷ > Ph > Me > *i*-Pr ≃ *s*-Bu in both solvents with low and high viscosity nearly in accordance with our prediction shown in Table 1 except for $\mathbf{12}_{R=Me}$. This discrepancy (R = Me) may be due to the difference between a very intimate *t*-butoxy–acetoxy ($\mathbf{14}_{R=Me}$) radical pair generated by simple O—O bond homolysis of *t*-butyl peracetal ($\mathbf{13}_{R=Me}$) and the rather loose acetoxy–*t*-butoxy radical pair ($\mathbf{14}_{R=Me}$) separated by a dinitrogen molecule formed from the hyponitrite (**12**) (see p. 605–606). Since recombination would be slow in such a separated pair, k_2 becomes more important in determining f_r in such a case than in the very intimate radical pair (**14**) of perester origin. Therefore, the yield of perester from **12** seems to depend substantially on the lifetime of the acyloxy radical before it recombines with a *t*-butoxy radical; the acyloxy radicals with relatively slow rates of decarboxylation are considered to have a better chance or recombination with *t*-butoxy radicals to form the perester than those which undergo very fast decarboxylation in thermolysis of **12**.

The data in Table 2 show that the yield of perester (f_r) is much higher in nujol, a much more viscous medium than hexane. Since the lifetime of benzoyloxy radicals is longer than

TABLE 2. Thermal decomposition of hyponitrites **12** at 32°C[82b]

R	Case	Solvent	$f_r \times 100^a$	Scrambling of ^{18}O in **13**^{b} (%)	$k_2/k_{-1}{}^c$	Ref.
Ph	II	n-C_6H_{14}	9.0	98	0	24
		Nujol	38.0	—	—	
▷	I_A	n-C_6H_{14}	18.0	90	—	49
		Nujol	46.0			
Me	I_A	n-C_6H_{14}	2.7	93	4.2	23
		Nujol	10.2	—	—	
i-Pr	I_B	n-C_6H_{14}	0.1	—	—	84
		Nujol	—	—	—	
s-Bu	I_B	n-C_6H_{14}	0.1	—	—	84
		Nujol	—	—	—	

^{a}Yield of perester **13** (%).
^{b}The extent of scrambling of the original carbonyl-^{18}O label of the starting **12** in the cage product **13**.
^{c}Intercept of fluidity plot (equation 66).

that of aliphatic acyloxy radicals, the extent of cage recombination of Case II radical pairs ($\mathbf{14}_{R=Ph}$) should be, and indeed is, more sensitive to solvent viscosity than that of Case I radicals pairs ($\mathbf{14}_{R=alkyl}$). Recombination of a benzoyloxy radical with a t-butoxy radical would reach up to 100% even in an infinitely viscous medium due to very slow decarboxylation of the benzoyloxy radical (see Table 9), whereas recombination of short-lived acetoxy radicals with t-butoxy radicals would never be quantitative. In fact the intercept of the fluidity plots (Table 2) reveals that in infinitely viscous solvent, $\mathbf{14}_{R=Ph}$ recombines quantitatively while $\mathbf{14}_{R=Me}$ recombines to give t-butyl peracetate in 20% yield only upon calculation with equation (73) using the value $k_2/k_{-1} = 4.2$ in Table 2. The original ^{18}O label of the carbonyl group of **12** is nearly completely randomized in the perester $\mathbf{13}_{R=Ph}$; the other results, although possibly within experimental error, show that the extent of ^{18}O scrambling in the recombination of the Case II radical pair, $\mathbf{14}_{R=Ph}$, is greater than that of the Case I radical pairs, $\mathbf{14}_{R=alkyl}$, which are short-lived[23,24,49].

c. Thermal decomposition of t-*butyl peresters.* Taylor and Martin have observed ^{18}O scrambling in the thermolysis of carbonyl ^{18}O-labelled acetyl peroxide[88]; Koenig and his coworkers have observed also ^{18}O scrambling in peresters recovered after the partial decomposition of the latter in nonpolar solvents, and suggested the radical mechanism shown in Scheme 1[24,46,89]. Obviously, the reaction is initiated by one-bond fission. Goldstein and Judson[37] have postulated [1,3]-sigmatropy also to be responsible for the oxygen scrambling (equation 69). The references pertaining to the ^{18}O scrambling data are listed in Table 3.

$$\mathrm{R\overset{\overset{*O}{\|}}{C}OO{+}} \rightleftharpoons \left[\mathrm{RC}\overset{\cdot\cdot\,{}^{*}O\,\cdot\cdot}{\underset{\cdot\cdot\,O\,\cdot\cdot}{}}\mathrm{O{+}}\right]^{\ddagger} \rightleftharpoons \mathrm{R\overset{\overset{O}{\|}}{C}{}^{*}OO{+}} \qquad (69)$$

According to the radical-pair mechanism shown in Scheme 1, the rate constants for ^{18}O scrambling, k_s, and decomposition, k_d, and the fraction of cage return, f_r, in the formation of ^{18}O-scrambled perester from **14** can be expressed by equations (70)–(73). F in Koenig's

TABLE 3. Thermal decomposition of carbonyl-^{18}O-labelled *t*-butyl perester $R-\overset{*O}{\overset{\|}{C}}OOBu\text{-}t$ (**13**)

R	Case	Temperature (°C)	Solvent	$k_d \times 10^4$ (s^{-1})	$k_s \times 10^4$ (s^{-1})	$(k_d + k_s) \times 10^4$ (s^{-1})	k_d/k_s	$f_r \times 100$[a]	n[b]	Ref.
Ph	II	130	$n\text{-}C_6H_{14}$	3.72	0.16	3.88	23.3	4	2	24
			$i\text{-}C_8H_{18}$	3.36	0.20	3.56	16.8	5.6	2	24
Me	I_A	130	$n\text{-}C_6H_{14}$	5.08	1.31	6.39	3.89	20.4	2	46
			60% $n\text{-}C_6H_{14}$–nujol	4.43	1.66	6.09	2.56	28.1	2	46
			Nujol	3.13	2.08	5.21	1.54	39.4	2	46
i-Pr	I_B	103	$i\text{-}C_8H_{18}$	2.16	0.006	2.17	360	0.3	1	89
		50.6	$i\text{-}C_8H_{18}$					0.5	1	89
		103	Nujol	2.23	0.01	2.24	223	0.4	1	89
Ph	II	105.5	Cumene	0.175 ± 0.002	0.042 ± 0.002	0.217	4.17	19	9	37
Me	I_A	105.5	Cumene	0.237 ± 0.004	0.17 ± 0.8	0.407	1.39	42	8	37
t-Bu	I_B	50.5	Cumene	0.096 ± 0.002	0.0025 ± 0.0011	0.0985	38.4	2.5	6	37

[a] The f_rs have been calculated assuming that the [18]O scrambling took place exclusively via **14** as shown in Scheme 1 and that the [18]O label was completely scrambled in the acyloxy radical before recombination to **13**.
[b] n is the number of kinetic points determined.

cage equation (equation 66) can be changed into equation (74), in which only k_3 is presumed to depend on the viscosity of the medium[82b].

$$k_s = \frac{k_1 \cdot k_{-1}}{k_{-1} + k_2 + k_3} = k_1 \cdot f_r \tag{70}$$

$$k_d = \frac{k_1(k_2 + k_3)}{k_{-1} + k_2 + k_3} \tag{71}$$

$$k_1 = k_s + k_d \tag{72}$$

$$f_r = \frac{k_{-1}}{k_{-1} + k_2 + k_3} = \frac{1}{(k_d/k_s) + 1} = 1 - \frac{k_d}{k_1} \tag{73}$$

$$F = \frac{1}{f_r} - 1 = \frac{k_3}{k_{-1}} + \frac{k_2}{k_{-1}} = \frac{1}{(k_1/k_d) - 1} = \frac{k_1}{k_s} - 1 = \frac{k_d}{k_s} \tag{74}$$

Koenig and his coworkers[46,47] and Wolf and coworkers[48] have measured secondary hydrogen–deuterium kinetic isotope effects on the rates of thermal decomposition of various peresters by two methods. One method is a direct measurement of rates of decomposition of normal and deuterated peresters by following the decrease of the carbonyl stretching vibration absorption or the fading of galvinoxyl added in the system. The other method is to measure the mass peak height ratio of *m/e* 46/44, i.e. $C^{16}O^{18}O/C^{16}O_2$, generated in the thermal decomposition of a mixture of $CD_3C(O)OOBu$-*t* and $CH_3C(^{18}O)OOBu$-*t* (see Table 4).

Pryor and coworkers have proposed a very simplified expression (equation 75) to correlate the rate of unimolecular decomposition of one-bond fission radical initiators with solvent viscosity assuming that only the rate of the diffusion process of caged germinate radical pairs, k_3, is affected by the viscosity of the solvent[90,91]. In this equation η

$$\frac{1}{k_d} = \frac{1}{k_1} + \frac{k_{-1}}{k_1 A_d}\left(\frac{\eta}{A_v}\right)^{\alpha} \tag{75}$$

denotes the viscosity of the solvent, α is a constant (0.5–0.7) in the linear relationship between the activation energy for the self-diffusive flow of solvent (E_v) and the energy barrier for the diffusion of the radical pair (E_d) ($E_d = \alpha E_v$) and A_d and A_v are the frequency factors of the self-diffusive flow of the solvent and of the diffusion of the radical pair, respectively. The rate constant, k_1, can be obtained from the intercept of linear reciprocal plots of k_ds measured in various *n*-alkanes of different viscosities against $(\eta/A_v)^{\alpha}$ values[90]. The extent of recombination of **14**, f_r, can be calculated by equation (73) using the values of k_1 and k_d, measured directly by kinetic experiments.

The f_r values for **14** can also be calculated from k_d and k_s values measured directly (equation 73), assuming that ^{18}O scrambling in **13** results exclusively from recombination of the radical pair **14** (see Table 3). Since k_1 depends on the nature of the solvent, as described later, the ^{18}O scrambling experiment of **13** may be more accurate in monitoring the bond-breaking–bond-forming phenomenon of the perester than the influence of viscosity on k_d[46,82b,95]. The f_r values listed in Table 3 fall in the order of R = Me > Ph > *i*-Pr,*t*-Bu. This order fits that observed for **14** generated by homolytic cleavage of **12**, namely R = ▷ > Ph > Me > *i*-Pr,*s*-Bu except for R = Me, suggesting that the radical mechanism shown in Scheme 1 is responsible for the ^{18}O scrambling in **13**, with the exception of $\mathbf{14}_{R=Me}$ which has already been discussed (see p. 602). Peresters **13** can

TABLE 4. Secondary deuterium kinetic isotope effects (α and β) on thermal decomposition of perester **13**

R	Case	Temperature (°C)	Solvent	(α) k_H/k_D[a]	(β) k_H/k_D[a]	Ref.
Me	I_A	130.1	i-C_8H_{18}	1.000 ± 0.007	—	46
			Nujol	1.007	—	46
$PhCH_2$—	I_B	84.98	i-C_8H_{18}	1.066	—	47
			Nujol	1.060	—	47
Ph—CH(Me)—	I_B	73.99	i-C_8H_{18}	1.046	1.020	47
Ph—C(Me)(Me)—	I_B	60.56	i-C_8H_{18}	—	1.018	47
t-Bu	I_B	60.56	i-C_8H_{18}	—	1.016	47
			90% dioxane–H_2O	—	1.014	47
p-$MeOC_6H_4CH_2$—	I_B	60.46	i-C_8H_{18}	1.034	—	46
			Nujol	1.048	—	46
p-$O_2NC_6H_4CH_2$—	I_B	85.1	PhCl	1.048	—	46
		84.98	Nujol	1.049	—	46
18$_3$[c]	I_A	102.6	C_8H_{18}[b]	1.006 ± 0.005	—	48
18$_4$[c]	I_A	102.6	C_8H_{18}	1.006 ± 0.009	—	48
18$_5$[c]	I_B	102.6	C_8H_{18}	1.049 ± 0.009	—	48
18$_6$[c]	I_B	102.6	C_8H_{18}	1.050 ± 0.009	—	48

[a] Per d_1.
[b] 2,2,4-Trimethylpentane.
[c] See Section III.A.3.

also be classified into Case I_A, Case I_B and Case II, as already described. The mode of thermolysis of the peresters of each case is explained in the same manner as for the ^{18}O scrambling in diacyl peroxides. Indeed, the structural effects postulated in Table 1 are in good accordance with the experimental results shown in Table 3.

(*i*) *Case II peresters.* The mechanism of decomposition of Case II peresters is quite simple. The first-order rate constant of decarboxylation of the benzoyloxy radical (see Table 9) is of the same order of magnitude as that for the β-scission of t-butoxy radicals to form methyl radical and acetone (equation 67) but several orders of magnitude smaller than the rate for diffusion in usual solvents. Therefore, there is no possibility of decomposition of **14**$_{R=Ph}$ in the solvent cage. This fits with the result that the linear fluidity $(1/f_r - 1)$ plots (equation 66) for both **12** and **13** (R = Ph) cross the ordinate at zero[24]. In other words, f_r is small in the usual solvents of low viscosity, but increases as the solvent viscosity increases and eventually reaches unity at infinite viscosity.

The value of f_r(0.19) for t-butyl perbenzoate calculated from k_d and k_s in cumene at 105.5°C[37] is in good agreement with the value of f_r (0.11 in hexane) calculated by equation (75) based on k_d values obtained in n-alkanes of different viscosities in sealed tubes at 130.1°C[91]. Thus the radical-pair mechanism is suggested as being responsible for the ^{18}O scrambling in t-butyl perbenzoate. The large activation volume ($\Delta V^{\neq} = 10.4\,cc\,mol^{-1}$) for the thermal decomposition of t-butyl perbenzoate observed by Neuman and Behar appears to indicate that the decomposition of the perester involves a peroxide bond-breaking–bond-forming equilibrium[92,93].

In order to see the effect of spin multiplicity and also the effect of the intervening nitrogen molecule on the efficiency of recombination of $\mathbf{14}_{R=Ph}$, Koenig and Hoobler[94] photolysed ^{18}O-labelled *t*-butyl perbenzoate both in the absence and presence of a sensitizer under the same conditions as used for $\mathbf{12}_{R=Ph}$. The f_r of the singlet radical pair $\mathbf{13}_{R=Ph}$ generated from the perester was found to be 0.177, i.e. about twice the f_r (0.09) of the radical pair separated by N_2 formed in the cleavage of $\mathbf{12}_{R=Ph}$. The triplet pair of $\mathbf{14}_{R=Ph}$ is found to recombine only 8% prior to the diffusion out of cage. Thus, *t*-butyl perbenzoate (a Case II perester) is considered to undergo a typical one-bond fission and the radical-pair mechanism seems to be responsible for the ^{18}O scrambling[94].

(*ii*) *Case I_A peresters.* The kinetic data in Tables 3 and 4 suggest that *t*-butyl peracetate ($\mathbf{13}_{R=Me}$) undergoes one-bond fission, with the ^{18}O scramgling being a radical-pair process (Scheme 1)[46]. In this thermolysis, there is no α-secondary hydrogen kinetic isotope effect (Table 4); and k_s is of the same order of magnitude as k_d, but k_s and k_d are affected markedly in opposite directions, by the change of solvent viscosity (Table 3).

According to equation (74), plots of k_d/k_s and of $(k_1/k_d - 1)^{-1}$ against $\eta^{-0.5}$ should fall on the same line. However, this is not found to be the case for $\mathbf{13}_{R=Me}$ due mainly to two reasons[46]. (1) The ^{18}O scrambling in **14** may not only be incomplete before coupling to form **13**, but may also vary with the viscosity of the medium. The kinetic data measured[46] would suggest that the fraction of recombination with complete ^{18}O scrambling for $\mathbf{14}_{R=Me}$ varies from 1.00 to 0.65 (35% of ^{18}O label remaining unchanged in the recombination product) with the change of solvent from *n*-hexane to paraffin oil. (2) The value of k_1 is dependent on the nature of *n*-alkane solvents.

Koenig and coworkers, however, prefer the view that k_1 depends on the internal pressure of the solvent[93b,c] but not on its viscosity[46,82b,95]. Based on the value $\Delta V^{\neq} = +12\,cc\,mol^{-1}$ for $\mathbf{13}_{R=Ph}$, which was determined by the external pressure dependency of k_d [93a], Owens and Koenig[95] have estimated a set of differential pressures of hydrocarbon solvents using k_ds of the same perester. The activation volume for k_1 of *t*-butyl peracetate ($\mathbf{13}_{R=Me}$) is calculated to be $5\,cc\,mol^{-1}$ at 130°C from the slope of $\ln(k_d + k_s)$ against the differential solvent pressure. $\Delta V^{\neq} = 8\,cc\,mol^{-1}$ for k_1 of $\mathbf{13}_{R=Ph}$. Therefore, the ^{18}O scrambling may be more sensitive in estimating the extent of cage recombination of **14** than the effect of solvent viscosity on k_d. Thus, the intercept of the fluidity plot (equation 66) for $\mathbf{13}_{R=Me}$ using k_d/k_s, which is independent of k_1, is found to be 0.8, suggesting that only 56% of $\mathbf{14}_{R=Me}$ upon calculation with equations (**66**) and (**73**) recombines to the original perester ($\mathbf{13}_{R=Me}$) even at infinite viscosity at 130°C[46].

t-Butyl cycloalkaneperformates with strained small rings ($\mathbf{18}_{3,4}$) (see Section III.A.3) are also assigned to Case I_A rather than Case I_B because of the instability of the strained cycloalkyl radicals.

(*iii*) *Case I_B peresters.* The secondary H–D kinetic isotope effect found in the thermolysis of *t*-butyl perpivalate indicates that the perester undergoes two-bond fission[47,96], as was suggested earlier[44]. However, ^{18}O scrambling was detected[37] in the decomposition of carbonyl-^{18}O-labelled *t*-butyl perpivalate ($\mathbf{13}_{R=t\text{-}Bu}$). Since only the radical-pair mechanism is responsible for the ^{18}O scrambling ($f_r = 0.025$)[95], the major path of concerted two-bond fission decomposition may be concurrently accompanied by a one-bond fission process in the thermolysis of this perester. The extremely small f_r value suggests that the decomposition of Case I_B peresters proceeds almost exclusively via a concerted two-bond fission as shown in equation (65) ($k_1 \ll k_{conc}$ in Scheme 1). The relatively large β-secondary H–D kinetic isotope effect found in the thermolysis of *t*-butyl *para*-substituted phenylperacetate is also in accordance with the concerted two-bond fission mechanism suggested earlier[43,45]. There is a substantial effect of solvent viscosity on k_d of $\mathbf{13}_{R=p\text{-}NO_2C_6H_4}$, but because of the lack of ^{18}O tracer experiments we cannot discuss further the thermolysis of this perester. Case I_B peresters (R = PhC(Me)$_2$—, PhCHMe—)

which undergo very fast decomposition show no solvent effect but give reasonably large k_H/k_D values, as expected from the two-bond fission mechanism (Table 4)[41]. Unstrained *t*-butyl cycloalkaneperformates (**18**$_{5,6}$) also undergo two-bond fission in thermolysis[48].

d. [1,3]-*Sigmatropic mechanism for* ^{18}O *scrambling*. Goldstein and Judson[37] have proposed a [1,3]-sigmatropic mechanism for ^{18}O scrambling in peresters, since scrambling is observable even in the thermolysis of *t*-butyl perpivalate which is considered to undergo typical two-bond fission. Also, the f_r value for the perbenzoate is lower than that for the peracetate, despite the much slower rate of decarboxylation of the benzoyioxy radical compared with that of the acetoxy radical[37]. Thus, the mechanism of ^{18}O scrambling in **13** cannot be said to be clear until the viscosity dependence of the [1,3]-sigmatropic shift is well understood[37]. Goldstein has given a warning on the routine use of ^{18}O scrambling in **13** as the measure of recombination[37]. However, in view of the rather minor contribution of the sigmatropic shift to the ^{18}O scrambling in the thermolysis of diacyl peroxides as described later, ^{18}O scrambling is still one of the best ways to estimate the extent of recombination.

2. *Diacyl peroxides*

Both the rate and the mode of decomposition of diacyl peroxides (**15**) are influenced by the change of R more than with peresters. For example, when R and R′ are primary, decomposition of **15** gives mainly radicals, while if R is secondary or tertiary, the decomposition is ionic even in nonpolar solvents at low temperatures. There are the following three modes (equations 76, 77 and 78) in the thermal decomposition of **15**, depending on the stabilities of R· and R′·, whereas the carboxy inversion, a heterolytic 1,2-rearrangement of R and R′CO$_2$ groups (equation 79) is the major mechanistic path for the decomposition when R$^+$ is stable, as described later.

$$\mathrm{RC(=O)OOC(=O)R'} \longrightarrow [\mathrm{RC(=O)O\cdots OC(=O)R'}]^{\ddagger} \longrightarrow \mathrm{RC(=O)O\cdot \ \ \cdot OC(=O)R'} \tag{76}$$

$$\mathrm{RC(=O)OOC(=O)R'} \longrightarrow [\mathrm{R\cdots C(=O)\cdots O\cdots OC(=O)R'}]^{\ddagger} \longrightarrow \mathrm{R\cdot + CO_2 + \cdot OC(=O)R'} \tag{77}$$

$$\mathrm{RC(=O)OOC(=O)R'} \longrightarrow [\mathrm{R\cdots C(=O)\overset{..}{=}O\cdots O\overset{..}{=}C(=O)\cdots R'}]^{\ddagger} \longrightarrow \mathrm{R\cdot + 2CO_2 + R'\cdot} \tag{78}$$

$$\mathrm{RC(=O)OOC(=O)R'} \longrightarrow [\overset{+}{\mathrm{R}}\cdots\mathrm{C(=O)(O)}\cdots\overset{-}{\mathrm{O}}-\mathrm{C(=O)R'}] \longrightarrow \mathrm{ROC(=O)OC(=O)R'} \tag{79}$$

Szwarc and coworkers[97,98] have suggested that the decomposition of acetyl peroxide proceeds via one-bond fission (equation 76) and the acetoxy radical pair (**16**$_{R=Me}$) should diffuse out from the solvent cage prior to decomposition, based on their observation that the decomposition of acetyl peroxide in solvent requires more activation energy than that in the gas phase. The nearly identical activation energies required for decomposition of **15**$_{R,R'=Me,Et,Pr}$ seem to suggest that these primary alkyl diacyl peroxides decompose by one-bond fission (equation 76)[99]. On the other hand, the thermal decomposition of diacyl peroxides (**15**$_{R=s\text{-alkyl}}$) requires much smaller activation energies and may proceed via a two-bond fission process (equation 77)[100,101]. ^{18}O scrambling in diacyl peroxides should be observable during one-bond fission thermolysis. However, Szwarc failed to detect any ^{18}O scrambling in the thermolysis of carbonyl-^{18}O-labelled acetyl peroxide[102]. Meanwhile, Braun, Rajbenbach and Eirich found a substantial solvent viscosity effect on the rate of decomposition of acetyl peroxide[87], seemingly supporting Szwarc's initial prediction. Taylor and Martin eventually observed the ^{18}O scrambling in acetyl peroxide[88] (Scheme 2) and the contradiction was thus solved. However, in 1970, Goldstein and coworkers suggested sigmatropic shifts for ^{18}O scrambling of acetyl peroxide in solution[103] and indeed found it in the gas-phase decomposition[104].

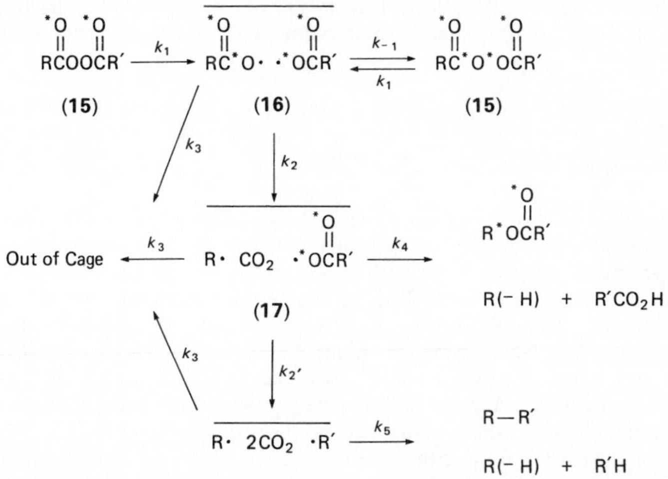

$$\mathrm{RC(={}^{*}O)OOC(={}^{*}O)R'}\ (\mathbf{15}) \xrightarrow{k_1} [\mathrm{RC(={}^{*}O)O\cdot\ \cdot{}^{*}OC(=O)R'}]\ (\mathbf{16}) \underset{k_1}{\overset{k_{-1}}{\rightleftharpoons}} \mathrm{RC(={}^{*}O)O{}^{*}OC(=O)R'}\ (\mathbf{15})$$

$$(\mathbf{16}) \xrightarrow{k_2} [\mathrm{R\cdot\ CO_2\ \cdot{}^{*}OC(={}^{*}O)R'}]\ (\mathbf{17}) \xrightarrow{k_4} \mathrm{R{}^{*}OC(={}^{*}O)R'};\ \mathrm{R(-H)} + \mathrm{R'CO_2H}$$

$$(\mathbf{16}),\ (\mathbf{17}) \xrightarrow{k_3} \text{Out of Cage}$$

$$(\mathbf{17}) \xrightarrow{k_{2'}} [\mathrm{R\cdot\ 2CO_2\ \cdot R'}] \xrightarrow{k_5} \mathrm{R{-}R'};\ \mathrm{R(-H)} + \mathrm{R'H}$$

SCHEME 2. Homolytic decomposition of diacyl peroxides and the acyloxy radical-pair mechanism for ^{18}O scrambling.

According to Goldstein and coworkers in Scheme 3 there are two experimentally observable rate constants for ^{18}O scrambling, i.e. total scrambling (k_{ts}) and random scrambling (k_{rs}) (equations 80 and 81)[103,104]. Equations (66) and (70)–(75) mentioned above can be again applied in the decomposition of **15**. Both rate constants for oxygen scrambling can be determined by mass spectral analysis of O_2 derived from the two peroxidic oxygens in the residue, after partial decomposition[103,104]. The rate constant of

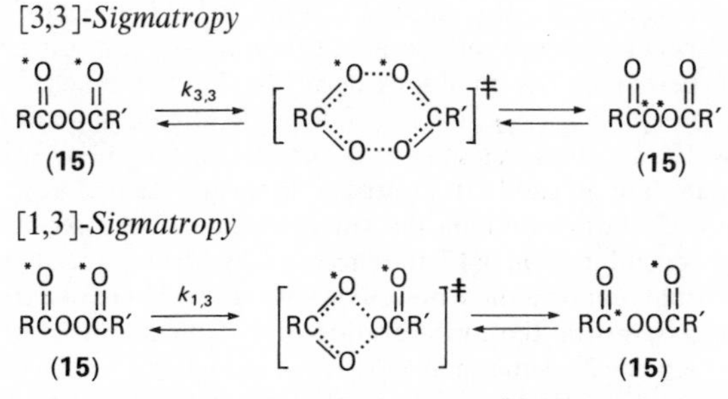

[3,3]-*Sigmatropy*

$$\mathrm{RC(={}^{*}O)OOC(={}^{*}O)R'}\ (\mathbf{15}) \underset{}{\overset{k_{3,3}}{\rightleftharpoons}} [\text{cyclic TS}]^{\ddagger} \rightleftharpoons \mathrm{RC(=O){}^{*}O{}^{*}OC(=O)R'}\ (\mathbf{15})$$

[1,3]-*Sigmatropy*

$$\mathrm{RC(={}^{*}O)OOC(={}^{*}O)R'}\ (\mathbf{15}) \underset{}{\overset{k_{1,3}}{\rightleftharpoons}} [\text{cyclic TS}]^{\ddagger} \rightleftharpoons \mathrm{RC(=O){}^{*}OOC(={}^{*}O)R'}\ (\mathbf{15})$$

SCHEME 3. Sigmatropic mechanisms for ^{18}O scrambling in diacyl peroxides.

^{18}O scrambling via the radical-pair mechanism can be expressed as $k_1 f_r$, assuming that the ^{18}O label is completely scrambled in the acyloxy radical pair. This assumption could be supported by two observations. Firstly, the ^{18}O label is completely scrambled in methyl acetate formed by the thermolysis of labelled acetyl peroxide in solution[39,88] and secondly in the photolysis of benzoyl-^{18}O-labelled acetyl benzoyl peroxide in ethanol at 0°C the two oxygens in the benzoyloxy radical are also completely equilibrated before recombination with methyl radical to form methyl benzoate. However, in the photodecomposition of acetyl benzoyl peroxide labelled with ^{18}O in the peroxidic oxygens both in crystalline state and in ethanol matrix at 77 K, the ether oxygen of methyl benzoate retains 62% and 72% of the original ^{18}O label[69]. Also in the thermolysis of diacyl peroxides, the extent of ^{18}O scrambling and the amount of cage recombination product seem to vary with the acyloxy radical.

$$k_{ts} = k_1 f_r + 2(k_{3,3} + k_{1,3}) \tag{80}$$

$$k_{rs} = k_1 f_r + 4k_{1,3} \tag{81}$$

a. Classification of diacyl peroxides and acyloxy radical pairs. If decomposition of a diacyl peroxide proceeds by one-bond fission (equation 76), the reactions after the O—O bond cleavage would vary markedly with the acyloxy radical pair (**16**)[40]. The cage return to generate peroxide, f_r, is controlled by the relative rates of recombination, k_{-1}, decarboxylation, k_2, and diffusion, k_3, of the respective acyloxy radicals in equation (73). Diacyl peroxides have been classified by us into Case I (R,R′ = alkyl), Case II (R = alkyl, R′ = aryl) and Case III (R,R′ = aryl) assuming that the energy barrier for the recombination of aromatic acyloxy radical pairs is greater than that of aliphatic ones[40], so that the rate of recombination, k_{-1}, decreases in the order of Case I > Case II > Case III. Since k_{-1} of acetoxy radical pairs is of the same order of magnitude as that of decarboxylation of acetoxy radicals ($1.6 \times 10^9\,s^{-1}$ at 60°C[87,88]), the rate of combination of acyloxy radical pairs of Case I is approximately $10^9\,s^{-1}$ or more, while k_{-1} for benzoyloxy radical pairs is lower than k_3, the rate of diffusion (about $10^9\,s^{-1}$) in usual solvents[86], as is exemplified in the very small f_r value for benzoyl peroxide in *i*-octane and the rate of recombination shown in equation (84). Since k_2 of aliphatic acyloxy radicals varies with the structure of R, Case I and Case II are subdivided into A (R = primary alkyl, strained cycloalkyl) and B (R = *s*- and *t*-alkyl, benzyl), expanding our original classification[40]. Thus, the rate of decarboxylation, k_2, decreases in the order of Case I_B > Case II_B > Case I_A > Case II_A ≫ Case III (see Table 9). However, the rate of diffusion, k_3, is considered to be affected by changes of the R group in **16** to a much smaller extent than k_{-1} and k_{-2}.

(*i*) *In usual solvents of low viscosity*, since k_3 is about $10^9\,s^{-1}$ [86], ecombination of Case I_A acyloxy radical pairs can compete with decarboxylation and diffusion, giving a fairly large f_r value. In Case I_B peroxides, f_r is small since decarboxylation is much faster than diffusion and recombination ($k_2 \gg k_3, k_{-1}$). In Case II, the rate of recombination, k_{-1}, is smaller than in Case I, and f_r becomes less than in the corresponding Case I peroxide. Case III acyloxy radical pairs are so stable that most of them diffuse out from the solvent cage without decarboxylation or recombination, thus giving small values of f_r. Upon the rate of ester formation by recombination of **17**, the effect of structural change would be very small, since the process would require almost no activation energy. Therefore, the yield of the ester formed would depend only on the concentration of the precursors, i.e. acyloxy–alkyl/aryl radical pairs (**17**) (Scheme 2). Thus, in an acyloxy radical pair in which decarboxylation of one radical is facile and the other radical is stable, formation of the ester from the second is favoured. Hence, the ester yields are in the order Case II > Case I > Case III[40].

(*ii*) *In an infinitely viscous solvent*, there is no diffusion ($k_3 = 0$), and hence cage return (f_r) for Case III peroxides becomes 1 ($k_{-1} \gg k_{-2}$). The extent of f_r for Case I and Case II peroxides may increase with the increase of solvent viscosity but to a much less extent than for Case III peroxides. The effect of structural changes in the thermolysis of diacyl peroxides are summarized in Table 5.

The peroxidic linkage is highly energetic because of unfavourable lone-pair–lone-pair repulsion between the vicinally situated oxygen atoms. Therefore, the recombination of two oxy radicals to form a peroxidic linkage against such unfavourable electronic repulsion may involve appreciable activation energy[40]. A recent MO calculation[106] seems to support our earlier assumption: when the O—O bond of acetyl peroxide is stretched, the energy maximum appears at 50% homolysis and further stretching to give the acetoxy radical pair somewhat lowers the energy of the system. The benzoyloxy radical is believed to be a σ radical at the ground state[71,106] and MINDO calculations suggest that the dihedral angle between the phenyl ring and the carbonyl plane is 90° [106]. Since the dihedral angle in benzoyl peroxide is nearly zero[107] the recombination of benzoyloxy radical pairs must change the conformation of benzoyloxy radical. If the energy required to change the conformation cannot be compensated by O—O bond formation, the recombination of $\mathbf{16}_{R=Ph}$ requires a somewhat greater energy of activation than that for aliphatic ones.

A few experimental data seem to support this hypothesis. The value calculated by Koenig[82b] for the rate constant of recombination involving his cage theory is much greater for ether formation (equation 82) than for peroxide formation (equations 83 and 84). Koenig observed a positive ρ value (+0.4) for the recombination of $\mathbf{14}_{R=Ph}$ generated from the hyponitrite (**12**) to form *t*-butyl esters of substituted perbenzoic acids (equation 85), suggesting that the recombination of *t*-butoxy–benzoyloxy radical pairs ($\mathbf{14}_{R=XC_6H_4}$[108]) requires some activation energy. This fits the Hammett ρ value of −0.7 obtained in the thermal decomposition of *t*-butyl peresters of substituted perbenzoic acids[109].

$$s\text{-Bu}\cdot CO_2\ N_2\ \cdot\text{OBu-}t \xrightarrow[32^\circ C]{3.9 \times 10^{10}\ s^{-1}} s\text{-BuOBu-}t \quad (82)$$

$$t\text{-BuO}\cdot\ N_2 \cdot\text{OBu-}t \xrightarrow[45^\circ C]{2 \times 10^{10}\ s^{-1}} t\text{-BuOOBu-}t \quad (83)$$

$$PhC(=O)O\cdot\ N_2\ \cdot\text{OBu-}t \xrightarrow[45^\circ C]{7 \times 10^{8}\ s^{-1}} PhC(=O)OOBu\text{-}t \quad (84)$$

$$X\text{-}C_6H_4\text{-}C(=O)\text{-}O\text{-}N{=}N\text{-}O\text{-}Bu\text{-}t\ (\mathbf{12}_{R=XC_6H_4}) \longrightarrow [X\text{-}C_6H_4\text{-}C(=O)\text{-}O\cdot\ N_2\ \cdot O\text{-}Bu\text{-}t]\ (\mathbf{14}_{R=XC_6H_4}) \longrightarrow X\text{-}C_6H_4\text{-}C(=O)\text{-}O\text{-}O\text{-}Bu\text{-}t\ (\mathbf{13}_{R=XC_6H_4}) + N_2 \quad (85)$$

In keeping with our *a priori* precept (Table 5), the amount of cage return (f_r) decreases in the order of cyclopropaneformyl peroxide (Case I) > benzoyl cyclopropaneformyl peroxide (Case II) > benzoyl peroxide (Case III) (see Table 6 which lists the effects of various groups on the oxygen scrambling in the thermolysis of diacyl peroxides[40,56] and in the Cope rearrangement[110]). The effect of the phenyl group on the oxygen scrambling in diacyl peroxides is exactly opposite to that on the rates of the Cope rearrangement[40,56], a

TABLE 5. Structural effects on the thermal decomposition of diacyl peroxides, RC(=O)OOC(=O)R′ (**15**)

Case	R	R′	k_d	k_d/k_{ts}	f_r[a]	f_r[b]	Ester yield by radical process	Contribution of ionic path
I	Alkyl	Alkyl						
A	primary alkyl, strained cycloalkyl		Small	Small	Large	Large	Low	Very small
B	*s*- and *t*-alkyl, benzyl		Large	Large	Small	Small	Low	Large
II	Alkyl	Aryl						
A	primary alkyl, strained cycloalkyl		Small	Large	Small	Small	High	Very small
B	*s*- and *t*-alkyl, benzyl		Large	Large	Very small	Very small	Very high	Large
III	Aryl	Aryl	Small	Large	Very small	Very large	Very low	Small

[a] In usual solvents of low viscosity.
[b] In infinitely viscous solvent.

TABLE 6. Comparison of oxygen scrambling in diacyl peroxides (**15**) with Cope rearrangement of 1,5-hexadiene-1-d_2

	1,5-hexadiene	2-Ph-1,5-hexadiene	2,5-Ph$_2$-1,5-hexadiene
k^{rel}_{Cope} [a]	5×10^{-4}	2×10^{-2}	1
	(cyclopropyl-CO-O)$_2$	Ph-CO-O-O-CO-cyclopropyl	(Ph-CO-O)$_2$
k^{rel}_{ts} [e]	11[b]	4[b,c]	1[d]
$(k_d/k_{ts})^{rel}$	6.58	1.89	1
f^{rel}_r [f]	5.2	1.8	1

[a]Relative rate of the Cope rearrangement of 1,5-hexadienes, Ref. 110.
[b]in CCl_4 at 80°C, Ref. 56.
[c]Cyclopropyl-side carbonyl-^{18}O-labelled peroxide was used to determine k_{ts}.
[d]In isooctane at 80°C, Ref. 105.
[e]Relative rate of total scrambling of carbonyl-^{18}O in diacyl peroxide at 80°C.
[f]Relative value of fraction of cage return of **16** to generate **15**.

typical [3,3]-sigmatropic reaction (equation 86)[111]. Phenyl substituents at 2- and/or 5-position of 1,5-hexadiene increase the rate of the Cope rearrangement by stabilizing its transition state[112]. On the other hand, phenyl groups stabilize the intermediary (equation 86) acyloxy radicals (**16**) in the oxygen scrambling in diacyl peroxides (Scheme 2) and thus retard the rate of recombination, resulting in low f_r and decreasing the extent of ^{18}O scrambling [40,56]. The transition state of the concerted[111,112] Cope rearrangement of 1,5-hexadiene-1-d_2 is believed to be stabilized by 28.7 kcal mol^{-1} for the chair form[113] and by 17.5 kcal mol^{-1} for the boat form[114] more than the potential energy required for homolytic cleavage of the C(3)—C(4) bond to generate an allyl radical pair[113]. Lewis and Newman[115] have indicated that the transition state of the 'dioxa-Cope rearrangement' of allyl-3-d_2 trifluoroacetate (equation 87) in the gas phase is markedly stabilized due to the concerted nature of the reaction. On the other hand, activation enthalpies of oxygen scrambling in diacyl peroxides are usually nearly identical to those of homolytic cleavage of the O—O bond to generate **16**. For example, the activation enthalpies of the total and of

$$R^1\text{-C(=CD}_2\text{)CH}_2\text{CH}_2\text{C(R}^2\text{)=CH}_2 \rightleftharpoons [\text{TS}]^{\ddagger} \rightleftharpoons R^1\text{-C(=CH}_2\text{)CH}_2\text{CH}_2\text{C(R}^2\text{)=CD}_2 \quad (86)$$

$$CF_3\text{-C(=O)-O-CH}_2\text{CH=CD}_2 \rightleftharpoons [CF_3\text{-C}(\overset{\delta-}{\text{O}})_2 \cdots \overset{\delta+}{\text{C}_3\text{H}_3\text{D}_2}] \rightleftharpoons CF_3\text{-C(=O)-O-CD}_2\text{CH=CH}_2 \quad (87)$$

the random scrambling of cyclopropaneformyl peroxide are only 2.0 and 2.8 kcal mol^{-1} less, respectively, than that of homolytic fission of the O—O bond to generate a cyclopropaneformyloxy radical pair[56,117]. These small $\Delta\Delta H^{\neq}$, (homolytic decomposition–oxygen scrambling) values are nearly identical to the energy barrier for the diffusion of the intermediary acyloxy radical pair[116]. Even with acetyl peroxide in the gas phase the activation enthalpies for both modes of oxygen scrambling are only 1.0–1.5 kcal mol^{-1} smaller than that of decomposition[104], i.e. almost within the experimental error. In contrast to the negative activation entropies of the Cope rearrangement, i.e. −13.8 and −3.0 gibbs for 1,5-hexadiene in the chair-[113] and the boat-[114] form transition states, respectively, and −9.8 gibbs for allyl trifluoroacetate[115], the activation entropies observed for the oxygen scrambling in diacyl peroxides are positive, suggesting that this transition state is much less rigid, and does not seem to be concerted. While the driving force for the Cope rearrangement is the formation of the partial C(1)–C(6) bond which stabilizes the transition state, the driving force of the oxygen scrambling in diacyl peroxides[53,117] seems to be the release of the electronic repulsion due to the lone-pair–lone-pair interaction by cleavage of the peroxidic bond.

Table 7 summarizes the rates of ^{18}O scrambling (k_{ts}) of various carbonyl ^{18}O-labelled diacyl peroxides, rates of decomposition (k_d) and f_r values calculated from k_d and k_{ts}.

b. Case I$_A$ diacyl peroxides. (*i*) *Acetyl peroxide* has been considered to undergo one-bond fission, although acetoxy radicals could not be trapped in earlier experiments[118]. Later, however, Shine's group succeeded in trapping acetoxy radicals, by cyclohexene solvent, obtaining CO_2 in 65–75% yield and 1,1′-dicyclohexyl acetate and a mixture of 3-cyclohexenyl and cyclohexyl acetates in about 20% yield[119,120]. Furthermore, addition of I_2–H_2O, 9,10-dihydroanthracene or galvinoxyl reduced the yield of CO_2, without affecting the rate of decomposition of acetyl peroxide[121]. The ^{18}O tracer experiments[122] in the same system supported Shine's conclusion that cyclohexene reacts with acetoxy radicals but not with the peroxide, by showing that the original ^{18}O label in the carbonyl of the peroxide was scrambled completely in the products. These observations suggested that acetyl peroxide undergoes degradation by one-bond fission (equation 76).

This was further confirmed by Taylor and Martin[88] who have measured the k_{ts} of ^{18}O-labelled acetyl peroxide. They calculated that 38% of the acetoxy radical pairs recombines back to the ^{18}O-scrambled peroxide, assuming that the acyloxy radical-pair mechanism is responsible for the oxygen scrambling (equation 73, Table 7, Scheme 2). Martin and Dombchik[123] have examined the solvent viscosity effect on k_{ts}, and found that it increases with the increase of solvent viscosity; e.g. $k_{ts} \times 10^5$ (s^{-1}) at 80°C = 4.00 (isooctane), 5.25 (octadecane), 6.37 (nujol). This also fits earlier observations[87] clearly supporting the mechanism shown in Scheme 2.

The rather small secondary deuterium kinetic isotope effects $k_H/k_D = 1.039$ in *i*-octane and 1.057/d$_3$ in nujol at 80°C found with acetyl and trideuterioacetyl peroxides by Koenig and Cruthoff[66] also shows that the reaction is a one-bond fission process (Scheme 2), while the viscosity-dependent value of k_H/k_D in the rate of decomposition suggests the decomposition to be somewhat reversible, since the rate constant of the rate of overall decomposition, k_d, includes the rate constant of decarboxylation, k_2, which displays a substantial kinetic isotope effect. The value of k_H/k_D for the decarboxylation of acetoxy radical was obtained as 1.09/d$_3$ at 80°C in *i*-octane by measurements of the total yields of methyl acetate and the ratio of CH_3COOCD_3 vs. CD_3COOCH_3 in the thermal decomposition of $CD_3CO_2O_2CCH_3$. From this value, together with the value of k_2/k_{-1}, obtained from a fluidity plot (equation 66), the value of k_H/k_D in the decomposition was estimated as 1.009/d$_3$ in *i*-octane and 1.030/d$_3$ in nujol at 80°C. These values are substantially smaller than those determined by direct decomposition kinetics mentioned

TABLE 7. Rates of decomposition, k_d, rates of oxygen scrambling, k_{ts}, and extent of cage return of **16**, f_r, in the thermal decomposition of diacyl peroxides, $\mathrm{R\overset{O}{\overset{\|}{C}}O O\overset{O}{\overset{\|}{C}}R'}$ (**15**)

Case	R	R′	Solvent	Temp. (°C)	$k_{ts} \times 10^6$ (s^{-1})	$k_d \times 10^6$ (s^{-1})	$f_r \times 100$	Ref.
I_A		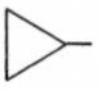	CCl_4	80	14.6	46.2	24	56, 117
I_A	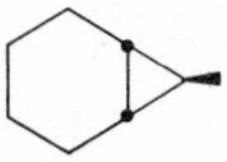	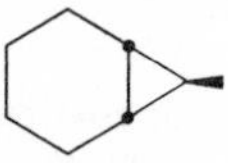	CCl_4	80	14.8	89.6	14.2	56
I_B	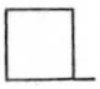		CCl_4	80	32.6	1040	3.0	135
I_A	Me	Me	$i\text{-}C_8H_{18}$	80	44.0	72.0	38	88
I_A	Me	Et	$i\text{-}C_8H_{18}$	80	16.6	78.9	17	126
I_A	Et	Et	$i\text{-}C_8H_{18}$	80	8.10	78.9	5.1	126
I_B	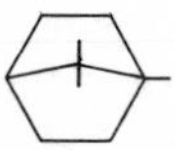	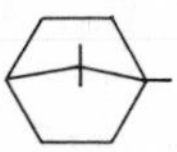	CCl_4	80	58.2	104	55	40
II_A		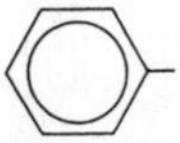	CCl_4	80	5.15	56.5	8.3	56, 117
II_B	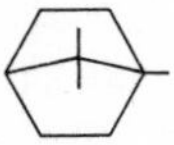	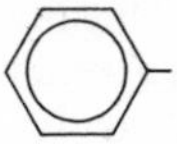	CCl_4	70	ca. 0.7	17.4	7	40
III	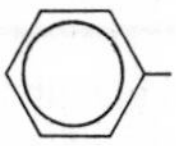	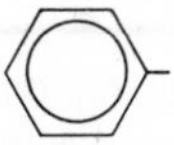	$i\text{-}C_8H_{18}$	80	1.30	27.0	4.6	105
I_A			CCl_4	45	0.129	0.201	39	56, 117
I_B	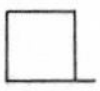	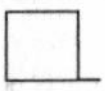	CCl_4	45	0.393	13.2	2.9	135
I_B	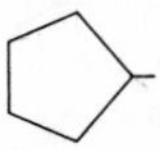	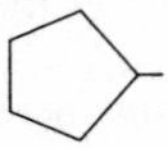	CCl_4	45	1.32	40.5		135
I_B	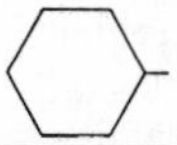	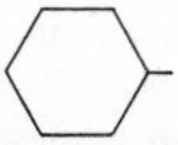	CCl_4	45	2.00	402		135

above. This discrepancy has been rationalized by assuming $k_H/k_D = 1.027/d_3$ for the O—O bond fission (equation 74); this suggests that some C—C bond reorganization is involved in the formation of acetoxy radicals from acetyl peroxide[66]. A similar rationalization has been made on the kinetic isotope effect, k_H/k_D, observed[48] in the decomposition of *t*-butyl cyclopropaneperformate (**18**$_3$).

Observations of CIDNP of the thermal decompositions of both unlabelled and labelled acetyl peroxides in hexachloroacetone at 110°C support the mechanism shown in Scheme 2[72,124]. ^{1}H-NMR runs during the thermolysis of acetyl peroxide give net polarization signals of methoxy protons of methyl acetate and of ethane (memory effect) which originate in the singlet–triplet (t_0) transition in the methyl–acetoxy radical pair. Kaptein has summarized the CIDNP results as follows[124]. (*a*) The rates of combinations of methyl–methyl radicals and methyl–acetoxy radical pairs are comparable to that of the decarboxylation of acetoxy radicals and are about 2–3 × $10^9\,s^{-1}$ at 110°C. (*b*) The g-factor of the acetoxy radical is estimated as 2.0058 from the intensity ratio in the doublet CIDNP signal of $CH_3COO^{13}CH_3$. (*c*) The following reactions are possible, but not important in the decomposition of acetyl peroxide (equations 88 and 89).

$$CH_3^{\bullet} + (CH_3\overset{O}{\overset{\|}{C}}O)_2 \longrightarrow CH_3O\overset{O}{\overset{\|}{C}}CH_3 + CH_3COO^{\bullet} \tag{88}$$

$$CH_3^{\bullet} + CH_3COO^{\bullet} \longrightarrow C_2H_6 + CO_2 \tag{89}$$

Thermal decomposition of other Case I$_A$ peroxides (**15**$_{R,R'=n\text{-alkyl}}$) does not exhibit any net polarization in the NMR signals of the ester, suggesting that the rate constant of decarboxylation of ordinary *n*-alkaneformyloxy radical is greater than $10^{10}\,s^{-1}$, since the appearance of CIDNP signals requires a lifetime of more than ca. $10^{10}\,s^{-1}$ [124]. These Case I$_A$ peroxides decompose at nearly the same rate as acetyl peroxide and require nearly the same activation energy to decompose[99] despite the different lifetime of each acyloxy radical. This may be taken as evidence to support the one-bond fission process for Case I$_A$ peroxides (equation 76). Goldstein has suggested, however, that the decomposition of acetyl peroxide proceeds via fission of 2 or 3 bonds (equations 77 and 78) based on their experimentally observed values of kinetic isotope effects, i.e. $k_{16}/k_{18} = 1.035 \pm 0.002$ in cumene, 1.029 ± 0.002 in *i*-octane (oxygen) and $k_{12}/k_{13} = 1.019 \pm 0.001$ in cumene and 1.023 ± 0.002 in *i*-octane (carbonyl carbon) in the thermal decomposition at 45°C[125]. Assuming acetyl peroxide to undergo multibond fission, or $k_2 \gg k_{-1}$, Goldstein, and Judson[103] have calculated (equations 80 and 81) that the [3,3]-sigmatropic shift can account for 63–85% of the oxygen scrambling in acetyl peroxide in solution (cumene, *i*-octane). It has been found later that both total and random scramblings in acetyl peroxide proceed somewhat faster in the gas phase than in solution[104]. Since there is no cage return in the gas phase, the [3,3]-sigmatropic shift has been suggested to be responsible for the scrambling[104].

(*ii*) *Acetyl propionyl peroxide* (**15**$_{R=Me,R'=Et}$) and *propionyl peroxide* (**15**$_{R,R'=Et}$). Replacement of H atoms of acetyl peroxide by Me groups decreases the rate of oxygen scrambling (k_{ts}) i.e. 4.00 × 10^{-5} (**15**$_{R,R'=Me}$), 1.66 × 10^{-5} (**15**$_{R=Me,R'=Et}$), 0.816 × 10^{-5} (**15**$_{R,R'=Et}$) in *i*-octane at 80°C, while the rate of decomposition increases only from 7.28 × 10^{-5} (**15**$_{R,R'=Me}$) to 7.89 × $10^{-5}\,s^{-1}$ (**15**$_{R,R'Et}$)[126]. This may be explained by the mechanism shown in Scheme 2 and fits the CIDNP results which indicate that $EtCO_2^{\bullet}$ decomposes much more rapidly than $MeCO_2^{\bullet}$.[124].

(*iii*) *Cyclopropaneformyl peroxide* (**19**$_3$). Since the cyclopropyl radical is unstable, cyclopropaneformyl peroxide is expected to undergo one-bond fission. Indeed, it has been observed that when the solvent is changed from octane to nujol at 80°C, the rates of oxygen scrambling increases 1.87-fold (k_{ts}) and 1.72-fold (k_{rs}), while the rate of decomposition decreases from 6.02×10^{-5} to $5.15 \times 10^{-5}\,s^{-1}$ [53,117]. The fraction of cage return for cyclopropaneformyloxy radical pairs (f_r) calculated by the Pryor–Smith equation from the solvent-dependent k_d of **19**$_3$ is nearly identical to that calculated from k_d and the rates of oxygen scrambling (k_{ts} and k_{rs}), thus ruling out the sigmatropic paths[53,117].

c. Case I_B diacyl peroxides. (i) *Cycloalkaneformyl peroxides* (**19**$_{4-6}$) have been investigated with isotopically labelled peroxides[135].

(*ii*) 1-*Apocamphoryl peroxide* is interesting in that it undergoes both homolytic and ionic (carboxy-inversion) decompositions and also oxygen scrambling at similar rates in CCl_4 [40].

d. Case II diacyl peroxides. (*i*) 1-*Apocamphoryl benzoyl peroxide* also decomposes through both ionic and homolytic paths at almost the same rates, but here, the rate of oxygen scrambling is too small to be measured accurately, $f_r < 0.07$ at 70°C in CCl_4 [40]. The scrambling of the apocamphoryl ^{18}O label appears to take place more rapidly than that of the benzoyl ^{18}O label. The decomposition affords 1-apocamphoryl benzoate, presumably the radical product, in 20% yield which corresponds to 40% of radical decomposition path, while 1-apocamphoryl peroxide (Case I), gives the corresponding ester only in 3% yield[40].

(*ii*) *Acetyl benzoyl peroxide.* Kobayashi, Minato and Hisada[67] have prepared acetyl-^{18}O-labelled and benzoyl-^{18}O-labelled acetyl benzoyl peroxides, respectively, and subjected them to thermolysis, and observed that the rate of scrambling of acetyl ^{18}O label is greater than that of benzoyl ^{18}O label in the recovered acetyl benzoyl peroxide. Both oxygen scramblings are markedly slower than those of the corresponding Case I peroxide (acetyl peroxide) but faster than the corresponding Case III peroxide (benzoyl peroxide).

MO calculations for the acyloxy radical suggest that there are four electronic states, (one π radical, $^2A_2$, and three σ radicals $^2A_1$, $^2B_2$ and $^2A'$), nearly at the same energy levels as shown in Scheme 4[106,127]. Efforts have been made to determine theoretically[106,127,128] and experimentally[70,71,129] the electronic state of acyloxy radicals at ground state. McBride and Mernil[71] have determined experimentally the electronic state of benzoyloxy radicals in the photolysis of single crystals of benzoyl peroxide and acetyl-d_3 benzoyl peroxides labelled with ^{17}O at either carbonyl or peroxidic oxygens at 77 K[70,71]. They have concluded that the ground state of the benzoyloxy radical is the $^2B_2$ state, based on

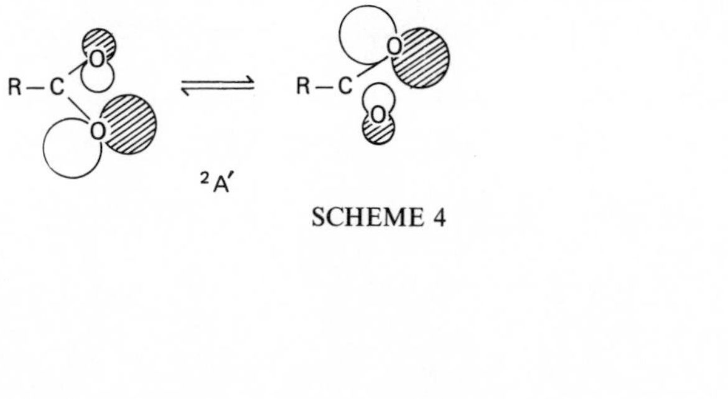

R—C $^2B_2$ R—C $^2A_1$ R—C $^2A_2$

R—C ⇌ R—C $^2A'$

SCHEME 4

the ^{17}O-hyperfine splittings of the radical pair $[CD_3\cdot \, \cdot O_2CPh]$ generated from either carbonyl-^{17}O-labelled or peroxidic-oxygen-^{17}O-labelled peroxide. There is no spin density on the phenyl ring, while the 2p orbitals of the two oxygens possess large and similar spin densities.

e. Case III diacyl peroxides. Hammond and Soffer[130] have found that benzoyloxy radicals can be intercepted quantitatively by I_2–H_2O to yield benzoic acid during the thermal decomposition of benzoyl peroxide, thus supporting clearly that this peroxide undergoes one-bond fission. However, the rate of decomposition, k_d, is found to be affected only a little by the change of viscosity of the solvent; the value of f_r calculated by the Pryor–Smith treatment is only 0.004 at 80°C in *i*-octane[91]. Oxygen scrambling in benzoyl peroxide is also very slow compared to decomposition and could not be measured by the earlier method[131]. Martin and Hargis[105] have determined the rate of oxygen scrambling by measuring the increase of ^{18}O in O_2 derived from peroxidic oxygens of the recovered peroxide during thermal decomposition of carbonyl-^{18}O-labelled peroxide and found that $k_{ts} = 2.89 \times 10^{-5}\,s^{-1}$ at 80°C in *i*-octane containing 0.2 M styrene ($f_r = 0.046$). The rate of ^{18}O scrambling is accelerated 4.4-fold by changing the solvent from *i*-octane to nujol, suggesting that the cage return of benzoyloxy radical pairs takes place during the decomposition. They have suggested three possibilities for this very small cage return: (*a*) an appreciable fraction of cage return without scrambling, (*b*) a high activation energy barrier for the recombination of $\mathbf{16}_{R,R'=Ph}$ and (*c*) an unfavourable $\Delta S^{\neq}$[105]. The possibility (*a*) can be ruled out, since dependency of k_d on solvent viscosity has revealed that f_r is also very small[91]. Thus (*b*) or (*c*) may be responsible for the small value of f_r. If one could measure the rate of oxygen scrambling in benzoyl peroxide at various temperatures, then a concrete answer might be obtained. However, since decarboxylation of benzoyloxy radicals requires a large energy (ca. 14 kcal mol^{-1})[141], in an infinitely viscous hypothetical solvent, f_r would become 1. Benzoyloxy radical pairs cannot diffuse out from the solvent cage in such a hypothetical solvent and hence the rate-determining step of the decomposition would become that of decarboxylation. Unfortunately this hypothesis cannot be confirmed directly. This difficulty was overcome by Fujimori and coworkers[117] by using carbonyl-^{18}O-labelled phthaloyl peroxide whose O—O fission gives phthaloyloxy radicals (an intramolecular Case III), in which there is no diffusion path. Therefore, if the hypothesis is correct, f_r of the phthaloyloxy radical should be very close to 1[40].

Earlier, Greene reported that there was no oxygen scrambling even after one day of refluxing with carbonyl-^{18}O-labelled phthaloyl peroxide in CCl_4[62]. However, we have found that oxygen scrambling does take place in CCl_4 rather quickly as compared to the decomposition[117], as shown in Table 8, thus substantiating our hypothesis. Actually, 98% of the phthaloyloxy radicals are found to recombine back to the original peroxide in which oxygens are scrambled (equation 90). This is the first example which reveals a high

k_1 ; k_1 / k_{-1} ; k_2 → Products (90)

TABLE 8. Thermal decomposition of phthaloyl peroxide and of benzoyl peroxide at 80°C

Peroxide	Solvent	$k_{ts} \times 10^7$ (s^{-1})	$k_d \times 10^7$ (s^{-1})	k_{ts}/k_d	f_r
Phthaloyl peroxide[a]	CCl_4	130 ± 3.23	3.12 ± 0.31	41.2	0.976
Benzoyl peroxide[b]	i-C_8H_{18}	13.0	270	0.05	0.046

[a]Refs. 63 and 117.
[b]Ref. 105.

preference of recombination of acyloxy radical pair over the decomposition[117]. The rate of decomposition, k_d, of benzoyl peroxide is roughly identical to that of the O—O bond fission of benzoyl peroxide (k_1) while the rate of oxygen scrambling (k_{ts}) in phthaloyl peroxide is considered to be roughly identical to the value of k_1 of phthaloyl peroxide. Hence k_d for benzoyl peroxide being about twice the value of k_{ts} for pthaloyl peroxide may mean that the rates of O—O bond fission of both peroxides are very similar. Thus, the cyclic structure of phthaloyl peroxide has no special effect on the rate of homolytic cleavage of the O—O bond as advocated earlier by Greene[62].

Since benzoyloxy radical pairs formed from benzoyl peroxide diffuse out predominantly, both ester formation (3%) and cage return of benzoyloxy radical pairs ($f_r = 0.04$) are quite low.

However, both direct and singlet-sensitized photolyses of benzoyl peroxide proceed via two-bond fission, generating directly phenyl–benzoyloxy radical pairs which subsequently give phenyl benzoate intramolecularly, in 15–20% yield[59]. Thus the mechanism shown in equation (91) has been suggested[59] on the basis of the cross-over experiments and CIDNP studies using 2,2′,4,4′,6,6′-hexadeuterio- and perdeuterio-benzoyl peroxides.

$$PhC(=O)OOC(=O)Ph \longrightarrow [Ph\cdot CO_2\ \cdot OC(=O)Ph] \longrightarrow PhOC(=O)Ph \qquad (91)$$

The photodecomposition of carbonyl-^{18}O-labelled benzoyl peroxide in benzene gives phenyl benzoate in which the ^{18}O-label is equilibrated[132]. The authors have concluded that the reaction is unimolecular decomposition of the excited singlet state, produced by energy transfer from excited singlet benzene to the peroxide as predicted earlier[133].

3. *Thermal decomposition of* t-*butyl cycloalkaneperformates and cycloalkaneformyl peroxides*

Rüchardt[134] has compared the thermal decomposition of t-butyl cycloalkaneperformates (**18**$_{4-8}$) with that of biscycloalkanediazenes (**20**$_n$) and claimed that the former reaction is much less endothermic than the latter. Therefore, the transition state of the former reaction, in which the O—O bond is essentially broken but the R—CO bond stretches only a little, comes in a relatively early stage, while that of the latter reaction, in which substantial C—N reorganization takes place, comes at a later stage in the reaction

$$(CH_2)_{n-1}C(H(D))C(=O)OO\text{-}t\text{-Bu} \qquad [(CH_2)_{n-1}CHC(=O)O\text{—}]_2 \qquad [(CH_2)_{n-1}CHN=]_2$$

(**18**$_n$) (**19**$_n$) (**20**$_n$)

coordinate. He considers that the rate of **18** is controlled mainly by the polar effect of the ring in the ground state, assuming the single-step decomposition path, based on the linear dependency of log k_d on the $J_{^{13}C-H}$ of the corresponding cycloalkanes and on other data.

Recently, Wolf and coworkers[48] have reinvestigated the mechanism of thermal decomposition of **18** and found that the secondary α-hydrogen–deuterium kinetic isotope effect plays an important role in clarifying the mechanism. Thermal decomposition of $\mathbf{18}_3$ and $\mathbf{18}_4$ exhibits no secondary α-deuterium isotope effect (Table 4) and the large activation enthalpies, 34.9 ± 0.5 ($\mathbf{18}_3$) and 35.9 ± 0.4 ($\mathbf{18}_4$) kcal mol^{-1} observed in octane suggest one-bond fission thermolysis. The solvent viscosity effect also supports this, and k_d of $\mathbf{18}_3$ is more dependent on solvent viscosity $[k_d(C_6H_{14})/k_d(C_{16}H_{34}) = 1.30 \pm 0.04]$ than k_d of $\mathbf{18}_4$ $[k_d(C_6H_{14})/k_d(C_{16}H_{34}) = 1.12 \pm 0.02]$ at 102.6°C[48]. Based on these and Koenig's results, Wolf and coworkers have proposed a genuine one-bond fission mechanism for $\mathbf{18}_3$, and a merged mechanism for $\mathbf{18}_4$ (equation 92), rather than a mixture of one-bond and two-bond fission mechanisms.

Merged mechanism

$$-\overset{|}{\underset{|}{C}}-\overset{O}{\overset{\|}{C}}-O-O-C(CH_3)_3 \underset{k_1}{\overset{k_{-1}}{\rightleftharpoons}} \left[-\overset{|}{\underset{|}{C}}\cdots\overset{O}{\overset{\|}{C}}-O\cdots O-C(CH_3)_3\right]^{\ddagger} \longrightarrow -\overset{|}{\underset{|}{C}}\cdots\overset{O}{\overset{\|}{C}}-O\cdot + \cdot O-C(CH_3)_3 \xrightarrow{k_2} -\overset{|}{\underset{|}{C}}\cdot + CO_2 + \cdot O-C(CH_3)_3 \quad (92)$$

where $k_{-1} < k_d$ and $k_{-1} < k_2$, k_d is given by equation (71)

Secondary α-deuterium kinetic isotope effects observed in the decomposition of $\mathbf{18}_5$ and $\mathbf{18}_6$ are substantial (Table 4), but are smaller than $k_H/k_D = 1.0931/d_1$, observed in the thermal decomposition of 1,1′-diphenylazoethane[96b], showing undoubtedly that the thermolysis of the peresters undergoes two-bond fission at the transition state which comes somewhat earlier than thermolysis of 1,1′-diphenylazomethane. The values of k_ds of these two peresters are not affected by the change of viscosity of medium. Activation enthalpies for $\mathbf{18}_5$ are (33.6 ± 0.8 kcal mol^{-1}) and for $\mathbf{18}_6$ (32.4 ± 1.3 kcal mol^{-1})[48].

The kinetics of decomposition and ^{18}O scrambling of $\mathbf{19}_n$ have been studied in detail by us[55,135]. The plot of lgoarithms of k_d and k_{ts} against $J_{^{13}C-H}$ of the cycloalkanes exhibits a curve (Figure 1). Similar plots for the peresters also give a curve rather than a straight line. Rüchardt has reported that the plot of log k_d of $\mathbf{18}_n$ whose ring sizes are greater than four against $J_{^{13}C-H}$ value give approximately a straight line[134]. The slope of the line for $\mathbf{19}_n$ is approximately twice of that for $\mathbf{18}_n$. The log–log plot of k_d for $\mathbf{18}_n$ vs. that for $\mathbf{19}_n$ gives a straight line with a slope of about 0.4, suggesting that the ring-size effect on k_d for $\mathbf{18}_n$ is about 40% of that for $\mathbf{19}_n$[55,56]. This is in keeping with the argument that the decomposition of the perester proceeds through a somewhat polar transition state, while in the thermal decomposition of $\mathbf{19}_n$ the contribution of the heterolytic carboxy–inversion increases with the increase of the ring size. Thus $\mathbf{19}_3$ decomposes exclusively by homolytic one-bond fission, while the thermal decomposition of $\mathbf{19}_4$ involves a contribution of ca. 5% by the carboxy-inversion process and $\Delta H^{\neq}$ for ^{18}O scrambling is somewhat (though within experimental error) greater than $\Delta H^{\neq}$ for decomposition, revealing that the major path of decomposition involves a concerted two-bond fission[55]. Decomposition of $\mathbf{19}_6$ proceeds via carboxy-inversion which would involve the same transition state as the homolytic decomposition, implying that $\mathbf{19}_6$ is a typical Case I_B diacyl peroxide which decomposes through two-bond fission. However, even in $\mathbf{19}_6$, ^{18}O scrambling was detectable in the recovered peroxide, suggesting a very small contribution of a one-bond fission process which leads to oxygen scrambling[55,135]. Stabilities of both cycloalkyl radicals and cycloalkyl cations increase with the ring size, changing the mechanism from

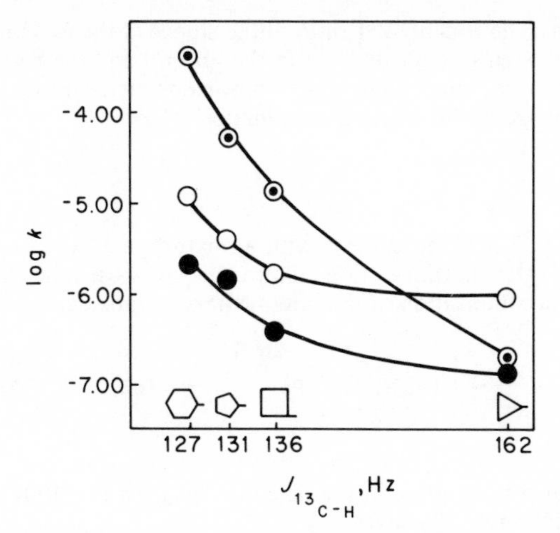

FIGURE 1. Ring-size effects on rates of decomposition of **18** in C_6H_{14} at 80°C, and oxygen scrambling and decomposition of **19** in CCl_4 at 45°C[135]. ○———○: k_d for **18**, ◉———◉: k_d for **19**, ●———●: k_{ts} for **18**.

one- to two-bond fission. If one assumes that the acyloxy radical energy level is perturbed to a lesser extent than the ground state and the transition state of the two-bond fission, the slope of the line for oxygen scrambling in Figure 2 reflects the ring-size effect on the energy level of the ground state of the peroxide. Meanwhile, the slope of the line for decomposition may indicate the ring-size effect on both the ground and the transition states of the major

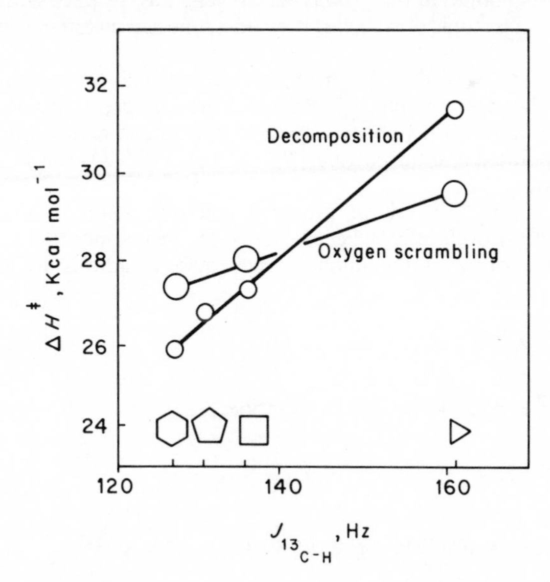

FIGURE 2. Ring-size effects on activation enthalpies of decomposition and oxygen scrambling of **19** in CCl_4 [55,135].

decomposition process. The roughly 2:1 ratio of the slopes of the two lines in Figure 2 implies that the ring-size effect appears both in the ground and the transition state of decomposition roughly to the same extent, suggesting again that the transition state comes at a relatively earlier stage of the reaction coordinate[56,135].

4. *Polyoxides*

Bartlett and Traylor[136] oxidized cumene with a mixture of $^{16}O{-}^{16}O$ and $^{18}O{-}^{18}O$. Since they detected $^{16}O{-}^{18}O$ in the reaction mixture, they proposed the scheme shown in equation (93) for the decomposition of the intermediary tetraoxide.

$$\mathrm{PhC^{*}O^{*}O\cdot} + \mathrm{PhCCO\cdot} \longrightarrow \left[\mathrm{PhC^{*}O^{*}OOOCPh}\right] \longrightarrow \mathrm{PhC^{*}O} + {}^{*}\mathrm{O{=}O} + \mathrm{OCPh} \tag{93}$$

A similar reaction also takes place in the reactions between two ROO· and radicals (R = *n*-, *s*- and *t*-alkyl, acetyl, 2-pyridyl)[137].

Intervention of diacetyl tetraoxide has also been confirmed using ^{18}O tracer in the autoxidation of acetaldehyde and in the induced decomposition of peracetic acid with *t*-butoxy radical[138].

5. *Polymerization initiated by free radicals generated by unimolecular decomposition of organic peroxides*

Polymerization of olefinic monomers is initiated either by oxy radicals formed directly by homolysis of O—O bonds or by carbon radicals generated by β scission of the primary oxy radicals. By using isotopically labelled peroxides, one can investigate which species is involved[139].

Yoshida found in 1950 that fragments of deuterated benzoyl peroxide were incorporated into polymer terminals[58]. Russian chemists used carbonyl-^{14}C-labelled benzoyl peroxide to initiate the polymerization of styrene and revealed that benzoyloxy groups were incorporated into the polymer[140]. The yield of $^{14}CO_2$ was found to increase at higher temperatures[140].

Bevington and Toole[141–144] have provided extensive data on the initiation of polymerization with isotopically labelled peroxides (equations 95 and 96). They determined the fraction χ of the initiating benzoyloxy radicals by measuring the

$$\mathrm{ArCO_2\cdot} \xrightarrow{k_2} \mathrm{Ph\cdot} + \mathrm{CO_2} \tag{94}$$

$$\mathrm{ArCO_2\cdot} + \mathrm{M} \xrightarrow{k_p} \mathrm{ArCOOM\cdot} \tag{95}$$

$$\mathrm{Ar\cdot} + \mathrm{M} \longrightarrow \mathrm{ArM\cdot} \tag{96}$$

$$\chi = k_p[\mathrm{PhCOO\cdot}][\mathrm{M}]/\{k_2[\mathrm{PhCOO\cdot}] + k_p[\mathrm{PhCOO\cdot}][\mathrm{M}]\} \tag{97}$$

$$\frac{1}{\chi} = 1 + \frac{k_1}{k_2}\frac{1}{[\mathrm{M}]} \tag{98}$$

radioactivity of polymers initiated by ring- or substituent-^{14}C-labelled benzoyl peroxides, before and after the removal of aroyl groups by hydrolysis (equation 97). A reciprocal plot of χ against the monomer concentration gave a straight line with a slope k_1/k_2 (equation 98). Some of their data are summarized in Table 9. The electron-releasing group in the *para*-position markedly stabilizes the benzoyloxy radical, while an electron-withdrawing substituent, e.g. the MeO group at the *meta*-position, accelerates the decarboxylation of benzoyloxy radicals. Since Arrhenius plots of $\log(k_1/k_2)$, for the thermolysis of the benzoyl peroxide–styrene system at 60°C and 80°C and photolysis at 25°C and 40°C, fall on the same line, benzoyloxy radicals formed by photolysis do not seem to be in the excited state. If one assumes the activation energy for the reaction (95) to be 7 kcal mol^{-1}, the activation energy of decarboxylation (equation 94) of aroyloxy radicals can be estimated as shown in Table 9[141–143].

The rates of addition of benzoyloxy radicals to olefins decrease in the order of 2,5-dimethylstyrene > styrene > vinyl acetate > methyl methacrylate ≫ acrylonitrile, revealing that the electron-donating ability of the olefin controls its reactivity toward the benzoyloxy radical[139].

Bevington and Allen have also found that polymerization of styrene is initiated mainly by *t*-butoxy radicals when methyl-^{14}C-labelled *t*-butyl peroxides are used at 80° and 130°C[25,144]. At the higher temperature the fraction of β scission of *t*-butoxy radicals (equation 67) increases. Dilution with a hydrocarbon solvent increases the fraction of hydrogen abstraction from the solvent by *t*-butoxy radicals and decreases the efficiency of polymerization. The rate of ^{12}C–^{14}C bond fission is found to be approximately 7% less than that of ^{12}C–^{12}C bond fission (equation 67).

B. Unimolecular Heterolytic Decomposition

1. Peresters

Transformation of *trans*-9-decalyl perbenzoate to 1-benzoyloxy-1,6-epoxycyclodecane was shown to be an intramolecular heterolytic rearrangement, first by Criegee and Kaspar[145], and later by Bartlett and Kice[146a] and Goering and Olson[146b], on the basis of solvent, salt and substituent effects. The elegant ^{18}O tracer work by Denney and Denney clearly reveals that in the rearrangement the ^{18}O label in the carbonyl group is nearly completely retained as shown in equation (99), pointing conclusively to a route through an intimate ion pair (**21**)[36].

AcOH—MeOH

(**21**)

(99)

2. Diacyl peroxides

The thermal decomposition of *p*-methoxy-*p*′-nitrobenzoyl peroxide in polar media has been suggested by Leffler to proceed mainly through heterolytic carboxy-inversion (equation 79), since both the rate of decomposition and the amount of carboxy-inversion product increases more in a more polar solvent[147]. Denney and his coworkers have carried out ^{18}O tracer studies using peroxides in which both (α and δ) carbonyl oxygen atoms

TABLE 9. Reactivities of aroyloxy radicals in polymerization of styrene

	Temp. (°C)	$PhCO_2\cdot$[a]	p-$MeOC_6H_4CO_2\cdot$[b]	m-$MeOC_6H_4CO_2\cdot$[c]	$CH_3CO_2\cdot$
k_1/k_2 (mol l^{-1})[f]	60	0.4	0.022	0.3	—
	80	0.7	0.053	0.6	—
$E_2 - E_p$ (kcal mol^{-1})		6.6	10.5	8.3	—
E_2 (kcal mol^{-1})[g]		ca. 14	ca. 17	ca. 15	ca. 6.6[d]
k_2 (s^{-1})	60	7,400	44		1.9×10^9[d]
	80	24000	200		—
	110	—	—		2.5×10^9[e]

[a] Ref. 141.
[b] Ref. 142.
[c] Ref. 143
[d] Ref. 87.
[e] Refs. 72 and 124.
[f] E_2 and E_p are the activation energies for the reactions in equations (94) and (95), respectively.
[g] k_1 and k_2 denote the rate constants for the homolysis of the O—O bond of $(ArCO_2)_2$ and decarboxylation of acyloxy radicals, respectively.

were ^{18}O-labelled[36,73]. In the carboxy-inversion products, obtained in thionyl chloride, the a-oxygen contained no ^{18}O while the d-oxygen retained only 66% of the original ^{18}O label. Without further examination of the distribution of the remaining ^{18}O in the other oxygen atoms, they postulated the ionic mechanism shown in equation (100)[36,73].

$$CH_3O-C_6H_4-\overset{\alpha}{\overset{*}{O}}=C-\overset{\beta}{O}-\overset{\gamma}{O}-C(=\overset{\delta}{\overset{*}{O}})-C_6H_4-NO_2 \longrightarrow [\ldots] \longrightarrow CH_3O-C_6H_4-\overset{a}{O}-C(=\overset{b}{O})-\overset{c}{O}-C(=\overset{d}{\overset{*}{O}})-C_6H_4-NO_2 \quad (100)$$

Treatment of benzoyl peroxide with $SbCl_5$ at room temperature gives phenyl benzoate. Here again, Denney's group have shown that the original carbonyl ^{18}O label does not appear in the ether oxygen of the phenyl benzoate and have suggested the mechanism shown in equation (101)[148]. The carboxy-inversion product has often been detected during thermal decomposition of symmetrical secondary and tertiary alkaneformyl peroxides[75,149]. Thus the carboxy-inversion has been found to be the main route of many thermal decompositions, especially of the highly polarized Case III peroxide, but also of Case I_B and Case II_B peroxides, even in non-polar solvents. These inversion products, acyl alkyl carbonates, usually decompose to the corresponding esters and carbon dioxide under the reaction conditions, and the yield of the esters is always high[39,149].

$$Ph-C(=\overset{*}{O})-O-O-C(=\overset{*}{O})-Ph \xrightarrow{SbCl_5} [\ldots] \longrightarrow [\ldots] \longrightarrow Ph-C(=\overset{*}{O}^{+}\bar{S}bCl_5)-O-C(=\overset{*}{O})-O-Ph \longrightarrow [\ldots] \longrightarrow Ph-C(=\overset{*}{O})-O-Ph + CO_2 + SbCl_5 \quad (101)$$

Thermal decomposition of optically active β-phenylisobutyryl peroxide in CCl_4 affords 60% of 1-phenyl-2-propyl β-phenylisobutyrate in which the alkyl group retains 75% of the optical activity. DeTar and Weis[150] suggested a free-radical mechanism in which the recombination of the optically active 2-phenylisopropyl and β-phenylisobutyloxy radicals in the solvent cage is so rapid that the stereointegrity of the alkyl group is retained in the resulting ester. The reaction was, however, later found to proceed mainly through carboxy-inversion, as shown by Kashiwagi, Kozuka and Oae[39] who carried out the decomposition with ^{18}O in the carbonyl group and found that ^{18}O label in the resulting optically active ester was retained to nearly the same extent (79–89%) as the optical activity (92%) (equation 102). Homolytic oxygen scrambling in the starting peroxide and that in the intermediary carbonate could be responsible for the small difference between the two values. Other secondary diacyl peroxides have been shown to decompose similarly, whereas in primary diacyl peroxides the major path involves homolytic O—O bond cleavage and the corresponding esters are obtained in only 16% for acetyl peroxide and 30% for δ-phenylvaleryl peroxide[39]. Thus Case I_B and Case II_B diacyl peroxides always give esters in high yields, while Case I_A and Case III diacyl peroxides afford the esters in poor yields[40].

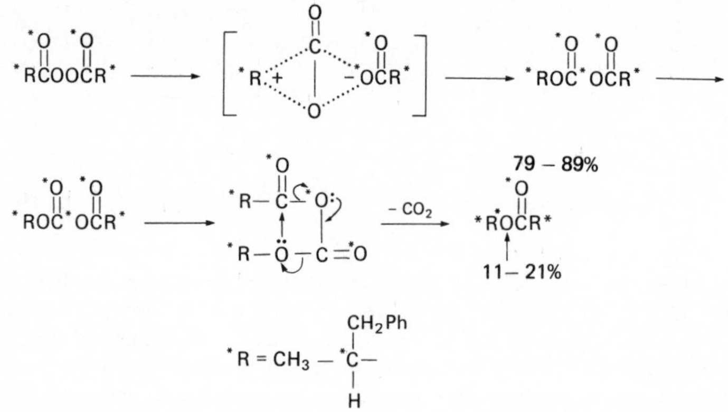

*RC(=*O)OOC(=*O)R* ⟶ [*R·+ C(=O)(O) ⁻*OC(=O)R*] ⟶ *ROC(=*O)*OC(=*O)R* ⟶

*ROC(=*O)*OC(=*O)R* ⟶ *R—C(=*O)—O: / *R—Ö—C=*O ⟶ (− CO_2) *ROC(=*O)R* (79 – 89%; 11 – 21%) (102)

*R = CH_3—*C(H)(CH_2Ph)—

In the Lewis acid-catalysed decomposition, e.g. with optically active ^{18}O-labelled β-phenylisobutyryl peroxide in the presence of $SbCl_5$ at room temperature in light petroleum, the carbonyl-^{18}O of the peroxide is retained 83–89% in the carbonyl group of the resulting ester while the alcohol portion of the ester shows 91% retention of configuration. Thus the thermolysis clearly proceeds through the carboxy-inversion process[151]. In the Baeyer–Villiger reaction of optically active 2-phenyl-2-propyl methyl ketone, the alcohol moiety of the resulting ester retains its original configuration nearly completely[152,153]. Oae, Fujimori and Kozuka[74] carried out thermolysis of 1-apocamphoryl benzoyl peroxide labelled with ^{18}O at either of the two carbonyl groups separately and found that ^{18}O in both cases was distributed equally into three oxygens, i.e. b, c and d of the carboxy-inversion product (equation 103). This observation appears to cast doubt on the mechanism[36,73] shown in equation (100). However, when 1-apocamphyl benzoyl carbonate labelled with ^{18}O either at c- or at d-oxygens was heated under the same conditions, in both cases ^{18}O was found to be equilibrated completely into the b-, c- and d-oxygens, proving complete scrambling in the carboxy-inversion product[154]. Thus, it was necessary to find a diacyl peroxide which undergoes thermolysis much faster than the oxygen scrambling in the carboxy-inversion product[55]. Decomposition of benzoyl cyclohexaneformyl peroxide in 2 M sulpholane–CCl_4, a fairly polar but aprotic medium, at 45°C, was found to be such a case. Here, the carboxy-inversion (equation 104) proceeds at a rate of $4.25 \times 10^{-4} s^{-1}$, while the separate experiment using benzoyl cyclohexyl carbonate labelled with ^{18}O at c- and d-oxygens reveals that the ^{18}O equilibration

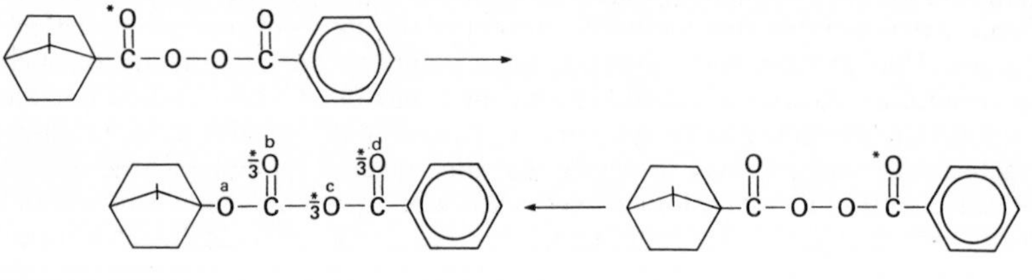

(apocamphyl)—C(=*O)—O—O—C(=O)—Ph ⟶

(apocamphyl)—$\overset{a}{O}$—C(=$\overset{b}{O}$, *⅓)—$\overset{c}{O}$(*⅓)—C(=$\overset{d}{O}$, *⅓)—Ph ⟵ (apocamphyl)—C(=O)—O—O—C(=*O)—Ph (103)

between b- and c-oxygens in benzoyl cyclohexyl carbonate (equation 105) takes place much more slowly ($k_e = 3.39 \times 10^{-5}\,s^{-1}$) than the decomposition of the peroxide; however, neither the decomposition of the carbonate nor equilibration of d-oxygen into the other two oxygens occurs at 45°C. At 75°C, oxygen scrambling among b-, c- and d-oxygens takes place smoothly; however, a-oxygen is not equilibrated with the other three oxygens (equation 105). ^{18}O distribution in the inversion products obtained from α and δ ^{18}O-labelled benzoyl cyclohexaneformyl peroxides, respectively, are shown in equations (106) and (107)[55]. The results clearly show that the carboxy-inversion proceeds by the path shown in equation (108), accordingly the Denney's hypothesis of the carboxy-inversion is still the most plausible[55].

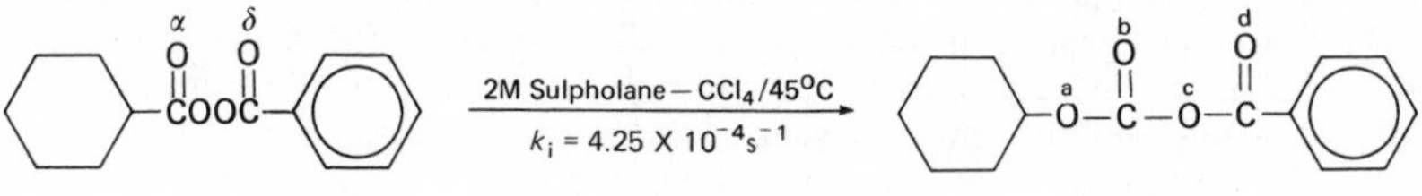

2M Sulpholane—CCl_4/45°C, $k_i = 4.25 \times 10^{-4}\,s^{-1}$ (104)

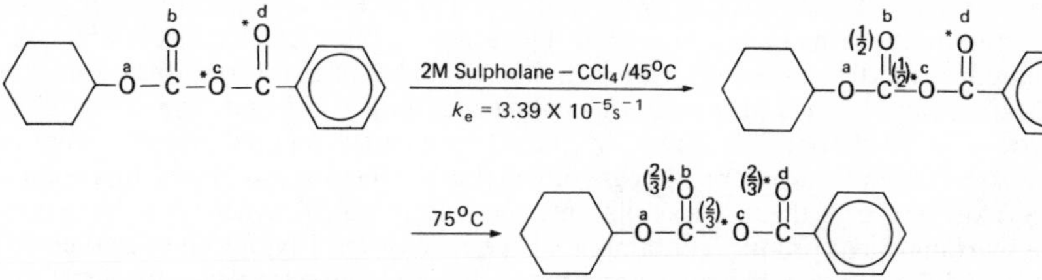

2M Sulpholane—CCl_4/45°C, $k_e = 3.39 \times 10^{-5}\,s^{-1}$; 75°C (105)

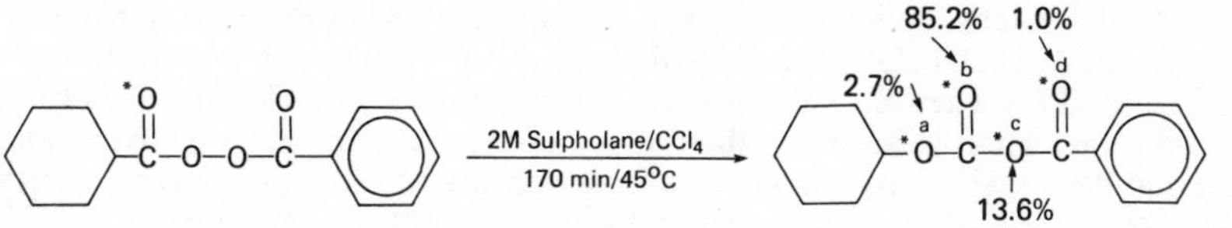

2M Sulpholane/CCl_4, 170 min/45°C: 85.2% (b), 1.0% (d), 2.7% (a), 13.6% (c) (106)

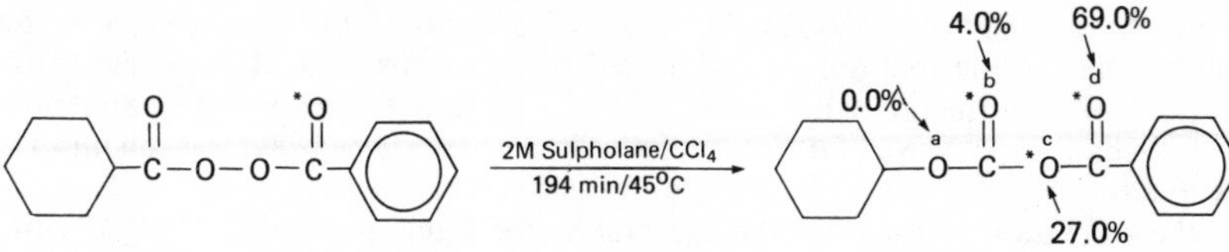

2M Sulpholane/CCl_4, 194 min/45°C: 4.0% (b), 69.0% (d), 0.0% (a), 27.0% (c) (107)

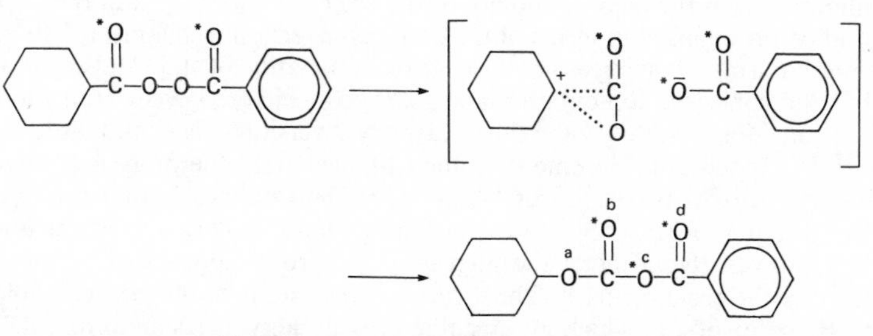

(108)

The amount of ^{18}O retention in the b-oxygen was found to be 85.6% (equation 106) instead of 100% as expected. However, since ^{18}O-scrambling takes place between the b- and c-oxygens at a fairly low temperature and among the b-, c- and d-oxygens at higher temperatures[165] (equation 105), one has to be careful in the evaluation of ^{18}O tracer experimental data in the mechanistic diagnosis of the thermolysis of diacyl peroxides[55]. Denney and coworkers[36,73] have been quite fortunate in that there was not much oxygen equilibration between the b- and c-oxygens in the carboxy-inversion product, as they did not analyse the ^{18}O content in the b- and c-oxygens.

One of the most important results in these experiments is that there is no ^{18}O-equilibration between the a-oxygen and the remaining three b-, c- and d-oxygens of the carbonate, once the carbonate has been formed[154]. Therefore, if there is any ^{18}O in the a-oxygen in the inversion product of a carbonyl ^{18}O-labelled peroxide, the oxygen scrambling must have taken place during the thermolysis of the peroxide, very probably within the diacyl peroxide itself and not after the rearrangement. Indeed, oxygen scrambling within diacyl peroxide is quite common and there are numerous examples in which ^{18}O of a carbonyl group is incorporated into the a-oxygen of the inversion product, as shown in Table 10. The data clearly show that incorporation of ^{18}O from carbonyl oxygen of the starting peroxide into a-oxygen of the inversion product is the highest with Case I peroxides, decreasing with Case II peroxides and further decreasing with Case III peroxides. One peculiar case is dicyclopropaneacetyl peroxide[41], which shows very little oxygen scrambling during thermolysis and yet gives a rearrangement product in which the a-oxygen contains 20% of the original carbonyl ^{18}O label of the starting peroxide.

Walling's group[155] have proposed a mechanism shown in Scheme 5, in which both homolytic and heterolytic cleavages of the O—O bond proceed via a common intermediate. Oae and coworkers[41] have assumed that the thermolysis fo dicyclopropane-acetyl peroxide proceeds through a similar intermediate, in which oxygen scrambling may give the rearrangement product containing labelled a-oxygen. Taylor and coworkers[156] have obtained the ring-contracted cyclopropylmethyl cyclobutaneformyl carbonate and the ring-opened 3-butenyl cyclobutaneformyl carbonate, as well as the normal inversion product, cyclobutyl cyclobutaneformyl carbonate, in the thermolysis of cyclobutane-formyl peroxide (**19**$_4$). They have also isolated the ring-expanded cyclobutyl derivative, though in a rather small yield, in the thermolysis of cyclopropaneacetyl peroxide, and suggested that these rearranged products are formed through the carboxy-inversion after the Wagner–Meerwein rearrangement within the ion pair (**22**) as shown in Scheme 5[156].

This mechanism may also explain the ^{18}O incorporation into the a-oxygen in the inversion product in the thermolysis of labelled dicyclopropaneacetyl peroxide. More investigations without the use of isotopes[149a,156–159] have attempted to substantiate the proposed mechanism; however, further studies will be necessary before this mechanism is fully accepted.

Thus, the formation of esters in the thermolysis of diacyl peroxides[150,160–163], the retention of configuration in the alcohol moiety of the ester[150,160,162,163] and the ^{18}O tracer data may all be rationalized in terms of the carboxy-inversion mechanism[39,41,74], and rule out the mechanisms involving either the cage recombination of alkyl and acyloxy radicals as well as the concerted ionic path via a polar six-membered cyclic transition state[164]. The transition state of the carboxy-inversion is undoubtedly polar[39,147,148,151–159]. Indeed, the Hammett ρ value, obtained in the thermolysis of ring-substituted isobutyryl benzoyl peroxide in CCl_4 is $+1.0$, showing clearly that the rate increases as the leaving $R'COO^-$ group becomes more polar and electron-withdrawing[158a]. However, the effect of the migrating R^+ group appears to be more decisive in controlling the reaction path. Thus, carboxy-inversion should be negligible with Case I_a and II_a peroxides, in which an unstable primary alkyl carbenium ion, R^+, would be involved in the ion-radical pair intermediate, while with Case I_B and II_B peroxides the secondary alkyl groups would stabilize the ion-radical intermediate substantially, presumably increasing the amount of carboxy-inversion products. With tertiary R^+ ions, the intermediate would be so polar that it would undergo facile

TABLE 10. Incorporation of ^{18}O into the a-oxygen of the inversion product and k_d/k_{ts} values in the thermal decomposition of carbonyl-^{18}O-labelled diacyl peroxides, $RC(=O)OOC(=O)R'$

R	R′	Case	Solvent	Temp. (°C)	^{18}O incorporation into the a-oxygen of the inversion product	k_d/k_{ts} [a]	Ref.
	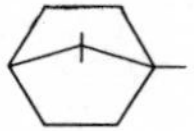	I_B	CCl_4	80	22	1.79	40
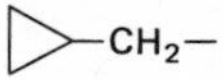 $—CH_2—$	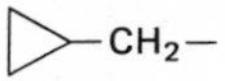 $—CH_2—$	I_B	CCl_4	45	20	∞	41
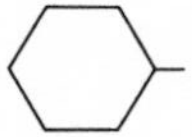	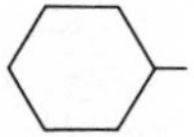	I_B	CCl_4 2M sulpholane–CCl_4	45 45	4 5	101 >25	217
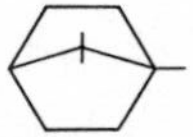	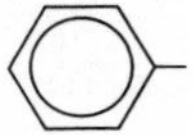	II_B	CCl_4	80	3	—	74
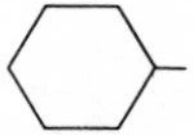	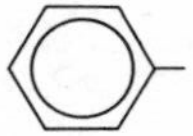	II_B	CCl_4 Nujol	45 45	2 2.5	— —	55, 217
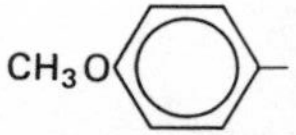 CH_3O	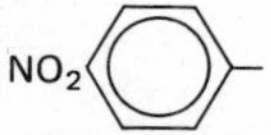 NO_2	III	$SOCl_2$	Reflux	0	—	73

[a] k_d is the rate constant for total disappearance of diacyl peroxide and k_{ts} is the rate constant for total oxygen scrambling in the remaining diacyl peroxide during thermolysis of carbonyl-^{18}O-labelled peroxide (see equation 74).

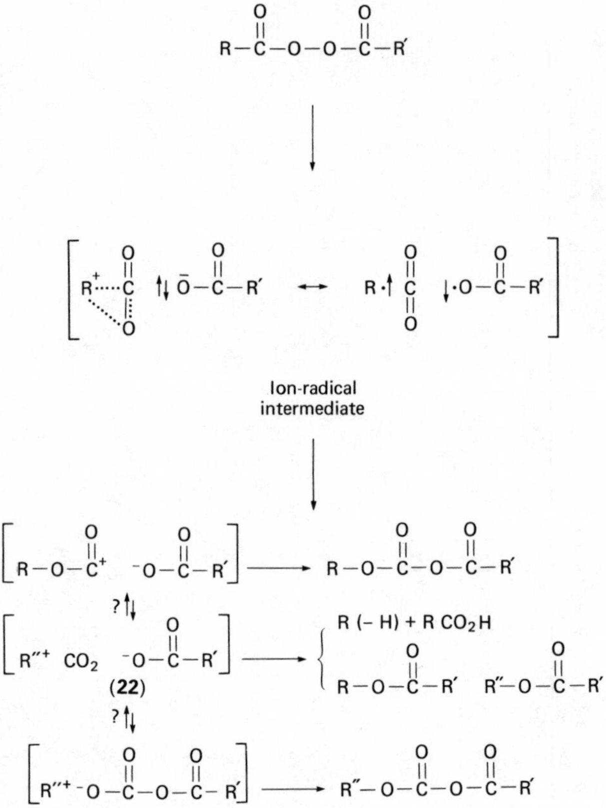

$$R-\overset{O}{\overset{\|}{C}}-O-O-\overset{O}{\overset{\|}{C}}-R'$$

$$\downarrow$$

$$\left[R^{+}\cdots C(=O)\cdots O \;\; \uparrow\downarrow \; \bar{O}-\overset{O}{\overset{\|}{C}}-R' \;\; \longleftrightarrow \;\; R\cdot\uparrow \; O=C=O \; \downarrow\cdot O-\overset{O}{\overset{\|}{C}}-R' \right]$$

Ion-radical
intermediate

$$\downarrow$$

$$\left[R-O-\overset{O}{\overset{\|}{C}}{}^{+} \;\; {}^{-}O-\overset{O}{\overset{\|}{C}}-R' \right] \longrightarrow R-O-\overset{O}{\overset{\|}{C}}-O-\overset{O}{\overset{\|}{C}}-R'$$

$$?\uparrow\downarrow$$

$$\underset{(\mathbf{22})}{\left[R''^{+} \;\; CO_2 \;\; {}^{-}O-\overset{O}{\overset{\|}{C}}-R' \right]} \longrightarrow \begin{cases} R(-H) + RCO_2H \\ R-O-\overset{O}{\overset{\|}{C}}-R' \quad R''-O-\overset{O}{\overset{\|}{C}}-R' \end{cases}$$

$$?\uparrow\downarrow$$

$$\left[R''^{+} \; {}^{-}O-\overset{O}{\overset{\|}{C}}-O-\overset{O}{\overset{\|}{C}}-R' \right] \longrightarrow R''-O-\overset{O}{\overset{\|}{C}}-O-\overset{O}{\overset{\|}{C}}-R'$$

SCHEME 5

heterolysis to yield carbon dioxide and a stable carbenium ion which would give either a mixture of olefin and acid or an ester. Therefore, the yield of the rearranged product would be small, as indeed was found to be the case. With Case III peroxides, only highly polarized ones, such as *p*-methoxy-*p'*-nitrobenzoyl peroxide, give the carboxy-inversion product. Benzoyl *p*-toluenesulphonyl peroxide, labelled either in the benzoyl or the sulphonyl group, has also been shown[76] to undergo thermolysis through carboxy-inversion to afford phenyl *p*-toluenesulphonyl carbonate.

The ^{18}O equilibration in the resulting acyl carbonate, described in equation (105)[165] is a kind of degenerated acyl migration. Cyclohexyl *para*-substituted-benzoyl carbonate labelled with ^{18}O at the c- and d-oxygens undergoes ^{18}O equilibration readily with the b- and c-oxygens at 45°C in CCl_4 and very slowly with the d-oxygen. However, at 70–75°C, all three b-, c- and d-oxygens are completely equilibrated. The ^{18}O equilibration between the b- and c-oxygens has been shown to be an intramolecular migration by a cross-over experiment, while the effect of polar substituents (X) is nearly nil. An ionic mechanism involving a zwitterion intermediate (**23**) was once suggested by us[165]; however, the lack of

(**23**) X = CH_3, H, CN

effect of polar substituents (X) and the very small effect of solvent polarity on the rate of ^{18}O-scrambling between b- and c-oxygens of cyclohexyl *para*-substituted benzoyl carbonate have led us to reconsider and adopt a new concerted mechanism, illustrated in Figure 3[166]. In this mechanism for the degenerated acyl migration of alkyl aroyl carbonate, the following three sets of σ–π reorganizations take place *concertedly*, even though the interacting orbitals are in a perpendicular orientation to each other, e.g. the orbital of the attacking n electrons of O_b and the breaking σ-bonding orbital (C_B—O_c) are in a perpendicular position to each other: (*i*) The n electrons (lone pair) of O_b attack nucleophilically the carbonyl carbon, C_B, interacting with π orbital (C_B—O_d) to form a new σ bond ($O_{b'}$—$C_{B'}$) and a lone pair on $O_{d'}$. (*ii*) The n orbital of O_d interacts with the anti-σ- bonding orbital (C_B—O_c) to form a new π-bonding orbital ($O_{d'}$—$C_{B'}$), which is in perpendicular orientation to the old π-bonding orbital (C_B—O_d) to be cleaved, with

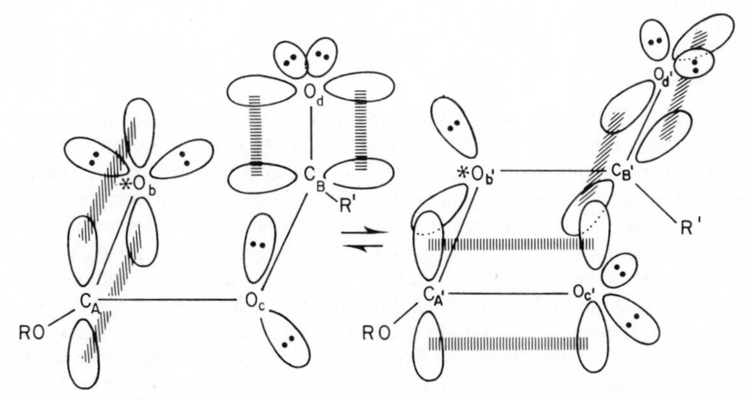

FIGURE 3. Stereoelectronic course of the concerted ^{18}O scrambling in alkyl aroyl carbonate[166].

breaking of the σ bonding ($C_B—O_c$). In this process the old σ-bond ($C_B—O_c$) electrons become n electrons of $O_{c'}$. (*iii*) The n electrons of O_c interact with the anti-π-bonding ($C_A—O_b$) electrons to form a new π bond ($O_{c'}—C_{A'}$) with a conversion of π electrons ($C_A—O_b$) to n electrons of $O_{b'}$.

A similar concerted σ–π reorganization was suggested earlier by Oae and coworkers[167] in the concerted nucleophilic substitution on the sp^2 nitrogen atom of alkyl nitrites (RON=O), which is still bearing an electron pair on the nitroso-oxygen atom and can form a new π bond when the alkoxy group leaves at a quasi-perpendicular position to both the old nitroso π orbital and the attacking lone pair of the nucleophile[167]. The MINDO MO calculation performed by Kikuchi also seems to support this concept[168]. A similar acyl migration following the stereoelectronic course shown in Figure 3 may be found in the rearrangement of *N*-nitroso-*N*-acyl-*O*-*t*-butylhydroxylamine to *O*-acyl *O'*-*t*-butyl hyponitrite[82] and the rearrangement of *N*-nitrosoamine to an acyloxy alkyldiazene shown in equation (109)[169].

$$R-C(=O)-N(-N=O)-R' \longrightarrow R-C(=O)-O-N=N-R' \quad (109)$$

R = alkyl, aryl; R′ = alkyl, *t*-BuO

A few other similar acyl migrations have also been reported by Curtin and Miller[170] and McCarthy and Hegarty[171]. In the rearrangement shown in equation (110), the rate of the rearrangement was measured with various R groups and the Hammett ρ_X values obtained were as small as $\rho_X = +0.59$ (R = *p*-nitrophenyl) and $+0.65$ (R = PhO), while the rate was quite insensitive to the polarity of the medium used. This may be another example of concerted nucleophilic substitution process in which the forming new bond and the cleaving old bond are oriented at a quasi-perpendicular position.

$$X-C_6H_4-C(=O)(-O-C(Ar)=N-R) \longrightarrow X-C_6H_4-C(=O)-N(R)-C(=O)-Ar \quad (110)$$

A few other intramolecular rearrangements also seem to proceed through a similar stereoelectronic course[172].

3. Ozonides

Ozonization of olefins has been investigated extensively by Criegee who suggested in 1949 his famous mechanism[173,174]. The mechanism involves the initial formation of a ‘primary ozonide’ (**24**) which decomposes to aldehyde and carbonyl oxide. The latter two recombine to form the ‘secondary ozonide’ (**25**). This mechanism was confirmed by a cross-over experiment which revealed that foreign aldehyde could be incorporated into the secondary ozonide. In 1966 Story, Murray and Youssefyeh (SMY) proposed a different mechanism for the formation of ozonides[175,176]. Details of the two arguments are well documented[174] and beyond the scope of this chapter, and we shall only discuss studies with ^{18}O tracers. Scheme 6 shows that using ^{18}O-labelled aldehyde, Criegee's mechanism demands ^{18}O incorporation into the ether oxygen of the ozonide, while in the SMY mechanism ^{18}O appears in the peroxidic oxygen. However, the experiment is by no means simple[174] due partly to the lack of an adequate method for the differential analysis of the

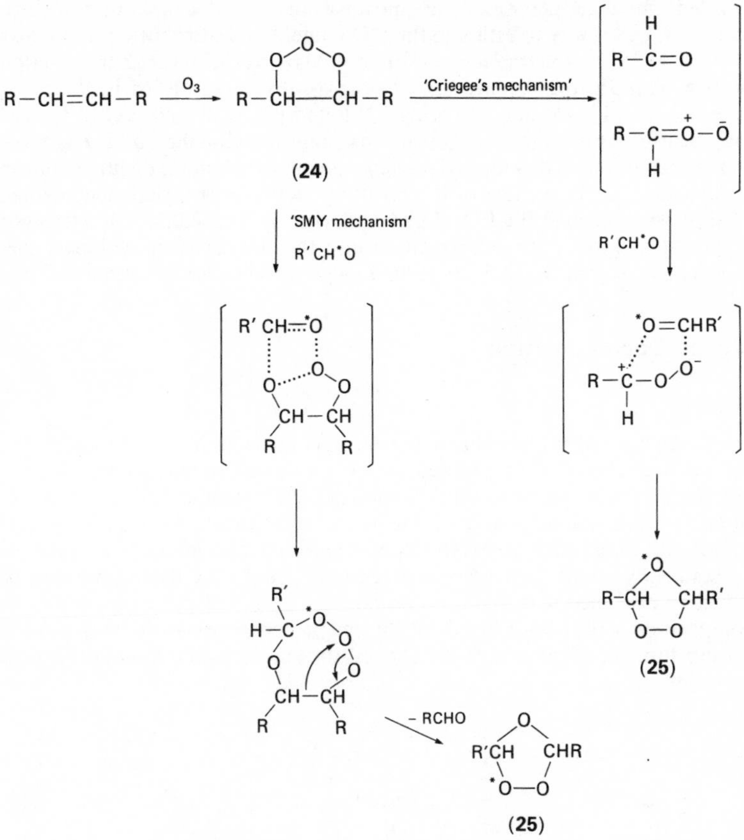

SCHEME 6. Mechanisms for ozonide formation[174].

^{18}O content of ether and peroxide oxygens, and partly to the occurrence of ^{18}O exchange between aldehyde and ozonide[177] as well as the facile oxygen exchange of the carbonyl function with various oxy and oxo functions. At first, Story and coworkers[178] carried out ozonolysis of *trans*-1,2-diisopropylethylene in the presence of acetaldehyde-^{18}O, and reduced the resulting labelled methylisopropylethylene ozonide with $LiAlH_4$ or $LiCH_3$ to ethanol and isobutyl alcohol, both of which were analysed by mass spectrometry. From these data they calculated that 68–77% of the ozonide was formed by a path which placed ^{18}O at the peroxide site. Later, Murray and Hagen[179] reinvestigated this ozonolysis in the presence of isobutyraldehyde-^{18}O and reported that incorporation of ^{18}O into the ether and peroxide sites varied with the temperature, e.g. the SMY mechanism contributed 40% at −120°C and 20% at 0°C. Unfortunately these values were later found to be unreliable due to ^{18}O interchange between aldehydes[175,176] and to the enrichment of ^{18}O in ozone by exchange with ^{18}O-aldehyde[177], etc. Thus, the direct mass spectroscopic analysis of the ozonide was performed[180]. Lattimer and Kuczkowski carried out ozonolysis of *cis*- and *trans*-diisopropylethylene in the presence of ^{18}O-labelled acetaldehyde and found that an upper limit of 10% of the original ^{18}O in acetaldehyde added was estimated to be incorporated into the peroxide site, most of the ^{18}O being found in the ether fragment[180]. A better way to estimate the ^{18}O content and distribution in the ozonide is the use of Ph_3P[181,182] which selectively attacks the peroxidic oxygens, the reaction being quantitative. Thus, in the ozonization of phenyl-substituted olefins in the presence of benzaldehyde-^{18}O, Fliszár and coworkers[183] found practically no ^{18}O in the peroxidic oxygens, clearly supporting the Criegee mechanism. Gallaher and Kuczkowski[184] repeated the ozonolysis of diisopropylethylene, as a model of simple olefins, in the presence of acetaldehyde-^{18}O. Treatment of the resulting methylisopropylethylene ozonide-^{18}O with Ph_3P gave unlabelled Ph_3PO. Higley and Murray[185] used the same technique and finding almost exclusive ^{18}O incorporation into the ether oxygen, withdrew their own SMY mechanism so that only the original Criegee mechanism has remained valid.

C. Induced Decomposition

1. *Peresters*

Decomposition of *t*-butyl perbenzoate bearing a nucleophilic substituent at the *ortho* position has been shown[186] to proceed by several powers of ten faster than that of the unsubstituted *t*-butyl perbenzoate. Obviously the decomposition is induced by neighbouring-group participation of the *ortho* substituent. An especially high rate acceleration was found with the *ortho* sulphenyl group which interacts with the peroxide bond. These early studies are well-documented[187], and only those involving isotopic tracers will be discussed here.

Decomposition of the *t*-butyl *ortho*-substituted perbenzoate is 167 times faster than that of the unsubstituted derivative at 60°C. The resulting cyclic ester is found to retain 88% of

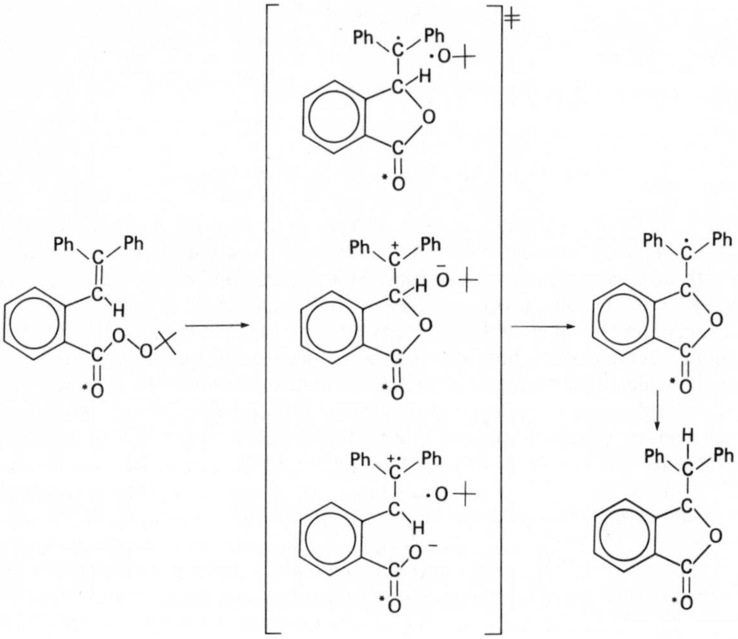

(111)

^{18}O at the carbonyl group, suggesting clearly that the homolysis of the O—O bond is induced by neighbouring-group participation involving the double bond, in the transition state as illustrated by the contributing resonance structures in equation (111)[186].

The decompositions of *o*-methanesulphenyl and *o*-benzenesulphenyl derivatives are 2–3 × 10^4-fold faster than that of the parent compound and are considered to proceed through a polar transition state forming sulphuranyl and *t*-butoxy radicals which eventually give the cyclic esters[188]. The formation of the sulphuranyl radical has been confirmed by CIDNP spectra[189] in the thermolysis of the peroxide and also by ESR signals[190] of deuterated sulphuranyl radicals in the decomposition of *t*-butyl deuterated perbenzoates. The radical formed is a σ radical with spin density mainly concentrated at the sulphur atom as shown below. In a further experiment with ^{17}O-labelled *t*-butyl *o*-methanesulphenyl perbenzoate, the distribution of ^{17}O in the resulting ester was determined by ^{17}O-NMR and the neighbouring-group participation of the *o*-sulphenyl group was confirmed (equation 112)[50].

Decomposition of ring-substituted *t*-butyl perbenzoates induced by triphenylphosphine is a bimolecular reaction with a Hammett ρ value of $+1.24$. The experiment with carbonyl-^{18}O-labelled *t*-butyl perbenzoate shown in equation (113) and the rate-enhancing effect of polar solvents was taken to suggest the mechanism shown, which involves nucleophilic attack of the trivalent phosphine on the peroxidic oxygen[191].

(112)

$$XC_6H_4\overset{\overset{*}{O}}{\overset{\|}{C}}-O-O-CMe_3 + Ph_3P \longrightarrow Ph_3P\begin{cases}O-CMe_3\\ O-\underset{\underset{O}{\|}}{C}C_6H_4X\end{cases} \longrightarrow Ph_3P\cdots\begin{matrix}O\cdots CMe_3\\ O\\ |\\ C(=O)C_6H_4X\end{matrix} \tag{113}$$

$$\longrightarrow Ph_3P{=}O + Ph-\overset{\overset{*}{O}\ (60-80\%)}{\overset{\|}{C}}-{}^{*}O-CMe_3\ (20-40\%)$$

In the reaction between *t*-butyl performate and base, Pinock found that there is a concerted decarboxylative 1,2-elimination in which k_H/k_D was found to be 4.1 and the rate was enhanced in polar media (equation 114)[192].

$$B{:}\frown \underset{(D)}{H}-\overset{\overset{O}{\|}}{C}-O-O-CMe_3 \longrightarrow \left[\underset{(D)}{BH^+}\quad CO_2\quad {}^{-}O-CMe_3\right] \longrightarrow B{:} + CO_2 + \underset{(D)}{HO}-CMe_3 \tag{114}$$

2. *Diacyl peroxides*

Diacyl peroxides undergo facile decomposition, induced by various free radicals and also by nucleophiles. For example, decomposition of diacyl peroxide is known to be induced by free radicals generated during the decomposition of themselves[193], and also by nucleophiles such as amines[194], sulphides[195], trivalent phosphorus compounds[196], phenols[197], olefins or other electron-rich compounds bearing π or σ lone-pair orbitals. The decomposition is usually a bimolecular reaction and takes place at relatively low temperatures. The groups of Horner[194–196], Bartlett[193], Greene[198] and Walling[197,199] have studied this interesting induced decomposition which proceeds through a transition state having both ionic and free-radical nature. Only studies involving isotopes are discussed in this section.

Denney and Freig[200] and Doering and coworkers[54] have independently reported that the attack of carbon radicals on carbonyl ^{18}O-labelled benzoyl peroxide takes place at the peroxide site, resulting in the formation of ^{18}O-scrambled ester, presumably via intervention of either a benzoyloxy radical or a benzoate anion as an incipient intermediate in the induced decomposition of benzoyl peroxide (equation 115).

$$(Ph\overset{\overset{*}{O}}{\overset{\|}{C}}O-)_2 + \cdot\overset{\overset{Me}{|}}{C}HOEt \longrightarrow \left[Ph\overset{\overset{*}{O}}{\overset{\|}{C}}O^{-}\ {}^{+}\overset{\overset{Me}{|}}{C}HOEt \quad Ph\overset{\overset{*}{O}}{\overset{\|}{C}}O\cdot\ \overset{\overset{Me}{|}}{C}H{=}\overset{+}{O}Et\right] \longrightarrow Ph\overset{\overset{*}{O}\ (70-80\%)}{\overset{\|}{C}}{}^{*}O\overset{\overset{Me}{|}}{C}HOEt\ (20-30\%) \tag{115}$$

$$\left[{}^{*}O{=}C(Ph)-\overset{*}{O}\cdot \longleftrightarrow {}^{*}O{=}C(Ph)-O^{-}\right]$$

In the induced decomposition of carbonyl-^{18}O-labelled benzoyl peroxide with triphenylmethyl radical, the resulting ester retains 85% of the original ^{18}O label, while with 3-cyclohexenyl radical, 70% of the ^{18}O is found in the carbonyl oxygen of the ester formed[54].

Denney and Denney[201] carried out the reaction between carbonyl ^{18}O-labelled benzoyl peroxide and dibenzylamine and obtained *O*-benzoyl-*N*,*N*-diphenylhydroxylamine whose carbonyl oxygen retained all the original ^{18}O label, suggesting that the reaction proceeded via nucleophilic attack by the amine on the peroxidic oxygen of benzoyl peroxide. However, the reaction between the same peroxide and diphenylamine gave *N*-phenyl-*N*-*o*-hydroxyphenylbenzamide in which ^{18}O was distributed 55% in the carbonyl oxygen and 45% in the hydroxyl group. They proposed a mechanism involving *O*-benzoyl-*N*,*N*-diphenylhydroxylamine as the incipient intermediate, formed by attack of the amine nitrogen at the peroxidic oxygen of the peroxide, yielding the product through heterolytic rearrangement (equation 116).

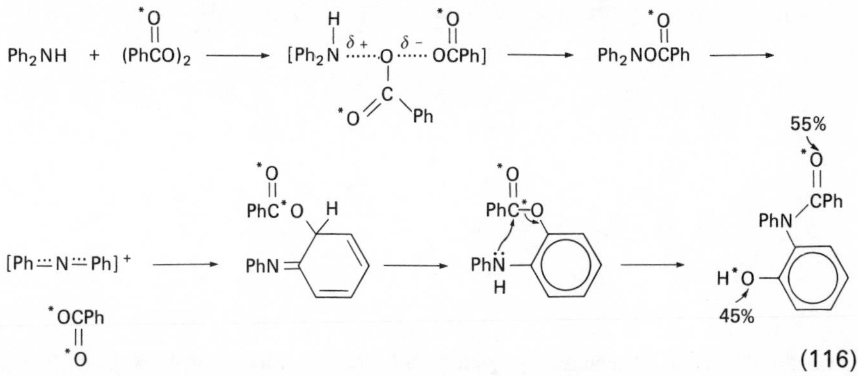

(116)

The induced decomposition of carbonyl-^{18}O-labelled benzoyl peroxide with *p*-cresol in benzene gave 2-hydroxy-4-methylphenyl benzoate in which the carbonyl oxygen retained 87% of the ^{18}O label while the 4-methylcatechol formed by the acid hydrolysis of the product contained 13% of the ^{18}O label. The ionic mechanism shown in equation (117) has been proposed[201]. Carbonyl-^{18}O-labelled benzoyl peroxide and triphenylphosphine gave unlabelled triphenylphosphine oxide and benzoic anhydride containing 75% of the ^{18}O at the carbonyl group[202]. Reaction of ring-substituted carbonyl-^{18}O-labelled benzoyl peroxides with triphenylphosphine and the ^{18}O distribution in the resulting substituted benzoic anhydride[203] showed clearly that triphenylphosphine attacked preferentially the electron-deficient peroxidic oxygen as shown in **26–28**. Apparently, relative

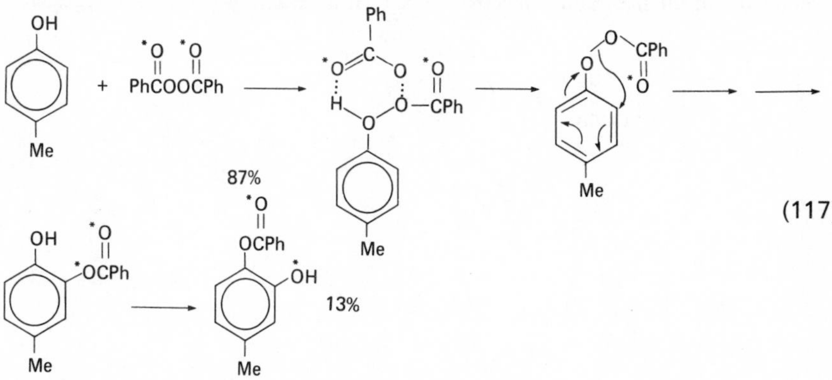

(117)

magnitudes of partial positive charge in the peroxidic oxygens of the peroxides play an important role in orienting the nucleophilic attacks of the phosphines in these S_N2 reactions on the peroxidic oxygens in which the transition state may lie at an early stage of the reaction coordinate[203]. One drawback of this ^{18}O tracer experiment is that no attention was paid to possible ^{18}O scrambling in the resulting acid anhydride, which was found later by us in the mixed anhydride of acyl alkyl carbonate, as illustrated in Figure 3. If there is any ^{18}O scrambling either in the peroxide and/or in the anhydride, all ^{18}O tracer data have to be reassessed.

$$p\text{-MeOC}_6\text{H}_4\text{C(=O)}-\overset{0\%}{\text{O}}-\overset{*}{\text{O}}^{100\%}-\text{C(=O)C}_6\text{H}_4\text{NO}_2\text{-}p$$

(26)

$$\text{C}_6\text{H}_5\text{C(=O)}-\overset{*}{\text{O}}^{15\%}-\overset{*}{\text{O}}^{85\%}-\text{C(=O)C}_6\text{H}_4\text{NO}_2\text{-}p$$

(27)

$$p\text{-PhC}_6\text{H}_4\text{C(=O)}-\overset{*}{\text{O}}^{50\%}-\overset{*}{\text{O}}^{50\%}-\text{C(=O)C}_6\text{H}_5$$

(28)

Carbonyl-^{18}O-labelled benzoyl peroxide readily reacts with sodium dialkyl malonate at the peroxidic site to afford dialkyl α-benzoyloxymalonate in which essentially all the ^{18}O originally present in one carbonyl group is retained (equation 118)[204].

$$\text{PhC}(=\overset{*}{\text{O}})\text{OOC}(=\overset{*}{\text{O}})\text{Ph} + \text{NaC(Ph)(CO}_2\text{Et)}_2 \longrightarrow \text{PhC}(=\overset{*}{\text{O}})\text{OC(Ph)(CO}_2\text{Et)}_2 + \text{PhC}^{*}\text{O}_2\text{Na} \tag{118}$$

The ^{18}O label in the addition product formed from carbonyl-^{18}O-*m,m'*-dibromobenzoyl peroxide with *trans-p,p'*-dimethoxystilbene was found to be completely scrambled and the free-radical mechanism shown in equation (119) was suggested[205].

$$\text{ArC}(=\overset{*}{\text{O}})\text{OOC}(=\overset{*}{\text{O}})\text{Ar} + \text{Ar'(H)C=C(H)Ar'} \longrightarrow \left[\text{ArC}(=\overset{*}{\text{O}})\overset{*}{\text{O}}-\text{C(Ar')(H)}-\dot{\text{C}}\text{(H)Ar'} \quad \cdot\overset{*}{\text{O}}\text{C}(=\overset{*}{\text{O}})\text{Ar}\right] \longrightarrow \text{ArC}(=\overset{*}{\text{O}})\overset{*}{\text{O}}-\text{C(Ar')(H)}-\text{C(H)(Ar')}-\overset{*}{\text{O}}\text{C}(=\overset{*}{\text{O}})\text{Ar} \tag{119}$$

The reaction between carbonyl-^{18}O-labelled phthaloyl peroxide and *trans*-stilbene gave two different products (equation 120)[62]. When **29** was hydrolysed, the resulting diol was found to contain 11% of the ^{18}O label. This is in contrast with the complete ^{18}O scrambling in the open-chain diester. Considering that oxygen scrambling is 42 times faster than thermal decomposition in phthaloyl peroxide, the rather small, i.e. 22%, ^{18}O scrambling may mean that the addition reaction is highly stereospecific and concerted. However, the initial step of the reaction might be a one-electron transfer from the olefin to the peroxide[206]. Greene[207] also postulated a polar mechanism for this addition, on the basis of substantial effects of both polar substituents and polar solvents.

(120)

(29)

In the reaction of carbonyl-^{18}O-labelled *m,m'*-dibromodibenzoyl peroxide with tetraethylethylene, however, Greene and Adam obtained unlabelled tetraethylene oxide[208], suggesting clearly that the reaction is ionic, as illustrated by equation (121).

Benzene 45°C

(75%)

(72%)

(121)

Hisada and coworkers similarly found that triphenylphosphine oxide formed in the reaction of sulphonyl-^{18}O-labelled *m,m'*-dinitrobenzenesulphonyl peroxide with triphenylphosphine did not contain any ^{18}O[209]. In the reaction of sulphonyl-^{18}O-labelled benzoyl *p*-toluenesulphonyl peroxide with various nucleophiles, the attacking site was found to vary with the nucleophile used. Thus, hydrazine and methoxide anion attack carbonyl carbon (site a), phosphine and sulphide attack the peroxidic b-oxygen while Grignard reagents react at the peroxidic c-oxygen of **30** as shown in equation (122)[209].

Ph—C(=O)—O—O—S(=O)(=O)—Tol-*p*

(30)

$$\text{PhC(=O)OS(=}^{*}\text{O)}_2\text{Tol-}p + p\text{-TolMgBr} \longrightarrow \left[\text{Ph—C(=O}\cdots\overset{+}{\text{Mg}}\text{Br)—O—O}\cdots\text{(}p\text{-Tol}^{-}\text{)}\text{—S}^{*}\text{O}_2\text{Tol-}p \right] \longrightarrow \text{Ph—C(=O)OMgBr} + p\text{-Tol—O—S}^{*}\text{O}_2\text{—Tol-}p \qquad (122)$$

In the decomposition of carbonyl-^{18}O-labelled benzoyl *p*-toluenesulphonyl peroxide in anisole, the ^{18}O label was found to be equally distributed in the carbonyl and ether oxygens of the resulting anisyl benzoates. The mechanism shown in equation (123) has been proposed by Kobayashi and coworkers[210].

$$\text{PhC(=}^{*}\text{O)OOTs} \longrightarrow \left[\text{PhC(=}^{*}\text{O)O}^{\delta-}\text{—OTs} \cdots \text{anisole}^{\delta+} \right] \longrightarrow \left[\text{PhC(=}^{*}\text{O)}^{*}\text{O}\cdot\ \bar{\text{O}}\text{Ts}\ \ \text{anisole}^{+\cdot} \right] \longrightarrow \left[\text{PhC(=}^{*}\text{O)}^{*}\text{O}\cdot\ \ \text{(Ts)(H)C}_6\text{H}_5^{\cdot}\text{—OMe} \right]$$

$$\left[\text{PhC(=}^{*}\text{O)}^{*}\text{O}\cdot\ \bar{\text{O}}\text{Ts}\ \ \text{anisole}^{+\cdot} \right] \longrightarrow \text{PhC(=}^{*}\text{O)}^{*}\text{O(H)C}_6\text{H}_5^{+}\text{—OMe} \longrightarrow \text{PhC(=}^{*}\text{O)}^{*}\text{O—C}_6\text{H}_4^{+}\text{—OMe} + \text{HOTs}$$

$$\left[\text{PhC(=}^{*}\text{O)}^{*}\text{O}\cdot\ \ \text{(Ts)(H)C}_6\text{H}_5^{\cdot}\text{—OMe} \right] \longrightarrow \text{TsO—C}_6\text{H}_4\text{—OMe} + \text{PhCO}_2\text{H} \qquad (123)$$

In the decomposition of sulphonyl-^{18}O-labelled *m,m'*-dinitrobenzenesulphonyl peroxide in benzene, however, the ether oxygen in the resulting phenyl *m*-nitrobenzenesulphonate contained only 35–36% of the original ^{18}O label while 64–65% was retained in the sulphonyl group. This led Kobayashi and coworkers to propose a mechanism involving the incipient formation of a loose π-complex[64] (equation 124).

$$\text{ArS(=}^{*}\text{O)}_2\text{OOS(=}^{*}\text{O)}_2\text{Ar} + \text{C}_6\text{H}_6 \longrightarrow \left[\text{ArS(=}^{*}\text{O)}_2^{*}\text{O} \cdots \text{C}_6\text{H}_6^{+} \cdots {}^{*}\text{OS(=}^{*}\text{O)}_2\text{Ar} \right]_{\pi\ \text{Complex}} \longrightarrow \text{ArS(=}^{*}\text{O)}_2^{*}\text{O(H)C}_6\text{H}_6^{+}\ \ \text{Ar—S}^{*}\text{O}_3^{-} \longrightarrow \text{ArS(=}^{*}\text{O)}_2^{*}\text{O—C}_6\text{H}_5 + \text{ArS}^{*}\text{O}_3\text{H} \qquad (124)$$

3. *Miscellaneous*

Edwards and coworkers[211] have investigated the mechanism of the decomposition of peracetic acid labelled with ^{18}O at both peroxidic oxygens. The pH–rate profile and the formations of $^{36}O_2$ and $^{32}O_2$ but not $^{34}O_2$ in the decomposition of a mixture of labelled and unlabelled acids led them to postulate the mechanistic scheme shown in equation (125).

$$CH_3\overset{O}{\overset{\|}{C}}OOH + CH_3\overset{O}{\overset{\|}{C}}{}^{*}O^{*}OH \xrightarrow{-H^+} \left[CH_3C(O^-)(^{*}O^{*}O\cdots H)\text{—}O\text{—}O\text{—}C(=O)CH_3 \right] \longrightarrow \begin{cases} ^{*}O{=}^{*}O \\ CH_3CO_2^- \\ CH_3CO_2H \end{cases} \quad (125)$$

The reaction of perbenzoic acid labelled at both peroxidic oxygens with hydrogen peroxide has been found by Akiba and Simamura[3] to give only $^{32}O_2$ when the reaction is carried out in methanol, suggesting the occurrence of reaction (126); however it gives $^{32}O_2 + {}^{36}O_2$ when water is used as solvent. This is apparently due to an equilibrium between perbenzoic acid and hydrogen peroxide as shown in equation (127).

$$PhC(=O){}^{*}O\text{—}{}^{*}O^- + H\text{—}O\text{—}O\text{—}H \longrightarrow PhC(OH)(={}^{*}O) + O{=}O + {}^{*}\bar{O}H \longrightarrow PhC^{*}O_2H + O_2 + {}^{*}OH^- \quad (126)$$

$$PhC(=O){}^{*}O\text{—}{}^{*}O^- + HOO^- \longrightarrow \left[PhC(O^-)(OO^-)({}^{*}O^{*}OH) \right] \longrightarrow Ph\overset{O}{\overset{\|}{C}}OO^- + H^{*}O^{*}O^- \quad (127)$$

The reaction of *trans*-9-decalyl hydroperoxide with tri-*n*-butylphosphine in ^{18}O-enriched water–ethanol gave *trans*-9-decalol and non-^{18}O-labelled tri-*n*-butylphosphine oxide (equation 128)[212], i.e. the configuration of the resulting alcohol was retained in the induced decomposition in keeping with the result observed by Davies and Feld[213] that an optically active α-phenethyl hydroperoxide was reduced by Ph_3P with retention of configuration of α-phenethyl alcohol. Treatment of cumyl hydroperoxide with Ph_3P in ^{18}O-enriched water–ethanol also gave unlabelled $Ph_3P{=}O$, clearly showing that Ph_3P attacks the terminal peroxidic oxygen linked to hydrogen of the peroxide[212].

$$\text{(trans-9-decalyl)}\text{—}O\text{—}O\text{—}H + {:}P(Bu\text{-}n)_3 \xrightarrow{H_2{}^{*}O/EtOH} \text{(trans-9-decalyl)}\text{—}OH + O{=}P(Bu\text{-}n)_3 \quad (128)$$

The phenol-forming reaction between ^{18}O-labelled phenylboric acid and hydrogen peroxide is an interesting modification of the Baeyer–Villiger reaction (equation 129). The phenol formed contains practically no ^{18}O label[214].

$$\mathrm{Ph{-}B({}^{*}OH)_2 + HOO^- \longrightarrow \left[H{}^{*}O{-}\underset{{}^{*}OH}{\overset{Ph}{B}}{-}OOH \right]^- \xrightarrow{-OH} H{}^{*}O{-}\underset{{}^{*}OH}{B}{-}OPh} \qquad (129)$$

In the Baeyer–Villiger reaction of ^{18}O-labelled benzophenone with perbenzoic acid the carbonyl oxygen is retained completely in the resulting phenyl benzoate (equation 130)[214]. Other experiments using optically active ketones have shown that the migrating groups retain their configuration during the rearrangement[215].

$$\mathrm{Ph{-}\overset{{}^{*}O}{\overset{\|}{C}}{-}Ph + Ph{-}\overset{O}{\overset{\|}{C}}{-}OOH \longrightarrow [\text{cyclic transition state}] \longrightarrow Ph{-}\overset{{}^{*}O}{\overset{\|}{C}}{-}OPh + PhCO_2H} \qquad (130)$$

A similar mechanism has also been suggested for the cleavage of benzil with alkaline hydrogen peroxide (equation 131)[4].

$$\mathrm{Ar{-}\overset{O}{\overset{\|}{C}}{-}\overset{O}{\overset{\|}{C}}{-}Ar + H{}^{*}O{}^{*}O^- \longrightarrow \left[Ar{-}\overset{O}{\overset{\|}{C}}{-}\underset{{}^{*}O{-}{}^{*}OH}{\overset{{}^{*}O^-}{C}}{-}Ar \right] \xrightarrow{-H\overset{*}{O}^-} Ar{-}\overset{O}{\overset{\|}{C}}{-}{}^{*}O{-}\overset{O}{\overset{\|}{C}}{-}Ar} \qquad (131)$$

IV. CONCLUDING REMARKS

The thermolysis of peroxides and their reactions with various nucleophiles as studied by the aid of isotopes and other means may be summarized as follows:

(1) In most cases, the reaction, which appears to be ionic, is accompanied by formation of free-radical species, while the transition state of what appears to be a typical free-radical reaction is very often highly polar in nature. Each of the two peroxidic oxygens has two unshared electron pairs which repulse each other, and both heterolysis and homolysis of the O—O bond should be facile and require similar free energies of activation.

(2) The main route for the thermolysis of aliphatic acyl peroxides with primary R groups involves homolysis of the O—O bond and oxygen scrambling in the acyloxy group, while the extent of oxygen scrambling varies with the change of R group and also with viscosity of the medium. On the other hand, carboxy-inversion, a heterolytic rearrangement, is the main path for the thermolysis of acyl peroxides with secondary and tertiary R groups which can stabilize the R^+ group.

(3) In rearrangements involving peroxidic oxygens, the group usually migrates to that peroxidic oxygen which is relatively less electron-rich. Very often the rearrangement proceeds through heterolytic and homolytic paths concurrently, e.g. in most unimolecular and induced decompositions of peroxides.

(4) Soft nucleophiles attack peroxidic oxygens due to the facile mixing of the LUMO of a relatively low energy level in the O—O bond and the HOMO of the soft nucleophile, which lies in a relatively high energy level.

(5) Of the two peroxidic oxygens of the peroxide, soft nucleophiles preferentially attack the oxygen which has a lower electron density in the ground state.

(6) In the induced decomposition of peroxides with nucleophiles or free radicals, both activation enthalpy and activation entropy are markedly small as compared to those of unimolecular decompositions.

V. REFERENCES

1. D. Staschewski, *Angew. Chem.* (*Intern. Ed.*), **13**, 357 (1974).
2. M. Ambar, *J. Amer. Chem. Soc.*, **83**, 2031 (1961).
3. K. Akiba and O. Simamura, *Tetrahedron*, **26**, 2527 (1970).
4. Y. Sawaki and C. S. Foote, *J. Amer. Chem. Soc.*, **101**, 6292 (1979).
5. (a) R. E. Ball, J. O. Edwards and P. Jones, *J. Inorg. Nucl. Chem.*, **28**, 2458 (1966).
   (b) P. A. Giguere, E. A. Secco and R. S. Eaton, *Discuss. Faraday Soc.*, **14**, 104 (1953).
6. F. A. Cotton and G. Wilkinson, *Advanced Inorganic Chemistry*, 4th ed., John Wiley & Sons, New York–London, 1980, p. 495.
7. R. Hiatt in *Organic Peroxides*, Vol. II (Ed. D. Swern), Wiley–Interscience, New York–London, 1970, p. 1.
8. M. R. Leonov, B. A. Redoshkin and V. A. Shushonov, *Zh. Obshch. Khim.*, **32**, 3959 (1962).
9. (a) K. H. Pausacker, *J. Chem. Soc.*, 3478 (1950).
   (b) T. Tezuka and N. Narita, *J. Amer. Chem. Soc.*, **101**, 7413 (1979).
   (c) T. Tezuka, N. Narita, W. Ando and S. Oae, *J. Amer. Chem. Soc.*, **103**, 3045 (1981).
10. (a) A. Schönberg, G. O. Schenk and O. A. Neumüllar, *Preparative Organic Photochemistry*, Springer-Verlag, New York, 1968, p. 373.
    (b) R. W. Denny and A. Nickon, *Org. Reactions*, **20**, 133 (1973).
    (c) K. Gollnick and H. J. Kuhn in *Singlet Oxygen* (Eds. H. H. Wasserman and R. W. Murray), Academic Press, New York, 1979, p. 287.
11. (a) G. O. Schenk, K. Gollnick and O. A. Neumüller, *Justus Liebigs Ann. Chem.*, **603**, 46 (1957).
    (b) G. O. Schenk, *Angew. Chem.*, **69**, 579 (1957).
    (c) G. O. Schenk and K.-H. Schlute-Elte, *Justus Liebigs Ann. Chem.*, **618**, 185 (1958).
    (d) L. M. Stephenson, D. E. McClure and P. K. Sysak, *J. Amer. Chem. Soc.*, **95**, 7888 (1973).
12. (a) R. W. Murray in *Singlet Oxygen*, (Eds. H. H. Wasserman and R. W. Murray), Academic Press, New York, 1979, p. 59.
    (b) P. D. Bartlett and A. P. Schaap, *J. Amer. Chem. Soc.*, **92**, 6055 (1970).
13. (a) C. S. Foote and S. Wexler, *J. Amer. Chem. Soc.*, **86**, 3879 (1964).
    (b) C. S. Foote, S. Wexler and W. Ando, *Tetrahedron Letters*, 4111 (1965).
    (c) C. S. Foote, S. Wexler, W. Ando and R. Higgins, *J. Amer. Chem. Soc.*, **90**, 975 (1968).
    (d) C. S. Foote, *Accounts Chem. Res.*, **1**, 104 (1968).
14. (a) A. E. Cahill and H. Taube, *J. Amer. Chem. Soc.*, **74**, 2312 (1952).
    (b) M. Kasha and A. Kahn, *Ann. N. Y. Acad. Sci.*, **171**, 5 (1970).
15. H. H. Wasserman, F. J. Vinick and Y. C. Chang, *J. Amer. Chem. Soc.*, **94**, 7180 (1972).
16. K. R. Kopecky and J. H. van de Sande, *Can. J. Chem.*, **50**, 4034 (1972).
17. N. Furutachi, Y. Nakadaira and K. Nakanishi, *J. Chem. Soc., Chem. Commun.* 1625 (1968).
18. A. A. Frimer, P. D. Bartlett, A. F. Boschung and J. G. Jewett, *J. Amer. Chem. Soc.*, **99**, 7977 (1977).
19. (a) A. G. Davies, *Organic Peroxides*, Butterworth, London, 1961.
    (b) A. G. Davies, *J. Chem. Soc.*, 3474 (1958); 4288 (1962).

(c) M. Bassey, C. A. Bunton, A. G. Davies, T. A. Lewis and D. R. Llewellyn, *J. Chem. Soc.*, 2471 (1955).
20. H. R. Williams and H. S. Mosher, *J. Amer. Chem. Soc.*, **76**, 3495 (1954).
21 (a) A. Rieche, E. Schmitz and E. Beyer, *Chem. Ber.*, **91** 1942 (1958).
(b) A. Rieche and C. Bischoff, *Chem. Ber.*, **94**, 2722 (1961).
22. K. R. Kopecky, J. H. van de Sande and C. Mumford, *Can. J. Chem.*, **46**, 25 (1968).
23. T. Koenig and M. Deinzer, *J. Amer. Chem. Soc.*, **90**, 7014 (1968).
24. T. Koenig, M. Deinzer and J. A. Hoobler, *J. Amer. Chem. Soc.*, **93**, 938 (1971).
25. J. K. Allen and J. C. Bevington, *Proc. Roy. Soc.* (*London*), **A 262**, 271 (1961).
26. K. R. Kopecky and C. Mumford, *Can. J. Chem.*, **47**, 709 (1969).
27. (a) A. P. Schaap and K. A. Zaklika in *Singlet Oxygen*, (Eds. H. H. Wasserman and R. W. Murray), Academic Press, New York, 1979, p. 173.
(b) P. D. Bartlett and M. E. Landis in *Singlet Oxygen*, (Eds. H. H. Wasserman and R. W. Murray), Academic Press, New York, 1979, p. 243.
28. P. D. Bartlett and A. P. Schaap, *J. Amer. Chem. Soc.*, **92**, 3233 (1970).
29. C. S. Foote and J. W.-P. Lin, *Tetrahedron Letters*, 3267 (1968).
30. J. H. Wieringa, J. Strating, H. Wynberg and W. Adam, *Tetrahedron Letters*, 169 (1972).
31. P. D. Bartlett and M. S. Ho, *J. Amer. Chem. Soc.*, **96**, 627 (1974).
32. K. Gollnick and G. O. Schenck in *1,4-Cycloaddition Reactions*, (Ed. J. Hamer) Academic Press New York, 1967, p. 255.
33. D. Swern in *Organic Peroxides*, Vol. 1 (Ed. D. Swern), Wiley–Interscience, New York–London, 1970, p. 313.
34. M. Kobayashi, H. Minato and R. Hisada, *Bull. Chem. Soc. Japan*, **44**, 2271 (1971).
35. L. Ponticorbo and D. Rittenberg, *J. Amer. Chem. Soc.*, **76**, 1705 (1954).
36. D. B. Denney and D. G. Denney, *J. Amer. Chem. Soc.*, **79**, 4806 (1957).
37. M. J. Goldstein and H. A. Judson, *J. Amer. Chem. Soc.*, **92**, 4120 (1970).
38. S. Oae, T. Kitao and Y. Kitaoka, *J. Amer. Chem. Soc.*, **84**, 3359 (1962).
39. T. Kashiwagi, S. Kozuka and S. Oae, *Tetrahedron*, **26**, 3619 (1970).
40. K. Fujimori and S. Oae, *Tetrahedron*, **29**, 65 (1973).
41. S. Oae, K. Fujimori, S. Kozuka and Y. Uchida, *J. Chem. Soc., Perkin Trans.2*, 1844 (1974).
42. R. Criegee, *Chem. Ber.*, **77**, 22 (1944).
43. P. D. Bartlett and R. R. Hiatt, *J. Amer. Chem. Soc.*, **80**, 1398 (1958).
44. L. P. Lorand and P. D. Bartlett, *J. Amer. Chem. Soc.*, **88**, 3294 (1966).
45. (a) R. Hecht and C. Ruchardt, *Chem. Ber.*, **96**, 1281 (1963).
(b) H. A. Staab, W. Rohr and F. Graf, *Chem. Ber.*, **98**, 1122 (1965).
46. T. Koenig, J. Huntington and R. Cruthoff, *J. Amer. Chem. Soc.*, **92**, 5413 (1970).
47. T. Koenig and R. Wolf, *J. Amer. Chem. Soc.*, **91**, 2574 (1969).
48. R. A. Wolf, M. J. Migliore, P. H. Fuery, P. G. Gagnier, I. C. Sabeta and R. J. Trocino, *J. Amer. Chem. Soc.*, **100**, 7967 (1978).
49. T. Koenig, J. A. Hoobler and W. R. Mabey, *J. Amer. Chem. Soc.*, **94**, 2514 (1972).
50. W. Nakanishi, T. Jo, K. Miura, Y. Ikeda, T. Sugawara, Y. Kawada and H. Iwamura, *Chem. Letters*, 387 (1981).
51. C. C. Price and H. Morita, *J. Amer. Chem. Soc.*, **75**, 3685 (1953).
52. L. S. Silbert and D. Swern, *J. Amer. Chem. Soc.*, **81**, 2364 (1959).
53. J. C. Martin and E. H. Drew, *J. Amer. Chem. Soc.*, **83**, 1232 (1961).
54. W. von E. Doering, K. Okamoto and H. Krauch, *J. Amer. Chem. Soc.*, **82**, 3579 (1960).
55. S. Oae and K. Fujimori, *Proceedings of2nd International Symposium on Organic Free Radicals at Aix-en-Provence, France* 1977, p. 349.
56. K. Fujimori, Y. Hirose and S. Oae, *20th Symposium on Organic Free Radical Reactions at Fukuoka*, Abstract, 1979, p. 4.
57. D. B. Denney, R. L. Ellsworth and D. Z. Denney, *J. Amer. Chem. Soc.*, **86**, 1116 (1964).
58. T. Yoshida, *Bull. Chem. Soc. Japan*, **23**, 209 (1950).
59. A. Kitamura, H. Sakuragi, M. Yoshida and K. Tokumaru, *Bull. Chem. Soc. Japan*, **53**, 1393 (1980).
60. J. C. Bevington and C. S. Brooks, *J. Polym. Sci.*, **22**, 257 (1956).
61. J. C. Bevington, J. Toole and E. Trossarelli, *Trans. Faraday Soc.*, **54**, 863 (1958).
62. F. D. Greene, *J. Amer. Chem. Soc.*, **81**, 1503 (1959).

63. (a) K. Fujimori, Y. Oshibe and S. Oae, *43rd Annual Meeting of Chemical Society of Japan*, Abstracts II, 1981, p. 1189.
(b) K. E. Russell, *J. Amer. Chem. Soc.*, **77**, 4814 (1955).
64. Y. Yokoyama, H. Wada, M. Kobayashi and H. Minato, *Bull. Chem. Soc. Japan*, **44**, 2479 (1971).
65. D. F. DeTar and L. A. Carpino, *J. Amer. Chem. Soc.*, **77**, 6370 (1955).
66. T. Koenig and R. Cruthoff, *J. Amer. Chem. Soc.*, **91**, 2562 (1969).
67. M. Kobayashi, H. Minato and R. Hisada, *Bull. Chem. Soc. Japan*, **44**, 2271 (1971).
68. F. Juračka and R. Chromeček, *Chem. Prumysl.*, **6**, 27 (1956).
69. N. J. Karch and J. M. McBride, *J. Amer. Chem. Soc.*, **94**, 5092 (1972).
70. N. J. Karch, E. T. Koh, B. L. Whitsel and J. M. McBride, *J. Amer. Chem. Soc.*, **97**, 6729 (1975).
71. J. M. McBride and R. A. Merrill, *J. Amer. Chem. Soc.*, **102**, 1723 (1980).
72. R. Kaptein, J. Brokken-Zijp and E. J. J. de Kanter, J.Amer. Chem. Soc., **94**, 6280 (1972).
73. D. B. Denney, *J. Amer. Chem. Soc.*, **78**, 590 (1956).
74. S. Oae, K. Fujimori and S. Kozuka, *Tetrahedron*, **28**, 5327 (1972).
75. F. D. Greene, H. P. Stein, C.-C. Chu and F. M. Vane, *J. Amer. Chem. Soc.*, **86**, 2080 (1964).
76. R. Hisada, H. Minato and M. Kobayashi, *Bull. Chem. Soc. Japan*, **45**, 2814 (1972).
77. (a) F. D. Greene and J. Kazan, *J. Org. Chem.*, **28**, 2168 (1963).
(b) D. F. DeTar and R. Silberstein, *J. Amer. Chem. Soc.*, **88**, 1013 (1966).
78. J. K. Kochi, A. Bemis and C. L. Jenkins, *J. Amer. Chem. Soc.*, **90**, 4616 (1968).
79. L. A. Singer in *Organic Peroxides*, Vol. I, (Ed. D. Swern), Wiley–Interscience, New York–London, 1970, p. 265.
80. W. A. Pryor and K. Smith, *Intern. J. Chem. Kinetics.*, **3**, 387 (1971).
81. P. D. Bartlett and C. Rüchardt, *J. Amer. Chem. Soc.*, **82**, 1756 (1960).
82. (a) T. Koenig and M. Deinger, *J. Amer. Chem. Soc.*, **88**, 4518 (1966).
(b) T. Koenig in *Organic Free Radicals* (Ed. W. A. Pryor), ACS Symposium Series, Vol. 69, 1978, p. 134.
83. T. Koenig, *J. Amer. Chem. Soc.*, **91**, 2558 (1969).
84. T. Koenig and W. R. Mabey, *J. Amer. Chem. Soc.*, **92**, 3804 (1970).
85. (a) C. Walling and A. Padwa, *J. Amer. Chem. Soc.*, **85**, 1593 (1963).
(b) J. A. Howard, *Advan. Free Radical Chem.*, **4**, 49 (1971).
86. K. U. Ingold in *Free Radicals*, Vol. I, (Ed. J. K. Kochi), John Wiley and Sons, New York–London, 1973, p. 37.
87. W. Braun, L. Rajbenbach and F. R. Eirich, *J. Phys. Chem.*, **66**, 1591 (1962).
88. J. W. Taylor and J. C. Martin, *J. Amer. Chem. Soc.*, **88**, 3650 (1966); **89**, 6904 (1967).
89. T. Koenig and J. G. Huntington, *J. Amer. Chem. Soc.*, **96**, 592 (1974).
90. W. A. Pryor and K. Smith, *J. Amer. Chem. Soc.*, **92**, 5403 (1970).
91. W. A. Pryor, E. H. Morkved and H. T. Bickley, *J. Org. Chem.*, **37**, 1999 (1972).
92. R. C. Neuman, Jr. and J. V. Behar, *J. Amer. Chem. Soc.*, **91**, 6024 (1969).
93. (a) R. C. Neuman, Jr., *Acc. Chem. Res.*, **5**, 381 (1972).
(b) R. C. Neuman, Jr., *J. Org. Chem.*, **37**, 495 (1972).
(c) Internal pressure of solvent varies with the kind of solvent and temperature. For recent review, see A. F. M. Barton, *J. Chem. Educ.*, **48**, 156 (1971).
94. T. Koenig and J. A. Hoobler, *Tetrahedron Letters*, 1803 (1972).
95. J. Owens and T. Koenig, *J. Org. Chem.*, **39**, 3153 (1974).
96. Secondary α-deuterium/hydrogen kinetic isotope effect on the thermolysis of 1,1′-diphenylazoethane is reported to be $k_H/k_D = 1.12^a$ and 1.093^b per deuterium.
(a) S. Seltzer, *J. Amer. Chem. Soc.*, **83**, 2625 (1961).
(b) Ref. 48.
97. A. Rembaum and M. Szwarc, *J. Amer. Chem. Soc.*, **76**, 5975 (1954).
98. H. Levy, M. Steinberg and M. Szwarc, *J. Amer. Chem. Soc.*, **76**, 5978 (1954).
99. J. Smid, A. Rembaum and M. Szwarc, *J. Amer. Chem. Soc.*, **78**, 3315 (1956).
100. J. Smid and M. Szwarc, *J. Chem. Phys.*, **29**, 432 (1958).
101. M. Szwarc and L. Herk, *J. Chem. Phys.*, **29**, 438 (1958).
102. L. Herk, M. Feld and M. Szwarc, *J. Amer. Chem. Soc.*, **83**, 2998 (1961).
103. M. J. Goldstein and H. A. Judson, *J. Amer. Chem. Soc.*, **92**, 4119 (1970).
104. M. J. Goldstein and W. A. Haiby, *J. Amer. Chem. Soc.*, **96**, 7358 (1974).
105. J. C. Martin and J. H. Hargis, *J. Amer. Chem. Soc.*, **91**, 5399 (1969).

106. O. Kikuchi, A. Hiyama, H. Yoshida and K. Suzuki, *Bull. Chem. Soc. Japan*, **51**, 11 (1978).
107. (a) M. Sax and R. K. McMullan, *Acta Cryst.*, **22**, 281 (1967).
(b) S. Caticha-Ellis and S. C. Abrahams, *Acta Cryst.*, **B24**, 277 (1968).
108. T. Koenig, *Tetrahedron Letters*, 3487 (1973).
109. A. T. Blomquist and J. Bernstein, *J. Amer. Chem. Soc.*, **73**,
110. M. J. S. Dewar and L. E. Wade, Jr., *J. Amer. Chem. Soc.*, **95**, 290 (1973); **99**, 4417 (1977).
111. R. Hoffmann and R. B. Woodward, *J. Amer. Chem. Soc.*, **87**, 4389 (1965); *Accounts Chem. Res.*, **1**, 17 (1968).
112. (a) J. J. Gajewski and N. D. Conrad, *J. Amer. Chem. Soc.*, **100**, 6269 (1978).
(b) J. J. Gajewski, *Accounts Chem. Res.*, **13**, 142 (1980).
113. W. von E. Doering, V. G. Toscano and G. H. Beasley, *Tetrahedron*, **27**, 5299 (1971).
114. M. J. Goldstein and M. S. Benzon, *J. Amer. Chem. Soc.*, **94**, 7147 (1972).
115. E. S. Lewis, J. T. Hill and E. R. Newman, *J.* Amer. Chem. Soc., **90**, 662 (1968).
116. (a) R. C. Lamb and J. G. Pacifici, *J. Amer. Chem. Soc.*, **86**, 914 (1964).
(b) G. Houghton, *J. Chem. Phys.*, **40**, 1628 (1964).
(c) O. Dobis, J. M. Pearson and M. Szwarc, *J. Amer. Chem. Soc.*, **90**, 278 (1968).
117. K. Fujimori, Y. Hirose, Y. Oshibe and S. Oae, *3rd International Symposium on Organic Free Radical Chemistry at Freiburg*, Abstract, 1981, p. 16.
118. C. Walling and R. B. Hodgdon, Jr., *J. Amer. Chem. Soc.*, **80**, 288 (1958).
119. H. J. Shine and D. M. Hoffman, *J. Amer. Chem. Soc.*, **83**, 2782 (1961).
120. H. J. Shine and J. R. Slagle, *J. Amer. Chem. Soc.*, **81**, 6309 (1959).
121. H. J. Shine, J. A. Waters and D. M. Hoffman, *J. Amer. Chem. Soc.*, **85**, 3613 (1963).
122. J. C. Martin, J. W. Taylor and D. H. Drew, *J. Amer. Chem. Soc.*, **89**, 129 (1967).
123. J. C. Martin and S. A. Dombchik in *Oxidation of Organic Compounds*, Advances in Chemistry Series, Vol. 75, 1968, p. 264.
124. R. Kaptein in *Advances in Free Radical Chemistry*, Vol. 5 (Ed. G. H. Smith), Academic Press, New York, p. 319.
125. (a) M. J. Goldstein, *Tetrahedron Letters*, 1601 (1964).
(b) M. J. Goldstein, H. A. Judson and M. Yoshida, *J. Amer. Chem. Soc.*, **92**, 4122 (1970).
126. S. A. Dombchik, *Ph.D. Thesis*, University of Illinois, Urbana U.S.A., 1968.
127. T. Koenig, R. A. Wielesek and J. G. Huntington, *Tetrahedron Letters*, 2283 (1974).
128. (a) O. Kikuchi, *Tetrahedron Letters*, 2421 (1977).
(b) O. Kikuchi, K. Utsumi and K. Suzuki, *Bull. Chem. Soc. Japan*, **50**, 1339 (1977).
(c) M. B. Yim, O. Kikuchi and D. E. Wood, *J. Amer. Chem. Soc.*, **100**, 1869 (1978).
129. (a) M. Iwasaki, K. Minakata and K. Toriyama, *J. Amer. Chem. Soc.*, **93**, 3533 (1971).
(b) K. Toriyama, M. Iwasaki, S. Noda and B. Eda, *J. Amer. Chem. Soc.*, **93**, 6415 (1971).
(c) B. Eda and M. Iwasaki, *J. Chem. Phys.*, **55**, 3442 (1971).
(d) K. Minakata and M. Iwasaki, *J. Chem. Phys.*, **57**, 4758 (1972).
(e) B. Eda and M. Iwasaki, *J. Chem. Phys.*, **57**, 1653 (1972).
130. G. S. Hammond and L. M. Soffer, *J. Amer. Chem. Soc.*, **72**, 4711 (1950).
131. M. Kobayashi, H. Minato and Y. Ogi, *Bull. Chem. Soc. Japan*, **41**, 2822 (1969).
132. M. Kobayashi, H. Minato and Y. Ogi, *Bull. Chem. Soc. Japan*, **42**, 2737 (1969).
133. Y. Nakata, K. Tokumaru and O. Simamura, *Tetrahedron Letters*, 3303 (1967).
134. C. Rüchardt, *Angew. Chem.* (*Intern. Ed.*), **9**, 830 (1970).
135. K. Fujimori, Y. Hirose and S. Oae, *Annual Meeting of Chemical Society of Japan*, Abstract II, 1980, p. 924.
136. P. D. Bartlett and T. G. Traylor, *Tetrahedron Letters*, 30 (1960); *J. Amer. Chem. Soc.*, **85**, 2407 (1963).
137. J. E. Bennett and J. A. Howard, *J. Amer. Chem. Soc.*, **95**, 4008 (1973).
138. (a) R. A. Kenley and T. G. Traylor, *J. Amer. Chem. Soc.*, **97**, 4700 (1975).
(b) N. A. Clinton, R. A. Kenley and T. G. Traylor, *J. Amer. Chem. Soc.*, **97**, 3746 (1975).
139. G. Ayrey, *Chem. Rev.*, **63**, 645 (1963).
140. M. M. Koton, T. M. Kiseleva and M. I. Bessenov, *Zh. Fiz. Khim.*, **28**, 2137 (1954); *Dokl. Akad. Nauk SSSR*, **96**, 85 (1954).
141. J. C. Bevington and J. Toole, *J. Polym. Sci.*, **28**, 413 (1958).
142. J. C. Bevington, J. Toole and L. Torossarelli, *Trans. Faraday Soc.*, **54**, 863 (1958).

143. J. C. Bevington, J. Toole and L. Torossarelli, *Makromol. Chem.*, **28**, 237 (1958).
144. J. C. Bevington, *Makromol. Chem.*, **34**, 152 (1959).
145. R. Criegee and Kaspar, *Ann.*, **560**, 127 (1948).
146. (a) P. D. Bartlett and J. L. Kice, *J. Amer. Chem. Soc.*, **75**, 5591 (1953).
(b) H. L. Goering and A. C. Olson, *J. Amer. Chem. Soc.*, **75**, 5853 (1953).
147. (a) J. E. Leffler, *J. Amer. Chem. Soc.*, **72**, 67 (1950).
(b) J. E. Leffler and C. C. Petropoulos, *J. Amer. Chem. Soc.*, **79**, 3068 (1957).
148. D. Z. Denney, T. M. Valega and D. B. Denney, *J. Amer. Chem. Soc.*, **86**, 46 (1964).
149. (a) R. C. Lamb, J. G. Pacifici and P. W. Ayers, *J. Amer. Chem. Soc.*, **87**, 3928 (1965).
(b) P. D. Bartlett and F. D. Greene, *J. Amer. Chem. Soc.*, **76**, 1088 (1954).
(c) D. B. Denney and N. Sherman, *J. Org. Chem.*, **30**, 3760 (1965).
150. D. F. DeTar and C. Weis, *J. Amer. Chem. Soc.*, **79**, 3045 (1957).
151. T. Kashiwagi and S. Oae, *Tetrahedron*, **26**, 3631 (1970).
152. T. Kashiwagi, K. Fujimori, S. Kozuka and S. Oae, *Tetrahedron*, **26**, 3647 (1970).
153. P. A. Smith in *Molecular Rearrangement* (Ed. P. de Mayo), Wiley–Interscience, New York–London, 1963, Chap. 8.
154. S. Oae, K. Fujimori and Y. Uchida, *Tetrahedron*, **28**, 5321 (1972).
155. C. Walling, H. P. Waits, J. Milovanovic and C. G. Pappiannou, *J. Amer. Chem. Soc.*, **92**, 4927 (1970).
156. K. G. Taylor, C. K. Govindan and M. S. Kaelin, *J. Amer. Chem. Soc.*, **101**, 2091 (1979).
157. (a) J. E. Leffler, *US–Japan Peroxide Symposium, Boulder, Colorado, U.S.A.*, August 1976.
(b) J. E. Leffler, *26th International Congress Pure and Applied Chemistry, Tokyo*, Abstract Session I, 1977, p. 47.
(c) J. E. Leffler and A. A. More, *J. Amer. Chem. Soc.*, **94**, 2484 (1972).
158. (a) R. C. Lamb and J. R. Sanderson, *J. Amer. Chem. Soc.*, **91**, 5034 (1969).
(b) R. C. Lamb, L. L. Vestal, G. R. Cipau and S. Debnath, *J. Org. Chem.*, **39**, 2096 (1974).
159. C. Walling and J. P. Sloan, *J. Amer. Chem. Soc.*, **101**, 7679 (1979).
160. M. S. Kharasch, J. Kuderna and W. Nudenberg, *J. Org. Chem.*, **19**, 1283 (1954).
161. M. S. Kharasch, F. Engelmann and W. H. Urry, *J. Amer. Chem. Soc.*, **65**, 2428 (1943).
162. F. D. Greene, *J. Amer. Chem. Soc.*, **77**, 4869 (1955).
163. H. H. Lau and H. Hart, *J. Amer. Chem. Soc.*, **81**, 4897 (1959).
164. (a) H. Hart and D. P. Wyman, *J. Amer. Chem. Soc.*, **81**, 4891 (1959).
(b) H. Hart and R. A. Cipriani, *J. Amer. Chem. Soc.*, **84**, 3697 (1962).
165. S. Oae, Y. Uchida, K. Fujimori and S. Kozuka, *Bull. Chem. Soc. Japan*, **46**, 1741 (1973).
166. K. Fujimori and S. Oae, *36th Annual Meeting of Chemical Society of Japan at Osaka*, Abstract II, 1977, p. 845.
167. S. Oae, N. Asai and K. Fujimori, *J. Chem. Soc. Perkin Trans 2*, 1124 (1978).
168. O. Kikuchi, *Tetrahedron Letters*, **21**, 1055 (1980).
169. E. H. White and D. J. Woodcock in *The Chemistry of the Amino Group*, (Ed. S. Patai), John Wiley and Sons Ltd., London–New York, 1958, p. 440.
170. D. Y. Curtin and L. L. Miller, *J. Amer. Chem. Soc.*, **89**, 637 (1967).
171. D. J. McCarthy and A. F. Hegarty, *J. Chem. Soc., Perkin Trans. 2*, 1085 (1977).
172. D. Spinelli, V. Frenna, A. Corrao and N. Vivona, *J. Chem. Soc., Perkin Trans. 2*, 19 (1978) and references cited therein.
173. R. Criegee and G. Wenner, *Ann.*, **564**, 9 (1949).
174. P. S. Bailey, *Ozonization in Organic Chemistry*, Vol. I, Academic Press, New York, 1978.
175. P. R. Story, P. N. Murray and R. D. Youssefyeh, *J. Amer. Chem. Soc.*, **88**, 3144 (1966).
176. R. W. Murray, R. D. Youssefyeh and R. S. Story, *J. Amer. Chem. Soc.*, **89**, 2429 (1967).
177. G. Klopman and C. M. Joiner, *J. Amer. Chem. Soc.*, **97**, 5287 (1975).
178. P. R. Story, C. E. Bishop, J. R. Burgess, R. W. Murray and R. D. Youssefyeh, *J. Amer. Chem. Soc.*, **90**, 1907 (1968).
179. R. W. Murray and R. Hagen, *J. Org. Chem.*, **36**, 1103 (1971).
180. R. P. Lattimer and R. L. Kuczkowski, *J. Amer. Chem. Soc.*, **96**, 6205 (1974).
181. (a) O. Lorenz and C. P. Parks, *J. Org. Chem.*, **30**, 1976 (1965).
(b) O. Lorenz, *Anal. Chem.*, **37**, 101 (1965).
182. J. Carles and S. Fliszár, *Can. J. Chem.*, **47**, 1113 (1969).

183. (a) S. Fliszár, J. Carles and J. Renard, *J. Amer. Chem. Soc.*, **90**, 1364 (1968).
(b) S. Fliszár and J. Carles, *J. Amer. Chem. Soc.*, **91**, 2637 (1969).
(c) J. Castonguay, M. Bertand, J. Carles, S. Fliszár and Y. Rousseau, *Can. J. Chem.*, **47**, 919 (1969).
184. K. L. Gallaher and R. L. Kuczkowski, *J. Org. Chem.*, **41**, 892 (1976).
185. D. P. Higley and R. W. Murray, *J. Amer. Chem. Soc.*, **98**, 4526 (1976).
186. J. C. Martin and T. W. Koenig, *J. Amer. Chem. Soc.*, **86**, 1771 (1964).
187. J. C. Martin, *Organic Free Radicals* (Ed. W. A. Pryor) ACS Symposium Series, Vol. 69, 1978, p. 71.
188. W. G. Bentrude and J. C. Martin, *J. Amer. Chem. Soc.*, **84**, 1561 (1962).
189. W. Nakanishi, S. Koike, M. Inoue, Y. Ikeda, H. Iwamura, H. Imahashi, H. Kihara and M. Iwai, *Tetrahedron Letters*, 81 (1977).
190. C. W. Perkins, J. C. Martin, A. J. Arduengo, W. Lau, A. Alegria and J. K. Kochi, *J. Amer. Chem. Soc.*, **102**, 7753 (1980).
191. D. B. Denney, W. F. Goodyear and B. Goldstein, *J. Amer. Chem. Soc.*, **83**, 1726 (1961).
192. R. E. Pincock, *J. Amer. Chem. Soc.*, **86**, 1820 (1964); **87**, 1274 (1965).
193. K. Nozaki and P. D. Bartlett, *J. Amer. Chem. Soc.*, **68**, 1686 (1946).
194. L. Horner and E. Schwenk, *Ann.*, **566**, 69 (1950).
195. L. Horner and E. Jürgens, *Ann.* **602**, 135 (1952).
196. L. Horner, W. Jurgeleit and K. Klüpfel, *Ann.*, **591**, 108 (1955).
197. L. Horner and W. Jurgeleit, *Ann.*, **591**, 138 (1955).
198. F. D. Greene, *J. Amer. Chem. Soc.*, **78**, 2250 (1956); F. D. Greene and W. W. Rees, *J. Amer. Chem. Soc.*, **80**, 3432 (1958); **82**, 890 (1960); F. D. Greene, W. Adam and J. E. Cantrill, *J. Amer. Chem. Soc.*, **83**, 3461 (1961).
199. C. Walling, *J. Amer. Chem. Soc.*, **102**, 6854 (1980).
200. D. B. Denney and G. Feig, *J. Amer. Chem. Soc.*, **81**, 5322 (1959).
201. D. B. Denney and D. Z. Denney, *J. Amer. Chem. Soc.*, **82**, 1389 (1960).
202. M. A. Greenebaum, D. B. Denney and A. K. Hoffmann, *J. Amer. Chem. Soc.*, **78**, 2563 (1956).
203. D. B. Denney and M. A. Greenebaum, *J. Amer. Chem. Soc.*, **79**, 979 (1957).
204. S.-O. Lawesson, C. Frisell, D. Z. Denney and D. B. Denney, *Tetrahedron*, **19**, 1229 (1963).
205. D. B. Denney and D. Z. Denney, *J. Amer. Chem. Soc.*, **82**, 1389 (1960).
206. G. B. Schuster, *Accounts Chem. Res.*, **12**, 366 (1979).
207. F. D. Greene and W. W. Rees, *J. Amer. Chem. Soc.*, **80**, 3432 (1958).
208. F. D. Greene and W. Adam, *J. Org. Chem.*, **29**, 136 (1964).
209. R. Hisada, M. Kobayashi and H. Minato, *Bull. Chem. Soc. Japan*, **45**, 2035 (1972).
210. R. Hisada, M. Kobayashi and H. Minato, *Bull. Chem. Soc. Japan*, **45**, 2902 (1972).
211. E. Koubek, M. L. Haggett, C. J. Battaglia, K. M. Ibne-Rasa, H. Y. Ryun and J. O. Edwards, *J. Amer. Chem. Soc.*, **85**, 2263 (1963).
212. D. B. Denney, W. F. Goodyear and B. Goldstein, *J. Amer. Chem. Soc.*, **82**, 1393 (1960).
213. A. G. Davies and R. Feld, *J. Chem. Soc.*, 4637 (1958).
214. A. G. Davies and R. B. Moodie, *J. Chem. Soc.*, 2372 (1958).
215. W. von E. Doering and E. Dorfman, *J. Amer. Chem. Soc.*, **75**, 5595 (1953).
216. K. Mislow and J. Brenner, *J. Amer. Chem. Soc.*, **75**, 2318 (1953); R. B. Turner, *J. Amer. Chem. Soc.*, **72**, 878 (1950); J. A. Berson and S. Suzuki, *J. Amer. Chem. Soc.*, **81**, 4088 (1959); T. Kashiwagi, K. Fujimori, S. Kozuka and S. Oae, *Tetrahedron*, **26**, 3647 (1970).
217. K. Fujimori and S. Oae, unpublished results.

The Chemistry of Functional Groups, Peroxides
Edited by S. Patai

CHAPTER **19**

Ozonation of single bonds

EHUD KEINAN and HAIM T. VARKONY

Department of Organic Chemistry, The Weizmann Institute of Science, Rehovot, 76100, Israel

I. INTRODUCTION 649
II. ELECTROPHILIC REACTIONS WITH MULTIPLE BONDS 650
III. ELECTROPHILIC REACTIONS WITH NUCLEOPHILIC HETEROATOMS . 651
IV. ELECTROPHILIC REACTIONS WITH SINGLE BONDS 651
A. Ozonation of Ethers and Acetals 652
B. Ozonation of Aldehydes 653
C. Ozonation of Hydrosilanes 653
D. Ozonation of Anthrones 654
E. Ozonation of Amines 654
F. Ozonation of Saturated Hydrocarbons 655
G. Dry Ozonation 658
H. Ozonation of Carbon–Carbon Single Bonds 671
I. Ozonation Reactions in Strong Acidic Media 675
V. REFERENCES 681

I. INTRODUCTION

Ozone, one of the most commonly used oxidizing agents, especially in organic chemistry has been known for more than 140 years[1]. Having a considerable high oxidation potential (−2.07 eV) it is capable of reacting with almost any organic molecule. Yet, techniques and procedures which have been developed within the last two decades allow easy control in order to carry out selective synthetic transformation.

Being an endothermic allotrope of oxygen, ozone may serve as a precursor for reactive oxygen species such as oxygen atoms, singlet oxygen molecules etc. Based on a convenient description of the ozone molecule as a resonance hybrid of four canonical forms, one could predict that ozone should be able to function as an electrophile or as a nucleophile:

$$\overset{+}{O}{-}O{-}\overset{-}{O} \longleftrightarrow \overset{-}{O}{-}O{-}\overset{+}{O} \longleftrightarrow O{=}\overset{+}{O}{-}\overset{-}{O} \longleftrightarrow \overset{-}{O}{-}\overset{+}{O}{=}O$$

However, there are few reports on ozone behaving as a nucleophile. A rare example is the oxidation of 9,10-dibromoanthracene to anthraquinone[2], in which one molecule of ozone acts as an electrophile and a second as a nucleophile. The vast majority of ozonation processes involve initial electrophilic attack by ozone at an electron-rich centre in the substrate molecule to form a 1:1 ozone–substrate adduct.

Most organic compounds which react with ozone may be divided into three classes, based on the nature of their reacting nucleophilic function:

(*a*) Multiple bonds having an electron rich π system as the nucleophile.
(*b*) Heteroatoms having a filled nonbonding orbital as a nucleophile.
(*c*) Saturated compounds having an electron-rich σ bond as a nucleophile.

Ozonation processes of classes (*a*) and (*b*) are the most common ones and they are satisfactorily covered by a number of monographs and review articles[3,4]. Therefore, the present account will focus attention on ozonations belonging to class (*c*), namely ozonations of single bonds in relatively less reactive compounds. Special emphasis will be given to dry ozonation reactions and their applications. Classes (*a*) and (*b*) are briefly mentioned for the sake of drawing analogies and common features to class (*c*).

II. ELECTROPHILIC REACTIONS WITH MULTIPLE BONDS

Most of the known unsaturated systems such as olefins, acetylenes, aromatic compounds and carbon–heteroatom multiple bonds are prone to direct electrophilic attack by ozone[3–5]. The mechanism of ozonolysis reactions, especially those of olefins, is now well established and well documented (an excellent review was provided by Bailey[4]).

The first step in the electrophilic reaction of ozone with double bonds is the formation of a π complex. This complex may undergo a 1,3-dipolar addition to form a primary ozonide (1,2,3-trioxolane) followed by a series of rearrangements, as suggested by Criegee[3m] (Scheme 1). Only in special cases of highly hindered olefins, will the electrophilic attack by ozone result in direct formation of epoxides.

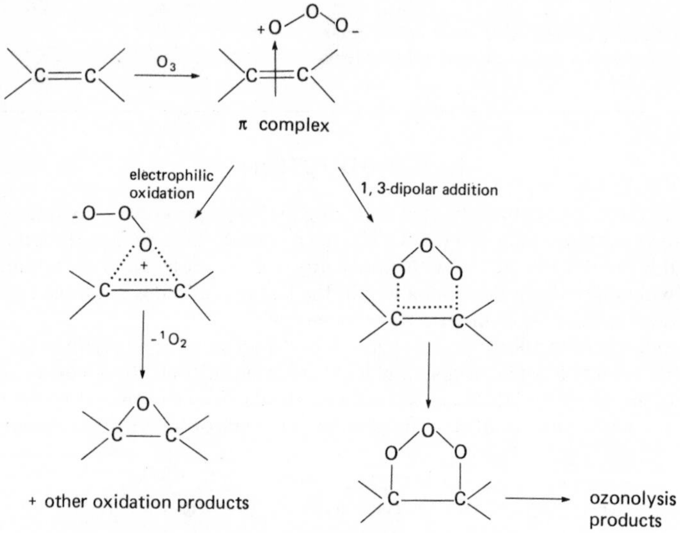

O_3

π complex

electrophilic oxidation

1, 3-dipolar addition

$-{}^1O_2$

+ other oxidation products

ozonolysis products

SCHEME 1

III. ELECTROPHILIC REACTIONS WITH NUCLEOPHILIC HETEROATOMS

The reaction of ozone with heteroatom nucleophiles involves primary interaction between ozone and a filled, nonbonding orbital of the heteroatom (equation 1). The resulting reactive zwitterion may undergo a large variety of inter- and intra-molecular processes, leading to a large number of possible products. This behaviour is typical for compounds containing nitrogen[6–10], phosphorus, arsenic[11], sulphur[12] and selenium[13], etc. at various oxidation levels.

$$R_nX + O_3 \longrightarrow R_n\overset{+}{X}{-}O{-}O{-}O^- \longrightarrow R_nX{=}O + {}^1O_2 \qquad (1)$$

X = N, P, As, S, Se

The most common mode in which the intermediate trioxide decomposes is by loss of singlet molecular oxygen yielding the corresponding oxide. For example, ozonation of sulphides results in sulphoxides which may undergo further ozonation to give sulphones (equation 2).

$$R_2S + O_3 \longrightarrow R_2\overset{+}{S}{-}O{-}O{-}O^- \longrightarrow R_2S{=}O + {}^1O_2$$

$$R_2S{=}O \xrightarrow{O_3} R_2\overset{+}{S}(=O){-}O{-}O{-}O^- \longrightarrow R_2S(=O){=}O + {}^1O_2 \qquad (2)$$

The reaction of triphenylphosphine with ozone produces a relatively stable adduct[14], which may decompose thermally to singlet molecular oxygen and triphenylphosphine oxide (equation 3).

$$(PhO)_3P \xrightarrow{O_3} (PhO)_3P\langle O_3 \rangle \xrightarrow{\Delta} (PhO)_3P{=}O + {}^1O_2 \qquad (3)$$

This controlled release of $^1O_2$ was found to be synthetically useful and has been extensively used as a method for convenient production of singlet molecular oxygen for various purposes[15].

The reactions of ozone with nitrogen-containing compounds such as primary, secondary and tertiary amines[6,7], azo compounds[8], azines[9], imines[10], oximes[10a], and nitroso compounds[10b] have been thoroughly investigated, mainly by Bailey and his coworkers.

IV. ELECTROPHILIC REACTIONS WITH SINGLE BONDS

Ozone attacks C—H bonds in saturated compounds via a 1,3-dipolar insertion to form an unstable hydrotrioxide. C—H bonds possessing high electron density are attacked preferentially.

A. Ozonation of Ethers and Acetals

The reaction of ozone with ethers was recognized as early as 1855[1c], but the first extensive investigation was carried out only in 1964, by Price and Tumolo[16]. They found that only C—H bonds α to the etheric oxygen were attacked by ozone. High preference for ozonation of the less acidic C—H was observed. For example, di-*t*-butyl ether was inert to oxidation by ozone. The tertiary C—H bond of propyl isopropyl ether was found to be, on a statistical basis, 1.7 times more reactive than the secondary C—H bond. The ozonation reactions were carried out at room temperature with ozone in an oxygen stream. These results[16] suggest a 1,3-dipolar insertion mechanism of ozone into the C—H bond. Since the transition state (**II**) (equation 4) for such a process has carbonium ion character, it is stabilized by electron-donating groups such as alkyl groups and neighbouring oxygen atoms.

$$\text{R—CH(H)—}\ddot{\text{O}}\text{—R}^1 + {}^{-}\text{O—O}^{+}\text{=O}\ \textbf{(I)} \rightleftharpoons \text{RCH}^{\delta+}\cdots\text{O—R}^1\ \ (\delta^-\text{O}\cdots\text{H}\delta^-\ \text{O}\cdots\text{O}^{\delta+})\ \textbf{(II)} \longrightarrow \text{R(H)C(OR}^1\text{)—O—O—O—H}\ \textbf{(III)} \qquad (4)$$

The same transition state was also proposed by Deslongchamps and coworkers[17] for similar ozonations of acetals. They showed that the only reactive conformations of the acetals were those in which the nonbonding orbitals of the oxygen atoms were oriented antiperiplanar to the C—H bond.

Erickson and collaborators[18] studied low-temperature ozonations of ethers using ozone–oxygen and ozone–nitrogen mixtures. They found that the presence of oxygen affects product distribution, deuterium isotope effects and the relative rates of ozonation of various ethers. They propose the 1,3-dipolar insertion mechanism (equation 4) to be predominant at −78°C and a competing radical-chain mechanism, involving both oxygen and ozone, at higher temperatures.

The formation of a hydrotrioxide intermediate (**III** in equation 4) via 1,3-dipolar insertion, is evident from low-temperature (−78°C) NMR studies[19]. Murray's group[19c] have shown that hydrotrioxides are stable at low temperatures due to internal hydrogen bonding. At higher temperatures they undergo thermal decomposition to give the corresponding alcohol and singlet molecular oxygen (equation 5).

$$\text{R(H)C(OR}^1\text{)—O—O—O—H} \xrightarrow{\Delta} \text{R}^1\text{—OH} + \text{R—CH=O} + {}^1\text{O}_2 \qquad (5)$$

Bailey and Lerdal[20] investigated the ozonation of four ethers, in Freon 11 at 0°C, −30°C and −78°C using ozone in a nitrogen stream. They found that ozone attacks ethyl isopropyl ether largely at the tertiary hydrogen, while 4-oxa-2-heptanone is principally attacked at the more acidic methylene group. Hydrotrioxide intermediates are also observed in low-temperature ozonations of aliphatic and aromatic acetals[21], the NMR spectra of which show two OOOH absorptions at ca. 13 ppm downfield from Me_4Si.

Taillefer and coworkers[22] systematically investigated the reaction between ozone and acetals. They found that the stoichiometry of the reaction is 1:1 in each reactant and that the reaction is also first order in each reactant. Substituent effects measured in a variety of systems and under several conditions of temperature and solvents were found to be small. Solvent polarity was also found to have little effect on the rate of the reaction.

The reaction of unsymmetric 1,3-dioxacyclanes with ozone[23] (equation 6) gives a mixture of ring-cleavage products. In all cases the reaction proceeds to yield the more substituted alcohol.

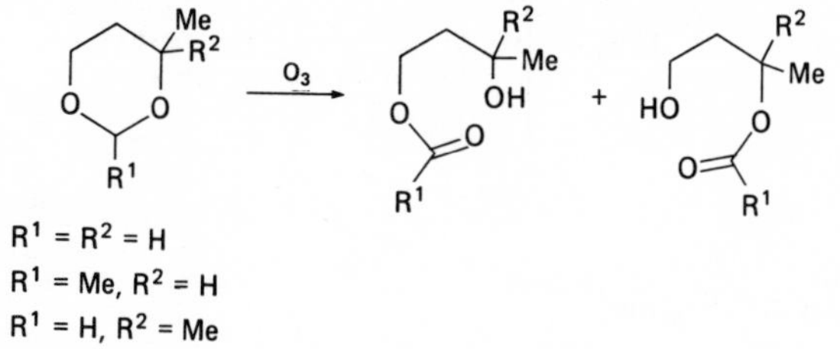

$R^1 = R^2 = H$
$R^1 = Me, R^2 = H$
$R^1 = H, R^2 = Me$

(6)

B. Ozonation of Aldehydes

Aldehydes react with ozone to produce acids, esters and peracids[24]. White and Bailey[25] have investigated the mechanism of these reactions and suggested a 1,3-dipolar insertion of ozone into the aldehydic C—H bond as the first step of the reaction (equation 7).

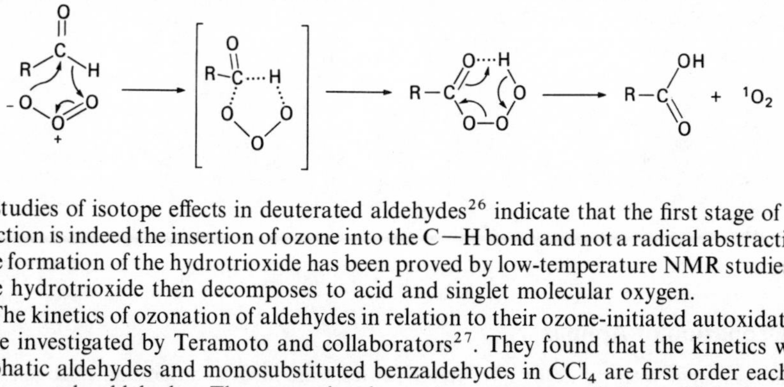

(7)

Studies of isotope effects in deuterated aldehydes[26] indicate that the first stage of the reaction is indeed the insertion of ozone into the C—H bond and not a radical abstraction. The formation of the hydrotrioxide has been proved by low-temperature NMR studies[19]. The hydrotrioxide then decomposes to acid and singlet molecular oxygen.

The kinetics of ozonation of aldehydes in relation to their ozone-initiated autoxidation were investigated by Teramoto and collaborators[27]. They found that the kinetics with aliphatic aldehydes and monosubstituted benzaldehydes in CCl_4 are first order each in ozone and aldehyde. The second-order rate constants of butyraldehyde and isobutyraldehyde are larger than those of the substituted benzaldehydes. The observed order of the reactivity of benzaldehydes was: *p*-$MeOC_6H_4CHO$ > *p*-MeC_6H_4CHO- > PhCHO > *p*- and *m*-ClC_6H_4CHO > *p*- and *m*-$NO_2C_6H_4CHO$. The rates of ozonization were linearly related to ρ values.

C. Ozonation of Hydrosilanes

Spialter and coworkers[28] investigated the ozonation of hydrosilanes, and found that the first stage of the reaction involves the reversible formation of a complex between ozone and the silane which is followed by a 1,3-dipolar insertion of ozone into the Si—H bond to form a hydrotrioxide. The latter decomposes with retention of configuration to form the corresponding silanol (equation 8).

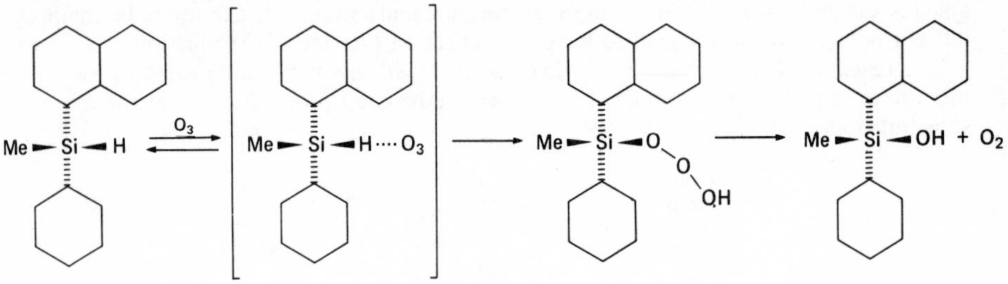

$$\text{Me}-\text{Si}-\text{H} \underset{}{\overset{O_3}{\rightleftharpoons}} [\text{Me}-\text{Si}-\text{H}\cdots O_3] \longrightarrow \text{Me}-\text{Si}-\text{O}-\text{O}-\text{OH} \longrightarrow \text{Me}-\text{Si}-\text{OH} + O_2 \quad (8)$$

D. Ozonation of Anthrones

Ozonation of anthrones was investigated by Batterbee and Bailey[29]. They suggested a dipolar insertion of ozone into the C—H bond to explain the formation of anthraquinone (equation 9).

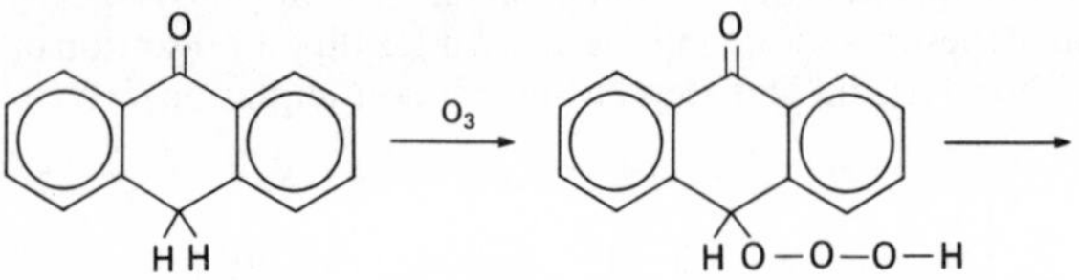

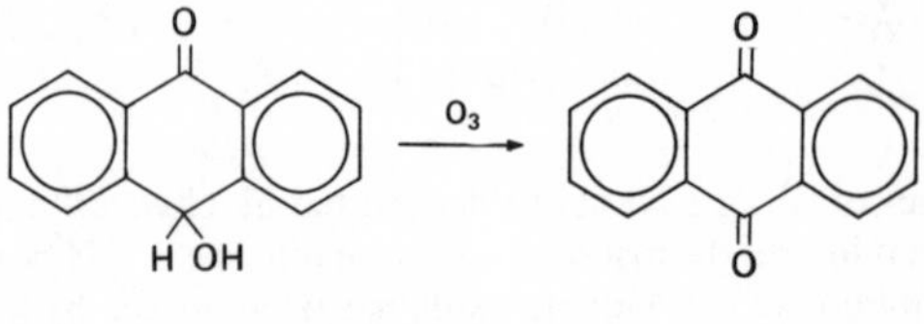

$$\text{anthrone (H H)} \xrightarrow{O_3} \text{(H O-O-O-H)} \longrightarrow \text{(H OH)} \xrightarrow{O_3} \text{anthraquinone} \quad (9)$$

E. Ozonation of Amines

Amines react with ozone by two major routes. One involves attack on nitrogen and formation of *N*-oxidation products such as nitroxides and ammonium salts[6,7]. The other involves side-chain oxidation[30,31]. The studies of different alkyl substituted primary, secondary and tertiary amines have led to the proposal of four competing reactions (a–d in Scheme 2), for the initially formed ozone–amine adduct (**I** in Scheme 2).

The first step is an electrophilic attack of ozone on the amine and formation of an amine–ozone adduct (**I**). This can further react by four different routes (Scheme 2): (*a*) amine oxide (**II**) formation, (*b*) formation of a radical-ion pair (**III**), (*c*) intramolecular side-chain oxidation via **IV** or (*d*) abstraction of hydrogen (in primary and secondary amines) and formation of *N*-oxides and ammonium salts.

An alternative mechanism, 1,3-dipolar insertion of ozone into a C—H bond α to nitrogen, has been suggested by Lerdal and Bailey[20]. This mechanism is similar to the mechanism of ozonation of ethers (equation 10).

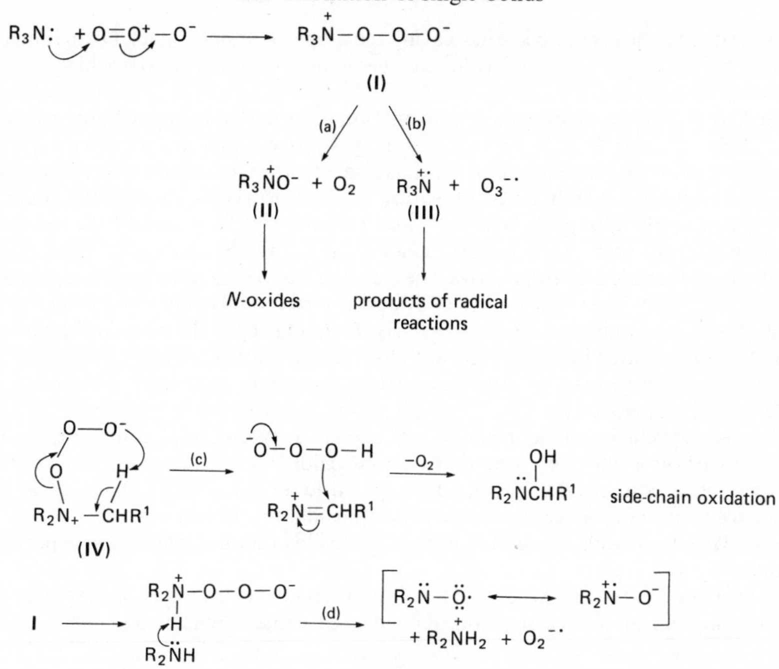

$$R_3N\!: + O{=}O^+{-}O^- \longrightarrow R_3\overset{+}{N}{-}O{-}O{-}O^- \;(\mathbf{I})$$

(a) $R_3\overset{+}{N}O^- + O_2$ (II) → N-oxides

(b) $R_3\overset{+\cdot}{N} + O_3^{-\cdot}$ (III) → products of radical reactions

(IV) $\xrightarrow{(c)}$ $^-O{-}O{-}O{-}H$, $R_2\overset{+}{N}{=}CHR^1$ $\xrightarrow{-O_2}$ $R_2\ddot{N}CHR^1(OH)$ side-chain oxidation

$$\mathbf{I} \longrightarrow R_2\overset{+}{N}(H){-}O{-}O{-}O^- \;(R_2\ddot{N}H) \xrightarrow{(d)} [R_2N{-}\ddot{O}\cdot \longleftrightarrow R_2\overset{+\cdot}{N}{-}O^-] + R_2\overset{+}{N}H_2 + O_2^{-\cdot}$$

SCHEME 2

$$R_2N{-}CHR^1(H) + O_3 \longrightarrow R_2N{-}CHR^1(H\cdots{}^-O{-}\overset{+}{O}{=}O) \longrightarrow R_2N{-}CH{-}R^1(H{-}O{-}O{-}O) \longrightarrow R_2N{-}CHR^1(OH) + O_2 \quad (10)$$

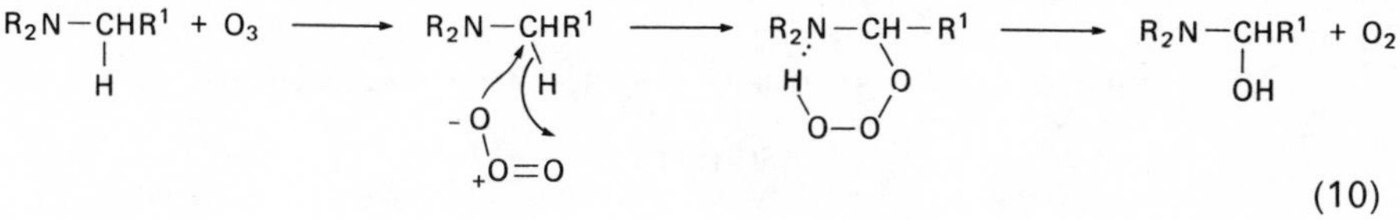

The nitroxide pathway is found to be the major ozonation route for secondary amines having tertiary or secondary alkyl groups. With those bearing primary alkyl groups, however, side-chain oxidation is predominant in most solvents. The most important processes in the ozonation of tertiary amines having primary or secondary alkyl groups are side-chain attack and amine oxide formation. Side-chain attack may proceed either by 1,3-dipolar insertion (predominant for secondary alkyl groups) or by intramolecular proton abstraction (predominant for primary alkyl groups). Amine oxide formation is a minor pathway except for ozonations of tertiary amines with primary alkyl groups in a protic solvent.

F. Ozonation of Saturated Hydrocarbons

The earliest (1898) report on ozonation of saturated hydrocarbons by Otto[32], described the oxidation of methane to formaldehyde. Later, Durland and Adkins[33] reported the formation of alcohols, ketones and acids upon ozonation of saturated hydrocarbons. Surprisingly, ozonation of 9,10-dihydrophenanthrene resulted in oxidation of the aliphatic methylene group to ketone leaving the aromatic system intact. Ozonation of *cis*-decalin produced 78% *cis*-9-decalol as the major product, accompanied by other oxidation products.

The first systematic investigation of ozonation of saturated hydrocarbons, was carried out by Schubert and Pease[34], who studied the gas-phase ozonation of methane, propane, butane and pentane. They suggested a mechanism of radical oxidation where ozone acts as a radical initiator. Ozonations of saturated hydrocarbons in solution were first investigated by the groups of Whiting[35] and Hamilton[36]. They studied the reactions of ozone with *cis*- and *trans*-decalin, adamantane, cyclohexane, *cis*- and *trans*-1,2-dimethylcyclohexane and isopentane. The reaction was found to be highly regioselective and stereospecific. The relative rates of attack at tertiary, secondary and primary C—H bonds were found to be 110:13:1 respectively. A similar regioselectivity in the gas–phase ozonation of saturated hydrocarbons was observed by Williamson and Cvetanovic[37] who reported the relative rates of attack at tertiary:secondary:primary C—H bonds to be 30,000:300:1, respectively.

The stereospecificity of the reaction was shown by retention of configuration at the oxidized carbon atom (e.g. formation of *cis*-decalol from *cis*-decalin), and the higher reactivity of an equatorial C—H bond compared to an axial one (7:1) as well as by the higher reactivity of *cis*-decalin compared to *trans*-decalin (5.6:1).

Several reaction mechanisms have been proposed in order to explain these experimental results.

Whiting and coworkers[35] suggested a radical mechanism for the gas-phase reaction and an ionic mechanism for the reaction in condensed phases (Scheme 3).

$$RH + O_3 \longrightarrow R^{\cdot} + {}^{\cdot}O_3H \longrightarrow RO^{\cdot} + {}^{\cdot}O_2H \quad \text{(gas phase)}$$

$$RH + O_3 \longrightarrow R^{+}\ {}^{-}O_3H \longrightarrow ROH + O_2 \quad \text{(condensed phase)}$$

SCHEME 3

Hamilton and coworkers[36] found the regioselectivity and stereospecificity in the ozonation of saturated hydrocarbons in liquid phase to be similar to those found in insertion reactions by highly reactive carbenes and nitrenes.

Hellmann and Hamilton[38] investigated the ozonation of liquid alkanes at ambient temperatures. The following characteristics were found to be independent of the ozone's carrier gas (oxygen or nitrogen): (*a*) The stereospecificity (60–70% net retention of configuration) in which tertiary alcohols were formed. (*b*) The relative reactivities (1:13:110) of primary, secondary and tertiary C—H bonds. (*c*) The ratio (0.3) of ketone to alcohol products formed from cyclohexane. The stereochemical results were unchanged by performing the ozonation in more polar solvents, but a somewhat higher (82%) net retention of configuration was observed in an alkane solvent (octadecane) of higher viscosity. Lower retentions were observed when additives or solvents with which O_3 reacts were present. When the ozonations were performed in the presence of $FeCl_3$, alcohols were formed with 100% retention of configuration together with alkyl halides with essentially 100% inversion of configuration. The Hammet ρ value for the oxidation of substituted toluenes was found to be -2.07. These data were rationalized by processes outlined in Scheme 4.

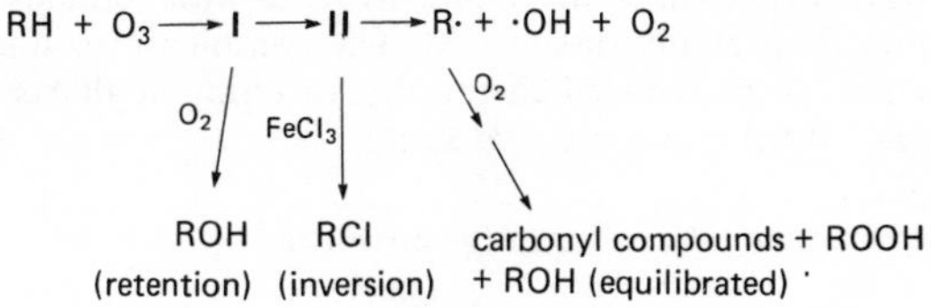

$$\text{RH} + \text{O}_3 \longrightarrow \text{I} \longrightarrow \text{II} \longrightarrow \text{R}\cdot + \cdot\text{OH} + \text{O}_2$$

$\text{I} \xrightarrow{O_2}$ ROH (retention); $\text{II} \xrightarrow{FeCl_3}$ RCl (inversion); $\text{R}\cdot \xrightarrow{O_2}$ carbonyl compounds + ROOH + ROH (equilibrated)

SCHEME 4

Ozone and alkane react to give either singlet O_2 and alcohol with retention of configuration or triplet O_2 and a triplet solvent-caged radical pair (**Ia** in Scheme 5). The radical pair is either trapped by $FeCl_3$ to give alkyl halide, or the radicals diffuse apart and react further to give the other observed oxidation products. **I** and **II** are intermediates and **I** exists in equilibrium between radical (**a**) and ionic (**b**) forms in a solvent cage (Scheme 5).

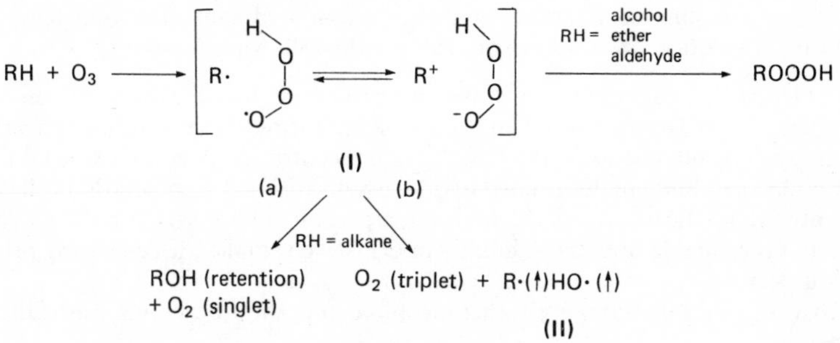

$$\text{RH} + \text{O}_3 \longrightarrow [\text{R}\cdot \ \cdot\text{O}{-}\text{O}{-}\text{O}{-}\text{H} \rightleftharpoons \text{R}^+ \ {}^-\text{O}{-}\text{O}{-}\text{O}{-}\text{H}] \ (\text{I}) \xrightarrow{\text{RH = alcohol, ether, aldehyde}} \text{ROOOH}$$

(a) RH = alkane: ROH (retention) + O_2 (singlet)

(b) RH = alkane: O_2 (triplet) + R·(↑)HO·(↑) (II)

SCHEME 5

Benson[39] has calculated the energy involved in the different suggested mechanisms. From thermodynamical considerations he has concluded that the radical mechanism ($O_3 + RH \rightarrow {}^{\cdot}O_3H + R^{\cdot}$) is too endothermic to occur at -78°C. He has proposed that the ozonation reaction of saturated compounds proceeds by hydride ion transfer (equation 11). He has calculated solvation enthalpies from the Kirkwood formula for dipole solvation and showed that these enthalpies lead to reasonable activation energies for the ozonation of alkanes, alcohols and acetals, all of which produce hydrotrioxides[40].

$$\text{RH} + \text{O}_3 \longrightarrow [\text{R}^+ + {}^-\text{O}_3\text{H}] \longrightarrow \text{products} \quad (11)$$

Mazur's group[41–44] have reinvestigated the low-temperature ozonation of saturated hydrocarbons in order to make it a synthetically useful process. Ozone is found to react with most organic solvents, thus interfering with ozonation of hydrocarbons. Therefore, ozonation reactions are carried out in neat liquids or in hydrocarbon solutions. It is found that at -80°C and below, stable ozone solutions are formed. Upon treatment of these solutions with a reducing agent (such as Ph_3P, $NaHSO_3$, KI) or on irradiation with visible light, or warming up to room temperature, smooth conversion to alcohols occurs, via net oxygen insertion into C—H bonds. However, no reaction products are observed when ozone is swept out of the cold solution by argon prior to any of the above treatments. These data suggest reversible formation of an ozone–hydrocarbon complex (Scheme 6)[45], in which the C—H bond is partially dissociated. The complex may undergo thermal or photochemical reactions or be reduced chemically, resulting in all cases in the same net insertion of an oxygen atom into the C—H bond.

$$\text{hydrocarbon} + O_3 \rightleftharpoons [\text{hydrocarbon}-O_3] \text{ complex}$$

$$\xrightarrow{T > -60^{\circ}C} \quad \xrightarrow{h\nu,\ \lambda > 334} \quad \xrightarrow{Ph_3P,\ NaHSO_3,\ KI}$$

$$\text{alcohols and ketones} + O_2$$

SCHEME 6

Oxidation of unactivated hydrocarbons by ozone either in the gas phase or in solution has clearly demonstrated the synthetic potential of these processes. However, neither of these methods allow general and practical applications in organic synthesis. Gas-phase reactions are very limited in terms of scale, volatility of substrates and temperature. Ozonations in solution are also impractical for the following reasons:

(*a*) Ozone is essentially insoluble in saturated hydrocarbons. Even at low temperatures (−78°C and below), at which ozone forms stable solutions in saturated hydrocarbons and no reaction takes place[42], its concentration does not exceed 0.1–0.2 %. Slightly higher solubility in fluorinated hydrocarbons (0.2–0.3 % at −80°C) still does not allow synthetic applications. At the higher temperatures necessary for the reactions to proceed at a reasonable rate, the solubility of O_3 is even smaller, necessitating prolonged ozonation periods.

(*b*) Most organic substrates are rather insoluble in perfluorocarbons, especially at low temperatures.

(*c*) Most organic solvents react with ozone[43] to form undersirable products, and even more so, reactive intermediates. This is true for ethyl acetate, various alcohols (including *t*-butanol), chloroform, carbon tetrachloride and even CCl_2F_2 and CCl_3F.

These severe limitations associated with the reaction matrix have led to an extensive search for alternatives to accommodate both ozone and substrate at any desired ratio. This search has resulted in the development of the dry ozonation method[46] which utilizes dry silica gel as a very convenient reaction matrix.

G. Dry Ozonation

Ozone is selectively adsorbed on silica gel in considerable quantities depending on its partial pressure and temperature[47] (Figure 1). The interaction between silica gel and ozone stabilizes the system even at high concentrations. Therefore, silica gel is commonly used in order to separate pure ozone from its carrier gas.

A typical output from a Welsbach ozonizer is 2 % ozone in oxygen, which means partial ozone pressures of 10–16 mm Hg. A regular silica gel may thus be loaded with ozone up to 4.5 % (w/w) at −78°C (Figure 2)[47].

The dry ozonation technique takes advantage of the availability of a high concentration of ozone. It is simple and easy to perform[46]: the silica gel is precoated with the organic substrate, loaded with ozone at −78°C, warmed to room temperature and the product eluted with an organic solvent.

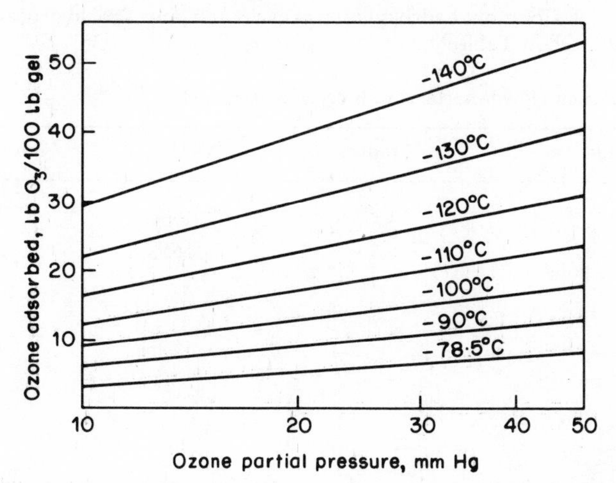

FIGURE 1. Smoothed adsorption isotherms for ozone in oxygen on Davison silica gel at temperatures ranging from −78.5° to −140°C. Total pressure 1 atm.

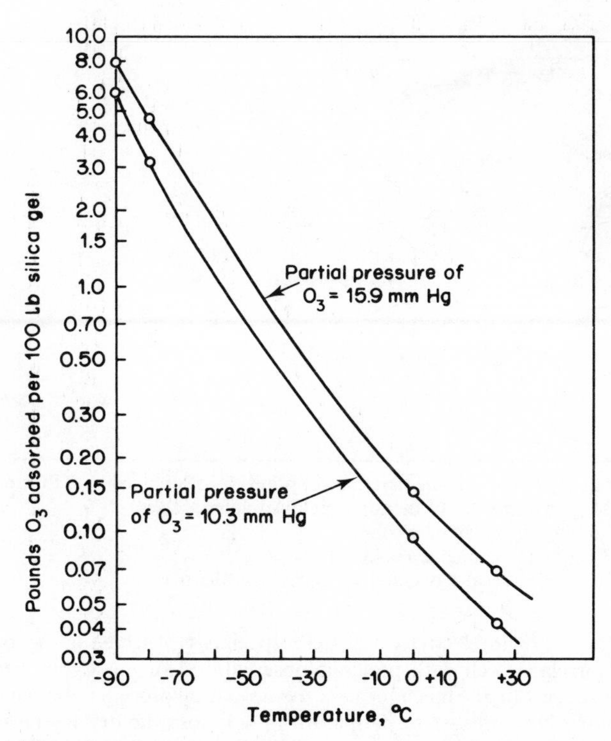

FIGURE 2. Adsorption of ozone on silica gel at a concentration of 2 weight % in oxygen. Total pressure 1 atm.

Typical examples for monohydroxylation of cyclic and polycyclic hydrocarbons via dry ozonation are listed in Table 1[46a].

TABLE 1. Reaction of hydrocarbons with ozone on silica gel

Substrate	Products	Yield (%)[a]	Conversion (%)[a]
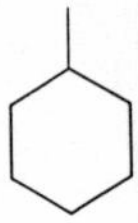	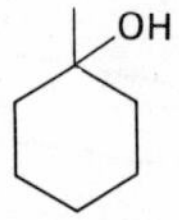 OH	65[b]	>99.5
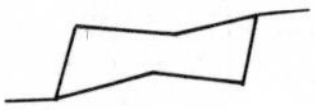	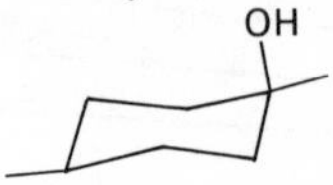 OH	79[c]	72
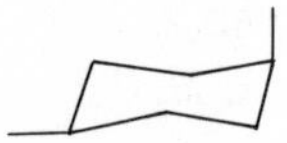	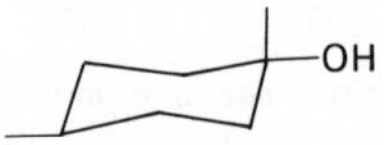 OH	76[d]	92
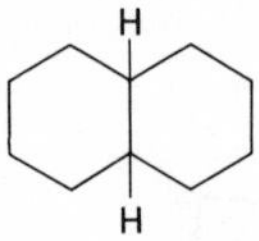 H H	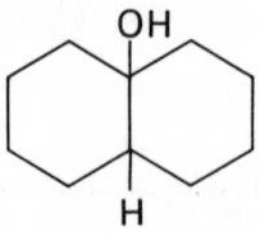 OH H	99	>99.5
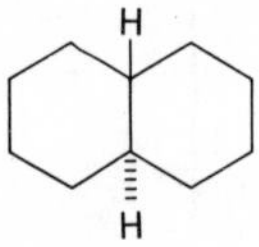 H H	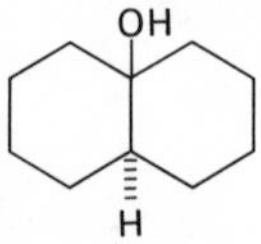 OH H	72[e]	88
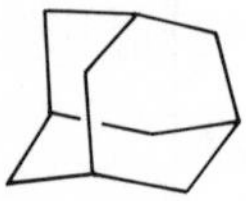	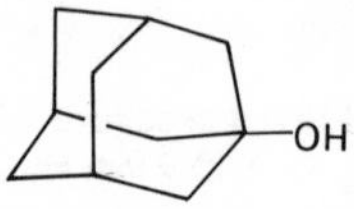 OH	99	>99.5
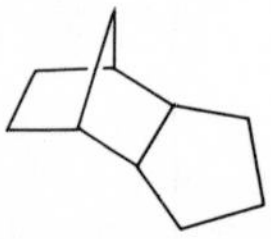	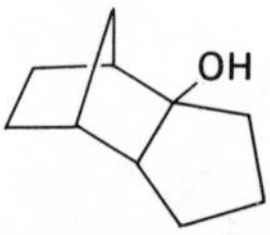 OH	90	>99.5

[a]Based on the quantity of the starting material consumed as determined by VPC analysis.
[b]In addition to 34 % of a mixture of the three methylcyclohexanones.
[c]In addition to 0.6 % of the epimeric alcohol.
[d]In addition to 3.5 % of the epimeric alcohol.
[e]In addition to 10 % *trans*-decal-1-one and 16 % *trans*-decal-2-one.

The high degree of regioselectivity and stereospecificity observed in dry ozonation of hydrocarbons correlates well with previous observations for analogous ozonations in solution[42], thus suggesting the mechanism of the ozonation process to be independent on the reaction matrix. In contrast to ozonation in solution, the dry ozonation is not a continuous process. This allows a controllable stepwise polyhydroxylation as demonstrated in Table 2[46a].

TABLE 2. Reaction of substituted hydrocarbons with ozone on silica gel

Substrate	Product	Yield (%)[a]	Conversion (%)[a]
 OH	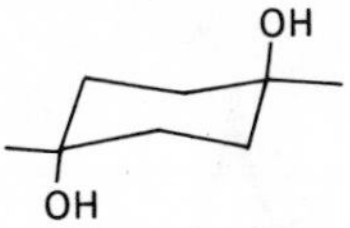 OH, OH	>99.5	95
 OH	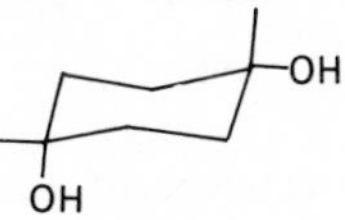 OH, OH	76[b]	97
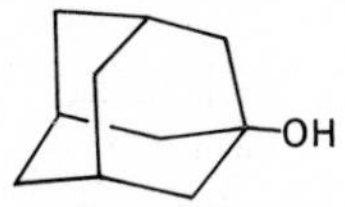 OH	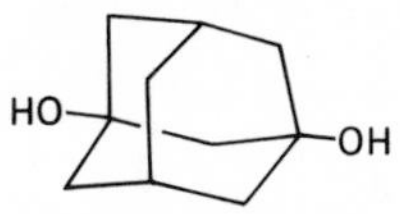 HO, OH	99	43
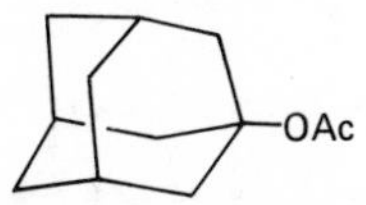 OAc	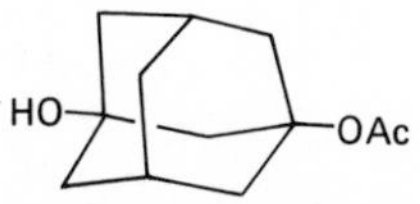 HO, OAc	86	50
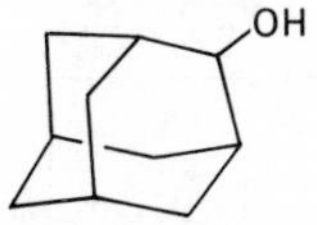 OH	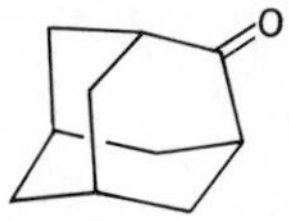 O	>99.5	>99.5

[a] Based on the starting material consumed.
[b] In addition to 5 % of the *trans* diol.

Tertiary carbon atoms in strained polycyclic molecules such as norbornane are rather inert towards dry ozonation conditions[48]. In such cases oxidation may be directed to secondary carbon atoms to form ketones (equations 12 and 13).

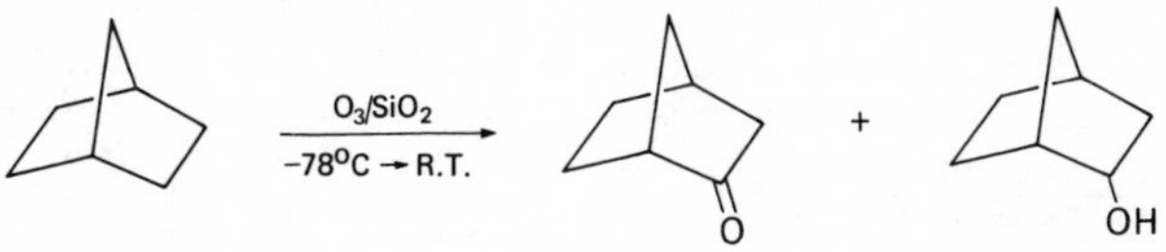

O_3/SiO_2, $-78^{o}C \rightarrow$ R.T.; products: O + OH (12)

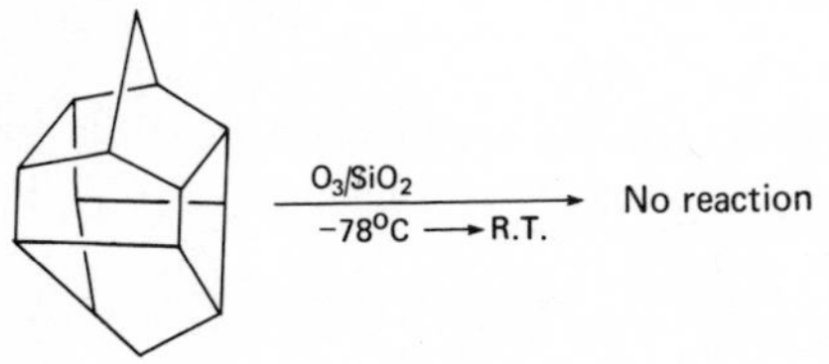

O_3/SiO_2, $-78^{o}C \rightarrow$ R.T.; No reaction (Ref. 49) (13)

Further examples for the inertness of strained carbon centres towards ozonation are given by various model compounds containing a cyclopropane ring[50]. However, the position α to the cyclopropyl ring, even if it is a secondary carbon, is electron-rich enough to react with ozone faster than a tertiary centre (equation 14).

O_3/SiO_2 87% 9% (14)

This oxidation α to the cyclopropyl unit has been found to be a quite general reaction as depicted in equations (15)–(23).

dry ozonation 95% conversion (15)

R	Yield (%)
Me	95
n-Pr	87

$(CH_2)_n$ dry ozonation 98% conversion (16)

n	Yield (%)
2	93
3	95
4	85

dry ozonation 97% conversion 88% + 7% + 2% (17)

dry ozonation 90% conversion 40% + 44% + 11% + 4% (18)

dry ozonation, 91% conversion → 70% + 29% (19)

dry ozonation, 55% conversion → 42% + 33% + 25% (20)

dry ozonation → 96% (21)

dry ozonation → 73% (22)

dry ozonation, −78°C → R.T. → dry ozonation, −78°C → R.T. → 61% (23)

The relative reactivity of tertiary sp^3 carbon atoms and aromatic sp^2 carbon atoms towards dry ozonation is exemplified by equations (24) and (25)[51].

NO_2 → O_3/SiO_2 → OH, NO_2 20% (24)

O_3/SiO_2 → CO_2H 80–90% (25)

A variety of functional groups can survive in the dry ozonation conditions (*vide infra*), in which tertiary carbon atoms undergo hydroxylation. These include tertiary alcohols and acetates, primary and secondary acetates, ketones and also bromides[52,53] (equation 26).

O_3/SiO_2, 78°C → R.T.; 35% conversion; 85% (26)

This observation allows significant expansion of the applications of the method. Saturated tertiary centres in unsaturated substrates can be selectively oxidized while the olefinic function is protected in the dibromide form as shown in equations (27) and (28)[53], and in the synthesis of 1α,25-dihydroxy vitamin D_3 (equation 39).

Br_2 → dry ozonation → Zn/AcOH (27)

77% O.P. — Br_2 → dry ozonation → Zn/AcOH, overall 10% — linalool 43% O.P. (28)

The synthesis of optically active linalool via dry ozonation of an optically active hydrocarbon related to citronellol illustrates a useful synthetic approach to chiral tertiary alcohols. The extent of linalool's optical purity represents 80 % retention of the configuration at the oxidized carbon atom. This observation supports the suggested mechanism[42] for ozonation of saturated hydrocarbons.

The regioselective oxidation at positions remote from the functional groups was also observed by Beckwith and his coworkers[54] in a number of model compounds (equations 29 and 30).

O_3/SiO_2, − 78°C → R.T.; 15% + 27% + 9% (29)

dry ozonation

AcO 1 2 3 4 5 6 7 8 9 10 11 12 O

(30)

Loading w/w (%)	Total yield (%)	% Oxidation at position:						
		5	6	7	8	9	10	11
1	20	4	18	41	18	10	4	5
1[a]	38	9	16	15	15	11	9	25
1[b]	45	6	14	10	10	9	9	42

[a]Substrate was precoated on 20 % of the silica gel then mixed with the untreated silica gel.
[b]As in *a*; substrate was precoated on 10 % of the silica gel.

Marked preference for attack at the penultimate position was observed for high loading rates. This was explained[54a] by the formation of a close-packed adsorbed monolayer leaving the hydrophobic part of the molecule remote from the silica surface, prone to attack by gaseous ozone.

An interesting similarity was found[54a] between the results of dry ozonation and biological oxidation of similar monofunctional compounds, where oxidation is directed to positions remote from the binding site. Further examples are given in equation (31).

R^1 R^2 dry ozonation OH + OH + O O

R^1 = OAc, R^2 = H	35%	25%	20%
R^1, R^2 = O	18%	12%	18%

(31)

An additional example for such a regioselectivity is given in equation (32)[54b,c].

AcO dry ozonation AcO OH + AcO + AcO OH + AcO OH OH (32)

SiO_2/Al_2O_3 5:1	72%	18%	0.7%	0.1%
SiO_2	61%	25%	0.6%	8.5%

Since silica gel is slightly acidic, it seems likely that in addition to acting as the reaction matrix it may play a role in the activation of ozone and the enhancement of its electrophilicity.

A much more significant increase in the electrophilicity of ozone can be achieved with the direct involvement of Lewis acids (as reported for ozonation of aromatic compounds[55]) or by using a highly acidic reaction matrix as studied by Olah and his coworkers[66–71] (*vide infra*).

Based on this approach, dry ozonation reactions of sluggishly reacting hydrocarbons such as norbonane were carried out with silica gel containing up to 10 % of various Lewis acids[48]. This technique gave quite substantial increases in conversion rates as illustrated in Table 3.

TABLE 3. Dry ozonation of norbornane on silica gel containing 10 % Lewis acid

Norbornane $\xrightarrow[O_3,\ -78°C]{MCl_n/SiO_2}$ 2-norbornanol (OH) + 2-norbornanone (O) + 2-chloronorbornane (Cl)

	OH	O	Cl	Conversion
$FeCl_3/SiO_2$	12 %	35 %	24 %	78 %
$TiCl_4/SiO_2$	2 %	5 %	53 %	77 %
$AlCl_3/SiO_2$	1 %	9 %	31 %	76 %
$SbCl_5/SiO_2$	10 %	14 %	29 %	59 %
$ZnCl_2/SiO_2$	2 %	41 %	6 %	49 %
$CuCl_2/SiO_2$	2 %	32 %	10 %	46 %
$CaCl_2/SiO_2$	3 %	38 %	—	41 %
$SbCl_3/SiO_2$	16 %	10 %	10 %	36 %
$CdCl_2/SiO_2$	3 %	31 %	—	34 %
HCl/SiO_2	—	1 %	18 %	21 %
SiO_2	2 %	32 %	—	34 %

The formation of chlorinated products accompanying the oxygenated ones was also reported for ozonation of hydrocarbons in solution in the presence of $FeCl_3$[38] (equation 33).

$$\text{cis-1,4-dimethylcyclohexane} \xrightarrow{O_3/FeCl_3} \text{(OH product)} + \text{(Cl product)} \qquad (33)$$

Dry ozonation of relatively large molecules such as steroids and triterpenes may lead to complex mixtures of polyoxygenated products. Therefore, it is recommended that in order to keep to low conversion (<20 %) rates the excess ozone should be desorbed from the silica gel at low temperatures by sweeping it with nitrogen.

As demonstrated in equations (34)–(38), steroids can be selectively ozonized[52c] to give oxidation products which are otherwise difficult to obtain.

O_3/SiO_2
$-78^oC \longrightarrow$ R.T.
74% conversion

AcO

OH

AcO

+

AcO

OH

+ other products

8%

7%

(34)

AcO

OH

dry ozonation
86% conversion

OH

AcO

OH

11%

(35)

AcO

Br Br

dry ozonation
74% conversion

OH

AcO

Br Br

11%

(36)

AcO

Br

Br

O_3/SiO_2
-65^oC
50% conversion

OH

AcO

Br

Br

32%

(37)

Vitamin D_3 $\xrightarrow{\text{3 steps}}$ OAc $\xrightarrow[\text{37\% conversion}]{\text{dry ozonation}}$ OH, OAc 15% $\xrightarrow{\text{3 steps}}$ OH

(38)

The presence of polar functional groups in the substrate molecule dictates a high degree of regioselectivity by directing the oxidation to positions remote from the polar groups. This selectivity may be rationalized either by assuming a specific interaction between the functional groups and the silica surface (*vide supra*) or by an unproductive reversible complexation of ozone to these functional groups.

The key step in the total synthesis of 1α,25-dihydroxy vitamin D_3[52] (equation 39) was based on this effect of regiocontrol by functional groups, as well as on olefin protection by dibromination.

By using similar conditions, friedelane (**1**) was oxidized to the monooxygenated products **2**–**7** (equation 40)[56].

The main product (**3**) was a secondary oxidation product resulting from initial hydroxylation at position 18 followed by dehydration to give the corresponding C(18)–C(19) olefin, which was subsequently epoxidized by ozone (equation 41)[56].

(**1**) $\xrightarrow{O_3/SiO_2}$ OH $\longrightarrow$ $\xrightarrow{O_3}$ O (**3**)

(41)

Dry ozonation of friedelin (**2**) afforded an analogous mixture of products (equation 42)[56].

5 steps from cholesterol

AcO AcO Br Br

dry ozonation 11% conversion

AcO AcO Br Br OH

51%

4 steps

HO OH OH

1α, 25-dihydroxy vitamin D_3 (39)

(1)

O_3/SiO_2 $-78^{o} \longrightarrow -60^{o}C$ 18% conversion

(3) 48% + (4) 2% + (5) 11% + (6) 11%

(7) 1% + Friedelin (2) 0.8% (40)

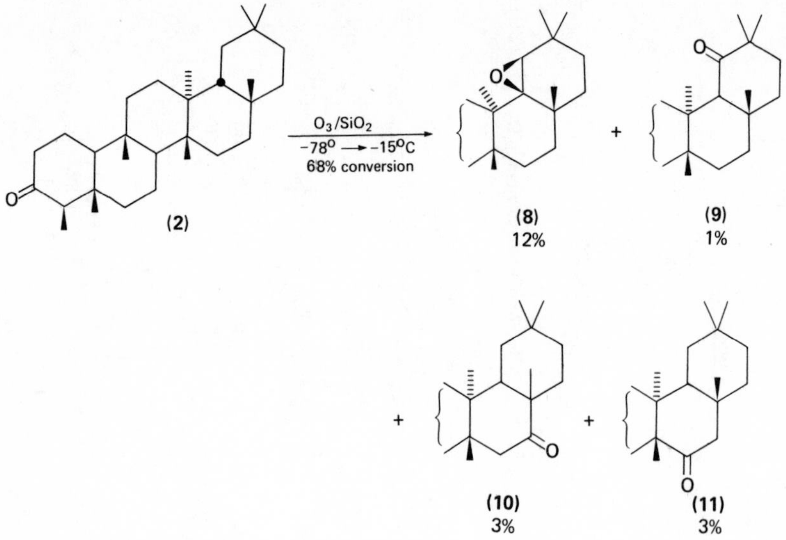

O_3/SiO_2, $-78° \rightarrow -15°C$, 68% conversion

(2) → (8) 12% + (9) 1% + (10) 3% + (11) 3% (42)

Similarly, dry ozonation of 3,28-diacetoxylupane (**12**) under low conversion rate (10 %) yielded only one product (**13**) (equation 43)[57].

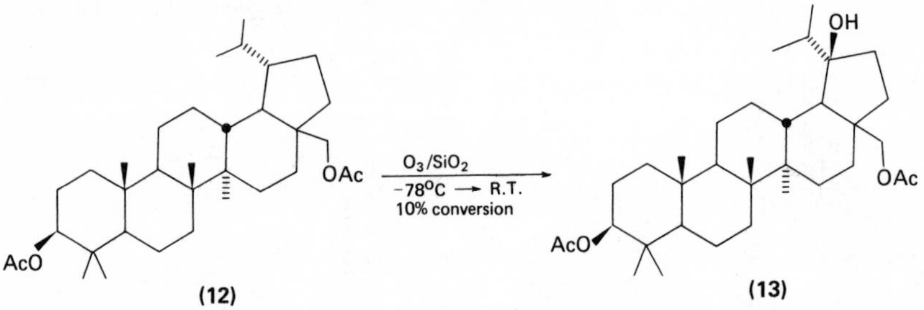

O_3/SiO_2, $-78°C \rightarrow$ R.T., 10% conversion

(12) → (13) (43)

Dry ozonation of naturally occurring sesquiterpenes was found to be highly efficient as studied independently by two different groups[48,58] (equations 44–47).

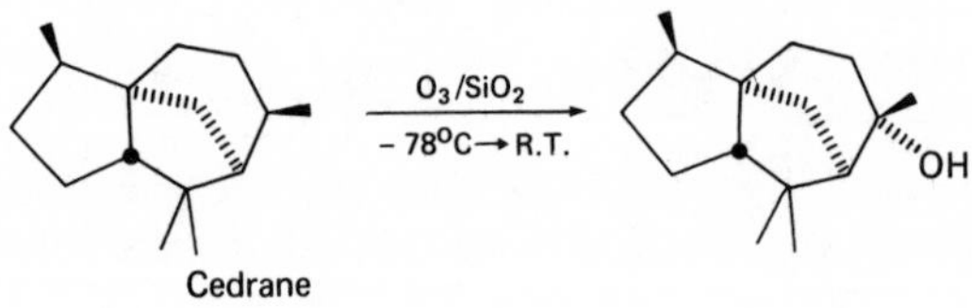

O_3/SiO_2, $-78°C \rightarrow$ R.T.

Cedrane → (44)

Cedrol (R = H)

Cedryl acetate (R = Ac)

O_3/SiO_2, $-78^{\circ}C \rightarrow R.T.$

50%

(45)

Cedrane oxide

O_3/SiO_2, $-78^{\circ}C \rightarrow R.T.$

30% + 10%

(46)

Patchoulol (R = H)

Patchoulyl acetate (R = Ac)

O_3/SiO_2, $-78^{\circ}C \rightarrow R.T.$

(47)

H. Ozonation of Carbon–Carbon Single Bonds

Oxidation products arising from carbon–carbon bond cleavage have been observed in many dry ozonation reactions. In most cases they were minor side-products accompanying the major route of ozone insertion into C—H bonds. In some cases considerable oxidative cleavage of single C—C bonds took place and yielded substantial quantities of the corresponding ketones[54,59,60] (equations 48–50).

dry ozonation, 100% conversion

66% + 19% + 15%

(Refs. 46, 60) (48)

dry ozonation

AcO ... 25% + AcO ... CO_2H 5% + tertiary alcohols 70%

(Refs. 54b, c) (49)

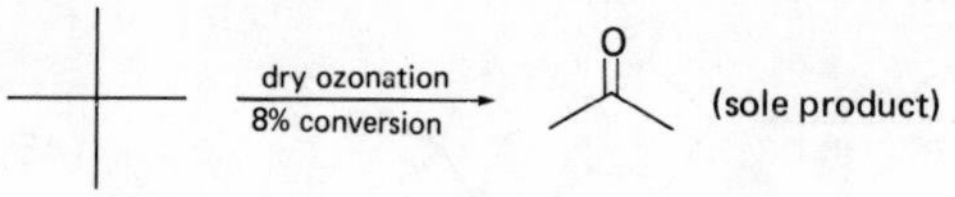

dry ozonation / 8% conversion (sole product) (Refs. 59, 60) (50)

A special case of C—C bond cleavage was observed when bicyclo[2.1.0]hydrocarbons were subjected to dry ozonation[61]. The strained central bond was efficiently cleaved when ozonized at −50° to −30°C (equation 51).

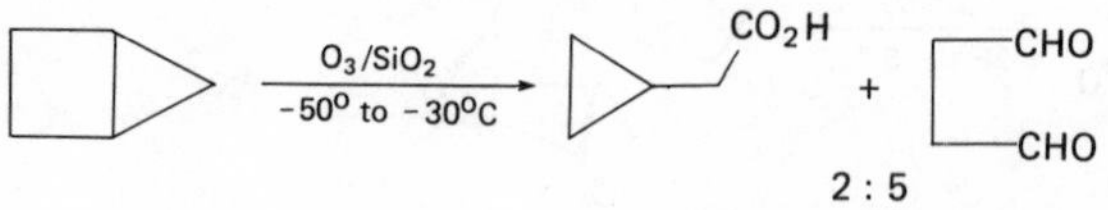

O_3/SiO_2, −50° to −30°C; CO_2H + CHO, CHO; 2 : 5 (51)

Based on deuterium labeling, a plausible mechanism for this cleavage has been suggested[61]. Initial 1,3-dipolar addition of ozone to the central C—C bond is followed by rearrangements which are analogous to those of the primary ozonides formed in ozonolysis of olefins (Scheme 7).

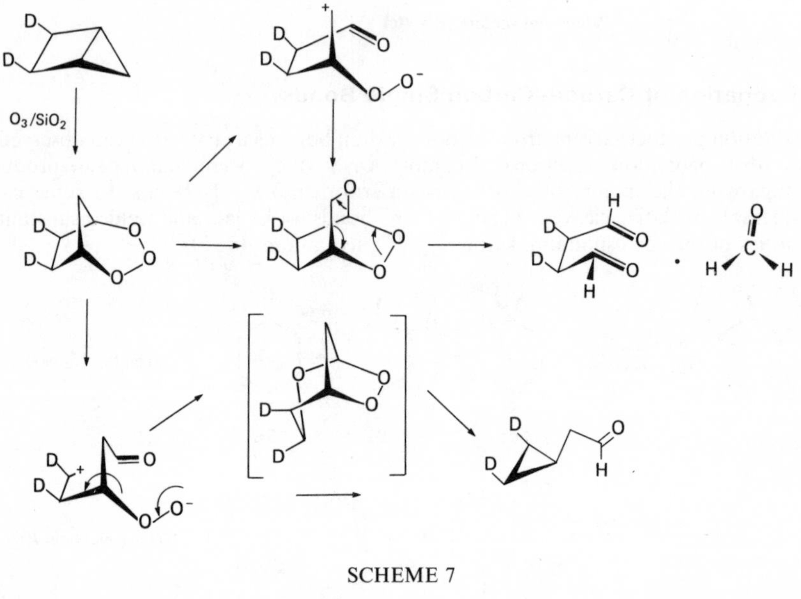

SCHEME 7

Further examples for ozonation of similar systems are depicted in quations (52)–(54)[61].

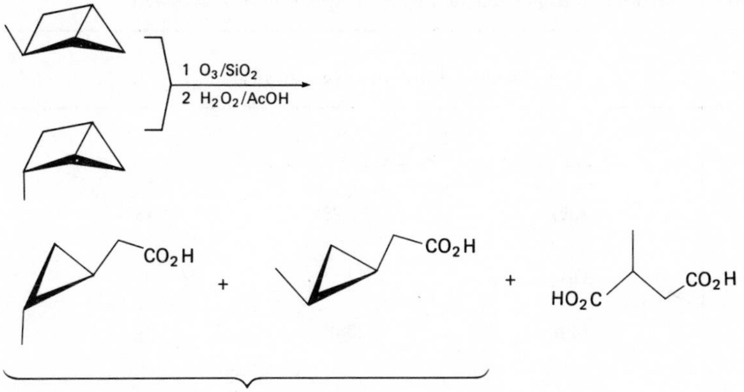

$$\xrightarrow[2\ H_2O_2/AcOH]{1\ O_3/SiO_2} \quad \text{(5)} : \text{(7)} \tag{52}$$

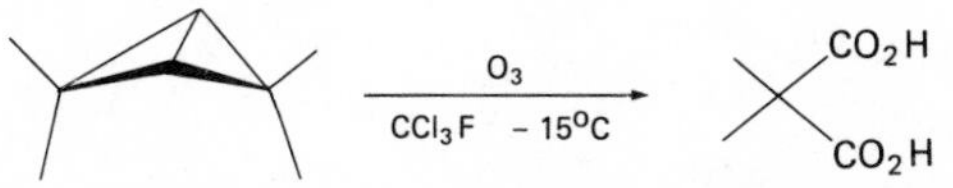

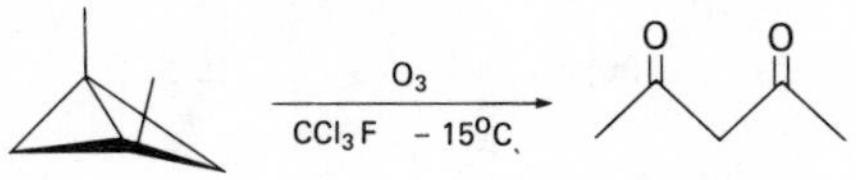

CO_2H + CO_2H + HO_2C CO_2H

5 : 7

$$\xrightarrow[CCl_3F\ \ -15^{o}C]{O_3} \tag{53}$$

CO_2H CO_2H

21%

$$\xrightarrow[CCl_3F\ \ -15^{o}C]{O_3} \tag{54}$$

O O

14%

High yields of C—C bond cleavage were obtained when dry ozonation was carried out continuously at $-45°C$[60,61]. Interestingly, the relative yields of ketones arising from cleavage of alkyl groups were found to be considerably higher in acyclic hydrocarbons than in cyclic ones as demonstrated in Table 4[59]

The proposed mechanism[59,60] involves a direct insertion of ozone into the C—C bond, leading to a dialkyl trioxide which in turn decomposes to give the observed cleavage products. The transition state for the insertion step (equation 55) is assumed to be analogous to that proposed by Olah[66] for the ozonation of alkanes in superacids (*vide infra*) and similar to the corresponding insertion of ozone into C—H bonds (equation 56)[42].

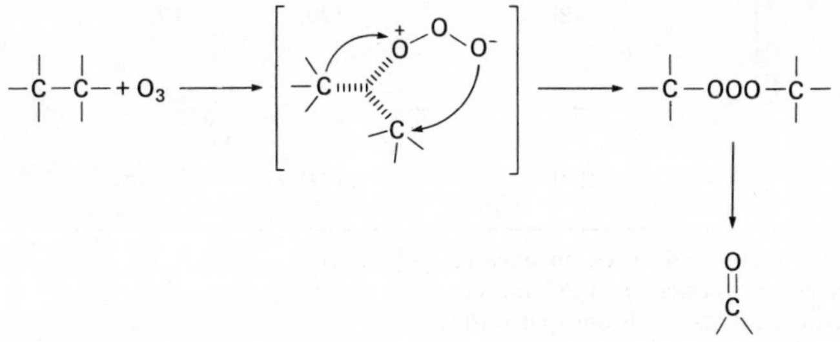

$$-\overset{|}{\underset{|}{C}}-\overset{|}{\underset{|}{C}}- + O_3 \longrightarrow [\text{transition state}] \longrightarrow -\overset{|}{\underset{|}{C}}-OOO-\overset{|}{\underset{|}{C}}- \longrightarrow \ >C{=}O \tag{55}$$

TABLE 4. Product distribution from ozonation of some hydrocarbons[a] at $-45°C$

Starting Material	Product distribution (molar yield[b], %)				
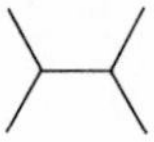	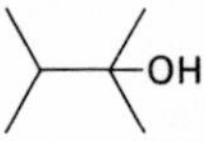 (35)	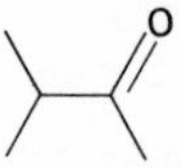 (28)	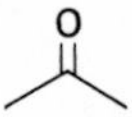 (37)		
	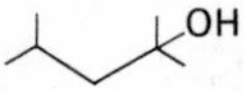 (43)	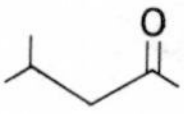 (38)[c]	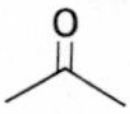 (19)		
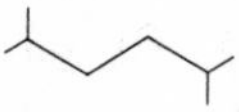	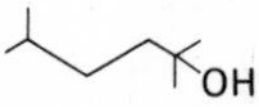 (63)	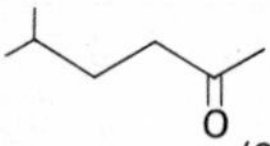 (22)	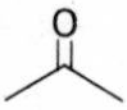 (15)		
	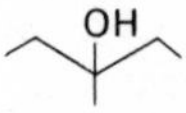 (34)	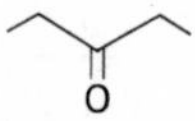 (19)	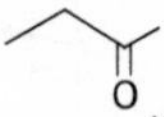 (31)	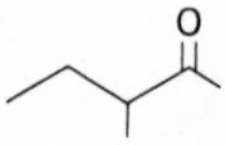 (16)	
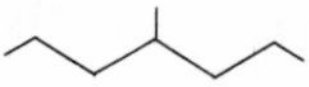	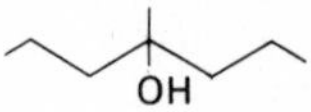 (30)	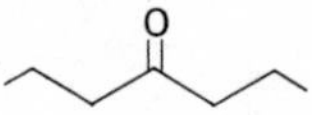 (19)	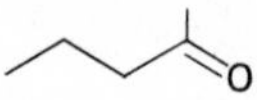 (51)		
	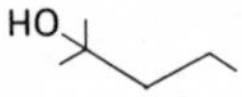 (21)	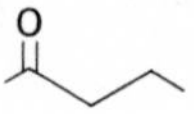 (25)	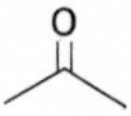 (52)	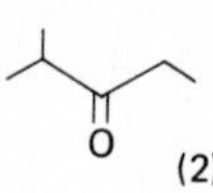 (2)	
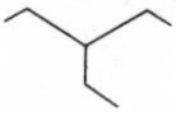	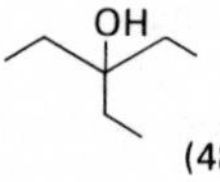 (48)	 (42)	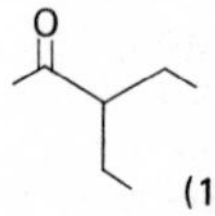 (10)		
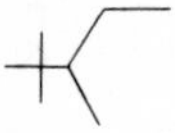	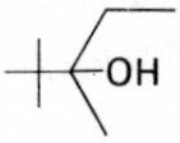 (48)	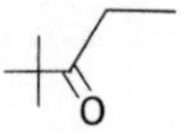 (20)	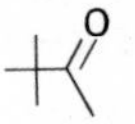 (10)	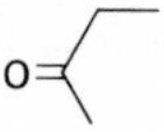 (14)	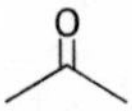 (8)
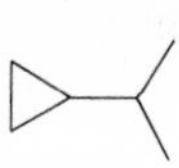	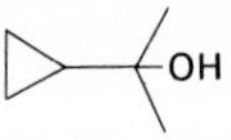 (57)	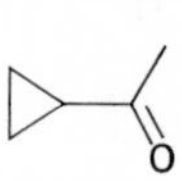 (37)	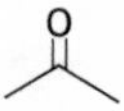 (6)		

[a]The substrates were preadsorbed on silica gel (1 % w/w).
[b]Based on detected products by VPC analysis.
[c]Included small amounts of diisopropyl ketone.

$$-\overset{|}{\underset{|}{C}}-H + O_3 \longrightarrow \left[-\overset{|}{\underset{|}{C}}\cdots\overset{+}{O}(-O-O^-)\cdots H \right] \longrightarrow -\overset{|}{\underset{|}{C}}-OH + O_2 \qquad (56)$$

This proposed mechanism is supported by the fact that the yields of ketones formed by the cleavage of either primary, secondary or tertiary alkyl groups are similar in magnitude and not substantially different (e.g. Table 4, penultimate entry). Also, the constant product distribution (Figure 3)[59,60] throughout the reaction suggests the formation of one common intermediate.

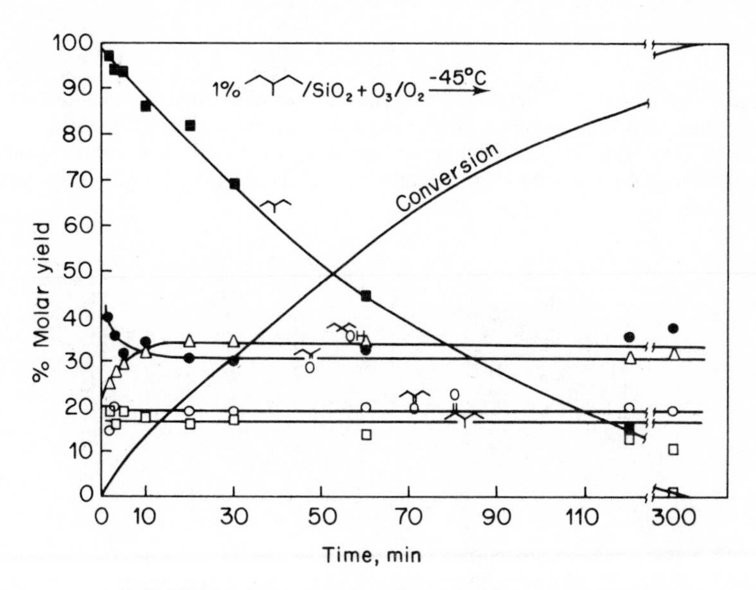

FIGURE 3. Dry ozonation of 3-methylpentane at −45°C.

When ozonation was carried out at temperatures higher than −20°C (Table 5)[59,60] yields of the respective ketones originating from cleavage of the more substituted alkyl groups increased, mainly at the expense of the tertiary alcohols. It may be assumed that at higher temperatures the fragmentation of alkanes into ketones occurs mainly by an alternative mechanism involving tertiary alkoxy radicals generated in the cleavage of tertiary C—H bonds by ozone.

Dry ozonation was also found to be an efficient method for the ozonation of a number of functional groups such as amines[62,63] and alkenes[64,65].

I. Ozonation Reactions in Strong Acidic Media

The reactions of ozone in highly acidic solvents such as FSO_3H–SbF_5–SO_2, FSO_3H–SbF_5–SO_2ClF and HF–SbF_5–SO_2ClF have been thoroughly investigated by

TABLE 5. Product distribution[a] from ozonation of 3-methylpentane at various temperatures

Temp. (°C)	OH	O	O	O
−45	32	15	20	33
−23	30	15	20	35
0	22	13	20	44
30	8	10	23	60

[a] In molar yield (%), and based on detected products by VPC analysis.

Olah and his coworkers[66]. They found that the electrophilic ozonation of alkanes occurs readily in these superacidic media to give oxygenated products resulting from C—H or C—C bond cleavage. The results were rationalized by initial electrophilic attack by protonated ozone on the σ bonds of alkanes through pentacoordinated carbonium ions (equation 57).

$$\mathrm{Me_3C{-}H} \xrightarrow{^{+}O_3H} \left[\mathrm{Me_3C}\cdots\begin{smallmatrix}\mathrm{O{-}O{-}OH}\\ \mathrm{H}\end{smallmatrix}\right]^{+} \xrightarrow{-\mathrm{H^+}} \left[\mathrm{Me_3C{-}O{-}O{-}OH}\right] \qquad (57)$$

$$\xrightarrow[-\mathrm{H_2O_2}]{\mathrm{H^+}} \left[\mathrm{Me_3C{-}O^{+}}\right] \longrightarrow \mathrm{Me_2C{=}\overset{+}{O}{-}Me} \xrightarrow{\mathrm{H_2O}} \mathrm{Me_2C{=}O}$$

Protonated ozone, O_3H^+, should have a much higher affinity (i.e. be a more powerful electrophile) for σ-donor single bonds in alkanes than neutral ozone. Therefore O_3H^+ initiates ozonolysis of alkanes in a similar manner to nitrolysis by the nitronium ion NO_2^+ [67], chlorolysis by chloronium ion Cl^+ [68] or protolysis by superacids[69]. However, the relative order of reactivity of σ bonds in alkanes with protonated ozone was found to be generally the following[66]: tertiary C—H > secondary, primary C—H > C—C; whereas the more usual order of reactivity towards a large variety of electrophilic reagents is: tertiary C—H > C—C > secondary C—H > primary C—H.

Several examples for ozonation of branched and linear alkanes with protonated ozone[70] are given in Tables 6 and 7.

TABLE 6. Ozonation of branched alkanes in magic acid at −78°C

Alkane	Major products			
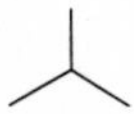	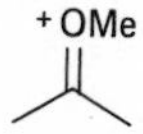 $^{+}$OMe	(~100%)		
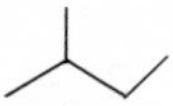	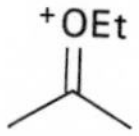 $^{+}$OEt	(~100%)		
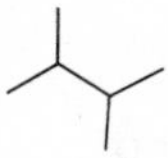	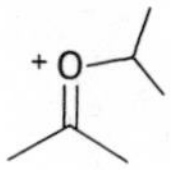 $^{+}$O	(60%)	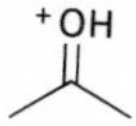 $^{+}$OH	(40%)
	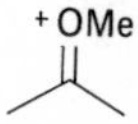 $^{+}$OMe	(50%)	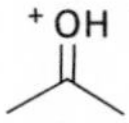 $^{+}$OH	(50%)

TABLE 7. Ozonation of straight-chain alkanes in magic-acid–SO_2ClF at −78°C

Alkane	Major products
CH_4 (at −50°C)	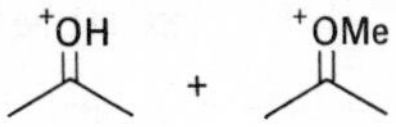 $^{+}$OH + $^{+}$OMe
C_2H_6	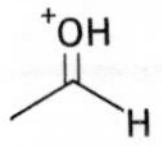 $^{+}$OH, H
C_3H_8	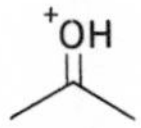 $^{+}$OH
C_4H_{10}	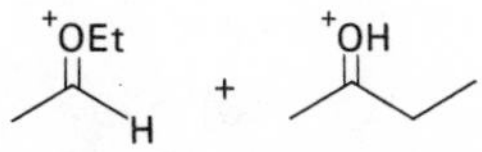 $^{+}$OEt, H + $^{+}$OH
C_5H_{12}	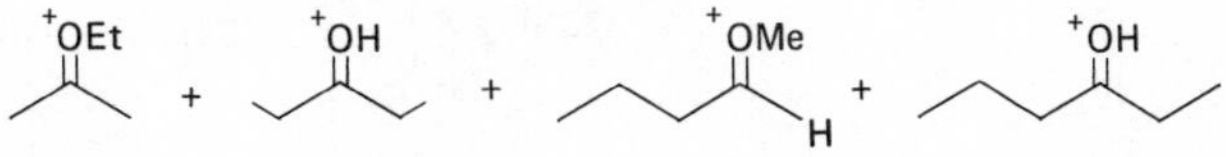 $^{+}$OEt + $^{+}$OH + $^{+}$OMe, H + $^{+}$OH

Reactions of cycloalkanes with ozone in superacid media proceed in a similar manner[66] (Scheme 8) with the formation of protonated cycloketones and cyclic carboxonium ions. Several examples are given in Table 8.

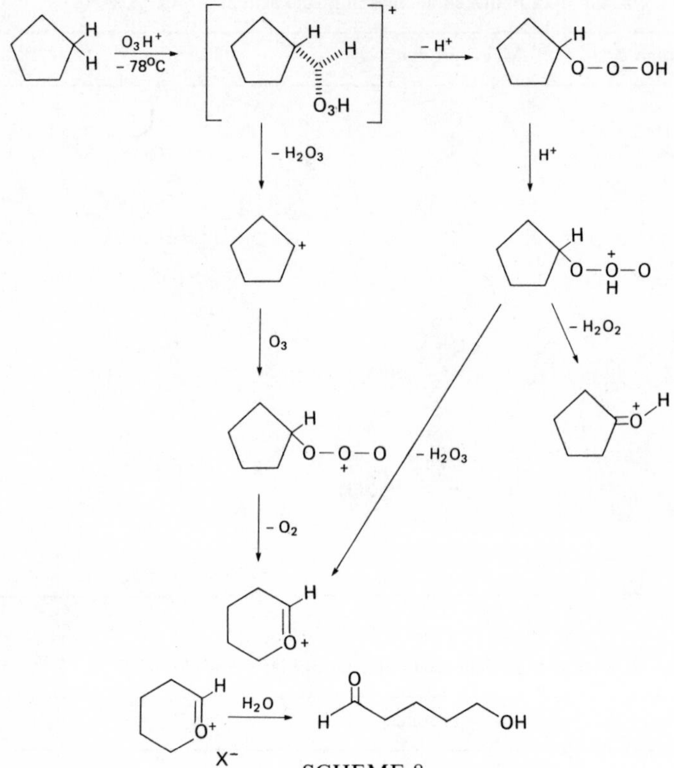

H H $\xrightarrow[-78°C]{O_3H^+}$ [H H O_3H]$^+$ $\xrightarrow{-H^+}$ H O—O—OH

$-H_2O_3$ ↓ ↓ H^+

$^+$ H O—$\overset{+}{O}$—O H

O_3 ↓ ↘ $-H_2O_2$

H O—O—O + $-H_2O_3$ $\overset{+}{O}$ H

$-O_2$ ↓

H O+

H O+ X$^-$ $\xrightarrow{H_2O}$ H O OH

SCHEME 8

TABLE 8. Ozonation of cycloalkanes in magic-acid–SO_2ClF at $-78°C$

Alkane	Major products		
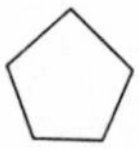	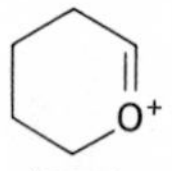 O+ (20%)	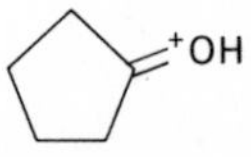 $^+$OH (80%)	
	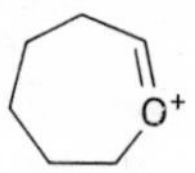 O+ (20%)	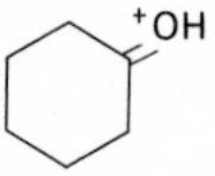 $^+$OH (65%)	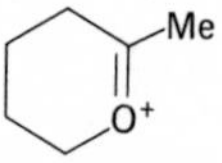 Me O+ (15%)
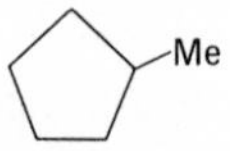 Me	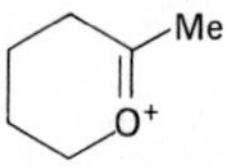 Me O+		
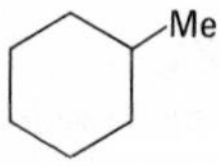 Me	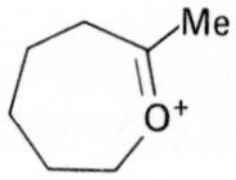 Me O+		

An interesting and synthetically useful regiocontrol by functional groups has been found by Olah and his coworkers[71]. They observed that methyl alcohol, acetone or acetaldehyde are not oxidized by ozone in magic-acid–SO_2ClF solution. In the superacid system these oxygenated compounds are present as completely protonated species, i.e. as Me^+OH_2, $Me_2C{=}O^+H$ and $Me(H)C{=}O^+H$; thus protonated ozone does not readily react with σ bonds located in the proximity of the carboxonium centre in the protonated substrates. However, reactions can occur at σ bonds which are located sufficiently far away from the charged carboxonium centre. Thus, the oxygenation of functionalized compounds in superacid media takes place at positions γ or further removed from the oxonium centre. These reactions allow the preparation of bifunctional oxygenated derivatives, as depicted in Table 9[66].

TABLE 9. Ozonation of alcohols, ketones and aldehydes in magic acid

Substrate	Reaction temp. (°C)	Conversion (%)	Product
OH	−40	30	$^+$OH $^+$OH$_2$
OH	−40	100	$^+$OMe
OH	−40	60	$^+$O $^+$OH$_2$
OH	−78	100	$^+$OH $^+$OH$_2$
O	−40	no reaction	
O	−78	100	$^+$OH $^+$OH
CHO	−78	80	$^+$OH H $^+$OH

Ozonation in superacid medium has been utilized for selective oxyfunctionalization of steroids[72]. Regiocontrol was achieved by existing keto functions as illustrated by equations (58)–(60).

$\xrightarrow[-20^{o}C,\ 1.5\ h]{O_3,\ HF\text{-}SbF_5}$ 15% + 50% (58)

$\xrightarrow[-20^{o}C,\ 1\ h]{O_3,\ HF\ SbF_5}$ 28% + 38% (59)

$\xrightarrow[-35^{o}C,\ 1\ h]{O_3\ \ HF\text{-}SBF_5}$ 12% + 22% + 6% (60)

V. REFERENCES

1. (a) C. F. Schoenbein, *Compt. Rend.*, **10**, 706 (1840).
   (b) C. F. Schoenbein, *J. Prakt. Chem.*, **52**, 135 (1851).
   (c) C. F. Schoenbein, *J. Prakt. Chem.*, **66**, 273 (1855).
2. P. Kolsaker, P. S. Bailey, F. Dobinson and N. B. Kumar, *J. Org. Chem.*, **29**, 1409 (1964).
3. (a) R. E. Oesper, *J. Chem. Ed.*, **6**, 432, 677 (1929).
   (b) R. Criegee, *Record Chem. Progr.*, **18**, 111 (1957).
   (c) P. S. Bailey, *Chem. Rev.*, **58**, 925 (1958).
   (d) R. Criegee, *Advan. Chem. Ser.*, **21**, 133 (1959).
   (e) R. Criegee in *Peroxide Reaction Mechanisms* (Ed. J. O. Edwards), John Wiley and Sons, New York–London, 1962, pp. 29–39.
   (f) R. W. Murray, *Trans. N.Y. Acad. Sci.*, [2] **29**, 854 (1967).
   (g) R. W. Murray, *Acc. Chem. Res.*, **1**, 313 (1968).
   (h) R. Criegee, *Chimia*, **22**, 392 (1968).
   (i) R. W. Murray, *Tech. Methods Org. Organomet. Chem.*, **1**, (1969).
   (j) J. S. Below in *Oxidation* (Ed. R. L. Augustine), Vol. 1, Marcel Dekker, New York, 1969, pp. 259–335.
   (k) E. H. Pryde and J. C. Cowan, *Top. Lipid Chem.*, **2**, 1 (1971).
   (l) J. S. Murphy and J. R. Orr (eds.), *Ozone Chemistry and Technology. A Review of the Literature:* 1961–1974. Franklin Institute Press, Philadelphia, Penn., 1975.
   (m) R. Criegee, *Angew. Chem.* (*Intern. Ed.*), **14**, 745 (1975).
   (n) R. Criegee, *Chemiker Ztg.*, **99**, 138 (1975).
4. P. S. Bailey, *Ozonation in Organic Chemistry*, Vol. 1, Academic Press, New York, 1978.
5. P. S. Bailey, *Ozonation in Organic Chemistry*, Vol. 2. Academic Press, New York, 1982.
6. (a) A. Maggiolo and S. J. Niegowski, *Advan. Chem. Ser.*, **21**, 202 (1959).
   (b) L. Horner, H. Schaefer and W. Ludwig, *Chem. Ber.*, **91**, 75 (1958).
   (c) H. B. Henbest and M. J. W. Stratford, *J. Chem. Soc.*, 711 (1964).
   (d) G. P. Shulman, *Can. J. Chem.* **43**, 3069 (1965).
7. (a) P. S. Bailey and J. E. Keller, *J. Org. Chem.*, **33**, 2680 (1968).
   (b) P. S. Bailey, T. P. Carter and L. M. Southwick, *J. Org. Chem.*, **37**, 2997 (1972).
   (c) P. S. Bailey, J. E. Keller, D. A. Mitchard and H. M. White, *Advan. Chem. Ser.*, **77**, 58 (1968).
   (d) S. Bailey, D. A. Mitchard and A. Y. Khashab, *J. Org. Chem.*, **33**, 2675 (1968).
   (e) P. S. Bailey, J. E. Keller and T. P. Carter, *J. Org. Chem.*, **35**, 2777 (1970).
   (f) P. S. Bailey and J. E. Keller, *J. Org. Chem.*, **35**, 2782 (1970).
8. J. Van Alpher, *Rec. Trav. Chim.*, **64**, 305 (1945).
9. R. E. Miller, *J. Org. Chem.*, **26**, 2327 (1961).
10. (a) R. E. Erickson, P. J. Andrulis, Jr., J. C. Collins, M. L. Lungle and G. D. Mercer, *J. Org. Chem.*, **34**, 2961 (1969).
    (b) A. H. Riebel, R. E. Erickson, C. J. Abshire and P. S. Bailey, *J. Amer. Chem. Soc.*, **82**, 1801 (1960).
11. L. Horner, H. Schaefer and W. Ludwig, *Chem. Ber.*, **91**, 75 (1958).
12. A. Maggiolo and E. A. Blair, *Advan. Chem. Ser.*, **21**, 200 (1959).
13. G. Ayreg, D. Barnard and D. T. Woodbridge, *J. Chem. Soc.*, 2089 (1962).
14. Q. E. Thompson, *J. Amer. Chem. Soc.*, **83**, 845 (1961).
15. R. W. Murray and M. L. Kaplan, *J. Amer. Chem. Soc.*, **91**, 5358 (1969).
16. C. C. Price and A. L. Tumolo, *J. Amer. Chem. Soc.*, **86**, 4691 (1964).
17. P. Deslongchamps, P. Atlani, D. Frehel, A. Malaval and C. Moreau, *Can. J. Chem.* **52**, 3651 (1974), and references cited therein.
18. R. E. Erickson, R. T. Hansen and J. Harkins, *J. Amer. Chem. Soc.*, **90**, 6777 (1968).
19. (a) R. W. Murray, W. C. Lumma, Jr. and J. W. P. Lin, *J. Amer. Chem. Soc.*, **92**, 3205 (1970).
    (b) F. E. Stray, D. E. Emge and R. W. Murray, *J. Amer. Chem. Soc.*, **96**, 5671 (1974).
    (c) F. E. Stray, D. E. Emge and R. W. Murray, *J. Amer. Chem. Soc.*, **98**, 1880 (1976).
20. (a) P. S. Bailey and D. A. Lerdal, *J. Amer. Chem. Soc.*, **100**, 5820 (1978).
    (b) D. Lerdal, *Ph.D. Dissertation*, The University of Texas at Austin, 1971.
21. F. Kovac and B. Plesnicar, *J. Chem. Soc. Chem. Commun.*, 122 (1978).
22. R. J. Taillefer, S. E. Thomas, Y. Nadeau, S. Fliszar and H. Henry, *Can. J. Chem.*, **58**, 1138 (1980).

23. B. M. Brudnik, L. V. Spirikhin, E. M. Kuramshin, U. B. Imashev, S. S. Zlofskii and D. L. Rakhmankulov, *Zh. Org. Khim.*, **16**(6), 1281 (1980).
24. (a) W. Lohringer and J. Sixt, *German Patent*, No. 1793 569 (1972); *Chem. Abstr.*, **78**, 29246f (1973).
(b) V. A. Bryukhovetskii, S. S. Levush and V. U. Shevchuk, *Zh. Prikl. Khim.*, **49**(2), 473 (1976).
(c) P. Sundararaman, E. C. Walker and C. Djerassi, *Tetrahedron Letters*, 1627 (1978).
25. H. M. White and P. S. Bailey, *J. Org. Chem.*, **30**, 3037 (1965).
26. R. E. Erickson, D. Bakalik, C. Richards, M. Scanlon and G. Huddleston, *J. Org. Chem.*, **31**, 461 (1966).
27. M. Teramoto, T. Ito and H. Teranishi, *J. Chem. Eng. Japan*, **10**(3), 218 (1977).
28. (a) L. Spialter and J. D. Austin, *J. Amer. Chem. Soc.*, **87**, 4406 (1965).
(b) J. D. Austin and L. Spialter, *Advan. Chem. Ser.*, **77**, 26 (1968).
(c) L. Spialter, L. Pazdernik, S. Bernstein, W. A. Swansiger, G. R. Buell and M. E. Freeburger, *Advan. Chem. Ser.*, **112**, 65 (1972).
29. J. E. Batterbee and P. S. Bailey, *J. Org. Chem.*, **32**, 3899 (1967).
30. P. S. Bailey, L. M. Southwick and T. P. Carter, Jr., *J. Org. Chem.*, **43**, 2657 (1978), and references cited therein.
31. P. S. Bailey, D. A. Lerdal and T. P. Carter, Jr., *J. Org. Chem.*, **43**, 2663 (1978).
32. M. Otto, *Ann. Chem. Phys.* (*VII*), **13**, 109 (1898).
33. J. Durland and H. Adkins, *J. Amer. Chem. Soc.*, **61**, 429 (1939).
34. (a) C. C. Schubert and R. N. Pease, *J. Chem. Phys.*, **24**, 919 (1956).
(b) C. C. Schubert and R. N. Pease, *J. Amer. Chem. Soc.*, **78**, 2044 (1956).
35. M. C. Whiting, A. J. N. Bolt and J. H. Parish, *Advan. Chem. Ser.*, **77**, 4 (1968).
36. G. A. Hamilton, B. S. Ribner and T. M. Hellman, *Advan. Chem. Ser.*, **77**, 15 (1968).
37. D. G. Williamson and R. J. Cvetanovic, *J. Amer. Chem. Soc.*, **92**, 2949 (1970).
38. T. M. Hellman and G. A. Hamilton, *J. Amer. Chem. Soc.*, **96**, 1530 (1974).
39. S. W. Benson, *Advan. Chem. Ser.*, **77**, 74 (1968).
40. P. S. Nangia and S. W. Benson, *J. Amer. Chem. Soc.*, **102**, 3105 (1980).
41. Y. Mazur, *Pure Appl. Chem.*, **41**, 145 (1975).
42. T. H. Varkony, S. Pass and Y. Mazur, *J. Chem. Soc., Chem. Commun.*, 437, (1974).
43. T. H. Varkony, *Ph.D. Thesis*, Weizmann Institute of Science, Israel, 1975.
44. S. Pass, *M.Sc. Thesis*, Weizmann Institute of Science, Israel, 1974.
45. T. H. Varkony, S. Pass and Y. Mazur, *J. Chem. Soc., Chem. Commun.*, 709 (1975).
46. (a) Z. Cohen, E. Keinan, Y. Mazur and T. H. Varkony, *J. Org. Chem.*, **40**, 2141 (1975).
(b) Y. Mazur, Z. Cohen, E. Keinan and T. H. Varkony, *Israel Patent*, No. 47344 (1975).
(c) Z. Cohen, T. H. Varkony, E. Keinan and Y. Mazur, *Org. Synth.* **59**, 176 (1980).
47. G. A. Cook, A. D. Kiffer, C. V. Klupp, A. H. Malik and L. A. Spencer, *Advan. Chem. Ser.*, **21**, 44 (1959).
48. E. Keinan, *Ph.D. Thesis*, Weizmann Institute of Science, Israel, 1977.
49. The polycyclic hydrocarbon ($C_{14}H_{16}$) was kindly provided by Prof. P. v. R. Schleyer of Princeton University.
50. (a) E. Proksch and A. de Meijere, *Angew. Chem.* (*Intern. Ed.*), **15**, 761 (1976).
(b) E. Proksch and A. de Meijere, *Tetrahedron Letters*, 4851 (1976).
51. H. Klein and A. Steinmetz, *Tetrahedron Letters*, 4249 (1975).
52. (a) Z. Cohen, E. Keinan, Y. Mazur and A. Ulman, *J. Org. Chem.*, **41**, 2651 (1976).
(b) Y. Mazur, Z. Cohen and E. Keinan, *Israel Patent*, No. 49287 (1976); *Chem. Abstr.*, **92**, 164178q (1980).
(c) Z. Cohen, *Ph.D. Thesis*, Weizmann Institute of Science, Israel, 1979.
53. E. Keinan and Y. Mazur, *Synthesis*, 523 (1976).
54. (a) A. L. J. Beckwith and T. Duong, *J. Chem. Soc., Chem. Commun.*, 413 (1978).
(b) A. L. J. Beckwith, C. L. Bodkin and T. Duong, *Chem. Letters*, 425 (1977).
(c) A. L. J. Beckwith, C. L. Bodkin and T. Duong, *Australian J. Chem.*, **30**, 2177 (1977).
55. (a) J. P. Wibaut, F. L. J. Sixma, L. W. F. Kampschmidt and H. Boer *Rec. Trav. Chim.*, **69**, 1355 (1950).
(b) F. L. J. Sixma, H. Boer and J. P. Wibaut, *Rec. Trav. Chim.*, **70**, 1005 (1951).
56. (a) E. Akiyama, M. Tada, T. Tsuyuki and T. Takahashi, *Bull. Chem. Soc. Japan*, **52**, 164 (1979).
(b) E. Akiyama, M. Tada, T. Tsuyuki and T. Takahashi, *Chem. Letters*, 305 (1978).

57. E. Suokas and T. Hase, *Acta Chem. Scand.*, **B32**, 623 (1978).
58. (a) E. Trifilieff, L. Bang and G. Ourisson, *Tetrahedron Letters*, 2991 (1977).
(b) E. Trifilieff, L. Bang, A. S. Narula and G. Ourisson, *J. Chem. Res.*, 64(s), 601(M) (1978).
59. D. Tal, E. Keinan and Y. Mazur, *J. Amer. Chem. Soc.*, **101**, 502 (1979).
60. D. Tal, *Ph.D. Thesis*, Weizmann Institute of Science, Israel, 1981.
61. T. Preub, E. Proksch and A. de Meijere, *Tetrahedron Letters*, 833 (1978).
62. E. Keinan and Y. Mazur, *J. Org. Chem.*, **42**, 844 (1977).
63. E. Keinan and Y. Mazur, *J. Amer. Chem. Soc.*, **99**, 3861 (1977).
64. I. E. D. Besten and T. H. Kinstle, *J. Amer. Chem. Soc.*, **102**, 5968 (1980).
65. J. P. Desvergne and H. Bouas-Laurent, *J. Catal.*, **51**, 126 (1977).
66. G. A. Olah, D. G. Parker and N. Yoneda, *Angew. Chem. (Intern. Ed.)*, **17**, 909 (1978).
67. G. A. Olah, A. Germain, H. C. Lin and D. A. Forsyth, *J. Amer. Chem. Soc.*, **97**, 2928 (1975).
68. G. A. Olah, R. Renner, P. Schilling and Y. K. Mo, *J. Amer. Chem. Soc.*, **95**, 7686 (1973).
69. G. A. Olah, Y. Halpern, J. Shen and Y. K. Mo, *J. Amer. Chem. Soc.*, **95**, 4960 (1973).
70. G. A. Olah, N. Yoneda and D. G. Parker, *J. Amer. Chem. Soc.*, **98**, 5261 (1976).
71. G. A. Olah, N. Yoneda and R. Ohnishi, *J. Amer. Chem. Soc.*, **98**, 7341 (1976).
72. J. C. Jacquesy and J. F. Patoiseau, *Tetrahedron Letters*, 1499 (1977).

Note added in proof

After completion of this manuscript, there appeared the second volume of *Ozonation in Organic Chemistry* by P. S. Bailey, published by Academic Press. The reader is referred to this book, which covers much of the material discussed here.

The Chemistry of Functional Groups, Peroxides
Edited by S. Patai

CHAPTER **20**

Pyrolysis of peroxides in the gas phase

LESLIE BATT

Department of Chemistry, University of Aberdeen, Aberdeen AB9 2UE, Scotland

MICHAEL T. H. LIU

Department of Chemistry, University of Prince Edward Island, Charlottetown, Prince Edward Island, C1A 4P3, Canada

I. INTRODUCTION 685
II. DIALKYL PEROXIDES 687
III. PEROXYNITRATES 697
IV. OZONIDES 700
V. DIOXETANES 704
VI. ACKNOWLEDGEMENT 708
VII. REFERENCES 708

I. INTRODUCTION

Our brief from the editor was to survey critically the mechanisms for the pyrolysis of peroxides in the gas phase. This subject matter includes peroxides, hydroperoxides, acyl peroxides, peroxy acids, peroxy esters, ozonides and dioxetanes. The years up to 1970 have been covered comprehensively by the three excellent reviews of Benson and Shaw[1], Richardson and O'Neal[2] and Cubbon[27]. This review is therefore confined to the period

1970–1980. We have not been able to find any new work on hydroperoxides, acyl peroxides and peroxy acids. However, there are two important comments to make about hydroperoxides (ROOH). First, as reported by Richardson and O'Neal[2], gas-phase kinetic results are very poor (Table 1). Observed activation energies are as much as 6 kcal mol^{-1}* lower than ΔH_1^0:

$$\text{ROOH} \longrightarrow \text{R}\dot{\text{O}} + \dot{\text{O}}\text{H} \quad (1)$$

This is probably because decomposition of the hydroperoxide is subject either to surface catalysis or to chain processes, such that reaction (1) was never isolated. However, Benson and Spokes[5] used the very-low-pressure pyrolysis technique (VLPP) to study the decomposition of *t*-butyl hydroperoxide in order to isolate step (1). By using the Rice, Ramsperger, Kassel, Marcus (RRKM)[6] model they were able to extrapolate these results to high-pressure conditions in order to obtain values for A_1 and E_1 (Table 1). Second, in relation to the oxidation of methyl radicals, the mechanism at low temperatures (<1000 K) involves the methylperoxy radical:

$$\dot{\text{C}}\text{H}_3 + \text{O}_2 + \text{M} \rightleftharpoons \text{CH}_3\text{O}_2^{\cdot} + \text{M} \quad (2)$$

TABLE 1. Pyrolysis of hydroperoxides: ROOH → R$\dot{\text{O}}$ + $\dot{\text{O}}$H (1)

R	Temp. range (K)	$\log A_1$ (s^{-1})	E_1 (kcal mol^{-1})	ΔH_1^0 (298) (kcal mol^{-1})	Reference
Me	565–651	11	32 ± 5	43.9	3
Et	553–653	13.4	37.7	43.6	4
i-Pr	553–653	15.2	40.0	43.6	4
		14.5	40.7	—	4
t-Bu	553–653	13.7	37.8	43.2	4
	573–1323	15.6 ± 0.5	42.2 ± 2	—	5

This mechanism (see Section II) becomes important over a wider temperature range, because of a new stronger CH_3—O_2 bond of 31 kcal mol^{-1} [7] which depends upon the latest value for ΔH_f^0 (HO_2) = 2.5 ± 0.6 kcal mol^{-1} [8] since D(HO_2—H) is assumed equal to D(RO_2—H)[1]. This now takes the value of 87.2 kcal mol^{-1}. Batt has briefly covered the work on peroxynitrates[9], but they deserve a proper review here, especially in view of their possible importance in relation to atmospheric chemistry. In connection with the latter, it is also important to consider the reaction of olefins with ozone[10,11] (equation 3):

$$\text{C}_2\text{H}_4 + \text{O}_3 \longrightarrow \text{H}_2\text{C}(\text{O})(\text{O—O})\text{CH}_2 \quad (3)$$

R¹—C(R²)(O—O)—C(R³)—R⁴

(1)

R^1, R^2, R^3, R^4 = alkyl

* 1 thermochemical calorie = 4.185 J.

and the chemistry of dioxetanes (**1**). The first of these cyclic four-membered ring peroxides, trimethyl-1,2-dioxetane, was prepared by Kopecky and Mumford[12]. Before this time, dioxetanes were thought to be intermediates only. Since this first paper, much work has gone into the synthesis and particularly the study of the thermal decomposition of these products. In all cases of substituted, 1,2-dioxetanes, the energetic quantities are such that the heat of reaction (ΔH_{RXN}) and the activation energy (E_a) combined exceed the energy required for excitation (E^*). The observed chemiluminescence accompanying thermal decomposition has been the subject of active research and reviews[13–18].

II. DIALKYL PEROXIDES

Dialkyl peroxides are important, clean, thermal sources of alkoxy radicals. They may also be used to generate these radicals photolytically, but since the peroxides absorb in the far-UV, the alkoxy radicals carry considerable excess energy. Thus photolysis at 253.7 nm leaves the two alkoxy radicals with some 75 kcal mol^{-1} (of peroxide) since the O—O bond dissociation energy is only 38 kcal mol^{-1} [1,19]. Large pressures of diluent gas are required for thermal equilibration[20].

Dimethyl peroxide (DMP) assumes a special significance because the reactions of methoxy radicals are important in relation to:

(*a*) the chemistry of the upper atmosphere, where they are formed via the photochemical oxidation of methane;
(*b*) the oxidation or combustion of organic compounds, in particular hydrocarbons; and
(*c*) the concomitant pollution of the lower atmosphere.

At low temperatures, the oxidation of methyl radicals appears to proceed via the formation of methylperoxy radicals which subsequently react with either a free radical (e.g. $CH_3O_2^{\bullet}$, $\cdot CH_3$ or $HO_2^{\bullet}$) or with organic compounds (RH):

$$\dot{C}H_3 + O_2 + M \rightleftharpoons CH_3O_2^{\bullet} + M$$

$$CH_3O_2^{\bullet} + HO_2^{\bullet} \longrightarrow CH_3O_2H + O_2 \quad (4)$$

$$CH_3O_2^{\bullet} + \dot{C}H_3 \longrightarrow (CH_3O)_2 \longrightarrow 2\,CH_3\dot{O} \quad (5)$$

$$CH_3O_2^{\bullet} + CH_3O_2^{\bullet} \longrightarrow (CH_3O_4CH_3) \longrightarrow 2\,CH_3\dot{O} + O_2 \quad (6)$$

$$CH_3O_2^{\bullet} + RH \longrightarrow CH_3O_2H + \dot{R} \quad (7)$$

$$CH_3O_2H \longrightarrow CH_3\dot{O} + \dot{O}H \quad (8)$$

As mentioned in the introduction, the new and stronger, reported CH_3—O_2 bond strength now extends the temperature range over which this mechanism is important. At still higher temperatures (>2000 K), the oxidation of methyl radicals may involve the direct interaction of the radicals with oxygen:

$$\dot{C}H_3 + O_2 \rightleftharpoons CH_3O_2^{\bullet} \longrightarrow CH_2O + \dot{O}H \quad (9)$$

$$\dot{C}H_3 + O_2 \longrightarrow CH_3\dot{O} + \dot{O} \quad (10)$$

Dimethyl peroxide needs careful handling because the liquid is shock-sensitive[21] and the vapour subject to explosive decomposition. Two explosions have been encountered by one of the present authors[19,22,23]. Takezaki and Takeuchi pyrolysed dimethyl peroxide in the presence of excess methanol[24]. They were able to extract information about the initial bond-breaking step (11) albeit from a rather complicated scheme (Table 2):

$$\text{DMP} \longrightarrow 2\,\text{CH}_3\dot{\text{O}} \tag{11}$$

Hanst and Calvert[25] concluded that the rate of decomposition of pure dimethyl peroxide was first order with respect to the peroxide and that the main products were methanol and carbon monoxide according to the stoichiometric equation (12):

$$\text{DMP} \longrightarrow 3/2\,\text{CH}_3\text{OH} + \tfrac{1}{2}\,\text{CO} \tag{12}$$

TABLE 2. Arrhenius parameters for the decomposition of dimethyl peroxide: DMP → S$CH_3\dot{O}$ (11)

$\log(A_{11}/\text{s}^{-1})$	E_{11}(kcal mol^{-1})	Reference
15.6	36.9	24
15.2	35.3	25
15.2 ± 0.3	36.4 ± 0.6	19[a]
15.7 ± 0.5	37.1 ± 0.9	26
16.1 ± 0.5	38.0 ± 0.9	22

[a] Arrhenius parameters recalculated.

They concluded that the mechanism consisted of the following steps:

$$\text{DMP} \longrightarrow 2\,\text{CH}_3\dot{\text{O}} \tag{11}$$

$$2\,\text{CH}_3\dot{\text{O}} \longrightarrow \text{CH}_3\text{OH} + \text{CH}_2\text{O} \tag{13}$$

$$\text{CH}_3\dot{\text{O}} + \text{CH}_2\text{O} \longrightarrow \text{CH}_3\text{OH} + \dot{\text{C}}\text{HO} \tag{14}$$

$$\text{CH}_3\dot{\text{O}} + \dot{\text{C}}\text{HO} \longrightarrow \text{CH}_3\text{OH} + \text{CO} \tag{15}$$

However, their determined value for E_{11} is lower than that predicted by thermochemical calculations. Both Benson and Shaw[1] and O'Neal and Richardson[2] attributed this to the participation of a chain-reaction which could amount to 10–13 % at 400 K. This was subsequently verified by Barker, Benson and Golden[26]. Since this reaction is very exothermic, spurious results may also be observed due to the onset of thermal gradients in the reaction vessel[28]. Batt and McCulloch[19] studied the decomposition of dimethyl peroxide in the presence of excess isobutane (t-BuH) in order, amongst other things, to avoid the onset of thermal gradients. They were able to show that for complete reaction 96 ± 4 % of the available methoxy radicals were converted to methanol. This verifies that dimethyl peroxide is a clean source of methoxy radicals. Their mechanism for the decomposition in the presence of isobutane is the following:

$$\text{DMP} \longrightarrow 2\,\text{CH}_3\dot{\text{O}} \tag{11}$$

$$\text{CH}_3\dot{\text{O}} + t\text{-BuH} \longrightarrow \text{CH}_3\text{OH} + t\text{-}\dot{\text{B}}\text{u} \tag{16}$$

$$2\,t\text{-}\dot{B}u \longrightarrow (t\text{-Bu})_2 \qquad (17a)$$

$$2\,t\text{-}\dot{B}u \longrightarrow t\text{-BuH} + (CH_3)_2C{=}CH_2 \qquad (17b)$$

The Aberdeen group have also determined the rate of decomposition of dimethyl peroxide in the presence of either nitric oxide or nitrogen dioxide[22]. In these cases the step (11) is followed either by (18) or (19):

$$CH_3\dot{O} + NO \longrightarrow CH_3ONO \qquad (18)$$

$$CH_3\dot{O} + NO_2 \longrightarrow CH_3ONO_2 \qquad (19)$$

It is concluded that the most reliable value for the Arrhenius parameters of reaction (11) (Table 2) are given from the study of the decomposition in the presence of nitrogen dioxide. Here the system is completely devoid of any surface effects. However, there is reasonable accord between several studies (Table 2 and Figures 1 and 2). The preferred result leads to a value of $\Delta H_f^0\,(CH_3\dot{O})$ of $+4.0 \pm 0.5\,\text{kcal mol}^{-1}$ [29]. One other step has been proposed in the decomposition of dimethyl peroxide[30,31]. This is the isomerization of the methoxy radical (20):

$$CH_3\dot{O} + M \longrightarrow \dot{C}H_2OH + M \qquad (20)$$

In the gas phase, this step will almost certainly be pressure-dependent at normal pressures (~1 atm). Christe and Pilipovich[30] discarded this step because no ethylene glycol was observed whereas Batt, Burrows and Robinson[29] argued against this step on thermochemical kinetic grounds. This reaction may be important at 900 K where Baldwin and coworkers have studied the oxidation of methane[32] and acetaldehyde[33]. It should also be pointed out that the decomposition of the methoxy radical (reaction 21) is too slow[34],

$$CH_3\dot{O} + M \longrightarrow CH_2O + \dot{H} + M \qquad (21)$$

under the conditions where the dimethyl peroxide pyrolysis has been studied, to compete with the other reactions of the methoxy radical. Under Baldwin's[32,33] conditions, however, decomposition as opposed to isomerization is the major process[29].

It is interesting to compare the decomposition of dimethyl peroxide with that of the fully fluorinated compound bis(trifluoromethyl) peroxide [$(CF_3O)_2$]. One difference in the experimental conditions is that monel[35], aluminium[36,37], magnesium[37] and nickel[38,39] reactors were used. In connection with the stability of bis(trifluoromethyl) peroxide, Levy and Kennedy[35] first studied the equilibrium between the peroxide and carbonyl fluoride and trifluoromethyl hypofluorite:

$$(CF_3O)_2 \rightleftharpoons CF_2O + CF_3OF \qquad (22)$$

Porter and Cady[40] had previously reported that the two products in (22) reacted together to give the peroxide. From $\Delta H_{22}^0\,(298) = 24.5 \pm 0.7\,\text{kcal mol}^{-1}$, Kennedy and Levy deduced that $\Delta H_f^0\,(CF_3OOCF_3) = -360.2 \pm 3\,\text{kcal mol}^{-1}$. They subsequently studied the kinetics of both the forward and reverse steps[36]. There were no side- or surface-reactions and the stoichiometric reaction (22) indicated an increase in pressure upon reaction, so they measured the rate of reaction manometrically. Since the reaction reaches equilibrium, kinetic measurements were confined to initial rates. The order of the reaction changed from first order at low pressures to half order at high pressures and whereas

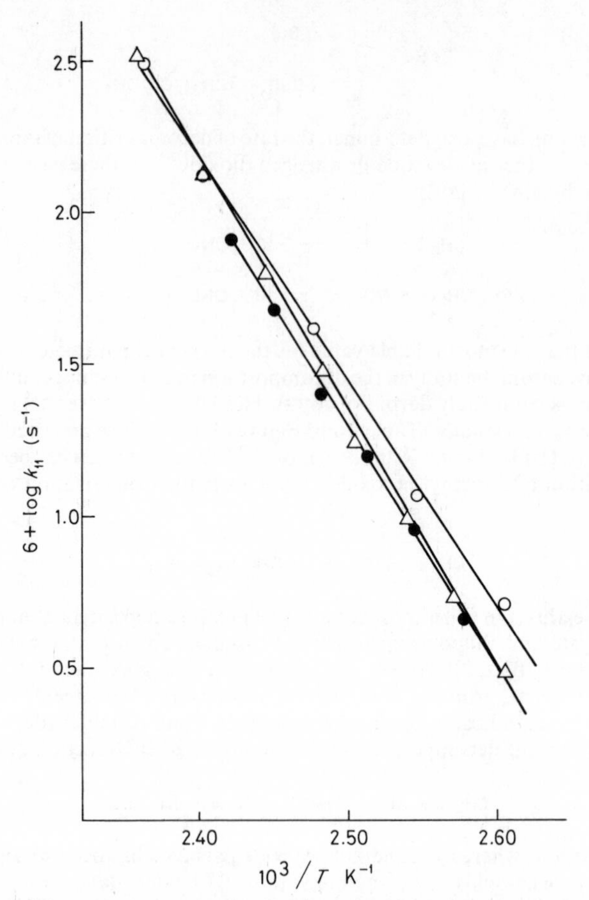

FIGURE 1. Arrhenius plots for the decomposition of dimethyl peroxide: DMP → $2CH_3\dot{O}$ (11) in the presence of ●: *t*-BuH, △: NO_2 and ○: NO. Reprinted with permission from L. Batt and G. N. Rattray, *Intern. J. Chem. Kinet.*, **11**, 1183 (1979).

carbonyl fluoride inhibited the rate of reaction, trifluoromethyl hypofluorite and inert gases had no effect. They proposed the following mechanism:

$$(CF_3O)_2 \longrightarrow 2CF_3\dot{O} \quad (23)$$

$$2CF_3\dot{O} \longrightarrow (CF_3O)_2 \quad (24)$$

$$CF_3\dot{O} \longrightarrow COF_2 + F \quad (25)$$

$$CF_3\dot{O} + F \longrightarrow CF_3OF \quad (26)$$

$$CF_3OF \longrightarrow CF_3\dot{O} + F \quad (27)$$

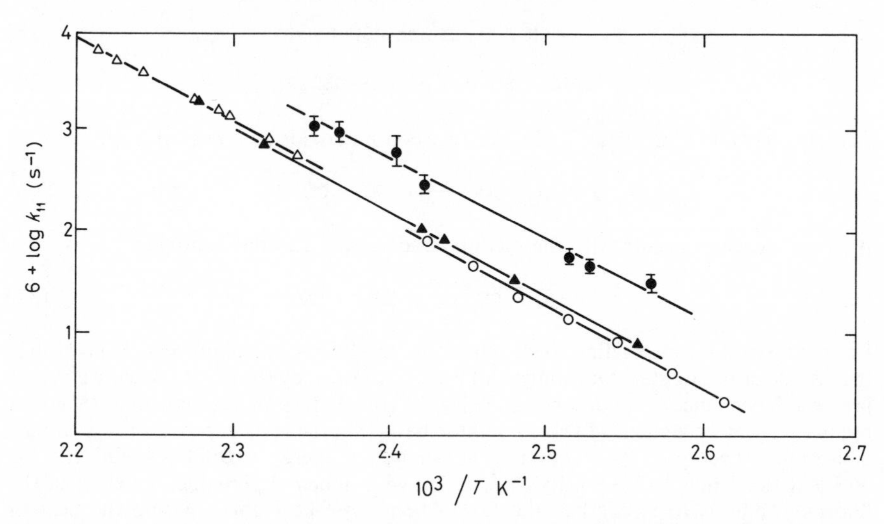

FIGURE 2. Arrhenius plots for the decomposition of dimethyl peroxide: DMP → $2CH_3\dot{O}$ (11) in the presence of ○: *t*-BuH (from Ref. 19), ▲: CH_3OH (from Ref. 24); △: from Ref. 25 and 26; ●: from Ref. 26. Reprinted with permission from J. R. Barker, S. W. Benson and D. M. Golden, *Intern. J. Chem. Kinet.*, **9**, 31 (1977).

At low pressures of peroxide, the rate-determining step is reaction (25), so that the rate of reaction was given by:

$$-d[(CF_3O)_2]/dt = k_{25}(k_{23}/k_{24})^{\frac{1}{2}}[(CF_3O)_2]^{\frac{1}{2}}$$

Kennedy and Levy appreciated that the conditions for the above relationship were not quite fully met by their experimental conditions, but were nevertheless able to interpolate a value for k_{25}. This was given by:

$$k_{25} = 10^{14.5 \pm 0.2} 10^{-31.0 \pm 0.5/\theta} \mathrm{s}^{-1} *$$

Kennedy and Levy also measured initial rates for the reverse of reaction (22). They concluded that k_{27} was given by:

$$k_{27} = 10^{14.5} 10^{-43.5/\theta} \mathrm{s}^{-1}$$

Czarnowski and Schumacher[41] had come to the same conclusion about the mechanism with rate data for the reverse of reaction (22) over the temperature range 223–233°C. They also showed that k_{27} was given by:

$$k_{27}(\infty) = 10^{14.3} 10^{-43.3 \pm 0.5/\theta} \mathrm{s}^{-1}$$

This is in good agreement with the previous result. Both Czarnowski and Schumacher[37] and Descamps and Forst[38] simplified the mechanism for the decomposition of the peroxide by the addition of fluorosulphate dimer. Under these conditions, the trifluoromethoxy radicals were irreversibly removed by addition to the fluorosulphate free radical:

---

* $\theta = 2.303\,RT$ kcal mol^{-1}.

$$(SO_3F)_2 \rightleftharpoons 2\dot{S}O_3F \quad (28)$$

$$CF_3\dot{O} + \dot{S}O_3F \longrightarrow CF_3OSO_3F \quad (29)$$

Czarnowski and Schumacher[37] also used carbon monoxide as a radical trap:

$$CF_3\dot{O} + CO \longrightarrow \dot{C}F_3 + CO_2 \quad (30)$$

A similar reaction occurs with the methoxy radical and carbon monoxide[42]:

$$CH_3\dot{O} + CO \longrightarrow \dot{C}H_3 + CO_2 \quad (31)$$

For the bis(trifluoromethyl)peroxide, under the conditions stated, process (23) is clearly the rate-determining step. Descamps and Forst[39] showed, by the Forst procedure[43], that pressure dependence of k_{23} occurs only below 10 Torr. Values for k_{23} were obtained which are in excellent agreement (Table 3). On the basis that the reverse process (24) has zero activation energy, the bond dissociation energy $D(CF_3O{-}OCF_3)$ is $46.8 \pm 0.5\,\text{kcal mol}^{-1}$ (298 K). By comparison with dimethyl peroxide, the effect of the fluorine atoms is to strengthen the O—O bond by $9\,\text{kcal mol}^{-1}$. Using the heat of formation for the fluorinated peroxide cited earlier, a value for the heat of formation of the trifluoromethoxy radical may be determined since:

$$\Delta H_4^0(CF_3\dot{O}) = \tfrac{1}{2}[\Delta H_{23}^0 + \Delta H_f^0(CF_3OOCF_3)]$$
$$= -156.7 \pm 1.5\,\text{kcal mol}^{-1}$$

TABLE 3. Arrhenius parameters for the decomposition of bis(trifluoromethyl) peroxide: $(CF_3O)_2 \rightarrow 2CF_3\dot{O}$ (25)

$\log(A_{23}/s^{-1})$	$E_{23}(\text{kcal mol}^{-1})$	Reference
15.2 ± 0.1	46.2 ± 0.33	36
16.1 ± 0.15	46.3 ± 0.4	37
15.9 ± 0.23	46.2 ± 0.5	38

Using this value and the heats of formation of carbonyl fluoride[44] and the fluorine atom[44], one may determine ΔH_{25}^0 to be $22.9 \pm 1.6\,\text{kcal mol}^{-1}$, very close to $\Delta H_{21}^0 = 22.1 \pm 0.6\,\text{kcal mol}^{-1}$ for the similar process for the methoxy radical[34].

The entropy of the trifluoromethoxy radical may be determined by comparison with that of carbon tetrafluoride[44] using the method of O'Neal and Benson[45]. Taking into account changes in mass, symmetry and electron degeneracy, $S^0(CF_3\dot{O}) = 66.3$ e.u. For comparison the values for the entropies of carbon tetrafluoride and the methoxy radical are 62.5 e.u.[44] and 54.3 e.u.[19] respectively. The standard entropy for bis(trifluoromethyl) peroxide is 97 ± 1.2 e.u.[35]. Hence the standard entropy change for reaction (23) is given by:

$$\Delta S_{23}^0 = 35.6\ \text{e.u.}$$

The preexponential factors for reactions (23) and (24) are related by the expression[46]:

$$\ln(A_{23}/A_{24}) = \Delta S_{23}^0/R) - (1 + \ln R'T)$$

where $R = 1.987\,\text{cal}\,\text{K}^{-1}\,\text{mol}^{-1}$, $R' = 0.082\,\text{litre-atm}\,\text{K}^{-1}\,\text{mol}^{-1}$ and T is the absolute mean temperature of the experiments. Since E_{23} may be taken to be zero, $A_{24} = k_{24} = 10^{10.1}\,\text{M}^{-1}\,\text{s}^{-1}$. A similar calculation may be made to estimate k_{26}. This is given by $k_{26} = 10^{9.4}\,\text{M}^{-1}\,\text{s}^{-1}$. Kennedy and Levy[36] found a value of $10^{9.5}\,\text{M}^{-1}\,\text{s}^{-1}$ in good agreement, but their value for k_{24} is half a power of ten less.

Descamps and Forst[39] found that in the decomposition of bis(trifluoromethyl) peroxide in the absence of a radical trap, carbonyl fluoride inhibited the rate of reaction in agreement with Kennedy and Levy[36]. This provided good evidence for the participation of the reverse of reaction (25). Unlike Kennedy and Levy[36], they found that the rate depended upon the concentration of added inert gases. However Kennedy and Levy were working at much higher pressures. In neat peroxide, Descamps and Forst found that the order of the reaction varied between 0.89 and 1.04 over the temperature range 509–545 K. They considered two mechanisms, one of which was identical to that of Kennedy and Levy[36] and the other involved the thermoneutral chain step in the place of reaction (26):

$$F + CF_3OOCF_3 \longrightarrow CF_3OF + CF_3\dot{O} \tag{32}$$

followed by:

$$CF_3\dot{O} \longrightarrow CF_2O + F \tag{25}$$

The chain step is a displacement process which therefore probably has a high activation energy. The chain length appears to be only 0.5[38], i.e. 1.5 molecules of peroxide decompose for each initiation step, so that it is difficult to differentiate between the two mechanisms. However, the nonchain mechanism does seem the more plausible. As discussed previously, analysis of the first mechanism leads to the result for the rate of reaction[38]:

$$-\mathrm{d}[CF_3OOCF_3]/\mathrm{d}t = k_{25}(k_{23}/k_{24})^{\frac{1}{2}}[CF_3OOCF_3]^{\frac{1}{2}}$$

Since reaction (25) is almost certainly pressure-dependent under these conditions, this would raise the order dependence of the peroxide in agreement with the observed value of 0.9[38]. An extrapolation is required[38] in order to obtain $k_{25}(\infty)$. Using the calculated value for k_{24} of $10^{10.1}\,\text{M}^{-1}\,\text{s}^{-1}$ and $10^{16}\,\text{s}^{-1}$ for A_{23} (Table 3) leads to the result:

$$A_{25} = 10^{12.8}\,\text{s}^{-1}$$

Also, since $E_{\text{obs}} = E_{25} + \frac{1}{2}(E_{23} - E_{24}) = 49.7 \pm 1.4\,\text{kcal}\,\text{mol}^{-1}$ [38] and with $E_{24} = 0$ and $E_{23} = 46.2\,\text{kcal}\,\text{mol}^{-1}$ (Table 3), $E_{25} = 26.6\,\text{kcal}\,\text{mol}^{-1}$. These Arrhenius parameters do not agree very well with those of Kennedy and Levy[36] given by:

$$k_{25} = 10^{14.5}10^{-31/\theta}\,\text{s}^{-1}$$

although at 550 K the rate constants k_{25} calculated from the two expressions are almost identical. It is well known that the Lindemann extrapolation procedure used by Descamps and Forst[38] leads to too low values of k_{25} and therefore A_{25} and E_{25}. One would expect the Arrhenius parameters to lie in between these two sets of reported results[47]. This makes the rate constant for the decomposition of the trifluoromethoxy radical similar but somewhat smaller than that for the methoxy radical reaction[34].

Both Kennedy and Levy's[36] and Descamps and Forst's[39] results preclude reaction (33):

$$2\,CF_3\dot{O} \longrightarrow CF_3OF + CF_2O \tag{33}$$

On this basis it may be concluded that the ratio of the two rates $R_{23}/R_{33} = k_{23}/k_{33}[CF_3\dot{O}] < 10^{-2}$. Analysis of the mechanism under average conditions leads to the result $[CF_3\dot{O}] = 10^{-7.5}$ M. Hence $k_{33} < 10^8\,M^{-1}\,s^{-1}$. The main difference between the decompositions of the two peroxides is the slowness of step (33) compared to the very fast step (13)[19]:

$$2\,CH_3\dot{O} \longrightarrow CH_3OH + CH_2O \tag{13}$$

Also the stoichiometric equation (22) is endothermic by 24.5 kcal mol^{-1} whereas the stoichiometric equation for dimethyl peroxide[19,25] (equation 12) is exothermic to the extent of 55.2 kcal mol^{-1}.

$$(CH_3O)_2 \longrightarrow 3/2\,CH_3OH + \tfrac{1}{2}\,CO \tag{12}$$

It has been shown that for dialkyl peroxides the bond dissociation energy $D(RO{-}OR) = 38$ kcal mol^{-1} [1,19] independent of the nature of R. This rules out the possibility of alkyl-group: oxygen-atom *gauche* interactions for these peroxides[19]. East and Phillips[47] studied the decomposition of di-*n*-propyl peroxide $(n\text{-PrO})_2$ in the presence of nitric oxide. Under these conditions no sensitized decomposition of the peroxide would be possible. They proposed the following mechanism to account for the products of the reaction:

$$(n\text{-PrO})_2 \longrightarrow 2\,n\text{-PrO} \tag{34}$$

$$n\text{-Pr}\dot{O} + NO \longrightarrow n\text{-PrONO} \tag{35}$$

$$n\text{-Pr}\dot{O} + NO \longrightarrow C_2H_5CHO + HNO \tag{36}$$

$$n\text{-Pr}\dot{O} + HNO \longrightarrow n\text{-PrOH} + NO \tag{37}$$

They concluded that the rate constant for reaction (34) was given by:

$$k_{34} = 10^{14.5}\,10^{-34.5/\theta}\,s^{-1}$$

In terms of the previous argument E_{34} is too low by 3 kcal mol^{-1}. The discrepancy may be due to the errors involved in estimating the yields of products from reactions (35), (36) and (37) or their further reaction. Perona and Golden[48] studied the decomposition of di-*t*-amyl peroxide (DTAP) using the VLPP technique. Under these conditions, since the reactant concentration is so low, sensitized decomposition involving hydrogen abstraction reactions from the peroxide should be absent. The activation energy obtained by Perona and Golden for reaction (38)—the rate constant k_{38} is given by $k_{38} = 10^{15.8}\,10^{-36.4/\theta}\,s^{-1}$—is lower than that expected in terms of the above argument. However, the fit between the experimental VLPP data and the calculated RRKM[6] curve is not significantly altered by varying $E_{38}(\infty)$ by $+0.6$ kcal mol^{-1}. One other important result from this study is that the tertiary amyloxy radical (t-Am$\dot{O}$) only decomposes to give acetone and an ethyl radical reaction (39a). A specific search was made for methyl ethyl ketone, but none was detected. This is in agreement with the results of Batt, Islam and Rattray[49] although earlier results are conflicting.

$$DTAP \longrightarrow 2\,t\text{-Am}\dot{O} \tag{38}$$

$$t\text{-Am}\dot{O} \longrightarrow \dot{C}_2H_5 + CH_3COCH_3 \quad (39a)$$

$$t\text{-Am}\dot{O} \longrightarrow \dot{C}H_3 + CH_3COCH_2CH_3 \quad (39b)$$

The pyrolysis of di-*t*-butyl peroxide (DTBP) must be a candidate for the record of the most studied reaction! It had been observed by earlier work of Hinshelwood and his coworkers[50] that the addition of certain gases accelerated the rate of decomposition of di-*t*-butyl peroxide. In some cases this acceleration in the rate was interpreted in terms of an extension to current theories of unimolecular reactions. However they realized that some of these accelerations were due to chemical sensitization. Batt and Cruickshank[51] suggested that in the presence of sulphur hexafluoride, the acceleration in the rate was due to a fluorine-atom-sensitized decomposition of di-*t*-butyl peroxide:

$$\text{DTBP} \longrightarrow 2\,t\text{-Bu}\dot{O} \quad (40)$$

$$t\text{-Bu}\dot{O} \longrightarrow \dot{C}H_3 + CH_3COCH_3 \quad (41)$$

$$\dot{C}H_3 + SF_6 \longrightarrow CH_3F + \dot{S}F_5 \quad (42)$$

$$SF_5 \longrightarrow SF_4 + F \quad (43)$$

$$F + \text{DTBP} \longrightarrow HF + \text{DT}\dot{B}\text{P}_{-H} \quad (44)$$

$$\text{DT}\dot{B}\text{P}_{-H} \longrightarrow t\text{-Bu}\dot{O} + (CH_3)_2C\overset{O}{\frown}CH_2 \quad (45)$$

$$\dot{C}H_3 + \dot{S}F_5 \longrightarrow CH_3F + SF_4 \quad (46)$$

(However this interpretation has been disputed[52].) Shaw and Pritchard[53] also refuted any extension to the theories of unimolecular reactions and showed that high pressures of added carbon dioxide (15 atm) had no effect on the rate. They also noted that the Arrhenius plot for reaction (40) was linear over the temperature range 363–623 K. The Arrhenius parameters are also independent of the nature of the alkyl group in the peroxide[17]. Mention must also be made of the continued use of di-*t*-butyl peroxide for sensitizing the decomposition of other compounds, in particular in the production of other alkoxy radicals. Loucks, Liu and Hooper[54] generated a trifluoroisopropoxy radical via the sensitized decomposition of trifluoroacetaldehyde (equations 40, 41 and 47). The

$$\text{DTBP} \longrightarrow 2\,t\text{-Bu}\dot{O} \quad (40)$$

$$t\text{-Bu}\dot{O} \longrightarrow \dot{C}H_3 + CH_3COCH_3 \quad (41)$$

$$\dot{C}H_3 + CF_3CHO \longrightarrow (CF_3)(CH_3)CH\dot{O} \quad (47)$$

trifluoroisopropoxy radical abstracted a hydrogen atom from an organic compound (RH) (equation 48). Decomposition of the radical resulted only in the reverse of reaction (47) and the production of trifluoroacetone (equation 49c) but none of the expected acetaldehyde (equation 49b). The heats of reaction were calculated using Group

$$(CF_3)(CH_3)CH\dot{O} + RH \longrightarrow (CF_3)(CH_3)CHOH + \dot{R} \tag{48}$$

		ΔH^0(kcal mol^{-1})
$(CF_3)(CH_3)CH\dot{O} \longrightarrow \dot{C}H_3 + CF_3CHO$	(49a)	11.4
$(CF_3)(CH_3)CH\dot{O} \longrightarrow \dot{C}F_3 + CH_3CHO$	(49b)	8.5
$(CF_3)(CH_3)CH\dot{O} \longrightarrow \dot{H} + CF_3COCH_3$	(49c)	20.2

$$(CF_3)_3C{-}\dot{O} \longrightarrow CF_3COCF_3 + \dot{C}F_3 \tag{50}$$

Additivity Rules[10]. The group value $\Delta H_f^0[C(F)_3(CO)] = -151$ kcal mol^{-1} was generated from the result that $\Delta H_{50}^0 = +21.5 \pm 1.3$ kcal mol^{-1} [55] (see equation 50). The production of trifluoroacetone from reaction (49c) could not be differentiated from that produced via a disproportionation with a radical $\dot{R}'$ (equation 51). However, the yield of hydrogen was

$$(CF_3)(CH_3)CH\dot{O} + \dot{R}' \longrightarrow CF_3COCH_3 + R'H \tag{51}$$

much less than that of the trifluoroacetone. Loucks, Liu and Hooper[54] also found evidence for the addition of trifluoromethyl radicals to trifluoroacetaldehyde (equations 52 and 53) (other relevant work is given in Reference 57). Finally, after many studies, the pressure-

$$\dot{C}F_3 + CF_3CHO \rightleftharpoons (CF_3)_2CH\dot{O} \longrightarrow CF_3COCF_3 + \dot{H} \tag{52}$$

$$(CF_3)_2CH\dot{O} + RH \longrightarrow (CF_3)_2CHOH + \dot{R} \tag{53}$$

dependent decomposition of the *t*-butoxy radical, generated from the pyrolysis of di-*t*-butyl peroxide, has been unequivocally demonstrated[34]. When the results were subjected to a RRKM[6] analysis the rate constant was given by $k_{41}(\infty) = 10^{14.6}\ 10^{-15.9}\ \theta$ s^{-1} [56].

Ireton, Gordon and Tardy studied the decomposition of perfluorodi-*t*-butyl peroxide (PFDTBP)[58]. Over the temperature range of 381–422 K and 5–600 Torr they found that the decomposition was homogeneous, first order with respect to peroxide and free from any chain-sensitized decomposition. Products were analysed by GLC. Only three temperatures were investigated so that errors would be expected to be relatively large The mechanism is given by:

$$\text{PFDTBP} \longrightarrow 2\,(CF_3)_3C\dot{O} \tag{54}$$

$$(CF_3)_3C\dot{O} \longrightarrow \dot{C}F_3 + CF_3COCF_3 \tag{55}$$

Step (55) is followed by the combination of the trifluoromethyl radicals. The rate-determining step is (54) given by:

$$k_{54} = 10^{16.2 \pm 1.2} : 10^{-35.5 \pm 1}/\theta \text{ s}^{-1}$$

They concluded that the presence of the fluorine atoms had no effect on the O—O bond strength just like the dialkyl peroxides but unlike trifluoromethyl peroxide mentioned earlier.

III. PEROXYNITRATES

Peroxynitrates are of considerable current interest because of their possible influence on the chemistry of the upper and lower atmosphere. In the upper atmosphere, the parent compound, peroxynitric acid, represents a sink for both hydroperoxy radicals and nitrogen dioxide:

$$HO_2^{\cdot} + NO_2 + M \longrightarrow HO_2NO_2 + M \tag{56}$$

Both precursors of peroxynitric acid affect the ozone cycle, referred to as the Chapman mechanism[111]:

$$O_2 + h\nu \longrightarrow 2O \tag{57}$$

$$O + O_2 + M \longrightarrow O_3 + M \tag{58}$$

$$O_3 + h\nu \longrightarrow O_2 + O \tag{59}$$

$$O + O_3 \longrightarrow 2O_2 \tag{60}$$

Peroxynitric acid may also play a role in photochemical smog formation[59]. Peroxynitrates may be formed via the photolysis of freons:

$$CX_3Cl + h\nu \longrightarrow \dot{C}X_3 + Cl \tag{61}$$

$$\dot{C}X_3 + O_2 \longrightarrow CX_3O_2^{\cdot} \tag{62}$$

$$CX_3O_2^{\cdot} + NO_2 \longrightarrow CX_3O_2NO_2 \tag{63}$$

However, other fates may await the halogenotrimethylperoxy radical[7,60]. Equally, the nonhalogenated species may also play a role in the chemistry of the atmosphere. Peroxyacyl nitrates (PANS) are also formed in photochemical smog and are believed to be very toxic to vegetation[61]. PANS are powerful eye irritants. Peroxybenzoyl nitrate is about two hundred times more irritating than formaldehyde and about one hundred times more irritating than peroxyacetyl nitrate (PAN)[61]. PAN is an extremely explosive compound[62].

Cox, Derwent and Hutton photolysed nitrous acid in the presence of synthetic air (1 atm) and carbon monoxide using a flow system[63]. The photolytic process generated hydroperoxy radicals and nitrogen dioxide according to the chain mechanism below:

$$HONO + h\nu \longrightarrow H\dot{O} + NO \tag{64}$$

$$H\dot{O} + CO \longrightarrow CO_2 + \dot{H} \tag{65}$$

$$H + O_2 + M \longrightarrow HO_2^{\cdot} + M \tag{66}$$

$$HO_2 + NO \longrightarrow \dot{H}O + NO_2 \tag{67}$$

$$HO_2^{\cdot} + NO_2 + M \longrightarrow HO_2NO_2 + M \tag{56}$$

Nitric oxide, nitrogen dioxide and nitrous acid were monitored by chemiluminescent techniques. By modelling the reaction, the data could be used to determine the rate constant k_{68}:

$$HO_2NO_2 \longrightarrow HO_2 + NO_2 \tag{68}$$

Simonaitis and Heicklen photolysed nitrous oxide in the presence of oxygen, hydrogen (1 atm) and small amounts of nitric oxide and nitrogen dioxide[64]:

$$N_2O + h\nu \longrightarrow N_2 + O \tag{69}$$

$$O + H_2 \longrightarrow \dot{O}H + H \tag{70}$$

$$H + O_2 + M \longrightarrow HO_2^{\cdot} + M \tag{66}$$

$$HO_2^{\cdot} + NO \longrightarrow \dot{H}O + NO_2 \tag{67}$$

$$HO_2^{\cdot} + NO_2 + M \longrightarrow HO_2NO_2 + M \tag{56}$$

$$HO_2NO_2 \longrightarrow HO_2 + NO_2 \tag{68}$$

By a judicious choice of experimental conditions, data for k_{68} could be extracted. Graham, Winer and Pitts monitored the decay of peroxynitric acid directly in the presence of excess nitric oxide using Fourier transform infrared (FTIR) spectroscopy to determine k_{68} [65]. The Arrhenius parameters are given in Table 4. Baldwin and Golden calculated a value for k_{68} based upon an estimated value for k_{56} of $10^{9.5}\,M^{-1}\,s^{-1}$ and the thermochemistry for reaction (68)[66]. They also applied RRKM theory[6] using a hindered rotational Gorin model transition state. Their result was in good agreement with Cox, Derwent and Hutton[63], although this may be fortuitous. This is because Graham, Winer and Pitts have demonstrated the pressure dependence of k_{68} [67]. Thus all the experimental data for k_{68} may well refer to pressure-dependent conditions. In fact the most reliable data could be determined from the data in Reference 67 using RRKM theory[6]. The results in Table 4 show that peroxynitric acid is thermally unstable at room temperature in the troposphere. However in the stratosphere the molecule will have considerable thermal stability, reinforced by the pressure-dependent decomposition at the low pressures encountered there. The molecule may also have a considerable photolytic lifetime[67,68].

Duynstee and coworkers have synthesized *t*-butyl peroxynitrate by the reaction of the peroxy acid with dinitrogen pentoxide[69]:

$$t\text{-}BuO_2H + N_2O_5 \longrightarrow t\text{-}BuO_2NO_2 + HNO_3$$

Attempts to synthesize methyl peroxynitrate this way did not meet with the same success. Niki and coworkers observed the formation of peroxynitrates (RO_2NO_2), where $R = C_1–C_6$, via FTIR spectroscopy[70]. The method was to photolyse chlorine

(300–400 nm) in the presence of the appropriate hydrocarbon, nitrogen dioxide and air (1 atm):

TABLE 4. Arrhenius parameters for the decomposition of peroxynitric acid at 1 atm: $HO_2NO_2 \rightarrow HO_2^{\cdot} + NO_2$ (68)

$\log(A_{68}/s^{-1})$	E_9(kcal mol^{-1})	Excess gas	Reference
16.1 ± 1.6	23.2 ± 2	air	63
17.8	26	H_2	64
14.2	20.7 ± 0.5	N_2	65
16.4	23.0	—	66

$$Cl_2 + h\nu \longrightarrow 2\,Cl \tag{71a}$$

$$Cl + RH \longrightarrow \dot{R} + HCl \tag{71b}$$

$$\dot{R} + O_2 \longrightarrow RO_2^{\cdot} \tag{72}$$

$$RO_2^{\cdot} + NO_2 \longrightarrow RO_2NO_2 \tag{73}$$

Difficulties were encountered with methyl peroxynitrate. In this case synthesis was achieved via the hydroperoxide:

$$CH_3O_2H + Cl \longrightarrow CH_3O_2^{\cdot} + HCl \tag{74}$$

$$CH_3O_2^{\cdot} + NO_2 \longrightarrow CH_3O_2NO_2 \tag{75}$$

Similar results were found for halogenated methyl peroxynitrates (R = CCl_3, $CFCl_2$, CF_2Cl)[71]. Heicklen and coworkers have studied the decomposition of two halogenated methyl peroxynitrates[72,73] using a technique similar to that used by them for peroxynitric acid[64]. They monitored the concentration of nitric oxide as a function of time in order to determine k_{76} and k_{77}:

$$CCl_3O_2NO_2 \longrightarrow CCl_3O_2^{\cdot} + NO_2 \tag{76}$$

$$CCl_2FO_2NO_2 \longrightarrow CCl_2FO_2^{\cdot} + NO_2 \tag{77}$$

The results are shown in Table 5. Both of these nitrates have thermal stabilities similar to that of peroxynitric acid (Table 4).

Cox and Roffey[74], Hendry and Kenley[75] and Schurath and Wipprecht[76] studied the decomposition of PAN in the presence of excess nitric oxide:

$$CH_3CO_3NO_2 \longrightarrow CH_3CO_3 + NO_2 \tag{78}$$

$$CH_3CO_3 + NO \longrightarrow CH_3CO_2 + NO_2 \tag{79}$$

The rate of removal of the nitrate was monitored by IR spectroscopy[75] or GLC[75,76]. PAN has a similar stability to that of the other peroxynitrates (Tables 4 and 5) but as noted previously has been shown to be subject to explosive decomposition. A similar study was made on peroxypropionyl nitrate[76] (Table 5):

$$C_2H_5CO_3NO_2 \longrightarrow C_2H_5CO_3 + NO_2 \qquad (80)$$

In comparison with the other peroxynitrates, A (Tables 4 and 5) appears to be too high by 1–2 powers of ten. Finally Spence, Edney and Hanst studied the decomposition of peroxychloroformyl nitrate using the same technique[77]:

$$ClC(O)O_2NO_2 \longrightarrow ClC(O)O_2^{\cdot} + NO_2 \qquad (81)$$

$$ClC(O)O_2^{\cdot} + NO \longrightarrow ClC(O)O + NO_2 \qquad (82)$$

Similar Arrhenius parameters were obtained (Table 5).

TABLE 5. Arrhenius parameters for the decomposition of some peroxynitrates: $RO_2NO_2 \rightarrow RO_2^{\cdot} + NO_2$

R	log (A/s^{-1})	E_A(kcal mol^{-1})	Reference
CH_3CO	14.9	24.9	74
	16.3	26.9	75
	15.4	25.7	76
C_2H_5CO	18.0	29.2	76
CCl_3	15.6 ± 1	21.9	72
CCl_2F	16.6 ± 1	24.4	73
ClCO	16.8	27.7	77

IV. OZONIDES

The mechanism for the pyrolysis of ozonides in the gas phase is important in relation to the reaction of ozone with olefins and photochemical smog formation. In solution the Criegee mechanism[78] is well established[9,79]. NMR results show that the initial olefin adduct (**A**) (equation 83) can only react one way[80–84]. Decomposition of (**A**) results in the formation

$$O_3 + (H)(R^1)C{=}C(R^2)(CH_2CH_2R^3) \longrightarrow \text{(A)}: (H)(R^1)C{-}C(R^2)(CH_2CH_2R^3) \text{ bridged by } {-}O{-}O{-}O{-} \qquad (83)$$

(A)

of carbonyl products and two zwitterions or biradicals **Z** and **Z′** (equations 84 and 85). **Z**

$$A \rightarrow R^1CHO + R^2(CH_2CH_2R^3)\overset{+}{C}OO^- \quad \textbf{(Z)} \qquad (84)$$

$$A \rightarrow R^2COCH_2CH_2R^3 + R^1\overset{+}{C}HOO^- \quad \textbf{(Z′)} \qquad (85)$$

and **Z′** subsequently decompose to give acids, esters, alcohols, ketones, alkanes, carbon monoxide, carbon dioxide and water (equations 86–93). Secondary reactions include

$$Z \rightarrow R^2CH_2CH_2R^3 + CO_2 \qquad (86)$$

$$Z \rightarrow R^2COOCH_2CH_2R^3 \qquad (87)$$

$$Z \rightarrow R^3CH_2CH_2COOR^2 \qquad (88)$$

$$Z' \longrightarrow R^1H + CO_2 \qquad (89)$$

$$Z' \longrightarrow R^1COOH \qquad (90)$$

$$Z' \longrightarrow HCOOR^1 \qquad (91)$$

$$Z' \longrightarrow R^1OH + CO \qquad (92)$$

$$Z' \longrightarrow R^1(H)C{=}O + H_2O \qquad (93)$$

reaction of a zwitterion, e.g. **Z′**, with aldehydes or alcohols (equations 94 and 95) (in

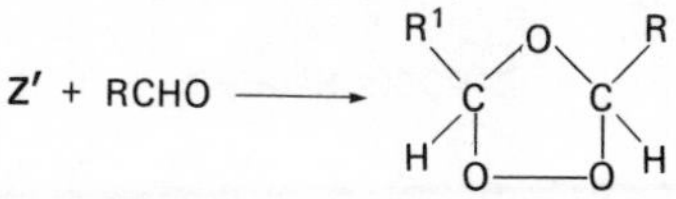

$$Z' + RCHO \longrightarrow \text{R}^1(\text{H})\text{C}\langle\text{O}\rangle\text{C}(\text{H})\text{R (ring: C–O–C, C–O–O–C)} \qquad (94)$$

$$Z' + ROH \longrightarrow H{-}\underset{OR}{\overset{R^1}{C}}{-}OOH \qquad (95)$$

reaction 94 the true ozonide is formed) and dimerization of, for example, **Z** (equation 96). As this mechanism suggests, this sequence is very complex. Leighton assumed that the same mechanism operated in the gas phase[85]. In 1973 O'Neal and Blumstein[11] proposed a new mechanism in terms of biradial intermediate equilibria (equations 97 and 98) of the molozonide formed in reaction (83).

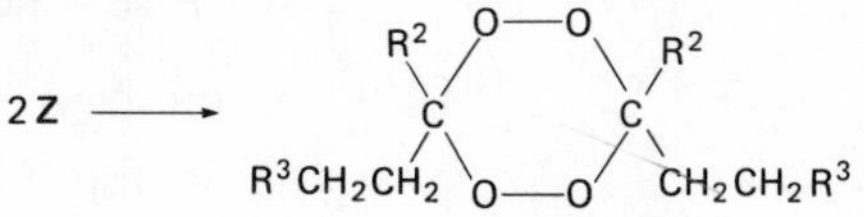

$$2Z \longrightarrow \text{cyclic dimer: } (R^2)(R^3CH_2CH_2)C\langle{}^{O-O}_{O-O}\rangle C(R^2)(CH_2CH_2R^3) \qquad (96)$$

$$\mathrm{A} \rightleftharpoons \mathrm{R^1{-}C(H)(O^{\bullet}){-}C(R^2)(O{-}\dot{O}){-}CH_2CH_2R^3} \quad \textbf{(B)} \tag{97}$$

$$\mathrm{A} \rightleftharpoons \mathrm{R^1{-}C(H)(O{-}\dot{O}){-}C(R^2)(O^{\bullet}){-}CH_2CH_2R^3} \quad \textbf{(B')} \tag{98}$$

Both **B** and **B′** can take part in three hydrogen-atom abstraction paths and a Criegee split. For **B** we have:

α-H abstraction

$$\mathrm{B} \longrightarrow \left(\mathrm{R^1{-}\overset{\dot{O}}{\overset{|}{C}}(H\cdots O{-}O){-}C(R^2)CH_2CH_2R^3}\right) \longrightarrow \mathrm{R^1{-}\overset{O}{\overset{\|}{C}}{-}C(R^2)(OOH)CH_2CH_2R^3} \quad \textbf{(C)} \tag{99}$$

Since the rearrangement is very exothermic, **C** carries excess energy.

$$\mathrm{C} \longrightarrow \mathrm{R^1COOH + R^2COCH_2CH_2R^3} \tag{100}$$

$$\mathrm{C} \longrightarrow \mathrm{R^1COCOCH_2CH_2R^3 + R^2OH} \tag{101}$$

$$\mathrm{C} \longrightarrow \mathrm{R^1COC(R^2)(O^{\bullet})\;CH_2CH_2R^3 + OH} \tag{102}$$

β-H abstraction

$$\mathrm{B} \longrightarrow \left(\mathrm{R^1{-}C(H)(O\cdots H){-}C(R^2)(O{-}\dot{O})CHCH_2R^3}\right) \longrightarrow \mathrm{R^1CH{-}C(R^2)(O{-}\dot{O}){-}\dot{C}HCH_2R^3} \longrightarrow \mathrm{R^1CH(OH){-}C(R^2){-}CHCH_2R^3\ (cyclic\ O{-}O\ bridge)} \quad \textbf{(D)} \tag{103}$$

D is a dioxetane and its decomposition results in fluorescence (see later):

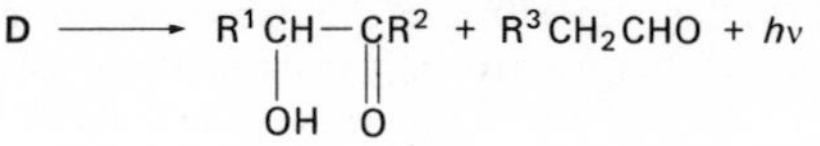

$$\mathbf{D} \longrightarrow R^1CH(OH){-}C({=}O)R^2 + R^3CH_2CHO + h\nu \qquad (104)$$

γ-H abstraction

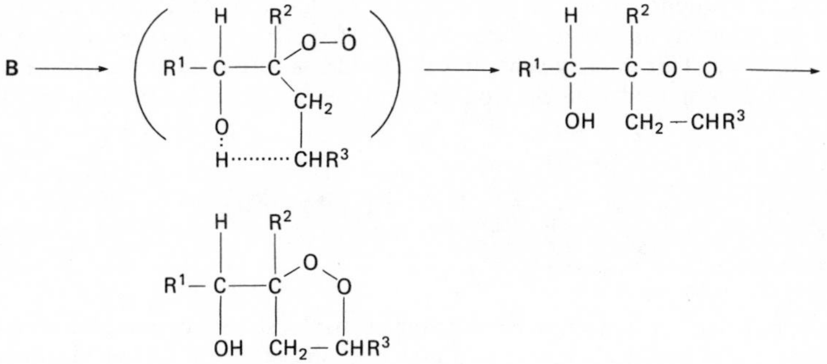

$$\mathbf{B} \longrightarrow \left(R^1{-}C(H)(O{\cdots}H){-}C(R^2)(O{-}\dot{O})CH_2{-}CHR^3 \cdots H \right) \longrightarrow R^1{-}C(H)(OH){-}C(R^2)(O{-}O)(CH_2{-}CHR^3) \longrightarrow$$

$$R^1{-}C(H)(OH){-}\overline{C(R^2){-}O{-}O{-}CHR^3{-}CH_2}\qquad (105)$$

This rearrangement produces a five-membered ring dioxetane.

Criegee split

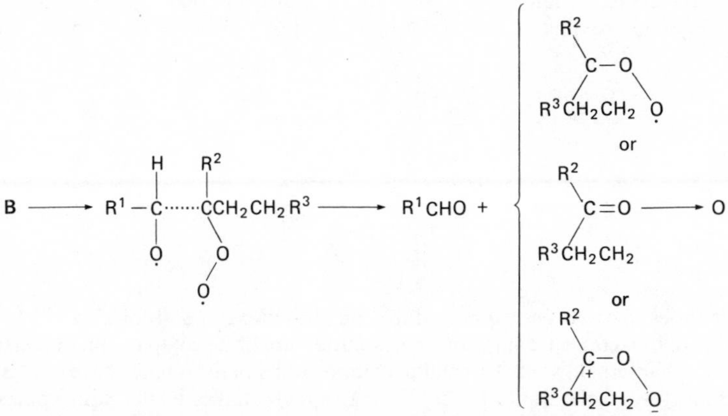

$$\mathbf{B} \longrightarrow R^1{-}C(H)(\dot{O}){\cdots}C(R^2)(O{-}\dot{O})CH_2CH_2R^3 \longrightarrow R^1CHO + \begin{cases} R^2(R^3CH_2CH_2)\dot{C}{-}O{-}\dot{O} \\ \text{or} \\ R^2(R^3CH_2CH_2)C{=}O \longrightarrow O \\ \text{or} \\ R^2(R^3CH_2CH_2)\overset{+}{C}{-}O{-}O^- \end{cases} \qquad (106)$$

This mechanism satisfactorily explains the products formed and, because many intermediates contain excess energy, explains the observed fluorescence[86] and the production of the excited hydroxyl radicals observed by Pitts and coworkers[86]. Pitts and coworkers analysed their data on ozone + olefin reactions[87] and concluded that the O'Neal–Blumstein mechanism gave a very satisfactory explanation of their results. However, it must be noted that Herron and Huie interpreted their results on the addition of ozone to ethene[88], propene and isobutene[89] in terms of the Criegee mechanism.

V. DIOXETANES

It is clear that studies of the decomposition of dioxetanes bear some relation to that of the ozonides. Like the ozonides, much controversy still remains over the actual mechanism of the decomposition[90–98]. However some general conclusions have been made:

(1) Dioxetanes fragment quantitatively into two carbonyl derivatives, with one being electronically excited—$<1\%$ (S_1) and $<50\%$ (T_1).

(2) Decomposition activation energies are in the range 25 ± 3 kcal mol^{-1}. These data have led to two proposed mechanisms. The first, suggested by McCapra[90] and Kearns[91] involves a concerted, simultaneous cleavage of the oxygen–oxygen and carbon–carbon bonds in the ring. This directly generates the products with no intermediates (equation 107). The asterisk indicates an electronically excited species. This mechanism was later

$$R^1R^2C(-O-O-)CR^3R^4 \longrightarrow R^1COR^{2*} + R^3COR^4 \qquad (107)$$

advanced by Turro[92] who suggested that singlet–triplet surface crossing occurred along with the cleavage. The mechanism could now explain the high triplet yields observed in experimental studies. The second mechanism suggested by O'Neal and Richardson[93,97,99] involves two steps. First, the oxygen–oxygen bond cleaves on the strained ring, producing a 1,4-diradical intermediate. At this point, the singlet unpaired oxygen electrons have very little repulsive interaction in the triplet state and almost no bonding energy. Therefore singlet and triplet states will be quite similar in energy, allowing transitions. Partitioning then occurs between singlet and triplet excited carbonyl products and ground-state carbonyl products (equation 108). O'Neal and Richardson explain the high triplet yield via an intersystem crossing to the triplet biradical just before the C—C bond breaks.

$$\text{dioxetane} \longrightarrow {}^{\uparrow}\dot{O}-C-C-\dot{O}^{\downarrow} \rightleftharpoons {}^{\uparrow}\dot{O}-C-C-\dot{O}^{\uparrow}$$
$$\downarrow \qquad\qquad\qquad \downarrow \qquad\qquad (108)$$
$$C{=}O + C{=}O^{*1} \qquad\qquad C{=}O + C{=}O^{*3}$$

Both theories have been supported by further dioxetane studies[94,95,98,100]. The mechanisms of dioxetane decomposition in solution are quite complex but lie outside the scope of this chapter. However thermal decompositions of dioxetanes have been studied almost exclusively in the liquid phase[13–18]. Although studies in the liquid phase have allowed the successful determination of relative amounts of excited singlet and triplet products, thermochemical values and factors involving chemiluminescence efficiency, important information has been lost. Studies of gas-phase dioxetane decomposition at low pressures minimize complications due to excited state quenching and energy transfer so that high-resolution emission spectra can be obtained. Reactions in the gas phase have revealed amounts of vibrational energy in the products and have provided an easy identification of primary excited products in unsymmetrical dioxetanes.

The first gas-phase luminescence experiment was reported by White and coworkers[101]. Upon warming trimethyl-1,2-dioxetane above 50°C and at a vapour pressure of 5–15 Torr, a faint blue luminescence was observed. A low-resolution emission spectrum with a broad band extending from 320 to 580 nm with a maximum at ~430 nm was obtained. Approximately 85 % of the emission was quenched by the addition of air from 5 to 13 Torr. The oxygen quenching suggests that the majority of the emission was probably phosphorescence. The pulsed CO_2 laser-enhanced decomposition of gas-phase tetramethyl-1,2-dioxetane employing methyl fluoride as a sensitizer was carried out by Flynn and Turro's group[102]. The methyl fluoride was laser-pumped and the energy was transferred to tetramethyl-1,2-dioxetane to initiate decomposition. The experiments were performed in a static cell with one Torr of tetramethyl-1,2-dioxetane and 2–30 Torr of methyl fluoride. These workers found quantitative formation of acetone and the induced decomposition of tetramethyl-1,2-dioxetane was accompanied by the emission of blue light ($\lambda_{max} \sim 410$ nm). The spectrum observed was similar to that obtained under thermal conditions. Direct excitation of tetramethyl-1,2-dioxetane by IR laser was carried out by Haas and Yahav[103,104]. They found that chemiluminescence occurred following absorption of the laser pulse. Their studies indicated that the excited tetramethyl-1,2-dioxetane decomposed by two processes, a unimolecular process and a bimolecular collision-induced process. The chemiluminescence observed at 420 nm did not match the fluorescence or the phosphorescence of acetone and the possibility of a mixture of both emissions was suggested.

Most gas-phase work on dioxetane decomposition has been conducted in a flow discharge system by Bogan and coworkers[105–110] under the conditions of ~4 Torr, 400–800 K and flow velocities of $1.5\,m\,s^{-1}$. A generalized coordinate diagram for dioxetane formation from $O_2(^1\Delta)$ plus olefin and decomposition to ground and excited carbonyl products is shown in Figure 3. Although the diagram is only a generalized one, in all cases the activation energy (12–13 $kcal\,mol^{-1}$ for ethyl, methyl, *n*-butyl and vinyl ethers[107]) for conversion of reactants to complex is lower than the activation energies (~25 $kcal\,mol^{-1}$) of dioxetane decomposition. Therefore, in the complete process shown in reaction (109), the rate of product formation is dependent on reaction (109a). The primary aim of these gas-phase studies was to provide high-resolution spectra of excited-

$$O_2(^1\Delta) + R^1R^2C{=}CR^3R^4 \longrightarrow R^1R^2C(-O-O-)CR^3R^4 \qquad (109a)$$

$$R^1R^2C(-O-O-)CR^3R^4 \longrightarrow R^1COR^{\bullet} + R^3COR^4 \qquad (109b)$$

state products. It was deduced from the thermochemistry and the measured E_{109a} that the dioxetane intermediate contains at least 45 $kcal\,mol^{-1}$ of excess vibrational energy[105,106,108]. This is an important difference between gas- and liquid-phase addition of $O_2(^1\Delta)$ to olefins. Experiments have shown that there is no evidence of collisional stabilization of the dioxetane, pointing to a lifetime of $< 10^{-8}$ s. The excess energy has allowed a further investigation into partitioning of vibronic energy of the formaldehyde

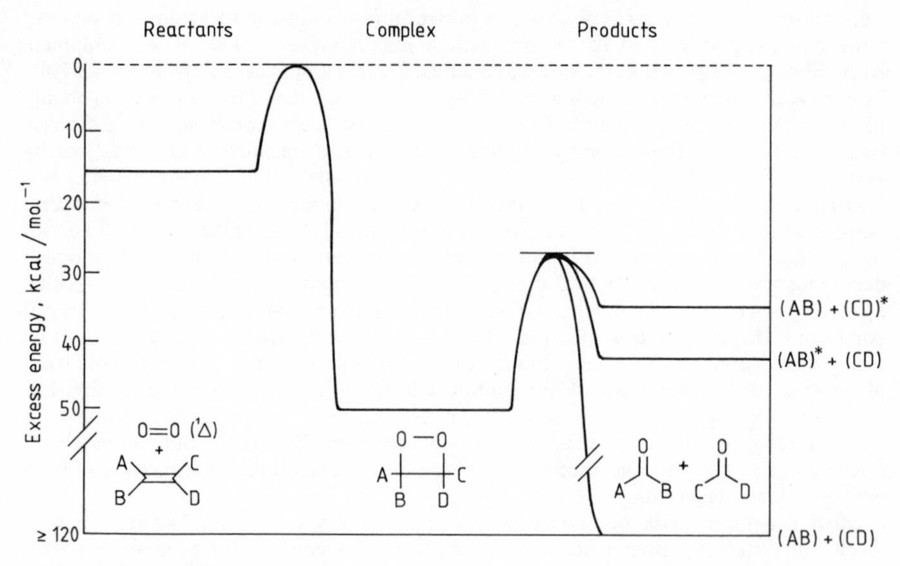

FIGURE 3. Generalized reaction coordinate diagram. Excess energy denotes the energy available for partitioning among the vibrational, rotational and translational motion, at any point on the reaction coordinate. In all cases studied, the entrance barrier, located between Reactants and Complex lies higher in energy than the exit barrier lying between Complex and Products. The product channel (AB) + (CD)* in some cases (vinyl ethers) may be higher than the exit barrier, but is always lower than the entrance barrier. The figure is a schematic and does not exactly conform to the reaction coordinate for any of the reactions studied, however, it bears a close resemblance to the reaction coordinate for all but *N,N*-dimethylisobutenylamine. Reprinted with permission from D. T. Bogan, J. L. Durant, R. S. Sheinson and F. W. Williams, *Photochem. Photo. Biol.*, **30**, 3 (1979). Copyright 1979, Pergamon Press Ltd.

product. The available energy for partitioning among vibrational, rotational and translational degrees of freedom (E_{VRT}) is:

$$E_{VRT} = E_a - H_{RXN} - 81\,\text{kcal mol}^{-1}$$

since only $H_2CO(A^1A_2 \rightarrow X^1A_1)$ transitions are observed. Here, 81 kcal mol^{-1} is the excitation energy of singlet excited formaldehyde[110]. Several factors were tested for their correlation with the hot-band activity of the formaldehyde spectra. These are shown in Table 6. Some discussion of the possible involvement of these factors was given and it was concluded that there is an important fundamental relationship between high quantum yields of S_1 formaldehyde and high hot-band activity.

Gas-phase studies have added valuable information to dioxetane chemistry, by complementing information from liquid-phase studies and, in some cases, providing more information. It remains an important method for probing reaction coordinates of isolated molecules. It also provides a method free of solvent effects and a need to synthesize dioxetanes.

TABLE 6. Correlation of hot-band activity and quantum yield with molecular parameters for alkene + O_2 ($^1\Delta$) reactions[105 a]

Alkene	E_{VRT}	Number of vibrational modes			Ring-carbon hybridization	Hot-band activity	Φ_{rel}
		Dioxetane	H_2CO	Other product			
Ketene	264	15	6	4	sp^3, sp^2	10	26.0
1,1-Difluoroethylene	238	18	6	6	sp^3, sp^3	6	42.0
Ethylene	130	18	6	6	sp^3, sp^3	1.5	14.0
Ethyl vinyl ether	163	39	6	27	sp^3, sp^3	1	1.0
Allene	205	24	6	9	sp^3, sp^2	10	2.7

[a]Reprinted with permission from D. J. Bogan, J. L. Durant, R. S. Sheinson and F. W. Williams, *Photochem. Biol.*, **30**, 3 (1979). Copyright 1979, Pergamon Press Ltd.

VI. ACKNOWLEDGEMENT

The authors acknowledge discussion with D. B. Smith and R. Walsh, and U. Schurath for the communication of results prior to publication and are grateful to J. C. Batt for kinetic calculations involving di-*n*-propyl peroxide.

VII. REFERENCES

1. S. W. Benson and R. Shaw, *Organic Peroxides*, Vol. 1 (Ed. D. Swern), John Wiley and Sons, New York, 1970, pp. 105–140.
2. W. H. Richardson and H. E. O'Neal, *Comprehensive Chemical Kinetics*, Vol. 5 (Eds. C. H. Bamford and C. F. H. Tipper), Elsevier, Amsterdam, 1972, pp. 381–565.
3. A. D. Kirk, *Can. J. Chem.*, **43**, 2236 (1965).
4. A. D. Kirk and J. H. Knox, *Trans. Faraday Soc.*, **56**, 1296 (1960); *Proc. Chem. Soc.*, 384 (1959).
5. S. W. Benson and G. N. Spokes, *J. Phys. Chem.*, **72**, 1182 (1968).
6. P. J. Robinson and K. A. Holbrook, *Unimolecular Reactions*, John Wiley and Sons Ltd., London, 1972.
7. L. Batt and R. Walsh, *Intern. J. Chem. Kinet.*, **14**, 933 (1982).
8. C. J. Howard, *Sixth International Symposium on Gas Kinetics, Southampton*, 1980.
9. L. Batt, *The Chemistry of Amino, Nitroso and Nitro Compounds and their Derivatives* (Eds. S. Patai and Z. Rappoport), John Wiley and Sons Ltd., Chichester, 1982, pp. 417–458.
10. S. W. Benson, *Thermochemical Kinetics*, 2nd ed., John Wiley and Sons, New York, 1976.
11. H. E. O'Neal and C. Blumstein, *Intern. J. Chem. Kinet.*, **5**, 397 (1973).
12. K. R. Kopecky and C. Mumford, *Can. J. Chem.*, **47**, 709 (1969).
13. W. Adam, *Advan. Heterocyclic Chem.*, **21**, 437 (1977).
14. K. A. Horn, J. Y. Koo, S. P. Schmidt and G. B. Schuster, *Molecular Photochemistry*, **9**, 1 (1978–9).
15. N. J. Turro, P. Lechtken, N. E. Schore, G. Schuster, H.-C. Steinmetzer and A. Yekta, *Acc. Chem. Res.*, **7**, 97 (1974).
16. F. McCapra, *Acc. Chem. Res.*, **9**, 21 (1976).
17. T. Wilson, *Intern. Rev. Sci.: Phys. Chem. Ser. II*, **9**, 265 (1976).
18. E. H. White, J. D. Miano, C. J. Watkins and E. J. Breaux, *Angew. Chem. (Intern. Ed. Engl.)*, **13**, 229 (1974).
19. L. Batt and R. D. McCulloch, *Intern. J. Chem. Kinet.*, **8**, 491 (1976).
20. C. K. Yip and H. O. Pritchard, *Can. J. Chem.*, **50**, 1531 (1972).
21. G. Baker, R. Pape and R. Shaw, *Chem. Ind. (London)*, **48**, 1988 (1964).
22. L. Batt and G. N. Rattray, *Intern. J. Chem. Kinet.*, **11**, 1183 (1979).
23. L. Batt and G. N. Robinson, *Intern. J. Chem. Kinet.*, **11**, 1045 (1979).
24. Y. Takezaki and C. Takeuchi, *J. Chem. Phys.*, **22**, 1527 (1954).
25. P. L. Hanst and J. G. Clavert, *J. Phys. Chem.*, **63**, 104 (1958).
26. J. R. Barker, S. W. Benson and D. M. Golden, *Intern. J. Chem. Kinet.*, **9**, 31 (1977).
27. R. C. P. Cubbon, *Progress in Reaction Kinetics* (Ed. G. Porter), Pergamon Press, Oxford, 1970, pp. 29–111.
28. L. Batt and S. W. Benson, *J. Chem. Phys.*, **36**, 895 (1962); **38**, 3031 (1963).
29. L. Batt, J. P. Burrows and G. N. Robinson, *Chem. Phys. Letters*, **78**, 467 (1981).
30. K. O. Christe and D. Pilipovich, *J. Amer. Chem. Soc.*, **923**, 51 (1971).
31. H. E. Radford, *Chem. Phys. Letters*, **71**, 195 (1980).
32. R. R. Baldwin, D. E. Hopkins, A. C. Norris and R. W. Walker, *Combust. Flame*, **15**, 33 (1970).
33. R. R. Baldwin, M. J. Matcham and R. W. Walker, *Combust. Flame*, **15**, 109 (1970).
34. L. Batt, *Intern. J. Chem. Kinet.*, **11**, 977 (1979).
35. J. B. Levy and R. C. Kennedy, *J. Amer. Chem. Soc.*, **94**, 3302 (1972).
36. R. C. Kennedy and J. B. Levy, *J. Phys. Chem.*, **76**, 3480 (1972).
37. J. Czarnowski and H. J. Schumacher, *Z. Phys. Chem. (Frankfurt)*, **92**, 329 (1974).
38. B. Descamps and W. Forst, *Can. J. Chem.*, **53**, 1442 (1975).
39. B. Descamps and W. Forst, *J. Phys. Chem.*, **80**, 933 (1976).
40. R. S. Porter and G. H. Cady, *J. Amer. Chem. Soc.*, **79**, 5628 (1957).

41. J. Czarnowski and H. J. Schumacher, *Z. Phys. Chem. (Frankfurt)*, **73**, 68 (1970).
42. E. A. Liesi, G. Maseiff and A. E. Villa, *J. Chem. Soc., Faraday 1*, **69**, 346 (1973).
43. W. Forst, *Theory of Unimolecular Reactions*, Academic Press, New York, 1973.
44. D. R. Stull and H. Prophet *et al.*, *JANAF Thermochemical Tables*, 2nd ed., John Wiley and Sons, New York, 1976.
45. H. E. O'Neal and S. W. Benson, *Intern. J. Chem. Kinet.*, **1**, 221 (1969).
46. D. M. Golden, *J. Chem. Ed.*, **48**, 235 (1971).
47. R. L. East and L. Phillips, *J. Chem. Soc. (A)*, 331 (1970).
48. M. J. Perona and D. M. Golden, *Intern. J. Chem. Kinet.*, **5**, 55 (1973).
49. L. Batt, T. S. A. Islam and G. N. Rattray, *Intern. J. Chem. Kinet.*, **10**, 931 (1978).
50. F. W. Birss, *Proc. Roy. Soc. (London)*, **A247**, 381 (1958); A. N. Bose and C. N. Hinshelwood, *Proc. Roy. Soc. (London)*, **A249**, 173 (1958); G. Archer and C. N. Hinshelwood, *Proc. Roy. Soc. (London)*, **A261**, 293 (1961).
51. L. Batt and F. R. Cruickshank, *J. Phys. Chem.*, **70**, 723 (1966).
52. H. F. LeFevre, J. D. Kale and R. B. Timmons, *J. Phys. Chem.*, **73**, 1614 (1969).
53. D. H. Shaw and H. O. Pritchard, *J. Can. Chem.*, **46**, 2721 (1968); see also C. K. Yip and H. O. Pritchard, *Can. J. Chem.*, **47**, 4708 (1969).
54. L. F. Loucks,, M. T. H. Liu and D. G. Hooper, *Can. J. Chem.*, **57**, 2201 (1979).
55. A. S. Gordon, *J. Chem. Phys.*, **36**, 1330 (1962).
56. L. Batt and G. N. Robinson, unpublished results.
57. K. J. Laidler and M. T. H. Liu, *Proc. Roy. Soc. (London)*, **A297**, 365 (1967); M. T. H. Liu, and K. J. Laidler, *Can. J. Chem.*, **46**, 479 (1968); L. Batt, *J. Chem. Phys.*, **47**, 3674 (1967); *J. Phys. Chem.*, **73**, 2091 (1969); M. T. H. Liu, L. F. Loucks and R. C. Michaelson, *Can. J. Chem.*, **51**, 2292 (1973); M. T. H. Liu, L. F. Loucks and D. G. Hooper, *Intern. J. Chem. Kinet.*, **9**, 589 (1977).
58. R. Ireton, A. S. Gordon and D. C. Tardy, *Intern. J. Chem. Kinet.*, **9**, 769 (1977).
59. S. Z. Levine, W. M. Uselman, W. H. Chan, J. G. Calvert and J. H. Shaw, *Chem. Phys. Letters*, **48**, 528 (1977).
60. R. Atkinson, K. R. Darnall, A. C. Lloyd, A. M. Winer and J. N. Pitts, Jr., *Advances in Photochemistry* (Eds. J. N. Pitts Jr., G. S. Hammond, K. Gollnick and D. Grosjean), Vol. 11, John Wiley and Sons, New York, 1979, pp. 375–488.
61. E. R. Stephens and M. A. Price, *J. Chem. Ed.*, **50**, 351 (1973).
62. E. R. Stephens, F. R. Burleson and K. M. Holtsclaw, *J. Air. Pollut. Ass.*, **19**, 261 (1969).
63. R. A. Cox, R. G. Derwent and A. J. L. Hutton, *Nature (London)*, **270**, 328 (1977).
64. R. Simonaitis and J. Heicklen, *Intern. J. Chem. Kinet.*, **10**, 67 (1978).
65. R. A. Graham, A. M. Winer and J. N. Pitts, Jr., *Chem. Phys. Letters*, **51**, 215 (1977).
66. A. C. Baldwin and D. M. Golden, *J. Phys. Chem.*, **82**, 644 (1978).
67. R. A. Graham, A. M. Winer and J. N. Pitts, Jr., *J. Chem. Phys.*, **68**, 4505 (1978).
68. R. A. Cox and K. G. Patrick, *Intern. J. Chem. Kinet.*, **11**, 635 (1979).
69. E. F. J. Duynstee, J. G. H. M. Housmans, J. Vlengels and W. Vaskuil, *Tetrahedron Letters*, 2275 (1973).
70. H. Niki, P. D. Maker, C. M. Savage and L. P. Breitenbach, *Chem. Phys. Letters*, **55**, 289 (1978).
71. H. Niki, P. D. Maker, C. M. Savage and L. P. Breitenbach, *Chem. Phys. Letters*, **61**, 100 (1978).
72. R. Simonaitis and J. Heicklen, *Chem. Phys. Letters*, **62**, 473 (1979).
73. R. Simonaitis, S. Glavas and J. Heicklen, *Geophys. Res. Letters*, **6**, 385 (1979).
74. R. A. Cox and M. J. Roffey, *Environ. Sci. Technol.*, **11**, 900 (1976).
75. D. G. Hendry and R. A. Kenley, *J. Amer. Chem. Soc.*, **99**, 3198 (1977).
76. U. Schurath and V. Wipprecht, *Symposium on Atmospheric Pollution, Ispra*, October 1979 and private communication.
77. J. W. Spence, E. O. Edney and P. L. Hanst, *Chem. Phys. Letters*, **56**, 478 (1978).
78. R. Criegee, *Record. Chem. Progr.*, **18**, 111 (1957).
79. For a review see P. S. Bailey, *Chem. Rev.*, **58**, 925 (1958).
80. R. Criegee and G. Schroder, *Chem. Ber.*, **93**, 689 (1960).
81. F. L. Greenwood, *J. Org. Chem.*, **29**, 1321 (1964).
82. F. L. Greenwood, *J. Org. Chem.*, **30**, 3108 (1965).
83. P. S. Bailey, J. A. Thompson and B. A. Shoulders, *J. Amer. Chem. Soc.*, **88**, 4098 (1966).
84. L. J. Durham and F. L. Greenwood, *Chem. Commun.*, 843 (1967).

85. P. A. Leighton, *The Photochemistry of Air Pollution*, Academic Press, New York, 1961.
86. B. J. Finlayson, J. N. Pitts, Jr. and H. Akimoto, *Chem. Phys. Letters*, **12**, 495 (1972).
87. B. J. Finlayson, J. N. Pitts, Jr. and R. Atkinson, *J. Amer. Chem. Soc.*, **96**, 5356 (1974).
88. J. T. Herron and R. E. Huie, *J. Amer. Chem. Soc.*, **99**, 5430 (1977).
89. J. T. Herron and R. E. Huie, *Intern. J. Chem. Kinet.*, **10**, 1019 (1978).
90. F. McCapra, *Chem. Commun.*, 155 (1968).
91. D. R. Kearns, *J. Amer. Chem..Soc.*, **91**, 6554 (1969).
92. N. J. Turro and P. Lechtken, *J. Amer. Chem. Soc.*, **95**, 264 (1973).
93. W. H. Richardson, F. C. Montgomery, M. B. Yelvington and H. E. O'Neal, *J. Amer. Chem. Soc.*, **96**, 7525 (1974).
94. L. B. Hardin and W. A. Goddard, III, *J. Amer. Chem. Soc.*, **99**, 4520 (1977).
95. D. R. Roberts, *Chem. Commun.*, 683 (1974).
96. G. Varnett, *Can. J. Chem.*, **52**, 3837 (1974).
97. W. H. Richardson and H. E. O'Neal, *J. Amer. Chem. Soc.*, **94**, 8665 (1972).
98. J.-Y. Koo and G. B. Schuster, *J. Amer. Chem. Soc.*, **99**, 5403 (1977).
99. H. E. O'Neal and W. H. Richardson, *J. Amer. Chem. Soc.*, **92**, 6553 (1970).
100. T. Wilson, D. E. Golan, M. S. Harris and A. L. Baumstark, *J. Amer. Chem. Soc.*, **98**, 1086 (1976).
101. E. H. White, P. D. Wildes, J. Wiecko, H. Doshan and C. C. Wei, *J. Amer. Chem. Soc.*, **95**, 7050 (1973).
102. W. E. Farneth, G. Flynn, R. Slater and N. J. Turro, *J. Amer. Chem. Soc.*, **98**, 7877 (1976).
103. Y. Haas and G. Yahav, *Chem. Phys. Letters*, **48**, 63 (1977).
104. Y. Haas and G. Yahav, *J. Amer. Chem. Soc.*, **100**, 4885 (1978).
105. D. J. Bogan, R. S. Sheinson, R. G. Gann and F. W. Williams, *J. Amer. Chem. Soc.*, **97**, 2560 (1975).
106. D. J. Bogan, R. S. Sheinson and F. W. Williams, *J. Amer. Chem. Soc.*, **98**, 1034 (1976).
107. D. J. Bogan, R. S. Sheinson and F. W. Williams, *J. Photochem.*, **7**, 156 (1976).
108. D. J. Bogan and J. L. Durant, *NBS Special Publication*, Vol. 526, US Department of Commerce, 1978.
109. D. J. Bogan, R. S. Sheinson and F. W. Williams, *ACS Symposium Series*, **56**, 127 (1977).
110. D. J. Bogan, J. L. Durant, R. S. Sheinson and F. W. Williams, *Photochem. Photobiol.*, **30**, 3 (1979).
111. S. Chapman, *Proc. Roy. Soc.* (*London*), **A132**, 353 (1931).

The Chemistry of Functional Groups, Peroxides
Edited by S. Patai

CHAPTER 21

Photochemistry and radiation chemistry of peroxides

YOSHIRO OGATA, KOHTARO TOMIZAWA AND KYOJI FURUTA

Department of Applied Chemistry, Faculty of Engineering, Nagoya University, Nagoya, Japan

---

I. INTRODUCTION . . . 712
A. Comparison of Thermolysis and Photolysis of Peroxides . . . 716
1. Excited radicals . . . 716
2. Excited substrates . . . 718
B. Reactions of Peroxide-derived Radicals with Organic Compounds . . . 719
II. HYDROGEN PEROXIDE . . . 720
A. General Considerations in the Photolysis of Hydrogen Peroxide . . . 720
B. Oxidation of Alcohols and Ethers . . . 721
C. Oxidation of Carboxylic Acids and Esters . . . 722
D. Oxidation of Nitrogen Compounds . . . 724
E. Oxidation of Aromatic Rings . . . 725
III. HYDROPEROXIDES . . . 727
A. Alkyl Hydroperoxides . . . 727
B. α-Ketohydroperoxides . . . 730
C. α-Azohydroperoxides . . . 730
D. Alkenyl Hydroperoxides . . . 731
E. Polymeric Hydroperoxides . . . 731
IV. DIALKYL PEROXIDES . . . 731
A. Di-*t*-butyl Peroxide . . . 731
B. Dimethyl Peroxide . . . 733
C. Diethyl Peroxide . . . 735
D. Bis(trifluoromethyl) Peroxide . . . 735
E. Cyclic Peroxides . . . 735
F. Other Peroxides . . . 737
V. DIACYL PEROXIDES . . . 738
A. Aliphatic Diacyl Peroxides . . . 738

B. Diaroyl Peroxides . . . 739
C. Cyclic Diacyl Peroxides . . . 742
D. Other Diacyl Peroxides . . . 743

VI. PEROXYCARBOXYLIC ACIDS . . . 743
A. Aliphatic Peroxycarboxylic Acids . . . 743
1. Cycloalkanes . . . 743
2. Aromatic compounds . . . 744
a. Sensitized decomposition . . . 744
b. Direct photolysis . . . 745
B. Aromatic Peroxycarboxylic Acids . . . 747

VII. PEROXYCARBOXYLIC ESTERS . . . 748
A. Simple Photolysis of Peroxycarboxylic Esters . . . 748
1. Aliphatic peroxy acid esters . . . 748
2. Aromatic peroxy acid esters . . . 752
3. Peroxylactones . . . 753
B. α-Acyloxylation . . . 755
C. Intramolecular Oxidation of Unreactive C—H Bonds . . . 755

VIII. MISCELLANEOUS PEROXIDES . . . 757
A. Peroxymono- and Peroxydi-sulphates . . . 757
1. Simple photolysis of peroxysulphates . . . 757
2. Reaction of $SO_4^{\cdot -}$ . . . 757
B. Phosphorus-containing Peroxides . . . 758
1. Peroxydiphosphate . . . 758
2. Peroxyphosphates and peroxyphosphonates . . . 759
3. Bisdiphenylphosphinic peroxide . . . 759
C. Aromatic Sulphoxyperoxides . . . 760
D. Silicon- and Boron-containing Peroxides . . . 760

IX. RADIATION CHEMISTRY . . . 761
A. Introduction . . . 761
B. Radiolysis of Hydrogen Peroxide . . . 761
C. Radiolysis of Alkyl Hydroperoxides . . . 763
D. Radiolysis of Dialkyl Peroxides . . . 764
E. Radiolysis of Diacyl Peroxides . . . 765
F. Radiolysis of Peroxycarboxylic Acids . . . 765
G. Radiolysis of Peroxydisulphate and Peroxydiphosphate . . . 767

X. ACKNOWLEDGEMENTS . . . 770

XI. REFERENCES . . . 771

I. INTRODUCTION

The homolysis of the O—O bond in peroxides may take place either by thermolysis or by photolysis.

In general, peroxides absorb in the ultraviolet region, although the absorption curves for most of them do not form a peak. For example, diacyl peroxides have continuous and weak absorption bands commencing at ca. 280 nm as shown in Table 1 and Figure 1[1]. Also, dialkyl peroxides[2–4] as well as alkyl hydroperoxides[5] and hydrogen peroxide[6] have UV spectral curves increasing continuously from 300 to 200 nm, i.e., at longer wavelengths than ethers and alcohols (<200 nm). Diaroyl peroxides[2,7], however, have primary and secondary absorption peaks (230 nm and 273 nm) close to those of benzoic acid or benzoic anhydride.

TABLE 1. UV absorption of diacyl peroxides in *n*-hexane[d]

Peroxide	$\varepsilon(\text{M}^{-1}\,\text{cm}^{-1})$[a]
Acetyl	47[b]
Propionyl	50
n-Butyryl	59
Isobutyryl	54
2-Methylbutyryl	62
H_2O_2	23[c]

[a]At 2537 Å.
[b]In cyclohexane.
[c]In water.
[d]Reprinted with permission from R. A. Sheldon and J. K. Kochi, *J. Amer. Chem. Soc.*, **92**, 4395 (1970). Copyright 1970 American Chemical Society.

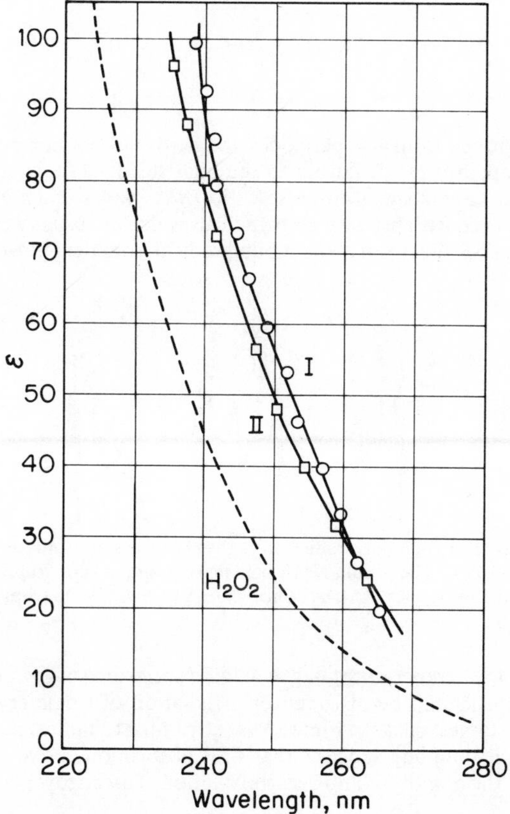

FIGURE 1. Absorption of acetyl peroxide in cyclohexane (I) and in alcohol (II) together with H_2O_2. Reprinted with permission from O. J. Walker and G. L. E. Wild, *J. Chem. Soc.*, 1132 (1937). Copyright 1937 Royal Society of Chemistry.

When light energy is absorbed in a peroxide molecule, the molecule is excited until the energy is removed. In most cases, the energy is used for homolytic dissociation of the O—O bond to give oxygen radicals. The photodissociation of some peroxides is possible even with 290 nm light, which is predominant in the Pyrex glass-filtered light obtained by a high-pressure mercury lamp.

The photolysis can be used for production of hydroxyl radicals from hydrogen peroxide (equation 1), alkoxy radicals from dialkyl peroxides (equation 2), acyloxy and alkyl radicals from diacyl peroxides (equation 3), and also from peroxy esters, and, finally, alkyl and hydroxyl radicals from peroxy acids (equation 4a), and alkyl, alkoxy and acyloxy radicals from peroxy esters (equation 4b).

$$\mathrm{HO{-}OH} \longrightarrow \mathrm{2HO^{\bullet}} \tag{1}$$

$$\mathrm{RO{-}OR^1} \longrightarrow \mathrm{RO^{\bullet}} + \mathrm{{}^{\bullet}OR^1} \tag{2}$$

$$\begin{aligned}\mathrm{RCOO{-}OCOR^1} &\longrightarrow \mathrm{RCOO^{\bullet}} + \mathrm{R^1COO^{\bullet}} \\ &\longrightarrow \mathrm{R^{\bullet}} + \mathrm{2CO_2} + \mathrm{R^{1\bullet}}\end{aligned} \tag{3}$$

$$\mathrm{RCOO{-}OH} \longrightarrow \mathrm{RCOO^{\bullet}} + \mathrm{{}^{\bullet}OH} \longrightarrow \mathrm{R^{\bullet}} + \mathrm{CO_2} + \mathrm{{}^{\bullet}OH} \tag{4a}$$

$$\mathrm{RCOO{-}OR^1} \longrightarrow \mathrm{RCO_2^{\bullet}} + \mathrm{{}^{\bullet}OR^1} \longrightarrow \mathrm{R^{\bullet}} + \mathrm{CO_2} + \mathrm{{}^{\bullet}OR^1} \tag{4b}$$

Alkyl radicals, formed from acyl peroxides and peroxy esters, can be identified by ESR spectra at low temperatures[1a,b]. Owing to the high steady-state concentration of alkyl radicals and the excellent signal-to-noise ratio (S/N ratio), intense ESR spectra are generally observed even with primary alkyl radicals which are usually unstable (Table 2)[8]. Figure 2 shows the ESR spectrum of the *n*-butyl radical from the photolysis of di-*n*-valeryl

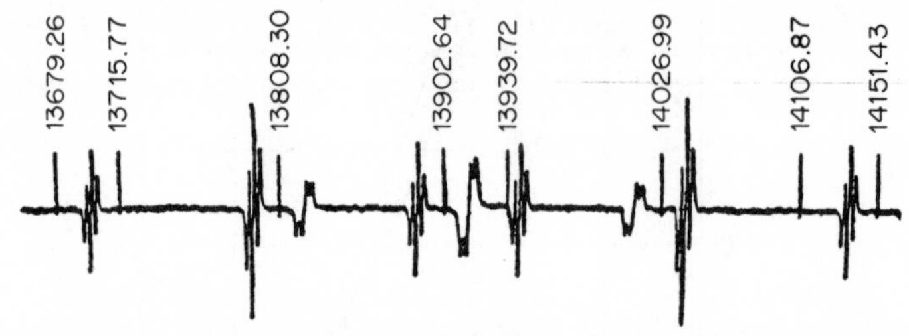

FIGURE 2. ESR spectrum of *n*-butyl radical from the photolysis of a solution of *n*-valeryl peroxide in cyclopropane at −105°C. The proton NMR field markers are in kcps. Reprinted with permission from J. K. Kochi and P. J. Krusic, *J. Amer. Chem. Soc.*, **91**, 3940 (1969). Copyright 1969 American Chemical Society.

peroxide[8]. At very low temperatures, say at 4.2 K, a spectrum suggesting the presence of a pair of phenyl radicals can be observed on irradiation of single crystals of dibenzoyl peroxide[9]. The two largest peaks in Figure 3 are attributed to radical pairs. The magnitude of the dipole–dipole coupling suggests that ESR absorption is due to a pair of phenyl radicals formed by the homolysis and decarboxylation of dibenzoyl peroxide (equation 5).

$$\mathrm{PhC(=O){-}O{-}O{-}C(=O){-}Ph} \xrightarrow{h\nu} \mathrm{2Ph^{\bullet}} + \mathrm{2CO_2} \tag{5}$$

TABLE 2. Hyperfine coupling constants of alkyl radicals from acyl peroxides[a,h]

		Hyperfine coupling constants (G)		
Alkyl radical	T(°C)	a_α	a_β	a_γ
$CH_3^{\cdot}$	−97	22.83		
$CH_3CH_2^{\cdot}$	−85	22.30	26.81	
$CH_3CH_2CH_2^{\cdot}$	−105	22.14	30.33	0.27
$(CH_3)_2CH^{\cdot}$	−105	22.06	24.74	
$CH_3CH_2CH_2CH_2^{\cdot}$	−106	22.12	29.07	0.71
$(CH_3)_2CHCH_2^{\cdot}$	−57	21.93	30.02	*b*
$(CH_3)_3CCH_2^{\cdot}$	−58	21.81		1.00
$CH_2{=}CH(CH_2)_2^{\cdot}$	−105	22.17	28.53	0.61, 0.35[c]
$CH_2{=}CH(CH_2)_3^{\cdot}$	−63	22.13	28.84	0.59
$CH_2{=}CH(CH_2)_4^{\cdot}$	−82	21.93	28.58	0.69
c-$C_5H_9CH_2^{\cdot}$	−90	21.3[d]	21.3[d]	*e*
$C_6H_5CH_2CH_2^{\cdot}$	−33	22.00	29.27	
p-$CH_3OC_6H_4CH_2CH_2^{\cdot}$	−20	22.04	29.50	
m-$CH_3OC_6H_4CH_2CH_2^{\cdot}$	−37	22.2[f]	29.5[f]	
p-$CH_3C_6H_4CH_2CH_2^{\cdot}$	−25	22.00	29.35	
$C_6H_5CH_2CH_2CH_2CH_2^{\cdot}$	−75	22.06	28.66	0.78
$C_6H_5CH_2CH_2CH_2CH_2CH_2^{\cdot}$	−44	22.07	28.12	0.69
$C_6H_5C(CH_3)_2C(CH_3)_2^{\cdot}$[g]	−88		22.61	*e*

[a]Solutions ca. 0.1–1 M peroxide in cyclopropane or cyclopentane solvent.
[b]Unresolved.
[c]One δ hydrogen.
[d]Near-degenerate coupling constants for α and β protons leading to broadened quartet. Envelope of each multiplet ca. 4 G wide, due also to unresolved γ-hyperfine interactions.
[e]Unresolved.
[f]Approximate value due to low solubility.
[g]From photolysis of di-*t*-butyl peroxide in 2,3-dimethyl-2-phenylbutane.
[h]Reprinted with permission from J. K. Kochi and P. J. Krusic, *J. Amer. Chem. Soc.*, **91**, 3940 (1969). Copyright 1969 American Chemical Society.

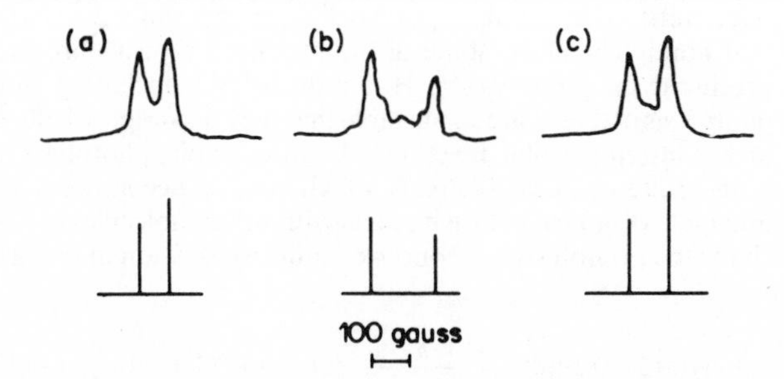

FIGURE 3. ESR absorption spectra obtained from dibenzoyl peroxide with magnetic field applied parallel to the a, b and c crystal axes. The spectra were obtained from an oscillographic display of the absorption at 4.2 K. Magnetic field strength increases to the right. Stick diagrams show calculated intensities of transitions. Reprinted with permission from H. C. Box, E. E. Budzinski and H. G. Freund, *J. Amer. Chem. Soc.*, **92**, 5305 (1970). Copyright 1970 American Chemical Society.

It is of interest to note that no radical pairs of (PhCOO$^{\cdot}$ + PhCOO$^{\cdot}$) or (PhCOO$^{\cdot}$ + Ph$^{\cdot}$) are observed, which shows that photoinduced benzoyloxy radicals are very unstable and rapidly eliminate CO_2 affording Ph$^{\cdot}$ radicals[9].

The O—O bond fission occurs with alkyl hydroperoxide ROOH to give RO$^{\cdot}$ and $^{\cdot}$OH, but the radicals detected are usually ROO$^{\cdot}$ because of the facile hydrogen atom abstraction from ROOH.

Quantum yields for the photolysis are high. For example, peresters RCOOOR^1 on irradiation with 254 nm light cleanly liberate 1 mol of CO_2 in a quantum yield of unity, when irradiated in a pentane, decalin or acetonitrile solution at 30°C (equation 6)[10].

$$RCO_2{-}OR^1 \xrightarrow{h\nu} R^{\cdot} + CO_2 + {}^{\cdot}OR^1 \qquad (6)$$

Similarly, the UV irradiation of diacyl peroxides[1] liberating 2 mol of CO_2 shows a quantum yield of 2.

A number of sensitizers for the photolysis of diacyl peroxides, peresters and hydroperoxides with >305 nm light have been postulated[11,12]. They include aromatic ketones (e.g. benzophenone and acetophenone) and aromatic hydrocarbons (e.g. anthracene). Sensitizers absorb light energy of longer wavelength and transfer it to peroxides. Sensitization by ketones, which is readily quenched by triplet quenchers, occurs via the excited triplet state of the sensitizer, while the sensitization by hydrocarbons may involve excited singlet states[11,12]. The rate constant for the energy transfer from benzophenone to benzoyl peroxide is very large ($3.2 \times 10^6\,\text{M}^{-1}\,\text{s}^{-1}$), but only 25 % of the excited peroxide decomposes because of deactivation or cage recombination[12].

A. Comparison of Thermolysis and Photolysis of Peroxides

Thermal and photochemical decompositions of peroxides are similar, since in both the primary reaction is the homolytic fission of peroxide linkage (O—O), giving oxygen radicals (RO$^{\cdot}$, RCOO$^{\cdot}$, HO$^{\cdot}$ etc). But there are several points in which their behaviour differs.

1. Excited radicals

The photolysis is initiated by the absorption of light energy which excites the peroxide to cleave the O—O bond. Therefore, some of the produced radicals are excited; i.e. 'hot radicals' are produced on photolysis[13]. Hot radicals[13,14], including those which are formed in the photolysis of esters, are so unstable that they decompose before being able to take part in further intermolecular reactions. As an example, photolysis of diisopropyl peroxide gives an excited radical (*i*-PrO$^{\cdot *}$) which readily decomposes to give mostly acetaldehyde and methyl radical before it reacts with other molecules (equation 7)[15]. An alternative path for the photolysis is a concerted multiple scission in the primary process (equation 8a).

$$(CH_3)_2CHO{-}OCH(CH_3)_2 \xrightarrow{h\nu} (CH_3)_2CHO^{\cdot *} + (CH_3)_2CHO^{\cdot} \qquad (7a)$$

$$(CH_3)_2CHO^{\cdot *} \longrightarrow CH_3^{\cdot} + CH_3CHO \qquad (7b)$$

$$(CH_3)_2CHOOCH(CH_3)_2 \xrightarrow{h\nu} (CH_3)_2CHO^{\cdot} + CH_3CHO + CH_3^{\cdot} \qquad (8a)$$

$$\longrightarrow 2CH_3^{\cdot} + 2CH_3CHO \qquad (8b)$$

Ground-state (ordinary) alkoxy radicals formed by thermolysis can be trapped completely by scavengers such as nitric oxide; e.g. isopropoxy radical gives isopropyl nitrite and then acetone (equation 9). However, excited alkoxy radicals are not trapped by nitric oxide, but decompose according to equation (7b)[15].

$$(CH_3)_2CHO^{\bullet} + NO \longrightarrow (CH_3)_2CHONO \longrightarrow CH_3COCH_3 + HNO \qquad (9)$$

Therefore, the molar ratio of excited and ground-state alkoxide, α, can be measured by the trapping experiments; e.g. in the case of isopropoxy radical, α can be expressed as:

$$\alpha = \phi\,(CH_3CHO)\Big/\Big[\phi\,(CH_3CHO) + \phi\,(i\text{-PrONO}) + \phi\,(CH_3COCH_3)\Big]$$

Here ϕ means the quantum yield of the product in parentheses. The α values for isopropoxy and *t*-butoxy radicals derived from the photolysis of the corresponding peroxides are shown in Table 3[16], which shows that more than half of the alkoxy radicals are excited.

TABLE 3. α Values of alkoxy radicals formed by photolysis of some dialkyl peroxides[a]. Peroxide pressure 23.5 mm, NO pressure 15 mm, λ 2537 Å at 26°C

Peroxide	$\alpha(i\text{-PrO}^{\bullet})$	$\alpha(t\text{-BuO}^{\bullet})$
Diisopropyl	0.60	—
Di-*t*-butyl	—	0.53
Isopropyl *t*-butyl	0.72	0.64

[a]Reprinted with permission from G. R. McMillan, *J. Amer. Chem. Soc.*, **84**, 2514 (1962). Copyright 1962 American Chemical Society.

Photolysis of dicumyl peroxide in *n*-hexane with 313 nm light gives mainly (95–100 %) α,α'-dimethylbenzyl alcohol by hydrogen abstraction of cumyloxy radical from the solvent RH, while the photolysis in carbon tetrachloride with the same 313 nm light gives mainly (95–100 %) acetophenone and some acetone, which are the products of C—C fission of the hot cumyloxy radical (equation 10)[17a].

$$Ph(CH_3)_2C{-}OO{-}C(CH_3)_2Ph \xrightarrow{h\nu} 2Ph(CH_3)_2C{-}O^{\bullet} \qquad (10a)$$

$$2Ph(CH_3)_2C{-}O^{\bullet} + 2RH \longrightarrow 2Ph(CH_3)_2C{-}OH + 2R^{\bullet} \qquad (10b)$$

$$2Ph(CH_3)_2C{-}O^{\bullet *} \longrightarrow PhCOCH_3\ [+ Ph^{\bullet} + CH_3^{\bullet} + (CH_3)_2CO] \qquad (10c)$$

Interestingly, small amounts of aromatic compounds accelerate the hydrogen atom transfer from solvent to oxyl radical via rapid addition of $R^{\bullet}$ radical to arenes followed by hydrogen abstraction from the adducts[17b].

In the case of di-*t*-butyl peroxide, the decomposition of the excited radical gives acetone (equation 11).

$$(CH_3)_3CO^{\bullet *} \longrightarrow CH_3^{\bullet} + CH_3COCH_3 \qquad (11)$$

2. *Excited substrates*

Excited substrates may show behaviour different from that of the ground state, as has been observed in the light-induced hydrolysis of esters of *p*- and *m*-nitrophenols[18,19]. Similarly abnormal orientation in ionic substitution was observed in the solvolysis of methoxybenzyl acetate in which nucleophilic attack is enhanced by a *meta*-methoxy group[20] and in the H–D exchange reaction with toluene and anisole, giving *ortho* and *meta* preference[21].

Abnormal orientation was observed in the photochemical reaction of toluene with ethyl chloroacetate, which should be an attack by the electrophilic radical $^{\cdot}CH_2COOEt$ on the aromatic ring of toluene. Contrary to the normal isomer distribution of electrophilic attacks (*para, ortho* ≫ *meta*), the reaction gave 5.2% *ortho*, 9.6% *meta* and 0.9% *para* ethyl tolylacetate as in equation (12)[22]. This abnormal behaviour can be explained by a preferential attack of radicals on an excited aromatic molecule, in which the electron density is different from that of the ground-state molecule (Table 4)[23].

$$C_6H_5CH_3 + ClCH_2CO_2Et \xrightarrow{h\nu,\ N_2} CH_3C_6H_4CH_2CO_2Et \left(+ \begin{array}{c} CH_2COOEt \\ | \\ CH_2COOEt \\ \text{trace} \end{array} \right) \quad (12)$$

Analogously, the methyl radical derived by photolysis of acetyl peroxide with 254 nm light gives xylenes, in which the *ortho*:*meta*:*para* ratio produced at 30°C is 51:35:14[24] (Table 4). Hence, the partial rate factors, in which the statistical factor of 2 for *ortho* and *meta* positions is taken into account, for the *ortho*:*meta*:*para* positions are 45:31:24, which is significantly different from the ratio 48:23:29 observed in the thermal reaction of toluene with methyl radical formed by thermolysis of acetyl peroxide[11]. This trend of increasing *meta*-methylation is even more remarkable at lower temperatures (Table 4). The abnormal behaviour of photochemical methylation is explained by the participation of excited toluene, which has different electron densities from the ground state (frontier

TABLE 4. Frontier electron densities of excited arenes, PhR, at first excited state and product distribution in radical methylation with acetyl peroxide and ethoxycarbonylmethylation (equation 12)[e]

	Lowest excited state frontier electron density			Substitution with					
				$Me^{\cdot}$			$EtO_2CCH_2^{\cdot}$		
R	*o*	*m*	*p*	*o*	*m*	*p*	*o*	*m*	*p*
Me	0.340	0.321	0.309	51	35	14[a]	33	61	6[c]
				37	51	12[b]	—	—	—
MeO	0.350	0.314	0.267	68	21	11[c]	59	28	13[c]
				54	32	14[d]	—	—	—
i-Pr	0.335	0.311	0.287	—	—	—	17	46	37[c]
t-Bu	0.327	0.295	0.258	—	—	—	0	65	35

[a] At 30°C.
[b] At −70°C.
[c] At 20°C.
[d] At −60°C.
[e] Reprinted with permission from Y. Ogata, E. Hayashi and H. Kato, *Bull. Chem. Soc. Japan*, **51**, 3657 (1978). Copyright 1978 Chemical Society of Japan.

electron density of first excited state: *ortho* > *meta* > *para*), and by enhanced steric hindrance at the *ortho* position at low temperature. Analogous behaviour is observed in the methylation of anisole. The reaction via exciplex (excited arene-peroxide) before O—O cleavage of peroxide is conceivable.

In the substitution of toluene with methyl radicals formed by photolysis of peracetic acid (254 nm light) the partial rate factors are *ortho*:*meta*:*para* = 30:49:21[25]. Attack by hydroxyl radicals, formed simultaneously by photolysis of peracetic acid, gives cresols, but the isomer distribution (*ortho* > *para* > *meta*) is analogous to that of the thermal reaction. The different behaviour of HO˙ radicals is attributable to the more random attack of the HO˙ radicals which are more reactive than the $CH_3^{\cdot}$ radicals[25].

The difference of behaviour between photochemical and thermal radical reactions was also observed for the reaction of toluene with $^{\cdot}CMe_2CN$ radicals derived from α,α′-azobis(isobutyronitrile); the thermal reaction gives only a methyl substitution, while photochemical reaction gives both methyl and ring substitution products[26].

B. Reactions of Peroxide-derived Radicals with Organic Compounds

As stated above, photolysis of peroxide gives oxy radicals by primary homolysis of the O—O bond. The radicals may be cleaved further and may give other radicals as well as their coupling products.

The simplest case is the formation of hydroxyl radical from hydrogen peroxide by photolysis. The hydroxyl radical thus formed may react in the following processes: (*a*) hydrogen abstraction from organic compounds or from H_2O_2 itself, (*b*) addition to multiple bonds, and (*c*) coupling and yielding again H_2O_2. Processes *a* and *b* then produce radicals again, and thus the radical chain may continue producing a variety of compounds. For example, when an aqueous solution of benzoic acid and H_2O_2 is photolysed, the initial process is the addition of HO˙ to the aromatic ring of benzoic acid, followed by dehydrogenation by HO˙ to give hydroxybenzoic acid and oxdative decarboxylation to give phenyl radical or phenol (equation 13)[27].

$$C_6H_5CO_2H + HO^{\cdot} \longrightarrow {}^{\cdot}C_6H_5(OH)CO_2H \xrightarrow{HO^{\cdot}} HOC_6H_4CO_2H + H_2O \quad (13a)$$

$$^{\cdot}C_6H_5(OH)CO_2H + HO^{\cdot} \longrightarrow C_6H_5OH + H_2O + CO_2 \quad (13b)$$

$$C_6H_5CO_2H + HO^{\cdot} \xrightarrow{-H_2O} C_6H_5CO_2^{\cdot} \longrightarrow Ph^{\cdot} + CO_2 \quad (13c)$$

Further reaction of the products, including cleavage of the benzene ring, gives a variety of secondary products: CO_2, HCO_2H, CH_3CO_2H, $HOCH_2CO_2H$, HOOCCOOH, $HOOCCH_2COOH$, HOOCCH=CHCOOH, HOOCCH=CH—CH=CHCOOH (muconic acid), HOC_6H_4OH, PhC_6H_4COOH, $PhCOOC_6H_4COOH$, etc.

Furthermore, photochemical reactions of the starting material and products may occur in the absence of peroxide, as well as dark reactions of the starting materials and products with peroxide such as perbenzoic acid formation from benzoic acid and H_2O_2, although these are fortunately very slow compared with the photolysis.

II. HYDROGEN PEROXIDE

A. General Considerations in the Photolysis of Hydrogen Peroxide

Hydrogen peroxide shows, due to the O—O and O—H bonds, continuous UV absorption without peak increasing with a decrease of wavelength (see Figure 1). Therefore, H_2O_2 can dissociate by absorbing UV light. The quantum yields for some types of this photolysis depend on the wavelength as shown in Table 5[28].

TABLE 5. Summary of primary processes in the photolysis of hydrogen peroxide[c]

	Quantum yield	
Primary process	1236 Å	1470 Å, 2537 Å
$H_2O_2 \rightarrow H^{\bullet} + HO_2^{\bullet}$	≥0.25[a]	<0.01
$H_2O_2 \rightarrow \begin{cases} H_2 + O_2 \\ H_2 + 2O(^3P) \end{cases}$	0.25	<0.01
$H_2O_2 \rightarrow H_2O + O$ (singlet)	<0.01	<0.01
$H_2O_2 \rightarrow 2OH^{\bullet}$	≤0.50[b]	~1.0

[a]This value is based on the reduction in H_2 yield upon addition of C_2D_4 and represents a minimum since reaction of H with H_2O_2 does not always lead to H_2 formation.
[b]This assumes that all quanta absorbed lead to decomposition and that OH formation accounts for all decomposition other than H or H_2 formation.
[c]Reprinted with permission from L. J. Stief and V. J. DeCarlo, *J. Chem. Phys.*, **50**, 1234 (1969). Copyright 1969 American Institute of Physics.

Under acidic and neutral conditions and UV light of a low-pressure Hg lamp (254 nm), the primary photolysis of H_2O_2 gives almost completely hydroxyl radical $HO^{\bullet}$. The latter reacts with H_2O_2 to give $HO_2^{\bullet}$, which can be observed with ESR at high concentrations of H_2O_2 at 77 K[29,30]. Therefore, the photolysis has a chain mechanism analogous to the Fenton's reagent decomposition in the dark, postulated by Haber–Weiss[28,31,32] (equation 14). In contrast, in alkaline media, H_2O_2 is photolysed to give ozone via a transient ozonide[33].

$$\text{Initiation} \quad H_2O_2 \xrightarrow{h\nu} 2HO^{\bullet} \tag{14a}$$

$$\text{Propagation} \quad HO^{\bullet} + H_2O_2 \longrightarrow H_2O + HO_2^{\bullet} \tag{14b}$$

$$HO_2^{\bullet} + H_2O_2 \longrightarrow H_2O + O_2 + HO^{\bullet} \tag{14c}$$

$$\text{Termination} \quad 2HO_2^{\bullet} \longrightarrow H_2O_2 + O_2 \tag{14d}$$

The photodecomposition of H_2O_2 is markedly accelerated by the presence of carbon monoxide, which reduces $HO^{\bullet}$ to hydrogen in a chain mechanism (equation 15)[34–36].

$$HO^{\bullet} + CO \longrightarrow CO_2 + H^{\bullet} \tag{15a}$$

$$H^{\bullet} + H_2O_2 \longrightarrow HO^{\bullet} + H_2O \text{ (or } H_2 + HO_2^{\bullet}) \tag{15b}$$

In the presence of organic compounds, the excitation of H_2O_2 occurs by the photoenergy transfer from excited organic compounds as well as by the direct absorption of photon by H_2O_2; thus H_2O_2 decomposes to give $HO^{\bullet}$ radicals, which in turn react to abstract hydrogen atoms or add to any multiple bonds if present.

Photolysis of H_2O_2 for organic oxidation has the following advantages: (*i*) the generation of $HO^{\bullet}$ radicals without contamination by metallic ions such as Fe^{2+} or Fe^{3+}; (*ii*) controllable concentration of $HO^{\bullet}$ by the change of wavelength and intensity of light; (*iii*) possibility of reaction at low temperatures even in the solid state; (*iv*) possibility of peculiar reactions.

B. Oxidation of Alcohols and Ethers

Primary and secondary alcohols react with $HO^{\bullet}$ generated by photolysis of H_2O_2, giving aldehydes and ketones along with dimeric substrates (equation 16)[37–39]. For example, methanol gives formaldehyde and ethylene glycol[40], and isopropanol gives acetone and pinacol[41,42]. The rates of oxidation and product yields depend on the light intensity, temperature and concentration of reactants.

$$RR^1CH{-}OH \xrightarrow[H_2O_2]{h\nu} RR^1\dot{C}{-}OH \xrightarrow{HO^{\bullet}} RR^1C{=}O \qquad (16a)$$

$$2\,RR^1\dot{C}{-}OH \longrightarrow RR^1C(OH){-}CRR^1(OH) \qquad (16b)$$

Glycols and ethers are similarly oxidized by initial hydrogen atom abstraction with $HO^{\bullet}$. If H_2O_2 is in excess, the products are further oxidized as shown in Tables 6 and 7 for ethylene glycol dimethyl ether (EDE) and ethylene glycol monomethyl ether (EME) (equation 17). Since the concentration of H_2O_2 is high in these cases, peroxy radical $HO_2^{\bullet}$ as well as $HO^{\bullet}$ may participate in these oxidations[43].

TABLE 6. Yields of products in photooxidation of 0.223 M EDE with 0.909 M H_2O_2 at 20°C[a]

Irrad. time (min)	Dec $[H_2O_2]$ (%)	Dec [EDE] (%)	CH_3OH (%)	HCO_2CH_3 (%)	CH_3OCH_2CHO (%)	EME (%)	$CH_3OCH_2CO_2H$ (%)
5	2.8	11.2	19.4	8.5	trace	25.8	2.9
10	7.4	24.3	15.7	6.9	2.5	22.9	3.1
15	11.6	35.8	14.5	7.8	2.1	19.6	2.0
20	16.5	42.3	14.6	7.3	3.7	16.2	1.8
30	27.3	57.1	14.3	6.0	4.1	16.2	1.3

[a]Reprinted with permission from Y. Ogata, K. Tomizawa and K. Fujii, *Bull. Chem. Soc. Japan*, **51**, 2628 (1978). Copyright 1978 Chemical Society of Japan.

$$\underset{\text{(EME)}}{HOCH_2CH_2OCH_3} \xrightarrow[\text{aq. } H_2O_2]{h\nu} CH_3OH,\ CH_3OCH_2CHO,\ HCO_2CH_3, CH_3CO_2H \qquad (17a)$$

$$\underset{\text{(EDE)}}{CH_3OCH_2CH_2OCH_3} \xrightarrow[\text{aq. } H_2O_2]{h\nu} CH_3OH,\ HOCH_2CH_2OCH_3, CH_3CH_2CHO,\ CH_3OCH_2CO_2H,\ HCO_2CH_3 \qquad (17b)$$

TABLE 7. Yields of products in photooxidation of 0.167 M EME with 0.919 M H_2O_2 at 20°C[a]

Irrad. time (min)	Dec [H_2O_2] (%)	Dec [EME] (%)	CH_3OH (%)	HCO_2H (%)	HCO_2CH_3 (%)	CH_3CO_2H (%)	CH_3OCH_2CHO (%)
5	2.9	5.4	30.2	—	trace	—	11.8
10	4.6	13.6	18.8	trace	3.1	—	8.0
15	7.3	13.8	24.5	0.1	4.5	trace	9.2
20	7.8	20.6	19.5	0.1	3.2	0.1	8.0
30	13.8	40.1	16.1	0.2	2.4	0.2	6.6

[a]Reprinted with permission from Y. Ogata, K. Tomizawa and K. Fujii, *Bull. Chem. Soc. Japan*, **51**, 2628 (1978). Copyright 1978 Chemical Society of Japan.

The products from EME may be formed by the scheme shown in equation (18) initiated by hydrogen atom abstraction by HO˙ radicals[43].

$$\text{EME} \xrightarrow{\text{HO}^\bullet} \text{CH}_3\text{OCH}_2\dot{\text{C}}\text{HOH} \xrightarrow{\beta\text{-fiss}} \text{CH}_3\text{O}^\bullet + \text{CH}_3\text{CHO} \quad (18a)$$

$$\text{CH}_3\text{OCH}_2\dot{\text{C}}\text{HOH} \xrightarrow{\text{HO}^\bullet \text{ or } \text{H}_2\text{O}_2} \text{CH}_3\text{OCH}_2\text{CHO} \quad (18b)$$

$$\text{CH}_3\text{OCH}_2\dot{\text{C}}\text{HOH} \xrightarrow{\text{H}^+} \text{CH}_3\text{OH} + {}^\bullet\text{CH}_2\text{CH}{=}\overset{+}{\text{O}}\text{H} \quad (18c)$$

$$\text{CH}_3\text{CHO} \xrightarrow{\text{O}} \text{CH}_3\text{COOH} \quad (18d)$$

$$\text{EME} \xrightarrow{\text{HO}^\bullet} \text{CH}_3\text{O}\dot{\text{C}}\text{HCH}_2\text{OH} \xrightarrow{\text{HO}_2^\bullet} \text{CH}_3\text{OCH(OOH)CH}_2\text{OH} \xrightarrow{h\nu} \text{CH}_3\text{OCH}{=}\text{O} + \text{HCHO} + \text{H}_2\text{O} \quad (18e)$$

C. Oxidation of Carboxylic Acids and Esters

Alkyl chains of carboxylic acids are usually subject to hydrogen atom abstraction with HO˙. For example, acetic acid gives ˙CH_2COOH, which is identified by ESR[38,44] while a γ-hydrogen atom is preferentially abstracted in the case of longer chain aliphatic acids because of the repulsion between the electrophilic carboxyl group and the electrophilic HO· radical (equation 19)[45].

$$\text{CH}_3\text{CH}_2\text{CH}_2\text{CO}_2\text{H} \xrightarrow[\text{aq. H}_2\text{O}_2]{h\nu} \text{HO}_2\text{CCH}_2\text{CH}_2\text{CO}_2\text{H} \quad (19a)$$

$$\text{CH}_3\text{CH}_2\text{CH}_2\text{CH}_2\text{CO}_2\text{H} \xrightarrow[\text{aq. H}_2\text{O}_2]{h\nu} \text{CH}_3\text{C(=O)CH}_2\text{CH}_2\text{CO}_2\text{H} \quad (19b)$$

$$\text{CH}_3\text{CH}_2\text{CH}_2\text{CH}_2\text{CH}_2\text{CO}_2\text{H} \xrightarrow[\text{aq. H}_2\text{O}_2]{h\nu} \text{CH}_3\text{CH}_2\text{C(=O)CH}_2\text{CH}_2\text{CO}_2\text{H} \quad (19c)$$

In the photolysis of carboxylic acids in the presence of excess H_2O_2, the subsequent oxidation gives a variety of products which eventually lead to carbon dioxide and water[46]. The oxidizability decreases in the order: $CH_3CH_2CO_2H > HCO_2H > CH_3CO_2H$. At high concentration of H_2O_2, the hydrogen atom of COOH may be abstracted. This process is assumed based on the formation of methane and ethane from acetic acid presumably by the scheme shown in equation (20)[47].

$$\text{CH}_3\text{CO}_2\text{H} \xrightarrow{\text{HO}^\bullet} \text{CH}_3\text{CO}_2^\bullet \longrightarrow \text{CH}_3^\bullet + \text{CO}_2; \quad \text{CH}_3^\bullet \longrightarrow \text{CH}_3\text{—CH}_3; \quad \text{CH}_3^\bullet \xrightarrow{\text{CH}_3\text{CO}_2\text{H}} \text{CH}_4 \quad (20)$$

Carboxylic acids bearing α-OH groups are more active than the parent acids toward H_2O_2, and their α-hydrogen atoms are abstracted to give α-keto acids (equation 21).

$$RCH(OH)CO_2H \xrightarrow[H_2O_2]{h\nu} RCOCO_2H \tag{21}$$

Aromatic carboxylic acids are attacked mainly by the initial addition of HO˙ to a ring carbon forming phenolcarboxylic acids, e.g. equation (22).

$$PhCO_2H \xrightarrow[H_2O_2]{h\nu} HOC_6H_4CO_2H + PhOH + PhH \tag{22}$$

The formation of phenol and benzene may be the result of hydrogen atom abstraction from the COOH group, followed by decarboxylation (cf. equation 20)[47]. In the case of a large excess of H_2O_2, a variety of carboxylic acids are formed and consumed as the reaction proceeds (Figure 4)[27].

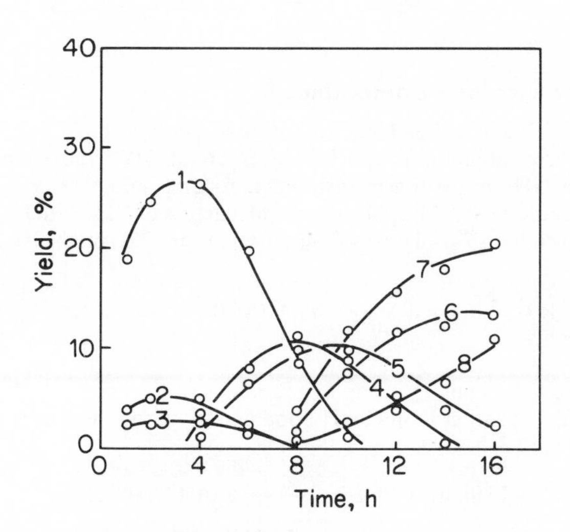

FIGURE 4. Time dependence of product yields in photolysis of $PhCO_2H$ in aqueous H_2O_2. Initial concentration $[PhCO_2H]_0$, 1.72×10^{-2} M and $[H_2O_2]_0$, 4.22×10^{-1} M ($[H_2O_2]_0/[PhCO_2H]_0$ = 24.5). Here $[\,]_0$ means initial concentration.

$$\text{Yield } (\%) = \frac{\text{carbon number of product}}{\text{carbon number of substrate}} \times \frac{[\text{Product}]\ (\text{M})}{[\text{PhCO}_2\text{H decomposed}]\ (\text{M})} \times 100$$

except for aromatic products whose yields are based on the decomposed $PhCO_2H$. Curve 1, $HOC_6H_4CO_2H$; curve 2, PhOH; curve 3, PhH; curve 4, $HO_2CCH{=}CHCO_2H$; curve 5, $HO_2CCH_2CO_2H$; curve 6, HO_2CCO_2H; curve 7, CH_3CO_2H; curve 8, HCO_2H. Reprinted with permission from Y. Ogata, K. Tomizawa and Y. Yamashita, *J. Chem. Soc., Perkin Trans. 2*, 616 (1980). Copyright 1980 Royal Society of Chemistry.

Phenols are oxidized further to give *o*-quinones and then muconic acids (equation 23)[27].

(23)

In the photolysis of esters in the presence of H_2O_2, the alkyl group is preferentially attacked; e.g. a methyl hydrogen atom of methyl formate is abstracted (equation 24)[44].

$$HCO_2CH_3 \xrightarrow[H_2O_2]{h\nu} HCO_2CH_2^{\bullet} \quad (24)$$

D. Oxidation of Nitrogen Compounds

Ammonia (pK_a 9.3) existing as NH_4^+ in acidic solution is stable, but in a neutral or alkaline solution, free ammonia is attacked by H_2O_2 on UV irradiation. The primary attack of $HO^{\bullet}$ on NH_3 is hydrogen abstraction to give amino radical, which is also observed in pulse radiolysis[48]. The photochemical reaction of NH_3 with H_2O_2 in aqueous solution gives the products[49] and yields shown in equation (25a) on the basis of consumed ammonia:

$$\underset{}{NH_3 + H_2O_2} \xrightarrow[\text{pH 10.2}]{h\nu} \underset{23\%}{N_2} + \underset{61\%}{NH_2OH} + \underset{15\%}{NO_2^-} + \underset{1.7\%}{NO_3^-} \quad (25a)$$

Primary and secondary aliphatic amines are photooxidized with H_2O_2, i.e. with $HO^{\bullet}$, via N—H bond fission rather than C—H bond fission, giving the corresponding amino radicals (equation 25b)[50].

$$2RR^1NH + H_2O_2 \xrightarrow{h\nu} 2RR^1N^{\bullet} + 2H_2O \quad (25b)$$

On the other hand, peptides are attacked by $HO^{\bullet}$ at the carbon between the NH and CO_2^- groups and by C—H bond fission to give carbon radicals characterized by ESR (Figure 5)[51,52].

Amino acids are subject to oxidative cleavage. The presence of an aromatic group facilitates the reaction by the energy transfer from the excited aromatic ring to H_2O_2. The reaction of phenylalanine is shown in equation (26)[53].

$$PhCH_2CH(NH_2)CO_2H \xrightarrow[H_2O_2]{h\nu\ (254\ nm)} HO_2CCH_2CH(NH_2)CO_2H + HOCH_2CH(NH_2)CO_2H + CH_3CH(NH_2)CO_2H + NH_3 + H_2N(CH_2)_4CH(NH_2)CO_2H \quad (26)$$

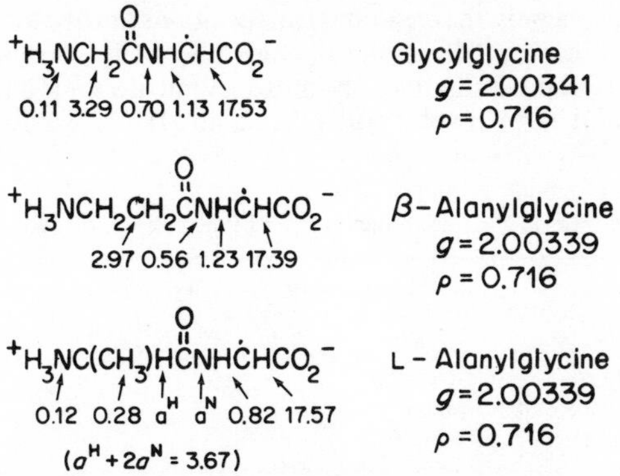

$^{+}H_3NCH_2\overset{O}{\overset{\|}{C}}NH\dot{C}HCO_2^-$ — Glycylglycine, $g = 2.00341$, $\rho = 0.716$; coupling constants: 0.11 3.29 0.70 1.13 17.53

$^{+}H_3NCH_2CH_2\overset{O}{\overset{\|}{C}}NH\dot{C}HCO_2^-$ — β-Alanylglycine, $g = 2.00339$, $\rho = 0.716$; coupling constants: 2.97 0.56 1.23 17.39

$^{+}H_3NC(CH_3)H\overset{O}{\overset{\|}{C}}NH\dot{C}HCO_2^-$ — L-Alanylglycine, $g = 2.00339$, $\rho = 0.716$; coupling constants: 0.12 0.28 a^H a^N 0.82 17.57 ($a^H + 2a^N = 3.67$)

FIGURE 5. Formulae of radicals studied in photolysed aqueous solutions containing the indicated dipeptide and hydrogen peroxide. Numbers below the formulae are hyperfine coupling constants in gauss for the indicated nuclei; *g* values and spin densities are also listed. Reprinted with permission from R. Livingston, D. G. Doherty and H. Zeldes, *J. Amer. Chem. Soc.*, **97**, 3198 (1975). Copyright 1975 American Chemical Society.

This kind of energy transfer is confirmed by quenching the fluorescence of tryptophan **(1)** by H_2O_2 [54].

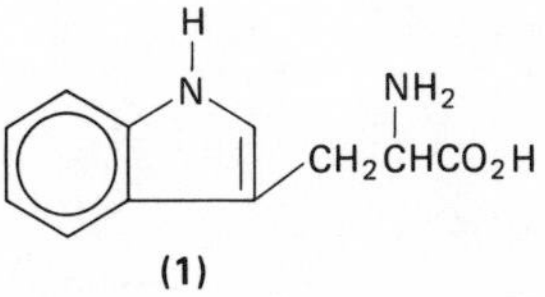

(1)

Cyclic amides such as uracil are degraded via oxidative decarboxylation; e.g. equation (27)[55].

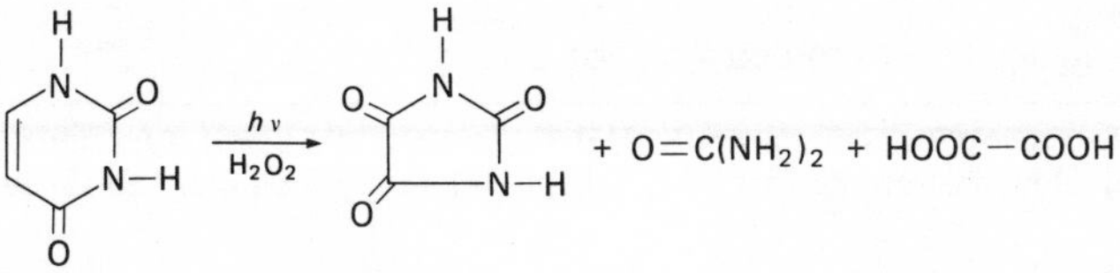

$$\text{uracil} \xrightarrow[H_2O_2]{h\nu} \text{(N-H imidazolidinetrione)} + O{=}C(NH_2)_2 + HOOC{-}COOH \qquad (27)$$

E. Oxidation of Aromatic Rings

As stated above, hydroxyl radicals add to aromatic rings yielding phenols after elimination of hydrogen from the adduct. Hence photochemical reaction of H_2O_2 can be used to prepare phenols; however the high reactivity of phenols toward the oxidant reduces the yield of phenol, since it leads to further oxidation and ring-fission products, e.g. equation (28)[56,57].

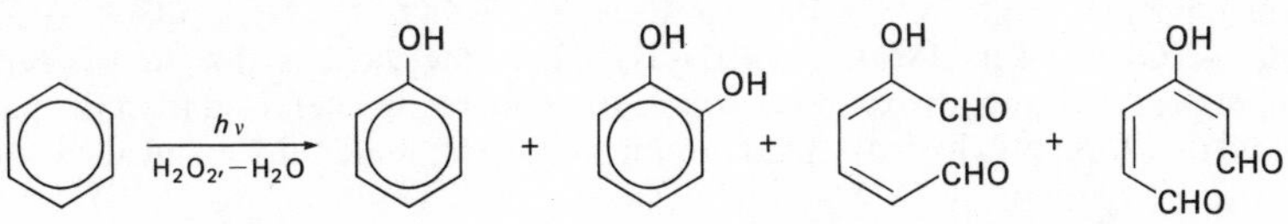

$$C_6H_6 \xrightarrow[H_2O_2,\,-H_2O]{h\nu} C_6H_5OH + C_6H_4(OH)_2 + \text{OH-CHO-CHO dienes} \qquad (28)$$

In the oxidation of phenols, the *ortho* and *para* positions are hydroxylated because of the electrophilic nature of the HO· radical, with the *ortho* position being preferred (equation 29)[58–60]. The most suitable solvent is acetonitrile, since it is most stable against HO· radical attack and it can dissolve both phenols and H_2O_2. Table 8 lists some products

TABLE 8. Hydroxylation of phenols in acetonitrile by photolysis of hydrogen peroxide[a,b,g]

X in (X-substituted phenol) (mmol)	H_2O_2 decomposed (mmol)	Conversion of phenol (mmol[%])	(HO)₂-catechol product (%)[c]	Hydroquinone product (%)[c]	Other products (%)[c]
H(30)	65	26 (89)	14 (26[d])	5 (14[d])	
p-Me (37)	32	20 (53)	18 (25–29[d])	—	
p-Ph (29)	[e]	23 (80)	14	—	2,4′-Dihydroxy-diphenyl (5.5) 4,4′-Dihydroxy-diphenyl (>1.0)
p-*t*-Bu (30)	61	21 (69)	24	—	
p-Cl (31)	[e]	14 (46)	<4	—	Phenol (9.0) Hydroquinone (trace)
p-Ac (29)	[e]	12 (42)	28	—	
p-COOH (29)	[e]	18 (61)	29 (38[d])	—	
m-COOH (30)	20	12 (41)	31[f]	6.4	
p-CN (30)	48	18 (59)	19	—	Hydroquinone (trace) *p*-Hydroxy-benzoic acid (trace)
o-NO_2 (31)	[e]	8.4 (27)	16	9.8	
p-NO_2 (31)	20	6.4 (21)	40	—	

[a]A solution of a phenol (20–37 mmol) and 35% H_2O_2 aq. (30 ml, 0.31 mol) in acetonitrile (200 ml) was irradiated with a low-pressure mercury lamp with quartz housing (mainly 253 nm) under bubbling N_2 at ca. 40–45°C.
[b]Irradiated for 3 h.
[c]Yields are based on the consumed phenol.
[d]Yield obtained in the aqueous solution.
[e]Not determined.
[f]2,3-Dihydroxybenzoic acid (20%) and 3,4-dihydroxybenzoic acid (11%).
[g]Reprinted with permission from K. Omura and T. Matsuura, *Tetrahedron*, **26**, 255 (1970).

obtained by the photooxidation of phenols bearing various substituents[59]. The apparent order of reactivity is as follows[30]: *p*-Ph > *p*-Ac, *p*-Cl > *p*-CO_2H, *o*-NO_2, *p*-CN, *p*-*t*-Bu > *m*-CO_2H, *p*-NO_2 > 2,4-$(COMe)_2$ > 2,4-$(NO_2)_2$. The low yield is due to the further oxidation of produced dihydroxyarenes which are more readily oxidized than the parent phenols and thus form polyhydroxylated products or a tarry material, the so-called 'humic acid'.

$$\text{C}_6\text{H}_4(\text{OH})\text{X} \longrightarrow \text{C}_6\text{H}_3(\text{OH})_2\text{X} + \text{C}_6\text{H}_3(\text{OH})_2\text{X} \quad (29)$$

III. HYDROPEROXIDES

Alkyl hydroperoxides have generally UV absorption commencing at 340 nm, so that they are photolysed by a mercury lamp light of 313 nm or 254 nm, resulting in O—O bond fission (equation 30). The radicals thus formed react further with peroxide itself, solvent molecules and/or other radicals; they are also cleaved at the C—C bonds to give smaller radicals and stable oxygen-containing (carbonyl) compounds. The C—C fission at the β position is favoured in a solvent which resists the hydrogen abstraction in the following order of leaving radical:

$$\text{H} \ll \text{Ph} < p\text{-NO}_2\text{C}_6\text{H}_4 \ll \text{Me} < t\text{-BuCH}_2 < n\text{-Pr} < \text{Et} < i\text{-Pr} < t\text{-Bu} < \text{PhCH}_2.$$

For example, the cumyloxy radical generated from cumyl hydroperoxide is unstable, releasing a methyl radical (equation 31)[61]. Photolysis products from hydroperoxides are usually analogous to the thermolysis products, but sometimes the product distribution is different owing to the excited nature of the radicals (see Introduction).

$$\text{RO—OH} \xrightarrow{h\nu} \text{RO}^\bullet + {}^\bullet\text{OH} \quad (30)$$

$$\text{PhCMe}_2\text{OOH} \xrightarrow{h\nu} \text{PhCMe}_2\text{O}^\bullet \longrightarrow \text{PhCOMe} + \text{Me}^\bullet \quad (31)$$

A. Alkyl Hydroperoxides

Ultraviolet irradiation of alkyl hydroperoxides gives as the initial step alkoxy and hydroxyl radicals by O—O fission. In the photolysis of *t*-butyl hydroperoxide, the *t*-butylperoxy radical has been identified by ESR[62], since both the *t*-BuO$^\bullet$ and HO$^\bullet$ radicals are more reactive than *t*-BuOO$^\bullet$ and abstract hydrogen rapidly from the original hydroperoxide, e.g. equation (32a). The *t*-butylperoxy radical either reacts with solvent or yields di-*t*-butyl peroxide (equation 32b).

$$t\text{-BuOOH} + \text{HO}^\bullet \ (\text{or } t\text{-BuO}^\bullet) \longrightarrow t\text{-BuOO}^\bullet + \text{H}_2\text{O} \ (\text{or } t\text{-BuOH}) \quad (32a)$$

$$2t\text{-BuOO}^\bullet \longrightarrow t\text{-Bu}_2\text{O}_2 + \text{O}_2 \quad (32b)$$

Alkoxy radicals abstract hydrogen atoms from the solvent. For example, *t*-butoxy radical in isopropanol abstracts a tertiary hydrogen atom from the solvent and the 2-hydroxypropyl radical thus formed reacts according to the scheme shown in equation (33), giving the enol **2** (and then acetone) and hemiketal **3** which is then also converted to acetone. Intermediates **2** and **3** have been identified by means of the ^{13}C CIDNP technique[63,64]. Table 9 shows the products identified during the photolysis of *t*-butyl hydroperoxide in various alcohols, and Figure 6 illustrates the ^{13}C spectrum recorded during the same process in ethanol.

TABLE 9. Reaction products identified from carbon-13 CIDNP signals observed during photolysis of *t*-BuOOH in alcohols[a]

Parent alcohol ROH	$R^1C(O)R^2$		t-BuOO–$C(R^3)(R^4)$–OH		t-BuOOC(O)R^5	$R^7R^8C{=}C(OH)R^6$		
R	R^1	R^2	R^3	R^4	R^5	R^6	R^7	R^8
CH_3			H	H				
CH_3CH_2	CH_3	H	CH_3	H	CH_3	H	H	H
$CH_3CH_2CH_2$	CH_3CH_2	H	CH_3CH_2	H	CH_3CH_2	H	CH_3	H
$CH_3(CH_2)_3$					$CH_3CH_2CH_2$			
$(CH_3)_2CHCH_2$					$(CH_3)_2CH$			
$CH_3CH_2(CH_3)CHCH_2$					$CH_3CH_2(CH_3)CH$			
$(CH_3)_2CH$	CH_3	CH_3	CH_3	CH_3		CH_3	H	H
$CH_3CH_2(CH_3)CH$	CH_3CH_2	CH_3	CH_3CH_2	CH_3	CH_3	(a) CH_3	CH_3	H
						(b) CH_3CH_2	H	H
$CH_3CH_2CH_2(CH_3)CH$	$CH_3CH_2CH_2$	CH_3			CH_3	(a) CH_3	CH_3CH_2	H
						(b) $CH_3CH_2CH_2$	H	H
$CH_3(CH_2)_3CHCH_3$	$CH_3(CH_2)_3$	CH_3				$CH_3(CH_2)_3$	H	H
$(CH_3CH_2)_2CH$	CH_3CH_2	CH_3CH_2	CH_3CH_2	CH_3CH_2	CH_3CH_2			
$(CH_3)_2CH(CH_3)CH$	$(CH_3)_2CH$	CH_3	$(CH_3)_2CH$	CH_3	CH_3			
$(CH_3)_3C(CH_3)CH$	$(CH_3)_3C$	CH_3			CH_3			
$PhCHCH_3$	Ph	CH_3						
C_6H_{11}						$(CH_2)_2$	$(CH_2)_2$	H

[a]Reprinted with permission from W. B. Moniz, S. A. Sojka, C. F. Poranski, Jr. and D. L. Birkle, *J. Amer. Chem. Soc.*, **100**, 7940 (1978). Copyright 1978 American Chemical Society.)

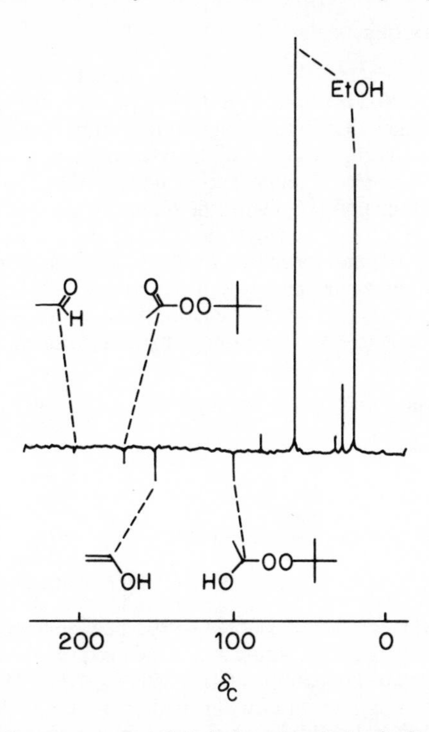

FIGURE 6. Carbon-13 spectrum recorded during photolysis of *t*-BuOOH in ethanol. Starting concentration of *t*-BuOOH was 10 % (v/v). Reprinted with permission from W. B. Moniz, S. A. Sojka, C. F. Poranski, Jr. and D. L. Birkle, *J. Amer. Chem. Soc.*, **100**, 7940 (1978). Copyright 1978 American Chemical Society.

$$t\text{-BuO}^{\bullet} + \text{Me}_2\text{CHOH} \longrightarrow t\text{-BuOH} + \text{Me}_2\dot{\text{C}}\text{OH} \quad (33a)$$

$$\text{Me}_2\dot{\text{C}}\text{OH} \longrightarrow \underset{(\mathbf{2})}{\text{CH}_2{=}\text{C}(\text{CH}_3){-}\text{OH}} \longrightarrow (\text{CH}_3)_2\text{C}{=}\text{O} \quad (33b)$$

$$t\text{-BuOO}^{\bullet} + \text{Me}_2\dot{\text{C}}\text{OH} \longrightarrow \underset{(\mathbf{3})}{t\text{-BuOOCMe}_2\text{OH}} \longrightarrow \text{MeC}{=}\text{O} \quad (33c)$$

The product ratio of hydrogen atom abstraction vs. C—C cleavage in alkoxy radicals was estimated for the photolysis of *t*-amyl hydroperoxide (equation 34) in 2,4-dimethylpentane at 50°C by measuring the ratio of [amyl alcohol]/[acetone] [solvent] in a dilute solution[65]. The ratio was ca. 0.22, which was analogous to that in pyrolysis.

$$\text{CH}_3{-}\text{CH}_2{-}\text{C}(\text{CH}_3)_2{-}\text{O}^{\bullet} \xrightarrow{\text{H abstr.}} \text{CH}_3\text{CH}_2\text{C}(\text{CH}_3)_2\text{OH} \quad (34a)$$

$$\text{CH}_3{-}\text{CH}_2{-}\text{C}(\text{CH}_3)_2{-}\text{O}^{\bullet} \xrightarrow{\text{C--C fission}} (\text{CH}_3)_2\text{C}{=}\text{O} + \text{CH}_3\text{CH}_2^{\bullet} \quad (34b)$$

B. α-Ketohydroperoxides

Ultraviolet irradiation of some α-hydroperoxyketones $R^1CO-CR^2R^3OOH$ gives a carboxylic acid R^1COOH and a ketone $R^2R^3C{=}O$ via a radical scheme. The quantum yield is usually much higher than unity, suggesting a radical-chain mechanism[66]. For example, photolysis of α-hydroperoxy-α,α-diphenylacetophenone, $PhCOCPh_2OOH$, in benzene affords benzoic acid (65%) and benzophenone (90%), along with benzaldehyde (4%), phenol (3%) and biphenyl (1%) with the quantum yield of 2.5.

The intermediary formation of benzoyl radical was demonstrated by formation of PhCOCl on addition of CCl_4 and formation of $PhCO_3H$ on addition of O_2. These facts together with facile intramolecular energy transfer from $C{=}O^*$ to O—O suggest the mechanism shown in equation (35), for the reaction, which involves the O—O fission of excited hydroperoxy groups to give acyl radical, hydroxyl radical and ketone.

$$\underset{\|\ \ \ \ |}{\overset{}{PhC-CPh_2}}\ \ldots\ PhC(=O)-CPh_2(OOH) \xrightarrow{h\nu} PhC(=O)-CPh_2(O\dot{O}H) \longrightarrow PhC(=O)^{\bullet} + (C{=}O)Ph_2 + {}^{\bullet}OH \qquad (35a)$$

$$PhC(=O)^{\bullet} + PhCOCPh_2OOH \longrightarrow PhCOOH + PhC(=O)-CPh_2(O^{\bullet}) \longrightarrow \text{chain reaction} \qquad (35b)$$

The reaction is sensitized with various ketones such as benzophenone, phenyl α-naphthyl ketone and fluorenone, but not with anthracene which retards the decomposition as an inner filter. The ketone-sensitized decomposition occurs regardless of its triplet energy (E_T), configuration (π–π^* or n–π^*) or ability for hydrogen atom abstraction. Other α-ketohydroperoxides are photolysed by an analogous chain mechanism.

C. α-Azohydroperoxides

α-Azohydroperoxides are prepared by autoxidation of hydrazones and are known as HO˙ radical sources effective for aromatic hydroxylation in anhydrous media[67–69].

The hydroperoxide **4a** possessing absorption maxima at 285 and 413 nm is photolysed by UV irradiation in a benzene solution, resulting in the O—O fission followed by β scission (equation 36)[69]. Subsequently the solvent benzene reacts with HO˙ to give phenol (23%) and biphenyl (3%), while aryl radicals Ar˙ give ArPh (86%) and ArH (3%) and also ArOH. The orientation of aromatic hydroxylation in this reaction is similar to that with Fenton's reagent; e.g. anisole with **4a** affords hydroxyanisole in a ratio of *ortho*:*meta*:*para* = 76:0:24, compared to Fenton's reagent giving 84:0:16[70]. The addition of molecular oxygen increases the yield of phenol (up to 52%) due to the acceleration of dehydrogenation of intermediary cyclohexadienyl radicals **5**.

$$R^2-C(R^1)(OOH)-N{=}N-C_6H_4-X \longrightarrow {}^{\bullet}OH + {}^{\bullet}C_6H_4-X + R^1R^2C{=}O + N_2 \qquad (36)$$

(4)

(a) R^1 = Ph, R^2 = H, X = Br
(b) R^1 = R^2 = Me, X = Br

HO H (5)

When *t*-butanol is used as a solvent, the photolysis of **4a** gives benzaldehyde mainly (32 %) together with a small amount of *p*-bromophenol, $(CH_3)_2C(OH)CH_2OH$ and $(CH_3)_2C(OH)CH_2CH_2C(OH)(CH_3)_2$.

D. Alkenyl Hydroperoxides

Alkenyl hydroperoxides give cyclic peroxides by acetophenone-sensitized photolysis in Freon 11 solvent[71]. The reaction is initiated by the abstraction of hydroperoxy hydrogen by excited acetophenone[12,72] followed by an intramolecular attack of peroxy radical on the double bond leading to cyclic peroxide (equation 37).

HOO(alkenyl, Me, H, H) + PhCOMe $\xrightarrow[-\text{Ph}\dot{\text{C}}(\text{OH})\text{Me}]{h\nu}$ ·OO(alkenyl, Me, H, H) $\longrightarrow$ cyclic O—O (Me, $CH_2^{\cdot}$) (37)

$\xrightarrow{O_2}$ cyclic O—O (Me, $CH_2OO^{\cdot}$) $\xrightarrow[-R^{\cdot}]{RH}$ cyclic O—O (Me, CH_2OOH) $\xrightarrow{Ph_3P}$ cyclic O—O (Me, CH_2OH)

E. Polymeric Hydroperoxides

In connection to the autoxidation of vinyl polymers, photolysis of polymeric hydroperoxides has been studied[65,73–75]. For example, irradiation of polystyrene hydroperoxide excites the phenyl group and then the energy is transferred to the hydroperoxy group which is cleaved at the O—O bond, initiating photooxidation[74]. The photolysis of *cis*-1,4-poly(isoprene hydroperoxide)[75] has a primary quantum yield of 0.8, while the subsequent radical-chain decomposition increases the overall quantum yield with 313 nm light. The ratio of quantum yields for O—OH fission vs. polymer C—C chain scission is 71.4.

IV. DIALKYL PEROXIDES

Dialkyl peroxides have generally absorption at UV region shorter than 340 nm (Figure 7)[3,76], so that they are decomposed by a Hg lamp. The primary reaction is again the O—O bond fission, giving two alkoxy radicals which are apt to react further. Since the behaviour of these alkoxy radicals is often similar to those produced by thermolysis, the special features of photolysis will be presented in the following.

A. Di-*t*-butyl Peroxide

Di-*t*-butyl peroxide is the most common and stable dialkyl peroxide available, and it gives *t*-BuO· radicals on photolysis[77] (equation 38a). *t*-BuO· radical decomposes by itself to give Me· and acetone[78] (equation 38b) or abstracts an atom from solvent to give the solvent radical[79]. The *t*-BuO· radical, formed by photolysis, has excess energy, since the

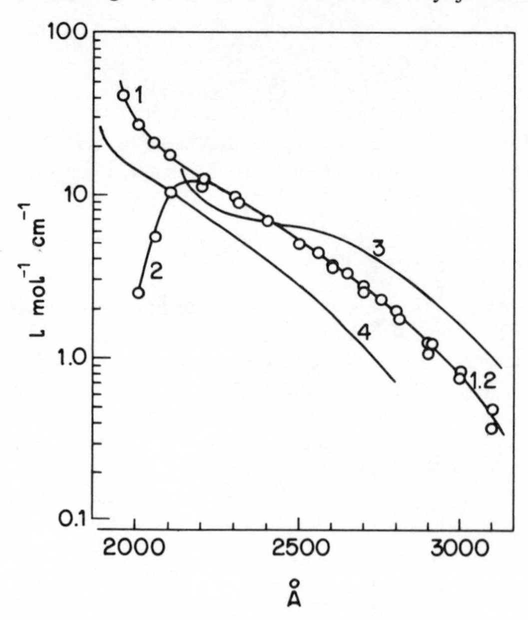

FIGURE 7. UV absorption spectra of MeOOMe (1)[3] and (2)[76], *t*-BuOO-*t*-Bu (3) and H_2O_2 (4)[3], reduced by factor of 10 [$\varepsilon = (\log Io/I)C^{-1}L^{-1}\,1\,\text{mol}^{-1}\,\text{cm}^{-1}$], Curve 2 may be the results of measurements extended beyond the reliable range of their spectrometer. Reprinted with permission from L. M. Toth and H. S. Johnston, *J. Amer. Chem. Soc.*, **91**, 1276 (1969). Copyright 1969 American Chemical Society.

O—O bond energy is below 40 kcal[80], much lower than the ca 110 kcal of the 254 nm light energy. This excited radical, ('hot radical') is subject to direct decomposition (equation 38b) or deactivated to ground state. This 'hot radical' is unimportant in the liquid phase because of its easy deactivation by collision with solvent molecules[81].

$$t\text{-BuO—OBu-}t \xrightarrow{h\nu} 2t\text{-BuO}^{\bullet} \tag{38a}$$

$$t\text{-BuO}^{\bullet} \longrightarrow \text{MeCOMe} + \text{Me}^{\bullet} \tag{38b}$$

On irradiation of a gaseous mixture of di-*t*-butyl peroxide and a hydrocarbon, the quantum yield of peroxide decomposition depends on the concentration of the hydrocarbon, since the photoexcited peroxide is deactivated by collision with the hydrocarbon; i.e., the quantum yield of decomposition ($\phi = 2$) in the absence of hydrocarbon decreases with increasing concentration of hydrocarbon[82].

On irradiation of t-Bu_2O_2, the radicals *t*-BuO$^{\bullet}$, Me$^{\bullet}$ and t-BuOOCMe$_2$CH$_2^{\bullet}$ were identified by ESR as expected by equations (38a) and (38b)[83,84]. Irradiation of a mixture of di-*t*-butyl peroxide and *t*-butyl hydroperoxide gave *t*-butylperoxy radicals (ESR) (equation 39)[85,86]. Since the deactivation rate constant of *t*-BuO$^{\bullet}$ ($1.3 \times 10^9\,\text{M}^{-1}\text{s}^{-1}$) is much larger than that of *t*-BuOO$^{\bullet}$ ($3.0 \times 10^5\,\text{M}^{-1}\text{s}^{-1}$)[83], ESR can measure only *t*-BuOO$^{\bullet}$. To identify unstable radicals formed by photolysis, some radical scavengers such as nitroso compounds[87–89], trialkylarsines and trialkylphosphines[90,91] are used. This technique is called 'spin trapping'. For example, nitrosodurene is used for the ESR detection of *t*-butoxy and methyl radicals (equation 40).

$$t\text{-BuO}^\bullet + t\text{-BuOOH} \longrightarrow t\text{-BuOH} + t\text{-BuOO}^\bullet \tag{39}$$

$$R^\bullet + \text{ON}\!-\!C_6(\text{Me})_4H \xrightarrow{h\nu} R\!-\!N(O^\bullet)\!-\!C_6(\text{Me})_4H \tag{40}$$

t-BuO$^\bullet$ radicals, formed by thermolysis, can completely (>99 %) be trapped by NO, while 'hot' *t*-BuO$^\bullet$ radicals, formed by gaseous photolysis, decompose to acetone and methyl radical as stated in equation (38b)[16]. The α value (= [acetone]/[acetone] + [*t*-BuONO]), which is a measure of hot radical content, is estimated to be 0.5–0.6[16].

The *t*-BuO$^\bullet$ radical can abstract hydrogen from hydrocarbons, alcohols and nitriles, forming carbon radicals, and similarly gives PhO$^\bullet$ from PhOH and $Me_3Si^\bullet$ from Me_3SiH[79,92]. Irradiation of an oxygen-saturated alkane (RH) solution of di-*t*-butyl peroxide gives alkylperoxy radicals, ROO$^\bullet$, via alkyl radicals formed by hydrogen abstraction from the alkane by *t*-BuO$^\bullet$ [93].

t-BuO$^\bullet$ radicals can even abstract highly polarized hydrogen atoms from HCl, affording atomic chlorine, which in turn abstracts hydrogen from peroxide as shown in equation (41)[94].

$$Cl^\bullet + t\text{-BuOOBu-}t \xrightarrow[-\text{HCl}]{} \text{Me}_2C(^\bullet CH_2)OOCMe_3 \longrightarrow \text{Me}_2\overline{C\!-\!O\!-\!CH_2} + Me_3CO^\bullet \tag{41a}$$

$$\text{Me}_2\overline{C\!-\!O\!-\!CH_2} \xrightarrow[(-\text{H}^\bullet)]{Cl^\bullet \text{ or } t\text{-BuO}^\bullet} \text{Me}_2\overline{C\!-\!O\!-\!CH^\bullet} \longrightarrow \text{Me}_2\dot{C}\!-\!CH\!=\!O \tag{41b}$$

In contrast, *t*-BuO$^\bullet$ radicals seem to abstract protons from trifluoroacetic acid to form the cation radical, $t\text{-BuOH}^{+\bullet}$, which abstracts hydrogen from alkanes and adds to alkenes[95] (equation 42) because of its strong electrophilicity, although *t*-BuO$^\bullet$ itself tends only to abstract hydrogen from alkene.

$$t\text{-BuOH}^{+\bullet} + R_2C\!=\!CR_2 \longrightarrow t\text{-Bu}\overset{+}{O}(H)\!-\!R_2C\!-\!\dot{C}R_2 \tag{42}$$

A strong singlet ($g = 2.0091$) of ESR was observed at low temperature[19]; this signal may be assigned to $t\text{-Bu}_2O_2^{+\bullet}$ formed by electron abstraction from $t\text{-Bu}_2O_2$ by $t\text{-BuOH}^{+\bullet}$, but $t\text{-Bu}_2O_2^{+\bullet}$ decomposes to give Me$^\bullet$ radical at over −80°C[96].

There are many examples which use $t\text{-Bu}_2O_2$ as a radical polymerization initiator[97,98].

B. Dimethyl Peroxide

Gaseous photolysis (240–340 nm light) of dimethyl peroxide gives methanol, formaldehyde, CO and a trace of H_2[3]. The quantum yields of methanol (10 → 3) and formaldehyde ($10 \rightarrow 0$) decrease rapidly with reaction time, while that of CO ($0 \rightarrow 1$) increases and that of total carbon compounds ($13 \rightarrow 3.5$) decreases. These facts suggest a chain mechanism (equation 43a)[99].

$$CH_3O^{\bullet} + CH_3O_2CH_3 \xrightarrow{-CH_3OH} {}^{\bullet}CH_2O_2CH_3 \longrightarrow HCHO + CH_3O^{\bullet} \qquad (43a)$$

There is a competitive reaction (equation 43b) with this chain as formaldehyde builds up, thus decreasing the quantum yields of CH_3OH and HCHO (Figure 8)[3].

$$CH_3O^{\bullet} + HCHO \longrightarrow CH_3OH + H\dot{C}O \qquad (43b)$$

The chain scheme with $CH_3O^{\bullet}$ as a carrier is terminated by coupling [$2\,CH_3O^{\bullet} \rightarrow (CH_3)_2O_2$] and disproportionation [$2\,CH_3O^{\bullet} \rightarrow HCHO + CH_3OH$] at first, or reaction with formyl radical [$CH_3O^{\bullet} + H\dot{C}O \rightarrow CH_3OH + CO$] at a later stage. The side-reaction is the formation of molecular hydrogen (equation 43c). Formaldehyde acts as a chain terminator or polymerizes by itself, giving $CH_3O(CH_2O)_nR$, where $R^{\bullet}$ is any radical.

$$HCHO^{*} \longrightarrow H^{\bullet} + H\dot{C}O \longrightarrow H_2 + CO \qquad (43c)$$

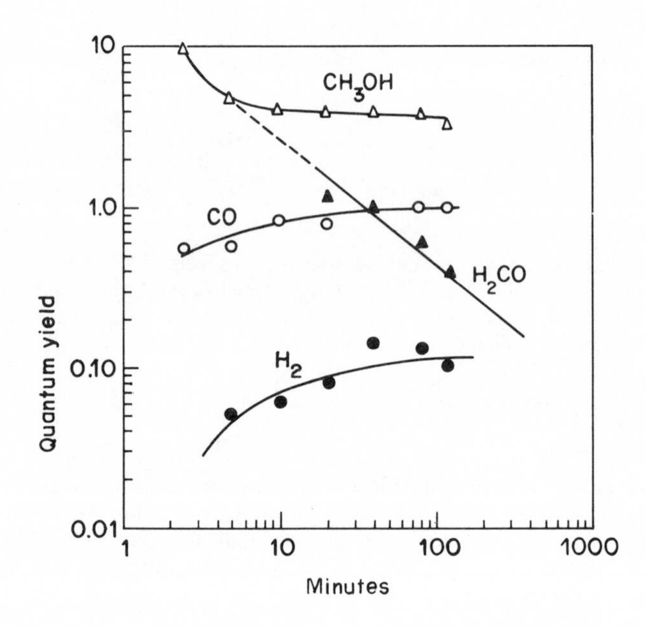

FIGURE 8. Integrated quantum yields for different products in the photolysis of $CH_3O_2CH_3$; 60°C, 2537 Å radiation, 30 torr of $CH_3O_2CH_3$. Reprinted with permission from L. M. Toth and H. S. Johnston, *J. Amer. Chem. Soc.*, **91**, 1276 (1969). Copyright 1969 American Chemical Society.

C. Diethyl Peroxide

Gaseous photolysis (313 nm) at 30°C of diethyl peroxide gives CO, CH_4, HCHO, CH_3CHO, ethanol, acetone and biacetyl[100]. Photolyses carried out in CCl_4, cyclohexane and water solutions give ethanol and acetaldehyde as the sole products[101]. The quantum yields in CCl_4 and cyclohexane are 2, while that in water is over 5, suggesting catalysis with a trace of some catalysts. In water, Cu^{2+} ion is a good catalyst for the decomposition, and no effect of acid and temperature is observed. These facts suggest the mechanism shown in equations (44)–(46) for the photolysis in water. The mechanism for catalysis leads to a quantum yield expression: $\phi = k_1[Cu^{2+}]k_0^{1/2}/(k_4A)^{1/2}$, which is in accord with the observation.

$$\mathrm{EtOOEt} \xrightarrow[k_0A]{h\nu} 2\mathrm{EtO^{\bullet}} \qquad (44)$$

A = absorbed light energy

Uncatalysed reaction

$$\mathrm{EtO^{\bullet}} + \mathrm{EtOOEt} \longrightarrow \mathrm{EtOH} + \mathrm{EtOO\dot{C}HCH_3} \qquad (45a)$$

$$\mathrm{EtOO\dot{C}HCH_3} \longrightarrow \mathrm{EtO^{\bullet}} + \mathrm{CH_3CHO} \qquad (45b)$$

$$2\mathrm{EtO^{\bullet}} \xrightarrow{k_4} \mathrm{CH_3CHO} + \mathrm{EtOH} \qquad (45c)$$

Cu^{2+} catalysed reaction

$$\mathrm{EtO^{\bullet}} + \mathrm{Cu^{2+}} \xrightarrow{k_1} \mathrm{Cu^{+}} + \mathrm{H^{+}} + \mathrm{CH_3CHO} \qquad (46a)$$

$$\mathrm{Cu^{+}} + \mathrm{EtOOEt} \xrightarrow{k_2} \mathrm{Cu^{2+}} + \mathrm{EtO^{-}} + \mathrm{EtO^{\bullet}} \qquad (46b)$$

D. Bis(trifluoromethyl) Peroxide

The UV photolysis of CF_3OOCF_3 gives CO_2 and COF_2, which have been identified by IR at 8 K[102]. The trifluoromethylperoxy radical $CF_3OO^{\bullet}$ has been identified by ESR at 103–74 K during the photolysis of CF_3OOCF_3 as well as CF_3OF, but $CF_3O^{\bullet}$ is observed at room temperature[103]. This is ascribed to the rapid reaction of $CF_3O^{\bullet}$ initially formed by O—O fission with a trace of oxygen in the system, since the dissolution of oxygen at low temperature renders the complete removal of oxygen difficult[102].

$$\mathrm{CF_3OO^{\bullet}} \xrightarrow{-\mathrm{F^{\bullet}}} \mathrm{F_2C{<}(O{-}O)} \longrightarrow \begin{cases} \mathrm{COF_2} + \mathrm{O} & (47a) \\ \mathrm{CO_2} + 2\mathrm{F} & (47b) \end{cases}$$

The photolysis of CF_3OOCF_3 under oxygen at -190°C gives $CF_3OOO^{\bullet}$ radical by the reaction of $CF_3O^{\bullet}$ and O_2 (ESR)[104].

The photolysis (<300 nm) of CF_3OOOCF_3 gives CF_4, CF_3OCF_3, COF_2 and $CF_3OO^{\bullet}$ as products[105].

E. Cyclic Peroxides

Tetramethyl-1,2-dioxane is photolysed with 310–350 nm light to give acetone (99 %) and ethylene (96 %) probably via an oxygen 1,6-biradical (equation 48).

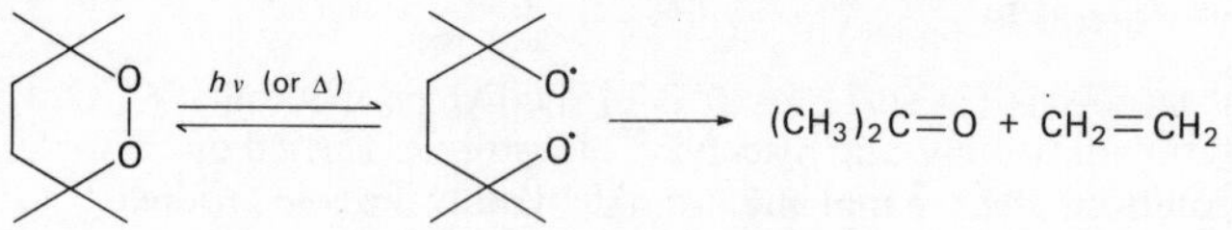

$$(CH_3)_2C{=}O + CH_2{=}CH_2 \quad (48)$$

Photolysis of 1,2-dioxolane gives, via a 1,3-diradical, epoxide and ketone, the latter being a rearrangement product (equation 49)[106]. The leaving ability of the ketone $R_2C{=}O$ is in the order: $Ph_2C{=}O(25) > MePhC{=}O\ (10) > Me_2C{=}O(1)$[107].

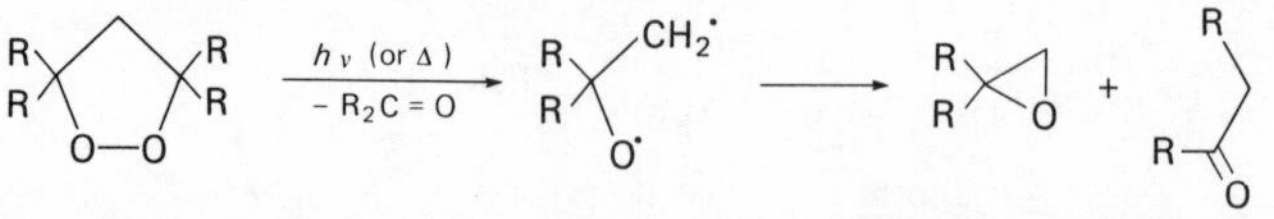

(49)

Cyclohexane diperoxide is photolysed to give interesting ring-enlarged products: cyclodecane (14 %) (**6**), undecalactone (9 %) (**7**), as well as cyclohexanone (~20 %); thermolysis gives the same products in 44, 23 and 21 % yields, respectively[108] (equation 50). Cycloheptane diperoxide behaves analogously. The ring-enlargement is explained by the intermediacy of a macrocyclic diacyl peroxide.

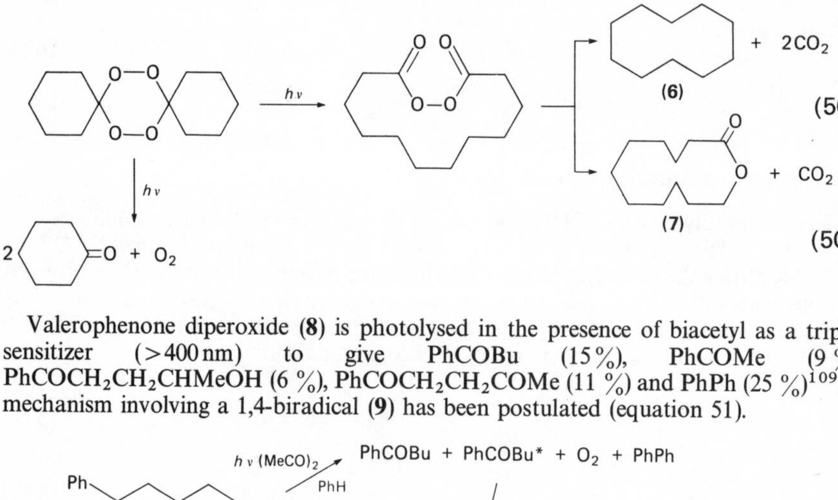

(50a)

(50b)

Valerophenone diperoxide (**8**) is photolysed in the presence of biacetyl as a triplet sensitizer (>400 nm) to give PhCOBu (15 %), PhCOMe (9 %), $PhCOCH_2CH_2CHMeOH$ (6 %), $PhCOCH_2CH_2COMe$ (11 %) and PhPh (25 %)[109]. A mechanism involving a 1,4-biradical (**9**) has been postulated (equation 51).

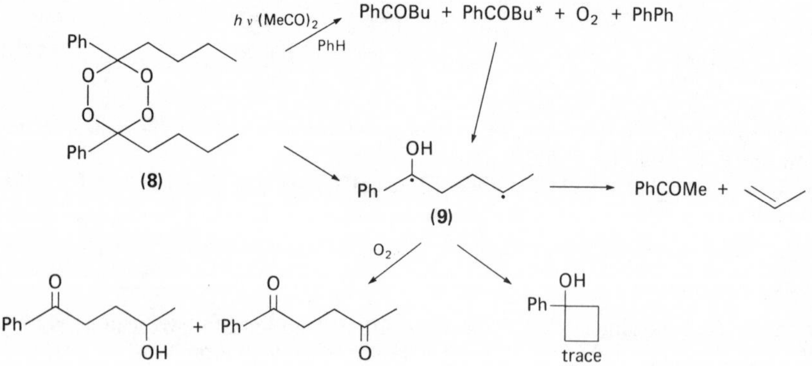

(51)

F. Other Peroxides

Gaseous photolyses in the presence of NO of some dialkyl peroxides $ROOR^1$ ($R = R^1 = i$-Pr; $R = t$-Bu, $R^1 = i$-Pr; $R =$ Et, $R^1 = t$-Am) have been studied[16]. At 254 nm, most hot alkoxy radicals except $EtO^{\cdot *}$ are not completely trapped by NO, but expel $Me^{\cdot}$ or $Et^{\cdot}$ to give ketones. In contrast, the radicals are trapped by NO in longer wavelength (313 nm) photolysis and in thermolysis, where most radicals have no excess energy. t-$AmO^{\cdot}$ radical decomposes in two ways (equation 52) with a preference for equation (52a).

$$Et\text{-}C(Me)_2\text{-}O^{\cdot} \xrightarrow{k_1} Et^{\cdot} + Me_2C{=}O \qquad (52a)$$

$$Et\text{-}C(Me)_2\text{-}O^{\cdot} \xrightarrow{k_2} Me^{\cdot} + MeCOEt \qquad (52b)$$

Light of a shorter wavelength tends to decrease the k_1/k_2 ratio, which is observed to be 16–22 for thermolysis, ~16 for 313 nm light and ~10 for 254 nm light.

Dicumyl peroxide $(PhCMe_2O\text{+}_2$ is photolysed to give acetophenone[110]. In photolysis of a mixture of dicumyl peroxide and cumyl hydroperoxide the only detectable radical is $PhCMe_2OO^{\cdot}$ as observed in the other cases (equation 39)[86].

Cyclohexadienone t-butyl peroxide (**10**) is photolysed by sunlight to give a ring-contracted ketone (**12**) via the cyclopentenone radical (**11**) which is identified by ESR[111].

(**10**) $\xrightarrow{h\nu}$ (**11**) $\longrightarrow$ (**12**) (53)

UV photolysis of the same peroxide **10** in methanol gives two cyclopentenone derivatives, **13** and **14**, which are probably formed by the scheme shown in equation (54)[112]. However, in benzene solution, more complicated products are formed.

(**10**) $\xrightarrow{h\nu}$ $\longrightarrow$ $\xrightarrow{-t\text{-}BuO^{\cdot}}$ $\xrightarrow[\text{(from MeOH)}]{^{\cdot}CH_2OH}$ (**13**) $\longrightarrow$ (**14**) (54)

Photolyses of di-t-amyl peroxide[65] and N,N-dialkylaminomethyl alkyl peroxides[113] have been reported.

V. DIACYL PEROXIDES

A. Aliphatic Diacyl Peroxides

Diacyl peroxides have continuous and weak absorption in the UV region shorter than 350 nm (see Table 1)[1b]. Hence UV light of 250–350 nm can cleave their O—O bonds and then rapid decarboxylation occurs, giving alkyl radicals (equation 55)[114,115].

$$(RCOO-)_2 \xrightarrow{h\nu} RCOO^{\cdot} \longrightarrow R^{\cdot} + CO_2 \quad (55)$$

The ESR spectrum at low temperature indicates the presence of alkyl radical R$^{\cdot}$, but not RCOO$^{\cdot}$ because of its instability[18,116]. The rate of loss of radicals follows second-order kinetics, which points to recombination of the radicals; this rate decreases with increasing chain length of R[117]. A stable radical pair is observed in some cases[118a].

Photolysate radicals couple with each other, disproportionate or isomerize. In the presence of solvent, they attack the solvent, abstracting hydrogen atoms and generating solvent radicals which react further, or add to unsaturated bonds. Primary diacyl peroxides $(RCOO-)_2$ in *n*-pentane with 254 nm light generally give alkanes RH, alkene R(−H) and dimer R—R in a molar ratio of 9:3:4, but the yield of ester RCO_2R is below 1 %[1b]. For example, neat acetyl peroxide is photolysed to give ethane and methane (equation 56a); On the other hand, the reactions (56b) and (56c) occur in the presence of solvent RH.

$$(CH_3COO-)_2 \xrightarrow{h\nu} CH_3CH_3 + CH_4 + CO_2 \quad (56a)$$

$$(CH_3COO-)_2 \xrightarrow[RH]{h\nu} CH_3COOR + CH_4 + CO_2 \quad (56b)$$

$$(CH_3COO-)_2 \xrightarrow[RH]{h\nu} CH_3COOH + CH_3R + CO_2 \quad (56c)$$

Products obtained by photolysis of acetyl peroxide in solid, liquid and solution (cyclohexane) states are given in Table 10[1a], which shows the highest yield of ethane in solid-state photolysis because of the lower probability of hydrogen abstraction by Me$^{\cdot}$ radical and the higher concentration of free Me$^{\cdot}$ radical. It is of interest to note that the photolysis of unsymmetrical diacyl peroxide $RCOO_2COR^1$ at low temperature (−78°C) gives unsymmetrical dimeric hydrocarbon RR^1 in a good yield (75 %)[118b].

Photolysis (254 nm) of acetyl peroxide in the presence of aromatic compounds affords methylarenes, with the orientation different from that of thermolysis; e.g. photolysis in

TABLE 10. Photodecomposition of acetyl peroxide[a]

		Composition (vol. %)						
State/solvent	Temp. (°C)	CO_2	O_2	Unsat.	CO	C_2H_6	CH_4	C_2H_6/CH_4
Solid	16–18	67.4	0.4	1.0	1.3	25.6	5.1	4.9
Liquid	30	61.5	—	—	1.8	17.5	19.5	0.90
Cyclohexane	16–18	59.7	0.9	0.9	1.3	8.4	28.9	0.29
Toluene	80	44.6	1.4	—	2.0	1.4	50.7	0.03

[a]Reprinted with permission from O. J. Walker and G. L. E. Wild, *J. Chem. Soc.*, 1132 (1937). Copyright 1937 Royal Society of Chemistry.

toluene gives at $-70°C$ xylenes $[C_6H_5CH_3 + CH_3^{\cdot} \rightarrow C_6H_4(CH_3)_2]$ in a ratio of *ortho*:*meta*:*para* = 37:51:12, in contrast to the thermolysis ratio of 56:27:17. At higher temperatures the photolysis and thermolysis isomer ratios converge[24]. Photolysis of propionyl peroxide gives ethyl radicals, which form butane by combination, and ethane and ethylene by disproportionation[50,51].

Secondary and tertiary diacyl peroxides give by disproportionation some isomerized alkenes; e.g. bis(2-methylbutyryl)peroxide affords a mixture of 1-butene, *trans*- and *cis*-2-butenes in the ratio of 5:3:1[19], in which 1-butene is favoured by statistical rather than thermodynamic considerations[1b].

Photolysis of cyclobutylcarbonyl and cyclopropylacetyl peroxides gives the products in equation (57)[1b].

$$(c\text{-}C_4H_7COO)_2 \xrightarrow{-CO_2} c\text{-}C_4H_8 + c\text{-}C_4H_6 + c\text{-}C_4H_7\text{-}c\text{-}C_4H_7 \qquad (57a)$$

$$(c\text{-}C_3H_5CH_2COO)_2 \xrightarrow{-CO_2} \text{[alkene and cyclopropyl products]} \qquad (57b)$$

If the peroxide contains a C=C double bond, cyclization occurs by intramolecular attack of the radical on the double bond, e.g. equation (58)[119].

$$(CH_2{=}CH(CH_2)_3CH_2{-}COO)_2 \xrightarrow{h\nu} CH_2{=}CH(CH_2)_3CH_2^{\cdot} \longrightarrow c\text{-}C_5H_9CH_2^{\cdot} \qquad (58)$$

In the photolysis of $C{=}^{18}O$-labelled acyl peroxide, the ester formed is partially (15 %) scrambled, and in the photolysis of optically active peroxides the alkyl group in the ester retains 68 % of its configuration. Hence, most of the reaction may go through a four-membered cyclic transition state (equation 59)[26].

$$R{-}C({=}^{18}O){-}O\cdots O\cdots C({=}^{18}O)\cdots R \longrightarrow R{-}C({=}^{18}O){-}O{-}R + C^{18}O_2 \qquad (59)$$

B. Diaroyl Peroxides

Benzoyl peroxide has UV absorption at 231 nm (ε 3.98×10^4) and 274 nm (ε 2.16×10^3)[7], so that it is excited by UV irradiation to a singlet state followed by O—O fission to give benzoyloxy radical and then phenyl radical by decarboxylation. The benzoyloxy radical is more stable than simple acyloxy radicals; hence it is often trapped by efficient scavengers such as vinyl monomers[120] and iodine[121]. In the photolysis of aroyl peroxides, a radical pair is formed and then reacts as shown in equation (60)[122].

$$(ArCOO)_2 \longrightarrow ArCOO^{\cdot} \longrightarrow Ar^{\cdot} + CO_2 \qquad (60a)$$

$$[ArCOO^{\bullet}\ {}^{\bullet}Ar^1] \xrightarrow{\text{recomb.}} ArCOOAr^1 \quad (60b)$$

$$[ArCOO^{\bullet}\ {}^{\bullet}Ar^1] \xrightarrow{-CO_2} [Ar^{\bullet}\ {}^{\bullet}Ar^1] \longrightarrow ArAr^1 \quad (60c)$$

$$[ArCOO^{\bullet}\ {}^{\bullet}Ar^1] \xrightarrow{\text{escape}} Ar^{\bullet} + {}^{\bullet}Ar^1 + CO_2 \quad (60d)$$

However, the radical pair ($PhCO_2^{\bullet}\ {}^{\bullet}Ph$) is not observed in the photolysis of solid benzoyl peroxide[9,123]. The effect of an external magnetic field on the cage and escape reactions has been measured with benzoyl peroxide in toluene[124,125], where a decrease (8 %) of cage product (PhCOOPh) and an increase (2 %) of radical products are observed[125]. This phenomenon can be explained by Kaplan's theory[126]. UV irradiation of benzoyl peroxide in benzene under N_2 affords the products shown in equation (61)[127].

$$(PhCOO-)_2 \xrightarrow[\text{PhH, r. t.}]{h\nu\ (N_2)} \underset{13\%}{PhCO_2Ph},\ \underset{33\%}{PhPh},\ \underset{30\%}{PhCO_2H},\ \underset{135\%}{CO_2} \quad (61)$$

The yield of $PhCO_2Ph$ is higher than that in thermolysis or triplet-sensitized photolysis (3 %). Further, similar high yields are obtained without benzene and in the presence of radical scavengers such as iodine[121] or *p*-xylene[127]. These facts suggest that phenyl benzoate ($PhCO_2Ph$) is a product in a solvent cage ($PhCO_2^{\bullet} + Ph^{\bullet} + CO_2$)[128]. The different spin states of radical pairs explain the different yields of Ph CO_2Ph; i.e., the singlet radical pair can easily couple, but the triplet pair escapes from the cage[129].

The yield of $PhCO_2Ph$ increases under oxygen atmosphere owing to the acceleration of dehydrogenation from the cyclohexadienyl radical (equation 62)[34,35].

$$PhCO_2^{\bullet} + PhH \rightleftharpoons \text{(PhCO}_2\text{)(H)C}_6\text{H}_5^{\bullet} \xrightarrow{O_2} PhCO_2Ph + HO_2 \quad (62)$$

An alternative mechanism for the formation of $PhCO_2Ph$ could be a direct attack of excited benzene on benzoyl peroxide, but this is ruled out by a tracer experiment with $[PhC(={}^{18}O)O-]_2$, which gives scrambled $PhC^{18}O_2Ph$[130].

In the presence of vinyl monomers, $PhCOO^{\bullet}$ adds to the double bond to initiate vinyl polymerization; no addition of $Ph\dot{C}O$ is observed, since no deoxygenation reaction occurs from benzoyl peroxide $[(PhCOO-)_2 \nrightarrow 2\,Ph\dot{C}O + O_2]$[131].

Decarboxylation of $PhCO_2^{\bullet}$ gives $Ph^{\bullet}$, which reacts with solvent benzene giving biphenyl, while the formation of biphenyl by coupling of $Ph^{\bullet}$ radicals is unimportant. Photolysis (254 nm) of benzoyl peroxide in toluene gives methylbiphenyls in an isomeric ratio of *ortho*:*meta*:*para* = 64:21:15 together with benzene and other products expected from equation (61). The ratio is little affected by temperature and analogous to the ratio in thermolysis[24].

$Ph^{\bullet}$ radical abstracts atoms from solvent; e.g. PhCl is formed in CCl_4 [132,133] and alkyl radical is formed from alkyl halide[134].

The kinetics of the photolysis of benzoyl peroxide in chloroform has been studied by means of ^{13}C CIDNP[135], the products being shown in Table 11. In Scheme 1[135], the rate constant k_c (i.e. $k_1 + k_2$) for the cage collapse process is nearly equal to $k_1 (10^{10}\ s^{-1})$, i.e. the rate constant for formation of $PhCO_2Ph$, since the rate for decarboxylation of $PhCO_2^{\bullet}$, k_2, is negligibly small (103–107 s^{-1}) by comparison to k_1. The cage escape is a diffusion-limited (very fast) process, $k_e \approx k_{diff}[CHCl_3] = 1.2 \times 10^{11}\ s^{-1}$. The rate constants, k_3 and k_4, for hydrogen abstraction with $Ph^{\bullet}$ and $PhCOO^{\bullet}$ from $CHCl_3$ are 3.7×10^7 and $1.5 \times 10^7\ s^{-1}$, respectively.

TABLE 11. Product distribution from photolysis of 0.83 M benzoyl peroxide in chloroform[e]

Product	Yield[a]	
	First period[b]	Subsequent period[c]
PhH	1.0	6.8
PhPh	0.16	0.08
$PhCO_2H$	0.3	3.2
$PhCO_2Ph$	1.0	1.0
$PhCCl_3$	0.05	0.2
PhCl	0.01	0.01
C_2Cl_4	0.01	0.01
C_2Cl_5H	0.01	0.01
C_2Cl_6[d]	0.7	2.9

[a]Normalized to the yield of phenyl benzoate.
[b]0–1300 s.
[c]1300–6400 s.
[d]Irradiation of $CHCl_3$ produced no detectable C_2Cl_6.
[e]Reprinted with permission from C. F. Poranski, Jr., W. B. Moniz and S. A. Sojka, *J. Amer. Chem. Soc.*, **97**, 4275 (1975). Copyright 1975 American Chemical Society.

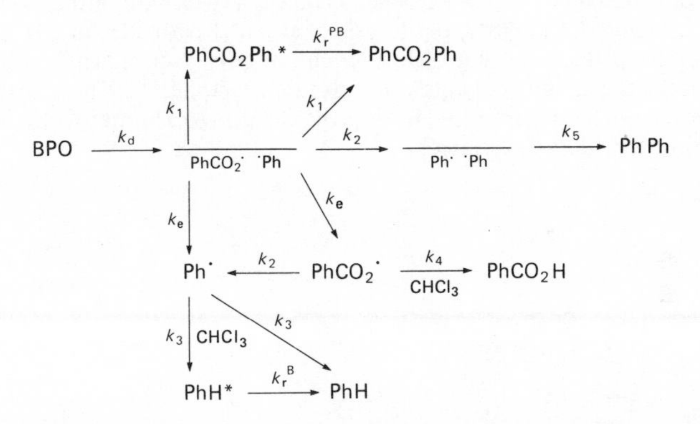

SCHEME 1. Reprinted with permission from C. F. Poranski, Jr., W. B. Moniz and S. A. Sojka, *J. Amer. Chem. Soc.*, **97**, 4275 (1975). Copyright 1975 American Chemical Society.

Diaroyl peroxides are subject to photosensitized decomposition. Sensitized photolysis in benzene shows that aromatic ketones possessing triplet energy of over 59 kcal mol^{-1} (e.g. benzophenone) act as triplet-excited (n–π*) sensitizers, while aromatic hydrocarbons (e.g. naphthalene) act in their singlet-excited (π–π*) state[12,136]. The excited sensitizer transfers its energy to the peroxide to cleave the O—O bond. The sensitization is suppressed by molecular oxygen because of the deactivation of the excited aromatic electron donor[137].

Evidence for the triplet ketone-sensitized decomposition are the suppression by triplet quencher, 1,3-pentadiene, and also the enhanced absorption of the CIDNP spectrum of formed benzene[138]. On the other hand, the singlet decomposition is supported by the lack of reduction of the quantum yield on addition of a triplet quencher, by quenching the fluorescence of sensitizers by benzoyl peroxide, by analysis of Stern–Volmer plots[138c] and by CIDNP analysis[138b]. The singlet-excited sensitizer forms an exciplex with ground-state peroxide which then decomposes, but the exciplex is usually so unstable that it is undetectable.

Introduction of a benzoyl group into the ring of benzoyl peroxide is effective (with $\phi \simeq 1$) for intramolecular sensitization[139]. A *p*-PhCO group is more efficient than a *meta* one for the energy transfer.

C. Cyclic Diacyl Peroxides

Photolysis of phthaloyl peroxide at room temperature gives benzyne[140], while at 8 K it affords a ketoketene and benzopropiolactone as primary products (equation 63)[141].

hν, r. t. (63a)

hν, 8 K; C=O; + (63b)

The ratio of ketoketene vs. lactone depends on the wavelength of the light, because of their interconversion[55]. Similarly, photolysis of succinyl peroxides in CH_2Cl_2 gives the corresponding propiolactones and olefins (equation 64)[142] and phenylmaleoyl peroxide affords phenylacetylene and polymeric unsaturated lactones[143]. These reactions occur through O—O bond fission, followed by decarboxylation to a biradical **(15)** which in turn is cyclized or further decarboxylated.

R, R; hν; $-CO_2$; **(15)** (64a)

$-CO_2$; R, R (64b)

α-Lactone is formed as an intermediate from di-*n*-butylmalonyl peroxide (equation 65)[144].

n-Bu, *n*-Bu; $-CO_2$; *n*-Bu, *n*-Bu; ⟶ polyester etc. (65)

D. Other Diacyl Peroxides

Benzoyl acyl peroxides are photolysed at low temperature to give alkyl benzoates and alkylbenzenes; e.g. acetyl benzoyl peroxide gives on photolysis methyl benzoate and toluene in a molar ratio of 1.8:1 (equation 66)[145]. The reaction occurs in a solvent cage and Me$^{\bullet}$ has more chance to couple with peroxy oxygen, so that only 55 % scrambling is observed with ^{18}O-labelled peroxide in glassy ethanol at 77 K (equation 67). Phenyl and methyl radicals can be detected at very low temperatures by means of matrix isolation IR spectroscopy[147–149]. The [PhCOO$^{\bullet}$ $^{\bullet}$Me] radical pair is also observed by low-temperature ESR[146].

$$PhCOO_2COMe \xrightarrow[-70^{\circ}C,\ neat]{254\ nm} PhCO_2Me + PhCH_3 \tag{66}$$

$$Ph{-}\overset{O}{\overset{\|}{C}}{-}^{18}O{-}^{18}O{-}\overset{O}{\overset{\|}{C}}{-}Me \xrightarrow{h\nu\ (-CO_2)} [Ph{-}C(=O){-}^{18}O^{\bullet}\ \ ^{\bullet}Me]_{cage} \tag{67}$$

$$Ph{-}CO{-}^{18}O{-}^{18}O{-}CO{-}Me \xrightarrow{h\nu\ (-2CO_2)} [Ph^{\bullet}\ \ ^{\bullet}Me]_{cage} \longrightarrow PhMe$$

$$[Ph{-}C(=O){-}^{18}O^{\bullet}\ \ ^{\bullet}Me]_{cage} \xrightarrow{h\nu\ (-CO_2)} [Ph^{\bullet}\ \ ^{\bullet}Me]_{cage}$$

$$[Ph{-}C(=O){-}^{18}O^{\bullet}\ \ ^{\bullet}Me]_{cage} \longrightarrow PhCO{-}^{18}OMe$$

VI. PEROXYCARBOXYLIC ACIDS

A. Aliphatic Peroxycarboxylic Acids

Aliphatic peroxycarboxylic acids (RCO_2OH) have no characteristic UV absorption at 230–300 nm, although there is a continuous increase in absorption with a decrease of wavelength, so that the O—O bond is easily cleaved by UV light as well as by heating. The formed carbonyloxy radical is rapidly decarboxylated to give the alkyl radical; i.e. alkylperoxy acids are photolysed to give CO_2 quantitatively (equation 68)[47]. The radicals thus formed react with organic compounds via hydrogen atom abstraction or addition to unsaturated bonds. Thus far, peracetic acid is the only aliphatic peracid used for the systematic photolysis studies.

$$RCO_2OH \xrightarrow{h\nu} RCO_2^{\bullet} + HO^{\bullet} \longrightarrow R^{\bullet} + CO_2 + HO^{\bullet} \tag{68}$$

1. Cycloalkanes

The photolysis of peracetic acid in cyclohexane gives cyclohexanol (90.2 %) and cyclohexanone (6.3 %) (equation 69)[150] but no methylation product is observed. This hydroxylation involves a chain mechanism of induced decomposition of peracetic acid with cyclohexyl radical R$^{\bullet}$ as a carrier (equation 70a–d).

$$C_6H_{12} + CH_3CO_3H \xrightarrow{h\nu} C_6H_{11}{-}OH + C_6H_{10}{=}O \tag{69}$$

$$CH_3CO_3H \xrightarrow{h\nu} CH_3CO_2^{\bullet} + HO^{\bullet} \tag{70a}$$

$$CH_3CO_2^{\bullet} \longrightarrow CH_3^{\bullet} + CO_2 \quad (70b)$$

$$CH_3^{\bullet}\ (\text{or } HO^{\bullet}) + RH \longrightarrow CH_4\ (\text{or } H_2O) + R^{\bullet} \quad (70c)$$

$$R^{\bullet} + CH_3CO_3H \longrightarrow ROH + CH_3CO_2^{\bullet} \quad (70d)$$

2. *Aromatic compounds*

Peracetic acid dissolved in aromatic compounds is photolyzed by excitation via an energy transfer from aromatic compounds or by direct excitation of peracetic acid.

a. Sensitized decomposition. Since arenes have generally higher absorption coefficients than peracids, the incident light of 254 nm from a low-pressure Hg lamp is absorbed and excites the former to [ArH]*. Then the energy is transferred to the peracid which is cleaved at the O—O bond (equation 71).

$$[ArH]^* + CH_3CO_3H \longrightarrow ArH + [CH_3CO_3H]^* \quad (71a)$$

$$[CH_3CO_3H]^* \longrightarrow CH_3CO_2^{\bullet} + HO^{\bullet} \quad (71b)$$

$$CH_3CO_2^{\bullet} \longrightarrow CH_3^{\bullet} + CO_2 \quad (71c)$$

Since the fluorescence of excited toluene is quenched by peracetic acid, as shown in the Stern–Volmer plots of Figure 9, this sensitized decomposition proceeds by singlet energy transfer from excited aromatics[25].

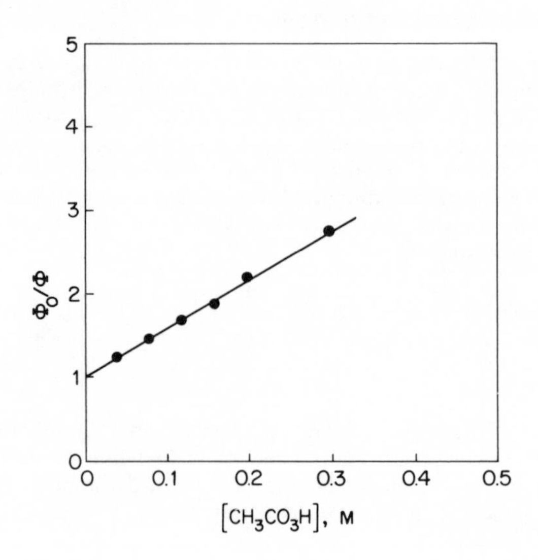

FIGURE 9. Stern–Volmer plot for quenching of the singlet state of toluene by peracetic acid. Slope $(k_q\tau) = 5.83$, $k_q = 1.71 \times 10^8\ \text{M}^{-1}\,\text{s}^{-1}$, where k_q is the quenching rate constant and τ is the lifetime of lowest singlet state. The τ value of $(3.4 \times 10^{-8}\,\text{s})$ is quoted from S. L. Murov, *Handbook of Photochemistry*, Marcel Dekker, New York, 1973. Reprinted with permission from Y. Ogata and K. Tomizawa, *J. Org. Chem.*, **43**, 261 (1978). Copyright 1978 American Chemical Society.

b. Direct photolysis. Simple alkylbenzenes such as toluene and ethylbenzene absorb little light at over 290 nm, so that peracetic acid is photolysed by direct absorption of light energy. Sensitized (254 nm) and direct (>290 nm) photolyses result in different product distributions and isomeric compositions of the methylated arenes.

Irradiation by >290 nm light favours hydroxylation of the side-chain more than irradiation by 254 nm light[151–153]. For example, photolysis of peracetic acid in ethylbenzene gives α-phenethyl alcohol in a yield of 40–50 % at >290 nm, but only 15–20 % at 254 nm. The product distribution is almost identical when using different intensities with the same wavelength (300 W and 1 kW high-pressure Hg lamps).

In contrast, the methylation of a side-chain with Me˙ radical is favoured with 254 nm light rather than >290 nm light. Thus the photolysis of peracetic acid in toluene with 254 nm light gives higher yields of ethylbenzene (10–13 %) and xylene (10–15 %) compared with lower yields of ethylbenzene (4–7 %) and xylene (8–6 %) at >290 nm[25]. Also the photolysis of peracetic acid in ethylbenzene gives higher yields of propylbenzene (25–30 %) and ethyltoluene (4–5 %) at 254 nm than their yields at >290 nm (5 % and 2 %, respectively)[152].

The wavelength of light also affects the methylation of ring carbons, i.e. the isomeric distribution of xylenes produced from toluene (Table 12)[25]. The order of isomer distribution of xylenes at 254 nm is *meta* > *ortho* > *para*, while at >290 nm it is *ortho* > *meta* > *para*. The latter distribution is in accord with those observed in the thermolysis of acetyl peroxide in toluene[24].

TABLE 12. The isomer distribution of xylene in the photochemical reaction of toluene with peracetic acid[c]

$[CH_3CO_3H]^a$ × 10 (M)	$[CH_3CO_3H$ decomposed] × 10^4 (mol)	Xylene × 10^4 mol (orientation %) *o-*	*m-*	*p-*	$[CH_3CO_3H]^b$ × 10 (M)	$[CH_3CO_3H$ decomposed] × 10^4 (mol)	Xylene × 10^4 mol (orientation %) *o-*	*m-*	*p-*
3.51	54.4	2.64	4.19	0.66	4.71	51.3	1.36	0.78	0.25
		(35)	(56)	(9)			(57)	(32)	(11)
2.00	43.7	1.95	3.34	0.70	3.77	45.0	0.95	0.53	0.18
		(33)	(56)	(11)			(57)	(32)	(11)
1.81	30.5	1.15	2.08	0.47	2.74	34.2	0.84	0.38	0.18
		(31)	(56)	(13)			(60)	(27)	(13)
0.65	13.8	0.53	0.94	0.24	1.87	23.4	0.56	0.32	0.14
		(31)	(55)	(14)			(55)	(31)	(14)
0.37	7.23	0.27	0.40	0.08	1.06	11.6	0.34	0.24	0.12
		(36)	(53)	(11)			(49)	(34)	(17)

[a]Irradiation with 2537 Å light.
[b]Irradiation with >2900 Å light.
[c]Reprinted with permission from Y. Ogata and K. Tomizawa, *J. Org. Chem.*, **43**, 261 (1978). Copyright 1978 American Chemical Society.

The photolysis of peracetic acid in xylene gives the aromatic products shown in Scheme 2 and Tables 13 and 14[153].

The effect of wavelength on the product yields is similar to the photolysis in toluene; i.e., methylation of side-chain and benzene ring is favoured with 254 nm light, while hydroxylation of side-chain is favoured with >290 nm light. Further, there is a large difference in the orientations of methylation with 254 and >290 nm lights (Tables 13 and 14[153]). The intensity of light, however, does not affect the product distribution.

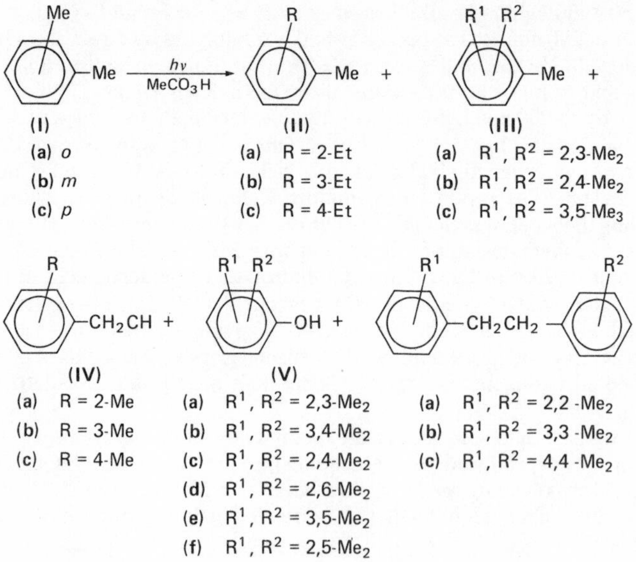

(I) (a) *o* (b) *m* (c) *p* —(*hν*, $MeCO_3H$)→ (II) + (III) +

(II) (a) R = 2-Et (b) R = 3-Et (c) R = 4-Et

(III) (a) R^1, R^2 = 2,3-Me_2 (b) R^1, R^2 = 2,4-Me_2 (c) R^1, R^2 = 3,5-Me_3

(IV) $RC_6H_4CH_2CH$ + (V) $R^1R^2C_6H_3OH$ + (VI) $R^1C_6H_4CH_2CH_2C_6H_4R^2$

(IV) (a) R = 2-Me (b) R = 3-Me (c) R = 4-Me

(V) (a) R^1, R^2 = 2,3-Me_2 (b) R^1, R^2 = 3,4-Me_2 (c) R^1, R^2 = 2,4-Me_2 (d) R^1, R^2 = 2,6-Me_2 (e) R^1, R^2 = 3,5-Me_2 (f) R^1, R^2 = 2,5-Me_2

(a) R^1, R^2 = 2,2-Me_2 (b) R^1, R^2 = 3,3-Me_2 (c) R^1, R^2 = 4,4-Me_2

SCHEME 2. Reprinted with permission from Y. Ogata and K. Tomizawa, *J. Org. Chem.*, **45**, 785 (1980). Copyright 1980 American Chemical Society.

TABLE 13. Product yields in photolysis of peracetic acid in xylenes[a]

Xylene	$[AcO_2H] \times 10^2$ (M)	Light (Å)	Product (%) EtC_6H_4Me (II)	$Me_3C_6H_3$ (III)	$MeC_6H_4CH_2OH$ (IV)	$Me_2C_6H_3OH$ (V)	$(MeC_6H_4CH_2-)_2$ (VI)
ortho (**Ia**)	3.4–30.7	2537	24.3	8.9	15.1	6.0	17.0
ortho (**Ia**)	3.4–30.7	>2900	9.3	6.2	36.9		45.1
meta (**Ib**)	5.8–34.2	2537	20.3	9.1	10.4	6.3	18.9
meta (**Ib**)	5.8–34.2	>2900	7.4	6.9	30.0		46.6
para (**Ic**)	6.0–33.5	2537	23.5	8.4	13.5	6.1	16.7
para (**Ic**)	6.0–33.5	>2900	10.2	5.9	32.3		43.4

[a] Reprinted with permission from Y. Ogata and K. Tomizawa, *J. Org. Chem.*, **45**, 785 (1980). Copyright 1980 American Chemical Society.

These facts are explained as follows. The excited arene at 254 nm is highly reactive and attacks peracid yielding methylated products probably via exciplex $[ArCH_3-CH_3CO_3H]^*$, while no excitation of arene occurs at >290 nm and peracid decomposes to $Me^\cdot$, CO_2, and $HO^\cdot$ which can abstract hydrogen. Therefore, the concentration of $ArCH_2^\cdot$ radical is higher and the induced decomposition of peracid, leading to hydroxylation is accelerated at >290 nm as shown in equation (72)[153].

$$ArCH_2^\cdot + CH_3CO_3H \longrightarrow ArCH_2OH + CH_3CO_2^\cdot \quad (72a)$$

$$ArCH_2^\cdot + HO^\cdot \longrightarrow ArCH_2OH \quad (72b)$$

TABLE 14. Isomer distribution in photochemical methylation of xylene ring[a]

Substrate	$[AcO_2H] \times 10^2$ (M)	Light (Å)	Distribution of $C_6H_3Me_3$ (%)		
			IIIa 1,2,3-Me_3	**IIIb** 1,2,4-Me_3	**IIIc** 1,3,5-Me_3
Ia	3.4–30.7	2537	46	54	
Ia	3.4–30.7	2900	63	37	
Ib	5.8–34.2	2537	31	46	25
Ib	5.8–34.2	2900	41	46	13
Ic	6.0–33.5	2537		100	
Ic	6.0–33.5	2900		100	

[a]Reprinted with permission from Y. Ogata and K. Tomizawa, *J. Org. Chem.*, **45**, 785 (1980). Copyright 1980 American Chemical Society.

Hence the methylation with 254 nm light occurs not only by simple radical coupling and radical addition to aromatic rings at their ground state, but also (*i*) by reaction between radical and photoexcited arene, which has a different electron distribution from that of the ground-state arene, (*ii*) by reaction via an exciplex $[ArH{-}CH_3CO_3H]^*$ in which an electron is transferred from arene to peracid, giving thus a different electron distribution in the arene, and (*iii*) *ortho* attack by $HO^{\bullet}$ forming a cyclohexadienyl radical, followed by attack of $Me^{\bullet}$ radical.

B. Aromatic Peroxycarboxylic Acids

Aromatic peroxycarboxylic acids are photolysed by UV irradiation to cleave their O—O bond, giving arylcarbonyloxy radicals, $ArCO_2^{\bullet}$, which are generally more stable than alkylcarbonyloxy radicals; hence decarboxylation and hydrogen atom abstraction compete. For example, photolysis of *m*-chloroperbenzoic acid affords chlorobenzene and *m*-chlorobenzoic acid (equation 73a)[154], and *o*-methoxycarbonylperbenzoic acid is also photolysed to give methyl benzoate and monomethyl phthalate (equation 73b)[154].

$$m\text{-}ClC_6H_4CO_3H \xrightarrow{h\nu} C_6H_5Cl + m\text{-}ClC_6H_4CO_2H \qquad (73a)$$

$$o\text{-}CH_3O_2CC_6H_4CO_3H \xrightarrow{h\nu} C_6H_5CO_2CH_3 + o\text{-}CH_3O_2CC_6H_4CO_2H \qquad (73b)$$

The photolysis of the latter peracid in the presence of olefins results in the epoxidation of olefins followed by opening of the epoxide ring by alcohol or water present, e.g. equation (74)[154].

$$o\text{-}CH_3O_2CC_6H_4CO_3H \xrightarrow[\text{MeOH}]{h\nu,\ MeCH{=}CMe_2} o\text{-}CH_3O_2CC_6H_4CO_2H + Me(HO)CH{-}CMe_2(OMe) \qquad (74)$$

VII. PEROXYCARBOXYLIC ESTERS

A. Simple Photolysis of Peroxycarboxylic Esters

1. *Aliphatic peroxy acid esters*

Aliphatic peroxy acid esters have in general weak and continuous UV absorption beginning around 280 nm due to n–π* transition in the C=O group and the peroxide continuum, which is analogous to diacyl peroxides (Table 15)[8,10,155]. Hence, peroxy esters RCO_2OR^1 are decomposed by UV irradiation via O—O bond fission, giving primarily $RCO_2^{\cdot}$ and $R^1O^{\cdot}$ radicals.

TABLE 15. Ultraviolet absorption of peresters at 2537 Å in *n*-hexane[a]

Perester $RCO_2OC(CH_3)_2R^1$			Perester $RCO_2OC(CH_3)_2R^1$		
R	R^1	ε ($M^{-1} cm^{-1}$)	R	R^1	ε ($M^{-1} cm^{-1}$)
CH_3	CH_3	21	cyclopropyl—CH_2	CH_3	20
CH_3	CH_2CH_3	20	cyclobutyl	CH_3	30
CH_3	$CH(CH_3)_2$	23	$(CH_3)_3C$	CH_3	38
CH_3CH_2	CH_3	28	$(CH_3)_3C$	CH_2CH_3	43
CH_3CH_2	CH_2CH_3	27	$(CH_3)_3C$	$CH(CH_2)_3$	41
$(CH_3)_2CH$	CH_3	26			

[a]Reprinted with permission from R. A. Sheldon and J. K. Kochi, *J. Amer. Chem. Soc.*, **92**, 5175 (1970). Copyright 1970 American Chemical Society.

The ESR spectra and product analyses show that the carbonyloxy radical $RCOO^{\cdot}$ formed is subject to very rapid decarboxylation (rather than hydrogen atom abstraction) to give alkyl radical $R^{\cdot}$, with very little formation of aliphatic acid RCOOH and carboxy inversion product $ROCOR^1$, Thus photolysis of peroxy esters is a good way for obtaining alkyl radicals in organic solvents, in which the peresters are usually soluble. However, aromatic peroxy acid esters, yield $ArCOO^{\cdot}$ radicals which are decarboxylated only slowly.

As is apparent from Table 16[10], the decarboxylation of methyl, ethyl and isopropyl esters of peroxyacids occurs in high yield (over 90 %), irrespectively of the nature of the alkyl group, with a quantum yield of unity (equation 75a). The paraffins and olefins produced, may contain also some products, such as methane, derived from β-C—C fission of the radicals[10].

$$RCO_2OCMe_2R^1 \xrightarrow{h\nu} [RCO_2^{\cdot} + {}^{\cdot}OCMe_2R^1] \longrightarrow [R^{\cdot} + CO_2 + {}^{\cdot}OCMe_2R^1]$$

$$\longrightarrow RH,\ R(-H),\ ROCMe_2R^1,\ CO_2,\ R^1H,\ R^1(-H),\ Me_2C{=}O \quad (75a)$$

[a trace of R—R, R—R^1, R^1—R^1, R^1Me_2COH, etc.]

TABLE 16. Photolysis of peresters $RCO_2OC(CH_3)_2R^1$ in decalin solutions at 2537 Å[a,i]

Perester														
R	R^1	mmol	CO_2[b]	RH	R(−H)	R^1H	$R^1(-H)$	R—R	$R^1—R^1$	$R—R^1$	Alcohol[c]	Ether[d]	Me_2CO	ΣR[f]/CO_2
CH_3	CH_3	0.54	0.52 (96)	0.26						t[g]	0.25	0.23	t	0.94
CH_3CH_2	CH_3	0.51	0.48 (94)	0.21	0.06	t		t		t	0.27	0.19	t	0.96
$(CH_3)_2CH$	CH_3	0.51	0.50 (98)	0.11	0.24	t		~0.01		t	0.37	0.10	t	0.90
$(CH_3)_3C$	CH_3	0.52	0.50 (96)	0.065	0.37	t		~0.01			0.45	0.02	t	0.91
CH_3	CH_2CH_3	0.51	0.49 (96)	0.24		0.03	None	None		~0.01	0.22	0.21	0.04	0.92
CH_3CH_2	CH_2CH_3	0.54	0.50 (92)	0.25	0.06			t		~0.01	0.28	0.18	0.04	1.00
$(CH_3)_2CH$	CH_2CH_3	0.52	0.51 (98)	0.10	0.18	0.03	None	~0.01		t	0.36	0.10	0.04	0.77
$(CH_3)_3C$	CH_2CH_3	0.50	0.45 (90)	0.08	0.27	0.03	None	~0.01	t	t	0.37	~0.01	0.03	0.82
CH_3	$CH(CH_3)_2$	0.52	0.48 (92)	0.28		0.13	0.06	None	~0.01	0.03	0.09	0.14	0.22	0.94
CH_3CH_2	$CH(CH_3)_2$	0.53	0.52 (98)	0.29	0.03	0.14	0.06	t	~0.01	0.04	0.11	0.12	0.22	0.92
$(CH_3)_2CH$	$CH(CH_3)_2$	0.51	0.50 (98)	0.23	0.16					0.08	0.22	0.06	0.18	1.06
$(CH_3)_3C$	$CH(CH_3)_2$	0.50	0.46 (92)	0.10	0.23	0.10	0.05	t	~0.01	~0.01	0.27	~0.01	0.16	0.77
Ph	CH_3	0.50	0.32[h] (64)	0.17				None	None	None	0.34	0.14	None	

[a] In solutions containing 5 ml of decalin and approximately 0.5 mmol/ester at 30°C. All runs in duplicate. All yields expressed as mmoles; t = traces found.
[b] Number in parentheses is yield of CO_2 based on perester charged.
[c] $R^1C(CH_3)_2OH$.
[d] $R^1C(CH_3)_2OR$.
[e] Acetone.
[f] ΣR includes all products derived from R·.
[g] Traces ($\lesssim$0.01 mmol).
[h] Benzoic acid (0.15 mm) was formed.
[i] Reprinted with permission from R. A. Sheldon and J. K. Kochi, *J. Amer. Chem. Soc.*, **92**, 5175 (1970). Copyright 1970 American Chemical Society.

The photolysis of a mixture of *t*-butyl peracetate and *t*-amyl perpropionate gives *t*-BuOMe and *t*-AmOEt, but neither of the crossed ethers, i.e. *t*-BuOEt nor *t*-AmOMe. Hence coupling and disproportionation of alkyl–alkoxy radical pairs only occur within the solvent cage[10].

The yield of the cage product *t*-BuOMe from the photolysis of *t*-butyl peracetate (*t*-$BuOO_2CMe$) is higher than that of thermolysis; for example, the photolysis yields are 27 % in pentane (viscosity 0.21 cP), 44 % in decalin (viscosity 1.55 cP), 75 % in mineral oil (viscosity 80 cP), while thermolysis yield is 17 % in decalin (viscosity 0.59 cP) at 115°C. These facts suggest that low temperature and higher viscosity tend to keep the radical pairs inside of the cage.

The yield ratio of alcohol to ether, $R^1Me_2COH:R^1Me_2COR$, increases in the series: R:Me < Et < *i*-Pr < *t*-Bu in the photolysis of $RCO_3CMe_2R^1$. This is ascribed to an increase in the ratio of (disproportionation + hydrogen abstraction):(coupling) in the $R^1Me_2C{-}O^{\bullet}$ radical[10].

In the photolysis of peroxyformate (R = H in equation 75a), hydrogen gas (46 %) is generated as a result of hydrogen atom abstraction or disproportionation, while the formation of *t*-BuOH is 90 % and formic acid is only 4 % (equation 75b)[156].

$$HCO_2OBu\text{-}t \xrightarrow{h\nu} CO_2 + H_2 + t\text{-}BuOH + (Me_2CO, CH_4, HCO_2H)_{trace} \quad (75b)$$

The photolysis of *t*-butyl performate in deuterated solvents (acetone-d_6, benzene-d_6, $CDCl_3$) gives DH, while thermolysis at 100°C gives only a trace amount of DH. The hydrogen atom abstraction rate estimated from this method is comparable to the other methods (e.g., photolysis of RSH or radiolysis of water). These facts suggest that the photolysis of HCO_2O-Bu-*t* is a convenient method for generation of hydrogen atoms[156].

The possibility of rearrangement of alkyl radicals generated from peroxy esters has been examined. Allylacetyl and cyclobutanecarbonyl peroxy esters afford radicals which cannot isomerize (equations 76 a and b), whereas, cyclopropylacetyl peroxy ester gives an alkyl radical which can isomerize by ring-opening (equation 76c)[10].

C(=O)–O–OBu-*t* $\xrightarrow{h\nu}$ CO_2 + [22% , trace , 55% OBu-*t*, *t*-BuOH 33%] (76a)

C(=O)–O–OBu-*t* $\xrightarrow{h\nu}$ CO_2 + [17% , 22% , 65% OBu-*t*, *t*-BuOH 24%] (76b)

C(=O)–O–OBu-*t* $\xrightarrow{h\nu}$ CO_2 + [20% , trace , 43% OBu-*t*, OBu-*t* 33%] (76c)

The facile and quantitative formation of alkyl radicals by the photolysis of *t*-butyl peroxycarboxylates is often used for the study of radicals. For example, the photolysis of norbornenyl peresters proves the rapid equilibrium between *syn* and *anti* isomers, since both *syn* and *anti* 7-deuterated 7-norbornenyl peresters give the same mixture of *syn*- and *anti*-7-

deuterionorbornene in which the ratio of *syn* to *anti* isomer is ca. 2 according to the NMR analysis after separation with preparative GLC (equation 77)[157,158].

t-BuO_3C D —$h\nu$, $-CO_2$, t-BuOH→ D (radical) ⇅ D CO_3Bu-t —$h\nu$→ D (radical) (77)

The photolysis of *t*-butyl peracetate was employed as a source of methyl radical to methylate purine derivatives in relation to the carcinogenesis. Purine derivatives such as caffeine (**16**), adenine (**17**), and guanine (**18**) are methylated at the 8-position, and the methylation rate varies with the pH of the solution (Figure 10)[159].

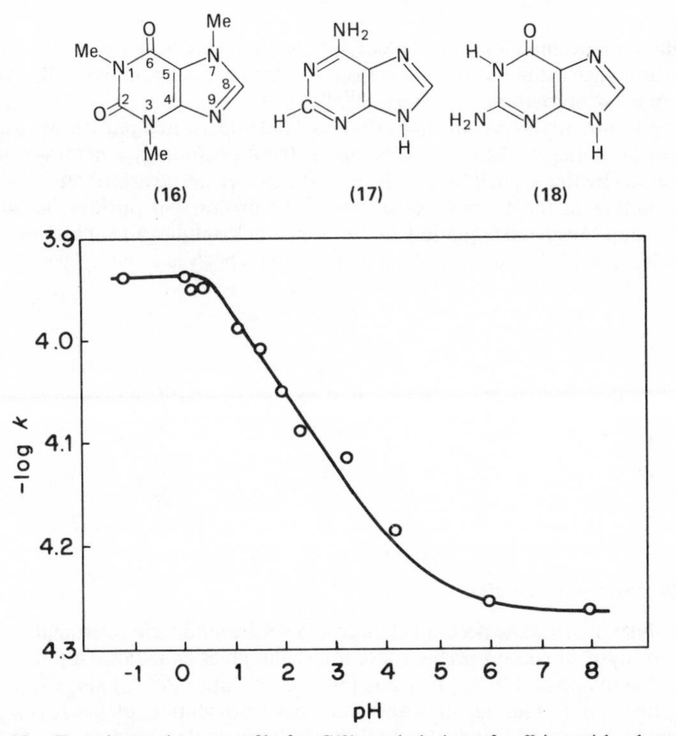

FIGURE 10. Experimental rate profile for C(8) methylation of caffeine with *t*-butyl peracetate under irradiation by a 1200 W mercury lamp with Pyrex filter at 25°C. Reprinted with permission from M. F. Zady and J. L. Wong, *J. Amer. Chem. Soc.*, **99**, 5096 (1977). Copyright 1977 American Chemical Society.

As apparent from the figure, the rate constant with caffeine increases with a decrease of pH, suggesting a Me˙ radical attack on neutral amino nitrogen at pH > 6, (equation 78a) along with a faster attack on protonated amine which leads to a charge-transfer complex at pH < 6 (equation 78b)[159a].

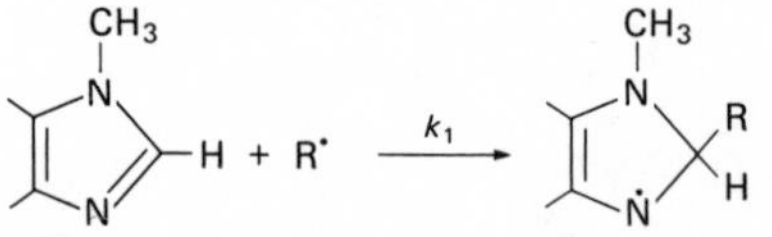

(78a)

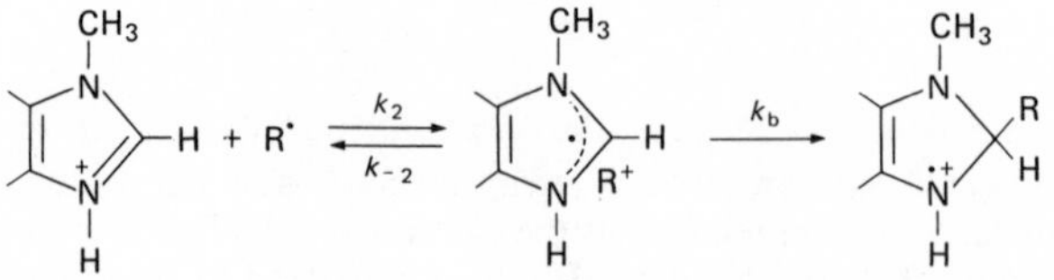

(78b)

The similar enhancement of radical methylation with decreasing pH is also observed in the methylation of pyridines at the 4-position[159a]; the reactivity order of alkyl radical is: $CH_3^{\cdot}$ < primary < secondary < tertiary R˙[160].

Interestingly, 3-methylhypoxanthine (**19**) and 3,6-disubstituted purines are methylated at the 2-position under acidic conditions, but at the 8-position in a neutral solution[161]. This is explained by the localization of the positive charge in the neighbourhood of the 2-position, as well as in the 1- and 3-positions of the protonated purines, because of the electron-releasing Me group (equation 79); thus the nucleophilic radical Me˙ attacks the 2-position, yielding a Me^+–purine radical complex which gives 2-methylation[161].

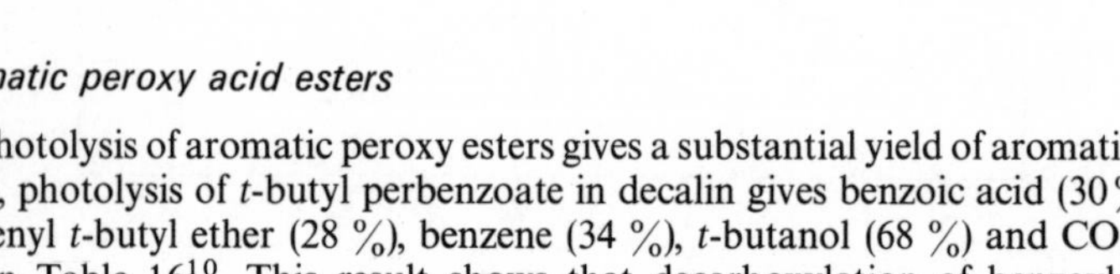

(79)

(**19**)

2. Aromatic peroxy acid esters

The photolysis of aromatic peroxy esters gives a substantial yield of aromatic acids. For example, photolysis of *t*-butyl perbenzoate in decalin gives benzoic acid (30%) together with phenyl *t*-butyl ether (28 %), benzene (34 %), *t*-butanol (68 %) and CO_2 (64 %) as shown in Table 16[10]. This result shows that decarboxylation of benzoyloxy radical giving Ph˙ radical proceeds more slowly than aliphatic acyloxy radical ($RCO_2^{\cdot}$).

The lower quantum yield for the CO_2 formation (0.59) compared with that in aliphatic peresters indicates recombination of PhCOO˙ with *t*-BuO˙ to the original perester as well as slow decarboxylation of PhCOO˙ (equation 80)[10].

$$PhCO_2OBu\text{-}t \rightleftharpoons [PhCO_2^{\bullet} + t\text{-}BuO^{\bullet}] \xrightarrow{(30\%)} PhCO_2H + t\text{-}BuOH$$
$$\downarrow (64\%)$$
$$[Ph^{\bullet} + CO_2 + t\text{-}BuO^{\bullet}]$$
$$\downarrow$$
$$PhH(34\%)\ t\text{-}BuOH(34\%)\ PhOBu\text{-}t(28\%) \quad (80)$$

3. Peroxylactones

Cyclic peroxycarboxylic esters, i.e., peroxylactones, are cleaved at the O—O bond by heat or light, affording biradicals which are then decarboxylated to ketones and epoxides[162–166]. For example, β-peroxylactone (**20**) is subject to photolysis and thermolysis to give ketones (**21** and **21**′) and epoxides (**22**).

(81)

	(21)	(21′)	(22)
Photolysis (%)	7.1	7.6	49.5
Thermolysis (%)	13.8	77.4	0.5

The epoxide is a main product in photolysis, while ketones are the main products in thermolysis[164]. In the ketone formation, the migratory aptitudes for Me:Ph, i.e. the ratio of Me migration product (ketone **21**) vs. Ph migration product (ketone **21**′), is 0.94 for the photolysis and 5.6 for the thermolysis.

A difference in products of photolysis and thermolysis is observed in the decomposition of tetramethyl-β-peroxylactone (**23**) (Table 17)[165]. The table shows that the thermolysis

TABLE 17. Product and quantum yields for the decarboxylation of β-peroxylactones[c]

		Yields (%)			
Process	Conditions	Ketone	Epoxide	Acetone	Total
Thermolysis	125°C, c-C_6H_{12}	100 ± 0.5	0.0	0.0	100 ± 0.5
TMD energized[a]	60°C, n-C_6H_{14}	0.9 ± 0.1	97 ± 2	*b*	98 ± 2
Photolysis (direct)	355 nm, n-C_6H_{14}	49 ± 3	22 ± 1	26 ± 4	97 ± 4
Photolysis (sensitized)	313 nm, acetone	32 ± 1	44 ± 1	*b*	76 ± 2
Photolysis (1.0 M Pip)	355 nm, n-C_6H_{14}	50 ± 2	20 ± 2	25 ± 5	95 ± 5

[a]Tetramethyl-1,2-dioxetane (TMD) chemienergization.
[b]Not determined.
[c]Reprinted with permission from W. Adam, O. Cueto and L. N. Guedes, *J. Amer. Chem. Soc.*, **102**, 2106 (1980). Copyright 1980 American Chemical Society).

gives ketone alone, and the direct photolysis gives a mixture of ketone, epoxide and acetone (equation 82), while tetramethyl-1,2-dioxetane-energized decomposition (via energy transfer from chemically excited ketone) gives epoxide alone.

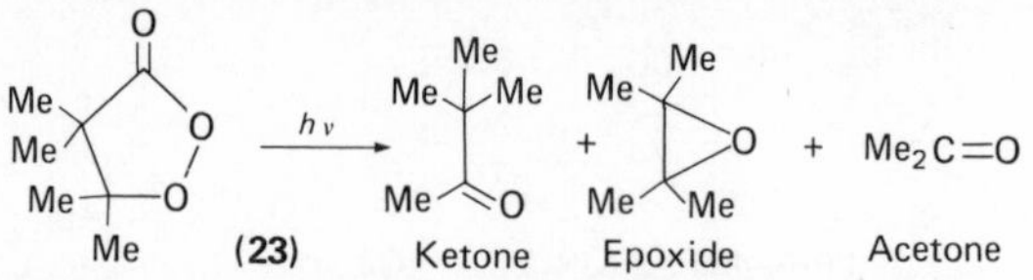

(23) $\xrightarrow{h\nu}$ Ketone + Epoxide + $Me_2C{=}O$ (Acetone) (82)

As shown in Scheme 3[165], the O—O bond shearing in thermolysis occurs, involving overlap of the odd electron orbital with the carbonyl bond in generating a $^{S}\pi$-type 1,5-diradical, which is resistant to decarboxylation, thus leading to ketone. On the other hand, the photolysis proceeds via n–π^* excitation to $^{S}\sigma$-type 1,5-dioxyl radical leading to epoxide by facile decarboxylation. Here S and T mean singlet and triplet states, respectively.

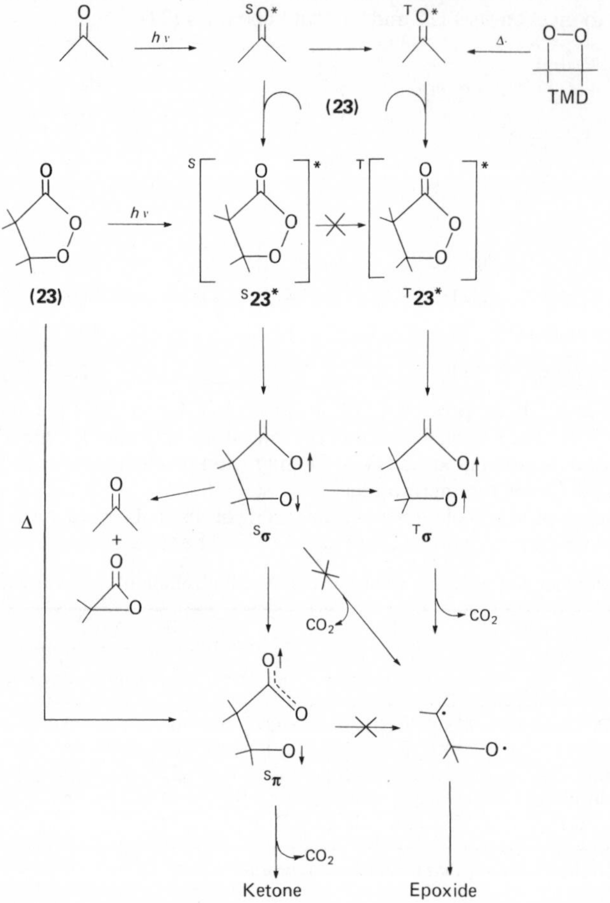

SCHEME 3. Reprinted with permission from W. Adam, C. Cueto and L. N. Guedes, *J. Amer. Chem. Soc.*, **102**, 2106 (1980). Copyright 1980 American Chemical Society.

The photolysis of γ-methyl-γ-peroxyvalerolactone gives 2,2-dimethyloxetane in a higher yield (16 %)[166] than the thermolysis (5.8 %) (equation 83).

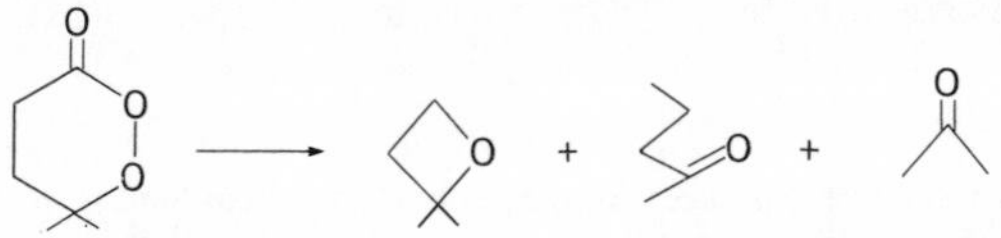

Photolysis	16%	0.5%	73.7%
Thermolysis	5.8%	0.5%	87.7%

B. α-Acyloxylation

Thermolysis of peroxy esters at 65–115°C in the presence of cuprous salt gives acyloxy radical which can react with ethers and olefins leading to α-acyloxylation; this method is not suitable for heat-sensitive acyloxy products and low-boiling substrates must be treated under pressure. Photochemical reaction is appropriate for these cases, since it gives 50–80 % of α-acyloxylation[167], e.g. equation (84). The reaction, if it is catalysed by light or Cu^+ only, proceeds at a much slower rate.

$$C_2H_5OC_2H_5 + CH_3CO_3Bu\text{-}t \xrightarrow[75\%]{CuBr,\ h\nu} C_2H_5OCH(CH_3)OCOCH_3 \qquad (84a)$$

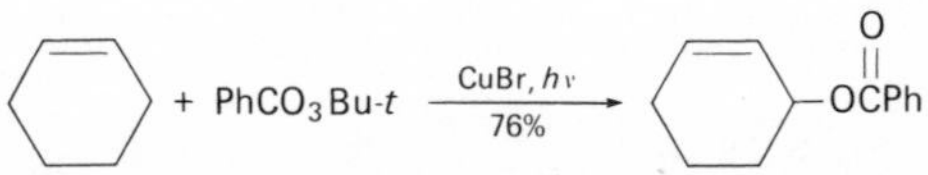

$$\text{cyclohexene} + PhCO_3Bu\text{-}t \xrightarrow[76\%]{CuBr,\ h\nu} \text{cyclohexenyl}\text{-}OCPh(=O) \qquad (84b)$$

In the absence of Cu^+ ion, acyloxy radicals formed by photolysis are easily decarboxylated, especially in the case of aliphatic ones ($RCO_2^{\cdot} \rightarrow R^{\cdot} + CO_2$); hence no acyloxylation occurs, but coupling products between the substrate radicals formed by hydrogen atom abstraction are obtained, e.g. equation (85)[168].

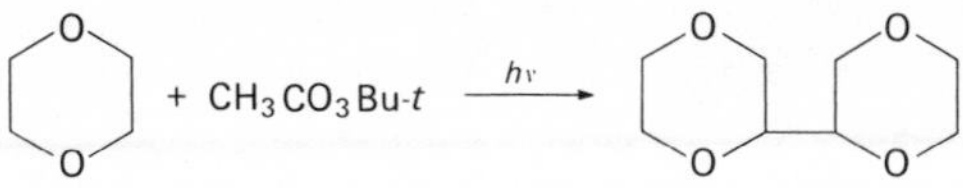

$$\text{dioxane} + CH_3CO_3Bu\text{-}t \xrightarrow{h\nu} \text{2,2'-bi(1,4-dioxanyl)} \qquad (85)$$

C. Intramolecular Oxidation of Unreactive C—H Bonds

Photolysis of α-acetylperoxyacetonitriles in benzene or *t*-butanol enables the effective and successive oxidation of unreactive C—H bonds with an efficient regioselectivity. For example, the photolysis of peroxyester **24** gives δ-ketonitrile **25** via O—O bond fission followed by δ-hydrogen abstraction, ring-formation and ring-opening[169].

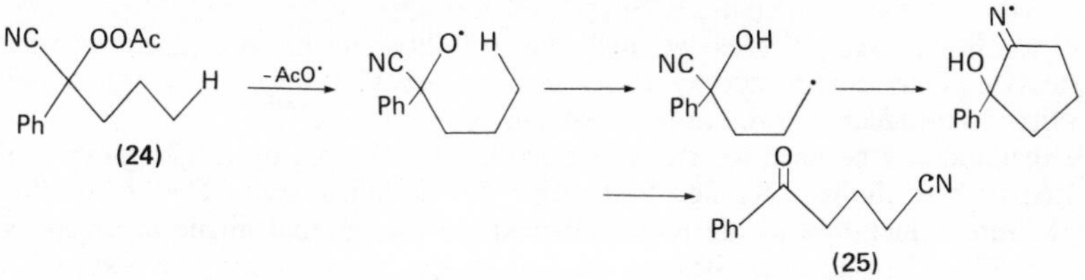

(86): **(24)** $\xrightarrow{-AcO^{\cdot}}$ ⟶ ⟶ ⟶ **(25)**

The starting peroxy ester can be prepared by the autoxidation of anions of secondary nitriles in alkaline solution to hydroperoxides followed by acetylation with acetyl chloride in 50–90 % yield (Table 18)[169].

TABLE 18. The photolysis of α-peracetoxynitriles in 0.25 M benzene for one hour[a]

α-Peracetoxynitrile	δ-Ketonitrile	Isolated yield (%)
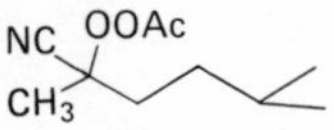 NC OOAc CH_3	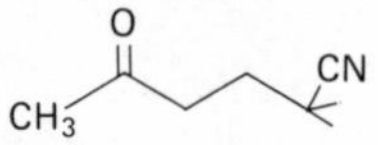 O CH_3 CN	50
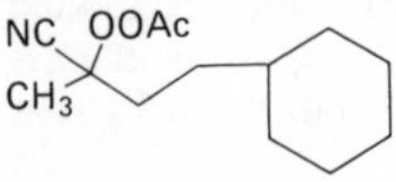 NC OOAc CH_3	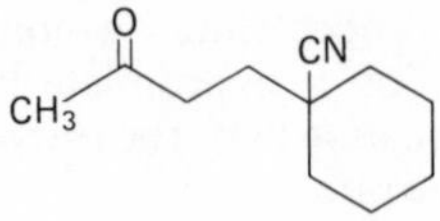 O CH_3 CN	47
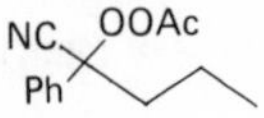 NC OOAc Ph	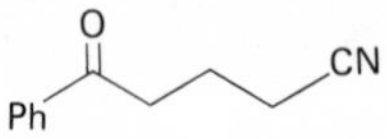 O Ph CN	15
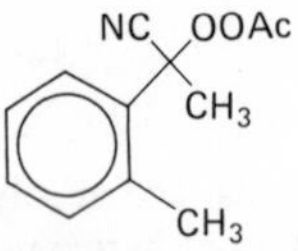 NC OOAc CH_3 CH_3	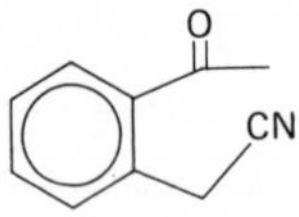 O CN	18
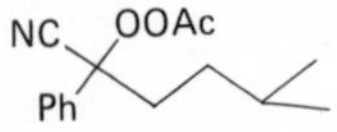 NC OOAc Ph	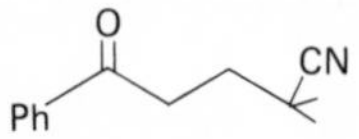 O Ph CN	52
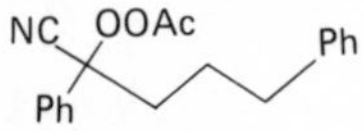 NC OOAc Ph Ph	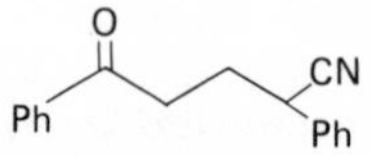 O Ph CN Ph	10

[a]Reprinted with permission from D. S. Watt, *J. Amer. Chem. Soc.*, **98**, 271 (1976). Copyright 1976 American Chemical Society.

The yields of δ-ketonitriles parallel the stability of the presumed free-radical intermediate. Thus α-acetylperoxynitriles bearing benzylic δ-hydrogen atoms suffer a competitive photodecarboxylation to yield α-methoxynitriles[169].

The reaction can be used for successive oxidation followed by nitrile group shift in saturated carbon chains, so-called 'reiterative functionalization of C—H bonds'. For example, monofunctional nitrile is transformed to trifunctional nitrile in several steps (equation 87). α-Peroxy esters bearing no δ-hydrogen atom cannot give δ-ketonitrile.

H CN Ph — 1. LiN(i-Pr)$_2$ 2. O$_2$ 3. AcCl → AcOO CN Ph

$h\nu$, – AcOH → Ph, O, CN — HO OH, – H$_2$O, H$^+$ → O O, Ph, CN (87)

→ O O, Ph, AcOO CN — $h\nu$, – AcOH → O O, Ph, O, CN

VIII. MISCELLANEOUS PEROXIDES

A. Peroxymono- and Peroxydi-sulphates

1. Simple photolysis of peroxysulphates

Peroxymonosulphate (HSO_5^-) and peroxydisulphate ($S_2O_8^{2-}$) are cleaved at the O—O bond on UV irradiation, forming sulphate anion radical $SO_4^{\cdot-}$ (equation 88)[170–172]. The anion radical $SO_4^{\cdot-}$ has an absorption maximum at 455 nm ($\varepsilon = 460\,\mathrm{M}^{-1}\,\mathrm{cm}^{-1}$) and a half-life of 300 μs[171]. In the absence of inorganic or organic reactants, $SO_4^{\cdot-}$ couples very rapidly to reform $S_2O_8^{2-}$ with the second-order rate constant of $3.7 \times 10^8\,\mathrm{M}^{-1}\,\mathrm{s}^{-1}$. $SO_4^{\cdot-}$ is stable in acidic and neutral solutions, but very unstable in alkaline solution because of fast reaction with hydroxide ions with a large rate constant $7.3 \times 10^7\,\mathrm{M}^{-1}\,\mathrm{s}^{-1}$ (equation 89).

$$HSO_5^- \xrightarrow{h\nu} SO_4^{\cdot-} + HO^{\cdot} \qquad (88a)$$

$$S_2O_8^{2-} \xrightarrow{h\nu} 2SO_4^{\cdot-} \qquad (88b)$$

$$SO_4^{\cdot-} + HO^- \longrightarrow SO_4^{2-} + HO^{\cdot} \qquad (89)$$

Irradiation of $S_2O_8^{2-}$ in the presence of a ruthenium complex such as $Ru(bipyridyl)_3^{2+}$ gives electron transfer from the complex to $S_2O_8^{2-}$ with a high quantum yield of ca. unity and a high rate constant of $5 \times 10^8\,\mathrm{M}^{-1}\,\mathrm{s}^{-1}$ (equation 90)[173].

$$S_2O_8^{2-} + e^- \longrightarrow SO_4^{2-} + SO_4^{\cdot-} \qquad (90)$$

2. Reaction of $SO_4^{\cdot-}$

$SO_4^{\cdot-}$ is generated by photolysis and radiolysis of persulphates as well as by the redox reaction of peroxydisulphate with metallic ions; e.g. equation (91). Photolysis is more suitable for the study of the reactions of $SO_4^{\cdot-}$ since no complications are caused by the metallic ions.

The anion radical $SO_4^{\cdot -}$ easily abstracts hydrogen atoms from organic and inorganic substrates, generating various radicals, such as $CO_2^{\cdot -}$ from formate, $CO_3^{\cdot -}$ from bicarbonate, $HO_2^{\cdot}$ from hydrogen peroxide and $NH_2O^{\cdot}$ from hydroxylamine[172].

Double-bonded compounds, e.g. fumarate ion, react with $SO_4^{\cdot -}$ to give an addition radical (ESR technique) $^{-}O_2CCH(OSO_3^{-})\dot{C}HCO_2^{-}$ [171]. The anion of nitromethane reacts with $SO_4^{\cdot -}$ to give an addition compound and a dimeric derivative. These reactions are used for trapping $SO_4^{\cdot -}$ (equations 92a–c)[171].

$$S_2O_8^{2-} \xrightarrow{h\nu} 2SO_4^{\cdot -} \tag{91a}$$

$$S_2O_8^{2-} + Fe^{2+} \longrightarrow Fe^{3+} + SO_4^{\cdot -} + SO_4^{2-} \quad \text{(Ref. 174)} \tag{91b}$$

$$SO_4^{\cdot -} + CH_2{=}NO_2^{-} \longrightarrow {}^{-}O_3SOCH_2\dot{N}O_2^{-} \tag{92a}$$

$$SO_4^{\cdot -} + CH_2{=}NO_2^{-} \longrightarrow {}^{\cdot}CH_2NO_2 + SO_4^{2-} \tag{92b}$$

$$^{\cdot}CH_2NO_2 + CH_2{=}NO_2^{-} \longrightarrow O_2NCH_2CH_2NO_2^{\cdot -} \tag{92c}$$

Carboxylate ions, e.g. $CH_3CO_2^{-}$ and $^{-}O_2CCH_2CH_2CO_2^{-}$, are decarboxylated, giving alkyl radicals (equation 92d)[172]. In this case, $SO_4^{\cdot -}$ acts as an electron abstractor rather than as a hydrogen atom abstractor.

$$CH_3CO_2^{-} + SO_4^{\cdot -} \longrightarrow CH_3^{\cdot} + CO_2 + SO_4^{2-} \tag{92d}$$

Since there are many studies on radiolytic formation of $SO_4^{\cdot -}$, the properties, reactivities and reaction schemes of $SO_4^{\cdot -}$ will be discussed in Section IX.G.

B. Phosphorus-containing Peroxides

1. *Peroxydiphosphate*

Irradiation of aqueous peroxydiphosphate by 254 nm light generates the phosphate radical ion[175], which can behave as both acid and base as shown in equations (93a) and (93b)[176a].

$$P_2O_8^{4-} \xrightarrow{h\nu} 2PO_4^{2\cdot -} \tag{93a}$$

$$H_2PO_4^{\cdot} \underset{pK_a = 5.7}{\overset{-H^+}{\rightleftharpoons}} HPO_4^{\cdot -} \underset{pK_a = 8.9}{\overset{-H^+}{\rightleftharpoons}} PO_4^{2\cdot -} \tag{93b}$$

As in the case of $SO_4^{\cdot -}$, phosphate radical can add to fumarate or maleate anions forming $^{-}O_2CCH(OPO_3^{-})\dot{C}HCO_2^{-}$, and also to the anion of the *aci* form of nitromethane ($CH_2{=}NO_2^{-}$) giving $^{2-}O_3POCH_2\dot{N}O_2^{-}$. The products and reactivities of phosphate radicals are analogous to those of $SO_4^{\cdot -}$; e.g. hydrogen atom abstraction to hydroxylalkyl radicals from aliphatic alcohols, $CO_3^{\cdot -}$ from bicarbonate ion and $PO_4^{2\cdot -}$ from HPO_3^{2-} [176b].

On the other hand, aliphatic carboxylic acids react with $PO_4^{2\cdot -}$ to afford α-carbon radicals by hydrogen atom abstraction (equation 93c), whereas $SO_4^{\cdot -}$ gives decarboxylation products via one-electron oxidation. The reactivity of phosphate radicals

$$CH_3CO_2^- + PO_4^{2\cdot-} \longrightarrow {}^{\cdot}CH_2CO_2^- + HPO_4^{2-} \quad (93c)$$

varies with their extent of protonation. In general, the electrophilic reactivity decreases in the order: $H_2PO_4^{\cdot} > HPO_4^{\cdot-} > PO_4^{2\cdot-}$ and the reactivity of $H_2PO_4^{\cdot}$ is comparable to that of $SO_4^{\cdot-}$. This will be discussed in Section IX.

2. *Peroxyphosphates and peroxyphosphonates*

Photochemical Cu^+-catalysed acyloxylation of olefins is possible by *t*-butyl peroxyphosphate **(26)** and *t*-butyl peroxyphosphonate **(27)** in analogy to the acyloxylation with percarboxylic esters (Section VII). For example, cyclohexene is acyloxylated in the presence of CuBr to give 50–70 % yield of phosphates (equation 94a) and phosphonates (equation 94b)[177].

$$(RO)_2P(=O)OOBu\text{-}t\ \textbf{(26)} + \text{cyclohexene} \xrightarrow[\text{in PhH}]{\text{CuBr, }h\nu} (RO)_2P(=O)\text{—O—(cyclohex-2-enyl)} + t\text{-BuOH} \quad (94a)$$

$$(RO)R^1P(=O)OOBu\text{-}t\ \textbf{(27)} + \text{cyclohexene} \xrightarrow[\text{in PhH}]{\text{CuBr, }h\nu} (RO)R^1\text{—}P(=O)O\text{—(cyclohex-2-enyl)} + t\text{-BuOH} \quad (94b)$$

R = Et, Pr, *i*-Pr, *t*-Bu; R^1 = Me, Et

This type of phosphorylation does not occur thermally even in the presence of Cu^+. The reaction is accelerated by the triplet sensitizer, e.g. benzophenone or acetone, and is retarded by radical inhibitors, but is not affected by a singlet sensitizer, e.g. Eosin-Y. Therefore, the peroxide is excited to the triplet state and gives acyloxy radicals which may be stabilized by complexing with Cu^+ and react at allylic positions of olefins.

3. *Bisdiphenylphosphinic peroxide*

The decomposition of bisdiphenylphosphinic peroxide to its rearrangement product, an unsymmetrical anhydride, is much accelerated by UV (260 nm) light, i.e. the half-life of the reactant is shortened from 10 h to 7 min by UV irradiation (equation 95)[178].

$$Ph_2P(=O)\text{—O—O—}P(=O)Ph_2 \xrightarrow{h\nu} \left[Ph_2P\dot{\ }(O)O \quad \cdot P(Ph)_2(O)O \right] \textbf{(28)} \longrightarrow Ph_2P(=O)\text{—O—}P(=O)(Ph)\text{—OPh} \quad (95)$$

In spite of the identical product, the mechanism of photolysis is different from that of thermolysis. The photolysis of the oxo-^{18}O-labelled peroxide gives the completely scrambled anhydride, while the thermolysis gives the oxo-^{18}O-labelled anhydride. Hence photolysis seems to occur by way of a free-radical intermediate (**28**) as shown in equation (95), but the thermolysis may proceed via a concerted process or via an intimate radical-pair intermediate which cannot scramble[179].

C. Aromatic Sulphoxyperoxides

m-Nitrobenzenesulphonyl peroxide **29** decomposes in aromatic compounds, sulphoxylating arenes, but the yields from aromatic compounds bearing electron-attracting groups are generally poor. Irradiation accelerates the reaction. For example, the photochemical and thermal reactions in nitrobenzene give the products shown in equation (96) with yields in mol per mol peroxide. The isomer distribution and yields shown in equation (96) differ considerably in the dark and photochemical reactions; e.g. the *ortho*:*meta*:*para* composition of sulphonate **30** is 21:58:21 for the photolysis and 0:100:0 for the dark reaction[180].

$$(m\text{-}O_2NC_6H_4SO_3)_2 \ (\mathbf{29}) \xrightarrow[h\nu]{C_6H_5X} m\text{-}O_2NC_6H_4SO_3H + m\text{-}O_2NC_6H_4SO_3C_6H_5 + m\text{-}O_2NC_6H_4SO_3C_6H_4X \ (\mathbf{30}) \quad (96)$$

X = NO_2			
$h\nu$	1.5	0.28	0.047
Δ	0.9	0	0.35

On the other hand, the reaction of **29** with benzonitrile (X = CN) gives a similar isomer distribution for **30**; i.e., *ortho*:*meta*:*para* = 43:35:22 for the photolysis and 39:42:19 for the dark reaction. These facts are explicable by assuming an additional *ipso* attack by the sulphoxy radical $ArSO_3^{\bullet}$ on nitrobenzene, leading to elimination of NO_2 (equation 97)[180].

$$ArSO_3^{\bullet} + C_6H_5NO_2 \longrightarrow \begin{cases} ArSO_3(H)\dot{C}_6H_5NO_2 \xrightarrow{ArSO_3^{\bullet}} ArSO_3H + ArSO_3C_6H_4NO_2 \\ ArSO_3(O_2N)\dot{C}_6H_5 \longrightarrow ArSO_3Ph + NO_2 \end{cases} \quad (97)$$

D. Silicon- and Boron-containing Peroxides

Peroxides containing Si or B can also be photolysed by O—O bond fission to give oxygen radicals (equation 98 and 99). These react further by hydrogen atom abstraction and addition to C—C double bonds as observed in the case of other peroxides.

$$R_3SiOOX \xrightarrow{h\nu} R_3SiO^{\bullet} + {}^{\bullet}OX \quad \text{(Ref. 181)} \quad (98)$$

R = Me, Ph; X = H, *t*-Bu

$$RCMe_2OOB(OBu)_2 \xrightarrow{h\nu} RCMe_2O^{\bullet} + {}^{\bullet}OB(OBu)_2 \quad \text{(Ref. 182)} \quad (99)$$

R = Me, Ph

The hydrogen atom abstraction occurs not only from the solvent, but also from peroxide itself via both intermolecular and intramolecular processes; e.g. (equation 100)[181c]. Here $\dot{R}(-H)$ means $\dot{C}H_2$ or $\dot{C}HCH_3$ for R = Me or Et, respectively. These radicals have been identified by ESR spectra. The ESR study also shows that the photolysis of bis(trialkylsilyl) peroxide $R_3SiOOSiR_3$ affords alkyl radical $R^{\bullet}$ by C—Si fission together with the above radicals as observed with di-*t*-butyl peroxide[181c].

$$R_3SiO^{\bullet} \xrightarrow{XOOSiR_3} XOOSiR_2\dot{R}(-H) \qquad R = Me, Et \qquad (100a)$$

$$R_3SiO^{\bullet} \xrightarrow{1,5\text{-H shift}} R_2Si(OH)-CH_2CH_2CH_2^{\bullet} \qquad R = Pr \qquad (100b)$$

IX. RADIATION CHEMISTRY

A. Introduction

For the radiolysis of organic and inorganic compounds[183] α-, β-, γ- and X-rays, and accelerated ion beams may be used. In general, γ-rays from ^{60}Co and ^{137}Cs and pulse radiolysis techniques are used, which employ a high-energy electron beam impulse followed by photoflash for spectral analysis just as in flash photolysis.

The rays used for the radiolysis have generally much higher (over 10^6 times) energy than light energy. The marked difference between photolysis and radiolysis is that the radiolysis is always initiated by high-energy particles resulting in the ionization of molecules, while photolysis occurs by absorbing light quanta by definite molecules at a definite absorption band without any ionization of molecules except in the case of very short wavelength UV. In radiolysis the energy is absorbed by a number of molecules, leading to nonselective ionization and excitation of a number of molecules along the track of charged particles. The photochemical reaction prohibits the direct transition to a triplet state from a singlet ground state, but the radiochemical reaction permits direct transition; e.g. *cis* olefin can be transformed directly to *trans* isomer via a triplet state.

In radiolysis the *G* value is the number of molecules reacted or produced per 100 eV of absorbed energy. For example, the radiolysis of water at neutral pH affords hydrated electron (e_{aq}^-), hydroxyl radical, proton, hydrogen atom and hydrogen peroxide with *G* values shown in parentheses[183c]:

$$H_2O \rightsquigarrow e_{aq}^-\ (2.7),\quad HO^{\bullet}\ (2.7),\quad H^+\ (2.7),\quad H^{\bullet}\ (0.6),\quad H_2O_2\ (0.7)$$

In a N_2O-saturated aqueous solution, e_{aq}^- reacts with N_2O to form $HO^{\bullet}$ radical (equation 101). Hence N_2O is often used for scavenging electrons in order to simplify the reaction system. Since $HO^{\bullet}$ is about 10 times more reactive than $H^{\bullet}$[184a], reaction in the presence of N_2O occurs mainly by way of $HO^{\bullet}$ [184b].

$$e_{aq}^- + N_2O + H_2O \longrightarrow N_2 + HO^- + HO^{\bullet} \qquad (101)$$

B. Radiolysis of Hydrogen Peroxide

The radiolysis of aqueous hydrogen peroxide is not initiated by the direct radiolysis of H_2O_2 ($H_2O_2^+ \rightarrow HO_2^{\cdot} + H^+$), but by radiolysis of water, generating e_{aq}^-, $H^{\cdot}$ and $HO^{\cdot}$. These active species then react with H_2O_2 (equation 102)[185].

$$e_{aq}^- + H_2O_2 \longrightarrow HO^- + HO^{\cdot} \quad (102a)$$

$$H^{\cdot} + H_2O_2 \longrightarrow H_2O + HO^{\cdot} \quad (102b)$$

$$HO^{\cdot} + H_2O_2 \longrightarrow H_2O + HO_2^{\cdot} \quad (102c)$$

These radicals are the same as those observed in the photolysis of H_2O_2. Here they are formed simultaneously, so that their concentration is higher than in photolysis of the same solution. Hydroxyl radical initiates the decomposition of H_2O_2 according to the Haber–Weiss chain mechanism[186] (cf. equations 14b and c).

Hydroperoxy radical $HO_2^{\cdot}$, which is less reactive than e_{aq}^-, $H^{\cdot}$ and $HO^{\cdot}$, reacts further as shown below, with the rate constants given in parentheses at 26–28°C. The reaction between neutral and anionic species is the most rapid one[185]:

$$HOO^{\cdot} + HOO^{\cdot} \longrightarrow H_2O_2 + O_2 \quad (7.6 \times 10^5\ \text{M}^{-1}\,\text{s}^{-1})$$

$$HOO^{\cdot} + O_2^{\overline{\cdot}} \xrightarrow{H^+} H_2O_2 + O_2 \quad (8.5 \times 10^7\ \text{M}^{-1}\,\text{s}^{-1})$$

$$O_2^{\overline{\cdot}} + O_2^{\overline{\cdot}} \xrightarrow{2H^+} H_2O_2 + O_2 \quad (\leq 100\ \text{M}^{-1}\,\text{s}^{-1})$$

The radiolysis of organic compounds in the presence of H_2O_2 is generally initiated through an attack by the $HO^{\cdot}$ radical (equations 102a and b), hence the products in radiolysis resemble those of photolysis. The radiolysis of an aqueous H_2O_2 solution of isopropanol is initiated by hydrogen atom abstraction with $HO^{\cdot}$, yielding acetone with a G value of 36–70[187]. The yield ($G = 0.30$) in the radiolysis of aqueous benzene under argon atmosphere to form phenol is increased by the presence of H_2O_2 ($G = 0.65$). The preparation of phenols by radiolysis of aqueous benzene alone or in the presence of metallic ions has been studied by several workers[188]. Since water can generate $HO^{\cdot}$ during radiolysis, the source of $HO^{\cdot}$ in aqueous saturated benzene oxidation was examined by using $H_2{}^{18}O_2$. The labelled ratio of phenol, $Ph^{18}OH/Ph^{16}OH$, was 40/60, which means that 40 % of phenol is formed from H_2O_2[189]. The increase of yield in the presence of H_2O_2 is ascribed to the increase of $HO^{\cdot}$ concentration owing to the reaction of H_2O_2 with e_{aq}^- (equation 102a).

Radiolysis of cysteine in the presence of H_2O_2 gives cystine ($G = 3.0$–3.4) (equation 103a)[190], and radiolysis of tryptophan gives alanine and glycine (equation 103b)[191]. The yield of the decomposition of tryptophan under N_2 ($G = 0.23$) is lower than the yield ($G = 1.8$) under O_2[190].

$$HSCH_2CH(NH_2)CO_2H \xrightarrow[H_2O_2]{\gamma\text{-ray}} HO_2C-CH(NH_2)-CH_2SSCH_2-CH(NH_2)CO_2H \quad (103a)$$

$$\text{(indol-3-yl)}-CH_2CH(NH_2)CO_2H \xrightarrow[H_2O_2]{\gamma\text{-ray}} CH_3CH(NH_2)CO_2H,\ NH_2CH_2CO_2H \quad (103b)$$

C. Radiolysis of Alkyl Hydroperoxides

In general, the radiolysis of alkyl hydroperoxides gives the same products as photolysis. The primary process may be the fission of O—O bond of the OOH group, but the products are more various and an electron (e_{aq}^-) participates in the primary process.

Radiolysis of neat *t*-butyl hydroperoxide gives mainly O_2 and small amounts of hydrogen, methane, ethane, ethylene, isobutylene and carbon monoxide as gaseous products along with liquid products containing water, *t*-butanol, di-*t*-butyl peroxide, acetone and a trace amount of *t*-butyl methyl ether and *t*-butyl methyl peroxide[192]. The nature and yields of these products resemble those in the photolysis, suggesting analogous initiation of the reaction by O—O fission of the excited OOH group (equation 104a). (*G* value for consumption of *t*-BuO_2H is 33.7 at 20°C)[192b].

$$t\text{-BuOOH}^* \longrightarrow t\text{-BuO}^\cdot + {}^\cdot\text{OH} \quad (104a)$$

$$t\text{-BuOOH} + {}^\cdot\text{OH} \longrightarrow t\text{-BuOO}^\cdot + \text{H}_2\text{O} \quad (104b)$$

Here, the primary processes should involve ionic species such as those in H_2O radiolysis (cf. Section IX.A). However, the yields are not influenced by the presence of N_2O, an efficient scavenger for electrons, and also the usual colour of electron-scavenged peroxide is not observed (Section IX.D). Therefore, the ionic species produced must rapidly undergo neutralization, presumably within the spurs, e.g. cationic hydroperoxide is neutralized by the hydroperoxide and electrons (equation 105) and the hydroperoxide reacting with the electrons rapidly expels hydroxide ion which is also involved in the neutralization (equation 106)[192b].

$$t\text{-BuOOH} \longrightarrow t\text{-BuOOH}^{+\cdot} + e^- \quad (105a)$$

$$t\text{-BuOOH}^{+\cdot} + t\text{-BuOOH} \longrightarrow t\text{-BuOOH}_2^+ + t\text{-BuO}_2^\cdot \quad (105b)$$

$$t\text{-BuOOH}_2^+ + e^- \longrightarrow t\text{-BuO}^\cdot + \text{H}_2\text{O} \quad (105c)$$

$$t\text{-BuOOH} + e^- \longrightarrow t\text{-BuO}^\cdot + \text{OH}^- \quad (106a)$$

$$t\text{-BuOOH}^{+\cdot} + \text{OH}^- \longrightarrow t\text{-BuO}_2^\cdot + \text{H}_2\text{O} \quad 106b)$$

$$t\text{-BuOOH}_2^+ + \text{OH}^- \longrightarrow t\text{-BuOOH} + \text{H}_2\text{O} \quad (106c)$$

The radicals thus formed react further according to the scheme shown in equation (107), which is well-known for photolysis or thermolysis.

$$2t\text{-BuOO}^\cdot \longrightarrow 2t\text{-BuO}^\cdot + \text{O}_2 \text{ (or } t\text{-Bu}_2\text{O}_2 + \text{O}_2) \quad (107a)$$

$$t\text{-BuO}^\cdot + t\text{-BuOOH} \longrightarrow t\text{-BuOH} + t\text{-BuO}_2^\cdot \quad (107b)$$

$$t\text{-BuO}^\cdot \longrightarrow \text{CH}_3\text{COCH}_3 + \text{CH}_3^\cdot \text{ etc.} \quad (107c)$$

Aliphatic and alicyclic hydroperoxides, e.g. *t*-butyl, cumyl, α-phenethyl, cyclohexyl and cyclopentyl hydroperoxides are radiolysed (^{60}Co γ-ray) in organic solvents to the corresponding ketones and alcohols[193,194].

Hydroperoxides bearing higher alkyl chains such as $(CH_3)_3C-CH_2C(CH_3)_2OOH$ and n-$C_5H_{11}CH(CH_3)OOH$ in hydrocarbon solvents, e.g. octane, are radiolysed to give dialkyl peroxides and alcohols, but not carbonyl compounds[195]. The reaction does not involve a chain mechanism. These facts are probably due to the stable nature of higher aliphatic peroxides and to the facile hydrogen abstraction by alkoxyl groups in the presence of a high concentration of hydrogen source such as the aliphatic chain. This radiolysis has a low energy of activation ($\sim 0.7\,kcal\,mol^{-1}$)[195].

The radiolysis at low peroxide concentrations at low temperatures may induce a branched-chain reaction which results in a very rapid rise of radical concentration and radiation yield. For example, radiolysis of cumyl hydroperoxide in cumene at $-170°C$ at low concentration (<10 mol %) gives rise to a branched-chain reaction via energy transfer from excited solvent to peroxide, and it is assumed that the O—H bond of the hydroperoxide is cleaved (equation 108)[196].

$$RO_2H^{\cdot} \longrightarrow RO_2^{\cdot} + H^{\cdot} \qquad (108)$$

D. Radiolysis of Dialkyl Peroxides

Radiolysis (1 MeV electrons at 80 K) of di-t-butyl peroxide affords the products shown in equation (109) with the G values given[197].

$$t\text{-BuOOBu-}t \longrightarrow t\text{-BuOH},\ Me_2CO,\ Me_2\overset{O}{C-CH_2},\ RCH_2^{\cdot} \qquad (109)$$

$$G = \quad 10 \quad 1.5 \quad 4 \quad 2\text{–}3$$

Radiolysis (γ-ray) of di-t-butyl peroxide in rigid matrices of an amine, ether or hydrocarbon at $-196°C$ gives a coloured peroxide anion with λ_{max} at ca. 560 nm by scavenging electrons delivered from the matrix molecules[198]. If the matrix is alcohol, it supplies an α-hydrogen atom to convert the peroxide anion to the corresponding alcohol; e.g. equations (110) and (111)[198].

$$t\text{-Bu}_2O_2 \xrightarrow{e^-} t\text{-BuOOBu-}t^- \xrightarrow{CH_3OH} t\text{-BuO}^{\cdot} + t\text{-BuOH} + CH_3O^- \qquad (110)$$

$$t\text{-BuO}^{\cdot} + CH_3OH \longrightarrow t\text{-BuOH} + {}^{\cdot}CH_2OH \qquad (111)$$

The positively charged matrix molecule M^+, which is produced by delivering an electron on radiolysis, may transfer its positive charge to the peroxide, if M^+ is mobile in the matrix. Thus the peroxide may be both positively and negatively charged on radiolysis. On irradiation by visible light in this system, the peroxide anion emits an electron, which induces reaction (112).

$$e^- + t\text{-BuOOBu-}t^+ \longrightarrow t\text{-BuO}^{\cdot} + t\text{-BuO}^{\cdot *} \qquad (112a)$$

$$t\text{-BuO}^{\cdot *} \longrightarrow CH_3^{\cdot} + CH_3COCH_3 \qquad (112b)$$

An ESR study of the radiolysis of $RMe_2COOCMe_2R$ (R = Ph, Me, Et) in frozen toluene or polystyrene identified the ion and radical pairs[199].

E. Radiolysis of Diacyl Peroxides

γ-Ray radiolysis of some diacyl peroxides $(RCO_2)_2$ [R = Ph, Me, $Me(CH_2)_{10}$, $HO_2C(CH_2)_2$, $HO_2C(CH_2)_3$, $MeO_2C(CH_2)_2$] in the neat state or in aromatic hydrocarbons at −196°C suggests an ionic mechanism[200]. The radiolysis of benzoyl peroxide at −160 to −170°C gives primarily CO_2 and phenyl radical which then couples to biphenyl[201]. Mixed crystals of benzoyl peroxide with deuterated dibenzoyl disulphide give on X-ray irradiation at 4.2 K primarily a radical pair $(PhCO_2^{\cdot}\,{}^{\cdot}O_2CPh)$ in contrast to the radical pair $(Ph^{\cdot}\,{}^{\cdot}Ph)$ observed in UV photolysis at 4.2 K[202].

The *G* values of diaroyl peroxide decomposition in dilute solutions are relatively high (often over 40), probably because of the energy transfer from the excited solvent benzene to the peroxide, or owing to a chain mechanism in solvent cyclohexane[192b].

F. Radiolysis of Peroxycarboxylic Acids

The radiolysis of aqueous peroxycarboxylic acids proceeds mainly by the reduction of the acid by electrons formed in the radiolysis of water to produce the anion of the parent acid and a hydroxyl radical, via a radical anion intermediate (equation 113).

$$RCO_2{-}OH + e_{aq}^- \xrightarrow{a} [RCO_2{-}OH]^{\dot{-}} \xrightarrow{b} RCO_2^- + HO^{\cdot} \qquad (113)$$

The pulse radiolysis of aqueous *p*-nitroperbenzoic acid has been studied by means of spectrophotometric and conductivity measurements and the acid found to decompose according to equation (113)[184a,203]. The overall scheme including the rate constants of the respective steps and the acidity constants (pK_a) of the participating acids is shown in Scheme 4[203a].

The scheme shows that the leaving of the $HO^{\cdot}$ radical is the rate-determining step. The constancy of conductivity during the decomposition of the radical anion excludes the possibility of elimination of HO^- ($[RCO_2OH]^{\dot{-}} \nrightarrow RCO_2^{\cdot} + HO^-$).

Figure 11 shows the dependence of the *G* value for the radiolysis on the acidity of

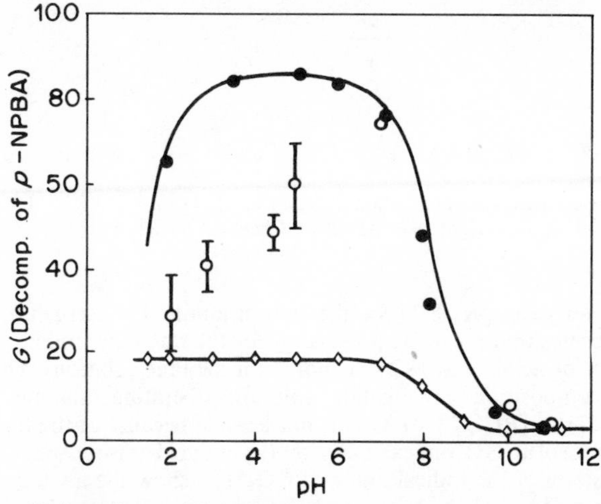

FIGURE 11. The pH dependence of the *G* value for the decomposition of *p*-nitroperoxybenzoic acid (1.6×10^{-4} M) in N_2O saturated aqueous solutions of 8×10^{-3} M (◇) 2-propanol and 8×10^{-3} M (○) and 8×10^{-2} M (●) formic acid at 21 ± 1°C. Dose rate = 3.1×10^{18} eV g^{-1} min^{-1}. Reprinted with permission from E. Heckel, *J. Phys. Chem.*, **80**, 1274 (1976). Copyright 1976 American Chemical Society.

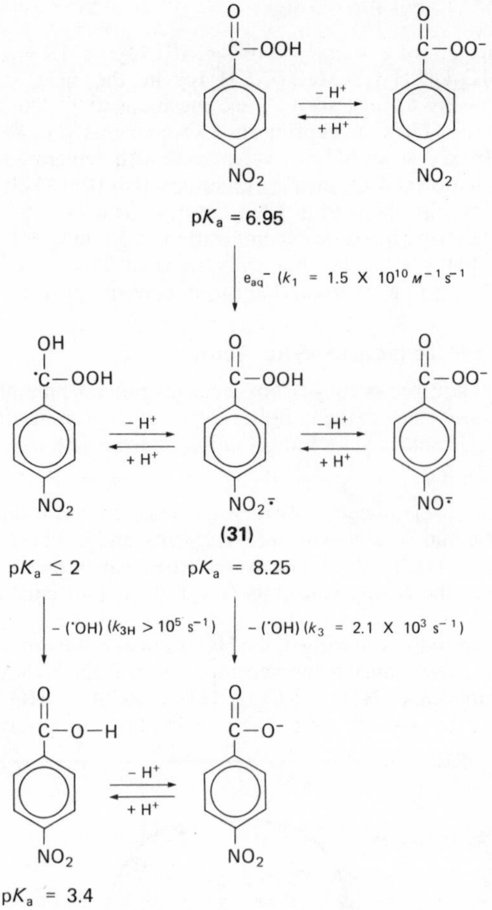

p-$O_2NC_6H_4C(O)OOH$ $\underset{+H^+}{\overset{-H^+}{\rightleftharpoons}}$ p-$O_2NC_6H_4C(O)OO^-$

$pK_a = 6.95$

e_{aq}^- ($k_1 = 1.5 \times 10^{10}\ M^{-1}s^{-1}$)

p-$O_2NC_6H_4\dot{C}(OH)OOH$ $\underset{+H^+}{\overset{-H^+}{\rightleftharpoons}}$ p-$O_2N^{\cdot -}C_6H_4C(O)OOH$ (31) $\underset{+H^+}{\overset{-H^+}{\rightleftharpoons}}$ p-$NO^{\cdot -}C_6H_4C(O)OO^-$

$pK_a \leq 2$ $\quad$ $pK_a = 8.25$

$-(^{\cdot}OH)$ ($k_{3H} > 10^5\ s^{-1}$) $\quad$ $-(^{\cdot}OH)$ ($k_3 = 2.1 \times 10^3\ s^{-1}$)

p-$O_2NC_6H_4C(O){-}O{-}H$ $\underset{+H^+}{\overset{-H^+}{\rightleftharpoons}}$ p-$O_2NC_6H_4C(O){-}O^-$

$pK_a = 3.4$

SCHEME 4. Reprinted with permission from J. Lilie, E. Heckel and R. C. Lamb, *J. Amer. Chem. Soc.*, **96**, 5543 (1974). Copyright 1974 American Chemical Society.

solution, and suggests pK_a 8.25 for the radical anion. The Arrhenius diagram of the radiolysis indicates that the activation energy for the rate-determining step ($HO^{\cdot}$ loss) is 2.7 kcal mol^{-1} for acidic and 18.2 kcal mol^{-1} for alkaline solutions. These facts suggest that the decomposition in alkaline solution liberating anionic atomic oxygen ($[RCO_2{-}O^-]^{\cdot -} \rightarrow [RCO_2^{\cdot}]^{\cdot -} + O^{\cdot -}$) is not feasible because of the high energy of $O^{\cdot -}$.

In the presence of an $HO^{\cdot}$ radical scavenger (formic acid or isopropanol) the G values for the decomposition of the radical anion $[RCO_3H]^{\cdot -}$ show the analogous pH–G profile (Figure 11). In these cases, the radicals $^{\cdot}CO_2^-$ from formate and $Me_2\dot{C}OH$ from isopropanol act as reducing agents for peroxy acid; e.g. equation (114).

$$Me_2\dot{C}OH + RCO_2{-}OH \longrightarrow [RCO_2{-}OH]^{\cdot -} + Me_2C{=}O + H^+ \qquad (114)$$

This carbon radical reduction (equation 114) is more efficient than electron reduction (equation 113) with the G value twice as great. Equation (114) combined with equation (113) and hydrogen abstraction from isopropanol constitutes a chain process.

As shown in Figure 11, the G value in the presence of formic acid decreases at pH < 3. This is ascribed to the lower reducing power of the radical $^{\bullet}CO_2H$ than its anion $^{\bullet}CO_2^-$.

G. Radiolysis of Peroxydisulphate and Peroxydiphosphate

Peroxydisulphate $S_2O_8^{2-}$ is reduced by electrons derived from radiolysis of water at a diffusion-controlled rate ($k = 1.1 \times 10^{10}\,M^{-1}\,s^{-1}$ at pH 5) to form sulphate radical anion (equation 115)[204–207].

$$S_2O_8^{2-} + e_{aq}^- \text{ (or } e^-\text{)} \longrightarrow SO_4^{\bar{\bullet}} + SO_4^{2-} \qquad (115)$$

The sulphate radical $SO_4^{\bar{\bullet}}$ can oxidize rapidly both inorganic and organic substances. For example, chloride ion is oxidized to $Cl_2^{\bar{\bullet}}$ at a very rapid rate ($k = 1.3 \times 10^8\,M^{-1}\,s^{-1}$) and a G value higher than that for the formation of $SO_4^{\bar{\bullet}}$ by excited Cl^{-*} ion.

The net reaction of a carboxylic acid with radiolytic $SO_4^{\bar{\bullet}}$ is decarboxylation giving carbon radicals (equation 116) (R = alkyl or aryl). The reaction (116a) is an electron transfer from carboxylate to sulphate radical[208].

$$RCO_2^- + SO_4^{\bar{\bullet}} \longrightarrow RCO_2^{\bullet} + SO_4^{2-} \qquad (116a)$$

$$RCO_2^{\bullet} \longrightarrow R^{\bullet} + CO_2 \qquad (116b)$$

The absolute rates for the reaction of substituted arenes with $SO_4^{\bar{\bullet}}$ generating the cation radical **32** have been determined by pulse radiolysis (equation 117)[205].

$$X\text{-}C_6H_5 + SO_4^{\bar{\bullet}} \xrightarrow[-SO_4^{2-}]{} [X\text{-}C_6H_5]^{+\bullet}\ \textbf{(32)} \xrightarrow{H_2O} X\text{-}C_6H_4\text{-}OH \qquad (117)$$

High rates are observed; e.g. $k = 5 \times 10^9\,M^{-1}\,s^{-1}$ for anisole and $3 \times 10^9\,M^{-1}\,s^{-1}$ for benzene at pH 7, which are comparable to those of $HO^{\bullet}$ and $H^{\bullet}$ ($k \sim 10^9\,M^{-1}\,s^{-1}$), but with higher selectivity. The Hammett plots with σ for substituted benzenes and benzoic acids (Figure 12) gives a negative ρ value of -2.4; i.e. an electron-releasing group facilitates the electron transfer to $SO_4^{\bar{\bullet}}$, which means that $SO_4^{\bar{\bullet}}$ has an electrophilic nature. The corresponding ρ value for $HO^{\bullet}$ or $H^{\bullet}$ addition is -0.5. The higher selectivity of $SO_4^{\bar{\bullet}}$ is ascribed to an electron-transfer mechanism to form radical cation **32** with no formation of adducts with $SO_4^{\bar{\bullet}}$, unlike $HO^{\bullet}$ and $H^{\bullet}$. Then the radical cation of arene reacts with water to form hydroxycyclohexadienyl radical (equation 118) which is identified by its characteristic UV peak of 310 nm[208].

$$[X\text{-}C_6H_5]^{+\bullet} + H_2O \xrightarrow{-H^+} X\text{-}C_6H_5(H)(OH)^{\bullet} \xrightarrow{-H^{\bullet}} X\text{-}C_6H_4\text{-}OH \qquad (118)$$

An analogous one-electron transfer is observed for the oxidation of cyanamide with $SO_4^{\bar{\bullet}}$ to form $NH_2CN^{\dot{+}}$ [204].

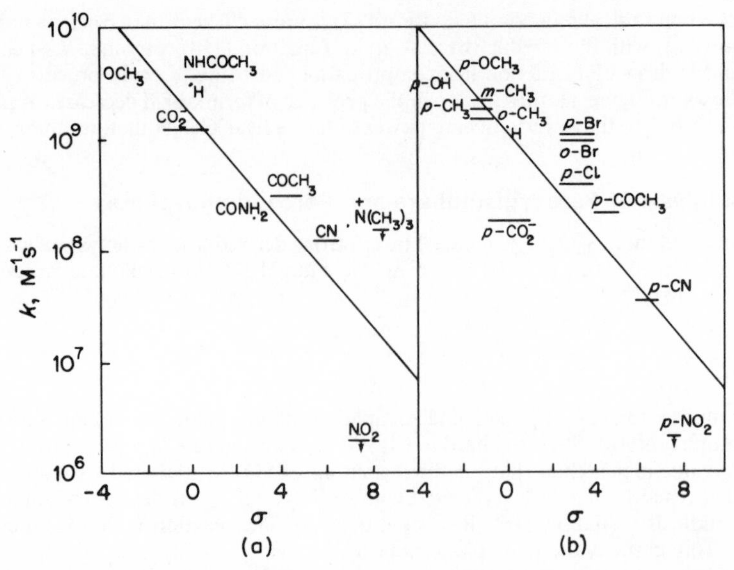

FIGURE 12. A correlation of rate constants for reaction of $SO_4^{\cdot-}$ with aromatic compounds with the substituent constants σ: (A) substituted benzenes; (B) substituted benzoate ions. The values of σ were taken from C. D. Ritchie and W. F. Sager, *Prog. Phys. Org. Chem.*, **2**, 323 (1964). In most cases, except for OH and OCH_3, both σ_{meta} and σ_{para} were used as indicated by the horizontal lines. The arrow for $^+N(CH_3)_3$ indicates that the rate constant should be decreased by the contribution from the reaction of $SO_4^{\cdot-}$ with the methyl groups. The arrow for p-CO_2^- indicates that the rate constant is lowered by the extra negative charge. Reprinted with permission from P. Neta, V. Madhavan, H. Zemel, and R. W. Fessenden, *J. Amer. Chem. Soc.*, **99**, 163 (1977). Copyright 1977 American Chemical Society.

On the other hand, the reactions of peroxydiphosphoric acid $H_4P_2O_8$ with e_{aq}^- have different rate constants based on the different dissociation degrees of the acid, where the constants $K_1 > 40$, $K_2 = 40$, $K_3 = 6.6 \times 10^{-6}$ and $K_4 = 2.1 \times 10^{-8}$ are measured for successive dissociations of the acid[209]. For example, second-order rate constants are: $5.3 \times 10^9\,M^{-1}\,s^{-1}$ for $H_2P_2O_8^{2-}$ at pH 4.23, $1.6 \times 10^9\,M^{-1}\,s^{-1}$ for $HP_2O_8^{3-}$ at pH 6.54 and $1.8 \times 10^8\,M^{-1}\,s^{-1}$ for $P_2O_8^{4-}$ at pH 10.5[207], which are smaller than those of H_2O_2 and $S_2O_8^{2-}$ ($\sim 10^{10}\,M^{-1}\,s^{-1}$).

The dissociation constants for phosphate radicals, formed by radiolysis and measured by the absorbance dependence on pH, are shown in Figure 13 and equation (119)[176a].

$$H_2PO_4^{\cdot} \underset{pK_a\,5.7}{\overset{-H^+}{\rightleftharpoons}} HPO_4^{\cdot-} \underset{pK_a\,8.9}{\overset{-H^+}{\rightleftharpoons}} PO_4^{2\cdot-} \qquad (119)$$

$\lambda_{max}(\varepsilon)$ 520nm (1850) 510nm (1550) 530nm (2150)

The oxidizing power of phosphate radicals is generally lower than that of $SO_4^{\cdot-}$. For example, the rate constants for the oxidation of Cl^- are $2.2 \times 10^6\,M^{-1}\,s^{-1}$ with $H_2PO_4^{\cdot}$ and $<10^4\,M^{-1}\,s^{-1}$ with $HPO_4^{\cdot-}$, while the rate constant with $SO_4^{\cdot-}$ is $3 \times 10^8\,M^{-1}\,s^{-1}$. Thus the order of oxidizing rates is: $SO_4^{\cdot-} > H_2PO_4^{\cdot} > HPO_4^{\cdot-} > PO_4^{\cdot}$.

The reaction rates for some organic compounds with $H_2PO_4^{\cdot}$, $SO_4^{\cdot-}$ and $HO^{\cdot}$ are listed in Table 19, which shows that the oxidative powers of $H_2PO_4^{\cdot}$ and $SO_4^{\cdot-}$ are similar, while

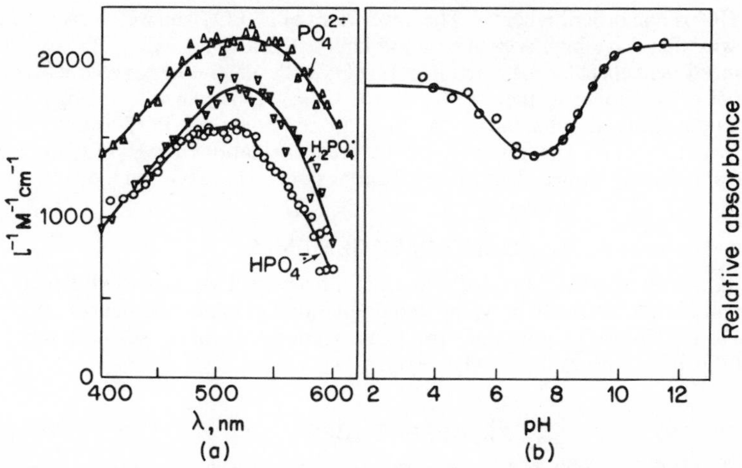

FIGURE 13. Absorption spectra and dissociation constants of the phosphate radicals. (a) Spectra of the three acid–base forms of the phosphate radical. Recorded using aqueous solutions of 2.5×10^{-2} M peroxodiphosphate ions at pH 4 (▽), pH 7 (○), and pH 11 (△). The transient absorption was monitored 2–10 μs after the pulse, after the formation was complete, and in the case of pH 4 it was extrapolated to time zero because some decay was observable. (b) Effect of pH on the absorbance at 540 nm. The curve was calculated from the plateau values using $pK_1 = 5.7$ and $pK_2 = 8.9$. Reprinted with permission from P. Maruthamuthu and P. Neta, *J. Phys. Chem.*, **82**, 710 (1978). Copyright 1978 American Chemical Society.

TABLE 19. A comparison of the rate constants of $H_2PO_4\cdot$ with those of $SO_4^{\cdot-}$ and $HO\cdot$[f]

Compound	$H_2PO_4\cdot$[a]	$SO_4^{\cdot-}$[b]	$HO\cdot$[c]
Methanol	4.1×10^7	1.1×10^7	9×10^8
Ethanol	7.7×10^7	3.4×10^7	1.8×10^9
2-Propanol	1.4×10^8	8.0×10^7	2.2×10^9
2-Methyl-2-propanol	3.9×10^6	8.0×10^5	5×10^8
Formate ion	1.5×10^8	1.7×10^8	3×10^9
Acetic acid	3.4×10^5	9×10^4	2×10^7
Propionic acid	4.2×10^6	(4.6×10^6)[d]	5×10^8
Malonic acid	1.8×10^5	(5.5×10^6)[d]	2×10^7
Succinic acid	1.6×10^6	(7.1×10^6)[d]	1×10^8
Glycine	$\leqslant 10^5$	9×10^6	6×10^6
Alanine	(1.6×10^7)[e]	1.0×10^7	7×10^7
Acetone	3.3×10^5		9×10^7
Fumaric acid	1.5×10^7	(1.6×10^7)[d]	1×10^9
Maleic acid	3.1×10^7		5×10^8

[a] Determined at pH 3.2–4.6.
[b] Determined at pH 7 by J. L. Redpath and R. L. Willson, *Intern. J. Radiat. Biol.*, **27**, 389 (1975), except for ethanol and acetic acid which were measured by L. Dogliotti and E. Hayon, *J. Phys. Chem.*, **71**, 3802 (1967), and except as noted.
[c] Average literature values from the compilation by Farhataziz and A. Ross, National Bureau of Standards, Report No. NSRDS-NBS 59.
[d] Determined in the present work for reaction with the anion.
[e] For reaction of PO_4^{2-} at pH 12.
[f] Reprinted with permission from P. Maruthamuthu and P. Neta, *J. Phys. Chem.*, **81**, 1622 (1977). Copyright 1978 American Chemical Society.

that of $HO^{\cdot}$ is one order higher[210]. The lower rates of $H_2PO_4^{\cdot}$ and $SO_4^{\cdot-}$ seem to be the reason why they have high selectivities and high values of ρ.

The substituent effect for the reaction of $H_2PO_4^{\cdot}$ with substituted benzoic acids shows a ρ value of -1.8, similar to that of $SO_4^{\cdot-}$ ($\rho = -2$)[210]. This suggests analogous radical cation intermediates **(32)** for the oxidation of aromatics with $H_2PO_4^{\cdot}$ as described with $SO_4^{\cdot-}$ (equation 117). The products of the arene oxidations by $H_2PO_4^{\cdot}$ in aqueous solution are phenols formed via nucleophilic attack of H_2O or HO^- on the radical cation.

X. ACKNOWLEDGEMENTS

The authors wish to thank Mr. Morio Inaishi for his helpful assistance in the preparation of the manuscript, especially in typing and preparation of tables and figures. Also we are grateful to Dr. Toyoaki Kimura of Prof. Fueki's laboratory for his advice in reading the draft of the section on radiation chemistry.

XI. REFERENCES

1. (a) O. J. Walker and G. L. E. Wild, *J. Chem. Soc.*, 1132 (1937).
   (b) R. A. Sheldon and J. K. Kochi, *J. Amer. Chem. Soc.*, **92**, 4395 (1970).
   (c) B. C. Gilbert and A. J. Dobbs in *Organic Peroxides*, Vol. 3 (Ed. D. Swern), Wiley–Interscience, New York, 1971, p. 271.
2. L. S. Silbert in *Organic Peroxides*, Vol. 2 (Ed. D. Swern) Wiley–Interscience, New York, 1971, pp. 678, 811.
3. L. M. Toth and H. S. Johnston, *J. Amer. Chem. Soc.*, **91**, 1276 (1969).
4. A. Rieche, *Alkylperoxyde und Ozonide*, Steinkopff, Dresden, 1931.
5. A. J. Everett and G. J. Minkoff, *Trans. Faraday Soc.*, **49**, 410 (1953).
6. W. C. Schumb, C. N. Satterfield and R. L. Wentworth, *Hydrogen Peroxide*, Reinhold, New York, 1955.
7. J. W. Breitenbach and J. Derkosch, *Monatsh. Chem.*, **81**, 530 (1950).
8. J. K. Kochi and P. J. Krusic, *J. Amer. Chem. Soc.*, **91**, 3940 (1969).
9. H. C. Box, E. E. Budzinski and H. G. Freund, *J. Amer. Chem. Soc.*, **92**, 5305 (1970).
10. R. A. Sheldon and J. K. Kochi, *J. Amer. Chem. Soc.*, **92**, 5175 (1970).
11. T. Nakata, K. Tokumaru and O. Simamura, *Tetrahedron Letters*, 3303 (1967).
12. C. Walling and M. J. Gibian, *J. Amer. Chem. Soc.*, **87**, 3413 (1965).
13. (a) M. H. J. Wijnen, *J. Chem. Phys.*, **27**, 710 (1957).
    (b) M. H. J. Wijnen, *J. Chem. Phys.*, **28**, 271 (1958).
    (c) M. H. J. Wijnen, *J. Amer. Chem. Soc.*, **82**, 1847 (1960).
14. (a) G. F. Sheats and W. A. Noyes, Jr., *J. Amer. Chem. Soc.*, **77**, 1421 (1955).
    (b) J. E. Jolley, *J. Amer. Chem. Soc.*, **79**, 1537 (1957).
    (c) F. P. Lossing, *Can. J. Chem.*, **35**, 305 (1957).
15. G. R. McMillan, *J. Amer. Chem. Soc.*, **83**, 3018 (1961).
16. G. R. McMillan, *J. Amer. Chem. Soc.*, **84**, 2514 (1962).
17. (a) R. G. W. Norrish and M. H. Searby, *Proc. Roy. Soc. (London)*, **A237**, 464 (1956).
    (b) I. Chodak, D. Sakos and V. Mihalov, *J. Chem. Soc., Perkin Trans. 2*, 1457 (1980).
18. E. Havinga, R. O. de Jongh and W. Dorst, *Rec. Trav. Chim.*, **75**, 378 (1956).
19. R. L. Letsinger, O. B. Ramsay and J. H. McCain, *J. Amer. Chem. Soc.*, **87**, 2945 (1965).
20. H. E. Zimmerman and V. R. Sandel, *J. Amer. Chem. Soc.*, **85**, 915 (1963).
21. D. A. de Bie and E. Havinga, *Tetrahedron*, **21**, 2359 (1965).
22. Y. Ogata and E. Hayashi, *Bull. Chem. Soc. Japan*, **50**, 323 (1977).
23. Y. Ogata, E. Hayashi and H. Kato, *Bull. Chem. Soc. Japan*, **51**, 3657 (1978).
24. Y. Ogata, K. Tomizawa, K. Furuta and H. Kato, *J. Chem. Soc., Perkin Trans. 2*, 110(1981); Y. Ogata, K. Tomizawa and K. Furuta, *Memoir Fac. Eng. Nagoya Univ.*, **33**, No. 2, 209 (1981).
25. Y. Ogata and K. Tomizawa, *J. Org. Chem.*, **43**, 261 (1978).
26. Y. Ogata, K. Takagi and E. Hayashi, *J. Org. Chem.*, **44**, 856 (1979).

27. Y. Ogata, K. Tomizawa and Y. Yamashita, *J. Chem. Soc. Perkin Trans. 2*, 616 (1980).
28. L. J. Stief and V. J. DeCarlo, *J. Chem. Phys.*, **50**, 1234 (1969).
29. R. C. Smith and S. J. Wyard, *Nature*, **191**, 896 (1961).
30. M. A. Pospelova, V. G. Belokon and V. S. Gurman, *Dokl. Akad. Nauk SSSR*, **212**, 414 (1973); *Chem. Abstr.*, **80**, 8965r (1974).
31. W. H. Koppenol, J. Butler and J. W. van Leeuwen, *Photochem. Photobiol.*, **28**, 655 (1978); *Chem. Abstr.*, **90**, 175279h (1979).
32. (a) F. Haber and J. Weiss, *Naturwiss.*, **20**, 948 (1932).
(b) F. Haber and J. Weiss, *Proc. Roy. Soc.* (*London*), **A147**, 332 (1934).
(c) Y. Ogata, *Oxidation and Reduction of Organic Compounds*, Nankodo, Tokyo, 1963, p. 225.
33. V. R. Landi and L. J. Heidt, *J. Phys. Chem.*, **73**, 2361 (1969).
34. R. A. Gorse and D. H. Volman, *J. Photochem.*, **1**, 1 (1972).
35. J. F. Meagher and J. Heicklen, *J. Photochem.*, **3**, 455 (1975).
36. G. Buxton and W. K. Wilmarth, *J. Phys. Chem.*, **67**, 2835 (1963).
37. R. Livingston and H. Zeldes, *J. Chem. Phys.*, **44**, 1245 (1966).
38. R. Livingston, J. K. Dohrmann and H. Zeldes, *J. Chem. Phys.*, **53**, 2448 (1970).
39. Y. Odaira, S. Morimoto, H. Yamamoto and S. Tsutsumi, *Kogyo Kagaku Zasshi*, **64**, 457 (1961).
40. B. Y. Ladygin, *Khim. Vys. Energy*, **5**, 170 (1971); *Chem. Abstr.*, **75**, 135790b (1971).
41. K. Takeda, H. Yoshida, K. Hayashi and S. Okamura, *Bull. Chem. Soc. Japan*, **39**, 1632 (1966).
42. S. Paszyc and W. Augustyniak, *Photochem. Photobiol.*, **8**, 235 (1968); *Chem. Abstr.*, **69**, 101727x (1968).
43. Y. Ogata, K. Tomizawa and K. Fujii, *Bull. Chem. Soc. Japan*, **51**, 2628 (1978).
44. V. I. Mal'tsev, B. N. Shelimov and A. A. Petrov, *Zh. Org. Khim.*, **4**, 545 (1968); *Chem. Abstr.*, **69**, 229lu (1968).
45. V. E. Shurygin, L. S. Boguslavskaya, Z. P. Volkova and G. A. Razuvaev, *Zh. Org. Khim.*, **2**, 2007 (1966); *Chem. Abstr.*, **66**, 75494k (1967).
46. Y. Ogata, K. Tomizawa and K. Takagi, *Can. J. Chem.*, **59**, 14 (1981).
47. W. Braun, L. Rajbenbach and F. R. Eirich, *J. Phys. Chem.*, **66**, 1591 (1962).
48. P. Neta, P. Maruthamuthu, P. M. Carton and R. W. Fessenden, *J. Phys. Chem.*, **82**, 1875 (1978).
49. Y. Ogata, K. Tomizawa and K. Adachi, *Memoirs Fac. Eng. Nagoya Univ.*, **33**, No. 1, 58 (1981).
50. V. I. Mal'tsev and A. A. Petrov, *Zh. Org. Khim.*, **3**, 216 (1967); *Chem. Abstr.*, **66**, 85278g (1967).
51. R. Livingston, D. G. Doherty and H. Zeldes, *J. Amer. Chem. Soc.*, **97**, 3198 (1975).
52. D. G. Doherty, R. Livingston and H. Zeldes, *J. Amer. Chem. Soc.*, **98**, 7717 (1976).
53. A. S. Ansari, S. Tahib and R. Ali, *Experientia*, **32**, 573 (1976).
54. P. Cavatora, R. Favilla and A. Mazzini, *Biochem. Biophys. Acta*, **578**, 541 (1979); *Chem. Abstr.*, **91**, 70358a (1979).
55. H. Ochiai, *Agr. Biol. Chem.*, **35**, 622 (1971); *Chem. Abstr.*, **75**, 34132h (1971).
56. N. Jacob, I. Balakrishnan and M. P. Reddy, *J. Phys. Chem.*, **81**, 17 (1977).
57. D. I. Dimitrov, B. K. Kumanova and N. Bisolnakov, *Zh. Prikl. Khim.*, **46**, 1803 (1973); *Chem. Abstr.*, **79**, 126006f (1973).
58. T. Matsuura and K. Omura, *Chem. Commun.*, 127 (1966).
59. K. Omura and T. Matsuura, *Tetrahedron*, **26**, 255 (1970).
60. S. N. Massie, *U.S. Patent*, No. 3600287 (1971); *Chem. Abstr.*, **75**, 110050x (1971).
61. J. J. Zwolenik, *J. Phys. Chem.*, **71**, 2464 (1967).
62. W. J. Maguire and R. C. Pink, *Trans. Faraday Soc.*, **63**, 1097 (1967).
63. S. A. Sojka, C. F. Poranski, Jr. and W. B. Moniz, *J. Amer. Chem. Soc.*, **97**, 5953 (1975).
64. W. B. Moniz, S. A. Sojka, C. F. Poranski, Jr. and D. L. Birkle, *J. Amer. Chem. Soc.*, **100**, 7940 (1978).
65. T. Mill, H. Richardson and F. R. Mayo, *J. Polym. Sci., Polym. Chem. Ed.*, **11**, 2899 (1973).
66. Y. Sawaki and Y. Ogata, *J. Amer. Chem. Soc.*, **98**, 7324 (1976).
67. W. Busch and W. Diez, *Chem. Ber.*, **47**, 3277 (1914); K. H. Pausacker, *J. Chem. Soc.*, 3478 (1950).
68. R. Criegee and G. Lohaus, *Chem. Ber.*, **84**, 219 (1951).
69. T. Tezuka and N. Narita, *J. Amer. Chem. Soc.*, **101**, 7413 (1979).
70. R. O. C. Norman and G. K. Radda, *Proc. Chem. Soc. London*, 138 (1962).
71. N. A. Porter, M. O. Funk, D. Gilmore, R. Isaac and J. Nixon, *J. Amer. Chem. Soc.*, **98**, 6000 (1976).

72. W. H. Richardson, G. Ranney and F. C. Montgomery, *J. Amer. Chem. Soc.*, **96**, 4688 (1974).
73. D. J. Carlsson and D. M. Wiles, *J. Polym. Sci. Polym. Chem. Ed.*, **12**, 2217 (1974).
74. G. Geuskens, D. Baeyens-Volant, G. Delaunois, Q. L. Vinh, W. Piret and C. David, *Eur. Polym. J.*, **14**, 299 (1978).
75. H. C. Ng and J. E. Guillet, *Macromol.*, **11**, 929 (1978).
76. Y. Takezaki, T. Miyazaki and N. Nakahara, *J. Chem. Phys.*, **25**, 536 (1956).
77. L. M. Dorfman and Z. W. Salsburg, *J. Amer. Chem. Soc.*, **73**, 255 (1951).
78. H. Paul, R. D. Small, Jr. and J. C. Scaiano, *J. Amer. Chem. Soc.*, **100**, 4520 (1978).
79. P. J. Krusic and J. K. Kochi, *J. Amer. Chem. Soc.*, **90**, 7155 (1968).
80. (a) Y. Ogata, *Chemistry of Organic Peroxides*, Nankodo, Tokyo, 1971, p. 155, compiles these O—O dissociation energies of peroxides.
(b) K. S. Pitzer, *J. Amer. Chem. Soc.*, **70**, 2140 (1948).
(c) D. Swern and L. S. Silbert, *Anal. Chem.*, **35**, 880 (1963).
81. F. H. Dorer and S. N. Johnson, *J. Phys. Chem.*, **75**, 3651 (1971).
82. C. K. Yip and H. O. Pritchard, *Can. J. Chem.*, **50**, 1531 (1972).
83. S. Weiner and G. S. Hammond, *J. Amer. Chem. Soc.*, **91**, 2182 (1969).
84. P. Svejda and D. H. Volman, *J. Phys. Chem.*, **73**, 4417 (1969).
85. K. U. Ingold and J. R. Morton, *J. Amer. Chem. Soc.*, **86**, 3400 (1964).
86. J. C. W. Chien and C. R. Boss, *J. Amer. Chem. Soc.*, **89**, 571 (1967).
87. H. D. Rehorek, *Z. Chem.*, **19**, 65 (1979); *Chem. Abstr.*, **90**, 203169t (1979).
88. B. B. Adeleke, S. Wong and J. K. S. Wan, *Can. J. Chem.*, **52**, 2901 (1974).
89. R. G. Gasanov, I. I. Kandror and R. K. Freidlina, *Tetrahedron Letters*, 1485 (1975).
90. A. G. Davies, D. Griller and B. P. Roberts, *J. Organometal. Chem.*, **38**, C8 (1972).
91. (a) E. Furimsky, J. A. Howard and J. R. Morton, *J. Amer. Chem. Soc.*, **95**, 6574 (1973).
(b) A. G. Davies, D. Griller and B. P. Roberts, *Angew. Chem.* (*Intern. Ed. Engl.*), **10**, 738 (1971).
92. (a) D. Griller, K. U. Ingold and J. C. Scaiano, *J. Mag. Res.*, **38**, 169 (1980).
(b) K. Y. Choo and P. P. Gaspar, *Taehan Hwahak Hoechi*, **21**, 270 (1977); *Chem. Abstr.*, **87**, 200475s (1977).
93. J. E. Bennett, D. M. Brown and B. Mile, *Trans. Faraday Soc.*, **66**, 386 (1970).
94. (a) M. Flowers, L. Batt and S. W. Benson, *J. Chem. Phys.*, **37**, 2662 (1962).
(b) D. J. Edge and J. K. Kochi, *J. Amer. Chem. Soc.*, **94**, 6485 (1972).
95. P. G. Cookson, A. G. Davies, B. P. Roberts and M. Tse, *J. Chem. Soc., Chem. Commun.*, 937 (1976).
96. *T. Gillbro, Chem. Phys.*, **4**, 476 (1974).
97. A. A. Frimer, A. Havron, D. Leonov, J. Sperling and D. Elad, *J. Amer. Chem. Soc.*, **98**, 6026 (1976).
98. (a) A. Erndt, A. Kostuch and A. Para, *Rocz. Chem.*, **50**, 769 (1976).
(b) Z. Lorberbaum, J. Sperling and D. Elad, *Photochem. Photobiol.*, **24**, 389 (1976); *Chem. Abstr.*, **86**, 51714u (1977).
99. P. L. Hanst and J. G. Galvert, *J. Phys. Chem.*, **63**, 104 (1959).
100. (a) A. Mortlock, *Ph. D. Thesis*, University of London, 1952.
(b) J. Boulton, *Ph.D. Thesis*, University of London, 1956; cited from Reference 24.
101. G. A. Holder, *J. Chem. Soc., Perkin Trans. 2*, 1089 (1972).
102. K. O. Christie and D. Pilipovich, *J. Amer. Chem. Soc.*, **93**, 51 (1971).
103. N. Vanderkooi, Jr. and W. B. Fox, *J. Chem. Phys.*, **47**, 3634 (1967).
104. R. W. Fessenden, *J. Chem. Phys.*, **48**, 3725 (1968).
105. R. R. Smardzewski, R. A. de Marco and W. B. Fox, *J. Chem. Phys.*, **63**, 1083 (1975).
106. W. Adam and J. Sanabia, *J. Amer. Chem. Soc.*, **99**, 2735 (1977).
107. W. Adam and N. Durán, *J. Amer. Chem. Soc.*, **99**, 2729 (1977).
108. P. R. Story, D. D. Denson, C. E. Bishop, B. C. Clark, Jr. and J.-C. Farine, *J. Amer. Chem. Soc.*, **90**, 817 (1968).
109. Y. Ito, T. Matsuura and H. Yokoya, *J. Amer. Chem. Soc.*, **101**, 4010 (1979).
110. J. Kowal and B. Waligóra, *Makromol. Chem.*, **179**, 707 (1978).
111. (a) H. Lind and H. Loeliger, *Tetrahedron Letters*, 2569 (1976).
(b) J. Lerchová, L. Kotulák, J. Rotschová, J. Pilár and J. Pospíšil, *J. Polym. Sci., Polym. Symp.*, **57**, 229 (1976).

112. H. Lind, T. Winkler and H. Loeliger, *J. Polym. Sci., Polym. Symp.*, **57**, 225 (1976).
113. R. V. Kucher, A. A. Turovskii, L. V. Lukyanenko and N. V. Dzumedzei, *Teor. Eksp. Khim.*, **12**, 41 (1976); *Chem. Abstr.*, **85**, 114618t (1976).
114. R. Hiatt in *Organic Peroxides*, Vol. 2 (Ed. D. Swern), Wiley–Interscience, New York, 1971, Chap. 8, p. 799.
115. Y. Ogata, Y. Sawaki, K. Sakanishi and T. Morimoto, *Chemistry of Organic Peroxides*, Nankodo, Tokyo, 1971, Chap. 6, p. 170.
116. S. S. Ivanchev, A. I. Yurzhenko, A. F. Lukovnilov, S. I. Peredereeva and Y. V. Gak, *Dokl. Akad. Nauk, SSSR*, **171**, 894 (1966); *Chem. Abstr.*, **66**, 60584r (1967).
117. S. S. Ivanchev, A. I. Yurzhenko, A. F. Lukovnikov, Y. N. Gak and S. M. Kvasha, *Teor. Eksp. Khim.*, **4**, 780 (1968); *Chem. Abstr.*, **70**, 52977p (1969).
118. (a) A. T. Koritskii, A. V. Zubkov and Y. S. Lebedev, *Khim. Vys. Energ.*, **3**, 387 (1969); *Chem. Abstr.*, **72**, 2840j (1970).
(b) M. Feldhues and H. J. Schäffer, *Tetrahedron Letters*, **22**, 433 (1981).
119. T. Kashiwagi, K. Fujimori, S. Kozuka and S. Oae, *Tetrahedron*, **26**, 3639 (1970).
120. P. D. Bartlett and S. G. Cohen, *J. Amer. Chem. Soc.*, **65**, 543 (1943).
121. G. S. Hammond and L. M. Soffer, *J. Amer. Chem. Soc.*, **72**, 4711 (1950).
122. V. I. Barchuk, A. A. Dubinskii, O. Y. Grinsberg and Y. S. Lebedev, *Chem. Phys. Letters*, **34**, 476 (1975).
123. M. C. R. Symons and M. G. Townsend, *J. Chem. Soc.*, 263 (1959).
124. Y. Sakaguchi, H. Hayashi and S. Nagakura, *Bull. Chem. Soc. Japan*, **53**, 39 (1980).
125. H. Sakuragi, M. Sakuragi, T. Mishima, S. Watanabe, M. Hasegawa and K. Tokumaru, *Chem. Letters*, 231 (1975).
126. Y. Tanimoto, H. Hayashi, S. Nagakura, H. Sakuragi and K. Tokumaru, *Chem. Phys. Letters*, **41**, 267 (1976).
127. J. Saltiel and H. C. Curtis, *J. Amer. Chem. Soc.*, **93**, 2056 (1971).
128. J. D. Bradley and A. P. Roth, *Tetrahedron Letters*, 3907 (1971).
129. K. Tokumaru, A. Ohshima, T. Nakata, H. Sakuragi and T. Mishima, *Chem. Letters*, 571 (1974).
130. M. Kobayashi, H. Minato and Y. Ogi, *Bull. Chem. Soc. Japan*, **42**, 2737 (1969).
131. P. K. SenGupta and J. C. Bevington, *Polymer*, **14**, 527 (1973).
132. J. A. den Hollander and J. P. M. van der Ploeg, *Tetrahedron*, **32**, 2433 (1976).
133. H. D. Roth and M. L. Kaplan, *J. Amer. Chem. Soc.*, **95**, 262 (1973).
134. M. Lehnig and H. Fischer, *Z. Naturforsch.*, **A25**, 1963 (1970); *Chem. Abstr.*, **74**, 47416h (1971).
135. C. F. Poranski, Jr., W. B. Moniz and S. A. Sojka, *J. Amer. Chem. Soc.*, **97**, 4275 (1975).
136. W. F. Smith, Jr., *Tetrahedron*, **25**, 2071 (1969).
137. P. Lebourgeois, R. Arnaud and J. Lemaire, *J. Chim. Phys. Physicochim. Biol.*, **69**, 1637 (1972); *Chem. Abstr.*, **78**, 36195s (1973).
138. (a) R. Kaptein, J. A. den Hollander, D. Antheunis and L. J. Oosterhoff, *J. Chem. Soc. (D)*, 1687 (1970).
(b) S. R. Fahrenholtz and A. M. Trozzolo, *J. Amer. Chem. Soc.*, **93**, 251 (1971).
(c) T. Nakata and K. Tokumaru, *Bull. Chem. Soc. Japan*, **43**, 3315 (1970).
139. J. W. Miley, *Diss. Abstr. Intern.*, **B32**, 177 (1971).
(b) J. E. Leffler and J. W. Miley, *J. Am. Chem. Soc.*, **93**, 7005 (1971).
140. M. Jones, Jr. and M. R. DeCamp, *J. Org. Chem.*, **36**, 1536 (1971).
141. (a) V. Dvořáh, J. Kolc and J. Michl, *Tetrahedron Letters*, 3443 (1972).
(b) O. L. Chapman, C. L. McIntosh and J. Pacansky, *J. Amer. Chem. Soc.*, **95**, 4061 (1973).
142. P. B. Dervan and C. R. Jones, *J. Org. Chem.*, **44**, 2116 (1979).
143. M. M. Martin and J. M. King, *J. Org. Chem.*, **38**, 1588 (1973).
144. W. Adams and R. Rucktäschel, *J. Org. Chem.*, **43**, 3886 (1978).
145. N. J. Karch and J. M. McBride, *J. Amer. Chem. Soc.*, **94**, 5092 (1972).
146. N. J. Karch, E. T. Koh, B. L. Whitsel and J. M. McBride, *J. Amer. Chem. Soc.*, **97**, 6729 (1975).
147. J. Pacansky and J. Bargon, *J. Amer. Chem. Soc.*, **97**, 6896 (1975).
148. J. Pacansky, G. P. Gardini and J. Bargon, *J. Amer. Chem. Soc.*, **98**, 2665 (1976).
149. J. Pacansky, G. P. Gardini and J. Bargon, *Ber. Bunsenges. Phys. Chem.*, **82**, 19 (1978).
150. D. L. Heywood, B. Phillips and H. A. Stansburg, Jr., *J. Org. Chem.*, **26**, 281 (1961).
151. Y. Ogata and K. Tomizawa, *J. Org. Chem.*, **44**, 2770 (1979).

152. Y. Ogata and K. Tomizawa, *Bull. Chem. Soc. Japan*, **53**, 2419 (1980).
153. Y. Ogata and K. Tomizawa, *J. Org. Chem.*, **45**, 785 (1980).
154. M. R. DeCamp and M. Jones, Jr., *J. Org. Chem.*, **37**, 3942 (1972).
155. J. J. Davis and J. G. Miller, *J. Amer. Chem. Soc.*, **99**, 245 (1977).
156. W. A. Pryor and R. W. Henderson, *J. Amer. Chem. Soc.*, **92**, 7234 (1970).
157. P. Bakuzis, J. K. Kochi and P. J. Krusic, *J. Amer. Chem. Soc.*, **92**, 1434 (1970).
158. J. K. Kochi, P. Bakuzis and P. J. Krusic, *J. Amer. Chem. Soc.*, **95**, 1516 (1973).
159. (a) M. F. Zady and J. L. Wong, *J. Amer. Chem. Soc.*, **99**, 5096 (1977).
(b) M. F. Zady and J. L. Wong, *Nucleic Acid Chem.*, **1**, 29 (1978); *Chem. Abstr.*, **89**, 197474z (1978).
160. (a) F. Minisci, R. Mondelli, G. P. Gardini and O. Porta, *Tetrahedron*, **28**, 2403 (1972).
(b) A. Clerici, F. Minisci and O. Porta, *Tetrahedron*, **29**, 2775 (1973).
(c) A. Clerici, F. Minisci and O. Porta, *J. Chem. Soc. Perkin Trans. 2*, 1699 (1974).
(d) A. Clerici, F. Minisci and O. Porta, *Tetrahedron*, **30**, 4201 (1974).
161. M. F. Zady and J. L. Wong, *J. Org. Chem.*, **44**, 1450 (1979).
162. F. D. Greene, W. Adam and G. A. Knudsen, Jr., *J. Org. Chem.*, **31**, 2087 (1966).
163. (a) W. Adam and Y. M. Cheng, *J. Amer. Chem. Soc.*, **91**, 2109 (1969).
(b) W. Adam, Y. M. Cheng, C. Wilkerson and W. A. Zaidi, *J. Amer. Cem. Soc.*, **91**, 2111 (1969).
(c) W. Adam and G. S. Aponte, *J. Amer. Chem. Soc.*, **93**, 4300 (1971).
164. W. Adam, O. Cueto, L. N. Guedes and L. O. Rodriguez, *J. Org. Chem.*, **43**, 1466 (1978).
165. W. Adam, O. Cueto and L. N. Guedes, *J. Amer. Chem. Soc.*, **102**, 2106 (1980).
166. W. Adam and L. Szendrey, *J. Chem. Soc. (D)*, 1299 (1971).
167. G. Sosnovsky, *J. Org. Chem.*, **28**, 2934 (1963).
168. G. Sosnovsky, *Tetrahedron*, **21**, 871 (1965).
169. D. S. Watt, *J. Amer. Chem. Soc.*, **98**, 271 (1976).
170. V. D. McGinniss and F. A. Kah, *J. Coat. Technol.*, **51**, 81 (1979); *Chem. Abstr.*, **91**, 124047u (1979).
171. (a) O. P. Chawla and R. W. Fessenden, *J. Phys. Chem.*, **79**, 2693 (1975).
(b) R. O. C. Norman, P. M. Storey and P. R. West, *J. Chem. Soc. (B)*, 1087 (1970).
172. L. Dogliotti and E. Hayon, *J. Phys. Chem.*, **71**, 2511 (1967).
173. F. Bolletta, A. Juris, M. Maestri and D. Sandrini, *Inorg. Chim. Acta*, **44**, 1175 (1980); *Chem. Abstr.*, **93**, 85029j (1980).
174. I. M. Kolthoff, A. I. Medalia and H. P. Raaen, *J. Amer. Chem. Soc.*, **73**, 1733 (1951).
175. R. J. Lussier, *Diss. Abstr. Int.* **B31**, 103 (1970); *Chem. Abstr.*, **74**, 106669t (1971).
176. (a) P. Maruthamuthu and P. Neta, *J. Phys. Chem.*, **82**, 710 (1978).
(b) P. Maruthamuthu and H. Taniguchi, *J. Phys. Chem.*, **81**, 1944 (1977).
177. G. Sosnovsky and G. A. Karas, *Z. Naturforsch.*, **33B**, 1177 (1978).
178. R. L. Dannley and K. R. Kabre, *J. Amer. Chem. Soc.*, **87**, 4805 (1965).
179. R. L. Dannley, R. L. Waller, R. V. Hoffman and R. F. Hudson, *J. Org. Chem.*, **37**, 418 (1972).
180. M. Kobayashi, M. Sekiguchi and H. Minato, *Chem. Letters*, 393 (1972).
181. (a) R. L. Dannley and G. Jalics, *J. Org. Chem.*, **30**, 3848 (1965).
(b) D. J. Edge and J. K. Kochi, *J. Chem. Soc., Perkin Trans. 2*, 182 (1973).
(c) P. G. Cookson, A. G. Davies, N. A. Fazal and B. P. Roberts, *J. Amer. Chem. Soc.*, **98**, 616 (1976).
182. G. B. Sadikov, V. P. Maslennikov and G. I. Makin, *Zh. Obshch. Khim.*, **42**, 1571 (1972); *Chem. Abstr.*, **77**, 107666p (1972).
183. (a) P. Ausloos, *Fundamental Processes in Radiation Chemistry*, Wiley–Interscience, New York–London, 1968.
(b) J. W. T. Spinks and R. J. Wood, *An Introduction to Radiation Chemistry*, Wiley–Interscience, New York–London, 1964.
(c) A. J. Swallow, *Radiation Chemistry*, Longman Group Ltd., London, 1973.
(d) I. G. Draganić and Z. D. Draganić, *Radiation Chemistry of Water*, Academic Press, New York, 1971.
184. (a) E. Heckel, *J. Phys. Chem.*, **80**, 1274 (1976).
(b) K.-D. Asmus, H. Möckel and A. Henglein, *J. Phys. Chem.*, **77**, 1218 (1973).
185. D. Behar, G. Czapski, L. M. Dorfman, J. Rabani and H. A. Schwarz, *J. Phys. Chem.*, **74**, 3209 (1970).

186. C. Ferradini, J. Foos, C. Houee and J. Pucheault, *Photochem. Photobiol.*, **28**, 697 (1978); *Chem. Abstr.*, **90**, 175281c (1979).
187. C. E. Burchill and I. S. Ginns, *Can. J. Chem.*, **48**, 1232 (1970).
188. (a) P. V. Phung and M. Burton, *Radiat. Res.*, **7**, 177 (1957).
(b) H. Hotta and A. Terakawa, *Bull. Chem. Soc. Japan*, **33**, 335 (1960).
189. V. S. Zhikharev and N. A. Vysotskaya, *Khim. Vys. Energ.*, **2**, 249 (1968); *Chem. Abstr.*, **69**, 31997j (1968).
190. M. Lal, *Proc. Chem. Symp., 1st*, **2**, 219 (1969); *Chem. Abstr.*, **74**, 105628s (1971).
191. S. L. Kul'chitskaya, T. V. Maksimuk, L. N. Zhigunova and E. P. Petryaev, *Vesti Akad. Navuk Belarus. SSR, Ser. Khim. Navuk*, 46 (1974); *Chem. Abstr.*, **81**, 136469q (1974).
192. (a) D. Verdin, *Proc. Tihany Symp. Radiat. Chem., 2nd. Tihany, Hung.*, 265 (1966); *Chem. Abstr.*, **67**, 59549x (1967).
(b) D. Verdin, *Trans. Faraday Soc.*, **65**, 2438 (1969).
193. B. Striefler and F. Boberg, *Erdöl, Kohle, Erdgas, Petrochem., Brennst.-Chem.*, **25**, 725 (1972); *Chem. Abstr.*, **79**, 72208c (1973).
194. F. Boberg, A. Oezkan and R. Voss, *Erdöl, Kohle, Erdgas, Petrochem., Brennst.-Chem.*, **30**, 369 (1977); *Chem. Abstr.*, **88**, 22171a (1978).
195. M. F. Romantsev, V. V. Saraeva and O. A. Mishchenko, *Zh. Fiz. Khim.*, **39**, 2599 (1965); *Chem. Abstr.*, **64**, 5987e (1966).
197. A. T. Koritskii, A. A. Karatun, F. F. Sukhov and N. A. Slovokhotova, *Khim. Vys. Energ.*, **10**, 406 (1976); *Chem. Abstr.*, **86**, 98934f (1977).
198. T. Shida, *J. Phys. Chem.*, **72**, 723 (1968).
199. A. T. Koritskii and A. V. Zubkov, *Khim. Vys. Energ.*, **2**, 544 (1968); *Chem. Abstr.*, **70**, 33211t (1968).
200. A. T. Koritskii, A. V. Zubkov, E. K. Starostin and B. A. Golovin, *Khim. Vys. Energ.*, **5**, 166 (1971); *Chem. Abstr.*, **75**, 13474f (1971).
201. I. I. Chkheidze, V. I. Trofimov and A. T. Koritskii, *Kinet. Katal.*, **8**, 453 (1967); *Chem. Abstr.*, **67**, 77843t (1967).
202. H. C. Box, *J. Phys. Chem.*, **75**, 3426 (1971).
203. (a) J. Lilie, E. Heckel and R. C. Lamb, *J. Amer. Chem. Soc.*, **96**, 5543 (1974).
(b) E. Heckel, *J. Phys. Chem.*, **80**, 1274 (1976).
204. I. G. Draganić, Z. D. Draganić and K. Sehested, *J. Phys. Chem.*, **82**, 757 (1978).
205. P. Neta, V. Madhavan, H. Zemel and R. W. Fessenden, *J. Amer. Chem. Soc.*, **99**, 163 (1977).
206. K.-J. Kim and W. H. Hamill, *J. Phys. Chem.*, **80**, 2325 (1976).
207. G. Levey and E. J. Hart, *J. Phys. Chem.*, **79**, 1642 (1975).
208. V. Madhavan, H. Levanon and P. Neta, *Radiat. Res.*, **76**, 15 (1978); *Chem. Abstr.*, **90**, 38287v (1979).
209. L. M. Bharadwaj, D. N. Sharma and Y. K. Gupta, *Inorg. Chem.*, **15**, 1695 (1976).
210. P. Maruthamuthu and P. Neta, *J. Phys. Chem.*, **81**, 1622 (1977).

The Chemistry of Functional Groups, Peroxides
Edited by S. Patai

CHAPTER 22

Solid-state reactions of peroxides

M. LAZÁR

Polymer Institute of the Slovak Academy of Sciences, 80934 Bratislava, Czechoslovakia

I. INTRODUCTION . . . 778
II. ISOMERIZATION REACTIONS . . . 778
A. Topotactic Isomerization of 2-Iododibenzoyl Peroxide Derivatives . . . 779
B. Oxygen-18 Scrambling in Acyl Peroxides . . . 780
C. Carboxy Inversion . . . 780
III. EVOLUTION OF OXYGEN FROM ENDOPEROXIDES OF AROMATIC COMPOUNDS . . . 781
A. 9,10-Dioxy-9,10-diphenylanthracene . . . 781
B. Other Endoperoxides . . . 783
IV. DECOMPOSITION OF PEROXIDES . . . 783
A. 9,10-Dioxyanthracene . . . 783
1. Topotaxy of the reaction . . . 783
2. Kinetic features of the decomposition . . . 784
3. Decomposition products and by-products . . . 786
4. Mechanism of decomposition . . . 787
B. Dibenzoyl Peroxide and its Derivatives . . . 788
1. Topochemical control . . . 788
2. Cause of autocatalysis . . . 788
3. Kinetic studies . . . 790
4. Decomposition in the presence of admixtures . . . 791
C. Cyclic Alkylidene Peroxides . . . 792
D. Hydroperoxides . . . 793
E. Peroxy Acids . . . 794
1. Autocatalysis . . . 794
2. Stoichiometry of the decomposition . . . 795
F. Polymer Hydroperoxides . . . 797
1. Influence of the aggregation of hydroperoxide groups on the decomposition rate . . . 797
2. Mobility and reactivity . . . 798

V. DECOMPOSITION OF LOW MOLECULAR WEIGHT PEROXIDES IN POLYMERIC MEDIA 799
A. Kinetic Parameters 799
1. Spontaneous and chain decomposition 799
2. Activation energy of the dissociation. 801
3. Cage effect in the polymer matrix 802
B. Microheterogeneity of the Polymer Medium 803
VI. CONCLUSION 803
VII. REFERENCES 804

I. INTRODUCTION

Interest in reactions of organic compounds in the solid phase has been motivated mostly by the attraction of unknown phenomena. At present it seems that in this field the main aim is the enhancement of specificity and stereoselectivity in chemical transformations, since the reactant and the reaction product may maintain the same crystalline arrangement during some solid-state reactions (topotaxy)[1]. The chemical change of a crystalline compound usually begins in defective centres. Therefore the defect concentration will be regarded as the most important parameter for describing the state of a solid organic reactant, although unfortunately its magnitude is seldom known. This leads to poor reproducibility of experiments and often discourages us from studying these reactions.

Fluctuations in the structure of solids are also evident in reactions in the glassy state[2]. In the condensed phase, each particle interacts with its surroundings which vary with time. Kinetic laws depend on the rate of structural changes of the reacting system and on the lifetime of an intermediate which mediates chemical reaction. In liquid-state reactions, the lifetime of the intermediate is much longer than the time taken for structural change of the surroundings. The different arrangements of the surroundings average the reactivity of the intermediate and the liquid medium appears to be a homogeneous phase. A different situation arises in the glassy state where changes in the physical structure of the medium are slower than the lifetime of the intermediate. This results in the structural nonequivalency of microregions where chemical reaction can take place, and causes deviations of the reaction kinetics from 'classical' laws.

In solid-state reactions of peroxides, much attention has been devoted to decomposition reactions, primarily in order to solve the problem of safety during handling in chemical technology and in research practice. Another important aspect is the study of peroxide decomposition in solid polymers. This has been investigated together with various reactions of primary radicals, in processes such as crosslinking, grafting, selective degradation, bleaching, polymerization of residual monomer, etc.

Isomerization of peroxides has been less examined, although these reactions are probably most affected by crystalline structure and may account for the rate laws observed in the solid phase. Even less is known about bimolecular solid-state reactions of peroxides.

II. ISOMERIZATION REACTIONS

Solid-state isomerization is certainly more frequent than it would seem from the published studies. The outcome of peroxide isomerization is often determined only from the structure of decomposition products. Most isomerizations have so far been studied only in solution[3].

A. Topotactic Isomerization of 2-Iododibenzoyl Peroxide Derivatives

It has long been known that 2,2′-diododibenzoyl peroxide (**1**) is unstable[4]. The solid-phase transformation of **1** into a heterocyclic compound containing trivalent iodine (**2**) takes place at room temperature and is essentially complete in six weeks. This rate is three to four orders slower than the isomerization in liquid phase. Further substitution in phenyl rings has only little influence on the rate of isomerization[5].

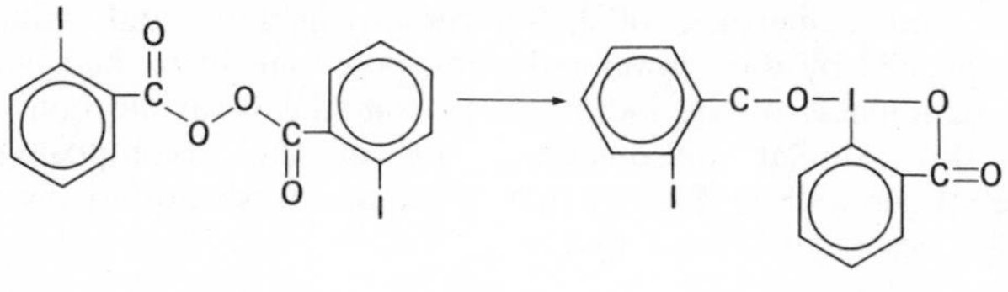

(1) (2)

UV- or X-irradiation and higher temperatures have an accelerating effect on the solid-state isomerization of **1**[6,7].

The isomerization of **1** under various conditions yields the monoclinic crystal structure of **2**, which can also be obtained by crystallization from solvents. During isomerization the external crystal shape of **1** is maintained, but the clear colourless crystals gradually become opaque, with numerous microscopic defects. The change of **1** to **2** has steric requirements too great to be accommodated by the volume available in the crystal lattice of **1**[6]. One phenyl ring must flip through ca. 180° about an axis parallel to the layers of the peroxide molecules. The reaction probably begins at defect nucleation sites and spreads through the crystal, forming a remarkably well ordered product phase.

2-Iodo-2′-chlorodibenzoyl peroxide (**3**), despite the similarity of its crystal structure to **1**, exhibits different patterns of behaviour[8]. On thermal isomerization the product is formed in the monoclinic crystal phase. In contrast, on exposure to X-rays or UV light, **3** is transformed to an isomer with a triclinic structure, which is in good accord with cell parameters of the parent **3** lattice. The triclinic phase could not be recovered by recrystallization of the product obtained from **3** either by UV or by X-ray isomerization. After the recrystallization the isomer is obtained in the usual monoclinic structure. It seems that the triclinic phase of the isomer is a polymorph, formed only through solid-state topotactic isomerization owing to a template effect.

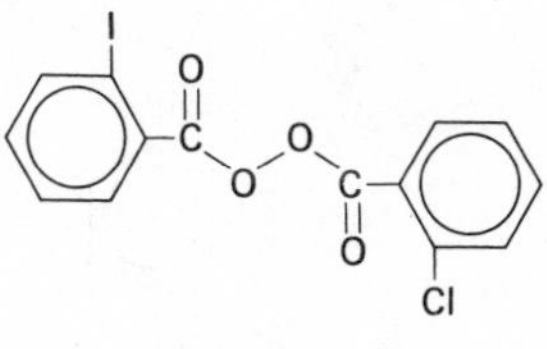

(3)

The monoclinic crystal phase is not homogeneous. Since the symmetry axes of the monoclinic reactant peroxide and the isomeric product are not aligned, a twinning accompanies topotactic reaction. (The term twinning is used for the formation of two or more crystalline phases which are identical, except for their orientation relative to the reactant crystal lattice.)

Triclinic and monoclinic modifications of isomers are formed through two distinct topotactic pathways of rearrangement. Under photochemical initiation the solid-state isomerization proceeds probably by an intermolecular mechanism whereas during the thermal decomposition an intramolecular rearrangement takes place[8].

Isomerization has also been observed with other substituted derivatives of *o*-iododibenzoyl peroxide[6,9].

The decomposition of **1** and some related iodo-substituted compounds is several thousand times faster than the decomposition of other *ortho*-substituted dibenzoyl peroxides. The relative inertness of 2,2′-dibromo-, dichloro- and difluoro-dibenzoyl peroxides is explained by their lower reduction potential. In *m*- and *p*-diiodobenzoyl peroxides the remoteness of the iodine atom from the peroxidic bond prevents the isomerization. For similar reasons the rate of the decomposition of bis(*o*-iodophenylacetyl) peroxide is close to that of the decomposition of the unsubstituted bis(phenylacetyl) peroxide[10].

B. Oxygen-18 Scrambling in Acyl Peroxides

A special type of isomerization of acyl peroxides is the mutual exchange of oxygen in peroxidic and carbonyl groups. Although no chemical transformation occurs in the process, the interchange of oxygen atoms can yield valuable information about motion in the cage of a radical pair[11]. Oxygen-18 scrambling has been mainly studied and discussed[12a] for the decomposition of acyl peroxides in solution. However, it also takes place in the solid state[11b] and future studies might contribute to the elucidation of the elementary steps in solid-state reactions of peroxides.

C. Carboxy Inversion

An alternative mode of isomerization is the loss of the peroxidic arrangement (equation 1). Isomerization of this type in solution is enhanced by polar media, electron asymmetry of the acyl peroxy compound and nucleophilicity of the group Z[12]. Some studies are available on carboxy inversion in the solid phase of benzoyl cumyl peroxide (equation 2)[13] and of *m*-trifluoromethylbenzene sulphonyl peroxide[14].

$$\text{X–O–O–Y–Z} \longrightarrow \text{X–O–Y–O–Z} \qquad (1)$$

$$\text{C}_6\text{H}_5\text{C(CH}_3)_2\text{–O–O–C(=O)C}_6\text{H}_5 \longrightarrow [\text{cyclic transition state}] \longrightarrow (\text{CH}_3)_2\text{C(OC(=O)C}_6\text{H}_5)(\text{OC}_6\text{H}_5) \qquad (2)$$

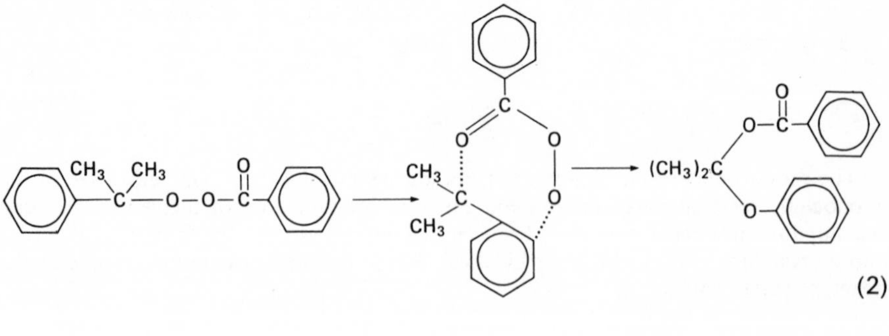

A similar reaction is expected in the decomposition of crystalline *t*-butyl *N*-methyl-*N*-(*p*-nitrophenyl)peroxycarbamate **(4)**. The half-time of decomposition of **4** at 30°C is about 2 days. In compounds where hydrogen instead of methyl is bound to nitrogen, the half-time of the decomposition is 600 days. This different reactivity in the solid phase is accounted for by the assumption[15] that a six-membered ring:

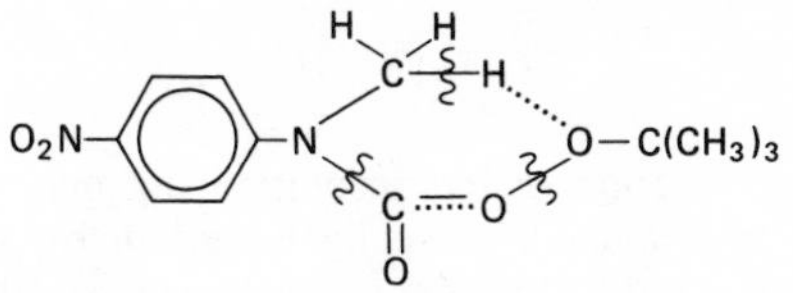

(**4**)

is more effective for decomposition than a five-membered ring in the second case:

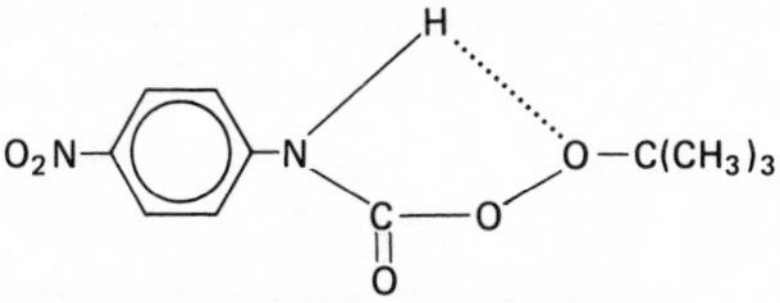

On the other hand the higher decomposition rate of the methyl derivative may be connected with greater mobility of its molecules in the crystalline lattice, shown by its lower m.p. (m.p. **4** = 66°C; m.p. hydrogen derivative 93°C). A similar influence of the melting point on the decomposition rate is observable in several other cases.

Among other types of isomerization, the rearrangement of endoperoxides into diepoxides has often been studied[16]. We failed to find any report on this isomerization in the solid phase, although, as it will be seen later, some results on the thermal decomposition of 9,10-dioxyanthracene indicate that such a reaction is probable.

III. EVOLUTION OF OXYGEN FROM ENDOPEROXIDES OF AROMATIC COMPOUNDS

Many aromatic endoperoxides release oxygen upon thermolysis with regeneration of the parent aromatic hydrocarbons[17]. The ease and the yield of oxygen depend primarily on the nature of the aromatic system. It appears that peroxides in the anthracene series give the highest yields of oxygen and that aryl substituents in the *meso* positions lead to increased oxygen release relative to alkyl or hydrogen[18]. The oxygen evolution from endoperoxides is usually followed in solution. Experiments in the solid phase are rare and have been formed under not very well defined conditions.

A. 9,10-Dioxy-9,10-diphenylanthracene

When the endoperoxide **5** is heated to 180°C in vacuo, it liberates 96% of its oxygen[19]. The differential enthalpic decomposition curve of **5** (Figure 1) shows that the reaction consists of small endo- and exo-thermal parts[20]. The two effects are very similar, each being about 20 kJ mol^{-1}. A similar value of the heat of reaction has also been reported elsewhere[21]. It is possible that the melting of **5** crystals contributes to the observed endothermal first peak. The heat of crystallization of 9,10-diphenylanthracene and

Ph

O

O

Ph

(5)

probably also the heat of reaction of the accompanying minor decomposition of the peroxide, where oxygen is retained chemically bound in by-products, belong to the exothermal part of the curve. On further temperature increase, the melting region of the formed 9,10-diphenylanthracene is reached.

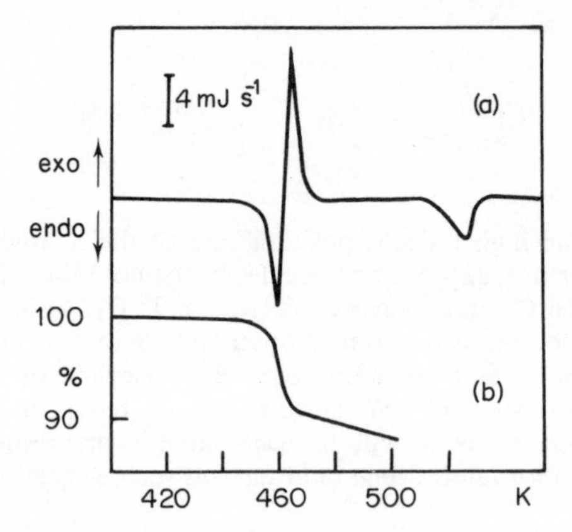

FIGURE 1. Calorimetric (a) and thermogravimetric (b) study[20] of the decomposition of 9,10-dioxy-9,10-diphenylanthracene at a heating rate of 16 K/min. Sample weight = 2 mg.

The thermogravimetric dependence shows that the weight loss (~8%) occurs in both the endo- and exo-thermal region. The result could indicate that the processes mentioned are superimposed upon the approximately thermoneutral reaction of oxygen release. The thermoneutrality of the decomposition follows from the thermochemical characteristics of the reaction components[22].

Originally it was expected that pyrolysis of pure **5** in an evacuated tube would provide a convenient source of singlet oxygen in the gas phase. However, no allylic hydroperoxide was formed with added gaseous tetramethylethylene[23a], showing that there was no singlet oxygen present. In contrast, singlet oxygen is formed during the decomposition of **5** in solution[24,25].

A study of the thermolysis of **5** has shown[26] that in solution this reaction proceeds via two pathways[27], one being a concerted mechanism in which $^1O_2$ is produced, and the second a biradical mechanism in which $^3O_2$ is produced.

In the light of the results, the decomposition of **5** in the solid state proceeds through the relatively long-lived alkylperoxy biradical.

B. Other Endoperoxides

In the case of 9,10-diphenylanthracene derivatives oxygen can be bound either to the 9,10- or to the 1,4- position. The latter, when activated by methoxy groups, releases oxygen at lower temperatures (at 80°C in one hour)[28].

Rubrene peroxide recrystallized from carbon disulphide gives oxygen in 80% yield at 140–150°C[29]. It is interesting to note that the same peroxide recrystallized from petroleum ether decomposes at a higher temperature (~170°C)[25].

IV. DECOMPOSITION OF PEROXIDES

A. 9,10-Dioxyanthracene

One of the first examples of an organic topotactic reaction is the conversion of anthracene peroxide **(6)** to anthraquinone **(7)** and anthrone **(8)**. Thermal decomposition of **6** in the solid state had long been known[30,31], but a better insight into the reaction was possible after an X-ray study[32] of a single crystal of **6**. This had begun as a straight single-crystal structure determination, without expecting any changes in **6** at room temperatures. Surprisingly spots and new layer lines appeared on the X-ray photographs after prolonged irradiation. Upon continued exposure to X-ray irradiation at room temperature or slow heating of the sample to 80–120°C, the crystals of **6** were transformed, without any change in the external shape, into mixed crystals of reaction products. On quick heating to 140°C, the single crystal of **6** decomposed into a powder, in which only diffraction patterns of **7** and **8** were found. In the Laue photograph there was much small-angle scattering, indicating that by-product molecules were present in a disordered state.

(6) → (7) + (8) + H_2O

(6)	(7)	(8)
m.p. 120–166°C	284°C	155°C

1. *Topotaxy of the reaction*

There is a clear general resemblance between the monoclinic structure of **6** and **7** or pseudoorthorhombic **8** in one direction only, namely the '*a*' lengths are almost equal. The other repeat distances of the molecules are different. Hence the transformation of **6** into **7** or **8** requires movements of molecules. However, owing to the relationship between the repeat distances in the crystal lattices ($a\ 6 \doteq 2c\ 7$), there is sufficient resemblance in the structures for the one to change into the other.

The product phase appears in two preferred orientations as a result of two different geometrical fits of the reactant and product lattices (Figure 2). Therefore the new crystal is never really single, but is always twinned on two twin planes. The two orientations are present in unequal amounts, being dependent on the orientation of newly generated nucleation centres, where the formation of the product phase begins. The orientation of crystalline product phases is due to the topochemical control of the regular molecular

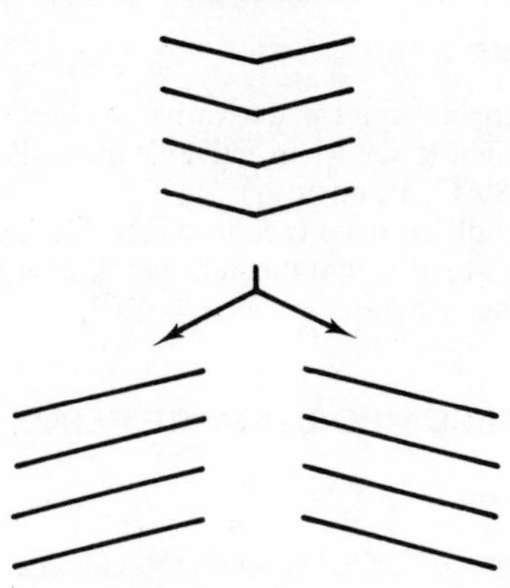

FIGURE 2. Scheme for the transformation of the bent anthracene peroxide molecule into the two alternative strings of the planar anthraquinone molecule.

arrangement of **6** upon which the formed **7** or **8** molecules may orient and further molecules then crystallize on product crystallites.

The reaction proceeds via an intermediate stage in which some break-up of the original **6** structure into partially disordered crystallites takes place. It may be that in this early stage of the reaction, arrays of **6** molecules are transformed into isomers which lose their sideways periodicity, while still remaining parallel with the main structure. During the decomposition of the labile isomers into **7** and **8** recrystallization of the reaction products begins.

2. *Kinetic features of the decomposition*

The time of decomposition of **6** subjected to X-rays at room temperature is variable; sometimes the reaction is apparently complete after 60 h irradiation, while sometimes it is not quite complete after 485 h irradiation[32].

The thermal stability of polycrystalline **6** isolated by different procedures was also found to be different[20,33] (Figure 3). The endothermic stage of melting could not be recorded.

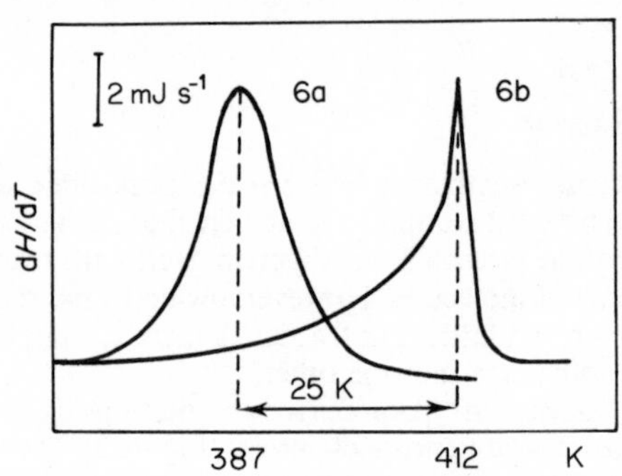

FIGURE 3. Dependence of the evolution of heat on decomposition of **6** purified in different ways: (**6a**) by freezing out from carbon disulphide solution at $-25°C$, (**6b**) by precipitation from a benzene solution with light petroleum. Programmed temperature increase = 4 K/min, sample weight = 1 mg[20].

Largely varying melting points, or more correctly, decomposition temperatures, are quoted in the literature: 120°C[30], 139–143°C[34], 147°C[35], 160°C[31] or 166°C[32]. The highest thermal stability is reached for single crystals, i.e. those containing the least defects and dislocations.

Under isothermal conditions, an autocatalytic decomposition is observed immediately after heating the samples up and no induction period can be detected. Isothermal records of the heat evolution from polycrystalline **6** exhibit irregularities (Table 1). More regular curves are obtained if 10% mixture of **6b** with Al_2O_3 or with anthracene is heated after careful mixing (Figure 4)[20]. In the presence of anthracene the decomposition is retarded, while in the presence of Al_2O_3 it is accelerated. These influences on the reaction rate are connected with the great sensitivity of the chain-decomposition of **6**[36].

TABLE 1. Decomposition rates, v_x (% min^{-1}), and time intervals, τ_x (s), needed to attain the relative maximum rate v_x (peak of the heat evolution at the given temperature, T) of anthracene peroxide (**6a** and **6b**) and the activation energies (E) of the decomposition processes[20]

Sample	T(°C)	τ_1	τ_2	τ_3	v_1	v_2	v_3
6a	92	625	4470	—	1.62	0.90	—
	94	501	3340	—	2.88	1.44	—
	99	322	1930	—	5.94	1.80	—
	102	249	1500	—	4.32	1.62	—
	105	198	670	—	8.28	4.08	—
	112	134	500	—	12.78	5.70	—
	E(kJ mol^{-1})	90	134	—	111	103	—
6b	100	614	2360	4000	1.08	0.90	0.72
	110	210	710	1370	2.70	2.52	2.52
	120	94	283	540	5.04	4.86	4.86
	125	66	183	580	8.10	8.82	7.56
	130	42	110	200	11.16	13.32	13.32
	135	28	87	212	16.56	18.36	21.96
	140	14	42	130	15.48	22.86	21.96
	E(kJ mol^{-1})	115	123	108	90	105	111

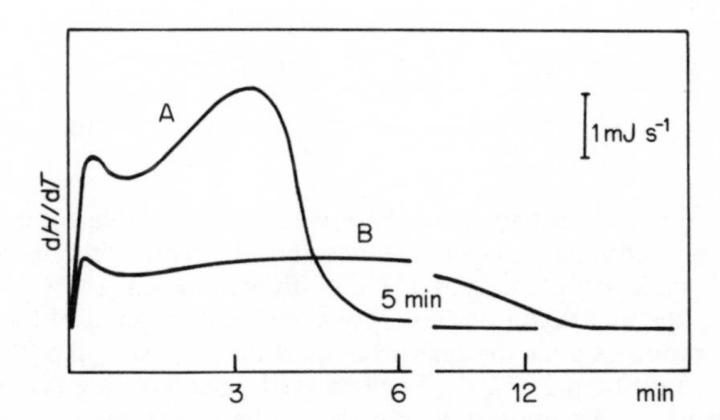

FIGURE 4. The rate of release of the heat of reaction in isothermal (393 K) decomposition of 1 mg of **6b** mixed either with aluminium oxide (A) or with anthracene (B). Total sample weight = 10 mg.

The apparent activation energy of the decomposition of solid **6** is nearly the same as in solution ($125\,kJ\,mol^{-1}$). However, the rate decomposition is slower in the solid state. At room temperature only about 1% of **6** decomposes during one month, while in a solvent[33,36,37] at the same temperature, 50% is already decomposed in two days.

3. Decomposition products and by-products

We may infer from mass spectra of the reaction products and model compounds[32] that in addition to **7**, **8** and H_2O, 9-fluorenone, 9-fluorene, anthracene, biphenylene, C_6H_4 (benzyne), CO_2, CO and H_2 are also formed. Chromatographic analysis[20] shows that in thermal decomposition (220°C, 40 kPa, N_2), 2.7 wt.% water and 0.03% hydrogen are released. Assuming that the water is formed only in the indicated reaction, accompanying the formation of **7** and **8**, this type of reaction constitutes about 63% of the decomposition of **6**.

The overall weight loss of the sample after 10 min at 100–140°C, (programmed temperature increase 4°/min) is 6.5%. Accordingly, in addition to water and hydrogen, other gaseous products must also be formed. If we consider carbon monoxide as an additional weight loss ($6.5 - 2.7 = 3.8$) accompanying formation of 9-fluorenone, then this mode of decomposition takes about 26%. However, this value is higher than experimental, because fluorene and biphenylene are also among the reaction products and during decomposition not only CO but also CO_2 and low molecular unsaturated hydrocarbons, partially included in the weight loss, are formed.

Small amounts of oxygen are only sporadically detected in the decomposition products of **6** by mass spectrometry. Since a relatively high (ca. 10%) amount of anthracene is observed in the decomposition products, the released oxygen either immediately reacts with other decomposition products or, more probably, the anthracene is not formed by evolution of molecular oxygen from **6**.

Another interesting observation in the mass spectrometric studies is that one or more by-products have the same molecular weight as the original peroxide **6**. These may have the structure of a bicyclic acetal **(9)** or a cyclobutane derivative **(10)**. Similar results are obtained[38] in the thermal decomposition of 9-phenyl-10-methyl-9,10-dioxyanthracene and of 9,10-dimethyl-9,10-dioxyanthracene in boiling chlorobenzene or *o*-dichlorobenzene.

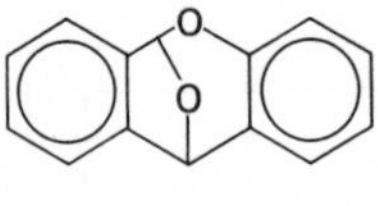

(9) 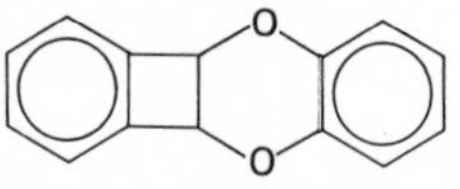 **(10)**

In thermal decomposition in the solid phase, the content of stable isomers is greater than in irradiated samples. Similarly, as for other decomposition products, on irradiation, more water is formed and less fluorene (m.p. 114°C), 9-fluorenone (m.p. 85°C) and biphenylene (110°C) than on thermal decomposition. Thus, the melting point of the mixture of the decomposition products after thermal decomposition is lower (m.p. 248°C) than after irradiation by X-rays (m.p. 256°C)[32]. The observed differences are probably due to the lower temperature of decomposition during irradiation compared to the thermally initiated decomposition. The role of X-rays consists in initiation of a chain-reaction and not of a simple decomposition.

In addition to stable isomers, some unstable isomers may appear in the reaction systems, such as 10-hydroxyanthrone, 9,10-dihydroxyanthracene or anthracene 9-hydroperoxide, which may arise from **6** by exothermal, probably chain, reaction. Such unstable isomers could also explain the formation of the mesomorphic phase in the initial stage of X-ray irradiation of **6**[32]. The molecules in the mesomorphic phase become sideways and lengthways displaced in comparison to the original structure though they are still parallel with respect to their neighbours. The molecules remain in a set of parallel equally spaced planes.

4. Mechanism of decomposition

The occurrence of a chain-reaction in the decomposition of **6** follows from kinetic features and also from the observed crystalline regions in the products.

The formation of hydroxyanthrone can be described by a set of chain-propagation reactions. After hydrogen abstraction from **6**, the free radical obtained is isomerized by β fragmentation of the peroxidic bond (equation 3). Hydrogen abstraction from a neighbouring molecule gives hydroxyanthrone and the cycle of propagating reactions may be rapidly continued in a series of ordered molecules, so that one primary radical may cause the isomerization of a whole series of **6** molecules to hydroxyanthrone. In a similar hydrogen-transfer reaction a hydroxyanthranyl radical is formed from hydroxyanthrone, followed by fragmentation to **7** and a hydrogen atom (equation 4). Then the chain-reaction is carried on by the hydrogen atom; the reaction is, however, less specific because hydrogen is much more mobile than the radicals participating in the change of **6** to hydroxyanthrone. The hydrogen atom may form a new hydroxyanthranyl radical by transfer reaction or may be added to hydroxyanthrone, yielding a dihydroxyanthranyl radical which in turn is fragmented to **8** and hydroxyl radical (equation 5). The chain-reaction continues, with the hydroxyl radical giving water and a radical derived from **6** or reaction products.

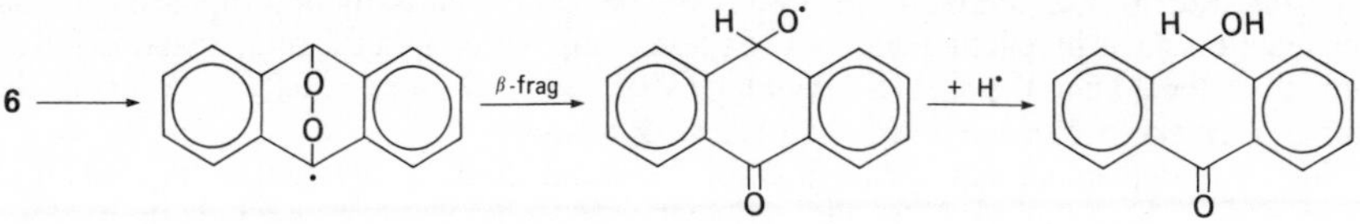

6 → [structure] —β-frag→ [structure: H, O•; O] —+ H•→ [structure: H, OH; O] (3)

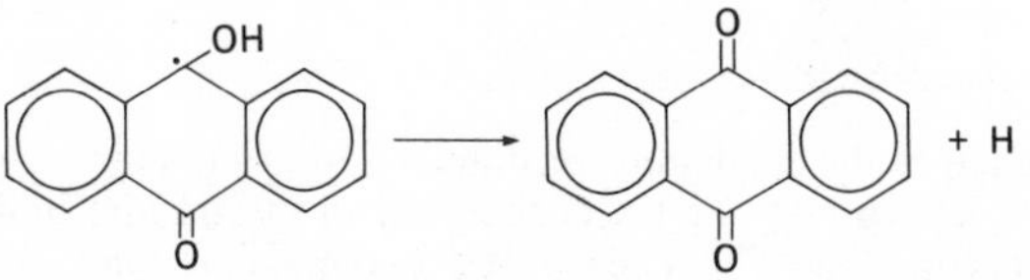

[structure: •, OH; O] → [structure: O; O] + H (4)

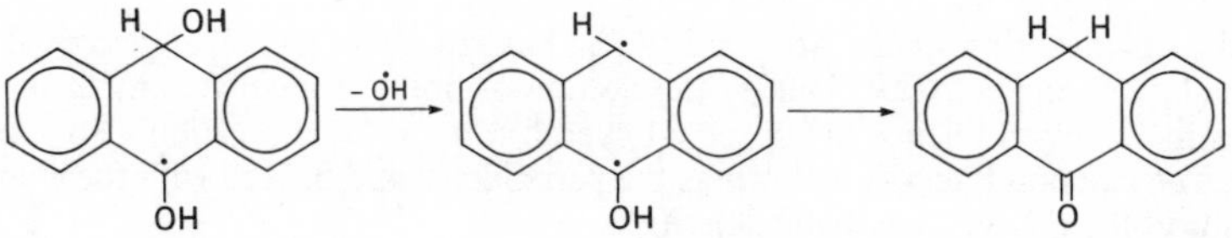

[structure: H, OH; •, OH] —- •OH→ [structure: H, •; •, OH] → [structure: H, H; O] (5)

By-products in which the anthracene skeleton has been destroyed arise by fragmentation of oxygen biradicals formed from **6** (equation 6). The stable isomers **9** and **10** may also arise by primary splitting of the peroxide bond, with subsequent isomerization of the biradical.

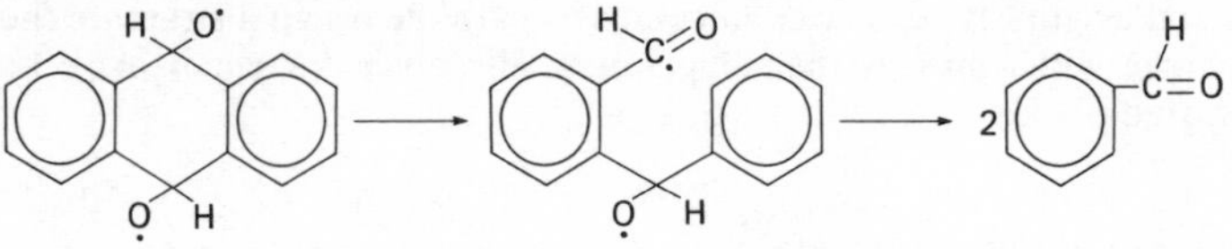

(6)

The proposed chain mechanism of the decomposition of **6** also accounts for the great sensitivity of the process to different reaction conditions in the solid state. Small retardation of the spontaneous decomposition or deactivation of the primary radicals may influence strongly the subsequent stages of the decomposition. This may be the reason for the relatively large scattering of the decomposition temperatures obtained for **6**.

The mechanism for the decomposition of **6** is rather complex and still uncertain owing to inadequate experimental data. Nondestructive and unambiguous analysis of the reaction products at various time intervals in single crystals under various decomposition conditions would be necessary in order to give a better foundation for mechanistic proposals.

B. Dibenzoyl Peroxide and its Derivatives

1. *Topochemical control*

Visual examination of dibenzoyl peroxide **(11)** partially decomposed in the solid state shows that its single crystals retain their shape but with a decrease in the thickness in one direction. By optical microscopy[39] the development of nuclei and of a 'river-line pattern' (with line-widths of about 0.001 mm) was detected on the partially decomposed crystals The reaction starts in reactive centres and the decomposition in the formed reaction nuclei is much faster than in the regular crystalline structure. Enhanced reactivity occurs at the points of emergence of both edge and screw dislocations[39].

The large dependence of peroxide stability on the state of aggregation is evident in differences of the stability of long-chain diacyl peroxides. Dilauroyl peroxide (m.p. 55°C) showed no decomposition at room temperature after 18 months, whereas the liquid dipelargonyl peroxide (m.p. 13°C) lost 80% of its peroxide oxygen in 7 months at 25°C[40].

2. *Cause of autocatalysis*

The autocatalysis of the solid-state decomposition of **11** is observable on the sigmoid pressure/time curves representing the dependence of the amount of the released carbon dioxide on the reaction time[42,43]. The varying decomposition rate is better seen on a plot of the released heat of reaction against the time on isothermal heating of **11**[44] or its derivatives[41] (Figure 5).

A trivial explanation of the autocatalysis is based on the idea that decomposition products liquefy the peroxide. During the decomposition the relative number of liquid domains increases and hence the observed overall rate of decomposition also increases rapidly. The reaction then slows down as the peroxide is used up and with the change of composition the mixture gradually solidifies.

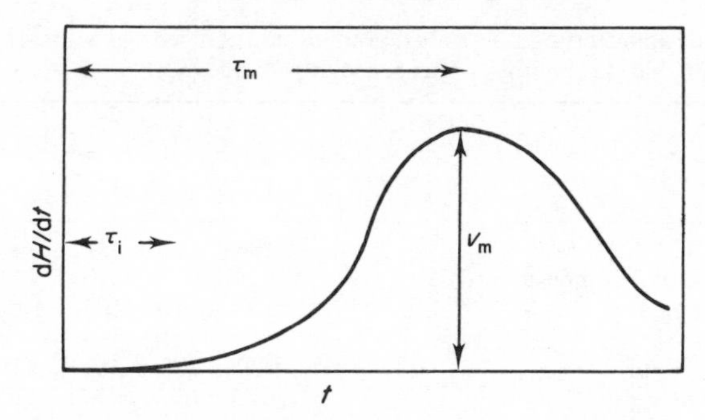

FIGURE 5. Typical course of the release of the heat of reaction of dibenzoyl peroxide (**11**) below its melting temperature; τ_i is the induction period, τ_m is the time on reaching the maximum rate of decomposition; v_m is the maximum rate of decomposition.

To compare the theory with reality we have to know the influence of the reaction products on the melting temperature of the mixture. The phase diagram of **11** and its main decomposition products[45,46] shows that the latter remarkably decrease the melting temperatures (Table 2). If we realize that in the decomposition of **11** a small amount of benzene is also formed, then the pure solid-state decomposition of **11** may only be considered for the initial phase of the reaction.

However, the autocatalytic decomposition of bis(*p*-nitrobenzoyl) peroxide (**13**) does not support the above theory, since here the melting points of the major reaction products are by about 100° higher than the temperature of peroxide decomposition.

TABLE 2. Eutectics of dibenzoyl peroxide (**11**, m.p. 105°C) with its main decomposition products

Compound	m.p. (°C)	Eutectic temperature (°C)	Mol. fraction of **11**
Biphenyl	69	55	0.16
Phenyl benzoate	71	56	0.24
Benzoic acid	122	87	0.5

The shapes of the thermochemical isotherms of **11** and its derivatives seem to be similar, although small differences are observable on more detailed analysis. Different amounts of the different peroxides decompose before reaching the maximum rate of decomposition, and these amounts decrease with increasing temperature for **11** and for *p*-methyl-*p'*-nitrodibenzoyl peroxide **12**, but they are not temperature-dependent for **13** (Table 3).

As mentioned above, the decomposition of **13** proceeds autocatalytically even though the products have much higher melting points than the reaction temperature. Hence, in the absence of liquefaction, gradual increase in the number and size of reaction centres may be also responsible for the acceleration in the decomposition of solid peroxides.

The approaches for the explanation of the autocatalysis, involving reaction centres on the one hand and eutectic liquefaction on the other, supplement rather than contradict each other.

TABLE 3. Kinetic parameters of solid-state decomposition of dibenzoyl peroxide (**11**), *p*-methyl-*p*′-nitrodibenzoyl peroxide (**12**) and bis(*p*-nitrobenzoyl) peroxide (**13**)

Property/parameter[a]	Peroxide					
	11		**12**		**13**	
Melting point (°C)	106		134		158	
Temperature range of decomposition (°C)	90–102		110–115		140–145	
Induction period $\tau_i(s)^b$	1570	10	505	105	640	120
Induction period $\tau_m(s)^b$	4300	110	1630	300	1615	675
Decomposition rate, $v_m(\%\ s^{-1})^b$	0.09	0.58	0.32	0.70	0.45	1.23
Decomposed peroxide at $v_m(\%)^b$	55	20	36	20	59	59
$E(\mathrm{kJ\ mol^{-1}})^c$ from temperature dependence of						
τ_i	543		401		464	
τ_m	418		353		259	
v_m	188		238		260	

[a] τ_i, τ_m, v_m are defined in Figure 5.
[b] The parameter at the lowest and the highest decomposition temperature of the temperature range.
[c] Apparent activation energy of decomposition; the value was determined from at least six experimental points[41,44].

3. Kinetic studies

The decomposition of **11** proceeds as a chain process initiated by spontaneous dissociation of the peroxide bond. The total rate of decomposition is approximately described by the known Bartlett–Nozaki relation:

$$-\frac{d[\mathrm{BP}]}{dt} = k_s[\mathrm{BP}] + k_i[\mathrm{BP}]^n$$

where [BP] is the concentration of (**11**), k_s is the rate constant for spontaneous decomposition and the second term expresses the rate of induced decomposition. The exponent n (0.5, 1, 1.5, 2) depends on the mechanism of the chain-reaction. The relation for the value n is simplified, because in the model scheme only one chain-carrying and one chain-terminating step is considered.

The initiation of the decomposition is very sensitive to temperature changes as shown by the rapid decrease in the induction periods with increasing temperature and from the high apparent activation energies in the solid state (Table 3) compared with those in solution. In inert solvents the activation energy of the decomposition of **11** and its derivatives is 125 kJ mol^{-1}, which is about a third or even a fourth of the value which should be assigned to the decomposition of crystalline samples.

During the induction period reactive centres are probably formed, where an induced chain-decomposition begins. The high apparent activation energy of the spontaneous decomposition is probably connected with the necessity of simultaneous decomposition of several molecules of **11**–**13** close to each other, thus forming a nucleus for the chain-reaction. In another explanation, we have to assume temperature changes of the physical (crystalline) peroxide structure determining the extent of the peroxide decomposition.

The values of E determined from τ_i are higher than those from τ_m; this follows from a higher degree of the overall decomposition at τ_m and hence more reaction centres in the crystalline peroxide. The activation energy calculated directly from the values of the maximum decomposition rate of **11** gives 272 kJ mol^{-1}. Taking into account the fact that with increasing temperature less undecomposed peroxide is present when reaching the maximum decomposition rate, the activation energy decreases to 188 kJ mol^{-1}. This corrected value is still about 50% higher than that obtained in solution. For the nitro derivatives of **11**, the values are even larger. A rise in the activation energy calculated from v_m for **12** and **13** is probably due to the lower contribution of liquefaction of these peroxides by reaction products.

For elimination product–reactant interactions in kinetic studies a high vacuum of 10^{-4} Pa was used[39]. In this case the reaction products quickly evaporate from the decomposing **11**, which should therefore not be liquefied. Despite this, autocatalysis of the decomposition was observed although not to the extent like at normal pressure.

The kinetics of the isothermal decomposition of single crystals of **11** in high vacuo have been studied by microgravimetry. In these experiments, after a minor acceleratory stage preceded by a very short induction time, the rate of the weight loss becomes slower.

High vacuum will enhance the fragmentation of benzoyloxy radicals to phenyl radicals and carbon dioxide, reducing the recombination of oxy radicals to **11**. The decomposition of **11** will therefore be faster in high vacuo than at normal pressure. A retardation effect of high pressure on the decomposition of **11** is also observed in solution[47]. The higher yield of phenyl radicals in the primary decomposition of **11** will reflect an increase in the fraction of its induced decomposition, which ought to decrease the total activation energy for decomposition reaction and shorten the induction time. Much longer induction times of **11** and its derivatives have been observed by calorimetry[41,44] and are probably connected to the presence of oxygen which decreases the fraction of induced decomposition and thus also the total peroxide decomposition rate. Retardation by oxygen of the decomposition of **11** in solution has been proved by several methods and in various systems.

For the decomposition in high vacuo at 343–373 K, activation energies of 45 kJ mol^{-1} and 72 kJ mol^{-1} have been determined for the nucleation stage and for the main reaction, respectively. The former is attributed[39] to the energy required to break the O—O bond of molecules situated at dislocations, and the latter is assigned to the induced decomposition of **11** in perfect regions of the crystal as a result of the attack of phenyl radicals generated in the nucleation process. The low activation energy of the spontaneous decomposition of **11** is explained by an increased dihedral angle of the peroxide, due to distortions in the vicinity of a dislocation. The magnitude of the dihedral angle is important, since any change from its optimal value will lead to a lessening of the O—O bond energy. This is qualitatively valid, but according to theoretical calculations[48] the increase the angle cannot lead to such a large decrease (125–45 kJ mol^{-1}) of the dissociation energy of the O—O bond.

Discussing the dissociation energy of the peroxidic bond we should note that the method used incorporates into the result the activation energy of the induced decomposition of **11**. Moreover, the low values may be connected with some sublimation of undecomposed **11** and of the reaction products. The heat of sublimation of **11** and the products will range between 50 and 80 kJ mol^{-1} [49].

4. *Decomposition in the presence of admixtures*

The presence of carbon black or of activated charcoal decreases the decomposition temperature of **11** in the solid state to 60°C[50a]. The shape of the isothermal autocatalytic curve is the same as in the absence of admixtures but the induction period is shortened[50b]. For all types of carbon black the shortening of the induction period is relatively more than acceleration in the region of the maximum decomposition rate. The presence of an admixture facilitates primary formation of radicals and reaction centres in the system. The activation energy determined from τ_m is still very high (380 kJ mol^{-1}). This indicates again how difficult is spontaneous dissociation of most peroxide molecules in the crystalline phase. Autocatalysis is much more affected by the area of the surface of the admixtures than by the concentration of unpaired spins in the admixture[50b]. A similar accelerating influence of carbon black is also observed when **11** is dissolved in benzene[51] or methyl methacrylate[52]. The redox decomposition of **11** on the surface of carbon black is able to initiate polymerization as well as induce decomposition of peroxide. In the catalysed decomposition, hydrogen is transferred from carbon black to peroxide. The decomposition of **11** may also be initiated by other organic and inorganic admixtures[53].

Water retards the decomposition of **11** in the solid phase. One of the explanations is that water, owing to its high heat of evaporation, cools **11** and thus retards its decomposition. However, the induction period of the decomposition releases only a little heat which should be transferred the surroundings without difficulty, and hence the mechanism of cooling must only be considered in the region of high decomposition rate, when it is rather difficult to stop the reaction. This is supported by experience of an explosion of a concentrated solution of **11** in chloroform at about 65°C.

A more probable mechanism of retardation may be based on adsorption of water on primary defects of the crystalline structure of **11**, which might lead to their partial deactivation.

Chemical interaction of a polarized radical pair with water is not excluded either. In this way the accumulation of free radicals during the induction period of the decomposition would be diminished.

C. Cyclic Alkylidene Peroxides

Dimeric peroxides of the tetraoxane type:

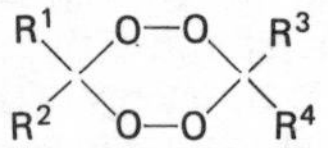

R¹ O—O R³
R² O—O R⁴

are highly explosive when the substituents have small. On the other hand, some cyclic dimeric peroxides with larger substituents are stable up to 200°C[23b].

A comparison[54] of the kinetic parameters of the decomposition of molten 3,6-bis(4-methoxycarbonylbutyl)-1,2,4,5-tetraoxane **(14)** with the solid-state decomposition of 3,6-bis(4-aminocarbonyl-butyl)-1,2,4,5-tetraoxane **(15)** is interesting (Table 4). The decomposition of molten **14** has very similar kinetic parameters to those of liquid alkyl peroxides[55]. A similar behaviour has also been found for **15** dissolved in dimethyl formamide ($A = 3 \times 10^{13}\, s^{-1}$, $E = 138\, kJ\, mol^{-1}$). In the absence of solvent, the decomposition of **15** shows unusually high values for both the apparent activation energy and for the frequency factor. These very high Arrhenius parameters are due to the stable aggregation of **15** in the crystalline state.

The decomposition in the crystalline state proceeds with a large autocatalytic acceleration and the exponent n in the relationship $\alpha = kt^n$ is 3–3.5. Here α is the fraction of the peroxide which reacts in time t and is a temperature-dependent constant. The products formed do not melt below 300°C. The reason for autocatalysis in the decomposition of **15** is mainly a gradual increase in the size of reaction nuclei.

An analysis of the reaction products and a study of the changes of the crystal structure of the decomposing cyclic peroxides would be desirable.

TABLE 4. A comparison of characteristic features of two cyclic peroxides **14** and **15**

Property	$RC(O)(CH_2)_4CH\langle O{-}O, O{-}O\rangle CH(CH_2)_4C(O)R$	
	R = CH_3O— **(14)**	R = H_2N— **(15)**
Melting point (°C)	82	207
Temperature at the beginning of decomposition (°C)[a]	159	200
Temperature at the maximal rate of decomposition (°C)[a]	199	214
Maximal rate of decomposition ($\%\,s^{-1}$)	0.28	3.29
$A(s^{-1})$[b]	5.8×10^{14} [d]	6.3×10^{74} [c]
E(kJ mol^{-1}) b	151[d]	710[c]
E(kJ mol^{-1}) f	—[g]	230[e]

[a]Rate of heating = 8°/min.
[b]From the differential enthalpic curves calculated according to Ref. 56.
[c]For the temperature range 207–225°C in the solid state.
[d]For the temperature range 161–221°C in the molten peroxide.
[e]For the temperature range 190–202°C in the solid state.
[f]From the parameters τ_m and v_m of isothermal decomposition; τ_m and v_m are defined in Figure 5.
[g]Decomposition of the molten peroxide without autocatalysis.

D. Hydroperoxides

Although several hundreds of hydroperoxides have been synthesized, little is known about their solid-state reactions, since most of them are liquid and stable[57]. Hydroperoxides with high melting points are often stated as decomposing at higher temperatures.

Mesityl hydroperoxide **(16)** decomposes at 105°C with a half-time of about one hour[58]. There are at least three pathways, two of which have been experimentally verified (equation 7).

$$2,4,6\text{-}(CH_3)_3C_6H_2COCH_2OOH \ \textbf{(16)}\ (\text{m.p. } 134^\circ C) \xrightarrow{105^\circ C} \begin{cases} \xrightarrow{49\%} 2,4,6\text{-}(CH_3)_3C_6H_2COOH + C_6H_5CHO \\ \xrightarrow{11\%} 2,4,6\text{-}(CH_3)_3C_6H_2COCO\text{-}C_6H_5 + H_2O \\ \xrightarrow{40\%\,?} \end{cases} \quad (7)$$

The decomposition of hydroperoxides is very sensitive to the catalytic effect of various compounds, and hence it is not surprising that the decomposition of **16** also takes place at room temperature. An interesting aspect of the slow room-temperature decomposition is that only the hydroperoxide in contact with the glass container turns yellow, while the material in the interior or on top of the sample does not change. Decomposition is evidently catalysed by contact with the glass surface.

The large sensitivity of hydroperoxides to catalytic decomposition could possibly be used for studying bimolecular solid-state reactions.

E. Peroxy Acids

The decomposition yields two basic products: aliphatic peroxy acids yield alcohols by loss of CO_2 and aromatic peroxy acids decompose into acids. Among the products released O_2, CO_2 and H_2O have been detected. The decomposition is enhanced by lowering of the overall gas pressure in the reaction system. In inert solvents peroxy acids are decomposed by a bimolecular mechanism with activation energies of about 71 kJ mol^{-1}. The stability of these compounds increases in the solid state by a factor of 10^3–10^6, probably due to the decrease in the mobility of molecules in the crystalline state. This is supported by the fact that the order of stability of peroxy acids in the solid state roughly corresponds to their melting points[40] (Table 5), although the rate is also influenced by the chemical structure. Similar effects of substituents have also been observed in the decomposition of peroxy acids in solution[59].

1. Autocatalysis

The decomposition of *p*-nitroperoxybenzoic acid **(17)** at 85–100°C is described by the relation:

$$\alpha = kt^{3/2}$$

where α is the fraction of decomposed **17** in time t and k is the temperature-dependent constant. The activation energy is 160 kJ mol^{-1}, more than twice the value in solution[60].

In agreement with the autocatalytic character of the decomposition, *p*-nitrobenzoic acid added to **17** accelerates the reaction, by producing defects in the crystal structure. Faster decompositions of aged peroxy acids can be explained similarly as due to the small nuclei of *p*-nitrobenzoic acid formed by a very slow decomposition at room temperature. Glass wool catalyses the initial stages of the decomposition as it does with alkyl hydroperoxides, while grinding of the crystals or the presence of the product gases has in this case no significant effect on the rate of decomposition. Oxalic acid inhibits the decomposition by deactivation of primary reaction nuclei, either by adsorption on the active centres of the crystal lattice or by chemical inhibition of the decomposition chain-reaction.

The decomposition of *m*-nitroperoxybenzoic acid **(18)** is faster than that of **17** at the same temperature, but acceleration occurs only to a lower percentage of decomposition[61]. The mathematical description of this reaction is given by the relation:

$$\log\left(\frac{\alpha}{1-\alpha}\right) = k't + \text{const.}$$

The lower activation energy of **18** may be due to the decomposition temperature being closer (2–25°C) to the melting point of the crystals. The value (125 kJ mol^{-1}) is similar to that of the radical decomposition of **18** in a solution of benzene and 2,2-diphenyl-1-picryl

TABLE 5. Melting points and approximate half-times of decomposition of some peroxy acids

Peroxy acid	m.p. (°C)	$\tau/2$ at 25°C (weeks)
$O_2N-C_6H_4-CO_3H$ (para)	138	>2700[a], 4500[b]
$H_3CO-C_6H_4-CO_3H$ (para)	110	105[a]
$C_6H_4(NO_2)CO_3H$ (ortho)	95	550[a]
$C_6H_4(NO_2)CO_3H$ (meta)	92	95[c]
$(CH_3)_3C-C_6H_4-CO_3H$ (para)	82	550[a]
$CH_3(CH_2)_{10}CO_3H$	50	115[a]
$C_6H_5CO_3H$	42	2.9
$CH_3(CH_2)_{15}CHBrCO_3H$	41	1.5

[a]Half-times are calculated from initial decomposition[40]. Regarding autocatalysis in decomposition reaction, half-times may be by 10–30% shorter.
[b]Value extrapolated from the temperature range 90–105°C[60].
[c]Value extrapolated from the temperature range 67–90°C[61].

hydrazyl[62]. In the solid-state decomposition, however, it is the activation energy of the total decomposition, but in solution it is only of a minor process, because the main reaction is bimolecular with a lower value of E (71 kJ mol^{-1}).

2. *Stoichiometry of the decomposition*

The decomposition of **17** and **18** conforms[60,61] to equation (8). This equation seems to show that in the decomposition some external oxygen was also consumed. Considering the analytical inaccuracies the question need not be analysed any more. It is, however, substantial that apart from a large portion (ca. 90 mol %) of the decomposition, where only oxygen atom is released, in the rest of the molecules almost all the carbon and

$$25\,NO_2C_6H_4COOOH \longrightarrow 23\,NO_2C_6H_4COOH + 14\,CO_2 + 5\,H_2O + N_2 \qquad (8)$$

hydrogen atoms are totally oxidized to give H_2O and CO_2. The causes are not clear. We can, however, assume that the normal decomposition may lead to microexplosions. During the slow decomposition, oxygen accumulates in the vacancies of the crystal structure of the relatively stable nitrobenzoic acids being formed. On reaching a critical concentration of the liberated oxygen, explosive combustion of the surrounding small part of peroxy acid will occur in this region. The explosive reaction may be initiated by reactive intermediates; the released heat of reaction supports the decomposition. The interruption of the explosive decomposition causes consumption of available oxygen in the microregion of the fast reaction. The microexplosion is thus stopped.

With increasing temperature, the frequency of microexplosions in peroxy acid will increase until it turns into an explosion of the whole sample. The stoichiometric shortage of oxygen will appear as production of soot and hydrogen.

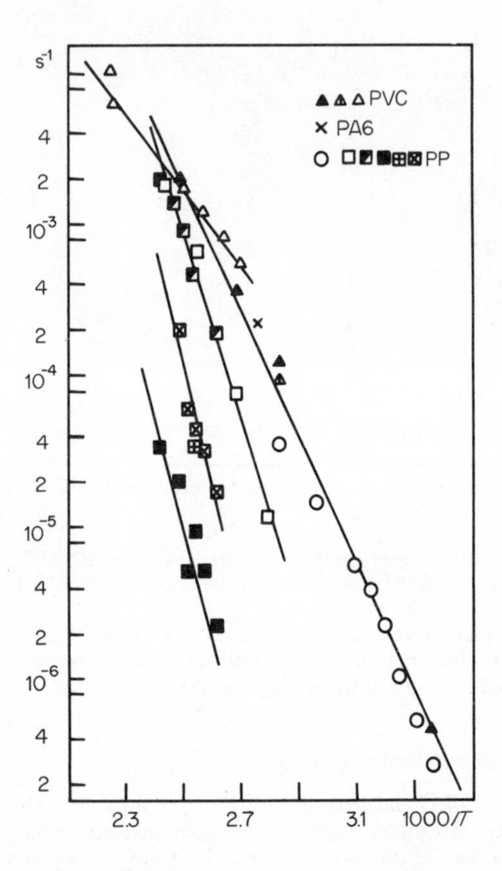

FIGURE 6. Dependence of the rate constants of decomposition on the reciprocal values of the absolute temperature for various polymer hydroperoxides in the solid phase: Polymer hydroperoxides of poly(vinyl chloride) ▲[63], ◭[64], △[65], polycaprolactam ×[66] and polypropylene (PP) ○[67] were prepared by oxidation in the solid state. Further PP hydroperoxides were obtained by oxidation in cumene ■[70], trichlorobenzene ◩[69] ⊠[70], chlorobenzene □[68] and benzene ⊞[71].

F. Polymer Hydroperoxides

Although many papers have dealt with models of ageing, oxidation and combustion of polymers which considered hydroperoxides, only a few quantitative results are available on the solid-state decomposition and kinetics of polymeric hydroperoxides.

Even in these cases the structure of the polymeric hydroperoxides was not determined unambiguously. This is mainly due to the complicated structure of the polymers and the nonspecific methods for their preparation.

The rate of decomposition is more affected by the method of preparation than by the type of the polymer (Figure 6). This conclusion cannot be now generalized for all kinds of polymers; but it is noteworthy that the same temperature dependence of the decomposition rate constant is valid for hydroperoxides of different polymers [poly(vinyl chloride), polycaprolactam, polypropylene] if they are prepared by oxidation of the polymer at about 20°C in the solid state. On the other hand, hydroperoxides of the same polymer prepared by oxidation in solution in various types of solvent and different oxidation temperatures, decompose under identical conditions at rates differing by two to three orders of magnitude (Table 6).

TABLE 6. Kinetic parameters of the decomposition of polymer hydroperoxides in the absence of solvent

Hydroperoxide	A (s^{-1})	E ($kJ\ mol^{-1}$)	k at 100°C (s^{-1})	Reference
Polypropylene[a]	9.6×10^{9}	114	8.8×10^{-7}	70
[b]	2.4×10^{6}	79	1.8×10^{-5}	70
[b]	5.4×10^{10}	106	6.6×10^{-5}	69
[c]	2.3×10^{13}	125	6.0×10^{-5}	68
[d]	9.2×10^{6}	76	1.9×10^{-4}	67
Poly(vinyl[d]	1.2×10^{8}	87	6.9×10^{-5}	65
chloride)[d]	1.6×10^{8}	83	3.4×10^{-4}	63
Polycaprolactam[d]	1.9×10^{12}	110	6.4×10^{-4}	66
Polyethylene[c]	1.6×10^{14}	146	4.6×10^{-7}	70
	1.5×10^{10}	105	2.5×10^{-5}	72
Polystyrene[c]	4.3×10^{14}	136	3.2×10^{-5}	74
	9.0×10^{6}	82	2.6×10^{-5}	75
t-Butyl hydroperoxide[e]	3.0×10^{15}	155	4.7×10^{-7}	73

[a–d]Preparation of polymer hydroperoxides by oxidation in: (*a*) cumene, (*b*) trichlorobenzene, (*c*) chlorobenzene, (*d*) solid state.
[e]Decomposition in low molecular weight solvents.

1. Influence of the aggregation of hydroperoxide groups on the decomposition rate

The great influence of the method of preparation of hydroperoxides, especially of polypropylene, on their stability is connected with the distribution of unstable hydroperoxide groups on a polymer chain. Owing to the considerable sensitivity of hydroperoxide groups both to radical-induced and to bimolecular decomposition, the relative rates are higher when the local concentration of hydroperoxide groups is high.

It seems therefore logical that hydroperoxide groups prepared by oxidation of polymers in the solid state decompose most rapidly. This method will produce hydroperoxide groups at the surface of the polymer particles close to each other. Oxidation of polypropylene in a dilute solution of an inert solvent (e.g. trichlorobenzene) gives hydroperoxide groups dispersed more regularly, but association of hydroperoxide groups within one macromolecule is observable by infrared spectroscopy[76]. More regular attack on the polypropylene chain is achieved by applying a reactive solvent which is also oxidized (e.g. cumene), when isolated hydroperoxide groups are formed on the polypropylene[70]. This type of polypropylene hydroperoxide has the highest thermal stability.

We can see the explanation of the solvent effect on hydroperoxidation of polypropylene in the scheme of propagation reactions in an inert solvent (equation 9) and in the reactive solvent SH (equation 10).

$$-CH(CH_3)-CH_2-C(CH_3)(O-O\cdot)- \longrightarrow -\dot{C}(CH_3)-CH_2-C(CH_3)(O-O-H)- \xrightarrow{O_2} -C(CH_3)(O-O\cdot)-CH_2-C(CH_3)(O-O-H) \quad (9)$$

$$-CH(CH_3)-CH_2-C(CH_3)(O-O\cdot)- + SH \longrightarrow -CH(CH_3)-CH_2-C(CH_3)(OOH)- + S^{\cdot} \quad (10)$$

The behaviour of associated OOH groups during decomposition in solution and in the solid phase is comparable[68]. In the polymer, the distance among OOH groups may be similar in solution and in the solid phase.

The large effect of the local concentration of OOH groups generally leads to lower activation energies of decomposition and lower frequency factors than in the decomposition of low molecular weight hydroperoxides. The lowering of the activation energy and of the frequency factor is connected with the fact that the decomposition of polymeric hydroperoxides is not a monomolecular dissociation of the peroxide bond, but a much more complex process.

There is a relation between the frequency factor A and the activation energy E in the modified Arrhenius equation:

$$\ln A = \ln k + E/RT$$

The temperature of 178°C can be calculated from the slope, which is probably the limit above which the decomposition of various hydroperoxides proceeds with the similar rate constant. It follows from the prevalence of spontaneous monomolecular dissociation over induced decomposition of hydroperoxy groups in the higher temperature range.

2. *Mobility and reactivity*

The decomposition of polymer hydroperoxides may also be induced by semistable radicals. The bimolecular hydrogen-transfer reaction of phenoxy radicals with the hydroperoxide of isotactic polypropylene (HOOPP) in the solid state (equation 11) is

$$R-C_6H_2(C(CH_3)_3)_2-O^{\cdot} + HOOPP \longrightarrow R-C_6H_2(C(CH_3)_3)_2-OH + {}^{\cdot}OOPP \quad (11)$$

slower by 2–3 orders of magnitude than with low molecular weight hydroperoxides in the liquid phase[77]. Such decreases in the rate constants are observed even in relatively slow reactions which should not be limited by diffusion of the reacting particles.

The structure of a semistable radical on the rate constant with a solid polymeric hydroperoxide has a much smaller effect than with low molecular weight liquid hydroperoxides. Several models have been proposed for the explanation of the levelling of the reactivity in various systems[2]. The lengthening of the time of contact of the reactant in the cage of the polymer medium seems to be decisive in reducing the differences in the reactivity. When the contact is long enough, even very slightly reactive reactants will react, whereas in the course of short collisions only very reactive particles may do the same.

In further investigations it would be interesting to examine reactions of peroxide groups in the polymer backbone. Reactions of high molecular weight peroxides ought to be sensitive to the mobility of the polymer chain, to mechanical deformation of the solid samples, and to changes in the physical structure of the polymer.

V. DECOMPOSITION OF LOW MOLECULAR WEIGHT PEROXIDES IN POLYMERIC MEDIA

The specific influence of polymers on the decomposition of peroxides is determined by the small mobility and relatively great microheterogeneity of the reaction medium. The effect of mobility governs the kinetic parameters of the reaction. The influence of microheterogeneity has not been studied sufficiently up to now, although it too seems to be an important factor.

A. Kinetic Parameters

1. Spontaneous and chain decomposition

Since the polymer's great viscosity retards diffusion of primarily formed radicals, it might seem surprising that the rate constants for the spontaneous decomposition of dibenzoyl peroxide (**11**) in polymers even in the glassy state are only slightly lower than in solvents (Table 7). This phenomenon seems to be connected with several possible reactions of **11**. These include splitting of **11** into benzoyloxy radicals ($R^{\bullet}$), reversible dimerization of the latter and their diffusion into the reaction medium (equation 12). In addition, primary radicals may fragment in a cage (k_{β}) and form a modified pair of radicals, the recombination of which may lead to formation of stable products. Fragmentation and combination reactions are thus mainly responsible for the relatively fast decomposition of **11** in a solid polymer medium. Slow diffusion and cage reactions lower the yield of free radicals in the reaction medium. This is the inherent reason for the smaller extent of induced decomposition of **11** in polymers, compared with solvents. For instance, the ratio

$$\mathbf{11} \underset{k_r}{\overset{k}{\rightleftharpoons}} [R^{\bullet}\ {}^{\bullet}R] \xrightarrow{k_D} R^{\bullet} + R^{\bullet}$$

$$[R^{\bullet}\ {}^{\bullet}R] \xrightarrow{k_{\beta}} [R^{\bullet} + r^{\bullet}] + CO_2$$

$$\text{Unreactive products} \xleftarrow{k_t} [R^{\bullet} + r^{\bullet}] \xrightarrow{k_D} R^{\bullet} + r^{\bullet} \qquad (12)$$

TABLE 7. Rate constants for the decomposition of dibenzoyl peroxide (**11**) at 75°C and activation energies of the spontaneous homolysis of its peroxidic bond in various polymeric media

Medium	Atmosphere	k_s at 75°C ($10^6\,s^{-1}$)	T(°C)	E (kJ mol^{-1})	Ref.
Poly(vinyl chloride)	O_2	3.7	65–77	171	82
			77–83	162	
			83–99	136	
Polyethylene	O_2	3.4	80–90	173	83
	N_2	5.6	60–89	211	84
Polystyrene	O_2	5.3	70–80	149	82
			80–90	124	
	O_2	6.3	56–83	151	85
			83–91	126	
	O_2	4.9	80	150[a]	86
	N_2	16	80	150[a]	86
Triacetyl cellulose	O_2	4.3	92–115	117[b]	87
	Vacuum	6.1	80–98	117[b]	88
Polyisobutylene	N_2	6.1	65–95	148	89
Polycarbonate[c]	CO_2	7.9	86–100	125	90
Polypropylene (isotactic)	Vacuum	9.2	80	125[a]	91
Polypropylene (atactic)	O_2	18	71–92	124	92
	N_2	15	65–87	125	93
	Vacuum	37	80	125	91
Poly(1-butene)[d]	O_2	19	71–92	124	92
Poly(4-methyl-1-pentene)[d]	O_2	25	71–92	123	92
Polyamide[e]	He	23	75–98	138	94
	O_2	10	80–98	125	95
Poly(ethyl acrylate)	O_2	18	85–110	124	96
Poly(methyl methacrylate)	O_2	6.0	100	125[b]	97
	N_2	7.2	80	125[b]	98
	N_2	25	72–105	169	99
			105–125	134	
Polyformaldehyde	O_2	12	80–100	124	100
	Vacuum	53	80–100	124	100
Benzene	N_2	17.5[f] ±3.8	50–100	125 ±4	101

[a] Value used for conversion of measured k to k (75°C).
[b] Determined with 1% of **11** in the polymer; with 10% of **11** the value of E is 155 kJ mol^{-1}.
[c] Reaction product of phosgene with 2,2′-bis(4-hydroxyphenyl)propane.
[d] The polymer contained 2,6-di-*t*-butyl-*p*-cresol (for elimination of the induced decomposition).
[e] Condensation product of hexamethylenediamine with adipic acid.
[f] Average from values of seven literature sources.

of the total decomposition rate constant at 5% of **11** and at an extrapolated zero peroxide concentration is 1.86 in polystyrene[86] but is 6.19 in isopropylbenzene[78]. The induced decomposition should be affected by the slower diffusion in the polymer medium, so that a generally higher activation energy should be expected for the process than in solvents. This has indeed been observed experimentally[84,88–90,93,100]. Chain-decomposition of **11** in polymer media is slower in the presence of oxygen[79].

A slight decrease in the rate of the spontaneous decomposition in polymeric media is also observed for dicumyl peroxide. At 110°C, the rate constants in isotactic polypropylene[70] are 2–3 times lower than in benzene. However, the viscosity of the polymers has an almost negligible effect on the rate of spontaneous peroxide decompositions in the rubber-like state of polymers as is observed with dimethyldi(*t*-butylperoxy)silane in polyethylene and in polystyrene at temperatures of 170–210°C[81]. The reason for the more enhanced decrease of the rate constant of the spontaneous decomposition in glassy or crystalline polymers compared to rubber-like media involves the slower dissociation of the peroxide and not only the reversible recombination of the primary radicals, as commonly assumed for the small retardation of peroxide decomposition in low molecular weight solvents with increasing viscosity.

2. *Activation energy of the dissociation*

If decomposition of **11** takes place above the glass transition temperature (T_g) of the amorphous polymer, the activation energy is similar to the values found in solvents. This is probably due to a low modulus of elasticity of the polymer matrix in the rubber-like state[102].

Dissociations producing two radicals by a simple bond fission should be characterized by a volume increase in the transition state of about $\Delta V^{\neq} \approx +10\,cm^3\,mol^{-1}$. The bond lengthening calculated from this value is about 50 pm[103]. During the decomposition of **11** in poly(methyl methacrylate)[97] $\Delta V^{\neq}$ is $7.5\,cm^3\,mol^{-1}$ and in polyethylene[104] $9.1\,cm^3\,mol^{-1}$. The energy needed for pushing away the surrounding molecules can be calculated from a product of the modulus of elasticity of a polymer matrix and the activation volume of the dissociation. If we take a modulus of elasticity of 5×10^3 MPa, a typical value for polymer glasses, and an activation volume of $8\,cm^3\,mol^{-1}$, then the activation energy for the decomposition in such a matrix will be higher by $40\,kJ\,mol^{-1}$. This value agrees quite well with the observed increase in the activation energy for the decomposition of **11** in the glassy state of poly(vinyl chloride), polystyrene and poly(methyl methacrylate). The zero or small increase in the activation energy of the dissociation of **11** in polycarbonate is surprising. T_g of this polymer is 150°C whereas the three preceding polymers had a T_g of about 100°C. Hence we must assume that polycarbonate has free space, allowing formation of the activated complex without having to push away polymer chains. The modulus of volume elasticity of gases is 4–5 orders of magnitude smaller than that of polymer glasses. Then the contribution to the increase of activation energy for decomposition in microvoids of polymer, filled with gaseous molecules, is negligible.

Another exception is the high activation energy for spontaneous decomposition in polyethylene. Its cause does not seem to be the high viscosity of the reaction medium as had earlier been assumed[84], but rather the limited solubility of **11** in polyethylene. With increasing temperature the amount of dissolved **11** increases and the fraction of the crystalline phase of polyethylene decreases. Since the decomposition of dissolved peroxide is faster than of crystalline **11**, a rise in the amorphous phase of the polymer leads to an increase in the reaction rate, which is in turn reflected in the apparent activation energy for dissociation.

A similar situation arises in triacetylcellulose at higher **11** concentration (see footnote *b* to Table 7).

A high activation energy for the spontaneous decomposition has also been observed for dicumyl peroxide in isotactic polypropylene. The decomposition below the melting point of the crystallites of the polymer has the activation energy of $162\,kJ\,mol^{-1}$[70] or $166\,kJ\,mol^{-1}$[80], while in benzene it is $144\,kJ\,mol^{-1}$[105] or $146\,kJ\,mol^{-1}$[106].

3. Cage effect in the polymer matrix

In radical dissociation of molecules in the condensed phase, a radical pair is formed, which exists for a certain time during which it may diffuse into the surroundings or may react to form molecules.

The fraction f of radicals which will escape from the cage in a polymer is evidently smaller than in a low molecular weight solvent because of slower diffusion of radicals in the polymer medium. The greater the viscosity of the medium, and the larger the radicals formed, the smaller is f. This qualitative tendency is observed (Table 8) if we realize that the value of f in low molecular weight solvents varies between 0.5 and 0.7.

TABLE 8. Fraction of radical escape (f) from polymer cage for some peroxides

Peroxide	Polymer	T(°C)	f	Reference
Dibenzoyl peroxide	Polypropylene[a]	80	0.15[g]	107
		85	0.09[h]	92
		130	0.10[h]	108
	Polypropylene[b]	70	0.1[h]	92
		80	0.5[h]	92
	Poly(1-butene)[c]	70	0.05[h]	92
		80	0.07[h]	92
		90	0.1[h]	92
		95	0.5[h]	92
	Poly(4-methyl-1-	85	0.01[c]	92
	pentene)[d]	95	0.05[c]	92
	Poly(butanediol dimethacrylate)	85	0.17	109
Dilauroyl peroxide	Polystyrene[e]	80	0.33[i]	110
Dicumyl peroxide	Polypropylene[a]	120	0.058[g,j]	80
		140	0.098[g,j]	80

[a] Isotactic polymer, m.p. = 165°C.
[b] Atactic polymer, $T_g = -10$°C.
[c] Isotactic polymer, m.p. = 112°C.
[d] Isotactic polymer, m.p. = 240°C.
[e] Atactic polymer, $T_g = 95$°C.
[g] From chromatographic analysis of the reaction products.
[h] Escape of radicals capable of initiation reaction in polymer oxidation.
[i] Fraction of radicals reacting with present α-naphthol.
[j] Escape of primary cumyloxy radicals from the cage.

The quantitative evaluation of the effect of the viscosity of the polymer medium on the diffusion of free radicals shows[108] that the fraction of radicals released from the cage is greater by 3–4 orders of magnitude than the theoretically estimated value of f. The degree of escape from the cage depends more on the reactivity of the radicals than on their diffusion coefficient. It is therefore justifiably assumed that the escape of radicals from the cage proceeds via competitive chemical reaction of the original radicals with the polymer medium. However, quantitative analysis does not agree with this as the only additional mechanism besides diffusion. Lower mobility of radicals in the polymer cage may thus also retard their reversible recombination.

The problem may be clarified by a study of oxygen scrambling and carboxy inversion of peroxy compounds dissolved in solid polymers.

B. Microheterogeneity of the Polymer Medium

The glassy and crystalline polymer media have frozen fluctuations presenting nonequivalent positions for the decomposing peroxide. Peroxide molecules fragment more easily, e.g. at sites where there is more free space than in the close surroundings of rigid polymer chains. The few available experimental data support this idea. The decomposition of **11** in polypropylene shows[91] that the overall rate constant for the decomposition and the efficiency of the escape of free radicals from the cage depends on the peroxide concentration and on the amount of additives, such as phenyl benzoate **(19)**. Addition of **19** accelerates the decomposition of **11** in an inert atmosphere, and usually increases the rate of polymer oxidation (in the presence of O_2) but decreases the velocity of polypropylene degradation. Dissolution of both **11** ($0.01\ \mathrm{mol\,kg^{-1}}$) and **19** ($0.1\ \mathrm{mol\,kg^{-1}}$) causes only slight changes in the arrangement of the polypropylene molecules. All the facts indicate that the low molecular weight compounds preferentially dissolve in defects or free spaces and that chemical reactions take place in these microregions.

The effect of local surroundings or the physical state of the sample on the decomposition of **11** can be illustrated by a change in the reaction rate as a function of the degree of mechanical deformation of the medium[111]. In deformed polycarbonate containing **11** as much as a tenfold acceleration of the initial decomposition rate of the peroxide is observed, compared to samples without mechanical extension. Since in the motion of macromolecules structural microheterogeneity is developed, **11** will probably be displaced into regions where the decomposition is faster. Both spontaneous and induced peroxide decomposition may be accelerated. Any detailed discussion is, however, untimely due to insufficient experimental data.

VI. CONCLUSION

Reactions of organic peroxides in the solid phase have not been a frequent subject of investigation. Achievements in the study of these compounds in the liquid phase will probably stimulate efforts to compare reactions under less common conditions. Recently developed simpler and faster methods for determining the structure of crystalline substances enable analysis of the arrangement of reacting groups in reactants and products.

At present the investigation of peroxide reactions in the solid state does not offer many practical examples, but we may expect that future research will be directed toward reactions interesting from the synthetic point of view (e.g. decomposition of cyclic peroxides to yield cyclic hydrocarbons, etc.). The different conformations of molecules in the crystal and in solution will probably influence the selectivity of the reaction as will also changes of mobility in the crystal. The latter may be realized by examining reaction products in the crystalline phase at temperatures both considerably below and approaching the melting temperature of the reactant.

Reactions of peroxides in the solid state are often more sensitive to temperature changes than in the liquid phase. Kinetic parameters obtained may be influenced by structural changes in the reactant or in the solid reaction medium. The values for apparent activation energies may therefore be different for various temperature regions and their physical interpretation is not always the same. Obviously, the cage effect is more prominent in solid-state reactions than in solution.

The kinetic courses of peroxide decomposition in the solid phase support the theory[112] that these reactions start in defects of the crystalline structure and many of them take place at the boundary of the old and new crystalline phases. Depending on the rate of formation of reaction centres (nucleation), the reaction may proceed either with or without an induction period. The reaction is autocatalytic under isothermal conditions. When the heat dispersion is insufficient, explosions may occur.

VII. REFERENCES

1. D. Y. Curtin, J. C. Paul, E. N. Duesler, T. W. Lewis, B. J. Mann and W. Shiau, *Mol. Cryst. Liquid Cryst.*, **50**, 25 (1979); J. Z. Gougoutas, *Israel J. Chem.*, **10**, 395 (1972); M. D. Cohen and B. S. Green, *Chem. Brit.*, **9**, 490 (1973).
2. O. N. Karpukhin, *Usp. Khim.*, **47**, 1119 (1979).
3. V. A. Yablokov, *Usp. Khim.*, **49**, 1711 (1980).
4. J. E. Leffler, R. D. Faulkner and C. C. Petropoulos, *J. Amer. Chem. Soc.*, **80**, 5435 (1958).
5. W. Honsberg and J. E. Leffler, *J. Org. Chem.*, **26**, 733 (1961).
6. J. Z. Gougoutas, *Pure Appl. Chem.*, **27**, 305 (1971).
7. M. C. Etter, *J. Amer. Chem. Soc.*, **98**, 5326, 5331 (1976).
8. J. Z. Gougoutas and D. G. Naae, *J. Phys. Chem.*, **82** 393 (1978).
9. J. Z. Gougoutas and L. Lessinger, *J. Solid State Chem.*, **12**, 51 (1975).
10. J. E. Leffler and A. F. Wilson, *J. Org. Chem.*, **25**, 424 (1960).
11. (a) J. W. Taylor and J. C. Martin, *J. Amer. Chem. Soc.*, **88**, 3650 (1966); **89**, 6904 (1967). (b) N. J. Karch and J. M. McBride, *J. Amer. Chem. Soc.*, **94**, 5092 (1972).
12. (a) J. Connor, 'The thermal decomposition of organic peroxides', *R.A.R.D.E. Memorandum*, 15 (1974).
    (b) J. Leffler, *J. Amer. Chem. Soc.*, **72**, 67 (1950); F. D. Greene, H. P. Stein, C. C. Chu and F. M. Vane, *J. Amer. Chem. Soc.*, **86**, 2080 (1964); T. Kashiwagi, S. Kozuka and S. Oae, *Tetrahedron*, **26**, 3619 (1970); C. Walling, H. P. Waits, J. Milanovic and C. G. Pappiaonnou, *J. Amer. Chem. Soc.*, **94**, 2483 (1972); S. Oae, K. Fujimori, S. Kozuka and Y. Uchida, *J. Chem. Soc., Perkin Trans. 2*, 1844 (1974); A. Blaschette and D. Brendes, *Chem. Z.* **99**, 125 (1975); K. Taylor, C. K. Govindan and M. S. Kaelin, *J. Amer. Chem. Soc.*, **101**, 2091 (1979).
13. H. Hock and H. Kropf, *Ber. Deut. Chem. Ges.*, **88**, 1544 (1955).
14. R. L. Dannley and P. K. Tornstrom, *J. Org. Chem.*, **40**, 2278 (1975).
15. C. J. Pedersen, *J. Org. Chem.*, **23**, 255 (1958).
16. W. Adam, *Angew Chem.* (*Intern. Ed. Engl.*), **13**, 616 (1974).
17. J. Rigaudy, *Pure Appl. Chem.*, **16**, 169 (1968); M. Sasaoka and H. Hart, *J. Org. Chem.*, **44**, 368 (1979).
18. K. Gollnick and G. O. Schenck in *1,4-Cycloaddition Reactions* (Ed. J. Hamer), Academic Press, New York, 1967, pp. 255–345.
19. C. Dufraisse and L. Enderlin, *Compt. Rend.*, **191**, 1321 (1930); C. Dufraisse and J. Le Bras, *Bull. Soc. Chim. Fr.*, **4**, 349, 1037 (1937); C. Dufraisse and A. Etienne, *Compt. Rend.*, **201**, 280 (1935).
20. S. Markuš, *Thesis*, Bratislava, 1975.
21. N. J. Turro, M. F. Chow and J. Rigaudy, *J. Amer. Chem. Soc.*, **101**, 1300 (1979).
22. B. Stevens and R. D. Small, Jr., *J. Phys. Chem.*, **81**, 1605 (1977).
23. M. Schulz and K. Kirshke, in *Organic Peroxides, III* (Ed. D. Swern) Wiley–Interscience, London–New York, 1972; (a) p. 133, (b) p. 67.
24. H. Wasserman and J. R. Scheffer, *J. Amer. Chem. Soc.*, **89**, 3173 (1967).
25. H. Wasserman, J. R. Scheffer and J. L. Cooper, *J. Amer. Chem. Soc.*, 4991 (1972).
26. N. J. Turro, M. F. Chow and J. Rigaudy, *J. Amer. Chem. Soc.*, **101**, 1300, 3701 (1979).
27. N. J. Turro and B. Kraentler, *Acc. Chem. Res.*, **13**, 369 (1980).
28. C. Dufraisse, J. Rigaudy, J. J. Basselier and N. K. Cuong, *Compt. Rend.*, **260**, 5031 (1965); J. Rigaudy, F. Gobert and N. K. Cuong, *Compt. Rend.*, **274**, 541 (1972).
29. C. Moreau, C. Dufraisse and P. M. Dean, *Compt. Rend.*, **182**, 1440 (1584 (1926).
30. C. Dufraisse, *Bull. Soc. Chim. Fr.*, **4**, 2052 (1937).
31. P. Bender and J. Faber, *J. Amer. Chem. Soc.*, **74**, 1450 (1952).
32. K. Lonsdale, E. Nave and J. F. Stephens, *Phil. Trans. Roy. Soc.* (*London*), **A261**, 1 (1966).
33. S. Markuš and M. Lazár, *Collect. Czech. Chem. Commun.*, **40**, 2469 (1975).
34. C. S. Foote, S. Wexler, W. Ando and R. Higgins, *J. Amer. Chem. Soc.*, **90**, 975 (1968).
35. N. Sugiyama, M. Iwata, M. Yoshioka, K. Yamada and H. Acyama, *Bull. Chem. Soc. Japan*, **42**, 1377 (1969).

36. M. Lazár, J. Rychlý and S. Markuš, *Chem. Zvesti*, **33**, 81 (1979).
37. J. W. Breitenbach and A. Kastell, *Monatsh. Chem.*, **85**, 676 (1954).
38. J. Rigaudy, C. Moreau and N. K. Cuong, *Compt. Rend.*, **274**, 1589 (1972).
39. S. E. Morsi, J. M. Thomas and J. O. Williams, *J. Chem. Soc., Faraday Trans. 1*, **71**, 1857 (1975).
40. L. S. Silbert, E. Siegel and D. Swern, *J. Org. Chem.*, **27**, 1336 (1962).
41. M. Lazár, P. Ambrovič, J. Mikovič and E. Borsig, *Thermochim. Acta*, **6**, 481 (1973).
42. R. C. Farmer, *J. Soc. Chem. Ind. (London)*, **40**, 84 (1921).
43. P. C. Bowes, *Combust. Flame*, **12**, 289 (1968).
44. P. Ambrovič and M. Lazár, *Europ. Polym. Suppl.*, 331 (1969).
45. H. Rheinboldt, *J. Prakt. Chem.*, **111**, 242 (1925).
46. D. H. Fine and P. Gray, *Combust. Flame*, **11**, 71 (1967).
47. Ch. Walling, H. N. Moulden, J. H. Waters and R. C. Neuman, *J. Amer. Chem. Soc.*, **87**, 518 (1965).
48. O. Kikuchi, A. Hiyama, H. Yoshida and K. Suzuki, *Bull. Chem. Soc. Japan*, **51**, 11 (1978).
49. A. S. Carson, P. G. Laye and H. Moris, *J. Chem. Thermodyn.*, **7**, 993 (1975).
50. J. Mikovič and M. Lazár, *Thermochim. Acta*, (a) **2**, 321 (1971); (b) **2**, 429 (1971).
51. B. R. Puri, K. V. Sud and K. C. Kalra, *Indian J. Chem.*, **9**, 966 (1971).
52. K. Ohkita, M. Uchiyama and M. Shimomura, *Kobushi Ronbushu*, **36**, 465 (1979).
53. D. H. Solomon, J. D. Swift, G. O'Leary and I. G. Treely, *J. Macromol. Sci. Chem.*, **A5**, 995 (1971); J. Rychlý, L. Matisová-Rychlá and M. Lazár, *Collect. Czech. Chem. Commun.*, **40**, 865 (1975); M. Lazár and P. Ambrovič, *Chem. Zvesti*, **23**, 881 (1969).
54. M. Lazár, P. Ambrovič and E. Borsig, *Thermochim. Acta*, **10**, 55 (1974).
55. P. Molyneux, *Tetrahedron*, **22**, 2929 (1966).
56. K. E. Barret, *J. Appl. Polym. Sci.*, **11**, 1167 (1967).
57. R. Hiatt in *Organic Peroxides, II* (Ed. D. Swern), Wiley–Interscience, New York–London, 1971, p. 40.
58. A. G. Pinkus, M. Z. Haq and J. G. Lindberg, *J. Org. Chem.*, **35**, 2557 (1970).
59. D. Swern in *Organic Peroxides, I*, (Ed. D. Swern), Wiley–Interscience, New York–London, 1970, p. 420.
60. D. F. Debenham, A. J. Owen and E. F. Pembridge, *J. Chem. Soc. (B)*, 213 (1966).
61. D. F. Debenham and A. J. Owen, *J. Chem. Soc. (B)*, 675 (1966).
62. S. R. Cohen and J. O. Edwards, *J. Phys. Chem.*, **64**, 1086 (1960).
63. J. Landler and P. Lebel, *J. Polym. Sci.*, **48**, 477 (1960).
64. A. Michel, E. Castaneda and A. Guyot, *J. Macromol. Sci. Chem.*, **A12**, 227 (1978).
65. L. Matisová-Rychlá, J. Rychlý, A. Michel, P. Ambrovič and I. Paška, *J. Luminiscence*, **17**, 73 (1978).
66. L. Sawtchenko, L. Dunsch and K. Jobst, *Acta Polymerica*, **31**, 206 (1980).
67. P. Citovický, D. Mikulášová and V. Chrástová, *Europ. Polymer J.*, **12**, 627 (1976).
68. J. C. Chien and H. Jabloner, *J. Polym. Sci.*, **A1,6**, 393 (1968).
69. L. Matisová-Rychlá, Z. Fodor, J. Rychlý and M. Ihring, *14th French–Czechoslovak Meeting on Oxidative Ageing and Combustion of Polymers, October, 1980.*
70. N. V. Zolotova and E. T. Denisov, *J. Polym. Sci.*, **A1,9**, 3311 (1971).
71. D. E. Van Sicle, *J. Polym. Sci.*, **A1,10**, 355 (1972).
72. J. C. Chien, *J. Polym. Sci.*, **A1,6**, 375 (1968).
73. L. Batt and S. W. Benson, *J. Chem. Phys.*, **30**, 895 (1962).
74. T. V. Pokholok, O. N. Karpukhin and V. Ya. Shlyapintokh, *J. Polym. Sci., Chem. Ed.*, **13**, 525 (1975).
75. L. Dulog and K. H. David, *Makromol. Chem.*, **145**, 67 (1971).
76. R. C. Boss, H. Jabloner and E. J. Vendenberg, *Polym. Letters*, **10**, 915 (1972).
77. V. I. Rubcov, V. A. Roginskij and V. B. Miller, *Vysokomol. Soedin.*, **A22**, 2506 (1980).
78. W. R. Foster and G. H. Williams, *J. Chem. Soc.*, 2862 (1962).
79. R. Rado, *Chem. Listy*, **61**, 785 (1967).
80. A. M. Tolks and V. S. Pudov, *Izv. Akad. Nauk Latv. SSR, Ser. Khim.*, 373 (1973).
81. N. P. Sluchevskaya, N. V. Yablokova, V. A. Yablokov and Yu. A. Alexandrov, *Zh. Obshch. Khim.*, **48**, 2511 (1978).
82. H. C. Haas, *J. Polym. Sci.*, **55**, 33 (1961).
83. R. Rado and M. Lazár, *J. Polym. Sci.*, **45**, 257 (1960).

84. R. Rado and M. Lazár, *Vysokomol. Soedin.*, **3**, 310 (1961).
85. H. C. Haas, *J. Polym. Sci.*, **39**, 493 (1959).
86. R. Rado and M. Lazár, *J. Polym. Sci.*, **62**, S167 (1962).
87. L. N. Guseva, Yu. A. Mikheyev and D. Ya. Toptygin, *Vysokomol. Soedin.*, **A20**, 2006 (1978).
88. L. S. Rogova, L. N. Guseva, Yu. A. Mikheyev and D. Ya. Toptygin, *Vysokomol. Soedin.*, **A21**, 1373 (1979).
89. R. Rado and D. Shimunkova, *Vysokomol. Soedin.*, **3**, 1277 (1961).
90. Yu. A. Mikheyev, O. A. Ledneva and D. Ya. Toptygin, *Vysokomol. Soedin.*, **A13**, 931 (1971); O. A. Ledneva, Yu. A. Mikheyev and D. Ya. Toptygin, *Izv. Akad. Nauk SSSR, Ser. Khim.*, **1**, 66 (1977).
91. A. P. Marin and Yu. A. Shlyapnikov, *Dokl. Akad. Nauk SSSR*, **215**, 1160.
92. J. C. W. Chien and D. S. T. Wang, *Macromolecules*, **8**, 920 (1975).
93. R. Rado, D. Shimunkova and L. Malak, *Vysokomol. Soedin.*, **4**, 304 (1962).
94. G. G. Makarov, L. M. Postnikov, G. B. Pariskii, Yu. A. Mikheyev and D. Ya. Toptygin, *Vysokomol. Soedin.*, **A20**, 2567 (1978).
95. G. G. Makarov, L. M. Postnikov, G. B. Pariskii, Yu. A. Mikheyev and D. Ya. Toptygin, *Vysokomol. Soedin.*, **A22**, 314 (1980).
96. A. V. Tobolsky, P. M. Norling, N. H. Frick and H. Ju, *J. Amer. Chem. Soc.*, **86**, 3925 (1964).
97. R. Rado, F. Szöcs, M. Vinkovičová and J. Plaček, *Chem. Zvesti*, **27**, 796 (1973).
98. R. Rado, *Chem. Zvesti*, **19**, 46 (1965).
99. L. Matisová-Rychlá, J. Rychlý and M. Lazár, *Europ. Polym. J.*, **8**, 655 (1972).
100. E. G. Atovmyan and A. F. Lukovnikov, *Kinet. Katal.*, **5**, 1340 (1970).
101. E. Brandrup and E. H. Immergut (Eds.), *Polymer Handbook*, Wiley–Interscience, New York–London, 1975, p. 19.
102. V. P. Roshchupkin and U. G. Gafurov, *Dokl. Akad. Nauk SSSR*, **235**, 869 (1977).
103. W. J. Le Noble in *High Pressure Chemistry* (Ed. H. Kelm), Reidel, Dordrecht–Boston, 1978, p. 325.
104. E. Borsig and F. Szöcs, *Polymer*, **22**, 1400 (1981).
105. H. C. Bailey and G. W. Godin, *Trans. Faraday Soc.*, **52**, 68 (1956).
106. S. M. Kabun and A. L. Buchachenko, *Izv. Akad. Nauk SSSR, Ser. Khim.*, 1483 (1966).
107. A. A. Tatarenko and V. S. Pudov, *Vysokomol. Soedin.*, **B9**, 287 (1967).
108. B. R. Smirnov and V. D. Sukhov, *Vysokomol. Soedin.*, **A19**, 236 (1977).
109. G. V. Kolcev, B. R. Smirnov, S. G. Bashkirova and A. A. Berlin, *Vysokomol. Soedin.*, **6**, 1256 (1964).
110. E. T. Denisov, S. S. Ivanchev, L. A. Zborshchik and N. V. Zolotova, *Izv. Akad. Nauk SSSR, Ser. Khim.*, 1500 (1968).
111. O. A. Ledneva, Yu. A. Mikheyev, D. Ya. Toptygin, L. B. Gavrilov and M. S. Akutin, *Vysokomol. Soedin.*, **21**, 1432 (1979).
112. G. Pannetier and P. Souchay, *Chemical Kinetics*, Elsevier, Amsterdam, 1967, pp. 386–418.

The Chemistry of Functional Groups, Peroxides
Edited by S. Patai

CHAPTER 23

Organometallic peroxides

P. B. BRINDLEY

Kingston Polytechnic, Kingston upon Thames KT1 2EE, England

I. INTRODUCTION . . . 807
A. General Properties . . . 808
B. General Methods of Preparation . . . 809

II. ORGANOMETALLIC PEROXIDES OF GROUP II METALS . . . 810
A. Beryllium and Magnesium . . . 810
B. Zinc, Cadmium and Mercury . . . 810

III. ORGANOMETALLIC PEROXIDES OF GROUP III METALS . . . 812
A. Boron . . . 812
B. Aluminium . . . 816
C. Gallium, Indium and Thallium . . . 817

IV. ORGANOMETALLIC PEROXIDES OF GROUP IV METALS . . . 818
A. Titanium, Zirconium and Hafnium . . . 818
B. Tin and Lead . . . 820

V. ORGANOMETALLIC PEROXIDES OF GROUP V METALS . . . 822
A. Antimony, Arsenic and Bismuth . . . 823

VI. PEROXY DERIVATIVES OF GROUP VI AND VIII ORGANOMETALLICS . . . 824
A. Group VI . . . 824
B. Group VIII . . . 825

VII. REFERENCES . . . 826

I. INTRODUCTION

Organometallic peroxides are defined for the purposes of this chapter as compounds with organic carbon bonded directly to the metal as well as having dioxygen bonded to the metal. The latter may be an aryl-, alkyl- or acyl-peroxy (μ-peroxo) group as in R_nMOOR[1] or a hydroperoxy group R_nMOOH. In these cases both the carbon of the organic group R and the oxygen are bonded to the metal by σ bonds. Also included is another type of organometallic peroxide which results when the organic carbon is π-bonded to the metal, as one finds in olefin or cyclopentadienyl complexes of transition metals e.g. $(C_5H_5)Zr(Y)OOR^1$. Metal peroxo compounds, which are well established for those transition-metal complexes having suitable ligands, are also to be found with organometallic bonded carbon, e.g. **1**[59].

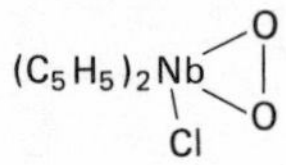

$$(C_5H_5)_2Nb(Cl)(\eta^2\text{-}O_2)$$

(1)

It has been pointed out by Sheldon and Kochi[1] that metals in low oxidation states will have relatively expanded 'd' orbitals and will favour π-bonded dioxygen, whereas high oxidation state metals will have 'd' orbitals less available for back-bonding and are more likely to form μ-peroxy links with dioxygen. If a peroxy group is not directly attached to the metal but is part of an organic ligand then it generally behaves like an ordinary organic hydroperoxide or dialkyl peroxide. Reference to this type of organometallic peroxide will be reserved for those cases where special properties result or there is topical interest.

In the early 1970s a number of reviews[2–4] appeared following publications of work which had provided some understanding of what had previously been a little studied area. In many instances the organometallic peroxides are labile compounds which sometimes undergo explosive decomposition, yet in other instances they are stable up to 100°C. In the past decade many examples of preparation, and in some cases isolation, of organometallic peroxides have been reported and the mechanisms of decomposition have been studied and classified. Organometallic peroxides have been used in the syntheses of organic hydroperoxides and more recently for aldehyde and ketone syntheses[35]. Interest has also been partly stimulated by the application of the organometallic peroxides to commercial processes although, apart from the Halcon processes for the epoxidation of olefins[5] and their possible use in polymerization reactions, their exploitation has been limited.

A. General Properties

The thermal homolysis of the oxygen–oxygen bond, well known in the case of organic hydroperoxides and dialkyl peroxides, is frequently demonstrated in the case of organometallic peroxides (equations 1 and 2). However, the presence of the metal can significantly modify the oxygen bond strengths. It should be noted that in many cases the metal provides additional routes for removal of the peroxy group and frequently these relatively low-energy pathways will convert the peroxy group to the alkoxy group and are the reason why isolation of the organometallic peroxide is sometimes difficult or impossible. The rearrangement of the peroxy group can be either intermolecular (equation 3) or intramolecular (equation 4) depending on the nature of the metal and other ligands that are present. It is frequently the case that an alkoxy group or halogen directly attached to the metal of an organometallic peroxide confers hydrolytic instability on the compound.

$$R_nMOOR^1 \longrightarrow R_nMO^\bullet + {}^\bullet OR \quad (1)$$

$$R_nMOOR^1 \longrightarrow R_nM^\bullet + {}^\bullet OOR^1 \quad (2)$$

$$R_nMOOR + R_nM \longrightarrow 2R_nMOR \quad (3)$$

$$R_nMOOR \longrightarrow R_{n=1}MOR_2 \quad (4)$$

B. General Methods of Preparation

One can divide the methods used to prepare organometallic peroxides into two broad classes. The first is nucleophilic substitution of halogen or other good leaving group, by hydroperoxide (equation 5) or the sodium salt of the hydroperoxide (equation 6). When X is a low molecular weight alkoxy group or oxygen it is often advantageous to remove the by-product alcohol or water by azeotropic distillation and in this way high yields of the more thermally stable peroxides can be obtained. Alternatively the sodium salt of the hydroperoxide can be used for the more unstable systems as in equation (6). The second is the insertion of dioxygen, derived from the air, into a carbon metal bond (equation 7). When studied in detail with individual organometallics, this autoxidation turns out to be a free-radical chain reaction. This was first recognized in the case of alkyl-boron[6], -zinc, -cadmium and -aluminium[7] and the sequence of reactions involving initiation (equation 8), followed by the chain propagation steps (equations 9 and 10) is now widely applicable. The spontaneous inflammability of some low molecular weight metal alkyls reported by chemists in the previous century may be attributed to this rapid, free-radical chain sequence. Equation (10) is an example of bimolecular homolytic substitution (S_H2). In the case of the more reactive organometallics such as trialkylboranes or homoleptic* transition-metal alkyls the initiation step (8) appears to be spontaneous and the initial attack by molecular oxygen is followed by rapid subsequent stages giving rise to the chain-propagating alkyl radical[8].

$$R_nMX_m + {}_m R^1OOH \longrightarrow R_nM(OOR^1)_m + {}_m HX \tag{5}$$

$$R_nMX_m + {}_m R^1OONa \longrightarrow R_nM(OOR^1)_m + {}_m NaX \tag{6}$$

$$R{-}M + O_2 \longrightarrow MOOR \tag{7}$$

$$RM + O_2 \longrightarrow R^{\cdot} \tag{8}$$

$$R^{\cdot} + O_2 \longrightarrow RO_2^{\cdot} \tag{9}$$

$$RO_2^{\cdot} + RM \longrightarrow RO_2M + R^{\cdot} \tag{10}$$

The exact nature of the initial dioxygen–metalalkyl complex is open to conjecture, bearing in mind that since dioxygen has a triplet ground state it can readily interact with metals in lower oxidation states. It may be of the superoxo type $R_nMO{-}O$ (similar to the reversible enzymatic oxygen adducts), which could eliminate alkyl radicals, recombination of radicals being spin forbidden (equation 11). Alternatively it could be a peroxy complex (**2**) where the oxygens are π-bonded in a manner similar to metal–olefin complexes. Insertion by alkyl groups (derived from metal–olefin complexes) into transition-metal complex peroxy compounds has been reviewed by Mimoun[9] and can equally well apply to certain organometallics (equation 12).

$$R_nMOO \longrightarrow R_{n-1}MOO^{\cdot} + R^{\cdot} \tag{11}$$

$$R_nM\langle{}^{O}_{O}| \longrightarrow R_{n-1}MOOR \tag{12}$$

(**2**)

* Metal alkyls which have only metal–carbon σ bonds but which may also have metal–metal bonds[85].

II. ORGANOMETALLIC PEROXIDES OF GROUP II METALS

A. Beryllium and Magnesium

Organometallic peroxides of beryllium or magnesium, ROOMR, have not been clearly identified, although the related peroxides derived from the Grignard reagents, ROOMgX, are well established. Walling and Buckler[10] used the autoxidation method at $-75°C$ to prepare the alkylperoxymagnesium halide *in situ*, and then isolated hydroperoxides in high yield after treatment with acid (equation 13). The use of dilute ether solutions of oxygen at $-75°C$ minimized the conversion of peroxide to alkoxide[10,11]. However, one should bear in mind that alkylmagnesium halides in ether solution are solvated and are in equilibrium with the dialkylmagnesium compounds[12]. Pure dialkylmagnesium compounds are known and they are frequently polymeric, although dineopentylmagnesium is trimeric in benzene[13]. One might ask to what extent the Schenk equilibrium results in compounds of the type ROOMgR, contributing to the overall hydroperoxide yield. It is unlikely that alkylperoxymagnesium alkyls could be isolated since they would be expected to undergo rapid reduction in a manner analogous to the related magnesium peroxides ROOMgX (Section I.A) whose reduction to alkoxide is well established (equation 14)[14]. Autoxidation of dimethylberyllium in diethyl ether gave beryllium methoxide consistent with a free-radical mechanism involving methylberyllium methyl peroxide[3,66].

$$\mathrm{RMgX + O_2 \xrightarrow{ether} ROOMgX \xrightarrow{H^+X^-} ROOH + MgX_2} \qquad (13)$$

$$\mathrm{RMgX + \mathit{t}\text{-}BuOOH \longrightarrow \mathit{t}\text{-}BuOOMgX \xrightarrow{RMgX} ROMgX + \mathit{t}\text{-}BuOMgX} \qquad (14)$$

B. Zinc, Cadmium and Mercury

As one progresses down Group IIB from zinc to mercury, so the organometallic peroxides become increasingly more stable, the rearrangement reactions (equations 3 and 4) become less facile and in some cases the organometallic peroxides are insolable. By applying their low-temperature, dilute-solution, autoxidation method to di-*n*-butylzinc, Walling and Buckler were able to prepare a solution of butylperoxyzinc-butyl[10] which at ambient temperature and with excess oxygen forms dibutyl peroxyzinc (equation 15, $R = n$-Bu)[15]; reduction by excess dibutylzinc gave butoxyzinc (equation 16, $R = n$-Bu). Previous to this work the vapour-phase autoxidation of dimethylzinc had been shown to be a chain-reaction[16]. Subsequently, autoxidation of diethylzinc in anisole solvent was confirmed as a free-radical chain process (equations 8–10) and since the total amount of oxygen absorbed was 1.5 mol then the overall reaction was as shown in equation (17). The first peroxyradical substitution could not be inhibited by the free-radical inhibitor galvinoxyl (1.3 mol %), but the second substitution yielding the diperoxide was completely inhibited[17].

$$\mathrm{R_2Zn + O_2 \longrightarrow ROOZnR \xrightarrow{O_2} (ROO)_2Zn} \qquad (15)$$

$$\mathrm{(ROO)_2Zn + R_2Zn \longrightarrow 2(RO)_2Zn \quad R = Et, \mathit{n}\text{-}Bu} \qquad (16)$$

$$\mathrm{Et_2Zn + 1.5O_2 \longrightarrow EtOZnOOEt} \qquad (17)$$

$$\mathrm{(CH_3)_2Cd + ROOH \longrightarrow CH_3CdOOR \longrightarrow CH_3OCdOR} \qquad (18)$$

Generally, dialkylcadmium compounds when interacted with alkylhydroperoxides according to equation (5) give dialkylperoxycadmiums, but it was reported that the less reactive dimethylcadmium with decahydro-9-naphthyl hydroperoxide (decalyl OOH) yielded the organometallic peroxide as did 2-phenylpropyl-2-hydroperoxide (cumyl OOH); the latter was stable at 20°C rearranging to the mixed alkoxide at 50°C (equation 18)[17,87]. Syntheses by the autoxidation route, with one molecule of oxygen, also gave the organocadmium peroxide, the reaction being slower in diethyl ether than in anisole. The second molecule of oxygen gave a white precipitate of the dialkyl peroxide (equation 19). In the absence of solvent, dimethylcadmium with air gave a glass which exploded when touched. Both stages of the free-radical chain substitution reaction could be inhibited by galvinoxyl in contrast to the zinc case, indicating slower initiation and/or propagation stages (equations 8 and/or 10) in the case of cadmium[17,18].

$$R_2Cd + O_2 \longrightarrow ROOCdR \xrightarrow{O_2} (ROO)_2Cd \qquad (19)$$

$$R = Me, Et, n\text{-}Bu$$

Increased stability of the peroxides as one progresses down the subgroup is clearly demonstrated with alkylperoxymercury alkyls. Preparation can be conveniently carried out by the hydroperoxide substitution methods. Heating dialkylmercury compounds with a hydroperoxide gives the mercury peroxide, but the elevated temperature then gives rise to rearrangement of the mercury peroxides as well as to their homolyses[19], e.g. equation (20). The use of the sodium salt of the hydroperoxide at a lower temperature is a more satisfactory synthesis allowing the preparation of cumylperoxymercury-phenyl[20], $PhC(CH_3)_2OOHgPh$. The slow autoxidation of dialkylmercury compounds follows a free-radical chain mechanism. Razuvaev and coworkers[67] concluded that diisopropylmercury and oxygen at 50°C first formed an intermediate oxygen complex which rearranged to alkylmercuric alkyl peroxide. This then either interacted with unchanged dialkylmercury to form isopropylmercuric isopropoxide or underwent mercury–oxygen homolysis (equations 21 and 22, R = *i*-Pr). By-products—acetone and isopropyl alcohol—were observed in the case of diisopropylmercury autoxidation. Similar products were observed by later workers[68], who using di-*s*-butylmercury obtained a homolytic displacement at mercury of *s*-butyl radicals (equation 23, R = *s*-Bu). It seems probable that equations (8)–(10) are applicable.

$$i\text{-}Pr_2Hg + t\text{-}BuOOH \longrightarrow i\text{-}PrH + t\text{-}BuOOHgPr\text{-}i \xrightarrow{i\text{-}Pr_2Hg}$$

$$\longrightarrow i\text{-}PrHgOBu\text{-}t + i\text{-}PrOHgPr\text{-}i \qquad (20)$$

$$R_2Hg + O_2 \longrightarrow [R_2Hg\,O_2{}_{-}] \longrightarrow [ROOHgR] \qquad (21)$$

$$[ROOHgR] \begin{cases} \xrightarrow{R_2Hg} 2\,ROHgR \\ \longrightarrow RO_2^{\cdot} + {}^{\cdot}HgR \end{cases} \qquad (22)$$

$$R = i\text{-}Pr, t\text{-}amyl \text{ or benzyl}$$

$$ROO^{\cdot} + R_2Hg \xrightarrow{S_H2} ROOHgR + R^{\cdot} \qquad (23)$$

It is worth noting in passing that peroxymercuration of alkenes, which gives organometallic compounds where the organic peroxy group is β to the mercury, forms a compound **(3)** where there is intramolecular coordination between the mercury and the peroxy oxygen[21]. Mimoun and coworkers[65] have used this type of stable peroxymercurial to prepare the labile palladium peroxy intermediate **4**, which rapidly converts to acetophenone and *t*-butoxypalladium.

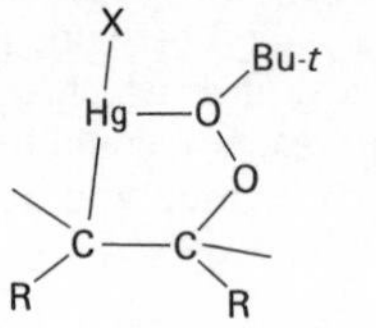

X = Br or trifluroacetate

(3)

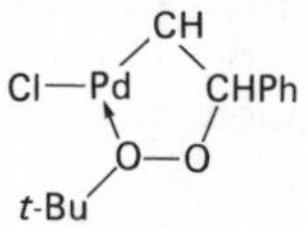

(4)

III. ORGANOMETALLIC PEROXIDES OF GROUP III METALS

With an increase in valency of the metal, so the variety of peroxides derived from the metal increases and $(ROO)_nMR_{3-n}$ ($n = 1$ or 2) and ROOM(R)OR or ROOM(R)–O–M(R)OOR become possible in Group III.

A. Boron

It was in the 1950s that organoboron μ-peroxides of the type $(ROO)_2MR_{3-n}$ were first detected[22,23] and subsequent work has resulted in these compounds being studied in considerable detail, although their isolation and purification has been hampered by the rapidity with which the peroxide rearranges to alkoxide (equations 3 and 4). The best method for the preparation of $(ROO)_nMR_{3-n}$ and ROOB(OR)R in solution has proved to be the autoxidation method carried out in an inert solvent, under carefully controlled conditions. The kinetic measurements carried out by Davies, Ingold and coworkers[24,25] have substantiated the earlier discovery that the process was a free-radical reaction of long chain length[26,27] (equations 24 and 25). Their work provided quantitative comparisons of the ease of propagation, the rate-determining step being the rapid S_H2 displacement of alkyl radicals from the boron atom (equation 25); k_p is the rate constant for the S_H2 process. Propagation was terminated by two alkylperoxy radicals combining to give stable organic molecules (equation 26). The quantity $k_p/(2k_t)^{1/2}$ has been referred to as the 'oxidizability' and is highly dependent on the structure of the organoboron compound. An understanding of the factors involved is necessary if good yields of organoboron peroxide are required. The autoxidations are usually inhibited by water and organic bases[26]. The conversion of diisopinocampheylbutyl boranes to the mono- (and di-) peroxides at 25°C in benzene showed steric retardation which was most pronounced in the case of the *s*-butyl isomer[27] (*s*-butyl isomer autoxidation to diisopinocampheyl-*s*-butylperoxyborane—$t_{1/2}$ ca. 4 s). However, tris-*exo*-2-norbornylborane showed no steric retardation when oxidized to the monoperoxide, rather it was comparable to tri-*s*-butylborane in behaviour.

$$R^{\bullet} + O_2 \xrightarrow{\text{very fast}} ROO^{\bullet} \tag{24}$$

$$ROO + R{-}B{-}R \xrightarrow[\text{fast}]{k_p} ROOB{-}R + R^{\bullet} \tag{25}$$

$$ROO^{\cdot} + ROO^{\cdot} \xrightarrow{2k_t} \text{stable products} \qquad (26)$$

In the case of trialkyl- or triaryl-boranes the rate-determining propagation steps are close to diffusion control, excess oxygen and highly efficient stirring being needed in the above experiments. At temperatures as low as -74°C there was no detectable induction period. For example, tri-*n*- -*s*- and -*i*-butylboranes (2 millimolar in ether at -74°C) absorbed the first molecule of oxygen to form $BuOOBBu_2$ with half-lives *ca.* 20 s, reaction being complete in *ca.* 3 min, whereas the second molecule of oxygen was absorbed much more slowly to form the diperoxide: $BuOOBBu_2 \rightarrow (BuOO)_2BBu$ ($t_{1/2}$ at 25°C, 8 mM in benzene—*n*-butyl = 5 min, *s*-butyl = 18 s, *i*-butyl = 31 min)[27]. It was by carrying out the autoxidations of trialkylboranes in the presence of pyridine that Ingold and coworkers were able to take advantage of the equilibrium (27) to calculate the values given in Table 1, the pyridine–borane complex not being subject to autoxidation.

$$R_3B{\cdot}C_5H_5N \rightleftharpoons R_3B + C_5H_5N \qquad (27)$$

TABLE 1. Oxidation of organoboron to peroxyboron compounds at 30°C in isooctane[24,25]

Starting compound	Product	Oxidizability[a] ($M^{-\frac{1}{2}}s^{-\frac{1}{2}}$)	k_p ($M^{-1}s^{-1}$)
$(n\text{-Bu})_3B$	$n\text{-BuOOB(Bu-}n)_2$	300[b]	2×10^{6} [b]
$(s\text{-Bu})_3B$	$s\text{-BuOOB(Bu-}s)_2$	70[b]	8×10^{4} [b]
(*exo*-2-Norbornyl)$_3$B	Similar to $(s\text{-Bu})_3B$		—
Ph_3B	Similar to $(s\text{-Bu})_3B$		—
$(PhCH_2)_3B$	$PhCH_2OOB(CH_2Ph)_2$	3×10^2	5×10^{6} [c,d]
$(PhCH_2)_2BOOCH_2Ph$	Indistinguishable from	$(PhCH_2)_3B$[c]	—
$(n\text{-Bu})_2BOEt$	$n\text{-BuOOB}(n\text{-Bu})OEt$	0.8	5×10^3
$(s\text{-Bu})_2BOBu\text{-}n$	$s\text{-BuOOB}(s\text{-Bu})OBu\text{-}n$	1.6	2×10^3
$(n\text{-Bu})_2BOOBu\text{-}n$	$(n\text{-BuOO})_2BBu\text{-}n$	4.4	3×10^4
$(s\text{-Bu})_2BOOBu\text{-}s$	$(s\text{-BuOO})_2BBu\text{-}s$	11	1×10^4
$(n\text{-Bu})_2BOB(Bu\text{-}n)_2$		43	3×10^5
$(s\text{-Bu})_2BOB(Bu\text{-}s)_2$		56	7×10^4

[a] $k_p/(2k_t)^{\frac{1}{2}}$.
[b] In isooctane/pyridine mixture.
[c] In benzene.
[d] In benzene with a small amount of pyridine.

It is significant that when an alkoxy group is attached to boron then autoxidation is much slower, yielding peroxides of the type ROOB(R)OR. This is partly because of a slower rate of self-initiation, partly a result of a slower S_H2 propagation. The retarding effect of oxygen-bound groups whether alkoxy or peroxy, is attributed to π back-bonding from oxygen to boron:

$$\rangle B{-}OR \longleftrightarrow \rangle \bar{B}{=}\overset{+}{O}R$$

which limits attack of the peroxy radical onto the boron, whereas in the case of trialkylboranes the boron has a vacant p orbital available to the substituting radical. In some instances the autoxidation of trialkylboranes, at low temperature in hydrocarbon solvent stops sharply after formation of the monoperoxide, and if the residual oxygen gas is rapidly displaced by nitrogen a solution of the pure peroxide is obtained (equation 28). In a similar manner the pure diperoxyboranes can be obtained at ambient temperatures[8,35]. Self-initiation of trialkylboranes was found to be first order in borane and zero order with respect to oxygen and this was explained in terms of equations (29) and (30)[8,28]. The nature of the complex was not discussed by these authors, the rate of initiation being suggested as a measure of the rearrangement of the dioxygen–borane complex to the μ-peroxoborane:

$$(n\text{-Bu})_3B + O_2 \xrightarrow[-74^\circ C]{\text{isooctane}} (n\text{-BuOO})B(n\text{-Bu})_2 \quad (28)$$

$$R_3B + O_2 \underset{}{\overset{\text{fast}}{\rightleftharpoons}} R_3B\cdot O_2 \xrightarrow{\text{slow}} ROOBR_2 \quad (29)$$

$$ROOBR_2 + R_3B \xrightarrow{\text{fast}} R^{\bullet} + R_2BO^{\bullet} + \text{stable products} \quad (30)$$

Rate constants for trialkylborane autoxidation self-initiation in benzene at 25°C.

$1.7 \times 10^{-5}\,s^{-1}$	for tri-*s*-butylborane
$0.9 \times 10^{-5}\,s^{-1}$	for tri-*i*-butylborane
$6.8 \times 10^{-5}\,s^{-1}$	for tricyclohexylborane

If a mixed organoborane such as $(i\text{-Bu})_2BBu\text{-}t$ is autoxidized the *t*-butyl group peroxidizes preferentially and 2 mol of oxygen gives almost pure di-isobutylperoxy-*t*-butylperoxyborane[30].

When nucleophilic substitution by hydroperoxide was used as the preparative method for organoboron peroxides the presence of the organometallic peroxide could be deduced from identification of the rearrangement products. Cumylperoxydiphenylborane was reported as being unstable, rearranging rapidly to the mixed alkoxy–phenoxy compound; this was deduced by relating it to its hydrolysis products; phenylboronic acid, phenol and 2-phenylpropan-2-ol (equations 31 and 32)[29]. Hydrogen peroxide has been used as well as hydroperoxides, thus with hydrogen peroxide and the mixed borane, diisobutyl-*t*-butylborane rearrangement followed initial reaction (equation 33). It was demonstrated by kinetics[31] and oxygen-18 labelling studies[32] that the labile phenylboronic peroxy anion, which is formed from phenylboronic acid and aqueous hydrogen peroxide, intramolecularly rearranges to the boric acid ester; the ester being moisture-sensitive, gives phenol whose oxygen is derived from the peroxy group (equations 34 and 35).

$$Ph_2BCl + \text{cumylOONa} \xrightarrow[\text{benzene}]{-50^\circ C} Ph_2BOOcumyl + NaCl \quad (31)$$

$$Ph_2BOOcumyl \longrightarrow PhOB(Ph)Ocumyl \xrightarrow{H_2O} PhB(OH)_2 + PhOH + PhC(CH_3)_2OH \quad (32)$$

$$(i\text{-Bu})_2BBu\text{-}t + HOOH \longrightarrow \left[(i\text{-Bu})_2BCBu\text{-}t\right] \xrightarrow{H_2O} t\text{-BuOH} + i\text{-Bu}_2BOH \quad (33)$$

$$PhB(^{18}OH)_2 + ROOH \longrightarrow [PhB(^{18}OH)_2OOR]^- \longrightarrow H^{18}OB(OPh)(OR) + {}^{18}OH^- \quad (34)$$

$$H^{18}OB(OPh)(OR) + 2H_2{}^{18}O \longrightarrow (H^{18}O)_3B + PhOH + ROH \quad (35)$$

R = H or tetrahydronaphthyl

π Back-bonding from oxygen is the stabilizing factor in *s*-butoxy-*s*-butyl-*s*-butylperoxyborane, *s*-BuOB(Bu-*s*)OOBu-*s*, which was only 10% decomposed after 24 hours at 50°C (equation 36) Here it was the difficulty in purifying the starting chloroborane which was the most significant factor leading to the 75% yield of organoboron peroxide[8].

$$s\text{-BuOB}(s\text{-Bu})\text{Cl} + s\text{-BuOOH} \xrightarrow[0^{\circ}\text{C}]{\text{pentane}} \text{HCl} + s\text{-BuOB}(s\text{-Bu})\text{OOBu-}s \quad (36)$$

The relative importance of the rearrangement reactions and homolysis of the peroxy groups have been studied quantitatively for butylboranes. Using pure samples of peroxyboranes in isooctane a synchronous, intramolecular rearrangement (equation 37), involving a cyclic transition state, was inferred from enthalpies and entropies of activation. A consideration of rate constants for the intramolecular rearrangement and the homolytic fission of peroxy groups showed that the latter play only a minor role at temperatures below 100°C (Table 2). On the other hand the intermolecular decomposition of the monoperoxide and trialkylborane is very rapid at ambient temperature and complete in less than 30 s (equations 38 and 39)[8].

$$\text{BuOOB(Bu)(Z)} \longrightarrow [\text{BuO}\cdots\overset{\delta+}{\text{O}}\cdots\text{Bu}\cdots\overset{\delta-}{\text{B}}(\text{Z})] \longrightarrow \text{BuOB(OBu)(Z)} \quad (37)$$

Z = Bu or OOBu

$$(n\text{-Bu})_2\text{BOOBu-}n + (n\text{-Bu})_3\text{B} \longrightarrow (n\text{-Bu})_2\text{BOBu-}n + n\text{-Bu}_2\text{BO}^{\bullet} + n\text{-Bu}^{\bullet} \quad (38)^*$$

$$(n\text{-Bu})_2\text{BO}^{\bullet} + (n\text{-Bu})_3\text{B} \longrightarrow (n\text{-Bu})_2\text{BOB(Bu-}n)_2 + n\text{-Bu}^{\bullet} \quad (39)$$

TABLE 2. Rate constants for decomposition of butylperoxyboranes

Compound	Rearrangement in isooctane, $k(s^{-1})$	Homolysis in k vinyl acetate, $k(s^{-1})$
$(n\text{-BuOO})_2\text{BBu-}n$	3.78×10^{-5} at 71°C	1.5×10^{-8} at 61°C
$(s\text{-BuOO})_2\text{BBu-}s$	4.80×10^{-5} at 71°C	2.6×10^{-8} at 61°C
$n\text{-BuOOB(Bu-}n)_2$	4.68×10^{-5} at 25°C	2.6×10^{-7} at 25°C

It has been suggested that organoboron peroxides could be used as low-temperature, free-radical polymerization initiators. Whether polymerization of vinyl monomers is as a result of chain transfer from the trialkylborane autoxidation or from intermolecular boronperoxide homolysis (equations (38) and (39) or a combination of both is not always

* Further confirmation of the free-radical nature equation (38) follows as a result of a recent CIDNP study[86].

clear[33,34]. A similar type of initiation supplies the alkyl free radical necessary to initiate the oxygen-induced S_H2 borane substitution by α,β-unsaturated carbonyl compounds to yield ketones or aldehydes, (equations 40 and 41)[35].

$$R^{\bullet} + {>}C{=}C{-}C{=}O \longrightarrow R{-}C{-}\dot{C}{-}C{=}O \longleftrightarrow R{-}C{-}C{=}C{-}\dot{O}$$

$$\xrightarrow[S_H2]{R_3B} R{-}C{-}C{=}C{-}OBR_2 + R^{\bullet} \qquad (40)$$

$$R{-}C{-}C{=}C{-}OB{<} + H_2O_2 + NaOH \longrightarrow R{-}C{-}C{=}C{-}OH + \text{etc.}$$

$$\rightleftharpoons R{-}C{-}\underset{H}{C}{-}C{=}O \qquad (41)$$

An important application of di-alkylperoxy-alkylboranes is their use to prepare and isolate in high yield the corresponding alkyl hydroperoxide. Hydroboration of alkene to trialkylborane is followed by controlled addition of oxygen, then the reaction mixture is treated with aqueous hydrogen peroxide, all operations being carried out consecutively in the same flask (equation 42)[36]. The hydroperoxide is separated by extraction with aqueous potassium hydroxide.

$$R_3B + 2O_2 \longrightarrow (ROO)_2BR \xrightarrow[0^{\circ}C]{H_2O_2} ROOH + ROH + (HO)_3B \qquad (42)$$

The use of di-butylperoxy-butylborane in the place of *t*-butyl hydroperoxide in an attempt to bring about Halcon-type epoxidation[5] (discussed in Section VI) of cyclohexene in the presence of molybdenum hexacarbonyl gave, after work-up, low yields of trans-1,2-diol. Similarly indene gave a low yield of the indanone[63]. In both cases the products can be related to the epoxide. The low yields were largely attributed to the marked increase in rate of decomposition of di-butylperoxy-butylborane in the presence of the molybdenum catalyst and it is more likely that epoxide formation was as a result of homolysis of the peroxide to butoxy radicals[64] rather than a complex consisting of olefin–molybdenum and peroxide[9].

B. Aluminium

Alkylperoxyaluminium alkyls have not been studied in anything like the detail of the corresponding boron compounds but it seems that they are essentially similar. Autoxidation of aluminium alkyls at low temperature first forms the alkylperoxy-aluminium compound but only in low yields; the ease of rearrangement of the peroxide is once again a major factor (equations 43 and 44). Triphenylaluminium autoxidizes less readily than the methyl or ethyl compounds; the identification of biphenyl in the products further substantiates the free-radical nature of the autoxidation[17,37]. A similar situation

$$R_3Al + O_2 \longrightarrow R_2AlOOR \nearrow RAl(OR)_2 \quad (43)$$

$$R_2AlOOR \xrightarrow{RAl<} R_2AlOR \quad (44)$$

R = Me, Et or Ph

applies to the autoxidation of butylaluminium dichloride at −50°C, only 10 % of aluminium peroxide being detected[38].

The 'Alfol' process is a method for manufacturing long-chain alcohols and involves the formation of intermediate alkylperoxyaluminium alkyls. The process is more than 20 years old and converts ethylene into aluminium alkyls having a wide range of molecular weights. The aluminium alkyls are then autoxidized and following rearrangement according to equations (3) and (4) the aluminium alkoxides are hydrolysed to a range of primary alcohols and aluminium hydroxide[72].

C. Gallium, Indium and Thallium

The previously noted increase in stability of the Group IIB peroxides compared with their lower atomic number elements is equally noticeable with Group IIIB. Treatment of the metal alkyl with peroxide gives good yields of the monoperoxide (equation 45). Also trimethylgallium or trimethylindium with dioxygen gives the monoperoxide in quantitative yield. At elevated temperature (120°C for 3 h) these two monoperoxides undergo simultaneous homolysis and rearrangement, the major products being the alkoxides $Me_nM(OMe)_{3-n}$ ($n = 1$ or 2) with some radical attack by $Me_2GaO^{\bullet}$ on the nonane solvent giving 5 % of Me_2GaOR[39].

$$R_3M + ROOH \longrightarrow R_2MOOR \quad (45)$$

M = Ga, In, Tl; R = Me, Et

The stable diethylthallium peroxides[40] have been known for some years; the triphenylgermylperoxy derivative is particularly stable, being light-sensitive yellow crystals, and can be prepared in 80 % yield in high purity (equation 46)[41]. At temperatures above 55°C it rearranges over a number of hours by a bimolecular process (equation 47)[42]. The corresponding triphenylsilylperoxythallium compound cannot be isolated because it too readily rearranges to the diphenylphenoxy-siloxy-thallium-diethyl, $Et_2TlOSi(Ph)_2OPh$. The trialkyls of gallium, indium and thallium undergo autoxidation and it seems reasonable that a mechanism similar to that for trialkylborons involving the organometallic peroxides is involved[69].

$$Et_3Tl + Ph_3GeOOH \xrightarrow{C_6H_6 \text{ at } 0^{\circ}C} Et_2TlOOGePh_3 + EtH \quad (46)$$

$$2Et_2TlOOGePh_3 \xrightarrow{C_6H_6 \text{ at } 60^{\circ}C} 2Et_2TlOGe(Ph)_2OPh \quad (47)$$

IV. ORGANOMETALLIC PEROXIDES OF GROUP IV METALS

The peroxides throughout this group show great variety and complexity; there is a considerable amount of published work dealing with silicon and germanium but they are not dealt with in this chapter, which is concerned only with metal peroxides.

A. Titanium, Zirconium and Hafnium

The known organometallic peroxides of this subgroup have only transient existence; the peroxides were prepared following the discovery that the previously unknown tetraalkyls of the early transition metals could be isolated if β-hydrogen elimination to yield alkenes was not possible.

Autoxidation of the tetraalkyls, in hydrocarbon solvent at 20°, was close to diffusion control, and was not significantly altered by lowering the temperature to −74°C. Peroxide formation was at first inferred because retardation resulted from the inclusion of free-radical inhibitors. It was possible at the lower temperature to detect the organometallic peroxides in low yield, but the final products from the interaction with hydrocarbon solutions of oxygen were the metal alkoxides (equation 48)[43,44]. Tetrabenzylzirconium gave the most clear-cut results; the half-life for the oxygen absorption, at −74°C, in toluene solvent, was 2 min and the peroxide yield was 30 % based on equation (49).

$$R_4M + 2O_2 \longrightarrow R_2M(OOR)_2 \longrightarrow (RO)_4M \qquad (48)$$

M = Ti, Zr, Hf, R = benzyl, neopentyl or trimethylsilylmethyl

Hydrolysis of the toluene solution gave benzyl hydroperoxide and benzyl alcohol. The benzaldehyde present was attributed in part to the termination reaction (50) and partly to further hydrolysis of the hydroperoxide. The half-life for the rearrangement of the organozirconium peroxide to alkoxide at 20°C was 50 min. In comparison to the other zirconium metal alkyls, tetrabenzylzirconium oxidized the least vigorously and gave the best yield of peroxide—a characteristic which was attributed to π interaction between the aromatic rings* and the low-lying vacant 'd' orbitals of the metal[43]. Dibenzylperoxy-hafnium-dibenzyl was prepared *in situ* under conditions similar to those for the zirconium compound. However, a significant difference was observed in the presence of pyridine. Due to complex formation, the initial rate of peroxidation was reduced and only the monoperoxide was formed (equation 51)[45]. On raising the temperature from −74°C to 20°C a second molecule of oxygen was absorbed suggesting that at the higher temperature the pyridine complex dissociated to allow formation of benzylperoxy-dibenzyloxy-hafnium-benzyl (equation 52). It seems from the earlier reports of the spontaneous inflammability of Me_4Ti, Ph_4Ti and $CpTiMe_3$ (Cp–cyclopentadienyl) that the organometallic peroxides are involved in the highly reactive autoxidations of the organometallics.

$$(PhCH_2)_4Zr + 2O_2 \longrightarrow (PhCH_2OO)_2Zr(CH_2Ph)_2 \qquad (49)$$

$$2PhCH_2OO^{\bullet} \longrightarrow PhCHO + O_2 + PhCH_2OH \qquad (50)$$

* It has been shown by X-ray structure analysis that tetrabenzylzirconium and tetrabenzylhafnium have distorted tetrahedral metal σ bonds and that the aromatic rings are closer to the metal atoms than in the undisturbed compounds such as tetrabenzyltin[46]. A weak interaction between the aromatic rings and the transition-metal centre contributes to the folding of the rings around the central metal.

$$(PHCH_2)_4Hf{\cdot}C_5H_5N + O_2 \longrightarrow PhCH_2OOHf(CH_2Ph)_3C_5H_5N \longrightarrow (PhCH_2O)_2Hf(CH_2Ph)_2{\cdot}C_5H_5N \quad (51)$$

$$(PhCH_2O)_2Hf(CH_2Ph)_2{\cdot}C_5H_5N \longrightarrow (PhCH_2O)_2Hf(CH_2Ph)_2 + C_5H_5N \xrightarrow{O_2} (PhCH_2)HfOOCH_2Ph(OCH_2Ph)_2 \quad (52)$$

$$Cp_2Zr(CH_2Ph)_2 + O_2 \longrightarrow Cp_2Zr(OOCH_2Ph)CH_2Ph \longrightarrow Cp_2Zr(OCH_2Ph)_2 \quad (53)$$

It was by reducing the five vacant orbitals in tetraalkylzirconiums to one in dicyclopentadienylzirconium dialkyls that *in situ* formation of the organozirconium peroxides became more manageable at ambient temperatures (equation 53). It was demonstrated that there was rapid intramolecular rearrangement of the zirconocene peroxide to the dialkoxyzirconocene, and it was this which prevented either spectroscopic identification of the peroxide, or its isolation. Kinetic studies involving free-radical inhibition and initiation on a range of substituted dibenzylzirconocenes fully established the free-radical substitution of alkyl radicals from the metal centre of the zirconocene, viz. an S_H2 propagation stage, involving the alkylperoxy radical:

Spontaneous initiation

$$Cp_2ZrR_2 + O_2 \longrightarrow Cp_2Zr(R)O_2 + R^{\bullet} \quad (54)$$

Propagation

$$R^{\bullet} + O_2 \xrightarrow{\text{fast}} ROO^{\bullet} \quad (55)$$

$$ROO^{\bullet} + Cp_2ZrR_2 \xrightarrow{\text{rate-determining}} Cp_2Zr(R)OOR + R^{\bullet} \quad (56)$$

$$Cp_2Zr(R)OOR \xrightarrow{\text{rapid}} Cp_2Zr(OR)_2 \quad (57)$$

Autoxidation of benzylzirconocene chloride or bromide again yielded by a similar route an unstable peroxide which, in this case, rearranged intermolecularly to a mixture of the alkoxide and the oxygen-bridged species $Cp_2Zr(X)O(X)ZrCp_2$ (equations 58 and 59)[45,47].

$$Cp_2Zr(CH_2Ph)X + O_2 \longrightarrow Cp_2Zr(OOCH_2Ph)X \quad (58)$$

$$Cp_2Zr(CH_2Ph)X + Cp_2Zr(OOCH_2Ph)X \longrightarrow 2Cp_2Zr(OCH_2Ph)X \qquad X = Cl \text{ or } Br \quad (59)$$

Acidolysis of dibenzylzirconocene with *t*-butyl hydroperoxide at room temperature, although first forming the *t*-butyl peroxy derivatives of the metal, rapidly rearranged to the mixed alkoxide which was identified by ^{1}H-NMR in 70 % yield (equations 60 and 61). Acidolysis of the less reactive benzyl-benzyloxyzirconocene under similar conditions still gave the mixed alkoxide (equation 62).

$$Cp_2Zr(CH_2Ph)_2 + t\text{-}BuOOH \longrightarrow Cp_2Zr(CH_2Ph)OOBu\text{-}t + PhCH_3 \quad (60)$$

$$Cp_2Zr(CH_2Ph)OOBu\text{-}t \longrightarrow \left[Cp_2Zr \cdots (CH_2Ph) \cdots O \cdots O(t\text{-}Bu) \right]$$

activated complex

$$\longrightarrow Cp_2Zr(OCH_2Ph)OBu\text{-}t \quad (61)$$

$$Cp_2Zr(OCH_2Ph)CH_2Ph + t\text{-}BuOOH \longrightarrow Cp_2Zr(OCH_2Ph)OBu\text{-}t + PhCH_2OH \quad (62)$$

B. Tin and Lead

1. Tin

This type of tin compound has been studied in detail over a number of years[4]. Peroxides of varying complexity have been prepared by the nucleophilic substitution method (equation 63)[48–51], and good yields have been achieved using the sodium salt of the hydroperoxide (equation 64) and similarly equations (65)[51] and (66)[49b]. Although the butyl isomer is unstable in air its thermal stability is comparable to that of di-*t*-butyl peroxide, i.e. well in excess of 100°C. In the presence of tetraethyltin, triethyltin-alkyl peroxide undergoes homolysis (equation 67)[50]; in some respects this can be compared with the interaction of tetraethyltin and di-*t*-butyl-peroxide. Trimethyltin-*t*-butyl peroxide forms a stable complex with *t*-butyl hydroperoxide. Crystalline $Me_3SnOOBu$-*t*·HOOBu-*t* complex (m.p. 28°C) sublimes *in vacuo* whereas the free tin peroxide $Me_3SnOOBu$-*t* is a mobile liquid[49b].

$$R_{4-n}SnX_n + nR^1OOH \longrightarrow R_{4-n}Sn(OOR^1)_n + nHX \quad (63)$$

$n = 1$ or 2; X = Cl, Br, CN, OR, NEt_2, OH

$$R_3SnCl + R^1OONa \longrightarrow R_3SnOOR^1 + NaCl \quad (64)$$

$$R_3SnOR^1 + R^2OOH \longrightarrow R_3SnOOR^2 + R^1OH \quad (65)$$

R^1 = Me, Bu-*t*

$$(Bu_3Sn)_2O + 2t\text{-}BuOOH \longrightarrow 2Bu_3SnOOBu\text{-}t + H_2O \quad (66)$$

$$Et_3SnOOBu\text{-}t + Et_4Sn \xrightarrow[\text{solvent}]{\text{undecane}} Et_3SnO^{\bullet} + t\text{-}BuO^{\bullet} \quad (67)$$

It has been reported by Aleksandrov[57] that tin peroxides of the type R_3SnOOR^1 ($R = Me, Et; R^1 = t$-Bu, cumyl) form 1:1 complexes with alkenes (and amines) and this is followed by homolytic first-order decomposition. The tin peroxides of the form $R_{4-n}Sn(OOR^1)_n$ are readily hydrolysed and when $n = 3$ or 4 their involatility, coupled with their thermal instability, has prevented their isolation[49].

The interaction of dialkyltin methoxides with 98% hydrogen peroxide gives stable, insoluble polymeric peroxides having the formulae $\mathrm{-\!\!\!+}R_2SnOO\mathrm{+\!\!\!-}_n$ (R = Me, Et, Bu) which decompose slowly at 20°C over a period of weeks. When the preparation is carried out in the presence of an aldehyde or ketone the polymeric peroxides have the formulae $(-R_2SnOO\text{-}CR^1R^2\text{-}O-)_n$. The structure and thermal decomposition of these compounds has been reported in detail by Dannley, Aue and Shubber[54b].

With triethyltin oxide and hydrogen peroxide it was reported that the unstable peroxide $Et_3SnOOSnEt_3$ was a green oil which exploded on warming[53a]. On the other hand the triisopropyl analogue formed stable crystals and the *n*-butyl isomer has been used for the synthesis of strained, bicyclic endoperoxides and for prostaglandin peroxide synthesis[55]. It was pointed out by Dannley and Aue[54] that the stability of the organotin hydroperoxides R_3SnOOH was sensitive to impurities that and this probably also applies to $R_3SnOOSnR_3$. Aleksandrov and coworkers[52] had previously demonstrated that the ditin peroxides $R_3SnOOSnR_3$ (R = Et, Ph) underwent homolysis over the temperature range 80–120°C to form the radical $R_3SnO\cdot$ which ultimately formed $R_3SnOSn(OR)R_2$ in a manner analogous to the corresponding germanium compounds. They later reported the preparation of a triethyltin hydroperoxide $(Et_3SnOOH)_2 \cdot H_2O_2$. Dannley's method of synthesis was to remove the water of reaction as the water–toluene azeotrope (equation 68). Whereas the trimethyl compound melted at 97°C with decomposition[54], triphenyltin hydroperoxide was reported as exploding at 75°C. The azeotropic distillation, previously used for the synthesis of organolead peroxide was also used by Davies and coworkers to prepare R_3SnOOR^1 ($R = Ph, R^1 = t$-Bu, amyl) using excess *t*-butyl hydroperoxide with the trialkyltin oxide and this gave very high yields. Dialkyltin oxides and *t*-butyl hydroperoxide formed another type of tin peroxide having tin–oxygen–tin linkages (equation 69)[49]. These are dimeric in benzene solution and therefore resemble dialkyltin oxychlorides and carboxylates.

$$Me_3SnOH + H_2O_2 \longrightarrow Me_3SnOOH + H_2O \tag{68}$$

$$R_2SnO + 2t\text{-}BuOOH \longrightarrow \underset{OOBu\text{-}t}{R_2Sn}-O-\underset{OOBu\text{-}t}{SnR_2} + H_2O \tag{69}$$

$$R = Et, Bu$$

Acylperoxytin alkyls have been prepared by Razuvaev[56] and Davies[49] and their coworkers who have reported that they undergo spontaneous and rapid, first-order, rearrangement (equations (70 and 71). The extent of the alternative mode of decomposition which liberates dioxygen depends on the nature of the group attached to the tin, e.g. equation (72).

$$(PhCOO)_2 + Et_3SnOR \longrightarrow [Et_3SnOOCOPh] + PhCOOR$$
$$[Et_3SnOOCOPh] \longrightarrow Et_2Sn\begin{matrix} OEt \\ OCOPh \end{matrix} \tag{70}$$

$$C_9H_{19}CO_3H + Ph_3SnOH \longrightarrow [Ph_3SnOOCOC_9H_{19}] + H_2O \longrightarrow Ph_2Sn(OPh)(OCOC_9H_{19}) \quad (71)$$

$$Et_3SnOOCOMe \longrightarrow Et_3SnOCOMe + \tfrac{1}{2}O_2 \quad (72)$$

Autoxidation of tetraethyltin or hexaethyldistannane at temperatures below ambient, often with ultraviolet light initiation, give as primary products peroxides such as $Et_3SnOOEt$ and $Et_3SnOOSnEt_3$ respectively, but as the temperature is raised free-radical derived decomposition products are observed[4].

2. Lead

Although similar to the preparation of the organotin peroxides the nucleophilic substitution method for preparing organolead peroxides of the type R_3PbOOR^1 is more limited in scope. Alkyl hydroperoxides do not interact sufficiently with alkyllead halides, therefore elimination of sodium halide using the sodium salt of the alkyl hydroperoxide is used (R = Me, Ph; R^1 = *t*-Bu, cumyl, Ph_3C). On the other hand, compounds of the general formulae R_3PbZ, where Z may be $-OH$, $-OR$ or $-OPbR_3$, can be substituted by the free hydroperoxide[4,53]. Triethyllead oxide reacts rapidly at room temperature (equation 73)[58]. In some circumstances low thermal stability of the organolead alkyl peroxides makes them difficult to isolate but this might be due to contamination; for example hexaalkyldilead rapidly converts the trialkyllead-alkyl peroxide to the corresponding alkoxide, e.g. equation (74).

$$Et_3PbOPbEt_3 + ROOH \longrightarrow Et_3PbOOR + Et_3PbOH \quad (73)$$

ROOH = *t*-Bu, cumyl, 1,4-diisopropylbenzene hydroperoxide

$$Et_3PbPbEt_3 + Et_3PbOOcumyl \longrightarrow Et_3PbOPbEt_3 + Et_3PbOcumyl \quad (74)$$

On heating the lead peroxides undergo homolysis. However, generally the stability of the lead peroxides is reasonably good and melting points are often reported for these crystalline compounds[4] with the exception of triethyllead-alkyl peroxides of low molecular weights. Trialkylleadhydroxides interact with peroxycarboxylic acids, but in common with tin, germanium and silicon the trialkylleadacyl peroxides are unstable, readily rearranging intramolecularly (compare equation 71), which is accompanied by dioxygen elimination decomposition (compare equation 72).

Photooxidation of tetraethyllead or hexaethyldilead is not a good method for preparing peroxides such as $Et_3PbOOEt$ or $Et_3PbOOPbEt$; as in the case of tin the yields are low. Autoxidation of both tin and lead alkyls are radical chain processes[2b].

V. ORGANOMETALLIC PEROXIDES OF GROUP V METALS

The organometallic peroxides of the vanadium subgroup have been but little studied. The interaction of dicyclopentadienylniobium dichloride with 30 % hydrogen peroxide in methylene dichloride formed in 80 % yield the yellow dioxygen complex (equation 75) which was soluble in a number of polar solvents, and was stable in air over a period of days yet was reported as being explosive[59]. Its structure was suggested from a consideration of its infrared spectrum.

$$(C_5H_5)_2NbCl_2 + HOOH \longrightarrow (C_5H_5)_2Nb(O_2)Cl \quad (75)$$

A. Antimony, Arsenic and Bismuth

Stable peroxides of the type $Ph_3Sb(OOR)Z$ are known where Z = Ph, Et, OH, OR or Br, and are readily prepared from alkyl hydroperoxides[60,61] or their alkali-metal salts. Displacement of halide or alkoxide from the metal is the preferred method of preparation (equations 76 and 77). In one case displacement of a phenyl group from pentaphenylantimony by an alkyl hydroperoxide has been reported (equation 78). The alkyl hydroperoxides were usually *t*-butyl hydroperoxide or cumyl hydroperoxide, although Razuvaev and coworkers found that triphenylsilyl or germyl hydroperoxide formed a stable metal peroxide (equation 79). The bis-μ-peroxy compounds can similarly be prepared using two equivalents of hydroperoxide or its sodium derivative (equation 80). All these antimony peroxides are only slowly decomposed or rearranged above 100°C (equation 81), and they are hydrolytically stable unless Z is alkoxide or halogen.

$$Ph_3Sb(Et)OR^1 \xrightarrow{t\text{-BuOONa}} Ph_3Sb(Et)OOBu\text{-}t + NaBr \quad (76)$$

$$Ph_3Sb(Et)Br \xrightarrow[t\text{-BuOOH}]{} Ph_3Sb(Br)OOBu\text{-}t + R^1OH \quad (77)$$

$$Ph_5Sb + ROOH \longrightarrow Ph_4SbOOR + PhH \quad (78)$$

$$Ph_4SbBr + Ph_3MOOH \xrightarrow{Et_3N} Ph_4SbOOMPh_3 \quad (79)$$

$$M = \text{Si or Ge}$$

$$R_3SbX_2 + 2R^1OOH \xrightarrow{\text{base}} R_3Sb(OOR^1)_2 + 2HX \quad (80)$$

$$Ph_4SbOOR \longrightarrow Ph_3Sb(OPh)OR \quad (81)$$

Hydroperoxides of antimony have been prepared using hydrogen peroxide (equation 82)[53b].

$$R_3Sb(OR^1)_2 + H_2O_2 \longrightarrow R_3Sb(OOH)_2 + 2R^1OH \quad (82)$$

The diperoxide of several organoarsenic compounds has been prepared[60] in the same way as the antimony diperoxide (equation 83).

$$R_3AsX_2 + 2R^1OOH \longrightarrow R_3As(OOR^1)_2 + 2HX \quad (83)$$

$$R = \text{Me, Ph};\ R^1 = t\text{-Bu, cumyl}$$

The arsenic peroxy compounds, like the antimony, are solids melting above 50°C. Both antimony and arsenic form polymeric peroxides of the type $[R_3MOO-]_n$. Studies on the action of air on trialkylstibines or trialkylbismuthines indicate that a free-radical chain-reaction occurs although no clear-cut product is formed. Alkylperoxystibines and bismuthines are formed, but react further with the trialkylmetal and their reduction results in insoluble polymeric metal oxides which separate out from the benzene solutions. It had been suggested earlier on that triethylbismuthine when autoxidized at −50°C gave an unstable peroxidic compound with the structure $Et_3\overline{Bi.OO}$ [70]. However, Davies and Hook point out that trialkylstibines and bismuthines when autoxidized have a lot in common with trialkylphosphine autoxidation[71] (compare equation 85) and they favour the mechanism shown in equations (84)–(87). Equations (84) and (85) are believed to be in competition.

$$ROO^{\cdot} + R_3M \longrightarrow ROOMR_2 + R^{\cdot} \quad (84)$$

$$ROO^{\cdot} + R_3M \longrightarrow RO^{\cdot} + OMR_3 \quad (85)$$

$$RO^{\cdot} + R_3M \longrightarrow ROMR_2 + R^{\cdot} \quad (86)$$

$$R^{\cdot} + O_2 \longrightarrow RO_2^{\cdot} \quad (87)$$

VI. PEROXY DERIVATIVES OF GROUPS VI AND VIII ORGANOMETALLICS

A. Group VI

The metals of Group VIA having organic groups with carbon–metal σ bonds have only been prepared in recent years and consequently little is known about their interaction with molecular oxygen. During the autoxidation of hexaneopentylbimolybdenum or hexabenzylbitungsten, the presence of alkyl peroxides of these organometallics was inferred by analogy with tetraalkylzirconium autoxidations because the reaction could be retarded by known antioxidants, thereby implying that a free-radical chain-process was involved. Also small amounts of peroxide were detected in the final solution[43].

Epoxidation of carbon–carbon double bonds using alkyl hydroperoxides and early transition-metal, homogeneous catalysts (such as molybdenum, tungsten, vanadium, niobium, tantalum or titanium, in high oxidation states) (equation 88) is the basis of the Halcon process for the production of propylene oxide on a large tonnage scale[5], as well as stereospecific epoxidations of certain steroids and terpenes. It has been established that the epoxy oxygen is derived from a molybdenum– or vanadium–alkyl hydroperoxide complex[73]. In the earlier stages it is envisaged that the olefin is π-bonded to the metal complex. It has been suggested by Mimoun[9], from a consideration of the various reaction characteristics, that the π-olefin complex converts into a σ-bonded group and this then permits the formation of a five-membered peroxo metallocyclic intermediate which can rearrange to the epoxide (equation 89).

$$ROOH + CH_3CH{=}CH_2 \xrightarrow[\text{catalyst}]{\text{soluble metal}} ROH + CH_3\overset{O}{\overbrace{CH{-}CH_2}} \quad (88)$$

$$R = t\text{-Bu or } PhCHCH_3$$

$$\text{C=C} + [\text{Mo}]\text{—OOR} \longrightarrow [\text{Mo}]\text{(C–C–O–O–R cyclic)} \longrightarrow [\text{Mo}]\text{—OR} + \text{epoxide} \qquad (89)$$

In so far as elemental tellurium can be considered to have metallic characteristics, it is worth bearing in mind that the rearrangement of 2-chlorocyclohexyltellurium trihalide in the presence of *t*-butyl hydroperoxide may well involve an intermediate cyclohexylalkyl peroxide (equation 90)[74].

$$\text{2-Cl-C}_6\text{H}_{10}\text{TeBr}_3 + 2t\text{-BuOOH} \longrightarrow \left[\text{2-Cl-C}_6\text{H}_{10}\text{Te}\text{Br}_{(3-n)}(\text{OOBu-}t)_n\right] \longrightarrow \text{2-Cl-C}_6\text{H}_{10}\text{Br} \qquad (90)$$

B. Group VIII

Organometallic peroxides of this group have as yet not been systematically studied although peroxy compounds of the Group VIII metals having a variety of ligands attached to the transition metals are of considerable interest[9], particularly with reference to enzymatic reactions involving molecular oxygen[75].

A thorough study involving the preparation of alkylperoxy(pyridinato)cobaloximes (equation 91) has been carried out by Giannotti and coworkers[76]. Benzyl- and substituted benzyl-(pyridinato)cobaloximes were found to insert dioxygen into the cobalt–carbon bond at ambient temperature to form the μ-peroxo compound (equation 92)[76]. Photolysis was necessary to bring about the reaction when simple alkyl groups were attached to the cobalt and a cobaloxime–molecular oxygen complex was inferred as a result of ESR studies, which detected the cobalt(III) superoxide anion. The final stage was considered to be the homolysis of the cobalt–alkyl bond, which resulted in the formation of the stable alkylperoxycobalt compound (equation 93)[77]. The alkylperoxycobaloxime could be prepared in high yield using *t*-butyl hydroperoxide or cumyl hydroperoxide (equation 94)[78,84].

$$\text{R–Co(dmgH)}_2\text{–pyr} \equiv \text{R–[Co]–pyr} \qquad (91)$$

(CH₃C=N–O···H···O–N=CCH₃ bridged glyoxime ligands around Co, with axial R and pyr)

$$\text{R—[Co]— pyr} + \text{O}_2 \longrightarrow \text{ROO—[Co]— pyr} \qquad (92)$$

$$\text{R—}[\text{Co}^{\text{II}}] + \text{O=O} \longrightarrow \underset{\text{Superoxide}}{\text{R—}[\text{C}^{\text{III}}]\text{—O—O}^{\cdot-}} \longrightarrow [\text{Co}^{\text{III}}]\text{—O—OR} \qquad (93)$$

$$\text{R—[Co]— pyr} + \text{R}^1\text{OOH} \xrightarrow{h\nu} \text{R—[Co(OOR}^1)]\text{— pyr} + \text{ROOR}^1 \qquad (94)$$

In a few cases the organometallic ligand is olefinic in character; for example, the iridium–ethylene complex having peroxo oxygen has been characterized **(5)**[79]. More recently the four-coordinate 1,5-cyclooctadiene (COD), phenanthroline (phen) complex was reported as forming a peroxy five-coordinate iridium complex[80]. The molecular oxygen reacted more readily with an intermediate five-coordinate iodine complex, [(COD)Ir(phen)I], than with the starting compound to form the five-coordinate iridium(II) superoxo complex or the six-coordinate iridium(III) peroxo complex (equation 95). A cyclooctene–rhodium–oxygen complex was suggested by James and Ochiai who believed that a hydroperoxy intermediate was involved in its conversion to ketone (equation 96)[81]. It was later shown that the final products were equimolar amounts of 2-cyclooctene-1-one, cyclooctanone and water[83]. More recently a dirhodium–dioxygen complex having the formula **6** was reported (equation 97). It liberated hydrogen peroxide on treatment with ROH where R = H, Me, Et or CH_3CO[82]. Similar compounds were obtained with dicyclopentadiene– or norbornadiene–rhodium complexes. The same authors had previously described a π-olefin-complexed palladium–dioxygen complex.

Cl CH_2 O Ir CH_2 O PPh_3 PPh_3

(5)

$$[(COD)Ir(phen)]Cl + NaI + O_2 \longrightarrow [(COD)Ir(phen)O_2]I \qquad (95)$$

H O_2 Cl Rh ⟶ O_2H Cl Rh ⟶ O OH Cl Rh (96)

$$(CODRhCl)_2 + 2KO_2 \xrightarrow[20^{\circ}C]{CH_2Cl_2} CODRh(O_2)RhCOD + 2KCl \qquad (97)$$

(6)

The potential use of transition-metal organometallic peroxides, particularly olefin–transition-metal π complexes, is currently stimulating interest since they may offer the opportunity to selectively oxidize olefins to a range of organic compounds. It is in this area of homogeneous catalysis that future developments are likely to take place.

VII. REFERENCES

1. R. Sheldon and J. K. Kochi, *Advan. Catal.*, **25**, 272 (1976).
2. (a) G. A. Razuvaev, V. A. Sushunov, V. A. Dodonov and T. G. Brilkine, *Organic Peroxides*, Vol. III (Ed. D. Swern), Wiley–Interscience, New York–London, 1972.
   (b) A. G. Davies, *Organic Peroxides*, Vol. II (Ed. D. Swern), Wiley–Interscience, New York–London, 1971.
3. T. G. Brilkina and V. A. Syshunov, *Reactions of Organometallic Compounds with Oxygen and Peroxides* (Ed. A. G. Davies), Iliffe Books, London, 1969.
4. D. Brandes and A. Blaschette, *J. Organometal. Chem.*, **78**, 1 (1974).
5. R. A. Sheldon, *J. Mol. Catal.*, **7**, 107 (1980).
6. A. G. Davies and B. P. Roberts, *J. Chem. Soc., Chem. Commun.*, 298 (1966).
7. A. G. Davies and B. P. Roberts, *J. Chem. Soc. (B)*, 1074 (1968).

8. P. B. Brindley, J. C. Hodgson and M. J. Scotton, *J. Chem. Soc., Perkin Trans. 2*, 45 (1979).
9. H. Mimoun, *J. Mol. Catal.*, **7**, 1 (1980).
10. C. Walling and S. A. Buckler, *J. Amer. Chem. Soc.*, **75**, 4372 (1953); **77**, 6032 (1955).
11. C. Walling and A. Cioffari, *J. Amer. Chem. Soc.*, **92**, 6609 (1970).
12. B. J. Wakefield, *Chem. Ind. (London)*, 450 (1972).
13. R. A. Anderson and G. Wilkinson, *J. Chem. Soc., Dalton Trans.* 809 (1977).
14. S. O. Lawesson and N. C. Yang, *J. Amer. Chem. Soc.*, **81**, 4230 (1959).
15. M. H. Abraham, *J. Chem. Soc.*, 4130 (1960).
16. C. H. Bamford and D. M. Newitt, *J. Chem. Soc.*, 688 (1946).
17. A. G. Davies and B. P. Roberts, *J. Chem. Soc. (B)*, 1074 (1968).
18. A. G. Davies and J. E. Packer, *J. Chem. Soc.*, 3164 (1959).
19. G. A. Razuvaev, S. F. Zhiltsov, O. N. Druzhov and G. G. Petukhov, *Zh. Obsch. Khim.*, **36**, 258 (1966); **37**, 2018 (1967).
20. (a) G. A. Razuvaev and E. I. Fedotova, *Dokl. Akad. Nauk SSSR*, **169**, 355 (1966).
(b) G. A. Razuvaev and N. S. Vasileyskaya, *Dokl. Akad. Nauk SSSR*, **74**, 279 (1950).
21. (a) J. Halfpenny and R. W. H. Small, *J. Chem. Soc., Chem. Commun.*, 879 (1979).
(b) A. J. Bloodworth, *J. Chem. Soc., Perkin Trans. 2*, 531 (1975).
22. R. C. Petry and F. H. Verhoek, *J. Amer. Chem. Soc.*, **78**, 6416 (1956).
23. M. H. Abraham and A. G. Davies, *Chem. Ind. (London)*, 1622 (1957); *J. Chem. Soc.*, 429 (1959).
24. A. G. Davies, K. Ingold, B. P. Roberts and R. Tudor, *J. Chem. Soc., (B)*, 698 (1971).
25. S. Korcek, G. B. Watts and K. U. Ingold, *J. Chem. Soc. Perkin Trans. 2*, 242 (1972).
26. A. G. Davies and B. P. Roberts, *J. Chem. Soc. (B)*, 17 (1967); 311, 317 (1969).
27. P. G. Allies and P. B. Brindley, *Chem. Ind. (London)*, 319 (1967); 1439 (1968); *J. Chem. Soc. (B)*, 1126 (1969).
28. J. Grotewold, J. Hernandez and E. A. Lissi, *J. Chem. Soc. (B)*, 182 (1971).
29. G. P. Gerbert, V. P. Maslyanikov, V. A. Shushunov and L. N. Jankitova, *Tr. po. Khim. i. Khim. Tekhnol (Gorky)*, **3**, (21) 41 (1968).
30. A. G. Davies, D. G. Hare and R. F. M. White, *J. Chem. Soc.*, 341 (1961).
31. H. G. Kuivila and A. G. Armour, *J. Amer. Chem. Soc.*, **79**, 5659 (1957).
32. A. G. Davies and R. B. Moodie, *J. Chem. Soc.*, 2372 (1958).
33. R. R. Hamann and R. L. Hansen, *J. Phys. Chem.*, **67**, 2868 (1963).
34. P. B. Brindley and R. G. Pearson, *Polymer Letters*, **6**, 831 (1968).
35. H. C. Brown and G. W. Kabalka, *J. Amer. Chem. Soc.*, **92**, 714 (1970).
36. H. C. Brown and M. M. Midland, *J. Amer. Chem. Soc.*, **93**, 4078 (1971).
37. A. G. Davies and C. D. Hall, *J. Chem. Soc.*, 1192 (1963).
38. H. Hock and F. Ernst, *Chem. Ber.*, **92**, 2716 (1959).
39. Y. A. Aleksandrov, N. V. Chikinova, G. I. Makin, N. V. Kornilova and V. I. Bregadze, *Zh. Obschch. Khim.*, **48**, (2), 467 (1978).
40. G. A. Razuvaev, E. V. Mitrofanova and V. A. Dodonov, *Dokl. Akad. Nauk SSSR*, **182**, 1342 (1968).
41. G. A. Razuvaev, E. V. Mitrofanova and V. A. Dodonov, *Dokl. Akad. Nauk SSSR*, **187**, 5, 1072 (1969).
42. G. A. Razuvaev, V. A. Dodonov, T. J. Starostina and T. A. Ivanova, *J. Organometal. Chem.*, **37**, 233 (1972).
43. P. B. Brindley and J. C. Hodgson, *J. Organometal. Chem.*, **65**, 57 (1974).
44. P. J. Davidson, M. F. Lappert and R. Pearce, *J. Organometal. Chem.*, **57**, 269 (1973).
45. P. B. Brindley and M. J. Scotton, *J. Chem. Soc. Perkin Trans. 2*, 419 (1981).
46. G. R. Davies, J. A. J. Jarvis and B. T. Kilbourn, *J. Chem. Soc., Chem. Commun.*, 1511 (1971).
47. T. F. Blackburn, J. A. Labinger and J. Schwartz, *Tetrahedron Letters*, 3041 (1975).
48. A. Rieche and T. Bertz, *Angew. Chem.*, **70**, 507 (1958).
49. (a) A. G. Davies and I. F. Graham, *Chem. Ind. (London)*, 1622 (1963).
(b) A. J. Bloodworth, A. G. Davies and I. F. Graham, *Organometal. Chem.*, **13**, 351 (1968).
50. G. A. Razuvaev and E. V. Mitrofanova, *Zh. Obshch. Khim.*, **38**, 249 (1968).
51. D. L. Alleston and A. G. Davies, *J. Chem. Soc.*, 2465 (1962).
52. Y. A. Aleksandrov and V. A. Schuschunov, *Dokl. Akad. Nauk SSSR*, **140**, 565 (1961); *Zh. Obshch. Khim.*, **35**, 115 (1965).
53. (a) A. Rieche and J. Dahmann, *Annalen*, **675**, 19 (1964);
(b) A. Rieche and J. Dahmann, *Ber.*, **100**, 1544 (1967).

54. (a) R. L. Dannley and W. A. Aue, *J. Org. Chem.*, **30**, 3845 (1965).
(b) R. L. Dannley, W. A. Aue and A. K. Shubber, *J. Organometal. Chem.*, **38**, 281 (1972).
55. M. F. Salomen and R. G. Salomen, *J. Amer. Chem. Soc.*, **99**, 3500 (1977).
56. Reference 2, pp. 155 and 224; Reference 4 and references therein.
57. Y. A. Aleksandrov, V. V. Gorbatov, N. V. Yablokova and V. G. Tsvetkov, *J. Organometal. Chem.*, **157**, 267 (1978).
58. Y. A. Aleksandrov, T. G. Brilkina and V. A. Shushunov, *Tr. Khim. Teknol.*, **3**, 381 (1960).
59. J. Sala-Pala, J. Roue and J. E. Guerchais, *J. Mol. Catal.*, 141 (1980).
60. A. Reiche, J. Dahlmann and D. List, *Angew. Chem.*, **73**, 494 (1961); *Annalen*, **678**, 167 (1964).
61. G. A. Razuvaev, T. I. Zinovjeva and T. G. Brilkina, *Izv. Akad. Nauk SSSR, Ser. Khim*, 2007 (1969); *Dokl. Akad. Nauk SSSR*, **188**, No. 4, (1969).
62. S. Venura and S. Fukizawa, *J. Chem. Soc., Chem. Commun.*, 1033 (1980).
63. J. C. Hodgson, *Ph.D. Thesis*, Kingston Polytechnic, C.N.A.A., 1973.
64. R. A. Sheldon and J. A. Van Dorn, *J. Catal.*, **31**, 427 (1973).
65. H. Mimoun, R. Charpentier, A. Mitschler, J. Fisher and R. Weiss, *J. Amer. Chem. Soc.*, **102**, 1047 (1980).
66. R. Masthoff, *Z. Anorg. Allgem. Chem.*, **336**, 252 (1965).
67. Y. A. Alexsandrov, G. A. Razuvaev, S. F. Zhil'tov, O. N. Druzhkov and G. G. Petukhov, *Dokl. Akad. Nauk SSSR*, **152**, 633 (1963); **157**, 1395 (1964); *Zh. Obschch. Khim.*, **35**, 1152, 1440 (1965).
68. F. R. Jensen and J. D. Heyman, *J. Amer. Chem. Soc.*, **88**, 3438 (1966).
69. K. U. Ingold and B. P. Roberts, *Free Radical Substitution Reactions* Wiley–Interscience, London–New York, 1971, p. 69 and references therein.
70. G. Calingaert, H. Soroos and W. Hnizda, *J. Amer. Chem. Soc.*, **64**, 392 (1942).
71. A. G. Davies and S. C. W. Hook, *J. Chem. Soc.* (*C*), 1660 (1971).
72. *Hydrocarbon Processing*, Gulf Publishing Company, Houston, Texas, 1977, p. 126.
73. A. O. Chong and K. B. Sharpless, *J. Org. Chem.*, **42**, 1587 (1977).
74. S. Vemura and S. Fukuzawa, *J. Chem. Soc., Chem. Commun.*, 1033 (1980).
75. T. Matsuura, *Tetrahedron*, **33**, 2869 (1977).
76. C. Fontaine, K. N. V. Duong, C. Merienne, A. Gaudemer and C. Giannotti, *J. Organometal. Chem.*, **38**, 167 (1972).
77. C. Giannotti, G. Merle and J. R. Bolton, *J. Organometal. Chem.*, **99**, 145 (1975).
78. C. Giannotti, C. Fontaine, C. Chiavoni and A. Riche, *J. Organometal. Chem.*, **113**, 57 (1976).
79. A. Van der Ent and A. L. Onderdelinden, *Inorg. Chim. Acta*, **7**, 203 (1973).
80. W. J. Louw, T. I. A. Gerber and D. J. A. de Waal, *J. Chem. Soc., Chem. Commun.*, 760 (1980).
81. B. R. James and E. Ochiai, *Can. J. Chem.*, **49**, 975 (1971).
82. F. Sakurai, H. Suzuki, Y. Moro-oka and T. Ikawa, *J. Amer. Chem. Soc.*, **102**, 1749 (1980); **101**, 748 (1979).
83. J. Farmer, D. Holland and D. J. Milner, *J. Chem. Soc., Dalton Trans.*, 815 (1975).
84. J. H. Espenson and J. T. Chen, *J. Amer. Chem. Soc.*, **101**, 2036 (1981); J. Denian and A. Gaudemer, *J. Organometal. Chem.*, **191**, C1 (1980).
85. F. A. Cotton and G. Wilkinson, *Advanced Inorganic Chemistry*, 4th ed., Wiley–Interscience, London–New York, 19**00**, p. 1123.
86. H. Friebolin, R. Huschens and R. Rensch, *Chem. Ber.*, **114**, 3581 (1981).
87. Yu A. Aleksandrov *et al.*, *Zh. Obshch. Khim*, **51**, (7), 1586 (1981).

The Chemistry of Functional Groups, Peroxides
Edited by S. Patai

CHAPTER **24**

Four-membered ring peroxides: 1,2-dioxetanes and α-peroxylactones

WALDEMAR ADAM

Institut für Organische Chemie, Universität Würzburg, Am Hubland, D-8700 Würzburg, BRD and Departamento de Quimica, Universidad de Puerto Rico, Rio Piedras, Puerto Rico 00931, U.S.A.

I. INTRODUCTION . . . 830
A. Nomenclature . . . 830
B. History . . . 831
C. Literature . . . 844
II. PHYSICAL ASPECTS . . . 844
A. Structural Properties . . . 844
B. Spectral Properties . . . 846
1. Nuclear magnetic resonance spectra . . . 846
2. Infrared spectra . . . 846
3. Electronic spectra . . . 847
III. SYNTHESIS . . . 847
A. 1,2-Dioxetanes . . . 848
1. The Kopecky method . . . 848
2. Singlet oxygenation . . . 849
3. Miscellaneous methods . . . 850
B. α-Peroxylactones . . . 852
1. The α-hydroperoxy acid route . . . 852
a. Preparation of α-hydroperoxy acids . . . 852
b. Cyclization of α-hydroperoxy acids . . . 855
2. The singlet oxygen route . . . 857

* This chapter is dedicated to Professor P. D. Bartlett (Texas Christian University), a scholar, a gentleman and a friend.

C. Purification . . . 858
D. Characterization . . . 859
1. Physical methods . . . 859
2. Chemical methods . . . 859
IV. 1,2-DIOXETANES AS REACTION INTERMEDIATES . . . 860
A. Singlet Oxygenations . . . 861
1. Alkenes and cycloalkenes . . . 861
2. Enol ethers, enols and ketene acetals. . . . 864
3. Enamines . . . 866
4. Heteroarenes . . . 871
B. Autoxidations . . . 873
1. Thermal and photochemical oxidations . . . 873
2. Base-catalysed oxidations . . . 874
C. Superoxide Cleavages . . . 878
D. Miscellaneous . . . 880
V. CHEMILUMINESCENCE . . . 880
A. Energy Sufficiency . . . 880
B. Unimolecular Decomposition Mechanisms . . . 882
1. Diradicals as intermediates . . . 882
2. Concerted decomposition . . . 891
3. Theoretical work . . . 893
C. Catalytic Decomposition . . . 894
1. Intermolecular electron exchange . . . 894
2. Intramolecular electron exchange . . . 895
D. Excitation Parameters . . . 897
1. Photophysical methods . . . 897
a. Direct chemiluminescence . . . 901
b. Energy-transfer chemiluminescence . . . 902
2. Photochemical methods . . . 905
a. Intramolecular transformations . . . 905
b. Intermolecular transformations . . . 906
VI. CHEMICAL TRANSFORMATIONS . . . 909
A. Nucleophiles . . . 910
B. Electrophiles . . . 912
VII. ACKNOWLEDGEMENTS . . . 912
VIII. REFERENCES . . . 913

I. INTRODUCTION

A. Nomenclature

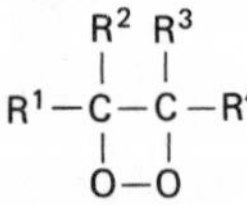

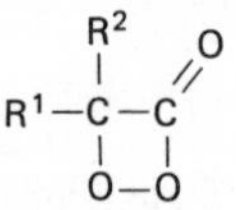

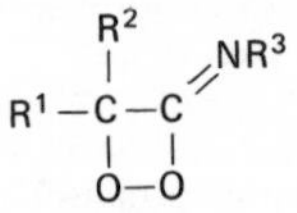

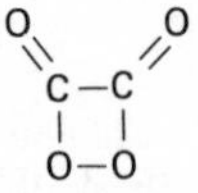

Four-membered ring peroxides are cyclobutane derivatives in which two adjacent methylene units have been replaced by oxygen atoms. Stable entities of this type include the 1,2-dioxetanes **1**, the 1,2-dioxetanones **2**, also known as α-peroxylactones, and the imino-1,2-dioxetanes **3**. The 1,2-dioxetanedione **4** or carbon dioxide dimer, has been postulated as reaction intermediate[1], but no spectral nor chemical evidence exists in support of this structure[2]. Consequently, this chapter will deal mainly with the chemistry of the four-membered ring peroxides **1** and **2**, with some mention of **3**, but not **4**.

B. History

Although 1,2-dioxetanes **1** were postulated as reaction intermediates over 80 years ago[3], these elusive materials were isolated as stable entities only as late as 1969. Thus, Kopecky and Mumford[4] prepared the first example, namely trimethyl-1,2-dioxetane (**5**), via base-catalysed dehydrobromination of β-bromohydroperoxides (equation 1). Earlier, the autoxidation of 1,2,3,3-tetraphenyl-1-propanone into benzoic acid and benzhydryl phenyl ketone was interpreted[5] as involving dioxetane **6**. Indeed, very recent work[6] on authentic α-hydroperoxyketones has shown that they decompose via their cyclic tautomers, i.e. the hydroxydioxetanes such as **6**, into the observed products. However, the formation of dioxetane **7** in the photooxygenation of ergosterol[7] was shown to be incorrect since the endoperoxide **8** was produced[8]. Similarly, in the autoxidation of cyclohexene, instead of the initially proposed[9] dioxetane **9**, the correct structure[10] was the allylic hydroperoxide **10**.

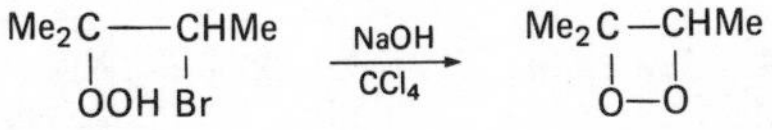

$$Me_2C(OOH)-CHMe(Br) \xrightarrow[CCl_4]{NaOH} Me_2C-CHMe \text{ (O–O ring)} \quad (1)$$

(**5**)

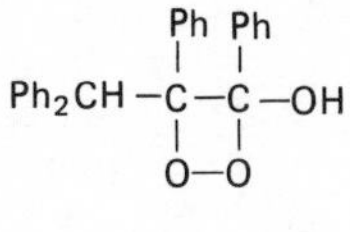

(**6**) 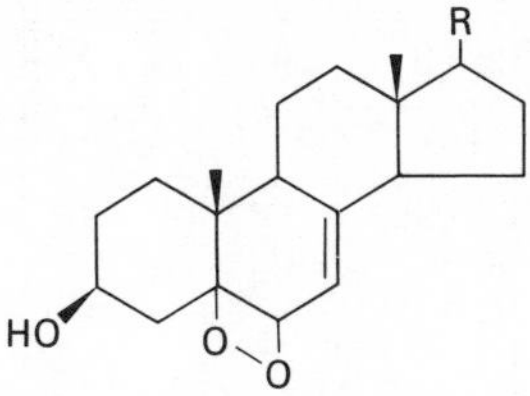 (**7**) 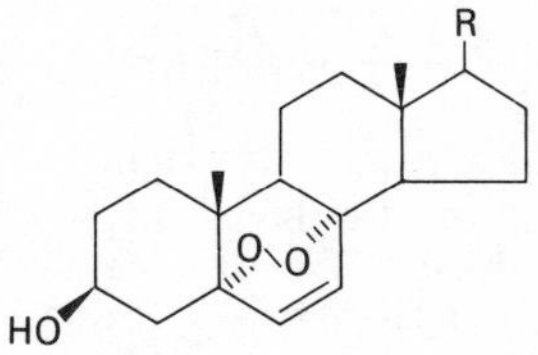 (**8**)

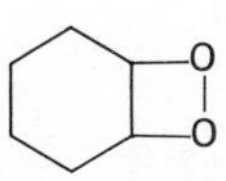

(**9**) 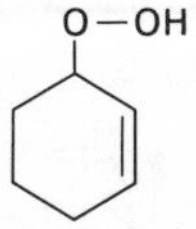 (**10**) 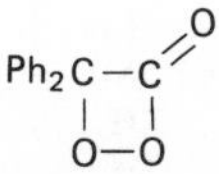 (**11**)

Since then well over a hundred stable derivatives of **1** have been isolated and characterized, of which the more typical ones are listed in Table 1. These include alkyl-, alkenyl-, aryl-, alkoxy-, thio-, amino-, bromoalkyl- and carbonyl-substituted derivatives of the monocyclic, fused bicyclic and spirobicyclic type.

One of the earliest suggestions for the intermediacy of α-peroxylactones stems from Staudinger and colleagues[11], who proposed derivative **11** in the autoxidation of diphenylketene. Indeed, recently this claim has been substantiated in the singlet oxygenation of ketenes[12]. However, the first stable α-peroxylactone that was isolated and characterized[13] was the *t*-butyl derivative **12**, prepared via dicyclohexylcarbodiimide

(DCC) dehydration of the corresponding α-hydroperoxy acid (equation 2). Other derivatives are shown in Table 2.

$$t\text{-BuCH(OOH)}-CO_2H \xrightarrow[-40^\circ C]{DCC} t\text{-BuCH}-C(=O) \text{ (O—O ring) } \quad \textbf{(12)} \tag{2}$$

TABLE 1. 1,2-Dioxetanes

R^1	R^2	R^3	R^4	Method[a]	^{1}H-NMR δ (ppm)[b] ^{13}C-NMR δ (ppm)	Ref.
$-O-CH_2-O$		H	H	A[c]	6.68	220
$-O-CH_2CH_2-O-$		H	H	A	6.22	220, 221
Me	H	H	Me	B	4.90	215, 222
Me	H	Me	H	B[c]	5.1–5.2	2
Me	Me	H	H	B	4.7	2
$-CH_2CH_2CH_2-$		H	H	B	5.5	38, 158
Me	Me	Me	H	B	5.7	27
(tricyclic dioxetane structure)				A	5.2	71
CH_2Br	Me	Me	CH_2Br	D	—	24, 45, 223
CH_2Br	CH_2Br	Me	Me	C	—	45, 223
$-CH_2CH_2CH_2-$		Me	H	B[c]	—	156
$(-CH_2-)_4$		H	H	B[c]	5.35	38, 158
MeC=O	H	Me	Me	B	4.92	35
CH_2Br	Me	Me	Me	D	—	45, 223
CH_2Cl	Me	Me	Me	B, D	—	46
Me	Me	Me	Me	C	89.4	27, 157
Et	H	Me	Me	B[c]	5.0	2
Et	Me	H	Me	B[c]	5.04	2
EtO	EtO	H	H	A	5.87	39
EtO	H	EtO	H	A	5.67	39
MeO	MeO	MeO	MeO	A	—	225, 221
Me_2N	Me	Me	H	A[c]	4.9	226
(cycloheptadiene-fused dioxetane structure)				A	5.18–5.45 6.15–6.35 81.96; 90.73	227
$(-CH_2-)_5$		H	H	B	5.35	38, 158
$(-CH_2-)_4$		Me	H	B	5.1	156
$(-CH_2-)_4$		MeO	H	A[c]	—	81

TABLE 1. *continued*

R^1	R^2	R^3	R^4	Method[a]	^1H-NMR δ (ppm)[b] ^{13}C-NMR δ (ppm)	Ref.
	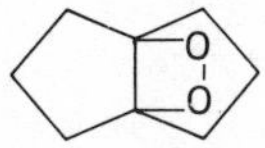 [structure]			C	—	159
	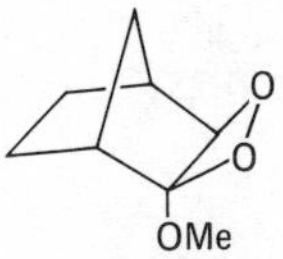 [structure, OMe]			A	5.08	228
$(-CH_2-)_6$		H	H	B[c]	—	158b
$(-CH_2-)_4$		Me	Me	B	—	27
	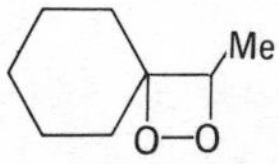 [structure, Me, O–O]			B	5.09	27
$Me_2C{=}CH$	Me	Me	Me	A	5.89	229
$(-CH_5-)_5$		Me	H	B[c]	—	156
	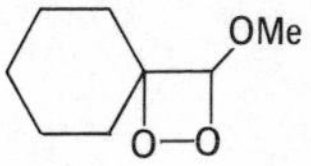 [structure, OMe, O–O]			A	5.08 109	230
Me_3SiO	Me	Me	Me	D[c]	91.8; 109.4	51
Ph	H	H	Me	B[c]	5.00	145b, 231
	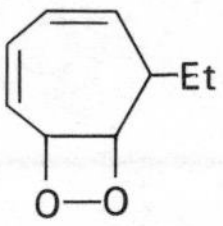 [structure, Et, O–O]			A	4.93–5.15; 6.20–6.34 81.83; 92.55	227
	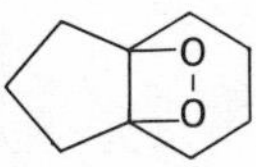 [structure]			C	—	159
n-Bu	Me	Me	Me	C[c]	—	156
	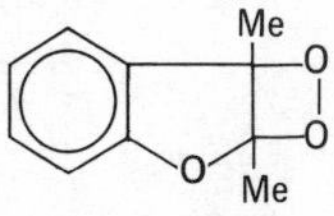 [structure, Me, O, Me]			A	—	232
	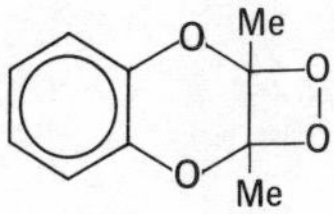 [structure, O, Me, O, Me]			A	—	233

TABLE 1. *continued*

R^1	R^2	R^3	R^4	Method[a]	^{1}H-NMR δ (ppm)[b] ^{13}C-NMR δ (ppm)	Ref.
PhO	EtO	H	H	A[c]	6.07; 6.35	234
PhO	—	EtO	H	A[c]	5.89; 6.12	234
	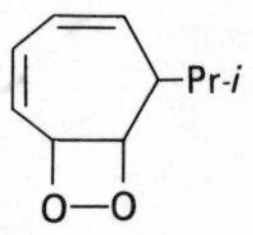			A	5.25–5.50 6.15–6.40 81.70; 91.25	227
	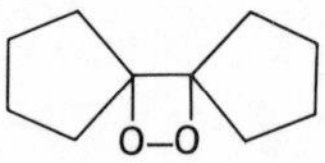			C	—	199
	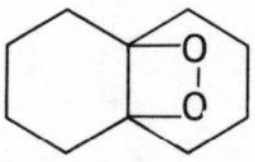			C	87.99	27
	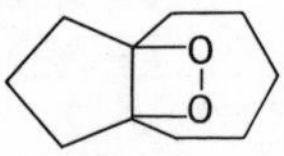			C	—	159
	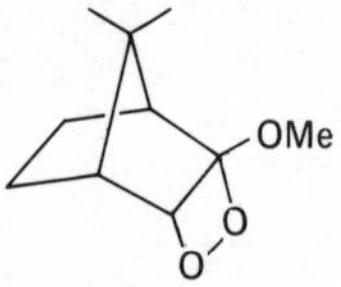			A	5.68	228
	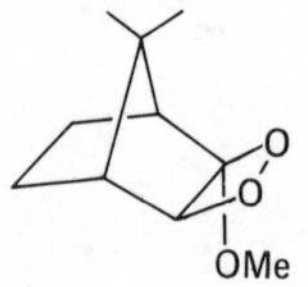			A	5.20	228
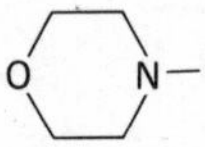	H	$(-CH_2-)_4$		A	—	60
Et	Et	Et	Et	C[c]	—	160
n-BuO	*n*-BuO	H	H	A[c]	5.95	44, 235
n-BuO	H	*n*-BuO	H	A[c]	5.65	44, 235
	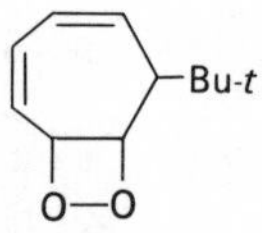			A	5.7–6.4	227

TABLE 1. *continued*

R^1	R^2	R^3	R^4	Method[a]	^{1}H-NMR δ (ppm)[b] ^{13}C-NMR δ (ppm)	Ref.
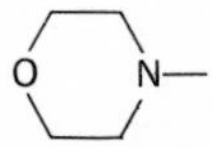 O N— (morpholino)	H	$(-CH_2-)_5$		A	—	60
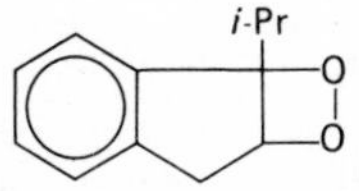	[structure: i-Pr, O, O]			A	5.88	237
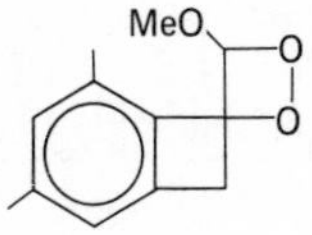	[structure: MeO, O, O]			A	6.05	238
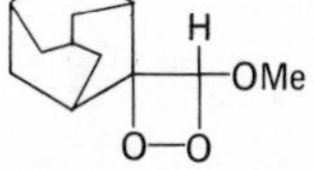	[structure: H, OMe, O—O]			A[c]	5.43	239
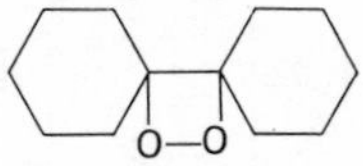	[structure: O—O]			C[c]	—	160
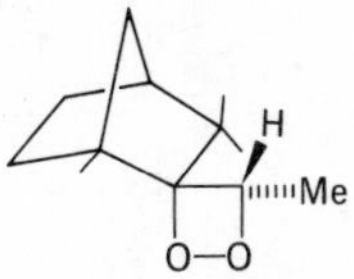	[structure: H, Me, O—O]			A[c]	6.08	239
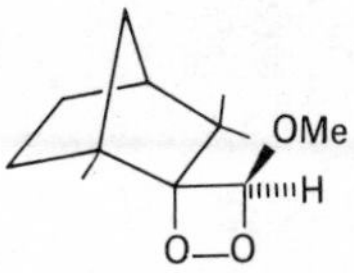	[structure: OMe, H, O—O]			A[c]	5.46; 5.86	239
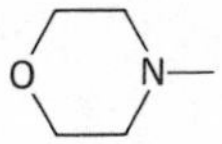 O N— (morpholino)	H	$(-CH_2-)_6$		A	—	60
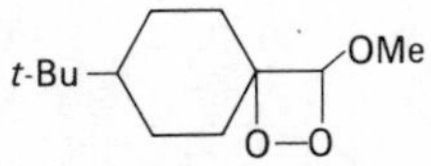	[structure: t-Bu, OMe, O—O]			A	5.08 89.5; 109	230
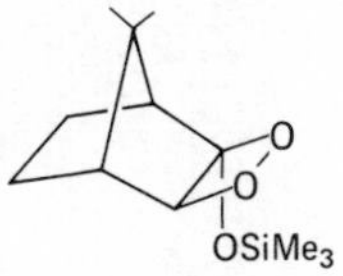	[structure: O, O, OSiMe$_3$]			A	5.58	228

TABLE 1. *continued*

R^1	R^2	R^3	R^4	Method[a]	^{1}H-NMR δ (ppm)[b] ^{13}C-NMR δ (ppm)	Ref.
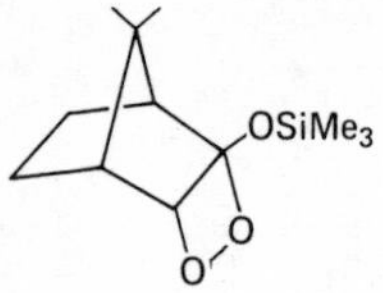				A	4.84	228
n-Bu	*n*-Bu	Me	Me	C[c]	—	200, 240
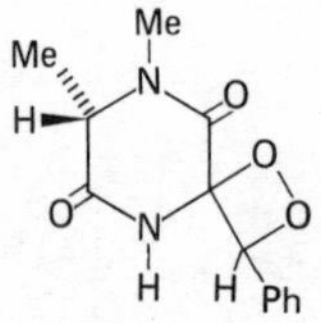				A[c]	5.2	241
$PhCH_2CH_2$	Me	Me	Me	C[c]	—	156
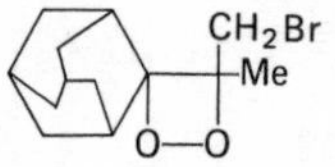				C	88.1; 92.6	47
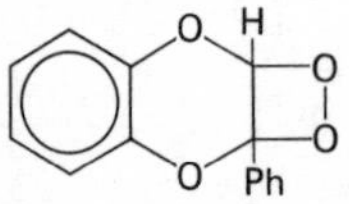				A	6.33	233
Ph	H	Ph	D	C[c]	6.34	146
Ph	H	Ph	H	C	6.4	146, 211c
Ph	H	H	Ph	C	5.63	145a
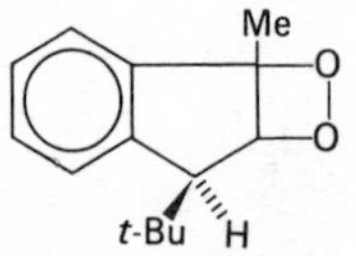				A	5.88	237
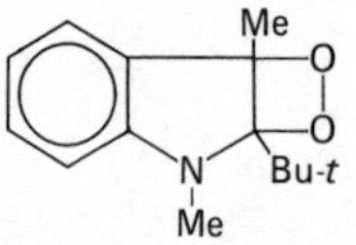				A[c]	96.1; 100	103
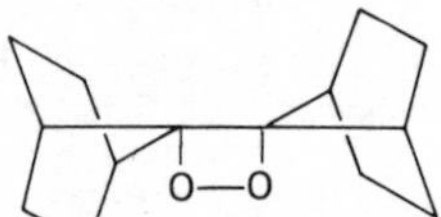				A	—	43

TABLE 1. *continued*

R^1	R^2	R^3	R^4	Method[a]	^{1}H-NMR δ (ppm)[b] ^{13}C-NMR δ (ppm)	Ref.
				A	96.6; 109.6	25, 47
				A	—	230
				C[c]	5.92	224
				A	—	233
		H	H	B[c]	4.65; 5.19	242
	H		H	B[c]	5.15; 5.32	242
Ph	MeO	H	Ph	C[c]	6.06	224
p-BrC_6H_4	t-Bu	H	Me_3SiO	A	4.8	54
PhO	t-Bu	D	Me_3SiO	A	—	54
PhO	t-Bu	H	Me_3SiO	A	4.7	54
				A	—	237

TABLE 1. *continued*

R^1	R^2	R^3	R^4	Method[a]	^{1}H-NMR δ (ppm)[b] ^{13}C-NMR δ (ppm)	Ref.
				A	109	31
$PhCH_2$	H	H	$PhCH_2$	B	4.85	145a
Ph	Me	Ph	Me	C	—	145b
p-$MeOC_6H_4$	H	H	p-$MeOC_6H_4$	B[c]	5.68	145d
Ph	Ph	MeO	MeO	A	—	244
Ph	MeO	Ph	MeO	A	—	244
p-MeC_6H_4	t-Bu	H	Me_3SiO	A	4.75	54
				A	5.7	47
				A	6.16 93.1; 100.2	25, 47
$-CH_2CH_2-$	Me	Me	Me	C	—	156
				A	5.8	47
				A	5.96	31, 245
N—	Ph	H	Ph	A	5	226
PhO	t-Bu	H	PhO	A	4.8	246
				A	5.45	47

TABLE 1. *continued*

R^1	R^2	R^3	R^4	Method[a]	^{1}H-NMR δ (ppm)[b] ^{13}C-NMR δ (ppm)	Ref.
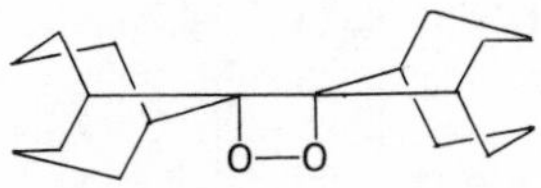 [structure]				A	—	44
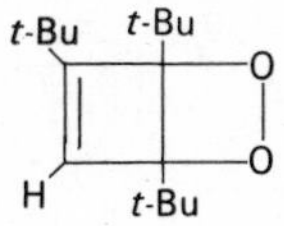 [structure: t-Bu, t-Bu, H, t-Bu, O—O]				A[c]	—	243a
PhO	t-Bu	H	Me_2SiO— (t-Bu)	A	4.5	246
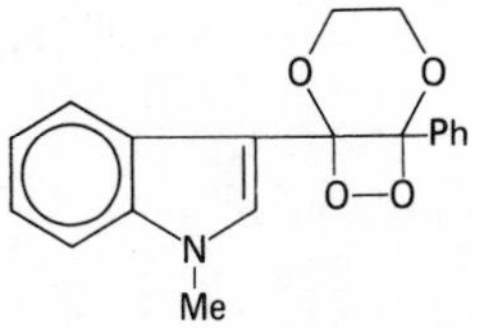 [structure: N-Me, O, O, Ph, O—O]				A[c]	—	30
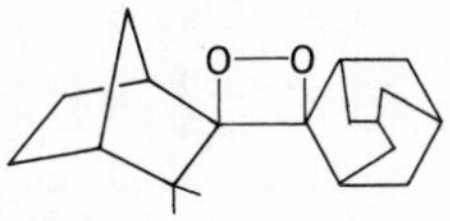 [structure: O—O]				A	—	250
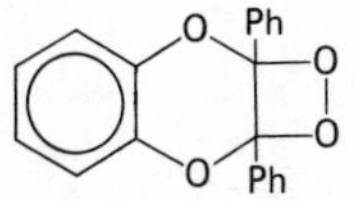 [structure: O, Ph, O, Ph, O—O]				A	113.4	23
Ph	Ph	H	Ph	B	6.04	145c
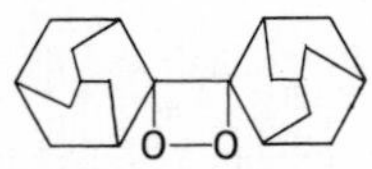 [structure: O—O]				A	95	42, 157
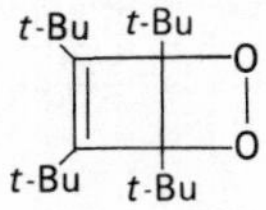 [structure: t-Bu, t-Bu, t-Bu, t-Bu, O—O]				A	108.7	243b
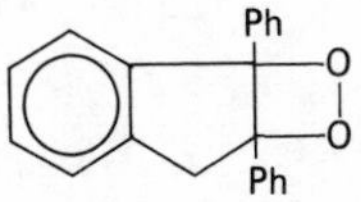 [structure: Ph, Ph, O—O]				A	—	237

TABLE 1. *continued*

R^1	R^2	R^3	R^4	Method[a]	^1H-NMR δ (ppm)[b] ^{13}C-NMR δ (ppm)	Ref.
Me–N … O–O … Ph, H	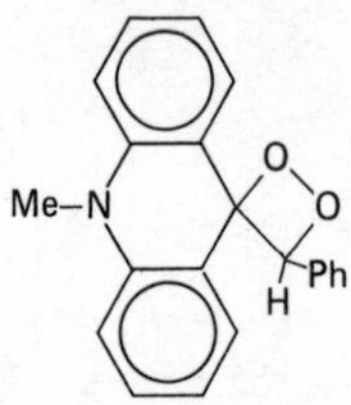			A, C[c]	6.44	180c
Ph, Ph … Me, Me … O–O	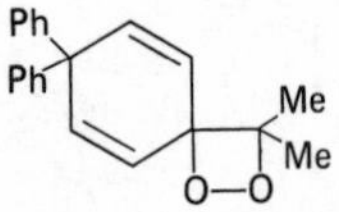			A	—	202
O–O	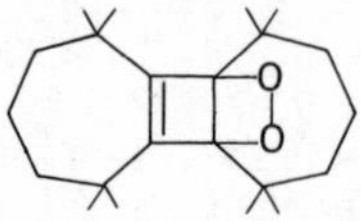			A	103.6	22
O … O–O	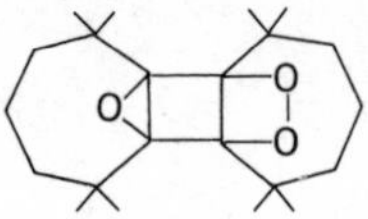			D	104.8	22
O–O	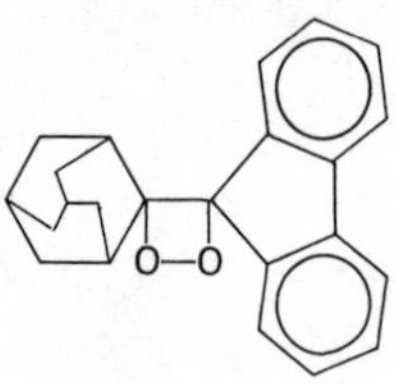			A	94.5; 97.7	47
OPh, OPh … O–O	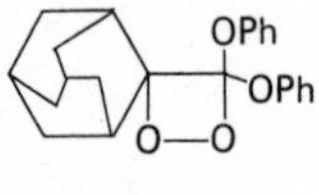			A	98.8; 117.7	47
Me–N … O–O	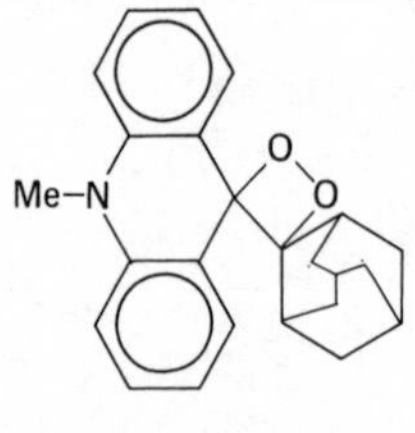			A[c]	—	180

TABLE 1. *continued*

R^1	R^2	R^3	R^4	Method[a]	^{1}H-NMR δ (ppm)[b] ^{13}C-NMR δ (ppm)	Ref.
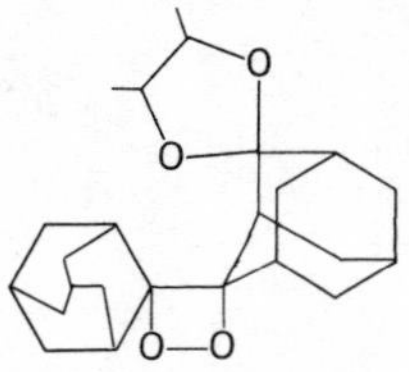				A	96.9; 97.6	248
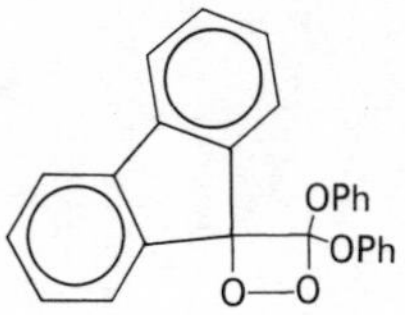				A	—	256
Ph	PhO	PhO	Ph	A	—	256
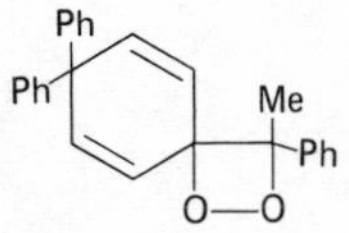				A	—	202
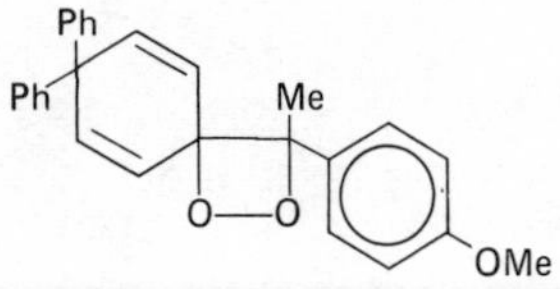				A	—	202
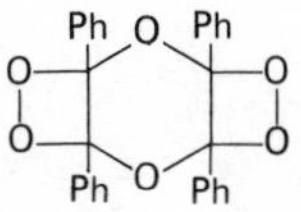				A	—	86
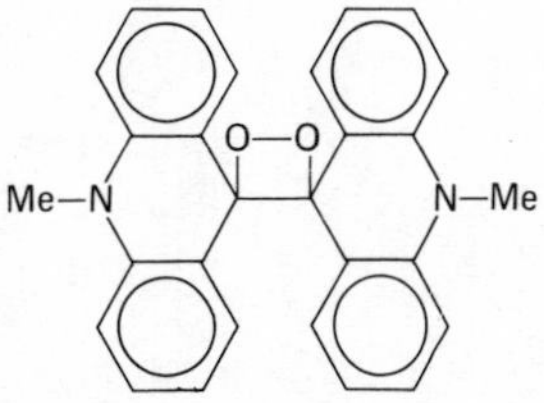				A[c]	—	180c

TABLE 1. *continued*

R^1 R^2 R^3 R^4	Method[a]	^{1}H-NMR δ (ppm)[b] ^{13}C-NMR δ (ppm)	Ref.
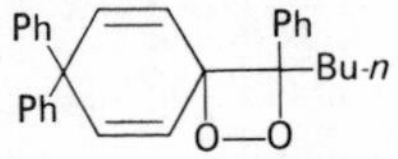	A	—	202
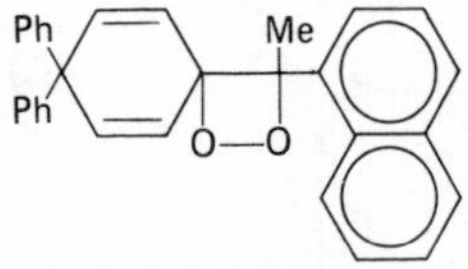	A	—	202
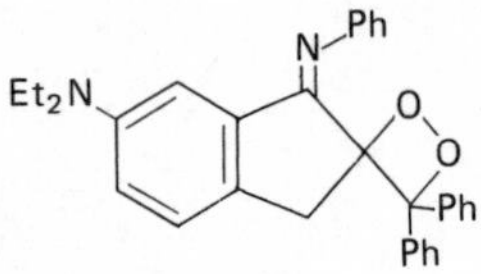	A	—	249
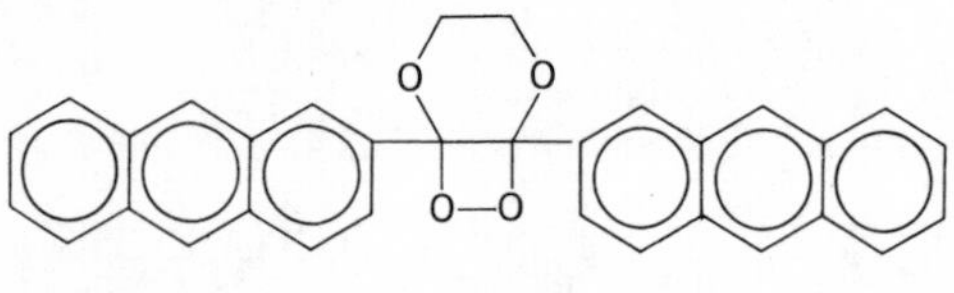	A	—	31, 48
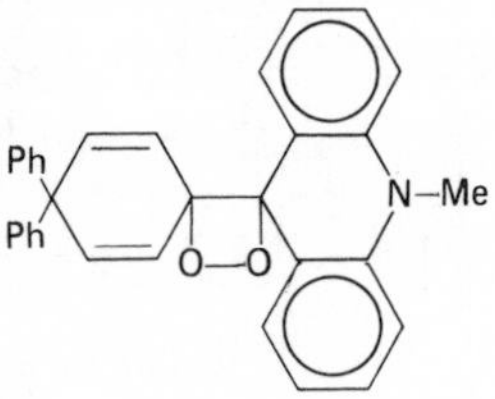	A[c]	—	180c
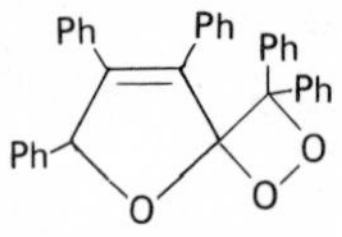	A		251a
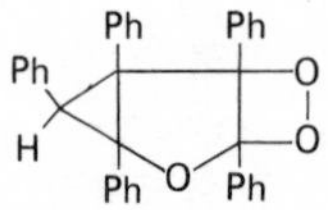	A	—	251b

TABLE 1. *continued*

R^1	R^2	R^3	R^4	Method[a]	^{1}H-NMR δ (ppm)[b] ^{13}C-NMR δ (ppm)	Ref.
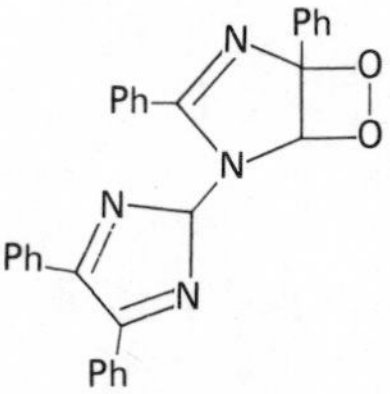				A	—	252

[a] Method A: singlet oxygenation; Method B: base-catalysed dehydrobromination; Method C: Silver ion-catalysed dehydrobromination; Method D: miscellaneous.
[b] Dioxetane ring proton or carbon.
[c] Not isolated.

TABLE 2. α-Peroxylactones and 2-imino-1,2-dioxetanes

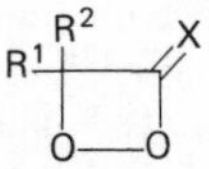

X	R^1	R^2	Method[a]	^{1}H-NMR (ppm)[b] IR, ν(cm^{-1})[c]	Reference
O	Me	Me	A, B	1874	12, 26, 201
O	CF_3	CF_3	A	1940	12
O	*t*-Bu	H	A, B	5.48, 1870	12, 26
O	*n*-Pr	Me	A	1870	12
t-BuN	Me	Me	A	—	57
i-PrN	Et	Et	A	—	57b
O	*t*-Bu	*t*-Bu	B	1855	26
c-C_6H_{11}N	Et	Et	A	—	57b
t-BuN	Et	Et	A	—	57b
t-BuN	*t*-Bu	H	A	5.59	58
O	Ph	*n*-Bu	A	1860	12, 201
O	1-Adamantyl	H	B	1875	253
t-BuN	Ph	Me	A	—	57b
O	Ph	Ph	A	1870	12, 201
t-BuN	Ph	*n*-Bu	A	—	57b

[a] Method A: singlet oxygenation; Method B: dicyclohexylcarbodiimide cyclization.
[b] Dioxetane ring protons.
[c] Carbonyl stretching frequency.

In the 1960s α-peroxylactone **13** was postulated as the energizing intermediate in the bioluminescence of fireflies[14,15]. While early ^{18}O-labelling experiments refuted this claim[16], careful recent studies[17] demonstrated that α-peroxylactones are indeed *bona fide* reaction intermediates in luciferin-type bioluminescence.

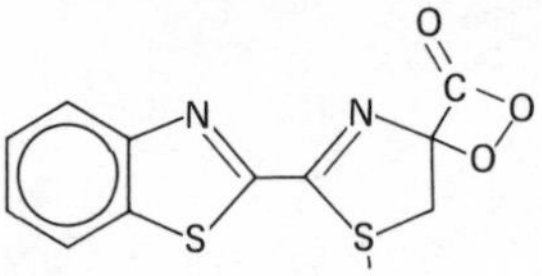

(13)

C. Literature

Since the discovery of stable four-membered ring peroxides, 1,2-dioxetanes[4] in 1969 and α-peroxylactones[13] in 1972, these unusual molecules have been studied with great intensity and interest. Well over a couple of hundred publications on this subject have accumulated during the last decade, either claiming these elusive materials as reaction intermediates or isolating them as stable compounds. Fortunately this rapidly expanding novel area of organic peroxide chemistry has been well reviewed in the literature, either as chapters in monographs or reviews in journals[18]. An international conference[19] on the chemical and biological aspects was held in 1978 and a multiauthored treatise[20] has appeared.

General reviews, which cover the synthesis, spectroscopy, chemiluminescence, mechanism of decomposition, chemical and photochemical transformations and biological implications, include those by Mumford[18e], Bartlett[18h], Schuster and students[18j], Bartlett and Landis[18m] and Adam[18i,o]. General reviews on singlet oxygen which cover dioxetane chemistry are those of Matsuura and Saito[18q], Schaap and Zaklika[18r], Gorman and Rodgers[18u], Frimer[18v], George and Bhat[18w] and Adam[36]. Specific reviews on chemiluminescence properties have been published by Turro and coworkers[18c], Gundermann[18d], Wilson[18g], Schuster[18k] and Adam[18p]. The biological aspects, both of bioluminescent and dark enzymatic processes, are covered in reviews by Goto and Kishi[18a], McCapra[18b], Cilento[18n] White and colleagues[18s], Hastings and Wilson[18t], Ward[18y] and Adam[18f], and in a book on bioluminescence edited by Herring[18x]. Since the older literature is well reviewed, we shall concentrate on the recently published material.

II. PHYSICAL ASPECTS

A. Structural Properties

The determination of the crystal structures of dioxetanes by X-ray diffraction is difficult, because in most instances these thermally and photolytically labile materials do not survive the irradiation. Moreover, it is quite cumbersome to grow crystals with good diffraction properties. Nevertheless, crystal structures have become available on a few stable derivates, i.e. the dioxetanes **14**–**17**. Their structural properties of the dioxetane moiety are summarized in Table 3, listing the O—O, C—O and C—C single-bond distances and the degree of puckering of the dioxetane ring.

The first crystal structure to be determined[21] is that of the dispiroadamantane-1,2-dioxetane **14**. This dioxetane shows the greatest degree of puckering, ca. 21°. Dreiding

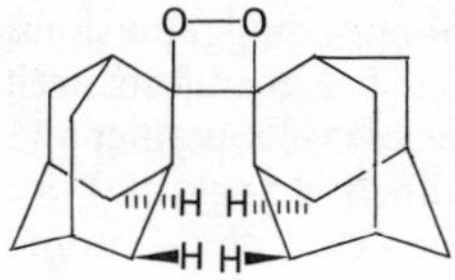

(14) 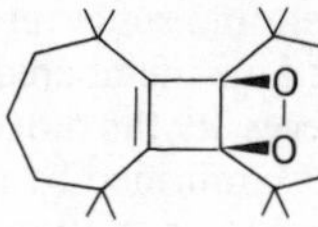 (15) 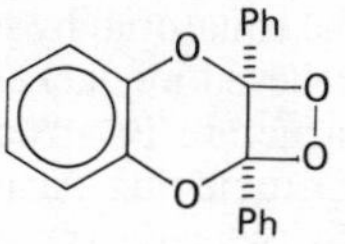 (16) 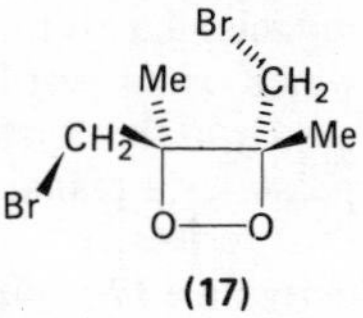 (17)

TABLE 3. X-Ray structural parameters of 1,2-dioxetanes

Dioxetane	Bond distances (pm)[a]				Torsional angle (deg.)[b]	Reference
	O(1)–O(2)	O(2)–C(3)	C(3)–C(4)	C(4)–O(1)		
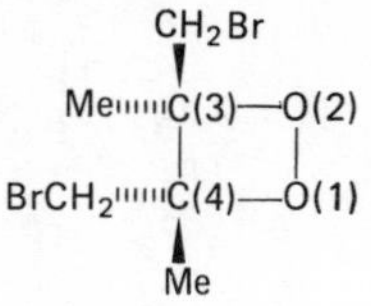	144(3)	148(2)	151(4)	149(2)	15.3	24
A	155.2(8)	144.3(9)	161(2)	144.6(9)	0.8	23
B	152.1(7)	143.8(8)	158(1)	146.8(8)		
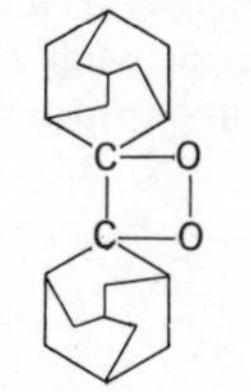	148.0	147.5	154.9	147.5	21.3	21
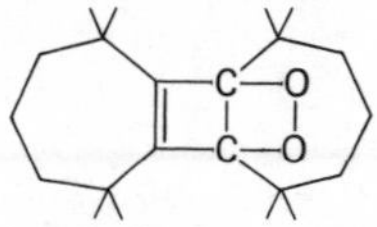	148.8	149.0	155.2	149.0	0	22
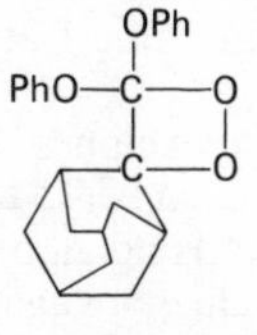	147(2)	148(2)	148(2)	151(2)	11.7	24
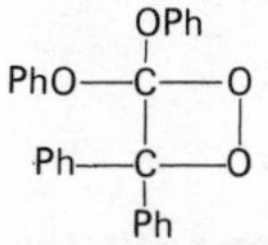	150.5(3)	148.1(3)	155.0(4)	143.2(3)	9.6	24

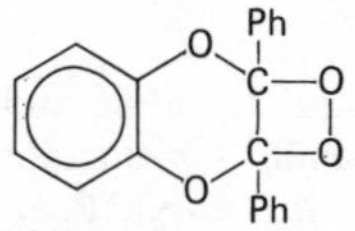

[a] Standard deviations in parentheses.
[b] O(2)C(3)C(4)O(1) angle.

models suggest that the equatorial hydrogens should suffer large nonbonding repulsions, which could best be relieved by puckering the four-membered ring. This manifests itself clearly in the crystal structure. By structural necessity the dioxetane of cyclobutadiene **15** possesses a planar dioxetane ring. Again this is confirmed by the dihedral angle of 0° for this derivative[22]. Also the *p*-dioxin dioxetane **16** is planar[23]; however, the dibromo derivative **17** is puckered, exhibiting a dihedral angle of 13°[24]. This latter example is the simplest of the four, suggesting that such monocyclic dioxetanes should all be puckered. Since in the *d,l* isomer **17** the two bromine atoms are diametrically arranged, no special steric effect by the voluminous bromines can be responsible for the puckering. It is truly unfortunate that tetramethyl-1,2-dioxetane **18** crystallizes in a too disordered form to permit structure determination by X-ray diffraction.

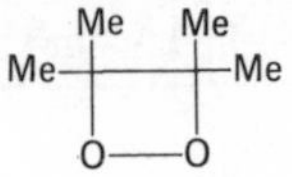

Me Me
Me— —Me
O—O

(18)

B. Spectral Properties

1. Nuclear magnetic resonance spectra

^{1}H- and ^{13}C-NMR methods are undoubtedly the most convenient and definitive tools for the identification of 1,2-dioxetanes and α-peroxylactones, especially for unstable derivatives which cannot be isolated. For example, with the help of ^{13}C-NMR the extremely labile sulphur-substituted dioxetanes **19** and **20** can be detected at low temperatures[25]. Thus, **19** shows a singlet at 93.1 ppm and a doublet at 100.2 ppm, while **20** shows two singlets at 96.6 and 109.6 ppm. These were the first sulphur-substituted dioxetanes to be characterized.

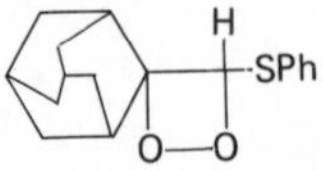

H
SPh
O—O

S
S
O—O

(19) 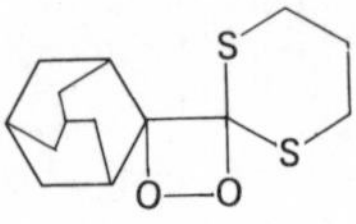 **(20)**

In general the dioxetane ring carbons, if not substituted by heteroatoms, are located characteristically at δ(TMS) 88–90 ppm. Heteroatom substitution shifts this carbon resonance in expected amounts to lower fields. Specific values are listed in Table 1 and have been invaluable in the identification of dioxetanes.

The dioxetane ring protons are also valuable for the identification of dioxetanes. In general these proton resonances exhibit characteristic chemical shifts at δ(TMS) 4.9–5.2 ppm, provided that the dioxetanes are alkyl-substituted. Olefinic, aryl and alkoxy substituents shift the dioxetane protons to δ(TMS) 5.6–5.9 ppm. Specific values are again listed in Table 1.

2. Infrared spectra

Except for the α-peroxylactones **2**, which display characteristic carbonyl stretching frequencies at 1850–1870 cm^{-1} (for specific values consult Table 2)[12,26], infrared spectra are relatively uninformative for the characterization of dioxetanes.

3. *Electronic spectra*

Most dioxetanes are yellow coloured. The λ_{max} is located at ca. 280 nm ($\varepsilon \sim 20$)[27] with a tail-end absorption reaching as far as 450 nm, hence the yellow colour. This characteristic yellow colour is assigned from photoelectron spectra[28] to $\pi_y^* \rightarrow \pi_z$ excitation of the oxygen–oxygen bond. The vertical ionization potentials were found to be 8.98, 10.98, 11.41 and 12.09 eV for tetramethyl-1,2-dioxetane (**18**) by photoelectron spectroscopy[28].

A characteristic property of 1,2-dioxetanes and α-peroxylactones is their ability to luminesce on heating. In fluid media and in the presence of oxygen gas the observed chemiluminescence is usually due to the fluorescence emitted by the electronically excited carbonyl product. It is located at 420 ± 10 nm and corresponds to $\pi^* \rightarrow n$ deexcitation[18c,29] of the carbonyl product. Probably the shortest wavelength fluorescence has been observed for the indole dioxetane **21**, which emits at 320 nm[30]. Recently also π,π^* fluorescence has been reported[31] for dioxetane **22**. Furthermore the emission of transient 1,2-dioxetanes in the gas phase has been actively investigated and matches the n,π^* fluorescence of the carbonyl fragments[32].

Emission of phosphorescence is rare, but under ideal conditions can be occasionally observed in liquid media. Thus, the chemiluminescence observed in the thermolysis of tetramethyl-1,2-dioxetane (**18**)[33] or dimethyl-α-peroxylactone (**23**)[34] at 430 nm in deaerated acetonitrile has been assigned to the n,π^* phosphorescence of acetone. For the acetyl derivative **24** both n,π^* fluorescence and phosphorescence of methylglyoxal was observed simultaneously[35].

(21) (22) (23) (24)

III. SYNTHESIS

The preparation of carbocyclic and heterocyclic four-membered rings is usually not a simple task; but when such strained rings contain a weak peroxide linkage as in 1,2-dioxetanes and α-peroxylactones, the synthetic problem is expected to be still more difficult. Indeed, the methods of preparing these thermally labile molecules are rather limited. Only two procedures possess general applicability and then with severe limitations. These are the hydroperoxide cyclization and singlet oxygen cycloaddition (equation 3).

For the dioxetanes the first route can be executed by base (HO^-, RO^-) or by metal ion (Ag^+, Hg^{+2}) catalysis; but for the considerably more labile α-peroxylactones neutral agents such as the mild carbodiimides are essential. Even then, the yields are usually very low (<30 %), as a result of thermal decomposition of the dioxetane products or their destruction by the catalysts that are essential for cyclization.

$$R_2C(OOH)-CR_2X \xrightarrow{(-HX)} R_2C(O)-CR_2(O)\ [\text{dioxetane}] \xleftarrow{^1O_2} R_2C{=}CR_2 \qquad (3)$$

The singlet oxygenation technique can give much higher yields, provided the olefinic substrates, i.e. alkenes for the dioxetanes and ketenes for the α-peroxylactones, are sufficiently reactive. Singlet oxygen is mildly electrophilic and requires electron-rich substrates. Even then, ene reaction with substrates bearing allylic hydrogens and [4 + 2]cycloaddition with substrates bearing olefinic or even aromatic substituents can be menacing side-reactions (equation 4)[36,37]. These shortcomings cannot usually be circumvented. We shall now outline these synthetic methods separately for the 1,2-dioxetanes and α-peroxylactones.

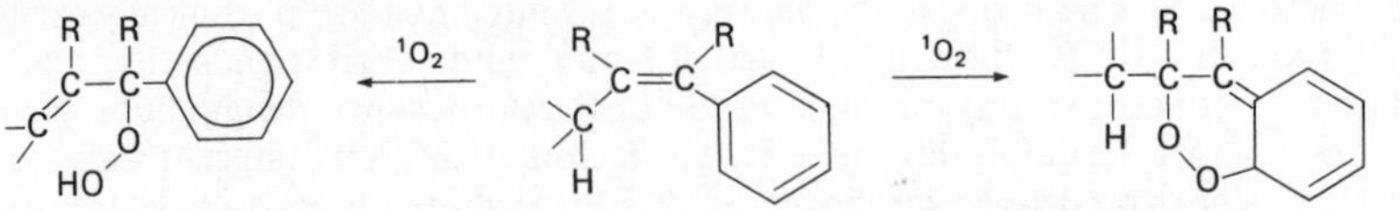

(4)

A. 1,2-Dioxetanes

1. The Kopecky method

As already illustrated in equation (1), the dehydrobromination of β-bromohydroperoxides constituted the very first synthesis of a stable dioxetane, namely the trimethyl derivative **5**[4]. We shall refer to this synthetic route as the Kopecky method[27], which is given in general terms in equation (5).

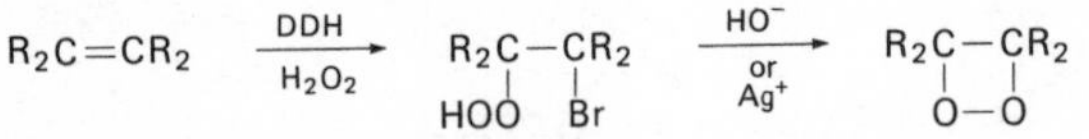

$$R_2C{=}CR_2 \xrightarrow[H_2O_2]{DDH} \underset{(25)}{R_2C(OOH){-}CR_2Br} \xrightarrow[\text{or } Ag^+]{HO^-} R_2C{-}CR_2 \,(\text{O{-}O ring}) \qquad (5)$$

The β-bromohydroperoxides **25** are prepared by electrophilic addition of bromine, using most commonly 1,3-dibromo-5,5-dimethylhydantoin (DDH) as carrier of Br^+, in the presence of concentrated (85–95 %) hydrogen peroxide (*CAUTION! All safety precautions must be taken when working with this dangerous chemical; the active oxygen content should always be below 10 % to avoid explosion hazards!*). Other bromine carriers, e.g. *N*-bromosuccinimide (NBS), can be used as well. The yields of β-bromohydroperoxides **25** are normally quite high (>60 %), but purification can be problematic because these hydroperoxides are quite unstable. Solid derivatives can be purified by recrystallization, but most often they are obtained as viscous oils. Low-temperature column chromatography on silica gel, eluting with methylene chloride, can be quite effective. Although formation of solid complexes with 1,4-diazobicyclo[2.2.2]octane (DABCO) has been recommended[27] for the purification of β-bromohydroperoxides **25**, we have found this method of very limited value. The corresponding β-chloro- and β-iodohydroperoxides can be prepared analogously by using 1,3-dichloro- or 1,3-diiodo-5,5-dimethylhydantoin as Cl^+ or I^+ carriers, respectively[27].

The critical preparative step in equation (5) is the cyclization. For primary and secondary bromides base catalysis is required, while for tertiary bromides silver acetate or silver oxide are more effective as cyclization catalysts. In the case of tertiary β-bromohydroperoxides dehydrobromination is a serious side-reaction and for this reason silver ion catalysis is essential. The silver catalyst must be free of metallic silver by preparing the silver salt freshly and recrystallizing it in the dark, because silver metal particles efficiently catalyse the destruction of dioxetanes. In this way tetramethyl-1,2-dioxetane (**18**) can be prepared in 5–20 % yields. The β-chloro- and β-iodo-hydroperoxides do not give higher yields.

In view of the labile nature of the weak peroxide bond, the cyclization must be performed at low temperatures, typically −30 to +10°C. However, too low temperatures usually result in lack of reactivity of the β-bromohydroperoxide substrate towards cyclization. In the case of relatively stable dioxetanes, the cyclization can be executed at room temperature (up to 30°C).

Protic solvents such as water and/or methanol are advantageous in the base-catalysed cyclizations; but heterogeneous solvent systems such as aqueous methanol–pentane can also be useful. Recently we[38] found phase-transfer catalysis with tetraalkylammonium salts convenient in the base-catalysed preparation of the bicyclic dioxetanes **26** and **27**. These latter conditions minimize base-catalysed destruction of the dioxetane product. In the silver-ion-catalysed cyclizations solvents such as pentane, hexane, cyclohexane, methylene chloride, dichlorodifluoromethane, chloroform and carbon tetrachloride are usually employed.

(26) (27)

2. *Singlet oxygenation*

Electron-rich olefins, especially vinyl ethers, ketene acetals, enamines[181] and more recently thio-substituted olefins[25,47], react readily with singlet oxygen to give dioxetane products (equation 3). Under carefully controlled temperature conditions and with filtering of the detrimental ultraviolet radiation, the dioxetane product accumulates sufficiently for isolation. The first dioxetanes prepared in this manner were the *cis* and *trans* isomers **28** and **29**, respectively[39].

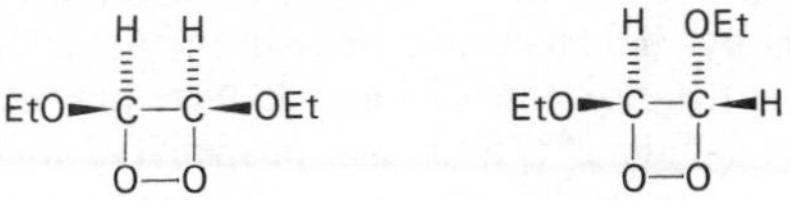

(28) (29)

The singlet oxygen can be generated either via photosensitization or chemically. The former is the more convenient method. Frequently used sensitizer–solvent combinations are tetraphenylporphyrin (TPP) in methylene chloride, Rose Bengal (RB) in acetone and methylene blue (MB) in methanol[36]. In certain applications the polymer-bound Rose Bengal sensitizers can be advantageous[40], especially when the photooxygenated product mixture is to be used directly for spectroscopic detection. In that case the sensitizer can be removed by simple filtration. However, the reaction times are significantly longer in such heterogeneous reactions.

As light source in the photosensitized singlet oxygenation we recommend the use of sodium lamps. These are easily available commercially in the form of street lamps, have a relatively high monochromatic light output in the spectral region for photosensitization, produce relatively little ultraviolet radiation and heat, and have a long life.

Excellent chemical sources of singlet oxygen are triphenylphosphite ozonate (**30**) and 1,4-dimethylnaphthalene-1*H*-peroxide (**31**)[41]. These singlet oxygen carriers liberate singlet oxygen at sufficiently low temperatures to preserve the dioxetane product.

$[PhO]_3P{\cdot}O_3$

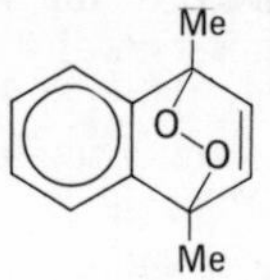

(30) **(31)**

The most serious side-reactions are ene reaction and [4 + 2]cycloaddition (equation 4). Thus, alkyl-substituted olefins, which are otherwise quite reactive towards singlet oxygen, cannot be used for the preparation of dioxetane via singlet oxygenation. However, when the allylic hydrogens are positioned at bridgeheads in the olefin, then singlet oxygen is obliged to undergo [2 + 2]cycloaddition, as was the case in the formation of dioxetanes **14**, **32** and **33**, prepared respectively from bisadamantylidene[42], 7,7-bisnorbornylidene[43] and 9,9-bisbicyclo[3.3.1]nonylidene[44] via photosensitized singlet oxygenation.

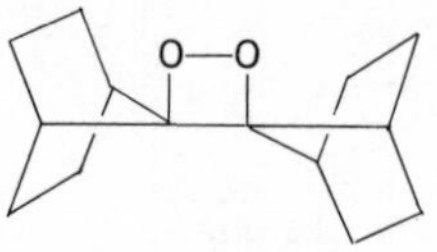

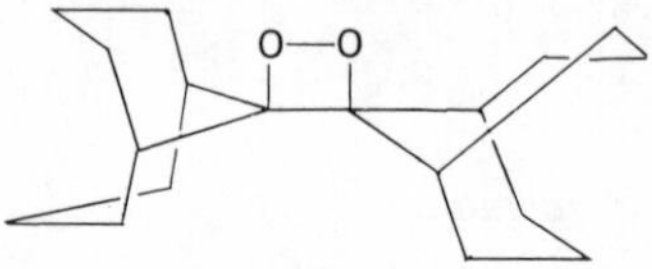

(32) **(33)**

3. Miscellaneous methods

A few alternative methods to the Kopecky route (equation 5) and singlet oxygenation (equation 3) are available for the preparation of 1,2-dioxetanes. Although none of these are general, it is worthwhile to mention them. For example, intramolecular peroxymercuration, followed by halogenation (equation 6) affords halogen-substituted dioxetanes[45]. In this way the dioxetanes **34** and **35** were prepared, respectively from 2,3-dimethyl-3-hydroperoxy-1-butene[46] and 2-(2-propenyl)-2-hydroperoxyadamantane[47]. This method has synthetic potential, especially if electrophiles other than halogens could be employed for the preparation of functionalized dioxetanes.

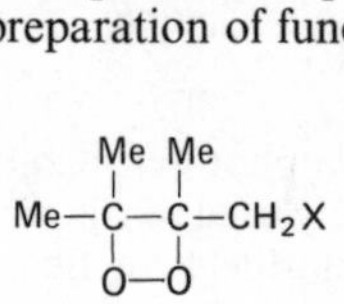

X = Br, Cl

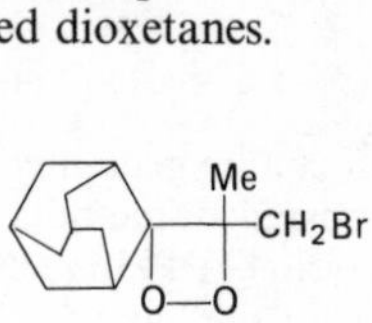

(34) **(35)**

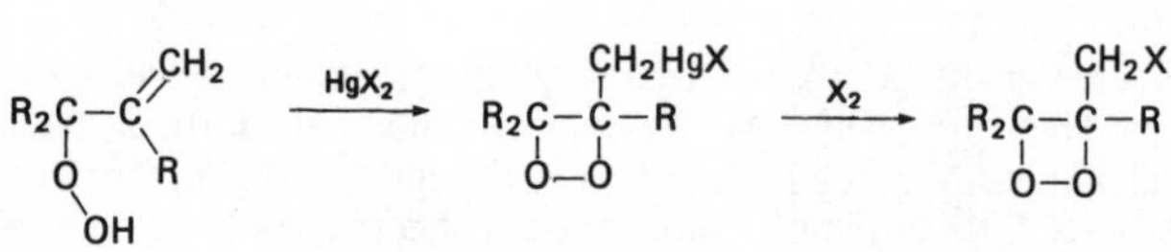

$$R_2C(OOH)-C(R)=CH_2 \xrightarrow{HgX_2} R_2C(O\text{-}O)C(R)CH_2HgX \xrightarrow{X_2} R_2C(O\text{-}O)C(R)CH_2X \quad (6)$$

Another interesting method, but of limited scope, is the silica-gel-catalysed rearrangement of endoperoxides into dioxetanes (equation 7)[48]. Thus, in the singlet oxygenation, first the endoperoxide **36** was formed, which on attempted silica gel chromatography gave dioxetane **37**. Similarly, but not requiring silica gel catalysis, the singlet oxygenation of the 3-indolyl-1,2-dioxene (**38**) gave at −70°C first the endoperoxide **39**, which on warming to −46°C rearranged into the thermally labile dioxetane **21** (equation 8)[30]. Surely with time other examples like these will become known, but this route will be limited to aryl-substituted, electron-rich substrates.

Of interest is the recent report[49] that electrochemical oxidation of diadamantylidene in the presence of molecular oxygen afforded dioxetane **14** (equation 9). Similarly, one-electron oxidants such as $NO_2^+PF_6^-$ in the presence of molecular oxygen also converted diadamantylidene into **14**[50]. The success of this novel method undoubtedly resides to a great extent in the high stability and inertness of dioxetane **14**. However, if this convenient and novel method could be generalized, it would constitute a valuable synthetic tool for the preparation of dioxetanes.

$^1O_2$ → **(36)**

silica gel → **(37)** (7)

(38) $\xrightarrow{^1O_2,\ -70°C}$ **(39)** $\xrightarrow{-46°C}$ **21** (8)

O, O, Ph, N, Me

$\xrightarrow[(n\text{-Bu})_4N^+ClO_4^-]{\text{current; } ^3O_2}$ **14** (9)

The last method worthy of mention entails the observation[51] that the ozonization of vinylsilanes gives silyloxy-substituted dioxetanes (equation 10). In this unusual manner dioxetane **40** was obtained. In view of the mildness of the reaction conditions, this method might constitute a convenient entry into hydroxy-substituted dioxetanes.

$$Me_2C{=}C(Me)SiMe_3 \xrightarrow{O_3} Me_2C{-}C(Me)(OSiMe_3){-}O{-}O \text{ (cyclic)} \quad (10)$$

(40)

B. α-Peroxylactones

In view of the increased strain energy of the α-peroxylactones that is caused by the exocyclic double bond and the larger exothermicity of their decomposition as the result of decarboxylation, it is logical to expect that these 'high-energy' molecules should be still more difficult to prepare than the 1,2-dioxetanes. As we shall see, this expectation has certainly become fact. Yet, since nature has succeeded in efficiently producing such labile molecules in the phenomenon of bioluminescence, it should be possible to prepare them through chemical means. An obvious approach is to mimic the biosynthetic route which is outlined in equation (11). We shall, therefore, first take up the synthetic strategy outlined in equation (11). The key synthetic intermediate is the α-hydroperoxy acid and consequently we shall term this approach the α-hydroperoxy acid route. Subsequently we shall consider the singlet oxygen route, in which α-peroxylactones are prepared by singlet oxygenation of ketenes (equation 12). This latter route is also the only access to the imino-1,2-dioxetanes **3**, for which ketenimines are used in equation (12) instead of ketenes.

$$R_2CH{-}C(=O)OH \xrightarrow[\text{enzyme}]{^3O_2} R_2C(OOH){-}C(=O)OH \xrightarrow{(-H_2O)} R_2C{-}C(=O) \text{ with } O{-}O \text{ ring} \quad (11)$$

(41) **(2)**

$$R_2C{=}C{=}O \xrightarrow{^1O_2} R_2C{-}C(=O) \text{ with } O{-}O \text{ ring} \quad (12)$$

(2)

1. The α-hydroperoxy acid route

Since the preparative problems of the two steps in equation (11) are so distinct, it is convenient to consider each step separately.

a. Preparation of α-hydroperoxy acids. In view of the ease of acid- and base-catalysed decarboxylation of the α-hydroperoxy acids **41** via Grob fragmentations (equation 13), it was clear that perfectly neutral or buffered conditions had to be developed which would permit accumulation of the labile α-hydroperoxy acids **41**.

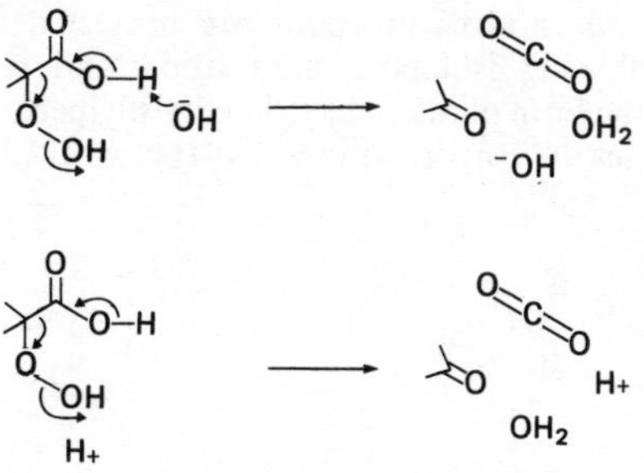

(13)

In equation (14) are outlined the three successful methods for the synthesis of α-hydroperoxy acids[26]. Of these, the first two, i.e. the ketene bis(trimethylsilyl)acetal (equation 14, top) and the α-lactone (equation 14, centre) routes occur under strictly neutral conditions. The third method, i.e. the carbanion route (equation 14, bottom), takes place under highly basic conditions; but when the necessary precautions are taken, this method can be one of the most general and convenient.

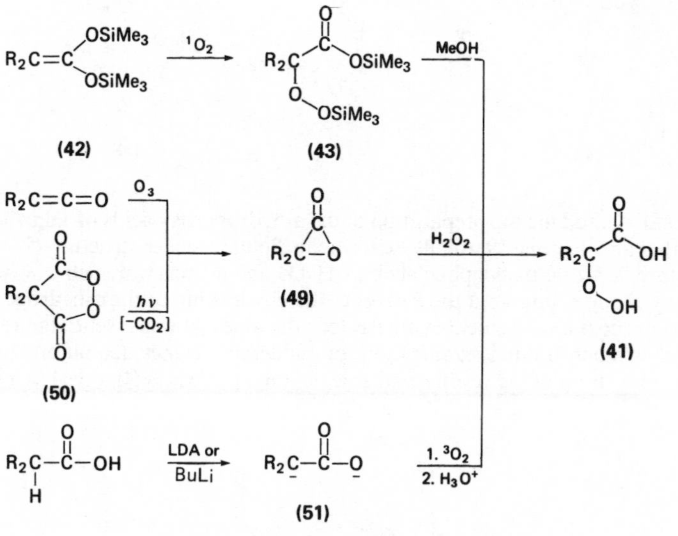

(14)

First we shall consider the ketene bis(trimethylsilyl)acetal route. The timely observation[52] that silyl enol ethers react with singlet oxygen to afford α-trimethylsilylperoxyketones opened up a convenient entry for the α-hydroperoxy acids **41**. Using ketene bis(trimethylsilyl)acetals **42** as substrates, singlet oxygen generates the trimethylsilylperoxy ester **43** which on desilylation affords the desired α-hydroperoxy acid **41**[26]. This amazing singlet oxygenation fulfills several important requirements. Firstly, and most importantly, through this silatropic shift a peroxy functionality is introduced next to an ester carbonyl. Secondly, and from the practical point of view equally important, the peroxy and carboxy functions are protected with readily removable trimethylsilyl groups, allowing isolation, purification and storage of the labile α-hydroperoxy acids **41** in the form of the stable trimethylsilylperoxy esters **43**. Desilylation of **43** with methanol releases the α-hydroperoxy acid **41** at will. All operations are perfectly neutral.

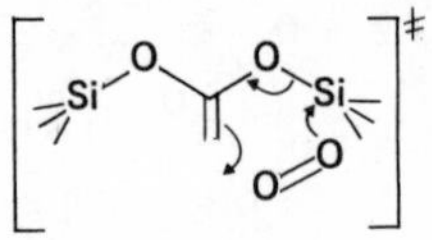

(44) 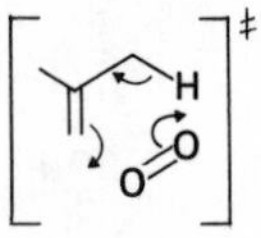 **(45)**

Although one can write the silatropic transition state **44**, which is analogous to the classical prototropic transition state **45** for the singlet oxygen ene reaction[36,37,53], like the latter, the mechanism is much more complex. For example, we recently observed[54] that in the singlet oxygenation of ketene aryltrimethylsilylacetals **46** the 1,2-dioxetane **48** was formed, besides the expected trimethylsilylperoxy ester **47** (equation 15). Clearly, such a product composition demands the intervention of an intermediate as branching point[54].

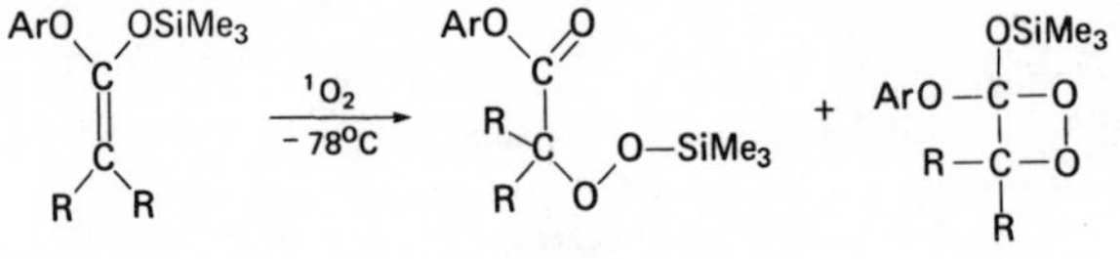

$$R_2C{=}C(OAr)(OSiMe_3) \xrightarrow[-78^\circ C]{^1O_2} ArO{-}C(=O){-}CR_2{-}O{-}O{-}SiMe_3 + \text{1,2-dioxetane} \quad (15)$$

(46) **(47)** **(48)**

The second method for the preparation of the α-hydroperoxy acids **41** takes advantage of the fact that an α-lactone **49** usually exists in the dipolar valence structure **49a** (equation 16). Therefore, a protic nucleophile such as H_2O_2 should react in such a way that the hydroperoxy group is bound at the α carbon, thereby leading to the α-hydroperoxy acid **41**. For this purpose a convenient synthesis for α-lactones **49** was essential to replace the preparatively rather limited ozonization of hindered ketenes (equation 14)[55]. The photodecarboxylation of the readily available malonyl peroxides **50** served as an effective method for the *in situ* formation of α-lactones[56].

$$R_2C{-}O{-}C{=}O \rightleftharpoons R_2C^{+}{-}C(=O){-}O^{-} \quad (16)$$

(49) **(49a)**

The third synthetic method engaged the α carbanion **51** as synthons (equation 14). In fact, this was our first synthetic approach for the preparation of α-hydroperoxy acids **41**. However, our early attempts always led to extensive decomposition of the α-hydroperoxy acids because of their labile nature towards base- or acid-catalysed decomposition (equation 13). Yet, when the oxygenation reaction of the α carbanion **51** and the protonation of the resulting α-peroxycarboxylate dianion are conducted at −78°C or lower and the α-hydroperoxy acid isolated and purified immediately at low temperatures, respectable yields of the desired product can be achieved[56]. Indeed, this method is to date the most convenient and general of the three described in equation (14). Fortunately, with these three preparative methods a large variety of α-hydroperoxy acids **41** can be prepared (Table 4).

TABLE 4. α-Hydroperoxy acids

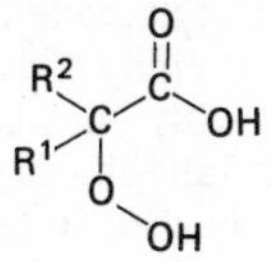

R^1	R^2	Method[a]	m.p. (°C)	Reference
Me	Me	A	44–46	56a
t-Bu	H	A, B	69–70	26, 56a
t-Bu	*t*-Bu	C	81–83	26
1-Ad	H	B	114–115	253
Ph	H	A	96–97	56b
Ph	Ph	A	100–102	56b
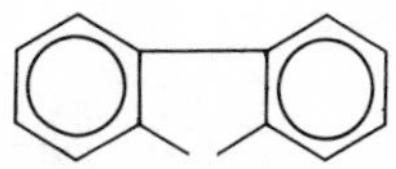		A	126–128	56b
n-Pr	Me	A	—	247
$CH_3(CH_3)_5$—	H	A	67–68	247
$CH_3(CH_2)_7$—	Me	A	45–46	247
$CH_3(CH_2)_7$—	Et	A	—	247
$CH_3(CH_2)_7$—	*n*-C_7H_{15}	A	—	247
$CH_3(CH_2)_{11}$—	H	A	89–90	247
$CH_3(CH_2)_{11}$—	Me	A	57–58	247
$CH_3(CH_2)_{11}$—	Et	A	—	247
$CH_3(CH_2)_{11}$—	*n*-Pr	A	—	247
cis-$CH_3CH_2CH{=}CH$—	H	A	—	247
cis-$CH_3(CH_2)_7CH{=}CH(CH_2)_6$—	H	A	—	247
trans-$CH_3(CH_2)_7CH{=}CH(CH_2)_6$—	H	A	79–80	247
—CO_2H	H	D	120	254

[a]Method A: direct oxygenation of enolate; Method B: methanolysis of α-silylperoxy ester; Method C: hydroperoxylation of α-lactone; Method D: ozonolysis.

b. Cyclization of α-hydroperoxy acids. The cyclization of the α-hydroperoxy acids **41** into α-peroxylactones **2** required an effective and convenient reagent. With hindsight we now know that in view of the delicate nature of the α-peroxylactones as well as that of the α-hydroperoxy acids, the requisites and conditions placed on the reagent are extremely stringent. For example, the reagent must be nonacidic, nonbasic, nonnucleophilic, nonelectrophilic and nonparamagnetic since all these properties would cause destruction of the α-peroxylactone product and/or its α-hydroperoxy acid precursor. In addition, the reagent must possess low-temperature reactivity so that the α-peroxylactone can accumulate and not suffer thermal decarboxylation. Finally, the reagent must allow easy isolation and purification of the α-peroxylactone product. Obviously, to meet all these stringent requisites, such an exigent reagent is yet to be discovered.

Of all the dehydrating agents that were tried (Table 5), the one that stood out by a large margin in effectiveness, convenience, and availability was the dicyclohexylcarbodiimide (DCC)[26]. The cyclization is executed by mixing solutions of the α-hydroperoxy acid in CH_2Cl_2 and DCC in CH_2Cl_2 at −78°C and allowing to warm up to ca. −40°C. At this point the urea by-product precipitates instantaneously. This is a most fortunate occurrence for two reasons. Firstly, it permits separation of the α-peroxylactone solution from the urea by low-temperature (ca. −78°C) decantation and thereby aids in the isolation of the labile product. Secondly, and still more importantly, the catalytic decomposition of the α-peroxylactone by the urea by-product is substantially supressed. Final purification of the volatile α-peroxylactones is achieved by flash distillation at reduced pressure and low temperature. Quantitative IR analysis of the 1850–1870 cm^{-1} carbonyl band or iodometric titration can be used to determine the α-peroxylactone content.

TABLE 5. Dehydrating agents[255]

$PhSO_2Cl$	$Ph_2C{=}C{=}N{-}Ph$
$[CF_3CO]_2O$	$t\text{-BuN}{=}C{=}N^+(Me)\text{Bu-}t\ BF_4^-$
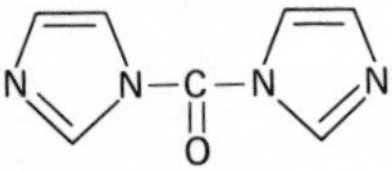 imidazolyl–C(=O)–imidazolyl (N–C–N, O)	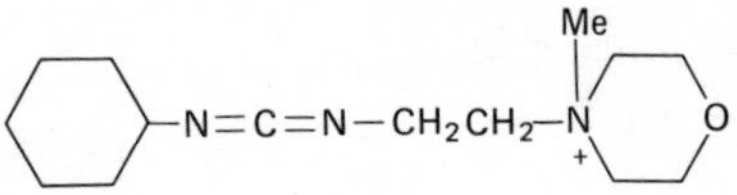 cyclohexyl–N=C=N–CH_2CH_2–N^+(Me) morpholino (O)
CF_3COCF_3	cyclohexyl–N=C=N–cyclohexyl
$[EtO]_2P(O)CN$	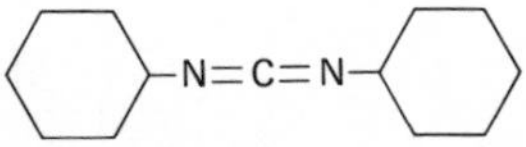 (P)–C_6H_4–CH_2–N=C=N–Pr-*i*
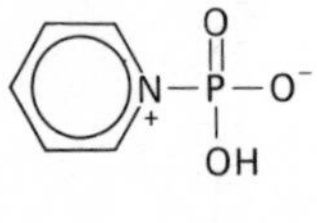 pyridinium N^+–P(=O)(OH)–O^-	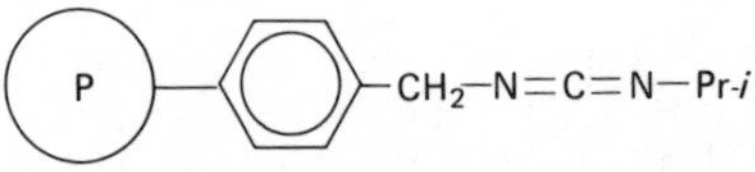
$H{-}C{\equiv}C{-}OEt$	
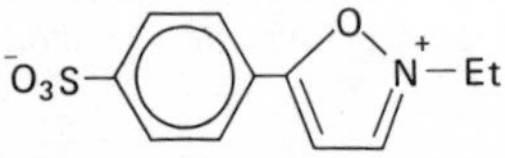 ^-O_3S–C_6H_4–isoxazolium (O, N^+–Et)	

Why is DCC so particularly effective compared to other dehydrating agents in the cyclization of α-hydroperoxy acids into the α-peroxylactones? Some plausible reasons are apparent in the dehydration mechanism postulated in equation (17). Even at −78°C initial protonation of DCC must be fast, thereby electrophilically activating the DCC towards nucleophilic attack by the carboxylate site. The adduct apparently requires some thermal activation towards cyclization since only on warm-up to ca. −40°C does precipitation of the urea by-product take place. Possibly internal proton transfer activates the nucleophilicity of the hydroperoxy group and at the same time enhances the leaving ability of the urea. It appears, therefore, that in the carbodiimide reagent we encounter an optimal sequence of nucleophilic and electrophilic activation steps by timely proton transfers.

In fact, it is hard to improve on the effectiveness of DCC. For example, if the carbodiimide is activated by conversion to its carbodiimidinium cation, the latter causes decomposition of the α-hydroperoxy acid instead of the desired α-peroxylactone cyclization (equation 18)[26]. Presumably the already electrophilically activated carbodiimide is attacked by the more nucleophilic hydroperoxy function, forming an adduct which is ideally set up for Grob fragmentation. This dead-end route is of course avoided with DCC because the first proton transfer (equation 17) enhances the nucleophilicity of the carboxyl over the hydroperoxy site, but at the same time activates the carbodiimide towards nucleophilic attack. In Table 2 are summarized the α-peroxylactones **2** that have been prepared via the α-hydroperoxy acid route.

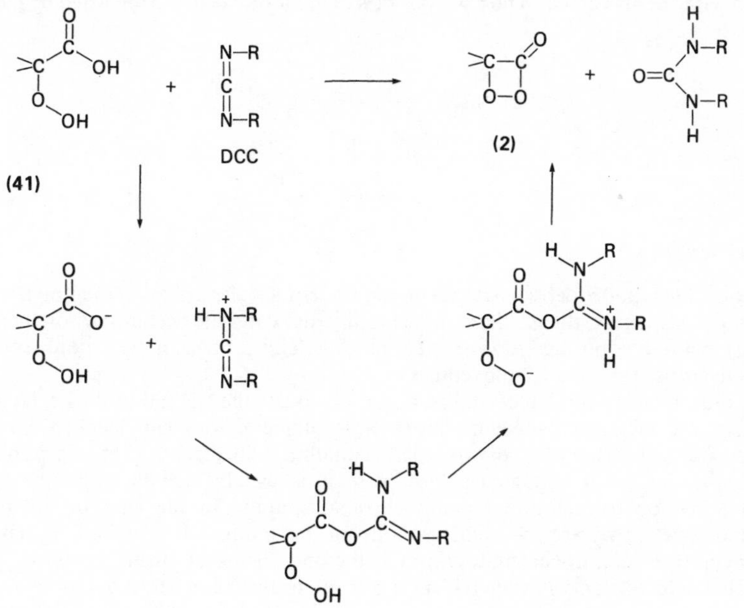

(41) + DCC ⟶ **(2)** + $O{=}C(NHR)_2$ (17)

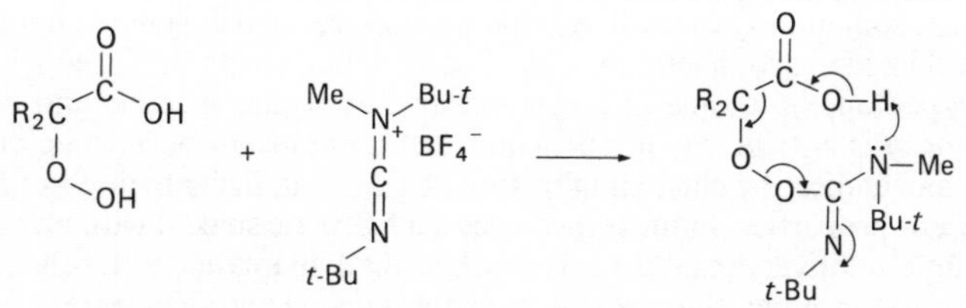

$R_2C(OOH)COOH$ + $[Me(t\text{-}Bu)N^+{=}C{=}N\text{-}t\text{-}Bu]\,BF_4^-$ ⟶ (18)

2. *The singlet oxygen route*

The most direct method for the preparation of α-peroxylactones is via singlet oxygenation of ketenes[12] (equation 12). The few α-peroxylactones that have been prepared by this method are listed in Table 2. Apparently ene reaction and [4 + 2]cycloaddition (equation 4) are not serious side-reactions since ketenes bearing methyl and phenyl substituents give good yields of α-peroxylactones on singlet oxygenation. Since ketenes are prone to autoxidation[11], it is sometimes essential to use chemically generated[41] instead of photosensitized singlet oxygen. An additional shortcoming of this method is the fact that sterically hindered ketenes, e.g. di-*t*-butylketene, and electron-poor ketenes, e.g. bis(trifluoromethyl)ketene, react quite sluggishly with singlet oxygen.

Singlet oxygenation of ketenimines constitutes the only method for the preparation of imino-1,2-dioxetanes **3** (equation 19)[57,58]. The iminodioxetanes **3** are thermally still more labile than the α-peroxylactones **2**, so that the singlet oxygenations must be performed at very low temperatures. Thus the iminodioxetanes cannot usually be isolated and are used directly in solution as obtained in the singlet oxygenation. As with the ketenes, the ketenimines may bear alkyl and aryl groups, since [2 + 2]cycloáddition to give iminodioxetanes predominates over ene reaction with the alkyl group or [4 + 2]cycloaddition with the aryl group. The few examples of iminodioxetanes **3** are listed in Table 2.

$$R_2C{=}C{=}NR \xrightarrow{^1O_2} \underset{\text{(3)}}{R_2C(-O-O-)C{=}NR} \qquad (19)$$

C. Purification

1,2-Dioxetanes and α-peroxylactones are unusual molecules in view of their high energy content. Consequently, due to thermal instability and catalytic decomposition, many of the usual methods of purification are not applicable. This problem is even more severe for α-peroxylactones than for 1,2-dioxetanes.

If the dioxetane is a solid, recrystallization is obviously the method of choice. Here it is critical to use metal-free solvents since traces of metal ions can lead to extensive decomposition. In the case of volatile, crystalline dioxetanes, prepurification via sublimation can be advantageous. Otherwise it is usually helpful to prepurify the dioxetane by low-temperature column chromatography. In the case of liquid 1,2-dioxetanes, unless they are sufficiently volatile for low-temperature distillation, repeated low-temperature column chromatography is the only means of purification.

Silylated or low-activity (Grade III) silica gel are effective adsorbants, but these also can frequently cause decomposition of the very sensitive dioxetanes even at −50°C. In such cases Florisil can sometimes be useful. As eluants mixtures of halogenated hydrocarbons, e.g. methylene chloride, carbon tetrachloride, fluorotrichloromethane (Freon-11), etc. and alkanes, e.g. *n*-pentane, *n*-hexane and cyclohexane, are quite effective and convenient. Again, their purity is critical, e.g. in particular metal impurities have to be absent.

To spot the dioxetanes, the eluate is monitored by TLC, utilizing their peroxidic and/or chemiluminescent properties. In most instances a TLC plate soaked with an aqueous KI solution will do; but for very resistant cases such as diadamantylidene-1,2-dioxetane (**14**), ferrous sulphate–ammonium thiocyanate and concentrated hydrochloric acid should not fail. In the case of detection via chemiluminescence, the TLC plate is sprayed with a 9,10-dibromoanthracene (DBA) or 9,10-diphenylanthracene (DPA) solution and heated in the dark. The dioxetane spot glows bright blue.

We have made the interesting observation that most dioxetanes bleach iodine when the latter is used as spotting agent. Thus, a bright white spot remains where a dioxetane is located, while the rest of the TLC plate turns yellowish on exposure to iodine vapours. The combination of potassium iodide detection (brown spot) and iodine detection (white spot) can be quite definitive for the presence of dioxetanes. However, if these tests are negative, it does not necessarily mean that dioxetanes are not present.

D. Characterization

Once a four-membered ring peroxide is reasonably pure (≥90 %), a number of physical, chemical and spectroscopic methods can be employed for their identification. The characteristic spectral properties, i.e. ^{1}H- and ^{13}C-NMR, infrared and electronic spectra, of the excited carbonyl decomposition products are discussed in Section II.B and summarized in Tables 1 and 2. Thus, we shall deal here only with the physical and chemical properties that are helpful in the characterization of four-membered ring peroxides.

1. Physical methods

To differentiate monomeric, dimeric and polymeric products, a molecular weight determination is essential. The usual cryoscopic and osmometric techniques are applicable[2,27,42]. In fact, these methods have been used for impure dioxetanes[59] and at low temperatures[59,60]. For example, singlet oxygenation of **52** gave instead of the expected monomeric dioxetane the dimer **53** (equation 20), as confirmed by cryoscopy[61].

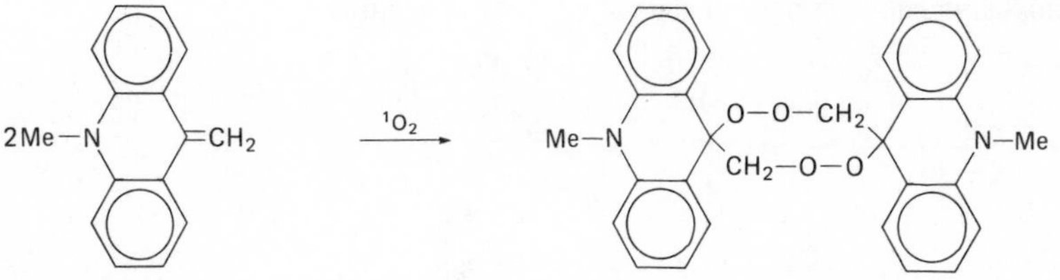

2 Me—N ... =CH_2 $\xrightarrow{^{1}O_2}$ Me—N ... O—O—CH_2 / CH_2—O—O ... N—Me (20)

(52) **(53)**

Melting points of crystalline and stable dioxetanes have served for identification purposes. However, these melting points are frequently decomposition temperatures, which limits their use.

2. Chemical methods

Whenever feasible, a combustion analysis is an essential method of characterization by establishing the elemental composition. All precautions should be exercised in view of the explosive nature of 1,2-dioxetanes.

As already mentioned in connection with the determination of purity, iodometric titration is useful in the identification of dioxetanes[27]. Unfortunately this simple and convenient method is not specific since most peroxides release iodine from acidic potassium iodide. If the necessary care is taken, iodometry is quantitative and thus an excellent purity criterion.

Catalytic reduction over platinum or palladium, which is usually a quantitative method for the identification of organic peroxides, is problematic. Little of the expected 1,2-diol is obtained because the dioxetane fragments into its carbonyl products due to metal catalysis. However, lithium aluminium hydride reduction at low temperatures[27] affords the expected 1,2-diol quantitatively. Again, the sterically hindered dioxetane **14** is an exception, but zinc in acetic acid proved successful[42].

A convenient and quantitative method is thermal decomposition into the carbonyl fragments, which can be easily characterized by IR and NMR.

IV. 1,2-DIOXETANES AS REACTION INTERMEDIATES

In Tables 1 and 2, some thermally labile cases were included which could not be isolated, but some sort of spectral evidence, other than the observation of chemiluminescence, i.e. infrared, ^{1}H- and ^{13}C-NMR, etc., was on hand to substantiate the claim of the four-membered ring peroxide.

The danger of merely relying on chemiluminescence as the only evidence for the intermediacy of a 1,2-dioxetane is well illustrated in the autoxidation of the imine **54**, generated *in situ* from 9-anthrylamine and isobutyraldehyde (equation 21). Initially it was claimed[62a] that dioxetane **55** was the direct product of this autoxidation, which on heating should give the observed chemiluminescence. However, the unusual chemical and physical properties of the isolated dioxetane **55** provoked several sceptics to reinvestigate this reaction. Indeed, it was shown[63] that the initial peroxide was the 1,2,4-trioxane **56**, which on base treatment led to efficient light emission via dioxetane **57**[64]. An X-ray crystal structure[62b] confirmed the 1,2,4-trioxane structure beyond doubt. Clearly, it is easy to become a victim of such 'chemiluminonsence', and to avoid it, all powers of criteria should be employed when claiming a dioxetane or α-peroxylactone.

(54) $\xrightarrow[i\text{-PrCHO}]{^{3}O_2}$ (56) $\xrightarrow[[-i\text{-Pr(OH)}]]{\Delta}$ (57) $\xrightarrow{[-Me_2C{=}O]}$ 9-(HN–CHO)anthracene + $h\nu$

(54) $\not\rightarrow$ ($^{3}O_2$, *i*-PrCHO) (55)

(21)

Nevertheless, numerous examples abound in the recent literature which claim the formation of four-membered ring peroxides as reaction intermediates. In some the observed chemiluminescence is taken as back-up for the intervention of these hyperenergetic species, but in most they are postulated as a mechanistic convenience (equation 22). That is to say, the fact that the carbon–carbon double bond has on oxidation led to the carbonyl products is taken as evidence for the intermediary dioxetane.

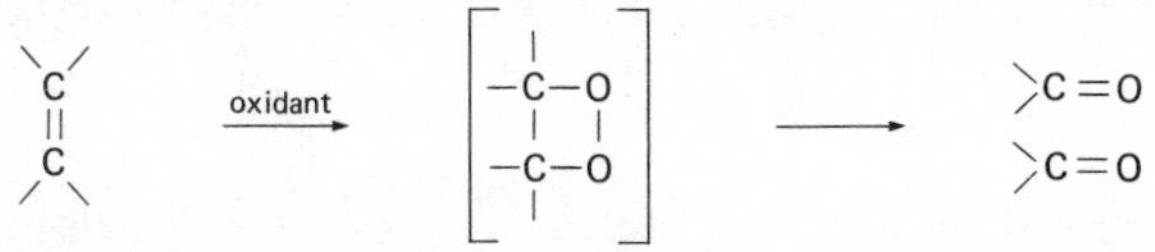

$$>C{=}C< \xrightarrow{\text{oxidant}} \left[\begin{array}{c} -C-O \\ | \quad | \\ -C-O \end{array} \right] \longrightarrow \begin{array}{c} >C{=}O \\ >C{=}O \end{array} \qquad (22)$$

Although the authenticity of such claims can be argued, it is still useful to enumerate recent examples. No doubt many of these present food for thought and demand mechanistic confirmation. Interesting examples since our last review[18i] on this subject will be featured from the literature of the last five years. The present coverage is by no means exhaustive, but we hope to have focused on representative cases involving singlet oxygen, triplet oxygen, superoxide ion, hydrogen peroxide, peroxides, etc. in order to stimulate mechanistic activity in this fascinating area.

A. Singlet Oxygenations

1. Alkenes and cycloalkenes

When the carbon–carbon double bond is activated through electron donation or strain, such substrates are cleaved with singlet oxygen, presumably via dioxetane intermediates. For example, under basic conditions **58** is converted into vanillin in which dioxetane **59** is proposed[65] as intermediate (equation 23). Similarly, the small amounts of fluorenone as cleavage product in the reaction of fluorene derivatives **60** (equation 24) with $^1O_2$ have been interpreted[66] in terms of dioxetane **61**. The formation of the keto acetal **65** in the cycloaddition of singlet oxygen to the activated and strained double bond of **62** in ethanol (equation 25) has been accounted for[67] in terms of dioxetane **63** via cleavage into **64** and subsequent acetalation with ethanol.

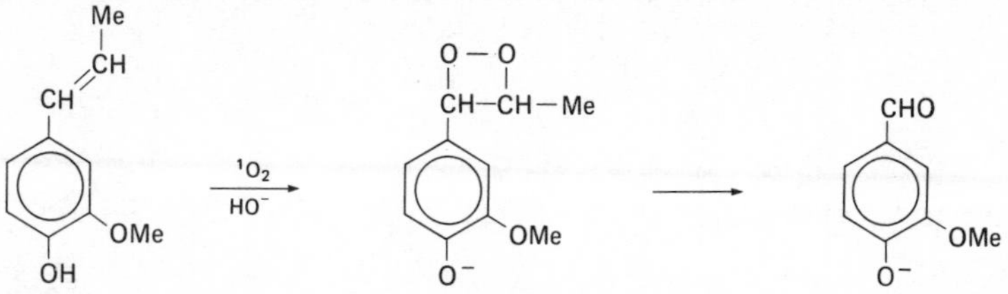

Me, CH, CH, OMe, OH **(58)** $\xrightarrow[HO^-]{^1O_2}$ O—O, CH—CH—Me, OMe, O^- **(59)** $\longrightarrow$ CHO, OMe, O^- (23)

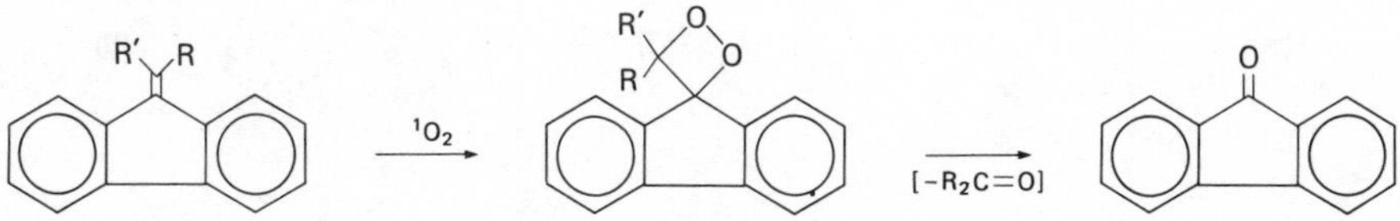

R′, R **(60)** $\xrightarrow{^1O_2}$ R′, R, O, O **(61)** $\xrightarrow[{[-R_2C{=}O]}]{}$ O (24)

R′ = R = Me

R′ = H; R = Me

R′ + R = $\{CH_2\}_5$
(60)

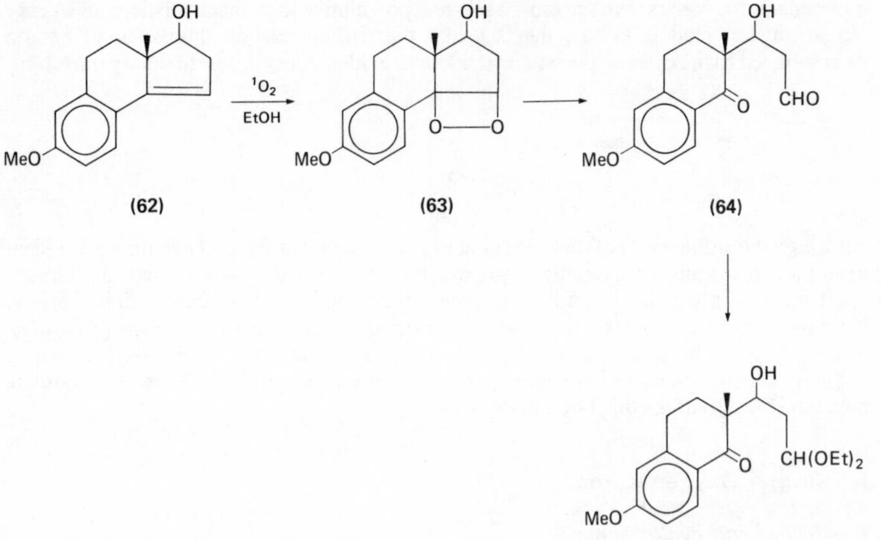

(62) (63) (64)

(65) (25)

The reaction of the dimer **66** of spiro[2.4]hepta-2,4-diene with $^1O_2$ under Rose Bengal photosensitization in methanol and pyridine mixture is unusual in that the cleavage product **68** and reduction product **69** are formed from the intermediary dioxetane **67** (equation 26)[68]. The photoreduction of dioxetane **67** to the diol **69** is surprising since normally dioxetanes cleave to carbonyl products on photolysis[18i]. However, an analogous photoreduction of dioxetane **70** into *cis*-2,3-dihydroxyindane was observed[68] in the photosensitized oxygenation of indene. Presumably methanol is oxidized into formaldehyde. This unprecedented photochemical behaviour of dioxetanes merits further scrutiny.

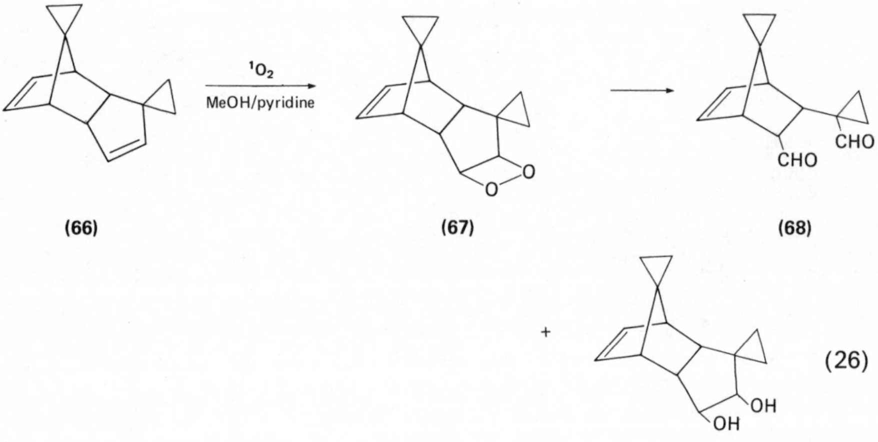

(66) (67) (68)

(69) (26)

The cleavage of 2-phenyl-2-norbornene into *cis*-1-benzoyl-3-formylcyclopentane with $^1O_2$ involves dioxetane **71** as intermediate[69]. Similarly, the singlet oxygenation of *trans*-cyclooctene[70] gives the labile dioxetane **72**, which cleaves into the corresponding dialdehyde even at $-78°C$. However, the still more strained dioxetane **73** can be spectroscopically observed[71a] (Table 1) and reduced with triphenylarsine to the corresponding diol[71b].

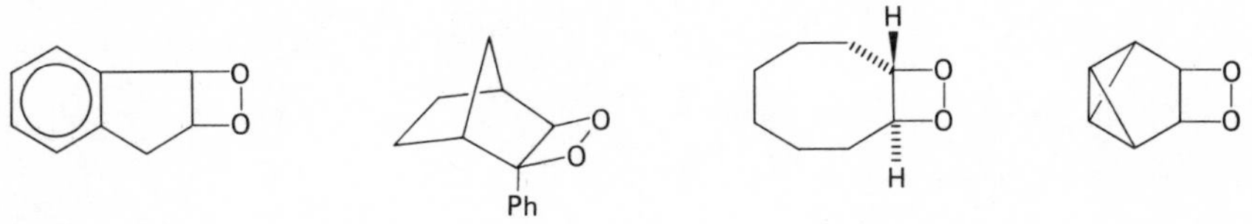

(70) (71) (72) (73)

Strained cycloalkenes of three- and four-membered rings have received considerable attention. For example, the formation of dione **76** in the singlet oxygenation of tetraphenylcyclopropene has been proposed[72] to involve dioxetanes **74** and **75** (equation 27). In this connection it would have been of interest to attempt trapping the carbonyl oxide intermediate **77** with a dipolarophile[73].

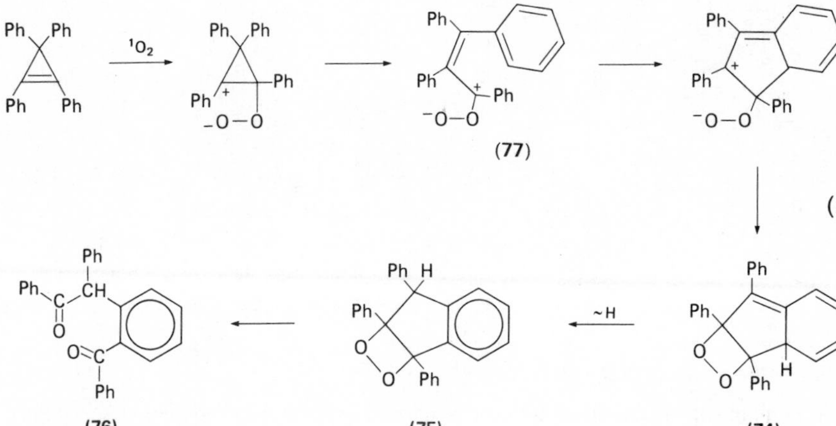

(27)

(77) (76) (75) (74)

The cleavage of methylenecyclopropanes[74,75] by singlet oxygen has been suggested as engaging dioxetane **78** (equation 28). Trapping agents such as Ph_3P[27] or Ph_2S[60] divert the dioxetane **78** to the corresponding cyclobutanone product. On the other hand, such trapping experiments suggest[76] that dioxetanes **80** are not precursors to the observed cleavage products in the reaction of dicyclopropylethylenes **79** with $^1O_2$ (equation 29). Finally, in the photosensitized singlet oxygenation of 1-phenylcyclobutene[77,78] the dioxetane **81** figures as intermediate to the corresponding cleavage product (equation 30).

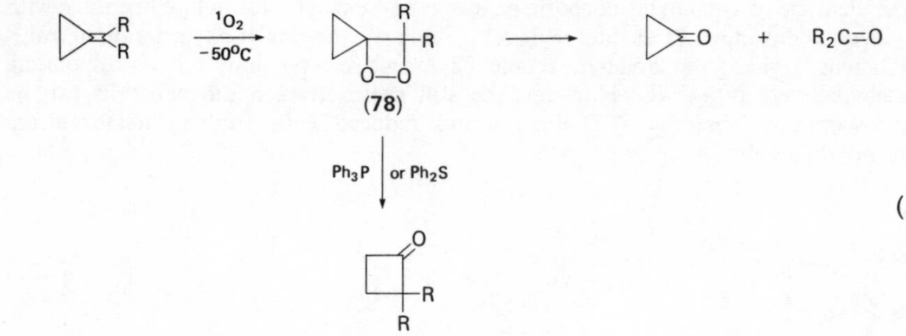

(28)

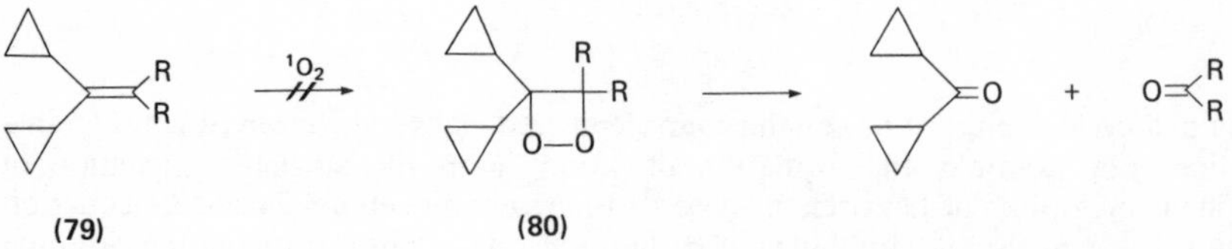

(78)

(29)

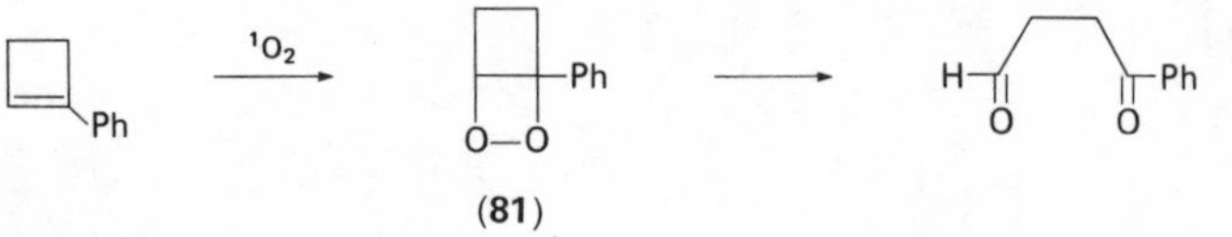

(79) (80)

(30)

(81)

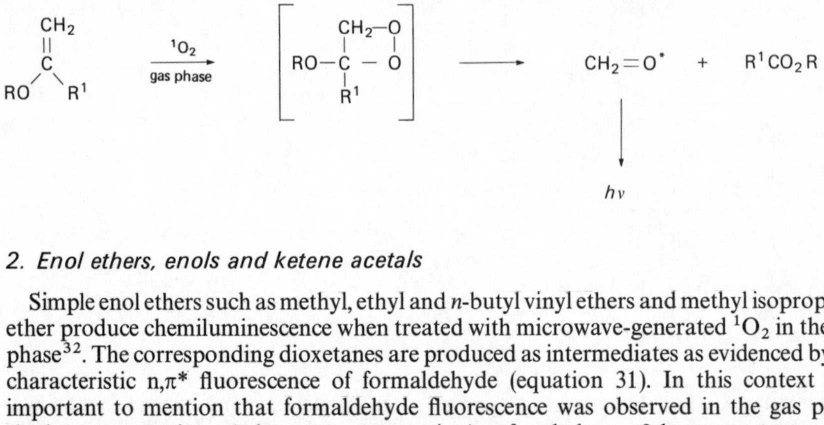

(31)

2. *Enol ethers, enols and ketene acetals*

Simple enol ethers such as methyl, ethyl and *n*-butyl vinyl ethers and methyl isopropenyl ether produce chemiluminescence when treated with microwave-generated $^1O_2$ in the gas phase[32]. The corresponding dioxetanes are produced as intermediates as evidenced by the characteristic n,π* fluorescence of formaldehyde (equation 31). In this context it is important to mention that formaldehyde fluorescence was observed in the gas phase singlet oxygenation (microwave generation) of ethylene, 2-bromopropene, 1,2-difluoroethylene, allene, 1,1-dimethylallene and ketene[32]. Our efforts[79] to isolate the dioxetane from ethylene in liquid-phase singlet oxygenations have failed so far. Gas-phase singlet oxygenation (laser-generated)[80] of dimethylketene dimethylacetal did afford acetone and dimethyl carbonate, presumably via the dioxetane **82**. In this study an effort was made to differentiate between the chemical behaviour of the $^1\Delta_g$ and the versus $^1\Sigma_g$

forms of singlet oxygen, but it was concluded that both undergo [2 + 2]cycloaddition affording dioxetane **82**.

Among the cyclic enol ethers, 1-methoxycyclohexene was shown[81] to give the expected dioxetane **83**, besides other products, in dye-sensitized singlet oxygenation in solution. Dioxetane **83** gave on heating the expected 5-carbomethoxypentanal. Dioxetane **84** was proposed[82] as transient intermediate in the photosensitized singlet oxygenation of the corresponding enol ether. Similarly, the cleavage of 4,5-diethylidene-2,2-dimethyl-1,3-dioxolane into 5-ethylidene-2,2-dimethyl-1,3-dioxolan-4-one and acetaldehyde during singlet oxygenation was interpreted[83] in terms of dioxetane **85**.

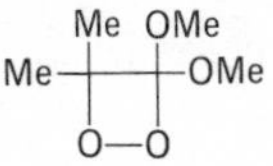

(82)  (83) 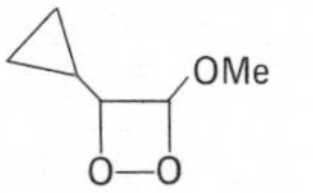 (84) 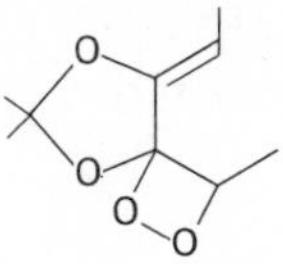 (85) 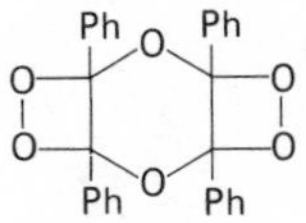 (87)

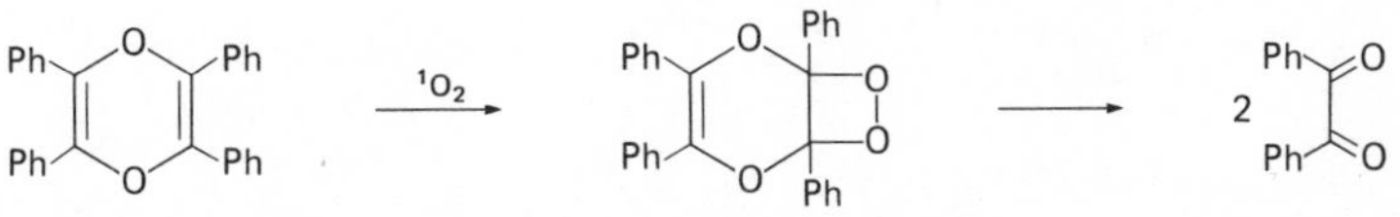

(86) (32)

The suggestion of the formation of benzil in the self-sensitized singlet oxygenation[84] of tetraphenyl-1,4-dioxin (equation 32) via dioxetan **86** is surprising. A related cleavage of dioxetanes has been observed in the singlet oxygenation of tetrathioethylenes[85] (equation 33), but not for enol ethers. In fact, in the case of tetraphenyl-1,4-dioxin it was shown[86] that the bis-dioxetane **87** is formed as a stable and isolable entity. In the conversion of 3-hydroxybisflavenylidenes **89** into lactone **92** (equation 34) with $^1O_2$ [87] it was proposed that the stable enol **89** cycloadds $^1O_2$ to give first dioxetane **90**. The latter cleaves into the hydroxy acid **91**, which dehydrates into lactone **92**.

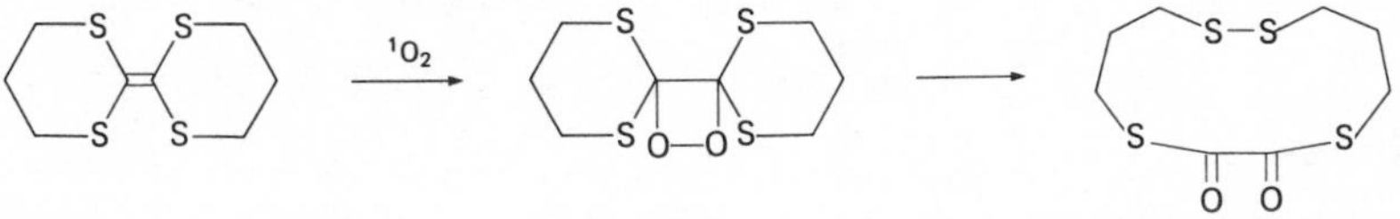

(88) (33)

The cyclic imino ether **93** apparently reacts with $^1O_2$ at the enol ether site to give dioxetane **94** (equation 35)[88]. Dioxetane cleavage, formyl migration and tautomerization affords **95** as the final product. Dioxetane **97** was proposed[89] as intermediate in the singlet oxygenation of the heterocycle **96** (equation 36). Fragmentation of **97** with decarboxylation was proposed to give phenylbenzimide (**98**), which via homolysis into the corresponding radical pair leads to the observed typical radical products.

(89) R=H, MeO

(90)

(91)

(92)

(34)

(93)

(94)

(95)

(35)

3. Enamines

The structures **93** and **96**, which have been featured under enol ethers, also possess the enamine functionality. Since enamines have been abundantly singlet-oxygenated, it is of interest to cite a few examples at this point. Thus, singlet oxygen reacts with the enamine **99**, present in the mixture as the tautomer of the corresponding imine, to give the typical cleavage products of dioxetane **100** (equation 37)[90]. An astonishing example was reported[91] in the singlet oxygenation of the steroid **101** which contains an enamine function (equation 38). The formation of the stable dioxetane **102** was proposed, which contrary to normal thermolytic behaviour was supposed to isomerize into hydroxyketone **104** either on heating or on SiO_2 chromatography. However, no chemiluminescence could be observed during the thermolysis of dioxetane **104**. With HCl in ethanol **102** gave **105**

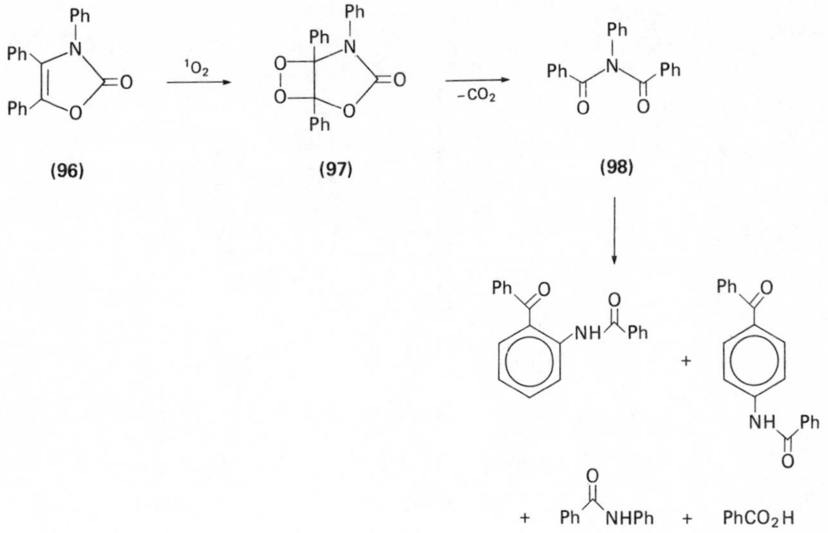

(96) (97) (98)

(36)

and **106**. In view of this unusual dioxetane chemistry it would be of interest to confirm the structure of **102** by X-ray analysis. A related case was observed[92] in the singlet oxygenation of the heterocycle **107** (equation 39) in MeOH. The formation of acetanilide via cleavage of dioxetane **108** and subsequent solvolysis is normal; however, formation of **111** via dioxetane **110** is unusual and analogous to the isomerization of **102** → **104**.

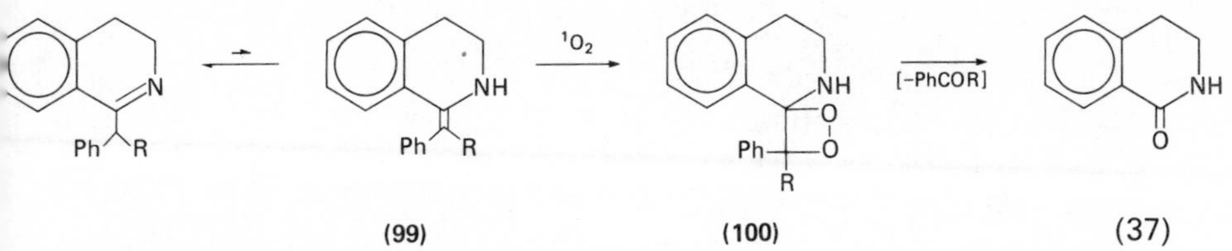

(99) (100) (37)

On the other hand, the dioxetane **113** that is produced in the singlet oxygenation of the alkaloid 3-oxovincadifformine (**112**) behaves normally and cleaves into **114** (equation 40)[93]. Similarly, but in a more complex manner, the alkaloid **115** affords on reaction with $^1O_2$ the spirolactone **118** via the dioxetanes **116** and **117** (equation 41)[94]. The rearrangement of dioxetane **116** into **117** is remarkable, but dioxetane–dioxetane rearrangements are known, e.g. dioxetane **119** derived from tetra-*t*-butylcyclobutadiene[243] exhibits a single ^{13}C resonance at room temperature as a result of interconverting dioxetane structures (equation 42). In this case the interconverting dioxetanes are structurally identical, but structurally distinct dioxetanes are formed in the rearrangement shown in equation (43)[22].

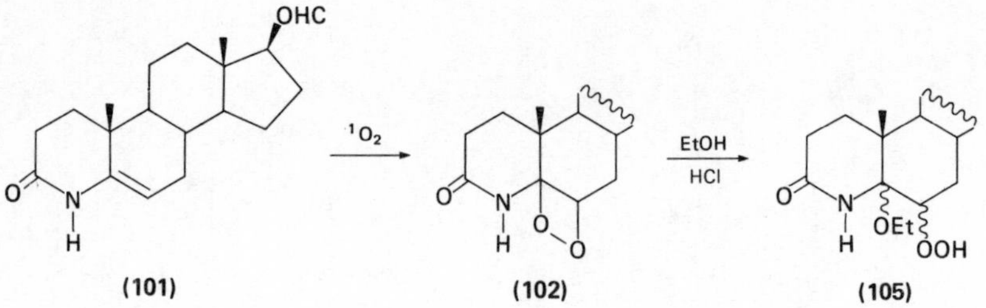

Examples are known in which a reactive enamine moiety is generated *in situ*. For example, the singlet oxygenation of the heterocycle **120** leads to benzimide and benzamide, presumably via dioxetane **121** (equation 44)[95]. Finally, in the photochromism of the spirocycle **122** the formation of pyridine *N*-oxide and fulvene were rationalized in terms of

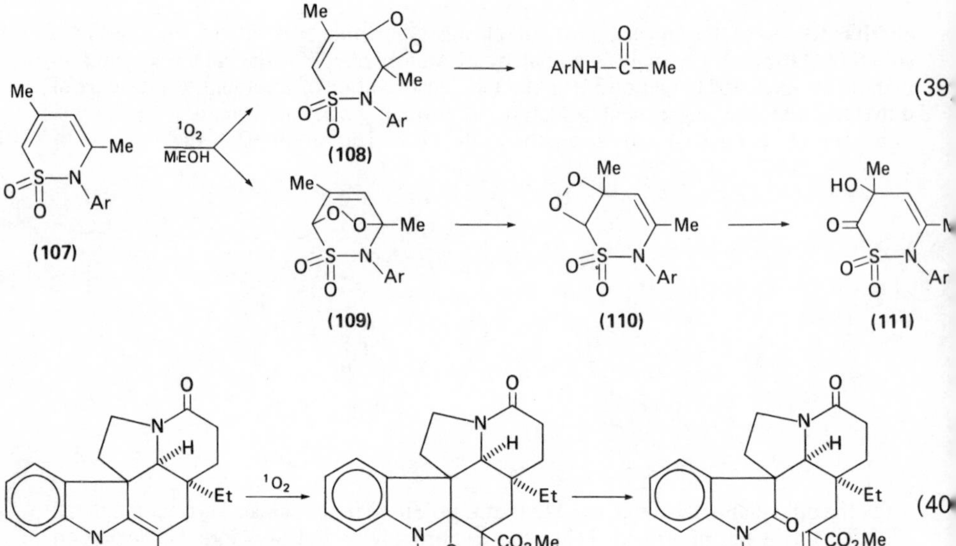

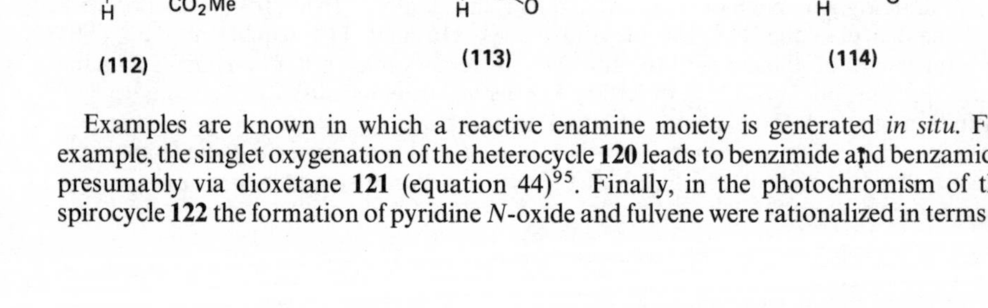

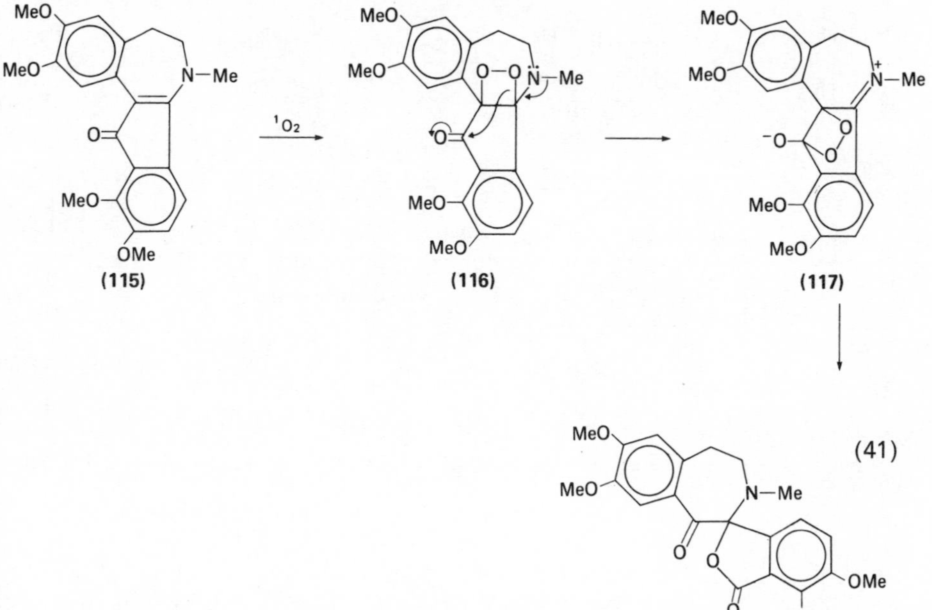

MeO, MeO, N—Me, O, MeO, OMe (115) $\xrightarrow{^1O_2}$ (116) → (117) → (118) (41)

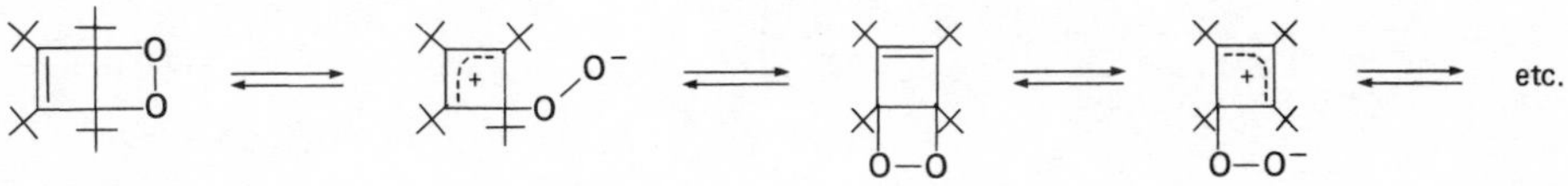

(119) ⇌ ... ⇌ etc. (42)

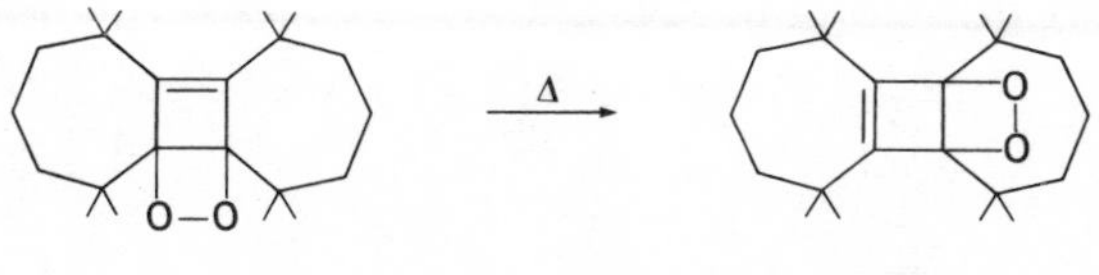

$\xrightarrow{\Delta}$ (15) (43)

dioxetane **124**, produced by self-sensitized singlet oxygenation of the betain **123** (equation 45)[96]. Again, instead of conventional peroxide bond fragmentation, the dioxetane **124** was proposed as undergoing ring-opening, in order to explain the observed fragmentation products.

Of interest is the observation that 3-(*N*-methyl)indolylstyrene (**125**) affords with $^1O_2$ *N*-methyl-3-indolecarboxaldehyde and benzaldehyde under light emission[97]. Presumably the thermally relatively stable endoperoxide **126**, formed via [4 + 2]cycloaddition of $^1O_2$

(120)

(121)

(44)

(122)

(123)

(124)

(45)

(125)

(126)

(127)

(46)

and **125** rearranges into the dioxetane **127** (equation 46), which fragments. Another example is dioxetane **21**[30].

4. Heteroarenes

Pyrroles, indoles, imidazoles and other related nitrogen-containing heteroarenes have always been favoured substrates for singlet oxygenation[18q,18r,98]. In view of the biological importance of these substrates, the involvement of dioxetanes in the oxidative degradation has been a central issue. For example, the reaction of 2,3,4-trimethylpyrrole affords the pyrrolinone **130**[99], presumably via dioxetane **128** (equation 47). Methanolysis of dioxetane **128** into the hydroperoxide **129** instead of cleavage to the dicarbonyl product is again unusual dioxetane behaviour, but we have seen it already in equation (38). Similar chemistry is proposed for the dioxetane **131** derived from 3-methylpyrrole[100], but the dioxetane **132** derived from *N*-*t*-butylpyrrole[101] gives the expected cleavage product. Also the cleavage of tetraphenylporphyrin has been interpreted as involving dioxetanes[102].

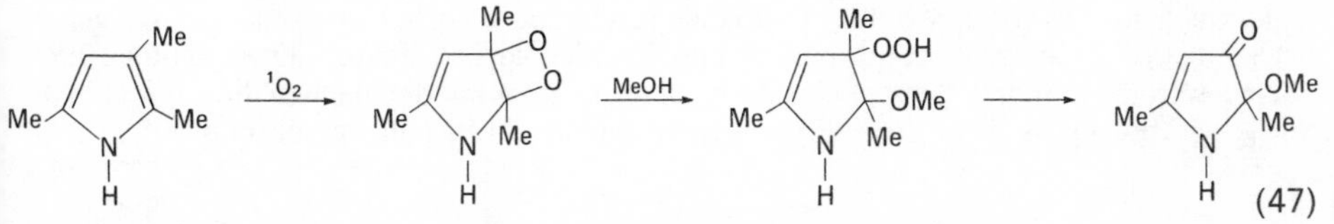

$^1O_2$ MeOH (47)

(128) **(129)** **(130)**

The singlet oxygenation of indoles has been particularly intensively investigated and a number of dioxetane intermediates have been postulated. In fact, the stable indole-derived dioxetane **133** produced in the singlet oxygenation of *N*-methyl-2-*t*-butyl-3-methylindole[103] has been isolated and fully characterized. In this case fragmentation into the dicarbonyl product is accompanied by chemiluminescence. Furthermore, as already pointed out in Section II, the indole-derived dioxetane **21** is sufficiently stable at low temperatures to be detected by NMR (see Table 1)[30].

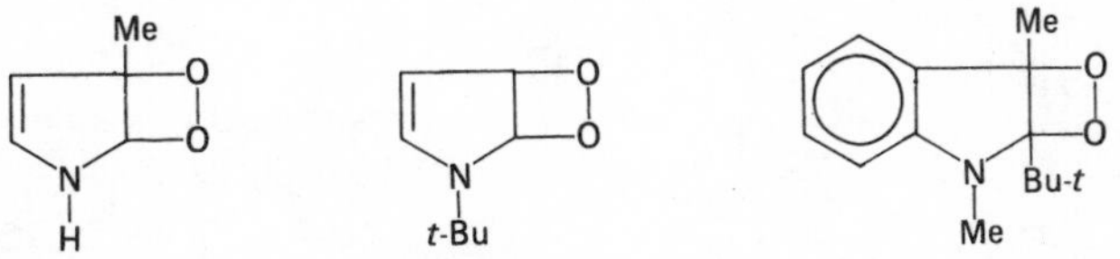

(131) **(132)** **(133)**

Dioxetane **134** has been postulated as a reaction intermediate in the singlet oxygenation shown in equation (48)[104].

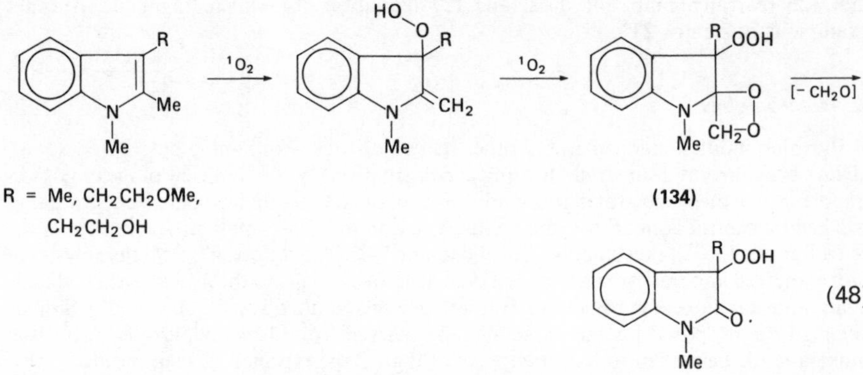

R = Me, CH_2CH_2OMe, CH_2CH_2OH

(134)

(48)

Particularly interesting from the mechanistic point of view is *N*-methyl-3-(2-hydroxyethyl)indole as substrate, since the competition between intramolecular trapping to give **136** and dioxetane (**137**) formation suggests[105] the intervention of the dipolar intermediate **135** (equation 49). Analogous results are reported on the singlet oxygen cleavage of a variety of tryptophanes into kyunrenine derivatives **139** (equation 50). However, the authors[106] express caution concerning the intermediacy of dioxetane **138**. Clearly this process is of great importance in the metabolic pathway of indoles[107].

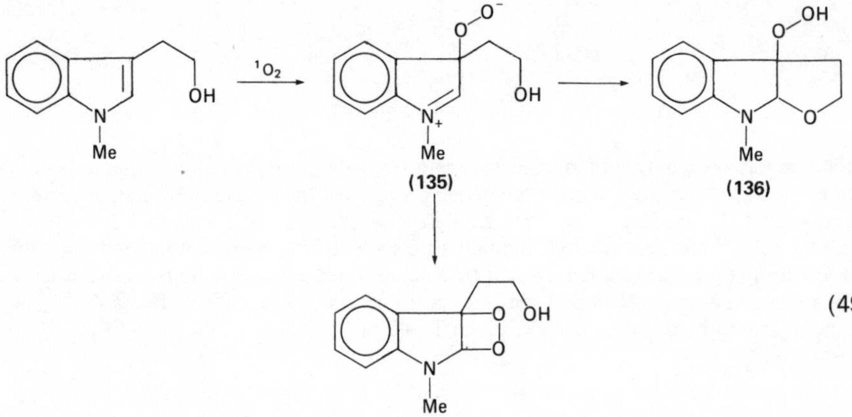

(135) (136)

(49)

(137)

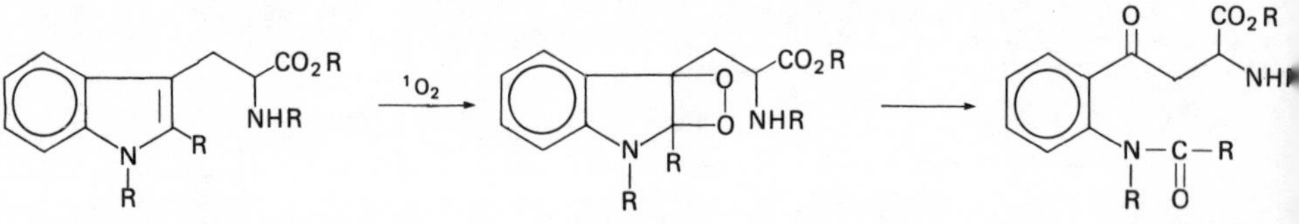

(138) (139)

(50

Other heteroaromatic cases that have been reported[108] include the betain **140**, which on singlet oxygenation affords lactone **142** via dioxetane **141** (equation 51). Furthermore, the ring-cleavage of 2-methyl-4,5-diphenylimidazole[109] and 1,2,4,5-tetraphenylimidazole[110] has been interpreted as proceeding via dioxetanes **143** and **144**, respectively. Finally, the conversion of thiazoles into nitriles and α-diones on singlet oxygenation has been postulated[111] as involving dioxetane **145**. Here again we are dealing with an unusual fragmentation in that the dioxetane carbon–carbon bond is preserved.

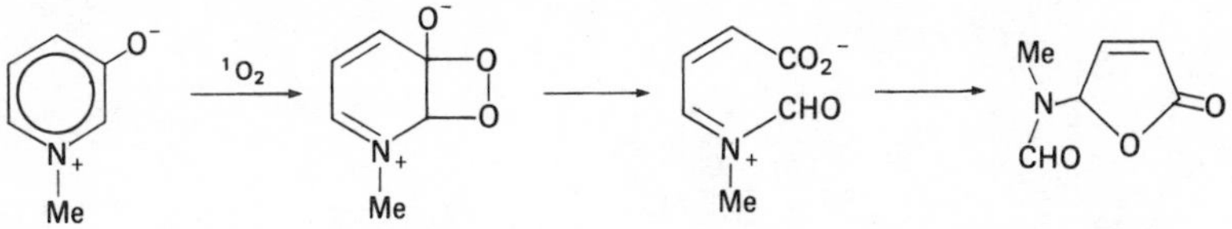

(51)

(140) **(141)** **(142)**

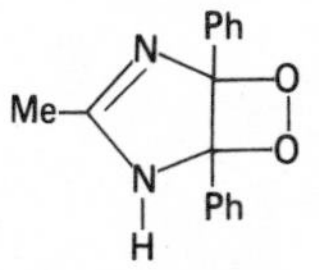

(143) 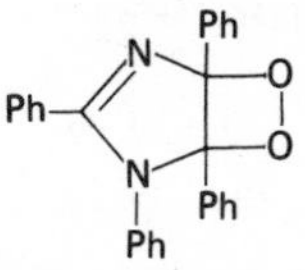 **(144)** 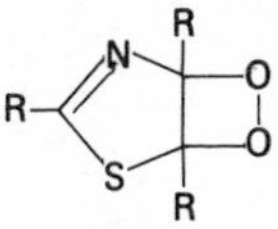 **(145)**

B. Autoxidations

In this section we shall discuss reactions of triplet oxygen ($^3O_2$), i.e. autoxidations, which lead to cleavage of a carbon–carbon double bond, presumably via a dioxetane intermediate. In some cases detailed mechanistic studies of the accompanying chemiluminescence suggests that such hyperenergetic intermediates intervene, but in most cases the only evidence is the fact that carbonyl products are produced.

1. Thermal and photochemical oxidations

A surprising example is the very recent observation of the high-temperature (200–250°C) chemiluminescent autoxidation of bisadamantylidene into adamantanone (equation 52)[112]. Of course, under these conditions even the very stable dioxetane **14** cannot survive and cleaves into adamantanone. Also the diadamantylidene epoxide is formed, but it is not clear how dioxetane **14** is generated. Whether at these high temperatures the diadamantylidene catalyses the spin inversion of $^3O_2$ into $^1O_2$, like strained acetylenes[113] and ketenes[114] do, is mechanistic conjecture at this stage, but such a process would be a convenient route to the dioxetane **14**. Other substrates that undergo this cleavage are the enol ethers methoxymethyleneadamantane, 4-*t*-butylmethoxymethylenecyclohexane and methoxymethylenedodecane, all giving the corresponding ketones and methyl formate under light emission.

(52)

(14)

Such cleavage can also be promoted on photolysis of olefins in the presence of molecular oxygen. Thus, on irradiation at 300 nm of 1,1-dichloro-2,2-di(*p*-chlorophenyl)ethylene[115] in the presence of $^3O_2$ the double bond is cleaved into phosgene and *p,p'*-dichlorobenzophenone (equation 53). Dioxetane **146** is postulated as intermediate, but not formed via singlet oxygenation. Quite analogously, the heterocycle **147** affords *N*-alkylbenzamide and methyl benzoate via dioxetane **148** (equation 54) during photolysis[116] in methanol.

$$\xrightarrow[300\ nm]{^3O_2} \textbf{(146)} \longrightarrow Ar_2C{=}O + Cl_2C{=}O \qquad (53)$$

$$\textbf{(147)} \xrightarrow[MeOH]{^3O_2/h\nu} \textbf{(148)} \xrightarrow{[CO_2]} Ph{-}C({=}O){-}NH{-}R + Ph{-}C({=}O){-}OMe \qquad (54)$$

A rather extraordinary case involves the phenanthrene **149**, which on photolysis in the presence of $^3O_2$ gives the lactone **153** (equation 55)[117]. Indeed, some rather unconventional chemistry is depicted here. It is proposed that the excited **149** yields the highly strained allene **150** which in turn reacts with $^3O_2$ to give dioxetane **151**. The latter rearranges via ketene **152** into the final product **153**.

Rather interesting results have been reported[118,119] on the 9,10-dicyanoanthracene (DCA)-sensitized cleavage of olefins in the presence of molecular oxygen in which dioxetane intermediates are involved. For example, stilbene, 1,1-diphenylethylene, triphenylethylene and tetraphenylethylene are all cleaved into the corresponding carbonyl products in this way (equation 56) via dioxetanes[118]. The proposed mechanism involves electron transfer between the excited DCA, triplet oxygen and the substrate as shown in equation (57). Similar results have been obtained for 2,3-diaryl-1,4-dioxene[119]. In fact, in this case it was possible to show that potassium superoxide and tris(*p*-bromophenyl-ammonium tetrafluoroborate led to the same oxidation products of the olefinic substrate.

2. *Base-catalysed oxidations*

It is known[120] that enolates react with molecular oxygen to give the next lower homologous carbonyl product. There is evidence[121] that an α-peroxy anion intermediate is formed which decomposes via the α-peroxylactone into the observed products (equation 58). By using the α-hydroperoxy esters it was shown[121] through chemiluminescence

(149) $\xrightarrow{h\nu}$ 149* $\longrightarrow$ (150)

(150) $\xrightarrow{^3O_2}$ (151) $\longrightarrow$ (152) $\longrightarrow$ (153) (55)

$$R_2C{=}CR^1_2 \xrightarrow[\text{DCA/MeCN}]{400\text{ nm};\ ^3O_2} R_2C(-O-O-)CR^1_2 \longrightarrow R_2C{=}O + R^1_2C{=}O \quad (56)$$

$h\nu$: DCA → DCA*; DCA* + $^3O_2$ → DCA$^{+\cdot}$ + $O_2^{\cdot-}$; DCA$^{+\cdot}$ + $R_2C{=}CR_2$ → DCA + $R_2\overset{+}{C}-\dot{C}R_2$; $R_2\overset{+}{C}-\dot{C}R_2$ + $O_2^{\cdot-}$ → $R_2C(-O-O-)CR_2$

$$R_2C(-O-O-)CR_2 \longrightarrow 2R_2C{=}O \quad (57)$$

$$R_2CH-C(=O)OR^1 \xrightarrow{B^-} R_2C^{-}-C(=O)OR^1 \xrightarrow{^3O_2} R_2C(OO^-)-C(=O)OR^1 \xleftarrow{B^-} R_2C(OOH)-C(=O)OR^1$$

$$R_2C(OO^-)-C(=O)OR^1 \xrightarrow{[-R^1O^-]} R_2C(-O-O-)C{=}O \longrightarrow R_2C{=}O + CO_2 \quad (58)$$

studies that indeed the α-peroxylactones are plausible intermediates. Also the base-catalysed autoxidations of nitriles[122] and aromatic ketones[123] are postulated as involving the dioxetanes **154** and **155**, respectively.

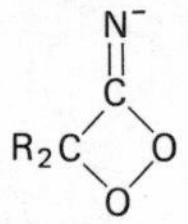

(154) 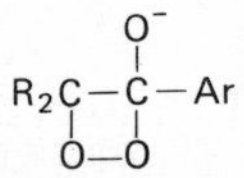 (155)

Dioxetanes similar to **155** are also invoked in the base-catalysed autoxidation of aryl-substituted pyruvic acids (equation 59)[124]. Again, the enolate of **156** leads to the dioxetane **157** on oxygenation, and fragmentation leads to the observed products. In this context it is of interest to mention that the base-catalysed hydrogen peroxide cleavage of benzil does not involve a dioxetane as intermediate, but a Baeyer–Villiger rearrangement takes place instead (equation 60), as confirmed by ^{18}O-labelling experiments[125].

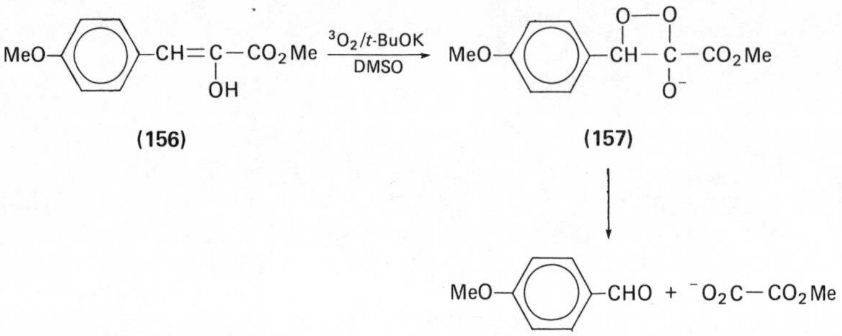

MeO–C6H4–CH=C(OH)–CO2Me (156) —$^3O_2$/t-BuOK, DMSO→ dioxetane (157) → MeO–C6H4–CHO + ^-O_2C–CO2Me (59)

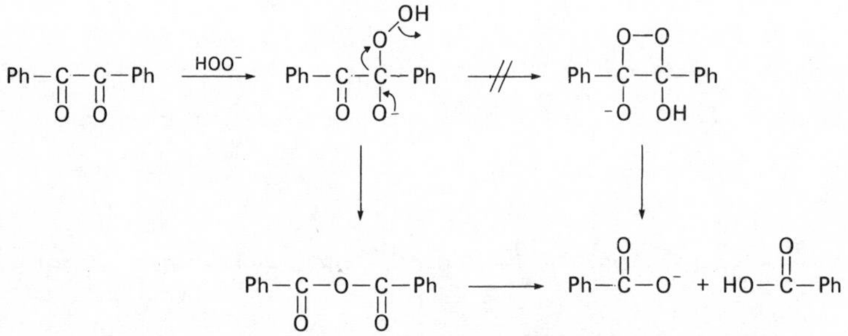

Ph–CO–CO–Ph —HOO⁻→ ... (60)

The benzofuranone **158** was converted into the oxazindione **160** under chemiluminescence when treated with molecular oxygen and base (equation 61)[126]. The α-peroxylactone **159** was postulated as intermediate in this chemiluminescent process.

In the base-catalysed autoxidation of the tryptamine derivative **161** the cleavage product **163** is formed via the dioxetane **162** (equation 62)[127]. In a related study[128] the base-catalysed decomposition of the hydroperoxide **164** was investigated. It was argued

(158) → (159) → (160) (61)

(161) → (162) → (163) → etc. (62)

(164) → (165) → [ ]* → $h\nu$; (164) → (166) → (63)

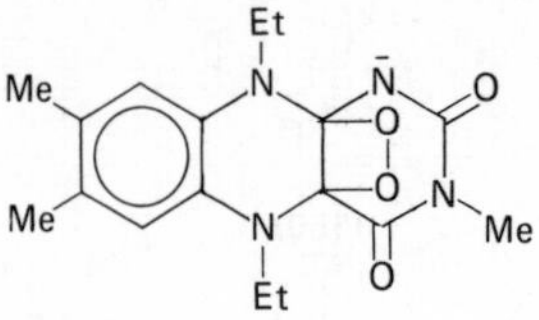

(167) 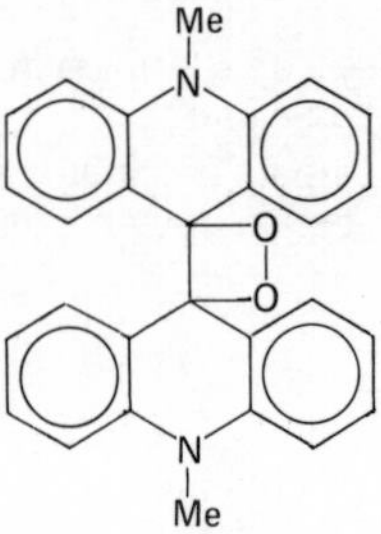 (168) 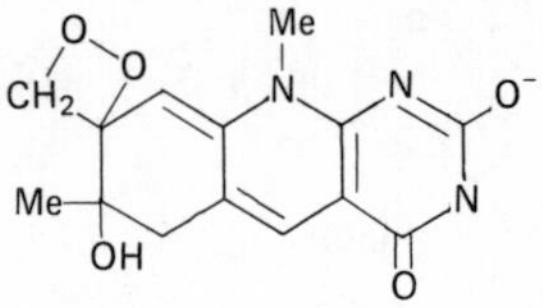 (169)

that the path via dioxetane **165** is of minor importance compared to the pathway (Hock–Criegee rearrangement) via **166** (equation 63). However, it was recently suggested that the Hock–Criegee rearrangement of **164** into **165** is unlikely[129].

In model studies of the transfer of oxygen to phenolates[130] and mandelic acid derivatives[131] by flavoenzymes the possible involvement of dioxetanes has been pointed out. For example, dioxetane **167** has been proposed[130]. The base-catalysed cleavage of lucigenin[132] by hydrogen peroxide or molecular oxygen could involve dioxetane **168**[133].

It was also proposed that the autoxidation of 7-hydroxy-6,7-dihydrolumiflavin involves dioxetane **169**. A rather complex mechanism involving dioxetane **171** has been suggested[134] in the Co(Salpr)-catalysed autoxidation of the Schiff base **170** (equation 64).

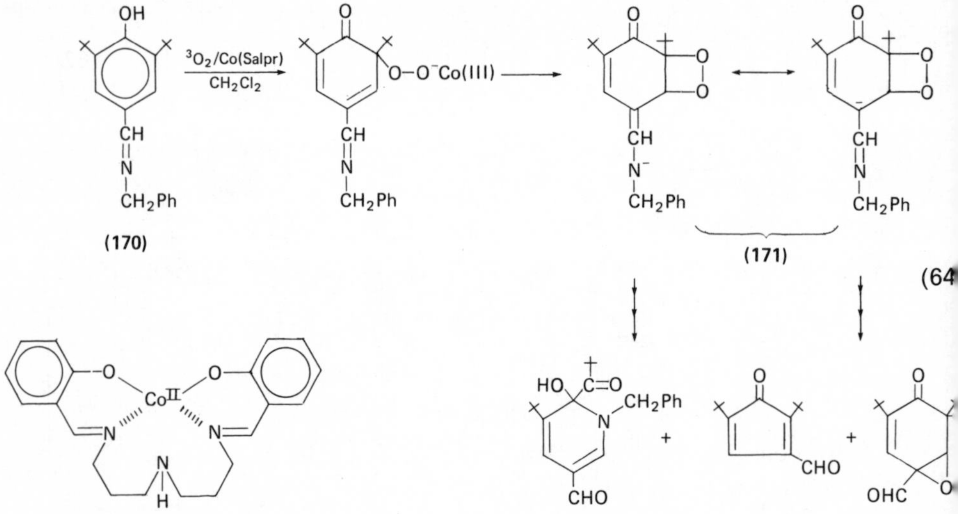

$^3O_2$/Co(Salpr), CH_2Cl_2

(170) (171) (64)

Co(Salpr)

C. Superoxide Cleavages

It has been proposed that the cleavage of electron-poor ethylenes by superoxide ion involves dioxetane intermediates. For example, chalcones[135] are cleaved into the corresponding carboxylic acids according to the mechanism shown in equation (65).

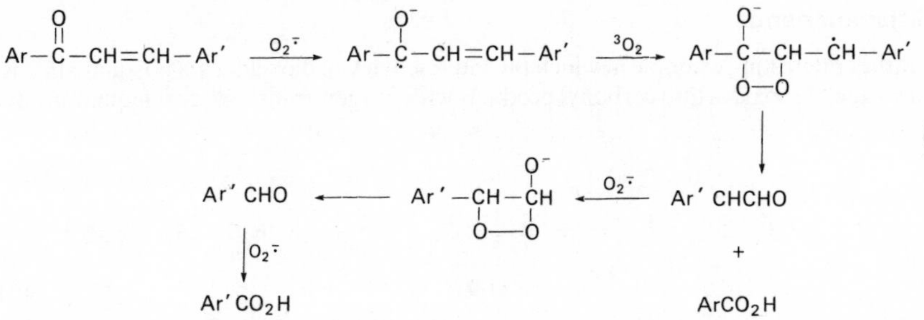

$$Ar-\overset{O}{\overset{\|}{C}}-CH{=}CH-Ar' \xrightarrow{O_2^{\cdot-}} Ar-\overset{O^-}{\overset{|}{\dot{C}}}-CH{=}CH-Ar' \xrightarrow{^3O_2} Ar-\overset{O^-}{\overset{|}{C}}(O{-}O)-CH-\dot{C}H-Ar' \longrightarrow Ar'CHCHO + ArCO_2H \quad (65)$$

$$Ar'CHCHO \xrightarrow{O_2^{\cdot-}} Ar'-CH-\overset{O^-}{\overset{|}{C}H}(O{-}O) \longrightarrow Ar'CHO \xrightarrow{O_2^{\cdot-}} Ar'CO_2H$$

Similarly, the reaction of 1,1-dicyano-, 1,1-dinitro- and mononitro-ethylenes, etc. with superoxide ion leads to cleavage, presumably via dioxetane intermediates[136]. In fact, the conversion of benzyl cyanide into benzoic acid with superoxide ion has been suggested[137] as proceeding via dioxetane-type intermediates. Furthermore, the oxidative degradation of paraquat (**172**) by electrochemically generated superoxide ion has been interpreted in terms of dioxetane **173** as intermediate (equation 66)[138]. Finally, the conversion of 3-hydroxyflavones **174** into **176** by superoxide ion[139] could involve dioxetane **175**, as shown in equation (67).

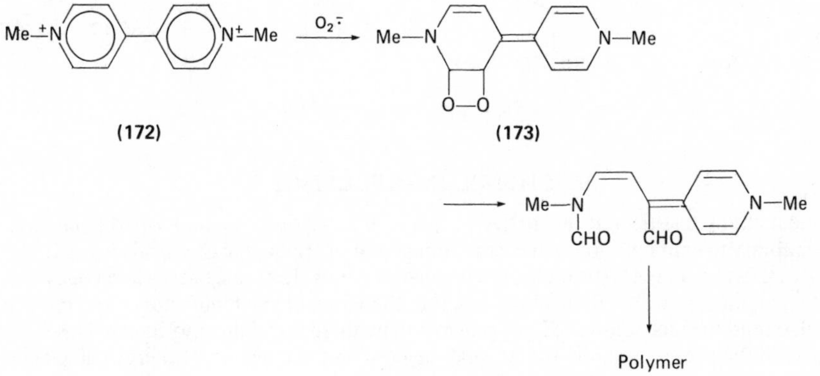

Me–⁺N(C₅H₄)–(C₅H₄)N⁺–Me $\xrightarrow{O_2^{\cdot-}}$ Me–N ... O–O ... N–Me

(172) **(173)**

$\longrightarrow$ Me–N(CHO) ... CHO ... N–Me $\longrightarrow$ Polymer (66)

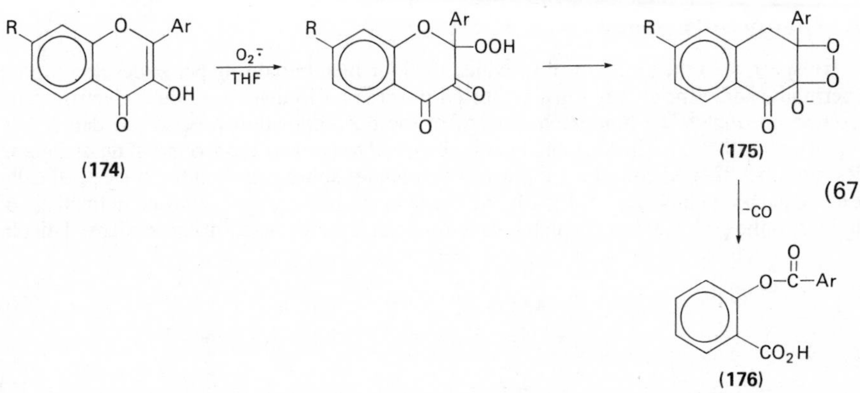

R ... O, Ar, OH, O $\xrightarrow[THF]{O_2^{\cdot-}}$ R ... O, Ar, OOH, O, O $\longrightarrow$ R ... Ar, O–O, O⁻, O $\xrightarrow{-CO}$ O–C(=O)–Ar, CO_2H (67)

(174) **(175)** **(176)**

D. Miscellaneous

A rather interesting example in which the intermediacy of dioxetanes is postulated[140] is the cleavage of epoxides into carbonyl products with oxygen-transfer agents (equation 68).

$$R_2C(O)CR_2 \xrightarrow[\text{or } O_3]{C_5H_5N{-}O/h\nu} R_2C(O^+{-}O^-)CR_2 \; (177) \longrightarrow R_2C(O{-}O)CR_2 \longrightarrow 2R_2C{=}O \qquad (68)$$

Thus, an oxygen atom from pyridine oxide or from ozone is transferred to the epoxide oxygen, forming first a perepoxide intermediate **177**, which is transformed into the carbonyl fragments via the dioxetane. Finally, in the cleavage of the labile dinitrites **178** of *vic* diols, it is proposed that the aldehyde products are derived from the diradical **179** (equation 69)[141]. Diradicals such as **179** are postulated in dioxetane decompositions.

$$RCH(OH){-}CH(OH)R \longrightarrow RCH(O{-}N{=}O){-}CH(O{-}N{=}O)R \; (178) \longrightarrow RCH(O\cdot){-}CH(O\cdot)R \; (179) \longrightarrow 2RCHO \qquad (69)$$

V. CHEMILUMINESCENCE

The most characteristic and distinctive property of 1,2-dioxetanes and α-peroxylactones is their ability to emit light. In such chemiluminescent processes the chemical energy that is stored in these hyperenergetic molecules is converted into electronic energy and released in the form of photons rather than vibrations. It is, therefore, of great importance and interest to understand the mechanism of light generation by these intriguing molecules. This is the purpose of the present section and we shall begin with the question of energy sufficiency.

A. Energy Sufficiency

Irrespective of the mechanism by which the four-membered ring peroxides decompose thermally, sufficient energy must be stored in these molecules to produce electronically excited products. The fundamental steps of the decomposition process are detailed in equation (70)[18o,p]. In the first step heat is absorbed to convert the ground-state dioxetane R_0 into an activated complex ($\neq$), which dissociates subsequently into an electronically excited carbonyl product P^*. Finally the excess excitation energy is emitted in the form of light. It is the $(\neq) \rightarrow P^*$ step which is so unusual, at least for reactions in condensed media

$$\begin{aligned} R_0 + \Delta &\longrightarrow (\neq) \\ (\neq) &\longrightarrow P^* \\ P^* &\longrightarrow P_0 + h\nu \end{aligned} \qquad (70)$$

$$\Delta H^{\neq} + (-\Delta H_0) \gg E^* \qquad (71)$$

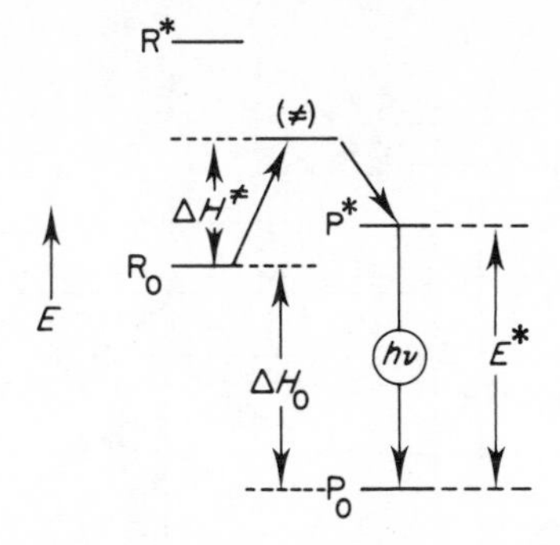

FIGURE 1. Energetics of chemiluminescent reactions.

since a crossover from a ground-state surface to an excited-state surface, i.e. a nonadiabatic path, has taken place. The reaction profile for this typical chemiluminescent reaction is schematized in Figure 1.

Clearly, it is the favourable exothermicity of the chemiluminescent process which leads to an electronically excited product P* rather than to a vibrationally excited product $P^{\neq}$. Consequently, the most important criterion that allows a molecule to decompose on heating into an electronically excited product is that of energy sufficiency. The sum of its activation enthalpy ($\Delta H^{\neq}$) and reaction enthalpy (ΔH_0) must be greater than the excitation energy (E^*) of the electronically excited product (equation 71). A substance that conforms to this energy balance we designate as *hyperenergetic*. Recently this oversimplified criterion of chemienergization has been justly criticized because free energies should be employed instead of enthalpies in order to deal properly with the entropic factor[142].

The energy sufficiency criterion for the two hyperenergetic molecules under discussion, i.e. the 1,2-dioxetanes and α-peroxylactones, is convincingly demonstrated in Figure 2, which summarizes the energetics of their thermal decomposition in a heat of formation (ΔH_f) diagram. Typically, the activation energies, determined from isothermal kinetic methods, range between 20 and 30 kcal mol^{-1} and the heats of reaction, estimated from thermochemical calculations[143], vary around 69–90 kcal mol^{-1}.

Since the lowest excited states (n, π*) of simple carbonyl compounds such as aldehydes and ketones are in the range ca. 75–85 kcal mol^{-1} for singlet excitation and ca. 70–80 kcal mol^{-1} for triplet excitation, the energy sufficiency criterion (equation 71) demands that at least 70–85 kcal mol^{-1} of energy must be made available at the transition state in the dioxetane decomposition to create one of the carbonyl fragments in its n,π* excited state.

For the specific case of tetramethyl-1,2-dioxetane, the heat of reaction was determined experimentally by means of differential scanning calorimetry (DSC) to be $\Delta H_0 = -61$ kcal mol^{-1} [144]. Its activation enthalpy is ca. 25 kcal mol^{-1} [27]. Thus, a total of ca. 86 kcal mol^{-1} of energy is available in the activated complex of tetramethyl-1,2-dioxetane. This is sufficient to produce one of the acetone molecules in its n,π* excited state, since the singlet and triplet energies are respectively at 84 and 78 kcal mol^{-1}. Note, however, that the corresponding π,π* states of acetone cannot be chemienergized by tetramethyl-1,2-dioxetane since these excited states lie well above 86 kcal mol^{-1}. Thus, on the basis of energy balance, we can expect to observe fluorescence or phosphorescence from the n,π* excited carbonyl fragment that is generated in the thermolysis of 1,2-dioxetanes.

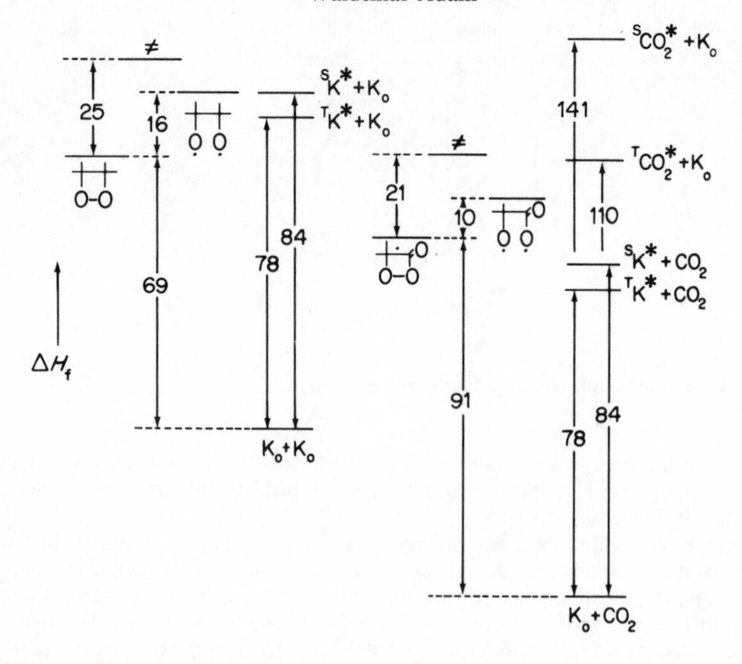

FIGURE 2. Enthalpy of formation (ΔH_f) diagram for tetramethyl-1,2-dioxetane and dimethyl-1,2-dioxetanone thermolysis. Numerical values in kcal mol^{-1} energy units. ^{S}K refers to singlet, ^{T}K to triplet, K* to excited and K_0 to ground-state ketone product.

Comparing the dimethyl-1,2-dioxetanone with the tetramethyl-1,2-dioxetane (Figure 2), we note that although the same electronically excited species is produced, i.e. singlet or triplet acetone, the decomposition of dioxetanones is considerably more exothermic. This is not surprising, because the incorporation of an sp^2 carbon into an already strained ring must increase the exothermicity of the reaction. However, the more interesting fact concerns the chemienergization of the carbon dioxide product. We see in Figure 2 that both the triplet and singlet excited states of CO_2 lie too high in energy to become electronically excited during the thermolysis of α-peroxylactones. But it should be evident that both 1,2-dioxetanes and α-peroxylactones fulfill the energy sufficiency criterion of equation (71).

B. Unimolecular Decomposition Mechanisms

Two mechanisms have been put forward over the years concerning the thermal decomposition of 1,2-dioxetanes leading to electronically excited (n,π*) carbonyl products. These are the diradical mechanism, suggested by Richardson and O'Neal[143], and the concerted mechanism, suggested by Turro and coworkers[18c].

1. *Diradicals as intermediates*

In the diradical mechanism the dissociation of the dioxetane ring proceeds in a stepwise fashion (equation 72). Stretching of the oxygen–oxygen bond (step k_r) leads to the singlet-state diradical (^{S}DR). This singlet diradical has three options: (*a*) to recyclize via step k_{-r} to the dioxetane, (*b*) to disengage the carbon–carbon bond via step k_S into ground-state carbonyl product and singlet excited (^{S}n, π*) carbonyl product, or (*c*) to intersystem-cross via step k_{isc} to the triplet-state diradical (^{T}DR). The latter can either reverse intersystem-cross via step k_{-isc} or fragment via k_T into ground-state singlet carbonyl product and excited triplet (^{T}n, π*) carbonyl product. Finally, deexcitation of the ^{S}n, π* and ^{T}n, π* carbonyl products by the usual photophysical and photochemical ways, e.g. fluorescence via step k_{fl} and phosphorescence via step k_{ph}, respectively, affords ground-state carbonyl products. To account for the high spin-state selection of triplet product, all one would need to impose is the condition that $k_T > k_{isc} > k_S$. This is not unreasonable, as suggested by most of the experimental evidence.

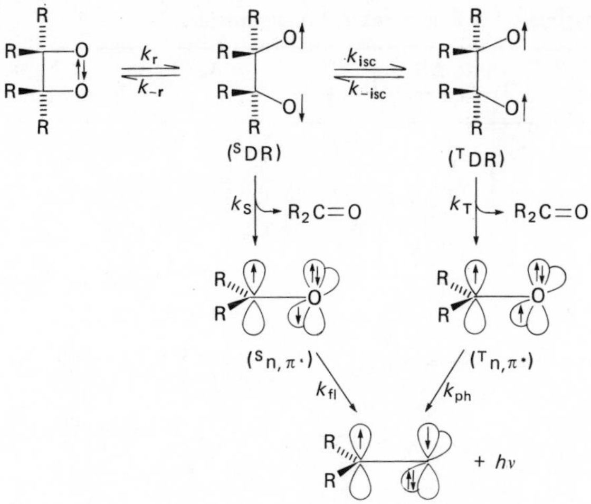

The very earliest experimental evidence in support of the diradical mechanism[143] rests on the fact that alkyl and phenyl substitution does not significantly alter the activation parameters (Table 6) for dioxetane decomposition[145]. It was argued that if carbon–carbon bond cleavage occurs simultaneously with oxygen–oxygen bond cleavage, the incipient carbonyl group in the activated complex should be stabilized in the relative order phenyl > alkyl > hydrogen. Thus, the activation energies should obey the relative order $E_a(Ph) < E_a(R) < E_a(H)$, i.e. lowest for phenyl-substituted dioxetanes. Since this expectation was not borne out by the experimental data[145] (Table 6), the diradicals ^{S}DR and ^{T}DR were proposed as intermediates. As additional support for the diradical mechanism it was shown that the 3,4-diethoxy-1,2-dioxetane (**29**) and the *p*-dioxene-1,2-dioxetane (**180**) had identical activation energies[221] (Table 6; entries 12 and 25, respectively), implying that the carbon–carbon bond is not significantly stretched in the activated complex.

The strongest support in favour of the diradical mechanism is the lack of a deuterium isotope effect in the thermal decomposition of *trans*-3,4-diphenyl-1,2-dioxetane (**181**)[146]. In the concerted mechanism the ring carbon of the dioxetane changes its hybridization state from sp^3 to sp^2 in the activated complex and an inverse secondary isotope effect (k_H/k_D) would be expected[147]. Consequently, a diradical mechanism was argued to accommodate these results. Similarly, in the thermal decarboxylation of the dimethyl α-peroxylactone a negligible ($k_H/k_D = 1.06 \pm 0.04$) secondary isotope effect was

TABLE 6. Activation parameters of four-membered ring peroxides

Dioxetane	E_a (kcal mol^{-1})	$\Delta H^{\neq}$ (kcal mol^{-1})	log A	$\Delta S^{\neq}$ (e.u.)	$\Delta G^{\neq}$ (293.2 K)[a] (kcal mol^{-1})	Ref.
1. 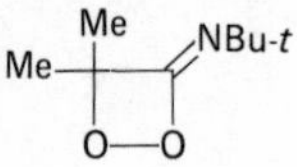	17.0 ± 2	17.1 ± 2	13.8	2.7 ± 1.0	16.3	57
2. 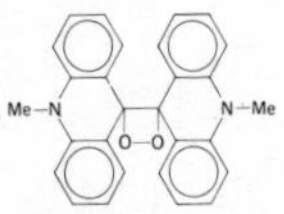	18.7 ± 0.3	18.2 ± 0.3	—	−1.4 ± 1.2	18.6	180a, b
3. 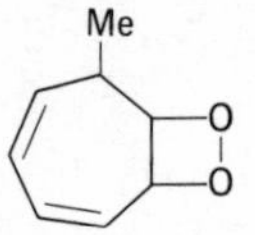	—	21.1 ± 0.6	—	8 ± 1	18.8	227
4. 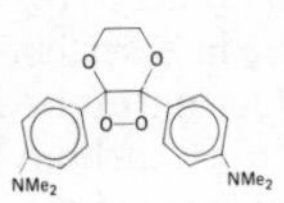	19.7	19.1	11.2	−7.2	21.2	31
5. 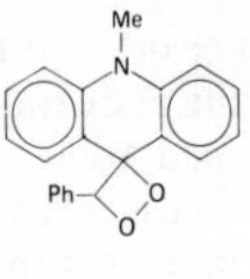	19.7	—	11.6	—	21.3	180c
6. 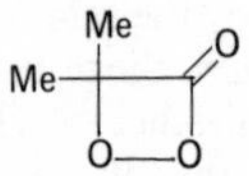	19.8	19.2 ± 0.2	—	−8.2 ± 0.5	21.6	172
7. 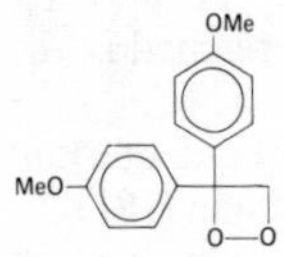	20.9 ± 0.3	20.2 ± 0.3	11.8 ± 0.2	−6.8 ± 0.8	22.2	145d
8. 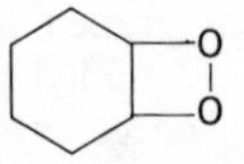	22.5 ± 0.3	—	12.8	—	22.5	158

TABLE 6. *continued*

Dioxetane	E_a (kcal mol^{-1})	$\Delta H^{\neq}$ (kcal mol^{-1})	log A	$\Delta S^{\neq}$ (e.u.)	$\Delta G^{\neq}$ (293.2 K)[a] (kcal mol^{-1})	Ref.
9. 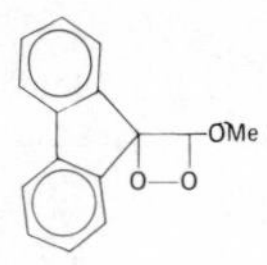 OMe, O—O	21.0 ± 1	—	11.6 ± 0.6	—	22.6	224
10. 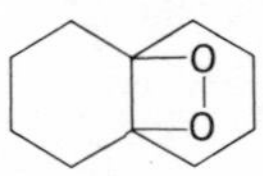 O—O	22.7 ± 0.9	—	—	−3 ± 3	23.0	27, 159
11. 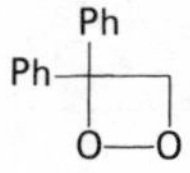 Ph, Ph, O—O	22.7 ± 0.1	22.0 ± 0.1	12.4	−4.1 ± 0.3	23.2	145a
12. 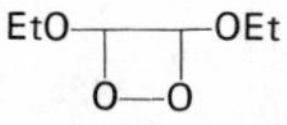 EtO, OEt, O—O	23.6	—	13.1	—	23.2	168, 221
13. 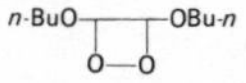 *n*-BuO, OBu-*n*, O—O	20.0 ± 0.5	—	10.3	—	23.2	235
14. 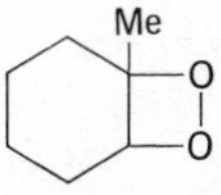 Me, O—O	23.7 ± 0.5	—	13.0	—	23.4	156
15. 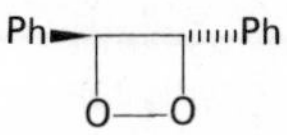 Ph, Ph, O—O	23.6 ± 1.6	—	12.9	—	23.5	146
16. 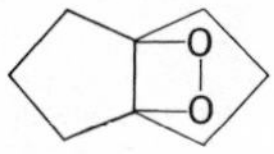 O—O	25.6 ± 0.6	—	—	4 ± 2	23.8	159
17. 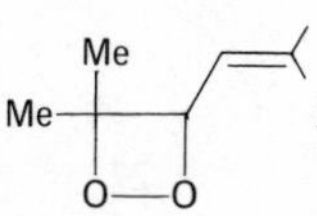 Me, Me, O—O	22.3 ± 0.3	21.6 ± 0.6	—	−7.6 ± 0.9	23.8 ± 0.8	229b
18. 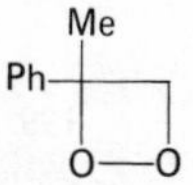 Me, Ph, O—O	22.9	—	12.1	−5.3	23.9	145a

TABLE 6. *continued*

Dioxetane	E_a (kcal mol^{-1})	$\Delta H^{\neq}$ (kcal mol^{-1})	log A	$\Delta S^{\neq}$ (e.u.)	$\Delta G^{\neq}$ (293.2 K)[a] (kcal mol^{-1})	Ref.
19. 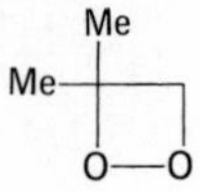	23.0	—	12.2	−5.0	23.9	145a
20. 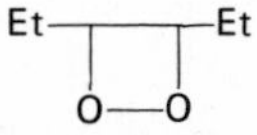	24.5 ± 0.3	—	13.1	—	24.1	158
21. 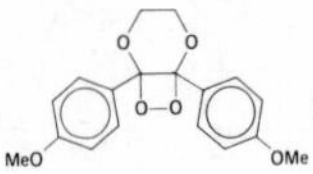	24.0	23.5	12.4	−1.9	24.1	31
22. 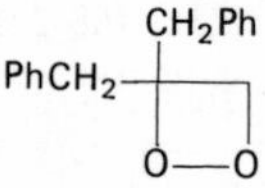	24.3 ± 0.1	23.6 ± 0.1	12.8	−2.0 ± 0.5	24.2	145a
23. 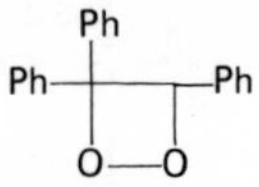	23.3 ± 0.3	22.6 ± 0.3	12.0 ± 0.2	−5.6 ± 0.9	24.2	145c
24. 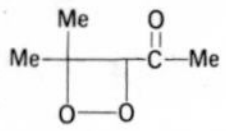	26.4 ± 1.0	25.5 ± 1.0	14.2	4.0 ± 2.0	24.3	35
25. 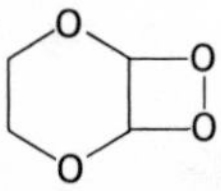	24.6 ± 1	—	13.0	—	24.3	221
26. 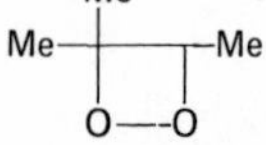	23.5 ± 0.5	—	—	−5 ± 2	24.4	27
27. 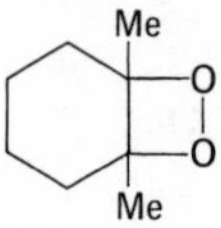	25.7 ± 0.7	—	—	2 ± 2	24.5	27
28. 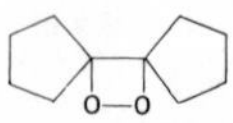	25.3 ± 0.2	—	—	0 ± 2	24.7	199

TABLE 6. *continued*

Dioxetane	E_a (kcal mol^{-1})	$\Delta H^{\neq}$ (kcal mol^{-1})	log A	$\Delta S^{\neq}$ (e.u.)	$\Delta G^{\neq}$ (293.2 K)[a] (kcal mol^{-1})	Ref.
29. 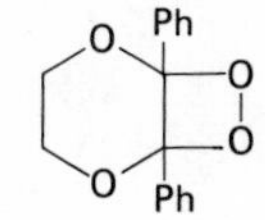	24.8	24.2	12.4	−1.8	24.7	31
30. 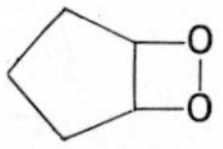	26.0 ± 0.3	—	13.6	1.6	24.7	158a
31. 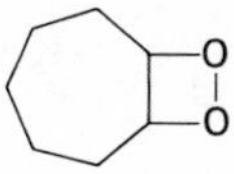	25.9 ± 0.3	—	13.6	1.9	24.7	158a
32. 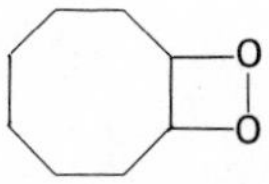	24.8 ± 0.3	—	13.0	−1.6	24.7	158a
33. 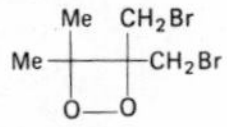 Me, Me, CH_2Br, CH_2Br	28.0 ± 0.2	27.3 ± 0.2	—	7.7 ± 0.4	25.0	45
34. 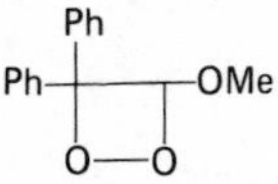 Ph, Ph, OMe	26.1 ± 1	—	13.5 ± 0.6	—	25.1	224
35. 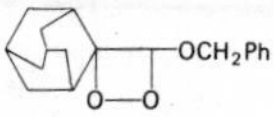 OCH_2Ph	24.5 ± 0.2	23.4 ± 0.6	—	−6 ± 0.4	25.2	256
36. 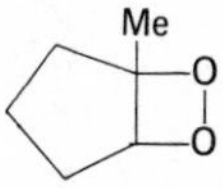 Me	25.2 ± 0.5	—	12.7	—	25.3	156
37. 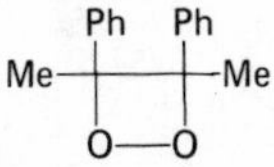 Me, Ph, Ph, Me	25 ± 1	—	—	−3 ± 3	25.3	231
38. 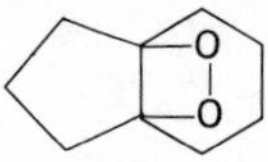	26.3 ± 0.5	—	—	1 ± 2	25.4	159

TABLE 6. *continued*

Dioxetane	E_a (kcal mol^{-1})	$\Delta H^{\neq}$ (kcal mol^{-1})	log A	$\Delta S^{\neq}$ (e.u.)	$\Delta G^{\neq}$ (293.2 K)[a] (kcal mol^{-1})	Ref.
39. 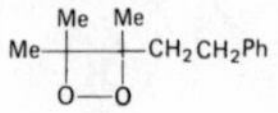 Me, Me, Me, CH_2CH_2Ph; O—O	25.5 ± 0.5	—	12.9	—	25.4	156
40. 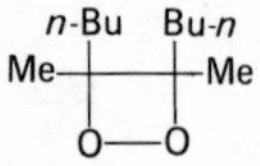 *n*-Bu, Bu-*n*, Me, Me; O—O	25.2 ± 1	—	12.5	—	25.6	240
41. 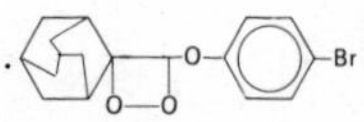 (adamantane)–O–C_6H_4–Br; O—O	25.1 ± 0.2	24.6 ± 0.3	—	−3.5 ± 0.3	25.6	256
42. 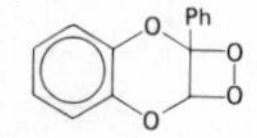 (benzodioxin), Ph; O—O	24.4 ± 1	23.8 ± 1	—	−5.1 ± 2	25.7	233
43. 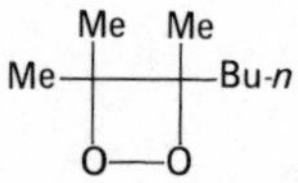 Me, Me, Me, Bu-*n*; O—O	25.3 ± 0.5	—	12.5	—	25.7	156
44. 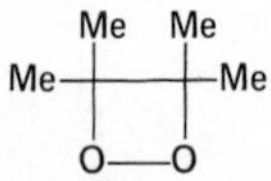 Me, Me, Me, Me; O—O	27.6 ± 0.1	26.9 ± 0.1	—	3.7 ± 0.3	25.8	45
45. 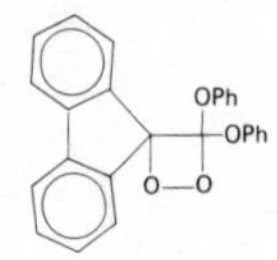 (fluorenylidene), OPh, OPh; O—O	—	25.0 ± 0.4	—	−2.8 ± 0.5	25.8	256
46. 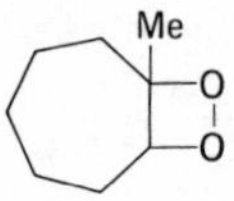 (cycloheptane), Me; O—O	25.9	—	12.8	—	25.9	156
47. 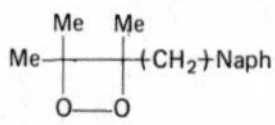 Me, Me, Me, (CH_2)Naph; O—O	25.7 ± 0.5	—	12.5	—	26.1	156
48. 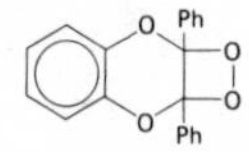 (benzodioxin), Ph, Ph; O—O	—	24.5 ± 0.4	—	−5.4 ± 1	26.1	236

TABLE 6. *continued*

Dioxetane	E_a (kcal mol^{-1})	$\Delta H^{\neq}$ (kcal mol^{-1})	log A	$\Delta S^{\neq}$ (e.u.)	$\Delta G^{\neq}$ (293.2 K)[a] (kcal mol^{-1})	Ref.
49. 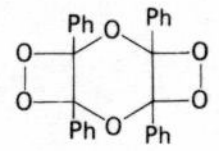	26 ± 1	—	12.7	−2.4	26.1	86
50. 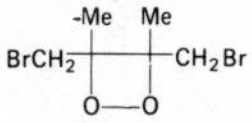	27.6 ± 0.2	26.9 ± 0.2	—	2.5 ± 1.0	26.2	45
51. 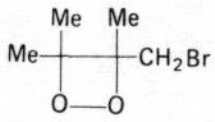	28.4 ± 0.1	27.7 ± 0.1	—	5.2 ± 0.4	26.2	45
52. 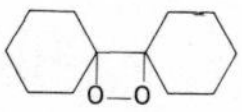	27.7	—	13.8	—	26.3	160
53. 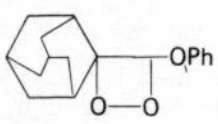	25.9 ± 0.2	25.0 ± 0.3	—	−4.7 ± 0.3	26.4	256
54. 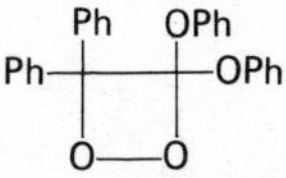	—	24.1 ± 0.4	—	−9.1 ± 1	26.7	256
55. 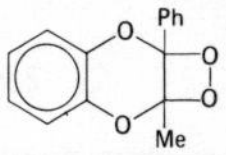	25.8 ± 1.5	25.1 ± 1.0	—	−5.7 ± 2	26.9	233
56. 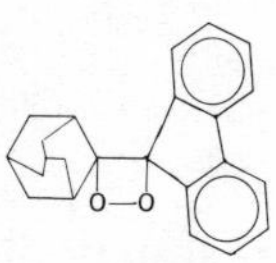	—	25.8 ± 0.4	—	−3.7 ± 0.9	26.9	256
57. 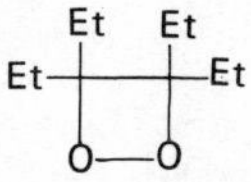	30.0 ± 1	—	14.8	—	27.3	160, 240
58. 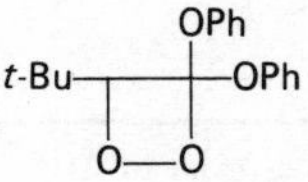	26.7 ± 0.1	25.8 ± 0.9	14.1 ± 0.1	−5.2 ± 2.5	27.3	246

TABLE 6. *continued*

Dioxetane	E_a (kcal mol^{-1})	$\Delta H^{\neq}$ (kcal mol^{-1})	log A	$\Delta S^{\neq}$ (e.u.)	$\Delta G^{\neq}$ (93.2 K)[a] (kcal mol^{-1})	Ref.
59. 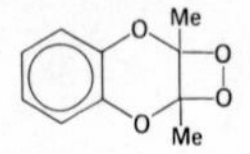	27.0	26.2 ± 1	12.5	−3.7 ± 1.5	27.3	233
60. 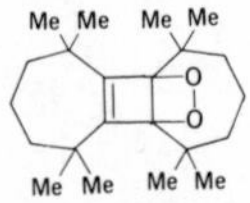	29.7 ± 0.6	29.0 ± 0.6	14.6	5.7 ± 1	27.3	179
61. 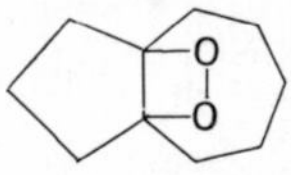	29.8 ± 0.4	—	—	4 ± 2	28.0	159
62. 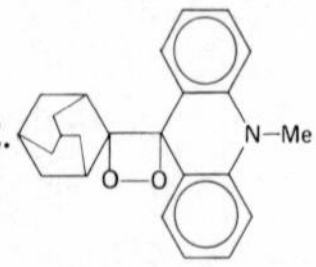	26.3	—	11.5	—	28.0	180c
63. 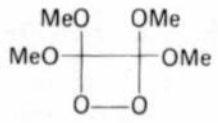	28.6 ± 1	—	12.9	—	28.5	27
64. 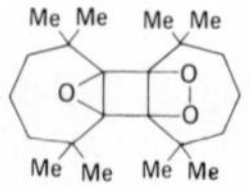	27.5 ± 0.8	27.0 ± 0.8	12.0	−5.1 ± 1	28.5	22, 179
65. 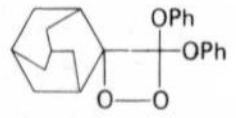	—	28.0 ± 0.2	—	−4.0 ± 0.6	29.2	256
66. 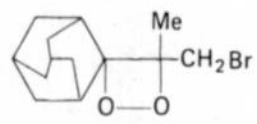	—	28.4 ± 0.3	—	−3.5 ± 0.9	29.4	256
67. 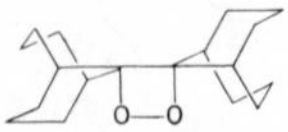	35.6 ± 0.5	—	14.8	—	32.9	235
68. 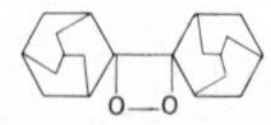	34.6 ± 1.5	33.8	—	2.9 ± 2	32.9	157

[a]Estimated from E_a and log A or $\Delta H^{\neq}$ and $\Delta S^{\neq}$ for 293.2 K; arranged in ascending order of $\Delta G^{\neq}$.

observed[148]. Presumably in the α-peroxylactone decomposition a diradical mechanism similar to that of dioxetanes (equation 72) also operates.

A dramatic solvent effect in the thermolysis of dioxetanes[149] was taken as evidence against the diradical mechanism. However, it was shortly thereafter reported[150] that the solvent effect in methanol was the result of catalysis by transition-metal ion impurities, and could be suppressed in the presence of metal ion complexing agents such as EDTA or Chelex 100. The utmost care that must be taken in measuring reliable kinetic parameters in 1,2-dioxetane decomposition cannot be overemphasized.

Trapping experiments would constitute the most unequivocal proof for the intervention of diradical intermediates in the decomposition of 1,2-dioxetanes. Although such experiments have not been reported to date, the interesting observation[151] that tri-*t*-butylphenol extinguished the trimethyldioxetane chemiluminescence more efficiently than piperylene, was construed as evidence that the phenol scavenged a relatively long-lived precursor to the electronically excited product, presumably a diradical. Tentative evidence in favour of the diradical mechanism has been claimed[57b] in the chemiluminescence of the imino-1,2-dioxetane **182** versus **183**.

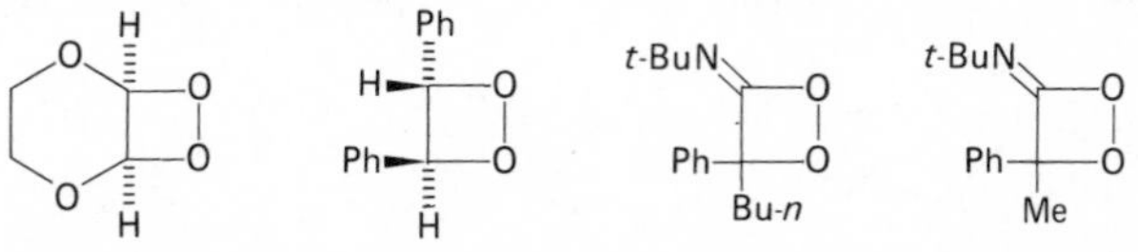

(180) (181) (182) (183)

Direct spectroscopic observation of the postulated diradical intermediates has not been possible so far. Thus, multiphoton infrared laser excitation of tetramethyldioxetane in the gas phase failed to detect diradical intermediates with lifetimes greater than ca. 5 ns[152]. Picosecond spectroscopy limited the lifetime of a diradical intermediate, if formed, to less than ca. 10 ps in the 264 nm pulsed photolysis of tetramethyldioxetane in acetonitrile, using a mode-locked neodymium phosphate laser[153]. However, a diradical intermediate of this lifetime has been proposed in the photolysis of tetramethyl-1,2-dioxetane[154]. Thus, we must conclude that the diradical mechanism (equation 72) of dioxetane decomposition remains uncertain.

2. Concerted decomposition

In the concerted mechanism (equation 73) oxygen–oxygen and carbon–carbon bonds are disengaged simultaneously[149]. Vibrational deformation leading to a puckered four-membered ring transition state aligns the orbitals (hatched) optimally to create an n, π* excited state of the carbonyl product. It is argued[155] that during this puckering act electron density is displaced in such a way as to promote spin-orbital coupling at the oxygen atom and thereby generating preferentially ^{T}n, π* carbonyl product.

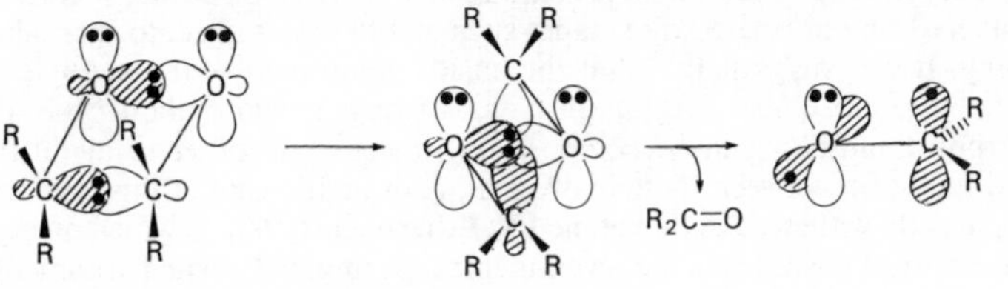

$R_2C{=}O$ (73)

The unusual stability of the diadamantylidene-1,2-dioxetane **14** (Table 6; entry 68) has been interpreted in terms of a concerted mechanism[21,156], but a diradical mechanism has also been proposed[157]. Thus, on the basis of the puckered structure of **14**, it was rationalized[21] that the inertial mass of the rigid and bulky spiroadamantane moieties was impeding further twisting of the four-membered ring in the postulated unscrewing mode in the activated complex of the concerted mechanism (equation 73). On the other hand, not aware of the puckered structure, an alternative argument postulated[155] that the rigid planar four-membered ring resists the necessary twisting action of the concerted mechanism. However, inspection of Dreiding models reveals that the four equatorial methylenic hydrogens in **14** would prevent twisting about the carbon–carbon dioxetane bond in the diradical as well as in the concerted decomposition mode.

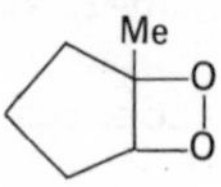

(184) 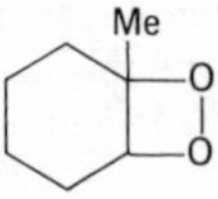 (185) 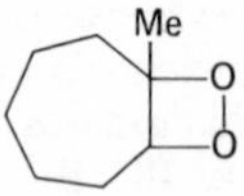 (186) 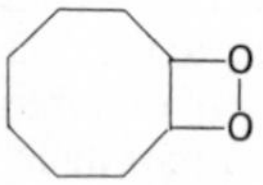 (187)

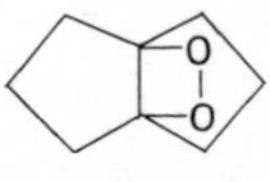

(188) 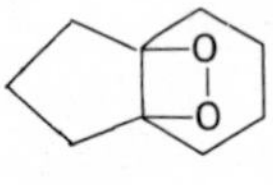 (189) 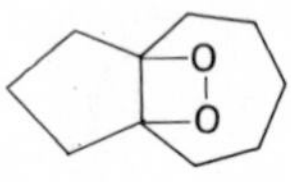 (190) 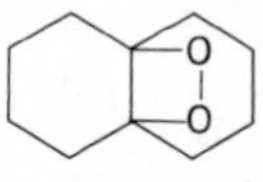 (191) 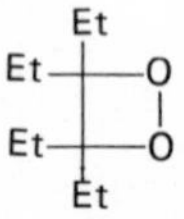 (192)

More convincing evidence in support of the concerted mechanism is the thermal stability of the dioxetane series **184** through **186**[156] (Table 6; entries 36, 14 and 46, respectively). The fact that the fused six-membered ring dioxetane **185** is considerably less stable than either **184** or **186** suggests that carbon–carbon fission occurs concurrently with oxygen–oxygen bond rupture. Another series of this type are the homologous unsubstituted dioxetanes **26**, **9**, **27** and **187** (Table 6; entries 30, 8, 31 and 32, respectively), for which a concerted mechanism is also argued[38,158]. Inspection of Dreiding models implies that dioxetanes with annelated six-membered rings must have puckered dioxetane rings, while those with annelated five-, seven- and eight-membered rings can be planar. Therefore, the twisted transition state of the concerted mechanism is in part realized in **9** and **185** even in their ground states in view of their puckered structures compared to the presumably planar dioxetanes **26**, **27**, **184** and **186**. Consequently, dioxetanes **9** and **185** decompose more easily. However, the series of dioxetanes **188**–**191** exhibits the same stability trends[159] (Table 6; entries 16, 38, 61 and 10, respectively), i.e. **188** and **189** are of normal stability (like tetramethyl-1,2-dioxetane), **190** is unusually stable and **191** unusually labile. This has been rationalized in terms of the diradical mechanism. It seems advisable to determine the crystal structures of the two extreme cases, i.e. **190** (the most stable) and **191** (the most labile), to confirm the degree of puckering of these structures.

Be this as it may, the unusual stability of dioxetane **190** requires an additional explanation beyond the puckering argument. In fact, there is something special about the seven-membered ring-annelated dioxetanes such as **190** with respect to their abnormally high stability. It was suggested[159] that the 'inside' hydrogens of the methylene groups adjacent to the dioxetane ring require a twisting motion about the dioxetane carbon–carbon bond along the decomposition trajectory in order to avoid the severe nonbonded repulsion between these hydrogens. A dramatic case of this conformational effect is apparently witnessed in dioxetane **15** (Table 6; entry 60), in which methyl groups occupy these critical positions in the seven-membered ring[22]. A related argument of steric crowding was forwarded[160] to explain the unusual stability (Table 6, entry 57) of tetraethyl-1,2-dioxetane (**192**). This apparently simple dioxetane is one of the more blatant failures of the diradical theory of dioxetane decomposition. It was argued that a 'stable' diradical intermediate, i.e. one that lies in an energy well of at least 5 kcal mol^{-1} depth, cannot be involved. Finally, a pressure dependence study of the energy-transfer chemiluminescence of tetramethyldioxetane[161] suggests that the activation volumes are more readily reconciled in terms of the concerted mechanism.

3. Theoretical work

In view of the two proposed mechanisms for the thermal decomposition of dioxetanes, we shall briefly review the theoretical contributions towards the understanding of this challenging problem.

The earliest theoretical thoughts on the chemiluminescent decomposition of 1,2-dioxetanes were based on orbital symmetry arguments[15]. It was predicted that the suprafacial [2 + 2]retrocyclization must lead to an electronically excited carbonyl product. However, such a retrocyclic process should chemienergize the π, π^* state, which is inaccessible on grounds of energy balance (cf. Section V.A). A more thorough orbital and state symmetry analysis of the concerted decomposition revealed[162] that a n, π^* excited carbonyl product should be formed. More important, this intuitive and penetrating study predicted well before it was confirmed experimentally[18c] that a triplet excited n, π^* product should be formed preferentially. Specifically it was noticed that the energy surface for the less energetic $^{T}n, n^*$ carbonyl product intersects that of the singlet excited n, π^* carbonyl fragment along the out-of-plane bending vibrational mode.

Numerous semiempirical calculations have been carried out on the dioxetane decomposition, including CNDO/2 calculations with and without configuration interaction[163] and MINDO/3 with configuration interaction[164]. Although these differ in their opinions of whether oxygen–oxygen bond homolysis occurs first, leading to a diradical intermediate via **193**, followed by fast carbon–carbon bond cleavage, or whether both bonds are cleaved simultaneously via **194**, these calculations all agree that the decomposition mode engages twisting of the dioxetane ring, as shown in the respective activated complexes **193** and **194**. Clearly, puckered transition states are involved in both, but in the diradical case **193** the carbon–carbon bond is still intact.

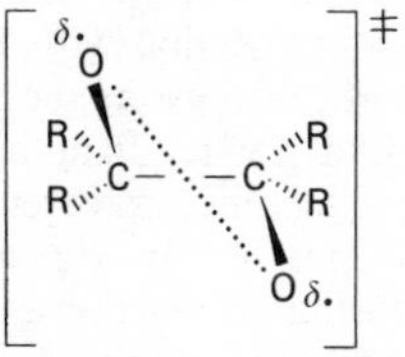

(193) 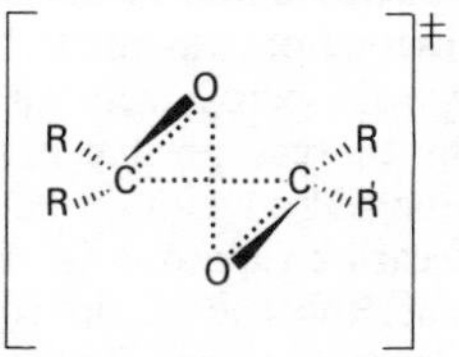 **(194)**

On the basis of qualitative considerations it was argued[165] that a crossover of the diradical path to the triplet excited product path, prior to reaching a *bona fide* stable diradical intermediate, is feasible. This attractive alternative represents a merger between the diradical (equation 72) and the concerted (equation 73) decomposition routes. This merged mechanism was also adopted to explain the tetraethyldioxetane **192** case[160].

The most ambitious theoretical investigation on this problem has employed the nonempirical GVB method[166]. It was concluded that a 1,4-diradical is an intermediate, resulting from oxygen–oxygen bond cleavage but leaving the carbon–carbon bond intact. The eight possible singlet and triplet state electronic configurations of the diradical all lie within a narrow bond of 3 kcal mol^{-1} and correlate with singlet and triplet excited carbonyl product and ground-state carbonyl product. What specific spin state of a particular excited state is chemienergized depends on its energy relative to the 1,4-diradical intermediate. For example, for the unsubstituted 1,2-dioxetane the energy of the surface crossing point of the $^{T}n,\pi^*$ state of formaldehyde lies ca. 8 kcal mol^{-1} lower than its $^{S}n,\pi^*$ state relative to the 1,4-diradical. Consequently, triplet excited formaldehyde is preferentially chemienergized, as observed experimentally[32]. The quantitative results on diradical stability obtained from the *ab initio* GVB calculations match well those derived from thermochemical estimations[143] (cf. Figure 2).

The most recent theoretical study on this subject concerns the parent α-peroxylactone[167]. These SCF calculations suggest that a diradical mechanism is involved, since stretching of the peroxide bond does not cause significant loosening of the carbon–carbon bond. The proportionally higher yield of singlet excited product for the α-peroxylactones compared to dioxetanes is explained in the greater rate of decarboxylation versus intersystem crossing, i.e. steps k_2 versus k_{isc} in equation (72), but involving α-peroxylactones instead of dioxetanes.

C. Catalytic Decomposition

The mechanisms discussed in Section V.B concern the unimolecular thermolysis of 1,2-dioxetanes and α-peroxylactones. However, such molecules also undergo catalytic decompositions involving electron-transfer processes, which we shall discuss presently.

1. *Intermolecular electron exchange*

The fact that hyperenergetic molecules such as the 1,2-dioxetanes should be prone to catalytic decomposition is not surprising. Early examples include the protecting effect of molecular oxygen on the thermal decomposition of 3,4-diethoxydioxetane[168], the efficient catalytic decomposition of this dioxetane by amines[169], and of alkyl-substituted dioxetanes by transition-metal ion impurities[150]. However, all these are competing dark reactions, which greatly diminish the chemiluminescence of the dioxetanes.

An unusual observation was made by us[170] in connection with the energy-transfer chemiluminescence of α-peroxylactones with aromatic fluorescers such as rubrene. Under similar conditions rubrene produced about 50-fold greater light intensity than 9,10-diphenylanthracene (DPA) with dimethyl α-peroxylactone; however, the rate of the α-peroxylactone decarboxylation was significantly greater for rubrene than for 9,10-diphenylanthracene. Rubrene was not consumed during the reaction and served, therefore, as catalyst in the decomposition of the α-peroxylactone, by enhancing the efficiency of light emission of the system. In other words, the reaction between rubrene and peroxylactones represented one of the few examples of a catalytic, but chemiluminescent, reaction of dioxetanes. Another such reaction, known for even longer[1], is that of aryl oxalates with hydrogen peroxide and aromatic fluorescers[171].

As soon as the phenomenon of *C*hemically *I*nduced *E*lectron *E*xchange *C*hemiluminescence (CIEEL) exhibited by peroxides and easily oxidized fluorescers (Fl) had been recognized[18k], it was obvious that the α-peroxylactone–rubrene case belonged to this category of chemiluminescence reactions. The electron-exchange mechanism shown in equation (74) was proposed[172] to account for the facts (*a*) that the easily oxidized fluorescers catalyse the decomposition of the α-peroxylactones and (*b*) that the rate of catalysis is proportional to the ease of oxidation of the fluorescer. During the slow step an electron is transferred from the fluorescer to the α-peroxylactone, producing a fluorescer radical cation–peroxide radical anion pair. Decarboxylation affords a fluorescer radical cation–ketyl radical anion pair, which can either diffuse to free ion radicals or back-exchange the electron to generate electronically excited fluorescer. The latter emits light. This process constitutes a chemical equivalent of the well-known phenomenon of electrochemiluminescence[173].

slow; $Fl^{+\cdot}$; CO_2; diffusion; free ion radicals; $Fl^{+\cdot}$; + Fl*; $Fl + h\nu$ (74)

Recently[174] the electron-transfer theory[173,175] has been extended to incorporate the slow and reversible chemically induced electron-exchange reactions as observed for the fluorescer-catalysed chemiluminescent decomposition of α-peroxylactones[172]. It was argued that the electron transfer is complete in the transition state for such a slow and irreversible endergonic electron-transfer reaction, but that the typically flat slopes ($-\alpha/RT$ where α ca. 0.3) of the ln (intensity) vs. oxidation potential plot was due to the fact that only a fraction (α) of the total free-energy change manifested itself in the activation energy. The recent criticism[176] of this treatment has been refuted[177].

It is noteworthy that tetramethyl-1,2-dioxetane does not exhibit electron-exchange chemiluminescence[172]; however, tetramethoxy-1,2-dioxetane does[178]. In fact, in this interesting case both electron donors such as DPA as well as electron acceptors such as 9,10-dicyanoanthracene give rise to such chemiluminescent catalytic decomposition. Furthermore, we have observed[179] that the dioxetane **195** exhibits electron-exchange chemiluminiescence. In this context it is of interest to mention that the related dioxetane **15** is inactive towards electron exchange even with rubrene.

2. *Intramolecular electron exchange*

1,2-Dioxetanes possessing easily oxidized substituents behave quite distinctly in the chemiluminescent properties when compared to tetramethyldioxetane. Examples are 1,2-dioxetanes substituted with *N*-methyllindolyl (**21**)[30], *N,N*-dimethylaminophenyl (**22**)[31] and *N*-methylacridanyl (**196**)[180] groups. First of all, these dioxetanes are relatively thermally unstable (Table 6) and second they give relatively high yields of singlet excited

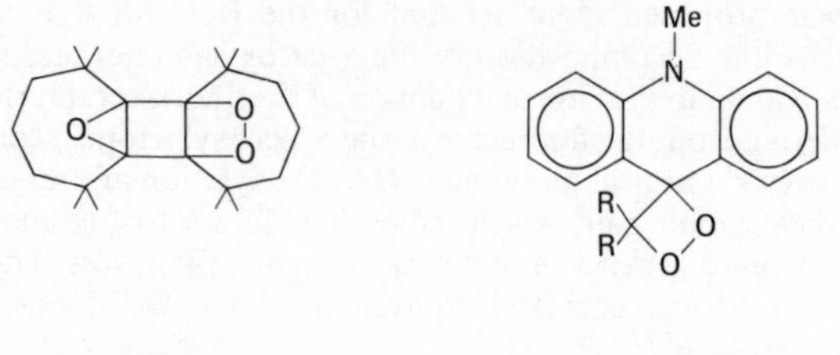

(195) (196)

carbonyl product (Table 7). For such systems an intramolecular electron-exchange chemiluminescence mechanism has been proposed, as illustrated in general terms in equation (75). When X is a substituent such as those in the dioxetanes **21**, **22** and **196**, internal electron transfer from X to the peroxide bond affords a dipolar intermediate such as **197**. Cleavage of the dioxetane carbon–carbon bond affords the radical-anion–radical-cation pair and back-transfer of an electron leads finally to singlet excited carbonyl product and chemiluminescence.

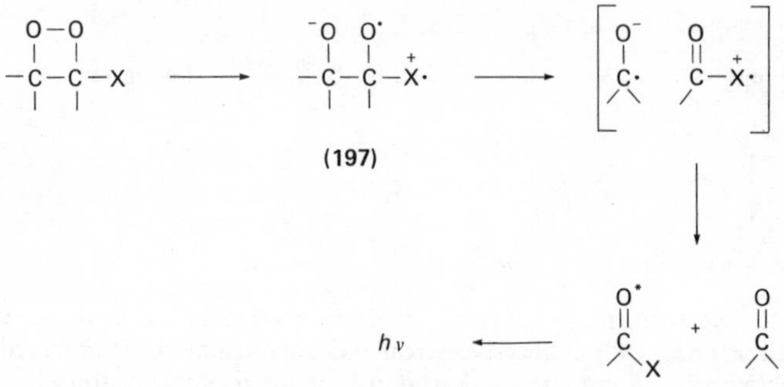

(197)

(75)

$h\nu$

An unusual chemiluminescent catalysis of this type has been observed on silica gel for dioxetanes **22**[31], **36**[48] and **196**[180c]. The much enhanced chemiluminescence efficiency was attributed to promotion of intramolecular electron exchange analogous to equation (75). The acidic silica gel supposedly complexes with the peroxide bond of the dioxetane, thereby lowering its reduction potential and thus enhancing the efficiency of internal electron transfer.

Of considerable interest are the implications of this intramolecular electron-exchange mechanism in bioluminescence[181]. For example, the high yield of singlet excitation that was observed in bioluminescent systems, especially of the firefly[182], was not understood. In fact, a key feature that had escaped the mechanistic chemist for a long while was the fact that methylation of the phenol group of the firefly luciferin extinguished almost completely the bioluminescence[183]. Presumably the phenolate moiety is essential for efficient light production. Consequently, as shown in equation (76), the mechanism of efficient bioluminescence requires intramolecular electron transfer from the electron-rich phenolate moiety to the α-peroxylactone ring[181]. After decarboxylation and electron back-transfer an electronically excited singlet-state oxyluciferin **198** is formed, which then emits the greenish-yellow bioluminescence.

A detailed theoretical analysis[184] suggests that the fluorescence of the oxyluciferin anion is derived from a low-lying π,π^* singlet state with substantial charge transfer from

(76)

(198)

the benzenethiazolyl group to the oxythiazoline chromophore. Subsequent charge annihilation leads to the observed emission.

D. Excitation Parameters

As already pointed out on several occasions, the unique property of dioxetanes is to generate electronically excited states on thermolysis, which then manifest themselves by light emission (equation 70). The total yield of excited states (equation 77), i.e. the sum of the singlet excitation yield (ϕ^S) and triple excitation yield (ϕ^T) and the spin-state selectivity (equation 78), i.e. the ratio of the triplet and singlet yields, are parameters that characterize a particular dioxetane. Typical values are compiled form the literature in Table 7. A number of methods are known to determine the ϕ^S and ϕ^T quantities. For simplicity and clarity, the methods for the determination of these excitation yields are classified into photophysical and photochemical techniques. This is warranted in view of the distinct experimental methodologies involved.

$$\text{Total excitation yield} \equiv \phi^{T+S} = \phi^T + \phi^S \tag{77}$$

$$\text{Spin-state selectivity} \equiv \phi^T / \phi^S \tag{78}$$

1. Photophysical methods

In the photophysical techniques for determining excitation yields (ϕ^*) of chemiluminescent processes the physical properties of the electronically excited state are utilized, specifically their luminescent properties. Thus, the observed chemiluminescence, i.e. fluorescence in the case of singlet states and phosphorescence in the case of triplet states, of the chemienergized process is related to the photoluminescence of the electronically excited product. For convenience we shall distinguish between direct chemiluminescence (DC), in which the chemienergized product K* directly exhibits chemiluminescence (fluorescence and/or phosphorescence), and energy-transfer chemiluminescence (ETC) or enhanced chemiluminescence (EC), in which the chemienergized product K* first transfers its excitation energy to a suitable luminescer Lu_0 (fluorescer

TABLE 7. Excitation parameters of four-membered ring peroxides

Dioxetane	Excitation parameters[a,b]				Ref.
	ϕ^{S}(%)[c]	ϕ^{T}(%)[d]	ϕ^{T+S}(%)	ϕ^{T}/ϕ^{S}	
1. 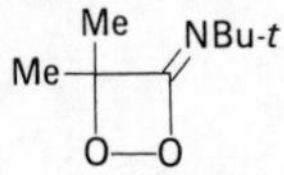 Me, Me, NBu-*t*, O—O	5.0×10^{-6} (EC)	3.4×10^{-3} (EC)	0.003	700	57
2. 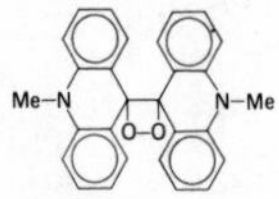 Me—N, N—Me, O—O	3.6 (DC)	4.0 (CT)	8	1	180a, b
3. 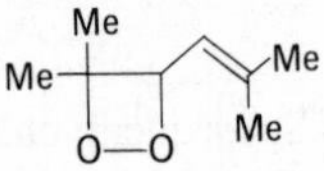 Me, Me, Me, Me, O—O	3.9×10^{-2} (EC)	0.28 (EC)	0.3		229b, 70
4. 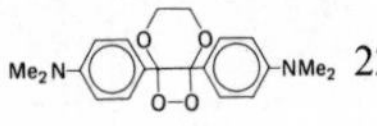 Me_2N, NMe_2, O—O	22 (DC)	*e*	22	—	31
5. 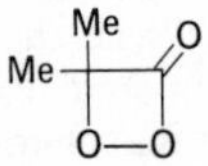 Me, Me, O, O—O	0.1(DC)	1.5 (DC, CT)	2	16	207, 257
6. 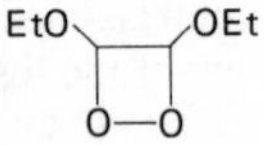 EtO, OEt, O—O	$<10^{-2}$ (EC)	20 (EC)	20	>2000	221
7. 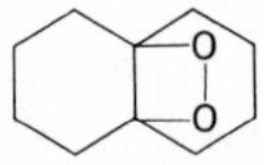	4.8×10^{-4} (EC)	1.1 (EC)	1	2000	159
8. 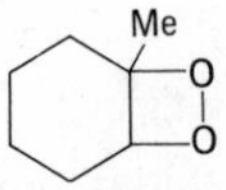 Me	*e*	*e*	81	150	156
9. 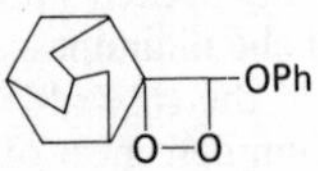 OPh, O—O	1.2 (EC)	5.0 (EC)	6	4	256
10. 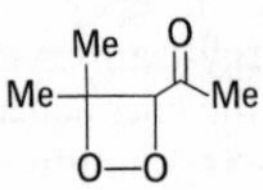 Me, Me, O, Me, O—O	1.6 (EC)	15 (EC, DC)	17	10	35

TABLE 7. *continued*

Dioxetane	Excitation parameters[a,b]				Ref.
	ϕ^{S}(%)[c]	ϕ^{T}(%)[d]	ϕ^{T+S}(%)	ϕ^{T}/ϕ^{S}	
11. 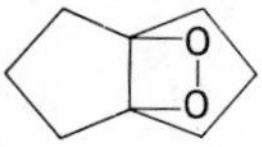	1.1 (EC)	10 (EC)	11	10	159
12. 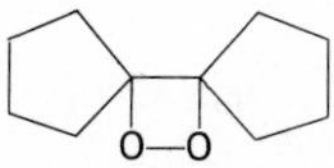	1×10^{-4} (EC)	8.7 (EC)	9	1100	199
13. 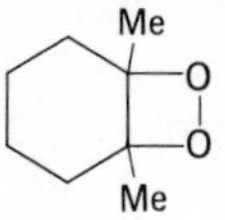	9×10^{-2} (EC)	23 (EC)	23	250	159b
14. 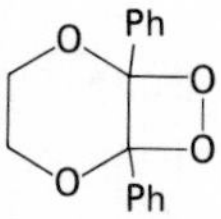	1.9×10^{-2} (EC)	6.8 (EC)	7	340	31
15. 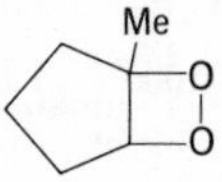	*e*	*e*	10	10	156
16. 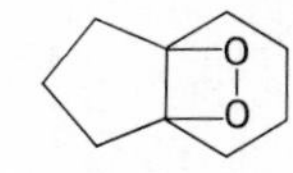	0.13 (EC)	11 (EC)	11	100	159a
17. 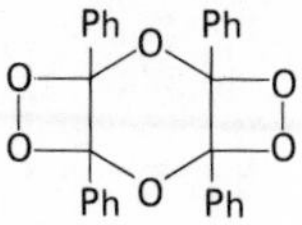	1×10^{-2} (EC)	22 (EC, CT)	22	2200	86
18. 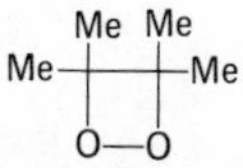	0.25 (DC, CT)	30 (EC, CT)	30	120	159b, 207, 208, 223, 79
19. 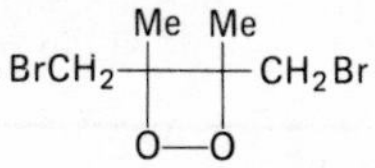	5.7×10^{-3} (EC)	7.1×10^{-3} (EC)	0.01	1	223
20. 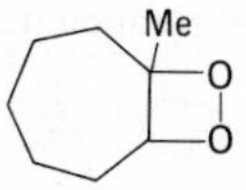	*e*	*e*	30	120	156

TABLE 7. *continued*

Dioxetane	Excitation parameters[a,b] $\phi^S(\%)^c$	$\phi^T(\%)^d$	$\phi^{T+S}(\%)$	ϕ^T/ϕ^S	Ref.
21. O—O	4×10^{-2} (EC)	30 (EC)	30	750	160
22. O, O	5.7×10^{-3} (EC)	0.37 (EC)	0.4	60	179
23. Ph, Me, O, O, O, O	1.6×10^{-3} (EC)	3.5 (EC)	4	2000	233
24. Et, Et, Et, Et, O—O	0.2 (EC)	60 (EC)	60	300	160
25. Ph, Ph, O, O, O, O	1.7×10^{-3} (EC)	1.6 (EC)	2	1000	236
26. O, O	0.19 (EC)	18 (EC)	18	90	159a
27. MeO, OMe, MeO, OMe, O—O	1.0 (EC)	10 (EC)	11	10	221
28. OPh, *t*-Bu, OPh, O—O	3.1×10^{-2} (EC)	9.5×10^{-2} (EC)	0.1	3	246
29. O—O	2.0 (CT)	15 (CT)	17	8	157

[a] The errors are usually 30–50% so that we have rounded the values up or off.
[b] Enhanced chemiluminescence (EC), direct chemiluminescence (DC), chemical titration (CT).
[c] 9,10-Diphenylanthracene was used in EC.
[d] 9,10-Dibromoanthracene was used in EC.
[e] Values were not reported.

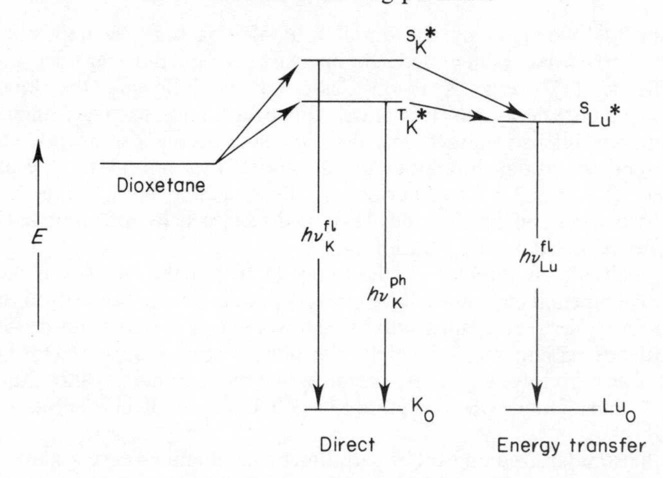

FIGURE 3. Energy diagram for direct (DC) and energy-transfer (ETC) chemiluminescence.

and/or phosphorescer) and the electronically excited luminescer Lu* emits, giving rise to the observed enhanced chemiluminescence. The two events are illustrated in terms of an energy diagram in Figure 3.

a. Direct chemiluminescence. In this method the luminescence of the chemienergized carbonyl product K* is directly detected and measured. It is essential that the electronically excited product K* be known and characterized, which is usually confirmed through photoluminescence with the authentic material, fluorescence in the case of singlet-state $^{S}K^{*}$ and phosphorescence in the case of triplet-state $^{T}K^{*}$. Since phosphorescence is usually difficult to detect in solution at ambient conditions in the presence of molecular oxygen, the direct chemiluminescence technique is essentially restricted to the determination of singlet excitation yields (ϕ^{S}). Our discussion focuses on the latter; however, it should be clear that in principle the same methodology applies to the determination of triplet excitation yields (ϕ^{T}), except that instead of the fluorescence the phosphorescence is measured. Excellent reviews on the instrumentation and calibration of light emission required to measure chemiluminescence quantum yields have appeared recently[185].

The direct chemiluminescence quantum yield (ϕ^{DC}) is given by equation (79), where ϕ^{S} is

$$\phi^{DC} = \phi^{S} \cdot \phi_{k}^{fl} \quad (79)$$

the singlet excitation quantum yield and ϕ_{K}^{fl} is the fluorescence quantum yield of the singlet excited carbonyl product $^{S}K^{*}$. The latter is directly responsible for the observed chemiluminescence. If ϕ_{K}^{fl} is known from photoluminescence work, determination of ϕ^{DC} allows us to calculate the desired ϕ^{S} parameter. Typical values are given in Table 7. Frequently ϕ_{K}^{fl} is not known and it is necessary to measure it, using routine fluorescence techniques[170,185b].

For the experimental determination of the ϕ^{DC} it is necessary to measure the light output of the direct chemiluminescent process. The experimental definition of the direct chemiluminescence quantum yield is given by equation (80), i.e. the initial rate of photon

$$\phi^{DC} = I_{0}^{DC}/k_{D}[D]_{0} \quad (80)$$

production (I_0^{DC}) divided by the initial rate of dioxetane decomposition ($k_D[D]_0$). Alternatively the total or integrated light intensity per total dioxetane decomposed can be used. The $k_D[D]_0$ term is readily assessed by following the kinetics of the chemiluminescence decay, which is usually first order. Thus, from a semilogarithmic plot of the emission intensity versus time the dioxetane decomposition rate constant k_D is obtained and the initial dioxetane concentration $[D]_0$ is known[170], especially if the dioxetane has been isolated and purified. In those cases in which the dioxetanes are too labile for isolation and purification, $[D]_0$ is determined by quantitative spectroscopic measurements or iodometric titration.

With a suitable photometer[185a] the initial or total light intensity is measured. The detailed experimental techniques have been published[170,185a]. It is critical to standardize the photomultiplier tube against suitable light standards and to calibrate for wavelength response if the emission intensities of the chemiluminescent process and the light standard occur at different wavelength. In recent years the luminol[186] and the 'scintillation cocktail'[187] have found wide acceptance for standardizing light intensities in dioxetane work[18g].

Once the standardized and calibrated direct chemiluminescence quantum yield (ϕ^{DC}) has been acquired experimentally, the singlet excitation yield (ϕ^{S}) can be calculated for the chemienergized process from equation (79). However, as already stated, this requires that the fluorescence quantum yield (ϕ_K^{fl}) be known under the same experimental conditions at which ϕ^{DC} was determined. This is not always the case and the disadvantage of the direct chemiluminescence method is that ϕ_K^{fl} may have to be determined. For very weakly chemiluminescing systems that can be a difficult task because the chemienergized emitter may not be defined.

b. Energy-transfer chemiluminescence. By far the most popular photophysical technique to count chemienergized singlets and triplets is via energy transfer to suitable luminescent acceptors (Figure 3). Usually the fluorescence of the acceptor is chemienergized; but in principle the phosphorescence can also be stimulated, provided the acceptor exhibits measurable phosphorescence under the conditions of dioxetane decomposition, i.e. in solution, at ambient temperatures and in the presence of molecular oxygen, as in the case of biacetyl[18c]. Here we shall consider the case of chemienergized fluorescence by energy transfer, although the same treatment also applies to phosphorescence.

In the case of fluorescence that is chemienergized by energy transfer, an energy acceptor is chosen which exhibits efficient fluorescence, e.g. polycyclic aromatic hydrocarbons and particularly 9,10-disubstituted anthracene derivatives[188]. Consequently, in the presence of such fluorescers (Fl) the feeble direct chemiluminescence emission intensity is significantly enhanced. Such a phenomenon is commonly referred to as enhanced chemiluminescence (EC), or energy transfer chemiluminescence (ETC)*.

(*i*) *Singlet–singlet energy transfer.* We shall distinguish between enhanced chemiluminescence chemienergized by singlet–singlet (SS) energy transfer and triplet–singlet (TS) energy transfer. The former permits us to determine singlet excitation yields (ϕ^{S}), the latter triplet excitation yields (ϕ^{T}). In the singlet–singlet energy-transfer process the fluorescer of choice is 9,10-diphenylanthracene (DPA)[18c,155c,188] since it is readily available commercially, easily purified and has a high quantum yield of fluorescence.

By means of steady-state kinetics the relationship in equation (81) for the DPA-

$$\phi_{DPA}^{EC} = \phi^{S} \cdot \phi_{ET}^{SS} \cdot \phi_{DPA}^{fl} \tag{81}$$

enhanced chemiluminescence quantum yield (ϕ_{DPA}^{EC}) is derived in terms of the singlet excitation yield (ϕ^{S}), the efficiency of singlet–singlet energy transfer (ϕ_{ET}^{SS}) and the DPA fluorescence quantum yield (ϕ_{DPA}^{fl}). The ϕ^{S} parameter can be readily assessed once the remaining terms are known. Typical values are collected in Table 7.

*Although we prefer the ETC designation, the use of EC is so engrained in the literature that we will continue to employ it here.

The DPA fluorescence quantum yield is essentially unity and relatively insensitive to temperature and solvent[168]. However, if the DPA-enhanced chemiluminescence is run under drastically different conditions, it would be essential to determine the DPA fluorescence yield under such conditions. This can be readily achieved by measuring the relative quantum yields under the two sets of conditions[185b] and making the necessary corrections.

The energy-transfer term ϕ_{ET}^{SS} is unity under conditions of infinite DPA concentration. Typically, one measures the DPA-enhanced chemiluminescence intensity (I_{DPA}^{EC}) as a function of DPA concentration and constructs a plot of $1/I_{DPA}^{EC}$ versus 1/[DPA]. The intercept of such a double reciprocal plot represents the DPA-enhanced chemiluminescence intensity at infinite DPA concentration, i.e. $I_{[DPA]_\infty}^{EC}$. The DPA-enhanced chemiluminescence quantum yield that is calculated from this emission intensity, i.e. $\phi_{[DPA]_\infty}^{EC}$, represents complete singlet–singlet energy transfer, i.e. ϕ_{ET}^{SS} is unity.

The experimental procedure for the determination of ϕ_{DPA}^{EC} is analogous to that discussed for ϕ^{DC}. The experimental definition is given by equation (82), in which all the

$$\phi_{[DPA]_\infty}^{EC} = I_{[DPA]_\infty}^{EC} / k_D\ [D]_0 \qquad (82)$$

terms have been already defined. Again the dioxetane decomposition rate constant k_D is determined by following the first-order kinetics of the DPA-enhanced chemiluminescence decay. The initial or total DPA fluorescence intensity is standardized with a suitable light standard, usually with luminol[186] or the 'scintillation cocktail'[187]. The photomultiplier tube should be corrected for wavelength response[185b].

To avoid reabsorption problems, the fluorescer concentration should not exceed 10^{-3} M. Typically the fluorescer concentration is taken between 10^{-5} and 10^{-3} M for the double reciprocal plot. Should it be necessary to work at much higher fluorescer concentration, correction for reabsorption is essential. This is readily done by measuring the fluorescer emission intensity as a function of path length[189]. From a plot of fluorescer emission intensity versus path length one extrapolates I_{fl}^{EC} at zero path length and applies the necessary corrections.

Another potential complication with fluorescer-enhanced chemiluminescence concerns electron-exchange chemiluminescence[18k]. While this is usually of little importance for simple 1,2-dioxetanes[179], it can be the dominant mechanism for the α-peroxylactones[172]. Furthermore, for readily oxidized fluorescers like rubrene such electron exchange is considerably more likely than for DPA. It is therefore essential, especially for new dioxetanes, to test for electron-exchange chemiluminescence. A simple and convenient diagnosis is to measure under identical conditions the relative enhanced chemiluminescence intensities of DPA versus rubrene, chemienergized by the dioxetane in question. Since the fluorescence quantum yields of DPA and rubrene are both essentially unity, the enhanced intensities should be approximately equal. If rubrene gives rise to a much larger enhanced intensity (at least by one magnitude), then the electron-exchange mechanism probably operates. Under such circumstances the singlet excitation yield derived from fluorescer-enhanced chemiluminescence will be erroneous and a different counting technique must be sought.

Finally, although DPA is a most favoured fluorescer for enhanced chemiluminescence, one of its inherent disadvantages is its high singlet-state energy, $E_S = 70\,\text{kcal mol}^{-1}$[190]. As should be evident from Figure 3, chemienergized carbonyl products with singlet-state energies lower than $70\,\text{kcal mol}^{-1}$ will go undetected by DPA. While this is no problem for simple aliphatic carbonyl products, since their singlet-state energies are normally in considerable excess of $70\,\text{kcal mol}^{-1}$, for aromatic carbonyl products such as fluorenone ($E_S = 63.2\,\text{kcal mol}^{-1}$) DPA is ineffective. In such cases rubrene could be used, for which

$E_S = 55\,kcal\,mol^{-1}$[191]. For excited states with E_S values below $50\,kcal\,mol^{-1}$ it would be difficult to employ the enhanced chemiluminescence technique to determine the singlet excitation yields.

(*ii*) *Triplet–singlet energy transfer*. In view of the fact that triplet excited states do not generally phosphoresce in solution, neither the direct chemiluminescence nor the enhanced chemiluminescence (via triplet–triplet energy transfer) techniques are of much help in counting chemienergized triplets. In fact, usually it is quite difficult to determine triplet excitation yields by photophysical methods.

Fortunately, molecules with heavy atoms such as 9,10-dibromoanthracene (DBA)[192] or europium tris(thenolytrifluoroacetonate)-1,10-phenanthroline[2] are capable of accepting the excitation energy of a chemically generated triplet carbonyl product ($^TK^*$) and release it in the form of fluorescence. The mechanism of this overall triplet–singlet energy transfer appears to be first spin-allowed triplet–triplet energy transfer from the first excited triplet state of $^TK^*$ to the T_2 state of the Fl (equation 83). Subsequently the second excited triplet state of the fluorescer undergoes spin-forbidden internal conversion to the first excited singlet state of the fluorescer, which is promoted via spin-orbital coupling by the heavy-atom substituent[193]. Although the mechanistic details of this energy transfer appear to be complex, for our purposes we will consider it as an overall triplet–singlet energy transfer, leading to the observed fluorescence.

$$\begin{aligned} {}^TK^* + Fl_0 &\longrightarrow K_0 + {}^{T_2}Fl^* \\ {}^{T_2}Fl^* &\longrightarrow {}^SFl^* \\ {}^SFl^* &\longrightarrow Fl_0 + h\nu \end{aligned} \tag{83}$$

The fluorescer of choice for counting chemienergized triplet states via triplet–singlet energy-transfer chemiluminescence has been DBA. Like DPA, it is readily available and easily purified; however, unlike DPA it has a relatively low fluorescence quantum yield, i.e. ϕ_{DBA}^{fl} ca. 0.10, and is temperature- and solvent-dependent[194]. For reliable triplet yields, the fluorescence quantum yields of DBA should be measured under the conditions at which the chemienergized carbonyl product $^TK^*$ is generated.

Steady state kinetics afford the expression given in equation (84) for the DBA-enhanced chemiluminescence quantum yield (ϕ_{DBA}^{EC}). The critical term is ϕ_{ET}^{TS} which designates the efficiency of triplet–singlet energy transfer from $^TK^*$ affording $^SDBA^*$, which is defined in equation (85), where k_{ET}^{TS} and k_{ET}^{TT} are respectively triplet–singlet and triplet–triplet energy-

$$\phi_{DBA}^{EC} = \phi_{ET}^{T} \cdot \phi^{TS} \cdot \phi_{DBA}^{fl} \tag{84}$$

$$\phi_{ET}^{TS} = k_{ET}^{TS} \Big/ \left(k_{ET}^{TS} + k_{ET}^{TT}\right) \tag{85}$$

transfer steps. In the latter case, DBA molecules in their first triplet excited state are energized, and do not luminesce.

As in the case of DPA-enhanced chemiluminescence, to ensure that all chemienergized carbonyl triplets $^TK^*$ are intercepted by DBA via triplet–triplet and triplet–singlet energy transfer, the DBA-enhanced chemiluminescence intensity I_{DBA}^{EC} is determined at infinite DBA concentration. Thus, the initial or total I_{DBA}^{EC} values are measured as a function of DBA concentration, and $I_{[DBA]\infty}^{DC}$ is extrapolated as the intercept of a double reciprocal plot of $1/I_{DBA}^{DC}$ versus $1/[DBA]$. Under these conditions of 100 % energy transfer, the energy-transfer parameter ϕ^{TS}_{DBA} takes values between 0.20 and 0.30. These values have been determined by a variety of techniques, including triplet–singlet energy transfer by chemienergized triplet-state cyclohexanone (from autoxidation) to DBA[192], by photoenergized triplet-state acetophenone to DBA[18g,195], and by chemienergized triplet acetone (from tetramethyl-1,2-dioxetane)[189]. However, a recent investigation has shown[196] that the triplet–singlet energy-transfer parameter ϕ^{TS}_{ET} for DBA (equation 85) is dependent on solvent and excited ketone donor. If these findings are general, the value of the DBA photoluminescence method is severely limited.

In practical terms, the enhance chemiluminescence intensity of DBA ($I^{EC}_{[DBA]_\infty}$) is only ca. 1/40 that of DPA ($I^{EC}_{[DPA]_\infty}$) because one has $\phi^{SS}_{ET} \sim 4 \times \phi^{TS}_{ET}$ and $\phi^{fl}_{DPA} \sim 10 \times \phi^{fl}_{DBA}$. However, it must again be emphasized and cautioned that equation (84) is only valid when $\phi^{T}/\phi^{S} > 10$, because this expression for the DBA-enhanced chemiluminescence quantum yield (ϕ^{EC}_{DBA}) was derived neglecting the formation of $^{T}K^*$ by intersystem crossing from chemienergized $^{S}K^*$. When $\phi^{T}/\phi^{S} < 10$, it is essential to assess the relative contribution to the DBA-enhanced chemiluminescence intensity $I^{EC}_{[DBA]_\infty}$ from triplet–singlet energy transfer via chemienergized $^{S}K^*$. This can be attempted by measuring $I^{EC}_{[DBA]_\infty}$ in the presence and absence of triplet quenchers, for example piperylene. However, for $\phi^{T}/\phi^{S} \sim 1$ the interpolated I^{EC}_{DBA} values (from a double reciprocal plot) for the piperylene quenched and unquenched DBA-enhanced chemiluminescence are indistinguishable within the experimental error, so that the determination of triplet yields by the DBA-enhanced chemiluminescence technique is problematic in such cases[246].

$$\phi^{ET}_{[DBA]_\infty} = I^{EC}_{[DBA]} / k_D [D]_0 \tag{86}$$

The experimental procedure to measure the DBA chemiluminescence yield ($\phi^{EC}_{[DBA]_\infty}$) follows that outlined for DPA, using equation (86). Again the DBA-enhanced chemiluminescence intensity must be standardized against a reliable light standard and calibrated for wavelength response. Precautions must be taken against reabsorption problems at high fluorescer concentrations. Contributions from electron-exchange chemiluminescence are usually not important in view of the relatively high oxidation potential of DBA. As already stated, the DBA fluorescence quantum yield is low and temperature- and solvent-dependent[194], and corrections should be applied for changes in reaction conditions. Even with all these shortcomings, DBA-enhanced chemiluminescence is to date still the most used technique for counting chemienergized triplets derived from 1,2-dioxetanes. Typical values are given in Table 7.

2. *Photochemical methods*

Over the last two decades a large body of valuable photochemical data has been accumulated on the chemical titration[197] of chemienergized singlet and triplet excited carbonyl products. For convenience we shall distinguish between intramolecular and intermolecular chemienergized photochemical transformations. In the intramolecular chemienergization the electronically excited carbonyl product K* undergoes directly a given photochemical change. In the intermolecular case, however, the chemienergized product K* first transfers its excitation energy to a suitable photochemically active acceptor. The electronically excited acceptor subsequently undergoes a given photochemical transformation.

a. Intramolecular transformations. The basic relationship, derived by steady-state kinetics, allows us to determine the excitation yield (equation 87). ϕ_{CHEM} represents the chemienergized photochemical quantum yield, ϕ_{PHOTO} is the photoenergized photochemical quantum yield and ϕ^* the excitation yield of the 1,2-dioxetane. The latter may represent the singlet (ϕ^S) or triplet (ϕ^T) excitation yields, depending on whether the singlet ($^SK^*$) or triplet ($^TK^*$) excited carbonyl products are photochemically active. This must be assessed previously by photoenergizing the carbonyl product K_0 via the usual photomechanistic techniques. If both the singlet and triplet excited carbonyl products are photochemically active, then ϕ^* represents the total excitation yield (ϕ^{T+S}).

$$\phi_{CHEM} = \phi^* \cdot \phi_{PHOTO} \quad (87)$$

If ϕ_{PHOTO} is not known under the conditions of the chemienergization, it is necessary to measure it by the usual photomechanistic methods. Frequently it suffices to measure a relative ϕ_{PHOTO} value between the established and new conditions and apply the necessary correction.

The chemienergized quantum yield ϕ_{CHEM} is experimentally defined in equation (88) as the concentration of photochemical product P formed per concentration of dioxetane D decomposed in a given time interval. Thus, to determine ϕ_{CHEM}, a known amount of dioxetane is completely thermally decomposed under the same conditions at which the ϕ_{PHOTO} has been determined and the amount of photochemical product P is determined by the usual spectroscopic and/or chromatographic methods. With ϕ_{CHEM} and ϕ_{PHOTO} available, the excitation yield ϕ^* is readily calculated.

$$\phi_{CHEM} = [P]/[D] \quad (88)$$

The intramolecular chemical titration is conceptually and experimentally simple and convenient; but it requires a particular dioxetane which chemienergizes the photochemically active carbonyl product K*. This is usually a formidable and challenging synthetic problem.

Representative intramolecularly chemienergized photochemical transformations include Norrish Type I cleavage[198] (equation 89) of dioxetane **199**. Dioxetane **200** gives 4-pentenal as photoproduct as the result of α cleavage of the chemienergized cyclopentanone (equation 90)[199]. Norrish Type II cleavage[200] (equation 91) of dioxetane **201** gives cyclobutanol and acetone as photoproducts. Such cleavages have also been reported for the α-peroxylactone **202**[201] and the imino-1,2-dioxetane **182**[57b], both affording acetophenone. Rather extensive use has been made of the cyclohexadienone rearrangement with dioxetane **203**[202]. The photoproduct is the rearranged enone **204**. Similarly, dioxetane **205** cleaves thermally to produce the rearranged photoproducts **206** and **207**[203]. The excitation parameters of these dioxetanes and others are summarized in Table 7.

$$PhCH_2-C(PhCH_2)(-O-O-)CH_2 \ (\mathbf{199}) \xrightarrow[{[-CH_2O]}]{\Delta} [PhCH_2-C(=O)-CH_2Ph]^* \longrightarrow [PhCH_2-\dot{C}(=O) \ \cdot CH_2Ph] \xrightarrow{[-CO]} PhCH_2-CH_2Ph \quad (89)$$

b. Intermolecular transformations. The more widely employed chemical titration technique for chemienergized carbonyl products is via intermolecular photochemical transformations. From steady-state kinetics the expression in equation (92) can be derived. As in the intramolecular case, ϕ^* may represent the singlet (ϕ^S) or triplet (ϕ^T)

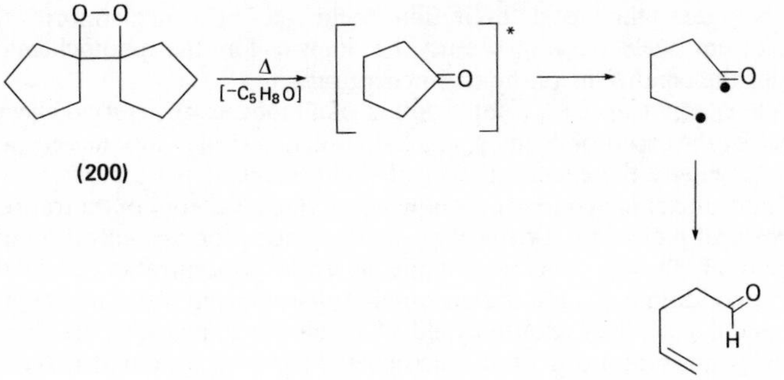

(200)

(90)

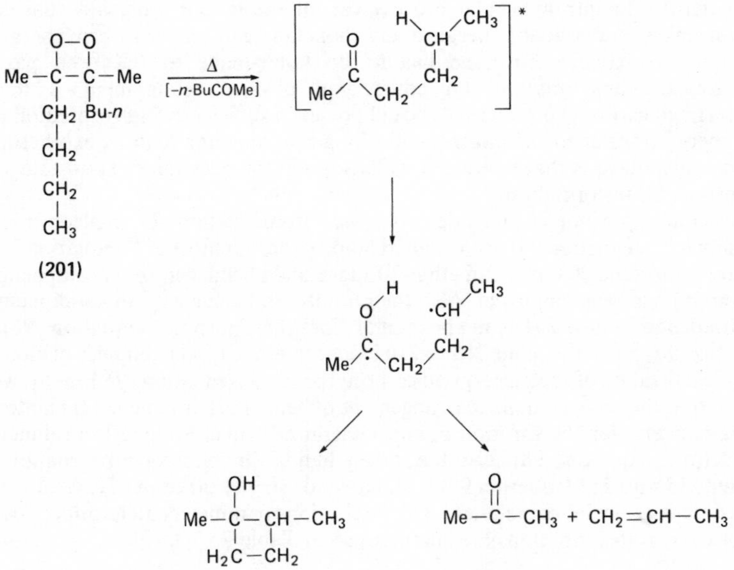

(201)

(91)

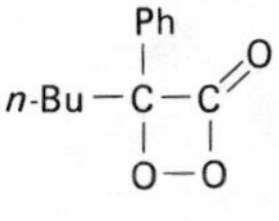

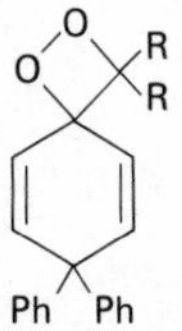

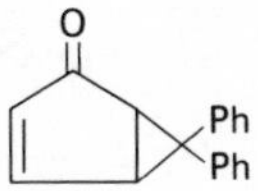

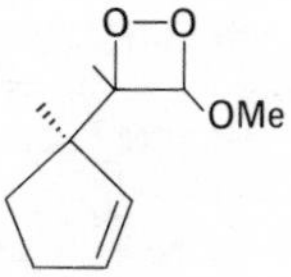

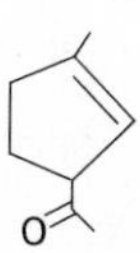

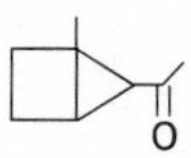

(202) (203) (204) (205) (206) (207)

$$\phi_{CHEM} = \phi^{*} \cdot \phi_{ET} \cdot \phi_{PHOTO} \quad (92)$$

excitation yields if the acceptor A undergoes photochemical transformations from its singlet state $^{S}A^{*}$ or triplet state $^{T}A^{*}$. This must be assessed previously from photomechanistic work. If both the $^{S}A^{*}$ and $^{T}A^{*}$ give rise to the same photoproduct P, then ϕ^* represents again the total excitation yield (ϕ^{T+S}). Furthermore, the photoenergized quantum yield (ϕ_{PHOTO}) must be known for the photochemical transformation under the conditions of the chemienergization.

To assure complete energy transfer, i.e. interception of all the chemienergized excited carbonyl product K* by the acceptor A, the ϕ_{CHEM} term is determined at infinite acceptor concentration. Experimentally the chemical yield of chemienergized photoproduct P is determined at constant dioxetane concentration, with varying acceptor concentrations. From a double reciprocal plot of the chemical yield versus acceptor concentration one interpolates the yield of photoproduct at infinite acceptor concentration from the intercept. Under these conditions ϕ_{ET} is unity since all K* molecules have been intercepted by A. From ϕ_{CHEM}^{∞} and ϕ_{PHOTO} the excitation yield ϕ^* is calculated, provided the chemi- and photo-energized transformations of the acceptor have been run under similar experimental conditions.

Compared to the intramolecular process, the intermolecular process is considerably more convenient and valuable because any dioxetane can serve as chemienergization source and no specific dioxetane has to be tailor-made to release a particular photochemical transformation. Of course, an obvious requirement is that the chemienergized carbonyl product K* should possess a sufficiently high excitation energy for the energy transfer to the photochemically active acceptor A to be exothermic. An additional advantage is that one has a wider margin for selection of spin-state specific photochemical transformations.

Representative examples include *cis–trans* isomerization of maleonitrile into fumaronitrile[197] (equation 93) or the dimerization of acenaphthene[204] (equation 94). The formation of oxetane **208** from 2-methyl-2-butene and chemienergized acetophenone[205] (equation 95) has been employed. Also the photorearrangement[204] of 4,4-diphenyl-2,5-cyclohexadienone into **204** can be useful for this purpose (equation 96). The photocyclization[206] of the dione **209** to **210** by tetramethyl-1,2-dioxetane is of interest for the chemical titration of chemienergized carbonyl products (equation 97). Finally, we have shown[207] that the di-π-methane rearrangement of benzonorbornadiene (**211**) into **212** is particularly useful for the titration of chemienergized triplet acetone from dimethyl-α-peroxylactone (equation 98). On the other hand, the photodenitrogenation[208] of azoalkane **213** into **212** (equation 99) is valuable for singlet titration of excited carbonyl products derived from dioxetanes. The excitation parameters determined by such intermolecular transformations are summarized in Table 7.

(93)

(94)

(95)

(208)

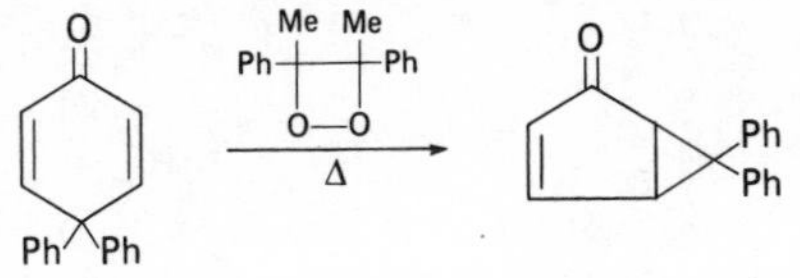

(96)

(204)

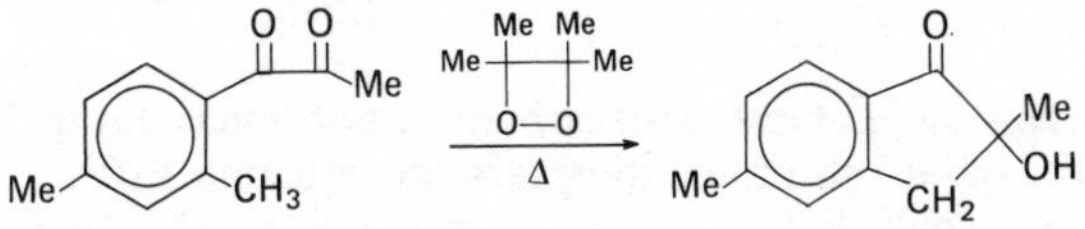

(97)

(209) (210)

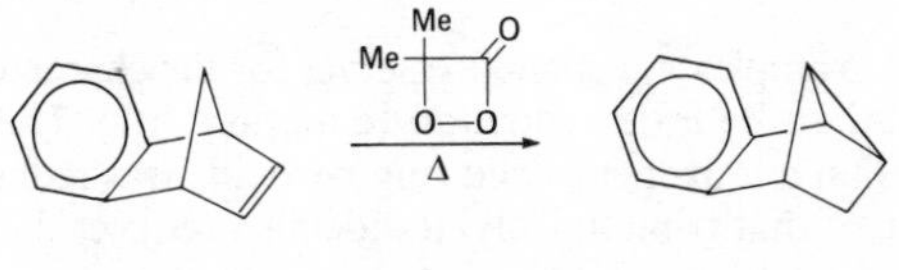

(98)

(211) (212)

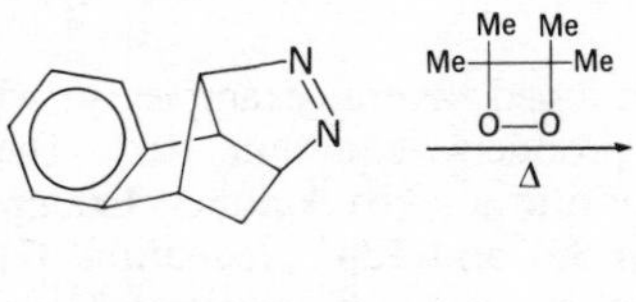

(99)

(213)

VI. CHEMICAL TRANSFORMATIONS

As we have seen in Section IV, the most common transformation of four-membered ring peroxides is their thermal and photochemical cleavage into two carbonyl fragments (equation 22). In fact, this characteristic chemical behaviour of dioxetanes and α-peroxylactones serves for identification purposes. Only a few exceptions have been claimed to this cleavage. For example, besides cleavage the dioxetane **214** (equation 100) eliminates morpholine to afford α-hydroxyketone and α-diketone products[209], in which the dioxetane carbon–carbon bond has survived while carbon–nitrogen bond cleavage has taken place. Similarly, the dioxetanes **215** and **216** give the respective α-diones on

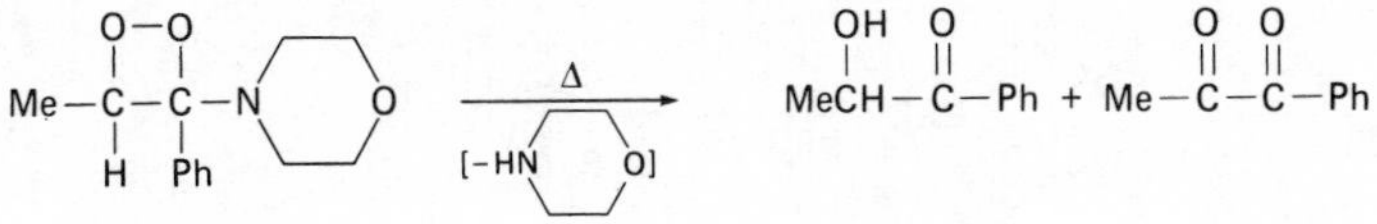

(100)

(214)

(215) (216)

thermolysis[60]. Carbon–sulphur instead of carbon–carbon bond cleavage has already been mentioned for the sulphur-substituted dioxetane **88** (equation 33)[85].

In this section we shall discuss dioxetane reactions which avoid cleavage of the dioxetane carbon–carbon bond, and which involve nucleophiles and electrophiles.

A. Nucleophiles

One of the earliest examples[27], which also serves for the characterization of dioxetanes, is the lithium aluminium hydride reduction (equation 101). The fact that a 1,2-diol is formed is indicative that a four-membered ring peroxide was reduced by the hydride. We have recently shown[210] that trimethylsilyl iodide also reduces dioxetanes efficiently into 1,2-diols.

$$\begin{matrix} R_2C-O \\ | \quad | \\ R_2C-O \end{matrix} \xrightarrow[-78^{\circ}C]{LiAlH_4} \xrightarrow{H_3O^+} \begin{matrix} R_2C-OH \\ | \\ R_2C-OH \end{matrix} \qquad (101)$$

In Section IV we have mentioned several examples in which alcohols transform dioxetanes into ring-opened products (equation 102). The labile α-alkoxyhydroperoxides **217** further transform into α-alkoxyketones. Examples are dioxetanes **102**[91] (equation 38), **110**[92] (equation 39) and **128**[99] (equation 47). Surprising, as already mentioned, is the observation that dioxetanes **67** (equation 26) and **70**[68] give the respective 1,2-diols during their photosensitized generation in methanol and Rose Bengal as sensitizer.

$$\begin{matrix} R_2C-O \\ | \quad | \\ R_2C-O \end{matrix} \xrightarrow{R'OH} \begin{matrix} R_2C-OOH \\ | \\ R_2C-OR' \end{matrix} \longrightarrow \begin{matrix} R-C{=}O \\ | \\ R_2C-OR' \end{matrix} \qquad (102)$$

(217)

The reaction of dioxetanes with trivalent phosphorus nucleophiles has been extensively investigated[27,211,212]. Under carefully controlled conditions the dioxaphospholanes **218** (equation 103) can be detected as intermediates. On warming, deoxygenation occurs with

$$\begin{matrix} R_2C-O \\ | \quad | \\ R_2C-O \end{matrix} \xrightarrow{R'_3P} \begin{matrix} R_2C-O \\ | \quad\; PR'_3 \\ R_2C-O \end{matrix} \xrightarrow{[-R'_3P=O]} \begin{matrix} R_2C \\ | \;\; \rangle O \\ R_2C \end{matrix} + \begin{matrix} R-C{=}O \\ | \\ R_2C-R \end{matrix} \qquad (103)$$

(218)

formation of epoxides and rearranged ketones. A recent example, the formation of the indolinone **219**[103], is shown in equation (104). The formation of such rearrangement products has been construed as evidence for the intermediacy of dioxetanes in singlet oxygenations, e.g. equation (28)[74,75]. On reaction with triphenylphosphine α-peroxylactones give α-lactones (equation 105), which polymerize to the corresponding polyester[57]. On the other hand, imino-1,2-dioxetanes are cleaved on deoxygenation with triphenylphosphine into acetone and isocyanide[57] (equation 106).

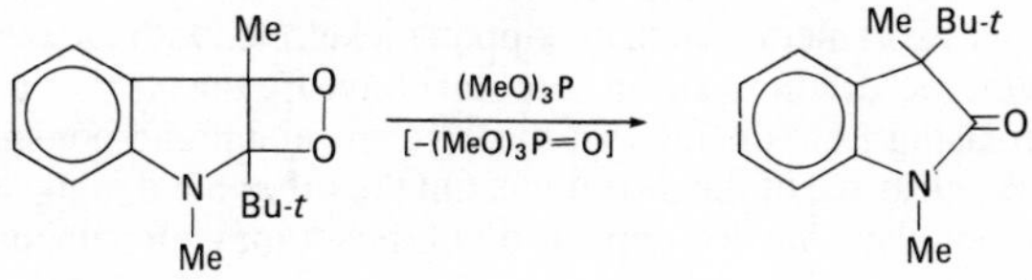

$$\mathbf{(133)} \xrightarrow[\text{[-(MeO)}_3\text{P=O]}]{(MeO)_3P} \mathbf{(219)} \qquad (104)$$

$$Me_2C(O{-}O)C{=}O \xrightarrow[\text{[-Ph}_3\text{PO]}]{Ph_3P} \left[Me_2C(O)C{=}O \right] \longrightarrow (-Me_2C-\overset{O}{\overset{\|}{C}}-O-)_n \qquad (105)$$

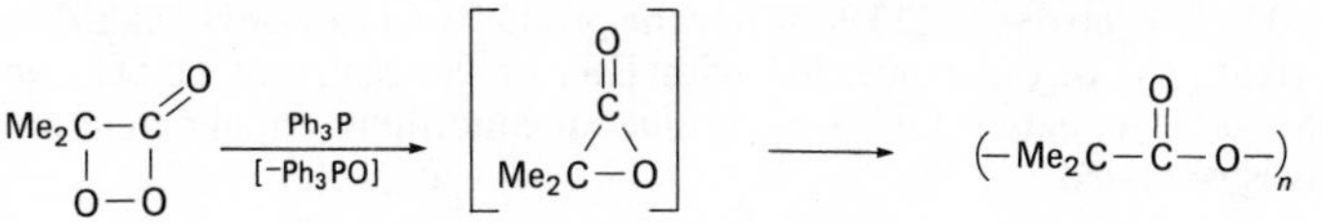

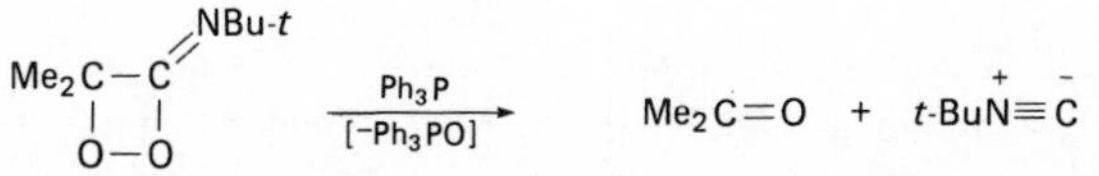

$$Me_2C(O{-}O)C{=}NBu\text{-}t \xrightarrow[\text{[-Ph}_3\text{PO]}]{Ph_3P} Me_2C{=}O + t\text{-}Bu\overset{+}{N}{\equiv}\overset{-}{C} \qquad (106)$$

Recently the reaction of triphenylarsine and triphenylantimony with dioxetanes has been investigated[213]. Stable dioxarsolane **220** and dioxastilbolane **221** were obtained,

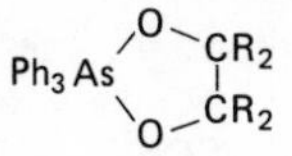

$$Ph_3As(OCR_2)_2 \quad \mathbf{(220)} \qquad Ph_3Sb(OCR_2)_2 \quad \mathbf{(221)}$$

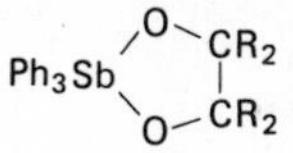

respectively. The relative reaction rates with tetramethyl-1,2-dioxetane were $Ph_3P \geq Ph_3Sb > Ph_3As$, indicating the biphilic nature of this reaction.

Sulphur nucleophiles also react quite readily with dioxetanes and can be a diagnostic test for dioxetane intermediates in singlet oxygen reactions[74,75] (equation 28). One of the earliest observations[214] of this type of dioxetane reactivity was the formation of epoxide **222** and the rearranged ketone **223** (equation 107). This reaction is analogous to that observed with phosphines (equation 103). However, when sulphoxylate is used as nucleophile, the stable sulphurane **224** is produced[215] (equation 108).

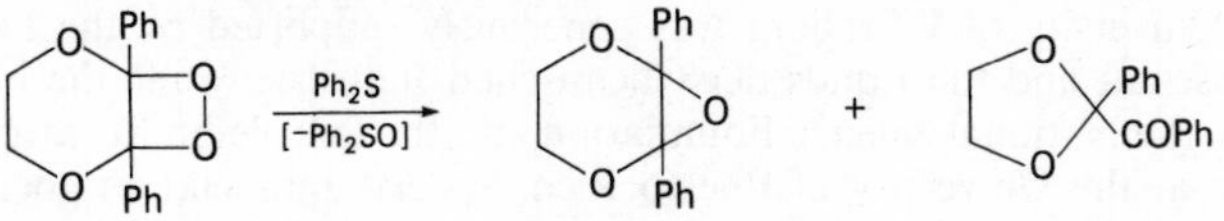

$$\xrightarrow[\text{[-Ph}_2\text{SO]}]{Ph_2S} \mathbf{(222)} + \mathbf{(223)} \qquad (107)$$

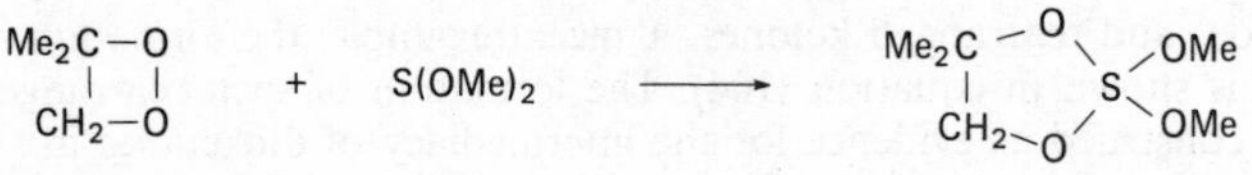

$$\begin{matrix} Me_2C-O \\ | \quad | \\ CH_2-O \end{matrix} + S(OMe)_2 \longrightarrow \text{(224)} \quad (108)$$

(224)

B. Electrophiles

The reaction of divalent metals, such as copper, nickel, etc., with dioxetanes in methanol leads to clean catalytic decomposition into carbonyl fragments[216]. The reaction rates increase with increasing Lewis acidity of the divalent metal and constitute, therefore, a typical electrophilic cleavage of the dioxetane. On the other hand, univalent rhodium and iridium complexes catalyse the decomposition of dioxetanes into carbonyl fragments via oxidative addition[217].

Simple Lewis acids such as boron trifluoride or protons, which only have a single site for coordination, promote completely different chemistry[218]. For example, the reaction of tetramethyldioxetane with BF_3 gives acetone, pinacolone and the 1,2,4,5-tetroxane **225** (equation 109). The tetroxane **225** is clearly the product of a carbonyl oxide dimerization. Similarly, treatment of dioxetane **203** with BF_3 or concentrated H_2SO_4 affords 3,4-diphenylphenol[219] (equation 110). Under aqueous conditions a rather complex spectrum of products is obtained.

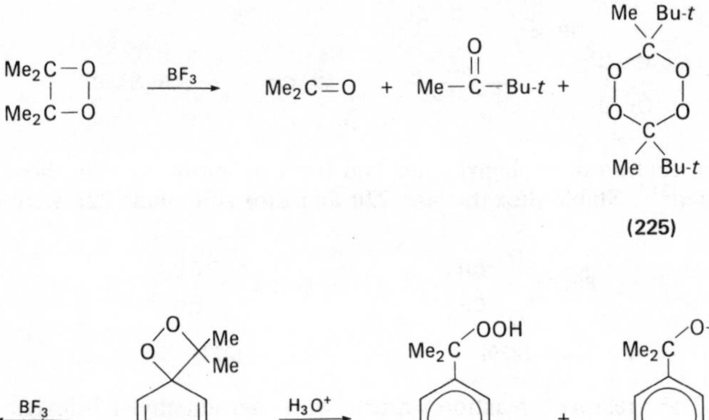

$$\begin{matrix} Me_2C-O \\ | \quad | \\ Me_2C-O \end{matrix} \xrightarrow{BF_3} Me_2C{=}O + Me-\overset{O}{\overset{\|}{C}}-Bu\text{-}t + \text{(225)} \quad (109)$$

(225)

$$\text{3,4-diphenylphenol} \xleftarrow[H_2SO_4]{BF_3} \text{(203)} \xrightarrow{H_3O^+} \text{products} \quad (110)$$

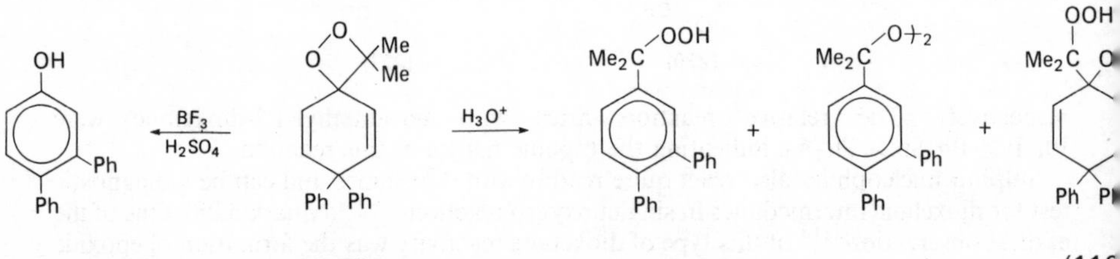

(203)

VII. ACKNOWLEDGEMENTS

The work at the University of Würzburg was generously supported by the Deutsche Forschungsgemeinschaft and the Fonds der Chemischen Industrie, while the National Institute of Health, the National Science Foundation and the Petroleum Research Fund financed our work at the University of Puerto Rico. Special appreciation goes to my students for their enthusiastic, diligent and exciting collaboration. Their names are cited in the individual references.

VIII. REFERENCES

1. M. M. Rauhut, *Acc. Chem. Res.*, **2**, 80 (1969).
2. E. H. White, P. D. Wildes, J. Wiecko, H. Doshan and C. C. Wei, *J. Amer. Chem. Soc.*, **95**, 7050 (1973).
3. H. Blitz, *Justus Liebigs Ann. Chem.*, **296**, 219 (1897).
4. K. R. Kopecky and C. Mumford, *Can. J. Chem.*, **47**, 709 (1969).
5. E. P. Kohler and R. B. Thompson, *J. Amer. Chem. Soc.*, **59**, 887 (1937).
6. (a) W. H. Richardson, V. F. Hodge, D. L. Stiggall, M. B. Yelvington and F. C. Montgomery, *J. Amer. Chem. Soc.*, **96**, 6652 (1974).
   (b) W. H. Richardson, G. Ranney and F. C. Montgomery, *J. Amer. Chem. Soc.*, **96**, 4688 (1974).
7. M. Müller, *Z. Physiol. Chem.*, **231**, 75 (1935).
8. W. Bergmann, F. Hirschmann and E. L. Skau, *J. Org. Chem.*, **4**, 29 (1939).
9. H. Hock and O. Schrader, *Angew. Chem.*, **49**, 565 (1936).
10. (a) R. Criegee, *Justus Liebigs Ann. Chem.*, **522**, 75 (1936).
    (b) R. Criegee, H. Pilz and H. Flygare, *Chem. Ber.*, **72**, 1799 (1939).
11. H. Staudinger, K. Dyckerhoff, H. W. Klever and L. Ruzicka, *Chem. Ber.*, **58**, 1079 (1925).
12. N. J. Turro, Y. Ito, M.-F. Chow, W. Adam, O. Rodriguez and F. Yany, *J. Amer. Chem. Soc.*, **99**, 5836 (1977).
13. W. Adam and J.-C. Liu, *J. Amer. Chem. Soc.*, **94**, 2894 (1972).
14. Th. A. Hopkins, H. H. Seliger, E. H. White and M. W. Cass, *J. Amer. Chem. Soc.*, **89**, 7148 (1967).
15. F. McCapra, *J. Chem. Soc., Chem. Commun.*, 155 (1968).
16. M. DeLuca and M. E. Dempsey, *Biochem. Biophys. Res. Commun.*, **40**, 117 (1970).
17. (a) O. Shimomura, F. H. Johnson and T. Goto, *Proc. Natl. Acad. Sci. (USA)*, **74**, 2799 (1977).
    (b) J. Wannlund, M. DeLuca, K. Stempel and P. D. Boyer, *Biochem. Biophys. Res. Commun.*, **81**, 987 (1978).
    (c) E. H. White, M. G. Steinmetz, J. D. Miano, P. D. Wildes and R. Morland, *J. Amer. Chem. Soc.*, **102**, 3199 (1980).
18. (a) T. Goto and Y. Kishi, *Angew. Chem. (Intern. Ed. Engl.)*, **7**, 407 (1968).
    (b) F. McCapra, *Endeavour*, **32**, 139 (1973).
    (c) N. J. Turro, P. Lechtken, N. E. Schore, G. Schuster, H.-C. Steinmetzer and A. Yekta, *Acc. Chem. Res.*, **7**, 97 (1974).
    (d) K.-D. Gundermann, *Top. Curr. Chem.*, **46**, 63 (1974).
    (e) C. Mumford, *Chem. Brit.*, **11**, 402 (1975).
    (f) W. Adam, *J. Chem. Educ.*, **52**, 138 (1975).
    (g) T. Wilson, *Intern. Rev. Sci. Phys. Chem. Ser. Two*, **9**, 265 (1976).
    (h) P. D. Bartlett, *Chem. Soc. Rev.*, **5**, 149 (1976).
    (i) W. Adam, *Advan. Heterocycl. Chem.*, **21**, 437 (1977).
    (j) K. A. Horn, J. Y. Koo, S. P. Schmidt and G. B. Schuster, *Mol. Photochem.*, **9**, 1 (1978–79).
    (k) G. B. Schuster, *Acc. Chem. Res.*, **12**, 366 (1979).
    (m) P. D. Bartlett and M. E. Landis in *Singlet Oxygen*, (Eds. H. H. Wasserman and R. W. Murray), Academic Press, New York, 1979, p. 243.
    (n) G. Cilento, *Acc. Chem. Res.*, **13**, 225 (1980).
    (o) W. Adam, *Pure Appl. Chem.*, **52**, 2591 (1980).
    (p) W. Adam, *Chemie in unserer Zeit*, **14**, 44 (1980).
    (q) T. Matsuura and I. Saito in *Photochemistry of Heterocyclic Compounds* (Ed. O. Buchardt), Wiley–Interscience, New York–London, p. 480 (1976).
    (r) A. P. Schaap and K. A. Zaklika in *Singlet Oxygen*, (Eds. H. H. Wasserman and R. W. Murray), Academic Press, New York, 1979, p. 173.
    (s) E. H. White, J. D. Miano, C. J. Watkins and E. J. Breaux, *Angew. Chem.*, **86**, 292 (1974).
    (t) J. W. Hastings and T. Wilson, *Photochem. Photobiol.*, **23**, 461 (1976).
    (u) A. A. Gorman and M. A. J. Rodgers, *Chem. Soc. Rev.*, **10**, 205 (1981).
    (v) A. A. Frimer, *Chem. Rev.*, **79**, 359 (1979).
    (w) M. V. George and V. Bhat, *Chem. Rev.*, **79**, 447 (1979).
    (x) P. J. Herring (Ed.), *Bioluminescence in Action*, Academic Press, New York, 1978.
    (y) W. W. Ward, *Photochem. Photobiol.*, **33**, 965 (1981).
19. W. Adam and G. Cilento, *Photochem. Photobiol.*, **30**, 1 (1979).

20. W. Adam and G. Cilento (Eds.), *Chemical and Biological Generation of Electronically Excited States*, Academic Press, New York, (1982).
21. H. Numan, J. H. Wieringa, H. Wynberg, J. Hess and A. Vos, *J. Chem. Soc., Chem. Commun.*, 591 (1977).
    (b) J. Hess and A. Vos, *Acta Cryst.*, **333B**, 3527 (1977).
22. (a) A. Krebs, H. Schmalstieg, O. Jarchow and K.-H. Klaska, *Tetrahedron Letters*, 3171 (1980).
    (b) H. Schmalstieg, *Dissertation*, University of Hamburg, July 1981.
23. W. Adam, E. Schmidt, K. Peters, E.-M. Peters and H. G. von Schnering, unpublished results.
24. W. Adam, L. A. Arias, A. Zahn, K. Zinner, K. Peters, E.-M. Peters and H. G. von Schnering, *Tetrahedron Letters*, **23**, 3251 (1982).
25. W. Adam, L. A. Arias and D. Scheutzow, *Tetrahedron Letters*, **23**, 2835 (1982).
26. W. Adam, A. Alzerreca, J.-C. Liu and F. Yany, *J. Amer. Chem. Soc.*, **99**, 5768 (1977).
27. K. R. Kopecky, J. E. Filby, C. Mumford, P. A. Lockwood and J. Y. Ding, *Can. J. Chem.*, **53**, 1103 (1975).
28. R. S. Brown, *Can. J. Chem.*, **53**, 3439 (1975).
29. K. R. Kopecky, P. A. Lockwood, J. E. Filby and R. W. Reid, *Can. J. Chem.*, **51**, 468 (1973).
30. (a) H. Nakamura and T. Goto, *Photochem. Photobiol.*, **30**, 27 (1979).
    (b) H. Nakamura and T. Goto, *Heterocycles*, **15**, 1119 (1981).
31. K. A. Zaklika, T. Kissel, A. L. Thayer, P. A. Burns and A. P. Schaap, *Photochem. Photobiol.*, **30**, 35 (1979).
32. D. J. Bogan, J. L. Durant, Jr., R. S. Sheinson and F. W. Williams, *Photochem. Photobiol.*, **30**, 3 (1979).
33. N. J. Turro, H.-C. Steinmetzer and A. Yekta, *J. Amer. Chem. Soc.*, **95**, 6468 (1973).
34. S. P. Schmidt and G. B. Schuster, *J. Amer. Chem. Soc.*, **102**, 306 (1980).
35. K. A. Horn and G. B. Schuster, *J. Amer. Chem. Soc.*, **100**, 6649 (1978).
36. W. Adam, *Chem. Z.*, **99**, 142 (1975).
37. L. M. Stephenson, M. J. Grdina and M. Orfanopoulos, *Acc. Chem. Res.*, **13**, 419 (1980).
38. C. Babatskiko, *Diplomarbeit*, University of Würzburg, September 1981.
39. P. D. Bartlett and A. P. Schaap, *J. Amer. Chem. Soc.*, **92**, 3223, 6055 (1970).
40. A. P. Schaap, A. L. Thayer, E. C. Blossey and D. C. Neckers, *J. Amer. Chem. Soc.*, **97**, 3741 (1975).
41. R. W. Murray in *Singlet Oxygen* (Eds. H. H. Wasserman and R. W. Murray), Academic Press, New York, 1979.
42. J. H. Wieringa, J. Strating, H. Wynberg and W. Adam, *Tetrahedron Letters*, 169 (1972).
43. P. D. Bartlett and M. S. Ho, *J. Amer. Chem. Soc.*, **96**, 627 (1974).
44. J. Keul, *Chem. Ber.*, **108**, 1198, 1207 (1975).
45. W. Adam and K. Sakanishi, *J. Amer. Chem. Soc.*, **100**, 3935 (1978).
46. A. Zahn, *Diplomarbeit*, University of Würzburg, September 1981.
47. W. Adam and L. A. Arias, *Chem. Ber.*, **115**, 2592 (1982).
48. K. A. Zaklika, P. A. Burns and A. P. Schaap, *J. Amer. Chem. Soc.*, **100**, 318 (1978).
49. E. L. Clennan, W. Simmons and C. W. Almgren, *J. Amer. Chem. Soc.*, **103**, 2098 (1981).
50. S. F. Nelson and R. Akaba, *J. Amer. Chem. Soc.*, **103**, 2096 (1981).
51. G. Büchi and H. Wüest, *J. Amer. Chem. Soc.*, **100**, 294 (1978).
52. G. M. Rubottom and M. J. Lopez Nieves, *Tetrahedron Letters*, 2423 (1972).
53. K. Yamaguchi, S. Yabushita, T. Fueno and K. N. Houk, *J. Amer. Chem. Soc.*, **103**, 5043 (1981).
54. W. Adam, J. del Fierro, F. Quiroz and F. Yany, *J. Amer. Chem. Soc.*, **102**, 2127 (1980).
55. R. Wheland and P. D. Bartlett, *J. Amer. Chem. Soc.*, **92**, 6057 (1970).
56. (a) W. Adam, O. Cueto and V. Ehrig, *J. Org. Chem.*, **41**, 370 (1976).
    (b) W. Adam and O. Cueto, *J. Org. Chem.*, **42**, 38 (1977).
57. (a) Y. Ito, T. Matsuura and H. Kondo, *J. Amer. Chem. Soc.*, **101**, 7105 (1979).
    (b) Y. Ito, H. Yokoya, K. Kyono, S. Yamamura, Y. Yamada and T. Matsuura, *J. Chem. Soc., Chem. Commun.*, 898 (1980).
58. W. Adam, O. De Lucchi, H. Quast, R. Recktenwald and F. Yany, *Angew. Chem.*, **91**, 855 (1979).
59. C. S. Foote, A. A. Dzakpasu and J. W.-P. Lin, *Tetrahedron Letters*, 1247 (1975).
60. H. H. Wasserman and S. Terao, *Tetrahedron Letters*, 1735 (1975).
61. E. H. White, N. Suzuki and W. H. Hendrickson, *Chem. Letters*, 1491 (1979).
62. (a) M. Akutagawa, H. Aoyama, Y. Omote and H. Yamamoto, *J. Chem. Soc., Chem. Commun.*, 180 (1976).

(b) H. Yamamoto, H. Aoyama, Y. Omote, M. Akutagawa, A. Takenaka and Y. Sasada, *J. Chem. Soc., Chem. Commun.*, 63 (1977).
63. T. Goto and H. Nakamura, *Tetrahedron Letters*, 4627 (1976).
64. (a) F. McCapra and A. Burford, *J. Chem. Soc., Chem. Commun.*, 874 (1977); 607 (1976).
(b) F. McCapra, Y. C. Chang and A. Burford, *J. Chem. Soc., Chem. Commun.*, 608 (1976).
65. K. Eskins, *Photochem. Photobiol.*, **29**, 609 (1979).
66. W. Ando and S. Kohmoto, *J. Chem. Soc., Chem. Commun.*, 120 (1978).
67. A. W. Nowicki and A. B. Turner, *J. Chem. Res., Synop.*, 110 (1981).
68. (a) H. Takeshita and T. Hatsui, *J. Org. Chem.*, **43**, 3080 (1978).
(b) T. Hatsui and H. Takeshita, *Bull. Chem. Soc. (Japan)*, **53**, 2655 (1980).
69. C. W. Jefford, A. F. Boschung and C. G. Rimbault, *Helv. Chim. Acta*, **59**, 2542 (1976).
70. Y. Inoue and N. J. Turro, *Tetrahedron Letters*, 4327 (1980).
71. (a) H. Leininger and M. Christl, *Angew. Chem.*, **92**, 466 (1980).
(b) Private communication.
72. G. W. Griffin, I. R. Politzer, K. Ishikawa, N. J. Turro and M.-F. Chow, *Tetrahedron Letters*, 1287 (1977).
73. W. Adam and A. Rodriguez, *Tetrahedron Letters*, **22**, 3505 (1981).
74. (a) G. Rousseau, P. Le Perchec and J. M. Conia, *Tetrahedron*, **32**, 2533 (1976).
(b) G. Rousseau, P. Le Perchec and J. M. Conia, *Tetrahedron*, **34**, 3475 (1978).
75. A. A. Frimer, T. Farkash and M. Sprecher, *J. Org. Chem.*, **44**, 989 (1979).
76. A. A. Frimer and D. Roth, *J. Org. Chem.*, **44**, 3882 (1979).
77. C. W. Jefford and C. G. Rimbault, *Tetrahedron Letters*, 2479 (1976).
78. M. Sakuragi and H. Sakuragi, *Chem. Letters*, 1017 (1980).
79. W. Adam and W. Baader, unpublished results.
80. W. B. Hammond, *Tetrahedron Letters*, 2309 (1979).
81. P. D. Bartlett and A. A. Frimer, *Heterocycles*, **11**, 419 (1978).
82. G. Rousseau, A. Lechevallier, F. Huet and J. M. Conia, *Tetrahedron Letters*, 3287 (1978).
83. H.-D. Scharf, H. Plum, J. Fleischhauer and W. Schleker, *Chem. Ber.*, **112**, 862 (1979).
84. M. V. George, C. V. Kumar and J. C. Scaiano, *J. Phys. Chem.*, **83**, 2452 (1979).
85. W. Adam and J.-C. Liu, *J. Chem. Soc., Chem. Commun.*, 73 (1972).
86. W. Adam, C.-C. Cheng, O. Cueto, I. Erden and K. Zinner, *J. Amer. Chem. Soc.*, **101**, 4735 (1979).
87. R. J. Molyneux, H. Aft and P. Loveland, *Chem. Ind. (London)*, 68 (1976).
88. O. Seshimoto, T. Tezuka and T. Mukai, *Chem. Letters*, 793 (1976).
89. O. Tsuge, K. Oe and Y. Ueyama, *Chem. Letters*, 425 (1976).
90. N. H. Martin, S. L. Champion and P. B. Belt, *Tetrahedron Letters*, **21**, 2613 (1980).
91. F. Abello, J. Boix, J. Gomez, J. Morell and J.-J. Bonet, *Helv. Chim. Acta*, **58**, 2549 (1975).
92. E. Fanghänel, E. D. Nascimento, R. Radeglia, G. Lutze and H. Bärhold, *Z. Chem.*, **17**, 135 (1977).
93. G. Hugel, J.-Y. Laronze, J. Laronze and J. Levy, *Heterocycles*, **16**, 581 (1981).
94. K. Orito, R. H. Manske and R. Rodrigo, *J. Amer. Chem. Soc.*, **96**, 1944 (1974).
95. V. Bhat and M. V. George, *J. Org. Chem.*, **44**, 3288 (1979).
96. H. Groß and H. Dürr, *Tetrahedron Letters*, **22**, 4679 (1981).
97. M. Matsumoto and K. Kondo, *J. Amer. Chem. Soc.*, **99**, 2393 (1977).
98. H. H. Wasserman and J. L. Ives, *Tetrahedron*, **37**, 1825 (1981).
99. D. A. Lightner and L. K. Low, *J. Heterocycl. Chem.*, **12**, 793 (1975).
100. D. A. Lightner, R. D. Norris, D. I. Kirk and R. M. Key, *Experientia*, **30**, 587 (1974).
101. D. A. Lightner and C.-S. Pak, *J. Org. Chem.*, **40**, 2724 (1975).
102. K. M. Smith, S. B. Brown, R. F. Troxler and J.-J. Lai, *Tetrahedron Letters*, **21**, 2763 (1980).
103. I. Saito, S. Matsugo and T. Matsuura, *J. Amer. Chem. Soc.*, **101**, 4757 (1979).
104. I. Saito, S. Matsugo and T. Matsuura, *J. Amer. Chem. Soc.*, **101**, 7332 (1979).
105. (a) I. Saito, M. Imuta, S. Matsugo and T. Matsuura, *J. Amer. Chem. Soc.*, **97**, 7191 (1975).
(b) I. Saito, M. Imuta, Y. Takahashi, S. Matsugo and T. Matsuura, *J. Amer. Chem. Soc.*, **99**, 2005 (1977).
106. (a) M. Nakagawa, K. Yoshikawa and T. Hino, *J. Amer. Chem. Soc.*, **97**, 6496 (1975).
(b) I. Saito, Y. Takahashi, M. Imuta, S. Matsugo, H. Kaguchi and T. Matsuura, *Heterocycles*, **5**, 53 (1976).
(c) I. Saito, T. Matsuura, M. Nakagawa and T. Hino, *Acc. Chem. Res.*, **10**, 346 (1977).
(d) M. Nakagawa, H. Watanabe, S. Kodato, H. Okajima, T. Hino, J. L. Flippen and B. Witkop,

Proc. Natl. Acad. Sci., USA, **74**, 4730 (1977).
(e) M. Nakagawa, S. Kato, S. Kataoka and T. Hino, *J. Amer. Chem. Soc.*, **101**, 3136 (1979).
107. O. Hayaishi (Ed.), *Molecular Mechanisms of Oxygen Activation*, Academic Press, New York, 1974.
108. A. Mori, S. Ohta and H. Takeshita, *Bull. Chem. Soc. Japan*, **47**, 2437 (1974).
109. M. L. Graziano, M. R. Iesce and R. Scarpati, *J. Chem. Soc., Chem. Commun.*, 7 (1979).
110. H. H. Wasserman, M. S. Wolff, K. Stiller, I. Saito and J. E. Pickett, *Tetrahedron* (*Suppl. 9*), **37**, 191 (1981).
111. G. Vernin, S. Treppendahl and J. Metzger, *Helv. Chim. Acta*, **60**, 284 (1977).
112. E. W. Meijer and H. Wynberg, *Tetrahedron Letters*, 785 (1981).
113. N. J. Turro, V. Ramamurthy, K.-C. Liu, A. Krebs and R. Kemper, *J. Amer. Chem. Soc.*, **98**, 6758 (1976).
114. N. J. Turro, M.-F. Chow and Y. Ito, *J. Amer. Chem. Soc.*, **100**, 5580 (1978).
115. C. L. Pedersen and C. Lohse, *Tetrahedron Letters*, 3141 (1978).
116. F. S. Guziec, Jr. and E. T. Tewes, *J. Heterocycl. Chem.*, **17**, 1807 (1980).
117. R. J. F. M. van Arendonk and W. H. Laarhoven, *Recueil*, **100**, 263 (1981); *Tetrahedron Letters*, 2629 (1977).
118. (a) J. Eriksen and C. S. Foote, *J. Amer. Chem. Soc.*, **102**, 6083 (1980).
(b) J. Eriksen, C. S. Foote and T. L. Parker, *J. Amer. Chem. Soc.*, **99**, 6455 (1977).
119. A. P. Schaap, K. A. Zaklika, B. Kaskar and L. W.-M. Fung, *J. Amer. Chem. Soc.*, **102**, 389 (1980).
120. H. H. Wasserman, B. H. Lipshutz, A. W. Tremper and J. S. Wu, *J. Org. Chem.*, **46**, 2991 (1981).
121. Y. Sawaki and Y. Ogata, *J. Org. Chem.*, **42**, 40 (1977).
122. (a) Y. Sawaki and Y. Ogata, *J. Amer. Chem. Soc.*, **99**, 6313 (1977).
(b) A. Donetti, O. Boniardi and A. Ezhaya, *Synthesis*, 1009 (1980).
123. I. Kamiya and T. Sugimoto, *Photochem. Photobiol.*, **30**, 49 (1979).
124. C. W. Jefford, W. Knöpfel and P. A. Cadby, *J. Amer. Chem. Soc.*, **100**, 6432 (1978).
125. Y. Sawaki and C. S. Foote, *J. Amer. Chem. Soc.*, **101**, 6292 (1979).
126. G. J. Lofthouse, H. Suschitzky, B. J. Wakefield and R. A. Whittaker, *J. Chem. Soc., Perkin Trans. 1*, 1634 (1979).
127. M. Nakagawa, T. Maruyama, K. Hirakoso and T. Hino, *Tetrahedron Letters*, **21**, 4839 (1980).
128. S. Muto and T. C. Bruice, *J. Amer. Chem. Soc.*, **102**, 7379 (1980).
129. F. McCapra and P. V. Long, *Tetrahedron Letters*, **22**, 3009 (1981).
130. (a) S. Muto and T. C. Bruice, *J. Amer. Chem. Soc.*, **102**, 4472 (1980).
(b) T. C. Bruice, *Acc. Chem. Res.*, **13**, 256 (1980).
131. M. Novak and T. C. Bruice, *J. Chem. Soc., Chem. Commun.*, 372 (1980).
132. R. Maskiewicz, D. Sogah and T. C. Bruice, *J. Amer. Chem. Soc.*, **101**, 5347, 5355 (1979).
133. R. Addink and W. Berends, *Tetrahedron*, **37**, 833 (1981).
134. A. Nishinaga, T. Shimizu and T. Matsuura, *Tetrahedron Letters*, **21**, 4097 (1980).
135. (a) A. A. Frimer and I. Rosenthal, *Photochem. Photobiol.*, **28**, 711 (1978).
(b) I. Rosenthal and A. Frimer, *Tetrahedron Letters*, 2805 (1976).
136. A. A. Frimer, I. Rosenthal and S. Hoz, *Tetrahedron Letters*, 4631 (1977).
137. M. Tezuka, H. Hamada, Y. Ohkatsu and T. Osa, *Denki Kagaku, Kogyo Butsumikagaku*, **44**, 17 (1976).
138. E. J. Nanni, Jr., C. T. Angelis, J. Dickson and D. T. Sawyer, *J. Amer. Chem. Soc.*, **103**, 4268 (1981).
139. M. M. A. El-Sukkary and G. Speier, *J. Chem. Soc., Chem. Commun.*, 745 (1981).
140. Y. Ito and T. Matsuura, *J. Chem. Soc. Perkin Trans. 1*, 1871 (1981).
141. N. Suzuki, *Angew. Chem.*, **91**, 852 (1979).
142. (a) C. L. Perrin, *J. Amer. Chem. Soc.*, **97**, 4419 (1975).
(b) E. Lissi, *J. Amer. Chem. Soc.*, **98**, 3386 (1976).
(c) E. B. Wilson, *J. Amer. Chem. Soc.*, **98**, 3387 (1976).
143. (a) H. E. O'Neal and W. H. Richardson, *J. Amer. Chem. Soc.*, **92**, 6553 (1970), and corrections *J. Amer. Chem. Soc.*, **93**, 1828 (1971).
(b) W. H. Richardson and H. E. O'Neal, *J. Amer. Chem. Soc.*, **94**, 8665 (1972).
144. P. Lechtken and G. Höhne, *Angew. Chem.*, **85**, 822 (1973).
145. (a) W. H. Richardson, F. C. Montgomery, M. B. Yelvington and H. E. O'Neal, *J. Amer. Chem. Soc.*, **96**, 7525 (1974).
(b) W. H. Richardson, M. B. Yelvington and H. E. O'Neal, *J. Amer. Chem. Soc.*, **94**, 1619 (1972).

(c) W. H. Richardson, J. H. Anderegg, M. E. Price, W. A. Tappen and H. E. O'Neal, *J. Org. Chem.*, **43**, 2236 (1978).
(d) W. H. Richardson, J. H. Anderegg, M. E. Price and R. Crawford, *J. Org. Chem.*, **43**, 4045 (1978).
146. J.-Y. Koo and G. B. Schuster, *J. Amer. Chem. Soc.*, **99**, 5403 (1977).
147. E. K. Thornton and E. R. Thornton in *Isotope Effects in Chemical Reactions* (Eds. C. J. Collins and M. S. Bouman), Van Nostrand–Reinhold, New York, 1970, p. 213.
148. W. Adam and F. Yany, *Photochem. Photobiol.*, **31**, 267 (1980).
149. N. J. Turro and P. Lechtken, *Pure Appl. Chem.*, **33**, 363 (1973).
150. (a) T. Wilson, M. E. Landis, A. L. Baumstark and P. D. Bartlett, *J. Amer. Chem. Soc.*, **95**, 4765 (1973).
(b) W. H. Richardson, F. C. Montgomery, P. Slusser and M. B. Yelvington, *J. Amer. Chem. Soc.*, **97**, 2819 (1975).
151. (a) I. Simo and J. Stauff, *Chem. Phys. Letters*, **34**, 326 (1975).
(b) C. Neidl and J. Stauff, *Z. Naturforsch.*, **33B**, 763 (1978).
152. (a) Y. Haas and G. Yahav, *J. Amer. Chem. Soc.*, **100**, 4885 (1978).
(b) Y. Haas and G. Yahav, *Chem. Phys. Letters*, **48**, 63 (1977).
(c) W. E. Farneth, G. Flynn, R. Slater and N. J. Turro, *J. Amer. Chem. Soc.*, **98**, 7877 (1976).
153. (a) K. K. Smith, J.-Y. Koo, G. B. Schuster and K. J. Kaufmann, *J. Phys. Chem.*, **82**, 2291 (1978).
(b) D. C. Doetschman, J. L. Fish, P. Lechtken and D. Negus, *Chem. Phys.* **51**, 89 (1980).
154. P. Lechtken and H.-C. Steinmetzer, *Chem. Ber.*, **108**, 3159 (1975).
155. (a) N. J. Turro and P. Lechtken, *J. Amer. Chem. Soc.*, **95**, 264 (1973).
(b) H.-C. Steinmetzer, A. Yekta and N. J. Turro, *J. Amer. Chem. Soc.*, **96**, 282 (1974).
(c) H.-C. Steinmetzer, P. Lechtken and N. J. Turro, *Justus Liebigs Ann. Chem.*, 1984 (1973).
156. P. Lechtken, G. Reissenweber and P. Grubmüller, *Tetrahedron Letters*, 2881 (1977).
157. G. B. Schuster, N. J. Turro, H.-C. Steinmetzer, A. P. Schaap, G. Faler, W. Adam and J.-C. Liu, *J. Amer. Chem. Soc.*, **97**, 7110 (1975).
158. (a) A. L. Baumstark and C. E. Wilson, *Tetrahedron Letters*, **22**, 4363 (1981).
(b) A. L. Baumstark and C. E. Wilson, *Tetrahedron Letters*, 2569 (1979).
159. (a) K. R. Kopecky, P. A. Lockwood, R. R. Gomez and J.-Y. Ding, *Can. J. Chem.*, **59**, 851 (1981).
(b) K. R. Kopecky and J. E. Filby, *Can. J. Chem.*, **57**, 283 (1979).
160. E. J. H. Bechara and T. Wilson, *J. Org. Chem.*, **45**, 5261 (1980).
161. R. Schmidt, H.-C. Steinmetzer, H.-D. Brauer and H. Kelm, *J. Amer. Chem. Soc.*, **98**, 8181 (1976).
162. D. R. Kearns, *J. Amer. Chem. Soc.*, **91**, 6554 (1969).
163. (a) K. Yamaguchi, T. Fueno and H. Fukutome, *Chem. Phys. Letters*, **22**, 461 (1973).
(b) G. Barnett, *Can. J. Chem.*, **52**, 3837 (1974).
(c) T. Aoyama, H. Yamakawa, K. Akiba and N. Inamoto, *Chem. Phys. Letters*, **42**, 347 (1976).
(d) C. W. Eaker and J. Hinze, *Theoret. Chim. Acta*, **40**, 113 (1975).
(e) E. A. Halevi and C. Trindle, *Israel J. Chem.*, **16**, 283 (1977).
(f) E. A. Halevi, *Intern. J. Quantum Chem.*, **12**, (Suppl. 1) 289 (1978).
(g) P. Lechtken, *Chem. Ber.*, **111**, 1413 (1978).
(h) E. M. Evleth and G. Feler, *Chem. Phys. Letters*, **22**, 499 (1973).
(i) D. R. Roberts, *J. Chem. Soc., Chem. Commun.*, 683 (1974).
164. (a) M. J. S. Dewar and S. Kirschner, *J. Amer. Chem. Soc.*, **96**, 7578 (1974).
(b) M. J. S. Dewar, S. Kirschner and H. W. Kollmar, *J. Amer. Chem. Soc.*, **96**, 7579 (1974).
165. N. J. Turro and A. Devaquet, *J. Amer. Chem. Soc.*, **97**, 3859 (1975).
166. L. B. Harding and W. A. Goddard, III, *J. Amer. Chem. Soc.*, **99**, 4520 (1977).
167. S. P. Schmidt, M. A. Vincent, C. E. Dykstra and G. B. Schuster, *J. Amer. Chem. Soc.*, **103**, 1292 (1981).
168. T. Wilson and A. P. Schaap, *J. Amer. Chem. Soc.*, **93**, 4126 (1971).
169. D. C.-S. Lee and T. Wilson in *Chemiluminescence and Bioluminescence* (Eds. M. J. Cormier, D. M. Hercules and J. Lee), Plenum Press, New York, 1973, p. 265.
170. W. Adam, G. A. Simpson and F. Yany, *J. Phys. Chem.*, **78**, 2559 (1974).
171. P. Lechtken and N. J. Turro, *Mol. Photochem.*, **6**, 95 (1974).
172. (a) W. Adam, O. Cueto and F. Yany, *J. Amer. Chem. Soc.*, **100**, 2587 (1978); W. Adam and O. Cueto, *J. Amer. Chem. Soc.*, **101**, 6511 (1979).
(b) S. P. Schmidt and G. B. Schuster, *J. Amer. Chem. Soc.*, **102**, 306, 7100 (1980).

173. A. Weller and K. A. Zachariasse in *Chemiluminescence and Bioluminescence* (Eds. M. Cormier, D. M. Hercules and J. Lee), Plenum Press, New York, 1973, p. 169.
174. G. B. Schuster, *J. Amer. Chem. Soc.*, **101**, 5851 (1979).
175. R. A. Marcus, *J. Chem. Phys.*, **24**, 966 (1956).
176. C. Walling, *J. Amer. Chem. Soc.*, **102**, 6854 (1980).
177. F. Scandola, V. Balzani and G. B. Schuster, *J. Amer. Chem. Soc.*, **103**, 2519 (1981).
178. T. Wilson, *Photochem. Photobiol.*, **30**, 177 (1979).
179. W. Adam, K. Zinner, A. Krebs and H. Schmalstieg, *Tetrahedron Letters*, **22**, 4567 (1981).
180. (a) C. Lee and L. A. Singer, *J. Amer. Chem. Soc.*, **102**, 3823 (1980).
(b) K. W. Lee and L. A. Singer, *J. Org. Chem.*, **41**, 2685 (1976).
(c) F. McCapra, I. Beheshti, A. Burford, R. A. Hann and K. A. Zaklika, *J. Chem. Soc., Chem. Commun.*, 944 (1977).
181. (a) F. McCapra, *J. Chem. Soc., Chem. Commun.*, 946 (1977).
(b) J.-Y. Koo, S. P. Schmidt and G. B. Schuster, *Procl. Natl. Acad. Sci. (USA)*, **75**, 30 (1978).
182. H. H. Seliger and W. D. McElroy, *Arch. Biochem. Biophys.*, **88**, 136 (1960).
183. (a) E. H. White, H. Wörther, H. H. Seliger and W. D. McElroy, *J. Amer. Chem. Soc.*, **88**, 2015 (1966).
(b) E. H. White, E. Rapaport, H. H. Seliger and T. A. Hopkins, *Bioorg. Chem.*, **1**, 92 (1971).
184. J. Jung, C.-A. Chin and P.-S. Song, *J. Amer. Chem. Soc.*, **98**, 3949 (1976).
185. (a) J. E. Anderson, G. J. Faini and J. E. Wampler in *Methods in Enzymology*, Vol. LVII, Academic Press, New York, 1978, pp. 529–549.
(b) H. H. Seliger in *Methods in Enzymology*, Vol. LVII, Academic Press, New York, 1978, pp. 560–600.
186. J. Lee and H. H. Seliger, *Photochem. Photobiol.*, **4**, 1015 (1965).
187. J. W. Hastings and G. Weber, *J. Opt. Soc. Amer.*, **53**, 1410 (1963).
188. N. J. Turro, P. Lechtken, G. Schuster, J. Orell, H.-C. Steinmetzer and W. Adam, *J. Amer. Chem. Soc.*, **96**, 1627 (1974).
189. (a) W. Adam, E. M. Cancio and O. Rodriguez, *Photochem. Photobiol.*, **27**, 617 (1978).
(b) W. Adam, O. Rodriguez and K. Zinner, *J. Org. Chem.*, **43**, 4495 (1978).
190. P. S. Engel and B. M. Monroe, *Advan. Photochem.*, **8**, 245 (1971).
191. T. D. Santa Cruz, D. L. Akins and R. L. Birke, *J. Amer. Chem. Soc.*, **98**, 1677 (1976).
192. V. A. Belyakov and R. F. Vasil'ev, *Photochem. Photobiol.*, **11**, 179 (1970).
193. (a) M. A. El-Sayed and M. L. Bhaumik, *J. Chem. Phys.*, **39**, 2391 (1963).
(b) W. Adam, G. Cilento and K. Zinner, *Photochem. Photobiol.*, **32**, 87 (1980).
(c) R. Schmidt, H. Kelm and H.-D. Brauer, *J. Photochem.*, **14**, 261 (1980).
194. K.-C. Wu and W. R. Ware, *J. Amer. Chem. Soc.*, **101**, 5906 (1979).
195. V. M. Berenfel'd, E. V. Chumaevskii, M. P. Grinev, Yu. I. Kuryatnikow, E. T. Artm'ev and R. V. Dzhagatspanyan, *Bull. Acad. Sci. USSR Ser. Phys.*, **3**, 587 (1970).
196. (a) T. Wilson and A. M. Halpern, *J. Amer. Chem. Soc.*, **102**, 7272, 7279 (1980).
(b) E. J. H. Bechara, private communication.
197. N. J. Turro and P. Lechtken, *J. Amer. Chem. Soc.*, **94**, 2886 (1972).
198. W. H. Richardson, J. H. Burns, M. E. Price, R. Crawford, M. Foster, P. Slusser and J. H. Anderegg, *J. Amer. Chem. Soc.*, **100**, 7596 (1978).
199. K. R. Kopecky and J. A. Lopez Sastre, *Can. J. Chem.*, **58**, 2089 (1980).
200. T. R. Darling and C. S. Foote, *J. Amer. Chem. Soc.*, **96**, 1625 (1974).
201. N. J. Turro and M.-F. Chow, *J. Amer. Chem. Soc.*, **102**, 5058 (1980).
202. H. E. Zimmerman, G. E. Keck and J. L. Pflederer, *J. Amer. Chem. Soc.*, **98**, 5574 (1976).
203. M. J. Mirbach, A. Henne and K. Schaffner, *J. Amer. Chem. Soc.*, **100**, 7127 (1978).
204. E. H. White, J. Wiecko and C. C. Wei, *J. Amer. Chem. Soc.*, **92**, 2167 (1970).
205. W. H. Richardson, M. B. Lovett, M. E. Price and J. H. Anderegg, *J. Amer. Chem. Soc.*, **101**, 4683 (1979).
206. K. R. Kopecky and J. E. Filby, private communication.
207. W. Adam, C.-C. Cheng, O. Cueto, K. Sakanishi and K. Zinner, *J. Amer. Chem. Soc.*, **101**, 1324 (1979).
208. W. Adam and K. Hannemann, *J. Amer. Chem. Soc.*, (in press).
209. W. Ando, T. Saiki and T. Migita, *J. Amer. Chem. Soc.*, **97**, 5028 (1975).
210. W. Adam and S. Chandrasekhar, unpublished results.

211. (a) P. D. Bartlett, A. L. Baumstark and M. E. Landis, *J. Amer. Chem. Soc.*, **95**, 6486 (1973).
(b) P. D. Bartlett, A. L. Baumstark, M. E. Landis and C. L. Lerman, *J. Amer. Chem. Soc.*, **96**, 5267 (1974).
(c) P. D. Bartlett, M. E. Landis and M. J. Shapiro, *J. Org. Chem.*, **42**, 1661 (1977).
(d) A. L. Baumstark, C. J. McCloskey, T. E. Williams and D. R. Chrisope, *J. Org. Chem.*, **45**, 3593 (1980).
212. (a) B. S. Campbell, N. J. De'Ath, D. B. Denney, D. Z. Denney, I. S. Kipnis and T. B. Min, *J. Amer. Chem. Soc.*, **98**, 2924 (1976).
(b) B. C. Campbell, D. B. Denney, D. Z. Denney and L. S. Shih, *J. Chem. Soc., Chem. Commun.*, 854 (1978).
213. A. L. Baumstark, M. E. Landis and P. J. Brooks, *J. Org. Chem.*, **44**, 4251 (1979).
214. H. H. Wasserman and I. Saito, *J. Amer. Chem. Soc.*, **97**, 905 (1975).
215. B. S. Campbell, D. B. Denney, D. Z. Denney and L. S. Shih, *J. Amer. Chem. Soc.*, **97**, 3850 (1975).
216. P. D. Bartlett, A. L. Baumstark and M. E. Landis, *J. Amer. Chem. Soc.*, **96**, 5557 (1974).
217. P. D. Bartlett and J. S. McKennis, *J. Amer. Chem. Soc.*, **99**, 5334 (1977).
218. P. D. Bartlett, A. L. Baumstark and M. E. Landis, *J. Amer. Chem. Soc.*, **99**, 1890 (1977).
219. A. Kawamoto, H. Uda and N. Harada, *Bull. Chem. Soc., Japan*, **53**, 3279 (1980).
220. A. P. Schaap, *Tetrahedron Letters*, 1757 (1971).
221. T. Wilson, D. E. Golan, M. S. Harris and A. L. Baumstark, *J. Amer. Chem. Soc.*, **98**, 1086 (1976).
222. W. H. Richardson and V. F. Hodge, *J. Amer. Chem. Soc.*, **93**, 3996 (1971).
223. C.-C. Cheng, *M.Sc. Thesis*, University of Puerto Rico, August 1979.
224. A. L. Baumstark, T. Wilson, M. E. Landis and P. D. Bartlett, *Tetrahedron Letters*, 2397 (1976).
225. S. Mazur and C. S. Foote, *J. Amer. Chem. Soc.*, **92**, 3225 (1970).
226. C. S. Foote, A. Dzakpasu and J. W. P. Lin, *Tetrahedron Letters*, 1247 (1975).
227. W. Adam and H. Rebollo, *Tetrahedron Letters*, **22**, 3049 (1981).
228. C. W. Jefford and C. G. Rimbault, *J. Amer. Chem. Soc.*, **100**, 295, 6437 (1978).
229. (a) N. M. Hasty and D. R. Kearns, *J. Amer. Chem. Soc.*, **95**, 3380 (1973).
(b) W. Adam and O. Cueto, *Z. Naturforsch.*, **36b**, 1653 (1981).
230. E. W. H. Asveld and R. M. Kellogg, *J. Amer. Chem. Soc.*, **102**, 3644 (1980).
231. M. A. Umbreit and E. H. White, *J. Org. Chem.*, **41**, 479 (1976).
232. J. J. Basselier, J. C. Cherton and J. Caille, *Compt. Rend.*, **273C**, 514 (1971).
233. W. Adam, O. Cueto, E. Schmidt and K. Takayama, *Angew. Chem.*, **93**, 1100 (1981).
234. A. P. Schaap and N. Tontapanish, *J. Chem. Soc., Chem. Commun.*, 490 (1972).
235. P. Lechtken, *Chem. Ber.*, **109**, 2862 (1976).
236. W. Adam and E. Schmidt, unpublished results.
237. P. A. Burns and C. S. Foote, *J. Amer. Chem. Soc.*, **96**, 4339 (1974).
238. G. Rousseau, P. Le Perchec and J. M. Conia, *Synthesis*, 67 (1978).
239. E. W. Meijer and H. Wynberg, *Tetrahedron Letters*, 3997 (1979).
240. E. J. H. Bechara, A. L. Baumstark and T. Wilson, *J. Amer. Chem. Soc.*, **98**, 4648 (1976).
241. S. A. Matlin and P. G. Sammes, *J. Chem. Soc., Perkin Trans. 1*, 624 (1978).
242. (a) E. Wong and J. M. Wilson, *Phytochem.*, **15**, 1325 (1976).
(b) J. M. Wilson and E. Wong, *Phytochem.*, **15**, 1333 (1976).
243. (a) G. Maier, *Angew. Chem.*, **86**, 491 (1974).
(b) H. Irngartinger, N. Riegler, K.-D. Malsch, K.-A. Schneider and G. Maier, *Angew. Chem.*, **92**, 214 (1980).
244. (a) G. Rio and J. Berthelot, *Bull. Soc. Chim. Fr.*, 3555 (1971).
(b) G. Rio and J. Berthelot, *Bull. Soc. Chim. Fr.*, 822 (1972).
245. A. P. Schaap, P. A. Burns and K. A. Zaklika, *J. Amer. Chem. Soc.*, **99**, 1270 (1977).
246. W. Adam and J. del Fierro, unpublished results.
247. D. A. Konen, L. S. Silbert and P. E. Pfeffer, *J. Org. Chem.*, **40**, 3253 (1975).
248. (a) H. Wynberg and H. Numan, *J. Amer. Chem. Soc.*, **99**, 603 (1977).
(b) H. Wynberg, H. Numan and H. P. J. M. Dekkers, *J. Amer. Chem. Soc.*, **99**, 3870 (1977).
249. M. Muramatsu, N. Obata and T. Takizawa, *Tetrahedron Letters*, 2133 (1973).
250. (a) F. McCapra and I. Beheshti, *J. Chem. Soc., Chem. Commun.*, 517 (1977).
(b) P. B. Hitchcock and I. Beheshti, *J. Chem. Soc., Perkin Trans. 2*, 126 (1979).
251. (a) J.-J. Basselier and J.-P. Le Roux, *Bull. Soc. Chim. Fr.*, 4443 (1971).
(b) J.-P. Le Roux, G. Letertre, P.-L. Desbene and J.-J. Basselier, *Bull. Soc. Chim. Fr.*, 4059 (1971).

252. G. Rio and B. Serkiz, *J. Chem. Soc., Chem. Commun.*, 849 (1975).
253. W. Adam and H.-C. Steinmetzer, *Angew. Chem.* (*Intern. Ed. Engl.*), **11**, 540 (1972).
254. E. Bernatek, T. Ledaal and S. Asen, *Acta Chem. Scand.*, **18**, 1317 (1964).
255. F. Yany, *Doctoral Dissertation*, University of Puerto Rico, August 1977.
256. W. Adam, L. A. Arias and K. Zinner, *Chem. Ber.*, (in press).
257. S. P. Schmidt and G. B. Schuster, *J. Amer. Chem. Soc.*, **100**, 5559 (1978).

Author index

This author index is designed to enable the reader to locate an author's name and work with the aid of the reference numbers appearing in the text. The page numbers are printed in normal type in ascending numerical order, followed by the reference numbers in parentheses. The numbers in *italics* refer to the pages on which the references are actually listed.

Abakerli, R. A. 456 (212), *461*
Abe, S. 211 (82), *231*
Abe, T. 451 (179, 180), *460*
Abello, F. 227 (208), *234*, 866, 910 (91), *915*
Ablezova, K. 91 (85), *95*
Abraham, M. H. 810 (15), 812 (23), *827*
Abrahams, S. C. 86 (10), 88 (10, 81), 91 (81), *94*, *95*, 376 (3), 377 (20), 407 (130), *412*, *415*, 611 (107b), *646*
Abramov, A. F. 137 (44), *158*
Abramovitch, R. A. 105 (1), *126*
Abshire, C. J. 651 (10b), *681*
Achari, B. 312 (9b), *370*
Acher, A. J. 207 (43), *230*
Achon, M. A. 424 (60), *427*
Achord, J. M. 436 (48), 437 (55), *457*
Achrem, A. A. 465, 470 (65), *480*
Acott, B. 174 (53), *197*
Acyama, H. 785 (35), *804*
Adachi, K. 724 (49), *771*
Adam, E. 317 (42), *370*
Adam, H. K. 312 (6), *369*
Adam, W. 2 (16), 47, 48 (174), *78*, *81*, 110 (9–11), 119 (25), *126*, *127*, 204 (5), 210 (66f, 68a, 68b), 213 (94a, 94b, 97), 214 (66f, 97), 216 (94a, 94b, 123a, 123b), 226 (179, 180), 228 (213), *229–234*, 236, 255 (7), *256*, 285 (66), *305*, 314 (27), 315 (30), 320 (55, 57), 321 (55, 58–61), 322 (55, 62–65), 323 (66–72), 335 (116), 337 (132), 338 (135), 342 (72), 343 (59–61), 344 (66), 345 (162), 346, 349 (64), 350 (62, 70, 174), 352 (180), 359 (196), *370–373*, 382 (32), 385 (43), *412*, *413*, 546 (120, 127), *581*, 592 (30), 636 (198), 639 (208), *644*, *648*, 667, 684 (13), *688*, 736 (106, 107), 753 (162, 163a–c, 164–166), 754 (165), 755 (166), *772*, *774*, 781 (16), *804*, 831 (12, 13), 832 (24, 45, 157, 227), 833 (227, 229b, 233), 834 (227), 836 (47, 233), 837 (25, 47, 54, 233), 838 (25, 47, 54, 246), 839 (23, 42, 157, 246), 840 (47), 841 (86, 256), 843 (12, 26, 58, 253), 844 (13, 18f, 18i, 18o, 18p, 19, 20, 36), 845 (23, 24), 846 (12, 23–26), 848 (36), 849 (18i, 25, 36, 47), 850 (42, 45, 47), 853 (26), 854 (36, 54, 56a, 56b), 855 (26, 56a, 56b, 253), 856 (26), 857 (12, 26), 858 (58), 859, 860 (42), 861, 862 (18i), 863 (73), 864 (79), 865 (85, 86), 880 (18o, 18p), 884 (172a, 227), 885 (229b), 887 (45, 256), 888 (45, 233, 236, 256), 889 (45, 86, 233, 246, 256), 890 (157, 179, 233, 256), 891 (148), 892 (157), 894 (170), 895 (172a, 179), 898 (207, 229b, 256), 899 (79, 86, 207, 208), 900 (157, 179, 233, 236, 246), 901 (170), 902 (170, 188), 903 (172a, 179, 188, 189a, 189b), 904 (193b), 905 (189a, 189b, 246), 908 (207, 208), 910 (85, 210), *913–915*, *917–919*
Adamic, K. 247 (42), *257*, 487 (39, 40), 488–490 (40), *518*
Adams, J. M. 51 (204–206), *82*, 86, 88 (14–16), *94*, 376 (10–13), *412*
Adams, W. 742 (144), *773*
Adams, W. R. 205 (8), *229*
Addink, R. 878 (133), *916*
Adeleke, B. B. 732 (88), *772*
Adkins, H. 513 (112), *520*, 655 (33), *682*
Adler, M. 281 (42, 43), 284 (43, 62), *305*
Adman, E. 386, 387 (44), *413*
Ad Yaasuren, P. 396, 397 (87b), *414*
Afanas'ev, I. B. 432 (26, 27), *457*
Aft, H. 865 (87), *915*
Agouri, E. 420 (12), *427*
Ahearn, G. P. 561 (198), *583*
Aher, W. A. 366 (207), *374*
Ahmed, F. U. 274 (63), *277*
Aida, T. 544 (98), *580*
Ainalem, I. B. 469 (47), *480*
Aiura, M. 285 (65), *305*
Akaba, R. 341 (151), *373*, 446 (151), *459*, 851 (50), *914*

Akashi, K. 190 (128, 129), *199*
Akatsuka, A. 226 (183a, 185c), *233*
Akiba, K. 587, 593, 641 (3), *643*, 893 (163c), *917*
Akimoto, H. 683 (86), *690*
Akins, D. L. 904 (191), *918*
Akiyama, E. 668 (56a, 56b), *682*
Aksenov, V. S. 525 (9), *578*
Akutagawa, M. 393 (71), *414*, 860 (62a, 62b), *914*
Akutin, M. S. 803 (111), *806*
Albericci, M. 384 (38f), *413*
Albert, A. 133 (11a), *158*
Alberts, A. H. 281 (37), *305*
Alcock, N. W. 474 (138), *482*
Alcock, W. G. 246 (40), *257*
Alder, K. 209 (53), *230*
Alder, R. W. 454 (201), *460*
Alegria, A. 254 (70), *258*, 635 (190), *648*
Aleksandrov, A. L. 145, 147, 149 (67), *159*
Aleksandrov, Y. A. 817 (39), 821 (52, 57), 822 (58), *827*, *828*
Aleksandrov, Yu. A. 144 (64), 157 (91, 92), *159*, *160*, 811 (87), *828*
Aleksejeva, E. 574 (236), *584*
Alexandrov, Yu. A. 801 (81), *805*
Alexksandrov, Yu. A. 98, 99 (15), *104*
Alexsandrov, Y. A. 811 (67), *828*
Alford, J. A. 193 (148), *199*, 228 (216), *234*
Ali, A. E. 316 (39), *370*
Ali, R. 724 (53), *771*
Aljadeff, G. 435 (37a, 37b), 436 (50), 437 (37a, 37b, 50), 438 (37a, 37b), 440, 445 (50), *457*
Allara, D. L. 164 (11), *196*
Allavena, M. 54 (218), 63 (275), *82*, *83*
Allen, A. O. 430, 454 (12b), *456*
Allen, J. K. 591, 623 (25), *644*
Allen, L. C. 36 (89, 96, 108, 117), 37 (117), 38 (89, 134), 53 (96), *79*, *80*, 87, 88 (41, 42), 90 (42), *95*
Allen, P. 419 (5), *426*
Allen, R. M. 384 (38e), *413*
Alleston, D. L. 820 (51), *827*
Allies, P. G. 812 (27), *827*
Allinger, N. L. 67 (291), *83*
Almenningen, A. 387 (54), *413*
Almgren, C. W. 446 (152), *459*, 851 (49), *914*
Aloyan, A. G. 144, 149 (60), *159*
Alscher, A. 330, 331, 354 (103), *372*
Altenbach, H. J. 213, 216 (105), *231*
Altenbach, H.-J. 324 (74), *371*
Altona, C. 38 (137), *80*
Alvin, A. N. 312 (7a), *369*
Alzerreca, A. 843, 846, 853, 855–857 (26), *914*
Amako, Y. 87, 88 (36), *95*
Ambar, M. 587 (2), *643*
Ambrovič, P. 788, 790, 791 (41, 44), 792 (53, 54), 796, 797 (65), *805*
Amma, E. L. 465 (95), 466 (110), 472 (95), 473 (110), *481*
Ammers, M. van 297 (185), *308*
Andal, R. K. 471 (89), *481*
Anderegg, J. H. 244 (30), *257*, 838 (145d), 839 (145c), 883 (145c, 145d), 884 (145d), 886 (145c), 906 (198), 908 (205), *916*, *918*
Anders, B. 271 (48), *277*, 287 (82), *305*
Andersen, A. 21, 58 (40), *78*
Anderson, G. H. 554, 556, 561 (148), *582*
Anderson, J. E. 901, 902 (185a), *918*
Anderson, L. R. 246 (39), *257*, 496 (68, 71), *519*
Anderson, R. A. 810 (13), *827*
Anderson, S. G. 55 (221), *82*
Andersson, M. 282 (56), *305*
Ando, W. 206 (23), 228 (209–211, 212a, 212b), *229*, *234*, 445 (124), 446 (124, 127–129), 449 (127), 454, 456 (124), *459*, 588 (9c, 13b, 13c), *643*, 785 (34), *804*, 861 (66), 909 (209), *915*, *918*
Andrews, L. 52 (209), 54 (209, 214), 61, 62 (214, 242, 243), *82*
Andrews, L. J. 138 (50), *159*
Andreyeva, T. P. 193 (146), *199*
Andrist, A. H. 394 (77), *414*
Andrulis, P. J., Jr. 651 (10a), *681*
Anet, E. F. 244 (29), *257*
Anet, F. A. L. 396 (87a), *414*
Angelis, C. T. 447, 454 (155), *459*, 879 (138), *916*
Anisimov, Yr. N. 140, 146 (46), *159*
Ans, J. d' 287 (86), 288 (91), *305*, *306*
Ansari, A. S. 724 (53), *771*
Antheunis, D. 742 (138a), *773*
Antonovskii, V. L. 142, 144 (53), 145, 146 (66), 149 (53, 66), 150, 153, 154 (66), *159*
Aoki, K. 546 (106), *580*
Aoyagi, A. 226 (178), *233*
Aoyama, H. 393 (71), *414*, 860 (62a, 62b), *914*
Aoyama, T. 893 (163c), *917*
Aponte, G. S. 753 (163c), *774*
Appel, H. R. 280 (32), *304*
Appelbaum, A. 301 (222), *309*
Appelman, E. H. 49 (195), *82*
ApSimon, J. W. 271 (51), *277*
Arai, H. 336 (126), *372*
Arai, T. 228 (210), *234*
Arakane, K. 443, 444 (121), 446 (121, 126b), *459*
Arakawa, H. 440, 454 (74), *458*, 465, 470 (64), *480*

Aranda, G. 111 (13b), *126*
Arata, Y. 366 (205a), *374*
Aratani, M. 291 (141), *307*
Arbuzov, Y. A. 210 (66a), *230*
Archer, G. 675 (50), *689*
Arduengo, A. J. 254(70), *258*, 635 (190), *648*
Arendonk, R. J. F. M. van 874 (117), *916*
Arias, L. A. 832 (24), 836 (47), 837, 838 (25, 47), 840 (47), 841 (256), 845 (24), 846 (24, 25), 849 (25, 47), 850 (47), 887–890, 898 (256), *914*, *919*
Aristoff, P. A. 223, 225 (156), *233*
Arkell, A. 61 (253), *83*
Armour, A. G. 814 (31), *827*
Armstrong, D. A. 435 (43), *457*
Arnau, J. L. 485 (19), *518*
Arnaud, R. 741 (137), *773*
Arndt, R. W. 152 (82), *159*
Arnold, E. 320 (53), *371*
Arnold, S. J. 203, 204 (2b), *229*
Aroney, M. J. 88 (62), 89 (62, 64), *95*
Artm'ev, E. T. 905 (195), *918*
Arudi, R. L. 441 (101), 456 (210), *459*, *461*
Aruga, S. 263 (28), *277*
Asai, N. 632 (167), *647*
Asao, T. 331 (106), *372*
Åsbrink, L. 44, 45 (159), *81*
Asen, S. 855 (254), *919*
Ashford, R. D. 208 (51), 209 (63), 210 (51), 212 (63), *230*
Ashmore, F. S. 46, 47 (171), *81*
Asinger, F. 525 (14), *578*
Asmus, K.-D. 761 (184b), *774*
Assada, K. 435 (42), *457*
Aston, J. G. 63 (279), *83*
Asveld, E. W. H. 332 (109d), *372*, 833, 835, 837 (230), *919*
Atal, C. K. 312 (7b), *370*
Atkhodzhaev, V. I. 423 (46), *427*
Atkins, R. L. 259, 260 (3), *276*
Atkinson, R. 677 (60), 683 (87), *689*, *690*
Atlani, P. 501, 507 (97, 98), *519*, 652 (17), *681*
Atovmyan, E. G. 800 (100), *806*
Atwood, J. L. 65 (287), *83*
Aubry, J. M. 456 (211), *461*
Audeh, C. A. 271, 272, (54), *277*
Audier, H. E. 111 (13a, 13b), *126*
Aue, W. A. 821 (54a, 54b), *828*
Augustine, R. L. 283 (59), *305*
Augustyniak, H. 113 (16), *126*
Augustyniak, W. 721 (42), *771*
Aune, J. P. 174 (59), *197*
Ausloos, P. 761 (183a), *774*
Austin, J. D. 501 (91), *519*, 653 (28a, 28b), *682*
Avdeef, A. 377 (24), *412*
Axworthy, A. E., Jr. 25 (63), *79*
Ayer, W. A. 216 (121), *232*, 319 (51), *371*
Ayers, P. W. 625, 628 (149a), *647*
Ayreg, G. 651 (13), *681*
Ayrey, G. 622, 623 (139), *646*
Azar, J. C. 253 (68), *258*
Azman, A. 289 (112), *306*
Ažman, A. 63 (272, 277), 67 (272), 68 (272, 277, 297, 298), 69 (272), *83*, 88, 92, 93 (103), *96*, 485 (18, 21), 486 (18), *518*, 523, 524 (3), 530 (3, 49), 531 (3), 573 (233), *578*, *579*, *583*

Baader, W. 864, 899, (79), *915*
Babatskiko, C. 832, 849, 892 (38), *914*
Bach, R. D. 88, 90, 92 (102), 93 (102, 106), *96*, 287 (89), 291 (137), *306*, *307*, 530 (51, 52), *579*
Bachi, G. 228 (215), *234*
Bachietto, G. 226 (186), *233*
Bachmann, W. E. 271 (49), *277*
Baczynskyj, L. 316 (37), *370*
Bader, R. F. W. 55 (221–225), 57 (223), *82*, 455 (204), *461*
Badger, R. M. 135, 138, 142, 148 (34a, 34b), *158*
Badhuri, S. 474, 475 (133), *481*
Baer, Y. 44, 45 (158), *81*
Baeyens-Volant, D. 731 (74), *772*
Baeyer, A. 280 (25, 31), 295 (160), *304*, *307*
Baeyer, A. von 559 (181), *582*
Bagus, P. S. 44, 45 (161), *81*
Bailey, H. C. 801 (105), *806*
Bailey, P. S. 2, 61, 63, 73 (12), *78*, 168 (39), *197*, 501 (87, 90, 92–95), 504 (109), 509 (87), *519*, 632, 633 (174), *647*, 650 (2, 3c, 4, 5), 651 (7a–f, 10b), 652 (20a), 653 (25), 654 (7a–f, 20a, 29–31), *681*, *682*, 680 (79, 83), *689*
Bailey, S. M. 40 (139), *80*
Bailey, S. W. 377 (22), *412*
Bakalik, D. 501 (88), *519*, 653 (26), *682*
Baker, G. 98 (12), *104*, 668 (21), *688*
Baker, K. A. 53 (217), *82*
Baker, R. 454 (201), *460*
Bakh, A. N. 484 (4), *517*
Bakuzis, P. 751 (157, 158), *774*
Balakrishnan, I. 725 (56), *771*
Balci, M. 2, 65 (10), *78*, 210 (66g, 68a), 212 (66g), 213 (97), 214 (66g, 97), *231*, 323 (66, 68–72), 342 (72, 152), 344 (66), 345 (162), 348 (152), 350 (70), 352 (180), *371*, *373*
Baldwin, A. C. 99 (33), *104*, 678, 679 (66), *689*
Baldwin, J. E. 337 (128), 354 (183), *372*, *373*, 465 , 469 (51), *480*

Baldwin, R. R. 669 (32, 33), *688*
Ball, M. 206 (6), *276*
Ball, R. E. 568 (223b), *583*, 587 (5a), *643*
Ball, S. 540 (85, 86), 543 (96), *580*
Ballard, D. H. 116 (21), *126*
Ballinger, P. 133 (15), *158*
Balogh-Hergovich, E. 436, 437 (51), *457*
Balzani, V. 550 (132), *581*, 895 (177), *918*
Bamford, C. H. 810 (16), *827*
Banderet, A. 419 (9), *426*
Banerji, J. 336 (123), *372*
Baney, R. H. 142 (57), *159*
Bang, L. 670 (58a, 58b), *683*
Banks, D. F. 546 (102), *580*
Barbara, P. F. 557, 559 (179, 180), *582*
Barbe, A. 48, 49, 54 (184), *81*
Barborak, J. C. 501 (95), *519*
Barchuk, V. I. 739 (122), *773*
Barciszewski, J. 113 (16), *126*
Bard, A. J. 451 (171, 172), 456 (172), *460*
Bard, E. J. 446, 454, 456 (153), *459*
Bargon, J. 743 (147–149), *773*
Bärhold, H. 866, 867, 910 (92), *915*
Barker, J. R. 668, 671 (26), *688*
Barlett, P. D. 101 (36), *104*, 245 (35), *257*
Barnard, D. 145, 147, 150 (68), *159*, 423 (50), *427*, 651 (13), *681*
Barnett, G. 893 (163b), *917*
Barnette, W. E. 313, 339, 359, (16b), *370*
Bar-nun, A. 69 (308), *84*
Barret, K. E. 793 (56), *805*
Barrett, H. C. 366 (204), *374*
Barriel, J. M. 59, 60 (236), *82*
Bartell, L. S. 37 (124, 125), 67 (124), *80*, 88 (69, 71), 89 (69), 90 (71), *95*, 381 (29, 30), *412*
Bartlett, P. A. 185 (103), *198*
Bartlett, P. D. 205 (7), 206 (29–31), 207 (34), 210 (65c), 222 (148), 225 (164), 226 (65c, 190), 227 (65c, 197), 228 (197, 219–221), *229*, *230*, *232–234*, 247 (44), *257*, 262 (18), 265 (33), *276*, *277*, 288 (109), 299 (200, 202), *306*, *308*, 445 (125), *459*, 486 (27, 28, 32), 494 (32, 66, 67), *518*, *519*, 528 (40), 544 (98), 552 (135), 553 (142, 143), 555 (157), *579–582*, 588 (12b), 589 (18), 592 (27b, 28, 31), 594 (43, 44), 600 (43, 44, 81), 607 (43, 44), 622 (136), 623 (146a), 625 (149b), 636 (193), *643–648*, 739 (120), *773*, 832 (39, 81), 836 (43, 211c), 837 (224), 844 (18h, 18m), 849 (39), 850 (43), 854 (55), 865 (81), 885, 887 (224), 891, 894 (150a), 910 (211a–c), 912 (216–218), *913–915*, *917–919*
Bartoli, G. 451 (176b), *460*
Barton, A. F. M. 606, 607 (93c), *645*
Barton, D. H. R. 341 (145–147), *372*, 388 (48), *413*
Bartsch, R. A. 271 (49), *277*
Basch, H. 45 (165), 46 (172), 53, 57 (215), *81*, *82*
Bashkirov, A. N. 193 (146), *199*
Bashkirova, S. G. 802 (109), *806*
Basolo, F. 2 (21), *78*, 464, 465 (12, 14), 471 (12, 81), 475 (12), *479*, *480*
Basselier, J. J. 351 (176), *373*, 783 (28), *804*, 833 (232), 842 (251a, 251b), *919*
Bassey, M. 554 (149), *582*, 590 (19c), *644*
Basson, H. H. 354 (183), *373*
Bateman, L. 224 (162), *233*, 486 (23), *518*
Bateman, R. 243 (28), *257*
Bates, J. S. 425 (70), *428*
Bates, R. G. 133 (11b), *158*
Batich, C. 47, 48 (174), *81*, 382 (32), *412*
Bats, J. P. 388 (50), *413*
Batt, L. 99 (19, 21), *104*, 666 (7, 9), 667 (19), 668 (19, 22, 23, 28), 669 (22, 29, 34), 671 (19), 672 (19, 34), 673 (34), 674 (19, 49), 675 (51), 676 (34, 56, 57), 677 (7), 680 (9), *688*, *689*, 733 (94a), *772*, 797 (73), *805*
Battaglia, A. 297 (181a, 181b), *308*, 545 (99, 100), *580*
Battaglia, C. J. 130 (1c), *157*, 641 (211), *648*
Batten, J. J. 285 (69), *305*
Batterbee, J. E. 501 (90), *519*, 654 (29), *682*
Bauder, A. 88, 92 (95), *96*
Bauer, S. H. 37, 67 (127, 130), 69 (307), *80*, *84*, 88, 90 (72), *95*, 135, 138, 142, 148 (34a), *158*
Baumstark, A. L. 227 (195), 228 (219–221), *234*, 684 (100), *690*, 832 (158a, 158b, 221), 833 (158b), 836 (240), 837 (224), 883 (221), 884 (158a, 158b), 885 (221, 224), 886 (158a, 158b, 221), 887 (158a, 224), 888, 889 (240), 891 (150a), 892 (158a, 158b), 894 (150a), 898, 900 (221), 910 (211a, 211b, 211d), 911 (213), 912 (216, 218), *917–919*
Bauslaugh, G. 312 (3), *369*
Bawn, C. E. H. 280 (7), *304*, 564 (210), *583*
Bayer, E. 464, 465 (5), *479*
Bayerlein, F. 271 (48), *277*
Baysal, B. M. 419 (10), *426*
Bazzi, A. A. 62, 63 (267), *83*
Beagley, B. 37, 67 (129), *80*
Beasley, G. H. 613, 614 (113), *646*
Beati, E. 422 (27), *427*
Beaulieu, W. B. 466 (117), 467 (24), 473 (117), *479*, *481*
Bechara, E. J. H. 834, 835 (160), 836, 888 (240), 889 (160, 240), 893, 900 (160), 905 (196b), *917–919*
Becker, J. 224 (157b), *233*

Becker, K. H. 58, 59 (233), *82*
Becker, M. 166 (24), *196*
Becker, N. 289 (114), *306*
Beckey, H. D. 106 (4), *126*
Beckwith, A. L. J. 174 (53), *197*, 301 (220), *309*, 338 (136), *372*, 498 (78), *519*, 664, 665, 671 (54a–c), *682*
Beebe, N. H. F. 21, 58 (40), *78*
Beers, Y. 25, 33 (82), *79*
Behar, D. 451, 452 (184a), *460*, 761, 762 (185), *774*
Behar, J. V. 606 (92), *645*
Beheshti, I. 71 (312), *84*, 226 (171), *233*, 384 (39, 40), 386 (40), *413*, 839 (250a, 250b), 840–842, 884, 890, 895, 896 (180c), *918*, *919*
Behn, N. S. 292 (145), *307*
Behrman, E. J. 568 (223b), *583*
Beierbeck, H. 501, 502 (105), *519*
Beileryan, N. M. 144 (61, 62), *159*
Beisler, J. A. 561 (198), *583*
Belew, J. S. 389 (60), 390 (62b), *413*
Belfoure, E. L. 262 (13), 272 (58), 273 (61), *276*, *277*, 546, 549 (118b), *581*
Belitskus, D. 88, 92 (96), *96*, 134 (18), *158*, 288 (94), *306*, 405 (107), *414*
Bell, E. R. 164 (10), *196*
Bell, R. P. 573 (235), *584*
Bellet, J. 33, 48, 49, 62 (70), *79*
Bellič, I. 118 (23), *126*
Bellucci, G. 526 (22), *579*
Belokon, V. G. 720, 726 (30), *771*
Below, J. S. 650 (3j), *681*
Belt, P. B. 866 (90), *915*
Belyaev, V. A. 136 (35), 137 (35, 41), 139 (35, 52), 140, 141, 143, 149 (35), *158*, *159*
Belyakov, V. A. 904, 905 (192), *918*
Belyayev, V. M. 425 (68), *428*
Bemis, A. 599 (78), *645*
Bemis, A. G. 437, 438, 440, 454 (60, 62), *457*, *458*
Bender, C. F. 25, 33 (71), 61 (245, 260), 62 (245), 68 (301), *79*, *82–84*
Bender, P. 783, 785 (31), *804*
Benioff, P. A. 61 (261), *83*
Bennett, J. E. 245 (36), 247 (41), *257*, 487 (36–38), 488, 489 (37), 490 (50), 491 (37, 50, 53), 492 (37, 53), 497 (74), *518*, *519*, 622 (137), *646*, 733 (93), *772*
Bennett, M. J. 466 (100), 472 (100, 101), *481*
Bennett, R. A. 48, 49 (179), *81*
Beno, M. A. 526, 527 (32), *579*
Benson, S. W. 2 (13), 25 (63), 40 (140–142), 41 (142–148), 42 (143), 43 (141–145, 148), 44 (143, 146, 147), 61 (13), *78–81*, 97 (1–3), 98 (6), 100 (3, 6, 18), 101 (6, 37), 102 (3, 6), *102*, *104*, 134 (31), 153 (85), *158*, *159*, 237, 239 (11), 247 (46), 248 (11), 252, 253 (66), *256–258*, 485 (14, 15a, 15b), 489, 490 (48), 491 (59), 492 (64a), 496 (14, 15b), 501, 507, 509, 514 (106), 517 (117), *518–520*, 657 (39, 40), *682*, 665 (1), 666 (5, 10), 668 (26, 28), 671 (26), 672 (45), 676 (10), *688*, *689*, 733 (94a), *772*, 797 (73), *805*
Bentrude, W. G. 546 (125), *581*, 635 (188), *648*
Benzon, M. S. 613, 614 (114), *646*
Beolyaev, V. A. 143, 144 (58), *159*
Berchthold, G. H. 324 (73), *371*
Berchtold, G. A. 213 (103, 104), 214 (104), *231*, 345 (161), *373*, 526 (25), *579*
Bercovici, T. 449 (163), *460*
Berends, W. 878 (133), *916*
Berenfel'd, V. M. 905 (195), *918*
Berenjian, N. 446 (145), *459*
Berger, M. 303 (233a), *309*
Berger, M. G. (240), *584*
Berglund, C. 282 (56), 302 (226, 228, 230, 231), 303 (231, 236, 239), *305*, *309*
Bergmann, W. 831 (8), *913*
Bergmark, T. 44, 45 (158), *81*
Berguer, Y. 432 (18, 19), 441 (18, 19, 88), 442 (19, 88), 443 (88), 444 (19, 88), 449, 452, 454 (18, 19), *457*, *459*
Beringer, F. M. 299 (209a, 209b), *308*
Berlin, A. A. 802 (109), *806*
Bernatek, E. 855 (254), *919*
Bernheim, M. 272 (60), *277*
Bernstein, J. 611 (109), *646*
Bernstein, S. 653 (28c), *682*
Berson, J. A. 560 (193), *583*, (216), *648*
Bertand, M. 634 (183c), *647*
Berthelot, J. 210 (73), *231*, 319 (50), *371*, 838 (244a, 244b), *919*
Berthelot, M. 484 (2), *517*
Berthet, D. 215 (115), *232*, 318 (47), 365 (200), *371*, *374*
Berti, G. 525 (20), 526 (22), 528, 530 (38), *579*
Bertrand, M. 121, 122 (33), 123 (36), *127*
Bertz, T. 820 (48), *827*
Bessenov, M. I. 622 (140), *646*
Besten, I. E. D. 675 (64), *683*
Bethke, H. 105 (2), *126*
Betzel, C. 546, 547 (111c), *581*
Beugelmans-Verrier, M. 295 (169, 170), *307*
Beurskens, P. 88, 92 (97), *96*, 134 (17), *158*, 405 (108), *414*
Beuttner, G. R. 446 (134), *459*
Bevington, J. C. 591 (25), 596 (60), 597 (60, 61), 618 (141), 622 (141–144), 623 (25, 141–144), 624 (141–143), *644*, *646*, *647*, 740 (131), *773*

Bex, J. P. 422 (31), *427*
Beyer, E. 590 (21a), *644*
Bharadwaj, L. M. 768 (209), *775*
Bhat, H. K. 312 (7b), *370*
Bhat, V. 211 (86), *231*, 332 (108b), *372*, 844 (18w), 867 (95), *913*, *915*
Bhatia, V. P. 286 (72), *305*, 546 (130), *581*
Bhattacharya, R. G. 466, 474 (137), *482*
Bhaumik, M. L. 904 (193a), *918*
Bianco, D. R. 57 (227), *82*, 87 (28), *94*
Bickley, H. T. 546 (128a), *581*, 605, 618 (91), *645*
Bie, D. A. de 718 (21), *770*
Biegler-König, F. W. 55 (224), *82*
Bielski, B. H. J. 430 (12a, 12b), 452 (187), 454 (12b), *456*, *460*, 484 (8, 9), *517*
Billingsley, F. B. 61, 62 (244), *82*
Billmers, R. 571 (232), *583*
Biloski, A. J. 194 (152, 153), *199*
Bilyk, I. 291 (138), *307*
Bingham, K. D. 289 (113), *306*, 525 (17), 529 (44), *579*
Bingham, R. C. 88, 89 (67), *95*
Bird, C. W. 405 (120), *415*
Birke, A. 315 (30), *370*
Birke, R. L. 904 (191), *918*
Birkle, D. L. 727–729 (64), *771*
Birss, F. W. 675 (50), *689*
Bisacchi, G. S. 227 (206), *234*, 334, 335 (120), *372*
Bischoff, C. 590 (21b), *644*
Bishop, C. E. 633 (178), *647*, 736 (108), *772*
Bishop, R. D. 268 (44, 46), 269 (44), *277*
Bisolnakov, N. 725 (57), *771*
Bissinger, W. E. 281 (44), *305*
Bitman, G. L. 138–142, 149, 150 (47), *159*
Bixon, M. 396 (85), *414*
Bkouche-Waksman, A. 465, 469 (40), *480*
Black, J. F. 251 (60), *257*
Blackburn, T. F. 819 (47), *827*
Blair, E. A. 651 (12), *681*
Blaisten-Barojas, E. 63 (275), *83*
Blanchard, H. S. 486 (24), *518*
Bland, J. 429 (3), *456*
Blank, F. 312 (3), *369*
Blaschette, A. 37, 67 (123), *80*, 86–89 (21), *94*, 147 (69), *159*, 382 (31), *412*, 780 (12b), *804*, 808 (4), 821 (56), 822 (4), *826*, *828*
Blattner, R. 281 (39, 40), 282, 283 (39), *305*
Blint, R. J. 25 (73, 78), 33 (78), 54, 61, 63 (73), *79*, 485, 499 (20), *518*
Blitz, H. 831 (3), *913*
Block, E. 62, 63 (627), *83*
Blomquist, A. T. 611 (109), *646*
Bloodworth, A. J. 116 (21), 119 (24a), *126*, 216 (123a), *232*, 314 (27), 315 (32), 316 (40, 41), 317 (41–43), *370*, 812 (21b), 820, 821 (49b), *827*
Blossey, E. C. 205 (11, 13), *229*, 849 (40), *914*
Blount, J. F. 356 (188), *373*
Blumberger, J. S. P. 290 (125), *306*
Blumbergs, J. H. 288, 289 (106), *306*
Blumstein, C. 666, 681 (11), *688*
Blustin, P. H. 67 (290), *83*
Boarland, V. 290 (117a), *306*
Boberg, F. 763 (193, 194), *775*
Bobrowicz, F. W. 21, 58 (39), *78*
Bobylev, B. A. 180 (81), *198*
Bobylev, B. N. 178 (74), 180 (81), *198*
Bobyleva, L. I. 180 (81), *198*
Boche, G. 272 (60), *277*
Boche, J. 213 (91a), *231*
Bock, C. W. 68 (293), *83*
Bockelheide, V. 331 (107), *372*
Boden, R. M. 205 (9), *229*
Bodkin, C. L. 498 (78), *519*, 664, 665, 671 (54b, 54c), *682*
Boekelheide, V. 326, 353 (86), *371*
Boer, H. 666 (55a, 55b), *682*
Boer, Th. J. de 211 (83), *231*
Boerseken, J. 290 (125), *306*
Boeseken, J. 525 (8), *578*
Boeyens, J. C. A. 475 (158), *482*
Bogan, D. J. 685 (105–110), 686 (110), 687 (105), *690*, 847, 864, 894 (32), *914*
Boggs, J. E. 67 (289), *83*
Boguslavskaya, L. S. 722 (45), *771*
Bohlmann, F. 105 (2), *126*
Böhm, M. 526, 527 (42), *579*
Böhm, M. C. 526, 527 (32), *579*
Bois, C. 465, 469 (40), *480*
Bois, J. 227 (208), *234*
Boix, J. 866, 910 (91), *915*
Bolletta, F. 757 (173), *774*
Bollyky, L. J. 226 (177), *233*
Bolotin, A. B. 87, 89, 92 (54), *95*
Bolt, A. J. N. 501, 513, 514 (89), *519*, 656 (35), *682*
Bolte, J. 261, 262 (8), 263 (20), 268 (8, 43), *276*, *277*
Bolton, J. R. 825 (77), *828*
Boness, M. J. W. 48, 49 (178), *81*
Bonet, J. J. 227 (208), *234*
Bonet, J.-J. 866, 910 (91), *915*
Bonetti, G. A. 187 (117), *199*
Boniardi, O. 876 (122b), *916*
Bontempelli, G. 442–444 (119), *459*
Booth, B. 475, 476 (146), *482*
Borders, D. B. 213 (100, 101), *231*, 314 (25), *370*
Bordner, J. 394 (76), *414*
Boreham, C. J. 465, 469 (33), *479*
Borgers, T. R. 388 (46), *413*

Bors, W. 453 (194), *460*
Borsig, E. 788, 790, 791 (41), 792 (54), 801 (104), *805*, *806*
Borstnik, B. 289 (112), *306*
Borštnik, B. 530 (49), *579*
Bortolini, O. 469 (44), 471 (77), *480*
Boschung, A. F. 205 (7), 226 (173a, 173b), *229*, *233*, 453 (195), *460*, 589 (18), *643*, 863 (69), *915*
Bose, A. N. 675 (50), *689*
Bosnich, B. 465, 467, 472 (22), *479*
Boss, C. R. 732, 737 (86), *772*
Boss, R. C. 798 (76), *805*
Bossert, W. 88, 92 (95), *96*
Botschwina, P. 36, 52, 53 (102), *80*
Bottari, F. 528, 530 (38), *579*
Bouas-Laurent, H. 675 (65), *683*
Bouchard, J. M. 469 (58), *480*
Bouillon, G. 281 (38, 40, 41, 47), 282 (38), 284 (38, 41), *305*
Boujlel, K. 448, 454 (160), *460*
Boulch, N. L. 292 (144), *307*
Boulton, J. 735 (100b), *772*
Bovey, F. A. 422 (34), *427*
Bowden, K. 133 (14), *158*
Bowes, P. C. 788 (43), *805*
Bowman, D. F. 491, 492 (60), *518*
Box, H. C. 714–716, 740 (9), 765 (202), *770*, *775*
Boyd, D. B. 67 (291), *83*
Boyd, D. R. 570, 572 (229), *583*
Boyd, J. D. 210 (70, 71), 213, 214 (93), *231*, 330 (101), 357 (191), *371*, *373*
Boyd, M. E. 25 (84), *79*
Boyer, P. D. 844 (17b), *913*
Boyle, J. W. 491 (54), *518*
Bozik, J. E. 166 (25), *196*
Bozzini, S. 88, 91 (78), *95*, 401 (98), *414*
Brabham, D. E. 10 (28), *78*
Bradley, D. C. 377 (26), *412*, 475 (159), *482*
Bradley, J. D. 740 (128), *773*
Braekman, J. C. 384 (38f), *413*
Brambilla, R. 388 (52), *413*
Brandes, D. 37, 67 (123), *80*, 86–89 (21), *94*, 382 (31), *412*, 808 (4), 821 (56), 822 (4), *826*, *828*
Brandrup, E. 800 (101), *806*
Brant, P. 68 (303), *84*
Bratož, S. 54 (218), *82*
Bratring, K. 297 (182a), *308*
Brauer, H.-D. 331 (105), *372*, 893 (161), 904 (193c), *917*, *918*
Brauman, J. I. 61, 62 (262), *83*
Braun, G. 282, 287 (50), *305*
Braun, W. 609, 610, 624 (87), *645*, 722, 723, 743 (47), *771*
Breaux, E. J. 667, 684 (18), *688*, 844 (18s), *913*
Bregadze, V. I. 817 (39), *827*
Brégeault, J. M. 475, 476, 478 (145), *482*
Breitenbach, J. W. 712 (7), *770*, 786 (37), *805*
Breitenbach, L. P. 69 (305), *84*, 678 (70), 679 (71), *689*
Brelière, C. 351 (177b), 355 (185), *373*
Brendes, D. 780 (12b), *804*
Brennan, M. E. 207 (35a), *230*
Brennan, R. L. 280 (10), *304*
Brenner, J. 560 (193), *583*, (216), *648*
Bressel, B. 147 (69), *159*
Brewer, G. A. 470 (75), *480*
Bricout, D. 328 (95), *371*
Bridger, R. F. 451 (182), *460*
Brilkina, T. G. 356 (186), *373*, 808, 810 (3), 822 (58), 823 (61), *826*, *828*
Brilkine, T. G. 808 (2a), 821 (56), *826*, *828*
Brill, W. F. 162 (2), *196*, 224 (159), *233*
Brindley, P. B. 809 (8), 812 (27), 814, 815 (8), 816 (34), 818 (43, 45), 819 (45), 824 (43), *827*
Brittain, M. J. 285 (63), *305*
Brock, N. 397 (88), *414*
Brocker, U. 216 (124a), *232*, 344 (159), *373*
Brocker, V. 325 (83), *371*
Brokken-Zijp, J. 598, 616, 624 (72), *645*
Broman, R. F. 432 (22), *457*
Brooks, C. S. 596, 597 (60), *644*
Brooks, J. R. 166 (19), *196*
Brooks, P. J. 911 (213), *919*
Brooks, W. F. 88, 92 (94a), *96*
Brossas, J. 420, 422 (11), *427*
Brown, D. 178 (70), *197*, 478 (154), *482*
Brown, D. M. 247 (41), *257*, 487 (36, 37), 488, 491, 492 (37), *518*, 733 (93), *772*
Brown, H. C. 267 (37), *277*, 526 (28, 29), *579*, 808, 814 (35), 816 (35, 36), *827*
Brown, J. H. 259 (1b), *276*
Brown, J. M. 454 (201), *460*
Brown, J. N. 394 (77), *414*
Brown, K. H. 350 (174), *373*
Brown, O. 356 (189), *373*
Brown, P. 110 (12). *126*
Brown, R. S. 46 (170), 47, 48 (170, 175), *81*, 120 (27), *127*, 216 (122), *232*, 847 (28), *914*
Brown, S. B. 871 (102), *915*
Browne, L. M. 216 (121), *232*, 319 (51), 366 (207), *371*, *374*
Brown-Winsley, K. A. 446 (146), *459*
Brudkin, B. M. 501 (120), *520*
Brudnik, B. M. 653 (23), *682*
Bruice, T. 195 (158), *200*
Bruice, T. C. 195 (156, 157), *200*, 540 (85, 86), 543 (95, 96), 566 (213), *580*, *583*, 876 (128), 878 (130a, 130b, 131, 132), *916*

Brunck, T. K. 38 (133), *80*, 89 (56), *95*
Brundle, C. R. 44, 45 (163), 46 (172), *81*
Brunken, J. 312 (1), *369*
Brunton, G. 245 (36), *257*, 497 (74), *519*
Bryan, R. F. 213 (102), *231*
Bryukhovetskii, V. A. 653 (24b), *682*
Bucciarelli, M. 570, 572 (230), *583*
Buchachenko, A. L. 801 (106), *806*
Buchanan, J. D. 435 (43), *457*
Buchanchenko, A. L. 148, 149 (70), *159*
Büchi, G. 366 (204), *374*, 833, 851 (51), *914*
Buckler, S. A. 168 (34, 35), *197*, 810 (10), *827*
Budai, S. I. 280 (29b), *304*
Budtov, V. P. 425 (68), *428*
Budzinski, E. E. 714–716, 740 (9), *770*
Buehler, C. A. 298 (193a), *308*
Buell, G. R. 653 (28c), *682*
Buenker, R. J. 13 (30), 24 (52, 53, 60), 25 (60, 75, 77), 33 (52, 60, 75), 45 (53, 77), 46 (77), 58 (53, 75, 77, 230), 59 (53, 77), *78*, *79*, *82*
Bugg, C. E. 397 (92), *414*
Buhlmayer, P. 560 (197), *583*
Bukowick, P. A. 540, 541 (88), *580*
Buloichik, Zh. J. 280 (27, 28b), *304*
Bunce, N. J. 280 (33), *304*
Bunnett, J. F. 435 (44), 451 (177, 183), *457*, *460*
Bunton, C. A. 554 (149), 566 (212), 567 (212, 215), 568 (212), *582*, *583*, 590 (19c), *644*
Burchill, C. E. 762 (187), *775*
Burford, A. 840–842 (180c), 860 (64a, 64b), 884, 890, 895, 896 (180c), *915*, *918*
Burgess, A. R. 46, 47 (171), *81*, 113 (14a, 14b), *126*
Burgess, J. R. 633 (178), *647*
Burka, L. T. 546 (109), *581*
Burkholder, W. J. 166 (23), *196*
Burlant, W. J. 418, 426 (3), *426*
Burleson, F. R. 677 (62), *689*
Burns, B. A. 327, 328 (90), *371*
Burns, J. H. 906 (198), *918*
Burns, P. A. 120 (30), *127*, 210 (69, 72), 227 (193, 198), *231*, *234*, 329 (96, 98), 330 (100), 355 (98), *371*, 835–837 (237), 838 (31, 245), 839 (237), 842 (31, 48), 847 (31), 851 (48), 884, 886, 887, 895 (31), 896 (31, 48), 898, 899 (31), *914*, *919*
Burrows, J. P. 669 (29), *688*
Burstain, I. G. 211, 217 (87a), *231*
Burt, J. G. 222 (146), *232*
Burton, M. 762 (188a), *775*
Burton, P. G. 24, 33 (56), 36 (105), *79*, *80*, 87, 90 (51), *95*
Burtzlaff, G. 554 (150), *582*
Burykina, I. K. 280 (28d, 28f), *304*
Busch, W. 730 (67), *771*
Busing, W. R. 49, 51 (191), *81*, 86, 88 (9), *94*, 376, 377, 382 (4), *412*
Buss, J. H. 97 (1), *102*
Busse, W. 287 (86), *305*
Butler, J. 452 (190), 456 (190, 208), *460*, *461*, 720 (31), *771*
Buxton, G. 720 (36), *771*
Buys, H. R. 38 (137), *80*
Buytenhek, M. 339 (137), *372*
Byers, A. 190 (127), *199*
Byers, J. D. 316 (38, 39), *370*
Bylina, G. S. 280 (26, 27, 28b, 28d–f, 30a, 30b), *304*
Byrne, D. R. 280 (10), *304*
Bystrov, V. F. 139 (52), *159*

Cadby, P. A. 876 (124), *916*
Cade, P. E. 455 (204), *461*
Cadena, R. 272 (55, 56), *277*, 546, 549 (118a), *581*
Cádiz, C. 315 (30), *370*
Cady, G. H. 262 (11), *276*, 669 (40), *688*
Cahill, A. E. 206 (20), *229*, 588 (14a), *643*
Caille, J. 833 (232), *919*
Cais, R. E. 422 (34), *427*
Calder, G. V. 405 (114), *415*
Calingaert, G. 824 (70), *828*
Calligaris, M. 88, 91 (78), *95*, 401 (98), *414*
Calvert, J. G. 2 (17), 43 (154), 60 (238), 68 (17, 302), *78*, *81*, *82*, *84*, 491 (57), 492 (61, 62), *518*, 677 (59), *689*
Camaioni, D. M. 298 (193c), *308*
Cambie, R. C. 546 (107), *581*
Camerman, A. 400 (96, 97), *414*
Camerman, N. 400 (96, 97), *414*
Campbell, B. C. 910 (212b), *919*
Campbell, B. S. 229 (222), *234*, 832 (215), 910 (212a), 911 (215), *919*
Campbell, I. M. 312 (6), *369*
Campbell, R. A. 292 (150), *307*, 536 (73), *580*
Campbell, T. W. 88–90 (60), *95*
Campos, O. 189 (124), *199*
Cancio, E. M. 903, 905 (189a), *918*
Candlish, E. M. C. 471 (85), *480*
Cantrill, J. E. 285 (66), *305*, 546 (120), *581*, 636 (198), *648*
Capiton, G. A. 114 (18c), *126*
Capparella, G. 471 (91), *481*
Cariati, F. 474 (127), *481*
Carles, J. 123 (36), *127*, 634 (182, 183a–c), *647*
Carlsen, N. R. 36, 57, 58 (106), *80*, 87, 88 (49), *95*
Carlsmith, A. 297 (184), *308*
Carlson, R. G. 292 (145), *307*
Carlsson, D. J. 731 (73), *772*
Caroll, H. F. 69 (307), *84*

Carotenuto, A. 334 (115), *372*
Carpino, L. A. 597 (65), *645*
Carr, R. V. C. 226 (172), *233*, 320 (53), *371*
Carreri, D. 466, 468, 474 (30), *479*
Carrondo, M. A. A. F. deC. T. 86, 88 (13), *94*
Carson, A. S. 791 (49), *805*
Carter, F. L. 68 (303), *84*
Carter, T. P. 651, 654 (7b, 7e), *681*
Carter, T. P., Jr. 501 (94), *519*, 654 (30, 31), *682*
Carton, P. M. 724 (48), *771*
Casabianca, F. 465, 467–469 (26), *479*
Casella, L. 474, 475 (133), *481*
Caserio, M. J. 436 (46), *457*
Caspar, R. 551 (134), *581*
Cass, M. W. 844 (14), *913*
Cass, W. E. 287 (80), *305*
Castaneda, E. 796 (64), *805*
Castellano, E. E. 388 (47), *413*
Castonguay, J. 123 (36), *127*, 634 (183c), *647*
Catala, J. M. 420, 422 (11), *427*
Caticha-Ellis, S. 88, 91 (81), *95*, 407 (130), *415*, 611 (107b), *646*
Caton, M. P. L. 438, 446 (68), *458*
Cava, M. P. 271 (49), *277*
Cavatora, P. 725 (54), *771*
Cavit, B. E. 465, 471 (88), *481*
Cecere, M. 176 (65), *197*
Cederbaum, L. S. 45 (166), *81*
Cekovic, Z. 174 (55, 56), *197*
Cella, J. A. 295 (159a, 159b), *307*
Celotta, R. J. 48, 49 (179), *81*
Cenci, S. 186 (109), *198*
Cenini, S. 471 (91), *481*
Cerc, G. 296 (174), *308*
Cerefice, S. A. 526 (26), *579*
Ceresa, R. J. 418 (1, 2), 423 (41, 42a), 424 (42b, 56), 426 (1, 2), *426*, *427*
Cerniani, A. 296 (173a), *308*, 544, 545 (97a), *580*
Cha, D. Y. 190 (132), *199*
Chabaud, B. 189 (125), *199*
Chaltykyan, O. A. 144, 149 (60), *159*
Chalvet, O. 324 (78), *371*
Chamberlain, P. 526, 532 (24), *579*
Chambers, R. D. 194 (150), *199*, 534 (64), *580*
Champion, S. L. 866 (90), *915*
Chan, H. W. S. 465, 469 (51), *480*
Chan, S. I. 388 (46), *413*
Chan, T. W. 195 (158), *200*, 543 (95), *580*
Chan, W. H. 677 (59), *689*
Chan, Y. M. 114 (18a, 18c), *126*
Chandrasekhar, S. 910 (210), *918*
Chang, C.-C. 898, 899, 908 (207), *918*
Chang, H.-M. 67 (291), *83*
Chang, K. H. 91 (83), *95*, 407 (135, 136, 138), *415*
Chang, Y. C. 333 (111), *372*, 588, 592, 593 (15), *643*, 860 (64b), *915*
Chan-Leuthauser, S. W. H. 446 (134), *459*
Chapiro, A. 424 (51–54), *427*
Chapman, O. L. 405 (114), *415*, 742 (141b), *773*
Chapman, S. 677 (111), *690*
Charpentier, R. 187 (119), *199*, 383 (38b), *413*, 475, 476 (144), 477 (148), *482*, 812 (65), *828*
Chauhan, J. S. 285 (71), *305*
Chautemps, P. 532 (54), *579*
Chawla, O. P. 757, 758 (171a), *774*
Chedekel, M. R. 446 (133), *459*
Chen, J. S. 571 (232), *583*
Chen, J. T. 825 (84), *828*
Chen, L. C. 465, 472 (95), *481*
Chen, M. J. W. 478 (150), *482*
Chen, M. M. L. 61 (249), *83*
Chen, Y. Y. 296 (175), *308*
Chenevert, R. 501, 507 (99), *519*
Cheng, C.-C. 832 (223), 841, 865, 889 (86), 899 (86, 223), *915*, *919*
Cheng, Y. M. 110 (10), *126*, 753 (163a, 163b), *774*
Chenier, J. H. B. 248 (49–52), 249 (57), 250 (49), *257*, 489 (46), *518*
Cherdnickenko, V. M. 134 (23), *158*
Cherest, M. 446 (130), *459*
Chern, C. I. 436, 437 (52), 438 (67), 441 (67, 87, 89), 448 (67), 453, 454 (87), *457–459*
Chern, C.-I. 442–445 (117), *459*
Cherton, J. C. 833 (232), *919*
Chevalet, J. 448 (157), *460*
Chevrier, B. 465, 470 (76), *480*
Chiang, W. 291 (138), *307*
Chiaroni, A. 382 (36, 37), *413*, 475 (156), *482*
Chiasson, B. A. 526 (25), *579*
Chiavoni, C. 825 (78), *828*
Chien, J. C. 796 (68), 797 (68, 72), 798 (68), *805*
Chien, J. C. W. 732, 737 (86), *772*, 800, 802 (92), *806*
Chikinova, N. V. 817 (39), *827*
Chin, C.-A. 896 (184), *918*
Chin, D. 484 (10), *517*
Chin, E. 560 (197), *583*
Chinuki, T. 167 (29), *197*
Chio, H. 89 (64), *95*
Chirko, A. I. 137 (44), 151 (74), *158*, *159*
Chkheidze, I. I. 765 (201), *775*
Cho, B. R. 271 (49), *277*
Chodak, I. 717 (17b), *770*
Choi, C. S. 49, 51 (193), *81*, 86, 88 (11), *94*
Chong, A. O. 171 (44), *197*, 824 (73), *828*

Choo, K. Y. 733 (92b), *772*
Chopra, C. L. 312 (7b), *370*
Chow, M. F. 347 (167, 168a, 168b), 348 (168a, 168b), *373*, 781 (21), 782 (26), *804*
Chow, M.-F. 226 (180), *233*, 831 (12), 843 (12, 201), 846, 857 (12), 863 (72), 873 (114), 906 (201), *913*, *915*, *916*, *918*
Choy, V. J. 464, 465, 473 (8), *479*
Chrástová, V. 796, 797 (67), *805*
Chrisope, D. R. 910 (211d), *919*
Christe, K. O. 669 (30), *688*
Christen, D. 49, 50, 67 (197), *82*, 87, 89 (22), *94*
Christian, D. F. 471 (87), *480*
Christiansen, P. A. 36, 38 (112), *80*
Christie, K. 99 (21), *104*
Christie, K. O. 735 (102), *772*
Christl, M. 832, 863 (71a), *915*
Christoffersen, R. E. 53 (217), *82*
Chromeček, R. 598 (69), *645*
Chu, C. C. 557 (171), *582*, 780 (12b), *804*
Chu, C.-C. 599, 625 (75), *645*
Chu, H.-K. 206 (31), *230*, 544 (98), *580*
Chu, S. 88, 92 (97), *96*, 134 (17), *158*, 405 (108), *414*
Chu, S.-C. 405 (109), *414*
Chuang, V. T. 219 (130), *232*
Chuiko, L. S. 425 (65, 72), 426 (72), *427*, *428*
Chumaevskii, E. V. 905 (195), *918*
Chung, P. J. 440, 454 (74), *458*
Church, D. F. 546, 547 (110b), *581*
Ciabattoni, J. 292 (150), 298 (192), *307*, *308*, 536 (73, 74), 538 (79), *580*
Ciannotti, C. 382 (37), *413*
Cilento, G. 844 (18n, 19, 20), 904 (193b), *913*, *914*, *918*
Cinquini, M. 296 (172), *307*
Cioffari, A. 810 (11), *827*
Cipau, G. R. 628 (158b), *647*
Cipriani, R. A. 557 (170), *582*, 628 (164b), *647*
Citovický, P. 796, 797 (67), *805*
Clapp, L. B. 570 (227), *583*
Clardy, J. 314 (22, 23), 320 (53), *370*, *371*, 396 (80, 81), 400 (94), *414*
Clardy, J. C. 91 (82), *95*, 407 (131), *415*
Clark, B. C., Jr. 736 (108), *772*
Clark, H. C. 474 (134), *482*
Clark, J. H. 215 (112b), *232*, 441 (98), *459*
Clark, J. T. 260 (4), *276*
Clark, M. 194 (150), *199*
Clarke, J. T. 280 (9), *304*
Clarke, S. M. 312 (5), *369*
Clavert, J. G. 668, 671, 674 (25), *688*
Clements, A. H. 134 (28), *158*
Clennan, E. L. 446 (152), *459*, 851 (49), *914*
Clerici, A. 752 (160b–d), *774*
Clinton, N. A. 240, 247 (19), *257*, 499, 500 (81–83), *519*, 622 (138b), *646*
Clipsham, J. 249, 250 (58), *257*
Clough, S. A. 57 (228), *82*
Coates, R. M. 287 (88), *305*
Coburn, W. C., Jr. 398 (93), *414*
Coggon, P. 213 (98), *231*, 314 (24), *370*
Cohen, M. D. 778 (1), *804*
Cohen, N. C. 349 (171), *373*
Cohen, S. G. 739 (120), *773*
Cohen, S. R. 795 (62), *805*
Cohen, Z. 658 (46a–c), 660 (46a), 664 (52a–c), 666 (52c), 668 (52a–c), 671 (46a–c), *682*
Cole, R. J. 314 (22, 23), *370*, 396 (80, 81), *414*
Collart-Lempereur, M. 384 (38f), *413*
Collin, R. L. 86, 88 (10), *94*, 376 (3), *412*
Collins, C. J. 556 (164), *582*
Collins, J. C. 651 (10a), *681*
Collman, J. P. 466 (112, 116), 471 (82), 473 (112, 116), 474 (126, 130), *480*, *481*
Concannon, P. W. 292 (150), 298 (192), *307*, *308*, 536 (73, 74), *580*
Conia, J. M. 221, 225 (136), *232*, 835 (238), 863 (74a, 74b), 865 (82), 911 (74a, 74b), *915*, *919*
Connick, R. E. 206 (19), *229*
Connor, J. 780 (12a), *804*
Connor, J. A. 464, 465, 468, 469, 471, 474 (15), *479*
Conover, W. W. 294 (153), *307*
Conrad, N. D. 613 (112a), *646*
Cook, C. D. 466, 474 (128), *481*
Cook, G. A. 658 (47), *682*
Cook, J. M. 189 (124), *199*
Cookson, P. G. 314 (29), *370*, 733 (95), 760, 761 (181c), *772*, *774*
Coon, C. L. 259, 260 (3), *276*
Coon, M. J. 471 (83), *480*
Cooper, D. W. 166 (23), *196*
Cooper, J. L. 207 (38), *230*, 782, 783 (25), *804*
Cooper, W. 280 (17), *304*
Cope, A. C. 290 (118a–c), *306*
Coppens, P. 55 (219), *82*
Corbett, G. E. 261, 262, 266 (9a), *276*
Corey, E. J. 316 (36), *370*, 438 (65), 441 (65, 109), *458*, *459*
Corrao, A. 632 (172), *647*
Cosijn, A. H. M. 153 (83), *159*
Cotton, F. A. 587 (6), *643*, 809 (85), *828*
Coughlin, D. F. 360, 362 (197), *373*
Coughlin, D. J. 120 (27), *127*, 216 (122), *232*, 320 (54), 350 (172), 360, 363 (54), *371*, *373*, 384 (38h), *413*
Court, W. A. 213 (102), *231*
Courtney, A. 468 (31), *479*

Cowan, J. C. 650 (3k), *681*
Cowles, C. 292 (145), *307*
Cox, J. D. 40 (138), *80*, 98, 99 (7), *104*
Cox, J. K. 178 (78), *198*
Cox, R. A. 677 (63, 678 (63, 68), 679 (63, 74), 680 (74), *689*
Cragg, L. 420 (15), *427*
Craig, J. C. 297 (183), *308*
Crandall, J. K. 294 (152, 153), 297 (189), *307*, *308*
Crank, G. 435, 436 (41), *457*
Crawford, R. 838, 883, 884 (145d), 906 (198), *916*, *918*
Crawley, L. C. 297 (189), *308*
Creig, J. A. 425 (76), *428*
Cremer, D. 13, 21 (32), 24 (58, 59), 25 (58, 59, 87), 33 (58), 36 (87, 107), 37, 38 (107), 41, 42 (149), 45 (59), 46 (107), 49, 50 (107, 197), 51 (107, 199), 53, 54, 57 (107), 61 (58), 62 (58, 266, 267), 63 (266, 267, 269, 270, 273), 64–66 (266), 67 (197, 266), 68 (273), 69 (266, 273, 304), 71 (58, 269, 270, 313, 316–318), 72 (269, 317, 318), 73 (319, 320), *78–84*, 87 (22, 47), 88 (47), 89 (22, 47), 90 (47), *94*, *95*, 485, 486 (22), *518*
Criegee, R. 165 (18), *196*, 299 (198), *308*, 492 (64b), *519*, 551 (133, 134), 559 (183), 574 (239), *581*, *582*, *584*, 594 (42), 623 (145), 632 (173), *664*, *647*, 650 (3b, 3d, 3e, 3h, 3m, 3n), *681*, 680, (78, 80), *689*, 730 (68), *771*, 831 (10a, 10b), *913*
Cristoph, G. G. 526, 527 (32), *579*
Cross, P. C. 36, 38, 49, 50 (119), *80*, 87–90 (24), *94*
Crovatt, L. W. 263, 266 (21), *276*
Crow, W. D. 313 (20), *370*, 394 (74), *414*
Cruickshanck, F. R. 40 (140), *80*
Cruickshank, F. R. 675 (51), *689*
Cruikshank, D. W. J. 395 (79), *414*
Cruikshank, F. R. 97 (2), *102*
Cruthoff, R. 594 (46), 597 (66), 603–607 (46), 614, 616 (66), *644*, *645*
Csizmadia, I. G. 41, 48, 49 (151), *81*, 538 (77a), 562 (202), *580*, *583*
Cubbon, R. C. P. 665 (27), *688*
Cudd, M. A. 384 (38d), *413*
Cueto, O. 322 (63), *371*, 753 (164, 165), 754 (165), *774*, 833 (229b, 233), 836, 837 (233), 841 (86), 854, 855 (56a, 56b), 865 (86), 884 (172a), 885 (229b), 888 (233), 889 (86, 233), 890 (233), 895 (172a), 898 (207, 229b), 899 (86, 207), 900 (233), 903 (172a), 908 (207), *914*, *915*, *917–919*
Cugley, J. 88, 92 (94b), *96*
Cugley, J. A. 88, 92 (95), *96*
Cumbo, C. C. 290 (116), *306*
Cummins, R. W. 542, 543 (92), *580*
Cuong, N. K. 349 (171), *373*, 783 (28), 786 (38), *804*, *805*
Cuppers, H. G. M. 473 (118), *481*
Curci, R. 2 (14), *78*, 186 (107–109, 112), *198*, 288 (100), 296 (176), 298 (191, 192), *306*, *308*, 444, 453, 454 (122), *459*, 522, 523, 525 (1c), 533 (59), 542 (91, 93), 543 (93, 94), 544 (93, 97b), 545 (97b), 563 (204), 568 (1c), *578–580*, *583*
Curtin, D. Y. 632 (170), *647*, 778 (1), *804*
Curtis, A. B. 430 (9), *456*
Curtis, H. C. 740 (127), *773*
Curtiss, L. A. 25, 33, 45 (74), *79*
Cutting, J. D. 182 (86), *198*
Cvetanovic, R. J. 656 (37), *682*
Czapski, G. 452 (189), *460*, 484 (8), *517*, 761, 762 (185), *774*
Czarnowski, J. 99 (29–32), 101 (31, 32), *104*, 497 (72, 73), *519*, 669 (37), 671 (37, 41), 672 (37), *688*, *689*

Dahlmann, J. 823 (60) *828*
Dahmann, J. 821 (53a), 822 (53a, 53b), 823 (53b), *827*
Dailey, R. G., Jr. 213 (102), *231*
Dakin, H. D. 567 (216), *583*
Dalcze, D. 384 (38f), *413*
Dale, J. 396 (86), *414*
Dalton, A. J. 425 (62), *427*
Dandel, R. 324 (78), *371*
Danen, W. C. 441 (100, 101), 452 (100), 456 (210), *459*, *461*
Daniel, R. B. 181, 192 (85), *198*
Daniels, K. 532–535 (53), *579*
Daniels, P. J. L. 219 (130), *232*
Dankleft, M. A. P. 543 (94), *580*
Dannley, R. L. 261, (9a, 9b), 262 (9a, 12), 266 (9a, 9b, 12, 35, 38), 267 (9b, 35, 38, 39), 268 (12, 35), 269 (9b), 274 (64), 275 (65), *276*, *277*, 759 (178, 179), 760 (181a), *774*, 780 (14), *804*, 821 (54a, 54b), *828*
D'Ans, J. 134 (21a), *158*
D'Antonio, P. 88, 89 (70), *95*
Dao, L. H. 441, 453 (106), *459*
Darling, T. R. 836, 906 (200), *918*
Darnall, K. R. 677 (60), *689*
Darnbrough, G. 438 446 (68), *458*
DaRooge, M. A. 249 (56), *257*
Das, G. 21 (43, 44), 61 (261), *79*, *83*
Dauben, W. G. 60 (239), *82*
Daudel, R. 54 (218), *82*
Dauphin, G. 263 (20), *276*
David, C. 731 (74), *772*
David, K. H. 797 (75), *805*
Davidson, A. J. 534 (66), *580*
Davidson, P. J. 818 (44), *827*

Davidson, R. B. 36, 53 (96), *79*, 87, 88, 90 (42), *95*
Davidson, R. S. 208, 222 (46), *230*
Davies, A. G. 2 (4), *78*, 168 (33), 193 (144), *197*, *199*, 236 (2), *256*, 280 (19), 299 (212), *304*, *308*, 314 (29), *370*, 522, 525 (1a), 554 (149), 556 (167), *578*, *582*, 590 (19a–c), 641 (213, 214), 642 (214), *643*, *644*, *648*, 733 (90, 91b, 95), 760, 761 (181c), *772*, *774*, 808 (2b), 809 (6, 7), 810 (17), 811 (17, 18), 812 (23, 24, 26), 813 (24), 814 (30, 32), 816 (17, 37), 820 (49a, 49b, 51), 821 (49a, 49b, 56), 822 (2b), 824 (71), *826–828*
Davies, D. I. 287 (83), *305*
Davies, D. W. 47 (173), *81*
Davies, G. R. 818 (46), *827*
Davis, B. T. 356 (189), *373*
Davis, D. W. 44, 45 (161), *81*
Davis, F. A. 571 (232), *583*
Davis, G. T. 271, 272 (54), *277*
Davis, J. J. (155), *774*
Davis, R. C. 441 (90), *459*
Davis, W. H. 130 (5), *157*
Davison, W. H. T. 92 (91), *96*, 134 (25), *158*, 514, 515 (115), *520*
Dayagi, S. 271 (49), *277*
De, B. R. 25 (83), *79*
Dean, L. B. 87, 88 (31), *94*
Dean, P. M. 207, 211 (36), *230*, 783 (29), *804*
De'Ath, N. J. 910 (212a), *919*
Debenham, D. F. 794, 795 (60, 61), *805*
Debnath, S. 628 (158b), *647*
De Boer, T. J. 326 (87), *371*
DeCamp, M. R. 742 (140), 747 (154), *773*, *774*
De Carlo, V. J. 720 (28), *771*
Declercq, J. P. 384 (38f), 400 (95), *413*, *414*
Defion, A. 351 (177a), *373*
Degani, Y. 271 (49), *277*
Deglise, X. 485 (13), *517*
Degtayareva, T. G. 422 (29), *427*
Deguchi, S. 424 (57), *427*
Dehmlow, E. V. 437, 438 (66), *458*
Dehnel, R. B. 534 (62), *579*
Deinger, M. 600, 632 (82a), *645*
Deinzer, M. 591 (23, 24), 594 (24), 595, 603 (23, 24), 604, 606 (24), *644*
De Jesus, R. 441, 453, 454 (87), *459*
Dekerk, J.-P. 400 (95), *414*
Dekkers, H. P. J. M. 841 (248b), *919*
De La Mare, H. E. 214, 221, 225, 228 (107), *232*, 573 (234), *583*
De La Mare, H. E. 175 (61), *197*, 453 (196), *460*
Delaunois, G. 731 (74), *772*
Deletang, C. 351 (176), *373*
De Luca, H. F. 189 (124), *199*
DeLuca, M. 844 (16, 17b), *913*
De Lucci, O. 322 (63), *371*, 843, 858 (58), *914*
Delaunois, G. 731 (74), *772*
Delzenne, G. 425 (74, 75), *428*
De Marco, R. 68 (303), *84*
de Meijere, A. 498 (77), *519*
Demerjian, K. L. 2, 68 (17), *78*
Demole, C. 215 (115), *232*, 318 (47), 365 (200), *371*, *374*
Demole, E. 215 (115), *232*, 318 (47), 365 (200), *371*, *374*
Dempsey, M. E. 844 (16), *913*
Demuth, M. R. 367 (208), *374*
Denes, A. S. 538 (77a), *580*
Denisev, Ye. T. 422 (30), *427*
Denison, E. T. 134, 143 (16), *158*
Denisov, E. T. 796–798, 801 (70), 802 (110), *805*, *806*
Denney, D. B. 229 (222), *234*, 264 (30), 265 (30, 32), 271 (48), *277*, 282 (57), 300 (217), 301 (219, 222), *305*, *309*, 453 (193), *460*, 546 (113, 124a–c), 547, 548 (113), 552 (138, 139), *581*, 593, 594 (36), 596 (57), 599 (36, 73), 623 (36), 625 (36, 73, 148, 149c), 626 (36, 73), 628 (36, 73, 148), 629 (73), 635 (191), 636 (200), 637 (201–203), 638 (203–205), 641 (212), *644*, *645*, *647*, *648*, 832 (215), 910 (212a, 212b), 911 (215), *919*
Denney, D. G. 593, 594, 599, 623, 625, 626, 628 (36), *644*
Denney, D. Z. 229 (222), *234*, 265 (32), 271 (48), *277*, 282 (57), 301 (219, 222), *305*, *309*, 546–548 (113), 552 (139), *581*, 596 (57), 625, 628 (148), 637 (201), 638 (204, 205), *644*, *647*, *648*, 832 (215), 910 (212a, 212b), 911 (215), *919*
Denning, P. H. 292 (142a), *307*
Dennis, N. 336 (123), *372*
Denny, R. 211, 217 (87a), *231*
Denny, R. W. 210 (65a, 66c), 212 (66c), 219 (130, 133), *230*, *232*, 318, 324 (45d), *370*, 588, 592 (10b), *643*
Deno, N. C. 298 (195), *308*
Denson, D. D. 736 (108), *772*
Denton, W. I. 178 (75), *198*
Depannemaecker, J. C. 33, 48, 49, 62 (70), *79*
De Pooter, H. 282 (58), *305*
Derkosch, J. 712 (7), *770*
Dervan, P. B. 742 (142), *773*
Derwent, R. G. 677–679 (63), *689*
Desbene, P.-L. 842 (251b), *919*
Descamps, B. 669 (38, 39), 671 (38), *672*, *673* (38, 39), *688*
Deslongchamps, P. 501, 507 (96–99), *519*, 652 (17), *681*

DesMarteau, D. D. 37, 67 (124), *80*, 88, 90 (71), *95*, 381 (29), *412*, 496 (70, 71), *519*
Desvergne, J. P. 675 (65), *683*
Detar, D. F. 597 (65), *645*
De Tar, D. F. 599 (77b), 625, 628 (150), *645*, *647*
Detitta, G. T. 384 (38h), *413*
Devaquet, A. 24, 58 (54), *79*, 893 (165), *917*
Dever, D. F. 492 (61), *518*
Dewar, M. J. S. 24, 45, 58 (62), 65 (282), *79*, *83*, 88, 89 (67), *95*, 209, 212 (64), *230*, 611, 613 (110), *646*, 893 (164a, 164b), *917*
Dewey, R. S. 320 (56), *371*
De Witt, B. J. 281 (44), *305*
Dhar, K. L. 312 (7b), *370*
Diak, J. 312 (9a), *370*
Dial, W. R. 281 (44), *305*
Dibiase, S. A. 437 (59), *457*
DiBiase, S. A. 440 (79), *459*
Dickey, F. H. 164 (10), *196*
Dickson, J. 447, 454 (155), *459*, 879 (138), *916*
Di Cosimo, R. 441, 453, 454 (87), *459*
Diebold, Th. 465, 470 (76), *480*
Dietz, R. 432 (23), 441 (86), 448 (158), 449, 450, 453 (86), *457*, *459*, *460*
Diez, W. 730 (67), *771*
DiFuria, F. 186 (107–109, 112), *198*, 298 (191, 192), *308*
Di Furia, F. 533 (59), 542 (91), 563 (204), *579*, *580*, *583*
DiGiorgio, J. B. 219 (130), *232*
Diki, M. A. 150 (73), *159*
Dimitrijevic, J. 174 (56), *197*
Dimitrov, D. I. 725 (57), *771*
Ding, J. Y. 227–229 (191), *234*, 314 (28), *370*, 832–834, 847, 848, 859, 881, 885, 886, 890, 910 (27), *914*
Ding, J.-Y. 156 (88b), *160*, 833, 834, 885, 887, 890, 892, 898–900 (159a), *917*
Diodati, F. P. 37 (125), *80*, 88, 89 (69), *95*, 381 (30), *412*
Diprete, R. A. 288 (100), *306*
Di Prete, R. A. 542–544 (93), 580
Dishong, D. M. 437 (59), *457*
Dittmann, W. 290 (135), *307*
Divisek, J. 441 (110), *459*
Dix, T. A. 225 (167b), *233*
Dixon, B. G. 577 (247), *584*
Dixon, J. E. 566 (213), *583*
Dixon, W. T. 57 (229), *82*
Djerassi, C. 110 (12), *126*, 312 (10b), *370*, 653 (24c), *682*
Djokic, G. 174 (56), *197*
Djokic, S. M. 494, 495 (65), *519*
Djordjevic, C. 465 (37), 469 (37, 43), *479*, *480*
Dmitruk, A. F. 90 (73), *95*
Dobashi, S. 210 (74, 75, 79), *231*, 328 (91), 329 (99), 368 (210), *371*, *374*
Dobbs, A. J. 712 (1c), *770*
Dobinson, F. 650 (2), *681*
Dobis, O. 614 (116c), *646*
Dobrynina, T. A. 86, 88 (12, 19), 89 (12), *94*
Dodonov, V. A. 356 (186), *373*, 808 (2a), 817 (40–42), 821 (56), *826–828*
Doedens, R. J. 466, 473 (111), *481*
Doering, W. von E. 559, 560 (185, 187), *582*, 596 (54), 613, 614 (113), 636 (54), 642 (215), *644*, *646*, *648*
Doetschman, D. C. 891 (153b), *917*
Dogliotti, L. 757, 758 (172), *774*
Doherty, D. G. 724 (51, 52), 739 (51), *771*
Dohgane, I. 167 (29), *197*
Dohrmann, J. K. 721, 722 (38), *771*
Dolphin, D. H. 471 (93), *481*
Domagala, J. M. 530 (51), *579*
Dombchik, S. A. 614 (123), 615, 616 (126), *646*
Domcke, W. 45 (166), *81*
Donaldson, P. B. 466 (100), 472 (100, 101), *481*
Dondoni, A. 297 (181a, 181b), *308*, 539 (82), 545 (99, 100), *580*
Donetti, A. 876 (122b), *916*
Doorn, J. A. van 177 (67), *197*
Dorer, F. H. 732 (81), *772*
Dorfman, E. 559, 560 (185), *582*, 642 (215), *648*
Dorfman, L. M. 731 (77), 761, 762 (185), *772*, *774*
Dörhöfer, G. 204 (6), *229*
Dorp, D. A. van 339 (137), 359 (195), *372*, *373*
Dorst, W. 718, 738 (18), *770*
Doshan, H. 684 (101), *690*, 831, 832, 859 (2), *913*
Doskotch, R. W. 114 (18a–c), *126*
Doughtery, R. C. 451 (174), *460*
Downer, J. 419 (5), *426*
Draganic, I. G. 761 (183d), *774*
Draganić, I. G. 767 (204), *775*
Draganic, Z. D. 761 (183d), *774*
Draganić, Z. D. 767 (204), *775*
Drago, R. S. 135, 138, 142, 148 (34c), *158*
Dreiding, A. S. 271 (49), *277*
Drens, W. 331 (105), *372*
Drew, D. H. 614 (122), *646*
Drew, E. H. 282 (48), *305*, 596, 617 (53), *644*
Drew, R. E. 465, 469 (34, 36), *479*
Drozdova, T. I. 145, 147, 149 (67), *159*
Druckrey, E. 333 (113), *372*
Druzhkov, O. N. 811 (67), *828*
Druzhov, O. N. 811 (19), *827*

Dryle, M. P. 286 (74), *305*
Dryuk, V. G. 288 (105), *306*, 525, 528 (5c), *578*
Dubinskii, A. A. 739 (122), *773*
Dudarev, V. Ya. 86, 88 (12, 19), 89 (12), *94*
Dudley, C. 472 (105), *481*
Duesler, E. N. 778 (1), *804*
Duffraisse, M. C. 349 (171), *373*
Dufraisee, C. 207, 211 (36), *230*
Dufraisse, C. 781 (19), 783 (28–30), 785 (30), *804*
Duke, A. J. 55 (221), *82*
Dulog, L. 797 (75), *805*
Duncan, J. L. 89 (63), *95*
Dunlap, D. E. 319, 343 (48), *371*
Dunning, T. H. 87–90 (46), *95*
Dunning, T. H., Jr. 7 (26), 24 (48–50), 25 (26, 49, 81), 33 (26, 50, 81), 36 (99, 101), 37, 38 (101), 45 (49), 52 (81), 53, 55, 56 (101), 57 (49), 58 (48–50), *78–80*
Dunsch, L. 796, 797 (66), *805*
Duong, K. N. V. 825 (76), *828*
Duong, T. 498 (78), *519*, 664, 665, 671 (54a–c), *682*
Dupin, J. F. 111 (13a), *126*
Dupuis, M. 55 (220), *82*
Duran, J. L., Jr. 847, 864, 894 (32), *914*
Duran, N. 226 (179), *233*, 359 (196), *373*
Durán, N. 736 (107), *772*
Durant, J. L. 685 (108, 110), 686 (110), *690*
Durham, D. L. 207 (34), *230*
Durham, L. J. 680 (84), *689*
Durland, J. 655 (33), *682*
Durland, J. D. 513 (112), *520*
Dürr, H. 869 (96), *915*
Duynstee, E. F. J. 678 (69), *689*
Dvořáh, V. 742 (141a), *773*
Dyckerhoff, K. 831, 858 (11), *913*
Dygos, J.-H. 526 (30), *579*
Dyke, J. M. 44, 45 (164), *81*
Dykstra, C. E. 538 (77c), *580*, 894 (167), *917*
Dzakpasu, A. 832, 838 (226), *919*
Dzakpasu, A. A. 859 (59), *914*
Dzhagatspanyan, R. V. 905 (195), *918*
Dzhemilev, U. M. 178, 179 (73), 184 (93, 94, 97, 98), 185 (97), 186 (110, 111, 114), *197–199*
Dzumedzei, N. V. 90 (73), *95*, 737 (113), *773*

Eaker, C. W. 893 (163d), *917*
East, R. L. 673, 674 (47), *689*
Eaton, R. S. 587 (5b), *643*
Ebsworth, E. A. V. 464, 465, 468, 469, 471, 474 (15), *479*
Eda, B. 617 (129b, 129c, 129e), *646*
Eddy, C. R. 134 (24), *158*, 403 (103), *414*
Edge, D. J. 733 (94b), 760 (181b), *772*, *774*
Edney, E. O. 680 (77), *689*
Edqvist, O. 44, 45 (159), *81*
Edward, J. T. 286 (77), *305*
Edwards, F. G. 281 (46), *305*
Edwards, J. 288 (100), *306*
Edwards, J. O. 130 (1c), 131 (3), *157*, 186 (109), *198*, 441 (102a, 102b), 444, 453, 454 (122), *459*, 522, 523, 525 (1b, 1c), 538 (79), 539 (83), 540 (84), 542 (93), 543 (93, 94), 544 (84, 93), 563 (204), 566 (213), 568 (1c), *578*, *580*, *583*, 587 (5a), 641 (211), *643*, *648*, 795 (62), *805*
Edwards, O. E. 2 (14), *78*, 271 (51), *277*
Egerton, A. C. 134 (29), *158*
Eggelte, H. J. 213 (94a, 94b), 216 (94a, 94b, 123a), *231*, *232*, 317 (42), 320 (55, 57), 321 (55, 58, 60, 61), 322 (55), 343 (60, 61), *370*, *371*
Egorochkin, A. N. 142, 147, 150, 151 (56), *159*
Ehrig, V. 854, 855 (56a), *914*
Eickhoff, D. J. 532–535 (53), *579*
Eickman, N. 314 (23), *370*, 396 (81), *414*
Eilers, J. E. 38 (135), *80*
Einspahr, H. M. 397 (92), *414*
Einstein, F. W. B. 465 (34, 36, 79), 469 (34, 36), 471 (79), *479*, *480*
Eirich, F. R. 609, 610, 624 (87), *645*, 722, 723, 743 (47), *771*
Elad, D. 733 (97, 98b), *772*
Elden, E. J. 340 (142), *372*
Elder, R. C. 393 (72), *414*
El-Feraly, F. S. 114 (18a–c), *126*
Eling, J. E. 316 (39), *370*
Elishewitz, S. 166 (21), *196*
Ellam, R. M. 574 (238, 241), *584*
Elling, W. 47, 48, 71 (176), *81*, 88–90 (65), *95*
Ellsworth, R. L. 596 (57), *644*
El'nitskii, A. P. 280 (26, 28a, 28c, 29a, 29b, 30a–c), *304*
El-Sayed, M. A. 904 (193a), *918*
Elstner, E. F. 433 (33), *457*
El-Sukkary, M. M. 879 (139), *916*
Elyashberg, M. E. 138–142, 149, 150 (47), *159*
Emanuel, N. M. 134 (23), *158*, 288 (96), *306*
Emge, D. E. 501, 503, 507 (101), *519*, 652, 653 (19b, 19c), *681*
Emmons, W. D. 297 (186a, 186b, 187, 188), *308*, 540 (87), 559 (189, 192), 560 (189), 561 (192), 570 (226), *580*, *582*, *583*
Emte, W. 134 (29), *158*
Enderlin, L. 781 (19), *804*
Engberts, J. B. F. N. 89 (58), *95*
Engel, P. S. 903 (190), *918*
Engelking, P. C. 48, 49, 54 (186), *81*
Engelmann, F. 628 (161), *647*

England, W. 36, 38 (111), *80*, 87, 88, 90 (33), *94*
Engler, D. A. 470 (68), *480*
Engström, S. 24 (57), 61 (258), *79*, *83*
Eppner, J. B. 303 (233b), *309*
Erden, I. 216 (123b), *232*, 321 (59), 322 (62, 64, 65), 337 (132), 343 (59), 346, 349 (64), 350 (62), *371*, *372*, 841, 865, 889, 899 (86), *915*
Erickson, R. E. 501 (86, 88), 503 (86), *519*, 651 (10a, 10b), 652 (18), 653 (26), *681*, *682*
Eriksen, J. 341 (150), *373*, 446 (137, 138, 140, 141), 450 (137), *459*, 874 (118a, 118b), *916*
Erman, M. G. 384 (38h), *413*
Erndt, A. 733 (98a), *772*
Ernst, F. 817 (38), *827*
Erskine, R. W. 464, 465, 473 (11), *479*
Erykalova, M. F. 157 (97), *160*
Eschinasi, E. H. 178 (76), *198*
Eskins, K. 861 (65), *915*
Espenson, J. H. 825 (84), *828*
Ester, W. 165 (16), *196*
Etienne, A. 781 (19), *804*
Etlis, V. S. 262 (17), 263 (25–27), 264, 265, (25, 31), *276*, *277*
Etter, M. C. 407 (135), *415*, 779 (7), *804*
Evan, C. A. 446 (131), *459*
Evans, C. A. 243 (26), *257*
Evans, G. W. 301 (220), *309*
Evans, R. F. 297 (185), *308*
Evans, W. H. 40 (139), *80*
Everett, A. J. 131, 132, 147 (2), *157*, 712 (5), *770*
Evleth, E. M. 60 (240, 241), *82*, 893 (163h), *917*
Evnin, A. B. 174 (52), *197*
Ewig, C. S. 36–38, 67 (122), *80*, 89, 90 (59), *95*
Exner, O. 87 (30), 88 (87, 110), 89 (58), 91 (30, 86–88), 92 (100, 101), 93 (100), 94 (88, 110), *94–96*, 408 (139), *415*
Ezhaya, A. 876 (122b), *916*

Faber, J. 783 , 785 (31), *804*
Fahrenholtz, S. R. 207 (41), *230*, 742 (138b), *773*
Faini, G. J. 901, 902 (185a), *918*
Fairchild, E. H. 114 (18a–c), *126*
Faler, G. 832, 839, 890, 892, 900 (157), *917*
Fang, T.-S. 544 (98), *580*
Fanghänel, E. 866, 867, 910, (92), *915*
Farberov, M. I. 178 (72, 74), 180 (81), 185 (100), *197*, *198*
Farid, S. 446 (144, 146), *459*
Farine, J.-C. 736 (108), *772*
Farkash, T. 220, 222, 223 (134), *232*, 555 (156), *582*, 863, 911 (75), *915*
Farmer, E. H. 222 (143), *232*, 555, 556 (158), *582*
Farmer, J. 826 (83), *828*
Farmer, R. C. 788 (42), *805*
Farneth, W. E. 685 (102), *690*, 891 (152c), *917*
Farnham, N. 559 (184), *582*
Farr, M. K. 49, 51 (193), *81*, 86, 88 (11), *94*
Farr, R. 223, 225 (156), *233*
Farrell, N. P. 471 (93), *481*
Fasiska, E. J. 387, 388 (51), *413*
Fattorusso, E. 312 (10a), *370*
Faulkner, D. J. 313 (18), 314 (21), *370*
Faulkner, R. D. 779 (4), *804*
Favilla, R. 725 (54), *771*
Fayat, C. 91 (79), *95*
Fayos, J. 314 (22), *370*, 396 (80), *414*
Fazal, N. A. 760, 761 (181c), *774*
Feder, R. L. 166 (21), *196*
Fedotova, E. I. 811 (20a), *827*
Fee, J. A. 448, 452 (156), 455 (205), *460*, *461*
Feig, G. 301 (219), *309*, 636 (200), *648*
Feinberg, A. M. 113 (16), *126*
Felber, U. 554 (150), *582*
Feld, M. 609 (102), *645*
Feld, R. 641 (213), *648*
Feldhues, M. 738 (118b), *773*
Feler, G. 893 (163h), *917*
Fell, B. 525 (14), *578*
Felton, R. H. 45 (167), *81*
Felzmann, J.-H. 291 (140), *307*
Fenical, W. 313 (19), *370*, 400 (94), *414*
Fenton, S. W. 290 (118a), *306*
Feringa, B. L. 335 (118), *372*
Feroci, G. 441 (85), 449 (168), *459*, *460*
Ferradini, C. 762 (186), *775*
Ferradni, C. 452 (186), *460*
Ferrandini, C. 456 (211), *461*
Fessenden, R. W. 724 (48), 735 (104), 757, 758 (171a), 767, 768 (205), *771*, *772*, *774*, *775*
Fétizon, M. 111 (13a, 13b), *126*
Fiatto, R. A. 525 (19), *579*
Fiedler, R. 298 (194a), *308*
Field, B. O. 464, 465, 473 (11), *479*
Field, G. S. P. 187 (116), *199*
Fields, E. K. 526 (26), *579*
Fierro, J. del 837 (54), 838 (54, 246), 839 (246), 854 (54), 889, 900, 905 (246), *914*, *919*
Filby, J. E. 156 (88b), *160*, 227–229 (191), *234*, 314 (28), *370*, 832 (27), 833, 834, (27, 159b), 847 (27, 29), 848, 859, 881 (27), 885 (27, 159b), 886 (27), 887 (159b), 890 (27, 159b), 892, 898, 899 (159b), 908 (206), 910 (27), *914*, *917*, *918*
Filliatre, C. 193 (145), *199*

Fina, N. J. 566 (213), *583*
Findley, T. W. 291 (136), *307*
Fine, D. H. 789 (46), *805*
Finer, J. 400 (94), *414*
Fink, E. H. 58, 59 (233), *82*
Fink, W. H. 36, 38 (89, 109), *79*, *80*, 87, 88 (41), *95*
Finkelstein, E. 446 (132), *459*
Finlayson, B. J. 683 (86, 87), *690*
Fischer, H. 740 (134), *773*
Fischer, J. 187 (119), *199*, 383 (38b), *413*, 475, 476 (144), *482*
Fish, J. L. 891 (153b), *917*
Fisher, J. 465, 467–469 (26), *479*, 812 (65), *828*
Fitzgerald, P. 562 (203), *583*
Fleet, G. W. J. 465, 469 (50), *480*
Fleischhauer, J. 865 (83), *915*
Flippen, J. L. 872 (106d), *915*
Fliszár, S. 634 (182, 183a–c), *647*
Fliszar, S. 121, 122 (33), 123 (36), *127*, 501 (104), *519*, 653 (22), *681*
Florian, L. R. 340 (142), *372*, 393 (72), *414*
Flowers, M. 733 (94a), *772*
Floyd, M. B. 333 (110), *372*
Flygare, H. 831 (10b), *913*
Flynn, G. 685 (102), *690*, 891 (152c), *917*
Fodor, Z. 796, 797 (69), *805*
Folli, U. 288 (92), *306*
Fölsing, R. 301, 302 (224), *309*
Foner, S. N. 25, 45, 46 (85), *79*, 98 (9), *104*
Fono, A. 173 (49), *197*
Fontaine, C. 382 (37), *413*, 475 (156), *482*, 825 (76, 78), *828*
Foos, J. 452 (186), *460*, 762 (186), *775*
Foote, C. S. 120 (30), *127*, 202 (1), 206 (22, 23), 208 (47), 210 (69–72, 77), 211 (87a), 213, 214 (93), 217 (87a), 219 (133), 226 (188), 227 (193), 229 (1, 223b), *229–234*, 318 (45b), 327 (90), 328 (90, 92), 329 (96), 330 (100, 101), 335 (121), 341 (150), 357 (191), *370–373*, 385 (42), *413*, 430 (11), 432, 433 (24), 446 (137–142), 448 (162), 450 (137), 453 (11), 454 (24), 456 (212), *456*, *457*, *459–461*, 568 (218), *583*, 587 (4), 588 (13a–d), 592 (29), 642 (4), *643*, *644*, 785 (34), *804*, 832 (225, 226), 835 (237), 836 (200, 237), 837 (237), 838 (226), 839 (237), 859 (59), 874 (118a, 118b), 876 (125), 906 (200), *914*, *916*, *918*, *919*
Föppl, H. 48, 49 (180), *81*
Formanek, K. 174 (59), *197*
Forni, A. 570, 572 (230), *583*
Forno, A. E. J. 432 (23), 441, 449, 450, 453 (86), *457*, *459*
Forst, W. 669, (38, 39), 671 (38), 672 (38, 39, 43), 673 (38, 39), *688*, *689*
Forster, C. H. 345 (161), *373*
Forsyth, D. A. 666, 676 (67), *683*
Fossey, J. 252 (65), *258*
Foster, C. H. 213 (103, 104), 214 (104), *231*, 324 (73), *371*
Foster, M. 906 (198), *918*
Foster, R. V. 556 (167), *582*
Foster, W. R. 800 (78), *805*
Fournier, A. 290 (118b), *306*
Fox, J. E. 223 (153), *233*
Fox, W. B. 68 (303), *84*, 246 (39), *257*, 496 (68, 71), *519*, 735 (103, 105), *772*
Fracheboud, M. 224 (157a), *233*
Franchini, P. F. 36, 53, 57 (92), *79*
Franchuk, I. F. 136–138, 150 (42), *158*
Frank, C. E. 224 (161), *233*
Franken, T. 204 (6), *229*
Frankenburg, P. E. 295 (164), *307*, 559, 560 (186), *582*
Frantzen, V. 292 (149a), *307*
Franzen, G. R. 290 (127), *306*
Franzen, V. 536, 537 (70), *580*
Fraser, R. T. M. 116 (19), *126*
Frauenglass, E. 561 (198), *583*
Freeburger, M. E. 653 (28c), *682*
Freed, K. F. 36 (118), *80*
Freeman, F. 529 (45), *579*
Frehel, D. 501, 507 (97–99), *519*, 652 (17), *681*
Freidlina, R. K. 732 (89), *772*
Freiesleben, W. 187 (120), *199*
Frenna, V. 632 (172), *647*
Freppel, C. 526 (21), *579*
Freund, H. G. 714–716, 740 (9), *770*
Frey, W. 288 (91), *306*
Frick, N. H. 800 (96), *806*
Fridovich, I. 430 (4, 5), 437 (53), *456*, *457*
Friebolin, H. 815 (86), *828*
Friedlander, B. F. 526 (31), *579*
Friedman, R. L. 263 (28), *277*
Friedrich, L. E. 525 (19), *579*
Fries, S. L. 559 (184), *582*
Friess, S. L. 295 (163, 164), *307*, 559, 560 (186), *582*
Frimer, A. 440 (80, 82), 445 (82), 449, 451 (80), *459*, 555 (157), *582*
Frimer, A. A. 2, 65 (10), *78*, 205 (7), 209, 212, 217 (55), 219 (131, 132), 220 (55, 131, 134), 222 (55, 132, 134, 148), 223 (134), 225 (163, 164, 169), 226 (55, 131), 229 (55), *229*, *230*, *232*, *233*, 433 (34), 435 (35, 36, 37a, 37b), 436 (50), 437 (37a, 36b, 50), 438 (35, 36, 37a, 37b, 64), 439 (35, 36, 70), 440 (50, 83), 441 (91), 444 (64), 445 (35, 50, 83), 449 (34, 165, 166), 451 (166, 169), 456 (91), *457–460*, 469, 470 (52), *480*, 555 (152, 156), *582*, 589 (18), *643*, 733 (97),

772, 832 (81), 844 (18v), 863 (75, 76), 865 (81), 878 (135a, 135b), 879 (136), 911 (75), *913*, *915*, *916*
Frisell, C. 282 (55, 57), *305*, 638 (204), *648*
Fritzche, M. 204 (3a), *229*
Frommer, U. 295 (155), *307*, 536 (68), *580*
Frost, D. C. 44, 45 (162), *81*
Frostick, F. C., Jr. 564 (207), *583*
Fry, A. 559, 561, 562 (191), *582*
Fueno, T. 25 (65), 61 (257, 259, 263), 65 (284–286), 71 (285), *79*, *83*, 209, 220 (61), *230*, 332 (109b), *372*, 466, 474 (122), *481*, 854 (53), 893 (163a), *914*, *917*
Fuery, P. H. 594, 605, 606 (48), 607 (96b), 608, 616 (48), 620 (48, 96b), *644*, *645*
Fuhrmann, R. 166 (21), *196*
Fujihara, M. 451 (175), *460*
Fujihira, M. 440 (72, 73), 441 (84), *458*, *459*
Fujii, H. 263 (28), *277*
Fujii, K. 721, 722 (43), *771*
Fujimori, K. 557 (174), *582*, 594 (40, 41), 596 (55, 56), 597 (63a), 599 (55, 56, 74), 601, 610 (40), 611, 613 (40, 56), 614 (56, 117), 615 (40, 56, 117, 135), 617, 618 (40, 117), 619 (63a, 117), 620 (55, 56, 135), 621 (55, 135), 622 (56, 135), 626 (40, 55, 74, 152, 154), 627 (55), 628 (41, 55, 74, 154, 165), 629 (40, 41, 55, 74, 217), 631 (165, 166), 632 (167), (216), *644–648*, 739 (119), *773*, 780 (12b), *804*
Fujimoto, H. 69 (306), *84*
Fujita, T. 61 (247), *83*, 87–89 (35), *95*
Fukizawa, S. (62), *828*
Fukui, K. 69 (306), *84*
Fukutome, H. 893 (163a), *917*
Fukuyama, T. 291 (141), *307*
Fukuzawa, S. 825 (74), *828*
Fukuzumi, S. 489 (44), 490, 491 (49), *518*
Füllbier, B. 288, 289 (108), *306*
Fung, L. W.-M. 446 (143), *459*, 874 (119), *916*
Fung, S. 216 (121), *232*, 319 (51), 366 (207), *371*, *374*
Funk, M. O. 317 (44), 339 (44, 139), *370*, *372*, 731 (71), *771*
Furia, F. di 469 (44), 471 (77), *480*
Furimsky, E. 247 (43), *257*, 733 (91a), *772*
Furuta, K. 718, 739, 740, 745 (24), *770*
Furutachi, N. 589 (17), *643*
Fusi, A. 466, 468 (30), 471 (91), 474 (30), *479*, *481*
Futrell, J. H. 48, 49, 54 (186), *81*
Fykin, L. E. 86, 88, 89 (12), *94*

Gabrische, A., Jr. 343 (155), *373*
Gadola, M. 348 (169), *373*
Gafurov, U. G. 801 (102), *806*
Gagen, J. E. 261 (9b), 266, 267 (9b, 38), 269 (9b), *276*, *277*
Gagnier, P. G. 594, 605, 606 (48), 607 (96b), 608, 616 (48), 620 (48, 96b), *644*, *645*
Gailyunas, I. A. 187 (118), *199*
Gajewski, J. J. 613 (112a, 112b), *646*
Gak, Y. N. 738 (117), *773*
Gak, Y. V. 738 (116), *773*
Gallagher, J. J. 57 (228), *82*
Gallagher, R. B. 425 (64), *427*
Gallagher, T. F. 560 (193), *583*
Gallaher, K. L. 69, 71 (314), *84*, 634 (184), *648*
Galli, R. 175 (63), 176 (65), *197*
Galliani, G. 438, 440, 445 (63), *458*
Galliard, T. 223 (154a–c), *233*
Gallopo, A. R. 540, 544 (84), *580*
Galvert, J. G. 734 (99), *772*
Gambale, R. J. 369 (212, 213), *374*
Gambarjan, S. 286 (76), *305*
Gancher, E. 303 (233a), *309*
Ganem, B. 194 (151–153), *199*, 295 (158b), *307*, 367 (209), *374*
Ganicke, K. 222 (140b), *232*
Gann, R. G. 685, 687 (105), *690*
Ganschow, S. 293 (146), *307*
Ganyushkin, A. V. 396, 397 (87b, 87c), *414*
Garatt, D. G. 525 (5d), *578*
Gardini, G. P. 743 (148, 149), 752 (160a), *773*, *774*
Gardner, E. J. 393 (72), *414*
Gardner, H. W. 225 (167a), *233*
Gardner, J. L. 44 (160), *81*
Gareil, M. 452 (184b), *460*
Garret, P. E. 367 (208), *374*
Garst, J. F. 454 (200), *460*
Gasanov, R. G. 732 (89), *772*
Gasi, G. P. 313, 339, 359 (16b), *370*
Gaspar, P. P. 733 (92b), *772*
Gassman, P. G. 272 (59), *277*, 526 (30), *579*
Gaudemar, A. 825 (76), *828*
Gaultier, J. 406 (125), *415*
Gavrilov, L. B. 803 (111), *806*
Gaylord, N. G. 421 (19), *427*
Geels, E. J. 437, 438, 440, 454 (60, 62), *457*, *458*
Geisen, F. 339 (140), *372*
Geisler, G. 136, 137 (38), *158*
Geldard, J. F. 465, 472 (97), *481*
Gelius, U. 44, 45 (158), *81*
George, G. 88, 89 (70), *95*
George, M. V. 211 (86), *231*, 332 (108b), *372*, 844 (18w), 865 (84), 867 (95), *913*, *915*
George, P. 68 (293), *83*
Gerber, T. I. A. 826 (80), *828*
Gerbert, G. P. 814 (29), *827*
Gerchikova, F. G. 184 (98), *198*

Gerkin, R. M. 178 (77), *198*
Germain, A. 666, 676 (67), *683*
Germain, G. 384 (38f), 400 (95), *413*, *414*
Gershanov, F. B. 184 (94, 96), 185 (96), *198*
Gerstein, J. 566 (213), *583*
Geuskens, G. 731 (74), *772*
Ghormley, J. A. 58, 59 (235), *82*, 491 (54), *518*
Ghotra, I. C. 475 (159), *482*
Ghotra, J. S. 377 (26), *412*
Giannotti, C. 825 (76–78), *828*
Gianotti, C. 475 (156), *482*
Gibian, M. J. 430 (6, 8), 431 (6, 14), 432 (14), 437, 440 (8), 441 (105), 442 (105, 115), 443 (6, 8), 445 (8, 105, 115), 448 (6), 450 (8), 453 (105, 115), *456*, *459*, 570 (225), *583*, 716, 731, 741 (12), *770*
Gibson, K. H. 212 (89), *231*, 313, 339, 359 (16a), *370*
Gigere, P. A. 288 (97), *306*
Giguere, P. A. 101 (35), *104*, 134 (22), *158*, 587 (5b), *643*
Giguère, P. A. 49 (190), 50 (198), 51, 52 (208), *81*, *82*, 86 (3, 5), 87 (3, 36), 88 (5, 36), 92 (92), *94–96*, 376 (14), 403 (105), *412*, *414*, 484 (1b, 1c, 10–12), 485 (12, 13, 19), *517*, *518*
Gilbert, B. C. 712 (1c), *770*
Gilberz, A. 260 (6), *276*
Gilinsky, P. 435, 438, 439 (35, 36), 445 (35), *457*
Gilinsky-Sharon, P. 435, 437, 438 (37a, 37b), *457*
Gill, N. 525 (18), *579*
Gillan, T. 491, 492 (60), *518*
Gillbro, T. 733 (96), *772*
Gillies, C. W. 71 (314), *84*, 90 (75), *95*, 125 (38), *127*, 389 (56, 57), *413*
Gilmore, C. J. 384 (38e), *413*
Gilmore, D. 317, 339 (44), *370*, 731 (71), *771*
Gilmore, D. W. 243, 244 (27), *257*, 315 (31), 316 (38), *370*
Gilmore, J. 213 (102), *231*
Gimarc, B. M. 13 (29, 31), 16 (29), 20 (34), 21 (35), *78*
Ginns, I. S. 762 (187), *775*
Ginsberg, D. 319 (52), *371*
Giorgianni, P. 297 (181a), *308*, 545 (99), *580*
Giovine, A. 296 (176), *308*, 544, 545 (97b), *580*
Glavas, S. 679, 680 (73), *689*
Gleason, E. 421 (25), *427*
Gleim, R. D. 116 (22), *126*, 315 (33), *370*
Gleiter, R. 526, 527 (32), *579*
Glidewell, C. 88, 90 (74), *95*, 382 (34, 35), *413*
Glover, L. C. 577 (246), *584*
Goasdoue, C. 355 (184), *373*
Gobert, F. 783 (28), *804*
Goddard, W. A., III 7 (26), 21 (38, 39), 24 (48, 49, 51), 25 (26, 49, 69), 33 (26, 51), 43 (69), 45, 57 (49), 58 (38, 39, 48, 49), 61, 63 (254), 65 (283), 68 (301), *78*, *79*, *83*, *84*, 209, 220, 226 (58), *230*, 680 (84), *689*, 894 (166), *917*
Godin, G. W. 801 (105), *806*
Goel, A. B. 474 (134), *482*
Goering, H. L. 552 (136), *581*, 623 (146b), *647*
Goggin, P. 534 (64), *580*
Gokel, G. W. 437 (59), 440 (79), *457*, *459*
Golan, D. E. 227 (195), *234*, 684 (100), *690*, 832, 883, 885, 886, 898, 900 (221), *919*
Golden, D. M. 40 (140), *80*, 97 (2), 99 (24, 33), *102*, *104*, 668, 671 (26), 672 (46), 674 (48), 678, 679 (66), *688*, *689*
Goldstein, B. 546 (124b), *581*, 635 (191), 641 (212), *648*
Goldstein, M. J. 593, 595, 603, 604, 608 (37), 609 (103, 104), 613 (114), 614 (104, 114), 616 (103, 104, 125a, 125b), *644–646*
Gole, J. L. 25 (72), 61 (251), *79*, *83*
Golebiewski, A. 36 (115), *80*
Gollnick, K. 167 (31), *197*, 204 (4b, 6), 209 (56), 210, 212 (66b), 220 (56), 225 (166), *229*, *230*, *233*, 318, 319 (45a), *370*, 588 (10c, 11a), 592 (32), *643*, *644*, 781 (18), *804*
Golob, L. 44, 45 (164), *81*
Golovin, B. A. 765 (200), *775*
Golubović, A. 152 (82), *159*
Gomez, J. 227 (208), *234*, 866, 910 (91), *915*
Gomez, R. R. 833, 834, 885, 887, 890, 892, 898–900 (159a), *917*
Goodall, B. L. 465, 472 (98), *481*
Goodgame, M. M. 25, 33, 52 (81), *79*
Goodman, J. F. 130, 131 (1b), *157*, 568 (222), *583*
Goodyear, W. F. 546 (124b), *581*, 635 (191), 641 (212), *648*
Goolsby, A. D. 431 (15), *456*
Gorbatov, V. V. 821 (57), *828*
Gorden, M. S. 36, 38 (111), *80*
Gordon, A. S. 99 (27), *104*, 676 (55, 58), *689*
Gordon, G. 474 (136), *482*
Gordon, M. S. 87, 88, 90 (32, 33), *94*
Gordy, W. 36 (121), 37 (128), 38, 50 (121), 67 (128), *80*, 87, 88, (23), *94*
Gorelik, A. V. 262 (17), *276*
Gorman, A. A. 844 (18u), *913*
Gorman, R. R. 316 (37), *370*
Gormish, J. F. 268 (41), *277*
Gorse, R. A. 720, 740 (34), *771*
Gosavi, R. K. 538 (77a, 77b), *580*

Goto, 217 (126), *232*, 291 (141), *307*, 839 (30a, 30b), 844, (17a, 18a), 847, 851 (30a, 30b), 860 (63), 871, 895 (30a, 30b), *913*, *914*
Gough, S. T. D. 546 (124c), *581*
Gougoutas, J. Z. 91 (82, 83), *95*, 406 (123, 126, 127), 407 (131, 133–135, 137, 138), 408 (133), *415*, 778 (1), 779 (6, 8), 780 (6, 8, 9), *804*
Gould, D. E. 496 (71), *519*
Govindan, C. K. 557, 558 (177), *582*, 628 (156), *647*, 780 (12b), *804*
Graf, F. 299 (204), *308*, 594, 596, 607 (45b), *644*
Graf, R. 262 (16), *276*
Graham, B. W. 465, 471 (86, 90), *480*, *481*
Graham, D. M. 546–548 (114a), *581*
Graham, I. F. 820, 821, (49a, 49b), *827*
Graham, R. A. 678 (65, 67), 679 (65), *689*
Grakovich, L. K. 280 (27, 28b), *304*
Grandjean, D. 469 (56), *480*
Grane, H. R. 162 (12), *196*
Gratton, S. 88, 91 (78), *95*, 401 (98), *414*
Gray, H. B. 2 (23), *78*, 472 (96), *481*
Gray, P. 25 (86), *79*, 789 (46), *805*
Gray, R. 326, 353 (86), *371*
Graziano, M. L. 334 (115), *372*, 873 (109), *916*
Grdina, M. J. 168 (32), *197*, 209, 218, 220 (60), *230*, 848, 854 (37), *914*
Greco, A. 296 (173b), *308*
Greed, D. 340 (144), *372*
Green, B. S. 778 (1), *804*
Green, F. D. 285 (66, 67), *305*
Green, F. R. 185 (103), *198*
Green, M. M. 174 (55), *197*
Greenbaum, M. A. 300 (217), *309*, 546 (124a), *581*
Greene, F. D. 280, 281 (14a–c), *304*, 405 (121), *415*, 546 (119a–c, 120), 549 (119a–c), 557 (171), 568 (220), *581–583*, 597 (62), 599 (75, 77a), 618, 619 (62), 625 (75, 149b), 628 (162), 636 (198), 639 (62, 207, 208), *644*, *645*, *647*, *648*, 753 (162), *774*, 780 (12b), *804*
Greenebaum, M. A. 637 (202, 203), 638 (203), *648*
Greenwood, F. L. 680 (81, 82, 84), *689*
Gregorijan, S. K. 144 (61), *159*
Greibrokk, T. 226 (176), *233*
Greig, J. A. 419, 426 (4), *426*
Greigger, B. A. 298 (195), *308*
Greist, L. 448 (157), *460*
Grekoc, A. P. 566 (213), *583*
Grekov, A. P. 441 (103), *459*
Grevels, F. W. 222 (150), *232*
Grieder, A. 224 (157a), *233*
Griesbaum, K. 157 (93), *160*
Griesshammer, R. 223 (155b), *233*
Griffin, G. W. 863 (72), *915*
Griffin, I. M. 316 (40), *370*
Griffith, W. P. 86, 88 (13), *94*, 190 (133), *199*
Griller, D. 237 (13), 241 (23), 242 (24), *257*, 733 (90, 91b, 92a), *772*
Grimbert, D. 24, 58 (54), *79*
Grinenko, N. M. 425 (65, 72), 426 (72), *427*, *428*
Grinev, M. P. 905 (195), *918*
Grinsberg, O. Y. 739 (122), *773*
Grobe, K. H. 165 (15), *196*
Grönwall, S. 302 (226), *309*
Gropen, O. 63 (276), *83*
Gross, H. 869 (96), *915*
Grosse, A. V. 63 (279), *83*
Grossman, S. 440, 445 (82, 83), *459*
Grotewold, J. 814 (28), *827*
Groth, P. 88, 91 (77), *95*, 389 (58), 391 (63–65, 67), 392 (63), 396 (63, 82), 397 (63), 402 (99, 100), 403 (101), *413*, *414*
Groves, J. T. 191 (134, 135, 137, 138), *199*, 471 (84), *480*
Grow, D. T. 61–63 (246), *82*
Grubmüller, P. 832, 833, 836, 838, 885, 887, 888, 892, 898, 899 (156), *917*
Gruen, F. M. 280 (10), *304*
Grundry, K. R. 465, 471 (88), *481*
Grunwald, E. 130 (6), *158*
Guaraldi, G. 247 (44), *257*, 486, 494 (32), *518*
Guberman, S. L. 21 (41), *78*
Guedes, L. N. 753 (164, 165), 754 (165), *774*
Guerchais, J. E. 465, 469 (38–40, 54), *479*, *480*, 808, 822 (59), *828*
Guidotti, C. 36, 57, 58 (98), *79*, 87, 88, 90 (45), *95*
Guillet, J. E. 731 (75), *772*
Guilmet, E. 191 (136), *199*
Gundermann, K.-D. 844 (18d), *913*
Gunning, H. E. 538 (77b), *580*
Günthard, H. H. 88, 92 (94b, 95), *96*, 290 (124), *306*
Günther, P. 101 (36), *104*, 494 (66), *519*
Gupta, Y. K. 768 (209), *775*
Gurman, V. S. 720, 726 (30), *771*
Guseva, L. N. 800 (87, 88), *806*
Guyot, 796 (64), *805*
Guziec, F. S., Jr. 874 (116), *916*
Gwinn, W. D. 388 (46), *413*

Haas, C. M. 88, 92 (94a), *96*
Haas, H. C. 280 (16), *304*, 800 (82, 85), *805*, *806*
Haas, Y. 685, (103, 104), *690*, 891 (152a, 152b), *917*
Haber, F. 452 (185), *460*, 720 (32a, 32b), *771*

Hadik, G. 525 (14), *578*
Hadži, D. 88, 92 (93), *96*, 134 (26, 27), 155, 156 (86), *158*, *159*, 514, 515 (115), *520*
Haegele, R. 475 (158), *482*
Hagen, R. 633 (179), *647*
Hagenbuch, J. P. 215 (111, 112a), *232*, 353 (182), *373*
Haggett, M. L. 130 (1c), *157*, 641 (211), *648*
Hagihara, N. 466, 474 (123), *481*
Hagstrom, S. 36 (103), *80*, 87, 90 (53), *95*
Hahn, W. 420 (16), *427*
Haiby, W. A. 609, 614, 616 (104), *645*
Haines, L. M. 466, 472 (103), 473 (113), *481*
Halevi, E. A. 893 (163e, 163f), *917*
Halfpenny, J. 384 (38c), *413*, 812 (21a), *827*
Hall, C. D. 816 (37), *827*
Hall, N. F. 156, 157 (94), *160*
Halow, I. 40 (139), *80*
Halpern, A. 312 (12), *370*
Halpern, A. M. 905 (196a), *918*
Halpern, J. 465 (98), 466 (124), 472 (98), 474 (124), *481*
Halpern, Y. 666, 676 (69), *683*
Halpert, T. R. 471 (82), *480*
Halsall, T. G. 356 (189), *373*
Hamada, H. 437 (58), *457*, 879 (137), *916*
Hamanaka, N. 358 (192a, 192b, 193), 363 (192a, 192b), 364 (193), *373*
Hamann, K. 290 (135), *307*
Hamann, R. R. 816 (33), *827*
Hamberg, M. 313 (14, 15), *370*
Hameiri, J. 435, 437, 438 (37b), 439 (70), *457*, *458*
Hamer, D. E. 189 (124), *199*
Hamill, W. H. 767 (206), *775*
Hamilton, G. A. 513 (113, 114), *520*, 656 (36, 38), 666 (38), *682*
Hammond, G. S. 472 (96), *481*, 618 (130), *646*, 732 (83), 739, 740 (121), *772*, *773*
Hammond, L. M. 526 (31), *579*
Hammond, W. B. 864 (80), *915*
Hampel, M. 288, 289 (108), *306*
Hamrin, K. 44, 45 (158), *81*
Hanafusa, M. 183 (92), *198*
Hanaki, A. 432 (30, 31), *457*
Hanaoka, M. 366 (205a), *374*
Hancock, R. D. 419 (4), 425 (76), 426 (4), *426*, *428*
Hands, A. P. 356 (189), *373*
Hangen, G. R. 40 (140), *80*
Hann, R. A. 840–842, 884, 890, 895, 896 (180c), *918*
Hanna, R. 435 (39), *457*
Hannemann, K. 899, 908 (208), *918*
Hansen, R. L. 816 (33), *827*
Hansen, R. T. 501 503 (86), *519*, 652 (18), *681*
Hanst, P. L. 110 (8), *126*, 668, 671, 674 (25), 680 (77), *688*, *689*, 734 (99), *772*
Hanzlik, R. P. 530, 532 (47), *579*
Haq, M. Z. 703 (58), *805*
Haque, F. 469 (58), *480*
Harada, N. 912 (219), *919*
Hardin, L. B. 684 (94), *690*
Harding, L. B. 24 (51), 25 (69), 33 (51), 43 (69), 65 (283), *79*, *83*, 209, 220, 226 (58), *230*, 894 (166), *917*
Harding, M. J. C. 152 (77), *159*
Hare, D. G. 814 (30), *827*
Hargis, E. 282 (49), *305*
Hargis, J. H. 613, 615, 618, 619 (105), *645*
Hargrave, K. R. 145, 147, 150 (68), *159*
Hariharan, P. C. 61 (264), *83*
Harirchia, B. 215 (119), *232*
Harkins, J. 501, 503 (86), *519*, 652 (18), *681*
Harlow, R. L. 336 (123), *372*, 394 (75), *414*
Harman, D. 299 (207), *308*
Harmony, J. A. K. 242 (25), *257*
Harris, D. O. 36–38, 67 (122), *80*, 89, 90 (59), *95*
Harris, H. P. 441 (99), *459*
Harris, M. S. 227 (195), *234*, 684 (100), *690*, 832, 883, 885, 886, 898, 900 (221), *919*
Harris, S. A. 162 (1), *196*
Harrison, C. R. 297 (177), *308*
Hart, E. J. 767, 768 (207), *775*
Hart, F. A. 377 (26), *412*, 475 (159), *482*
Hart, H. 120 (28), *127*, 298 (193a), *308*, 324 (80), *371*, 534 (65), 557 (170), *580*, *582*, 628 (163, 164a, 164b), *647*, 781 (17), *804*
Hartman, G. D. 272 (59), *277*
Harvey, R. B. 37, 67 (127), *80*, 88, 90 (72), *95*
Harvey, R. J. 164 (13), *196*
Hase, T. 670 (57), *683*
Hasegawa, H. 336 (126), *372*, 546–548 (114d), *581*
Hasegawa, M. 740 (125), *773*
Hasenfratz, H. 281–283 (39), *305*
Hashmall, J. A. 68 (303), *84*
Hasman, J. S. 430, 450 (10), *456*
Hassall, C. H. 295 (162), *307*, 559 (182), *582*
Hastings, G. 419 (5), *426*
Hastings, J. W. 205 (17), 227 (196), *229*, *234*, 844 (18t), 902, 903 (187), *913*, *918*
Hasty, N. M. 833 (229a), *919*
Haszeldine, R. 475, 476 (146), *482*
Hatsui, T. 211 (88a–c), 226 (170), 228 (217, 218), *231*, *233*, *234*, 343 (157), *373*, 862, 910 (68a, 68b), *915*
Hatzenbuhler, D. A. 61, 62 (242), *82*
Haugen, G. R. 97 (2), *102*
Haushalter, R. C. 191 (135), *199*
Hauw, C. 406 (125), *415*

Havinga, E. 38 (137), *80*, 718 (18, 21), 738 (18), *770*
Havron, A. 733 (97), *772*
Hawkins, E. G. E. 2 (5), *78*, 174 (57), *197*, 236 (3, 5), *256*, 263 (24a), *276*, 280 (20), 295 (168), *304*, *307*
Hawthorne, J. O. 440 (78), *459*
Hawthorne, M. F. 559 (189, 192), 560 (189), 561 (192), *582*, *583*
Hay, P. J. 7 (26), 24 (48–50), 25 (26, 49), 33 (26, 50), 45, 57 (49), 58 (48–50), *78*, *79*
Hay, R. K. M. 89 (63), *95*
Hayaishi, O. 464 (1), *479*, 872 (107), *915*
Hayashi, E. 718 (22, 23), 719, 739 (26), *770*
Hayashi, H. 740 (124, 126), *773*
Hayashi, K. 721 (41), *771*
Hayes, E. F. 25 (64, 72), 33 (64), 61 (251), *79*, *83*
Haynes, R. K. 341 (147, 148), *372*
Hayon, E. 757, 758 (172), *774*
Hays, R. C. 166 (23), *196*
Hayward, P. J. 474 (131), *481*
Haywood–Farmer, J. H. 526 (31), *579*
Hazeldine, R. W. 261, 262, 268 (10a), *276*
Hearne, G. W. 164 (9), *196*
Heathcock, C. H. 437 (57), *457*
Heaton, L. 136–138, 140, 141, 149, 150, 152 (39), *158*
Hecht, K. T. 36, 38 (120), *80*, 87–90 (25), *94*, 376 (2), *412*
Hecht, R. 299 (203), *308*, 594, 596, 607 (45a), *644*
Heckel, E. 761 (184a), 765 (184a, 203a, 203b), 766 (203a), *774*, *775*
Hedaya, E. 553, 561 (144), *582*
Heden, P. F. 44, 45 (158), *81*
Hedman, J. 44, 45 (158), *81*
Hegarty, A. F. 280, 288, 299 (3), *303*, 632 (171), *647*
Heggs, R. O. 194 (151), *199*
Heggs, R. P. 194 (152, 153), *199*
Hehre, W. J. 25, 33 (74), 36, 38, 40 (100), 41 (150), 45 (74), 61 (264), 63 (100, 150, 271), 65, 66 (150), 67 (100), *79–81*, *83*, 87, 88, 90 (43), *95*, 485 (16, 17), *518*
Heicklen, J. 491 (56), 492, 495 (63), *518*, 678 (64), 679 (64, 72, 73), 680 (72, 73), *689*, 720, 740 (35), *771*
Heidt, L. J. 720 (33), *771*
Heimbach, P. 474 (119), *481*
Hell, J. L. 48, 49 (179), *81*
Hellman, T. M. 513 (113, 114), *520*, 656 (36, 38), 666, (38), *682*
Hemingway, R. J. 213 (98, 99), *231*, 314 (24), *370*
Henbest, H. B. 290 (120), *306*, 526, 532 (23a, 23b), *579*, 651, 654 (6c), *681*
Henderson, R. W. 750 (156), *774*
Hendrickson, J. B. 396 (83), *414*
Hendrickson, W. H. 546 (128b, 128c), *581*, 859 (61), *914*
Hendrickson, W. H., Jr. 271 (53), *277*, 303 (237), *309*
Hendry, D. G. 99 (34), *104*, 164 (11), *196*, 679, 680 (75), *689*
Henglein, A. 761 (184b), *774*
Henke, H. 389 (61), *413*
Henne, A. 906 (203), *918*
Henrici Olivé, G. 464, 465 (9), *479*
Henry, H. 501 (104), *519*, 653 (22), *681*
Henry, Y. 111 (13b), *126*
Herk, L. 608 (101), 609 (102), *645*
Herman, K. 101 (35), *104*, 484 (11, 12), 485 (12), *517*
Hernandez, J. 814 (28), *827*
Herring, J. W. 226 (184), *233*
Herring, P. J. 844 (18x), *913*
Herron, J. T. 683 (88, 89), *690*
Hertel, L. W. 526, 527 (32), *579*
Hertog, H. J. den 297 (185), *308*
Herz, W. 213, 214 (106), 215, 219 (113), 222 (147), *231*, *232*, 356 (187, 188, 190), *373*, 555 (155), *582*
Herzberg, G. 54 (213), *82*
Heslop, R. B. 261, 262, 268 (10a), *276*
Hess, J. 71 (311), *84*, 385 (41), *413*, 844, 845, 892 (21a, 21b), *914*
Hewat, A. W. 51 (204), *82*
Hey, D. H. 287 (83), *305*
Heyman, J. D. 811 (68), *828*
Heywood, D. L. 743 (150), *773*
Hiatt, R. 133 (12), 144 (65), *158*, *159*, 163 (7), *196*, 214 (108), 221 (108, 137), 224 (158), 225 (108), *232*, *233*, 239 (17), 240, 242 (20), 249 (53–55, 58), 250 (54, 58), *257*, 260, 263 (5), 271 (48), *276*, *277*, 280–282, 286 (1), *303*, 453, 454 (197), *460*, 486 (30), *518*, 546, 547 (110a), 554 (146), 556 (165), 577 (246), *581*, *582*, *584*, 587, 590–592 (7), *643*, 738 (114), *773*, 793 (57), *805*
Hiatt, R. R. 185 (101), *198*, 236 (6), *256*, 299 (200), *308*, 594, 600, 607 (43), *644*
Hiberty, P. C. 25, 33 (67), 61 (67, 255), *79*, *83*
Hickinbottom, W. J. 190 (127), *199*
Higgens, R. 206 (23), *229*
Higgins, G. M. C. 145, 147, 150 (68), *159*
Higgins, R. 588 (13c), *643*, 785 (34), *804*
Higgs, M. D. 313 (18), *370*
Higley, D. P. 634 (185), *648*
Hill, D. W. 397 (89), *414*
Hill, J. T. 613, 614 (115), *646*
Hiller, J. F. 48, 49 (188), *81*
Hillier, I. H. 36 (95), *79*

Hillman, J. J. 52 (211), *82*
Hinchliffe, A. 65, 66 (281), *83*
Hindenlang, D. M. 366 (205b), *374*
Hinder, M. 222 (151), *232*, 555 (154), *582*
Hino, T. 872 (106a, 106c–e), 876 (127), *915*, *916*
Hinshelwood, C. N. 675 (50), *689*
Hinze, J. 893 (163d), *917*
Hirakoso, K. 876 (127), *916*
Hirao, K. 69 (310), *84*
Hiraoka, K. 69 (309), *84*
Hirobe, M. 443, 444 (121), 446 (121, 126b), *459*
Hirose, Y. 596, 599, 611, 613 (56), 614 (56, 117), 615 (56, 117, 135), 617–619 (117), 620 (56, 135), 621 (135), 622 (56, 135), *644*, *646*
Hirota, E. 67 (292), *83*
Hirotsu, K. 400 (94), *414*, 475, 476 (147), *482*
Hirschhausen, H. v. 41 (153), *81*
Hirschmann, F. 831 (8), *913*
Hisada, R. 262 (14, 15), 263 (14, 15, 22a, 23), 264 (14, 23), 265 (14), 273 (62), *276*, *277*, 593 (34), 597 (67), 598 (76), 617 (67), 631 (76), 639 (209), 640 (210), *644*, *645*, *648*
Hisatome, M. 119 (26), *127*
Hitchcock, P. B. 71 (312), *84*, 384, 386 (40), *413*, 839 (250b), *919*
Hitz, F. 299 (211), *308*
Hiyama, A. 611, 617 (106), *646*, 791 (48), *805*
Hjelmeland, L. M. 68 (294, 295), *83*, 88, 90, 92, 93 (105), *96*, 288 (95), *306*, 523, 524 (2), 530 (50), *578*, *579*
Hnizda, W. 824 (70), *828*
Ho, J. C. J. 152 (82), *159*
Ho, M. S. 210, 226, 227 (65c), *230*, 592 (31), *644*, 836, 850 (43), *914*
Hochanadel, C. J. 58, 59 (235), *82*, 491 (54), *518*
Hock, H. 166 (20), 193 (143), *196*, *199*, 222 (140a, 140b, 141a, 141b, 142), *232*, 555 (153), *582*, 780 (13), *804*, 817 (38), *827*, 831 (9), *913*
Hodder, O. J. R. 388 (47), *413*
Hodgdon, R. B. 285 (70), *305*, 546, 547 (112c), *581*
Hodgdon, R. B., Jr. 614 (118), *646*
Hodge, P. 297 (177), *308*, 419 (78), *428*
Hodge, V. F. 130 (1a), 132 (1a, 9), 133, 147, 148 (9), *157*, *158*, 555, 556, 575 (159), *582*, 831 (6a), 832 (222), *913*, *919*
Hodgson, J. C. 809, 814, 815 (8), 816 (63), 818, 824 (43), *827*, *828*
Hoff, W. S. 297 (179), *308*
Hoffman, A. S. 418, 426 (3), *426*
Hoffman, B. M. 464, 465 (14), 471 (81), *479*, *480*
Hoffman, D. H. 529 (42), *579*
Hoffman, D. M. 289 (111), *306*, 614 (119, 121), *646*
Hoffman, H. M. R. 209 (54), *230*
Hoffman, J. 177 (69), *197*, 221 (139), *232*
Hoffman, J. M. 337 (129), *372*
Hoffman, R. V. 262 (13), 267 (39), 268 (44, 46), 269 (44), 270 (47), 272 (55–58), 275 (65), *276*, *277*, 546 (118a, 118b), *581*, 759 (179), *774*
Hoffmann, A. K. 300 (217), *309*, 637 (202), *648*
Hoffmann, R. 613 (111), *646*
Hoffmann, R. W. 405 (115), *415*
Hoffsommer, J. C. 499 (79), *519*
Hofland, A. 211 (83), *231*
Hofmann, R. 574 (237), *584*
Hoft, E. 571 (231), *583*
Höft, E. 293 (146), *307*
Hohne, E. 71 (315), *84*, 391 (66), *413*
Höhne, G. 881 (144), *916*
Hohorst, F. A. 496 (71), *519*
Hohorst, H. J. 114 (17), *126*, 397 (88), *414*
Holan, G. 388 (49), *413*
Holbert, G. W. 367 (209), *374*
Holbrook, K. A. 666, 678 (6), *688*
Holcomb, A. G. 268 (45), *277*
Holden, D. A. 248 (51), *257*
Holder, G. A. 735 (101), *772*
Holland, D. 826 (83), *828*
Hollander, J. A. den 740 (132), *742* (138a), *773*
Holtsclaw, K. M. 677 (62), *689*
Holubka, J. W. 287 (89), 291 (137), *306*, *307*
Honsberg, W. 779 (5), *804*
Hoobler, J. A. 595, 603 (49), 607 (94), *644*, *645*
Hoobler, J. O.-A. 591, 594, 595, 603, 604, 606 (24), *644*
Hook, S. C. W. 824 (71), *828*
Hooper, D. G. 675 (54), 676 (54, 57), *689*
Hopkins, D. E. 669 (32), *688*
Hopkins, T. A. 896 (183b), *918*
Hopkins, Th. A. 844 (14), *913*
Hopkinson, A. C. 41, 48, 49 (151), *81*, 441, 453 (106), *459*
Hopper, D. G. 24 (61), *79*
Hoppilliard, Y. 111 (13a), *126*
Hordvik, A. 377 (27), *412*
Hori, K. 188 (121), *199*
Horn, K. A. 227 (200a, 200b), *234*, 405 (118, 119), *415*, 546–548, 550 (116), *581*, 667, 684 (14), *688*, 832 (35), 844 (18j), 847, 886, 898 (35), *913*, *914*
Horn, R. W. 466, 473 (116), *481*
Horner, C. L. 271 (48), *277*
Horner, L. 287 (81, 82), *305*, 546 (111a–e,

123), 547 (111a–e), *581*, 636 (194–197), *648*, 651 (6b, 11), 654 (6b), *681*
Hosking, J. W. 474 (130), *481*
Hosokawa, T. 435 (40), *457*
Hospital, M. 388 (50), *413*
Hotta, H. 762 (188b), *775*
Hotta, K. 226 (182, 185c), *233*
Houce, C. 452 (186), *460*
Houee, C. 762 (186), *775*
Houghton, G. 614 (116b), *646*
Houk, K. N. 65 (286), *83*, 209, 220 (61, 62), *230*, 854 (53), *914*
House, H. O. 178 (77), *198*, 288 (102), *306*, 522, 525 (1e), 567 (214), *578*, *583*
Housmans, J. G. H. M. 678 (69), *689*
Howard, B. M. 400 (94), *414*
Howard, C. J. 25, 33 (82), *79*, 98 (10), *104*, 666 (8), *688*
Howard, J. A. 164 (8), 167 (27), *196*, 236, 237 (8, 9), 238, 245 (8), 246 (37), 247 (8, 9, 42, 43), 248 (8, 49–52), 249 (57), 250 (9, 49, 59), 251 (62, 63), *256–258*, 487 (39, 40, 42), 488 (40, 43), 489 (40, 45–47), 490 (40, 47, 50, 52), 491 (50, 52, 53), 492 (53), 499, 500 (80), *518*, *519*, 602 (85b), 622 (137), *645*, *646*, 733 (91a), *772*
Howard, R. E. 36 (103), *80*, 87, 90 (53), *95*
Howe, G. R. 185 (101), *198*
Hoz, S. 438, 444 (64), *458*, 879 (136), *916*
Hrncir, D. C. 65 (287), *83*
Hsu, C. K. 285 (68), *305*
Huang, C. T. 114 (18b), *126*
Huber, F. 86, 88 (18), *94*
Hubner, H. 554 (150), 574 (237), *582*, *584*
Huddleston, D. 501 (88), *519*
Huddleston, G. 653 (26), *682*
Hudee, J. 213, 216 (91b), *231*
Hudson, B. E. 157 (93), *160*
Hudson, R. F. 275 (65), *277*, 759 (179), *774*
Hudson, R. L. 25, 45, 46 (85), *79*, 98 (9), *104*
Huet, F. 865 (82), *915*
Hugel, G. 867 (93), *915*
Hughes, E. W. 86, 88 (5), *94*, 376 (14), *412*
Hughes, H. 224 (162), *233*
Hughes, R. H. 377 (19), *412*
Huie, R. E. 683 (88, 89), *690*
Huisgen, R. 271 (48), *277*, 286 (75), *305*, 529 (46), *579*
Hull, C. A. 61 (256), *83*
Hull, L. A. 271, 272 (54), *277*
Hunt, R. H. 36, 38 (120), *80*, 87–90 (25, 26), *94*, 376 (2), *412*
Hunt, W. J. 7, 25, 33 (26), *78*
Huntington, J. 594, 693–607 (46), *644*
Huntington, J. G. 603, 604 (89), 617 (127), *645*, *646*
Hunziker, H. E. 58, 59 (232), *82*
Hurdlik, A. M. 560 (197), *583*
Hurdlik, P. F. 560 (197), *583*
Hurley, A. C. 41 (152), *81*
Hurst, J. R. 281 (35), *305*
Hursthouse, M. B. 377 (26), *412*, 475 (159), *482*
Huschens, R. 815 (86), *828*
Hussey, C. L. 436 (48), 437 (55), *457*
Hussman, G. 296 (175), *308*
Husthouse, M. B. 474, 475 (133), *481*
Huthmacher, K. 223, 225 (156), *233*
Hutton, A. J. L. 677–679 (63), *689*

Ibers, J. A. 466 (111), 467 (21), 473 (111), *479*, *481*
Ibne-Rasa, K. B. 539 (83), 540, 544 (84), *580*
Ibne-Rasa, K. M. 130 (1c), *157*, 538 (79), *580*, 641 (211), *648*
Iesce, M. R. 334 (115), *372*, 873 (109), *916*
Igersheim, F. 466 (106), 467 (25), 472, 473 (106), 474, 477 (135), *479*, *481*, *482*
Igeta, H. 336 (126), *372*
Ihring, M. 796, 797 (69), *805*
Iitaka, Y. 384 (38g), 389 (59), *413*
Ikawa, T. 440, 454 (74), *458*, 475 (140, 141), *482*, 826 (82), *828*
Ikeda, M. 272 (60), *277*
Ikeda, T. 528 (39), *579*
Ikeda, Y. 596 (50), 635 (50, 189), *644*, *648*
Ikegami, S. 526 (28), *579*
Ikegami, Y. 451 (179, 180), *460*
Ilan, Y. A. 452 (189), *460*
Imagawa, D. K. 210 (71), 213, 214, (93), *231*, 357 (191), *373*
Imahashi, H. 635 (189), *648*
Imai, J. 366 (205c, 206), *374*
Imamura, J. 167 (28), *196*
Imashev, U. B. 501 (120), *520*, 653 (23), *682*
Immergut, E. H. 800 (101), *806*
Imoto, M. 546–548 (114b), *581*
Imuta, M. 211 (81), *231*, 291 (139), *307*, 326 (85), *371*, 872 (105a, 105b, 106b), *915*
Inaba, H. 445, 446, 454, 456 (124), *459*, 491 (58), *518*
Inamoto, N. 893 (163c), *917*
Inamoto, Y. 525 (10), *578*
Indest, K. 280 (23), *304*
Indictor, N. 162 (2), *196*, 271 (48), *277*, 546, 547 (112b), *581*
Ingold, C. K. 451 (176a), *460*
Ingold, K. 812, 813 (24), *827*
Ingold, K. U. 167 (27), *196*, 208 (52), *230*, 237 (14), 238–240 (16), 241 (16, 23), 242 (24), 247 (42), 250 (59), 254 (16), *257*, 487 (39–41), 488 (40), 489 (40, 45, 46), 490 (40, 52), 491 (52, 60), 492 (60), 499, 500

(80), *518*, *519*, 602, 610 (86), *645*, 732 (85), 733 (92a), *772*, 812, 813 (25), 817 (69), *827*, *828*
Ingrosso, G. 526 (22), *579*
Inoue, H. 366 (205c, 206), *374*, 538 (78), *580*
Inoue, K. 207 (42), *230*, 347 (165), *373*
Inoue, M. 635 (189), *648*
Inoue, Y. 226 (175), *233*, 863 (70), *915*
Insinger, T. H. 166 (21), *196*
Inukai, T. 287 (87), *305*
Ireland, R. E. 223, 225 (156), *233*
Ireton, R. 99 (27), *104*, 676 (58), *689*
Irngartinger, H. 839, 867 (243b), *919*
Isaac, R. 317, 339 (44), *370*, 731 (71), *771*
Ishibe, N. 226 (185b), *233*
Ishihara, T. 133 (13), *158*
Ishii, Y. 525 (10), *578*
Ishikawa, K. 863 (72), *915*
Ishizumi, K. 367 (209), *374*
Islam, T. S. A. 674 (49), *689*
Israili, Z. H. 560, 561 (194), *583*
Itahara, T. 451, 454, 456 (170), *460*
Itai, A. 384 (38g), *413*
Ito, T. 653 (27), *682*
Ito, Y. 122 (35), *127*, 216 (125), 226 (180), *232*, *233*, 344 (158), *373*, 736 (109), *772*, 831 (12), 843 (12, 57a, 57b), 846, 857 (12) 858 (57a, 57b), 873 (114), 880 (140), 884 (57a, 57b), 891 (57b), 898 (57a, 57b), 906 (57b), 911 (57a, 57b), *913*, *914*, *916*
Itoh, K. 151 (75), *159*
Itoh, S. 211 (88b), *231*
Itoh, T. 181, 192 (85), *198*, 534 (61), *579*
Itokawa, H. 389 (59), *413*
Itskovich, V. A. 136, 137, 143, 144 (37), *158*
Ivanchenko, P. A. 422 (30), *427*
Ivanchev, S. S. 140, 146 (46), *159*, 425 (66, 68), *428*, 738 (116, 117), *773*, 802 (110), *806*
Ivanova, T. A. 817 (42), *827*
Ives, J. L. 210, 212, 214, 221 (66e), 227 (203–205), *230*, *234*, 342, 348, 365 (153), *373*, 871 (98), *915*
Iwabuchi, R. 211 (88b), *231*
Iwai, M. 635 (189), *648*
Iwamura, H. 303 (238), *309*, 596 (50), 635 (50, 189), *644*, *648*
Iwasaki, M. 617 (129a–e), *646*
Iwata, M. 785 (35), *804*
Iwata, S. 61 (248), *83*
Iwata, T. 397 (90, 91), *414*
Izumi, M. 297 (182c), *308*

Jabloner, H. 796, 797 (68), 798 (68, 76), *805*
Jackson, R. H. 37, 67 (126), *80*, 86, 88, 90 (20), *94*, 377 (21), *412*
Jackson, W. G. 465, 467, 472 (22), *479*
Jacob, N. 725 (56), *771*
Jacobs, D. 557, 559 (179), *582*
Jacobson, S. E. 465 (53, 70), 469 (53), 470 (69, 70), 471 (53, 70), *480*
Jacquesy, J. C. 679 (72), *683*
Jaffé, J. 98, 99 (13), *104*
Jaffe, R. L. 25 (79, 80), 33 (79), 52, 57 (80), 58, 59 (79), *79*
Jahn, A. 421 (20), *427*
Jakobsen, H. J. 283 (60, 61), *305*
Jalics, G. 760 (181a), *774*
James, B. R. 471 (93, 94), 478 (94), *481*, 826 (81), *828*
James, H. J. 432 (22), *457*
Jang, N. C. 291 (138), *307*
Jankitova, L. N. 814 (29), *827*
Janoschek, R. 53 (216), *82*
Janzen, E. G. 243 (26), *257*, 437, 438, 440, 454 (60, 62), *457*, *458*
Jaoven, G. 440 (77), *459*
Jarchow, O. 386, 388 (45b), *413*, 840, 845, 846, 867, 890, 892 (22a), *914*
Jaroslavskii, N. G. 288 (96), *306*
Jarossi, D. 288 (92), *306*
Jarvis, J. A. J. 818 (46), *827*
Jauhal, G. S. 466, 474 (128), *481*
Jayathirtha, Rao, V. 226 (183b, 185c), *233*
Jefford, C. W. 209 (59), 220 (59, 135), 225 (165), 226 (173a, 173b, 186), *230*, *232*, *233*, 332 (109c), *372*, 453 (195), *460*, 833–836 (228), 863 (69, 77), 876 (124), *915*, *916*, *919*
Jeffrey, A. M. 223 (152), *233*
Jeffrey, G. A. 88, 92 (96, 98), *96*, 288 (94), *306*, 395 (78), 405 (107, 109), *414*
Jeffry, G. A. 134 (18), *158*
Jehlička, V. 88, 91 (87), 92 (101), *96*, 408 (139), *415*
Jemilev, U. M. 184, 185 (96), *198*
Jencks, W. P. 566 (213), *583*
Jendrychowska-Bonamour, A.-M. 424 (54, 55), *427*
Jenkins, A. D. 98 (14), *104*
Jenkins, C. L. 599 (78), *645*
Jenkins, R., Jr. 571 (232), *583*
Jensen, F. R. 811 (68), *828*
Jerchel, D. 297 (182b), *308*
Jerina, D. M. 223 (152), *233*
Jewett, J. E. 205 (7), *229*
Jewett, J. G. 589 (18), *643*
Jinnai, O. 226 (170), *233*
Jira, R. 187 (120), *199*
Jitsukawa, K. 181, 192 (85), *198*, 534 (61), *579*
Jo, T. 596, 635 (50), *644*
Jobst, K. 796, 797 (66), *805*
Johansen, H. 87 (48), *95*
Johansen, J. 36 (97), *79*

Johansson, G. 44, 45 (158), *81*
Johns, J. W. 52 (210), *82*
Johnson, F. H. 844 (17a), *913*
Johnson, G. J. 298 (193b), *308*
Johnson, M. R. 526, 532, 533 (27), *579*
Johnson, R. A. 316 (37), *370*, 441 (104, 108, 111), 442 (116), *459*
Johnson, S. N. 732 (81), *772*
Johnson, V. L. 465, 472 (97), *481*
Johnston, H. S. 58, 59 (234), *82*, 712, 731–734 (3), *770*
Joiner, C. M. 633 (177), *647*
Jolley, J. E. 716 (14b), *770*
Jonathan, N. 44, 45 (164), *81*
Jones, C. R. 742 (142), *773*
Jones, M., Jr. 742 (140), 747 (154), *773*, *774*
Jones, P. 587 (5a), *643*
Jones, P. D. 86, 88 (13), *94*
Jones, P. L. 48, 49, 54 (186), *81*
Jones, R. D. 2 (21), *78*, 464, 465 (12, 14), 471, 475 (12), *479*
Jones, R. G. 484 (1a), *517*
Jongh, R. O. de 718, 738 (18), *770*
Jönsson, B. 24 (57), 61 (258), *79*, *83*
Jönsson, P. G. 282 (53), *305*
Jorgensen, W. L. 8 (27), *78*
Jouffret, M. 174 (59), *197*
Jouve, P. 48, 49, 54 (184), *81*
Ju, H. 800 (96), *806*
Judson, H. A. 593, 595, 603, 604, 608 (37), 609 (103), 616 (103, 125b), *644–646*
Jung, G. 297 (182b), *308*
Jung, J. 896 (184), *918*
Junkerman, H. 546, 547 (111e), *581*
Juračka, F. 598 (68), *645*
Jurgeleit, W. 546 (123), *581*, 636 (196, 197), *648*
Jürgens, E. 287 (81), *305*, 636 (195), *648*
Juris, A. 757 (173), *774*
Jurjev, V. P. 184, 185 (96), *198*
Just, G. 312 (3), *369*, 574 (237), *584*

Kabalka, G. W. 808, 814, 816 (35), *827*
Kabayashi, S. 445, 446, 454, 456 (124), *459*
Kabe, Y. 445, 446, 454, 456 (124), *459*
Kabre, K. 274 (64), *277*
Kabre, K. R. 759 (178), *774*
Kabun, S. M. 801 (106), *806*
Kadokura, A. 125 (39), *127*
Kaelin, M. S. 557, 558 (177), *582*, 628 (156), *647*, 780 (12b), *804*
Kagan, H. B. 465, 470 (66), *480*
Kaguchi, H. 872 (106b), *915*
Kah, F. A. 757 (170), *774*
Kahn, A. 588 (14b), *643*
Kahn, A. U. 455 (206), 456 (206, 207, 209), *461*
Kaiser, J. K. 225, 228 (168), *233*
Kaiser, S. 63, 67–69 (272), *83*, 485, 486 (18), *518*
Kakudo, M. 466, 474 (123), *481*
Kaldor, U. 36 (88), *79*, 87, 88 (37), *95*
Kale, J. D. 675 (52), *689*
Kalinina, L. I. 136–138, 150 (42), *158*
Kalnajs, J. 377 (20), *412*
Kalra, K. C. 792 (51), *805*
Kalyanaraman, B. 446 (135), *459*
Kamano, Y. 389 (59), *413*
Kamigata, N. 263 (22a), *276*
Kamiya, I. 876 (123), *916*
Kamiya, K. 471 (92), *481*
Kamiya, Y. 193 (147), *199*, 251 (61), *258*, 474, 476 (132), *481*
Kamp, D. 331 (107), *372*
Kampschmidt, L. W. F. 666 (55a), *682*
Kamzolkin, V. V. 193 (146), *199*
Kan, C. S. 491 (57), *518*
Kanamori, H. 211 (88c), *231*, 343 (157), *373*
Kanaoka, Y. 285 (65), *305*
Kandror, I. I. 732 (89), *772*
Kaneda, K. 181, 192 (85), *198*, 534 (61), *579*
Kanematsu, S. 435 (42), *457*
Kanoko, C. 343, 344 (156), *373*
Kanter, E. J. J. de 598, 616, 624 (72), *645*
Kantyukova, R. G. 186 (110), *198*
Kao, J. 63 (274), 67 (291), *83*
Kaplan, M. L. 2 (19), *78*, 206 (26–28), *229*, *230*, 391 (68), *414*, 651 (15), *681*, 740 (133), *773*
Kaptein, R. 598 (72), 616 (72, 124), 624 (72), *645*, *646*, 742 (138a), *773*
Karaban, A. A. 280 (29a), *304*
Karas, G. A. 759 (177), *774*
Karasev, Yu. Z. 138–142, 149, 150 (47), *159*
Karatun, A. A. 764 (197), *775*
Karban, J. 389 (60), 390 (62b), *413*
Karch, N. J. 88, 91 (84), *95*, 406, 407 (124), *415*, 598 (69, 70), 610 (69), 617 (70), *645*, 743 (145, 146), *773*, 780 (11b), *804*
Kargin, V. A. 423 (45–47), *427*
Karle, I. L. 377 (18), *412*
Karlström, G. 24 (57), 61 (258), *79*, *83*
Karplus, M. 38 (131), *80*
Karpukhin, O. N. 778 (2), 797 (74), 799 (2), *804*, *805*
Kasai, N. 466, 474 (123), *481*
Kasatochkin, V. 91 (85), *95*
Kasha, M. 10 (28), *78*, 206 (21), 229 (227), *229*, *234*, 588 (14b), *643*
Kashar, B. 446 (143), *459*
Kashigawi, T. 466, 474 (123), *481*
Kashino, S. 546–548 (114d), *581*
Kashiwagi, T. 557 (173), *582*, 594, 596, 610, 625 (39), 626 (39, 151, 152), 628 (39),

(216), *644*, *647*, *648*, 739 (119), *773*, 780 (12b), *804*
Kaskar, B. 874 (119), *916*
Kasper, 623 (145), *647*
Käss, D. 37, 67 (123), *80*, 86–89 (21), *94*, 382 (31), *412*
Kassab, E. 60 (241), *82*
Kastelic-Suhadok, T. 118 (23), *126*
Kastell, A. 786 (37), *805*
Kastening, B. 441 (110), *459*
Katagiri, K. 397 (90, 91), *414*
Kataoka, S. 872 (106e), *915*
Kato, H. 88, 90–93 (89), *96*, 530 (48), 544 (97c), *579*, *580*, 718 (23, 24), 739, 740, 745 (24), *770*
Kato, S. 61 (259), *83*, 133 (13), *158*, 326 (84), *371*, 872 (106e), *915*
Katritzky, A. R. 336 (123), *372*
Katsuki, T. 182 (90), 183 (91), *198*, 479 (160), *482*
Katsumura, A. 326 (88), *371*
Kaufmann, K. J. 891 (153a), *917*
Kavčič, R. 88, 92 (93), *96*, 118 (23), *126*, 134 (26, 27), 135 (33), 155, 156 (86), *158*, *159*, 514, 515 (115), *520*, 527, 528 (36), *579*
Kavic, R. 288 (99), *306*
Kawabe, N. 287 (90), *306*
Kawada, Y. 596, 635 (50), *644*
Kawai, W. 425 (63), *427*
Kawakami, J. H. 526 (28, 29), *579*
Kawamoto, A. 912 (219), *919*
Kayama, Y. 216 (124b), *232*, 352 (179), *373*
Kaye, I. A. 556 (164), *582*
Kaynes, R. K. 454 (202), *461*
Kazan, J. 568 (220), *583*, 599 (77a), *645*
Kealy, T. J. 411 (140), *415*
Kearns, D. R. 2, 65 (10), *78*, 204 (2c), 205 (12), 226 (189), *229*, *233*, 318, 348 (45c), 350 (173), *370*, *373*, 451, 456 (173), *460*, 684 (91), *690*, 833 (229a), 893 (162), *917*, *919*
Keaugh, A. H. 290 (118c), *306*
Keavenay, W. P. 303 (233a), *309*
Keaveney, P. (240), *584*
Kebarle, P. 69 (309), *84*, 441 (93), *459*
Keck, G. E. 226 (187), *233*, 840–842, 906 (202), *918*
Keefer, R. M. 138 (50), *159*
Keehn, P. M. 325 (82), *371*
Keeney, F. N. 266, 267 (36), *277*
Kees, K. 207 (35b), *230*
Keinan, E. 498 (75), *519*, 658 (46a–c), 660 (46a), 661 (48), 668 (52a, 52b, 53), 666 (48), 668 (52a, 52b), 670 (48), 671 (46a–c, 59), 672, 673 (59), 675 (59, 62, 63), *682*, *683*
Kell, R. W. 280 (5), *304*
Keller, J. E. 501 (93), *519*, 651, 654 (7a, 7c, 7e, 7f), *681*
Kelley, J. A. 295 (159b), *307*
Kelley, R. S. A. 213, 216 (91b), *231*
Kellog, R. M. 225, 228 (168), *233*
Kellogg, K. M. 332 (109d), *372*
Kellogg, R. M. 833, 837 (230), *919*
Kelly, R. C. 190 (132), *199*
Kelm, H. 893 (161), 904 (193c), *917*, *918*
Kemal, C. 195 (158), *200*, 543 (95), *580*
Kemper, R. 226 (174), *233*, 873 (113), *916*
Kenehan, E. F. 295 (159b), *307*
Kenley, R. 99 (34), *104*
Kenley, R. A. 240 (19), 247 (19, 48), 252 (48), *257*, 499 (81–84), 500 (81–83), 501 (84), *519*, 622 (138a, 138b), *646*, 679, 680 (75), *689*
Kennedy, R. C. 99 (28), *104*, 669, 672 (35, 36), 673 (36), *688*
Kereselidze, R. V. 564 (209), *583*
Kergoat, R. 465, 469 (54), *480*
Kergomard, A. 261, 262 (8), 263 (20), 268 (8, 43), *276*, *277*
Kern, C. W. 38 (131), *80*
Kern, W. 424 (60), *427*
Kerr, J. A. 2, 68 (17), *78*, 99 (25), *104*
Kesarev, S. A. 178 (74), *198*
Keul, H. 389 (61), *413*
Keul, J. 834, 839, 850 (44), *914*
Key, R. M. 871 (100), *915*
Khachkuruzov, G. A. 49, 50 (196), 51 (207), 53, 54 (212), *82*, 87–90 (27), *94*
Khalil, M. M. 288, 289 (107), *306*, 525 (12), *578*
Khan, A. U. 206 (21), *229*
Khan, J. A. 317 (43), *370*
Kharasch, M. S. 173 (49, 50), 175 (60), *197*, 222 (146), *232*, 301 (218), *309*, 628 (160, 161), *647*
Khare, G. P. 465, 472 (98), *481*
Kharitonov, V. V. 422 (29, 30), *427*
Khashab, A.-I. 501 (92), *519*
Khashab, A. Y. 651, 654 (7d), *681*
Khomenko, B. A. 140, 146 (46), *159*
Khorshev, S. Ya. 142, 147, 150, 151 (56), *159*
Kice, J. L. 260 (7), *276*, 552 (135), 566 (213), *581*, *583*, 623 (146a), *647*
Kiffer, A. D. 658 (47), *682*
Kihara, H. 635 (189), *648*
Kikuchi, O. 68 (299), *83*, 611 (106), 617 (106, 128a–c), 632 (168), *646*, *647*, 791 (48), *805*
Kilbourn, B. T. 818 (46), *827*
Kim, B. 291 (138), *307*
Kim, H. 49 (195), *82*
Kim, H. S. 88, 92 (98), *96*, 405 (109), *414*
Kim, K.-J. 767 (206), *775*
Kim, L. 194 (151), *199*
Kim, S. S. 298 (193c), *308*

Kim, Y. H. 445 (126a, 126c), 446 (126c), *459*
Kimura, K. 46 (169), 47, 48 (177), *81*
Kimura, M. 571 (232), *583*
King, G. S. 107 (5), *126*
King, J. M. 742 (143), *773*
Kinkel, K. G. 213, 216 (96), *231*
Kinstle, T. H. 675 (64), *683*
Kiparisova, E. G. 98, 99 (15), *104*
Kipnis, I. S. 910 (212a), *919*
Kirby, G. W. 384 (38e), *413*
Kirk, A. D. 666 (3, 4), *688*
Kirk, D. I. 871 (100), *915*
Kirksey, J. W. 314 (22, 23), *370*, 396 (80, 81), *414*
Kirkwood, J. G. 44 (155), *81*
Kirshcke, K. 71 (315), *84*, 391 (66), 405 (113), *413*, *415*
Kirschner, S. 893 (164a, 164b), *917*
Kirshenbaum, A. D. 63 (279), *83*
Kirshke, K. 782, 792 (23), *804*
Kiseleva, T. M. 622 (140), *646*
Kishi, J. 291 (141), *307*
Kishi, Y. 526, 532, 533 (27), *579*, 844 (18a), *913*
Kislyak, G. S. 280 (28d), *304*
Kissel, T. 838, 842, 847, 884, 886, 887, 895, 896, 898, 899 (31), *914*
Kitagawa, Y. 365 (199), *373*
Kitahara, Y. 216 (124b, 125), *232*, 344 (158), 346 (163), *373*
Kitaigorodsky, A. I. 386, 408 (45a), *413*
Kitamura, A. 596, 619 (59), *644*
Kitao, T. 593, 595, 603, 604, 608 (37), *644*
Kitaoka, Y. 593, 595, 603, 604, 608 (37), *644*
Kithara, Y. 352 (179, 181), *373*
Kiva, E. A. 139 (48), 140 (49), 141 (48, 49), 142 (48, 54, 55), 145 (54), 148, 149 (54, 55), 150 (48, 55), *159*
Klaska, K.-H. 386, 388 (45b), *413*, 840, 845, 846, 867, 890, 892 (22a), *914*
Kleemann, A. 290 (128), *306*
Klein, E. 222 (149), *232*
Klein, H. 663 (51), *682*
Klein, M. W. 291 (137), *307*
Klein, R. J. 299 (199b), *308*
Klein, W. M. 287 (89), *306*
Kleinfeller, H. 280, 281 (13), *304*
Klever, H. W. 831, 858 (11), *913*
Klimchuk, M. A. 139, 141, 142 (48), 150 (48, 72), *159*
Kline, M. 557, 559 (180), *582*
Klopman, G. 633 (177), *647*
Klotz, I. M. 65 (288), *83*
Klug, H.-H. 330 (104), *372*
Klug, J. T. 215 (114), *232*
Kluger, E. W. 571 (232), *583*
Klüpfel, K. 636 (196), *648*
Klupp, C. V. 658 (47), *682*
Kneip, A. 134 (21a), *158*
Knesel, G. A. 290 (127), *306*
Knipple, W. R. 266–268 (35), *277*
Knöpfel, W. 876 (124), *916*
Knorre, D. G. 134 (23), *158*
Knox, B. E. 431 (16a), *457*
Knox, J. H. 666 (4), *688*
Knudsen, G. A., Jr. 753 (162), *774*
Kobayashi, H. 560 (196), *583*
Kobayashi, M. 262 (14, 15), 263 (14, 15, 19, 22a, 22b, 23), 264 (14, 23), 265 (14), 267, 269 (19), 273 (62), *276*, *277*, 546 (122), *581*, 593 (34), 597 (64, 67), 598 (76), 617 (67), 618 (131), 619 (132), 631 (76), 639 (209), 640 (64, 210), *644–646*, *648*, 740 (130), 760 (180), *773*, *774*
Kobayashi, S. 446 (127–129), 449 (127), *459*
Kobyakov, A. K. 143, 144 (58), *159*
Kobylinski, T. P. 190 (131), *199*
Koch, D. 223 (155b), *233*
Koch, E. 204 (4a), 206 (32, 33), 208, 210 (48–50), *229*, *230*
Kochi, J. K. 168 (38), 169 (40), 170 (40, 41), 171 (40), 174 (54), 175 (61, 62), 176 (62), *197*, 241 (22), 254 (70), *257*, *258*, 301 (221, 223), *309*, 478 (150), *482*, 533 (58), *579*, 599 (78), 635 (190), *645*, *648*, 712, 713 (1b), 714 (1b, 8), 715 (8), 716 (10), 731 (79), 733 (79, 94b), 738, 739 (1b), 748–750 (10), 751 (157, 158), 752 (10), 760 (181b), *770*, *772*, *774*, 808 (1), *826*
Kocjan, D. 63, 68 (277), *83*, 485 (21), *518*
Kodato, S. 872 (106d), *915*
Koenig, T. 591 (23, 24), 594 (24, 46, 47), 595 (23, 24, 49), 597 (66), 600 (82a, 82b, 83), 602 (82b), 603 (23, 24, 46, 49, 82b, 84, 89), 604 (24, 46, 89), 605 (46, 47, 82b, 95), 606 (24, 46, 47), 607 (46, 47, 82b, 94, 95), 611 (82b, 108), 614, 616 (66), 617 (127), 632 (82a, 82b), *644–646*
Koenig, T. W. 546 (121), *581*, 634, 635 (186), *648*
Koerner von Gustorf, E. 222 (150), *232*
Koh, E. T. 88, 91 (84), *95*, 406, 407 (124), *415*, 598, 617 (70), *645*, 743 (146), *773*
Kohler, E. P. 831 (5), *913*
Kohmoto, S. 861 (66), *915*
Koike, S. 635 (189), *648*
Kokuka, S. (216), *648*
Kolbeck, W. 286 (75), *305*
Kolc, J. 742 (141a), *773*
Kolcev, G. V. 802 (109), *806*
Kolesinski, H. S. 280 (16), *304*
Kollar, J. 162 (3), 185 (99), *196*, *198*
Koller, J. 68 (298), *83*, 538 (80a), 573 (233), *580*, *583*

Kollmar, H. W. 893 (164b), *917*
Kolsaker, P. 387 (54), *413*, 501 (95), *519*, 650 (2), *681*
Kolthoff, I. M. 132, 147 (10), *158*, (174), *774*
Komazawa, K. 451 (175), *460*
Komin, J. B. 294 (153), *307*
Komornicki, A. 25, 52, 57 (80), *79*
Kondo, H. 843, 858, 884, 898, 911 (57a), *914*
Kondo, K. 120 (29), *127*, 210 (74, 78, 79), 215 (116–118), *231*, *232*, 318 (46), 319 (46, 49), 327 (89), 328 (91), 329 (97, 99), 336 (127), 365 (201–203), *371*, *372*, *374*, 869 (97), *915*
Kondo, Y. 366 (205c, 206), *374*
Konen, D. A. 288 (93), *306*, 855 (247), *919*
Konieczny, M. 259 (1b), *276*
Konishi, M. 122 (35), *127*
Konoreva, I. I. 423 (45), *427*
Konrad, F. M. 166 (19), *196*
Kontoyiannidou, V. 225 (169), *233*
Koo, J. Y. 667, 684 (14), *688*, 844 (18j), *913*
Koo, J.-Y. 648 (98), *690*, 836, 883, 885 (146), 891 (153a), 896 (181b), *916–918*
Koopmans, T. 44 (156), *81*
Kooyman, E. C. 554, 561 (147), *582*
Kopecky, K. R. 156 (88a, 88b), *160*, 227–229 (191), *234*, 314 (28), *370*, 589 (16), 591 (16, 22), 592 (26), *643*, *644*, 667 (12), *688*, 831 (4), 832 (27), 833 (27, 159a, 159b), 834 (27, 159a, 159b, 199), 844 (4), 847 (27, 29), 848 (4, 27), 859, 881 (27), 885 (27, 159a, 159b), 886 (27, 199), 887 (159a, 159b), 890 (27, 159a, 159b), 892, 898 (159a, 159b), 899 (159a, 159b, 199), 900 (159a), 906 (199), 908 (206), 910 (27), *913*, *914*, *917*, *918*
Koppenol, W. H. 452 (190), 456 (190, 208), *460*, *461*, 720 (31), *771*
Korber, H. 574 (239), *584*
Korcek, S. 489 (46), *518*, 812, 813 (25), *827*
Koritskii, A. T. 738 (118a), 764 (196, 197, 199), 765 (200, 201), *773*, *775*
Kornblum, N. 214, 221, 225, 228 (107), *232*, 352 (178), *373*, 435 (45), 442, 445 (118), 453 (196), *457*, *459*, *460*, 573 (234), *583*
Kornilova, N. V. 817 (39), *827*
Korp, J. 389 (60), *413*
Korshak, V. V. 423 (48, 49), *427*
Koshel, G. N. 136 (45), *159*, 185 (100), *198*
Kosser, G. F. 546 (103), *580*
Kosswig, K. 185 (102), *198*
Kost, M. T. 540, 544 (84), *580*
Kostuch, A. 733 (98a), *772*
Kotera, N. 167 (29), *197*
Koton, M. M. 622 (140), *646*
Kotsuki, H. 210 (80), *231*, 328 (93), 330 (102), *371*
Kotulák, L. 737 (111b), *772*
Koubek, E. 130 (1c), 131 (3, 4), *157*, 641 (211), *648*
Kovac, F. 574 (242), *584*, 652 (21), *681*
Kovač, F. 501 (102, 103), 504 (110a, 110b, 111), 507 (110a, 110b), 509, 510 (102, 103), 511 (103), 516 (103, 110a, 110b), *519*
Kovacic, P. 268 (41), *277*, 285 (63), *305*
Kovelan, M. J. 394 (77), *414*
Kowal, J. 737 (110), *772*
Kowala, C. 388 (49), *413*
Kozlov, P. V. 423 (45), *427*
Kozolov, N. A. 99 (16), *104*
Kozuka, S. 226 (183a, 185c), *233*, 557 (173, 174), *582*, 594 (39, 41), 596 (39), 599 (74), 610, 625 (39), 626 (39, 74, 152), 628 (39, 41, 74, 165), 629 (41, 74), 631 (165), *644*, *645*, *647*, 739 (119), *773*, 780 (12b), *804*
Kraentler, B. 782 (27), *804*
Kralj, B. 118 (23), *126*
Kramer, R. 433 (33), *457*
Kramer, V. 118 (23), *126*
Kramer, W. P. 45 (166), *81*
Krat, A. V. 425, 426 (72), *428*
Kratzsch, L. 574 (237), *584*
Kratzsch, M. 291 (140), *307*
Krauch, H. 288 (103), *306*, 596, 636 (54), *644*
Krauss, H. 354 (183), *373*
Krebs, A. 226 (174), *233*, 386, 388 (45b), *413*, 840, 845, 846, 867 (22a), 873 (113), 890 (22a, 179), 892 (22a), 895, 900, 903 (179), *914*, *916*, *918*
Krebs, E. 280 (5, 6), *304*
Kresze, G. 296 (174), *308*
Kretschmer, G. 526, 527 (33), *579*
Krinsky, N. I. 229 (226), *234*
Kritchevsky, T. H. 560 (193), *583*
Kropf, H. 222 (142), *232*, 260 (6), *276*, 301, 302 (224), *309*, 526, 527 (34), 555 (153), *579*, *582*, 780 (13), *804*
Krull, I. S. 106 (3a–c), *126*
Krupenie, P. H. 21, 24, 48, 49, 58 (36), *78*
Kruper, W. J. 191 (135, 137), *199*
Krushinskaya, G. A. 185 (100), *198*
Krusic, P. J. 241 (22), *257*, 714, 715 (8), 731, 733 (79), 751 (157, 158), *770*, *772*, *774*
Kryszewski, M. 426 (77), *428*
Kubo, M. 203, 204 (2b), *229*
Kubota, H. 421 (21–24), 425 (71), *427*, *428*
Kubota, M. 474 (130), *481*
Kucher, R. V. 90 (73), *95*, 737 (113), *773*
Kuczkowski, R. L. 65, 67 (280), 71 (314), *83*, *84*, 90 (75), *95*, 125 (37), *127*, 389 (56, 57), *413*, 633 (180), 634 (180, 184), *647*, *648*
Kuderna, J. 628 (160), *647*
Kudryavstev, A. B. 143, 144, 149 (59), *159*
Kuhn, H. J. 209, 220 (56), *230*, 588 (10c), *643*

Kuhnen, L. 180 (80), 186 (105, 106), *198*
Kuivila, H. G. 814 (31), *827*
Kukovinets, O. S. 120, 121 (32), *127*
Kul'chitskaya, S. L. 762 (191), *775*
Kulig, M. J. 440, 449 (81), *459*
Kumanova, B. K. 725 (57), *771*
Kumar, C. V. 865 (84), *915*
Kumar, N. B. 650 (2), *681*
Kung, W. 290 (117b), *306*
Küng, W. 525 (13), *578*
Kunz, W. 288 (103), *306*
Kunze, K. L. 536, 537 (76), *580*
Kupchan, S. M. 213 (98, 99, 102), *231*, 314 (24), *370*
Kuramshin, E. M. 501 (120), *520*, 653 (23), *682*
Kurnath, N. B. 266, 268 (40), *277*
Kuroda, K. 210 (75, 76a, 76b), *231*, 328 (94), 368 (210), *371*, *374*
Kurtz, D. M., Jr. 65 (288), *83*
Kuryatnikow, Yu. I. 905 (195), *918*
Kurz, M. E. 298 (193b), *308*, 517 (119), *520*
Kuschke, K. 245 (34), *257*
Kvasha, S. M. 738 (117), *773*
Kwan, C.-Y. 134 (20), *158*
Kwart, H. 289 (111), *306*, 529 (42, 43), 564 (43), *579*
Kyla, E. P. 105 (1), *126*
Kyono, K. 843, 858, 884, 891, 898, 906, 911 (57b), *914*
Laarhoven, W. H. 874 (117), *916*
L'abbé, G. 400 (95), *414*
Labinger, J. A. 819 (47), *827*
Lacombe, M. 328 (95), *371*
Ladygin, B. Y. 721 (40), *771*
Lahav, M. 245 (35), *257*, 494 (67), *519*
Lai, J.-J. 871 (102), *915*
Laidig, W. D. 25, 33 (68), *79*
Laidler, K. J. 676 (57), *689*
Laing, K. R. 465, 471 (86, 90), *480*, *481*
Laing, M. 377 (25), *412*, 466 (104, 114), 472 (104), 473 (104, 114, 115), *481*
Laker, T. M. 436 (48), 437 (55), *457*
Lal, M. 762 (190), *775*
Lalande, R. 193 (145), *199*
Lam, A. V. 174 (52), *197*
Lam, Y.-S. 86, 88 (17), *94*, 376 (16), *412*
Lamanna, U. 36, 57, 58 (98), *79*
Lamanna, V. 87, 88, 90 (45), *95*
Lamb, R. C. 557 (172), *582*, 614 (116a), 625 (149a), 628 (149a, 158a, 158b), *646*, *647*, 765, 766 (203a), *775*
Lambert, J. B. 268 (45), *277*
Lambert, J. P. 448 (157), *460*
Lamendola, J., Jr. 571 (232), *583*
Lancaster, J. E. 213 (100, 101), *231*, 314 (25), *370*
Landau, R. 178 (70), *197*, 478 (154), *482*
Landheer, I. 319 (52), *371*
Landi, V. R. 720 (33), *771*
Landis, M. E. 227 (197), 228 (197, 219, 221), *234*, 446 (150), *459*, 592 (27b), *644*, 836 (211c), 837 (224), 844 (18m), 885, 887 (224), 891, 894 (150a), 910 (211a–c), 911 (213), 912 (216, 218), *913*, *917–919*
Landler, J. 796, 797 (63), *805*
Landler, Y. 423 (43, 44), *427*
Lane, R. D. G. 113 (14b), *126*
Lang, S. 166 (20), 193 (143), *196*, *199*, 222 (141a, 141b), *232*
Lang, T. J. 88, 90, 92 (102), 93 (102, 106), *96*, 530 (52), *579*
Langen, P. 58, 59 (233), *82*
Langhoff, S. R. 25, 33, 58, 59 (79), *79*
Langs, D. A. 384 (38h), *413*
Langston, J. H. 281 (44), *305*
Lappert, M. F. 818 (44), *827*
Lapshin, N. M. 157 (97), *160*
Lapuka, L. F. 136 (45), *159*
Laputte, R. 420 (12), *427*
Larcombe, B. E. 432 (23), 441, 449, 450, 453 (86), *457*, *459*
Laronze, J. 867 (93), *915*
Laronze, J.-Y. 867 (93), *915*
Larsen, D. L. 207 (39), *230*, 324, 325, 347 (79, 81), 348 (81), *371*
Larsen, E. H. 283 (60, 61), *305*
Larson, E. H. 282 (54), *305*
Lassettre, F. N. 87, 88 (31), *94*
Laster, W. R., Jr. 397 (89), 398 (93), *414*
Lastochkina, N. S. 137 (41), *158*
Lastukhin, Yu. A. 425 (67), *428*
Lathan, W. A. 25, 33, 45 (74), 61 (264), *79*, *83*
Latour, J. M. 465, 469 (33), *479*
Lattimer, R. O. 125 (37), *127*
Lattimer, R. P. 389 (57), *413*
Lau, H. H. 628 (163), *647*
Lau, W. 254 (70), *258*, 635 (190), *648*
Lawesson, S. O. 546 (129), *581*, 810 (14), *827*
Lawesson, S.-O. 282 (53–57), 283 (60, 61), 300 (214, 215), 301 (214), 302 (226, 228, 230, 231), 303 (214, 231, 236, 239), *305*, *309*, 638 (204), *648*
Lawler, R. G. 557, 559 (179, 180), *582*
Lawson, D. E. 405 (116), *415*
Laye, P. G. 791 (49), *805*
Lazár, M. 784 (33), 785 (36), 786 (33, 36), 788, 790 (41, 44), 791 (41, 44, 50), 792 (50, 53, 54), 800 (83, 84, 86, 99), 801 (84), *804–806*
Lazurin, E. A. 139, 142, 149 (51), *159*
Leacock, R. A. 36, 38 (120), *80*, 87–90 (25, 26), *94*, 376 (2), *412*

Lebedev, N. N. 143, 144, 149 (59), *159*, 528, 530 (38), *579*
Lebedev, Y. S. 738 (118a), 739 (122), *773*
Lebel, P. 423 (43, 44), *427*, 796, 797 (63), *805*
Le Berre, A. 432 (18, 19), 441 (18, 19, 88), 442 (19, 88), 442 (19, 88), 442 (88), 444 (18, 19, 88), 449, 452, 454 (18, 19), *457*, *459*
Lebourgeois, P. 741 (137), *773*
Le Bras, J. 781 (19), *804*
Leca, J.-P. 424 (54), *427*
Le Carpentier, J. M. 465, 469 (61), 475 (155), *480*, *482*
Lechevallier, A. 865 (82), *915*
Lechtenbohmer, H. 420 (16), *427*
Lechtken, P. 2, 21 (15), *78*, 667 (15), 684 (15, 92), *688*, *690*, 832, 833 (156), 834 (235), 836, 838 (156), 844, 847 (18c), 881, (144), 882 (18c), 885 (156, 235), 887, 888 (156), 890 (235), 891 (149, 153b, 154, 155a, 155c), 892 (155a, 155c, 156), 893 (18c, 163g), 894 (171), 898, 899 (156), 902 (18c, 155c, 188), 903 (188), 905, 908 (197), *913*, *916–919*
LeClerc, G. 341 (145, 146), *372*
Lecoq, J. C. 471 (78), *480*
Ledaal, T. 122 (34), *127*, 855 (254), *919*
Leddy, B. P. 315 (32), *370*
Ledneva, O. A. 800 (90), 803 (111), *806*
Lee, C. 840, 884, 895, 898 (180a), *918*
Lee, D. C. 227 (199), *234*
Lee, D. C.-S. 894 (169), *917*
Lee, J. 902, 903 (186), *918*
Lee, J. B. 221 (138), *232*, 522, 525, 546 (1d), *578*
Lee, K. W. 840, 884, 895, 898 (180b), *918*
Lee, L. C. 48 (187), *81*
Lee, R. A. 438 (69), *458*
Lee, S. T. 44, 45 (162), *81*
Lee-Ruff, E. 432 (25), 440 (76), 441 (106), 448 (159), 453 (106), *457*, *459*, *460*
Leeuwen, J. W. van 720 (31), *771*
Le Fevre, H. F. 675 (52), *689*
Le Fèvre, R. J. W. 88 (62), 89 (62, 64), *95*
Leffler, J. 780 (12b), *804*
Leffler, J. E. 130 (6), *158*, 264 (29), *277*, 552 (141), 557 (168, 169, 178), 558 (178), *581*, *582*, 623 (147a, 147b), 628 (157a–c), *647*, 742 (139b), *773*, 779 (4, 5), 780 (10), *804*
Leforestier, C. 25, 33, 61 (67), *79*
Lefort, D. 252 (65), *258*, 287 (85), *305*
Lefort, J. B. 351 (177a), *373*
Lehmann, M. S. 49, 51, 55 (192), *81*
Lehnig, M. 740 (134), *773*
Leighton, P. A. 681 (85), *690*
Leininger, H. 832, 863 (71a), *915*
Lemaire, J. 564 (208), *583*, 741 (137), *773*
Lendal, D. 120 (30), *127*
Le Noble, W. J. 801 (103), *806*
Lenssen, U. 114 (17), *126*
Lenz, G. R. 333 (112), *372*
Leonov, D. 291 (138), *307*, 733 (97), *772*
Leonov, M. R. 588 (8), *643*
Le Perchec, P. 221, 225 (136), *232*, 835 (238), 863, 911 (74a, 74b), *915*, *919*
Lerchová, J. 737 (111b), *772*
Lerdal, D. 327 (90), 328 (90, 92), *371*
Lerdal, D. A. 504 (109), *519*, 652 (20a, 20b), 654 (20a, 20b, 31), *681*, *682*
Lerman, C. L. 228 (221), *234*, 910 (211b), *918*
Le Roux, J. P. 355 (184), *373*
Le Roux, J.-P. 842 (251a, 251b), *919*
Leroy, F. 388 (50), *413*
Leshem, Y. 440, 445 (82), *459*
Leshem, Y. Y. 440, 445 (83), *459*
Lessinger, L. 403 (104), 407 (104, 137), *414*, *415*, 780 (9), *804*
Letertre, G. 842 (251b), *919*
Lethbridge, J. W. 261, 262, 268 (10a), *276*
Letsinger, R. L. 718, 733, 739 (19), *770*
Levanon, H. 767 (208), *775*
Lever, A. B. P. 2 (23), *78*, 432 (25), *457*
Lever, D. W., Jr. 337 (128), *372*
Levey, G. 767, 768 (207), *775*
Levi, E. M. 268 (41), *277*
Levine, J. 48, 49 (179), *81*
Levine, S. G. 394 (76), *414*
Levine, S. Z. 677 (59), *689*
Levison, J. J. 466, 474 (129), *481*
Levonowich, P. F. 451 (174), *460*
Levush, S. S. 653 (24b), *682*
Levy, H. 608 (98), *645*
Levy, H. A. 49 (191, 194), 51 (191), *81*, 86, 88 (9), *94*, 376, 377, 382 (4), *412*
Levy, J. 867 (93), *915*
Levy, J. B. 99 (28), *104*, 669, 672 (35, 36), 673 (36), *688*
Levy, M. 36 (103), *80*, 87, 90 (53), *95*
Lewis, C. 205 (14), *229*
Lewis, D. K. 99 (23), *104*
Lewis, E. S. 613, 614 (115), *646*
Lewis, J. 469 (59), *480*
Lewis, R. N. 263 (28), *277*
Lewis, S. N. 290 (129a), *307*
Lewis, T. A. 554 (149), *582*, 590 (19c), *644*
Lewis, T. W. 778 (1), *804*
Li, J. P. 560, 561 (194), *583*
Liberles, A. 38 (135), *80*
Lichtenberger, J. 280 (15), *304*
Lichtenstein, M. 57 (228), *82*
Lick, C. 303 (233c), *309*
Liehr, J. E. van 113 (15a–c), *126*
Liesi, E. A. 672 (42), *689*

Lifshitz, A. 69 (308), *84*
Lifson, S. 396 (85), *414*
Liftman, Y. 440, 445 (83), *459*
Light, R. E., Jr. 425 (64), *427*
Lightner, D. A. 227 (206, 207), 228 (214), *234*, 334 (120, 130), 335 (120), *372*, 871 (99–101), 910 (99), *915*
Ligon, R. C. 356 (188), *373*
Likhterov, V. R. 262 (17), 263 (25–27), 264, 265 (25, 31), *276*, *277*
Liles, D. C. 88, 90 (74), *95*, 382 (34, 35), *413*
Lilie, J. 765, 766 (203a), *775*
Lim, H. S. 465, 472 (98), *481*
Lin, H. C. 666, 676 (67), *683*
Lin, J. W. P. 652, 653 (19a), *681*, 832, 838 (226), *919*
Lin, J. W.-P. 501 (100), *519*, 859 (59), *914*
Lind, H. 737 (111a, 112), *772*, *773*
Lindberg, B. 392 (69), *414*
Lindberg, J. G. 793 (58), *805*
Lindblom, L. A. 475 (157), *482*
Lindholm, E. 44, 45 (159), *81*
Lindsay, B. G. 546 (107), *581*
Lineberger, W. C. 48, 49, 54 (186), *81*
Linnett, J. W. 87 (50), 89 (57), *95*
Linton, E. P. 87 (29), *94*
Liotta, C. L. 441 (99), *459*
Liotta, D. C. 219 (129), *232*
Liotta, R. 297 (179), *308*
Lippmaa, E. T. 557 (250), *584*
Lipscomb, W. N. 86, 88 (10), *94*, 376 (3), *412*
Lipshutz, B. H. 211 (85), *231*, 332 (108c), *372*, 874 (120), *916*
Liskow, D. H. 25, 33 (71), *79*
Lisle, J. B. 25, 33, 45 (74), *79*
Lissel, M. 437, 438 (66), *458*
Lissi, E. 881 (142b), *916*
Lissi, E. A. 814 (28), *827*
List, D. 823 (60), *828*
List, F. 180 (80), *198*
List, W. F. 186 (106), *198*
Litinskii, A. O. 87, 89, 92 (54), *95*
Litkowez, A. K. 574 (237), *584*
Little, W. 465, 469 (50), *480*
Littlefair, J. H. 98 (12), *104*
Liu, B. 21, 58 (42), *78*
Liu, J.-C. 228 (213), *234*, 831 (13), 832, 839 (157), 843 (26), 844 (13), 846, 853, 855–857 (26), 865 (85), 890, 892, 900 (157), 910 (85), *913–915*, *917*
Liu, J. W.-P. 592 (29), *644*
Liu, K.-C. 226 (174), *233*, 873 (113), *916*
Liu, K.-T. 526 (29), *579*
Liu, M. T. H. 675 (54), 676 (54, 57), *689*
Livingston, R. 721 (37, 38), 722 (38), 724 (51, 52), 739 (51), *771*
Llewellyn, D. 554 (149), *582*
Llewellyn, D. R. 590 (19c), *644*
Lloyd, A. C. 677 (60), *689*
Lo, D. H. 88, 89 (67), *95*
Lo, S. T. D. 465, 467, 472 (22), *479*
Lobunez, W. 88 (61, 99), 89–91 (61), 92, 93 (99), *95*, *96*, 134 (30), *158*, 382 (33), 405 (106), *412*, *414*
Lockwood, P. A. 156 (88b), *160*, 227–229 (191), *234*, 314 (28), *370*, 832 (27), 833, 834 (27, 159a), 847 (27, 29), 848, 859, 881 (27), 885 (27, 159a), 886 (27), 887 (159a), 890 (27, 159a), 892, 898–900 (159a), 910 (27), *914*, *917*
Lodygina, V. P. 145, 147, 149 (67), *159*
Leoliger, H. 737 (111a, 112), *772*, *773*
Loew, G. 68 (295), *83*, 88, 90, 93 (105), *96*, 288 (95), *306*, 523, 524 (2), *578*
Loew, G. H. 68 (294), *83*, 530 (50), *579*
Lofthouse, G. J. 876 (126), *916*
Logan, J. 225 (169), *233*
Lohaus, G. 730 (68), *771*
Lohringer, W. 653 (24a), *682*
Lohse, C. 874 (115), *916*
Lokensgard, D. 314 (22), *370*, 396 (80), *414*
Lombardi, E. 36 (113), *80*
Long, F. A. 133 (15), *158*
Long, P. V. 878 (129), *916*
Lonsdale, K. 381 (28), *412*, 783–787 (32), *804*
Lopata, A. 125 (37), *127*
Lopez Nieves, M. J. 853 (52), *914*
Lopez Sastre, J. A. 834, 886, 899, 906 (199), *918*
Lorand, J. P. 299 (202), *308*
Lorand, L. P. 594, 600, 607 (44), *644*
Lorberbaum, Z. 733 (98b), *772*
Lorenz, O. 634 (181a, 181b), *647*
Lossing, F. P. 716 (14c), *770*
Loucks, L. F. 675 (54), 676 (54, 57), *689*
Louw, W. J. 826 (80), *828*
Lovas, F. J. 13, 61, 62, 69, 71 (33), *78*
Loveitt, M. E. 119 (24a), *126*, 216 (123a), *232*, 316 (41), 317 (41, 42), *370*
Loveland, P. 865 (87), *915*
Lovett, M. B. 908 (205), *918*
Lovtsova, A. N. 557 (250), *584*
Low, L. K. 228 (214), *234*, 871, 910 (99), *915*
Lowrey, A. H. 88, 89 (70), *95*
Lu, C. S. 86, 88 (5), *94*
Lu, C.-S. 376 (14), *412*
Lucchese, R. R. 24, 33 (55), *79*
Luchese, R. R. 63, 67 (278), *83*
Ludwig, P. 165 (18), *196*
Ludwig, W. 651 (6b, 11), 654 (6b), *681*
Luibrand, R. T. 405 (115), *415*
Lukovnikov, A. F. 738 (117), *773*, 800 (100), *806*
Lukovnilov, A. F. 738 (116), *773*

Lukyanenko, L. V. 737 (113), *773*
Lumma, W. C., Jr. 501 (100), *519*, 652, 653 (19a), *681*
Lumpkin, O. 57 (229), *82*
Lund, H. 440 (72, 73), 441 (84), 451 (175), *458–460*
Lungle, M. L. 651 (10a), *681*
Luo, Y. R. 517 (118), *520*
Luong, T. M. 134 (28), *158*
Lusinchi, X. 446 (130), *459*
Lussier, R. J. 758 (175), *774*
Lutes, H. D. 111 (13b), *126*
Lutze, G. 866, 867, 910 (92), *915*
Lynch, B. M. 288, 290 (110), *306*, 525 (7), *578*
Lyons, J. E. 181 (83, 84), *198*, 464 (3, 18), 474, 478 (18), *479*

Maass, O. 87 (29), *94*
Mabey, W. R. 595 (49), 603 (49, 84), *644*, *645*
Maccagnani, G. 297 (181a, 181b), *308*, 545 (99, 100), *580*
Machida, Y. 316 (36), *370*, 438, 441 (65), *458*
Machleder, W. H. 294 (152, 153), *307*
Machlin, L. J. 432 (28), *457*
Madan, V. 570 (227), *583*
Madhavan, V. 767 (205, 208), 768 (205), *775*
Madoux, D. C. 446 (150), *459*
Madsen, P. 283 (61), *305*
Maestri, M. 757 (173), *774*
Maestro, M. 36, 57, 58 (98), *79*, 87, 88, 90 (45), *95*
Mageli, O. L. 2 (6), *78*, 152 (82), *159*, 299 (205), *308*, 425 (64), *427*
Maggiolo, A. 651 (6a, 12), 654 (6a), *681*
Magnasco, V. 36 (110), *80*
Magno, F. 442–444 (119), *459*
Magno, S. 312 (10a), *370*
Magnus, P. D. 215 (119), *232*, 341 (145, 146), *372*
Magnus, P. G. 107 (5), *126*
Maguire, W. J. 727 (62), *771*
Magyar, E. S. 268 (45), *277*
Maheshwari, K. K. 213 (92), *231* 350 (175), *373*
Mahoney, L. R. 249 (56), *257*
Maier, G. 839, 867 (243a, 243b), *919*
Maiolli, L. 543 (94), *580*
Maizus, Z. K. 167 (26), 172 (46), *196*, *197*
Maîzus, Z. K. 134 (23), *158*
Majumder, P. C. 312 (9b), *370*
Mak, S. 437, 438, 440, 454 (60), *457*
Mak, T. C. W. 86, 88 (17), *94*, 376 (16), *412*
Makarov, G. G. 800 (94, 95), *806*
Maker, P. D. 69 (305), *84*, 678 (70), 679 (71), *689*
Makin, G. I. 760 (182), *774*, 817 (39), *827*
Makin, M. I. H. 435, 436 (41), *457*
Maksimuk, T. V. 762 (191), *775*
Malak, L. 800 (93), *806*
Malatesta, V. 405 (116), *415*
Malaval, A. 652 (17), *681*
Malik, A. H. 658 (47), *682*
Malley, M. 406 (123), *415*
Malsch, K.-D. 839, 867 (243b), *919*
Maltes, S. L. 446 (144, 146), *459*
Mal'tsev, V. I. 722 (44), 724 (44, 50), 739 (50), *771*
Mamchur, L. P. 150 (73), *159*
Mameniskis, W. A. 190 (130), *199*
Mandelbaum, A. 106 (3b, 3c), *126*
Manecke, G. 425 (73), *428*
Mangiaracina, P. 388 (52), *413*
Mann, B. J. 778 (1), *804*
Mann, C. K. 271, 272 (54), *277*
Manne, R. 44, 45 (158), *81*
Manner, J. A. 259, 260 (2b), 263 (28), *276*, *277*
Manoharan, P. T. 471 (89), *481*
Manske, R. H. 867 (94), *915*
Manson, J. 420 (15), *427*
Mao, M. 560 (197), *583*
Marangelli, V. 542, 543 (92), *580*
March, J. A. 266 (34), *277*
Marchon, J. C. 465, 469 (33), *479*
Marcinko, R. W. 47, 48 (175), *81*
Marco, R. A. de 735 (105), *772*
Marcus, R. A. 895 (175), *918*
Marcuzzi, F. 298 (191), 300 (216), *308*, *309*, 542 (91), *580*
Mare, H. E. de la 352 (178), *373*
Mares, F. 341 (149), *372*, 446 (149), *459*, 465 (53, 70), 469 (53), 470 (69, 70), 471 (53, 70), 472 (107), *480*, *481*
Margaryan, Sh. A 144 (61), *159*
Margrave, J. L. 377 (22), *412*
Margulis, T. N. 386, 387 (44), *413*
Marin, A. P. 800, 803 (91), *806*
Mark, C. 465, 470 (66, 67), *480*
Mark, H. W. 268 (45), *277*
Markarov, M. G. 143, 144, 149 (59), *159*
Markey, B. R. 36 (105), *80*, 87, 90 (51), *95*
Markham, J. L. 336 (124), *372*
Markuš, S. 781, 782 (20), 784 (20, 33), 785 (20, 36), 786 (20, 33, 36), *804*, *805*
Marloni, A. 312 (10c), *370*
Marnett, L. J. 225 (167b), *233*
Marsden, C. J. 37 (124, 125), 67 (124), *80*, 88 (69, 71), 89 (69), 90 (71), *95*, 381 (29, 30), *412*
Marsel, J. 118 (23), *126*
Marsh, R. E. 475 (157), *482*
Martell, A. E. 464, 465, 475 (13), *479*

Martin, J. C. 254 (69, 70), *258*, 282 (48, 49), *305*, 546 (121, 125), *581*, 596 (53), 603, 609, 610 (88), 613 (105), 614 (88, 122, 123), 615 (88, 105), 617 (53), 618, 619 (105), 634 (186, 187), 635 (186, 188, 190), *644–646*, *648*, 780 (11a), *804*
Martin, M. M. 742 (143), *773*
Martin, N. H. 866 (90), *915*
Martinez, F. 484 (7), *517*
Martirosyan, A. A. 144 (62), *159*
Maruthamuthu, P. 724 (48), 758 (176a, 176b), 768 (176a), 769 (176a, 210), 770 (210), *771*, *774*, *775*
Maruyama, T. 876 (127), *916*
Maseiff, G. 672 (42), *689*
Mashiko, T. 182 (89), *198*
Maskiewicz, R. 878 (132), *916*
Maslenikov, V. P. 556 (166), *582*
Maslennikov, V. P. 760 (182), *774*
Maslyanikov, V. P. 814 (29), *827*
Mason, P. P. 446 (135), *459*
Mason, R. 474 (127), *481*
Massey, J. T. 57 (227), *82*, 87 (28), *94*
Massie, S. N. 726 (60), *771*
Masthoff, R. 810 (66), *828*
Mastrorilli, E. 526 (22), *579*
Masuo, F. 133 (13), *158*
Matcham, M. J. 669 (33), *688*
Matheos, J. L. 559 (190), *582*
Mathern, G. 465, 469 (41, 42), *480*
Mathew, C. T. 166 (21), *196*
Mathur, K. B. L. 285 (71), 286 (72, 73), *305*, 546 (130), *581*
Matienko, L. I. 172 (46), *197*
Matisová-Rychlá, L. 792 (53), 796, 797 (65, 69), 800 (99), *805*, *806*
Matlin, S. A. 470 (74), *480*, 836 (241), *919*
Matsugo, S. 332 (109a), *372*, 836 (103), 871 (103, 104), 872 (105a, 105b, 106b), 911 (103), *915*
Matsumoto, M. 120 (29), *127*, 210 (74, 75, 76a, 76b, 78, 79), 215 (116–118), *231*, *232*, 318 (46), 319 (46, 49), 327 (89), 328 (91, 94), 329 (97, 99), 336 (127), 365 (201–203), 368 (210), *371*, *372*, *374*, 466, 474 (121, 122), *481*, 869 (97), *915*
Matsumoto, S. 397 (90, 91), *414*, 432 (29a–c, 30, 31), 433 (29a–c), *457*
Matsumura, N. 293 (147), *307*
Matsuo, M. 432 (29a–c, 30, 31), 433 (29a–c), *457*
Matsushita, S. 435, 438 (38), *457*
Matsuura, T. 65 (284), *83*, 122 (35), *127*, 151 (80), *159*, 207 (42), 209 (61), 210 (80), 211 (81, 82), 220 (61), 223 (155a, 155b), *230*, *231*, *233*, 324 (76), 326 (84, 85, 88), 328 (93), 330 (102), 332 (108a, 109a, 109b), 335 (117), 338 (134), 347 (76, 165, 166), 348 (76), *371–373*, 446 (148, 154), 449 (167), 451, 454, 456 (154, 170), *459*, *460*, 475, 476 (147), *482*, 726 (58, 59), 736 (109), *771*, *772*, 825 (75), *828*, 836 (103), 843 (57a, 57b), 844 (18q), 858 (57a, 57b), 871 (18q, 103, 104), 872 (105a, 105b, 106b, 106c), 878 (134), 880 (140), 884 (57a, 57b), 891 (57b), 898 (57a, 57b), 906 (57b), 911 (57a, 57b, 103), *913–916*
Matthew, J. A. 223 (154b, 154c), *233*
Mattner, J. 287 (86), *305*
Matuselli, E. 426 (77), *428*
Mauclaire, G. H. 48, 49, 67 (185), *81*
Maurer, B. 224 (157a), *233*
Mauring, L. E. 446 (141), *459*
Maurya, R. C. 469 (60), *480*
May, D. H. 166 (23), *196*
Mayeda, E. A. 446 (153), 451 (171, 172), 454 (153), 456 (153, 172), *459*, *460*
Mayer, N. 272 (60), *277*
Mayo, F. R. 164 (11, 14), 167 (14), *196*, 237 (10), *256*, 281 (46), *305*, 422 (32, 33), 423 (35), *427*, 729, 731, 737 (65), *771*
Mayo, P. de 213 (92), *231*, 350 (175), *373*, 446 (145), *459*
Mazel, K. S. 557 (250), *584*
Mazur, S. 120 (30), *127*, 210 (69), *231*, 327, 328 (90), 330 (100), *371*, 385 (42), *413*, 832 (225), *919*
Mazur, U. 69, 71 (314), *84*
Mazur, W. 125 (37), *127*
Mazur, Y. 295 (154), *307*, 394 (76), *414*, 498 (75), 503 (108), *519*, 657 (41, 42), 658 (42, 45, 46a–c), 660 (42, 46a), 664 (42, 52a, 52b, 53), 668 (52a, 52b), 671 (46a–c, 59), 672 (59), 673 (42, 59), 675 (59, 62, 63), *682*, *683*
Mazzanti, G. 297 (181a, 181b), *308*, 545 (99, 100), *580*
Mazzini, A. 725 (54), *771*
McAlloon, K. T. 37, 67 (129), *80*
McAndrews, C. 438 (69), *458*
McAtee, J. L., Jr. 389 (60), 390 (62b), *413*
McBride, J. M. 88, 91 (84), *95*, 406 (124), 407 (124, 129, 132, 141), *415*, 546, 548 (117), *581*, 598 (69–71), 610 (69), 611 (71), 617 (70, 71), *645*, 743 (145, 146), *773*, 780 (11b), *804*
McCain, D. C. 48, 49, (189), 61 (189, 252), *81*, *83*
McCain, J. H. 718, 733, 739 (19), *770*
McCallum, K. S. 559, 560 (189), *582*
McCapra, F. 226 (171), *233*, 384 (39), *413*, 667 (16), 684 (16, 90), *688*, *690*, 839 (250a), 840–842 (180c), 844 (15, 18b), 860 (64a, 64b), 878 (129), 884, 890 (180c), 893

(15), 895 (180c), 896 (180c, 181a), *913*, *915*, *916*, *918*, *919*
McCarrick, T. 249, 250 (54), *257*
McCarthy, D. J. 632 (171), *647*
McCloskey, C. J. 910 (211d), *919*
McClure, D. E. 588 (11d), *643*
McCorkindale, N. J. 312 (6), *369*
McCready, R. 194 (154, 155), *199*, 533 (56), *579*
McCulloch, R. D. 667, 668, 671, 672, 674 (19), *688*
McCullough, K. J. 245 (32, 33), *257*
McCullough, R. D. 99 (19), *104*
McDonald, R. N. 292 (149b), *307*, 536, 537 (71), *580*
Mcdonald, T. L. 546 (109), *581*
McDougall, A. O. 573 (235), *584*
McDougall, D. J. 384 (38e), *413*
McDowell, C. A. 44, 45 (162), *81*
McElroy, A. D. 430, 450 (10), *456*
McElroy, W. D. 896 (182, 183a), *918*
McGandy, E. L. 387, 388 (51), *413*
McGinetty, J. A. 464, 465 (7), 466 (111), 467 (21), 473 (111), *479*, *481*
McGinness, R. 244 (31), *257*
McGinniss, V. D. 757 (170), *774*
McGlinckey, M. 440 (77), *459*
McGrath, J. P. 295 (159a), *307*
McIntosh, C. L. 405 (114), *415*, 742 (141b), *773*
McIsaac, J. E., Jr. 568 (223b), *583*
McKean, D. C. 89 (63), *95*
McKee, R. L. 263, 266 (21), *276*
McKellar, A. R. W. 52 (210), *82*
McKennis, J. S. 912 (217), *919*
McKenzie, M. 312 (5), *369*
McKeown, E. 206 (24), *229*
McLaren, J. W. 465, 467, 472 (22), *479*
McLendon, G. 464, 465, 475 (13), *479*
McMillan, G. R. 68 (302), *84*, 716 (15), 717 (15, 16), 733, 737 (16), *770*
McMullan, R. K. 88, 89 (70), *95*, 407 (128), *415*, 611 (107a), *646*
McPhail, A. J. 314 (24), *370*
McPhail, A. T. 213 (98), *231*, 339 (138), *372*, 384 (38d), 388 (48, 52), *413*
McQuigg, R. D. 491 (57), *518*
McReady, R. 533 (55), *579*
Meagher, J. F. 720, 740 (35), *771*
Meakins, G. D. 289 (113), *306*, 525 (17), 529 (44), *579*
Mebane, R. C. 316 (38), *370*
Medalia, A. I. 132, 147 (10), *158*, (174), *774*
Medvedev, S. S. 574 (236), *584*
Mehl, K. S. 453, 456 (191), *460*
Meier, H. 536 (75), *580*
Meijer, E. W. 835 (239), 873 (112), *916*, *919*
Meijere, A. de 662 (50a, 50b), 672, 673 (61), *682*, *683*
Meinwald, J. 561 (198), *583*
Meisenheimer, J. 297 (182a), *308*
Melhuish, W. H. 452 (188), *460*
Melius, C. F. 25, 33 (78), *79*
Mellish, S. F. 280 (7), *304*
Melloni, G. 300 (216), *309*, 538 (80b), *580*
Melnik, L. V. 180 (81), *198*
Melrose, M. P. 36 (108), *80*
Melville, H. 419 (5), *426*
Melville, H. W. 423 (40), *427*
Menart, V. 574 (242), *584*
Menchaca, H. 559 (190), *582*
Mendeleev, D. I. 484 (3), *517*
Mendenhall, G. D. 206 (29), 207 (34), *230*
Menyailo, A. T. 168 (39), *197*
Menzies, I. D. 341 (145, 146), *372*
Mercer, G. D. 466 (117), 467 (24), 473 (117), *479*, *481*, 651 (10a), *681*
Merienne, C. 825 (76), *828*
Merkel, P. B. 204 (2c), *229*
Merle, G. 825 (77), *828*
Merrill, R. A. 546, 548 (117), *581*, 598, 611, 617 (71), *645*
Merrit, M. V. 441 (111), *459*
Merrit, M. W. 441 (104), *459*
Merritt, M. V. 441 (107), *459*
Mertens, H. 208, 210 (49), *230*
Mertens, H. J. 213, 216 (96), *231*
Mesrobian, R. 420 (14), *427*
Mesrobian, R. B. 2 (3), *78*, 280 (21), 299 (209a, 209b), *304*, *308*, 546–548 (114a), *581*
Messer, L. A. 298 (195), *308*
Messmer, R. P. 45 (167), 58 (231), *81*, *82*
Metelitza, D. I. 465, 470 (65), *480*
Metz, D. 420 (14), *427*
Metzger, J. 174 (59), *197*, 873 (111), *916*
Meunier, B. 191 (136), *199*
Meyer, M. D. 298 (195), *308*
Meyer, R. 88, 92 (94b), *96*
Meyer, W. 36, 52, 53 (102), *80*
Miano, J. D. 667, 684 (18), *688*, 844 (17c, 18s), *913*
Michaelson, R. C. 181 (82), 182 (86, 88), *198*, 676 (57), *689*
Michel, A. 796 (64, 65), 797 (65), *805*
Michel, C. 453 (194), *460*
Michelin, R. A. 475, 476 (142), *482*
Michl, J. 742 (141a), *773*
Michot, C. 424 (55), *427*
Midland, M. M. 470 (73), *480*, 816 (36), *827*
Migita, T. 228 (209–211, 212a, 212b), *234*, 491 (58), *518*, 909 (209), *918*
Migliore, M. J. 594, 605, 606 (48), 607 (96b), 608, 616 (48), 620 (48, 96b), *644*, *645*

Mihalov, V. 717 (17b), *770*
Mihelich, E. D. 182 (87), *198*, 532–535 (53), *579*
Mikhailova, I. P. 151 (74), *159*
Mikheyev, Yu. A. 800 (87, 88, 90, 94, 95), 803 (111), *806*
Miki, M. 467 (28), *479*
Mikovič, J. 788, 790 (41), 791 (41, 50), 792 (50), *805*
Miksztal, A. M. 471 (85), *480*
Mikulášová, D. 796, 797 (67), *805*
Milanovic, J. 780 (12b), *804*
Milas, N. A. 152 (82), *159*, 162 (1), 168 (36, 37), *196*, *197*, 299 (199a, 199b), *308*, 487 (33, 34), 494 (33, 34, 65), 495 (65), *518*, *519*
Mile, B. 246 (40), 247 (41), *257*, 487 (36–38), 488, 489, 491, 492 (37), *518*, 733 (93), *772*
Miley, J. W. 742 (139a, 139b), *773*
Milkie, T. H. 423 (37), *427*
Mill, T. 164 (11), *196*, 247 (45), *257*, 487, 495 (35), *518*, 729, 731, 737 (65), *771*
Miller, A. A. 422 (32), *427*
Miller, J. 382 (33), *412*
Miller, J. A. 406 (126), *415*
Miller, J. G. 88 (61, 99, 109), 89–91 (61), 92, 93 (99), 94 (109), *95*, *96*, 134 (30), *158*, 405 (106, 112), *414*, *415*, (155), *774*
Miller, L. L. 632 (170), *647*
Miller, R. E. 651 (9), *681*
Miller, R. S. 88, 93 (108), *96*
Miller, R. W. 384 (38d), *413*
Miller, V. B. 799 (77), *805*
Miller, W. H. 465, 472 (95), *481*
Milliet, P. 446 (130), *459*
Milligan, W. O. 389 (60), 390 (62b), *413*
Milne, R. T. 99 (21), *104*
Milner, D. J. 826 (83), *828*
Milovanovic, J. 557 (175), *582*, 628 (155), *647*
Mimoun, H. 171, 172 (45), 187 (45, 119), *197*, *199*, 383 (38b), *413*, 464 (17, 19, 20), 465 (26, 57, 62, 66, 80), 466 (106), 467 (20, 25, 26), 468 (20, 26, 29), 469 (26, 57), 470 (62, 66), 471 (57, 80), 472 (106), 473 (29, 106, 108), 474 (135), 475 (144, 145), 476 (20, 29, 144, 145), 477 (135, 148, 149), 478 (20, 29, 145), *479–482*, 533 (60), *579*, 809 (9), 812 (65), 816, 824, 825 (9), *827*, *828*
Min, T. B. 910 (212a), *919*
Minabe, M. 125 (39), *127*
Minakata, K. 617 (129a, 129d), *646*
Minale, L. 312 (10c), *370*
Minato, H. 262 (14, 15), 263 (14, 15, 19, 22a, 22b, 23), 264 (14, 23), 265 (14), 267, 269 (19), 273 (62), *276*, *277*, 593 (34), 597 (64, 67), 598 (76), 617 (67), 618 (131), 619 (132), 631 (76), 639 (209), 640 (64, 210), *644–646*, *648*, 740 (130), 760 (180), *773*, *774*
Minato, T. 69 (306), *84*
Minisci, F. 175 (63), 176 (64, 65), *197*, 752 (160a–d), *774*
Minkoff, G. J. 131, 132 (2), 134 (29), 147 (2), *157*, *158*, 712 (5), *770*
Minkoff, G. O. 567 (215), *583*
Mirbach, M. J. 906 (203), *918*
Mischler, A. 475 (144, 155), 476 (144), *482*
Mishchenko, O. A. 764 (195), *775*
Mishima, T. 740 (125, 129), *773*
Miskowski, V. M. 472 (96), *481*
Mislow, K. 560 (193), *583*, (216), *648*
Misra, H. P. 437 (53), *457*
Mitchard, D. A. 501 (92), *519*, 651, 654 (7c, 7d), *681*
Mitchell, P. A. 209, 220 (62), *230*
Mitrofanova, E. V. 817 (40, 41), 820 (50), *827*
Mitschler, A. 187 (119), *199*, 383 (38b), *413*, 465 (26, 61), 467, 468 (26), 469 (26, 61), *479*, *480*, 812 (65), *828*
Mitsuhashi, T. 541 (89), 542 (90), 562 (201), *580*, *583*
Mitsui, E. 441 (112), *459*
Miura, K. 596, 635 (50), *644*
Miyadera, H. 562 (201), *583*
Miyamoto, R. 344 (160), *373*
Miyazaki, H. 352 (179), *373*
Miyazaki, T. 731 (76), *772*
Mizutani, K. 475 (141), *482*
Mo, Y. K. 666, 676 (68, 69), *683*
Mobius, L. 529 (46), *579*
Moccia, R. 36, 57, 58 (98), *79*, 87, 88, 90 (45), *95*
Möckel, H. 761 (184b), *774*
Modena, G. 186 (107–109, 112), *198*, 288 (100), 296 (173a, 173b, 176), *306*, *308*, 469 (44), 471 (77), *480*, 533 (59), 538 (80b), 539 (82), 542 (92, 93), 543 (92–94), 544 (93, 97a, 97b), 545 (97a, 97b), *579*, *580*
Molyneux, P. 419 (5), *426*, 792 (55), *805*
Molyneux, R. J. 865 (87), *915*
Mondelli, R. 752 (160a), *774*
Moniot, J. L. 366 (205b), *374*
Moniz, W. B. 727 (63, 64), 728, 729 (64), 740, 741 (*771*, *773*
Monroe, B. M. 903 (190), *918*
Montanari, F. 296 (172), *307*
Montgomery, F. C. 227 (192), *234*, 555, 556, 575 (159), *582*, 684 (93), *690*, 731 (72), *772*, 831 (6a, 6b), 836, 838, 883, 885, 886 (145a), 891, 894 (150b), *913*, *916*, *917*
Moodie, R. B. 641, 642 (214), *648*, 814 (32), *827*
More, A. A. 557, 558 (178), *582*, 628 (157c), *647*

Moreau, C. 207, 211 (36), *230*, 501, 507 (96–99), *519*, 652 (17), *681*, 783 (29), 786 (38), *804*, *805*
Morell, J. 227 (208), *234*, 866, 910 (91), *915*
Moretti, I. 570, 572 (230), *583*
Morgan, A. R. 245 (32, 33), *257*
Morgan, M. S. 440 (78), *459*
Mori, A. 873 (108), *915*
Morimoto, S. 721 (39), *771*
Morimoto, T. 738 (115), *773*
Morino, Y. 48 (183), *81*
Moris, H. 791 (49), *805*
Morita, H. 596, 598 (51), *644*
Morita, N. 331 (106), *372*
Moritz, H. 280 (11), *304*
Morkved, E. H. 605, 618 (91), *645*
Morland, R. 844 (17c), *913*
Moro-Aka, Y. 465, 470 (64), *480*
Morokuma, K. 61 (259), *83*, 87, 88 (40), *95*
Moro-oka, Y. 440, 454 (74), *458*, 826 (82), *828*
Moro-Oka, Y. 432, 433, 454 (24), *457*, 475 (140, 141), *482*
Morozov, O. S. 144 (63, 64), 156 (90), 157 (90–92, 95), *159*, *160*
Morris, A. 44, 45 (164), *81*
Morris, G. W. 246 (40), *257*
Morrison, M. M. 430 (8), 431, 432 (14), 437 440, 443, 445, 450 (8), *456*
Morrow, L. R. 394 (76), *414*
Morsi, S. E. 788, 791 (39), *805*
Mortlock, A. 735 (100a), *772*
Morton, J. 388 (48), *413*
Morton, J. R. 732 (85), 733 (91a), *772*
Morukuma, K. 36 (90), *79*
Mosher, H. S. 297 (184), *308*, 568 (219), 577 (246), *583*, *584*, 590 (20), *644*
Moss, B. J. 21, 58 (38, 39), *78*
Mossman, A. B. 293 (148), *307*
Moulden, H. N. 791 (47), *805*
Moulines, J. 388 (50), *413*
Moye, A. J. 437, 438, 440, 454 (60, 62), *457*, *458*
Muccigrosso, D. A. 465, 470, 471 (70), *480*
Mukai, C. 366 (205a), *374*
Mukai, T. 344 (160), *373*, 865 (88), *915*
Mukai, Y. 546–548 (114d), *581*
Mulcahy, M. F. R. 285 (69), *305*
Mullan, L. F. 566 (213), *583*
Müller, G. 529 (46), *579*
Muller, K. A. 99 (17), *104*
Müller, M. 831 (7), *913*
Muller, W. 208, 210 (49), *230*
Müller, W. 295 (156, 157), *307*, 534 (65), *580*
Mullica, D. F. 389 (60), 390 (62b), *413*
Mullier, M. 422 (31), *427*
Mumford, C. 156 (88b), *160*, 227–229 (191), *234*, 314 (28), *370*, 454 (198), *460*, 591 (22), 592 (26), *644*, 667 (12), *688*, 831 (4), 832–834 (27), 844 (4, 18e), 847 (27), 848 (4, 27), 859, 881, 885, 886, 890, 910 (27), *913*, *914*
Murahashi, S.-I. 435 (40), *457*
Muramatsu, M. 842 (249), *919*
Muramikava, J. 272 (60), *277*
Muratake, H. 335, 359 (119), *372*
Murovec, S. 63, 68 (277), *83*, 485 (21), *518*
Murphy, J. S. 650 (36), *681*
Murray, R. N. 632, 633 (175), *647*
Murray, R. W. 2, 65 (11), *78*, 206 (26–28), *229*, *230*, 318 (45e), *370*, 391 (68), *414*, 501 (100, 101), 503, 507 (101), *519*, 563 (205), *583*, 588 (12a), 632 (176), 633 (176, 178, 179), 634 (185), *643*, *647*, *648*, 650 (3f, 3g, 3i), 651 (15), 652, 653 (19a–c), *681*, 850, 858 (41), *914*
Musgova, K. K. 423 (48), *427*
Musgrave, O. C. 287 (79), *305*
Musgrave, W. K. R. 534 (64), *580*
Musso, G. F. 36 (110), *80*
Musso, H. 454 (202), *461*
Muto, S. 471 (92), 474, 476 (132), *481*, 876 (128), 878 (130a), *916*
Myall, C. J. 261, 262 (10b), *276*
Myers, R. S. 191 (134), *199*

Naae, D. G. 407, 408 (133), *415*, 779, 780 (8), *804*
Nadeau, Y. 501 (104, 105), 502 (105), *519*, 653 (22), *681*
Nadir, U. 571 (232), *583*
Nagakura, S. 740 (124, 126), *773*
Nagano, T. 443, 444 (121), 446 (121, 126b), *459*
Nagashima, H. 188 (121), *199*
Nagata, R. 330 (102), *347* (166), *371*, *373*
Nagendrappa, G. 560 (197), *583*
Nair, V. G. K. 239 (17), *257*
Nakadaira, Y. 589 (17), *643*
Nakagawa, M. 872 (106a, 106c–e), 876 (127), *915*, *916*
Nakahara, N. 731 (76), *772*
Nakajima, N. 336 (125), *372*
Nakamura, A. 466 (102, 120, 122, 125), 467 (28), 472 (102), 474 (120, 122, 125), *479*, *481*
Nakamura, H. 217 (126), *232*, 839, 847, 851 (30a, 30b), 860 (63), 871, 895 (30a, 30b), *914*
Nakamura, K. 223 (155a, 155b), *233*
Nakamura, M. 226 (183a, 185c), *233*
Nakanishi, K. 113 (16), *126*, 589 (17), *643*
Nakanishi, W. 596 (50), 635 (50, 189), *644*, *648*

Nakano, H. 441 (112), *459*
Nakano, M. 441 (112), *459*, 491 (58), *518*
Nakata, A. 335 (117), 338 (134), *372*
Nakata, T. 716, 718 (11), 740 (129), 742 (138c), *770*, *773*
Nakata, Y. 619 (133), *646*
Nakatsu, K. 466, 474 (122), *481*
Nakatsu, N. 466, 474 (121), *481*
Nakayama, K. 324 (75), *371*, 384 (38g), *413*
Nalewajski, R. F. 87, 88 (34), *94*
Nangia, P. S. 2 (13), 41 (143–147), 42 (143), 43 (143–145), 44 (143, 146, 147), 61 (13), *78*, *80*, *81*, 101 (37), *104*, 247 (46), 252, 253 (66), *257*, *258*, 485 (14), 489, 490 (48), 492 (64a), 496 (14), 501, 507, 509, 514 (106), *518*, *519*, 657 (40), *682*
Nanni, E. J. 432, 433, 435 (21), 436, 437 (47), *457*
Nanni, E. J., Jr. 431 (13), 447 (155), 448 (13), 454 (155), *456*, *459*, 879 (138), *916*
Nappa, M. 471 (85), *480*
Narasimhan, N. 546 (109), *581*
Nardin, G. 88, 91 (78), *95*, 401 (98), *414*
Narita, N. 588 (9b, 9c), *643*, 730 (69), *771*
Narula, A. S. 670 (58b), *683*
Nascimento, E. D. 866, 867, 910 (92), *915*
Natsume, H. 335, 359 (119), *372*
Natsume, M. 335 (122), 359 (122, 194), 365 (199), *372*, *373*
Natta, G. 420 (13), 422 (27), *427*
Nave, E. 381 (28), *412*, 783–787 (32), *804*
Naylor, R. 299 (206), *308*
Neary, N. 472 (107), *481*
Neckers, D. C. 205 (11, 13), *229*, 849 (40), *914*
Nedelec, J. Y. 287 (85), *305*
Nefedova, A. I. 136 (45), *159*
Negus, D. 891 (153b), *917*
Neidl, C. 891 (151b), *917*
Neill, D. C. 570, 572 (229), *583*
Neiman, M. B. 295 (158a), *307*
Nekrasov, L. I. 484 (6), *517*
Nelsen, S. F. 341 (151), *373*, 446 (151), *459*
Nelson, E. K. 312 (11), *370*
Nelson, S. F. 851 (50), *914*
Nelyubin, V. I. 464, 465 (10), *479*
Nemo, T. E. 191 (134), *199*
Neporent, B. S. 134 (23), *158*, 288 (96), *306*
Nery, R. 556 (167), *582*
Neta, P. 451, 452 (184a), *460*, 724 (48), 758 (176a), 767 (205, 208), 768 (176a, 205), 769 (176a, 210), 770 (210), *771*, *774*, *775*
Neuman, R. C. 791 (47), *805*
Neuman, R. C., Jr. 606 (92, 93a, 93b), 607 (93a, 93b), *645*
Neumüllar, O. A. 588, 592 (10a), *643*
Neumüller, O. A. 588 (11a), *643*
Neuss, G. 475, 476 (146), *482*
Newitt, D. M. 810 (16), *827*
Newman, E. R. 613, 614 (115), *646*
Newman, M. S. 525 (18), *579*
Newton, M. D. 25, 54, 61, 63 (73), *79*, 485, 499 (20), *518*
Ng, H. C. 731 (75), *772*
Nguyen-Dang, T. T. 55 (222), *82*
Nicholls, W. 313 (20), *370*, 394 (74), *414*
Nichon, A. 318, 324 (45d), *370*
Nickon, A. 210 (65a, 66c), 212 (66c), 219 (130), *230*, *232*, 588, 592 (10b), *643*
Niclause, M. 564 (208), *583*
Nicolaou, K. C. 313 (16b), 316 (36), 339, 359 (16b), *370*, 438 (65), 441 (65, 109), *458*, *459*
Nicolova-Nankova, Z. 424 (61), *427*
Nidy, E. G. 316 (37), *370*, 441 (104, 108), *459*
Niegowski, S. J. 651, 654 (6a), *681*
Nielsen, A. T. 259, 260 (3), *276*
Nielson, A. J. 190 (133), *199*
Niessen, W. v. 45 (166), *81*
Niki, H. 69 (305), *84*, 678 (70), 679 (71), *689*
Niki, M. 466, 474 (125), *481*
Nikishin, G. I. 564 (209), *583*
Nikulina, N. R. 157 (97), *160*
Nilsson, R. 205 (12), *229*, 451, 456 (173), *460*
Nishi, Y. 243 (26), *257*
Nishikimi, M. 432 (28), 433 (32), *457*
Nishinaga, A. 151 (80), *159*, 223 (155a, 155b), *233*, 446 (154), 451, 454, 456 (154, 170), *459*, *460*, 475, 476 (147), *482*, 878 (134), *916*
Nishio, T. 285 (64), *305*, 336 (125), *372*
Nishizawa, K. 451, 454, 456 (170), *460*, 475, 476 (147), *482*
Nixon, J. 317, 339 (44), *370*, 731 (71), *771*
Nixon, J. R. 243, 244 (27), *257*, 316 (38), *370*
Noble, P. N. 61 (250), *83*
Noda, S. 617 (129b), *646*
Noel, F. 157 (89), *160*
Noftle, R. E. 262 (11), *276*
Noland, W. E. 185 (104), *198*
Nolte, M. J. 377 (25), *412*, 466 (104, 114), 472 (104), 473 (104, 114, 115), *481*
Nonengraaf, C. J. A. 525 (8), *578*
Nonhebel, D. C. 245 (32, 33), *257*
Nord, A. G. 392 (69), *414*
Nordling, C. 44, 45 (158), *81*
Norling, P. M. 800 (96), *806*
Norman, R. O. C. 534 (66), *580*, 730 (70), 757, 758 (171b), *771*, *774*
Norris, A. C. 669 (32), *688*
Norris, J. E. 422 (26), *427*
Norris, R. D. 227 (206), *234*, 334, 335 (120), *372*, 871 (100), *915*
Norrish, R. G. W. 717 (17a), *770*

Northington, D. J. 290 (127), *306*
Novak, M. 878 (131), *916*
Novick, S. E. 48, 49, 54 (186), *81*
Novis, W. P. 259, 260 (3), *276*
Novitskaya, N. N. 186 (110, 111), *198*
Nowicki, A. W. 861 (67), *915*
Noyes, W. A., Jr. 716 (14a), *770*
Noyori, R. 358 (192a, 192b, 193), 363 (192a, 192b), 364 (193), *373*, 560 (196), *583*
Nozaki, A. 465, 470 (64), *480*
Nozaki, H. 182 (86), *198*, 469, 471 (55), *480*
Nozaki, K. 636 (193), *648*
Nozaki, T. 421 (18), *427*
Nudenberg, W. 173 (50), 175 (60), *197*, 628 (160), *647*
Nugteren, D. H. 339 (137), *372*
Nukui, M. 423 (36), *427*
Numan, H. 841 (248a, 248b), 844, 845, 892 (21a), *914*, *919*
Nyman, C. J. 474 (131), *481*

Oae, S. 271 (50), *277*, 443, 444 (120), 445 (120, 126a, 126c), 446 (120, 126c), 453 (120), *459*, 557 (173, 174), *582*, 588 (9c), 593 (38), 594 (39–41), 596 (39, 55, 56), 597 (63a), 599 (55, 56, 74), 601 (40), 610 (39, 40), 611, 613 (40, 56), 614 (56, 117), 615 (40, 56, 117, 135), 617, 618 (40, 117), 619 (63a, 117), 620 (55, 56, 135), 621 (55, 135), 622 (56, 135), 625 (39), 626 (39, 40, 55, 74, 151, 152, 154), 627 (55), 628 (39, 41, 55, 74, 154, 165), 629 (40, 41, 55, 74, 217), 631 (165, 166), 632 (167), (216), *643–648*, 739 (119), *773*, 780 (12b), *804*
Oars, J. A. 350 (174), *373*
Obata, N. 842 (249), *919*
Oberhammer, H. 37 (123), 67 (123, 289), *80*, *83*, 86–89 (21), *94*, 382 (31), *412*
Oberley, L. W. 446 (134), *459*
O'Brien, E. L. 299 (209a, 209b), *308*
Ochiai, E. 826 (81), *828*
Ochiai, H. 725, 742 (55), *771*
Ochrymowycz, L. A. 297 (178), *308*
O'Connor, C. J. 464 (8), 465 (8, 86, 90), 471 (86, 90), 473 (8), *479–481*
Oda, M. 216 (124b, 125), *232*, 344 (158), 346 (163), 352 (179, 181), *373*
Oda, R. 421 (17, 18), *427*
Odaira, Y. 721 (39), *771*
Odani, M. 226 (185b), *233*
Odinokov, V. N. 120, 121 (32), *127*
Oe, K. 865 (89), *915*
Oelfke, W. C. 36, 38, 50 (121), *80*, 87, 88 (23), *94*
Oesper, R. E. 650 (3a), *681*
Oezkan, A. 763 (194), *775*
Offermanns, H. 290 (128), *306*
Ogata, H. 474, 476 (132), *481*
Ogata, Y. 252 (64), *258*, 445 (123), 448 (161), *459*, *460*, 525 (11, 15), 528 (39), 535 (67), 538 (78), 539 (81), 544 (97c), 546 (106), 555 (160, 161), 556 (160–163), 561 (199), 562 (200), 564 (163), 566 (163, 211), 567 (217), 568 (221, 224), 569 (224), 570, 571 (228), 575 (243), 576 (243, 244), *578–580*, *582–584*, 718 (22–24), 719 (25–27), 720 (32c), 721 (43), 722 (43, 46), 723 (27), 724 (27, 49), 730 (66), 732 (80a), 738 (115), 739 (24, 26), 740 (24), 744 (25), 745 (24, 25, 151–153), 746, 747 (153), *770–774*, 874 (121), 876 (122a), *916*
Ogawa, M. 359 (194), *373*
Ogi, Y. 618 (131), 619 (132), *646*, 740 (130), *773*
Ogiwara, Y. 421 (21–24), 425 (71), *427–428*
Ogrem, P. J. 58, 59 (235), *82*
Ogren, P. J. 491 (54), *518*
Ogryzlo, E. A. 203, 204 (2b), 208 (44, 51), 209 (63), 210 (51), 212 (63), *229*, *230*
Ohara, E. 484 (5), *517*
Ohkatsu, Y. 423 (36), *427*, 437 (58), 440 (75), *457*, *459*, 879 (137), *916*
Ohkita, K. 792 (52), *805*
Ohkubo, K. 61 (247), 68 (300), *83*, *84*, 87–89 (35), *95*
Ohloff, G. 210 (66d), 211 (87b), 212 (66d), 214 (66d, 109), 216 (66d), 217 (87b), 221 (66d), 222 (66d, 149, 151), 224 (157a, 157b), *230–233*, 348 (169), *373*, 555 (154), *582*
Ohmo, M. 287 (90), *306*
Ohnishi, R. 666, 679 (71), *683*
Ohno, T. 535 (67), 538 (78), *580*
Öhrn, Y. 44, 45 (157), *81*
Ohshima, A. 740 (129), *773*
Ohta, H. 546–548 (114c), *581*
Ohta, K. 25 (65), *61* (257), *79*, *83*
Ohta, N. 193 (147), *199*
Ohta, S. 873 (108), *915*
Ohta, T. 435 (40), *457*
Ojala, W. H. 406 (126), *415*
Okada, K. 287 (90), *306*
Okajima, H. 872 (106d), *915*
Okamoto, K. 596, 636 (54), *644*
Okamoto, Y. 435 (40), *457*
Okamura, S. 721 (41), *771*
Oki, M. 303 (238), *309*
Oku, A. 324 (80), *371*
Okuda, M. 44, 45 (164), *81*
Olah, G. A. 474 (138), *482*, 498 (76), 503 (107), *519*, 536 (69), 555 (151), *580*, *582*, 666 (66–71), 673 (66), 676 (66–70), 677 (66), 679 (66, 71), *683*
Olbertz, B. 470 (72), *480*

Ol'dekop, Yu. A. 280 (26, 27, 28a–f, 29a, 29b, 30a–d), *304*
O'Leary, B. 38 (135), *80*
O'Leary, G. 792 (53), *805*
Olivé, S. 464, 465 (9), *479*
Olivella, S. 24, 45, 58 (62), *79*
Oliver, J. D. 390 (62b), *413*
Olmos, A. W. 92 (92), *96*, 134 (22), *158*, 288 (97), *306*
Olovsson, I. 86, 88 (7), *94*, 376 (5), *412*
Olsen, A. C. 552 (136), *581*
Olsen, J. F. 40, 63, 67 (136), *80*, 88, 90 (76), *95*
Olson, A. C. 623 (146b), *647*
Olson, W. B. 36, 38, 49, 50 (119), *80*, 87–90 (24), *94*
Omote, Y. 285 (64), *305*, 336 (125), *372*, 393 (71), *414*, 860 (62a, 62b), *914*
Omura, K. 726 (58, 59), *771*
Onan, K. D. 388 (48, 52), *413*
Onderdelinden, A. L. 826 (79), *828*
O'Neal, H. E. 40 (140), *80*, 97 (2), 100 (18), *102*, *104*, 153 (85), *159*, 227 (192), *234*, 244 (30, 31), *257*, 665 (2), 666 (2, 11), 668 (2), 672 (45), 681 (11), 684 (93, 97, 99), *688–690*, 833 (145b), 836 (145a), 838 (145a, 145b), 839 (145c), 881, 882 (143a, 143b), 883 (143a, 143b, 145a–c), 885 (145a), 886 (145a, 145c), 894 (143a, 143b), *916*
O'Neil, S. V. 61, 62 (245), *82*
Ono, Y. 489 (44), 490, 491 (49), *518*
Onodera, K. 211 (84), *231*
Ooi, S. 475, 476 (147), *482*
Oosterhoff, L. J. 742 (138a), *773*
Orell, J. 902, 903 (188), *918*
Orfanopoulos, M. 168 (32), *197*, 209, 218, 220 (60), *230*, 848, 854 (37), *914*
Orhan, E. H. 419 (10), *426*
Orito, K. 867 (94), *915*
Orphanos, D. G. 299 (199b), *308*
Orr, G. 405 (114), *415*
Orr, J. R. 650 (36), *681*
Orr, R. 419 (6, 7), *426*
Ortiz de Montellano, P. R. 285 (68), *305*, 536, 537 (76), *580*
Orton, G. 205 (10), *229*
Osa, T. 437 (58), 440 (72, 73, 75), 441 (84), 451 (175), *457–460*, 879 (137), *916*
Osafune, K. 46 (169), 47, 48 (177), *81*
Osawa, Z. 441 (112), *459*
Oshibe, Y. 597 (63a), 614, 615, 617, 618 (117), 619 (63a, 117), *645*, *646*
Oshima, K. 469, 471 (55), *480*
Ossewold, M. G. J. 153 (83), *159*
Oswald, A. A. 157 (89, 93), *160*
Otradina, G. A. 425 (68), *428*
Otsuka, S. 183 (92), *198*, 466 (102, 120, 125), 467 (27, 28), 472 (102), 474 (27, 120, 122, 125), 475, 476, 478 (27), *479*, *481*
Otsuki, T. 449 (167), *460*
Otto, M. 655 (32), *682*
Ouannes, C. 208 (45), *230*
Ouderaa, F. J. van der 339 (137), *372*
Ourisson, G. 292 (144), *307*, 435 (39), *457*, 670 (58a, 58b), *683*
Overberger, C. G. 542, 543 (92), *580*
Overbergh, N. 424 (59), *427*
Owen, A. J. 794, 795 (60, 61), *805*
Owens, J. 605, 607 (95), *645*
Ozawa, T. 432 (30, 31), *457*

Paaren, H. E. 189 (124), *199*
Pacansky, J. 405 (114), *415*, 742 (141b), 743 (147–149), *773*
Pacifici, J. G. 614 (116a), 625, 628 (149a), *646*, *647*
Packer, J. E. 811 (18), *827*
Padbury, J. M. 574 (238, 241), *584*
Padwa, A. 602 (85a), *645*
Pagano, A. S. 297 (187), *308*, 540 (87), *580*
Page, G. A. 298 (194b), *308*
Pak, C.-S. 227 (207), *234*, 871 (101), *915*
Pakiari, A. H. 87 (50), *95*
Pakrashi, S. C. 312 (9b), *370*
Palacin, F. 424 (61), *427*
Palermo, R. E. 182 (88), 190 (129), *198*, *199*
Palet, K. B. 449 (164), *460*
Palke, W. E. 36 (91, 112), 38 (112), 48, 49 (189), 61 (189, 252), *79–81*, *83*, 87, 88, 90 (38), *95*
Palmer, B. M. 559, 561, 562 (191), *582*
Palmer, H. B. 431 (16a), *457*
Palumbo, R. 426 (77), *428*
Pannetier, G. 803 (112), *806*
Pant, L. M. 405, (110, 111), *414*
Panunto, T. W. 571 (232), *583*
Pape, R. 668 (21), *688*
Pappas, J. J. 303 (233a), *309*, (240), *584*
Pappiannou, C. G. 628 (155), *647*
Pappiaonnou, C. G. 557 (175), *582*, 780 (12b), *804*
Paquette, L. A. 219 (129), 226 (172), *232*, *233*, 320 (53), *371*, 526, 527 (32, 33), *579*
Para, A. 733 (98a), *772*
Parczewski, A. 36 (115), *80*
Parish, J. H. 501, 513, 514 (89), *519*, 656 (35), *682*
Pariskii, G. B. 800 (94, 95), *806*
Parker, A. J. 441 (94–96), *459*
Parker, D. G. 503 (107), *519*, 536 (69), 555 (151), *580*, *582*, 666 (66, 70), 673 (66), 676 (66, 70), 677, 679 (66), *683*
Parker, L. 429 (2), *456*

Parker, T. 438 446 (68), *458*
Parker, T. L. 341 (150), *373*, 446, 450 (137), *459*, 874 (118b), *916*
Parker, V. B. 40 (139), *80*
Parker, W. E. 134 (24), *158*, 403 (103), *414*
Parkes, D. A. 491 (55), *518*
Parks, C. P. 634 (181a), *647*
Parr, R. G. 36 (108), *80*
Pascard-Billy, C. 382 (36), *413*
Pascher, F. 209 (53), *230*
Paška, I. 796, 797 (65), *805*
Pass, S. 503 (108), *519*, 657 (42, 44), 658 (42, 45), 660, 664, 673 (42), *682*
Pasto, D. J. 290 (116), *306*
Paszyc, S. 721 (42), *771*
Pate, C. T. 110 (7), *126*
Patel, K. M. 438 (69), *458*
Pater, R. H. 538 (79), 563 (204), *580*, *583*
Patlyakevich, D. D. 143, 144, 149 (59), *159*
Patoiseau, J. F. 679 (72), *683*
Patrick, K. G. 678 (68), *689*
Patrie, W. J. 286 (74), *305*
Patterson, D. P. 343 (155), *373*
Paukert, T. T. 58, 59 (234), *82*
Paul, H. 237, 249 (12), *256*, 731 (78), *772*
Paul, J. C. 778 (1), *804*
Paul, K. 193 (148), *199*
Paul, N. C. 116 (19), *126*
Pauling, L. 2, 49 (25), *78*
Pausacker, K. H. 288, 290 (110), *306*, 525 (7), *578*, 588 (9a), *643*
Pauson, P. 173 (50), *197*
Pauson, P. L. 245 (32, 33), *257*
Pavelich, W. A. 561 (248), *584*
Pavlovskaya, T. E. 134 (23), *158*, 288 (96), *306*
Payne, G. B. 292 (142a, 142b), *307*
Payne, N. C. 467 (21), *479*
Payne, P. W. 36, 37 (117), *80*
Payzant, J. D. 441 (93), *459*
Pazdernik, L. 653 (28c), *682*
Peake, S. L. 298 (197), *308*, 546 (108), *581*
Pearce, R. 818 (44), *827*
Pearson, E. F. 49 (195), *82*
Pearson, G. O. 441 (102b), *459*
Pearson, J. M. 614 (116c), *646*
Pearson, P. K. 61, 62 (262), *83*
Pearson, R. G. 566 (213), *583*, 816 (34), *827*
Pease, R. N. 656 (34a, 34b), *682*
Pechmann, H. von 282 (51), *305*
Pedersen, B. 51 (202), *82*, 86, 88 (8), *94*, 376 (6), *412*
Pedersen, B. F. 51 (200–203), *82*, 86, 88 (6, 8), *94*, 376 (6–9, 15), 377 (9), 398 (15), *412*
Pedersen, C. J. 781 (15), *804*
Pedersen, C. L. 874 (115), *916*
Pedersen, L. 36 (90), *79*, 87, 88 (40), *95*
Pederson, J. 299 (208), *308*
Pedley, J. B. 98 (8), *104*
Pekhk, T. I. 557 (250), *584*
Pelizza, F. 555 (151), *582*
Pelletier, J. 455 (204), *461*
Pembridge, E. F. 794, 795 (60), *805*
Penfold, P. R. 465, 471 (79), *480*
Penn, R. E. 62, 63 (267), *83*
Penney, W. G. 86–88 (4), *94*, 376 (1), *412*
Peover, M. E. 432 (23), 441 (86), 448 (158), 449, 450, 453 (86), *457*, *459*, *460*
Perchugov, G. Ya. 157 (95, 96), *160*
Perederееva, S. I. 738 (116), *773*
Perez-Machirant, M. M. 473 (108), *481*
Perez-Reyes, E. 446 (135), *459*
Perkins, C. W. 254 (70), *258*, 635 (190), *648*
Perkle, W. H. 441 (113), *459*
Perlina, S. 91 (85), *95*
Perona, M. J. 99 (24), *104*, 674 (48), *689*
Perrin, C. L. 881 (142a), *916*
Peruzzotti, G. 214 (110), *232*
Peslak, J., Jr. 63 (268), *83*
Peter, G. 114 (17), *126*
Peters, C. W. 36, 38 (120), *80*, 87–90 (25), *94*, 376 (2), *412*
Peters, E.-M. 832 (24), 839 (23), 845, 846 (23, 24), *914*
Peters, J. A. M. 223 (155c), *233*
Peters, J. W. 430, 453 (11), *456*
Peters, K. 832 (24), 839 (23), 845 , 846 (23, 24), *914*
Peterson, P. E. 290 (118c), *306*
Peterson, S. W. 49 (194), *81*
Petrashkevich, S. F. 280 (28e, 29a), *304*
Petrongolo, C. 68 (296), *83*, 88, 90, 92, 93 (104), *96*
Petropoulos, C. C. 623 (147b), *647*, 779 (4), *804*
Petropoulos, C. D. 557 (169), *582*
Petrov, A. A. 722 (44), 724 (44, 50), 739 (50), *771*
Petrukhov, G. G. 811 (67), *828*
Petry, R. C. 812 (22), *827*
Petryaev, E. P. 762 (191), *775*
Petukhov, G. G. 811 (19), *827*
Peyerimhoff, S. D. 13 (30), 24 (52, 53), 25 (75–77), 33 (52, 75), 45 (53, 77), 46 (77), 58 (53, 75, 77, 230), 59 (53, 77), *78*, *79*, *82*
Pfeffer, P. E. 855 (247), *919*
Pflederer, J. L. 840–842, 906 (202), *918*
Philardeau, Y. 420 (12), *427*
Philippi, K. 289 (114), *306*
Phillips, B. 564 (207), *583*, 743 (150), *773*
Phillips, D. 223 (154a), *233*
Phillips, L. 99 (22), *104*, 116 (19), *126*, 673, 674 (47), *689*
Phillips, S. E. V. 388 (53), *413*

Phoenix, F. H. 446 (145), *459*
Phung, P. V. 762 (188a), *775*
Pickett, J. E. 873 (110), *916*
Pickett, J. F. 333 (114), *372*
Picot, A. 446 (130), *459*
Pierens, R. K. 88, 89 (62), *95*
Pierre, J.-L. 532 (54), *579*
Pierson, G. O. 213, 216, 221, 223 (90), *231*, 346 (164), *373*
Pietrzak, B. 323 (68, 69), 345 (162), *371*, *373*
Pilár, J. 737 (111b), *772*
Pilcher, G. 40 (138), *80*, 98, 99 (7) *104*
Pilipovich, D. 669 (30), *668*, 735 (102), *772*
Pilz, H. 831 (10b), *913*
Pimentel, G. C. 61 (250, 265), *83*
Pincock, R. E. 577 (245), *584*, 636 (192), *648*
Pink, R. C. 727 (62), *771*
Pinkus, A. G. 793 (58), *805*
Pinson, J. 452 (184b), *460*
Piret, W. 731 (74), *772*
Pirola, L. 36 (113), *80*
Pisanchijn, J. 166 (21), *196*
Pitts, J. N. 229 (225), *234*
Pitts, J. N., Jr. 43 (154), 60 (238), *81*, *82*, 110 (7), *126*, 677 (60), 678 (65, 67), 679 (65), 683 (86, 87), *689*, *690*
Pitzer, K. S. 732 (80b), *772*
Pitzer, R. M. 36 (91), 38 (131), 48, 49, 54 (182), 61 (182, 246), 62 (246), 63 (182, 246), *79–82*, 87, 88, 90 (38), *95*
Plaček, J. 800, 801 (97), *806*
Plas, H. C. van der 451, 452 (181), *460*
Platé, N. A. 423 (45, 47), *427*
Plesnicar, B. 288 (99), 289 (112), 290 (119), *306*, 652 (21), *681*
Plesničar, B. 63 (272, 277), 67 (272), 68 (272, 277, 297, 298), 69 (272), *83*, 88 (87, 93, 103, 110), 91 (87), 92 (93, 100, 101, 103), 93 (100, 103), 94 (110), *96*, 134 (26, 27), 135 (32, 33), 155, 156 (86), *158*, *159*, 408 (139), *415*, 485 (18, 21), 486 (18), 487, 494 (33, 34), 501 (102, 103), 504 (110a, 110b, 111), 507 (110a, 110b), 509, 510 (102, 103), 511 (103), 514, 515 (115), 516 (103, 110a, 110b), 517 (116), *518–520*, 522 (1f), 523, 524 (3), 525 (1f), 527 (36), 528 (36, 37), 530 (3, 49), 531 (3), 538 (80a), 543 (37, 249), 546 (104, 105), 573 (233), 574 (242), *578–580*, *583*, *584*
Pletcher, D. 261, 262 (10b), *276*
Ploeg, J. P. M. van der 740 (132), *773*
Ploetz, T. 280 (23), *304*
Plum, H. 865 (83), *915*
Pluth, J. J. 465, 472 (98), *481*
Podgornova, V. A. 180 (81), *198*
Podosenova, N. G. 425 (68), *428*
Poelkler, D. J. 272 (56, 57), *277*
Pokholok, T. V. 797 (74), *805*
Pokrovskaya, I. E. 382 (38a), *413*
Pokrovskaya, Z. A. 139 (52), *159*
Politzer, I. R. 863 (72), *915*
Polovsky, S. 388 (48), *413*
Polozova, N. I. 432 (26, 27), *457*
Ponticorbo, L. 593 (35), *644*
Poot, A. 422 (31), *427*
Pople, J. A. 25, 33 (74), 36 (100, 114), 38 (100), 40 (100, 114), 41 (150), 45 (74), 61 (264), 63 (100, 150, 271), 65, 66 (150), 67 (100), 69 (304), *79–81*, *83*, *84*, 395 (78), *414*, 485 (16, 17), *518*
Pople, L. A. 87, 88, 90 (43), *95*
Poranski, C. F., Jr. 727 (63, 64), 728, 729 (64), 740, 741 (135), *771*, *773*
Porta, O. 752 (160a–d), *774*
Porter, N. A. 225 (169), *233*, 243, 244 (27), *257*, 315 (31), 316 (38, 39), 317 (44), 339 (44, 138, 139), 342, 364 (154), *370*, *372*, *373*, 384 (38d), *413*, 731 (71), *771*
Porter, R. S. 669 (40), *688*
Pospelov, M. V. 168 (39), *197*
Pospelova, M. A. 720, 726 (30), *771*
Pospišil, J. 737 (111b), *772*
Postel, M. 465, 467–469 (26), *479*
Postnikov, L. M. 800 (94, 95), *806*
Potehkin, V. M. 137 (43), *158*
Potekchin, V. M. 136, 137, 143, 144 (37), *158*
Pouchot, O. 152 (78), *159*
Poupko, R. 436, 449, 456 (49), *457*
Powell, D. L. 142 (57), *159*
Powers, E. L. 2 (20), *78*
Pozdeeva, A. A. 184 (98), *198*
Pralus, M. 471 (78), *480*
Prelog, V. 290 (117a–c, 124), *306*, 312 (2), *369*, 525 (13), *578*
Preston, S. B. 470 (73), *480*
Preub, T. 672, 673 (61), *683*
Preuss, T. 498 (77), *519*
Preut, H. 86, 88 (18), *94*
Price, C. C. 280 (5, 6, 11), *304*, 501, 503 (85), *519*, 596, 598 (51), *644*, 652 (16), *681*
Price, M. A. 677 (61), *689*
Price, M. E. 244 (30), *257*, 838 (145d), 839 (145c), 883 (145c, 145d), 884 (145d), 886 (145c), 906 (198), 908 (205), *916*, *918*
Price, P. 214 (110), *232*
Prilezhaev, E. N. 288 (104), *306*
Prilezhaev, N. 525 (4), *578*
Prilezhaeva, E. N. 525 (5b), *578*
Prince, E. 86, 88 (11), *94*
Pritchard, H. O. 99 (20), *104*, 667 (20), 675 (53), *688*, *689*, 732 (82), *772*
Pritchard, R. G. 51 (205), *82*, 86, 88 (15, 16), *94*, 376 (10), *412*
Pritzkow, W. 99 (17), *104*, 165 (15), *196*, 288,

289 (107, 108), *306*, 525 (12), 554 (150), 574 (237), *578*, *582*, *584*
Prokopchuk, S. P. 425 (67), *428*
Proksch, E. 498 (77), *519*, 662 (50a, 50b), 672, 673 (61), *682*, *683*
Prophet, H. 672 (44), *689*
Prosen, E. J. 98, 99 (13), *104*
Proskuryakov, V. A. 136 (37), 137 (37, 43), 143, 144 (37), *158*
Pryde, E. H. 650 (3k), *681*
Pryor, W. A. 130 (5), *157*, 271 (53), *277*, 303 (237), *309*, 429 (1), 453 (191, 192), 456 (191), *456*, *460*, 517 (119), *520*, 546 (110b, 128a–c), 547 (110b), *581*, 600 (80), 605 (90, 91), 618 (91), *645*, 750 (156), *774*
Przhevalskii, I. N. 49, 50 (196), 51 (207), 53, 54 (212), *82*, 87–90 (27), *94*
Przybylski, M. 114 (17), *126*
Pucheault, J. 452 (186), 456 (211), *460*, *461*, 762 (186), *775*
Puchin, V. A. 425 (67), *428*
Pudov, V. S. 801 (80), 802 (80, 107), *805*, *806*
Pulwer, M. J. 369 (212), *374*
Punderson, J. O. 174 (58), *197*
Purcell, K. F. 135, 138, 142, 148 (34c), *158*
Puri, B. R. 792 (51), *805*
Puring, M. N. 136, 137, 143, 144 (37), *158*
Purushothaman, K. K. 297 (183), *308*
Purvis, G. D. 44, 45 (157), *81*
Pyun, H. Y. 130 (1c), *157*, 543 (94), *580*

Qazi, G. N. 312 (7b), *370*
Quast, H. 843, 858 (58), *914*
Quiroz, F. 837, 838, 854 (54), *914*
Quistad, G. B. 334 (130), *372*

Raab, O. 204 (3b), *229*
Raaen, H. P. (174), *774*
Rabani, J. 761, 762 (185), *774*
Rabeck, J. F. 229 (224a, 224b), *234*
Rabinovich, I. B. 98 (15), 99 (15, 16), *104*
Rabjohn, N. 189 (122), *199*
Radda, G. K. 730 (70), *771*
Radeglia, R. 866, 867, 910 (92), *915*
Rademacher, P. 47, 48, 71 (176), *81*, 88–90 (65), *95*
Radford, D. V. 89 (64), *95*
Radford, H. E. 669 (31), *688*
Rado, R. 800 (79, 83, 84, 86, 89, 93, 97, 98), 801 (84, 97), *805*, *806*
Radom, L. 36 (100, 106), 38, 40 (100), 41 (150), 57, 58 (106), 61 (264), 63 (100, 150, 271, 278), 65, 66 (150), 67 (100, 278), *80*, *81*, *83*, 87, 88 (43, 49), 90 (43), *95*, 395 (78), *414*, 485 (16, 17), *518*
Rafalski, A. J. 113 (16), *126*
Rafikov, S. R. 184 (94, 96), 185 (96), *198*
Rainey, W. J. 556 (164), *582*
Raison, J. K. 446 (136), *459*
Raithby, P. R. 377 (26), *412*, 475 (159), *482*
Rajbenbach, L. 609, 610, 624 (87), *645*, 722, 723 (47), *771*
Rajee, R. 226 (185a), *233*
Rakhimov, A. I. 87, 89, 92 (54), *95*
Rakhmankulov, D. L. 501 (120), *520*, 653 (23), *682*
Raley, J. H. 164 (10), *196*
Ramachandran, V. 563 (205), *583*
Ramamurthy, V. 226 (174, 183b, 185a, 185c), *233*, 873 (113), *916*
Ramasseul, R. 151, 152 (79), *159*
Ramdas, V. 51 (204, 206), *82*, 86, 88 (14), *94*, 376 (11–13), *412*
Ramsay, O. B. 718, 733, 739 (19), *770*
Ranby, B. 229 (224a, 224b), *234*
Ranck, J. P. 36 (97), *79*, 87 (48), *95*
Ranjon, A. 152 (78), *159*
Ranney, G. 731 (72), *772*, 831 (6b), *913*
Raoul, J. 292 (144), *307*
Rapaport, E. 896 (183b), *918*
Rassat, A. 151, 152 (79), *159*, 561 (198), *583*
Rastädter, K. 280, 281 (13), *304*
Rasuwajew, G. A. 264, 265 (31), *277*
Rattimer, R. P. 633, 634 (180), *647*
Rattray, G. N. 668, 669 (22), 674 (49), *688*, *689*
Rauchmann, E. J. 446 (132), *459*
Rauhut, M. M. 831 (1), *913*
Rauk, A. 59, 60 (236), *82*
Raviola, F. 424 (61), *427*
Rawlinson, D. I. 454 (199), *460*
Rawlinson, D. J. 173 (47, 48), *197*, 240, 241 (18), *257*, 280, 300, 302, 303 (2), *303*, 546 (126), *581*
Razuvaev, G. A. 262 (17), 263 (25–27), 264, 265 (25), *276*, *277*, 280 (24), *304*, 356 (186), *373*, 722 (45), *771*, 808 (2a), 811 (19, 20a, 20b, 67), 817 (40–42), 820 (50), 821 (56), 823 (61), *826–828*
Read, G. 472 (105), *481*
Reader, A. M. 501 (95), *519*
Rebek, J. 194 (154, 155), *199*
Rebek, J., Jr. 293 (148), *307*, 533 (55, 56), *579*
Rebollo, H. 210 (68b), *231*, 323 (67, 69), *371*, 832–834, 884 (227), *919*
Recktenwald, R. 843, 858 (58), *914*
Reddy, M. P. 725 (56), *771*
Redington, R. L. 36, 38, 49, 50 (119), *80*, 87–90 (24), *94*
Redoshkin, B. A. 588 (8), *643*
Rees, M. W. 636 (198), *648*
Rees, W. W. 285 (67), *305*, 546, 549 (119b), *581*, 639 (207), *648*

Regen, S. L. 470 (71), *480*
Regent, A. 135 (32), *158*, 517 (116), *520*, 528, 543 (37), *579*
Rehorek, H. D. 732 (87), *772*
Reich, H. J. 298 (197), *308*, 546 (108), *581*
Reiche, A. 823 (60), *828*
Reichle, W. T. 166 (19), *196*
Reid, C. G. 285 (63), *305*
Reid, R. W. 847 (29), *914*
Reissenweber, G. 832, 833, 836, 838, 885, 887, 888, 892, 898, 899 (156), *917*
Rekers, J. W. 339, 362 (141), *372*
Rembaum, A. 99 (26), *104*, 608 (97, 99), 616 (99), *645*
Renaldi, P. L. 441 (113), *459*
Renard, J. 634 (183a), *647*
Renner, C. A. 292 (150), *307*, 536 (73), *580*
Renner, R. 666, 676 (68), *683*
Renolen, P. 527 (35), *579*
Rensch, R. 815 (86), *828*
Reusch, W. 438 (69), *458*
Reuter, P. 425 (73), *428*
Reutov, O. A. 557 (250), *584*
Rheenen, V. van 190 (132), *199*
Rheinboldt, H. 789 (45), *805*
Ribner, B. S. 513 (113), *520*, 656 (36), *682*
Riccio, R. 312 (10c), *370*
Ricciuti, C. 88, 93 (107), *96*
Richards, C. 501 (88), *519*, 653 (26), *682*
Richardson, H. 729, 731, 737 (65), *771*
Richardson, J. H. 61 (260, 262), 62 (262), *83*
Richardson, W. H. 130 (1a), 132 (1a, 9), 133, 147, 148 (9), 151 (81), 153, 154 (84), *157–159*, 227 (192), *234*, 244 (30, 31), *257*, 555, 556, 575 (159), *582*, 665, 666, 668 (2), 684 (93, 97, 99), *688*, *690*, 731 (72), *772*, 831 (6a, 6b), 832 (222), 833 (145b), 836 (145a), 838 (145a, 145b, 145d), 839 (145c), 881, 882 (143a, 143b), 883 (143a, 143b, 145a–d), 884 (145d), 885 (145a), 886 (145a, 145c), 891 (150b), 894 (143a, 143b, 150b), 906 (198), 908 (205), *913*, *916–919*
Riche, A. 825 (78), *828*
Riche, C. 382 (37), *413*, 475 (156), *482*
Richter, J. C. 526 (21), *579*
Rickborn, B. 178 (77), *198*
Rideau, J. 420 (12), *427*
Riebel, A. H. 651 (10b), *681*
Rieche, A. 299 (211), *308*, 571 (231), *583*, 590 (21a, 21b), *644*, 712 (4), *770*, 820 (48), 821 (53a), 822 (53a, 53b), 823 (53b), *827*
Riegler, N. 839, 867 (243b), *919*
Rieker, A. 223 (155b), *233*
Riess, G. 419 (9), 420, 422 (11), 424 (61), *426*, *427*
Rigaudy, J. 217 (127), *232*, 324 (77, 78), 347 (167), 349 (77, 170, 171), 351 (176, 177a, 177b), 355 (185), *371*, *373*, 432 (25), 441, 453 (106), 456 (211), *457*, *459*, *461*, 781 (17, 21), 782 (26), 783 (28), 786 (38), *804*, *805*
Riggin, M. 52 (210), *82*
Rimbault, C. G. 209 (59), 220 (59, 135), 225 (165), *230*, *232*, *233*, 332 (109c), *372*, 833–836 (228), 863 (69, 77), *915*, *919*
Rindone, B. 438, 440, 445 (63), *458*
Rinquist, O. A. 346 (164), *373*
Rinsdorf, H. 114 (17), *126*
Rio, G. 152 (78), *159*, 210 (73), *231*, 319 (50), 328 (95), 338 (133), *371*, *372*, 838 (244a, 244b), 843 (252), *919*
Rios, A. 546 (127), *581*
Risaliti, A. 88, 91 (78), *95*, 401 (98), *414*
Rittenberg, D. 593 (35), *644*
Rittenhouse, J. 382 (33), *412*
Rittenhouse, J. R. 88 (61, 99), 89–91 (61), 92, 93 (99), *95*, *96*, 134 (30), *158*, 405 (106), *414*
Rivera, J. 352 (180), *373*
Ro, R. S. 567 (214), *583*
Robbins, J. L. 472 (96), *481*
Roberts, B. P. 238–241, 254 (16), *257*, 314 (29), *370*, 733 (90, 91b, 95), 760, 761 (181c), *772*, *774*, 809 (6, 7), 810, 811 (17), 812 (24, 26), 813 (24), 816 (17), 817 (69), *826–828*
Roberts, D. R. 684 (95), *690*, 893 (163i), *917*
Roberts, H. L. 475, 476 (143), *482*
Roberts, J. D. 436 (46), *457*
Roberts, J. L., Jr. 441 (114), *459*
Roberts, M. L. 526, 532 (24), *579*
Robertson, A. J. B. 46 (168), *81*
Robertson, A. P. 395 (79), *414*
Robertson, G. B. 474 (127), *481*
Robertson, J. C. 563 (206), *583*
Robin, M. B. 46 (172), *81*
Robinson, G. C. 552 (140), *581*
Robinson, G. N. 668 (23), 669 (29), 676 (56), *688*, *689*
Robinson, P. J. 666, 678 (6), *688*
Robinson, S. D. 466, 474 (129), *481*
Robson, J. F. 568 (222), *583*
Robson, P. 130, 131 (1b), *157*
Rodgers, A. S. 40 (140), *80*, 97 (2), *102*
Rodgers, M. A. J. 2 (20), *78*, 844 (18u), *913*
Rodionova, N. M. 136 (45), 139, 142, 149 (51), *159*
Rodrigo, R. 867 (94), *915*
Rodriguez, A. 863 (73), *915*
Rodriguez, L. O. 753 (164), *774*
Rodriguez, O. 226 (180), *233*, 831, 843, 846, 857 (12), 903, 905 (189a, 189b), *913*, *918*
Rodriquez, A. 315 (30), 321 (61), 338, (135), 343 (61), *370–372*

Rodwell, W. R. 36, 57, 58 (106), 63, 67 (278), *80*, *83*, 87, 88 (49), *95*
Roe, A. N. 339 (138), *372*
Roffey, M. J. 679, 680 (74), *689*
Roffia, S. 441 (85), 449 (168), *459*, *460*
Rogers, J. D. 52 (211), *82*
Rogers, M. T. 88–90 (60), *95*
Rogers, R. D. 65 (287), *83*
Rogers, T. E. 260 (7), *276*
Roginskij, V. A. 799 (77), *805*
Rogova, L. S. 800 (88), *806*
Rohr, W. 299 (204), *308*, 594, 596, 607 (45b), *644*
Rolle, W. 554 (150), 574 (237), *582*, *584*
Romano, L. J. 442–445 (117), *459*
Romanova, O. S. 425 (66, 68), *428*
Romantsev, M. F. 764 (195), *775*
Romantsova, O. N. 425 (66, 68), *428*
Rombusch, K. 290 (126), *306*
Romers, C. 38 (137), *80*
Roper, W. R. 465 (86, 88, 90), 471 (86–88, 90), *480*, *481*
Ros, R. 475, 476 (142), *482*
Rosen, G. M. 446 (132), *459*
Rosen, J. D. 453 (193), *460*
Rosenblatt, D. H. 271, 272 (54), *277*
Rosenthal, I. 207 (43), *230*, 436 (49), 438 (64), 440 (80), 441 (91), 444 (64), 449 (49, 80, 163, 165, 166), 451 (80, 166, 169, 178), 456 (49, 91), *457–460*, 878 (135a, 135b), 879 (136), *916*
Rosenthal, R. 187 (117), *199*
Roshchupkin, V. P. 801 (102), *806*
Rossiter, B. E. 182 (86), 183 (91), *198*, 534, 535 (63), *580*
Roth, A. P. 740 (128), *773*
Roth, D. 219, 222 (132), *232*, 863 (76), *915*
Roth, H. D. 740 (133), *773*
Rothbaum, P. 448 (158), *460*
Rothenberg, S. 24, 57 (45), *79*
Rotman, A. 295 (154), *307*
Rotschová, J. 737 (111b), *772*
Roue, J. 465, 469 (39), *479*, 808, 822 (59), *828*
Rouelle, F. 448 (157), *460*
Roundhill, D. M. 466 (117), 467 (24), 473 (117), *479*, *481*
Rouse, R. A. 390 (62a), *413*
Rousseau, G. 221, 225 (136), *232*, 835 (238), 863 (74a, 74b), 865 (82), 911 (74a, 74b), *915*, *919*
Rousseau, Y. 121, 122 (33), 123 (36), *127*, 634 (183c), *647*
Roussel, M. 477 (148, 149), *482*
Routhby, R. 474, 475 (133), *481*
Roux, M. 54 (218), *82*
Roy, R. B. 286 (78), *305*
Roy, R. G. 271 (52), *277*
Royer, J. 295 (169, 170), *307*
Rozantzer, E. G. 295 (158a), *307*
Rubcov, V. I. 799 (77), *805*
Rubottom, G. M. 853 (52), *914*
Ruchardt, C. 594, 596, 607 (45a), *644*
Rüchardt, C. 299 (203, 210), 300 (210), *308*, 554 (145), *582*, 600 (81), 619, 620 (134), *645*, *646*
Rucktäschel, R. 742 (144), *773*
Rudnevskii, N. K. 137 (40), 144 (63), 150 (71), *158*, *159*
Rudolph, H. 281 (44), *305*
Ruedenberg, K. 38 (132), *80*
Runquist, O. A. 213 (90, 91a), 216, 221, 223 (90), *231*
Russel, G. A. 437, 438, 440 (60–62), 451 (182), 454 (60–62), *457*, *458*, *460*
Russell, G. A. 247 (47), *257*, 280 (12), 297 (178), *304*, *308*, 405 (116), *415*, 486 (25, 26), 490, 493 (25), *518*, 546 (104), *580*
Russell, J. L. 185 (99), *198*
Russell, J. R. 296 (171), *307*
Russell, J. W. 388 (46), *413*
Russell, K. E. 281 (36), *305*, 405 (122), *415*, 597, 619 (63b), *645*
Rust, F. F. 164 (10), 165 (17), 175 (61, 62), 176 (62), *196*, *197*
Rutledge, P. C. 546 (107), *581*
Ruzicka, L. 295 (165), *307*, 831, 858 (11), *913*
Ryan, P. B. 36 (104), *80*, 87 (52), *95*
Ryang, H.-S. 229 (223b), *234*, 335 (121), *372*
Rychlý, J. 785, 786 (36), 792 (53), 796, 797 (65, 69), 800 (99), *805*, *806*
Rylance, J. 98 (8), *104*
Ryntz, R. A. 287 (89), 291 (137), *306*, *307*
Rys, J. 55 (220), *82*
Ryun, H. Y. 641 (211), *648*
Rzepa, H. S. 24, 45, 58 (62), *79*, 107 (5), *126*

Sabeta, I. C. 594, 605, 606 (48), 607 (96b), 608, 616 (48), 620 (48, 96b), *644*, *645*
Sadikov, G. B. 760 (182), *774*
Segae, H. 440 (72, 73), 441 (84), 451 (175), *458–460*
Sager, W. F. 499 (79), *519*
Saigusa, T. 421 (17, 18), *427*
Saiki, T. 228 (209), *234*, 909 (209), *918*
Saito, I. 65 (284), *83*, 207 (42), 209 (61), 210 (80), 211 (81, 82), 220 (61), 229 (223a), *230*, *231*, *234*, 324 (76), 326 (84, 85, 88), 328 (93), 330 (102), 332 (108a, 109a, 109b), 335 (117), 338 (134), 347 (76, 165, 166), 348 (76), *371–373*, 446 (148), 449 (167), *459*, *460*, 836 (103), 844 (18q), 871 (18q, 103, 104), 872 (105a, 105b, 106b, 106c), 873 (110), 911 (103, 214), *913*, *915*, *916*, *919*

Sajus, L. 465 (57, 62), 469 (57), 470 (62), 471 (57), *480*
Sakaguchi, H. 193 (147), *199*
Sakaguchi, Y. 740 (124), *773*
Sakanishi, K. 738 (115), *773*, 832, 850, 887–889 (45), 898, 899, 908 (207), *914*, *918*
Sakore, T. D. 405 (110, 111), *414*
Sakos, D. 717 (17b), *770*
Saksena, A. K. 388 (52), *413*
Sakuragi, H. 211 (84), *231*, 596, 619 (59), *644*, 740 (125, 126, 129), *773*, 863 (78), *915*
Sakuragi, M. 740 (125), *773*, 863 (78), *915*
Sakurai, F. 475 (140), *482*, 826 (82), *828*
Sakurai, T. 271 (50), *277*
Salahub, D. R. 45 (167), 58 (231), *81*, *82*
Sala-Pala, J. 465, 469 (38–40), *479*, *480*, 808, 822 (59), *828*
Salem, L. 8 (27), 60 (239), *78*, *82*
Salomen, M. F. 821 (55), *828*
Salomen, R. G. 821 (55), *828*
Salomon, M. F. 116 (22), 119 (24b), *126*, 315 (33–35), 360 (197), 361 (35), 362 (197), *370*, *373*
Salomon, R. G. 116 (22), 119 (24b), 120 (27), *126*, *127*, 216 (122), *232*, 315 (33–35), 320 (54), 350 (172), 360 (54, 197, 198), 361 (35, 198), 362 (197), 363 (54), *370*, *371*, *373*, 384 (38h), *413*
Salsburg, Z. W. 731 (77), *772*
Saltiel, J. 740 (127), *773*
Saluja, P. P. S. 69 (309), *84*
Salzman, H. 298 (196), *308*
Sambe, H. 45 (167), *81*
Sammes, M. P. 394 (75), *414*
Sammes, P. G. 336 (124), *372*, 470 (74), *480*, 836 (241), *919*
Samson, J. A. R. 44 (160), *81*
Samuelson, B. 313 (13–15), *370*
Sanabia, J. 736 (106), *772*
Sande, J. H. van de 589 (16), 591 (16, 22), *643*, *644*
Sandel, V. R. 718 (20), *770*
Sanderson, D. G. 446 (133), *459*
Sanderson, J. R. 193 (148), *199*, 557 (172), *582*, 628 (158a), *647*
Sanderson, R. T. 2, 3 (24), *78*
Sandrini, D. 757 (173), *774*
San Filippo, J., Jr. 436, 437 (52), 441 (87, 89), 442–445 (117), 453, 454 (87), *457*, *459*
Sanhueza, E. 491 (56), *518*
Sannograhi, A. B. 25 (83), *79*
Santacrose, C. 312 (10a), *370*
Santa Cruz, T. D. 904 (191), *918*
Santiago, G. 110 (11), *126*
Sapunov, V. N. 528, 530 (38), *579*
Saraeva, V. V. 764 (195), *775*
Saran, M. 453 (194), *460*
Sarin, V. A. 86, 88 (12, 19), 89 (12), *94*
Sarkar, S. 469 (60), *480*
Sasada, Y. 393 (70, 71), *414*, 860 (62b), *914*
Sasaoka, M. 781 (17), *804*
Sasavka, M. 120 (28), *127*
Sastre, J. A. L. 156 (88a), *160*
Sato, H. 61 (247), *83*, 87–89 (35), *95*
Sato, J. 68 (300), *84*
Sato, T. 546–548 (114b), 560 (196), *581*, *583*
Satterfield, C. N. 2 (1), *78*, 86 (2), *94*, 712 (6), *770*
Sauers, R. R. 561 (198), *583*
Saunders, V. R. 36 (95), *79*
Saunders, W. H. 559, 560, 564 (188), *582*
Savage, C. M. 69 (305), *84*, 678 (70), 679 (71), *689*
Savariault, J. M. 49, 51, 55 (192), *81*
Savéant, J. M. 452 (184b), *460*
Savitskii, A. V. 464, 465 (10), *479*
Sawaki, Y. 445 (123), 448 (161, 162), *459*, *460*, 525 (11), 535 (67), 538 (78), 544 (97c), 555 (160, 161), 556 (160–163), 561 (199), 564 (163), 566 (163, 211), 568 (218, 221, 224), 569 (224), 570, 571 (228), 575 (243), 576 (243, 244), *578*, *580*, *582–584*, 587, 642 (4), *643*, 730 (66), 738 (115), *771*, *773*, 874 (121), 876 (122a, 125), *916*
Sawhney, K. N. 286 (73), *305*
Sawtchenko, L. 796, 797 (66), *805*
Sawyer, D. T. 430 (6–8), 431 (6, 7 13–15), 432 (14, 21), 433, 435 (21), 436 (47), 437 (8, 47, 54), 440 (8), 441 (107, 114), 443 (6–8), 445 (8), 447 (155), 448 (6, 13), 450 (8), 454 (155), 456 (213), *456*, *457*, *459*, *461*, 879 (138), *916*
Sax, M. 88 (80, 97), 91, (80), 92 (97), *95*, *96*, 134 (17), *158*, 405 (108), 407 (128), *414*, *415*, 611 (107a), *646*
Saxton, R. P. 21, 58 (42), *78*
Scaiano, J. C. 237 (12, 13), 249 (12), *256*, *257*, 731 (78), 733 (92a), *772*, 865 (84), *915*
Scandola, F. 550 (132), *581*, 895 (177), *918*
Scanion, M. 501 (88), *519*
Scanlan, J. T. 291 (136), *307*
Scanlon, M. 653 (26), *682*
Scardellato, C. 471 (77), *480*
Scarpati, R. 334 (115), *372*, 873 (109), *916*
Schaade, G. 204 (6), *229*
Schaap, A. P. 205 (11, 13, 15), 206 (30), 207 (35b), 209, 226 (57), *229*, *230*, 329, 355 (98), *371*, 446 (143), *459*, 588 (12b), 592 (27a, 28), *643*, *644*, 832 (39, 157, 220), 834 (234), 838 (31, 245), 839 (157), 842 (31, 48), 844 (18r), 847 (31), 849 (39, 40), 851

(48), 871 (18r), 874 (119), 884 (31), 885 (168), 886, 887 (31), 890, 892 (157), 894 (168), 895 (31), 896 (31, 48), 898, 899 (31), 900 (157), 903 (168), *913*, *914*, *916*, *917*, *919*
Schaap, P. A. 227 (198), *234*
Schaefer, H. 651 (6b, 11), 654 (6b), *681*
Schaefer, H. F., III 21 (37), 24 (45, 55), 25 (71), 33 (55, 71), 44, 45 (161), 57 (45), 61 (245, 249, 260, 262), 62 (245, 262), *78*, *79*, *81–83*
Schaefer, J. F., III 25, 33 (68), 63, 67 (278), *79*, *83*
Schaefer, W. P. 377 (24), *412*
Schäfer-Ridder, M. 325 (83), 330 (104), 344 (159), *371–373*
Schafer-Rider, M. 216 (124a), *232*
Schäffer, H. J. 738 (118b), *773*
Schaffner, K. 906 (203), *918*
Schamma, M. 366 (205b), *374*
Schamp, N. 282 (58), *305*
Schank, K. 281 (38–43, 47), 282 (38, 39), 283 (39), 284 (38, 41, 43, 62), 291 (140), 292 (143), 303 (233c), *305*, *307*, *309*
Scharf, H.-D. 302 (232), *309*, 865 (83), *915*
Schearer, G. O. 530, 532 (47), *579*
Scheffer, J. R. 207 (37, 38), *230*, 782 (24, 25), 783 (25), *804*
Scheiraenz, G. P. 208, 210 (49), *230*
Schenck, G. O. 167 (30, 31), *197*, 204 (4a), 208 (48, 49), 210 (48, 49, 66b, 67a, 67b), 211 (87a), 212 (66b), 213 (96), 216 (96, 120), 217 (87a), 222 (144, 145, 150), *229–232*, 319, 343 (48), *371*, 592 (32), *644*, 781 (18), *804*
Schenk, G. D. 552 (137), *581*
Schenk, G. O. 588 (10a, 11a–c), 592 (10a), *643*
Schenker, K. 290 (117b, 124), *306*, 525 (13), *578*
Scheutzow, D. 837, 838, 846, 849 (25), *914*
Schildcrout, S. M. 343 (155), *373*
Schilling, P. 666, 676 (68), *683*
Schleker, W. 865 (83), *915*
Schlessinger, R. H. 337 (129), *372*
Schleyer, P. v. R. 661 (49), *682*
Schmalstieg, H. 386, 388 (45b), *413*, 840, 845, 846, 867 (22a, 22b), 890 (22a, 22b, 179), 892 (22a, 22b), 895, 900, 903 (179), *914*, *918*
Schmid, G. H. 525 (5d), *578*
Schmid, G. M. 324 (78), *371*
Schmid, P. 241 (23), *257*
Schmidt, D. D. 472 (99), *481*
Schmidt, E. 833, 836, 837 (233), 839, 845, 846 (23), 888 (233, 236), 889, 890 (233), 900 (233, 236), *914*, *919*
Schmidt, R. 331 (105), *372*, 893 (161), 904 (193c), *917*, *918*
Schmidt, S. P. 667, 684 (14), *688*, 844 (18j), 847 (34), 884 (172b), 894 (167), 895 (172b), 896 (181b), 898 (257), 903 (172b), *913*, *914*, *917–919*
Schmidt, V. 281 (39, 42, 43), 282, 283 (39), 284 (43), *305*
Schmitt-Josten, R. 290 (130), *307*
Schmitt, G. 470 (72), *480*
Schmitz, A. 209 (53), *230*
Schmitz, E. 590 (21a), *644*
Schneider, H.-J. 289 (114), 295 (157), *306*, *307*, 534 (65), *580*
Schneider, K.-A. 839, 867 (243b), *919*
Schnering, H.-G. von 832 (24), 839 (23), 845, 846 (23, 24), *914*
Schnur, R. C. 299 (213), *309*
Schnurpfeil, D. 290 (115), *306*
Schoenbein, C. F. 649 (1a–c), 652 (1c), *681*
Schoes, H. K. 189 (124), *199*
Scholl, M. J. 338 (133), *372*
Scholl, M.-J. 152 (78), *159*
Schomaker, V. 49 (190), *81*
Schönberg, A. 588, 592 (10a), *643*
Schore, N. E. 667, 684 (15), *688*, 844, 847, 882, 893, 902 (18c), *913*
Schott, H. 474 (119), *481*
Schowalter, K. A. 440 (78), *459*
Schrade, O. 555 (153), *582*
Schrader, O. 222 (140a), *232*, 831 (9), *913*
Schramm, W. 296 (174), *308*
Schreiber, S. L. 176 (66), *197*
Schrenck, M. 44, 45 (161), *81*
Schroder, G. 424 (60), *427*
Schroder, M. 189 (126), 190 (126, 133), *199*
Schröder, R. 301, 302 (224), *309*
Schubert, C. C. 656 (34a, 34b), *682*
Schuetz, R. D. 280 (10), *304*
Schuler, N. W. 280 (16), *304*
Schulte-Elte, K. H. 211 (87a, 87b), 214 (109), 217 (87a, 87b), 222 (145), 224 (157b), *231–233*, 348 (169), *373*, 552 (137), *581*
Schulte-Elte, K.-H. 588 (11c), *643*
Schulz, G. J. 48, 49 (178), *81*
Schulz, M. 71 (315), *84*, 391 (66), 405 (113), *413*, *415*, 782, 792 (23), *804*
Schulz, R. 424 (60), *427*
Schumacher, H. J. 99 (29–32), 101 (31, 32), *104*, 497 (72, 73), *519*, 669 (37), 671 (37, 41), 672 (37), *688*, *689*
Schumann, D. 105 (2), *126*
Schumb, W. C. 2 (1), *78*, 86 (2), *94*, 712 (6), *770*
Schumm, R. H. 40 (139), *80*
Schurath, U. 58, 59 (233), *82*, 679, 680 (76), *689*

Schürgers, M. 60 (237), *82*
Schurig, V. 465, 470 (67), *480*
Schurig, V. S. 465, 470 (66), *480*
Schuschunov, V. A. 821 (52), *827*
Schushuhov, V. A. 356 (186), *373*
Schuster, G. 337 (131), *372*, 667, 684 (15), *688*, 844, 847, 882, 893 (18c), 902 (18c, 188), 903 (188), *913*, *918*
Schuster, G. A. 227 (200a, 200b, 201), *234*
Schuster, G. B. 281 (35), *305*, 405 (117–119), *415*, 546–548 (115, 116), 550 (115, 116, 132), 577 (247), *581*, *584*, 639 (206), *648*, 667 (14), 684 (14, 98), *688*, *690*, 832 (35, 157), 836 (146), 839 (157), 844 (18j, 18k), 847 (34, 35), 883 (146), 884 (172b), 885 (146), 886 (35), 890 (157), 891 (153a), 892 (157), 894 (18k, 167), 895 (172b, 174, 177), 896 (181b), 898 (35, 257), 900 (157), 903 (18k, 172b, *913*, *914*, *916–919*
Schutz, M. 245 (34), *257*
Schwab, P. A. 292 (149b), *307*, 536, 537 (71), *580*
Schwager, I. 61 (253), *83*
Schwarts, D. R. 194 (152), *199*
Schwartz, H. A. 484 (9), *517*
Schwartz, J. 819 (47), *827*
Schwartz, N. N. 288, 289 (106), *306*
Schwarz, H. 116 (20a–d), *126*
Schwarz, H. A. 761, 762 (185), *774*
Schwarzenbach, D. 465, 468 (32), *479*
Schwenk, E. 546, 547 (111a, 111b), *581*, 636 (194), *648*
Scott, A. I. 223 (153), *233*
Scotton, M. J. 809, 814, 815 (8), 818, 819 (45), *827*
Scouten, W. H. 205 (14), *229*
Scribe, P. 351 (177b), *373*
Scrimin, P. 469 (44), 471 (77), *480*
Scriven, E. F. V. 105 (1), *126*
Scrivener, F. E., Jr. 552 (141), *581*
Scroder, G. 680 (80), *689*
Scully, F. E., Jr. 441 (90), *459*
Sczeimies, C. 529 (46), *579*
Searby, M. H. 717 (17a), *770*
Secco, E. A. 587 (5b), *643*
Secroun, C. 48, 49, 54 (184), *81*
Sedergran, T. C. 571 (232), *583*
Sehested, K. 767 (204), *775*
Seinfeld, J. H. 2, 68 (18), *78*
Seip, H. M. 387 (54, 55), *413*
Sekiguchi, M. 263 (22b), *276*, 760 (180), *774*
Sekine, Y. 335 (122), 359 (122, 194), *372*, *373*
Seliger, H. H. 205 (16), *229*, 844 (14), 896 (182, 183a, 183b), 901 (185b), 902 (186), 903 (185b, 186), *913*, *918*
Selin, L. E. 44, 45 (159), *81*
Seltzer, S. 607 (96a), *645*
Selva, A. 176 (65), *197*
Selwyn, J. 247 (43), *257*
Semkov, A. M. 89 (57), *95*
Semkow, A. M. 36, 52, 53 (102), *80*
Sen, A. 466, 474 (124), *481*
SenGupta, P. K. 740 (131), *773*
Sen Sharma, D. K. 113 (14a, 14b), *126*
Seo, E. T. 431, 432 (14), *456*
Sérée de Roch, I. 465 (57, 62), 469 (57), 470 (62), 471 (57), 473 (108), *480*, *481*
Sergeeva, V. P. 556 (166), *582*
Serjeant, E. P. 133 (11a), *158*
Serkiz, B. 843 (252), *919*
Seshimoto, O. 865 (88), *915*
Sevchenko, A. N. 280 (26, 30a–c), *304*
Severini, F. 422 (27), *427*
Shabalin, I. I. 139 (48), 140 (49), 141 (48, 49), 142 (48, 54, 55), 145 (54), 148, 149 (54, 55), 150 (48, 55, 72), *159*
Shaefer, W. P. 475 (157), *482*
Shakhashiri, B. Z. 205 (18), *229*
Shamshev, V. N. 764 (196), *775*
Shand, W. 88, 89 (68), *95*
Shani, A. 215 (114), *232*
Shanks, R. A. 425 (70), *428*
Shapiro, E. L. 388 (48), *413*
Shapiro, M. J. 228 (220), *234*, 836, 910 (211c), *918*
Sharefkin, J. G. 298 (196), *308*
Sharma, D. N. 768 (209), *775*
Sharpless, K. B. 162 (5), 171 (44), 179, 180 (5), 181 (5, 82), 182 (86, 88, 90), 183 (91), 189 (123, 125), 190 (128, 129), *196–199*, 465, 470 (63), 478 (152), 479 (160), *480*, *482*, 532, 533 (53), 534, 535 (53, 63), *579*, *580*, 824 (73), *828*
Sharts, C. M. 441 (97), *459*
Shatkina, T. N. 557 (250), *584*
Shatkovskaya, D. B. 87, 89, 92 (54), *95*
Shavitt, I. 36 (88), *79*, 87, 88 (37), *95*
Shaw, D. H. 99 (20), *104*, 675 (53), *689*
Shaw, J. H. 677 (59), *689*
Shaw, R. 40 (140, 142), 41, 43 (142), *80*, 97 (2, 4, 5), 98 (6, 12), 100–102 (6), *102*, *104*, 237, 239, 248 (11), *256*, 485, 496 (15b), *518*, 665 (1), 668 (21), *688*
Shchegoleva, T. M. 382 (38a), *413*
Sheats, G. F. 716 (14a), *770*
Shei, J. 296 (175), *308*
Sheik, Y. M. 312 (10b), *370*
Sheinson, R. S. 685 (105–107, 109, 110), 686 (110), 687 (105), *690*, 847, 864, 894 (32), *914*
Sheldon, R. 808 (1), *826*
Sheldon, R. A. 169, 170 (40), 171 (40, 42, 43), 177 (42, 43, 67), 178 (43), 179 (43, 79), 180, 181 (42, 43), 192 (140), *197–199*, 466,

467, 474 (23), 478 (151), *479*, *482*, 533 (57, 58), *579*, 712–714 (1b), 716 (10), 738, 739 (1b), 748–750, 752 (10), *770*, 808 (5), 816 (5, 64), 824 (5), *826*, *828*
Sheldrick, G. M. 88, 90 (74), *95*, 382 (34), *413*
Shelimov, B. N. 722, 724 (44), *771*
Shen, J. 666, 676 (69), *683*
Sheng, M. N. 162, 179, 180 (4), 184 (95), 186 (113), 190 (130), *196*, *198*, *199*
Sheppard, C. S. 2 (6), *78*, 299 (205), *308*
Sheppard, W. A. 441 (97), *459*
Sherington, D. 425 (76), *428*
Sherman, N. 625 (149c), *647*
Sherrington, D. C. 419 426 (4), *426*
Sherrinton, D. C. 419 (78), *428*
Shevchuk, V. U. 653 (24b), *682*
Shiau, W. 778 (1), *804*
Shibasaki, M. 316 (36), *370*, 438 (65), 441 (65, 109), *458*, *459*
Shida, T. 764 (198), *775*
Shih, L. S. 229 (222), *234*, 832 (215), 910 (212b), 911 (215), *919*
Shih, S. 24 (52, 60), 25 (60), 33 (52, 60), *79*
Shih, S. K. 25 (76, 77), 45, 46, 58, 59 (77), *79*
Shimizu, T. 151 (80), *159*, 446 (154), 451, 454, 456 (154, 170), *459*, *460*, 878 (134), *916*
Shimizu, Y. 491 (58), *518*
Shimomura, M. 792 (52), *805*
Shimomura, O. 844 (17a), *913*
Shimooda, I. 228 (218), *234*
Shimunkova, D. 800 (89, 93), *806*
Shine, H. J. 280 (8), *304*, 614 (119–121), *646*
Shiner, C. S. 316 (36), *370*, 438, 441 (65), *458*
Shingel, I. A. 151 (74), *159*
Shiratori, O. 397 (91), *414*
Shirley, D. A. 44, 45 (161), *81*
Shirmann, J. P. 471 (78), *480*
Shiroyama, M. 448 (161), *460*
Shklover, V. E. 396, 397 (87b, 87c), *414*
Shkolina, M. A. 423 (49), *427*
Shlyapintokh, V. Ya. 797 (74), *805*
Shlyapnikov, Yu. A. 800, 803 (91), *806*
Shook, F. C. 456 (212), *461*
Shortridge, R. 492 , 495 (63), *518*
Shoulders, B. A. 680 (83), *689*
Showell, J. S. 296 (171), *307*
Shriver, D. F. 65 (288), *83*
Shu, P. 213 (100), *231*, 314 (25), *370*
Shubber, A. K. 821 (54b), *828*
Shuckler, H. 222 (151), *232*
Shull, H. 36 (103), *80*, 87, 90 (53), *95*
Shulman, G. P. 651, 654 (6d), *681*
Shurygin, V. E. 722 (45), *771*
Shushonov, V. A. 588 (8), *643*
Shushunov, V. A. 556 (166), *582*, 814 (29), 822 (58), *827*, *828*
Shybnev, V. P. 423 (47), *427*
Sica, D. 312 (10a), *370*
Siegbahn, K. 44, 45 (158), *81*
Siegel, E. 403 (102), *414*, 788, 794, 795 (40), *805*
Siegel, M. W. 48, 49 (179), *81*
Sigolaev, Yu. F. 136, 137, 143, 144 (37), *158*
Sih, C. J. 214 (110), *232*
Silberstein, R. 599 (77b), *645*
Silbert, L. S. 86 (1), 88 (90, 107), 90, 91 (1), 92 (1, 90), 93 (107), *94*, *96*, 134 (19), 136 (36), *158*, 280 (18a, 18b), 282 (52), 288 (93), 299 (201), *304–306*, *308*, 403 (102), *414*, 596 (52), *644*, 712 (2), 732 (80c), *770*, *772*, 788, 794, 795 (40), *805*, 855 (247), *919*
Silvi, B. 63 (275), *83*
Sim, G. A. 213 (98), *231*, 314 (24), *370*
Simamura, O. 541 (89), 542 (90), 562 (200, 201), *580*, *583*, 587, 593 (3), 619 (133), 641 (3), *643*, *646*, 716, 718 (11), *770*
Simmons, H. E. 290 (118b, 118c), *306*
Simmons, W. 446 (152), *459*, 851 (49), *914*
Simo, I. 891 (151a), *917*
Simon, A. W. 440 (78), *459*
Simonaitis, R. 491 (56), *518*, 678 (64), 679 (64, 72, 73), 680 (72, 73), *689*
Simonet, J. 448, 454 (160), *460*
Simonsen, S. H. 336 (123), *372*, 394 (75), *414*
Simpson, G. A. 226 (179), *233*, 894, 901, 902 (170), *917*
Singaram, S. 442, 445 (118), *459*
Singer, L. A. 253, 255 (67), *258*, 600 (79), *645*, 840, 884, 895, 898 (180a, 180b), *918*
Singleton, E. 377 (25), *412*, 466 (104, 114), 472 (104), 473 (104, 113–115), *481*
Sinn, E. 470 (75), *480*
Sitzmann, M. E. 259, 260 (3), *276*
Siu, A. K. Q. 25, 33 (64), *79*
Sixma, F. L. J. 666 (55a, 55b), *682*
Sixt, J. 653 (24a), *682*
Skapski, A. C. 86, 88 (13), *94*
Skau, E. L. 831 (8), *913*
Škerjanc, J. 135 (32), *158*, 517 (116), *520*, 528 (37), 543 (37, 249), *579*, *584*
Skibida, I. P. 172 (46), *197*
Skobeleva, S. E. 142, 147, 150, 151 (56), *159*
Slagle, J. R. 280 (8), *304*, 614 (120), *646*
Slater, J. C. 202 (2a), *229*, 455 (203), *461*
Slater, R. 685 (102), *690*, 891 (152c), *917*
Slaugh, L. H. 187 (115), *199*
Sloan, J. P. 271 (53), *277*, 557 (176), *582*, 628 (159), *647*
Slovokhotova, N. A. 764 (197), *775*
Sluchevskaya, N. P. 801 (81), *805*

Slusser, P. 891, 894 (150b), 906 (198), *917*, *918*
Small, R. D., Jr. 237, 249 (12), *256*, 731 (78), *772*, 782 (22), *804*
Small, R. W. H. 384 (38c), *413*, 812 (21a), *827*
Smardzewski, R. R. 61, 62 (243), *82*, 735 (105), *772*
Smets, G. 422 (31), 423 (38, 39), 424 (57–59), 425 (74, 75), *427*, *428*
Smid, J. 608 (99, 100), 616 (99), *645*
Smirnov, B. R. 802 (108, 109), *806*
Smirnov, P. A. 137 (43), *158*
Smissman, E. E. 560, 561 (194), *583*
Smit, W. C. 290 (121), *306*
Smith, D. E. 472 (107), *481*
Smith, D. W. 52, 54 (209), *82*
Smith, G. P. 48 (187), *81*
Smith, H. W. 400 (96, 97), *414*
Smith, J. G. 554, 556, 561 (148), *582*
Smith, J. P. 337 (131), *372*
Smith, J. R. L. 271, 272 (54), *277*
Smith, K. 600 (80), 605 (90), *645*
Smith, K. K. 891 (153a), *917*
Smith, K. M. 871 (102), *915*
Smith, L. L. 113 (15a–c), *126*, 440, 449 (81), *459*
Smith, P. A. 626 (153), *647*
Smith, P. A. S. 295 (161), *307*, 559 (182), *582*
Smith, R. C. 720 (29), *771*
Smith, R. M. 213 (99), *231*
Smith, R. S. 153, 154 (84), *159*
Smith, W. B. 556 (164), *582*
Smith, W. F., Jr. 741 (136), *773*
Smoliková, J. 92 (101), *96*
Snyder, L. C. 53 (215), 57 (215, 226), *82*
Sobczak, J. 478 (153), *482*
Soffer, L. M. 618 (130), *646*, 739, 740 (121), *773*
Sogah, D. 878 (132), *916*
Sojka, S. A. 294 (152), *307*, 727 (63, 64), 728, 729 (64), 740, 741 (135), *771*, *773*
Sokhar, B. C. 422 (26), *427*
Sokolov, N. A. 156 (90), 157 (90, 95–97), *160*
Sokolova, R. F. 280 (30d), *304*
Sokova, K. M. 193 (146), *199*
Solo'eva, E. I. 144 (64), *159*
Solomon, D. H. 792 (53), *805*
Solomon, R. G. 214 (110), *232*
Solov'eva, E. I. 157 (91, 92), *160*
Soloway, A. H. 559, 560 (186), *582*
Sommer, A. 165 (16), *196*
Sommerfeld, C. D. 213, 216 (105), *231*, 324 (74), *371*
Song, P.-S. 896 (184), *918*
Sonnenberg, J. 152 (76), *159*
Sonoda, N. 293 (147), *307*
Sood, R. 214 (110), *232*
Sorba, J. 287 (85), *305*
Sorkina, M. V. 151 (74), *159*
Soroos, H. 824 (70), *828*
Sosnovskii, G. M. 137 (44), *158*
Sosnovsky, G. 173 (47, 48), *197*, 240, 241 (18), *257*, 259 (1b), *276*, 280 (2), 300 (2, 214), 301 (214, 218), 302 (2, 225, 227), 303 (2, 214, 229, 234, 235, 240), *303*, *309*, 440 (71), 454 (199), *458*, *460*, 546 (126), *581*, 755 (167, 168), 759 (177), *774*
Souchay, P. 803 (112), *806*
Southwick, L. M. 501 (94), *519*, 651 (7b), 654 (7b, 30), *681*, *682*
Sovers, O. J. 38 (131), *80*
Soyagimi, H. 335, 359 (122), *372*
Spada, L. T. 446 (139), *459*
Sparfel, D. 217 (127), *232*, 349 (170), *373*
Speakman, P. R. H. 280, 287 (22), *304*
Speck, M. 290 (117c), *306*
Speers, L. 559, 560 (187), *582*
Speier, G. 436, 437 (51), *457*, 879 (139), *916*
Spence, J. W. 680 (77), *689*
Spencer, C. F. 290 (118a), *306*
Spencer, L. A. 658 (47), *682*
Sperling, J. 733 (97, 98b), *772*
Spialter, L. 501 (91), *519*, 653 (28a–c), *682*
Spinelli, D. 632 (172), *647*
Spinks, J. W. T. 761 (183b), *774*
Spirikhin, L. V. 653 (23), *682*
Spiro, T. G. 471 (85), *480*
Spokes, G. N. 666 (5), *688*
Spratley, R. D. 61 (265), *83*
Sprecher, C. M. 343 (155), *373*
Sprecher, M. 220, 222, 223 (134), *232*, 555 (156), *582*, 863, 911 (75), *915*
Sprenger, A. Q. 471 (85), *480*
Sprinkle, M. R. 156, 157 (94), *160*
Sprinzak, Y. 156 (87), *160*
Sprung, J. L. 110 (7), *126*
Squire, R. H. 340 (142), *372*, 393 (72), *414*
Squires, T. G. 296 (175), *308*
Srinivasan, R. 350 (174), *373*
Srinivasan, T. K. K. 50 (198), 51, 52 (208), *82*, 86, 87 (3), *94*
Srivastava, R. B. 267 (39), *277*
Srnic, T. 174 (56), *197*
Staab, H. A. 299 (204), *308*, 594, 596, 607 (45b), *644*
Stallings, M. D. 432, 433, 435 (21), *457*
Stangl, H. 529 (46), *579*
Stanley, J. P. 432 (20), 437 (56), 442–445, 453 (20), *457*
Stannett, V. 421 (25), *427*
Stansburg, H. A., Jr. 743 (150), *773*
Starcher, P. S. 529 (43), 564 (43, 207), *579*, *583*

Starikova, Z. A. 382 (38a), *413*
Starostin, E. K. 765 (200), *775*
Starostina, T. J. 817 (42), *827*
Stary, F. E. 501, 503, 507 (101), *519*
Staschewski, D. 587 (1), *643*
Staudinger, H. 831, 858 (11), *913*
Stauff, J. 891 (151a, 151b), *917*
Stautzenberger, A. L. 191 (139), *199*
Steed, R. F. 151 (81), *159*
Steffan, G. 525 (14), *578*
Steg, R. 405 (116), *415*
Steichen, D. S. 210 (77), *231*, 446 (142), *459*
Stein, H. P. 557 (171), *582*, 599, 625 (75), *645*, 780 (12b), *804*
Steinberg, H. 211 (83), *231*, 326 (87), *371*
Steinberg, M. 608 (98), *645*
Steinmetz, A. 663 (51), *682*
Steinmetz, M. G. 844 (17c), *913*
Steinmetzer, H.-C. 667, 684 (15), *688*, 832, 839 (157), 843 (253), 844 (18c), 847 (18c, 33), 855 (253), 882 (18c), 890 (157), 891 (154, 155b, 155c), 892 (155b, 155c, 157), 893 (18c, 161), 900 (157), 902 (18c, 155c, 188), 903 (188), *913*, *914*, *917–919*
Stempel, K. 884 (17b), *913*
Stener, A. 88, 91 (78), *95*, 401 (98), *414*
Stephans, E. R. 110 (6), *126*
Stephens, E. R. 677 (61, 62), *689*
Stephens, J. F. 381 (28), *412*, 783–787 (32), *804*
Stephenson, A. J. 157 (89), *160*
Stephenson, L. M. 61, 62 (262), *83*, 168 (32), *197*, 209 (60), 218 (60, 128), 220 (60), *230*, *232*, 588 (11d), *643*, 848, 854 (37), *914*
Sternglanz, H. 397 (92), *414*
Sterns, M. 313 (20), *370*, 393 (73), 394 (74), *414*
Stevens, B. 782 (22), *804*
Stevens, E. D. 55 (219), *82*
Stevens, H. C. 281 (44), *305*
Stevens, R. M. 36 (94), *79*, 87, 88 (39), *95*
Stevenson, R. 226 (178), *233*
Stewart, O. J. 261, 266, 267, 269 (9b), *276*
Stief, L. J. 720 (28), *771*
Stierle, D. B. 314 (21), *370*
Stiggall, D. L. 555, 556, 575 (159), *582*, 831 (6a), *913*
Stille, J. K. 292 149c), *307*, 536, 537 (72), *580*
Stiller, K. 873 (110), *916*
Stock, L. M. 267 (37), *277*
Stockmayer, W. H. 260 (4), *276*
Stockmayer, W. T. 280 (9), *304*
Stoll, M. 295 (165), *307*
Stomberg, R. 465 (46, 49), 469 (46–49), *480*
Storey, B. T. 262 (18), *276*, 553 (142), *581*
Storey, P. M. 757, 758 (171b), *774*
Story, P. R. 228 (216), *234*, 632 (175), 633 (175, 178), *647*, 736 (108), *772*
Story, R. S. 632, 633 (176), *647*
Stotter, P. L. 303 (233b), *309*
Stoute, V. 560, 561 (195), 562 (202, 203), *583*
Strachan, W. M. J. 249 (55), *257*
Strain, F. 281 (44), *305*
Stratford, M. J. W. 651, 654 (6c), *681*
Strating, J. 119 (25), *127*, 210 (65b), 226 (65b, 181), 227 (65b), *230*, *233*, 281 (37), 297 (180a), *305*, *308*, 385 (43), *413*, 592 (30), *644*, 839, 850, 859, 860 (42), *914*
Strauss, H. L. 388 (46), *413*
Strausz, O. P. 538 (77a, 77b), *580*
Stray, F. E. 652, 653 (19b, 19c), *681*
Streitweiser, A., Jr. 437 (57), *457*
Stretzmann, P. 464, 465 (5), *479*
Strickler, H. 555 (154), *582*
Striefler, B. 763 (193), *775*
Strigina, L. I. 312 (8), *370*
Stringham, R. S. 247 (45), *257*, 487, 495 (35), *518*
Strom, E. T. 340 (144), *372*, 437, 438, 440, 454 (60), *457*
Strong, J. D. 471 (85), *480*
Stroud, S. G. 298 (195), *308*
Struchkov, Yu. T. 396, 397 (87b, 87c), *414*
Struck, R. F. 397 (89), 398 (93), *414*
Strukul, G. 475, 476 (142), *482*
Stull, D. R. 672 (44), *689*
Stumpf, W. 290 (126), *306*
Stuurman, J. 290 (123), *306*
Stwalley, W. C. 21 (44), *79*
Style, D. W. G. 98 (14), *104*
Subbaratnam, N. R. 492 (62), *518*
Subbotina, I. V. 180 (81), *198*
Sud, K. V. 792 (51), *805*
Suda, H. 167 (29), *197*
Suenram, R. D. 13, 61, 62, 69, 71 (33), *78*
Sugawara, T. 303 (238), *309*, 596, 635 (50), *644*
Sugimoto, A. 343, 344 (156), *373*
Sugimoto, T. 876 (123), *916*
Sugiyama, N. 785 (35), *804*
Sukhov, F. F. 764 (197), *775*
Sukhov, V. D. 802 (108), *806*
Sul'din, B. V. 144 (64), 157 (91, 92), *159*, *160*
Sullivan, G. A. 178 (70), *197*, 478 (154), *482*
Sulton, H. C. 452 (188), *460*
Sum, F. W. 189 (124), *199*
Summers, A. J. 99 (21), *104*
Summers, B. 287 (83), *305*
Summers, N. L. 48 (181), *81*
Summers, R. 245 (36), *257*, 497 (74), *519*
Summerville, D. A. 464, 465 (12, 14), 471, 475 (12), *479*

Sumerville, D. D. 2 (21), *78*
Sunami, M. 226 (185b), *233*
Sundararaman, P. 653 (24c), *682*
Sundberg, R. J. 540, 541 (88), *580*
Sundralingam, A. 222 (143), *232*, 555, 556 (158), *582*
Sung, H.-N. 134 (20), *158*
Suokas, E. 670 (57), *683*
Surgenor, D. M. 168 (36, 37), *197*
Surgenov, D. M. 299 (199a), *308*
Suschitzky, H. 876 (126), *916*
Sushunov, V. A. 808 (2a), 821 (56), *826*, *828*
Suslick, K. S. 471 (82), *480*
Sutherland, G. B. B. M. 86–88 (4), *94*, 376 (1), *412*
Sutton, L. E. 377 (17), *412*
Suzuki, H. 475 (140, 141), *482*, 826 (82), *828*
Suzuki, J. 228 (210, 211), *234*
Suzuki, K. 68 (299), *83*, 125 (39), *127*, 611 (106), 617 (106, 128b), *646*, 791 (48), *805*
Suzuki, M. 358 (192a, 192b, 193), 363 (192a, 192b), 364 (193), *373*
Suzuki, N. 859 (61), 880 (141), *914*, *916*
Suzuki, S. 560 (193), *583*, (216), *648*
Suzuki, Y. 210 (76b), *231*, 328 (94), *371*
Svejda, P. 732 (84), *772*
Svenson, J. 313 (15), *370*
Sviridonov, V. N. 312 (8), *370*
Svitych, R. B. 148, 149 (70), *159*
Swain, C. G. 260 (4), *276*, 280 (9), *304*
Swain, H. A. 134 (20), *158*
Swallow, A. J. 761 (183c), *774*
Swallow, J. C. 465, 469 (51), *480*
Swan, G. A. 286 (78), *305*
Swann, G. A. 271 (52), *277*
Swansiger, W. A. 653 (28c), *682*
Swelin, A. 563 (206), *583*
Swern, D. 2 (7), *78*, 88 (90, 99, 107), 92 (90, 99), 93 (99, 107), *96*, 134 (21b, 24, 28, 30), *158*, 236 (4), *256*, 259–261 (2a), *276*, 280 (18a, 18b), 282 (52), 287 (84), 288 (98, 101), 290 (84, 122, 132), 291 (136), 293 (84), 296 (171), 299 (201), *304–308*, 314 (26), *370*, 403 (102, 103), 405 (106), *414*, 525 (5a, 6), 546 (101), *578*, *580*, 593 (33), 596 (52), *644*, 732 (80c), *772*, 788 (40), 794 (40, 59), 795 (40), *805*
Swift, H. E. 166 (25), *196*
Swift, J. D. 792 (53), *805*
Symes, W. R. 475, 476 (143), *482*
Symons, M. C. R. 740 (123), *773*
Syroezhko, A. M. 137 (43), *158*
Sysak, P. K. 588 (11d), *643*
Syshunov, V. A. 808, 810 (3), *826*
Szendrey, L. 753, 755 (166), *774*
Szmant, H. H. 312 (12), *370*
Szöcs, F. 800 (97), 801 (97, 104), *806*
Szwarc, M. 98 (13), 99 (13, 26), *104*, 263 (24b), *276*, 608 (97–101), 609 (102), 614 (116c), 616 (99), *645*, *646*

Tabushi, A. 567 (217), *583*
Tabushi, I. 525 (15), 539 (81), *578*, *580*
Tachi, Y. 389 (59), *413*
Tada, M. 668 (56a, 56b), *682*
Taft, R. W. 561 (248), *584*
Tahib, S. 724 (53), *771*
Taillefer, R. J. 501 (104, 105), 502 (105), *519*, 653 (22), *681*
Taipale, J. 282 (53), *305*
Tait, J. C. 488 (43), *518*
Takagi, K. 719 (26), 722 (46), 739 (26), *770*, *771*
Takagi, T. 425 (69), *428*
Takahashi, S. 466, 474 (123), *481*
Takahashi, T. 226 (178), *233*, 312 (4), *369*, 668 (56a, 56b), *682*
Takahashi, Y. 211 (82), *231*, 872 (105b, 106b), *915*
Takai, K. 469, 471 (55), *480*
Takamizawa, A. 397 (90, 91), *414*
Takata, T. 443, 444 (120), 445 (120, 126a, 126c), 446 (120, 126c), 453 (120), *459*
Takayama, H. 324 (75), *371*, 384 (38g), *413*
Takayama, K. 335 (116), *372*, 491 (58), *518*, 833, 836, 837, 888–890, 900 (233), *919*
Takeda, K. 721 (41), *771*
Takemoto, K. 546–548 (114b), *581*
Takenaka, A. 393 (70, 71), *414*, 860 (62b), *914*
Takeshita, H. 211 (88a–c), 226 (170), 228 (217, 218), *231*, *233*, *234*, 343 (157), *373*, 862 (68a, 68b), 873 (108), 910 (68a, 68b), *915*
Takeuchi, C. 668, 671 (24), *688*
Takezaki, Y. 668, 671 (24), *688*, 731 (76), *772*
Takizawa, T. 842 (249), *919*
Takula, A. 435, 438 (38), *457*
Takyu, C. 445, 446, 454, 456 (124), *459*
Tal, D. 498 (75), *519*, 671–673, 675 (59, 60), *683*
Tal, Y. 55 (222, 224), *82*
Tallman, R. L. 377 (22), *412*
Tamagaki, S. 226 (182, 183a, 185c), *233*
Tamelen, E. E. van 320 (56), 369 (211), *371*, *374*
Tamoto, K. 326 (88), *371*, 446 (148), *459*
Tamura, Y. 272 (60), *277*
Tanahashi, Y. 312 (4), *369*
Tanaka, K. 167 (28), *196*, 538 (77d), *580*
Tanaka, S. 182 (86), *198*, 343, 344 (156), *373*
Tanaka, T. 48 (183), *81*
Tang, R. 341 (149), *372*, 446 (149), *459*, 465, 469 (53), 470 (69), 471 (53), 472 (107), *480*, *481*

Tang, T. H. 55 (224), *82*
Tangpoonpholvivat, R. 430, 437, 440, 443, 445, 450 (8), *456*
Tani, K. 183 (92), *198*
Taniguchi, H. 758 (176b), *774*
Tanimoto, K. 167 (29), *197*
Tanimoto, Y. 740 (126), *773*
Tanino, H. 291 (141), *307*
Tannenbaum, H. P. 451 (174), *460*
Tanner, D. D. 280 (33), *304*
Tappen, W. A. 244 (30), *257*, 839, 883, 886 (145c), *916*
Taqui Khan, M. M. 471 (89), *481*
Tarabarina, A. P. 142, 147, 150, 151 (56), *159*
Tarantini, G. 36 (113), *80*
Tarbell, D. S. 298 (194b), *308*
Tardy, D. C. 99 (27), *104*, 676 (58), *689*
Tasevski, M. 68 (297), *83*, 88, 92, 93 (103), *96*, 523, 524, 530, 531 (3), *578*
Tatarenko, A. A. 802 (107), *806*
Tatsumi, K. 466, 474 (122), *481*
Tatsuno, S. 466, 474 (120), *481*
Tatsuno, Y. 466 (102, 125), 467 (27, 28), 472 (102), 474 (27, 125), 475, 476, 478 (27), *479*, *481*
Taube, H. 206 (20), *229*, 474 (136), *482*, 588 (14a), *643*
Tavale, S. S. 405 (110), *414*
Taylor, E. G. 369 (211), *374*
Taylor, I. F. 466, 473 (110), *481*
Taylor, J. W. 603, 609, 610 (88), 614, (88, 122), 615 (88), *645*, *646*, 780 (11a), *804*
Taylor, K. 780 (12b), *804*
Taylor, K. G. 557, 558 (177), *582*, 628 (156), *647*
Teif, Zh. D. 280 (27, 28b), *304*
Templeton, D. H. 86, 88 (7), *94*, 376 (5), *412*
Teng, J. I. 440, 449 (81), *459*
Terakawa, A. 762 (188b), *775*
Teramoto, M. 653 (27), *682*
Teranishi, H. 653 (27), *682*
Teranishi, S. 188, 192 (85), *198*, 534 (61), *579*
Terao, S. 227 (202), *234*, 834, 835, 859, 863, 910 (60), *914*
Terashima, S. 182 (89), *198*
Terent'ev, V. A. 142, 144 (53), 145, 146 (66), 149 (53, 66), 150, 153, 154 (66), *159*
Terman, L. M. 263 (26), *276*
Terry, N. W. 465, 472 (95), *481*
Testi, R. 186 (107), *198*
Tewes, E. T. 874 (116), *916*
Tezuka, M. 437 (58), 440 (75), *457*, *459*, 879 (137), *916*
Tezuka, T. 344 (160), *373*, 446 (127, 128), 449 (127), *459*, 588 (9b, 9c), *643*, 730 (69), *771*, 865 (88), *915*
Tezuka, Y. 541 (89), *580*
Thayer, A. L. 205 (11, 13, 15), 207 (35b), *229*, *230*, 838, 842, 847 (31), 849 (40), 884, 886, 887, 895, 896, 898, 899 (31), *914*
Thiel, W. 65 (282), *83*, 209, 212 (64), *230*
Thierbach, D. 86, 88 (18), *94*
Thomas, A. 487 (38), *518*
Thomas, A. F. 224 (157c), *233*
Thomas, J. M. 788, 791 (39), *805*
Thomas, J. R. 486 (31), 490 (51), *518*
Thomas, M. J. 453, 456 (191), *460*
Thomas, S. E. 501 (104, 105), 502 (105), *519*, 653 (22), *681*
Thompson, J. A. 680 (83), *689*
Thompson, P. G. 246 (38), *257*, 496 (69), *519*
Thompson, Q. E. 206 (25), *229*, 651 (14), *681*
Thompson, R. B. 831 (5), *913*
Thomson, D. W. 525 (18), *579*
Thornton, E. K. 883 (147), *917*
Thornton, E. R. 883 (147), *917*
Thorpe, M. C. 398 (93), *414*
Thulstrup, E. W. 21, 58 (40), *78*
Thunemann, K. H. 24, 45, 58, 59 (53), *79*
Thynne, J. C. J. 98 (12), *104*
Tidwell, T. T. 425 (62), *427*
Timm, U. 536 (75), *580*
Timmons, R. B. 675 (52), *689*
Timmons, R. J. 320 (56), *371*
Timms, N. 440 (76), *459*
Timofeeva, G. Ya. 134 (23), *158*
Timofeeva, T. N. 137 (43), *158*
Timoschtschuck, T. A. 465, 470 (65), *480*
Tinsley, S. W. 529, 564 (43), *579*
Tischenko, I. G. 151 (74), *159*
Tischenko, I. G. 137 (44), *158*
Titaka, Y. 432 (31), *457*
Tobolsky, A. V. 2 (3), *78*, 280 (21), *304*, 800 (96), *806*
Tochino, Y. 397 (90, 91), *414*
Todd, H. D. 87 (52), *95*
Todd, J. D. 36 (104), *80*
Todesco, P. E. 296 (173b), *308*, 451 (176b), *460*, 539 (82), 542, 543 (92), *580*
Tokumaru, K. 68 (299), *83*, 211 (84), *231*, 546–548 (114c), *581*, 596 (59), 619 (59, 133), *644*, *646*, 716, 718 (11), 740 (125, 126, 129), 742 (138c), *770*, *773*
Tolks, A. M. 801, 802 (80), *805*
Tolschow, J. E. 470 (68), *480*
Tolstikov, G. A. 120, 121 (32), *127*, 178, 179 (73), 184 (93, 94, 96–98), 185 (96, 97), 186 (110, 111, 114), 187 (118), *197–199*
Tomioka, H. 469, 471 (55), *480*
Tomita, H. 451, 454, 456 (170), *460*, 475, 476 (147), *482*
Tomizava, K. 562 (200), *583*
Tomizawa, K. 252 (64), *258*, 528 (39), 535 (67), *579*, *580*, 718 (24), 719 (25, 27), 721

(43), 722 (43, 46), 723 (27), 724 (27, 49), 739, 740 (24), 744 (25), 745 (24, 25, 151–153), 746, 747 (153), *770*, *771*, *773*, *774*
Tonelatto, U. 538 (80b), *580*
Tong, S. B. 248, 250 (49), 251 (62), *257*, *258*
Tontapanish, N. 834 (234), *919*
Toole, J. 597 (61), 618 (141), 622–624 (141–143), *644*, *646*, *647*
Top, S. 440 (77), *459*
Toppet, S. 400 (95), *414*
Toptygin, D. Ya. 800 (87, 88, 90, 94, 95), 803 (111), *806*
Toriyama, K. 617 (129a, 129b), *646*
Tornstrom, P. K. 262 (12), 266 (12, 42), 267 (39), 268 (12, 42), *276*, *277*, 780 (14), *804*
Torossarelli, L. 622–624 (142, 143), *646*, *647*
Torre, G. 570, 572 (229, 230), *583*
Torsellini, P. 36 (113), *80*
Toscano, V. G. 613, 614 (113), *646*
Toth, L. M. 712, 731–734 (3), *770*
Towns, R. L. R. 394 (77), *414*
Townsend, J. M. 465, 470 (63), *480*
Townsend, M. G. 740 (123), *773*
Trachtman, M. 68 (293), *83*
Traylor, T. G. 240 (19), 247 (19, 48), 252 (48), *257*, 265 (33), *277*, 486 (27–30), 499 (81–84), 500 (81–83), 501 (84), *518*, *519*, 553 (143), *581*, 622 (136, 138a, 138b), *646*
Traynham, J. G. 290 (127), *306*
Treely, I. G. 792 (53), *805*
Tremper, A. W. 874 (120), *916*
Treppendahl, S. 873 (111), *916*
Trethewey, K. R. 208, 222 (46), *230*
Trevino, S. F. 49, 51 (193), *81*, 86, 88 (11), *94*
Trifilieff, E. 670 (58a, 58b), *683*
Trindle, C. 61, 62 (244), *82*, 893 (163e), *917*
Trocino, R. J. 594, 605, 606 (48), 607 (96b), 608, 616 (48), 620 (48, 96b), *644*, *645*
Trofimov, V. I. 765 (201), *775*
Trofimova, N. F. 422 (29), *427*
Trossarelli, E. 597 (61), *644*
Trost, B. M. 560 (197), *583*
Trotter, J. 388 (53), *413*
Troxler, R. F. 871 (102), *915*
Trozzolo, A. M. 2 (19), *78*, 207 (41), *230*, 742 (138b), *773*
Trunov, V. K. 382 (38a), *413*
Tsai, R. S.-C. 110 (9), *126*
Tse, M. 733 (95), *772*
Tsinker, I. 396, 397 (87b), *414*
Tsuchiya, T. 336 (126), *372*
Tsuge, O. 865 (89), *915*
Tsuji, J. 188 (121), *199*
Tsuji, Y. 491 (58), *518*
Tsukamoto, Y. 566 (211), *583*
Tsunetsugu, J. 219 (130), *232*
Tsuruta, T. 423 (36), *427*
Tsutsui, M. 468 (31), *479*
Tsutsumi, S. 293 (147), *307*, 425 (63), *427*, 721 (39), *771*
Tsuyuki, T. 226 (178), *233*, 668 (56a, 56b), *682*
Tsvetkov, V. G. 821 (57), *828*
Tudor, R. 812, 813 (24), *827*
Tullen, D. L. 546 (125), *581*
Tumolo, A. L. 501, 503 (85), *519*, 652 (16), *681*
Turchi, I. J. 571 (232), *583*
Turner, A. B. 861 (67), *915*
Turner, J. A. 213, 214 (106), 215, 219 (113), 222 (147), *231*, *232*, 356 (187, 188, 190), *373*, 555 (155), *582*
Turner, L. 297 (184), *308*
Turner, R. B. (215), *648*
Turovskii, A. A. 88, 89 (66), 90 (66, 73), *95*, 737 (113), *773*
Turovskii, N. A. 88, 89 (66), 90 (66, 73), *95*
Turro, N. J. 2, 21 (15), 60 (239), *78*, *82*, 226 (174, 175, 180), 227 (194), *233*, *234*, 347 (167, 168a, 168b), 348 (168a, 168b), *373*, 667 (15), 684 (15, 92), 685 (102), *688*, *690*, 781 (21), 782 (26, 27), *804*, 831 (12), 832, 833, 836, 838 (156), 843 (12, 201), 844 (18c), 846 (12), 847 (18c, 33), 857 (12), 863 (70, 72), 873 (113, 114), 882 (18c), 885, 887, 888 (156), 891 (149, 152c, 155a–c), 892 (155a–c, 156), 893 (18c, 165), 894 (171), 898 (156), 899 (156), 902 (18c, 155c, 188), 903 (188), 905 (197), 906 (201), 908 (197), *913–918*
Tursch, B. 384 (38f), *413*
Tyrrell, J. 48 (181), *81*

Uchida, Y. 594 (41), 626 (154), 628 (41, 154, 165), 629 (41), 631 (165), *644*, *647*, 780 (12b), *804*
Uchiyama, M. 792 (52), *805*
Uda, H. 912 (219), *919*
Ueyama, Y. 865 (89), *915*
Uff, B. C. 221 (138), *232*, 522, 525, 546 (1d), *578*
Ugelstad, J. 527 (35), *579*
Ugo, R. 466, 468 (30), 474 (30, 127, 133), 475 (133), *479*, *481*
Ullrich, V. 295 (155), *307*, 464 (2), *479*, 536 (68), *580*
Ulman, A. 664, 668 (52a), *682*
Umasaka, T. 421 (24), *427*
Umbreit, M. A. 189 (123), *199*, 833, 887 (231), *919*
Unger, L. R. 205 (10), *229*
Ungerman, T. 441 (105), 442, 445, 453 (105, 115), *459*

Ungermann, T. 430, 437, 440, 443, 445, 450 (8), *456*, 570 (225), *583*
Unwin, J. R. 419 (8), *426*
Unwin, T. 419 (5), *426*
Upton, R. M. 470 (74), *480*
Urry, W. H. 628 (161), *647*
Uselman, W. M. 677 (59), *689*
Ushakova, T. B. 136 (45), 139, 142, 149 (51), *159*
Usova, L. G. 156, 157 (90), *160*
Ustanov, Kh. V. 423 (46), *427*
Utaka, M. 435, 438 (38), *457*
Utsumi, K. 617 (128b), *646*

Valega, T. M. 625, 628 (148), *647*
Valenti, P. C. 205 (15), *229*
Valentine, D. 466, 473 (112), *481*
Valentine, J. 466, 473 (112), *481*
Valentine, J. S. 430 (9), 432 (16b), 438 (67), 441 (67, 89, 92), 442–445 (117), 448 (67, 92, 156), 452 (156), 455 (205), *456–461*, 464, 465 (6), 471 (85), 473 (6), *479*, *480*
Van Alpher, J. 651 (8), *681*
Van Den Heuvel, C. J. M. 326 (87), *371*
Van den Heuvel, C. J. M. 211 (83), *231*
Van der Ent, A. 473 (118), *481*, 826 (79), *828*
Vanderkooi, N., Jr. 735 (103), *772*
Van Doorn, J. A. 466, 467, 474 (23), *479*
Van Dorn, J. A. 816 (64), *828*
Vane, F. M. 557 (171), *582*, 599, 625 (75), *645*, 780 (12b), *804*
Van Gaal, H. 473 (118), *481*
Van Ginkel, G. 446 (136), *459*
Vanino, L. 282 (51), *305*
Van Leeuwen, J. W. 452, 456 (190), *460*
Van Meerssche, M. 384 (38f), 400 (95), *413*, *414*
Van Sickle, D. E. 280 (4), *304*
Van Sicle, D. E. 796 (71), *805*
Van Stevenick, A. W. 554, 561 (147), *582*
Varkony, H. 503 (108), *519*
Varkony, T. H. 657 (42, 43), 658 (42, 43, 45, 46a–c), 660 (42, 46a), 664 (42), 671 (46a–c), 673 (42), *682*
Varlamov, V. T. 145, 147, 149 (67), *159*
Varnett, G. 684 (96), *690*
Vartapetyan, O. A. 144, 149 (60), *159*
Vartapotyan, O. A. 144 (62), *159*
Vary, M. W. 407 (129), *415*
Vasil'ev, R. F. 904, 905 (192), *918*
Vasileyskaya, N. S. 811 (20b), *827*
Vaska, L. 2 (22), *78*, 377 (23), *412*, 464 (4), 465 (4, 95, 95), 467 (4), 472 (95, 95), 473 (4, 109), *479*, *481*
Vaskuil, W. 678 (69), *689*
Vaughan, W. E. 164 (10), *196*
Vedejs, E. 470 (68), *480*
Veillard, A. 36 (93, 116), 38 (116), 53, 54 (93), *79*, *80*, 87, 88, 90 (44), *95*
Vemura, S. 825 (74), *828*
Vendenberg, E. J. 798 (76), *805*
Venier, C. G. 296 (175), *308*
Venugopalan, M. 484 (1a), *517*
Venura, S. (62), *828*
Verbruggen, A. 400 (95), *414*
Verderame, F. D. 88, 94 (109), *96*, 405 (112), *415*
Verdin, D. 763 (192a, 192b), 765 (192b), *775*
Vergani, C. 36, 53, 57 (92), *79*
Verhoek, F. H. 812 (22), *827*
Verhoeven, T. R. 162, 179–181 (5), 182 (86), *196*, *198*, 478 (152), *482*, 532, 533 (53), 534, 535 (53, 63), *579*, *580*
Vernin, G. 873 (111), *916*
Vernon, J. 529 (46), *579*
Veselov, V. Y. 441 (103), *459*
Veselov, V. Ya. 566 (213), *583*
Vestal, L. L. 628 (158b), *647*
Vestal, M. L. 48, 49 (185, 188), 67 (185), *81*
Vilhuber, H. G. 219 (130), *232*
Vilkas, M. 525 (16), *578*
Villa, A. E. 672 (42), *689*
Villiger, V. 280 (25, 31), 295 (160), *304*, *307*, 559 (181), *582*
Vincent, M. A. 894 (167), *917*
Vincent, S. 261, 262 (8), 268 (8, 43), *276*, *277*
Vinh, Q. L. 731 (74), *772*
Vinick, F. J. 333 (111, 114), *372*, 588, 592, 593 (15), *643*
Vinkovičová, M. 800, 801 (97), *806*
Vinogradov, A. N. 136, 137, 139–141, 143, 149 (35), *158*
Vinogradov, M. G. 564 (209), *583*
Visser, T. 249, 250 (58), *257*
Vivona, N. 632 (172), *647*
Vlengels, J. 678 (69), *689*
Vliet, N. P. van 223 (155c), *233*
Vocelle, D. 271 (51), *277*
Voelcker, G. 114 (17), *126*
Vogel, E. 213 (105), 216 (105, 124a), *231*, *232*, 324 (74), 325 (83), 330 (103, 104), 331 (103), 344 (159), 354 (103), *371–373*
Vogel, P. 215 (111, 112a), *232*, 353 (182), *373*
Volkova, Z. P. 722 (45), *771*
Volman, D. H. 720 (34), 732 (84), 740 (34), *771*, *772*
Volnov, I. I. 464, 465 (16), *479*
Voronekov, V. V. 139, 142, 149 (51), *159*
Voronenkov, V. V. 139 (52), *159*
Voronov, S. A. 425 (67), *428*
Vos, A. 385 (41), *413*, 844, 845, 892 (21a, 21b), *914*
Voss, A. 71 (311), *84*
Voss, R. 763 (194), *775*

Vuletic, N. 465 (37), 469 (37, 43), *479*, *480*
Vyazankin, N. S. 142, 147, 150, 151 (56), *159*
Vyshinskii, N. N. 144 (63, 64), 156 (90), 157 (90–92), *159*, *160*
Vysotskaya, N. A. 762 (189), *775*

Waal, D. J. A. de 826 (80), *828*
Wacek, A. 298 (194a), *308*
Wada, H. 263, 267, 269 (19), *276*, 597, 640 (64), *645*
Waddel, W. H. 227 (194), *234*
Wade, L. E., Jr. 611, 613 (110), *646*
Wadt, W. R. 61, 63 (254), *83*
Wagenaar, A. 226 (181), *233*
Wagman, D. D. 40 (139), *80*
Wagner, J. J. 396 (87a), *414*
Wagner, K. 272 (60), *277*
Wagner, R. D. 338 (136), *372*
Wagner, T. 114 (17), *126*
Wahl, A. C. 21 (43), 61 (261), *79*, *83*
Wahren, M. 574 (237), *584*
Waits, H. P. 557 (175), *582*, 628 (155), *647*, 780 (12b), *804*
Wakabayashi, T. 313 (15), *370*
Wakefield, B. J. 810 (12), *827*, 876 (126), *916*
Walch, S. P. 25, 33, 52 (81), *79*
Waligóra, B. 737 (110), *772*
Walker, E. C. 653 (24c), *682*
Walker, O. J. 712–714, 738 (1a), *770*
Walker, P. J. C. 472 (105), *481*
Walker, R. F. 99 (22), *104*
Walker, R. L. 267 (39), *277*
Walker, R. W. 669 (32, 33), *688*
Wallace, J. G. 290 (129b), *307*
Wallach, O. 213 (95), *231*
Waller, R. L. 275 (65), *277*, 759 (179), *774*
Walling, C. 136–138, 140, 141, 149, 150, 152 (39), *158*, 168 (34, 35), 173 (51), *197*, 253 (68), *258*, 271 (48, 53), *277*, 281 (45), 285 (70), 298 (193c), *305*, *308*, 546 (112a–c, 131), 547 (112a–c), 550 (131), 557 (175, 176), *581*, *582*, 602 (85a), 614 (118), 628 (155, 159), 636 (199), *645–648*, 716, 731, 741 (12), *770*, 780 (12b), *804*, 810 (10, 11), *827*, 895 (176), *918*
Walling, Ch. 791 (47), *805*
Walsh, A. D. 224 (160), *233*
Walsh, R. 40 (140), *80*, 97 (2), *102*, 666, 677 (7), *688*
Walter, D. W. 407 (141), *415*
Walton, D. J. 88, 90 (74), *95*, 382 (34), *413*
Walton, J. 429 (2), *456*
Wampler, J. E. 901, 902 (185a), *918*
Wamser, C. C. 226 (184), *233*
Wan, J. K. S. 732 (88), *772*
Wang, D. S. T. 800, 802 (92), *806*
Wannlund, J. 844 (17b), *913*
Ward, B. 487 (38), *518*
Ward, H. R. 241 (21), *257*
Ward, W. W. 844 (18y), *913*
Wardale, D. A. 223 (154b), *233*
Ware, W. R. 904, 905 (194), *918*
Warheit, A. C. 260 (7), *276*
Warner, R. J. 441 (100, 101), 453 (100), *459*
Wasserman, E. 206 (27), *230*
Wasserman, H. 782 (24, 25), 783 (25), *804*
Wasserman, H. H. 2, 65 (11), *78*, 207 (37–39), 210 (66e), 211 (85), 212, 214, 221 (66e), 227 (202–204), 229 (223a), *230*, *231*, *234*, 318 (45e), 324 (79), 325 (79, 82), 332 (108c), 333 (110–114), 342 (153), 347 (79), 348, 365 (153), 369 (212, 213), *370–374*, 588, 592, 593 (15), *643*, 834, 835, 859, 863 (60), 871 (98), 873 (110), 874 (120), 910 (60), 911 (214), *914–916*, *919*
Watanabe, H. 872 (106d), *915*
Watanabe, K. 228 (211, 212a, 212b), *234*
Watanabe, S. 740 (125), *773*
Waters, J. 614 (121), *646*
Waters, J. H. 791 (47), *805*
Waters, W. A. 2 (8), *78*, 206 (24), *229*, 528 (41), *579*
Watkins, C. J. 667, 684 (18), *688*, 844 (18s), *913*
Watson, C. G. 570, 572 (229), *583*
Watson, W. F. 423 (40), *427*
Watson, W. H. 571 (232), *583*
Watt, D. S. 755, 756 (169), *774*
Watts, G. B. 812, 813 (25), *827*
Webb, T. R. 185 (103), *198*
Weber, G. 902, 903 (187), *918*
Weber, L. 388 (48), *413*
Webrin, H. 340 (144), *372*
Wechsler, C. J. 471 (81), *480*
Weedon, A. C. 446 (145), *459*
Wegler, R. 290 (130), *307*
Wei, C. C. 684 (101), *690*, 831, 832, 859 (2), 908 (204), *913*, *918*
Weiberg, O. 290 (131), *307*
Weigert, W. 290 (128), *306*
Weigert, W. M. 2 (2), *78*
Weiler, L. 189 (124), *199*
Weinand, G. 424 (57–59), *427*
Weiner, S. 732 (83), *772*
Weingartshofer Olmos, A. 403 (105), *414*
Weinhold, F. 38 (133), *80*, 89 (56), *95*
Weininger, M. S. 466, 473 (110), *481*
Weinstein, J. 452 (187), *460*
Weir, N. A. 423 (37), *427*
Weis, C. 625, 628 (150), *647*
Weiss, F. 280 (15), 295 (167b), *304*, *307*
Weiss, J. 452 (185), *460*, 720 (32a, 32b), *771*
Weiss, L. B. 367 (209), *374*

Weiss, R. 187 (119), *199*, 383 (38b), *413*, 465 (41, 42, 61, 76), 469 (41, 42, 56, 61), 470 (76), 475 (144, 155), 476 (144), *480*, *482*, 812 (65), *828*
Weissberger, E. 466, 473 (116), *481*
Welch, J. 474 (138), *482*
Welch, J. E. 131 (4), *157*
Welge, K. H. 60 (237), *82*
Weller, A. 895 (173), *917*
Wells, R. J. 120 (31), *127*, 313 (17), *370*
Wendt, H. R. 58, 59 (232), *82*
Wenner, G. 632 (173), *647*
Wentworth, R. L. 2 (1), *78*, 86 (2), *94*, 712 (6), *770*
Werme, L. O. 44, 45 (158), *81*
Werstiuk, E. 219 (130), *232*
Wessling, D. 292 (143), *307*
West, P. R. 757, 758 (171b), *774*
West, R. 142 (57), *159*
Westlake, D. 465 , 469 (54), *480*
Westland, A. D. 469 (58), *480*
Wetmore, R. W. 61 (249), *83*
Wettach, R. H. 546 (103), *580*
Wexler, S. 206 (22, 23), 211, 217 (87a), *229*, *231*, 588 (13a–c), *643*, 785 (34), *804*
Wheland, R. 854 (55), *914*
Whilten, D. G. 446 (147), *459*
White, D. M. 152 (76), *159*
White, E. H. 152 (77), *159*, 632 (169), *647*, 667 (18), 684 (18, 101), *688*, *690*, 831, 832 (2), 833 (231), 844 (14, 17c, 18s), 859 (2, 61), 887 (231), 896 (183a, 183b), 908 (204), *913*, *914*, *918*, *919*
White, H. M. 501 (87, 95), 509 (87), *519*, 651 (7c), 653 (25), 654 (7c), *681*, *682*
White, J. D. 367 (208), *374*
White, L. S. 350 (174), *373*
White, M. J. D. 113 (14a), *126*
White, R. E. 471 (83), *480*
White, R. F. M. 814 (30), *827*
White, R. W. 297 (186b), *308*
Whitebeck, M. R. 491 (57), *518*
Whited, E. A. 228 (216), *234*
Whitehurst, D. D. 292 (149c), *307*, 536, 537 (72), *580*
Whitesides, G. M. 470 (71), *480*
Whitham, G. H. 181, 192 (85), *198*, 525 (17), 526 (24), 529 (44), 532 (24), 534 (62), *579*
Whiting, M. C. 501, 513, 514 (89), *519*, 656 (35), *682*
Whitman, D. R. 38 (135), *80*
Whitsal, B. L. 598, 617 (70), *645*
Whitsel, B. L. 88, 91 (84), *95*, 406, 407 (124), *415*, 743 (146), *773*
Whittaker, R. A. 876 (126), *916*
Whyman, R. 469 (59), *480*
Wibaut, J. P. 666 (55a, 55b), *682*
Wiberg, K. B. 2 (9), *78*, 396 (84), *414*, 568, 569 (223a), *583*
Wiecko, J. 684 (101), *690*, 831, 832, 859 (2), 908 (204), *913*, *918*
Wiede, O. F. 469 (45), *480*
Wiegand, D. 213 (92), *231*, 350 (175), *373*
Wieghardt, K. 465, 469 (35), *479*
Wieland, H. 238 (15), *257*, 280 (23, 24), *304*
Wieland, P. 312 (2), *369*
Wielesek, R. A. 617 (127), *646*
Wiemiorowski, M. 113 (16), *126*
Wieringa, J. H. 119 (25), *127*, 210, 226, 227 (65b), *230*, 385 (43), *413*, 592 (30), *644*, 839 (42), 844, 845 (21a), 850, 859, 860 (42), 892 (21a), *914*
Wijnen, M. H. J. 716 (13a–c), *770*
Wild, G. L. E. 712–714, 738 (1a), *770*
Wildes, P. D. 684 (101), *690*, 831, 832 (2), 844 (17c), 859 (2), *913*
Wiles, D. M. 731 (73), *772*
Wilke, G. 474 (119), *481*
Wilkerson, C. 753 (163b), *774*
Wilkinson, G. 587 (6), *643*, 809 (85), 810 (13), *827*, *828*
Willadsen, T. 387 (54), *413*
Willhalm, B. 211 (87b), 214 (109), 217 (87b), 222 (151), *231*, *232*, 555 (154), *582*
Williams, D. R. 465, 470 (63), *480*
Williams, F. W. 685 (105–107, 109, 110), 686 (110), 687 (105), *690*, 847, 864, 894 (32), *914*
Williams, G. H. 800 (78), *805*
Williams, H. R. 568 (219), *583*, 590 (20), *644*
Williams, J. C., Jr. 209, 220 (62), *230*
Williams, J. O. 788, 791 (39), *805*
Williams, J. R. 205 (10), *229*
Williams, J. W. 287 (88), *305*
Williams, L. G. 205 (18), *229*
Williams, P. H. 292 (142a), *307*
Williams, S. B. 286 (74), *305*
Williams, S. H. 419 (6, 7), *426*
Williams, T. E. 910 (211d), *919*
Williamson, D. G. 656 (37), *682*
Williamson, J. B. 564 (210), *583*
Willis, C. L. 88, 90, 92, 93 (102), *96*, 187 (115), *199*, 530 (51), *579*
Wilmarth, W. K. 720 (36), *771*
Wilms, K. 330, 331, 354 (103), *372*
Wilshire, J. 430, 431, 443 (7), *456*
Wilson, A. F. 780 (10), *804*
Wilson, C. E. 832 (158a, 158b), 833 (158b), 884, 886 (158a, 158b), 887 (158a), 892 (158a, 158b), *917*
Wilson, C. W., Jr. 24 (61), *79*
Wilson, E. B. 881 (142c), *916*
Wilson, E. R. 130, 131 (1b), *157*, 568 (222), *583*

Wilson, J. M. 837 (242a, 242b), *919*
Wilson, R. A. L. 290 (120), *306*
Wilson, R. L. 449 (164), *460*
Wilson, R. M. 339 (140, 141), 340 (142, 143), 362 (141), *372*, 393 (72), *414*
Wilson, T. 205 (17), 207 (40), 208 (45), 227 (195, 196, 199), *229*, *230*, *234*, 667, 675 (17), 684 (17, 100), *688*, *690*, 832 (221), 834, 835 (160), 836 (240), 837 (224), 844 (18g, 18t), 883 (221), 885 (168, 221, 224), 886 (221), 887 (224), 888 (240), 889 (160, 240), 891 (150a), 893 (160), 894 (150a, 168, 169), 895 (178), 898 (221), 900 (160, 221), 902 (18g), 903 (168), 905 (18g, 196a), *913*, *917–919*
Wilt, M. H. 440 (78), *459*
Wilterdink, R. J. 193 (148), *199*
Windaus, A. 312 (1), *369*
Winer, A. M. 677 (60), 678 (65, 67), 679 (65), *689*
Winkelmann, E. 436, 437 (51), *457*
Winkler, D. E. 164 (9), *196*
Winkler, T. 737 (112), *773*
Winnewisser, G. 37, 67 (128), *80*
Winnewisser, M. 37, 67 (128), *80*
Winnik, M. A. 560, 561 (195), 562 (202, 203), *583*
Winstein, S. 552 (140), 553, 561 (144), *581*, *582*
Winter, N. W. 36 (99, 101), 37, 38, 53, 55, 56 (101), 68 (301), *80*, *84*, 87–90 (46), *95*
Winter, W. 465, 470 (67), *480*
Wipprecht, V. 679, 680 (76), *689*
Wistuba, E. 554 (145), *582*
Witham, G. H. 289 (113), *306*
Witkop, B. 872 (106d), *915*
Witnauer, L. P. 88, 93 (107), *96*, 134 (24), *158*, 288 (98), *306*, 403 (103), *414*
Wojtowicz, J. A. 484 (7), *517*
Wolak, R. 194 (154), *199*, 533 (55), *579*
Wolak, R. P., Jr. 437 (59), *457*
Wolber, G. J. 93 (106), *96*, 530 (52), *579*
Wolf, J. F. 341 (149), *372*, 446 (149), *459*
Wolf, R. 131 (8), *158*, 594, 605–607 (47), *644*
Wolf, R. A. 594, 605, 606 (48), 607 (96b), 608, 616 (48), 620 (48, 96b), *644*, *645*
Wolf, S. F. 293 (148), *307*
Wolfe, S. 89 (55), *95*
Wolff, M. E. 271 (51), *277*
Wolff, M. S. 873 (110), *916*
Wolters, E. 302 (232), *309*
Wong, C. S. 474 (134), *482*
Wong, E. 837 (242a, 242b), *919*
Wong, J. L. 751 (159a, 159b), 752 (161), *774*
Wong, P. C. 237 (13), *257*
Wong, S. 732 (88), *772*
Wood, D. E. 617 (128c), *646*
Wood, G. W. 290 (118c), *306*
Wood, R. J. 761 (183b), *774*
Woodbridge, D. T. 194 (149), *199*, 651 (13), *681*
Woodcock, D. J. 632 (169), *647*
Woodgate, P. D. 546 (107), *581*
Woodward, A. E. 423 (38, 39), *427*
Woodward, R. B. 613 (111), *646*
Wörther, H. 896 (183a), *918*
Wright, J. S. 24 (46, 47, 60), 25, 33 (60), 69 (47), *79*
Wu, C. Y. 166 (25), 190 (131), *196*, *199*
Wu, J. S. 874 (120), *916*
Wu, K.-C. 904, 905 (194), *918*
Wu, T.-T. 366 (205b), *374*
Wüest, H. 228 (215), *234*, 833, 851 (51), *914*
Wuesthoff, M. T. 211, 217 (87a), *231*
Wunderlich, J. A. 388 (49), *413*
Wunderly, S. W. 340 (143), *372*
Wyard, S. J. 720 (29), *771*
Wyatt, J. F. 36 (95), *79*
Wyman, D. P. 628 (164a), *647*
Wynberg, H. 119 (25), *127*, 210, 226, 227, (65b), *230*, 281 (37), *305*, 385 (43), *413*, 592 (30), *644*, 835 (239), 839 (42), 841 (248a, 248b), 844, 845 (21a), 850, 859, 860 (42), 873 (112), 892 (21a), *914*, *916*, *919*

Ya, G. 134 (23), *158*
Yablokov, V. A. 142, 147, 150, 151 (56), *159*, 396, 397 (87b), *414*, 778 (3), 801 (81), *804*, *805*
Yablokova, N. V. 801 (81), *805*, 821 (57), *828*
Yablonskii, O. P. 136 (35, 45), 137 (35, 41), 139 (35, 52), 140, 141 (35), 143 (35, 58), 144 (58), 148 (70), 149 (35, 70), *158*, *159*, 564 (209), *583*
Yabushita, S. 61 (257, 259, 263), 65 (285, 286), 71 (285), *83*, 854 (53), *914*
Yager, W. A. 206 (27), *230*
Yagodovskaya, T. V. 484 (6), *517*
Yahav, G. 685 (103, 104), *690*, 891 (152a, 152b), *917*
Yamabe, S. 69 (306, 310), *84*
Yamada, K. 785 (35), *804*
Yamada, S. 182 (89), *198*, 324 (75), *371*, 384 (38g), *413*
Yamada, T. 251 (61), *258*
Yamada, Y. 843, 858, 884, 891, 898, 906, 911 (57b), *914*
Yamadagni, R. 441 (93), *459*
Yamagishi, A. 445, 446, 454, 456 (124), *459*
Yamaguchi, K. 25 (65, 66), 33 (66), 61 (248, 257, 259, 263), 65 (284–286), 71 (285), *79*, *83*, 209, 220 (61, 62), *230*, 332 (109b), *372*, 397 (90, 91), *414*, 854 (53), 893 (163a), *914*, *917*

Yamaguchi, T. 451, 452 (181), *460*
Yamakawa, H. 893 (163c), *917*
Yamakawa, K. 119 (26), *127*
Yamamoto, H. 182 (86), *198*, 393 (70, 71), *414*, 721 (39), *771*, 860 (62a, 62b), *914*
Yamamoto, O. 88, 90–93 (89), *96*, 530 (48), *579*
Yamamoto, S. 167 (29), *197*
Yamamura, S. 843, 858, 884, 891, 898, 906, 911 (57b), *914*
Yamasaki, H. 435, 438 (38), *457*
Yamashita, Y. 719, 723, 724 (27), *771*
Yang, F. 226 (180), *233*
Yang, N. C. 300 (215), 301 (218), 303 (240), *309*, 546 (129), *581*, 810 (14), *827*
Yany, F. 831 (12), 837, 838 (54), 843 (12, 26, 58), 846 (12, 26), 853 (26), 854 (54), 855 (26), 856 (26, 255), 857 (12, 26), 858 (58), 884 (172a), 891 (148), 894 (170), 895 (172a), 901, 902 (170), 903 (172a), *913*, *914*, *917*, *919*
Yaroslavskii, N. G. 134 (23), *158*
Yasuoka, N. 466, 474 (123), *481*
Yates, J. H. 48, 49, 54, 61, 63 (182), *81*
Yates, K. 41, 48, 49 (151), *81*
Yazdanbachsch, M. R. 526, 527 (34), *579*
Yekta, A. 667, 684 (15), *688*, 844 (18c), 847 (18c, 33), 882 (18c), 891, 892 (155b), 893, 902 (18c), *913*, *914*, *917*
Yelvington, M. B. 227 (192), *234*, 555, 556, 575 (159), *582*, 684 (93), *690*, 831 (6a), 833 (145b), 836 (145a), 838, 883 (145a, 145b), 885, 886 (145a), 891, 894 (150b), *913*, *916*, *917*
Yilgor, I. 419 (10), *426*
Yim, M. B. 617 (128c), *646*
Yimenu, T. 560 (197), *583*
Yip, C. K. 667 (20), 675 (53), *688*, *689*, 732 (82), *772*
Yoke, J. T. 472 (99), *481*
Yokoya, H. 736 (109), *772*, 843, 858, 884, 891, 898, 906, 911 (57b), *914*
Yokoyama, Y. 263, 267, 269 (19), *276*, 597, 640 (64), *645*
Yokozeki, A. 37, 67 (130), *80*
Yoneda, N. 498 (76), 503 (107), *519*, 536 (69), 555 (151), *580*, *582*, 666 (66, 70, 71), 673 (66), 676 (66, 70), 677 (66), 679 (66, 71), *683*
Yonezava, T. 530 (48), *579*
Yonezawa, T. 88, 90–93 (89), *96*
Yoshida, H. 611, 617 (106), *646*, 721 (41), *771*, 791 (48), *805*
Yoshida, M. 125 (39), *127*, 596 (59), 616 (125b), 619 (59), *644*, *646*
Yoshida, T. 596, 622 (58), *644*
Yoshida, Y. 466, 474 (122), *481*
Yoshikawa, K. 872 (106a), *915*
Yoshimine, M. 538 (77d), *580*
Yoshino, K. 423 (36), *427*
Yoshioka, M. 785 (35), *804*
Yoshizawa, R. 287 (87), *305*
Young, D. P. 174 (57), *197*
Young, P. W. 223 (153), *233*
Youssefyeh, R. D. 632 (175, 176), 633 (175, 176, 178), *647*
Yue, H. J. 341 (149), *372*, 446 (149), *459*
Yukawa, Y. 130, 131, 144 (7), *158*
Yurev, V. P. 178, 179 (73), 184 (93, 94, 97, 98), 185 (97), 186 (110, 111, 114), 187 (118), *197–199*
Yurzhenko, A. I. 738 (116, 117), *773*
Yurzhenko, T. I. 425 (65), *427*
Yuyama, M. 285 (64), *305*

Zachariasse, K. A. 895 (173), *917*
Zady, M. F. 751 (159a, 159b), 752 (161), *774*
Zagorski, M. G. 360, 361 (198), *373*
Zahn, A. 832 (24, 46), 845, 846 (24), 850 (46), *914*
Zaidi, W. A. 753 (163b), *774*
Zaikov, G. E. 167 (26), *196*, 499, 500 (80), *519*
Zajacek, J. G. 162, 179, 180 (4), 184 (95), 186 (113), *196*, *198*
Zak, K. 266, 267 (38), *277*
Zaklika, K. A. 205 (15), 209, 226 (57), 227 (198), *229*, *230*, *234*, 329, 355 (98), *371*, 446 (143), *459*, 592 (27a), *644*, 838 (31, 245), 840, 841 (180c), 842 (31, 48, 180c), 844 (18r), 847 (31), 851 (48), 871 (18r), 874 (119), 884 (31, 180c), 886, 887 (31), 890 (180c), 895 (31, 180c), 896 (31, 48, 180c), 898, 899 (31), *913*, *914*, *916*, *918*, *919*
Zalygin, L. L. 185 (100), *198*
Zanderighi, G. M. 466, 468, 474 (30), *479*
Zaret, E. H. 440 (71), *458*
Zarkov, V. V. 150 (71), *159*
Zaslowski, J. A. 484 (7), *517*
Zavitsas, A. 173 (51), *197*
Zavodnik, V. E. 86, 88 (12, 19), 89 (12), *94*
Zborshchik, L. A. 802 (110), *806*
Zdero, C. 105 (2), *126*
Zeelen, F. J. 223 (155c), *233*
Zeldes, H. 721 (37, 38), 722 (38), 724 (51, 52), 739 (51), *771*
Zeller, K. P. 536 (75), *580*
Zeller, K.-P. 292 (151), *307*
Zellers, E. T. 560 (197), *583*
Zemel, H. 767, 768 (205), *775*
Zemke, W. T. 21 (43, 44), *79*
Zharkov, V. V. 137 (40), *158*
Zhigunova, L. N. 762 (191), *775*

Zhikharev, V. S. 762 (189), *775*
Zhil'tov, S. F. 811 (67), *828*
Zhiltsov, S. F. 811 (19), *827*
Ziegler, K. 210 (67a), *231*
Ziemann, A. 397 (88), *414*
Ziffer, H. 291 (139), *307*
Zimmerman, H. E. 226 (187), *233*, 718 (20), *770*, 840–842, 906 (202), *918*
Zimmermann, H. 136, 147 (38), *158*
Zinner, G. 271 (48), *277*
Zinner, K. 832 (24), 841 (86, 256), 845, 846 (24), 865 (86), 887, 888 (256), 889 (86, 256), 890 (179, 256), 895 (179), 898 (207, 256), 899 (86, 207), 900 (179), 903 (179, 189b), 904 (193b), 905 (189b), 908 (207), *914*, *915*, *918*, *919*
Zinovjeva, T. I. 823 (61), *828*
Ziolkowski, J. J. 478 (153), *482*
Ziv, J. 440, 445 (82), *459*
Ziv, Y. 436, 437, 440, 445 (50), *457*
Zlofskii, S. S. 653 (23), *682*
Zlotski, S. S. 501 (120), *520*
Zolotova, N. V. 796–798, 801 (70), 802 (110), *805*, *806*
Zubkov, A. V. 738 (118a), 764 (199), 765 (200), *773*, *775*
Zuccaro, C. 474, 475 (133), *481*
Zupancic, J. J. 227 (200b), *234*, 405 (119), *415*, 546–548, 550, (116), *581*
Zvat'kov, I. P. 280 (26), *304*
Zwanenburg, B. 226 (181), *233*, 297 (180a, 180b), *308*
Zwolenik, J. J. 727 (61), *771*
Zyat'kov, I. P. 150 (73), *159*, 280 (30a–c), *304*

Subject index

Acenaphthene,
  dimerization of 908
  reactions of 440
Acetaldehyde,
  as photolytic product 735
  autoxidation of 622
Acetal hydrotrioxides 507
  decomposition of 509–512
Acetals
  oxidation of 295, 296
  ozonation of 501, 652, 653
Acetanilide 867, 868
Acetic acid, photooxidation of 722
Acetone,
  as photolytic product 727, 735, 754, 762
  oxidation of 563
Acetophenone 717
  as photolytic product 737
  as photosensitizer 716
  chemienergized 908
Acetophenones, oxidation of 561, 562
Acetoxy radicals, decarboxylation of 610
Acetyl benzoyl peroxide,
  acetyl^{18}O-labelled 598
  benzoyl-^{18}O-labelled 593, 598
  decomposition of 617
  structure of 406, 407
Acetyl-d_3 benzoyl peroxides, ^{17}O-labelled 617
Acetylene, oxidation of 538, 539
Acetyl peroxide,
  carbonyl-^{18}O-labelled 597, 609
  decomposition of,
    photolytic 718, 738
    radiolytic 765
    thermolytic 609, 614
  methyl-^{13}C-labelled 598
  UV absorption of 713
Acetyl peroxide-d_6 597
α-Acetylperoxyacetonitriles, photolysis of 755
Acetylperoxyl, reactions of 240
Acetyl propionyl peroxide, decomposition of 616
Acid–base interactions 61
Activation parameters, of four-membered ring peroxides 883–890
O-Acyl *O'*-*t*-butyl hyponitrites, decomposition of 600–603
Acyl chlorides,
  ^{18}O-labelled 594
  reaction with $O_2^{\cdot -}$ 442
Acyl migration 631, 632
N-Acyl-*N*-nitroso-*O*-*t*-butylhydroxylamine, carbonyl-^{18}O-labelled, thermolysis of 591
α-Acyloxylation 755
Acyloxy radicals 714
Acyl peroxides—*see* Diacyl peroxides
Acyl radical 730
Adamantanone 873
Adamantylideneadamantane, reactions of 592
Adamantylideneadamantanedioxetane, thermal stability of 227
Adenine, reactions of 437, 751
α-Agarofuran, synthesis of 366
Alanine, as radiolytic product 762
Alcohol hydrotrioxides 513–515
Alcohols,
  as alkyl hydroperoxide reduction products 193, 194
  as $O_2^{\cdot -}$ disproportionation catalysts 432
  complexes with hydroperoxides 143
  oxidation of 186, 187, 192, 193, 295, 721
  ozonation of 501, 514, 515, 661, 679
Aldehyde hydrotrioxides 509, 512, 513
Aldehydes,
  aromatic—*see* Aromatic aldehydes
  as alcohol photooxidation products 721
  oxidation of 295, 296, 444, 564–566, 593, 597, 598, 622
  ozonation of 653, 679
  α,β-unsaturated—*see* α,β-Unsaturated aldehydes
Alkanes,
  autoxidation of 163–165
  hydroxylation of 295
  ozonation of 655–658, 673–678
Alkenes,
  complexes with hydroperoxides 139, 142
  epoxidation of 177–184, 191, 192, 288–292, 525–534

hydroxyketonization of 184
ketonization of 187, 188
oxidation of 167, 168, 472, 473, 477, 525–534, 549–551, 874
with $O_2^{-\cdot}$ 449
with $^1O_2$ 209, 218–220, 226, 588–590, 592, 593, 861–864
ozonation of 574, 632–634
reaction of,
with diacyl peroxides 285
with H_2O_2 591
vicinal dihydroxylation of 189, 190
Alkenyl hydroperoxides, photolysis of 731
Alkenylperoxy radical cyclization 339
α-Alkoxyalkyl hydroperoxides, synthesis of 168
α-Alkoxybenzyl hydroperoxides, decomposition of 574
α-Alkoxyhydroperoxides 910
α-Alkoxyketones 910
Alkoxyls, cage disproportionation of 237
Alkoxy radicals 714, 717, 727
thermal sources of 687
N-Alkylbenzamide 874
Alkyl benzoates, as photolytic products 743
Alkyl halides,
formation in ozonation of alkanes 656, 657
reaction with $O_2^{-\cdot}$ 441, 442
Alkyl hydroperoxides—*see also* Hydroperoxides
as oxidizing agents 188–193, 543
base-catalysed decomposition of 574, 575
conformation of 90, 91
heterolysis of 169–172
acid-catalysed 176–177
metal-catalysed 177–188
mass spectra of 113–116
metal-catalysed homolysis of 169, 170
intermolecular processes in 173, 174
intramolecular processes in 174–176
photolysis of 727, 729
radiolysis of 763, 764
reaction of 194, 195
with $O_2^{-\cdot}$ 452, 453
reduction to alcohols 193, 194
α-substituted 194, 195
synthesis of 163–168, 816
thermal stability of 162
UV absorption of 727
Alkyl hydrotrioxides 501–517
β-Alkyl loss, in mass spectral fragmentation of β-peroxylactones 110
Alkyl peroxide complexes, of transition metals 464, 476
Alkyl peroxides—*see* Dialkyl peroxides
Alkylperoxyaluminium alkyls 816, 817
Alkylperoxycobaltoximes, structure of 382, 383
t-Alkylperoxy esters, rearrangement of 552
Alkylperoxyls, reactions of 236
4-Alkylperoxy-4-methyl-2,6-di-*t*-butyl-2,5-cyclohexadiene-1-one 236
Alkylperoxy(pyridinato)cobaloximes 825
Alkylperoxy radicals, self-reaction of 490–493
t-Alkylperoxy radical–tetroxide equilibria 486–488
thermodynamic parameters for 488
Alkyl radicals 714, 718, 733, 739, 743, 751, 758
rearrangement of 750
Alkyl trifluoromethanesulphonates 315
Alkynes,
complexes with hydroperoxides 139, 142
epoxidation of 292–294, 536–539
vicinal dihydroxylation of 189–191
Allene, reaction with $^1O_2$ 864
Allenes,
as intermediates in photolysis of phenanthrenes 874, 875
[2 + 2] cycloaddition of $^1O_2$ to 226
epoxidation of 292–294
Allylacetyl peroxy ester, photolysis of 750
Allyl *t*-butyl peroxide, decomposition of 239
Allylic alcohols,
epoxidation of 180–183, 532, 534
synthesis of 220, 221
Allylic alkoxy radicals, as intermediates in reactions of allylic hydroperoxides 224, 225
Allylic hydroperoxides 202, 831
decomposition of 249
reactions of 220–225, 555, 556
synthesis of 209
Amide hydroperoxides, as epoxidizing agents 533
Amides,
complexes with hydroperoxides 145, 147
reaction with $O_2^{-\cdot}$ 445
Amine oxides 655
Amines,
aromatic—*see* Aromatic amines
complexes with hydroperoxides 144, 156, 157
oxidation of 270–273, 539, 540, 547–549, 724
ozonation of 501, 654, 655
Amino acids, photooxidation of 724
Aminobenzaldehydes, oxidation of 567
Amino radical 724
Ammonia, photooxidation of 724
t-Amyl hydroperoxide, photolysis of 729
Anchimeric assistance, in decomposition of peroxides 242, 254

Anhydrides, reaction with $O_2^{-\cdot}$ 442, 443
Aniline, reaction with $O_2^{-\cdot}$ 436
Anisole, reaction with radiolytic $SO_4^{-\cdot}$ 767
Anisyl peroxides, methyl-^{14}C-labelled 597
Annulenes, photooxidation of 330, 331
Anomeric effect 38
Anthracene, reactions of 451
Anthracene photoxide 381
Anthraquinone 654
Anthrone, reactions of 440
Anthrones, ozonation of 654
9-Anthrylamine 860
1-Apocamphoryl benzoyl peroxide, decomposition of 617
1-Apocamphoryl peroxide, decomposition of 617
1-Apocamphyl benzoyl carbonate, decomposition of 626
Arene dioxides, synthesis of 213
Arenes,
  complexes with hydroperoxides 138–141
  photoreaction with peracetic acid 744–747
  reaction with $O_2^{-\cdot}$ 451, 452
Arenesulphonyl radicals 263
Arene trioxides, synthesis of 213
Aromatic aldehydes, ozonation of 501, 509, 513
Aromatic amines, reaction with $O_2^{-\cdot}$ 436, 437
Aromatic carboxylic acids, photooxidation of 723
Aromatic hydrocarbons, reaction with diacyl peroxides 285
Aromatic ketones, autoxidation of 876
Aromatic peroxy acids, photolysis of 747
Aromatic peroxy esters, photolysis of 752
Aromatic rings, photooxidation of 725–727
Arrhenius parameters, for thermolysis of acyl and aroyl peroxides 240
Arylacetic acid ester 718
Aryl migration, in oxidation of aldehydes 564
Aryl radicals 730
Arylsulphonyl peroxides, as oxidizing agents 549
Ascaridole 312
  photolysis of 350
  reactions of 213
Ascorbic acid, reaction with $O_2^{-\cdot}$ 433
Asymmetry parameter 57
Atomic charges 53, 62, 63
Atomic oxygen, anionic 766
Atomization energy 3
Autoxidation 236
  base-catalysed 435, 436, 438, 440, 587, 874–878
  of aldehydes 240, 252, 499, 500, 593, 597, 598, 622
  of enols 435
  of enones 440
  of hydrocarbons 163–168, 588, 873, 874
  of imines 860
  of ketenes 858
  of ketones 831, 876
  of organometallic compounds 168
  of polyunsaturated compounds 243, 731
  photolytic 873, 874
  thermolytic 873
AZBN 419
Azoalkanes, photodenitrogenation of 908, 909
α-Azoalkyl hydroperoxides, synthesis of 588
trans-Azobenzene, oxidation of 541
α,α′-Azobis(isobutyronitrile) 719
Azo compounds, oxidation of 541
Azocyclohexylnitrile, photolysis of 487
α-Azohydroperoxides, photolysis of 730
Azoisobutane, photolysis of 487
Azoisobutyronitrile, photolysis of 487

Badger–Bauer relationship 135, 138, 142, 148
Baeyer–Villiger oxidation 559–566
Baeyer–Villiger reaction 626, 641, 642
Baeyer–Villiger rearrangement 876
Barrier for internal rotation,
  of H_2O_2 36
  of X_2O_2 67
  of X_2S_2 67
  origin of 37
Barrier for inversion, of Li_2S_2 63
Basis set,
  augmented 36, 51
  double-zeta 51
  minimal 51
Basis set error 50
Benson mechanism, for self-reaction of alkylperoxy radicals 492, 493
Bent bond 57
Benzaldehyde 869
  as mass spectral product 110
  as photolytic product 730
  autoxidation of 593
  ozonation of 513
Benzaldehydes,
  oxidation of 564, 566
  ozonation of 653
Benzene,
  as benzoic acid photolytic product 723
  radiolysis of 762, 767
  reactions of 451
anti-Benzene dioxide 345
Benzene-1,4-endoperoxide 326
Benzene oxide 324
Benzene ring, cleavage of 719
Benzenethiazolyl group 897

Benzene trioxide, synthesis of 324
trans-Benzene trioxide 213
Benzhydrol, autoxidation of 587
Benzhydryl hydroperoxides, decomposition of 554
Benzil,
  cleavage of 876
  oxidation of 568
Benzofuranones, autoxidation of 876, 877
Benzoic acid,
  as photolytic product 730, 752
  photoreaction of 719, 723
Benzonitrile, photoreaction of 760
Benzonorbornadiene, rearrangement of 908, 909
Benzophenone,
  as photolytic product 730
  as photosensitizer 716, 730
  ^{18}O-labelled 642
  oxidation of 560
Benzopropiolactone, as photolytic product 742
Benzoylacetonitrile, reaction with $O_2^{\cdot-}$ 437
Benzoyl acyl peroxides, photolysis of 743
Benzoyl cumyl peroxide, carbonyl inversion in the solid phase 780
Benzoyl cyclohexaneformyl peroxide, decomposition of 626
Benzoyl cyclohexyl carbonate, ^{18}O-labelled, ^{18}O equilibration in 626, 627
cis-1-Benzoyl-3-formylcyclopentane 863
Benzoyloxy radical 739
Benzoyl peroxide,
  as oxidizing agent 548
  carbonyl-^{14}C-labelled 596, 622
  carbonyl-^{18}O-labelled 594, 636–638
  conformation of 91
  decomposition of,
    photolytic 714, 715, 741
    radiolytic 765
    solid-state 788–792, 800, 801
    thermolytic 242
  phenyl-^{14}C-labelled 596
  reactions of 283
  UV absorption of 739
Benzoyl peroxide-d_2, synthesis of 596
Benzoyl peroxide-2,2′4,4′,6,6′-d_6, synthesis of 596
Benzoyl peroxides, ring- or substituent-^{14}C-labelled 623
Benzoylperoxyl 248
Benzoyl radical 730
Benzoyl *p*-toluenesulphonyl peroxide,
  benzoyl-^{18}O-labelled 598
  carbonyl-^{18}O-labelled 640
  carboxy inversion in 264
  decomposition of 631, 639, 640
  sulphonyl-^{18}O-labelled 598, 599, 639
N-Benzyladenine, reaction with $O_2^{\cdot-}$ 437
Benzylamines, oxidation of 440, 549
Benzyl cyanide, reaction with $O_2^{\cdot-}$ 879
Benzylidenefluorene, reactions of 450
Berberal, synthesis of 366
Berberine alkaloids 366
Beryllium methoxide 810
Betains, reaction with $^1O_2$ 873
$BH_2O_2BH_2$ 63
BH_2O_2H 63
Biacetyl,
  as photolytic product 735
  as photosensitizer 736
Biadamantylidene dioxetane, reactions of 453
Bicyclic endoperoxides, as intermediates in prostaglandin biosynthesis 244
Bicyclic ketones, oxidation of 561
Bicyclic peroxides, mass spectra of 120, 121
Bicyclo[2.2.1]endoperoxides, synthesis of 339
Bicyclo[3.2.1]endoperoxides, synthesis of 339
Bilirubin, photooxidation of 334
Bioluminescence 844, 852, 896
Biphenyl,
  as photolytic product 730
  as radiolytic product 765
Biphenylacetylene, oxidation of 537
Biradicals, as intermediates,
  in bicyclic endoperoxide synthesis 339
  in cleavage of dioxetane ring 882–894
  in [2 + 2]cycloaddition of $^1O_2$ to double bonds 226
  in $^1O_2$ ene reaction 220
  in photolysis of cyclic peroxides 736, 742, 753, 754
Biradical states,
  of O_2 7
  of O_3 25
  of XO_2 7
Bisadamantylidene,
  autoxidation of 873
  reaction with $^1O_2$ 850
3,6-Bis(4-aminocarbonylbutyl)-1,2,4,5-tetraoxane, decomposition of 792, 793
9,9-Bisbicyclo[3.3.1]nonylidene, reaction with $^1O_2$ 850
1,3-Bis(*t*-butyl) peroxide of cyclopentane, mass spectrum of 117
Biscycloalkanediazenes, decomposition of 619
Bis(dimethylbenzylsilyl) peroxide, structure of 397
Bis-dioxetanes 865
Bis-diphenylphosphinic peroxide, photolysis of 759
Bis-diphenylphosphinyl peroxide 259, 260
  reactions of 275, 276
  synthesis of 274

Bis-hydroperoxides, structure of 402
Bis(2-methylbutyryl) peroxide, photolysis of 739
2,3-Bis(methylene)-7-oxanorbornane 353
Bis(*p*-nitrobenzoyl) peroxide, decomposition of 789, 790
7,7-Bisnorbornylidene, reaction with $^1O_2$ 850
Bis(pentafluorosulphur) trioxide 497
Bis-sulphonyl peroxides 259
reactions of 261, 266–274
synthesis of 261, 262
(1,4-β)-(2,3-β)-Bis(1,1,4,4-tetramethylcycloheptane)-5,6-dioxabicyclo[2.2.0]hex-2-ene 386
Bis(trialkylsilyl) peroxide, photolysis of 761
Bis(tributylstannyl) peroxide 315
Bis(trifluoromethyl)ketene, reaction with $^1O_2$ 858
Bis(trifluoromethyl) peroxide, decomposition of 689–693
photolytic 735
Bis(trifluoromethyl) trioxide 246, 496
IR spectrum of 499
Bis(triphenylmethyl) peroxide, crystal structure of 382
Block copolymers 418–426
Bond angles, ∠COO 378–381
Bond energy,
of OO bond 2
of XO bond 2
of XX bond 3
Bonding, in XO_2 61, 62
Bonding overlap 13, 20
Bond lengths,
C—O 377–380
O—O 377–380
Bond order 49
Bond separation energy 65
Bond staggering 38
Bond strengths 99, 100
Boron-containing peroxides, photolysis of 760
Bridged form, of X_2O_2 compounds 13, 25, 63
Bromobenzene, reactions of 451
β-Bromohydroperoxides, dehydrobromination of 831–840, 848, 849
threo-1-Bromomercuri-2-*t*-butylperoxy-1,2-diphenylethane 383
Bromonitrobenzenes, reactions of 451
o-Bromoperoxybenzoic acid, topotactic decomposition of 377
2-Bromopropene, reaction with $^1O_2$ 864
Bromostilbene, reactions of 441
N-Bromosuccinimide, as bromine carrier 848
Butane, as photolytic product 739
t-Butanol, as photolytic product 750
Butene, as photolytic product 739
trans-2-Butene ozonide, microwave study of 389
t-Butoxyl, formation of 237
t-Butoxy radical 600–602, 727, 731, 733, 763
t-Butyl arylperoxysulphonates, decomposition of 553
t-Butyl cycloalkaneperformates, decomposition of 619, 620
ring-size effects on 621
3-*t*-Butylcyclohexene, epoxidation of 525, 526
t-Butyl hydroperoxide,
as epoxidizing agent 534
complexes of,
with alkynes 142
with aniline 144
with arenes 138
with diethyl ether 144
with β-ethanolamine 144
with pyridine 144
decomposition of 249, 250
photolytic 727
radiolytic 763
thermolytic 686
^{18}O-labelled 595
oxidation of 486, 487, 494
photoreactions of 727–729, 732
synthesis of 163, 164
thermal stability of 162
use in organic synthesis 162, 163, 172, 173, 178–192
t-Butyl hydroperoxide sodium salt, as oxidizing agent 567
t-Butyl *o*-methanesulphenyl perbenzoate, ^{17}O-labelled 635
2-*t*-Butylmethoxymethylenecyclohexane, autoxidation of 873
t-Butyl-*N*-methyl-*N*-(*p*-nitrophenyl)peroxycarbamate, carboxy inversion in the solid phase 781
t-Butyl 2-methyl-2-phenylperacetate-2-d_1, synthesis of 594
t-Butyl 2-methylthioperbenzoate, carbonyl ^{17}O-labelled 596
t-Butyl *p*-nitroperbenzoate, ^{18}O-labelled 595
t-Butyl peracetate,
decomposition of 253
photolysis of 750, 751
t-Butyl perbenzoate,
decomposition of 606, 607, 635
photolytic 752, 753
mass spectrum of 106
^{18}O-labelled 607, 635
t-Butyl perbenzoates, *ortho*-substituted, decomposition of 634, 635
t-Butyl percarboxylates, deuterated alicyclic 594

t-Butyl performate,
  photolysis of 750
  reaction with base 636
t-Butyl peresters 601
  decomposition of 603, 604, 606, 607, 634–636
  deuterated 594
  ^{17}O-labelled 596, 635
  ^{18}O-labelled 595, 607, 635
t-Butyl peroxide—*see* Di-*t*-butyl peroxide
t-Butyl peroxides, methyl-^{14}C-labelled 623
Butyl peroxyacetates, decomposition of 577
t-Butyl peroxybenzoates, decomposition of 254
t-Butyl peroxycarboxylates, reactions of 300–303
t-Butyl peroxyformate, decomposition of 577
t-Butylperoxyl 248, 250
t-Butylperoxyls, self-reactions of 249
t-Butyl peroxynitrate, synthesis of 698
t-Butyl peroxyphosphate, photoreaction of 759
t-Butyl peroxyphosphonate, photoreaction of 759
t-Butylperoxy radical 727, 732
  equilibrium between di-*t*-butyl tetroxide and 486
Butylperoxyzincbutyl 810
t-Butyl 5-phenylisoxazole-3-peroxycarboxylate, mass spectrum of 107
t-Butyl 5-phenyl-Δ^2-isoxazoline-3-peroxycarboxylate, mass spectrum of 107
t-Butyl phenylperacetate-2-d_2, synthesis of 594
t-Butyl *o*-phenylthioperbenzoate, decomposition of 254
N-*t*-Butylpyrrole, reaction with $^1O_2$ 871
n-Butyl radical, ESR spectrum of 714

Caffeine, photomethylation of 751, 752
Caged geminate radical pair 600, 601
Cage escape 740
Cage reactions 799, 802
Carbanions, reactions with diacyl peroxides 282, 283
α-Carbanions, as synthons for α-hydroperoxy acids 854
Carbohydrates, anomeric effects in 377, 395
5-Carbomethoxypentanal 865
Carbon–metal bond cleavage 273, 274
Carbon monoxide, in photodecomposition of H_2O_2 720
Carbon trioxide, loss of, in mass spectral fragmentation of β-peroxylactones 110
Carbonyl compounds, reaction of,
  with H_2O_2 591
  with $O_2^{\cdot-}$ 437, 438, 444, 445
Carbonyl-forming eliminations 573–578
Carbonyl oxide 61, 63
Carbonyl oxides, as intermediates in Baeyer–Villiger oxidations 563
Carbonyloxy radical 748
Carboxy inversion 264, 608, 623, 625, 748
  in solid-state reactions of acyl peroxides 780, 781
Carboxylic acids,
  aromatic—*see* Aromatic carboxylic acids
  complexes with hydroperoxides 143
  photooxidation of 722–724
Carboxylic esters, photooxidation of 724
Cation radicals,
  radiolytic formation of 767
  reaction with $O_2^{\cdot-}$ 446
Cations, reaction with $O_2^{\cdot-}$ 446
Cedrane oxide, ozonation of 671
Cedrol, ozonation of 671
Cedryl acetate, ozonation of 671
Centrosymmetric conformations 391, 408
Chalcones, reactions of 449, 878, 879
Charge transfer 7, 25
  π overlap 7
  σ, π promotion 7
C—H bonds, reaction with $O_2^{\cdot-}$ 437
Chemical ionization mass spectrometry 110, 114
Chemically induced dynamic nuclear polarization 241
Chemically induced electron exchange chemiluminescence 894
Chemienergization 753, 905–909
Chemiluminescence 205, 209, 227, 844, 847, 860, 861, 864, 866, 871, 873, 874, 876, 880–909
  activation parameters for 883–890
  direct 901, 902
  electron-exchange 894–897, 903
  energy sufficiency for 880–882
  energy-transfer 902–905
  excitation parameters for 897–909
  mechanisms for 882–894
N-Chloramines, reactions of 441
Chlorine ion, oxidation of 767, 768
Chlorobenzene,
  as photolytic product 747
  reactions of 451
m-Chlorobenzoic acid, as photolytic product 747
β-Chlorohydroperoxides 848, 849
m-Chloroperbenzoic acid, photolysis of 747
m-Chloroperoxybenzoic acid,
  as oxidizing agent 530, 538, 546
  structure of 403
o-Chloroperoxybenzoic acid, topotactic

decomposition of 377
$CH_3O_2CH_3$ 63
CH_3O_2H 63, 67
Cholest-4-en-3-one, reaction with $O_2^{\cdot-}$ 439
Cholesterol, reactions of 440, 449
Chondrilline 313
Chromium peroxo complexes 469
CIDNP spectroscopy 616, 619, 635, 727, 728, 740–742
Cis-directing groups 534
Cis effect 63
CNDO/2 calculations, for dioxetane decomposition 893
Cobalt(III) peroxo complexes 472
Cobalt *meso*-tetraphenylporphine 357
π Complexation 267
Complex spin orbitals, of O_2 10
Concerted decomposition, of four-membered ring peroxides 891–893
Configuration interaction 13, 25, 58, 59
Configuration space,
of XO_2 compounds 5
of X_2O_2 compounds 5
Conformation,
of cyclic peroxides 47
of cyclic polyoxides 71
of diacyl peroxides 91
of dialkyl peroxides 89–91
of hydrogen peroxide 20, 36–40, 86–89
of ozonides 71
of peroxy acids 66–68, 92, 93
of polyoxides 68–70
Conjugated ketones, reaction with $O_2^{\cdot-}$ 448
Cope rearrangement 611, 613
activation entropies for 614
Correlation effects 36, 44, 53, 57
Correlation error 50
Coupling constants, of alkyl radicals 615
Covalent radius, of O 3
Criegee adducts 599–562
Criegee mechanism 632–634, 700–703
Criegee rearrangement 551–554
Cross-linking 426
Crotepoxide 314
synthesis of 367, 368
Crystal decomposition 385
anisotropic 395, 396
Crystal instability 377
Crystallographic isomorphism 409
Cumene, autoxidation of 166
Cumyl hydroperoxide,
benzylic-^{14}C-labelled 588
complexes of,
with aniline 144
with carbonyl compounds 144, 145
with phenol 143
radiolysis of 763, 764
synthesis of 163, 166, 249, 588
α-Cumyl hydroperoxides, decomposition of 554
Cumyloxy radicals 727
Cumylperoxydiphenylborane 814
Cumylperoxyl 248
Cyanamide, reaction with radiolytic $SO_4^{\cdot-}$ 767
ω-Cyanocarboxylic acids 333
Cyano olefins, oxidative cleavage of 450
Cybullol, synthesis of 366
Cyclic bonds, elongation of 386
Cyclic diacyl peroxides,
bond angles in 381
structure of 405, 406
Cyclic dialkyl peroxides, structure of 384–387
Cyclic peroxides 47, 69, 71–73, 731
free-radical reaction mechanisms involving 243–245, 250
mass spectra of 119–123
photolysis of 735, 742
solid-state decomposition of 792, 793
Cyclization,
intramolecular 739
of β-halohydroperoxides 848, 849
of α-hydroperoxy acids 843, 855–857
[2 + 2]Cycloaddition 858, 865
of $^1O_2$ to double bonds 209, 226–229, 592
[4 + 2]Cycloaddition, as side-reaction in synthesis of four-membered ring peroxides 848, 850, 857, 858
Cycloalkaneformyl peroxides 617
thermolysis of 619
ring-size effects on 621
Cycloalkanes,
ozonation of 662, 677, 678
photooxidation of 743, 744
Cycloalkenes,
bridged, epoxidation of 526, 527
reaction with $^1O_2$ 861–864
Cyclobutadiene dioxetane 846
Cyclobutanecarbonyl peroxy ester, photolysis of 750
Cyclobutaneformyl peroxide, thermolysis of 628
Cyclobutylcarbonyl peroxide, photolysis of 739
Cyclodecane, as photolytic product 736
Cycloheptane diperoxide, photolysis of 736
1,3,5-Cycloheptatriene, photooxidation of 323
Cyclohexadiene endoperoxide, photolysis of 350
Cyclohexadienes, reaction with $O_2^{\cdot-}$ 440
Cyclohexadienone *t*-butyl peroxide, photolysis of 737

Cyclohexadienyl radical 740, 767
1,3-Cyclohexadione, reaction with $O_2^{-\cdot}$ 437
Cyclohexane, photoreaction of 743
Cyclohexane diperoxide, photolysis of 736
Cyclohexane trioxide, decomposition of 252, 253
Cyclohexanol, as photolytic product 743
Cyclohexanone, as photolytic product 736, 743
Cyclohexanone oximes, oxidation of 540
Cyclohexene,
  autoxidation of 831
  epoxidation of 538
  photoreaction of 755, 759
  reaction with $O_2^{-\cdot}$ 440, 449
2-Cyclohexene-1-ol, epoxidation of 526
Cyclohex-2-en-1-ones, reaction with $O_2^{-\cdot}$ 438, 439
Cyclohexyl benzoyl carbonate, *para*-substituted ^{18}O-labelled, ^{18}O equilibration in 631
Cyclohexyl hydroperoxide, radiolysis of 763
Cyclohexyl peroxyacetate, decomposition of 577
Cyclohexyl radical 743
trans-Cyclooctene, reaction with $^1O_2$ 863
Cyclopentadiene, reaction with $O_2^{-\cdot}$ 437
Cyclopentenones 737
Cyclopentyl hydroperoxide, radiolysis of 763
[2,2,2,2]-(1,2,2,5)Cyclophane, reaction with $^1O_2$ 326
Cyclophosphamide 397
Cyclophosphamides, mass spectra of 114, 115
Cyclopropaneformyl peroxide, decomposition of 616, 617
Cyclopropanes, ozonation of 662
Cyclopropylacetyl peroxide, photolysis of 739
Cyclopropylacetyl peroxy ester, photolysis of 750
Cycloreversion 119, 121
Cysteine, radiolysis of 762

Dakin reaction 567
β-Damascenone 348
Deamination, oxidative 270
trans-9-Decalyl hydroperoxide, reactions of 641
trans-9-Decalyl hydroperoxide esters, rearrangement of 551
Decarboxylation 865, 867
  activation energy of 623
  photolytic 110, 719, 723, 725, 738–740, 742, 743, 748, 752–756, 758, 854
  radiolytic 110, 767
  thermolytic 855
n-Decyl hydroperoxide, complex with caproic acid 143
Deformation density 54
Delocalization,
  of lone-pair electrons 3, 38, 57, 66, 68, 71
  of π electrons 7, 57, 66
Delocalization energy, of O_3 43
Deuterium isotope effects,
  in decomposition reactions,
    of *t*-butyl cycloalkaneperformates 620
    of diacyl peroxides 614
    of peresters 605–607, 636
  in ene reactions 589
Deuterium labelling, in ozonation of C—C bonds 672
3,28-Diacetoxylupane, ozonation of 670
(Diacetyloxy)iodobenzene, reactions of 487, 494
Diacetyl peroxide—*see* Acetyl peroxide
Diacyl peroxides,
  as oxidizing agents 546–551
  carbonyl-^{18}O-labelled 739
  conformation of 91, 92
  cyclic—*see* Cyclic diacyl peroxides
  decomposition of 608–619, 623–632, 636–640
    photolytic 714–716, 738–743
    radiolytic 765
    structural effects on 611, 612
    thermolytic 557–559, 685, 686
  free-radical reaction mechanisms involving 236, 240–242
  isotopically labelled,
    synthesis of 596–599
    uses of 594, 608–619, 623–632, 636–640
  macrocyclic 736
  reaction of,
    with nucleophiles 282–287
    with $O_2^{-\cdot}$ 443, 444, 456
  stability of 281, 282
  structure of 405–409
  substituent effects in 68
  synthesis of 280, 281, 568
  UV absorption of 712, 713, 738
Diacyl tetroxides 247, 499–501
Diadamantylidene, electrooxidation of 851
Diadamantylidene-1,2-dioxetane 858, 892
Diadamantylidene epoxide 873
Di-*t*-alkyl acyl peroxides, decomposition of 240
Dialkylamines, reaction with $O_2^{-\cdot}$ 436
N,N-Dialkylaminomethyl alkyl peroxides, photolysis of 737
2,5-Dialkyl-1,4-dimethoxybenzene radical cations, reactions of 446, 456
Dialkyldithiocarbamates, reactions of 251
Dialkyldithiophosphates, reactions of 251

Dialkyl peroxides,
  acid-catalysed reactions of 556, 557
  as radiolytic products 764
  conformation of 89–91
  cyclic—*see* Cyclic dialkyl peroxides
  decomposition of 687–697
    base-catalysed 573
    photolytic 714, 716, 717, 731–737
    radiolytic 764
  free-radical reaction mechanisms involving 236–240
  heterolysis of 169, 170
    homolysis of 169, 170, 173
    isotopically labelled 591–593
    mass spectra of 116, 117
    reaction with $O_2^{\cdot-}$ 453
    structure of 381–387
    synthesis of 168, 169
    thermal stability of 162
  UV absorption of 731
Dialkylperoxycadmiums 811
Dialkyl polyoxides 486–499
  structure of 499
Dialkyl tetroxides, free-radical reaction mechanisms involving 247
Di-*t*-alkyl trioxides, free-radical reaction mechanisms involving 245
Di-*t*-amyl peroxide, photolysis of 737
Diaroyl peroxides,
  free-radical reaction mechanisms involving 242, 243
  photolysis of 714, 715, 739–742
  radiolysis of 765
  UV absorption of 712
2,3-Diaryl-1,4-dioxene, cleavage of 874
Diarylmethanes, reaction with $O_2^{\cdot-}$ 440
1,4-Diazabicyclo[2.2.2]octane (DABCO) 361
  in photooxidation of oxazoles 334
  in quenching of $^1O_2$ 208
Diazanaphthaquinone 411
Diazoaminobenzenes, oxidation of 542
Diazo compounds,
  oxidation of 541, 542
  ozonation of 501
Diazodiphenylmethane, oxidation of 542
Dibenzo[*b,e*]oxepin 355
Dibenzoylmethane, reaction with $O_2^{\cdot-}$ 437
Dibenzoyl peroxide—*see* Benzoyl peroxide
Dibenzylamine, oxidation of 548
9,10-Dibromoanthracene, as fluorescer 904
m,m'-Dibromodibenzoyl peroxide, carbonyl-^{18}O-labelled 638, 639
1,3-Dibromo-5,5-dimethylhydantoin, as bromine carrier 848
Di-*t*-Butyl disulphide, reaction with $O_2^{\cdot-}$ 445
1,3-Di-*t*-butylisobenzofuran, reaction with $^1O_2$ 338
Di-*t*-butylketene, reaction with $^1O_2$ 858
Di-*n*-butylmalonyl peroxide, photolysis of 742
2,6-Di-*t*-butyl-4-methylphenol 236
2,6-Di-*t*-butyl-4-methylphenoxyl, reactions of 236
Di-*t*-butyl peroxide 727
  t-carbon-^{14}C- labelled 591
  conformation of 89
  decomposition of 237
    photolytic 717, 731–733
    radiolytic 764
    thermolytic 695
  gas-phase structure of 382
  methyl-^{14}C- labelled 591
  reaction with $O_2^{\cdot-}$ 453
  UV spectrum of 732
Di-*t*-butyl peroxyoxalate 249, 255
2,6-Di-*t*-butylphenol, as free-radical inhibitor 208
Di-*t*-butyl sulphide, reaction with $O_2^{\cdot-}$ 445
Di-*t*-butyl tetroxide 247, 486, 487, 494
Di-*t*-butyl trioxide 494–496
  decomposition of 245
p,p'-Dichlorobenzophenone 874
1,1-Dichloro-2,2-di(*p*-chlorophenyl)ethylene, autoxidation of 874
1,3-Dichlorodimethylhydantoin, as chlorine carrier 848
Dicumyl peroxide,
  decomposition in polypropylene 801
  photolysis of 717, 737
Dicumyl trioxide 494–496
  decomposition of 245
9,10-Dicyanoanthracene 874, 895
1,1-Dicyanoethylene, superoxide cleavage of 879
Dicyclohexylcarbodiimide, as dehydrating agent 831, 832, 843, 856, 857
Dicyclopropaneacetyl peroxide, thermolysis of 628
Dicyclopropylethylenes, reaction with $^1O_2$ 863
Dideuterioacetyl peroxide 598
[2 + 4]Diels–Alder addition, of $^1O_2$ to dienes 209–217, 317–338
  mechanism of 212
  nature of the substrate in 210–212
1,3-Dienes, reaction with $^1O_2$ 209–212, 317–338, 592, 593
Diepoxides, as endoperoxide rearrangement products 213, 214, 216
3,4-Diethoxydioxetane, decomposition of 883, 894
4,5-Diethylidene-2,2-dimethyl-1,3-dioxolane 865
Diethyl malonate, reaction with $O_2^{\cdot-}$ 437

Diethyl peroxide, decomposition of 240
photolytic 735
1,2-Difluoroethylene, reaction with $^1O_2$ 864
Di-(6-heptenoyl) peroxide 241
9,10-Dihydroanthracene, reaction with $O_2^{\cdot-}$ 440
2,24-Dihydro-4,24-dihydroxysigmosceptillin A, conformation of 384
1,2-Dihydronaphthalenes, reaction with $^1O_2$ 210
Dihydroperoxydialkyl peroxides, mass spectra of 118
9,10-Dihydrophenanthrene, reactions of 440
Dihydroxyarenes 726
reaction with $O_2^{\cdot-}$ 432, 433
1,1′-Dihydroxydicyclohexyl peroxide, mass spectrum of 121
cis-2,3-Dihydroxyindane 862
1α,25-Dihydroxy vitamin D_3, synthesis of 664, 668
Diimide, in selective reduction of unsaturated endoperoxides 320–323, 342, 343
2,2′-Diiododibenzoyl peroxide, topotactic iosmerization of 779
1,3-Diiodo-5,5-dimethylhydantoin, as iodine carrier 848
m-Diisopropylbenzene monohydroperoxide 419
Diisopropylbenzenes, autoxidation of 166
Diisopropyl peroxide, decomposition of 716, 717
α-Diketones,
as dioxetane rearrangement products 227
oxidative cleavage of 445, 448
Dilauroyl peroxide, decomposition of 788
Dimedone, reaction with dibenzoyl peroxide 283
1,4-Dimethoxyanthracene 1,4-endoperoxide 354
3,8-Dimethoxy-4,5,6,7-dibenzo-1,2-dioxacyclooctane, structure of 394
1,1-Dimethylallene, reaction with $^1O_2$ 864
Dimethylaminobenzenes, reaction with $^1O_2$ 211
N,*N*-Dimethylaminophenyl-1,2-dioxetane 895
α,α′-Dimethylbenzyl alcohol 717
Di-*O*-methylcurvularin, synthesis of 369
15,16-Dimethyldihydropyrene, reaction with $^1O_2$ 331
1,5-Dimethyl-6,7-dioxabicyclo[3.2.1]octane, decomposition of 362
Dimethyl-1,2-dioxetanone, enthalpies of formation for thermolysis of 882
2,3-Dimethyl-3-hydroperoxy-1-butene 850
2,4-Dimethyl-2-hydroperoxy-3-pentanone, decomposition of 575
Dimethylketene dimethylacetal, reaction with $^1O_2$ 864
1,4-Dimethylnaphthalene, photooxidation of 324, 325
1,4-Dimethylnaphthalene-1*H*-peroxide, as $^1O_2$ carrier 850
2,3-Dimethyl-1,4-naphthaquinone, reactions of 449
7,7-Dimethylnorbornene, epoxidation of 526
Dimethyl peroxide,
decomposition of 687–694
mass spectrum of 116
photolysis of 733, 734
UV spectrum of 732
Dimethyl-α-peroxylactone 908
decomposition of 847, 883, 894
Dimethyl trioxide, structure of 499
Dinitrites, cleavage of 880
Dinitrobenzenes, reactions of 451
m,*m*′-Dinitrobenzenesulphonyl peroxide, sulphonyl-^{18}O- labelled 639, 640
1,1-Dinitroethylene, superoxide cleavage of 879
1,2-Diols 910
1,3-Dioxabicyclo[1.1.0]butanes 537
2,3-Dioxabicyclo[2.2.1]heptane,
decomposition of 360, 362
synthesis of 315
1,3-Dioxacyclanes, ozonation of 653
1,2-Dioxacyclobutanes 209
Dioxacyclopentane, synthesis of 315
1,2-Dioxacyclopentanes, synthesis of 316
Dioxan, photoreaction of 755
1,2-Dioxanes, structure of 384
p-Dioxanyl hydroperoxide, structure of 392, 400
Dioxaphospholanes 910
Dioxarsolane 911
Dioxastilbolane 911
p-Dioxene-1,2-dioxetane, decomposition of 883
1,2-Dioxenes 318
Dioxetane 21, 71
1,2-Dioxetane, mass spectrum of 119
1,2-Dioxetanedione 830
1,2-Dioxetanes 202, 830–843
as endoperoxide rearrangement products 217
as reaction intermediates 860–880
as side-products 854
characterization of 859, 860
decomposition of 227, 704–707, 880–909
halogen-substituted 850
hydroxy-substituted 851
photoreduction of 862
purification of 858, 859
reaction of,

with electrophiles 912
with nucleophiles 228, 910–912
rearrangement of 227, 228
silyloxy-substituted 851
spectral properties of 832–843, 846, 847, 859, 860
structure of 844–846
sulphur-substituted 846
synthesis of 592, 847–852
by singlet oxygenation 209, 226, 849, 850
by the Kopecky method 848, 849
1,2-Dioxetanones—*see* α-Peroxylactones
1,2-Dioxetans,
O—O bond length in 377
structure of 384, 385
p-Dioxin dioxetane 846
1,2-Dioxin ring, structure of 384, 406
Dioxirane 13, 61, 71
Dioxiranes, as intermediates, in Baeyer–Villiger oxidations 563
1,2-Dioxolane, photolysis of 736
1,2-Dioxolans, structure of 384, 385
9,10-Dioxyanthracene, decomposition of 783–788
9,10-Dioxy-9,10-diphenylanthracene, decomposition of 781
Dioxygen–metal complexes 463–479
Dioxygen–organometallic complexes, X-ray analysis of 377
9,10-Diphenylanthracene, as fluorescer 894, 902, 903
9,10-Diphenylanthracene endoperoxide 207, 453
9,10-Diphenylanthracene 9,10-endoperoxide 355
4.4-Diphenyl-2,5-cyclohexadienone, photorearrangement of 908, 909
1,4-Diphenyl-2,3-dioxabicyclo [2.2.1]heptane, structure of 384
trans-3,4-Diphenyl-1,2-dioxetane, decomposition of 883
1,1-Diphenylethylene, cleavage of 874
2,5-Diphenylfuran, reactions of 451
2,3-Diphenyl-2-hydroperoxyvalerophenone, decomposition of 575
2,3-Diphenylindene, photooxidation of 327
1,3-Diphenylisobenzofuran,
as singlet oxygen trap 338
reactions of 451
1,3-Diphenylisobenzofuran cation radicals, reaction with $O_2^{\cdot-}$ 446, 456
Diphenylmethane, reaction with $O_2^{\cdot-}$ 440
1,1-Diphenyl-2-methoxyethylene, photooxidation of 327
Diphenylmethylcarbinyl hydroperoxide, complexes with aniline 144
3,4-Diphenylphenol 912
Diphenylphosphinyl chloride, reactions of 274
Diphenyl sulphide, reaction of,
with dioxetanes 229
with $O_2^{\cdot-}$ 445
1,3-Dipolar insertion 652–655
Dipole–dipole interactions 38, 66
Dipole moment,
of HO_2 57, 58
of H_2O_2 57
of hydrogen bonds 92, 93
of O_3 57
of peroxybenzoic acid 92, 93
Di-*n*-propyl peroxide, decomposition of 694
Dirhodium–dioxygen complex 826
Disordered molecules 385
Disproportionation 419, 739, 750
acid-catalysed, of $O_2^{\cdot-}$ 430, 431, 456
Dissociation,
of HO_2 25
of H_2O_2 60
of XO_2 compounds 5, 63
of X_2O_2 compounds 5
Dissociation energy,
of F_2O_2 63
of O_3 49, 58
of O_2 and its ions 49
Dissociation enthalpy,
of CO bonds 43
of OH bonds 42
of OO bonds 41
Distyrylmethane, reaction with $O_2^{\cdot-}$ 440
Disulphides 411
mass spectra of 116
reaction with $O_2^{\cdot-}$ 445
S—S bond length in 377
Di-*n*-valeryl peroxide, photolysis of 714
Divinyl ethers, formation in cleavage of allylic hydroperoxides 221–224
Donor–acceptor interactions 61, 66
Dreiding models 892

Eclipsed peroxide torsional geometry 384, 390
E2 eliminations, $O_2^{\cdot-}$-induced 441
Electric field gradient 57
Electrochemiluminescence 895
Electron affinity 44
Electron configuration, of O_2 9
Electron density 53, 55
bond critical point of 55
gradient of 55
Electronegativity, of O 2
Electronic spectra, of dioxetanes 847, 859
Electron spin resonance spectroscopy (ESR), in detection of radicals 241, 254, 714, 715,

720, 722, 727, 732, 733, 735, 738, 743, 748, 758, 761, 764
in detection of tetroxides 486, 487, 491
in detection of trioxides 497
Electron trapping 763, 764
Electrophilic aromatic substitution, by sulphonyl peroxides 266–268
Electrophilic behaviour of peroxides 522–566
β-Elimination mechanism, for base-catalysed rearrangement of peroxides 352
Enamines,
photooxidation of 228
reaction of,
with diacyl peroxides 283, 284
with $^1O_2$ 226, 849, 866–871
Endoperoxides 202, 831
bicyclic—*see* Bicyclic endoperoxides
formation in [4 + 2]cycloadditions 869
naturally occurring 312–314
^{18}O-labelled 592, 593
reactions of 212–217, 342–369
silica-gel-catalysed rearrangement of 851
synthesis of,
by nucleophilic displacement reaction 314–317
by $^1O_2$ reaction with 1,3-dienes 209–212, 317–338
by oxidation with triplet oxygen 338–342
thermal decomposition of,
as a method of preparing singlet oxygen 207, 208, 346–348
in the solid phase 781–783
thermal stability of 207
Ene reaction 209, 218–225, 588–590
as side-reaction in synthesis of four-membered ring peroxides 848, 850, 857, 858
mechanism of 219, 220
nature of the substrate in 218, 219
Energy,
correlation 40
experimental 40
Hartree–Fock limit 40
relativistic 40
theoretical 40
vibrational 40
zero-point 40
Enol ethers,
autoxidation of 873
reaction with $^1O_2$ 226, 864, 865
Enols,
formation in cleavage of allylic hydroperoxides 224
reaction of 269
with $O_2^{\cdot-}$ 435, 438
with $^1O_2$ 864–866
Enones,
formation in cleavage of allylic hydroperoxides 224
reaction with $O_2^{\cdot-}$ 438
Entropy values, estimation of 102, 103
Epidioxides—*see* Endoperoxides
8α,10α-Epidioxy-8,14-dihydro-14β-nitrothebaine, structure of 384
(6*R*)-6,19-Epidioxy-9,10-*seco*-5(10),7,22-ergostatriene-3β-ol benzoate 384
3-Epiuleine, synthesis of 365
Epoxidation 68, 816
of alkenes 177–184, 191, 192, 288–292, 525–534, 747
mechanism of 288–290
molybdenum-catalysed 470, 472, 478, 824, 825
transition state in 530, 531
vanadium-catalysed 534
of alkynes 292–294
of allenes 292–294
Epoxides 650
as photolytic products 753, 754
cleavage of 880
formation of 288–294
from dioxetanes 910, 911
in cleavage of allylic hydroperoxides 224
in oxidation of alkenes 525–534
in reactions of $O_2^{\cdot-}$ 439, 443, 446
Epoxy aldehydes, as endoperoxide rearrangement products 216, 217
Ergosterol, photooxidation of 831
Ergosterol endoperoxide 312
Ester hydroperoxides, as epoxidizing agents 533
Esters,
complexes with hydroperoxides 145, 146
reaction with $O_2^{\cdot-}$ 442, 443
α,β-unsaturated—*see* α,β-Unsaturated esters
Ethane, as photolytic product 722, 738, 739
Ethanol, as photolytic product 735
Ether hydrotrioxides 503–507
decomposition of 504, 507, 508
Ethers,
complexes with hydroperoxides 142, 143
ozonation of 501, 503–506, 652
photoreaction of 721, 722, 755
Ethoxycarbonylmethylation 718
Ethylbenzene,
autoxidation of 166
photoreaction of 166
Ethylbenzene hydroperoxide,
synthesis of 163, 166, 167
use in organic synthesis 166, 178, 190
Ethyl chloroacetate, photoreactions of 718
Ethylene,
as photolytic product 735, 739

reaction with $^1O_2$ 864
Ethylene glycol dimethyl ether, photooxidation of 721
Ethylene glycol monomethyl ether, photooxidation of 721, 722
Ethylene ozonide, structure of 387
5-Ethylidene-2,2-dimethyl-1,3-dioxolan-4-one 865
Ethyl radical 739
Euler angles 58
Europium tris(thenolytrifluoroacetonate)-1,10-phenanthroline, as fluorescer 904
Excitation parameters, of four-membered ring peroxides 897–909
Excited radicals 716, 717, 732, 733, 737
Excited states,
of HO_2 58
of H_2O_2 59
of O_2 7, 13, 21
of O_3 58
of XO_2 7
of X_2O_2 8
Excited substrates 718, 719, 747

Fatty acids, photooxidation of 722
Fenton reagent 720
Ferrocene cation radicals, reaction with $O_2^{\cdot-}$ 446, 456
Ferrocenophane cyclic peroxides, mass spectra of 119, 120
Field latex 422
Flavins, reaction with $O_2^{\cdot-}$ 436
Flavoenzymes 878
Flip–flop rotation 69
Fluorene, reaction with $O_2^{\cdot-}$ 440
Fluorenes, reaction with $^1O_2$ 861
Fluorenone 861, 903
as photosensitizer 730
Fluorescence 847, 881, 883, 897, 901–904
Fluorescers 894, 895, 897, 902–904
F_2O_2 21, 63, 66
FO_2H 63
Force constants,
of H_2O_2 52
cubic 53
quadratic 53
of H_2O_3 53
of O_3 53
of O_2 and its ions 53
of XO_2 61
Formaldehyde,
as photolytic product 733–735
n,π* fluorescence of 864, 894
Formic acid, as photolytic product 750
Formyl radical 734
Fourier expansion 38
Free radicals—*see* Radicals
Friedelane, ozonation of 668
Friedelin, ozonation of 668, 670
F_2S_2 65, 66
Fumarate ion, photoreaction of 758
Fumaronitrile 908
Fumitremorgin A 314, 396
Functionalization, reiterative 756
Furanoterpenes, synthesis of 215
Furans,
reaction with $^1O_2$ 211, 332
synthesis of 214, 215
Furfurylamines, reaction with $O_2^{\cdot-}$ 440

Gauche effect 89
Geometry,
of cyclic peroxides 71
of cyclic polyoxides 71
of F_2O_2 66
of F_2S_2 66
of HO_2 25
of H_2O_2 36, 49, 50, 51
of H_2O_3 69
of LiO_2 62
of O_2 48
of O_3 25, 48
of O_2 ions 48
of ozonides 71
of XO_2 62
of X_2O_2 66
Gilvanol 389
Glycine, as radiolytic product 762
Glycol ethers, photooxidation of 721
Graft copolymers 418–426
Grob fragmentation 852, 853, 857
Group additivity 97, 100–102
Guanine, photomethylation of 751

Haber–Weiss chain mechanism 720, 762
Haber–Weiss reaction 452, 453
Hafnium peroxo complexes 468, 469
o-Halodibenzoyl peroxides, structure of 407
1-Halo-2,4-dinitrobenzenes, reactions of 451
Halogenoquinolines, reaction with $O_2^{\cdot-}$ 452
α-Haloketones, oxidative cleavage of 448
Hammett correlation, of pK_a values 131
Hammett equation, in reaction of *t*-butylperoxyl with ring-substituted toluenes 248
Hammett plots 767, 768
Hammett ρ values 611
$HBeO_2BeH$ 63
$HBeO_2H$ 63
Heat capacities, estimation of 102, 103
Heats of formation 40, 98–102
Helix conformation, of H_2O_n 69
Hemiketals 727

Heptyl hydroperoxides, self-association in 136, 138
Heteroarenes, reaction with $^1O_2$ 871–873
Heterocoerdianthrone, photooxidation of 331
Heterocyclic compounds, reaction with $^1O_2$ 211, 332–338
Heterolytic cleavage, of OO bond 43
Hexadeuterioacetyl peroxide 598
2,2′,4,4′,6,6′-Hexadeuteriobenzoyl peroxide 619
trans-Hexahydrophthaloyl peroxide, carbonyl-^{18}O-labelled 597
Hexamethylbenzene, reaction with $^1O_2$ 326
Hexamethylcyclotrisilaperoxane, structure of 397
1,2,4,5,7,8-Hexaoxacyclononanes, structure of 396
HNO_2 62
HO_2 7, 13, 25, 45, 52, 57, 58, 69
  cyclic form of 25
  linear form of 25
  self-reaction of 69
HO_2^+ 13, 45
HO_{2n}^+ 69
H_2O_n 68, 69
H_2O_2—*see* Hydrogen peroxide
H_2O_3 63, 69
H_2O_4 63, 69
Hock cleavage 555
  of allylic hydroperoxides 221–225
  vs. dioxetane cleavage 222, 229
Hock–Criegee rearrangement 878
Hock dehydration 224, 225
Hole state 44
Homo-Diels–Alder reaction 211
Homolytic cleavage, of OO bond 41, 173–176, 224, 281, 282
Humic acid 726
Hydrazines, reaction with $O_2^{-\cdot}$ 436, 437
Hydride migration, in peroxy acid oxidation of aldehydes 565
Hydrocarbon hydrotrioxides 513, 514
Hydrocarbons,
  aromatic—*see* Aromatic hydrocarbons
  autoxidation of 163–168, 588
    rate constants for 164
  hydroxylation of 534–536
  ozonation of 501, 513, 514, 655–680
  polynuclear aromatic reaction with $^1O_2$ 207, 211
Hydrocinnamoyl-β,β-d_2 peroxide 599
Hydrogen atoms, photolytic generation of 750
Hydrogen bonding 376
  in bis(cyclophosphamide) peroxide 397, 398
  in hydroperoxides 135–155, 381, 400–403
  in hydrotrioxides 515–517
  in peroxy acids 92, 93, 134, 135, 403, 404
Hydrogen migration, intramolecular, in peroxy diesters 107
Hydrogen pentoxide 485
Hydrogen peroxide,
  as hydroxylating agent 536
  as oxidizing agent 543, 567–570
  bridged form of 13, 25
  concentrated 848
  conformation of 20, 36–40, 86–89
  ^{17}O- or ^{18}O-labelled 587
  oxidation of, as a method of preparing $^1O_2$ 205, 206
  photolysis of 714, 719–727
  physical properties of 25, 36–40, 46, 47, 49–60
  radiolysis of 761, 762
  reaction of,
    with acylating agents 593, 594
    with alkenes 591
    with alkylating agents 590, 591
    with carbonyl compounds 591, 592
    with imines 591
    with α-lactones 854
    with $O_2^{-\cdot}$ 452
  thermochemical data for 98–100
  UV spectrum of 720, 732
  Y form of 13, 25
Hydrogen polyoxides 484–486
Hydrogen tetroxide 484–486
Hydrogen trioxide 484–486
Hydroperoxides—*see also* Alkenyl hydroperoxides, Alkoxyalkyl hydroperoxides, Alkoxyhydroperoxides, Alkyl hydroperoxides, Allylic hydroperoxides, Amide hydroperoxides, Nitrile hydroperoxides
  acid-catalysed cleavage of 554, 555
  acidity of 131–133
  as epoxidizing agents 533, 534
  as $O_2^{-\cdot}$ disproportionation catalysts 432
  complexes of,
    with amines 156, 157
    with ammonium compounds 157
  decomposition of 685, 686
  formation in reaction of heteroarenes with $^1O_2$ 871
  free-radical reaction mechanisms involving 235, 247–251
  intermolecular association of 138–150
    with alcohols and phenols 143
    with alkenes and alkynes 139, 142
    with amines 144
    with arenes 138–141
    with carbonyls and carboxylic acid

derivatives 144, 145
with carboxylic acids 143
with sulphoxides 145–148
intramolecular association of 150–155
isotopically labelled 587–591
of transition metals 464, 476
polymeric 419–425
reaction of,
with alkylating agents 591, 592
with carbonyl compounds 592
self-association in 135–138
steric effects in dimer/trimer equilibrium of 136
structure of 398–403
torsional angles in 381
α-Hydroperoxy acids,
cyclization of 832, 843, 855–847
decarboxylation of 852, 854
synthesis of 852–855
α-Hydroperoxyalkyl radical 250
Hydroperoxy anion, as oxidizing agent 545
4-Hydroperoxycyclophosphamide, structure of 400
3-Hydroperoxy-2,3-dimethylbut-1-ene, radical-induced decomposition of 250
α-Hydroperoxy-α,α-diphenylacetophenone, photolysis of 730
α-Hydroperoxy-α,α-diphenylacetophenones, decomposition of 576
α-Hydroperoxyethyl methyl ketone, synthesis of 249
4a-Hydroperoxyflavin, as oxidizing agent 540, 543
4-Hydroperoxyisophosphamide, structure of 400, 401
α-Hydroperoxyketones,
decomposition of 831
acid-catalysed 556
base-catalysed 575
photolysis of 730
Hydroperoxy radical 113, 762
pK_a of 431
α-Hydroperoxytetrahydrofuran, synthesis of 249
4-Hydroperoxytrophosphamide, structure of 400
Hydrophenazines, reaction with $O_2^{\cdot-}$ 436
Hydroquinone, manufacture of 166
o-Hydroquinone, reaction with $O_2^{\cdot-}$ 433
Hydrosilanes, ozonation of 653, 654
Hydrotrioxides 501–517
decomposition of 252, 253, 503–515
formation in ozonation of single bonds 651–657
mechanism of formation of 501–503
structure of 515–517
Hydroxyacetophenones, oxidation of 561
α-Hydroxy acids, photooxidation of 723
Hydroxybenzaldehydes, oxidation of 567
Hydroxybenzoic acid 719
Hydroxybisflavenylidenes, reaction with $^1O_2$ 865
2-Hydroxycholesta-1,4-dien-3-one, reaction with $O_2^{\cdot-}$ 433
6-Hydroxychroman compounds, reaction with $O_2^{\cdot-}$ 432
4-Hydroxycyclophosphamide, structure of 397
2-Hydroxy-1,2-dehydrotestosterone, reaction with $O_2^{\cdot-}$ 433
Hydroxydioxetanes 831
Hydroxyenones, as endoperoxide rearrangement products 216
4-Hydroxy-2-en-1-ones, synthesis of 214, 215
3-Hydroxyflavones, reaction with $O_2^{\cdot-}$ 879
α-Hydroxyketones,
as dioxetane rearrangement products 227, 228
oxidative cleavage of 445, 448
Hydroxylamine, reaction with $O_2^{\cdot-}$ 436
Hydroxylation 743, 745
Hydroxyl radical 60, 714, 719, 720, 727, 730, 762
3-Hydroxy-3-phenylpropionitrile, as mass spectral product 110

Imidazoles, reaction with $^1O_2$ 211, 332, 871, 873
Imines, reactions of 570–573, 591, 860
1,6-Imino[10]annulene, reaction with $^1O_2$ 325
Imino-1,2-dioxetanes,
decomposition of 891, 906
nomenclature of 830, 843
reaction with nucleophiles 911
synthesis of 852, 858
Imino ethers, reaction with $^1O_2$ 865
Immonium cations, reaction with $O_2^{\cdot-}$ 446
Indenes, reaction with $^1O_2$ 210, 327
Indole dioxetane 847
Indoles, reaction with $^1O_2$ 211, 871, 872
Indolinones 911
3-Indolyl-1,2-dioxene, reaction with $^1O_2$ 851
Infrared intensities, of HO_2 52
Infrared spectroscopy, of α-peroxylactones 843, 846, 859
Intermolecular substitution at the peroxide oxygen 522–551
general characteristics of 522–524
Intramolecular sensitization 742
Inversion symmetry in the crystal 382
Iodides, oxidation of 546
2-Iodo-2′-chlorodibenzoyl peroxide, topotactic isomerization of 779

2-Iododibenzoyl peroxides, structure of 408
β-Iodohydroperoxides 848, 849
Iodosobenzene, reactions of 487, 494
Iodosyl compounds, oxidation of 546
Ionicity, of XO bond 2
Ionic states,
 of HO_2 25, 58
 of O_3 25
 of XO_2 7
 of X_2O_2 10
Ionization potential 43, 44, 59
 of cyclic peroxides 47
 of HO_2 45
 of H_2O_2 46
 of O_2 44
 of O_3 44
 of organic peroxides 47
Ion pair 44
Ion-radical pair intermediates, in thermolysis of diacyl peroxides 628, 630
Iridium hydroperoxidic species 478
Iridium peroxo complexes 473
Iron peroxo complexes 471
Isoascaridole 350
Isobutane, autoxidation of 164
Isobutyraldehyde 860
Isopropanol,
 photooxidation of 721
 radiolysis of 762
Isopropenyl ethers, reaction with $^1O_2$ 864
Isopropoxy radicals 717
Isopropyl ether, ozonation of 503
9-Isopropylidenebenzonorbornenes, epoxidation of 527
Isopropyltoluenes, autoxidation of 167
Isosteric molecules 411
Isotopic labelling in mass spectra,
 of ozonides 123, 125
 of β-peroxylactones 110

Ketene, reaction with $^1O_2$ 864
Ketene acetals, reaction with $^1O_2$ 849, 853, 854, 864
Ketene bis(trimethylsilyl)acetals, reaction with $^1O_2$ 853, 854
Ketenes,
 as intermediates in photolysis of phenanthrenes 874, 875
 ozonation of 854, 855
 reaction with $^1O_2$ 226, 831, 843, 852, 857, 858
Ketenimines, reaction with $^1O_2$ 852, 858
Keto acetals 861, 862
α-Keto acids,
 as photooxidation products 723
 oxidative cleavage of 445, 448
α-Ketohydroperoxides, photolysis of 730
Ketoketenes, as photolytic products 742
Ketone peroxides,
 dimeric 391, 397
 trimeric 396, 397
Ketones,
 aromatic—*see* Aromatic ketones
 as photolytic products 721, 730, 753, 754
 complexes with hydroperoxides 144, 145
 conjugated—*see* Conjugated ketones
 oxidation of 187, 295
  with $O_2^{\cdot-}$ 438, 444, 445, 448, 449
  with peroxy acids 559–564
 ozonation of 679
δ-Ketonitriles, as photolytic products 755, 756
Kharasch–Sosnovsky reaction 300, 301
Kirkwood's formula 44
Koopmans' theorem 44, 59
Kopecky method 831–840, 848, 849
Kornblum–DeLaMare reaction 214, 221, 225, 227, 453
Kyunrenines 872

α-Lactone elimination, in mass spectral fragmentation of β-peroxylactones 110, 112
Lactones, as photolytic products 742
α-Lactones 911
 reaction with H_2O_2 853–855
Lead tetraacetate,
 as oxidizing agent 487, 44
 reaction with alkyl hydroperoxides 251
Leffler carboxy inversion 557
Levopimaric acid methyl ester, photolysis of 350
Lewis acids,
 as catalysts,
  for decomposition of diacyl peroxides 626
  for oxidation of cyclic conjugated dienes 341
 reaction with 1,2-dioxetanes 912
Linalool, synthesis of 664
Linear free energy correlation, for hydroperoxides 138, 147
Li_2O_2 63
LiO_2H 63
Lithium aluminium hydride, as reducing agent,
 for 1,2-dioxetanes 910
 for endoperoxides 342–344
Lone pair–lone pair repulsion 3, 7, 8, 69, 71
Lucigenin, cleavage of 878

Macrocyclic peroxide ethers, mass spectra of 121, 122
Magic acid 677–679

Maleic acid, photoreaction of 758
Maleonitrile, isomerization of 908
Malononitrile, reaction with $O_2^{\cdot -}$ 437
Malonyl peroxides, photodecarboxylation of 853, 854
Mandelic acids 878
Markownikoff orientation, in reaction of sulphonyl peroxides with olefins 268
Mechanochemical degradation 424
Mesitylene, hydroxylation of 534, 535
Mesityl hydroperoxide, decomposition of 793
Metal ions, as catalysts,
  in dialkyl peroxide decomposition 240
  in endoperoxide decomposition 356–359, 362–364
  in hydroperoxide decomposition 251
Metal–peroxide bonding 382, 383
Metastable ions 110, 116, 113
Methane, as photolytic product 722, 735, 738
Methanediol model 395
Methanol,
  as photolytic product 733
  photooxidation of 721
1,6-Methanol[10]annulene, photooxidation of 330
Methoxybenzenes, reaction with $^1O_2$ 211
α-Methoxybenzyl hydroperoxide, decomposition of 574, 575
o-Methoxycarbonylperbenzoic acid, photolysis of 747
1-Methoxycyclohexene, reaction with $^1O_2$ 865
Methoxymethyleneadamantane, autoxidation of 873
Methoxymethylenedodecane, autoxidation of 873
α-Methoxynitriles, as photolytic products 756
p-Methoxy-*p'*-nitrobenzoyl peroxide, thermolysis of 623, 631
Methoxy radical 734
N-Methylacridanyl-1,2-dioxetane 895
Methylation, photochemical,
  of purines 751
  of side-chains 745
  of toluene 745
  of xylene 746, 747
  orientation in 718, 738, 745, 747, 752
  pH effect on 751, 752
Methyl benzoate 874
α-Methylbenzyl alkyl ether hydrotrioxides, decomposition of 504, 507
α-Methylbenzyl alkyl ethers, ozonation of 504
α-Methylbenzyl *t*-butyl peroxide, decomposition of 573
α-Methylbenzyl hydroperoxide, decomposition of 574
2-Methyl-2-butene 908
Methyl *t*-butyl trioxide 497, 498
1-Methylcyclohexyl hydroperoxide, complex with hexanoic acid 143
4-Methyl-2,3-dihydro-γ-pyrane, ene reaction of 589
2-Methyl-4,5-diphenylimidazole, ring-cleavage of 873
Methylene blue, as photosensitizer 849
Methylenecyclopropanes, reaction with $^1O_2$ 863
Methylglyoxal 847
N-Methyl-3-(2-hydroxyethyl)indole, reaction with $^1O_2$ 872
Methyl(5*Z*,8*E*,10*E*,12*S*)-12-hydroxy-5,8,10-heptadecatrienoate 363
3-Methylhypoxanthine, photomethylation of 752
N-Methyl-3-indolecarboxaldehyde 869
Methylindoles, reaction with $^1O_2$ 871
N-Methylindolyl-1,2-dioxetane 895
3-(*N*-Methyl)indolylstyrene, reaction with $^1O_2$ 869
Methyl isopropyl ether, ozonation of 503
Methyl ketones, formation of 472, 473, 477
p-Methyl-*p'*-nitrodibenzoyl peroxide, decomposition of 789, 790
2-Methyl-2-pentene, reactions of 440, 449
Methylperoxyl, reactions of 240
Methyl peroxynitrate 698, 699
Methyl peroxynitrates, halogenated 699
γ-Methyl-γ-peroxyvalerolactone, photolysis of 755
4-Methylpyridine *N*-oxide, complex with *p*-nitroperoxybenzoic acid 155
Methylpyridine-*N*-oxides, reaction with $O_2^{\cdot -}$ 441
Methylpyridines, reaction with $O_2^{\cdot -}$ 441
2-Methylpyrrole, reaction with $^1O_2$ 871
Methyl radical 718, 733, 743, 751
2-Methyltetrahydrofuran, ozonation of 503
Methyl viologen, reactions of 447
Microwave discharge, as a method of preparing $^1O_2$ 208
MINDO/3 calculations, for dioxetane decomposition 893
Molecular orbital description,
  of F_2O_2 21
  of HO_2 13
  of H_2O_2 13
  of O_2 8
  of O_3 13
Molecular oxygen 2, 6–8, 21, 48, 53, 69, 201–229
  electronic states of 202, 203
  ground-state 202
  molecular orbitals of 203
  singlet—*see* Singlet oxygen

Molozonides 701
Molybdenum peroxo complexes 469–472, 478
Mononitroethylene, superoxide cleavage of 879
Muconic acid 724
Mulliken population analysis 53

Naphthalene, reactions of 451
Naphthalene 1,4-endoperoxide, synthesis of 325
Naphthalene endoperoxides, thermolysis of 347
Neighbouring-group participation, in decomposition of peresters 634
Neoconcinndiol hydroperoxide, structure of 400
N—H bonds, reaction with $O_2^{\overline{\cdot}}$ 436, 437
NH_2O_2H 63, 66
$NH_2O_2NH_2$ 63
Nickel peroxo complexes 474
Niobium–dioxygen complex 822, 823
Niobium peroxo complexes 469
Nitric oxide, as radical scavenger 717, 737
Nitrile hydroperoxides, as epoxidizing agents 533
Nitriles, oxidation of 445, 568, 569, 876
Nitroanilines, reaction with $O_2^{\overline{\cdot}}$ 437
Nitroaromatic amines, reaction with $O_2^{\overline{\cdot}}$ 437
Nitrobenzenes, reaction with $O_2^{\overline{\cdot}}$ 451
m-Nitrobenzenesulphonyloxyperoxide, photolysis of 760
m-Nitrobenzenesulphonyl peroxide, sulphonyl-^{18}O- labelled 597
Nitrogen compounds, oxidation of 184–186, 297, 298, 724, 725
Nitromethane, photoreaction of 758
Nitrones, reaction with $O_2^{\overline{\cdot}}$ 446
Nitro olefins, oxidative cleavage of 450
p-Nitroperbenzoic acid, radiolysis of 765, 766
Nitroperoxybenzoic acid, structure of 405
m-Nitroperoxybenzoic acid, decomposition of 794, 795
o-Nitroperoxybenzoic acid, X-ray study of 134
*p*Nitroperoxybenzoic acid, decomposition of 794, 795
N-Nitroso-*N*-acyl-*O*-butylhydroxylamine, rearrangement of 632
N-Nitrosoamine, rearrangement of 632
Nitroso compounds,
  as radical scavengers 732, 733
  oxidation of 540
Nitrosodurene, as radical scavenger 733
Nitrotoluenes, reaction with $O_2^{\overline{\cdot}}$ 440
Nitrous oxide, as electron scavenger 763
Nitroxides 654, 655
NO_2^- 62
Norbornane, ozonation of 661, 666
Norbornene, epoxidation of 526
7,7′-Norbornylidene, reactions of 592
Norbornylidenenorbornanedioxetane, thermal stability of 227
Norbornyl peresters, photolysis of 750
Norcaradiene 323
Norrish Type I cleavage, of dioxetanes 906
Nuclear magnetic resonance (NMR) spectroscopy,
  in identification of hydrotrioxides 503, 504, 509, 513
  in peroxide decomposition 241
  of 1,2-dioxetanes 832–843, 846, 859, 860
  of α-peroxylactones 843, 846, 859
  ^{17}O-NMR 635
Nucleophilic addition, to dioxetanes 228
Nucleophilic behaviour of peroxides 566–573
Nucleophilic displacement,
  intramolecular, of hydroperoxides 592
  on aliphatic carbon by H_2O_2 590
  on electrophilic oxygen 522–551

O_2^+ 44, 48, 61
$O_2^{\overline{\cdot}}$—*see* superoxide anion radical
O_2 ($^1\Delta_g$)—*see* Singlet oxygen
$^1O_2$—*see* Singlet oxygen
$^3O_2$—*see* Triplet oxygen
O_4 69
O_5 69
4-Octyne, oxidation of 538
O—H bonds, reaction with $O_2^{\overline{\cdot}}$ 432–435
Olefinic monomers, polymerization of 622–624
Olefins—*see* Alkenes
Oligomers 417, 425
One-electron properties 57
Orbital energy,
  of HO_2 45
  of H_2O_2 38, 46
  of O_2 44
  of O_3 44
Organoaluminium peroxides 816, 817
Organoantimony peroxides 823
Organoarsenic peroxides 823
Organoberyllium peroxides 810
Organobismuth peroxides 823, 824
Organoboron peroxides 812–816
Organocobalt peroxides 825
Organogallium peroxides 817
Organohafnium peroxides 818
Organoindium peroxides 817
Organoiridium peroxides 826
Organolead compounds 274
Organolead peroxides 822
Organomagnesium peroxides 810

Organomercury compounds, reaction with methanesulphonyl peroxide 274
Organomercury peroxides 810–812
Organometallic compounds,
 autoxidation of 168
 structure of 382, 383
Organomolybdenum peroxides 824, 825
Organorhodium peroxides 826
Organosilicon compounds 274
Organothallium peroxides 817
Organotin peroxides 820–822
Organotitanium peroxides 818
Organotungsten peroxides 824, 825
Organozirconium peroxides 818, 819
Oscillator strength 58, 60
Osmium peroxo complexes 471
Overlap 20, 21
Overlap population 53
9-Oxabicyclo [3.2.1]octan-3-ones, oxidation of 560
[Oxabis(*t*-butylperoxytriphenylantimony)], structure of 382
2-Oxa-Δ^4-steroids, synthesis of 438
Oxaziridines 570
Oxazoles, reaction with $^1O_2$ 211, 332, 333
Oxetanes, structure of 387, 388
Oxidation, controlled thermal 426
Oxidative fragmentation, of peroxy esters 107
1,6-Oxido [10]annulene, photooxidation of 330
Oximes 411
 oxidation of 226, 540, 541
Oximino ethers 411
Oxiranes—*see* Epoxides
Oxirenes 292, 537, 538
Oxonium cations, reaction with $O_2^{\cdot-}$ 446
3-Oxovincadifformine 867
Oxygen,
 atomic 6
 molecular—*see* Molecular oxygen
Oxygen isotope effects, in decomposition of diacyl peroxides 616
Oxygen-18 labelling 876
Oxygen scrambling,
 in cycloalkaneformyl peroxides 620, 621
 in decomposition of diacyl peroxides 780, 781
 in diacyl peroxides 615, 617, 618, 627, 628, 631
  activation entropies for 614
  effect of the phenyl group on 611, 613
  mechanisms for 609, 610
 in peresters 603, 605, 607
  mechanism for 608
Oxygen transfer, mechanism of 171
Oxyluciferin anion 896, 897
Oxythiazoline chromophore 897
Ozonation 423, 507, 509
 dry 658–675
 in strongly acidic media 675–680
 low-temperature 501–506, 513–515
 of acetals 652, 653
 of aldehydes 653, 679
 of alkenes 632–634, 650
 of amines 654, 655
 of anthrones 654
 of ethers 652
 of heteroatom nucleophiles 651
 of hydrosilanes 653, 654
 of ketenes 854
 of saturated hydrocarbons 655–680
Ozone 13, 24, 44, 53, 57, 58
 biradical character of 25, 43
 cyclic form of 24
 decomposition of 24
 linear form of 25
Ozonides 650
 bond angles in 381
 C—O bond length in 377
 decomposition of 700–703
 isotopically labelled 632–634
 mass spectra of 123–125
 naturally occurring 389
 reactions of 206, 207
 substituent effects in 69, 71–73
Ozonization—*see* Ozonation
Ozonolysis 61, 73

Palladium(II) *t*-butyl peroxide carboxylates 476, 477
Palladium–dioxygen complex 826
Palladium peroxo complexes 474
[2,2]Paracyclophane, naphthalene analogue of, photooxidation of 325
Paraquat, oxidative degradation of 879
Pariser–Parr–Pople SCF-MO method 530
Patchoulol, ozonation of 671
Patchoulyl acetate, ozonation of 671
1,3-Pentadiene, as triplet quencher 741
Peptides, photooxidation of 724, 725
Peracetic acid,
 ^{18}O-labelled, decomposition of 641
 photoreactions with 743–747
Peracids,
 free-radical reaction mechanisms involving 236, 252
 isotopically labelled 593, 594
 reaction with acylating agents 598, 599
Perbenzoic acid, ^{18}O-labelled 593, 594
 hydrolysis of 587
Perdeuteriobenzoul peroxide 619
Perepoxide mechanism 589
Perepoxides, as intermediates,

in [2 + 2]cycloaddition of $^1O_2$ to double bonds 226
in $^1O_2$ Diels–Alder reaction 212
in $^1O_2$ ene reaction 220
Peresters,
decomposition of 600–608, 634–636
photolytic 716
isotopically labelled,
synthesis of 594–596
uses of 600–608, 623, 634–636
rearrangement of 623
Perfluorodi-*t*-butyl peroxide, decomposition of 696, 697
Performic acid 66
Perhydrates 376
Peroxides—*see also* Diacyl peroxides, Dialkyl peroxides, Diaroyl peroxides, Dihydroperoxydialkyl peroxides, Ketone peroxides
bicyclic—*see* Bicyclic peroxides
cyclic—*see* Cyclic peroxides
substituted 61, 63
transannular—*see* Endoperoxides
Peroxirane 65
Peroxiranes—*see* Perepoxides
Peroxo complexes 465–474
reaction of,
with electrophiles 467
with nucleophiles 467, 468
synthesis of 466, 467
μ-Peroxo complexes 475–478
Peroxy acetals, structure of 387–398
Peroxyacetic acid,
as oxidizing agent 563
decomposition of 501
MO calculations for 523, 524, 530
α-Peroxyacetoxynitriles, photolysis of 756
Peroxyacetyl nitrate 697
mass spectrum of 110
Peroxy acids,
acidity of 130, 131
aromatic—*see* Aromatic peroxy acids
as oxidizing agents 523–546, 559–566, 570–573
conformation of 66–68, 92, 93
decomposition of 685, 686
photolytic 714
radiolytic 765–767
solid-state 794–796
hydrogen bonding in 92, 93, 134, 135
hydrolysis of 568
photoreactions of 743–748
reactions of 288–299
with alkenes 288–292
with alkynes and allenes 292–294
stability and structure of 288
structure of 381, 403–405
substituent effects in 66
synthesis of 287, 288, 568
UV absorption of 743
Peroxy acid–solvent interactions 538
Peroxyacyl nitrates 697
Peroxyalkylpalladium(II) 172
Peroxybenzoic acid,
as oxidizing agent 538, 541, 545, 561, 564
dipole moment of 92, 93
Peroxybenzoic acids,
complexes of 155, 156
Hammett correlation for 134
Peroxybenzoyl nitrate 697
Peroxycarboximidic acid 568
Peroxycarboxylates—*see* Peroxy esters
Peroxycarboxylic acids—*see* Peroxy acids
Peroxycarboxylic esters—*see* Peroxy esters
Peroxychloroformyl nitrate, decomposition of 700
Peroxydiphosphates,
photolysis of 758
radiolysis of 767–770
Peroxydisulphates, radiolysis of 767–770
Peroxy esters,
aromatic—*see* Aromatic peroxy esters
conformation of 93, 94
decomposition of 685
base-catalysed 577, 578
photolytic 714, 748–757
free-radical reaction mechanisms involving 236, 253–255
mass spectra of 106–110
reactions of 300–303
rearrangement of 551–554
stability of 299
structure of 405
synthesis of 299
UV absorption of 748
Peroxyformate, photolysis of 750
Peroxyformic acid, as oxidizing agent, theoretical study of 530, 531, 538, 539
Peroxyketals, structure of 387–398
Peroxylactones,
mass spectra of 110–113
photolysis of 753–755
α-Peroxylactones 830–832, 843, 844
as reaction intermediates 874–876
decomposition of 855, 880–909
purification of 858
reaction with nucleophiles 911
spectral properties of 843, 846, 847, 859
synthesis of 847, 848, 852–858
by α-hydroperoxy acid route 852–857
by singlet oxygen route 857, 858
β-Peroxylactones, decomposition of 255, 256
Peroxymercuration 316

Peroxymetalation,
cyclic 468
pseudocyclic 477
Peroxymonophosphoric acid,
as hydroxylating agent 535
as oxidizing agent 562, 566
Peroxynitrates 686, 697–700
decomposition of 699, 700
Peroxynitric acid 697
decomposition of 698, 699
Peroxypelargonic acid,
structure of 403
X-ray study of 134
Peroxyphosphates, photoreaction of 759
Peroxyphosphonates, photoreaction of 759
Peroxypropionyl nitrate, decomposition of 700
Peroxysulphates, photolysis of 757
Peroxysulphuric acid, as oxidizing agent 563
Peroxytrifluoroacetic acid, MO calculations for 530
PGE_2 313, 359
PGG_2 313, 359
PGH_2 313, 359
Phenafulvenes, photooxidation of 331
Phenanthrenes, autoxidation of 874
α-Phenethyl alcohol, as photolytic product 745
α-Phenethyl hydroperoxide, radiolysis of 763
Phenol,
as photolytic product 719, 723, 730
as radiolytic product 762
Phenolates 878
Phenols,
as photooxidation products 725
complexes with hydroperoxides 143
hydroxylation of 726
oxidation of 432, 724
Phenylacetylene, as photolytic product 742
Phenylacetylenes, oxidation of 537, 538
Phenylalanine, photooxidation of 724
Phenylation, photolytic, orientation in 740
Phenylbenzimide 865, 867
Phenyl benzoate, as photolytic product 740
Phenylboronic peroxy anion 814
Phenylboric acid, ^{18}O-labelled 641
Phenyl *t*-butyl ether, as photolytic product 752
1-Phenylcyclobutene, reaction with $^1O_2$ 863
o-Phenylenediamine, reaction with $O_2^{\cdot -}$ 436
1-Phenylethyl peroxyacetate, decomposition of 577
1-Phenylethylperoxyl 248
β-Phenylisobutyryl peroxide, optically active 625
Phenylmaleoyl peroxide, photolysis of 742
Phenyl migration, from phosphorus to oxygen 275
Phenyl α-naphthyl ketone, as photosensitizer 730
2-Phenyl-2-norbornene, reaction with $^1O_2$ 863
Phenyl radical 719, 740, 743, 765
ESR spectrum of 714
p-Phenylstyrene, epoxidation of 530
δ-Phenylvaleryl peroxide 626
Phosgene 874
Phosphate radical 769
Phosphate radical anion 758, 768
Phosphates, as photolytic products 759
Phosphonates, as photolytic products 759
Phosphorescence 847, 881, 883, 897, 901, 902
Phosphorus nucleophiles, reaction with 1,2-dioxetanes 910, 911
Phosphorus ylides, [2 + 2]cycloaddition of $^1O_2$ to 226
Phosphorylation 759
Photochromism 867
Photodynamic action 202
Photo-endoperoxides 588
Photooxidation 201–229, 446, 447, 831–844, 849–852, 857, 858, 861–873
classes of 204
of alkenes 588–590, 592
Photosensitization 741, 744, 849, 862, 865
as a method of preparing $^1O_2$ 204, 205
rate of reaction and product distribution in 209
Photosensitized oxidation 201–229
Photosensitizers,
polymer-based or adsorbant-bound 205
solubility properties of 205
triplet energy of 204
Phthaloyl peroxide,
as oxidizing agent 549, 550
carbonyl-^{18}O-labelled 597, 618, 639
photolysis of 742
structure of 405
Pinacol 721
Plakortin 313
Plastoquinone-1, photooxidation of 340
Platinum hydroperoxidic complexes 478
Platinum peroxo complexes 474
Polar effects, in hydroperoxide-forming reactions 248
Polluted atmospheres, chemistry of 68
1,4-Poly(isoprene hydroperoxide), photolysis of 731
Polymeric hydroperoxides,
photolysis of 731
solid-state decomposition of 797–799
Polymeric initiators 420
Polymeric peroxides 417–426
Polymerization, of O_2 3
Polymers, irradiation of 423

Polymethylbenzenes, reaction with $^1O_2$ 211
Polymethylnaphthalene 1,4-endoperoxides, thermolysis of 347
Polyoxides 483–517
decomposition of 622
heats of formation for 40, 101
substituent effects in 68–71
Polyperoxides, synthesis of 236
Polystyrene hydroperoxide, photolysis of 731
Polystyrene peroxide 422
Potassium superoxide, in endoperoxide synthesis 316
Potential hypersurface,
of HO_2 25
of H_2O_2 25
of O_3 24
Prilezhaev reaction 525–533
catalysis of 528
kinetics and substituent effects in 525
mechanism of 528–538
solvent effects in 527, 528
stereochemistry of 525–527
theoretical studies of 530–533
2-(2-Propenyl)-2-hydroperoxy-adamantane 850
Propiolactones, as photolytic products 742
Propionyl peroxide, decomposition of 616
photolytic 739
Propylene ozonide, microwave study of 389
2-Propyl *p*-nitroperoxybenzoates, decomposition of 553
Prostacycline 313, 359, 364
Prostaglandin endoperoxides 313, 359–365
Prostaglandin G_2, synthesis of 316
Prostaglandins 313, 316, 359–365
synthesis of 212, 214, 216, 244
Pryor–Smith equation 617, 618
Pseudo-halogens 266
Pulse radiolysis 767
Purines,
photomethylation of 751, 752
reaction of,
with $O_2^{\cdot-}$ 437
with $^1O_2$ 211, 332
Pyridine, reactions of 451
Pyridine-*N*-oxide 867
complexes with peroxybenzoic acids 155
Pyridines, photomethylation of 752
Pyrroles, reaction with $^1O_2$ 211, 322, 871
Pyrrolinones 871
Pyruvic acids, autoxidation of 876

Quadrupole coupling constant 57
Quadrupole moment tensor 57
Quaternary ammonium salts, reactions of 441
o-Quinone 724

Radiation lifetime 58
Radical anions, as intermediates in radiolysis of peroxy acids 765, 766
Radical-induced decomposition,
of dialkyl peroxides 238, 239
of hydroperoxides 249, 250
Radical-pair mechanism 603
Radical pairs 739, 740, 743, 759, 765
acyloxy 609–614
acyloxy-*t*-butoxy 600–602
Radicals—*see also* Alkyl radical, etc.
$HC(O)O_2\cdot$ 68
reaction with $O_2^{\cdot-}$ 454
RO_3 247
Radical trapping 717, 732, 733, 737, 740, 758, 766
Rate constants,
for hydroperoxide-forming reaction in hydrocarbon autoxidation 248
for thermolysis of acyl and aroyl peroxides 240
Rearrangement,
Baeyer–Villiger 876
dioxetane–dioxetane 867
Hock–Criegee 878
intramolecular nucleophilic 551–573
of dioxetanes 227, 228
of endoperoxides 213, 214, 216, 217, 349, 352–355
of epoxides 851
of organophosphorus peroxides 275
of peroxy group 808
of sulphonyl peroxides 264, 265
Redox initiation 420
Reduction,
of allylic hydroperoxides 220, 221
of dioxetanes 228, 229, 910
of endoperoxides 216, 342–346
Reduction potential, for oxygen in aprotic solvents 448
Relaxation effect 44
Resorcinol, manufacture of 166
[2 + 2]Retrocyclization 893
Retro–Diels–Alder processes 120, 346, 350
'Retro' 1,3-dipolar processes 110
Rhodium(I) π-olefinic complexes 473
Rhodium(III) peroxo complexes 472, 473
Ring-contraction 737
Ring-expansion 736
Ring-inversion 69, 71, 72
Ring-opening 755, 869, 910
of peroxy esters 106
Ring-pseudorotation 69, 72
Ring-puckering 36, 63, 71, 72
Ring-substitution, influence on stability of aroyl peroxides 242

Rose Bengal, as photosensitizer 849, 862, 910
Rubrene,
as fluorescer 894, 903
reactions of 451
Rubrene peroxide, thermal decomposition in the solid phase 783
Russell mechanism 247
for self-reaction of alkylperoxy radicals 490, 491
Ruthenium peroxo complexes 471

Schiff bases, oxidation of 570–573, 878
Secondary overlap effect 66, 68
Second moment 57
Senepoxide, synthesis of 367, 368
Sesquiterpenes, ozonation of 670, 671
S—H bonds, reaction with $O_2^{-\cdot}$ 435
[1,3]-Sigmatropy 603, 608, 609
[3,3]-Sigmatropy 609, 613, 616
Silanes, ozonation of 501
Silatropic shift 853, 854
Silicon-containing peroxides, photolysis of 760, 761
Silyl enol ethers, reaction with $^1O_2$ 853, 855
Singlet oxygen 201–229
activation energies for reactions of 208, 210
diagnostic tests for 208, 209
direction of attack 210
formation of 204–208, 346–348, 588, 651
from $O_2^{-\cdot}$ 455
in decomposition of hydrotrioxides 504, 508, 509, 511
in self-reaction of peroxy radicals 491
rate of reaction of 210
reaction of 226, 831–844, 849–851
with alkenes 218–220, 588–590, 861–864
with 1,3-dienes 210–212, 317–338, 592, 593
with enamines 226, 592, 849, 866–871
with enol ethers 226, 592, 849, 864–866
with enols 864–866
with heteroarenes 871–873
with ketene acetals 849, 864
with ketenes 226, 843, 852, 857, 858
with vinyl sulphides 226
Slater determinant, of O_2 states 10
S_H2 mechanism, for decomposition of peroxides 240, 241
SMY mechanism, for ozonation of olefins 632–634
Sodium peroxide 274
Solanofuran, synthesis of 365
Solid-state isomerization 778
Solvation energy 44
Solvent cage 750
Solvolysis,
of diacyl peroxides 282
of dioxetanes 228
of endoperoxides 217
Spiro[2.4]hepta-2,4-diene, reaction with $^1O_2$ 862
Spiro[2.4]hepta-4,6-diene, reaction with $^1O_2$ 343
Steglitz rearrangement 272
Stemolide, synthesis of 369
Steric effects,
in decomposition of dialkyl peroxides 240
in hydroperoxide-forming reactions 248
in peroxy acid oxidation of ketones 561
Steric strain, relief of 248
Stern–Volmer plots 742, 744
Steroids, ozonation of 666, 667, 679, 680
trans-Stilbene, oxidation of 549
Strain 57
Structure,
of superoxides 62
of XO_2 compounds 5, 13, 61
of X_2O_2 compounds 5, 13, 21
Styrene, reactions of 449
Styrenes, substituted, reaction with $^1O_2$ 210
Substituent effects,
in mass spectra of β-peroxylactones 110
in peroxy acid oxidation of carbonyl compounds 560, 566
Substitutional disorder 376
Succinyl peroxides, photolysis of 742
Sulphate radical anion 757, 758, 767, 768
Sulphenic esters 411
Sulphides, oxidation of 445, 542, 543
Sulphines, oxidation of 226, 545
Sulphinyl chlorides, reaction with $O_2^{-\cdot}$ 446
Sulphonates, reaction with $O_2^{-\cdot}$ 441, 442
Sulphonic peracids 259
Sulphonoxy bridging 269
Sulphonoxy radicals 263
Sulphonyl acyl peroxides 259, 260
reactions of 263–265, 273
synthesis of 261, 262
Sulphonyl alkyl peroxides 259, 260
reactions of 265
synthesis of 261, 262
Sulphonyl chlorides, reaction with $O_2^{-\cdot}$ 446
O-Sulphonylhydroxylamine 270
Sulphoxides,
complexes with hydroperoxides 145, 147, 148
oxidation of 544, 545
Sulphoxilation 760
Sulphoxyperoxides, photolysis of 760
Sulphoxy radical 760
Sulphuranes 911, 912
Sulphur compounds, oxidation of 186, 296, 297, 445, 446

Sulphur nucleophiles, reaction with 1,2-dioxetanes 911
Superacids 673
Supernucleophilicity, of $O_2^{-\cdot}$ 441
Superoxide anion radical,
basic modes of action of 430
deprotonation vs. hydrogen atom abstraction by 430–441
effective basicity of 431
electron transfer from 448
hydration affinities of 441
methods for generating 430
nucleophilic reactions by 441–447
one-electron reductions by 448–456
reactions of, involving dioxetane intermediates 878–880
Superoxo complexes 377
Syn-directing groups 526, 532

Taft correlation, of pK_a values 130–133
Tantalum peroxo complexes 469
Temperature effects, in mass spectra of cyclic peroxides 119
Template effect 779
Tertiary amines, reaction with $O_2^{-\cdot}$ 436
Testosterone, reaction with $O_2^{-\cdot}$ 439
Tetraalkyltins, reactions of 274
Tetracyclone, reactions of 449
Tetraethyl-1,2-dioxetane 893
Tetralin, reactions of 440
Tetralin hydroperoxide, decomposition of 574
Tetramethoxy-1,2-dioxetane, decomposition of 895
Tetramethyl-1,2-dioxane, photolysis of 735
Tetramethyldioxetane 891, 892, 895, 912
Tetramethyl-1,2-dioxetane,
decomposition of 705
PE spectrum of 847
synthesis of 849
thermolysis of 847
enthalpies of formation for 882
3,3,5,5-Tetramethyl-1,2-dioxolane, decomposition of 244
Tetramethylethylene, reactions of 440, 449
Tetramethyl-β-peroxylactone, decomposition of 753
1,2,4,5-Tetraoxanes, structure of 391
Tetraphenylcyclopropene, reaction with $^1O_2$ 863
Tetraphenyl-1,4-dioxin, reaction with $^1O_2$ 865
Tetraphenylethylene, reactions of 449, 874
1,2,4,5-Tetraphenylimidazole, ring-cleavage of 873
Tetraphenylporphine 330
Tetraphenylporphyrin,
as photosensitizer 849
cleavage of 871
1,2,3,3-Tetraphenyl-1-propanone, autoxidation of 831
Tetrathioethylenes, reaction with $^1O_2$ 865
Tetra-μ-(trichloroaceto)tetra-μ-(*t*-butylperoxy)tetrapalladium 383
1,2,4,5-Tetroxane 912
Tetroxides,
diacyl—*see* Diacyl tetroxides
free-radical reaction mechanisms involving 236, 247
primary and secondary 490–493
tertiary 486–490
activation parameters for irreversible decomposition of 488, 489
equilibrium between *t*-alkylperoxy radicals and 486–488
Thianthrene, reactions of 445
Thianthrene cation radical, reactions of 447, 456
Thiazoles, reaction with $^1O_2$ 211
Thioalkenes, reaction with $^1O_2$ 849
Thiobenzophenones, oxidation of 545
Thioketenes, [2 + 2]cycloaddition of $^1O_2$ to 226
Thiolates, reaction with $O_2^{-\cdot}$ 445
Thiols, reaction with $O_2^{-\cdot}$ 435
Thiones, [2 + 2]cycloaddition of $^1O_2$ to 226
Thiophenes, reaction with $^1O_2$ 211, 332
Thiosulphinates, reaction with $O_2^{-\cdot}$ 445
Thiosulphonates, reaction with $O_2^{-\cdot}$ 445, 446
Thiourea, as reducing agent for endoperoxides 342–344
Thrombaxane 313, 359
Titanium peroxo complexes 468, 469
α-Tocopherol, reaction with $O_2^{-\cdot}$ 432
α-Tocopherol model compounds, oxidation of 433, 434
Toluene,
excited, fluorescence of 744
photoreaction of 718, 744, 745
Topology,
of XO_2 compounds 4
of X_2O_2 compounds 4
Topotactic solid-state behaviour 381, 408
Torsional angles 376, 378–381
Torsional potential function 376
Tosyloxy radicals 263
Transition energies,
of HO_2 58
of O_3 58
Transition-metal peroxides 463–479
Transition state, four-membered cyclic 739
Triacylamides 333
Trialkylarsines, as radical scavengers 732
Trialkylborane, autoxidation of 814

Trialkylphosphines, as radical scavengers 732
Trialkyl phosphite ozonides, decomposition of, as a method of preparing $^1O_2$ 206, 207
Trialkyltin radicals, reactions of 240
Triaryl phosphite ozonides, decomposition of, as a method of preparing $^1O_2$ 206, 207
Trifluoromethyl hydroperoxide, structure of 381
Trifluoromethylperoxy radical 735
Trifluoroperoxyacetic acid,
  ab initio MO study of 524
  as oxidizing agent 534, 540, 541
Trimethylbenzene, as photolytic product 746, 747
Trimethyldioxetane 891
Trimethyl-1,2-dioxetane,
  decomposition of 704, 705
  synthesis of 687, 831
Trimethylethylene, reactions of 440, 449
2,3,4-Trimethylpyrrole, reaction with $^1O_2$ 871
Trimethylsilyl iodide, reaction with 1,2-dioxetanes 910
β-Trimethylsilyl ketones, oxidation of 560
Trimethylsilylperoxy esters, desilylation of 853–855
Trimethylsilylperoxyketones 853
1,2,4-Trioxane 860
Trioxanes, synthesis of 340
1,2,4-Trioxanes, structure of 393
Trioxides 494–499
  as intermediates 651, 673
  free-radical reaction mechanisms involving 235, 245, 246
1,2,3-Trioxolanes 246
1,2,4-Trioxolanes 246
Triphenylantimony, reaction with 1,2-dioxetanes 911
Triphenylarsine, reaction with 1,2-dioxetanes 863, 911
Triphenylarsine oxide, complexes with peroxybenzoic acids 155
Triphenylethylene, cleavage of 874
Triphenylmethoxyl, rearrangement of 238
Triphenylmethyl, reactions of 236
Triphenylmethyl peroxide, synthesis of 236
Triphenylphosphine,
  as reducing agent for endoperoxides 342, 344–346
  reaction with α-peroxylactones 911
Triphenylphosphine oxide 651
  complexes with peroxybenzoic acids 155
Triphenyl phosphite, as reducing agent for endoperoxides 342, 344, 345
Triphenylphosphite ozonate, as singlet oxygen carrier 850
Triphenyl phosphite–ozone adduct 588
Triphenyl phosphite ozonide, direct reaction with reactive substrates 206
Triplet biradicals 339
Triplet oxygen 202
Triplet quenchers 741
Triterpenes, ozonation of 666
Tritium isotope effects, in ene reactions 589, 590
Trityl hydroperoxide, complexes with aniline 144
Tropilidene 323
Tropolones 344
Tropone, reactions of 449, 450
Tropylium ion, reaction with $O_2^{\cdot-}$ 446
Tryptamines, autoxidation of 876
Tryptophan 725
  radiolysis of 762
Tryptophanes, reaction with $^1O_2$ 872
Tungsten peroxo complexes 469, 471

Ultraviolet (UV) spectroscopy,
  of alkyl hydroperoxides 727
  of diacyl peroxides 712, 713, 738, 739
  of dialkyl peroxides 731, 732
  of HO_2 59
  of hydrogen peroxide 60, 720, 732
  of peroxy acids 743
  of peroxy esters 748
Undecalactone, as photolytic product 736
α,β,-Unsaturated aldehydes,
  selective epoxidation of 567
  synthesis of 221
α,β-Unsaturated esters, synthesis of 221
α,β-Unsaturated ketones, selective epoxidation of 567
Uracil, photooxidation of 725
Urea, as by-product, in cyclization of α-hydroperoxy acids 856

Valence bond description,
  of O_2 6
  of XO_2 7
  of X_2O_2 8
Valerophenone diperoxide, photolysis of 736
Vanadium peroxo complexes 469
Vanillin 861
Verruculogen 314, 396
Vibrational frequencies,
  of HO_2 52
  of H_2O_2 51
  of LiO_2 62
  of O_2 and its ions 49
Vinylarenes, reaction with $^1O_2$ 210, 327–330
Vinylcyclopropanes, [2+2]cycloaddition of $^1O_2$ to 226
Vinyl ethers, reaction with $^1O_2$ 849, 864
Vinylnaphthalenes, reaction with $^1O_2$ 210

Vinyl polymers 740
  autoxidation of 731
Vinylsilanes, ozonation of 851
Vinyl sulphides, photooxidation of 226, 228
2-Vinylthiophenes, reaction with $^1O_2$ 210
Vitamin-K-related compounds, reactions of 449

Wagner–Meerwein rearrangement 628
Water,
  ^{18}O- enriched, electronic discharge of 587
  radiolysis of 761

Xanthine, reaction with $O_2^{\cdot-}$ 440
Xylene,
  as photolytic product 745
  photoreaction of 746, 747
o-Xylylene peroxide 337

Y form,
  of F_2S_2 65
  of H_2S_2 65
  of peroxides 13, 25, 65

Zinc peroxides 810
Zirconium peroxo complexes 468, 469
Zirconocene peroxide 819
Zwitterions, as intermediates, in addition of $^1O_2$ to double-bond systems 220, 226, 332